BIOLOGY

HELENA CURTIS

BIOLOGY

Third Edition

BIOLOGY, THIRD EDITION

PRINTED IN THE UNITED STATES OF AMERICA

LIBRARY OF CONGRESS CATALOG CARD NO. 78-68582

ISBN: 0-87901-100-9

SECOND PRINTING, AUGUST 1979

EDITORS: SALLY ANDERSON

SUE BARNES

ILLUSTRATOR: SHIRLEY BATY

PICTURE EDITOR: ANNE FELDMAN

PRODUCTION: GEORGE TOULOUMES

DESIGN: MALCOLM GREAR DESIGNERS

TYPOGRAPHER: NEW ENGLAND TYPOGRAPHIC SERVICE, INC.

PRINTING AND BINDING: R. R. DONNELLEY & SONS

WORTH PUBLISHERS, INC.

444 PARK AVENUE SOUTH

NEW YORK, NEW YORK 10016

This book is dedicated to C. P. Rhoads, M.D.

PREFACE

As students of biology—whether professionals or amateurs—we are particularly fortunate to be living at a time when almost every aspect of the life sciences brims with new insights and discoveries. I feel privileged to be able to introduce newcomers to this subject, and I have tried to convey some of my own enthusiasm and to share some of the excitement that is so much a part of contemporary biology.

I find it remarkable that each new edition of the book requires more work than the preceding one. The entire book must be reconsidered, sentence by sentence, rewritten to a large extent, and also, to some degree, reorganized. The same questions must be answered with each edition: What is modern biology? What do students *have* to know? What should they remember ten years from now? What will still be true?

The most important undertaking of this revision has been to modernize the basic chemistry in the early chapters and to show throughout the remainder of the book how this knowledge of events at the molecular level deepens our understanding of biological phenomena. The goal has been to make the chemistry more coherent, more relevant to biology, but no more difficult. (Some reviewers have, indeed, found it to be easier.)

The Introduction, as before, deals with evolution, the major unifying principle of the life sciences. The historical background and essential ideas of Darwinian evolution provide a perspective that will be useful to students throughout their study of biology. No matter what organization is followed in a particular course, I hope the Introduction will be considered essential first reading.

I am reluctant to march you through a chapter-by-chapter description of the book, even though that seems to be the rule for textbook prefaces. For those who are not familiar with previous editions, such a thumbnail sketch is certainly inadequate, and, after all, the book is right in front of you. For those who are acquainted with the second edition, let me quickly mention the sources of information that guided us in this revision.

First of all, a number of you wrote with constructive criticisms of the book. Second, many teachers prepared detailed reviews of the second edition. A third source of information was the reporting by Worth representatives of conversations with teachers who had used the text. And finally, a number of ideas for improvement had been brewing in my own mind, many of them prompted by my work on new editions of both *Biology of Plants,* Second Edition, coauthored with Peter H. Raven and Ray F. Evert, and *Invitation to Biology,* Second Edition. So a plan emerged that involved a total rethinking of the chemistry, additional emphasis throughout the book on energetics, and a modernized ecology section, with less description and more analysis. And, of course, everything had to be brought up to date.

The overall structure of the book remains basically the same. The levels-of-biological-organization approach is again followed, beginning with atoms and moving to cells, Part I; organisms, Part II; and populations, Part III. For this edition, energetics now appears in a separate section, and the order of the ecology and evolution sections has been reversed. It is within the chapters themselves that a significant amount of restructuring has occurred, resulting, I believe, in a book that is more coherent and more intellectually satisfying.

Acknowledgments

Once again, I have had the benefit of a symbiotic relationship (I hope mutualistic, but perhaps commensalistic and, in a few instances, parasitic) with reviewers and consultants. Robert S. Edgar, University of California, Santa Cruz, and Abraham Flexer have been a continual source of guidance and wise counsel throughout the revision. I am also deeply indebted to Daryl Sweeney, University of Illinois, Urbana-Champaign, for his excellent suggestions and encouragement.

Although it is woefully inadequate recognition for their help, I am, as before, listing those who provided reviews for this edition, together with my thanks for their erudition and generosity:

ROSLYN B. ALFIN-SLATER, University of California, Los Angeles
GEORGE T. BARTHALMUS, North Carolina State University
JOHN W. BATEY, Seton Hall University
DOROTHY B. BERNER, Temple University
ANTONIE W. BLACKLER, Cornell University
WALTER F. BODMER, University of Oxford
RICHARD K. BOOHAR, University of Nebraska, Lincoln
CHARLES BROWN, Santa Rosa Junior College
MAREN H. BROWN, Onondaga Community College
EDWIN BURLING, DeAnza College
JOHN CORLISS, Oregon State University
MARY D. COYNE, Wellesley College
MEL CUNDIFF, University of Colorado, Boulder
CAROL DIAKOW, Adelphi University
LAWRENCE S. DILLON, Texas A & M University
MARK W. DUBIN, University of Colorado, Boulder
DAVID EPEL, Stanford University, Hopkins Marine Station
RAY F. EVERT, University of Wisconsin, Madison
BEN T. FEESE, Centre College of Kentucky
HERBERT FRIEDMANN, University of Chicago
JOHN GAPTER, University of Northern Colorado
MICHAEL T. GHISELIN, University of California, Bodega Marine Laboratory
FRANK H. GLEASON, Santa Rosa Junior College
MALCOLM S. GORDON, University of California, Los Angeles
JAMES L. GOULD, Princeton University
GOVINDJEE, University of Illinois, Urbana-Champaign
MICHAEL C. GRANT, University of Colorado, Boulder
PAUL B. GREEN, Stanford University
ALAN D. GRINNELL, University of California, Los Angeles
GUIDO GUIDOTTI, Harvard University

BURTON S. GUTTMAN, Evergreen State College
PENNY HANCHEY-BAUER, Colorado State University
JEAN B. HARRISON, University of California, Los Angeles
BRIAN A. HAZLETT, University of Michigan
ALVIN E. HIXON, Daytona Beach Community College
CHARLES E. HOLT, Massachusetts Institute of Technology
GEORGE B. JOHNSON, Washington University
RUSSELL L. JONES, University of California, Berkeley
JOHN KIRSCH, Yale University
RICHARD M. KLEIN, University of Vermont
PAUL KUGRENS, Colorado State University
KENNETH O. LLOYD, Memorial Sloan-Kettering Cancer Center
JANE LUBCHENCO, Oregon State University
DOROTHY S. LUCIANO, formerly of the University of Michigan
R. DAVIS MANNING, University of Mississippi Medical Center
EUGENE R. MEYER, New York University
DOUGLAS W. MORRISON, Rutgers University, Newark
TODD NEWBERRY, University of California, Santa Cruz
JOHN G. NICHOLLS, Stanford University
GARTH L. NICOLSON, University of California, Irvine
H. FREDERIK NIJHOUT, Duke University
R. D. O'BRIEN, Cornell University
JOHN H. OSTROM, Yale University
JANE A. PETERSON, University of California, Los Angeles
KEITH R. PORTER, University of Colorado, Boulder
GENE A. PRATT, University of Wyoming
CARL A. PRICE, Rutgers University, New Brunswick
JONATHAN REISKIND, University of Florida, Gainesville
EDWARD S. ROSS, California Academy of Sciences
CARL SAGAN, Cornell University
THOMAS K. SCOTT, University of North Carolina, Chapel Hill
DAVID G. SHAPPIRIO, University of Michigan
LINCOLN TAIZ, University of California, Santa Cruz
ROBERT M. THORNTON, University of California, Davis
JOSEPH VARNER, Washington University
JOHN L. YARNALL, Humboldt State University
NORTON D. ZINDER, Rockefeller University

I would like to call your attention to the excellent Study Guide prepared by Vivian Null to accompany the text. And with this edition we now for the first time have a Laboratory Manual to be used with the text. It was written by Ray F. Evert and Susan E. Eichhorn of the University of Wisconsin, Madison, and Barbara Saigo of the University of Wisconsin, Eau Claire. Transparency Masters of some of the illustrations in the book are also available for the first time to adopters.

I feel particularly privileged to be writing for young people; I like their curiosity, their energies, their imaginativeness, and their dislike of the pompous and pedantic. I hope I serve them well.

East Hampton, New York
January, 1979

Helen Curtis

CONTENTS IN BRIEF

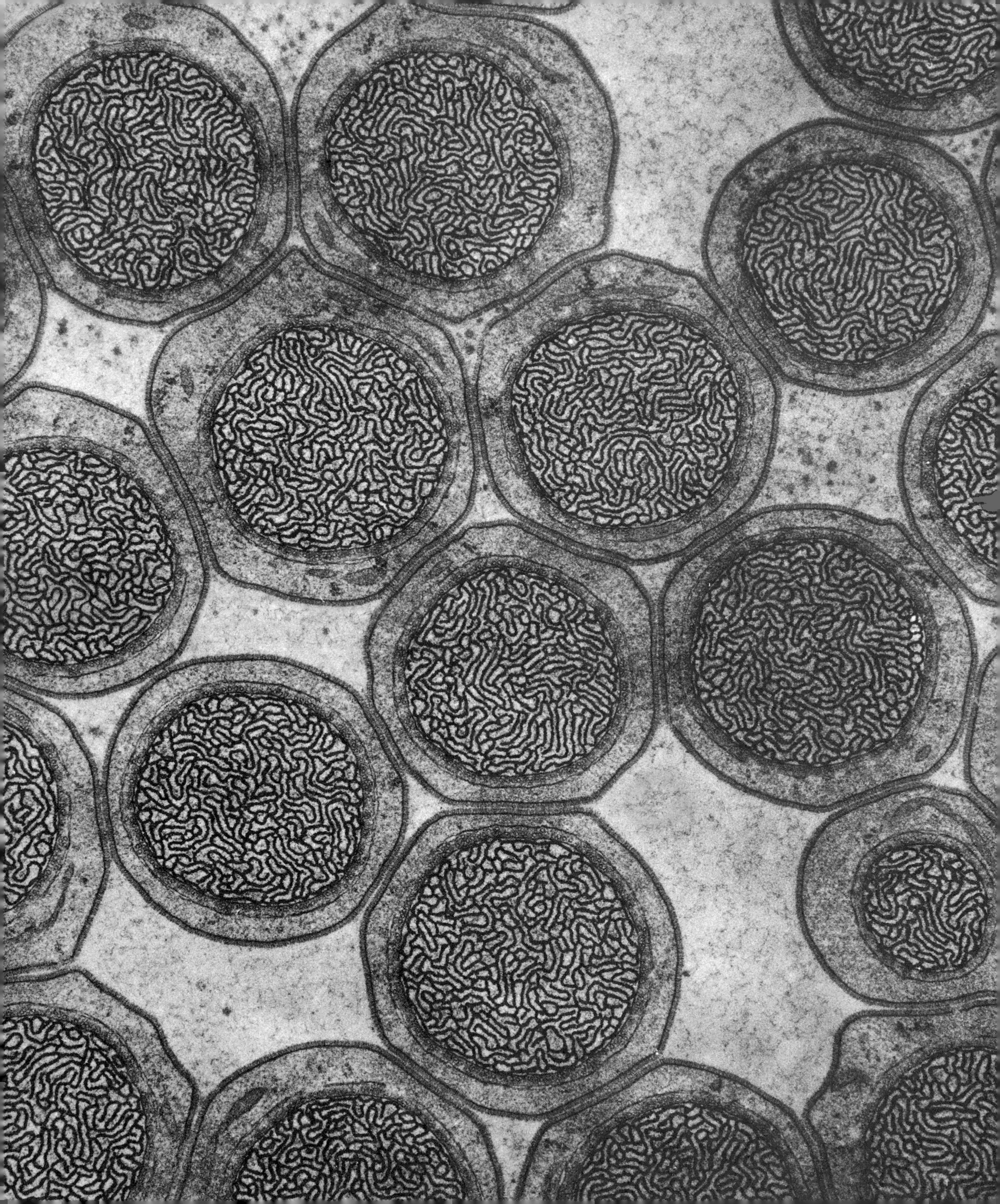

CONTENTS

Salamander larvae

Harvest mice

Meiosis (anaphase II)

Phage surrounded by its DNA

Cougar

Diatom

Hippopotamuses

Water hemlock

Orangutans

Lung tissue from a bird

Penguins

Leaflike insect (*Anaea*)

Wolves

Chimpanzees

BIOLOGY

I–1
When Darwin visited the Galapagos archipelago, he found that each major island had its own variety of tortoise, so distinct from the others that it was easily recognized by local sailors and fishermen. This was one of the clues that led him to the formulation of the theory of evolution. The Galapagos consists of 13 volcanic islands that pushed up from the sea more than a million years ago. The major vegetation is thornbush and cactus, and the original black basaltic lava is often visible, as it is beneath the lumbering feet of this tortoise on Hood Island—"what we might imagine the cultivated parts of the Infernal regions to be," young Darwin wrote in his diary.

Introduction

In 1831, the young Charles Darwin set sail from England on what was to prove the most consequential voyage in the history of biology. Not yet 23, Darwin had already abandoned a proposed career in medicine—he describes himself as fleeing a surgical theater in which an operation was being performed on an unanesthetized child—and was a reluctant candidate for the clergy, a profession deemed more suitable for the younger son of an English gentleman. An indifferent student, Darwin was an ardent hunter and fisherman, a collector of beetles, mollusks, and shells, and an amateur botanist and geologist. When the captain of the surveying ship H.M.S. *Beagle,* himself only a little older than Darwin, offered passage to any young man who would volunteer to go without pay as a naturalist, Darwin eagerly seized the opportunity to escape from Cambridge. This voyage, which lasted five years, shaped the course of Darwin's future work. He returned to an inherited fortune, an estate in the English countryside, and a lifetime of independent work and study that radically changed mankind's view of life and of our place in the living world.

THE ROAD TO EVOLUTIONARY THEORY

That Darwin was the founder of the modern theory of evolution is well known. In order to understand the meaning of his theory, however, it is useful to look briefly at the intellectual climate in which it was formulated. Aristotle (384–322 B.C.), the first great biologist, believed that all living things could be arranged in a hierarchy. This hierarchy became known as the *Scala Naturae,* or ladder of nature, in which the simplest creatures had a humble position on the bottommost rung, mankind occupied the top, and all other organisms had their proper places between. Until the end of the last century, many biologists believed in such a natural hierarchy. But whereas to Aristotle living organisms had always existed, the later biologists (at least those of the Occidental world) believed, in harmony with the teachings of the Old Testament, that all living things were the products of a divine creation. They believed, moreover, that most were created for the service or pleasure of mankind. Indeed, it was pointed out, even the lengths of day and night were planned to coincide with the human need for sleep.

I–2
Adaptations. The polar bear is suited to the snowy realms of the Arctic, the bat to hunting in the dark, and the monkey to the treetops, all as if by careful design.

That each type of living thing came into existence in its present form—specially and specifically created—was a compelling idea. How else could one explain the astonishing extent to which every living thing was adapted to its environment and to its role in nature? It was not only the authority of the church, but also the evidence before one's own eyes that gave such strength to the concept of special creation.

Among those who believed in divine creation was Carolus Linnaeus (1707–1778), the great eighteenth-century Swedish naturalist, who devised our present system of nomenclature for biological species. All the time that Linnaeus was at work on his encyclopedic *Systema Naturae* and *Species Plantarum,* explorers of Africa and the New World were continuing to return to Europe with new species of plants and animals and even, apparently, new kinds of human beings. Linnaeus revised edition after edition to accommodate these findings, but he did not change his opinion that all species now in existence were created on the sixth day of God's labor and have remained fixed ever since. During Linnaeus's time, however, it became clear that the pattern of creation was far more complex than had been originally envisioned.

Evolution before Darwin

The French scientist Georges-Louis Leclerc de Buffon (1707–1788) was among the first to suggest that species might undergo some changes in the course of time. Buffon believed that these changes took place by a process of degeneration. He suggested that, in addition to the numerous creatures that were produced by divine creation at the beginning of the world, "there are lesser families conceived by Nature and produced by Time." In fact, as he summed it up, ". . . improvement and degeneration are the same thing, for both imply an alteration of the original constitution." Buffon's hypothesis, although vague as to mechanism, did attempt to explain the bewildering variety of creatures in the modern world.

Another early doubter of the fixity of species was Erasmus Darwin (1731–1802), Charles Darwin's grandfather. Erasmus Darwin was a physician, a gentleman naturalist, and a prolific writer, often in verse, on both botany and zoology. Erasmus Darwin suggested, largely in asides and footnotes, that species have historical connections with one another, that competition plays a role in the development of different species, that animals may change in response to their environment, and that their offspring may inherit these changes. He maintained, for instance, that a polar bear is an "ordinary" bear that, by living in the Arctic, becomes modified and passes the modifications along to its cubs. These ideas were never clearly formulated but are interesting because of their possible effects on Charles Darwin, although the latter, born after his grandfather died, did not profess to hold his grandfather's views in high esteem.

The Age of the Earth

It was geologists, more than biologists, who paved the way for evolutionary theory. One of the most influential of these was James Hutton (1726–1797). Hutton proposed that the earth had been molded not by sudden, violent events but by slow and gradual processes—wind, weather, and the flow of water—the same processes that can be seen at work in the world today. This theory of Hutton's, which was known as uniformitarianism, was important for three reasons. First, it implied that the earth has a living history, and a long one. This was a new idea. Christian theologians, by counting the successive generations since Adam (as recorded in the Bible), had calculated the maximum age of the earth at about 6,000 years. No one, as far as we know, had ever thought in terms of a longer period. And, as we shall see in Chapter 41, 6,000 years is not enough time for such major evolutionary changes as the formation of new species to have taken place. Second, the theory of uniformitarianism stated that change is the *normal* course of events, as opposed to the concept of a normally static system interrupted by an occasional unusual event, such as a flood. Third, although this was never explicit, uniformitarianism suggested that there might be interpretations of the Bible other than the literal one.

The Fossil Record

During the latter part of the eighteenth century, there was a revival of interest in fossils. In previous centuries, fossils had been collected as curiosities, but they had generally been regarded either as accidents of nature—stones that somehow looked like shells—or as evidence of great natural catastrophes, such as the Flood. The English surveyor William Smith (1769–1839) was the first to make a systematic study of fossils. Whenever his work took him down into a mine or along canals or cross-country, he carefully noted the order of the different layers of rock, which are called strata, and collected the fossils from each layer. He eventually established that each stratum, no matter where he came across it in England, contained a characteristic group of specimens and that these fossils were actually the best way to identify a particular stratum. (The use of fossils to identify strata is still widely practiced, for instance, by geologists looking for oil.) Smith did not interpret his findings, but the implication that the present surface of the earth had been formed layer by layer over the course of time was an unavoidable one.

Like Hutton's world, the world seen and reported by William Smith was clearly a very ancient one. A revolution in geology was beginning; earth science was

(a)

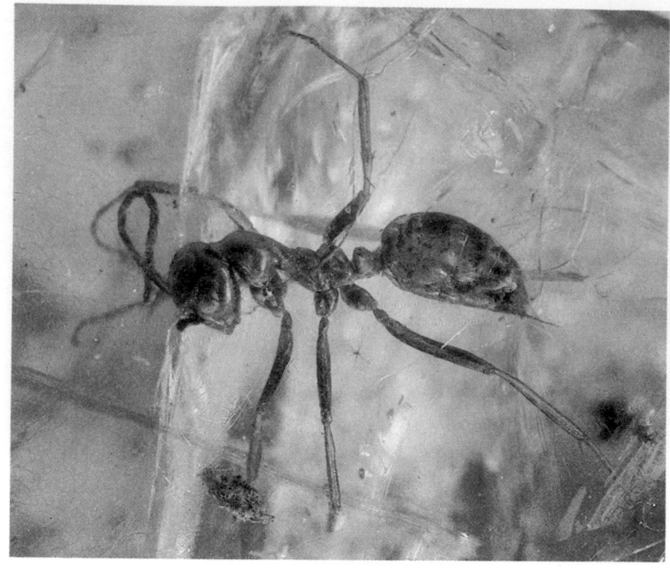

(b)

(c)

I–3

A fossil is a remnant or trace of a once-living organism. (a) *Minerals may fill the hollows left by the decay of soft tissues, as, for example, in this fossilized ginseng leaf.* (b) *A primitive wasplike ant caught in amber formed from the resin of a tree that lived some 100 million years ago, during the Cretaceous period. The ant is a worker (a sterile female), indicating that insect societies had evolved by this time.* (c) *Among the most common early fossils are the discarded exoskeletons of trilobites, a type of arthropod that flourished during the Cambrian period, which began 500 to 600 million years ago.*

becoming a study of time and change rather than a mere cataloging of types of rocks. As a consequence, the history of the earth became inseparable from the history of living organisms, as revealed in the fossil record.

Catastrophism

Although the way was being prepared by the revolution in geology, the time was not yet ripe for a parallel revolution in biology. The dominating force in European science in the early nineteenth century was Georges Cuvier (1769–1832). Cuvier was the founder of vertebrate paleontology, the scientific study of the fossil record. An expert in anatomy and zoology, he applied his special knowledge of the way in which animals are constructed to the study of fossil animals, and he was able to make brilliant deductions about the form of an entire animal from a few fragments of bone. We think of paleontology and evolution as so closely connected that it is surprising to learn that Cuvier was a staunch and powerful opponent of evolutionary theories. He recognized the fact that many species that had once existed no longer did. (In fact, according to modern estimates, considerably less than one percent of all species that have ever lived are represented on the earth today.) Cuvier explained the extinction of species by postulating a series of catastrophes. After each catastrophe, the most recent of which was the Flood, new species filled the vacancies.

Cuvier hedged somewhat on the source of the new animals and plants that appeared after the extinction of older forms; he was inclined to believe they moved in from parts unknown. A contemporary and supporter of his, the first professor of paleontology at the Paris museum, was more straightforward: He taught that all species were wiped out at each catastrophe and wholly new ones were specially created to take their places. (One symptom of the terminal illness of a scientific theory is the accumulation of new hypotheses required to bolster its shaking foundations.)

I–4
A drawing by Georges Cuvier of a mastodon. Although Cuvier was one of the world's experts in reconstructing extinct animals from their fossil remains, he was a powerful opponent of evolutionary theories.

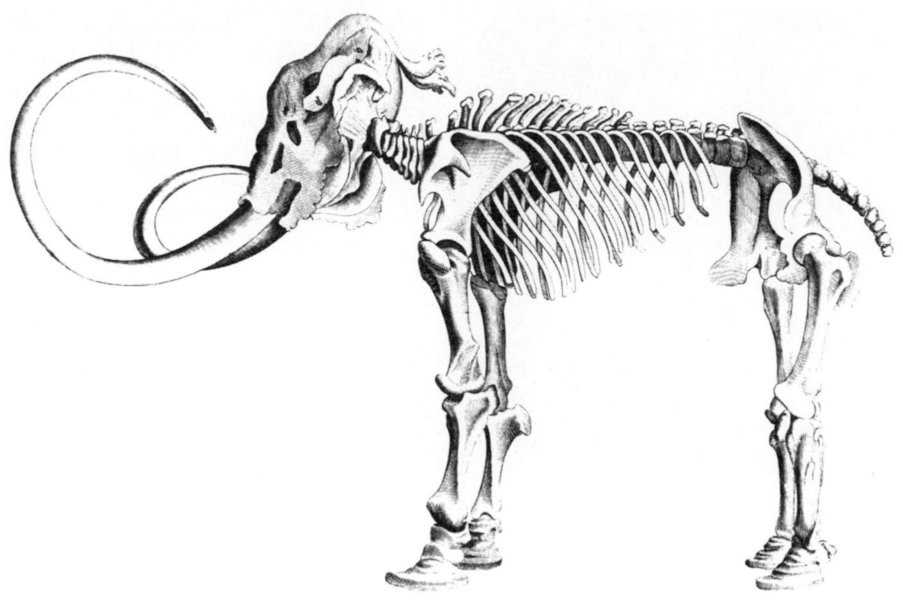

The Theories of Lamarck

The first scientist to work out a systematic theory of evolution was Jean Baptiste Lamarck (1744–1829). "This justly celebrated naturalist," as Darwin himself referred to him, boldly proposed in 1801 that all species, including *Homo sapiens*, are descended from other species. Lamarck, unlike most of the other zoologists of his time, was particularly interested in one-celled organisms and invertebrates. Undoubtedly it was his long study of these forms of life that led him to think of living things in terms of constantly increasing complexity, each species derived from an earlier, less complex one.

Like Cuvier and others, Lamarck noted that older rocks generally contained fossils of simpler forms of life. Unlike Cuvier, however, Lamarck interpreted this as meaning that the higher forms had risen from the simpler forms by a kind of progression. According to his hypothesis, this progression, or "evolution," to use the modern term, is dependent on two main forces. The first is the inheritance of acquired characteristics. Organs in animals become stronger or weaker, more or less important, through use or disuse, and these changes, according to Lamarck's theory, are transmitted from the parents to the progeny. His most famous example, and the one that Cuvier used most often to ridicule him, was that of the giraffe, which stretched its neck longer and longer to reach leaves on higher branches and transmitted this longer neck to its offspring, which stretched its neck even longer, and so on.

The second important factor in Lamarck's theory of evolution was a universal creative principle, an unconscious striving upward on the *Scala Naturae* that moved every living creature toward greater complexity. Every amoeba was on its way to man. Some might get waylaid—the orangutan, for instance, by being caught in an unfavorable environment had been diverted off its course—but the will was always present. Life in its simplest forms was constantly emerging by spontaneous generation to fill the void left at the bottom of the ladder. In Lamarck's formulation, the ladder of nature of Aristotle had been transformed into a steadily ascending escalator powered by a universal will.

Lamarck's concept of the inheritance of acquired characteristics is an attractive one. Darwin borrowed from it consciously and unconsciously. In fact, successive editions of *The Origin of Species* show that Darwin was maneuvered into a Lamarckian point of view by some of his critics and by his own inability (owing to the primitive state of the science of genetics at the time) to explain some of the ways in which animals changed. Belief in the inheritance of acquired characteristics persisted into the twentieth century in the now discredited work of the Russian biologist Trofim Lysenko.

Lamarck's contemporaries did not object to his ideas about the inheritance of acquired characteristics, as we would today with our more advanced knowledge of genetics. Nor did they criticize his belief in a metaphysical force, which was actually a common element in many of the theories of the time. But these vague, untestable postulates provided a very shaky foundation for the radical proposal that higher forms evolved from simpler forms, and Lamarck personally was no match for the brilliant and witty Cuvier. As a result, Lamarck's career was ruined, and both scientists and the public became even less prepared to accept any evolutionary doctrine.

I-5
According to Lamarck's hypothesis, the necks of giraffes became longer when they stretched to reach high branches, and this acquired characteristic was transmitted to their offspring.

DEVELOPMENT OF DARWIN'S THEORY

The Earth Has a History

The person who most influenced Darwin, it is generally agreed, was Charles Lyell (1797–1875), a geologist who was Darwin's senior by 12 years. One of the few books that Darwin took with him on his voyage was the first volume of Lyell's newly published *Principles of Geology,* and the second volume was sent to him while he was on the *Beagle.* On the basis of his own observations and those of his predecessors, Lyell opposed the theory of catastrophes. Instead, he produced new evidence in support of Hutton's earlier theory of uniformitarianism. According to Lyell, the slow, steady, and cumulative effect of natural forces had produced continuous change in the course of the earth's history. Since this process is demonstrably slow, its results being barely visible in a single lifetime, it must have been going on for a very long time. In his earlier works, Lyell did not discuss the biological implications of his theory, but apparently they were clear to Darwin. If the earth had a long continuous history and if no forces other than well known, natural ones were needed to explain the events as they were recorded in the geologic record, might not living organisms have had a similar history? What Darwin needed for the development of his theory was time, and it was time that Lyell gave him.

The Voyage of the Beagle

This, then, was the intellectual equipment with which Charles Darwin set sail from England. As the *Beagle* moved down the Atlantic coast of South America, through the Straits of Magellan, and up the Pacific coast, Darwin traveled the interior, fished, hunted, and rode horseback. He explored the rich fossil beds of South America (with the theories of Lyell fresh in his mind) and collected specimens of

I–6
"Afterwards, on becoming very intimate with Fitz Roy [the captain of the Beagle*], I heard that I had run a very narrow risk of being rejected on account of the shape of my nose! He . . . was convinced that he could judge of a man's character by the outline of his features; and he doubted whether anyone with my nose could possess sufficient energy and determination for the voyage. But I think he was afterwards well satisfied that my nose had spoken falsely." (Charles Darwin,* The Voyage of the Beagle.*)*

the many new kinds of plant and animal life he encountered. He was impressed most strongly during his long, slow trip down the coast and up again by the constantly changing varieties of organisms he saw. The birds and other animals on the west coast, for example, were very different from those on the east coast, and even as he moved slowly up the western coast, one species would give way to another.

Most interesting to Darwin were the animals and plants that inhabited a small, barren group of islands, the Galapagos, which lie some 950 kilometers off the coast of Ecuador. The Galapagos were named after the islands' most striking inhabitants, the tortoises (*galápagos* in Spanish), some of which weigh 100 kilograms or more. Each island has its own type of tortoise; sailors who took these tortoises on board and kept them as convenient sources of fresh meat on their sea voyages could readily tell which island any particular tortoise had come from. Then there was a group of finchlike birds, 13 species in all, that differed from one another in the sizes and shapes of their bodies and beaks, and particularly in the type of food they ate. In fact, although clearly finches, they had many characteristics seen only in completely different types of birds on the mainland. One finch, for example, feeds by routing insects out of the bark of trees. It is not fully equipped for this, however, lacking the long tongue with which the woodpecker flicks out insects from under the bark. Instead, the woodpecker finch uses a small stick or cactus thorn to pry the insects loose.

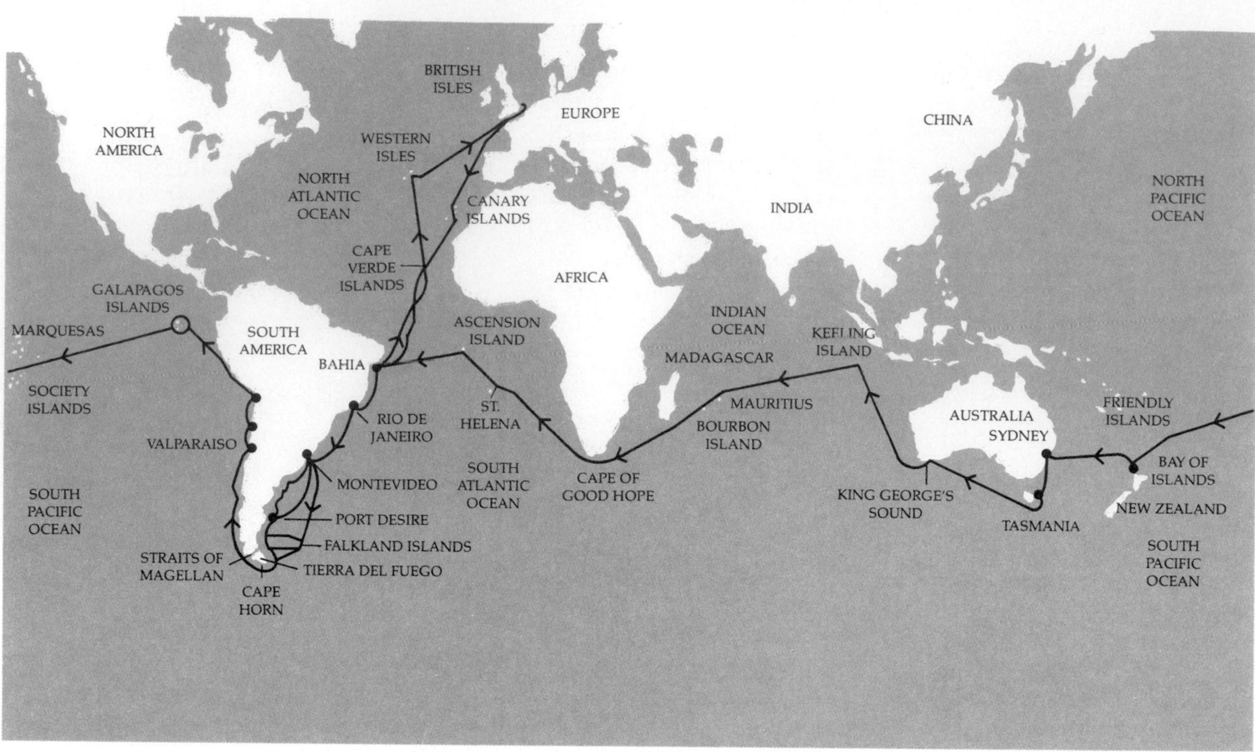

I–7
The Beagle's *voyage.*

From his knowledge of geology, Darwin knew that these islands, clearly of volcanic origin, were much younger than the mainland. Yet the plants and animals of the islands were different from those of the mainland, and in fact the inhabitants of different islands in the archipelago differed from one another. Were the living things on each island the product of a separate special creation? "One might really fancy," Darwin mused at a later date, "that from an original paucity of birds in this archipelago one species had been taken and modified for different ends." For years after his return, this problem continued, in his own word, to "haunt" him.

The Darwinian Theory

Not long after Darwin's return, he came across a book by the Reverend Thomas Malthus that had first appeared in 1798. In this book, Malthus warned, as economists have warned frequently ever since, that the human population was increasing so rapidly that it not only would soon outstrip the food supply, but also would leave "standing room only" on earth. Darwin saw that Malthus's conclusion—that food supply and other factors hold populations in check—is true for all species, not just the human one. For example, a single breeding pair of elephants, which are the slowest breeders of all animals, would, if all their progeny lived and reproduced the normal number of offspring over a normal life span, produce 19 million elephants in 750 years, yet the average number of elephants generally remains the same over the years. Where there might have been 19 million elephants in theory, there are, in fact, only two. But why those particular two? The process by which the two survivors are "chosen" Darwin called natural selection.

He saw it as a process analogous to the type of selection exercised by breeders of cattle, horses, or dogs—with which, as a country squire, he was very familiar. In the case of artificial selection, we humans choose individuals for breeding on the basis of characteristics that seem to us to be desirable. In the case of natural selection, environmental conditions are the principal forces that operate on the variations among individuals continually produced in all species of living organisms. As these forces select for some varieties and eliminate others, the characteristics of the population very slowly change. For example, if some individual horses were swifter than others, these individuals would be more likely to survive and their progeny, in turn, might be swifter, which is exactly what happened over many thousands of years.

Where do the variations come from? According to Darwin's theory, variations occur absolutely at random. They are not produced by the environment, by a "creative force," or by the unconscious striving of the organism. In themselves, they have no goal or direction. It is the operation of natural selection over a series of generations that gives direction to evolution. A variation that gives an animal even a slight advantage makes that animal more likely to leave surviving offspring. Thus, to return to Lamarck's giraffe, an animal with a slightly longer neck has an advantage in feeding and so is apt to leave more offspring than one with a shorter neck. If the longer neck is an inherited trait, some of these offspring will also have long necks, and, if the long-necked animals in this generation have an advantage, members of the next generation will include more long-necked individuals. Finally, the population of short-necked giraffes will give way to a population of long-necked ones.

I-8
The Beagle, *at anchor in Sydney Harbor. Only 28 meters in length, this "good little vessel" set sail on its five-year voyage with 74 people aboard. Darwin shared the poop cabin with a midshipman and 22 chronometers belonging to Captain Fitz Roy, who had a passion for exactness. His sleeping space was so confined that he had to remove a drawer from a locker to make room for his feet.*

DARWIN'S LONG DELAY

Darwin returned to England with the Beagle *in 1836. Two years later, he read the essay by Malthus, and in 1842 he wrote a preliminary sketch of his theory, which he revised in 1844. On completing the revision, he wrote a formal letter to his wife requesting her in the event of his death to publish the manuscript (which was some 230 pages long). Then, with the manuscript and letter in safekeeping, he turned to other work, including a four-volume treatise on the classification of barnacles. For more than 20 years following his return from the Galapagos, Darwin mentioned his ideas on evolution only in his private notebooks and in letters to his scientific colleagues.*

In 1856, urged on by his friends Charles Lyell and botanist Joseph Hooker, Darwin set slowly to work preparing a manuscript for publication. In 1858, some 10 chapters later, Darwin received a letter from another English naturalist, Alfred Russel Wallace, presenting a theory of evolution that exactly paralleled Darwin's own. Like Darwin, Wallace had traveled extensively, both in South America (in the Amazon basin) and in the Malay archipelago, from which his letter had been mailed. He also had read Malthus's essay, and he had corresponded with Darwin on several previous occasions. Wallace, tossing in bed one night with a fever, had a sudden flash of insight. "Then I saw at once," Wallace recollected, "that the ever present variability of all living things would furnish the material from which, by the mere weeding out of those less adapted to the actual conditions, the fittest alone would continue the race." Within two days, Wallace's 20-page manuscript was completed and in the mail.

When Darwin received Wallace's letter, he turned to his friends for advice, and Lyell and Hooker, taking matters into their own hands, presented the theory of Darwin and Wallace at a scientific meeting just one month later. (Darwin described Wallace as "noble and generous," as indeed he was.) Lyell and Hooker read four papers from Darwin's notes of 1844, excerpts from two letters written by Darwin, and Wallace's manuscript. Their presentation received little attention, but for Darwin the floodgates were opened. He finished his long treatise in little more than a year, and the book was finally published. The first printing was a mere 1,250 copies, but they were sold out the same day.

Why Darwin's long delay? His own writings, voluminous though they are, shed little light on this question. But perhaps his background does. He came from a very devout family, he himself had been a divinity student, and his wife, to whom he was deeply devoted, was extremely religious. It is difficult to avoid the speculation that Darwin, as has been the case with others, found the implications of his theory difficult to confront.

Alfred Russel Wallace (1823–1913).

I-9
(a) *A view of the universe, first proposed by the early Greeks and accepted throughout the Middle Ages. In this drawing, dated 1528, earth is in the center of the universe, surrounded by concentric circles representing spheres of air, fire, the moon, the sun, five planets, and fixed stars.* (b) *The solar system, as proposed by Nicholas Copernicus. In 1543, Copernicus set forth in* De Revolutionibus *the new concept that the sun, not the earth, is the center of the solar system. His theory was supported by the German astronomer Johannes Kepler (1571–1630), who discovered the laws of planetary motion, and by the Italian Galileo Galilei (1564–1642). The latter spent his last ten years confined to his home for heresy because of his advocacy of Copernican beliefs.*

(a)

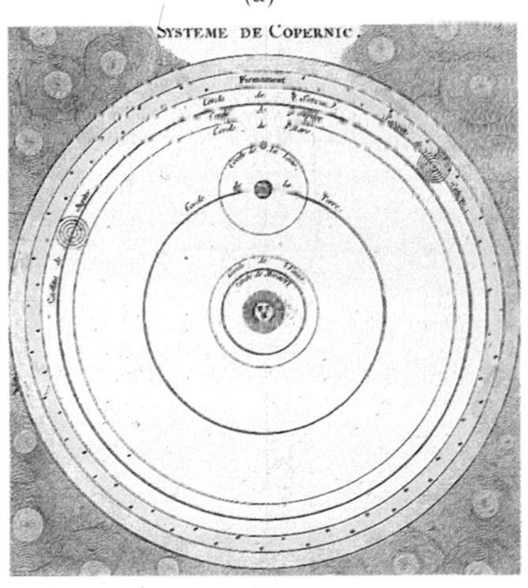

(b)

As you can see, the essential difference between Darwin's formulation and that of any of his predecessors is the central role he gave to variation. Others had thought of variations as mere disturbances in the overall design, whereas Darwin saw that variations among individuals are the real fabric of the evolutionary process. Species arise, he saw, when differences among individuals within a group are gradually converted into differences between groups as the groups become separated in space and time.

The Origin of Species, which Darwin pondered for more than 20 years before its publication in 1859 (see essay), is, in his own words, "one long argument." No experiments are performed. No new form of evidence is revealed. Fact after fact, observation after observation, culled from the most remote Pacific island to a neighbor's pasture, is recorded, analyzed, and commented upon. Every objection is anticipated and countered. Because the process of evolution is so slow, Darwin did not believe that direct proof of his theory was possible. However, as we shall see, the twentieth century has produced clear evidence of evolution in progress. Few scientists now doubt that species have originated in the past and are still originating, that species have become extinct in the past and are still becoming extinct, and that the different species living today share an ancestral species in the past.

The real difficulty in accepting Darwin's theory has always been that it seems to diminish our significance. The new astronomy had made it clear that the earth is not the center of the universe or even of our own solar system. Now the new biology asked us to accept the proposition that, like all other organisms, we too are the products of a random process and that, as far as science can show, we are not created for any special purpose or as a part of any universal design.

Importance of the Theory in Modern Biology

What is the importance of the theory of evolution to modern biologists, many of whom are concerned with such areas of investigation as the chemistry of heredity—a phrase that would have been meaningless in Darwin's time—or the interpretation of subcellular structures newly revealed by the electron microscope, or the tracing of a radioactive substance through the components of an ecosystem? In the words of Ernst Mayr of Harvard University:*

> The theory of evolution is quite rightly called the greatest unifying theory in biology. The diversity of organisms, similarities and differences between kinds of organisms, patterns of distribution and behavior, adaptation and interaction, all this was merely a bewildering chaos of facts until given meaning by the evolutionary theory. There is no area of biology in which that theory has not served as an ordering principle.

Because they are of particular importance in this book, we shall single out for mention three of the many ways in which Darwin's theory has "served as an ordering principle" in biology. The first involves the kinds of living organisms that fill the world about us. On the basis of evolutionary theory, we can understand, as Linnaeus could not, both the resemblances and differences among the various groups of plants and animals and other living things. And, we can correlate these differences, to some extent, with where and how these organisms live and with their history.

* Ernst Mayr, *Animal Species and Evolution,* Harvard University Press, Cambridge, Mass., 1963.

I–10
Darwin accurately predicted the discovery of an insect with a tongue 28 centimeters long in Madagascar, because this length would be required to reach the nectar of a species of orchid that blooms there. A similar, though slightly less startling, relationship exists between the sphinx moth and the tobacco blossom shown here.

The theory of evolution also helps us understand the seeming purposefulness of living things and their activities. Why do flowers often have sweet odors and bright colors? Why does our heart pump blood through our lungs? If we answer that the heart pumps blood through our lungs in order to supply it with oxygen, we imply that the heart has a purpose, that it knows what it is doing—which is, of course, not true. Yet in another sense, this sort of explanation is the correct one. Flowers and their insect pollinators evolved together; brightly colored flowers were more apt to be pollinated and, therefore, left more offspring. As you can see, within the framework of evolutionary theory, "why" and "what for" questions become interesting and meaningful.

The third point, closely related to the first two, is that evolutionary theory emphasizes the dynamic relationship between structure and function. Anatomy, on the one hand, and physiology or behavior, on the other, cannot logically be studied or understood apart from one another. The two have necessarily evolved together, the function shaping the structure and the structure directing and guiding the activity.

As you will see, we can say with equal truthfulness that we can use our hands for grasping because we have an opposable thumb and that we have an opposable thumb because we use our hands for grasping. This does not imply perfection; the wisdom of nature is a sentimental notion. Your body will automatically reject a skin graft from another person—unless the person happens to be your identical twin—even if you are dying of an extensive burn and the graft could save your life. Evolution has imposed a number of clumsy burdens. The human vertebral column is a good example. We are the only primate to walk upright. The resulting alterations in skeletal structure carry a heavy price—in slipped discs, other back ailments, and hernias. An engineer could devise a better structural support for us if he or she were permitted to start from the beginning. The point is, however, that our backbone and our pelvis have been molded by time—by variation and natural selection—into their present forms, and their structure and function are entirely inseparable.

This, then, is the background for the study of biology, "the science of life." The scope of biology extends from the organization of the atoms and molecules that make up living matter through the interactions of whole organisms and groups of organisms with one another and with the environment in which they live. This is also the scope of this text, which begins with the submicroscopic world within the cell and ends with a view of the biosphere, the entire world of living things.

THE NATURE OF SCIENCE

Science, biological and other, is a way of seeking principles of order in the natural world. Art is another, as are religion and philosophy. Science has two separate components: (1) objective evidence based on observation or experiment or a combination of the two—the data—and (2) the structuring of data by making meaningful connections among them. The great discoveries in science are not merely the addition of new facts but the perception of new relationships among the available facts, in other words, the development of new concepts and new ideas.

The ideas of science are categorized in ascending order of validity as hypotheses, theories, and principles or laws. Lower on the scale than the hypothesis is the hunch, or educated guess, which is how most hypotheses begin. A hunch becomes a hypothesis when it is stated in such a way that it can be tested. (As we shall see, a hypothesis may be stated verbally, mathematically, or in the form of a model.) Until it can be so stated, it is useless, scientifically speaking. When a hypothesis has survived a number of independent tests, it becomes, strictly speaking, a theory. A theory that has withstood repeated testing over a period of time becomes elevated to the status of a law or principle, although not always identified as such. The "theory" of evolution is an example. It is not "just a theory," in the common usage of the term, nor has it been for almost a hundred years. As far as scientists are concerned, it is a principle, just as is the cell "theory" (page 103). However, many of the details of cellular structure and of the evolutionary process are still in the stage of theory, or even hunches.

However, this emphasis on theories is not to say that the facts are not important. It is the facts that endure, that are passed from one worker to the next, from one scientific generation to its successor. It is for this reason that scientists stress objectivity; in all science, observations and experiments must always be reported in such a way that they can be repeated and verified, and must always be repeated and verified before they are incorporated into the body of knowledge. The data are the unyielding building blocks, the structural elements, stubborn and concrete, with which theories are erected and against which dreams are shattered.

CAUTIONARY NOTES

Most texts, and this one is no exception, tend to stress what is known at the present time, rather than what is not known or how we came to know what we do. This tendency, although understandable, somewhat distorts the nature of biology, and indeed of science in general. A modern science is not a static accumulation of facts organized in a particular way, but a somewhat amorphous body of knowledge that not only constantly grows, developing new bulges and unpredictable appendages, but also may suddenly change its entire shape (as biology did in the nineteenth

SOME COMMENTS ON SCIENCE AND SCIENTISTS

Art and Science

The act of discovery in science engages the imagination (first of the man who makes it, and then of the man who appreciates it) as truly as does the act of creation in the arts.

J. Bronowski, Identity of Man, *rev. ed., Natural History Press, Garden City, N.Y., 1971.*

Science and Theory

The power of theories is that they combine many generalizations and other theories into networks of interlocking ideas that point to the future.

J. Bronowski, The Common Sense of Science, *Random House, New York, 1959.*

On principle, it is quite wrong to try founding a theory on observable magnitudes alone. It is the theory which decides what we can observe.

A. Einstein, from J. Bernstein, "The Secrets of the Old Ones, II," The New Yorker, *March 17, 1973.*

The Scientific Method

Indeed, scientists are in the position of a primitive tribe which has undertaken to duplicate the Empire State Building, room for room, without ever seeing the original building or even a photograph. Their own working plans, of necessity, are only a crude approximation of the real thing, conceived on the basis of miscellaneous reports volunteered by interested travelers and often in apparent conflict on points of detail. In order to start the building at all, some information must be ignored as erroneous or impossible, and the first constructions are little more than large grass shacks. Increasing sophistication, combined with methodical accumulation of data, make it necessary to tear down the earlier replicas (each time after violent arguments), replacing them successively with more up-to-date versions. We may easily doubt that the version current after only 300 years of effort is a very adequate restoration of the Empire State Building; yet, in the absence of clear knowledge to the contrary, the tribe must regard it as such (and ignore odd travelers' tales that cannot be made to fit).

E. J. DuPraw, Cell and Molecular Biology, *Academic Press, Inc., New York, 1968.*

Scientists at Work

Scientists at work have the look of creatures following genetic instructions; they seem to be under the influence of a deeply placed human instinct. They are, despite their efforts at dignity, rather like young animals engaged in savage play. When they are near to an answer their hair stands on end, they sweat, they are awash in their own adrenalin. To grab the answer, and grab it first, is for them a more powerful drive than feeding or breeding or protecting themselves against the elements.

It sometimes looks like a solitary activity, but it is as much the opposite of solitary as human behavior can be. There is nothing so social, so communal, so interdependent. An active field of science is like an immense intellectual anthill; the individual almost vanishes into the mass of minds tumbling over each other, carrying information from place to place, passing it around at the speed of light.

There are special kinds of information that seem to be chemotactic. As soon as a trace is released, receptors at the back of the neck are caused to tremble, there is a massive convergence of motile minds flying upwind on a gradient of surprise, crowding around the source. It is an infiltration of intellects, an inflammation.

There is nothing to touch the spectacle. In the midst of what seems a collective derangement of minds in total disorder, with bits of information being scattered about, torn to shreds, disintegrated, reconstituted, engulfed, in a kind of activity that seems as random and agitated as that of bees in a disturbed part of the hive, there suddenly emerges, with the purity of a slow phrase of music, a single new piece of truth about nature. . . .

There is something like aggression in the activity, but it differs from other forms of aggressive behavior in having no sort of destruction as the objective. While it is going on, it looks and feels like aggression: get at it, uncover it, bring it out, grab it, halloo! It is like a primitive running hunt, but there is nothing at the end of it to be injured. More probably, the end is a sigh. But then, if the air is right and the science is going well, the sigh is immediately interrupted, there is a yawping new question, and the wild, tumbling activity begins once more, out of control all over again.

Lewis Thomas, "Notes of a Biology-Watcher," New England Journal of Medicine, *vol. 288, pages 307–308, February 8, 1973.*

century with the acceptance of the theory of evolution). Science is dynamic, not static; consequently it cannot be contained within textbooks or libraries or information retrieval centers, but rather it is a process taking place in the minds of living scientists. In our enthusiasm for telling you all that biology has discovered, do not let us convince you that all is known. Many questions are still unanswered. More important, many good questions have not yet been asked. Perhaps you may be the one to ask them.

You may have been persuaded to study biology because of the environmental problems now confronting us or because of a desire to know more about the mechanisms of your own body or an interest in the "green revolution" or genetic engineering or a career in medicine—in short, because it is "relevant." The study of biology is, indeed, pertinent to many aspects of our day-to-day existence, but do not make this your main reason for the study of biology. Above all other considerations, study biology because it is "irrelevant"—that is, study it for its own sake, because, like art and music and literature, it is an adventure for the mind and nourishment for the spirit.

QUESTIONS

1. What is the essential difference between Darwin's theory of evolution and that of Lamarck?

2. Lyell, like Darwin, was a collector, especially of insects. In what way might this pursuit have affected the formulation of the theory of evolution?

3. The chief predator of an English species of snail is the song thrush. Snails that inhabit woodland floors have dark shells, whereas those that live on grass have yellow shells, which are less clearly visible against the lighter background. Explain, in terms of Darwinian principles.

4. The phrase "chance and necessity" has been applied to the process by which species develop. Relate this to the fact that snails living on grass do not have green shells but there are, for example, green frogs and green insects.

5. In what ways are the organisms in Figure I–2 adapted to their environment?

SUGGESTIONS FOR FURTHER READING

BATES, MARSTON, and PHILIP S. HUMPHREY, eds.: *The Darwin Reader*, Charles Scribner's Sons, New York, 1956.*

A collection of Darwin's writings, including The Autobiography, *and excerpts from* The Voyage of the Beagle, The Origin of Species, The Descent of Man, *and* The Expression of the Emotions. *Darwin was a fine writer, and you can discover here the wide range of his interests and concerns at different periods of his life.*

BRONOWSKI, J.: *The Ascent of Man*, Little, Brown and Company, Boston, 1973.*

An informal and illuminating history of the sciences, originally prepared as a television series. The emphasis is on science's relation to human culture. Well designed and illustrated.

* Available in paperback.

DARWIN, CHARLES: *The Origin of Species by Means of Natural Selection, or The Preservation of Favored Races in the Struggle for Life,* Doubleday & Company, Inc., Garden City, N.Y., 1960.*

Darwin's "long argument." Every student of biology should, at the very least, browse through this book to catch its special flavor and to begin to understand its extraordinary force.

DARWIN, CHARLES: *The Voyage of the Beagle,* Natural History Press, Garden City, N.Y., 1962.*

Darwin's own chronicle of the expedition on which he made the discoveries and observations that eventually led him to his theory of evolution. The sensitive, eager young Darwin that emerges from these pages is very unlike the image many of us have formed of him from his later portraits.

MOOREHEAD, ALAN: *Darwin and the Beagle,* Harper & Row, Publishers, Inc., New York, 1969.

A delightful narrative of Darwin's journey, beautifully illustrated with contemporary or near contemporary drawings, paintings, and lithographs.

TOULMIN, STEPHEN, and JUNE GOODFIELD: *The Discovery of Time,* Harper & Row, Publishers, Inc., New York, 1965.*

The historical development of our concepts of time as they relate to nature, human nature, and human society.

---

* Available in paperback.

PART I

Biology of Cells

SECTION 1

THE UNITY OF LIFE

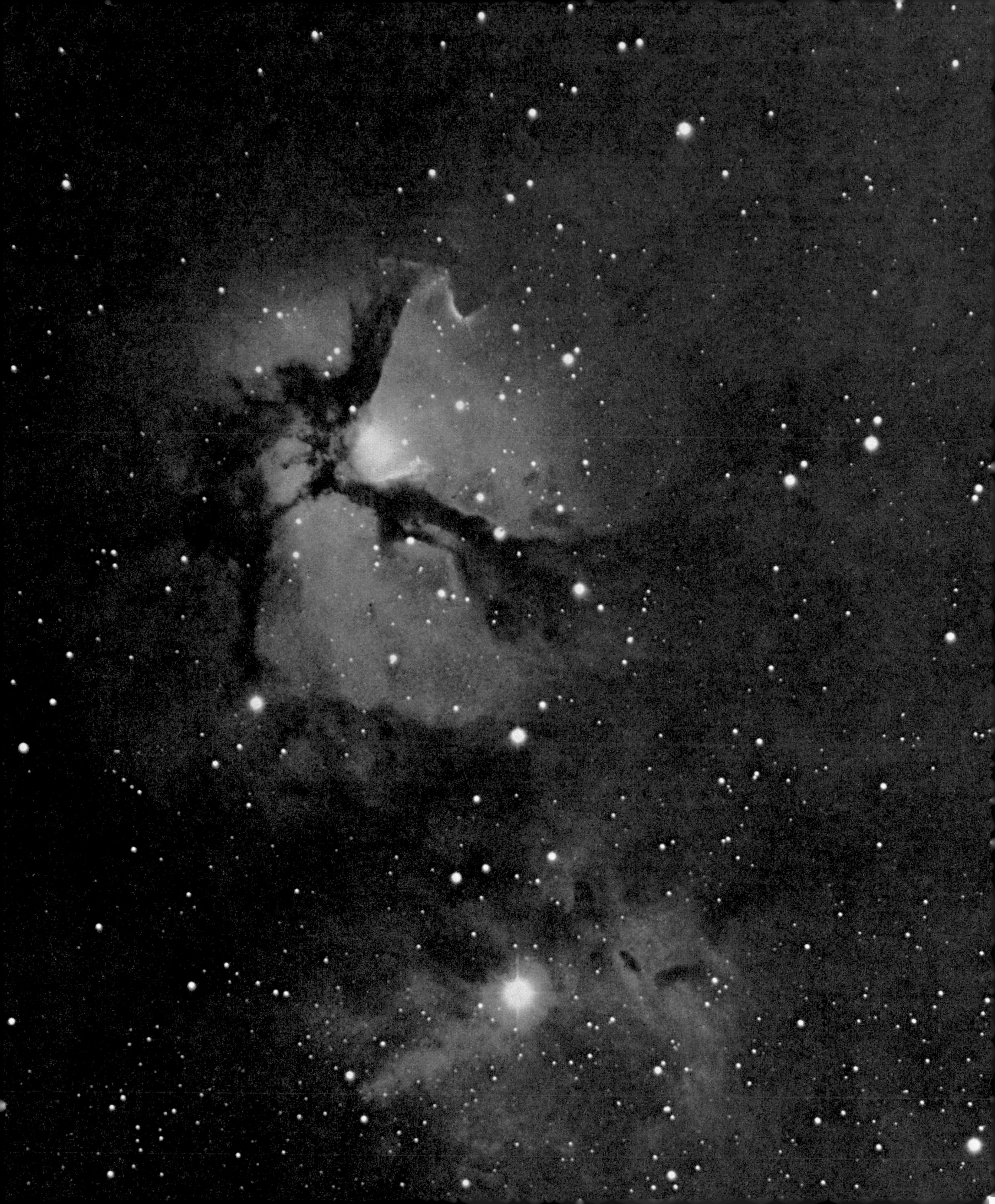

CHAPTER 1 *Atoms and Molecules*

The universe began, astronomers tell us, with an explosion that filled all space, with every particle of matter hurled away from every other particle. The temperature at the time of the explosion—some 20 billion years ago—was about 100,000,000,000 degrees Celsius (10^{11} °C). At this temperature, not even atoms could hold together; all matter was in the form of subatomic, elementary particles. Moving at enormous velocities, even these particles had fleeting lives. Colliding with great force, they annihilated one another, creating new particles and releasing more energy.

As the universe cooled, two types of stable particles, previously present only in relatively small amounts, began to assemble themselves. (By this time, 100,000 years after the "big bang" is believed to have taken place, the temperature had dropped to a mere 2500°C, about the temperature of a white-hot wire in an incandescent light bulb.) These particles—protons and neutrons—are very heavy as subatomic particles go. Held together by forces that are still little understood, they formed the central core, or nuclei, of atoms.

These nuclei, with their positively charged protons, attracted small, light, negatively charged particles—electrons—which moved rapidly around them. Thus, atoms came into being. (Perhaps atoms existed before the explosion, but how will one ever know?)

According to current theory, it is from these atoms—blown apart, formed, and re-formed over 20 billion years—that all the stars and planets of the universe are formed, including our particular star and planet. And it is from the atoms present on this planet that living systems assembled themselves and evolved. Each atom in our own bodies had its origins in this explosion of some 20 billion years ago. You and I are flesh and blood, but we are also stardust.

This text begins where life begins, with the atom. At first, the universe aside, it might appear that lifeless atoms have little to do with biology. Bear with us, however. A closer look reveals that the activities we associate with being alive depend on combinations and exchanges between atoms, and the force that binds the electron to the atomic nucleus stores the energy that drives living systems.

1-1
Clouds of gases and dust in the constellation Sagittarius. The stars and planets of the universe have their origins in nebulae such as these.

THE SIGNS OF LIFE

What do we mean when we speak of "the evolution of life," or "life on other planets," or "when life begins"? Actually, there is no simple definition. Life does not exist in the abstract; there is no "life," only living things. We recognize a system as being alive when it has certain properties that are more common among animate objects than inanimate ones. Let us take a look at some of these properties.

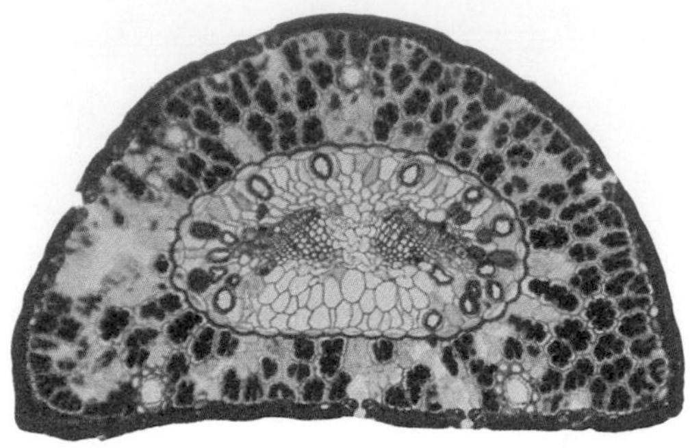

Living organisms are highly organized, as in this cross section of a pine needle. It reflects the complicated organization of many different kinds of atoms into molecules and of molecules into complex structures. Such complexity of form is never found in inanimate objects.

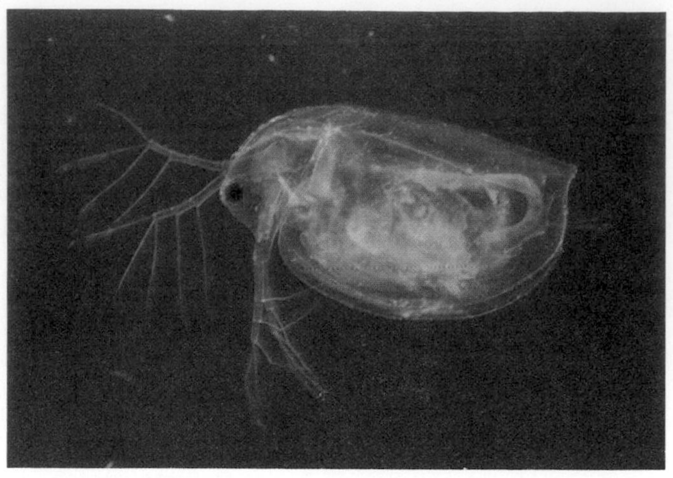

Living things are homeostatic, which means simply "staying the same." Even though they constantly exchange materials with the outside world, they maintain a relatively stable internal environment quite unlike that of their surroundings. Even this tiny, apparently fragile animal, a water flea, has a constant chemical composition that differs from its changing environment.

Living organisms take energy from the environment and change it from one form to another. They are highly specialized at energy conversion. Here a saw-whet owl is converting chemical energy to kinetic energy, thereby procuring a new source of chemical energy, in this case, a white-footed mouse.

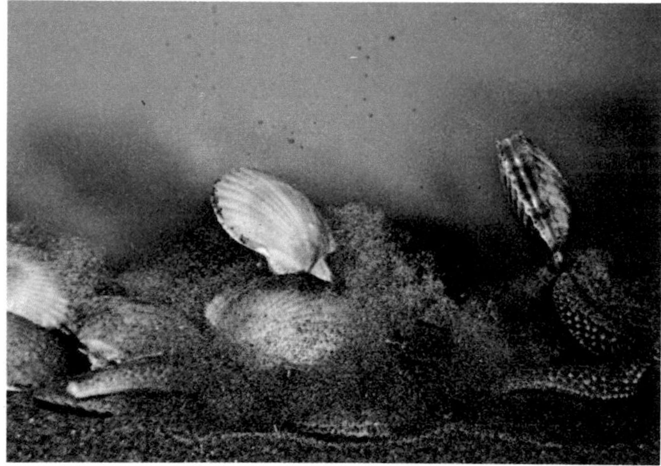

Living things respond to stimuli. Here scallops, sensing an approaching starfish, leap to safety.

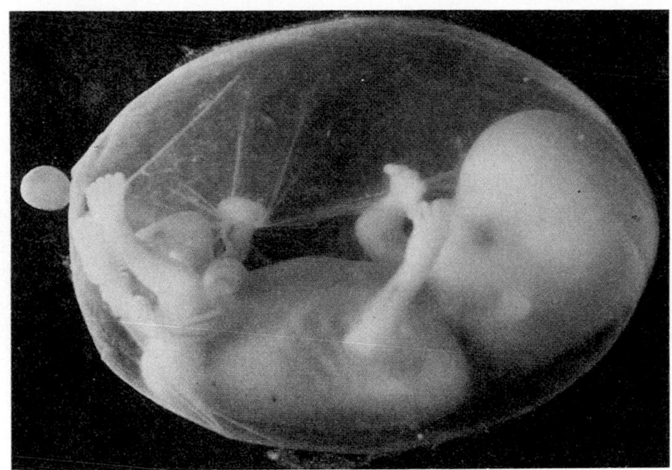

Living things grow and develop. Growth and development are the processes by which, for example, a single living cell, the fertilized egg, becomes a tree, or an elephant, or as shown here, a human infant.

Living things are adapted. Moles, for instance, live underground in tunnels shoveled out by their large forepaws. Their eyes are small and almost sightless. Their noses, with which they sense the worms and other small invertebrates that make up their diet, are fleshy and enlarged.

Living things reproduce themselves. They make more of themselves, copy after copy after copy, with astonishing fidelity (and yet, as we shall see, with just enough variation to provide the raw material for evolution).

Table 1-1 *Atomic Numbers and Weights of Some Common Atoms*

ATOM	ATOMIC NUMBER	ATOMIC WEIGHT*
Hydrogen (H)	1	1
Helium (He)	2	4
Carbon (C)	6	12
Nitrogen (N)	7	14
Oxygen (O)	8	16
Sodium (Na)	11	23
Phosphorus (P)	15	30
Sulfur (S)	16	32
Chlorine (Cl)	17	35
Calcium (Ca)	20	40

* For most common isotope.

ATOMS

There are 92 naturally occurring chemical elements, each differing from the others in the structure of its atoms. Each different type of atom has a different number of protons in its nucleus, ranging from the lightest, hydrogen, which has 1 proton, to the heaviest, uranium, which has 92. The number of protons in the nucleus of a particular atom is called the *atomic number*. Outside the nucleus of the atom are electrons, attracted by the positive charge of the protons. The electrons determine the chemical properties of atoms, and chemical reactions involve changes in the numbers and relative positions of these electrons.

Atoms also contain neutrons, which are uncharged particles about the same weight as protons. These, too, are found in the nucleus of the atom, where they seem to have a stabilizing effect. The *atomic weight* of an element is essentially equal to the number of protons and neutrons in the nucleus. (Electrons are so light by comparison that their weight is usually disregarded. When you weigh yourself, only about 30 grams—approximately 1 ounce—of your total weight is made up of electrons.)

Isotopes

Atoms with the same number of protons but different numbers of neutrons are known as *isotopes*; they differ from one another in their atomic weights but not in their atomic numbers. For example, the common form of hydrogen, with its one proton, has an atomic weight of 1 and is symbolized as ^{1}H, or simply H. Deuterium, ^{2}H, is an isotope of hydrogen that contains one proton and one neutron and so has an atomic weight of 2. Tritium, ^{3}H, the third isotope of hydrogen, has one proton and two neutrons and so has an atomic weight of 3. Like many, but not all of the less common isotopes, tritium is radioactive, which means that its nucleus is unstable and emits energy when changing into a more stable form.

The next heaviest atom is helium (He), which has an atomic number of 2 (two protons) and an atomic weight of 4 (two protons and two neutrons). Hydrogen, deuterium, and tritium are all similar in their chemical properties because each can attract only a single electron, but helium with its two electrons (attracted by its two protons) is very different chemically from any of the isotopes of hydrogen. Thus, to repeat, the protons determine the number of electrons attracted to an atomic nucleus and those electrons determine the chemical properties of the atom.

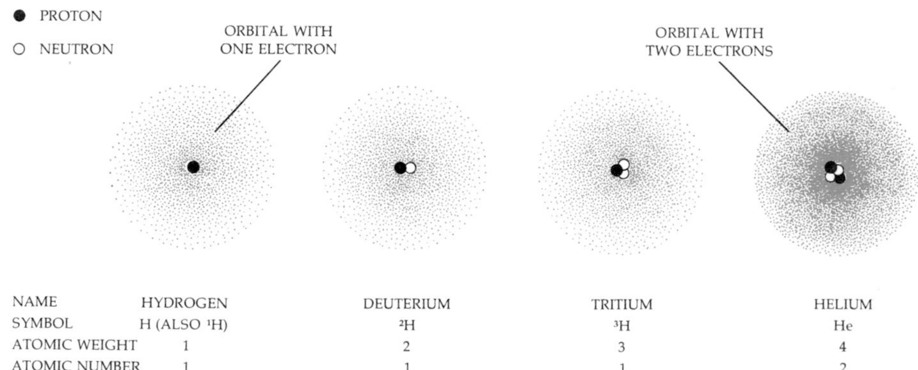

1-2
Diagrams of atoms of hydrogen, deuterium, tritium, and helium. Note that hydrogen, deuterium, and tritium each have only a single proton and a single electron, and so they behave similarly chemically although they differ in the number of neutrons. Helium, which has one more proton and one more electron, is chemically very different from hydrogen and its isotopes.

1–3
Models of the carbon atom: (a) *a planetary model and* (b) *a Bohr model.*

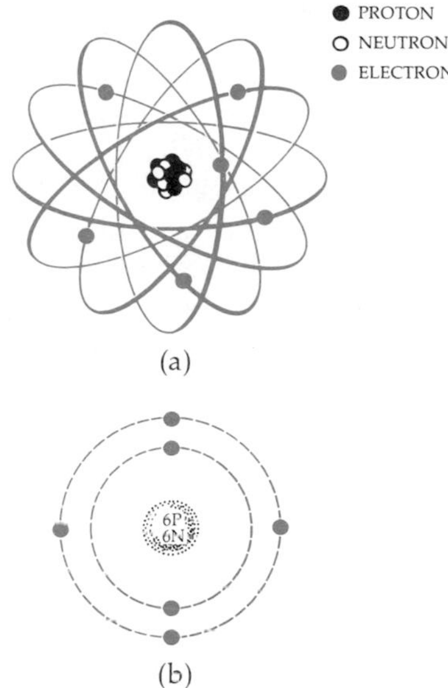

Electrons and Orbitals

The concept of the atom as the indivisible unit of the chemical elements is almost 200 years old; however, our ideas about its structure have undergone many changes. These ideas, or hypotheses, are usually presented in the forms of models, as are many scientific hypotheses.

The earliest model, emphasizing the indivisibility of the atom, resembled a billiard ball. As it came to be realized that electrons could be removed from the atom, the billiard ball gave way to the plum-pudding model, in which the atom was represented as a solid, positively charged mass with negatively charged particles, the electrons, embedded in it. Subsequently, however, physicists found that an atom is, in fact, mostly open space. The distance from electron to nucleus, experiments indicated, is about 1,000 times the diameter of the nucleus; the electrons are so exceedingly small that the space is almost entirely empty. Thus the more familiar planetary model of the atom came into being, in which the electrons were depicted as moving in orbits around the nucleus (Figure 1–3a). More recently, the Bohr model (named after physicist Niels Bohr) became the most popular one (Figure 1–3b). It emphasizes the fact that electrons are found outside the nucleus at different energy levels; this concept of energy levels is, as you will see, of great importance in the chemistry of living systems.

The current model of electron configurations is quite different from all previous ones and more accurate in terms of what we know about the behavior of electrons. According to this model, the electron moves unpredictably around the nucleus, and its position at any given moment cannot be known with certainty. For convenience, its pattern of motion is defined as the volume in which the electron will be found 90 percent of the time. This volume is known as the electron's *orbital*. Each orbital can hold a maximum of two electrons.

Orbitals vary in shape. The first two electrons occupy a single spherical orbital. (Thus, for instance, hydrogen's single electron moves about the nucleus—90 percent of the time—within a single spherical orbital, and so do the two electrons of helium.) This single spherical orbital, with its maximum of two electrons, makes up the first energy level. At the second energy level, there are four orbitals, each of which, as we noted, can hold two electrons. One of these orbitals is spherical, and the other three are dumbbell-shaped. The axes of the dumbbell orbitals are perpendicular to one another (Figure 1–4). The spherical orbitals are filled first, and then the dumbbell-shaped ones. The second energy level can hold a total of eight electrons, and so can the third.

1–4
Orbital models. (a) *The two electrons at the first energy level of an atom occupy a single, spherical orbital.* (b) *At the second energy level, there are four orbitals, each containing two electrons. One of these orbitals is spherical and the other three are dumbbell-shaped. The axes of the dumbbell-shaped orbitals (indicated by the lines) are perpendicular to one another. The nucleus is at the intersection of the axes.*

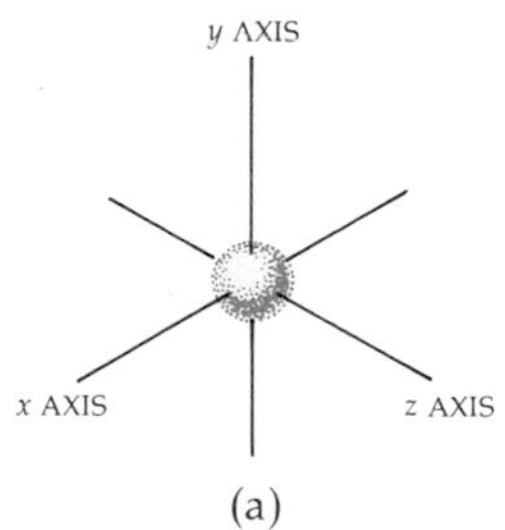

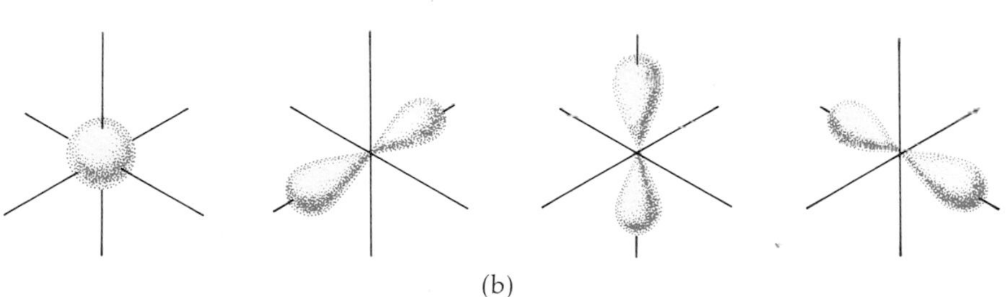

Atoms tend to complete their energy levels, and the chemical behavior of atoms is governed by this tendency. For instance, helium (atomic number 2), neon (atomic number 10), and argon (atomic number 18) all have completely filled outer energy levels and so tend to be unreactive; they are thus called the "noble" gases because of their disdain for reacting with other elements. Hydrogen (atomic number 1), lithium (atomic number 3), sodium (atomic number 11), and potassium (atomic number 19) each has a single electron in its outermost energy level, and each tends to lose this electron. As a consequence of such a loss, each has one more proton than electron and therefore acquires a positive charge: H^+, Li^+, Na^+, K^+. Fluorine and chlorine, by contrast, with atomic numbers of 9 and 17, tend to gain an electron in order to complete an outer energy level and so become negatively charged: F^- and Cl^-. Similarly, an atom with two electrons in its outer energy level may lose both of them, acquiring a double positive charge. Magnesium (atomic number 12) and calcium (atomic number 20) become Mg^{2+} and Ca^{2+}, and so on. Such charged atoms are known as *ions*.

Ions make up less than 1 percent of the weight of most living matter, but they play crucial roles. K^+ is the principal positively charged ion in most cells, and many essential biological reactions do not proceed in its absence. Both Na^+ and K^+ are involved in the production and propagation of the nerve impulse. Ca^{2+} is required for the contraction of muscles and Mg^{2+} forms a part of the chlorophyll molecule, the molecule that traps the energy from the sun. Some of the roles of ions in living systems will be discussed in future chapters.

Electrons and Energy

As we noted previously, electrons, which are negatively charged, are attracted to the atomic nucleus because of the positive charge of the protons. The orbital occupied by an electron is determined by the amount of energy of the electron. This energy is in the form of potential energy. An analogy may be useful. A rock on flat ground may be said to have no energy. If you push it up a hill, you give it energy—potential energy. So long as it sits on the peak of the hill, it neither gains nor loses energy. If it rolls down the hill, however, it loses its potential energy as it rolls back toward its original level ground. Similarly, water that has been pumped up to a water tank for storage has potential energy that will be released when the water runs back down.

The electron is like the boulder, or the water, in that an input of energy can raise it to a higher energy level—farther away from the nucleus. As long as it remains at

1–5
(a) *The energy used to push a boulder to the top of a hill (less the heat energy produced by friction between the boulder and hill) becomes potential energy, stored in the boulder as it rests at the top of the hill. This potential energy is converted to kinetic energy (energy of motion) as the boulder rolls downhill.* (b) *When an input of energy—such as light energy—boosts an electron to a high energy level, the electron, like the boulder, possesses potential energy that is released when the electron returns to its previous energy level.*

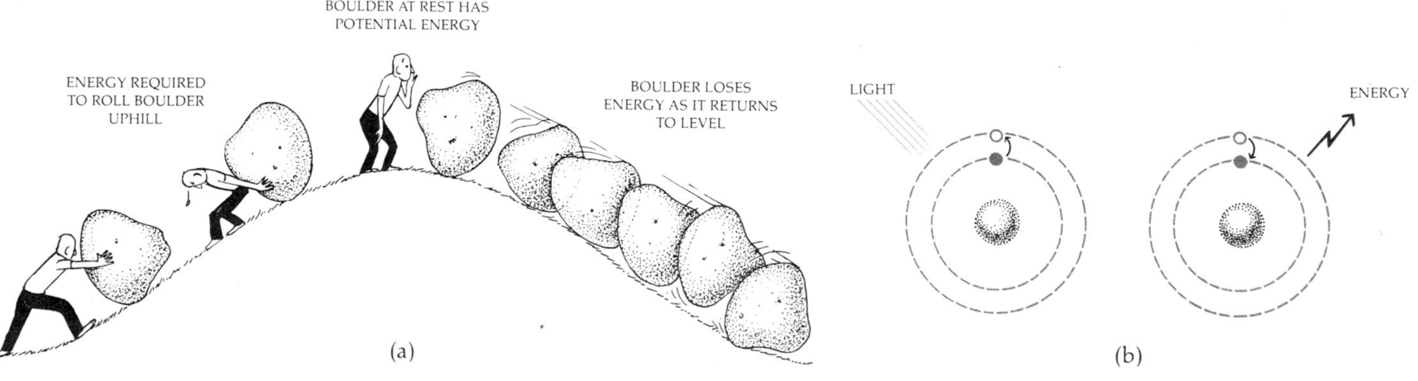

this higher level, it possesses the added energy. Also, like the boulder or water, the electron tends to go to its lowest possible energy level, just as the rock rolls downhill, and not up.

In a given atom, the first spherical orbital is the lowest energy level—the ground state. The four orbitals of the second level are occupied by electrons with more energy, and so on. It takes energy to move a negatively charged electron farther away from the positively charged nucleus, just as it takes energy to position a boulder at the top of a hill. However, unlike the boulder on the hill, the electron cannot be pushed part way up. With an input of energy, electrons may move from a lower energy level to any one of several higher ones, but they cannot move to somewhere in between. For an electron to move from one level to another, the atom must absorb a discrete packet of energy, known as a quantum, which contains just the precise energy needed for the transition and no more or no less. Thus the study of electron movements is known as quantum mechanics, and the term quantum jump, which has invaded our everyday discourse, refers to an abrupt, discontinuous movement from one level to another.

Electronegativity

Table 1–2 *Electronegativity of Some Common Atoms (On a Relative Scale of 0 to 4)*

Oxygen (O)	3.5
Nitrogen (N)	3.0
Chlorine (Cl)	3.0
Carbon (C)	2.5
Sulfur (S)	2.5
Hydrogen (H)	2.1
Phosphorus (P)	2.1
Sodium (Na)	0.9

For the atoms of any particular element, all the electrons at any given energy level have about equal amounts of energy. However, the atomic nuclei of different elements have various degrees of attraction for electrons. The strength of the attraction depends on the number of protons in the nucleus, the closeness of the electrons to the nucleus, and the number of electrons outside the nucleus. The affinity of an atom for electrons is called *electronegativity*. Electronegativity is expressed on a scale of 0 to 4. Helium and the other unreactive noble gases have electronegativities of 0. At the other end of the scale is fluorine with an electronegativity of 4. Oxygen, the next most electronegative, is 3.5. Electronegativity values for some of the elements are given in Table 1–2.

When an electron moves from an atom that is less electronegative to one that is more electronegative, it moves downhill, energetically speaking, and energy is released, as it is released when the boulder moves downhill.

In the green cells of plants and algae, the radiant energy of sunlight raises electrons to a higher energy level. In the course of a series of electron-transfer reactions, which will be described in Section 2, the electrons are passed slowly downhill until they are finally accepted by oxygen. During this transition, the radiant energy of sunlight is changed to the chemical energy on which all life on earth depends.

1–6
The green in the leaves of these corn plants is the green of the chlorophyll molecule. When a particle of light—a photon—strikes a molecule of chlorophyll, electrons in the molecule are raised to higher energy levels. As they return to their previous energy level, the energy they release is captured in the bonds of carbon-containing compounds.

1–7
Sodium (Na), which has only one electron in its outer energy level, becomes stable if it loses the electron. Chlorine (Cl), which has seven electrons in its outer energy level, becomes stable if it gains one. When sodium and chlorine interact, sodium loses its single electron and chlorine gains it; following this transaction, sodium has a positive charge and chlorine a negative one. Such charged atoms are called ions. (a) *Oppositely charged ions attract one another. Table salt is crystalline NaCl, a latticework of alternating Na^+ and Cl^- ions held together by their opposite charges. Such bonds between oppositely charged ions are known as ionic bonds.* (b) *The regularity of the latticework is reflected in the structure of salt crystals.*

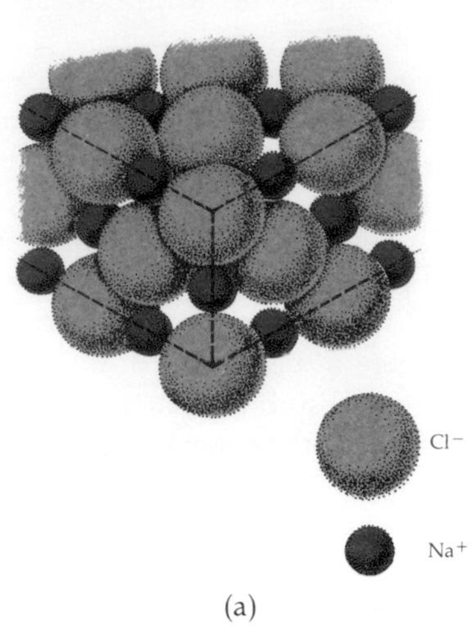

(a)

(b)

BONDS AND MOLECULES

The ways in which atoms combine to form molecules are dictated by the number and arrangement of their electrons. The forces that hold molecules together are known as *bonds*. There are two principal types of bonds: ionic and covalent.

Ionic Bonds

Positive and negative ions attract one another. Thus the sodium ion (Na^+) with its single positive charge is attracted to the chloride ion (Cl^-) with its single negative charge. The calcium ion (Ca^{2+}) with its double positive charge can attract and hold two Cl^- ions, forming $CaCl_2$, with the subscript 2 indicating that two atoms of chlorine are present for every one atom of calcium.* Such bonds, involving the mutual attraction of ions of opposite charge, are known as *ionic bonds.*

Covalent Bonds

Another way for an atom to complete its outer energy level is by sharing electrons with another atom. Bonds formed by shared pairs of electrons are known as *covalent bonds.* In a covalent bond, the shared pair of electrons forms a new orbital that encompasses both atoms. In such a bond, each electron spends part of its time around one nucleus and part of its time around the other. Thus the electron sharing both completes the outer energy level and neutralizes the nuclear charge. Covalent bonds are stronger and more stable than ionic bonds.

Carbon, which has an atomic number of 6, needs either to gain or lose four electrons to achieve a filled and therefore stable outer energy level, and so is able to form covalent bonds with as many as four other atoms. When these bonds are formed, the electron pairs assume new orbitals. These orbitals, as shown in Figure

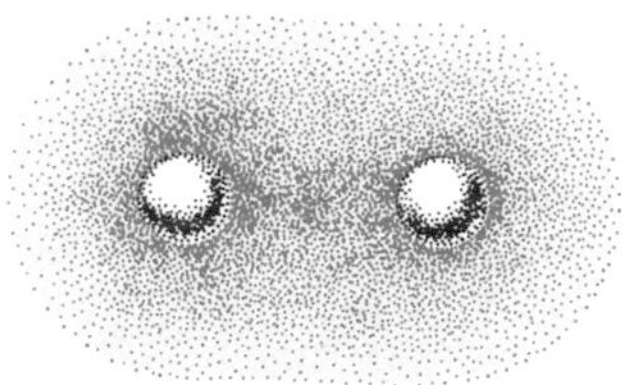

1–8
In a molecule of hydrogen, each atom shares its single first orbital electron with the other atom. As a result, both atoms effectively have a filled first orbital, containing two electrons—a highly stable configuration. The orbital shown here—a molecular orbital—indicates the volume in which the two electrons can be found 90 percent of the time.

* The "combining power" of an element is called its valence. It is a number that indicates how many electrons a particular atom will readily give up or acquire in order to stabilize its electron configuration. Sodium has a valence of 1+, calcium of 2+, chlorine of 1−. Thus, calcium will combine with chlorine in a ratio of 1 to 2.

1-9
(a) When carbon forms covalent bonds with four other atoms, new orbitals are formed. These new orbitals, which are all the same shape, are oriented toward the four corners of a tetrahedron. Thus the four orbitals are separated as far as possible. (b) When a carbon atom reacts with hydrogen to form methane, the four unpaired electrons of the carbon atom form pairs with each of the unpaired electrons in four hydrogen atoms, producing a methane molecule (c). Each pair of electrons moves in a new, molecular orbital. The molecule has the shape of a tetrahedron.

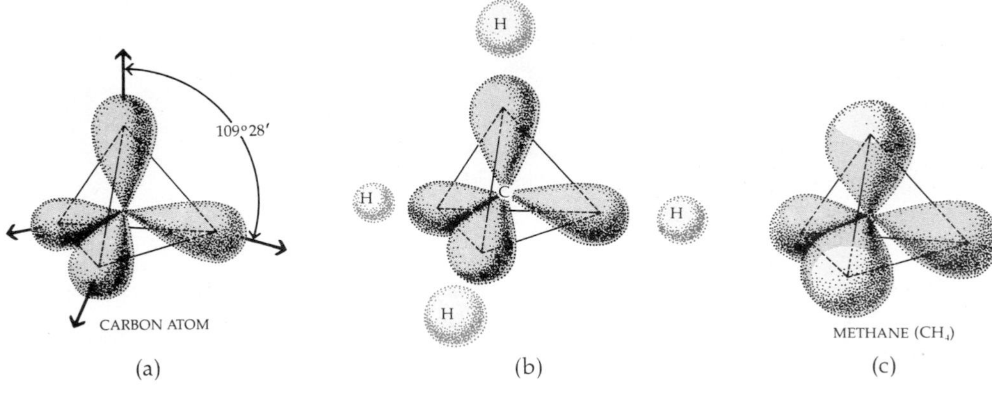

1-9, are distributed symmetrically in space, forming a regular tetrahedron. Carbon's capacity to form four covalent bonds is crucial to its central role in the chemistry of living systems.

Double and Triple Bonds

There are various ways in which atoms can participate in covalent bonds and satisfy their valence requirements. Oxygen, with two unpaired electrons in its outer electron orbitals, has a valence of two and can thus form a simple compound with carbon, carbon dioxide (CO_2), in which each oxygen atom shares two pairs of electrons with a central carbon atom. Such bonds in which two atoms are held together by two pairs of electrons (four electrons) are called double bonds. They are symbolized in a structural formula by two lines connecting the atomic symbols: C=C. Carbon atoms can form double or even triple bonds (in which three pairs of electrons are shared) with each other as well as with other atoms, and so the variety of kinds of molecules that carbon can form is very large.

Electrons shared in double and triple bonds form orbitals that differ in shape from the orbitals filled by single electron pairs. For instance, when four bonds satisfy the electron requirements of carbon, they will be directed toward the four corners of a tetrahedron that has the carbon atom at its center, as we saw in Figure 1-9. When two bonds are replaced by a double bond, the remaining single bonds form the arms of a Y with the double bond as its leg (Figure 1-10). When two double bonds are made by a single carbon atom, as in carbon dioxide, the three bonded atoms lie in a straight line.

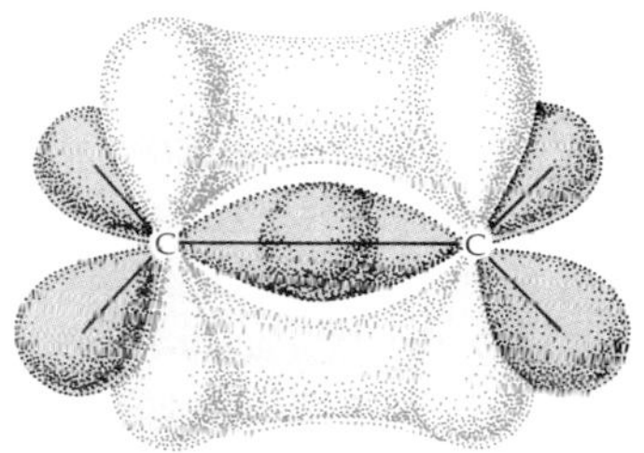

1-10
An orbital model of a carbon-carbon double bond. One pair of electrons occupies the inner orbital between the two carbon atoms. The other pair of electrons occupies the outer orbital, which has two phases, one above the plane and one below. This creates a rigid bond, about which the atoms cannot rotate. The two smaller orbitals extending from each carbon atom each contain one electron and can form covalent bonds with other atoms.

Single bonds are flexible, leaving atoms free to rotate in relation to one another. Double and triple bonds hold the atoms relatively rigid in relation to one another. The presence of double bonds in a molecule can make a significant difference in its properties. For example, both fats and oils are composed of carbon and hydrogen atoms covalently bonded to one another, but in oils some of the bonds are double, and in fats the bonds are all single. As a consequence, fats are solid at room temperature and oils are liquid.

LIVING SYSTEMS OBEY THE LAWS OF PHYSICS AND CHEMISTRY

Notice that by beginning the study of biology with an examination of how atoms and molecules are put together and how they interact, we are making a very fundamental assumption. We are saying, in effect, that living things are made up of the same chemical and physical components as nonliving things and that they obey the same chemical and physical laws. This is, in fact, one of the major ordering principles in the study of biology. It is a comparatively new concept. In fact, until fairly recently, many prominent biologists believed that living systems are qualitatively different from nonliving ones, containing within them a "vital spirit" that enables them to perform activities that cannot be carried on outside the living organism. This concept is known as vitalism and its proponents as vitalists. (Do not dismiss this "foolish" idea too rapidly; you may discover you too are a vitalist.)

In the seventeenth century, the vitalists were opposed by a group known as the mechanists. The French philosopher René Descartes (1596–1650) was a leading proponent of this group. The mechanists set about proving that the body worked essentially like a machine; the arms and legs move like levers, the heart like a pump, the lungs like a bellows, and the stomach like a mortar and pestle. By the nineteenth century, such simple mechanical models of living organisms had been abandoned, and the argument now centered on whether or not the chemistry of living organisms was governed by the same principles as the chemistry performed in the laboratory. The vitalists claimed that the chemical operations performed by living tissues could not be carried out experimentally in the laboratory, categorizing reactions as either "chemical" or "vital." The reductionists, as their opponents were now called (since they believed that the complex operations of living systems could be reduced to simpler and more readily understandable ones), achieved a partial victory when the German chemist Friedrich Wöhler (1800–1882) converted an "inorganic" substance (ammonium cyanate) into a familiar organic substance (urea). On the other hand, the claims of the vitalists were supported by the fact that, as chemical knowledge improved, many new compounds were found in living tissues that were never seen in the nonliving, or inorganic, world.

In the late 1800s, the leading vitalist was Louis Pasteur, who claimed that the marvelous changes that took place when fruit juice was transformed to wine were "vital" and could be carried out only by living cells—the cells of yeast. In spite of many advances in chemistry, this phase of the controversy lasted until almost the turn of the century.

This laboratory at Giessen, where Wöhler was a student, was one of the first where practical work in chemistry could be done.

However, in 1898, the German chemists Eduard and Hans Büchner showed that a substance extracted from the yeast cells could produce fermentation outside the living cell. (This substance was given the name enzyme, from zyme, *the Greek word meaning "yeast" or "ferment.") A "vital" reaction was proved to be a chemical one, and the subject was eventually laid to rest. Today it is generally accepted that living systems "obey the rules" of chemistry and physics, and modern biologists no longer believe in a "vital principle."*

Perhaps the greatest test for this principle of modern biology came about 25 years ago. As you know, one of the outstanding characteristics of living things is their capacity to reproduce, to generate faithful copies of themselves. By about 1950, this capacity had been shown to reside in a single type of chemical molecule, deoxyribonucleic acid (DNA). The race to discover the structure of this molecule began, and the question in everyone's mind was whether or not the structure of this one "simple" molecule could possibly explain the centuries-old question of the mysteries of heredity. As it turned out, and as we shall discuss further in this text, it could.

The next test for this concept seems to lie in the mechanisms of the brain. Even those of us who now willingly admit that the nature of heredity—so long a mystery—resides in a single molecule, an evolutionary accumulation of atoms organized in a particular way, are perhaps less willing to admit that our dreams and our ideals and our deepest emotions are based on "nothing but" the chemistry and organization of our brain cells.

1-11
In a polar molecule, such as hydrogen chloride (HCl), the shared electrons tend to spend more time around the more electronegative atom, in this case, the chlorine atom. As a result, the more electronegative atom (chlorine) has a slightly negative charge, and the less electronegative atom (hydrogen) has a slightly positive charge.

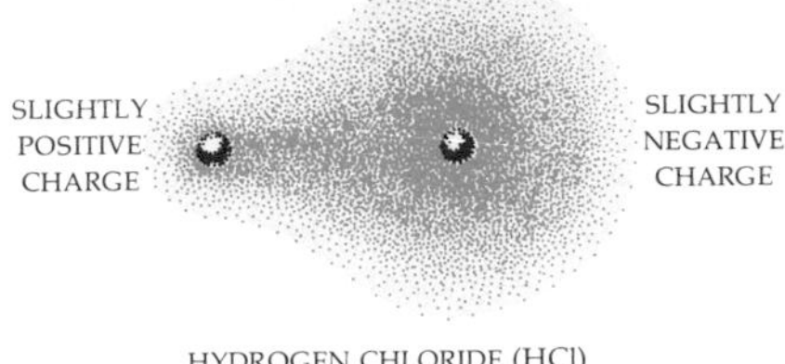

HYDROGEN CHLORIDE (HCl)

Polar Covalent Bonds

The electrons in covalent bonds are not always shared equally between the two atoms involved. Some atoms have a greater attractive force for the electrons than others—that is, they are more electronegative. As a consequence of this differential, the shared electrons tend to spend more time around the more electronegative atom. As a result, the more electronegative atom in the molecule will have a slightly negative charge, and the less electronegative atom will have a slightly positive one, since its nuclear charge is not entirely neutralized. The polar properties of some covalent bonds have very important consequences for living things. For example, many of the special properties of water, upon which life depends, derive largely from its polar nature, as we shall see in the next chapter.

Ionic, polar covalent, and covalent bonds actually may be considered different versions of the same type of bond. The differences depend on the differences in electronegativity among the combining atoms. In a wholly nonpolar covalent bond, the electrons are shared equally; such bonds can exist only between identical atoms, H_2, Cl_2, O_2, and N_2, for example. In polar covalent bonds, electrons are shared unequally, and in ionic bonds, there is an electrostatic attraction between the negatively and positively charged ions as a result of their having gained or lost electrons.

Chemical Equations

Chemical reactions—exchanges of electrons among atoms—can be compactly described by chemical equations. The equation for the formation of sodium chloride is

$$Na^+ + Cl^- \longrightarrow NaCl$$

The arrow in the equation designates "forms" or "yields" and shows the direction of chemical change. Like algebraic equations, chemical equations "balance"; that is, there are the same number of atoms in the products of the reaction as in the original reactants. For instance, hydrogen gas can combine with oxygen gas to produce water. Hydrogen gas is H_2, and oxygen gas is O_2. However, we know that each molecule of water contains two atoms of hydrogen and one of oxygen, and therefore the proportions must be two to one:

$$2H_2 + O_2 \longrightarrow 2H_2O$$

Two molecules of H_2 plus one molecule of O_2 yield two molecules of water. The equation for a chemical reaction thus tells the kinds of atoms that are present, their proportions, and the direction of the reaction.

Table 1-3 *Atomic Composition of Three Representative Organisms*

ELEMENT	MAN	ALFALFA	BACTERIUM
Carbon	19.37	11.34	12.14
Hydrogen	9.31	8.72	9.94
Nitrogen	5.14	0.83	3.04
Oxygen	62.81	77.90	73.68
Phosphorus	0.63	0.71	0.60
Sulfur	0.64	0.10	0.32
CHNOPS total:	97.90	99.60	99.72

THE BIOLOGICALLY IMPORTANT ELEMENTS

Of the 92 naturally occurring elements, only six make up some 99 percent of all living tissue (Table 1-3). These six are carbon, hydrogen, nitrogen, oxygen, phosphorus, and sulfur, conveniently remembered as CHNOPS. These are not the most abundant of the elements of the earth's surface. Why, as life assembled itself from stardust, were these chosen? One clue is that they are all small, light atoms and such atoms, in which electrons are held more closely to the nucleus, form tighter, more stable bonds. Another is that they all form covalent bonds; and finally, with the exception of hydrogen, they can all form bonds with two or more atoms.

In Chapter 2, we shall be looking at one particular combination of these atoms—water.

SUMMARY

Living and nonliving matter are composed of atoms, the smallest units of the chemical elements. Atoms are made up of smaller particles. The nucleus of an atom contains protons, positively charged particles, and (except for hydrogen) neutrons, which have no charge. The atomic number of an atom is equal to the number of protons in its nucleus. The atomic weight of an atom is the sum of the number of protons and neutrons in its nucleus. The chemical properties of an atom are determined by the configurations of its electrons—small, negatively charged particles found outside the nucleus.

Electrons move about the nucleus in orbitals, which are defined as the volume in which the electron is found 90 percent of the time. Only two electrons can occupy any given orbital. The number of electrons in an electrically neutral atom is equal to the number of protons and thus to the atomic number. Electron orbitals are at various energy levels. The first orbital, containing two electrons, is spherical and is the only orbital at the first energy level. There are four orbitals at the second energy level, with a total of eight electrons. One of these orbitals is spherical and three are shaped like dumbbells. Larger atoms have additional energy levels with electrons in additional orbitals.

Electrons at the lower energy levels have less energy than those at higher energy levels; hence the lower levels are always filled first (just as water tends to run downhill). With an input of energy, an electron may be boosted to a higher energy level. In doing so it absorbs a discrete unit—a quantum—of energy. When an electron falls to a lower energy level, energy is released. Different kinds of atoms differ in the strength with which the nucleus attracts electrons. This strength is expressed in terms of electronegativity. Oxygen is a highly electronegative atom that tends to gain electrons from other atoms.

Atoms that have two electrons at the first energy level, or eight at subsequent energy levels, are in a stable state. Chemical reactions between atoms result from the tendency of atoms to reach this stable outer electron configuration. Two common types of bonds are covalent and ionic. In covalent bonds, pairs of electrons are shared equally by atoms; in polar covalent bonds, pairs of electrons are shared unequally, giving the molecule areas of positive and negative charge (polarity). Ionic bonds are formed by the mutual attraction of oppositely charged ions. Covalent bonds are strong, stable, and of great importance in living systems.

QUESTIONS

1. Distinguish between the following: neutron/proton; deuterium/tritium; atomic weight/atomic number; ionic bond/covalent bond; orbital/energy level.

2. Determine the number of protons, the number of neutrons, the number of energy levels, and the valence in each of the following atoms: oxygen, nitrogen, carbon, sulfur, phosphorus, chlorine, calcium.

3. Magnesium has an atomic number of 12. How would you expect magnesium and chlorine to interact? Write the formula for magnesium chloride.

4. Knowing that chemical reactions have to be balanced, fill the appropriate numbers into the underlined spaces (*hint:* from 1 to 6 in all cases):

(a) ___H_2CO_3 (CARBONIC ACID) $\longrightarrow$ ___H_2O + ___CO_2

(b) ___H_2 + ___N_2 $\longrightarrow$ ___NH_3 (AMMONIA)

(c) ___$NaOH$ (SODIUM HYDROXIDE) + ___H_2CO_3 $\longrightarrow$ ___Na_2CO_3 + ___H_2O

(d) ___CH_3OH (METHYL ALCOHOL) + ___O_2 $\longrightarrow$ ___CO_2 + ___H_2O

(e) ___O_2 + ___$C_6H_{12}O_6$ (SUGAR) $\longrightarrow$ ___H_2O + ___CO_2

5. For each of the following isotopes, determine the number of protons and neutrons in the nucleus.

(a) ^{11}C, ^{12}C, ^{14}C

(b) ^{31}P, ^{32}P, ^{33}P

(c) ^{32}S, ^{35}S, ^{38}S

CHAPTER 2

Water

Life on this planet began in water, and today, wherever liquid water is found, life is also present. There are one-celled organisms that eke out their entire existence in no more water than can cling to a grain of sand. Some kinds of algae are found only on the melting undersurfaces of polar ice floes. Certain bacteria can tolerate the near-boiling water of hot springs. In the desert, plants race through an entire life cycle—seed to flower to seed—following a single rainfall. In the jungle, the water cupped in the leaves of a tropical plant forms a microcosm in which a myriad of small organisms are born, spawn, and die.

Observations of the polar ice caps on Mars and of its "canals," first seen more than 100 years ago, kindled hope that life might be present there also. However, no liquid water was detected in the soil or atmosphere at the site of the first spacecraft landings. Indeed, it was discovered that the air surrounding Mars is so dry and cold that liquid water could not exist on the planet. Nor was life discovered.

2–1
The first living systems came into being, according to present hypotheses, in the warm primitive seas and for many organisms, ourselves included, each new individual begins life bathed and cradled in water. These are salamander larvae.

2–2
Photograph of Mars taken on July 21, 1976, the day following Viking I's successful landing on the planet. The dusty, windblown soil is littered with rocks and boulders. The spacecraft landed with one foot on such a rock, leaving its spidery body tilted at an 8° angle. The red color of the surface of Mars may be due to iron oxides.

On earth, by contrast, water is a common liquid. Three-quarters of the earth's surface is covered by water. In fact, if the earth's surface were absolutely smooth, all of it would be 2.5 kilometers under water. Water makes up 50 to 95 percent of the weight of any functioning living system. If you weigh 60 kilograms, about 50 kilograms of that is water.

But "common" is not the same as "ordinary." Water is not in the least an ordinary liquid. Compared with other liquids it is, in fact, quite extraordinary. If it were not, it is highly probable that life could never have evolved.

WATER AND THE HYDROGEN BOND

In order to understand why water is so extraordinary and how, as a consequence, it can play its unique and crucial role in relation to living systems, we have to look again at its molecular structure. As we noted in the previous chapter, each water molecule is made up of two atoms of hydrogen and one of oxygen held together by two covalent bonds.

The water molecule as a whole is neutral in charge, having an equal number of electrons and protons. However, the molecule is electrically asymmetric—that is, polar. This polarity occurs because oxygen is more electronegative than hydrogen. The paired electrons in the outer orbitals spend more time around the oxygen nucleus than they do around the hydrogen nuclei. As a consequence, the region near the oxygen nucleus has two weakly negative zones, and each of the regions near the hydrogen nuclei has a weakly positive zone. Thus, the water molecule, in terms of its polarity, is four-cornered, with two positively charged "corners" and two negatively charged ones (Figure 2–3a).

When any of these charged regions comes close to the oppositely charged region of another water molecule, the force of the attraction forms a bond between them, which is known as a *hydrogen bond.* Hydrogen bonds are not found only in water. A hydrogen bond can form between any hydrogen atom that is covalently bonded to an electronegative atom—usually oxygen or nitrogen—and the oxygen or nitrogen atom of another molecule. In water, hydrogen bonds form between the negative "corners" of one water molecule and the positive "corners" of another. As a consequence, every water molecule can establish hydrogen bonds with four other water molecules. Liquid water is made up of water molecules bound together in this way, as shown in Figure 2–3.

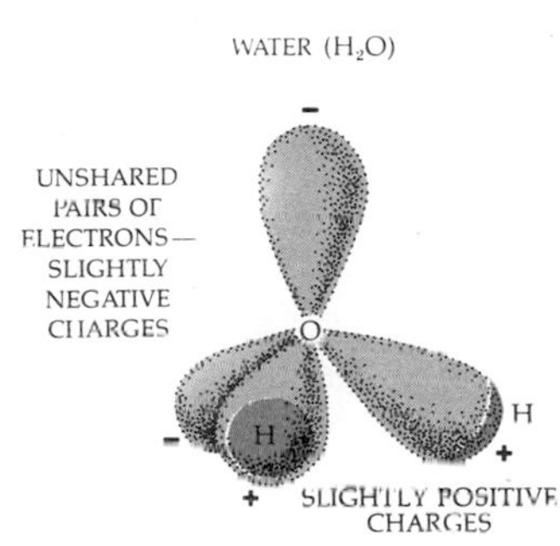

(a)

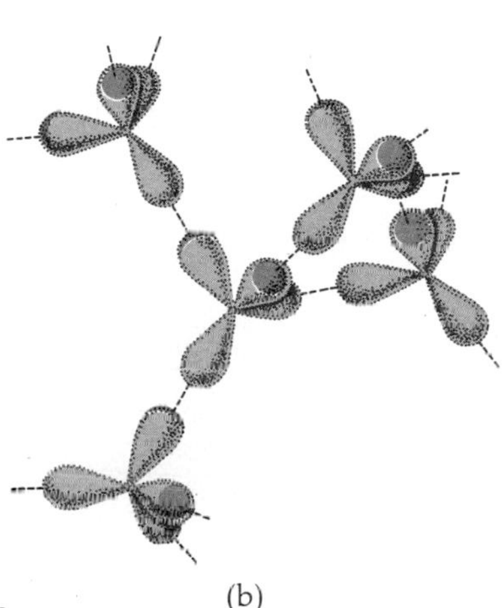

(b)

2–3
Water. (a) *Orbital diagram of a single molecule. Note the four-cornered distribution of positive and negative charges.* (b) *As a result of these positive and negative charges, each water molecule can form hydrogen bonds (dashed lines) with four other water molecules.*

2–4
The formation of a water droplet.

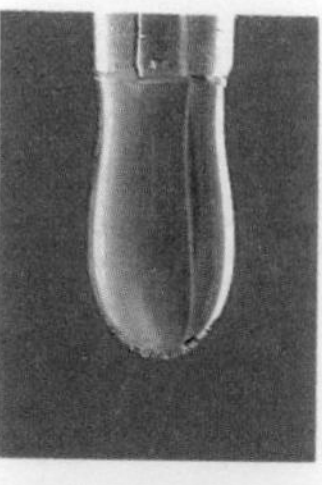

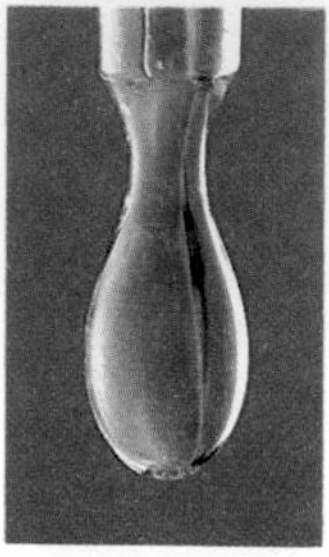

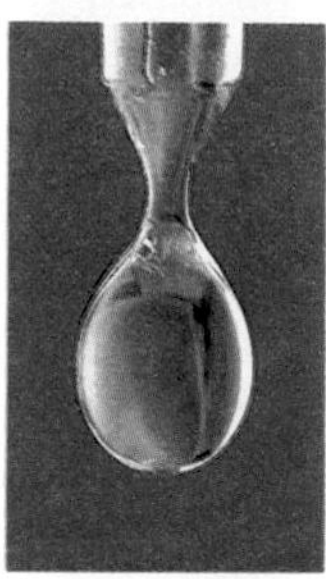

Any single hydrogen bond is relatively weak and has an exceedingly short lifetime; on an average, each such bond lasts about 1/100,000,000,000th of a second. But, as one is broken, another is made. All together, the hydrogen bonds have considerable strength, making water both liquid and stable under ordinary conditions of pressure and temperature.

Now let us look at some of the consequences of these attractions among water molecules, especially as they affect living systems.

Surface Tension

Look at water dripping from a faucet. Each drop clings to the rim and dangles for a moment by a thread of water; then just as the tug of gravity breaks it loose, its outer surface is drawn taut, to form a sphere as the drop falls free. Gently place a needle or a razor blade flat on the surface of the water in a glass. Although the metal is denser than water, it floats. Look at a pond in spring or summer; you will see water striders and other insects walking on its surface almost as if it were solid.

These phenomena are all the result of *surface tension.* Surface tension is the result of the cohesion or clinging together, of the water molecules. (*Cohesion* is, by definition, the holding together of like substances. *Adhesion* is the holding together of unlike substances.) The only liquid with a surface tension greater than that of water is mercury. Atoms of mercury are so greatly attracted to one another that they tend not to adhere to anything else.

Water, because of its negative and positive charges, also adheres strongly to any other charged molecules and to charged surfaces. The "wetting" capacity of water—that is, its ability to coat a surface—reflects its polar structure, as does its cohesiveness.

2–5
A water strider's legs dimple the surface of a pond. The water strider, which lives on this surface, has specialized hairs on its first and third pairs of legs that enable it to rest upon the surface film, depressing it but not penetrating. The second pair of legs, which penetrate the film, serve as propelling oars.

2–6
The germination of seeds begins with changes in the seed coat that permit a massive uptake of water. The embryo and surrounding structures then swell and burst the seed coat. In this acorn, photographed on the forest floor, the embryonic root has emerged through the tough outer layer of the fruit.

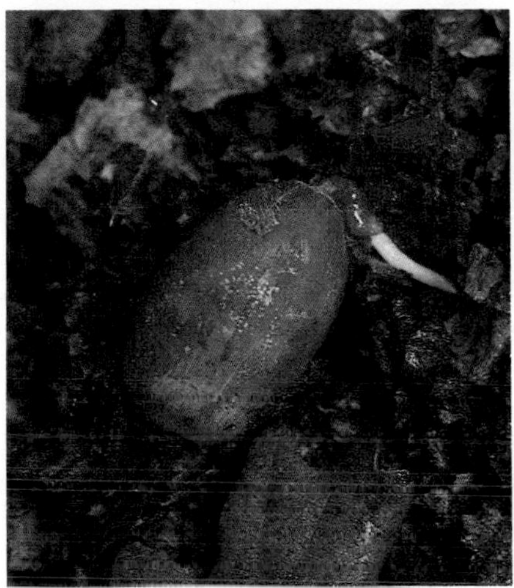

Capillary Action and Imbibition

If you hold two dry glass slides together and dip one corner in water, the combination of cohesion and adhesion will cause water to spread upward between the two slides. This is *capillary action.* Capillary action similarly causes water to rise in very fine glass tubes, to creep up a piece of blotting paper, or to move slowly through the micropores of the soil and so become available to the roots of plants.

Imbibition ("drinking up") is the capillary movement of water molecules due to adhesion into substances such as wood or gelatin, which swell as a result. The pressures developed by imbibition can be astonishingly great. It is said that stone for the ancient Egyptian pyramids was quarried by driving wooden pegs into holes drilled in the rock face and then soaking the pegs with water. The swelling of the wood created a force great enough to break the stone slab free. Seeds imbibe water as they begin to germinate, swelling and bursting their seed coats (Figure 2–6).

WATER AND TEMPERATURE

Specific Heat of Water

The amount of heat a given amount of a substance requires for a given increase in temperature is its *specific heat* (also called heat capacity). One calorie* is defined as the amount of heat that will raise the temperature of 1 gram (1 milliliter or 1 cubic centimeter) of water 1°C. The specific heat of water is about twice the specific heat of oil or alcohol; that is, approximately 0.5 calorie will raise the temperature of 1 gram of oil or alcohol 1°C. It is four times the specific heat of air or aluminum and 10 times that of iron. Only liquid ammonia has a higher specific heat (Table 2–1). In other words, it requires a high input of energy to raise the temperature of water.

The high specific heat of water is also a consequence of hydrogen bonding. Heat is a form of energy—the kinetic energy, or energy of movement, of molecules. In order for the kinetic energy of water molecules to increase sufficiently for the temperature to rise 1°C, it is necessary first to rupture a number of the hydrogen bonds holding the molecules together. The reason that it takes more heat to warm a gram of water to a given temperature than to warm a gram of almost any other liquid is that so much of the heat energy added to the water is used in breaking the hydrogen bonds between the water molecules. Only a relatively small amount of heat energy is therefore available to increase molecular movement.

Note that heat and temperature are not the same. A lake may have a lower temperature than does a bird flying over it, but the lake contains more heat because it comprises many more molecules than the bird; therefore it has more molecular motion. Temperature is measured in degrees and reflects the average kinetic energy of the molecules. Heat is measured in calories and reflects both molecular movement and the mass and number of the molecules present.

What does the high specific heat of water mean in biological terms? It means that for a given rate of heat input, the temperature of water will rise more slowly than the temperature of almost any other material. Conversely, the temperature will drop more slowly as heat is removed. Because so much heat input or heat loss is required to raise or lower the temperature of water, organisms that live in the oceans or large bodies of fresh water live in an environment where the temperature

Table 2–1 *Comparative Specific Heats (The Quantity of Heat, in calories, Required to Raise the Temperature of 1 Gram through 1°C)*

SUBSTANCE	SPECIFIC HEAT
Water	1.00
Lead	0.03
Iron	0.10
Salt (NaCl)	0.21
Glass	0.20
Sugar (sucrose)	0.30
Liquid ammonia	1.23
Chloroform	0.24
Ethyl alcohol (ethanol)	0.60

* Dietary calories are actually kilocalories (kcal); 1,000 calories equal 1 kilocalorie.

Table 2-2 *Comparative Heats of Vaporization (The Quantity of Heat, in calories, Required to Convert 1 Gram of Liquid to 1 Gram of Gas)*

LIQUID	HEAT REQUIRED
Water (at 0°C)	596
Water (at 100°C)	540
Ammonia	295
Chlorine	67.4
Hydrofluoric acid	360
Nitric acid	115
Carbon dioxide	72.2
Ethyl alcohol (ethanol)	236.5
Ether	9.4

is relatively constant. Also, the high water content of terrestrial plants and animals helps them to maintain a relatively constant internal temperature. This constancy of temperature is important because, as we shall see, biologically important chemical reactions take place only within a narrow temperature range.

Heat of Vaporization

Vaporization—or evaporation, as it is more commonly called—is the change from a liquid to a gas. Water has a high *heat of vaporization.* It takes more than 500 calories to change a gram of liquid water into vapor, fifty times as much as for ether and almost twice as much as for ammonia.

Hydrogen bonding is also responsible for water's high heat of vaporization. Vaporization comes about because some of the more rapidly moving molecules of a liquid break loose from the surface and enter the air. The hotter the liquid, the more rapid the movement of its molecules and, hence, the more rapid the rate of evaporation. But, whatever the temperature, so long as a liquid is exposed to air that is less than 100 percent saturated with the vapor of that liquid, evaporation will take place, right down to the last drop.

In order for a water molecule to break loose from its fellow molecules—that is, to vaporize—the hydrogen bonds have to be broken. This requires heat energy. At water's boiling point (100°C at a pressure of 1 atmosphere), 540 calories are needed to change 1 gram of liquid water into vapor. As a consequence, when water evaporates, as from the surface of your skin or a leaf, the escaping molecules carry a great deal of heat away with them. Thus evaporation has a cooling effect. Evaporation from the surface of a land-dwelling plant or animal is one of the principal ways in which these organisms "unload" excess heat and so stabilize their temperatures.

"Ammonia! Ammonia!"

[*Drawing by R. Grossman, © 1962 The New Yorker Magazine, Inc.*]

UNSHARED PAIR OF ELECTRONS—SLIGHTLY NEGATIVE CHARGE

−

N

H + H + H +

SLIGHTLY POSITIVE CHARGES

AMMONIA (NH_3)

2–7
Ammonia is very similar to water in its chemical structure, and biologists have speculated about whether it might substitute for water in life processes. The ammonia molecule, like the water molecule, is made up of hydrogen atoms covalently bonded to nitrogen, which, like the oxygen in the water molecule, retains a slight negative charge. But because there are three hydrogens with slight positive charges to one nitrogen, ammonia does not have the cohesive power of water and evaporates much more quickly. Perhaps this is why no form of life based on ammonia has been found, although NH_3 *was very common in the primitive atmosphere.*

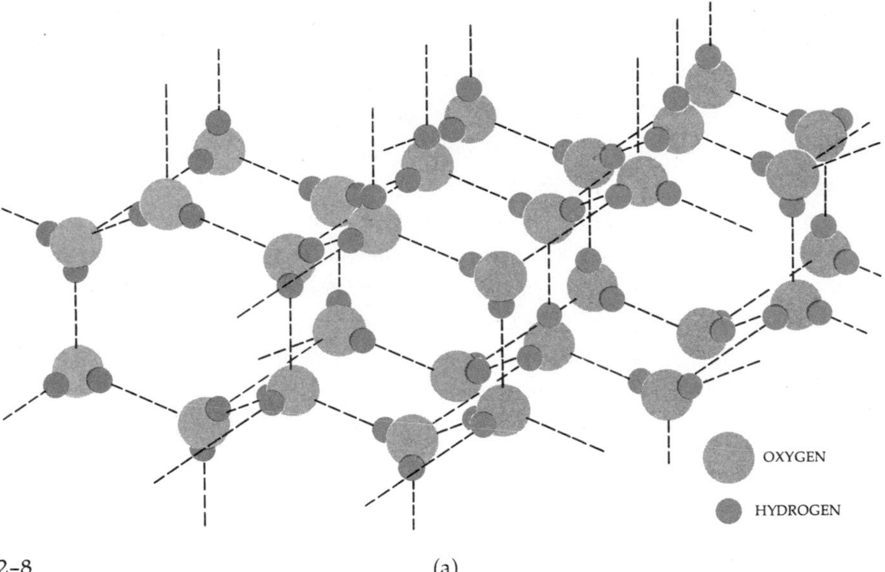

2–8 (a) (b)

(a) *In ice, each water molecule is hydrogen bonded to four other water molecules in an open latticework. In this schematic diagram, the large spheres represent the oxygen atoms, the small spheres represent the hydrogen atoms, and the dashed lines represent the hydrogen bonds.* (b) *Depending on atmospheric conditions of temperature and humidity, ice can form strikingly beautiful crystal patterns, as on these sedges.*

Freezing: The Heat of Fusion

The freezing point of water is 0°C. So is the melting point. To make the transition from a solid to a liquid, water requires 79.7 calories per gram. As water melts in an icebox or other closed, insulated container, it draws this much heat from the contents of the container. Conversely, as water freezes, it releases the same amount of heat into its surroundings. This 79.7 cal/gram is known as the *heat of fusion*. In this way, ice and snow also serve as temperature stabilizers, particularly during the temperate zone transition periods of fall and spring. Moderation of sudden changes in temperature gives organisms time to make seasonal adjustments essential to survival.

Water exhibits another peculiarity as it undergoes the transition from a liquid to a solid. As the temperature of a liquid drops, its density tends to increase. This increase occurs because the individual molecules are moving more slowly and so the spaces between them decrease. The density of water also increases, until the temperature nears 4°C. Then the water molecules come so close together that every one of them becomes hydrogen-bonded to four others, forming an open latticework. In the course of this bonding process, water expands again, so that water as a solid takes up more volume than water as a liquid.

This increase in volume has occasional disastrous effects on water pipes, but, on the whole, turns out to be so beneficial for life forms that it would seem to be almost miraculously devised. If water contracted as it froze, not only would ice cubes fail to tinkle in our glasses, but also lakes and ponds and other bodies of water would freeze from the bottom up. Once ice began to accumulate on the bottom, it would tend not to melt, season after season, and eventually the pond would freeze solid and life in the pond would be destroyed. By contrast, a layer of ice tends to protect the life of the pond, keeping the liquid water beneath it at 0°C or above.

THE WATER CYCLE

The earth's supply of water is the permanent possession of this planet, held to its surface by the force of gravity. It collects on the lowest surfaces of the earth's crust, forming puddles and pools, the largest of which are the oceans. Of all the water on the planet, 97.4 percent is the salt water of the oceans. Of the other 2.6 percent, some is locked in ice, some is in the form of water vapor in the atmosphere, and some, a very small proportion of the total water—1/20th of 1 percent—is on the land in a form available to terrestrial organisms.

Land organisms owe their water to the energy of the sun. Solar energy evaporates water from the oceans, leaving the salt behind. Water is also evaporated, but in much smaller amounts, from moist soil surfaces, from the leaves of plants, and from the bodies of other living organisms. These molecules are carried up into the atmosphere by air currents as water vapor and eventually fall to the earth's surface again as snow or rain. Most of the water falls on the oceans, since these cover most of the earth's surface. The water that falls on land is pulled back to the oceans by the force of gravity. Some of it, reaching low ground, emerges as ponds or lakes or streams, which pour water into the oceans.

Some of the water that falls on the land percolates down through the soil until it reaches a zone of saturation. In the zone of saturation, all pores and cracks in the rock are filled with water. The upper surface of this zone of saturation is known as the water table. Below the zone of saturation is solid rock, through which the water cannot penetrate. The deep groundwater, moving extremely slowly, eventually also reaches the ocean floor, thereby completing the water cycle.

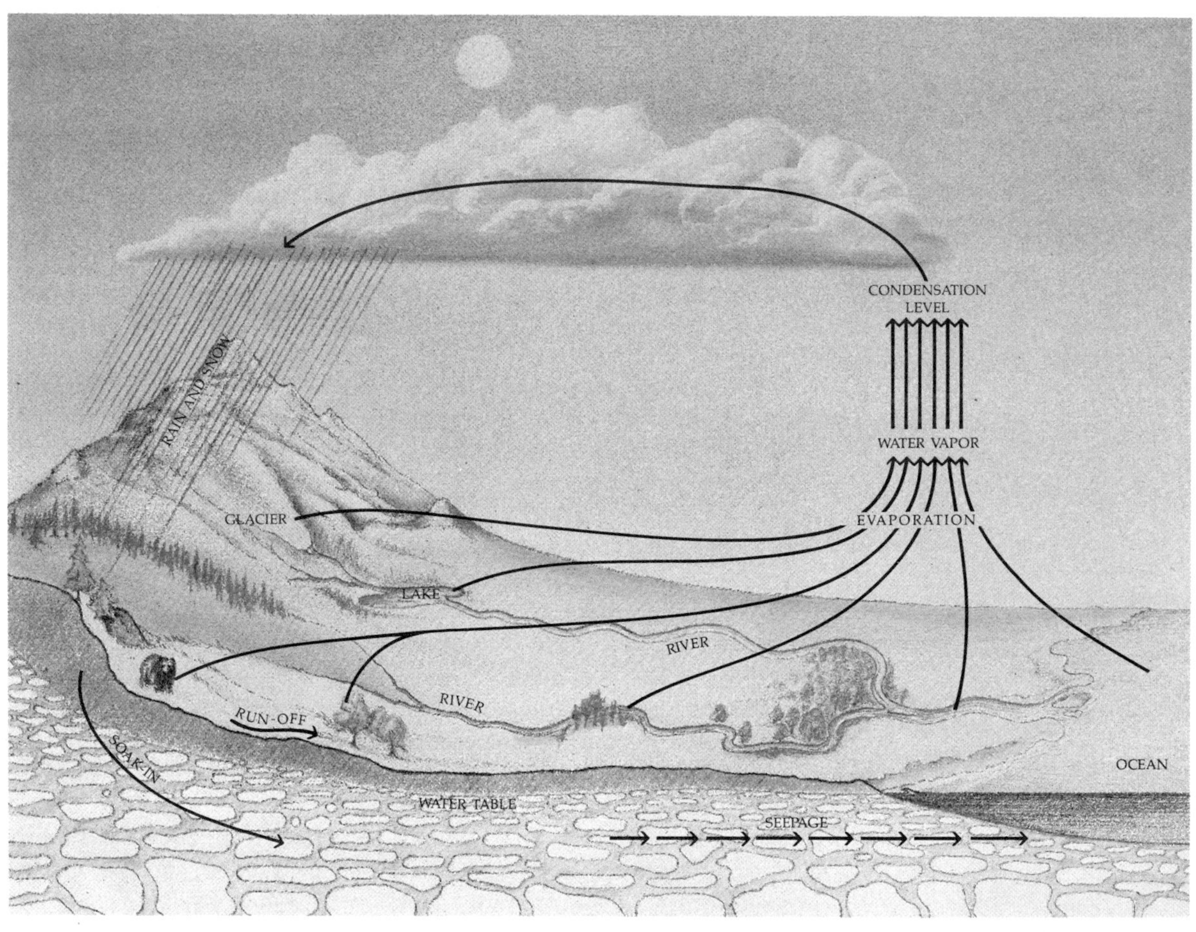

The presence of dissolved substances in water reduces the temperature at which water freezes, which is why salt is thrown on icy sidewalks and used in ice-cream freezers. The "hardening" process in several species of winter-hardy plants, by which they prepare themselves for cold weather, includes the breakdown of starch (which is insoluble) into simple sugars (which are soluble). Freshwater fish, whose body fluids are salty compared to the pond or lake in which they live, do not freeze when the temperature of the water is at or near 0°C. However, logically speaking, saltwater fish, whose body fluids are less salty than the ocean water surrounding them, should freeze at the below-zero temperatures of Arctic water. They do not, however, and animal physiologists investigating this phenomenon have discovered that at least one species, ghost fish, produces a complex protein named, appropriately, antifreeze protein. This protein, which is secreted into the bloodstream, appears, like the antifreeze in your car radiator, to prevent water molecules from crystallizing.

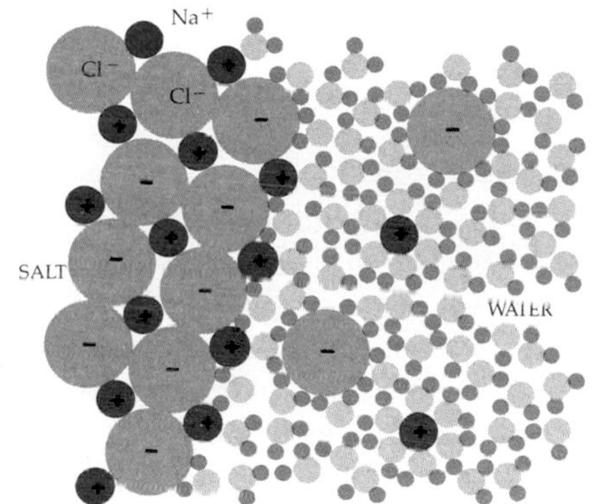

2–9
Because of the polarity of water molecules, water can serve as a solvent for any charged or polar atoms or molecules. This diagram shows sodium chloride (NaCl) dissolving in water as the water molecules cluster around the individual ions, separating them from each other.

WATER AS A SOLVENT

Many substances within living systems are found in solution. (A *solution* is a uniform mixture of the molecules of two or more substances; the substance present in the greatest amount—usually a liquid—is called the solvent, and the substances present in lesser amounts are called solutes.) The polarity of the water molecules is responsible for water's capacity as a solvent. The polar water molecules tend to separate substances such as NaCl into their constituent ions. Then, as shown in Figure 2–9, the water molecules cluster around and segregate the charged ions. Many of the molecules important in living systems—such as glucose (sugar)—also have areas of positive and negative charge. (Such polar regions of partial charge arise, as you might expect, in the neighborhood of covalently bonded atoms of unequal electronegativity.) These molecules therefore attract water molecules and so dissolve in water.

Molecules that readily dissolve in water are often called *hydrophilic* ("water-loving"). Such molecules slip into aqueous solution easily because their partially charged regions attract water molecules and so compete with the attraction between the water molecules themselves.

Molecules such as fats that lack polar, or charged, regions, tend to be very insoluble in water. The hydrogen bonding between the water molecules acts as a force to exclude the nonpolar molecules. As a result of this exclusion, nonpolar molecules tend to cluster together in water, just as droplets of fats tend to coalesce, as for example, on the surface of chicken soup. Such molecules are said to be *hydrophobic* ("water-fearing"), and the clusterings are known as hydrophobic interactions.

We will encounter these properties of hydrophobic and hydrophilic molecules again in later chapters. These weak forces—hydrogen bonds and hydrophobic forces—play very important roles in shaping the architecture of large biomolecules and, as a consequence, in determining their properties.

IONIZATION OF WATER

In liquid water, there is a slight tendency for a hydrogen atom to jump from the oxygen atom to which it is covalently bonded to the oxygen atom to which it is

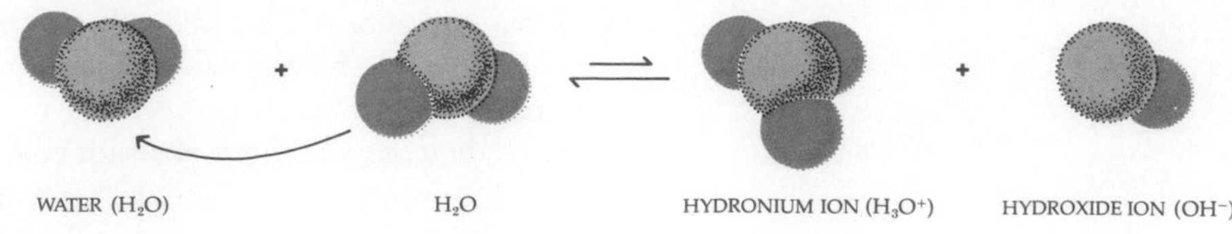

2–10
When water ionizes, a hydrogen atom shifts from the oxygen atom to which it is covalently bonded to the oxygen atom to which it is hydrogen-bonded. In this diagram, the large spheres represent oxygen and the small spheres represent hydrogen.

hydrogen-bonded. In this reaction, two ions are produced: the hydronium ion (H_3O^+) and the hydroxide ion (OH^-). In any given volume of pure water, a small but constant number of water molecules will be ionized in this way. The number is constant because the tendency of water to ionize is exactly offset by the tendency of the ions to reunite; thus even as some molecules are ionizing, an equal number of others are forming, a state known as equilibrium.

Although the hydrogen ion does not exist in a separate form, by convention the ionization of water is expressed by the equation:

$$HOH \rightleftharpoons H^+ + OH^-$$

The arrows indicate that the reaction can go in either direction. The fact that the arrow pointing toward HOH is longer indicates that, at equilibrium, most of the H_2O is not ionized. As a consequence, in any sample of pure water, only a small fraction exists in ionized form.

In pure water, the number of H^+ ions exactly equals the number of OH^- ions. This is necessarily the case since neither ion can be formed without the other when only H_2O molecules are present. A solution acquires the properties we recognize as acidic when the number of H^+ ions exceeds the number of OH^- ions; conversely, a solution is alkaline (basic) when the number of OH^- ions exceeds the number of H^+ ions.

The pH Scale

Chemists define degrees of acidity by means of the *pH scale*. In the expression "pH," the p stands for a negative power of 10, and the H stands for the concentration of hydrogen ions in moles per liter.

A mole is the amount of an element equivalent to its atomic weight expressed in grams, or the amount of a substance equivalent to its molecular weight expressed in grams. Thus, a mole of atomic hydrogen (atomic weight 1) is 1 gram of hydrogen atoms, a mole of atomic oxygen (atomic weight 16) is 16 grams of oxygen atoms, and a mole of water is 18 grams of water molecules. The most interesting thing about the mole is that a mole—of any substance—contains the same number of particles as any other mole. This number, known as Avogadro's number, is 6.02×10^{23}. Thus, a mole of water molecules (18 grams) contains exactly the same number of molecules as a mole of hydrochloric acid molecules (36.5 grams). Use of the mole in specifying quantities of substances involved in chemical reactions makes it possible for us to consider comparable numbers of reacting particles.

A liter of pure water contains 1/10,000,000 mole of hydrogen ions. For convenience, this is written exponentially as 10^{-7} mole per liter. In terms of the pH scale, it is referred to simply as pH 7 (see Table 2–3). At pH 7, the concentrations of free

Table 2-3 *The pH Scale*

	CONCENTRATION OF H^+ IONS (MOLES PER LITER)		pH	CONCENTRATION OF OH^- IONS (MOLES PER LITER)	
Acidic	1.0	$= 10^0$	0	10^{-14}	
	0.1	$= 10^{-1}$	1	10^{-13}	
	0.01	$= 10^{-2}$	2	10^{-12}	
	0.001	$= 10^{-3}$	3	10^{-11}	
	0.0001	$= 10^{-4}$	4	10^{-10}	
	0.00001	$= 10^{-5}$	5	10^{-9}	
	0.000001	$= 10^{-6}$	6	10^{-8}	
Neutral	0.0000001	$= 10^{-7}$	7	10^{-7}	$= 0.0000001$
Basic		10^{-8}	8	10^{-6}	$= 0.000001$
		10^{-9}	9	10^{-5}	$= 0.00001$
		10^{-10}	10	10^{-4}	$= 0.0001$
		10^{-11}	11	10^{-3}	$= 0.001$
		10^{-12}	12	10^{-2}	$= 0.01$
		10^{-13}	13	10^{-1}	$= 0.1$
		10^{-14}	14	10^0	$= 1.0$

2-11
Surface of the lining of the stomach, as shown in a scanning electron micrograph (magnified approximately 185 times). The numerous indentations are the openings into gastric pits, in which acid-secreting cells are located. Mucus, also secreted by cells of the stomach lining, coats the surface of the stomach and protects it from the acid.

H^+ and OH^- are exactly the same, and thus pure water is said to be neutral. Any pH below 7 is acidic, and any pH above 7 is basic. The lower the pH number, the higher the concentration of hydrogen ions. Thus pH 2 means 10^{-2} mole of hydrogen ions per liter of water, or 1/100 mole per liter—which is, of course, a much larger figure than 1/10,000,000. The difference of one pH unit represents a tenfold difference in the concentration of hydrogen ions.

We can now define "acid" and "base" more exactly:

1. An *acid* is a substance that donates H^+ ions to a solution, that is, a hydrogen-ion donor. A solution with a pH below 7 (with more than 10^{-7} mole of H^+ ions per liter) is acidic.
2. A *base* is a substance that decreases the number of H^+ ions, that is, a hydrogen-ion acceptor. A solution with a pH above 7 (with less than 10^{-7} mole of H^+ ions per liter) is basic.

In short, a solution that contains more H^+ ions than OH^- ions is acidic, and one that contains more OH^- ions than H^+ ions is basic.

Strong and Weak Acids and Bases

Hydrochloric acid (HCl) is an example of a common acid. It is a strong acid, meaning that it tends to be almost completely ionized in aqueous solution into H^+ and Cl^- ions. Sodium hydroxide (NaOH) is a common strong base; in aqueous solution, it exists almost entirely as Na^+ and OH^- ions. Weak acids and weak bases are those that ionize only slightly.

Most organic acids owe their acidic properties to a group of atoms called the carboxyl group, which includes one carbon, two oxygens, and a hydrogen atom (symbolized as —COOH). When an organic acid is dissolved in water, the carboxyl group dissociates to yield hydrogen ions:

$$\text{—COOH} \rightleftharpoons \text{—COO}^- + H^+$$

Thus any compound containing carboxyl is a hydrogen donor, or an acid. It is a weak acid, however, because, as indicated by the arrows, the —COOH ionizes only slightly.

Compounds that contain the amino group ($—NH_2$) act as weak bases since the $—NH_2$ has a weak tendency to accept hydrogen ions (protons), thereby forming $—NH_3^+$:

$$—NH_2 + H^+ \rightleftharpoons —NH_3^+$$

Because of the strong tendency of H^+ and OH^- to combine and the weak tendency of water to ionize, the concentration of H^+ ions will always decrease as the concentration of OH^- increases, and vice versa. If HCl is added to a solution containing NaOH, the following reaction will take place:

$$Na^+ + OH^- + H^+ + Cl^- \longrightarrow H_2O + Na^+ + Cl^-$$

In other words, if an acid and a base of comparable strengths are added in equivalent amounts, the solution will once more be neutral.

Buffers

Solutions more acidic than pH 1 or more basic than pH 14 are possible, but these are not included in the scale because they are almost never encountered in biological systems (Figure 2–12). In fact, almost all the chemistry of living things takes place at pH's between 6 and 8. Notable exceptions are the chemical processes in the stomach of humans and other animals, which take place at a pH of about 2. Human blood, for instance, maintains an almost constant pH of 7.4 despite the fact that it is the vehicle for a large number and variety of nutrients and other chemicals being delivered to the cells, as well as for the removal of wastes, many of which are acids or bases.

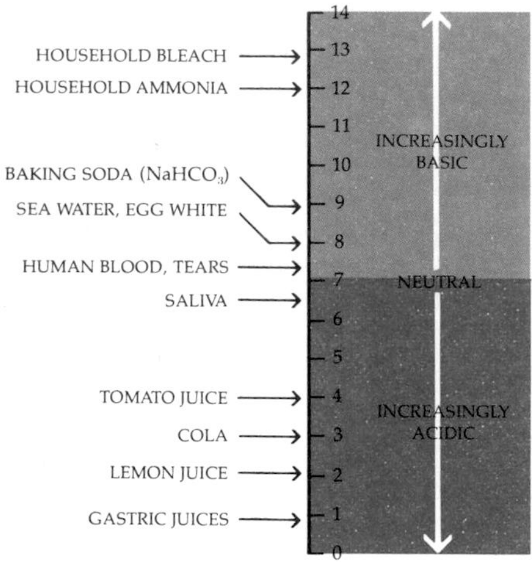

2–12
pH values of various common solutions. A difference of one pH unit reflects a tenfold difference in the H^+ ion concentration. Cola, for instance, is 10 times as acidic as tomato juice, and gastric juices are about 100 times more acidic than cola drinks.

The maintenance of a constant pH—an example of homeostasis (see page 20)—is important because the pH greatly influences the rate of chemical reactions. Organisms resist strong, sudden changes in the pH of blood and other fluids by means of *buffers*.

Buffers help maintain constant pH by their tendency to combine with H^+ ions and thus remove them from solution as the H^+ ion concentration begins to rise and to release them as it falls. Buffers are often combinations of H^+-donor and H^+-acceptor forms of weak acids or bases. The capacity of a buffer system to resist changes in pH is greatest when the concentrations of H^+ donor and H^+ acceptor are equal. As the ratio changes in either direction, the buffer becomes less effective.

The major buffer system in the human bloodstream is the acid-base pair H_2CO_3-HCO_3^-. The weak acid H_2CO_3 (carbonic acid) dissociates into H^+ and bicarbonate ions as shown in the equation below.

$$\underset{H^+\text{ DONOR}}{H_2CO_3} \rightleftharpoons H^+ + \underset{H^+\text{ ACCEPTOR}}{HCO_3^-}$$

The H_2CO_3-HCO_3^- buffer system resists the changes in pH that might result from the addition of small amounts of acid or base by "soaking up" the acid or base. For example, if a small amount of H^+ in the form of an organic acid is added to the system, it combines with the H^+ acceptor HCO_3^- to form H_2CO_3. This reaction removes the added H^+ and maintains the pH near its original value. If a small amount of OH^- is added, it combines with the H^+ to form H_2O; more H_2CO_3 tends to ionize to replace the H^+ as it is used.

Control of the pH of the blood is rendered even "tighter" by the fact that the H_2CO_3 is in equilibrium with dissolved carbon dioxide (CO_2) in the blood:

$$H_2O + CO_2 \rightleftharpoons H_2CO_3$$

As the arrows indicate, the two reactions are in equilibrium and the equilibrium favors the formation of CO_2; in fact, the ratio is about 100 to 1 in favor of CO_2 formation.

Dissolved CO_2 in the blood is, in turn, in equilibrium with the CO_2 in the lungs. By changing your rate of breathing, you can change the HCO_3^- concentration in the blood and thus adjust the pH of your internal fluids.

Obviously, if the blood should be flooded with a very large excess of acid or base, the buffer would fail, but normally it is able to adjust continuously and very rapidly to the constant small additions of acid or base that normally occur in body fluids.

SUMMARY

One of the chief components of living systems is water. Water, which is the most common liquid in the biosphere, has a number of remarkable properties that are responsible for its "fitness" for its role in living systems. These properties are a consequence of the molecular structure of water.

Water is made up of two hydrogen atoms and one oxygen atom held together by covalent bonds. Because of the electronegativity of the oxygen atom, the water molecule is polar, carrying two weak negative charges and two weak positive charges. As a consequence, weak bonds form between water molecules. Such bonds, which link a somewhat positively charged hydrogen atom that is part of one molecule to a somewhat negatively charged atom that is part of another molecule, are known as hydrogen bonds. Each water molecule can form hydrogen bonds with four other water molecules. Although individual bonds are weak and constantly shifting, the total strength of the bonds holding the molecules is very great.

Because of the hydrogen bonds holding the water molecules together (cohesion), water has a high surface tension and a high specific heat. (The specific heat of a substance is the amount of heat that a given amount of the substance requires for a given increase in temperature.) It also has a high heat of vaporization (the heat required to change a liquid to a gas) and a high heat of fusion (the heat required to change a solid to a liquid). Just before it freezes, water expands; thus ice has a lower density and a larger volume than liquid water.

The polarity of the water molecule (having zones of positive and negative charge) is responsible for water's adhesiveness and hence its tendency to capillary movement. Similarly, water's polarity makes it a good solvent for ions and other polar (charged) molecules. Molecules that dissolve readily in water are known as hydrophilic. Water molecules, as a consequence of their polarity, actively exclude uncharged molecules from solution. Molecules excluded from water solution are known as hydrophobic.

Water has a slight tendency to ionize, that is, to separate into H^+ ions (actually H_3O^+, hydronium ions) and OH^- ions. In pure water, the number of H^+ ions and OH^- ions is equal at 10^{-7} mole per liter. A solution that contains more H^+ ions than OH^- ions is acidic; one that contains more OH^- ions than H^+ ions is basic. The pH scale reflects the proportion of H^+ ions to OH^- ions. An acidic solution has more H^+ ions than OH^- ions and a pH less than 7; a basic solution has more OH^- ions than H^+ ions and a pH more than 7. Almost all of the chemical reactions of living systems take place within a narrow range of pH. Organisms maintain this narrow pH range by means of buffers.

QUESTIONS

1. (a) Sketch the water molecule and label the areas of positive and negative charge. (b) What are the major consequences of the polarity of the water molecule? (c) How are these effects important to living systems?

2. Can you explain maple sugar production in terms of its value to the sugar maple tree?

3. The trick with the razor blade works better if the blade is a little greasy. Why?

4. What is vaporization? Describe the changes that take place in water as it vaporizes. What is heat of vaporization? Why does water have an unusually high heat of vaporization?

5. Generally, coastal areas have more moderate temperatures (not as cold in the winter, nor as hot in the summer) than inland areas at the same latitude. What reasonable explanation can you give for this phenomenon?

6. Whereas pure (distilled) water has a pH of 7, water exposed for any time to the air dissolves some carbon dioxide gas (CO_2). As you would expect from the discussion of buffers, the following reactions occur:

$$\underset{\text{CARBON DIOXIDE}}{CO_2} + \underset{\text{WATER}}{H_2O} \rightleftharpoons \underset{\text{CARBONIC ACID}}{H_2CO_3}$$

$$\underset{\text{CARBONIC ACID}}{H_2CO_3} \rightleftharpoons \underset{\text{HYDROGEN ION}}{H^+} + \underset{\text{BICARBONATE ION}}{HCO_3^-}$$

What effect do you suppose this would have on the pH of the water? (The effect would certainly be seen in tap water.)

7. As we have seen, the digestive processes in the human stomach take place at a pH of about 2. When the food being digested reaches the small intestine, sodium bicarbonate ($NaHCO_3$) is released from the pancreas into the small intestine. What effect would you expect this to have on the pH of the food?

8. Write a short essay on the properties of water as exhibited and suggested by the photograph at the left.

CHAPTER 3

Organic Molecules

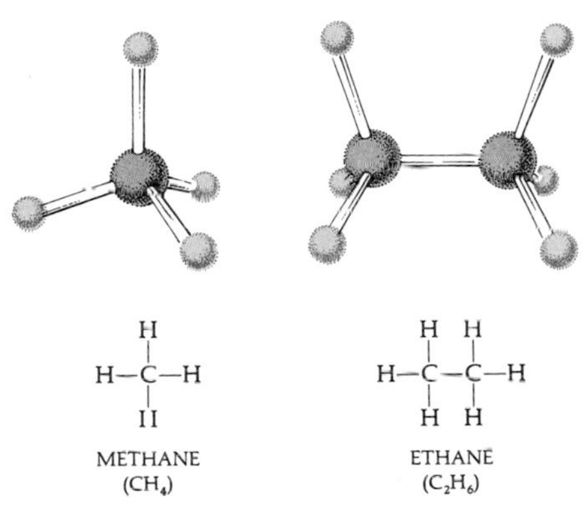

BUTANE
(C_4H_{10})

3–1
Structural formulas and ball-and-stick models of methane, ethane, and butane. In these models, the large black spheres represent carbon atoms and the small colored spheres represent hydrogen atoms.

In this chapter, we present some of the types of molecules that are found in living things. Consider this, if you will, an introduction to the principal characters; the plot begins to unfold in Chapter 4. As you will see, the molecular drama is a grand spectacular with, literally, a cast of thousands; a single bacterial cell contains some 5,000 different kinds of molecules, and a eukaryotic cell—the sort of cell that makes up the tissues of animals and plants—has about twice that many. These thousands of molecules, however, are composed of relatively few elements. Similarly, relatively few kinds of molecules play the major roles in living systems. If you bear with us, you will find that you readily learn to recognize the players and their roles and to distinguish the stars from the members of the chorus.

THE CENTRAL ROLE OF CARBON

As we noted previously, water makes up about 60 to 70 percent, on the average, of a living system, and ions such as K^+, Na^+, and Ca^{2+} account for no more than 1 percent. Almost all the rest, chemically speaking, is composed of organic molecules—molecules containing carbon.

In general, an organic molecule derives its overall shape from the arrangement of the carbon atoms that form the backbone, or skeleton, of the molecule. As you will recall from Chapter 1, a carbon atom has six protons and six electrons, two electrons in its innermost orbital and four in its outer orbitals. Thus carbon can form four covalent bonds with as many as four different atoms. Methane (CH_4), which is natural gas, is an example.

Even more important, in terms of carbon's biological role, carbon atoms can form bonds with each other. Ethane contains two carbons; propane, three; butane, four; and so on, forming long chains. In these compounds, every carbon bond that is not occupied by another carbon atom is taken up by a hydrogen atom. Compounds consisting of only carbon and hydrogen are known as *hydrocarbons*.

Chains containing oxygen atoms in addition to carbon and hydrogen are common in living systems. When these atoms occur in the ratio of one carbon to two hydrogens to one oxygen (CH_2O) the compounds they form are known as *carbohydrates*. Sugars are carbohydrates. The sugar glucose is a carbohydrate whose skeleton contains six carbon atoms. In solution, glucose forms a ring (Figure 3–2). Rings of carbon atoms are common types of biological molecules, so common, in fact, that the C's are usually omitted in representations of the ring form.

3-2
In living systems, alpha and beta forms of the ring-structured glucose molecule are in equilibrium. The molecules pass through the straight-chain form to get from one ring structure to the other.

ALPHA-GLUCOSE

GLUCOSE, STRAIGHT-CHAIN FORM

BETA-GLUCOSE

Functional Groups

The specific chemical properties of a molecule derive principally from groups of atoms, known as *functional groups*, that are attached to the carbon skeleton. An OH (hydroxyl) group is an example.* When one hydrogen and one oxygen are bonded covalently, one electron is left over for sharing. A compound with a hydroxyl group in place of one or more of the hydrogens in a hydrocarbon is known as an alcohol. Thus methane (CH_4), with the replacement of one hydrogen atom by a hydroxyl group, becomes methanol, or wood alcohol (CH_3OH), a pleasant-smelling, poisonous compound noted for its ability to cause blindness and death. Ethane similarly becomes ethanol, or grain alcohol (C_2H_5OH), which is present in all alcoholic beverages. Glycerol, $C_3H_5(OH)_3$, contains, as its formula indicates, three carbon atoms, five hydrogen atoms, and three hydroxyl groups. The carboxyl group (COOH), mentioned in the previous chapter, is a functional group that gives a compound the properties of an acid.

* —OH, the functional group, is called hydroxyl; OH^-, the ion, is called hydroxide.

Table 3-1 *Some Biologically Important Functional Groups*

GROUP	NAME	BIOLOGICAL SIGNIFICANCE
—OH	Hydroxyl	Polar, thus water soluble; forms hydrogen bonds
—C(=O)OH	Carboxyl	Usually negatively charged; a hydrogen donor (acid); most common acidic group —C(=O)O^- + H^+
—NH_2	Amino	Usually positively charged; a hydrogen acceptor (base) —NH_2—H^+
—C(H)=O	Aldehyde	Polar, thus soluble; characterizes a sugar
>C=O	Ketone (or carbonyl)	Polar, thus soluble; characterizes a sugar
—CH_3	Methyl	Hydrophobic (insoluble in water)
—P(=O)(OH)—OH	Phosphate	Usually negatively charged; a hydrogen donor (acid) —P(=O)(O^-)—O^-

Many functional groups are polar and so have a partial positive or negative charge in aqueous solution. Some of these tend to become fully ionized, depending on the pH of the solution. Thus these groups confer water-solubility and local electric charge to the molecules that contain them. Many functional groups participate directly in chemical reactions.

Functional groups are essential for recognizing a particular molecule and helping predict its properties. Alcohols, with polar hydroxyl groups, tend to be soluble in water, for instance, whereas hydrocarbons such as butane, with only nonpolar functional groups (such as methyl groups), are highly insoluble in water. Sugars contain hydroxyl groups plus an aldehyde group or a ketone group (Figure 3–6). Aldehyde groups are often associated with pungent odors and tastes. Smaller molecules with aldehyde groups, such as formaldehyde, have unpleasant odors, whereas larger ones, such as the chemicals that give vanilla, apples, cherries, and almonds their distinctive flavors, tend to be pleasing to the human sensory apparatus.

Four Representative Molecules

Figure 3–3 shows four kinds of molecules found frequently in living systems. The first (a), as you will recognize, has a long hydrocarbon chain; this chain, being nonpolar, is hydrophobic. Because of the COOH, you know that it is an acid. It is a *fatty acid*, a major component of fats, lipids, and waxes.

The second molecule (b) is a *sugar*. It is recognizable as a sugar because it consists of a carbon skeleton with hydrogen and oxygen atoms in the ratio of CH_2O, and it has a ketone group characteristic of a sugar. In aqueous solution, it forms a ring, as shown here. You can also tell that it is soluble because of the OH (alcohol) groups that confer solubility. This particular sugar is fructose.

The third molecule (c) is an *amino acid*. An amino acid consists of a carbon atom bonded to one hydrogen atom and to three functional groups. One of these groups is always a carboxyl group, which gives the molecule its acidic properties. Because of the amino group, an amino acid is also a weak base. The third functional group varies from amino acid to amino acid; in this case it is a methyl (CH_3) group, and the amino acid is alanine.

The fourth molecule (d) is a *nitrogenous base*. It is called nitrogenous because the rings of the molecule contain not only carbon but also nitrogen. It is a base because of the amino group. This nitrogenous base is called adenine and it is definitely one of the stars in the biological drama.

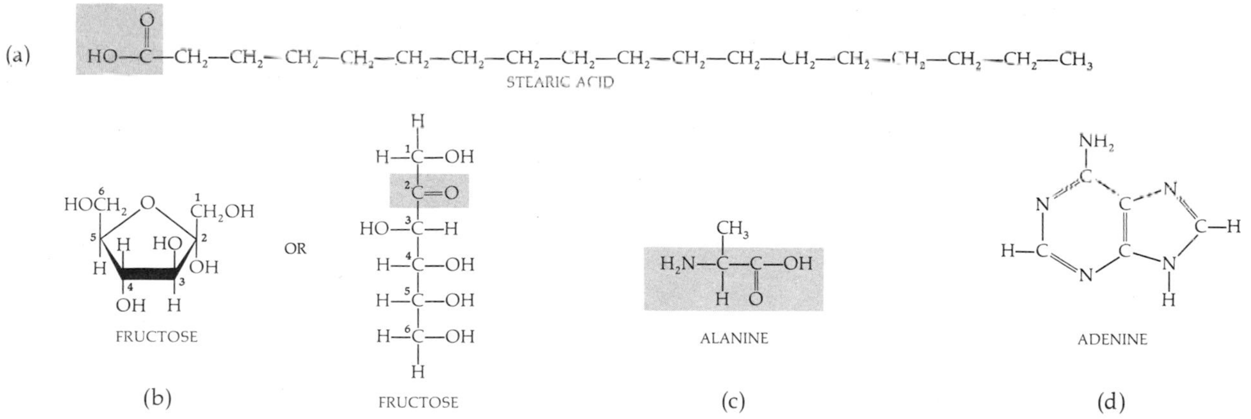

3–3
(a) *A fatty acid is a long hydrocarbon chain terminating in a carboxyl group (indicated in color). Fatty acids are major components of fats and lipids.* (b) *The sugar fructose. In the chain form, the ketone group is indicated by color. In aqueous solution, the molecule tends to assume the ring form.* (c) *Every amino acid contains an amino group (NH_2) and a carboxyl group (COOH) bonded to a central carbon atom. A hydrogen atom and a side group are also bonded to the same carbon atom. This basic structure is the same in all amino acids. In this amino acid, called alanine, the side group is CH_3 (methyl), one of the functional groups shown in Table 3–1.* (d) *A nitrogenous base, adenine. Adenine is an important component of DNA, the hereditary material, and of a number of other compounds as well. Note the two nitrogen-containing rings.*

(a) HO—C(=O)—CH₂—CH₂—CH₂—CH₂—CH₂—CH₂—CH₂—CH₂—CH₂—CH₂—CH₂—CH₂—CH₂—CH₂—CH₂—CH₂—CH₃

STEARIC ACID

(b) (c) (d)

WHY NOT SILICON?

Silicon, which has an atomic number of 14, is more abundant than carbon, which has an atomic number of 6. As you can tell from its atomic number, silicon also requires four electrons to complete its outer orbitals. Why then is it found so rarely in living systems? Because silicon is larger, the distance between two silicon atoms is much greater than the distance between two carbon atoms. As a result, the bonds between the more tightly held carbon atoms are almost twice as strong as those between silicon atoms. Thus carbon can form long stable chains and silicon cannot.

Carbon's capacity to form double bonds is also crucial to its central role in biology. Carbon combines with oxygen by means of two double bonds, and the carbon dioxide molecule, free and independent, all its electron requirements satisfied, floats in air as a gas. It also readily dissolves in water and so is available to living systems. In SiO_2*, by contrast, silicon is joined to oxygen by single bonds, leaving two unpaired electrons on the silicon atom and one on each oxygen. Unable to participate in multiple bonds, these unpaired electrons pair with the unpaired electrons on neighboring* SiO_2 *molecules, forming, eventually, grains of sand, rocks, or with biological intervention, the shells of microscopic marine organisms.*

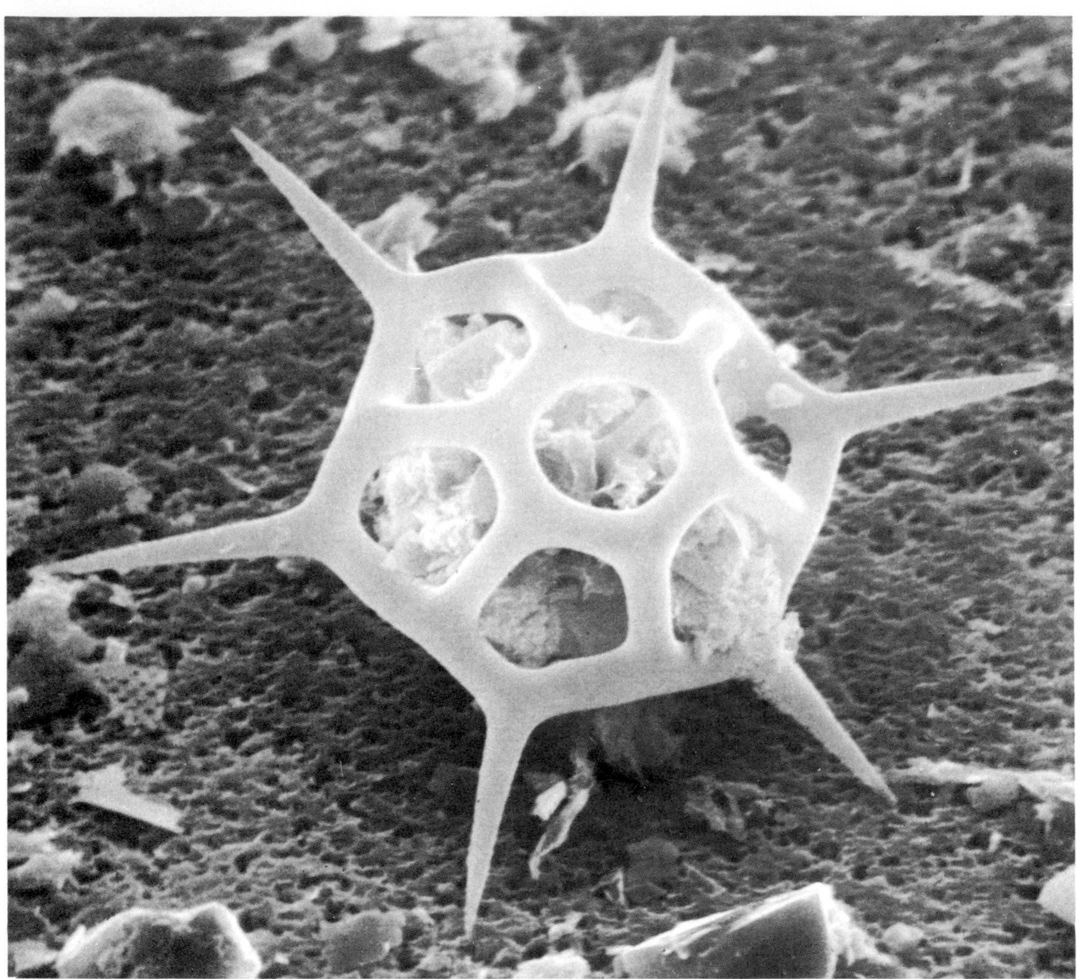

The delicate shell of this microorganism, from Narragansett Bay, Rhode Island, is composed of silicon dioxide. The material within the shell is organic debris.

3-4
A chemical bond is a force holding atoms together. The strength of the bond is measured in terms of the energy required to break it. The figures at the left indicate the number of kilocalories that will break the bonds between the pairs of atoms shown. The lines connecting the atoms represent the bonds, and the figures above them the characteristic center-to-center distances between them, expressed in nanometers (1 nanometer, abbreviated nm, equals 10^{-9} meter). Double lines indicate double bonds, which, as you can see, hold the atoms closer together and are stronger.

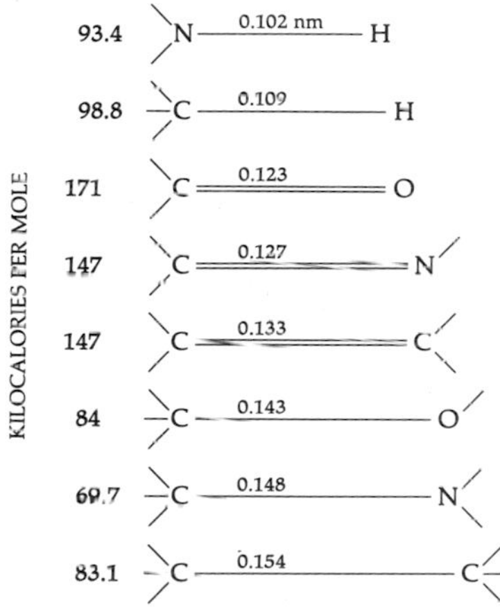

It has been said that it is necessary only to be able to recognize about 30 molecules for a working knowledge of the biochemistry of cells. One of these is a fatty acid; two others are the sugars glucose and ribose; five are nitrogenous bases such as adenine; and 20 are the biologically important amino acids. So, as you can see, the cast of characters is nearly complete.

The Energy Factor

Covalent bonds—the bonds commonly found in organic molecules—are strong, stable bonds consisting of electrons moving in orbitals about two or more different atomic nuclei. These bonds have different characteristic strengths, depending on the configuration of these orbitals. Bond strengths are conventionally measured in terms of kilocalories per mole (Figure 3-4). Chemical reactions always involve a change in electron configurations and therefore in bond strengths.

Consider, for example, the burning of methane, represented by the following equation:

$$CH_4 + 2O_2 \longrightarrow CO_2 + 2H_2O$$

This reaction can be set in motion by a spark, which is what causes explosions in coal mines, and it releases energy in the form of heat. The amount of energy released can be measured quite precisely (Figure 3-5). It turns out to be 213 kilocalories per mole of methane. This can be expressed by a simple equation:

$$\Delta H^\circ = -213 \text{ kcal}$$

The Greek letter delta (Δ) stands for change, *H* for heat, and the superscript ° indicates that the reaction occurs under certain standard conditions of temperature and pressure. The minus sign indicates that energy has been released.

Similarly, changes in energy occur in the chemical reactions that take place in living systems. However, living systems have evolved strategies for minimizing the energy required to set a reaction in motion and also the proportion of energy released as heat. (The word "strategy" in its ordinary meaning is a deliberate plan to achieve a specified goal. Biologists use it to mean a group of related traits evolved by natural selection to solve particular problems encountered by living systems.)

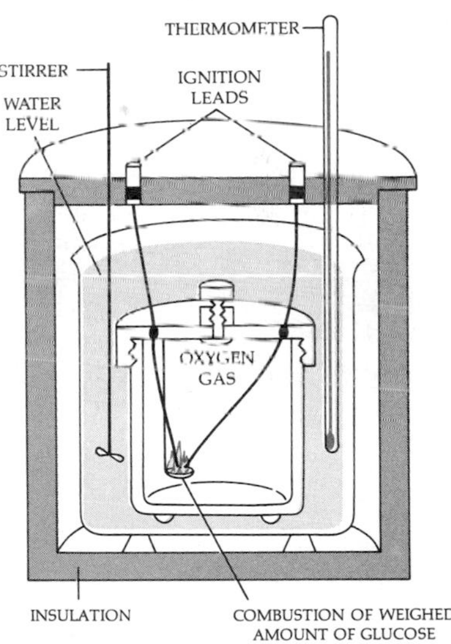

3-5
A calorimeter. A known quantity of glucose or some other material is ignited electrically. As it burns, the rise in the temperature of the water is measured. Based on the specific heat of water, one can then calculate the number of calories released by the burning of the sample.

MOLECULAR MODELS

Chemists have developed various ways of representing the structures of compounds. These representations differ in the degree to which they reflect the underlying architecture of the molecule. The simplest, and also the least informative, is the molecular formula, which indicates only the kind and number of atoms present. Glucose, for example, has 6 carbon atoms, 12 hydrogen atoms, and 6 oxygen atoms, so its molecular formula is $C_6H_{12}O_6$. However, fructose, a sugar with different properties, has the same molecular formula. The two can be distinguished by their structural formulas:

```
      H                    H
      |                    |
      C=O               H—C—OH
      |                    |
   H—C—OH                 C=O
      |                    |
  HO—C—H               HO—C—H
      |                    |
   H—C—OH               H—C—OH
      |                    |
   H—C—OH               H—C—OH
      |                    |
   H—C—OH               H—C—OH
      |                    |
      H                    H
```

GLUCOSE (CHAIN) FRUCTOSE (CHAIN)

Or, in the ring form:

GLUCOSE (RING): CH_2OH, H, O, H, H, OH, H, HO, OH, H, OH

FRUCTOSE (RING): $HOCH_2$, O, CH_2OH, H, HO, H, OH, OH, H

The lower edges of the rings are made thicker to hint at a three-dimensional structure. By convention, the carbon atoms at the intersections of the links in an organic ring structure are "understood" and not labeled.

These structural formulas are more informative because they indicate the way the atoms are joined together in the molecule. They are misleading, however, because they are essentially two-dimensional, and the properties of a molecule depend on its three-dimensional characteristics.

Still more informative are the ball-and-stick models of the kind shown in Figure 3–1. These emphasize three-dimensionality and, in particular, the bonds between atoms. But bonds are not sticks, any more than atoms are different-colored balls.

Another model is the space-filling one, in which each atom is represented by the edge of its outermost orbitals. This model is more "truthful" in its representation of the overall shape of the molecule and the nature of the bonds that hold it together.

Ball-and-stick and space-filling models are often used in the laboratory, but they are less useful on paper because it is necessary to see them from all angles to see all of the atoms and their bonds.

Space-filling models are also misleading in that molecules do not fill space in the same way that we think of a table or a rock as filling space. The atoms that make up molecules consist mostly of empty space. If the outer orbitals of the electrons in an oxygen atom were the size of the perimeter of the Astrodome in Houston, the nu-

Space-filling models of three fatty acids. Stearic acid is saturated and the molecule is quite straight. Oleic acid has one double bond in the hydrocarbon chain, and linoleic acid has two double bonds. Notice the kinks in their chains, caused by the double bonds. These models represent the ionized forms of the acids.

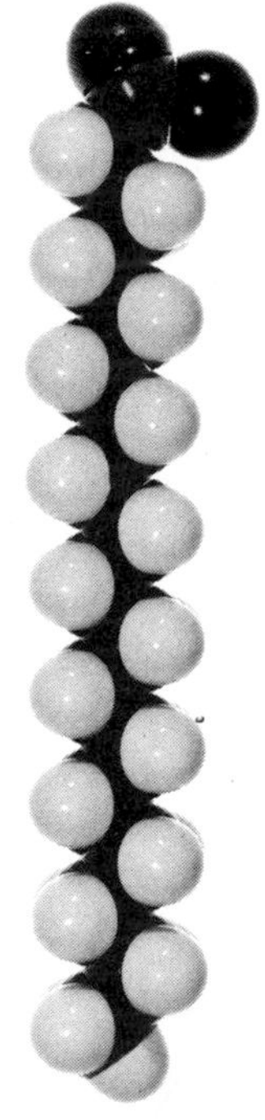

STEARIC ACID

cleus would be a ping-pong ball in the center of the stadium. What "fills" the space in molecules are areas of charge, produced by the movements of the electrons around the nuclei. One molecule "sees" another molecule in terms of these areas of charge. As a consequence, for instance, an enzyme that breaks down a glucose molecule will not have any effect on a fructose molecule because of the differences in shape. All the intricate biochemistry that goes on in the cell is based on this ability of molecules to "recognize" one another.

As is the case with atoms, none of these ways of showing molecules attempts to tell what a molecule actually looks like. Rather, they are models, ways of organizing a particular set of scientific data.

Probably the most famous molecular model is the model made of wire and pieces of tin that enabled Watson and Crick to figure out the structure of DNA. But that is another story.

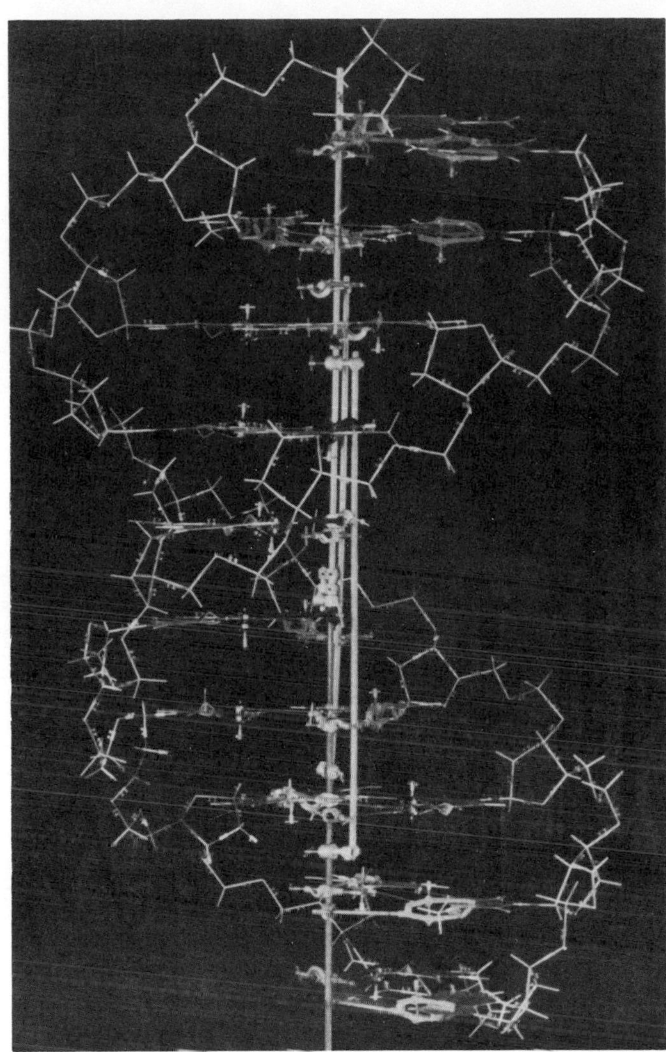

Wire and tin model of the DNA molecule.

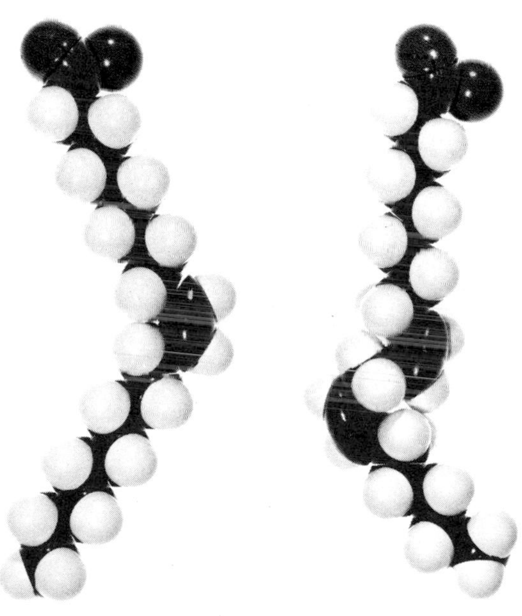

OLEIC ACID LINOLEIC ACID

CARBOHYDRATES: SUGARS AND POLYMERS OF SUGARS

Carbohydrates are the principal chemical energy sources of most living things. In addition, they form a variety of structural components of living cells; plant cell walls, for example, are mostly made of cellulose, which is the most common organic compound in the biosphere. Large molecules that are made up of similar or identical subunits are known as *polymers* ("many parts") and the subunits are called *monomers* ("single parts"). In the case of carbohydrates, the monomers are monosaccharides, single sugars, and the polymers are polysaccharides.

Monosaccharides: Ready Energy for Living Systems

As we noted previously, *monosaccharides* (sugars) can be described as $(CH_2O)_n$; n may be as small as 3, as in $C_3H_6O_3$, or as large as 8, $C_8H_{16}O_8$. As shown in Figure 3-6, they are characterized by hydroxyl groups and an aldehyde or a ketone group.

Like hydrocarbons, carbohydrates can be oxidized to yield carbon dioxide and water:

$$(CH_2O)_n + nO_2 \longrightarrow (CO_2)_n + (H_2O)_n$$

This reaction is also energy-yielding, and the amount of energy released as heat can be calculated by burning sugar molecules in a calorimeter. The same amount of energy is released when the equivalent amount of carbohydrate is oxidized in a living cell as when it is burned in a calorimeter.

This statement comparing the oxidation of food molecules with that of fuel molecules is not a metaphor; it is a fact. Look for instance at Table 3-2 in which

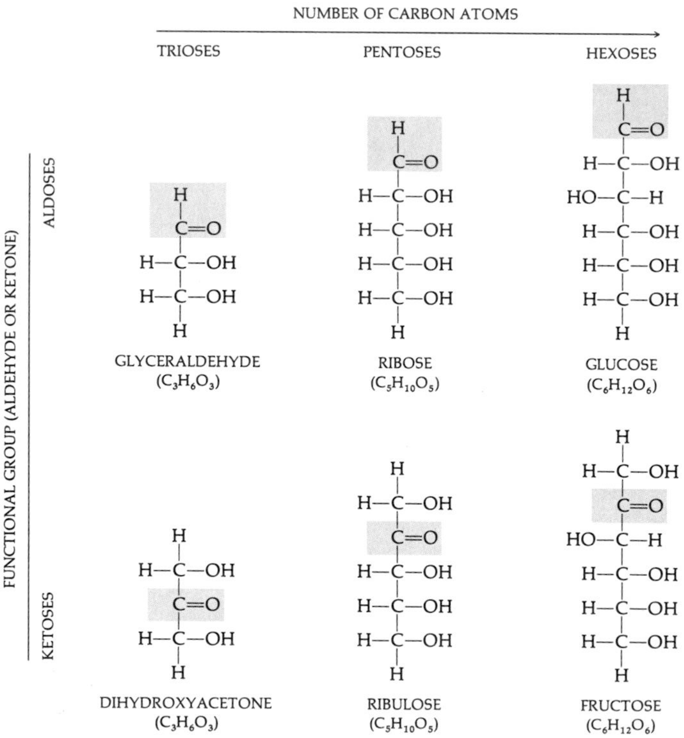

3-6
Two different ways of classifying monosaccharides: according to the number of carbon atoms and according to the functional groups (indicated in color). Glyceraldehyde, ribose, and glucose contain, in addition to hydroxyl groups, an aldehyde group. Dihydroxyacetone, ribulose, and fructose each contain a ketone group.

NUMBER OF CARBON ATOMS → TRIOSES, PENTOSES, HEXOSES
FUNCTIONAL GROUP (ALDEHYDE OR KETONE): ALDOSES, KETOSES

GLYCERALDEHYDE ($C_3H_6O_3$)
RIBOSE ($C_5H_{10}O_5$)
GLUCOSE ($C_6H_{12}O_6$)
DIHYDROXYACETONE ($C_3H_6O_3$)
RIBULOSE ($C_5H_{10}O_5$)
FRUCTOSE ($C_6H_{12}O_6$)

Table 3–2 *Cost of Transport*

ANIMAL OR VEHICLE	WEIGHT (kg)	SPEED (km/hr)	METABOLIC RATE (kcal/kg hr)	COST OF TRANSPORT (kcal/kg km)
Flyers				
Fruit fly *(Drosophila)*	2.0×10^{-6}	7.4	104	14.05
Honey bee *(Apis)*	1.0×10^{-4}	30.0	450	15.0
Hummingbird *(Calypte)*	0.003	49.0	197	4.02
Pigeon *(Columba)*	0.384	58.0	55.1	0.95
Walkers and Runners				
White-footed mouse *(Peromyscus)*	0.023	1.08	37.8	35.0
Ground squirrel *(Citellus)*	0.198	2.80	15.2	5.43
Lizard *(Iguana)*	0.90	0.370	2.31	6.24
Dog *(Canis)*	10.7	5.84	7.94	1.36
Man *(Homo)*	70.0	5.58	4.06	0.73
Horse *(Equus)*	707.0	5.00	2.61	0.52
Swimmer				
Salmon *(Oncorhynchus)*	0.19	2.8	1.1	0.39
Vehicles				
Cherokee airplane	978	200	240	1.2
F105F jet fighter	17,200	1,150	1,750	1.52
DC-8 jet	107,000	965	550	0.57
Volkswagen	1,010	81	59.6	0.74
Cadillac	2,210	119	98.3	0.83

From M. S. Gordon, *Animal Physiology: Principles and Adaptations*, 2d ed., The Macmillan Company, New York, 1972.

energy costs for running different kinds of machinery—biological and manmade—are compared, and note the remarkable similarities. (There are at least two other interesting generalizations that can be formulated from these data. Can you find them?)

A principal energy source for vertebrates is the monosaccharide glucose. It is in this form that sugar is generally transported in the animal body. A patient receiving an intravenous feeding in a hospital is getting glucose dissolved in water. This glucose is carried through the bloodstream to the cells of the body where the energy-yielding reactions are carried out. As measured in a calorimeter, the oxidation of a mole of glucose yields 673 kilocalories:

$$C_6H_{12}O_6 + 6O_2 \longrightarrow 6CO_2 + 6H_2O$$

$$\Delta H^\circ = -673 \text{ kcal}$$

Disaccharides: Transport Forms

Disaccharides consist of two single sugars (monosaccharides) covalently bonded together. Although glucose is the common transport sugar for vertebrates, sugars are often transported in other organisms as disaccharides. Sucrose, commonly called cane sugar, is the form in which sugar is transported in plants from the photosynthetic cells (mostly in the leaf), where it is produced, to other parts of the

3-7

Sucrose is a disaccharide made up of two monosaccharides, glucose and fructose. Note that the sucrose formed here is the beta form; the fructose has been rotated 180° and so can join to glucose in a 1→2 linkage. The formation of sucrose involves the removal of a molecule of water (condensation). Splitting sucrose requires, conversely, the addition of a water molecule (hydrolysis).

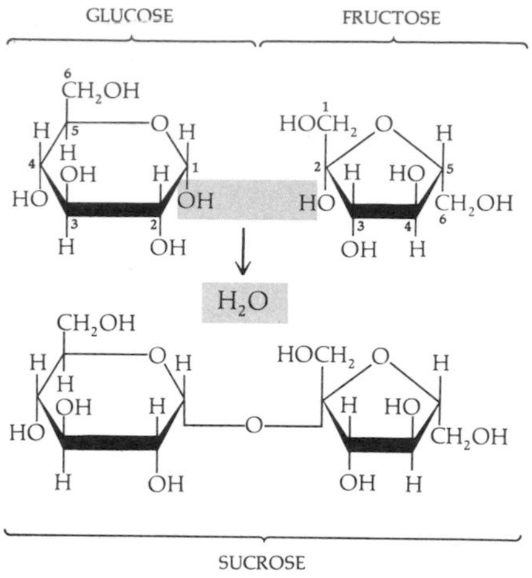

GLUCOSE FRUCTOSE

H_2O

SUCROSE

plant body. Sucrose is composed of the monosaccharides glucose and fructose. As is the case with all disaccharides and, indeed with most organic polymers, a molecule of water is removed in the course of bond formation, a reaction known as *condensation.* Thus, only the monomers of carbohydrates actually have a CH_2O ratio because of the removal of two atoms of hydrogen and one of oxygen every time such a bond is formed.

When the sucrose molecule is split into glucose and fructose, as it is when it is used as an energy source, the molecule of water is added again. This splitting requires the addition of a water molecule and so is known as *hydrolysis,* from *hydro,* meaning "water," and *lysis,* meaning "breaking apart."

Another common disaccharide is lactose, a sugar that occurs only in milk. Lactose is made up of glucose combined with another monosaccharide, galactose. Sugar is transported through the blood of many insects in the form of another disaccharide, trehalose, which consists of two glucose units linked together.

Polysaccharides: Sugars in Storage

Polysaccharides are made up of monosaccharides linked together in long chains. Some of them are storage forms of sugar. *Starch,* for instance, is the principal food storage form in most plants. A potato, for example, contains starch produced from the sugar formed in the green leaves of the plant; the sugar is transported underground and accumulated there in a form suitable for winter storage, after which it will provide for new growth in the spring. Starch occurs in two forms, amylose and amylopectin. Both consist of glucose units linked together (Figure 3-8).

Glycogen is the principal storage form for sugar in higher animals. Glycogen has a structure much like that of amylopectin except that it is more highly branched, with branches occurring every eight to ten glucose units. In vertebrates, glycogen is stored principally in the liver and in muscle tissue. When there is an excess of glucose in the bloodstream, the liver forms glycogen. When the concentration of glucose in the blood drops, the hormone glucagon, produced by the pancreas, is released into the bloodstream; glucagon stimulates the liver to hydrolyze glycogen to glucose, which then enters the bloodstream.

Formation of polysaccharides from monosaccharides requires energy. However, when the cell needs energy and these polysaccharides are hydrolyzed, the energy released is available for cellular work.

Structural Polysaccharides

A major function of molecules in living systems is to form the structural components of cells and tissues. The principal structural molecule in plants is *cellulose.* In fact, half of all of the organic carbon in the biosphere is incorporated into cellulose. Wood is about 50 percent cellulose, and cotton is nearly pure cellulose.

Cellulose molecules form the fibrous part of the plant cell wall. The cellulose fibers, cemented together by other kinds of polysaccharides, surround the plant cells, which, in effect, live in little cellulose boxes. When the cells are young, these boxes are flexible and stretch as the cell grows, but they become thicker and more rigid as the cell matures. In some plant tissues, such as the tissues that form wood and bark, the cells eventually die, leaving only their tough outer cellulose-containing walls.

3–8

In plants, sugars are stored in the form of starch. Starch is composed of two different types of polysaccharides, amylose (a) *and amylopectin* (b). *A single molecule of amylose may contain 1,000 or more glucose units with the first carbon of one glucose ring linked to the fourth carbon of the next in a long unbranched chain, which winds to form a helix* (c). *A molecule of amylopectin is made up of about 48 to 60 glucose units arranged in shorter, branched chains. Starch molecules, perhaps because of their helical nature, tend to cluster into granules.* (d) *In this scanning electron micrograph of a single storage cell of a potato, the spherical and egg-shaped objects are starch granules.* (e) *Glycogen, which is the common storage form for sugar in vertebrates, resembles amylopectin in its general structure except that each molecule contains only 16 to 24 glucose units. The dark granules in this liver cell are glycogen. When glucose is needed, it is provided by the conversion of glycogen.*

AMYLOSE

(a)

AMYLOPECTIN

(b)

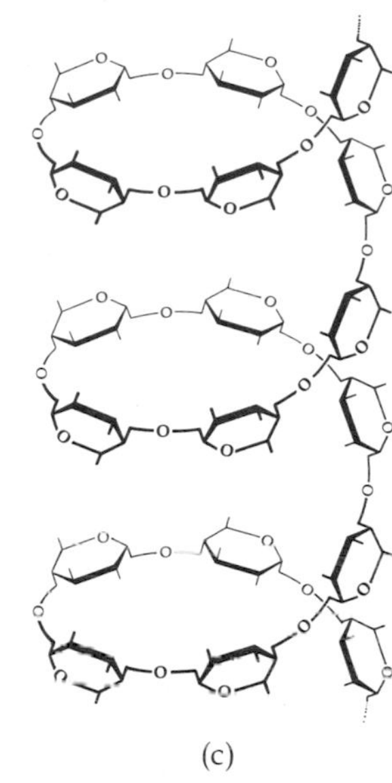

(c)

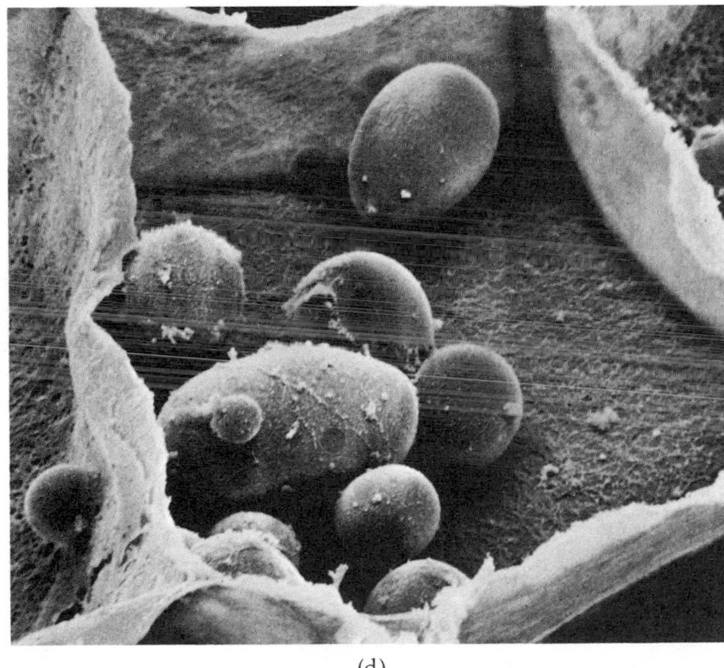

(d)

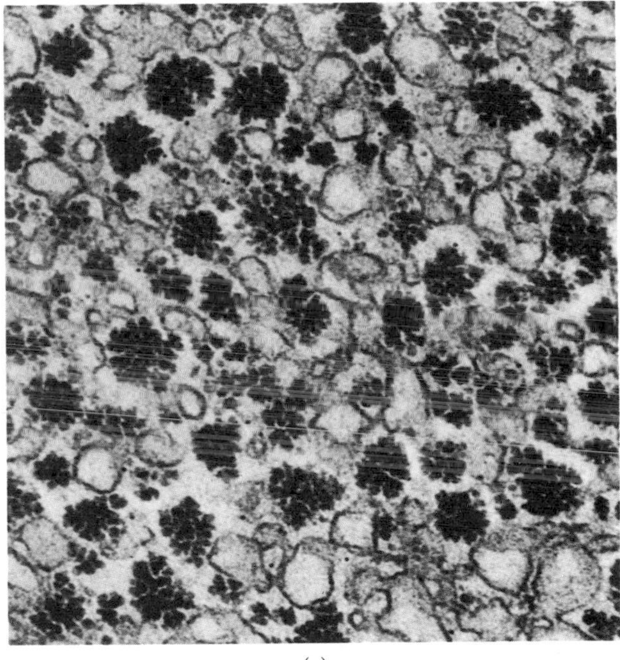

(e)

WHY IS SUGAR SWEET?

Actually it is not. Sugars have a number of objectively definable properties, such as molecular weight, melting point, and calorie content. Sweetness is not such a property. Just as beauty is in the eye of the beholder, sweetness resides not in the molecule itself but in the perception systems that have evolved to detect it.

Houseflies, for instance, have sugar detectors in their feet. If a housefly lights on a droplet of weak sugar solution, its proboscis will automatically be extended. This useful response evolved over the millennia as a food-detecting mechanism. We similarly have sensory receptors especially tuned for the detection of sugars, although our receptors, in keeping with our eating habits, are in our tongues. Because of sugar's value as an energy source, many other organisms have sugar-detecting mechanisms. A housefly reacts positively to about the same types of sugars that we do, although its detection mechanism, as gauged by the proboscis extension test, is about 10 million times more sensitive.

The capacity to detect sugars, and the fact that the sensation is pleasurable, probably evolved in insects and other animals because it promoted their ingestion of these energy-rich food molecules and, by extension, their survival. This taste for sugar has been exploited by plants, particularly the flowering plants. They have evolved nectaries, dripping with sugary syrup, by which they lure pollinators to their flowers, and fruits that turn sweet just as the seeds become mature and are ready to be transported to a germination site.

Today, our taste for sugar is being further exploited by manufacturers of prepared foods. Many dry cereals are more than 50 percent sugar, and sugar is an added hidden ingredient in countless other products. Sugar has thus become what behavioral scientists call a "supernormal stimulus." Baby cuckoos provide an example of a supernormal stimulus. Born in the nests of other birds, they gape so widely at the sight of food that the adult birds, all their parental instincts focused on this one oversized craw, let their own offspring starve. As another example, herring gulls abandon their own eggs for an artificial egg 1.5 times its size. And following a cue designed to lead us to a nutritious, vitamin-rich food supply, we let our children and ourselves grow fat, toothless, malnourished, and diabetic.

Cellulose is a polymer composed of monomers of glucose, just as starch and glycogen are. Starch and glycogen can be readily utilized as fuel by almost all kinds of living systems, but only a few microorganisms—certain bacteria, protozoa, and fungi—can hydrolyze cellulose. Cows and other ruminants, termites, and cockroaches can use cellulose for energy only because of microorganisms that inhabit their digestive tracts.

To understand the differences between structural polysaccharides, such as cellulose, and energy-storage polysaccharides, such as starch or glycogen, we have to look briefly once again at the glucose molecule. You will remember that the molecule is basically a chain of carbon atoms and that when it is in solution, as it is in the cell, it assumes a ring form. The ring may close in either of two ways (see Figure 3–2). One ring form is known as alpha, and the other as beta. The alpha and beta forms are in equilibrium, with a certain number of the molecules changing from one to the other all of the time, going through the open-chain structure to reach the other form. Starch and glycogen are both made up entirely of alpha units. Cellulose consists entirely of beta units (see Figure 3–9a). Because of this slight difference in structure, cellulose is impervious to the enzymes that so successfully break down the storage polysaccharides.

Chitin, which is a major component of the exoskeletons of insects and also of the cell walls of many fungi, is a tough, resistant, modified polysaccharide (Figure 3–10).

3–9

Starch (Figure 3–8) consists of alpha-glucose monomers. Cellulose (a) *consists of beta-glucose monomers. The OH groups (indicated in color), which project from both sides of the chain, form hydrogen bonds with neighboring OH groups, resulting in the formation of bundles of cross-linked parallel chains* (b). *In the starch molecule (Figure 3–8), most of the OH groups capable of forming hydrogen bonds face toward the exterior of the helix, making it more readily soluble in the surrounding water.* (c) *The plant cell wall is composed largely of cellulose. Each of the microfibrils you can see here is a bundle of hundreds of cellulose strands, and each strand is a chain of glucose units* (a). *The microfibrils, as strong as an equivalent amount of steel, are embedded in other polysaccharides, one of which is pectin.*

(a)

CELLULOSE MOLECULE

(b)

MODEL OF CROSS-LINKED CELLULOSE MOLECULES

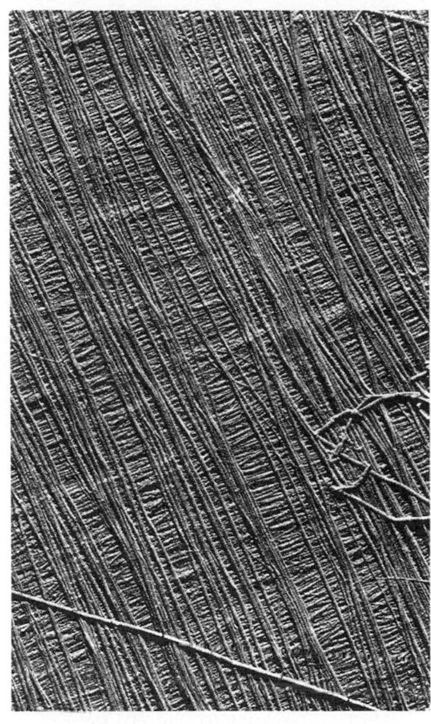

(c)

3–10

(a) *Chitin is a polymer consisting of repeated modified monosaccharides. As you can see, the monomer is a six-carbon sugar, like glucose, to which a nitrogen-containing group has been added.* (b) *Green darner molting. The relatively hard outer coverings, or exoskeletons, of insects contain chitin. Some types of insects, after molting, recycle their sugars and nitrogen by thriftily eating their discarded exoskeletons.*

CHITIN

(a)

(b)

LIPIDS

Lipids are a general group of substances, of which fats and waxes are particular examples. Unlike carbohydrates, lipids are hydrophobic. Many, although not all, contain fatty acids as major structural components. Lipids serve both as energy-storage forms—usually in the form of fats or oils—and for structural purposes, as in the case of phospholipids and waxes.

Fats: Energy in Storage

Unlike many plants, such as the potato, animals have only a limited capacity to store carbohydrates. In vertebrates, sugars in excess of what can be stored as glycogen are converted into fats. Some plants also store food energy as fats (oils), especially in seeds and fruits. Fats contain a higher proportion of energy-rich carbon-hydrogen bonds (see Figure 3–4) than carbohydrates do and, as a consequence, contain more chemical energy. On the average, fats yield about 9.3 kilocalories per gram* as compared to 3.79 kcal per gram of carbohydrate, or 3.12 kcal per gram of protein. Also, because they are hydrophobic, they do not attract water molecules, as glycogen does. Taking into account the water factor, fats store six times as much energy, gram for gram, as glycogen, which is undoubtedly why in the course of evolution they came to play a major role in energy storage.

For instance, a male ruby-throated hummingbird has a fat-free weight of 2.5 grams (about $\frac{1}{10}$ ounce). It migrates every fall from Florida to Yucatan, some 2,000 kilometers. Before doing so, it accumulates 2.0 grams of body fat, an amount almost equal to its original weight. However, if it were to carry the same energy reserves in the form of glycogen, it would have to carry 5 grams, twice its own fat-free weight. (Incidentally, as you will note in Table 3–2, flying is a more efficient method of travel for animals of comparable weight than walking.)

A neutral fat molecule consists of three molecules of fatty acid joined to one glycerol molecule. (Note the similarities of glycerol and glyceraldehyde, as shown in Figure 3–11a.) Fatty acids, which are seldom found in cells in a free state (that is, not as part of another molecule), typically consist of chains between 14 and 22 carbon atoms long. About 70 different fatty acids are known, which differ in their chain lengths and in whether the chain contains any double bonds (as in oleic acid) or not (as in stearic acid), and in the position in the chain of any double bonds

* 1,000 grams = 1 kilogram = 2.2 pounds, so oxidation of a pound of fat would yield about 4,200 kilocalories, more than the 24-hour requirement for moderately active adults.

3–11
(a) *Glycerol and glyceraldehyde differ by only two hydrogen atoms. The hydroxyl (OH) groups of glycerol are characteristic of an alcohol. Glycerol is one of the building blocks of fats.* (b) *A fat molecule consists of three fatty acids joined to a glycerol molecule. The long hydrocarbon chains of which the fatty acids are composed terminate in carboxyl groups, which become covalently bonded to the glycerol molecule. The bonds are formed by the removal of a molecule of water. The physical properties of the fat—such as its melting point—are determined by the lengths of the chains and by whether its component fatty acids are saturated or unsaturated. Three different fatty acids are shown here. Stearic acid and palmitic acid are saturated, and oleic acid is unsaturated, as you can see by the double bond in its structure.*

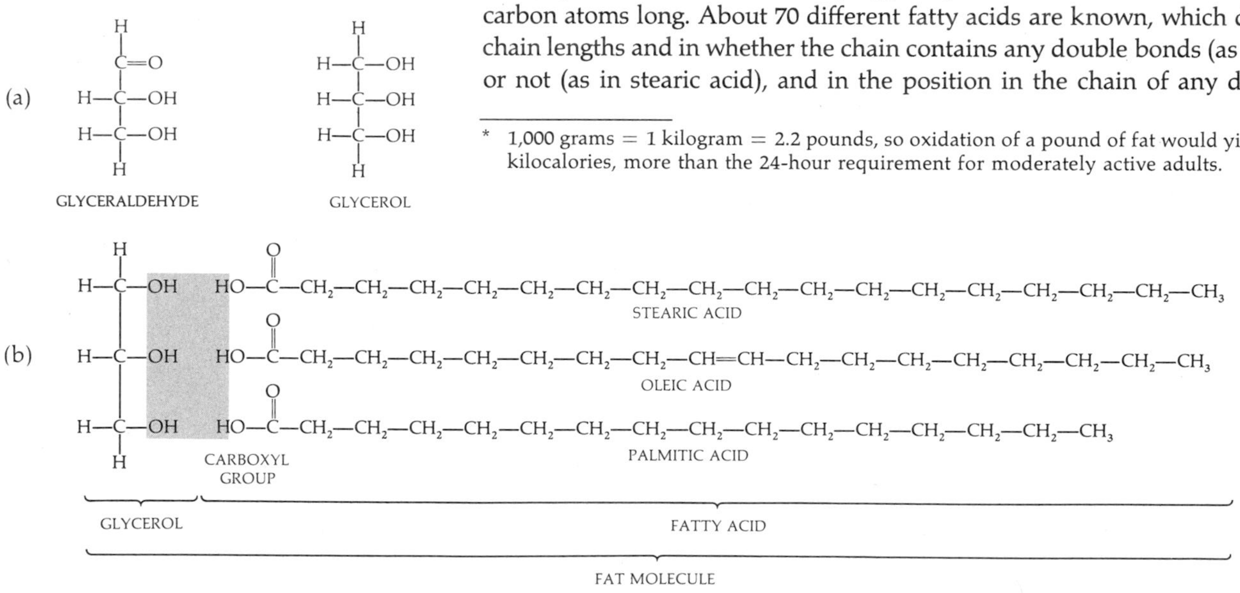

(a)

GLYCERALDEHYDE: H—C(=O)—C(H)(OH)—C(H)(OH)—H

GLYCEROL: H—C(H)(OH)—C(H)(OH)—C(H)(OH)—H

(b)

GLYCEROL: H—C—OH / H—C—OH / H—C—OH

STEARIC ACID: $HO{-}C({=}O){-}(CH_2)_{16}{-}CH_3$, i.e. HO—C(=O)—CH₂ repeated 16 times—CH₃

OLEIC ACID: $HO{-}C({=}O){-}(CH_2)_7{-}CH{=}CH{-}(CH_2)_7{-}CH_3$

PALMITIC ACID: $HO{-}C({=}O){-}(CH_2)_{14}{-}CH_3$

CARBOXYL GROUP

GLYCEROL — FATTY ACID

FAT MOLECULE

(Figure 3-11b). A fatty acid, such as stearic acid, in which there are no double bonds is known as a saturated fatty acid. A fatty acid, such as oleic acid, that contains carbon atoms joined by double bonds is said to be unsaturated. Unsaturated fats, which tend to be oily liquids, are more common in plants than in animals; examples are olive oil, peanut oil, and corn oil.

Sugars, Fats, and Calories

As we noted earlier, when carbohydrates are taken into the body in excess of the body's energy requirements, they are stored temporarily as glycogen or, more permanently, as fats. Conversely, when the energy requirements of the body are not met by its immediate intake of food, glycogen and, subsequently, fat are broken down to fill these requirements. Whether or not the body uses up its own storage molecules has nothing to do with the molecular form in which the energy comes into the body. It is simply a matter of whether these molecules, as they are broken down, release sufficient numbers of calories.

Structural Lipids

Insulators and Cushions

In general, fat stored in fat cells can be mobilized for energy when caloric intake is less than caloric expenditures. Some types of fat, however, seem to be protected from metabolic activity. Large masses of fatty tissue, for example, surround mammalian kidneys and serve to protect these precious organs from physical shock. For reasons that are not understood, these fat deposits remain intact even at times of starvation. Another mammalian characteristic is a layer of fat under the skin, which serves as thermal insulation. This layer is particularly well developed in sea-going mammals (Figure 3-12).

Among humans, females characteristically have a thicker layer of subdermal ("under-the-skin") fat than males. This capacity to store fat, although not much admired in our present culture, was undoubtedly very valuable 10,000 or more years ago. At that time, as far as we know, there was no other reserve food supply, and this extra fat not only nourished the woman but more important, the unborn child and the nursing infant, whose ability to fast without damage is much less than that of the adult. Thus many of us are strenuously dieting off what millennia of evolution have given us the capacity to accumulate.

3-12
This Weddell seal, enjoying the Antarctic spring, is well insulated by a thick layer of fat under the skin, which serves the same function that a wet suit serves for a diver.

POLAR HEAD — NONPOLAR TAILS

$R{-}O{-}\overset{O^-}{\underset{O}{P}}{-}O{-}{}^{3}CH_2$

$H{-}{}^{2}C{-}O{-}\overset{O}{\overset{\|}{C}}{-}CH_2CH_2CH_2CH_2CH_2CH_2CH_2CH{=}CHCH_2CH_2CH_2CH_2CH_2CH_2CH_2CH_3$

$H{-}{}^{1}C{-}O{-}\overset{O}{\overset{\|}{C}}{-}CH_2CH_2CH_2CH_2CH_2CH_2CH_2CH_2CH_2CH_2CH_2CH_2CH_2CH_2CH_2CH_2CH_3$

H

GLYCEROL

3–13
A phospholipid molecule consists of two fatty acids linked to a glycerol molecule, as in a fat, and a phosphate group (indicated by color) linked to the glycerol's third carbon. It also usually contains an additional chemical group, indicated by the letter R. The fatty acid "tails" are nonpolar and therefore insoluble in water (hydrophobic); the polar "head" containing the phosphate and R groups is soluble (hydrophilic).

Lipids and Membranes

Lipids also play extremely important structural roles. The lipids most important for structural purposes are phospholipids. Like fats, the phospholipids are composed of fatty acid chains attached to a glycerol backbone. In the phospholipids, however, the third carbon of the glycerol molecule is occupied not by a fatty acid but by a phosphate group (Figure 3–13) to which another polar group is usually attached. Phosphate groups are negatively charged. As a result, the phosphate end of the molecule is hydrophilic and the fatty acid portions are not. The consequences are shown in Figure 3–14. As we shall see, according to the present model, this arrangement of phospholipid molecules, with their hydrophilic heads extended and their hydrophobic tails clustered together, forms the structural basis of the cell membrane. We shall examine this structure further in Chapter 6.

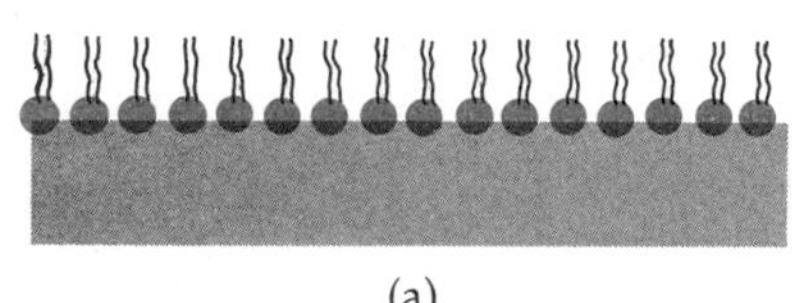

(a)

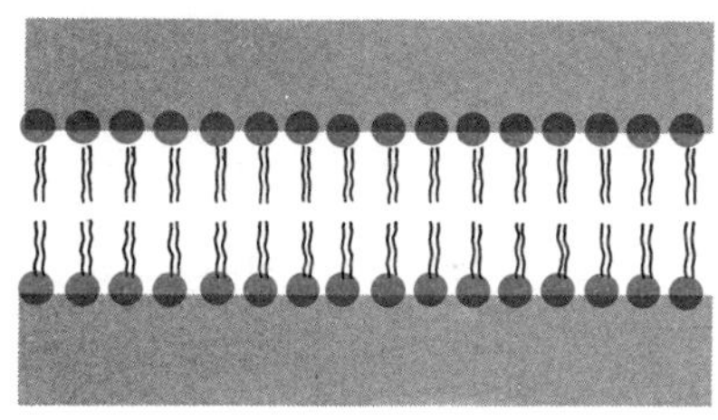

(b)

3–14
(a) *Because phospholipids have water-soluble heads and water-insoluble tails, they tend to form a thin film on a water surface with their tails extending above the water.* (b) *Surrounded by water, they spontaneously arrange themselves in two layers with their heads extending outward and their hydrophobic (water-fearing) tails inward. This arrangement is believed to be important in the structure of the cell membrane.*

Waxes

Waxes are also a form of structural lipid. They form protective coatings on skin, fur, feathers, on the leaves and fruits of higher plants, and on the exoskeletons of many insects.

Cholesterol

Cholesterol is found in cell membranes (with the exception of bacterial cells). About 25 percent (by dry weight) of the membranes of red blood cells is cholesterol; it is also a major component of the myelin sheath, the fatty membrane that wraps around fast-conducting nerve fibers, speeding the nerve impulse.

Cholesterol belongs to a group of compounds known as the *steroids*. All the steroids have four linked carbon rings, like cholesterol, and several of them, like cholesterol, have a tail. In addition, many of them have the OH functional group, which classifies them as alcohols. Although steroids do not resemble the other lipids structurally, they are grouped with the lipids because they are insoluble in water.

In some older people, cholesterol forms fatty deposits on the inner lining of the blood vessels. This blocks the vessels and decreases their elasticity, making such people more susceptible to high blood pressure, heart attacks, and strokes. Attempts have been made to reduce the incidence and extent of such deposits by reducing the dietary intake of high-cholesterol foods, such as eggs and cheese. But since many cells of the human body are able to synthesize cholesterol, it is not clear that a reduction of dietary cholesterol will solve the problem.

3–15
Waxes are also lipids. This electron micrograph shows waxy deposits on the upper surface of a eucalyptus leaf. Biosynthesis of waxes, which protect exposed plant surfaces from water loss, is a property of all groups of higher plants.

Regulatory Lipids

Sex hormones and the hormones of the adrenal cortex are also steroids; they are formed from cholesterol in the ovaries, testes, and other glands that produce them. Prostaglandins are a recently discovered group of chemicals with hormonelike actions, which are derived from fatty acids. Both the steroid hormones and the fatty acids will be discussed more fully in Section 6.

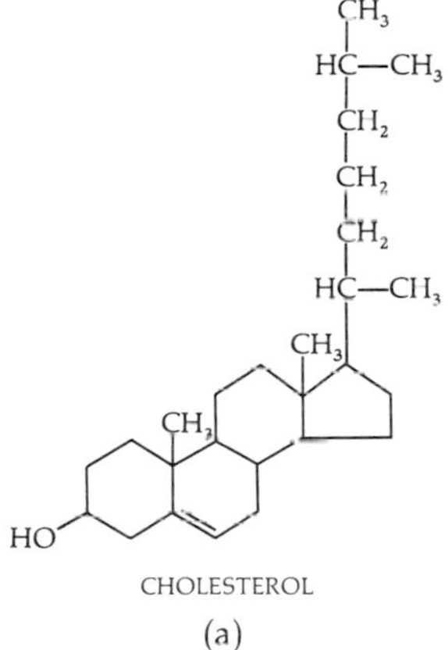

OH
CH_3
CH_3
O
TESTOSTERONE
(b)

3–16
(a) *The cholesterol molecule consists of four carbon rings and a hydrocarbon chain.* (b) *Testosterone, which is synthesized from cholesterol by cells in the testes, also has the characteristic four-ring structure but lacks the hydrocarbon tail.*

PROTEINS

Proteins are among the most abundant organic molecules; in most living systems they make up 50 percent or more of the dry weight. Only plants, with their high cellulose content, are less than half protein. There are many different protein molecules: enzymes; hormones; storage proteins, such as those in the eggs of birds and reptiles and in seeds; transport proteins, such as hemoglobin; contractile proteins of the sort found in muscle; immunoglobulins (antibodies); and many different types of structural proteins. In their functions, their diversity is overwhelming. In their structure, however, they all follow the same simple blueprint: They are all polymers of amino acids, arranged in a linear sequence.

Table 3–3 *Biological Functions of Proteins*

TYPES OF PROTEINS*	EXAMPLES
Structural proteins	Collagen, silk, virus coats, microtubules
Regulatory proteins	Insulin, ACTH, growth hormones
Contractile proteins	Contractile filaments in muscle cells
Transport proteins	Hemoglobin, myoglobin
Storage proteins	Egg white, seed protein
Protective proteins in vertebrate blood	Antibodies, complement
Membrane proteins	Membrane-transport proteins, antigens, permeases
Toxins	Botulism toxin, diphtheria toxin
Enzymes	Sucrase, pepsin

* Many of the proteins listed here will be discussed in other sections of the book, particularly Section 6.

3–17

(a) Every amino acid contains an amino group (NH_2) and a carboxyl group (COOH) bonded to a central carbon atom. A hydrogen atom and a side group are also bonded to the same carbon atom. This basic structure is the same in all amino acids. The "R" stands for the side group, which is different in each kind of amino acid. (b) The 20 kinds of amino acids used in making proteins. As you can see, their basic structures are the same, but they differ in their side groups. The side groups may be nonpolar (with no difference in charge between one zone and another), polar but with the two charges balancing one another out so that the side group as a whole is uncharged, negatively charged, or positively charged. The nonpolar side groups are not soluble in water, whereas the charged and polar side groups are water-soluble.

(a)

R
$H_2N-C-C-OH$
H O

(b)

NONPOLAR

ALANINE (ala) VALINE (val) LEUCINE (leu) ISOLEUCINE (ile)

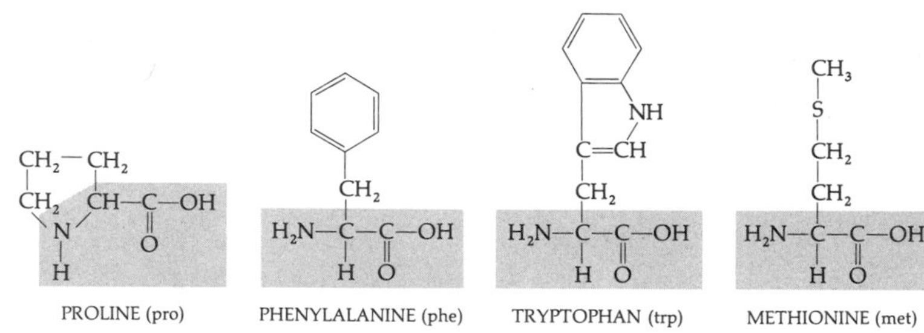

POLAR BUT UNCHARGED

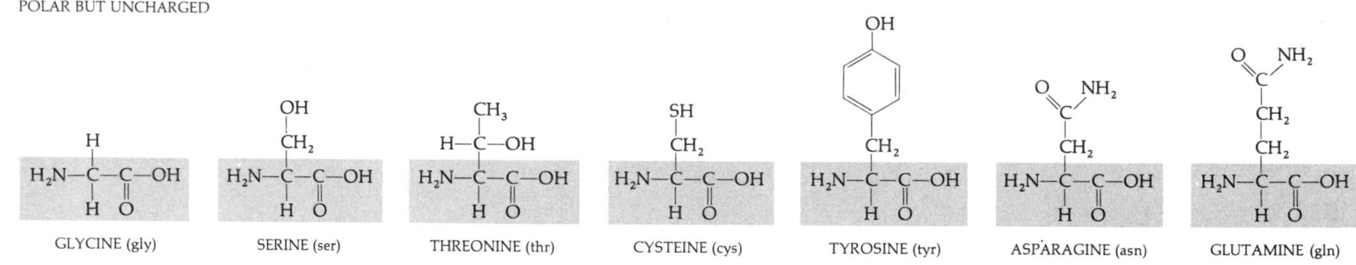

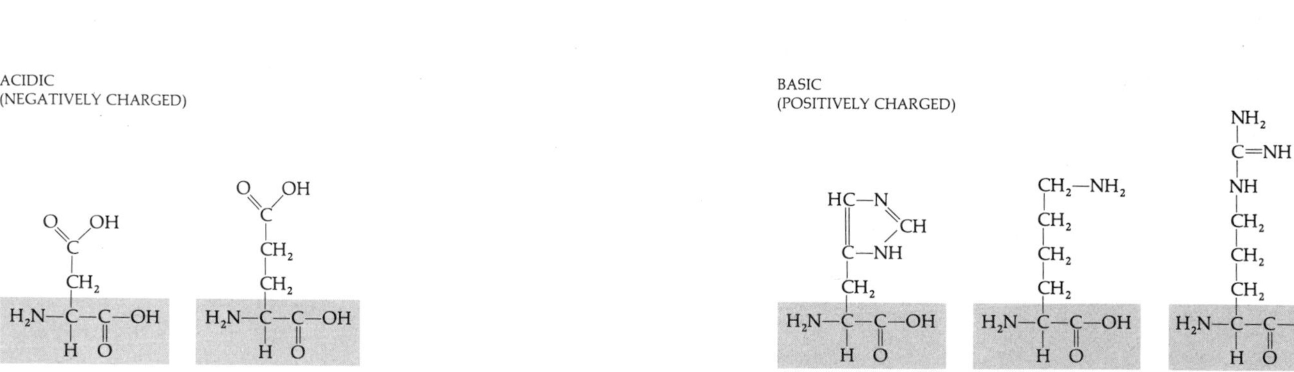

Amino Acids: The Building Blocks of Proteins

As we saw earlier, every amino acid has the same fundamental structure, which consists of a central carbon atom bonded to an amino group (NH_2), to a carboxyl group (COOH), and to a hydrogen atom. In every amino acid there is also another atom or group of atoms bonded to the central carbon. As shown in Figure 3–17, this side (R) group can be a hydrogen atom, in which case the amino acid is glycine; a CH_3 group, in which case the amino acid is alanine; and so on. The side group, depending on the atom or atoms that compose it, may have a positive charge, have a negative charge, be polar (with negative and positive zones), or have no charge at all (in which case it is hydrophobic).

A large variety of different amino acids is theoretically possible, but only about 20 are used to build proteins. And it is always the same 20, whether in a bacterial cell, a plant cell, or a cell in your own body.

In proteins, amino acids are joined together by peptide linkages (Figure 3–18). The sequence of amino acids in the chains determines the biological character of the protein molecule; even one small variation in the sequence may alter or destroy the way in which the protein functions. Protein molecules are large, often containing several hundred amino acids. Thus the number of different amino acid sequences, and therefore the possible variety of protein molecules, is enormous—about as enormous as the number of different sentences that can be written with our own 26-letter alphabet.

Organisms have only a very small fraction of the possible proteins. The single-celled bacterium *Escherichia coli*, for example, contains 600 to 800 different kinds of proteins at any one time, and the cell of a plant or animal has several times that number. In a complex organism, there are at least several thousand different proteins, each with a special function and each, by its unique chemical nature, specifically fitted for that function.

3–18
(a) *A peptide linkage is a covalent bond formed by condensation.* (b) *Polypeptides are polymers of amino acids linked together by peptide bonds, with the amino group of one acid joining the carboxyl group of its neighbor. The polypeptide chain shown here contains six different amino acids, but some chains may contain as many as 300 linked amino acid monomers.*

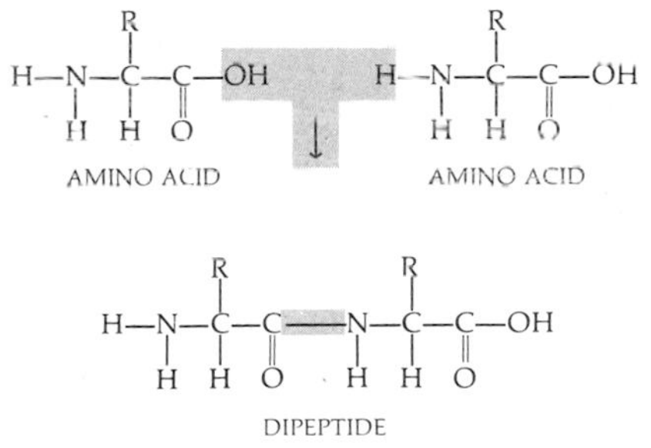

AMINO ACID AMINO ACID

DIPEPTIDE

(a)

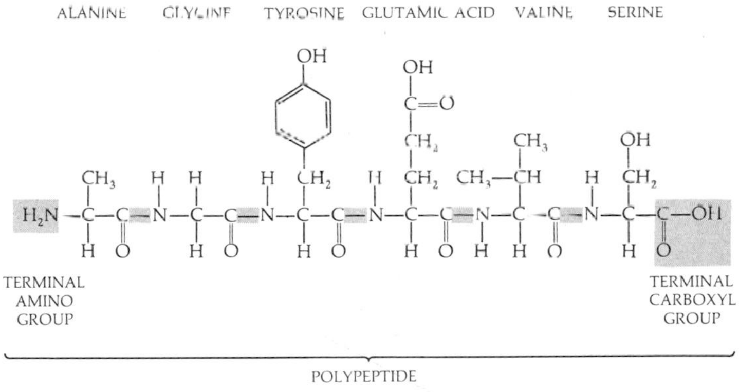

ALANINE GLYCINE TYROSINE GLUTAMIC ACID VALINE SERINE

TERMINAL AMINO GROUP

TERMINAL CARBOXYL GROUP

POLYPEPTIDE

(b)

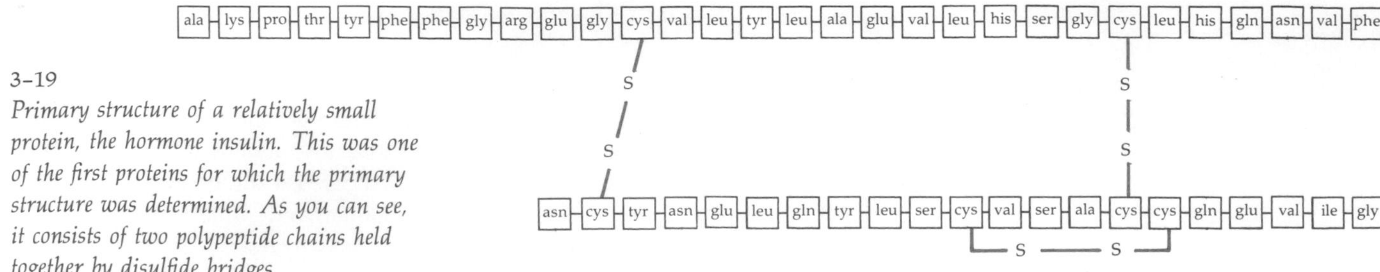

3–19
Primary structure of a relatively small protein, the hormone insulin. This was one of the first proteins for which the primary structure was determined. As you can see, it consists of two polypeptide chains held together by disulfide bridges.

The Levels of Protein Organization

Proteins are assembled in living systems with the amino group of one amino acid linked to the carbonyl* of another, like a line of boxcars. Such chains of covalently bonded amino acids may be called *polypeptides.* This linear sequence is known as the *primary structure* of the protein. Each different protein has a different primary structure. The primary structure of one protein is shown in Figure 3–19.

The protein is assembled in a long chain, one amino acid at a time. (This process will be described in some detail in Chapter 16.) As it is assembled, interactions begin to take place among the various amino acids along the chain. Linus Pauling and co-worker Robert Corey discovered that hydrogen bonds could form between the slightly positive amino hydrogen of one amino acid and the slightly negative carbonyl oxygen of another amino acid. They elucidated two structures that could result from these hydrogen bonds. One of these they called the alpha helix, because it was the first to be discovered, and the second, the beta pleated sheet. Proteins that have an extended form such as the helix or the pleated sheet are known as *fibrous proteins.* These structures are shown in Figure 3–21a and b. Biochemists refer to the regular, repeated configurations caused by hydrogen bonding between atoms of the polypeptide backbone as the *secondary structure* of a protein. Pauling won a Nobel Prize for his elucidation of the hydrogen bond and its role in protein structure.

3–20
Linus Pauling, who discovered the helical structure of proteins and the role of the hydrogen bond in maintaining it. On his right is a space-filling model of the alpha helix. Pauling won a Nobel Prize for his scientific accomplishments and a second Nobel Prize for his contributions to world peace, thus becoming the second person in history to win two Nobel Prizes. (Marie Curie was the first.)

Other forces, which involve the nature of the R groups, are also at work on the polypeptide chain, and these counteract the formation of the hydrogen bonds just described. For instance, an R group such as that of isoleucine is so bulky that it interrupts the turn of the helix, making the hydrogen bonding impossible. Wherever a cysteine encounters another cysteine, a covalent bond may form, making a disulfide bridge that locks the molecule in that position. R groups with unlike charges are attracted to each other and those with like charges are mutually repelled. As the molecule is twisted and turned in solution, the hydrophobic R groups become clustered together in the interior of the molecule and the hydrophilic groups extend outward into the aqueous solution. Hydrogen bonds form, linking together segments of the amino acid backbone. As a result of these interactions, many proteins take an intricately folded, globular shape, and these are known as *globular proteins.*

* In a peptide, the carboxyl group as such is no longer present—since the OH joined with an H of the amino group to form water—and so we refer to the C=O group as carbonyl.

AMINO ACIDS AND NITROGEN

Like fats, amino acids are formed within living cells using sugars as starting materials. But while fats are made up only of carbon, hydrogen, and oxygen atoms, all available in the sugar and water of the cell, amino acids also contain nitrogen. Most of the nitrogen in the biosphere exists in the form of gas in the atmosphere. Only a few organisms, all microscopic, are able to incorporate nitrogen from the air into compounds—nitrites and nitrates—that can be used by living systems. Hence the amount of nitrogen available to the living world is very small.

Plants incorporate the nitrogen in nitrites and nitrates into carbon-hydrogen compounds to form amino acids. Animals are able to synthesize some of their amino acids, using ammonia as a nitrogen source. The amino acids they cannot synthesize, the so-called essential amino acids, must be obtained either directly or indirectly from plants. For adult human beings, the essential amino acids are lysine, tryptophan, threonine, methionine, phenylalanine, leucine, valine, and isoleucine.

People who eat meat usually get enough protein and the correct balance of amino acids. People who are vegetarians, for either philosophical, esthetic, or economic reasons, have to be careful that they get enough protein and, in particular, the essential amino acids.

Until recently, agricultural scientists concerned with the world's hungry people concentrated on developing plants with a high caloric yield. Increasing recognition of the role of plants in supplying amino acids to the animal world has led to emphasis on the development of high-protein strains of food plants and of plants with essential amino acids, such as "high-lysine" corn.

Another approach to the right balance of amino acids is to combine certain foods. Beans, for instance, are likely to be deficient in tryptophan and in the sulfur-containing amino acids, but they are a good-to-excellent source of isoleucine and lysine. Rice is deficient in isoleucine and lysine but provides an adequate amount of the other essential amino acids. Thus rice and beans in combination make just about as perfect a protein menu as eggs or steak, as some non-scientists seem to have known for quite a long time.

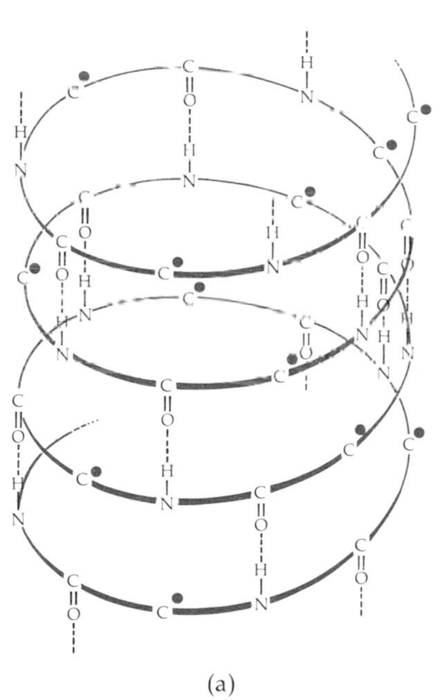

(a)

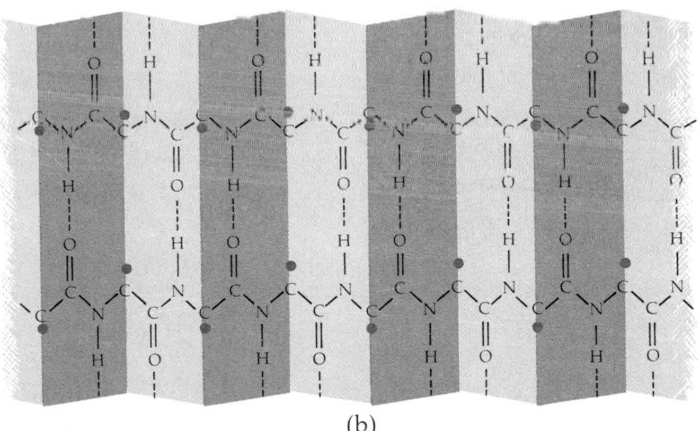

(b)

3–21
(a) *The alpha helix. The helix is held in shape by hydrogen bonds, indicated by the dashed lines. The bonds form between the oxygen atom in the carbonyl group in one amino acid and the amino group in another amino acid that occurs four amino acids farther along the chain. The R groups, not shown in this diagram, would be attached to the carbons indicated by the colored dots.* (b) *The beta pleated sheet structure of proteins. The pleats are formed by hydrogen bonding between atoms of the backbone of the polypeptide; the R groups, which would be attached to the carbons indicated by the colored dots, extend above and below the folds of the pleat.*

3–22
(a) *Bonds stabilizing the three-dimensional protein structure.* (b) *Space-filling model of a globular protein, chymotrypsin.*

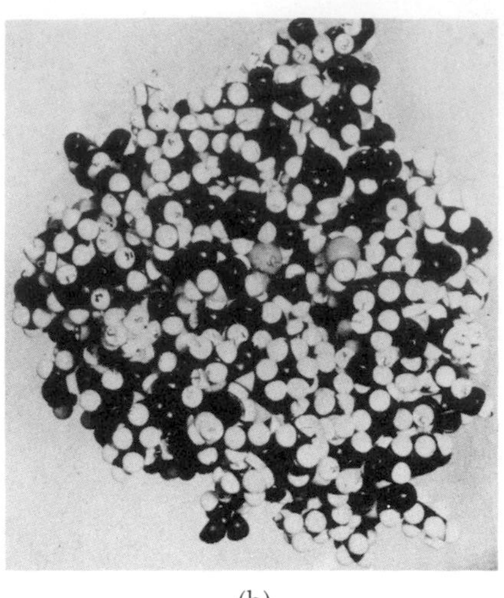

(b)

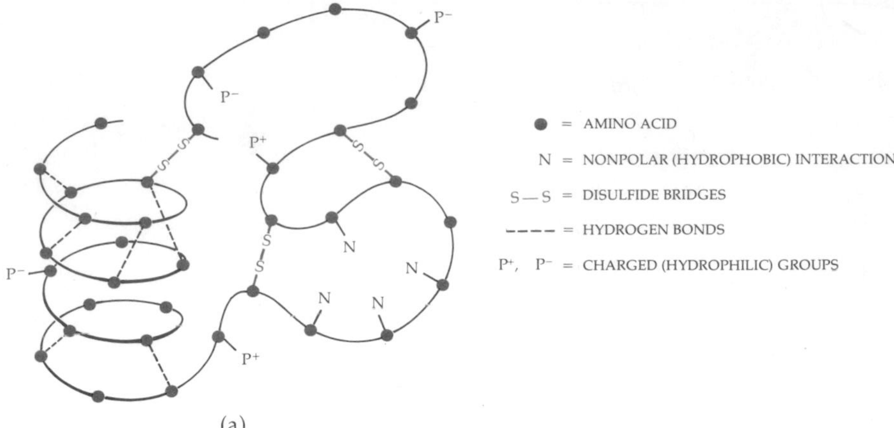

(a)

Figure 3–22 indicates the various types of bonds that are involved in a globular structure. The intricate three-dimensional structure involving interactions among R groups is often referred to as *tertiary structure.*

Many proteins are composed of more than one polypeptide chain. The polypeptide chains are held together by hydrogen bonds, disulfide bridges, attractions between positive and negative charges, and hydrophobic forces. Such proteins are often called multimeric; a protein containing four polypeptide chains is termed a tetramer. Often some of the polypetide chains within a multimeric protein are dissimilar. Hemoglobin, for example, is a tetramer; it is composed of four polypeptide chains, two of one kind (alpha) and two of another (beta). This level of organization of proteins, which involves interaction of two or more polypeptides, is called a *quaternary structure.*

Note that the tertiary and quaternary structures depend on the primary structure—the sequence of amino acids—and on the local chemical environment.

Structural Proteins

Fibrous Proteins

In general, fibrous proteins have a regular, repeated sequence of amino acids and so a regular, repetitious structure. An example is collagen, which makes up about one-third of all the protein in vertebrates. The basic collagen molecule is composed of three very long polymers of amino acids—about 1,000 amino acids per chain. These three polymers, which are made up of repeating groups of amino acids, are held together by hydrogen bonds linking amino acids of different chains in a tight coil (Figure 3–24). The molecules can coil so tightly because every third amino acid is glycine, the smallest of the amino acids.

Collagen performs many functions in the body. Consider a cow. Tendons, which link muscle to bone, are made up of collagen fibers in parallel bundles; thus arranged, they are very strong but do not stretch. The cow's hide, by contrast, is made up of collagen fibrils arranged in an interlacing network laid down in sheets. Even its corneas—the transparent coverings of the eyeballs—are composed of collagen. Boiling in water disperses the polymers of collagen into shorter chains, which we know as gelatin.

Other fibrous proteins include elastin, present in the elastic tissue of ligaments, silk, and keratin (Figure 3–25).

3–23
(a) *The primary structure of a protein is the linear sequence of its amino acids.* (b) *Because of interactions among these amino acids, the molecule spontaneously coils into a secondary structure, such as the alpha helix shown here, and* (c) *into a tertiary structure, such as a globule.* (d) *Many globular proteins, including hemoglobin and some enzymes, are made up of more than one amino acid chain. This structure is known as a quaternary structure.*

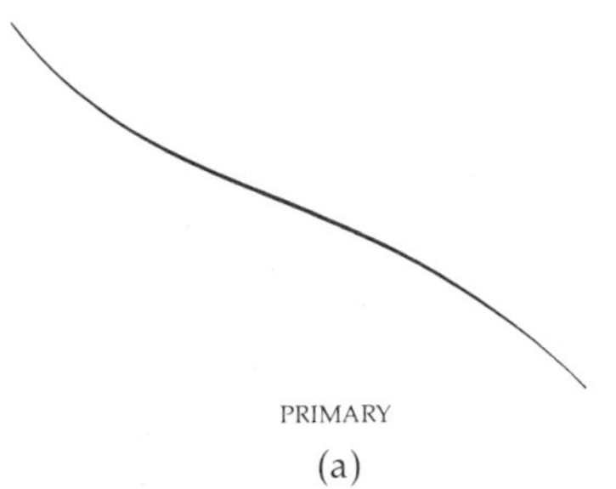

(a)

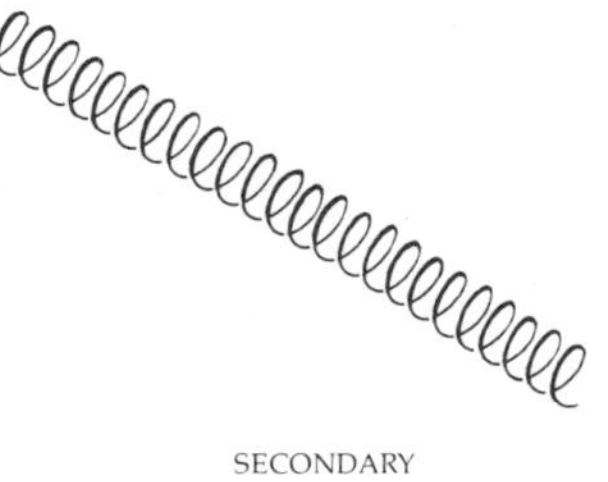

(b)

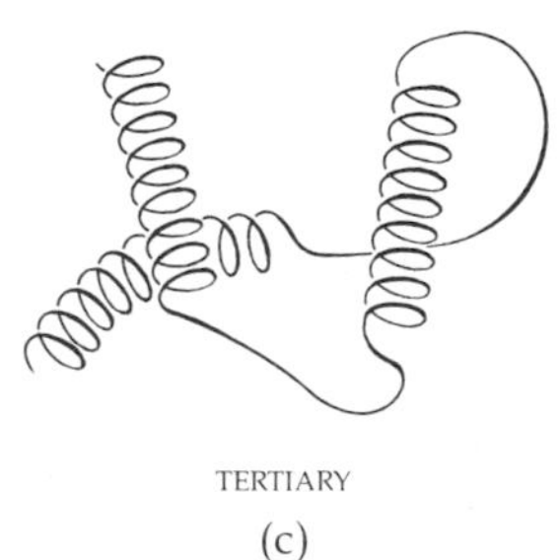

(c)

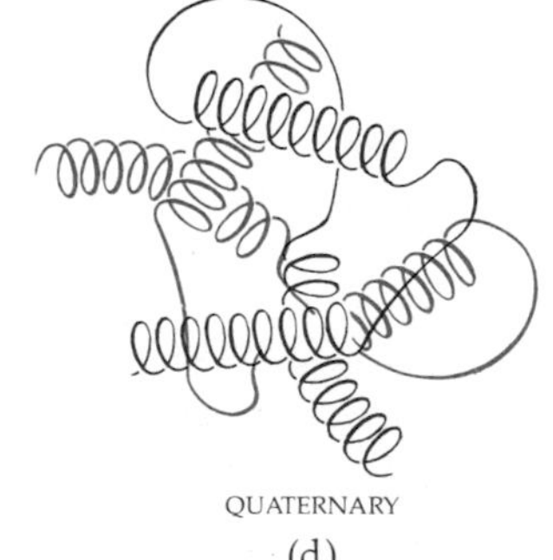

(d)

3–24
The structural protein collagen makes up about one-third of the protein in the human body and is probably the most abundant protein in the animal kingdom. Collagen is composed of long polypeptide chains woven together to form fibrils, such as those shown here. These fibers are a major constituent of skin, tendon, ligament, cartilage, and bone.

(a)

(b)

3–25
(a) *The structural protein keratin is found in all vertebrates. It is the chief component found in scales, horns, wool, nails, and feathers. These are cross sections of human hairs as seen by the scanning electron microscope.* (b) *Keratin also is the chief component of the formidable armor of this member of the class Reptilia.*

3–26

(a) *Microtubules are hollow tubes, so small that they cannot be visualized by a light microscope. They are composed of subunits, each of which is a globular protein. The subunits are of two types, alpha tubulin and beta tubulin, which first come together to form a soluble dimer. The dimers then self-assemble into insoluble hollow tubules.* (b) *Among their many functions, microtubules make up the internal structure of cilia, the small, hairlike appendages visible on this* Paramecium.

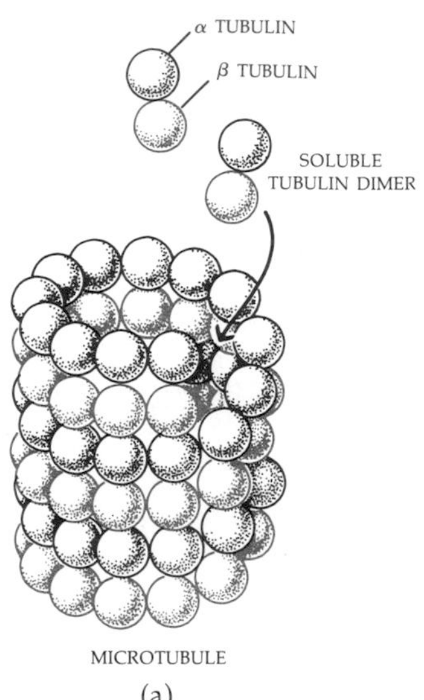

(a)

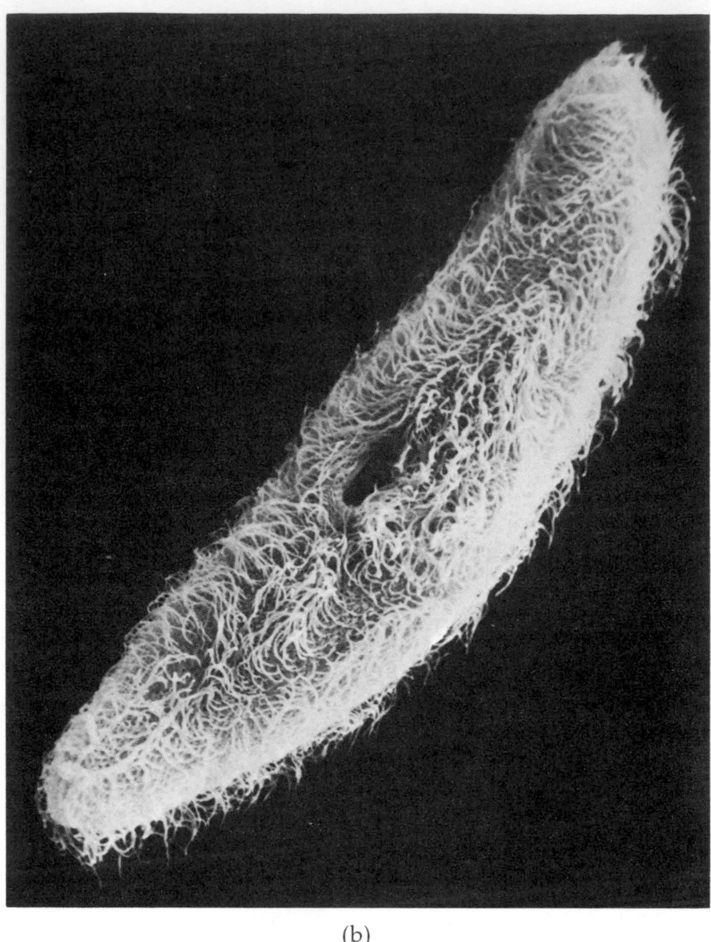
(b)

Structural Uses of Globular Proteins

Some structural proteins are globular. For example, microtubules, which function in a variety of ways inside the cell, are made up of globular proteins. They are long hollow tubes—so long that their entire length can seldom be traced in a single microscopic section. They apparently act as internal skeletons, stiffening parts of the cell body. They also may serve as tracks along which substances can move inside the cell. The formation of a new cellulose cell wall in a plant can be predicted by the appearance at the site of large numbers of microtubules; when cellulose fibrils are being laid down outside a plant cell membrane, as a cell wall forms or grows, it is possible to detect microtubules inside the cell aligned in the same direction as the fibrils outside.

Chemical analysis shows that each microtubule consists of a very large number of subunits, each of which is a globular protein made up of two polypeptide chains. The two polypeptides fit together because of their complementary configurations, forming approximately spherical subunits. The subunits assemble themselves into tubules (Figure 3–26), adding on length as required. When their job is over, they separate. Assembling requires an input of energy but just how it is triggered is not known.

3-27
The heme group of hemoglobin. It contains an iron atom (Fe) held in a porphyrin ring. The porphyrin ring consists of four nitrogen-containing rings, which are numbered in the diagram. Each heme group is attached to a long polypeptide chain that wraps around it. The oxygen molecule is held flat against the heme.

Hemoglobin: An Example of Specificity

Fibrous proteins, like structural polysaccharides, are usually molecules with a relatively small variety of monomers in a repetitive sequence. Many proteins, by contrast, have extremely complex, irregular amino acid sequences, as complex and irregular as the sequence of letters in a sentence on this page. Just as these sentences make sense (if they do) because the letters are the right ones and in the right order, the proteins make sense, biologically speaking, because their amino acids are the right ones in the right order.

Hemoglobin, for example, has a quaternary structure that consists of four polypeptide chains, each of which is combined with an iron-containing molecule known as *heme*. In heme, an iron atom is held in the center of a ring of nitrogen-containing atoms known as a porphyrin ring (Figure 3-27). Hemoglobin has two identical alpha chains and two identical beta chains, each about 150 amino acids long, for a total of about 600 amino acids in all. In man, hemoglobin molecules are manufactured and carried in the red blood cells. A mature red blood cell contains about 265 million molecules of hemoglobin. These molecules possess the special property of being able to combine loosely with oxygen so that they can collect oxygen in the lungs and release it in the tissues.

Sickle cell anemia is a disease in which the hemoglobin molecules are defective. When oxygen is removed, the defective molecules change shape and combine with one another to form stiffened rodlike structures. Red cells containing large proportions of these molecules become stiff and deformed, taking on the characteristic sickle shape. The deformed cells may clog the smallest blood vessels (capillaries), cutting off the local blood supply and resulting in anemia. The disease is usually painful and often fatal.

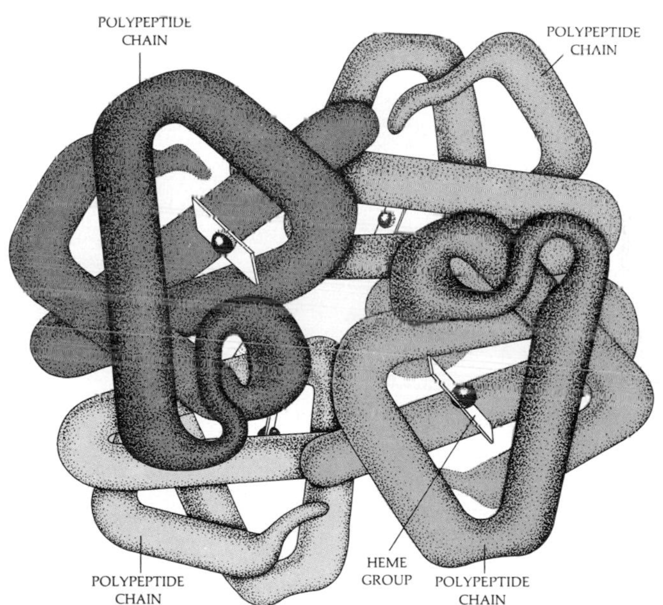

3-28
The hemoglobin molecule consists of four heme groups with their polypeptide chains intertwined in a quaternary structure. The outside of the molecule and the hole through the middle are lined by charged amino acids, and the uncharged amino acids are packed inside. Each molecule can hold up to four oxygen molecules. The sequence of the amino acids in each chain is its primary structure. The helical form assumed by any part of the chain as a consequence of hydrogen bonding between nearby C=O and NH groups is its secondary structure. The folding of the chains in three-dimensional shapes is the tertiary structure, and the combination of the four chains into a single functional molecule is the quaternary structure. (Adapted with permission from R. E. Dickerson and I. Geis, The Structure and Action of Proteins, *W. A. Benjamin, Inc., Menlo Park, Calif., 1969. Copyright 1969 by Dickerson and Geis.)*

3–29

An example of the remarkable precision of the "language" of proteins. Portions of the beta chains of the hemoglobin A (normal) molecule and the hemoglobin S (sickle cell) molecule are shown. The hemoglobin molecule is composed of two identical alpha chains and two identical beta chains, each chain consisting of about 150 amino acids, or a total of 600 amino acids in the molecule. The entire structural difference between the normal molecule and the sickle cell molecule (literally, a life-and-death difference) consists of one change in the sequence of each beta chain: One glutamic acid is replaced by one valine.

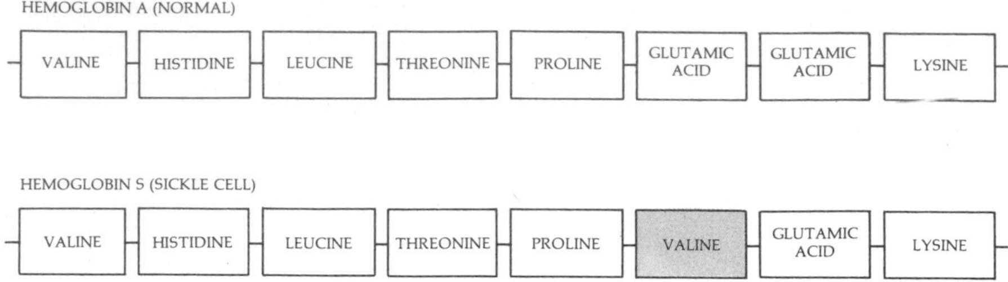

3–30

Scanning electron micrographs of (a) *red blood cell containing normal hemoglobin, and* (b) *red blood cells containing the abnormal hemoglobin associated with sickle cell anemia. In these sickle cells, so called because of their shape, the hemoglobin molecules stick together.*

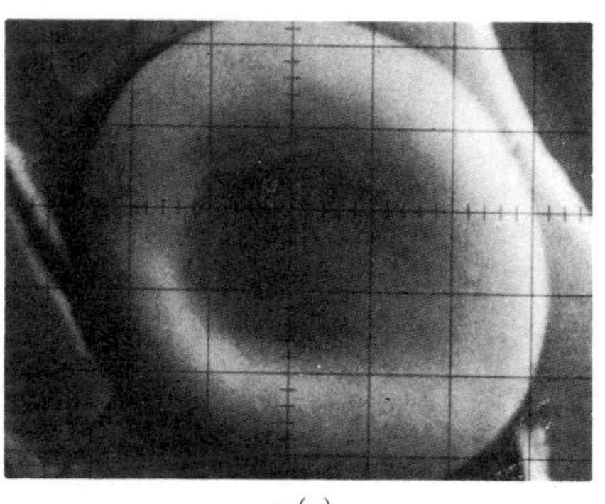

(a)

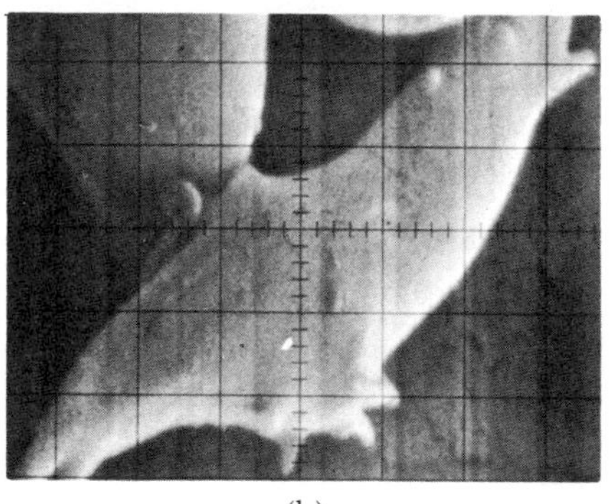

(b)

Analysis of the hemoglobin molecules of patients with sickle cell anemia reveals that the only difference between normal and sickle cell hemoglobin is that in a precise location in each beta chain, one glutamic acid is replaced by one valine.

When one considers that this difference of two amino acids in a total of almost 600 is actually the difference between life and death, one begins to get an idea of the meaning of specificity and of the precision and the importance of the arrangement of amino acids in a precise sequence in a protein.

LEVELS OF ORGANIZATION

In the last three chapters we have progressed from a brief consideration of subatomic particles to an examination of large proteins, among the most complex of all molecules. Although the neutrons, protons, and electrons of which atoms are composed are all the same, the elements differ greatly from one another. Mercury is a heavy metallic liquid, chlorine is a green gas, sulfur is a yellow powder, pure carbon can take the form of a hard solid—a diamond—and so on. The differences lie not in the nature of the subatomic particles, but rather in their number and arrangement—in their organization.

And just as subatomic particles combine to form atoms, atoms, as we have seen, combine to form molecules. At each level of organization, new properties appear. Water, for instance, as we have seen, is not the sum of the properties of hydrogen and oxygen; it is something more and also something different. In a molecule such as hemoglobin, we see how amino acids, with their diverse properties, become organized into a polypeptide and how polypeptide chains combine in a new level of organization, the quaternary structure of the complete molecule. Only at this level of organization do the complex properties of the molecule emerge, and only then can the molecule assume its function.

Living systems, as we noted earlier, obey the laws of physics and chemistry. Are we then "nothing but" a collection of atoms and molcules? There are recognizable differences between "life" and "nonlife." What is the basis of these differences? According to biologists, the differences are to be found not in the atoms and molecules themselves, but in their organization.

SUMMARY

The chemistry of living organisms is, in essence, the chemistry of compounds containing carbon. Carbon is uniquely suited to this central role by the fact that it is the lightest atom capable of forming four covalent bonds. Because of this capacity, carbon can combine with carbon and other atoms to form a great variety of strong and stable chain and ring compounds. Organic molecules derive their three-dimensional shapes primarily from their carbon skeletons. Many of their specific properties, however, are dependent on functional groups. Among the major types of organic molecules are (1) fatty acids, which have long hydrocarbon chains containing only carbon and hydrogen; (2) carbohydrates, whose monomers contain carbon, hydrogen, and oxygen in the ratio CH_2O; (3) amino acids; and (4) nitrogenous bases, such as adenine.

A general characteristic of all organic compounds is that they yield energy when oxidized. Sugars serve as a primary source of chemical energy for living systems. The simplest sugars are the monosaccharides ("single sugars"), such as glucose and fructose. Monosaccharides can be combined to form disaccharides ("two sugars"), such as sucrose, which is the form in which sugars are transported through the plant body.

Polysaccharides (chains of many monosaccharides), such as starch and glycogen, are storage forms for sugars. These molecules can be broken apart by hydrolysis. Polysaccharides also form important structural elements. The chief structural polysaccharide is cellulose, found in plant cell walls.

Lipids are hydrophobic organic molecules that, like carbohydrates, play important roles in energy storage and as structural components. Fats are the chief energy-storing lipids. A fat molecule consists of one molecule of glycerol bonded to three fatty acids. Fats are designated as saturated or unsaturated depending on whether or not their fatty acids contain any double bonds. Unsaturated fats, which tend to be oily liquids, are more commonly found in plants.

Phospholipids are major structural components of cellular membranes. Phospholipids consist of one unit of glycerol, two fatty acids (instead of the three fatty acids present in fats), and a phosphate group to which another polar group may be attached. Because of their hydrophilic "heads" and hydrophobic "tails," phospholipids spontaneously orient in water to form films and clusters that are the basis of the membrane structure.

Proteins are very large molecules composed of long chains of amino acids. The 20 different amino acids used in making proteins vary according to the properties of their side groups. From these an extremely large variety of different kinds of protein molecules can be synthesized, each of which has a highly specific function in living systems.

The sequence of the amino acids is known as the primary structure of the protein. Depending on the amino acid sequence, the molecule may take on any of a variety of forms. Hydrogen bonds between CO and NH groups tend to fold the chain into a repeating structure, including the alpha helix or the pleated sheet, sometimes referred to as the secondary structure, which, in turn, is further folded into an intricate, globular form, called a tertiary structure. Two or more polypeptide chains may interact to form a quaternary structure.

In fibrous proteins, the long molecules interact with other similar, or identical, long polypeptide chains to form cables or sheets. Collagen, hair, silk, wool, horns, nails, and feathers are all made up of fibrous structural proteins. Globular proteins may also serve structural purposes. Microtubules, which are important cell components, are composed of repeating units of globular proteins assembled helically into a hollow tubule.

Because of the variety of amino acids, proteins are able to have a high degree of specificity. An example is hemoglobin, the oxygen-carrying molecule of the blood, which is composed of four (two pairs of) protein chains, each attached to an iron-containing (heme) group. Substitution of one amino acid for another in one of the pair of chains results in a serious and sometimes fatal form of anemia known as sickle cell anemia.

QUESTIONS

1. Distinguish among the following: hydrocarbon/carbohydrate; glucose/fructose/sucrose; amylopectin/glucagon/glycogen; polysaccharide/polymer/polypeptide.

2. Identify the functional groups in the following compounds:

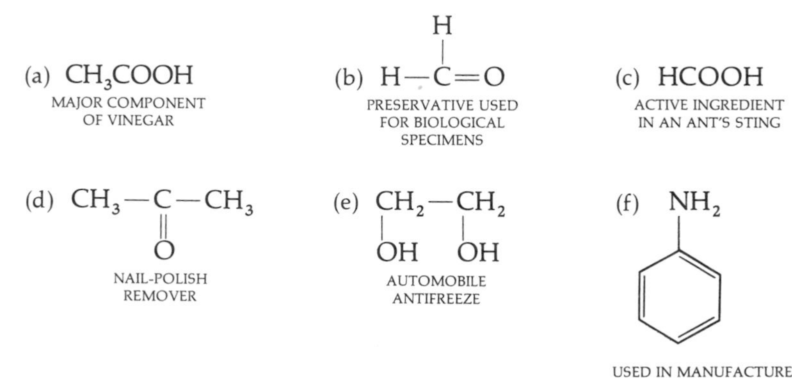

Which of these is hydrophilic? Hydrophobic?

3. Draw a structural formula for (a) a carbohydrate; (b) a fatty acid; (c) an amino acid.

4. Butyric acid, $CH_3CH_2CH_2COOH$, gives rancid butter its odor and flavor. Draw its structural formula.

5. Why do plants usually store energy reserves as polysaccharides, whereas, in most animals, lipids are the principal form of energy storage? What kinds of storage materials would you expect to find in seeds?

6. Why do higher animals have two forms of energy reserves? Why are oils a common form of energy storage in small aquatic organisms?

7. Both disaccharides, such as sucrose, and polysaccharides, such as starch and glycogen, are broken down by hydrolysis. What is hydrolysis? Considering the function of these reactions in the life of a cell, what are two advantages of hydrolysis?

8. What do we mean when we say that some polysaccharides are "energy-storage" molecules and that others are "structural" molecules? Give an example of each. In what sense should any polysaccharide be regarded as an "energy-storage" molecule?

9. In pioneer days, soap was made by boiling animal fat with lye (sodium hydroxide). The boiling hydrolyzed the bonds linking the fatty acids to the glycerol molecule, and the sodium hydroxide reacted with the fatty acid to produce soap. A typical soap is sodium stearate:

$$CH_3CH_2CH_2CH_2CH_2CH_2CH_2CH_2CH_2CH_2CH_2CH_2CH_2CH_2CH_2CH_2CH_2C\begin{matrix}\nearrow O \\ \searrow ONa\end{matrix}$$

Explain how soap functions to trap and remove particles of dirt and grease.

10. Silk is a protein in which polypeptide chains are arranged in a beta pleated sheet. In these chains, the peptide sequence glycine-serine-glycine-alanine-glycine-alanine occurs repeatedly. (a) Draw the structural formula for this hexapeptide, and show the peptide bonds in color. (b) Explain how a peptide bond is formed.

11. As the text indicated, several interesting generalizations can be formulated from the data in Table 3-2. Study the table and formulate two generalizations that you think are justified by the data.

CHAPTER 4

Chemical Reactions, Enzymes, and Living Systems

4–1
(a) *An individual hydrogen atom has one electron in its first orbital (that is, at the lowest energy level). As two hydrogen atoms approach one another, repulsive and attractive forces are set in motion.* (b) *When the nuclei of the two atoms are a certain distance apart, these forces are exactly balanced. The two electrons are shared equally between the two nuclei, giving each atom a completed first energy level and forming a strong bond.*

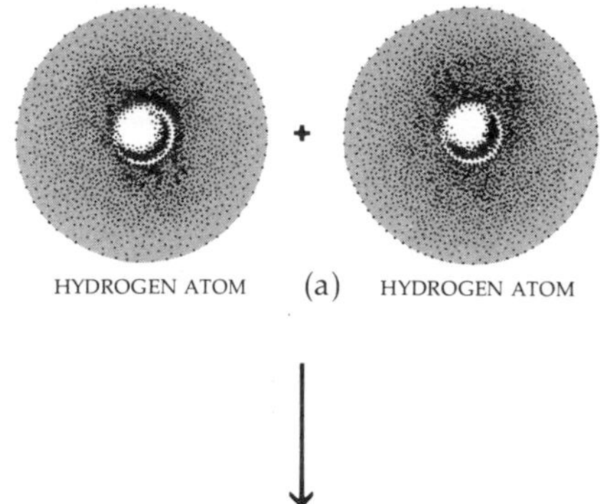

(a)

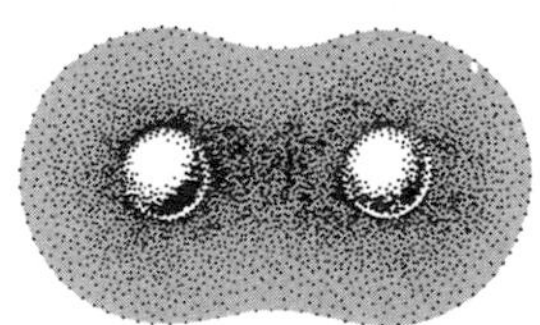

(b)

In the previous chapters, we have introduced you to the principal kinds of molecules found in living systems. In this chapter, we shall try to offer a glimpse of molecular dynamics. First we shall look at some general principles regarding the formation of molecules, and next we shall see how living systems build up and break down molecules, how they convert one kind to another, creating more of their own substance, and how they extract and harness the energy that drives these processes.

THE NATURE OF CHEMICAL REACTIONS

Let us look at a very simple example of a chemical reaction and consider what happens when two hydrogen atoms approach one another. For our purposes, a hydrogen atom is, simply, a positively charged proton and a negatively charged electron. When two hydrogen atoms approach one another, a variety of forces are set in motion. First, as they come closer, the negative charges of the electrons repel the atoms from one another. As they come closer still, the two atoms encounter one another's positive charges and are repelled. But there is a third force at work: The nucleus of the first atom attracts the electron of the second, and vice versa. When the two nuclei are a certain distance (about 7.4 nanometers or 0.74×10^{-8} centimeter—in other words, a little less than a hundred millionth of a centimeter) apart from one another, these forces of repulsion and attraction are equalized. At this distance, the two electrons are shared between the two nuclei, forming a strong covalent bond. This simple example illustrates two universal facts about chemical reactions:

1. All involve a change in the electron configurations of the atoms or molecules involved.
2. All require energy to make the transition between one stable electron configuration and another. This energy is known as the *energy of activation*.

Rates of Chemical Reactions

According to present concepts, atoms or molecules react with one another only when they collide with sufficient force to overcome the initial forces of repulsion.

The force required varies with the nature of the atoms or molecules; the more stable their initial state, the more forceful the collision must be. In any given group of atoms, it is likely that some proportion is moving with sufficient energy to cause a reaction to occur, but often this proportion is so small that the reaction, for all practical purposes, does not take place.

Reaction rates can be increased by increasing the likelihood of forceful collisions. One way to do this is to raise the temperature, thereby increasing the average velocity at which the atoms or molecules move and so increasing their likelihood of colliding with sufficient force. Sometimes, as in the case of natural gas, a spark is all that is needed. Once the reaction begins, it liberates heat that is transferred to the other CH_4 molecules until all are moving rapidly enough to react almost simultaneously with explosive force. Driving a reaction by heat is a method commonly used in both chemical laboratories (for example, the familiar Bunsen burner) and industry.

A second way to increase the rate of collisions is to increase the number of reacting molecules (or to decrease the volume in which they move). Laboratories and chemical factories usually work with pure chemicals in high concentrations.

A third way is to use catalysts. Catalysts lower the energy of activation of a reaction (see Figure 4–2). Metals, such as iron, nickel, and platinum, are commonly used as catalysts in industrial laboratories. Certain molecules apparently tend to cluster on the surfaces of such metals and so the likelihood of close encounters of the necessary kinds between the reactants is increased. The metals, although they participate in the reactions, are not used up by them, and so they can be used over and over again.

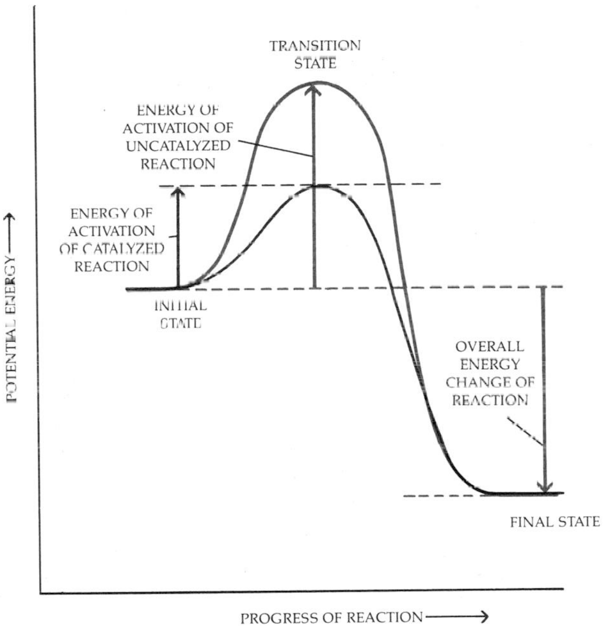

4–2
Chemical reactions require an input of energy to get them started—analogous to the spark that lights a fire or the push that sends a boulder on its downhill course. An uncatalyzed reaction requires more activation ("input") energy than a catalyzed one, such as an enzymatic reaction. The lower activation energy in the presence of the catalyst is often within the range of energy possessed by the molecules, and so the reaction can occur with little or no added energy. Note, however, that the overall energy change from the initial state to the final state is the same with and without the catalyst.

Types of Reactions

Chemical reactions can be classified into a few general types. One type can be represented by the expression:

$$A + B \longrightarrow AB$$

An example of this sort of reaction is the previously mentioned combination of hydrogen gas with oxygen gas to produce water:

$$2H_2 + O_2 \longrightarrow 2H_2O$$

A reaction may also take the form of a dissociation:

$$AB \longrightarrow A + B$$

For example, the equation above showing the formation of water can be reversed:

$$2H_2O \longrightarrow 2H_2 + O_2$$

This means that water molecules yield hydrogen and oxygen gases. A reaction may also involve an exchange, taking the form:

$$AB + CD \longrightarrow AD + CB$$

An example of such a reaction is the combination of hydrochloric acid (HCl) and sodium hydroxide (NaOH) to make table salt and water:

$$HCl + NaOH \longrightarrow NaCl + H_2O$$

Oxidation-Reduction

The passing of an electron from one molecule to another is known as an oxidation-reduction (or redox) reaction. The loss of an electron is known as *oxidation*, and the compound that loses the electron is said to be oxidized. The reason electron loss is called oxidation is that oxygen, with its high electronegativity, is often the electron acceptor.

Reduction is, conversely, the gain of an electron. Oxidation and reduction take place simultaneously because an electron that is lost by the oxidized atom is accepted by another atom, which is then reduced. (The fact that reduction means a gain of an electron seems paradoxical and may make it hard to remember the distinction between the two terms. It may help to recall that gain of an electron *reduces* the charge of an atom or molecule.)

Redox reactions may involve only a solitary electron, as for instance when sodium loses an electron and becomes oxidized to Na^+, and chlorine gains an electron, thereby reducing its charge (Cl^-). Often the electron travels with a proton, so oxidation involves removal of hydrogen atoms and reduction the gain of hydrogen atoms. When glucose is burned (oxidized), hydrogen atoms are lost by the glucose molecule and gained by oxygen:

$$C_6H_{12}O_6 + 6O_2 \longrightarrow 6CO_2 + 6H_2O$$

Chemical Equilibrium

Chemical reactions can go in either direction, as we have noted previously. When net change ceases, the reaction is said to be at equilibrium. In the reaction

$$A + B \rightleftharpoons C + D$$

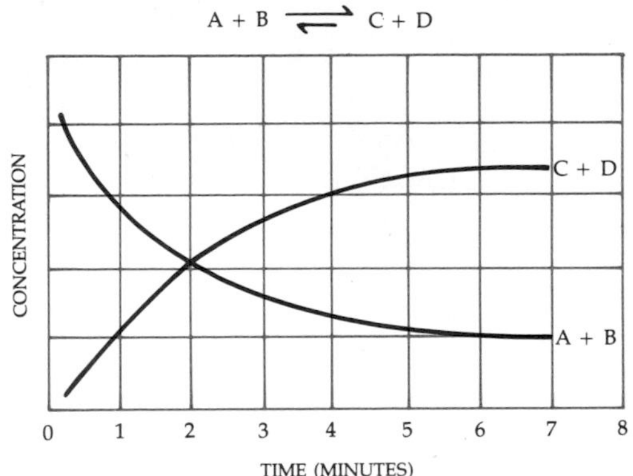

4-3
The changes in concentration of products and reactants in a reversible reaction. At first, only molecules of A and B are present. The reaction begins when A and B start to yield products C and D. At the end of two minutes, the concentrations of A + B and C + D are equal. As the reaction proceeds, the concentration of C + D will continue to increase to the point of chemical equilibrium (at about the sixth minute) and will thereafter remain greater than the concentration of A + B. This is the proportion at which the rates of forward and reverse reactions are the same.

the point of equilibrium is reached when as many molecules of C and D are being converted to molecules of A and B as molecules of A and B are being converted to molecules of C and D.

The concentration of reactants does *not* have to equal the concentration of products in order for equilibrium to be established; only the rates of the forward and reverse reactions must be the same. Consider the reaction shown at the bottom of the facing page. The different lengths of the arrows indicate that at equilibrium there is more C + D present than A + B. If only A and B molecules are present initially, the reaction occurs at first to the right, with A and B molecules converting into C and D molecules. Figure 4-3 shows the relative changes in concentration as the reaction continues. As C and D accumulate, the rate of the reverse reaction increases, and at the same time, the rate of the forward reaction decreases because of the decreasing concentrations of A and B. At about minute 6, the rates of the forward and reverse reactions equalize and no further changes in concentration take place. The proportions of A + B and C + D will remain the same, but there will always be more C and D molecules in the system.

METABOLISM

In any living system, thousands of different chemical reactions take place, many of them simultaneously. The sum of all these reactions is referred to as *metabolism* (from the Greek *metabole,* meaning "change"). If we were merely to list the individual chemical reactions, it would be difficult indeed to understand a cell's metabolic activities. Fortunately, there are some guiding principles that lead one through the maze of cell metabolism. First, virtually all the chemical reactions that take place in a cell involve *enzymes*—large protein molecules that play very specific roles. Second, biochemists are able to group these reactions in an ordered series of steps, commonly called a pathway; a pathway often has a dozen or more sequential reactions or steps. Each pathway serves a function in the overall life of the cell or organism. Furthermore, certain pathways have many steps in common, for instance, those that are concerned with the synthesis of the different amino acids or the various nitrogenous bases. Some pathways converge; for example, the pathway by which fats are broken down to yield energy leads to the pathway by which glucose is broken down to yield energy.

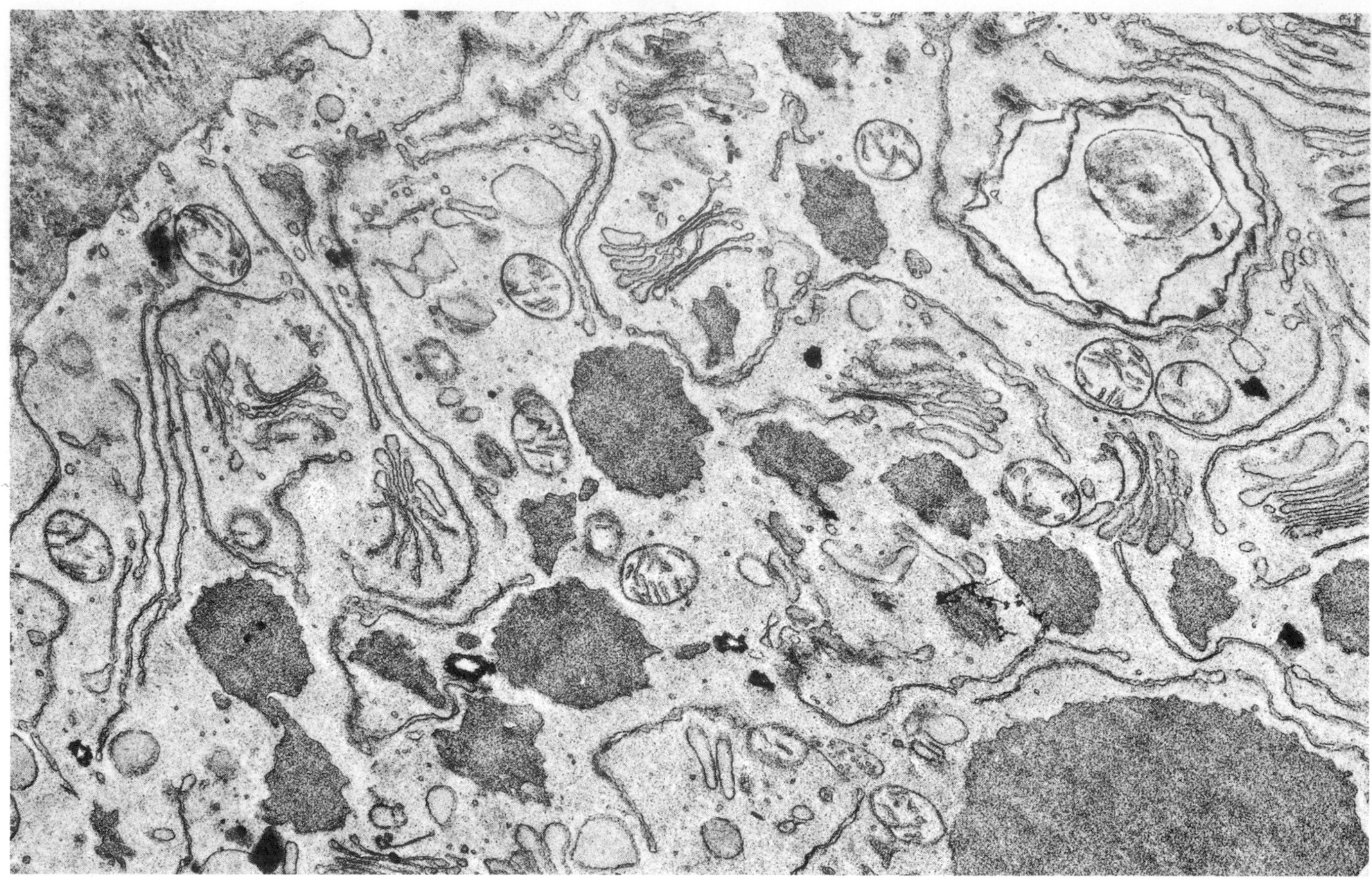

4–4
A chemical factory. Part of a cell from the root of a wheat plant. Cells such as this contain thousands of different organic molecules. These molecules form the many intricate structures of which the cell is composed and carry out the multitude of chemical reactions necessary for the life of the cell, its maintenance and growth, and its interactions with the cells around it. The chemical activities of the cells are carried out on the surface of the membranes, within many of the vesicles and other formed bodies, and in solution in the watery material surrounding them.

Many living systems have pathways unique to them. Plant cells expend much of their energy building their cellulose walls, an activity not engaged in by animal cells. Red blood cells specialize in the synthesis of hemoglobin molecules, not made anywhere else in the animal body. It is not surprising that the distinctive differences among cells and organisms are reflected not only in their forms and functions but also in their biochemistry. What is more surprising, however, is that much of the metabolism of even the most diverse of organisms is exceedingly similar; the differences in many of the metabolic pathways of humans, oak trees, mushrooms, and jellyfish are very slight. Some pathways are virtually universal, found in all living systems.

The magnitude of the chemical work carried out by a cell can be imagined if one recognizes that, for the most part, the thousands of different molecules, large and small, found within a cell are synthesized there. The total of chemical reactions involved in synthesis is called *anabolism*. Cells also are constantly involved in the breakdown of larger molecules; these activities are known, collectively, as *catabolism*. Catabolism serves two purposes: (1) It releases the energy for anabolism and other work of the cell, and (2) it serves as a source of raw materials for anabolic processes.

Not only do living systems carry out this multitude of chemical activities, but they do so under extraordinarily difficult conditions. Most of their chemical reactions are carried out within living cells (Figure 4–4), and in cells, not two or three but thousands of different kinds of molecules are intermingled. Moreover, temperatures are not high. (Otherwise, many of the fragile structures on which life depends would be destroyed.) How is all this complex chemical work accomplished? The question can be answered in a single word: enzymes. Enzymes are the catalysts and regulators for all metabolic processes of living systems. If it were not for enzymes, biochemical reactions would take place so slowly that, for all practical purposes, they would not occur at all, and the activities we associate with life would cease to exist.

SUBSTRATE (SUCROSE)

ACTIVE SITE

ENZYME MOLECULE (INVERTASE)

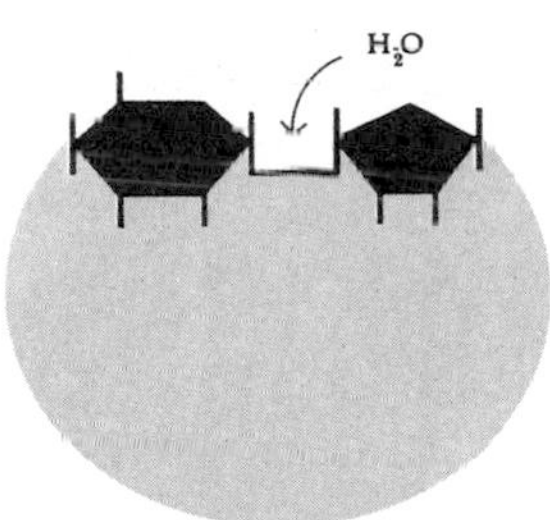

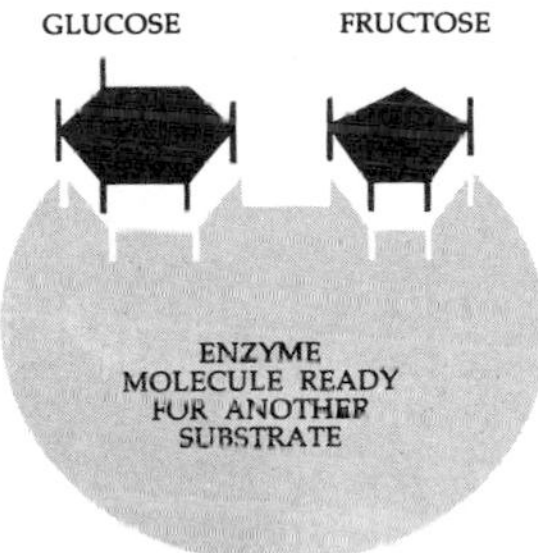

4–5

A model of the lock-and-key hypothesis of enzyme action. Sucrose is hydrolyzed to yield a molecule of glucose and a molecule of fructose. The enzyme involved in the reaction is specific for this process; as you can see, the active site of the enzyme exactly fits the opposing surface of the sucrose molecule.

ENZYMES AS CATALYSTS

Enzymes are the catalysts of biological reactions. They differ from the catalysts previously described in that they are very selective in their action. A particular enzyme can generally catalyze reactions only between a very specific group of reactants, known as its *substrate*. Otherwise, enzymes resemble the catalysts previously described in that they are not used up in the course of the reaction and so can be used over and over again. Enzymes also, as we shall see, appear to act in part by increasing the local concentrations of the reactants or substrate.

Enzymes enormously accelerate the rate at which reactions take place. For instance, the combination of carbon dioxide with water,

$$CO_2 + H_2O \rightleftharpoons \underset{\text{CARBONIC ACID}}{H_2CO_3}$$

can take place spontaneously, as it does in the oceans. In the human body, however, this reaction is catalyzed by an enzyme, carbonic anhydrase, which is one of the fastest enzymes known, each enzyme molecule producing 10^5 (100,000) molecules of carbonic acid per second. The catalyzed reaction is 10^7 times faster than the uncatalyzed one. In animals, this reaction is essential in the transfer of carbon dioxide from the cells, where it is produced, to the bloodstream, which transports it to the lungs. (Note also another characteristic of enzymes—they usually have names ending in *ase*.)

STRUCTURE OF ENZYMES

Enzymes are large to very large,* complex globular proteins consisting of one or more polypeptide chains. They are folded so as to form a groove or pocket into which the reacting molecule or molecules—the substrate—fit and where the reactions take place. This portion of the enzyme is known as the *active site*. The relationship between the active site and the substrate is very precise. Emil Fischer, who postulated the existence of active sites in 1894, compared the relationship to that of a lock and a key (Figure 4–5). The active site, we know now, is a result of the very exact folding of the polypeptide chain.

* Different enzymes vary in their molecular weight from about 12,000 to more than 1 million. (Molecular weight is simply the sum of the atomic weights of all the atoms in a given molecule.) Amino acids have an average molecular weight of about 120, and this figure is used to estimate the number of amino acids in a polypeptide of known molecular weight.

4–6

(a) *Primary structure of the enzyme lysozyme. This enzyme, which contains 129 amino acids, is found in many different types of animal cells. The sequence of the amino acids constitutes its primary structure. One consequential feature of the primary structure is the occurrence of the sulfur-containing amino acid cysteine. Covalent bonds form between the sulfur atoms, linking parts of the molecule together. Lysozyme breaks down the cell walls of many types of bacteria. It was first detected by Sir Alexander Fleming, better known for his discovery of penicillin, as a consequence of his having a bad cold. A droplet fell from his nose onto a culture dish containing bacteria, and he noticed that the bacteria in the vicinity of the droplet were destroyed. Lysozyme was subsequently discovered in tears, saliva, milk, and various other body fluids of many different animals. The molecule shown here was extracted from egg white. Lysozymes from other organisms might differ in some of their amino acids.* (b) *From this drawing you can gain a rough idea of the complex nature of the precise folding and coiling that make up the tertiary structure of a polypeptide. Again, it is the enzyme lysozyme that is shown. The crevice that forms the active site runs horizontally across the molecule. The substrate is shown in a darker color. (From R. E. Dickerson and I. Geis,* The Structure and Action of Proteins, *W. A. Benjamin, Inc., Menlo Park, Calif., 1969. Copyright 1969 by Dickerson and Geis.)*

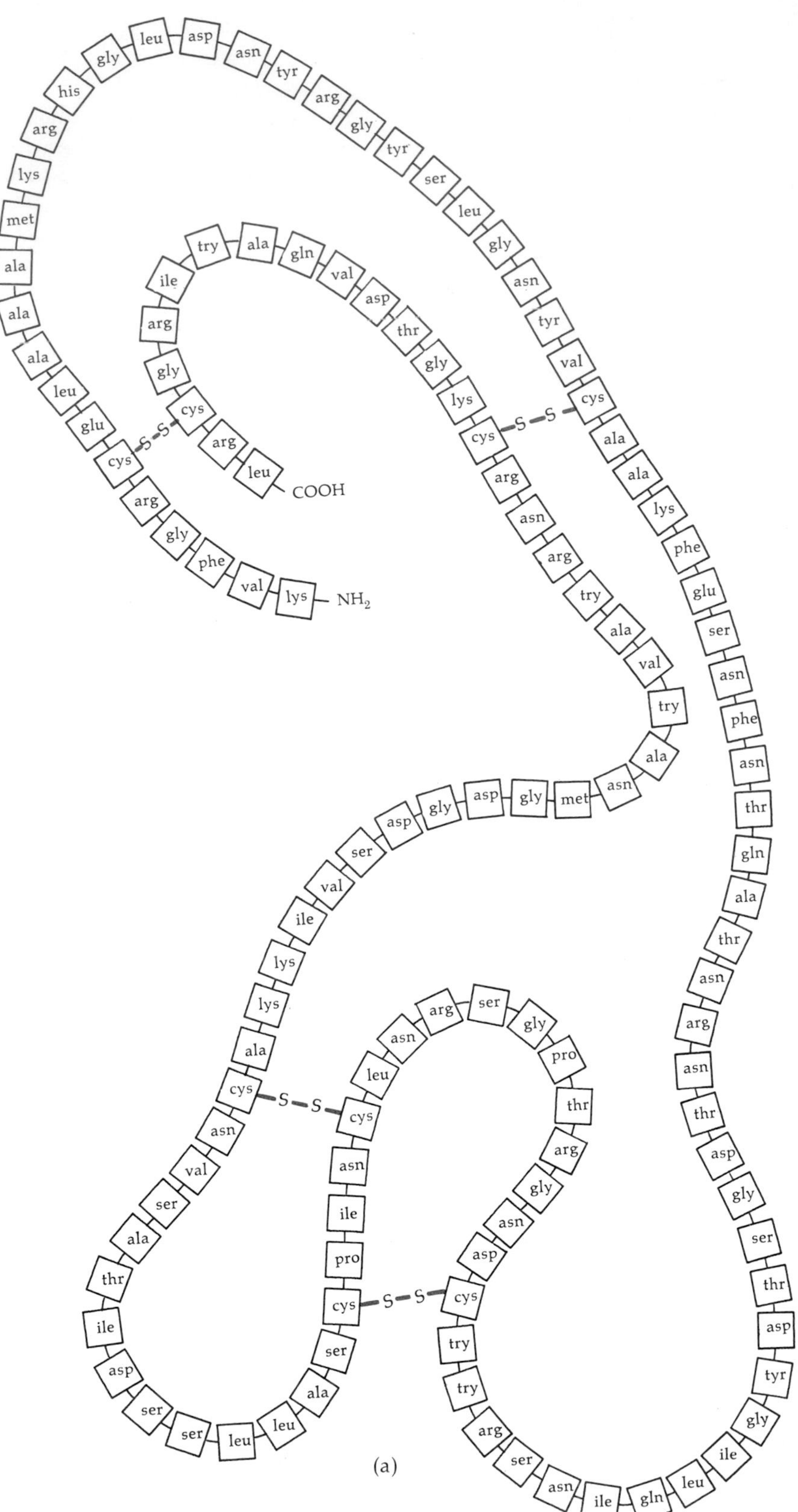

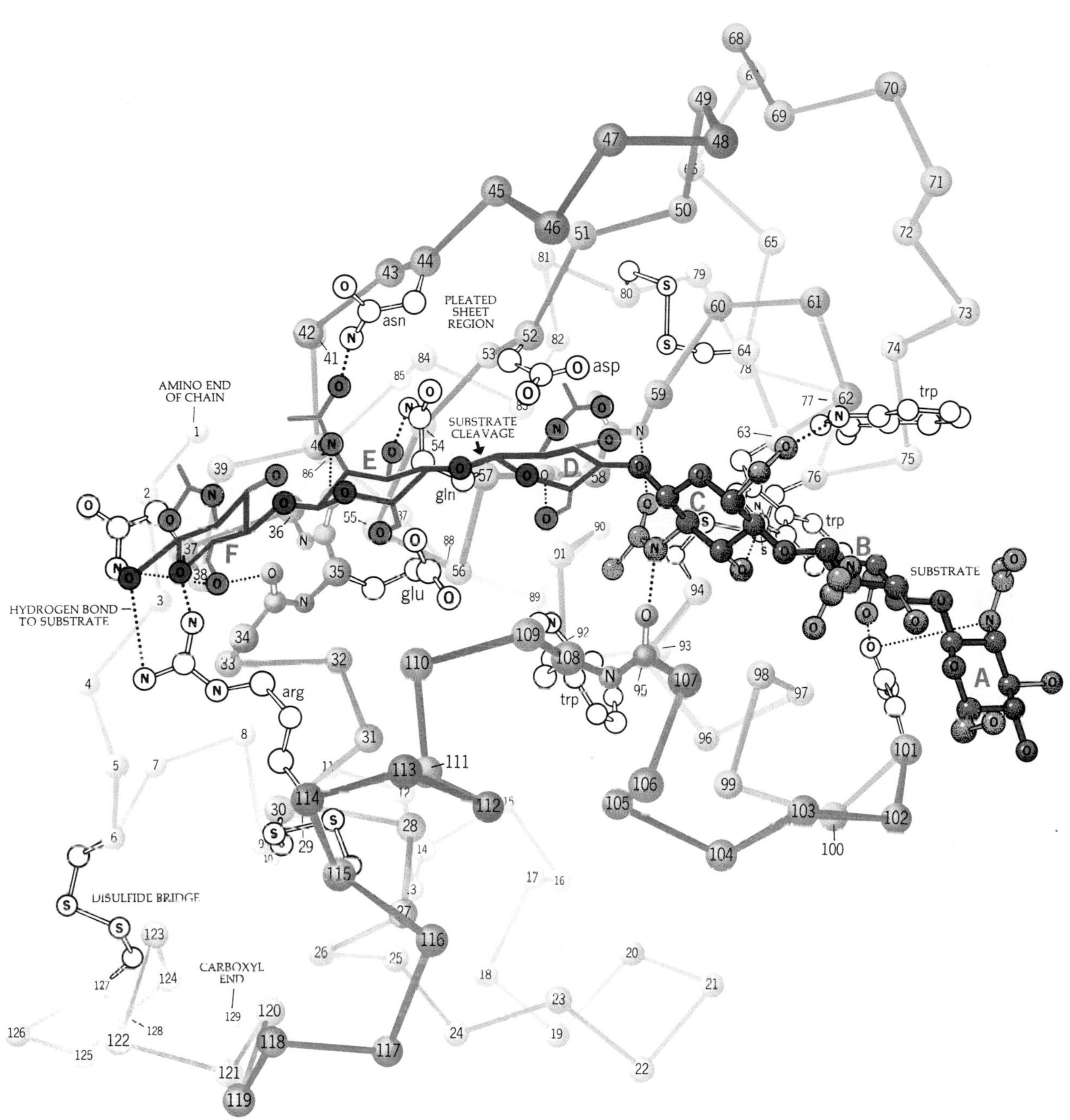

PLEATED
SHEET
REGION
AMINO END
OF CHAIN
SUBSTRATE
CLEAVAGE
HYDROGEN BOND
TO SUBSTRATE
SUBSTRATE
DISULFIDE BRIDGE
CARBOXYL
END
asn
asp
trp
gln
glu
arg
A
B
C
D
E
F

(b)

4-7
Model of an enzyme. This enzyme (the digestive enzyme chymotrypsin) is composed of three polypeptide chains. The amino (NH_2) and carboxyl (COOH) ends of each are labeled. The numbers represent the positions of particular amino acids in the chains. Five disulfide bridges connect amino acids 1 and 122, 42 and 58, 136 and 201, 168 and 182, and 191 and 220. The three-dimensional shape of the molecule is a result of a combination of disulfide bonds and of interactions among the chains and between the chains and the surrounding water molecules, based on the positive or negative charges or the polarity of the various amino acids. As a result of this bending and twisting of the polypeptide chains, particular amino acids come together in a highly specific configuration to form the active site of the enzyme. Two amino acids known to be part of the active site are shown in color.

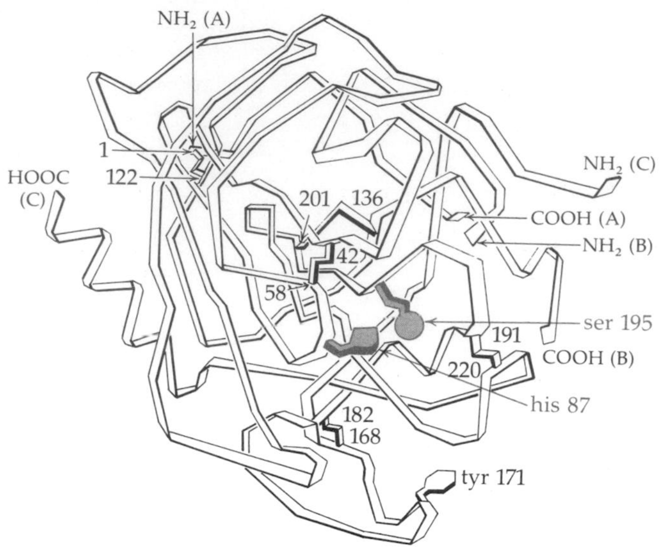

The active site not only has the right three-dimensional shape, but also it has exactly the right array of charged or uncharged, hydrophilic or hydrophobic areas on the binding surface. If a particular portion of the substrate has a negative charge, the corresponding feature on the active site has a positive charge, and so on. Thus the active site not only confines the substrate molecule but also orients it in the right direction.

The amino acids involved in the active site need not be adjacent to one another on the polypeptide chains. In fact, in an enzyme with quaternary structure, they may even be on different polypeptide chains, as shown in Figure 4-7. They are brought together at the active site by the precise folding of the molecule.

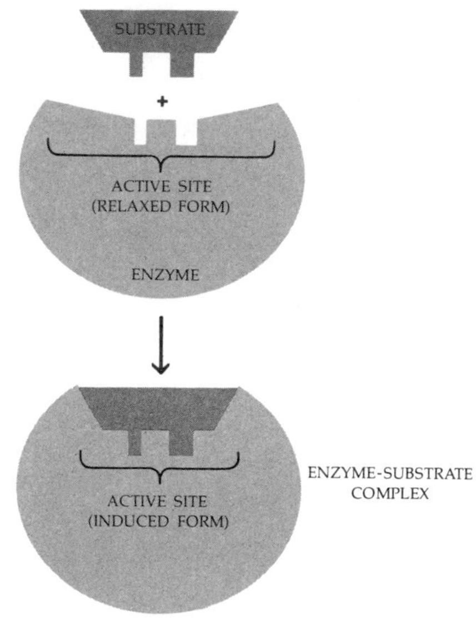

4-8
The induced-fit hypothesis. The active site is believed to be flexible and to adjust its conformation to that of the substrate molecule. This induces a close fit between the active site and the substrate.

The Induced-Fit Hypothesis

Within the last several years, studies of enzyme structure have suggested that the binding between enzyme and substrate alters the conformation of the enzyme, thus inducing the close fit between the active site and the reactants. It is believed that this induced fit may put some strain on the reacting molecules and so further facilitate the reaction (Figure 4-8).

COFACTORS IN ENZYME ACTION

The catalytic activity of some enzymes appears to depend only upon the physical and chemical interactions between the amino acids of the active site and the substrate. Many enzymes, however, require additional nonprotein low-molecular-weight substances in order to function. Such nonproteins that are essential for enzyme function are known as *cofactors*.

Ions as Cofactors

Certain ions are cofactors for particular enzymes. For example, the magnesium ion (Mg^{2+}) is required in all enzymatic reactions involving the transfer of a phosphate group from one molecule to another. Its two positive charges hold the negatively charged phosphate group in position. K^+, Ca^{2+}, and other ions play similar roles in other reactions. In some cases, ions serve to hold the protein together.

Coenzymes and Vitamins

Nonprotein organic molecules may also play a crucial role in enzyme-catalyzed reactions. Such molecules are called *coenzymes*. For example, in some oxidation-reduction reactions, electrons—often traveling as a hydrogen ion with a pair of electrons—are passed to a molecule that serves as an electron acceptor. There are several different electron acceptors in any given cell, each tailor-made to hold the electron at a slightly different energy level. As an example, let us look at just one, nicotinamide adenine dinucleotide (NAD^+), which is shown in Figure 4–9. At first glance, NAD^+ looks complex and unfamiliar, but if you look at it more closely, you will find that you recognize most of its component parts. The two units labeled ribose are five-carbon sugars. They are linked by two phosphate groups. One of the sugars is attached to the nitrogenous base adenine. The other is attached to another nitrogenous base, nicotinamide. (A nitrogenous base plus a sugar plus a phosphate is called a *nucleotide*, and a molecule that contains two of them is called a dinucleotide. Watch out for nucleotides; they will return.)

The nicotinamide ring is the business end of NAD^+, the part that accepts the electrons. Nicotinamide is a vitamin, niacin. Vitamins are compounds required in small quantities that humans and other animals cannot synthesize themselves and so must obtain in their diets. Thus we must eat foods containing niacin (which includes both nicotinamide and nicotinic acid but should not be confused with nicotine, found in tobacco). When nicotinamide is present, our cells can use it to make NAD^+. Many vitamins are coenzymes or parts of coenzymes.

Nicotinamide adenine dinucleotide, like other coenzymes, is recycled. That is, NAD^+ is regenerated when NADH passes its electrons on to another electron acceptor. Thus, although this coenzyme is involved in many cellular reactions, the actual number of NAD^+ molecules required is relatively small.

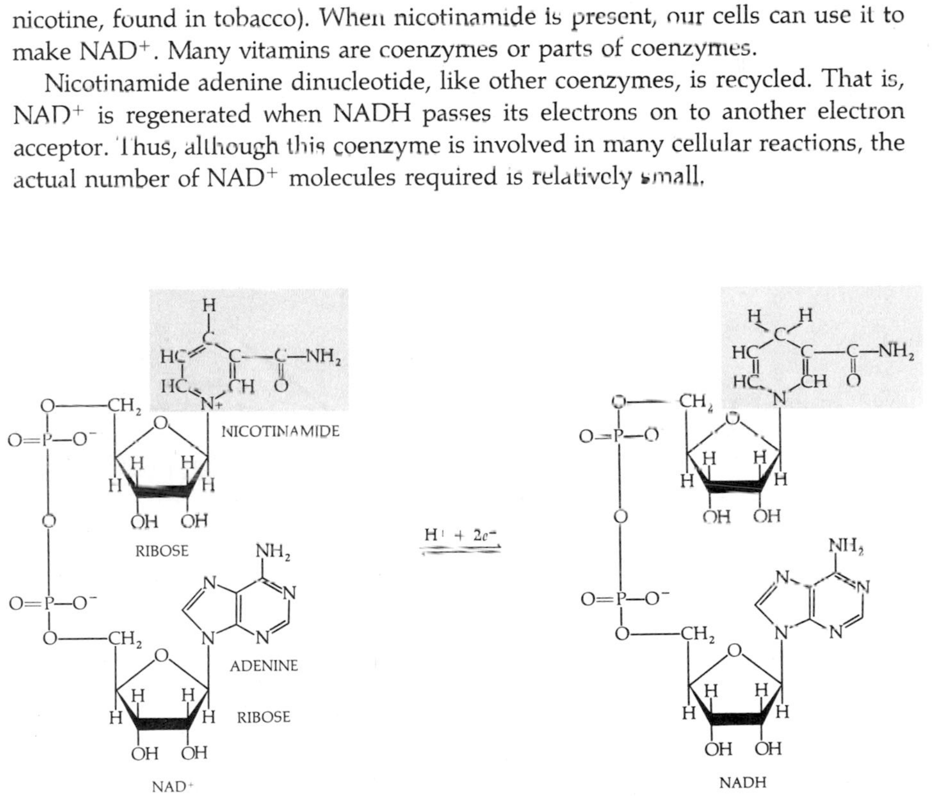

4–9
Nicotinamide adenine dinucleotide (NAD) in its oxidized form, NAD^+, and its reduced form, NADH.

NICOTINAMIDE
RIBOSE
ADENINE
RIBOSE
NAD^+
$H^+ + 2e^-$
NADH

4-10

The effect of temperature on the rate of an enzyme-controlled reaction. The concentrations of enzyme and reacting molecules (substrate) were kept constant. As you can see, the rate of the reaction, as in most chemical reactions, approximately doubles for every 10° C rise in temperature up to about 40° C. Above this temperature, it decreases, and at about 60° C it stops altogether, presumably because the enzyme is denatured.

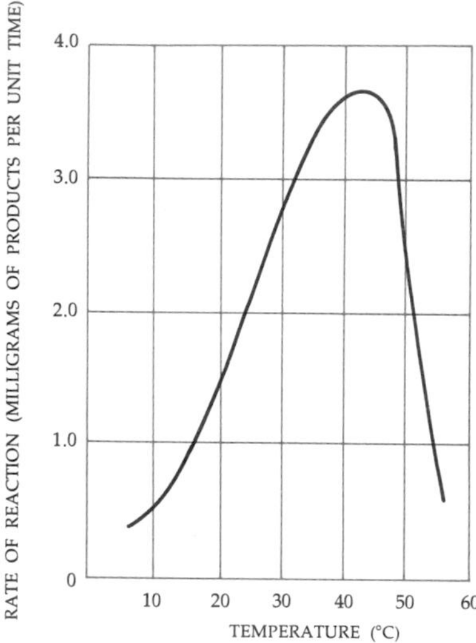

EFFECTS OF TEMPERATURE AND pH

As we noted earlier, an increase in temperature increases the rate of uncatalyzed chemical reactions. This temperature effect also holds true for enzyme-catalyzed reactions—but only up to a point. As you can see in Figure 4-10, the rate of most enzymatic reactions approximately doubles for each 10°C rise in temperature and then drops off very quickly at about 40°C. The increase in reaction rate occurs because of the increased energy of the reactants; the decrease in the reaction rate occurs as the enzyme molecule itself begins to move and vibrate, disrupting the hydrogen bonds and other relatively fragile forces that hold it together. A molecule that has lost its characteristic three-dimensional structure in this way is said to be *denatured*. Some denatured enzymes regain their activity on being cooled, indicating that their polypeptide chains have regained their necessary shape. Others, however, are permanently tangled and inactivated.

The pH of the surrounding solution also affects enzyme activity. The conformation of an enzyme depends, among other factors, on attractions and repulsions between negatively charged (acidic) and positively charged (basic) amino acids. As the pH changes, these charges change, and so the shape of the enzyme changes until it is so drastically altered that it is no longer functional. More important, probably, the charges of the active site and substrate are changed so that the binding capacity is affected. The optimum pH of one enzyme is not the same as that of another. The digestive enzyme pepsin, for example, works at the very low (highly acidic) pH of the stomach (page 42), in an environment where most other proteins would be permanently denatured.

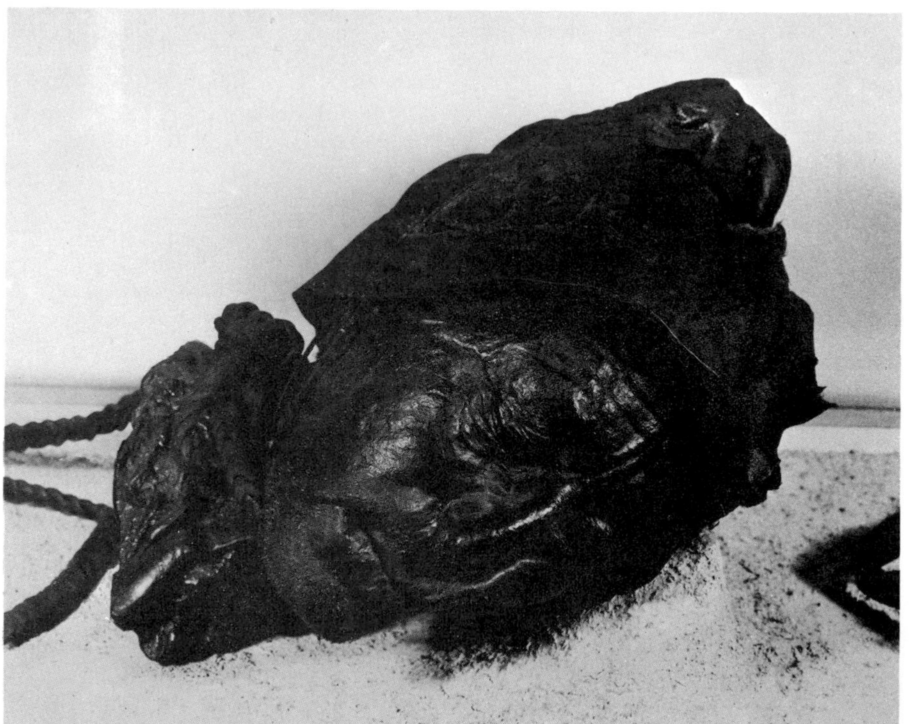

4-11

The body of a man with a noose around his neck was found in a Danish peat bog in 1950. He died some 2,000 years ago. The remarkable preservation of the body is the result of the extremely acidic pH of the peat bog, which almost completely inhibited the enzymatic activities of the microorganisms that customarily decompose organic molecules.

4–12
Particular sets of enzymes that carry out particular functions within the cell may be compartmentalized within membrane-bound structures, such as these mitochondria (striped structures), the powerhouses of the cell, and lysosomes (dense oval or irregularly shaped bodies), where destructive enzymes are segregated.

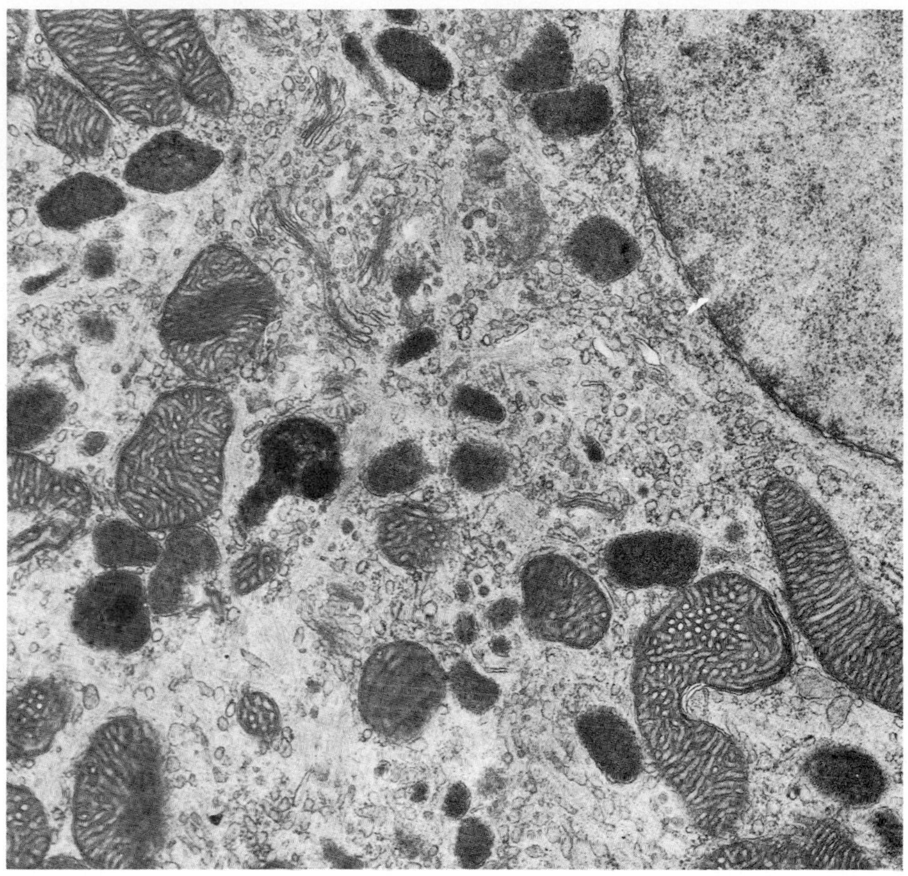

ENZYMATIC PATHWAYS

Enzymes work in series—the pathways we referred to earlier.

product 1 $\xrightarrow{\text{enzyme 1}}$ product 2 $\xrightarrow{\text{enzyme 2}}$ product 3 $\xrightarrow{\text{enzyme 3}}$
product 4 $\xrightarrow{\text{enzyme 4}}$ product 5 $\xrightarrow{\text{enzyme 5}}$ end product

As a consequence, compared to chemists in a laboratory, a living organism carries out its chemical activities with remarkable efficiency. First, there is little accumulation of waste products, since each tends to be used up in the next reaction along the pathway. A second advantage of such sequential reactions can be understood readily if you remember the principle of reversible reactions. If product 2, for example, is used up (by being converted into product 3) almost as rapidly as it is formed, the reaction product 1 $\longrightarrow$ product 2 can never reach equilibrium. If the eventual end product is also used up rapidly, the whole series of reactions will move toward completion.

A third advantage is that the groups of enzymes making up a common pathway can be segregated within the cell. Some are found in small vesicles, membrane-surrounded sacs in the cytoplasm. Others are embedded in the membranes of specialized organelles (Figure 4–12).

4-13
The characteristic color pattern of the Siamese cat is a result of a "mistake" in an enzyme affecting coat color. This enzyme, involved in controlling the synthesis of a dark pigment, is only active in the cooler, more peripheral areas of the body.

ENZYME DEFICIENCIES

As with the hemoglobin molecule described in the previous chapter, slight variations in the amino acid sequence of an enzyme can have drastic effects on the function of the enzyme. A "mistake" involving any of the amino acids directly involved in the active site renders the enzyme less efficient or even completely inactive. Changes in the rest of the molecule can be more or less important, depending on the extent to which they affect the tertiary or quaternary structure. For instance, one of the ways in which Siamese cats differ from their more common counterparts is that a particular enzyme, involved in controlling the synthesis of dark pigment for the coat, is especially sensitive to temperature. It functions adequately in the cooler, peripheral areas of the body, such as the ears, nose, paws, and tip of the tail, but it becomes inactive in the warmer areas of the body. Cat fanciers who want their Siamese to be perfect specimens do not let them out in the winter because exposure to cold can make the entire coat turn darker.

Enzymes and Human Diseases

Enzyme deficiencies are the cause of many human diseases. One such was described in 1649:

> The patient was a boy who passed black urine and who, at the age of fourteen years, was submitted to a drastic course of treatment which had for its aim the subduing of the fiery heat of his viscera, which was supposed to bring about the condition in question by charring and blackening his bile. Among the measures prescribed were bleedings, purgation, baths, a cold and watery diet and drugs galore. None of these had any obvious effect, and eventually the patient, who tired of the futile and superfluous therapy, resolved to let things take their natural course. None of the predicted evils ensued, he married, begat a large family, and lived a long and healthy life, always passing urine black as ink.

In 1902 the cause of this disease was identified by the physician Archibald Garrod. Alcaptonuria, as it came to be called, results from deficiencies in an en-

4–14

Four steps in the pathway for the breakdown of the amino acids phenylalanine and tyrosine. If the enzyme that catalyzes the conversion of phenylalanine to tyrosine is defective, the result is an accumulation of phenylalanine and the disease known as phenylketonuria. If the enzyme that catalyzes the conversion of homogentisate is missing, alcaptonuria ("black urine disease") results.

PHENYLALANINE

Step 1

TYROSINE

Step 2

p-HYDROXYPHENYLPYRUVATE

Step 3

HOMOGENTISATE

Step 4

4-MALEYLACETOACETATE

zyme that opens the ring structure (a benzene ring) at step 4 in the pathway of the breakdown of phenylalanine (see Figure 4–14). As a consequence, the ring-containing compound homogentisate accumulates in the urine and, on exposure to air, oxidizes, polymerizes, and turns black.

Unfortunately, not all enzyme deficiencies are so benign. Loss of function of another enzyme in this same pathway, the one that converts phenylalanine to tyrosine, produces phenylketonuria, which, if untreated, produces severe mental retardation and, in half of the cases, death by age 20.

The hereditary bases of alcaptonuria, phenylketonuria, and other diseases will be discussed in Chapter 18.

REGULATION OF ENZYME ACTIVITY

Another remarkable feature of the metabolic activity of cells is the extent to which each cell regulates the synthesis of the products necessary to its well-being in the amounts and at the rates required, while avoiding overproduction, which would waste both energy and raw materials. Temperature, as we have noted, affects enzymatic reactions, as does pH; some enzymes are usually found at a pH that is not their optimum, suggesting that this discrepancy may not be an evolutionary oversight but a way of damping enzyme activity. The availability of reactant molecules or of cofactors is a principal factor in limiting enzyme action, and most enzymes probably work at a rate well below their maximum for this reason.

Living systems also have more precise ways of turning enzyme activity on and off. Some enzymes are produced only in an inactive form and, just when they are needed, are activated, usually by another enzyme. Pepsin, a digestive enzyme, is controlled in this way. It is synthesized by cells in the lining of the stomach in the form of pepsinogen, which contains 42 additional amino acids on the amino end of the molecule and is inactive. After pepsinogen is released into the stomach, a particular enzyme snips off these last 42 amino acids, converting pepsinogen to pepsin, the active form. In this way, pepsin (and other digestive enzymes) are prevented from digesting the proteins in the cells in which they are synthesized. Once activated, these enzymes cannot again be deactivated, however.

Allosteric Interactions

An ingenious mechanism by which an enzyme may be temporarily activated or inactivated is known as *allosteric interaction.* Allosteric interactions occur among enzymes that have two binding sites, one the active site and another, into which a second molecule, known as an allosteric effector, fits. The binding of the effector changes the shape of the enzyme molecule and either activates or inactivates it.

4–15

An allosteric ("other shape") effector can bind to an enzyme and, by disrupting the bonds determining its tertiary structure, alter the active site. As a consequence, the enzyme may be altered so that it cannot interact with its substrate, as shown here, or it may be activated.

4–16

Feedback inhibition. (a) *In this series of reactions, each step (the black arrows) is catalyzed by a specific enzyme. Enzyme* E_1*, which converts A to B, is allosteric, and the allosteric inhibitor is the product F. Thus, enzyme* E_1 *will be more active when amounts of F are low.* (b) *A branched metabolic pathway, A being converted to B, B being converted to C, and C being converted to both D and X. In this case, the enzyme at the branch,* E_3*, is allosteric. As the quantity of the allosteric inhibitor F increases, the upper branch of the pathway will become less active, and the pathway will be primarily shunted to the production of Z.*

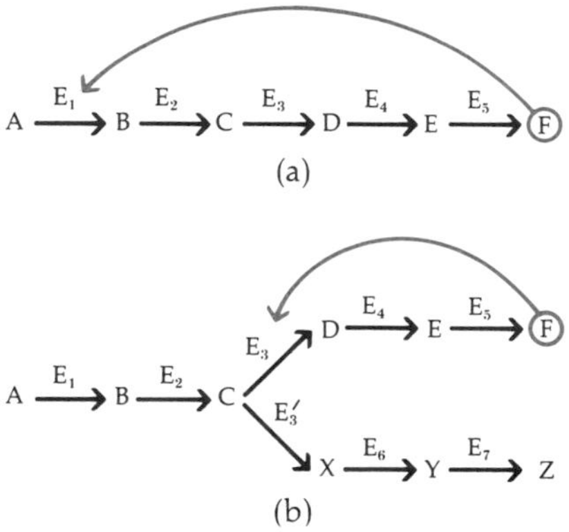

$$A \xrightarrow{E_1} B \xrightarrow{E_2} C \xrightarrow{E_3} D \xrightarrow{E_4} E \xrightarrow{E_5} (F)$$

(a)

(b)

Feedback Inhibition

Allosteric interactions are often involved in feedback inhibition, which is a common means of biological control. A familiar nonbiological example of feedback inhibition is a thermostat that turns off the furnace when the room temperature reaches a desired level. In feedback inhibition of enzymatic reactions, one of the products, usually the last in the series, acts as an effector, inhibiting the function of one of the enzymes, often the first in the series (Figure 4–16a). Or, in a reaction that may take one of two directions, the effector may act to shunt the reactions along another pathway (Figure 4–16b).

Competitive Inhibition

Some compounds inhibit enzyme activity by temporarily occupying the active site of the enzyme; regulation in this way is known as competitive inhibition, because the regulatory compound and the substrate compete for the active site. The result of the competition depends on how many of each kind of molecule are present. For example, in the reaction series

$$A \xrightarrow{E_1} B \xrightarrow{E_2} C \xrightarrow{E_3} D \xrightarrow{E_4} E \xrightarrow{E_5} F$$

the final product F might be rather similar in structure to product D. It could occupy the active site of enzyme E_4, preventing D, the normal substrate, from binding to the enzyme. As F was used up by the cell, the active site of enzyme E_4 would once more become available to D.

Competitive inhibition probably does not play a large role in the normal life of a cell, but it is an important mechanism in the action of some drugs. For instance, bacteria can make the vitamin folic acid, which animal cells do not make (animals obtain folic acid from their food). One of the compounds in the metabolic pathway leading to folic acid is para-aminobenzoic acid (Figure 4–17). Sulfanilamide, as you can see, has a structure very similar to PABA, so similar, in fact, that the enzyme involved in converting PABA to folic acid combines with the drug rather than with the PABA. Without folic acid, the bacterial cell dies, leaving the animal cell, which lacks this enzyme, unharmed.

Noncompetitive Inhibition

In noncompetitive inhibition, the inhibitory chemical, which need not resemble the substrate, binds with the enzyme either at the active site or elsewhere in the molecule and prevents its functioning. Lead, for instance, forms covalent bonds with sulfhydryl (SH) groups. Many enzymes contain cysteine, which has a sulfhydryl group. The binding of lead to such enzymes permanently deactivates them, producing the symptoms associated with lead poisoning.

The most potent poisons known, including arsenic, the cyanides, mercury, and the nerve gases, all combine with and prevent the functioning of enzymes necessary for survival, as do many useful drugs (Figure 4–18).

4–17

para-Aminobenzoic acid (PABA) is one of the compounds in the metabolic pathway to folic acid in bacterial cells. Sulfanilamide, a drug, has a similar structure. It can combine with the enzyme that converts PABA to folic acid, thereby blocking the synthesis of folic acid, without which the bacterial cell cannot live.

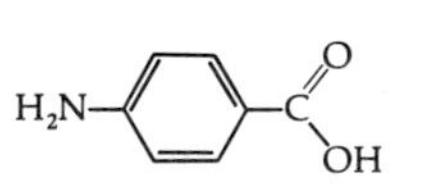

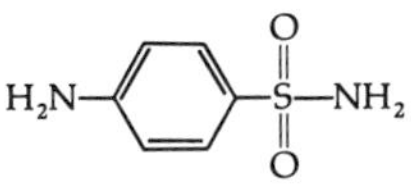

4–18
The bacterial cell wall has polysaccharide backbones formed by alternating molecules of two sugars, NAG and NAM. The backbones (indicated by the thick black lines) are crosslinked by short peptide chains. One of the amino acids in these chains, lysine (shown in color), forms three peptide bonds—one with the amino acid above it, one with the amino acid below it, and another with the adjacent amino acid. These bonds are formed in an unusual way—by the transfer of a peptide bond from one molecule to another. Penicillin blocks this transfer. Its structure mimics that of a dipeptide, and thus it can bind to the active site of the enzyme involved. Without crosslinks, the cell wall cannot hold the cell together. Thus penicillin acts specifically on growing bacterial cell walls.

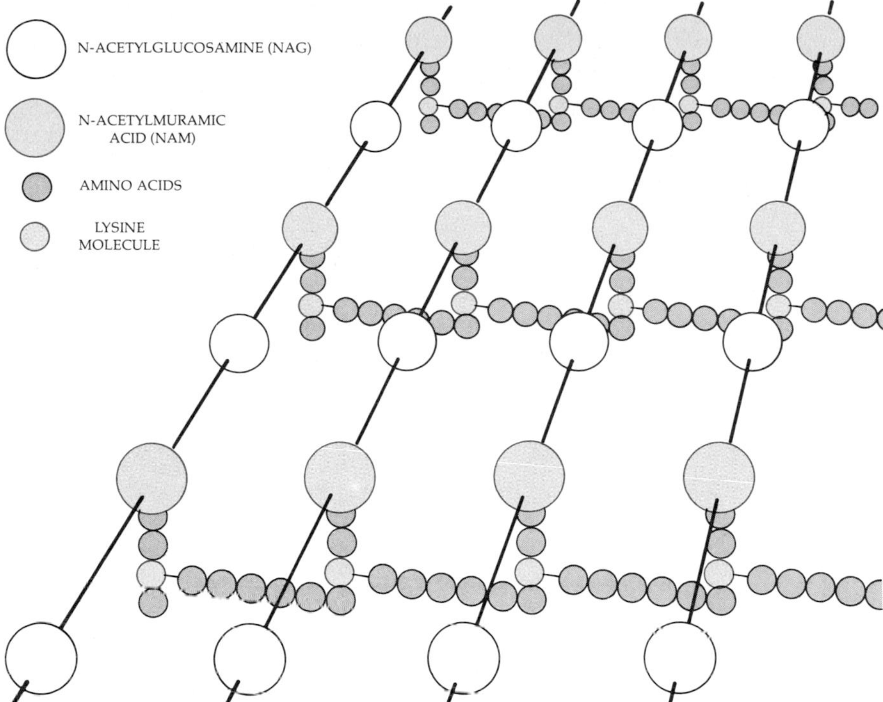

Regulation at the Source

Many enzymes are broken down rapidly, characteristically by other enzymes. For such enzymes, a highly efficient means of enzyme regulation is to produce these large protein molecules only when they are needed. The way in which bacterial cells turn on and off their enzyme production at the source is described in some detail in Chapter 17.

THE ENERGY FACTOR: ATP

All of the biosynthetic activities of the cell (and many other activities as well) require energy. A large proportion of this energy is supplied by a single molecule, *adenosine triphosphate,* or ATP. ATP is the cell's chief energy currency. Glucose, glycogen, and starch are like money in the bank. ATP is like change in your pocket or a subway token.

At first glance, ATP, as shown in Figure 4–19, appears to be a complex molecule. As with NAD^+, you will find, however, that its component parts are familiar to you. It is made up of adenine, a five-carbon sugar known as ribose, and three phosphate groups. These three phosphate groups with strong negative charges are covalently bonded to one another; this is an important feature in ATP function.

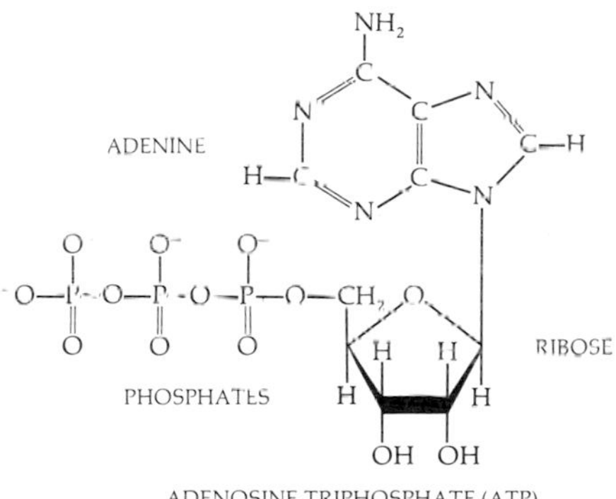

4–19
Adenosine triphosphate (ATP) is the cell's chief energy currency. The bonds between the three phosphate groups in the molecule are important in ATP function.

To understand the role of ATP, we have to return briefly to the concept of the chemical bond and bond energies. A chemical bond is the total of the forces that hold the constituent atoms together in a molecule. Because a bond is a stable configuration, it takes energy to break an old bond and form a new one. This energy is the energy of activation (Figure 4–2). Because of enzymes, which greatly reduce the energy of activation required, the reactions essential to life are able to proceed at an adequate rate. However, there is an important additional limitation on chemical reactions in living systems: The bond energies of the products must always be less than the bond energies of the reactants. (This is a variation on our previous statement that boulders will not roll uphill.) Following this statement to its logical conclusion, one might think that biosynthetic reactions could never take place. Not true. Cells circumvent this requirement by *coupled reactions* in which energy-requiring chemical reactions are linked to energy-releasing reactions. The molecule that participates most frequently in such coupled reactions is ATP.

The internal structure of the ATP molecule makes it unusually suited to this role in living systems. In the laboratory, energy is released from the ATP molecule when the third phosphate is removed by hydrolysis, leaving ADP (adenosine diphosphate) and a phosphate:

$$\text{ATP} + H_2O \longrightarrow \text{ADP} + \text{phosphate}$$

In the course of this reaction, about 7 kilocalories per mole of ATP are released, a relatively large amount of chemical energy. Removal of the second phosphate produces AMP (adenosine monophosphate) and releases an equivalent amount of chemical energy.

$$\text{ADP} + H_2O \longrightarrow \text{AMP} + \text{phosphate}$$

To indicate that relatively large amounts of chemical energy are involved, the covalent bonds linking these two phosphates to the rest of the molecule are often called high-energy bonds, symbolized by $\sim$. This term is somewhat misleading, however, because the energy released during the reaction does not arise entirely from the bond. The difference in energy between reactants and products is due only in part to the energies in the bonds. It is also a result of a rearrangement of the orbitals of the ATP or ADP molecules. The phosphate groups each carry negative charges and so tend to repel each other. When a phosphate group is removed, the molecule undergoes a change in electron configuration that results in a structure with less energy.

In living systems, ATP is also sometimes hydrolyzed to ADP; ATP hydrolysis provides, for example, a rapid means for producing heat, as in an animal waking up from hibernation. Usually, however, the terminal phosphate group is not simply removed but is transferred to another molecule. (This addition of a phosphate group is known as phosphorylation; enzymes that catalyze such transfers are known as kinases.) This reaction transfers some of the energy in the "high-energy" bond to the phosphorylated compound, which, thus energized, participates in the reaction.

For example, in the reaction,

$$A + B \longrightarrow C + D$$

if the bond energies of A plus the bond energies of B were less than those of C plus D, the reaction would not take place. Chemists could drive the reaction by supplying outside energy, probably in the form of heat. The cell might handle it this way:

$$A + \text{ATP} \longrightarrow \text{A-P} + \text{ADP}$$

The energy of the products is less than that of the reactants, so the reaction will take place. However, much of the ATP/ADP bond energy is conserved in the new compound A-phosphate, or A-P.

The next step in the reaction becomes:

$$\text{A-P} + \text{B} \longrightarrow \text{C} + \text{D} + \text{P}$$

With the release of the phosphate from A, this second reaction also becomes one in which the energy of the products is less than the energy of the reactants and which, therefore, can take place.

Take, for instance, the formation of sucrose in sugarcane.

$$\text{glucose} + \text{fructose} \longrightarrow \text{sucrose} + H_2O$$

In this reaction, the bond energy of the products is 5.5 kilocalories per mole greater than the bond energy of the reactants. However, the sugarcane plant couples this reaction to the breakdown of ATP:

$$\text{glucose} + \text{fructose} + 2\text{ATP} \longrightarrow \text{sucrose} + 2\text{ADP} + 2\text{P}$$

Since the bond energy of 2 ADPs is about 14 kilocalories per mole less than the bond energy of 2 ATPs, the overall difference in products and reactants becomes 8.5 kilocalories per mole. The coupling of the two reactions permits sugarcane to form sucrose.

Where does the ATP come from? As we shall see, energy released in the cell's catabolic reactions, such as the breakdown of glucose, is used to "recharge" the ADP molecule. Thus the ATP/ADP system serves as a universal energy-exchange system, shuttling between energy-releasing reactions and energy-requiring ones.

One marvels at the process of evolution as exemplified by the intricate flower of an orchid, the shell of a chambered nautilus, or the opposable thumb and forefinger of the human hand. Remember that NAD, ATP, and indeed the place of each amino acid in a polypeptide chain are also the products of evolution and also, you must admit, quite marvelous. Perhaps even beautiful.

SUMMARY

Chemical reactions involve a change in the electron configurations of the atoms or molecules involved, and require activation energy to make the transition between one stable electron configuration and another. Transfers of electrons from one atom or molecule to another are oxidation-reduction reactions. Oxidation is the loss of an electron, and reduction is the gain of an electron.

Metabolism is the total of all the chemical reactions that take place in cells. Reactions resulting in the breakdown or degradation of molecules are known, collectively, as catabolism. Biosynthetic reactions are called anabolism. Metabolic reactions take place in series, called pathways, each of which serves a particular function in a cell. Each step in the pathway is controlled by a particular enzyme.

Enzymes serve as catalysts, enormously increasing the rate at which reactions take place and remaining themselves unchanged in the process. They are large protein molecules folded in such a way that particular groups of amino acids form an active site. The reacting molecules, known as the substrate, fit precisely into this active site. Many enzymes require cofactors, which may be simple ions, such as Mg^{2+} or Ca^{2+}, or nonprotein organic molecules such as NAD. The latter are known as coenzymes; many vitamins are parts of coenzymes.

Enzyme-catalyzed reactions are under tight cellular control. The rate of enzymatic reactions is affected by temperature and pH, which affect the attractions among the amino acids of the protein molecule and also between the active site and substrate.

A principal means for enzyme control is allosteric interaction. Allosteric interaction occurs when a molecule other than the substrate combines with an enzyme at a site other than the active site and in so doing alters the shape of the active site to render it either functional or nonfunctional. Feedback inhibition occurs when the product of an enzymatic reaction at the end or at a branch of a particular pathway acts as an allosteric effector, temporarily inhibiting the activity of an enzyme earlier in the pathway and thus temporarily stopping the series of chemical reactions.

Enzymes may also be regulated by competitive inhibition, in which another molecule, similar to the normal substrate, competes for the active site.

ATP supplies the energy for most of the activities of the cell. The ATP molecule consists of a nitrogenous base, adenine; a five-carbon sugar, ribose; and three phosphate groups. The three phosphate groups are linked by two high-energy bonds—bonds that release a relatively large amount of energy when they are hydrolyzed. ATP participates as an energy carrier in most series of reactions that take place in living systems.

QUESTIONS

1. Distinguish among the following terms: metabolism/catabolism/anabolism; ATP/ADP/AMP; active site/substrate; allosteric interaction/competitive inhibition; lysosome/lysozyme.

2. Give another household example of feedback inhibition. If you cannot think of one, go flush the toilet.

3. Turn back to Figure 3–17, in which all the amino acids are shown, and try to make some educated guesses about which amino acids can substitute for one another in the structure of an enzyme, and what substitutions would have drastic effects.

4. In enzyme regulation by allosteric interaction, the inhibitor often works on the first enzyme of the series. In regulation by competitive inhibition, it often works on the last. Why this difference?

5. Most organisms cannot live at high temperatures. Explain at least one way in which high temperatures are harmful to organisms. Some bacteria and algae, however, live in hot springs, at temperatures far above those that can be tolerated by most organisms. How might such bacteria and algae differ from most other organisms?

6. When a plant does not have an adequate supply of an essential mineral, such as magnesium, it is likely to become sickly and may die. When an animal is deprived of a particular vitamin in its diet, it too is likely to become ill and may die. What is a likely explanation of such phenomena?

7. In one series of experiments with an enzyme *a* that catalyzes a reaction involving substrate *A*, it was found that a particular substance *X* inhibited the enzyme. When the concentration of *A* was high and the concentration of *X* was low, the reaction proceeded rapidly; as the concentration of *X* was increased and that of *A* was decreased, the reaction slowed down; when the concentration of *X* was high and that of *A* was low, the reaction stopped. If the concentration of *A* was again increased, the reaction resumed. How can you explain these results?

8. In another series of experiments, a substance *Y* was added to the enzyme-substrate mixture. As the concentration of *Y* was increased, the reaction slowed down and eventually stopped completely, regardless of the concentration of *A*. The investigators noticed that as *Y* was added, a solid material precipitated out of the solution. When more substrate was added, the reaction did not resume. How might these observations be explained? How does the inhibition by substance *Y* differ from that by substance *X* in Question 7? (*Hint:* Egg white is almost pure protein. Think about what happens when it is heated.)

9. When a sulfa drug, such as sulfanilamide, is prescribed for a bacterial infection, it is very important to remember to take the drug at the prescribed times and in the prescribed quantity. Why is this essential? Suppose you were instructed to take two tablets every three hours, and instead you took only one tablet every five hours. What do you think would happen?

CHAPTER 5

How Cells Are Organized

We have been pursuing our examination of living things by progressively noting the features that emerge as we move up the levels of organization. We have seen that atoms are composed of electrons, protons, and neutrons, and molecules, in turn, are made up of atoms. Large macromolecules are formed from smaller molecular building blocks. As we studied each level of organization we found that not all of the properties seen at that level could be fully explained in terms of the properties of its parts. Many properties of a system are determined by the way its components are organized. The three-dimensional structure of a protein molecule, for example, is in part a consequence of the properties of the individual amino acids, in part a consequence of their sequential arrangement, and in part a consequence of their environment.

Now we move to a level of biological organization higher than molecules, the cellular level. The characteristics of living systems, which we described in Chapter 1, do not emerge gradually as the degree of biochemical organization increases. They appear quite suddenly and specifically, in the form of the living cell.

PROKARYOTES AND EUKARYOTES

There are two fundamentally distinct kinds of cells, *prokaryotes* and *eukaryotes*. The prokaryotes include the bacteria and the blue-green algae (which many experts hold to be species of bacteria). Prokaryotic cells differ most notably from eukaryotic cells in that their nuclei are not surrounded by an outer membrane and their hereditary material is not organized into complex, protein-containing chromosomes (hence their name—*pro,* meaning "before," and *karyon,* meaning "nucleus" or "kernel").

All other organisms are eukaryotes ("true nucleus"). Eukaryotic cells have a well-defined nucleus, separated from the rest of the cell by a special membrane, the nuclear envelope. They have complex chromosomes, and they also have a variety of specialized structures not found in prokaryotes. Eukaryotic cells are nearly always larger than prokaryotic cells. The evolutionary relationship between the prokaryotes and the eukaryotes—quite an interesting subject—will be explored in Chapter 20.

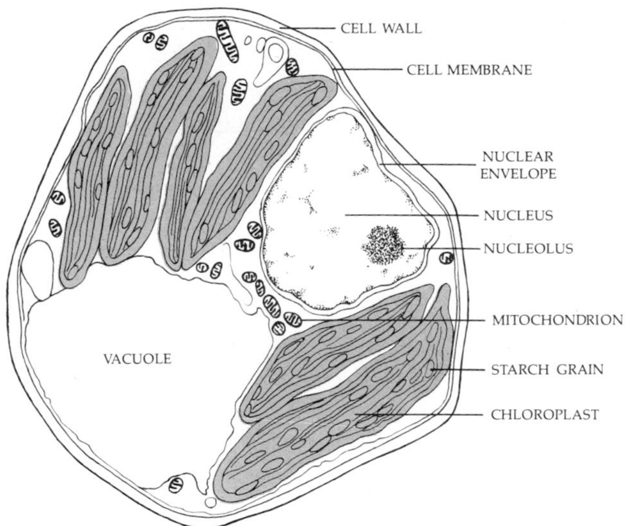

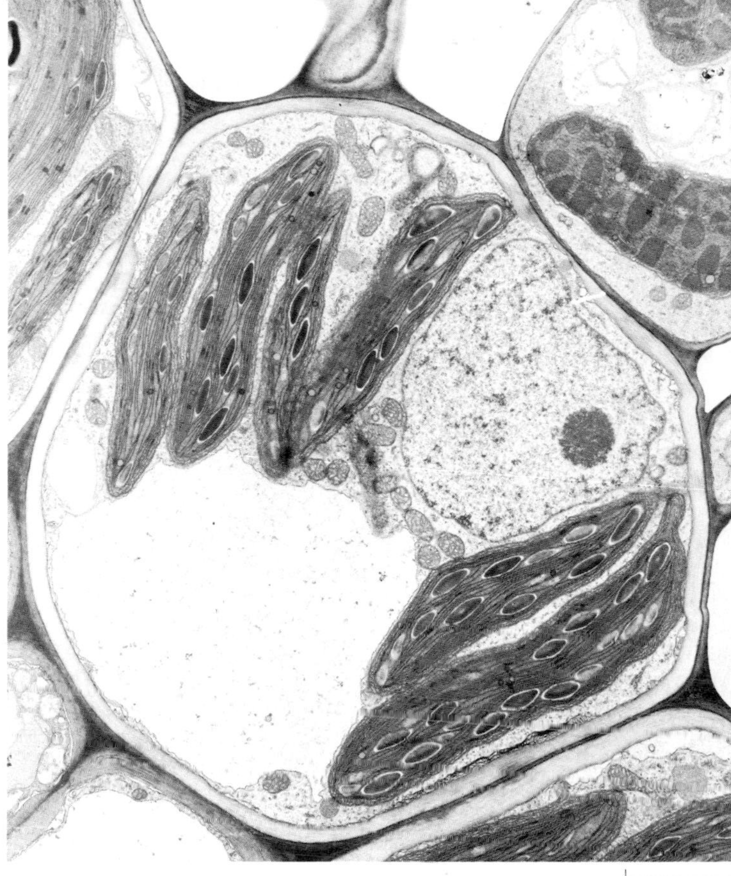

5–1
Electron micrograph of cells from the leaf of a corn plant. The nucleus can be seen on the right side of the central cell. The granular material within the nucleus is chromatin; it contains the hereditary material. Note the many mitochondria and chloroplasts, all enclosed by membranes. The vacuole and cell wall are characteristic of plant cells but are generally not found in animal cells. The short, straight line at the bottom of this micrograph and those that follow provides a reference for size; a micrometer, abbreviated μm, is 1/10,000 centimeter. The same system is used to indicate distances on a road map.

AUTOTROPHS AND HETEROTROPHS

A second fundamental distinction between cells concerns their energy source. *Autotrophs*, by definition, are "self-feeders." They require no organic sources of chemical energy. With the exception of certain bacteria, all autotrophs are also *phototrophs* ("light-eaters"), meaning that their energy source is the sun. *Heterotrophs*, by contrast, require energy-rich organic compounds, synthesized by other organisms, often phototrophs.

Five different kingdoms of organisms are recognized in this text: prokaryotes, protists (eukaryotic one-celled organisms and some simple multicellular ones), fungi, plants, and animals. Prokaryotes include both autotrophs, including phototrophs, and heterotrophs. Protists also include autotrophs and heterotrophs. All animals and fungi are heterotrophs. All plants, with a few curious exceptions (such as Indian pipe and dodder, which are parasites) are photosynthetic autotrophs. Note, however, that within a plant body, some of the cells are phototrophic, such as cells of a leaf, and some are heterotrophic, such as cells of a root. The organism as a whole, however, is a photosynthetic autotroph.

THREE CELLS

Figure 5-2 shows three different types of cells. The first is a prokaryote, the common bacterium *Escherichia coli*. You will be hearing a lot more about this particular bacterium in later chapters because it is the most intensively studied, best understood, of all living organisms. It is also at the center of the recombinant DNA controversy (see page 325), now, we hope, flickering to an end. Bacteria, like all cells, are surrounded by an outer cell membrane. Outside the cell membrane is a rigid cell wall.

The second cell is a protist, a single-celled alga (from the Latin *alga,* plural *algae,* meaning "seaweed"). Algae is a fairly general term applied to all single-celled photosynthetic organisms and many simple multicellular forms. Almost all are aquatic organisms. *Chlamydomonas* is a common inhabitant of freshwater ponds and aquariums. These organisms are small, bright green (because of their chlorophyll), and move very quickly with a characteristic darting motion. They are usually found near the water's surface. Unlike *E. coli,* Chlamydomonas* has a nucleus surrounded by a nuclear envelope. Prominent in this micrograph is a chloroplast, which is the site of photosynthesis, and a mitochondrion, which produces ATP. You will notice that the mitochondrion is found near the base of the flagella (singular, *flagellum*), the whiplike extensions at the anterior (forward end) of the cell. The mitochondrion provides energy for the flicking movements of the flagella that propel the cell through the water. Note also the starch granules; these are the organism's food reserves. The cell is surrounded by a cell membrane outside of which is a cell wall composed of cellulose and other polysaccharides. *Chlamydomonas* is believed to resemble the sort of cell from which the plants evolved.

* By convention, with the second mention of a scientific binomial ("two-name"), it is permissible to abbreviate the first name (which specifies the genus). This is fortunate, particularly when dealing with names such as *Escherichia.* It is also permissible to use the genus name alone, as in *Chlamydomonas,* when one is referring to all members of a particular genus. The second, species, name cannot stand alone, however; you may not, technically speaking, refer simply to *coli,* although it is sometimes done informally.

5-2
(a) *Cells of* Escherichia coli, *a prokaryote that is a common inhabitant of the human digestive tract. The hereditary material is in the less dense (lighter-appearing) material in the center of each cell. The most dense small bodies in the cytoplasm are ribosomes. The two cells in the center have just finished dividing and have not yet separated completely.* (b) *Electron micrograph of* Chlamydomonas, *a photosynthetic eukaryotic cell, which contains a membrane-bound ("true") nucleus and numerous organelles. A single, irregularly shaped chloroplast, surrounded by a double membrane, fills most of the cell. Numerous mitochondria are visible. The cytoplasm is surrounded by a cell membrane, outside of which is a cell wall.* (c) *Cells from the surface of the trachea (windpipe) of a bat. Numerous cilia are visible on the cell surfaces.*

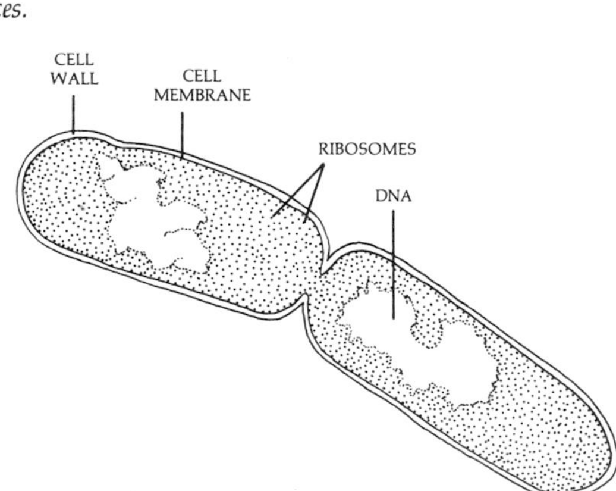

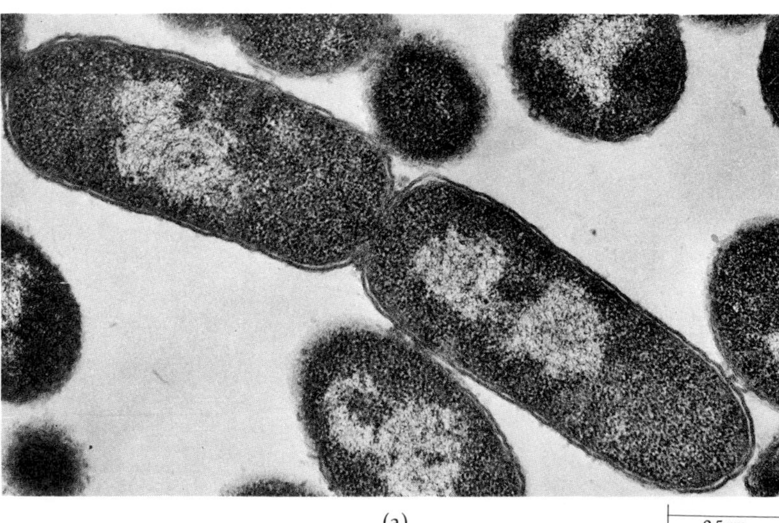

(a)

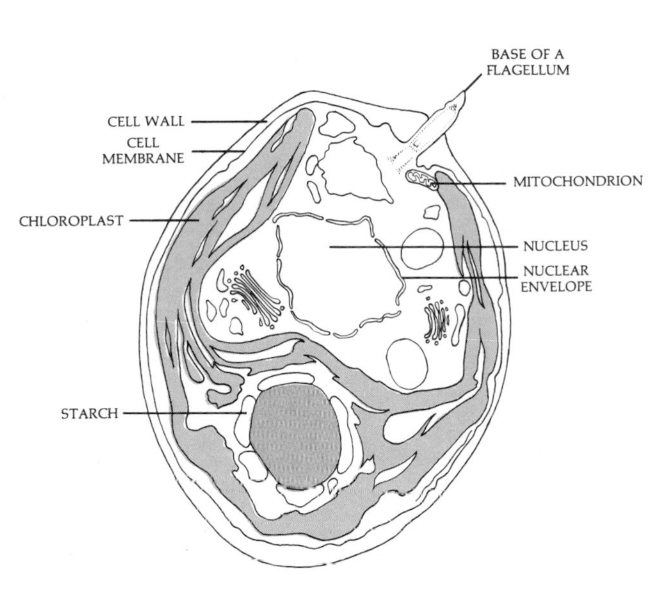
BASE OF A FLAGELLUM
CELL WALL
CELL MEMBRANE
MITOCHONDRION
CHLOROPLAST
NUCLEUS
NUCLEAR ENVELOPE
STARCH

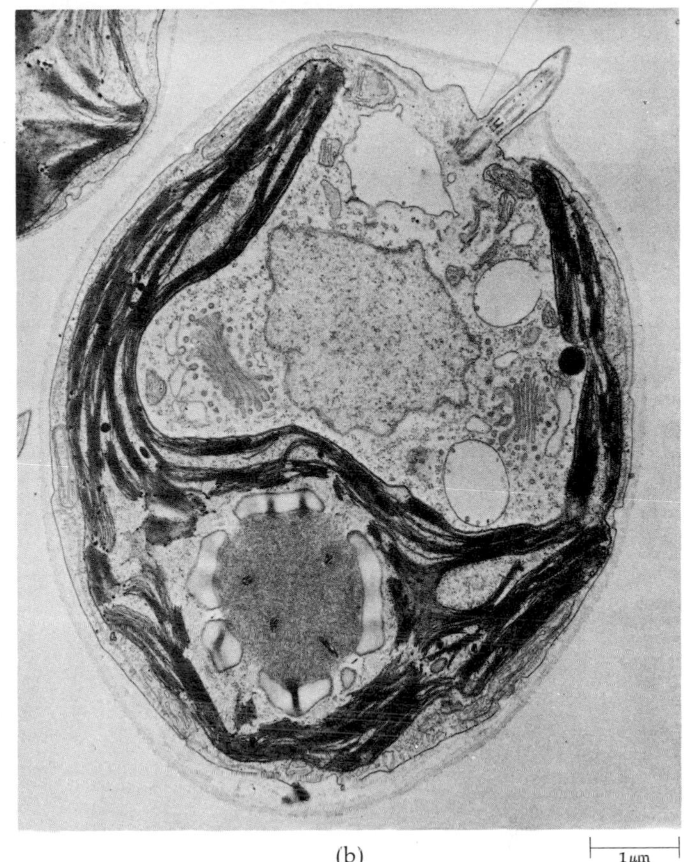

(b) 1 μm

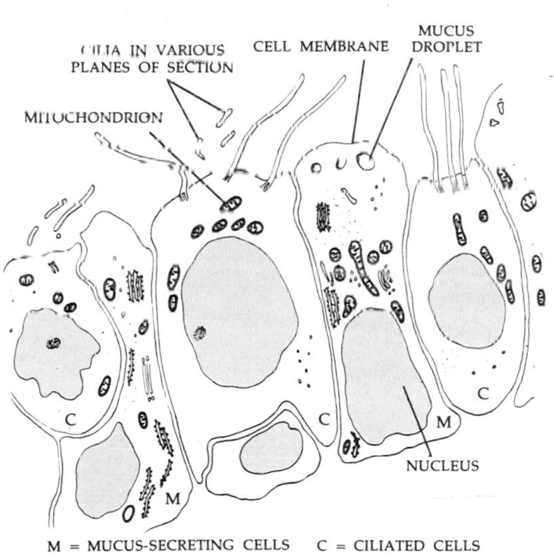
CILIA IN VARIOUS PLANES OF SECTION
CELL MEMBRANE
MUCUS DROPLET
MITOCHONDRION
C
M
NUCLEUS
M = MUCUS-SECRETING CELLS
C = CILIATED CELLS

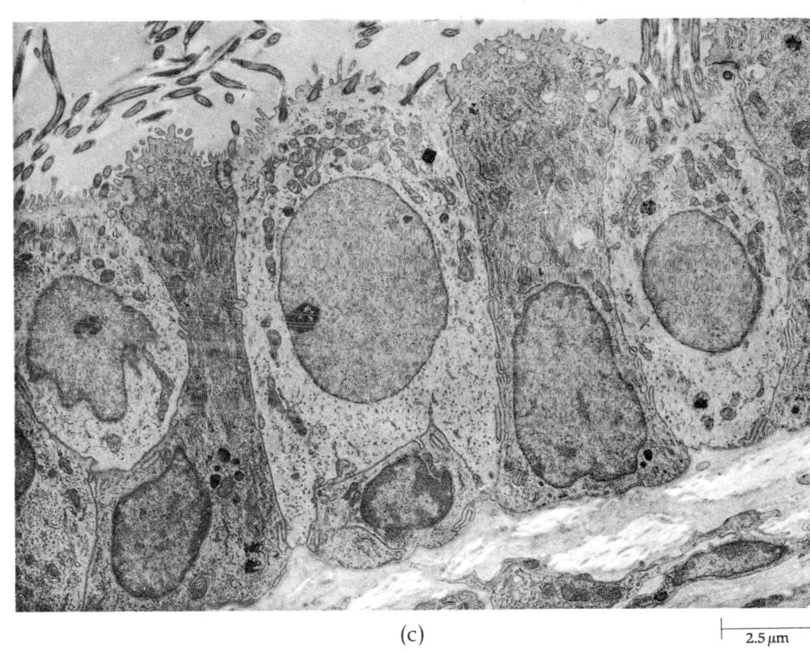

(c) 2.5 μm

Notice how similar *Chlamydomonas* is to the cell from the leaf of a corn plant shown in Figure 5–1. This plant cell is also photosynthetic, supplying its own energy needs from sunlight. However, unlike the alga, it is part of a multicellular organism and depends on other cells for water, minerals, protection from desiccation (drying out), and other necessities. In turn, it supplies other, heterotrophic cells of the plant with sucrose produced from photosynthesis.

The third micrograph shows cells from an animal trachea (windpipe). These cells are epithelial cells, the sort of cells that line all the internal and external body surfaces of animals. The free surface of the larger cell is covered with cilia. These cilia are exactly the same in structure as the flagella of *Chlamydomonas.* When they are fewer and longer, they are usually called flagella, whereas when shorter and more numerous, they are called cilia.

Next to the ciliated cell is a mucus-secreting cell. The white spots in the cell are mucus droplets; the mucus is released (secreted) onto the cell surface. Mucus currents, swept by cilia, remove foreign particles from the surface of the trachea. (One of the first visible effects of tobacco smoking on the human trachea is the destruction of these cilia.) Note that, as in *Chlamydomonas,* mitochondria are clustered at the base of the cilia. These epithelial cells of the trachea are part of an elaborate organ system involved in delivering oxygen to other cells in the body.

ORGANIZATION OF CELLS

There are many, many different kinds of eukaryotic cells. Within our own bodies are more than 100 different and distinct cell types. In a drop of pond water, you are likely to find several different kinds of protists, and in even a small pond, there are probably several hundred clearly different kinds. Plants are composed of cells superficially quite different from those of our own bodies, and insects have many cells of kinds not found either in plants or in vertebrates. Thus, the first remarkable fact about cells is their diversity.

The second, even more remarkable, fact is their similarity. Every cell is a self-contained and at least partially self-sufficient unit, surrounded by an outer membrane that controls the passage of materials into and out of the cell. This makes it possible for the cell to differ biochemically and structurally from its surroundings. All eukaryotic cells also, at least at some time in their existence, contain nuclei, their information and control centers. Many have a variety of internal structures, the organelles, which are similar or identical from one cell to another throughout a wide range of cell types. And, all are composed of the same remarkably few kinds of atoms and molecules.

Size and Shape

Most of the cells that make up a plant or animal body are within a size range of between 10 and 30 micrometers in diameter (see page 101). A principal restriction on cell size is the relationship between volume and surface area. As you can see in Figure 5–3, as volume increases, surface area decreases rapidly in proportion to volume. Materials entering and leaving the cell have to move through the surface membrane, and the more active a cell is, the more rapidly these materials must pass through. Also, oxygen, carbon dioxide, and other metabolically important molecules move into and out of the cell by diffusion, which, as we shall see in the next

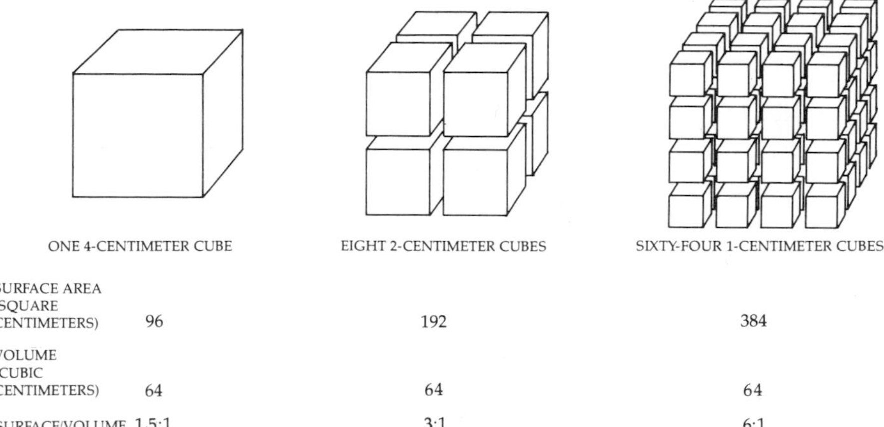

	ONE 4-CENTIMETER CUBE	EIGHT 2-CENTIMETER CUBES	SIXTY-FOUR 1-CENTIMETER CUBES
SURFACE AREA (SQUARE CENTIMETERS)	96	192	384
VOLUME (CUBIC CENTIMETERS)	64	64	64
SURFACE/VOLUME	1.5:1	3:1	6:1

5–3
The single 4-centimeter cube, the eight 2-centimeter cubes, and the sixty-four 1-centimeter cubes all have the same total volume; however, as the volume is divided up into smaller units, the amount of surface area increases (for example, the single 4-centimeter cube has only one-fourth the total surface area of the sixty-four 1-centimeter cubes). Similarly, smaller cells have more surface area (more membrane surface through which materials can diffuse into or out of the cell) and less volume (shorter diffusion distances within the cell).

chapter, is effective only over short distances. Materials can move faster into, out of, and through small cells.

A second limitation on cell size appears to involve the capacity of the nucleus, the cell's control center, to regulate the cellular activities of a large, metabolically active cell. The exceptions seem to "prove" the rule. In certain particularly large and complex one-celled organisms—the ciliates, of which *Paramecium* is an example—each cell has two or more nuclei, the additional ones apparently multiple copies of the original. Other organisms, such as the slime molds, are made up, in effect, of one giant cell but have thousands of nuclei. Such organisms are also, frequently, very thin and spread out, thereby avoiding the surface/volume problem.

It is not surprising, therefore, that the most metabolically active cells are usually small. The relationship between cell size and metabolic activity is nicely illustrated by egg cells. Many egg cells are very large. A frog's egg, for instance, is 1,500 micrometers in diameter. Some egg cells are several centimeters across. (Most of this is stored nutrients for the developing embryo.) When the egg cell is fertilized and begins to be metabolically active, many nuclear divisions occur and the cell divides many times before there is any increase in volume or mass, thus cutting the total mass into units small enough for efficient transfer and control processes.

Like drops of water and soap bubbles, cells tend to be spherical. Cells take other shapes because of cell walls, found in plants, fungi, and many one-celled organisms, because of attachments to and pressure from other neighboring cells or surfaces (as in intestinal epithelial cells), or because of arrays of microtubules and other structural filaments within the cell (Figure 5–4).

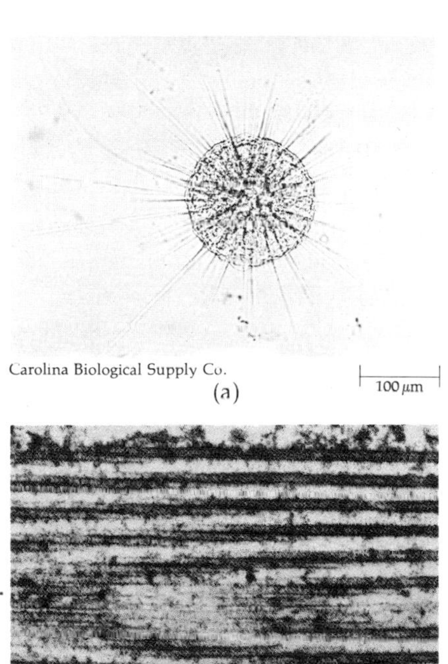

5–4
(a) *Numerous fine strands of cytoplasm (axopods) extend radially from the body of the protist* Actinospherium. (b) *Each axopod is bounded by an extension of the cell membrane and contains many microtubules, arranged longitudinally, which stiffen and extend the axopods. This array is from the base of an axopod.*

Subcellular Organization

Antony van Leeuwenhoek discovered the protists some 300 years ago. "This was for me," he wrote, "among all the marvels that I have discovered in nature, the most marvelous of all." As Leeuwenhoek and his successors observed the thousands of living creatures that they found "all alive in a drop of water," they were able to see, but only barely, structures within them, which they interpreted as being miniature hearts and stomachs and lungs, in other words, tiny organs, or organelles.

VIEWING THE MICROSCOPIC WORLD

Unaided, the human eye has a resolving power of about 1/10 millimeter, or 100 micrometers. This means that if you look at two lines that are less than 100 micrometers apart, they merge into a single line. Similarly, two dots less than 100 micrometers apart look like a single blurry dot. To separate structures closer than this, optical instruments such as microscopes are used. The best light microscope has a resolving power of 0.2 micrometer, or 200 nanometers, and so improves on the naked eye about 500 times. It is theoretically impossible to build a light microscope that will do better than this.

Notice that resolving power and magnification are two different things; if you take a picture through the best light microscope of two lines that are less than 0.2 micrometer, or 200 nanometers, apart, you can enlarge that photograph indefinitely, but the two lines will continue to blur together. By using more powerful lenses, you can increase magnification, but this will not improve resolution. The limiting factor is the wavelength of light.

With the electron microscope, resolving power has been increased about 400 times over that provided by the light microscope. This is achieved by using "illumination" of a much shorter wavelength, consisting of electron beams instead of light rays. Areas in the specimen that permit the transmission of more electrons—"electron-transparent" regions—show up bright, and areas that absorb or deflect electrons—"electron-dense" regions—are dark. Electron microscopy at present, under the very best conditions, affords a resolving power of about 0.5 nm, roughly 200,000 times greater than that of the human eye. (A hydrogen atom is about 0.1 nm in diameter.)

Electrons, which have a very small mass, can pass through specimens only when the specimens are exceedingly thin. As a consequence, living matter has to be fixed, dehydrated, embedded in hard materials, and sliced by special cutting instruments before it can be examined by transmission electron microscopy. Moreover, most specimens are electron-transparent and have to be stained with heavy metals that bind differentially to different subcellular components. Thus only dead cells can be studied, and moreover, it is sometimes difficult to determine whether or not the specimens have been changed by the preparation methods.

Recently, electron microscopy has been expanded in scope by the development of the scanning electron microscope. In scanning electron microscopy, the electrons whose imprints are recorded come from the surface of the specimen. The electron beam is focused into a fine probe, which is rapidly passed back and forth over the specimen. Complete scanning from top to bottom usually takes a few seconds.

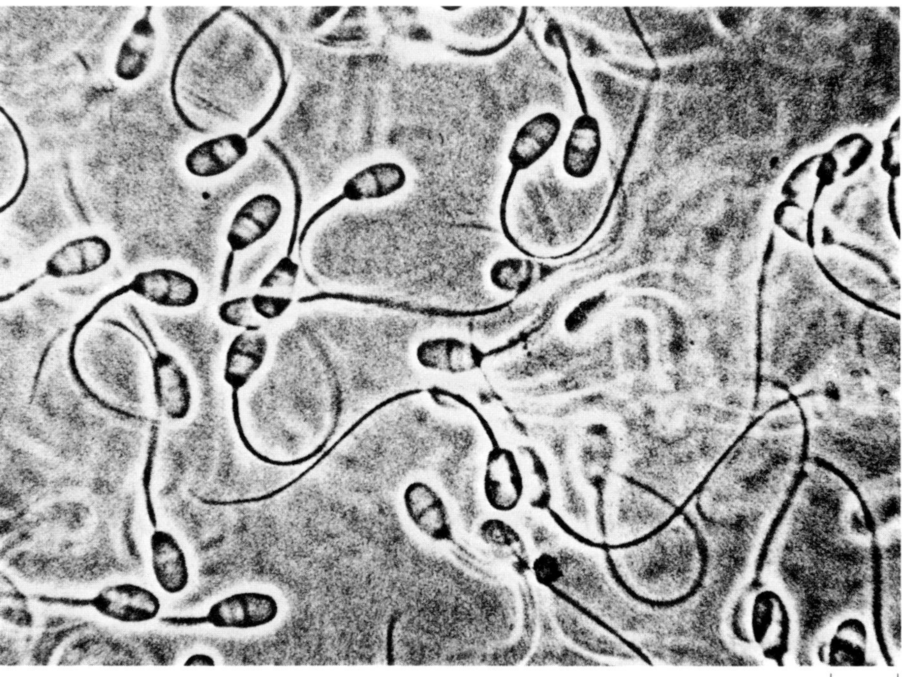
(a)

10 μm

Sperm cells in (a) *a light micrograph;* (b) *an electron micrograph;* (c) *a scanning electron micrograph.*

As a result of the electron bombardment from the probe, the specimen emits low-energy secondary electrons. Variations in the surface of the specimen alter the number of secondary electrons emitted. Holes and fissures appear dark, and knobs and ridges are light. Electrons scattered from the surface plus secondary electrons are amplified and transmitted to a screen, which is scanned in synchrony with the electron probe.

Table 5-1 *Measurements Used in Microscopy*

1 centimeter (cm) = 1/100 meter = 0.4 inch*
1 millimeter (mm) = 1/1,000 meter = 1/10 cm
1 micrometer (μm)† = 1/1,000,000 meter = 1/10,000 cm
1 nanometer (nm) = 1/1,000,000,000 meter = 1/10,000,000 cm
1 angstrom (Å) = 1/10,000,000,000 meter = 1/100,000,000 cm
or
1 meter = 10^2 cm = 10^3 mm = 10^6 μm = 10^9 nm = 10^{10} Å

* A metric-to-English conversion table is found in Appendix A.
† Micrometers were formerly known as microns (μ).

(c)

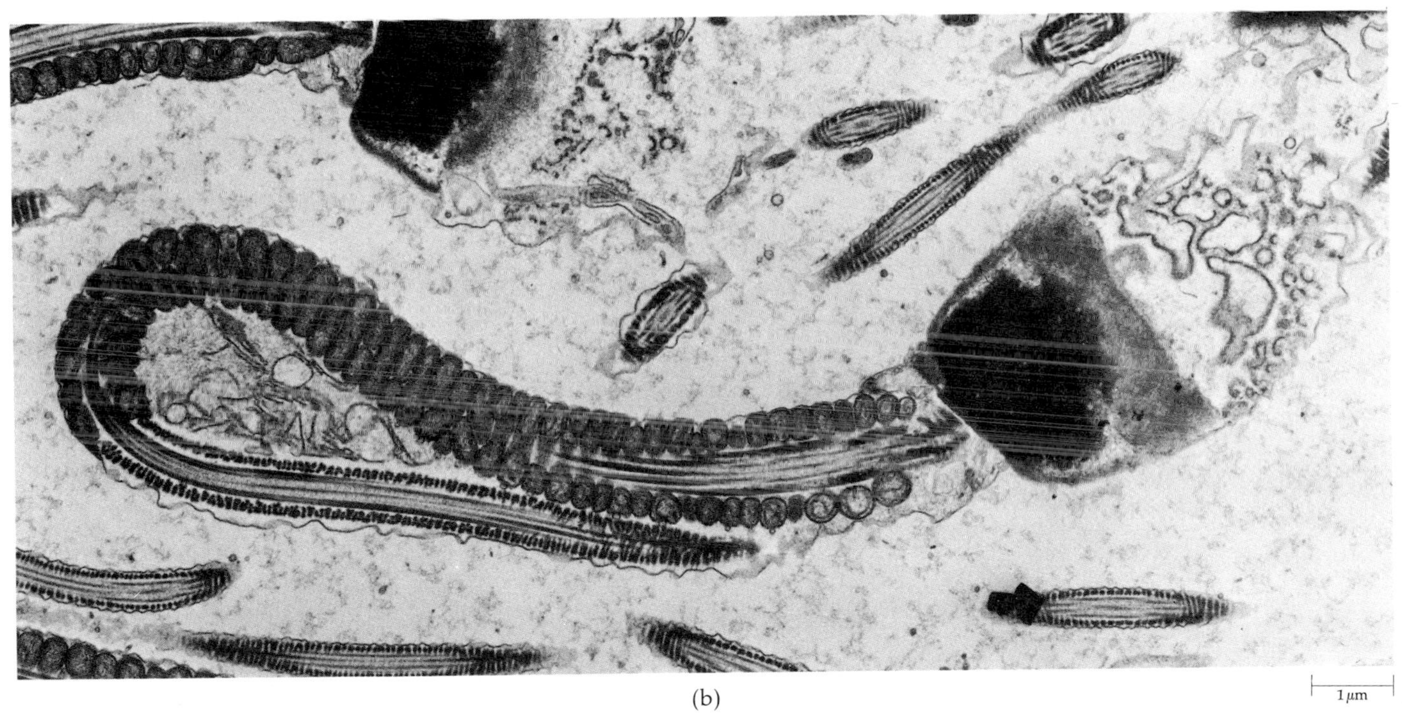

(b)

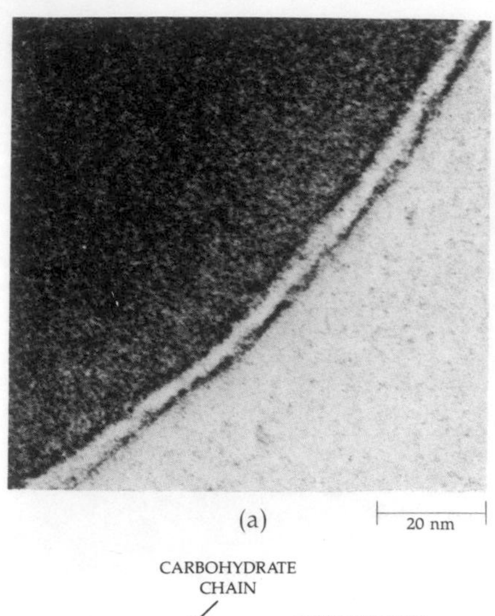

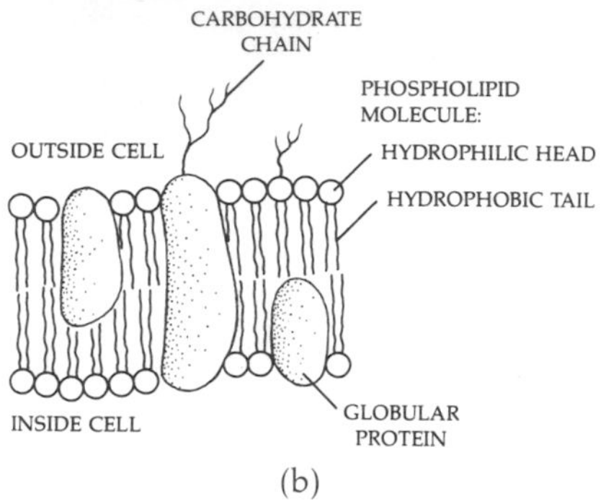

5–5
(a) *Electron micrograph showing a cross section of the cell membrane of a red blood cell. The "molecular sandwich" structure of the membrane is believed to consist of two electron-dense (dark) layers of phospholipid molecules arranged with their hydrophobic tails pointing inward, forming the electron-transparent (light) inner "filling," with globular proteins embedded throughout.* (b) *Model of the cell membrane showing the two layers of phospholipids within which the large globular proteins are embedded. Short carbohydrate chains are attached to some of the outside layer of phospholipid and protein molecules. These serve as receptors for hormones, antibodies, viruses, and are also involved with adhesion between cells.*

Modern microscopic techniques have confirmed that eukaryotic cells do indeed contain a multitude of structures. They are not, of course, organs such as those found in multicellular organisms, but they are in some ways comparable: They are specialized in form and function to carry out particular activities required in the cellular economy. Just as the organs of multicellular animals work together in organ systems, the organelles of cells engage in a number of cooperative and interdependent functions.

Every cell must carry out essentially the same processes—acquire and assimilate food, eliminate wastes, synthesize new cellular materials, and in many cases, be able to move and to reproduce. Just as the various organs of your body have a form or design that suits them for the specific functions they carry out (the kidney for elimination of wastes from the blood, the intestine for food absorption, and so on), so all cells have an internal architecture that includes organelles suited to the functions they perform. It is important to realize that a cell is not a random assortment of parts but a dynamic, integrated entity.

Also, although we can look at only one function at a time, remember that most activities of a cell go on simultaneously and influence one another. *Chlamydomonas,* for instance, is swimming, photosynthesizing, absorbing nutrients, building its cell wall, making enzymes, converting sugars to starch (or vice versa), and oxidizing food molecules for energy, all at the same time. It is also likely to be orienting itself in the sunlight, it is probably preparing to divide, it is possibly looking for a mate, and it is undoubtedly carrying out at least a dozen or more similarly important activities, many of which may be still unknown.

CELL STRUCTURE

All cells, as we stated previously, are basically very similar, containing many of the same structures, the same types of enzyme systems, and the same kind of genetic material. An outer membrane, apparently conforming to the same general biochemical design in all cells, surrounds the cell. The living materials bounded by the membrane consist, in eukaryotes, of the cytoplasm, which contains the organelles, and the nucleus.

The Cell Membrane

The outer *cell membrane* (sometimes called the plasma membrane or plasmalemma) is only about 7 to 9 nanometers thick and cannot be resolved in the light microscope. Now, with the electron microscope, it can be visualized as a continuous, thin, double line (Figure 5–5). According to current hypotheses, the eukaryotic cell membrane consists essentially of phospholipid and cholesterol molecules arranged so that their hydrophobic tails are clustered inward, forming the filling of a molecular sandwich, with globular proteins embedded throughout. All the membranes of a cell, including those surrounding the various organelles, have this same structure. There are, however, local differences in the types of lipids and, particularly, in the number and types of proteins. These differences undoubtedly reflect differences in membrane function. The cell membrane will be discussed in greater detail in Chapter 6.

The cell membrane of bacterial cells is much the same in basic composition as the membranes of eukaryotes, except that, with a few exceptions, bacterial cell membranes do not contain cholesterol.

ORIGIN OF THE CELL THEORY

In the seventeenth century, Robert Hooke, using a microscope of his own construction, noticed that cork and other plant tissues are made up of small cavities separated by walls. He called these cavities "cells," meaning "little rooms." The word did not take on its present meaning, however, for more than 150 years.

In 1838, Matthias Schleiden, a German botanist, came to the conclusion that all plant tissues are organized in the form of cells. In the following year, zoologist Theodor Schwann extended Schleiden's observation to animal tissues and proposed a cellular basis for all life. The cell theory is of tremendous and central importance to biology because it emphasizes the basic sameness of all living systems and so brings an underlying unity to widely varied studies involving many different kinds of organisms.

In 1858, the cell theory took on an even broader significance when the great pathologist Rudolf Virchow generalized that cells can arise only from preexisting cells: "Where a cell exists, there must have been a preexisting cell, just as the animal arises only from an animal and the plant only from a plant. . . . Throughout the whole series of living forms, whether entire animal or plant organisms or their component parts, there rules an eternal law of continuous development."

In the broad perspective of evolution Virchow's concept takes on an even larger significance. There is an unbroken continuity between modern cells—and the organisms that they compose—and the primitive cells that first appeared on earth more than three billion years ago.

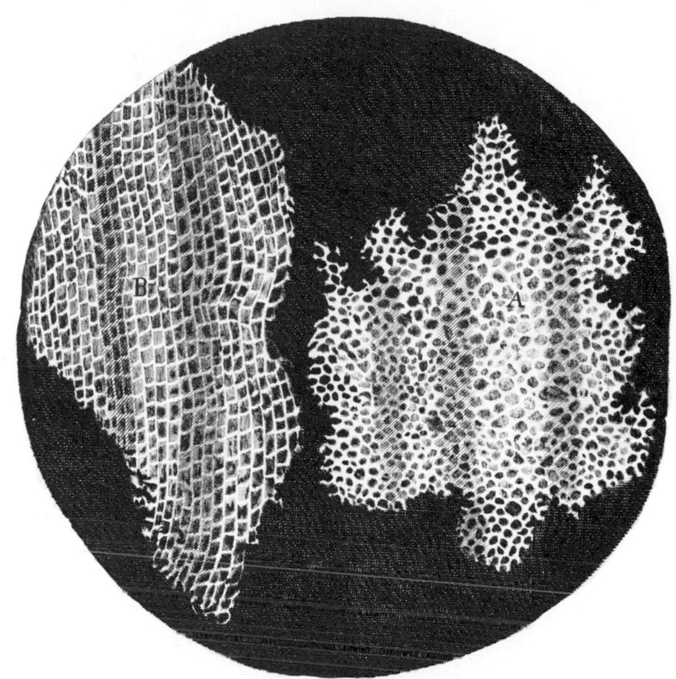

Robert Hooke's drawings of two slices of a piece of cork, reproduced from his Micrographia *(1665). Hooke was the first to use the word "cells" to describe these tiny compartments into which living organisms are organized.*

The Cell Wall

A principal distinction between plant and animal cells is that the former are surrounded by a *cell wall*. The wall is outside the membrane and is constructed by the cell. As plant cells divide, a thin layer of gluey material forms between them, which becomes the *middle lamella*. It is composed of pectins (the compounds that make jellies jell) and other polysaccharides. It holds the cells together. Next, under the middle lamella, the plant cell constructs its primary cell wall. This wall is composed, to a large extent, of cellulose molecules wound together like wires in a cable and laid down in a matrix of gluey polymers. As you can see in Figure 3–9c on page 57, the microfibrils are laid down with successive sheets oriented at right angles to one another. (Persons familiar with building materials will note that the cellulose cell wall thus combines the structural features of both fiber glass and plywood.)

In plants, growth takes place largely by cell elongation. Studies have shown that the cell adds new materials to its walls all during this elongation process. Notice,

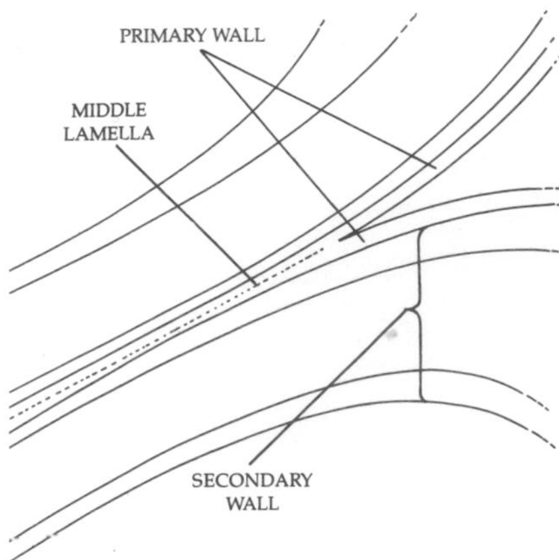

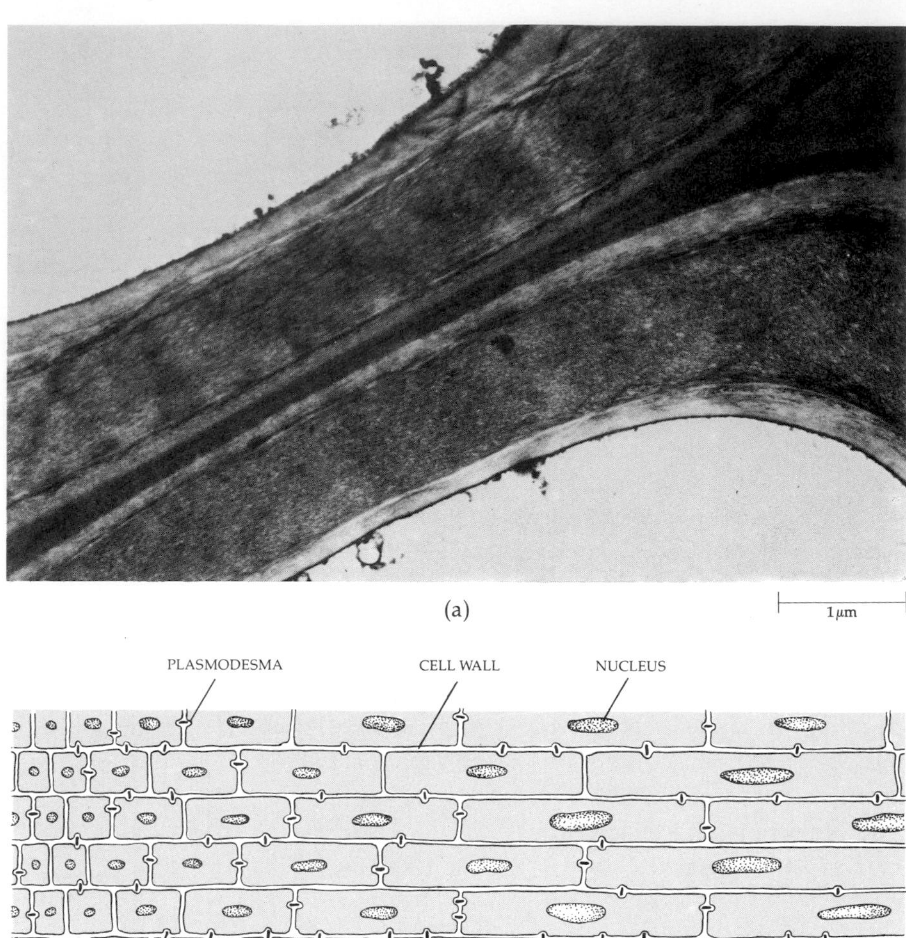

5–6
(a) *Electron micrograph and diagram of two adjacent cell walls of vessel elements, the cells that form the tubes through which water is conducted in plants. You can see the middle lamella, the primary walls, and the layered secondary walls, deposited inside the primary wall. The cells, which are from the wood of a black locust tree, have died.* (b) *Growth of plant cells is limited by the rate at which the cell walls expand. The walls control both the rate of growth and its direction; they do not expand in all directions but elongate in a single dimension. Cells at the left are newly formed; cells farther to the right are older and have started to elongate. Plasmodesmata are channels connecting adjacent cells.*

too, that the cell does not simply expand in all directions; its final shape is determined by the structure of its cell wall.

As the cell matures, a secondary cell wall may be constructed. This wall is not capable of expansion, as is the primary wall. It often contains other molecules, such as lignin, which have stiffening properties. In such cells, the living material of the cell often dies, leaving only the outer wall, a monument to the cell's architectural abilities.

Cell walls are also found in algae, fungi (see page 56), and prokaryotes.

The Cytoplasm

Not long ago, the cell was visualized as a bag of fluid containing enzymes and other dissolved molecules along with a few mitochondria and occasional other organelles that could be seen by special microscopic techniques. Within the last two or three decades, however, with the development of electron microscopy, an increasing number of structures have been identified within the cytoplasm, which is now known, as we saw in Figure 4–4, to be highly structured and crowded with organelles. A representative array of them is seen in Figure 5–7.

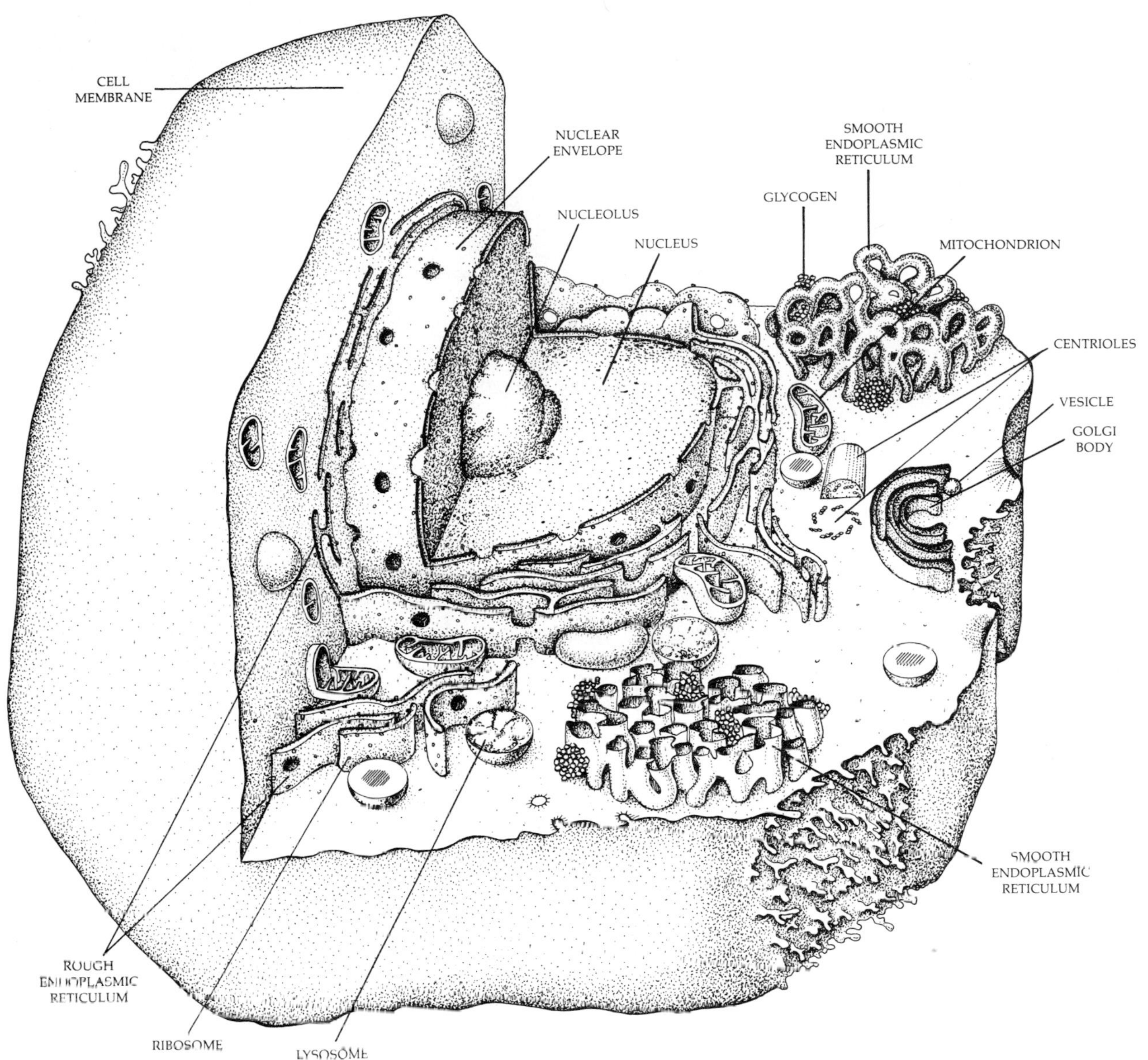

5–7

An animal cell, as interpreted from electron micrographs. Like all cells, this one is bounded by an outer cell membrane, which acts as a selectively permeable barrier to the surrounding environment. All materials that enter or leave the cell, including food, wastes, and chemical messages must pass through this barrier. Within the membrane is found the cytoplasm, which contains the enzymes and other solutes of the cell. The cytoplasm is traversed and subdivided by an elaborate system of membranes, the endoplasmic reticulum. In some areas, the endoplasmic reticulum is covered with ribosomes, the special structures on which amino acids are assembled into proteins. Ribosomes are also found free in the cytoplasm. Golgi bodies are packaging centers for molecules synthesized within the cell. The mitochondria are the chief sources of ATP production in the cell. The largest body in the cell is the nucleus, which is surrounded by a double membrane, the nuclear envelope, which is continuous with the endoplasmic reticulum. Within the nuclear envelope is a nucleolus, the site where the ribosomes are formed, and the chromatin, which is the material of the chromosomes in an extended form. These cellular structures are all described in further detail in the text.

5–8

(a) *Electron micrograph of a human fibroblast (a connective tissue cell) showing microtubules, microfilaments, and intermediate fibers.* (b) *An interpretation of the way these structures form an internal skeleton for the cell.*

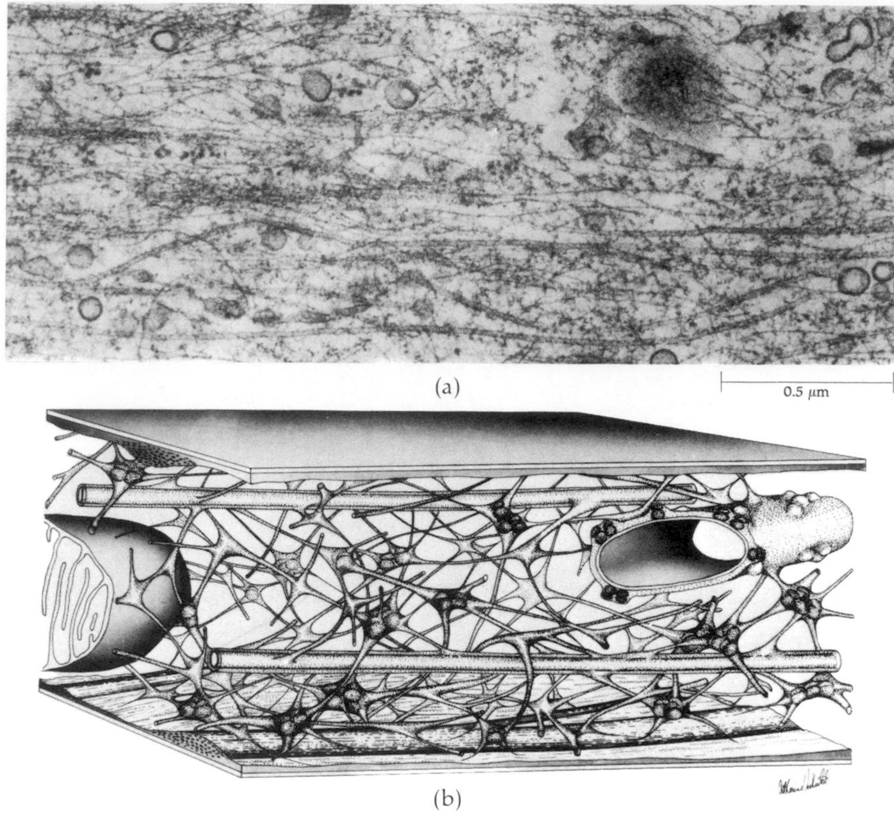

Microtubules, Microfilaments, and Intermediate Fibers

Most recently, new techniques that permit the interior of the cell to be visualized in three dimensions have shown additional, previously unsuspected structures in the cell. They seem to form an internal scaffolding that helps the cell keep its shape, anchor its organelles, and direct its traffic. These structures are of three types: *microtubules* (described on page 68), *microfilaments,* and *intermediate fibers* (Figure 5–8). Microtubules seem to extend out from the cell center ending near the cell surface. They are 20 to 25 nanometers in diameter. Microfilaments are fine protein threads, only 3 to 6 nanometers in diameter. They seem to be involved in cell motility. In cells that move, such as amoebas, they are found concentrated in bundles or a meshwork near the moving edge. Intermediate fibers, as their name implies, are intermediate in size between microtubules and microfilaments, with a diameter of 7 to 10 nanometers. They form a pattern of wavy intermingled lines throughout the cell. Their function is unknown.

Vacuoles

Many cells, especially plant cells, also contain *vacuoles.* A vacuole is a space in the cytoplasm filled with water and solutes, and surrounded by a single membrane. Immature plant cells characteristically have many vacuoles, but as the plant cells mature, the numerous smaller vacuoles coalesce into one large, central, fluid-filled vacuole that then becomes a major supporting element of the cell (see Figure 5–1).

Vesicles, more common in animal cells, have the same general structure as vacuoles. They are distinguished by size: Vesicles are usually less than 100 nanometers in diameter, whereas vacuoles are larger.

The Nucleus

In eukaryotic cells, the *nucleus* is a large, often spherical body, usually the most prominent structure within the cell. It is surrounded by two lipoprotein membranes, which together make up the *nuclear envelope* (Figure 5-9). These two membranes are fused together at frequent intervals to create pores 50 to 80 nanometers in diameter through which materials pass between the nucleus and cytoplasm. Whether or not these pores can be closed—and if so, by what—is a matter of current study.

The *chromosomes* are found within the nucleus. When the cell is not dividing, the chromosomes are visible only as a tangle of fine threads, called *chromatin*. The most conspicuous body within the nucleus is the *nucleolus*. There are usually two nucleoli per nucleus, although often only one is visible in a micrograph. The nucleolus is the site of assembly of the ribosomes (see page 278). Viewed with the electron microscope, the nucleolus appears to be a collection of fine granules and tiny fibers. These are thought to be parts of ribosomes.

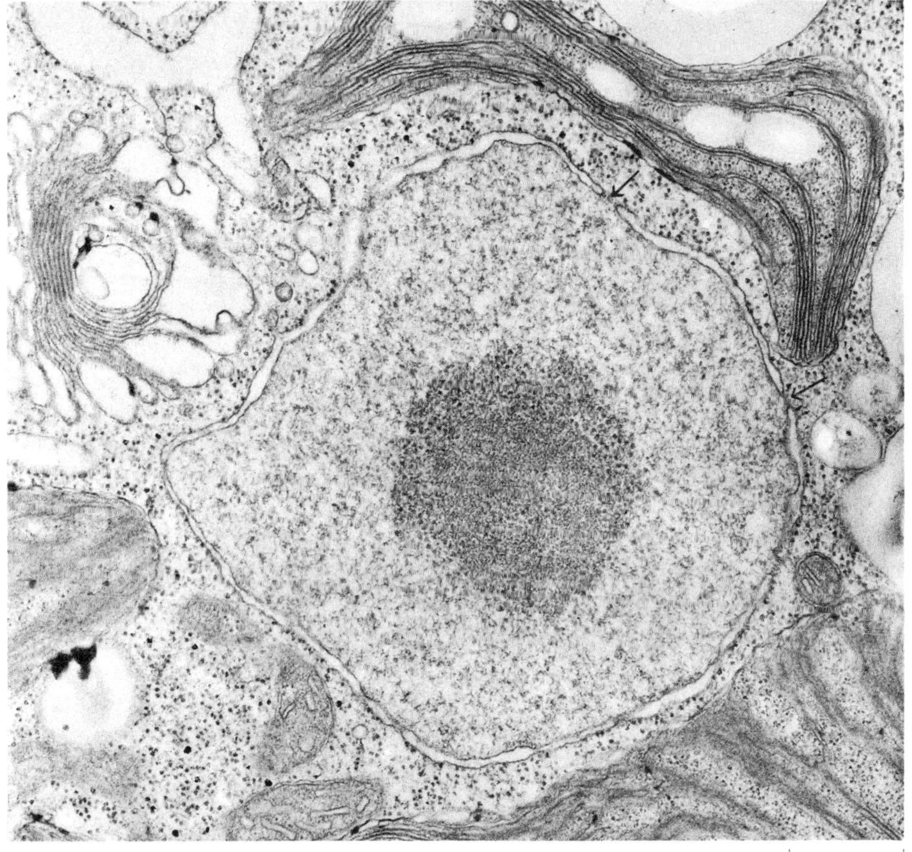

5-9
Electron micrograph of the nucleus of the alga Chlamydomonas. *The dark body in the center of the nucleus is the nucleolus. The nucleolus is the site of production of major components of ribosomes; you can see partially formed ribosomes around its periphery. Notice also the nuclear envelope with its many pores, two of which are indicated by arrows. Above the nucleus is a portion of a chloroplast containing starch grains. A Golgi body is to the left of the nucleus, and mitochondria are visible below.*

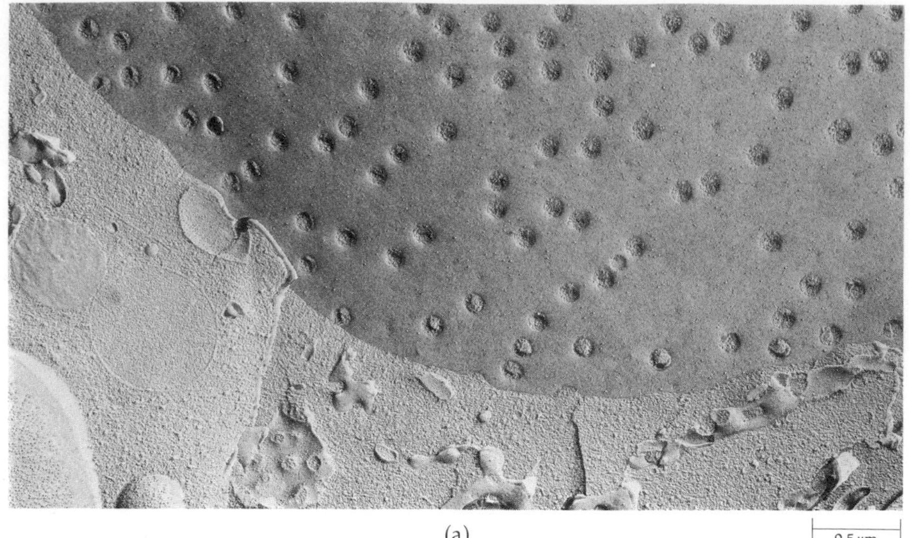

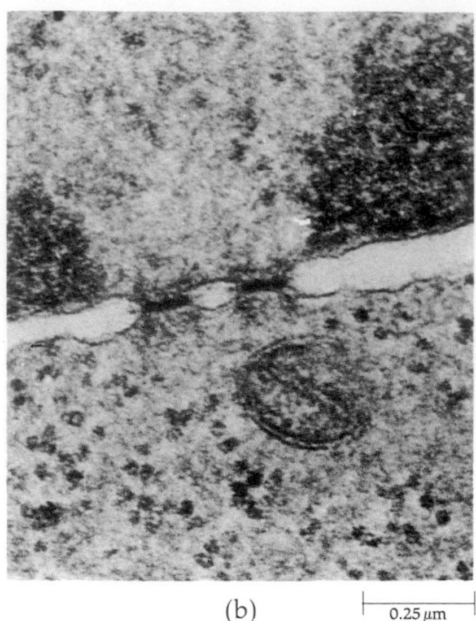

5–10
(a) *The cell nucleus fills the upper half of this electron micrograph. What you see here is the surface of the nuclear envelope. Clearly visible on this surface are pores, through which, it is believed, the nucleus and the cytoplasm are able to "communicate."* (b) *A single pore in the nuclear envelope of an immature red blood cell.*

The Functions of the Nucleus

The nucleus carries the hereditary information for the cell, the instructions that determine whether a particular organism will develop to become a *Paramecium*, oak tree, or human being—and not just any paramecium, oak tree, or human being, but one that resembles the parent or parents of that particular, unique organism. Each time a cell divides, this information is passed on to the two new cells. The nucleus exerts its influence by directing the ongoing activities of the cell, ensuring that the complex molecules the cell requires are synthesized in the number and of the kind needed. The way in which the nucleus performs these functions will be described in Section 3.

SOME PRINCIPAL ORGANELLES

Ribosomes

Ribosomes are the most numerous of the cell's many organelles. A rapidly growing *E. coli* cell has approximately 15,000 ribosomes, and a eukaryotic cell may have many times more. As we noted in Chapter 3, proteins are chains of amino acids, assembled in sequence. The ribosomes are the sites at which this assembly takes place, a process that will be explored in some detail in Chapter 16. The more protein a cell is making, the more ribosomes it has.

The way in which ribosomes are distributed in the cell seems to be related to the kinds of proteins it is making. Some proteins—enzymes or hemoglobin, for example—are used within the cell. Others, such as collagen, digestive enzymes, hormones, or mucus, are released outside the cell, sometimes carrying out their functions at a great distance, on a cellular scale, from their source. In cells that are making proteins for their own use, such as embryonic cells, ribosomes tend to be distributed in the cytoplasm. In cells that are making digestive enzymes or other proteins for export, ribosomes are also found attached to a complex system of internal membranes, the endoplasmic reticulum.

Endoplasmic Reticulum

The *endoplasmic reticulum* is a network of interconnecting flattened sacs, tubes, and channels found in eukaryotic cells. The amount of endoplasmic reticulum in a cell is not fixed but increases or decreases depending on the cell's activity.

There are two types of endoplasmic reticulum, rough (with ribosomes attached) and smooth (without ribosomes). Rough endoplasmic reticulum is found in cells making large amounts of proteins for export. It can sometimes be seen to be continuous with the outer layer of the nuclear envelope, which may also contain ribosomes. Rough endoplasmic reticulum often includes large, flattened sacs called cisternae. If cells engaged in protein synthesis are permitted to take up radioactive amino acids, the radioactive labels can be first detected at the surface of the rough endoplasmic reticulum. This lends support to the thesis that the rough endoplasmic reticulum is involved in the synthesis of proteins.

5–11

(a) *Rough endoplasmic reticulum, which fills most of this micrograph, is a system of membranes that separates the cell into channels and compartments and provides surfaces on which chemical activities take place. The dense objects on the membrane surfaces are ribosomes. This cell is from a pancreas, an organ extremely active in the synthesis of digestive enzymes, which are "exported" to the upper intestine, where most digestion takes place. In the lower right-hand corner of the micrograph are one mitochondrion and, above it, a portion of another.* (b) *Rough endoplasmic reticulum at higher magnification, showing the individual ribosomes.* (c) *An interpretation of the rough endoplasmic reticulum based on electron micrographs.*

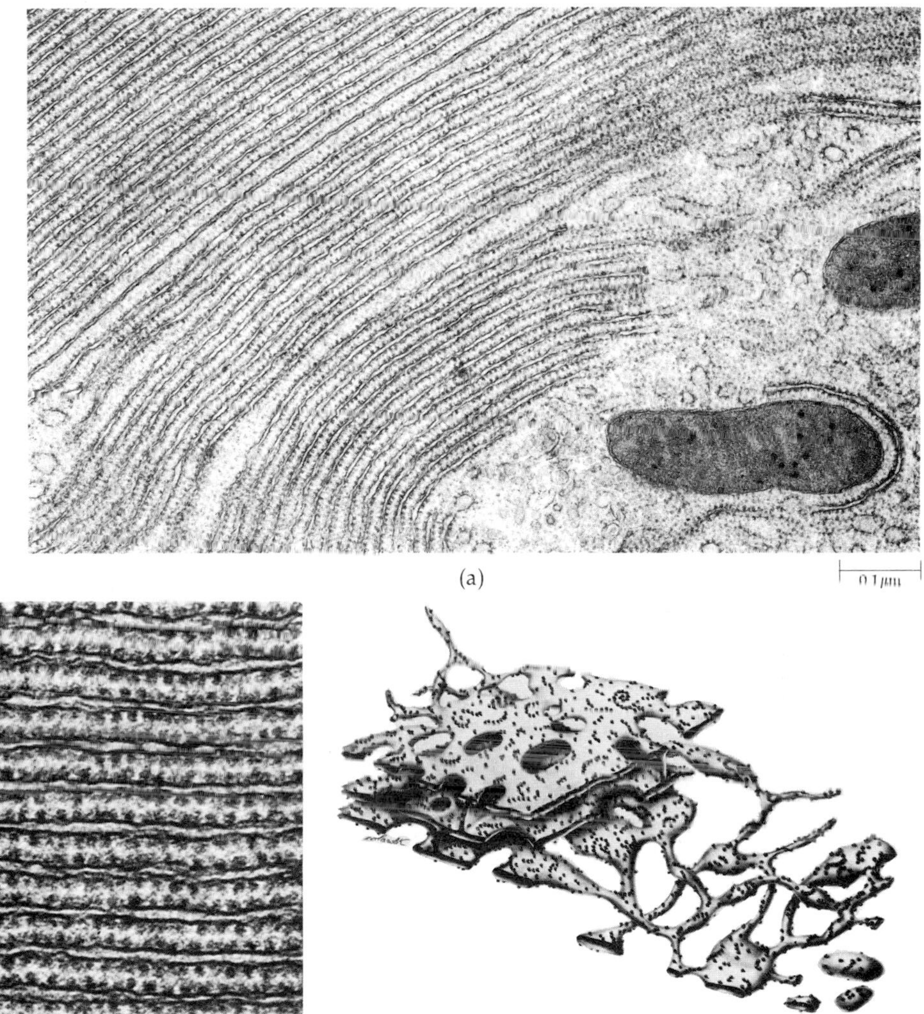

(a)

(b)

(c)

5–12
Smooth endoplasmic reticulum from the testicle of an opossum. These membranes participate in the synthesis of the hormone testosterone.

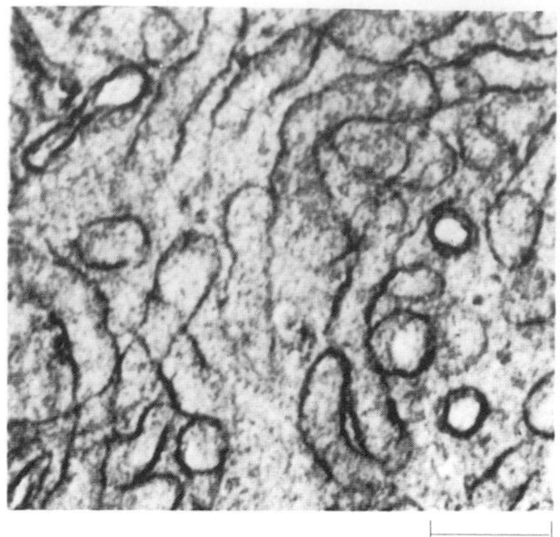

Cells concerned with the synthesis of lipids—such as the gland cells that make steroid hormones—have large amounts of smooth endoplasmic reticulum. Smooth endoplasmic reticulum is also found in liver cells where it appears to be involved in various detoxification processes (one of the many functions of the liver). For instance, in experimental animals fed large amounts of phenobarbital, the amount of smooth endoplasmic reticulum in the liver cells increases severalfold. Smooth endoplasmic reticulum also seems to be active in the liver's breakdown of glycogen to glucose. It appears, as well, to serve as a passageway for material moving from the rough endoplasmic reticulum to the Golgi bodies. As more of its functions are discovered, it seems likely that the smooth endoplasmic reticulum system actually represents a number of quite different kinds of endoplasmic reticulum, resembling one another only in their lack of ribosomes.

Golgi Bodies

Each *Golgi body* consists of a small group of flattened, membrane-bound sacs stacked loosely on one another and surrounded by tubules and vesicles. The function of the Golgi body is to accept vesicles from the endoplasmic reticulum, to modify the membranes of the vesicles and their contents, and to distribute the product to other parts of the cell and, especially, to the cell surface. Thus, they serve as packaging and distribution centers.

Combinations of sugars and proteins (glycoproteins) that are found on the surfaces of cell membranes are packed in the Golgi body. In plant cells, for example, Golgi bodies bring together some of the components of the cell walls and export them to the cell surface where they are assembled. Golgi bodies are found in most types of eukaryotic cells. Animal cells usually contain 10 to 20 Golgi bodies, and plant cells may have several hundred.

5–13
Graphic interpretation and electron micrograph of a Golgi body. Golgi bodies are composed of special arrangements of membranes. Materials are packaged in membrane-enclosed vesicles at the Golgi bodies and are distributed within the cell or shipped to the cell surface. Note the vesicles pinching off from the edges of the flattened sacs.

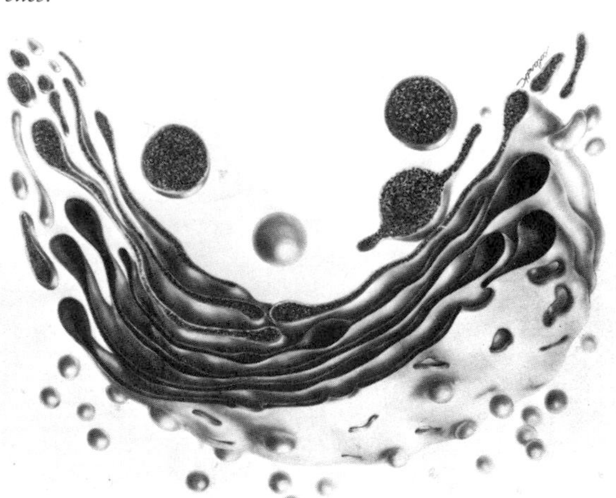

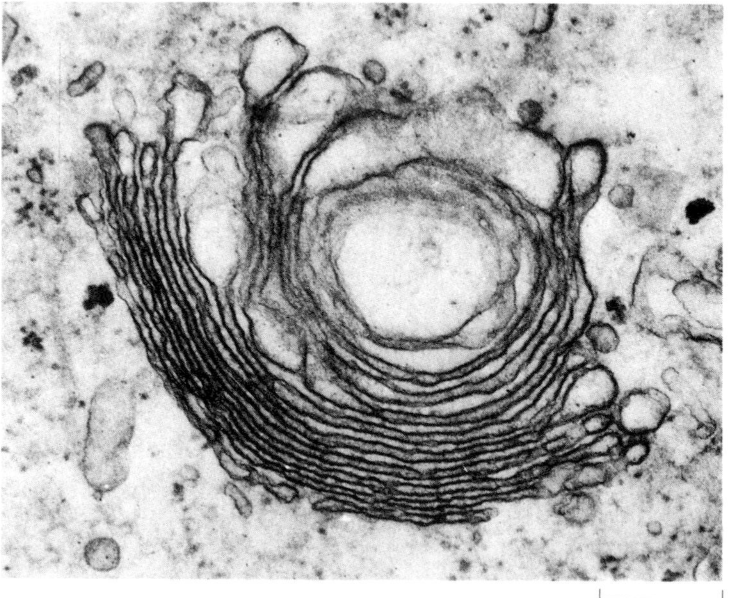

5–14
Diagram illustrating the interaction of cytoplasmic organelles in the synthesis of protein and in the packaging of macromolecules for export. Glycoproteins are combinations of sugars and proteins, and lipoproteins are combinations of fats and proteins. They are common components of membranes.

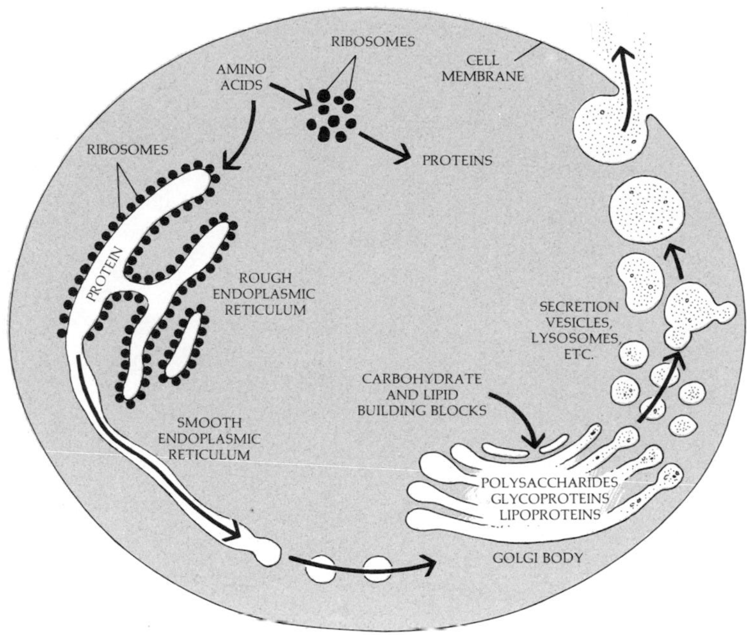

Figure 5–14 summarizes the ways in which the rough and smooth endoplasmic reticulum and the Golgi body and its vesicles may interact to produce macromolecules for export.

Lysosomes

One type of small vesicle commonly formed in the Golgi body is the *lysosome.* Lysosomes are essentially bags of destructive enzymes enclosed in membranes that separate them from the rest of the cell; they are involved in the catabolic activities of many types of cells. If the lysosomes break open, the cell itself is destroyed, since the enzymes they carry are capable of breaking down all the major compounds found in a living cell. The tenderness and inflammation associated with rheumatoid arthritis and gout appear to be related to the escape of hydrolytic enzymes from lysosomes.

An example of the function of lysosomes is given by white blood cells, which engulf bacteria in the human body. As the bacteria are taken up by the cell, they are wrapped in a membrane-enclosed sac, a vacuole. (This process is known as phagocytosis; we shall discuss it further in Chapter 6.) When this occurs, the lysosomes within the cell fuse with the vacuoles containing the bacteria and release their hydrolytic enzymes. The bacteria are quickly digested. In similar fashion, the lysosomes of single-celled organisms, such as *Paramecium, Didinium,* and amoebas, fuse with the phagocytic vacuoles containing food organisms. Hydrolytic enzymes released by the lysosomes into the vacuoles digest the contents. Why the enzymes do not destroy the membranes of the lysosomes that carry them is a pertinent question yet to be answered.

5-15
Mitochondria in heart muscle.

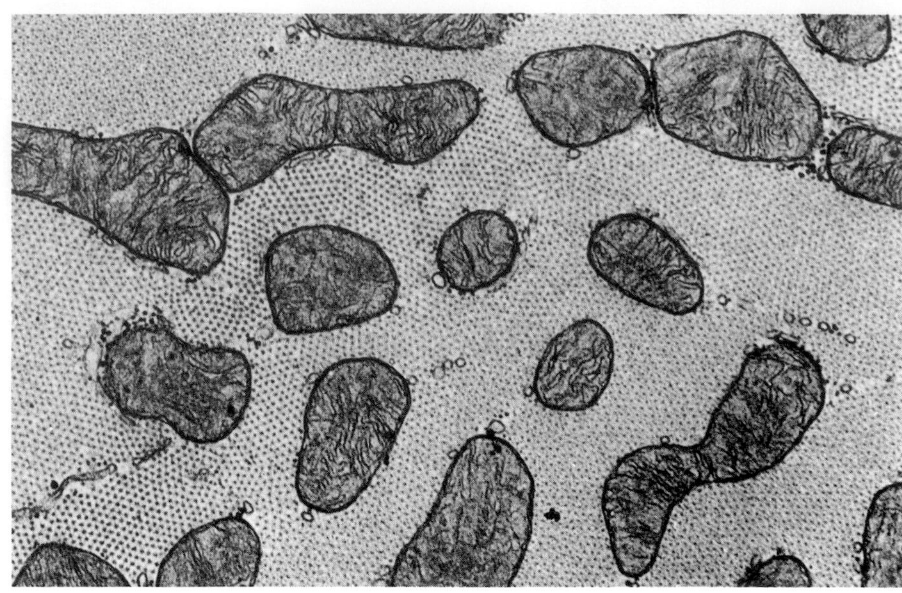

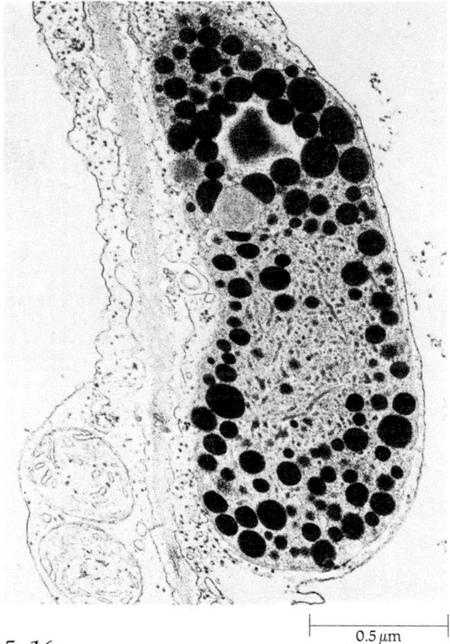

5-16
Chromoplast from a forsythia petal. The large, dark granules contain the orange and yellow pigments characteristic of certain flowers and fall leaves. To the left of the chromoplast are the faintly visible cell wall and two mitochondria. To the left and right of the micrograph are portions of two vacuoles.

Mitochondria

Mitochondria (singular, mitochondrion) are among the largest and often the most numerous organelles in a cell. In nonphotosynthetic cells, they are the principal sites of production of ATP, and the higher the energy requirements of a particular eukaryotic cell, the more or larger mitochondria it is likely to have. A liver cell, for example, which has modest energy requirements, has about 2,500, making up about 25 percent of the volume of the cell, whereas a heart muscle cell has several times as many very large mitochondria.

Mitochondria vary in shape from almost spherical, to potato-shaped, to greatly elongated cylinders. They are always surrounded by two membranes, the inner one of which folds inward; these folds, known as cristae, are working surfaces for enzymatic functions. The more active a mitochondrion, the more cristae it is likely to have. Mitochondria are often found clustered in areas in the cell where ATP requirements are high.

Plastids

Plastids are membrane-bound organelles found only in the cells of plants and algae. They are surrounded by two membranes, like mitochondria, and have an internal membrane system that may be folded intricately. Mature plastids are of three types: leucoplasts, chromoplasts, and chloroplasts.

Leucoplasts (*leuco* means "white") store starch or, sometimes, proteins or oils. Leucoplasts are likely to be numerous in storage organs such as roots, as in a turnip, or tubers, as in a potato (see page 55).

Chromoplasts (*chromo* means "color") contain pigments and are associated with the bright orange and yellow colors of fruits, flowers, fall leaves, and carrots.

Chloroplasts (*chloro* means "green") are the chlorophyll-containing plastids in which photosynthesis takes place. Their structure will be described in Chapter 9.

5–17

Chloroplast development. (a) *The immature plastid contains small crystalline structures (top).* (b) *In the presence of light, these structures begin to break up into elongated vesicles.* (c) *The vesicles flatten into stacks of membranes.* (d) *A mature chloroplast. Chromoplasts and leucoplasts develop from plastids similar to* (a).

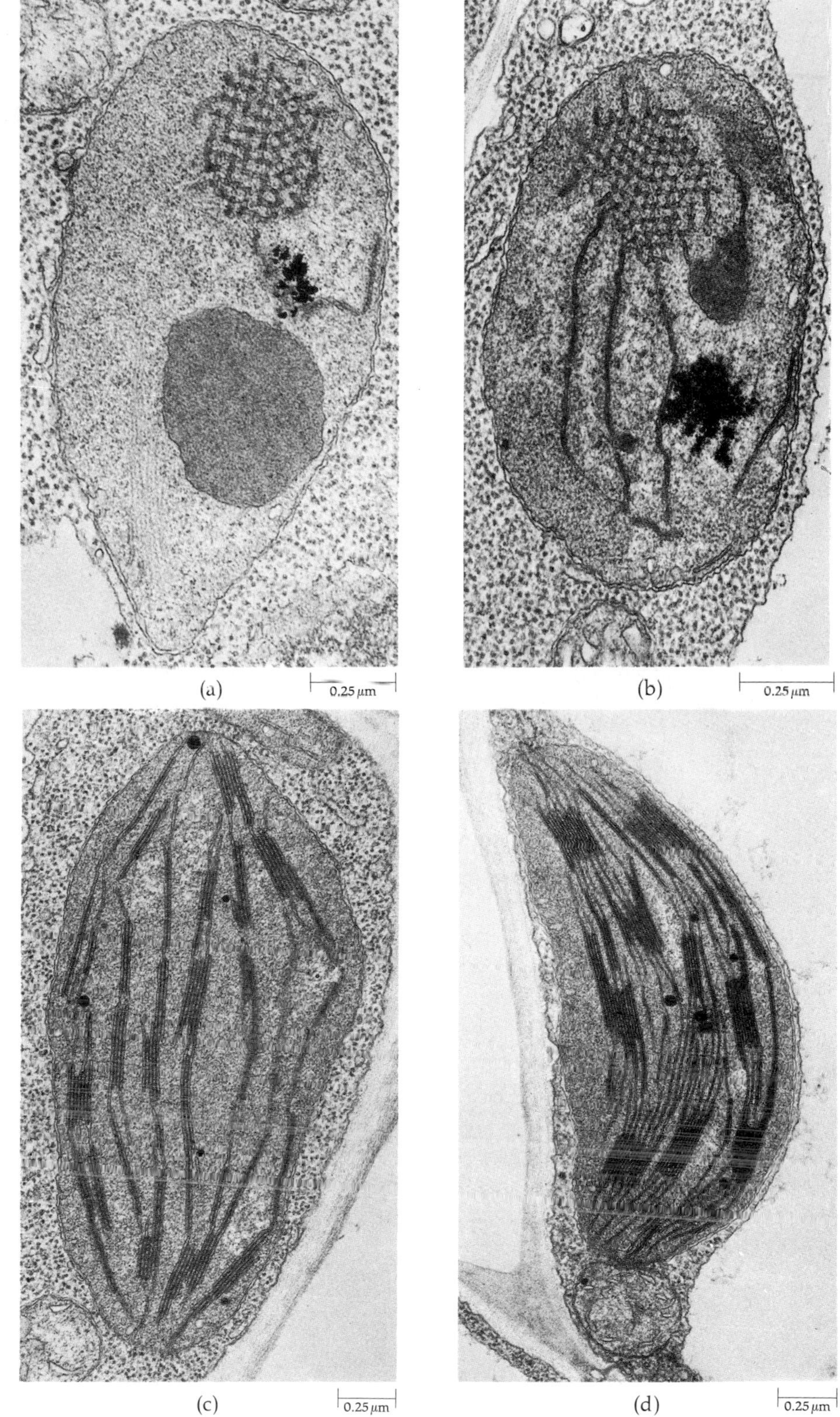

5-18

Contractile assemblies of fibrous proteins in a vertebrate skeletal muscle. Each unit is known as a sarcomere, approximately 14 of which are visible in this electron micrograph.

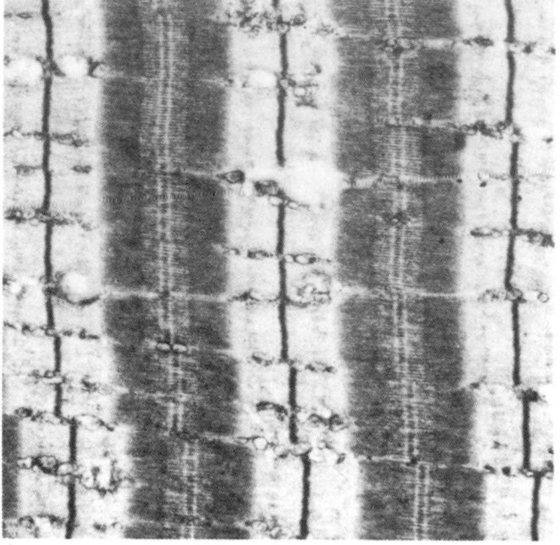

HOW CELLS MOVE

All cells exhibit some form of movement. Even plant cells, encased in a rigid cellulose cell wall, exhibit active cytoplasmic streaming (movement of cytoplasm within the cell) as well as chromosomal movements and changes in shape during cell division. As we saw in Figure 5-2c, cilia beat along the surface of the tracheal cells of animals. Embryonic cells migrate in the course of development. Differentiating and regenerating nerve cells send out axons, long slender extensions, which may be a meter or more in length. Amoebas pursue and engulf their prey. Even little *Chlamydomonas* cells dart toward a light source. Two different molecular mechanisms of cellular movement have been identified: (1) assemblies of fibrous proteins, usually referred to as muscle protein, and (2) the assemblies of microtubules in cilia and flagella.

Muscle Protein

The skeletal muscles of vertebrates contain elaborate contractile assemblies of fibrous proteins (Figure 5-18). The two principal types of protein involved are known as actin and myosin. (The contractile mechanism will be described in detail in Chapter 31.) These are commonly thought of as muscle proteins, since it is in muscle tissue that they were first identified and have been studied most intensively. However, it has now been found that actin, in particular (which is easier to isolate chemically), is present in a great variety of cells, including plant cells. It appears to be associated with internal movement within these cells.

For example, contractile proteins have been found in large, multinucleate organisms known as slime molds, which move like giant amoebas, with vigorous cytoplasmic streaming. Actin has also been found in a true amoeba, *Amoeba proteus.* In algal cells, microfilaments containing actin occur in bundles wherever cytoplasmic streaming is taking place. Such microfilaments have also been shown to act as a sort of "purse string" in animal cells during cytokinesis, pinching off the cytoplasm to separate the two cells produced in cell division. Cytologists are now coming to the conclusion that "muscle" proteins are common to all cells and that the elaborate contractile machinery of skeletal muscle cells is a recent evolutionary specialization. However, the way in which these proteins bring about cytoplasmic streaming and amoeboid movement is not yet understood.

5-19

Contractile proteins in nonmuscle cells as revealed by a special microscopic technique: (a) *actin in microfilament bundles and* (b) *myosin filaments, both from human fibroblast cells. These proteins are of the same type as found in the sarcomeres of skeletal muscle.*

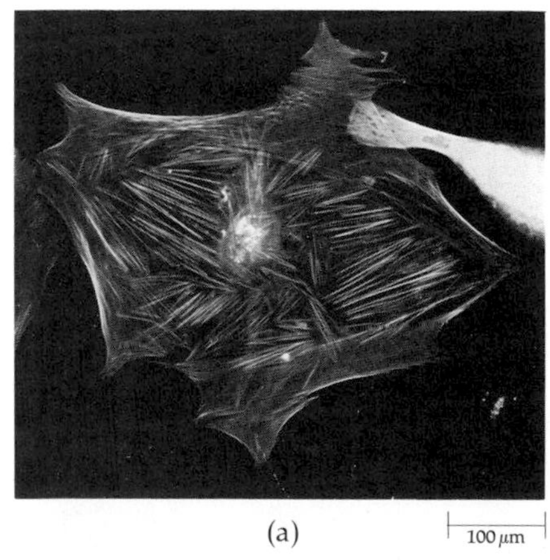

(a) 100 μm

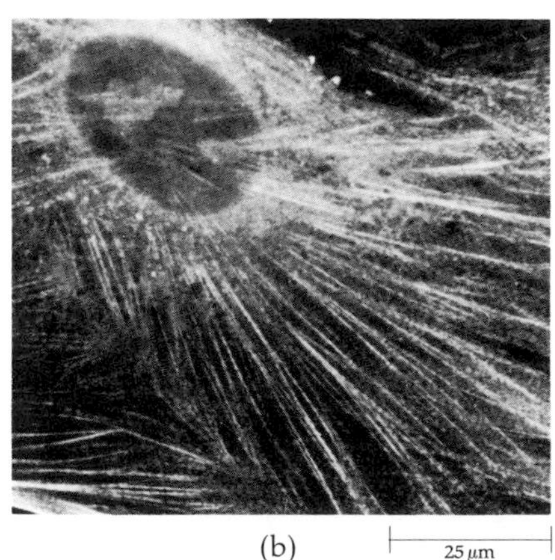

(b) 25 μm

5-20

Two ciliates, one-celled organisms distinguished by their many cilia. On the left, Paramecium; *on the right,* Didinium. Didinium *is stalking* Paramecium. Paramecium, *in defense, has discharged a barrage of barbs (visible as a cloud at the top of the micrograph).* Didinium *is about to eject a bundle of slender, poisonous strands (not visible), which will paralyze* Paramecium *in a matter of seconds. In* Paramecium, *the cilia are distributed fairly evenly over the cell surface. In* Didinium, *they form two wreaths that circle the organism's barrel-shaped body.*

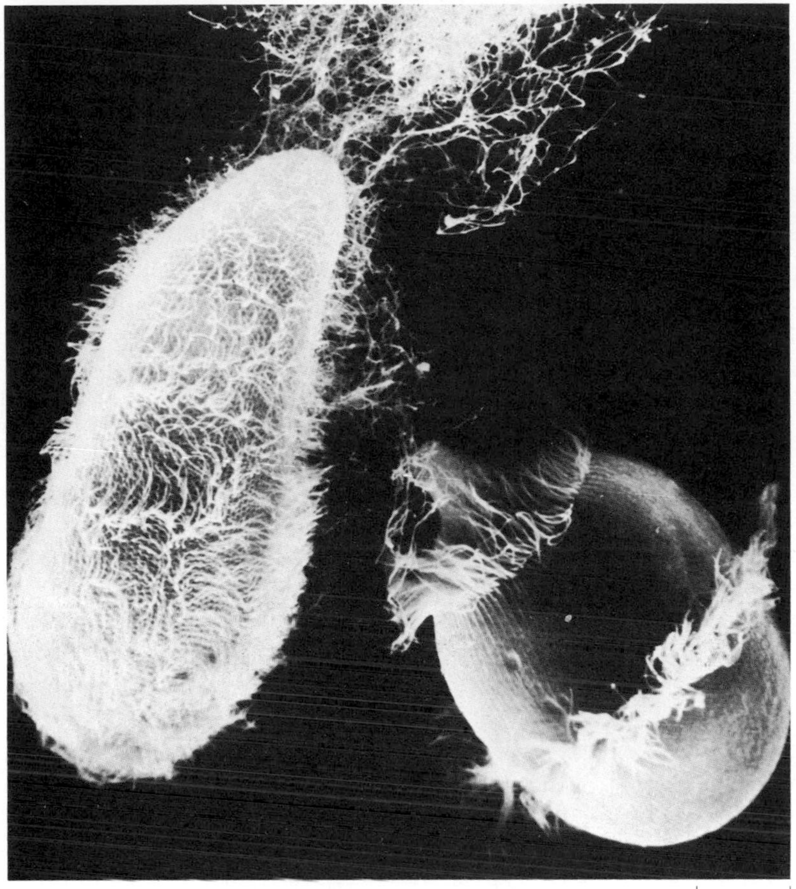

Cilia and Flagella

Cilia and *flagella* are long, thin (0.2 micrometer) structures extending from the surface of many types of eukaryotic cells. They are the same except for length; flagella are longer. (Prokaryotic cells also have flagella, but these are different in structure.) In one-celled organisms and some small animals (such as flatworms), cilia and flagella are associated with movement of the organism. For example, one type of *Paramecium* has approximately 17,000 cilia, each about 10 micrometers long, which propel it through the water by beating in a coordinated fashion. Other one-celled organisms, such as the members of the genus *Chlamydomonas,* have only two whiplike flagella, protruding from the anterior end of the organism, which move them through the water. The motile power of the human sperm cell comes from its single powerful flagellum, or "tail."

Many of the cells that line the surfaces within our bodies are also ciliated. These cilia do not move the cells, but rather serve to sweep substances from the environment across the cell surface. For example, cilia on the surface of cells of the respiratory tract beat upward, propelling a current of mucus that sweeps bits of soot, dust, pollen, tobacco tar—whatever foreign substances we have inhaled either accidentally or on purpose—to our throats, where they can be removed by swal-

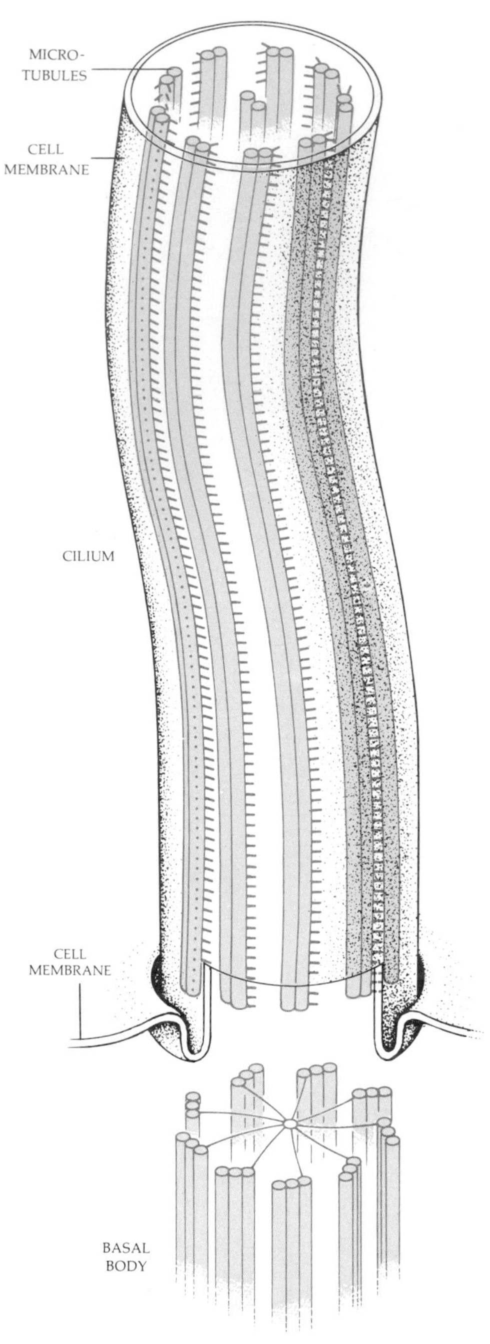

lowing. Human egg cells are propelled down the oviducts by the beating of cilia that line the inner surfaces of these tubes. Cilia and flagella are found extensively throughout the living world, on the cells of invertebrates, vertebrates, the sex cells of ferns and other plants, as well as on one-celled organisms. Only a few large groups of organisms, such as the red algae, higher fungi, and flowering plants, have no cilia or flagella in any cells.

All eukaryotic cilia and flagella, whether on a *Paramecium* or a human sperm cell, have the same structure: Nine fused pairs of microtubules form a ring that surrounds two additional, solitary microtubules in the center. (The microtubules themselves, you will recall from Chapter 3, are composed of identical globular protein units assembled in a hollow helix.) The movement of cilia and flagella comes from within the structures themselves; cilia removed from cells and exposed to ATP are capable of movement. The movement, according to one hypothesis, is caused by one outer pair moving tractor-fashion with respect to its nearest neighbor. The two "arms" that you can see on one of each pair of outer tubules (Figure 5-21a) have been shown to be enzymes involved in the splitting of ATP to ADP within the cilium, and so converting chemical energy to kinetic energy.

Basal Bodies and Centrioles

Underlying each cilium is a structure known as a basal body, which has the same diameter as a cilium, about 0.2 micrometer. It consists of microtubules arranged in nine triplets around the periphery. Unlike the cilium, it has no microtubules in the

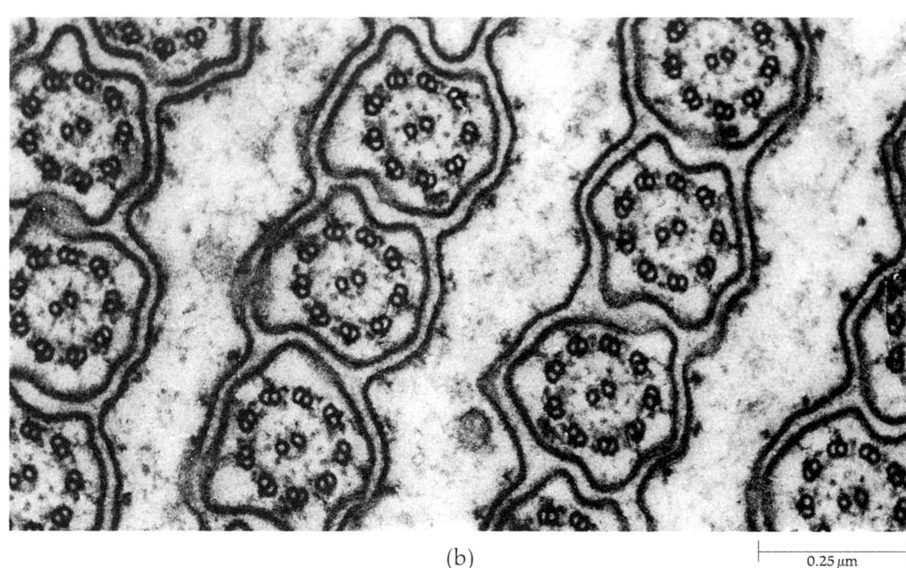

5-21
(a) *Diagram of a cilium with its underlying basal body. All cilia and flagella, whether they are found on one-celled organisms or on the surfaces of cells within our own bodies, contain this same internal structure, which consists of an outer ring of nine pairs of microtubules surrounding two additional microtubules in the center. The basal bodies from which they arise have nine outer triplets, with no microtubules in the center. The structural basis of the wheel-like formation in the basal body is not known. The "hub" of the wheel is not a microtubule, although it has about the same diameter.* (b) *Cross section of flagella. These are from* Trichonympha, *a one-celled organism that lives in the guts of termites where it breaks down cellulose.*

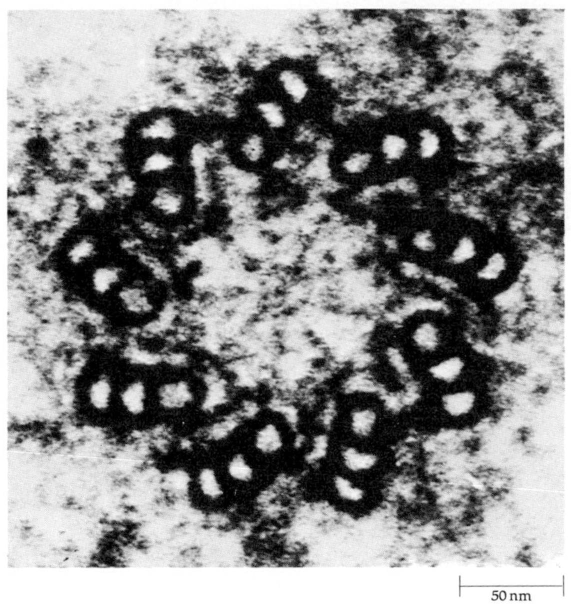

5–22
Centrioles are structurally identical to basal bodies. The one shown here in cross section is from a mouse embryo cell.

center, and none of the microtubules in the basal body has arms. Cilia and flagella arise from basal bodies. For instance, as a sperm cell takes form, a basal body moves near the cell membrane, and the sperm's flagellum arises from it through the assembly of microtubules. Basal bodies also apparently transmit ATP and perhaps other nutrients to cilia and flagella.

Many types of eukaryotic cells contain *centrioles*. Centrioles, which typically occur in pairs, are small cylinders, about 0.2 micrometer in diameter, containing nine microtubule triplets (Figure 5–22). Their structure is identical to that of basal bodies; however, their distribution in the cell is different. Until recently, it appeared that their function was also different, which is why they are called by different names even though electron microscopy has revealed their identical structures. Centrioles usually lie in pairs with their long axes at right angles to one another in the cytoplasm, near the nuclear envelope. They are found only in those groups of organisms that also have cilia or flagella (and, therefore, basal bodies). There is evidence that centrioles are important in organizing a structure known as the *spindle*, which appears at the time of cell division and is involved in chromosome movements. The spindle, it has been shown, also contains numerous microtubules; thus centrioles and basal bodies both appear to be organizers of microtubules. However, despite this evidence, cells without centrioles can, paradoxically, form spindles.

Bacterial Flagella

Some bacteria also move by means of flagella, although these flagella are so different in construction from those of eukaryotes that it would be useful if they had a different name. Each flagellum is made up of repeating units of a small globular protein, flagellin, assembled into chains that are wound in a triple helix (three chains) with a hollow core. Bacterial flagella grow from the tip; the individual flagellin molecules pass down a channel in the center of the helix and are assembled onto the end of the chains.

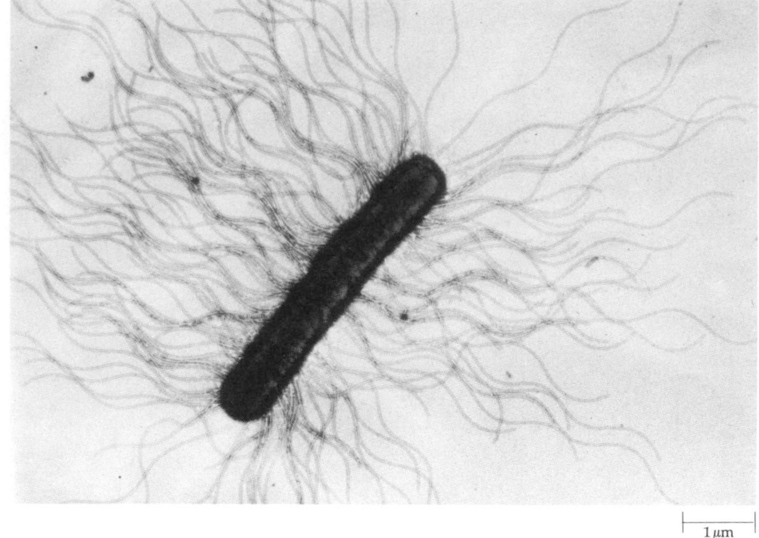

5–23
A bacterial cell (Proteus mirabilis) *with multiple flagella.*

The bacterial flagellum is not enclosed within the cell membrane, as eukaryotic flagella are, but protrudes from the cell as a naked, corkscrew-shaped protein filament. It is anchored in the cell membrane by a baseplate and rotates, at about 40 revolutions per second (the rate varies with the species). When it rotates in a counterclockwise direction, it pushes the cell forward. Clockwise rotation pulls the cell backward. The motor for this flagellar motion is not understood. *Escherichia coli* moves by means of six flagella. This bacterial cell, a heterotroph, can feed on glucose. If glucose is present, *E. coli* has no flagella. In the absence of glucose (or another abundant food source), the flagella regrow and propel the little cell into greener pastures.

HOW CELLS COMMUNICATE

In multicellular organisms a yet higher level of organization is found where the specialized cells cooperate to form a functional system. Cells are organized into tissues, groups of cells with common functions: muscle, nerve, connective, and epidermal (covering). Tissues are further organized in concert to form an organ, such as the heart, brain, or kidney, which, like the subcellular organelles, has a design that suits it for a specific function.

As you might imagine, in multicellular organisms it is essential that individual cells communicate with one another so that they can collaborate to create a harmonious tissue and organ. This communication sometimes occurs over long distances through the exchange of chemical messages (hormones), which may pass through the cell membrane or may bind to specific membrane protein receptors (page 131) at the cell surface. However, in many cases close and intimate contacts of various types occur between cells. Among plant cells, which are separated from one another by cell walls, cytoplasmic channels called *plasmodesmata* traverse the cell walls, connecting adjacent cells. Plasmodesmata, which are about 4 nanometers in diameter, appear to be lined by the cell membrane and to contain, on occasion, tubular extensions of the endoplasmic reticulum.

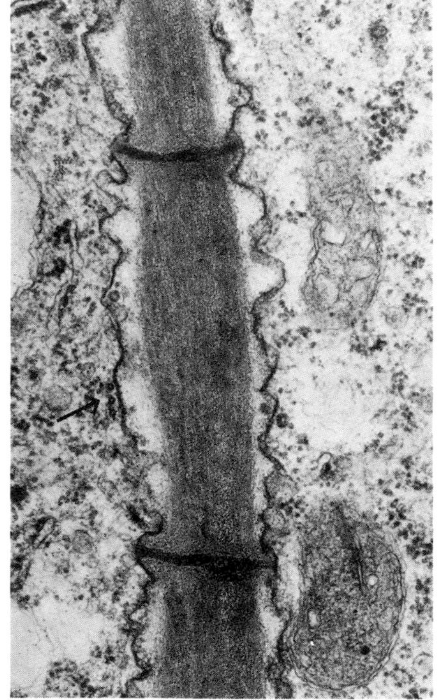

5–24
Electron micrograph of plasmodesmata, in longitudinal view, connecting two plant cells.

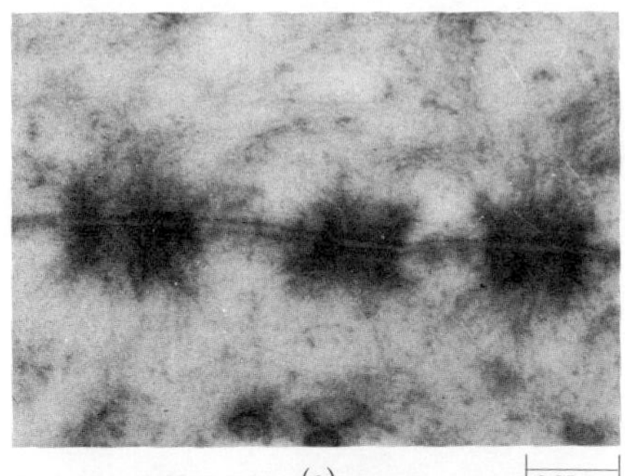

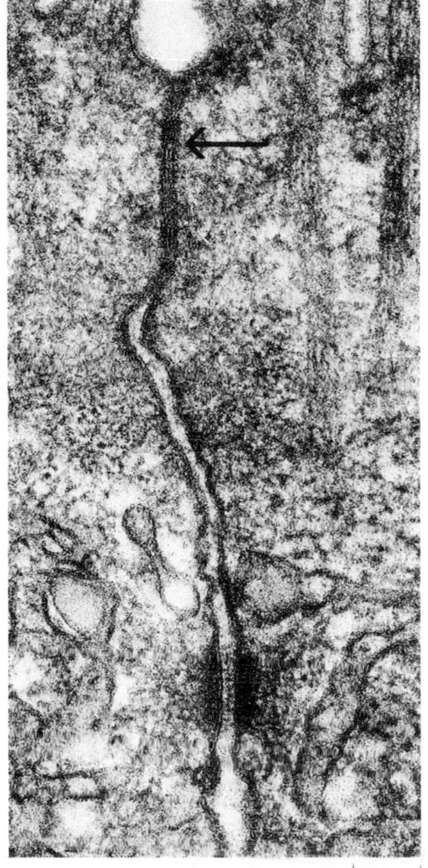

5-25
Junctions between cells. (a) Desmosomes cement cells to adjoining cells. Here three desmosomes connect the cell membranes of two adjacent skin cells of a salamander larva. (b) Tight junction (arrow) seals the spaces between cells. A desmosome is visible below the tight junction. These cells are from the surface layer (epithelium) of a rat intestine.

Table 5-2 *Comparison of Prokaryotic, Animal, and Plant Cells*

	PROKARYOTE	ANIMAL	PLANT
Cell membrane	Present	Present	Present
Cell wall	Present (noncellulose polysaccharide plus protein)	Absent	Present (cellulose)
Nucleus	No nuclear envelope	Surrounded by nuclear envelope	Surrounded by nuclear envelope
Chromosomes	Single, containing only DNA	Multiple, containing DNA and protein	Multiple, containing DNA and protein
Endoplasmic reticulum	Absent	Usually present	Usually present
Mitochondria	Absent	Present	Present
Plastids	Absent	Absent	Present in many cell types; chloroplasts in photosynthetic cells
Ribosomes	Present (smaller)	Present	Present
Golgi bodies	Absent	Present	Present
Lysosomes	Absent	Often present	Usually absent
Vacuoles	Absent	Small or absent	Usually large single vacuole in mature cell
9 + 2 cilia or flagella	Absent	Often present	Absent (in higher plants)
Centrioles	Absent	Present	Absent (in higher plants)

Cell-Cell Junctions

Three functional types of junctions between animal cells can now be distinguished by the electron microscope: (1) junctions that hold together adjacent cells in a tissue; these are called *desmosomes;* (2) junctions between cells that keep materials from leaking through tissues; these are called *tight junctions;* and (3) junctions through which cells can exchange nutrient molecules and molecular communications; these are called *gap junctions.*

Desmosomes have often been compared to spot welds between cells. They consist of plaques of dense fibrous material between cells with clusters of filaments from the cytoplasm of the neighboring cells looping in and out of them. They attach cells to one another and give tissues mechanical strength. They are found in especially large numbers in tissues subjected to severe mechanical stress, such as the skin.

Tight junctions form a continuous ring around each cell in a layer of tissue, preventing leakage between cells. For example, intestinal epithelial cells are surrounded by tight junctions that keep the intestinal contents from seeping between the cells. Similar arrangements are found in all other cells that form linings of internal organs, such as the bladder and the kidney. Tight junctions seem to involve the fusion of adjacent membranes.

Gap junctions permit the passage of materials between cells. They appear as clusters of very small channels (about 2 nanometers in diameter) surrounded by an ordered array of proteins. Experiments with labeled molecules have shown that messenger molecules can pass through these channels. The junctions also serve to transmit electrical signals. For example, gap junctions synchronize the contractions of muscle cells of the heart.

Chemical Communication

Cells send one another chemical messages. Nerve impulses are transmitted from neuron to neuron, or from neuron to muscle or gland, by means of chemical signals, as we shall see in Chapter 34. Cells in the body of a plant or animal release hormones that travel over a distance and affect other cells in the same organism. Embryonic cells in the course of development release chemicals that direct the differentiation of neighboring cells into organs and tissues. If normal cells from vertebrates (including humans) are isolated from one another and grown in a nutrient medium on a smooth glass surface, they move, amoebalike, ruffling their cell borders until they encounter another cell, at which time they stop, apparently in response to some signal. More significantly, they multiply until each cell is touching another cell; then they cease to multiply. This phenomenon is known as contact inhibition. (Cancer cells do not show contact inhibition; they pile on top of one another, moving, multiplying, crowding each other until all of the nutrients are used up.)

A cellular communication system of particular interest to biologists, because of the comparative ease with which it can be studied, is seen among a group of organisms known as the cellular slime molds. The slime mold *Dictyostelium discoideum* is an example. At one stage in its life cycle, it exists as a swarm of small

5-26
Skin cells growing on a glass plate. When the cells touch each other, they stop their amoebalike movement. They will multiply only until they fill the empty spaces.

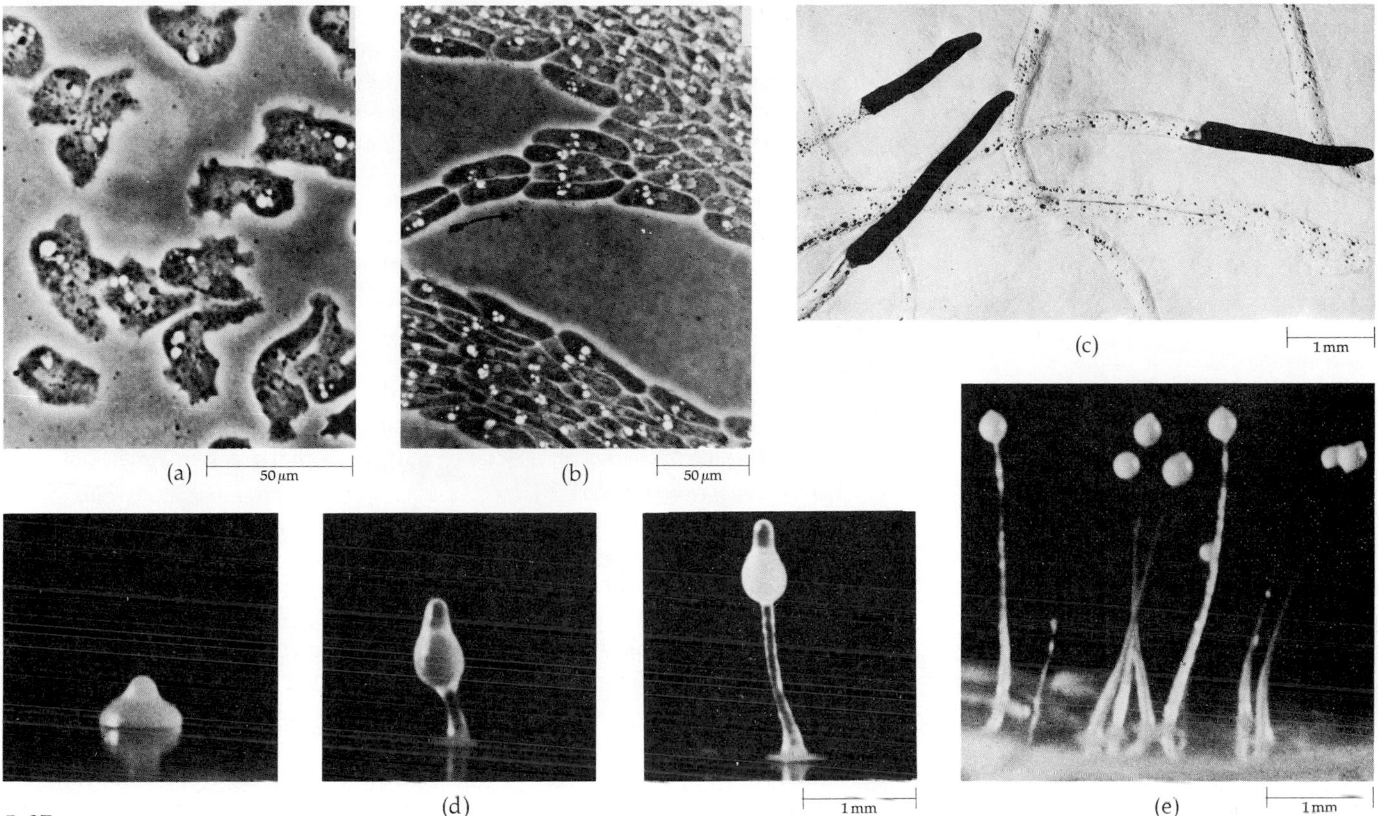

5–27
Life history of a cellular slime mold, Dictyostelium discoideum. (a) *Amoebas in feeding stage. The light gray area in the center of each cell is the nucleus, and the white areas are contractile vacuoles.* (b) *Amoebas aggregating. The direction in which the stream is moving is indicated by the arrow.* (c) *Migrating sluglike masses, each of which deposits a thick slime sheath that collapses behind it.* (d) *At the end of the migration, the mass gathers together and begins to rise vertically, differentiating into a stalk and fruiting body* (e).

individual amoebas, which divide and grow and feed, amoeba-fashion, until their food supply (mostly bacteria) gives out. At this point, the cells alter both their shape and behavior: They become sausage-shaped and begin to migrate toward the center of the group. Eventually, they pile up in a heap; the heap gradually takes on the form of a multicellular mass somewhat resembling a garden slug. The sluglike mass soon stops its migration, gathers itself into a mound and sends up a long stalk at the tip of which a small, shimmering fruiting body forms. The fruiting body matures and eventually bursts open, releasing a new swarm of tiny amoebas, and the cycle begins again. The chemical that initiates this remarkable sequence of events was first called acrasin, after Acrasia, the cruel witch in Spenser's *Faerie Queene* who attracted men and turned them into beasts. Acrasin was later identified as the chemical compound cyclic AMP. Cyclic AMP has now been shown to have an important role in human physiology, which will be described in Chapter 34.

SUMMARY

The properties associated with biological systems emerge at the cellular level of organization. All living organisms consist of one or more cells; the processes of metabolism, growth, response to stimuli, energy exchange, and reproduction occur at the cellular level.

There are two fundamentally distinct types of cellular organization—prokaryotes, which include only the bacteria and blue-green algae, and eukaryotes, which include everything else. Prokaryotic cells lack membrane-bound nuclei and most of the organelles found in eukaryotic cells. Cells also may be divided into two groups on the basis of their energy sources: Autotrophs are cells that get their energy from inorganic sources, whereas heterotrophs are dependent on organic compounds derived from other living organisms. With the exception of some prokaryotes, all autotrophs are photosynthetic.

Cells are separated from their environment by an outer cell membrane that restricts the passage of materials into and out of the cell and so protects the cell's structural and functional integrity. Plant cells are also usually enclosed in a cell wall. The size of cells is limited by proportions of surface to volume; the smaller a cell's volume in proportion to its surface area, the more readily substances can get into and out of the cell. Cell size is also limited by the capacity of the nucleus to regulate the cellular activities of very large cells. Cells that are active metabolically are likely to be small.

The cytoplasm of the cell is a concentrated aqueous solution containing enzymes, many other dissolved molecules and ions, and also, in the case of eukaryotic cells, a variety of membrane-bound organelles with specialized functions in the life of the cell. The eukaryotic cytoplasm has also been shown recently to contain microtubules, which appear to form a supporting cytoskeleton, and microfilaments, involved in various types of cell movement. Vacuoles and vesicles (small vacuoles) are also present in many cells, particularly plant cells. They are bounded by a single membrane.

In eukaryotic cells, the hereditary material is separated from the cytoplasm by the nuclear envelope, which contains pores through which molecules pass to and from the cytoplasm. The nucleus contains the chromosomes, which, when the cell is not dividing, exist in an extended form called chromatin. The nucleolus, visible within the nucleus, is the site of formation of the ribosomes.

Eukaryotic cells contain numerous organelles. The most numerous are the ribosomes, the sites of assembly of proteins. The endoplasmic reticulum is an interconnected system of membrane-lined sacs, tubes, and channels. It serves as a work surface for many of the cell's biochemical activities, and also, perhaps, as a system of passageways for moving materials through the cell. In eukaryotic cells, ribosomes may be bound to the surface of the endoplasmic reticulum or free in the cytoplasm. Rough endoplasmic reticulum (with ribosomes on it) is found characteristically in cells producing proteins for export. Smooth endoplasmic reticulum, which lacks ribosomes, is involved with lipid synthesis. Golgi bodies, also composed of membranes, are packaging centers for materials being moved through and out of the cell. Lysosomes, which contain destructive enzymes, are involved in intracellular digestive activities in some cells.

Mitochondria are membrane-bound organelles that are principal sites of ATP formation. Plastids are membrane-bound organelles found only in photosynthetic organisms. Leucoplasts are storage compartments, chromoplasts contain pigments, and chloroplasts are the sites of photosynthesis in eukaryotes.

Cells move either by means of specialized assemblies of "muscle" proteins or by means of cilia and flagella, hairlike appendages found on the surface of (and within the cell membrane) of many cell types. Cilia and flagella are associated with the movement of cells or the movement of materials across cell surfaces. They have a highly characteristic 9 + 2 structure, with nine pairs of microtubules forming a ring surrounding two central microtubules. One of each pair of outer microtubules contains enzymes that split ATP and provide the energy for ciliary motion.

Cilia and flagella arise from basal bodies, which are cylindrical structures containing nine microtubule triplets with no inner pair. Centrioles have the same internal structure as basal bodies and are found only in those groups of organisms that have cilia or flagella. They typically occur in pairs, lying near the nuclear envelope, and appear to play a role in the separation of the chromosomes at cell division.

Cells have a variety of systems for communicating with one another, including direct physical contacts, such as plasmodesmata and gap junctions. Cells also communicate with one another by chemical signals.

QUESTIONS

1. Distinguish among the following: prokaryote/eukaryote; autotroph/heterotroph/phototroph; cell membrane/cell wall; nucleus/nucleolus; cilia/flagella; basal body/centriole; smooth endoplasmic reticulum/rough endoplasmic reticulum; plasmodesmata/desmosomes; gap junctions/tight junctions; chloroplasts/mitochondria.

2. (a) Sketch an animal cell. Include the principal organelles and label them. (b) Prepare a similar, labeled sketch of a plant cell. (c) What are the major differences between the animal cell and the plant cell?

3. (a) Sketch a cross section of a cilium. (b) Sketch a cross section of the basal body of a cilium. (c) What are the differences between the two structures?

4. (a) Return to Chapter 2 and add approximate scale markers to Figures 2–1, 2–2, 2–4, 2–5, 2–6, and 2–11. (b) Use a ruler and the scale marker at the bottom of each micrograph on pages 102 and 116 to determine. (1) the thickness (roughly) of a cell membrane, (2) the diameter of a cilium, and (3) the diameter of a microtubule within a cilium. (This is how the sizes of cellular components are determined by microscopists.) (c) Would a cilium be resolvable in a light microscope (that is, is its diameter more than 0.2 μm)?

5. Explain the functions of each of the following structures: endoplasmic reticulum, ribosomes, Golgi bodies, lysosomes.

6. Why is the secondary cell wall of a plant *inside* the primary cell wall? Where is the cell membrane in relation to the two cell walls?

7. On the basis of what you know of the functions of each of the structures in Table 5–2 (page 119), what components would you expect to find most prominently in each of the following cell types: muscle cells, sperm cells, green leaf cells, red blood cells, white blood cells?

8. Two brothers were under medical treatment for infertility. Microscopic examination of their semen showed that the sperm were immotile and that the little "arms" were missing from the microtubular arrays. The brothers also had chronic bronchitis and other respiratory difficulties. Can you explain why?

CHAPTER 6

How Things Get into and out of Cells

Cells are able to regulate the passage of materials across cell membranes. This is an important capacity. One of the criteria by which we identify living systems is that living matter, although surrounded on all sides by nonliving matter, is different from it in the kinds and amounts of chemical substances it contains. Without this difference, of course, living matter would be unable to maintain the organization and structure on which its existence depends. The cell membrane is not simply an impenetrable barrier, however. Living matter constantly exchanges substances with the nonliving world around it. Control of these exchanges is essential in order to protect the cell's integrity and to maintain those very narrow conditions of pH and ionic concentrations at which metabolic activities can take place. Cell membranes thus have a complex double function of keeping things out and letting things in. Moreover, the outer membrane not only controls the passage of material from outside the cell, but internal membranes, such as those surrounding mitochondria, chloroplasts, and the nucleus, also regulate the passage of materials among intracellular compartments and so regulate the internal environment.

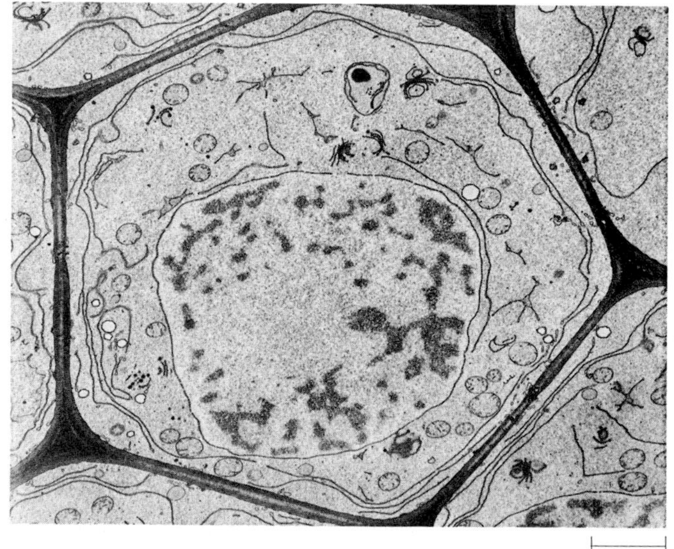

6–1
Electron micrograph of a cell from the root tip of a corn plant. The cell has a large, central nucleus with scattered chromatin, several clearly delineated Golgi bodies, many mitochondria, profiles of endoplasmic reticulum, small vacuoles, and numerous other cellular organelles. Not only is the cell itself surrounded by a membrane and the nucleus by a double membrane system (the nuclear envelope), but many of its organelles are also surrounded by membranes. The membranes of the endoplasmic reticulum further divide the cell into membrane-bound channels. These membranes regulate the movements of substances into and out of cells and restrict their passage from one part of the cell to another.

6–2
Diagram showing the relative concentrations of different ions in pond water (gray bars) and in the cytoplasm of the green alga Nitella *(colored bars). Differences such as these between cells and their surroundings indicate that cells regulate the passage of materials across their membranes.*

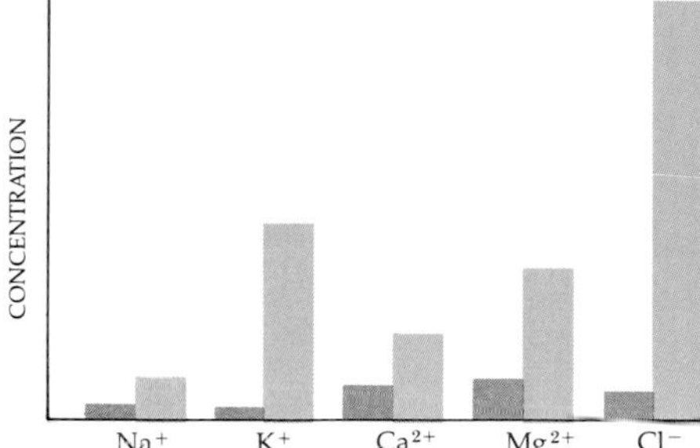

Control of these exchanges across membranes depends on the physical and chemical properties of the membranes and of the molecules that move through them. Of the many kinds of molecules moving into and out of cells, by far the most important is water. Let us therefore look again at water, focusing our attention this time on how water moves.

PRINCIPLES OF WATER MOVEMENT

We shall first discuss the movement of water in the inanimate world. Three mechanisms are involved in water movement: bulk flow, diffusion, and osmosis.

Bulk Flow

Bulk flow is the overall movement of water (or some other fluid). The molecules move all together and in the same direction. Bulk flow occurs in response to differences in the potential energy of water, usually referred to as the *water potential.*

A simple example of water that has potential energy is water at the top of a hill. As this water runs downhill, its potential energy can be converted to mechanical energy by a watermill or to electrical energy by a hydroelectric turbine.

Pressure is another source of water potential. If we fill a rubber bulb with water and squeeze, water will squirt out the nozzle. Like water at the top of a hill, this water has been given a high water potential and will move to a lower one. Can we make the water that is running downhill run uphill by means of pressure? Obviously we can. But only so long as the water potential produced by the pressure exceeds the water potential produced by gravity. Water moves from an area where water potential is greater to an area where water potential is lower, regardless of the reason for the water potential. Your heart pumps blood to your brain against gravity by creating a greater water (blood) potential.

The concept of water potential is a useful one because it enables us to predict the way that water will move under various combinations of circumstances. Mea-

6–3
Water at the top of a falls, like a boulder on a hilltop, has potential energy. The movement of water molecules as a group, as from the top of the falls to the bottom, is referred to as bulk flow.

surements of water potential are usually made in terms of the pressure required to stop the movement of water—that is, the hydrostatic (water-stopping) pressure—under the particular circumstances. The unit usually used to measure this pressure is the atmosphere. One atmosphere is the average pressure of the air at sea level, about 1 kg/cm^2 (or 15 lb/in^2).

Diffusion

Diffusion is a familiar phenomenon. If you sprinkle a few drops of perfume in one corner of a room, the scent will eventually permeate the entire room even if the air is still. If you put a few drops of dye in one end of a glass tank full of water, the dye molecules will slowly become distributed evenly throughout the tank. The process may take a day or more, depending on the size of the tank, the temperature, and the relative size of the molecules.

Why do the dye molecules move apart? If you could observe the individual dye molecules in the tank (Figure 6–4), you would see that each one of them moves individually and at random. Looking at any single molecule—at either its rate of motion or its direction of motion—gives you no clue at all as to where the molecule is located with respect to the others. So how do the molecules get from one side of the tank to the other? Imagine a thin slice through the tank, running from top to bottom. Dye molecules will move in and out of the slice, some moving in one direction, some moving in the other. But you will see more dye molecules moving from the side of greater dye concentration. Why? Simply because there are more dye molecules at that end of the tank. Since there are more dye molecules on the left, more dye molecules, moving at random, will move to the right, even though there is an equal probability that any one molecule of dye will move from right to left. Consequently, the overall (net) movement of dye molecules will be from left to right. Similarly, if you could see the movement of the individual water molecules in the tank, you would see that net movement is from right to left.

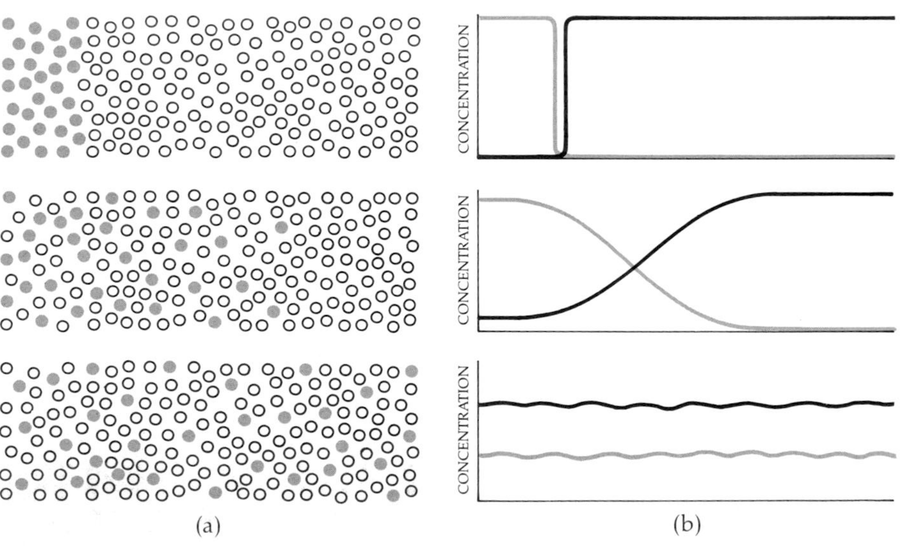

6–4
(a) *Diagram of the diffusion process. Diffusion is the result of the random movement of individual molecules, which produces a net movement from a more concentrated to a less concentrated area. Notice that as one type of molecule (indicated by color) diffuses to the right, the other diffuses in the opposite direction. The result will be an even distribution of both types of molecules. Can you see why the net movement of molecules will slow down as equilibrium is reached?* (b) *Graphs showing the concentration gradients of dye and water.*

What happens when all the molecules are distributed evenly throughout the tank? The even distribution does not affect the behavior of the molecules as individuals; they still move at random. And, since the movements are random, just as many molecules go to the left as to the right. But because there are now as many molecules of dye and as many molecules of water on one side of the tank as on the other, there is no net direction of motion. There is, however, just as much individual motion as before, provided the temperature has not changed.

Substances that are moving from a region of higher concentration of their own molecules to a region of lower concentration are said to be moving *down a gradient.* Diffusion only occurs along a gradient. A substance moving in the opposite direction, toward a higher concentration of its own molecules, would be moving *against a gradient,* which is analogous to being pushed uphill. The steeper the downhill gradient—that is, the larger the difference in concentration—the more rapid the net flow. Also, diffusion is more rapid in gases than in liquids and at higher rather than at lower temperatures. Can you explain why?

Notice that, in our imaginary tank, there are two gradients; the dye molecules are moving along one of them, and the water molecules are moving along the other in the opposite direction. In both cases, the movement is down the gradient. When the molecules have reached a state of equal distribution, that is, when there are no more gradients, they continue to move, but there is no net movement in either direction. This situation is similar to that which occurs in a reversible chemical reaction when equilibrium has been reached. (See page 76.)

The concept of water potential, introduced in the discussion of bulk flow, is also useful in understanding diffusion. A high solute concentration in one region, such as a corner of the tank, means low water concentration there and, thus, low water potential. Thus, if the pressure is everywhere equal, water molecules, as they move down the gradient, are moving from a region of higher water potential to a region of lower water potential. The region of the tank in which there is pure water has a greater water potential than the region containing water plus dye or some other dissolved substance (solute).

The essential characteristics of diffusion are (1) that each molecule moves independently of the others and (2) that these movements are random. The net result of diffusion is that the diffusing substance eventually becomes evenly distributed.

Cells and Diffusion

Diffusion is essentially a slow process, except over very short distances. It is efficient only if the concentration gradient is steep and the volume relatively small. For instance, the rapid spread of a scent, such as perfume, through the air is due not primarily to diffusion but rather to the circulation of air currents. Similarly, in many cells, the transport of materials is speeded by active streaming of the cytoplasm. Also cells hasten diffusion by their own metabolic activities. For example, oxygen is used up within the cell almost as rapidly as it enters, thereby keeping the gradient steep. Carbon dioxide is produced by the cell, and so a gradient from inside to outside is maintained.

Similarly, within a cell, materials are often produced at one place and used at another. Thus a concentration gradient is established between the two areas, and the material diffuses down the gradient from the site of production to the area of use.

Most organic molecules are polar and so cannot freely diffuse through the lipid barrier of the cell membrane. Carbon dioxide and oxygen, however, which are soluble in lipids, move freely through the membrane. Water also moves in and out freely. Water is not soluble in lipids (which is, of course, just another way of saying that lipids are not soluble in water), and so the fact that water moves freely has led biologists to postulate the presence of small pores in the membrane that permit the passage of water molecules (and also of some small ions).

Osmosis

While permitting the passage of water, cell membranes block the passage of most of the materials dissolved in it. Such a membrane is known as selectively permeable, and the movement of water through it is known as *osmosis.* Osmosis carries a net flow of water from a solution that has high water potential to a solution that has low water potential. In the absence of other factors that influence water potential (such as pressure), the movement of the water in osmosis will be from a region of lower solute concentration (and therefore of higher water concentration) into a region of higher solute concentration (lower water concentration). The presence of solute decreases the water potential and so creates a gradient of water potential along which water moves.

Osmosis can result in a buildup of pressure as water molecules continue to move across the membrane into regions of lower water concentration. The pressure exerted on the membrane as a result of the osmotic movement of water is known as the osmotic pressure and is another factor in determining water potential. As the increasing osmotic pressure raises the water potential, the net flow of water molecules will slow and then cease as the pressure gradient disappears.

The measurement of osmotic pressure is illustrated in Figure 6–5. The beaker contains distilled water, and within the tube is water plus a solute such as salt or sugar. Across the mouth of the tube is a selectively permeable membrane; such a membrane is freely permeable to water, but not to the solute. Water moving from the beaker through the membrane into the solution causes the solution level in the tube to rise until equilibrium is reached—that is, until the water potential on both sides of the membrane is equal.

The movement of water is not affected by what is dissolved in the water, only by how much solute it contains—how many particles of solute (molecules or ions). The word *isotonic* was coined to describe solutions that have equal numbers of dissolved

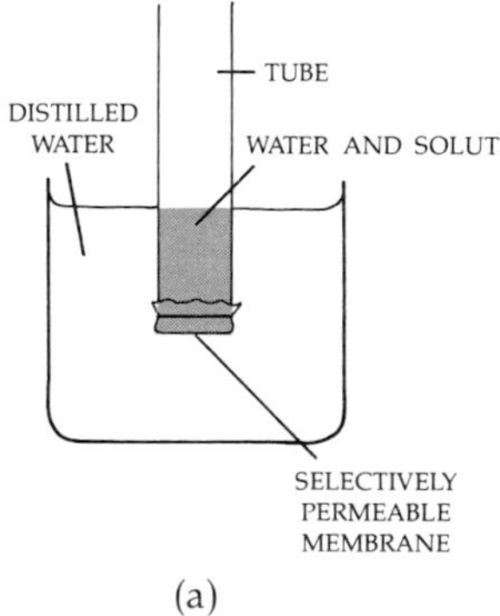

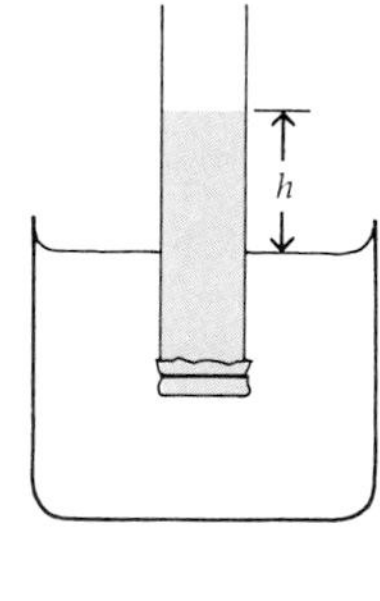

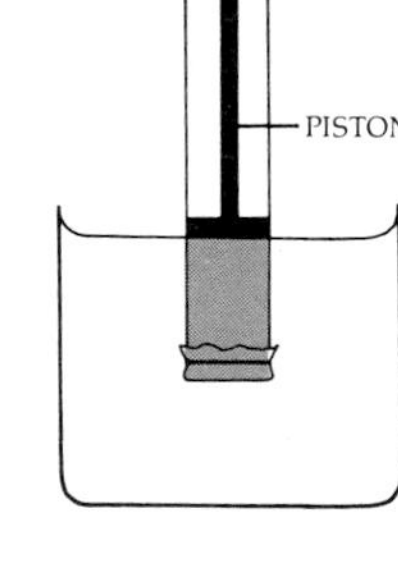

6–5
Osmosis and osmotic pressure. (a) *The tube contains a solution and the beaker contains distilled water.* (b) *The selectively permeable membrane permits the passage of water but not of solute. The movement of water into the solution causes the solution to rise in the tube until the osmotic pressure resulting from the tendency of water to move into a region of lower water concentration is counterbalanced by the height,* h, *and density of the column of solution.* (c) *The force that must be applied to the piston to oppose the rise in the tube of the solution is a measurement of the osmotic pressure. It is proportional to height and density.*

6–6
A Paramecium *is hypertonic in relation to its environment, and hence water tends to move into the cell by osmosis. Excess water is expelled through its contractile vacuoles.* (a) *Collecting tubules converge toward the vacuole, filling it.* (b) *Then it contracts, emptying outside the cell membrane by way of a small central pore.* (c) *Photomicrograph of a* Paramecium *showing the position of its two rosette-like contractile vacuoles.*

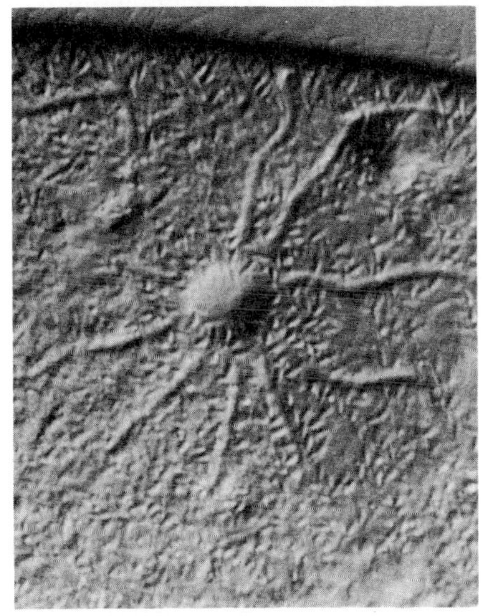

(a)

(b) 10 μm

particles and therefore exert equal osmotic pressures. There is no net movement of water across a membrane separating two solutions that are isotonic to one another, unless, of course, pressure is exerted on one side. In comparing solutions of different concentration, the solution that has less solute and therefore exerts a lower osmotic pressure is known as *hypotonic*, and the one that has more solute and exerts a higher osmotic pressure is known as *hypertonic*. (Note that *iso* means "the same"; *hyper* means "more"—in this case, more molecules of solute; and *hypo* means "less"—in this case, fewer particles of solute.)

Since solutes decrease water potential, a hypotonic solution has a higher water potential than a hypertonic one. In osmosis, water molecules move through a selectively permeable membrane into a hypertonic solution until the water potential is equal on both sides of the membrane.

Osmosis and Living Organisms

The movement of water across the cell membrane from a hypotonic to a hypertonic solution causes some crucial problems for living systems. These problems vary according to whether the cell or organism is hypotonic, isotonic, or hypertonic in relation to its environment. One-celled organisms that live in salt water, for example, are usually isotonic with the medium they inhabit, which is one way of solving the problem. Similarly, the cells of higher animals are isotonic with the blood and lymph that constitute the watery medium in which they live.

Many types of cells live in a hypotonic environment. In all single-celled organisms that live in fresh water, such as *Paramecium,* the interior of the cell is hypertonic to the surrounding water; consequently water tends to move into the cell by osmosis. If too much water were to move into the cell, it could dilute the cell contents to the point of interfering with function and could even eventually rupture the cell membrane. This is prevented by a specialized organelle known as a contractile vacuole, which collects water from various parts of the cell and pumps it out with rhythmic contractions. As you might expect, this bulk transport process requires energy.

Turgor

Plant cells expand as a result of the water pressure created by osmosis. Because plant cells are usually hypertonic to their surrounding environment, water tends to move into them. This movement of water into the cell creates pressure within the

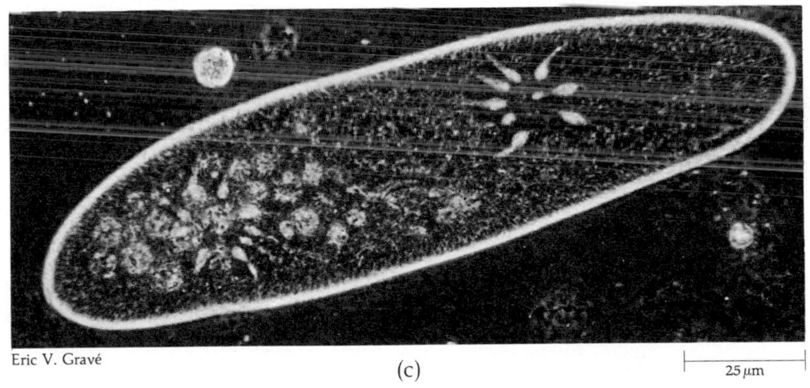

Eric V. Gravé

(c) 25 μm

6–7

(a) *A turgid plant cell. The central vacuole is hypertonic in relation to the fluid surrounding it and so gains water. The expansion of the cell is held in check by the cell wall.* (b) *A plant cell begins to wilt if it is placed in an isotonic solution, so that water pressure no longer builds up within the vacuole.* (c) *A plant cell in a hypertonic solution loses water to the surrounding fluid and so collapses, with its membrane pulling away from the cell wall. Such a cell is said to be plasmolyzed.*

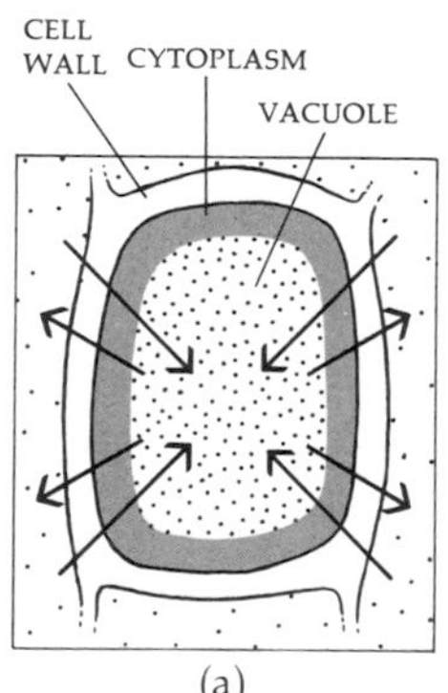

(a)

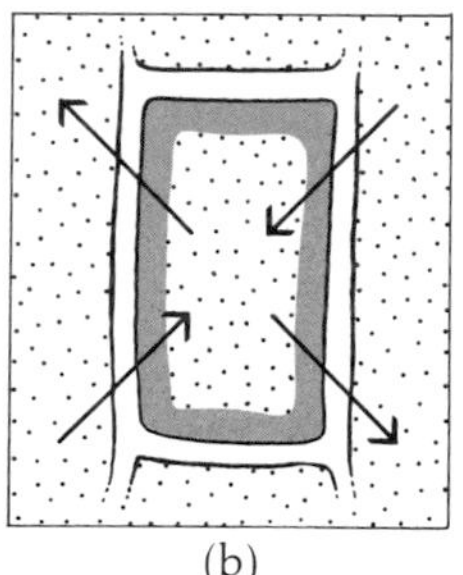

(b)

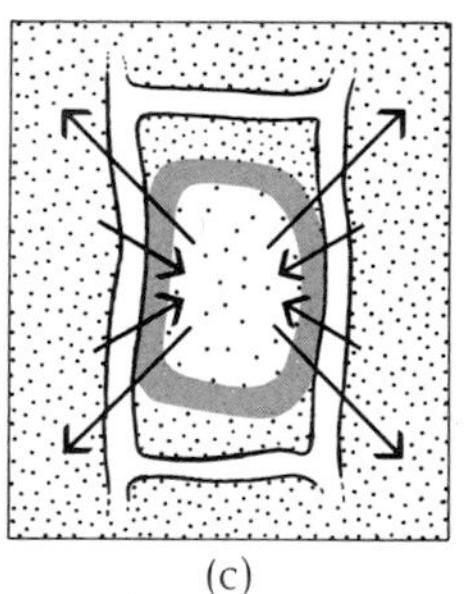

(c)

cell against the cell wall. The pressure causes the cell wall to expand and the cell to enlarge (Figure 5–6b). The elongation of a plant cell is a direct result of water uptake.

If the water potential on both sides of the plant cell membrane becomes equal, the net movement of water ceases. (We say "net movement" here because water molecules continue to diffuse back and forth across the membrane. However, these movements are in equilibrium; that is, as many water molecules are going in as are coming out.)

As the plant cell matures, the cell wall stops growing. Moreover, mature plant cells typically have large central vacuoles. These vacuoles often contain solutions of salts and other materials. (In citrus fruits, for example, they contain the acids that give the fruits their characteristic sour taste.) Because of these concentrated solutions, water continues to "try" to move into the cells. In the mature cell, however, the cell wall does not expand further, and equilibrium of salt concentration is not reached. As a consequence, the cell wall remains under constant pressure. This water pressure exerted on the cell wall from inside is known as *turgor*. Turgor, which keeps the cell walls stiff and the plant body crisp, results from an equilibrium of water potential. When turgor pressure is reduced, as a consequence of water loss, the plant wilts.

Osmotic pressure also plays an extremely important part in determining the composition of the urine excreted by the kidney, as we shall see in Chapter 38.

THE CELL MEMBRANE

Figure 6–8 shows the current model of the structure of cell membranes. The membrane is composed of a phospholipid bilayer (see Figure 3–14) in which are suspended numerous large globular proteins, floating like icebergs in the sea. Most membranes are about 40 percent lipid and 60 percent protein, though there is considerable variation. Some of the proteins protrude from both surfaces, but others are embedded in one or the other surface of the membrane, so the two surfaces differ considerably in chemical composition. The portion of the protein molecule embedded in the bilayer is hydrophobic. The structure of the bilayer is quite fluid, so the protein molecules can move within it, forming different patterns that vary from time to time and place to place. This widely accepted hypothesis of membrane structure is known as the *fluid-mosaic model*.

The structure of the outer membrane (cell membrane) of the cell has been studied most extensively, but this model is believed also to apply to all the cell's internal membranes.

Some of the proteins, and to a lesser extent, the lipids, have short chains of carbohydrates attached to their outer surfaces. These form a carbohydrate coat on the outer surface of the membranes of some eukaryotic cells. These carbohydrates appear to be involved with recognition and adhesion processes—between cells and cells, for instance, as well as cells and antibodies, and cells and viruses.

TRANSPORT ACROSS CELL MEMBRANES

As we noted in Chapter 2, water and other polar or charged (hydrophilic) molecules exclude lipids and other hydrophobic molecules. Conversely, hydrophobic molecules exclude hydrophilic ones. Thus the cell membrane's lipid bilayer would

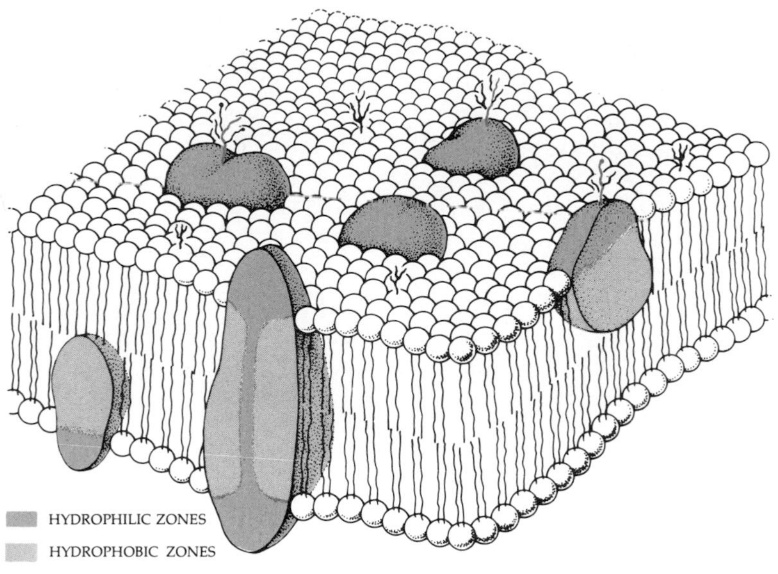

6–8
Model of a cell membrane, proposed by S. J. Singer and G. L. Nicolson, based on electron micrographs and biochemical data. The membrane is composed of phospholipid molecules, with their hydrophobic "tails" forming the inner layer, and of large protein molecules. The surface of the portion of the protein molecule embedded in the lipid bilayer is hydrophobic. The surface of the protein molecule exposed outside the lipid bilayer is hydrophilic, as are the surfaces of the pores that pass through some of the protein molecules. The short carbohydrate chains attached to the hydrophilic regions of the protein molecules are involved with cellular adhesion and recognition.

According to this model, the proteins vary from membrane to membrane, depending on cell function, and also from place to place on the same membrane. Some are enzymes and some are involved in cross-membrane transport and cell-cell contact. The whole structure is quite fluid, and the proteins can be thought of as floating in a lipid sea.

be expected to exclude a large number of ions and molecules essential to the cell's existence. (It was in fact the observation that hydrophobic molecules diffused readily across cell membranes that provided the first evidence of the lipid nature of the membrane.) A number of mechanisms apparently exist for getting hydrophilic molecules and ions through the membrane.

First, there are apparently apertures in the membrane through which water molecules can diffuse. (Water is the smallest polar molecule of biological importance.) These may be either structural pores or momentary openings resulting from the movements of the large molecules of which the membrane is composed. Other polar molecules, provided they are small enough, also pass through these openings. The permeability of the membrane to these solutes varies inversely with the size of the molecules, indicating that the apertures are small and that the membrane acts like a sieve in this respect.

Second, some of the proteins of the membrane act as carriers, ferrying back and forth molecules that could not readily cross the membrane by diffusion because of their size or polarity.

There are two basically different categories of carrier-assisted transport: facilitated diffusion and active transport. *Facilitated diffusion* is driven by the concentration gradient and so moves molecules from an area of higher concentration to an area of lower concentration. *Active transport* requires energy (often supplied by ATP) and can move molecules against the concentration gradient.

Carrier-assisted transport, whether facilitated diffusion or active transport, is highly selective; the transport protein may accept one molecule while it excludes a nearly identical one. Also, the transport molecule is not altered in the process. In this respect, carrier molecules are like enzymes and to emphasize this aspect of their function, transport proteins have been named permeases. However, they are unlike enzymes in the fact that they do not necessarily produce chemical change in the molecules with which they are temporarily bound.

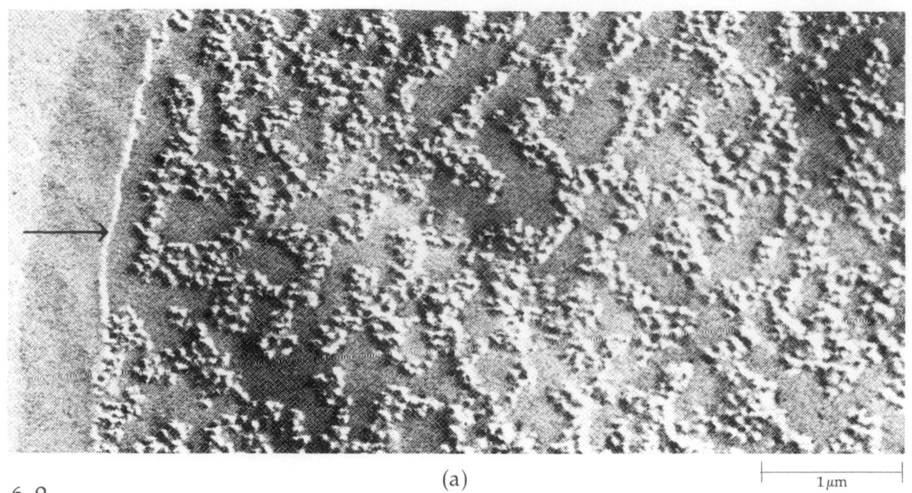

(a)

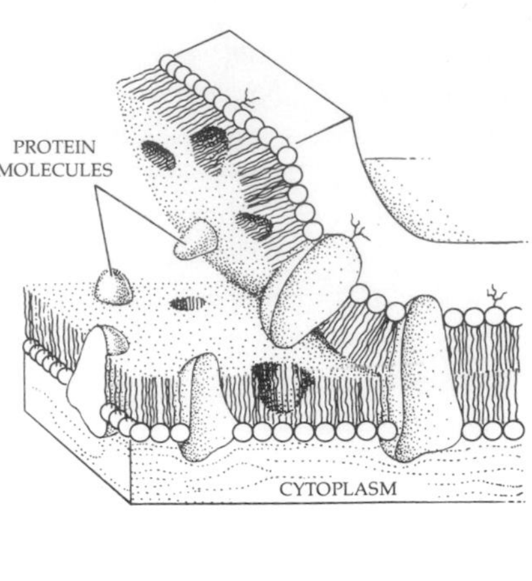

(b)

6–9
(a) *The internal surface of the membrane of a red blood cell, prepared by freeze-fracture technique.* (b) *In this procedure, the cell is frozen and then fractured with a sharp blow. The fracture line often runs through the hydrophobic layer of the cell membrane, revealing the proteins embedded throughout. The arrow in the micrograph indicates the edge of the fracture.*

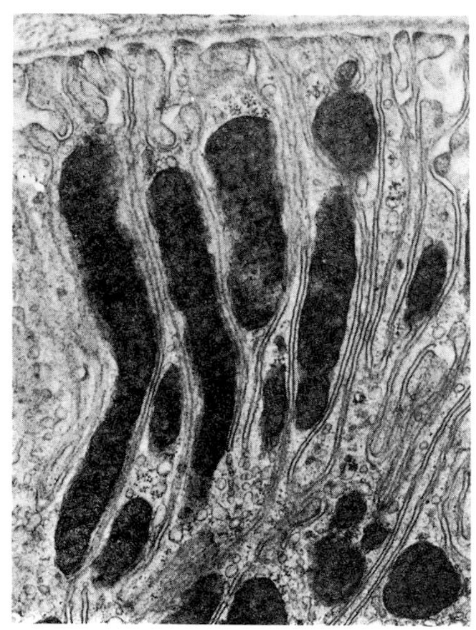

6–10
Mitochondria clustered near the surface of kidney cells. These cells are concerned with pumping out sodium against the concentration gradient. The mitochondria, presumably, provide the energy for this active-transport process.

Glucose is an example of a molecule whose entry into cells takes place by facilitated diffusion. There are several lines of evidence that support this conclusion. First, because glucose is rapidly broken down when it enters a cell, a steep concentration gradient is maintained. Second, if large numbers of glucose molecules are present outside the cell, the rate of entry into the cell does not increase as it would if glucose were moving by simple diffusion; it reaches a peak and then levels off. This limitation on the rate of entry is attributed to there being a limited number of carriers. Third, some molecules shaped almost like glucose compete with glucose for transport across the membrane. Thus, although the exact carrier molecule has not been isolated or identified, it is generally agreed that it exists.

The Sodium-Potassium Pump

One of the most important and best understood active-transport systems is the sodium-potassium pump. Most cells maintain a differential concentration gradient of sodium ions (Na^+) and potassium ions (K^+) across the cell membrane: Na^+ is maintained at a low concentration inside the cell, and K^+ is kept at a high concentration. This concentration gradient is exploited by nerve cells to propagate electrical impulses, as you will learn in Chapter 34. The sodium-potassium pump requires ATP. A measure of its importance to the organism is that more than a third of the ATP used by a resting animal is consumed by the sodium-potassium pump.

The transport of Na^+ and K^+ ions is accomplished by a carrier protein thought to exist in two alternative configurations. One has a cavity opening to the inside of the cell into which an Na^+ ion can fit; the other has a cavity opening to the outside, into which a K^+ ion fits. As shown in Figure 6–11, Na^+ binds to the transport

6–11

A model of the sodium-potassium pump. (a) Na^+ is bound to the carrier protein, and (b) the protein is phosphorylated; (c) the protein changes shape, and the Na^+ is released; (d, e) K^+ is bound to the carrier protein, which in this form provides a better fit for K^+ than for Na^+; (f) the protein is dephosphorylated, inducing conversion back to the other shape, and K^+ is released.

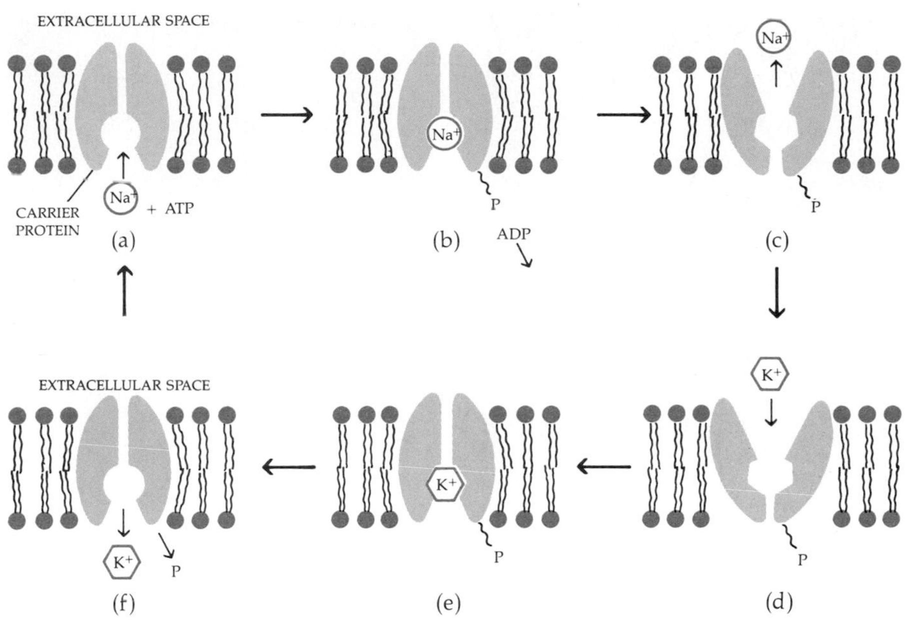

protein. ATP is broken down to ADP and the phosphate is attached to the protein (that is, the protein is phosphorylated). This triggers its shift to the alternative configuration and the release of the Na^+ to the outside of the membrane. The transport protein is now ready to pick up a K^+ ion, which results in a dephosphorylation of the protein, thus causing it to return to the first conformation and to release the K^+ ion to the inside of the cell. As you can see, this process will generate a gradient of Na^+ and K^+ ions across the membrane.

Many ingenious models have been proposed to show how carrier molecules might accept and eject their passengers. One of the first suggested that the carrier rotated, like a revolving door. A more recent model hypothesizes that the molecule has a hydrophilic core, which the transported molecule is squeezed through, propelled by changes in the configuration of the protein. It is likely that cell membranes contain a large variety of carrier proteins and that they employ a variety of different techniques for carrying out the transport process.

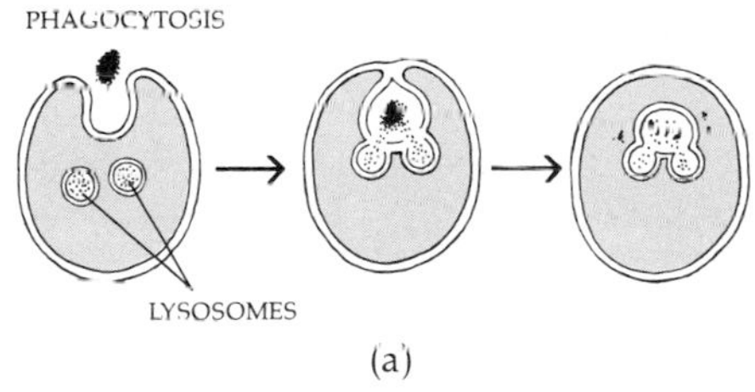

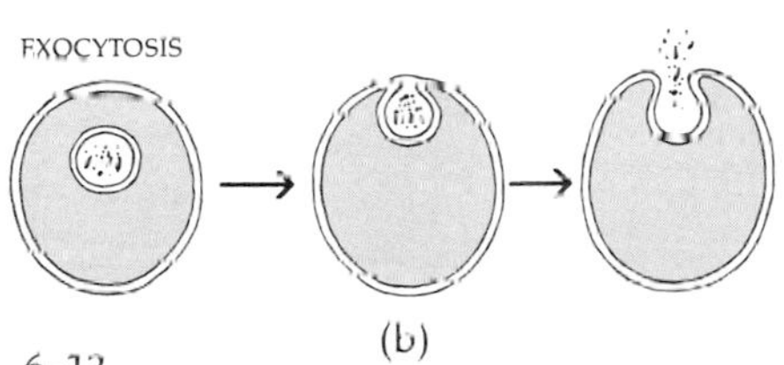

6–12

(a) Phagocytosis. A food particle is enveloped in a portion of the cell membrane, which pinches off to become a separate vacuole. Lysosomes spill their contents of lytic enzymes into the vacuole. (b) The indigestible remains are eliminated by exocytosis as the vacuole membrane fuses with the cell membrane.

ENDOCYTOSIS AND EXOCYTOSIS

In *endocytosis,* material to be taken into the cell attaches to special places on the cell membrane and induces the membrane to bulge inward, producing a little pouch or vacuole enclosing the substance. This vacuole is released into the cytoplasm.

When the substance to be taken in is a solid, such as a bacterial cell, the process is usually called *phagocytosis,* from the Greek word *phage,* "to eat." (See Figure 6–12.) Many one-celled organisms, such as amoebas, feed in this way, and white blood cells in our own bloodstreams engulf bacteria and other invaders in phagocytic vacuoles. Often lysosomes fuse with these vacuoles, emptying their enzymes into them and so digesting or destroying their contents.

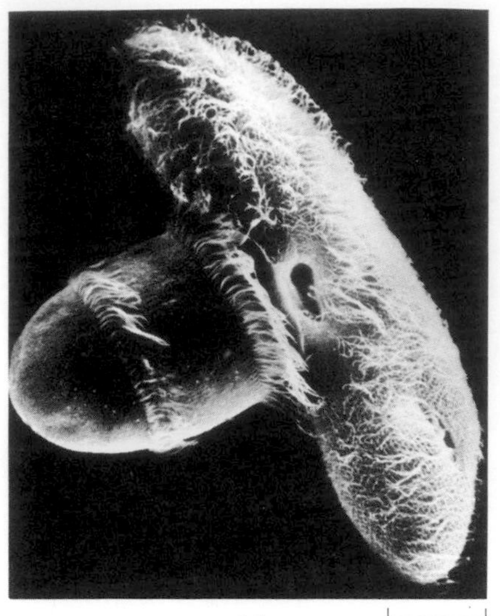

(a) 30 μm

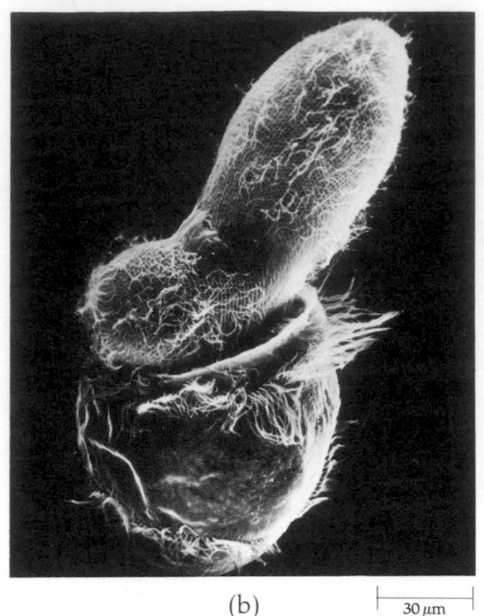

(b) 30 μm

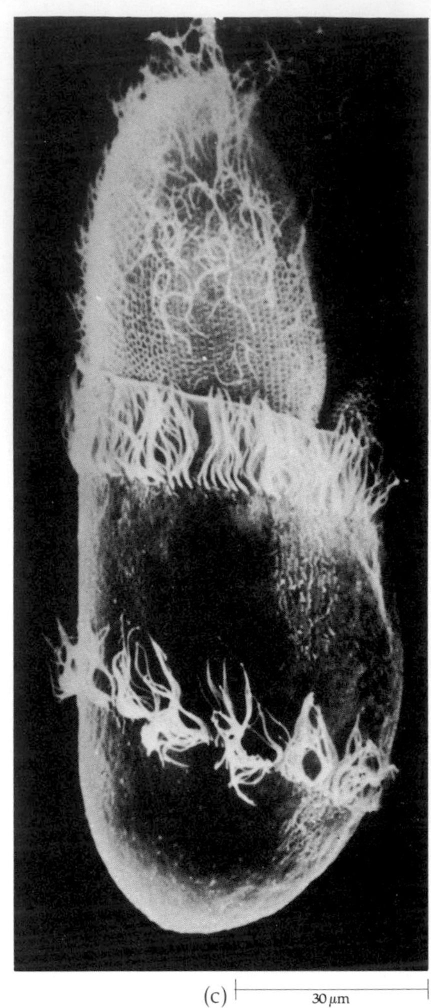

(c) 30 μm

6–13
Phagocytosis of a Paramecium *by a* Didinium. *(See Figure 5–20 for the preamble to their encounter.)* (a) *Ingestion of the* Paramecium *has begun. The concave area just above the oral rim of the* Didinium *is the oral groove of the* Paramecium. Paramecium, *a heterotroph, feeds largely on bacteria.* (b) *Because the* Paramecium *is larger than the* Didinium, *folding helps.* (c) *The* Paramecium *is half "swallowed"; the part that is within the* Didinium *is surrounded by a membrane composed of the cell membrane of the* Didinium. *The process of compression has begun, as you can see in the posterior tip protruding from the oral groove. This compression is largely a matter of squeezing out water. Once the* Paramecium *is completely inside, the cell membrane of the* Didinium *will fuse over it, forming a food vacuole. A* Didinium *can eat a dozen* Paramecium, *each larger than itself, in a single day. Moreover, the* Paramecium *must also provide the means for its own demolition: The* Didinium *apparently lacks a crucial digestive enzyme that the* Paramecium *supplies.*

The taking in of dissolved molecules, as distinct from particulate matter, is sometimes given the special name of *pinocytosis*, although it is the same in principle as phagocytosis. Pinocytosis occurs not only in single-celled organisms but also in multicellular animals. One type of cell in which it has been frequently observed is the human egg cell. As the egg cell matures in the ovary of the female, it is surrounded by "nurse cells," which apparently transmit nutrients to the egg cell, which takes them in by pinocytosis.

Phagocytosis and pinocytosis can also work in reverse. Many substances are exported from cells in vesicles or vacuoles. The vacuoles move to the surface of the cell. When the vacuole reaches the cell surface, its membrane fuses with the membrane of the cell, thus expelling its contents to the outside. This process is sometimes referred to as *exocytosis*.

Phagocytosis and pinocytosis appear superficially different from membrane-transport systems involving carrier molecules. They are fundamentally similar, however, in that all depend on the capacity of the membrane to "recognize"

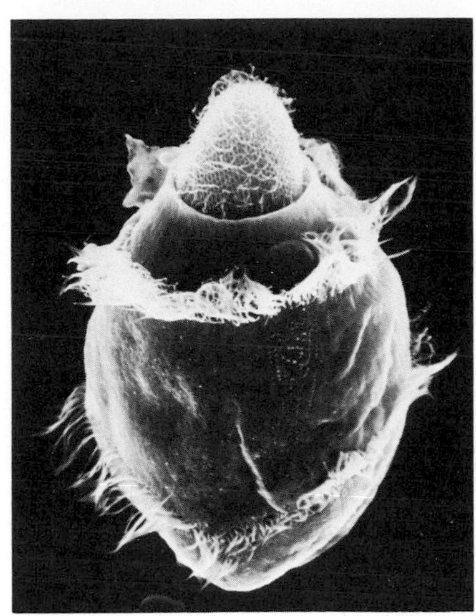

6–14
The end.

particular molecules. This capacity is, of course, the result of billions of years of an evolutionary process that began, as far as we are able to discern, with the formation of a fragile film around a few organic molecules, thus separating the molecules from their external environment and permitting them to maintain the particular kind of organization that we recognize as life.

SUMMARY

The cell membrane regulates the passage of materials into and out of the cell, a function that makes it possible for the cell to maintain its structural and functional integrity. This regulation depends on interaction between the membrane and the materials that pass through it.

One of the principal materials passing into and out of cells is water. Water moves by bulk flow, diffusion, and osmosis. Water potential determines the direction the water moves; that is, water movement takes place from where the water potential is greater to where it is lower. Bulk flow is the overall movement of water molecules as a group, as when water flows downhill or in response to pressure.

Diffusion involves the random movement of molecules and results in net movement down a concentration gradient. Carbon dioxide and oxygen are two important molecules that move into and out of cells by diffusion across the membrane. Diffusion is most efficient when the surface area is large in relation to volume, when the distance involved is small, and when the concentration gradient is steep. The rate of movement of substances within cells is increased by cytoplasmic streaming.

Osmosis is the movement of water through a membrane that permits the passage of water but inhibits the movement of solutes; such a membrane is called a selectively permeable membrane. In the absence of other forces, the movement of water in osmosis is from a region of lower solute concentration (a hypotonic medium), and therefore of higher water potential, to one of higher solute concentration (a hypertonic medium), and so of lower water potential. Turgor in plant cells is a consequence of osmosis.

According to the fluid-mosaic model of the membrane, cell membranes are lipid bilayers in which globular proteins are suspended. Some of these proteins act as carriers, ferrying molecules through the membrane. If the movement is driven by a concentration gradient, the process is known as facilitated diffusion. If the movement requires ATP (or other chemical energy), it is known as active transport. Active transport can move substances against a concentration gradient. One of the most important active-transport systems is the sodium-potassium pump, which maintains sodium ions at a low concentration and potassium ions at a high concentration in the cytoplasm.

Controlled movement into and out of a cell may also occur by endocytosis or exocytosis, in which substances are transported in vacuoles composed of portions of the cell membrane. Endocytosis of solids is called phagocytosis and of dissolved molecules, pinocytosis.

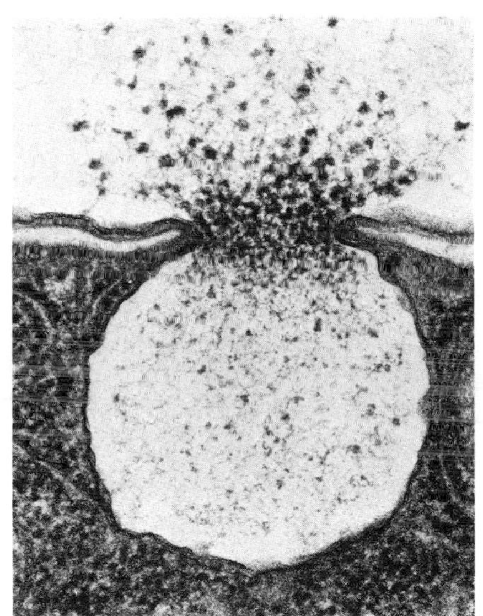

6–15
Exocytosis. A secretory vesicle, formed within the cell, discharges its contents.

QUESTIONS

1. Distinguish among the following: diffusion/osmosis; water potential/hydrostatic pressure/osmotic pressure; hypotonic/hypertonic/isotonic; pinocytosis/phagocytosis; exocytosis/endocytosis.

2. What is a concentration gradient? How does a concentration gradient affect diffusion? How does a concentration gradient affect osmosis?

3. In diffusion, random movement of molecules continues (as long as the temperature remains the same). However, net movement stops. How do you reconcile these two facts?

4. Why is diffusion more rapid in gases than in liquids? Why is it more rapid at higher temperatures than at lower temperatures?

5. Three funnels have been placed in a beaker containing a solution (see the figure below). What is the concentration of the solution? Explain your answer.

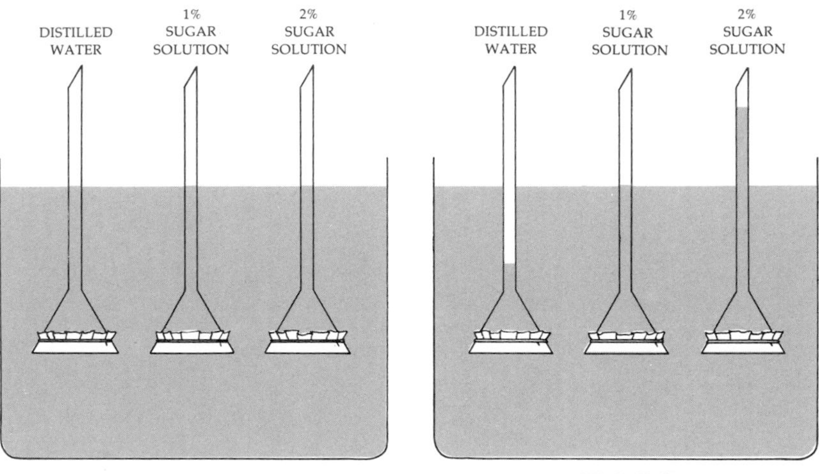

6. Imagine a pouch with a selectively permeable membrane containing a saltwater solution. It is immersed in a dish of fresh water. Which way will the water move? If you add salt to the water in the dish, how will this affect the water movement? What living systems exist under analogous conditions? How do you think they maintain water balance?

7. When you forget to water your houseplants, they wilt and the leaves (and sometimes the stems) become very limp. What has happened to the plants to cause this change in appearance and texture? Within a few hours after you remember to water your plants, they resume their normal, healthy appearance. What has occurred within the plants to cause this restoration? Sometimes, if you wait too long to water your plants, they never revive. What do you suppose has happened?

8. What limits the passage of water and other polar molecules and ions through the cell membrane? How do such molecules get into and out of the cell? Describe three possible routes.

9. Justify the conclusion that differences in ion concentration between cells and their surroundings (see Figure 6-2) indicate that cells regulate the passage of materials across membranes.

10. When you cut or scrape your skin, white blood cells rapidly converge on the site of the injury. What are they doing when they reach the site? Why is this important to your well-being?

11. In Figure 6–14, the *Paramecium* is sinking below the oral groove of *Didinium*. What will happen next? Complete the scenario, giving as many details as possible. (You might want to end your account with the fact that *Didinium* divides once for every two *Paramecium* consumed.)

SUGGESTIONS FOR FURTHER READING

Books

BAKER, J. J., and G. E. ALLEN: *Matter, Energy and Life: An Introduction for Biology Students*, 2d ed., Addison-Wesley Publishing Co., Inc., Reading, Mass., 1970.*

A book for students who have had no previous chemistry or physics, which deals with such topics as the structure of matter, the formation of molecules, the course and mechanism of chemical reactions, as well as with the chemistry of living systems.

BUECHE, F.: *Physical Science*, Worth Publishers, Inc., New York, 1972.

You may have remaining questions about the structure and properties of atoms. If so, the appropriate chapters in this text are likely to provide useful answers.

DYSON, ROBERT D.: *Cell Biology: A Molecular Approach*, 2d ed., Allyn & Bacon, Inc., Boston, 1978.

Comprehensive, clear, up-to-date account of cell structure and how it relates to metabolism and transport functions.

LEDBETTER, M. C., and KEITH R. PORTER: *Introduction to the Fine Structures of Plant Cells*, Springer-Verlag, New York, 1970.

An excellent atlas of electron micrographs of plant cells, with detailed explanations.

LEHNINGER, ALBERT L.: *Biochemistry*, 2d ed., Worth Publishers, Inc., New York, 1975.

This introductory text is outstanding both for its clarity and for its consistent focus on the living cell.

LOEWY, A. G., and P. SIEKEVITZ: *Cell Structure and Function*, 3d ed., Holt, Rinehart and Winston, Inc., New York, 1979.

An outstanding elementary text on cell structure and function.

PORTER, KEITH R., and MARY A. BONNEVILLE: *An Introduction to the Fine Structure of Cells and Tissues*, 3d ed., Lea & Febiger, Philadelphia, 1968.

An atlas of electron micrographs of animal cells; detailed commentaries accompany each. These are magnificent micrographs, and the commentaries describe not only what the pictures show but also the experimental foundations of our knowledge of cell ultrastructures.

STRYER, LUBERT: *Biochemistry*, W. H. Freeman and Company, San Francisco, 1975.

An introductory text, with many examples of medical applications of biochemistry. Handsomely illustrated.

* Available in paperback.

THOMAS, LEWIS: *The Lives of a Cell: Notes of a Biology Watcher*, Viking Press, Inc., New York, 1974.*

Thomas, a physician and medical researcher, reveals the extent to which science can tune our intellectual antennae, broaden our perceptions, and extend our appreciation of ourselves and of the world around us. Anyone who wants to refute the contention that science destroys human values need look no further than these short, sensitive essays.

WEINBERG, STEVEN: *The First Three Minutes: A Modern View of the Origin of the Universe*, Basic Books, Inc., New York, 1977.

A wonderful story, written for the intelligent nonscientist (characterized by the author as a smart old attorney who expects to hear some convincing arguments before he makes up his mind).

WHITE, EMIL H.: *Chemical Background for the Biological Sciences*, 2d ed., Prentice-Hall, Inc., Englewood Cliffs, N.J., 1970.

A short but quite rigorous introduction to modern chemistry, with particular emphasis on biochemistry and biology.

WOLFE, STEPHEN L.: *Biology of the Cell*, Wadsworth Publishing Co., Inc., Belmont, Calif., 1972.

An outstanding synthesis of cell structure and function.

Articles

ALBERSHEIM, PETER: "The Walls of Growing Plant Cells," *Scientific American*, April 1975, pages 81–95.

MILLER, JULIE ANN: "The Bone and Muscle of Cells," *Science News*, vol. 112, pages 250–253, October 15, 1977.

STAEHELIN, L. ANDREW, and BARBARA E. HULL, "Junctions between Living Cells," *Scientific American*, May 1978, pages 141–152.

TRIMBLE, VIRGINIA: "Cosmology: Man's Place in the Universe," *American Scientist*, vol. 65, pages 75–86, January–February 1977.

* Available in paperback.

SECTION 2 ENERGETICS

7–1
The flow of energy in a biological system. Radiant energy from the sun is transformed to chemical energy by the grasses of the African plains. The zebras, eating the grasses, use this chemical energy for growth and convert some of it to kinetic energy, which may help to keep them from participating in another energy transfer involving the waiting lions. Of the radiant energy falling on the grass, less than 10 percent is converted to chemical energy. Less than 10 percent of the chemical energy stored in the grass is converted to chemical energy stored in the zebra, and less than 10 percent of the zebra's stored energy is transferred to its predator, the lion.

CHAPTER 7

The Flow of Energy

Life here on earth depends on the flow of energy from the thermonuclear reactions taking place at the heart of the sun. The amount of energy delivered by the sun is 13×10^{23} (the number 13 followed by 23 zeros) calories per year. It is a difficult quantity to imagine. For example, the amount of energy striking the earth every day is the equivalent of about a million Hiroshima-sized A-bombs.

About one-third of this solar energy is immediately reflected back into space as light (as it is from the moon). Much of the remaining two-thirds is absorbed by the earth and converted to heat. Some of this absorbed heat energy serves to evaporate the waters of the oceans, producing the clouds that, in turn, produce rain and snow. Solar energy, in combination with other factors, is also responsible for the movements of air and of water that help set patterns of climate over the surface of the earth.

A small fraction—less than 1 percent—of the solar energy reaching the earth becomes, through a series of operations performed by the cells of plants and other photosynthetic organisms, the energy that drives all the processes of life. Living systems change energy from one form to another, transforming the radiant energy from the sun into the chemical and mechanical energy used by everything that is alive.

These concepts of a vital relationship between plants and animals and between energy and life are relatively recent ones. They form part of the study of *thermodynamics*—the science of energy exchanges. Energy is such a common word in our energy-conscious modern world that it is surprising to discover that the word itself did not come into existence until about 100 years ago. In the next few pages, we are going to trace briefly the growth of ideas about energy, especially as they relate to living systems.

PLANTS AND FIRE

Until about 350 years ago, observers of the biological world, noting that the life processes of animals were dependent on the food they ate, thought that plants derived their food from the soil in a similar way. This concept was widely accepted by those who thought about the problem, until the Belgian physician Jan Baptista van Helmont (1577–1644) offered the first experimental evidence to the contrary.

7-2
"I took an earthen vessel in which I put 200 pounds of earth that had been dried in a furnace, which I moistened with rainwater, and I implanted therein the trunk or stem of a willow tree, weighing five pounds. And at length, five years being finished, the tree sprung from thence did weigh 169 pounds and about three ounces. When there was need, I always moistened the earthen vessel with rainwater or distilled water, and the vessel was large and implanted in the earth. Lest the dust that flew about should be co-mingled with the earth, I covered the lip or mouth of the vessel with an iron plate covered with tin and easily passable with many holes. I computed not the weight of the leaves that fell off in the four autumns. At length, I again dried the earth of the vessel, and there was found the same 200 pounds, wanting about two ounces. Therefore 164 pounds of wood, bark and roots arose out of water only." (Jan Baptista van Helmont.)

Van Helmont grew a small willow tree in an earthenware pot for five years, adding only water to the pot. At the end of five years, the willow had increased in weight by 74 kilograms (164 pounds), while the earth in the pot had decreased in weight by only 57 grams (2 ounces). On the basis of these results, van Helmont concluded that all the substance of the plant was produced from the water and none from the soil. (This experiment is of general interest to those concerned with tracing the history of science, because it is one of the first carefully designed biological experiments ever reported. As we shall see, however, van Helmont's conclusions were too broad.)

The next stage in our knowledge of energy exchanges among organisms was the result of studies of combustion. Fire, one of the "elements" of the ancient Greeks, intrigued the early alchemists and the Renaissance chemists who were their successors. The question of what happens when something is consumed by fire was similarly a focus of study and debate among the chemists of the eighteenth century.

Like his contemporaries, Joseph Priestley (1733–1804) believed that the flames leaping away from a burning object, such as a candle, represented something escaping from it. This "something," the volatile constituent of all combustible substances, was called phlogiston. Air was needed to burn the candle because it was air that carried the phlogiston away. When the air became too saturated with phlogiston, combustion could no longer continue. And an intriguing additional observation was made: "Phlogisticated" air not only could not support burning, it could not support animal life. The phlogiston theory was the first theory in chemistry that began to deal with the unity of chemical and biological processes.

On August 17, 1771, Priestley "put a sprig of mint into air in which a wax candle had burned out and found that, on the 27th of the same month, another candle could be burned in this same air." Priestley believed, as he reported, that he had "accidentally hit upon a method of restoring air that has been injured by the burning of candles." The "restorative which nature employs for this purpose," he stated, is "vegetation." Priestley extended his observations and soon showed that air "restored" by vegetation was not "at all inconvenient to a mouse." Priestley's experiments offered the first logical explanation of how the air remained "pure" and able to support life despite the burning of countless fires and the breathing of many animals. When he was presented with a medal for his discovery, the citation read in part: "For these discoveries we are assured that no vegetable grows in vain . . . but cleanses and purifies our atmosphere."

Thus, it was beginning to be seen that plants, animals, and energy (as exemplified by fire) were somehow interconnected.

Only in the Light

Priestley's reports that plants purify the air were of great interest to his fellow chemists, but they soon attracted criticism because the experiments could not readily be confirmed. In fact, when Priestley tried to do the experiments again himself, he did not get the same results. (We think now that he may have moved his equipment to a dark corner of his laboratory.) It was a Dutch physician, Jan Ingenhousz (1730–1799), who was finally able to confirm Priestley's work with an important addition. He found that the purification takes place only in sunlight. Plants at night or in the shade, he reported, "contaminate the air which surrounds them, throwing out an air hurtful to animals." He also observed that only the green

7-3
Joseph Priestley—". . . no vegetable grows in vain. . . ."

parts of plants restore the air and, on the basis of control experiments, that "the sun by itself has no power to mend air without the concurrence of plants."

In 1782, Pastor Jean Senebier of Geneva showed that the amount of restored air produced by a plant is equal to the amount of "phlogisticated" air available to it. This was the beginning of the idea that carbon dioxide as well as light is required for photosynthesis.

Lavoisier and the Chemical Revolution

While Ingenhousz was performing his experiments on plants, Antoine Lavoisier (1743–1794), working in his private laboratory, was carrying out the experiments that put chemistry on an essentially modern basis. Lavoisier's work helped to establish the *law of conservation of matter*. This principle is one of the foundations of modern science and can be stated very simply: *Although matter can be transformed from one form to another, it is never lost.* We assume the law of conservation of matter every time we write a chemical equation. (See, however, the essay on page 150 for an exception that Lavoisier could not have been expected to anticipate.)

Most of Lavoisier's arguments were based on extremely careful measurements. The principles of accurate measurement that he established brought about a new era in chemistry. By the 1780s, Lavoisier's work brought him into direct confrontation with the widely accepted phlogiston theory. Lavoisier showed that a metal was heavier after combustion, a fact difficult to reconcile with the theory that it had lost phlogiston. He then proved that the gain in weight resulted from a combination with what he called "eminently respirable air." (He later came to call this air oxygen.) He showed that when charcoal was heated, it combined with oxygen to produce a gas that was a compound of carbon and oxygen. This compound we now know as carbon dioxide.

7-4
Lavoisier conducting an experiment on respiration. This drawing was made by Mme. Lavoisier, who pictured herself as an observer.

Among Lavoisier's many discoveries, those that are most directly connected with the story we are following here concern the exchanges of gases that take place when animals breathe. Working with the mathematician P. S. Laplace (1749–1827), Lavoisier confined a guinea pig for about 10 hours in a jar containing oxygen and measured the carbon dioxide produced. He also measured the amount of oxygen used by a man active and at rest. In these experiments, he was able to show that the combustion of carbon compounds with oxygen is the true source of animal heat and that oxygen consumption increases during physical work. "Respiration is merely a slow combustion of carbon and hydrogen, which is similar in every respect to that which occurs in a lighted lamp or candle, and, from this point of view, animals that breathe are really combustible bodies which burn and are consumed." In short:

$$(CH_2O) + O_2 \longrightarrow CO_2 + H_2O + \text{heat}$$

The work of Ingenhousz spanned the prematurely terminated career of Lavoisier, who was guillotined on May 8, 1794 during the French Revolution. (The judge presiding over the case is reported to have said, "The Republic has no need of savants.") Quick to adopt Lavoisier's ideas about gases, Ingenhousz hypothesized that the plant was not just exchanging "good air" for "bad" and so making the world habitable for animal life. In the sunshine, he suggested, a plant absorbs the carbon from carbon dioxide, "throwing out at that time the oxygen alone, and keeping the carbon to itself as nourishment."

Nicholas Theodore de Saussure (1767–1845) applied Lavoisier's principles of quantitative measurement and showed that equal volumes of CO_2 and O_2 are exchanged during photosynthesis and that the plant does indeed retain the carbon. He also showed that more weight was gained by the plant during photosynthesis than could be accounted for by the carbon taken in as carbon dioxide. In other words, the carbon in the dry matter of plants came from carbon dioxide, but equally important, the rest of the dry matter, with the exception of minerals from the soil, came from water.

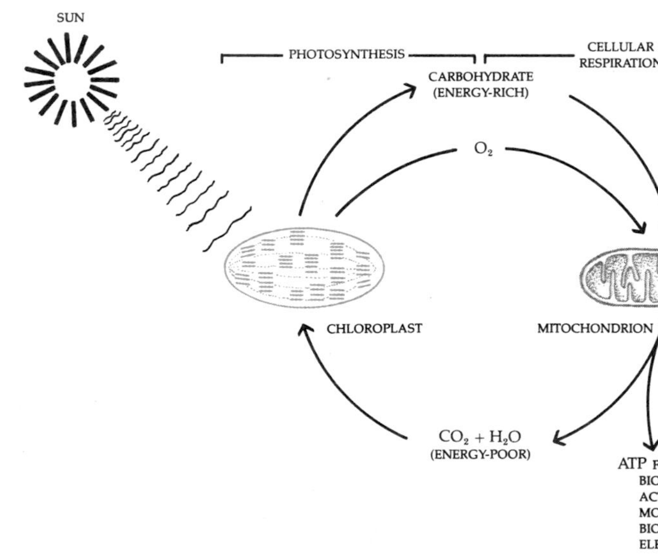

7–5
The flow of biological energy. The radiant energy of sunlight is produced by the fusion of hydrogen atoms to form helium. Chloroplasts, present in all photosynthetic eukaryotic cells, capture this energy and use it to convert water and carbon dioxide into carbohydrates, such as glucose, starch, and other foodstuff molecules. Oxygen is released into the air as a product of the photosynthetic reactions. Mitochondria, present in all eukaryotic cells, break down these carbohydrates and capture their stored energy in ATP molecules. This process, cellular respiration, consumes oxygen and produces carbon dioxide and water, completing the cycle.

THE CARBON CYCLE

By photosynthesis, living systems incorporate carbon dioxide from the atmosphere into organic, carbon-containing compounds. In respiration, these compounds are broken down again into carbon dioxide and water. These processes, viewed on a worldwide scale, result in the carbon cycle. The principal photosynthesizers in the cycle are plants and the phytoplankton, the marine algae. They synthesize carbohydrates from carbon dioxide and water and release oxygen into the atmosphere. About 75 billion metric tons of carbon per year are bound into carbon compounds by photosynthesis.

Some of the carbohydrates are used by the photosynthesizers themselves: Plants release carbon dioxide from their roots and leaves, and marine algae release it into the water where it maintains an equilibrium with the carbon dioxide of the air. Some 500 billion metric tons of carbon are "stored" in the seas, and some 700 billion metric tons in the atmosphere. Some of the carbohydrates are used by animals that feed on the live plants, on algae, and on one another, releasing carbon dioxide. The dead bodies of the plants and other organisms plus discarded leaves and shells, feces, and other waste materials settle into the soil or sink to the ocean floors where they are consumed by decomposers—small invertebrates, bacteria, and fungi. Carbon dioxide is also released by these processes into the reservoir of the air and oceans. Another, even larger store of carbon lies below the surface of the earth in the form of coal and oil, deposited there some 300 million years ago.

The natural processes of photosynthesis and respiration balance one another out. For many millions of years, the carbon dioxide of the atmosphere, as far as we can tell, has remained constant. By volume it is a very small proportion of the atmosphere, only about 0.03 percent. It is important, however, because carbon dioxide, unlike most other components of the atmosphere, absorbs heat from the sun's rays. Since 1850, carbon dioxide concentrations in the atmosphere have been increasing, owing in part to our use of fossil fuels, to our plowing of the soil, and to our destruction of forest land, particularly in the tropics. Some environmentalists predict that this increase in the carbon dioxide "blanket" will increase the temperature here on earth, with a consequential increase in the great deserts of the world. Others, looking on the sunnier side, foresee an increase in photosynthetic activity because of the increased carbon dioxide that will be available to plants and algae. Most, however, feel alarmed by the fact that although we do not know the consequences of what we are doing, we keep right on doing it.

The flow of energy in the living world parallels the flow of carbon.

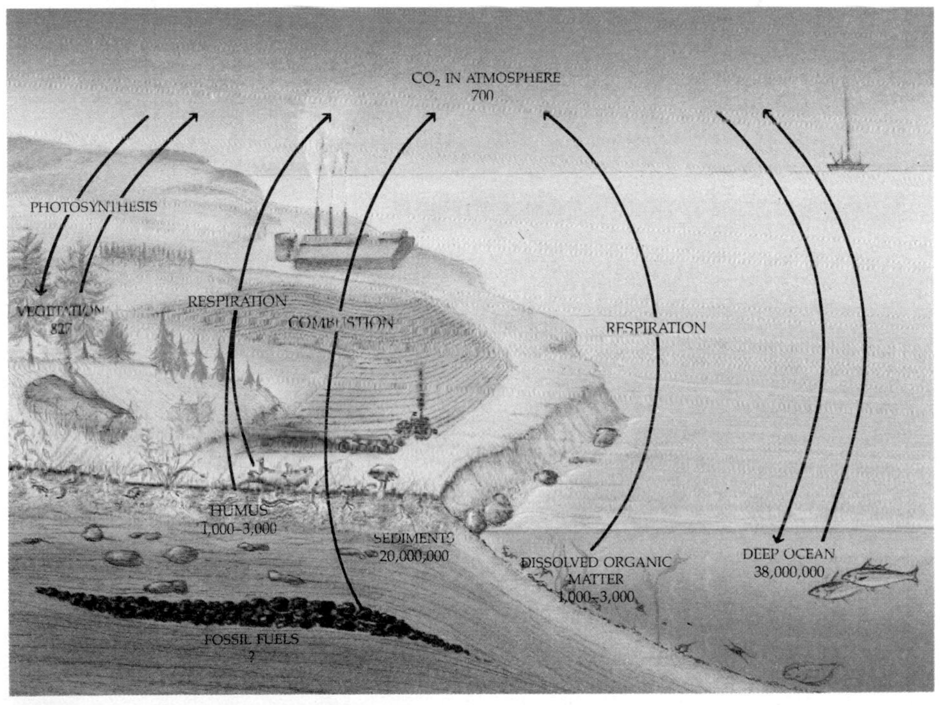

The carbon cycle. The arrows indicate the movement of carbon atoms. The numbers are all estimates of the amount of carbon stored, expressed in billions of metric tons. The amount of carbon released by respiration and combustion has, it is believed, begun to exceed the amount fixed by photosynthesis.

Thus, all the components were identified—carbon dioxide, water, and light—and it became possible to write the overall photosynthetic equation:

$$CO_2 + H_2O + \text{light} \longrightarrow (CH_2O) + O_2$$

If you compare this with the previous equation, you can see that photosynthesis is, in a sense, respiration (or combustion) in reverse. Although all the details were still not known, the outlines of the picture had come into focus.

THE LAWS OF THERMODYNAMICS

Energy is an elusive concept. It is usually defined operationally—by what it does rather than what it is. Until less than 200 years ago, heat—the form of energy most readily studied—was thought of as a separate, though weightless, substance called caloric. Even after the phlogiston theory had been upset, Lavoisier himself listed caloric among the simple substances, or elements. An object was hot or cold depending upon how much caloric it contained; when a cold object was placed next to a hot one, the caloric flowed from the hot object into the cold one; when metal was pounded with a hammer, the caloric was forced to the surface. As you can see, the concept is surprisingly useful.

The First Law

The rapid development of the steam engine in the latter part of the eighteenth century, more than any other single chain of events, changed scientific thinking about the nature of energy. Energy came to be associated with work, and heat and motion came to be seen as forms of energy. The way was opened for the formulation of the laws of thermodynamics. The *first law of thermodynamics* states, quite simply: *Energy can be changed from one form to another but it cannot be created or destroyed.*

In engines, for instance, chemical energy (in coal or gasoline, for example) is converted to heat, which is then partially converted to mechanical movements (kinetic energy). Some of the energy is converted back to heat by the friction of those movements, and some leaves the engine in the form of exhaust products. The

7–6
Electrical energy can be converted to light energy, as in these Las Vegas signs, for example. The energy emitted as an electron falls from one energy level to another is a discrete amount, characteristic for each atom. When electricity is passed through a tube of gas, the electrons are boosted to higher energy levels. As they fall back, light energy is emitted, producing, for example, the red glow characteristic of neon and the yellow glow characteristic of sodium vapor.

heat produced by friction and lost in the exhaust, unlike the heat in the engine or the boiler, cannot produce "work"—that is, it cannot turn the gears—because it is dissipated into the surroundings. But it is nevertheless part of the total equation. In fact, engineers calculate that most of the energy put into an engine is dissipated randomly as heat; most engines work with less than 25 percent efficiency.

In the course of such studies of engine efficiency, the notion of potential energy developed. A barrel of gasoline or a ton of coal could be assigned a certain amount of potential energy, expressed in terms of the amount of heat it would liberate when burned. The efficiency of the conversion of the potential energy to "useful" energy depended on the design of the system.

Although these concepts were formulated in terms of engines running on heat power, they apply to other systems as well. Returning to our earlier analogy, the boulder pushed to the top of the hill contains energy—potential energy. Given a little push (the energy of activation) it rolls down the hill again, converting the potential energy to the energy of motion and the heat produced by friction. Water, as we saw in Chapter 6, may also possess potential energy. As it moves by bulk flow from the top of a waterfall or over a dam, it can turn water wheels that turn gears and, for example, grind corn. Thus the potential energy of water, in this system, is converted to the kinetic energy of the wheels and gears and to heat, produced by the movement of the water itself and also by the turning wheels and gears.

Light is a form of energy, as is electricity. Light can be changed to electrical energy, and electrical energy can be changed to light (for example, by letting it flow through the tungsten wire in a light bulb).

The first law of thermodynamics states that in energy exchanges and conversions, wherever they take place and whatever they involve, the energy of the products of the reaction plus the energy released in the reaction is equal to the initial energy of the reactants.

The Second Law

The *second law of thermodynamics* is the more interesting one, biologically speaking. It predicts the direction of all events involving energy exchanges; thus it has been called "Time's Arrow." The second law states that in all energy exchanges and conversions, if no energy leaves or enters the system under study, the potential energy of the final state will always be less than the potential energy of the initial state. The second law is entirely in keeping with everyday experience. A boulder will run downhill but never uphill. Heat will flow from a hot object to a cold one and never the other way. Our cells can process glucose enzymatically to yield carbon dioxide and water but—unless we are plants and can capture the energy of the sun—our enzymatic processes will not produce glucose from carbon dioxide and water.

A process in which the potential energy of the final state is less than that of the initial state is one that yields energy (otherwise it would be in violation of the first law). An energy-yielding process is called an *exergonic* ("energy-out") reaction. As the second law predicts, only exergonic reactions can take place spontaneously. (Spontaneously, though the word has an explosive sound to it, says nothing about the rate of the reaction, just whether or not it can take place at all.) *Endergonic* reactions are energy-requiring reactions, and in order for them to proceed an input of energy is required that is greater than the difference in energy between product and reactants.

One important factor in determining whether or not a reaction is exergonic is already familiar to us—$\triangle H$. As we noted in Chapter 3, the energy change that takes place when glucose, for instance, is oxidized can be measured in a calorimeter and expressed in terms of $\triangle H$, the change in heat content of the system. The oxidation of a mole of glucose yields 673 kilocalories. Or,

$$C_6H_{12}O_6 + 6O_2 \longrightarrow 6CO_2 + 6H_2O$$
$$\triangle H = -673 \text{ kcal/mole}$$

Generally speaking, a chemical reaction with a negative $\triangle H$ is an exergonic reaction. However, there are exceptions. One of the most dramatic is found with a substance known as dinitrogen pentoxide, which decomposes spontaneously and with explosive force to nitrogen dioxide and oxygen, and in so doing, absorbs heat:

$$2N_2O_5 \longrightarrow 4NO_2 + O_2$$
$$\triangle H = +26.18 \text{ kcal/mole}$$

In short, another factor besides the gain or loss of heat can determine the direction of a process. This factor is given the scientific name of *entropy*, and it is a measurement of the disorder of a system.

Before we examine more closely why the decomposition of dinitrogen pentoxide is exergonic despite its positive $\triangle H$, let us return to the more familiar example of water. The change from ice to liquid water and the change from liquid water to water vapor are both endothermic processes—a considerable amount of heat is removed from the surrounding air as they take place. Yet, under the appropriate conditions, they proceed spontaneously. The key factor in all three of these examples is the increase in entropy. In the case of the dinitrogen pentoxide, a solid is being changed into two gases, and two molecules are being converted to five. In the case of ice and liquid water, a solid is being turned into a liquid, and some of the bonds that hold the water molecules together in a crystal (ice) are being broken. As the liquid water turns to vapor, the rest of the hydrogen bonds are ruptured as the individual water molecules dance off, one by one. In every case, the disorder of the system has increased.

The notion that there is more disorder associated with more numerous and smaller objects than with fewer, larger ones is in keeping with our everyday experience. If I have 20 papers on my desk, the possibilities for disorder are greater than if I have 2 or even 10. If I cut each of the 20 in half, the entropy of the system—the capacity for randomness—increases. Also, the relationship between entropy and energy is a commonplace idea. If you were to find your room tidied up and your books in alphabetical order on the shelf, you would recognize that someone had been at work—that energy had been expended. For me to organize the papers on my desk similarly requires that I expend energy. Furthermore, it would be feasible to measure the energy expenditure in calories.

Now let us return to the question of the energy changes that determine the course of chemical reactions. We have seen that both the change in the heat content of the system ($\triangle H$) and the change in the entropy (which is symbolized as $\triangle S$) contribute to the overall change in energy. This total change—the one that takes into account both heat and entropy—is called the *free energy change* and is symbolized as $\triangle G$, after the American physicist Josiah Willard Gibbs (1839–1903), who was one of the first to put all of these ideas together.

7-7
Some illustrations of the second law of thermodynamics. In nature, processes tend toward randomness, or disorder. Only an input of energy can reconstruct the initial state from the final state.

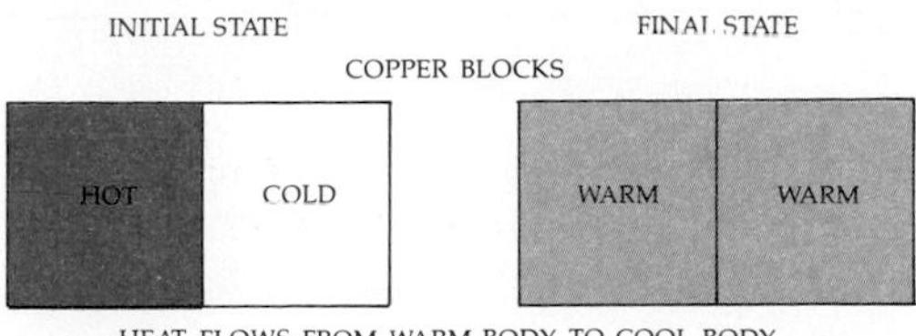

HEAT FLOWS FROM WARM BODY TO COOL BODY

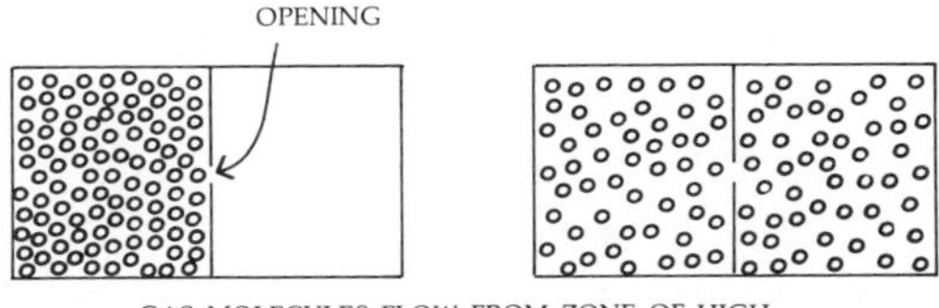

GAS MOLECULES FLOW FROM ZONE OF HIGH PRESSURE TO ZONE OF LOW PRESSURE

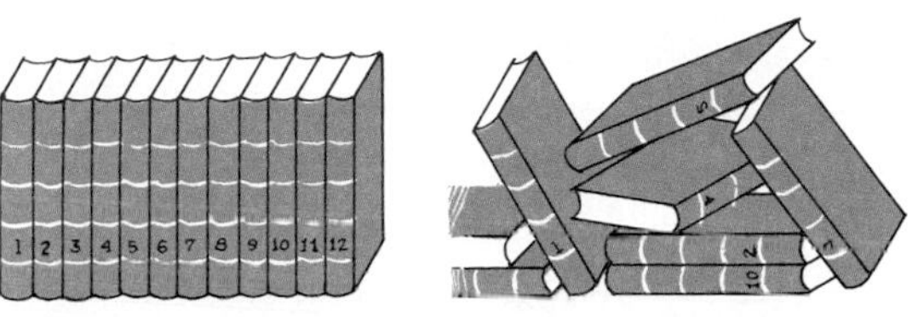

ORDER BECOMES DISORDER

With $\triangle G$ in mind, let us examine once more the combustion of glucose. The $\triangle H$ of that reaction is -673 kcal/mole. The $\triangle G$ is -686 kcal/mole. The entropy factor has contributed 13 kcal/mole to the free energy change of the process. Thus both the change in heat and the change in entropy contribute to the lower energy state of the products of the reaction.

$\triangle G$ can also enable one to predict processes that occur when $\triangle H$ is zero or even positive, as with dinitrogen pentoxide. For instance, it confirms our earlier observations that heat will flow from a warm object to a cold one, that dye molecules will diffuse in a beaker of water, or that my desk will revert to disorder. In each of these processes, the final state has more entropy—and therefore less potential energy—than the initial state.

The relationship between $\triangle G$, $\triangle H$, and entropy is given in the following equation:

$$\triangle G = \triangle H - T \triangle S$$

It states that the free energy change is equal to the change in heat (a minus figure in exothermic reactions, remember) minus the change in entropy, which varies according to the absolute temperature T. In exergonic reactions, $\triangle H$ may be zero or may even be positive, but $\triangle G$ is always negative.

As you can see in the equation, $T \triangle S$ is preceded by a minus sign. The greater the entropy, the more negative $\triangle G$ will be; that is, the more exergonic the reaction will be. Therefore it is possible to state the second law in another, simpler way: *All natural processes tend to proceed in such a direction that the disorder or randomness of the system increases.*

$E = mc^2$

Protons and neutrons, as we noted in Chapter 1, are arbitrarily assigned an atomic weight of 1. One would therefore expect that an element with, for example, twice as many protons and neutrons as another element would weigh twice as much. This assumption is true—almost. If the weights of nuclei are measured with great accuracy, as they can be by instruments developed by modern physics, small nuclei always have proportionately slightly greater weights than larger nuclei. For example, the most common isotope of carbon, as you know, has a combined total of 12 protons and neutrons, and carbon, by convention, is assigned an atomic weight of 12. The hydrogen atom, however, has an atomic weight not exactly of 1, as would be expected, but of 1.008. Helium has two protons and two neutrons. It does not have a weight of exactly 4, however, or of 4.032 (four times the weight of hydrogen); it weighs, in relation to carbon, 4.0026. Similarly, oxygen, with a combined total of 16 protons and neutrons, has an atomic weight, in relation to that of carbon, of 15.995. In short, when protons and neutrons are assembled into an atomic nucleus, there are slight changes in weight, which reflect changes in mass.

Albert Einstein in 1905, the year he published his paper on the theory of relativity. He was 26 years old and working at the Swiss Patent Office in Bern as a technical expert third class.

One of the oldest and most fundamental concepts of chemistry is the law of conservation of mass—that mass is never created or destroyed. Yet under conditions of extremely high temperature, atomic nuclei fuse to make new elements. What happens to the mass "lost" in the course of this fusion? This is the question answered by Einstein's fateful equation $E = mc^2$*.* E *stands for energy,* m *stands for mass, and* c *is a constant equal to the speed of light. Einstein's equation means simply that under certain extreme and unusual conditions, mass is turned into energy.*

The sun consists largely of hydrogen nuclei. At the extremely high temperatures at the core of the sun, hydrogen nuclei strike each other with enough velocity to fuse. In a series of steps, four hydrogen nuclei fuse to form one helium nucleus. In the course of these reactions, energy is released, enough to keep the fusion reaction going and to emit tremendous amounts of radiant energy into space. Life on this planet depends on energy emitted by the sun in the course of this reaction. This same reaction provides, of course—as Einstein foresaw—the energy of the hydrogen bomb.

A COSMIC VIEW

The universe, according to the present model, is an isolated system. The matter and energy present in it 20 billion years ago, at the time of the primordial explosion, are all the matter and energy it will ever have. Moreover, in every energy exchange, the final state has lower free energy and higher entropy than the initial one. In this view, of course, the universe is running down. The stars will flicker out, one by one; life—any form of life on any planet—will come to an end. Finally, even the motion of individual molecules will cease. However, even the most pessimistic among us do not believe this will occur for another 20 billion years or so. In the meantime, life can exist *because* the universe is running downhill. Life is one of the ways in which the universe is dying.

(a)

(b)

(c)

(d)

7–8
Organisms are experts in energy conversion. (a) Heat energy is being generated from chemical energy by these rapidly growing skunk cabbages. (b) Two harvest mice eating ripe wheat, in which is stored energy from the sun. (c) Garter snake and toad engaged in a chemical energy conversion. (d) A cheetah converting chemical energy to kinetic energy.

Biology and the Second Law

The laws of thermodynamics are of crucial importance to biology, as they are to physics and chemistry. As Gibbs saw, 100 years ago, they are organizing principles under which a number of different kinds of processes can be unified. Also, as we shall see in the following chapter, they permit biochemical bookkeeping. The difference of 13 kcal/mole due to the entropy factor in calculating the energy yield of glucose may not, for example, seem like a large amount, but it is almost enough to phosphorylate two ADPs to ATPs.

However, the most interesting implication of the second law, as far as biology is concerned, is the relationship between entropy on the one hand, and order, organization, and specificity on the other. Living systems are characterized, as we have seen, by nonrandomness. It is because of its nonrandomness that lysozyme, for example, is able to fill its biological function. Disorder, as in the case of the sickle cell hemoglobin molecule, can mean death. The order and organization found in every detail of a living cell is order produced by energy. Not only does energy run the machine, it creates and maintains it.

SUMMARY

Life on this planet is dependent on the flow of energy from the sun. A small fraction of this energy, captured in the process of photosynthesis, is converted to the energy that drives the many other metabolic reactions associated with living systems and from which living systems derive their order and organization.

The thermodynamic relationship between plants and animals—photosynthesizers and heterotrophs—has been recognized only recently. Stated in its simplest form: In the course of photosynthesis, the energy of the sun is used to forge high-energy carbon-carbon and carbon-hydrogen bonds; then, in the course of respiration, these bonds are once more broken down to carbon dioxide and water with the release of energy. As in machines, some useful energy is lost in each step of these energy conversions.

Living systems thus operate in accord with the laws of thermodynamics. The first law of thermodynamics states that energy can be changed from one form to another but is not created or destroyed. The potential energy of the initial state (or reactants) is equal to the potential energy of the final state (or products) plus the energy released in the process or reaction. The second law of thermodynamics states that in the course of energy conversions, the potential energy of the final state will always be less than the potential energy of the initial state. The difference in energy between the initial and final state is known as the free energy change and is symbolized as $\triangle G$. Exergonic (energy-yielding) reactions have a negative $\triangle G$. Factors that determine the $\triangle G$ include $\triangle H$, the change in heat content, and $\triangle S$, the change in entropy, which, multiplied by the absolute temperature (T), is a measure of randomness or disorder:

$$\triangle G = \triangle H - T \triangle S$$

QUESTIONS

1. Why, in Figure 7–1, are there more plants than zebras and more zebras than lions? (Explain in terms of thermodynamics.)

2. The laws of conservation of energy apply only to isolated systems, that is, to systems into which no energy is entering. Is an aquarium ordinarily an isolated system? Could you convert it to one? A spaceship may or may not be an isolated system, depending on certain features of its design. What would these features be? Is the earth an isolated system?

3. At present, at least four types of energy conversions are going on in your body. Name them.

4. Explain why it is that living systems, despite appearances, are not in violation of the second law of thermodynamics.

5. What is there about the orderliness of a living organism that most significantly distinguishes it from the orderliness of a machine, such as an IBM 370 computer or the Bell Telephone System?

6. Some human societies use the barter system for exchange of goods and services. However, all complex societies have some form of monetary exchange. What are the advantages of a monetary exchange? Relate your answer to the ADP/ATP system.

7. All natural processes proceed with an increase in entropy. How then do you explain the freezing of water?

CHAPTER 8

How Cells Make ATP: Glycolysis and Respiration

ATP is the principal energy carrier in living systems. As we have noted in previous chapters, it participates in a great variety of cellular events, from chemical biosyntheses, to the flick of a cilium, the twitch of a muscle, or the active transport of a molecule across a cell membrane. It is involved in the propagation of an electric impulse along a nerve or, in some remarkable organisms, the electrocution of prey (Figure 8–1). In the following pages, we shall show in some detail how a cell breaks down carbohydrates and captures and stores the released energy in the terminal phosphate bonds of ATP. The oxidation of glucose (or other carbohydrates) is complicated in detail—so go slowly—but simple in its overall design.

(a)

(b)

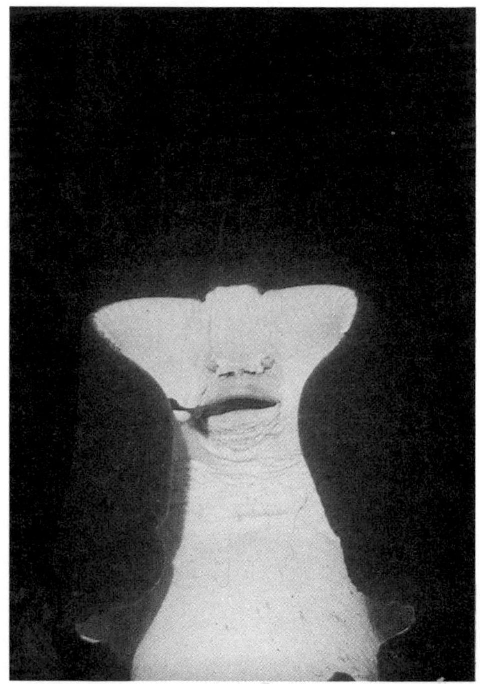

(c)

8–1
(a) *The electric ray converts chemical energy to electrical energy, stunning and immobilizing its prey with electric discharges.* (b, c) *The prey shown here, a small reef fish, is then moved to the mouth by the pectoral fins and is swallowed. The reef fish will be converted to chemical energy, which will, in turn, be converted to kinetic and electrical energy for the capture of additional prey. Only a fraction of the potential energy is transferred at each passage. These photographs were taken at night off Santa Barbara, California.*

OXIDATION-REDUCTION REVISITED

Oxidation, you will recall, is the loss of an electron. Reduction is the gain of an electron. Since, in spontaneous oxidation-reduction reactions, electrons go from higher to lower energy levels, a molecule usually releases energy as it is oxidized. In the oxidation of glucose, carbon-carbon bonds, carbon-hydrogen bonds, and oxygen-oxygen bonds are exchanged for carbon-oxygen and hydrogen-oxygen bonds, as the highly electronegative oxygen atom attracts and hoards electrons. The overall equation for this reaction is

$$\text{glucose} + \text{oxygen} \longrightarrow \text{carbon dioxide} + \text{water} + \text{energy}$$

Or,

$$C_6H_{12}O_6 + 6O_2 \longrightarrow 6CO_2 + 6H_2O$$

$$\triangle G = -686 \text{ kcal/mole}$$

Living systems are experts at energy conversions. They are organized to trap the free energy so that it will not be dissipated randomly but can be used to do the work of the cell. In living systems, the breakdown of glucose takes place in two stages. In the first, known as *glycolysis,* the six-carbon glucose molecule is split into two molecules of a three-carbon compound (pyruvic acid):

GLUCOSE ⟶ 2 PYRUVIC ACID

The electrons are accepted not by oxygen but by NAD (see page 83). The $\triangle G$ of this stage is -143 kcal/mole; this represents a relatively small proportion of the total energy of the glucose molecule.

In the second stage of the oxidation of glucose, which is called *respiration,* the electrons and protons removed from the carbon atoms are accepted by oxygen. For this series of reactions, $\triangle G$ is -543 kcal/mole, a comparatively large energy yield.

In the course of glycolysis and respiration, 36 molecules of ATP are regenerated from ADP in the breakdown of each molecule of glucose.

GLYCOLYSIS

Glycolysis—the lysis (splitting) of glucose—takes place in a series of nine steps, each catalyzed by a specific enzyme. Notice how the carbon skeleton of the molecule is dismembered and its atoms rearranged step by step. Note especially the formation of ATP from ADP and of NADH from NAD^+. ATP and NADH represent the cell's net energy harvest from this transaction.

8–2
The steps of glycolysis.

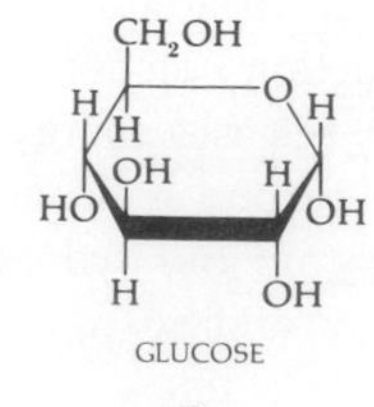

Step 1

HEXOKINASE

ATP

ADP

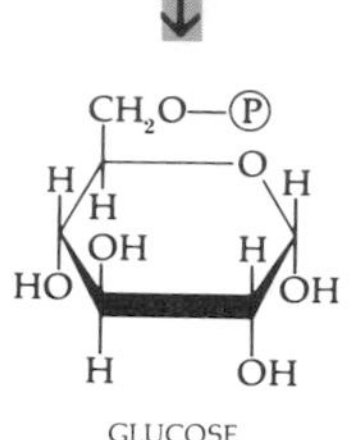

Step 2

PHOSPHOGLUCO-ISOMERASE

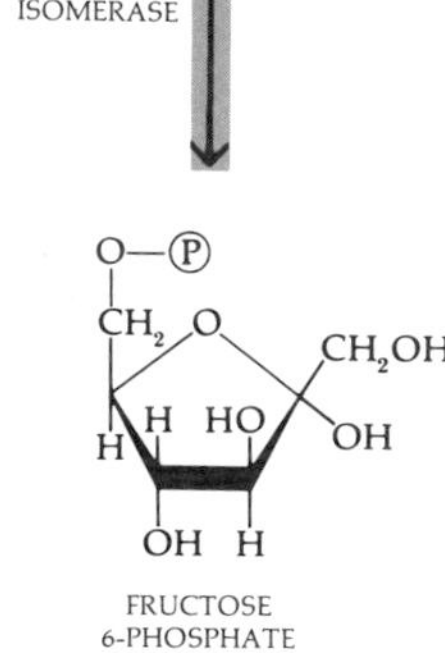

Step 3

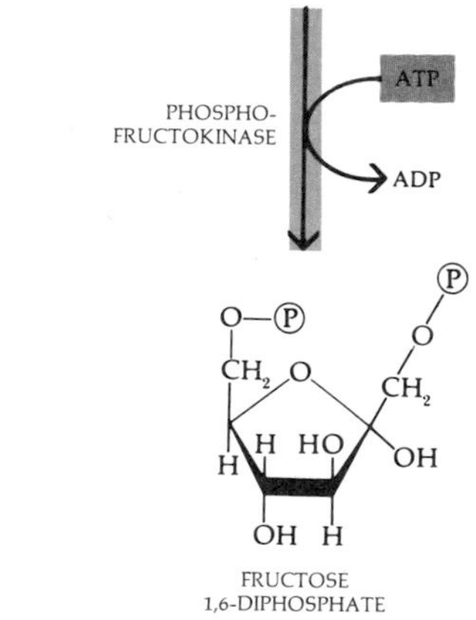

Step 1. The first steps in glycolysis require an input of energy, which is supplied by coupling these steps to the ATP/ADP system. The terminal phosphate group is transferred from an ATP molecule to the carbon in the sixth position of the glucose molecule, to make glucose 6-phosphate. The reaction of ATP with glucose to yield glucose 6-phosphate and ADP is an exergonic reaction. Some of the energy released is conserved in the chemical bond linking the phosphate to the sugar molecule, which then becomes energized. This reaction is catalyzed by a specific enzyme (hexokinase), and each of the reactions that follows is similarly regulated by a specific enzyme.

Step 2. The molecule is reorganized, again with the help of a particular enzyme. The six-sided ring characteristic of glucose becomes a five-sided fructose ring. (As you know, glucose and fructose both have the same number of atoms—$C_6H_{12}O_6$—and differ only in the arrangements of these atoms.) This reaction can proceed approximately equally well in either direction; it is pushed forward by the accumulation of glucose 6-phosphate and the removal of fructose 6-phosphate as the latter enters Step 3.

Step 3. In this step, which is similar to Step 1, fructose 6-phosphate gains a second phosphate by the investment of another ATP. The added phosphate is bonded to the first carbon, producing fructose 1,6-diphosphate, that is, fructose with phosphates in the 1 and 6 positions. Note that in the course of the reactions thus far, two molecules of ATP have been converted to ADP and no energy has been recovered.

The enzyme catalyzing this step, phosphofructokinase, is an allosteric enzyme, and ATP is an allosteric effector inhibiting this enzyme. The allosteric interaction between them is the chief regulatory mechanism of glycolysis. If ATP is present in adequate quantities for other purposes of the cell, ATP inhibits the activity of the enzyme and so ATP production ceases and glucose is conserved. As the cell uses up its supply of ATP, the enzyme is released from inhibition and the breakdown of glucose resumes. This is one of the major control points of ATP production.

Step 4. The six-carbon sugar molecule is split into two three-carbon molecules, dihydroxyacetone phosphate and glyceraldehyde phosphate. The two molecules are interconvertible by the enzyme isomerase. However, because the glyceralde-

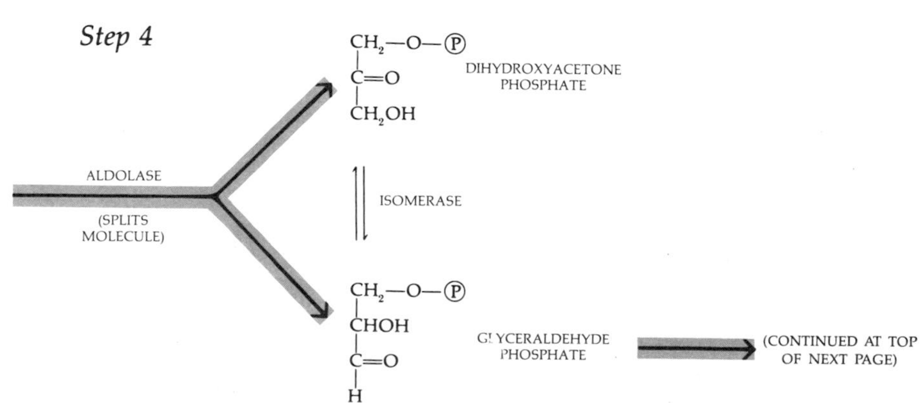

hyde phosphate is used up in subsequent reactions, all of the dihydroxyacetone phosphate is eventually converted to glyceraldehyde phosphate. Thus, all subsequent steps must be counted twice to account for the fate of one glucose molecule. With the completion of Step 4, the preparatory reactions are complete.

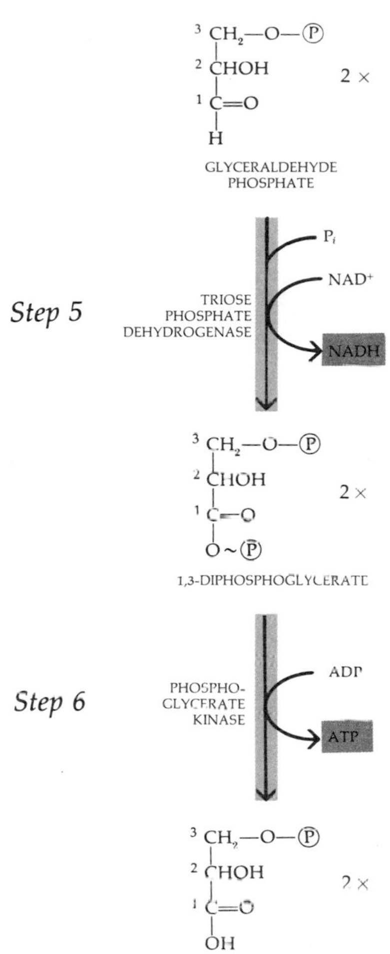

Step 5. Glyceraldehyde phosphate molecules are oxidized—that is, hydrogen atoms with their electrons are removed—and NAD^+ is reduced to NADH. This is the first reaction from which the cell harvests energy. Some of the energy from this oxidation reaction is also used to phosphorylate the glyceraldehyde molecules in what is now the 1 position of each molecule. (The designation P_i indicates inorganic phosphate available as a phosphate ion in solution in the cytoplasm.) Note that a "high-energy" bond is formed.

Step 6. The high-energy phosphate is transferred from each diphosphoglycerate molecule to a molecule of ADP (a total of two molecules of ATP per molecule of glucose). This is a highly exergonic reaction (the $\triangle G$ is negative and large) and so pulls all the preceding reactions forward.

Step 7. The remaining phosphate group is enzymatically transferred from the 3 position to the 2 position.

Step 8. In this step, a molecule of water is removed from the three-carbon compound. This internal rearrangement of the molecule concentrates energy in the vicinity of the phosphate group, forming a "high-energy" bond.

Step 9. The high-energy phosphate is transferred to a molecule of ADP, forming another molecule of ATP (again, a total of two molecules of ATP per molecule of glucose). This is also a highly exergonic reaction and thus pulls forward the preceding two reactions (Steps 7 and 8).

Summary of Glycolysis

The complete sequence begins with one molecule of glucose. Energy is invested at Steps 1 and 3 by the transfer of a phosphate group from an ATP molecule—one at each step—to the sugar molecule. The six-carbon molecule splits at Step 4, and

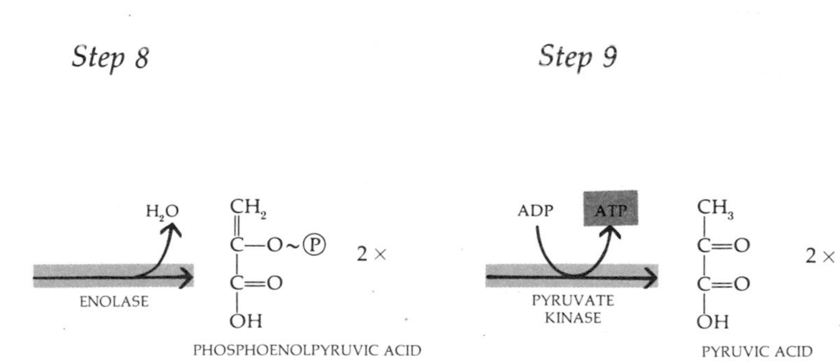

8–3

The two stages of glycolysis. Compounds other than glucose, such as galactose, mannose, pentoses, glycogen, and starch, can undergo glycolysis once they have been converted to glucose 6-phosphate.

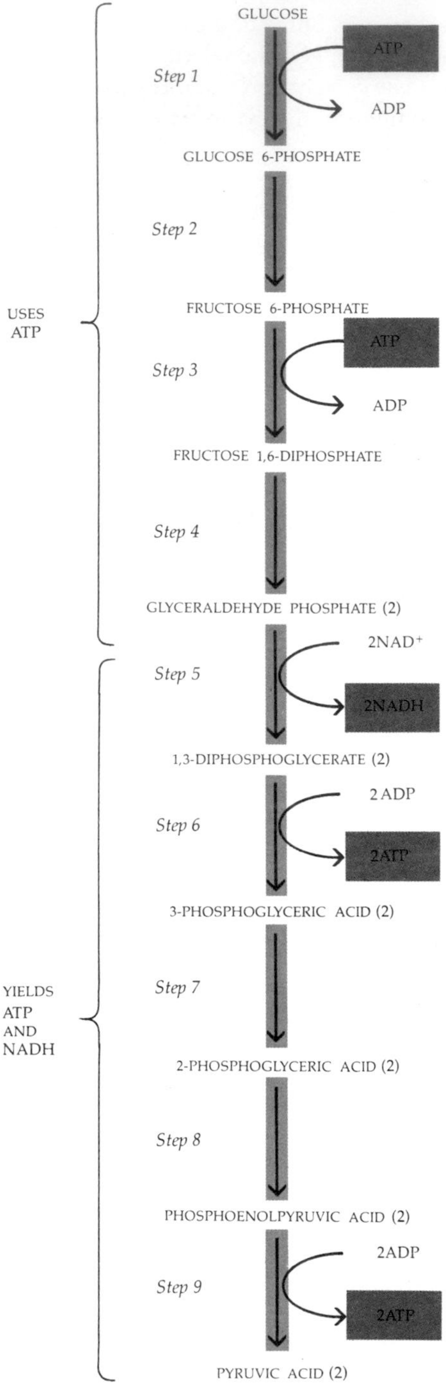

from this point onward, the sequence yields energy. At Step 5, a molecule of NAD^+, in being reduced to NADH, stores some of the energy from the oxidation of glyceraldehyde phosphate. At Steps 6 and 9, molecules of ADP take energy from the system, becoming phosphorylated to ATP. The energy from the phosphate bonds of two ATP molecules is needed to initiate the glycolytic sequence. Two NADH molecules are produced from two NAD^+ and four ATP molecules from four ADP molecules:

$$\text{glucose} + 2\text{ATP} + 4\text{ADP} + 2\text{P}_i + 2\text{NAD}^+ \longrightarrow$$
$$2 \text{ pyruvic acid} + 2\text{ADP} + 4\text{ATP} + 2\text{NADH} + 2\text{H}^+ + 2\text{H}_2\text{O}$$

Thus one glucose molecule has been converted to two molecules of pyruvic acid. The net harvest—the energy recovered—is two molecules of ATP and two molecules of NADH per molecule of glucose. The two molecules of pyruvic acid still contain a large amount of the potential energy that was stored in the original glucose molecule. This series of reactions is carried out by virtually all living cells—from prokaryotes to the eukaryotic cells of our own bodies.

AEROBIC AND ANAEROBIC PATHWAYS

Pyruvic acid can follow one of several pathways. Two pathways are anaerobic (without oxygen), and one is aerobic. We shall briefly discuss the anaerobic pathways and then follow the aerobic one, which is the principal pathway of energy metabolism for most cells in the presence of oxygen.

In the absence of oxygen, pyruvic acid can be converted either to ethanol (ethyl alcohol) or to lactic acid. Yeast cells, for example, can grow either with or without oxygen; without oxygen, they form alcohol from pyruvic acid. Yeast cells are present as a "bloom" on the skin of grapes. When the glucose-filled juices of grapes and other fruits are extracted and stored in airtight kegs, the yeast cells turn the fruit juice to wine by converting glucose into ethanol. Yeast, like all living things, have a limited tolerance for alcohol, and when a certain concentration (about 12 percent) is reached, the yeast cells cease to function.

Because of the economic importance of the wine industry, alcoholic fermentation was the first enzymatic process to be intensively studied. Louis Pasteur was the first to recognize the role of yeast cells in the process. The fermentation process was the setting for a phase of the vitalism-reductionism controversy (page 28) that came to an end around the beginning of the twentieth century. The formation of alcohol from sugar is called "fermentation," and for many years enzymes were commonly referred to as "ferments." The scope of their activities is now known to be so broad that this word is no longer appropriate.

Lactic acid is formed from pyruvic acid by a variety of microorganisms and also by some animal cells when O_2 is scarce or absent. It is produced, for example, in muscle cells during bursts of extra hard work, as by an athlete in a sprint. We breathe hard when we run fast, thereby increasing the supply of oxygen to the muscle cells, but the capacity of the muscles to do work is not limited by the amount of oxygen available to them at that moment. They can accumulate what is known as an oxygen debt by producing lactic acid from glucose (which is stored in the liver in the form of glycogen and released in the bloodstream as it is needed). The accumulated lactic acid lowers the pH of muscle and reduces the capacity of the muscle fibers to contract, producing the sensation of muscle fatigue. The lactic

8-4

(a) *The steps by which pyruvic acid, formed by glycolysis, is converted anaerobically to ethanol (ethyl alcohol). In the first step, carbon dioxide is released. In the second, NADH is oxidized, and acetaldehyde is reduced. Most of the energy of the glucose remains in the alcohol, which is the end product of the sequence. However, by regenerating NAD^+, these steps allow glycolysis to continue, with its small but sometimes vitally necessary yield of ATP.* (b) *Consequences of anaerobic glycolysis. Yeast cells visible on the grapes as dustlike "bloom," mix with the juice when the grapes are crushed. Storing the mixture under anaerobic conditions causes the yeast to break down the glucose in the grape juice to alcohol.*

CH_3—C(=O)—C(=O)—OH → (CO_2) → CH_3—C(=O)—H → (NADH → NAD^+) → CH_3—CH(OH)—H

PYRUVIC ACID (FROM GLYCOLYSIS) | ACETALDEHYDE | ETHANOL

(a)

(b)

8-5

The enzymatic reaction that produces lactic acid from pyruvic acid anaerobically in muscle cells. In the course of this reaction, NADH is oxidized and pyruvic acid is reduced. The NAD^+ molecules produced in this reaction and the one shown in Figure 8-4 are recycled in the glycolytic sequence. Without this recycling, glycolysis cannot proceed. Lactic acid accumulation results in muscle soreness and fatigue.

CH_3—C(=O)—C(=O)—OH → (NADH → NAD^+) → CH_3—CH(OH)—C(=O)—OH

PYRUVIC ACID (FROM GLYCOLYSIS) | LACTIC ACID

acid diffuses into the blood and is carried to the liver where it is resynthesized to pyruvic acid and back again to glucose or glycogen.

Why is pyruvic acid converted to lactic acid, only to be converted back again? The function of this reaction is simple: It regenerates the NAD^+ without which glycolysis cannot go forward. Even though this step seems to be wasteful in terms of energy consumption, it may be all-important in the economy of the organism, spelling the difference between life and death when an animal "out of breath" needs one last burst of ATP to escape from a predator or catch a prey.

An oxygen debt is usually accumulated during a short burst of intensive exercise, as during a rapid sprint. During sustained moderate exercise, the intake of oxygen by the lungs and the circulation of the blood supplying oxygen to the muscle tissues often catch up with the oxygen consumption in the muscle, thus producing the "second wind" phenomenon well known to runners.

The fact that glycolysis does not require oxygen suggests that the glycolytic sequence evolved early, before free oxygen was present in the atmosphere, and that primitive one-celled organisms used glycolysis (or something very much like it) to extract energy from the organic compounds that they absorbed from their watery surroundings.

RESPIRATION

In the presence of oxygen, the next stage in the breakdown of glucose involves the stepwise oxidation of pyruvic acid to carbon dioxide and water—the process known as *respiration.* Respiration has two meanings in biology. One is the breathing in of oxygen and breathing out of carbon dioxide; this is also the ordinary, nontechnical meaning of the word. The second meaning of respiration is the oxidation of food molecules by cells. This process, sometimes qualified as cellular respiration, is what we are concerned with here.

Cellular respiration takes place in two stages: the Krebs cycle and the electron transport system. In eukaryotic cells, these reactions take place within the mitochondria. Mitochondria, as we noted in Chapter 5, are surrounded by two membranes. The outer one is smooth and the inner one folds inward, forming cristae. Within the inner compartment, surrounding the cristae, is a dense solution (the matrix) containing enzymes, coenzymes, water, phosphates, and other molecules involved in respiration. The outer membrane is permeable to most small molecules, but the inner one permits the passage of only certain molecules, such as pyruvic acid and ATP, and restrains the passage of others. (As we shall see, this selective permeability of the inner membrane is critical to the ability of the mitochondria to harness the power of respiration to the production of ATP.)

Some of the enzymes of the Krebs cycle are in solution in the inner compartment. Other Krebs cycle enzymes and the enzymes and other components of the electron transport chains are built into the membrane of the cristae. These inner membranes of the mitochondria are about 80 percent protein. In the mitochondria, pyruvic acid from glycolysis is oxidized to carbon dioxide and water, completing the breakdown of the glucose molecule. Ninety-five percent of the ATP generated by heterotrophic cells is produced in the mitochondria.

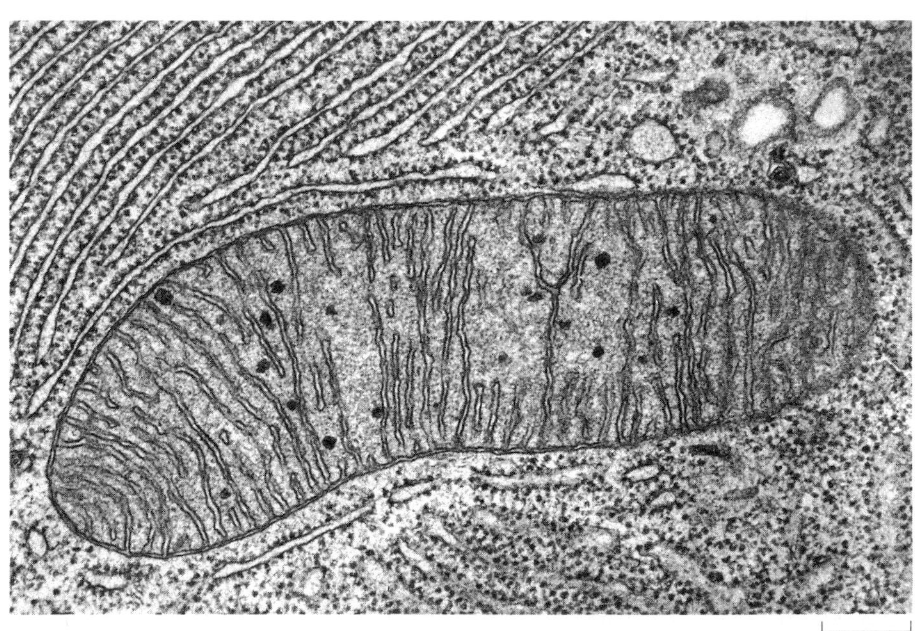

8–6
Mitochondria are surrounded by two membranes. The inner membrane folds inward to make a series of shelves, or cristae. The enzymes and electron carriers involved in the final stage of cellular respiration are built into these internal membranes.

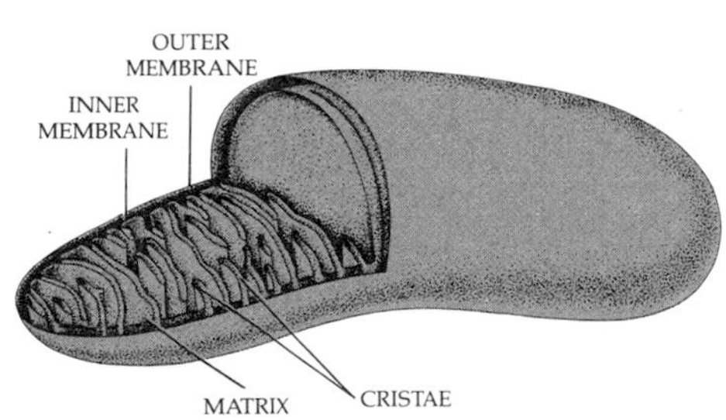

DISSECTING THE CELL

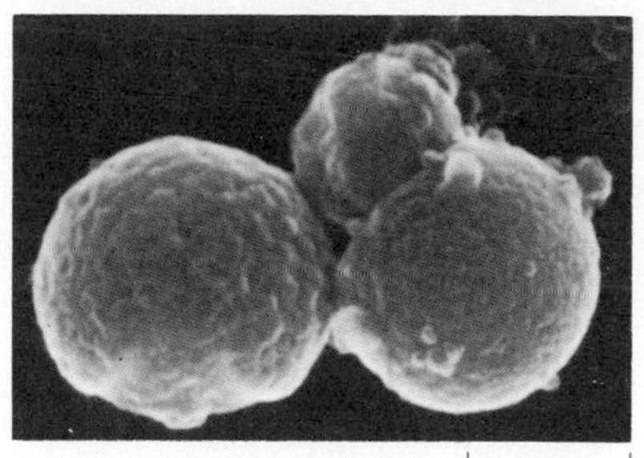

Scanning electron micrograph of isolated intact liver mitochondria.

In every living cell, many hundreds of chemical reactions proceed simultaneously. Molecules are continuously synthesized via certain enzymatic pathways, and molecules are broken down by other enzymatic pathways. Many of these reactions are mutually incompatible, as can be demonstrated by destroying the structure of cells and mixing their enzymes in a test tube. Chemical chaos results, and the enzymes are soon inactivated. In the cell, however, synthetic and degradative pathways operate in harmony because biochemical reactions are spatially localized and compartmentalized within specific subcellular organelles. A living cell is the most intensely concentrated set of chemical reactions known. A cell carries out many more chemical reactions than any apparatus devised by chemical engineers, and all within the space of a few cubic micrometers. The cell's unique chemical versatility results from the compartmentalization of biochemical pathways within organelles.

In order to study the specific functions of any organelle type, the organelle must be dissected free from all other cell structures and collected in large quantities. Mitochondria, lysosomes, and other organelles are, of course, far too small for hand dissection, but cell biologists can prepare pure samples of any organelle type by the technique of preparative centrifugation.

Small particles, ranging in size from cells to macromolecules, can be separated by centrifugation if the particle types differ in shape and density. Particles suspended in fluid and then subjected to strong gravitational force will move through the fluid at varying rates, the largest, densest particles settling most rapidly. Forces up to 400,000 times the force of gravity (400,000 g) can be generated in a test tube of suspended particles by rotating the tube at very high speeds in an ultracentrifuge. Thus, subcellular organelles, such as mitochondria, nuclei, and intracellular membranes, can be separated into purified fractions by spinning fragmented cells at appropriate centrifugal forces.

For example, in order to determine which enzymatic pathways are present in mitochondria, a tissue, such as rat liver, is minced into small pieces and homogenized—that is, the cells are gently broken up by grinding the tissue in a glass tube fitted with a Teflon pestle. The tube contains a sucrose solution that is isotonic with intracellular fluid. The resulting suspension of cell organelles is then placed in an unbreakable test tube and spun in the centrifuge at low speed (700 g) for 10 minutes, so as to drive the bulkiest organelles, the nuclei, to the bottom of the tube. All the lighter organelles remain suspended in the fluid, which is termed the supernatant. The supernatant is transferred to another centrifuge tube and spun at a higher speed (10,000 g for 20 minutes), which sediments particles such as mitochondria and lysosomes into a pellet at the bottom of the tube. The supernatant, containing ribosomes and various membranes, is discarded, and the pellet is retained.

At this point, the mitochondria have been partially purified by means of differential centrifugation. This type of centrifugation separates particles of quite different size and density properties by "spinning down" the larger particles to form a pellet. To separate particles of rather similar size, such as mitochondria and lysosomes, the more subtle technique of zonal centrifugation is used. To return to the experiment we have outlined, the pellet containing the mitochondria and lysosomes is resuspended and gently layered atop a sucrose density gradient in a centrifuge tube. A sucrose density gradient is prepared by layering increasingly dense sucrose solutions one beneath the other so that the densest solution is at the bottom of the tube and the least dense solution is at the top of the tube. When organelles are centrifuged in a density gradient (120,000 g for 8 hours), each organelle type moves through the gradient at a different rate, depending on the organelle's density. Following centrifugation, mitochondria will occupy one zone in the gradient, lysosomes another, and other organelles will be found in other zones. By puncturing the bottom of the tube and removing the contents drop by drop, we can collect a pure sample of mitochondria. The purity can be verified by the electron microscope. If the procedure has been performed correctly, the isolated mitochondria will emerge with membranes intact and all enzymatic pathways functioning. It is then possible to test for the activities of specific enzymes, thus determining which biochemical functions are compartmentalized within mitochondria. Similarly, other cell constituents can be isolated and their biochemical activities determined.

8-7
(a) *The three-carbon pyruvic acid molecule is oxidized to the two-carbon acetyl group, which is combined with coenzyme A to form acetyl CoA. The oxidation of the pyruvic acid molecule is coupled to the reduction of NAD^+. Acetyl CoA enters the Krebs cycle.* (b) *Electron micrograph showing the enzymes involved in the oxidation of pyruvic acid to acetyl CoA. Each of the complexes visible here represents multiple copies of three different enzymes.*

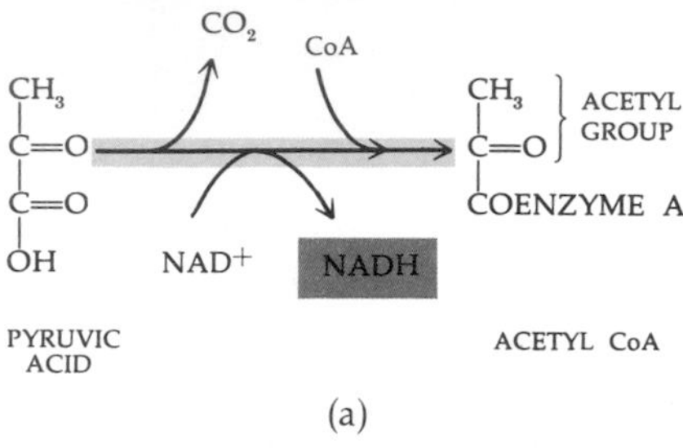

(a)

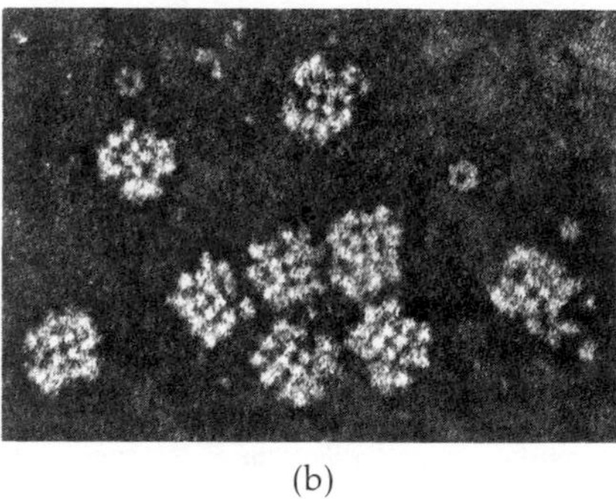

(b)

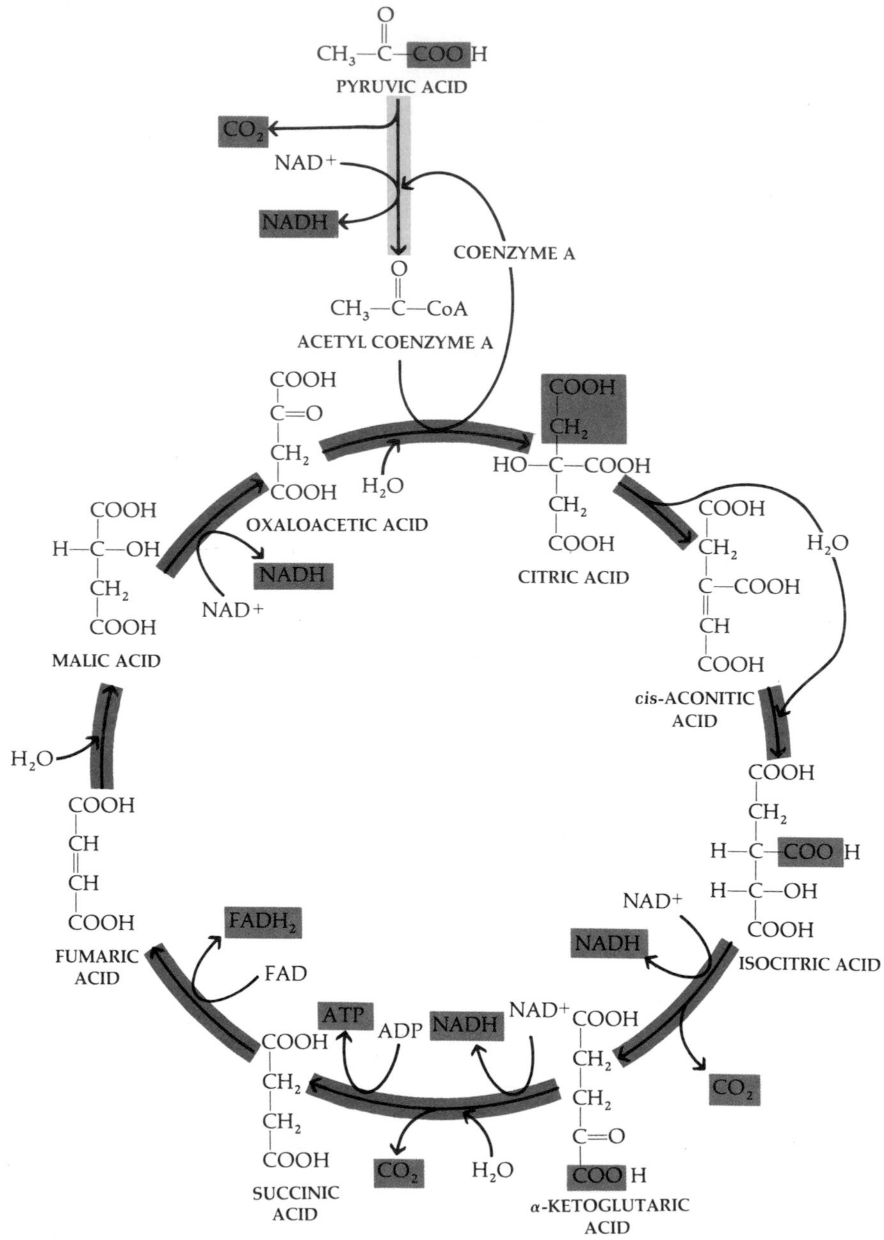

The Krebs Cycle

Pyruvic acid crosses the outer and inner membranes of the mitochondria. Before entering the Krebs cycle, the three-carbon pyruvic acid molecule is oxidized. The carbon of the carboxyl group is removed, forming carbon dioxide. In the course of this exergonic reaction, a molecule of NADH is formed from NAD^+. The original glucose molecule has now been oxidized to two CO_2 molecules and two acetyl (CH_3CO) groups, and, in addition, four NADH molecules have been formed. Each acetyl group is momentarily accepted by a coenzyme known as coenzyme A. Like the coenzymes we have examined previously, coenzyme A is a large molecule, a

PYRUVIC ACID
CO_2
NAD^+
NADH
COENZYME A
ACETYL COENZYME A
OXALOACETIC ACID
H_2O
CITRIC ACID
H_2O
cis-ACONITIC ACID
ISOCITRIC ACID
NAD^+
NADH
CO_2
α-KETOGLUTARIC ACID
NAD^+
NADH
ADP
ATP
H_2O
CO_2
SUCCINIC ACID
FAD
$FADH_2$
FUMARIC ACID
H_2O
MALIC ACID
NAD^+
NADH

8-8
The Krebs cycle. In the course of the cycle, the carbons donated by the acetyl group are oxidized to carbon dioxide, and the hydrogen atoms are passed to electron carriers. As in glycolysis, a specific enzyme is involved at each step.

8-9
Flavin adenine dinucleotide, an electron acceptor, in (a) its oxidized form (FAD) and (b) its reduced form ($FADH_2$). Riboflavin is a vitamin made by all plants and many microorganisms. It is also known as vitamin B_2. It is a pigment; in its oxidized form it is a bright yellow.

A related electron acceptor, flavin mononucleotide (FMN), consists of riboflavin and the first phosphate group shown here. It accepts electrons from NADH in the electron transport chain.

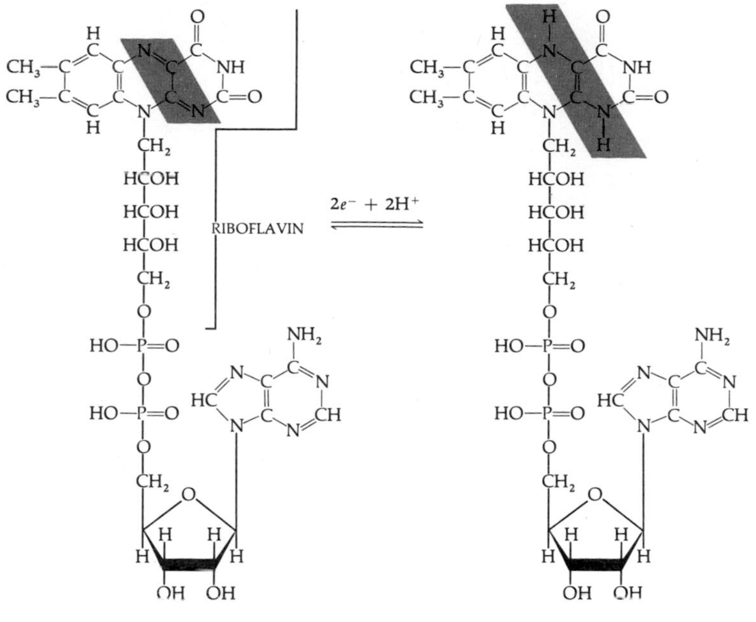

portion of which is a nucleotide and a portion of which is a vitamin (pantothenic acid, one of the B complex vitamins). The combination of the acetyl group and CoA is abbreviated as acetyl CoA. This reaction is the link between glycolysis and the Krebs cycle.

Upon entering the Krebs cycle, the two-carbon acetyl group is combined with a four-carbon compound (oxaloacetic acid) to produce a six-carbon compound (citric acid). In the course of the cycle, two of the six carbons are oxidized to CO_2, and oxaloacetic acid is regenerated—thus making this series literally a cycle. Each turn around the cycle uses up one acetyl group and regenerates a molecule of oxaloacetic acid, which is then ready to begin the sequence again. In the course of these steps, some of the energy released by the oxidation of the carbon-hydrogen and carbon-carbon bonds is used to convert ADP to ATP (one molecule per cycle), and some is used to produce NADH from NAD^+ (three molecules per cycle). In addition, some energy is used to reduce a second electron carrier, flavin adenine dinucleotide, abbreviated FAD (Figure 8-9). One molecule of $FADH_2$ is formed from FAD per turn of the cycle. No O_2 is required for the Krebs cycle; the electrons and protons removed in the oxidation of carbon are all accepted by NAD^+ and FAD.

$$\text{oxaloacetic acid} + \text{acetyl CoA} + \text{ADP} + P_i + 3NAD^+ + \text{FAD} \longrightarrow \text{oxaloacetic acid} + 2CO_2 + \text{CoA} + \text{ATP} + 3\text{NADH} + FADH_2 + 3H^+ + H_2O$$

Note that the oxaloacetic acid molecule the cycle ends with is not the same molecule with which the cycle began. If one begins with a glucose molecule in which the carbon atoms are radioactive, radioactive carbon atoms will appear among the four carbons of the oxaloacetic acid.

8-10
Summary of the Krebs cycle. One molecule of ATP, three molecules of NADH, and one molecule of $FADH_2$ represent the energy yield of the cycle.

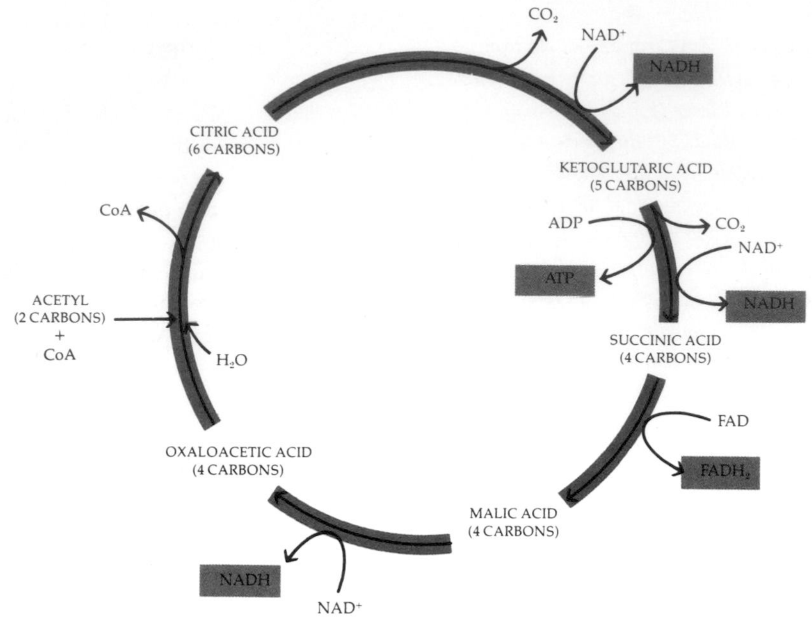

The Electron Transport Chain

The carbon atoms of the glucose molecule are now completely oxidized. Some of the energy of glucose has been used to produce ATP from ADP. Most of its energy, however, remains in electrons removed from the C—C and C—H bonds and passed to the electron carriers NAD^+ and FAD. These electrons are still at a high energy level.

In the final stage of the oxidation of glucose, these high-energy-level electrons are passed step-by-step to the low energy level of oxygen. The energy they yield in the course of this passage regenerates ATP from ADP. This step-by-step passage is made possible by a series of electron carriers, each of which holds the electrons at a slightly lower level.

These carriers make up what is known as an *electron transport chain.* At the top of the energy hill the electrons are held by NADH and $FADH_2$. Most of the energy of the glucose molecule now resides in these electron acceptors. The Krebs cycle yielded two molecules of $FADH_2$ and six molecules of NADH for each molecule of glucose. The oxidation of pyruvic acid to acetyl CoA yielded two molecules of NADH. Also, you will recall, two molecules of NADH were produced in glycolysis. In the presence of oxygen, the electrons held by these two NADH molecules are also transported into the mitochondrion where they are accepted by FMN (flavin mononucleotide; see Figure 8-9) and fed into the electron transport chain.

The principal components of the electron transport chain are molecules known as cytochromes (Figure 8-11). These molecules consist of a porphyrin ring, similar to that of heme, enclosing an atom of iron. Each iron atom alternately accepts and

8–11

(a) *Cytochromes are molecules in which a heme group is held in an intricate protein structure. In the cytochrome* c *structure shown here, the heme group is in color.*

(b) *The heme group of cytochrome* c. *In the cytochromes, an atom of iron (Fe) is held in a nitrogen-containing ring (a porphyrin ring). Cytochromes are involved in electron transfer. It is the iron within the molecule that actually combines with the electrons.*

releases an electron, passing it along to the next cytochrome at a slightly lower energy level until the electrons, their energy spent, are accepted by oxygen. The energy released in this downhill passage of electrons is used to form ATP molecules from ADP. Such ATP formation is known as *oxidative phosphorylation.* At the end of the chain, the electrons are accepted by oxygen, which then combines with protons (hydrogen ions) from the solution to produce water. Calculations show that each time two electrons pass from NADH to oxygen, three molecules of ATP are formed from ADP and phosphate. Each time a pair of electrons passes from $FADH_2$, which holds them at a slightly lower energy level than NADH, two molecules of ATP are formed. In oxidative phosphorylation the electron transfer potential of NADH and $FADH_2$ is converted to the phosphate transfer potential of ATP.

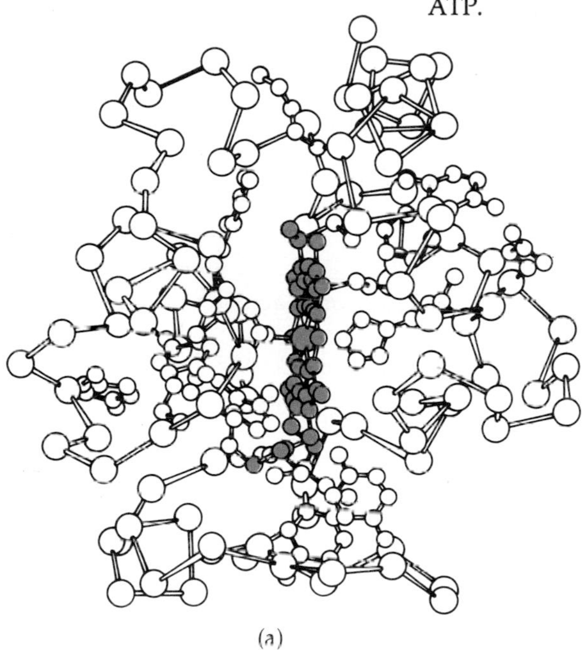

(a)

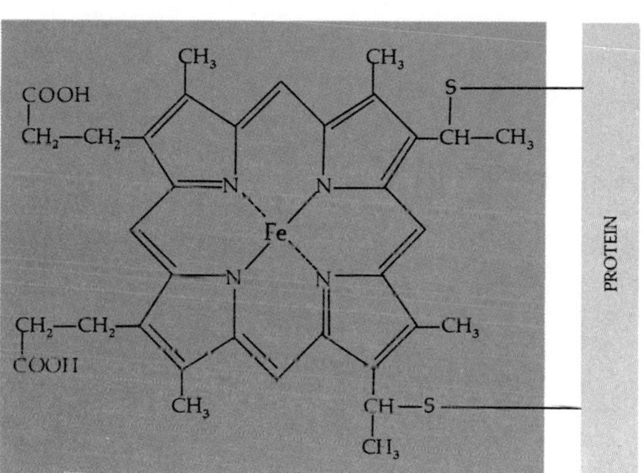

(b)

8–12

The electron transport chain Flavin mononucleotide (FMN) and coenzyme Q transfer electrons and protons. The cytochromes transfer only electrons. The electrons carried by NADH enter the chain when they are transferred to FMN; those carried by $FADH_2$ *enter the chain farther down the line at coenzyme Q. Each time two electrons from NADH pass down the chain, three molecules of ATP are formed from ADP; each time two electrons from* $FADH_2$ *pass down the chain, two molecules of ATP are formed.*

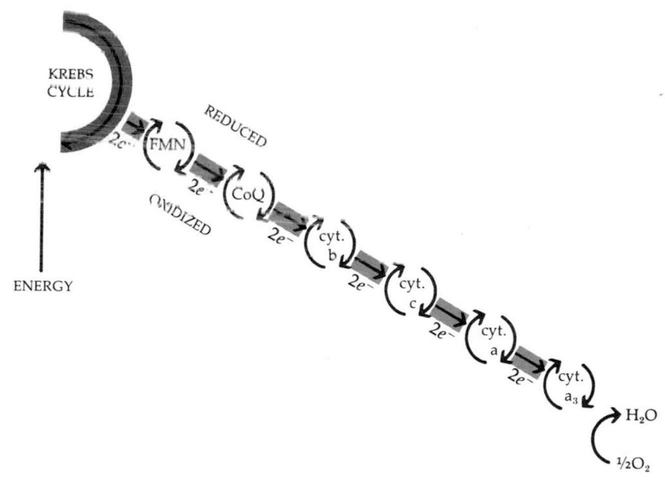

Mechanism of Oxidative Phosphorylation

The way in which the energy of the transferred electrons is harnessed to produce ATP from ADP is one of the most intriguing puzzles now facing biochemists. There are three current hypotheses: (1) chemical coupling, (2) chemiosmotic coupling, and (3) conformational change.

Chemical Coupling

According to this hypothesis, which is the oldest established one, the phosphorylation of ADP takes place enzymatically and is directly coupled to the oxidation of electron carriers at specific points in the electron transport chain. According to this model, then, oxidative phosphorylation is not very different from the phosphorylation steps of glycolysis. However, it has not been possible to verify this hypothesis experimentally.

Chemiosmotic Coupling

This model is the most popular one at the present time. It proposes that oxidative phosphorylation is driven not by chemical coupling between electron transport and phosphorylation reactions but by a proton gradient (differential H^+ concentration) produced between the two sides of the inner mitochondrial membrane during electron transport. According to this ingenious concept, first proposed in the 1960s by Peter Mitchell of the Glynn Research Laboratories in England, protons are pumped from the mitochondrial matrix to the outer mitochondrial compartment as electrons flow along the transport chain. Like the boulder at the top of the hill, the water at the edge of the dam, or the chemical energy in a stick of dynamite, this differential concentration of protons represents potential energy. The return of the protons across the membrane to the mitochondrial matrix releases the stored energy that, according to this theory, is harnessed by a membrane-bound enzyme that phosphorylates ADP to ATP (Figure 8–13).

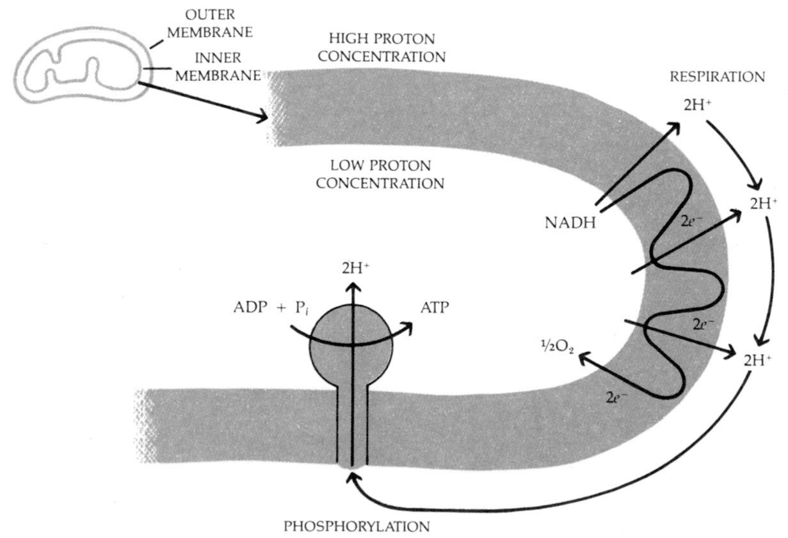

8–13
According to the chemiosmotic model for ATP production, protons are pumped out of the mitochondrial matrix as electrons are passed along the electron transport chain, which forms a part of the inner mitochondrial membrane. The movement of the protons back into the membrane provides the energy by which ATP is regenerated from ADP. Mitchell, who proposed this model, was awarded the Nobel Prize in 1978.

Conformational Change

The third and most recent hypothesis, proposed by David E. Green of the University of Wisconsin and Paul D. Boyer of the University of California, suggests that the energy released by electron transport produces transient changes in the shape of large molecules in the inner mitochondrial complex. The subsequent return of these molecules to their low-energy configuration releases sufficient energy, according to this hypothesis, to drive the phosphorylation of ADP. Support for this model comes from electron micrographs that show changes in mitochondria undergoing active transport.

Control of Oxidative Phosphorylation

Electrons continue to flow along the electron transport chain only if ADP is available to be converted to ATP. Thus oxidative phosphorylation is regulated by supply and demand. When the energy requirements of the cell decrease, fewer molecules of ATP are used, fewer molecules of ADP become available, and electron flow is decreased.

OVERALL ENERGY HARVEST

Glycolysis, in the presence of oxygen, yields two molecules of ATP directly and two molecules of NADH. The total gain, however, is not 8 ATP as you might calculate, but 6 ATP because there is a "cost" of 2 ATP to transport the electrons held by the two molecules of NADH across the mitochondrial membrane.

The conversion of pyruvic acid to acetyl CoA yields two molecules of NADH (inside the mitochondrion) for each molecule of glucose and so produces six molecules of ATP.

The Krebs cycle yields, for each molecule of glucose, two molecules of ATP, six of NADH, and two of $FADH_2$, or a total of 24 ATP.

8–14
Summary of glycolysis and respiration. Glucose is first broken down to pyruvic acid, with a yield of two ATP molecules and the reduction (dashed arrows) of two NAD^+ molecules. Pyruvic acid is oxidized to acetyl CoA, and one molecule of NAD^+ is reduced (note that this and subsequent reactions occur twice for each glucose molecule; this electron passage is indicated by solid arrows). In the Krebs cycle, the acetyl group is oxidized and the electron acceptors, NAD^+ and FAD, are reduced to NADH and $FADH_2$. The NADH and $FADH_2$ then transfer their electrons to the series of cytochromes and other electron carriers that make up the electron transport chain. As the electrons are passed "down-hill," the energy released is used to make ATP from ADP.

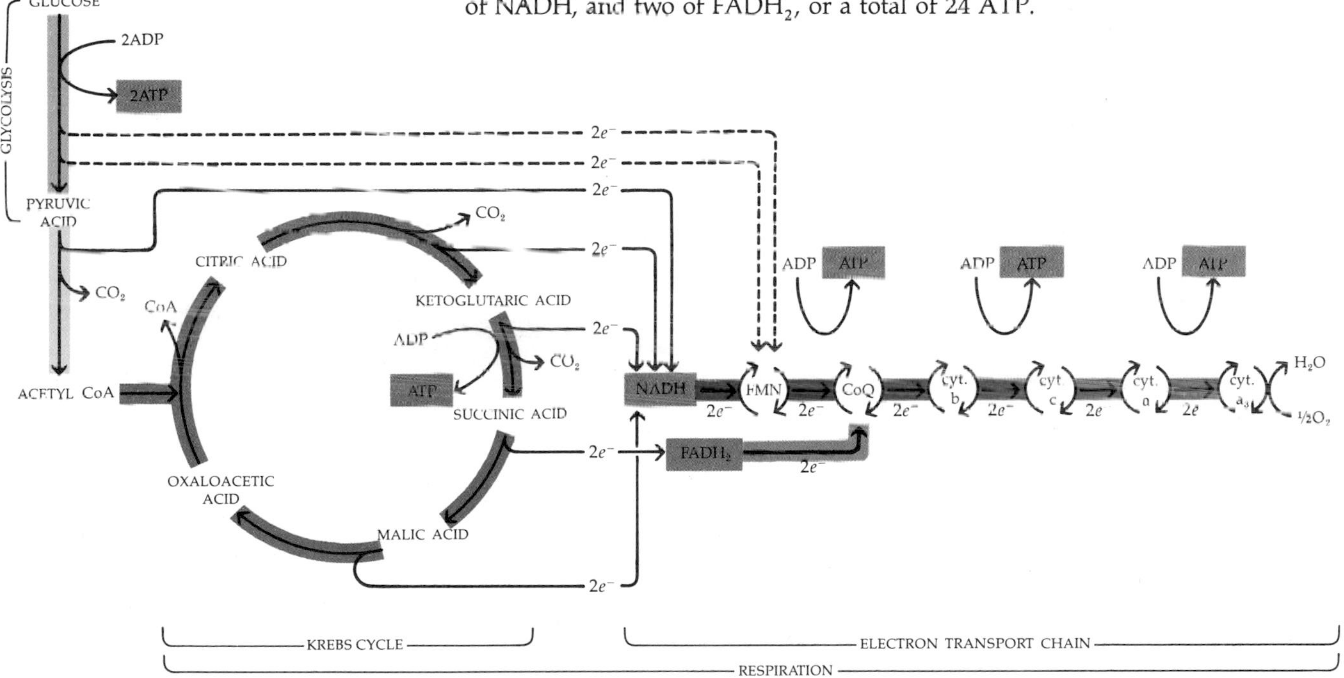

8-15
Energy changes in the oxidation of glucose. The complete respiratory sequence (glucose + $6O_2 \longrightarrow 6CO_2 + 6H_2O$) proceeds with an energy drop of 686 kcal/mole. Of this, almost 40 percent (252 kilocalories) is conserved in 36 ATP molecules. In anaerobic glycolysis (glucose ⟶ lactic acid), by contrast, only 2 ATP molecules are produced, representing only about 2 percent of the available energy of glucose.

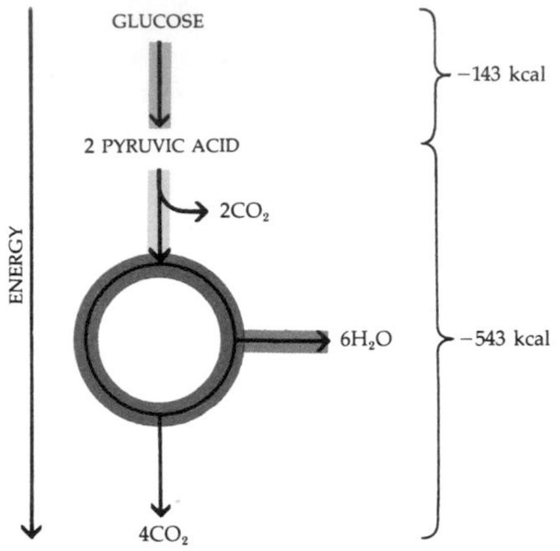

Table 8-1 *Summary of Energy Yield from One Molecule of Glucose*

Glycolysis:	2 NADH ⟶ 2 FMN	⟶ 2 ATP, 4 ATP		⟶	6 ATP
Respiration:					
Pyruvic acid ⟶ acetyl CoA:	1 NADH	⟶ 3 ATP	(× 2)	⟶	6 ATP
Krebs cycle:		1 ATP			
	3 NADH	⟶ 9 ATP	(× 2)	⟶	24 ATP
	1 $FADH_2$	⟶ 2 ATP			

As a balance sheet, Table 8-1, shows, the complete yield from a single molecule of glucose is 36 molecules of ATP. Note that all but two of the 36 molecules of ATP have come from reactions taking place in the mitochondrion, and all but four involve the oxidation of NADH or $FADH_2$ along the electron transport chain.

The free energy change (ΔG) of the reactions is −686 kilocalories per mole. About 252 (7 × 36) kilocalories per mole have been captured in the "high-energy" bonds of the ATP molecules, an efficiency of almost 40 percent.

The ATP molecules, once formed, are exported across the membrane of the mitochondrion by a shuttle system that simultaneously brings in one molecule of ADP for each ATP exported.

OTHER CATABOLIC PATHWAYS

Most organisms do not feed directly on glucose. How do they extract energy from, for example, fats or proteins? The answer lies in the fact that the Krebs cycle is a Grand Central Station for energy metabolism. Other foodstuffs are broken down and converted into molecules that can feed into this central pathway at some point or other.

Polysaccharides such as starch are broken down to their constituent monosaccharides and phosphorylated to glucose 6-phosphate and in this form enter the glycolytic pathway. Fats are first split into their glycerol and fatty acid components. These are then chopped up into two-carbon fragments and slipped into the Krebs cycle as acetyl CoA. Proteins are broken down into their constituent amino acids.

8-16
The strategy of energy metabolism. Organisms extract energy from compounds by oxidizing them to carbon dioxide and water. NAD^+ is the major oxidizing agent. The NADH gives up its high-energy electrons and associated protons to electron acceptors in the electron transport chain. The electrons are passed down the chain ultimately to oxygen. The energy released by the electron acceptors of the electron transport chain is used to phosphorylate ADP, converting it to ATP, which is used to drive the endergonic reactions of the organism. Note that both NADH and ATP are used in a cyclic fashion. Although they are critically important compounds, they are present in very small amounts, accomplishing their work as a result of constant and rapid turnover. The need for a constant supply of these molecules explains why a deprivation of oxygen brings about death in a very few minutes.

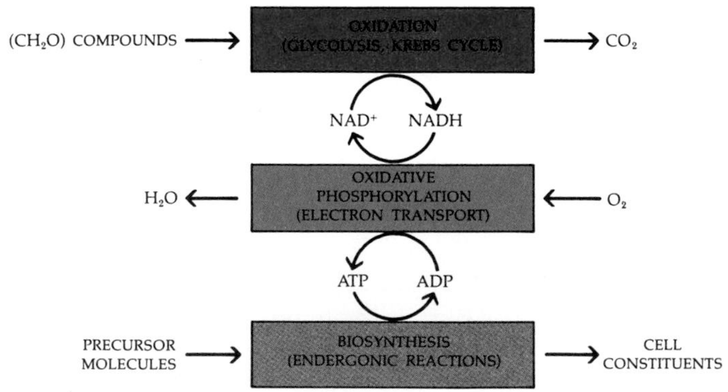

The amino acids are deaminated (the amino groups removed), and the residual carbon skeleton is either converted to an acetyl group or to one of the larger carbon compounds of the glycolytic pathway or the Krebs cycle so that it can be processed at this stage of the central pathway. The amino groups, if not reutilized, are eventually excreted as urea or other nitrogen-containing wastes. These various degradative pathways are, collectively, catabolism.

BIOSYNTHESIS

The pathways of glucose breakdown, central to catabolism, are also central to the biosynthetic or anabolic processes of life. These processes are the pathways of synthesis of the various molecules and macromolecules that make up an organism.

Since many of these substances, such as proteins and lipids, can be broken down and fed into the central pathway, you might guess that the reverse process can occur—namely, that the various intermediates of glycolysis and the Krebs cycle can serve as precursors for biosynthesis. This is in fact the case. However, the biosynthetic pathways, while similar to the catabolic ones, are distinctive. Different enzymes control the steps and the various critical steps of anabolism differ from the catabolic processes. These general pathways are sketched out in Figure 8–17.

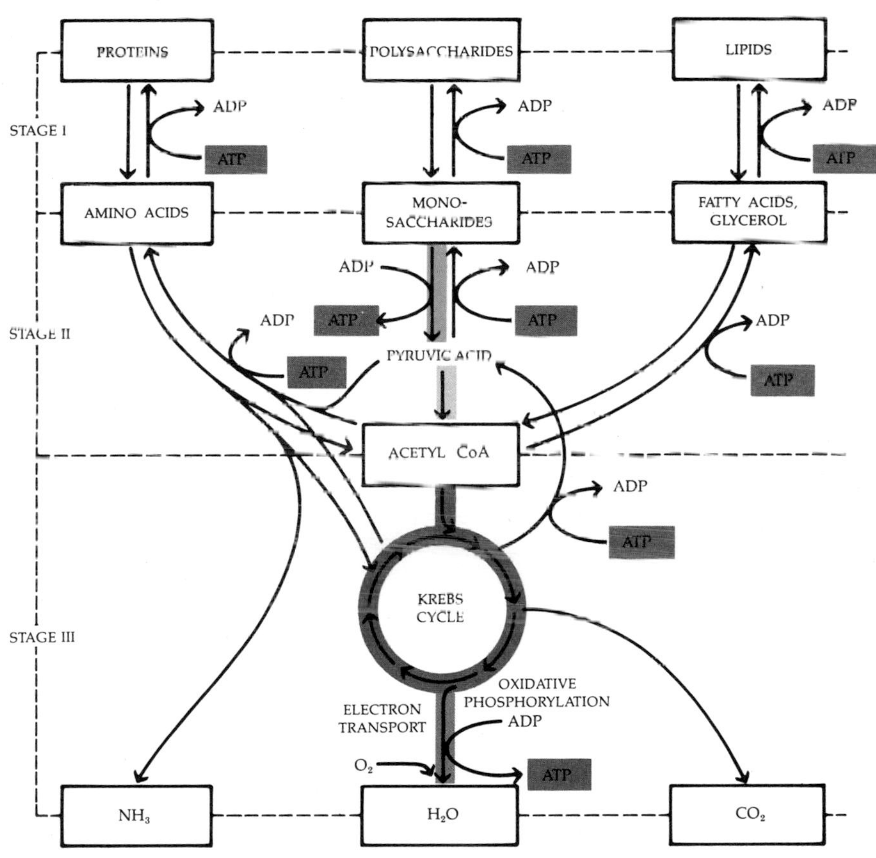

8–17
Major metabolic pathways in the cell.

SUMMARY

The oxidation of glucose is a chief source of energy in most cells. As the glucose is broken down in a series of small enzymatic steps, the energy in the molecule is packaged in the form of "high-energy" bonds in molecules of ATP.

The first phase in the breakdown of glucose is glycolysis, in which the six-carbon glucose molecule is split into two three-carbon molecules of pyruvic acid. A net yield of two molecules of ATP (from ADP) and two of NADH (from NAD^+) are formed in the process.

The second phase in the breakdown of glucose and other fuel molecules is respiration. It requires oxygen and, in eukaryotic cells, takes place in the mitochondria. It occurs in two stages: the Krebs cycle and the electron transport chain.

In the course of respiration, the three-carbon pyruvic acid molecules are broken down to two-carbon acetyl groups, which then enter the Krebs cycle. In a series of reactions in the Krebs cycle, the two-carbon acetyl group is oxidized completely to carbon dioxide. In the course of the oxidation of each acetyl group, four electron acceptors (three NAD^+ and one FAD) are reduced, and another molecule of ATP is formed.

The final stage of breakdown of the fuel molecule is the electron transport system, which involves a chain of electron carriers and enzymes embedded in the inner membranes of the mitochondrion. Along these chains of electron carriers, the high-energy electrons accepted by NADH and $FADH_2$ during glycolysis and the Krebs cycle pass downhill to oxygen. Each time a pair of electrons passes down the electron transport chain, ATP molecules are formed from ADP and phosphate. In the course of the breakdown of the glucose molecule, 36 molecules of ATP are formed. Of the free energy released through glucose oxidation, almost 40 percent of it is retained in the form of ATP.

Other food molecules, including fats, polysaccharides, and proteins, are utilized by being degraded to compounds that can slip into these central pathways at various steps. The biosynthesis of these substances also originates with precursor compounds derived from intermediates in the respiratory sequence and is driven by the energy derived from those processes.

QUESTIONS

1. Distinguish among the following: oxidation of glucose/glycolysis/respiration/fermentation; catabolism/anabolism; Krebs cycle/electron transport system; aerobic pathways/anaerobic pathways.

2. Describe the process of fermentation. What conditions are essential if fermentation is to occur? Why does fermentation stop when the alcohol concentration reaches 12 percent?

3. If oxygen-breathing organisms are so much more successful in converting energy than anaerobes, why are there any anaerobes left on this planet? Why haven't they all become extinct long ago?

4. Sketch the structure of a mitochondrion. Describe where the various stages in the breakdown of glucose take place in relation to mitochondrial structure. What molecules cross the mitochondrial membranes during these processes?

5. (a) As we have seen, a cell obtains 36 molecules of ATP from each molecule of glucose that is completely oxidized. Account for the production of each molecule of ATP. (b) In the course of glycolysis, the Krebs cycle, and the electron transport system, 40 molecules of ATP are actually formed. Why is the net yield for the cell only 36 molecules of ATP?

6. Cyanide can combine with—and deactivate—cytochrome *a* and cytochrome a_3. In our bodies, however, cyanide tends to react first with hemoglobin and to make it impossible for oxygen to bind to the hemoglobin. Either way, cyanide poisoning has the same effect: It inhibits the synthesis of ATP. Explain how this is so.

7. Describe how the processes of the cell are adapted to the efficient use of a variety of foodstuffs, and to the efficient production of the variety of materials that the cell needs to manufacture for its own use.

8. In terms of the cell's economy, what do anabolic processes provide for the cell? What do catabolic processes provide? How are they dependent on each other?

CHAPTER 9 Photosynthesis, Light, and Life

9–1
A cluster of cells of a green alga (Cosmarium botrytis) *that lives in fresh water. Each cell is an individual, self-sufficient photosynthetic organism. The green color is due to chlorophyll, in which the radiant energy of sunlight is converted to chemical energy.*

Over 300 years ago, the English physicist Sir Isaac Newton (1642–1727) separated visible light into a spectrum of colors by letting it pass through a prism (Figure 9–2). Then by passing the light through a second prism, he recombined the colors, producing white light once again. By this experiment, Newton showed that white light is actually made up of a number of different colors, ranging from violet at one end of the spectrum to red at the other. Their separation is possible because light of different colors is bent at different angles in passing through the prism. Newton believed that light was a stream of particles (or, as he termed them, "corpuscles"), in part, because of its tendency to travel in a straight line.

In the nineteenth century, through the genius of James Clerk Maxwell (1831–1879), it became known that what we experience as light is in truth a very small part of a vast continuous spectrum of radiation, the electromagnetic spectrum. As Maxwell showed, all the radiations included in this spectrum act as if they travel in waves. The wavelengths—that is, the distances from one peak to the next—range from those of gamma rays, which are measured in nanometers, to those of low-frequency radio waves, which are measured in kilometers. Within the spectrum of visible light, red light has the longest wavelength, violet the shortest. Another feature that these radiations have in common is that, in a vacuum, they all travel at the same speed—300,000 kilometers per second.

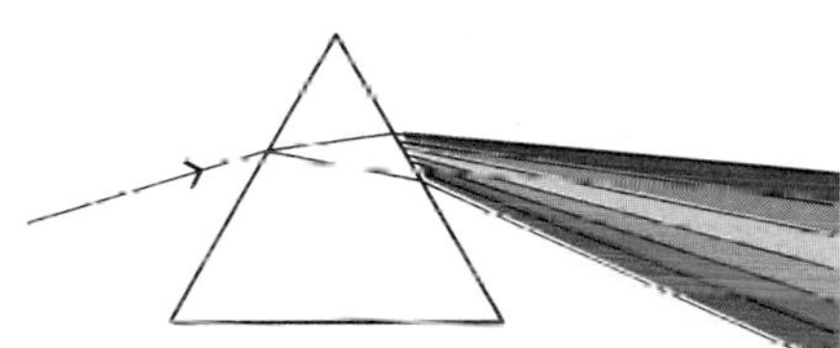

9–2
White light is actually a mixture of different colors, ranging from violet at one end of the spectrum to red at the other. It is separated into its component colors when it passes through a prism—"the celebrated phaenomena of colors," as Newton referred to it.

By 1900, it had become clear, however, that the wave model of light was not adequate. The key observation, a very simple one, was made in 1888: When a zinc plate is exposed to ultraviolet light, it acquires a positive charge. The metal, it was soon deduced, becomes positively charged because the light energy dislodges electrons, forcing them out of the metal atoms. Subsequently, it was discovered that this photoelectric effect, as it is called, can be produced in all metals. Every metal has a critical wavelength for the effect; the light (visible or invisible) must be of that wavelength or a shorter wavelength for the effect to occur.

With some metals, such as sodium, potassium, and selenium, the critical wavelength is within the spectrum of visible light, and as a consequence, visible light striking the metal can set up a moving stream of electrons (such a stream is an electric current). Burglar alarms, exposure meters, television cameras, and the electric eyes that open doors for you at supermarkets or airline terminals all operate on this principle of turning light energy into electrical energy.

THE NATURE OF LIGHT

Now here is a problem. The wave model of light would lead you to predict that the brighter the light—that is, the stronger the beam—the greater the force with which the electrons would be dislodged. But as we have already seen, whether or not light can eject the electrons of a particular metal depends not on the brightness of the light but on its wavelength. A very weak beam of the critical wavelength or a shorter wavelength is effective, while a stronger beam of a longer wavelength is not. Furthermore, as was shown in 1902, increasing the brightness of the light increases the number of electrons dislodged but not the velocity at which they are ejected from the metal. To increase the velocity, one must use a shorter wavelength of light. Nor is it necessary for energy to be accumulated in the metal. With even a dim beam of a critical wavelength, electrons may be emitted the instant the light hits the metal.

To explain such phenomena, the particle model of light was resurrected by Albert Einstein in 1905. According to this model, light is composed of particles of energy called *photons.* The energy of a photon is not the same for all kinds of light but is, in fact, inversely proportional to the wavelength—the longer the wavelength, the lower the energy. Photons of violet light, for example, have almost twice the energy of photons of red light, the longest visible wavelength.

The wave model of light permits physicists to describe certain aspects of its behavior mathematically, and the photon model permits another set of mathematical calculations and predictions. These two models are no longer regarded as opposed to one another; rather, they are complementary, in the sense that both—or a totally new model—are required for a complete description of the phenomenon we know as light.

The Fitness of Light

Light, as Maxwell showed, is only a tiny band in a continuous spectrum. From the physicist's point of view, the difference between radiations we can see and radia-

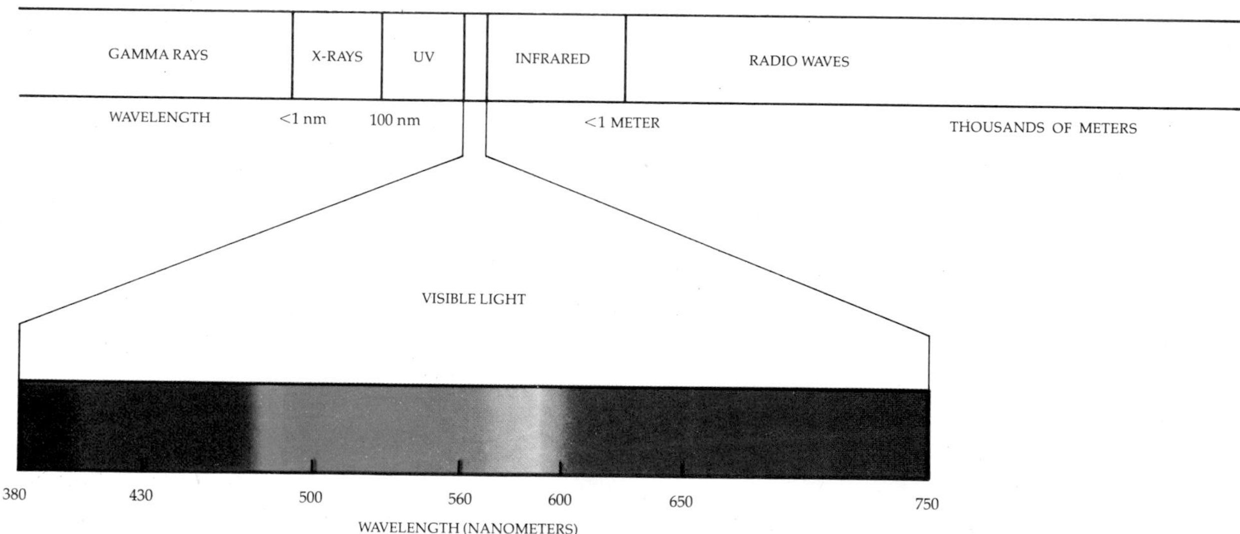

9–3
Visible light is only a small portion of the vast electromagnetic spectrum. For the human eye, the visible spectrum ranges from violet light, which is made up of comparatively short rays, to red light, the longest visible rays.

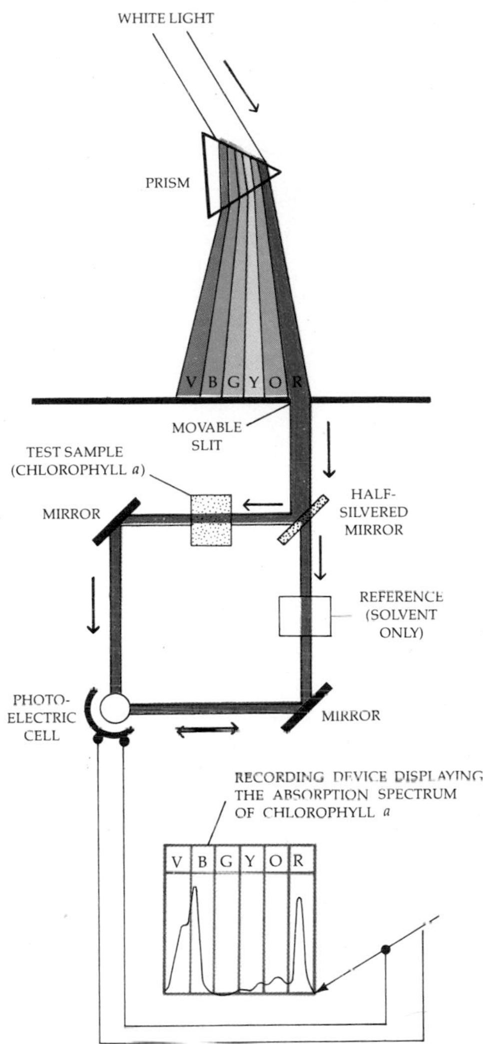

9–4
The absorption spectrum of a pigment is measured with a spectrophotometer. This device directs a beam of light of each wavelength at the object to be analyzed and records what percentage of light of each wavelength is absorbed by the pigment sample as compared to a reference sample. Because the mirror is lightly (half) silvered, half of the light is reflected and half is transmitted. The photoelectric cell is connected to an electronic device that automatically records the percentage absorption at each wavelength.

tions we cannot see—so dramatic to the human eye—is only a few nanometers of wavelength (Figure 9–3). Why does this particular small group of radiations, rather than some other, make the leaves grow and the flowers burst forth, cause the mating of fireflies and palolo worms, and when reflecting off the surface of the moon, excite the imagination of poets and lovers? Why is it that this tiny portion of the electromagnetic spectrum is responsible for vision, for the rhythmic, day-night regulation of many biological activities, for the bending of plants toward the light, and also for photosynthesis, on which all life depends? Is it an amazing coincidence that all these biological activities are dependent on these same wavelengths?

George Wald of Harvard, one of the greatest living experts on the subject of light and life, says no. He thinks that if life exists elsewhere in the universe, it is probably dependent on this same fragment of the vast spectrum. Wald bases this conjecture on two points. First, living things, as we have seen, are composed of large, complicated molecules held in special configurations and relationships to one another by hydrogen bonds and other weak bonds. Radiation of even slightly higher energies than the energy of violet light breaks these bonds and so disrupts the structure and function of the molecules. Radiations with wavelengths less than 200 nanometers drive electrons out of atoms. Light of wavelengths longer than those of the visible band is absorbed by water, which makes up the great bulk of all living things on earth. When this light reaches molecules, its lower energies cause them to increase their motion (increasing heat) but do not trigger changes in their electron configurations. Only those radiations within the range of visible light have the property of exciting molecules—that is, of moving electrons into higher-energy orbitals—and so of producing chemical and, ultimately, biological changes.

The second reason for the visible band of the electromagnetic spectrum being "chosen" by living things is that it, above all, is what is available. Most of the radiation reaching the earth from the sun is within this range. Higher-energy wavelengths are screened out by the oxygen and ozone high in the atmosphere. Much infrared radiation is screened out by water vapor and carbon dioxide before it reaches the earth's surface.

This is an example of what has been termed "the fitness of the environment"; the suitability of the environment for life and that of life for the physical world are exquisitely interrelated. If they were not, life could not, of course, exist.

CHLOROPHYLL AND OTHER PIGMENTS

In order for light energy to be used by living systems, it must first be absorbed. A pigment is any substance that absorbs light. Some pigments absorb all wavelengths of light and so appear black. Some absorb only certain wavelengths, transmitting or reflecting the wavelengths they do not absorb. Chlorophyll, the pigment that makes leaves green, absorbs light in the violet and blue wavelengths and also in the red; because it reflects green light, it appears green. Different pigments absorb light energy at different wavelengths. The absorption pattern of a pigment is known as the *absorption spectrum* of that substance (Figure 9–4).

Different groups of plants and algae use various pigments in photosynthesis. There are several different kinds of chlorophyll that vary slightly in their molecular structure. In plants, chlorophyll *a* is the pigment directly involved in the transformation of light energy to chemical energy. Most photosynthetic cells also contain a second type of chlorophyll—in plants, it is chlorophyll *b*—and a representative of

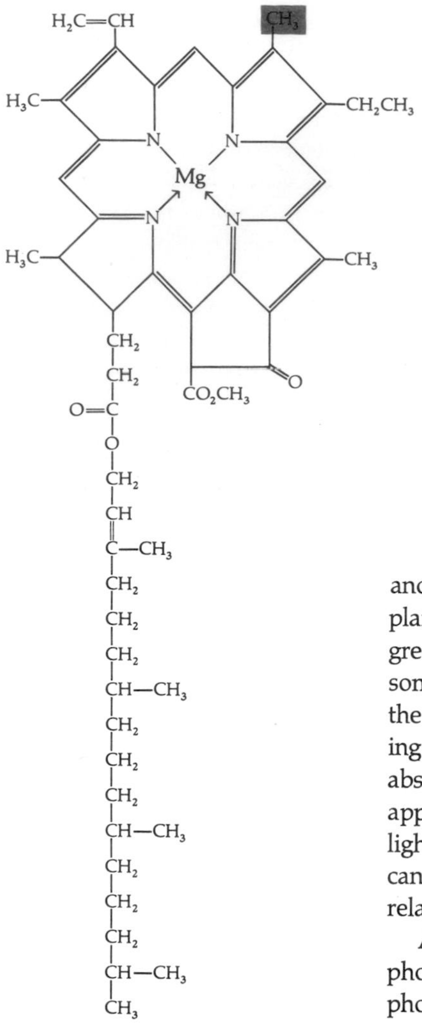

9-5
(a) *Chlorophyll* a *is a large molecule with a central atom of magnesium held in a porphyrin ring. Attached to the ring is a long, hydrophobic carbon-hydrogen chain that may help to anchor the molecule in the internal membranes of the chloroplast. Chlorophyll* b *differs from chlorophyll* a *in having an aldehyde (CHO) group in place of the* CH_3 *group indicated by color. Alternating single and double bonds, such as those in the chlorophylls, are common in pigments.* (b) *The estimated absorption spectra of chlorophyll* a *and chlorophyll* b. *(Prepared by Govindjee.)*

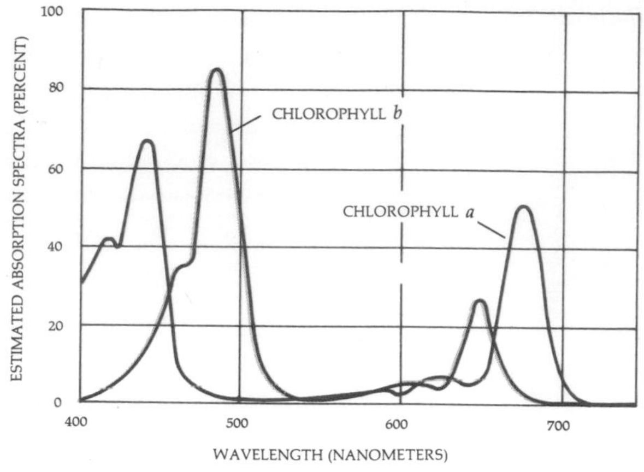

another group of pigments called the carotenoids. One of the carotenoids found in plants is beta-carotene. The carotenoids are red, orange, or yellow pigments. In the green leaf, their color is masked by the chlorophylls, which are more abundant. In some tissues, however, such as those of a ripe tomato or the root of a carrot plant, the carotenoid colors predominate, as they do also when leaf cells stop synthesizing chlorophyll in the fall. The other chlorophylls and the carotenoids are able to absorb light at wavelengths different from those absorbed by chlorophyll *a* and apparently can pass the energy on to chlorophyll *a*, thus extending the range of light available for photosynthesis (Figure 9-6). Whether or not a particular pigment can cause a chemical reaction depends not only on its structure but also on its relationship with neighboring molecules.

An *action spectrum* defines the relative effectiveness (per number of incident photons) of different wavelengths of light for light-requiring processes, such as photosynthesis, flowering, and phototropism (the bending of a plant toward light). Similarity between the absorption spectrum of a pigment and the action spectrum of a process is considered evidence that that particular pigment is responsible for that particular process (Figure 9-7).

When pigments absorb light, electrons are boosted to a higher energy level. Three of the possible consequences are: (1) The energy may be dissipated as heat; (2) it may be re-emitted immediately as light energy of a longer wavelength, a phenomenon known as fluorescence; (3) the energy may cause a chemical reaction, as happens in photosynthesis.

If chlorophyll molecules are isolated in a test tube and light is permitted to strike them, they fluoresce. In other words, the molecules absorb light energy, and the electrons are momentarily raised to a higher energy level and then fall back again to a lower one. As they fall to a lower energy level, they release much of this energy as light. None of the light absorbed by isolated chlorophyll molecules is converted to any form of energy useful to living systems. Chlorophyll can convert light energy to chemical energy only when it is associated with certain proteins and embedded in a specialized membrane.

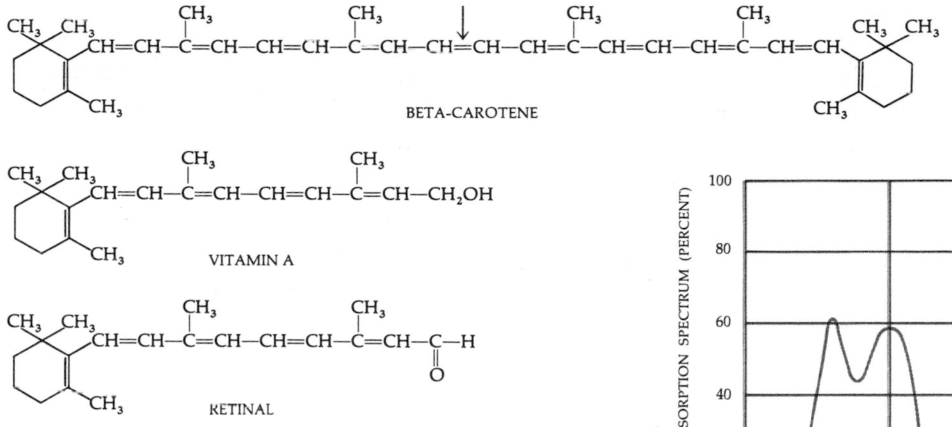

$CH_3 \quad CH_3 \qquad CH_3 \qquad CH_3 \downarrow \qquad CH_3 \qquad CH_3 \qquad CH_3 \quad CH_3$

$CH{=}CH{-}C{=}CH{-}CH{=}CH{-}C{=}CH{-}CH{=}CH{-}CH{=}C{-}CH{=}CH{-}CH{=}C{-}CH{=}CH$

$CH_3 \qquad CH_3$

BETA-CAROTENE

$CH_3 \quad CH_3 \qquad CH_3 \qquad CH_3$

$CH{=}CH{-}C{=}CH{-}CH{=}CH{-}C{=}CH{-}CH_2OH$

CH_3

VITAMIN A

$CH_3 \quad CH_3 \qquad CH_3 \qquad CH_3$

$CH{=}CH{-}C{=}CH{-}CH{=}CH{-}C{=}CH{-}C{-}H$ (with $C{=}O$)

CH_3

RETINAL

(a)

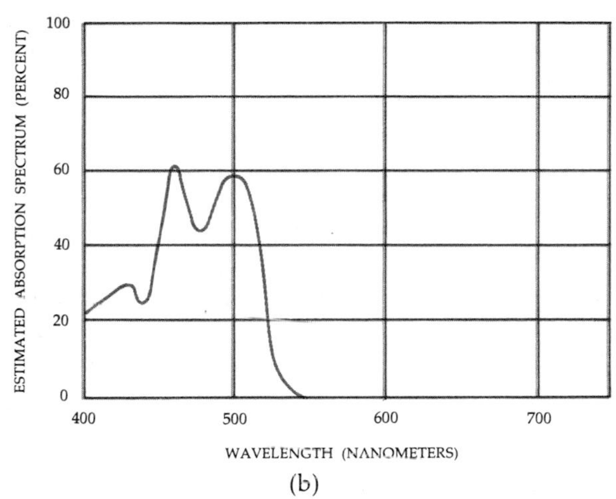

(b)

9–6
(a) Related carotenoids. Cleavage of the beta-carotene molecule at the point indicated by the arrow yields two molecules of vitamin A. Oxidation of vitamin A yields retinal, the pigment involved in vision. Absorption of light energy changes the electron configuration of retinal and triggers a nerve impulse in the retina. All of this explains why you were told to eat your carrots. (b) *The estimated absorption spectrum of carotenoids in the chloroplast. (Prepared by Govindjee.)*

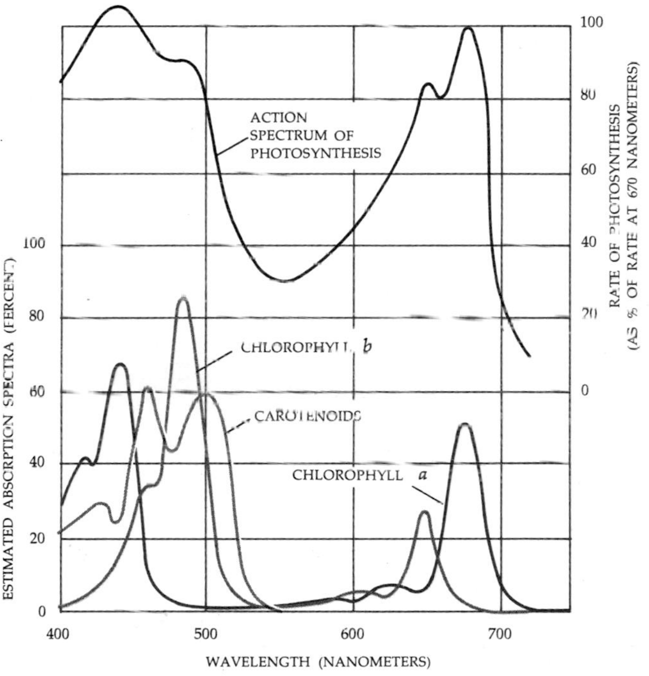

9–7
The upper curve shows the action spectrum for photosynthesis and the lower curves, absorption spectra for chlorophyll a, *chlorophyll* b, *and carotenoids in the chloroplast. Note that the action spectrum of photosynthesis indicates that chlorophyll* a, *chlorophyll* b, *and carotenoids all absorb light used in photosynthesis. (Prepared by Govindjee.)*

9–8
Results of an experiment performed in 1882 by T. W. Englemann revealing the action spectrum of photosynthesis in a filamentous alga. Like more recent investigators, Englemann used the rate of oxygen production to measure the rate of photosynthesis. Unlike his successors, however, he lacked sensitive devices for detecting oxygen. As his oxygen indicator, he chose bacteria that are attracted by oxygen. In place of the mirror and diaphragm usually used to illuminate objects under view in his microscope, he substituted a "microspectral apparatus" which, as its name implies, produced a tiny spectrum of colors that it projected upon the slide under the microscope. Then he arranged a filament of algal cells parallel to the spread of the spectrum. The oxygen-seeking bacteria congregated mostly in the areas where the violet and red wavelengths fell upon the algal filament. As you can see, the action spectrum for photosynthesis Englemann revealed in this elegant experiment paralleled the absorption spectrum of chlorophyll. He therefore concluded that photosynthesis depends on the light absorbed by chlorophyll.

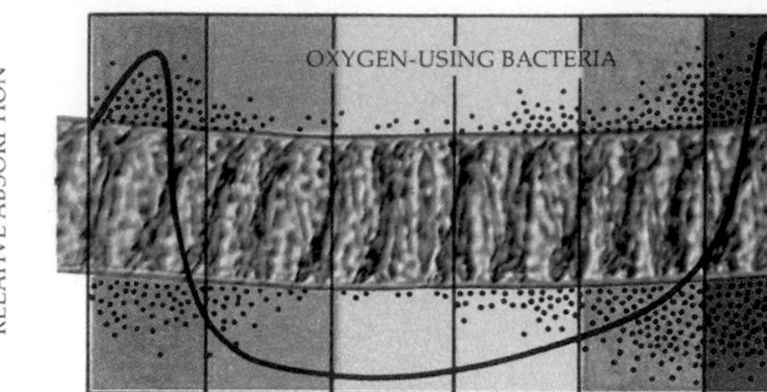

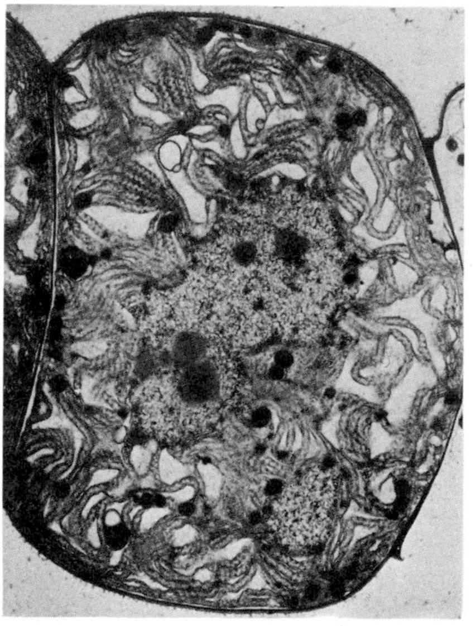

9–9
A photosynthetic blue-green alga. Photosynthesis takes place in the complex system of membranes that almost fills the cell.

PHOTOSYNTHETIC MEMBRANES: THE THYLAKOID

The structural unit of photosynthesis is the thylakoid, which usually takes the form of a flattened sac or vesicle. In the photosynthetic prokaryotes, thylakoids may form a part of the cell membrane, or they may occur singly in the cytoplasm, or, as in the blue-green algae, they may be part of an elaborate internal membrane structure (Figure 9–9).

Chloroplasts

In eukaryotes, the thylakoids form a part of the internal membrane structure of specialized organelles, the chloroplasts (Figure 9–10). The alga *Chlamydomonas,* for instance, has a single very large chloroplast; the cell of a leaf characteristically has 40 to 50 chloroplasts, and there are often 500,000 chloroplasts per square millimeter of leaf surface.

Chloroplasts, like mitochondria, are surrounded by two outer membranes. The interior of the chloroplast is filled with a dense solution, the stroma, which (like the matrix of the mitochondrion) is different in composition from the material surrounding the organelles in the cytoplasm. With the light microscope under high power, it is possible to see little spots of green within the chloroplasts of leaves. The early microscopists called these green specks grana ("grains"), and this term is still in use.

Under the electron microscope, it can be seen that the grana are stacks of thylakoids. Some of the thylakoid membranes have extensions that interconnect the grana through the stroma that separates them.

All the thylakoids in a chloroplast are oriented parallel to each other. Thus the whole chloroplast can swing toward the light, simultaneously aiming all of the millions of pigment molecules for optimum reception, as if they were a giant electromagnetic antenna (which, of course, they are).

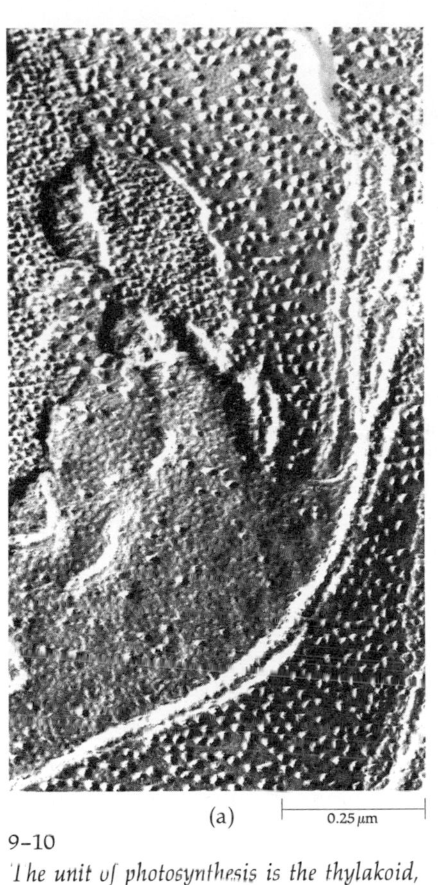

(a) 0.25 μm

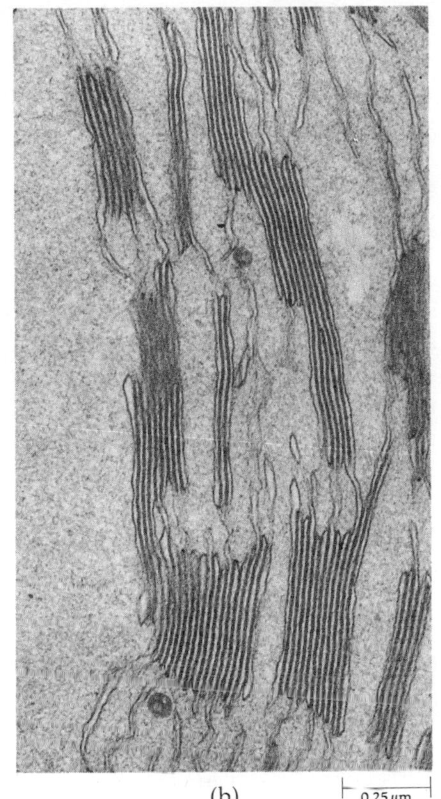

(b) 0.25 μm

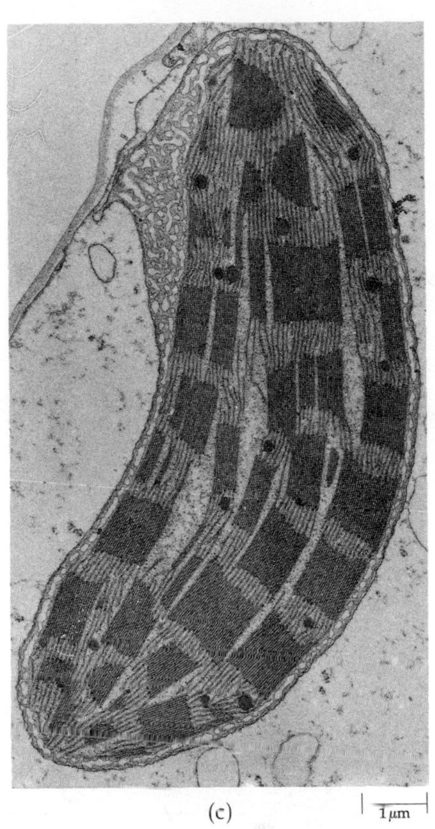

(c) 1 μm

9–10
The unit of photosynthesis is the thylakoid, a flattened sac, whose membranes contain chlorophyll and other pigments. In plants and algae, thylakoids are part of an elaborate membrane system enclosed in a special organelle, the chloroplast. (a) *Internal membrane surface of a thylakoid prepared by freeze-fracture technique (see page 132). The particles are believed to be enzymes involved in the light reactions.* (b) *A stack of thylakoids from a plant cell.* (c) *A chloroplast, showing the elaborate system of internal membranes comprising interconnected stacks of thylakoids.* (d) *A photosynthetic cell with eight chloroplasts visible. The center of the cell is filled with a large vacuole.*

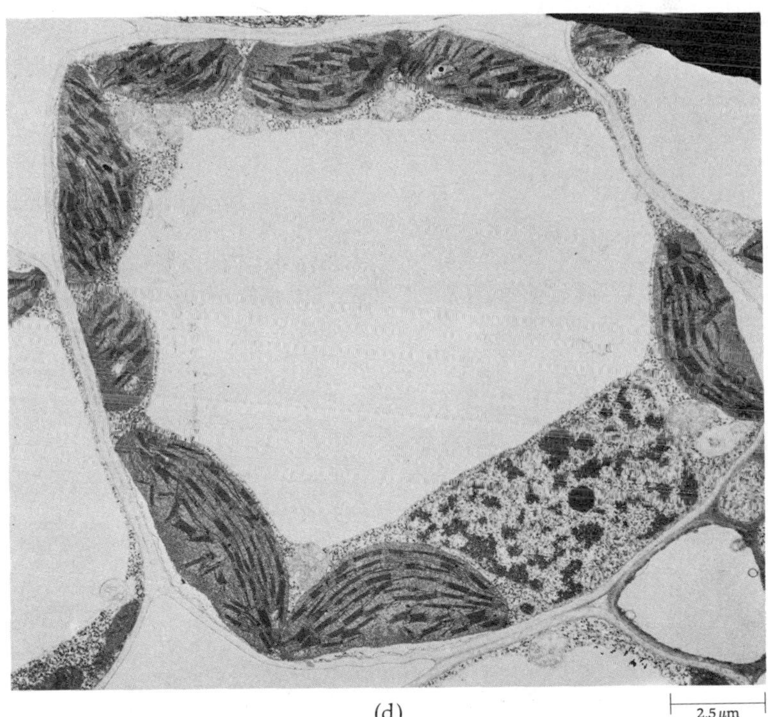

(d) 2.5 μm

THE LIGHT AND THE DARK REACTIONS

As we noted in Chapter 7, it was discovered about 200 years ago that light is required for the process we now call photosynthesis. It is now known that photosynthesis actually takes place in two stages, only one of which requires light. Evidence for this two-stage mechanism was first presented in 1905 by the English plant physiologist F. F. Blackman, as the result of experiments in which he measured the rate of photosynthesis under varying conditions.

Blackman first plotted the rate of photosynthesis at various light intensities. In dim to moderate light, increasing the light intensity increased the rate of photosynthesis, but at higher intensities, a further increase in light intensity had no effect. He then studied the combined effects of light and temperature on photosynthesis. In dim light, an increase in temperature had no effect. However, Blackman found that if he increased the light and also increased the temperature, the rate of photosynthesis was greatly accelerated. As the temperature increased above 30°C, the rate of photosynthesis slowed and finally the process ceased.

On the basis of these experiments, Blackman concluded that more than one set of reactions was involved in photosynthesis. First, there was a group of light-dependent reactions that were temperature-independent. The rate of these reactions could be accelerated in the dim-to-moderate light range by increasing the amount of light, but it was not accelerated by increases in temperature. Second, there was a group of reactions that were dependent not on light but rather on

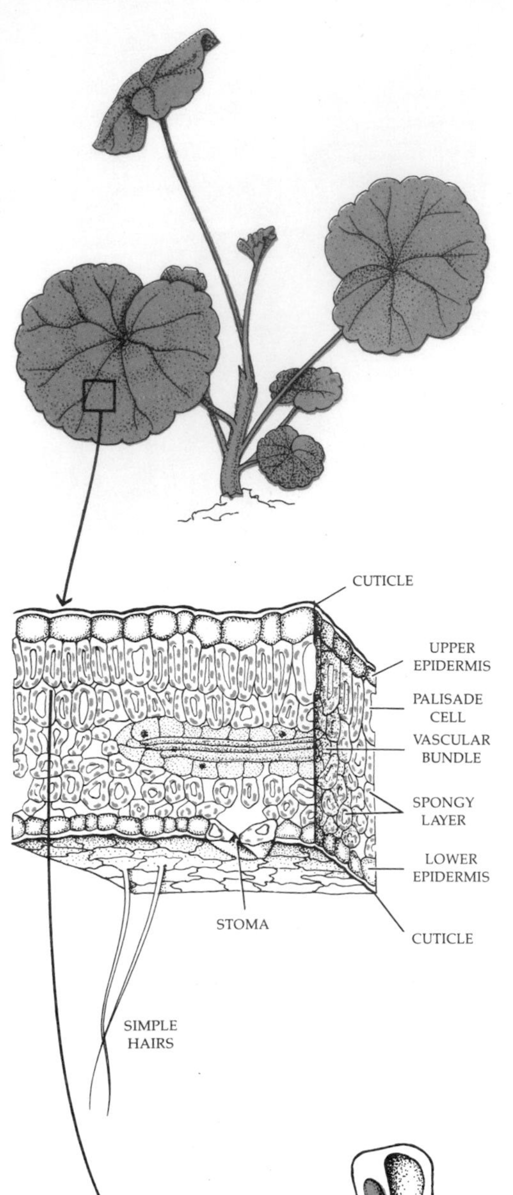

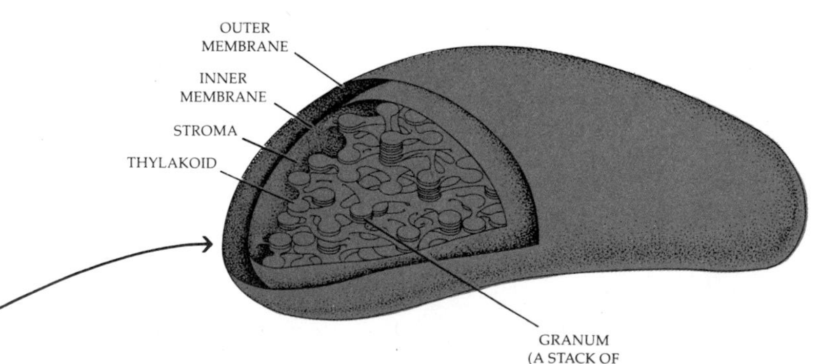

9–11

Journey into a chloroplast. The plant shown is a geranium, which you may recognize by the characteristic shape of its leaves. The inner tissues of the leaf are completely enclosed by transparent epidermal cells that are coated with a waxy layer, the cuticle. Oxygen, carbon dioxide, and other gases enter the leaf largely through special openings, the stomata (singular, stoma). Gases and water vapor fill the spaces between cells in the spongy layer, leaving and entering cells by diffusion. Water, taken up by the roots, enters the leaf by way of the vascular bundle, and sugars, the products of photosynthesis, leave the leaf by this route, traveling to nonphotosynthetic parts of the plant. Much of the photosynthesis takes place in the palisade cells, elongated cells directly beneath the upper epidermis. They have a large central vacuole and numerous chloroplasts that move within the cell, orienting themselves with respect to the light. Light is captured in the membranes of the disk-shaped thylakoids within the chloroplast.

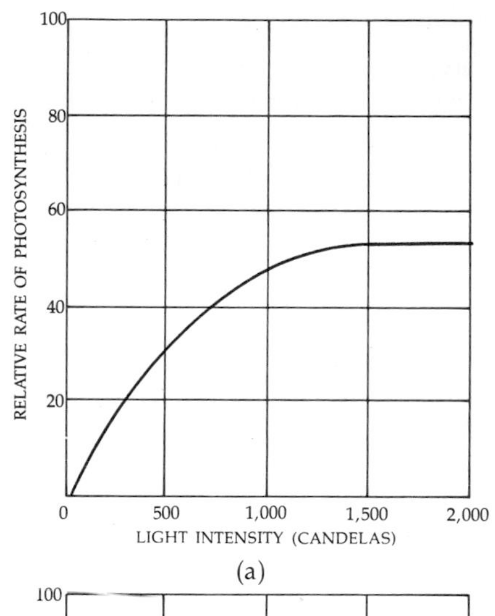

(a)

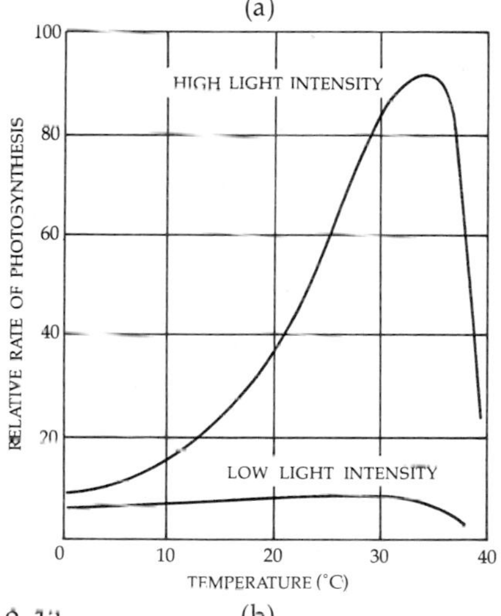

(b)

9–12
(a) *An increase in light intensity does not produce a corresponding increase in the rate of photosynthesis beyond about 1,200 candelas. A curve such as the one shown here indicates that some other factor—known as a rate-limiting factor—is involved in the process under study. Under field conditions* CO_2 *concentration is commonly the rate-limiting factor.* (b) *At a low intensity of light, an increase in temperature does not increase the rate of photosynthesis. At a high intensity, however, an increase in temperature has a very marked effect. Based on these data, Blackman concluded that photosynthesis included both light-dependent and light-independent reactions.*

temperature. Both sets of reactions seemed to be required for the process of photosynthesis. Increasing the rate of only one set of reactions increased the rate of the entire process only to the point at which the second set of reactions began to hold back the first (that is, it became rate-limiting). Then it was necessary to increase the rate of the second set of reactions in order for the first to proceed unimpeded.

Photosynthesis was thus shown to have both a light-dependent stage, the so-called "light reactions," and a light-independent stage, the "dark reactions." It is important to keep in mind that the dark reactions do not necessarily take place in the dark. They simply do not require light as such. (However, they do require the products of the light reactions.)

The "dark reactions" increased in rate as the temperature was increased, but only up to about 30°C, after which the rate began to decrease. From this evidence it was concluded that these reactions were controlled by enzymes, since this is the way enzymes are expected to respond to temperature (Figure 4–10). This conclusion has since proved to be correct.

In the first stage of photosynthesis—the light reactions—light energy is used to form ATP from ADP and to reduce electron carrier molecules. In the second stage of photosynthesis—the dark reactions—the energy products of the first stage are used to reduce carbon from carbon dioxide to a simple sugar, thus converting the chemical energy of the carrier molecules to forms suitable for transport and storage and, at the same time, forming a carbon skeleton on which other organic molecules can be built. This binding of CO_2 into organic compounds is known as the *fixation of carbon.*

THE LIGHT REACTIONS

In the thylakoids, chlorophyll and other molecules are, according to the present model, packed into photosynthetic units. Each unit contains from 250 to 400 molecules of pigment, which serve as light-trapping antennae. Once a quantum of energy is absorbed by one of the antenna pigments, it is bounced around (like a hot potato) until it reaches a special form of chlorophyll *a*, which is the reaction center. When this particular chlorophyll molecule absorbs the energy, an electron is boosted to a higher energy level from which it is transferred to another molecule, an electron acceptor. The chlorophyll molecule is thus oxidized (minus an electron) and positively charged.

The Photosystems

According to present evidence, there are two different kinds of photosynthetic units, each forming part of a different photosystem. In Photosystem I, the reactive chlorophyll *a* molecule is known as P_{700} because one of the peaks of its absorption spectrum is at 700 nanometers, a slightly longer wavelength than the usual chlorophyll *a* peak. When P_{700} (P is for pigment) is oxidized, it bleaches, which is how it was detected. No one has managed to isolate pure P_{700}. Recent evidence indicates that P_{700} is not an unusual kind of chlorophyll but rather a dimer ("two-part") of two chlorophyll *a* molecules that has unusual properties because of its association with special proteins in the membrane and its position in relation to other molecules. Photosystem II also contains a specialized chlorophyll *a* molecule, which passes its electron on to a different electron acceptor. The reactive chlorophyll *a* molecule of Photosystem II is P_{680}.

Model of the Light Reactions

The two photochemical systems probably evolved separately, with Photosystem I coming first. As we shall see, Photosystem I can operate independently. In general, however, the two systems work together simultaneously and continuously, as shown in Figure 9–13. According to this model, light energy enters Photosystem II where it is trapped by the reactive chlorophyll molecule P_{680}. An electron is boosted to a higher energy level from which it is transferred to an electron acceptor molecule. The electrons then pass downhill along an electron transport chain to Photosystem I. As the electrons pass along this transport chain, ATP is formed from ADP, probably via the same mechanism by which ATP is formed along the electron transport chain of the mitochondrion. This process is known as *photophosphorylation.*

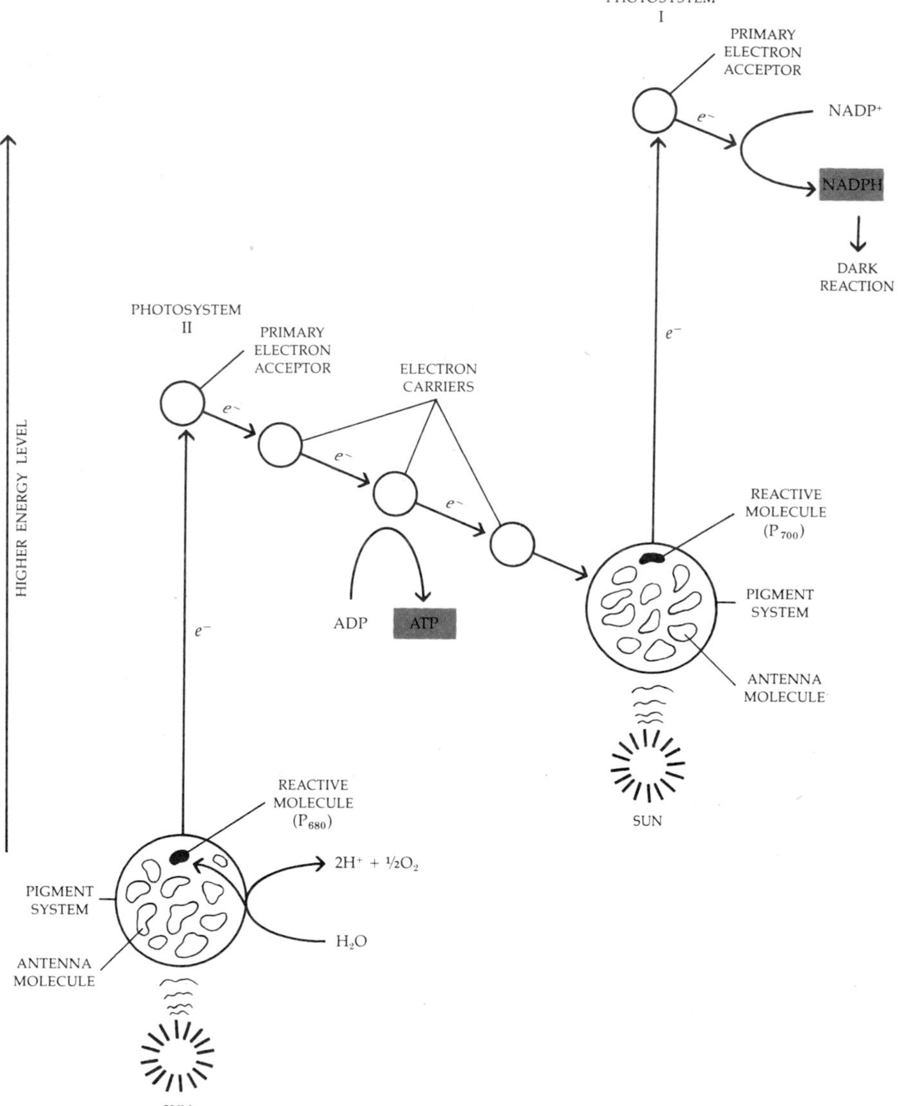

9–13

Light energy trapped in the reactive chlorophyll a molecule of Photosystem II boosts electrons to a higher energy level. These electrons are replaced by electrons pulled away from water molecules, releasing protons and oxygen gas. The electrons are passed from the electron acceptor along an electron transport chain to a lower energy level, the reaction center of Photosystem I. As they pass along this electron transport chain, some of their energy is packaged in the form of ATP. Light energy absorbed by Photosystem I boosts the electrons to another primary electron acceptor. From this acceptor, they are passed via other electron carriers to NADP+ to form NADPH. The electrons removed from Photosystem I are replaced by those from Photosystem II. ATP and NADPH represent the net gain from the light reactions.

9-14
Although NAD (Figure 4-9) and NADP resemble one another very closely, their biological roles are distinctly different. NADH transfers its electrons to other electron carriers, which continue to pass them on down to successively lower electron levels in discrete steps. In the course of this electron transfer, ATP molecules are formed. NADPH provides energy directly to biosynthetic processes of the cell that require large energy inputs.

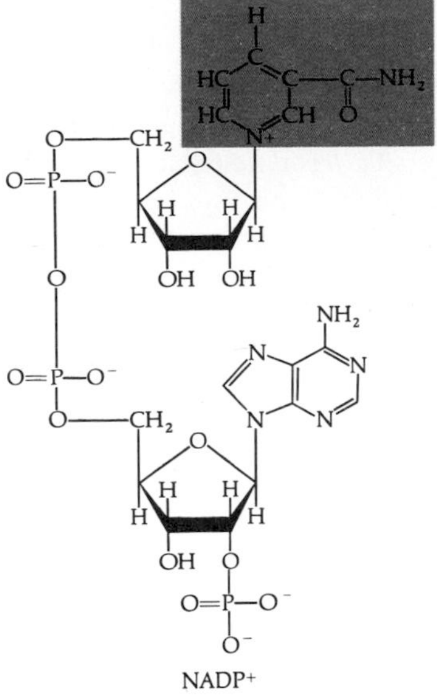

PHOTOSYSTEM I
ELECTRON ACCEPTOR
ADP
ATP
REACTIVE MOLECULE (P_{700})
PIGMENT SYSTEM
SUN

9-15
Cyclic electron flow bypasses Photosystem II and requires only Photosystem I. ATP is produced from ADP but oxygen is not released and NADP+ is not reduced.

Other events are taking place simultaneously:

1. The P_{680} chlorophyll molecule, as a result of having lost its electron, is avidly seeking a replacement. It finds it in the water molecule, which is thus dissociated into protons and oxygen gas.
2. Light energy is trapped in the reactive chlorophyll molecule (P_{700}) of Photosystem I. The molecule is oxidized, and an electron is passed to a primary electron acceptor from which it goes downhill to $NADP^+$ (Figure 9-14).
3. The electron removed from the P_{700} molecule is replaced by the electron from Photosystem II.

Thus in the light there is a continuous flow of electrons from water to Photosystem II to Photosystem I to $NADP^+$.

The energy harvest from these steps is represented by an ATP molecule (whose formation releases a water molecule) and NADPH, which then becomes the chief source of energy for the reduction of carbon. To generate one molecule of NADPH, four photons must be absorbed, two by Photosystem II and two by Photosystem I.

In the words of Nobel laureate Albert Szent-Györgyi: "What drives life is . . . a little electric current, kept up by the sunshine."

Cyclic Electron Flow

As we mentioned previously, there is also some evidence that Photosystem I can work independently. When this occurs, no NADPH is formed. In this process, called *cyclic electron flow*, electrons are boosted from P_{700} to an electron acceptor and from there pass downhill through a series of intermediates, including cytochromes, back into the reactive molecule. ATP is produced in the course of this passage. It is believed that the most primitive photosynthetic mechanisms did, indeed, work in this way, and this is apparently the way in which some prokaryotes carry out photosynthesis. Also, eukaryotic cells are able to synthesize ATP by cyclic electron flow in the absence of $NADP^+$. However, O_2 is produced and carbon is reduced only if both systems are in operation.

9–16
The chemiosmotic model of photophosphorylation. According to this model, the energy of light creates a proton gradient across the thylakoid membrane. (In the chloroplast, however, in contrast to the mitochondrion, the protons are pumped inward.) As the protons flow outward, down the gradient, ADP is phosphorylated to ATP.

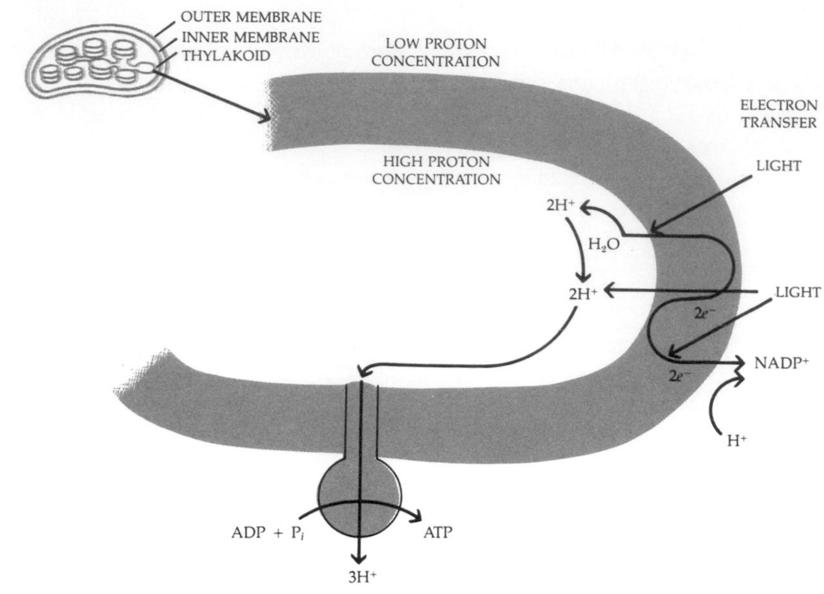

9–17
Scanning electron micrograph of open stomata on the lower surface of a leaf. The carbon dioxide used in photosynthesis reaches the photosynthetic cells through these openings.

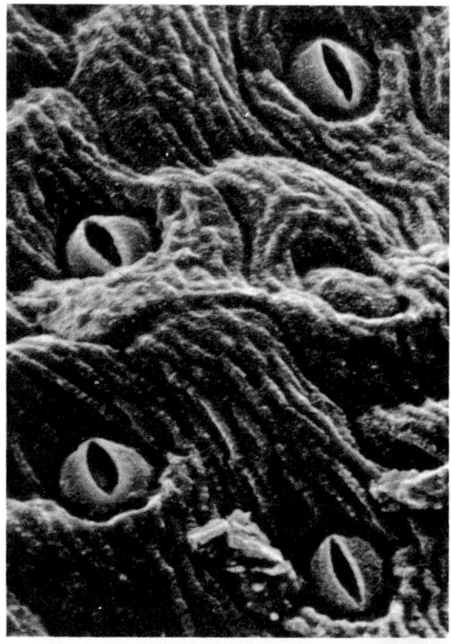

Photosynthetic Phosphorylation

We noted in the previous chapter that there are three current hypotheses for the mechanism of oxidative phosphorylation. These same three hypotheses also apply to photophosphorylation, and in the case of this latter process, there is much evidence supporting the chemiosmotic model. When isolated chloroplasts are illuminated, they absorb protons from the medium in which they are suspended; when the light is turned off and electron flow stops, the protons slowly return from the chloroplasts to the medium. Also, and most impressive, is the demonstration that an artificially produced proton gradient across the inner membranes of a chloroplast can drive the phosphorylation of ATP in the dark. The chemiosmotic model of photophosphorylation is shown in Figure 9–16.

Summary of the Light Reactions

The reactions that we have just described are the "light reactions" of photosynthesis. In the course of these reactions, as we saw, light energy is converted to electrical energy—the flow of electrons—and the electrical energy is converted to chemical energy stored in the bonds of NADPH and ATP.

THE DARK REACTIONS

In the second stage of photosynthesis, the energy generated by the light reactions is used to reduce carbon. Carbon is available to photosynthetic cells in the form of carbon dioxide. In algae, such as those seen in Figure 9–1, the carbon dioxide is dissolved in the surrounding water. In plants, carbon dioxide reaches the photosynthetic cells through specialized openings in leaves and green stems, called stomata (Figure 9–17).

VAN NIEL'S HYPOTHESIS

For more than 100 years, it was generally assumed that in the equation

$$CO_2 + H_2O + light \longrightarrow (CH_2O) + O_2$$

the carbohydrate (CH_2O) resulted from the combination of the carbon and water molecules and that the oxygen was released from the carbon dioxide molecule. This entirely reasonable hypothesis was widely accepted. But, as it turned out, it was wrong.

The investigator who upset this long-held theory was C. B. van Niel of Stanford University. Van Niel, then a graduate student, was investigating photosynthesis in different types of photosynthetic bacteria. In their photosynthetic reactions, bacteria reduce carbon to carbohydrates, but they do not release oxygen. Among the types of bacteria van Niel was studying were the purple sulfur bacteria, which require hydrogen sulfide for photosynthesis. In the course of photosynthesis, globules of sulfur (S) are excreted or accumulated inside the bacterial cells. In these bacteria, van Niel found that this reaction takes place during photosynthesis:

$$CO_2 + 2H_2S \xrightarrow{light} (CH_2O) + H_2O + 2S$$

This finding was simple enough and did not attract much attention until van Niel made a bold extrapolation. He proposed that the generalized equation for photosynthesis is

$$CO_2 + 2H_2A \xrightarrow{light} (CH_2O) + H_2O + 2A$$

In this equation, H_2A stands for some oxidizable substance such as hydrogen sulfide, free hydrogen, or any one of several other compounds used by photosynthetic bacteria—or water. In the photosynthetic algae and higher green plants, H_2A is water. In short, van Niel proposed that it was the water that was the source of oxygen in photosynthesis, not *the carbon dioxide.*

This brilliant speculation, first proposed in the early 1930s, was not proved until many years later. Eventually, investigators, using a heavy isotope of oxygen ($^{18}O_2$), traced the oxygen from water to oxygen gas:

$$CO_2 + 2H_2{}^{18}O \xrightarrow{light} (CH_2O) + H_2O + {}^{18}O_2$$

and confirmed van Niel's hypothesis. The overall concept of photosynthesis has remained unchanged from the time of van Niel's proposal. However, many of its details have subsequently been worked out, and more are still under active investigation.

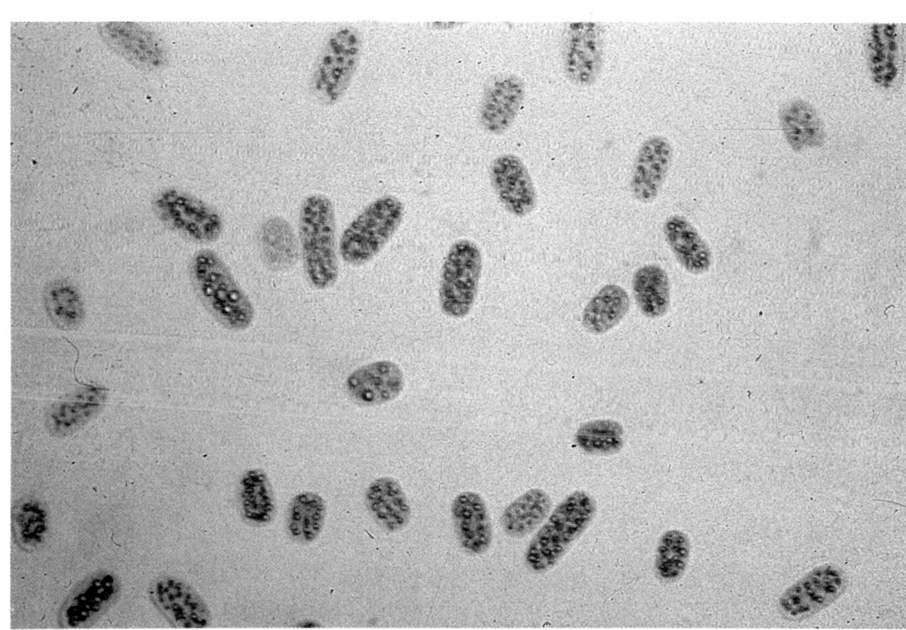

Purple sulfur bacteria. In these cells, hydrogen sulfide plays the same role as water does in the photosynthetic process of plants. The hydrogen sulfide (H_2S) is split, and the sulfur that is released accumulates as globules, visible within the cells.

PHOTOSYNTHESIS WITHOUT CHLOROPHYLL

Halobacteria are rod-shaped cells, quite similar in appearance to Escherichia coli. *They grow best in very salty water, about seven times as salty as sea water. If the salt concentration is reduced to only about three times that of sea water, the cell wall falls apart, and as it is reduced still further, the cell membrane begins to break up. Walther Stoeckenius, then at Rockefeller University, separated the membrane fragments by centrifugation. One of the fractions, it turned out, was purple, and this, though the investigators did not know it at the time, was the first clue to the energy source of the salt-loving bacteria. It has now been shown that these bacteria are photosynthetic and that their photosynthetic pigment is not a form of chlorophyll, as in all other photosynthetic organisms, but retinal, the visual pigment of the vertebrate eye (page 743). The membrane of the halobacteria contains molecules of retinal plus protein—the complex is called bacteriorhodopsin. When retinal is excited by light, a proton gradient is established across the membrane, and it is this proton gradient that drives the phosphorylation of ATP, thus providing additional support that the chemiosmotic mechanism is a universal one for the regeneration of ATP.*

Darwin himself confessed a certain uneasiness when called upon to explain how an organ as complex as the eye might have arisen by the slow, cumulative steps of evolution. Was retinal "invented" twice? Or do these purple fragments of membrane hold clues both to the mechanisms of human vision and also to its origins?

A freeze-fracture preparation of the membrane of a halobacterium. The fine-grained regions with a hexagonal pattern are patches of purple membrane.

The Calvin Cycle: The Three-Carbon Pathway

The reduction of carbon takes place in the stroma in a cycle called the Calvin cycle (named after its discoverer, Melvin Calvin). The Calvin cycle is analogous to the Krebs cycle (page 162) in that, in each turn of the cycle, the starting compound is again regenerated. The starting (and ending) compound is a five-carbon sugar with two phosphates attached, ribulose diphosphate (RuDP). The cycle begins when carbon dioxide enters the cycle and is bound to RuDP, which then splits to form two molecules of phosphoglycerate, or PGA (Figure 9–18). (Each PGA molecule contains three carbon atoms, hence the name, the three-carbon pathway.)

The enzyme catalyzing these crucial reactions (RuDP carboxylase) is very abundant in chloroplasts, making up more than 15 percent of the total chloroplast protein. (It is said to be the most abundant protein in the world.) This enzyme is located on the surface of the thylakoid membranes.

The complete cycle is diagrammed in Figure 9–19. As in the Krebs cycle, each step is regulated by a specific enzyme. At each full turn of the cycle, a molecule of carbon dioxide enters the cycle, is reduced, and a molecule of RuDP is regenerated.

9-18

Calvin and his collaborators briefly exposed photosynthesizing algae to radioactive carbon dioxide ($^{14}CO_2$). The radioactive carbon atom after 5 seconds of illumination was found almost entirely in the three-carbon compound phosphoglycerate. (The transient intermediate is inferred; it has not been isolated.) After 60 seconds, many compounds contained radioactive carbon.

$O{=}C{=}O$ + $CH_2OPO_3^{2-}$–$C{=}O$–H–C–OH–H–C–OH–$CH_2OPO_3^{2-}$

RIBULOSE 1,5-DIPHOSPHATE

↓

^{-}O–$C({=}O)$–$C(CH_2OPO_3^{2-})(OH)$–$C{=}O$–H–C–OH–$CH_2OPO_3^{2-}$

TRANSIENT INTERMEDIATE

↓ H_2O

$CH_2OPO_3^{2-}$–H–C–OH–COO^-

+

COO^-–H–C–OH–$CH_2OPO_3^{2-}$

TWO MOLECULES OF 3-PHOSPHOGLYCERATE

9-19

Summary of the Calvin cycle. At each full "turn" of the cycle, one molecule of carbon dioxide enters the cycle. Three turns are summarized here—the number required to make one molecule of glyceraldehyde phosphate. Three molecules of ribulose diphosphate (RuDP), a five-carbon compound, are combined with three molecules of carbon dioxide, yielding six molecules of phosphoglycerate, a three-carbon compound. These are converted to six molecules of glyceraldehyde phosphate. Five of these three-carbon molecules are combined and rearranged to form three five-carbon molecules of RuDP. The "extra" molecule of glyceraldehyde phosphate represents the net gain from the Calvin cycle. The energy that "drives" the Calvin cycle is in the form of ATP and NADPH, produced by the light reactions.

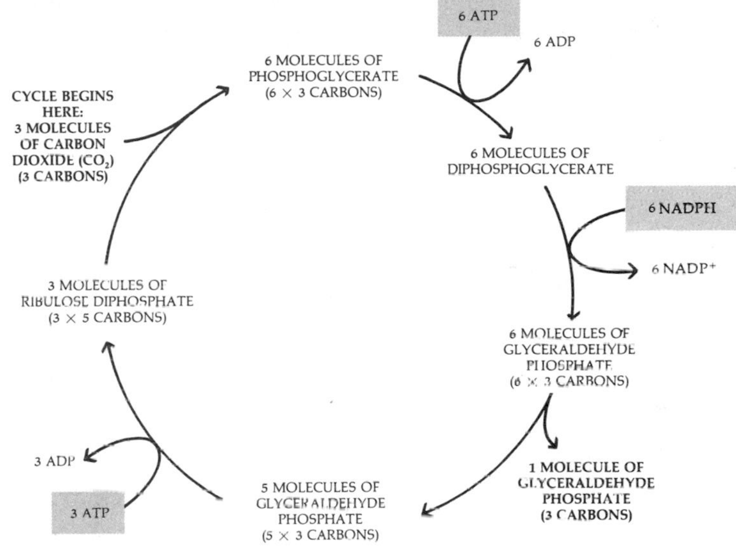

Six revolutions of the cycle, with the introduction of six atoms of carbon, are necessary to produce a six-carbon sugar, such as glucose. The overall equation is

$$6RuDP + 6CO_2 + 18ATP + 12NADPH + 12H^+ + 12H_2O \longrightarrow 6RuDP + \text{glucose} + 18P_i + 18ADP + 12NADP^+$$

The immediate product of the cycle itself is glyceraldehyde phosphate. This same sugar-phosphate is formed when the fructose diphosphate molecule is split at the fourth step in glycolysis (page 156).

The Four-Carbon Photosynthetic Pathway

The Calvin cycle is not the only carbon fixation pathway used in the dark reactions. In some plants, the first product of CO_2 fixation to be detected is not the three-carbon molecule phosphoglycerate, as it is in the Calvin cycle. It is the four-carbon compound oxaloacetic acid. (Oxaloacetic acid, you may recall, is also an intermediate in the Krebs cycle.) Plants that utilize this pathway, also known as the Hatch-Slack pathway, are commonly called C_4, or four-carbon plants, as distinct from the C_3 plants that use only the Calvin cycle.

Table 9-1 *Summary of the Stages of Photosynthesis*

	CONDITIONS	WHERE	WHAT APPEARS TO HAPPEN	RESULTS
Light reactions	Light	Thylakoids	Light striking Photosystem II boosts electrons uphill. (These electrons are replaced by electrons from water molecules, which release O_2.) Electrons then pass downhill along an electron transport chain, forming ATP, to Photosystem I. Light hits Photosystem I, boosts electrons uphill to an electron acceptor from which they are passed to $NADP^+$ to form NADPH. Electrons are replaced in Photosystem I by electrons from Photosystem II.	Energy of light is converted to chemical energy stored in bonds of ATP and NADPH.
Dark reactions	Do not require light	Stroma	Calvin cycle. NADPH and ATP formed in light reactions are used to reduce carbon dioxide. The cycle yields glyceraldehyde phosphate from which glucose or other organic compounds can be formed.	Chemical energy of light reactions stored in ATP and NADPH is used to incorporate carbon into organic molecules.

The oxaloacetic acid is formed when carbon dioxide is bound to a compound known as phosphoenolpyruvate (PEP). This reaction is catalyzed by the enzyme PEP carboxylase (Figure 9–20). The oxaloacetic acid is then reduced to malic acid or converted (with the addition of an amino group) to aspartic acid. These steps take place in mesophyll cells. The next step is a surprise: The malic acid (or aspartic acid, depending on the species) is transported to bundle-sheath cells where it is decarboxylated to yield CO_2 and pyruvic acid. The CO_2 then enters the Calvin cycle.

One might well ask why C_4 plants should have evolved such a seemingly clumsy and energetically expensive method of providing carbon dioxide to the Calvin cycle. This question can be answered only by considering the function of the leaf as a whole. Carbon dioxide is not continuously available to the photosynthesizing cells. It enters the leaf by way of the stomata, specialized pores that open and close depending on, among other things, water stress.

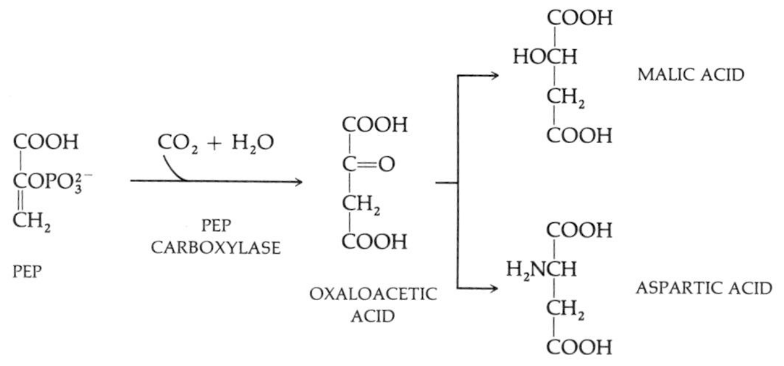

9–20
Carbon dioxide fixation by the C_4 pathway. Carbon dioxide is incorporated into phosphoenolpyruvate (PEP) by the enzyme PEP carboxylase. The resulting oxaloacetic acid is converted either to malic acid or aspartic acid. These steps later will be reversed, releasing carbon dioxide for use in the Calvin cycle.

PEP: COOH–$COPO_3^{2-}$=CH_2 → ($CO_2 + H_2O$, PEP CARBOXYLASE) → OXALOACETIC ACID: COOH–C=O–CH_2–COOH → MALIC ACID: COOH–HOCH–CH_2–COOH or ASPARTIC ACID: COOH–H_2NCH–CH_2–COOH

9–21
A pathway for carbon fixation in C_4 plants. CO_2 is first fixed in mesophyll cells as oxaloacetic acid. It is then transported to bundle-sheath cells, where the carbon dioxide is released. The CO_2 thus formed enters the Calvin cycle. Pyruvic acid returns to the mesophyll cell, where it is phosphorylated to PEP.

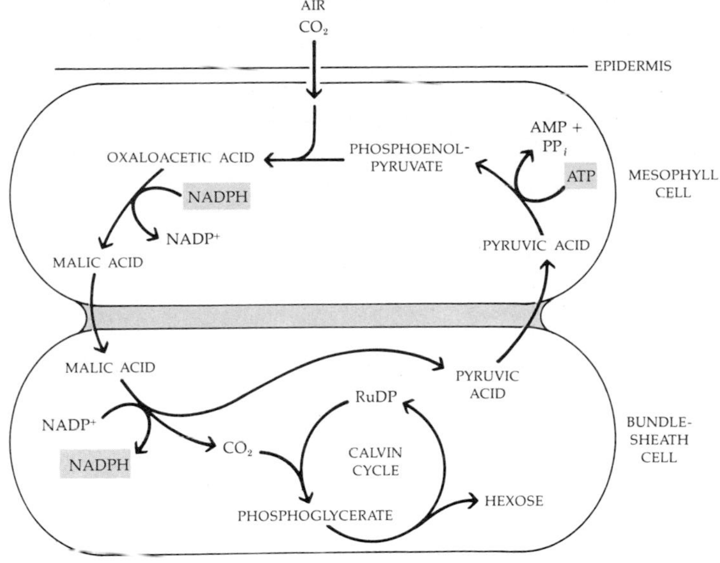

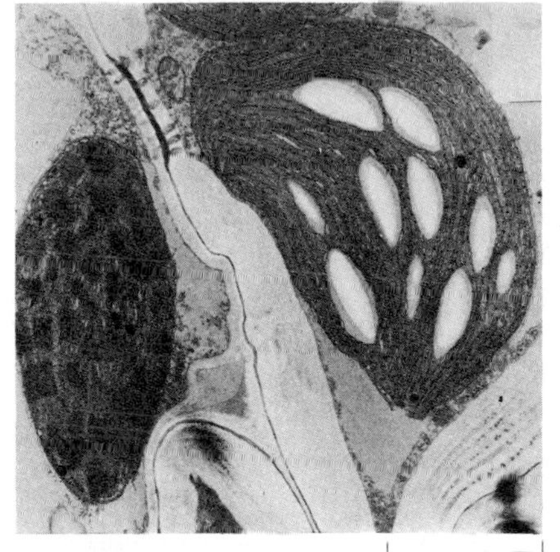

9–22
In C_4 plants, the chloroplasts of mesophyll cells differ from those of bundle-sheath cells. As shown in this micrograph, the bundle-sheath cells lack grana. Notice the plasmodesmata connecting the two cells and the large, white starch grains in the bundle-sheath cell.

PEP carboxylase has a higher CO_2 affinity than does RuDP carboxylase, so it keeps the CO_2 concentration lower within the leaf (that is, it fixes carbon dioxide faster, at low levels). This maximizes the gradient of carbon dioxide between the cells and the outside air. A higher gradient means the leaf will trap a larger fraction of the passing stream of carbon dioxide. If the stomata must be closed much of the time, as in a hot, dry climate, the plant with C_4 metabolism will take up more carbon dioxide with each gasp (so to speak) than the plant that has only C_3 metabolism. Hence, it is at a distinct advantage in drought-ridden areas.

The list of plants known to utilize the four-carbon pathway has grown to over 100 genera, at least a dozen of which have both C_3 and C_4 species. This pathway undoubtedly has arisen many times independently in the course of evolution. Sugarcane, corn, and sorghum are among the best-known C_4 plants.

Perhaps the most familiar example of the competitive capacity of C_4 plants is seen in lawns in the summertime. In most parts of the United States, lawns consist mainly of C_3 grasses such as Kentucky bluegrass and creeping bent. In the summer, these dark green, fine-leaved grasses are often overwhelmed by rapidly growing crabgrass, which, to the dismay of the suburbanite, disfigures the lawn as its yellowish-green, broader-leaved plants slowly take over. Crabgrass, you will not be surprised to hear, is a C_4 plant.

THE PRODUCTS OF PHOTOSYNTHESIS

The three-carbon sugar produced by the Calvin cycle may seem an insignificant reward, both for all the enzymatic activity on the part of the cell and for our own intellectual stress. However, this molecule and those derived from it provide (1) the energy source for all living systems, and (2) the basic carbon skeleton for all organic molecules. Carbon has been fixed—that is, it has been brought from the inorganic world into the organic one.

PHOTOSYNTHESIS AND THE ACCUMULATION OF OXYGEN

The first photosynthetic organism probably appeared almost 3 billion years ago. Before the evolution of photosynthesis, the physical characteristics of the planet itself were by far the most powerful forces in shaping the course of natural selection. But with the evolution of photosynthesis, organisms began to change the face of the planet, and they have continued to do so, at an ever-increasing rate, up to the present day.

The atmosphere in which the first cells evolved probably consisted largely of ammonia (NH_3) and methane (CH_4) and other carbon gases. Gradually, methane and ammonia were broken down, and their hydrogen atoms, which are very light, were released into space as hydrogen gas (H_2). These earliest forms of life were, of course, adjusted to living in an environment without free oxygen, and, in fact, oxygen, with its powerful electron-attracting capacities, would have been poisonous to them. Their energy probably came from glycolysis. Anaerobic glycolysis (page 159) might have resulted in the accumulation of carbon dioxide in the atmosphere. Thus carbon dioxide and nitrogen replaced methane and ammonia as significant components of the atmosphere.

Then, it is hypothesized, there slowly evolved photosynthetic organisms that used carbon dioxide as their carbon source and released oxygen, as do most modern photosynthetic forms. As these photosynthetic organisms multiplied, free oxygen began to accumulate. In response to the new selection pressures of these changing conditions, cell species arose for which oxygen was not a poison but a requirement of existence.

As we have seen, oxygen-consuming organisms have an advantage over those that do not use oxygen because a higher yield of energy can be extracted per molecule from the oxidation of carbon-containing compounds than from anaerobic fermentation processes, in which fuel molecules are not broken down completely. Energy resulting from the use of oxygen by cells made possible the development of increasingly active, increasingly complex organisms. Without oxygen, the higher forms of life that now exist on earth could not have evolved.

The planet continues to be dependent on photosynthesis for its oxygen. In fact, all the oxygen of the biosphere is renewed about every 2,000 years by photosynthetic cells. And the carbon dioxide produced in the course of respiration is the raw material on which photosynthesis depends.

Molecules of glyceraldehyde phosphate may flow into a variety of different metabolic pathways, depending on the activities and requirements of the cell. Often they are built up to glucose or fructose, following a sequence that is in many of its steps the reverse of the glycolysis sequence described in the previous chapter. (At some steps, the reactions are simply reversed and the enzymes are the same. Other steps—the highly exergonic ones of the downhill sequence—are bypassed.) Plant cells use these six-carbon sugars to make starch and cellulose for their own purposes and sucrose for export. Animal cells store them as glycogen. All cells use sugars, including glyceraldehyde phosphate and glucose, as the starting point for the manufacture of other carbohydrates, fats and other lipids, and, with the addition of nitrogen, amino acids and nitrogenous bases. Finally, as we saw in the preceding chapter, the carbon fixed in photosynthesis is the source of ATP energy for heterotrophic cells.

SUMMARY

In photosynthesis, light energy is converted to chemical energy, and carbon is fixed into organic compounds. The generalized equation for this reaction is

$$CO_2 + 2H_2A + \text{light energy} \longrightarrow (CH_2O) + H_2O + 2A$$

in which H_2A stands for water or some other substance from which electrons can be removed.

Photosynthesis takes place within cellular organelles known as chloroplasts. These organelles are surrounded by two membranes. Within the membranes is a solution of organic compounds and ions known as the stroma, and a complex internal membrane system consisting in plants of fused pairs of membranes that form sacs called thylakoids. The pigments and other molecules responsible for capturing light are located in and on these membranes.

Pigments are compounds that absorb light energy. The pigments involved in photosynthesis in eukaryotes include the chlorophylls and the carotenoids. Light absorbed by pigments boosts their electrons to a higher energy level. Because of the way the pigments are packed into the membranes, they are able to transfer this energy to reactive molecules, probably chlorophyll packed in a particular way. Two pigment systems have been identified in the thylakoids.

In the currently accepted model of the light reactions in photosynthesis, light energy strikes antennae pigments of Photosystem II, which contains several hundred molecules of chlorophyll *a* and chlorophyll *b*. Electrons are boosted uphill from the reactive chlorophyll molecule P_{680} to an electron acceptor. As the electrons are removed, they are replaced by electrons from water molecules, with the production of O_2. The electrons then pass downhill to Photosystem I along an electron transport chain, in the course of which ATP is generated (photophosphorylation). Light energy absorbed in antennae pigments of Photosystem I and passed to chlorophyll P_{700} results in the ejection of one electron. The electrons removed from P_{700} are replaced by the electrons from Photosystem II. The electrons are ultimately accepted by the electron carrier molecule, $NADP^+$. The energy yield from the light reactions is contained in the molecules of NADPH and in the ATP formed by photophosphorylation.

In the dark reactions, which take place in the stroma, the NADPH and ATP produced in the light reactions are used to reduce carbon dioxide to organic carbon. This is accomplished by means of the Calvin cycle. In the Calvin cycle, a molecule of carbon dioxide is combined with the starting material, a five-carbon sugar called ribulose diphosphate. At each turn of the cycle, one carbon atom enters the cycle. Three turns of the cycle produce a three-carbon molecule, glyceraldehyde phosphate. Two molecules of glyceraldehyde phosphate (six turns of the cycle) can combine to form a glucose molecule. At each turn of the cycle, RuDP is regenerated. The glyceraldehyde phosphate can also be used as starting material for other organic compounds needed by the cell.

In C_4 plants, carbon dioxide is initially accepted by a compound known as PEP (phosphoenolpyruvate) to yield the four-carbon oxaloacetic acid. The oxaloacetic acid is then reduced to malic acid, which is, in turn, oxidized, and carbon dioxide is transferred to the RuDP of the Calvin cycle. Under conditions of drought, C_4 plants are more efficient than C_3 plants.

QUESTIONS

1. Distinguish among the following: action spectrum/absorption spectrum; NAD^+/$NADP^+$; stroma/grana/thylakoid; oxidative phosphorylation/photophosphorylation; light reactions/dark reactions; C_3 photosynthesis/C_4 photosynthesis.

2. Sketch a chloroplast and label its structures. Compare with Figure 9–11.

3. Describe, in general terms, the events of photosynthesis. Compare your description with Table 9–1.

4. Trace a carbon atom through a series of biological events, such as those indicated in Figure 7–1 (page 140).

5. Given the scarcity of high-salt environments and the difficulties of surviving in them, explain in evolutionary terms why the halobacteria are found there.

6. Why is C_4 photosynthesis advantageous to those plants that have it? Many plants, however, do not have C_4 photosynthesis. Why is it advantageous to such plants *not* to have C_4 photosynthesis?

7. Why is it plausible to argue, as the distinguished physiologist and Nobel laureate George Wald does, that wherever in the universe we find living organisms, we will find them (or at least some of them) to be colored?

8. Return to Figure 9–21 and describe the biochemical events taking place in each of the chloroplasts.

SUGGESTIONS FOR FURTHER READING

Books

CONANT, JAMES BRYANT (ed.): *Harvard Case Histories in Experimental Science,* vol. 2, Harvard University Press, Cambridge, Mass., 1964.

Case # 5, Plants and the Atmosphere, *edited by Leonard K. Nash, describes the early work on photosynthesis, presented often in the words of the investigators themselves. The narrative illuminates the historical context in which the discoveries were made.*

LEHNINGER, ALBERT L.: *Biochemistry,* 2d ed., Worth Publishers, Inc., New York, 1975.

This introductory text is outstanding both for its clarity and for its consistent focus on the living cell.

LEHNINGER, ALBERT L.: *Bioenergetics: The Molecular Basis of Biological Energy Transformations,* 2d ed., The Benjamin/Cummings Publishing Company, Menlo Park, Calif., 1971.

A thorough account; Lehninger is one of the foremost experts on cellular energetics.

RABINOWITCH, EUGENE, and GOVINDJEE: *Photosynthesis,* John Wiley & Sons, Inc., New York, 1969.*

A lucid introduction, suitable for undergraduate students, to the processes of photosynthesis and to related physical and chemical concepts, such as entropy and free energy.

STRYER, LUBERT: *Biochemistry,* W. H. Freeman and Company, San Francisco, 1975.

A good introduction, handsomely illustrated, to cellular energetics.

Articles

GOVINDJEE, and RAJNI GOVINDJEE: "The Primary Events of Photosynthesis," *Scientific American,* December 1974, pages 68–82.

HINKLE, PETER C., and RICHARD E. MCCARTY: "How Cells Make ATP," *Scientific American,* March 1978, pages 104–123.

STOECKENIUS, WALTHER: "The Purple Membrane of Salt-loving Bacteria," *Scientific American,* June 1976, pages 38–46.

---

* Available in paperback.

SECTION 3 GENETICS

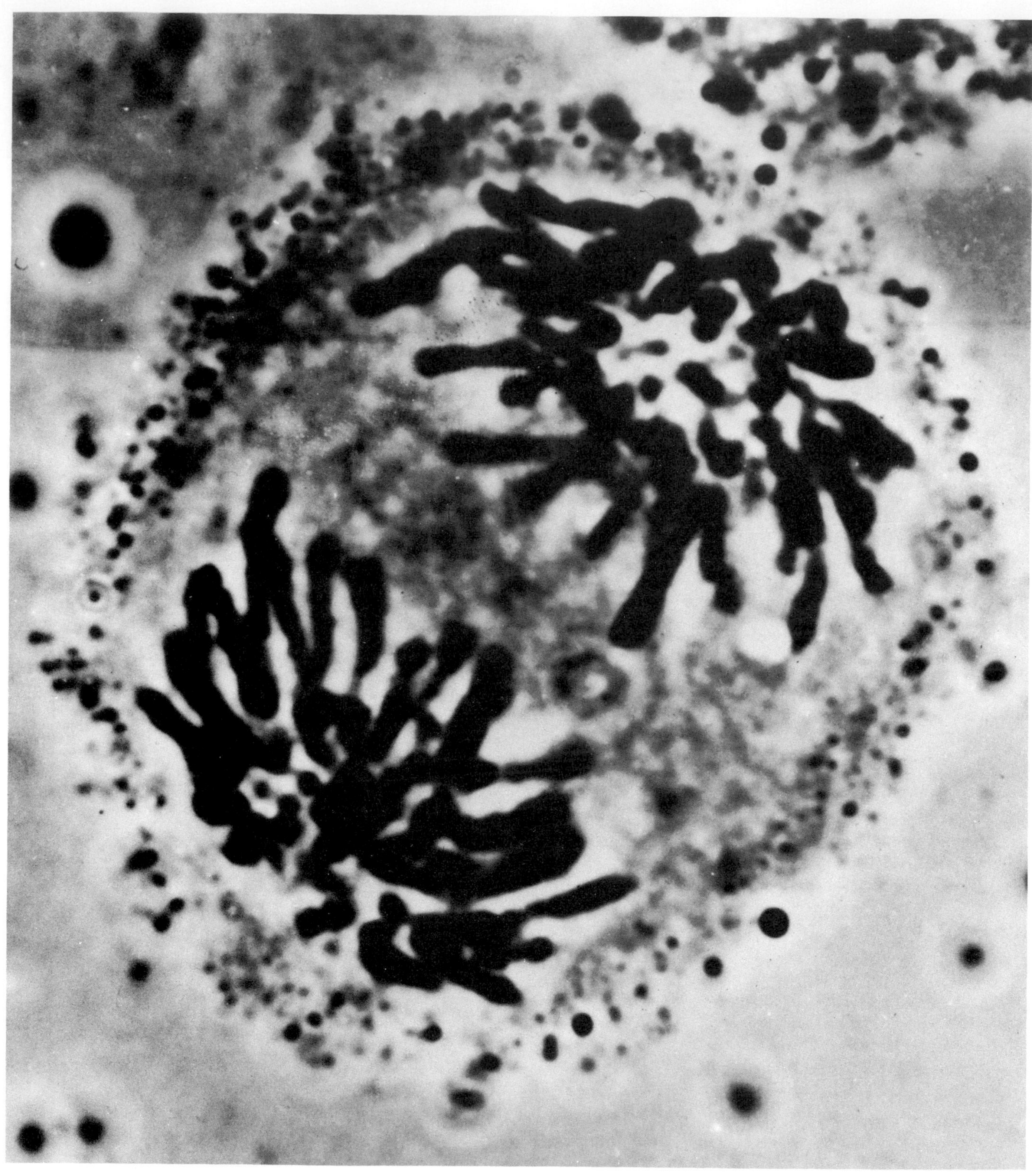

CHAPTER 10

The Nucleus and Mitosis

10–1
A human cell divides. The long, dark bodies are chromosomes, the carriers of the genetic information. The chromosomes have replicated and moved apart. Each set of chromosomes is an exact copy of the other. Thus the two new cells, the daughter cells, will contain the same hereditary material.

Genetics is the branch of biology that deals with heredity, the process by which traits are passed from parents to offspring. The capacity to reproduce is one of the special characteristics of living matter, and for more than 3 billion years living things have been reproducing themselves, generation after generation, with each parent passing on to its offspring all the biological instructions necessary for the offspring to develop into the same kind of organism as the parent. A primary question of genetics is how this hereditary information is transmitted by the parent to its young. Or to put it more simply, why do cats have kittens, dogs have puppies, and hens have chicks? Why do you resemble your father or mother? What is the nature of this hereditary information? How is it stored between generations? How is it passed from one generation to another?

Since every organism—whether a *Paramecium,* an oak, or a human—starts as a single cell, the hereditary information must be carried somewhere, somehow, in that single cell. More than a century ago, biologists began to suspect that the key to the mystery of heredity was to be found within the cell's nucleus, and as it turned out, they were quite right.

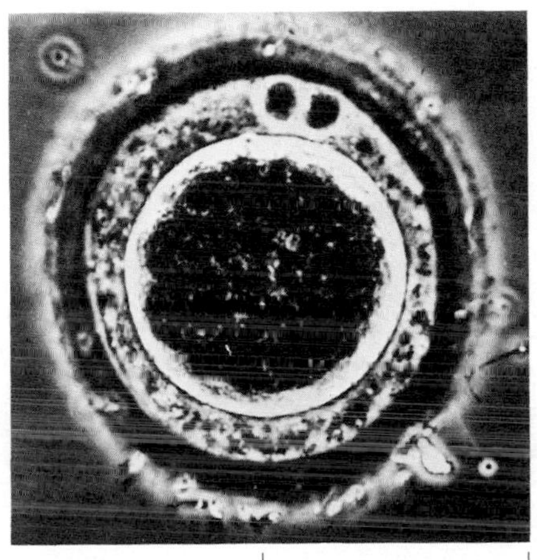

10–2
A human ovum surrounded by sperm cells. Note the great difference in size between the two types of cells. Each, however, contributes the same number of chromosomes to the fertilized egg from which the new individual will develop.

THE EUKARYOTIC NUCLEUS

The nucleus of the eukaryotic cell is a prominent, roughly spherical body. Unlike the cytoplasm, which contains many distinct structures, the nucleus contains little that can be seen with any kind of microscope. The only things usually visible are its outer membranes, and, within them, a tangle of threadlike material and one or two spherical bodies, the nucleoli. Yet despite its simple appearance, the nucleus plays a crucial role in the life of the organism, both as the repository of the hereditary information and as the control center of the individual cell.

Some Experimental Evidence

One of the most important early observations of the role of the nucleus in the life of the cell was made about a hundred years ago by a German embryologist, Oscar Hertwig, observing the eggs and sperm of a sea urchin. Sea urchins produce eggs and sperm in great numbers. The eggs are relatively large—about as large as a human ovum (Figure 10–2), which they closely resemble—and so are easy to observe. They are fertilized in the open water, rather than internally, as is the case with land-dwelling vertebrates such as ourselves. Watching the eggs being fertilized under his microscope, Hertwig observed that only a single sperm cell was required and that when the sperm cell penetrated the egg, its nucleus was released and fused with the nucleus of the egg. This observation, confirmed by other scientists and in other kinds of organisms, was important in establishing the fact that the nucleus is the carrier of heredity: The only link between father and offspring is the nucleus of the sperm.

Since Hertwig's time, a number of experiments have been performed concerning the role of the nucleus in the cell. In one simple experiment, the nucleus was removed from an amoeba by microsurgery. The protist stopped dividing and, in a few days, it died. If, however, a nucleus from another amoeba was implanted within 24 hours after the original one was removed, the cell survived and divided normally.

In the early 1930s, Joachim Hämmerling studied the comparative roles of the nucleus and the cytoplasm by taking advantage of some unusual properties of the marine alga *Acetabularia*. The body of *Acetabularia* consists of a single huge cell 2 to 5 centimeters in height. Individuals have a cap, a stalk, and a "foot," all of which are differentiated portions of the single cell. If the cap is removed, the cell will rapidly regenerate a new one. Different species of *Acetabularia* have different kinds of caps. *Acetabularia mediterranea,* for example, has a compact umbrella-shaped cap and *Acetabularia crenulata* has a cap of petal-like structures.

Hämmerling took the "foot," which contains the nucleus, from a cell of *A. crenulata* and grafted it onto a cell of *A. mediterranea,* from which he had first removed the "foot" and the cap. The cap that then formed had a shape intermediate between the two. When this cap was removed, the next cap that formed was completely characteristic of *A. crenulata* (Figure 10–3).

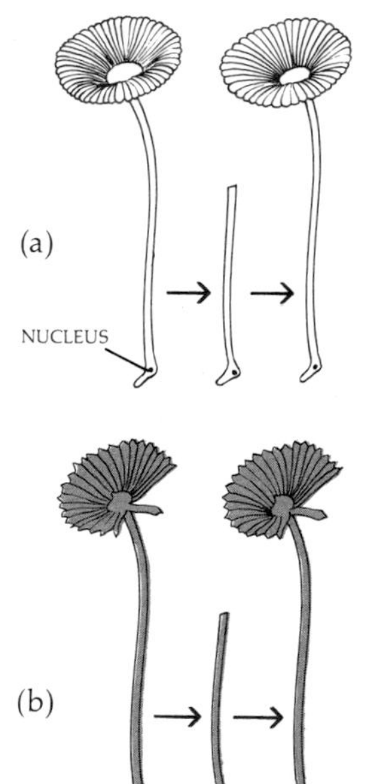

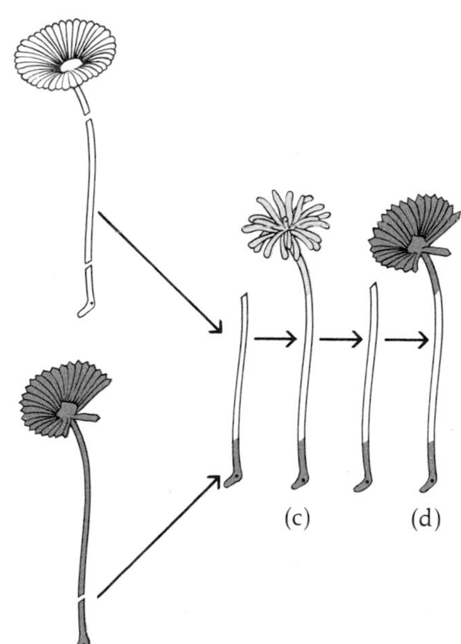

10–3
(a) *One species of* Acetabularia *has an umbrella-shaped cap, and* (b) *another has a ragged, petal-like cap. If the caps are removed, new caps form, similar in appearance to the amputated one. However, if the "foot" (containing the nucleus) is removed at the same time as the cap and a new nucleus from the other species is transplanted, the cap* (c) *that forms will have a structure with characteristics of both species. If this cap is removed, the next cap* (d) *that grows will be characteristic of the cell that donated the nucleus, not of the cell that donated the cytoplasm.*

Hämmerling interpreted these results as meaning that certain cap-determining substances are produced under the direction of the nucleus. These substances accumulate in the cytoplasm, which is why the first cap that formed after nuclear transplantation was of an intermediate type. By the time the second cap formed, however, the cap-determining substances present in the cytoplasm before the transplant had been exhausted, and the form of the cap was completely under the control of the new nucleus.

Conclusion: The Functions of the Nucleus

We can see from these experiments that the nucleus performs two crucial functions for the cell. First, it carries the hereditary information for the cell, the instructions that determine whether a particular organism will develop to become a *Paramecium,* an oak, or a human—and not just any *Paramecium,* oak, or human, but one that resembles the parent or parents of that particular, unique organism. Second, as Hämmerling's work indicated, the nucleus exerts a continuing influence over the ongoing activities of the cell, ensuring that the complex molecules that cells require are synthesized in the number and of the kind needed.

THE DIVISION OF THE NUCLEUS

Another clue to the importance of the nucleus came about as the result of the observations of Walther Flemming, also about 100 years ago. Flemming observed the "dance of the chromosomes" that takes place when eukaryotic cells divide, and he painstakingly pieced together the sequence of events. (The fact that Hertwig and Flemming made their observations at about the same time was no coincidence; enormous improvements had just occurred in the light microscope and in techniques of microscopy.) We know more than Flemming did about the structure of the chromosomes and their movements during nuclear division, but much still remains a mystery.

The Problem

Cell division is the means by which living organisms increase both in number and in size. In one-celled organisms, cells grow by assimilating materials from the environment and synthesizing these materials into new structural and functional molecules. When such a cell reaches a certain size, it divides. The two daughter cells, each of which has received about half of the mass of the parent cell, then begin growing again. A bacterial cell may divide every 20 minutes. In a one-celled organism such as a *Paramecium,* cell division may occur every few hours. In many-celled plants and animals, cell division is the means by which the organism grows, starting from one single cell, and also by which injured or worn-out tissues are replaced and repaired.

In all of these instances, the new cells produced are structurally and functionally similar both to the parent cell and to one another. They are similar, in part, because each new cell receives about half of the parent cell's cytoplasm and organelles. More important, in terms of structure and function, each new cell inherits an exact replica of the hereditary information of the parent cell. How cells accomplish this quite impressive feat is the subject of the rest of this chapter.

The solution is comparatively simple in prokaryotic cells. In such cells, the nuclear material is in the form of a single, long, circular molecule of DNA (deox-

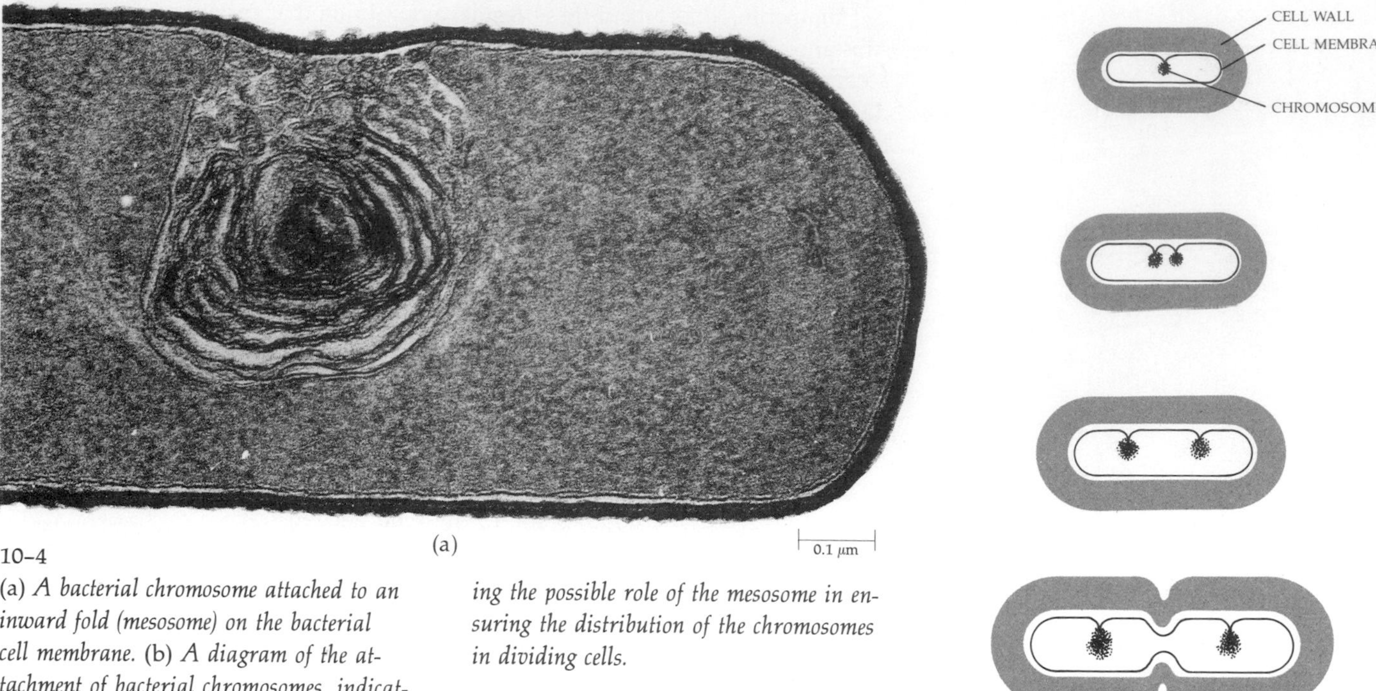

10–4
(a) *A bacterial chromosome attached to an inward fold (mesosome) on the bacterial cell membrane.* (b) *A diagram of the attachment of bacterial chromosomes, indicating the possible role of the mesosome in ensuring the distribution of the chromosomes in dividing cells.*

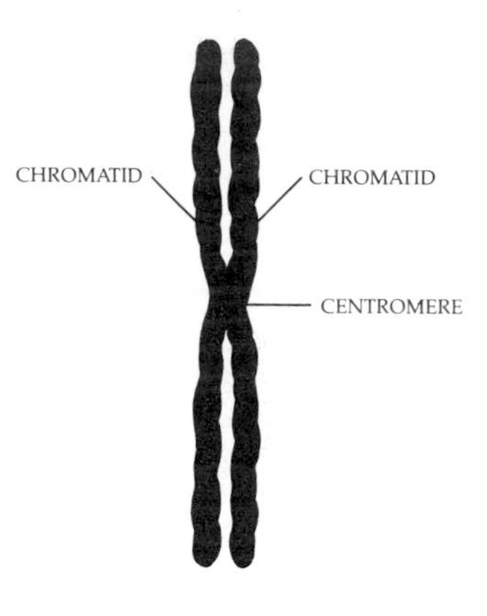

10–5
A chromosome at the beginning of mitosis. The chromosomal material has replicated, and each chromosome now consists of two identical parts, called chromatids. The centromere, the constricted area at the center, is the site of attachment of the two chromatids.

yribonucleic acid, about which we shall soon be hearing considerably more). This molecule, the cell's chromosome, is replicated before cell division. According to present evidence, each of the two daughter chromosomes is attached to a different spot on the interior of the cell membrane. As the membrane elongates, the chromosomes move apart (Figure 10–4). When the cell has approximately doubled in size and the chromosomes are separated, the cell membrane pinches inward, and a new cell wall forms that separates the two new cells and their chromosome replicas.

In eukaryotic cells, the problem is more complex. For instance, human cells have 46 chromosomes, each different from the others, and, at cell division, each daughter cell has to receive one, and only one, of each of the 46. The solution, as you will see, is ingenious and elaborate. The chromosomes are first replicated and then they are apportioned between the daughter cells by a series of steps called, collectively, *mitosis*.

The Chromosomes

When cells are not dividing, the chromosomal material is dispersed and is visible, if at all, only as threadlike strands, the chromatin. As we shall see in Chapter 17, it appears that the genetic material can play its role in controlling and determining cell functions only when it is in this extended form. The most prominent bodies in the nucleus at this time are the nucleoli (see page 107).

As mitosis begins, the chromatin slowly coils and condenses into a compact form; this condensation appears to be necessary for the complex movements and separation of the chromosomes.

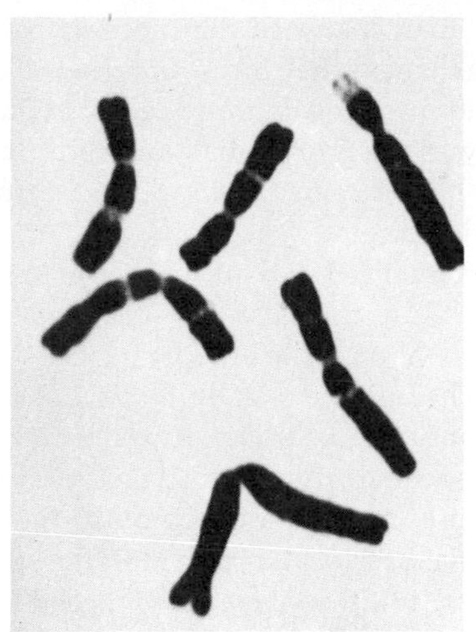

10–6
If you look carefully, you can see that each of these chromosomes actually consists of two longitudinal halves. Some chromosomes have several constrictions. The centromere, the area of attachment, is usually the constriction nearest the center.

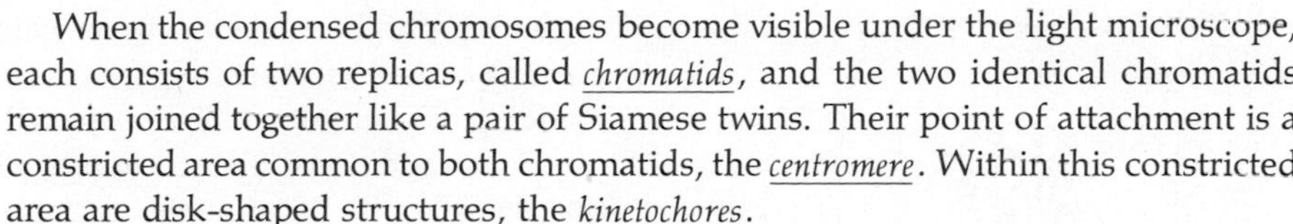

When the condensed chromosomes become visible under the light microscope, each consists of two replicas, called *chromatids*, and the two identical chromatids remain joined together like a pair of Siamese twins. Their point of attachment is a constricted area common to both chromatids, the *centromere*. Within this constricted area are disk-shaped structures, the *kinetochores*.

The Phases of Mitosis

The function of mitosis is to maneuver the replicated chromosomes so that each new cell gets a full complement—one of each.

The process of mitosis is conventionally divided into four phases: prophase, metaphase, anaphase, and telophase; of these, prophase is usually by far the longest. If a mitotic division takes ten minutes (which is about the minimum time required), during about six of these minutes the cell will be in prophase. The schematic drawings that follow show mitosis as it takes place in an animal cell.

Between divisions, the cell is in interphase. Although little can be seen in the nucleus during interphase, and it is not considered one of the stages of mitosis, it is an essential preparation for it, since it is during this period that replication of the chromosome occurs.

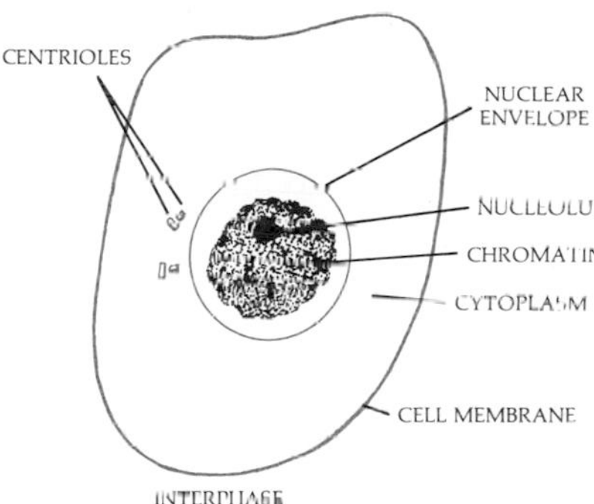

INTERPHASE

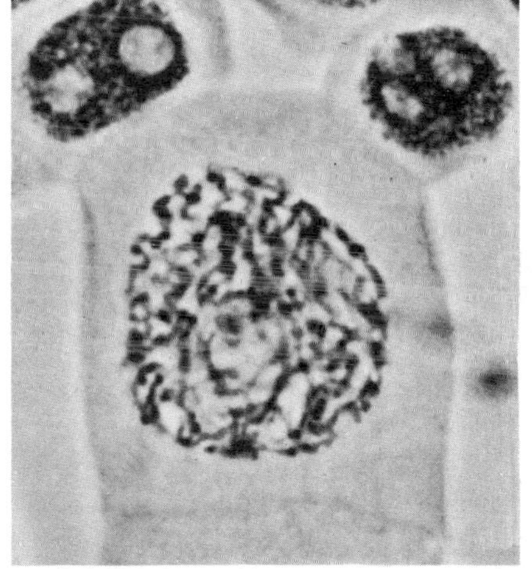

10–7
Early prophase in an onion cell. The chromosomes have begun to condense.

In early *prophase*, the chromatin condenses, and the individual chromosomes begin to become visible. If you look carefully, you can see that each chromosome consists of two duplicate chromatids pressed closely together longitudinally and connected at the centromere. In most cells (higher plants are the principal exception), two centriole pairs can be seen at one side of the nucleus, outside the nuclear envelope. Each pair consists of one mature centriole and a smaller, newly formed centriole lying at right angles to the first.

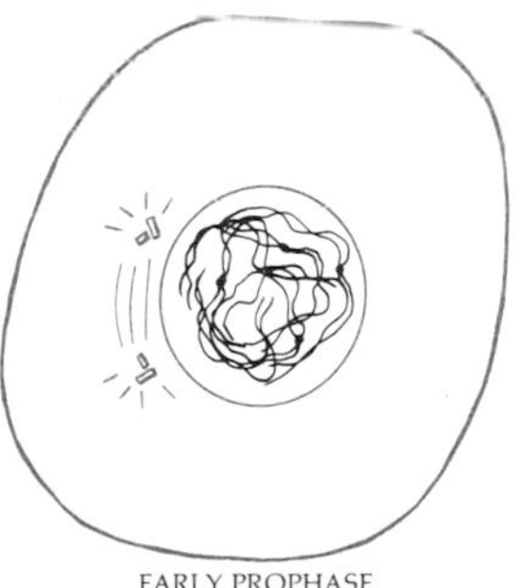

EARLY PROPHASE

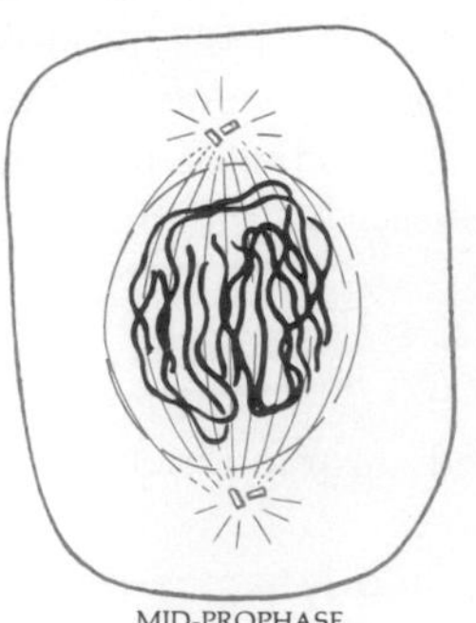

MID-PROPHASE

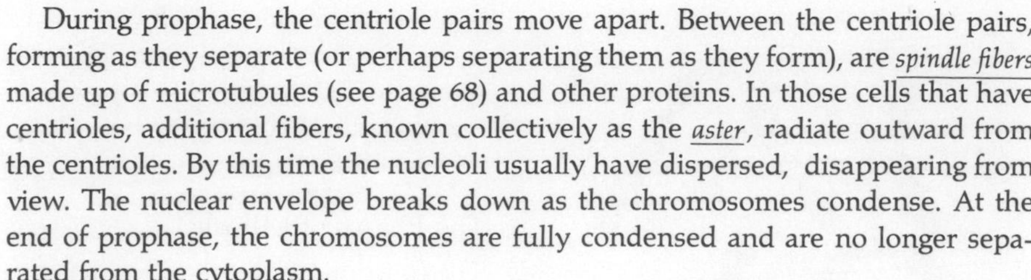

During prophase, the centriole pairs move apart. Between the centriole pairs, forming as they separate (or perhaps separating them as they form), are *spindle fibers* made up of microtubules (see page 68) and other proteins. In those cells that have centrioles, additional fibers, known collectively as the *aster*, radiate outward from the centrioles. By this time the nucleoli usually have dispersed, disappearing from view. The nuclear envelope breaks down as the chromosomes condense. At the end of prophase, the chromosomes are fully condensed and are no longer separated from the cytoplasm.

By the end of prophase, the centriole pairs are at opposite ends of the cell, and the members of each pair are of equal size. The spindle is fully formed. It is a three-dimensional football-shaped structure consisting of three groups of microtubules: (1) astral rays, (2) continuous fibers reaching from pole to pole, and (3) shorter fibers that are attached to the kinetochores of each sister chromatid.

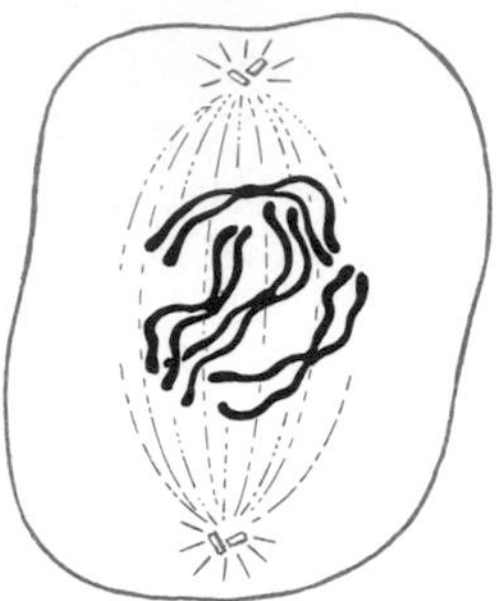

EARLY METAPHASE

During early *metaphase* the chromosomes move back and forth within the spindle, apparently maneuvered by the spindle fibers, as if they were being tugged first toward one pole and then the other. Finally they become arranged precisely at the midplane (equator) of the cell. This marks the end of metaphase.

At the beginning of *anaphase*, the chromatids separate into what are now called chromosomes. The two duplicate chromosomes move apart, apparently drawn toward the poles by the spindle fibers. The centromeres move first, while the arms of the chromosomes seem to drag behind.

As anaphase continues, the two identical sets of newly separated chromosomes move toward the opposite poles of the spindle. Anaphase is the most rapid portion of mitosis.

By the beginning of *telophase*, the chromosomes have reached the opposite poles and the spindle disperses into tubulin dimers, subunits of the globular protein building blocks that make up its microtubules (see Figure 3-26 on page 68).

During late telophase, nuclear envelopes form around the two sets of chromosomes, which once more become diffuse. Often, a new centriole forms adjacent to each of the previous ones, so that each daughter cell has a new centriole pair. In each nucleus, the nucleoli re-form.

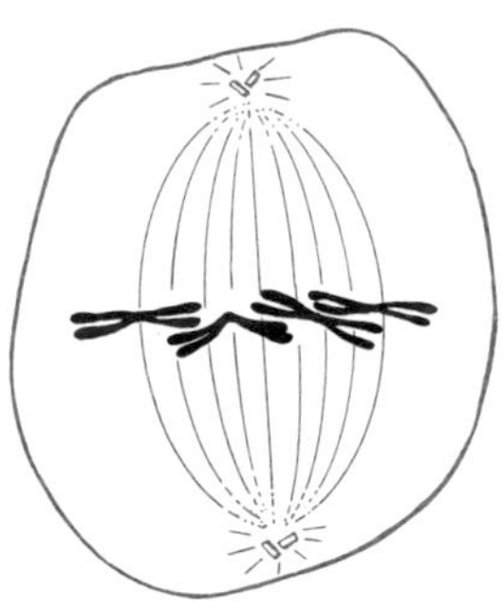

METAPHASE

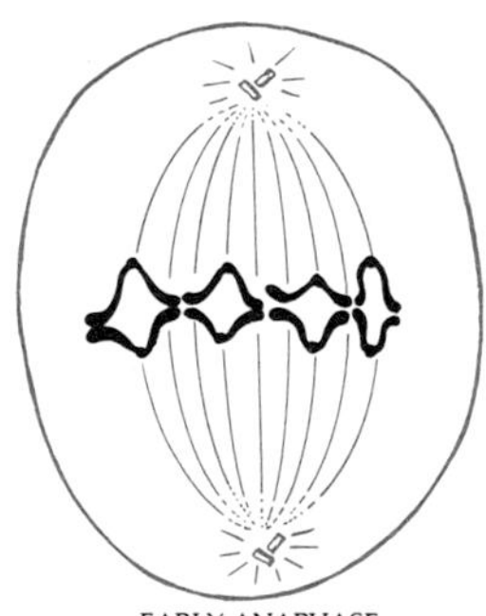

EARLY ANAPHASE

LATE ANAPHASE

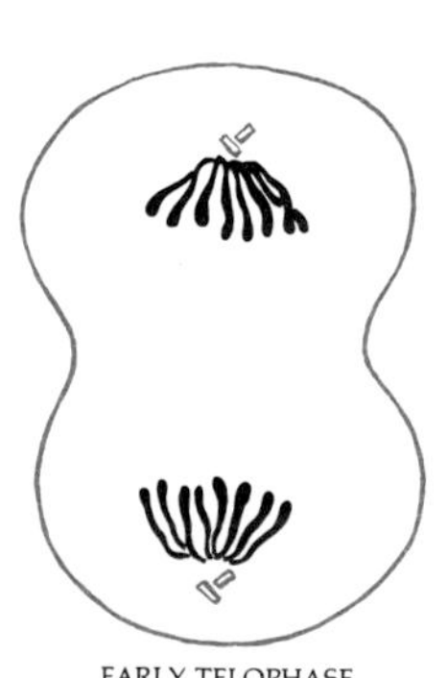

EARLY TELOPHASE

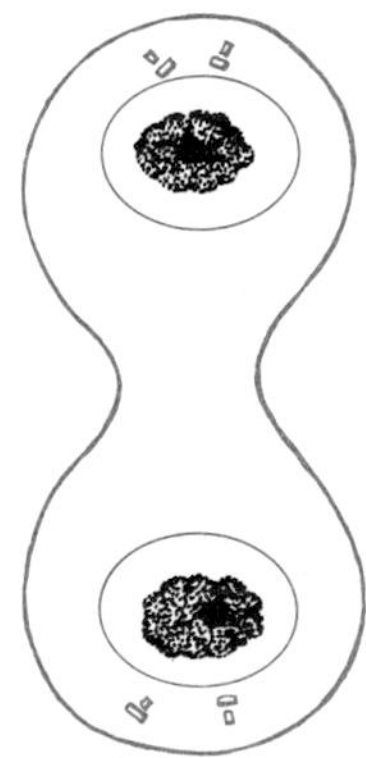

LATE TELOPHASE

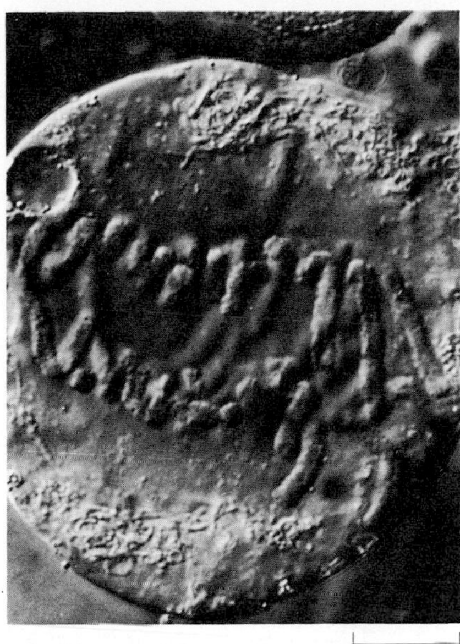

10-8
Middle anaphase in a cell from the seed of a South African lily. The chromosomes have moved halfway toward the poles.

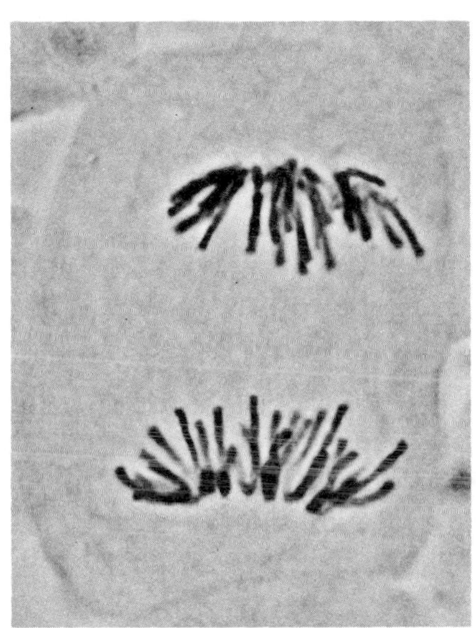

10-9
Early telophase in an onion cell.

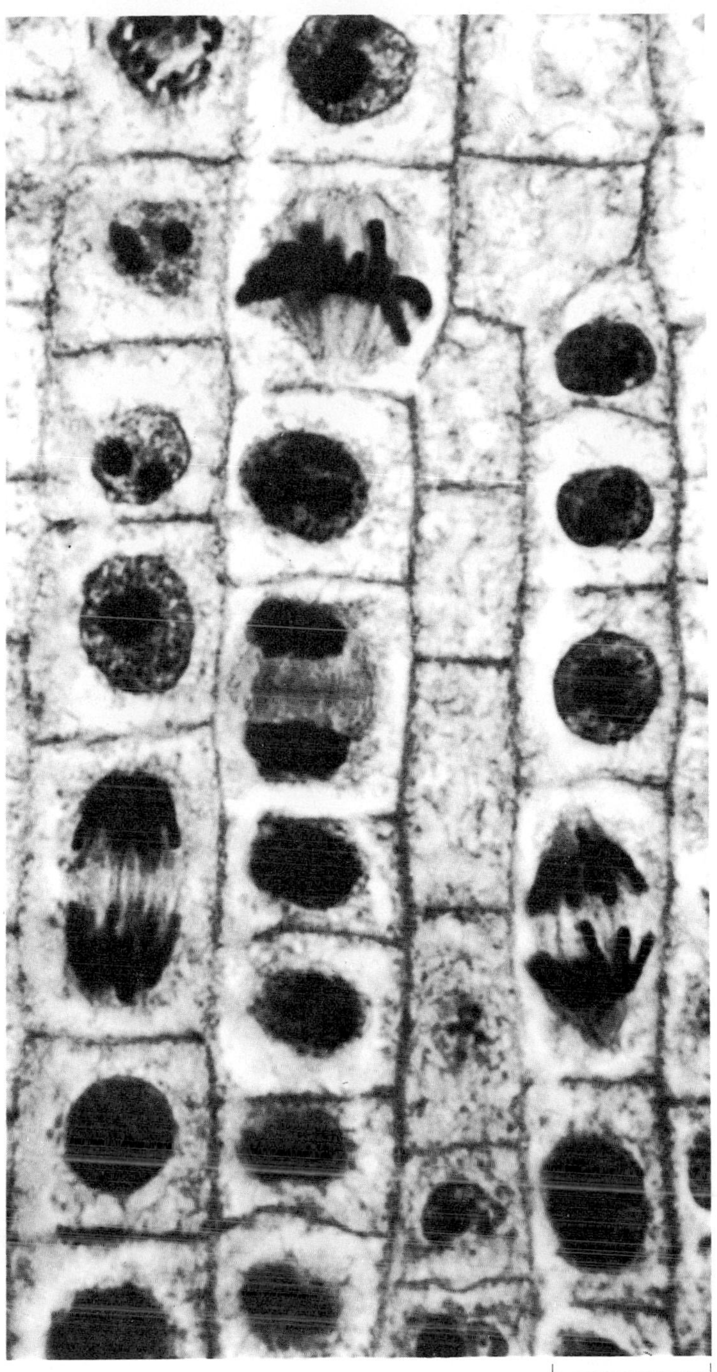

10-10
Cross section of the tip of an onion root. This is the area of the root where cells increase in number, and several shown here are in mitosis.

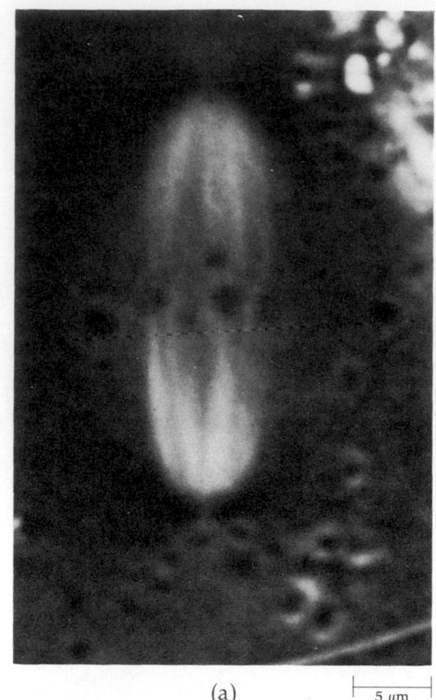

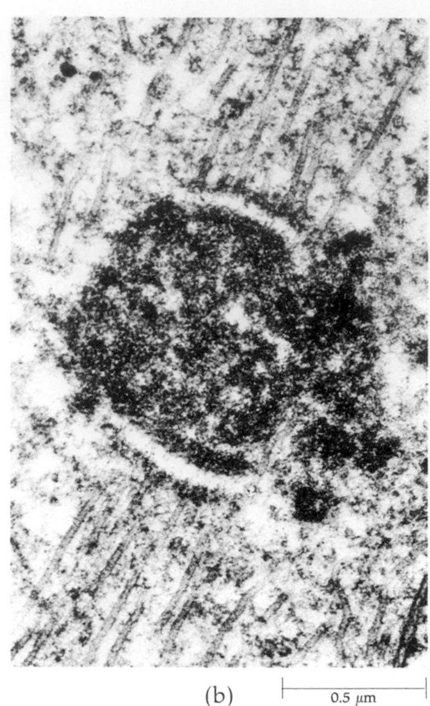

(a) (b)

10–11
(a) *This unusual micrograph of a spindle in an egg cell of a marine worm emphasizes the spindle's three-dimensional qualities. The bright streaks are the spindle fibers. The chromosomes appear as indistinct gray bodies near the equator of the spindle.* (b) *In this electron micrograph, spindle fibers can be seen extending to the kinetochores of a green alga. The dark material is the metaphase chromosome, most of which is out of the plane of the thin section prepared for this micrograph. The kinetochores are the two arc-shaped areas at either side of the chromosome.*

The Spindle

The spindle, as we noted previously, is composed largely of microtubules, some of which are attached to the chromosomes at their kinetochores (Figure 10–11) and some of which stretch from pole to pole. Although the spindle fibers lengthen during prophase and then shorten during anaphase, they do not appear to get thinner or thicker. This suggests that they do not contract but that material is added to or removed from the spindle fiber as the spindle changes shape. This hypothesis is supported by the observation that microtubules from other locations in the cell are disassembled to provide tubulin for the spindle. This borrowing from other microtubules is probably responsible for the characteristic rounded appearance of cells during mitosis.

The Centrioles

Basal bodies and centrioles are the same structure used, apparently, for different purposes. Basal bodies organize the microtubules of flagella and cilia, and it is tempting to think that the centrioles organize the microtubules of spindle fibers—spinning them out somewhat as a spider spins out silk. However, plant cells that do not have centrioles or basal bodies also form spindles with microtubules. No asters appear, but the spindles are as well-formed as those in animal cells. Also, in some animal cells, it is possible to remove the centrioles from the cells—yet spindle formation proceeds normally. It has been suggested that the spindles, instead of forming from the centrioles, serve to separate the centrioles, pushing them apart, and so ensuring that each daughter cell receives an adequate supply of basal bodies from which to construct flagella or cilia.

10–12
(a) Cross section of a centriole. Centrioles and basal bodies are structurally identical. It is hypothesized that basal bodies organize microtubules into cilia and flagella and that, similarly, centrioles organize the microtubules of spindle fibers. However, plant cells that do not have centrioles are able to form spindles. (b) Centriole pair. Centrioles are also usually—but not always—formed from preexisting centrioles, with the newly formed centriole appearing at right angles to the previously existing one.

10–13
Mitosis in the embryonic cells of a whitefish. (a) Prophase. The chromosomes have become visible, the nuclear envelope has broken down, and the spindle apparatus has formed. (b) Metaphase. The chromosomes, perhaps guided by the spindle fibers, are lined up at the equator of the cell. Some of them appear to have begun to separate. (c) Anaphase. The two sets of chromosomes are moving apart. (d) Telophase. The chromosomes are completely separated, the spindle apparatus is disappearing, and a new cell membrane is forming that will complete the separation of the two daughter cells.

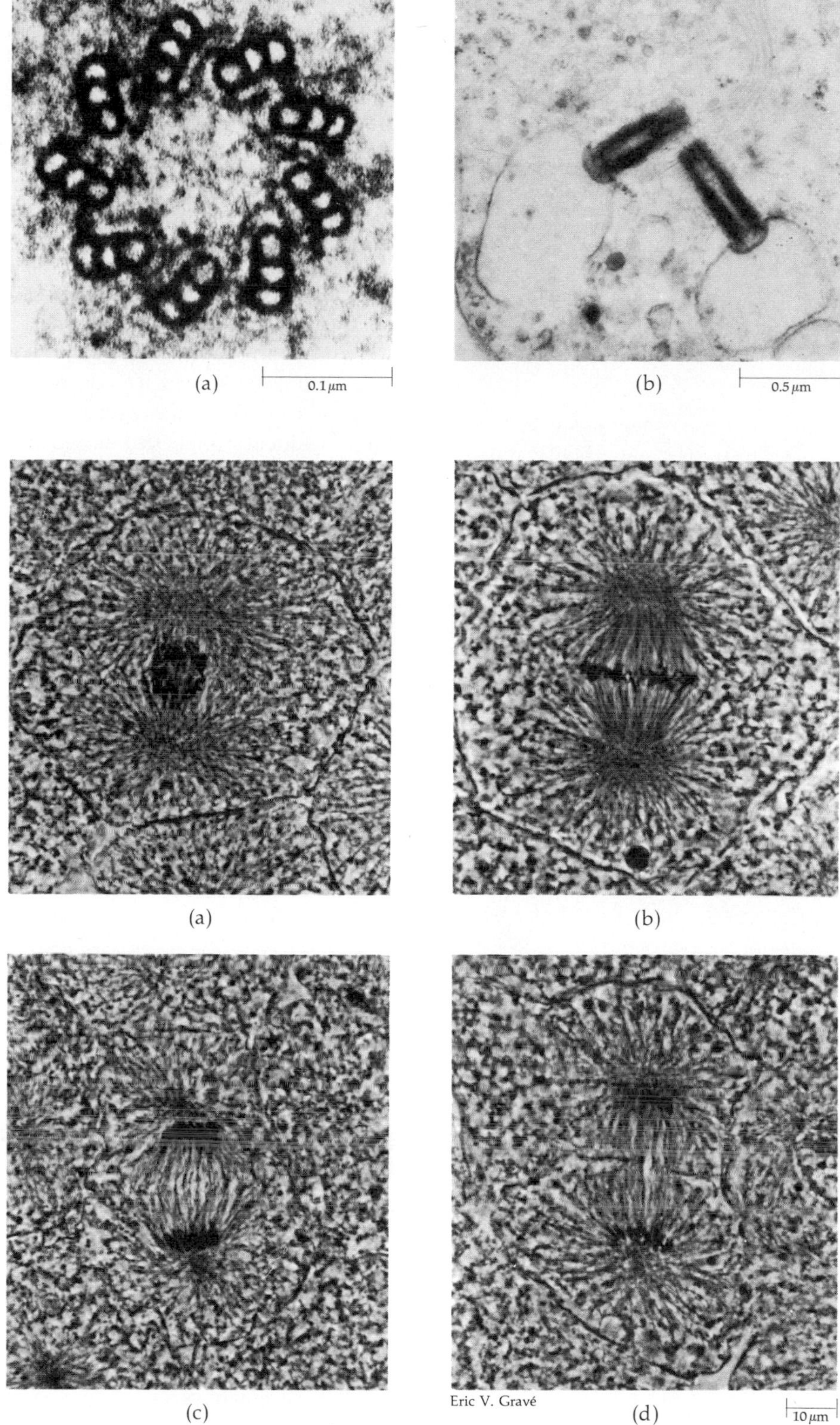

CYTOKINESIS

Cytokinesis, the division of the cytoplasm, usually but not always accompanies nuclear division. It is usually preceded by a doubling of the molecules that make up the cytoplasm, of the organelles, and of the size of the cell. Some of the cellular structures can be synthesized entirely *de novo* ("from scratch") by the cell; these include microtubules, microfilaments, and ribosomes, all of which are composed, at least in part, of protein subunits. Membranous structures, such as the nuclear envelope, Golgi bodies, lysosomes, vacuoles, and vesicles, are all apparently derived from the endoplasmic reticulum, which is renewed and enlarged by the addition of self-assembling molecules. These new structures are produced during interphase, prior to replication of the genetic material.

Mitochondria and chloroplasts are produced only from previously existing mitochondria and chloroplasts or proplastids. Each of these organelles also has its own chromosome, which is organized much like the single chromosome of the bacterial cell. These are two of the reasons that many biologists hypothesize that these organelles originated as separate organisms and then took up a symbiotic ("living together") relationship with early eukaryotes more than a billion years ago. This question will be discussed further in Chapter 20.

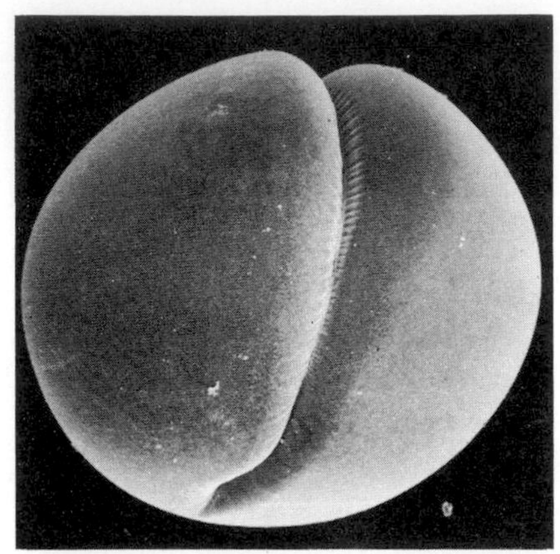

(a)

(b)

10–14
(a) *Cytokinesis in an animal cell, the egg of a frog.* (b) *Note the constriction furrows.*

The visible process of cytokinesis usually begins during telophase of mitosis. It usually divides the cell into two nearly equal parts. The cleavage always occurs at the midline of the spindle, although the spindle does not seem to be involved in cytoplasm division. For instance, if a spindle is pushed out of position at late metaphase, or even if it is removed from the cell, cytokinesis proceeds normally, with the cell dividing along the previously established equator.

Cytokinesis differs in some respects in plant and animal cells. In animal cells, during early telophase, the cell membrane begins to constrict along the circumference of the cell in the area where the equator of the spindle used to be. At first a furrow appears on the surface, and this gradually deepens into a groove. Eventually the connection between the daughter cells dwindles to a slender thread, which soon parts. Microfilaments, seen in large numbers near the furrows, are thought to play a role in the constriction. They are believed to act as a sort of "purse string" drawing the ends of the two daughter cells together and pinching them apart.

In plant cells, the cytoplasm is divided by the formation of a *cell plate*, which is formed from a series of vesicles produced from Golgi bodies. The vesicles eventually fuse to form a flat, membrane-bound space, the cell plate. As more vesicles fuse, the edges of the growing plate fuse with the membrane of the cell. In this way, a space is established between the two daughter cells, completing their separation. Cellulose is then laid down against the membranes, each new cell forming its own cell wall, and the space itself eventually is impregnated with pectins and becomes the middle lamella (see page 103).

When cell division is complete, two daughter cells are produced, indistinguishable from each other and from the parent cell.

THE CELL CYCLE

Dividing cells pass through a regular sequence of cell growth and division, known as the *cell cycle*. Completion of the cycle requires varying periods of time from a few hours to several days, depending on both the type of cell and external factors, such as temperature or available nutrients.

10-15
In plants, the final separation of the two cells is effected by the formation of a structure known as the cell plate. Vesicles appear across the equatorial plane of the cell and gradually fuse, forming a flat membrane-bound space, the cell plate, which extends outward until it reaches the wall of the dividing cell. The large, dark forms on either side of the micrograph are the chromosomes.

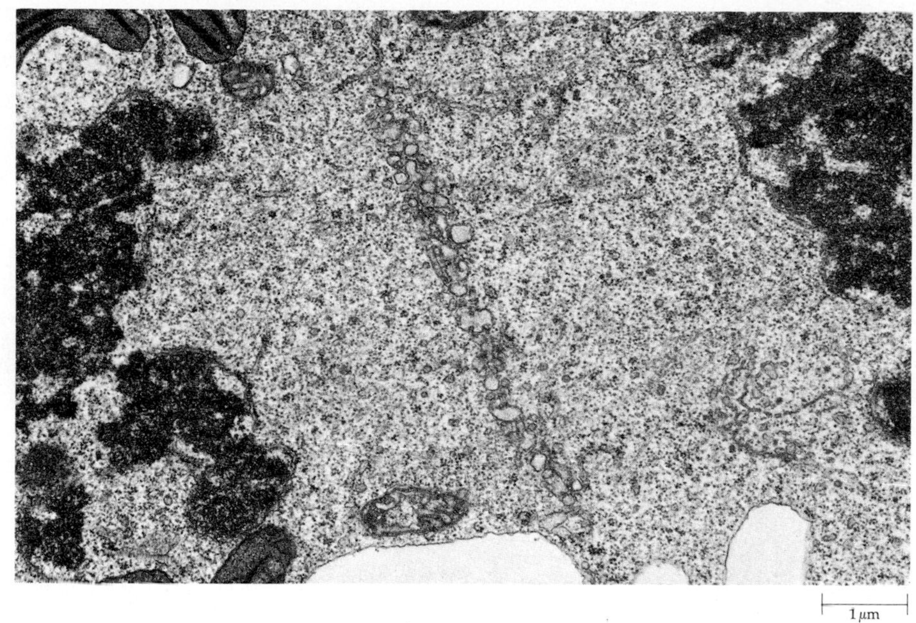

10-16
Mitosis in a plant cell with four chromosomes. Note that a spindle forms, although no centrioles or asters are visible. Cytokinesis takes place with the formation of a cell plate.

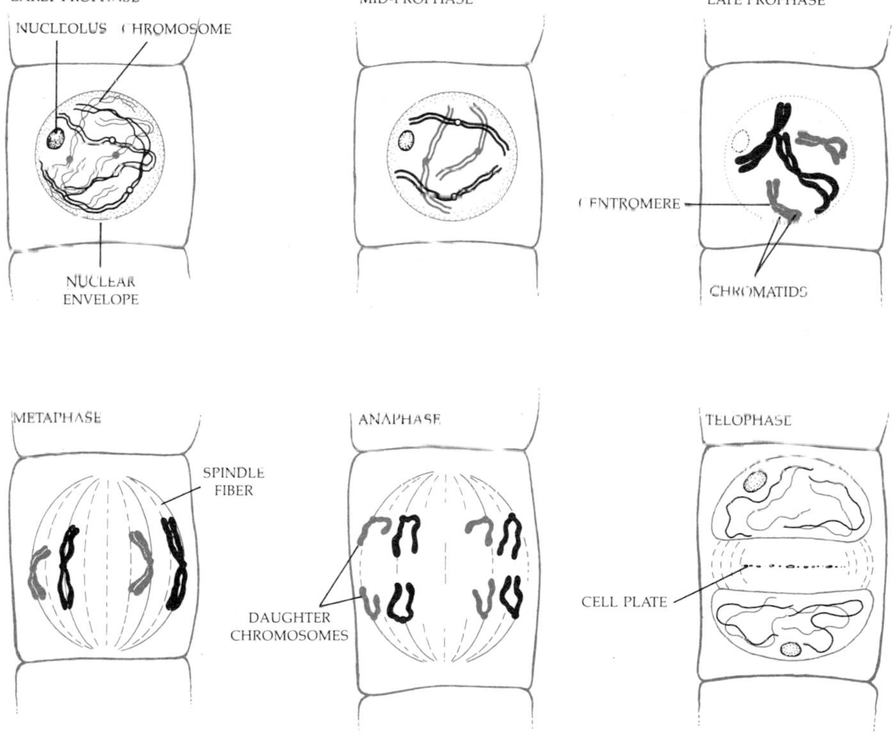

10–17
The cell cycle. Dividing cells go through four principal phases, including the stages of mitosis (shown in color) and the S (synthesis) phase, during which the genetic material is replicated. Separating mitosis and the S phase are two G phases. The first of these (G_1) is a period of general growth and replication of cytoplasmic organelles. During the second (G_2), structures directly associated with mitosis, such as the spindle fibers, are synthesized. After the G_2 phase comes mitosis, which is, in turn, divided into four stages. In cells of different species, the different phases occupy different proportions of the total cycle.

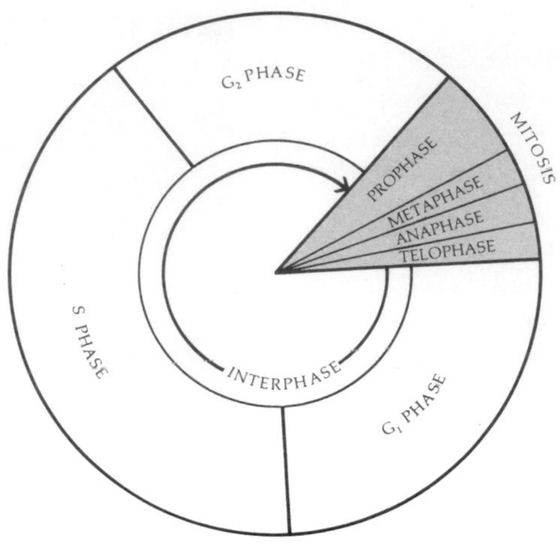

The S (synthesis) phase of the cell cycle is the period during which the genetic material (DNA) is replicated along with many of its associated proteins. G (gap) phases precede and follow the S phase. The G_1 period occurs after mitosis and precedes the S phase; the G_2 phase follows the S phase and occurs before mitosis. These are periods of intensive biochemical activity, during which the cell doubles in size and its enzymes, ribosomes, mitochondria, and other molecules and organelles approximately double in number. The G and S phases together are referred to as interphase. Mitosis occupies the rest of the cell cycle.

Dividing cells of different species show characteristic variations in the general pattern of the cell cycle. In the common bean, for example, the complete cycle requires about 19 hours, of which 7 hours are taken up by the S period; G_1 and G_2 are of equal length (about 5 hours each), and mitosis lasts 2 hours. By contrast, in mouse fibroblast cells, the cell cycle is 22 hours, of which mitosis is less than 1 hour, S is almost 10 hours, G_1 is 9 hours, and G_2 a little more than 2 hours.

When cells stop growing, as a result, for example, of depletion of nutrients or contact inhibition (page 120), they stop in the G_1 phase. Therefore, it is concluded that substances are synthesized during the G_1 phase that either inhibit or stimulate the S stage and the rest of the cycle, thus determining whether or not cell division will occur.

Some cells pass through successive cell cycles repeatedly. This group includes the one-celled organisms and certain cells in growth centers of both plants and animals. An example is the cells in the human bone marrow that give rise to red blood cells. The average red blood cell lives only about 120 days, and there are about 25 trillion (2.5×10^{13}) of them in an adult. To maintain this number, about 2.5 million new red blood cells must be produced by cell division each second. Some highly specialized cells, such as nerve cells, lose their capacity to replicate once they are mature. A third group of cells retains the capacity to divide but does so only under special circumstances. Cells in the human liver, for example, do not ordinarily divide, but if a portion of the liver is removed surgically, the remaining cells (even if only about a third of the total remain) continue to replicate themselves until the liver reaches its former size. Then they stop. About 2 trillion (2×10^{12})

cell divisions occur in an adult human every 24 hours, in other words, about 25 million per second. Further knowledge of the control mechanisms involved would not only be interesting biologically but also might be of great importance in the control of cancer. Cancer cells differ from normal cells largely in that they keep on dividing at the expense of the host tissues.

SUMMARY

The nucleus is the most prominent structure within most eukaryotic cells. It is surrounded by a double membrane, the nuclear envelope. Within it are the nucleolus and the chromosomes, the bodies that contain the genetic information of the cell.

Cell division in eukaryotes includes division of the nucleus (mitosis), during which the chromosomes are apportioned between the two daughter cells, and division of the cytoplasm (cytokinesis). When the cell is in interphase (not dividing), the chromosomes are visible only as thin strands of threadlike material (chromatin). During this period, if mitosis is to take place, the chromosomal material is replicated. As mitosis begins, the chromatin condenses into pairs of identical chromosomes held together at the centromere. At these stages the chromosomes are called chromatids. Next, the spindle forms. In animal cells, it forms between the centrioles as they separate. In both animal and plant cells, some spindle fibers stretch from pole to pole and some are attached to the chromosomes at their kinetochores, bodies adjacent to the centromeres. Prophase ends with the breakdown of the nuclear envelope and disappearance of the nucleoli. During metaphase, the chromosomes, apparently maneuvered by the spindle fibers, move toward the center of the cell. At the end of metaphase, the chromosomes are arranged on the equatorial plane. During anaphase, the sister chromatids separate, and each chromatid, now an independent chromosome, moves to an opposite pole. During telophase, a nuclear envelope forms around each group of chromosomes. The spindle breaks down, the chromosomes uncoil and once more become extended and diffuse, and the nucleoli re-form.

Cytokinesis in animal cells results from constrictions in the cytoplasm between the two nuclei. In plant cells, the cytoplasm is divided by the coalescing of vesicles to form the cell plate, within which the cell wall is subsequently laid down. In both cases, the result is the production of two new, separate cells. As a result of mitosis, each has received an exact copy of the enormous skein of genetic material of the parent cell, and, as a result of cytokinesis, approximately half of the cytoplasm and organelles.

Dividing cells pass through a regular sequence of cell growth and cell division known as the cell cycle. The cycle consists of a G_1 phase, during which the cytoplasmic material doubles; an S phase, during which DNA is synthesized; a G_2 phase, during which the tubulin of the spindle and other structures directly involved with mitosis are synthesized; and mitosis.

QUESTIONS

1. Distinguish among the following terms: mitosis/cytokinesis/cell division/cell cycle; centriole/centromere/kinetochore; chromatid/chromosome.

2. Identify each of the stages shown below and describe what is happening.

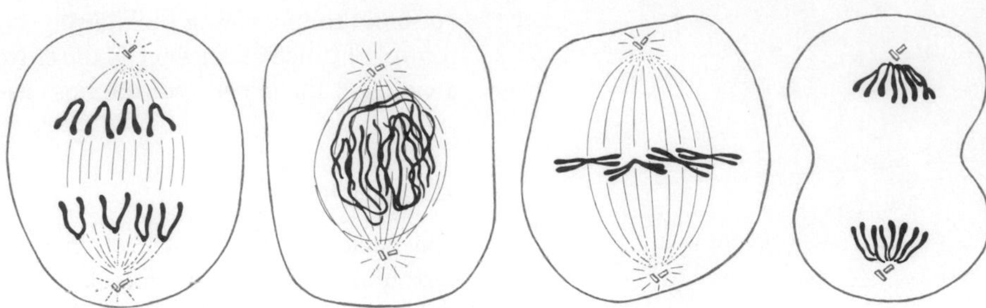

3. What is a chromosome? How is it related to chromatin?

4. Why do we often refer to chromatids as sister chromatids? When are sister chromatids formed? How? When do they first become visible under the microscope?

5. Look at Figure 10–10. Arrange the stages of mitosis you see there in the proper sequence. Are all the stages shown?

6. In what ways is cell division different in plant cells than in animal cells?

7. Why do cells divide? Suppose you, as an organism, were made up of a large single cell rather than several billion small ones. How would you differ from your present self?

CHAPTER 11

Meiosis and Sexual Reproduction

Mitosis, as we saw in the preceding chapter, is one way in which organisms reproduce. Most higher organisms, however, reproduce sexually. Sexual reproduction generally requires two parents, each of which contributes genetically to the new offspring. It always involves two events: *fertilization* and *meiosis.* Fertilization is the means by which the different genetic contributions of the two parents are brought together to form the new genetic identity (genome) of the offspring. Meiosis is a special kind of nuclear division that evolved from and uses much of the same machinery as mitosis. However, as you will see, it differs from it in some important respects.

HAPLOID AND DIPLOID

To understand meiosis, we must look again at the chromosomes. Every organism has a chromosome number characteristic of its particular species. A mosquito has 6 chromosomes per cell; a cabbage, 18; corn, 20; a sunflower, 34; a cat, 38; a human being, 46; a plum, 48; a dog, 78; and a goldfish, 94.

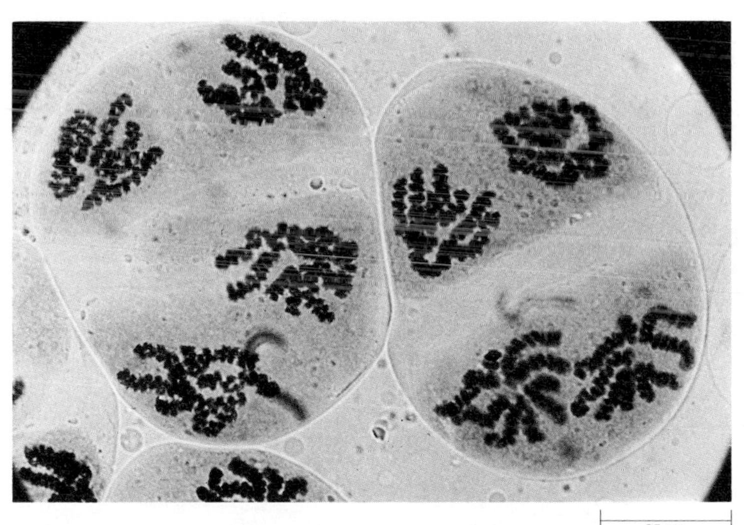

11–1
Meiosis in wake robin (Trillium erectum). *The clearly visible chromosomes are almost completely separated. Each diploid nucleus has divided to form four haploid sets of chromosomes.*

However, the sex cells—or *gametes* (from the Greek word *gamos,* "to marry")—have exactly half the number of chromosomes that is characteristic of the somatic (body) cells of an organism. The number of chromosomes in the gametes is referred to as the *haploid* ("single") number, and the number in the somatic cells as the *diploid* ("double") number. Cells that have more than one double set of chromosomes are known as *polyploid.*

For brevity, the haploid number is designated n and the diploid number as $2n$. In humans, for example, $n = 23$ and $2n = 46$. When a sperm fertilizes an egg, the two haploid nuclei fuse, $n + n = 2n$, and the diploid number is restored (Figure 11–2). The diploid cell produced by the fusion of two gametes is known as a *zygote.*

In every diploid cell, each chromosome has a partner. These pairs of chromosomes are known as homologous pairs, or *homologues.* The two resemble each other in size and shape and also, as we shall see, in the type of hereditary information each contains. Usually, one homologue contains genetic information from one parent, and its partner contains genetic information inherited from the other parent.

In the special kind of nuclear division called meiosis, the diploid number of chromosomes is reduced to the haploid number. If this reduction did not occur, the chromosome number would double each generation. Moreover, as we shall see, meiosis is in itself a source of new genetic combinations.

11–2
Sexual reproduction is characterized by two events: the coming together of the sex cells, or gametes (fertilization), and meiosis. Following meiosis, the number of the chromosomes is single, or haploid (n). *Following fertilization, the number is double, or diploid* (2n).

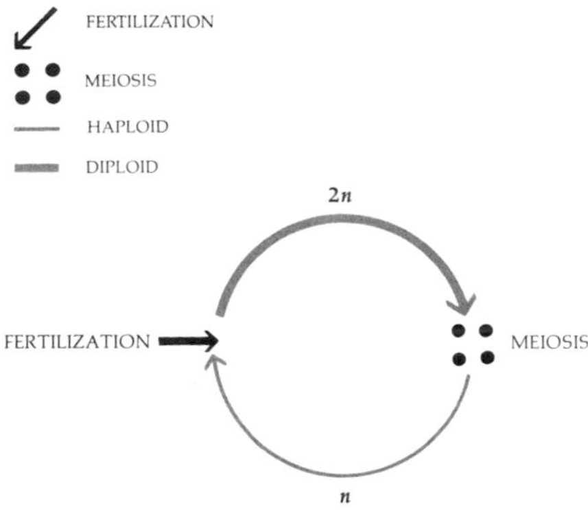

MEIOSIS AND THE LIFE CYCLE

Meiosis occurs at different times during the life cycle of different organisms (Figure 11–2). In the alga *Chlamydomonas* and the mold *Neurospora* it occurs immediately after fusion of the mating cells (Figure 11–3); the cells are ordinarily haploid, and meiosis restores the haploid number. In plants, such as ferns, a haploid phase typically alternates with a diploid phase (Figure 11–4). The common and conspicuous form of a fern is the sporophyte, the diploid organism. By meiosis, fern sporophytes produce spores, usually on the undersides of their fronds. These spores have only the haploid number of chromosomes. They germinate to form much smaller plants (gametophytes), often only a few cell layers thick. In these plants all the cells are haploid. The small, haploid plants produce gametes, which fuse and then develop into a new, diploid sporophyte. This process, in which the haploid stage is followed by the diploid stage and again by the haploid stage, is known as *alternation of generations.* As we shall see, alternation of generations occurs, although not always in the same form, in all plants.

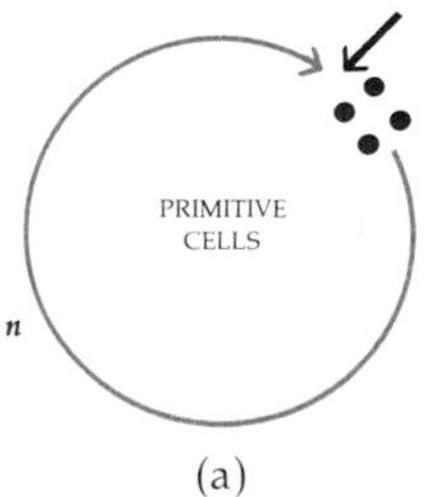

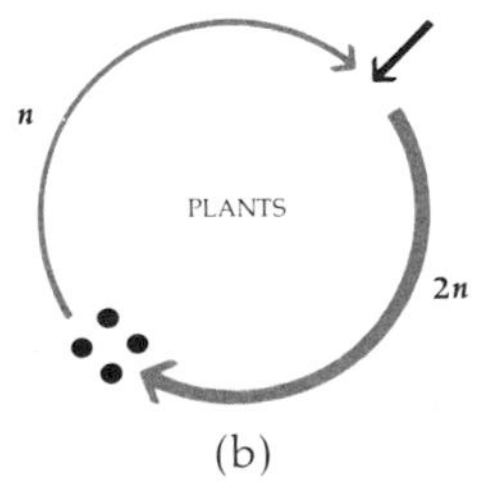

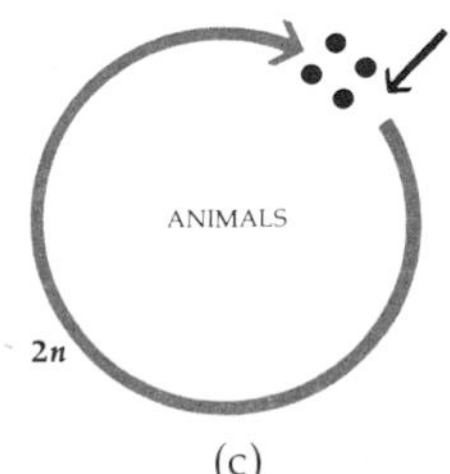

Fertilization and meiosis occur at different points in the life cycle of different organisms. (a) *In some one-celled organisms, meiosis occurs immediately after fertilization, and most of the life cycle is spent in the haploid state (signified by the thin line).* (b) *In plants, fertilization and meiosis are separated, and the organism characteristically has a diploid phase and a haploid phase.* (c) *In animals, meiosis is immediately followed by fertilization. As a consequence, during most of the life cycle the organism is diploid (signified by the thick line).*

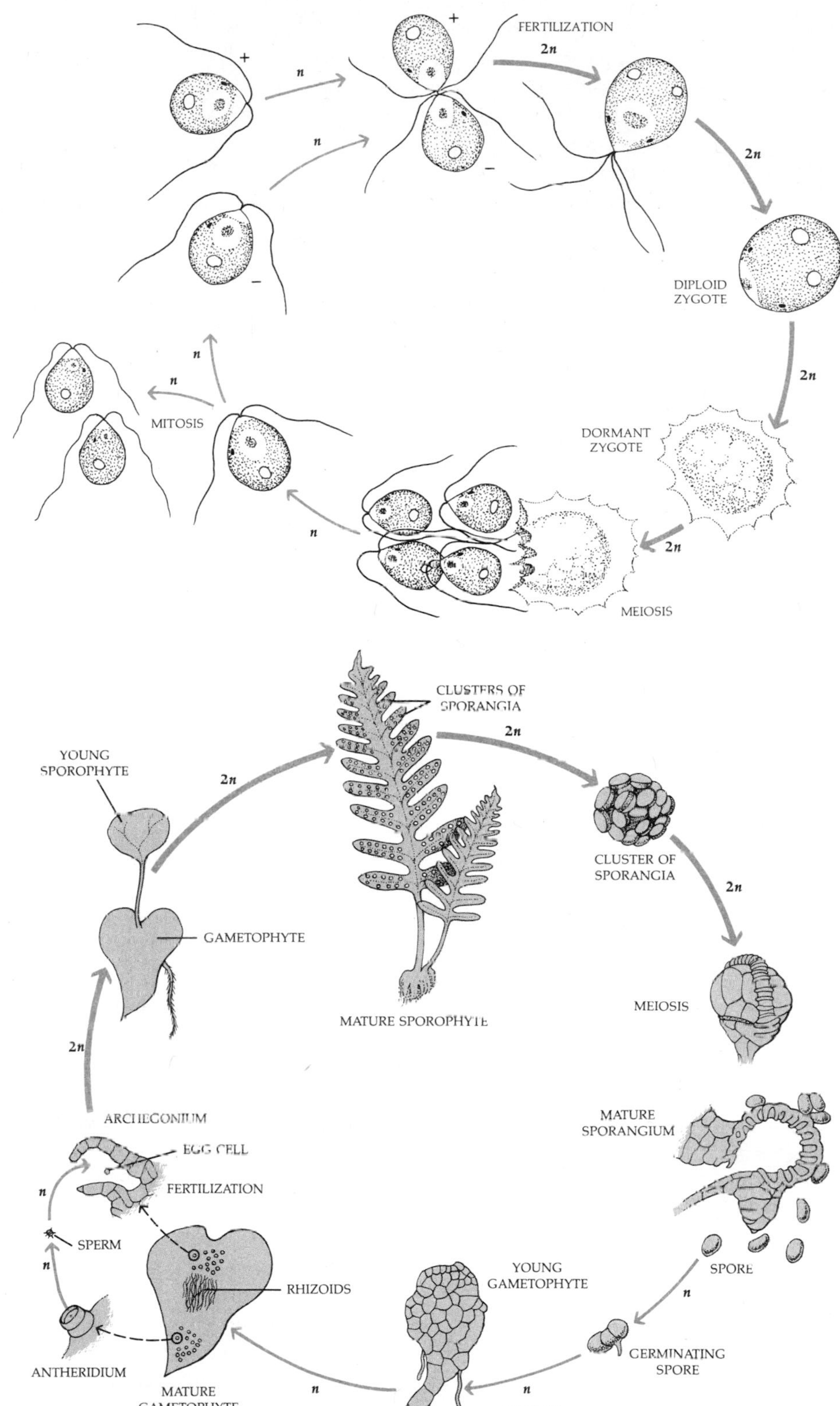

11–3

The life cycle of Chlamydomonas *is of the type shown in Figure 11–2a. The organism is haploid for most of its life cycle (thin arrows). Fertilization, the fusing of cells of different mating strains (indicated here by + and −), temporarily produces the diploid state (thick arrows). The diploid zygote divides meiotically, forming four new haploid cells, which will probably divide by mitosis (asexually) but may enter another sexual cycle.*

11–4

The life cycle of a fern is of the type shown in Figure 11–2b. Following meiosis, spores, which are haploid, are produced in the sporangia and then are shed (extreme right). The spores develop into haploid gametophytes. In many species, the gametophytes are only one layer of cells thick and are somewhat heart-shaped, as shown here (bottom). From the lower surface of the gametophyte, filaments, the rhizoids, extend downward into the soil.

On the lower surface of the gametophyte are borne the flask-shaped archegonia, which enclose the egg cells, and the antheridia, which enclose the sperm. When the sperm are mature and there is an adequate supply of water, the antheridia burst, and the sperm cells, which have numerous flagella, swim to the archegonia and fertilize the eggs. From the fertilized (2n) *egg, the zygote, the* 2n *sporophyte grows out of the archegonium within the gametophyte. After the young sporophyte becomes rooted in the soil, the gametophyte disintegrates. When it becomes mature, the sporophyte develops sporangia, in which meiosis occurs, and the cycle begins again.*

11-5
The life cycle of Homo sapiens. *Gametes, which are haploid, are produced by meiosis. At fertilization, they fuse, restoring the diploid number in the fertilized egg. The fertilized egg develops into a mature man or woman, who again produces haploid gametes. As is the case with most other animals, the life cycle is almost entirely diploid, the only exception being the gametes. This is the type of life cycle diagrammed in Figure 11-2c.*

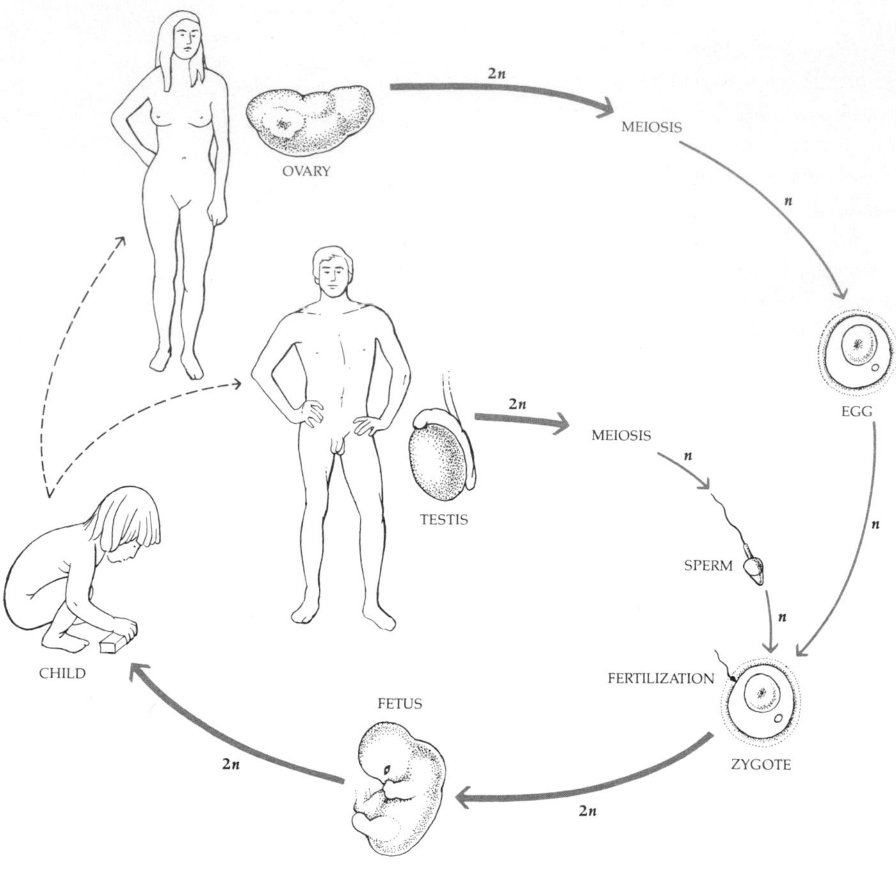

Human beings have the typical animal life cycle, in which meiosis immediately precedes fertilization and virtually all of the life cycle is spent in the diploid state (Figure 11-5).

MEIOSIS VS. MITOSIS

The events that take place during meiosis somewhat resemble those that take place during mitosis, and meiosis, which is believed to have evolved from mitosis, uses much of the same cellular machinery. But there are some important differences:

1. Two successive nuclear divisions occur, producing a total of four cells.
2. Each of the four cells contains half the number of chromosomes present in the original cell from which it was produced.
3. The haploid cells produced by meiosis contain new combinations of chromosomes. That is, the homologous chromosomes containing genetic information from each parent are assorted randomly among the four new haploid cells. (For example, whether a particular gamete produced by your body contains, for a particular hereditary characteristic, the genetic information provided by your mother or by your father is purely a matter of chance.) As a result, each cell produced by meiosis has a unique assortment of genetic information.

THE PHASES OF MEIOSIS

Meiosis consists of two successive nuclear divisions, conventionally designated meiosis I and meiosis II. In meiosis I, the homologues separate; in meiosis II, the chromatids separate. In the following discussion we shall describe meiosis in a plant cell in which the diploid number is 8 ($n = 4$).

Four of the eight chromosomes were inherited from one parent and four from the other parent. For each chromosome from one parent there is a homologous chromosome, or homologue, from the other parent—a chromosome of similar size and shape, which contains the other parent's genetic instructions for the same set of hereditary characteristics, for instance, flower color.

During interphase, the chromosomes are replicated, so that by the beginning of meiosis I each chromosome consists of two identical chromatids held together at the centromere as shown in Figure 11–6a. In the first prophase of meiosis, *prophase I*, the chromatin condenses into chromosomes and the nuclear envelope begins to dissolve.

During the first prophase, an event occurs for which the mechanism is completely unknown. The pairs of homologous chromosomes come together. Since each chromosome consists of two identical chromatids, the pairing of the homologous chromosomes actually involves four chromatids; this complex of paired homologous chromosomes is known as a *tetrad*. Once contact is made at any point between the two homologues, pairing extends, zipperlike, along the length of the chromatids. The alignment is very precise. *Crossing over*, the interchange of segments of one chromosome with corresponding segments from its homologue, takes place at this time (Figure 11–6b). Again, the mechanism by which such precise transfers take place is unknown. Such exchanges of chromosomal material are important sources of variation in the hereditary material. (We shall discuss crossing over in more detail in Chapter 13.)

If you remember the arrangement of the homologous chromosomes at this stage of meiosis, you will be able to remember all the subsequent events with little difficulty.

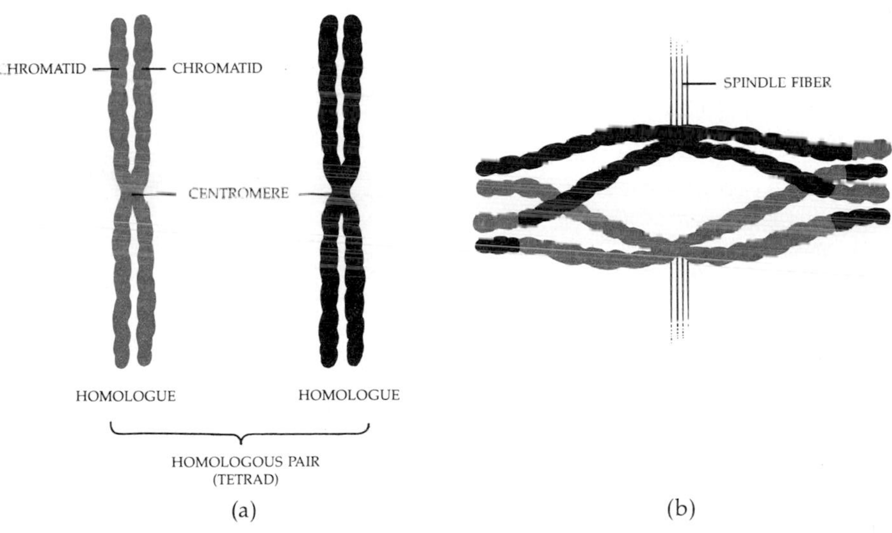

11–6
(a) *During prophase I of meiosis, chromosomes become arranged in homologous pairs. Each homologous pair consists of four chromatids and is therefore also known as a tetrad (from the Greek* tetra, *meaning "four").* (b) *Crossing over. During meiosis I, homologues, paired up as tetrads, connect at crossover points, where exchanges of segments of the chromosomes—crossing over—take place.*

LATE PROPHASE I

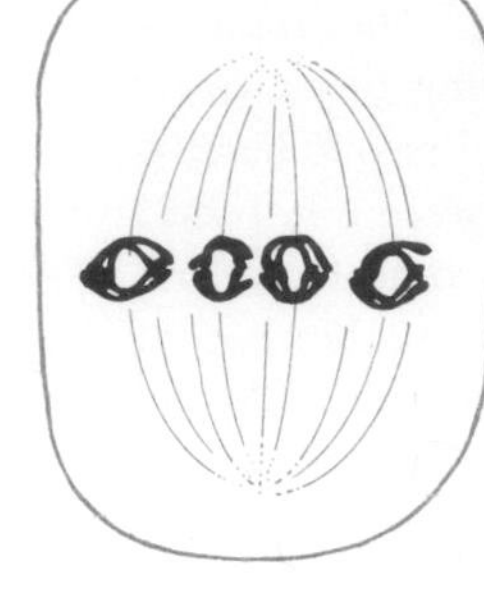

METAPHASE I

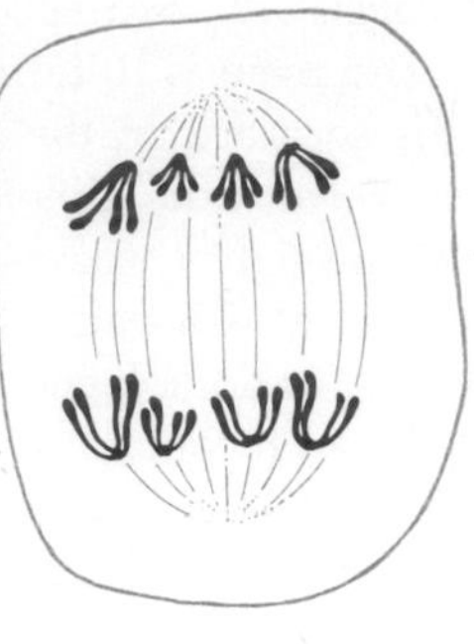

ANAPHASE I

TELOPHASE I

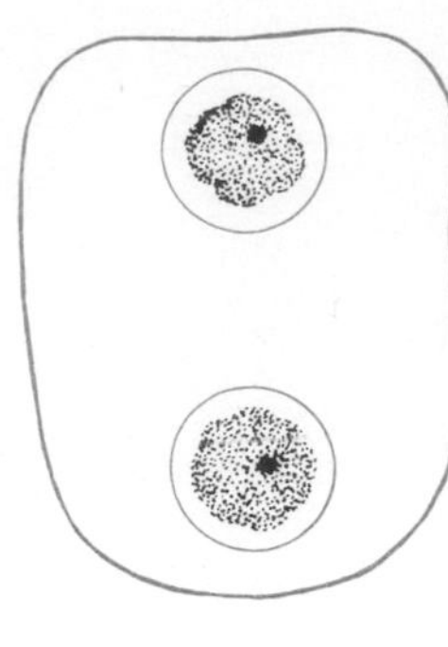

INTERPHASE II

Toward the end of prophase I, the spindle apparatus is formed, the nucleolus disperses, and the nuclear envelope breaks down.

In *metaphase I*, the homologous pairs (four in this example) line up along the equatorial plane of the cell. Each pair consists of four chromatids (two homologous chromosomes). The spindle has formed, and spindle fibers attach to each homologue at its kinetochore. If this were an animal cell, centrioles and asters would be present.

At *anaphase I*, the homologues, each consisting of two chromatids, separate, as if pulled apart by the spindle fibers attached to their kinetochores. However, the two sister chromatids of each chromosome do not separate as they did in mitosis.

By the end of the first meiotic division, *telophase I*, the homologues have moved to the poles. Each group now contains only half the number of chromosomes as the original nucleus.* Depending on the species, new nuclear envelopes may or may not form, and cytokinesis may or may not take place.

* In counting, it is often difficult to know whether to count a chromosome that has duplicated but has not divided as 1 or 2. It is customary to count such a chromosome as 1. The trick is to count centromeres.

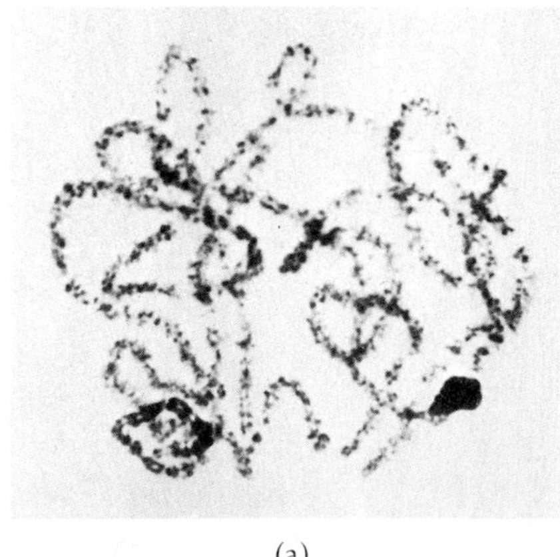

(a)

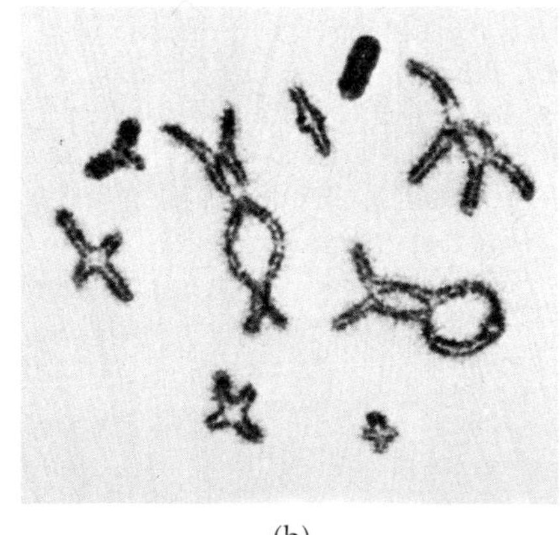

(b)

11–7
(a) *Early prophase I in the formation of a sperm cell in the grasshopper. The homologous chromosomes are now paired; the individual chromatids are not visible, however, so each chromosome still appears as a single structure, and the tetrads appear double-stranded (rather than four-stranded). The dark area in the lower right is a sex chromosome, which is very prominent in the grasshopper.* (b) *Late prophase I. All four chromatids can be seen in most of the tetrads. The spindle fibers are not visible.*

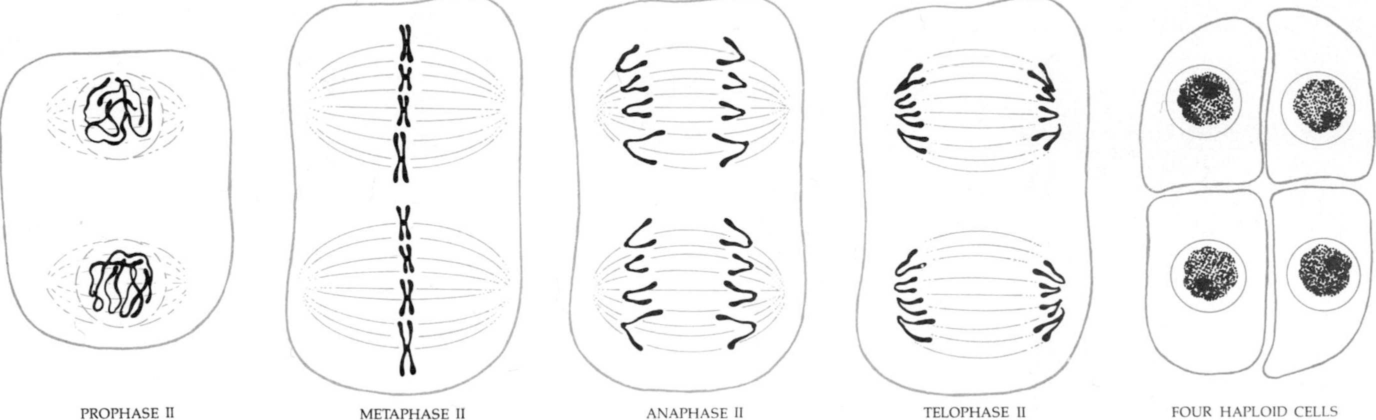

11–8
Metaphase II in crested wheat grass. The chromosomes (chromatid pairs) are approaching the equatorial plane.

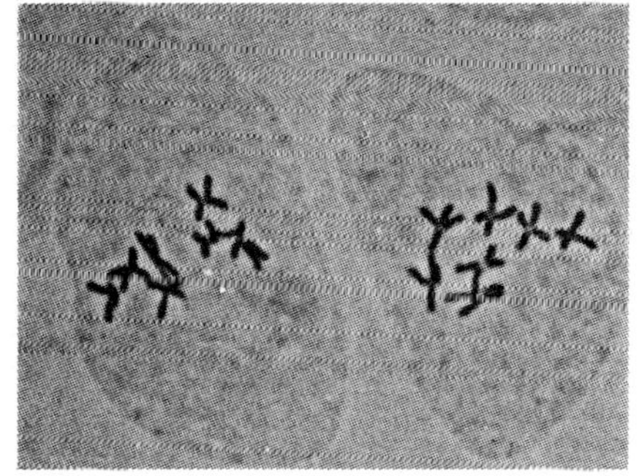

Meiosis II resembles mitosis except that it is not preceded by replication of the chromosomal material. At the beginning of the second meiotic division, the chromosomes, which may have dispersed somewhat, condense fully again. There are four in each nucleus (the haploid number), and they are still in the form of two chromatids held together at the centromere.

During *prophase II*, the nuclear envelopes, if present, dissolve, and new spindle fibers begin to appear.

During *metaphase II*, the four chromosomes in each nucleus line up on the equatorial plane. At *anaphase II*, as in mitosis, the sister chromatids separate, and each resulting chromosome moves toward one of the poles.

During *telophase II*, the spindles disappear and a nuclear envelope forms around each set of chromosomes. There are now four nuclei in all, each containing the haploid number of chromosomes. Cell division (cytokinesis) proceeds as it does following mitosis. Cell walls form, dividing the cytoplasm, and the cells begin to differentiate.

So, beginning with one cell containing eight chromosomes (four homologous pairs), we end with four cells, each with four chromosomes (no homologous pairs).

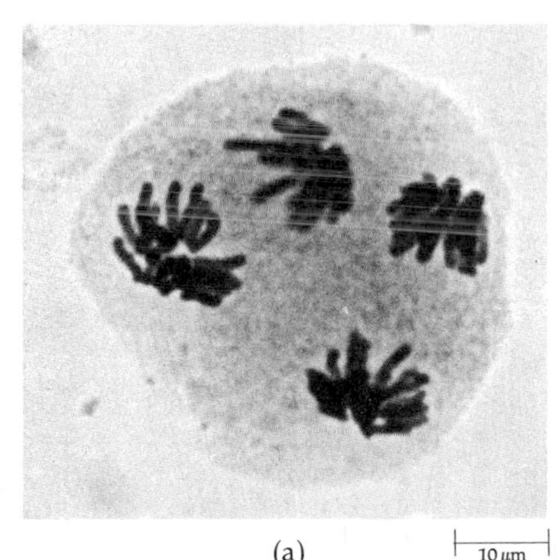

(a)

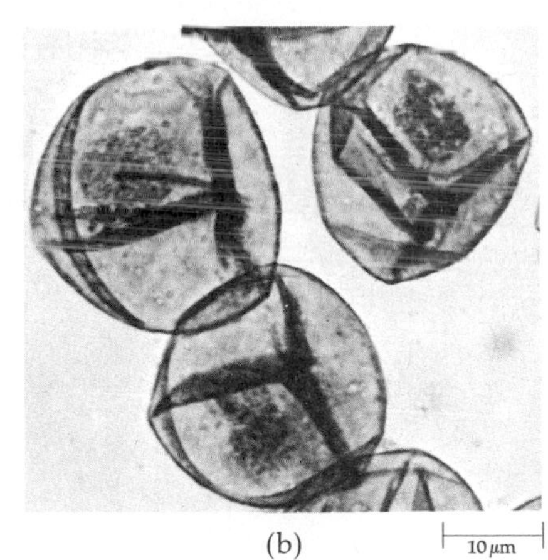

(b)

11–9
(a) *Anaphase II in the royal fern* Osmunda regalis. *The chromatids have separated, and the daughter chromosomes are moving to the opposite poles of the spindles.* (b) *The end of spore formation in* Osmunda regalis. *Each of these haploid cells can germinate to produce a gametophyte, the haploid phase in the life cycle.*

11-10
A comparison of mitosis and meiosis.

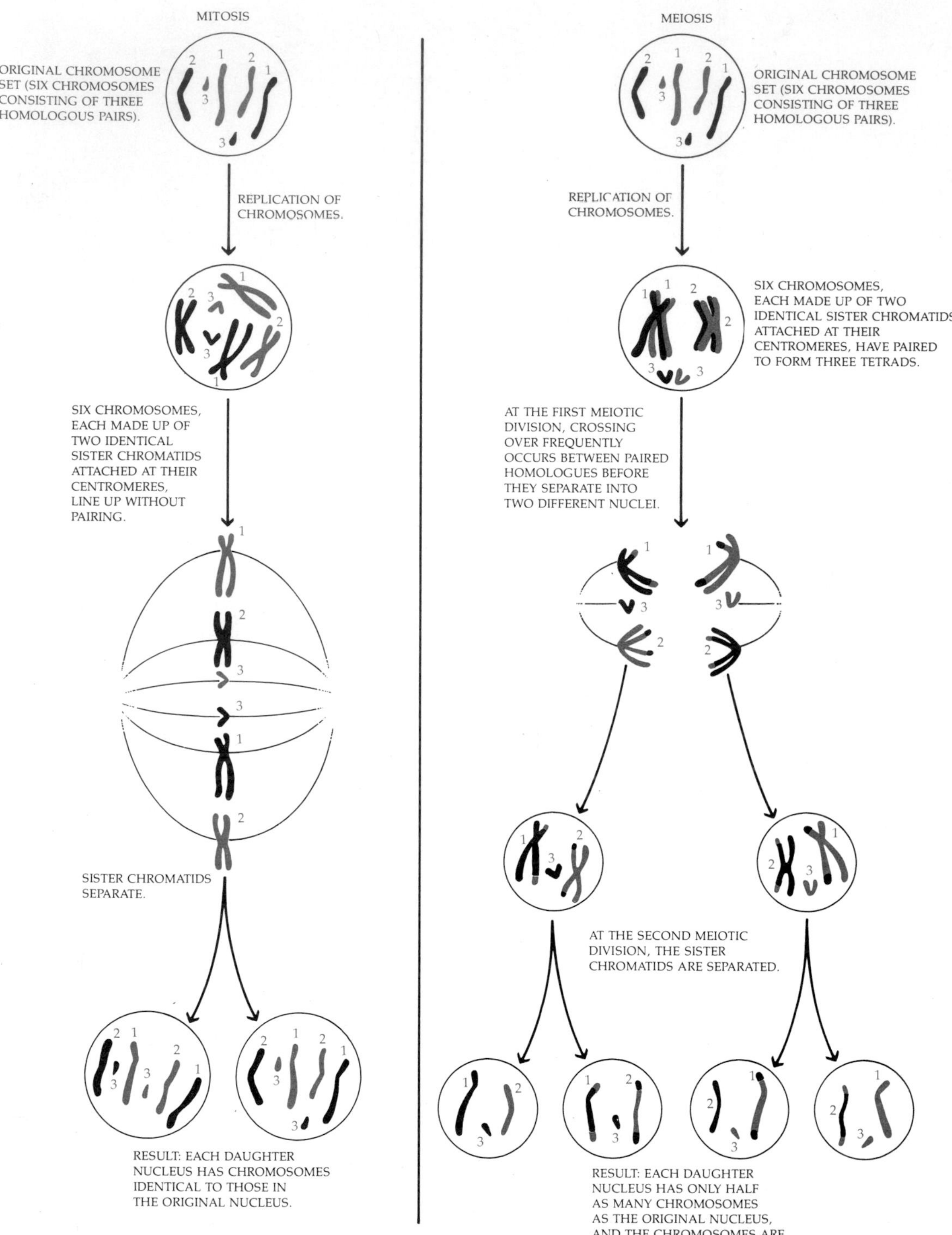

Meiosis in the Human Species

In all vertebrates, including humans, meiosis takes place in the reproductive organs, the testes of the male and the ovary of the female. In the male, a cell known as a primary spermatocyte undergoes the two divisions of meiosis to produce four spermatids, each of which then differentiates into sperm (Figure 11–11). The ejaculate of a normal human male contains 300 million sperm cells.

In females, the meiotic divisions are equal in terms of the genetic material but unequal in the way the cytoplasm is apportioned (Figure 11–12). One egg cell (ovum) is produced and two or three polar bodies, which contain the other postmeiotic nuclei. These then disintegrate. As a result of this unequal division, the ovum is well supplied with essential materials for the developing embryo. In Figure 10–2 on page 195, two polar bodies can be seen in the upper part of the micrograph.

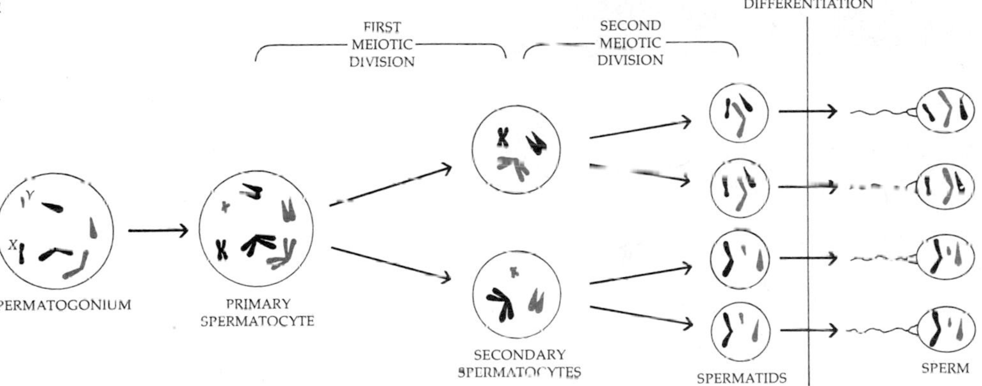

11–11
The series of changes resulting in the formation of sperm cells begins with the growth of spermatogonia into large cells known as primary spermatocytes. At the first meiotic division, each primary spermatocyte divides into two haploid secondary spermatocytes. The second meiotic division results in the formation of four haploid spermatids. The spermatids differentiate into functional sperm. For simplicity, only six (n = 3) *chromosomes are shown.*

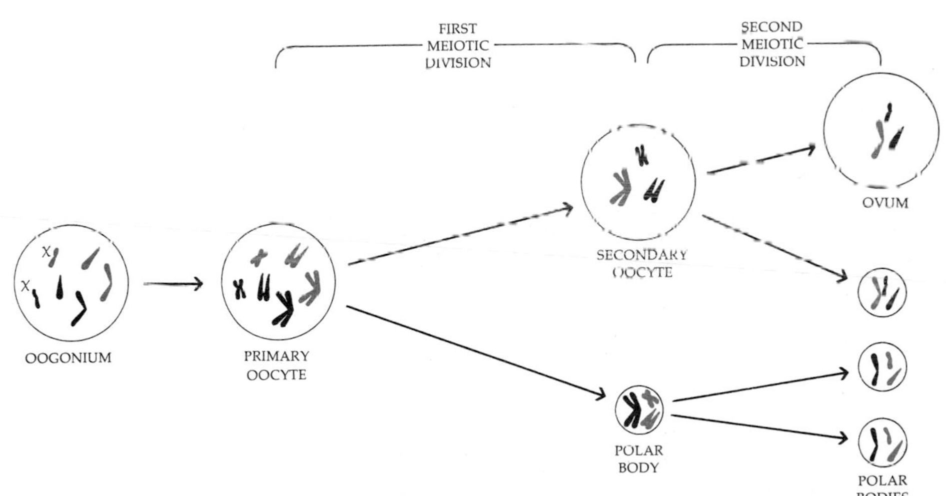

11–12
Formation of the ovum. A primary oocyte undergoes a meiotic division to produce a secondary oocyte and a polar body. This first meiotic division begins in the human female during the third month of fetal development and ends at ovulation, which may take place 50 years later. The second meiotic division, which produces the egg cell and a second polar body, does not take place until after fertilization. The first polar body may also divide.

11–13
In this multicellular animal, a Hydra, *a new organism is developing on the body of the parent. It will eventually break away and live independently. It is, of course, genetically identical to its parent.*

ASEXUAL VS. SEXUAL REPRODUCTION

Many organisms can reproduce both sexually and asexually (by mitosis). Most one-celled organisms have a life cycle similar to that of *Chlamydomonas* (Figure 11–3) in which either sexual or asexual reproduction may take place, often depending on environmental circumstances. Some one-celled animals, such as amoebas, reproduce only asexually. Many plants can also reproduce asexually; many grasses, for instance, spread by means of underground stems (rhizomes). In animals, asexual reproduction can take place by budding, as in *Hydra* (Figure 11–13), or by the breaking off of a fragment of the parent animal, as occurs in sponges, sea anemones, and certain types of worms. Because of the careful copying process of mitosis, asexually produced individuals are always genetically identical to their parents. Because of the reshuffling and crossing over of chromosomes that take place at meiosis, sexually produced individuals are different from their parents.

Sexual reproduction demands a tremendous expenditure of energy on the part of the organisms involved. Male animals often produce thousands or even millions of sperm cells for each one that reaches an egg. Plants cover themselves with flowers. Birds invest in bright, improbable plumage. Members of the human species write poems, sing songs, and renounce kingdoms, all in the cause of sexual reproduction.

11–14
Possible distributions of chromosomes at meiosis. The black chromosomes were originally of paternal origin, and the colored chromosomes of maternal origin. They were transmitted by replication and mitotic division to the cells from which the eggs and sperm were formed. In the course of meiosis, these chromosomes are sorted out among the haploid cells. As you can see, chromosomes of maternal or paternal origin do not stay together but are assorted independently. (a) *If the original number of chromosomes is 4 (n = 2), the number of possible combinations of chromosomes is 2^2, or 4.* (b) *If the original number is 6, the number of possible combinations is 2^3, or 8, and* (c) *if there are 8 chromosomes, 16 different combinations (2^4) are possible. Because maternal and paternal chromosomes differ in some of their genetic material, each of these cells is genetically different. A human male with his 46 chromosomes is capable of producing 2^{23} kinds of sperm cells—8,388,608 different combinations of chromosomes, equal in number to the population of New York City. Similarly, a human female is capable of producing 2^{23} kinds of egg cells—8,388,608 different combinations of chromosomes. And this does not take into account the additional variations introduced by crossing over, which we shall examine in Chapter 13.*

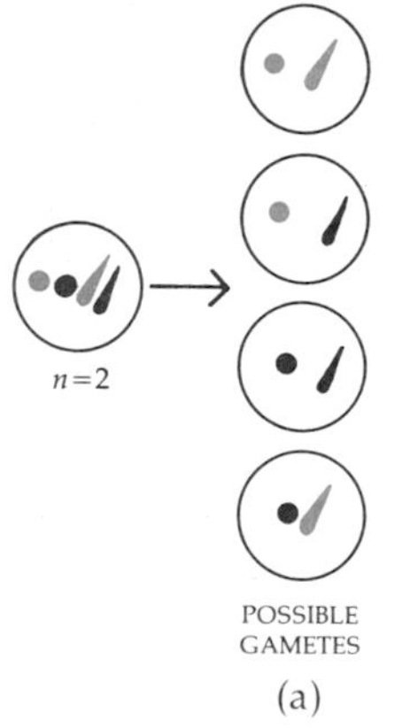

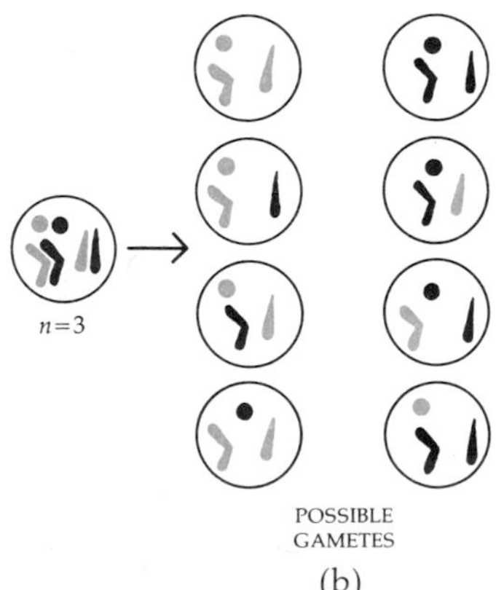

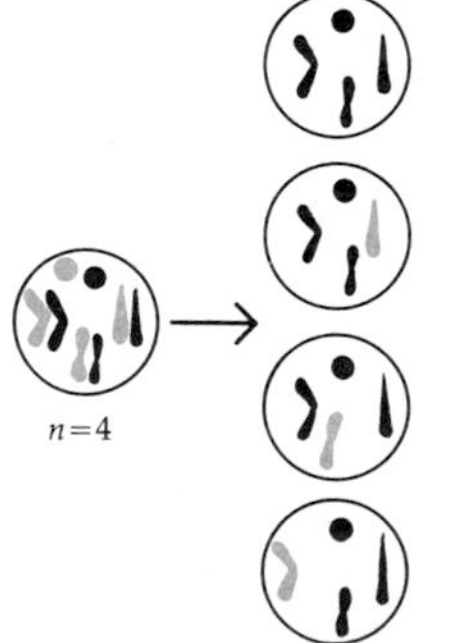
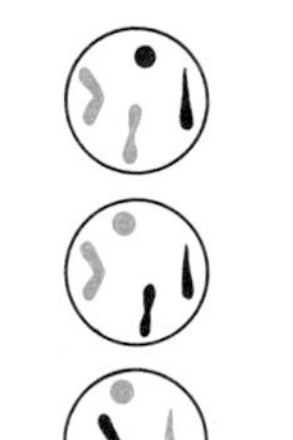
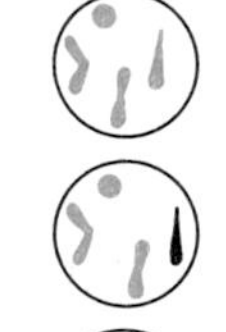

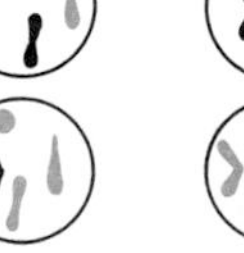

As in the case of other adaptations, biologists can appropriately ask the meaning of the phenomenon of sexual reproduction. The answer that biologists offer to this question is that sexual reproduction provides a source of variations in a population, and these variations provide the rich store of material upon which natural selection can operate. The fact that all higher animals—the most rapidly and recently evolved—reproduce sexually appears to confirm the evolutionary advantages of this method of reproduction.

Cloning

A clone, in its original usage, denoted any group of cells produced by mitosis from a single individual. It was usually used to describe colonies of bacteria or protists all descended from a single cell.

Some 25 years ago, botanist F. C. Steward of Cornell University produced a new kind of clone. He removed cells from mature carrot plants, placed them in a nutrient medium, and, by a manipulation of hormones and other chemicals, was able to coax individual cells to grow into new plants.

More recently, J. B. Gurdon of Oxford University performed a somewhat similar series of experiments with an animal. He removed nuclei from the intestinal cells of tadpoles and transplanted these nuclei into egg cells whose own nuclei had been destroyed. Some of these eggs developed normally and became mature frogs with the characteristics of the frogs that had donated the nuclei.

11-15
In experiments by J. B. Gurdon, the nucleus was removed from an intestinal cell of a tadpole and implanted into an egg cell in which the nucleus had been destroyed. In many cases, the egg developed normally, indicating that the intestinal cell nucleus contained all the information required for all the cells of the organism.

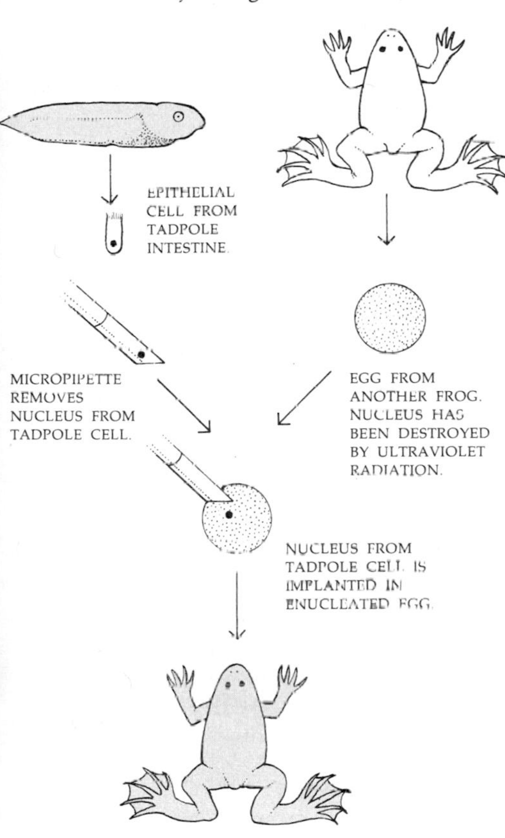

Artificial cloning of this sort is already proving a valuable technique for the production of uniform, disease-free varieties of plants. (Domestic plants have long been propagated, as you know, by cuttings, grafts, and other asexual means.) Its usefulness in the breeding of domestic animals is less clear. Animals in which cloning might be worthwhile, economically speaking, are those in which the fertilized egg, unlike the eggs of frogs, develops inside the body of the female. The technical problems of removing and reimplanting such an egg are soluble—as the recent birth of a test-tube baby in Britain has demonstrated. It is not clear, however, that the delicate procedures involved could be carried out on a large enough scale to make them economically feasible.

It has of course been suggested—although not, so far, in public by scientists in the field—that in this way a family could "reincarnate" a loved one simply by removing a cell from his or her body; a society could reproduce its leaders in politics and the arts; and a government could produce larger numbers of citizens of a proven useful type. So far, there have been great financial rewards from such experiments but only to writers of science fiction.

SUMMARY

Sexual reproduction involves a special kind of nuclear division called meiosis. Meiosis is the process by which the genetic material is reassorted and cells are produced that have the haploid chromosome number (n). The other principal component of sexual reproduction is fertilization, the coming together of haploid cells, which restores the diploid number ($2n$). There are characteristic differences among major groups of organisms as to where in the life cycle these events take place.

At the start of meiosis, the chromosomes arrange themselves in pairs, the members of which are known as homologues. Each homologue consists of two

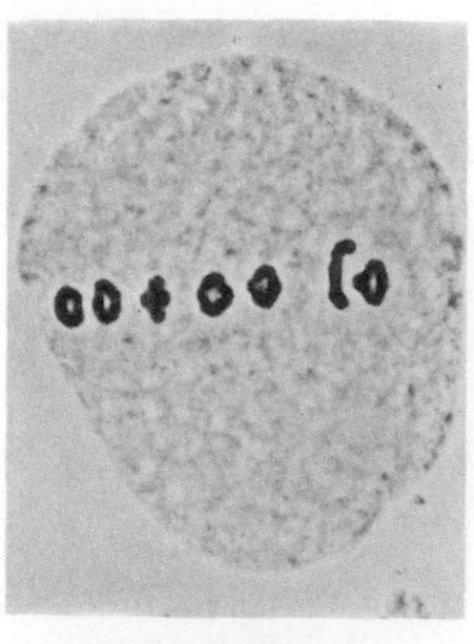
(a)

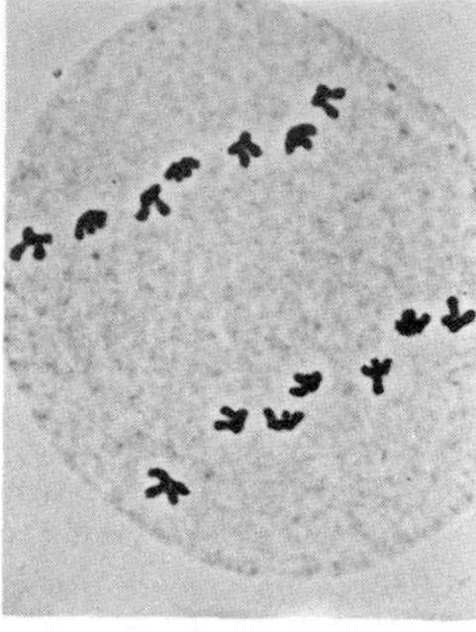
(b)

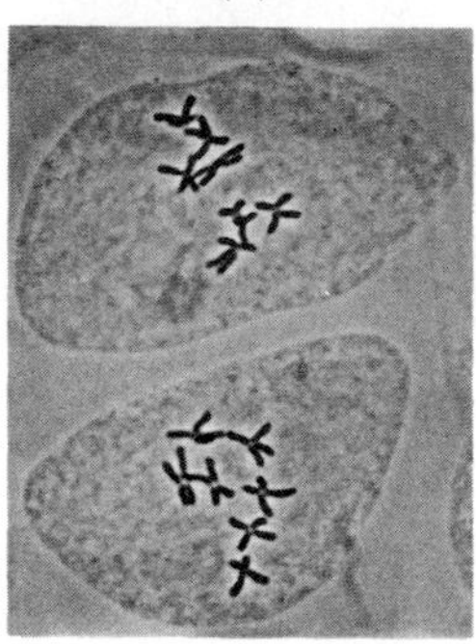
(c)

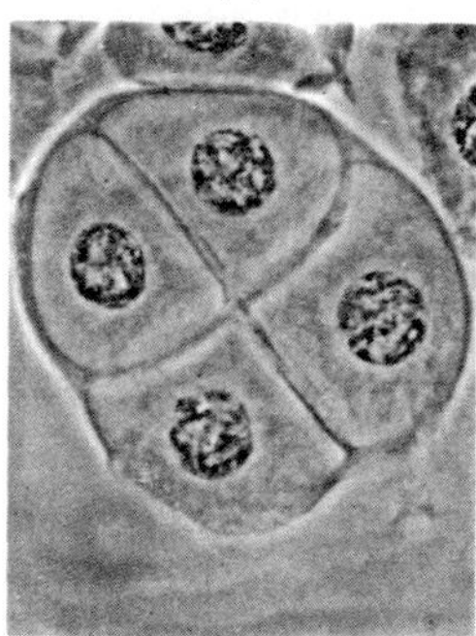
(d)

identical chromatids. Early in meiosis, crossing over occurs between homologues, resulting in exchanges of genetic material.

In the first stage of meiosis, the homologues are separated. Two nuclei are produced, each with a haploid number of chromosomes, each consisting of two chromatids. The cell enters interphase, but the chromosomal material is not replicated. In the second stage of meiosis, the sister chromatids separate as in mitosis. When the two new nuclei divide, four haploid cells result.

Meiosis results in haploid cells, each of which contains a unique assortment of genetic information due to crossing over and random assortment of homologues. Thus meiosis provides a source of variation and so makes it possible for evolution to take place.

QUESTIONS

1. Distinguish among the following: haploid/diploid/polyploid; sporophyte/gametophyte; gamete/zygote; meiosis I/meiosis II; homologue/tetrad.

2. Draw a diagram of a cell with six chromosomes ($n = 3$) at meiotic prophase I. Label each pair of chromosomes differently (for example, one pair labeled A^1 and A^2, and another B^1 and B^2, etc.).

3. Diagram the possible gametes resulting from meiosis in a plant cell with six chromosomes ($n = 3$).

4. Identify the stages of meiosis in crested wheat grass shown in the micrographs in the margin. What stage of meiosis is visible in Figure 11–1?

5. Compare and contrast the processes and the consequences of meiosis and mitosis.

6. In our bodies and in those of other higher animals, both mitosis and meiosis occur. Describe the functions of these two processes in our bodies. Why is it advantageous to us to have both processes?

7. When does sexual reproduction *not* require two parents? (If you can't answer this question, wait until you read the next chapter and try again.)

8. Suppose, as has been suggested in recent fiction, that a group of individuals were to be cloned from a person such as Adolf Hitler. In what respects do you think they would resemble their "parent"?

9. Recently an eminent scientist, in commenting on the possibility of a human clone, stated that in order to produce just one human clone, one would have to clone everybody in the world. Can you explain what he meant?

CHAPTER 12

From an Abbey Garden: The Beginning of Genetics

12-1
The protruding lower lip of the Hapsburgs is a famous example of an inherited trait. These portraits of members of the Hapsburg family encompass a period of about 300 years: (a) *Rudolph I (1218–1291), King of Germany,* (b) *Maximilian I (1460–1519), Holy Roman Emperor,* (c) *Charles V (1500–1558), Holy Roman Emperor,* (d) *Ferdinand I (1503–1564), Holy Roman Emperor.*

In this chapter and the one that follows, we are going to trace the early history of the science of genetics. Much of modern genetics and also of modern evolutionary theory is built upon these sturdy foundations. Do not dismiss these scientists because they knew less than you do today. In this account, you will become acquainted with some of the most brilliant scientists our civilization has ever known. As Isaac Newton observed, we can see as far as we do because we stand on the shoulders of giants.

Among all the symbols in biology, perhaps the most widely used and most ancient are the hand mirror of Venus (♀) and the shield and spear of Mars (♂), the biologists' shorthand for female and male. Ideas about the nature of biological inheritance—the role of male and female—are even older than these famous symbols. Very early, it must have been noticed that certain characteristics—hair color, for example, a large nose, or a small chin—were common to both parent and offspring. And throughout history, the concept of biological inheritance has been an important factor in the social organizations of mankind, determining the distribution of wealth, power, land, and royal privileges.

Sometimes a family trait is so distinctive that it can be traced through many generations. A famous example of such a characteristic is the Hapsburg lip (Figure 12-1), which has appeared in Hapsburg after Hapsburg, over and over again since at least the thirteenth century. Examples such as this have made it easy to accept the importance of inheritance in the formation of the individual, but it is only comparatively recently that we have begun to understand how this process works. In fact, the study of heredity as a science did not really begin until the second half of the nineteenth century. Yet the problems posed in this study are among the most fundamental in biology, since self-replication is the essence of the hereditary process and one of the principal properties of living systems.

EARLY IDEAS ABOUT HEREDITY

Far back in human history, people learned to improve domestic animals and crops by inbreeding and crossbreeding. In the case of date palms, male and female flowers are found on different trees, and artificial fertilization of the palm was well known to the ancient Babylonians and Egyptians. The nature of the difference between the two flowers was understood by Theophrastus (380–287 B.C.). "The males should be brought to the females," he wrote, "for the male makes them ripen and persist." The mule, a hybrid (a cross between a male donkey and a mare), was well known in the days of Homer. Both Plutarch and Lucretius noted in their writings that some children resemble their mothers, some resemble their fathers, and some even skip back a generation to resemble a grandparent; this fact, so easy to observe, continued to puzzle people for a very long time.

Since the nature of sexual reproduction was poorly understood, bizarre crossbreeds were the subject of many legends. The wife of Minos, according to Greek mythology, mated with a bull and produced the Minotaur. Folk heroes of Russia and Scandinavia were traditionally the sons of women who had been captured by bears, from which these men derived their great strength and so enriched the national stock. The camel and the leopard also crossbred from time to time, according to the early naturalists, who were otherwise unable—and it is hard to blame them—to explain an animal as improbable as the giraffe (the common giraffe still bears the species name of *camelopardalis*). Thus folklore reflected early and imperfect glimpses of the nature of hereditary relationships.

The first scientist known to have pondered the nature of heredity was Aristotle. He postulated that the male semen was made up of a number of imperfectly blended ingredients and that, at fertilization, it mixed with the "female semen," the menstrual fluid, giving form and power (*dynamis*) to this amorphous substance. No one had a better idea—or indeed many ideas at all—for two thousand years. Seventeenth-century medical texts show various stages in the coagulation of the embryo from the mixture of maternal and paternal semens. Indeed, many scientists as well as laymen did not believe that such mixtures were even always necessary; they held that life, at least the "simpler" forms of life, could arise by spontaneous generation. Worms, flies, and various crawling things, it was commonly believed, took shape from putrid substances, ooze, and mud, and a lady's hair dropped in a rain barrel could turn into a snake. Jan Baptista van Helmont, who also carried out experiments on the growth of plants (see page 142), published his personal recipe for the production of mice: One need only place a dirty shirt in an open pot containing a few grains of wheat, and in 21 days mice would appear. He had performed the experiment himself, he said. The mice would be adults, both male and female, he added, and would be able to produce more mice by mating.

12–2
Only fairly recently has it been realized that living things come only from other living things of the same species and never from another species or from lifeless matter. This picture from an old Turkish history of India shows a wakwak tree, which bears human fruit. According to the account, the tree is to be found on an island in the South Pacific.

THE FIRST EXPERIMENTS

In 1677, the Dutch lens maker Anton van Leeuwenhoek discovered living sperm—"animalcules," he called them—in the seminal fluid of various animals, including man. Enthusiastic followers peered through Leeuwenhoek's "magic looking glass" (his homemade microscope) and believed they saw within each human sperm a tiny creature—the homunculus, or "little man" (Figure 12–3). This little man was the future human being in miniature. Once implanted in the female womb, the future human being was nurtured there; the only contribution that the mother made was to serve as an incubator for the growing fetus. Any resemblance a child might have to its mother, these theorists held, was because of "prenatal influences."

During the very same decade (the 1670s) that van Leeuwenhoek first saw human sperm cells, another Dutchman, Régnier de Graaf, described for the first time the ovarian follicle, the structure in which the human egg cell forms. Although the actual human egg was not seen for another 150 years, the existence of a human egg was rapidly accepted. In fact, de Graaf attracted a school of followers, the ovists, who were as convinced of their opinions as the animalculists, or spermists, were of theirs and who soon contended openly with them. It was the female egg, the ovists said, that contained the future human being in miniature; the animalcules in the male seminal fluid merely stimulated the egg to grow. Ovists and animalculists alike carried the argument one logical step further. Each homunculus had within it another perfectly formed homunculus, and in that was still another one, and so on—children, grandchildren, and great-grandchildren, all stored away for future use. Some ovists even went so far as to say that Eve had contained within her body all the unborn generations yet to come, each egg fitting closely inside another like a child's hollow blocks. Each female generation since Eve has contained one less than the previous generation, they explained, and after 200 million generations, all the eggs will be spent and human life will come to an end.

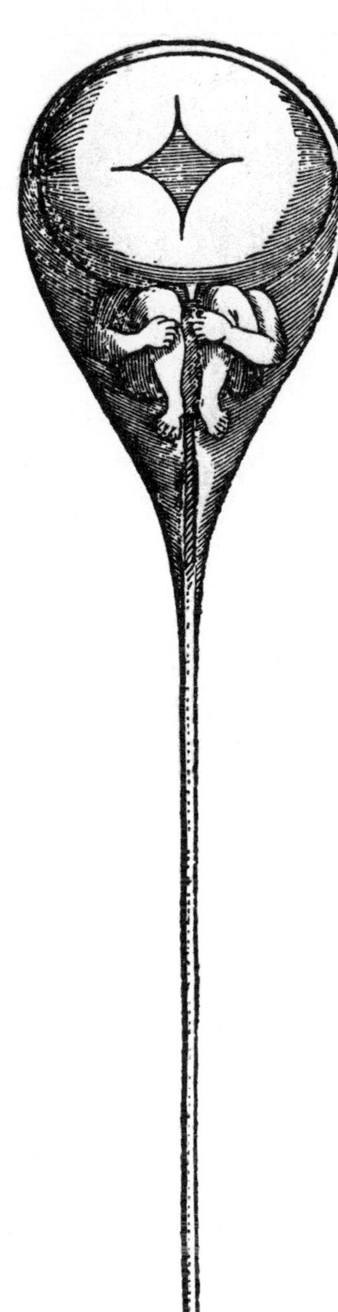

12–3
What the animalculists, or spermists, of the seventeenth and eighteenth centuries believed they saw when they looked through a microscope at sperm cells. This is a homunculus ("little man"), a future human being in miniature, in a sperm cell.

BLENDING INHERITANCE

By the middle of the nineteenth century, the concepts of the ovists and spermists began to yield to new data. The facts that challenged these earlier hypotheses came not so much from scientific experiments as from practical attempts by master gardeners to produce new ornamental plants. Artificial crossings of such plants showed that, in general, regardless of which plant supplied the pollen (which contains the male cells, or sperm) and which plant contributed the egg cells, or ova, both contributed to the characteristics of the new variety. But this conclusion raised even more puzzling questions: What exactly did each parent plant contribute? How did all the hundreds of characteristics of each plant get combined and packed into a single seed?

The most widely held hypothesis of the nineteenth century was blending inheritance. According to this hypothesis, when the sex cells (or gametes) combine, there is a mixing of hereditary material that results in a blend, analogous to a blend of two different-colored inks. On the basis of such a theory, one would predict that the offspring of a black animal and a white animal would be gray, and their offspring would also be gray because the black and white hereditary material, once blended, could never be separated again.

You can see why this concept was unsatisfactory. It ignored the phenomenon of traits skipping a generation, or even several generations, and then reappearing. To Charles Darwin and other proponents of the theory of evolution, it presented particular difficulties. Evolution, according to Darwin, as we noted in the Introduction (page 9), takes place as a result of random variations and natural selection. If the hypothesis of blending inheritance were valid, the small hereditary variations would disappear, like a single drop of ink in the many-colored mixture. Sexual reproduction would eventually result in complete uniformity, natural selection would have no raw material on which to act, and evolution would not occur.

THE CONTRIBUTIONS OF MENDEL

Gregor Mendel, who was born into a peasant family in 1822, entered a monastery in Brünn (now Brno, Czechoslovakia), where he was able to receive an education.

SPONTANEOUS GENERATION

Most of the early biologists, from the time of Aristotle, believed that simple things such as worms, beetles, frogs, and salamanders could originate spontaneously in dust or mud, that rodents formed from moist grain, and that plant lice condensed from a dewdrop. In the seventeenth century, Francesco Redi performed a famous experiment in which he put out decaying meat in a group of wide-mouthed jars—some with lids, some covered by a fine veil, and some open—and proved that maggots arose only where flies were able to lay their eggs.

By the nineteenth century, no scientist continued to believe that complex organisms arise spontaneously. With the advent of microscopy, however, belief in spontaneous generation was vigorously revived. It was necessary only to put decomposing substances in a warm place for a short time and tiny "live beasts" appeared under the lens, before one's very eyes. By 1860, the controversy had become so spirited that the Paris Academy of Sciences offered a prize for experiments that would throw new light on the question. The prize was claimed in 1864 by Louis Pasteur, who devised experiments to show that microorganisms appeared only as contaminants from the air and not "spontaneously," as his opponents claimed. In his experiments he used swan-necked flasks, which permitted the entrance of oxygen, thought to be necessary for life, but which, in their long, curving necks, trapped bacteria, fungal spores, and other microbial life and thereby protected the contents of the flask from contamination. He showed that if the liquid in the flask was boiled (which killed any microorganisms already present) and the neck of the flask was allowed to remain intact, no microorganism would appear. Some of his original flasks, still sterile, remain on display. Only if the neck of the flask was broken off, permitting air to enter the flask, did microorganisms appear.

"Life is a germ, and a germ is Life," Pasteur proclaimed at a brilliant "scientific evening" at the Sorbonne before the social elite of Paris. "Never will the doctrine of spontaneous generation recover from the mortal blow of this simple experiment!"

In retrospect, Pasteur's well-planned experiments were considered so decisive because the broad question of whether or not spontaneous generation ever occurred had been reduced to a simpler question of whether or not it occurred under the specific conditions claimed for it. In fact, today it is generally agreed that some form of "spontaneous generation" did indeed take place, although under quite different circumstances, when the earth was young.

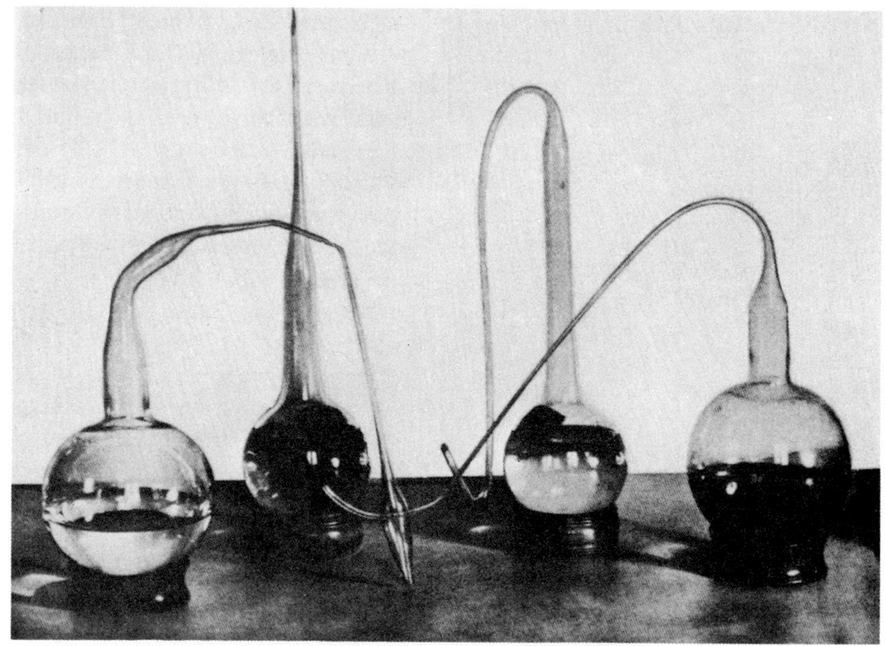

Pasteur's swan-necked flasks, which he used to counter the argument that spontaneous generation failed to occur in sealed vessels because air was excluded. These flasks permitted the entrance of oxygen, thought to be essential for life, but their long, curving necks trapped spores of microorganisms and thereby protected the liquids in the flasks from contamination.

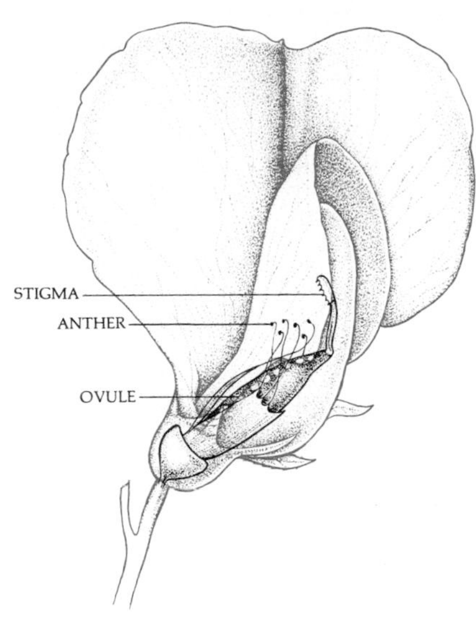

12–4
In a flower, pollen develops in the anther and the egg cells in the ovule. Pollination occurs when pollen grains, trapped on the stigma, germinate and grow down to the ovule, where they release sperm cells. The fertilized egg develops within the ovule. In the garden pea (Pisum sativum), *ovule and embryo form the peas (the seeds).*

Pollination in most species involves the pollen from one plant (often carried by an insect) being caught on the stigma of another plant. This is called cross-pollination.

In the pea flower, however, the stigma and anthers are completely enclosed by petals, and the flower, unlike most, does not open until after fertilization has taken place. Thus the plant normally self-pollinates. In his crossbreeding experiments, Mendel pried open the bud before the pollen matured and removed the anthers with tweezers. Then he artificially pollinated the flower by dusting the stigma with pollen collected from other plants.

He attended the University of Vienna for two years, pursuing studies in both mathematics and science. He failed his tests for the teaching certificate he was seeking and so retired to the monastery, of which he eventually became abbot. Mendel's work, carried on in a quiet monastery garden and ignored until after his death, marks the beginning of modern genetics.

Mendel's great contribution was to demonstrate that inherited characteristics are carried by factors that are discrete units and that are parceled out in different ways—reassorted—in each generation. These discrete units eventually came to be known as *genes.*

Mendel was successful where others were not, largely for three reasons. First, he planned his experiments carefully and imaginatively, choosing for study definite and measurable hereditary differences. Second, probably because of his background in physics and mathematics, he was one of the first to apply mathematics to the study of biology. Even though his mathematics was simple, the idea that mathematical laws could be applied to living organisms was startlingly new. Finally, he carried out his experiments in a thoroughly scientific way (as his predecessors had not). He tested a very specific model in a series of logically designed experiments.

Mendel's choice of pea plants for his experiments was a very deliberate one. The plants were commercially available, easy to cultivate, and grew rapidly. Different varieties had clearly different characteristics that "bred true," reappearing in crop after crop. Finally, the sexual organs of the pea flower are entirely enclosed by the petals, even when they are mature (see Figure 12–4). Consequently, the flower normally self-pollinates. Although the plants could be crossbred experimentally, accidental crossbreeding could not occur to confuse the experimental results. In Mendel's own words, quoted from his original paper, "The value and utility of any experiment are determined by the fitness of the material to the purpose for which it is used."

Mendel's choice of the pea plant for his experiments was not original. However, he performed two steps, both crucial, not carried out by his predecessors: (1) He studied the progeny of not only the first generation but also the second, and (2) he counted.

Principle of Segregation

Mendel began with 32 different types of pea plants, which he studied for two years to see which characteristics were clearly defined. As he said later in his report on this work, he did not want to experiment with traits in which the difference could be "of a 'more or less' nature, which is often difficult to define." As a result of these observations, he selected for study seven traits that appeared as conspicuously different characteristics in different varieties of plants. One variety of plant, for example, always produced yellow peas, or seeds, while another always produced green ones. In one variety, the seeds, when dried, had a wrinkled appearance, while in another variety they were smooth. The complete list of alternate traits is given in Table 12–1.

Mendel performed experimental crosses, removing the anthers from flowers and dusting the stigmas with pollen from a flower of another variety. He found that in every case in the first generation (now known as the F_1 in biological shorthand), one of the alternate traits disappeared completely without a sign. All the seeds

Table 12-1 *Results of Mendel's Experiments with Pea Plants*

	ORIGINAL CROSSES			(F_2) SECOND GENERATION		
TRAIT	DOMINANT	×	RECESSIVE	DOMINANT	RECESSIVE	TOTAL
Seed form	Round	×	Wrinkled	5,474	1,850	7,324
Seed color	Yellow	×	Green	6,022	2,001	8,023
Flower position	Axial	×	Terminal	651	207	858
Flower color	Purple	×	White	705	224	929
Pod form	Inflated	×	Constricted	882	299	1,181
Pod color	Green	×	Yellow	428	152	580
Stem length	Tall	×	Dwarf	787	277	1,064

12-5
Garden pea plant with seeds (peas) in the approximate ratio of three yellow seeds for each green seed, as observed by Mendel.

produced as a result of a cross between yellow-seeded plants and green-seeded plants were as yellow-seeded as the yellow-seeded parent. All of the flowers produced by plants resulting from a cross between a purple-flowered plant and a white-flowered plant were purple. These traits and other such traits Mendel called *dominant.*

The interesting question was: What had happened to the second trait—the whiteness of the flower or the greenness of the seed—which had been passed on so faithfully for generations by the parent stock? Mendel let the pea plant itself carry out the next stage of the experiment by permitting the F_1 plants to self-pollinate and produce another generation, the F_2. The traits that had disappeared in the first generation reappeared in the F_2. In Table 12-1 are the results of Mendel's actual counts. These traits, which appeared in the parent generation and reappeared in the F_2 generation, must also have been present somehow in the F_1 generation, although not apparent there. Mendel called these traits *recessive.*

If you analyze the results in Table 12-1 as Mendel did, you will notice that the dominant and recessive traits appear in the second, or F_2, generation in ratios of about 3:1. How do the recessives disappear so completely and then appear again, and always in such constant proportions? It was in answering this question that Mendel made his greatest contribution. He saw that the appearance and disappearance of traits and their constant proportions could be explained if hereditary characteristics are determined by discrete (separable) factors.* These factors, Mendel saw, must occur in the offspring as pairs, one factor inherited from each parent. These pairs of factors (genes) are separated again when the mature F_1 offspring produce sex cells, resulting in two kinds of gametes, with one gene of the pair in each.

The hypothesis that every individual carries pairs of factors for each trait that segregate during the formation of gametes is known as Mendel's first law, or the *principle of segregation.*

* Mendel called these factors *Elemente.* As we noted previously, they are now known as genes, and we shall use this term to refer to them.

MENDEL AND THE LAWS OF CHANCE

In applying mathematics to the study of genetics, Mendel was stating that the laws of chance apply to biology as they do to the physical sciences. Toss an imaginary coin. The chance that it will turn up heads is fifty-fifty, or $\frac{1}{2}$. *The chance that it will turn up tails is also fifty-fifty, or* $\frac{1}{2}$. *The chance that it will turn up one or the other is certain, or one chance in one. Now toss two imaginary coins. The chance that one will turn up heads is again* $\frac{1}{2}$. *The chance that the second will turn up heads is also* $\frac{1}{2}$. *The chance that both will turn up heads is* $\frac{1}{2} \times \frac{1}{2}$, or $\frac{1}{4}$. *The probability of two independent events occurring together is simply the probability of one occurring alone multiplied by the probability of the other occurring alone. The chance of both turning up tails is similarly* $\frac{1}{2} \times \frac{1}{2}$. *The chance of the first turning up tails and the second turning up heads is* $\frac{1}{2} \times \frac{1}{2}$, *and the chance of the second turning up tails and the first heads is* $\frac{1}{2} \times \frac{1}{2}$.

We can diagram this in a Punnett square (see figure), which indicates that the combination in each square has an equal chance of occurring. It was undoubtedly the observation that one-fourth of the offspring in the F_2 *generation showed the recessive phenotype that indicated to Mendel that he was dealing with a simple case of the laws of probability.*

If there were three coins involved, the probability of any given combination would be simply the product of all three fifty-fifty possibilities: $\frac{1}{2} \times \frac{1}{2} \times \frac{1}{2}$, or $\frac{1}{8}$. *Similarly, with four coins, the probability of any given combination is* $\frac{1}{2} \times \frac{1}{2} \times \frac{1}{2} \times \frac{1}{2}$, or $\frac{1}{16}$. *The Punnett square on page 230 expresses the probability (or chance) of each of any one of four possible combinations.*

Notice that in planning his experiments, Mendel made two assumptions: (1) that an equal number of male and female gametes are produced; and (2) that the gametes combined at random. Thus, the laws of probability could be employed.

If you toss two coins 4 times, it is unlikely that you will get the precise results diagrammed above. However, if you toss two coins 100 times, you will come close to the results predicted in the Punnett square, and if you toss two coins 1,000 times, you will be very close indeed. As Mendel knew, the relationship between dominants and recessives might well not have held true if he had been dealing with a small sample. The larger the sample, the more closely it will conform to results predicted by the laws of chance.

The two factors in a pair might be the same, in which case the self-pollinating plant would breed true. Or the two factors might be different; such different, or alternative, forms came to be known as *alleles*. Yellow-seededness and green-seededness, for instance, are determined by alleles, different forms of the gene (factor) for seed color. When the genes of a gene pair are the same, the organism is said to be *homozygous* for that particular trait; when the genes of a gene pair are different, the organism is *heterozygous* for that trait.

12–6
A pea plant homozygous for purple flowers is represented as WW in genetic shorthand. The gene for purple flowers is designated W because of a convention by which geneticists, in indicating a pair of alleles, use the first letter of the less common form (white). The capital indicates the dominant, the lowercase the recessive. A WW plant can produce gametes with only a purple-flower (W) gene. The female symbol ♀ indicates that this flower contributed the egg cells, or female gametes.

A white pea plant (ww) can produce gametes with only a white-flower (w) gene. The male symbol ♂ indicates that this flower contributed the sperm cells, or male gametes.

When a w sperm cell fertilizes a W egg cell, the result is a Ww pea plant, which, since the W gene is dominant, will produce purple flowers. However, this Ww plant can produce gametes with either a W or a w allele.

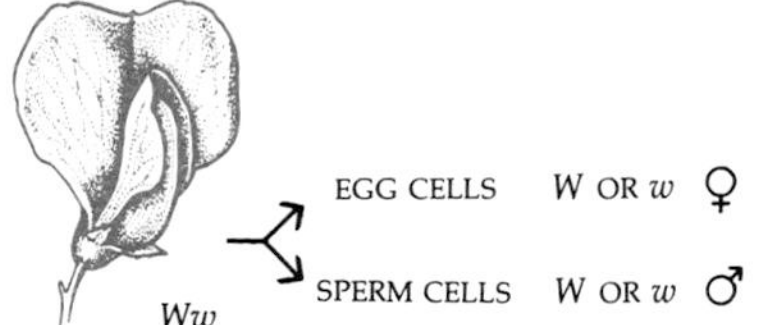

And so, if the plant self-pollinates, four possible crosses can occur:

♀ W × ♂ W → *purple flowers*
♀ W × ♂ w → *purple flowers*
♀ w × ♂ W → *purple flowers*
♀ w × ♂ w → *white flowers*

These results are summarized in Figure 12–7.

When gametes are formed, genes are passed on to them, but each gamete contains only one of two possible alleles for any given trait. When two gametes combine in the fertilized egg, the genes occur in matched pairs again. If the alleles are the same, both will be expressed. If the alleles are different, one may be dominant over the other; in this case, the organism will appear as if it had only this allele. This outward appearance is known as its *phenotype.* However, in its genetic makeup, or *genotype,* each allele still exists independently and as a discrete unit, even though it may not be visible in the phenotype. The recessive allele will separate from its dominant partner when gametes are again formed. Only if two recessive alleles come together—one from the female gamete and one from the male—will the phenotype then show the recessive trait. When pea plants homozygous for purple flowers are crossed with pea plants having white flowers, only pea plants with purple flowers are produced, although each plant in the F_1 generation will carry a gene for purple and a gene for white. Figure 12–6 shows what happens in the F_2 generation if the F_1 generation self-pollinates. Notice that the result would be the same if an F_1 individual is cross-fertilized with another F_1 individual, which is the way these experiments are performed with animals and with plants that are not self-pollinating.

In order to test his hypothesis (diagrammed in Figure 12–7), Mendel performed two additional experiments. He crossed white-flowering plants with white-flowering plants and confirmed that they bred true—that only white-flowering plants were produced. Next he crossed one of his F_1 individuals, the result of a cross between purple- and white-flowering plants, with a true-breeding white-flowering plant. To the outside observer, it would appear as if Mendel were simply repeating his first experiment, crossing plants having purple flowers with plants having white flowers. But he knew that if his hypothesis were correct, his results would be different from those of his first experiment. In fact, he actually predicted the results of such a cross before he made it. Can you? Stop a moment and think about it.

The easiest way to analyze the possible result of such a cross is to diagram it, as in Figure 12–8. This experiment, which reveals the genotypes, is known as a testcross. A testcross is an experimental cross between an individual with the dominant phenotype for a given trait with another individual that is homozygous recessive for the trait. The resulting ratio of offspring will indicate whether the individual with the dominant phenotype was homozygous or heterozygous for the trait being studied. In the testcross shown in Figure 12–8, the cross revealed that the genotype of the organism being tested was *Ww* rather than *WW.*

Principle of Independent Assortment

In a second series of experiments, Mendel studied crosses between pea plants that differed simultaneously in two characteristics; for example, one parent plant had peas that were round and yellow, and the other had peas that were wrinkled and green. The round and yellow traits, you will recall (see Table 12–1), are dominant, and the wrinkled and green are recessive. As you would expect, all the seeds produced by a cross between the parental types were round and yellow. When these seeds were planted and the flowers allowed to self-pollinate, 556 seeds were produced. Of these, 315 showed the two dominant characteristics, round and yellow, but only 32 combined the recessive traits, green and wrinkled. All the rest of the seeds were unlike either parent; 101 were wrinkled and yellow, and 108 were round and green. Totally new combinations of characteristics had appeared. This

12–7
A cross between a pea plant with two dominant genes for purple flowers (WW) and one with two recessive genes for white flowers (ww). The phenotype of the offspring in the F_1 generation is purple, but note that the genotype is Ww. The F_1 heterozygote self-pollinates, producing four kinds of gametes, ♀ W, ♂ W, ♀ w, and ♂ w, in equal proportions. The W and w sperm cells and eggs combine randomly to form, on the average, 1/4 WW (purple), 1/2 Ww (purple), and 1/4 ww (white) offspring. It is this underlying 1:2:1 genotypic ratio that accounts for the phenotypic ratio of 3 dominants (purple) to 1 recessive (white). The distribution of traits in the F_2 generation is shown by a Punnett square, named after the English geneticist who first used this sort of checkerboard diagram for the analysis of genetically determined traits.

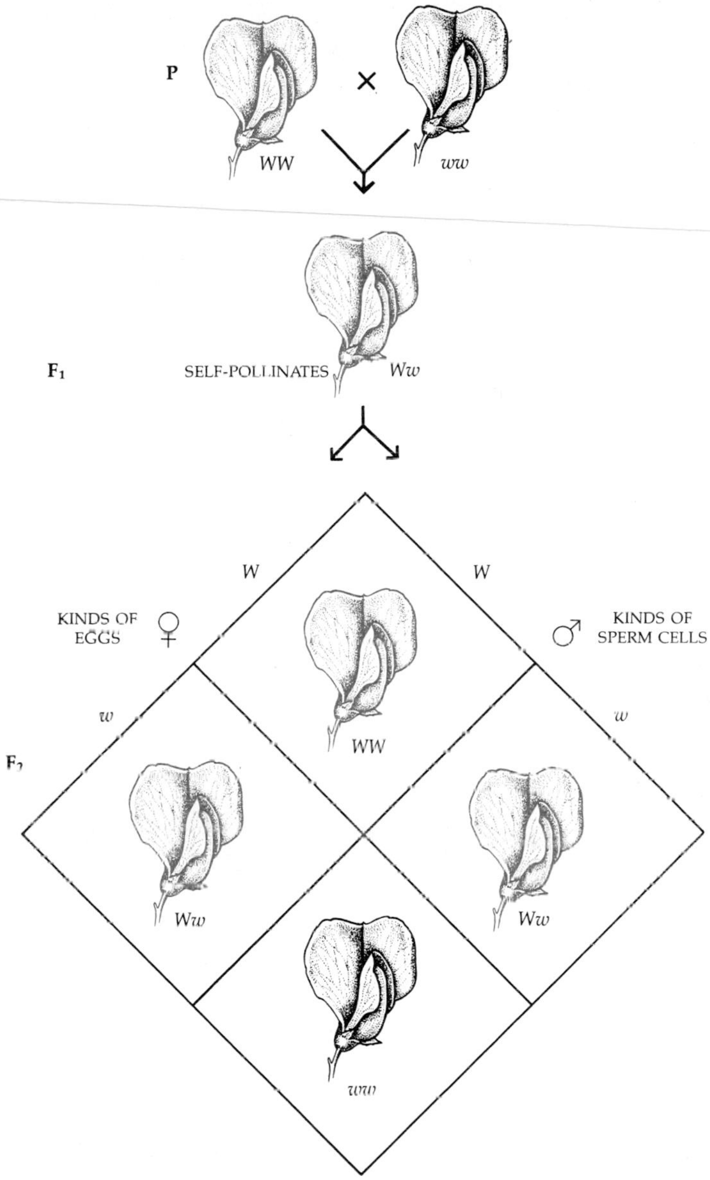

12–8
A testcross. In order for a pea flower to be white, the plant must be homozygous for the recessive gene (ww). But a purple pea flower can come from a plant with either a Ww or a WW genotype. How could you tell such plants apart? Mendel solved this problem by breeding such plants with homozygous recessives. This sort of experiment is known as a testcross. As shown here, a phenotypic ratio in the F_1 generation of 2 purple to 2 white indicates a heterozygous purple-flowering parent. What would have been the result if the plant being tested had been homozygous for purple flowers?

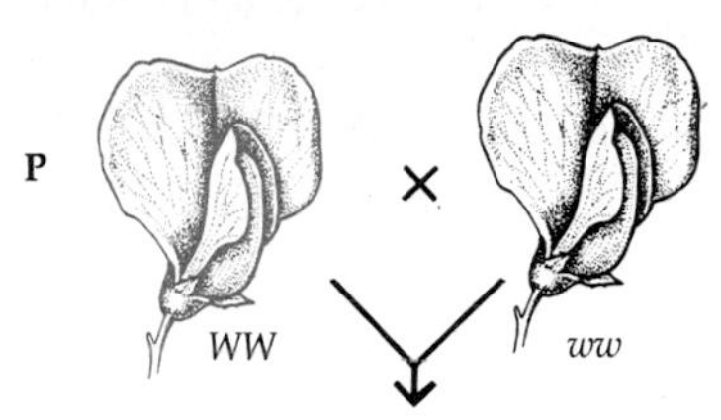

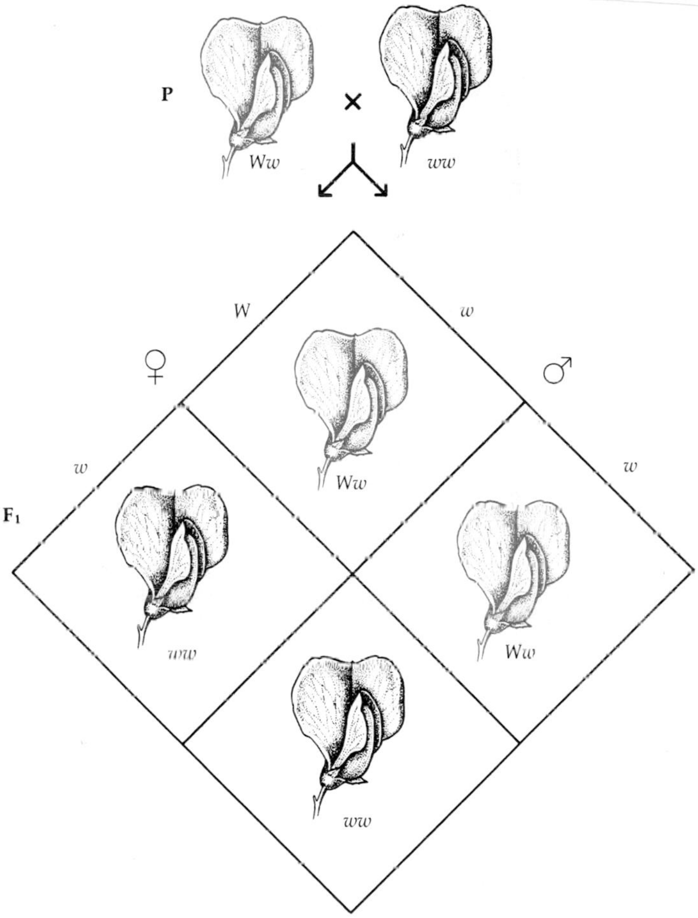

F₁
SELF-POLLINATES
Ww
W
W
KINDS OF EGGS ♀
♂ KINDS OF SPERM CELLS
w
w
WW
F₂
Ww
Ww
ww

P
×
Ww
ww
W
w
♀
♂
w
w
Ww
F₁
ww
Ww
ww

experiment did not contradict Mendel's previous results. Round and wrinkled still appeared in the same 3:1 proportion (423 round to 133 wrinkled), and so did yellow and green (416 yellow to 140 green). But the round and the yellow traits and the wrinkled and the green ones, which had originally combined in one plant, behaved as if they were entirely independent of one another. From this, Mendel formulated his second law, the *principle of independent assortment.* This principle states that members of each pair of traits are distributed independently when the gametes are formed.

Figure 12–9 diagrams these results and shows why, in a cross involving two gene pairs, each pair with one dominant and one recessive allele, the ratio of distribution will be, on the average, 9:3:3:1, with 9 representing the proportion of F_2 progeny that will show the two dominant traits, 1 the proportion that will show the two recessive traits, and 3 and 3 the proportions of the two alternative combinations of dominants and recessives. This is true when one of the original parents is homozygous for both recessive traits and the other homozygous for both dominant ones, as in the experiment just described (*RRYY* × *rryy*), as well as when each original parent is homozygous for one recessive and one dominant trait (*rrYY* × *RRyy*).

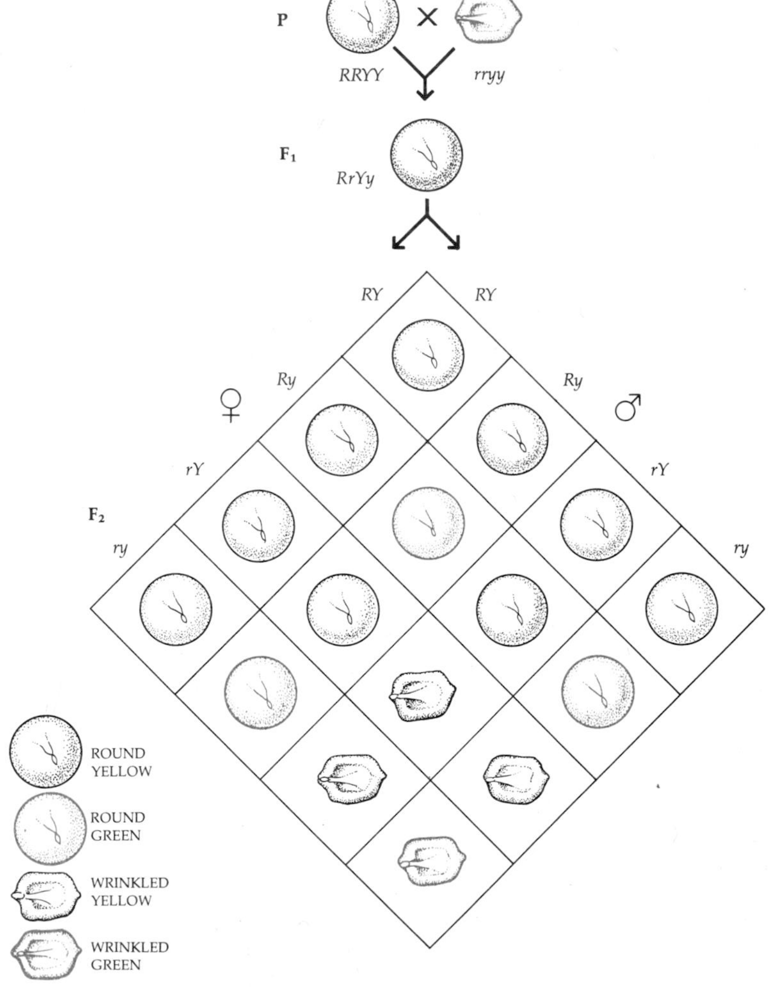

12–9

One of the experiments from which Mendel derived his principle of independent assortment. A plant homozygous for round (RR) *and yellow* (YY) *peas is crossed with a plant having wrinkled* (rr) *and green* (yy) *peas. The* F_1 *generation are all round and yellow, but notice how the traits will, on the average, appear in the* F_2 *generation. Of the 16 kinds of offspring, 9 show the two dominant traits* (RY), *3 show one combination of dominant and recessive* (Ry), *3 show the alternate combination* (rY), *and 1 shows the two recessives* (ry). *This 9:3:3:1 distribution is always the expected result from a cross involving two pairs of independent dominant-recessive alleles. (The letters* R *and* Y *are used because round and yellow are the less common forms in nature.)*

12–10
A cross involving two gene pairs.

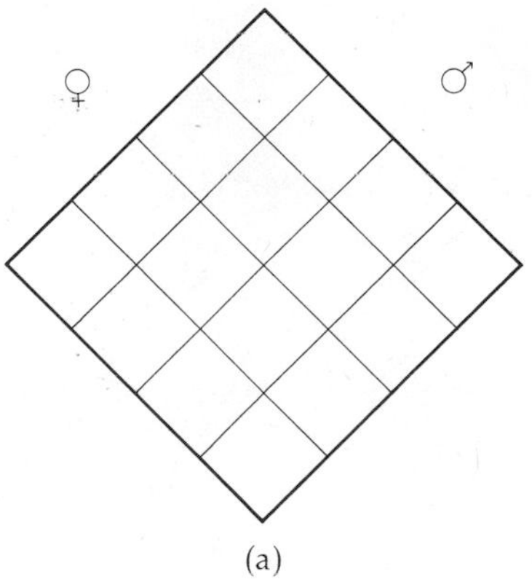

(a)

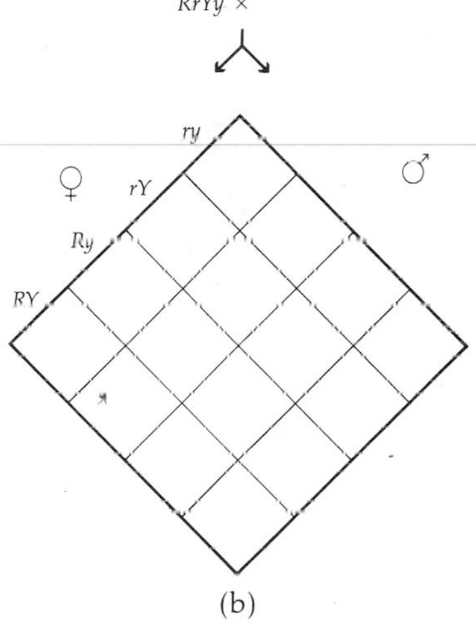

(b)

A Testcross

Can you predict the outcome of a cross between a homozygous recessive for the two genes and a heterozygote? Such a cross is similar to the one analyzed in Figure 12–8 but involves two gene pairs instead of one. For simplicity, let us again study the distribution of round versus wrinkled (*R* versus *r*) and green versus yellow (*Y* versus *y*) in a cross between a heterozygote and a homozygous recessive. Draw a Punnett square with 16 squares (Figure 12–10a). Put the female symbol on one side and the male on the other. Assume that the heterozygote contributes the female gametes. With a genotype of *RrYy*, the heterozygote can produce four kinds of gametes: *RY*, *Ry*, *rY*, and *ry*. At the head of each column at the left, where the female symbol is, put one of these possible combinations (Figure 12–10b). (Notice that in this step we are assuming, as Mendel did, that each of the possible kinds of gamete is produced in equal numbers.)

The homozygous recessive can produce only one type of gamete in terms of the traits being studied: *ry*. Put *ry* at the head of each column on the right (Figure 12–10c). The advantage of using a Punnett square is that it makes it impossible to overlook any combination of gametes.

Now, starting with the column on the far left, begin to fill in the squares. You are less likely to make a mistake if you fill in all the female gametes first (or all male gametes), a column at a time, than if you try to work with both male and female gametes at the same time. When you are halfway through, your square will look like the one in Figure 12–10d.

Next, fill in the symbols for the traits carried in the male gametes. Your square will then look like the one in Figure 12–10e.

Now count the phenotypes that this square predicts. Every capital *R* indicates a round seed (since *R* is dominant); there are eight capital *R*'s, and hence eight round seeds. Conversely you can count eight wrinkled (*rr*) seeds. Similarly there are eight yellow (*Y*) seeds and eight green (*yy*) seeds.

Notice that each of the possible combinations of traits, round green, round yellow, wrinkled green, and wrinkled yellow, appears in equal proportions.

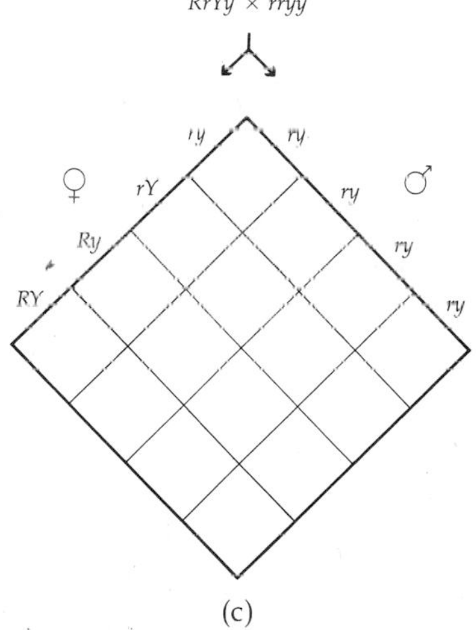

(c)

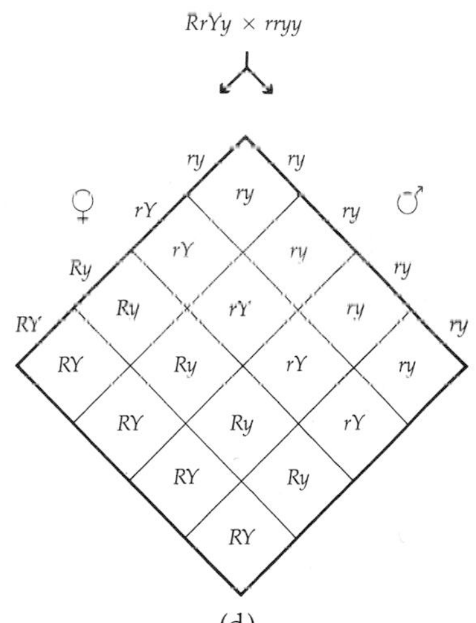

(d)

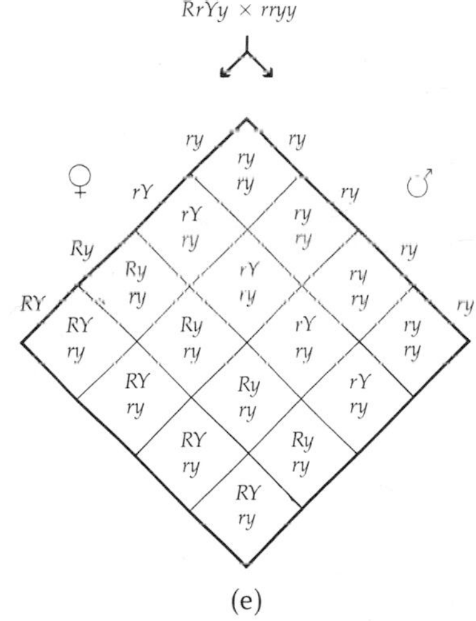

(e)

GENOTYPE AND PHENOTYPE

There is a crucial difference between genotype and phenotype. From the moment of its conception, every organism is acted upon by the environment, and the expression of any gene is always the result of the interaction of gene and environment. To take a simple, familiar example, a seedling may have the genetic capacity to be green, to flower, and to fruit, but it will never turn green if it is kept in the dark, and it may not flower and fruit unless certain precise environmental requirements are met. Himalayan rabbits are all white if they are raised at high temperatures (above 35°C). However, rabbits of the same genotype, when raised at room temperature, have black ears, forepaws, noses, and tails.

In humans, also, similar genotypes are expressed quite differently in different environments. A simple example is found in height, which is influenced by a group of genetic factors (see page 236). This influence is indicated by the fact that tall parents generally have tall children and short parents tend to have short children. On the other hand, adult height is also clearly influenced by environmental factors in childhood, such as diet, sunshine, and incidence of disease. The population of the United States grew taller with each generation for four generations, presumably as a result of general improvements in the standard of living and in the average diet (this trend has just come to an end; grown children in this country are no longer, on the average, taller than their parents).

As René Dubos reminds us, children are growing faster; a boy now reaches full height at about 19 years, but 50 years ago, maximum stature was not usually attained until age 29. Of more social consequence is the fact that puberty is also being reached earlier. In Norway, for example, the mean age of the onset of menstruation has fallen from 17 in 1850 to 13 in 1960. Historical evidence indicates, however, that also in Imperial Rome and Western Europe in Shakespeare's time, teen-agers reached puberty at an early age (Juliet, remember, was not yet 14). Apparently, the slowing of the growth and maturation rate of the general population was a consequence of the industrial revolution and increasing urbanization, which reduced the standards of nutrition and health. Thus we have a situation in which the genetic potential has remained apparently unaltered over the centuries, but the phenotypic expression has undergone fluctuations. A side effect of this change in phenotypic expression is that teen-agers are now reaching physical maturity early in a society in which childhood and dependency have become greatly prolonged.*

* *René Dubos,* So Human an Animal, *Charles Scribner's Sons, New York, 1968.*

The water buttercup, Ranunculus penicillatus, *grows with half the plant body submerged in water. Although the leaves are genetically identical, the broad, floating leaves differ markedly in both form and physiology from the finely divided leaves that develop under water. Such plants illustrate dramatically that the final form of a plant is the result of an interaction between heredity and environment.*

12–11
Gregor Mendel, holding a pea plant, is third from the right in this photograph of members of the Augustinian monastery in Brünn in the 1860s. In his experiments carried out in the monastery garden, Mendel showed that hereditary determinants are carried as separate units from generation to generation. His discoveries explained how inherited variations can persist for generation after generation.

The Influence of Mendel

Mendel's experiments were first reported in 1865 before a small group of people at a meeting of the Brünn Natural History Society. None of them, apparently, understood what Mendel was talking about. But his paper was published the following year in the *Proceedings* of the Society, a journal that was circulated to libraries all over Europe. In spite of this, his work was ignored for 35 years, during most of which he devoted himself to the administrative duties of an abbot, and he received no scientific recognition until after his death. (He was, to use DuPraw's phrase, page 14, an odd traveler whose tale could not be made to fit.)

It was not until 1900, 35 years after Mendel first reported his work, that biology was finally prepared to accept his findings. Within a single year, his paper was independently rediscovered by three scientists working in three different European countries. Each of them had done similar experiments and was searching the scientific literature to seek confirmation of his results. And each found, in Mendel's brilliant analysis, that much of their work had been anticipated.

12–12
Hugo de Vries is shown standing next to Amorphophallus titanum, *which has the largest inflorescence (flower cluster) of any of the flowering plants. De Vries, a Dutch botanist, was the first to recognize the nature of mutations and their role in hereditary processes.*

MENDEL'S LAWS AND THE THEORY OF EVOLUTION

Mendel's work filled an important gap in Darwin's evolutionary theory by explaining why small variations persisted in populations and were not lost by blending. However, Mendelian principles presented new problems to the early evolutionists, because they appeared to offer no possibility for major changes in the genetic makeup of organisms. Segregation of traits explained how variations were maintained from generation to generation. Independent assortment explained how individuals could have combinations of characteristics not present in either parent and so be better adapted, in evolutionary terms, than either parent. But if all hereditary variations were to be explained by the reshuffling process proposed by Mendel, there would be little or no opportunity for major changes in organisms. How, for instance, could new species come about? In fact, these same principles were used by opponents of evolution to demonstrate that evolution could not possibly have occurred.

Mutations

A solution to the problem was proposed by one of Mendel's rediscoverers, a Dutch botanist named Hugo de Vries. De Vries was studying genetics in the evening primrose. Heredity in the primrose, he found, was generally orderly and predictable, as in the garden pea, but occasionally a characteristic appeared that was not present in either parent or indeed anywhere in the lineage of that particular plant. De Vries hypothesized that this characteristic came about as the result of an abrupt change in a gene and that the trait produced by the changed gene was then passed along like any other hereditary trait. De Vries spoke of this hereditary change as a *mutation* and of the organism that carried it as a *mutant.* Different alleles of the same gene, de Vries proposed, arise as a result of mutations. Mutations are the source of new genetic characteristics, and these are the sources of the variations on which evolution depends.

As it turns out, only about 2 of some 2,000 changes in the evening primrose observed by de Vries were actually mutations. The rest were due to new combinations of genes rather than to actual changes in any particular gene. However, de Vries's definition of a mutant and his recognition of the importance of the concept of mutation are still valid, although most of his examples are not.

12–13
A cross between a red (W) snapdragon and a white (W′) snapdragon. This looks very much like the cross between a purple- and a white-flowering pea plant shown in Figure 12–7, but there is a significant difference because neither allele is dominant. The flower of the heterozygote is a blend of the two colors.

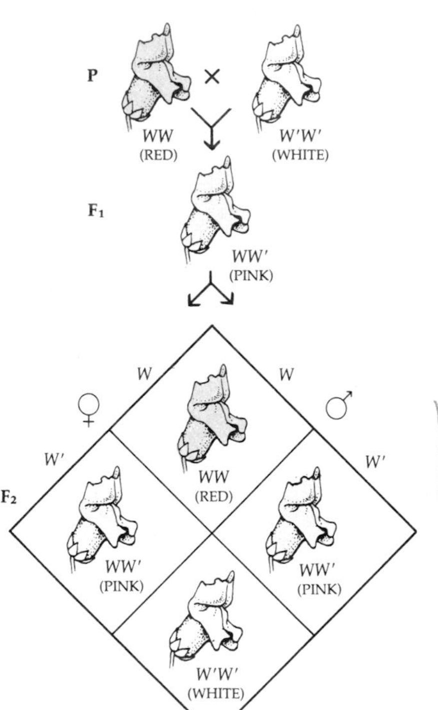

MODIFYING MENDEL

Incomplete Dominance

During the decade that followed the discovery of Mendel's work, many studies were carried out that, while confirming his work in principle, showed that the action of genes is more complex than it had at first appeared to be. An important example is *incomplete dominance.*

Dominant and recessive traits are not always so clear-cut as those selected by Mendel. Some traits do appear to blend. For instance, a cross between a red-flowering snapdragon and a white-flowering snapdragon produces heterozygotes that are pink (Figure 12–13). But when members of this generation self-pollinate, the traits begin to sort themselves out once again, showing that the alleles themselves, as Mendel had affirmed, remained discrete and unaltered. As we shall see, what occurs in the snapdragon and in similar crosses is actually a result of the combined effect of gene products.

Broadening the Concept of the Gene

Mendel's experiments seemed to suggest that each gene affects a single characteristic in a one-to-one relationship. It was soon discovered, however, that a single gene sometimes affects many traits in an organism (*pleiotropy*), and that, conversely, a single trait is often affected by many genes (*polygenic inheritance*).

Pleiotropy

One of the first investigators to demonstrate pleiotropy was Theodosius Dobzhansky. Dobzhansky arbitrarily selected 12 mutant female fruit flies, each with a single different mutation that changed a specific characteristic, such as eye or body color or wing shape. In each of these flies, he examined the shape of the spermatheca, a special sperm-storing organ associated with the female reproductive tract. Ten out of the twelve mutants showed a variation from normal in the size and

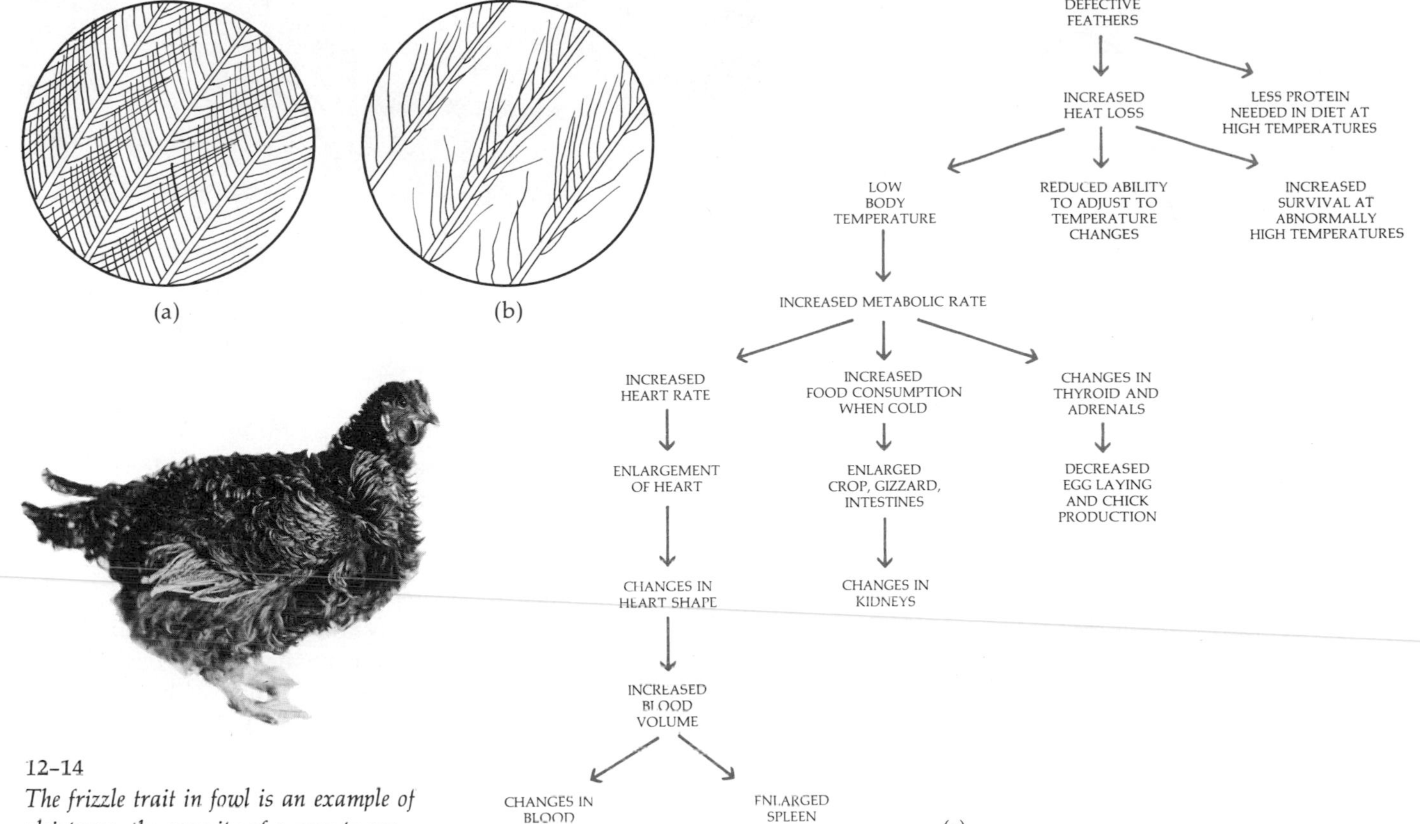

12–14
The frizzle trait in fowl is an example of pleiotropy, the capacity of a gene to produce a variety of phenotypic effects. "Frizzle" is manifested primarily by differences in the feathers. (a) *Under low magnification, feathers from normal birds show a closely interwebbed structure.* (b) *Frizzle feathers are weak and stringy and provide poor insulation. Some of the consequences of the manifestation of this single gene are shown in* (c). *Notice that the frizzle trait is a liability at low temperatures but may increase survival at high temperatures.*

shape of this particular organ, although each mutant fly was thought to differ from the norm in only a single obvious characteristic. Thus it was apparent that the genes affected more than one characteristic of the flies.

Sometime later, another investigator, Hans Gruneberg, approaching the problem from the other direction, studied a whole complex of congenital deformities in rats, including thickened ribs, a narrowing of the tracheal passage, a loss of elasticity of the lungs, hypertrophy of the heart, blocked nostrils, a blunt snout, and needless to say, a greatly increased mortality. All these changes, he was able to demonstrate, were caused by a single mutation, that is, a mutation involving only one gene. This particular gene produces a protein involved in the formation of cartilage, and since cartilage is one of the most common structural substances of the body, the widespread effects of such a gene are not difficult to understand. In fact, it is very likely that Mendel's allele for wrinkled, for example, affected other characteristics of the pea.

The frizzle trait in fowl (Figure 12–14) is another example of pleiotropy.

Polygenic Inheritance

A trait affected by a number of genes does not show a distinctive clear-cut difference between groups—such as the differences tabulated by Mendel—but rather shows a gradation of small differences, which is known as continuous variation. If

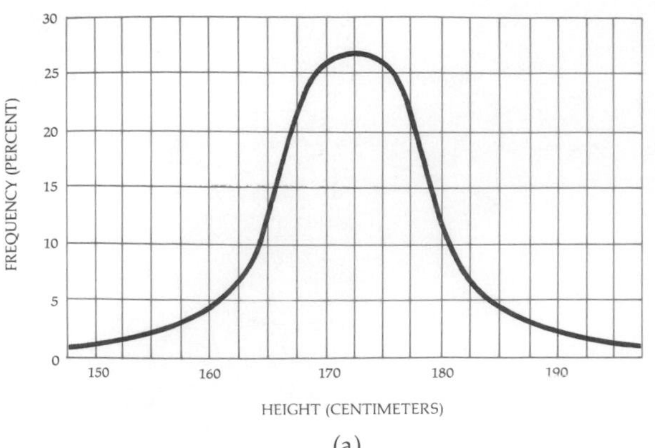

(a)

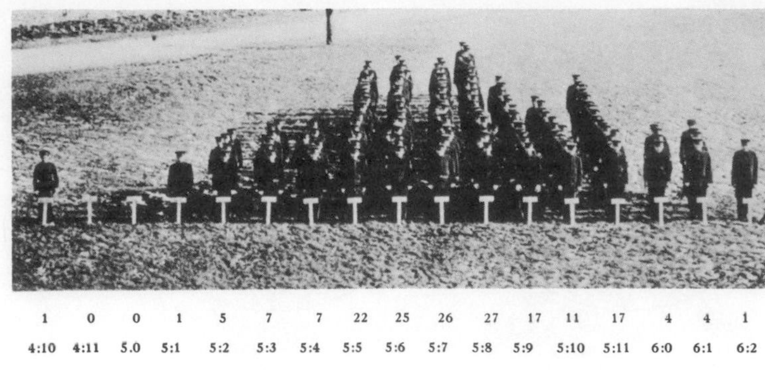

(b)

12–15
(a) *Height distribution of males in the United States. Height and weight are examples of polygenic inheritance. These genetic traits are characterized by small gradations of difference. A graph of the distribution of such traits always takes the form of a bell-shaped curve, as shown, with the mean, or average, falling in the center of the curve. The larger the number of genes involved, the smoother the curve.* (b) *A company of student recruits at the Connecticut Agricultural College about 65 years ago. The height (in feet and inches) of each group and the number of men in the group are shown below the photograph.*

you make a chart of genetic differences among individuals that are affected by a number of genes, you get a curve such as that shown in Figure 12–15.

Fifty years ago, the average height of males in the United States was less but the shape of the curve was the same; in other words, the great majority fell within the middle range and the extremes in height were represented by only a few individuals. Some of these height variations are produced by environmental factors, such as diet, but even if all the men in a population were maintained from birth on the same type of diet, there would still be a continuous variation in height in the population. This is due to genetically determined differences in hormone production, bone formation, and numerous other factors. A number of other human characteristics, notably skin color, are controlled by polygenic inheritance.

Table 12–2 illustrates a simple example of polygenic inheritance, color in wheat kernels, that is controlled by two pairs of genes. Human skin color is under a similar kind of genetic control, as are many other human characteristics.

Table 12–2 *The Genetic Control of Color in Wheat Kernels (Polygenic Inheritance)*

PARENTS:		$R_1R_1R_2R_2 \times r_1r_1r_2r_2$ (DARK RED) (WHITE)		
F_1:		$R_1r_1R_2r_2$ (MEDIUM RED)		
F_2:		GENOTYPE	PHENOTYPE	
1		$R_1R_1R_2R_2$	Dark red	15 red to 1 white
2	4	$R_1R_1R_2r_2$	Medium-dark red	
2		$R_1r_1R_2R_2$	Medium-dark red	
4	6	$R_1r_1R_2r_2$	Medium red	
1		$R_1R_1r_2r_2$	Medium red	
1		$r_1r_1R_2R_2$	Medium red	
2	4	$R_1r_1r_2r_2$	Light red	
2		$r_1r_1R_2r_2$	Light red	
1		$r_1r_1r_2r_2$	White	

SUMMARY

In this chapter, we started with the earliest ideas about inheritance and traced the gradual development of these ideas into a science. The first question with which this new science was concerned was the mechanics of inheritance. How are hereditary characteristics passed from one generation to another?

By the middle of the nineteenth century, it was recognized that ova and sperm are specialized cells and that the ova and sperm both contribute to the hereditary characteristics of the new individual. But how were these special cells, called gametes, able to pass on the many hundreds of characteristics involved in inheritance? One answer to this question was the theory of blending inheritance, which held that the traits of the parents blended in the offspring, like a mixture of two fluids.

The revolution in genetics came when the blending theory was replaced by a unit theory. According to Mendel's principle of segregation, hereditary characteristics are always determined by discrete factors (now called genes) that appear in pairs, one of each pair inherited from each parent.

The genetic makeup of an organism is known as its genotype. Its appearance, or set of outward characteristics, is its phenotype. Both genes in a pair may be alike (a homozygous condition), or they may be different (heterozygous). Different genes of a gene pair are called alleles (alternative forms). In a heterozygous pair, only one allele may be detected in the phenotype. An allele that is expressed in the phenotype to the exclusion of the other is a dominant allele; one that is concealed in the phenotype is a recessive allele. When two organisms that are homozygous for different alleles of the same gene are crossed, the ratio of dominant to recessive in the phenotype of the F_2 generation is 3:1.

Genes determine only potential capacities. Interactions of the genotype and the environment determine the phenotype.

Mendel's other great principle, the principle of independent assortment, applies to the behavior of two or more gene pairs. This law states that the members of each pair of genes segregate independently of one another. In crosses involving two independent gene pairs, the expected phenotypic ratio in the F_2 generation is 9:3:3:1. A testcross, in which an individual of the F_1 generation is crossed with a homozygous recessive, reveals the possible phenotypes in equal numbers.

Mutations are abrupt changes in the genotype. They are the ultimate source of the genetic variations that Mendel studied and that make possible the process of evolution.

Incomplete dominance is a situation in which the effects of the recessive allele are apparent in the heterozygote.

Genes may affect two or more superficially unrelated characteristics; this property of the gene is known as pleiotropy.

Many characteristics are under the control of a number of separate genes and are said to be polygenetically inherited. Traits under the control of a number of genes typically show continuous variation, as represented by a bell-shaped curve.

QUESTIONS

1. Distinguish between the following terms: gene/allele; dominant/recessive; homozygous/heterozygous; genotype/phenotype; the F_1/the F_2.

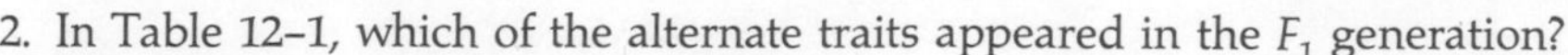

2. In Table 12–1, which of the alternate traits appeared in the F_1 generation?

3. Why is a homozygous recessive always used in a testcross?

4. (a) In a pea plant that breeds true for tall, what possible gametes can be produced? (Use the symbol *D* for tall, *d* for dwarf.) (b) In a pea plant that breeds true for dwarf, what possible gametes can be produced? (c) What will be the genotype of the F_1 generation produced by a cross between these two types? (d) What will be the phenotype of the F_1 generation produced by a cross between these two types? (e) What will be the probable distribution of traits in the F_2 generation? Illustrate with a Punnett square.

5. You have just flipped a coin five times and it has turned up heads every time. What are the chances that the next time you flip it, it will be tails?

6. The ability to taste a bitter chemical, phenylthiocarbamide (PTC), is due to a dominant gene. In terms of the tasting gene, what are the possible phenotypes of a man both of whose parents are tasters? What are his possible genotypes? (This is a complex question.)

7. If such a man marries a woman who is a nontaster, what proportion of their children will probably be tasters? Suppose one of the children is a nontaster. What would you know about his genotype? Explain your results by drawing Punnett squares.

8. Suppose this couple had four children, all of whom could taste PTC. What is the father's probable genotype? Is there another possibility?

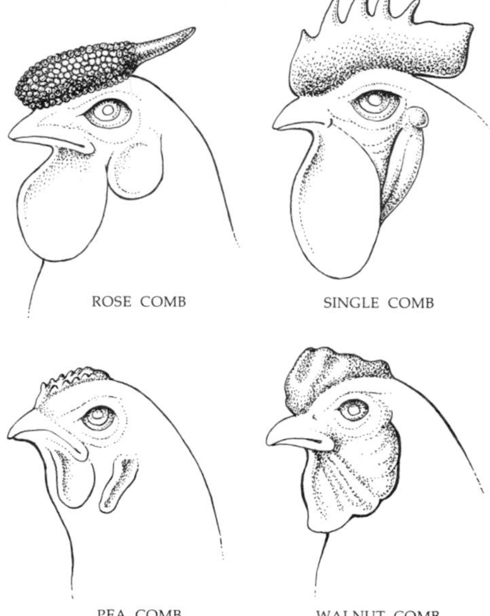

9. In chickens, two pairs of alleles determine comb shape, *R*, *r* and *P*, *p*. *RR* or *Rr* results in rose comb, whereas *rr* produces single comb. *PP* or *Pp* produces pea comb and *pp* produces single comb. When *R* and *P* occur together, they produce a new type of comb: walnut. What would be the genotype of the F_1 cross resulting from *RRpp* × *rrPP*? The phenotype? If F_1 hybrids were crossbred, what would be the probable distribution of genotypes? Of phenotypes? (Illustrate this cross with a Punnett square.)

10. The so-called "blue" (really gray) Andalusian variety of chicken is produced by a cross between the black and white varieties. What color chickens (and in what proportions) would you expect if you crossed two blues? If you crossed a blue and a black?

11. In one strain of mice, skin color is determined by five different pairs of alleles. The colors range from almost white to dark brown. Would it be possible for any given pair of mice to produce offspring darker or lighter than either parent? Explain.

12. Height and weight in animals follow a distribution similar to that shown in Figure 12–15. By inbreeding large animals, breeders are usually able to produce some increase in size among their stock. But after a few generations, increase in size characteristically stops. Why?

13. Mendel did not know of the existence of chromosomes. Had he known, what change might he have made in his second principle?

CHAPTER 13

The Physical Basis of Heredity

Between the publication of Mendel's experiments in 1865 and their rediscovery in 1900, some important observations were made, largely as a result of improvements in the light microscope and in techniques of preparing materials for microscopic observation. By Mendel's time, you will recall, microscopists had seen sperm and egg cells, and had recognized them to be the hereditary links between parents and offspring. In 1875, the German zoologist Oscar Hertwig witnessed the fertilization of the egg of a sea urchin and noted that the fertilized egg contained two nuclei, one from the sperm and one from the egg. At about this time, as a result of new staining techniques, chromosomes were first observed within cells. In 1882, the German cytologist Walther Flemming pieced together the separate steps in mitosis, the "dance of the chromosomes," which we described in Chapter 10.

CYTOLOGY AND GENETICS MEET: SUTTON'S HYPOTHESIS

In 1902 William S. Sutton, a graduate student at Columbia University, was studying the formation of sperm cells in male grasshoppers. In grasshoppers, as in man, the male gametes are formed from a special group of cells in the testes, the spermatogonia. Like other cells in the grasshopper body, spermatogonia contain the diploid number of chromosomes. At the time the gametes are formed, the spermatogonia undergo meiosis, producing four haploid cells that then develop into sperm.

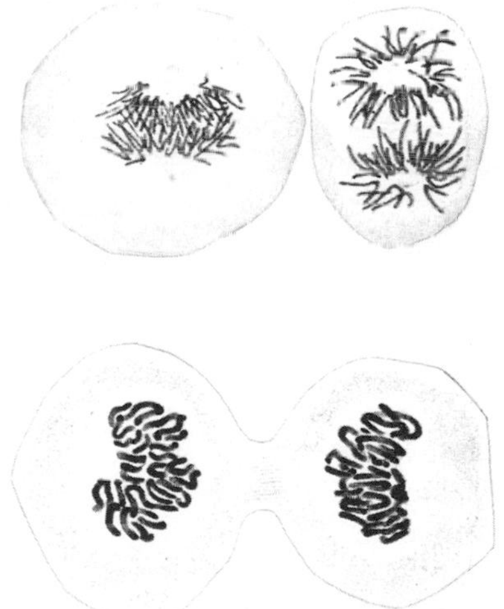

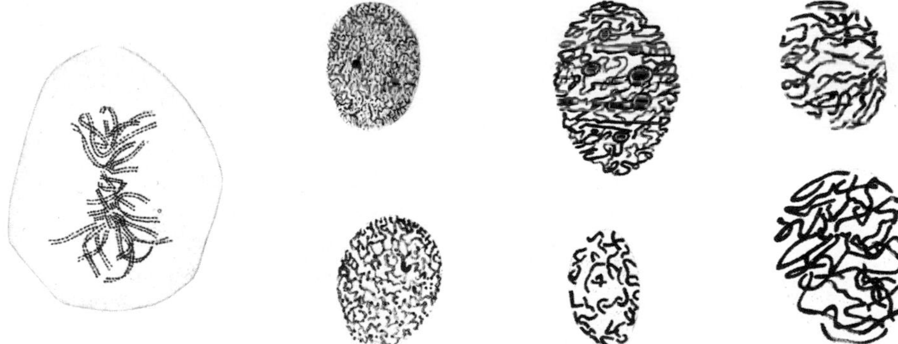

13–1
Drawings made by Walther Flemming in 1882 of chromosomes in dividing cells of salamander larvae.

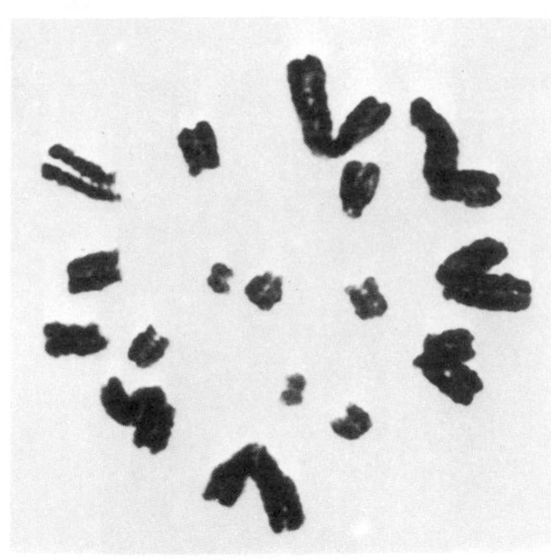

13–2
Chromosomes from a diploid cell of a grasshopper. Note that even though these chromosomes are not paired, it is possible to pick out some of the homologues. The observation that chromosomes come in homologous pairs was one of Sutton's clues to the meaning of meiosis.

Observing the process of meiosis, Sutton was struck by the fact that the chromosomes that paired with one another at the beginning of the first meiotic division had physical resemblances to one another. In diploid cells, he noted, chromosomes apparently came in pairs. The pairing was obvious only at meiosis, but the discerning eye could find the homologues in the unpaired chromosomes when they became visible at the time of mitosis.

Sutton postulated that the homologues observed undergoing separation in the formation of the sperm cells in the grasshopper were replicas of the homologous chromosomes that had come together in the fertilized egg, at the first moment in that particular grasshopper's biography. In other words, if one homologue of a chromosome pair came from the grasshopper's mother, the other homologue must have come from the father. The two had been copied faithfully in cell division after cell division through the grasshopper's several stages of development until the time of meiosis, when they were once again separated.

Suddenly the facts fell into place. Suppose chromosomes carried genes, the *Elemente* described by Mendel. This idea does not seem very startling to you now, with your superior knowledge, but remember that the gene was just an abstract idea or mathematical unit to the geneticist, and that the chromosome was just an unidentified colored body to the cytologist. Suppose, Sutton reasoned, alleles occurred on homologous chromosomes. Then the alleles could always remain independent and so could separate at meiosis, with new pairs of alleles forming when the gametes came together at fertilization. Mendel's law of the segregation of alleles could be explained by the segregation of the homologous chromosomes at meiosis.

Now, suppose that half of the chromosomes of the male grasshopper were inherited from his mother and half were inherited from his father. At the time of meiosis, a maternal chromosome could be separated from the paternal homologue. Or in terms of the garden pea, cross-fertilization of a plant having smooth and yellow seeds with a plant having green and wrinkled seeds could produce, in the F_2 generation, plants with smooth and green seeds and plants with yellow and wrinkled seeds, although neither parent plant had had this phenotype (Figure 13–3). Mendel's law of independent assortment could be explained on the basis of the independent movement of chromosomes at meiosis.

The arguments in support of Sutton's hypothesis that Mendel's factors, or genes, are carried on chromosomes can be summarized as follows:

1. All hereditary characteristics are carried in the sperm and egg cells since these cells are the only bridge from one generation to another.
2. Since sperm cells lose almost all their cytoplasm as they mature, the hereditary factors are probably carried in the nucleus.
3. The only visible parts of the nucleus that are accurately divided during cell division are the chromosomes. This suggests that the factors, or genes, must be carried on the chromosomes.
4. Chromosomes obey Mendel's laws.
   a. Chromosomes occur in pairs; so do Mendelian factors.
   b. Chromosomes segregate at meiosis; Mendelian factors segregate at the time of (or, in plants, before) gamete formation.
   c. The members of a chromosome pair appear to assort independently of other chromosome pairs; Mendelian factors assort independently.

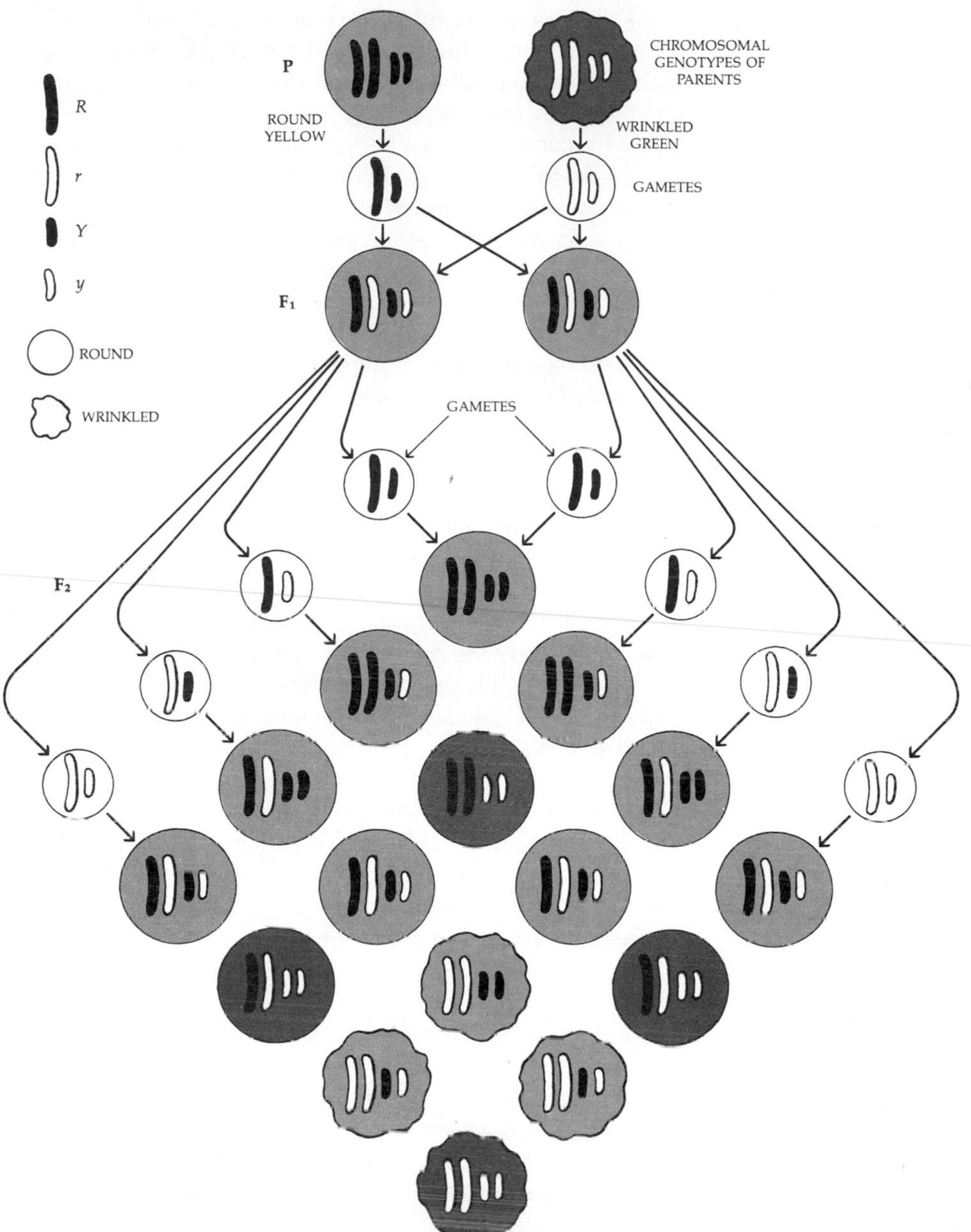

13–3

*The chromosome distributions in Mendel's cross of round yellow and wrinkled green peas, according to Sutton's hypothesis. Although the pea has 14 chromosomes (*n = *7), only 4 are shown here, the two carrying the genes for round or wrinkled and the two carrying the genes for yellow or green. (This is analogous to what Mendel did when he selected the two traits to study.) As you can see, one parent is homozygous for the recessives, one for the dominants. Therefore, the only gametes they can produce are* RY *and* ry. *(Remember,* R *now stands not just for the trait but also for the chromosome carrying the trait, as do the other letters.) The* F_1 *generation, therefore, must be* Rr *and* Yy. *When a mother cell of this generation undergoes meiosis,* R *is separated from* r *and* Y *from* y *when the respective homologues separate at anaphase I. Four different types of haploid egg cells are possible, as the diagram reminds us, and also four different types of sperm nuclei. These can combine in 4 × 4, or 16, different ways, as illustrated in the Punnett square. (Yellow is indicated by the color, green by gray.)*

As occurs often in the history of science, two other biologists recognized the correlation between the behavior of Mendel's *Elemente* and the observed movement of the chromosomes, but young Sutton's paper appeared first, and his presentation was by far the most convincing.

On the basis of Sutton's hypothesis, it was possible to make certain predictions. One of the first and most obvious concerned independent assortment. For example, the gametes of a pea plant have only seven chromosomes ($n = 7$), and they clearly carry more than seven genes. Thus, one is led to predict that many genes will be present on the same chromosome and that such genes may not follow Mendel's second law—that is, they may not assort independently. This prediction turned out to be true, and it is an important exception to Mendel's second law. Genes on the same chromosome are said to be linked, a subject we shall discuss more fully later in this chapter.

SEX DETERMINATION

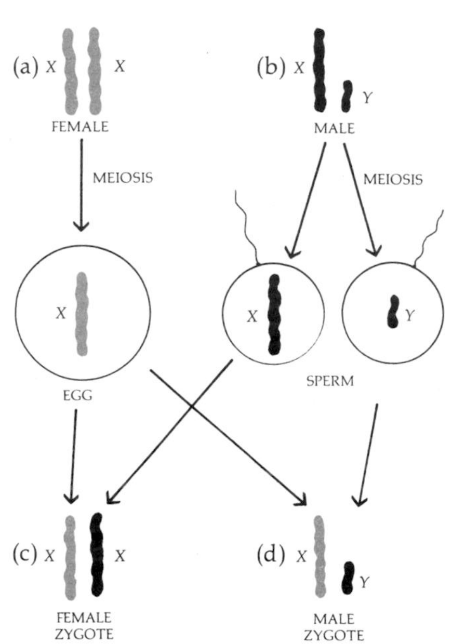

13–4
How the sperm cell determines the sex of human offspring. (a) *At meiosis, every egg cell receives an* X *chromosome from the mother.* (b) *A sperm cell may receive either an* X *chromosome or a* Y *chromosome.* (c) *If a sperm cell carrying an* X *chromosome fertilizes the egg, the offspring will be female* (XX). (d) *If a sperm cell carrying a* Y *chromosome fertilizes the egg, the offspring will be male* (XY).

Note that although Sutton showed that there are similarities between the behavior of the chromosomes at meiosis and the Mendelian laws, he did not actually prove that the genes, which were still abstract concepts, are really carried by the chromosomes. His closely argued thesis demonstrated only that whatever cellular organelles did carry the genes would have to behave like chromosomes. More convincing evidence for the physical location of the gene was to depend on another series of studies, which we shall now describe.

In the 1890s, microscopists noticed that male and female animals often show chromosomal differences, and they began to suspect that these differences were related to sex determination. One pair of chromosomes differs between the sexes, and these are known as the *sex chromosomes;* all the other chromosomes, which are the same whether the animal is male or female, are known as *autosomes.* In many animals, the two sex chromosomes are identical in the female but are dissimilar in the male, with one male sex chromosome resembling the female sex chromosomes and the other usually smaller and of a different shape. The sex chromosome that is similar in the cells of both males and females is called the *X* chromosome, and the unlike chromosome characteristic of the cells of males is called the *Y* chromosome. Thus we can characterize the two sexes as *XX* (female) and *XY* (male).

However, in some insects, including the grasshopper, which Sutton studied, the *Y* chromosome is missing entirely. In this case, we usually speak of *XX* females and *XO* males. In birds, moths, and butterflies (and in occasional species in other groups), the chromosomes are reversed; the male has the two *X* chromosomes, and the female only one. The *Y* chromosome may or may not be present.

Human beings have 22 pairs of autosomes, which are structurally the same in both sexes. Females have a twenty-third matching pair, the sex chromosomes, *XX*. Human males, as their twenty-third pair, have one *X* and one *Y*. During meiosis, as each diploid spermatocyte undergoes meiotic division into four haploid sperm cells, two of the sperm cells receive *X* chromosomes and two receive *Y* chromosomes. The ovum always contains an *X* chromosome, since a human female does not generally possess the *Y* in any of her cells. Thus the zygote will become *XX* or *XY*, depending on whether an *X*-bearing sperm or a *Y*-bearing sperm fertilizes the egg (Figure 13–4). It is in this way that the sperm cell determines the sex of the future offspring, and it is the process of meiosis that governs the almost equal

CALICO CATS, BARR BODIES, AND THE LYON HYPOTHESIS

A dark spot of chromatin—called a Barr body—can be seen at the outer edge of the nucleus of female mammalian cells in interphase. According to the Lyon hypothesis—named after Mary Lyon, who proposed it—this dark spot is an inactivated X chromosome. Early in embryonic life, according to Lyon, this inactivation occurs in one or the other X chromosome in each cell of the female mammal (except for those cells from which egg cells will form). Thus all the somatic cells of female mammals are not identical but are one of two types, depending on which of the X chromosomes is active.

Calico cats (also sometimes known as tortoise-shell cats) have coats that are both black and yellow (actually an orange-yellow). They are also almost always female. In cats, the alleles for black or yellow coat color are carried on the X chromosome, so calico cats neatly fit the Lyon hypothesis. There are, however, occasional male calicos. These are suspected of having an extra X chromosome, a supposition supported by the observation that they are almost always sterile.

production of male and female babies. (In actual fact, the ratio of human male to human female births is about 106 to 100. The reason for this is not known, but it has been suggested that the male-determining sperm may have an advantage in getting to the egg.)

The correlation of the appearance of chromosomes with a particular characteristic—sex—was strong evidence for Sutton's hypothesis. Stronger support was to come from a variety of studies carried out in what was known, because of the richness of the data it yielded, as the "Golden Age of Genetics." Much of the work of this Golden Age was carried out with the fruit fly, *Drosophila*

THE GOLDEN AGE OF DROSOPHILA

Early in the 1900s, Thomas Hunt Morgan began a study of genetics at Columbia University, founding what was to be the most important laboratory in the field for several decades. By a remarkable combination of insight and good fortune, he selected *Drosophila* as his experimental material.*

Drosophila means "lover of dew," although actually this useful little fly is not attracted by dew but feeds on the fermenting yeast that it finds in rotting fruit. The fruit fly was a likely choice for a geneticist since it is easy to breed and maintain. These tiny flies, each only 3 millimeters long, produce a new generation every 2

* Geneticists have often used for their experiments "insignificant" plants and animals—such as Mendel's pea plants, for instance, or Hertwig's sea urchins. Underlying this approach is the geneticist's assumption that genetic principles are universal, applying equally to all living things.

13–5

The fruit fly (Drosophila) *and its chromosomes. Fruit flies have only four pairs of chromosomes, a fact that simplified Morgan's experiments. Six of the chromosomes (three pairs) are autosomes (including the two small spherical chromosomes in the center) and two are sex chromosomes.*

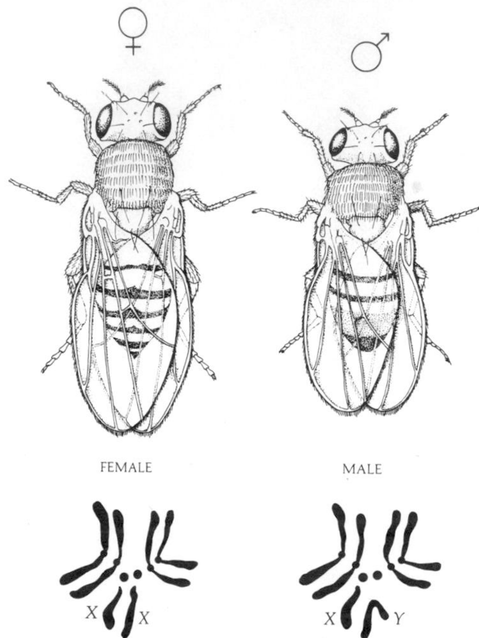

weeks. Each female lays hundreds of eggs at a time, and an entire population can be kept in a half-pint bottle, as they were in Morgan's laboratory.

Also, *Drosophila* has only four pairs of chromosomes, a feature that turned out to be particularly useful, although Morgan could not have foreseen that. Three pairs of these are autosomes, and the fourth is an *XX* pair in the female and an *XY* pair in the male (Figure 13–5).

Sex-Linked Characteristics

The aim in Morgan's laboratory was, initially, to do breeding experiments similar to those Mendel had carried out in the pea plant. Such experiments involved examining under a magnifying lens hundreds—eventually thousands—of individual fruit flies.

The investigators, at first, were looking for genetic differences between individual flies that they could study by interbreeding experiments. Shortly after Morgan established his colony, such a difference appeared. One of the prominent and readily visible characteristics of fruit flies is their brilliant red eyes. One day, a white-eyed fly, a mutant, appeared in the colony. The mutant, a male, was bred to a red-eyed female. All members of the first generation were red-eyed, indicating that the mutation was recessive. Then the members of the F_1 generation were interbred just as Mendel had done in his pea experiments, and this is what resulted:

Red-eyed females	2,459
White-eyed females	0
Red-eyed males	1,011
White-eyed males	782

All of the white-eyed F_2 were males. Why were there no white-eyed females? To explore the situation further, Morgan crossed the original white-eyed male with one of the F_1 (red-eyed) females. And these results were obtained:

Red-eyed females	129
White-eyed females	88
Red-eyed males	132
White-eyed males	86

In other words, females can be white-eyed; the trait behaves pretty much like a usual recessive one. So why were there no white-eyed females in the F_2 generation? Stop here a moment and see if you can answer this question. Morgan was able to.

Morgan and his co-workers examined these figures and on the basis of them came to an important conclusion: The gene for eye color is carried only on the *X* chromosome. (In fact, as it was later shown, the *Y* chromosome carries very little genetic information.) The allele for white eyes is recessive. Thus a heterozygous female would never have white eyes—which is why there were no white-eyed females in the F_2 generation. However, a male that received an *X* chromosome carrying the allele for white eyes would always be white-eyed since no other allele would be present.

Further experimental crosses proved Morgan's hypothesis to be right (Figure 13–6). They also showed that white-eyed fruit flies are more likely to die before they hatch than are red-eyed fruit flies, which explains the lower-than-expected numbers in the F_2 generation and the testcross.

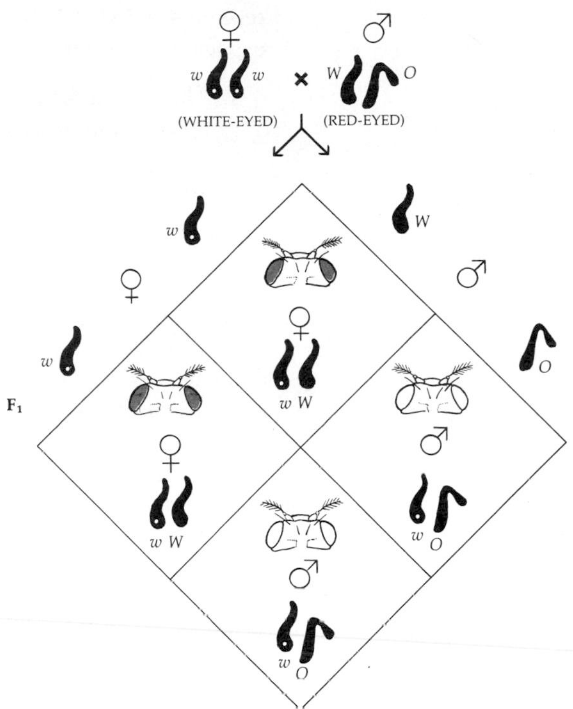

13–6
Offspring of a cross between a white-eyed female fruit fly and a red-eyed male fruit fly, illustrating what happens when a recessive gene, indicated by the white dot, is carried on an X chromosome. The F_1 females, with one X chromosome from the mother and one from the father, are heterozygous (Ww) and so will be red-eyed. But the F_1 males, with their single chromosome received from the mother carrying the recessive (w) trait, will be white-eyed because the Y chromosome carries no gene for eye color. Thus the recessive allele on the X chromosome inherited from the mother will be expressed.

These experiments introduced the concept of sex-linked traits, which are, as we shall see, important in the genetics of human beings as well as of fruit flies. More important, they convinced Morgan, and most other geneticists as well, that Sutton's hypothesis was right: Genes are on chromosomes.

Linkage

Mendel showed that certain pairs of alleles, such as those for round and wrinkled peas, assort independently of other pairs, such as those for yellow and green peas. However, as we noted previously, genes assort independently only if they are on different chromosomes. If two genes are on the same chromosome, they should both be transmitted to the same gamete at meiosis. The genes on any one chromosome are said to be in a *linkage group*.

As increasing numbers of mutants were found in Columbia University's *Drosophila* population, the mutations began to fall into four linkage groups, in accord with the four pairs of chromosomes visible in the cells. Indeed, in all organisms that have been studied in sufficient genetic detail, the number of linkage groups and the number of pairs of chromosomes have been the same, further supporting Sutton's hypothesis.

Crossing Over

Large-scale studies of linkage groups soon revealed some unexpected difficulties. For instance, most fruit flies have gray bodies and long wings. When these individuals were bred with mutant fruit flies having black bodies and short wings (both recessive traits), all the progeny had gray bodies and long wings, as would be expected. Then the F_1 generation was inbred. Two outcomes seemed possible:

1. The two recessives would be assorted independently and would appear in the 9:3:3:1 ratio, indicating that they were on different chromosomes.
2. The two recessives would be linked. In this case, 75 percent of the flies would be gray with long wings, and 25 percent, homozygous for both recessives, would be black with short wings.

In the case of these particular traits, the results most closely resembled the second possibility but they did not conform exactly. In a few of the offspring, the traits seemed to assort independently, not together; that is, some few flies appeared that were gray with short wings, and some that were black with long wings. How could this be? Somehow the alleles had moved out of their linkage groups.

To find out what was happening, Morgan tried a testcross, breeding a member of the F_1 generation with a homozygous recessive. If black and gray, long and short assorted independently—that is, if they were on different chromosomes—25 percent of the offspring of this cross should be black with long wings, 25 percent gray with long wings, 25 percent black with short wings, and 25 percent gray with short wings. On the other hand, if the two genes (color and wing size) were on the same chromosome and so moved together, half of the offspring of the testcross should be gray with long wings and half should be black with short wings. But actually, as it turned out, over and over, in counts of hundreds of fruit flies resulting from such crosses, 41.5 percent were gray with long wings, 41.5 percent were black with short wings, 8.5 percent were gray with short wings, and another 8.5 percent were black with long wings.

Morgan was convinced by this time that genes are located on chromosomes. It now seemed clear that the two traits were located on the same chromosome since they did not show up in the 25:25:25:25 percentage ratios of independently assorted gene pairs. The only way in which the observed figures could be explained was if one assumed (1) that the two alleles were on one chromosome and their two partners on the homologous chromosome, and (2) that sometimes alleles could be exchanged between homologous chromosomes.

13–7
Homologous chromosomes of a grasshopper, as seen in prophase I. All four chromatids are visible. Crossing over—the exchange of genetic material—has probably occurred at the point at which these chromatids intersect. The arrows indicate the centromeres.

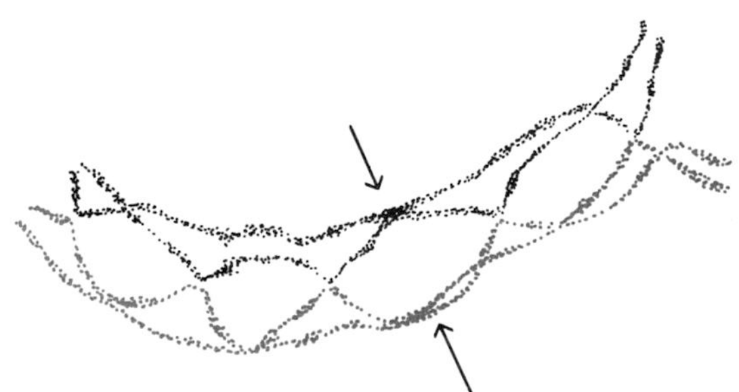

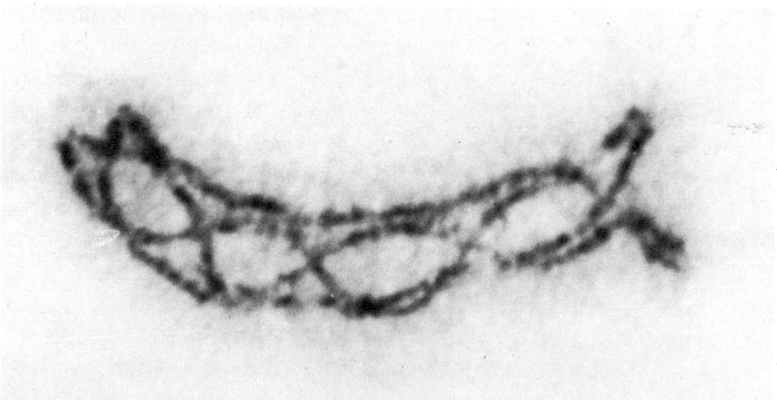

As we noted in Chapter 11, it now has been established that exchange of portions of homologous chromosomes—crossing over—takes place at the beginning of meiosis. Presumably, the actual exchange takes place when the homologues are closely grouped early in the first meiotic prophase. As the homologues begin to separate, chiasmata ("crosses") become visible where crossing over has occurred.

MAPPING THE CHROMOSOME

With the discovery of crossovers, it began to seem clear not only that the genes are carried on the chromosomes, as Sutton had hypothesized, but also that they must be positioned at particular spots, or *loci* (singular, *locus*), on the chromosomes. Furthermore, the alleles of any given gene must occupy corresponding loci on homologous chromosomes. Otherwise, exchange of sections of chromosomes would result in genetic chaos rather than an exact exchange of alleles.

As other traits were studied, it became clear that the percentage of separations, or crossovers, between any two genes, such as gray body and long wing, was different from the percentage of crossovers between two other genes, such as gray body and long leg. In addition, as Morgan's experiments had shown, these percentages were fixed and predictable. It occurred to A. H. Sturtevant, one of the many brilliant young geneticists attracted to Morgan's laboratory during these golden days of *Drosophila* genetics, that the percentage of crossovers probably had something to do with the distances between the genes, or in other words, with their spacing along the chromosome. This simple concept opened the way to the "mapping" of chromosomes.

Sturtevant postulated (1) that genes are arranged in a linear series on chromosomes; (2) that genes that are close together will be separated by crossing over less frequently than genes that are farther apart; and (3) that it should therefore be possible, by determining the frequencies of crossovers, to plot the sequence of the genes along the chromosome and the relative distances between them. In Figure 13–8, for example, you can see that in a crossover, the chances of a strand breaking and recombining with its homologous strand somewhere between *B* and *C* should be less likely than this happening somewhere between *A* and *C*.

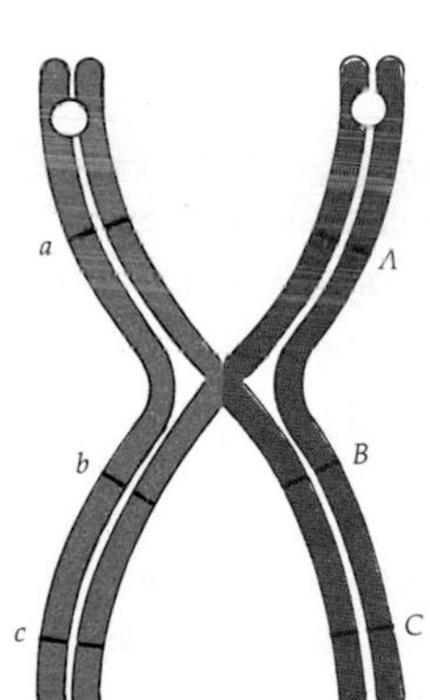

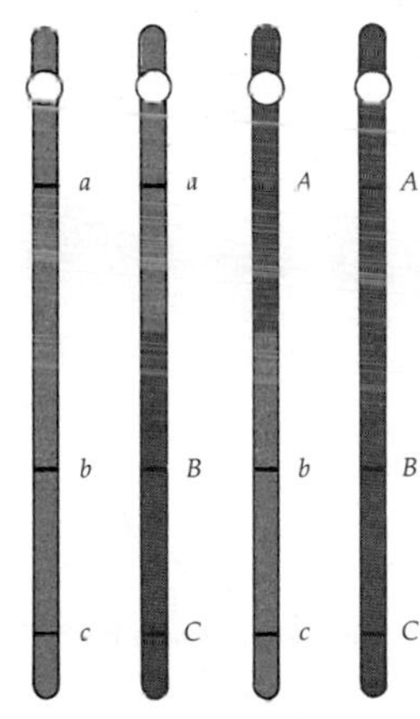

13–8
Crossing over takes place when breaks occur in homologous chromatids during prophase I, when the chromosomes are paired, and the broken end of each chromatid joins with the chromatid of a homologous chromosome. In this way, alleles are exchanged between chromosomes. The white circles symbolize centromeres.

13-9
How to "map" a chromosome. Genes A *and* B *recombine with* a *and* b *in 8 percent of the offspring, and genes* A *and* C *with* a *and* c *in 25 percent of the offspring. Use as a unit of measure the distance that will give (on the average) one recombinant per 100 fertilized eggs.* (a) *Start with the largest number and establish the relative positions of* A *and* C *on the chromosome.* (b) A *and* B*, as you know from the data, are 8 units apart. Thus* B *could theoretically be either to the left of* A *or to the right.* (c) *However, if it were to the left,* B *and* C *would be 33 units apart, a distance that does not conform to the data.* B *must therefore be between* A *and* C.

CROSS	OFFSPRING		
AB × *ab*	*AB* + *ab*	(PARENTAL)	92%
	Ab + *aB*	(RECOMBINANT)	8%
AC × *ac*	*AC* + *ac*	(PARENTAL)	75%
	Ac + *aC*	(RECOMBINANT)	25%
BC × *bc*	*BC* + *bc*	(PARENTAL)	83%
	Bc + *bC*	(RECOMBINANT)	17%

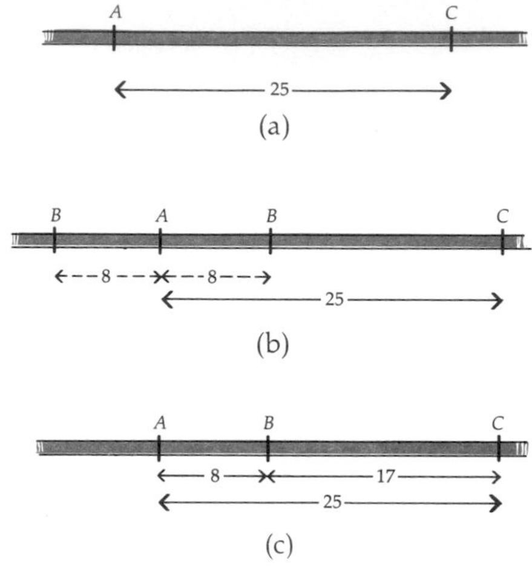

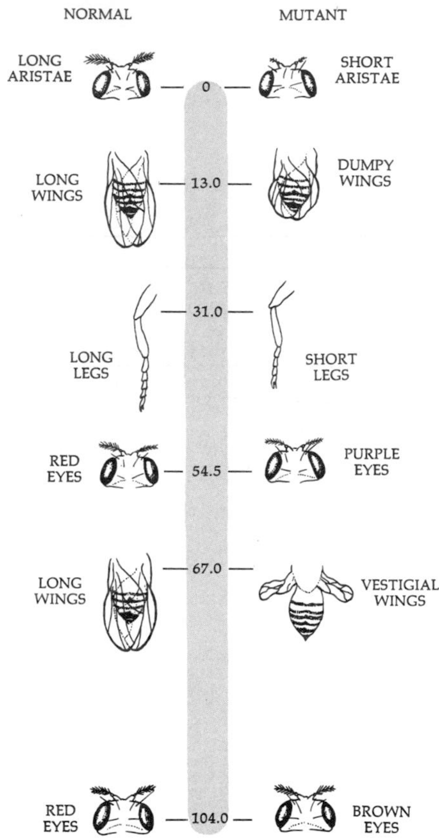

13-10
A portion of the genetic map of Drosophila melanogaster, *showing some of the genes on chromosome 2 and their relative positions, as calculated by the frequency of crossovers. As you can see, more than one gene may affect a single characteristic, such as eye color.*

In 1913, Sturtevant began constructing chromosome maps using data from crossover studies in fruit flies. As a standard unit of measure, he arbitrarily took the distance that would give (on the average) one crossover per 100 fertilized eggs. The genes with 10 percent crossover would be 10 units apart; those with 8 percent crossover would be 8 units apart. By this method, he and other geneticists constructed genetic maps locating a wide variety of genes and their mutants in *Drosophila* (Figure 13-10). The distances on such maps cannot be absolute because the distance unit is an arbitrary one. The relative distances may also be in error in some cases because breaking and rejoining of chromosome segments may be more likely to occur at some sites on the chromosomes than at others.

These important studies confirmed Morgan's hypothesis that genes are arranged linearly on chromosomes. However, there is still no explanation of how homologous chromosomes break in exactly corresponding sites and so exchange equal amounts of genetic material at crossing over.

GIANT CHROMOSOMES

In *Drosophila*, as in many other insects, certain cells do not divide during the larval (immature) stage of the insect. In such cells, however, the chromosomes continue to replicate, over and over again, but since daughter chromosomes do not separate from one another after replication, they simply become larger and larger. In 1933, such giant chromosomes were reported in the salivary glands of the larvae of

13–11
Chromosomes from the salivary gland of a Drosophila *larva. These chromosomes are 100 times larger than the chromosomes in ordinary body cells, and their details are therefore much easier to see. (This micrograph was taken with a light microscope, not an electron microscope.) Because of the distinctive banding patterns, it is possible in some cases to assign genes to specific bands on particular chromosomes.*

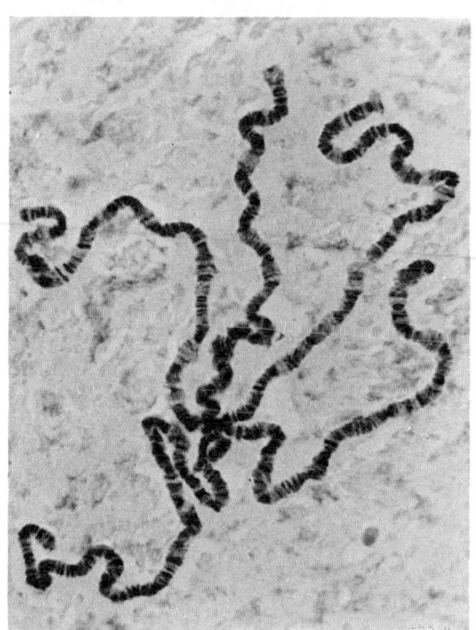

Drosophila. As you can see in Figure 13–11, these giant chromosomes when stained are marked by very distinctive dark and light bands. The bands are areas in which the chromatin is more tightly packed. Changes in these banding patterns were found to correlate with observed genetic changes in the flies. In addition, the results of crossovers and breaks could actually be *seen;* thus further confirmation of Sturtevant's brilliant hypothesis came 20 years later from an unexpected source.

SUMMARY

In the late nineteenth century, rapid advances in the tools and methods of microscopy led to the discovery of mitosis and meiosis.

Sutton was among the first to notice the analogy between the behavior of the chromosomes at meiosis and the segregation and assortment of genetic traits described by Mendel. On the basis of this observation, Sutton proposed that Mendelian factors (genes) are carried on chromosomes. Chromosomes appeared to come in pairs, and Sutton postulated that alleles occur on these homologous chromosomes. The chromosome pairs separate at meiosis, and new pairs form again when the egg is fertilized. Sutton's hypothesis led to the recognition of an important exception to Mendelian laws: Independent assortment may be modified if the genes involved are on the same chromosome.

Strong support for Sutton's hypothesis that genes are on the chromosomes came as a result of studies by Morgan and his group on the fruit fly *Drosophila. Drosophila,* because it is easy to breed and maintain, has been used in a wide variety of genetic studies. It has four pairs of chromosomes; three pairs are structurally the same in both sexes, but the fourth pairs, the sex chromosomes, are different. In fruit flies, as in many other species (including humans), the two sex chromosomes are *XX* in the females and *XY* in the males.

At the time of meiosis, the sex chromosomes are segregated. Each egg cell will receive an *X* chromosome, but half the sperm cells will receive *X* chromosomes and half will receive *Y* chromosomes. Thus it is the sperm cell that determines the sex of the embryo.

In the early 1900s, experiments with mutations in the fruit fly showed that certain characteristics are sex-linked, that is, their genes are carried on the sex chromosomes. Recessive genes carried on the *X* chromosomes (the *Y* chromosome carries less genetic information) appear in the phenotype far more often in males than in females; a female heterozygous for a sex-linked characteristic will show the dominant trait, whereas a single recessive gene in the male, if carried on the *X* chromosome, will result in a recessive phenotype since no other allele is present.

Some genes assort independently in breeding experiments, and others tend to remain together. Genes that tend to travel together are said to be in the same linkage groups.

Genes are sometimes exchanged between homologous chromatids at meiosis. Such crossovers could take place only if (1) the genes are arranged in a fixed linear array along the length of the chromosomes, and (2) the alleles are at corresponding sites (loci) on homologous chromosomes. On the basis of these assumptions, chromosome maps, showing the relative positions of gene loci along *Drosophila* chromosomes, were developed from crossover data provided by breeding experiments.

QUESTIONS

1. Draw a diagram similar to Figure 13–4 indicating sex determination in a robin.

2. A couple has three girls. What are the chances that the next child will be a boy?

3. Suppose you would like to have a family consisting of two girls and a boy. What are your chances, assuming you have no children now? If you already have one boy, what are your chances of completing your family as planned?

4. Segregation of alleles can occur at either of two stages of meiosis. Name the two stages and explain what happens in each of them.

5. An organism has 42 chromosomes. How many linkage groups does it have?

6. Does crossing over necessarily result in a recombination of genes? Explain your answer.

7. The first two pairs of genes studied simultaneously in crosses were *Rr* and *Yy* in pea plants. As it turned out, they were on different chromosomes. How do you think the development of the principles of genetics would have been affected had they been linked?

8. The genes for coat color in cats are carried on the *X* chromosome. Black (*b*) is the recessive and yellow (*B*) is the dominant. What coat colors would you expect in the offspring of a cross between a black female and a yellow male? What coat colors would you expect in the sons of a calico female, regardless of the coat color of the father?

9. In a series of breeding experiments, a linkage group composed of genes *A, B, C, D,* and *E* was found to show approximately the following crossover frequencies:

GENE	GENE: *A*	*B*	*C*	*D*	*E*
A	—	8	12	4	1
B	8	—	4	12	9
C	12	4	—	16	13
D	4	12	16	—	3
E	1	9	13	3	—

CROSSOVERS PER 100 FERTILIZED EGGS

Using Sturtevant's standard unit of measure, "map" the chromosome.

10. You and a geneticist are looking at a mahogany-colored Ayrshire cow with a newly born red calf. You wonder if it is male or female, and the geneticist says it is obvious from the color which sex the calf is. He explains that in Ayrshires the genotype *AA* is mahogany and *aa* is red, but the genotype *Aa* is mahogany in males and red in females. What is he trying to tell you—that is, what sex is the calf? What are the possible phenotypes of the calf's father?

CHAPTER 14

The Chemical Basis of Heredity I: The Search Begins

By the early 1940s the existence of genes and the fact that they were in chromosomes were no longer in doubt. But what were genes? What did they really do? What does biological inheritance mean? When we say, "She has her mother's eyes," we obviously mean something quite different from saying, "She has her mother's pocketbook." "He inherited his father's intelligence" is not at all the same as "He inherited his uncle's yacht." The microscopic fertilized egg contains neither eyes nor intelligence. All that can be inherited, biologically speaking, is a potentiality, an ability to develop in certain ways.

A turning point in genetics came when scientists began to focus on the question of how it was possible for these little lumps of matter—the chromosomes—to be the bearers of what they had come to realize must be an enormous amount of complex information. The chromosomes, like all the other parts of a living cell, are composed of atoms arranged into molecules. Some scientists, a number of them quite eminent in the field of genetics, thought it would be impossible to understand the complexities of genetics in terms of the structure of "lifeless" chemicals. (Although they did not use the term to describe themselves, they were, in fact, the vitalists* of the twentieth century.) Others thought that if the molecular structure of the chromosomes were understood, we could then come to understand how chromosomes can function as the bearers of genetic information. The work of the latter group is now termed "molecular genetics."

GENE-ENZYME RELATIONSHIPS

"Inborn Errors of Metabolism"

The first molecular geneticist—although he would not have used the term—was the English physician Sir Archibald Garrod. In 1908, Garrod presented a series of lectures in which he set forth a new concept of human diseases, one he called "inborn errors of metabolism." With a leap of the imagination that spanned almost half a century, Garrod postulated that certain diseases, such as alcaptonuria (page 86), that were hereditary in nature, were caused by the lack of a particular enzyme.

* See essay, page 28.

14-1
Asci of Neurospora. *Because of the shape of these asci, the sexual spores contained within the ascus are immobilized in the order in which they are produced by meiosis.* Neurospora *also produces fine, dust-like asexual spores known as conidia.*

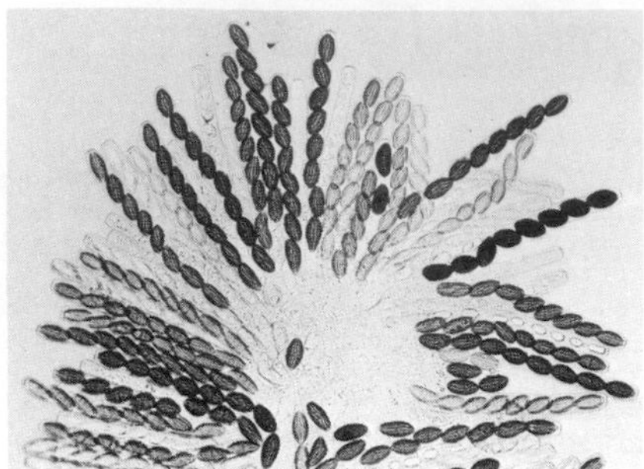

In postulating the existence of a relationship between genes and body chemistry, Garrod anticipated, by almost 30 years, the coming together of genetics and chemistry. An essential feature of modern biology is that it attempts to explain biological mechanisms in terms of the laws of physics and chemistry. It was not until the mid-1940s, however, that further research made it possible to begin to understand the role of genes in controlling enzyme production.

One Gene, One Enzyme

Based on observations such as Garrod's, the hypothesis—one of many—was formulated that genes act by influencing enzyme production. The problem was to devise a test for this hypothesis. Traits such as wrinkledness in peas or eye color and wing shape in *Drosophila*—or indeed the visible features of any organism—are characteristically the end product of a vast number of chemical reactions, most of which are still unknown.

In 1941, George Beadle and Edward L. Tatum decided to turn the problem around, and in so doing they changed the course of genetics and won a Nobel Prize. Instead of picking a genetic characteristic and working out its chemistry, they decided to begin with step-by-step chemical reactions—controlled by enzymes—and see if mutations affected these reactions. Their experiments are worth examining in some detail because of their influence on both the theories and the techniques of genetics in the next quarter century.

First, like Morgan and Mendel before them, they chose a suitable organism. For their purposes, it was the red bread mold *Neurospora crassa,* which has since become almost as famous a research tool in genetics as the fruit fly. This organism has several obvious advantages for genetic research:

1. Its life cycle is brief.
2. It can be grown in vast quantities in the laboratory.
3. Unlike most higher organisms, it is haploid; that is, during most of its life cycle it has only one set of chromosomes (seven in number) rather than two sets. As a consequence, when a mutation occurs, its effects are detectable immediately; the lack of a homologous chromosome rules out the masking of any mutation by a dominant allele.
4. Meiosis takes place in pod-shaped sacs known as asci (singular, *ascus*) that not only neatly package all the products of meiosis and line them up for inspection but even provide an extra copy (Figure 14-2).
5. Many mapping studies had already been done with *Neurospora,* which facilitated further genetic analyses.

Another important feature of *Neurospora,* from the point of view of the investigators, was the fact that it can be grown on a very simple medium containing any one of several sugars as a carbon and energy source, one vitamin (biotin), and a few minerals. This undemanding mold is able to make for itself all the amino acids, other vitamins, polysaccharides, and other substances essential for its growth and functioning.

The making of an amino acid or a vitamin requires, as we know, a series of reactions, each of which is mediated by a particular enzyme. If, as a result of mutation, *Neurospora* were to lose any one of the enzymes involved, for example, in making the amino acid arginine, it could no longer grow in the simple minimal

14-2
Meiosis in Neurospora crassa, *showing only one of its seven chromosomes. The zygote represents a cross of a mutant strain with a normal strain (lacking that particular mutation). As a result of meiosis and the single mitotic division that follows it, eight spores are produced, lined up in a single, narrow spore case, the ascus. Four will be normal and four mutants. Mutations in different chromosomes will assort independently. Mutations in the same chromosome will be separated by crossing over. Thus, by analyzing the products of meiosis, it is possible to demonstrate that a given biochemical change involves only a single gene.*

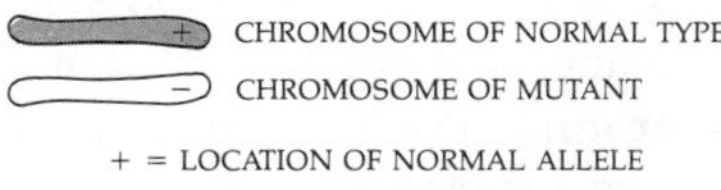

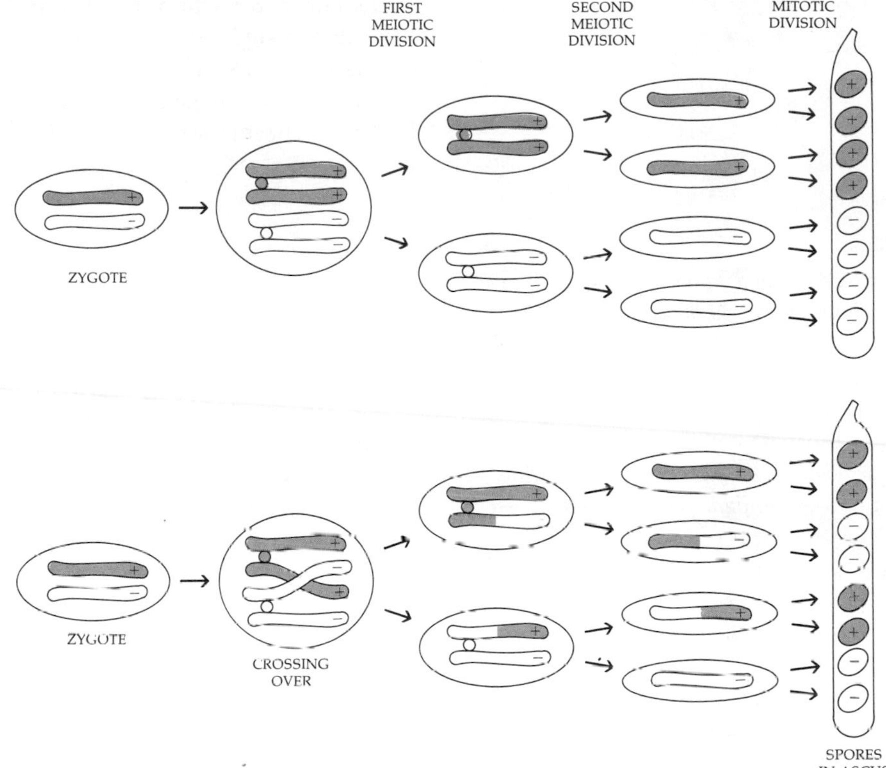

medium. The mutant could, however, survive on a medium supplemented with arginine. Thus geneticists began to study not just the distant effects of gene activity, such as white eyes and long wings, but also the most immediate products of gene function.

To understand the experiments, it is necessary first to know something about the life cycle of *Neurospora.* The mold produces tiny, dustlike spores known as conidia (singular, *conidium*); these are asexual spores, meaning that they are not produced by any mating processes. Conidia are carried in air currents, and when they land on a suitable medium, such as a loaf of bread, they germinate and grow to form a matlike structure, the mycelium. *Neurospora crassa* usually propagates in this rapid and efficient way. Sometimes, however, two of these fungal mycelia will encounter one another, and if they are of different mating strains, portions of the mycelia will fuse and specialized structures, fruiting bodies, are produced. These fruiting bodies are made up of the asci, in which meiosis takes place, producing sexual spores. These too can germinate, producing new mycelia. The entire cycle from sexual spore to sexual spore takes only about 10 days, which is another advantage of *Neurospora.*

Beadle and Tatum collected asexual spores (conidia) of *Neurospora,* x-rayed them to increase the mutation rate, then crossed these irradiated strains with normal strains. This step was taken to demonstrate that the changes in the x-rayed molds were really genetic changes. The sexual spores thus produced were planted in culture media containing all the nutrients *Neurospora* normally needs plus amino acids. This made it possible for the investigators to select mutants that could grow in the enriched medium but not in the minimal one. Subcultures of such molds were then tested individually to determine which nutrient they were unable to synthesize (Figure 14–3). By crossing *Neurospora* mutants with normal strains, it was possible to perform genetic analyses that, by segregating the alleles, correlated the loss of an enzyme with a single change in the chromosomes.

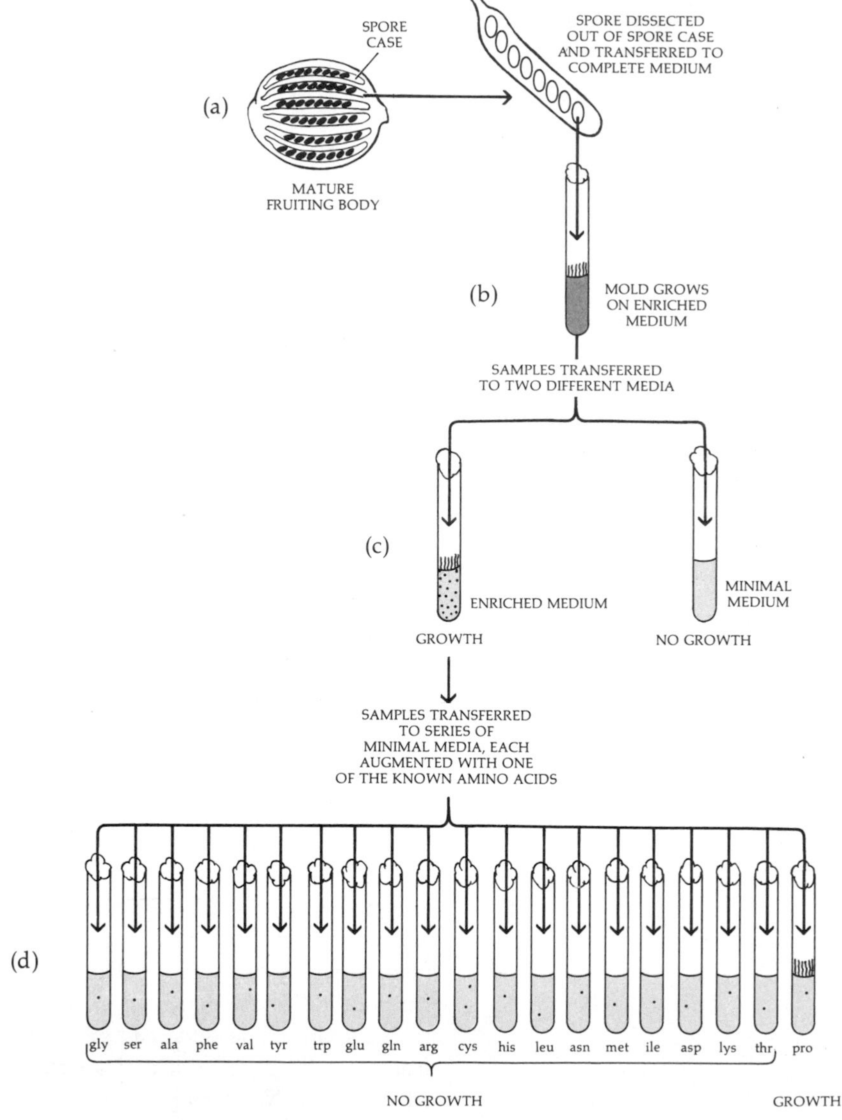

14–3

How Beadle and Tatum tested the mutants of Neurospora. *By these experiments, they were able to show that a change in a single gene results in a change in a single enzyme.* (a) *Asci are removed from fruiting bodies of* Neurospora *and the sexual spores dissected out.* (b) *Each spore is transferred to an enriched medium, containing all* Neurospora *normally needs for growth plus supplementary amino acids.* (c) *A fragment of the mycelium is tested for growth in the minimal medium. If no growth is observed on the minimal medium, it may mean that a mutation has occurred that renders this mutant incapable of making a particular amino acid, and so tests are continued.* (d) *Subcultures of mycelia that grow on the enriched medium but not on the minimal one are tested for their ability to grow in minimal media supplemented with only one of the amino acids. As in the example shown here, a mold that has lost its capacity to synthesize the amino acid proline is unable to survive in a medium that lacks that amino acid. Further tests are then made to discover, in each case, which enzymatic step has been impaired.*

At the time Beadle and Tatum's experiments were first performed, enzymatic pathways had been worked out for the synthesis of only a very few organic molecules in only a few species. Since that time, however, it has been found that a great variety of cells—including bacteria, yeast, and even the cells that make up the tissues of the human body—are very similar in their enzyme systems and in their stepwise synthesis of the various basic cellular nutrients, underlining once more the remarkable biochemical unity of living things.

One Gene, One Protein

The Beadle-Tatum hypothesis that a particular gene is responsible for the production of a particular enzyme was quick to gain acceptance. That enzymes are proteins had already been demonstrated in the 1930s. Not all proteins are enzymes, however; some, for instance, are hormones, like insulin. Others are structural proteins, like collagen. These proteins, too, are under gene control. This expansion of the original concept did not modify it in principle: "One gene, one enzyme," as the theory was first abbreviated, was simply amended to the less memorable but more precise "one gene, one polypeptide chain." In other words, enzymes and other protein molecules are the products of genes.

THE STRUCTURE OF HEMOGLOBIN

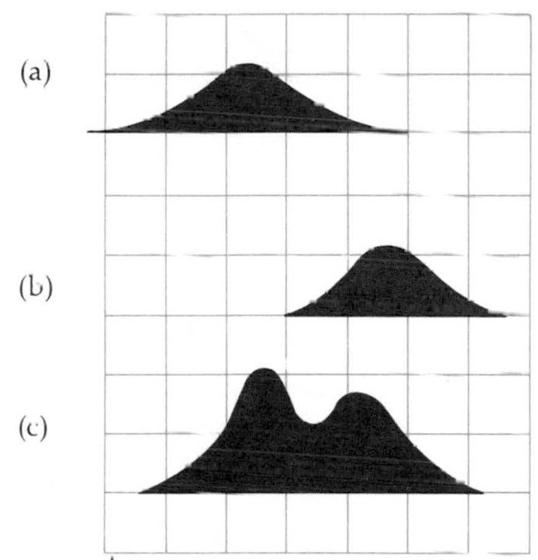

14-4
Results from electrophoresis of (a) *normal hemoglobin,* (b) *the hemoglobin of a person with sickle cell anemia, and* (c) *the hemoglobin of a person who is heterozygous for the sickle cell trait. Because of slight differences in electric charge, the normal and sickle cell hemoglobins move differently in an electric field. The normal hemoglobin is more negatively charged; hence, it is closer to the positive pole than the sickle cell hemoglobin. The hemoglobin of the heterozygote separates into the two different types.*

Linus Pauling was one of the first to see some of the implications of the work of Beadle and Tatum. Perhaps, Pauling reasoned, human diseases involving hemoglobin, such as sickle cell anemia, could be traced to a slight variation from normal in the protein structure of the hemoglobin molecule. To test this hypothesis—which, at this stage, was pure speculation—he took samples of hemoglobin from people with sickle cell anemia, from others heterozygous for the disease, and from others free of it. To try to detect differences in these proteins, he used a process known as electrophoresis, in which organic molecules are dissolved in a solution and exposed to a weak electric field.

As we noted in Chapter 3, individual amino acids may have a positive, a negative, or a neutral charge. Therefore, a mutation that causes the substitution of one amino acid for another may change the total charge of the protein molecule, and the normal and the mutant protein molecules will, as a result, move differently in an electric field.

Figure 14-4 shows the results of Pauling's experiment. A person who has sickle cell anemia makes a different sort of hemoglobin than a person who does not have the disease. A person who is heterozygous (carrying one copy of the recessive gene for sickling and one copy for normal hemoglobin) makes both kinds of hemoglobin molecules; however, enough normal molecules are produced to prevent anemia. (Notice that the terms "dominant" and "recessive" are beginning to take on a different meaning.)

A few years later, in 1959, Vernon Ingram, of Cambridge University in England, was able to show that the actual difference between the normal and the sickle cell hemoglobin molecules is one amino acid in 300, as we noted previously (page 70). The hemoglobin molecule is composed of four polypeptide chains—two identical alpha chains and two identical beta chains. Each chain consists of about 150 amino acids. In a precise location in each beta chain, where a glutamic acid is present in the normal hemoglobin, a valine appears in the sickle cell hemoglobin. This was the first demonstration that a protein has a unique series of amino acids.

Following Ingram's report, speculative thinkers in the field of biology were quick to see that the amino acids, the number of which was so provocatively close to the number of letters in our own alphabet, could be arranged in a variety of different ways. It appeared that these different arrangements might account for both the great diversity of enzymes and the great specificity of the biochemical reactions they mediate within the cell. The proteins were seen as making up a sort of language—"the language of life"—that spelled out the directions for all the many activities of the cell.

THE CHEMISTRY OF THE GENE: DNA VS. PROTEIN

During the 1930s and 1940s, scientists became more and more concerned with the question, "Exactly what is a gene?" The attempt to identify the chemical nature of the gene gave rise to two opposing schools of thought, and the controversy lasted for a number of years—until, as we shall see, it was conclusively resolved in 1953.

Chromosomes, as we mentioned previously, are composed of DNA and protein. Many prominent investigators, particularly those who had been studying proteins, believed that the genes themselves were proteins, that the chromosomes contained master models of all the proteins that would be required by the cell, and that enzymes and other proteins active in cellular life were copied from these master models. This was a logical hypothesis, but as it turned out, it was wrong.

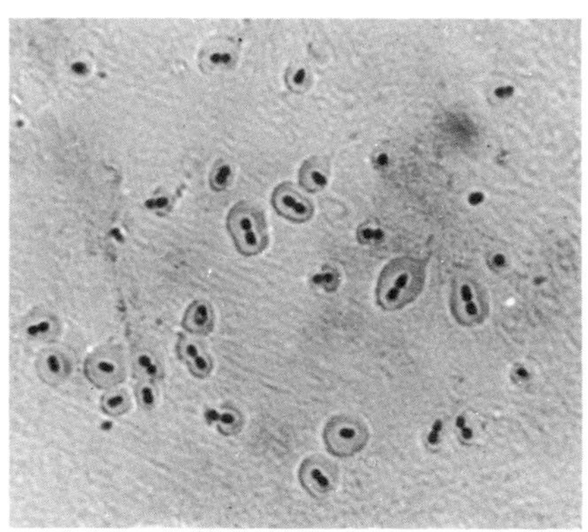

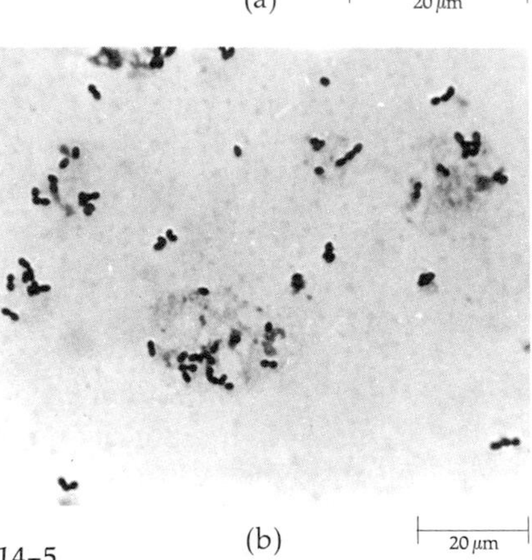

14–5
(a) *Encapsulated and* (b) *nonencapsulated forms of pneumococci. The encapsulated form, which is resistant to phagocytosis by white blood cells, produces pneumonia; the nonencapsulated form is harmless.*

The Transforming Factor

To trace the beginning of the other hypothesis—the one that proved right—it is necessary to go back to 1928 and pick up an important thread in modern biological history. In that year, an experiment was performed that seemed at the time very remote from either biochemistry or genetics. Frederick Griffith, a public health bacteriologist, was studying the possibility of developing vaccines against the bacterial cells, pneumococci, that cause one kind of pneumonia. In those days, before the development of modern antibiotics, bacterial pneumonia was a serious disease, the grim "captain of the men of death."

Pneumococci, as Griffith knew, come in either virulent (disease-causing) forms with capsules (polysaccharide coats) or nonvirulent (harmless) forms without capsules (Figure 14–5). Griffith was interested in finding out whether injections of heat-killed virulent pneumococci, which do not cause disease, could be used to vaccinate against pneumonia. In the course of various experiments, he performed one that gave him very puzzling results. He injected mice simultaneously with heat-killed virulent bacteria and with living but nonvirulent bacteria, both of which were harmless—but all the mice died. When Griffith performed autopsies on them, he found their bodies filled with living encapsulated (and therefore virulent) bacteria (Figure 14–6). Had the dead virulent cells come back to life or had something been passed from them to the living nonvirulent cells that endowed the living cells with the capacity to make capsules and therefore to be virulent?

Within the next few years it was shown that the same phenomenon could be reproduced in the test tube and these questions could be answered. It was found that extracts from the killed encapsulated bacteria, when added to the living

14-6
Discovery of the transforming factor, a substance that can transmit genetic characteristics from one cell to another, resulted from studies of pneumococci, pneumonia-causing bacteria. One strain of these bacteria has capsules (protective outer sheaths); another does not. The capacity to make capsules and cause disease is an inherited characteristic, passed from one bacterial generation to another as the cells divide. (a) *Injection into mice of encapsulated pneumococci killed the mice.* (b) *The nonencapsulated strain was harmless.* (c) *If the encapsulated strain was heat-killed before injection, it too was harmless.* (d) *If, however, heat-killed encapsulated bacteria were mixed with live nonencapsulated bacteria and the mixture was injected into mice, the mice died.* (e) *Blood samples from the dead mice revealed live encapsulated pneumococci. Something had been transferred from the dead bacteria to the live ones that endowed them with the capacity to make capsules and cause pneumonia. This "something" was later isolated and found to be DNA.*

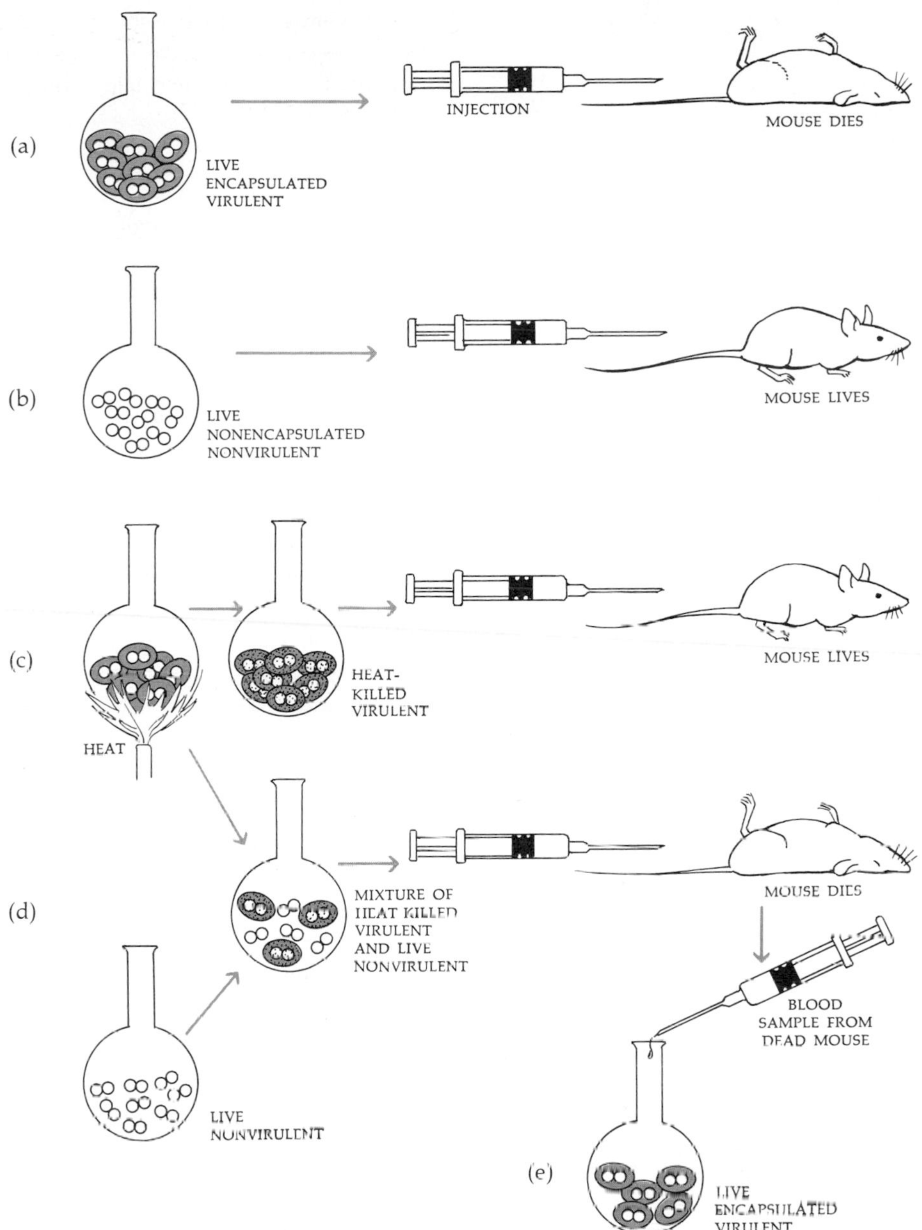

harmless bacteria, could convert them to the virulent type with the ability to make capsules. Furthermore, once converted, they could transmit this characteristic to their progeny.

One of the laboratories that worked on the nature of this *transforming factor,* as it came to be called, was that of O. T. Avery at Rockefeller University. After almost a decade of patient chemical isolation and analysis, Avery and his co-workers were convinced that the chemical substance in the cellular extracts of killed bacteria that transmitted the new genetic quality was the molecule known as *deoxyribonucleic acid.*

Subsequent experiments showed that a variety of genetic traits could be passed from one colony of bacterial cells to members of a similar colony by means of isolated deoxyribonucleic acid, which soon took on the now familiar abbreviation of DNA.

THE NATURE OF DNA

At this point, we shall begin to examine the chemical structure of DNA. It was largely because of the seeming simplicity of its structure, as compared to that of proteins, that Avery's experiment, although beautifully designed and executed, did not receive full recognition for almost a decade.

DNA was first discovered in that same remarkable decade in which Darwin published *The Origin of Species* and Mendel presented his results to an audience of forty at the Natural History Society in Brünn. In 1869, a German chemist, Friedrich Miescher, had extracted a substance from the nuclei of cells that was white, slightly acid, and contained phosphorus. He called it nucleic acid, which was later amended to deoxyribonucleic acid to distinguish it from a closely related chemical, ribonucleic acid (RNA), which subsequently was also isolated from cells.

In 1914, another German, Robert Feulgen, discovered that DNA had a remarkable attraction for a red dye called fuchsin, but he considered this finding so unimportant that he did not trouble to report it for a decade. Feulgen staining, as it was called when it finally made its way into use, revealed that DNA was present in all cells and was characteristically located in the chromosomes.

There was no particular interest in DNA for several decades since no role had been postulated for it in cellular metabolism. Most of the work on its chemistry was carried out by the great biochemist P. A. Levene. He showed that DNA could be broken down into four nitrogenous (nitrogen-containing) bases—adenine and guanine (the purines) and thymine and cytosine (the pyrimidines)—a five-carbon sugar, and a phosphate group. From the proportions of these components, he made two deductions, one correct and one incorrect:

1. Each nitrogenous base is attached to a molecule of sugar, which, in turn, is attached to a phosphate group to form a single molecule, a *nucleotide* (Figure 14–7). This hypothesis was right.
2. Since, in all the samples he measured, the proportions of the nitrogenous bases were approximately equal, all four nitrogenous bases must be present in nucleic acid in equal quantity. Furthermore, these molecules must be grouped in clusters of four—a tetranucleotide, he called it—that repeated over and over again along the length of the molecule. This hypothesis, which was wrong, dominated scientific thinking about the nature of DNA for more than a decade.

Because Levene's tetranucleotide theory was given great weight by his renown as a biochemist, biologists as a whole were slow to recognize the importance of Avery's experiment. Avery, like Mendel before him, was a traveler bearing an odd tale that did not fit. How could a chemical whose structure was "known" to be simple and repetitious carry the complicated and various hereditary information? For almost a decade, most influential biochemists continued to believe that proteins were the genetic material. In the next chapter, we shall review some of the evidence that made them change their minds.

14–7

(a) *A nucleotide is made up of three different parts: a nitrogenous base, a sugar, and a phosphate.* (b) *Each nucleotide in DNA contains one of the four possible nitrogenous bases, a deoxyribose sugar, and a phosphate.*

(a) PHOSPHATE — SUGAR — NITROGENOUS BASE

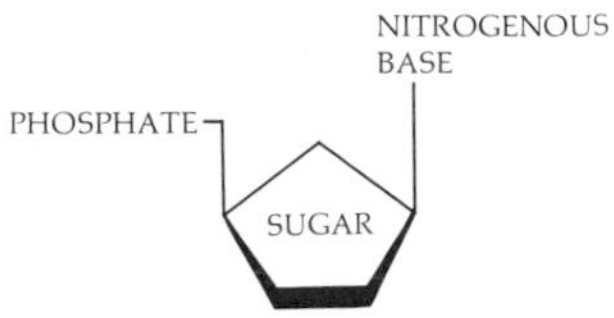

(b) PURINE-CONTAINING NUCLEOTIDES: ADENINE, GUANINE

PYRIMIDINE-CONTAINING NUCLEOTIDES: THYMINE, CYTOSINE

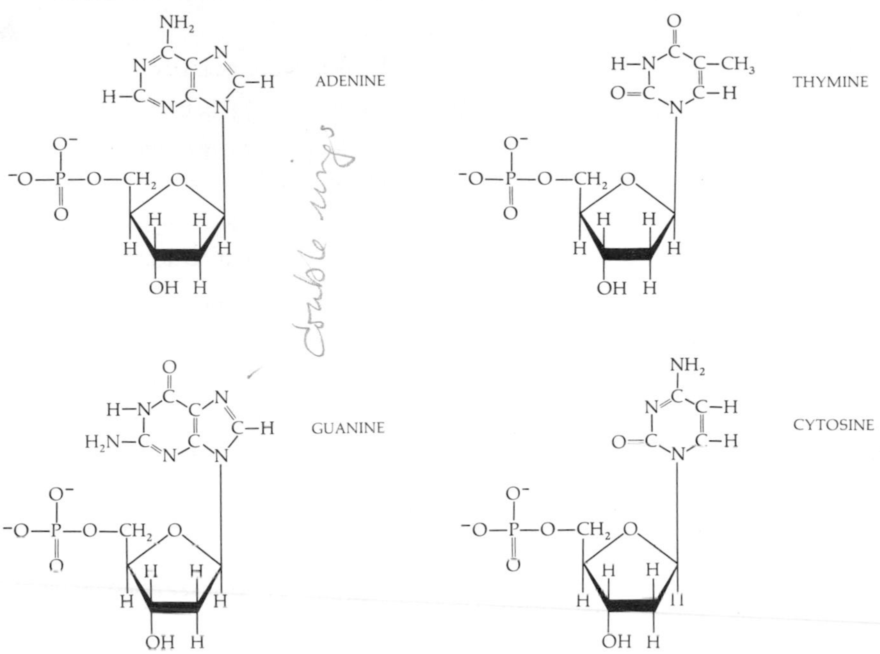

SUMMARY

Classical genetics had been concerned with the mechanics of inheritance—how the units of hereditary material were passed from one generation to the next and how changes in the hereditary material were expressed in individual organisms. In the late 1930s, a new question arose and geneticists began to explore the nature of the gene—its structure, composition, and properties—and its role in the internal chemistry of living organisms.

In 1941, working with the bread mold *Neurospora crassa,* Beadle and Tatum established the "one-gene-one-enzyme" principle. They were able to trace biochemical defects in *Neurospora* to the nonfunction of specific enzymes and to relate each of these defects to a mutation in a specific gene.

It was soon discovered that some enzymes are made up of more than one protein chain and that other proteins, such as some hormones and hemoglobin, are also under genetic control. The one-gene-one-enzyme principle was expanded to the "one-gene-one-polypeptide" principle.

Sickle cell anemia is inherited as a Mendelian recessive gene. Using electrophoresis, Linus Pauling demonstrated that normal hemoglobin is slightly different from the hemoglobin of a person with sickle cell anemia. Subsequently, it was shown that the actual difference lay in just two amino acid changes among the 600 amino acids making up the four polypeptide chains in the molecule.

During the 1940s, many investigators believed that genes were proteins, but others were convinced that the hereditary material was deoxyribonucleic acid (DNA). Important, although not widely accepted, evidence for the genetic role of DNA was presented by Avery in his experiments on the transforming factor of *Pneumococcus.*

QUESTIONS

1. Define a gene (a) in terms of Mendelian genetics, (b) in terms of the Beadle-Tatum hypothesis.

2. The bread mold with which Beadle and Tatum worked is haploid. Would haploid organisms follow Mendel's principles? Explain. Why would this feature make it easier and faster to conduct genetic experiments?

3. In what situation would segregation of a pair of alleles occur during the first meiotic division? During the second meiotic division?

4. What advantages does *Neurospora* have over *Drosophila* as an organism for genetic experiments?

5. In pea plants, a cross between a purple-flowered plant and a white-flowered plant produces a purple-flowered plant. In snapdragons, a cross between a red-flowered plant and a white-flowered plant produces a pink-flowered plant. Explain how the Beadle-Tatum hypothesis might account for these differences.

6. Although the person with sickle cell anemia is usually ill, the person who is simply a carrier for sickle cell anemia—that is, the heterozygote—rarely has any symptoms of disease. Explain, in terms of the examples of flower color mentioned above.

7. A new mutant in *Neurospora* requires chemical *X* for growth and accumulates chemical *Y*. What is the order of these substances in the biosynthetic sequence and where in this sequence does the gene that is deficient in the mutant normally act?

8. In *Neurospora,* mutant *x* will grow only if provided with cystathionine, homocysteine, or cysteine; mutant *r* will grow only if provided with homocysteine, but it accumulates cystathionine; mutant *w* will grow if provided with homocysteine or cystathionine, but not if only cysteine is available. Sketch the sequence of biosynthesis of these chemicals. What is the genetic defect in each of the mutants?

9. What are the steps by which Griffith demonstrated the existence of the transforming factor? Can you think of any implications of Griffith's discovery for modern medicine?

10. We have seen many examples of progress in the understanding of biological phenomena, in which new work was built upon previous work. But there are times when the results of previous work, or ignorance of such results, can block new understandings. Describe several examples of this situation in the history of genetics.

CHAPTER 15

The Chemical Basis of Heredity II: The Path to the Double Helix

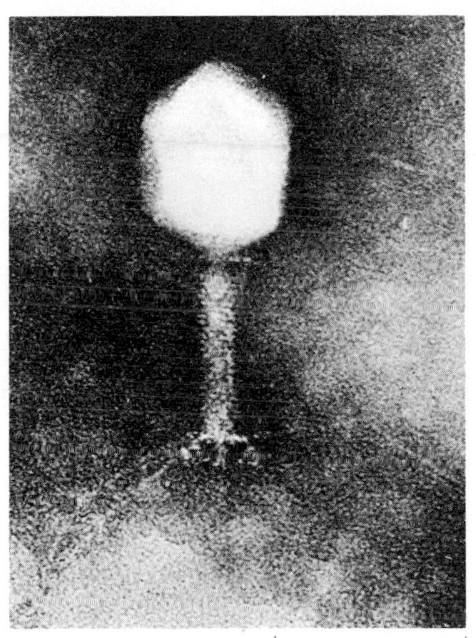

15–1
Electron micrograph of a T4 bacteriophage. Notice the highly distinctive "tadpole" shape. Each bacteriophage consists of a head, which appears hexagonal in electron micrographs, and a complex tail.

About the time that Beadle and Tatum began their studies with *Neurospora*, another pair of scientists, Max Delbrück and S. E. Luria, initiated a series of studies with another "fit material," destined to become as important to genetic research as the garden pea, the fruit fly, and the red bread mold. The fit material was a group of viruses that attacked bacterial cells and were therefore known as *bacteriophages*, "bacteria eaters." Every known type of bacterial cell is preyed upon by its own type of bacterial virus, and many bacteria are host to many different kinds of viruses. Delbrück, Luria, and the group that joined them in these studies agreed to concentrate on a series of seven related viruses that attacked *Escherichia coli*, a normal inhabitant of the human intestine. These viruses were numbered T1 through T7, with the T standing simply for "type." As it turned out, most of the early work was done on T2 and T4, which became known as the T-even bacteriophages.

These viruses were inexpensive, readily available, and demanded little space or equipment. Furthermore, they were phenomenal at reproducing themselves. Twenty-five minutes from the time a single virus infected a bacterial cell, that cell would burst open, releasing hundreds of new viruses, all exact copies of the original virus. Another advantage (which was not discovered until after the research was begun) was that this group of bacteriophages has a highly distinctive shape (Figure 15–1), and so can be readily identified with the electron microscope.

THE PHAGE EXPERIMENTS

According to electron-microscope studies of infected *E. coli* cells (broken open at regular intervals after infection), the bacteriophages do not multiply like bacteria. Except for a few fragments, they disappear the moment after infection, and for the first 10 to 11 minutes of the infectious cycle, not a single virus can be seen within the bacterial cell. Then, depending on when the cell is opened during the course of the infection, increasing numbers of completed bacteriophages can be seen and, mixed with them, odds and ends that resemble bits of incomplete phages.

Chemical analysis of the bacteriophages revealed that they consist quite simply of DNA and of protein, the two leading contenders in the 1940s for the role of the genetic material. The chemical simplicity of the bacteriophage offered geneticists a remarkable opportunity. The viral genes—the hereditary material by which new viruses are made within the bacterial cell—had to be carried either on the protein or on the DNA. If it could be determined which of the two it was, then the gene would be chemically identified. In 1952, a simple but ingenious experiment was carried out.

Alfred D. Hershey and his laboratory assistant, Martha Chase,* prepared two separate samples of viruses, one in which the DNA was labeled with the radioactive isotope ^{32}P and the other in which the protein was labeled with the radioactive isotope ^{35}S. These were made by growing the *E. coli* host on a medium that contained the radioactive isotope. After a cycle of multiplication the newly formed viruses all contained some of the radioactive isotope in place of the common

* We are including the names of the scientists involved in these experiments, not only to give credit where it is due, but also because the names have become synonymous with the work. What we are describing now are the Hershey-Chase experiments.

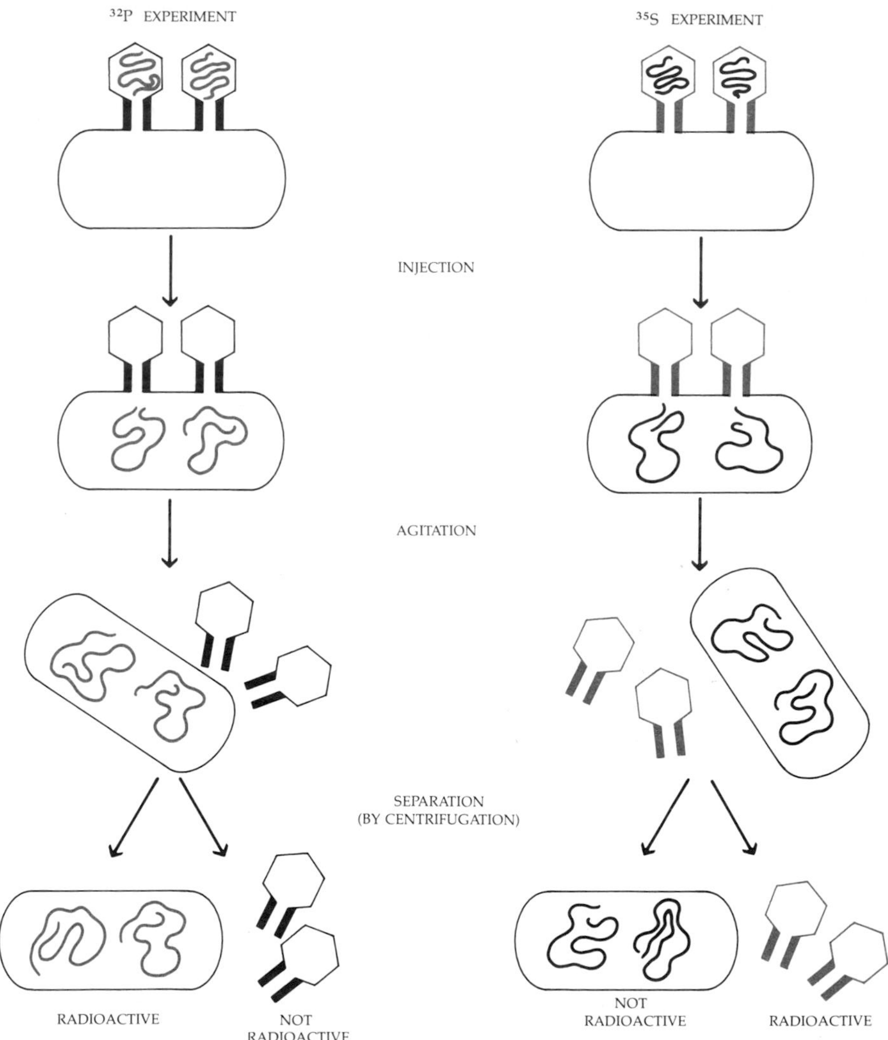

15–2
A summary of the Hershey-Chase experiments demonstrating that DNA is the hereditary material of a virus.

15–3
Max Delbrück and Salvador Luria at Cold Spring Harbor Laboratory of Quantitative Biology in 1953. They shared the Nobel Prize with A. D. Hershey in 1971 for "their discoveries concerning the replication mechanism and the genetic structure of viruses."

nonradioactive isotope. If you recall the chemical structure of nucleic acids and proteins, you will note that DNA contains phosphorus but no sulfur, while the amino acid components of proteins contain no phosphorus but two of them (methionine and cysteine) contain sulfur. Thus these two isotopes can serve as specific tags or labels that distinguish DNA from protein.

One culture of bacteria was infected with ^{32}P-labeled phage and another with ^{35}S-labeled phage (Figure 15–2). After the infectious cycle had begun, the cells were agitated in a blender to separate them from any viral material remaining outside the cells. The two samples—one containing extracellular material and the other intracellular material—were then tested for radioactivity. Hershey and Chase found that the ^{35}S had stayed outside the bacterial cells with the empty viral coats and the ^{32}P had entered the cells, infected them, and caused the production of new virus progeny. It was therefore concluded that the genetic material of the virus is DNA rather than protein.

Electron micrographs (Figure 15–4, for example) have now confirmed that the phage attaches to the bacterial cell wall by its tail. They also indicate that the phage injects its DNA into the cell, leaving the empty protein coat on the outside. In short, the protein is just a container for the bacteriophage DNA. It is the DNA of the bacteriophage that enters the cell and carries the complete hereditary message of the virus particle, directing formation of new viral DNA and new viral protein.

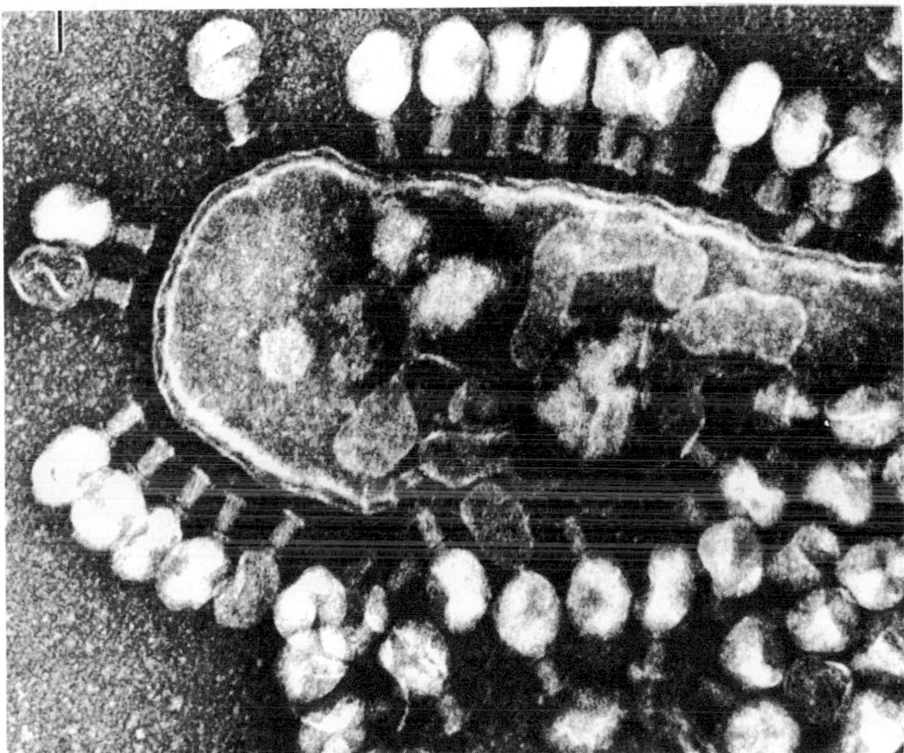

15–4
Electron micrograph of bacteriophages attacking a cell of E. coli. *The viruses are attached to the bacterial cell by their tails. Some of the viruses have apparently injected their DNA into the cell, as indicated by their empty heads. A complete cycle of virus infection takes only about 25 minutes. At the end of that period, several hundred new virus particles are released from the cell.*

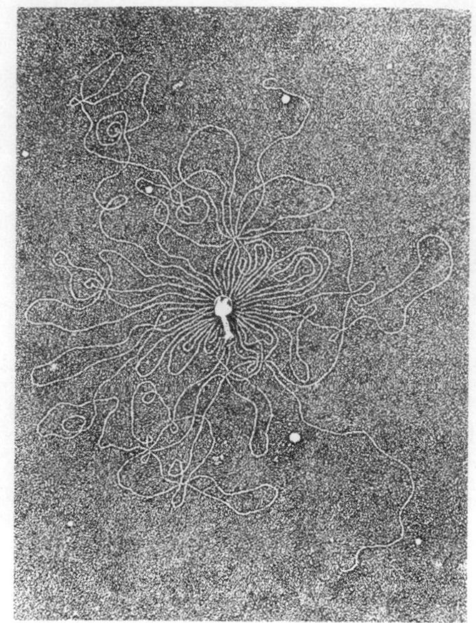

15-5
Electron micrograph of a bacteriophage surrounded by its single continuous molecule of DNA. The bacteriophage was burst open and the DNA was released by osmotic shock, produced by placing the bacteriophage in distilled water. The DNA was then allowed to expand on the water surface. The normally circular DNA has been broken at one point—note the two free ends at the top and bottom.

FURTHER EVIDENCE FOR DNA

The role of DNA in transformation and in viral replication formed very convincing evidence for believing that DNA was the chemical basis of the gene. Two other lines of experimental work also helped to lend weight to the argument. Alfred Mirsky, in a long series of careful studies conducted at Rockefeller University, showed that all the tissue cells of any given species contain equal amounts of DNA. The only exceptions are the gametes, which regularly contain just half as much DNA as the other cells of the same species, and certain unusual cells, such as polyploid cells and cells with giant chromosomes (see page 248).

Chargaff's Results

A second important series of contributions was made by Erwin Chargaff of Columbia University's College of Physicians and Surgeons. Chargaff analyzed the purine and pyrimidine content of the DNA of many different kinds of living things and found that, in contradiction to Levene's conclusions (see page 258), the nitrogenous bases do not always occur in equal proportions. The proportion of nitrogenous bases is the same in all cells of a given species but varies from one species to another. Therefore variations in base composition could very well provide a "language" in which the instructions controlling cell growth could be written. Some of Chargaff's results are reproduced in Table 15-1. Can you, by examining these figures, notice anything interesting about the proportions of purines and pyrimidines?

THE HYPOTHESIS IS CONFIRMED

The explanation of the way that the genetic information is contained in DNA was to be found in the structure of the DNA molecule. We have seen that the genetic material must meet at least four requirements:

1. It must carry genetic information from cell to cell and from generation to generation. Further, it must carry a great deal of information. Consider how many instructions must be contained in the set of genes that directs, for example, the development of an elephant, or a tree, or even a *Paramecium.*
2. It must contain information for producing a copy of itself, for it is copied with every cell division and with great precision.

Table 15-1 *Composition of DNA in Several Species*

	PURINES		PYRIMIDINES	
SOURCE	ADENINE	GUANINE	CYTOSINE	THYMINE
Human being	30.4%	19.6%	19.9%	30.1%
Ox	29.0	21.2	21.2	28.7
Salmon sperm	29.7	20.8	20.4	29.1
Wheat germ	28.1	21.8	22.7	27.4
E. coli	26.0	24.9	25.2	23.9
Sea urchin	32.8	17.7	17.3	32.1

3. On the other hand, it must sometimes mutate. When a gene changes, that is, when a "mistake" is made, the "mistake" must be copied as faithfully as was the original. This is a most important property, for without the capacity to replicate "errors," there could be no evolution by natural selection.
4. There must be some mechanism for decoding the stored information and translating it into action in the living cell.

It was when the DNA molecule was found to have the size, the configuration, and the complexity required to code the tremendous store of information needed by living things and to make exact copies of this code that DNA became widely accepted as the genetic material.

The scientists primarily responsible for working out the structure of the DNA molecule were James Watson and Francis Crick, and their feat is one of the milestones in the history of science.

THE WATSON-CRICK MODEL

In the early 1950s, a young American scientist, James Watson, went to Cambridge, England, on a research fellowship to study problems of molecular structure. There, at the Cavendish Laboratory, he met physicist Francis Crick. Both were interested in DNA, and they soon began to work together to solve the problem of its molecular structure. They did not do experiments in the usual sense but rather undertook to examine all the data about DNA and attempt to unify them into a meaningful whole.

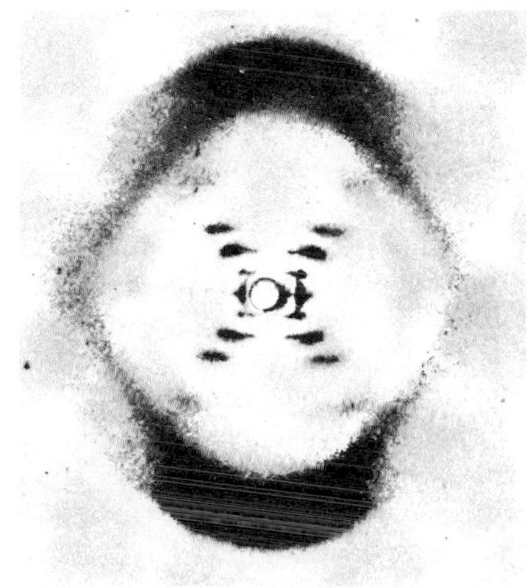

15–6
X-ray diffraction photograph of DNA taken by Rosalind Franklin in the laboratories of Maurice Wilkins, who shared the Nobel Prize with Watson and Crick. The reflections crossing in the middle indicate that the molecule is a helix. The heavy dark regions at the top and bottom are due to the closely stacked bases perpendicular to the axis of the helix.

The Known Data

By the time Watson and Crick began their studies, quite a lot of information on the subject had already accumulated:

1. The DNA molecule was known to be very large, and also very long and thin, and to be composed of nucleotides containing the nitrogenous bases adenine, guanine, thymine, and cytosine.
2. According to Levene's data, these nucleotides were assembled in repeating units of four.
3. Linus Pauling, in 1950, had shown that a protein's component chains of amino acids are often arranged in the shape of a helix and are held in that form by hydrogen bonds between successive turns of the helix. Pauling had suggested that the structure of DNA might be similar.
4. X-ray diffraction studies of DNA (Figure 15–6) from the laboratories of Maurice Wilkins and Rosalind Franklin at King's College, London, showed markings that almost certainly reflected the turns of a giant helix. (Neither Luria nor Pauling was permitted to visit England at that time and so did not see the x-ray diffraction photographs. The McCarthy era kept these two without passports. It has been suggested that the U.S. Passport Office may have determined the winners in the race for the double helix.)
5. Also crucial were the data of Chargaff indicating, as you perhaps noticed in Table 15–1, that (within experimental error) the amount of adenine is the same as the amount of thymine, and the amount of guanine is the same as the amount of cytosine: A = T and G = C.

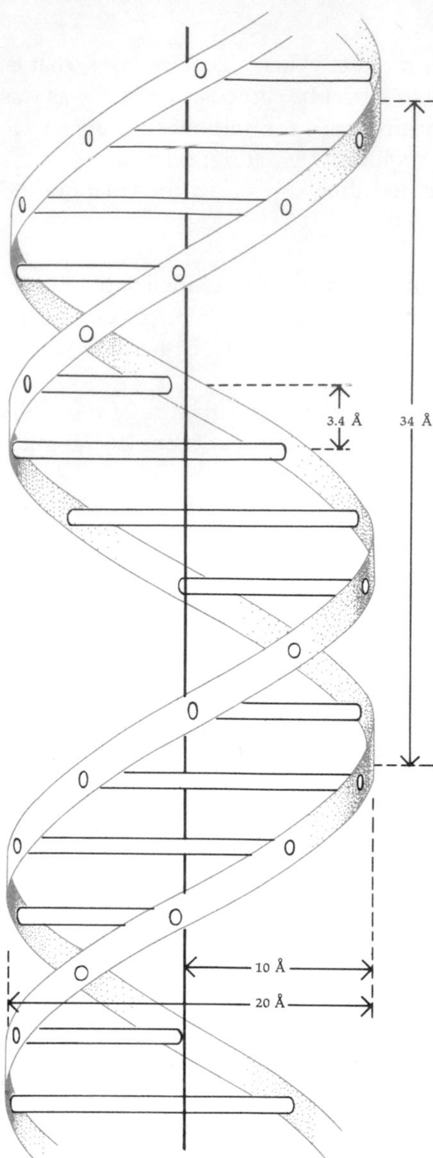

15-7
The double-stranded helical structure of DNA, as first presented in 1953 by Watson and Crick. The framework of the helix is composed of the sugar-phosphate units of the nucleotides. The rungs are formed by the four nitrogenous bases adenine and guanine (the purines) and thymine and cytosine (the pyrimidines). Each rung consists of two bases. Knowledge of the distances between the atoms, determined from x-ray diffraction pictures, was crucial in establishing the structure of the molecule.

Building the Model

From these data, some of them contradictory, Watson and Crick attempted to construct a model of DNA that would fit the known facts and explain the biological role of DNA. In order to carry the vast amount of genetic information, the molecules should be heterogeneous and varied. Also, there must be some way for them to replicate readily and with great precision so that faithful copies could be passed from cell to cell and from parent to offspring, generation after generation.

On the other hand, Watson and Crick could not be sure that the chemical structure of DNA actually would reflect its biological function. After all, this idea had never really been tested rigorously. Perhaps DNA was merely some sort of biological clay on which some outside "vital force" operated. "In pessimistic moods," Watson has recalled, "we often worried that the correct structure might be dull—that is, that it would suggest absolutely nothing."

It turned out, in fact, to be unbelievably "interesting." By piecing together the various data, they were able to deduce that DNA does not have a single-stranded helix structure, as do proteins, but is an exceedingly long, entwined double helix.

The banister of a spiral staircase forms a single helix. If you take a ladder and twist it so the rails become spirals, keeping the rungs perpendicular, this would form a crude model of the molecule (Figure 15-7). The two rails, or sides, of the ladder are made up of alternating sugar and phosphate molecules. The perpendicular rungs of the ladder are formed by the nitrogenous bases—adenine (A), thymine (T), guanine (G), and cytosine (C)—one base for each sugar-phosphate, as Levene had shown, and two bases forming each rung. The nucleotides along any one chain of the double helix can occur in any order: ATGCGTACATTGCCA, and so on (Figure 15-8). Since a DNA molecule may be many thousands of nucleotides long, there is a possibility for great variety, one of the primary requirements for the molecule. Note also that each phosphate group is attached to one sugar at the 5′ position (the fifth carbon in the ring) and the other at the 3′ position. Thus the chain has a 5′ end and a 3′ end.

The paired bases meet across the helix and are joined together by hydrogen bonds, the relatively weak, omnipresent bonds that Pauling had demonstrated in his studies of the secondary structures of proteins. The distance between the two sides, or railings, according to Wilkins' measurements, is 2 nanometers. Two purines in combination would take up more than 2 nanometers (since they have two nitrogen rings), and two pyrimidines would not reach all the way across (since they have only one nitrogen ring). But if a purine paired in each case with a pyrimidine, there would be a perfect fit. The paired bases—the "rungs" of the ladder—would therefore always be purine-pyrimidine combinations.

The most exciting discovery came, however, when they set out to match the pairs of bases. They encountered another interesting and important restriction. Not only could purines not pair with purines and pyrimidines not pair with pyrimidines, but because of the configurations of the molecules, adenine could pair only with thymine and guanine only with cytosine (Figure 15-8). Look at Chargaff's data (Table 15-1, page 264) again and see how well these chemical requirements confirm his data.

The double-stranded structure is shown in Figure 15-9. Note that the bridges between the nucleotides run in opposite directions along the two chains; that is, the direction from the 5′ end to the 3′ end of each chain is opposite. The chains are antiparallel.

15–8
(a) *The structure of a portion of one strand of a DNA molecule. Each nucleotide consists of a sugar, a phosphate group, and a purine or pyrimidine base. Note the repetitive sugar-phosphate-sugar-phosphate sequence that forms the backbone of the molecule. Each phosphate group is attached to the 5′-carbon of one deoxyribose sugar and to the 3′-carbon of the deoxyribose sugar in the adjacent nucleotide. The sequence of bases varies from one DNA molecule to another. In the figure, the order of nucleotides (from the top) is TTCAG.* (b) *Diagram and space-filling model of the hydrogen-bonded base pair adenine-thymine. The two hydrogen bonds are indicated by colored dotted lines.* (c) *The guanine-cytosine base pair, showing its three hydrogen bonds. Guanine-cytosine pairs are slightly closer together, more compact, and thus slightly denser than adenine-thymine pairs.*

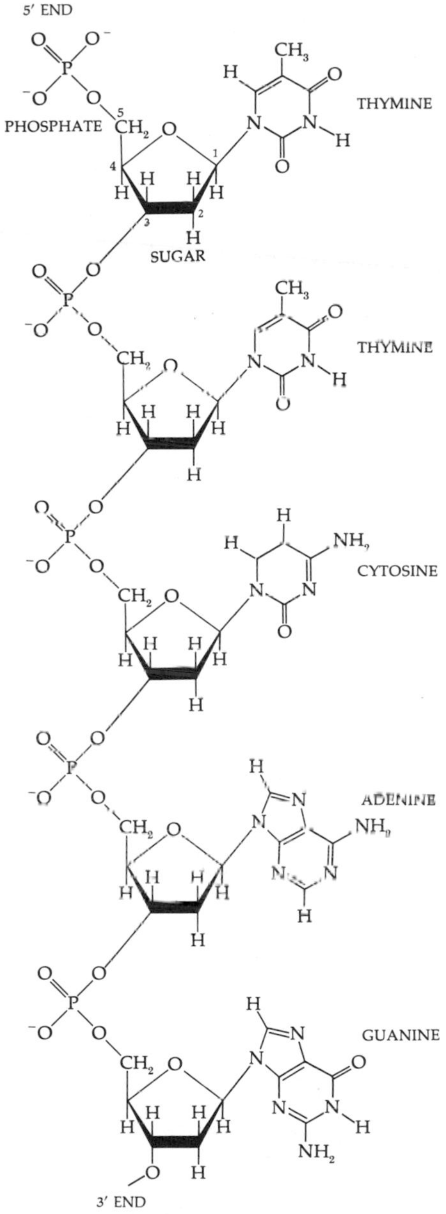

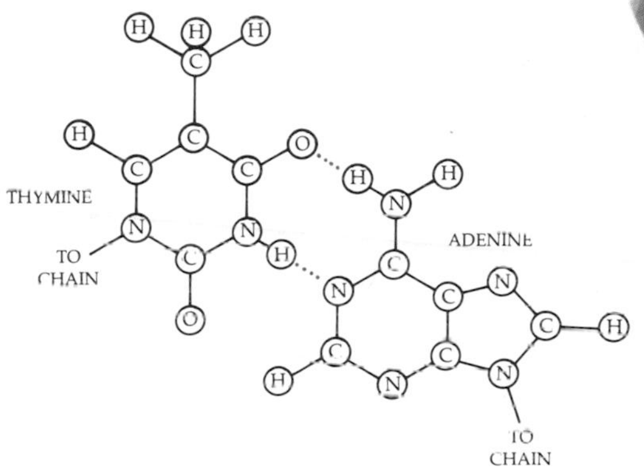

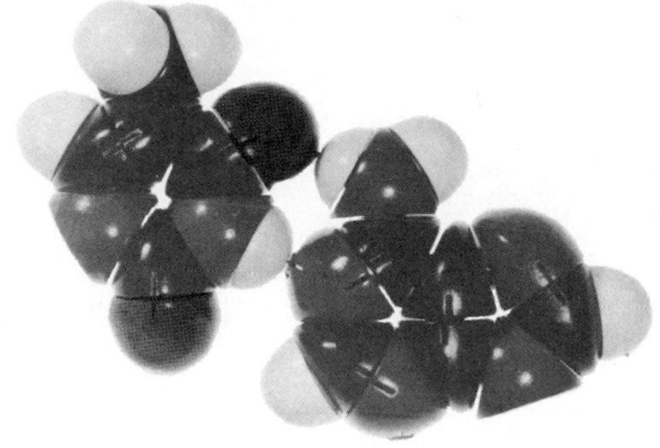

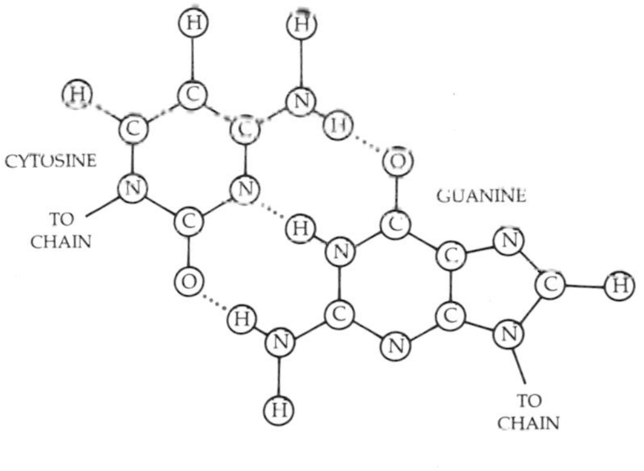

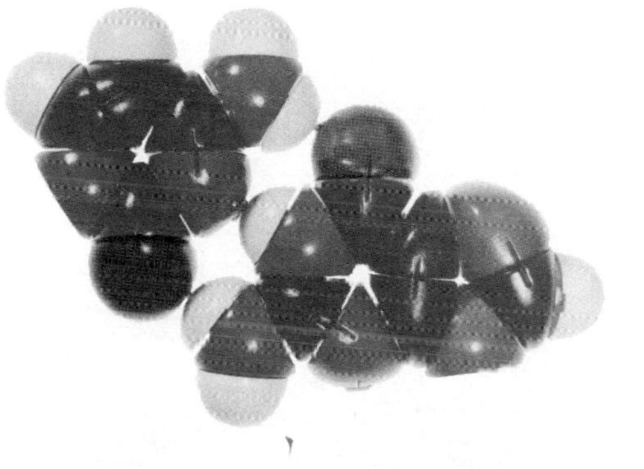

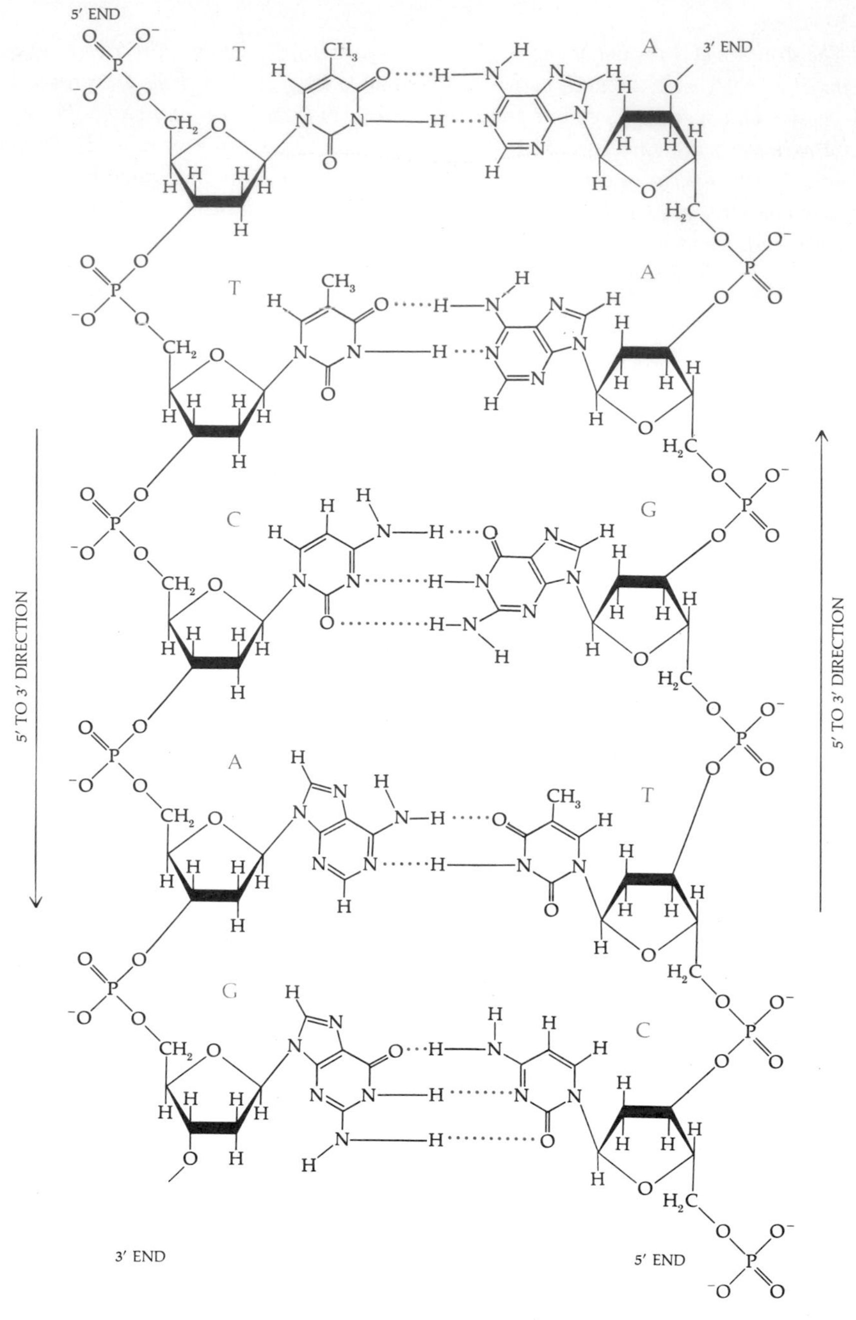

15-9

The double-stranded structure of a portion of the DNA molecule. Adenine can pair only with thymine and guanine only with cytosine. Thus the order of bases along one strand determines the order of bases along the other. Note that the strands run in opposite directions.

5′ END

3′ END

5′ TO 3′ DIRECTION

5′ TO 3′ DIRECTION

3′ END

5′ END

15-10

Watson (left) and Crick in 1953 with one of their models of DNA.

DNA as a Carrier of Information

You will recall that a necessary property of the genetic material is the ability to carry genetic information. The Watson-Crick model showed that the DNA molecule is able to do this. The information is carried in the sequence of the bases, and *any* sequence of bases is possible. Since the number of paired bases ranges from about 5,000 for the simplest known virus up to an estimated 5 billion in the 46

WHO MIGHT HAVE DISCOVERED IT?

Then there is the question, what would have happened if Watson and I had not put forward the DNA structure? This is "iffy" history which I am told is not in good repute with historians, though if a historian cannot give plausible answers to such questions I do not see what historical analysis is about. If Watson had been killed by a tennis ball I am reasonably sure I would not have solved the structure alone, but who would? Olby has recently addressed himself to this question. Watson and I always thought that Linus Pauling would be bound to have another shot at the structure once he had seen the King's College x-ray data, but he has recently stated that even though he immediately liked our structure it took him a little time to decide finally that his own was wrong. Without our model he might never have done so. Rosalind Franklin was only two steps away from the solution. She needed to realise that the two chains must run in opposite directions and that the bases, in their correct tautomeric forms, were paired together. She was, however, on the point of leaving King's College and DNA, to work instead on TMV [tobacco mosaic virus] with Bernal. Maurice Wilkins had announced to us, just before he knew of our structure, that he was going to work full time on the problem. Our persistent propaganda for model building had also had its effect (we had previously lent them our jigs to build models but they had not used them) and he proposed to give it a try. I doubt myself whether the discovery of the structure could have been delayed for more than two or three years.

There is a more general argument, however, recently proposed by Gunther Stent and supported by such a sophisticated thinker as Medawar. This is that if Watson and I had not discovered the structure, instead of being revealed with a flourish it would have trickled out and that its impact would have been far less. For this sort of reason Stent had argued that a scientific discovery is more akin to a work of art than is generally admitted. Style, he argues, is as important as content.

I am not completely convinced by this argument, at least in this case. Rather than believe that Watson and Crick made the DNA structure, I would rather stress that the structure made Watson and Crick. After all, I was almost totally unknown at the time and Watson was regarded, in most circles, as too bright to be really sound. But what I think is overlooked in such arguments is the intrinsic beauty of the DNA double helix. It is the molecule which has style, quite as much as scientists. The genetic code was not revealed all in one go but it did not lack for impact once it had been pieced together. I doubt if it made all that difference that it was Columbus who discovered America. What mattered much more was that people and money were available to exploit the discovery when it was made. It is this aspect of the history of the DNA structure which I think demands attention, rather than the personal elements in the act of discovery, however interesting they may be as an object lesson (good or bad) to other workers.

Francis Crick: "The Double Helix: A Personal View," Nature, *vol. 248, pages 766–769, 1974.*

human chromosomes, the possible variations are astronomical. The DNA from a single human cell which if extended in a single thread would be almost 2 meters long—can contain information equivalent to some 600,000 printed pages of 500 words each, or a library of about a thousand books. Obviously, the DNA structure can well account for the endless diversity among living things. We shall return to this subject of the genetic code in Chapter 16.

DNA REPLICATION

An essential property of the genetic material is the ability to provide for exact copies of itself. Does the Watson-Crick model satisfy this requirement? In their published account, Watson and Crick wrote, "It has not escaped our notice that the specific pairing we have postulated immediately suggests a possible copying mechanism for the genetic material." Implicit in the double and complementary

15-11
The DNA molecule shown here is in the process of replication, separating down the middle as the paired bases separate at the hydrogen bonds. (For clarity, the bases are shown out of plane.) Each of the original strands then serves as a template along which a new, complementary strand forms from nucleotides available in the cell.

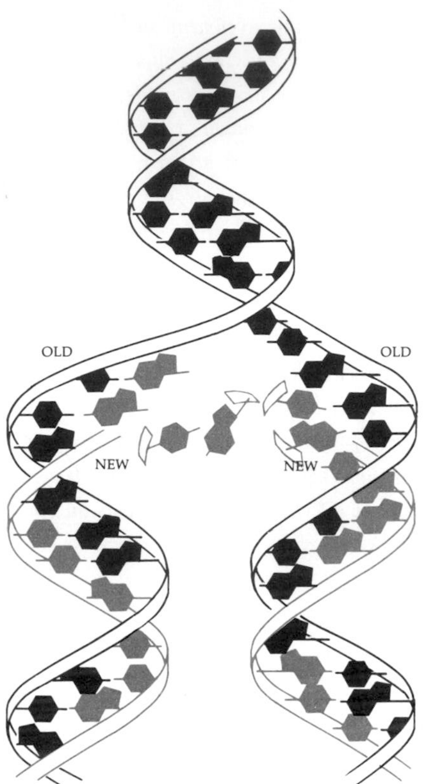

structure of the DNA helix is the method by which it reproduces itself. The molecule "unzips" down the middle, the paired bases separating at the hydrogen bonds. As the two strands separate, new strands form along each old one, using the raw materials in the cell. Each old strand forms a template, or guide, for the production of the new one. If a T is present on the old strand, only an A can fit into place in the new strand; a G will pair only with a C, and so on. In this way, each strand forms a copy of the original partner strand, and two exact replicas of the molecule are produced (Figure 15-11). The age-old question of how hereditary information is duplicated and passed on, duplicated and passed on, for generation after generation, had in principle been answered.

A Confirmation

The Watson-Crick hypothesis of DNA replication is not the only possible one. Matthew Meselson and Franklin W. Stahl, working at the California Institute of Technology, devised an elegant experiment to choose among three possible models (Figure 15-12).

In designing their experiment, they took advantage of the availability of a heavy isotope of nitrogen (^{15}N). To separate DNA molecules according to the relative amounts of ^{14}N and ^{15}N they contain, they are first dissolved in a solution of the salt cesium chloride (CsCl). The solution is then placed in a tube and centrifuged (subjected to a high gravitational field in an ultracentrifuge). The CsCl settles partly, so that it forms a density gradient—it becomes more concentrated in the bottom of the tube and less concentrated in the top of the tube. The DNA molecules dissolved in the cesium chloride seek their own density in this range of densities. Thus a molecule rich in ^{15}N is relatively dense and finds its place low in the tube; a molecule poor in ^{15}N is light and finds its place high in the tube. In this way, DNA molecules containing heavy nitrogen can be separated from one another on the basis of how much of the isotope they contain.

Meselson and Stahl grew *E. coli* for several generations in a medium containing the heavy isotope of nitrogen (^{15}N). At the end of this period virtually all of the ordinary isotope of nitrogen (^{14}N) had been replaced by heavy nitrogen. They isolated the DNA from some of these bacteria, put it in solution, spun it in an ultracentrifuge with cesium chloride, and recorded the position of the DNA (Figure 15-13a).

They then placed a sample of cells containing heavy nitrogen in a medium containing ^{14}N and left them there long enough for the DNA to replicate once (as determined by a doubling of the number of cells). A sample of this DNA was spun in the ultracentrifuge. A second generation was then grown in the ^{14}N medium. This DNA was also ultracentrifuged.

Each sample of DNA contained more light DNA, as could be expected, because newly formed DNA had to incorporate the available ^{14}N (Figure 15-13d and e). Moreover—and this was of crucial importance—the density of the first generation DNA was exactly halfway between that of heavy parent DNA and that of ordinary light DNA, as it should be if each molecule contained one old (heavy) strand and one new (light) strand, as predicted by Watson and Crick (Figure 15-12b). The second generation contained one-half half-heavy DNA and one-half light DNA, which again, exactly and ingeniously, confirmed the Watson-Crick hypothesis.

15–12
Three possible mechanisms of replication of DNA. Newly replicated strands are shown in color. (a) *Conservative replication. Each of the two strands of parent DNA is replicated, without strand separation. In the first generation, one daughter is all old DNA and one daughter is all new. The* F_2 *generation contains one helix composed of two old strands and three made up entirely of new strands.* (b) *Semiconservative replication of DNA. In the first generation, each daughter is half old and half new. The* F_2 *generation comprises two hybrid DNAs (half old, half new) and two new DNAs made up entirely of new strands.* (c) *Dispersive replication. During replication, parent chains break at intervals and replicated segments are combined into strands with segments from parent chains. All daughter helices are part old, part new. The Meselson-Stahl experiment (Figure 15–13) was undertaken to determine which of these three possibilities was correct. Watson and Crick had predicted* (b).

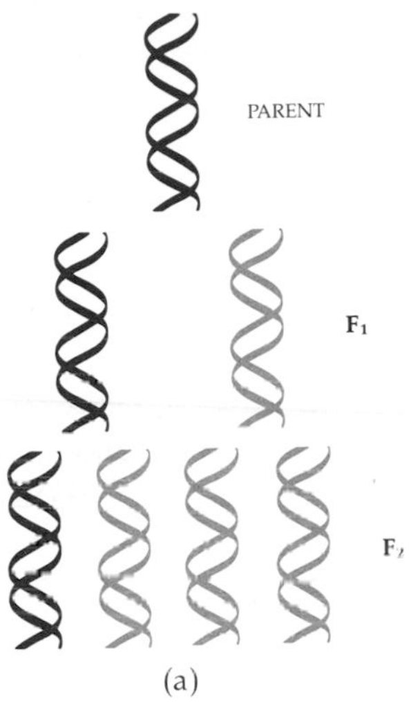
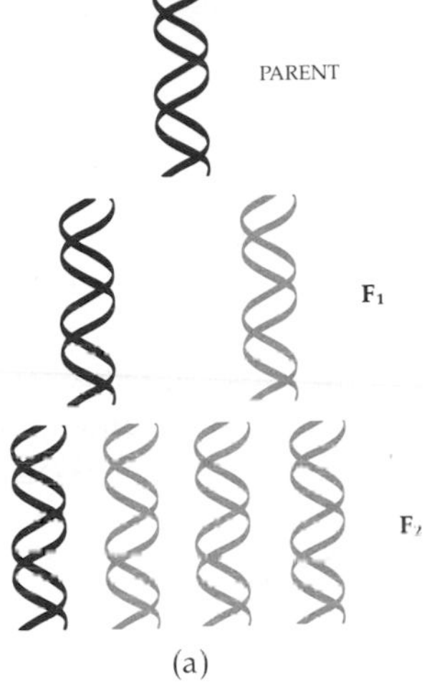

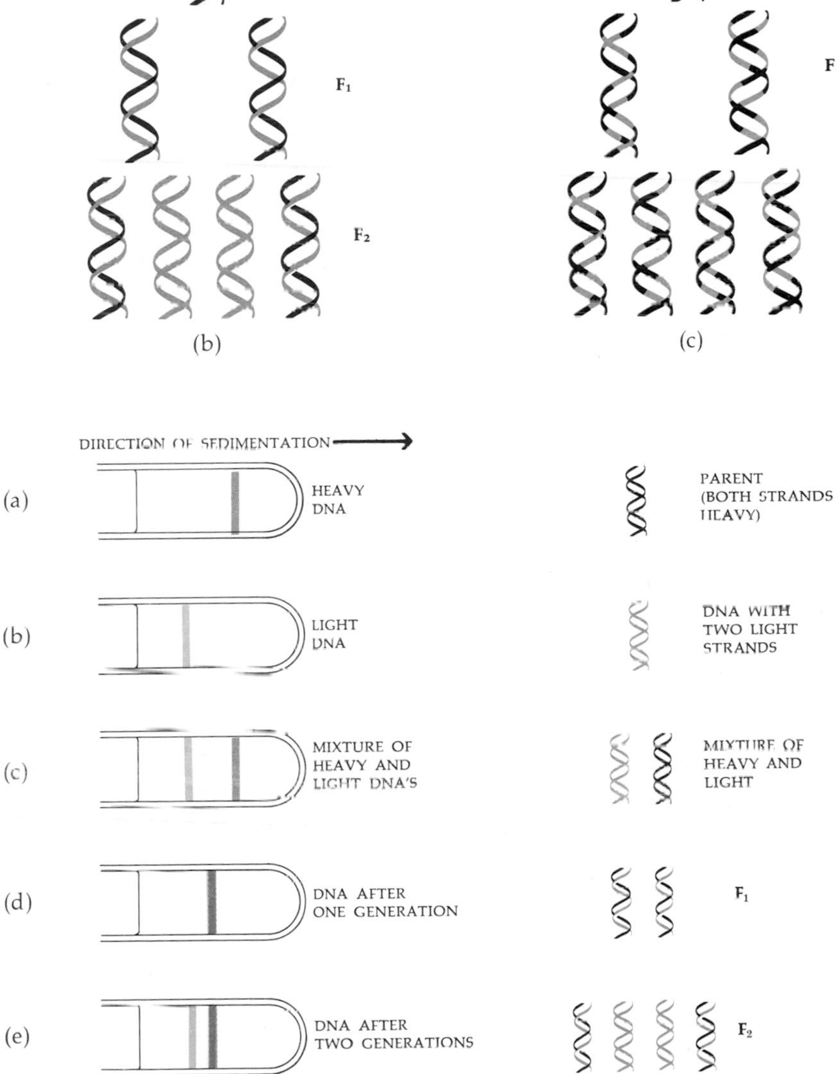

15–13
Meselson-Stahl experiment. Escherichia coli *cultured in a medium containing heavy nitrogen* (^{15}N) *accumulated a "heavy" DNA. Then the cells were removed from the* ^{15}N *medium and permitted to multiply on a "light"* ^{14}N *medium. Light and heavy DNAs were separated by ultracentrifuging the molecules in a cesium chloride gradient. The column on the left shows the results of the experiment; the column on the right shows the investigators' interpretation. As you can see, the experiment confirmed the Watson-Crick hypothesis. Where would the bands have appeared if DNA replication had been conservative (Figure 15–12a)? Dispersive (Figure 15–12c)?*

The Mechanics of DNA Replication

Although DNA is often referred to as a self-replicating molecule, this description is not precisely true. If DNA is placed in solution along with all the necessary components for new DNA, nothing happens. As in other biological reactions, a special enzyme is needed. This enzyme, DNA polymerase, links the nucleotides of the new DNA strand along the template of the old strand.

Along with DNA polymerase, there are a dozen or more different enzymes that specialize in performing various operations on DNA. Some seem to help unwind the DNA for replication; others fill in gaps that arise in its synthesis; and a third group seals together the broken ends of DNA strands. Probably some of these enzymes are also important in repairing damaged DNA. Such enzymes may also help to cause crossovers, thereby recombining DNA from different sources.

The Energetics of DNA Replication

The nucleotides required for DNA synthesis are assembled along biosynthetic pathways of the cell. They are not assembled in the form of the monophosphates shown in Figure 14-7, but rather as triphosphates—that is, adenine is provided as ATP, guanine as the analogous guanine triphosphate (GTP), and so forth. These "extra" P~P groups provide the energy that powers the reactions mediated by DNA polymerase. As the nucleotide is attached to the growing DNA strand, the two phosphates are removed and released. Almost immediately, another enzyme breaks the high-energy bond between the two phosphates, releasing them as inorganic phosphates.

Why does the cell do it this way? The release of the energy-rich P~P and the subsequent breaking of the high-energy bond seem like a waste of carefully stored chemical energy. Is the cell really so profligate?

Measurement of the energy changes involved reveals an interesting point. The reaction in which the activated nucleotide is attached to the growing DNA strand is only slightly exergonic. Therefore, it could tend to go in either direction. Under certain equilibrium conditions, the DNA strand could come apart about as fast as it was synthesized, with the enzyme working both ways. However, with immediate degradation of the P~P fragment, the reaction becomes highly exergonic. Thus the reverse reaction—which would necessitate reforging the P~P group—becomes highly endergonic and so, for all practical purposes, does not occur. This is another example of the way in which cells exercise control over their biochemical activities.

SUMMARY

DNA (deoxyribonucleic acid) is the genetic material of the cell. Investigations showing that both the transforming factor in bacteria and the carrier of the genetic information in bacteriophages are DNA provided some of the first evidence for this hypothesis.

Further support for the genetic role of DNA came from two more findings: (1) Almost all tissue cells of any given species contain equal amounts of DNA, and (2) the proportions of nitrogenous bases are the same in the DNA of all cells of a given species, but they vary in different species.

In 1953, Watson and Crick proposed a structure of DNA. The DNA molecule, according to their model, is a double-stranded helix, shaped somewhat like a twisted ladder. The two sides of the ladder are composed of repeating groups of a phosphate and a five-carbon sugar. The "rungs" are made up of paired bases, one purine base pairing with one pyrimidine base. There are four bases—adenine (A), guanine (G), thymine (T), and cytosine (C)—and A can pair only with T, and G only with C. The four bases are the four "letters" used to spell out the genetic message. The paired bases are joined by hydrogen bonds. On the basis of this structure, as revealed by Watson and Crick, the role of DNA as the carrier and transmitter of the genetic information became widely accepted.

When the DNA molecule replicates, the two strands come apart, breaking at the hydrogen bonds. Each strand forms a new complementary strand from nucleotides available in the cell. The semiconservative (one-half conserved) nature of this process was confirmed by studies using radioactive and heavy isotopes.

Replication of DNA is enzymatically mediated, and energy is supplied by high-energy phosphate bonds.

QUESTIONS

1. One of the chief arguments for the erroneous theory that proteins constitute the genetic material is that proteins are heterogeneous. Explain why the genetic material must have this property. What feature of the Watson-Crick DNA model is important in this respect?

2. What characteristics of bacteriophages made them such a useful experimental tool in resolving the question of whether the genetic material was DNA or protein?

3. When the structure of DNA was being worked out, it became apparent that one purine base must be paired with a pyrimidine base, and that the other purine base must be paired with the other pyrimidine base. The evidence for this requirement came from two types of data. What were the data, and how did they indicate this structural requirement? Further consideration of the structures of the four nitrogenous bases indicated that adenine could pair only with thymine, and cytosine only with guanine. What features of the structures of the bases imposed this requirement on the structure of the DNA molecule?

4. Suppose you are talking to someone who has never heard of DNA. How would you support an argument that DNA is the genetic material? List at least five of the strong points in such an argument.

5. Suppose the Meselson-Stahl experiment was extended to the third generation. What would be the proportion of light to heavy DNA?

6. Eukaryotic cells are grown in a medium containing thymine labeled with ^3H. Then they are removed from the radioactive medium, placed in an ordinary medium, and allowed to divide. Studies of the distribution of the radioactive isotope are made after each generation to determine whether or not a chromatid contains radioactive material. Before they are placed in the nonradioactive medium, all the chromatids contain ^3H. After one generation in the nonradioactive medium, the ^3H is still divided evenly among the chromatids. Assuming each chromatid contains a single DNA molecule, explain the results. Does this confirm the Watson-Crick hypothesis? What would be the distribution of the ^3H after two divisions in the nonradioactive medium? Why?

CHAPTER 16

The Code: Transcription and Translation

Like most important scientific discoveries, the Watson-Crick model raised more questions than it answered.

Given that genes are made of DNA and that the products of genes are specific proteins, what is the link between them? How does DNA influence protein synthesis? One early hypothesis was that the DNA somehow formed a template for protein production. But there was simply no way that a strand of protein could be matched up against a strand of DNA in a templatelike fashion. The relationship between DNA and protein had to be a more complicated one. If the proteins, with their 20 amino acids, were the "language of life," to extend the metaphor of the 1940s, the DNA molecule, with its four nitrogenous bases, could be envisioned as a sort of code for this language. So the term *genetic code* came into being.

THE TRIPLET CODE

As it turned out, the idea of a "code of life" was useful not only as a dramatic metaphor but also as a working analogy. Scientists, seeking to understand how the DNA so artfully stored in the nucleus could order the quite dissimilar structures of protein molecules, approached the problem with methods used by cryptographers in deciphering codes. There are 20 biologically important amino acids, and there are four different nucleotides. As George Gamow, a Nobel Prize–winning physicist, pointed out, if a single nucleotide "coded" one amino acid, only four amino acids could be provided for. If two nucleotides specified one amino acid, there could be a

16–1
Diagram of tobacco mosaic virus (TMV), representing about half of its total length. This virus has a central core of RNA and a protein coat composed of 2,200 identical protein molecules. If the RNA is removed from its protein coat and rubbed into scratches on a tobacco leaf, new TMV viruses are formed, complete with new protein coats. These results were among the first to indicate that RNA as well as DNA can carry genetic instructions.

16–2
Chemically, RNA is very similar to DNA, but there are two differences in its nucleotides. One difference is in the sugar component; instead of deoxyribose, RNA contains ribose, which has an additional oxygen atom. The other difference is that instead of thymine, RNA contains the closely related pyrimidine uracil (U). (A third, and very important, difference between the two nucleic acids is that most RNA does not possess a regular helical structure and is usually single-stranded.)

DEOXYRIBOSE RIBOSE

THYMINE URACIL

maximum number, using all possible arrangements, of 4^2, or 16—still not quite enough to code all 20 amino acids. Therefore, following the code analogy, at least three nucleotides in sequence must specify each amino acid. This would provide for 4^3, or 64, possible combinations, or codons. The codon was widely and immediately adopted as a working hypothesis, although its existence was not actually demonstrated until a decade after the Watson-Crick discovery. Proof depended on answering yet another question: How is the information encoded in the DNA molecule translated into protein?

THE RNAs

The search for the answer to this question led investigators to the examination of another kind of molecule, ribonucleic acid, a close chemical relative of DNA (Figure 16–2). There were several clues indicating that RNA might play a role in the translation of genetic information into a sequence of amino acids. First, there was some circumstantial evidence. Unlike DNA, which is found primarily in the nucleus, RNA is found mostly in the cytoplasm, and it is there that most protein synthesis takes place. Also, cells making large amounts of protein have numerous ribosomes. A rapidly growing *Escherichia coli* cell, for instance, contains about 15,000 ribosomes, constituting about one-half of the total mass of the cell. And ribosomes are rich in RNA.

Additional evidence came from experiments with viruses. When a bacterial cell is infected by a DNA-containing bacteriophage, a new RNA appears in the cell at the beginning of the infection cycle, before protein synthesis begins. Also, some viruses contain no DNA—only RNA and protein; tobacco mosaic virus is an example. If the RNA is removed from TMV and rubbed onto a scratched tobacco leaf, infection results and new viruses are produced, protein coats and all. In other words, RNA as well as DNA seemed to contain information about proteins.

Many of the experiments to define the role of RNA in protein biosynthesis were carried out using cell-free extracts of *E. coli* and of animal liver cells. (Cell-free in this context means simply that the cells have been broken apart.) A principal advantage of using such extracts is that they can be separated into various fractions and the fractions studied separately.

As it turned out in the course of these studies, which occupied many different scientific groups for more than a decade, not one but three types of RNA are involved: messenger RNA, transfer RNA, and ribosomal RNA.

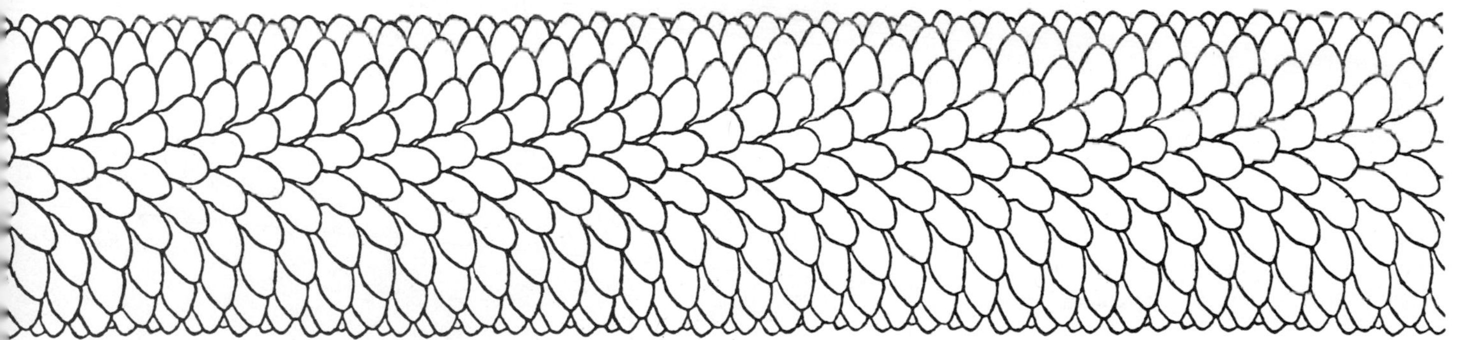

16-3

A schematic representation of RNA transcription. At the point of attachment of the enzyme RNA polymerase, the DNA opens up. RNA nucleotide building blocks are assembled using one strand of the DNA as a template. As the RNA polymerase moves along the DNA molecule, the hydrogen bonds between the two strands of the DNA re-form, squeezing out the newly formed RNA strand. Transfer RNA and ribosomal RNA are also formed by this same copying process from chromosomal DNA and nucleolar DNA, respectively.

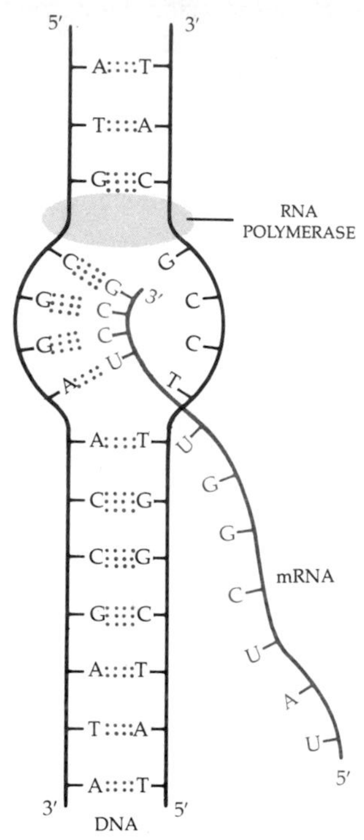

The messenger RNA copies the information from the DNA molecule. The process of making messenger RNA is known as *transcription*. Transfer RNA and ribosomal RNA, following the instructions coded in the messenger RNA molecule, line up the amino acids in the sequence dictated by the messenger RNA molecule. The process of converting information from the language of nucleic acids to the language of proteins is known as *translation*.

Messenger RNA

Messenger RNA (mRNA) is a long molecule consisting of a single strand of 1,000 to 10,000 nucleotides. This molecule forms along a strand of the DNA using it as a template (Figure 16–3), in a way similar in principle to that in which a new DNA strand forms. Like the DNA strand, the RNA molecule has a 3′ end and a 5′ end. The nucleotides are present in the cell as triphosphates and are added, one at a time, to the 3′ end of the growing RNA chain. The reactions are catalyzed by special enzymes, the RNA polymerases.

Transfer RNA

Transfer RNA (tRNA) molecules are relatively small, containing only about 80 nucleotides. Like other RNA molecules, tRNAs are formed in the nucleus along DNA templates.

All of the tRNA molecules are similar though not identical. One end (the 3′ end) always terminates in a CCA sequence; the other (the 5′ end) often terminates in a guanine nucleotide. The other nucleotides vary according to the particular type of tRNA. There are more than 20 kinds of tRNA molecules—at least one for each amino acid found in proteins—and the sequence of nucleotides in a number of these molecules has now been analyzed. All tRNA molecules appear to have the cloverleaf shape shown in Figure 16–5a; in addition, the cloverleaf is folded over on itself to form a three-dimensional structure (Figure 16–5b). Some of the bases, as indicated in the diagram, are hydrogen-bonded to one another (A with U, G with C). Transfer RNA molecules characteristically contain a number of unusual bases.

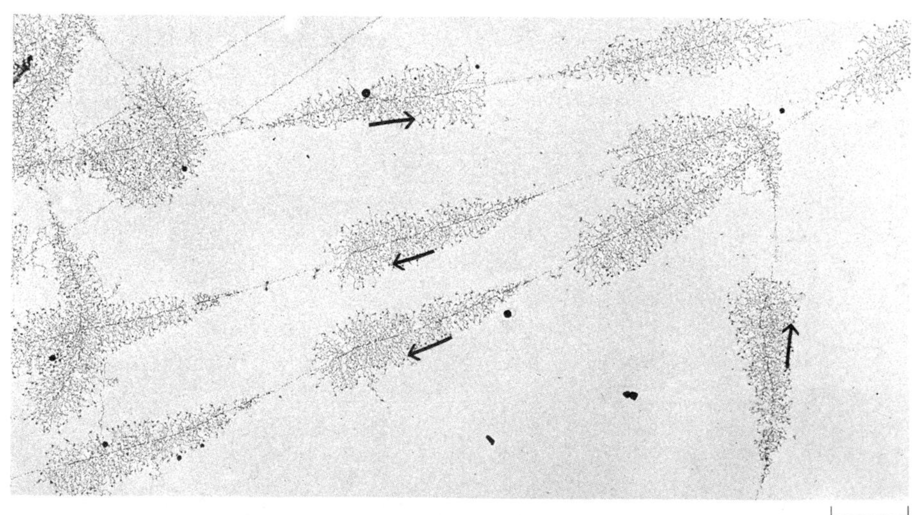

16–4

Transcription. This electron micrograph shows DNA from the nucleolus of an amphibian egg cell. The fine fibrils are RNA molecules that have formed along the DNA strand. The arrows indicate the direction in which the transcription is proceeding.

16-5
(a) *General structure of tRNA molecules. Such molecules consist of about 80 nucleotides linked together in a single chain. The chain always terminates in a CCA sequence. The amino acid is linked to the tRNA at this CCA end. The other nucleotides vary according to the particular tRNA. All tRNA molecules appear to have the same general configuration shown here; in some, however, there is an additional bulge, indicated by the dashed lines. Some of the bases are hydrogen-bonded to one another, as indicated by colored dots. The unpaired bases at the bottom of the diagram (indicated in color) serve as the anticodon and "plug in" the tRNA molecule to an mRNA codon.* (b) *Three-dimensional structure of a tRNA molecule.* (c) *Bonding between an mRNA codon and its corresponding tRNA anticodon.*

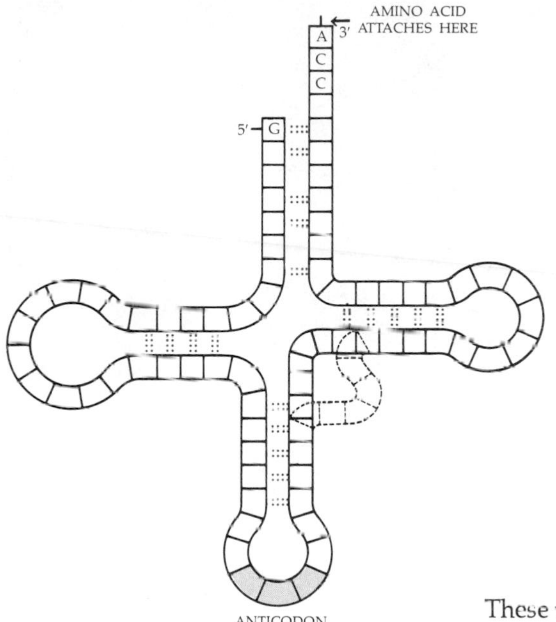

(a)

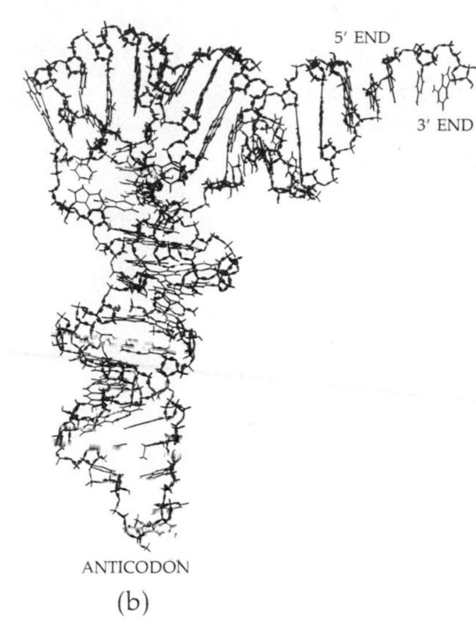

(b)

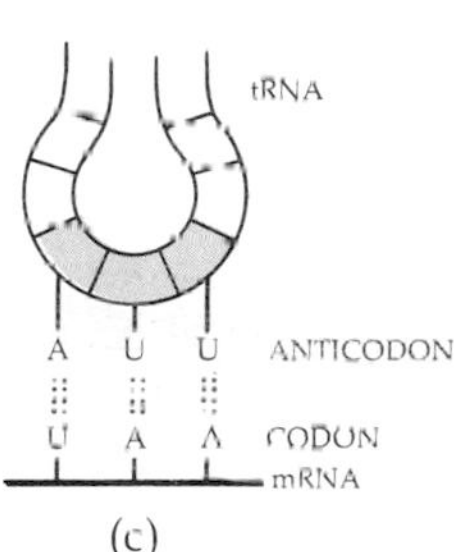

(c)

These unusual bases are formed after the molecule is assembled by the addition of extra chemical groups, such as the methyl (CH_3) group, mediated by special enzymes. The reason for these additional groups is not known, but it is hypothesized that their presence—by preventing base pairing within the molecule—maintains the tRNA in its particular three-dimensional configuration. It will soon be possible to produce tRNA molecules in the laboratory and so to compare the properties of the unmodified molecule with those of the modified one.

The function of the tRNA is to provide the translational link between the sequence of nucleotides in the DNA (and therefore in the messenger RNA) and the sequence of amino acids in the protein. The amino acid is bonded to the tRNA at the CCA end. For each amino acid there is at least one specific tRNA molecule to which it can be attached. The joining of the amino acid to its appropriate tRNA is accomplished by a special enzyme, and there is a different enzyme for each amino acid-tRNA combination.

The role of the enzyme is to unite the amino acid with the right tRNA molecule. (This is, in effect, the moment of translation.) Once attached to the appropriate tRNA molecule, the amino acid can be correctly positioned according to the instructions in the messenger RNA strand. Three unpaired bases in one of the loops of the tRNA molecule—the *anticodon*—specifically recognize and bond to three corresponding bases—the codon—on the mRNA molecule. In this way, only one kind of amino acid can be brought in at any one point (Figure 16-5c).

16–6
Electron micrograph of the nucleus of the single-celled alga Chlamydomonas. *The dark body in the center is the nucleolus, where the RNA of the ribosomes is synthesized and the ribosomal subunits assembled. Very fine RNA strands are barely visible, around which ribosomal subunits are grouped. There are usually two nucleoli per nucleus, each a part of and attached to a specific area on a specific chromosome. Notice also the nuclear envelope, with its many clearly visible pores. There is a portion of a chloroplast to the left of the nucleus and a Golgi body below the nucleus to the left. Ribosomal proteins are manufactured in the cytoplasm and migrate into the nucleus where the ribosomal subunits are assembled. The subunits then move into the cytoplasm where they combine to form the ribosome.*

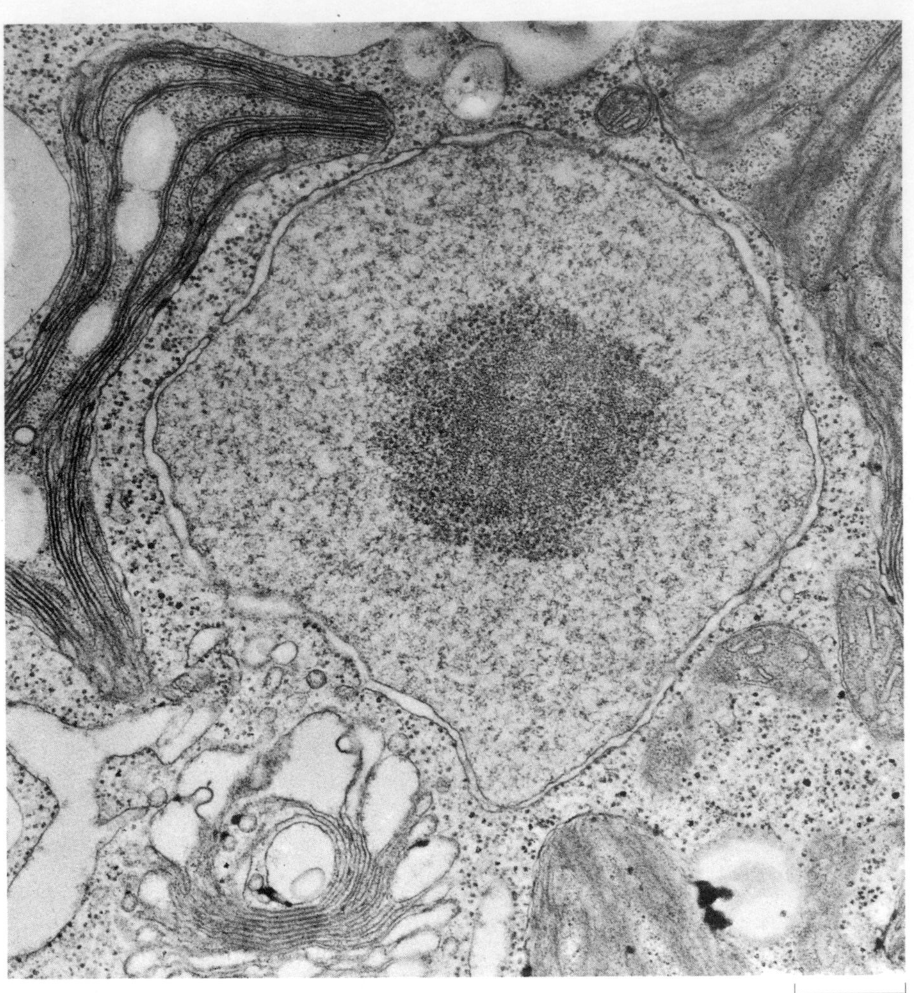

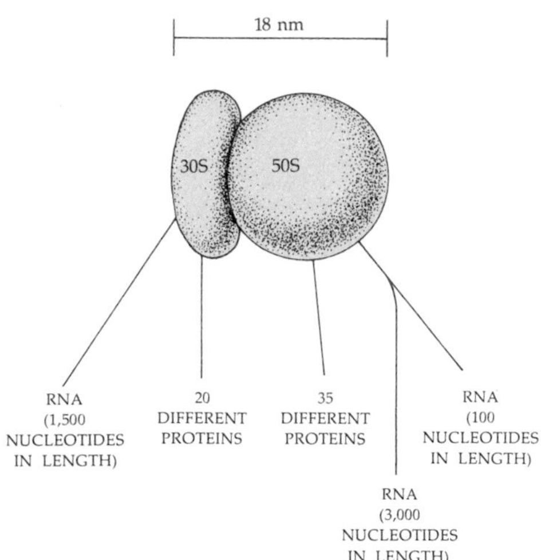

16–7
Diagram of a ribosome from a bacterial cell. As you can see, it consists of two subunits, one slightly larger than the other, and each composed of specific RNA and protein molecules. The terms 30S and 50S refer to relative densities of the two subunits, as indicated by sedimentation rates in the ultracentrifuge.

Ribosomal RNA

Ribosomes are about half RNA and half protein. The ribosomal RNAs are formed on the DNA of the nucleolus (Figure 16–6). Each ribosome is composed of two subunits, each with its characteristic RNAs and proteins (Figure 16–7). In the ribosomes of *E. coli,* which have been studied most extensively, the smaller subunit has one type of ribosomal RNA molecule, about 1,500 nucleotides in length, and about 20 different proteins. The larger subunit has two types of ribosomal RNA molecules, one about 100 nucleotides in length and the other about 3,000, and about 35 different proteins. Ribosomes of eukaryotic cells are somewhat larger but apparently structurally and functionally similar.

Messenger RNA and tRNA come together at the ribosome. The function of the ribosome, presumably, is to orient mRNA, tRNA, amino acids, and the growing protein in a precise relationship to one another. Exactly how the ribosome carries out this function is still an area of active exploration.

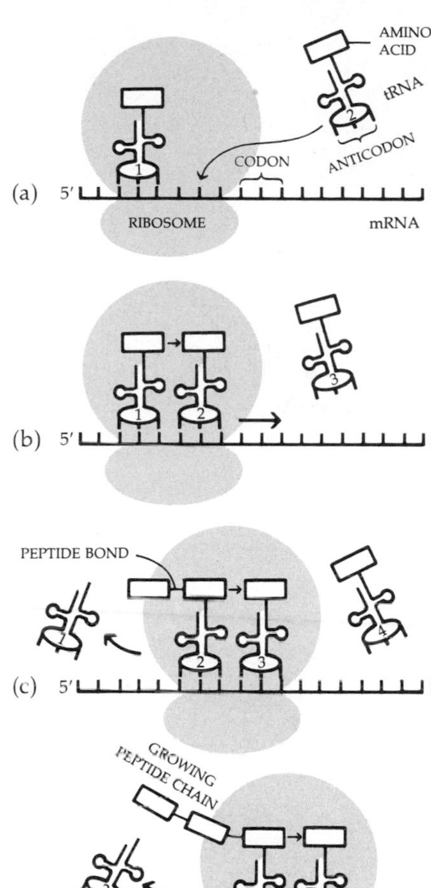

16–8
The process of chain elongation in protein synthesis. (a) *After the ribosome and the first tRNA have complexed with the mRNA,* (b) *the second tRNA with its amino acid associates with the complex.* (c) *A peptide bond is formed between the two amino acids and the first tRNA is released. The ribosome then is translocated along the messenger so that the next codon is exposed within the ribosome. The appropriate tRNA with its attached amino acid complexes with the exposed mRNA codon.* (d) *A new peptide bond is formed and the second tRNA is released. The ribosome is translocated once again as in* (c) *and the process continues.*

THE MAKING OF A PROTEIN

The manufacture of a protein is one of the most intricate molecular processes carried out by living organisms. The essential feature of the process is the faithful linking up of the amino acids attached to their specific tRNAs, guided by the messenger RNA molecule. The whole process occurs on the ribosome and takes place in three stages: *initiation*, *elongation*, and *termination*.

First, the smaller subunit (known as the 30S component) of the ribosome combines with a special tRNA molecule. Then the mRNA molecule joins the initiation complex, as it is called, and finally the larger (50S) subunit of the ribosome joins and completes the complex. The first codon in the mRNA is now hydrogen-bonded to the anticodon of the tRNA. A second tRNA then bonds to the adjacent (second) codon in the mRNA, bringing the first two amino acids—both still attached to their tRNAs—into juxtaposition. This is the beginning of the elongation process. The bond holding the first amino acid to its tRNA is then broken, and the amino acid immediately becomes attached through a peptide bond to the second amino acid still attached to its tRNA. The ribosome then moves along the mRNA until the next (third) codon in the mRNA is brought into the ribosome so that the third tRNA with its attached amino acid can bind to the next codon in the messenger. In this way, the third amino acid is brought into alignment with the dipeptide already formed. The bond between the dipeptide and the second tRNA is broken, and a peptide bond is immediately formed between the dipeptide and the amino acid attached to the newly arrived, third tRNA, forming a tripeptide (Figure 16–8).

This elongation process is repeated over and over as the ribosome moves along the messenger RNA, translating its information into the unique sequence of amino acids that make up the protein. The process terminates at a specific nucleotide sequence on the messenger, where the final tRNA is cleaved from the long chain of amino acids to which it has been joined. At this point, the protein and the last tRNA are liberated from the complex. The ribosome dissociates into its two subunits, falls off the messenger, and the process is over. It is, of course, a cyclic procedure, with the various elements of the system—the tRNAs, the mRNA, and the ribosomal subunits—participating in many rounds of protein synthesis. Further, a number of ribosomes can be moving along the mRNA molecule at the same time making several identical protein products.

The Energetics of Protein Biosynthesis

The activation energy for the incorporation of amino acids into proteins is supplied, again, by ATP, although the role of ATP in this transaction is slightly different from previous ones described. First, the amino acid reacts with an ATP molecule, and in the course of this reaction, the adenine-containing portion of the ATP molecule, rather than the P∼P, is linked to the amino acid:

$$\text{amino acid} + \text{ATP} \rightleftharpoons \text{amino acid}{\sim}\text{AMP} + \text{P}{\sim}\text{P}$$

The P∼P group is almost immediately hydrolyzed, thus making the overall reaction highly exergonic (see page 147). The enzymes that catalyze this reaction are called amino acid activating enzymes. There are at least 20 different kinds of activating enzymes, each one with specific binding sites for a particular amino acid and its matching tRNA molecule.

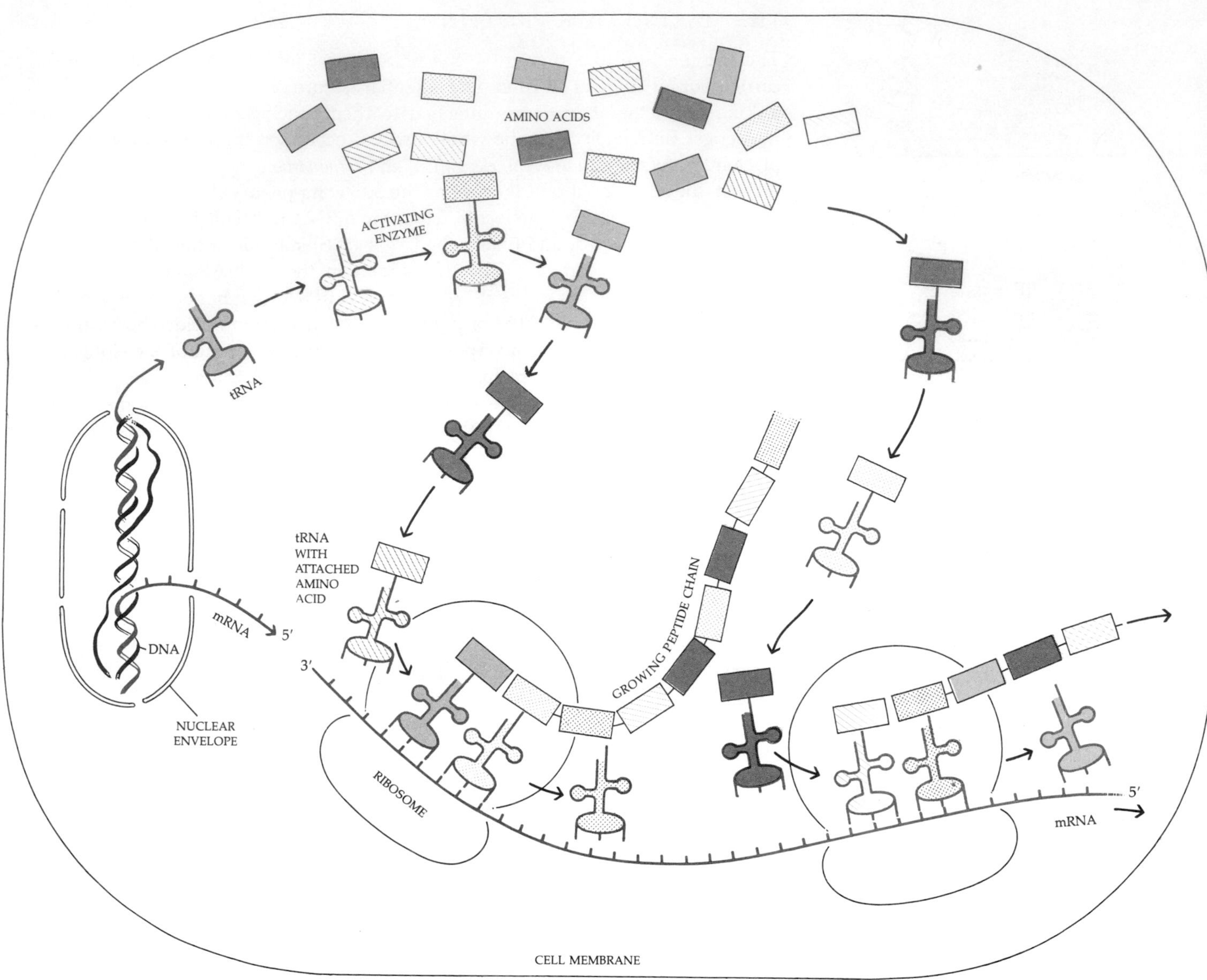

16–9

How a protein is made. At least 20 different kinds of tRNA molecules are formed on the DNA in the nucleus of the cell. These molecules are so structured that each can be attached (by a special enzyme) at one end to a specific amino acid. Each contains an anticodon that fits an mRNA codon for that particular amino acid.

The process begins when an mRNA strand is formed on the DNA template in the nucleus and travels to the cytoplasm. At the point of attachment to the ribosome, the matching tRNA molecule, with its amino acid, plugs in momentarily to the codon in the mRNA. As the ribosome moves along the mRNA strand, a tRNA linked to its particular amino acid fits into place and the first tRNA molecule is released, leaving behind its amino acid, which is bonded to the second amino acid by a peptide bond. As the process continues, the amino acids are brought into line one by one, following the exact order laid down by the DNA code.

16–10
Groups of ribosomes "reading" the same mRNA strand are called polyribosomes. These are from yeast cells.

The amino acid~AMP complex remains bound to its activating enzyme until it reacts with the tRNA specific for that particular amino acid. Then the activated amino acid is transferred from the AMP to its particular tRNA by the following reaction:

$$\text{amino acid}\sim\text{AMP} + \text{tRNA} \rightleftharpoons \text{amino acid}\sim\text{tRNA} + \text{AMP}$$

The amino acid~tRNA molecule, held together by a "high-energy" bond, is then released from the enzyme.

When the amino acid is added to the growing protein chain, the energy of the bond linking the amino acid with its tRNA molecule is used to make the peptide bond that adds the amino acid to the forming protein:

$$\text{growing peptide} + \text{amino acid}\sim\text{tRNA} \rightleftharpoons \text{larger peptide} + \text{tRNA}$$

BREAKING THE CODE

Long before all the details of transcription and translation became known, Marshall Nirenberg of the National Institutes of Health set out to test the messenger RNA hypothesis. He took cell-free extracts of *E. coli* and added to them radioactively labeled amino acids and crude samples of RNA from a variety of cell sources. All of the RNAs stimulated protein synthesis; the amounts of protein produced were small but definite. In other words, the cellular material would start producing protein molecules even when the RNA "orders" it received were from a "complete stranger." Even the RNA from tobacco mosaic virus, which naturally multiplies only in cells of the leaves of tobacco plants, could be read as an mRNA by the ribosomes and tRNA molecules of the bacterial cell.

16–11
A bacterial gene in action. In the micrograph, you can see molecules of RNA polymerase, the enzyme that regulates the transcription of RNA from DNA. The one at the far right is approximately at the point where transcription begins. Several different mRNA strands (shown in color in the diagram) are being formed simultaneously. The longest one, at the left, was the first one synthesized. As each mRNA strand peels off the DNA molecule, ribosomes attach to the RNA, translating it into protein. The protein molecules are not visible in the micrograph.

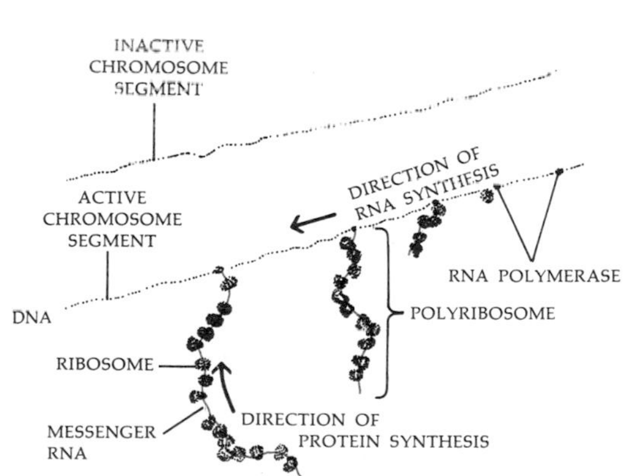

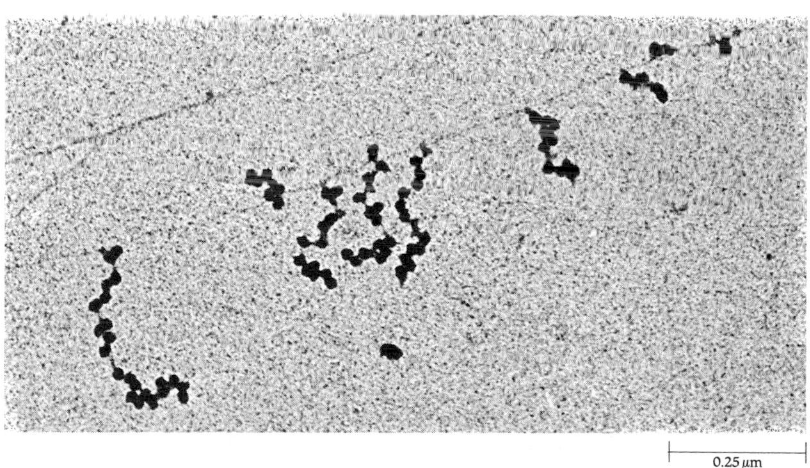

Nirenberg and Heinrich Matthaei next tried an artificial RNA. Perhaps if the cell extracts could read a foreign message and translate it into protein, they could read a totally synthetic message, one dictated by the scientists themselves. Severo Ochoa of New York University had developed an enzymatic process for linking ribonucleotides into a long strand of RNA and, by this process, had produced an RNA molecule that contained only one nitrogenous base, uracil, repeated over and over again. It was called "poly-U." What kind of protein would poly-U dictate?

Nirenberg and Matthaei prepared 20 different test tubes, each of which contained cellular extracts of *E. coli* with ribosomes, tRNA, ATP, the necessary enzymes, and all the amino acids. In each test tube, one of the amino acids, and only one, carried a radioactive label. Synthetic poly-U was added to each test tube. In 19 of the test tubes, nothing could be detected, but in the twentieth one, to which radioactive phenylalanine had been added, the investigators were able to detect newly formed, radioactive polypeptide chains. When the polypeptide was analyzed, it was found to consist only of phenylalanines, one after another. Nirenberg and Matthaei had dictated the message "uracil . . . uracil . . . uracil," and a clear answer had come back, "phenylalanine . . . phenylalanine . . . phenylalanine." The experiment not only defined the first code word (UUU = phe) but made available a method for defining the others.

H. G. Khorana at the University of Wisconsin succeeded in producing an artificial messenger in which two or three nucleotides were repeated over and over again in a known sequence: AGAGAGAGAG, UCUCUCUCUC, ACACACACAC, and UGUGUGUGUG. Each of these RNA chains, when used as a messenger in the cell-free system, produced polypeptide chains of alternating amino acids. Poly-AG produced arginine and glutamic acid over and over again; poly-UC, serine and leucine; poly-AC, threonine and histidine; and poly-UG, cysteine and valine. This is, of course, what you would expect from a triplet code, since the message would be read AGA . . . GAG . . . AGA. . . . These studies provided the first clear demonstration (1) that mRNA is read sequentially (that is, one codon after another); (2) that how it was read depended on the reading frame—that is, the nucleotide at which transcription starts; and (3) that the codon consists of an uneven number of nucleotides.

An artificial mRNA containing repeating triplets of three different nucleotides could produce three different polypeptides, depending on where the reading process began. Each would contain only one type of amino acid repeated over and over again.

mRNA BASE SEQUENCE	READ AS	AMINO ACID SEQUENCE OBTAINED
$(AG)_n$	··· AGA GAG AGA GAG ···	··· arg–glu–arg–glu ···
$(AGC)_n$	··· AGC AGC AGC ···	··· ser–ser–ser ···
	··· GCA GCA GCA ···	··· ala–ala–ala ···
	··· CAG CAG CAG ···	··· gln–gln–gln ···

16–12
In Khorana's experiments, an artificial mRNA in which two nucleotides were repeated over and over produced a polypeptide chain of alternating amino acids. Depending on where the reading process began, an artificial mRNA with three different nucleotides could produce three different polypeptides, each consisting of only one type of amino acid.

Artificial Triplets

In another approach to the problem, Nirenberg worked with tiny trinucleotide fragments. He found that when artificially produced RNA triplets were added to a cell-free extract, the triplets combined with ribosomes, tRNA, and their appropriate amino acid. These relatively large complexes could then be filtered out of the extract. Apparently the trinucleotides behaved as messenger codons and stimulated binding of the tRNA with its attached amino acid to a ribosome. Investigators quickly proceeded to carry out this procedure with the various trinucleotides to determine which amino acid–tRNA complex bound to a particular trinucleotide. As a result of these kinds of experiments carried out by Nirenberg, Khorana, and others, RNA codes for all of the amino acids have been worked out (Figure 16–13).

16–13

The genetic code, consisting of 64 triplet combinations (codons) and their corresponding amino acids (see page 62). Since 61 triplets code 20 amino acids, there are "synonyms," as many as six for leucine, for example. Most of the synonyms, as you can see, differ only in the third nucleotide. Of the 64 codons, only 61 specify particular amino acids. The other three codons are stop signals, which cause the chain to terminate. The code is shown here as it would appear in the mRNA molecule.

SECOND LETTER

FIRST LETTER (5′ END)	U		C		A		G		THIRD LETTER (3′ END)
U	UUU	phe	UCU	ser	UAU	tyr	UGU	cys	U
	UUC		UCC		UAC		UGC		C
	UUA	leu	UCA		UAA	stop	UGA	stop	A
	UUG		UCG		UAG	stop	UGG	trp	G
C	CUU	leu	CCU	pro	CAU	his	CGU	arg	U
	CUC		CCC		CAC		CGC		C
	CUA		CCA		CAA	gln	CGA		A
	CUG		CCG		CAG		CGG		G
A	AUU	ile	ACU	thr	AAU	asn	AGU	ser	U
	AUC		ACC		AAC		AGC		C
	AUA		ACA		AAA	lys	AGA	arg	A
	AUG	met	ACG		AAG		AGG		G
G	GUU	val	GCU	ala	GAU	asp	GGU	gly	U
	GUC		GCC		GAC		GGC		C
	GUA		GCA		GAA	glu	GGA		A
	GUG		GCG		GAG		GGG		G

Of the 64 possible triplet combinations, 61 of them specify particular amino acids. With 61 combinations coding for 20 amino acids, you can see that there must be more than one codon for many of the amino acids. As shown in Figure 16–13, codons specifying the same amino acid often differ only in the third nucleotide, leading to the speculation that the first two may be sufficient to hold the tRNA in most instances.

Hemoglobin Reexamined

Some of the biological implications of these findings are strikingly clear. Let us take another look at sickle cell anemia, for example, in the light of Figure 16–13. Normal hemoglobin contains glutamic acid; sickle cell hemoglobin contains valine. In mRNA, GAA or GAG specifies glutamic acid (glu), and GUU, GUC, GUA, or GUG specifies valine (val). So the difference between the two is merely the replacement of one adenine by one uracil in a molecule that, since it dictates a protein that contains more than 150 amino acids, must contain more than 450 bases. In other words, the tremendous functional difference—literally a matter of life and death—can be traced to a single "misprint" in over 450 nucleotides.

Punctuation

The genetic material, the DNA, is undifferentiated in form, being an enormously long sequence of nucleotides. Yet the information it contains is compartmentalized in the units we call genes. How is this accomplished? In a written language, punctuation takes the form of special symbols, such as spaces between words and a period at the end of a sentence. In the genetic language, punctuation takes the form of special nucleotide sequences other than the triplets that code for the 20 amino acids. As you have perhaps already realized, a number of different punctuation sequences are required. At what point in the DNA should the RNA polymerase

BACTERIOPHAGE ϕX174 BREAKS THE RULES

In 1977, the complete nucleotide sequence of the DNA of a small bacteriophage known as ϕX174 was worked out by Fred Sanger and his co-workers at the Medical Research Council in Cambridge, England. The DNA was known to code for nine proteins, whose sequences were also known. (Viruses borrow enzymes and other proteins from host cells, so they are able to get along with such a relatively small molecular arsenal.)

Even before the last details were worked out, however, the investigators began to realize they had a serious and interesting problem: The DNA (which contained some 5,375 nucleotides) was insufficient to code for the known proteins by the triplet code hypothesis. It simply was not long enough, a fact that caused some ripples of excitement and tension in laboratories all over the world as the news spread. Were the concepts established by a quarter-century of intensive effort to be overthrown by a submicroscopic particle?

When the nucleotide sequence became known, its secret was revealed. The investigators had originally assumed that each gene was physically separate along the DNA molecule. However, it turns out that there are pairs of overlapping genes. In other words, different genes in ϕX174 are coded by the same regions of DNA using different reading frames. Whether it is common among organisms to practice such economies is not yet known, but it does provide a warning that molecular geneticists may not yet become complacent in their knowledge.

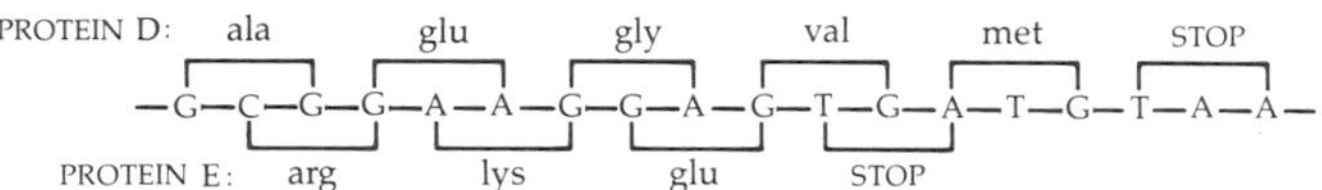

A segment of the DNA of ϕX174, showing a portion of the genes for proteins D and E. Both are coded by the same stretch of DNA; their reading frames, however, are different. The gene for protein E is completely contained within the gene for D, an entirely unrelated protein.

start synthesizing messenger RNA, and at what point should it stop? And once the messenger is formed, what determines the initiation and termination of protein synthesis?

The sequence of DNA to which RNA polymerase binds initially is called the *promoter;* most promoters that have been identified contain the sequence TATAATG.

The code AUG on mRNA means "start protein synthesis." AUG also specifies methionine, and the clue leading to the discovery of its other punctuation role is that many proteins begin with the amino acid methionine. (In others, it is removed after the protein is formed.) Other special codons before AUG in the mRNA are probably also involved. The tRNA that starts protein synthesis by bringing a methionine into place is different from the tRNA that positions methionine in the middle of the message.

Conversely, the codons UAA, UAG, and UGA mean "stop protein synthesis."

Editing

The RNAs may undergo extensive processing after transcription. The two larger ribosomal RNAs, for instance, are produced as one single molecule, which is then cleaved by special enzymes. This can be interpreted as an efficient way for a cell to produce the molecules in the proportions required.

Some editing is more mysterious as to its purpose. Each tRNA, for instance, is also produced as part of a larger molecule, much of which, apparently, is simply discarded. Then, as we noted earlier, as many as 10 of the approximately 80 nucleotides in the tRNA molecule are altered by enzymatic processes.

Messenger RNA is also edited. One common alteration found in eukaryotic cells is the addition of a segment of poly-A at the 3′ end. The poly-A consists of approximately 200 adenine nucleotides, added enzymatically, one by one. This poly-A tail may facilitate the migration of the mRNA strand out of the nucleus.

Early in 1978, as techniques improved for direct analysis of nucleic acids, news began to spread of a particularly baffling discovery. As more DNAs and more mRNAs were being analyzed, nucleotide after nucleotide, it became possible to make direct comparisons between chromosomal DNA and the corresponding mRNA. And a completely unexpected finding occurred. Right in the middle of a DNA coding for protein, a long sequence of bases—up to several hundred—would be discovered that was not accounted for in the mature RNA molecule. These intervening sequences, as they are called, may be several hundred nucleotides in length, comprising as much as a third of the original mRNA molecule. Before translation they are removed enzymatically.

Genes coding for such intervening sequences have now been found in *Drosophila,* yeast, chick, rabbit, mouse, and an animal DNA virus. They are not apparently present in prokaryotes. Given the extraordinary efficiency of the transcription-translation process, the fact that intervening sequences are so widespread leads one to believe that they have some meaning, either functionally or historically. But what could that meaning be? Their discovery comes at a time when some investigators had begun to fear that most of the excitement was over—which, once more, only goes to show that molecular biologists cannot take anything for granted.

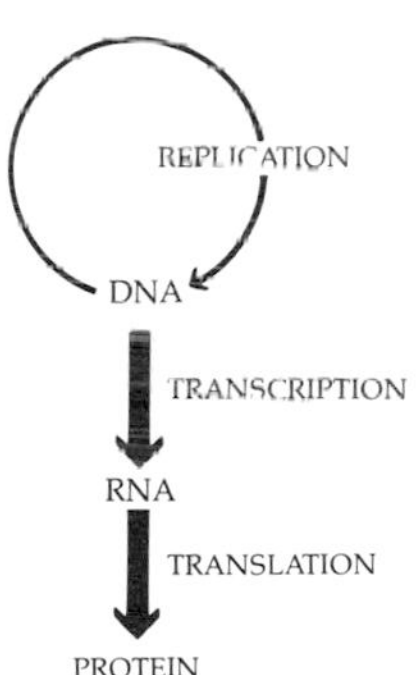

16–14
A representation of the information flow from DNA to protein.

SUMMARY

Genetic information is coded in molecules of DNA, and these, in turn, determine the sequence of amino acids in molecules of protein. A gene is a segment of a DNA molecule that specifies the complete sequence of one polypeptide.

The way in which the gene directs the production of a protein, according to current theory, is as follows. Each series of three nucleotides along a DNA strand is the DNA code for a particular amino acid. The information is transcribed from the DNA to a long, single strand of RNA (ribonucleic acid). This type of RNA molecule is known as messenger RNA, or mRNA. The mRNA forms along one of the strands of DNA, following the principles of base pairing first suggested by Watson and Crick. The mRNA therefore is complementary to the DNA strand.

The mRNA strand leaves the cell's nucleus and becomes attached to a ribosome. At the point where the strand of mRNA is in contact with the ribosome, small molecules of another type of RNA, known as transfer RNA (tRNA), which serve as adapters between the mRNA and the amino acids, are bound temporarily to the mRNA strand. This bonding takes place by complementary base pairing between the mRNA codon and the tRNA anticodon. Each tRNA molecule carries the specific amino acid called for by the mRNA codon to which the tRNA attaches. Thus, following the sequence originally dictated by the DNA, the amino acid units are brought into line one by one and are formed into a polypeptide chain.

The genetic code has now been "broken," that is, it is now known which amino acid is called for by a given codon. Of the 64 possible triplet combinations of the four-letter DNA code, 61 combinations have been identified with one of the 20 amino acids that make up protein molecules. The other three triplets serve as "punctuation marks," terminating protein synthesis.

RNA may undergo considerable modification after transcription. This may include the cleaving into separate molecules, as in ribosomal RNA, the alteration of specific bases, as in tRNA, or even the removal of long intervening sequences, as in mRNA.

QUESTIONS

1. Explain the term "genetic code." In what ways is it a useful analogy?

2. In the last 25 years, the concept of the gene has undergone a series of changes and modifications. How would you now define "gene?" (Note: There is no "right" answer to this question.)

3. Describe the structure and function of each of the three types of RNA involved in protein biosynthesis.

4. What kind of RNA is being transcribed in Figure 16–4? Where will it be translated?

5. Define and distinguish between transcription and translation.

6. In a hypothetical segment of a DNA molecule, the sequence of bases is (3′)-AAGTTTGGTTACTTG-(5′). What would be the sequence of bases in an mRNA strand transcribed from this DNA segment? What would be the sequence of amino acids coded by the mRNA? Does it matter where the transcription from DNA to mRNA begins? Show why.

7. Fill in the missing letters.

(5′)	T G T	__ __ __	__ __ __	(3′)	DNA
(3′)	__ __ A	C __ __	__ __ __	(5′)	
(5′)	U __ __	__ C A	__ __ __	(3′)	mRNA codons
	__ __ __	__ __ __	G C A		tRNA anticodons

Specify the amino acid that each triplet codes for.

CHAPTER 17

The Structure of the Chromosome and Regulation of Gene Expression

The genetic code is universal. It must have been frozen in its present form more than 3 billion years ago, before the appearance of the first cells. It is essentially the same for *Escherichia coli* and *Homo sapiens,* awesome evidence that all living things are descended from a common ancestor. Recent work in molecular genetics has emphasized, however, that, despite their basic similarities, there are important differences between the chromosomes of bacteria (with which much of the original work was done) and those of eukaryotes (which are now receiving more attention).

THE BACTERIAL CHROMOSOME

The bacterial chromosome is a single, circular loop of DNA (Figure 17–1) with little or no protein attached to it. Not all of a bacterium's genes are in its chromosome. In addition to the DNA in the chromosome, many bacteria also contain additional but much smaller DNA molecules, called *plasmids* (Figure 17–2), which replicate in synchrony with the cell. A plasmid may contain as few as two genes or as many as several hundred. Plasmids are passed from mother to daughter cells at cell division.

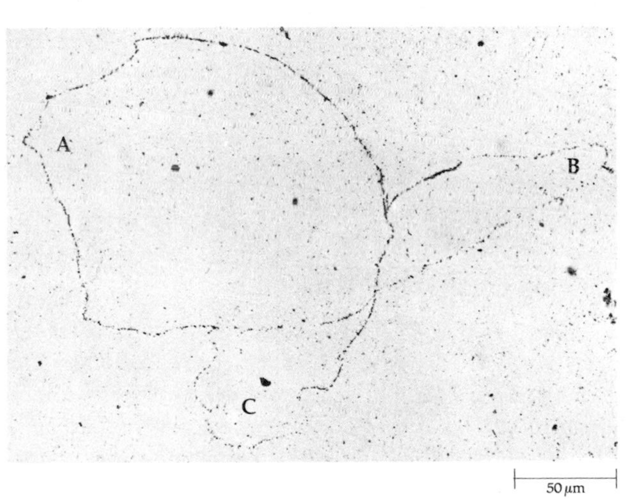

17–1
DNA replicating. To make this autoradiograph, an E. coli *cell was grown for a short time in a medium containing thymine labeled with a radioactive isotope (^3H). The cell was broken open and the DNA, which was allowed to spread out, was placed on a photographic plate for two months. During this period, decay of the ^3H left spots on the photographic emulsion. The chromosome had been caught in the process of replication. Replication is bidirectional, traveling out from the starting point in opposite directions around the circular chromosome. A and B are believed to represent already replicated portions and C the unreplicated portion. This micrograph, taken by John Cairns, established that the DNA of a bacterial cell is in the form of a circle—that is, of a single molecule with no end. The chromosome is about 1,100 micrometers long, or about 500 times as long as an* E. coli *cell.*

17–2
Plasmids from Neisseria gonorrhoeae, *the bacterium that causes gonorrhea. The two connected plasmids in the upper right-hand corner are probably replicating.*

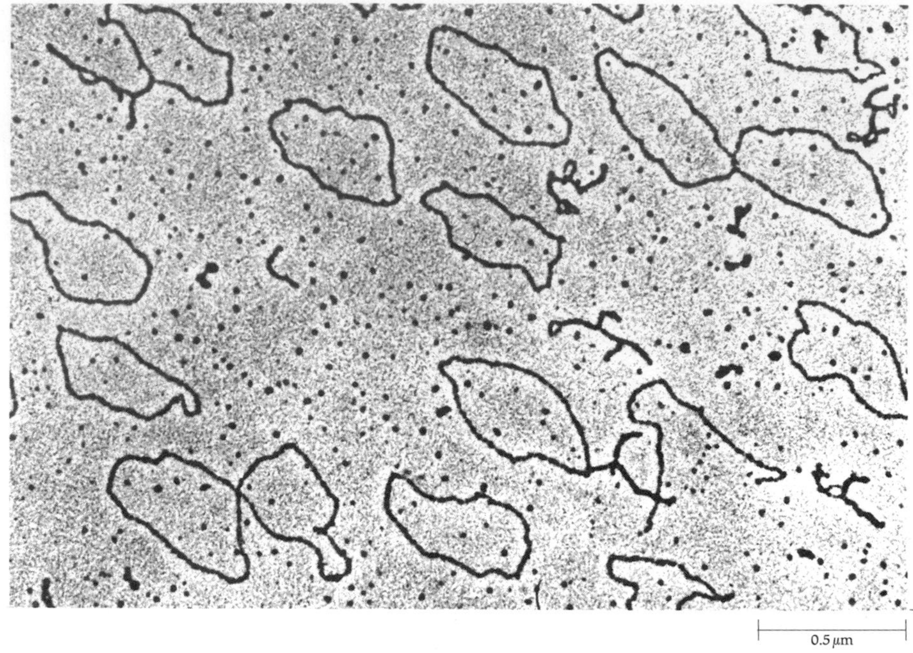

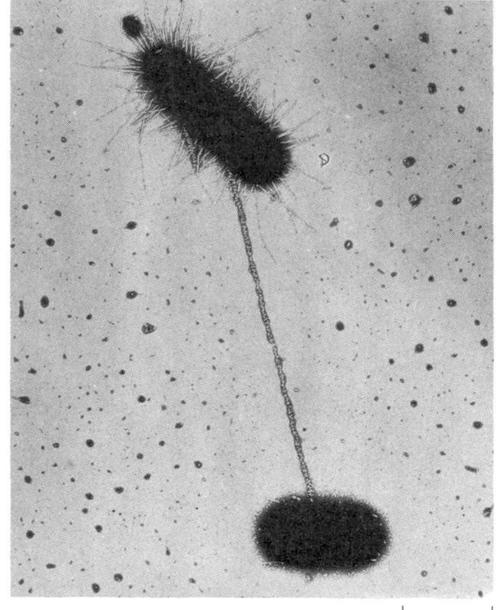

17–3
Electron micrograph of conjugating E. coli *cells. The F^+ cell at the top of the micrograph is connected to an F^- cell by a single long pilus, which is covered with bacteriophages, seen here as tiny spherical structures. Numerous shorter pili are visible on the F^+ cell.*

Many plasmids carry genes that confer resistance to various antibiotics.

Some plasmids (though not, apparently, those conferring antibiotic resistance) can become part of the main chromosome and, conversely, can escape from it. Such plasmids are known as *episomes.*

THE F FACTOR AND CONJUGATION

The first plasmid to be discovered was the F (for fertility) factor of *E. coli.* The F factor plasmid contains at least 15 genes, nine of which control the production of F pili, long appendages that extend from the surface of F^+ cells (cells containing the F factor). F^+ cells can attach themselves to F^- cells by their special pili and transfer the F factor to these cells, giving them the capacity to produce special pili and transfer the F factor.

The F factor is an episome. When it is attached to the bacterial chromosome, some or, rarely, all of the bacterial chromosome can be transferred to an F^- cell. This transfer of genetic material between prokaryotes is known as *conjugation.*

During conjugation, the circular DNA molecule opens and begins to replicate at the point of attachment of the F factor. As it replicates, the free end of the chromosome moves into the recipient cell (see Figure 17–4). Usually only part of the chromosome is transferred. The F factor, which would be the last portion of the chromosome to enter the cell, is therefore seldom transferred at conjugation.

Recombination by crossing over occurs between the chromosome of the recipient cell and the chromosome or chromosomal fragment of the donor cell. When the recipient and donor differ in certain characteristics—such as the ability to make a particular enzyme—these recombinations can be detected.

17–4

(a) The F factor is a plasmid that confers upon its host cell the capacity to be a donor cell and transfer an F factor to a recipient (F^-) cell. (b) When the F factor is part of the chromosome, the cell is referred to as HFr, for high frequency of recombination. From such donor cells, the chromosome itself, or a portion of it, enters the recipient cell. That portion may undergo recombination with the chromosome of the recipient cell, exchanging genes with it. The recipient cell usually remains an F^- cell because the F factor is not usually transferred.

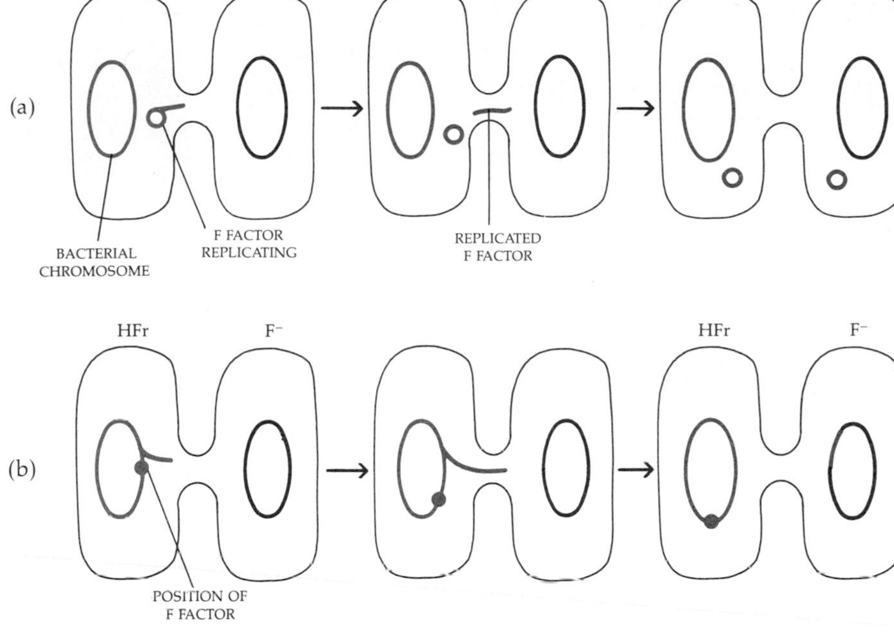

The conjugation process takes about 90 minutes (at 37°C). By separating the cells at various points in the process (which can be accomplished by whirling them at high speed in an electric blender and then analyzing which genes have been transferred), it is possible to construct a map of the chromosome (Figure 17–5). The mapping process confirms that the genes are carried in a linear array on the chromosome: *A* is followed by *B*, *B* by *C*, and so on, down to *XYZ*. It was these studies that also first gave a clue that the *E. coli* chromosome is circular; that is, *Y* followed *X*, and *Z* followed *Y*, but next would come *A* and then *B*.

17–5

In bacterial conjugation, genes are transferred from one cell to another in a definite order and at a constant rate. By employing mating strains with a variety of different mutations, it was possible to map the bacterial chromosome simply by observing the order in which genes entered the recipient cell. The letters on the chromosome represent genetic markers, such as the capacity to produce a particular enzyme or resistance to a particular antibiotic.

Transfer of Other Plasmids

Plasmids are very promiscuous and readily undergo recombination with other plasmids. Thus, for example, a single plasmid may gain the capacity to confer immunity to several antibiotics. Also a plasmid conferring resistance may simultaneously carry the capacity to make its host cell a donor cell and so to transfer the plasmids to a recipient cell (see page 290).

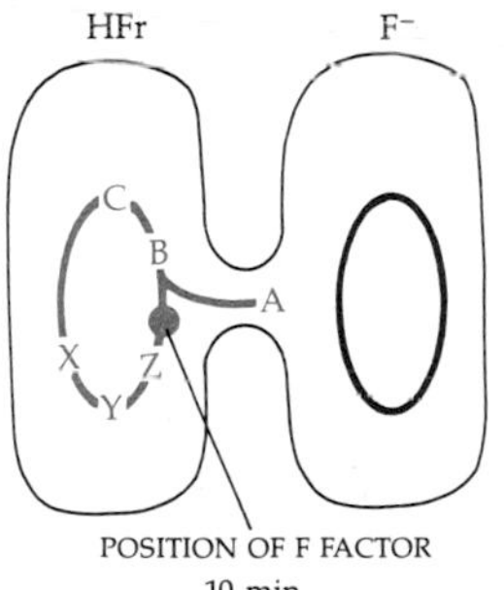

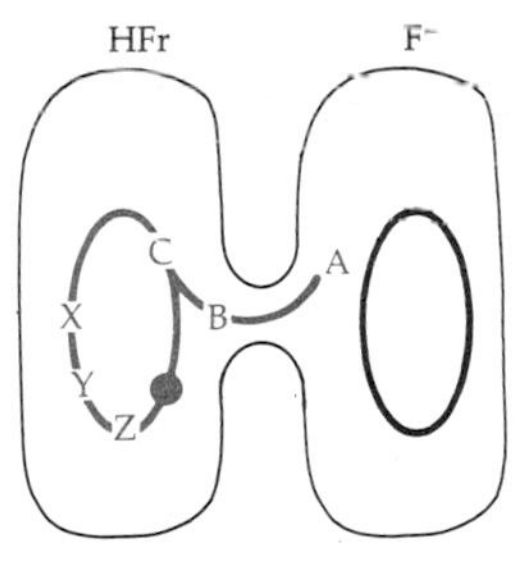

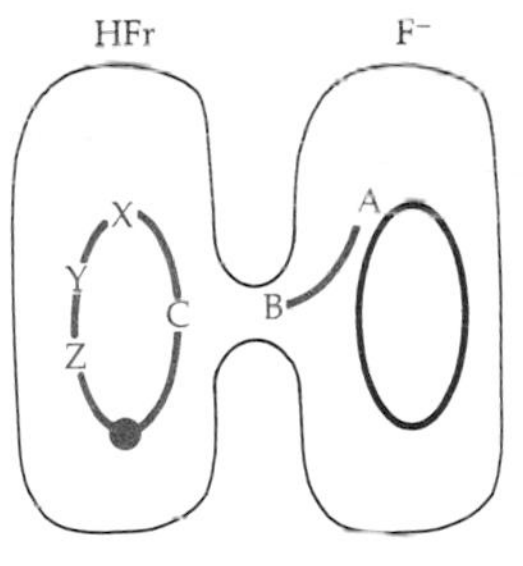

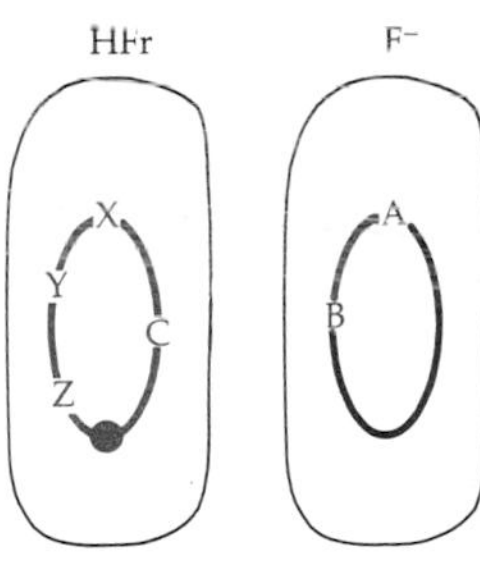

INFECTIOUS DRUG RESISTANCE

For many years, transfers of genetic information among bacterial cells have been regarded mainly as laboratory phenomena, chiefly of interest to research scientists engaged in studies such as gene mapping. However, it is now clear that such transfers also occur frequently in nature among certain groups of bacteria.

In a large bacterial population, there may be a few cells that, as a result of mutation, are resistant to a particular drug—streptomycin, for example. If the bacterial population is exposed to streptomycin—as when a patient is being treated with the drug—susceptible cells are destroyed and the resistant ones remain. These multiply and produce a population of bacterial cells made up entirely of resistant individuals. This is one reason why responsible physicians advocate restraint in the use of antibiotics. (A second reason is that with each treatment with antibiotics, the chance of the patient's developing an allergic reaction to that drug increases.)

It is now known that bacteria can become drug-resistant by the transfer of plasmids from resistant to nonresistant cells. Like F factors, these plasmids are readily passed from cell to cell. Under experimental conditions, 100 percent of a population of nonresistant cells can become resistant within an hour after being mixed with suitable resistant bacteria. Such transfers were first discovered in a strain of Shigella, *a bacterium that causes dysentery. Further studies showed that not only can* Shigella *transfer drug resistance to other* Shigella *organisms but also the innocuous* E. coli *can transfer resistance factors to this and other, unrelated groups of bacteria. Infectious resistance is now found among an increasing number of types of bacteria, including those that cause typhoid fever, gastroenteritis, and undulant fever as well as dysentery.*

These findings cast considerable doubt on the advisability of the current widespread use of antibiotics. For example, antibiotics are ineffective against viruses, yet many physicians (sometimes at their patients' insistence) routinely prescribe them for the common cold, a virus-caused disease. Antibiotics are routinely added to animal feed because they promote growth. In members of one farm family whose chickens were being fed antibiotics, 36 percent of the bacteria present in their intestines were resistant to three or more antibiotic compounds, compared with 6 percent for a neighboring control family.

The problem is, of course, that if these practices continue, an increasingly larger percentage of disease-causing bacteria will be resistant to any treatment now available for them. Thus, by our own lack of wisdom, we are threatening to negate some of the greatest medical advances of our century.

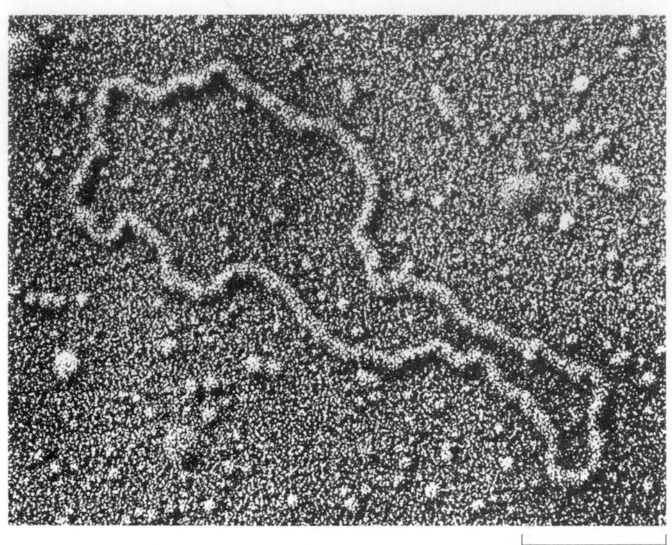

pSC101, a plasmid of E. coli *that confers resistance to tetracycline.*

CONTROL OF GENE EXPRESSION IN BACTERIA

The chromosome of *E. coli* is calculated, on the basis of its weight, to consist of about 4 million nucleotide pairs. If you estimate that the polypeptide chains of proteins average 300 amino acids, you can calculate that the chromosome could consist of more than 4,000 genes, and, indeed, as many as 3,000 different proteins have been isolated from *E. coli* cells.

Not all of a bacterial cell's genes are active at the same time. For example, cells of *E. coli* need the enzyme beta-galactosidase in order to split the disaccharide lactose into glucose and galactose, thereby enabling them to use lactose as an energy

17-6
Splitting of lactose (milk sugar) to galactose and glucose requires the enzyme beta-galactosidase. Beta-galactosidase is an inducible enzyme; that is, its production is regulated by an inducer—in this case, lactose.

LACTOSE

H_2O

BETA-GALACTOSIDASE

GLUCOSE

GALACTOSE

source (Figure 17-6). In cells growing on lactose, approximately 3,000 molecules of beta-galactosidase are present in every normal *E. coli* cell. This represents about 3 percent of all the protein in the cell. If lactose is not present, there is only one molecule of this enzyme per cell.

The capacity to produce or not produce the enzyme is genetically determined, as disclosed by the presence of mutants of *E. coli* that make beta-galactosidase even in the absence of lactose. These mutants are at a disadvantage compared to cells with a balanced protein synthesis, since by making an enzyme in the absence of its substrate they are using their energies and resources uneconomically.

Substances such as lactose that increase the amount of enzyme produced by a cell are known as *inducers*, and the enzymes they influence are known as *inducible enzymes* (Figure 17-7a).

Some substances act to repress enzyme production. For example, *E. coli* can make all its own amino acids. If a particular amino acid—histidine, for example—is present in the medium, however, the cell will then stop making all of the enzymes associated with the biosynthesis of histidine. Such enzymes are known as *repressible enzymes*, and the substances that inhibit their production are known as *corepressors* (Figure 17-7b).

The Operon

To explain how bacterial cells regulate enzyme biosynthesis, the French scientists François Jacob and Jacques Monod developed the hypothesis of the *operon*. An operon is a group of structural genes (genes that code for enzymes or other proteins) aligned along a single segment of DNA that are regulated as a unit. In the beta-galactosidase system, the operon includes the gene that codes for beta-galactosidase and two other genes, which also code for enzymes involved in lactose metabolism. These genes are adjacent to one another and are transcribed consecutively, one after the other along the DNA, forming a single mRNA molecule. The mRNA molecules produced are active for only a very short time, after which they are degraded. Thus, if a cell needs a continuous supply of enzymes, it must make a continuous supply of the specific mRNA.

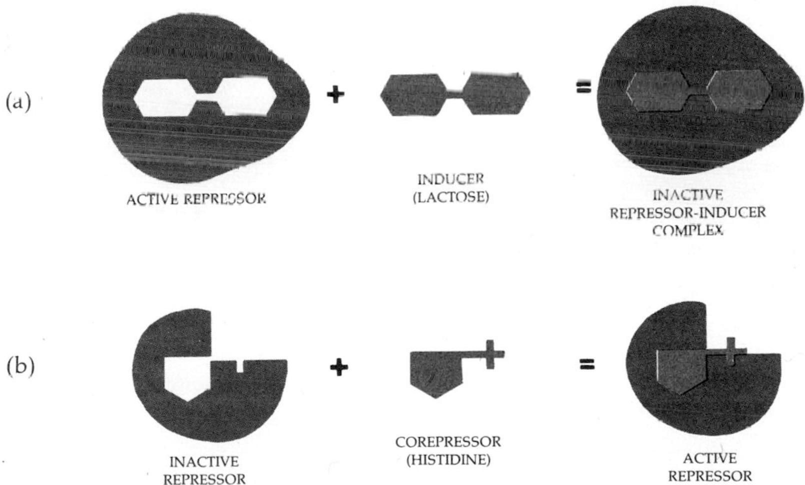

17-7
According to the operon theory, the synthesis of proteins is regulated by interactions involving either a repressor and inducer or a repressor and corepressor. (a) *In the case of inducible enzymes, such as beta-galactosidase, the repressor molecule is active until it combines with the inducer (in this case, lactose). It is then inactivated, and so the genes in the operon are no longer repressed.* (b) *In the case of a repressible enzyme, the repressor is not active until it combines with the corepressor. Thus, in the absence of the corepressor, the genes in the operon are active.*

17–8
(a) *An operon is a group of structural genes that code for proteins, often enzymes that work sequentially in a particular reaction pathway. Adjacent to the operon on the bacterial chromosome is a segment of DNA that contains the promoter and the operator. The operator slightly overlaps the promoter. The promoter is the site at which RNA polymerase attaches; this is the enzyme responsible for the transcription of mRNA. Another gene involved in operon function is the regulator. The regulator gene, which is not necessarily adjacent to the other genes, produces a regulatory protein known as a repressor. Some operons are inducible, others repressible.*

(b) *In operons activated by inducers, the regulator gene codes for a protein that binds to DNA at the operator and so prevents the RNA polymerase from initiating transcription. The inducer counteracts the effect of the repressor by binding with it and maintaining it in an inactive form. Thus, when the inducer is present, the repressor can no longer attach to the operator, and synthesis of mRNA proceeds.*

(c) *In operons regulated by corepressors, the repressor can bind to the operator only when it is combined with a corepressor. Thus transcription proceeds until a corepressor is produced.*

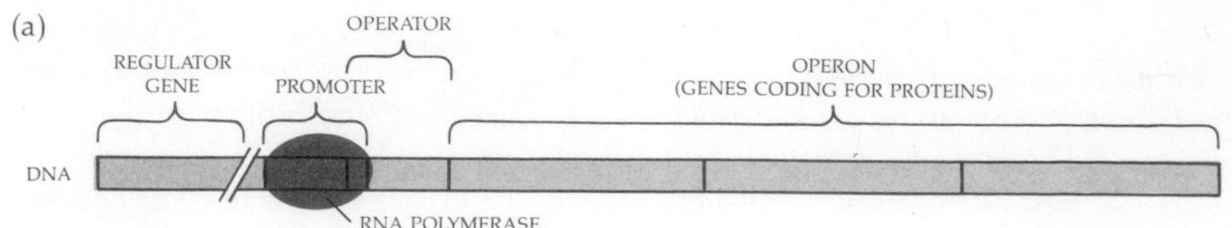

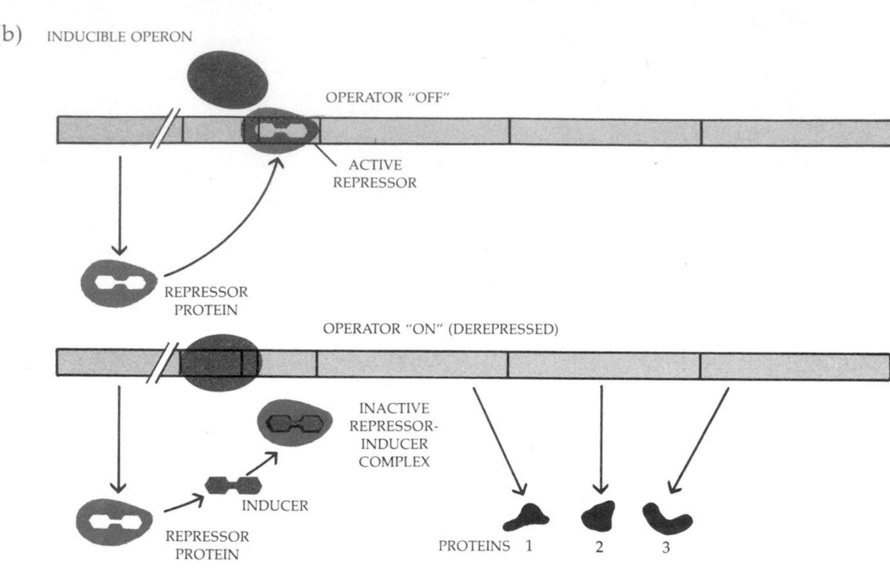

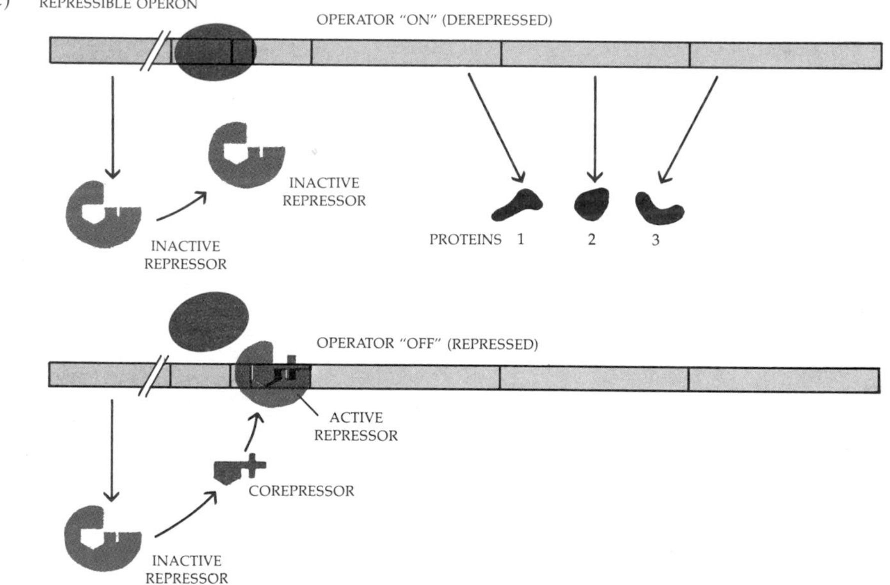

Adjacent to the operon, as shown in Figure 17-8, are (1) the *promoter*, a short segment of DNA to which RNA polymerase attaches at the initiation of transcription, and (2) the *operator*, which serves as the on-off switch for the operon. If the operator is "on," the mRNA polymerase can move down the DNA and transcribe a molecule of mRNA. If the operator is "off," the mRNA polymerase cannot move off the promoter and no mRNA will be made.

Whether the operator is on or off is controlled by yet another genetic element, (3) the *regulator*. The regulator codes for a specific protein known as the *repressor*, which may bind to the operator. If the repressor is bound to the operator, the operator is "off," and no mRNA can be produced by the operon.

The repressor, an allosteric protein, is controlled by another "signal" compound. In the case of inducible enzymes, this compound is the inducer. The inducer binds with the repressor molecule and changes its shape so that it can no longer attach itself to the operator. In the absence of the active form of the repressor, mRNA molecules are formed along the structural genes, and from these molecules proteins are produced. When the supply of the inducer is exhausted as a result of the enzymatic action of the proteins produced, the repressor, once again active, assumes control, and mRNA production and protein formation cease. In the beta-galactosidase system, the inducer is lactose or a closely related compound derived from lactose. The lactose or related compound binds with the repressor and keeps it in an inactive form, thus permitting enzyme biosynthesis to proceed.

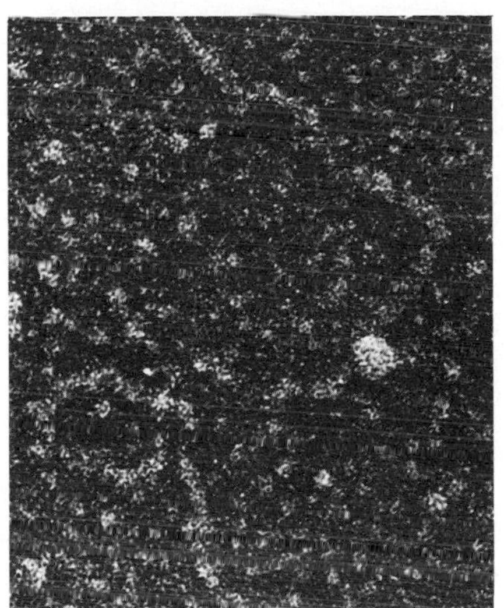

17-9
A repressor protein (the white spherical form at the middle right in the micrograph) attached to the operator of the lactose operon.

In the case of repressible enzymes, the "signal" compound is a corepressor. The repressor is active only when bound with the corepressor. In the histidine system, which involves some 10 different enzymes, histidine combined with its tRNA is the corepressor. Histidine-$tRNA_{his}$ combines with the repressor and activates it, thus halting mRNA synthesis and, ultimately, synthesis of histidine.

In further confirmation of the operon hypothesis, mapping studies show that genes for enzymes with related activities occur in clusters. The operon system is now widely accepted as a principal genetic control mechanism in bacteria. It is a simple and efficient means of coordinating the production of enzymes that the cell needs in the same amounts at the same time.

Feedback Control Systems

Note that these inducible and repressible enzyme systems are further examples of regulatory systems operating through feedback control. The key to the control of the system is the regulatory protein, the repressor, which is designed to complex not only with a specific sequence of nucleotides in the DNA—the operator—but with the inducer or corepressor as well. Thus repressors are allosteric proteins; they exert their regulatory influence in a manner not unlike that of the allosteric enzymes described on page 87. That is, they are stabilized in either an active or an inactive form by small molecules in the cellular environment, and these small molecules are often precursors or products of the processes they control.

THE EUKARYOTIC CHROMOSOME

The eukaryotic chromosome is far more complex in structure. Unlike the bacterial chromosome, it is more than half protein. These proteins are believed to play a major role in maintaining the structure of the chromosomes and in the events that take place during mitosis and meiosis. They are also believed to be involved in regulating gene expression in eukaryotic cells.

The regulation of gene expression is a much different problem for a multicellular eukaryote than for *E. coli.* A multicellular organism—a human being, for instance—usually starts life as a fertilized egg. The egg cell undergoes multiple mitotic divisions, producing many cells, and at some stage these cells begin to differentiate, becoming muscle cells, bone cells, blood cells, intestinal cells, and so forth. Each cell type, as it differentiates, begins to produce characteristically different proteins that distinguish it from other types of cells. This is nicely illustrated by the mammalian red blood cell, which first produces one type of fetal hemoglobin, then a second type, and then, sometime after the birth of the organism, begins producing the alpha and beta chains characteristic of adult hemoglobin. Thus the genes are expressed as a carefully controlled sequence, one after the other. The DNA segments that code for all these hemoglobin molecules are expressed only in the red blood cells.

There are two possible explanations for these changes that take place during development: (1) As cells differentiate, they lose some of their genetic material, or (2) the total genetic material remains the same in all the cells of an organism but a large portion of it is inactivated. Since the total DNA per diploid cell is the same in all cells of a multicellular eukaryote, the first hypothesis seems unlikely. Also, the cloning experiments of Steward and Gurdon (page 219) indicate that the genetic information remains constant, at least during the earlier stages of cell differentiation. On the basis of such evidence, it is generally believed that the genetic information originally in the fertilized egg cell, or zygote, is also present in every diploid cell of the organism.

In other words, the DNA segments that code for hemoglobin are present in skin cells and heart cells and liver cells and nerve cells and, indeed, in every one of the more than 100 different types of cells in the body. The DNA sequence that codes for insulin is present not only in special cells of the pancreas that manufacture insulin but also in all the other cells. Following this argument further, one is rapidly led to conclude that in any cell in any multicellular organism, most of the DNA has been selectively silenced.

Much evidence links this repression of DNA expression with the arrangement of DNA and proteins in the eukaryotic chromosome.

CHROMOSOMAL PROTEINS AND GENE REGULATION

In eukaryotic cells, hundreds of different proteins are bound to the chromosomes. Some proteins bind directly to the DNA by the attraction between the positively charged side groups of arginine, lysine, and histidine (page 62) and the negatively charged phosphate groups in the backbone of the DNA chains. Other proteins bind to the proteins that are bound to the DNA. Chromatin, the "colored threads" seen by earlier microscopists, is this complex of DNA and protein. Within the nucleus, chromatin may be found in a tightly packed, highly condensed form, called heterochromatin, or in a dispersed form, euchromatin. There are several lines of evidence that indicate that the tightly packed heterochromatin is genetically inactive and that RNA is transcribed only on euchromatin.

First, no new RNA is produced in mitotic chromosomes where the chromosomal material is tightly condensed. Transcription takes place only during interphase.

17-10
Observations of chromosome puffs support the concept that the DNA is somehow unwound to make it available for mRNA transcription. These puffs were observed in chromosomes of the Brazilian gnat, which, like the fruit fly, has giant chromosomes in some of its cells. Puffs occur normally but can also be induced experimentally. These puffs occurred in response to a hormone that causes molting. As the micrographs indicate, the puffs occurred sequentially along one chromosome.

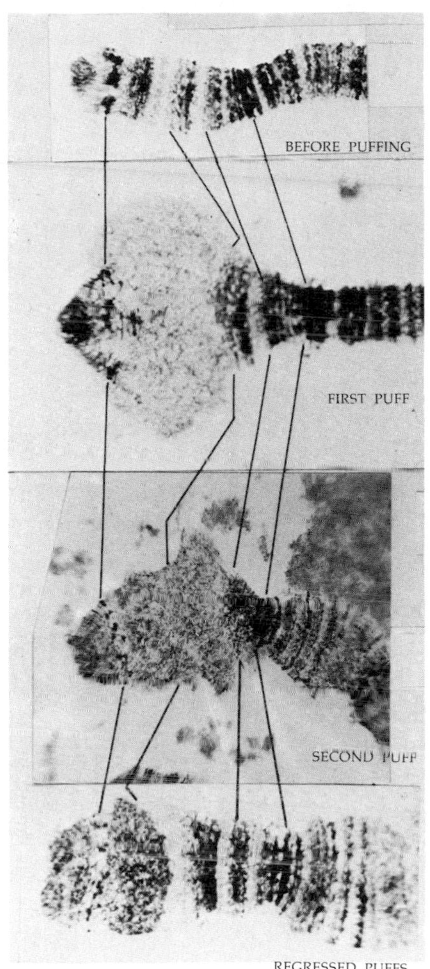

Second, no transcription takes place on the inactivated female chromosomes, the Barr bodies, which are tightly condensed (see essay, page 243).

A third type of evidence comes from studies of the giant chromosomes of insects (see page 248). At various stages of larval growth in insects, it is possible to observe diffuse thickenings, or "puffs," in various regions of these chromosomes (Figure 17-10). The puffs are open loops of DNA, and studies with radioactive isotopes indicate that these loops are sites of rapid RNA synthesis (transcription). When ecdysone, a hormone that produces molting in insects, is injected, the puffs occur in a definite sequence that can be related to the developmental stage of the animal. For example, in one species of *Drosophila,* ecdysone initiates three new puffs and causes increases in 18 other puffs within 20 minutes after it is injected; during this same period, 12 other puffs decrease in size. After 4 to 6 hours, five additional puffs can be seen. The looping out of the DNA occurs before RNA synthesis is initiated. The mechanism of this unraveling, or spinning out, of the chromosomal DNA is not known.

Somewhat similar observations have been made with another type of chromosomal material, lampbrush chromosomes, so called because they resemble the brushes used to clean kerosene lamps. They are found in the nuclei of egg cells of fish, birds, reptiles, and amphibians during the first prophase of meiosis. These cells are very actively engaged in the RNA and protein syntheses required for the rapid growth of the mature egg. Each lampbrush chromosome consists of two homologues held together by their centromeres and chiasmata (page 246), each homologue consisting of two sister chromatids. As you can see in Figure 17-11, each chromatid has a number of lateral loops, which branch out from the main axis. These loops appear to be reeled in and out from the body of the chromosome. Each of these loops consists of a thin fiber. If this thin fiber is exposed to an enzyme that digests away the associated protein, all that remains is a thin filament about the diameter of a single DNA molecule. These loops, like the chromosome puffs in giant chromosomes, are also the sites of active RNA synthesis.

Thus, although it may not be a cause-and-effect relationship, according to present information, tightly coiled DNA does not transcribe mRNA and, conversely, the unfolding of DNA is correlated with mRNA transcription.

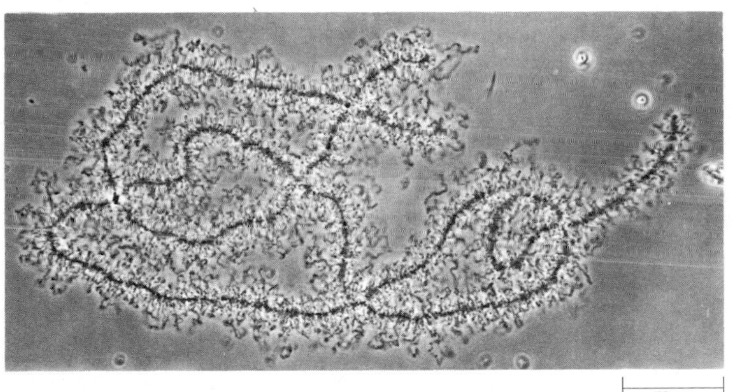

17-11
Two homologous lampbrush chromosomes held together by several chiasmata. The individual chromatids are closely paired and so cannot be distinguished.

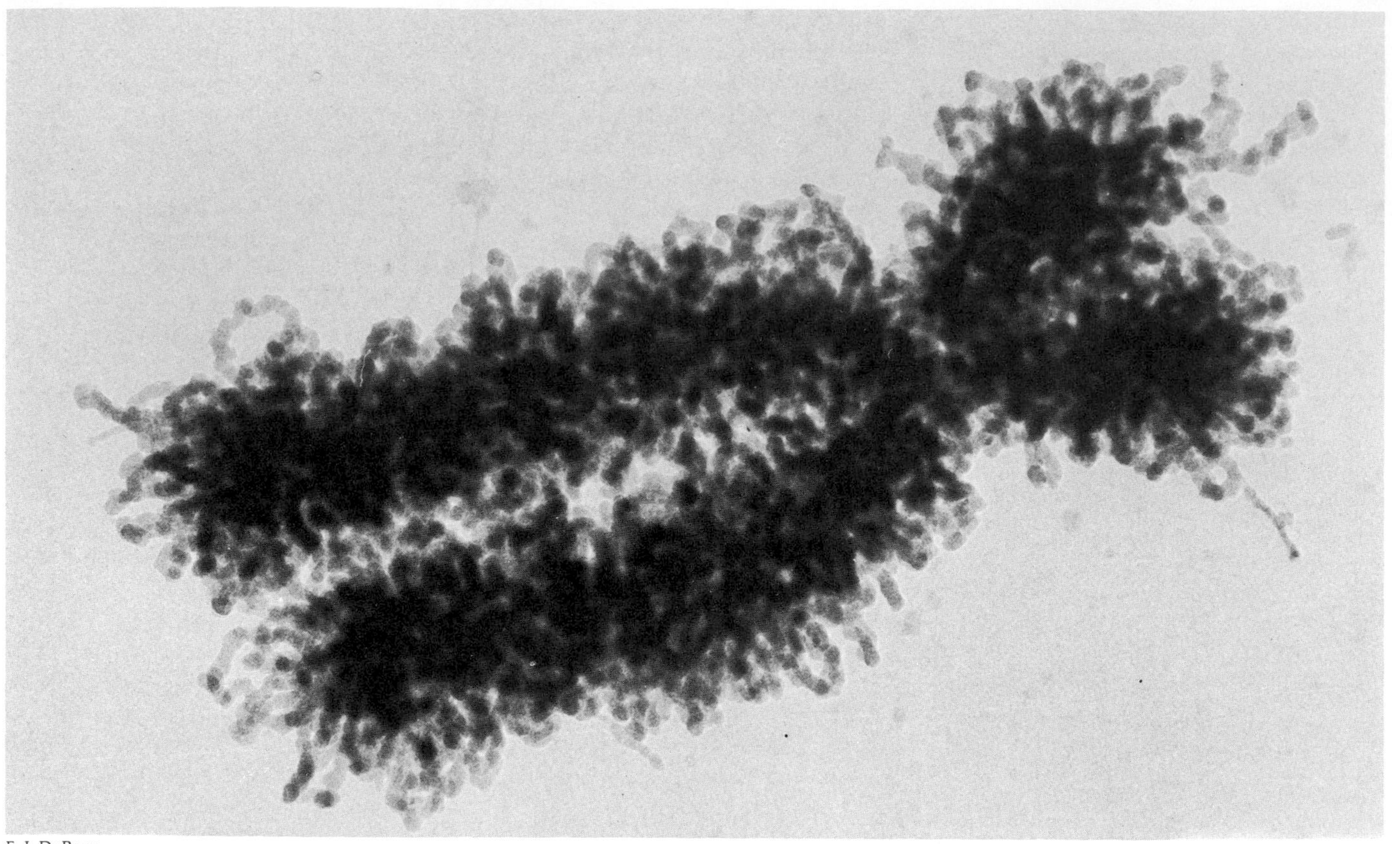

E. J. DuPraw

17–12

Human chromosome 12, shown at metaphase, before the chromatids have separated. The chromatin is highly condensed. Each chromatid, according to present evidence, contains a single DNA molecule.

THE PROTEINS OF THE CHROMOSOME

Histones

Two types of proteins are present in chromatin. One is a group of relatively small molecules known as histones; they are positively charged (basic) and so are attracted to the negatively charged (acidic) DNA. They are always present in chromatin, very similar from species to species, and are synthesized in large numbers during the S phase of the cell cycle (see page 204), when DNA is also being synthesized. There are at least five distinct types of histones. One type is invariably associated with the nuclei of eukaryotic cells and is virtually identical from species to species; it consists of 102 amino acids. The molecules of this histone found in calf nuclei differ from those found in pea seedlings by only two amino acids out of the 102. This constancy of structure throughout billions of years of evolution strongly suggests that the histones play some important role in the chromosome. The total weight of histone proteins is always about equal to the weight of DNA in the chromatin.

At one time, because of the obvious intimate relationship between DNA and the histone proteins, they were believed to be the regulatory proteins. However, because of their lack of heterogeneity (and therefore lack of specificity), this is no

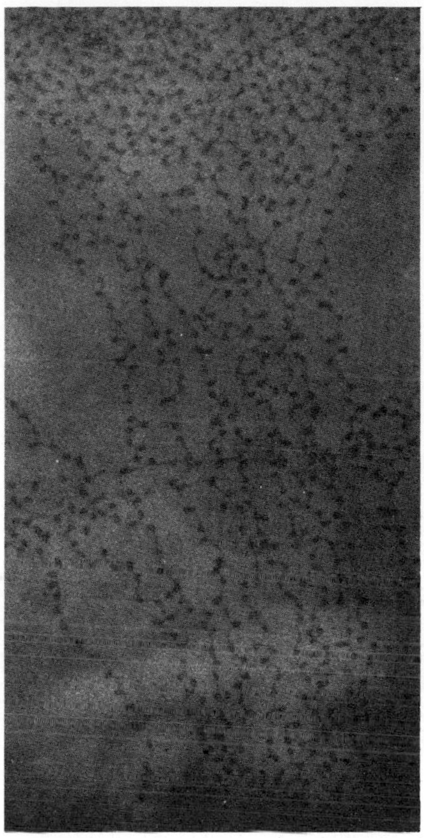

17–13
Chromatin from the red blood cell of a chicken. The distance between beads is about 14 nanometers, and the diameter of each bead is about 7 nanometers. Each bead is composed of about 140 base pairs of DNA and an assembly of histone molecules.

longer thought to be true, and they are regarded as probably structural rather than regulatory elements.

Much recent attention has been focused on the way in which DNA and histones might fit together. Until recently, it had been widely accepted that the DNA double helix was coated in some way with histones and twisted on itself in a large single helix, which investigators called the supercoil. Most recently the supercoil has been superseded, and a new and wholly unexpected structure has taken its place. This structure, which is based on electron micrographs (Figure 17–13) and is supported by biochemical evidence, resembles beads on a string. These beads have been given the name nucleosomes, or nu bodies.

Each nucleosome consists of four of the five different types of histones and DNA. The "string" in between is DNA plus the fifth type of histone. Each nucleosome contains about 140 pairs of nucleotides, and the strand between the particles contains another 30 to 60 base pairs. Thus an average gene is made up of three or four nucleosomes. When a fragment of DNA is tied up in a bead, it is about one-sixth the length it would be if it were fully extended. Presumably the beads and string are further folded in some way.

Nonhistone Proteins

The nonhistone proteins, which make up 50 to 70 percent of the total protein of chromatin, are varied and heterogeneous and, for this reason, are believed to play a regulatory role. They are difficult to analyze and, as yet, little is known about them.

DNA OF THE EUKARYOTIC CHROMOSOME

DNA is an "exquisitely thin filament," in the words of E. J. DuPraw, who calculates that a length sufficient to reach from the earth to the sun would weigh only half a gram. The DNA of each chromosome is believed to be in the form of a single molecule. In a human chromosome, each of these molecules is believed to be from 3 to 4 centimeters long. Thus each human cell contains between 1 and 2 meters of DNA, and the entire human body contains some 25 billion kilometers of DNA double helix.

The technical problems of investigating a molecule as large as the DNA of a eukaryotic cell are enormous. Because of the base-pairing properties of DNA, it is possible, however, to compare structures of different DNA molecules and different sequences on the same molecule without actually identifying them. If DNA is heated gently, the hydrogen bonds holding the two strands together are broken and the strands separate. (This process is called denaturation.) When the solution is cooled slightly the bonds re-form. The extent to which segments from two samples reassociate and the speed with which they do so provides an estimate of the similarity between the DNAs. Similarly, the extent and rate of combination of an mRNA molecule with a DNA molecule makes it possible to detect whether or not the mRNA was transcribed from that DNA.

Studies of this sort have revealed that the DNA of the eukaryotic cell is composed of three distinct kinds of sequences.

One type of DNA consists of gene-sized (that is, about 1,000 base pairs) segments that are unique sequences. About 70 percent of the total DNA of a eukaryotic cell is made up of this single-copy DNA. This much DNA could code for about half a million genes. The number of genes in a mammalian cell has been estimated to be about 50,000. Therefore, it would appear that the single-sequence DNA contains about 10 times more DNA than is required.

A second type of DNA consists of gene-sized sequences repeated over and over again, from a few times to hundreds of thousands of times.

The third type of DNA is made up of very short (about 100 base pairs), simple, and highly repetitive sequences. The sequence in a hermit crab, for instance, is ATCCATCCATCC; in a kangaroo rat, ACACAGCGGG; in a guinea pig, CCCTAA; and so on. Such sequences may be repeated as many as 10^7 times per cell and may make up as much as 10 percent of all the DNA. These repetitive sequences are likely to be found in the region of the centromere, suggesting that they have a structural role.

Restriction Enzymes

Another method for studying the composition of chromosomes has become available with the discovery of enzymes that cleave DNA at specific sites. One, for instance, known as Eco RI (isolated from a plasmid-containing *E. coli*), cleaves DNA only at the sequence GAATTC. These enzymes, which are known as *restriction enzymes*, apparently have the function of recognizing and destroying DNA that is foreign to the cell.

The analysis of large protein molecules was made possible to a great extent by the availability of digestive enzymes that hydrolyze the peptide bonds at specific points in the polypeptide chain, yielding relatively small, uniform fragments for analysis. Restriction enzymes are now serving as similar probes for DNA structure. They are also being used in a host of quite different investigations, generally described by the controversial terms "genetic engineering" and "recombinant DNA research," which will be described in Chapter 19.

SUMMARY

There are major structural differences between the prokaryotic and eukaryotic chromosome, primarily related to the differences in gene regulation in the two types of organisms.

The bacterial chromosome is a single, circular loop of DNA. In addition, a bacterial cell contains a number of much smaller, also circular, molecules of DNA called plasmids.

In the course of bacterial conjugation (a form of mating), the DNA of a bacterial cell replicates, and as it replicates, some or all of the chromosome is passed to a recipient cell. Donor cells are characterized by the presence of the F (fertility) factor. The F factor is a plasmid and, like many other plasmids, it can exist independently or as part of the host chromosome. An isolated F factor can be passed from one conjugating cell to another; the recipient cell then becomes a donor cell. When the F factor is integrated in the chromosome, the chromosome can later break at that point and replicate, and all or part of the daughter chromosome may be passed to a recipient cell.

A principal means of genetic regulation in bacteria is the operon system. This system is made up of a promoter, the site at which the synthesis of the mRNA (transcription) begins; an operator, which is the site of regulation; and the operon, one or more structural genes that code for polypeptide chains. The operator is under the influence of another gene, known as the regulator. The regulator codes for a protein, known as a repressor, that attaches to the DNA molecule at the operator and blocks mRNA transcription. The repressor does not act alone but rather in conjunction with another compound, either an initial reactant or an end product of the reaction series catalyzed by the enzymes coded by the structural genes of the operon. This compound may be an inducer, in which case it inactivates the repressor, or it may be a corepressor, in which case it enables the repressor to function.

The eukaryotic chromosome, which is far more complex in structure, consists of DNA complexed with two types of protein molecules. One is a group of very simple, basic (positively charged) proteins known as histones. Recent evidence indicates that the histone-DNA complexes resemble beads on a string. These beads are called nucleosomes. Other proteins are large and heterogeneous and are believed to have specific regulatory functions.

Each chromosome, according to present evidence, contains a single, very long DNA double helix. Three types of DNA sequences have been discovered. In one type, the sequences are short and highly repetitive. This type of DNA is found in the region of the centromere and is believed to serve a structural function. In the second type, the sequences are long and highly repetitive. In the third type, the sequences are moderately long, heterogeneous, and nonrepetitive. Only this third type, which comprises about 50 percent of the total DNA, is believed to contain the genetic message.

QUESTIONS

1. Distinguish among the following: plasmid/episome/F factor; regulator/promoter/operator/operon; inducer/repressor/corepressor; inducible enzyme/repressible enzyme/restriction enzyme; heterochromatin/euchromatin.

2. In what ways are the chromosomes of prokaryotes and eukaryotes similar? In what ways are they different?

3. Compounds can be used by cells in two different ways. One type is broken down (usually as an energy source). Another is used as a building block for a larger molecule. Which type of compound would you expect to function as a corepressor? As an inducer? Do the examples given in the text conform to these expectations?

4. Gene expression theoretically can be regulated at the level of transcription, translation, or activation of the protein. What would be the advantages of each, in terms of the cell? Under what circumstances might one type be more useful than another? Which is the more economical?

5. In what way does bacterial mating resemble fertilization? How does it differ from it?

6. As bacterial mating indicates, it is possible to separate the production of new genetic material from reproduction. Why do you think these two processes are combined in eukaryotic cells?

CHAPTER 18 Human Genetics

The principles of genetics are, of course, the same for humans as they are for any other species. There are some important differences, however, particularly in methodology. Breeding experiments, so readily performed with fruit flies, bread molds, and peas, are not possible with humans, and even if they were, the generation time is so long that one is likely to lose interest in the question before the answer comes along. Most data on patterns of human heredity are based on family pedigrees (Figure 18–1). With the exception of those of us who belong to royal families, few have knowledge about their forebears that extends over more than four generations. Do you, for instance, know the color of the eyes or skin of your various great-great-grandparents? More important, do you know what health problems they had and what caused their deaths?

Most information about human genetics comes from medical sources, and it is here that human genetics, as a science, has a great advantage. Large numbers of people are kept under careful scrutiny by a great variety of medical experts. As a result, a large amount of data has been accumulated on certain characteristics, such as the genetics and biochemistry of disorders involving hemoglobin.

SINGLE-FACTOR INHERITANCE

A number of diseases are known that are inherited as simple Mendelian recessives and therefore show a hereditary pattern similar to that diagrammed in Figure 18–1.

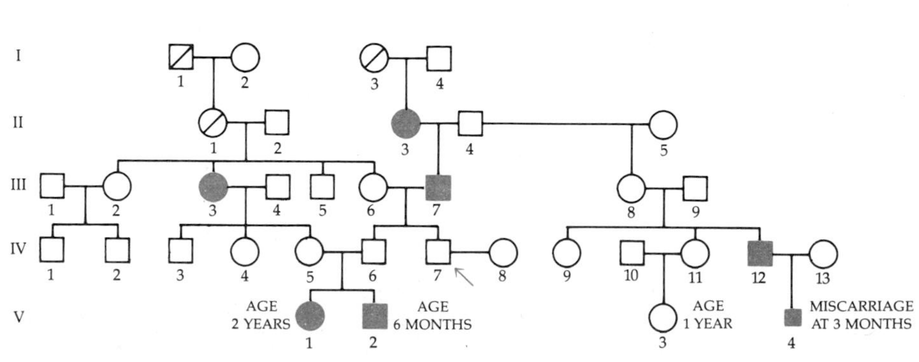

18–1
Pedigree showing inheritance of a genetic trait in five generations of a family, indicated by the roman numerals. By convention, squares represent males and circles, females. Each generation is on the same horizontal line. Siblings are indicated by a line above the squares or circles. Lines connecting a circle and a square indicate a marriage; vertical lines link parents and offspring. Individuals carrying the trait under study are indicated in color. Individuals believed but not proven to have the trait are designated by a line through the square or circle. The arrow indicates the person under study. From this pedigree, what can you deduce about the trait under study? If the trait was an undesirable one and you were a doctor, what advice would you give the patient?

For many of these diseases something is known of the biochemical basis of the defect as well.

Phenylketonuria

One of the most familiar examples of such a disease is phenylketonuria, or PKU, as it is commonly known. PKU is caused by lack of function of the enzyme that normally breaks down the amino acid phenylalanine. When this enzyme is missing or deficient, phenylalanine and its abnormal breakdown products accumulate in the bloodstream. Thus it resembles alcaptonuria (page 86), with the important difference that in phenylketonuria the breakdown products that accumulate are harmful to the developing cells of the brain and can result in mental retardation.

PKU is caused by a single recessive allele in the homozygous state. Infants with PKU usually appear healthy and normal at birth, but after the first few months the symptoms of the disease set in, and without treatment severe mental retardation usually results. Many never learn to walk or talk and are subject to periodic convulsions and seizures. Most afflicted individuals must be hospitalized for their entire lives, which, in untreated persons, is seldom more than 30 years.

It is not yet known how the high levels of phenylalanine and derivative compounds bring about the tragic mental symptoms. However, the knowledge we do have is enough to effectively treat the afflicted individuals and prevent the symptoms from appearing. Many states now require routine tests of all newborn babies in order to detect PKU homozygotes. Those identified at birth are simply put on a special diet containing low amounts of phenylalanine—enough to supply dietary needs but not enough to permit toxic accumulations. On the basis of 15 years' experience, PKU homozygotes develop normally on low phenylalanine diets. About 1 in every 15,000 infants born is homozygous for this recessive gene.

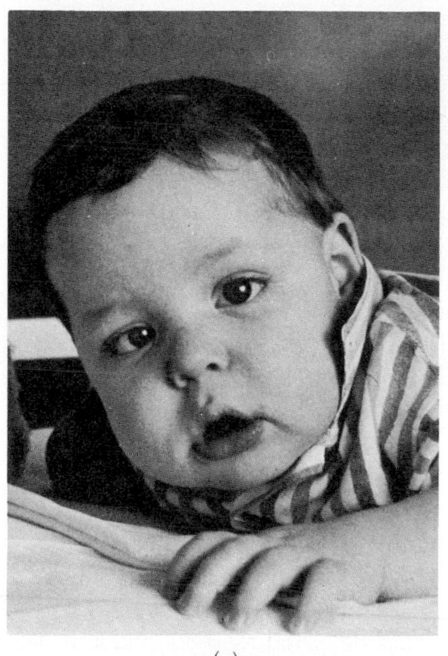

(a)

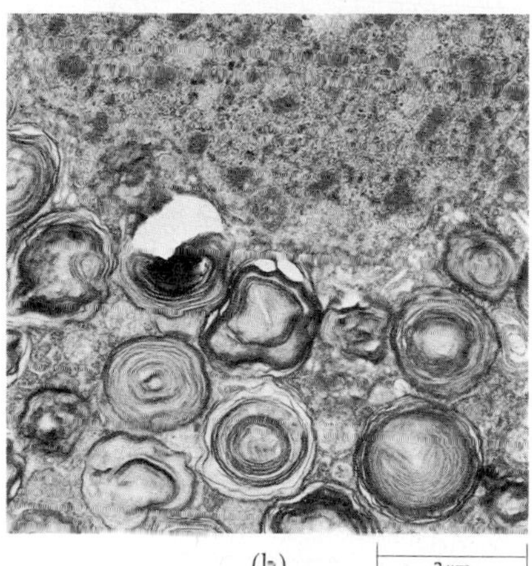

(b) 2 μm

18-2
(a) *In this one-year-old infant with Tay-Sachs disease, deterioration of the brain, already begun, will progress rapidly. The child will probably die before he is six years old.* (b) *The disease is caused by the absence of an enzyme involved with lipid metabolism. Without the enzyme, harmful fatty deposits accumulate in the lysosomes of brain cells, as shown in this micrograph. A portion of the nucleus can be seen at the top of the micrograph.*

Tay-Sachs Disease

Tay-Sachs disease is a recessive condition involving degeneration of the nervous system. As with PKU, Tay-Sachs homozygotes appear normal at birth and through the early months. However, by about eight months, symptoms of severe listlessness set in. Blindness usually occurs within the first year. Afflicted children rarely survive past their fifth year. The biochemical basis of this disease is now, at least in part, understood. Homozygous individuals lack an enzyme, ganglioside GM_2-hexosaminidase, which breaks down a special lipid found in brain cells. This enzyme is normally found in the lysosomes of brain cells and plays a crucial role in keeping the lipid from accumulating. In the child lacking the enzyme, the lysosomes of the brain cells fill with this lipid and swell, and the cells die. There is no therapy yet available for Tay-Sachs disease. While in the general population it is a rare disorder (1 in 300,000 births), it has a very high incidence among Jews of Central European extraction, Ashkenazic Jews, who make up more than 90 percent of the American Jewish population. The disease shows up once in every 3,600 births, which means that about 1 in 28 Ashkenazic Jews is a carrier (a heterozygote).

Sickle Cell Anemia and the Genetics of Hemoglobin

The early role that research on hemoglobin played in helping to clarify the relationship between a gene and the protein determined by that gene has been described. However, this is only one facet of the story. An enormous amount has also been learned about the genetics of hemoglobin, in large part for reasons more medical than biological: The blood is a most important bodily tissue, its analysis is central to many medical diagnoses, and it is readily obtainable in quite large amounts since it is quickly regenerated when lost.

While many different hereditary variants of hemoglobin have been discovered, the most serious one from a medical point of view is sickle cell anemia. This hereditary disease occurs with a very high frequency among blacks. In the United States about 14 percent of blacks are heterozygous for the sickle cell gene, and about 1 percent are homozygous and therefore have the symptoms of sickle cell anemia.

It was pointed out in Chapter 3 that the hemoglobin molecule in adult blood is a tetramer composed of two alpha chains and two beta chains. Sickle cell anemia is due to a genetic alteration in the beta chain, a single amino acid substitution. However, sickle cell anemia is not the only known genetic alteration of the hemoglobin molecule. More than 100 hereditary variants that are due to amino acid substitution in the beta chains have now been found, and a comparable number of alpha-chain alterations have also been detected. Unlike sickle cell anemia, most of these hereditary conditions are rare, and often the symptoms of the condition are mild or nonexistent. Families have been found with traits indicating alterations in both the alpha and beta chains, and genetic analysis of the segregation of the factors responsible has shown that the two genes determining the alpha and beta chains are on different chromosomes.

Heterozygotes for sickle cell anemia are symptomless; their "good" allele makes enough hemoglobin that the effects of the "bad" allele are not discernible. However, if blood samples are treated in ways that remove oxygen from all the hemoglobin molecules, some of the cells in the samples of a heterozygote will sickle (Figure 18-3). Thus it is possible to detect heterozygotes quite readily. If two heterozygotes marry, there is 1 chance in 4 of their having a child with sickle cell anemia, and a fifty-fifty chance of having a child who is, like themselves, a heterozygote and so a "carrier" of the disease.

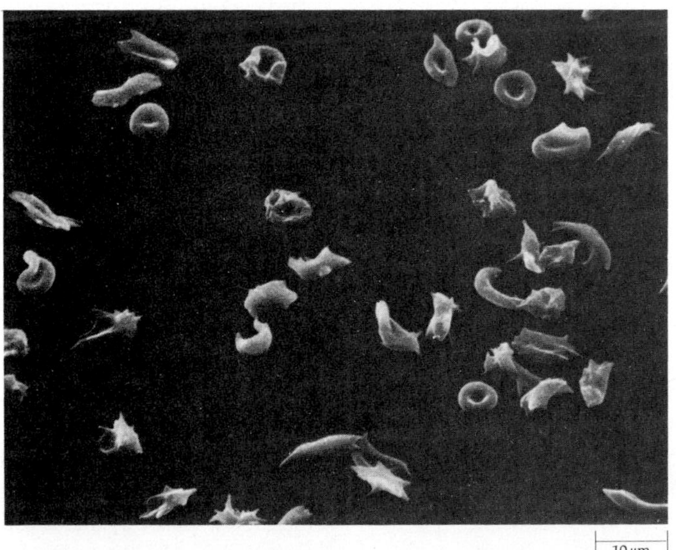

18-3
Scanning electron micrograph of blood from a heterozygote for sickle cell anemia. As you can see, most of the blood cells have sickled.

18-4
Many human characteristics are governed by simple Mendelian inheritance. One such characteristic is a cleft chin. This trait is due to a simple dominant allele.

Although sickle cell anemia has long been present among blacks and probably at a higher incidence than it is today, the disease rose to sociological prominence only in recent decades, paralleling the rise in political visibility of the black population. In a hasty effort to right old wrongs, programs for the detection of heterozygotes and for their genetic counseling were instituted, many of them among school children. The administrators of the test often were unable to distinguish between the meaning of heterozygote and homozygote, either in genetic, medical, or social terms, and the children, on the basis of the information given them, certainly could not. As a result, the screening programs were, generally speaking, quite predictably disastrous. At present, major efforts are confined to voluntary screening programs of adults (as are carried out with most other genetic diseases) and an intensified search for chemical agents that will prevent sickling in the cells of affected individuals.

TRANSMISSION BY DOMINANT GENES

The foregoing examples all involve recessive alleles. A number of human conditions are caused by simple dominants. Among these are cleft chin, polydactylism (extra fingers), and brown teeth. Huntington's disease (formerly called Huntington's chorea), the disease that afflicted Woody Guthrie, is caused by a dominant gene. As you can rapidly calculate, each child of a heterozygous parent with Huntington's disease has a fifty-fifty chance of inheriting the gene. The usual time of onset of the disease is between 35 and 45; by that time, individuals with the disease have usually had children, who must then spend their lives waiting to see whether or not the disease will occur. The disease is chronic and progressive and usually causes death 10 to 20 years after onset.

MULTIPLE ALLELE INHERITANCE: BLOOD GROUPS

Although a single individual can have only two possible alleles of any single gene, many alleles can exist within a population. Probably the most familiar characteristic in human beings that is determined by a single group of alleles is the ABO blood series. The existence of blood groups was discovered in 1900 by Karl Landsteiner. Mixing samples of blood taken from members of his laboratory staff, Landsteiner found that sometimes the red blood cells would clump together, or agglutinate, and sometimes they would not. From these experiments, he developed the theory that there were different categories of blood and that agglutination was caused by mixing blood of different categories. Soon after, the four major blood groups were determined: A, B, AB, and O. Persons who belong to blood group A, it was found, can receive blood from donors of blood group A or blood group O. Persons who belong to blood group B can receive blood from persons with blood group B or O. Persons with AB blood can receive blood from anyone but can give blood only to other persons with AB blood. Persons with blood type O can safely donate blood to anyone but can receive only type O blood. (These data are summarized in Figure 18-5.)

This account is beginning to sound like one of those logic puzzles found in *Scientific American*. However, it is far simpler than any of those. A, B, and O are all alleles of the same gene, and all affect the surface of the red blood cell. Individuals with allele A have a specific polysaccharide called A on the surface of their blood

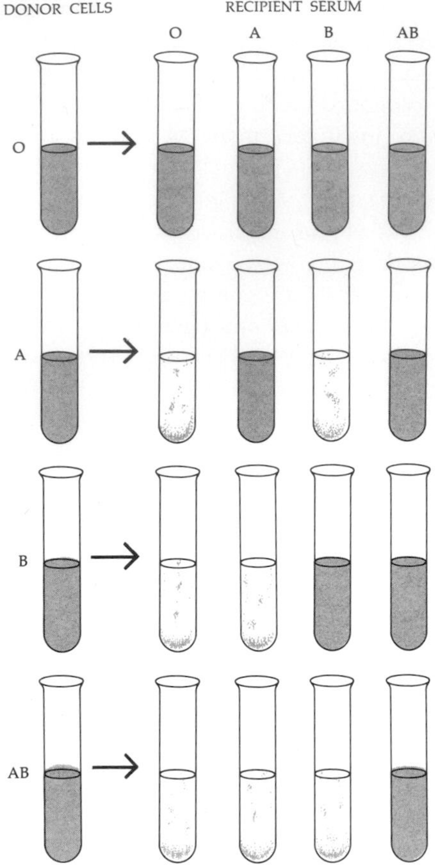

18–5
Severe and sometimes fatal reactions can occur following transfusions of blood of a different type from the recipient's. These reactions are the result of agglutination of the blood cells caused by antibodies present in the recipient's serum. Blood-group reactions to transfusions can be demonstrated equally well in test tubes, as shown here. The blood that is shown agglutinating has natural antibodies against the donor blood. Persons with type O blood used to be called universal donors and those with type AB blood, universal recipients. Now other factors are checked as well.

Table 18–1 *Blood Groups*

PHENOTYPE (POLYSACCHARIDES MADE)	GENOTYPE	REACTION WITH ANTIBODIES		ANTIBODIES IN BLOOD PLASMA
		ANTIBODY A	ANTIBODY B	
O	O/O	–	–	Antibody A, antibody B
A	A/A, O/A	+	–	Antibody B
B	B/B, O/B	–	+	Antibody A
AB	A/B	+	+	None

cells. Individuals with B have another polysaccharide, called B. ABs have both polysaccharides. Individuals with O have neither. If you receive a transfusion of blood cells containing foreign polysaccharide—that is, one not present in your own body—you will have an immune reaction against it. It is the immune reaction that causes the agglutination of the blood cells.

The genetic basis of the ABO blood types is shown in Table 18–1. A and B are codominants; the effects of one are not masked by the effects of the other.

If you have AB blood, it means that one of your parents is A or AB and the other is B or AB. If you have A blood, it means that you inherited a gene for A from one parent and either a gene for A or a gene for O from the other. If you have O-type blood, you must have had parents who each carried a gene for O, although they may have been, phenotypically, either A or B.

In the famous Charlie Chaplin paternity case in the 1940s, the baby's blood was B, the mother's A, and Chaplin's O. If you had been the judge, how would you have decided the case?*

On the basis of the ABO groups alone, it can never be proved that someone *is* the father of a particular child, but, as you can see in Table 18–2, it is possible to prove that someone could *not* be the father.

* As a matter of fact, Chaplin was judged guilty. Blood-group data are not admitted as evidence by some states in cases of disputed parentage.

Table 18–2 *Inheritance of the ABO Blood Types*

PHENOTYPES OF PARENTS	CHILDREN POSSIBLE	CHILDREN NOT POSSIBLE
A × A	A, O	AB, B
A × B	A, B, AB, O	–
A × AB	A, B, AB	O
A × O	A, O	AB, B
B × B	B, O	A, AB
B × AB	A, B, AB	O
B × O	B, O	A, AB
AB × AB	A, B, AB	O
AB × O	A, B	O, AB
O × O	O	A, B, AB

18–6
Color blindness in humans is determined by a recessive allele on the X chromosome. In the chart, the mother has inherited one normal and one defective allele. The normal allele will be dominant, and she will have normal color vision. However, half her eggs (on the average) will carry the defective allele and half will carry the normal allele—and it is a matter of chance which kind is fertilized. Since her husband's Y chromosome, the one that determines a son rather than a daughter, carries no gene for color discrimination, the single gene the wife contributes (even though it is a recessive gene) will determine whether or not the son is color-blind. Therefore, half her sons (on the average) will be color-blind. Assuming that her children marry individuals with X chromosomes with the normal alleles, the expected distribution of the trait among her grandchildren will be as shown on the chart.

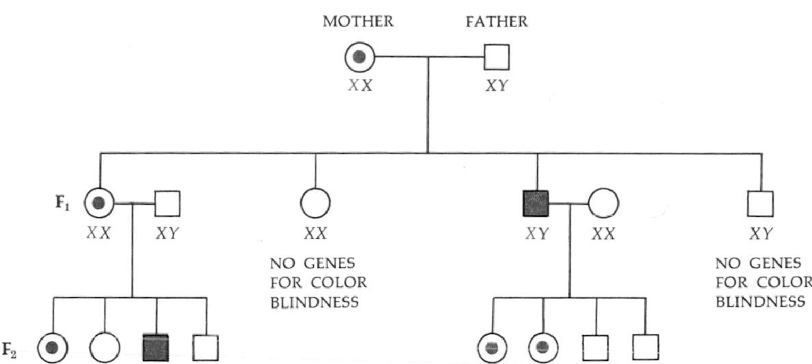

Persons with O-type blood used to be known as universal donors and those with AB-type blood as universal recipients. However, a number of additional blood factors have been found since Landsteiner's time, and they are also checked in modern blood banks to determine blood compatibility.

Discovery of the ABO blood groups, which has made blood transfusions safe and practical, ranks among the great medical advances of the twentieth century.

SEX-LINKED CHARACTERISTICS

Color Blindness

As in *Drosophila*, the *Y* chromosome of a human male carries less genetic information than the *X* chromosome. Genes for color vision, for example, are carried on the *X* chromosome in humans but not on the *Y* chromosome. Color blindness is produced by a recessive allele of the normal gene. The allele for complete color vision is dominant; a woman with one X chromosome with the normal allele and one X chromosome with the allele for color blindness will have normal color vision. If she transmits the X chromosome with the recessive allele to a daughter, the daughter will also have normal color vision if she receives an X chromosome with the normal allele from her father (that is, if he is not color-blind). If, however, the X chromosome with the recessive allele is transmitted from mother to son, he will be color-blind since, lacking a second X chromosome, he has only the recessive allele (Figure 18–6). Nonsexual characteristics, such as color blindness, that are controlled by alleles on the X chromosome are said to be *sex-linked*.

18-7

As this pedigree shows, Queen Victoria was the original carrier of the allele for hemophilia that has afflicted male members of the royal families of Europe since the nineteenth century. The British royal family escaped the disease because King Edward VII, and consequently all his progeny, did not inherit the defective gene.

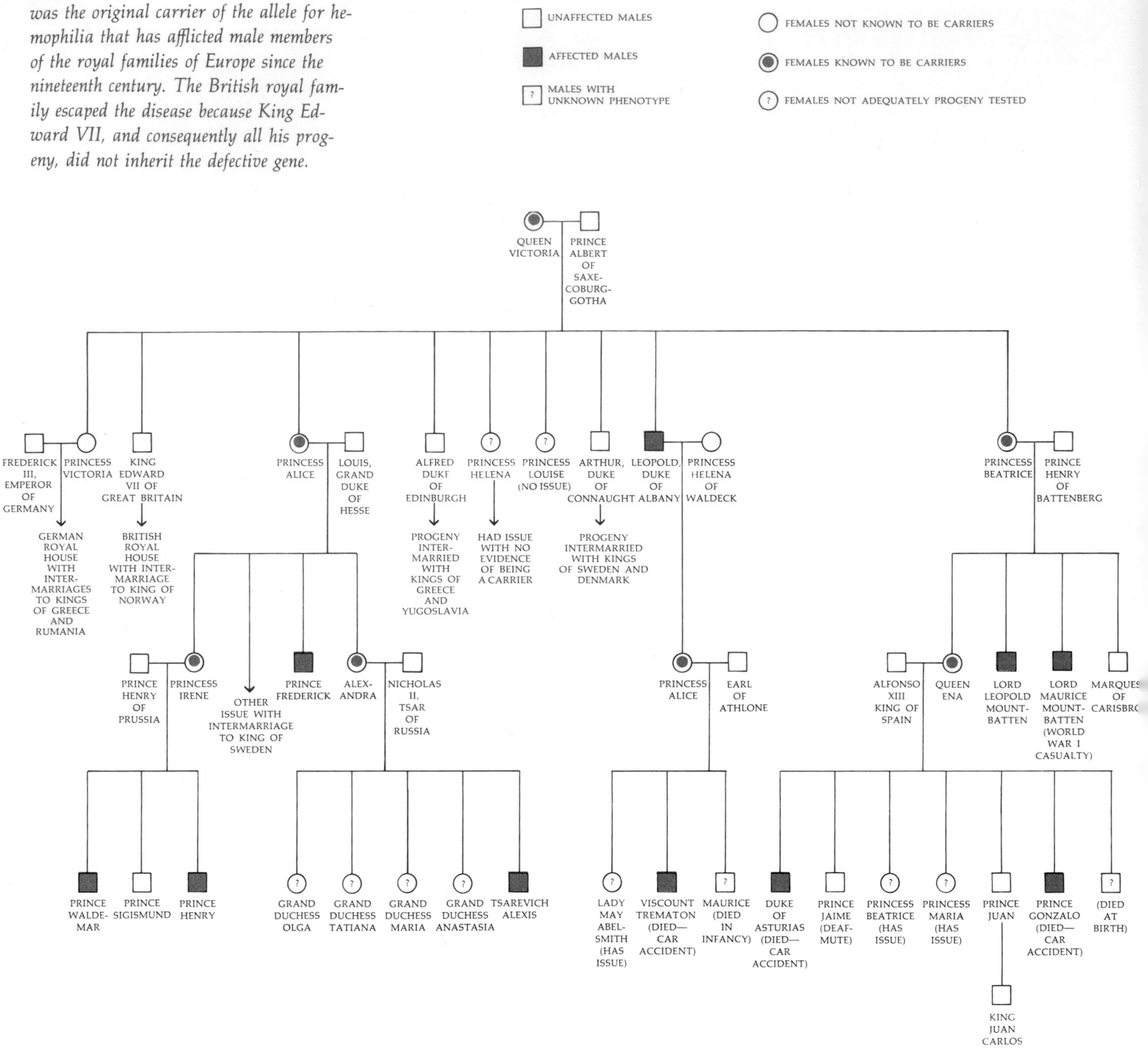

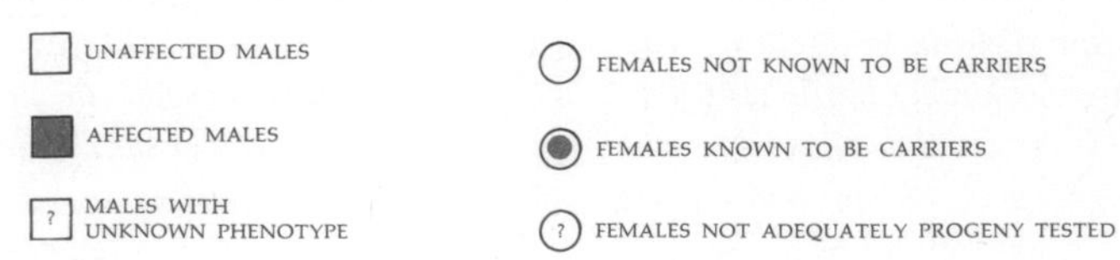

UNAFFECTED MALES
AFFECTED MALES
MALES WITH UNKNOWN PHENOTYPE
FEMALES NOT KNOWN TO BE CARRIERS
FEMALES KNOWN TO BE CARRIERS
FEMALES NOT ADEQUATELY PROGENY TESTED
QUEEN VICTORIA
PRINCE ALBERT OF SAXE-COBURG-GOTHA
FREDERICK III, EMPEROR OF GERMANY
PRINCESS VICTORIA
KING EDWARD VII OF GREAT BRITAIN
PRINCESS ALICE
LOUIS, GRAND DUKE OF HESSE
ALFRED DUKE OF EDINBURGH
PRINCESS HELENA
PRINCESS LOUISE (NO ISSUE)
ARTHUR, DUKE OF CONNAUGHT
LEOPOLD, DUKE OF ALBANY
PRINCESS HELENA OF WALDECK
PRINCESS BEATRICE
PRINCE HENRY OF BATTENBERG
GERMAN ROYAL HOUSE WITH INTER-MARRIAGES TO KINGS OF GREECE AND RUMANIA
BRITISH ROYAL HOUSE WITH INTER-MARRIAGE TO KING OF NORWAY
PROGENY INTER-MARRIED WITH KINGS OF GREECE AND YUGOSLAVIA
HAD ISSUE WITH NO EVIDENCE OF BEING A CARRIER
PROGENY INTERMARRIED WITH KINGS OF SWEDEN AND DENMARK
PRINCE HENRY OF PRUSSIA
PRINCESS IRENE
OTHER ISSUE WITH INTERMARRIAGE TO KING OF SWEDEN
PRINCE FREDERICK
ALEX-ANDRA
NICHOLAS II, TSAR OF RUSSIA
PRINCESS ALICE
EARL OF ATHLONE
ALFONSO XIII KING OF SPAIN
QUEEN ENA
LORD LEOPOLD MOUNT-BATTEN
LORD MAURICE MOUNT-BATTEN (WORLD WAR I CASUALTY)
MARQUES OF CARISBRC
PRINCE WALDE-MAR
PRINCE SIGISMUND
PRINCE HENRY
GRAND DUCHESS OLGA
GRAND DUCHESS TATIANA
GRAND DUCHESS MARIA
GRAND DUCHESS ANASTASIA
TSAREVICH ALEXIS
LADY MAY ABEL-SMITH (HAS ISSUE)
VISCOUNT TREMATON (DIED—CAR ACCIDENT)
MAURICE (DIED IN INFANCY)
DUKE OF ASTURIAS (DIED—CAR ACCIDENT)
PRINCE JAIME (DEAF-MUTE)
PRINCESS BEATRICE (HAS ISSUE)
PRINCESS MARIA (HAS ISSUE)
PRINCE JUAN
PRINCE GONZALO (DIED—CAR ACCIDENT)
(DIED AT BIRTH)
KING JUAN CARLOS

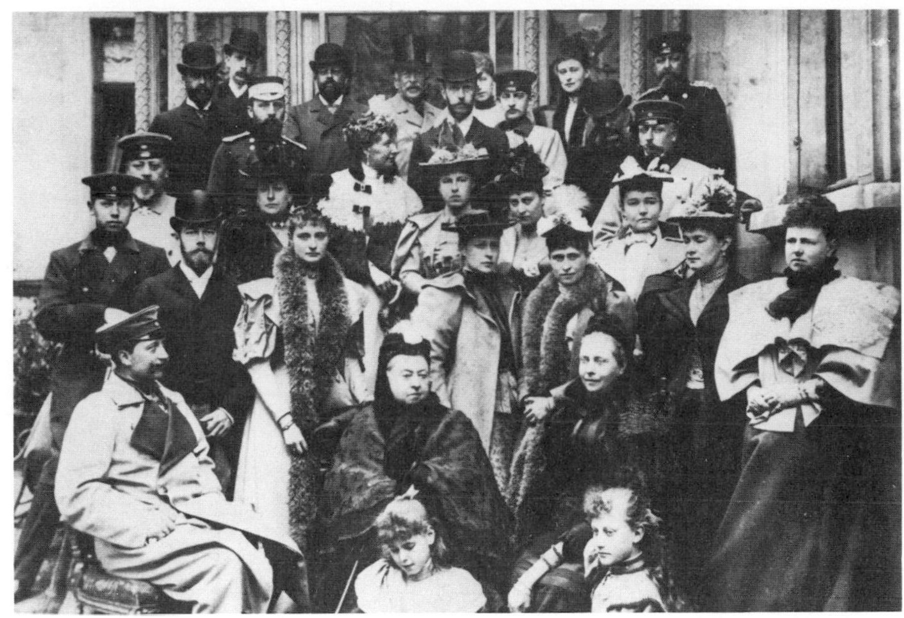

18–8
Queen Victoria (seated center) and some of her immediate family. Seventeen of the people in this photograph, which was taken in 1894, are her direct descendants. These include Princess Irene of Prussia, standing to the right of Victoria and wearing a feather boa, and, to the left of Victoria, Alexandra (also wearing a boa), the future Tsarina of Russia. Nicholas II, to become the last Tsar of Russia, is standing beside Alexandra. Both Irene and Alexandra were carriers of hemophilia.

18–9
Gregory Efimovitch Rasputin (1871–1916). Nicholas II and Alexandra of Russia fell under the sinister influence of Rasputin because of the hemophilia of their son Alexis, on whom Rasputin exerted apparently mystical healing powers. The gene for hemophilia had been inherited from Queen Victoria. Alexis did not die of his disease, but was executed with other members of the royal family in 1918.

Hemophilia

A classic example of a recessive allele transmitted on the *X* chromosome is the hemophilia that has afflicted some royal families of Europe since the nineteenth century. Hemophilia is a group of diseases in which the blood does not clot normally. In some kinds of hemophilia, even minor injuries carry the risk of the patient's bleeding to death. Queen Victoria was probably the original carrier in her family (Figure 18–7). Because none of her forebears or collateral relatives was affected, we conclude that the mutation occurred on an *X* chromosome in one of her parents or in the cell line from which her own eggs were formed. Prince Albert, Victoria's consort, could not have been responsible because male-to-male inheritance of the disease is impossible. (Why?) One of her sons, Leopold, Duke of Albany, died of hemophilia at the age of 31. At least two of Victoria's daughters were carriers, since a number of *their* descendants were hemophiliacs. And so, through various intermarriages, the disease spread from throne to throne across Europe. In the Tsarevitch, son of the last Tsar of Russia, the gene for hemophilia, inherited from Victoria, had considerable political consequences.

CHROMOSOME ABNORMALITIES

Certain genetic diseases are caused by gross abnormalities in the number or structure of chromosomes. For example, from time to time, usually because of "mistakes" at the time of meiosis, homologues may not separate. In this case, one of the sex cells has one too many chromosomes and the other, one too few. This phenomenon is known as *nondisjunction.* The cell with one too few (unless it is a *Y*) cannot produce a viable embryo, but the cell with one too many sometimes can. The result is an individual with an extra chromosome in every cell of his or her body.

PREPARATION OF A KARYOTYPE

Chromosome typing for the identification of hereditary defects is being carried out at an increasing number of genetic counseling centers throughout the United States. The result of the procedure is known as a karyotype. The chromosomes shown in a karyotype are metaphase chromosomes, each consisting of two sister chromatids held together at their centromeres. White blood cells in the process of dividing have been interrupted at metaphase by the addition of colchicine, which prevents the subsequent steps of mitosis from taking place. After treating and staining, the chromosomes are photographed, enlarged, cut out, and arranged according to size. Certain abnormalities, such as an extra chromosome or piece of a chromosome, can be detected.

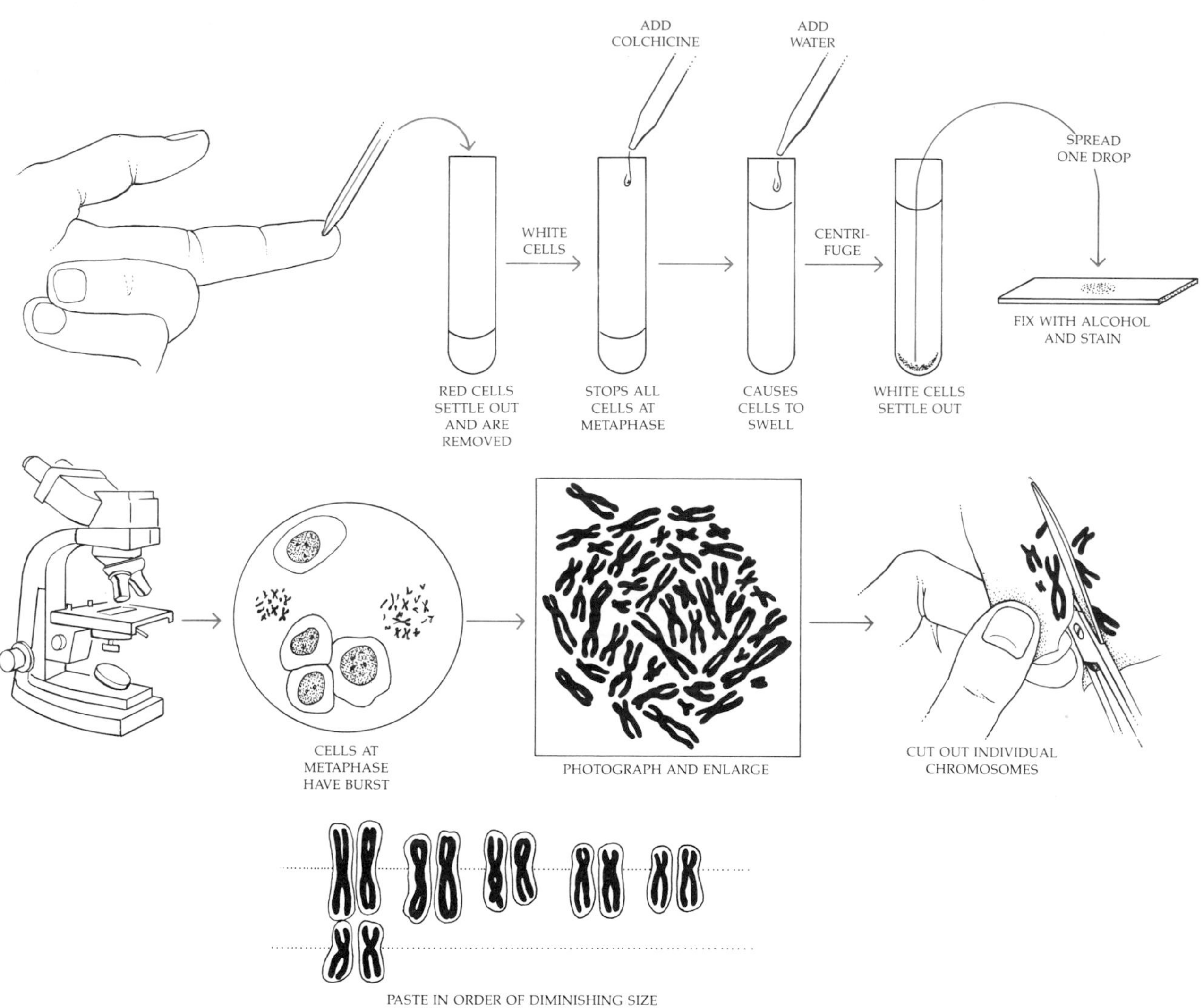

The presence of additional chromosomes often produces widespread abnormalities. Many infants with such abnormalities are stillborn. Among those who survive, many are mentally deficient. In fact, studies of abnormalities in human chromosome number among living subjects are often carried out on patients in mental hospitals. These patients frequently have abnormalities of the heart and other organs as well.

Down's Syndrome

One of the most familiar chromosomal abnormalities is the form of mental deficiency known as Down's syndrome, after the physician who first described it. (The disease is also sometimes called mongolism, a name that derives from the characteristic appearance of the eyefold in these patients, which makes them look "foreign," or mongoloid, to Europeans.) Down's syndrome usually involves more than one defect and so is referred to as a syndrome, a group of disorders that occur together. The syndrome includes, in most cases, not only mental deficiency but also a short, stocky body type with a thick neck and, often, abnormalities of other organs, especially the heart.

In about 95 percent of the cases of Down's syndrome, the cause of the genetic abnormality is nondisjunction of chromosome 21. This results in an extra chromosome 21 (Figure 18–10b) in the cells of the defective child.

18–10
(a) *The normal diploid chromosome number of a human being is 46, 22 pairs of autosomes and the 2 sex chromosomes. In the karyotype below, the autosomes are grouped by size* (A, B, C, *etc.*), *and then the probable homologues are paired. A normal woman has two* X *chromosomes and a normal man, shown here, an* X *and a* Y. (b) *The karyotype of a male patient with Down's syndrome caused by nondisjunction. Note that there are three chromosomes 21.*

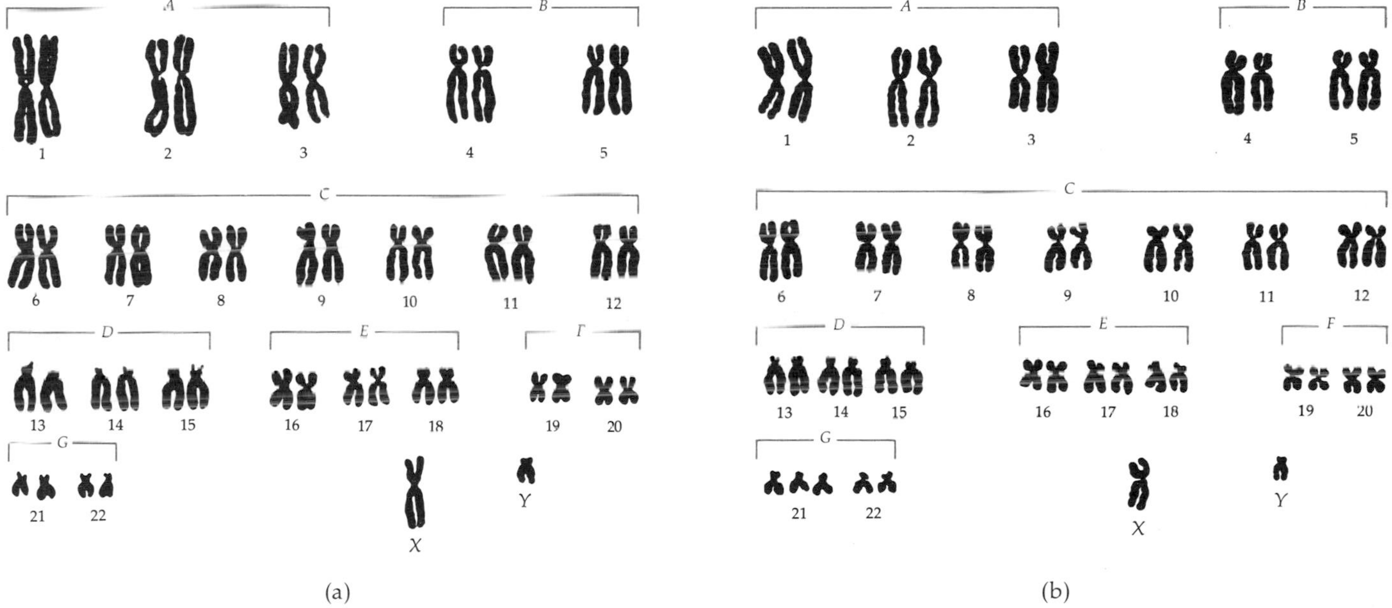

18–11
Transmission of translocation mongolism. The father, top row, has normal pairs of chromosomes 21 and 15, and each of his sperm cells will contain a normal 21 and a normal 15. The mother (more frequently, although not always, the translocation carrier) has one normal 15, one normal 21, and a translocation 15/21. She herself appears normal, but her chromosomes cannot pair normally at meiosis. There are six possibilities for the offspring of these parents; the infant will (1) die before birth (three of the six possibilities), (2) be mongoloid, (3) be a translocation carrier like the mother, or (4) be normal. Tests for the chromosomal abnormality can be made in prospective parents and in the fetus before birth.

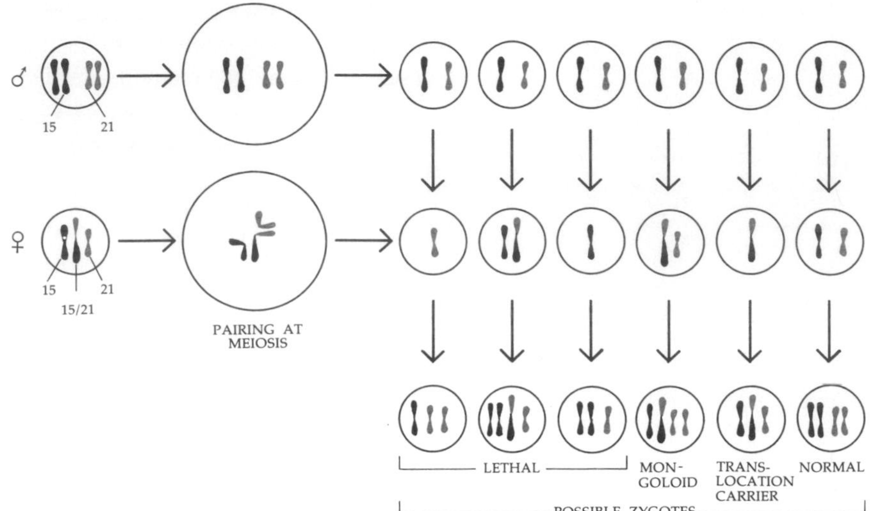

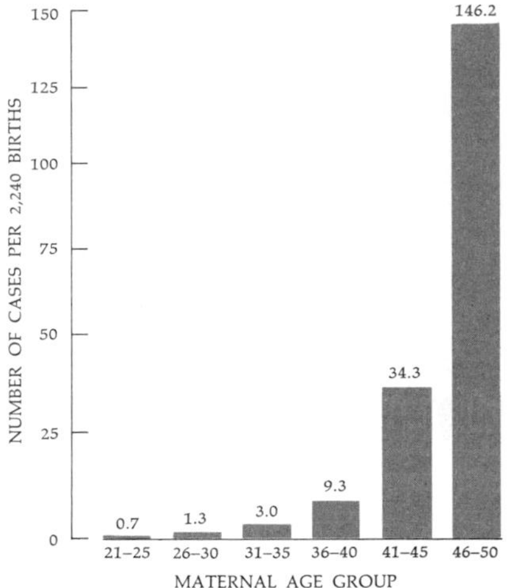

18–12
The frequencies of births of infants with Down's syndrome in relation to the ages of the mothers. The number of cases shown for each age group represents the occurrence of Down's syndrome in every 2,240 births by mothers in that group. As you can see, the risk of having a child with Down's syndrome increases rapidly after the mother's age exceeds 40.

Down's syndrome may also be the result of an abnormality, known as a *translocation,* in the chromosomes of one of the parents. Translocation occurs when a portion of a chromosome is broken off and becomes attached to another chromosome. The patient with translocation mongolism usually has a third chromosome 21 (or, at least, most of it) attached to a larger chromosome, such as 15. (Thus, in both cases, the patient has three chromosomes 21, or their equivalent.)

When cases of Down's syndrome due to translocation are studied, it is usually found that one parent, although phenotypically normal, has only 45 separate chromosomes—one chromosome being composed of most of chromosomes 15 and 21 joined together. The possible genetic makeups of the offspring of this parent are diagrammed in Figure 18–11. Three out of the six possible combinations are lethal. One of the remaining three will produce Down's syndrome, one will be normal, and one will be a carrier.

Thus parents who have had a child with Down's syndrome are advised to have their karyotypes prepared. If either parent shows an abnormality, they are warned that they may transmit the condition. If the karyotypes of both parents are normal, they do not run a greater-than-average risk of having another congenitally defective child. Down's syndrome and a number of other defects involving nondisjunction are more likely to occur among infants born to older women (Figure 18–12). The reasons for this are not known, but the formation of the egg cells is well under way in the human female before she is born, so the increasing incidence of abnormalities may be correlated in some way with the aging of the mother's reproductive cells.

Abnormalities in the Sex Chromosomes

Nondisjunction may also produce individuals with extra sex chromosomes. An *XY* combination in the twenty-third pair, as you know, produces maleness, but so do *XXY*, *XXXY*, and even *XXXXY*. These latter males are usually sexually underde-

QUESTIONS DISCUSSED AT A MEETING OF JURISTS AND PHYSICIANS ON "ETHICAL ISSUES IN GENETIC COUNSELING AND THE USE OF GENETIC KNOWLEDGE"

Is a genetic counselor's responsibility first to the couple involved, or to society?

What constitutes a "defective" fetus?

Does an unborn baby have "rights"—including even the right not to be conceived (by cloning) as an identical twin of many others, or of his father? Can a parent ethically give "consent" on behalf of an unborn child—for gene manipulation, for example?

What will be the effect on society, the family, and the individual himself of being able to choose in advance which sex a child will be?

Mass prenatal screening can detect some 130 biochemical abnormalities. Who should do such screening? Which defects should have priority in the search? Should it be voluntary or compulsory?

Who should have access to the information from prenatal tests? How should it be used?

Could a child with genetic defects arising from a rubella infection early in pregnancy sue his mother's obstetrician—or his parents—for "wrongful life" in failing to abort him?

What public policy can and should be worked out to prevent abuses of current and future advances in genetics?

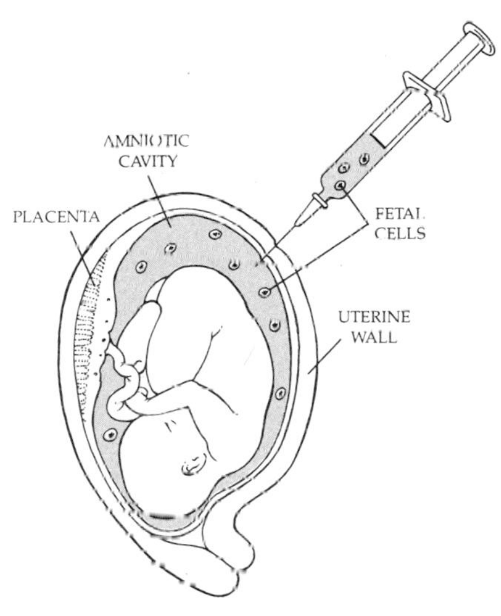

18–13
Amniocentesis. A sharp-pointed needle is inserted into the amniotic cavity, and fluid containing cells from the fetus is withdrawn into a syringe. The cells can then be analyzed for genetic defects. The procedure cannot be done until the fourth month of pregnancy.

veloped and sterile, however. *XXX* combinations sometimes produce normal females, but many of the *XXX* women and all *XO* women (women with only one *X* chromosome) are sterile. Many individuals with sex chromosome abnormalities are mentally retarded. In 1965 it was reported that there was a high incidence of males with an extra *Y* chromosome in institutions for the criminally insane. These men were said to be above average in height, to be below normal in intelligence, to have severe acne, and to be self-destructive and homicidal—"super males." Richard Speck, a mass murderer who killed eight nurses in Chicago in 1966, was reported by the media to be *XYY* and to be planning a defense based on genetic predetermination.

Recent studies, more carefully planned and carried out, have shown that *XYY* individuals are no more likely to be violent or aggressive than *XY*s or indeed than *XXY*s, and that many are of normal intelligence. Also, as it was later reported, Richard Speck, chromosomally speaking, was a normal *XY*.

AMNIOCENTESIS

A procedure known as amniocentesis is becoming a very powerful aid in diagnosing Down's syndrome, as well as Tay-Sachs disease and a variety of other genetic conditions in the fetus. A very thin needle is inserted through the mother's abdominal wall and through the membranes that enclose the fetus, and a sample of the amniotic fluid surrounding the fetus is withdrawn. While the procedure must be done with great care, it is simple, quick, and relatively harmless. This fluid will contain living cells sloughed off by the embryo. These cells, grown in tissue culture, can be tested for various enzymes and can provide mitotic cells from which a karyotype can be made.

There is no treatment for conditions such as Down's syndrome or Tay-Sachs disease, however. Amniocentesis merely provides the information on which the parents may base a decision as to whether to continue or terminate the pregnancy.

THE SOCIAL COST OF GENETIC DISEASE

As a conservative estimate, some 120,000 infants in the United States are born each year with genetic diseases. The financial cost to society is tremendous. The emotional cost is, of course, incalculable. At present, our capacity to reduce these costs is limited.

PKU is an example of a genetic disease that can be treated. If it is detected early and the child is kept on a carefully controlled diet during the first six years of life, he or she will develop into a normal adult. The only method of detection presently available is to routinely test the urine or blood of all newborn infants for abnormally high levels of phenylalanine. Although this is a very costly procedure, it is in fact now being done in most hospitals nationwide. Is it worth the cost? For this particular disease, it can be shown that it is indeed financially worth the expense, quite apart from humanitarian considerations.

A "cost-benefit" analysis shows that the costs of not *diagnosing and treating PKU children are greater than those of providing them with treatment. The test costs $1.25 per child. That comes to about $4,000,000 to test the 3,000,000 children born each year in the United States. Since the incidence of PKU is 1 in 15,000 births, the testing should uncover about 200 afflicted children each year who can then receive the special diet. The cost of administering the diet is about $1,000 a year per child. The cost of treatment is thus $1,200,000 (200 children at $1,000 a year for six years). The*

This little girl, age 1½, has phenylketonuria (PKU), which is caused by a defect in an enzyme that breaks down the amino acid phenylalanine. She has just had a blood test, part of a treatment program that includes a diet low in phenylalanine.

NATURE AND NURTURE

A little more than half a century ago, a leading American psychologist, John Watson, threw out a challenge: "Give me a dozen healthy infants, well-formed, and my own specified world to bring them up in, and I'll guarantee to take any one at random and train him to become any type of specialist I might select—doctor, lawyer, merchant, chief, and yes, even beggarman and thief, regardless of his talents, penchants, tendencies, abilities, vocations, and the race of his ancestors."*

The pendulum now seems to have swung entirely in the other direction, and no longer, in the popular view, is an infant completely plastic, capable of being molded into any shape by his or her environment. Rather, a member of the human species is seen as an untidy collection of behavioral souvenirs, acquired during our ascendancy from the apes (or perhaps even on the long route from the amoeba).

This change in attitude has little to do with studies actually carried out in humans, but is rather a product of a new and extremely interesting discipline known as ethology. Ethology is the study of animal behavior under natural conditions. Ethologists are particularly concerned with species differences in behavior

* J. B. Watson, *Behaviorism,* Peoples Institute, New York, 1925, p. 82.

total annual cost of this program nationwide thus comes to $5,200,000. This is a considerable sum to spend on 200 children, but what would be the cost to society if the program were not carried out? Each of those 200 children would have to be maintained in a mental institution for an average of approximately 30 years at a cost of about $5,000 per child per year, or a grand total of $30,000,000 per year. This simple type of cost-benefit analysis shows that carrying out the PKU program nationwide, while costing $5,200,000, actually saves almost $25,000,000 a year. Thus, for PKU at least, the effort is clearly worth the cost, and the same is probably true for other genetic disorders as well.

With other genetic diseases, it is possible to routinely screen prospective parents to detect "carriers." Such screening is now being carried out almost entirely on a voluntary basis, and the decisions made on the basis of the results are private ones. Few would argue against the expansion of such programs and of encouraging members of any "at risk" groups to take advantage of them. Sickle cell anemia, some types of Down's syndrome, and Tay-Sachs disease can be detected in the carriers, but, unfortunately, many hereditary diseases cannot.

For some of these diseases, amniocentesis can be used to detect afflicted fetuses. It is now becoming possible not only to detect conditions involving gross chromosomal abnormalities, such as Down's syndrome, but also diseases based on enzyme deficiencies, such as Tay-Sachs. However, the only "treatment" for diseases diagnosed by amniocentesis is abortion. Financial assistance is available in many states for women seeking abortion in order to avoid the birth of a handicapped child. In financial terms, such programs are undoubtedly cost effective, given the cost to society of caring, for example, for a person with Down's syndrome for a lifetime. Even those who would personally choose abortion under such circumstances would find it abhorrent, however, to insist—or even suggest—that all women elect this alternative.

Finally, and most tragic, are those diseases for which virtually nothing can be done at this time. In the case of Huntington's disease, for example, early detection is not possible by any known means and so it follows that it is impossible to predict who will or who will not transmit the disease. The costs of long-term care for Huntington's disease patients are so high that health agency personnel sometimes advise the healthy husband or wife to get a divorce so that the patient can become a ward of the state and so become eligible for medical benefits. There is no "cost benefit" here, to society, the patient, or the family. What are our responsibilities? Even were biologists able to trace the origins of the disease to faulty nucleotides, biology cannot answer these social and moral questions.

and their evolutionary origin (as contrasted, for example, with studying a solitary pigeon shut up in a box).

In animals, some patterns of behavior are clearly genetically determined. A spider, for instance, although raised in solitary confinement, can build an extremely complex web on its first try. Among vertebrates, many species of fish and birds have elaborate courtship and territory-defending rituals, whose basic patterns are part of their genetic equipment—that is, instinctive rather than learned.

Mammals also show inherited behavior. For example, the first time a flying squirrel raised in a bare laboratory cage isolated from other squirrels is given nuts or nutlike objects, it will "bury" them in the bare floor, making scratching movements as if to dig out the earth, pushing the nuts down into the "hole" with its nose, covering them over with imaginary earth, and stamping on them. Another example is domestic animals, which are bred for personality traits and disposition as well as for physical characteristics.

Is man naturally aggressive? Does he inherit genes for territoriality? Are some races more intelligent than others? At this time, very, very little is known about the answers to these questions. Schizophrenia has recently been shown to have a strong genetic basis, as indicated by studies of siblings separated at birth (by being adopted into different families). These studies revealed that if one sibling devel-

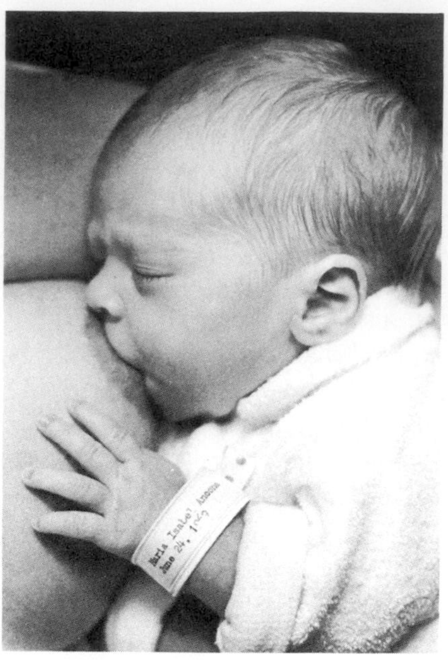

18–14
Very few patterns of human behavior are clearly based solely on instinct. One such, of obvious survival value, is the suckling response, present in infants at birth.

oped schizophrenia, the other was far more likely to be schizophrenic than an unrelated person of the same age. Some forms of mental deficiency have been associated with chromosomal abnormalities, as previously noted, and with genetic "errors," as in cases of PKU. Identical twins separated at birth have been found to have IQs more like one another than like other members of the household in which they are raised. However, recent studies purporting to show genetically based differences in IQ between races or ethnic groups have been under attack, largely because of environmental inequities or the cultural bias of the test itself. The single factor that correlates most consistently with low IQ is low birth weight—a subject explored further in Chapter 33.

Beyond such scraps of evidence as we have cited, there are very few data. The nature of one's nature is, of course, an interesting subject and generates argument and discussions in which anyone can participate, since, even if special knowledge is required, it is not available. Consequently, at present, arguments about these issues must be regarded as social, ethical, and political, rather than as scientific. However, because they have reverberating consequences, they may not be regarded as trivial.

SUMMARY

Our information about human genetic traits is based largely on observations accumulated in the course of diagnosis, treatment, and research involving conditions of medical concern.

Some human diseases are due to simple Mendelian recessives. In such diseases, the genetic trait appears in the phenotype of the offspring only if both parents carry the recessive allele. In cases in which both parents are heterozygotes, the probabilities are that, of four children, one will be homozygous for the recessive allele and so phenotypically abnormal, two will be heterozygotes, and one will be genotypically and phenotypically normal for the trait under study. PKU, Tay-Sachs disease, and sickle cell anemia are due to Mendelian recessives in the homozygous state. The biochemical nature of these diseases is known.

In diseases that are inherited as Mendelian dominants, such as Huntington's disease, there is a fifty-fifty chance that the child of any afflicted individual will develop the disease.

Some conditions, such as the ABO blood groups, are the result of multiple alleles of the same single gene.

Humans have two sex chromosomes (*X* and *Y*) and 22 pairs of autosomes. For study, the chromosomes are arranged in a karyotype.

In humans (and all other mammals), the *Y* chromosome determines maleness. The *Y* chromosome carries fewer genes than the *X* chromosome. As a consequence, if a male receives an *X* chromosome carrying recessive alleles, those alleles (which in the female would be dominated by the normal dominant alleles) will usually be expressed. Characteristics resulting from the expression of such genes are said to be sex-linked, or *X*-linked. Color blindness and hemophilia are examples of sex-linked characteristics that are carried by females but are seen chiefly among males.

Nondisjunction is the failure of two homologues to separate at the time of meiosis. As a consequence of nondisjunction, children may be born with an extra chromosome. The presence of an extra chromosome can produce abnormalities such as Down's syndrome. The presence of extra sex chromosomes is often but not always associated with sterility and mental retardation.

QUESTIONS

1. Construct a four-generation pedigree for single-factor inheritance involving a dominant gene.

2. Construct a similar pedigree for inheritance of ABO blood groups.

3. PKU, phenylketonuria, is a disease caused by the coming together of two recessive genes. If two healthy parents have a child with PKU, what are their genotypes with respect to PKU? What are their chances of having another child with the same disease?

4. If two healthy parents have a child with sickle cell anemia, what are their genotypes with respect to this allele? Having had one such child, what are their chances of having another child with the same disease?

5. Albinism results from the failure to make the pigment melanin, which is made from tyrosine. Like other enzyme deficiency diseases, it is transmitted as a recessive allele. In 1952 it was reported that two albinos, who had met at a school for the partially sighted, had married and had three children, all of whom had normal pigmentation. Assuming the children are not illegitimate, how do you explain that the children are normal? (*Hint:* Consider the possibility that more than one enzymatic step is required in the biosynthesis of melanin.)

6. If you are of blood type O and neither of your parents is, what is the probability that your brother is also of type O?

7. In a paternity suit, the baby has type O blood and the mother has type A blood. The man alleged to be the father has type B blood. His defense lawyer maintains that this man could not be the father of the child. Is the lawyer correct? Explain your answer.

8. Why isn't male-to-male inheritance of color blindness possible? Under what conditions would color blindness be found in a woman? If she married a man who was not color-blind, would her sons be color-blind? Her daughters?

9. A woman whose maternal grandfather was hemophiliac has parents who seem to be normal. She too seems normal, as does her husband. What are the chances that her first son will be normal?

10. Nondisjunction can occur at the first meiotic division or the second. How do the effects differ?

11. Describe the two types of chromosomal abnormality that can cause Down's syndrome. With which type is it possible to identify prospective parents who are at a higher-than-average risk? How is this done?

12. How would you answer the questions in the essay on page 311? Who do you think should determine such policies?

CHAPTER 19

Genetic Change: Spontaneous and Induced

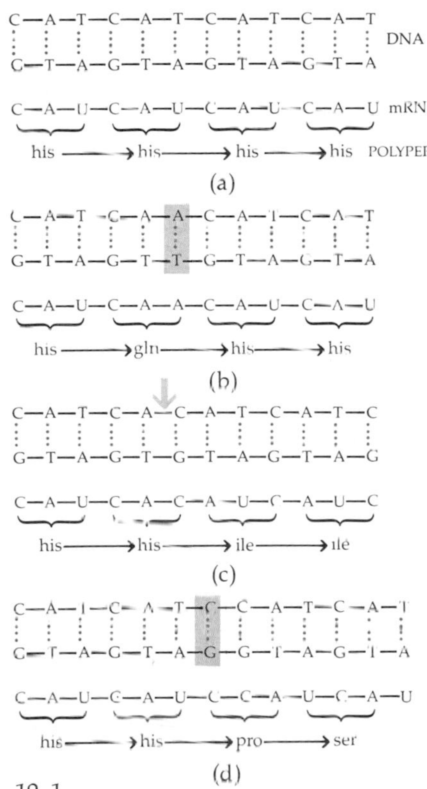

19–1
Types of point mutations: (a) *the original DNA molecule, the mRNA transcribed from it, and the polypeptide produced;* (b) *substitution of one base pair for another;* (c) *deletion of a base pair; and* (d) *addition of a base pair. You might wish to refer back to the mRNA code on page 283.*

In this chapter, we shall be concerned with changes in the genetic material. These include spontaneous changes (meaning, in this context, those that occur without human intervention), changes occurring as by-products of human activities, and, finally, changes produced deliberately. Broadly speaking, genetic changes take place in two ways: (1) mutation and (2) recombination.

MUTATION

De Vries, some 80 years ago, defined mutation in terms of traits appearing in the phenotype. The current definition is somewhat different: A mutation is a change in the sequence or number of nucleotides in a nucleic acid that can be transmitted from parent to offspring.

Mutations can occur in asexually reproducing cells of prokaryotes or eukaryotes. They can take place in the germ cell line, that is, in eggs or sperm, or the cells that give rise to them. Mutations may also occur in somatic cells, and there is some evidence that such mutations are a principal cause of cancer.

TYPES OF MUTATIONS

Mutations are often characterized as *point mutations*, those involving only a small number of base pairs, and *chromosomal aberrations*, those involving relatively large segments of a chromosome or chromosomes.

Point Mutations

Three kinds of changes can occur in a nucleotide sequence: (1) substitution of one base for another, (2) deletion of one or more bases, and (3) insertion of new bases (Figure 19–1). Substitutions involving even a single nucleotide may have large consequences. For example, a substitution may lead to the substitution of an amino acid with different properties than the one originally called for, as in hemoglobin, or the substitution of a stop signal leading to premature termination of a polypeptide chain.

Deletions and insertions of one or more nucleotides may lead to subtractions or additions of particular amino acids in the corresponding protein. However, since such deletions and insertions of nucleotides will probably not take place in threes, they are more likely to lead to frame shifts and so to the production of a totally new and almost certainly useless polypeptide.

Base substitutions can mutate in both directions, from mutant to normal as well as from normal to mutant (depending on which you have decided is which). However, additions or subtractions are far less likely to revert.

Chromosomal Aberrations

Abrupt changes in the genetic makeup of an organism also occur as a result of alterations involving comparatively large (sometimes, indeed, microscopically visible) alterations in chromosomes. Pieces of chromosomes may break off and be lost. They may become attached to other, nonhomologous chromosomes (translocation). If two breaks occur in the same chromosome, a broken piece may be removed, turned around, and replaced "backwards." This type of chromosomal aberration is called an *inversion*.

EFFECTS OF MUTATIONS

It is not surprising that mutations are usually detrimental to the organism and that, in fact, many are lethal. If one were to change a word at random in a Shakespearean sonnet or a wire at random in a television set, an improvement would be unlikely, and the results might well be disastrous. It is reasonable to assume that the genetic makeup of an individual is as precise in its engineering and as delicate in its balance as a sonnet or a television set. (Consider, for example, the complexity of structure of the overlapping genes in ϕX174 discussed on page 284.) However, as a result of mutations, there is a wide range of variability in a natural population, and, in a complex or shifting environment, a variation might give an individual or its progeny a slight edge. A simple example would be a series of base substitutions in a single enzyme that produced a series of variant enzymes that functioned more efficiently at one temperature than at another.

MUTATION RATES

Spontaneous mutation rates are low. For instance, in *E. coli,* in the course of cell division, 1 cell out of every 10^9 will mutate from streptomycin sensitivity to streptomycin resistance. In *Neurospora,* among asexual spores, 4 in every 10^8 will gain the capacity to synthesize adenine. Among fruit flies, about 4 sperm cells in every 10^5 will carry a mutation for red eye to white eye, or vice versa. In a study involving 7 loci and 688,921 male mice, the average mutation rate per locus was 8.1 per million (10^6) male gametes. In human beings, mutations for Huntington's disease occur in about 5 out of every 10^6 gametes. Note that in all these examples, the mutations that are being counted are those detectable in the phenotype. Many individual nucleotide substitutions, for example, would be overlooked.

CAUSES OF MUTATIONS

Most spontaneous mutations, it is believed, are due to mistakes in the processes of DNA replication and crossing over, and, at the chromosomal level, mistakes during

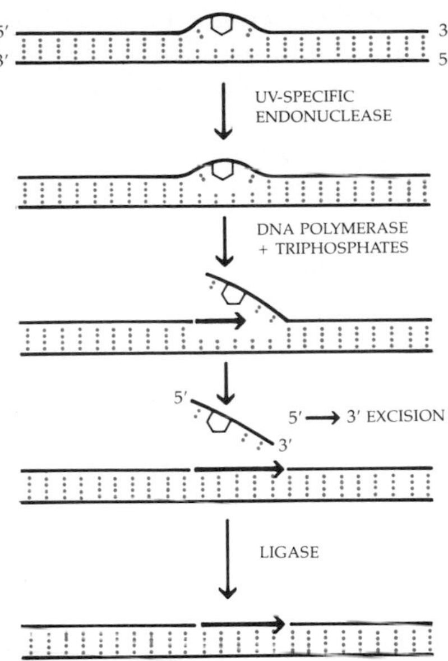

19–2

Ultraviolet light produces bonding between adjacent thymine nucleotides, distorting the DNA double helix. Repair enzymes—endonucleases—clip the DNA on the 5′ side of the distortion. New bases are inserted in the gap by another enzyme, DNA polymerase. A second repair enzyme then clips the DNA on the 3′ side of the distortion. And, finally, the new DNA patch is joined to the rest of the DNA strand by the enzyme ligase.

mitosis and meiosis. Just as the alleles themselves are under selection pressures, the spontaneous mutation rate in a given species is determined by evolutionary pressures. Too high a rate will endanger an organism's capacity to produce viable progeny, whereas too low a rate will prevent the appearance of new variants that can compete successfully in a varied or changing environment.

MUTAGENS

Mutagens are agents that increase the natural rate of mutations. Some mutagens, such as ultraviolet radiation and radioactivity from isotopes present in the earth, are part of the environment in which life evolved. It is roughly estimated that these known factors may account for 10 percent of spontaneous mutations. Other mutagens, particularly certain chemicals, have been introduced into the environment as a consequence of human activities and are, rightfully, a cause of current alarm.

Mutagenic Effects of Ultraviolet Light

We shall begin with a discussion of the mutagenic effect of ultraviolet light because it is among the best understood. Ultraviolet light increases the frequency of mutation in bacterial cells. One of the ways in which it damages DNA is by causing two adjacent thymine molecules to become covalently bonded to one another in what is known as a thymine dimer, producing a structural distortion in the DNA that may cause chromosome breaks.

Special enzymes have been isolated that repair those distortions, literally excising and replacing them (Figure 19–2). Apparently the effect of low exposure to ultraviolet can be minimized by these repair enzymes.

Ultraviolet exposure also causes cancer; the frequent development of skin cancers in people exposed over long periods of time to sunlight is well documented. Xeroderma pigmentosum is inherited, like phenylketonuria (PKU) and Tay-Sachs disease, as a homozygous recessive. It is associated with extreme sensitivity to sunlight and the development of multiple cancers. It has recently been discovered that this disease is due to the deficiency of one of the enzymes participating in the excision of the thymine dimer previously described.

Ionizing Radiations

X-radiation was the first mutagenic agent discovered; in the 1920s, H. J. Muller, working with fruit flies, showed that exposure to x-rays greatly increased the frequency of mutations as revealed by breeding experiments. Now it is known that a wide range of radiations are mutagenic, including cosmic rays, x rays, the emissions of radioactive isotopes, and, as we mentioned previously, ultraviolet light.

The primary chemical effect of x-radiation is the absorption of energy by an atom and the consequent release of electrons to produce an ion, thus the term ionizing radiation. Ionizing radiation is normally measured in roentgens (r), after the discoverer of x-rays, W. C. Roentgen. A roentgen is the amount of radiation that will produce 1.6×10^{12} ion pairs in 1 cubic centimeter of water. A dose of 100 r to a typical cell will produce about 10,000 ion pairs, with the result that about 1 percent of the protein and nucleic acid will be ionized. The instability of ions results in chemical reactions that may either directly affect the DNA or the molecules involved in its replication or repair.

19–3
An x-ray taken by W. C. Roentgen of the hand of his wife. Many of the early workers with x-rays developed cancers on their hands as a consequence of careless exposures. Roentgen, however, perhaps recognizing the possible hazards, was more cautious.

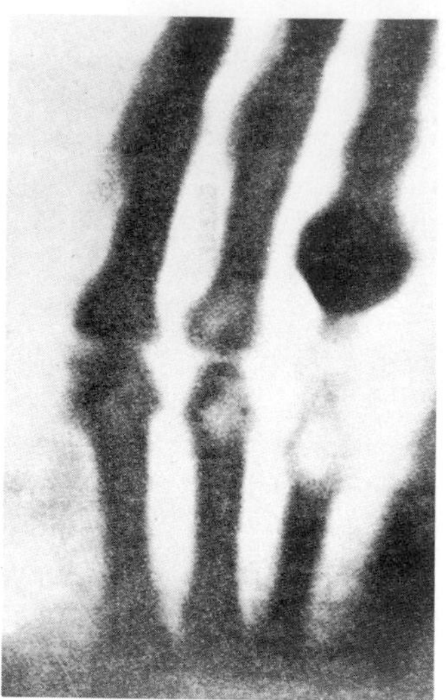

Ionizing radiation is ubiquitous. We are all exposed to some radiation from cosmic rays and naturally occurring radioisotopes. Furthermore, x-rays are used extensively for medical diagnostic work (chest x-rays, for example, and dental checkups) and also therapeutically for the destruction of cancer cells that cannot safely be removed through surgery. The spreading use of atomic power brings large amounts of radioactive isotopes into our communities. What is a safe level of exposure? The frequency of mutations or of chromosome breaks induced by radiation is directly and linearly proportional to the amount of radiation exposure (Figure 19–4). In other words, with regard to the genetic effects of radiation there is *no* safe level of exposure. That is, there is no threshold level of radiation below which no damage is caused to the genetic material.

Radiation and Cancer

It has long been known that exposure to ionizing radiation increases the incidence of cancer. Many of the early workers with x-rays developed cancers on their hands and other exposed parts of their bodies. Among the first cases studied of cancers due to environmental exposure were workers, mostly women, who put luminescent paint containing small amounts of radium on watch dials, pointing their brushes with their lips. The radioactive material accumulated in their bones and caused bone cancers. The incidence of leukemia among the "survivors" of the bombing of Hiroshima was 1 in 60 after 12 years, an increase of almost fifty-fold. Recently, a larger-than-expected number of women have developed breast cancer after x-ray screening procedures to detect breast cancer. Also, larger-than-expected numbers of thyroid cancers have been detected among persons receiving radiation to the head and face for dermatological conditions.

These considerations were important factors in bringing about the cessation in the 1950s of surface testing of nuclear weapons, which released high levels of radioisotopes in the atmosphere. They also led to the use of lower levels of radiation in diagnostic x-rays and to limitations on the use of x-rays for the treatment of trivial conditions, such as acne or warts.

Chemical Mutagens

A number of chemicals may also act as mutagens, as shown by studies in bacteria. One class of mutagens, which includes nitrous acid, acts directly on the bases to cause deamination, the removal of a nitrogen atom. Adenine deaminated, for instance, behaves like guanine so that when the DNA is next replicated, a cytosine will occur opposite the altered base instead of the thymine that should be there. One of the reasons for concern over the safety of the food preservative sodium nitrite (found in hot dogs, bacon, and luncheon meats) is its possible conversion to nitrous acid in the stomach.

19–4
The relationship between radiation dosage (measured in roentgens) and mutation rate, as indicated by studies in Drosophila.

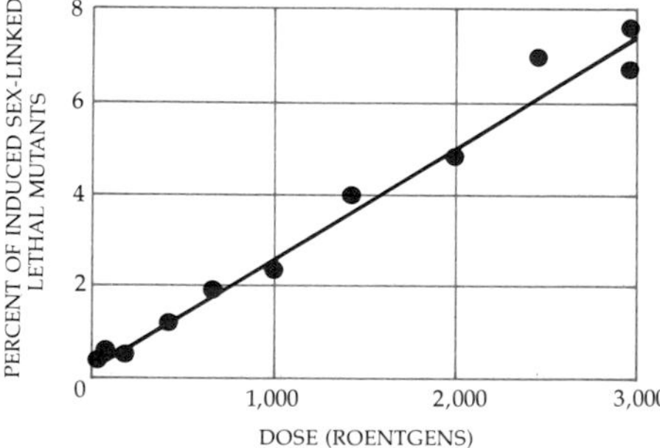

Another class of chemicals, called alkylating agents, reacts mainly with guanine, causing it to be released from DNA and, usually, replaced with another base. Mustard gas (nitrogen mustard), which was developed during World War I and was the first chemical mutagen to be discovered, acts in this way. Many insecticides (the organophosphates) are derived from nitrogen mustards.

A third group of mutagens leads to the presence of bases that become incorporated into the DNA in place of the normal purine or pyrimidine. One such analog is shown in Figure 19–6. 5-Bromouracil (5BU) is an unnatural pyrimidine that is similar in structure to thymine except that it has a bromine atom attached to

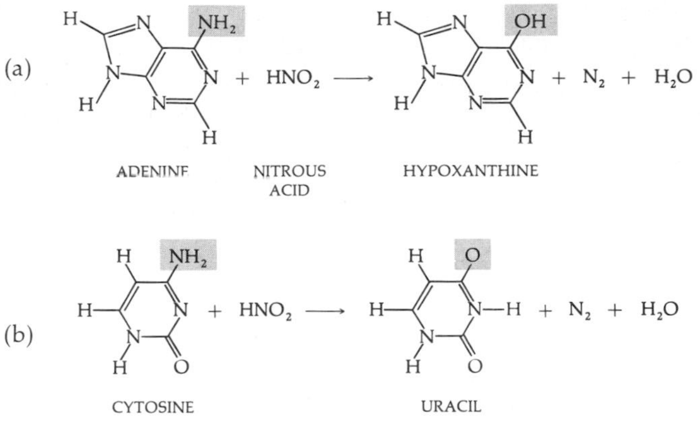

19-5
(a) When adenine is deaminated by nitrous acid, the product is hypoxanthine, which forms three hydrogen bonds as does guanine. When the DNA strand replicates, cytosine—and not thymine—pairs with the altered base. (b) Similarly, deamination of cytosine produces uracil, which behaves like thymine. When the DNA strand replicates, adenine—and not guanine—pairs with the altered base.

(a) ADENINE + NITROUS ACID (HNO_2) → HYPOXANTHINE + N_2 + H_2O

(b) CYTOSINE + HNO_2 → URACIL + N_2 + H_2O

THYMINE

5-BROMOURACIL

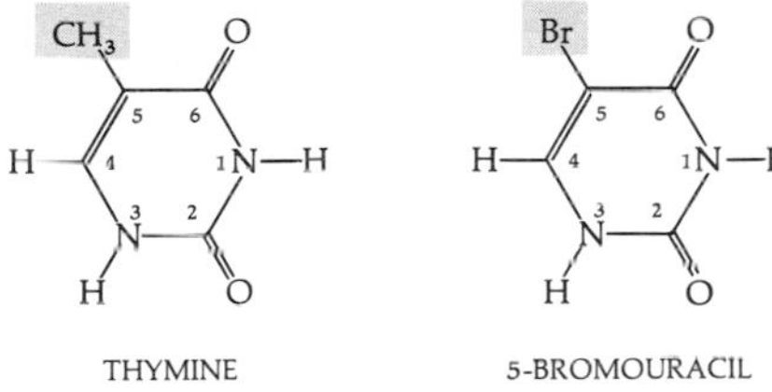

19-6
5-Bromouracil is similar in structure to thymine except for the replacement of the methyl (CH_3) group in position 5 with a bromine atom. When 5-bromouracil is incorporated into the DNA molecule in place of thymine, it may pair with guanine rather than adenine at the next replication of the molecule.

the ring rather than a methyl group. The cell is "tricked" by 5-bromouracil, which it will treat as a normal pyrimidine and incorporate into the DNA. When incorporated as thymine (resulting in an A-5BU base pair), the 5-bromouracil may, at the next replication of the DNA, accept guanine as its base-pairing partner rather than adenine. This will in further generations result in a permanent base-pair change from A-T to A-5BU to G-5BU to G-C. If the 5-bromouracil is incorporated as a pairing partner of guanine, the reverse change will take place, G-C to G-5BU to A-5BU to A-T.

Many of the chemicals found to be mutagenic in bacteria have been found to be carcinogens (cancer-causing agents). Because bacterial assays can be carried out much more rapidly and inexpensively than tests of carcinogenicity (which take many months, numerous laboratory animals, and expensive facilities), they are now being used in routine screening of the many thousands of chemicals in our environment.

GENETIC RECOMBINATIONS

Changes in the genotype of cells can also be produced by the insertion of new DNA containing new information into a recipient cell. New genetic information is, of course, inserted into a cell every time fertilization takes place. In previous chapters, we have mentioned three other ways that such genetic transfers can take place in nature: (1) the release of DNA (transforming factor) into the environment, where it can be taken up by other cells; (2) viral infection, in which DNA (or RNA) is injected into a host cell; and (3) bacterial conjugation. Here we shall describe a fourth, because it is pertinent to the discussions that follow.

TRANSDUCTION

Transduction is the transfer of genes from one cell to another by means of viruses. It has been observed with bacterial viruses, which, as we noted in Chapter 15, are composed of nucleic acid wrapped in a protein coat. Transduction occurs when some of the nucleic acid of the host cell becomes wrapped in the protein coat of a virus and is carried to a new host cell where it is incorporated by recombination into the host-cell chromosome.

Two forms of transduction are known: general and restricted. General transduction occurs when fragments of the DNA of a bacterial cell, broken down in the course of infection, are caught up in a bacteriophage head and are carried to a new host cell. These particles carry little or no viral DNA and so they cannot set up an active infection in the recipient cell. Instead, the portion of the bacterial chromosome carried into the new host may become a part of the host chromosome.

Restricted transduction occurs only with certain types of bacterial viruses. These bacteriophages do not necessarily set up an infective cycle when they enter a host cell but may instead exist as a molecule of naked DNA in the host cell—in other words, a plasmid. Like the F factor and some other plasmids, they may become part of the cell chromosome (Figure 19–7). They are then called *prophages* or *proviruses* and are duplicated faithfully each time the bacterial cell divides.

From time to time the relationship between a bacterial cell and its resident prophage may shift, and the viral DNA breaks loose from the chromosome and sets up an infective cycle. Exposure to x-rays, ultraviolet light, or certain chemicals (all generally the same agents that cause mutations) increases the rate of activation of viruses in host bacteria. Bacteria that harbor prophages are known as *lysogenic bacteria*. When a lysogenic bacterium releases a burst of newly synthesized viruses, the viruses infect and lyse other nearby bacterial cells.

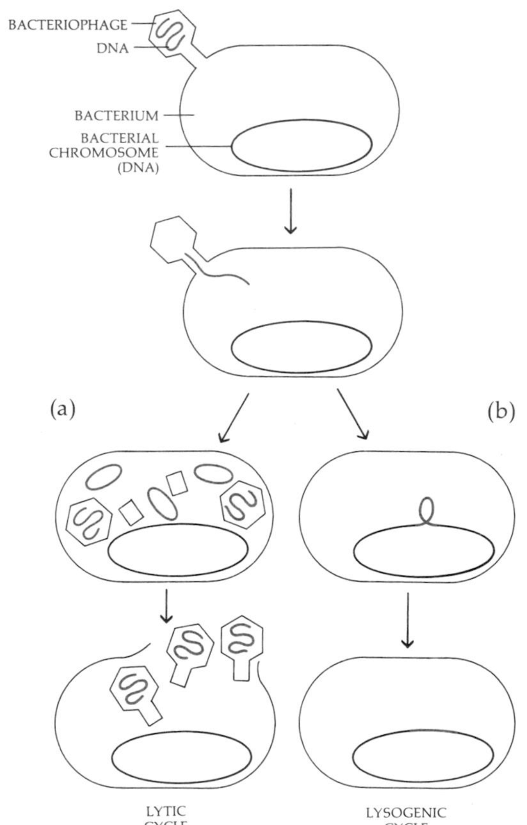

19–7
When certain types of bacterial viruses infect bacterial cells, one of two events may occur. (a) *The virus DNA may enter the cell and set up an infection, such as we described in Chapter 15, or* (b) *the DNA of the virus may simply lie latent in the cell. In the latter case it may become part of the bacterial chromosome, replicating with it. From time to time, such a virus becomes activated and sets up a new infective cycle. Bacteria harboring such viruses are known as lysogenic.*

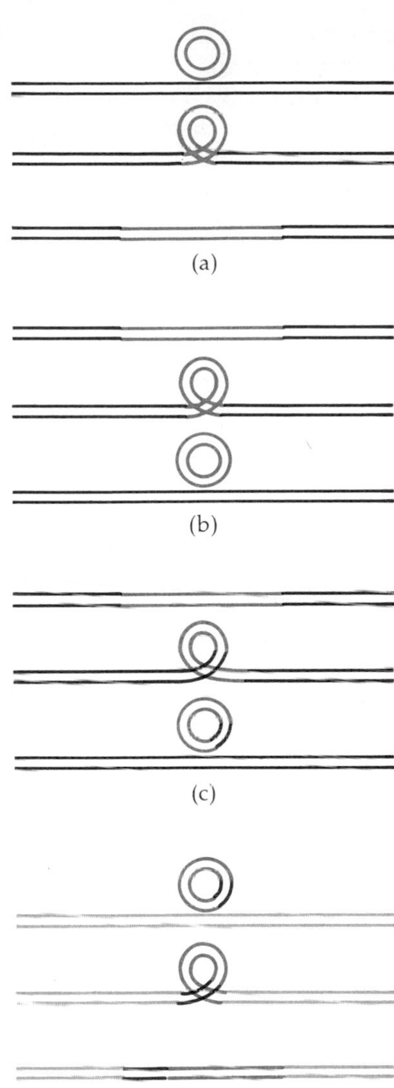

19–8
(a) *Like the F factor, a virus may become part of a bacterial chromosome.* (b) *As a result of x-rays, chemical treatment, or "spontaneously," the virus may break loose from the chromosome, setting up a new infectious cycle.* (c) *Sometimes the virus takes a fragment of adjacent host chromosome with it. This fragment may be large enough to include several host genes.* (d) *When and if the virus combines with the chromosome of a new host cell, the cell acquires the new genetic information. This process is known as transduction.*

When the viruses leave a host bacterial cell chromosome, they sometimes take a fragment of the chromosome with them (Figure 19–8). This fragment is replicated with them through any infectious cycle. If one of these viruses becomes a part of the chromosome of a new host, the genes from the previous host may be inserted into the new host's chromosome and become a part of its genetic equipment. For instance, the bacteriophage lambda can carry the bacterial genes that produce the enzymes concerned with the breakdown of galactose—in other words, the galactose operon. When a lambda provirus carrying the galactose genes becomes a provirus in a bacterial cell that cannot synthesize one or more of these enzymes, the infected cell gains the capacity to utilize galactose for growth.

It has long been suspected that certain viruses can enter into relationships with eukaryotic cells similar to those relationships between the lysogenic bacteria and their viruses. Herpes simplex, the virus that causes fever blisters, is an example. If you suffer from fever blisters or know someone who does, you know that the lesions tend to break out repeatedly in the same person and that they characteristically occur in times of stress, such as when that person has a fever or eats certain food or gets a sunburn (ultraviolet exposure). It is believed that certain cells constantly harbor the herpes simplex virus, that it multiplies when the cells multiply, and that it causes cell damage—the fever blister—only when it is activated.

GENES, VIRUSES, AND CANCER

Cancer is a disease, or group of diseases, in which particular cells in the body cease to respond to whatever controls growth under normal conditions. They multiply autonomously—crowding out, invading, and destroying other tissues. There have been many theories about the causes of cancer. One of the older hypotheses, now gaining renewed support, is that cancer is the result of a somatic mutation. A somatic mutation, as we noted previously, is simply a change in the genes of a body cell. Such a mutation would not necessarily affect the germ cells, but it would be passed on by mitosis to all the progeny of the malignant cell.

More recently, accumulating information about viruses led many scientists to believe that viruses are the cause of at least some of the forms of human cancer. One of the reasons for believing this is the fact that many different kinds of cancers in other animals are caused by viruses, including cancers in frogs and chickens and in mice, rats, and other mammals. It has been impossible to test conclusively the hypothesis that any human cancers are caused by viruses. First, the latent period between infection and the appearance of symptoms is so long that no infectious pattern emerges, as it does in influenza or measles, for example. Some of the animal viruses are passed from generation to generation—for instance, in the mother's milk—and may even skip a generation or two. Second, it is, of course, not permissible to inject human subjects with any virus or other substance suspected of causing cancer. Finally, although viruses can be detected in some electron micrographs of human cancer, there is no way to prove that such viruses are causative agents and not merely passengers.

With the discovery of viruses—and particularly of proviruses—the somatic theory of mutation and the virus theory appeared to merge. It has recently been discovered that at least one cancer-causing virus, the polyoma virus, does become part of the chromosomes of cells growing in tissue culture. When the viral DNA is inserted into the host chromosome, the cell becomes cancerous. Its offspring are

19-9
SV40 (simian virus 40), originally isolated from monkeys, has been shown to produce cancers in baby hamsters and other laboratory animals. It is small, even for a virus; each particle is only about 45 nanometers in diameter. It is composed of a molecule of DNA surrounded by an icosahedral (twenty-sided) protein coat. Although many viruses have now been shown to produce cancers in many different types of animals, there is no conclusive evidence that any human cancer is caused by a virus.

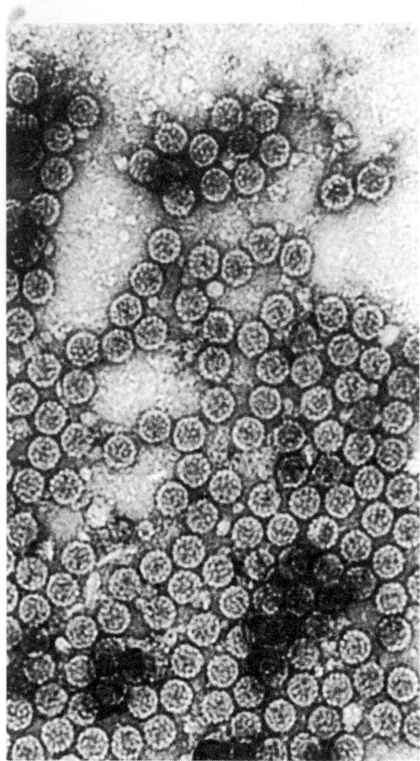

also cancerous by all criteria. They have the capacity to cause cancer in the animal or strain of animal from which the cells were originally taken.

One of the difficulties with the virus theory of cancer has been that many cancer viruses contain RNA instead of DNA. DNA viruses replicate themselves and produce mRNA, from which viral proteins are made. RNA viruses replicate themselves by producing a "negative" copy complementary to the original single RNA strand. This "negative" copy is used repeatedly to make large numbers of "positive" copies identical to the original viral RNA. Either the "positive" copies of the RNA or the "negative" complementary copies can serve as mRNA for viral protein production. But RNA cannot integrate itself into a chromosome in the way that DNA can. So the theory that cancer viruses act by becoming part of the genetic apparatus of the cell did not seem reconcilable with the existence of cancer-causing RNA viruses. Recently, this problem was resolved in a totally unexpected way: A group of virologists discovered that some RNA viruses make DNA. When these viruses infect a cell, they bring with them a new enzyme, reverse transcriptase, that makes viral DNA on the template of the viral RNA. Moreover, most of the cancer-producing RNA viruses tested so far have been found to produce this new enzyme and to make DNA, and reverse transcriptase is made by only one RNA virus not known to cause cancer.

Thus, it is at least possible that the cancer-causing RNA viruses operate by the same mechanism as the cancer-causing DNA viruses: They manufacture DNA that becomes part of the host chromosome and, as they do so, endow the host cell with the capacity to multiply, invade, and destroy.

Discovery of the cause of cancer, although a great intellectual achievement, would not necessarily mean control of the disease. We have, for instance, the example of sickle cell anemia, in which the cause is known to the last nucleotide, yet for which, at this writing, there is no known cure. Nor are there any cures for virus diseases, although some of them can be prevented very effectively before infection occurs. On the other hand, the cause of at least one very common type of cancer, cancer of the lung, is well established, but cigarette smoking has not stopped. Nevertheless, we continue to hope that such research will surely hasten the day when a control will be found for what is probably the most dreaded of human diseases.

GENETIC ENGINEERING

Transformation, bacterial conjugation, and transduction are not new discoveries. There are, however, new developments that make it possible to manipulate these phenomena to some extent. They involve principally the discovery of new enzymes that operate on DNA. One group of enzymes, known as DNA ligases, mend breaks in DNA molecules. A second group comprises the various DNA restriction enzymes, or endonucleases. Some of these enzymes cut through both strands at the same site. Others cleave the molecule by cutting through the two strands a few bases apart. This can have surprising consequences. Look again at the sequence at which Eco RI cleaves the molecule: GAATTC. Now visualize the double-stranded molecule. It would be

5′ . . . GAATTC . . . 3′
3′ . . . CTTAAG . . . 5′

19–10

An enzyme (Eco RI) has been found that opens DNA molecules at specific sites, leaving "sticky" ends exposed. These ends, consisting of TTAA and AATT sequences can rejoin or can join with any other fragment of DNA that has been cleaved by the same enzyme. Thus it is possible to splice together DNA from different sources, such as inserting a foreign gene into a bacterial plasmid, as shown here.

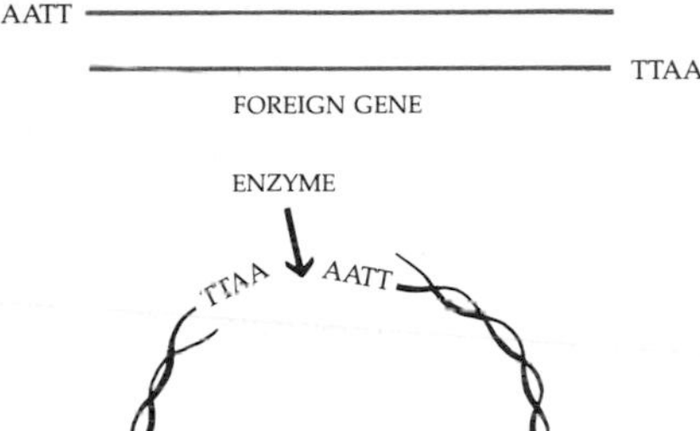

Eco RI cleaves each strand between the guanine and the adenine:

```
        ↓
5′ ... GAATTC ... 3′
3′ ... CTTAAG ... 5′
             ↑
```

What is left, as you will readily see, is

```
... G                AATTC ...
          and
... CTTAA                G ...
```

The protruding ends, . . . TTAA and AATT . . . , will thus join again with each other and also—and this is most important—with any other segment of DNA that has been cleaved by the same enzyme. Thus it is now possible to splice together segments of DNA from, apparently, a limitless variety of sources. (Such DNA molecules have been called chimeras after the mythological beast that was part lion, part goat, and part snake.) This is the work commonly referred to as "recombinant DNA" research.

These segments of DNA can be carried by plasmids (page 287). For instance, Stanley Cohen of Stanford University Medical School and a group of co-workers have isolated a small plasmid of *E. coli,* designated pSC101 (see page 290), that makes the bacteria resistant to the antibiotic tetracycline. pSC101 has only one GAATTC sequence in the entire molecule and, as a consequence, is cleaved at only one spot by Eco RI. These investigators have shown that the insertion of an additional segment of a DNA molecule into the plasmid (as shown in Figure 19–10) does not affect either the uptake of the plasmid by *E. coli,* its capacity to make the recipient cells tetracycline-resistant, or its ability to replicate.

The Possibilities

What can be done with the new technology? Some researchers, including Robert Sinsheimer of the University of California, Santa Cruz, have suggested that viruses might be tailor-made to carry particular genes to particular cells and so to cure diseases such as PKU or sickle cell anemia. Others pointed out that if, for example, the DNA segments that code for the polypeptide chains of insulin could be inserted into a bacterial plasmid, one would have a low-cost, virtually limitless source of insulin. At present, millions of diabetics depend on daily injections of insulin. The molecule is too complex to synthesize commercially. It is presently extracted from the pancreatic glands of pigs and cattle slaughtered for food, but the demand, due to the increasing number of diabetics, is threatening to exceed the supply. Moreover, many patients are allergic to these nonhuman proteins. Perhaps most important and intriguing of all is the possibility of being able to isolate a eukaryotic gene and study it outside of the enormously complex eukaryotic cell and in the well-understood and relatively simple *E. coli* host. A new field of research is now open to molecular biology. Another particularly important practical application might be the production of agriculturally important plants with combinations of new characteristics—such as the capacity both to carry out C_4 photosynthesis (page 187) and to fix nitrogen (page 351). Some investigators are working with four strains of the bacterium *Pseudomonas,* each of which has a capacity to break down a component of fuel oil, and are attempting to combine them genetically into one super oil-eater that, in powdered form, can simply be sprinkled on oil spills.

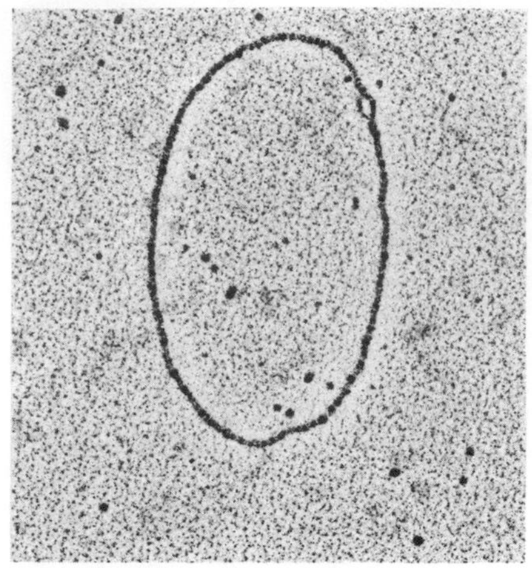

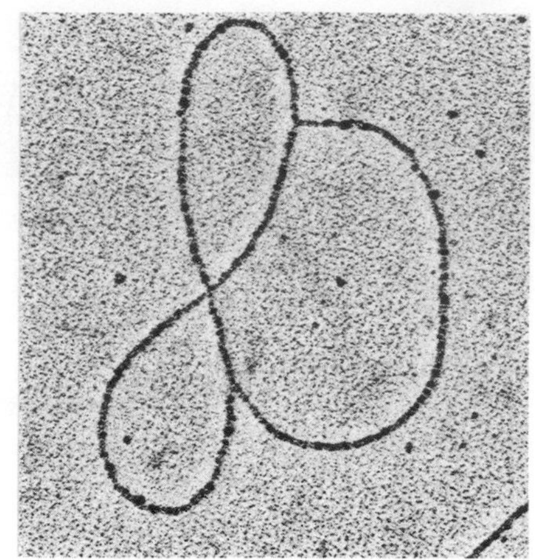

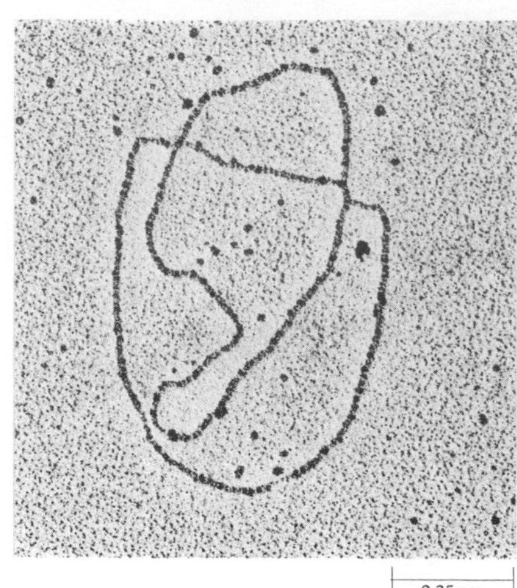

19–11
A plasmid replicating. At almost 2:00 in the chromosome in the micrograph at the left, you can see the two DNA double helices beginning to separate (each consisting, as you will remember, of one old strand and one new one). In the center micrograph, the replication process is more than half completed, and in the micrograph at the right, the two plasmids are almost at the point of separation.

However, despite the possibilities and the enthusiasm they engendered, the investigators realized that there were perhaps some hazards. *Escherichia coli* lives in the digestive tract of many animals, including human beings. Ideally, these studies should be made with a rare prokaryote, whose spread could be closely controlled. But *E. coli* is so well understood—probably the most extensively studied of any organism—that the scientists are understandably reluctant to abandon it. (Also, the notion of a rare prokaryote is a contradiction in terms; all prokaryotes multiply with great rapidity.) What would be the consequences of creating a new race of *E. coli* with the capacity to produce insulin or growth hormone or some other biologically potent protein? The fact that many plasmids confer drug resistance also bothers critics of the program. (pSC101 is so common in nature that its use is not considered dangerous, but others might well be.) Moreover, it is all very well to create a super oil-eater but alarming to think of it turned loose in the Standard Oil Refineries. Similarly, might it not be possible to produce—either accidentally or intentionally—an infectious organism of unheard-of virulence and unbreakable resistance to known antibiotic treatment?

In 1974, in a totally unprecedented move, a group of molecular biologists—all of whom were working actively in these investigations—called for a temporary ban on certain experiments involving DNA recombinations, and for almost two years the work was virtually at a standstill.

In June 1976, the National Institutes of Health issued guidelines governing research on recombinant DNA. Some kinds of experiments are prohibited, including recombinations that increase the virulence of a disease-causing organism or that might transfer drug resistance to an organism that does not acquire it naturally. Also, conditions are specified under which various categories of experiments may be carried out. These are designed to protect both laboratory personnel and the general public. For example, the *E. coli* used in these experiments will be "disabled"; that is, they will lack the capacity to synthesize their cell walls, for example, or to make thymine, and so will be able to survive only in an enriched laboratory medium, like the *Neurospora* strains of Beadle and Tatum.

Now, on the basis of additional experience, most molecular biologists believe that qualified investigators can carry out recombinant DNA research with little or no hazard to themselves or to the public at large.

19–12

How a bacterial plasmid was made to produce the hormone somatostatin.

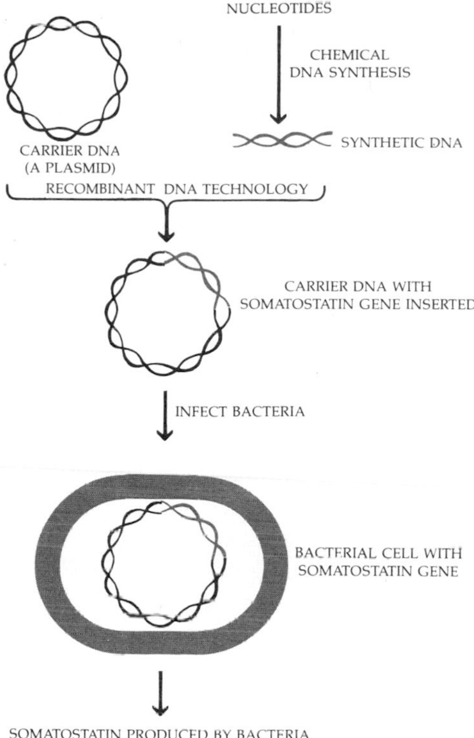

The Realities

We shall close this discussion by describing the work of one group of investigators, Keiichi Itakura and his associates at the City of Hope Medical Center. In 1976, the group constructed a plasmid containing the lactose operator gene, which, you will recall, is the on-off switch for the lactose operon (page 292). The next obvious step was to choose a gene for the operator gene to turn on and off. They selected the gene for the mammalian hormone somatostatin because (1) it was a small protein (only 14 amino acids), (2) it could be detected in very small amounts, and (3) it was a potentially useful compound.

The amino acid sequence of somatostatin is alanine-glycine-cysteine-lysine-asparagine-phenylalanine-phenylalanine-tryptophan-lysine-threonine-phenylalanine-threonine-serine-cysteine. Knowing this sequence, the investigators determined what the sequence of nucleotides in the DNA should be (you can do this, too) and synthesized the gene, bonding together the nucleotides one by one (you could not do this; it is technically very difficult and required 6 months). They spliced the somatostatin gene into the beta-galactosidase gene; beta-galactosidase is the first protein of the lactose operon (page 290). At the point of the splice, for reasons we shall see shortly, they inserted the codon for methionine. This plasmid, naked DNA, was then supplied to *E. coli* cells, following essentially the same techniques used by Avery many years ago (page 257). Only a few bacteria were transformed, but these multiplied, the plasmid along with them, until several clones of bacteria were obtained, all making the hybrid protein. The protein was then isolated and treated with the chemical cyanogen bromide, which cleaved it exactly at the methionine insert, releasing pure somatostatin. The somatostatin, tested in laboratory animals, was found to have the biological activity of the natural hormone.

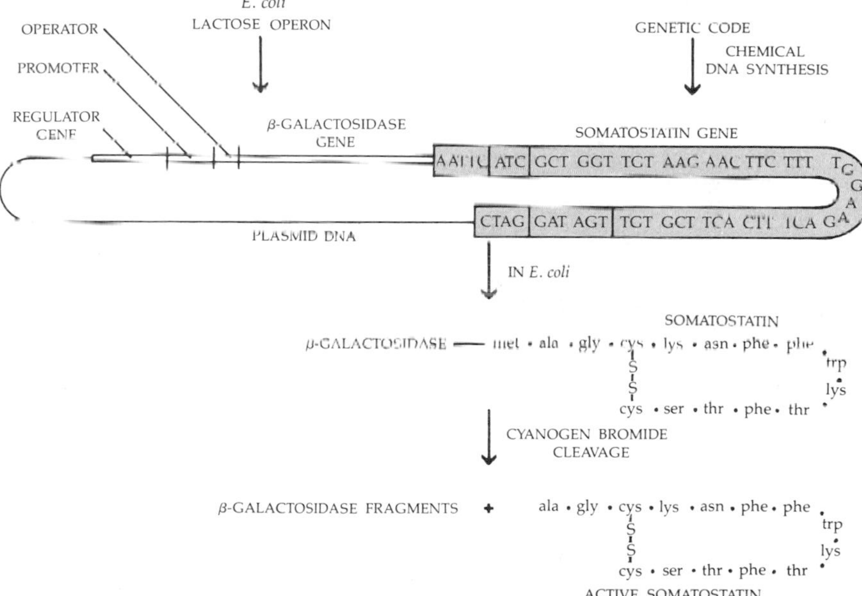

19–13

The somatostatin project. The gene for somatostatin, synthesized artificially, was fused to the beta-galactosidase gene in a bacterial plasmid. Introduced into E. coli*, the plasmid directs the synthesis of a protein that begins as beta-galactosidase but ends as somatostatin. Cyanogen bromide cleaves the protein at methionine, thus releasing the hormone.*

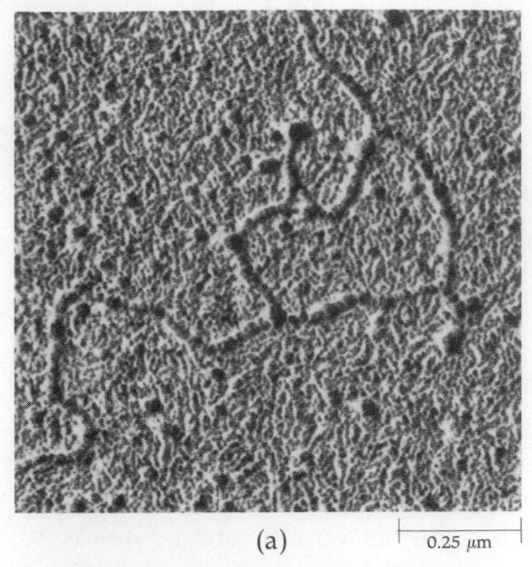

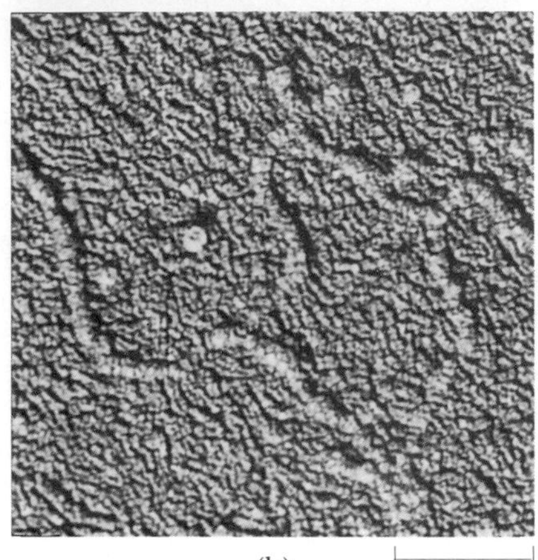

19–14
Recombinant DNA techniques have recently been used by scientists at the University of Michigan to isolate a human gene. The gene that codes for ribosomal RNA is shown here at two different magnifications and under different conditions. In each micrograph, the human gene is that region of the double-stranded DNA molecule where the strands have split apart to form a loop. The DNA extending from either side of this loop is bacteriophage DNA.

In other words, in the span of one scientific generation, molecular geneticists not only have identified the genetic material in chemical terms and demonstrated how it is replicated, transcribed, and translated, but they have also produced a gene of their own making and been able to make it function. Philip Handler, president of the National Academy of Sciences, described this "astonishing" accomplishment as "a scientific triumph of the first order." And so it is.

Most recently, these recombinant DNA techniques have made it possible to insert the gene for insulin into a plasmid. The insulin was made using the same techniques as in the somatostatin work. So far only very small amounts have been obtained, but it is a beginning.

SUMMARY

Mutations are defined as changes in the sequence or number of nucleotides in the nucleic acid of a cell or organism. Point mutations involve only a relatively few nucleotides. They may take the form of substitutions, deletions, or insertions. The latter two usually result in frame shifts. Mutations may also be in the form of chromosomal aberrations, including losses of sections of chromosomes, transposition of chromosome segments to nonhomologous chromosomes (translocation), and inversion. Spontaneous mutations are probably the result of mistakes in the replication of nucleic acids or in mitosis or meiosis. Mutagens are agents that increase the frequency of mutations. Known mutagens include ultraviolet radiation, ionizing radiations, and certain chemicals. Many mutagens are also carcinogens.

Genetic engineering is the term used to describe the introduction of new genetic material into a cell. In nature, DNA is passed between prokaryotic cells by transformation, conjugation, and transduction. The possibilities of exploiting one or more of these forms of genetic exchange has been increased by the discovery of enzymes that cleave the DNA molecule in such a way that new segments can be inserted into an existing bacterial or viral chromosome or into a plasmid (recombinant DNA). Using these new techniques developed by recombinant DNA research, a mammalian gene has been successfully incorporated into a bacterial cell.

QUESTIONS

1. Describe the possible outcomes of infection of a bacterial cell by a bacterial virus.

2. Three types of plasmid were mentioned in this chapter and Chapter 17. Describe them.

3. Obviously, experiments using a plasmid that confers resistance to an antibiotic has disadvantages. However, it also has an important advantage. What is it?

4. Given the fact that nearly all mutations are harmful, why would control systems that increase mutation rate be selected for? (If you have difficulty with this question, think in terms of an organism that produces large numbers of offspring—a pine tree, for instance, or a bacterial cell.)

SUGGESTIONS FOR FURTHER READING

Books

BODMER, WALTER F., and L. L. CAVALLI-SFORZA: *Genetics, Evolution, and Man,* W. H. Freeman and Company, San Francisco, 1976.

An excellent undergraduate text with emphasis on human genetics.

DYSON, ROBERT D.: *Cell Biology: A Molecular Approach,* 2d ed., Allyn & Bacon, Inc., Boston, 1978.

Although not primarily a genetics text, this book is a particularly good source of information on the structure of the eukaryotic chromosome and regulation of gene expression.

GOODENOUGH, URSULA: *Genetics,* 2d ed., Holt, Rinehart and Winston, Inc., New York, 1978.

An up-to-date, general introductory text with emphasis on molecular genetics.

JACOB, FRANÇOIS: *The Logic of Life: A History of Heredity,* Pantheon Books, New York, 1973.

Jacob's principal theme concerns the changes in the way people have looked at the nature of living beings. These changes, which are part of our total intellectual history, determine both the pace and direction of scientific investigation. The opening chapters are particularly brilliant.

LURIA, S. E.: *Life: The Unfinished Experiment,* Charles Scribner's Sons, New York, 1973.*

An interpretation of biology written for the layman by a Nobel laureate. A National Book Award winner.

OLBY, ROBERT: *The Path to the Double Helix,* University of Washington Press, Seattle, Wash., 1975.

An account, written by a professional historian of science, of twentieth-century genetics. Olby is interested not only in the scientific concepts and experiments but also in the various personalities involved and their effects on one another and on the course of scientific discovery.

PETERS, JAMES A. (ed.): *Classic Papers in Genetics,* Prentice-Hall, Inc., Englewood Cliffs, N.J., 1959.*

Includes papers by most of the scientists responsible for the important developments in genetics: Mendel, Sutton, Morgan, Beadle and Tatum, Watson and Crick, and so on. You should find this book very interesting; the authors are surprisingly readable, and the papers give a feeling of immediacy that no modern account can achieve.

STRICKBERGER, MONROE W.: *Genetics,* 3d ed., The Macmillan Company, New York, 1975.

An up-to date and cohesive account of the science, this book is much broader in coverage than most texts at this level.

WATSON, J. D.: *The Double Helix,* Atheneum Publishers, New York, 1968.*

"Making out" in molecular biology. A brash and lively book about how to become a Nobel laureate.

WATSON, J. D.: *Molecular Biology of the Gene,* 3d ed., The Benjamin/Cummings Publishing Company, Menlo Park, Calif., 1976.*

For the student who wants to go more deeply into the questions of molecular biology, this is a detailed and authoritative account.

WOLFE, STEPHEN: *Biology of the Cell,* 2d ed., Wadsworth Publishing Co., Inc., Belmont, Calif., 1977.

This excellent cytology text contains an unusually complete discussion of mitosis and meiosis.

---

* Available in paperback.

Articles

CAMPBELL, ALLAN M.: "How Viruses Insert Their DNA into the DNA of the Host Cell," *Scientific American,* December 1976, pages 102–113.

COHEN, STANLEY N.: "The Manipulation of Genes," *Scientific American,* July 1975, pages 25–33.

FIDDES, JOHN C.: "The Nucleotide Sequence of a Viral DNA," *Scientific American,* December 1977, pages 54–67.

PART II

Biology of Organisms

SECTION 4

THE DIVERSITY OF LIFE

20–1
Representatives of two major classes of organisms, Insecta and Angiospermae: African grasshoppers eating euphorbia.

CHAPTER 20

The Origin and Classification of Organisms

The universe was perhaps 15 billion years old, cosmologists calculate, when the star that is our sun came into being. According to current hypotheses, it formed, like other stars, from an accumulation of particles of dust and hydrogen and helium gases whirling in space among the older stars.

The immense cloud that was to become the sun condensed gradually as the hydrogen and helium atoms were pulled toward one another by the force of gravity, falling into the center of the cloud and gathering speed as they fell. As the cluster grew denser, the atoms moved faster and faster. More and more atoms collided with each other, and the gas in the cloud became hotter and hotter. As the temperature rose, the collisions became increasingly violent until the hydrogen atoms collided with such force that their nuclei fused, forming helium and releasing nuclear energy. This thermonuclear reaction is still going on at the heart of the sun and is the source of the energy radiated from its glowing surface.

The planets, according to current theory, formed from the remaining gas and dust moving around the newly formed star. At first, particles would have collected at random, but as each mass grew larger, other particles began to be attracted by the gravity of the largest masses. The whirling dust and forming spheres continued to revolve around the sun until finally each planet had swept its own path clean, picking up loose matter like a giant snowball. The orbit nearest the sun was swept by Mercury, the next by Venus, the third by earth, the fourth by Mars, and so on out to Neptune and Pluto, the most distant of the planets. The planets, including earth, are calculated to have come into being about 4.6 billion years ago.

THE EARTH AND THE BIOSPHERE

During the time earth and the other planets were being formed, the release of energy from radioactive materials kept their interiors very hot. When earth was still so hot that it was mostly liquid, the heavier materials collected in a dense core whose diameter is about half that of the planet. As soon as the supply of stellar dust, stones, and larger rocks was exhausted, the planet ceased to grow. As earth's surface cooled, an outer crust, a skin as thin by comparison as the skin of an apple, was formed. The oldest known rocks in this layer are about 3.98 billion years old.*

* The process by which these dates are estimated is described on page 804.

HELLO FROM THE CHILDREN OF THE PLANET EARTH

Sagittarius, the archer, is a constellation that is visible in our summer skies. If you look toward Sagittarius on a clear, moonless night, you are looking toward the center of the Milky Way, the galaxy of which our planet is a part. There are 250 billion (2.5 × 10^{10}) stars in our galaxy, and within the range of our most powerful telescopes there are about 10 billion more such galaxies.

Some biologists contend that given certain conditions—such as an energy source, water, and a temperature range in which water can exist in liquid form—plus a long enough time, the evolution of some form of life is inevitable. There are approximately 10^{20} (the number 1 followed by 20 zeros—a hundred billion billion) stars in the universe, like our own sun, that can provide energy for living things. At least 10 percent of these, according to astronomers, are likely to be surrounded by planetary systems such as our own. If only 1 percent were to have planets with environments roughly similar to those of earth, that would offer some 10^{18} possibilities for the existence of life in other solar systems.

What form would this life take? There is no reason to believe that the living organisms of another solar system would have any resemblance whatsoever to the organisms here on earth. However, even if the biology were not the same, the physics and chemistry would be: Radio wavelengths would be the same, the speed of light would be unchanging, the carbon atom would have the same properties, and so forth. Thus one would have both means of communication and something in common to talk about.

Radio signals would be one form of communication. It has been estimated that if a far-off planet were eavesdropping on our radio emissions, the first signals they would pick up would be those of the Ballistic Missile Early Warning Systems sweeping the local horizons with powerful radar beams. The next signals to be received, about 100 times less powerful, would be the video portion of television programs. Even if the content of individual programs could not be made out, TV monitoring could reveal a surprising amount of information, such as the location of our major population centers, the length of an earthly day, and the movement of the earth around the sun. NASA scientists have proposed Project Cyclops, an array of 1,500 radio antennas each 100 meters in diameter, each connected to one another and to a large computer system. Such a system would be capable of eavesdropping on radio communications of civilizations several hundred light-years away.

Within the last decade, we have, for the first time, succeeded in sending objects out of our own solar system. Pioneer 10 and Pioneer 11 spacecraft, launched in 1973, carry engraved metal plaques designed to last for billions of years (see figure). The second space messengers, Voyagers 1 and 2, launched in 1977, are carrying an audiovisual message in the form of a 30-centimeter copper disk sealed in an aluminum container with its own ceramic cartridge and needle and operating instructions. The disk contains, in a picture sequence, a rapid survey of our solar system and of basic biology, including a model of DNA, plus a message from President Carter. The second part of the disk contains, in audio form, greetings in 55 languages, beginning with Sanskrit, the world's oldest language, and ending with a small voice saying in English, "Hello from the children of the planet earth." The greetings are followed by a variety of sounds of the planet, including the song of the whale, the roar of a rocket, rain, laughter, and a sequence of music, which includes the first movement of the second Brandenburg Concerto and Louis Armstrong playing Melancholy Blues. *It may arrive in about 10 billion years. "We were here," it will say. "We cared."*

Bon voyage.

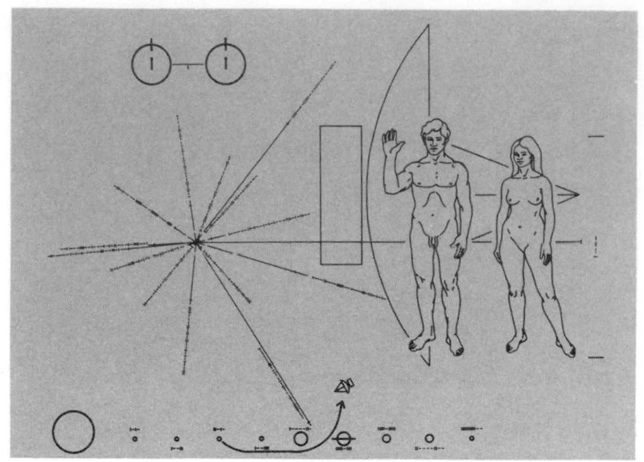

Metal plaque designed by Carl and Linda Sagan for the first spacecraft to leave our solar system. In this cosmic greeting card, the sun is located with respect to 14 pulsars, whose precise periods are specified in binary code. At the bottom, the earth is located with respect to the sun and other planets, and the path of the spacecraft out of the solar system is shown. The man's hand is raised in peaceful greeting.

20–2
These Greenland rocks, according to radioactive dating techniques, crystallized 3.75 billion years ago. The white strip in the center of the photo was produced by a lava flow 2.6 billion years ago.

Only 50 kilometers below its surface, the earth is still hot—a small fraction of it is even still molten. We see evidence of this in the occasional volcanic eruption that forces lava, which is molten rock, through weak points in the earth's skin, or in the geyser, which spews up boiling water that has trickled down to the earth's interior.

The biosphere is the part of the planet within which life exists. It forms a thin film on the outermost layer, extending only about 8 or 10 kilometers up into the atmosphere and about as far down into the depths of the sea.

Why on Earth?

In our solar system, earth among all the planets seems most favored for the production of life. A major factor is that earth is neither too close nor too distant from the sun. The chemical reactions on which life—any form of life—depends virtually cease at very low temperatures. At high temperatures, complex chemical compounds are too unstable for life to form or survive.

Earth's size and mass are also important factors. Planets much smaller than earth do not have enough gravitational pull to hold a protective atmosphere, and any planet much larger than earth is likely to have so dense an atmosphere that light from the sun cannot reach its surface.

THE BEGINNING OF LIFE

Darwin believed that it would probably be impossible ever to look back down that long corridor of time and reconstruct life's beginnings. Until very recently, the earliest fossils known were those of the Cambrian period, a mere 600 million years ago, and for a long time after the publication of *The Origin of Species,* biologists regarded the earliest events in the history of life as chapters that would probably remain forever closed.

Two scientific developments, however, have greatly improved our long-distance vision. One was the formulation of a testable hypothesis about the events preceding life's origins; the second was the discovery of cells more than 3 billion years old.

The testable hypothesis was offered by the Russian biochemist A. I. Oparin. According to Oparin, the appearance of life was preceded by a long period of what is sometimes called chemical evolution. During this period, which occupied almost a third of the earth's history, there was little or no free oxygen in the atmosphere surrounding the earth, and the oxygen that was present existed mainly in the form of water vapor. The atmosphere was a reducing one—an atmosphere that abounded in hydrogen gas, methane (CH_4), ammonia (NH_3), and other carbon-hydrogen gases. As we have seen, combinations of these four elements available in the primitive atmosphere—hydrogen, oxygen, carbon, and nitrogen—make up more than 95 percent of living tissues.

Energy was required to break apart the simple gases of the atmosphere and re-form them into organic molecules. And energy abounded on the young earth. There was heat energy, both boiling (moist) heat and baking (dry) heat. Water vapor spewed out of the primitive seas, cooled in the upper atmosphere, collected into clouds, fell back on the crust of the earth, and steamed up again. Violent rainstorms were accompanied by lightning, which provided electrical energy. The sun bombarded the earth's surface with high-energy particles and ultraviolet light, and radioactive elements within the earth released their energy into the atmosphere. Oparin hypothesized that under such conditions organic molecules were formed and collected in a thin soup in the earth's seas and lakes. Some of these molecules might have become locally more concentrated by the drying up of a lake or by the adhesion of the molecules to a solid surface. (Oparin published his small monograph in 1922, but at that time biochemists were so convinced by Pasteur's demonstrations disproving spontaneous generation—page 224—that the scientific community ignored his ideas.)

20-3
Bolts of lightning in the steam boiling up from a volcanic crater. Such sources of energy would have been present on the primitive earth and might have contributed to the formation of organic molecules. This photograph, taken in 1963, shows the birth of the island of Surtsey off the coast of Iceland.

20–4
Conditions believed to exist on the primitive earth are simulated in this apparatus. Methane and ammonia are continuously circulated between a lower "ocean," which is heated, and an upper "atmosphere," through which an electric discharge is transmitted. At the end of 24 hours, 45 percent of the carbon originally present in the methane gas is converted to amino acids and other organic molecules.

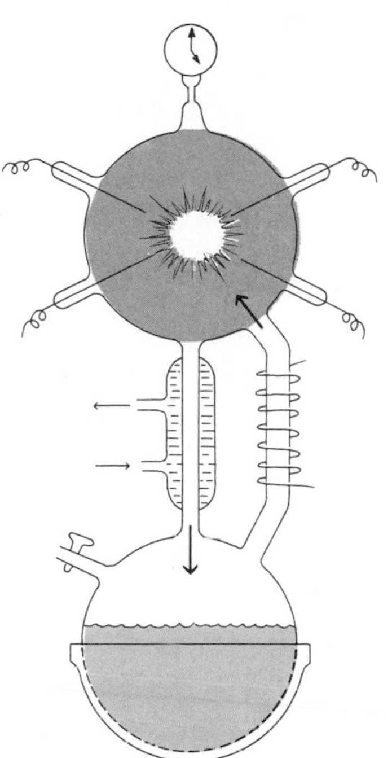

In the 1950s, Oparin's hypothesis was tested in the laboratory by Stanley Miller, then a graduate student at the University of Chicago. Miller showed that under conditions simulating those believed to have been present some 3 to 4 billion years ago, amino acids are produced from hydrogen gas, ammonia, methane, and water vapor (Figure 20–4). Miller's experiments have now been repeated many times in other laboratories. With various modifications in the experimental conditions, almost all of the common amino acids have been produced and also purines, pyrimidines, ribose, and nucleotides.

As the concentrations of such molecules increased, it is not difficult to imagine that they would have assembled themselves in aggregates, held together by hydrophobic interactions, hydrogen bonds, and other weak forces. The most stable of the chemical combinations would have tended to survive, and hence a form of natural selection played a role in chemical evolution as well as in the biological evolution that was to follow.

At some stage in chemical evolution, a boundary or membrane formed around such an aggregate of macromolecules, permitting it to lead a separate existence from its external environment. What was enclosed in that boundary membrane? A catalytic protein? A self-replicating nucleic acid? Such questions lead directly to the mystery of the origin of the genetic code. As far as we can see, there is no underlying molecular logic to the code, no reason why a particular triplet should specify one amino acid rather than another. Did the code come into being purely by chance? It would appear so.

All we know for sure is that all living systems are the descendants of these tiny metabolizing droplets.

20–5
(a) *Alga-like microfossils in thin sections and* (b) *bacteria-like microfossils in surface replicas. They were found in the same deposit of South African black chert and are about 3.4 billion years old, the oldest fossils now known.*

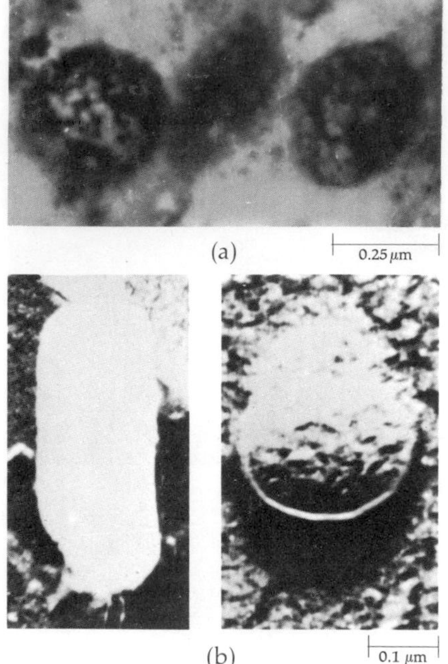

The First Cells

When did life begin? We will probably never be able to answer this question because it depends first on how we define life and, second, on finding chemical traces probably long since erased. We can establish some sort of time scale, however, owing to the discovery little more than a decade ago of microfossils in flintlike rocks, called chert, in South Africa. The fossils, whose structure is visible only by electron microscopy, resemble present-day bacteria and blue-green algae. The rocks in which they were found, according to radioactive dating, are 3.4 billion years old. Thus life began very early, perhaps in the first billion years of earth's history.

The Evolution of Energy Systems

Although both heterotrophs and autotrophs are represented among the earliest microfossils, it is logical to postulate that the first living cell was an extreme heterotroph. It probably possessed few enzymes, compared with a modern cell, and so was capable of only a few anabolic reactions. It would have had to find most of the molecules it required in Oparin's organic soup.

As the primitive prokaryotic heterotrophs increased in number, according to this hypothesis, they began to use up the complex molecules on which their existence depended and which had taken millions of years to accumulate. Competition began. Under the pressure of this competition, cells that could make efficient use of the limited energy sources now available were more likely to survive than cells that could not. In the course of time, cells evolved that were able to synthesize organic molecules out of simple inorganic materials. Without the evolution of these autotrophs, life on earth would soon have come to an end.

The most successful of the autotrophs were those that evolved a system for making direct use of the sun's energy—the process of photosynthesis. With the advent of photosynthesis, the flow of energy in the biosphere came to assume its modern form: Radiant energy channeled through photosynthetic autotrophs to other forms of life.

With the evolution of photosynthetic cells that produced oxygen, the atmosphere slowly began to change. As a result of photosynthesis, the concentration of oxygen in the atmosphere rose to 1 percent about 600 to 1,000 million years ago and to 10 percent only about 400 million years ago. The present concentration of 21 percent oxygen has been created and is maintained by photosynthesis. The accumulation of oxygen in the air made possible, in turn, the evolution of organisms that produced most of their chemical energy by cellular respiration.

According to this hypothesis, in terms of energy production, anaerobic glycolysis came first, followed by photosynthesis and finally, as oxygen accumulated, by respiration. The genetic code was determined and the pathways of glycolysis, respiration, and photosynthesis were established in nearly their present form long before the appearance of any eukaryotic cells.

THE EVOLUTION OF EUKARYOTES

The microfossil record indicates that the eukaryotes probably evolved about 1.6 billion years ago. Eukaryotes, as we saw in Section 1, are distinguished from prokaryotes by their larger size, the separation of nucleus from cytoplasm by a nuclear envelope, the organization of their DNA, and their complex organelles, among which are chloroplasts and mitochondria.

The step from prokaryotes to the first eukaryotes was one of the big evolutionary transitions, second only to the origin of life. The question of how it came about is a matter of current and lively discussion. One interesting hypothesis is that larger, more complex cells evolved partly as a result of certain prokaryotes' taking up residence inside other cells. (Such a close association between two organisms of different species is known as symbiosis, about which we shall have more to say on page 364.)

As oxygen gas began slowly to accumulate in the atmosphere, those bacteria that were able to convert to the use of oxygen in ATP production gained a strong advantage, and so such forms began to prosper and increase. Some of these evolved into modern forms of bacteria. Others, according to this theory, became symbionts within larger cells and evolved into mitochondria.

Several lines of evidence support the hypothesis that mitochondria are descended from specialized bacteria: Mitochondria contain their own DNA, and this DNA is present in a single, continuous ("circular") molecule, like the DNA of bacteria; many of the same enzymes contained in the cell membranes of bacteria are found in mitochondrial membranes; the ribosomes of mitochondria resemble those of bacteria both in their size (they are smaller than those of eukaryotic cells) and in some details of their chemical composition; mitochondria appear to be produced only by other mitochondria, which divide within their host cell. (However, to make the situation more complex, there is nowhere near enough DNA in mitochondria to code for all the mitochondrial proteins. Host-cell DNA is also required for mitochondrial structures.)

We know little about the original cells in which these bacteria first set up housekeeping—or, indeed, if they actually existed. But if they did exist, they probably had no means of using oxygen for cellular respiration and so were dependent entirely on anaerobic fermentation as an energy-releasing process, which, as we have seen, is relatively inefficient. Cells with respiratory assistants would have been more efficient than those lacking them and so would have outreproduced them.

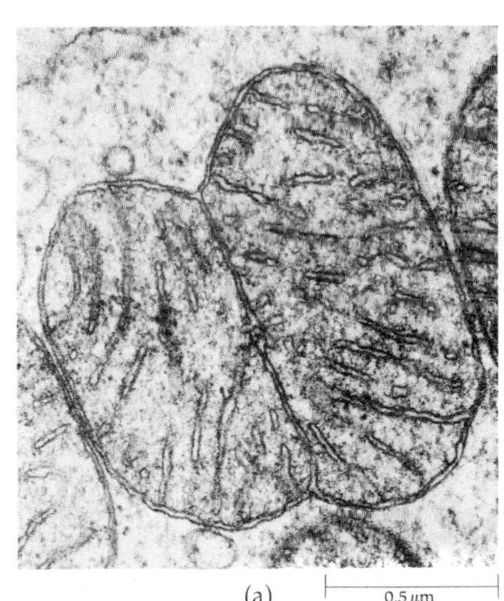

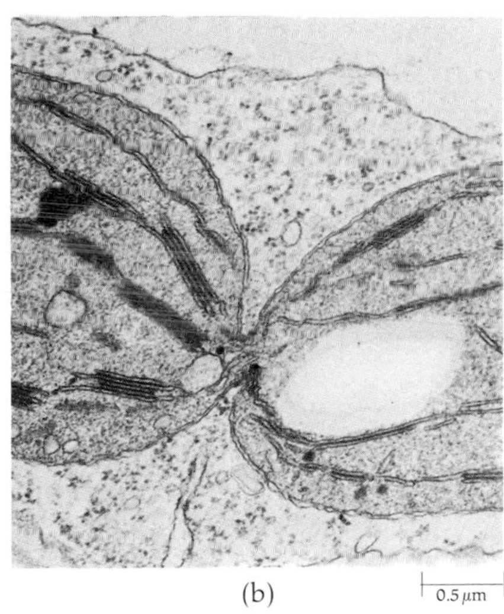

20-6
(a) Mitochondria and (b) chloroplasts may have originated as symbiotic prokaryotes. This hypothesis is supported by the facts that both contain their own DNA and both, as shown here, replicate by division.

20-7
(a) *Symbiotic bacteria from a flashlight fish. Luminescent bacteria such as these are collected in continuously glowing organs below the eyes of several species of fish. As many as 10 billion are found in a milliliter of fluid. The fish, which resemble small porgies, swim in reefs in the Indian Ocean and Red Sea. The French divers who first discovered the flashlight fish with its blinking headlights called it "le petit Peugeot."* (b) *A nudibranch* (Elysia elsiae), *a marine mollusk. These animals obtain their chloroplasts by eating certain green algae. The chloroplasts, which line the digestive tubules bordering the respiratory chamber of the nudibranch, carry on photosynthesis so efficiently that some individuals are reported to evolve oxygen more rapidly than it is consumed.*

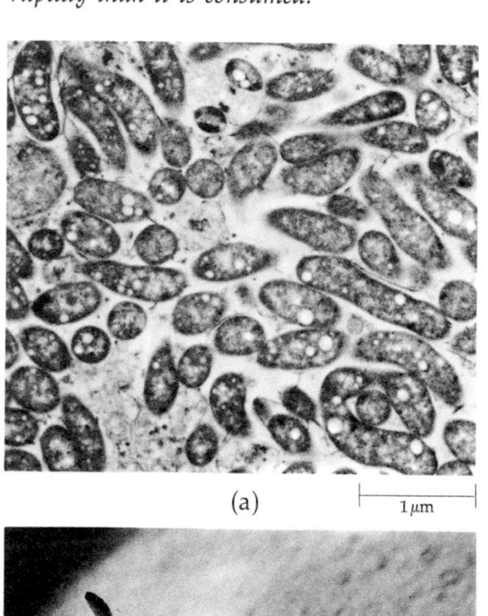

(a)

(b)

In an analogous fashion, photosynthetic prokaryotes ingested by larger, nonphotosynthetic cells are believed to be the forerunners of chloroplasts. By this symbiosis, the smaller cells gained nutrients and protection, and the larger cells were given a new energy source.

This hypothesis accounts for the presence in eukaryotic cells of complex organelles not found in the far simpler prokaryotes. It gains support from the fact that many modern organisms contain intracellular symbiotic bacteria and algae, indicating that such associations are not difficult to establish.

Modern one-celled eukaryotes—the protists—are a very diverse group, most of which probably bear little or no resemblance to the first eukaryotic cells. The multicellular organisms are all eukaryotes. The major groups—the fungi, the plants, and the animals—are believed each to have originated from one (and possibly more) separate types of one-celled eukaryotes. There is no fossil evidence of their origins. The first fossils of multicellular organisms are found in sediments a mere 0.6 billion years old, and their diversity and complexity indicate that eukaryotes had been in existence for millions of years. The history of evolution from this time forward is summarized in Table 26-2, on page 478. Some aspects will be covered in more detail in the chapters that follow.

As you can see in Figure 20-9, members of the human species appeared only very recently in the evolutionary timetable. If we measure out the history of the earth on a 24-hour time scale, the oldest known fossils appeared at 6 A.M., the oldest eukaryotes between 4 and 5 P.M., the first multicellular organisms between 8 and 9 P.M., and *Homo sapiens* in the last 30 seconds at 11:59:30 P.M.

20-8
Possible evolutionary relationships among the major groups of organisms. According to these hypotheses, the fungi, the various phyla of eukaryotic algae, and the modern protozoans evolved separately from single-celled eukaryotes. The origins of the higher plants have been traced to one group of algae. The animals are presumed to have arisen from a single-celled eukaryotic heterotroph, although direct evidence is lacking.

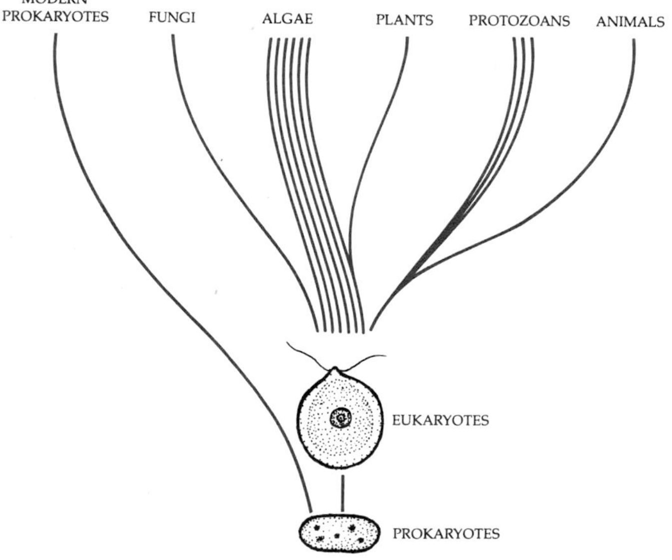

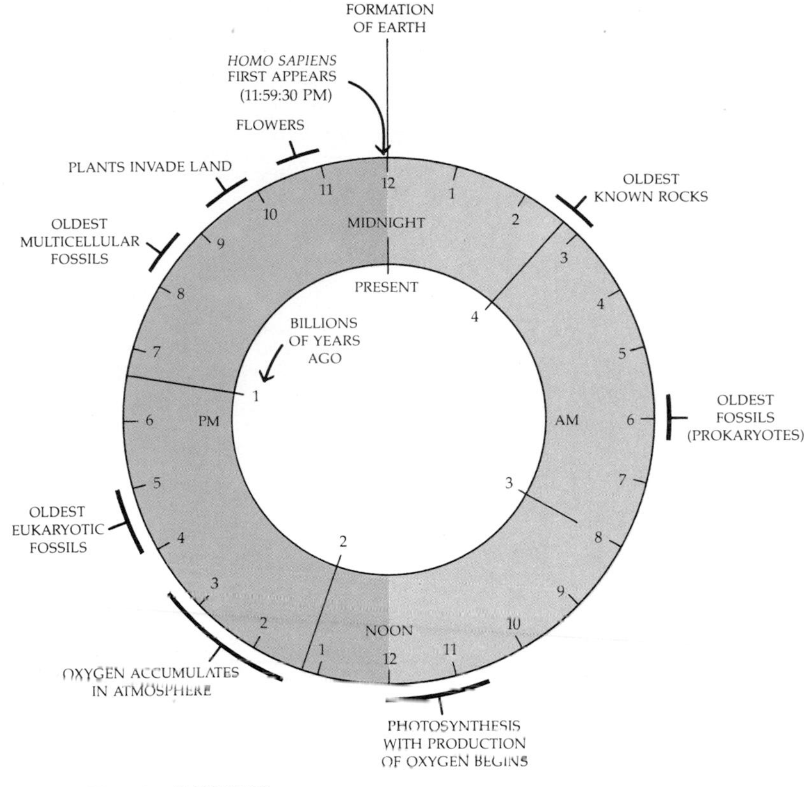

20–9
The clockface of biological time. Life first appears relatively early in the earth's history, before 7:00 A.M. *on a 24-hour time scale. The first multicellular organisms do not appear until the twilight of that 24-hour day, and* Homo sapiens *is a late arrival—at about 30 seconds to midnight.*

CLASSIFICATION OF ORGANISMS

As late arrivals on the evolutionary scene, we humans are confronted with a bewildering variety of living things. As many as 5 million different kinds of organisms share the biosphere. For thousands of years, we have been trying to find order and meaning in this diversity. Systematists are biologists who study this order, and taxonomists arrange organisms in systems of classification.

Aristotle is generally considered to be among the first to have tried to group organisms in ways that made sense. He recognized, for example, that you could not choose any single characteristic as the basis for grouping. For instance, suppose you decided to group all the animals into those that fly and those that do not. You would end up with one category that included most insects, birds, and bats, and you would have to separate some obviously close relations, such as the winged and the wingless ants that occur as members of the same colony. Aristotle recognized, as do modern taxonomists, that one had to look not just at a single feature but at many characteristics of a plant or animal. On the basis of overall similarities and differences, an organism could be placed with other forms in increasingly restricted categories: vertebrate, bird, duck, mallard, for example.

20–10

Representatives of the five kingdoms. (a) *Prokaryotes. Bacteria,* Neisseria gonorrhoeae *(dark spheres), the causative agent of gonorrhea, being ingested by white blood cells.* (b) *Protists.* Vorticella, *like most protists, is a single cell. It attaches itself to a substrate by a long stalk. A contractile fiber, visible in this micrograph, runs through the stalk.* (c) *Fungi. An inky cap mushroom. Fungi are characterized by a multicellular underground network and also spores and sporangia, which, in mushrooms, are borne in these familiar structures.* (d) *Plants. Marsh marigolds. The flowers of angiosperms, attractive to pollinators, are among the principal reasons for their evolutionary success.* (e) *Animals. A luna moth. A nervous system with a variety of sense organs is a chief characteristic of the animal kingdom.*

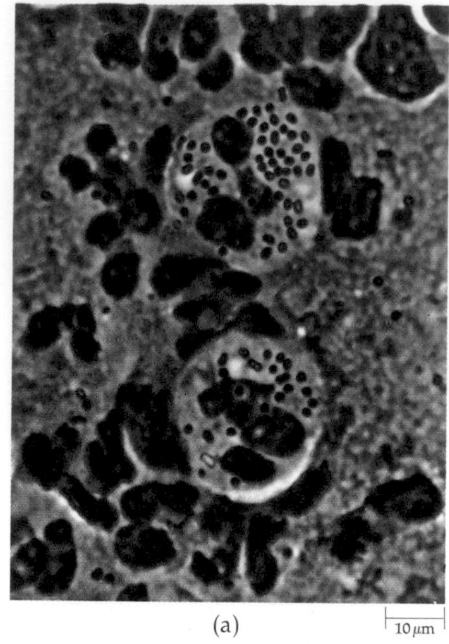

(a) 10 μm

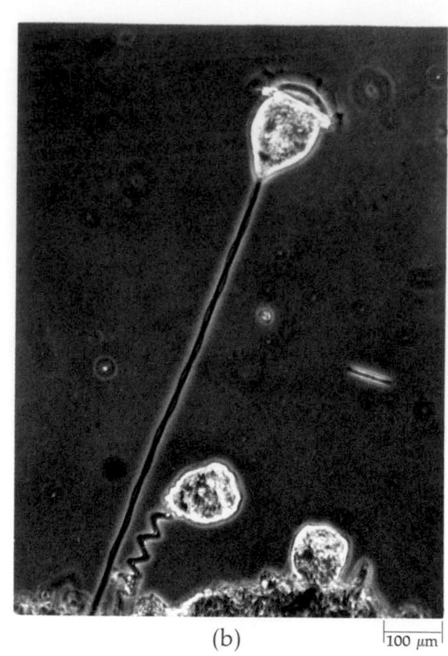

(b) 100 μm

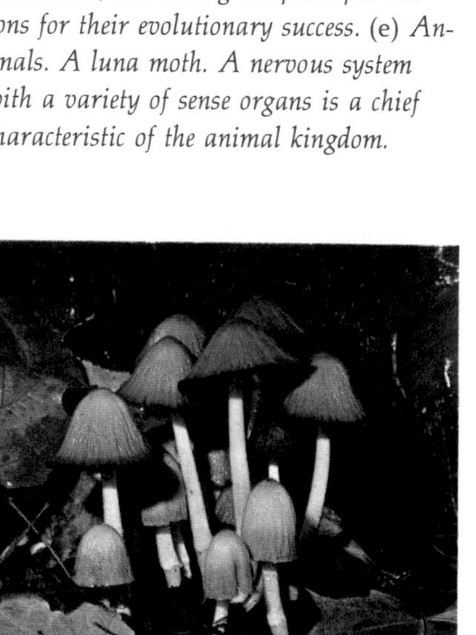

(c)

(e)

(d)

In the middle of the seventeenth century, an English clergyman, John Ray, set out to catalog all the organisms in the world and to arrange them systematically. He was the first to use the word *species* to describe a kind of organism. A species, according to Ray, was made up of organisms that were morphologically the same—from *morphe*, meaning "shape" in Greek—and that could reproduce their own kind.

Ray cataloged all the plants in the vicinity of Cambridge, the first complete catalog of one locality ever to be made. Then, traveling through England and the Continent, he worked his way, astonishingly, through all the plants that he could learn about (almost 19,000) and the birds, the fishes, and the four-footed animals.

At the time of Ray, and for a number of years thereafter, plants and animals were designated by cumbersome phrase names, or polynomials. These polynomials were brief descriptive phrases concerning the plant or animal to which they were applied. The first word in the polynomial had come, by the close of the seventeenth century, to designate the *genus* (plural, *genera*), an inclusive group of similar species. Thus, the numerous kinds of roses were grouped in the genus *Rosa*, many butterflies in the genus *Papilio*, and cats and catlike animals in the genus *Felis*.

THE BINOMIAL SYSTEM

20–11
Linnaeus, the originator of modern hierarchical classification and nomenclature. He is wearing his collector's outfit.

The system of naming living things was simplified by the eighteenth-century Swedish professor, physician, and naturalist, Carolus Linnaeus (1707–1778). (Linnaeus was born Carl von Linné but latinized his name in the scholarly fashion of his time.) His ambition was to classify all the known kinds of plants and animals according to their genera. In 1753, he published a two-volume work, *Species Plantarum* ("the kinds of plants"), which contained brief analytical descriptions of every species of plant known to European science. Although he used the polynomial designations and regarded them as the proper names for species, he made an important innovation. In the margin of his book, opposite the "proper" name of each species, Linnaeus entered a single word, which, together with the generic name, formed a convenient "shorthand" designation for the species. In the book, catnip, which had previously been designated by the polynomial *Nepeta floribus interrupte spicatus pedunculatis* ("*Nepeta* with flowers in an interrupted pedunculate spike"), was described under *Nepeta*, and *cataria* was put in the margin, making it *Nepeta cataria*, which is its name today.

The convenience of this system was quickly recognized, and Linnaeus and subsequent authors soon replaced all "proper" names with "shorthand" ones. This binomial ("two-name") system is still used today. A species name, as we have seen, consists of two parts—the generic name and the specific epithet (adjective or modifier). However, a genus name may be written alone when one is referring to members of the entire group of species making up that genus, such as *Drosophila* or *Paramecium*.

A specific epithet is meaningless when written alone, however, because many different species in different genera may have the same specific epithet. The domestic dog is *Canis familiaris*, for instance; *familiaris* is a commonly used specific epithet and by itself would not identify any organism. For this reason, the specific epithet is always preceded by the genus name, or, in a context where no ambiguity is possible, the genus name may be abbreviated to its initial letter. Thus *Canis familiaris* may be designated *C. familiaris*.

THE NAMING OF ORGANISMS

The use of Latin names for organisms—a practice that makes biology in general, and taxonomy in particular, seem so formidable a subject in terms of language alone—came about in medieval times, when Latin was the language of scholarship. Although the terms seem cumbersome, they are a necessary tool of science. Common names are, at best, inadequate. We tend to give names only to what is of interest to us. For example, gauchos, the cowboys of Argentina who are famous for their horsemanship, have some 200 names for different colors of horses but divide all the plants known to them into four groups: pasto, or fodder; paja, or bedding; cardo, wood; and yuyos, everything else. Moreover, even when common names do exist for a kind of organism, they may vary from place to place. A robin in North America is distinctly different from the English bird of the same name. "Mole" refers to a placental mammal in North America but to a marsupial in Australia. A yam in the southern United States is a totally different vegetable from a yam several hundred kilometers away in the West Indies. A sheepshead in Ohio is an ugly, unwanted, inedible cousin of the carp, but a sheepshead in Georgia is a delicate-fleshed saltwater fish. Or more than one name may be given the same animal, such as groundhog and woodchuck, gnu and wildebeest, pill bug and sow bug and wood louse. When different languages are involved, the problems of communication become virtually insurmountable. For this reason, biologists refer to organisms by Latin names, registered with and officially recognized by international organizations of botanists and zoologists.

American robin, Turdus migratorius.

European robin, Erithacus rubecula.

WHAT IS A SPECIES?

Species in Latin simply means "kind," and so species, in the simplest sense, are the different kinds of organisms. A more technical definition of species is "a group of interbreeding organisms that do not ordinarily breed with members of other groups." This definition conforms to common sense. If one species interbred freely with other species, it would no longer represent a distinct kind of organism. This definition, although useful for animals, is not so useful for other organisms, such as some bacteria and protists that do not breed at all, as far as is known, and for some plants in which fertile crossings can take place among very different kinds of plants.

(a)

(b)

(c)

20–12
Three examples of the genus Felis: (a) *a cougar,* Felis concolor, (b) *an ocelot,* Felis pardalis, *and* (c) *a domestic cat,* Felis cattus.

A profound change occurred in the concept of species with the acceptance of evolutionary theory. Rather than being the immutable, ideal form conceived by Linnaeus, a species is now seen to change constantly in both time and space. However, even though a precise definition of species is elusive, the term is understandable in a practical way. For instance, anthropologists point out that primitive peoples, particularly hunter-gatherers, are able to recognize, distinguish, and name great numbers of the organisms with which they come in contact. What these people, operating on a practical level, recognize as a "kind"—something different enough to have its own name—almost always coincides with what scientists recognize as a "species." So despite the inability of experts always to agree upon a definition (or the organism to conform to one), the species is a clear and useful reality.

OTHER TAXONOMIC GROUPS

Scientists before Linnaeus recognized the plant, animal, and mineral kingdoms, and the *kingdom* is still the major category used in biological classification. Between the level of genus and the level of kingdom, however, Linnaeus and subsequent taxonomists have added a number of categories. Thus, genera are grouped into *families,* families into *orders,* orders into *classes,* and classes into *phyla.* These categories may be subdivided or aggregated into a number of less important ones, such as subgenera and superfamilies.

The groups that are placed in the categories are known as *taxa* (singular, *taxon*). Thus, in Table 20–1, Plantae and Animalia are taxa within the category of kingdom, and Angiospermae and Mammalia are taxa having the categorical rank of class. By convention, generic and specific names are written in italics, while the names of families, orders, classes, and other taxa whose categories rank above the genus level are not, although they are capitalized.

Table 20–1 shows the classification of two different types of organisms. Note how much information we have about a plant or animal if we know its classification. Indexing this information is one of the chief functions of a classification scheme.

Table 20-1 *Biological Classifications*

Red Maple *(Acer rubrum)*

CATEGORY	TAXON	CHARACTERISTICS
Kingdom	Plantae	Organisms that usually have rigid cell walls and usually possess chlorophyll
Subkingdom	Embryophyta	Plants forming embryos
Phylum	Tracheophyta	Vascular plants
Subphylum	Pterophytina	Generally large, conspicuous leaves, complex vascular pattern
Class	Angiospermae	Flowering plants, seed enclosed in ovary
Subclass	Dicotyledoneae	Embryo with two seed leaves (cotyledons)
Order	Sapindales	Soapberry order; trees and shrubs
Family	Aceraceae	Maple family; trees of temperate regions
Genus	*Acer*	Maples and box elder
Species	*Acer rubrum*	Red maple

Man *(Homo sapiens)*

CATEGORY	TAXON	CHARACTERISTICS
Kingdom	Animalia	Multicellular organisms requiring organic plant and animal substances for food
Phylum	Chordata	Animals with notochord, dorsal hollow nerve cord, gills in pharynx at some stage of life cycle
Subphylum	Vertebrata	Spinal cord enclosed in a vertebral column, body basically segmented, skull enclosing brain
Superclass	Tetrapoda	Land vertebrates, four-limbed
Class	Mammalia	Young nourished by milk glands, skin with hair or fur, body cavity divided by diaphragm, red corpuscles without nuclei, high body temperature
Order	Primates	Tree dwellers or their descendants, usually with fingers and flat nails, sense of smell reduced
Family	Hominidae	Flat face, eyes forward, color vision, upright, bipedal locomotion, with hands and feet differently specialized
Genus	*Homo*	Large brain, speech, long childhood
Species	*Homo sapiens*	Prominent chin, high forehead, sparse body hair

TAXONOMY AND EVOLUTION

To Linnaeus and his immediate successors, taxonomic classification was a revelation of grand, unchanging design. When evolutionary theory came to be the dominant ordering principle in the biological sciences, systematics, as the science of taxonomy is called, took on a new meaning. The goal is to classify organisms in a way that reflects their evolutionary history. Thus species are groups that have diverged only recently, genera share more distant ancestors, and so on. A major problem is that all species alive today are equally modern; none is an ancestor of any other. So the degrees of similarity and differences among modern forms may or may not represent the closeness of relationship. Traditionally, taxonomy has depended on morphological criteria—the arrangement of parts and the external appearance of the organism. Recently, increased knowledge of cellular biochemistry is leading to greater reliance on biochemical criteria, such as the composition of the cell membrane and the cell wall or the use of certain anabolic or catabolic pathways. The bacteria are a particularly difficult group, taxonomically speaking, because their forms are variable and their evolutionary histories are, to a large extent, unknown. Now studies are being made using hybridizing techniques to reveal the extent of differences and similarities among the DNAs of certain bacte-

20–13
Three of the classification systems in common use. In (a), *all organisms are classified as either plants or animals. In* (b), *the prokaryotes are considered a third kingdom. In* (c), *five separate kingdoms are recognized.*

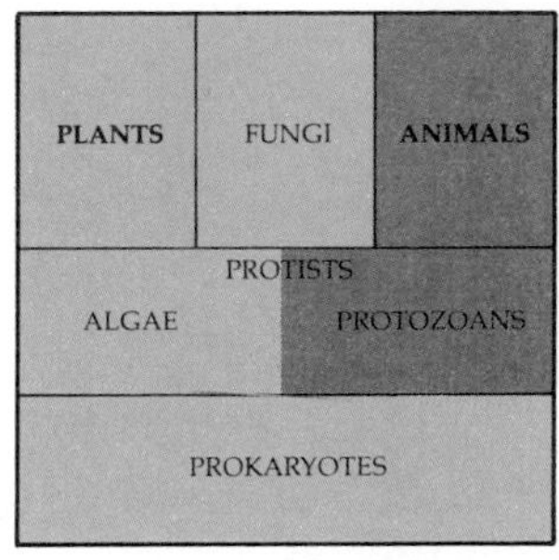

(a)

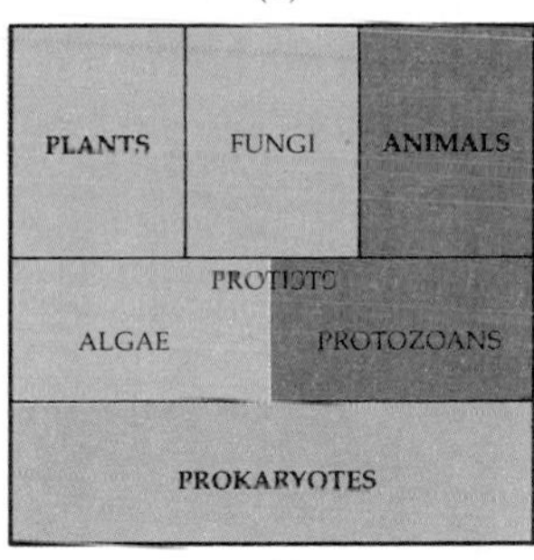

(b)

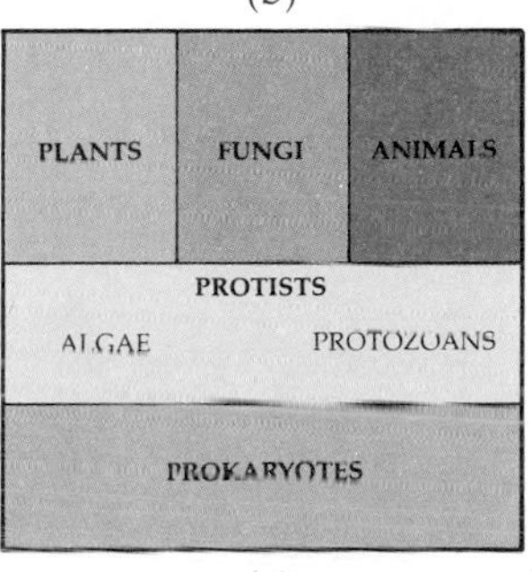

(c)

rial groups. The use of genotype rather than phenotype as a basis for classification probably represents the taxonomic ideal, not soon to be obtained but clearly the solution to many problems.

A QUESTION OF KINGDOMS

In Linnaeus's time, as we mentioned, it was thought that there were three kingdoms—animals, plants, and minerals—and until very recently it was popular to classify every living thing as either an animal or a plant. Animals were organisms that moved, ate things, and breathed, and their limbs and organs and bodies grew to a certain size and then stopped growing. Plants were organisms that did not move, eat, or breathe, and that grew indefinitely. Thus the fungi, algae, and bacteria were grouped with the plants, and the protozoans—one-celled organisms that ate and moved—were classified with the animals.

In the twentieth century, new data began to emerge. This was partly a result of improvements in the light microscope and, subsequently, the development of the electron microscope, and partly because of the application of biochemical techniques to studies of differences and similarities among organisms. As a result, the number of groups recognized as constituting different kingdoms has increased as biological knowledge has accumulated. The most recent proposals recommend five kingdoms: Prokaryotae, Protista, Fungi, Plantae, and Animalia. In Appendix C, the major groups of organisms are classified within these five kingdoms. Other systems group the eukaryotic algae and the fungi with the plants, ending up with three kingdoms: prokaryotes, plants, and animals. Some retain the two-kingdom classification system: plants and animals.

In truth, no system is really satisfactory. For instance, at the one-celled level of life, there are no useful criteria for separating plants from animals. Two species of single-celled, motile organisms may be almost identical in most respects, except that one has chloroplasts and the other does not. In some cases, the one that has chloroplasts can lose them from time to time and still continue to survive and reproduce indefinitely. Yet in a plant-animal division based on the capacity for photosynthesis, these two closely related forms are separated at the level of kingdom. On the other hand, as we shall see in Chapter 24, there is a clear evolutionary sequence, with modern, living representatives, leading from certain one-celled algae to the flowering plants. So if we group all of the one-celled eukaryotes together—the protists—as we do in this text, we break up what would appear to be a clear evolutionary line leading to the plants.

Whether organisms are classified in two, three, or five kingdoms, however, their species designations are not affected, nor are most of the other categories in which they are classified.

Table 20–2 summarizes some of the essential similarities and differences among these five kingdoms of organisms. We shall discuss each group in turn in the chapters that follow.

SUMMARY

The sun and its planets were formed 4.6 billion years ago—the sun probably from the condensation and contraction of a hydrogen gas cloud, and the planets as accumulations of interstellar debris. Of the nine planets in this solar system, only

Table 20-2 *Characteristics of Major Groups of Organisms*

	PROKARYOTES	PROTISTS	FUNGI	PLANTS	ANIMALS
Cell type	Prokaryotic	Eukaryotic	Eukaryotic	Eukaryotic	Eukaryotic
Chromosome	DNA (single circular molecule)	DNA plus protein	DNA plus protein	DNA plus protein	DNA plus protein
Nucleus	No nuclear envelope	Nuclear envelope present	Nuclear envelope present	Nuclear envelope present	Nuclear envelope present
Mitochondria	Absent	Present	Present	Present	Present
Chloroplasts	None (photosynthetic lamellae in some types)	Present (some forms)	Absent	Present	Absent
Cell wall	Noncellulose (polysaccharide plus amino acids)	Present in some forms, various types	Chitin and other noncellulose polysaccharides	Cellulose	Absent
Means of genetic recombination	Conjugation, transduction, transformation, or none	Fertilization (syngamy) and meiosis, conjugation, or none	Fertilization and meiosis, or none, dikaryosis (page 388)	Fertilization and meiosis	Fertilization and meiosis
Mode of nutrition	Autotrophic (chemosynthetic and photosynthetic) and heterotrophic (saprobic and parasitic)	Photosynthetic and heterotrophic, or combination of these	Heterotrophic (saprobic and parasitic), by absorption	Photosynthetic	Heterotrophic, by ingestion
Motility	Bacterial flagella, gliding, or nonmotile	9 + 2 cilia and flagella, amoeboid, contractile fibrils	9 + 2 cilia and flagella in some forms, none in most forms	9 + 2 cilia and flagella in lower forms and in some gametes, none in most forms	9 + 2 cilia and flagella, contractile fibrils
Multicellularity	Absent	Absent in most forms	Present in most forms, but limited	Present in all forms	Present in all forms
Nervous system	None	Primitive mechanisms for conducting stimuli in some forms	None	None or primitive (page 563)	Present, often complex

earth is known to support life, but there are likely to be other planets in the galaxy with some form of life.

The primitive atmosphere of earth held the chief raw materials of living matter—hydrogen, oxygen, carbon, and nitrogen—combined in water vapor and gases. The energy required to break apart the simple gases in the atmosphere and re-form them into more complex molecules was supplied by heat, lightning, radioactive elements, and high-energy radiation from the sun. Laboratory experiments have shown that the types of molecules characteristic of living organisms, including amino acids, proteins, and nucleotides, can be formed under such conditions.

These molecules, it is believed, gradually accumulated to form cells, the first living things.

The oldest fossils discovered so far were found in rocks 3.4 billion years old. They are microscopic in size and resemble modern prokaryotes. The first prokaryotes probably were heterotrophs, ingesting organic molecules from the waters in which they lived. Anaerobic glycolysis presumably evolved during this period. As the supply of organic molecules became increasingly depleted, natural selection favored cells with greater anabolic capacities. During this period, photosynthesis evolved and, as a consequence, oxygen began to accumulate in the atmosphere. The accumulation of oxygen set the stage for the evolution of cellular respiration. All of these events took place, it is believed, before the appearance of the first eukaryotes, some 1.6 billion years ago. The transition from prokaryotes to eukaryotes is a far larger and more significant evolutionary leap than any event thereafter. One factor in the evolution of eukaryotes may have been the establishment of symbiotic relationships with prokaryotic cells that eventually became specialized as mitochondria and chloroplasts. The one-celled eukaryotes are known as the protists, which is a large and varied group. All other organisms—including fungi, plants, and animals—are believed to have had their origins in various types of one-celled eukaryotes. The first fossil records of multicellular organisms are found in deposits 0.6 billion years old.

There are more than 5 million kinds of organisms in the biosphere. Scientists have sought order in this vast diversity of living things by classifying them, that is, by grouping them in meaningful ways. Evolutionary theory provides the intellectual framework for modern taxonomic systems. A species is the basic unit of classification. Ideally, species are groups of closely related individuals; genera are groups of species that share a common ancestor. Similarly, genera are grouped into families, families into orders, orders into classes, classes into phyla, and phyla into kingdoms. Depending on the classification system used, the phyla may be grouped into as few as two kingdoms (plants and animals) or as many as five (Prokaryotae, Protista, Fungi, Plantae, and Animalia). The latter system is followed in this text.

QUESTIONS

1. Why would energy sources have been necessary for the synthesis of simple organic molecules?

2. Why is Oparin's hypothesis of a reducing atmosphere so important? What properties of oxygen would have made chemical evolution unlikely in an atmosphere containing O_2?

3. What arguments can you advance for the hypothesis that nucleic acids preceded proteins in chemical evolution? What can you advance for the reverse?

4. As we noted, it is believed that many planets in our galaxy might contain some form of life. Suppose you were seeking such a planet, what characteristics would you look for?

5. Can you suggest a system of life different from that on earth? What might its features be?

6. Name the distinguishing features of the members of the five kingdoms.

CHAPTER 21

The Prokaryotes

In the next three chapters, we are going to describe the three kingdoms comprising the microorganisms: the prokaryotes, the protists, and the fungi. Although they are dissimilar in many important ways, and many algae and fungi are quite large, they are conventionally studied together in the science known as microbiology. Viruses will also be described briefly in this chapter. Although they do not fit comfortably in any of the taxa of living systems, they too are traditionally a part of microbiology.

The prokaryotes are the oldest and most abundant group of organisms in the world. About 1,600 species have been described. They are the smallest cellular organisms; a single gram (about $\frac{1}{28}$ of an ounce) of fertile soil can contain as many as $2\frac{1}{2}$ billion individuals.

The success of the prokaryotes, biologically speaking, is undoubtedly due to their rapid rate of cell division and their great metabolic versatility. Growing under optimum conditions, a population can double in size every 20 or 30 minutes. Prokaryotes can survive in many environments that support no other form of life. They have been found in the icy wastes of Antarctica, the near-boiling waters of natural hot springs, and even in the dark depths of the ocean. Some bacteria are among the very few modern organisms that can survive without free oxygen, obtaining their energy by anaerobic glycolysis (see page 158). Oxygen is lethal to some types (obligate anaerobes), whereas others can exist with or without oxygen (facultative anaerobes).

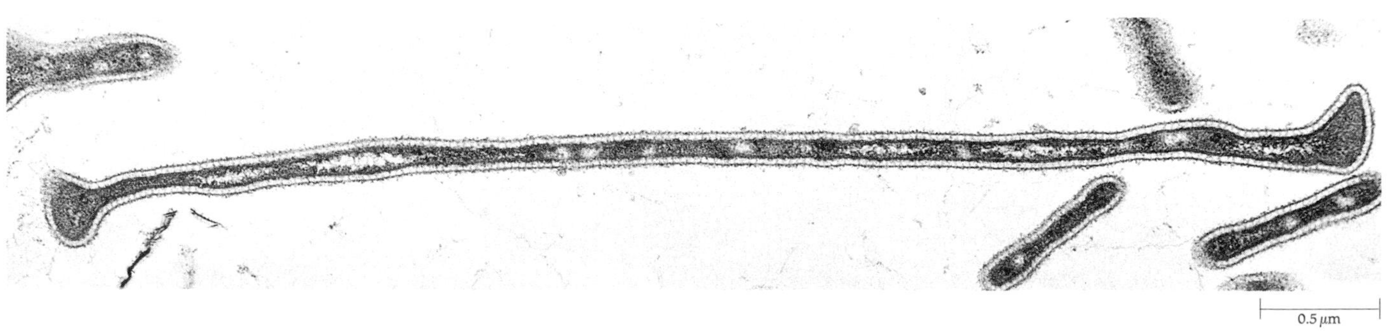

21-1
Thermophilic bacteria thrive at 92° C, a temperature close to the boiling point of water. This electron micrograph shows a single long filamentous type and several short rodlike types.

A few prokaryotes can form thick-walled, dry spores. These spores are inactive, resistant forms that enable the cells to survive for long periods of time without water or nutrients or in conditions of extreme heat or cold. They may stay dormant for years, and some remain viable even when boiled in water for as long as 2 hours.

From the ecological point of view, prokaryotes are important as decomposers, breaking down organic material to a form in which it can be used by plants. They also play a major role in the process known as nitrogen fixation, by which nitrogen gas (N_2) is reduced to ammonia (NH_3) or ammonium (NH_4^+). Although nitrogen is abundant in the atmosphere, no eukaryotes are able to use atmospheric nitrogen, and so the crucial first step in the incorporation of nitrogen into organic compounds depends largely on a few species of prokaryotes, some free-living and some in symbiotic association with plants. Some forms are photosynthetic, and some few are both photosynthetic and nitrogen-fixing.

THE PROKARYOTIC CELL

The essential features of a prokaryotic cell were outlined in Chapter 5 and are reviewed in Figure 21-2. The cell shown here is the familiar *Escherichia coli*, one of the eubacteria. The eubacteria, in general, and *E. coli*, in particular, are the best studied of the prokaryotes.

The Cytoplasm

The cytoplasm of prokaryotes is relatively unstructured. Except for the blue-green "algae," the cytoplasm is not divided or compartmentalized by membranes and does not contain any membrane-bound organelles. The most prominent feature within the cytoplasm is the chromosome. All prokaryotic chromosomes analyzed so far have proved to be one single, continuous ("circular") molecule of DNA. In addition, a prokaryotic cell may contain one or more plasmids (page 287).

The cytoplasm often has a fine granular appearance due to its many ribosomes. These are somewhat smaller than eukaryotic ribosomes but have the same general shape.

21-2
Cells of Escherichia coli, *a modern prokaryote that is a common inhabitant of the human digestive tract. The DNA is in the less dense (lighter-appearing) material in the center of each cell. The most dense small bodies in the cytoplasm are ribosomes. The two cells in the center have just finished dividing and have not yet separated completely.*

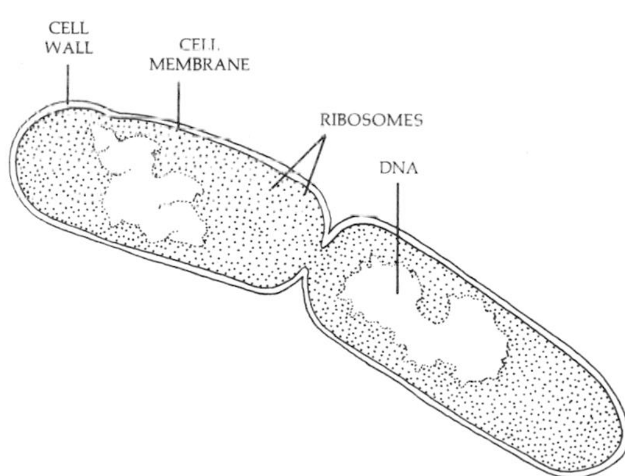

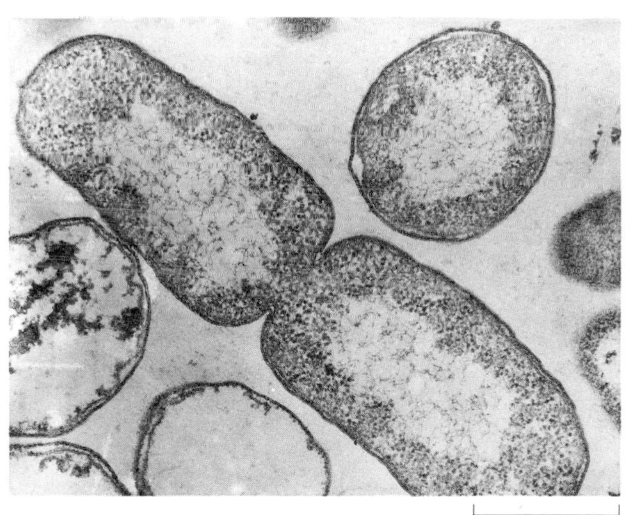

The Cell Membrane

The membrane surrounding a prokaryotic cell is similar in chemical composition to that of a eukaryotic cell (see page 102) but, with the exception of the mycoplasma (the smallest free-living cells known), it lacks cholesterol or other steroids. In the aerobic prokaryotes, the cell membrane incorporates the electron transport system found in the mitochondrial membrane of eukaryotic cells, and in one group of photosynthetic prokaryotes (the purple bacteria), the sites of photosynthesis are found in the outer membrane (page 186). In the purple bacteria and in aerobes with large energy requirements, the membrane is often extensively convoluted, with folds extending into the interior. These, of course, greatly increase the working surface of the membrane. Also, as we noted on page 198, the membrane appears to contain specific attachment sites (called mesosomes) for the DNA molecules; these sites are believed to play a role in ensuring the separation of the duplicate chromosomes at cell division.

The Cell Wall

Almost all prokaryotes are surrounded by a cell wall, which ranges from 5 to 80 nanometers in diameter and gives the different types their characteristic shapes. Many prokaryotes have rigid walls, some have flexible walls, and only the mycoplasma have no cell walls at all. Because most bacterial cells are hypertonic in relation to their environment, they would burst without their walls. (The mycoplasma live as intracellular parasites in an isotonic environment.)

21–3
A gram-stained smear of E. coli *and* Staphylococcus aureus *cells. The gram-positive* S. aureus *retain the gentian violet dye and appear as dark spheres. The rod-shaped* E. coli *cells, which are gram-negative and do not retain the violet dye, have been stained red.*

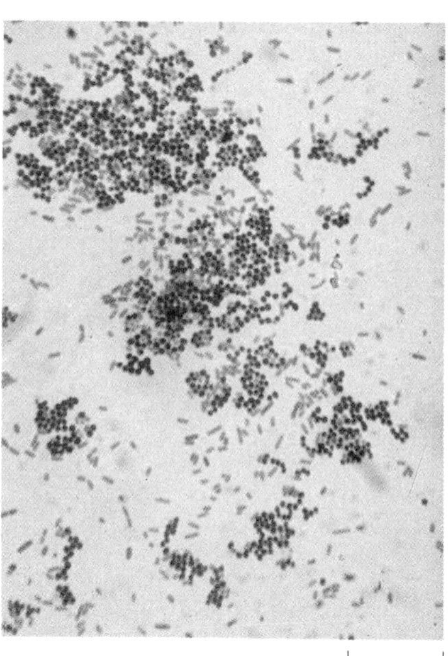

Chemical Structure of the Cell Wall

The cell walls of prokaryotes are complex and contain many kinds of molecules not present in eukaryotes. The walls of prokaryotic cells contain complex polymers known as peptidoglycans, which are primarily responsible for the mechanical strength of the wall (see pages 57 and 103). In addition, in some prokaryotes, large molecules of lipopolysaccharide are deposited over the peptidoglycan layer. Bacterial cell walls that lack the lipopolysaccharide layer combine firmly with such dyes as gentian violet, and those in which it is present do not. Those that combine with the dyes are known as gram-positive, whereas the others are gram-negative, after Hans Christian Gram, the Danish microbiologist who discovered this distinction. Gram staining is widely used as a basis for classifying bacteria, since it reflects a fundamental difference in the architecture of the cell wall. This architecture in turn affects various other characteristics of the bacteria, such as their patterns of susceptibility to antibiotics. Gram-positive bacteria are more susceptible to most antibiotics than are gram-negative bacteria. They are also more susceptible to lysozyme (page 80), an enzyme found in nasal secretions, saliva, and other body fluids, which digests the cell walls of bacteria.

In certain bacteria, a gluey polysaccharide capsule, which is secreted by the bacterium, is present outside the cell wall. The function of the capsule is not entirely clear, but its presence is associated with pathogenic activity in certain organisms. For example, as shown in Figure 14–5, the encapsulated form of *Diplococcus pneumoniae* is virulent, whereas the nonencapsulated form is generally nonvirulent. It appears that the capsule interferes with phagocytosis by host white blood cells.

21–4
(a) *A bacterial flagellum showing the basal end.* (b) *Diagram of a flagellum from* E. coli. *The basal body, which serves to anchor the flagellum to the cell wall and the cell membrane, consists of several rings surrounding a rod. The S and M rings are integrated into the cell membrane, the P ring is in the peptidoglycan layer of the cell wall, and the L ring is in the lipopolysaccharide layer. The filament is made up of several protein chains that form a helix with a hollow core.*

(a)

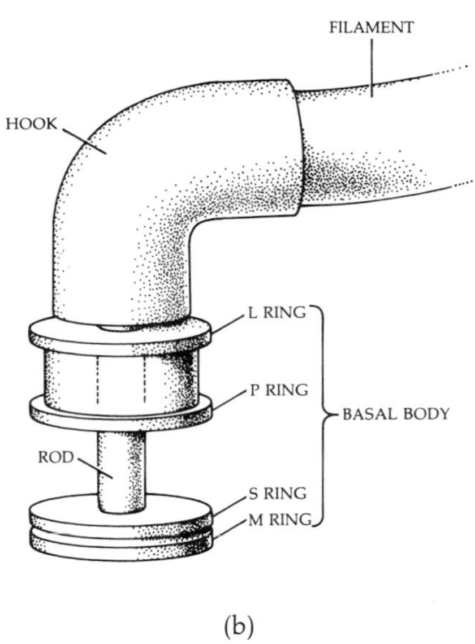

(b)

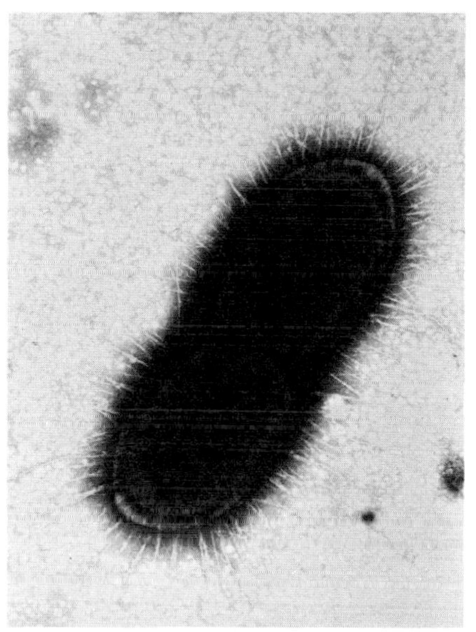

21–5
Pair of cells of Proteus mirabilis *stained to show the numerous pili.*

Flagella and Pili

Some types of bacteria have long, slender extensions, known as flagella and pili. Bacterial flagella, as we noted in Chapter 5, are made up of monomers of protein known as flagellin, which are assembled in chains and wound around a hollow central core. The flagella of different species differ slightly in diameter (12 to 18 nanometers), probably due to slight differences in the composition of their flagellin.

Electron microscopy has recently revealed that bacterial flagella are anchored into the cell wall and membranes by a fascinatingly complicated assembly (Figure 21–4). The filament (the helix of flagellin monomers with its hollow core) terminates in a hook made up of a different protein. The hook is inserted into a basal body that consists of a rod and, in gram-negative bacteria, two pairs of rings, one pair of which is embedded in the cell wall and one pair in the cell membrane. In gram-positive bacteria, only the inner pair of rings is present.

Pili (singular, pilus) are assembled from protein monomers in much the same way that the filaments of flagella are. (You will not be surprised to learn that the protein is called pilin.) They are rigid, cylindrical rods that extend out from the cell, sometimes to a considerable distance. They are shorter and thinner than flagella and are often present in large numbers (hundreds on a single cell). They serve to attach bacteria to a food source, to the surface of a liquid (where oxygen is present), or, in the case of conjugating bacteria, to one another.

DIVERSITY OF FORM

The oldest method of identifying microorganisms is by their physical appearance. The shape of individual prokaryotes is a result, as we noted previously, of their cell wall. In addition, different types of bacteria have characteristic patterns of growth,

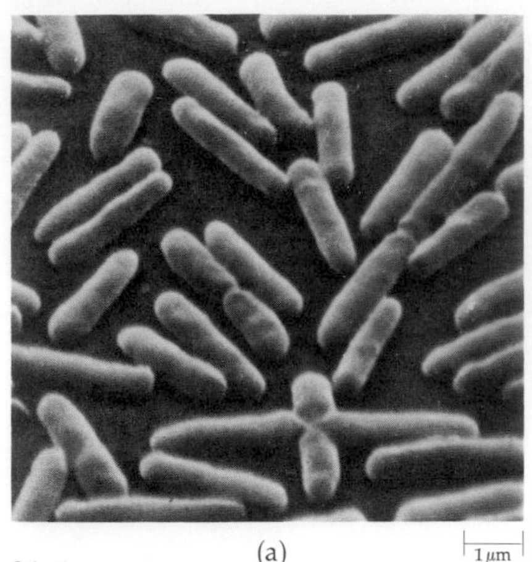

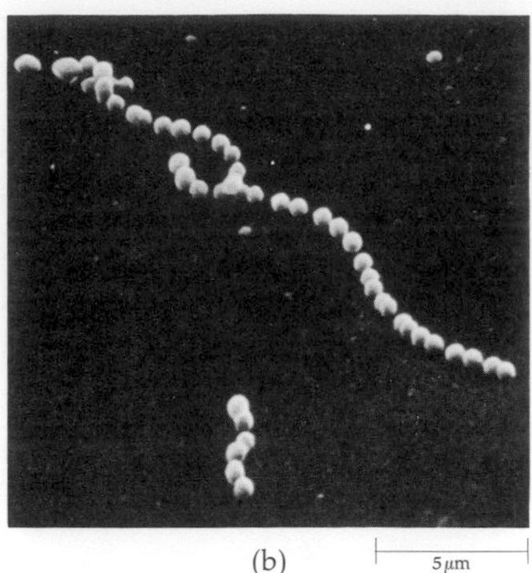

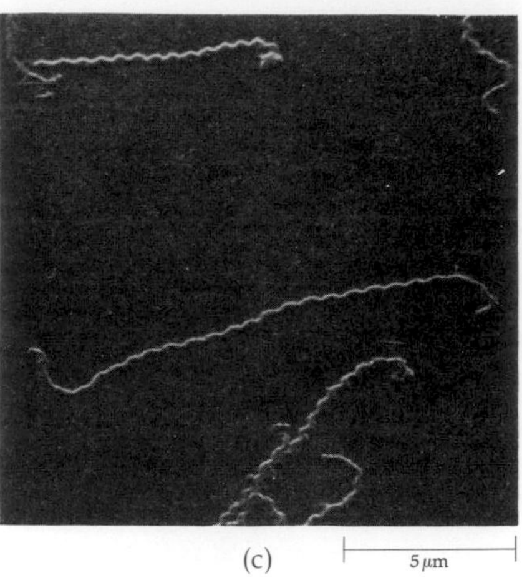

21–6

Three of the four major form-groups of eubacteria: (a) *bacilli,* (b) *cocci, and* (c) *spirilla. The rod-shaped bacteria (bacilli) include those microorganisms that cause lockjaw* (Clostridium tetani), *diphtheria* (Corynebacterium diphtheriae), *and tuberculosis* (Mycobacterium tuberculosis), *as well as the familiar* E. coli. *Among the cocci are* Diplococcus pneumoniae, *the cause of bacterial pneumonia,* Streptococcus lactis, *which is used in the commercial production of cheese, and* Nitrosococcus, *soil bacteria that oxidize ammonia to nitrates. The spirilla, which are less common, are helically coiled bacteria. Cell shape is a relatively constant feature in most species of bacteria.*

producing colonies that also have a distinctive shape, and many forms have sheaths or other coatings outside their cell walls. The eubacteria, or true bacteria, for example, may have any one of four forms. Straight, rod-shaped forms like *E. coli* are known as *bacilli;* spherical ones are called *cocci;* long, spiral rods are called *spirilla;* and short, curved rods, probably incomplete spirals, are called *vibrios.* Cocci may stick together in pairs after division (diplococci), they may occur in clusters (staphylococci), or they may form chains (streptococci). One bacterium that causes pneumonia is a diplococcus, while the staphylococci are responsible for many serious infections characterized by boils or abscesses.

Rod-shaped bacilli usually separate after cell division. When they do remain together, they spread out end to end in filaments, since they always divide in the same plane (transversely). Because these filaments are funguslike in appearance, the combining form *myco-* (from the Greek word for "fungus") is often a part of the name of these organisms. *Mycobacterium tuberculosis,* for example, the cause of tuberculosis, is a rod-shaped bacillus that forms a filamentous, funguslike growth.

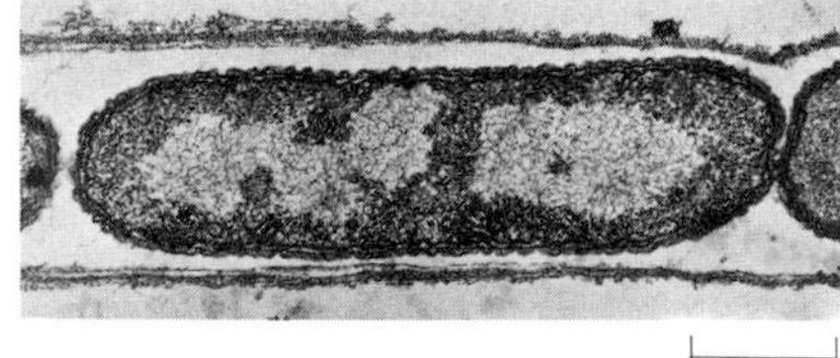

21–7

Sphaerotilus natans, *a filamentous bacterium enclosed in a sheath. Filaments of members of this genus are responsible for the brownish scum often seen on the surfaces of polluted streams.*

Spirochetes are among the easiest microorganisms to identify. They are very long (5 to 500 micrometers) and slender (about 0.5 micrometer in diameter) and have an unusual structure, known as an axial filament, attached at each end of the cell. Axial filaments have now been shown to contain two fibrils, identical to flagella, in their structure, and so they are recognized as modified flagella. They are wrapped around the cell between the cell membrane and the delicate wall, with the fibrils of each filament overlapping at the middle of the cell.

Rickettsia, another type of prokaryote distinguishable by its form, is shown in Figure 21–9.

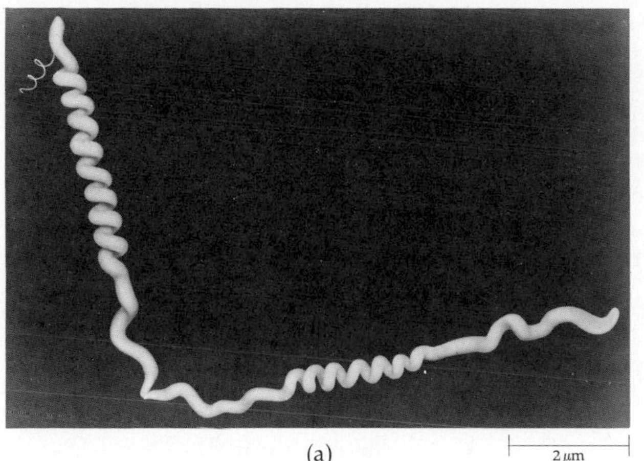

(a)

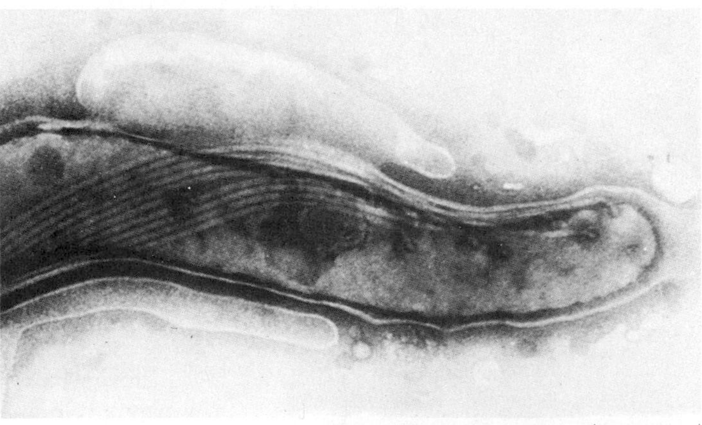

(b)

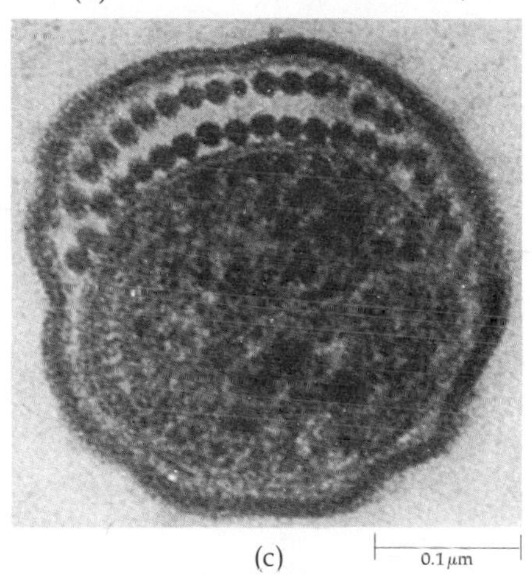

(c)

21–8
(a) *The spirochetes range in size up to 500 micrometers long, which is an enormous size for prokaryotes.* Treponema pallidum, *two of which are shown here, is the causative agent of syphilis.* (b) *The end of a spirochetal cell that has been stained to show the insertion points of the two fibrils of the axial filament (at the far right of the micrograph).* (c) *Cross section of a spirochete showing the fibrils of the axial filament between the cell membrane and the cell wall.*

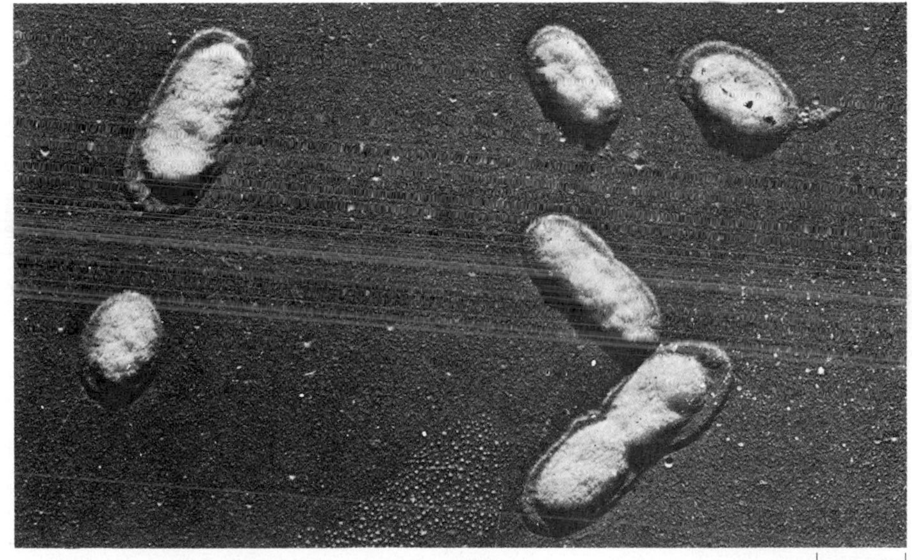

21–9
Rickettsiae are the smallest known cells. Typhus is caused by Rickettsia typhi, *shown here. It is spread from rats to humans by fleas. It can then be transmitted by body lice, and, under crowded conditions, large numbers of people can be infected in a very short time. More human lives have been taken by rickettsial diseases than by any other infection except malaria. At the siege of Granada in 1789, 17,000 Spanish soldiers were killed by typhus, 3,000 in combat. In the Thirty Years War, the Napoleonic campaigns, and the Serbian Campaign during World War I, typhus was also the decisive factor.*

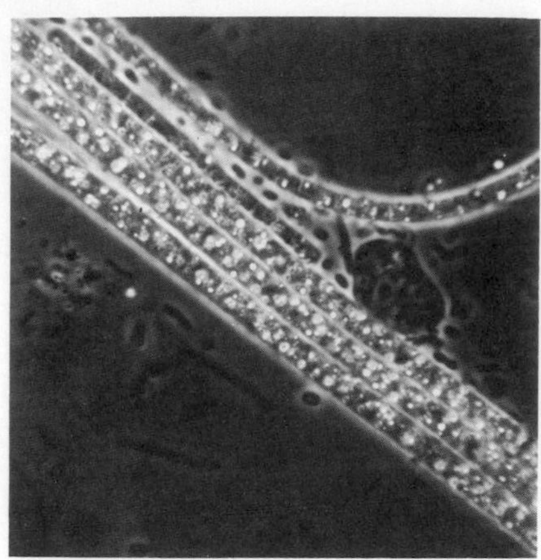

21–10
Beggiatoa, *gliding myxobacteria, are relatively large cells that form filaments and move in a "gliding" fashion similar to blue-green algae. Waves of contraction cause periodic changes in the form of the cell and bring about its movements.* Beggiatoa *is a chemosynthetic autotroph. The conspicuous granules in the cells are sulfur, produced by the oxidation of hydrogen sulfide—a process utilized by the bacteria for the production of energy. The gliding bacteria are essentially like colorless blue-green algae, a fact that underscores the fundamental similarity between these prokaryotes.*

TYPES OF MOTILITY

Prokaryotes move in a variety of ways, all of which are more or less mysterious. Flagella beat with a rotary movement. The flagella are so fine and the beat so fast, the motion of the flagellum itself cannot be seen. (The flagella of *Spirillum serpens,* for instance, are reported to have been clocked at 2,400 rpm.) Ingenious methods have been devised by which the cells can be tethered in place by the flagella. As a result, the cells rotate instead of the flagella, and the rotational movement can be observed. The basal end of the flagella, with its complex structure, apparently acts as a motor, perhaps running on chemiosmotic power (see page 166).

Another characteristic form of movement for some bacteria is gliding. Gliding requires attachment to a solid surface and the secretion of some sort of slime or mucus along which the cells slide. It has been suggested that cytoplasmic streaming may play a role in imparting motility to these cells, but so far the mechanism remains unknown.

Spirochetes move by a corkscrew motion caused, apparently, by contractions of the axial filament.

REPRODUCTION AND RESTING FORMS

Most prokaryotes reproduce by simple cell division (page 198), also called binary fission. In some forms, reproduction is by budding or by the breaking off of fragments of the cell. As they multiply, these prokaryotes, barring mutations, produce clones of genetically identical cells. Genetic recombinations take place as a result of conjugation, transformation, transduction, and exchanges of plasmids. It is not known how common such genetic recombinations are in nature or whether or not they occur in all types of prokaryotes.

Many prokaryotes have the capacity to form spores, which are dormant, resting cells. Again, the process has been studied most extensively in the rod-shaped eubacteria. It occurs characteristically when a population of cells, growing very rapidly, has begun to use up its food supply. Each cell, at the beginning of sporulation (spore formation), contains two duplicate chromosomes. A cell membrane grows around one of the chromosomes, separating it from the rest of the cell, which then engulfs the newly formed cell (as in phagocytosis). Thus the spore-to-be is now surrounded by two membranes, its own and that of the larger cell (this series of events is shown in Figure 21–11). A spore coat containing a unique peptidoglycan forms around this smaller cell. Covered by an outer layer of proteins composed largely of hydrophobic amino acids, this peptidoglycan is completely different from that present in the bacterial cell wall. The spore remains in this protected state until appropriate events trigger its germination. Germination takes place rapidly, with the uptake of water, the dissolution of the spore coat, and the formation of a new cell wall. Genetic studies of the spore-making *Bacillus subtilis* indicate that some 50 genes, clustered in about five segments of the chromosome, are involved in spore production.

The myxobacteria, a type of gliding bacteria, form fruiting bodies, which are brightly colored collections of spores and slime large enough to be seen by the unaided eye.

Spore formation greatly increases the capacity of prokaryotic cells to survive. The spores of *Clostridium botulinum,* for instance, the bacterium that causes botulism, are not destroyed by boiling for several hours.

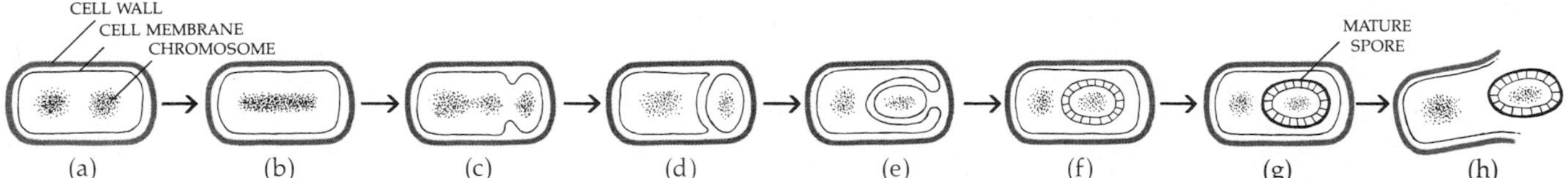

21–11
Endospore formation in Bacillus cereus. (a, b) *The two chromosomes within the vegetative cell have condensed into a rod-shaped form.* (c) *The transverse wall begins to form,* (d) *cutting off the spore material from the vegetative cell.* (e, f) *The vegetative cell grows around the spore, and the spore coat is formed.* (g, h) *The spore matures and is released from the cell. The micrograph shows an endospore of* Clostridium.

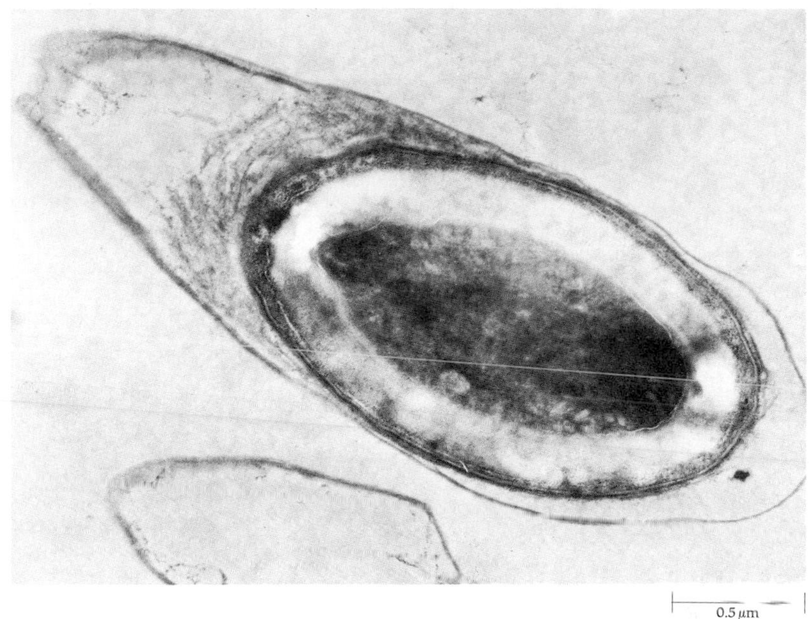

21–12
Scanning electron micrograph of the fruiting bodies of Chondromyces crocatus, *a myxobacterium. Each fruiting body may contain as many as 1 million cells.*

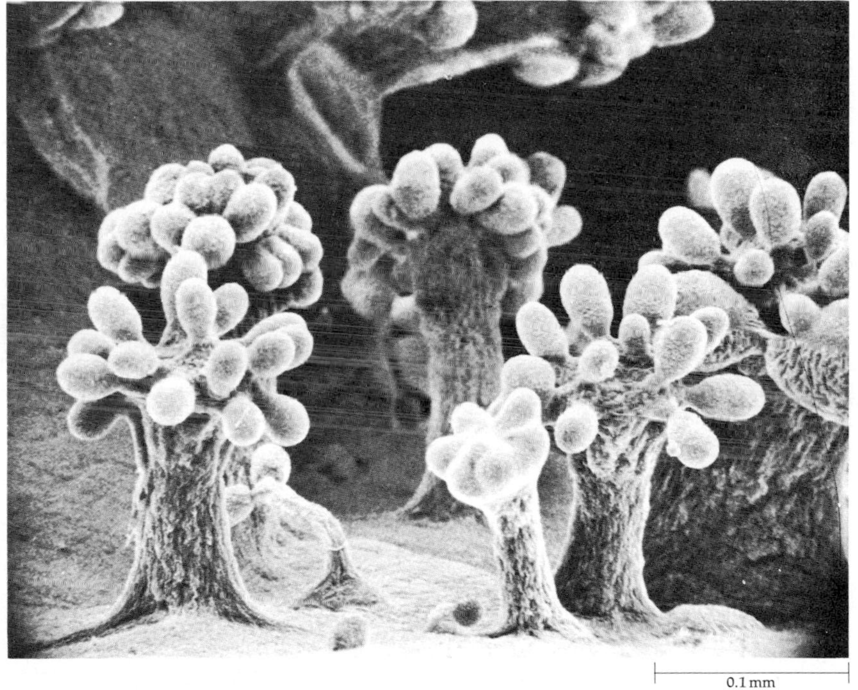

PROKARYOTIC NUTRITION

Heterotrophs

Most prokaryotes are heterotrophs; most types feed on dead organic matter. Bacteria and other microorganisms are responsible for the decay and recycling of organic material in the soil. Typically, different groups of bacteria play different, specific roles—such as the digestion of cellulose, starches, or other polysaccharides, or the hydrolysis of specific peptide bonds, or the breakdown of amino acids. Because of their high degree of nutritional specialization, prokaryotes are able to live in large numbers in the same small area with reduced competition and, indeed, with mutual assistance, as the activities of one group make food molecules available to another group. These combined activities release the nutrients and make them available to plants and, through plants, to animals. Thus, they are an essential part of ecological systems.

Some heterotrophic bacteria are symbionts. Some of these are parasites that break down organic material in the bodies of living organisms. The disease-causing (pathogenic) bacteria belong to this group, as do a number of other nonpathogenic forms. Some of the symbiotic bacteria have little effect on their hosts, and some are actually beneficial. Cows and other ruminants can digest cellulose only because their stomachs contain bacteria and certain symbiotic protozoans. Our own intestines contain a number of types of generally harmless bacteria (including *E. coli*). Some supply vitamin K, which is necessary for blood clotting. Others prevent us from developing serious infections. When the normal bacterial inhabitants of the human intestinal tract are destroyed—as can happen, for example, following prolonged antibiotic therapy—our tissues are much more vulnerable to disease-causing microorganisms. One group of very small bacteria, members of the genus *Bdellovibrio* (from *bdello,* the Greek word for leech) are parasites on other bacteria. They make a hole in the host cell wall and multiply between the wall and the membrane, digesting the host cell as they multiply.

Chemoautotrophs

Chemoautotrophic prokaryotes obtain their energy from the oxidation of inorganic compounds. Only prokaryotes are able to use inorganic compounds as an energy source.

Certain chemosynthetic bacteria are essential components of the nitrogen cycle, the process by which nitrogen compounds are cycled and recycled through ecosystems. One group oxidizes ammonia or ammonium (derived from the breakdown of organic materials, the activities of nitrogen-fixing prokaryotes, or, to a minor extent, from lightning or volcanic activities). The products of this reaction are nitrite (NO_2^-) and energy. Another group oxidizes nitrites, producing nitrate (NO_3^-) and energy. Nitrate is the form in which nitrogen moves from the soil into the roots of plants.

Sulfur is also required by plants for amino acid synthesis. Like nitrogen, it is converted to the form in which it is taken up by plant roots by the activities of chemoautotrophic bacteria that oxidize elemental sulfur to sulfate:

$$2S + 2H_2O + 3O_2 \longrightarrow 2H_2SO_4$$

Other sulfur bacteria, such as *Thiothrix* and *Beggiatoa,* obtain energy by oxidizing hydrogen sulfide.

A THIRD FORM OF LIFE

Recently a group of prokaryotes has been found to be so different from all the other forms that it has been proposed that they constitute a separate kingdom. They are called the methanogens ("methane-makers"), and, although their existence has long been known, they have not previously been studied extensively because they are obligate anaerobes—that is, they are poisoned by oxygen—and hence difficult to isolate and impossible to grow in ordinary culture media. The methanogens are the final participants in decomposing processes involving organic matter in an anaerobic environment, including marshes, lake sediments, and the digestive tracts of animals. They convert CO_2 and H_2 formed by other anaerobes by fermentation to methane gas (CH_4). They differ from all other prokaryotes in several significant ways. First, their cell walls do not contain peptidoglycans. Second, they do not use the Calvin cycle for carbon reduction. Third, they require several unique coenzymes that serve the function of NAD^+ and FAD as electron acceptors. Fourth, their mechanism for generating ATP is unusual and unknown. Finally, and perhaps most important, the RNA sequences of their transfer RNA and ribosomal RNA are markedly different from those in all other organisms.

A particularly interesting feature of the methanogens is that they would have been excellently suited to the conditions prevailing on earth during the earliest stages of biological evolution. They may also give exobiologists some clues as to what to look for in the search for life on other planets.

These important biogeochemical cycles and the roles of microorganisms in them will be discussed further in Chapter 44.

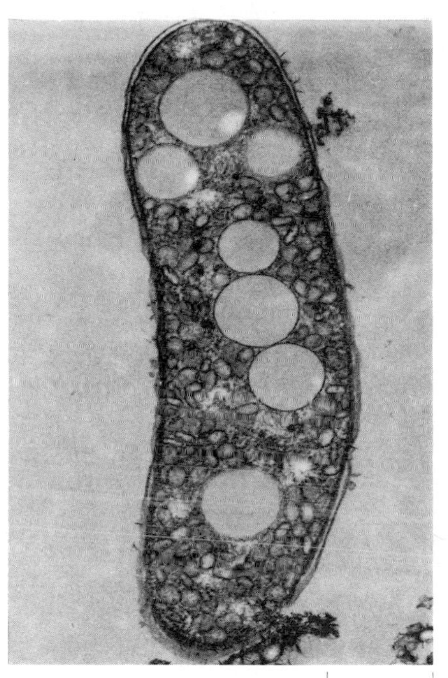

21-13
A photosynthetic bacterium, Rhodospirillum rubrum. *These bacteria have bacteriochlorophyll rather than chlorophyll* a *as a photosynthetic pigment. They do not lyse water and do not release oxygen.*

Photosynthetic Prokaryotes

Among the eubacteria are three photosynthetic forms: the green sulfur bacteria, the purple sulfur bacteria, and the purple nonsulfur bacteria. (The colors of the third group may actually range from purple to red or brown.) The chlorophyll found in the green sulfur bacteria, chlorobium chlorophyll, is chemically similar to chlorophyll *a*. The chlorophyll found in the two groups of purple bacteria is bacteriochlorophyll, which differs chemically in several details from chlorophyll *a* and is a pale blue-gray. The colors of the purple bacteria are due to the presence of several different yellow and red carotenoids, which function as accessory pigments.

In the photosynthetic sulfur bacteria, as we noted in Chapter 9, the sulfur compounds are the electron donors, playing the same role in bacterial photosynthesis that water does in photosynthesis in eukaryotes.

$$CO_2 + 2H_2S \xrightarrow{\text{light}} (CH_2O) + H_2O + 2S$$

Photosynthesis by eubacteria is carried out anaerobically, and it never results in the production of molecular oxygen (O_2). However, in all species except the methanogens (see essay), carbon is fixed by means of the Calvin cycle.

In the nonsulfur photosynthetic bacteria, other compounds, including alcohols, fatty acids, and a variety of other organic substances, serve as electron donors for the photosynthetic reaction.

THE BLUE-GREEN "ALGAE"

The blue-green "algae" are essentially bacteria, members of the gliding group. Unlike other prokaryotes, they contain chlorophyll *a*, which is also found in all photosynthetic eukaryotes. They have several kinds of accessory pigments, in-

21–14
Electron micrograph of the blue-green alga Anabaena cylindrica. *Blue-green algae have no chloroplasts and no membrane-bound nucleus, such as are found in eukaryotic algae and plant cells. Photosynthesis takes place in chlorophyll-containing membranes scattered throughout the cell, and the nucleus is a single molecule of DNA. The three-dimensional quality of this electron micrograph is due to freeze-fracturing (page 132).*

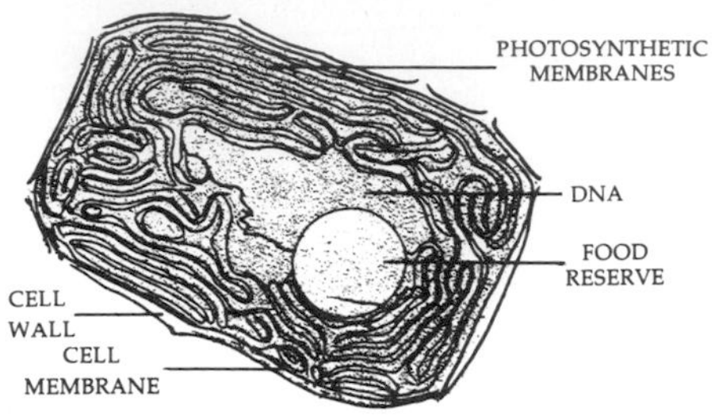

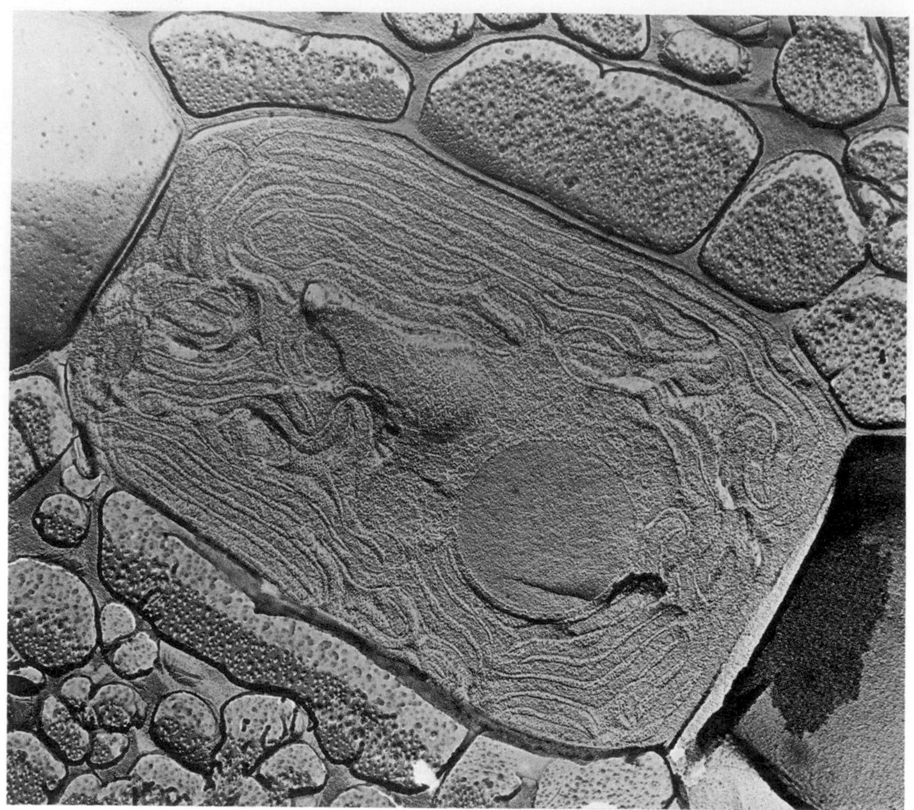

cluding xanthophyll, which is a yellow carotenoid, and several other carotenoids. (Carotenoids can also be found in photosynthetic eukaryotes and in some other bacteria.) The cells of blue-green algae may also contain one or two pigments known as phycobilins: phycocyanin, a blue pigment, which is always present, and phycoerythrin, a red one, which is often present. Chlorophyll and the accessory pigments are not enclosed in chloroplasts, as they are in plant cells, but are part of a membrane system distributed in the peripheral portion of the cell.

Cells of the blue-greens, as with many other prokaryotes, have an outer polysaccharide sheath, or coating. The outer sheath is often deeply pigmented, particularly in species that spread up onto the land; the colors include a light golden yellow, brown, red, emerald green, blue, violet, and blue-black. In addition, the carotenoids and phycobilins modify the color of the cells in which they occur. Thus, not only are they not algae, but only about half of the known species are actually blue-green. Indeed, the Red Sea was so named because of the dense concentrations, or "blooms," of red-pigmented blue-green algae that float on its surface.

Some species of blue-green algae are capable of nitrogen fixation. The photosynthetic, nitrogen-fixing blue-green algae have gone the farthest along the pathway to independence hypothesized for prokaryote evolution. They have the simplest nutritional requirements of any living thing, needing only N_2 and CO_2, which are always present in the atmosphere, a few minerals, and water.

Table 21-1 *Major Groups of Prokaryotes*

GROUP (EXAMPLES)	FORM	MODE OF MOVEMENT	MODE OF REPRODUCTION	MODE OF NUTRITION	ECOLOGICAL ROLE	OTHER DISTINGUISHING FEATURES
Eubacteria *(Escherichia coli, Streptococcus, Staphylococcus, Mycobacterium tuberculosis, Clostridium)*	Rod-shaped, spherical, spirillum, vibrio	Flagella or nonmotile	Binary fission	Chemoautotrophs, photosynthetic autotrophs, heterotrophs	Decomposers, symbionts, pathogens (lockjaw, diphtheria, tuberculosis)	Rigid cell wall; form endospores
Prosthecate and budding bacteria	Unusual shapes, appendages (prosthecae)	Flagella	Budding, binary fission (some members)	Most heterotrophic, some chemoautotrophs	Decomposers	
Gliding bacteria	Filaments; large and short rods	Gliding (mechanism unknown)	Short filaments of cell break off; binary fission	Heterotrophic, chemoautotrophic	Decomposers, especially of complex polysaccharides (myxobacteria)	Flexible cell wall; some form fruiting bodies
Filamentous (sheathed) bacteria	Individual cells enclosed in common sheath	Flagella	Binary fission	Heterotrophic		
Actinomycetes *(Streptomyces, Actinomyces)*	Branching multicellular filaments	Nonmotile	Spores, fragmentation, binary fission	Heterotrophic	Decomposers (degrading lipids and waxes), pathogenic (tuberculosis and leprosy)	Many are moldlike in appearance
Spirochetes *(Spirocheta, Treponema, Leptospira)*	Extremely long, helical, flexible	Twisting (axial filament)	Binary fission	Heterotrophic	Symbionts (in mollusks), decomposers, pathogens (syphilis)	Obligate anaerobes (many forms)
Rickettsiae; Chlamydiae (trachoma)	Small (0.5 × 1.1 micrometers), rigid cell wall	Nonmotile	Binary fission	Heterotrophic	Pathogens (typhus, spotted fever)	Very heterogeneous group; all obligate intracellular parasites
Mycoplasma	Smallest free-living cells, no cell walls	Nonmotile	Binary fission	Heterotrophic	Pathogens (mycoplasmal pneumonia)	Cell membranes contain cholesterol, many intracellular parasites
Blue-green "algae" *(Nostoc)*	Unicellular and filaments	Gliding and nonmotile	Binary fission	Photosynthetic autotrophs	Carbon and nitrogen fixation	Should be classified with gliding bacteria

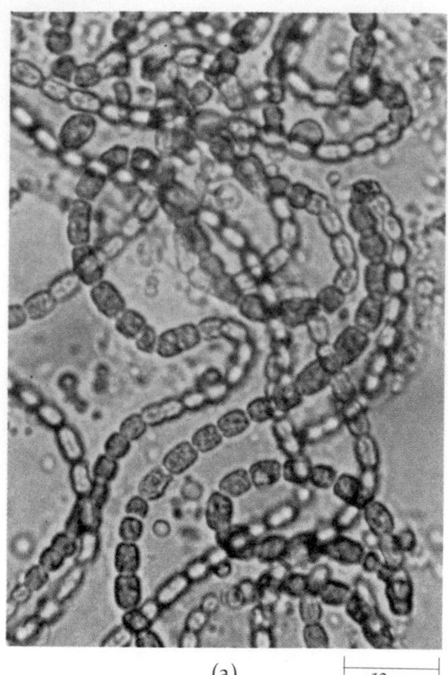

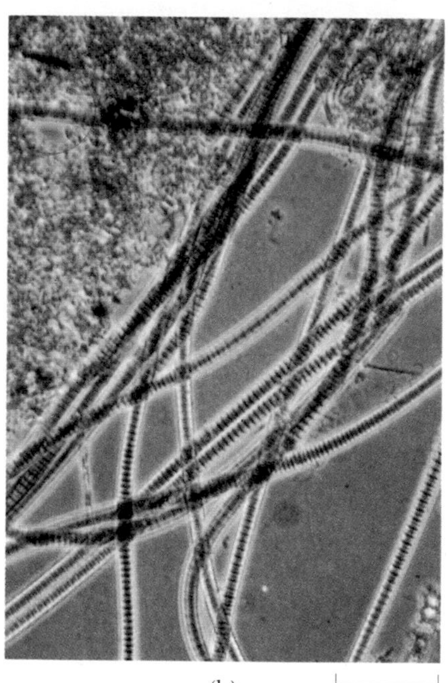

21–15
(a) *A gelatinous colony of the blue-green alga* Nostoc. (b) Oscillatoria, *a filamentous blue-green alga. These algae, now widely regarded as types of bacteria, have chlorophyll* a *and produce oxygen during photosynthesis.*

The ecological importance of the blue-green algae appears to be less than that of the nitrogen-fixing bacteria, at least for agriculture. However, in Southeast Asia, rice can be grown on the same land for years without the addition of fertilizers because of the rich growth of nitrogen-fixing blue-green algae in the rice paddies.

Because of their nutritional independence, the blue-green algae are able to colonize bare areas of rock and soil. A dramatic example of such colonization was seen on the island of Krakatoa in Indonesia, which was denuded of all visible plant life by its cataclysmic volcanic explosion of 1883. Filamentous blue-green algae were the first living things to appear on the pumice and volcanic ash; within a few years they had formed a dark-green gelatinous growth. The layer of blue-green algae eventually became thick enough to provide a substrate for the growth of higher plants. It is very probable that the blue-green algae were similarly the first colonizers of land in the course of biological evolution.

VIRUSES

Viruses do not fit easily into any of the traditional kingdoms of living organisms, and the problem of categorizing them is made even more difficult by the fact that there is doubt about whether or not they should be considered living.

Structure of Viruses

In size, viruses range from about 17 nanometers (a hemoglobin molecule is 6.4 nanometers in diameter) to about 300 nanometers, larger than small bacteria. The larger ones are at the limits of resolution of the light microscope.

Viruses are made up of nucleic acid, either DNA or RNA, enclosed in a protein coat. The protein coat determines the specificity of the virus; a cell can be infected by a virus only if that type of cell has a receptor site for the virus protein. Thus cold viruses infect cells in the mucous membranes of the respiratory tract; measles and chicken pox infect skin cells; and polio infects the upper respiratory tract, the intestinal lining, and sometimes nerve cells. Even bacteria, as we noted in Section 3, have their own set of specific viruses, the bacteriophages.

In some virus infections, the protein coat is left outside the cell (see Figure 15–4, page 263); in others, the intact virus enters the cell, but once inside, the protein is destroyed by enzymes. The DNA of a DNA-containing virus serves as a template for more viral DNA and also for messenger RNA, which codes for viral enzymes, viral coat protein, and probably, at least in some cases, repressors and other regulatory chemicals. The virus uses the equipment of the host cell, including ribosomes, transfer RNA molecules, amino acids, and nucleotides. Many viruses use host enzymes as well as those coded for by their own nucleic acids, and some break up host DNA and recycle the nucleotides as viral DNA. In the case of the RNA viruses, the viral RNA not only serves as a template for more RNA but also acts directly as messenger RNA. As we noted in Chapter 19, viral RNA can also serve as a template for viral DNA. This phenomenon of reverse transcription is observed almost exclusively with cancer-causing viruses.

Virus particles are assembled within the host cell. They then leave the cell, often lysing the membrane. Some viruses, such as influenza virus, bud off from the host cell membrane and, in so doing, become wrapped in fragments of it. Each new viral particle is capable of setting up a new infective cycle in an uninfected cell.

21–16
(a) *Adenovirus, one of the many viruses that cause colds in humans. This virus is an icosahedron. Each of its 20 sides is an equilateral triangle composed of identical protein subunits. Many viruses, and also Buckminster Fuller's geodesic domes, are constructed on this principle. There are 252 subunits in all. Within the icosahedron is a core of double-stranded DNA.* (b) *A model of the adenovirus, made up of 252 tennis balls.* (c, d) *Electron micrograph and diagram of influenza virus. The virus is composed of a core of RNA surrounded by a lipoprotein envelope through which protrude stubby protein spikes. The influenza virus, for reasons that are not understood, mutates frequently. The changes in its nucleic acid alter its proteins and, hence, previously formed antibodies no longer "recognize" it. New strains of influenza virus are likely to arise more rapidly than new vaccines can be produced to combat them.* (e, f) *Electron micrograph and diagram of a bacteriophage, showing the many different structural components of the protein coat. The DNA of the virus codes for all these structural proteins.*

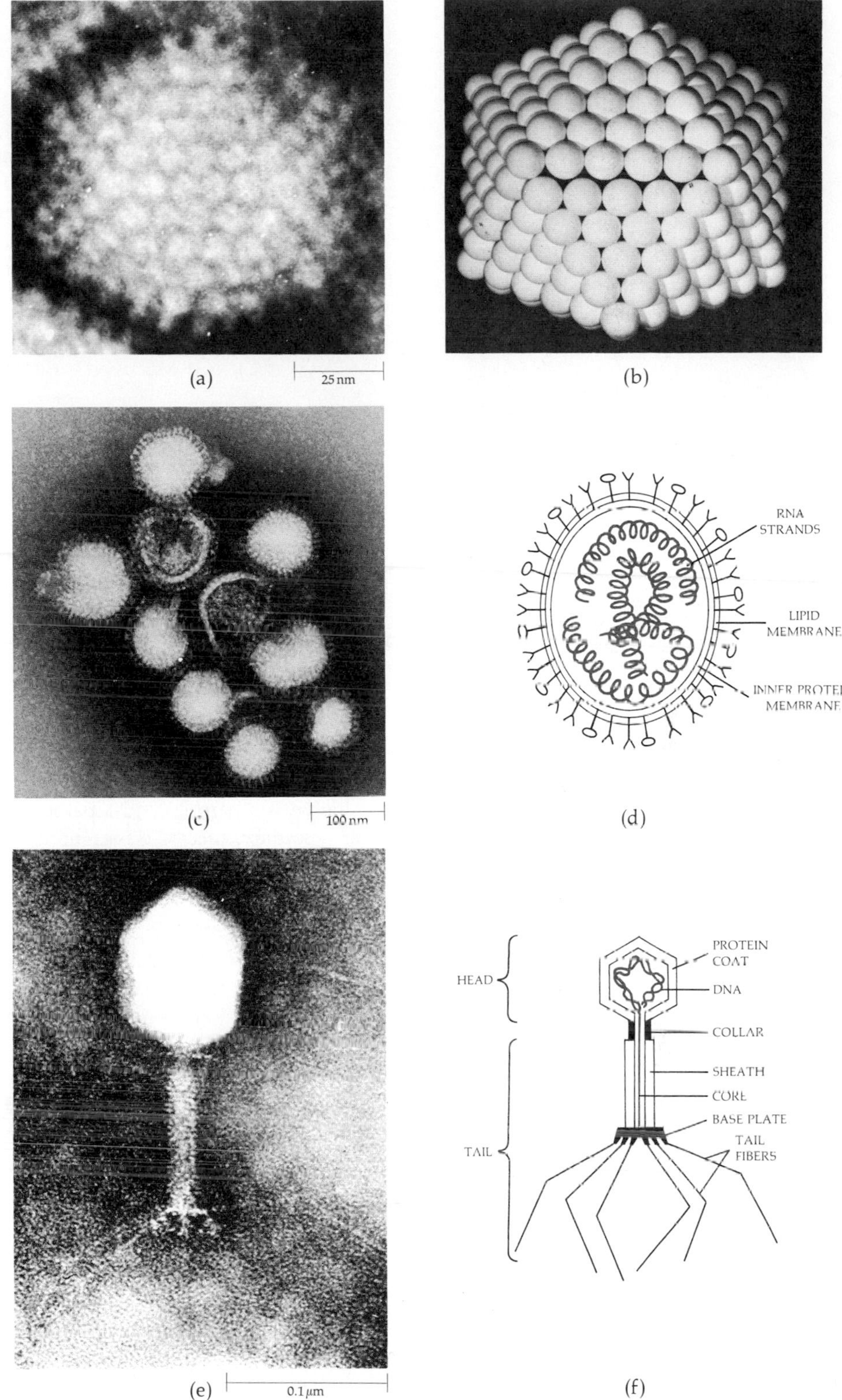

Origin of Viruses

Are viruses very simple forms of life? Can they give us clues to the nature of a precellular living system of which no fossil is likely to remain? As information increases on the nature of viruses and, in particular, of their relationships with host cells, this interpretation of viral origins seems less and less likely. Viruses are able to reproduce and to make their protein coats only because they are capable of commandeering the enzymes and other metabolic machinery of the host cell. Without this machinery, they are as inert as any other macromolecule, lifeless by many criteria. It seems more likely that viruses are cellular fragments that have set up a partially independent existence. S. E. Luria (page 263) has called viruses "bits of heredity looking for a chromosome."

MICROORGANISMS AND HUMAN ECOLOGY

Symbiosis

Symbiosis ("living together") is a close and permanent association between organisms of different species. There are three types of symbiotic relationships. If the relationship is beneficial to both species, it is called *mutualism*. If one species benefits from the association while one is neither harmed nor benefited, it is called *commensalism*. If one species benefits and the other is harmed, the relationship is known as *parasitism*. Microorganisms and human beings are often symbionts. When the microorganisms are parasites, they are said to be pathogenic—disease-causing. The lines of demarcation between the different categories of symbiosis are not clearcut. A bacterium such as *E. coli*, which lives in the lower intestine, may be commensal, depending on its host for food and shelter and neither helping nor harming. If it produces a needed vitamin or digestive enzyme, the relationship is then mutualistic. If it gets in the bloodstream and causes septicemia (blood poisoning), it is a parasite and a pathogen.

Evolutionary success is measured, as we noted, in terms of surviving progeny. A parasite that destroys its host is less likely to be successful by the evolutionary criterion than one that enjoys a long and comfortable relationship with its protector. Disease is likely to be the result of a sudden change in the parasite, in the host, or in their relationship. For instance, many persons harbor small numbers of *Mycobacterium tuberculosis* without any symptoms of disease; however, factors such as malnutrition, fatigue, or other diseases may weaken host defenses so that the signs of tuberculosis appear. Similarly, herpes simplex, the virus that causes "cold sores," may remain latent for months or years at a time, with the lesion appearing only in response to some change in the condition of the host.

How Microbes Cause Disease

The pathogenic effects of microbes are produced in a variety of ways. The viruses, as we have seen, enter particular types of cells and often destroy them. Bacteria may produce cell destruction also. Frequently, however, the effects we recognize as disease are caused not by the direct action of the pathogens but by toxins, or poisons, produced by them. For instance, diphtheria is caused by a bacillus, *Corynebacterium diphtheriae*. The organisms are inhaled and establish infection in the upper respiratory tract. They do not invade the bloodstream, but the powerful toxin they produce does. (This toxin is made only when the bacterium is harboring a particular prophage, see page 322.) The toxin is absorbed by the body cells. Under the influence of the toxin, an enzyme involved in the assembly of activated amino acids into polypeptides becomes irreversibly bound to a fragment of NAD^+. The cells can no longer make proteins and so can no longer function.

Some diseases are the result of the body's reaction to the pathogen. In pneumonia caused by *Streptococcus pneumoniae,* the infection causes a tremendous outpouring of fluid and cells into the air sacs of the lungs, thus interfering with respiration. The symptoms caused by fungus infections of the skin similarly result from inflammatory responses.

A single disease agent can cause a variety of diseases. Skin infections of *Streptococcus pyogenes* cause the disease known as impetigo. Throat infections by the same bacteria are the familiar strep throat. Throat infections with strains of the bacteria that produce toxins (again as a result of a bacteriophage) are known as scarlet fever. Among the persons with untreated strep throat or scarlet fever, about 0.5 percent develop rheumatic fever, which is characterized by inflammatory changes in the joints, heart, and other tissues, apparently as a result of reactions involving the body's own immune system. Conversely, many agents may cause the same symptoms; the "common cold" can result from infection with any one of a large number of viruses.

Prevention and Control of Infectious Disease

Although microorganisms were first seen and depicted with remarkable accuracy by Antony van Leeuwenhoek in the late seventeenth century, they were not generally recognized as a cause of disease until 100 years ago. The relationship was made clear by the work of Louis Pasteur on disease in silkworms and by Robert Koch, who studied anthrax in sheep and tuberculosis in humans.

Recognition of microbes as disease agents opened the way to control measures, among the most important of which were the introduction of sterile procedures in hospitals. Among the agents spreading bacterial infection were physicians, surgeons, nurses, and hospitals. Childbed fever, for example, once a leading cause of sickness and death among young women, was transmitted almost exclusively by physicians who carried streptococci from patient to patient on their hands or unsterilized instruments.

The battlefields of World War II were the proving grounds for new antimicrobial drugs, such as sulfa and penicillin. These drugs made possible not only treatment of wounds but also the widespread and often life-saving use of major surgery as a treatment for cancer and other diseases. Ironically, with the advent of strains of bacteria resistant to these drugs and endemic in hospitals, the latter are once more becoming reservoirs of serious bacterial diseases.

Even more important than improvement in medical practices was the institution of public health measures, for example, the eradication of fleas, lice, mosquitoes, and other agents that carry disease; disposal of sewage and other wastes; protection of public water supplies; pasteurization of milk; quarantine; and other procedures. Not long ago, for instance, infant mortality during the first years of life was often as high as 50 percent in some localities owing to infant diarrhea caused by contaminated milk and water.

Many infectious diseases can be prevented by immunization, a practice that dates back to 1796. Observing that persons who had cowpox, a mild disease, did not develop smallpox, Edward Jenner boldly undertook to inoculate healthy people with cowpox. (Cowpox is called vaccinia and hence immunization came to be called vaccination.) We know now that vaccination is effective because the protein coat of the vaccinia virus is similar enough to that of the smallpox virus to stimulate the production of antibodies effective against both viruses. (Antibodies and immunity will be discussed further in Chapter 35.) Immunization procedures have now been developed against a large number of diseases including mumps, measles, German measles, and polio (all virus-caused) and diphtheria, whooping cough, and tetanus (all bacteria-caused). They involve stimulation of the immune system either with a related microorganism (as in smallpox or tuberculosis), with a killed strain (as in the Salk vaccine or typhus vaccine), or with a strain attenuated (weakened) by adaptation to another host (oral polio vaccine, measles, German measles).

In 1935, sulfanilamide, the first of the new "wonder drugs" for the control of bacterial infection, was discovered in Germany, and in 1940 the effects of penicillin (first noted by Alexander Fleming in 1929) were reported from England. Penicillin was the first-discovered antibiotic—by definition, a chemical that is produced by a living organism and is capable of inhibiting the growth of microorganisms.

Many antibiotics are produced by bacteria, especially the actinomycetes; some are formed by fungi. Many, including penicillin, can now be synthesized in the laboratory.

The tremendous decrease in deaths from infectious disease over the last decades is a chief cause of the present population explosion.

21-17
Until the 1950s, poliomyelitis was one of the most feared of all childhood diseases. It was brought under control first by a killed-virus (Salk vaccine), which was subsequently replaced by a live-virus vaccine (Sabin vaccine), the one now in wide general use. The possible aftereffects of polio are depicted in this Egyptian hieroglyph, dating from the nineteenth dynasty (1320–1200 B.C.).

SUMMARY

The prokaryotes include the bacteria and the blue-green algae (now widely regarded as a class of bacteria). The prokaryotes are the only group of cells in which the DNA is not associated with protein and in which there is no membrane-bound nucleus or other membrane-bound organelle. Bacteria are characteristically surrounded by a cell wall containing peptidoglycans, amino sugars linked in a complex polymer. Within the wall is a cell membrane that does not contain cholesterol and that often incorporates enzyme systems such as—in aerobic forms—the electron transport chain. Many species also have long, slender extensions, flagella and pili. The varied forms of prokaryotes are imparted by the cell walls, which may be rigid or flexible.

Prokaryotes move by means of their flagella, which have a complex structure and a rotary movement; by gliding, which involves the secretion of a mucus or slime; or not at all. Reproduction is generally by binary fission, but some forms reproduce by budding or fragmentation. Genetic exchanges and recombinations take place by conjugation, transformation, and transduction, but the extent of these phenomena in nature is not known. Many types form tough, resistant spores.

Prokaryotes comprise both heterotrophs and autotrophs; the autotrophs include both chemosynthetic and photosynthetic forms. Chemosynthetic autotrophs are found only among prokaryotes. The photosynthetic eubacteria use a variety of compounds, including sulfur, as electron acceptors and do not produce oxygen gas. The blue-green "algae" have a type of photosynthesis essentially like that of plants. They lack chloroplasts and their photosynthetic pigments are embedded in internal membranes.

Viruses are combinations of nucleic acids (DNA or RNA) and proteins. They are obligate intracellular parasites. The nucleic acid replicates within the host cell and directs the formation of new protein coats, utilizing the host cell's enzymes and other metabolic equipment.

Microorganisms and human beings often live in symbiotic relationships. Symbiosis includes (1) commensalism, in which one member benefits, (2) mutualism, in which both benefit, and (3) parasitism, in which one is harmed. Pathogenic microorganisms cause disease by direct tissue destruction, by producing toxins, and by provoking host defenses. A symbiotic relationship may change from one type to another as a result of changes in the host or, sometimes, in the pathogen.

QUESTIONS

1. Label the drawings below.

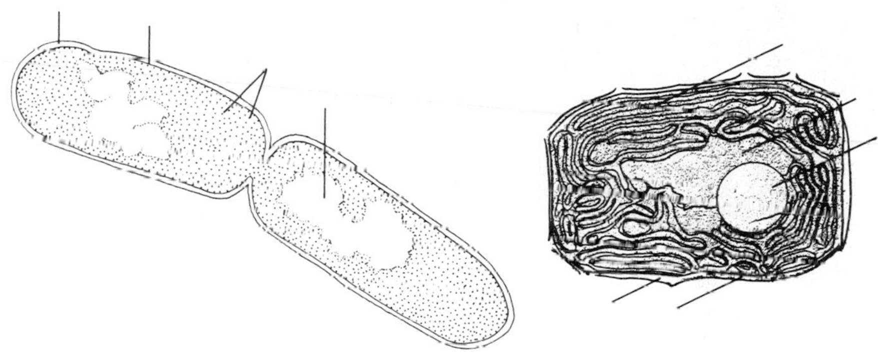

2. Many microorganisms produce antibiotics. What do you think their function might be?

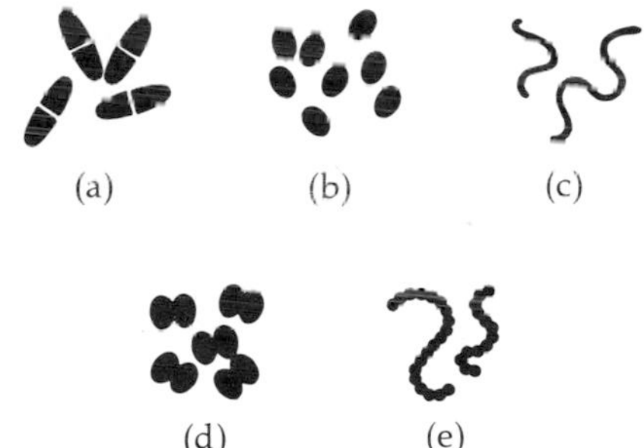

3. Identify the types of bacteria shown in the drawings on the left.

4. Describe the three types of symbiotic association that, in your opinion, would apply to the following relationships: dog and flea; human being and dog; maggot and wound; *E. coli* and human being; diphtherial microorganism and human being.

5. Before the structure of any virus was known, Crick and Watson predicted that the protein coats of viruses would prove to be made up of large numbers of identical subunits. Can you explain the basis of their prediction?

6. Some biologists consider viruses to be living organisms. By what criteria might viruses be considered as alive?

CHAPTER 22

The Protists

22–1
Stentor, *a protozoan. Extended, it looks like a trumpet (this genus is named after "bronze-voiced" Stentor of the* Iliad, *"who could cry out in as great a voice as 50 other men").* Stentor *is crowned with a wreath of membranelles. These beat in rhythm, creating a powerful vortex that draws edible particles up to and into the funnel-like groove leading to the cytostome ("mouth"). Elastic protein threads, myonemes, run the length of the body. When contracted,* Stentor *is an almost perfect sphere. The macronuclei are visible here.*

The protists are all eukaryotic. Most are unicellular, and those that are not have relatively simple multicellular structures. Apart from these two characteristics, they are an extremely diverse group of organisms. The kingdom includes heterotrophs, among them a number of parasitic forms; photosynthetic autotrophs; and a few versatile organisms that are both heterotrophic and photosynthetic.

Protists are far from simple. Indeed among one major phylum, the Protozoa, are found probably the most complex of all cells, with an astonishing variety of highly specialized structures. That single cells should be so complex is actually not surprising. Among the protists, each cell is a self-sufficient organism, as capable of meeting all of the requirements of living as a complex, multicellular organism.

As their complexity reminds us, modern protists—that is, all protists alive today—are not beginnings. They are in themselves the end products of millions of years of evolution. Those that have not changed much in that time—amoebas might be an example—have survived over the millennia because they are exquisitely adapted to the environment in which they live. It is all a question of how one measures evolutionary success.

There are eight major phyla of Protista (Table 22–1), six of which are algae. Biologists generally agree (1) that the protists represent a number of quite different evolutionary lines, and (2) that all other eukaryotic organisms—fungi, plants, and animals—originated from primitive protists.

PHYLUM PROTOZOA

The Protozoa—"first animals"—are one-celled heterotrophs. Classification of the Protozoa into three of their four major groups is based upon their characteristically different methods of locomotion: (1) by flagellar movement (the zooflagellates—"animal flagellates"—or mastigophores), (2) by pseudopodia (the sarcodines), and (3) by ciliary movements (the ciliates). The fourth major group, the sporozoans, are nonmotile during the major phases of their lives and all are parasites.

Protozoa usually reproduce asexually, by mitosis. Many also have sexual cycles, involving meiosis and the fusion of gametes, or *syngamy.* (Syngamy in animals is usually referred to as fertilization.) The fusion of gametes results in a diploid ($2n$) zygote. The zygote is often a thick-walled, resistant resting cell, formed during periods of cold or drought. Some protozoa, notably the ciliates, have a special type of sexuality involving exchanges of genetic material between individuals.

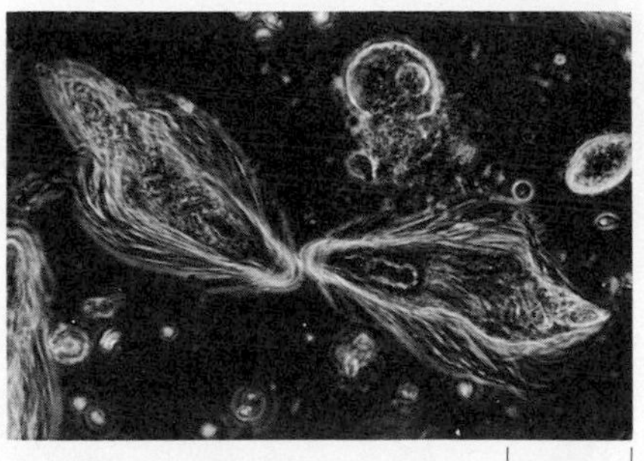

22–2
Two Trichonympha *cells. These zooflagellates, which break down cellulose, are entirely responsible for the well-known proficiency of their termite hosts to digest wood.*

Table 22–1 *The Kingdom Protista*

Phylum Protozoa	Unicellular heterotrophs, including flagellates (mastigophores), amoeba-like organisms (sarcodines), ciliates, and sporozoans
Phylum Euglenophyta	*Euglena* and related algae, all unicellular, mostly freshwater
Phylum Chrysophyta	Diatoms and related algae, all unicellular, mostly marine
Phylum Pyrrophyta	Dinoflagellates and related algae, all unicellular, marine and freshwater
Phylum Chlorophyta	Green algae, including unicellular, colonial, and multicellular forms; mostly freshwater
Phylum Phaeophyta	Brown algae, including the kelps, all multicellular, almost all marine
Phylum Rhodophyta	Red algae, all multicellular, mostly marine
Phylum Gymnomycota	Slime molds

Class Mastigophora

The zooflagellates are regarded as the most primitive of the Protozoa. They are thought to have been derived from photosynthetic flagellated cells, such as *Euglena,* that lost their chloroplasts. Almost all of the smaller cells have one or two flagella, and larger cells frequently have many flagella. These flagella have the characteristic 9 + 2 structure (page 116). The zooflagellates multiply asexually by mitosis and cell division, and, in some forms, sexually, by syngamy. They generally have no outer wall, and some are able to form pseudopodia, temporary extensions of the cell body that are used in locomotion and in engulfing food particles. Most are free-living, but some are parasitic. The latter include *Trypanosoma gambiense,* a flagellate that causes African sleeping sickness, and members of the genus *Trichonympha,* complex and beautiful flagellates that live as symbionts in the digestive tracts of termites, where they digest the wood ingested by the termite.

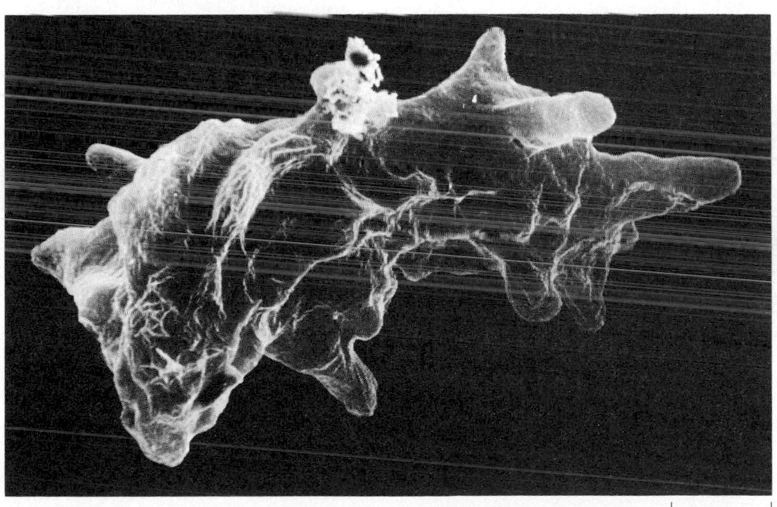

22–3
Scanning electron micrograph of Amoeba proteus, *a single-celled organism named for Proteus, a Greek god capable of changing his shape at will. Amoebas use their pseudopodia ("false feet") both for moving and for capturing prey.*

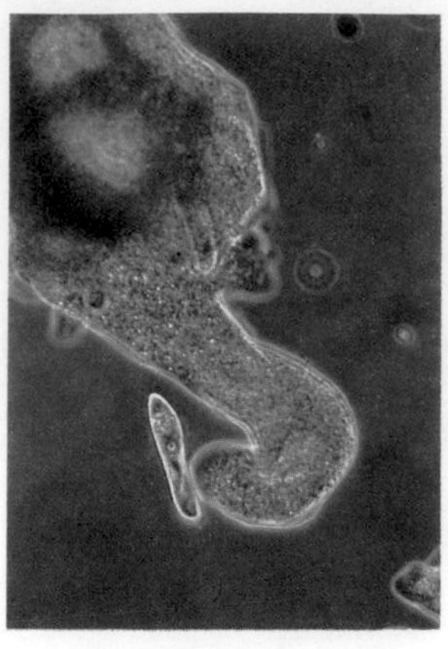
(a)

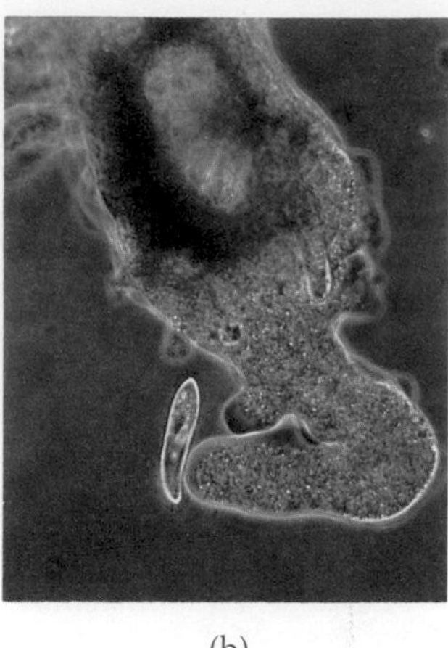
(b)

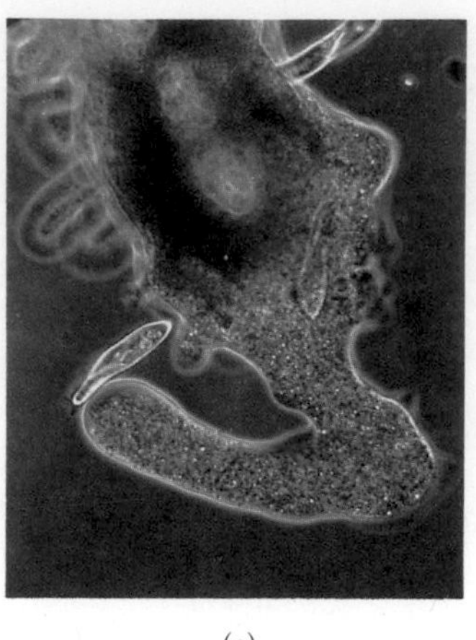
(c)

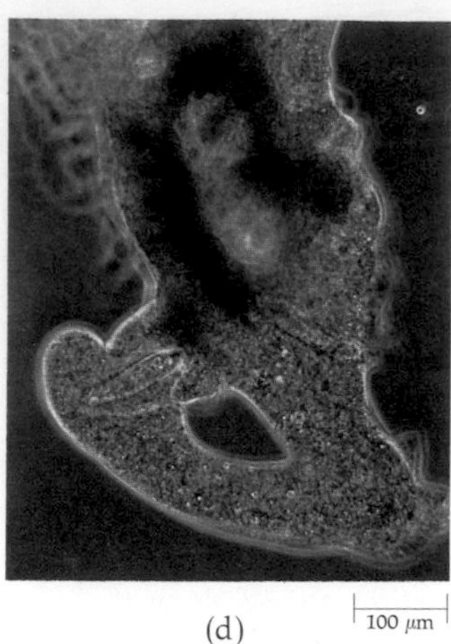
(d)

22–4
The giant amoeba Chaos chaos *captures a* Paramecium. *Although seemingly disorganized,* Chaos *is able to sense its prey, move toward it, and send out a pseudopod of the right size and shape to envelop it.*

Class Sarcodina

The sarcodines include amoebas and their relatives. They have no coat or wall outside their cell membranes and generally move and feed by the formation of pseudopodia. They take their name from the word "sarcode," coined in the early nineteenth century to describe the "simple, glutinous, and homogeneous jelly" of which, at one time, simple cells were thought to be composed. Despite their uncomplicated appearance, they are complex cells and are even capable of some complex behavior patterns, as when sensing and pursuing prey organisms (Figure 22–4).

Sarcodines are thought to have originated from mastigophores; some sarcodines may develop flagella during particular stages of their life cycle or under particular environmental conditions. They are found in both fresh water and salt water. Some are parasites, such as those that cause amoebic dysentery.

Reproduction may be asexual or sexual. Asexual reproduction takes place by cell division accompanied by mitosis in which the nuclear envelope usually does not break down. In sexual reproduction, the cells, which are diploid, undergo meiosis, forming gametes, which then fuse to form zygotes.

Many of the sarcodines have outer shells, often brightly colored. Some, like *Arcella,* secrete a hard, keratinlike material, while others, like *Difflugia,* exude a sticky organic substance upon which forms a sort of haphazard mosaic made up of specks of sand and bits of the discarded shells of other minute creatures. Such shells presumably evolved as defenses against predators. Some, the "sun animals," resemble pincushions, with fine pseudopodia stiffened with microtubules (page 99) radiating from their bodies. *Actinospherium,* the protozoan shown in Figure 5–4, may reach as much as a millimeter in diameter and may sometimes be seen as a tiny white speck floating on the surface of a pond. Other sarcodines, the Foraminifera, have snail-like shells and live in the sea. Their shells are made of calcium carbonate, extracted from the sea water. The white cliffs of Dover and similar chalky deposits throughout the world are the result of the long accumulation of these discarded shells. The shells of the Foraminifera have been accumulating on the ocean bottom for millions of years, and in many areas, as a result of geologic changes, thick deposits of their skeletons (the foraminiferan ooze) can be found on

22–5
Shell of a foraminiferan, a sarcodine.

the surface of the land or under later rock formations. Since the skeletons have evolved over this long period of time, it is possible to date a particular stratum by the type of Foraminifera that it contains, a fact that has proved of immense practical value in locating oil-bearing strata in Texas, Oklahoma, and many other oil-rich areas.

Class Ciliophora

The ciliophores, or ciliates, are the most highly specialized and complicated of the Protozoa, and indeed probably represent the most complex of all living cells. They are believed to have been derived from primitive mastigophores, which, traveling in the opposite evolutionary direction from the sarcodines, developed elaborate ciliary systems (Figure 22–6). They are characterized by 9 + 2 cilia. In some species, the cilia adhere to each other in rows, forming brushlike structures called membranelles or clumps of cilia called cirri, which can be used for walking or jumping. Cilia, membranelles, and cirri move in a coordinated fashion, although the way in which they are coordinated is not understood. Some ciliates also have myonemes, contractile threads. All have a complex skin, the cortex, which includes the cell membrane. In some groups, the cortex contains small barbs known as trichocysts, which are discharged when the cell is stimulated in certain ways.

The ciliates have another unusual feature: They have two kinds of nuclei, micronuclei and macronuclei. One or more of each kind is present in all cells. They also have a complex system for exchange of genetic information, in which cells conjugate and the micronuclei undergo meiosis. Cells then exchange haploid micronuclei that fuse, so that each cell has a new diploid micronucleus, which then divides. The old macronucleus dissolves, and a new macronucleus develops from one of the new micronuclei. The macronucleus in certain ciliates contains 50 to 100 times as much DNA as the micronucleus and so is believed to represent multiple

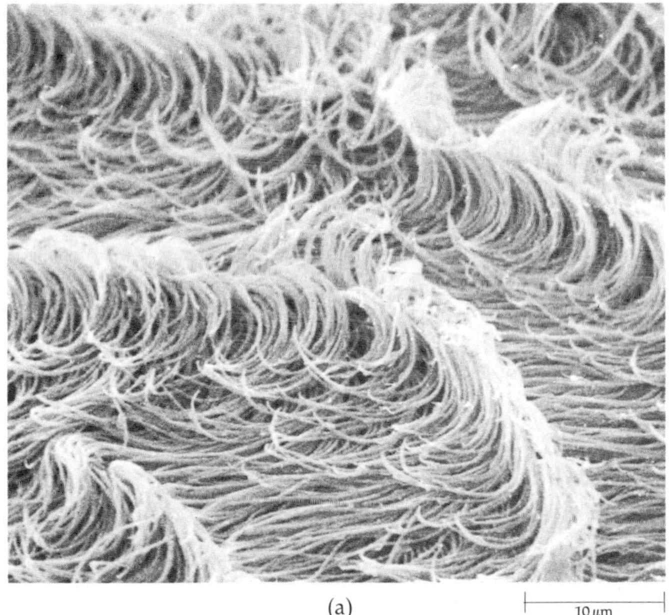

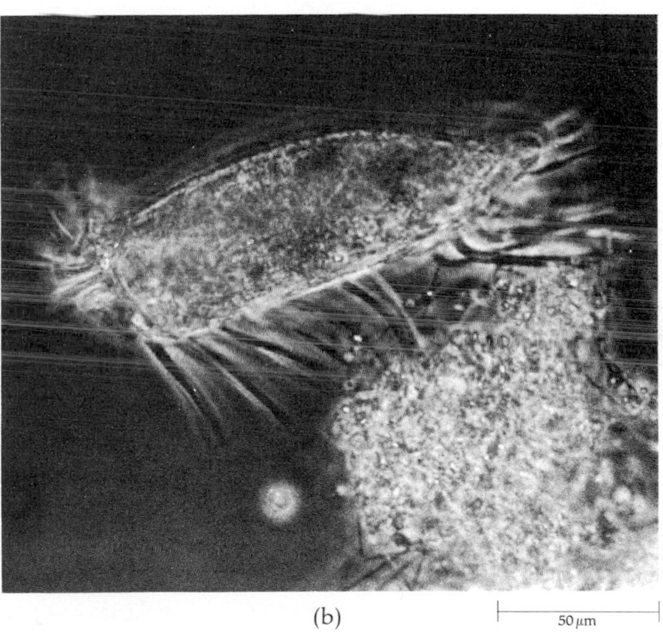

22–6
(a) *Cilia beat continuously, about 40 to 60 times per second. The beat is coordinated, producing synchronous waves of motion, as shown in this scanning electron micrograph of the surface of the ciliate* Opalina. (b) *Cilia can be fused to form cirri. In this protozoan,* Euplotes patella, *they move individually, propelling the cell in a jerky motion.*

22–7
Drawing of a Paramecium, *a ciliate. The body of this protist is completely covered by 9 + 2 cilia, although only a relative few are shown here. Like other ciliates, a* Paramecium *feeds largely on bacteria, smaller microorganisms, and other particulate matter. The beating of specialized cilia drive particles into the gullet, where they are formed into food vacuoles. The food is digested in the vacuoles, and the undigested matter, still in vacuoles, is emptied out through the anal pore.*

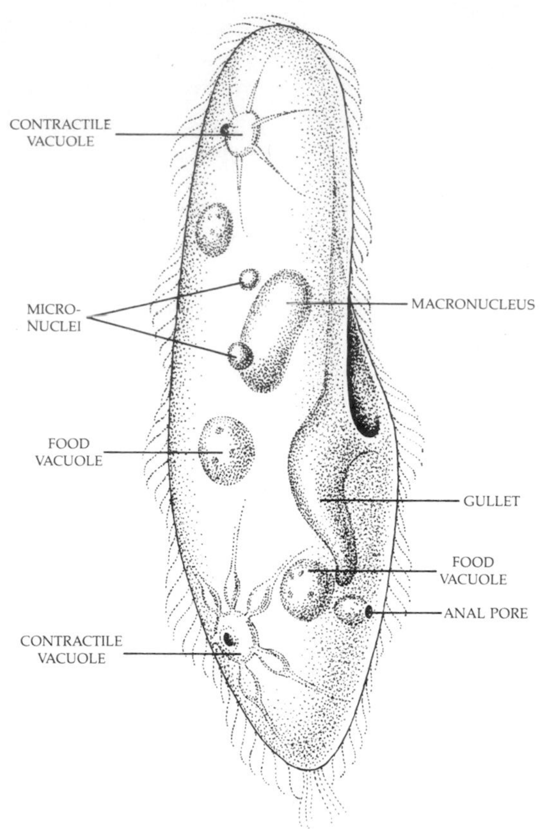

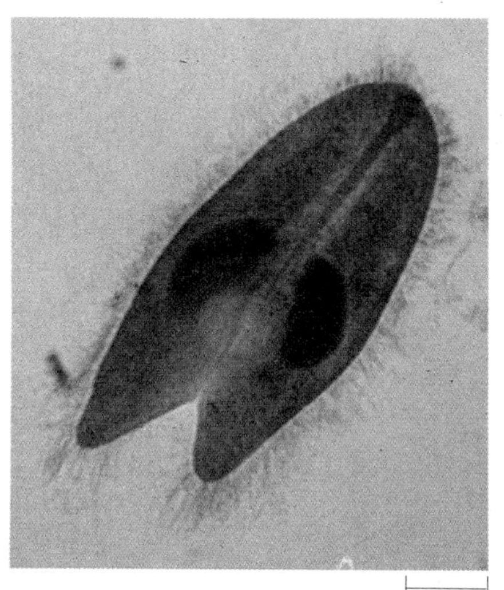

22–8
Paramecium *conjugating. The cells have been stained to show the macronuclei.*

copies of it. This view is supported by the fact that, in many ciliates, a cell can survive indefinitely without a micronucleus if even a small portion of a macronucleus is left, although the cell cannot conjugate. However, it cannot live without a macronucleus, even if it has a micronucleus. The macronucleus does not divide mitotically but is apportioned approximately equally between dividing cells as they constrict and separate.

The complexity of the conjugation process among ciliates reminds us again of the expenditure of energy and other resources involved in effecting exchanges of genetic information among individuals of the same species, from bacteria to *Homo sapiens.* The high biological cost of these activities is an indication of the great survival value, in evolutionary terms, of the genetic variability produced by these exchanges.

About 6,000 species of ciliates are known, both freshwater and saltwater forms. Almost all are free-living (nonparasitic).

Class Sporozoa

The fourth class of Protozoa is the Sporozoa, all of which are parasitic. They are characterized by the lack of cilia or flagella in adult forms and by their complex life cycles. A single sporozoan undergoes multiple fission, dividing into numerous smaller cells (the spores) simultaneously.

The best known sporozoans are members of the genus *Plasmodium,* which cause malaria. The *Plasmodium* life cycle is shown in Figure 22–9.

22–9
Life cycle of Plasmodium vivax, *the sporozoan that causes malaria in humans. The cycle begins* (a) *when a female* Anopheles *mosquito bites a person with malaria and, along with the blood, sucks up gametes* (b) *of the protozoan. In the mosquito's body, the gametes unite* (c) *and form a zygote* (d). *From the zygotes, multinucleate structures called oocysts develop* (e), *which, within a few days, divide into thousands of very small, spindle-shaped cells, sporozoites* (f), *which then migrate to the mosquito's salivary glands. When the mosquito bites another victim* (g), *she infects the person with the sporozoites. These first enter liver cells* (h), *where they undergo multiple division* (i). *The products of these divisions (merozoites) enter the red blood cells* (j), *where again they divide repeatedly* (k). *They break out of the blood cells* (l) *at regular intervals of about 48 or 72 hours, producing the recurring episodes of fever characteristic of the disease. After a period of asexual reproduction, some of these merozoites become gametes, and, if they are ingested by a mosquito at this stage, the cycle begins anew.*

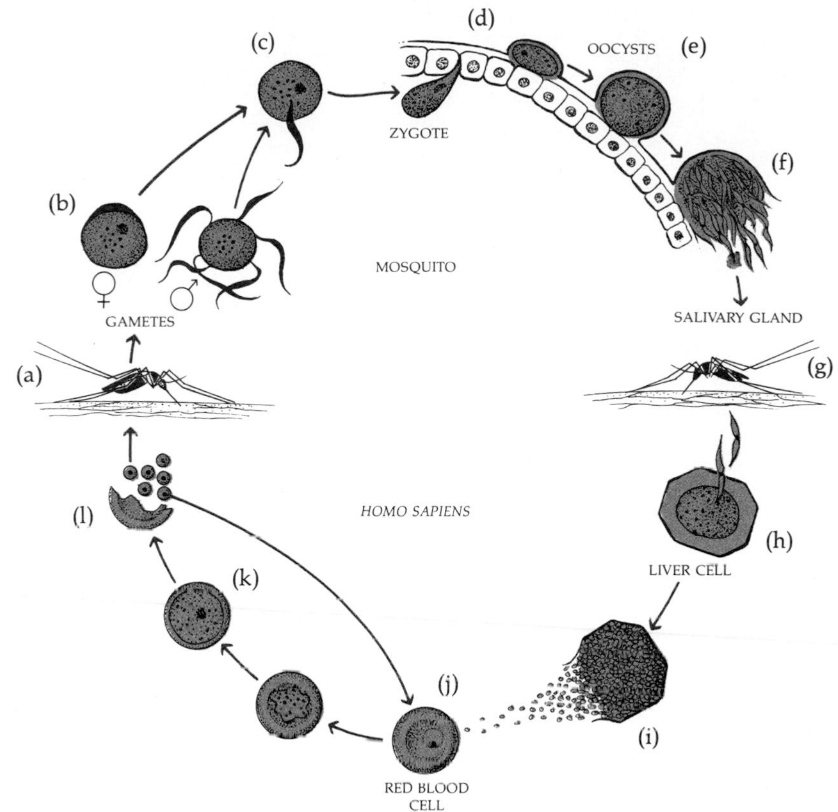

CHARACTERISTICS OF THE ALGAE

Among the eukaryotic algae, a number of distinct lines had evolved by the early Paleozoic era, more than 450 million years ago (see Table 26–2, page 478). The approximately 22,000 described species now in existence are grouped into six phyla. The term "algae" has been abandoned as a formal term in modern classification, because the various groups are not directly related to one another. Three of these phyla—the Euglenophyta, the Chrysophyta, and the Pyrrophyta—consist almost entirely of unicellular organisms. The other three phyla—Chlorophyta, Rhodophyta, and Phaeophyta—include groups that are multicellular.

Nearly all members of these phyla are photosynthetic. In general, they all have a relatively simple structure, which may be a single cell, a filament of cells, a plate of cells, or a solid body more or less comparable to that found in the vascular plants. The cell walls of algae, in general, have a cellulose matrix, often with massive amounts of other polysaccharides that give certain algae a mucilaginous consistency. When algal cells divide, the cell membranes generally pinch inward from the margin of the cell (furrowing), just as in animals, fungi, and protozoa. However, cell plates, like those of plants (see page 204), are formed during cell division in one brown alga and a few green algae. Most algal cells—except those of the red algae—have centrioles.

Table 22–2 *Comparative Summary of Characteristics in the Six Major Phyla of Eukaryotic Algae*

PHYLUM	NUMBER OF SPECIES	PHOTOSYNTHETIC PIGMENTS	FOOD RESERVE	FLAGELLA	CELL WALL COMPONENT	REMARKS
Euglenophyta (euglenoids)	800	Chlorophyll *a* and *b*, carotenoids	Paramylum and fats	1, 2, or 3 per cell, apical	No cell wall	All unicellular, most freshwater; sexual reproduction unknown
Chrysophyta (golden-brown algae and diatoms)	6,000–10,000	Chlorophyll *a*, and often *c*, carotenoids, including fucoxanthin	Leucosin and oils	1 or 2, apical, equal or unequal	Pectic compounds with siliceous material	Most marine
Pyrrophyta (dinoflagellates)	1,100	Chlorophyll *a* and *c*, carotenoids	Starch and oils	2, lateral	Cellulose	Marine and fresh-water; sexual reproduction rare
Chlorophyta (green algae)	7,000	Chlorophyll *a* and *b*, carotenoids	Starch	Usually 2 per cell, identical	Cellulose	Mostly fresh-water, but some marine
Phaeophyta (brown algae)	1,500	Chlorophyll *a* and *c*, carotenoids, including fucoxanthin	Laminarin and fats	2, lateral, in reproductive cells only	Cellulose and algin	Almost all marine, flourish in cold ocean waters
Rhodophyta (red algae)	4,000	Chlorophyll *a*, carotenoids, phycobilins, chlorophyll *d* in some	Floridean starch	None	Cellulose, pectic materials common, xylan in *Porphyra*	Mostly marine, some freshwater; complex sexual cycle; many species tropical

The phyla of algae differ from one another widely in the nature of their flagella (when present) and in their biochemical characteristics, especially with respect to differences in pigmentation, nature of reserve food, and cell wall components (Table 22–2). The names of some phyla are derived from the colors of the predominant accessory pigments, which mask the bright green of the chlorophylls. A wide variety of carotenoids is found in the algae. The xanthophylls are yellowish-brown carotenoids; the xanthophyll fucoxanthin gives the brown algae their characteristic color and name. It is also found in the golden-brown algae and diatoms (phylum Chrysophyta). The red algae (Rhodophyta) owe their colors to several kinds of phycobilins, accessory pigments that, unlike the carotenoids, are water-soluble. In the green algae the color of the chlorophylls is usually not masked by accessory pigments. The diversity of pigments present in the chloroplasts of the various phyla of algae suggests that (1) various types of oxygen-producing prokaryotes were in existence prior to the development of eukaryotic cells, and (2) the various phyla of algae may have evolved as a result of the establishment of symbiotic relationships with different photosynthetic prokaryotes, which then evolved into modern chloroplasts.

22–10
Laminarin, the principal storage product of brown algae. Like starch, it is made up of glucose residues but unlike starch, there are only about 15 to 30 glucose units per molecule, and their linkage is different (see page 55).

LAMINARIN

A rich diversity of storage products is found in the different phyla of algae, with most having distinctive carbohydrate food reserves, often in addition to lipids (see Table 22–2). The green algae and dinoflagellates store their carbohydrates as starch, as do vascular plants. In the brown algae, another glucose polymer, laminarin (Figure 22–10), takes the place of starch. The carbohydrate reserves of the red algae are biochemically similar to starch.

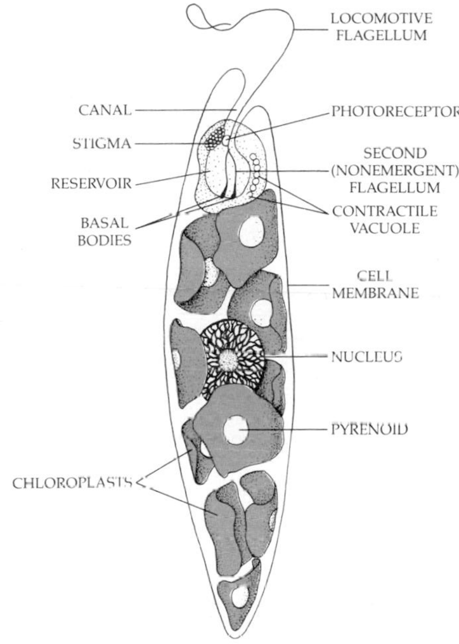

22–11
Euglena *is one of the most versatile of all one-celled organisms. Containing numerous chloroplasts, it is photosynthetic, but it can also absorb organic nutrients from the medium in which it lives. The pyrenoid is believed to be a food storage body.* Euglena *moves by the whiplike motion of a single flagellum.*

Unicellular Algae

These minute photosynthesizing cells are usually found floating near the surface of the oceans and inland waters. Together with small invertebrates and immature forms of larger animals they form the *plankton* (from *planktos,* the Greek word for wanderer). Each cell is a totally self-sufficient individual, dependent only on light from the sun and carbon dioxide and minerals from the water that surrounds it. These cells may live in colonies, and sometimes the cells of a single colony may be attached to one another as the gliding bacteria, for instance, are held together by such sticky secretions, but there is no division of labor among the cells. There are, as we noted, three principal phyla of these unicellular forms.

Phylum Euglenophyta

Euglenophyta is a small group of unicellular organisms (about 800 species), most of which are found in fresh water. They are named for the genus *Euglena,* the most common of the group. Characteristically, they have one or two flagella. Their chloroplasts contain chlorophylls *a* and *b* like those of plants. They store their food as paramylum, an unusual polysaccharide found almost exclusively in this group. The cells lack a cell wall but have a flexible series of protein strips, which make up the pellicle, inside the cell membrane. They reproduce asexually, dividing longitudinally to form two new cells that are mirror images of one another.

The cells are complex. *Euglena* is characteristically an elongated cell with a single nucleus, two flagella (one of which is small and inactive), and numerous small emerald-green chloroplasts that give the cell its bright color. The flagella are attached at the base of a flask-shaped opening, the reservoir, at the anterior end of the cell. Emptying into the reservoir is the contractile vacuole, which collects excess water from all parts of the cell and discharges it into the reservoir. Unlike the stiff walls of the cells of plants, the flexible pellicle permits *Euglena* to change its shape, providing an alternative means of locomotion—wriggling—for mud-dwelling forms. Only a few other photosynthetic organisms have contractile vacuoles, which are common features among the protozoans (see page 129).

22–12
A cell of Euglena, *broken open, showing the flexible protein strips that make up the pellicle. The large organelle at the upper right is a chloroplast. The spongy network below the chloroplast is endoplasmic reticulum. At the bottom left is a broken mitochondrion, identifiable by its characteristic cristae.*

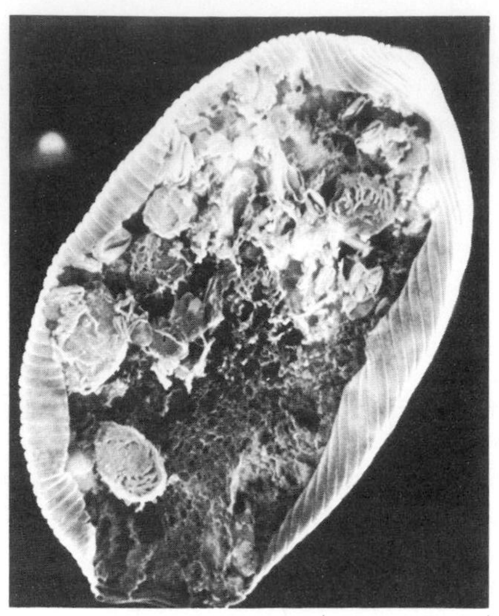

If one leaves a culture of *Euglena* near a sunny window, a clearly visible green cloud will form in the water, and this will move as the light changes. *Euglena* is probably able to orient with respect to light because of a pair of special structures: the stigma, or eyespot, which is a patch of pigment, and an underlying photoreceptor. The exact way in which these function is not known.

Some nonphotosynthetic protists—members of the genus *Peranema,* for example—closely resemble *Euglena* but lack chloroplasts. In fact, if some strains of *Euglena* are kept at an appropriate temperature and in a rich medium, the cells may replicate faster than the chloroplasts, producing nonphotosynthetic cells, which can survive indefinitely in a suitable medium containing a carbon source. It is tempting to speculate that some modern heterotrophs arose from autotrophs in an analogous manner early in evolutionary history.

Phylum Chrysophyta: "Golden" Algae

The chrysophytes are the "golden" algae. There are 6,000 to 10,000 species, most of which are diatoms. Members of this group have several identifying characteristics: (1) Their photosynthetic pigments are chlorophyll *a* and chlorophyll *c,* which closely resembles the chlorophyll *b* of plants; (2) they contain a yellow-brown carotenoid, fucoxanthin, which functions as an accessory pigment and gives them their characteristic color; (3) their cell walls, which contain no cellulose, are often impregnated with silicon compounds and thus are very rigid; (4) they store food in the form of oil rather than starch. Because of these oil reserves, fresh water containing large numbers of these algae may have an unpleasant oily taste, as may fish caught in such waters. It has, in fact, been suggested that golden algae might be a potential source of oil for fuel. They are a major component of plankton.

Diatoms are enclosed in a fine double shell, the two halves of which fit together, one on top of the other, like a carved pillbox (Figure 22–13). Electron microscopy has shown that the fine tracings in diatom shells actually represent minute, intricately shaped pores or passageways connecting the protoplasm within the cell to the exterior environment. The piled-up silicon shells of diatoms, which have col-

22–13
Diatoms, chrysophytes. (a) *Side view showing the characteristic intricately marked shell.* (b) *Pillbox type seen from above and from the side. Notice that one cell is dividing. Each new cell will get half of the "pillbox" and then construct the other half. When the parent cell divides, one half is always smaller than the other.*

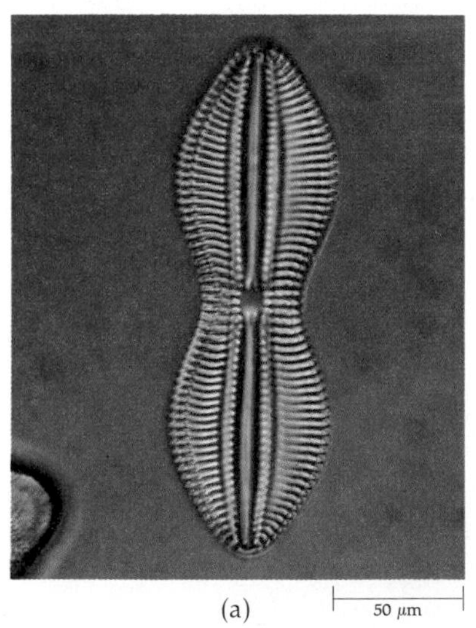
(a)

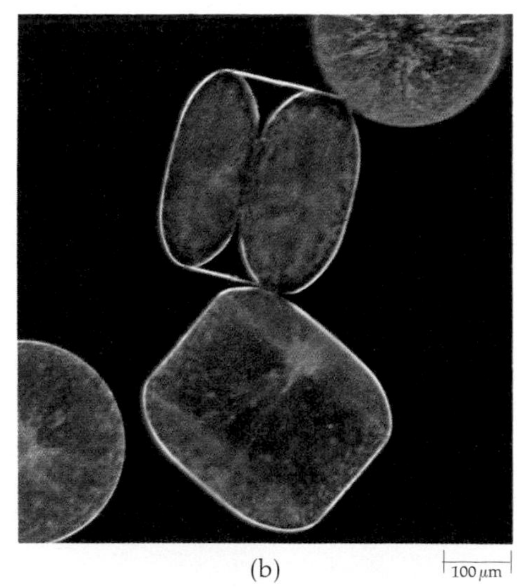
(b)

22-14
The silicon-containing walls of a diatom, as shown by the scanning electron microscope. Each species has its characteristic pattern of perforations in the walls. The delicate markings of these shells, by which the species are identified, were traditionally used by microscopists to test the resolving power of their lenses.

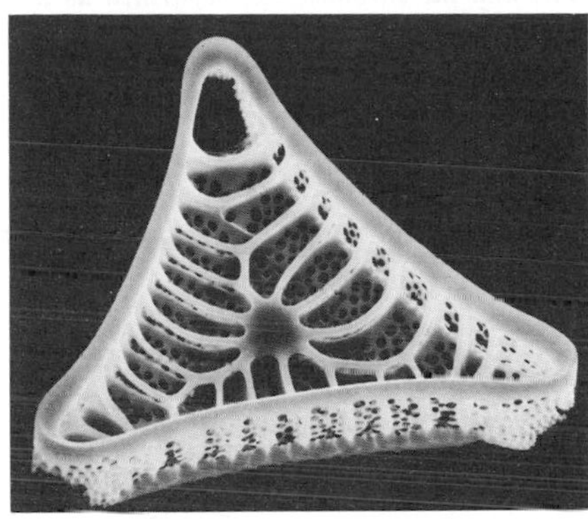

lected over millions of years, form the fine, crumbly substance known as "diatomaceous earth," used as an abrasive in silver polish and toothpaste and for filtering and insulating materials.

Some other members of the phylum lack cell walls and are amoeboid. Except for the presence of chloroplasts, the amoeboid cells are indistinguishable from amoeboid protozoans (the sarcodines), and the two groups may be closely related. Reproduction usually takes place by mitosis, but zygotes are sometimes formed. Diatoms are diploid, except for the gametes.

Phylum Pyrrophyta: "Fire" Algae

The Pyrrophyta is also largely composed of single-celled algae, the dinoflagellates ("spinning flagellates"), of which about 1,000 different species are known, almost all of them marine forms. Like the diatoms, they are important components of the phytoplankton. Other members of the phylum are heterotrophs clearly related to the photosynthetic forms.

The dinoflagellates usually have two flagella that beat within grooves: one encircling the body like a belt and a second lying perpendicular to the first. The beating of the flagella in their respective grooves causes the cells to spin like tops as they move through the water.

Many of the dinoflagellates are bizarre in appearance, with a stiff cellulose wall (theca), which often looks like a strange helmet or an ancient coat of armor.

These dinoflagellates are often red in color (hence the name of the phylum), and the infamous red tides, in which thousands of fish die, are caused by great blooms of red dinoflagellates. The poison in these red tides has been traced to one species of armored dinoflagellate, *Gonyaulax catanella.* It is such an extraordinarily powerful nerve toxin that 1 gram of it would be enough to kill 5 million mice in 15 minutes. Blooms of *Gonyaulax catanella* appear regularly on the Pacific Coast and the Gulf of Mexico and have recently been reported off the New England coast. Mussels ingest the algae and concentrate the poison; the mussels then become dangerous for consumption by vertebrates. Many dinoflagellates are bioluminescent.

22-15
Dinoflagellates. (a) Ceratium tripos, *an armored dinoflagellate. You can see one flagellum in motion.* (b) Noctiluca scintillans, *a bioluminescent marine dinoflagellate. Yellow-brown diatoms that have been ingested can be seen inside the cell.*

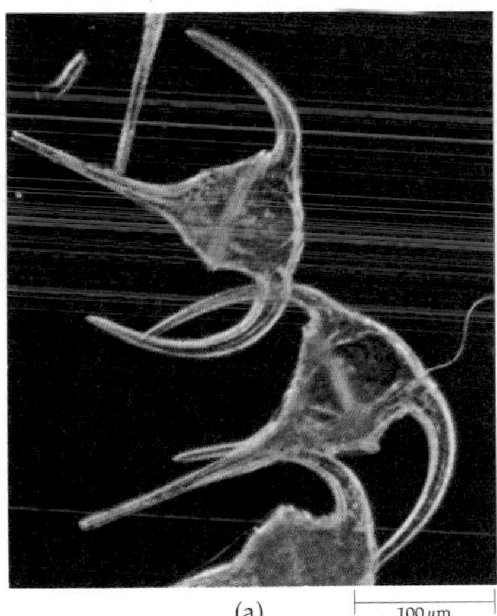

(a)

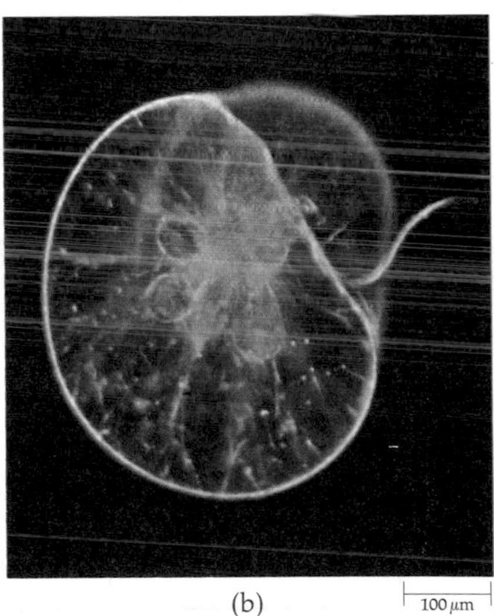

(b)

(a)

(b)

22–20
(a) *A brown alga,* Laminaria, *showing holdfasts, stipes, and blades.* (b) *Unlike the brown algae, most red algae are made up of filaments. The branched filaments of this red alga are hooked, enabling it to cling to other seaweeds.*

The brown algae are often very large, and many have a variety of specialized tissues. Some of the giant kelps are nearly 60 meters long; these are annuals, reaching their full size in a single season. The body of an alga (or a plant) that is multicellular, but relatively unspecialized, is known as a *thallus.* In some kelps, the thallus is well differentiated into *holdfast* ("root"), *stipe* ("stalk"), and *blade* ("leaf"). The words are put into quotation marks because, although the corresponding parts of the alga superficially resemble the organs of plants, they are not really comparable in their internal organization. However, the kelps have strands of elongated conducting cells in the center of the stipe that are similar to the cells that conduct sugars in the vascular plants. Carbohydrates produced in the blades, which are exposed to sunlight, are thus transported to the stipe and holdfast, which may be far below the surface of the water.

There are no modern unicellular forms in this group, except for the gametes. Sperm and spore cells are often flagellated.

Phylum Rhodophyta: Red Algae

The red algae are most commonly found in warm marine waters; they make up the other principal group of large seaweeds. The red algae contain chlorophylls *a* and *d,* carotenoids, and also certain phycobilins, which give them their distinctive colors. (Phycobilins and chlorophyll *a* are also found in the blue-green "algae," and it is believed that the modern blue-greens and the chloroplasts of the red algae are related.)

Most of the seaweeds of the world are red algae, of which there are some 4,000 species. Less than 2 percent are freshwater forms. Red algae usually grow attached to rocks and other algae; there are no large red algae capable of prolonged life in the floating state, like the sargasso weed of the Sargasso Sea. (They die when floating because the water must pass over them for adequate gas exchange; without it, they suffocate.) As you would expect from their red color, which indicates that they absorb blue light, some grow at greater depths than other algae; they have been found attached 175 meters below the ocean surface in the clear water of the tropics. Although some are several meters long, red algae never attain the size of the largest of the brown algae. They have a reproductive pattern in which neither of the gametes is motile; the male is carried to the fixed female reproductive cell by the movement of the water. They are one of the few groups of organisms that have no flagellated cells; they also lack centrioles.

PHYLUM GYMNOMYCOTA: SLIME MOLDS

The slime molds are a group of curious organisms classified with the protists—although some of them are multicellular in some stages—because of their similarity to protozoans, especially the amoebas. Two main groups (classes) are known, the plasmodial slime molds (Myxomycetes) and the cellular slime molds (Acrasiomycetes).

Most slime molds live in cool, shady, moist places in the woods—on decaying logs, dead leaves, or other damp organic matter. One of the common species *(Physarum cinereum)*, however, is sometimes found creeping across city lawns. The plasmodia come in a variety of colors and can be spectacularly beautiful. The function of the pigments is not known with certainty, but they are probably photoreceptors because only slime molds with pigmented plasmodia require light for spore production.

During their nonreproductive stages, the Myxomycetes, or "true" slime molds, are thin, streaming masses of protoplasm, which creep along in amoeboid fashion. As one of these plasmodia travels, it engulfs bacteria, yeast, fungal spores, and small particles of decayed plant and animal matter, which it digests. It may grow to weigh as much as 50 grams or more, and, since slime molds are spread thinly, this mass can cover an area more than a meter in diameter. The plasmodium is coenocytic, and as it grows, the nuclei divide repeatedly and, in the early stages, synchronously.

22-21
(a) *Plasmodium of a slime mold. Such a plasmodium can pass through a piece of silk or filter paper and come out the other side apparently unchanged.* (b) *Sporangia of a plasmodial slime mold on a rotting log.*

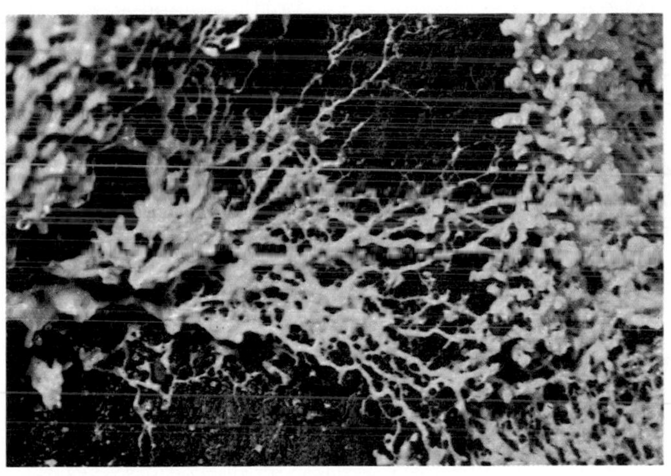

(a)

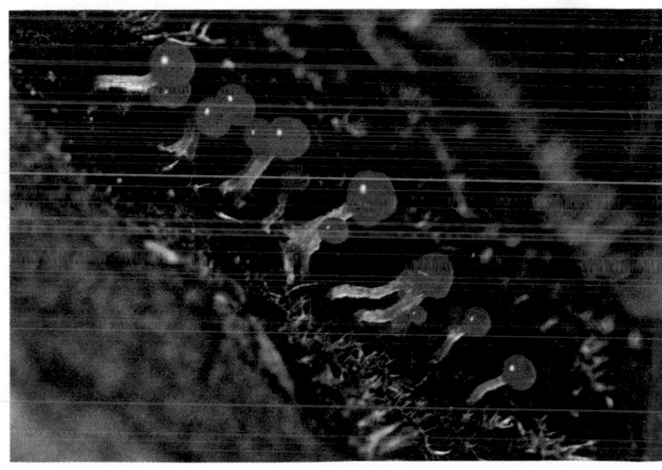

(b)

Plasmodial growth continues as long as an adequate food supply and moisture are available. When either of these is in short supply, the plasmodium separates into many mounds of protoplasm, each of which develops into a mature sporangium (a structure in which spores develop) borne at the tip of a stalk. Meiosis takes place and cell walls form around the individual haploid nuclei to produce spores, which are resistant resting forms.

The spores germinate under favorable conditions, and each spore, depending on the species, produces one to four haploid, flagellated cells. Some of these cells fuse to form a zygote, from which a new plasmodium develops.

The life cycle of an Acrasiomycetes, or cellular slime mold, is shown on page 121. These slime molds also begin as amoebalike organisms but differ from the plasmodial slime molds in that the amoebas, on swarming together, do not lose their cell membranes but retain their identity as individual cells.

SUMMARY

The kingdom Protista comprises a variety of eukaryotic organisms, mostly unicellular with some relatively simple multicellular forms. There are eight principal phyla, six of which are algae. Each of the various phyla of protists probably had a separate origin among the prokaryotes, and it appears likely that symbiotic relationships with prokaryotes gave rise to chloroplasts and mitochondria.

The Protozoa, the nonphotosynthetic protists, are subdivided into four classes: the Mastigophora (flagellates), the Sarcodina (amoebas), the Ciliophora (ciliates), and the Sporozoa. The first three classes may be identified on the basis of their locomotor structures. Members of the latter group, which is composed largely of parasitic forms, have no locomotor organelles. Among the protists are some of the largest known cells and also the most complex. The original protozoan was believed to have been a small, colorless flagellate.

Euglenophyta is a small group of unicellular algae, mostly found in fresh water. They contain chlorophylls *a* and *b* and store carbohydrates in an unusual starchlike substance, paramylum. The cells lack a cell wall but have a flexible series of protein strips, which make up the pellicle, inside the cell membrane. The cells are highly differentiated, containing chloroplasts, a contractile vacuole, eyespot, or stigma, and flagella. No sexual cycle is known. The group also contains nonphotosynthetic forms.

The Chrysophyta are important components of freshwater and marine phytoplankton. Most of the known species are unicellular organisms known as diatoms. Diatoms are characterized by fine, double silicon-containing shells. Abundant fossils of these shells have been found.

The Pyrrophyta are unicellular biflagellates, many of which are marine. This phylum includes the dinoflagellates, which are characterized by two flagella that beat in different planes, causing the organism to spin; dinoflagellates often have stiff, bizarrely shaped cellulose walls.

The Chlorophyta, the green algae, are hypothesized to be the ancestors of the plants, on the basis of similarities between the groups. These similarities include the fact that both groups have chlorophylls *a* and *b* and beta-carotene as their photosynthetic pigments. Both store food reserves in the form of starch, and the cells of both have cellulose cell walls. The widely diverse forms of green algae range from the single-celled *Chlamydomonas* to a variety of multicellular forms belonging to several different evolutionary lines.

The Phaeophyta (brown algae) and Rhodophyta (red algae) are the principal seaweeds. The brown algae, which include the kelps, are found more commonly in cooler water; the red algae, in the tropics. In some brown algae the thallus is differentiated into holdfast, stipe, and blade, analogous to the root, stem, and leaf of plants. The kelps have tissues specialized for the conduction of sugar from the blades to nonphotosynthetic parts of the thallus.

The slime molds, or Gymnomycota, are amoeba-like organisms. They resemble fungi in that they reproduce by the formation of spores and are heterotrophic. Unlike the fungi, however, they lack a cell wall, and they ingest their food as particles. There are two principal classes: Myxomycetes (plasmodial slime molds) and Acrasiomycetes (cellular slime molds).

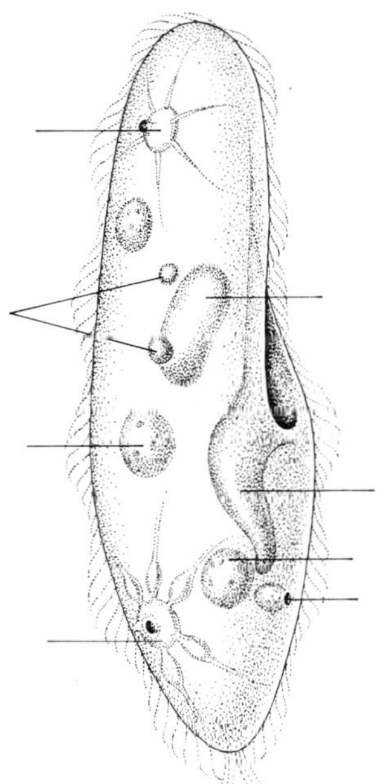

QUESTIONS

1. Label the drawing at the left.

2. Name the four classes of phylum Protozoa, and give the distinguishing characteristics of each.

3. Several different possible pathways to multicellularity have now been discussed. What are they?

4. Consider the life cycle of *Plasmodium.* At what stages in the cycle do its numbers increase? Why might a parasite that requires several hosts find it advantageous to evolve a life cycle in which its numbers increase at several stages? Why might it be advantageous to a parasite to have a second host, such as the mosquito?

5. Arrange in order of evolutionary development: shell or skeleton, phagocytosis, active transport, multicellularity. (Like the rest of us, you will only be guessing, but be prepared to defend your guess.)

CHAPTER 23

The Fungi

The fungi are a group of organisms so unlike any others that, although they were long classified with the plants, it has come to seem appropriate to assign them to a separate kingdom. Except for some one-celled forms, such as the yeasts, the fungi are basically coenocytic organisms composed of masses of filaments. A fungal filament is called a *hypha,* and all the hyphae of a single organism are collectively called a *mycelium.* The mycelium may appear as a mass on the surface of the nutrient or may be hidden beneath the surface. A fungus is essentially a multinucleate mass of cytoplasm enclosed within a rigid, much-branched system of tubes (the hyphae). The cytoplasm, organelles, and nuclei flow within those tubes.

A mycelium normally arises by the germination and outgrowth of a single cell, with growth taking place only at the tips of hyphae. The complex, spore-producing structures of fungi, such as mushrooms, are tightly packed hyphae. In most groups of fungi, the cell walls are composed primarily of chitin, a polysaccharide that is never found in the kingdom Plantae (it is, however, the principal component of the exoskeletons—the hard outer coverings—of insects).

23–1
Gill fungi on a dead quaking aspen.

23–2
Fungal cell walls characteristically contain chitin (a) *rather than cellulose* (b), *the polysaccharide found in the cell walls of plants. Chitin resembles cellulose in that it is tough, inflexible, and insoluble in water. As you can see, they are also structurally very similar; in chitin, the hydroxyl (OH) group in the 2 position is replaced by a nitrogen-containing group.*

(a) CHITIN

(b) CELLULOSE

23–3
The perforated cross wall of a fungus showing a nucleus in the perforation. The fungus is Neurospora crassa, *the red bread mold.*

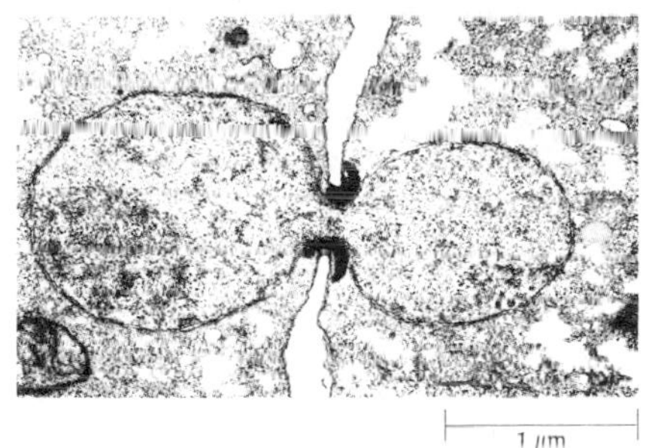

In some groups, the hyphae are septate—divided by cell walls—but the walls, or septa, are perforated, and the cytoplasm and even the nuclei (Figure 23–3) are able to flow through the septa. Only the reproductive structures are separated by cell membranes.

All fungi are heterotrophs with a highly characteristic means of nutrition. Because of their filamentous form, each fungal cell is no more than a few micrometers from the soil, water, or other substance in which the fungus lives, and is separated from it only by a thin cell wall. Because of their rigid walls, fungi are unable to engulf small microorganisms or other particles. They obtain food by absorbing dissolved inorganic and organic materials. Typically a fungus will secrete digestive enzymes onto a food source and then absorb the smaller molecules released.

The only motile cells of fungi are reproductive cells which may travel through water or air. Growth of the mycelium substitutes for motility, bringing the organism into contact with new food sources and different mating strains. Under favorable conditions, a fungus can expand very rapidly, as evidenced by the overnight

23–4
Mycelium of a fungus growing on wood.

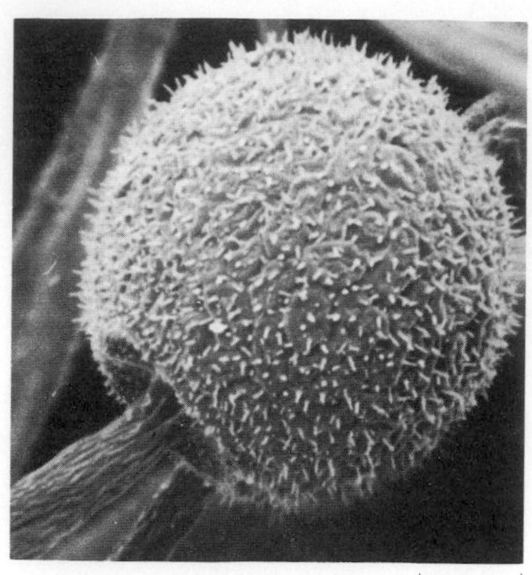

23–5
Reproductive structure of a fungus, a sporangium, from the black bread mold Rhizopus.

appearance of a lawnful of mushrooms, produced by the sudden transport of material from the underground mycelium into the fruiting bodies, or mushrooms.

The fungi, together with the bacteria, are the principal decomposers of the world. As we shall see in Section 8, their activities are as vital to the continued survival of higher forms of life as are those of the food producers. Some are also destructive; they may interfere with human activities by attacking our foodstuffs, our domestic plants and animals, our shelter, our clothing, and even our persons.

REPRODUCTION IN THE FUNGI

Fungi reproduce both asexually and sexually. Asexual reproduction takes place either by the fragmentation of the mycelium (with each fragment becoming a new individual) or by the production of spores. In some of the fungi, spores are produced in reproductive structures called *sporangia,* and are borne on specialized hyphae called *sporangiophores.* Spores are often, but not necessarily, resting forms, surrounded by a tough, resistant wall. They are able to survive during periods of lack of water and extreme temperatures. Some airborne spores are very small and so can remain suspended in the air for long periods and be carried for great distances. Often the sporangia are raised above the mycelium by the sporangiophores; thus the spores are easily caught up and transported by air currents. The bright colors and powdery textures associated with many particular types of molds are produced by the spores.

Sexual reproduction can occur in a variety of ways: (1) by fusion of gametes, (2) by penetration of a gamete into a specialized reproductive structure called a *gametangium,* or (3) by fusion of gametangia.

Sometimes the fusion of fungal hyphae is not followed immediately by the fusion of nuclei. Thus strains of fungi may exist with two genetically distinct kinds of nuclei operating simultaneously. Such a combination, known as *dikaryosis,* is found uniquely among the higher fungi (Ascomycetes and Basidiomycetes).

PHYLUM MYCOTA

The kingdom Fungi consists of only one phylum, Mycota, which is divided into five principal classes: Oomycetes, Zygomycetes, Ascomycetes, Basidiomycetes, and Fungi Imperfecti.

Class Oomycetes

The oomycetes are also known as the water molds because many of them are aquatic. They are the only group of fungi that produce flagellated, swimming spores. Even the terrestrial forms produce swimming spores that require free water.

Most oomycetes are saprobes, living on dead organic matter. Some forms are parasitic and pathogenic, however, and, as mycologist C. J. Alexopoulos has said, "At least two of them have had a hand—or should we say a hypha!—in shaping the economic history of an important portion of mankind." The first of these is *Phytophthora infestans (phytophthora* literally means "plant destroyer"), the cause of the "late blight" of potatoes, which produced the great potato famines in Ireland. The second economically important member of this group is *Plasmopara viticola,* the

23–6
Phytophthora infestans, *cause of potato blight. Infection begins when an airborne sporangium alights on a leaf, releasing spores that move about in the film of water on the leaf's surface. These spores germinate, producing hyphae that penetrate the epidermis and attack the mesophyll cells. Eventually aerial hyphae—sporangiophores—bear a sporangium from which a new generation of spores are released.*

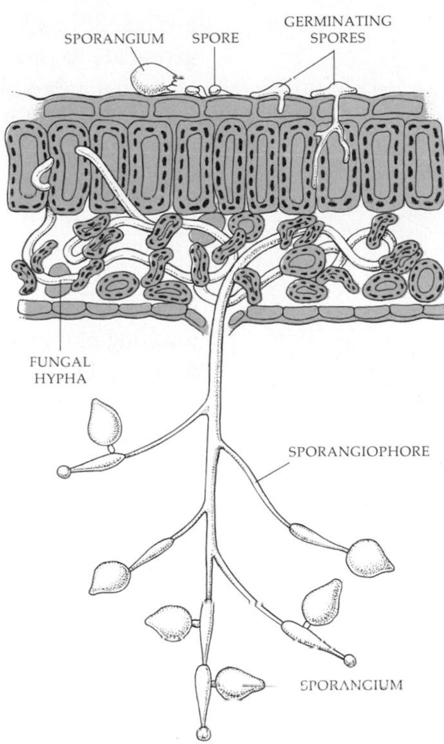

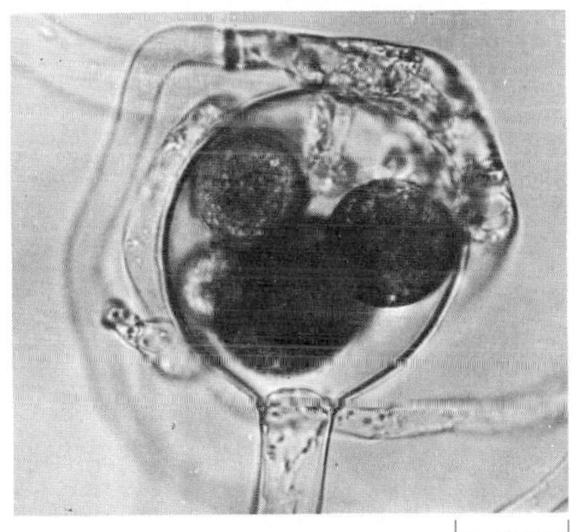

23–7
Mating in the fungus Achyla ambisexualis, *an oomycete. The large spherical structure is the female gametangium. The dark bodies within it contain eggs. Encircling the female gametangium is the male gametangium. Fertilization tubes extending from the male into the female gametangium are barely visible. Male nuclei pass through these tubes to the egg nuclei. Development of the male gametangia and their attraction to the female are controlled by the production of a steroid hormone remarkably similar in structure to the human sex hormones.*

cause of downy mildew of grapes. This mildew threatened the entire French wine industry during the latter part of the nineteenth century.

The oomycetes derive their name from *oion*, the Greek word for "egg." They are the only group of fungi in which the gametes are clearly male (sperm) and female (egg). They are also the only group of fungi with cellulose rather than chitin in their cell walls. Sperm and eggs, each of which are borne in their own type of gametangia, fuse to produce a zygote.

Class Zygomycetes

The zygomycetes are terrestrial fungi, most of which live in the soil, feeding on dead plant or animal matter. Some are parasites of plants, insects, or small soil animals. Unlike oomycetes, they produce no flagellated spores at any stage of the life cycle. In sexual reproduction, they produce zygospores, which are thick-walled, resistant spores that develop from a zygote.

One of the most common members of this class is *Rhizopus stolonifer*, the black bread mold. Infection begins when a spore germinates on the surface of bread, fruit, or some other organic matter and forms hyphae. Some of the hyphae extend rootlike anchoring structures, called rhizoids. The rhizoids secrete digestive enzymes and absorb dissolved organic materials. Other specialized hyphae, the sporangiophores, push up into the air, and sporangia form on their tips. As the sporangia mature, they become black, giving the mold its characteristic color. They eventually break open, releasing numerous airborne spores, each of which can germinate to produce a new mycelium.

Sexual reproduction in *Rhizopus* occurs when the specialized hyphae of two different mating strains meet and fuse, attracted toward one another by hormones that diffuse in the form of gases. The two strains are designated plus (+) and minus (−), since there are no morphological differences between them on which to base male and female designations. Septa, or cross walls, form behind the tips of the touching hyphae; the two tip cells thus formed are gametangia, one containing numerous + nuclei, the other containing numerous − nuclei. Two gametangia fuse, and the two types of nuclei then fuse, producing a diploid nucleus. The resulting multinucleate cell, the zygote, then forms a hard, warty wall and becomes dormant. During this dormant stage, it can survive periods of extreme heat or cold or desiccation. At the end of dormancy, only one diploid nucleus remains, and this undergoes meiosis when the zygote germinates. Only one of the four nuclei produced by meiosis generally survives. It commonly gives rise to a sporangiophore, from which airborne spores are released.

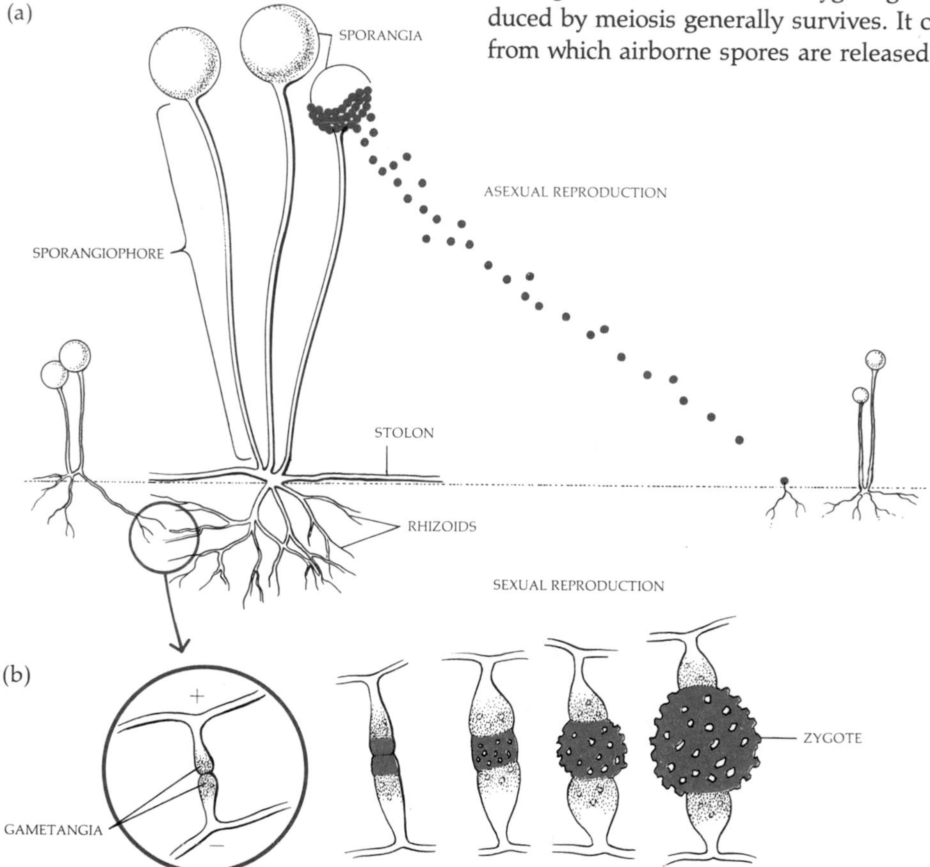

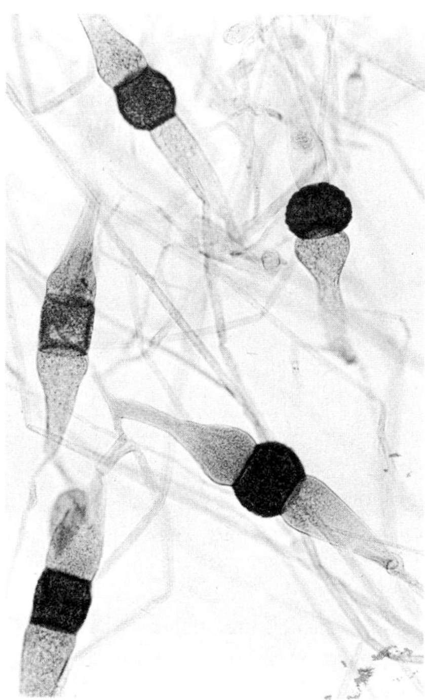

Carolina Biological Supply Co.

(c)

23–8
Asexual and sexual reproduction in Rhizopus. *The mold consists of branched hyphae, including rhizoids, which anchor the mycelium; stolons, which run aboveground; and sporangiophores.* (a) *At maturity, the fragile wall of the sporangium disintegrates, releasing the asexual spores, which are carried away by air currents. Under suitable conditions of warmth and moisture, the spores germinate, giving rise to new masses of hyphae.* (b) *Sexual reproduction occurs when two hyphae from different mating strains come together, forming gametangia, which fuse to form a thick-walled, resistant zygote, commonly called a zygospore* (c). *After a period of dormancy, the zygote undergoes meiosis and germinates, producing a new sporangium.*

23-9
Two ascomycetes. (a) *A common morel. These (and truffles) are among the most prized of the edible fungi.* (b) *Scarlet cup, a harbinger of spring in hardwood forests throughout the United States. It is usually found arising from a fallen branch.*

(a)

(b)

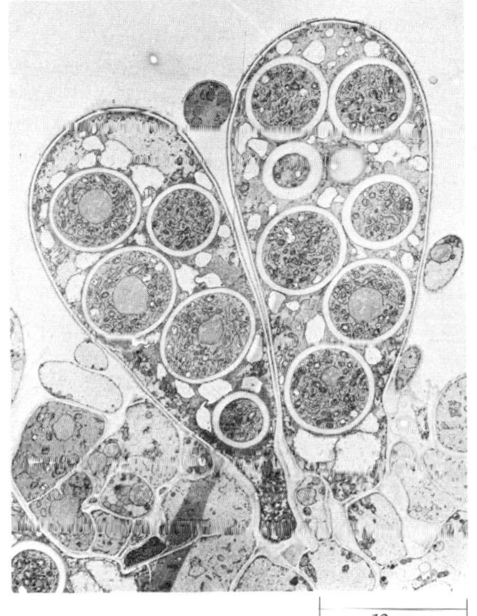

10 μm

23-10
Electron micrograph of two asci. The closed ascus within which the sexually produced spores develop is the "trademark" of the ascomycete.

Class Ascomycetes

The ascomycetes are the largest class of fungi (about 30,000 species), including the yeasts and powdery mildews, many of the common black and blue-green molds, and the morels and truffles prized by gourmets. Members of this group of fungi are the cause of many plant diseases, such as chestnut blight and Dutch elm disease, and are the source of many of the antibiotics. The red bread mold *Neurospora,* which played a major role in the history of modern genetics, is an ascomycete.

In ascomycetes, the hyphae are divided by cross walls, or septa, unlike the hyphae of the oomycetes and most zygomycetes. Each compartment generally contains a separate nucleus, but the septa have pores in them through which the cytoplasm and the nuclei can move. Spores are formed sexually and asexually. Asexual spores are formed either singly or in chains at the tip of a specialized hypha. They are characteristically very fine and so are often called *conidia,* from the Greek word for "dust."

Sexual reproduction always involves the formation of an *ascus* ("little sac"), a structure that is characteristic of the group. Ascus formation is preceded by the fusion of hyphae of different mating strains to form a dikaryon. The nuclei form pairs that divide synchronously as the hypha grows. Eventually some of the nuclei fuse; this is the only truly diploid stage in the life cycle. The diploid nuclei immediately undergo meiosis, producing four haploid nuclei, which then divide mitotically, producing eight haploid nuclei. Each of these nuclei becomes surrounded by a tough wall; a mature ascus contains eight of these spores (ascospores). In most ascomycetes, the ascus becomes turgid at maturity and finally bursts, releasing its ascospores explosively into the air.

Single-celled ascomycetes are known as yeasts. They are characteristically small, oval cells that reproduce by budding. Many yeasts are adapted to environments with high sugar content, such as the nectar of flowers or the surface of fruits and, as we noted in Chapter 8, they are responsible for the fermentation of fruit juice to wine.

Many members of this class are plant parasites. Ergot, for instance, one of the most famous fungus-produced diseases, is caused by *Claviceps purpurea*, a parasite of rye. Although ergot seldom causes serious damage to a crop of rye, it is dangerous because a small amount mixed with rye grains is enough to cause severe illness among domestic animals or among the people who eat bread made with the flour. Ergotism is often accompanied by gangrene, nervous spasms, psychotic delusions, and convulsions. It occurred frequently during the Middle Ages, when it was known as St. Anthony's fire. In one epidemic in the year 994, more than 40,000 people died. Ergot, which causes muscles to contract and blood vessels to constrict, has various medical uses. It is also the initial source for the psychedelic drug lysergic acid diethylamide (LSD).

Class Basidiomycetes

The most familiar basidiomycetes are mushrooms. The mushroom, which is the spore-producing body, is composed of masses of tightly packed hyphae. The mycelium from which the mushrooms are produced forms a diffuse mat, which may grow as large as 35 meters in diameter. Mushrooms usually form at the outer

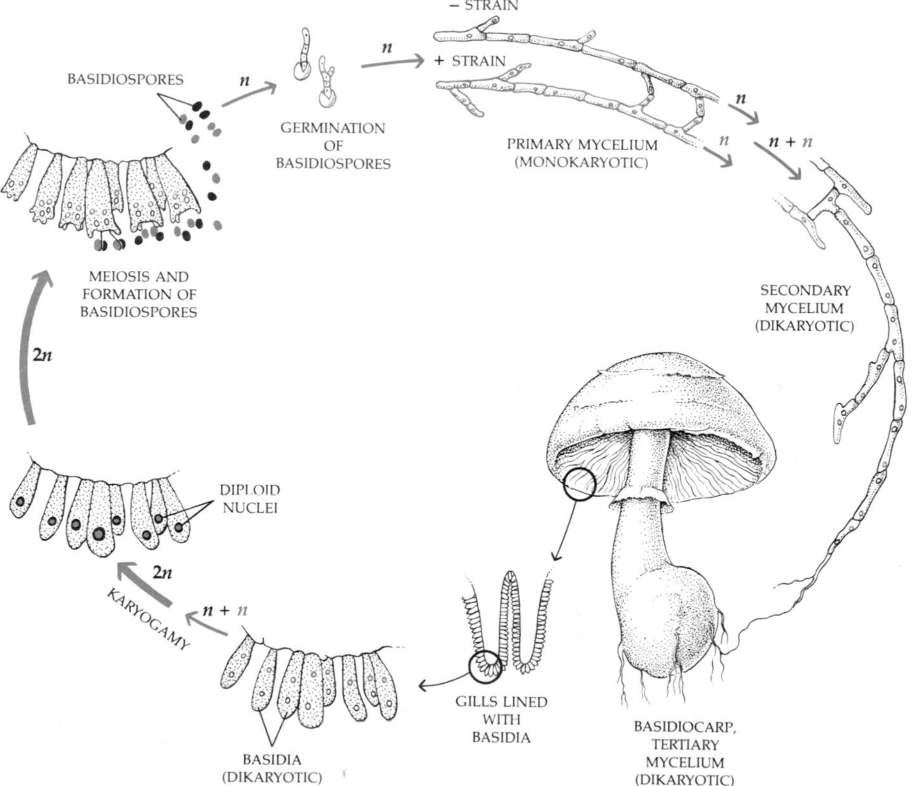

23–11
Life cycle of a basidiomycete. Basidiospores (upper left) germinate to produce primary monokaryotic mycelia. Secondary dikaryotic mycelia are formed by the fusion of hyphae from different mating types. The secondary mycelia divide and differentiate to form the reproductive structures (basidia). In mushrooms, the basidia form within the gill. After the basidium enlarges, the two nuclei, one from each mating strain, fuse. Meiosis follows almost immediately, resulting in the formation of four nuclei, from each of which a basidiospore develops.

edges, where the mycelium grows most actively, since this is the area in which there is the freshest nutritive material. As a consequence, the mushrooms will appear in rings, and as the mycelium grows, the ring becomes larger and larger in diameter. Such circles of mushrooms, which might appear in a meadow overnight, were known in European folk legends as "fairy rings." They can make such a rapid appearance because most of the production of new protoplasm takes place underground, in the mycelium. The protoplasm then streams into the new hyphae of the fruiting body as it forms aboveground.

The basidiomycetes have hyphae subdivided by perforated septa. Sexual reproduction is initiated by the fusion of haploid hyphae to form a dikaryon. The dikaryon may persist for years, forming an elaborate mycelium. Eventually, some of the nuclei fuse to form diploid nuclei that immediately undergo meiosis. Fusion and meiosis always take place in a specialized hypha called a *basidium* (from the Greek word for "club"). The spores (basidiospores) are formed externally on the basidium. Many of the larger basidiomycetes seem to have lost the capacity to produce asexual spores.

The best-known mushrooms belong to the group known as the gill fungi, which includes *Agaricus campestris*, the common field mushroom. Varieties of this species are the only mushrooms it has been possible to cultivate commercially. The gill fungi also include most of the known poisonous mushrooms. Mushrooms of the genus *Amanita* are the most highly poisonous of all mushrooms; even one bite of the white *Amanita verna*, the "destroying angel," can be fatal. Some species of toxic mushrooms, such as *Psilocibe mexicana* (the source of psilocybin), are eaten for their hallucinogenic effects.

The spores of the gill mushrooms are found in the furrows, or "gills," under the cap. If you cut off the cap of a mature mushroom and place it on a piece of white paper, it will release fine spores that will trace out a negative copy of the gill structure. The spores, which come in a variety of colors, are a useful means of identifying various mushrooms. Other types of basidiomycetes include puffballs (a few of which are a meter in diameter), earthstars, stinkhorns, and the parasitic rusts and smuts, some of which cause severe losses among cereal crops.

23–12
Three basidiomycetes. (a) *Corn smut, a common fungus disease of corn. The black, dusty-looking masses are spores.* (b) *Sulfur shelf fungi, which grow on decaying wood.* (c) Amanita muscaria. *Members of this genus include the most beautiful and also the most poisonous of the mushrooms. The ring near the top of the stalk is one of the identifying characteristics of the group. One mushroom has been picked to show the gills, on which the sexual spores are formed.*

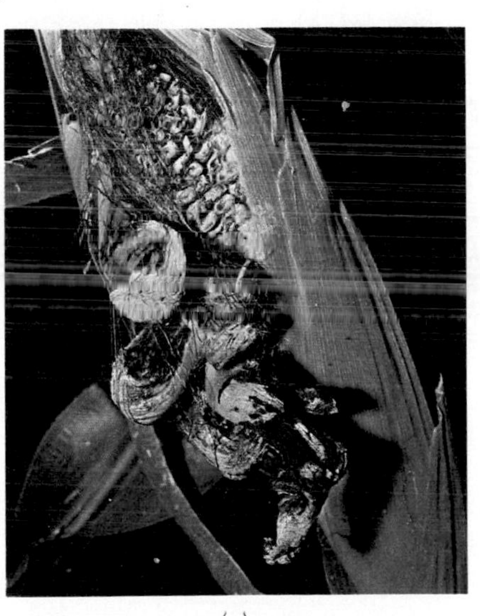

(a)

(b)

(c)

PREDACEOUS FUNGI

Among the most highly specialized of the fungi are the predaceous fungi that have developed a number of mechanisms for capturing small animals that they use as food. Some secrete a sticky substance on the surface of their hyphae in which passing protozoans, rotifers, small insects, or other animals become glued. More than 50 species of fungi capture small roundworms (nematodes) that abound in the soil. In the presence of a population of roundworms (or even of water in which the worms have been growing), the hyphae of the fungi produce loops that swell rapidly, closing the opening like a noose when a nematode rubs against its inner surface. Presumably the stimulation of the cell wall increases the amount of osmotically active material in the cell, causing water to enter the cells and rapidly increase their turgor. (a) *The predaceous imperfect fungus* Arthrobotrys dactyloides *has trapped a nematode. The traps consist of rings, each comprising three cells, which swell rapidly to about three times their original size and garrote the nematode. Once the worm has been trapped, fungal hyphae grow into its body and digest it. When triggered, the ring cells can expand completely in less than a tenth of a second.* (b) *Another nematode-trapping fungus,* Dactylella drechsleri. *This species traps the worms with small adhesive knobs.*

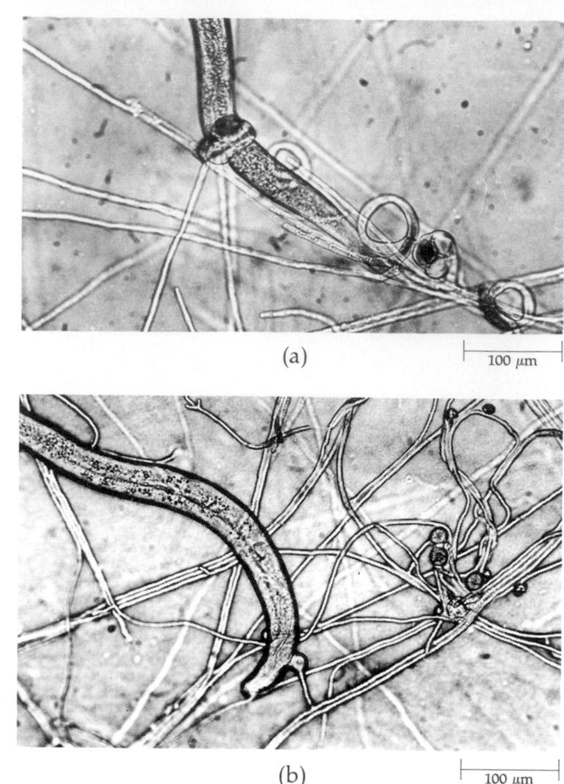

The Fungi Imperfecti

This class includes fungi in which sexual reproduction is unknown, either because it has disappeared or because it has not been observed. Also included in this class for convenience are certain other closely related fungi whose sexual stages are known. Taxonomists refer to groups like this as "waste baskets" because species are included here only because they do not fit in other taxa.

Among the Fungi Imperfecti are parasites that cause diseases of plants and animals. The most common human diseases caused by this group are infections of the skin and mucous membranes, such as ringworm (including "athlete's foot") and thrush. A few are of economic importance due to the part they play in the production of certain cheeses (Roquefort and Camembert, for example) and of antibiotics, including penicillin.

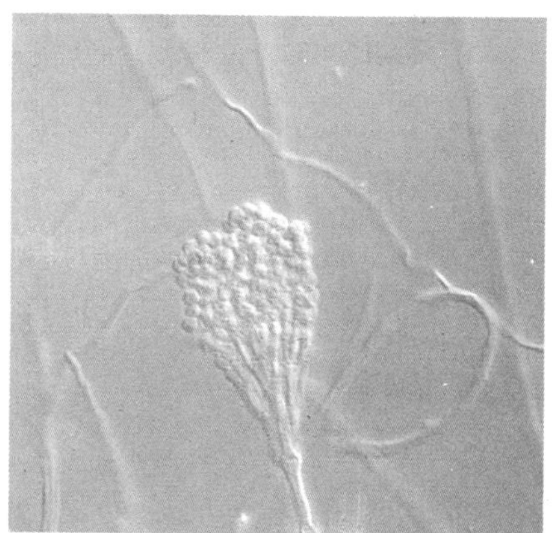

23–13
Penicillium, *an "imperfect fungus," showing conidiophores ("conidia-bearers"), which have formed at the tips of hyphae. Conidia are dust-fine asexual spores.*

THE LICHENS

A lichen is a combination of a specific fungus and an alga. The organisms resulting from these combinations are very different from either the alga or the fungus growing alone, as are the physiological conditions under which the lichen can

Table 23-1 *Major Classes of Fungi*

CLASS	NO. OF SPECIES	EXAMPLES	DISTINCTIVE CHARACTERISTICS	DISEASES	ECONOMIC USES
Oomycetes	Several hundred	Potato blight fungus	Some aquatic; flagellated spores; formation of eggs and sperm in special gametangia; cell walls contain cellulose	Blights and mildews of plants, fish infections	None
Zygomycetes	Several hundred	Black bread mold	Formation of zygospores (tough, resistant spores resulting from a fusion of gametangia); no flagellated cells	Few	None
Ascomycetes	30,000	*Neurospora*, yeasts, morels, truffles	Formation of fine asexual spores (conidia); sexual spores in asci; hyphae divided by perforated septa; dikaryons; no flagellated cells	Powdery mildews of fruits, chestnut blight, Dutch elm disease, ergot	Food (morels, truffles), wine, beer, bread-making (yeasts)
Basidiomycetes	25,000	Toadstools, mushrooms	Sexual spores in basidia; hyphae divided by perforated septa; dikaryons; no flagellated cells	Rusts, smuts	Food (mushrooms)
Fungi Imperfecti	25,000	*Penicillium*	Fungi with no known sexual cycles; no flagellated cells	Ringworm, thrush	Cheeses, antibiotics

survive. The lichens are widespread in nature. They occur from arid desert regions to the Arctic and grow on bare soil, tree trunks, sun-baked rocks, fence posts, and windswept alpine peaks all over the world. They are often the first colonists of bare rocky areas.

Lichens do not need an organic food source, as do their component fungi, and unlike free-living algae, they can remain alive even when desiccated. They require only light, air, and a few minerals. They apparently absorb some minerals from their substrate (this is suggested by the fact that particular species are characteristically found on specific kinds of rocks or soil or tree trunks), but minerals reach the plant primarily through the air and in rainfall. Because lichens rapidly absorb substances from rainwater, they are particularly susceptible to airborne toxic compounds. Thus, the presence or absence of lichens is a sensitive index of air pollution.

The algae found in lichens also occur as free-living species, whereas the lichen fungi are generally found in nature only in the lichens. For these reasons, lichens are generally classified according to the species of the fungal complement.

There are about 17,000 species of lichens—that is, of lichen fungi. Most are ascomycetes or Fungi Imperfecti, although a few are basidiomycetes. Algae from some 30 different genera are found in symbiotic combination with these fungi.

23–14
(a) *Lichen growing on a dead cedar in a North Carolina salt marsh.* (b) *British soldier lichen* (Cladonia). *Each soldier (so called because of the scarlet color) is about 3 millimeters tall.* (c) *Crustose lichens on a rock.*

(a)

(b)

(c)

Lichens reproduce most commonly by the breaking off of fragments containing both fungal hyphae and algae. New lichens are formed by the capture of an alga by a fungus. Sometimes the alga is destroyed by the fungus, in which case the fungus also dies. If it survives, a lichen is produced.

SUMMARY

The fungi are considered in this text to be a separate kingdom. They are coenocytic, consisting of cytoplasm moving in a system of tubes composed of chitin. They are heterotrophs, deriving their nutrition by absorption of organic compounds digested extracellularly by secreted enzymes. Their bodies are composed of masses of filaments called hyphae, sometimes with perforated septa (walls). The principal component of their cell walls is usually chitin. They form both asexual and sexual spores, although not all fungi form both kinds. The spores may be formed in sporangia. The sexual cycle is initiated by the fusion of hyphae of different mating strains. In some groups of fungi, the nuclei in the fused hyphae immediately combine, and a zygote is formed. In others (the ascomycetes and basidiomycetes), the two genetically distinct nuclei remain separate, forming pairs that divide synchronously, sometimes over prolonged periods. A cell or organism with such paired, genetically distinct nuclei is known as a dikaryon. Once the nuclei fuse, meiosis always follows immediately.

There are five principal classes of fungi: the Oomycetes, or water molds, which form flagellated spores; the Zygomycetes, which have nonmotile spores and whose zygotes become thick-walled zygospores; the Ascomycetes, in which the sexual spores develop within an ascus (sac); the Basidiomycetes, in which the sexual spores develop on a basidium (club); and the Fungi Imperfecti, most of which have no known sexual cycle.

Fungi have an important ecological role as decomposers of organic material. They also are major plant pathogens.

Lichens are combinations of fungi and algae that are morphologically and physiologically different from either organism as it exists separately. They are able to survive under adverse environmental conditions where neither partner could exist independently. The lichen represents a symbiotic relationship in which the fungus encloses the algal cells and is dependent on the algal cells for nourishment.

QUESTIONS

1. Distinguish between the following: hypha/mycelium; chitin/cellulose; sporangia/gametangia.

2. Give the distinguishing characteristics of the five major classes of fungi.

3. Coenocytic organisms—the fungi, for instance—show little differentiation. When differentiation does occur, as in basidium formation, it is preceded by construction of a septum. In your opinion, why?

4. As you can see from the last two chapters, our method of classification into kingdoms does not produce entirely satisfactory results. Can you suggest alternatives?

5. What type of symbiotic relationship would you say exists between the alga and the fungus of a lichen?

CHAPTER 24 *Plants and Their Ancestors*

24–1
Indian pipe, which lacks chlorophyll, was long thought to live on decaying organic matter in the soil. Recently it has been shown that it depends for its nutrition on a fungus that transports carbohydrates from a photosynthetic plant.

Plants are multicellular photosynthetic organisms adapted for life on land. As usual there are a few exceptions to this simple statement. Some plants—such as Indian pipe and dodder—in the course of evolution have become parasites and are no longer photosynthetic (Figure 24–1). Others, such as the water fern *Marsilea,* have returned to an aquatic existence (Figure 24–2); however, like whales and porpoises, these plants bear the unmistakable traces of an ancestral sojourn on the land.

As we have noted, all plants appear to have arisen from a group of the green algae (Chlorophyta). There are several lines of evidence leading to this conclusion. Some of the clues are biochemical: Like the plants, the green algae contain chlorophylls *a* and *b* as their photosynthetic pigments; they accumulate their food as starch; and their cell walls contain cellulose. Additional evidence is found in their pattern of cytokinesis. In almost all other organisms, division of the cytoplasm takes place by constriction and pinching off of the cell membrane. In plants and in a few species of green algae (Figure 24–3), the cytoplasm is divided by the formation of a cell plate at the equator of the spindle. (Cell-plate formation is also seen in one species of brown algae, but this alga does not contain chlorophyll *b* or store its food as starch.)

24–2
The water fern, Marsilea.

24–3
Cell division in Ulothrix, *a filamentous green alga. Formation of the cell plate is almost complete. Note the large chloroplasts and the two Golgi bodies just outside the nuclear envelopes.*

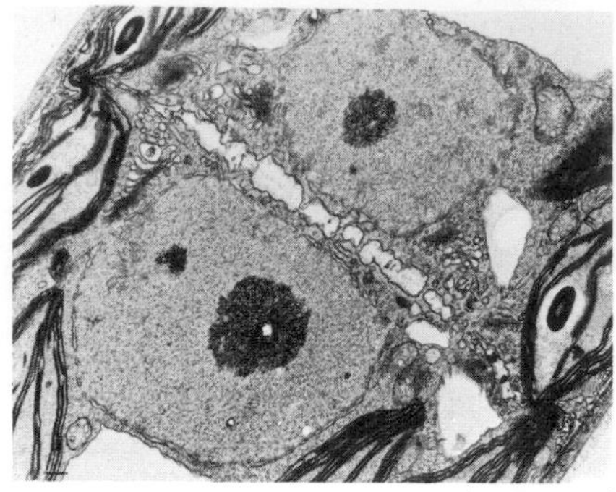

Finally, all multicellular green algae have a life cycle characterized by alternation of generations, which is also found in all plants. As we described in Chapter 11, in this type of life cycle, a diploid, spore-producing generation alternates with a haploid, gamete-producing generation. The plant that produces gametes is known as the *gametophyte,* the gamete-producing plant. The individual cells of this plant are haploid *(n)* and so are the gametes it produces. The gametes fuse to form a diploid (2*n*) zygote. The zygote develops into a plant in which all the cells are diploid. This plant is known as the *sporophyte,* the spore-producing plant. Spores, which are produced by meiosis, are always haploid.

Among plants, spores differ from gametes in that when a spore germinates it forms a new organism, while a gamete must first unite with another gamete before further development occurs. The gametophyte and the sporophyte are always genetically different, since one is composed of haploid cells and the other of diploid cells.

In some chlorophytes, such as the sea lettuce *Ulva* (Figure 24–4), the two generations look alike. In other green algae, the sporophyte and the gametophyte do not resemble one another, and the generations are said to be heteromorphic. In fact, in some cases, the two generations of the same alga were once considered to be two entirely different species until the organism was studied in the laboratory. Figure 24–5 shows such an alga—a large coenocytic alga like *Valonia*—which was called *Derbesia* in its sporophyte (2*n*) form and *Halicystis* in its gametophyte *(n)* form.

The gametes may be identical in size and structure, as they are in *Ulva;* this condition is known as *isogamy.* When one gamete is larger than another, but both are motile, the condition is known as *heterogamy,* as in *Derbesia* (Figure 24–5). When

24–4
In the sea lettuce, Ulva, *we can see the reproductive pattern known as alternation of generations, in which one generation produces spores, the other gametes. The haploid* (n) *gametophyte produces haploid isogametes, and the gametes fuse to form a diploid* (2n) *zygote. A sporophyte, a multicellular body in which all the cells are diploid, develops from the zygote. The sporophyte produces haploid spores by meiosis. The haploid spores develop into haploid gametophytes and the cycle begins again. The micrograph shows gametes of* Ulva *before cytoplasmic fusion.*

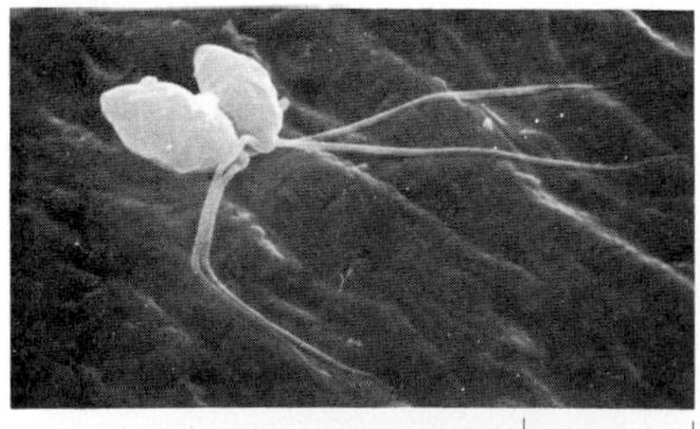

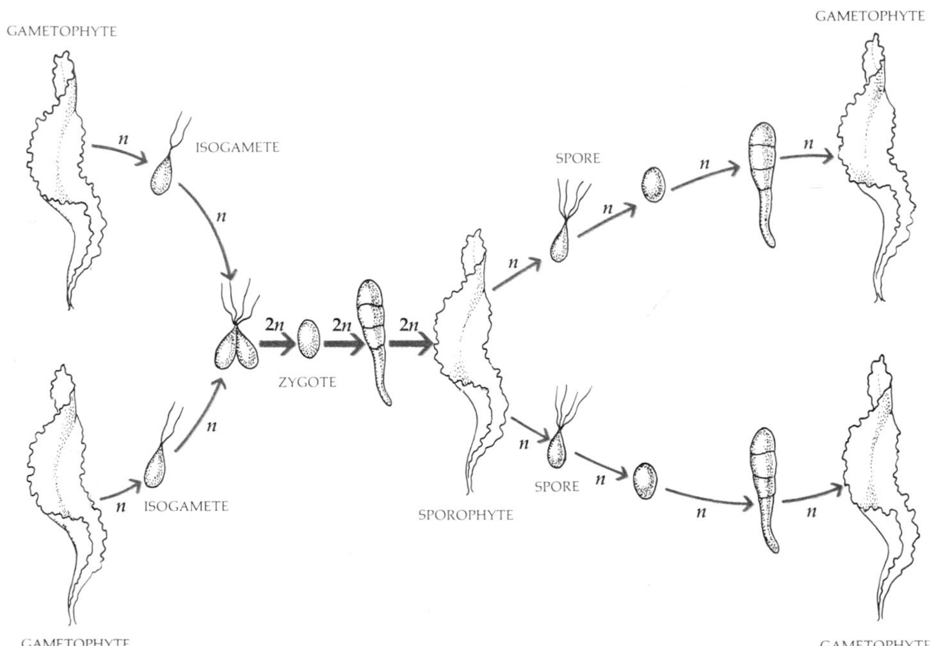

24-5
In some algae the alternating generations are so different that they once were believed to be different genera. At one time, the gametophyte of the Valonia-*like alga shown here was called* Halicystis *and the sporophyte* Derbesia.

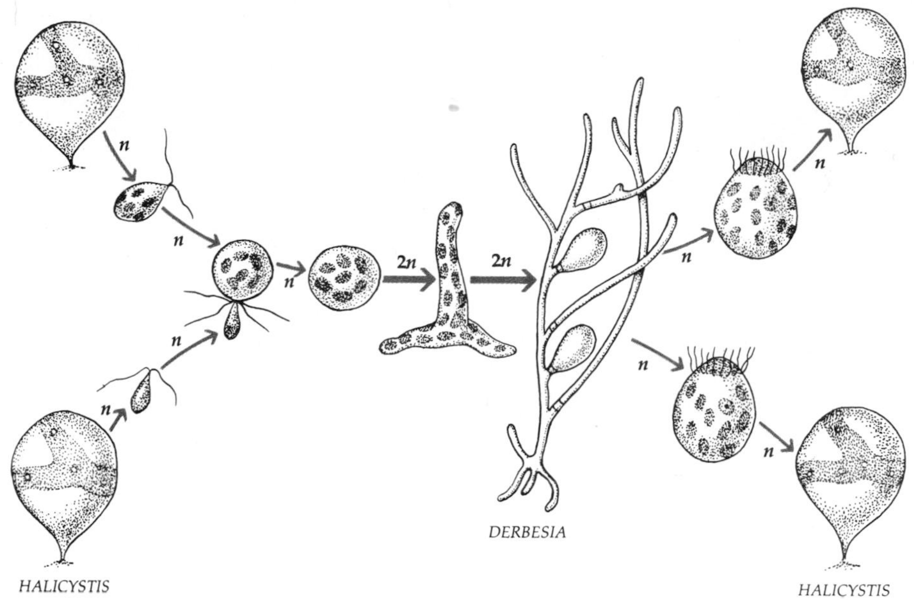

24-6
Evolution of sex. (a) *Isogamy. The gametes are similar in size and shape.* (b) *Heterogamy. One gamete, conventionally termed male, is smaller than the other.* (c) *Oogamy. One gamete, usually the smaller, is motile, and the other is not.*

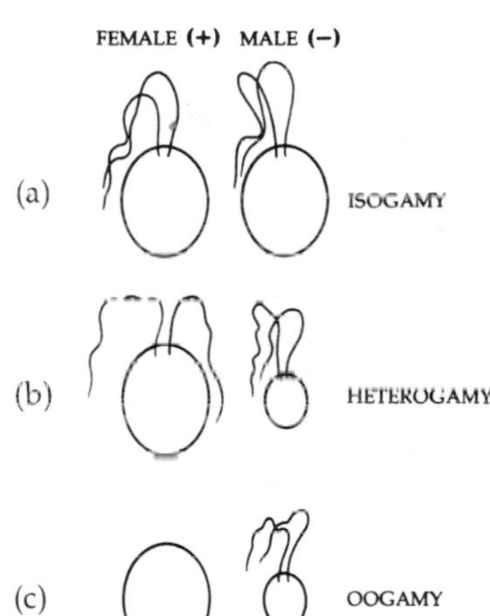

one gamete is motile and the other is not, the condition is *oogamy* (Figure 24–6c). An individual that produces gametes that are nonmotile and (usually, though not always) larger is known as female, and this is where sexual differentiation, with its mixed blessings and many complications, all began. All plants are oogamous and heteromorphic.

THE TRANSITION TO LAND

Water was the first life zone for living things. Land was the next life zone, and air was the third. We are now beginning the occupation of a possible fourth life zone, outer space.

The second life zone, the land, offered a wealth of advantages to plants. Light was abundant from daylight to dusk, unfiltered by the turbulent water. Carbon dioxide, needed for photosynthesis, was plentiful in the atmosphere, as was oxygen by this time, and both circulated much more freely in air than in water. And there was a great abundance of unoccupied space, as compared with the shorelines, now teeming with life.

Terrestrial existence presented some major difficulties, however. First and foremost—and still the principal problem for many plants—is water loss and desiccation. Water is freely available to every cell of a green alga, but the major source of water for plants is beneath the ground. No light is available there for photosynthesis, however, and so the belowground parts of a plant are, in essence, heterotrophic, depending on the aboveground photosynthetic plant parts for organic molecules. The modern plant represents a collection of solutions to and compromises with these problems.

ADAPTATIONS TO LAND

One of the principal adaptations to land, which apparently occurred early in plant evolution, was the development of a fatty protective cuticle (see Figure 3–15, page 61), which reduces water loss from the aboveground portions of the plant. The stomata, openings that permit the gas exchange necessary for photosynthesis, are a related adaptation.

Another adaptation related to the hazard of water loss was the development of multicellular reproductive organs with a protective layer of sterile (nonreproductive) cells around the gametangia. These gametangia are called *archegonia* if they give rise to egg cells and *antheridia* if they give rise to sperm (Figure 24–7).

Among algae the zygote leads an independent existence. In plants, the zygote and embryo are retained within the female gametangium, and the embryo and young sporophyte are protected during their early development.

The presence of lignin in the cell walls is another feature that appeared early in plant evolution. Lignin stiffens the cell walls so they are better able to support the plant body.

As past evolutionary events are reconstructed, it is apparent that these developments took place before the plants diverged into the two separate lines, represented today by: (1) Bryophyta, a small phylum that includes the mosses, hornworts, and liverworts, and (2) Tracheophyta, the vascular plants, a large phylum that includes all of the dominant modern plants. The principal difference between the bryophytes and the tracheophytes is that the tracheophytes have a well-developed vascular system that transports water, minerals, sugars, and other nutrients.

According to this evidence, the bryophytes and the vascular plants diverged long ago. The oldest fossils of vascular plants are from the Silurian period, which

24–7
Multicellular gametangia of Marchantia, *a bryophyte.* (a) *Female gametangia, or archegonia, in various stages of development. The archegonia are flask-shaped, with a single egg developing in the base of the flask.* (b) *A developing antheridium, containing future sperm cells. When mature, the sperm cells will swim to the egg through the neck canal of the female gametophyte.*

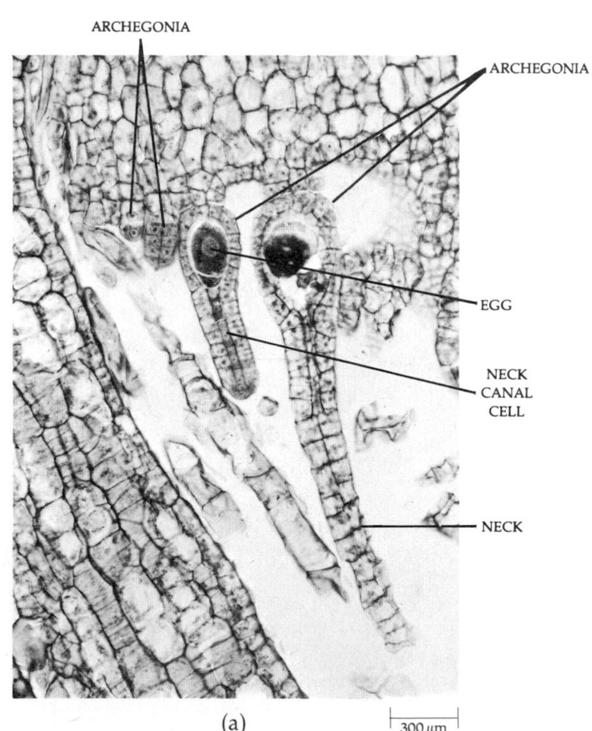

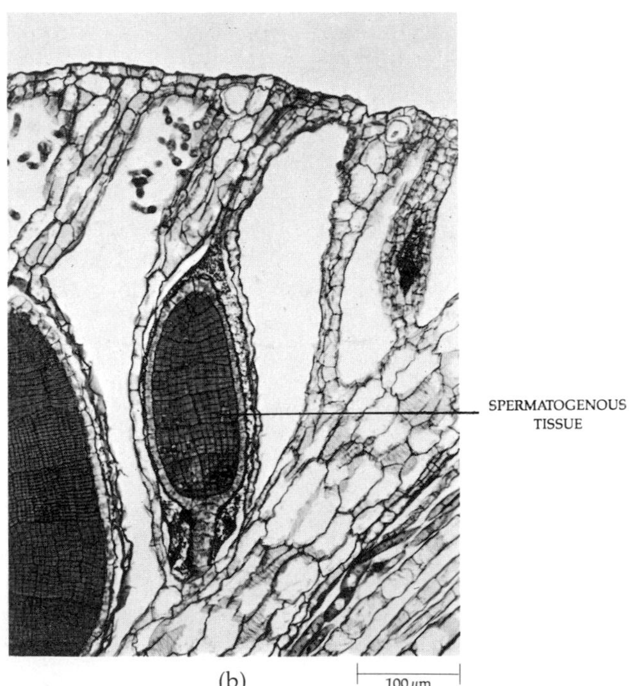

24-8
Evolution of plants.

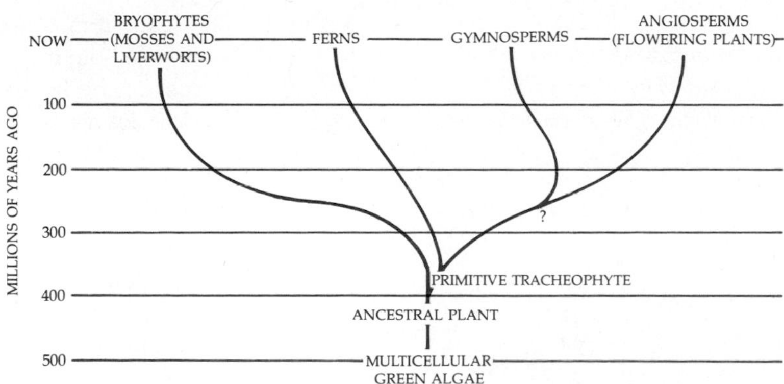

Table 24-1 *The Bryophytes (Land Plants without True Vascular Systems)*

CLASS	COMMON NAME	NO. OF SPECIES
Hepaticae	Liverworts	9,000
Anthocerotae	Hornworts	100
Musci	Mosses	14,500

ended some 400 million years ago. The bryophytes first appear in the fossil record in the Devonian period, more than 350 million years ago (see Table 26-2, page 478). These ancient fossils are quite similar to bryophytes living today.

PHYLUM BRYOPHYTA: MOSSES, HORNWORTS, AND LIVERWORTS

Lacking water-gathering roots and specialized vascular tissues, the bryophytes absorb moisture through aboveground structures. As a consequence, they grow most successfully in moist, shady places and in bogs. Some of them, like sphagnum (peat moss), are able to absorb and hold large amounts of water, so that they in effect maintain a watery existence even on land. Most bryophytes are tropical, but some species occur in temperate regions, and a few even reach the Arctic and Antarctic.

In the damp environments frequented by the bryophytes, individual cells can absorb water and nutrients directly from the air or by diffusion from nearby cells. (Thus, like lichens, they are sensitive indices of air pollution.) These plants do not have any stiff supporting structures to hold them erect, and, in general, they have no conducting cells to transport water up the plant body. As a consequence, many of them are very small, and even the largest reach little more than 30 centimeters above the ground.

Bryophytes have motile sperm; free water is necessary for the completion of their sexual cycle, because the sperm swim to the egg. Thus the bryophytes are not completely adapted to land.

The bryophytes do have some special structural adaptations to a terrestrial existence. They have small leaflike structures in which photosynthesis take place, and, as in the tracheophytes, the body of the plant is specialized for support and food storage. Thus the bryophytes resemble other plants far more than they do the algae, although they are not usually classified as "higher plants" and do not seem to have been their ancestors.

The bryophytes do not have roots. In the mosses, the individual plants (gametophytes) grow up like branches from a network of horizontal filaments known as protonemata (singular, protonema). Each moss plant is attached to the substrate by means of elongate single cells called rhizoids. A single moss plant may sprawl over

24–9
Protonema of a moss. Protonemata, which are young gametophytes, are characteristic of some bryophytes. They resemble filamentous green algae.

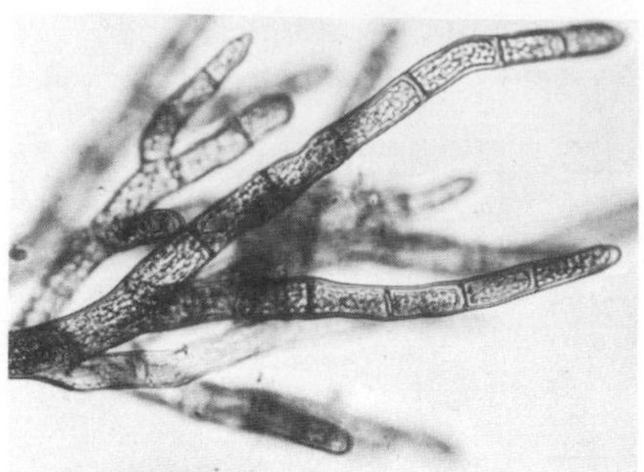

a considerable area, but most liverworts are so small that they are noticeable only to a keen observer. In both mosses and liverworts, small leaflike structures lack the specialized tissues of the leaves of higher plants and are only one or a few cell layers thick. Moss leaves are believed to have evolved separately from "true" leaves.

Bryophyte Reproduction

In the bryophytes, unlike any tracheophytes, the gametophyte (the haploid form, you will recall) is often larger than the sporophyte. The antheridia and archegonia are borne on the gametophyte. When sufficient moisture is present, the sperm, which are biflagellate, are released and swim to the archegonium, to which they are attracted chemically. Fusion of sperm and egg takes place within the archegonium.

Inside the archegonium, the zygote develops into a sporophyte, which remains attached to the gametophyte and is often nutritionally dependent on it. Typically the sporophyte has a single, large sporangium (or capsule) elevated on a stalk, from which the spores are discharged.

Asexual reproduction, often by fragmentation, is also common. A moss life cycle is shown in Figure 24–10.

24–10
In a representative moss life cycle, spores are released from a capsule, which opens when a small lid (operculum) bursts. The spore germinates to form a branched, filamentous protonema, from which a leafy gametophyte develops. Sperm cells, which are expelled from the mature antheridium, are attracted into the archegonium, where one fuses with the egg cell to produce the zygote. The zygote divides mitotically to form the sporophyte and, at the same time, the venter of the archegonium divides to form the protective calyptra. The sporophyte consists of a capsule, which may be raised on a stalk, also part of the sporophyte, and a foot. Meiosis occurs within the capsule, resulting in the formation of haploid spores.

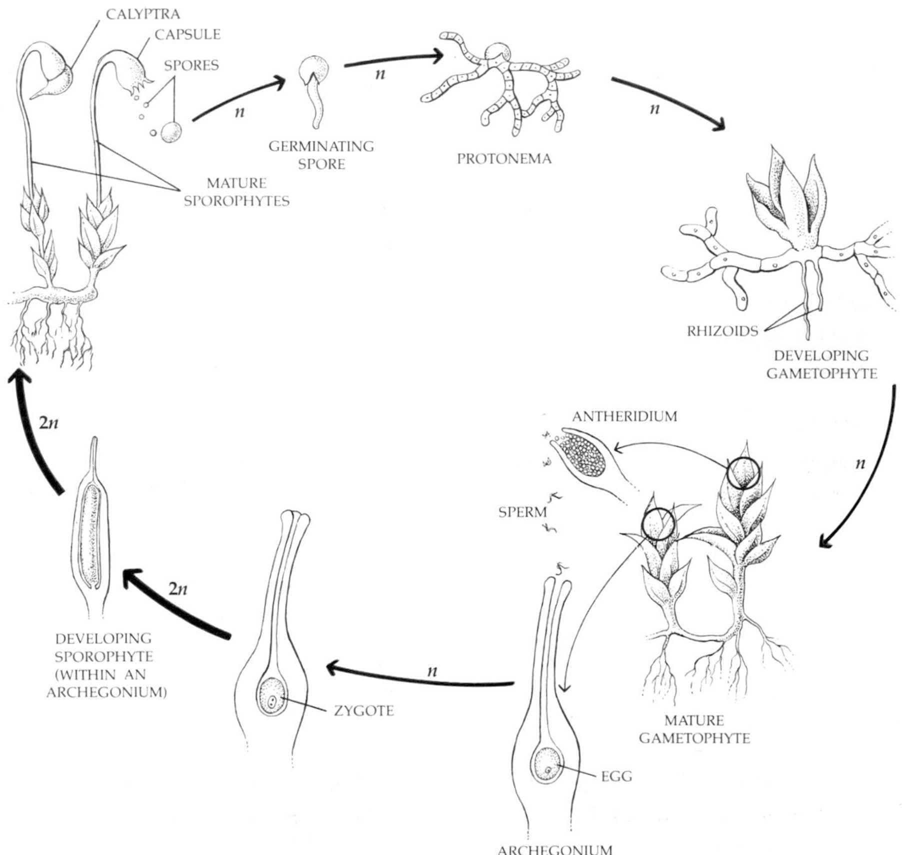

24–11
A haircap moss with spore capsules. The lower green structures are female gametophytes; the stalks and capsules are the sporophytes.

24–12
(a) *Young liverwort growing on a rock. In the liverwort* Marchantia, *the archegonia and the antheridia are formed on different plants, the female* (b) *and the male* (c) *gametophytes. The zygote, formed in the archegonium, develops into the sporophyte, which remains attached to the female gametophyte. The cuplike gemmae visible in* (c) *participate in asexual reproduction.*

(a)

(b)

(c)

24–13
Rhynia major, *one of the earliest known vascular plants. It lacked leaves and roots. Its aerial stems, which were photosynthetic, were attached to an underground stem, or rhizome. The aerial stems were covered with a cuticle and contained stomata. The dark structures at the tips of the stems are sporangia, which released the spores by splitting longitudinally.*

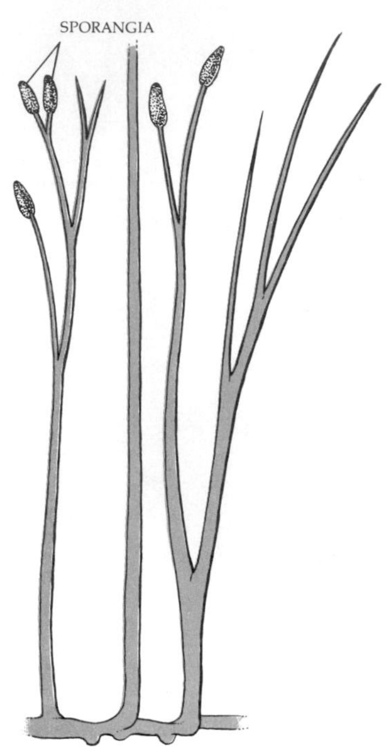

PHYLUM TRACHEOPHYTA: THE VASCULAR PLANTS

Rhynia major (Figure 24–13) is an example of the earliest known vascular plants, dating back some 400 million years ago. As you can see, it does not look much like any modern vascular plant and, indeed, is much more primitive in appearance than a bryophyte. However, it differs from all bryophytes in one important respect: Within its stem is a central cylinder of vascular tissue, specialized for conducting water up the plant body and products of photosynthesis down.

The vascular system in *Rhynia,* and in modern tracheophytes as well, consists of two distinct tissues: the *phloem,* which conducts sugars and other soluble organic molecules from the leaves to other parts of the plant body, and the *xylem,* which conducts water and minerals up from the roots. The conducting elements of the phloem are *sieve cells* or *sieve-tube elements,* and the conducting elements of the xylem are *tracheids* and *vessel elements.* In stems, the strands of xylem and phloem are side by side, either in vascular bundles or arranged in two concentric layers (cylinders), in which one tissue (typically the phloem) appears outside the other.

Rhynia major and its relatives are now extinct, as are some other subphyla of vascular plants.

Three groups of vascular plants still have living representatives. Two of the groups are small: the Lycophytina (club mosses) and the Sphenophytina (horsetails). The third group, the Pterophytina, a very large subphylum, includes all the major groups of plants: ferns, gymnosperms, and flowering plants.

Trends among Vascular Plants

Beginning with a very simple vascular plant, such as *Rhynia,* it is possible to trace some major evolutionary trends. One early development was the root, a structure specialized for the absorption of water. Another was the leaf. Two distinct types of leaves evolved, the microphyll and the megaphyll. The microphyll contains only a single strand of vascular tissue, whereas the megaphyll typically contains a complex system of veins. These two types of leaves seem clearly to have evolved in different ways (Figure 24–14). Of the three groups of modern plants, the Lycophytina and the Sphenophytina have microphylls and the Pterophytina have megaphylls.

A third evolutionary trend is the development of increasingly complex and efficient vascular systems.

24–14
According to one widely accepted theory, microphylls (left) evolved as outgrowths of the main axis of the plant. Megaphylls (right) evolved by fusion of branch systems.

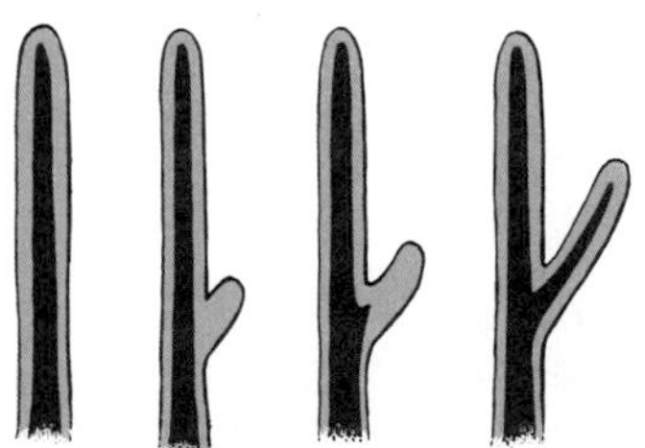

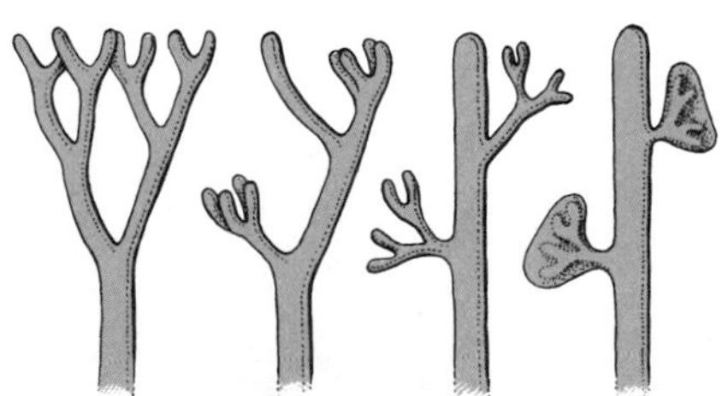

(a)

(b)

(c)

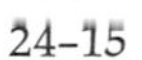

24-15
(a) *In the club mosses,* Lycopodium, *the sporangia are borne on specialized leaves, sporophylls, which are aggregated into a cone at the apex of the branches, as shown in this running ground pine,* Lycopodium complanatum. *The airborne, waxy spores give rise to small, independent, subterranean gametophytes. The sperm, which are biflagellated, swim to the archegonium, and the plant embryo develops there.* (b) *Tree club moss,* Lycopodium obscurum. (c) *The horsetails, of which there is only one living genus* (Equisetum), *are easily recognized by their jointed, finely ribbed stems, which contain silicon. At each node, there is a circle of small, scalelike leaves. Spore-bearing structures are clustered into a cone at the apex of the stem. The gametophytes are independent, and the sperm are coiled, with numerous flagella.*

A fourth trend is seen in the reduction of the gametophyte. In all vascular plants, the gametophyte is smaller than the sporophyte. However, in the more primitive tracheophytes, the gametophyte is separate and nutritionally independent, whereas in the more recently evolved forms, it has been reduced to microscopic size and dependency status.

Also related to the reproductive cycle is the trend toward heterospory. The earliest vascular plants produced only one kind of spore (homospory) in one kind of sporangium. Upon germination, such spores produce gametophytes on which both antheridia and archegonia form. In plants that are heterosporous, two different kinds of gametophytes develop; one bearing archegonia and the other, antheridia. (As the gametophytes become reduced, archegonia and antheridia decrease in size until, in the most highly evolved gymnosperms and in the angiosperms, they disappear.)

The final and perhaps the most significant innovation was the seed. The seed is, in effect, the sporangium, outside of which is one or two layers of tissue, the integument, and inside of which is the spore. The spore germinates to form the female gametophyte, in which an egg cell differentiates and becomes an embryonic sporophyte—"three generations under one roof." The sporangium with its integument and contents is called the <u>ovule</u>, and it is the mature ovule that is properly known as the seed. The earliest known seeds were fossilized in late Devonian deposits some 350 million years ago.

All of these developments will be discussed in greater detail in the pages that follow.

24–16
Fern spores develop on the sporophyte in sporangia, which are usually found in clusters (sori) on the underside of a leaf (sporophyll), as in the woodfern shown here.

Class Filicinae: Ferns

Ferns are vascular plants that can usually be distinguished from most other plants by their large, feathery leaves, which, in most species, unroll from base to tip during growth. The sporangia commonly are on the undersurface of the leaves or, sometimes, on specialized leaves. Spore-bearing leaves are called *sporophylls*. As we shall see, the carpels and stamens of flowers are also sporophylls.

According to the fossil record, ferns first appeared almost 400 million years ago, and they are still relatively abundant. Most of the 11,000 living species are found in the tropics, but some occur in temperate and even arid regions. Because they have flagellated sperm and need free water for fertilization, those species growing in arid regions exploit the seasonal occurrence of water for sexual reproduction.

The stems of ferns are anatomically simple compared with those of gymnosperms and angiosperms, and are often reduced to a creeping underground stem (rhizome). Although ferns do not exhibit secondary growth—the type of growth that results in increase of girth and formation of bark and woody tissue—some grow very tall. For instance, *Cyathea australis,* a tree fern found on Norfolk Island in the South Pacific, sometimes reaches 28 meters in height.

The leaves of ferns are usually finely divided into pinnae; these divided leaves spread widely and so collect more light, and apparently they are thus adapted to growing on the forest floor in diffused light. The sporangia develop on the lower

(a)

(b)

24–17
(a) *The immature sporophyte of many common ferns develops as a "fiddle head," which uncoils and spreads as it elongates.* (b) *A cinnamon fern. The sporangia are borne on separate stalks.*

24–18
Reconstruction of a Devonian forest showing western New York as it is believed to have looked 370 million years ago. The tall plant to the right of center is a relative of the club mosses. Although modern club mosses are typically only a few centimeters high, these primitive trees often grew to heights of 30 meters. The tall plants at the left and right are tree ferns. In the foreground at the right is a primitive leafless plant, Psilophyton, *related to the ferns. Several clumps of horsetails, identifiable by their whorled branches, are also visible.*

surface of sporophylls, which may resemble the other green leaves on the plant or may be nonphotosynthetic stalks (modified leaves). The sporangia commonly occur in small clusters known as sori (singular, sorus).

In the ferns, as in all the vascular plants, the predominant plant form is the sporophyte. The gametophyte of the homosporous ferns begins development as a small algalike filament of cells, each filled with chloroplasts, and then develops into a flat structure, often only one layer of cells in thickness. Although this gametophyte is small, it is nutritionally independent, as is the sporophyte. All but a few genera are homosporous, and the single gametophyte produces both antheridia and archegonia. The sperm are coiled and multiflagellate. The life cycle of a fern is shown on page 211.

24–19
Ponderosa pine (also called western yellow pine) in Colorado. Like most other conifers, pine trees have needlelike leaves and their gametophytes form on the scales of cones. As in other pines, the long, flexible needles are held together in a bunch by a papery sheath. Ponderosa pine, the main forest growth of the Rockies from Canada to Mexico, is a staple of the Northwest lumber industry.

Gymnosperms

The humid Paleozoic era was the age when the earth's coal deposits were formed from lush vegetation that sank so swiftly into the warm, marshy soil that there was no chance for much of it to decompose. At the close of this era, in the Permian period, 225 to 280 million years ago, there were worldwide changes of climate, with the advent of widespread glaciers and drought. The amphibians gave way to the scaly-skinned reptiles, which may have been better suited to the harsher climate. Under similar selective pressures, plants began to evolve specialized structures that enabled them to survive during periods when no water was available.

It was during the Permian period that the gymnosperms—the naked-seed plants—evolved. There are four groups of gymnosperms with living representatives, three small groups (the cycads, the ginkgos, and the Gnetinae) and one large and familiar group (the conifers).

The Seed

The seed is a protective structure in which the embryonic plant can lie dormant until conditions become favorable for its survival. Thus, in its function, it parallels the spores of bacteria or the resistant zygotes of the freshwater algae. In structure it is far more elaborate, however. A seed includes the embryo (the young, dormant sporophyte), a store of nutritive tissue, and an outer protective coat.

To understand the evolution of the seed it is necessary to return for a moment to the life cycle, or alternation of generations, found in all the vascular plants, and foreshadowed in the green algae. In the green algae, the two generations, sporo-

24-20
(a) *Male cones of jack pine* (Pinus banksiana) *shedding pollen. The pollen grains are immature male gametophytes, which complete their maturation when they reach the ovules, embedded in the female cones. There they release sperm cells that fuse with the egg cells on the scales of the female cone.* (b) *Female cone. The female gametophyte develops in an ovule on the base of a scale of the cone and is fertilized there. Each scale contains two ovules. When the seeds are mature, they drop from the cone.*

(a) 1 cm

(b) 2 mm

phyte and gametophyte, are independent. In most ferns, the gametophytes, although still independent, are usually smaller than the sporophytes. In the seed plants, the gametophytic generation is reduced still further and is totally dependent on the sporophyte. All gymnosperms are heterosporous, producing both male and female gametophytes.

The female gametophyte develops on the mother sporophyte within the ovule. Within the female gametophyte, one or more egg cells are formed. Sperm cells develop in another gametophyte, also protected and nourished by a sporophyte. They are carried to the female gametophyte and fertilize the eggs developing there. The zygote develops into an embryo. Following fertilization, the ovule enlarges and its outer surface hardens, forming a protective cover around the embryo and the gametophyte tissue in which the embryo is embedded. This complex is released as the seed.

Let us look at a specific example, the formation of a pine seed. A pine tree has two types of cones, which produce the two types of spores. The small "male" cones resemble those of club mosses, but on the large "female" cones, the scales that bear the ovules are much thicker and tougher than the sporophylls of the male cones. In the male cones, specialized cells inside the sporangia undergo meiosis to produce haploid spores. Each spore differentiates into a microscopic, windborne pollen grain, an immature male gametophyte. The wind is an unreliable messenger, disseminating the pollen grains at random, and wind-pollinated plants characteristically produce pollen in great quantities (Figure 24-20a).

Fertilization

Within the ovules of the female cones, a second type of spore is formed by meiosis. (Spores that give rise to female gametophytes are often referred to as megaspores, as distinct from microspores, from which male gametophytes arise.) Of the four spore cells produced within the ovule by each meiotic sequence, three disintegrate and the remaining one forms a tiny gametophyte. This haploid gametophyte grows within the cone and develops two or more archegonia, each of which contains a single egg cell. The development from the spore into the gametophyte with its egg cells may take many months—slightly more than a year, for example, in some common pines. The ripening ovule secretes a sticky liquid. When the cones become dusted with pollen, some of the pollen sifts down between the cone scales and comes into contact with the liquid. Pollen grains, caught in the sticky liquid, are drawn to the ovule as the liquid dries. Here the pollen grain develops into a mature gametophyte. The male gametophyte produces two nonmotile cells, the male gametes, or sperm. These are carried toward the egg within the pollen tube, which is produced by the male gametophyte and which grows through the tissues of the ovule.

Because the drought-resistant pollen is blown to the ovule of the female cones by the wind, and the sperm are carried to the egg by the pollen tube, the pines and other conifers are not dependent on free water for fertilization. Thus they are able to reproduce sexually when (and where) ferns and bryophytes cannot.

Following fertilization, the zygote begins to divide and forms the embryo. As the ovule matures, its outer walls harden into a seed coat, enclosing both the embryo and the female gametophyte (the latter provides food for the embryo when the seed germinates). After the cone matures, it releases its seeds. In most conifers the seeds are winged and are spread widely by the wind.

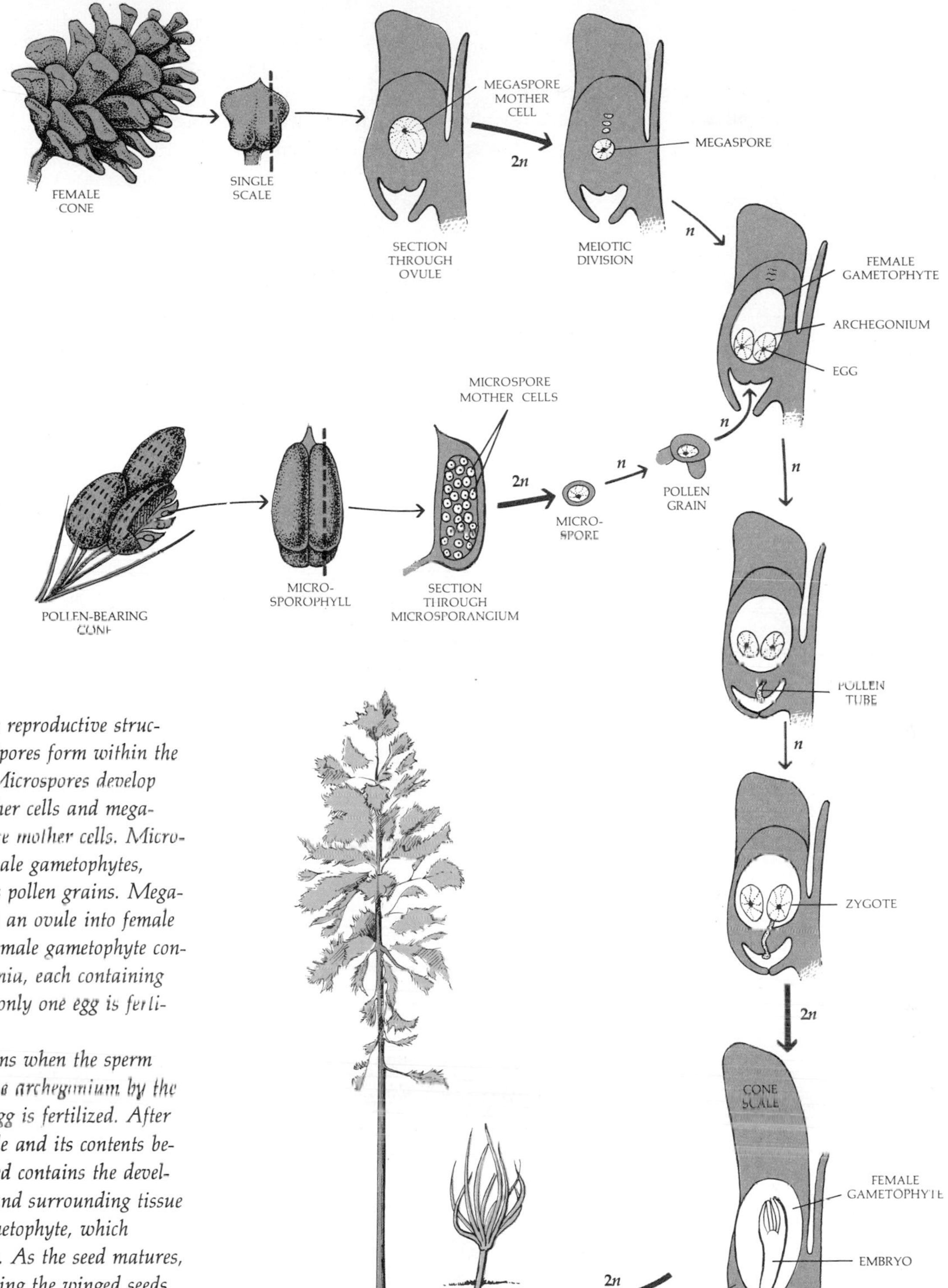

24–21

Pine life history. The reproductive structures are the cones. Spores form within the cones on the scales. Microspores develop from microspore mother cells and megaspores from megaspore mother cells. Microspores develop into male gametophytes, which are released as pollen grains. Megaspores develop within an ovule into female gametophytes; each female gametophyte contains several archegonia, each containing an egg cell. Usually only one egg is fertilized and develops.

The life cycle begins when the sperm cells are carried to the archegonium by the pollen tube and the egg is fertilized. After fertilization, the ovule and its contents become the seed; the seed contains the developing plant embryo and surrounding tissue derived from the gametophyte, which nourishes the embryo. As the seed matures, the cone opens, releasing the winged seeds, which germinate, producing the seedling (sporophyte). Both types of cones develop on the sporophyte.

24–22
Pine seed. The outer layers of the ovule have hardened into a seed coat, enclosing the female gametophyte and the embryo, which now consists of an embryonic root, the radicle, and a number of embryonic leaves, the cotyledons. When the seed germinates, the radicle will emerge from the seed coat and penetrate the soil. When the root takes in water, the tightly packed cotyledons will elongate and swell with the moisture, rising above ground on the lengthening stem and forcing off the seed coat.

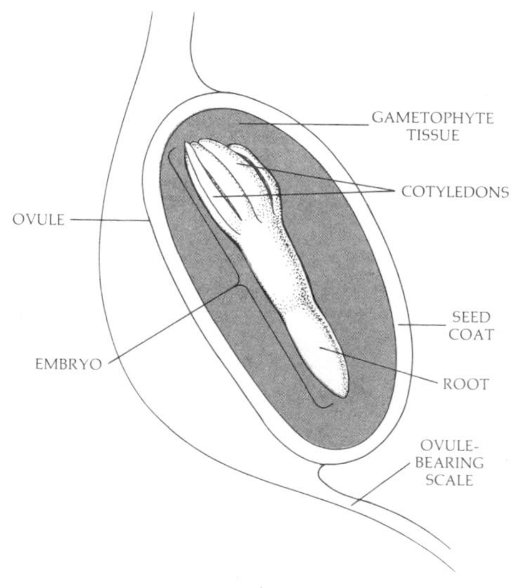

Figure 24–22 shows a cross section of a pine seed. The seed coat and the wing on which the seed is carried arise from the hardened outer layers of the ovule, derived from the mother sporophyte. The next layer represents the body of the gametophyte; swollen and packed with stored food reserves, it grows and displaces the original sporophyte tissue. The inner core is the embryo with its many *cotyledons,* the embryonic leaves, which will appear as the first leaves of the shoot of the new sporophyte when the seed germinates. The lower part of the embryo, the radicle, will form the root.

The seed was in existence by the close of the Carboniferous period, and according to the fossil record, some of the fernlike plants and even some of the club mosses had seedlike structures. But it was not until the close of the Permian period, at the end of the Paleozoic era (some 225 million years ago), when the land became colder and drier, that plants that had seeds gained a major evolutionary advantage and the seed plants became the dominant plants of the land.

The Conifer Leaf

Another feature commonly associated with conifers, although not with all gymnosperms, is their needlelike leaf. Figure 24–23 shows a cross section of a pine leaf, which may be 10 or more centimeters long but only 1 to 3 millimeters in diameter. In the center you can see veins, the vascular transport system, which carries water in one set of conducting cells (the tracheids) and sugars in another (the sieve cells). Outside the veins are parenchyma cells in which photosynthesis takes place. The ducts on the facing sides of the needles carry resin, a substance that is released if the plant is wounded and may serve to close the break. The outside layer of cells (epidermis) is porous but very hard.

The needlelike leaf, despite its slender form, is a megaphyll. It is well adapted to long periods of low humidity, such as in northern winters, and to moisture-losing, sandy soils. One or both of those are characteristic of many regions in which modern conifers are abundant. (In areas with more available water in the air or soil, they cannot usually compete with angiosperms.)

24–23
Cross section of a pine leaf stained to show the cell types. The hard outer covering and compact shape protect the leaf from water loss, an essential factor in the survival of these trees in areas in which there is little rainfall, or in which the water is locked in the ground as ice during many months of the year.

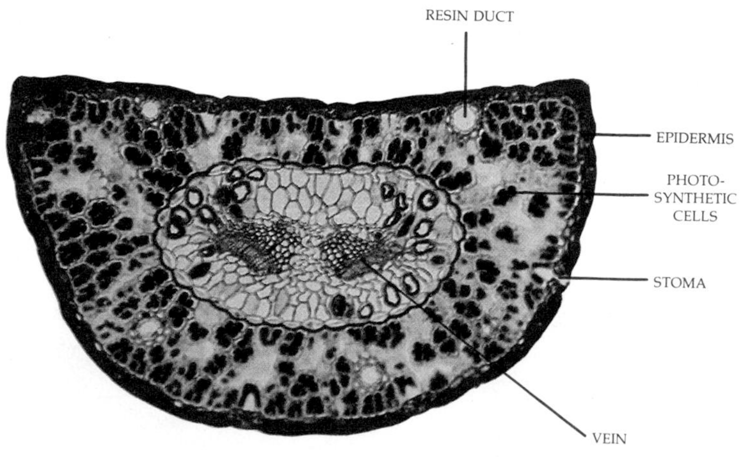

Table 24-2 *Vascular Plants: Phylum Tracheophyta*

		NO. OF SPECIES	LEAVES	GAMETOPHYTES	SPERM	SEED
	Subphylum Lycophytina (Club mosses)	1,000	Microphylls, in spirals	Independent	Biflagellate	Not present in modern forms
	Subphylum Sphenophytina (Horsetails)	12	Microphylls, scalelike, in whorls	Independent	Multiflagellate	Not present
	Subphylum Pterophytina	More than 260,000	Megaphylls	Small, mostly microscopic	Flagellated in some	Present in most forms
	Class Filicinae (Ferns)	11,000	Megaphylls	Independent in most	Flagellated	Not present in modern forms
Gymnosperms	Class Coniferinae	550	Megaphylls, often reduced to needles or scales	Dependent, reduced	Not flagellated, carried in pollen tubes	Present, naked seed
Gymnosperms	Class Cycadinae	100	Megaphylls, large	Dependent, reduced	Flagellated but carried in pollen tubes	Present, naked seed
Gymnosperms	Class Ginkgoidae	1	Megaphylls, fanlike	Dependent, reduced	Flagellated but carried in pollen tubes	Present, naked seed
Gymnosperms	Class Gnetinae	70	Megaphylls	Dependent, reduced	Not flagellated, carried in pollen tubes	Present, naked seed
	Class Angiospermae (Flowering plants)	About 250,000	Megaphylls	Dependent, greatly reduced	Not flagellated, carried in pollen tubes	Present, enclosed within ovary (fruit)

Class Angiospermae: Flowering Plants

It is believed that the angiosperms evolved from a now-extinct group of gymnosperms. They appear in the fossil record suddenly and in abundance during the Cretaceous period, about 100 million years ago, as the dinosaurs were vanishing. Numerous angiosperm genera appeared suddenly at that time, and many of these seem to have been very similar to our modern plants.

According to Daniel Axelrod, a paleobotanist at the University of California, angiosperms probably arose long before this time, during the Permian period. They are likely to have originated on the less fertile hills and uplands of tropical areas, the richer lowlands being crowded with ferns, gymnosperms, and lycopods.

24–24
Evolutionary relationships among some cells of the tracheophytes. Tracheids, which are water-conducting cells of the conifers, are believed to resemble the more primitive (that is, earlier evolved) cells. Tracheids are elongated cells with thin areas (pits) in their lateral walls through which water moves from one tracheid to another up the trunk from the roots. Tracheids also serve as mechanical supporting structures. From tracheids evolved wood fibers, specialized for support, and vessel elements, specialized for conducting water. In the most highly evolved vessels, the end walls of the individual cells (vessel elements) disintegrate during development and the elements are stacked on top of one another, leaving a continuous tube. Both vessels and wood fibers are presumed to have been derived from primitive tracheids, the water-conducting cells of primitive vascular plants.

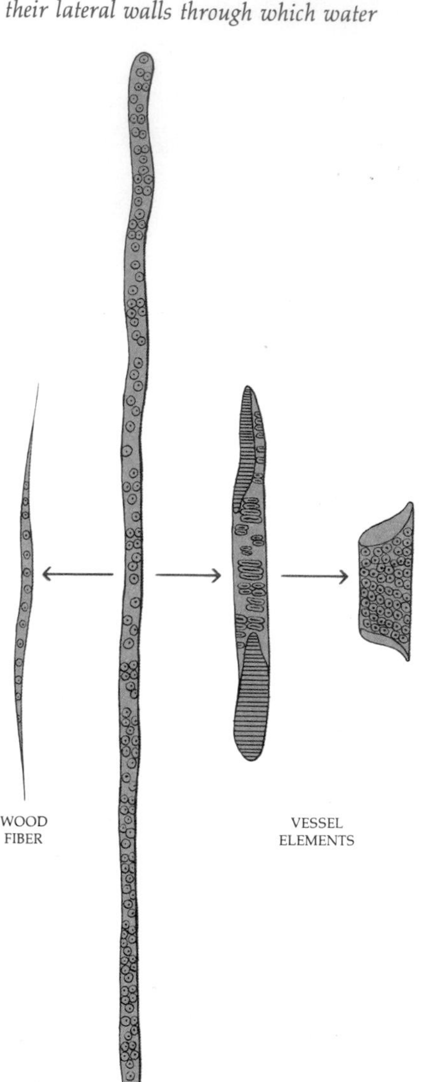

Once established, they spread into the lowlands, where they soon became the dominant plant forms and were deposited as fossils. During this mid-Cretaceous period, the climate of the earth was warmer and more uniform than it is at present, and by the end of the Cretaceous period, much of the land was covered with a rich forest of angiosperms, reaching almost as far north as the Arctic Circle.

About 250,000 different species of angiosperms are known. Just listing them, one after another, would require a book twice the length of this one. They dominate the tropical and temperate regions of the world. The angiosperms include not only the plants with conspicuous flowers but also most of the great trees, the oak, the willow, the elm, the maple, and the birch; all the fruits, vegetables, nuts, and herbs; the cactus and the coconut; and all the corn, wheat, rice, and other grains and grasses that are the staples of the human diet and the basis of agricultural economy all over the world.

Angiosperms, like other vascular plants, contain chlorophylls *a* and *b* and beta-carotene, and have megaphylls, stomata, and a cuticle impervious to water. The modern forms have a more highly evolved vascular system than is found in other groups (Figure 24–24). They also have two new, interrelated structures that distinguish them from all other plants: the flower and the fruit.

The Flower

Figure 24–25 is a diagram of a simple flower. The central structure is the *carpel,* the female reproductive structure.* The carpel is believed to be a modified leaf, which, in the course of evolution, enfolded in such a way that the ovule is attached to its inner surface (Figure 24–26a). A single carpel may contain one or more ovules, and a single flower one or more carpels. The carpels may be separate (blackberry) or fused (tomato and apple). The swollen base of the carpel or carpels is the *ovary.* Within the ovary is the ovule, or ovules, in which the female gametophyte develops. The tip of the carpel or carpels has become specialized as a *stigma,* a sticky surface to which pollen grains adhere. The stigma and ovary are connected by a slender column of tissue, the *style.*

The pollen grains (immature male gametophytes) develop in the *stamen,* which, like the carpel, is the evolutionary descendant of a sporophyll. It consists of the *anther,* which is a group of sporangia in which the male gametophytes develop, and a supporting *filament.*

* A single carpel or a group of fused carpels used to be known as the pistil because of their resemblance to an apothecary's pestle. This older term is now being abandoned, however.

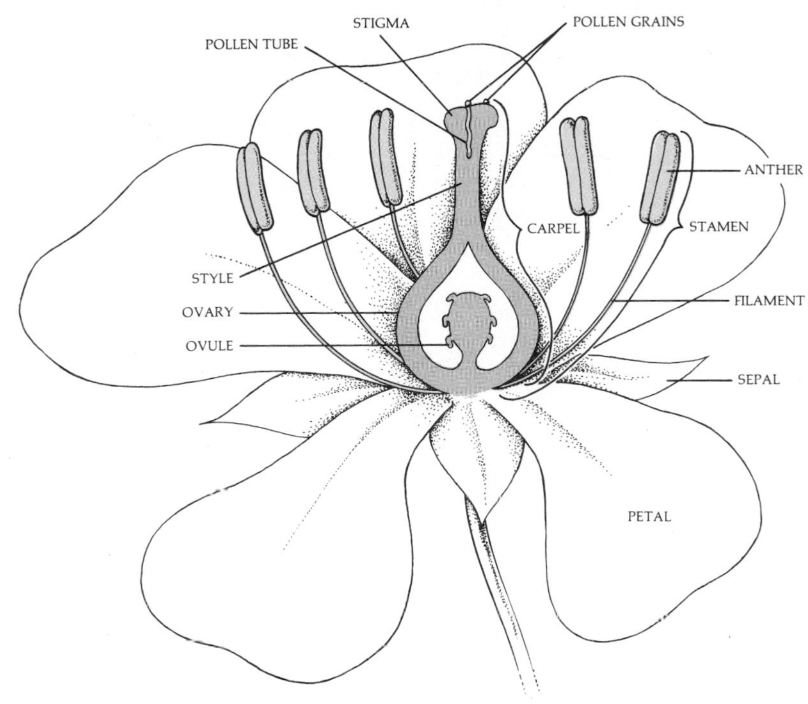

24–25
Structure of a flower. The sex organs are shown in color. Some flowers have only the male reproductive structures, some only the female. A flower that possesses both stamens and carpel, such as this flower, is known as a perfect flower.

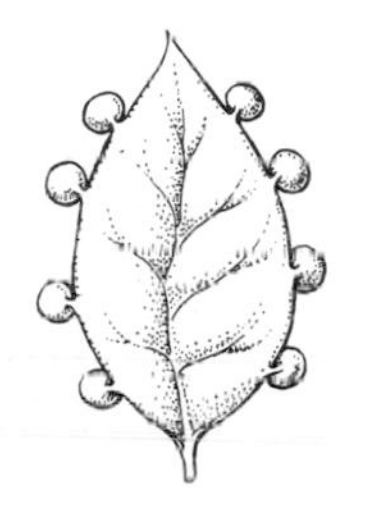

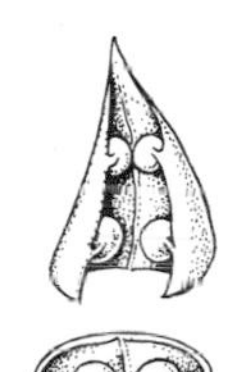

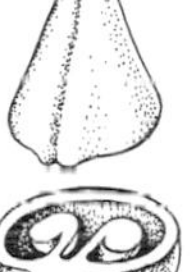

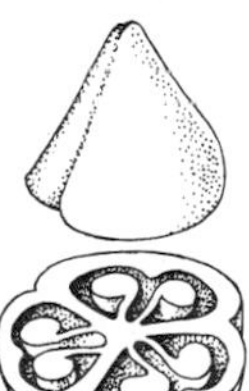
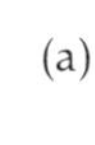

(a) (b)

24–26
(a) *Presumed evolutionary development of simple and compound ovaries. A leaf-shaped carpel with marginal ovules (upper left) folded in on itself, and the margins fused to form a simple ovary (upper right). Compound ovaries (lower) were formed by the fusing of separate infolded carpels.* (b) *Flower of a round-leaved hepatica* (Hepatica americana). *Notice the open, bowl shape and the numerous and separate floral parts. The insect is a pollen-eating beetle.*

The other parts of the flower are specialized to attract insects and other pollinators. Pollen grains produced in the anthers are carried to the stigma of (usually) another flower, where they germinate, developing pollen tubes that grow down through the style toward the ovule. The sperm cells are carried by the pollen tubes to the female gametophyte. This gametophyte typically consists of only eight cells, one of which is the egg cell. It is fertilized by a sperm cell, giving rise to an embryo sporophyte. (The extraordinary phenomenon of double fertilization is described in Section 5.)

Evolution of the Flower

The flower evolved as a device by which plants induce animals to transport their pollen, and hence sperm, to the egg cells. The most primitive flowers are believed to have resembled the modern hepatica, shown in Figure 24–26b, which has numerous floral parts, each clearly separate from the other. By comparing this type of flower with some of the more specialized ones, such as a composite (Figure 24–27) or an orchid (Figure 24–28), it is possible to see four main trends in flower evolution, as listed on the next page.

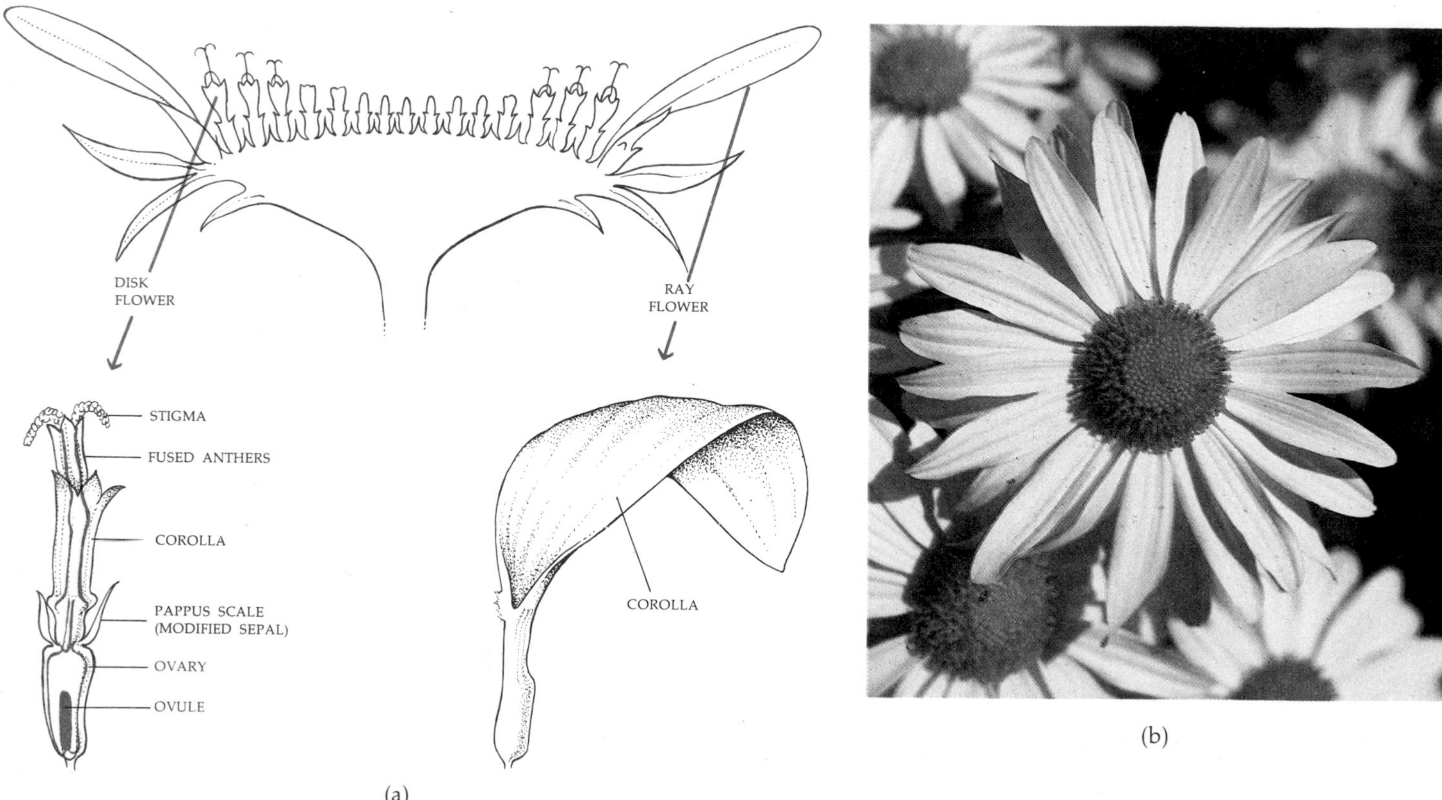

24–27
(a) *The organization of the head of a composite. The individual flowers are subordinated to the overall effect of the head, which acts as a large, single flower in attracting insects. The internal structure of the disk flower is indicated in color.* (b) *Ox-eye daisy* (Chrysanthemum leucanthemum)*, a representative composite. The ray flowers (with white petals) are often sterile.*

24–28

Orchidaceae, with about 20,000 species, is the largest family of flowering plants. (a) *The parts of an orchid flower. The lip is a modified petal that serves as a landing platform for insects.* (b) Epipactus gigantea, *stream orchids.*

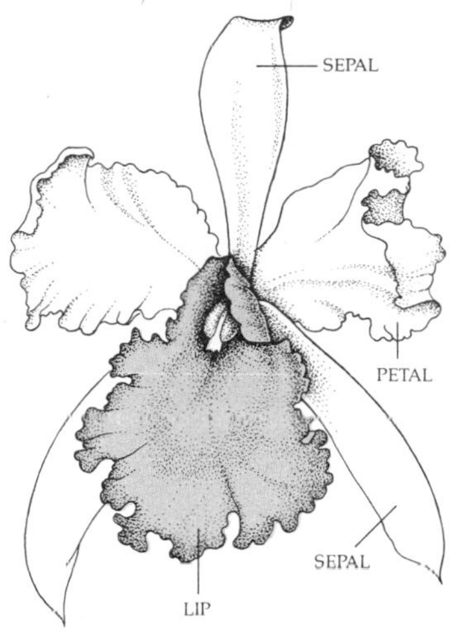

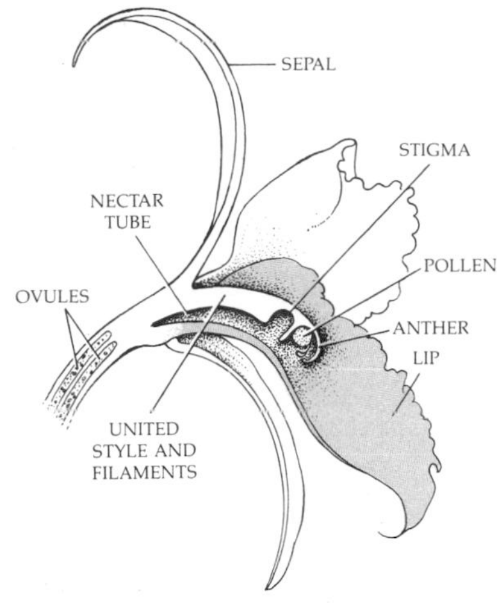

(a)

(b)

24–29

(a) *Superior and* (b) *inferior ovaries. In flowers with superior ovaries, the floral parts are attached below the ovaries. In flowers with inferior ovaries, the floral parts are attached above, and the ovaries are protected.*

(a)

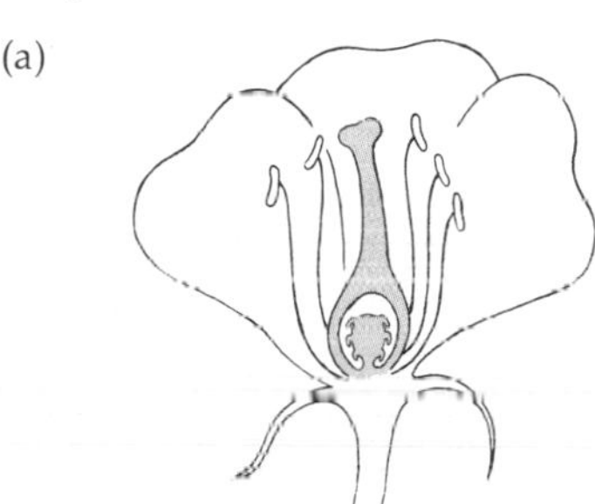

(b)

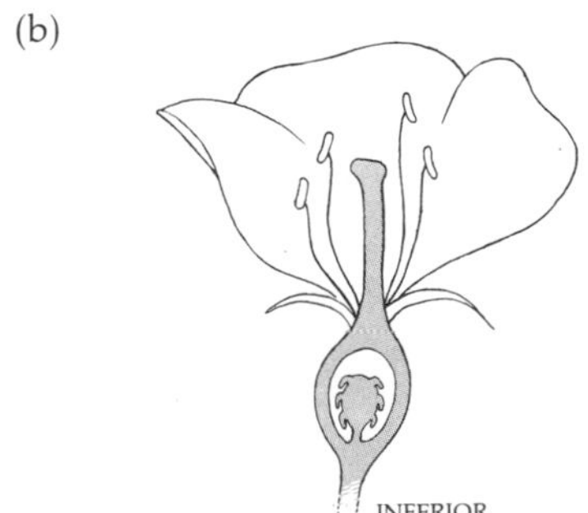

1. Reduction in number of floral parts. Most specialized flowers have few stamens and few carpels, and these are present in a definite number, depending on the species.
2. Fusion of floral parts. Carpels and petals, in particular, have become fused, sometimes elaborately so.
3. Elevation of free floral parts above the ovary. In the primitive flower, the floral parts arise at the base of the ovary (Figure 24–29a). Such ovaries are said to be superior. In the more advanced flowers, the free portions of the floral parts are above the ovary (Figure 24–29b); this is an important adaptation by which the ovules are protected from foraging insects. Such ovaries are said to be inferior.
4. Changes in symmetry. The radial symmetry of the primitive flower has given way, in more advanced flowers, to bilaterally symmetrical forms.

The Agents of Evolution

The early gymnosperms from which the angiosperms evolved were probably wind-pollinated, as are modern gymnosperms. And, as in the modern gymnosperms, the ovule probably exuded droplets of sticky sap in which pollen grains were caught and drawn to the female gametophytes. Insects, probably beetles, feeding on plants must have come across the protein-rich pollen grains and the sticky, sugary droplets. As they began to depend on these new-found food supplies, they inadvertently carried pollen from plant to plant.

Beetle pollination must have been more efficient than wind pollination for some plant species because, clearly, selection began to favor those plants that had insect pollinators. The more attractive the plants were to the beetles, the more frequently

24-30
Orchid pollen is bound together in a mass, called a pollinium, which contains a sticky extension that attaches the pollinium to the pollinating insect. In a bee-pollinated orchid, a bee with a pollinium on its back lands on the lip of the flower and crawls inside to gather the nectar. As it backs out, the flap separating the stigma from the pollen mass first scrapes the transported pollinium off its back onto the sticky surface of the stigma. Next, as it pushes the anther cap outward, the pollinium from the flower it is leaving becomes deposited on its back and is thus carried to another flower. Only a strong, sturdy insect can force its way into the flower to reach the nectar.

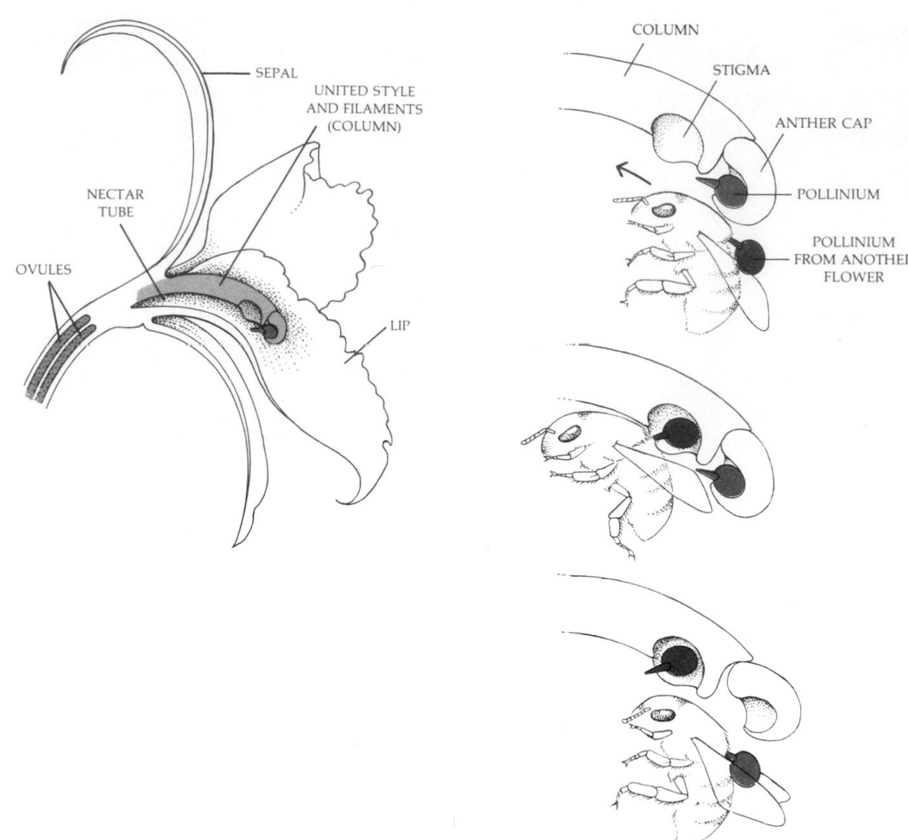

they would be visited and the more seeds they would produce. Any chance variations that made the visits more frequent or that made pollination more efficient thus offered immediate advantages; more seeds would be formed, and more offspring would survive. Nectaries (nectar-secreting organs) evolved, which lured the pollinators. Plants developed white or brightly colored flowers that called attention to the nectar and other food supplies. The carpel, originally a leaf-shaped structure, became folded on itself, enclosing and protecting the ovule from hungry pollinators. By the beginning of the Cenozoic era, some 65 million years ago, the first bees, wasps, butterflies, and moths had appeared. These are long-tongued insects for which flowers are often the only source of nutrition for the adult forms. From this time on, flowers and certain insect groups had a profound influence on one another's history, each shaping the other as they evolved together.

A flower that attracts only a few kinds of animal visitors and attracts them regularly has an advantage over flowers visited by more promiscuous pollinators: Its pollen is less likely to be wasted on a plant of another species. In turn, it is an advantage for the insect to have a "private" food supply that is relatively inaccessible to competing insects. Most of the distinctive features of modern flowers are special adaptations that encourage regular visits (constancy) by particular pollinators. The varied colors and odors allow sensory recognition by pollinators. The diverse shapes such as deep nectaries and complex landing platforms that are found, for example, in orchids, snapdragons, and irises, represent ways of excluding indiscriminate pollinators.

(a)

(b)

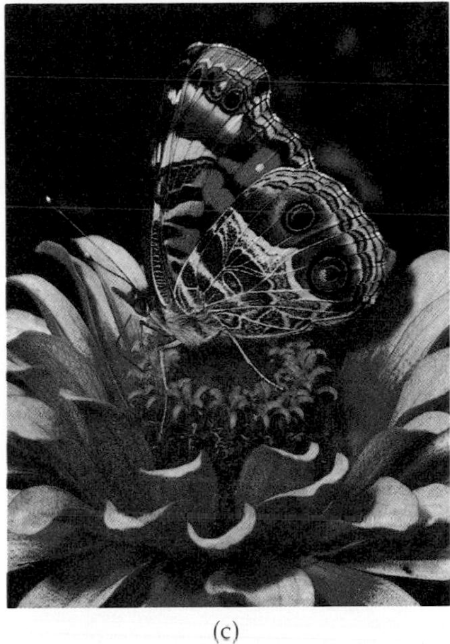

(c)

(d)

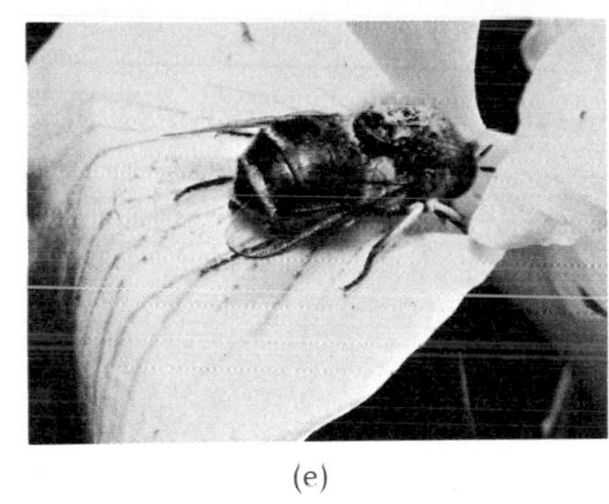

(e)

(f)

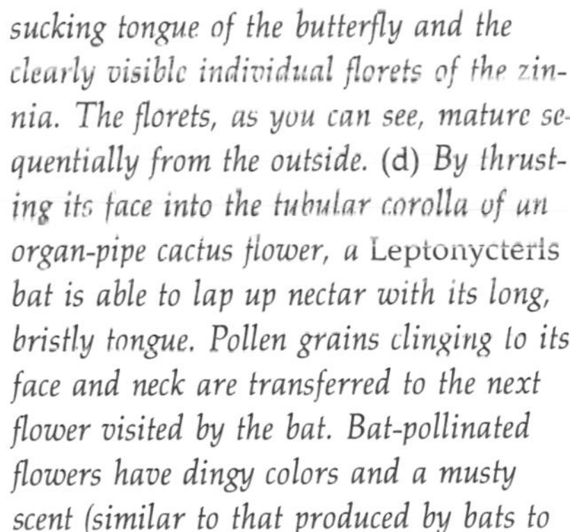

24-31
Pollinators. (a) *A honey bee foraging in a flower of* Salvia. *Notice that the anthers are depositing pollen grains on the top of the bee's thorax.* (b) *Female rufous hummingbird probing for nectar in a columbine. Her head is collecting pollen, which she will carry to another flower. Flowers pollinated by birds are scentless, bright red or orange, and have copious nectar to make the visit worthwhile.* (c) *An American painted lady butterfly sipping nectar in a zinnia, a composite. Notice the long, sucking tongue of the butterfly and the clearly visible individual florets of the zinnia. The florets, as you can see, mature sequentially from the outside.* (d) *By thrusting its face into the tubular corolla of an organ-pipe cactus flower, a* Leptonycteris *bat is able to lap up nectar with its long, bristly tongue. Pollen grains clinging to its face and neck are transferred to the next flower visited by the bat. Bat-pollinated flowers have dingy colors and a musty scent (similar to that produced by bats to attract one another), and they open at night.* (e) Acrocerid *("small headed") fly on the lip of an iris flower. The fly has pollen on its back.* (f) *Unlike most angiosperms, grasses and most common species of trees are wind-pollinated. The staminate flowers of the turkey oak* (Quercus laevis) *hang in catkins, which are flexible, thin tassels several centimeters long. These catkins are whipped by passing breezes, and the pollen, when ripe, is thrown out and caught by the wind.*

Dispersal of Fruits and Seeds

Following fertilization in the angiosperm, the ovary develops into the fruit. Fruits are adaptations for seed dispersal. Many edible fleshy fruits become brightly colored as they ripen, attracting the attention of birds and mammals. The seeds within the fruits pass through the digestive tract hours later and are often deposited some distance away. In some species the seed coat's exposure to digestive juices or bile is a prerequisite for germination. The seeds themselves may be toxic, as in apples, discouraging animals from grinding up and digesting the seeds.

In some dry fruits, the fruit bursts open, shooting out the seeds, as in impatiens or witchhazel. Some fruits, such as the samara of maples, carry wings; in others, the seeds themselves are winged or tufted, as in the milkweed. Some species of geranium send forth their seeds by a sort of slingshot. Burrs are dry fruits that adhere to fur, feathers, or clothing and so are carried by involuntary porters to far-off fields and meadows. In the tumbleweed, the whole plant is blown across open country, with fruits opening and scattering the seeds as they go. The many extraordinary means of seed dispersal are another major reason for the evolutionary success of angiosperms.

Angiosperms differ from the other vascular plants in their more efficient conducting systems, their flowers, which increase the efficiency of fertilization, and their fruits, which increase the efficiency of distribution. Angiosperms represent the most successful of all plants in terms of numbers of individuals, numbers of species, and their effects on the existence of other organisms. The anatomy and physiology of this dominant group of plants will be explored further in Section 5.

24-32

Development and structure of the pear. (a) *Flower of the pear.* (b) *Older flower, after petals have fallen.* (c) *Longitudinal section and* (d) *cross section of the mature fruit. The core of the pear is the ripened ovary wall. The fleshy, edible part develops from the floral tube.*

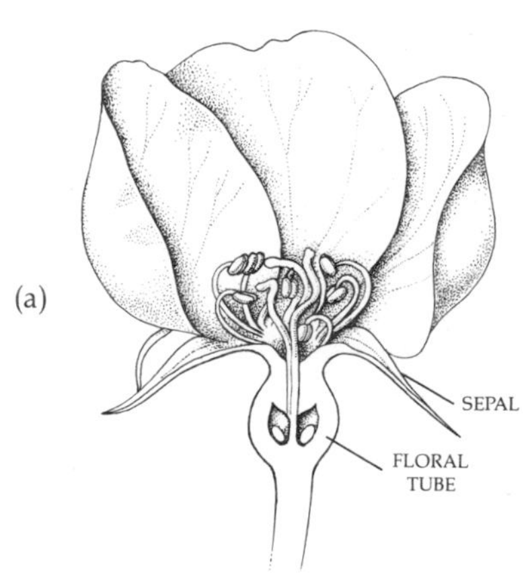

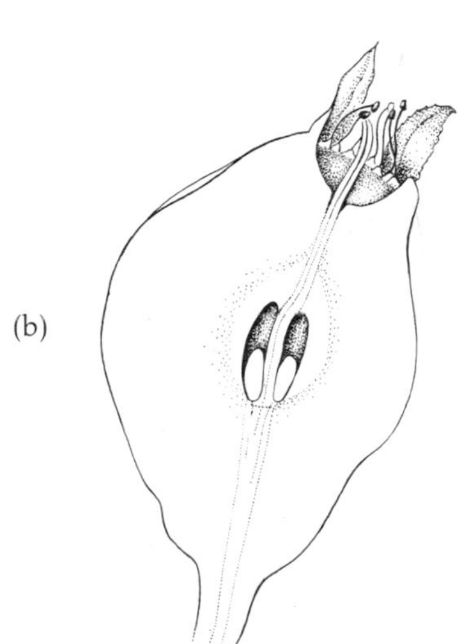

SUMMARY

Plants are multicellular photosynthetic organisms adapted for life on land. The plant kingdom comprises the Bryophyta (mosses, liverworts, and hornworts) and the Tracheophyta (vascular plants).

On the basis of similarities between the groups, the Chlorophyta, the green algae, are believed to be the ancestors of plants. These similarities include the fact that both groups have chlorophylls *a* and *b* and beta-carotene as their photosynthetic pigments. Both store food reserves in the form of starch, and the cells of both have cellulose cell walls.

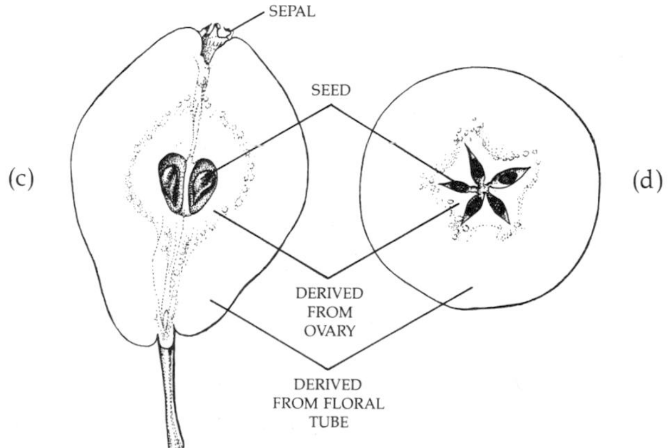

(a) (b) (c) (d)

24–33
Angiosperms are characterized by fruits. A fruit is a mature ovary, including the seed or seeds and, often, accessory parts of the flower. The fruit aids in dispersal of the seed. Some fruits are borne on the wind, some are carried from one place to another by animals, some float on water, and some are even forcibly ejected by the parent plant. (a) *In milkweed, the fruit bursts open when it is ripe, releasing seeds with tufts of silky hair that aid in their dispersal.* (b) *The goldfinch loosens the thistle's seeds. Some it eats, and others are carried off by the wind.* (c) *The tough seeds of these blackberries will pass unharmed through the digestive tract of the dormouse and germinate when they are eliminated.* (d) *A coconut, one of the largest of all fruits, is adapted for sea-going voyages and carries a food and water supply that lasts up to two years.*

Alternation of generations is found among the multicellular green algae and the plants. The term designates a reproductive cycle in which a haploid phase alternates with a diploid phase. In one generation, the plant body, known as the gametophyte, is haploid *(n)*. It produces haploid gametes, the sperm and egg cells. The gametes fuse to form the zygote, which develops into a diploid ($2n$) sporophyte. The sporophyte produces spores by meiotic division. A spore is a single cell that, unlike a gamete, can develop into an adult organism asexually, that is, without combining with another cell. In alternation of generations, the spore, which is haploid, produces the haploid gametophyte.

The mosses and other modern bryophytes are small plants found usually in moist locations. Most lack specialized vascular tissues and true leaves, although the plant body is differentiated into photosynthetic, food-storing, and anchoring tissues, as in higher plants. Unlike the algae, they have multicellular reproductive structures in which the gametes develop. The female gamete (egg) develops in the archegonium; the male gametes (sperm) develop in the antheridium. The sperm, which are flagellated, swim to the archegonium and fertilize the egg cell. The resultant zygote develops into an embryo within the archegonium. In the bryophytes—and in no other members of the plant kingdom—the gametophyte *(n)* is dominant and the sporophyte ($2n$) is smaller, attached, and often nutritionally dependent.

The important subdivisions of the true vascular plants, the tracheophytes, include the ferns, the gymnosperms, and the angiosperms, or flowering plants. Members of all these groups have megaphylls (leaves with complex venation), stomata, cuticles, and specialized vascular tissue.

Ferns are characterized by large, often finely dissected leaves called fronds. The sporophyte is the dominant generation, but in most ferns the gametophytes are independent. The sperm are flagellated, and free water is needed for fertilization. Sporangia are typically formed on the underside of leaves (sporophylls) of the sporophyte.

The conifers are the largest group of gymnosperms—naked-seed plants—so called in contrast to the angiosperms in which the seed is enclosed in an ovary, which develops to form a fruit. Among conifers, the male gametophyte is formed on modified leaves, the sporophylls, which form a cone. The gametophytes are released in the form of windblown pollen. The female gametophyte develops on scales of a separate, larger cone within an ovule composed of the tissue of the parent sporophyte. Within the female gametophyte, archegonia form. The male gametophyte germinates and produces a pollen tube through which male reproductive cells (the sperm) enter the archegonium and fertilize the egg cell. The seed consists of the ovule and its contents, including the young embryo and the gametophyte, which serves as nutritive tissue. The seed, which is shed from the female cone, can remain dormant for long periods of time and hence is adapted to withstanding cold and drought.

The angiosperms, of which there are about 250,000 species, are the dominant modern plants. Angiosperms are characterized by the flower and the fruit. The reproductive structures of the flower are the stamens, composed of filament and anther, and the carpels, composed of ovary, style, and stigma. The stamens and the carpels are highly specialized sporophylls. The ovule or ovules are enclosed in the ovary, the base of the carpel or group of fused carpels. Pollen grains, the immature male gametophytes, are formed in the anthers and germinate on the sticky surface of the stigma. The sperm cells are carried through a pollen tube, which grows down the style into the ovary and enters an ovule. As the embryo begins to develop, the ovary enlarges and the other parts of the flower, such as petals and anthers, often fall away. The ovary becomes the fruit, enclosing the mature ovules, the seeds. Flowers attract pollinators, and fruits aid in dispersal of the seeds.

The various shapes and colors of flowers evolved under selection pressures for more efficient pollinating mechanisms. Major trends in flower evolution include reduction and fusion of floral parts, a change in the position of the ovary relative to the other flower parts to a more protected (inferior) position, and a shift from radial to bilateral symmetry.

QUESTIONS

1. Distinguish among the following: moss/club moss; spore/sporangia/sporophyte/sporophyll; egg/sperm; gametophyte/antheridium/archegonium; microphyll/megaphyll; conifer/gymnosperm; carpel/sepal/petal.

2. Where did the liverworts get their name? (The answer is not in the text, but a little research should tell you.) What does this name tell you about the human perspective of the rest of the biosphere?

3. Describe the sporophylls of a club moss, a horsetail, a fern, and an angiosperm.

4. Sketch a flower and label the structures.

5. Sketch a pine seed, label it, and indicate the origin and ploidy of its components.

6. Bryophytes, among the plants, and amphibia, among the animals, often live in habitats intermediate between fresh water and dry land, rather than between the salt water and land. Propose a physiological argument (referring back to Chapter 6) to explain why invasions of the land were more likely by organisms previously adapted for life in fresh water.

7. Tall trees today are not appreciably taller than plants of the Devonian forests. What factors select for tallness in trees? What factors, by contrast, select against tallness? What kind of evolutionary innovation might be required to alter the optimal balance among these various factors?

8. In many areas on earth, large gymnosperms are either dominant or manage to coexist with angiosperms. List some advantages for a big plant's being a gymnosperm.

9. Plants are characterized by the presence of lignin, which stiffens and supports the plant body. What is the primary source of support for the multicellular seaweeds?

10. Compare and contrast the structure of a "primitive" flower, such as an hepatica, with that of a "specialized" one, such as a sunflower or an iris.

11. The dark spots on the foxglove flowers at the left are called honey guides. What do you suppose their function is?

CHAPTER 25

The Animal Kingdom I: Introducing the Invertebrates

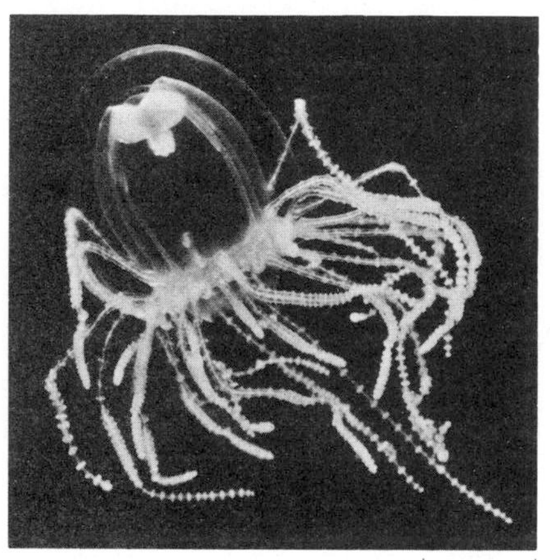

25–1
Animals are characterized by their mobility, usually a result of the contraction of assemblies of protein fibers within specialized (muscle) cells. Gonionemus murbachii, *a jellyfish, is shown here swimming actively, its bell contracted by muscles around its margin.*

Animals are many-celled heterotrophs. They depend directly or indirectly for their nourishment on photosynthetic autotrophs—algae or plants. Most digest their food in an internal cavity, and most store food as glycogen or fat. Their cells do not have walls. Most move by means of contractile cells (muscle cells) containing characteristic proteins. Reproduction is usually sexual. As adults, most are fixed in size and shape, in contrast to plants, in which growth often continues for the lifetime of the organism. The higher animals—the arthropods and the vertebrates—are the most complex of all organisms, with many kinds of specialized tissues, including elaborate sensory and neuromotor mechanisms not found in any of the other kingdoms.

For most of us, animal means mammal, and mammals are, in fact, the chief focus of attention in Section 6. However, the mammals, or even the vertebrates as a whole, represent only a small fraction of the animal kingdom. More than 90 percent of the different species of animals are invertebrates—that is, animals without backbones—and most of these are insects. Indeed, the enormous variety displayed by the invertebrates is partly why they are so endlessly fascinating to study. They are, in addition, of great ecological importance; the insects, for example, have long challenged human dominance of the earth.

Finally, and perhaps most important, the invertebrates, confronted with the same biological problems that we face, demonstrate a spectrum of ingenious solutions. In this way, they illuminate the essential nature of these problems and so help us to understand and evaluate the solutions arrived at by mammals.

THE SOURCES OF DIVERSITY

In Chapter 21, we commented upon the tremendous versatility of the prokaryotes, as exemplified by the wide range of environments they inhabit and the many ways in which they satisfy their energy requirements. Among the invertebrates, we see, on a slightly different scale, this same pattern of adaptation to many different ways of life. Thus, for instance, on a single coral head, only a meter or two in diameter, one finds a dazzling array of different forms—sponges, jellyfish, starfish, sea urchins, anemones, and the coral animals themselves. Similarly, to take terrestrial

25–2
Tidal pool on the Maine coast. Three different species of invertebrates are visible. Can you identify them?

examples, a single spadeful of soil turns up earthworms, pillbugs, spiders, nematodes, and various other tiny animals; and the branch of a single tree may harbor a dozen different kinds of insects. Given the relentless force of natural selection, why do not the larger ones crowd out the smaller? Why are not the "lower" animals replaced by "higher" animals with superior strength or intelligence? Darwin, again, offered the answer: The different organisms, he noted, "occupy different positions in the economy of nature." Each has been shaped by the long process of evolution to occupy a different niche in the environment. Natural selection has worked not to make one "superior" to the other but to continuously adapt the different forms to different ways of life. This process of adaptation, of course, continues. Every species, including our own, is a traveler through time, caught for only an instant in the present.

25–3
Animals are heterotrophs. These gooseneck barnacles, which are small crustaceans, sweep food particles into their mouths with their six pairs of hairy appendages that protrude from the shell.

THE SOURCES OF CONTINUITY

Through the patterns of diversity, there is a strong theme of continuity. One reason is simply that of "descent." We are all related; not only are we made of the same atoms and molecules and even macromolecules, but from *E. coli* to elephant, we even share many of the same enzymes. Although the evolutionary relationships between us and the invertebrates are obscure, we can read into them traces of our own biological beginnings. In the twitch of a tiny segment of an earthworm's artery, we sense the echo of our own heartbeat.

Second, all organisms face the same set of problems. These problems can best be defined by recalling that an organism is a cell or group of cells. A primary need is to supply the cell or cells with, first of all, a source of energy. Also, most cells—on this planet, at least—require water, oxygen, a source of nitrogen, fixed carbon (in the case of heterotrophs), and a few ions. Another requirement is to eliminate wastes, including excess carbon dioxide, nitrogenous wastes from the breakdown of amino acids, and, in some cases, excess water. Cells that live individual or colonial lives in a watery environment can solve these problems in relatively simple ways, but as organisms get larger, thicker, and more complex, the problem of servicing each individual cell becomes correspondingly more complicated.

25-4
Animals have a variety of sensory equipment. Even this small and relatively simple organism, a planarian, has two light-sensitive eyespots (ocelli) and a variety of chemoreceptors in the head region.

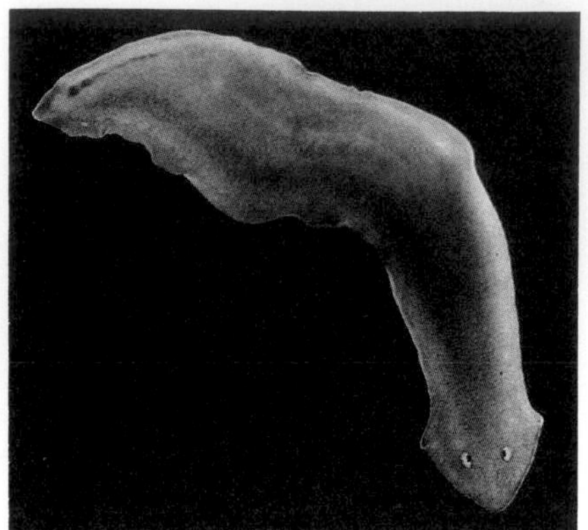

0.5 mm

Another set of problems that a multicellular organism must solve in order to exist arises from the fact that it is more than just a group of cells. It is, in fact, a complex society of cells, in which the needs of each individual cell are subordinated to the needs of the society. In a population of *Paramecium,* the organisms have common requirements, but each is in competition with the others. In a society of even a few thousand cells—a small crustacean, for instance—the individual cells are dependent on the existence of the group and are organized in a system of mutual cooperation. The second group of problems faced by organisms, therefore, relates to the organization or integration of activities. Hormones are one of the chief means of integration in both plants and animals. In the animals, another, more rapid integrating mechanism has evolved: the nervous system, by which the organism keeps in touch with its environment and coordinates its own activities.

A QUESTION OF SIZE

At this point, one might well ask: Considering the problems faced by larger organisms, why did larger animals evolve? What selective advantages do the multicellular, more complex animals have as compared with the smaller ones? Some answers to these questions are obvious and simple. Larger animals are, in general, more likely to eat than be eaten. Larger organisms, especially those that live underwater or on land, are generally able to travel faster and farther than small ones, and this is an advantage. A small ciliate, for instance, might starve only a few centimeters from a food supply. On the other hand, its requirements are very modest.

Perhaps even more important, however, than mobility and edibility is what the French physiologist Claude Bernard called the *milieu interieur,* the internal environment of the animal, as distinct from the external environment that surrounds it. A single-celled organism is as cold or as hot and as wet or as dry as its surroundings, whereas a larger animal is more independent and, to some extent, controls the environment in which its cellular society lives. Control of the internal environment is more readily achieved by the many-celled animal because of the simple surface-to-volume geometry we noted in Chapter 5. Exchanges between a cell and its surroundings take place across a cell's available surface area. This is a principal reason why a cell, which depends for its existence on the exchange of substances with its environment, cannot be very large. On the other hand, since it may be advantageous to conserve certain substances, such as water and heat, an organism may be better off, within limits of weight and mobility, if its relative surface area is reduced. One-celled animals can live successfully only in water or as parasites in the bodies of other organisms, which amounts to the same thing. Many-celled animals can live not only in water but also on land, in the sky, and even, as we are now beginning to discover, in outer space—which is a logical extension of an old evolutionary trend.

In this chapter, we shall discuss the so-called lower invertebrates and some that must clearly be considered higher, such as the clever and highly emotional octopus. The chapter that follows deals with the insects and other arthropods and, briefly, with the vertebrates.

There are almost 30 phyla of invertebrates. Of these, we are going to discuss relatively few, concentrating on the largest phyla and those of particular biological interest. A complete listing can be found in Appendix C.

25-5
A breadcrumb sponge showing some of the many openings (oscula) through which water leaves the animal. The oscula usually protrude above the rest of the animal. This arrangement allows the natural water flow in the habitat to draw water through the sponge.

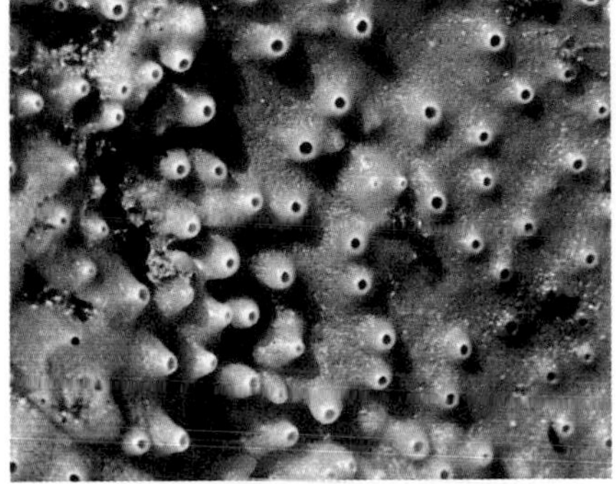

PHYLUM PORIFERA: SPONGES

Sponges seem to have had a different origin from other members of the animal kingdom and to have traveled a solitary evolutionary route. For this reason, they are often classified in a subkingdom of their own, the Parazoa ("alongside of animals"). In fact, until the eighteenth century, the sponges were classified as plant-animals ("zoophytes") since they are all sessile (attached to a substrate) during their adult life. Sponges are found on ocean floors throughout the world. Most live along the coasts in shallow water, but some, such as the fragile glass sponges, are found at great depths, where the water is almost motionless. A few types are found in fresh water.

A sponge is essentially a water-filtering system, made up of one or more chambers through which water is driven by the action of numerous flagellated cells. Sponges are made up of a relatively few cell types, the most characteristic of which are the choanocytes, or collar cells, the flagellated cells that line the interior cavity of the sponge (Figure 25-6). Similar cells, the choanoflagellates, are found among the ciliated protozoans, and it is possible that the sponges arose from such organisms. All of a sponge's digestive processes are carried out intracellularly; hence, even a giant sponge—and some stand taller than a person—can eat nothing larger than microscopic food particles.

25-6
The body of a simple sponge is dotted with tiny pores, from which the phylum derives its name (Porifera, or "pore bearer"). Water containing food particles is drawn into the internal cavity of the sponge through these pores and is exhaled out the osculum. The water is moved by the beating of the flagella protruding from the collars of choanocytes and by the sucking effect of flow of the local currents across the osculum. Each collar is made up of about 20 filaments, each of which is retractile. Each choanocyte has a single flagellum, the lashing of which directs a current of water through the filaments. Minute particles are filtered out and cling to one or more filaments and are then drawn into the cell and digested. The digested food is then shared by diffusion with other sponge cells. A sponge 10 centimeters high filters more than 20 liters of water a day.

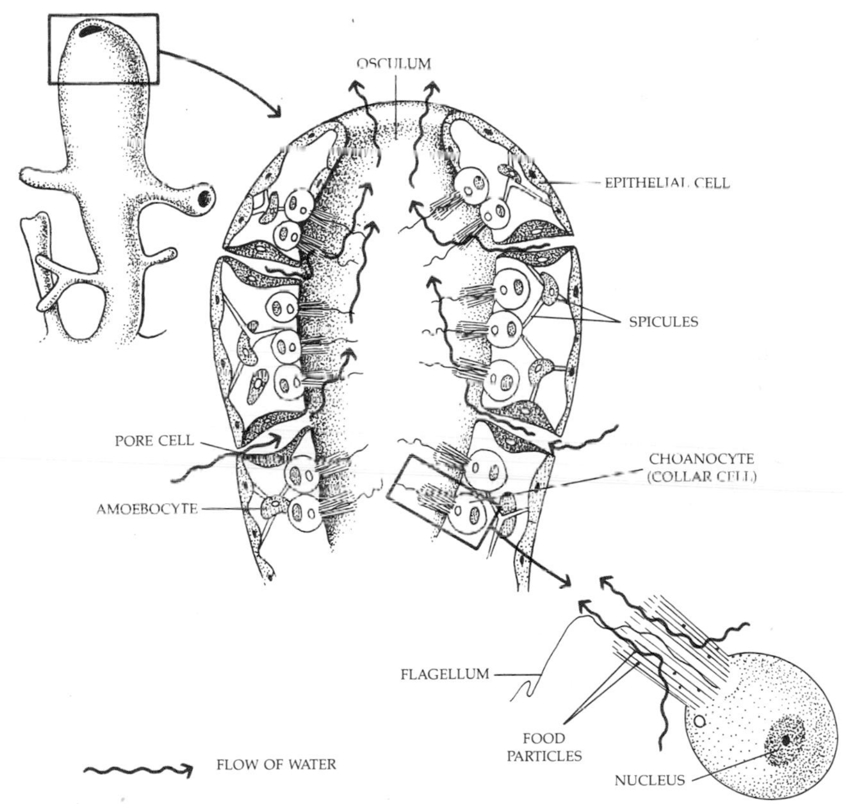

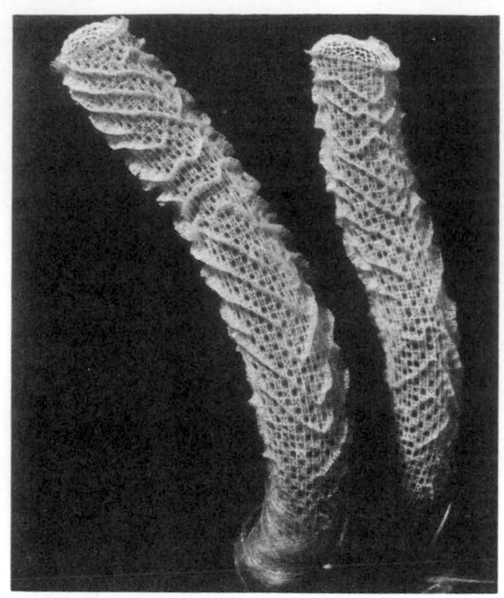

25-7
The skeleton of Euplectella speciosissima, *a species of glass sponge picturesquely known as Venus's flower basket. These fragile sponges, with their delicate silica-containing skeletons, are usually found at great depths. According to the fossil record, glass sponges were present in the Ordovician period, which began 500 million years ago.*

25-8
A cluster of sponges.

The outer surface of the sponge is covered with epithelial cells. Among these epithelial cells are cells that contract in response to touch or to irritating chemicals, and in so doing, close up pores and channels. Each cell acts as an individual, however; there is little coordination among them. Between the epithelial cells and the choanocytes is a middle, jellylike layer, and in this layer are amoebalike cells, amoebocytes, which carry out various functions. Some amoebocytes carry food particles from the choanocytes to the epithelial cells. Amoebocytes also secrete skeletal materials.

Sponges are grouped into three classes, according to their skeletal structure. In some species, the skeleton consists of individual spicules of calcium carbonate. Some, the glass sponges, have spicules of silica fused in a continuous and often very beautiful structure. The third and largest group has unfused silica spicules, or a tough, fibrous keratinlike protein called spongin, or a combination of the two. The skeleton of sponges serves only for protection, stiffening, and support, but not for locomotion, since the adult forms are sessile.

The sponge shown in Figure 25-6 is a small and simple one. In larger sponges, the body plan, although it is essentially the same, looks far more complex. These sponges have greatly increased feeding and filtering surfaces, owing to their highly folded body walls. We have already encountered this evolutionary strategem for increasing biological work surfaces at the cellular level—as in the inner membrane of the mitochondrion—and we shall be encountering it again as we examine the structure of gills and lungs.

Sponges are somewhere between a colony of cells and a true multicellular organism. The cells are not organized into tissues or organs; each leads an independent existence. Yet there is a form of recognition among the cells that holds them together and organizes them. If the sponge *Microciona prolifera* is squeezed through a fine sieve or a piece of cheesecloth, the body of the sponge is separated into individual cells and small clumps of cells. Within an hour, the isolated sponge cells begin to reaggregate, and as these aggregations get larger, canals, flagellated chambers, and other characteristics of the body organization of the sponge begin to appear. This phenomenon has been used as a model for the analysis of cell adhesion, recognition, and differentiation, all of which are basic biological features of development in higher organisms.

Most kinds of sponges are hermaphroditic; that is, they have male and female reproductive organs in the same individual. Gametes appear to arise from an enlarged amoebocyte, but there are reports that choanocytes can also form gametes. A sperm enters another sponge in a current of water. It is captured by a choanocyte and transferred to an amoebocyte, which then transfers it to a ripe egg (a method of fertilization unique to the sponges). The fertilized egg develops into a ciliated, free-swimming larva. After a short life among the plankton, the larva settles and becomes sessile.

Sponges also reproduce asexually, either by fragments that break off from the parent animal, or by gemmules, aggregations of amoebalike cells within a hard, protective outer layer. Production of such resistant forms is found, in general, only among freshwater organisms. In the ocean, conditions are relatively unchanging, but the freshwater environment is much harsher. Invertebrates that live in fresh water are more likely to have protected embryonic forms than even closely related marine species.

25–9
In organisms with radial symmetry, any plane through the animal that passes through the central axis divides the body into halves that are mirror images of one another.

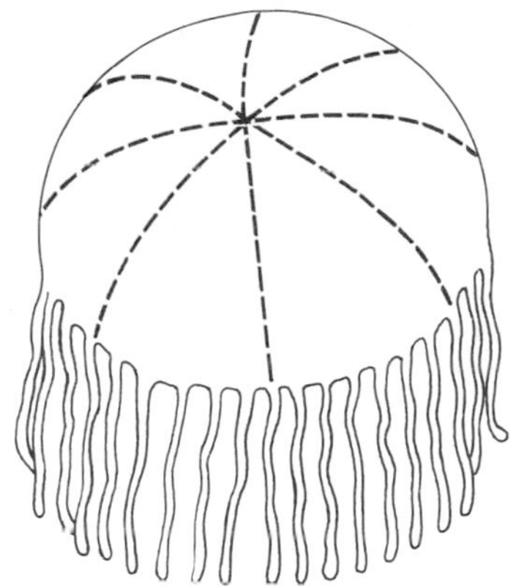

PHYLUM COELENTERATA: POLYPS AND JELLYFISH

The coelenterates are a large and often strikingly beautiful group of aquatic organisms. Their adult form is generally radially symmetrical; that is, their body parts are arranged around a central axis, like spokes around a hub (Figure 25–9). As you can see in Figure 25–10, the basic body plan is a simple one: The animal is essentially a hollow container, which may be either vase-shaped, the *polyp*, or bowl-shaped, the *medusa*. The polyp is usually sessile; the medusa, motile. Both consist of two layers of tissue: *ectoderm* and *endoderm* (from the Greek *ektos*, "outer," and *endon*, "inner," plus *derma*, "skin"). Between the two layers is a gelatinous filling, the *mesoglea* ("middle jelly"), which is made of a collagenlike material. In the polyp form, the mesoglea is sometimes very thin, but in the medusa, it often accounts for the major portion of the body substance.

Most coelenterates go through both a polyp and a medusa stage in their life cycles. In such species, polyps reproduce asexually and medusas sexually. This sort of life cycle, in which the sexually reproductive form is distinctly different from the asexual form, superficially resembles alternation of generations in plants. There is, however, no alternation between haploid and diploid forms as there is in plants; the only haploid forms are the gametes.

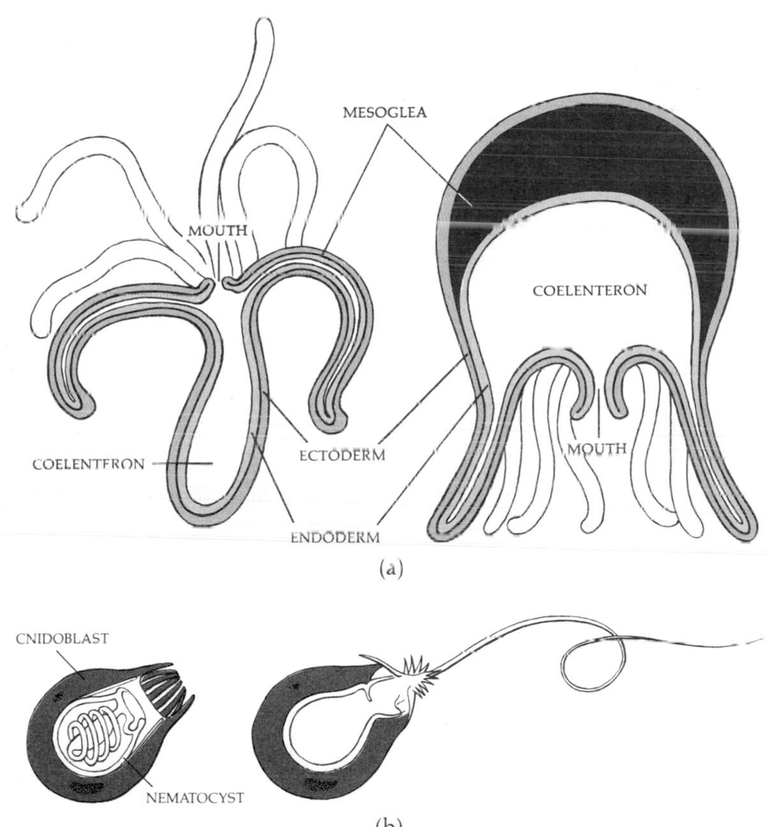

25–10
(a) *Among coelenterates, there are two basic body plans: the vase-shaped polyp (left) and the bowl-shaped medusa (right). The coelenteron, characteristic of the phylum, is a digestive cavity with a single opening. The coelenterate body has two tissue layers, ectoderm and endoderm, with gelatinous mesoglea between them.* (b) *Cnidoblasts, specialized cells located in the tentacles and body wall, are a distinguishing feature of coelenterates. The interior of the cnidoblast is filled by a nematocyst, which consists of a capsule containing a coiled tube, as shown on the left. A trigger on the cnidoblast, responding to chemical or mechanical stimuli, causes the tube to shoot out, as shown on the right. The capsule is forced open and the tube turns inside out, exploding to the outside. The cnidoblast cannot be "reloaded"; it is absorbed and a new cell grows to take its place.*

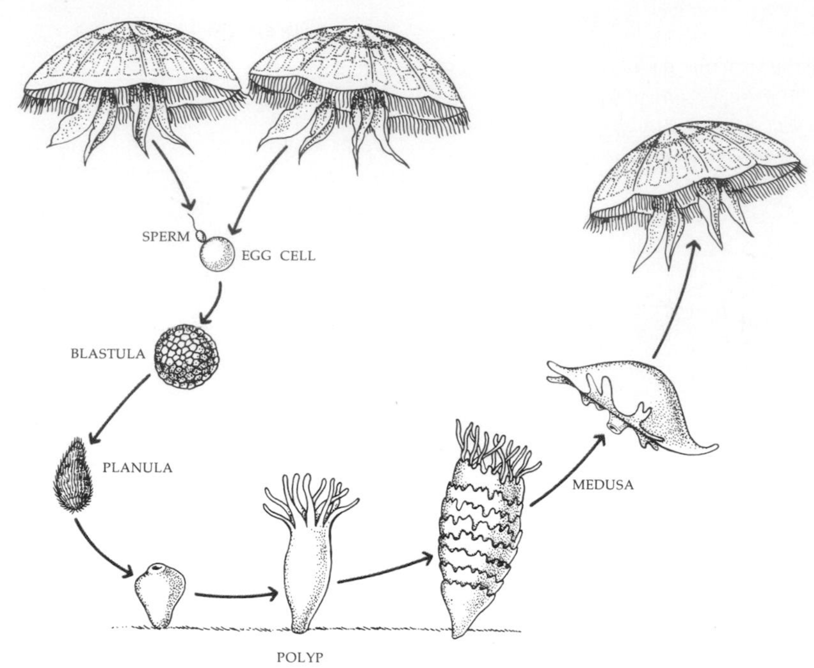

25–11
The life cycle of the coelenterate Aurelia. *Sperm and egg cells are released from adult medusas into the surrounding water. Fertilization takes place, and the resulting zygote develops first into a hollow sphere of cells, the blastula, and then elongates and becomes a ciliated larva called a planula. The planula eventually settles to the bottom, attaches by one end to some object, and develops a mouth and tentacles at the other end, thus transforming into the polyp stage. The body of the polyp grows and, as it grows, begins to form medusas, stacked upside down like saucers. These bud off, one by one, and grow into full-sized jellyfish.*

Another distinctive feature of the animals in this phylum is the *coelenteron*, a digestive cavity with only one opening. Within this cavity, enzymes are released that break down food, partially digesting it extracellularly, as our own food is digested within the stomach and intestinal tract. The food particles are then taken up by the cells lining the cavity; they complete the digestive process and pass the products on to the other cells of the animal. Inedible remains are ejected from the single opening.

A third distinctive feature of coelenterates is the *cnidoblast*. Coelenterates are carnivores. They capture their prey by means of tentacles that form a circle around the "mouth." These tentacles are armed with cnidoblasts, special cells that contain nematocysts (thread capsules). Nematocysts are discharged in response to a chemical stimulus or touch. The nematocyst threads, which are often poisonous and may be sticky or barbed, can lasso prey, harpoon it, or paralyze it—or some useful combination of all three. The toxin apparently produces paralysis by attacking the lipoproteins of the nerve cell membrane of the prey.

Cnidoblasts occur only in this phylum, with some interesting exceptions. Certain other invertebrates, including nudibranchs (a kind of mollusk) and flatworms, can eat coelenterates without triggering the nematocysts. The nematocysts then migrate to the surface of the predator and can be fired in their new host's defense.

Classes of Coelenterates

There are three major classes of coelenterates: Hydrozoa, in which the polyp is usually the dominant form; Scyphozoa, predominantly medusoid, exemplified by the common jellyfish; and Anthozoa, which includes the sea anemones and the reef-building corals, and has only the polyp.

25–12
Hydra has a nervous system that integrates the body into a functional whole, making possible a range of fairly complex activities. For instance, Hydra *may swim, glide on its base, or, as shown here, it may travel by a somersaulting motion.*

Class Hydrozoa: Hydra

One of the most thoroughly studied of the coelenterates is *Hydra,* which is a small, common freshwater form, convenient to keep in the laboratory. Figure 25–13 shows a small section of the body wall of *Hydra.* The ectoderm is composed largely of epitheliomuscular cells, which perform a covering, protective function and also serve as muscle tissue. Each cell has contractile fibers, myonemes, at its base and so can contract individually, like the contractile epithelial cells of the sponge. The endoderm is mostly made up of cells concerned with digestion; these cells also contain contractile fibers. In *Hydra,* as in other polyps, the contractile fibrils of the ectoderm attach lengthwise to the mesoglea and the fibrils of the endoderm cells attach transversely, so the body walls can stretch or bulge, depending on which group is stimulated.

In addition to cnidoblasts and epitheliomuscular cells, which are independent effectors—cells that both receive and respond to stimuli—*Hydra* contain two other types of nerve cells: sensory receptor cells and cells connected into a network, the nerve net. Sensory receptor cells are more sensitive than other epithelial cells to chemical and mechanical stimuli, and when stimulated they transmit their impulses to an adjacent cell or cells. The adjacent cell may be simply an epitheliomuscular cell, an effector, which then responds. Note that this system is one step more complicated than the epitheliomuscular cell or cnidoblast, which acts as both receptor and effector. The nerve net, a loose connection of nerve cells lying at the base of the epithelial layers, is the simplest example of a nervous system that links an entire organism into a functional whole. It coordinates the muscular contrac-

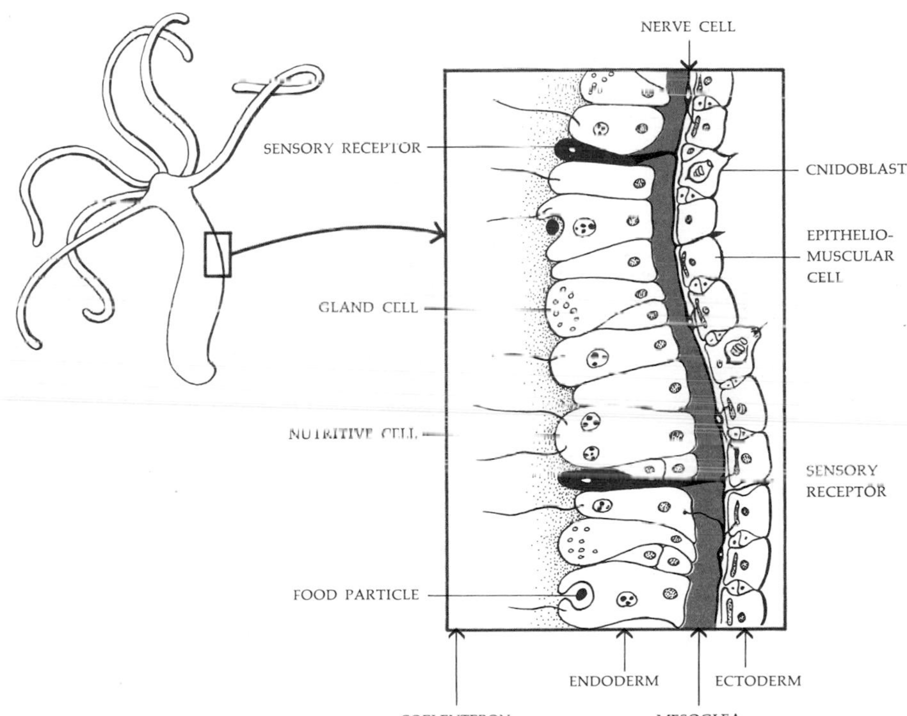

25–13
The structure of the body wall of Hydra. *The outer layer of cells, the ectoderm, is primarily for protection, while the inner layer, the endoderm, performs the digestive function. One type of digestive cell, the gland cell, secretes the digestive enzymes that are released into the coelenteron. Gland cells can also produce a gas bubble that enables the animal to float to the surface. The other type, the nutritive cell, using its flagella, mixes the food as it is being processed, and then extends pseudopodia that collect the food particles for further digestion.*

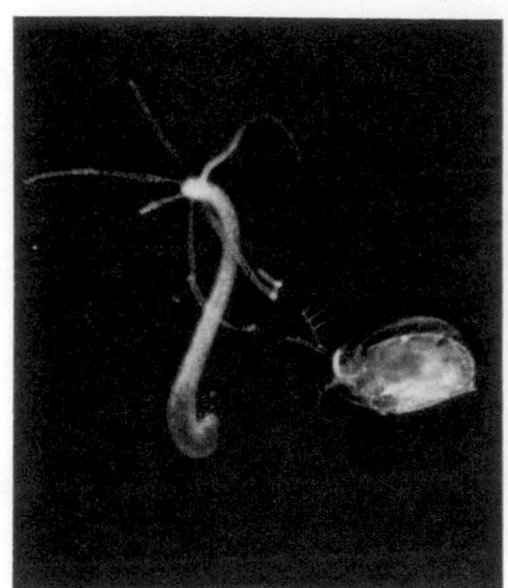
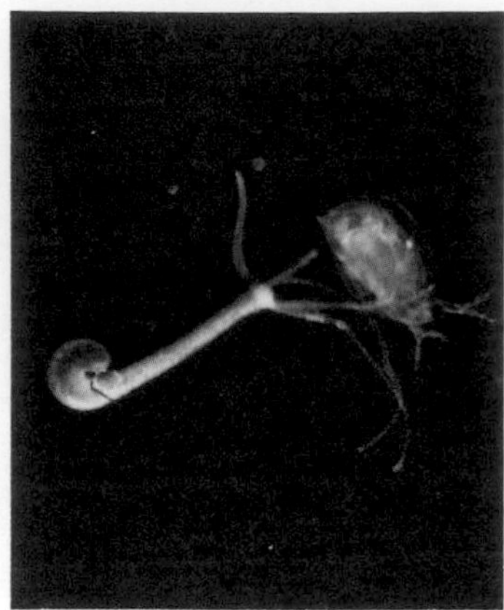
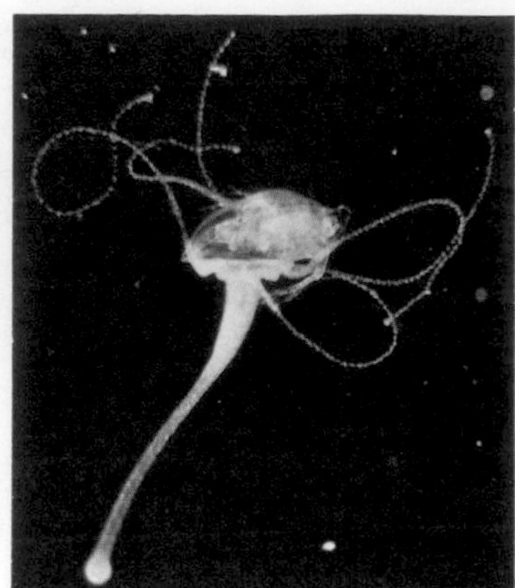
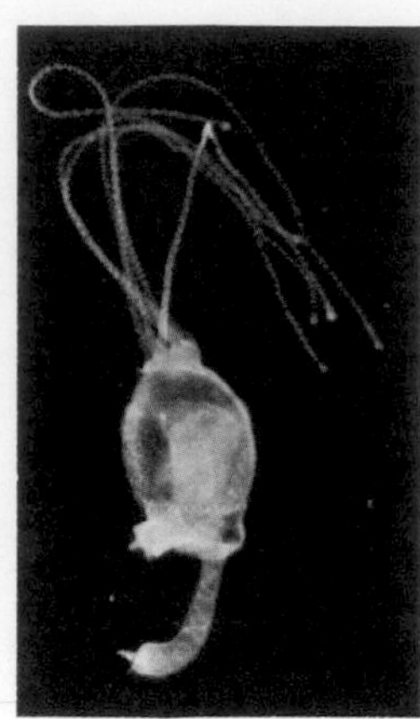

25–14
A Hydra *encounters a small crustacean* (Daphnia), *grasps it with its tentacles, engulfs it, and digests it.*

tions of *Hydra,* making possible a wide variety of activities (Figure 25–14). However, there is no center of operations for the nervous system. This type of conducting system occurs in *Hydra* and certain other coelenterates.

Although *Hydra* has only the polyp form, many hydrozoans have both hydroid (polyp) and medusoid forms at different times in their life cycles. Coelenterates of the genus *Obelia,* for example, spend most of their lives as colonial polyps. The colony arises from a single polyp, which multiplies by budding. The new polyps do not separate but remain interconnected so that their body cavities form a continuous channel, through which food particles are circulated. Within the colony are two types of polyps: feeding polyps with tentacles and cnidoblasts, and reproductive polyps from which tiny medusas bud off. These medusas produce eggs or sperm that are released into the water and fuse to form zygotes.

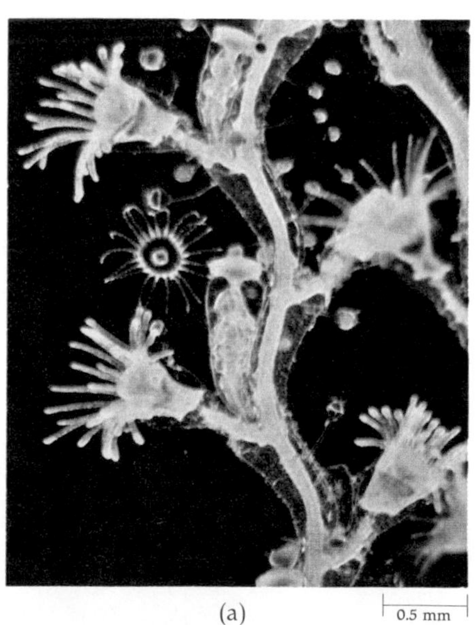
(a) 0.5 mm

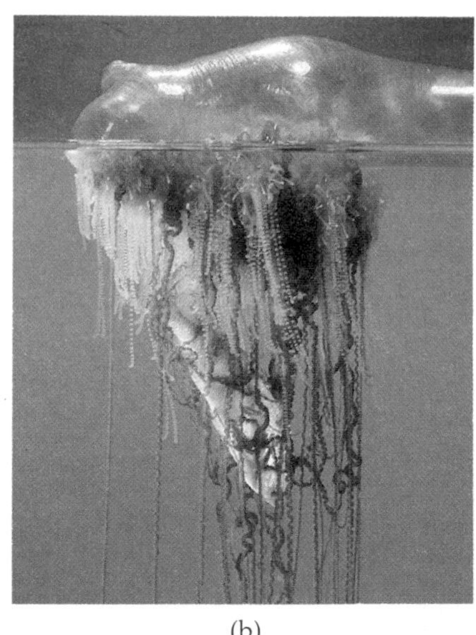
(b)

25–15
Colonial hydrozoans. (a) Obelia *is made up of two types of polyp, a* Hydra-*like form, shown here with tentacles extended, which is the feeding polyp, and a reproductive form, which lacks tentacles. You can see two of these reproductive forms in the axils of the branches. At the left is a newly released free-swimming medusa.* (b) *Coelenterates of the order Siphonophora are large, floating colonies made up of both polyps and medusas. The polyps are feeding forms and the medusas are reproductive forms; in some species, nonreproductive medusas are swimming bells. The colony produces a gas-filled float, or pontoon. In the Portuguese man-of-war shown here, the large float serves also as a sail. The blue strands are composed of reproductive and feeding individuals; the purple strands, which may grow as long as 15 meters, are made up of stinging, food-gathering polyps armed with nematocysts. A large colony can kill a human being.*

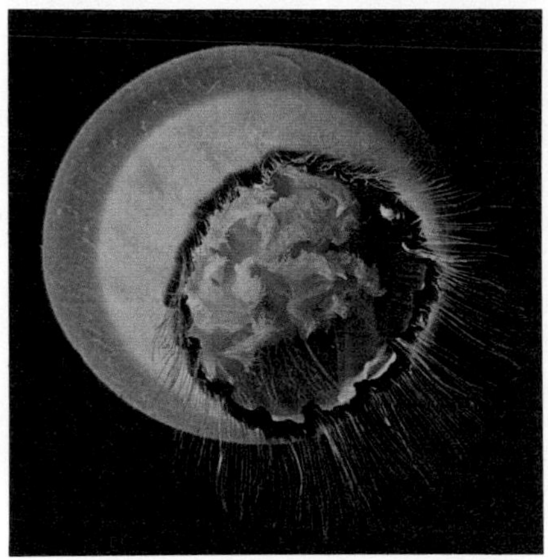

25–16
The moon jellyfish (Aurelia limbata) *has four frilly mouth lobes that gather minute organisms into the coelenteron. It has short tentacles and grows to only about 30 centimeters in diameter. This specimen was photographed in the cold waters of the Atlantic.*

Thus colonial polyps, with their division of labor between feeding and reproductive forms, are very like a single organism. Such a high degree of specialization of structure and function among social organisms is seen in other phyla only among the social insects.

Class Scyphozoa: Jellyfish

A second major class of coelenterates is Scyphozoa, or "cup animals," in which the medusa form is dominant. Scyphozoans, more commonly known as jellyfish, range in size from less than 2 centimeters in diameter up to animals 4 meters across and trailing tentacles 10 meters long. In the adult animal, the mesoglea is so firm that a large, freshly beached jellyfish can easily support the weight of a human being. The mesoglea of some jellyfish is filled with wandering, amoebalike cells, which serve to transport food from the nutritive cells of the endoderm. Unlike *Hydra*, scyphozoans have true muscle cells; these underlie the ectoderm, contracting rhythmically to propel the medusa through the water.

Nervous System of Medusa

In the medusa, there are concentrations of nerve cells in the margin of the bell. These nerve cells connect with fibers innervating (providing the nerve supply for) the tentacles, the musculature, and the sense organs.

The bell margin is liberally supplied with sensory receptor cells sensitive to mechanical and chemical stimuli. In addition, the jellyfish has two types of true sense organs: statocysts and light-sensitive ocelli. *Statocysts* are specialized receptor organs that provide information by which an animal can orient itself with respect to gravity (Figure 25–17a). The statocyst, which seems to have been one of the first special sensory organs to have appeared in the course of evolution, has persisted apparently unchanged to the present day, appearing in many animal phyla. *Ocelli*, which may have evolved even earlier, are groups of pigment cells and photoreceptor cells. They are typically located at the bases of the tentacles.

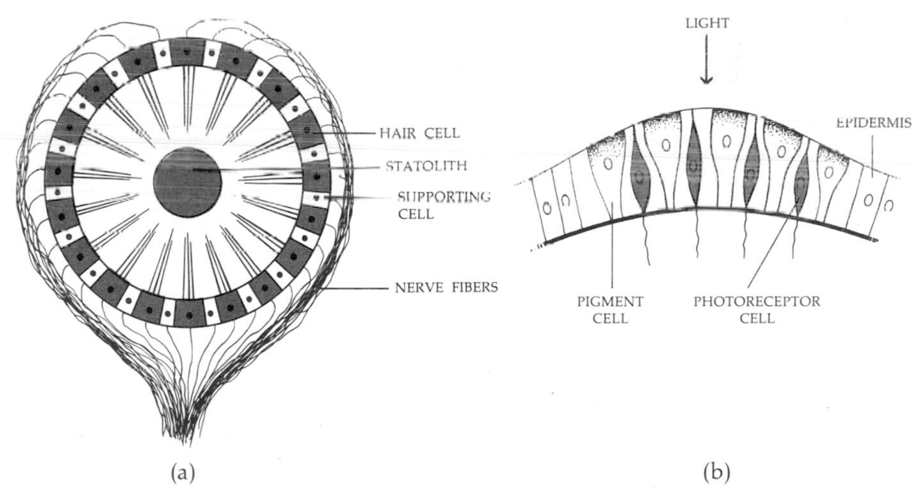

25–17
(a) *The statocyst is a specialized receptor organ that orients the jellyfish with respect to gravity. When the bell tilts, gravity pulls the statolith, a grain of hardened calcium salts, down against the hair cells. This stimulates the nerve fibers and signals the animal to right itself.* (b) *Eyespot (ocellus), the simplest type of photoreceptor organ. Ocelli of this sort are found among coelenterates.*

25-18
Burrowing sea anemones. The flowerlike appearance is deceptive; anemones are carnivorous animals belonging to a class of coelenterates that, like Hydra, *have dropped the medusa stage. In common with other coelenterates, their tentacles are equipped with stinging nematocysts. The tentacles move food into the coelenteron, which is divided longitudinally by partitions.*

Class Anthozoa

Anthozoans ("flower animals")—the corals and sea anemones—are members of a class of coelenterates that, like *Hydra,* have lost the medusa stage. They differ from *Hydra* in having a gullet lined with epidermis and a coelenteron divided by vertical partitions. In most corals, which are colonies of anthozoans, the epidermal cells secrete protective outer walls, usually of calcium carbonate (limestone), into which each delicate polyp can retreat. The coral-forming polyps are the most ecologically important of the coelenterates. The 2,000-kilometer-long Great Barrier Reef off the northeast shores of Australia, and the Marshall Islands in the Pacific are examples of coral-created land masses. A coral reef is composed primarily of the accumulated limestone skeletons of coral coelenterates, covered by a thin crust occupied by the living colonial animals. A reef is both the structural and nutritional basis of the complex coral reef community.

PHYLUM PLATYHELMINTHES: FLATWORMS

The flatworms are the simplest animals, in terms of body plan, to show bilateral symmetry. In bilaterally symmetrical animals, the body plan is organized along a longitudinal axis, with the right half an approximate mirror image of the left half. It also has a top and a bottom or, in more precise terms (applicable even when it is turned upside down or, as in the case of humans, standing upright), a dorsal and a ventral surface. Like most bilateral organisms, the flatworm also has a distinct "headness" and "tailness," anterior and posterior. Having one end that goes first—cephalization—is characteristic of actively moving animals. In such animals, many of the sensory cells are also collected into the anterior end. With the aggregation of sensory cells, there came a concomitant gathering of nerve cells; this gathering is a forerunner of the brain.

Flatworms have three distinct tissue layers—ectoderm, mesoderm, and endoderm—characteristic of all animals above the coelenterate level of organization.

25-19
In a bilaterally symmetrical organism, the right and left halves of the body are mirror images of one another. The upper and lower (or back and front) surfaces are known as dorsal and ventral. With some exceptions, the end that goes first is termed anterior and the rear, posterior.

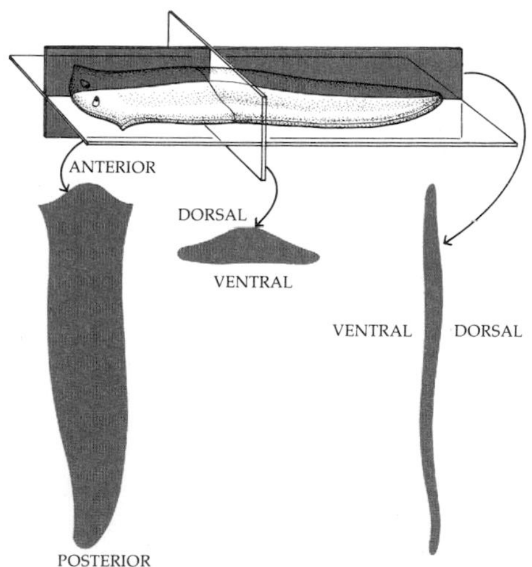

25–20
A polyclad flatworm. The many species belonging to this phylum, though widely various in color and shape, are nearly all small and flat-bodied, like a planarian (Figure 25–21).

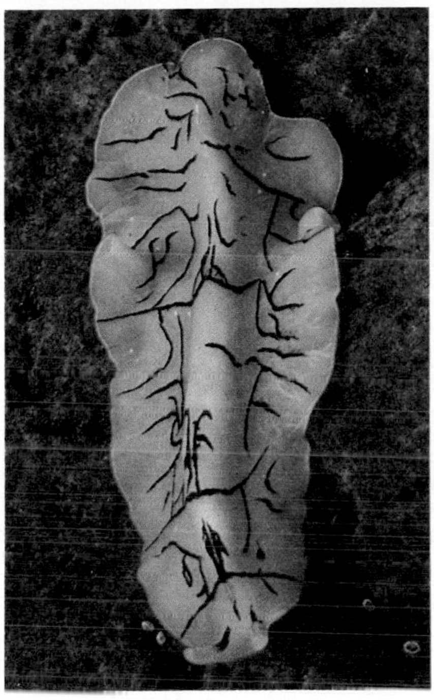

Moreover, not only are their tissues specialized for various functions, but also two or more types of tissue cells may combine to form organs—for example, the muscular pharynx (Figure 25–21). Thus, while sponges are made up of aggregations of cells and coelenterates are largely limited to the tissue level of organization, flatworms can be said to exemplify the organ level of complexity. There are three classes of flatworms: the free-living turbellaria and two parasitic forms, flukes and tapeworms.

The flatworms are believed by some zoologists to have evolved from the coelenterates (or, perhaps, vice versa), not by way of either adult form, however, but from the ciliated larval form. Others are persuaded that the flatworms had independent origins among the ciliates. Still another group maintains that they are degenerate annelids. It is unlikely that this matter will soon be resolved.

The Planarian

The free-living flatworms form a large and varied group, and we shall single out just one for examination, the freshwater planarian. The ectoderm of the planarian is made up of cuboid epithelial cells, many of which are ciliated, particularly those on the ventral surface. Ventral ectodermal cells secrete mucus, which provides traction for the planarian as it moves by means of its cilia along its own slime trail. Planarians are among the largest animals that can use cilia for locomotion. The cilia in larger species are usually employed for moving water or other substances along the surface of the animal, as in the human respiratory tract, rather than for propelling the animal.

The planarian has an endoderm composed largely of amoeboid cells, which, although they are phagocytic, are not wandering cells like the amoebocytes of the jellyfish. Between the ectoderm and the phagocytic endoderm is a *mesoderm*, or middle tissue layer. In planarians, as in all other groups to follow, the muscle cells and the principal organ systems are of mesodermal origin.

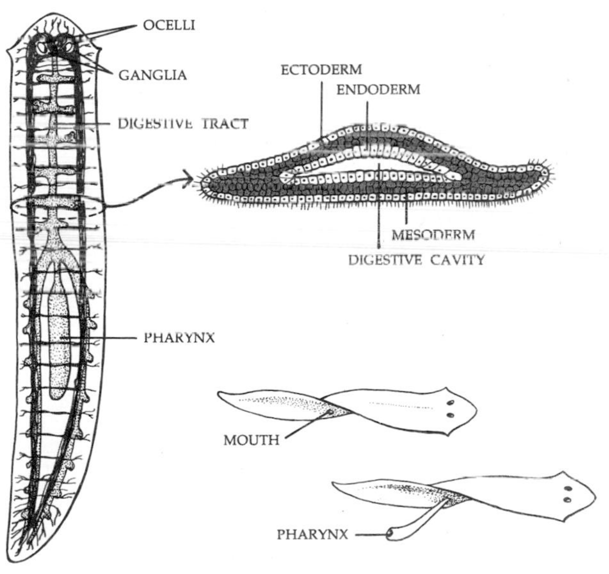

25–21
Example of a flatworm: the freshwater planarian. The planarian, which is carnivorous, feeds by means of its extensible pharynx. The nervous system is indicated in color in the left-hand diagram. Note that some of the fibers have been aggregated into two cords, one on each side of the body, and there is a cluster of nerve cells in the head, the beginnings of a brain. The ocelli are light-sensitive areas.

The planarian, like the coelenterates, has a digestive cavity (gut) with only one opening, located on the ventral surface. This digestive cavity has three main branches, which is why planarians are placed in the order of flatworms known as Tricladida.

Like other flatworms, the planarian is carnivorous. It eats either dead meat or slow-moving animals it can fasten itself to or subdue by sitting on, such as smaller planarians. It feeds by means of a muscular tube, the pharynx, which is free at one end. The free end can be stretched out through the mouth opening. Muscular contractions in the tube cause strong sucking movements, which tear the meat into microscopic bits and draw them into the internal cavity, where they are phagocytized by the endodermal cells.

Unlike the sponges or coelenterates, most flatworms have an excretory system (Figure 25–22). In the planarian, the system is a network of fine tubules that runs the length of the animal's body. Side branches of the tubules contain flame cells, each of which has a hollow center in which a tuft of cilia beats, flickering like a tongue of flame, moving water along the tubules to the exit pores between the epidermal cells. The flame-cell system appears to function largely to regulate water balance; most of the metabolic waste products probably diffuse out through the ectoderm or the endoderm.

Planarians have a complicated reproductive system. The eggs are fertilized internally. At mating, each partner deposits sperm in the copulatory sac of the other partner (Figure 25–23). These sperm then travel along special tubes, the *oviducts*, to fertilize the eggs as they become ripe. As we mentioned earlier, organisms like planarians, in which both types of gametes are produced by one individual, are known as hermaphrodites (from Hermes and Aphrodite). Solitary, slow-moving animals, such as earthworms and snails, are often hermaphrodites; these animals may seldom encounter another adult member of the species, but every such encounter can result in a mating. Some types of hermaphroditic animals can fertilize themselves, although they do not usually do so if another individual is present.

25–22
Flatworms have a tubular excretory system. The system usually consists of two or more branching tubules running the length of the body. In the planarian and its relatives, the tubules open to the body surface through a number of tiny pores. At the ends of the side branches are small bulblike structures known as flame cells. Within each of these cells, a tuft of cilia in constant motion resembles the flickering of a flame. Water and some waste materials from the tissue fluids are moved by the cilia through tubules to the excretory pores, where the collected liquid leaves the body.

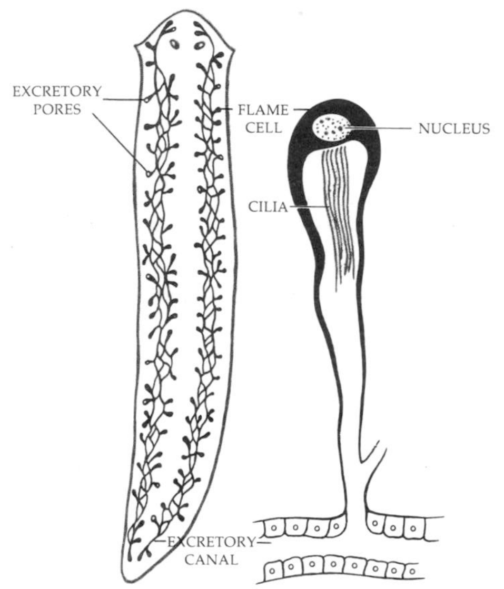

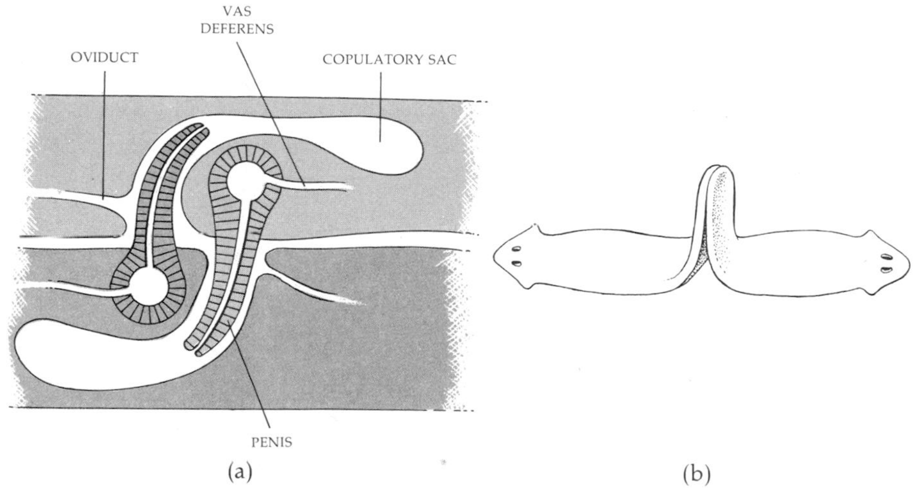

25–23
(a) *Planarians have both male and female reproductive structures, and mating involves a mutual exchange of sperm. The erect penis of each partner is inserted into the sperm receptacle of the other.* (b) *Planaria mating.*

Planaria may also reproduce asexually, by fragmentation or by fission. Fragments of the tail, for example, may regenerate new heads, and sometimes the animal divides longitudinally, forming two mirror images.

25-24

Ocellus of the planarian. The light stimulus is received by the ends of the photoreceptor cells adjacent to the pigment cup. Thus the light travels first through the fibers carrying the signals to the cerebral ganglia. In this inside-out organization, the planarian ocellus resembles the vertebrate eye.

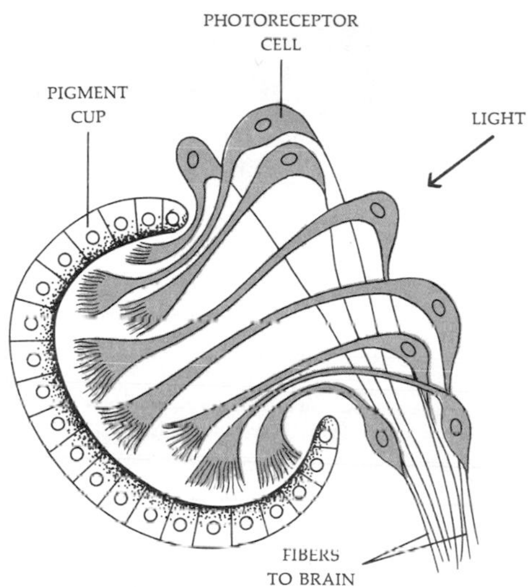

The Planarian Nervous System

The evolution of bilateral symmetry brought with it marked changes in the organization of the nervous system as well as of other systems. Even among primitive flatworms, the neurons (nerve cells) are not dispersed in a loose network, as in *Hydra,* but are instead condensed into longitudinal cords. In the planarians, this condensation is carried further, and there are only two main conducting channels, one on each side of the flat, ribbonlike body. These channels carry impulses to and from the aggregation of nerve cells in the anterior end of the body. (See Figure 25-21.) Such aggregations of nerve cell bodies are known as *ganglia* (singular, ganglion).

The ocelli of the planarian are usually inverted pigment cups (Figure 25-24). They have no lenses, and they cannot form an image. However, they can distinguish light from dark and can tell the direction from which the light is coming. Planarians are photonegative; if you shine a light on a dish of planarians from the side, they will move steadily away from the source of light.

Among the epithelial cells are receptor cells sensitive to certain chemicals and to touch. The head region in particular is rich in *chemoreceptors.* If you place a small piece of fresh liver in the culture water so that its juices diffuse through the medium, the planarians will raise their heads off the bottom and, if they have not eaten recently, will lope directly and rapidly (on a planarian scale) toward the meat, to which they then attach themselves to feed. The animal locates the food source by repeatedly turning toward the side on which it receives the stimulus more strongly until the stimulus is equal on both sides of its head. If the chemoreceptor cells are removed from one side of the head, the animal will turn constantly toward the intact side.

Planarians, with their simple nervous systems, their capacity to react to a variety of stimuli, and their powers of regeneration, have been the subject of a number of experiments. In one group, planarians trained to avoid electric shock were fed to untrained planarians. The result, it was claimed, was that the untrained planarians that had eaten the trained ones behaved as if they had been trained. These experiments, still not verified, on the "transfer of training by cannibalism" led to a great deal of controversy in the 1960s and also, as you might expect, to a number of suggestions for more meaningful student-teacher relationships.

Tapeworms and Flukes

Phylum Platyhelminthes includes also the tapeworms and the trematodes (flukes), parasitic forms that can cause serious and sometimes fatal diseases among vertebrates. Members of both of these parasitic classes have a tough outer layer of cells that is resistant to digestive fluids and, usually, suckers or hooks on their anterior ends by which they fasten to their victims. Trematodes feed through a mouth, but the tapeworms, which have no mouths, digestive cavities, or digestive enzymes, merely hang on and absorb predigested food molecules through their skin. Tapeworms are found in the intestines of many vertebrates, including humans, and may grow as long as 5 or 6 meters. They cause illness not only by encroaching on the food supply but also by producing wastes and by obstructing the intestinal tract. The most common human tapeworm, the beef tapeworm, infects people who eat the undercooked flesh of cattle that have eaten fodder contaminated by human feces containing tapeworm segments.

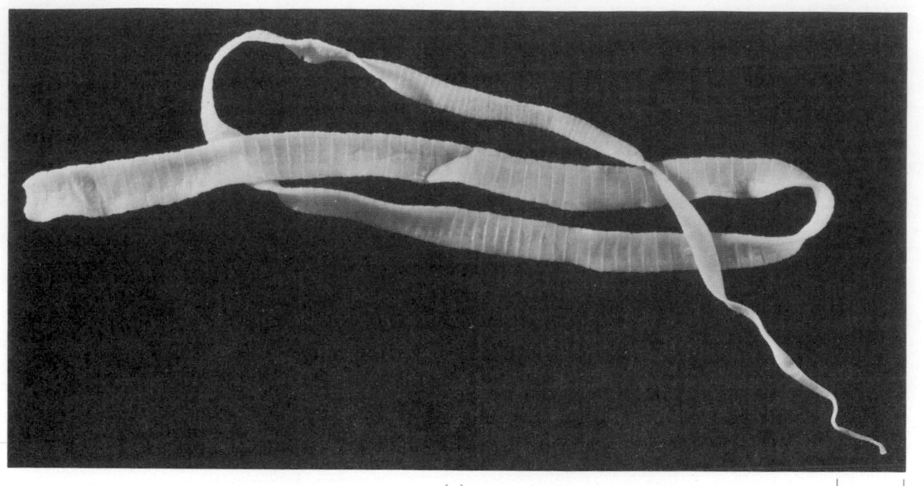

(a)

(b)

25–25
(a) *Human tapeworm,* Taenia solium. *Tapeworms are intestinal parasites that lack any digestive system of their own. They cling by their heads, which are equipped with hooks and suckers, and absorb the food molecules digested by their hosts through their body walls. Their bodies, posterior to the heads, are divided into segments, each of which is a sexually complete hermaphroditic unit* (b) *in which sperm and ova are produced. Segments break off when the ova are mature, and new segments are formed. Characteristically, the ova are eaten and develop to a larval form in one species and the adult worm parasitizes members of another species. The dark areas are the reproductive structures; the opening is the genital pore.*

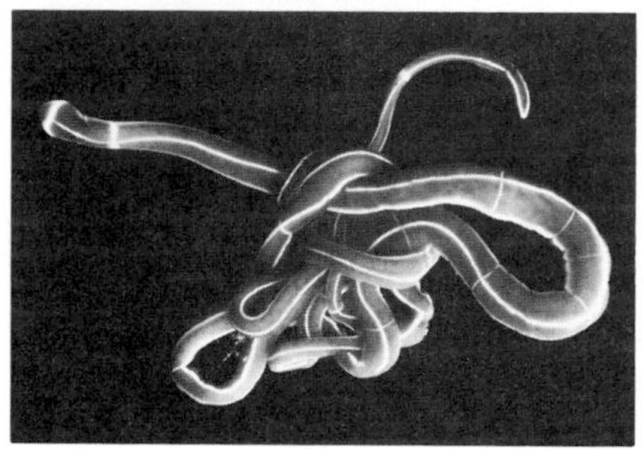

25–26
A ribbon worm in an embarrassing situation. Ribbon worms range in length from less than 2 centimeters to 30 meters and through virtually all the colors of the spectrum. All members of the species, however, have flat, thin (seldom more than ½ centimeter thick) bodies, a mesoderm, a mouth-to-anus digestive tract, and a long muscular tube that can be thrust out to grasp prey.

All parasites, including parasitic flatworms, are believed to have originated as free-living forms and to have lost certain tissues and organs (such as the digestive tract) as a secondary effect of their parasitic existence, while developing adaptations of advantage to the parasitic way of life. Such adaptations also often include a complex life cycle.

PHYLUM RHYNCHOCOELA: RIBBON WORMS

The ribbon worms (sometimes called nemertines), although a small phylum, are of special interest to biologists attempting to reconstruct the evolution of the invertebrates. They appear to be closely related to the flatworms, but with an important difference: They have a one-way digestive tract beginning with a mouth and ending with an anus. This is a far more efficient arrangement than the one-opening digestive system of the coelenterates and flatworms. In the one-way tract, food moves assembly-line fashion, with the consequent possibilities (1) that eating can be continuous and (2) that various segments of the tract can become specialized for different stages of digestion. The ribbon worms also have a circulatory system, usually consisting of one dorsal and two lateral blood vessels that carry the colorless blood.

This phylum is called Rhynchocoela ("beak" plus "hollow") because these worms are characterized by a long, retractile, slime-covered tube (proboscis). The proboscis, sometimes armed with a barb, seizes prey and draws it to the mouth where it is engulfed. Some inject a paralyzing poison into their prey.

PHYLUM NEMATODA: ROUNDWORMS

The number of species of roundworms (nematodes) has been variously estimated as low as 10,000 and as high as 400,000 to 500,000. Most are free-living, microscopic forms. It has been estimated that a spadeful of good garden soil usually contains about a million nematodes. Some are parasites; most species of plants and animals are parasitized by at least one species of nematodes. Humans are hosts to about 50 species, including hookworms, pinworms, and *Trichinella.* The latter causes trichinosis, which is transmitted by eating uncooked or undercooked pork, a single gram of which may contain 3,000 cysts (resting forms) of *Trichinella.* Ingestion of only a few hundred of these cysts can be fatal to human beings.

Nematodes have a three-layered body plan and a tubular gut with a mouth and an anus. They are unsegmented and are covered by a thick, continuous cuticle, which is molted periodically as they grow. An interesting, and unique, feature of nematode construction is the absence of circular muscles. The contraction of the longitudinal muscles acting against the tough, elastic cuticle gives the worm its characteristic whipping movement in water. The sexes are usually separate.

Nematodes may have evolved from early Platyhelminthes, possessing, as they do, a three-layered body plan without a true coelom (Figure 25–28c). They have, however, what is known as a pseudocoelom, a body cavity that is between the endoderm and the mesoderm and lacks the epithelial lining of a true coelom. Six other minor (in terms of species and numbers) phyla, mostly small, wormlike animals, have body plans based on the pseudocoelom, but Nematoda is the only major pseudocoelomate phylum.

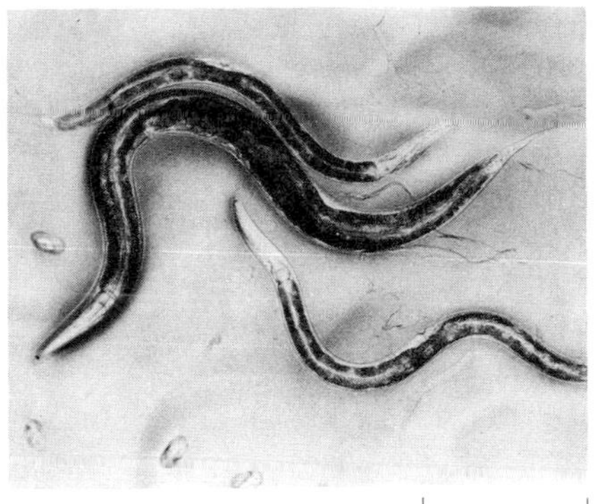

25–27
An adult and two larval nematodes, Caenorhabditis elegans. *Strands of shed cuticle can be seen extending from the right-hand side of the adult nematode. Four eggs are visible at the lower left of the micrograph.*

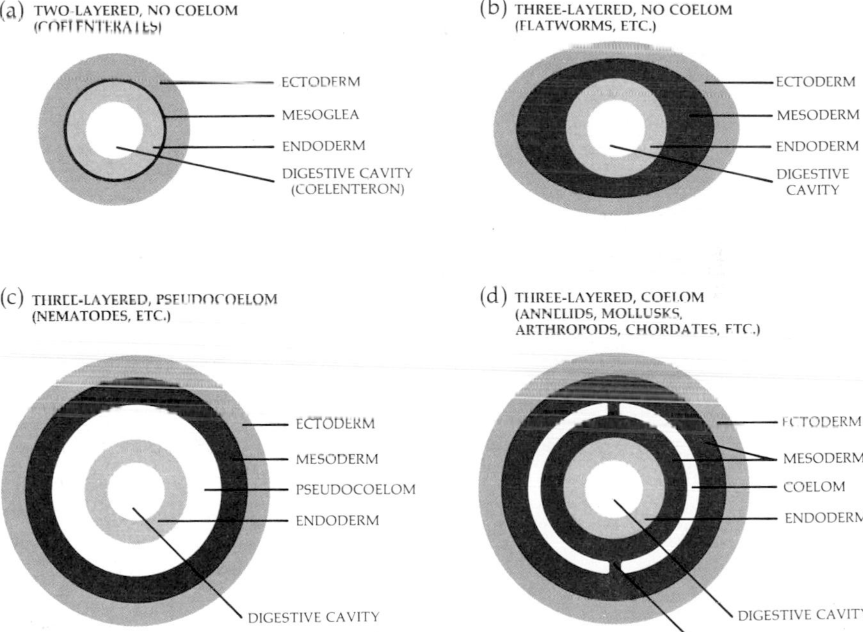

25–28
Basic body plans of the animal phyla, as shown in cross section. (a) *A body that consists of only two tissue layers is characteristic of coelenterates.* (b) *Flatworms and ribbon worms have three-layered bodies, with the layers closely packed on one another.* (c) *Nematodes have three-layered bodies with a pseudocoelom between the endoderm and mesoderm.* (d) *Annelids and most other animals, including vertebrates, have bodies that are three-layered with a cavity, the coelom, within the middle layer (mesoderm). The mesodermal mesenteries suspend the gut within the body wall.*

PHYLUM ANNELIDA: SEGMENTED WORMS

This phylum includes almost 9,000 different species of marine, freshwater, and soil worms, including the familiar earthworm. The term annelid means "ringed" and refers to the most distinctive feature of this group, which is the division of the body into segments, visible as rings on the outside and with partitions on the inside. This segmented pattern is found in a modified form in higher animals, too, such as dragonflies, millipedes, and lobsters, which are thought to have evolved from the same ancestors that gave rise to modern annelids.

The annelids have a three-layered body plan, a tubular gut, and a well-developed circulatory system that transports oxygen (diffused through the skin or through fleshy extensions of the skin) and food molecules (from the gut) to all parts of the body. The excretory system is made up of specialized paired tubules, nephridia, which occur in each segment of the body except the head. Annelids have a nervous system and a number of special sense cells, including touch cells, taste receptors, light-sensitive cells, and cells concerned with the detection of moisture. Some also have well-developed eyes and sensory antennae.

In the flatworms and ribbon worms, the mesoderm is packed solid with muscle and other tissues, but in the annelids there is a fluid-filled cavity, the *coelom*, (pronounced see-loam) in this middle layer (Figure 25-28d). (Note that the term coelom, although it sounds similar to coelenteron and comes from the same Greek root, meaning "cavity," refers quite specifically to a cavity *within* the mesoderm, whereas the coelenteron is a digestive cavity lined by endoderm.) Within the coelom, the gut—lined with an epithelium—is suspended by double layers of mesoderm known as mesenteries. The fluid in the coelom constitutes a hydrostatic skeleton for the annelid, stiffening the body in somewhat the same way water pressure stiffens and distends a fire hose. The muscles of the earthworm work against this hydrostatic skeleton much as our muscles work against our bony skeleton.

Although the opening of a cavity within the mesoderm may seem less dramatic than other evolutionary innovations, it is extremely important. Within such a space, organ systems can bend, twist, and fold back on themselves, increasing their functional surface areas and filling, emptying, and sliding past one another, surrounded by lubricating coelomic fluid. Consider the human lung, constantly expanding and contracting in the chest cavity, or the 6 or 7 meters of coiled human intestine; neither of these could have evolved until the coelom made room for them.

Earthworms

The earthworm is the most familiar of the annelids. Figure 25-29 shows a portion of the body of an earthworm. Note how the body is compartmented into regular segments. Most of these segments, particularly the central and posterior ones, are identical, each exactly like the one before and the one after. Each identical segment contains four pairs of bristles, or *setae;* two *nephridia,* excretory tubules that pick up waste materials from the body fluids and excrete them through pores on the ventral surface of the worm; four sets of nerves branching off from the central nerve cord running along the ventral surface; a portion of digestive tract; and a left and right coelomic cavity. The chief exceptions to this rule of segmented structure are found in the most forward segments. In these, sensory cells, a cluster of nerve cells

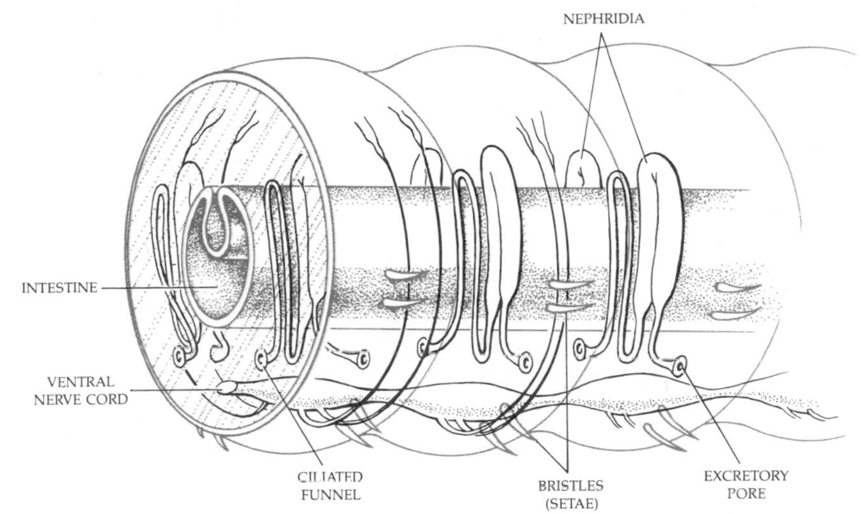

25-29
Three segments of the earthworm, an annelid. On each segment are four pairs of bristles, which are extended and retracted by special muscles. These are used by the worm to anchor one part of its body while it moves another part forward. Two excretory tubes, or nephridia, are in each segment (except the first three and the last). Each nephridium really occupies two segments since it opens externally by a pore in one segment and internally by a ciliated funnel in the segment immediately in front of it. The intestine, nephridia, and other internal organs are suspended in the large coelom, which also serves as a hydraulic skeleton.

(ganglia), and specialized areas of the digestive, circulatory, and reproductive systems are found.

The tubelike body is wrapped in two sets of muscles, one set running longitudinally and the other encircling the segments. When the earthworm moves, it anchors some of its segments by its setae, and the circular muscles of the segments anterior to the anchored segments contract, thus extending its body forward. Then its forward setae take hold, and the longitudinal muscles contract while the posterior anchor is released, drawing the posterior segments forward.

Digestion in Earthworms

The digestive tract of the earthworm (Figure 25-30) is a long, straight tube. The mouth leads into a strong, muscular pharynx, which acts like a suction pump, drawing in decaying leaves and other organic matter, as well as dirt, from which organic materials are extracted. The earthworm makes burrows in the earth by passing such material through its digestive tract and depositing it outside in the form of castings, a ceaseless activity that serves to break up, enrich, and aerate the soil. The narrow section of tube posterior to the pharynx, the *esophagus,* leads to the crop, where food is stored. In the gizzard, which has thick, muscular walls lined with protective cuticle, the food is ground up with the help of the ever-present soil particles. The rest of the digestive tract is made up of a long intestine, which has a large fold along its upper surface that increases its surface area. The intestinal epithelium consists of enzyme-secreting cells and ciliated absorptive cells.

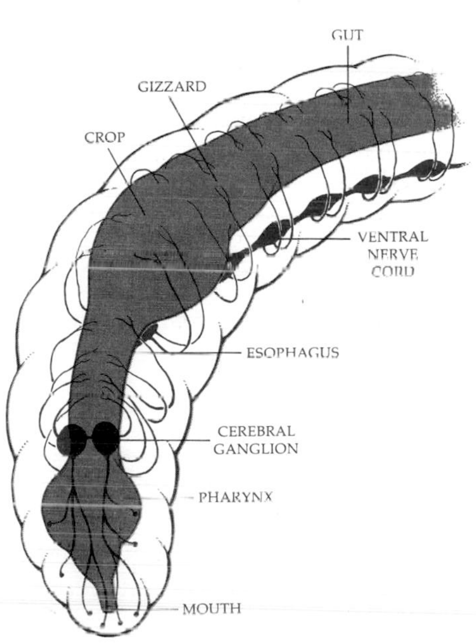

25-30
The digestive tract of an earthworm. The mouth leads into a muscular pharynx, which sucks in decaying vegetation and other material. These are stored in the crop and ground up in the gizzard with the help of soil particles. The rest of the tract is a long intestine (gut) in which food is digested and absorbed.

Circulation in Earthworms

In the protists and in the smaller and simpler animals, food molecules and oxygen are supplied to cells largely by diffusion, aided by the movement of external fluids and, sometimes, as we saw in coelenterates, by wandering, amoeboid cells. A circulatory system that propels extracellular fluid around the body solves the problem of providing each cell with a more direct and rapid line of supply.

The circulatory system of the earthworm (Figure 25-31) is composed of longitudinal vessels running the entire length of the worm, one dorsal and several

25-31
The circulatory system of the earthworm is made up of longitudinal vessels running the entire length of the animal, one dorsal and several ventral. Smaller vessels (the parietal vessels) in each segment collect the blood from the tissues and feed it into the muscular dorsal vessel, through which it is pumped forward. In the anterior segments are five pairs of hearts—muscular pumping areas in the blood vessels—whose irregular contractions force the blood downward through the parietal vessels to the ventral vessels, from which it returns to the posterior segments. The arrows indicate the direction of blood flow.

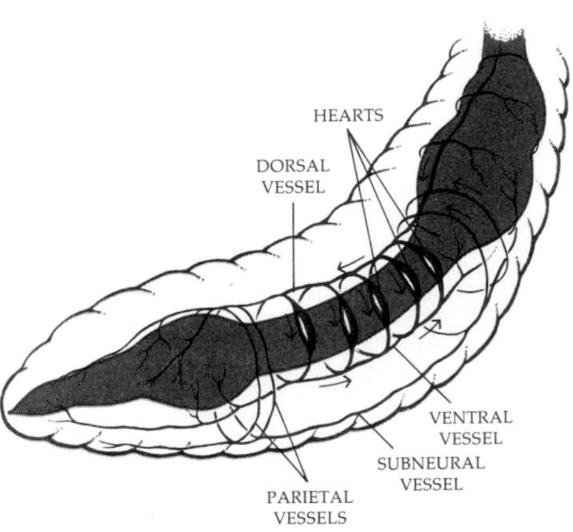

ventral. The largest ventral vessel underlies the intestinal tract, collecting nutrients from it and distributing them by means of many small branches to all the tissues of the body and to the three smaller ventral vessels that surround the nerve cord and nourish it. Numerous small capillaries in each segment carry blood from the ventral vessels through the tissues to the dorsal vessel. Also in each segment are larger parietal ("along the wall") vessels transporting blood from the subneural vessels to the dorsal vessel. Fluids collected in this way from all over the animal's body are fed into the muscular dorsal vessel, which propels the blood forward.

Connecting the dorsal and ventral vessels, and so completing the circuit, are five pairs of hearts, muscular pumping areas in the blood vessels. Their irregular contractions force the blood down to the ventral vessels and also forward to the vessels that supply the more anterior segments. Both the hearts and the dorsal vessel have valves that prevent backflow. Note that the blood flows entirely through vessels. Such a system is known as a closed circulatory system. Evolution of a closed circulatory system, in effect, added a new compartment to the body plan and, in so doing, made possible a degree of control, not previously feasible, over the content of the circulating body fluid.

Excretory System of Earthworms

The excretory system consists of pairs of tubules, the nephridia; one pair for each segment. Each nephridium consists of a long, convoluted tubule that terminates in a ciliated funnel opening into the coelomic cavity of the anteriorly adjacent segment. Coelomic fluid is carried into the funnel by the beating of the cilia and is excreted through an outer pore. As the fluid makes its way through the long tubule, water, sugar, salts, and other needed materials are returned to the coelomic fluid through the walls of the tubule, while other materials are absorbed into the tubule for excretion. Thus the excretory system is concerned not only with the problem of water balance, as are the contractile vacuoles of *Paramecium* and the flame cells of planarians, but also with the homeostatic regulation of the chemical composition of the body fluids. Thus, in effect, one segment monitors and regulates this aspect of the physiology of its neighboring segment—a neat trick for integrating the animal as a whole.

Respiration in Earthworms

The earthworm has no special respiratory organs; respiration takes place by simple diffusion through the body surface. The gases of the atmosphere dissolve in the liquid film on the surface of the earthworm's body, which is kept moist by secreted mucus and excreted water. Oxygen travels inward by diffusion, since the surface film, exposed to the oxygen-rich atmosphere, contains more oxygen than does the blood in the network of capillaries just underlying the body surface. The oxygen is consumed by body cells as the blood circulates. Carbon dioxide moves out to the surface film and then into the air by the same principle. In fact, all gas exchange in animals, whether the organism is land-dwelling or water-dwelling, takes place across moist membranes.

Nervous System in Earthworms

The earthworm has a variety of sensory cells. It has touch cells, or *mechanoreceptors.* These contain tactile hairs, which, when stimulated, trigger a nerve impulse. Patches of these hair cells are found on each segment of the earthworm. The hairs probably also respond to vibrations in the ground, to which the earthworm is very sensitive. The earthworm does not have ocelli—as one might expect, since it lives most of its life in complete darkness—but it does have light-sensitive cells. Such cells are more abundant in its anterior and posterior segments, the parts of its body most likely to be outside of the burrow. These cells are not responsive to light in the red portion of the spectrum, a fact exploited by anglers who search for worms in the dark using red-lensed flashlights.

Among the earthworm's most sensitive cells are those that detect moisture. The cells are located on its first few segments. If an earthworm emerging from its burrow encounters a dry spot, it swings from side to side until it finds dampness; failing that, it retreats. However, when the anterior segments are anesthetized, the earthworm will crawl over dry ground. The animal also appears to have taste cells. In the laboratory, worms can be shown to select, for example, celery in preference to cabbage leaves and cabbage leaves in preference to carrots.

Each segment of the worm is supplied by nerves that receive impulses from sensory cells and by nerves that cause muscles to contract. The cell bodies for these nerves are grouped together in clusters (ganglia). The movements of each segment are directed by a pair of ganglia. Movement in each segment is triggered by movement in the adjacent anterior segment; thus a headless earthworm can move in a coordinated manner. However, an earthworm without its cerebral ganglia moves ceaselessly; in other words, cephalic ganglia modulate activity.

There are also, as in planarians, conducting channels made up of nerve fibers bound together in bundles, like cables, which run lengthwise through the body. These nerve fibers are gathered together in a double nerve cord that runs along the ventral surface of the body. The nerve cords contain fast-conducting fibers that make it possible for the earthworm to contract its entire body very quickly, withdrawing into its burrow when disturbed.

Reproduction in Earthworms

Earthworms are hermaphrodites. Two earthworms, held together by mucous secretions from the clitellum (a special collection of glandular cells), exchange sperm and separate. Two or three days later, the clitellum forms a second mucous sheath surrounded by an outer, tougher protective layer of chitin. This sheath is pushed forward along the animal by muscular movements of its body. As it passes over the female gonopores, it picks up a collection of mature eggs, and then, continuing forward, it picks up the sperm deposited in the spermathecas. Once the mucous band is slipped over the head of the worm, its sides pinch together, enclosing the now fertilized eggs in a small capsule from which the infants hatch.

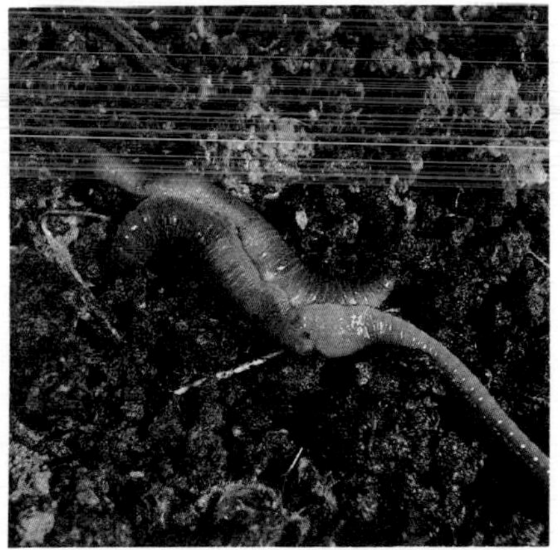

25–32
Earthworms mating. The worms' heads are facing in opposite directions and their ventral surfaces are in contact. The clitellum, a thickened band that surrounds the body of each, secretes mucus, which holds them together during copulation, which may take as long as 2 hours. Sperm cells are released through pores in specialized segments of one worm into the sperm receptacles of its partner. After the partners separate, the clitellum secretes a cocoon into which first the eggs and then the sperm are released. The eggs are fertilized within this cocoon.

25–33
This annelid, a polychaete worm, unlike the more familiar earthworm, has a well-differentiated head with sensory appendages and lateral parapodia ("side feet") with many setae.

25–34
Leeches are primarily blood suckers with digestive tracts specially adapted for storage of blood. They range in size from 1 to 30 centimeters and are found mostly in inland waters or in damp places on land, where they parasitize fish, turtles, and other vertebrates. A few species are predatory, feeding on worms and insects that they swallow whole.

Shown here is a medicinal leech, of the type commonly used for bloodletting. Bloodletting, one of the most common forms of medical treatment as late as the 1800s, was used for patients suffering from whooping cough, gout, drunkenness, rheumatism, sore throat, and asthma, among other ailments. Patients were commonly "bled to faintness," losing as much as 1.5 liters of blood.

Other Annelids

Other annelids resemble the earthworm in that they are cylindrical worms divided into a series of similar segments, and they have a complex circulatory system of blood vessels, a main ventral nerve trunk, a complete digestive tube, and a coelom. The phylum is usually divided into three principal classes: Oligochaeta, Polychaeta, and Hirudinea. Oligochaeta is the group that includes the earthworms and related freshwater species. The polychaetes, which are marine animals, differ from the earthworms and other oligochaetes in a number of ways. The most striking difference is that they typically have a variety of appendages, including tentacles, antennae, and specialized mouthparts. Each segment contains two fleshy extensions, parapodia, which function in locomotion and also, because they contain many blood vessels, are important in gas exchange. Many polychaetes live in elaborately fashioned tubes constructed in the mud or sand of the ocean bottom. Usually the sexes are separate, fertilization is external, and the larvae are free-swimming. Hirudinea are the leeches, which have flattened, often tapered, bodies with a sucker at each end. Bloodsucking leeches attach themselves to their hosts by their posterior sucker, and then, with their anterior sucker, either slit the host's skin with their sharp jaws or digest an opening through the skin by means of enzymes. Finally, they secrete a special chemical (hirudin) into the host's blood to prevent it from coagulating.

PHYLUM MOLLUSCA: MOLLUSKS

The mollusks constitute one of the largest phyla of animals, both in numbers of species and in numbers of individuals. They are characterized by soft bodies within a hard, calcium-containing shell, although in some forms the shell has been lost in the course of evolution, as in slugs and octopuses, or greatly reduced in size and internalized, as in squids. There are three major classes of mollusks: (1) the gastropods, such as the snails, whose shells are generally in one piece; (2) the bivalves, including the clams, oysters, and mussels, which have two shells joined by a hinge ligament; and (3) the cephalopods, the most active and most intelligent of the mollusks, including the cuttlefish, squids, and octopuses.

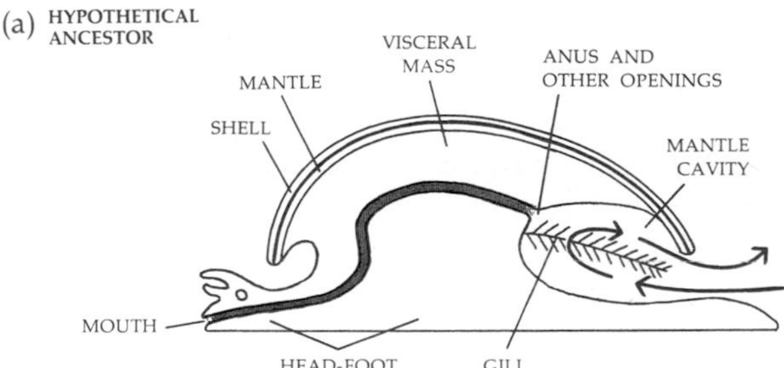

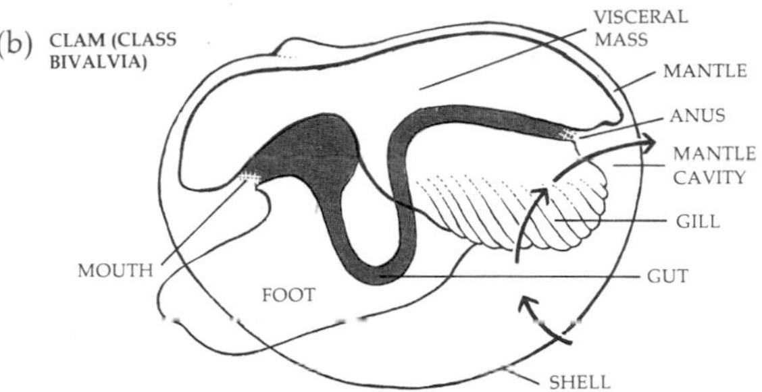

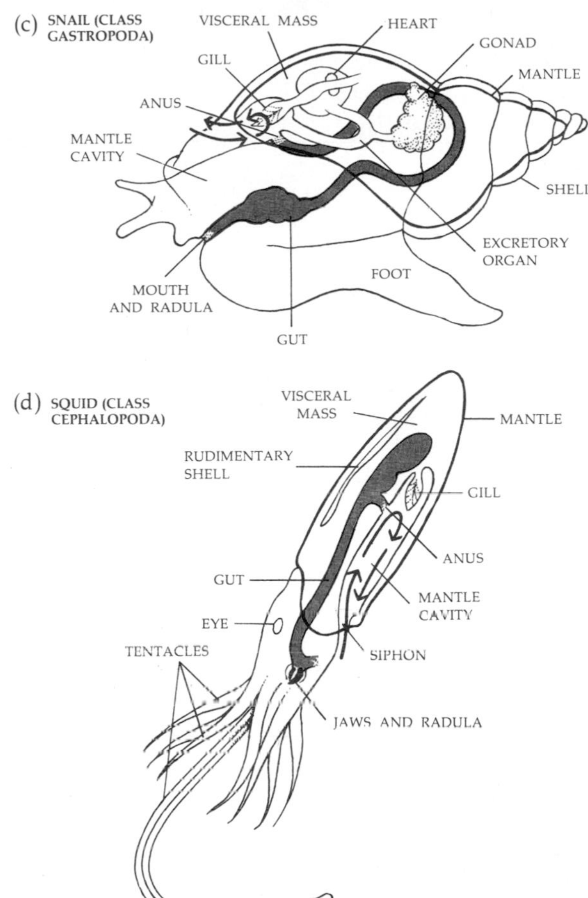

25–35
Mollusks are characterized by soft bodies composed of a head-foot, a visceral mass, and a mantle, which can secrete a shell. They breathe by gills, except for the land snails, in which the mantle cavity has been modified for air breathing. The hypothetical primitive mollusk is shown in (a). *The three major modern classes are the bivalves, such as the clam* (b), *which are generally sedentary and feed by filtering water currents, created by beating cilia, through large gills; the gastropods, exemplified by the snail* (c), *in which the visceral mass has become coiled upward and the gut turned back so that mouth, anus, and gills all share the same small aperture in the mantle; and the cephalopods, such as the squid* (d), *in which the head is modified into a circle of tentacles and part of the head-foot forms a tubelike siphon through which water can be forcibly expelled, providing for locomotion by jet propulsion. The arrows indicate the direction of water movement.*

The basic molluscan body plan is shown in Figure 25–35a. As you can see, this hypothetical animal was bilaterally symmetrical and segmented. Among modern mollusks, only the chitons, a relatively small group, bear any obvious resemblance to the archetypal model, but modern mollusks, although diverse in size and shape, all have the same fundamental body plan. There are three distinct body zones: a head-foot, which contains both the sensory and the motor organs; a visceral mass, which contains the organs of digestion, excretion, and reproduction; and a mantle, which hangs over and enfolds the visceral mass and which secretes the shell. The mantle cavity, a space between the mantle and the visceral mass, houses the gills; the digestive, excretory, and reproductive systems discharge into it. Water sweeps into the mantle cavity (propelled in the bivalves by cilia on the gills), passing through the gills and aerating them. It then passes by the nephridia, gonopores, and rectum, which are always downstream from the gills. Water leaving the mantle cavity carries excreta and, in season, gametes.

The digestive tract is far more convoluted and so provides more working surface than that of the annelids. In all mollusks, the digestive tract is extensively ciliated, with many different working areas. Food is taken up by the cells lining the digestive glands arising from the stomach and the anterior intestine, and then is passed into the blood.

25–36
A vertical section through the head of a snail to show the radula. The radula rasps and tears plant materials.

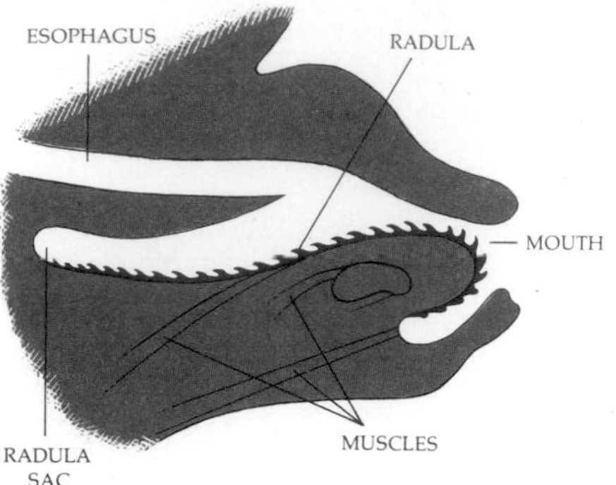

A characteristic organ of the mollusk, found only in this phylum, and in all classes except the bivalves, is the radula (Figure 25–36), a tooth-bearing strap of movable pieces of chitinous tissue covering the tongue. The radula apparatus, which operates with a rhythmic back-and-forth movement, serves both to scrape off algae and other food materials and also to convey them backward to the digestive tract. It is also used in combat.

Mollusks, as we noted previously, have gills. To understand the basic plan of gill structure and function, it is necessary only to recall the moist epidermis of the earthworm, through which oxygen diffuses, and the blood vessel lying close beneath it, which transports the oxygen to other parts of the body. A gill is a structure with an increased amount of surface area, through which gases can diffuse, and a rich blood supply for transport of these gases. Oxygen diffuses inward, along the gradient, because the cells of the animal have removed oxygen from the bloodstream by cellular respiration. Carbon dioxide, produced by cellular respiration, diffuses outward.

Mollusks have three-chambered hearts; two of the chambers (atria) collect oxygenated blood from the gills, and the third (the ventricle) pumps it to the oxygen-depleted tissue. Except for the cephalopods, mollusks have what is known as an open circulation; that is, the blood does not circulate entirely within vessels—as it does in the earthworm, for example—but is oxygenated, pumped through the heart, and released directly into spaces in the tissues from which it returns, deoxygenated, to the gills and then to the heart. Such a blood-filled space is known as a hemocoel ("blood cavity"). Cephalopods, which are extremely active animals, have accessory hearts that propel blood into the gills, and a closed circulatory system.

Class Bivalvia

In the bivalves, the two-shelled mollusks, the body has become flattened between the two shells, and "headness" has generally disappeared. The bivalves are sometimes called Pelecypoda—"hatchet foot"—because the muscular foot is often highly developed in this group. A clam, using its "hatchet foot," can dig itself into sand or mud with remarkable speed. However, the bivalves are largely sessile forms, and many of them secrete strong strands of protein by which they anchor themselves to rocks.

Most bivalves are filter-feeding herbivores; they live largely on microscopic algae. Their gills, which are large and elaborate, collect food particles. Water is circulated through the sievelike gills by the beating of gill cilia. Small organisms and particles of food are trapped in mucus on the gill surface and swept toward the mouth by the cilia; the gills also sort particles by size, rejecting sand and other larger particles. The shells are held together and opened at the hinge by a strong ligament and are drawn closed by one or two large muscles connecting the two shells.

Throughout the molluscan phylum, there is a wide range of development of the nervous system. The bivalves have three pairs of ganglia of approximately equal size—cerebral, visceral, and pedal (supplying the foot)—and two long pairs of nerve cords interconnecting them. They have statocysts, usually located near the pedal ganglia, and sensory cells for discrimination of touch, chemical changes, and light. The scallop has quite complex eyes; a single individual may have a hundred or more eyes located among the tentacles on the fringe of the mantle. The lens of this

(a)

(b)

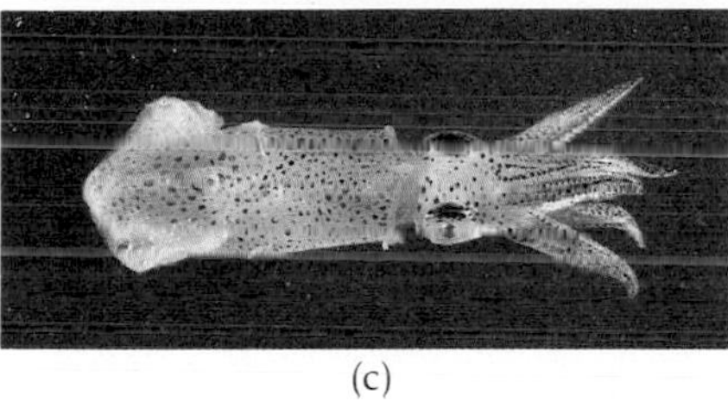

(c)

(d)

25–37
(a) *A bivalve. The blue eyes of this scallop are visible among its tentacles.* (b) *A land-dwelling gastropod. The shell, which is secreted by the mantle and grows as the soft body grows, covers and protects the visceral mass. The head contains sensory organs, including two eyes at the tips of the longer tentacles.* (c) *A squid. Jet-propelled, the squid is moving to the left. Note that its siphon is directed to point backward.* (d) Octopus macropus. *Note its well-developed eyes and the suckers on the undersurface of the arms.*

eye cannot focus on images, however, so it does not appear to serve for more than the detection of light and dark and movement.

Class Gastropoda

The gastropods, which include the snails, whelks, periwinkles, abalones, and slugs, are the largest group of mollusks. They have either a single shell or, as a secondary evolutionary development, no shell. Another feature common to all members of this group, as compared with the ancestral mollusk, is that all of them have undergone torsion. In other words, the internal organs, the shell, and the mantle have been twisted 180° so that in the modern animal, the mouth and anus and also the gills share the same comparatively small mantle cavity, now pointing forward instead of toward the rear. Third, the stomach and digestive gland have become twisted upward into a spirally coiled visceral mass. In response to this displacement and consequent crowding of the internal organs, the gill and nephridium of the right side have been lost in many species. In some close relatives of the snails, such as the slugs, the digestive tract has become straightened out again by another course of evolutionary events, in which the shell was lost but the missing organs were not regained.

Land-dwelling snails do not have gills but the area in their mantle cavities once occupied by gills is rich in blood vessels, and the snail's blood is oxygenated there.

BEHAVIOR IN THE OCTOPUS

Behaviorally, the octopus is the best studied of the cephalopods. The octopus is a sea dweller that creeps about actively on its "arms" or swims rapidly through the water by strong rhythmic muscular contractions that expel water from the mantle cavity. The octopus, like the other cephalopods, is carnivorous, living on smaller sea animals, usually crabs. It bites the crab, or other prey, with its parrotlike beak, injecting a toxin from its salivary glands. It then bundles up the paralyzed animal in its web and carries it home to eat. When it is not actively in pursuit of food, the octopus, lacking any protective shell, lives in small caves behind rocks or in reefs or wreckage.

Curious and able to use its arms with great dexterity, the octopus makes an extremely apt experimental subject. In a seawater tank, it will gather together bricks, shells, and any movable debris into a crude sort of house, and there it sits and watches, often bobbing its head up and down. Although the eye is remarkably similar to our own and equally acute, apparently the octopus does not have stereoscopic vision, and this head bobbing seems to be the way it estimates distance, fixing on an object from two points, just as surveyors triangulate a distant landmark.

The sight of a crab can so excite an octopus that its arms weave about and changes appear in its skin color. Unlike the response to threat of danger, in which the animal becomes lighter, the octopus darkens on seeing a crab, breaking out in patches of bright blue, pink, or purple, depending on the species. These skin changes are brought about by the contraction of muscles that draw out small sacs of pigment, the chromatophores, to form flat plates.

By using a system of rewards (crabs, for example) and punishments (mild electric shock), the investigator can readily teach the octopus to seize certain objects and not to seize others. Such experiments approximate what an octopus must learn in nature—that some objects are edible, some are not, and some may bite or sting. A small crab, for example, is a meal, but if it is carrying a sea anemone on its shell, it had better be left alone. Because of the comparative ease with which such tests can be performed, owing to the octopus's natural curiosity and appetite, the animal has been the subject of a great many experiments on vision, touch, and learning.

The manipulatory powers of the octopus are great, and the tentacles are very sensitive to texture and rich in chemoreceptors. However, the animal seems to have difficulty processing and coordinating sensory data. For example, if an octopus sees a crab behind a glass partition, it will rush directly toward it, flushed with excitement, and when it reaches the glass, will press itself against the pane,

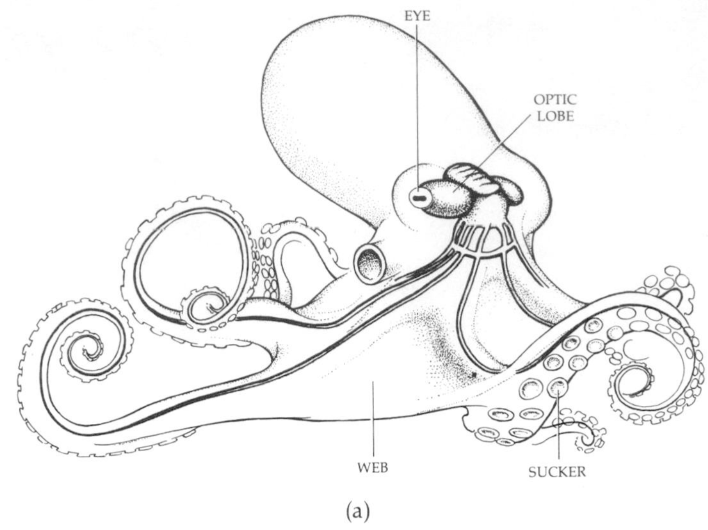

(a)

writhing its tentacles. One of the tentacles may chance over the top of the glass and reach the crab. In this case, the tentacle will close about the crab. The octopus, however, will continue to respond to the visual stimulation of the crab and press excitedly against the glass as if it were still in pursuit of the prey. Apparently the impulses received from the tactile stimulation of the arm are not integrated with those received from the eye, nor is the movement of the arm detectably influenced by the fact that the octopus can see its arm and the prey.

For similar reasons, the octopus is unable to discriminate between objects on the basis of their weight. The observer can tell which of two objects is the heavier by watching the strain in the animal's muscles, but the octopus lifting the objects cannot. As Martin Wells, of Churchill College, Cambridge, who has carried out many of the studies, points out, the problems of handling data received from eight extensible arms, each of which has several hundred suckers and can move separately and in any direction, would probably be insuperable. As we shall see in the following chapter, the severe limitations of movement provided by an articulated skeleton prove, in contrast, to be a great advantage for many types of specialized activities.

(b)

(a) *A simplified view of the brain and nervous system of the octopus. The relatively huge lobes behind the eyes are the optic lobes, concerned with collecting and analyzing data from the retina. The octopus often finds its prey by reaching its tentacles into a crevice into which it cannot see. The suckers on these tentacles are softer and more flexible than our fingertips and so can make fine distinctions among textures. They also contain chemoreceptors that can detect sugars, salts, and other chemicals in dilutions well below the range of discrimination of the human tongue.* (b) Octopus vulgaris *literally pales when confronted by an object larger than itself. It flattens and turns almost white except for a dark area around the eyes and a violet trim along the edges of its web. This makes the octopus seem larger than it is and probably serves to deter a would-be predator. The same response can be obtained in the octopus by stimulating its brain with electric currents.*

Some snails that were once land dwellers have returned to the water, but they have not regained gills. Instead, they must bob up to the surface at regular intervals to entrap a fresh bubble of air in their mantle cavities. Thus the mantle cavity has, in effect, become a lung. Moreover, as with all lungs, the opening is reduced to retard evaporation.

Gastropods, which lead a more mobile, active existence than bivalves, have a ganglionated nervous system with as many as six pairs of ganglia connected by nerve cords (Figure 25-38). There is a concentration of nerve cells at the anterior end of the animal, where the tentacles, which have chemoreceptors and touch receptors, and the eyes are located. In some of the animals, the eyes are quite highly developed in structure; they appear, however, to function largely in the detection of changes in light intensity, like the eyes of the scallop.

Class Cephalopoda

The cephalopods (the "head-foots") are the most highly developed mollusks. The large head has conspicuous eyes and a central mouth surrounded by arms, some 70 or 80 in the chambered nautilus, 10 in the squid, and 8 in the octopus. The nautilus, as the only modern shelled cephalopod, offers an indication of some of the steps by which this class disposed of the shell entirely. The animal occupies only the outermost portion of its elaborate and beautiful shell, the rest of which serves as a flotation chamber. In the squid and its relative, the cuttlefish, the shell has become an internal stiffening support, and in the octopus, it is lacking entirely.

The octopus body seldom reaches more than 30 centimeters in diameter (except on the late late show), but giant squids sometimes attain sea-monster proportions. One caught in the Atlantic some hundred years ago was 15 meters long, not counting the tentacles, and was estimated to weigh 2 tons.

Freedom from the external shell has given the mantle more flexibility. The most obvious effect of this is the jet propulsion by which cephalopods dart through the water. Usually, water taken into the mantle cavity bathes the gills and is then expelled slowly through a tube-shaped structure, the siphon; but when the cephalopod is hunting or being hunted, it can contract the mantle cavity forcibly and suddenly, thereby squirting out a sudden jet of water. Contraction of the mantle-cavity muscles usually shoots the animal backward, head last, but the squid and the octopus can turn the siphon in almost any direction they choose. In addition to the siphon, cephalopods have sacs from which they can release a dark fluid that forms a cloud, concealing their retreat and confusing their enemies. These colored fluids were at one time a chief source of commercial inks. *Sepia* is the name of the genus of cuttlefish from which a brown ink used to be obtained.

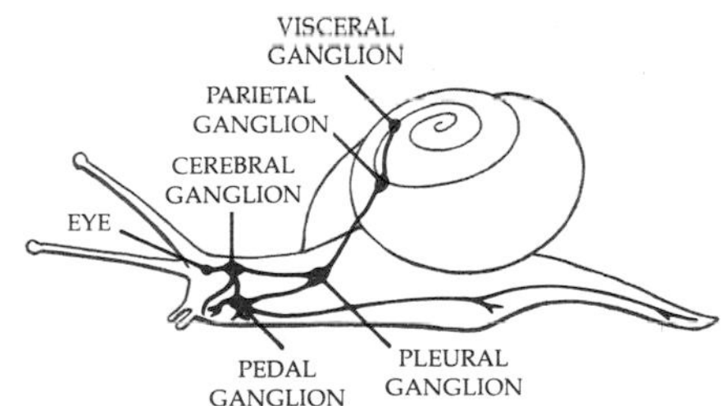

25-38
The ganglionated nervous system of a gastropod. Each dot represents a pair of ganglia, each innervating a different segment of the body. The cerebral ganglia supply the tentacles and eyes; the pleural ganglia, the mantle; the pedal ganglia, the foot muscles; and so on.

25-39
(a) The trochophore larva. Although their adult forms are very different, certain annelids and mollusks have larvae of this type. This particular larva will develop into a polychaete worm. (b) How the annelid trochophore develops into a segmented worm. The process begins with the elongation of the lower part of the trochophore. The elongated region then becomes constricted into segments, which soon develop bristles. The apical tuft disappears, and the upper part of the trochophore becomes the head. The worm will continue to grow throughout its lifetime by adding new segments just in front of its rear segment.

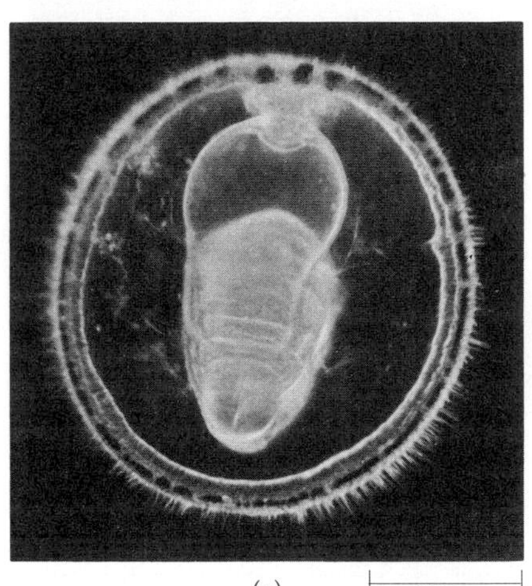

(a)

The cephalopods have well-developed brains, composed of many groups of ganglia, in keeping with their highly developed sensory systems and their lively, predatory behavior. These large brains are covered with cartilaginous cases.

The rapid responses of the cephalopods are made possible by a bundle of giant nerve fibers that control the muscles of the mantle. Many of the studies on conduction of the nerve impulse are made with the giant axon of the squid, which is large enough to permit the insertion of an electrode.

Evolutionary Affinities of the Mollusks

Although the annelids and the mollusks are quite different in their basic body plans, there are some similarities between them that seem to suggest evolutionary links. One of these is the trochophore larva (Figure 25-39). Many of the annelids (the oligochaetes and hirudineans excepted) have this very distinct larval form. Most marine mollusks (except the cephalopods) also pass through a trochophore stage in their development. Until fairly recently, the lack of unequivocal traces of segmentation in the mollusks seemed to argue against the evidence of close affinity provided by the trochophore. In the 1950s, however, 10 living specimens of *Neopilina,* a genus of mollusks previously known only from Cambrian fossils, were dredged from a deep ocean trench off the coast of Costa Rica. *Neopilina,* which is little more than 2.5 centimeters long, resembles a combination of gastropod and chiton, with a single large shell but five pairs of gills, five pairs of retractor muscles, and six pairs of nephridia, all arranged in what seems to be a distinctly segmental pattern. A third link between mollusks and annelids is the pattern of embryonic development—protostomes vs. deuterostomes—which we shall discuss further at the end of this chapter.

PHYLUM ECHINODERMATA: STARFISH

The starfish and its relatives are known as echinoderms, or "spiny skins." Adult echinoderms are radially symmetrical, like most coelenterates, although the symmetry is imperfect with some traces of bilaterality in the adults and with bilaterally symmetrical larvae.

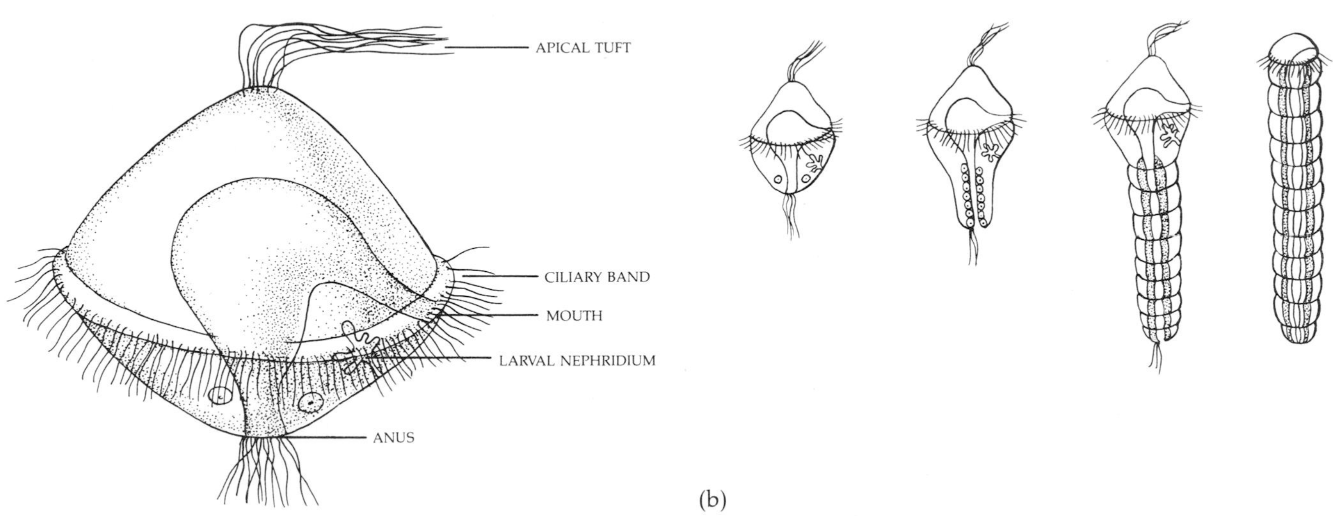

(b)

25–40
Some echinoderms: (a) *starfish,* (b) *sea urchins,* (c) *a sea cucumber,* (d) *sand dollar. Echinoderms are characterized by a calcium-containing skeleton of rods, plates, or spicules embedded just below the skin, a water vascular system of canals, and tube feet.*

Starfish

The most familiar of the echinoderms is the starfish, whose body consists of a central disk from which radiate a number of arms. Most starfish have five arms, which was the ancestral number, but some have more. A starfish has no head, and any arm may lead in its sluggish, creeping movements along the sea bottom. The central disk contains a mouth on the ventral surface, above which is the stomach. Like all echinoderms, the starfish has an interior skeleton that typically bears projecting spines, the characteristic from which the phylum derives its name. The skeleton is made up of tiny, separate calcium-containing plates held together by the skin tissues and by muscles. Each arm contains a pair of digestive glands and also a nerve cord, with an eyespot at the end. These latter are the only sensory organs, strictly speaking, of the starfish, but the epidermis contains thousands of neurosensory cells (as many as 70,000 per square millimeter) concerned with touch and chemoreception. Each arm also has its own pair of gonads, which open directly to the exterior through small pores.

The circulatory system consists of a series of channels within the coelomic cavity. Respiration is accomplished by many small fingerlike projections, the skin gills, which are protected by spines. Amoeboid cells circulate in the coelomic fluid, picking up the wastes and then escaping through the thin walls of the skin gills, where they are pinched off and ejected.

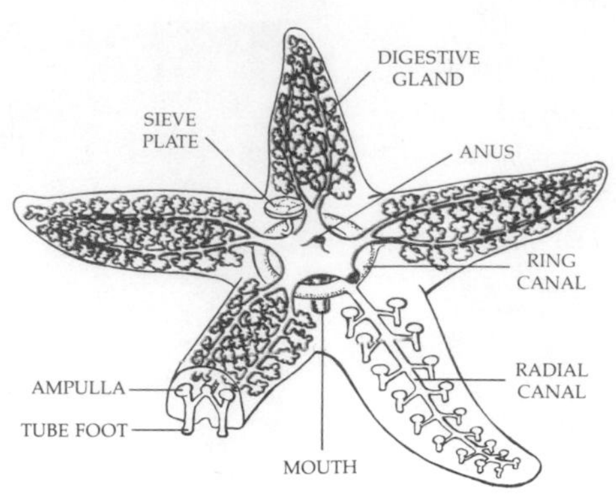

25–41
The water vascular system of the starfish is its means of locomotion. Water enters through minute openings in the sieve plate and is drawn, by ciliary action, down a tube to the ring canal. Five radial canals, one for each arm, connect the ring canal with many pairs of tube feet, which are hollow, thin-walled cylinders ending in suckers. Each tube foot connects with a rounded muscular sac, the ampulla. When the ampulla contracts, the water in it, prevented by a valve from flowing back into the radial canal, is forced under pressure into the tube foot. This stiffens the tube, making it rigid enough to walk on and extends the foot until it attaches to the substrate by its sucker. The longitudinal muscles of the foot then contract, forcing the water back into the ampulla and creating suction that holds the foot to the surface. If the tube feet are planted on a hard surface, such as a rock or a clamshell, the collection of tubes will exert enough force by suction to pull the starfish forward or to open the clam.

The water vascular, or hydraulic, system is a unique feature of the phylum. Each arm of a starfish contains two or more rows of water-filled tube feet (Figure 25–41). These tube feet are interconnected by a central ring and radial canals. Water filling the soft, hollow tubes makes them rigid enough to walk on. Each tube foot connects with a rounded muscular sac, the ampulla. When the ampulla contracts, the water is forced under pressure through a valve into the tube foot; this extends the foot, which attaches to the substrate by its sucker. When the muscles at the base of the tube feet contract, the animal is pulled forward.

If the tube feet are planted on a hard surface, such as a rock or a clam shell, the collection of tubes will exert enough suction to pull the starfish forward or to pull apart a bivalve mollusk, a feat that will be appreciated by anyone who has ever tried to open an oyster or a clam. When attacking bivalves, which are its staple diet, the starfish everts its stomach through its mouth opening and then squeezes the stomach tissue through the minute space that the starfish has made between the bivalve shells. The stomach tissues can insinuate themselves through a slit as narrow as 0.1 millimeter and begin to digest the soft tissue of the prey.

Echinoderm Evolution

The echinoderms are believed to have evolved from an ancestral, bilateral, mobile form that settled down to a sessile life, and then became radially symmetrical. The sea lilies represent this second hypothetical stage. In the third evolutionary stage, some of the animals, as represented by the starfish and sea urchins, became mobile again. Following this line of reasoning, one might expect an eventual return to bilateral symmetry in this group, and, in fact, this is seen to some extent in the soft, elongated bodies of sea cucumbers.

The ancestral bilateral form, like most other hypothetical ancestors, was wormlike. It had a coelom and a one-way digestive tract. However, it differed from the ancestor of the mollusks and annelids in what zoologists consider a very fundamental way, its early embryonic development. Among mollusks, annelids, and also arthropods, the early cell divisions of the zygote are spiral, occurring in a plane oblique to the long axis of the egg. In the echinoderms, the cleavage pattern is radial, parallel to and at right angles to the axis of the egg (Figure 25–42). The second difference appears when the embryo becomes a hollow sphere of cells. In the embryos of both groups, an opening, the blastopore, appears. Among the

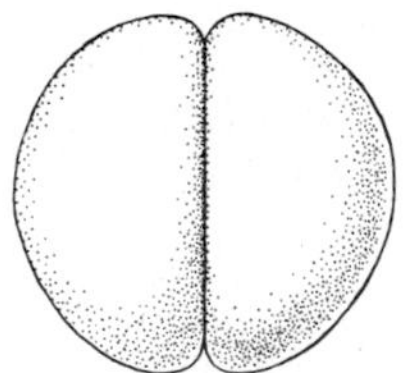

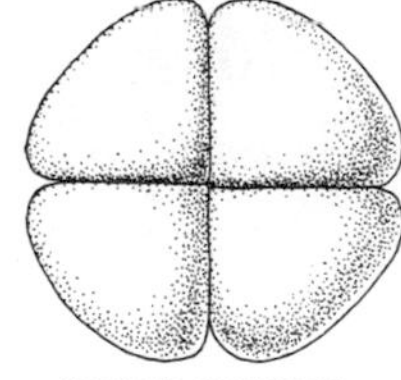

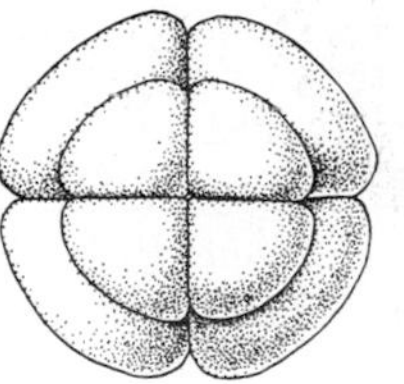

25–42
Egg cleavages, showing radial and spiral patterns at the third cleavage. Radial cleavage occurs in deuterostomes (echinoderms and chordates); spiral cleavage occurs in protostomes (mollusks, annelids, and arthropods).

mollusks, annelids, and arthropods, the mouth (stoma) of the animal develops at or near the blastopore, and this group is called the protostomes—"first the mouth." In the echinoderms, which are called deuterostomes, the anus forms at or near the blastopore and the mouth forms secondarily. The chordates, the phylum to which we vertebrates belong, share these characteristics with the echinoderms.

SUMMARY

Sponges, phylum Porifera, are composed of a number of different cell types, including choanocytes, or collar cells, which are the feeding cells of the sponge; epithelial cells, some of which are contractile; and amoebocytes, largely undifferentiated cells from which other cell types, such as eggs and sperm, may arise. In the sponge, there is little coordination among the various cells. The animal, although it may grow very large, is limited nutritionally to very small particles of a size that can be ingested by the choanocytes.

The distinctive features of the coelenterates, phylum Coelenterata, are (1) a two-layered body, in which the two layers, the ectoderm and the endoderm, are divided by a jellylike substance, the mesoglea; (2) the coelenteron, a cavity in which the food can be partially digested extracellularly; and (3) the cnidoblasts, special stinging cells found only rarely outside this phylum. The coelenterates may take the form of either the polyp or the medusa; in many species, both forms are seen in the course of each life cycle. *Hydra,* a common freshwater coelenterate, in which only the polyp form occurs, has a variety of sensory cells and a nerve net that coordinates the movements of the animal. *Obelia* is an example of a coelenterate in which the polyp form predominates. The polyps, which are joined, live in colonies in which some of the individuals feed and others are reproductive forms, giving rise to medusas that produce gametes. In *Aurelia,* the medusa form predominates. In addition to special sensory cells, two specialized sense organs, statocysts and ocelli, are found in this phylum.

The flatworms, phylum Platyhelminthes, are bilaterally symmetrical and elongated, with a distinct "headness" and "tailness" and a concomitant clustering of nerve cells in the anterior region. The flatworms have three distinct tissue layers—ectoderm, mesoderm, and endoderm—and also an excretory system, involving flame cells, that serves largely to maintain water balance.

Table 25-1 *Characteristics of Some Major Invertebrate Phyla*

PHYLUM	NO. OF SPECIES	BODY PLAN: CELL LAYERS	SYMMETRY	BODY CAVITIES	SEGMENTATION	ANUS	CHARACTERISTIC FEATURES	HYPOTHETICAL ANCESTOR
Porifera (sponges)	5,000	2, 3, or aggregate	None or radial	None	–	–	Unique body plan, choanocytes	Ciliated (choano-protozoan flagellates)
Coelenterata (jellyfish, corals, sea anemones)	9,000	2	Radial	None	–	–	Cnidoblasts, with nematocysts, mesoglea	Ciliated protozoan
Platyhelminthes (flatworms)	15,000	3	Bilateral	None	–	–	Flame cells	Ciliated protozoan, planula larva
Rhynchocoela (ribbon worms)	600	3	Bilateral	Circulatory, rhynchocoel	–	+	Retractile proboscis	Primitive flatworms
Nematoda (roundworms)	80,000	3	Bilateral	Pseudocoelom	–	+	Only longitudinal muscle, periodically molted cuticle	Primitive flatworms
Annelida (segmented worms)	8,800	3	Bilateral	Circulatory, coelom	+	+	Setae, segmentation	Segmented coelomate worm (protostomes)
Mollusca (snails, clams, octopuses)	128,000	3	Bilateral	Hemocoel, coelom	In some classes, perhaps	+	Body divided into head-foot, visceral mass, shell-secreting mantle, radula	Primitive annelid or flatworm
Echinodermata (starfish, sea urchins, sea lilies, sea cucumbers)	6,000	3	Larvae bilateral; adults secondarily pentaradial	Compartmentalized coelom, including water vascular system	–	+	Tube feet, internal dermal skeleton, pentaradial symmetry	Segmented coelomate worm (deuterostomes)
Arthropoda	765,000	3	Bilateral	Coelom	+	+	Exoskeleton	Primitive annelid
Chordata	45,000	3	Bilateral	Coelom	+	+	Endoskeleton	Segmented coelomate worm (deuterostomes)

The ribbon worms, phylum Rhynchocoela, are characterized by a retractile, prey-seizing proboscis, a one-way (mouth-to-anus) digestive tract, and a simple circulatory system.

The roundworms, or nematodes, phylum Nematoda, of which there are, conservatively, 80,000 species, are pseudocoelomates. A pseudocoelom is a cavity between the endoderm and the mesoderm.

Members of phylum Annelida—the earthworm is an example—have segmented bodies; coeloms, which are cavities within the mesoderm and within which internal organs are supported by mesenteries; one-way digestive tracts; and closed circulatory systems, often with contractile vessels. Excretion is accomplished by special organs, the nephridia, which are convoluted tubules. The nephridia collect fluids from the coelom and exchange salts and other substances with the body fluids as the urine passes along the tubules for excretion. The earthworm has a relatively complex nervous system, consisting of a ventral nerve cord, which divides to encircle the pharynx at the anterior end of the animal, and pairs of ganglia (clusters of nerve cells), one pair to each segment, which receive sensory impulses and trigger motor activities in each segment.

There are three main classes of phylum Mollusca: the bivalves (oysters and clams), the gastropods (snails), and the cephalopods (octopuses and their relatives). In each of these classes, the basic body plan is the same, but it has been modified in the course of adaptation to a particular environment. The body is always divided into a head-foot, visceral mass, and mantle. In most mollusks, respiration is carried out by means of a gill, a thin-walled structure that is an extension of the epidermis. It is richly endowed with blood vessels that serve as an area of gaseous exchange (respiration). Mollusks are also characterized by an efficient three-chambered heart and a toothed tongue, the radula. Shells are often present but may be reduced or absent. Nervous systems and behavior vary among the species, reaching a zenith of complexity in the brainy octopus.

Phylum Echinodermata includes the starfish, the sea urchins, and the sea lilies, all of which are radially symmetrical in their adult forms. An unusual characteristic of this phylum is the water vascular system, which provides a hydraulic skeleton for the animal and also provides suction for the clinging and pulling activities associated with moving and predatory behavior.

QUESTIONS

1. Distinguish among the following: coelenteron/coelom/pseudocoelom/hemocoel.

2. Compare the nervous systems of a hydra, a planarian, a clam, an annelid, and an octopus.

3. A coelom provides hydraulic support for an animal. Why does not a coelenteron fill this same function?

4. Describe respiration in a coelenterate, an earthworm, and an aquatic snail.

5. A larval or medusoid stage occurs in the life histories of many marine invertebrates. Yet in group after group of freshwater and terrestrial invertebrates, each supposedly separately evolved from marine ancestors, the free-living larva or medusa has been lost. What might be the selective factor underlying the repeated loss?

6. Nematodes have only longitudinal muscles in their body walls but have very high internal fluid pressures; earthworms, with both longitudinal and circular muscles, have low fluid pressures. Can you suggest a mechanism by which internal fluid at high pressure can circumvent the need for certain muscles?

7. In which phylum do you find each of the following and what is its function: cnidoblast, radula, flame cell, siphon, tube foot?

8. Many mollusks have lost or may be in the evolutionary process of losing their shells. What are the advantages to an organism of having a shell? Of losing one?

9. Smallness may also be an advantage to an organism. Name some advantages of smallness.

10. Label the drawing below.

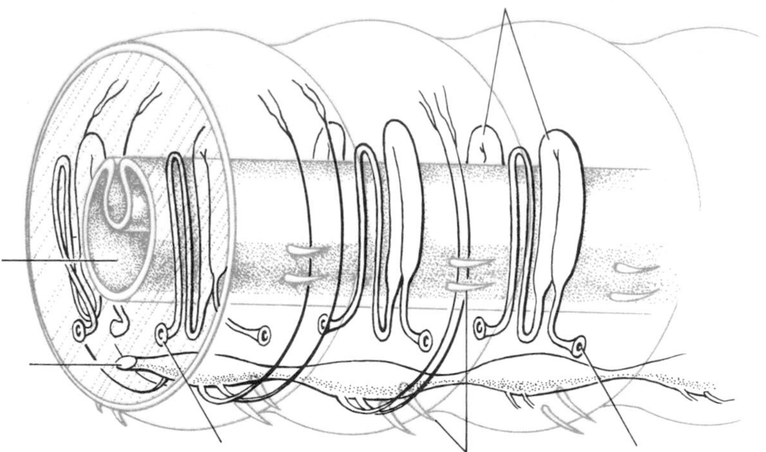

CHAPTER 26

The Animal Kingdom II: Arthropods and Chordates

26–1
Arthropods are characterized by a variety of specialized appendages. The antennae of this male Cecropia moth, for example, are receptors for alluring chemicals released by the females.

This chapter will be concerned chiefly with the arthropods, the largest by far of all the phyla. Almost 1 million species of insects and other arthropods have been classified to date, and estimates of the total number are as high as 10 million. As to the number of individuals, it has been calculated that, of insects alone, as many as 10^{18}—a billion billion—are alive at any one time. It has been estimated that over every square kilometer in the temperate zone there are, at certain seasons, some 20 million individual arthropods, layered in the atmosphere, like plankton.

At the end of this chapter, we shall discuss the evolution of the Chordata, a much smaller phylum. This latter discussion will serve as an introduction to Section 6, which deals principally with the human species and other chordates.

PHYLUM ARTHROPODA

The arthropods, "joint-footed" animals, include such creatures as the crustaceans (crabs and lobsters), arachnids (spiders, ticks, mites, and scorpions), and insects; the latter constitute by far the largest group. Despite the huge size and diversity of the arthropod phylum, there are a number of features shared by all the members of this group.

CHARACTERISTICS OF THE ARTHROPODS

The Exoskeleton

First, all arthropods have an articulated (jointed) exoskeleton. This exoskeleton, or cuticle, is secreted by the underlying epidermis and is attached to it; it is made up of an outer waxy layer, composed of lipoprotein, and a middle horny layer and an inner flexible one, both composed principally of chitin. The exoskeleton not only covers the surface of the animal but also extends inward at both ends of the digestive tract and, in insects, lines the tracheae (breathing tubes) as well. Muscles are attached to the various segments of the exoskeleton, just as they are attached to the various bones of the endoskeleton in vertebrates. When the muscles contract, the exoskeleton moves at its joints, rather than restricting the animal to moving inside of it, as with the shelled mollusks. The cuticle may form a veritable coat of

26–2
Some arthropod characteristics. (a) *A many-jointed skeleton, as in this South American katydid. The slits in each foreleg are the insect's ears.* (b) *Molting. An exoskeleton, once formed, does not grow and so must be periodically discarded. The old cuticle is at the top, and below is the newly emerged "soft-shelled" blue crab. The new exoskeleton, already formed, has expanded and hardened.* (c) *Segmentation. Arthropods are segmented, like the annelids from which they presumably arose. However, unlike annelids, adult arthropods, such as the millipede shown here, have rigid, jointed exoskeletons and appendages.*

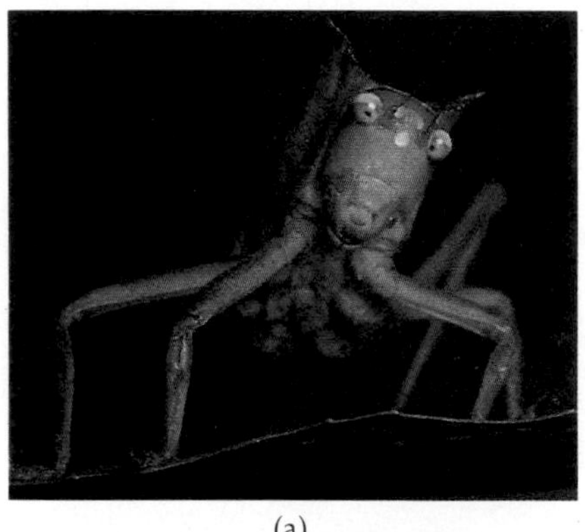

(a)

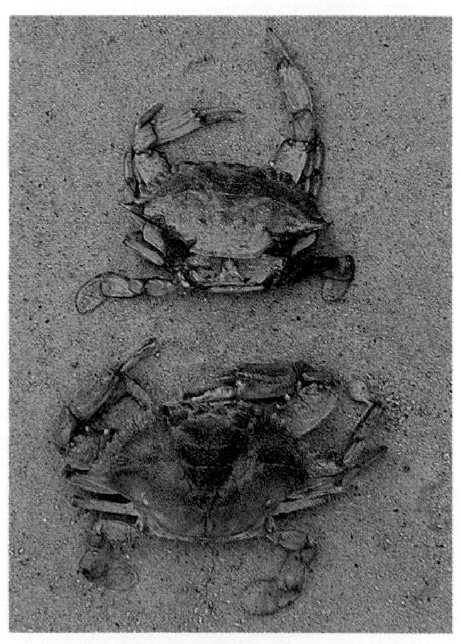

(b)

armor, as it does in beetles and in some of the crustaceans (in which it is often infiltrated with calcium salts), but at the joints it is flexible and thin, permitting free movement. It serves as protection against predators, and it is waterproof, keeping exterior water out and interior water in. It is used for food grinders in the foregut, for wings, and for tactile hairs. Cuticle even forms the lens of the arthropod eye.

The exoskeleton has certain disadvantages. It does not grow (as the bony vertebrate endoskeleton does) and so it must be discarded and re-formed many times as the animal grows and develops. Molting is dangerous; the newly molted animal is soft and hence particularly vulnerable to predators and, in the case of terrestrial forms, subject to water loss. Many arthropods go into hiding until their new cuticle has hardened. Molting is also costly in terms of metabolic expenditures (although a number of insects and some freshwater crustaceans limit their losses by thriftily eating the old exoskeleton).

The fact that the exoskeleton is waterproof made possible the evolution of terrestrial forms among the arthropods. The arthropods are the only major invertebrate phylum with many species adapted to life on land in nonmoist habitats.

Other Arthropod Characteristics

All arthropods are segmented, a characteristic that suggests a common ancestry with the annelids. In some of the arthropods the segments have become fused, forming a head, a thorax (sometimes fused with the head to form a cephalothorax), and an abdomen. But the basic segmented pattern is often still clearly evident in the immature stages (witness the caterpillar) and can be discerned in the adult by examination of the appendages, the musculature, and the nervous system.

Arthropods have a large number of jointed appendages. Especially among the insects and crustaceans, these appendages constitute a kit of highly specialized tools, including jaws, gills, poison fangs, tongs, egg depositors, sucking tubes, claws, antennae, paddles, and pincers.

The insects and some other terrestrial forms have unusual means of respiration, consisting of a system of cuticle-lined air ducts (tracheae) that pipe air directly into various parts of the body. Air flow is regulated by the opening and closing of special pores (spiracles) on the exoskeleton. Terrestrial arthropods, such as spiders,

(c)

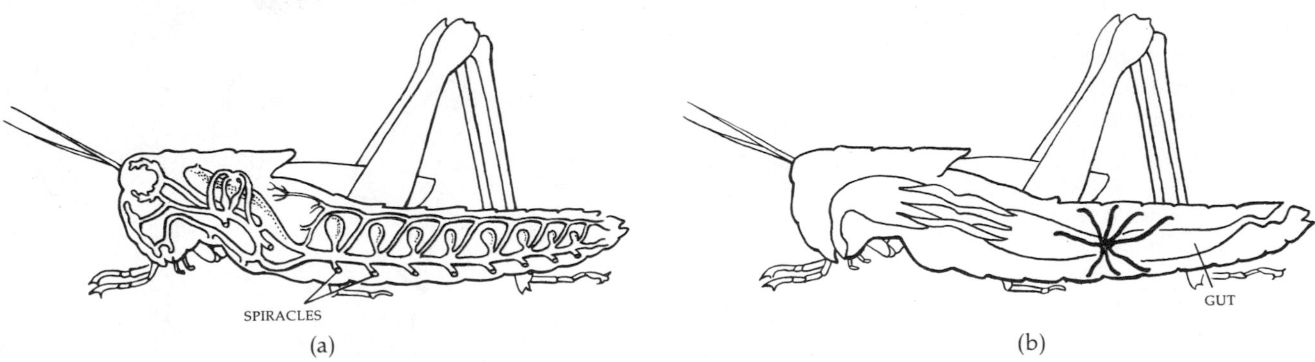

26–3
(a) *Respiration by means of a system of internal tubes (called tracheae) is found only among arthropods. Tracheae are usually branched, like those shown here, and open to the outside by spiracles that may be closed to conserve water. The tubes are lined with spiral rings and cuticle, which keeps them open. A tracheal system is one of the most efficient respiratory systems in the animal kingdom because it delivers oxygen directly to the cells.* (b) *Malpighian tubules represent another exclusively arthropod characteristic, although like tracheae, they are not found in all classes. These tubules collect water and nitrogenous wastes from the hemocoel and empty them into the gut. The wastes (in the form of uric acid) are excreted with the feces.*

that do not have tracheae have book lungs, structures that are also unique to this phylum. Excretion in terrestrial forms is by means of tubes (called Malpighian tubules) attached to and emptying into the midgut or hindgut. They absorb wastes from the body cavities. Respiration by tracheae, book lungs, or book gills and excretion by Malpighian tubules are found only in arthropods (although not all arthropods have these features).

Arthropods also share a number of characteristics, besides segmentation, with the annelids. Like them, they are bilaterally symmetrical, three-layered, and have a tubular mouth-to-anus gut. They have a coelom, as do the annelids, but the arthropod coelom is markedly reduced, consisting only of the cavities of the gonads and the nephridia. (As you will recall, the coelom serves as a hydrostatic skeleton in annelids, but the arthropods with their exoskeletons require less internal stiffening.)

The arthropods, like the bivalve mollusks, have an open circulatory system in which blood flows through free spaces among the tissues—the hemocoel—as well as through vessels. Blood returns from the hemocoel to the tubular heart through special valved openings. An open blood system creates little in the way of turgor pressure, which accounts for the general squashiness of the internal organs of insects compared with the firmness, for example, of the vertebrate kidney.

The premise of an evolutionary relationship between the annelids and the arthropods is strengthened by the existence of a group of caterpillar-like worms called Onychophora. These worms have nephridia and reproductive tracts that resemble those of annelids but a heart and hemocoel and jawlike appendages resembling those of arthropods. The body is covered with a thin cuticle of chitin; however, there are no joints in the cuticle, which is soft enough so that it bends as the animal moves.

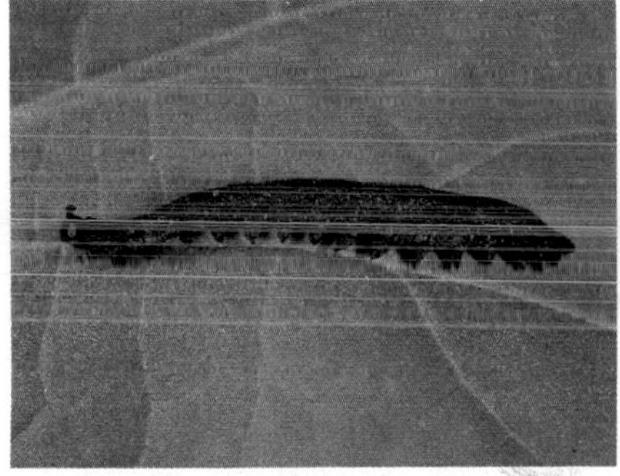

26–4
Peripatus, *an onychophoran, found living under a log in a rain forest in South America.*

The Arthropod Nervous System

Arthropods have a ladderlike nervous system, with a double chain of segmental ganglia running along the ventral surface. The double chains part to encircle the esophagus, ending in a pair of dorsal ganglia. The dorsal ganglia are enough larger than the other ganglia to be called a brain. However, many arthropod activities are controlled at the segmental level, as in annelids. For example, members of a number of species can move, eat, and carry on other functions normally, even after the brain is removed. In fact, in the arthropods generally, the brain appears to act not so much as a stimulator of the action of the animal but as an inhibitor, as in the earthworm. The grasshopper, for example, can walk, jump, or fly with its brain

26–5
Arthropod nervous systems: a bee, a crayfish, and an ant. The brain is a large dorsal pair of ganglia at the anterior end of a double chain of ganglia. The chains are interconnected by two bundles of nerve fibers running along the ventral surface. Because of the arthropod's "ladder" type of nervous system, many arthropod activities are controlled at a local level, and a number of species can carry on a few of their normal functions after the brain has been removed.

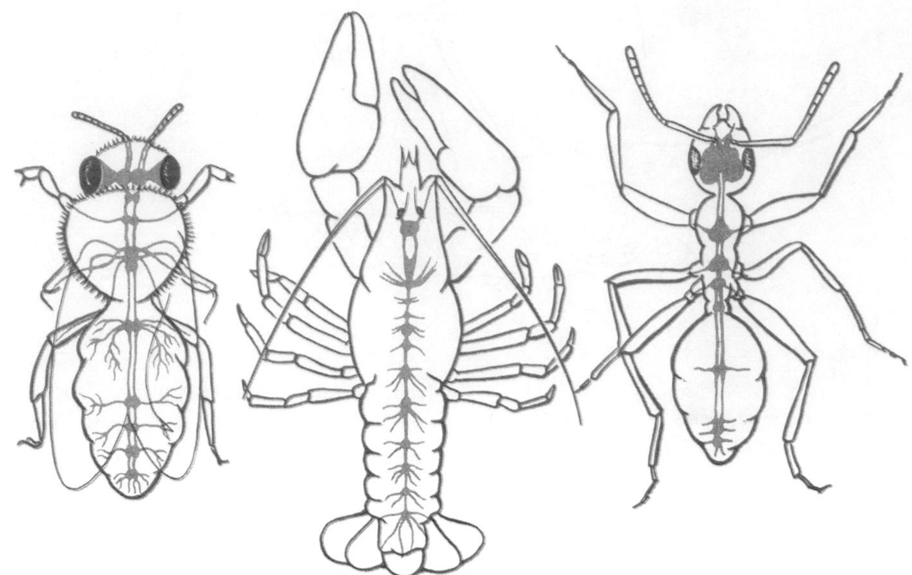

removed; indeed, the brainless grasshopper responds to the slightest stimulus by jumping or flying. The extreme consequences of this releasing of inhibition can be seen in the praying mantis. Mantises are carnivorous and cannibalistic, and the female, being larger than the male, frequently captures her mate and, grasping him with her forelegs, begins to eat him, head first. This decapitation results in the release of strong motor activities by which the headless male struggles loose from the grasp of the female, mounts her, and mates with her. The headless male is more likely to copulate than an intact male; investigators seeking to breed mantises have found that males of strains that do not mate readily in captivity will often do so after decapitation.

26–6
A female mantis will often decapitate a courting male. As with other arthropods, loss of the brain removes inhibitions to certain activities, and a headless male is more likely to copulate than an intact one. After mating, she may leisurely complete her meal.

SUBDIVISIONS OF THE PHYLUM

The arthropods are often classified in two subphyla, the Mandibulata and the Chelicerata, which can be clearly distinguished by even an inexperienced eye. There are conspicuous differences in the appendages. The most anterior appendages in the mandibulates (insects, crustaceans) are antennae, one or two pairs, and the next are mandibles (jaws). The chelicerates, which include horseshoe crabs, sea spiders, and arachnids (spiders, mites, scorpions, and their relatives), have no antennae and no true mandibles. Their first pair of appendages consist of chelicerae (singular, chelicera), which usually take the form of pincers or fangs. In spiders, ducts from a pair of poison glands lead through the chelicerae, which are sharp and pointed and are used for biting and paralyzing prey. Chelicerates may have book lungs or book gills; these structures, which are not present in mandibulates, derive their name from their resemblance to the leaves of a partially opened book.

Each of these subphyla are, in turn, divided into several classes. The three largest classes are the crustaceans, the insects (both mandibulates), and the arachnids (chelicerates). Smaller classes include the horseshoe crab and sea spiders (xiphosurans), which are chelicerates, and the centipedes and millipedes, which are mandibulates (Figure 26–2c).

Class Crustacea

The crustaceans include crabs, crayfish, lobsters, barnacles, shrimp, prawns, *Daphnia* (water fleas), and a number of smaller forms found largely in freshwater and marine plankton, as well as some terrestrial forms such as the familiar pillbugs or sowbugs. Crustaceans are mandibulates. They differ from the insects, which are also mandibulates, in that they have legs or leglike appendages on the abdomen as well as the thorax, and have two pairs of antennae as compared to the insects' one.

Lobster

Figure 26–8 shows the structure of a lobster, a representative crustacean. A crayfish, which is a freshwater form, differs morphologically from the lobster only in minor respects.

The lobster has 20 segments, the first 13 of which are united on the dorsal side in a combined head and thorax, or cephalothorax. A heavy shield, or carapace, arises from the head and covers the thorax. Carapaces are common among crustaceans. The abdomen consists of six distinct segments. (To lobster eaters, the abdomen is the "tail," but to biologists, the designation "tail" is usually reserved for areas of the body posterior to the anus.)

The various appendages have special functions. The antennae, of which there are two pairs, are sensory. The mandibles, or jaws, are used for crushing food; like all arthropod jaws, they move laterally, opening and closing from side to side like a pair of ice tongs rather than up and down like the jaws of vertebrates. The maxillae and the maxillipeds serve chiefly to collect food, mince it, and pass it on to the mouth. The claws are unequal in size in the full-grown lobster; the larger claw is used for defense and for crushing food, and the smaller claw, which has the sharper teeth, seizes and tears the prey. The first two pairs of walking legs have

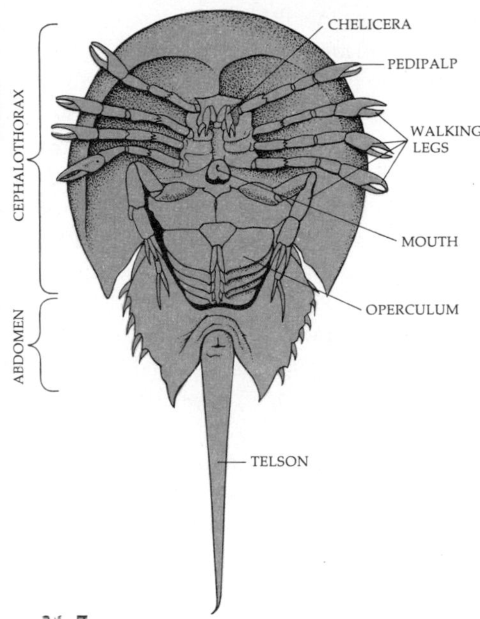

26–7
Because the body of the horseshoe crab is largely covered with a heavy shield, or carapace, its segmented body plan is only evident from the underside. The operculum is a flat, movable plate that covers and protects the book gills dorsal to it. Horseshoe crabs are often called living fossils because their remarkably similar ancestors date back to the Cambrian period. There are now only five living species.

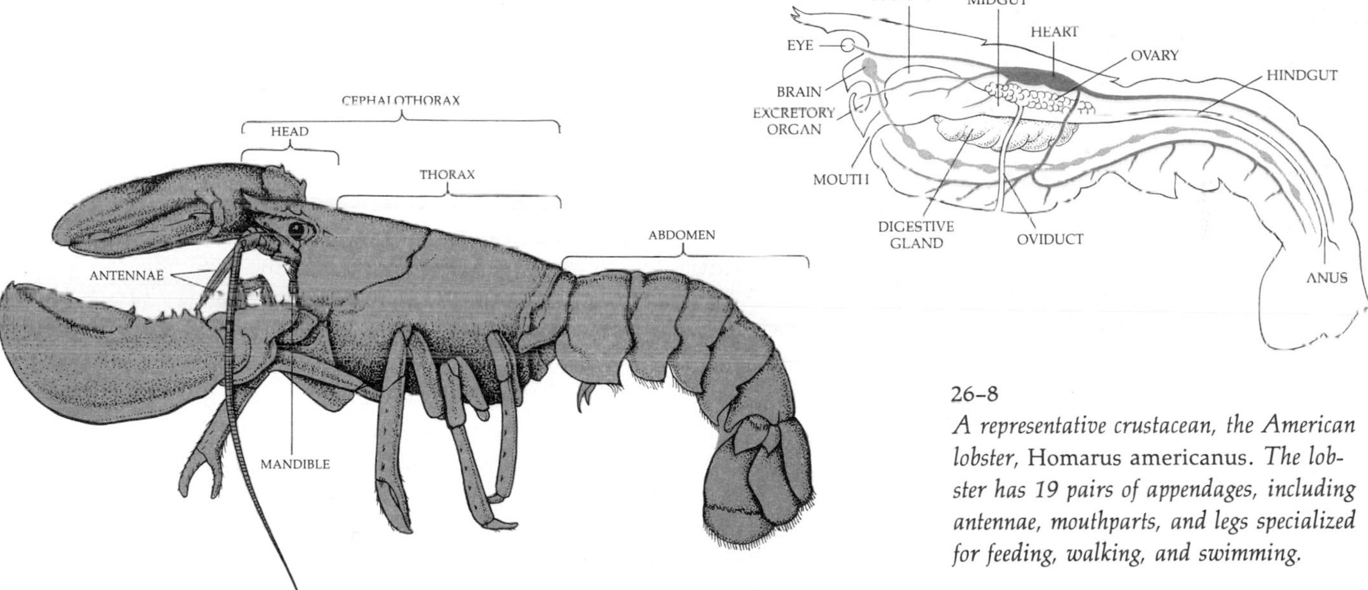

26–8
A representative crustacean, the American lobster, Homarus americanus. *The lobster has 19 pairs of appendages, including antennae, mouthparts, and legs specialized for feeding, walking, and swimming.*

(a)

(b)

(c)

26-9
Some crustaceans. (a) *Hermit crab, which houses itself in the abandoned shells of mollusks or other crustaceans;* (b) *pillbugs (also known as sowbugs or wood lice), terrestrial crustaceans found in damp places;* (c) *a painted lady shrimp.*

small pincers, which can also seize prey. The last pair of walking legs also serves to clean the abdominal appendages. The flattened posterior appendages are used, like flippers, for swimming. The claws and abdomen are filled almost entirely with the large striated muscles many of us enjoy eating. These muscles are extremely powerful. A lobster can snap its "tail" ventrally with enough force to shoot backward through the water, and a large lobster can shatter the shell of a clam or oyster with its crushing claw. Any one of its appendages can be regenerated over a series of molts if lost, and a lobster will often drop off (automize) a claw or leg that is held by a predator, in order to escape. Severely damaged legs are also automized, reducing blood loss.

The anterior and posterior regions of the digestive tract, the foregut and hindgut, are lined with cuticle. As a consequence, most of the food is absorbed through the midgut and through the cells of the large digestive gland, an appendage of the midgut.

Respiration is accomplished by the flow of water over 20 pairs of feathery gills attached to the bases of the legs. A pair of nephridia is located in the head; wastes extracted from the blood are collected into a bladder and excreted from a pore at the base of each of the second antennae. Fertilization is external; the sperm, which are deposited near the female gonopores, fertilize the eggs as they are laid, and the fertilized eggs cling by means of a sticky secretion to the swimmerets of the female until they hatch.

Terrestrial Crustaceans

Unlike other arthropod groups, almost all crustaceans are aquatic; some crabs, however, are amphibious or terrestrial. Amphibious crabs continue to breathe with gills, carrying around water in their thoracic cavities with which to keep the gills wet and aerating the water through holes in their exoskeletons. The true land crab has lost the gill structures and instead has an area of highly vascularized epithelial tissue through which oxygen is exchanged. The land snail, you will recall, solved the respiratory problems involved in the transition from water to land in an analogous way.

Class Arachnida

The arachnids, which include spiders, ticks, mites, scorpions, and daddy longlegs, are chelicerates. In this class, the first pair of appendages, the chelicerae, which are usually sharp and fanglike, are used for biting and paralyzing prey by the injection of a poison. The second pair are the pedipalps, which, in scorpions, for example, are used for handling and tearing food. Male spiders also use pedipalps to transfer semen to the female. Arachnids have four pairs of walking legs.

Spiders, like most arachnids, live on a completely liquid diet. All are predatory. The prey is bitten and often paralyzed by the chelicerae, and then enzymes from the midgut are poured out over the torn tissues to produce a partially digested broth. The liquefied tissues of the prey are pumped into the stomach by the muscular pharynx, where digestion is completed and the juices absorbed. Arachnids respire by means of tracheae or book lungs, or both. Book lungs are a series of leaflike plates within a chitin-lined chamber into which air is drawn and expelled by muscular action.

On the posterior portion of the spider's abdominal surface is a cluster of spinnerets, fingerlike organs from which a fluid protein exudes that polymerizes into

26–10

A spider, an arachnid. Ducts from the poison gland open at or near the tips of the chelicerae. The flow of poison is voluntarily controlled by the spider. Only a few spiders are dangerous to human beings; perhaps the most dangerous are members of the species shown here, the black widow.

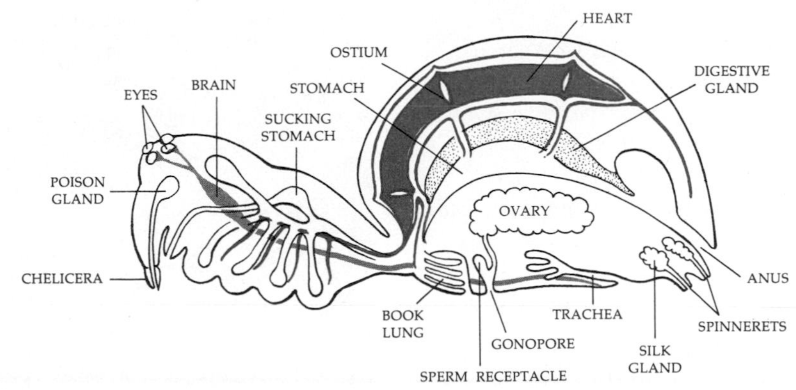

(a)

(b)

(c)

(d)

26–11

Some arachnids. (a) *A scorpion belonging to the genus* Centruroides, *a highly toxic genus. Scorpions sting in self-defense, releasing a venom that attacks the nervous systems of their victims.* (b) *A jumping spider from the Congo. These spiders are among the few with well-developed sight (note the large eyes). They stalk their prey, sometimes over considerable distances.* (c) *An adult red velvet mite of the genus* Trombidium. (d) *A banded garden spider* (Argiope trifasciata) *has caught a grasshopper in its web. The prey, immobilized by venom, is being wrapped in silk. Webs, which may be as much as a meter in diameter, are reconstructed daily.*

silk as it is exposed to air. Silk is used not only for the variety of prey-snaring webs made by the different species but for a number of other purposes as well, such as a drop line, on which the spider can make a defensive dive, draglines for marking a course, gossamer threads for ballooning, hinges for trap doors, an egg case, lining for a burrow, the shroud of a victim, or a wrapping for an edible offering presented to the female of certain species by the courting male. Most spiders can spin several kinds and thicknesses of silk. Webs are species-specific, and web-building is a genetically programmed behavior.

(a)

(b)

26–12
The late J. B. S. Haldane, noted for his agnosticism and crusty disposition, as well as for his scientific achievements, was once asked what his study of biology had revealed to him about the mind of God. "Madame," he replied, "only that He had an inordinate fondness for beetles." Shown here is everybody's favorite beetle, a ladybug (a), *accompanied by a stag beetle* (b), *so called because of the antlerlike mandibles, characteristic of the pugnacious males.*

Class Insecta

The insects constitute the largest class, by far, of the arthropods. In fact, there are more species of insects than of all other animals combined. More than 700,000 are known.

Insects are the only invertebrates capable of flight. When insects began to fly—more than 240 million years ago—they were able to move into and exploit a life zone almost totally unoccupied by any other form of animal life. In terms both of numbers of species and of individuals, they are the dominant organisms of this planet.

Insect Characteristics

Figure 26–13 shows a grasshopper. Here you can see many of the characteristic features of insects: three main divisions—the head, the thorax, and the abdomen; three pairs of legs; one pair of antennae; and a set of mouthparts similar to those of the lobster. In the more primitive insects, such as the grasshopper, the mouthparts are used for handling and masticating food, but in the more highly specialized groups, the mouthparts are often modified into sucking, piercing, slicing, or sponging organs. Some are exquisitely adapted to draw in nectar from the deep, tubular nectaries of specialized flowers.

Most adult insects have two pairs of wings made up of light, strong sheets of chitin; the veins in the wings are chitinous tubules that serve primarily for strengthening. In some entire orders, such as fleas and lice, wings are totally lost, returning the insect to the condition of its wingless ancestors. Other species may have short, nonfunctional wings in one or both sexes.

There are more than 20 orders of insects, of which we shall briefly describe four examples: Diptera, Lepidoptera, Hymenoptera, and Coleoptera. The Diptera ("two-winged") include the familiar flies, gnats, and mosquitoes. The Lepidoptera ("scale wings") are the moths and the butterflies. Hymenoptera ("membrane-winged") include ants, wasps, and bees, many species of which are social. The Coleoptera ("shield-winged") are the beetles, most of which have a pair of hard protective forewings, which pivot forward out of the way during flight, and a pair of membranous hindwings used for flying. Of the more than 700,000 classified insects, at least 275,000 species are beetles.

26–13
In the grasshopper, an insect, the head consists of six fused segments that have appendages specialized for tasting and biting. Each of the three segments of the thorax carries a pair of legs (three pairs in all), and two of them carry wings (in the grasshopper, the forewings are hardened as protective covers). The spiracles in the abdomen open into a network of chitin-lined tubules through which air circulates to various tissues of the body. This sort of tubular breathing system is found only among insects and some other land-dwelling arthropods. Excretion takes place by Malpighian tubules that empty into the hindgut.

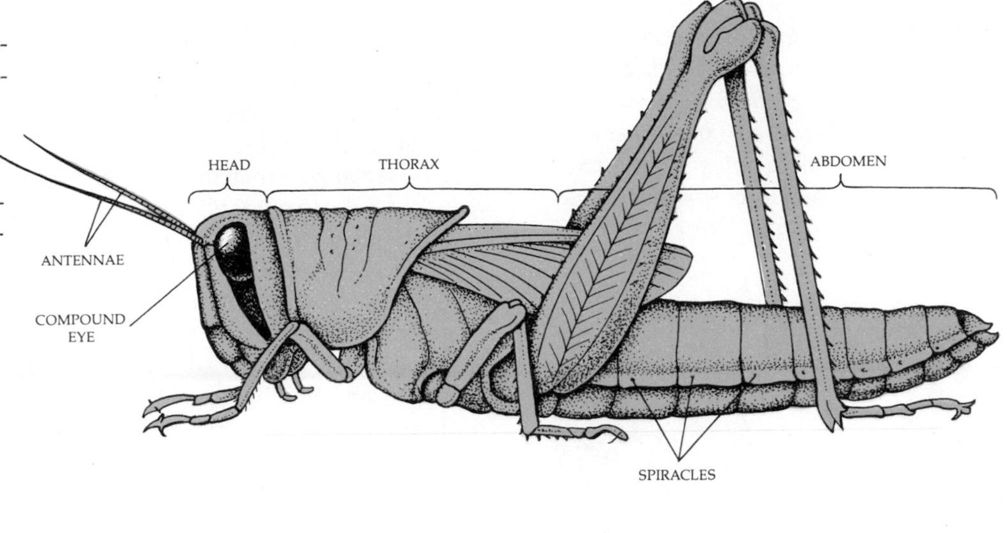

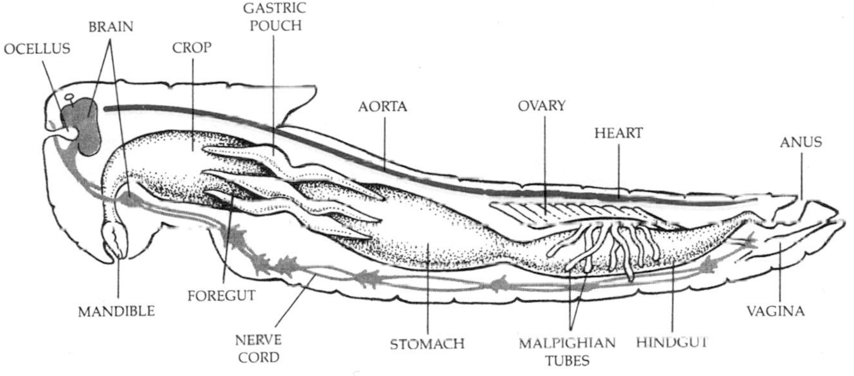

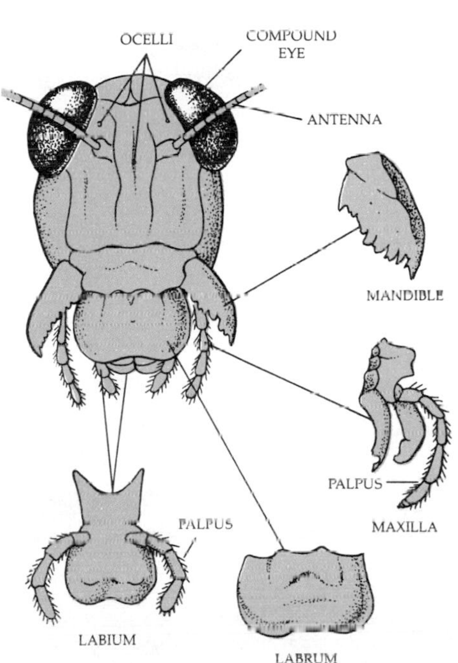

26–14
Mouthparts of a grasshopper. The mandibles are crushing jaws. The labium and the labrum are the lower and upper lip. The maxillae move food into the mouth, and the palpi assist in tasting.

Digestive and Respiratory Systems

The foregut and hindgut of the insect digestive tract are lined with chitin. Salivary gland fluids are carried with food into the crop, where digestion begins. The stomach, or midgut, which lies mainly in the abdomen, is the chief organ of absorption. Insects have digestive enzymes as specialized as their mouthparts; the structure of the enzymes depends on whether the insect dines on blood, seeds, other insects, eggs, flour, cereal, glue, wood, paper, or your woolen clothes. Excretion is carried out through Malpighian tubules. In the grasshopper and many other insects, the nitrogenous and other wastes are eliminated in the form of nearly dry crystals of uric acid, an adaptation that promotes water conservation.

The respiratory system consists of a network of chitin-lined tubules through which air circulates to the various tissues of the body, supplying each cell directly. Muscular movements of the animal's body improve the internal circulation of air. The amount of incoming air and also the degree of water loss is regulated by the opening and closing of the spiracles.

26-15
Some immature forms. (a) *Scarab beetle (white grub) larva in soil.* (b) *Black swallowtail larva (caterpillar).* (c) *Mosquito larvae and a pupa (on right). Mosquito larvae are aquatic, hanging to the undersurface of the water with respiratory tubes.* (d) *Tent caterpillars on their protective web.*

(a)

(b)

Metamorphosis

Growing insects change not only in size but often in form, a phenomenon known as metamorphosis. The extent of change varies. In some species, the young, although sexually immature, looks like a small adult; it grows larger by a series of molts until it reaches full size. In others, like the grasshopper, the newly hatched young are wingless but may gain wingpads in later immature stages; otherwise they are similar to the adult. These immature, nonreproductive forms are known as *nymphs*. Almost 90 percent of the insects, however, undergo a complete metamorphosis, in which adults are drastically different from their immature forms. The immature feeding forms are all correctly referred to as *larvae*, although they are also commonly known as caterpillars, grubs, or maggots, depending on the different species. Following the larval period, the insect undergoing complete metamorphosis enters an immobile pupal stage within which extensive remodeling of the organism occurs. The adult (sexually mature) insect emerges from the pupa. Both eggs and pupae (which are nonfeeding) can endure lengthy cold or dry seasons.

The insect that undergoes complete metamorphosis exists in four different forms in the course of its life history. The first form is the egg and the embryo. The second form is the larva, the animal that hatches from the egg; larvae eat and grow. In many larvae, such as those of flies, growth takes place not by an increase in the number of cells, as in most animals, but by an increase in the size of the cells, in somewhat the same way that growth takes place in certain plant tissues. During the course of its growth, the larva molts a characteristic number of times—twice in the fruit fly, for example. The stages between molts are known as *instars*. Then, when the larva is full-grown, it molts to form the pupa. During this outwardly lifeless pupal stage, many of the larval cells break down, and entirely new groups of cells, set aside in the embryo, begin to proliferate, using the degenerating larval tissue as a culture medium. These groups of cells are known as imaginal discs since they form the imago, the adult insect, which, according to Aristotle, is the perfect form or ideal image that the immature form is "seeking to express." These imaginal discs develop into the complicated structures of the adult.

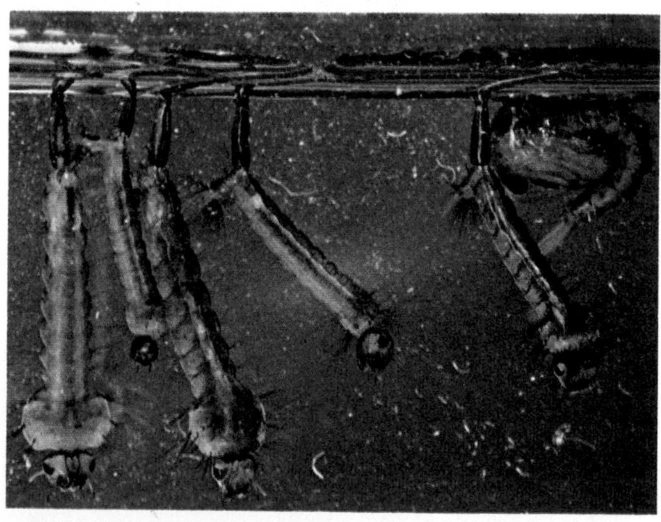

(c)

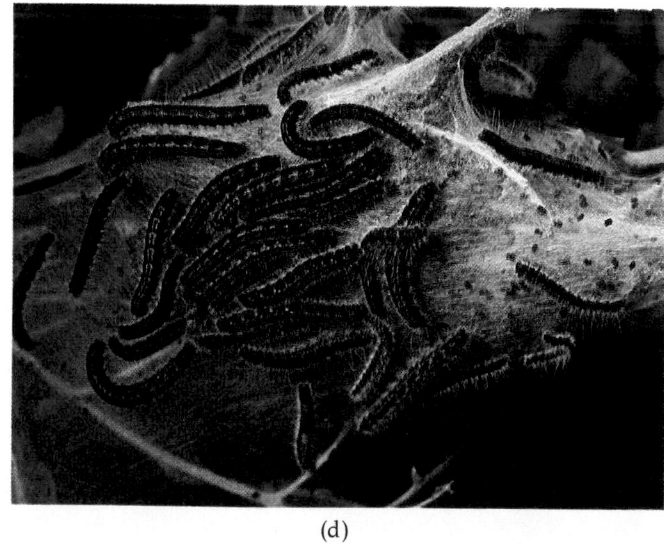

(d)

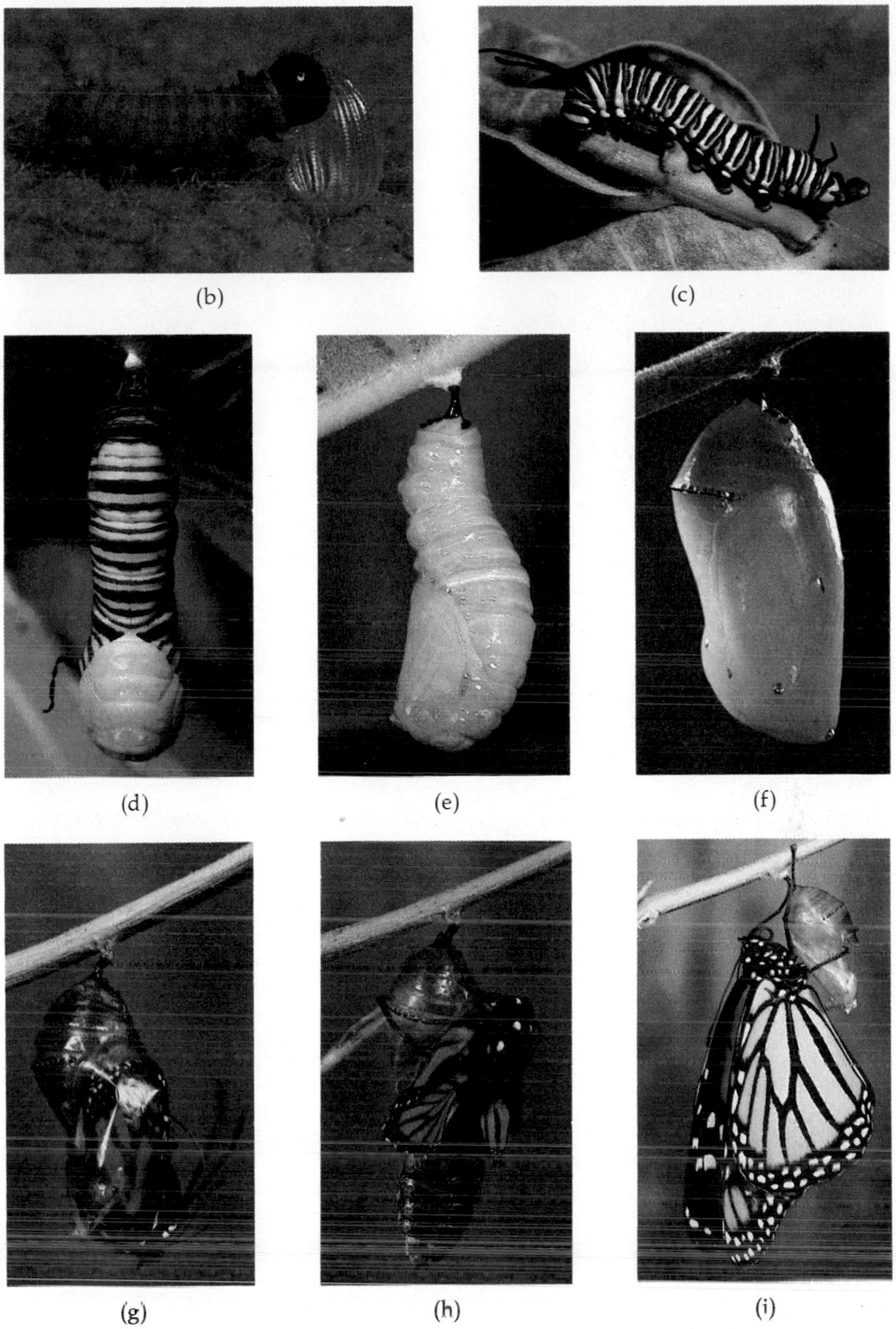

(a)

(b)

(c)

26–16

Developmental stages of a monarch butterfly. (a) *Larva hatching and* (b) *eating its rubbery egg shell.* (c) *A caterpillar at the fifth molt.* (d) *When ready to pupate, the caterpillar attaches itself to a leaf with a patch of silk. Its skin splits, revealing the pupa.* (e) *Inside the pupa, development of the butterfly begins by recycling the tissues of the caterpillar.* (f) *The pupating butterfly remains inside the gold-studded chrysalis for 2 to 3 weeks.* (g) *The butterfly forces its swelling thorax out of the pupa and* (h) *pulls free, forcing blood into its wings.* (i) *The butterfly rests until its wings harden and it can fly.*

(d)

(e)

(f)

(g)

(h)

(i)

26-17
Molting is under hormonal control. In insects, a hormone (brain hormone) produced by neurosecretory cells in the brain and released from corpora cardiaca stimulates the prothoracic gland, which, in turn, produces molting hormone (ecdysone). Although all molts require ecdysone, whether or not metamorphosis occurs depends on a third hormone, juvenile hormone, produced by the corpora allata. Continued presence of juvenile hormone at high concentrations ensures that larval molts occur during the first portion of the life history. In later larval life, production of juvenile hormone declines, permitting adult structures to develop.

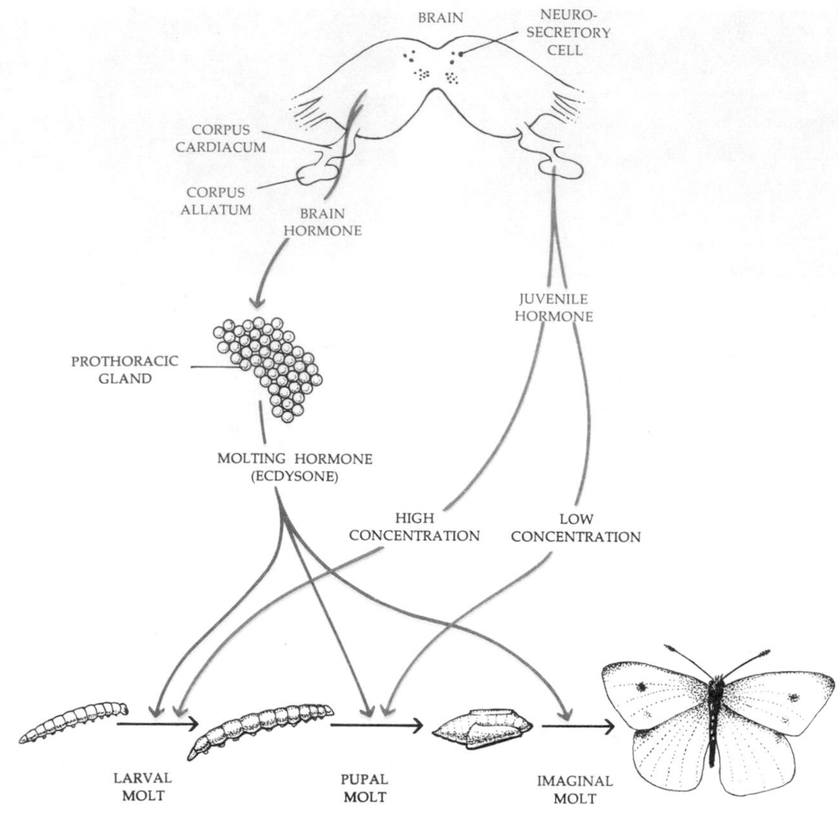

Molting and metamorphosis are under hormonal control. Like many hormone-controlled processes, they are the end result of an interplay of several hormones: brain hormone, molting hormone (ecdysone), and juvenile hormone. At intervals during larval growth, brain hormone, produced by neurosecretory cells in the brain, is released into the blood. It stimulates the release, in turn, of molting hormone from a gland in the thorax. The molting hormone stimulates not only molting but also the formation of a pupa and the development of adult structures. The latter are held in check, however, by a third hormone, the juvenile hormone. Only when the production of juvenile hormone declines, in later larval life, can metamorphosis to the adult form take place.

REASONS FOR ARTHROPOD SUCCESS

Among all the invertebrates, why are the arthropods, in general, and the insects, in particular, so spectacularly successful? One important reason is undoubtedly the nature of the exoskeleton, which waterproofs, provides protection, and makes possible the evolution of the many finely articulated appendages characteristic of this phylum.

A second reason, which applies especially to insects, is the high specificity of diet and other requirements of each species. As a consequence, many different species can live in a single small environment—in a few cubic centimeters of soil, on

26–18
(a) *Compound eye of* Drosophila, *as shown by the scanning electron microscope. Although insect eyes cannot change focus, they can define objects only a millimeter from the lens, a useful adaptation for an insect.* (b) *Structure of the compound eye. The eye is composed of a large number of structural and functional units called ommatidia. Each ommatidium has its own cornea, which forms one of the facets of the compound eye, and its own light-focusing lens. The light-sensitive part is the rhabdom, which is surrounded by the retinular cells, which transmit the stimulus. The ommatidium is surrounded by pigment cells that prevent light from traveling from one ommatidium to another.*

a small plant, or on or within the egg or body of a single animal—without competing with one another. The varied and highly specialized mouthparts are a reflection of this specificity of diet. Among the insects with complete metamorphosis, even the larvae of a species do not compete with the adults for food and for the same territory.

A third reason for the success of the insects, as we mentioned previously, is their capacity for flight, which gives them an extraordinary mobility, in three dimensions.

A fourth reason for success is undoubtedly the arthropod nervous system, with its fine control over the various appendages and the many extraordinarily sensitive sensory organs found in great diversity throughout the phylum.

The rest of this discussion deals with sensory perception among arthropods, especially insects, and some examples of arthropod behavior. The social behavior of bees is described in Chapter 48.

Arthropod Senses and Behavior

Vision: The Compound Eye

The most conspicuous sensory organ of the arthropods is the compound eye (Figure 26–18), which is an evolutionary development characteristic of this one phylum. The basic structural unit of this eye is the *ommatidium*. A dragonfly has some 30,000 ommatidia. Each ommatidium is covered by a cornea, usually with a round or hexagonal surface; these are visible under low-power magnification as individual facets of the eye. Underlying the cornea is a group of eight retinular cells surrounded by pigment cells. The light-sensitive portion of the ommatidium is the *rhabdom,* which is the central core of the ommatidium. Nerve fibers carry the stimulus from each ommatidium to the brain. The pigment cells prevent light from traveling from one ommatidium to another. An ommatidium is much larger than a vertebrate photoreceptor, and so there are far fewer in an equivalent space. Hence, the image has less resolution, like a newspaper picture under high magnification.

Although the compound eye is deficient in acuity, offering less detail than the vertebrate eye, it is better for detecting motion because each ommatidium is stimulated separately and so has a separate visual field. Also, each ommatidium responds to stimuli more rapidly than does a vertebrate photoreceptor. Ability to detect motion can be measured accurately in the laboratory by testing a phenome-

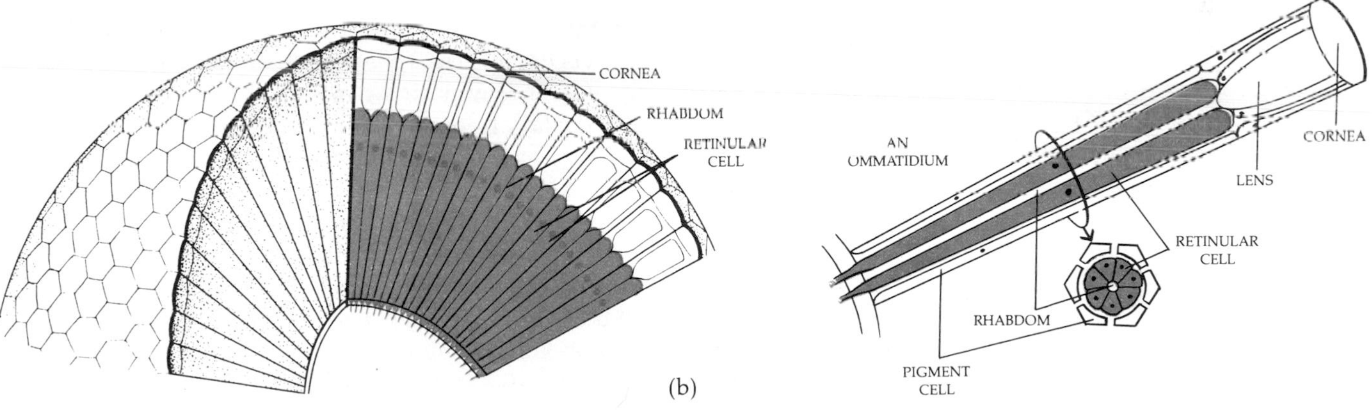

Table 26-1 *Major Classes of the Phylum Arthropoda*

	NO. OF SPECIES	BODY PARTS	HEAD APPENDAGES	LEGS	WINGS	EYES	RESPIRATION	DISTRIBUTION
Subphylum Chelicerata								
Class Arachnida (Scorpions, spiders, daddy longlegs, mites, ticks)	35,000	Two (cephalothorax, abdomen)	Chelicerae, pedipalps, no antennae	4 pairs	None	Simple	Book lungs, tracheae, or both	Terrestrial, some aquatic (secondarily)
Subphylum Mandibulata								
Class Crustacea (Lobsters, crabs, barnacles, *Daphnia*, copepods)	30,000	Two (cephalothorax, abdomen) or three (head, thorax, abdomen)	Antennae (2 pairs), mandibles, maxillae (2 pairs)	Numerous (on both thorax and abdomen)	None	Usually compound	Gills	Marine and freshwater, some terrestrial
Class Insecta	700,000	Three (head, thorax, abdomen)	Antennae (1 pair), mandibles, maxillae (2 pairs)	3 pairs (on thorax only)	2 pairs, 1 pair, or none	Compound and simple	Tracheae	Terrestrial, some aquatic (secondarily)

non known as flicker fusion. In this test, a light is flicked on and off with increasing rapidity until the observer sees the flicker as a continuous beam. The beam is perceived as continuous because stimulation of any retinal cell persists for a brief period even after the stimulus disappears. So, in effect, the flicker-fusion test is a measurement of how quickly the photoreceptor cell recovers from one stimulus and becomes sensitive to another. It is possible to test flicker-fusion rates in animals by training experiments in which the animal learns to associate a flickering light with a reward (usually food) and a steady beam with no reward, or vice versa. Such tests have proved that the compound eye greatly exceeds the camera eye of vertebrates in this respect. A bee would see in clear outline a moving figure that we would see as blurred, and if the bee went to the movies, the film seen by us as a continuous picture would jerk along from frame to frame for the bee. The ability to perceive motion is extremely important for an insect since it must be able to make out objects when it is flying at high speed (which, as far as the visual apparatus is concerned, presents the same problems as following a moving object).

In addition to, or instead of, compound eyes, many of the arthropods possess simple eyes, or ocelli, which seem generally to serve only for light detection. Most insects have two or three ocelli, and spiders, which do not have compound eyes, may have as many as eight ocelli, depending on species.

26-19
The hairs on the legs of this tiger beetle are touch receptors. At the base of the hairs are sensory cells. When a hair is touched or bent, nerve impulses are initiated.

Touch Receptors

The body surfaces of terrestrial arthropods are often covered with sensory-receptor units known as sensilla (singular, sensillum), or "little sense organs." Most of the sensilla take the form of fine spines, or setae, composed of hollow shafts of chitin. At the bases of these shafts are sensory cells. In their simplest form, the sensilla are touch receptors. In these, when the hair is touched or bent, the sensory cell responds and initiates nerve impulses. Such receptors are found, in particular, on the antennae and the legs. In addition to being stimulated by direct contact, they can also be stimulated by vibrations and air currents. A spider monitors what is going on in its web by sensing vibrations transmitted through the threads when the web is touched. Soldier termites of certain species strike the ground or the walls of their nest with their heads when threatened or disturbed; the vibrations they produce warn their colony mates. A fly perceives the air currents from the movement of a hand or fly swatter and so escapes; a fly in a glass jar is much less likely to be disturbed by such movements.

Proprioceptors

Proprioceptors are sensory receptors that provide information about the position of various parts of the body and the stresses and strains on them. A type common in the arthropods is the campaniform sensillum (Figure 26–20a). Campaniform sensilla are located in thin, stretchable areas of the cuticle. When the cells are twisted or stretched, a nerve fiber signals the central nervous system.

Touch receptors can also serve as proprioceptors. The praying mantis, for example, is capable of making a lightning-swift strike at a moving object. When it sights a potential victim, the insect moves its entire head to bring it into binocular range, since the eyes themselves do not move (Figure 26–20b). Movement of the head results in the stimulation of proprioceptive hairs on the head and thorax of the insect. On the basis of the impulses received from these hairs, the position of the prey and the movement of its own legs are automatically coordinated by the mantis. If these hairs are removed, the mantis can strike a moving object only if the object is directly in front of it.

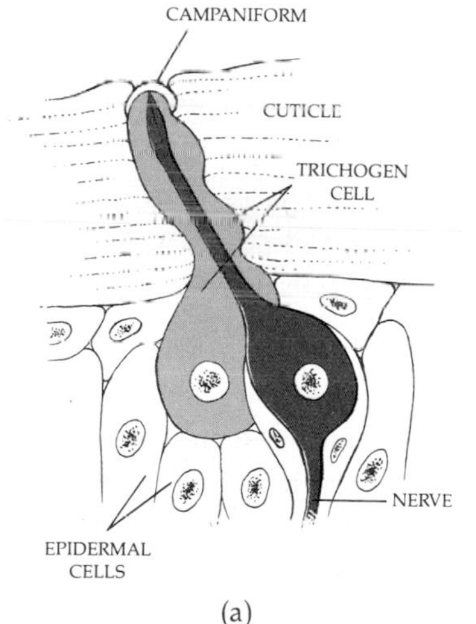

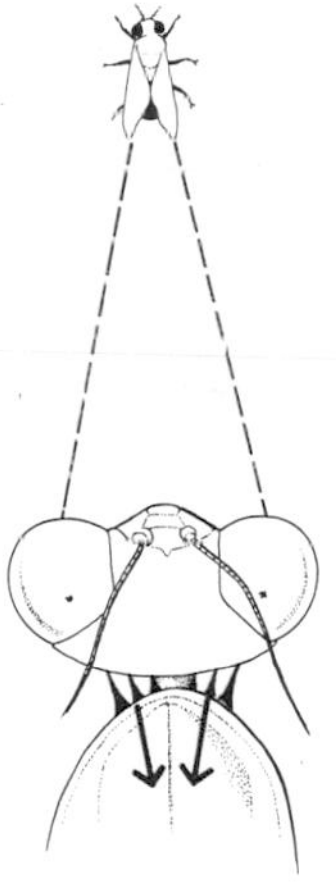

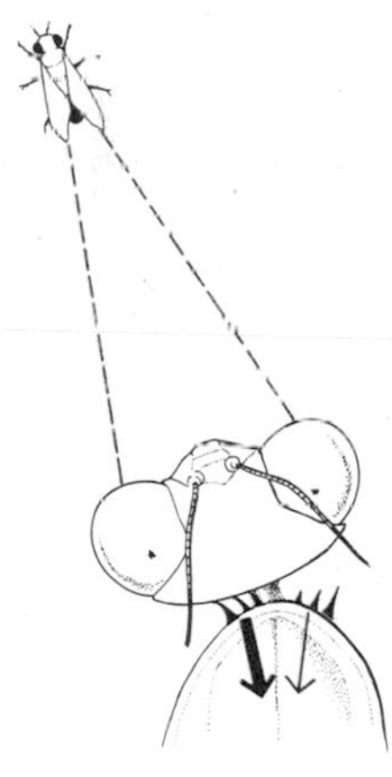

26-20
(a) Campaniform sensillum, a type of proprioceptor common in arthropods. (b) Since its eyes do not move, the praying mantis must move its entire head to bring its victim into binocular range. The movement of its head sends impulses through proprioceptive hairs on its head and thorax to its legs. The position of its prey and the movement of its legs are thus automatically coordinated, giving the mantis the ability to strike at a victim swiftly and at the proper range.

26–21
The most elaborate of the insect sound receptors is the tympanic organ. The tympanic air sacs are covered by a membranous drum, and the sensory cells are so arranged in the organ that they are stimulated by movements of the drum or the air-sac walls. Tympanic organs respond to vibrations in the air or in water. This is a tympanic organ on the inside front tibia of a katydid.

Sound Receptors

Arthropods have a variety of sound receptors. The simplest is a sensillum with a tactile hair that vibrates as a result of being "touched" by sound waves. The antennae of the male mosquito contain thousands of such hairs, which are responsive to the vibrations made by the wings of the female mosquito in flight and so serve to bring the sexes together. When the male mosquito first emerges from its pupal shell, it is sexually immature and also deaf, with its antennal hairs lying flat along the shafts. When the male matures sexually, some 24 hours later, the hairs almost simultaneously become erect and are now free to vibrate when a female approaches.

Other insects have developed special groups of cells for hearing; these are known as tympanic organs (Figure 26–21). In these organs, a fine membrane, the tympanum (or eardrum) is stretched across an air-filled cavity. The tympanic membrane vibrates in response to sounds of certain frequencies, and this vibration is transmitted to underlying receptor cells.

Communication

Communication by Sound

Arthropods, particularly insects, have developed complex forms of sensory communication. A number of species, such as the locusts, grasshoppers, and crickets, call to one another by sounds made by rubbing their legs or wings together or against their bodies. Five distinct types of calls are known: (1) calling by males and (2) calling by females, both of which are long-range sounds; (3) courtship sounds by males and (4) aggressive sounds by males, both of which are short-range; and (5) alarm sounds, which may be given either by males or by females. Recognition of and response to the sound seem to be based on pattern and on rhythm because insects apparently are not able to distinguish frequencies, or the differences between high and low notes, and so are essentially "tone deaf." The effectiveness of calling songs is often increased by group singing, such as the famous chorus of male seventeen-year cicadas, which can attract females from distances far greater than an individual "voice" would reach. Insects produce songs and respond to appropriate songs without ever having heard a song before.

26–22
The katydid produces its characteristic loud, shrill sounds by rubbing the scraper at the base of its right wing against the file at the base of its left wing.

Communication by Pheromones

The use of chemicals for communication is common among animals, and the substances employed range from the sex attractants of the little algal cell *Chlamydomonas* to Chanel No. 5. Many insects communicate by chemicals; such chemicals are known as *pheromones*. Pheromones are chemical messengers. They are usually produced in special glands and are discharged into the environment, where they act on other members of the same species.

Among the best studied of the pheromones are the mating substances of moths. One female gypsy moth, by the emission of minute amounts of a pheromone commonly known as gyplure, can attract male moths that are several kilometers downwind. In fact, a single female contains enough gyplure, somewhere around a millionth of a gram, to attract more than a billion males, supposing that it were distributed with maximum efficiency. Since the male can detect as little as a few hundred molecules per cubic centimeter of the attractant, gyplure is still potent even when it has become widely diffused.

26–23
A cricket. The two large appendages extending from either side of the head are palpi, which contain chemoreceptors (taste organs).

The male moth characteristically flies upwind, and the pheromone, of course, disperses downwind. Therefore, when a male moth detects the odor of a female of the species, he will fly toward the source. If he loses the scent, he flies about at random until he either picks it up again or abandons the search. It is not until he is quite close to the female that he can fly "up the gradient" and use the intensity of the odor as a locating device. Figure 26–1 shows the lavishly plumed antennae of a male Cecropia moth by which he detects the pheromone emitted by the female.

Insect Behavior

Complex patterns of unlearned, genetically transmitted behavior—such as web-building among spiders—are another arthropod characteristic. This rigid programming of behavior may be a necessary correlate to the shortness of the life span of the smaller arthropods. It provides an interesting contrast to the more flexible behavior patterns of the higher mammals, with their comparatively long life spans, long periods of learning, and much larger brains.

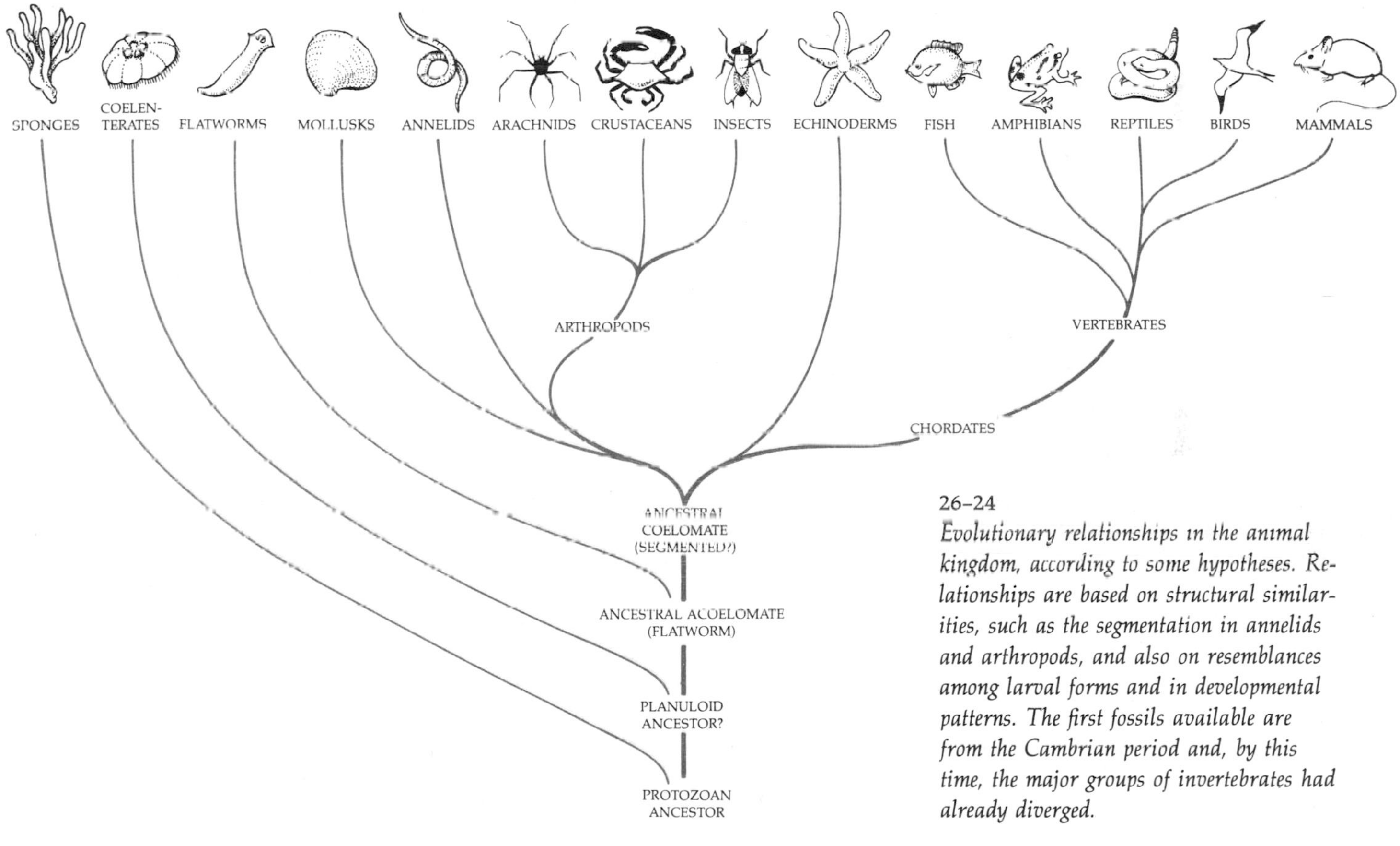

SPONGES
COELENTERATES
FLATWORMS
MOLLUSKS
ANNELIDS
ARACHNIDS
CRUSTACEANS
INSECTS
ECHINODERMS
FISH
AMPHIBIANS
REPTILES
BIRDS
MAMMALS
ARTHROPODS
VERTEBRATES
CHORDATES
ANCESTRAL COELOMATE (SEGMENTED?)
ANCESTRAL ACOELOMATE (FLATWORM)
PLANULOID ANCESTOR?
PROTOZOAN ANCESTOR

26–24
Evolutionary relationships in the animal kingdom, according to some hypotheses. Relationships are based on structural similarities, such as the segmentation in annelids and arthropods, and also on resemblances among larval forms and in developmental patterns. The first fossils available are from the Cambrian period and, by this time, the major groups of invertebrates had already diverged.

PHYLUM CHORDATA

The phylum Chordata comprises three subphyla: the Cephalochordata, or lancelets, which includes *Branchiostoma* (formerly called *Amphioxus*); the Tunicata, or tunicates, of which the most familiar are the sea squirts; and the Vertebrata, or vertebrates.

Branchiostoma (Figure 26–25) is a small, blade-shaped, semitransparent animal found in shallow marine waters all over the warmer parts of the world. Although it can swim very efficiently, it spends most of its time buried in the sandy bottom, with only its mouth protruding above the surface. This animal exemplifies all four of the salient features of the chordates. The first is the notochord, a rod that extends the length of the body and serves as a firm but flexible axis. The notochord is a structural support. Because of it, *Branchiostoma* can swim with strong undulatory motions that move it through the water with a speed unattainable by the flatworms or aquatic annelids.

The second chordate characteristic is the dorsal, hollow nerve cord, a tube that runs beneath the dorsal surface of the animal above the notochord. (The principal nerve cords in other phyla, as you will recall, are ventral.)

The third characteristic is a pharynx with gill slits. The pharyngeal gill slits become highly developed in fishes, in which they serve a respiratory function, and traces of them remain even in the human embryo. In *Branchiostoma,* they serve primarily for collecting food. The cilia around the mouth and at the opening of the pharynx pull in a steady current of water, which passes through the pharyngeal slits into a chamber known as the atrium and then exits through the atrial pore. Food particles are collected in the sievelike pharynx, mixed with mucus, and channeled along ciliated grooves to the intestine.

The fourth characteristic is a tail, posterior to the anus, consisting of blocks of muscle around an axial skeleton. Most of the body tissue of *Branchiostoma* is made up of blocks of muscles, the myotomes.

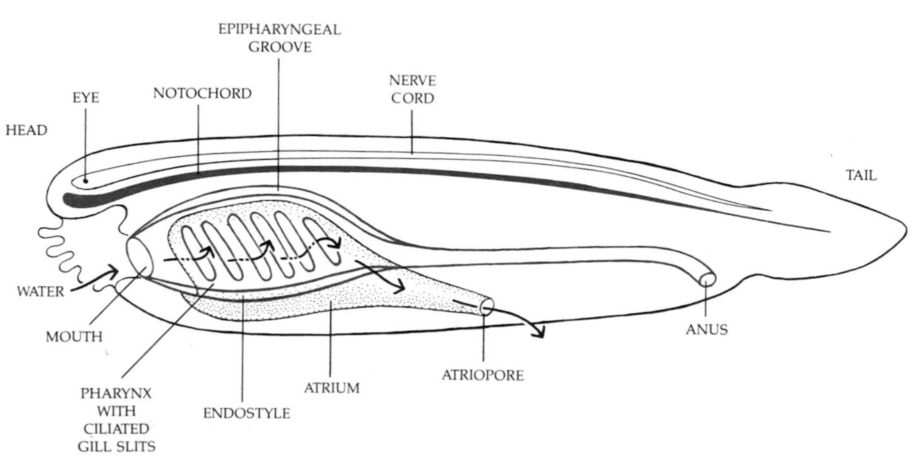

26–25
Branchiostoma, *a lancelet, exemplifies four distinctive chordate characteristics: (1) a notochord, the dorsal rod that extends the length of the body; (2) a dorsal, tubular nerve cord; (3) pharyngeal gill slits; and (4) a tail.*

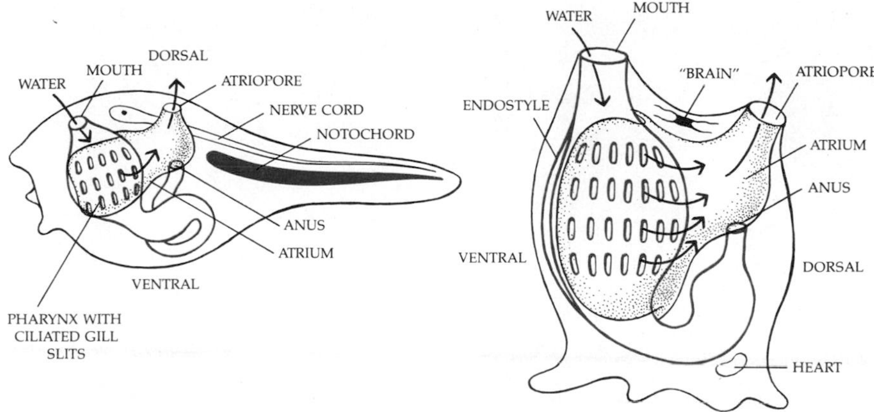

26–26
Two stages in the life of a tunicate. At the left is the larva; the adult form is shown on the right. After a brief free-swimming existence, the larva settles to the bottom and attaches at the anterior end. Metamorphosis then begins. The larval tail, with the notochord and dorsal nerve cord, disappears, and the animal's entire body is turned 180°. The mouth is carried backward to open at the end opposite that of attachment, and all the other internal organs are also rotated back. As some biologists reconstruct the past, tunicate larvae wriggled up the rivers, where they gave rise, after many generations and millions of years, to the ancestral vertebrates.

Although *Branchiostoma* usefully exemplifies the chordates, many biologists believe it is more likely to be a degenerate form of primitive fish rather than a truly primitive member of the phylum. A more probable candidate for the ancestral form is the tunicate. Although the adult form does not have all of the typical chordate features, the larva, which resembles *Branchiostoma,* is clearly a chordate possessing the four chordate characteristics (Figure 26–26).

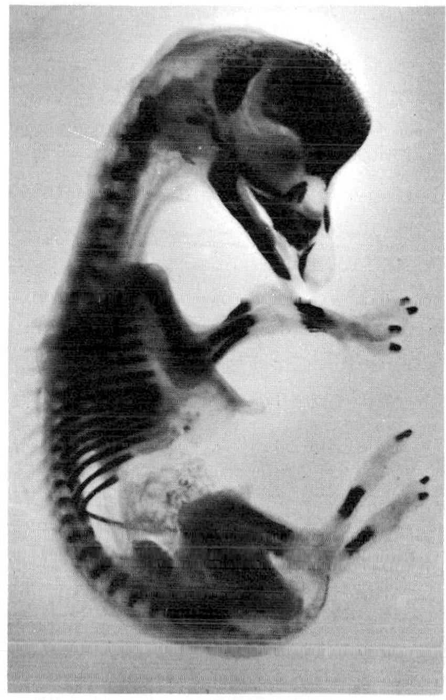

26–27
An elk fetus. The bones have been stained to show them more clearly, so that you can see the extent to which the skeleton is still cartilaginous. Notice the legs, for example. Only the dark areas are bone; these will gradually grow and replace the cartilage as the animal matures.

SUBPHYLUM VERTEBRATA

The vertebrates are a large (about 43,000 species) and familiar subphylum of the chordates. All vertebrates have a backbone, or vertebral column, as their structural axis—a flexible bony support that develops around the notochord, supplanting it entirely in most species. Dorsal projections of the vertebrae encircle the nerve cord along the length of the spine. The brain is similarly enclosed and protected by bony skull plates. Between the vertebrae are cartilaginous disks, which give the vertebral column its flexibility.

One of the great advantages of this bony endoskeleton, as compared with the exoskeletons of the invertebrates, is that it is composed of living tissue that can grow with the animal. In the developing embryo, the skeleton is largely cartilaginous, with bone gradually replacing cartilage in the course of maturation. Figure 26–27 shows the developing bone tissue of a vertebrate embryo. In the vertebrates, the growing parts of the bones remain cartilaginous until the animal reaches its full adulthood.

There are seven living classes of vertebrates: the fish (comprising three classes), the amphibians, the reptiles, the birds, and the mammals.

Classes Agnatha, Chondrichthyes, and Osteichthyes: Fish

The first fish were jawless and had a strong notochord running the length of their bodies. Today these jawless fish (class Agnatha), once a large and diverse group, are represented only by the hagfish and the lampreys. They have a notochord throughout their lives, like *Branchiostoma.* Although their ancestors had bony skeletons, modern agnaths have a cartilaginous skeleton. Lacking true bones, they are very flexible; a hagfish can actually tie itself in a knot. Many cyclostomes ("round mouths"), as they are called, are highly predatory, attaching to other fish by their

(a)

(b)

26–28
(a) *The heavily armored placoderm is the ancestor of two major classes of present-day fish—the Chondrichthyes (cartilaginous fish) and the Osteichthyes (bony fish).*
(b) *Rays, like skates and sharks, are cartilaginous fish that have existed in their present form for about 350 million years. Their flattened body is an adaptation to bottom living.*

suckerlike mouths and rasping through the skin into the viscera of their hosts. The juvenile lamprey, which resembles *Branchiostoma,* however, feeds by sucking up mud containing microorganisms and organic debris—as, most probably, did the primitive Agnatha.

The sharks (including the dogfish) and skates, the Chondrichthyes, the second major class of fish, also have a completely cartilaginous skeleton. Like agnaths, their ancestors were also bony animals. Their skin is covered with small, pointed teeth (denticles), which resemble vertebrate teeth structurally and give the skin the texture and abrasive quality of coarse sandpaper.

The third major class of fish includes those with bony skeletons, the Osteichthyes. This group includes the trout, bass, salmon, perch, and many others—most of the familiar freshwater and saltwater fish.

According to present evidence, fish evolved in fresh water. The chondrichthyans returned to the sea early in their evolution, while the bony fish went through most of their evolution in fresh water and spread to the seas at a much later period. Some still recapitulate in each lifetime this difficult physiological transition. Salmon, for example, return to fresh water to spawn, while eels leave the fresh waters of Europe and North America to return to the Sargasso Sea at breeding time, from which distant point the young begin the long, difficult journey, often lasting many years, back to the rivers and lakes.

The Transition to Land

Another characteristic of the bony fishes is that the early forms seem to have had lungs or lunglike structures, as well as gills. These lungs, however, were not efficient enough to serve as more than accessory structures to the gills. They were a special adaptation to fresh water, which, unlike ocean water, may become stagnant (depleted of oxygen) because of decay of organic matter or of algal bloom. Lunged fish were the most common fish in the later Devonian seas, apparently evolving independently several times. In most of them, the lung evolved into an air bladder, or swim bladder; many modern osteichthyans have gas-filled swim bladders that serve as flotation chambers or organs of sound production. A fish raises or lowers itself in the water by adding gases to or removing them from the air bladder via the

26–29
A modern lungfish. When the dry seasons come, members of this African species wriggle downward into the mud, which eventually hardens around them. Mucus glands under the skin secrete a watertight film around the body, preventing evaporation. Only the mouth is left exposed. During this period, they take a breath only about once every two hours.

bloodstream. Still other primitive fish evolved into the modern lungfish (Figure 26–29). These fish can live in water that does not have sufficient oxygen to support other fish life. Lungfish surface and gulp air into their lungs in much the same way that certain aquatic but air-breathing snails bob to the surface to fill their mantle cavities.

In yet others, skeletal supports evolved that served to prop up the thorax of the fish. These fish could gulp air even when their bodies were not supported by water. These osteichthyans could waddle, dragging their bellies on the ground, along the muddy bottom of a drying stream bed to seek deeper water or perhaps even make their way from one water source to another one nearby. Thus the transition to land began as an attempt to remain in the water.

Class Amphibia

Amphibians descended from air-breathing lunged fish. Modern amphibians include frogs and toads (which are tailless as adults) and salamanders (which have tails throughout their lives). They can readily be distinguished from the reptiles by their thin, usually scaleless skins, which serve as breathing organs. (Frogs have lungs as adults. Some salamanders have lungs, but others breathe entirely through their skins and the mucous membranes of their throats.) Because water evaporates rapidly through their skins, some amphibians can readily die of desiccation in a dry environment. Those found in deserts spend the drier times of the day far below the surface of the sand.

Most frogs in cold climates have two life stages (hence their name, from *amphi* and *bios*, meaning "two lives"). The eggs are laid in water and are fertilized externally. They hatch into gilled larvae (tadpoles). The tadpoles later develop into adults that lose their gills and develop lungs. The adults may live out of the water, at least in the summer. However, there are many variations on this theme. Some of the American salamanders fertilize their eggs on land; the males deposit sperm

26–30
Many frogs are clearly fishlike in their larval (tadpole) stages. As adults, they require water to reproduce and their moist skins are an important accessory breathing organ. Frogs, like all adult amphibians, are carnivores. This green frog, Rana clamatans, *has caught a grasshopper with a flick of its long tongue, which is attached at the front of the mouth and which has a sticky, flypaper-like surface.*

26–31
Many reptiles lay eggs in which the embryos can develop on land. These newly hatched black snakes are sunning themselves, a behavior characteristic of ectotherms—animals that take in heat from the environment. Black snakes, like all other snakes, are carnivorous, devouring their prey whole.

packets that are picked up by the females. Many amphibians are now known to skip the free-living larval stage. The eggs, which may be laid on land, in a hollow log or cupped leaf, or even carried by the parent, hatch into miniature versions of the adult. Some salamanders, such as the mud puppy and the axolotl, never complete their metamorphosis, remaining essentially aquatic larval forms. In some species, these larvalike forms can be induced to metamorphose into adult forms by administration of hormones, indicating that the genetic capacity for this later developmental stage has not been lost.

Class Reptilia

As you will recall, the vascular plants freed themselves from the water by the development of the seed. Analogously, the vertebrates became truly terrestrial with the evolution in the reptiles of the amniote egg, an egg that retains its own water supply and so can survive on land. The reptilian egg, which is much like the familiar hen's egg in basic design, contains a large yolk, the food supply for the developing embryo; abundant albumen; and a water supply. A membrane, the amnion, surrounds the developing embryo with a liquid-filled space that substitutes for the ancestral pond. The gilled stage is passed in a shelled egg or in the maternal oviduct or uterus. In mammals also, although their eggs typically develop internally, the embryos are enclosed in water within an enveloping membrane, the amnion, and pass through a gilled stage (although the gills are never functional) before birth.

Reptiles are characteristically four-legged, although the legs are absent in most snakes and some lizards. In keeping with their terrestrial existence, reptiles have a dry skin, usually covered with protective scales. Modern reptiles, of which there are 5,400 species, include lizards, snakes, turtles, and crocodiles.

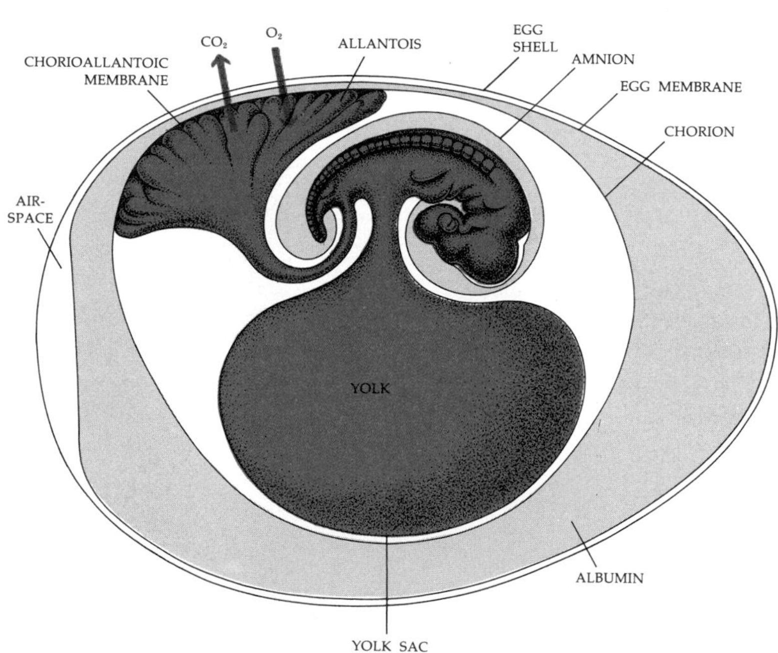

26–32
Amniote egg. The membranes, which are produced as outgrowths from the embryo as it develops, surround and protect the embryo and the yolk (its food supply). The egg shell and egg membrane, which are waterproof but permeable to gases, are added as the early embryo passes down the maternal reproductive tract.

(a)

(b)

26–33

Two members of the class Reptilia. (a) *A turtle, a river cooter.* (b) *Alligators and crocodiles, the largest modern reptiles, lay their eggs on land, and their skins are reinforced with epidermal horny scales. Crocodiles, which are essentially tropical animals, have more slender snouts than alligators. Alligators have jaws that are broader and more rounded anteriorly; they are also reported to be less aggressive. The animal shown here is an American alligator.*

Evolution of the Reptiles

By late in the Carboniferous period (see Table 26–2, page 478), the first reptiles had begun to evolve from closely similar amphibian ancestors. During the succeeding Permian period, there was an explosive increase in the number of reptilian species. (During this same period, conifers began to replace the ferns and other "amphibious" plants, suggesting that a drier climate may have been a primary selective force in both of these events.) During the Permian and much of Triassic time, the dominant land vertebrates were mammal-like reptiles, an abundant and diverse group, members of which were later to give rise to the mammals. Around the beginning of the Triassic period, several of the more specialized reptile groups arose, which include, among the surviving forms, turtles and lizards, and also, the thecodonts, ancestors of the Archosauria, or ruling reptiles, perhaps the most spectacular of all the land's inhabitants so far.

The origin and rise of the Archosauria was associated with improvements in reptilian locomotion. Vertebrates first came to the land with all four limbs sprawled far out to the side; turtles have retained this sprawling gait. Among the early Archosauria, there was a progressive tendency toward bipedalism, with a great increase in the strength and running capabilities of the hind legs, with the concomitant freeing of the front legs for other purposes—for example, flight. There were three major groups of archosaurs: the pterosaurs, or flying reptiles; the crocodilians, which reverting to four-legged posture, became our modern crocodiles and alligators; and the dinosaurs, a varied and splendid group of reptiles.

The largest of the dinosaurs, *Brachiosaurus,* was 25 meters in length and weighed, it is estimated, 50 tons, far larger than any land animal that has succeeded it. Throughout the long Mesozoic era, the dinosaurs dominated the life of the land, rulers of the earth for 150 million years. Then, they vanished; the reasons for their extinction are not known, but it did occur during a period of climatic change and may have been related to problems of temperature regulation or changes in the vegetation or the disappearance of swampy habitats. They left only a single line of descendants, the birds.

Table 26-2 *Major Physical and Biological Events in Geologic Time*

MILLIONS OF YEARS AGO	PERIOD	EPOCH	LIFE FORMS	CLIMATES AND MAJOR PHYSICAL EVENTS
Cenozoic Era				
	Quaternary	Recent Pleistocene	Planetary spread of *Homo sapiens;* extinction of many large mammals. Deserts on large scale.	Fluctuating cold to mild. Four glacial advances and retreats (Ice Age); uplift of Sierra Nevada.
1½–7	Tertiary	Pliocene	Large carnivores. First known appearance of hominids (manlike primates).	Cooler. Continued uplift and mountain building, with widespread extinction of many species.
7–26		Miocene	Whales, apes, grazing mammals. Spread of grasslands as forests contract.	Moderate uplift of Rockies.
26–38		Oligocene	Large, browsing mammals. Apes appear.	Rise of Alps and Himalayas. Lands generally low. Volcanoes in Rockies.
38–53		Eocene	Primitive horses, tiny camels, modern and giant types of birds.	Mild to very tropical. Many lakes in western North America.
53–65		Paleocene	First known primitive primates and carnivores.	Mild to cool. Wide, shallow continental seas largely disappear.
Mesozoic Era				
65–136	Cretaceous		Extinction of dinosaurs. Marsupials, insectivores, and angiosperms become abundant.	Lands low and extensive. Last widespread oceans. Elevation of Rockies at end of period.
136–195	Jurassic		Dinosaurs' zenith. Flying reptiles, small mammals. Birds appear. Gymnosperms, especially cycads, and ferns.	Mild. Continents low. Large areas in Europe covered by seas. Mountains rise from Alaska to Mexico.
195–225	Triassic		First dinosaurs. Primitive mammals appear. Forests of gymnosperms and ferns.	Continents mountainous. Large areas arid. Eruptions in eastern North America. Appalachians uplifted and broken into basins.
Paleozoic Era				
225–280	Permian		Reptiles evolve. Origin of conifers and possible origin of angiosperms; earlier forest types wane.	Extensive glaciation in southern hemisphere. Appalachians formed by end of Paleozoic; most of seas drain from continent.
280–345	Carboniferous Pennsylvanian Mississippian		Age of amphibians. First reptiles. Variety of insects. Sharks abundant. Great swamps, forests of ferns, gymnosperms, and horsetails.	Warm. Lands low, covered by shallow seas or great coal swamps. Mountain building in eastern U.S., Texas, Colorado. Moist, equable climate, conditions like those in temperate or subtropical zones, little seasonal variation, root patterns indicate water plentiful.
345–395	Devonian		Age of fish. Amphibians appear. Shellfish abundant. Lunged fish. Extinction of primitive vascular plants. Origin of modern subclasses of vascular plants.	Europe mountainous with arid basins. Mountains and volcanoes in eastern U.S. and Canada. Rest of North America low and flat. Sea covers most of land.
395–440	Silurian		Earliest vascular plants. Rise of fish and reef-building corals. Shell-forming sea animals abundant. Modern groups of algae and fungi.	Mild. Continents generally flat. Again flooded. Mountain building in Europe.
440–500	Ordovician		First primitive fish. Invertebrates dominant. Invasion of land by plants?	Mild. Shallow seas, continents low; sea covers U.S. Limestone deposits; microscopic plant life thriving.
500–600	Cambrian		Age of marine invertebrates. Shell animals.	Mild. Extensive seas. Seas spill over continents.
Precambrian Era				
Over 600			Earliest known fossils.	Dry and cold to warm and moist. Planet cools. Formation of earth's crust. Extensive mountain building. Shallow seas. Accumulation of free oxygen.

26-34
The oldest known fossil bird, Archaeopteryx, *dates from the middle Jurassic period, about 150 million years ago. It still has many reptilian characteristics. The teeth and the long, jointed tail are not found in modern birds. The clearly evident feathers may have been related as much to endothermy as to flight.*

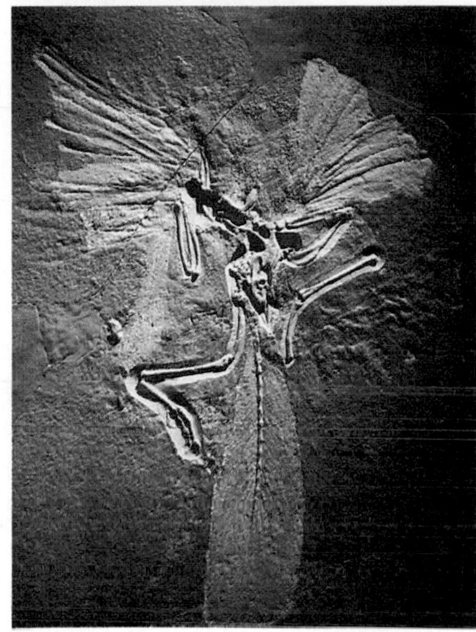

26-35
A red-shouldered hawk capturing a mouse.

Class Aves: Birds

Birds are essentially reptiles specialized for flight (Figure 26-34). Their bodies are lightened by air sacs and also by having hollow bones. The frigate bird, a large seagoing bird with a wingspread of more than 2 meters, has a skeleton that weighs only 110 grams (about 4 ounces). The most massive bone in the bird skeleton is the keel, or breastbone, to which are attached the huge muscles that operate the wings. Flying birds have jettisoned all extra weight; the female's reproductive system has been trimmed down to a single ovary, and even this becomes large enough to be functional only in the mating season.

Birds have feathers, which is their outstanding, unique physical characteristic, and they maintain a high and constant body temperature, which distinguishes them from most of the modern reptiles (although a few, such as leatherback turtles, show some degree of endothermy). In modern birds, feathers serve both to aid in flight and as insulation. (Only animals that are endothermic require insulation; animals that warm their bodies by exposure to the environment would find insulation a disadvantage.) Birds also have scales, a reminder of their reptilian ancestry. Many birds that are born at an immature stage require a long period of parental care.

Evolution of Flight

How did flight evolve? Biologists agree that evolution occurs by a series of small changes, each of which, to be conserved by natural selection, must be of survival value. Being able to fly not very well is a dubious advantage. Until recently, the most popular theory for the origin of flight has been that the ancestors of the birds were tree-dwelling reptiles and that flight evolved as a way to extend or brake jumps from branch to branch.

However, John Ostrom of Yale, having studied the anatomy of the five known specimens of *Archaeopteryx* and many related forms as well, has come to support a second theory: that the protobirds were ground-dwelling reptiles. *Archaeopteryx,* according to the fossil evidence, is a close relative of small, bipedal, carnivorous dinosaurs known as theropods. The only major distinctions are that *Archaeopteryx* has feathers and fused collar bones ("wish bone"), like modern birds. But why should feathers evolve in a ground-dwelling organism? According to Ostrom, feathers were originally an adaptation not for flight but for insulation. This raises an interesting question, because "cold-blooded" animals, such as reptiles, warm themselves from outside, and so, to these organisms, insulation is a disadvantage. Only the "warm-blooded" organisms, such as birds and mammals, require insulation as a help in conserving heat produced by a high metabolic rate. In other words, the theropods, according to this hypothesis, were warm-blooded.

Evidence that the original function of feathers was not flight is provided by anatomical studies showing that the wing feathers apparently were not attached to the bones of the "hand" in *Archaeopteryx,* as they are in modern birds, but were merely embedded in the skin. Also the breastbone and its keel are lacking, indicating that wing muscles were not well developed. *Archaeopteryx* was clearly not a good flyer, if indeed it could fly at all.

As Ostrom reconstructs it, the early stages in flight began with a feathered, warm-blooded, carnivorous dinosaur running after its prey, flapping its long feathered arms, and leaping. (The fact that the feathered forelimbs of *Archaeopteryx*

(a)

(b)

26–36
(a) *Marsupial infants are born at an immature stage and continue their development attached to a nipple in a special protective pouch of the mother. This tiny kangaroo accidentally became dislodged from its mother's pouch. As you can see, it is still attached to the nipple. After the picture was taken, the baby was restored to the pouch, with no apparent ill effects from its premature introduction to the outside world.* (b) *Opossum infants spend about two weeks in the womb and about three months in their mother's pouch, passively attached to a teat. A newborn opossum is much smaller than a honey bee.*

end in claws, as did the elongated arms of the theropods, lends support to this image.) The long "wing" feathers may have originated to serve as cagelike traps—natural nets—for capturing prey; some modern predatory birds use their wings in this way. Thus flight is seen as the culmination of a long, successful predatory leap.

Class Mammalia

Mammals also descended from the reptiles. Characteristics distinguishing mammals from other vertebrates are that mammals (1) have hair, (2) provide milk for their young from specialized glands (mammae), and (3) like birds, but unlike other vertebrates, maintain a high body temperature by generating metabolic heat. In nearly all mammalian species, the young are born alive, as they are in some fish and reptiles, which retain the eggs in their bodies until they hatch. Some very primitive mammals, however, the *monotremes,* such as the duckbilled platypus, lay eggs with shells but nurse their young after hatching. The *marsupials,* which include the opossum and the kangaroo, also bear their young alive, but they differ from the major group of mammals in that the infants are born at a tiny and immature stage and, in some species, are kept in a special protective pouch in which they suckle and continue their development. Most of the familiar mammals are *placentals,* so called because they have a more efficient nutritive connection, the placenta, between the uterus and the embryo. As a result, the young develop to a much more

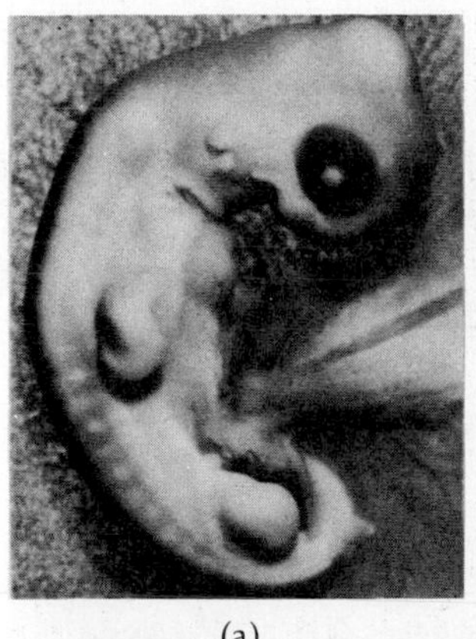
(a)

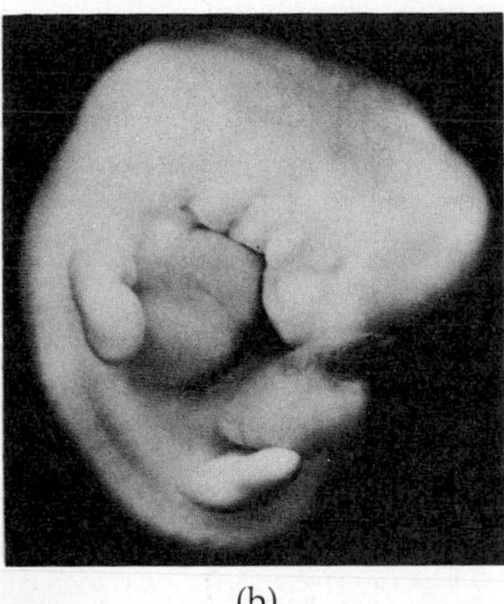
(b)

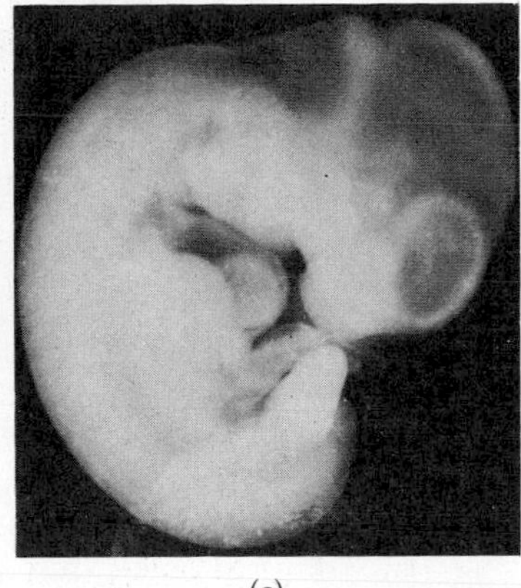
(c)

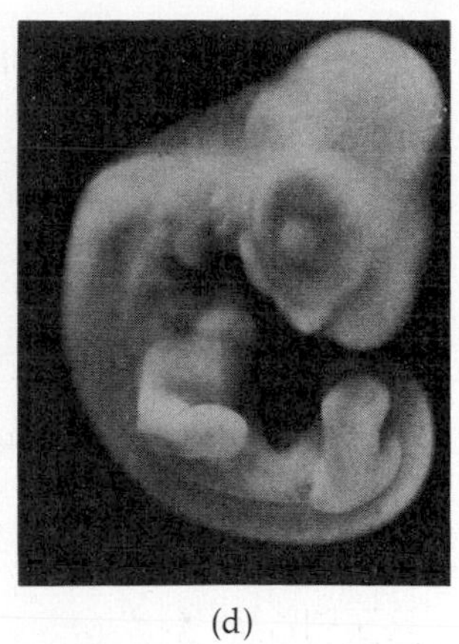
(d)

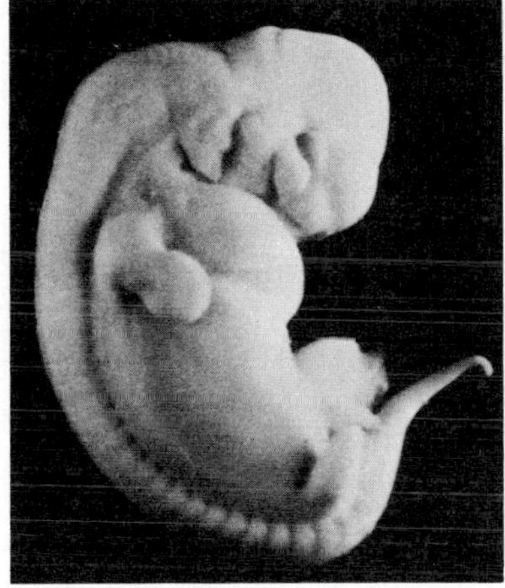
(e)

26–37
Which of these is human? As this comparison of the early fetuses of various vertebrates illustrates, the conservative nature of evolution is very evident in embryological development. The human embryo in its earlier stages greatly resembles the embryos of other vertebrates; for instance, at one stage we have gill slits and a tail. At one time, in fact, some students of evolution maintained that every embryo repeats the entire history of its evolution. Actually, what are retained are certain developmental pathways—ways of getting "from here to there." Embryo (c) *is human;* (a) *is a turtle,* (b) *a mouse,* (d) *a chick, and* (e) *a pig. (From Roberts Rugh,* Experimental Embryology: Techniques and Procedures, *Burgess Publishing Company, Minneapolis, 7th ed.)*

advanced stage before birth. Thus the young are afforded protection during their most vulnerable period. The earliest placentals were small, shy, and probably nocturnal, thus avoiding the carnivorous dinosaurs. They undoubtedly lived mostly on insects, grubs, worms, and eggs. Shrews, which are believed to closely resemble these primitive mammals, have retained their elusive habits.

Mammals have fewer, but larger, skull bones than the fish and reptiles, an example of the fact that "simpler" and "more primitive" may have quite opposite meanings. In the mammals, a bony platform or partition has developed that separates nasal and food passages far back in the throat, making it possible for the animal to breathe while eating. The lower mammalian jaw consists of a single bone. Mammals, unlike snakes or lizards, cannot move the upper jaw in relation to the brain case. They have lost the capacity of many reptiles to unhinge their jaws—an ability that makes it possible for a large anaconda, for example, to swallow a pig whole.

The major evolutionary lines of mammals are summarized in Figure 26–39.

The primates, the order to which we belong, are placental mammals characterized by three kinds of teeth (canines, incisors, and molars), opposable first digits (thumbs and usually toes), two pectoral mammae, expanded cerebral cortex, and a tendency toward single births. We are distinguished from the other primates by our upright posture, long legs and short arms, high forehead and small jaw, and lack of general body hair.

Among all the mammals, we are one of the least specialized. Humans are omnivores. Unlike the Carnivora, which are meat eaters, and the several orders of herbivores, we eat a wide variety of fruits, vegetables, and other animals. Our hands closely resemble those of a primitive reptile, in contrast to the highly specialized forelimbs developed by, for example, whales, bats, and horses. We cannot see as well as an eagle. Our sense of smell is much less keen than a dog's, and our sense of taste far less sensitive than a housefly's. Many animals can run faster, swim more powerfully, and climb trees with more agility (though few can do all three). Humans have, however, one area of extreme specialization: the brain. Because of our brain, we are unique among all the other animals in our capacity to reason, to speak, to plan, to learn, and so, to some extent, to control our own future and those of the other organisms with which we share this planet.

26–38
An assortment of mammals. (a) *Lagomorphs, such as the jack rabbit shown here, have two pairs of upper incisors whereas rodents, such as beavers* (b), *have only one. In rodents, the teeth grow continually. Carnivores, such as this nursing harbor seal* (c) *and lion* (d), *are adapted to hunt and kill for food. The lion's prey is a perissodactyl, an odd-toed ungulate.* (e) *Hippopotamuses, which are artiodactyls (two-toed ungulates), graze by night and spend most of the daylight hours resting in the water.* (f) *Elephants, the Proboscidea, are the largest land mammals living today; some reach a weight of 7.5 metric tons.*

26–39
There are about 4,500 species of mammals, divided into three subclasses, the monotremes, the marsupials, and the placentals. There are twelve orders of placentals.

SUMMARY

The arthropods constitute the largest animal phylum in both number of species and of organisms. Arthropods are segmented animals with jointed chitin exoskeletons and a variety of highly specialized appendages and sensory organs. In the higher forms, these segments are combined, forming a head, a thorax (sometimes combined with the head as a cephalothorax), and an abdomen. The arthropods are also characterized by an open circulatory system and a ladder-type nervous system, consisting of a series of ganglia, a pair per segment, interconnected by a double ventral nerve cord. Tracheae (chitin-lined breathing tubes) and Malpighian tubules (excretory ducts leading into the hindgut) are found exclusively among arthropods.

There are two subphyla of arthropods, the Mandibulata, characterized by mandibles (jaws) and antennae, and the Chelicerata with chelicerae (pincers) and cephalothorax but with no mandibles or antennae. The class Arachnida of the subphylum Chelicerata includes spiders, scorpions, mites, and ticks. Crustaceans, which are mandibulates, include lobsters, crabs, crayfish, barnacles, and shrimp. The class Insecta, largest of the arthropod classes, and also mandibulates, with more than 700,000 species, are terrestrial animals with a single pair of antennae (crustaceans have two, arachnids have none) and usually two pairs of wings.

In the course of their development, most insects pass through a complete metamorphosis. The stages are egg, larva (caterpillar, maggot, grub, etc.), pupa (often with a cocoon or some other protective covering), and adult. In a minority of species, including the grasshopper, the hatchling looks much like a miniature adult; in these, the immature form is known as a nymph. Metamorphosis is under hormonal control.

Arthropods, especially insects, have a variety of sensory receptors, one of which is the compound eye. The basic structural unit is the ommatidium, which consists of photoreceptor cells surrounding a light-conducting rhabdom and covered by a cornea. It is less acute than the vertebrate eye but more sensitive to motion.

Sensory receptors (sensilla) are found on the body surface of terrestrial arthropods. Most take the form of setae, hollow shafts of chitin associated with sensory cells that respond when the shafts are stimulated by touch, vibrations, or air currents.

Proprioceptors are sensory receptors that provide information about the position of various parts of the body and the stresses and strains on them. Arthropod proprioceptors include touch receptors and campaniform cells.

Arthropods communicate with members of the same species by sound and also by pheromones. Pheromones are chemicals released by one individual that affect the behavior or physiology of another. They serve as mating lures and also as trail-marking and alarm substances. Arthropod behavior is notable not only for its complexity and diversity but also for the extent to which it is programmed in the nervous system, that is, not learned.

The phylum Chordata comprises three subphyla: the Cephalochordata (lancelets), the Tunicata (tunicates), and the Vertebrata. The primary characteristics of the chordates are the notochord, a flexible longitudinal rod running just ventral to the nerve cord and serving as the structural axis of the body (present only in embryonic life in the vertebrates); the nerve cord, which is a hollow tube located dorsally; a pharynx with gill slits; and a tail. *Branchiostoma* illustrates the basic chordate body plan. It is hypothesized that the tunicate larva, which also shows these characteristics, resembles the primitive chordate from which the vertebrates evolved.

The vertebrates, the largest subphylum of the chordates, all have a vertebral column, a flexible bony support that develops around and supplants the notochord and encloses the nerve cord, and a cranium enclosing the brain. The vertebrates include the fish (three living classes), the amphibians, the reptiles, the birds, and the mammals.

QUESTIONS

1. Distinguish between the following: chelicerae/mandibles; exoskeleton/endoskeleton; tracheae/tracheid; Onycophora/Osteichthyes; larva/pupa; tympanic organ/ommatidium.

2. How would you distinguish an insect from an arachnid? An insect from a crustacean?

3. Name the identifying characteristics of the phylum Chordata and indicate the functional significance of each.

4. The arthropods, which include some of the most active animals, have open circulatory systems, often considered primitive or inefficient. Annelids, from which arthropods may have evolved, have closed circulatory systems. Presumably, half of the annelid system must have been lost. How could the acquisition of a relatively rigid exoskeleton make superfluous the vessels returning blood to the heart?

5. Consider the graceful swimming of a fish or squid. What do you think is the role of the semirigid beam running the length of the animal, whether the "pen" of a squid, the notochord of a *Branchiostoma,* or the vertebral column of a fish?

6. Note that only the shelled mollusks and the arthropods have open blood systems. Name an obvious feature shared by the members of these two groups. What is the correlation between these two structural similarities?

7. Why does any heart truly worthy of the name have two chambers?

8. Why do book lungs look like steam-heat radiators?

SUGGESTIONS FOR FURTHER READING

Books

AHMADJIAN, V.: *The Lichen Symbiosis,* Blaisdell Publishing Company, Waltham, Mass., 1967.

In this small but fascinating book, Ahmadjian describes the nature of the relationship between the fungal and algal components of a lichen.

ALEXOPOULOS, C. J.: *Introductory Mycology,* 2d ed., John Wiley & Sons, Inc., New York, 1962.

A general, brief introduction to the fungi.

BARNES, ROBERT D.: *Invertebrate Zoology,* 2d ed., W. B. Saunders Company, Philadelphia, 1968.

One of the best general introductions to protozoans and invertebrates.

BONNER, J. T.: *The Cellular Slime Molds,* 2d ed., Princeton University Press, Princeton, N.J., 1968.

A record of experimental work with a small but fascinating group of organisms.

BUCHSBAUM, RALPH, and LORUS J. MILNE: *The Lower Animals: Living Invertebrates of the World,* Doubleday & Company, Inc., Garden City, N.Y., 1962.

A collection of handsome photos of the invertebrates accompanied by a text prepared by two noted zoologists but directed toward the general reader.

CORNER, E. J. H.: *The Life of Plants,* Mentor Books, New American Library, Inc., New York, 1968.*

A renowned botanist with a flair for poetic prose describes the evolution of plant life, telling how plants modify their structures and functions to meet the challenge of a new environment as they invade the shore and spread across the land.

CURTIS, HELENA: *The Marvelous Animals,* Natural History Press, Garden City, N.Y., 1968.

An introduction to protozoans.

---

* Available in paperback.

CHAPTER 27

The Plant: An Introduction

For most of earth's history, the land was bare. A billion years ago, seaweeds clung to the shores at low tide and perhaps some gray-green lichens patched a few inland rocks, but had anyone been there to observe it, the earth's surface would generally have appeared as barren and forbidding as the surface of Mars does today. According to the fossil record, plants first began to invade the land a mere half billion years ago. Not until then did the earth's surface truly come to life. As a film of green spread from the edges of the waters, other forms of life, the heterotrophs, were able to follow. The shapes of these new forms and the ways in which they lived were determined by the plant life that preceded them. Plants supplied not only their food—their chemical energy—but also their nesting, hiding, stalking, and breeding places.

And so it is today. In all terrestrial communities except those of human creation, the character of the plants still determines the character of the animals and other forms of life that inhabit a particular area. Even we members of the human species, who have seemingly freed ourselves from the life of the land and even, on occasion, from the surface of the earth, are still dependent on the photosynthetic events that take place in the green leaves of plants.

Plants, as defined in this text (see page 397), are multicellular photosynthetic organisms adapted for life on land. In this section, we shall focus on the class of plants that evolved the most recently, the angiosperms. This class is characterized by specialized structures—flowers—in which sexual reproduction takes place, in which the seed is formed, and from which the fruit develops. The angiosperms are by far the most abundant class of plants, with about 235,000 species. The group is divided into two large subclasses, the dicots (170,000 species) and the monocots (65,000 species). The differences between the two groups are summarized in Figure 27-2 on the next page.

27-1
One of the most successful angiosperms, showing two characteristics of this large group. One is the flower, in which the seed develops. Note that here, as in other members of the sunflower family, many small flowers are grouped together to form a head. The other characteristic is the fruit, which, in this case, bears a plumelike structure that is caught by the wind and therefore aids in dispersal of the seed.

THE PLANT BODY AND EVOLUTION

As we noted in Chapter 24, the ancestor of the plants was probably a single-celled alga that floated on or just below the water's surface. Like modern plants, its photosynthetic pigments were chlorophylls *a* and *b* and carotenoids, especially beta-carotene, all of which were contained in chloroplasts. It had a membrane-bound nucleus and mitochondria and other cellular organelles, as well as an ex-

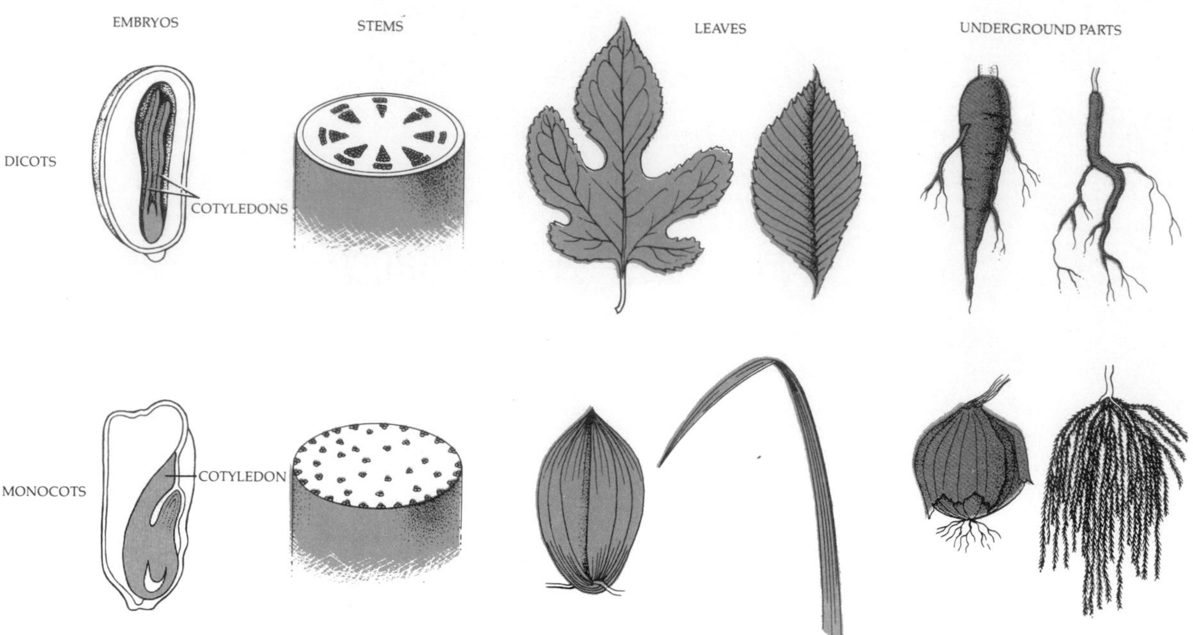

27–2
The angiosperms are divided into two broad groups: the dicots and the monocots. The names refer to the fact that the embryo in the dicots has two cotyledons ("seed leaves") and in the monocots has one. In the dicots, the vascular tissues are arranged around a central core in the stem; in monocots, they are scattered. The veins of dicot leaves are usually fanlike or featherlike; those of monocot leaves are usually parallel. Dicots characteristically have taproots, and monocot roots are often fibrous. These characteristics are discussed in more detail in the text.

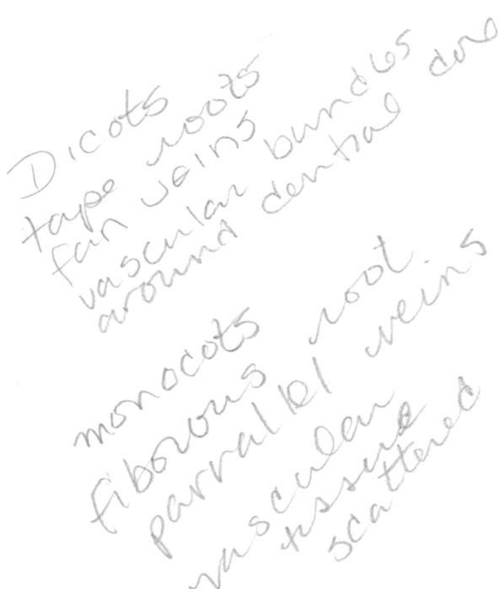

ternal cell wall of cellulose. Its energy source was sunlight, and it obtained oxygen, carbon dioxide, and the minerals it required from the waters in which it lived.

Plants have the same few and relatively simple requirements: light, water, oxygen, carbon dioxide, and certain minerals. From these simple materials they, like their ancestors, make the sugars, amino acids, and all other organic substances on which all plant and animal life depends. But there is an important difference. In the simplest photosynthesizing organism—the single green cell or filament of cells—each of the needed materials is immediately available to every cell. In a plant, however, the individual cells can no longer function autonomously but can survive only through cooperation and division of labor among the many cells and tissues forming the plant body. The reasons for this division of labor are easy to understand. The water and minerals needed by plants are found mostly below the surface of the earth; the development of a complex and extensive root system can be seen as a response to this selection pressure. Sunlight cannot reach these belowground structures, however, and photosynthesis is relegated to another part of the plant body. As the plants began to crowd each other and compete for light, selection pressures favored those with more extensive and efficient light-collecting surfaces—leaves.

The stem raises these photosynthesizing surfaces into the sunlight. Through the specialized vascular tissues of the stem, water and minerals from the ground travel to the leaves, and the products of photosynthesis formed in the leaves are transported to flowers, roots, and other nonphotosynthetic parts of the plant.

Figure 27–3 diagrams the external structure of a familiar plant, the geranium *(Pelargonium)*. It also indicates some of the terms used to describe parts of the plant body. In the rest of this chapter, we shall describe the three principal plant organs: leaves, stems, and roots.

27–3
The body plan of a flowering plant. The aboveground structures constitute the shoot, consisting of the stem, the leaves, whose primary function is photosynthesis, and the flowers, the reproductive structures. Leaves appear at regions on the stem known as nodes. The portions of the stem between nodes are called internodes. The belowground structures, the roots, supply water and minerals to the stem, leaves, and flowers.

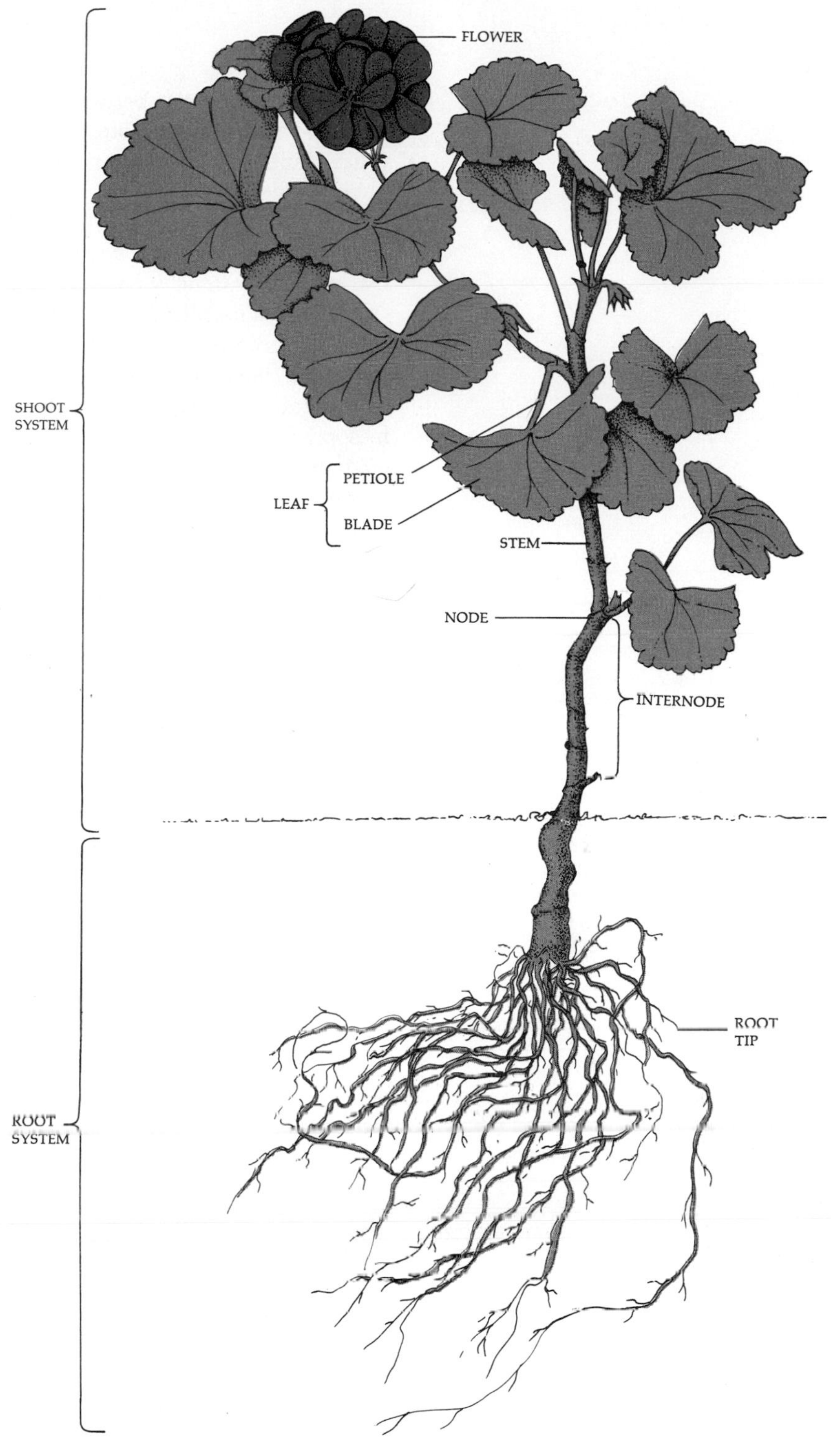

Leaf Structure

A leaf is a compromise between two conflicting evolutionary pressures. The first is to expose a maximum photosynthetic surface to sunlight; the second is to conserve water while, at the same time, providing for the exchange of gases necessary for photosynthesis.

The photosynthetic cells of leaves are of a general type known as *parenchyma*. They are many-sided cells with thin, flexible cell walls. In many leaves, there are two types of parenchyma cells: *palisade parenchyma*, consisting of long columnar cells in which most of the photosynthesis takes place, and *spongy parenchyma*, which consists of more rounded cells with larger air spaces surrounding them. These spaces are filled with gases, including water vapor.

Palisade and spongy parenchyma make up the *mesophyll*, or "middle leaf." The mesophyll is completely enclosed in an almost airtight wrapping made up of epidermal cells. These cells secrete a fatty substance called *cutin*, which forms a coating, the cuticle, over the outer surface of the epidermis. The epidermal cells and the cuticle are transparent, permitting light to penetrate to the photosynthetic cells of the mesophyll.

Materials move into and out of the leaf by two quite different routes: veins and stomata. Veins supply water and minerals to the leaf cells. The veins pass through the petioles (leaf stalks) and are continuous with the rest of the vascular system of the plant. (Note that they are called veins only when they are in the leaf; in the stem and root these same structures are called vascular bundles. It is correct to refer to veins as vascular bundles but not vice versa.) The tissue in vascular bundles that is concerned primarily with water transport is known as the *xylem*. Other groups of cells, the *phloem* (pronounced "flow-em"), transport sugars and other products from the photosynthetic (autotrophic) cells to the other (heterotrophic) cells of the plant.

Veins form distinctive patterns in leaf blades. There is a conspicuous difference between the veins of monocots and dicots: In monocots, the major veins are usually parallel; in dicots, the venation (vein pattern) is netted, either palmate (fanlike) or pinnate (featherlike), as shown in Figure 27–2.

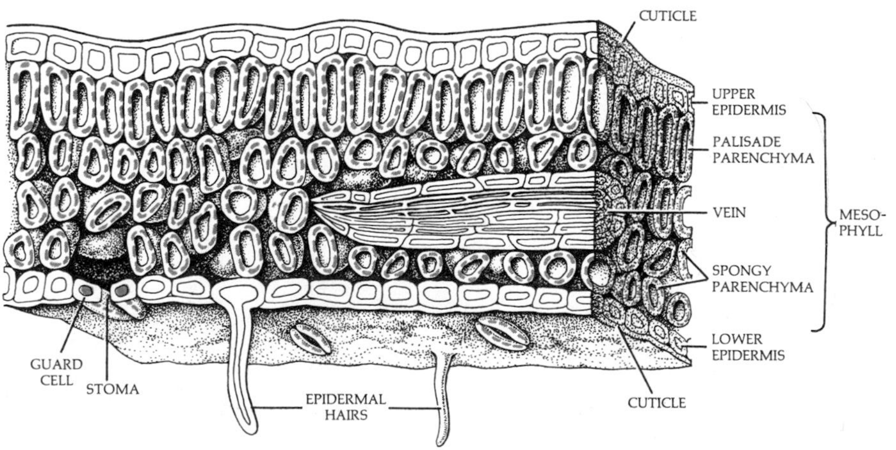

27–4
Diagram of the interior of a leaf. Chloroplasts, indicated in green, occur in the guard cells as well as in the cells of the mesophyll.

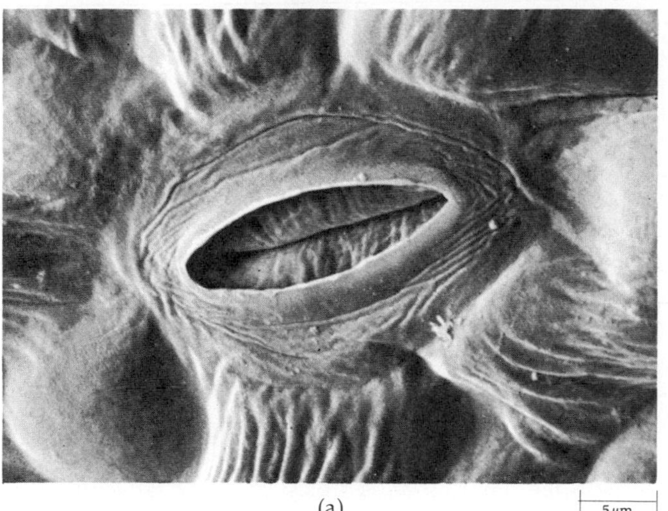

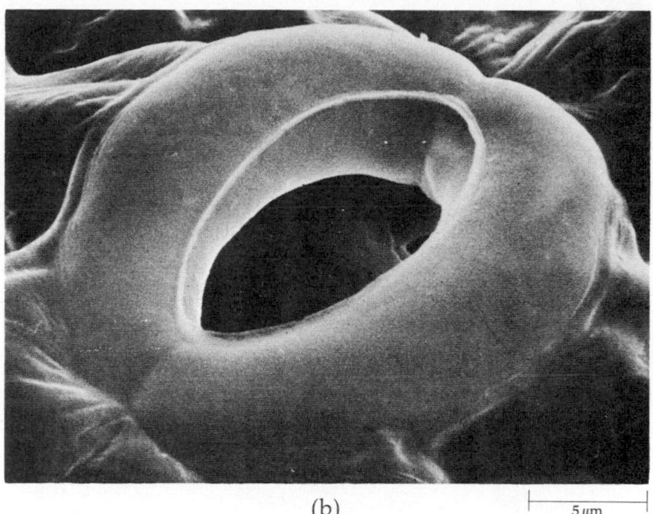

27-5
(a) *Portion of a parsley leaf showing a closed stoma.* (b) *View of an open stoma of cucumber. The stomata lead into a honeycomb of air spaces within the leaf that surround the thin-walled mesophyll cells. The air in these spaces, which make up 15 to 40 percent of the total volume of the leaf, is saturated with water vapor that has evaporated from the photosynthetic cells.*

Gases—oxygen and carbon dioxide—move into and out of the leaf by means of *stomata*. (A stoma, as we define it, includes both the opening, or pore, and the guard cells surrounding it; see Figure 27-5.) This exchange of gases is, as we saw in Chapter 9, necessary for photosynthesis. However, as gases are exchanged, water escapes from the leaf; 90 percent of the water that leaves the plant body is lost through the stomata. The stomata open and close in response to a variety of stimuli, both internal and external. This process will be described on page 533.

Stomata may be very numerous; for example, in tobacco leaves, there are 12,000 stomata per square centimeter of leaf surface. They are usually most abundant on the undersurface.

27-6
Cross section of the leaf of an oleander, a plant adapted to a dry climate. Note the very thick single-layered cuticle at the top covering the epidermis, which consists of four layers of cells. The stoma is contained within a stomatal crypt, which is lined with epidermal hairs.

100 μm

Leaf Shapes and Adaptations

Leaves come in all shapes and sizes, ranging from broad fronds to tiny scales. Some of these differences can be correlated with the environments in which the plants live. Large leaves with broad surfaces are often found in plants that grow under the canopy in a tropical rain forest, where water is plentiful but where there is intense competition for light. Leaves of such plants sometimes have "drip tips," which facilitate the runoff of rainwater. Leaves with small photosynthetic surfaces are associated with dry climates. In conifers, for example, the photosynthetic surfaces are greatly reduced and there is an extra-thick layer of epidermis and cuticle (see page 410). Similarly, angiosperms found in dry habitats often have small, leathery leaves. This reduction in leaf surface reaches its extreme in desert cacti, which have no leaves at all. In these plants, photosynthesis takes place in the fleshy stems, which are also water-storage organs.

Leaf Abscission

Another type of adaptation related to the conservation of water is found in the leaves of deciduous trees. Deciduous trees, which drop their leaves annually, are found in regions where there is a marked seasonal variation in available water. The fall of the leaf comes about when special enzymes dissolve the middle lamellae (see

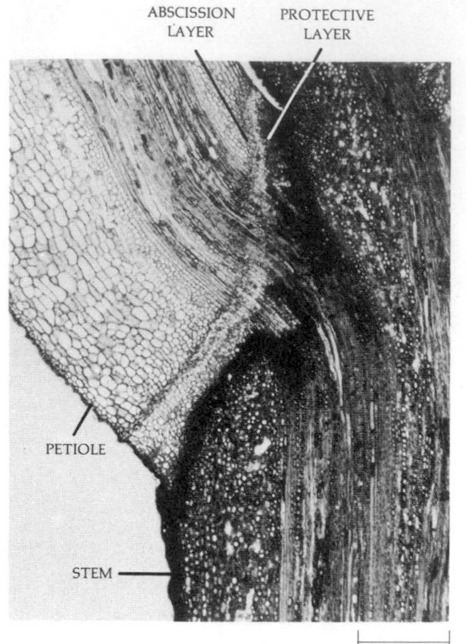

27–7
Abscission layer in a maple leaf, as seen in a longitudinal section through the base of the petiole.

page 103) in a layer of weak, thin-walled cells in the petiole. These cells compose the *abscission layer*, and the process is known as *abscission* (Figure 27–7). Deciduous trees are able, unlike the conifers, to present a broad, efficient light-collecting surface to the sun during favorable periods, while minimizing their water loss during dry periods. Thus deciduous trees are far better energy collectors. On the other hand, they must pay the relatively high price, energetically speaking, of replacing all their leaves annually.

Some Leaf Modifications

In some plants, leaves are modified as spines, which are hard, dry, and nonphotosynthetic. (The terms "spine" and "thorn" are often used interchangeably; however, thorns are technically modified branches.) In other plants, such as the garden pea, leaves are modified as tendrils. In many plants, leaves are succulent; that is, they are adapted for water storage. Other leaves are specialized for food storage. A bulb, such as the onion, is a large bud consisting of a short stem with many leaves modified for food storage. (In the onion, food is stored as sugar rather than as starch.) The "head" of a cabbage also consists of a compressed stem bearing numerous thick, overlapping leaves. In some plants, the petioles become thick and fleshy: Celery and rhubarb are two familiar examples.

In the flower-pot plant, which is an epiphyte (a plant that grows on other plants but is not parasitic on them), some of the leaves form tubes (Figure 27–8). Rainwater and debris collect in these tubes. Roots of the plant grow into the leaf tubes and collect water and minerals from them. In addition, ants colonize the tubes, adding to the nitrogen supply with their excreta and dead bodies.

Among the most spectacular modified leaves are those of the carnivorous plants, such as the pitcher plant, Venus flytrap, and sundew (Figure 27–9e). These plants capture insects by various means and then digest them with enzymes secreted by the leaf cells. The insects are an additional source of nitrogen for the plant.

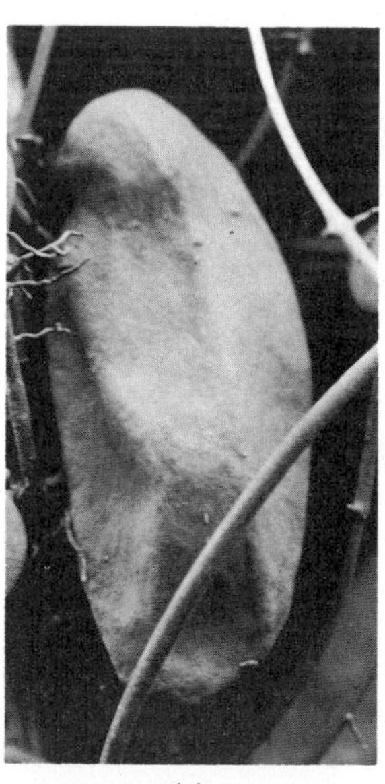

(a)

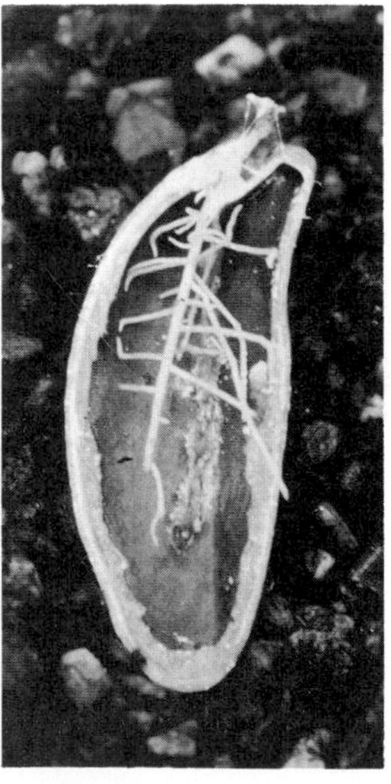

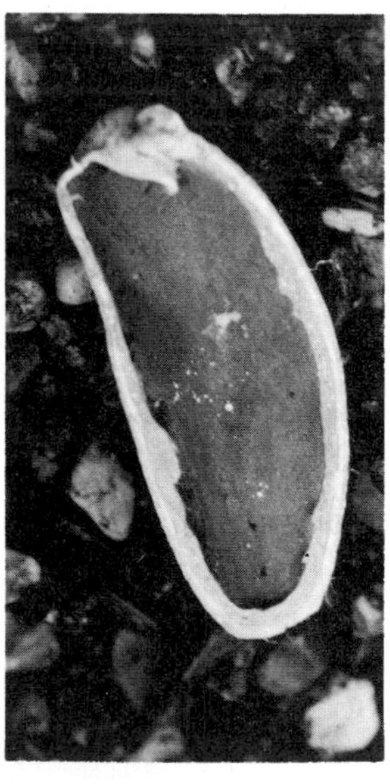

(b)

27–8
Dischidia rafflesiana, *the "flower-pot plant."* (a) *Tubular leaf, or "pot," which collects debris and rain runoff.* (b) *Leaf cut open to show roots that have grown down into it from node.*

(a)

(b)

(c)

(d)

(e)

27–9
Modified leaves. (a) *Leaves adapted for water storage* (Sedum). (b) *Tendrils of a pea plant.* (c) *An onion bulb is composed of leaves modified for storage.* (d) *Spines on a prickly-pear cactus.* (e) *Carnivorous leaves of the round-leaved sundew. The upper exposed surface of this leaf is covered with hairlike structures, the tentacles, each about 3 millimeters long, enlarged at the tip. A droplet of sticky secretion surrounds each enlargement. These droplets trap insects or other comparably small animals, and then the tentacles secrete enzymes that digest the prey.*

THE STEM

The Structure of the Stem

The stem holds the leaves up to the light and provides for the transportation of substances to and from the leaves. The outer surface of a young green stem, like that of the leaf and the root, is made of epidermal cells. Like the leaf, the green stem is covered with a fatty cuticle, contains stomata, and is photosynthetic.

Ground Tissue

The bulk of the tissue of a young stem is known as the *ground tissue*. Like the mesophyll of the leaf, ground tissue is composed mostly of parenchyma cells. The turgor (page 129) of these cells provides the chief support for young green stems.

Stems also may contain specialized supporting tissues formed from *collenchyma* or *sclerenchyma* cells. Collenchyma cells (Figure 27–10a) differ from the thin-walled parenchyma cells of the stem in having primary cellulose walls that are thickened at the corners or in some other uneven fashion. Their name derives from the Greek word *colla*, meaning "glue," which refers to their characteristic thick, glistening walls. Collenchyma cells are often located just inside the epidermis, forming either a continuous cylinder or distinct vertical strips of supporting tissue.

Sclerenchyma cells are of two types: fibers and sclereids. The name is derived from the Greek *skleros*, meaning "hard." Fibers, which are elongated, somewhat elastic cells (Figure 27–10b), typically occur in strands or bundles arranged in patterns characteristic of the plant. These supporting cells are often associated with the vascular tissues. Plant fibers such as flax, hemp, jute, sisal, and raffia have long been used in human artifacts, including baskets, rope, and cloth. Sclereids, or stone cells (Figure 27–10c), which are variable in form, are less common in stems than are fibers. Layers of sclereids are found more frequently in seeds, nuts, and fruit stones, where they form the hard outer coverings.

27–10
Some types of cells found in the ground tissue of stems. (a) *Collenchyma cell. Its irregularly thickened cellulose walls contain pectin and large amounts of water; they are plastic, so they permit growth while providing support.* (b) *Fiber. Its specialization is thickened, often lignified cell walls that give it strength and rigidity. Many fibers, but not all, are dead at maturity, like the cell shown here.* (c) *Sclereid. This type of cell, which has very thick lignified walls, is often found in seeds and fruit, as well as in stems. This sclereid is from a pear; sclereids give the fruit its characteristic gritty texture. This cell is a living cell with slender branches of cytoplasm—plasmodesmata—extending through the cell walls.*

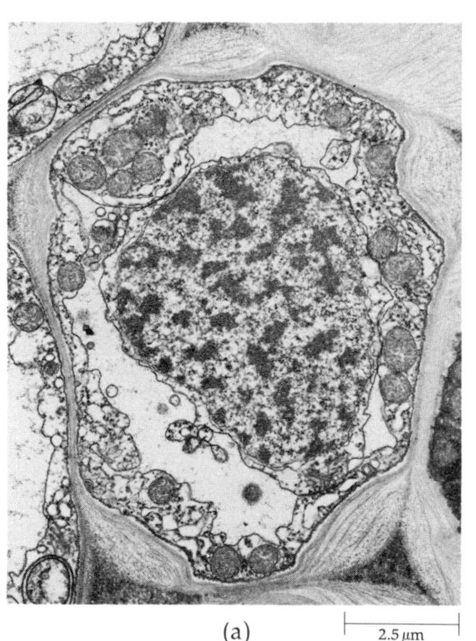

(a)

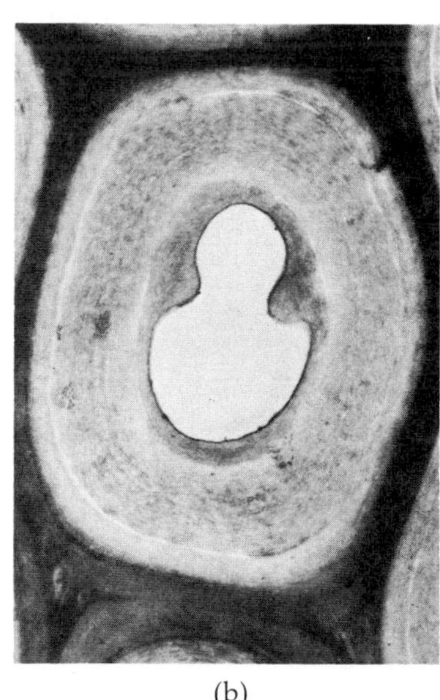

(b)

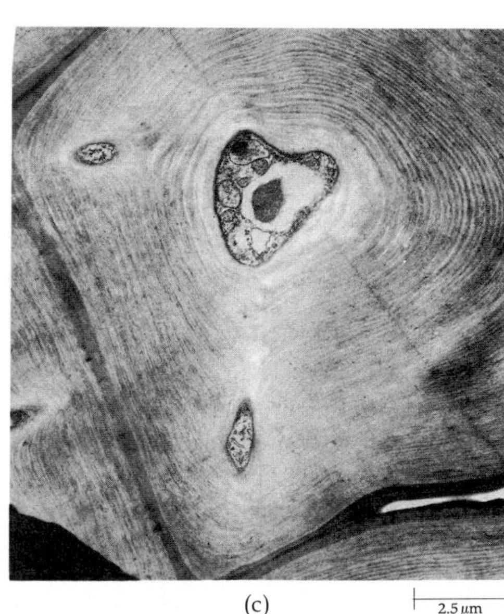

(c)

Sclerenchyma cells differ from collenchyma in three respects: (1) They have secondary walls (see page 104); (2) the walls often contain lignin, a complex macromolecule that impregnates the cellulose and toughens and hardens it; and (3) the cells are sometimes dead at maturity, with only their cell walls remaining.

Conducting Tissues

Phloem, as we noted previously, conducts the products of photosynthesis, chiefly in the form of sucrose, from the leaves to the nonphotosynthetic cells of the plant. In angiosperms, the conducting cells of the phloem are called *sieve-tube elements* (Figure 27–11). A *sieve tube* is a vertical column of sieve-tube elements joined by their end walls. These end walls, called *sieve plates*, have openings leading from one sieve-tube element to the next.

The sieve-tube elements, which are alive at maturity, are filled largely with a watery cytoplasm. They are characterized by the presence of a protein-containing substance called P-protein (or sometimes, slime). In mature sieve-tube elements, P-protein lies along the inner surfaces of the cell. It is continuous from one cell to the next through the sieve plates. The nucleus of a sieve-tube element disintegrates as the cell matures, as do many of the organelles. Sieve-tube elements are characteristically associated with *companion cells*, which are thought to provide nuclear control and energy for the sieve tubes. A sieve-tube element can function only if its cell membrane is intact, and it is likely that the companion cell helps to maintain the membrane.

Xylem conducts water and minerals from the roots to other parts of the plant body. It is customary to think of xylem as transporting water up and phloem as transporting sugars down, but if you think of the various shapes of plants you can see that water must also often be transported laterally, as along a tendril, or even down, as to the branches of a weeping willow. Conversely, sugars must often go upward, as into a flower or fruit.

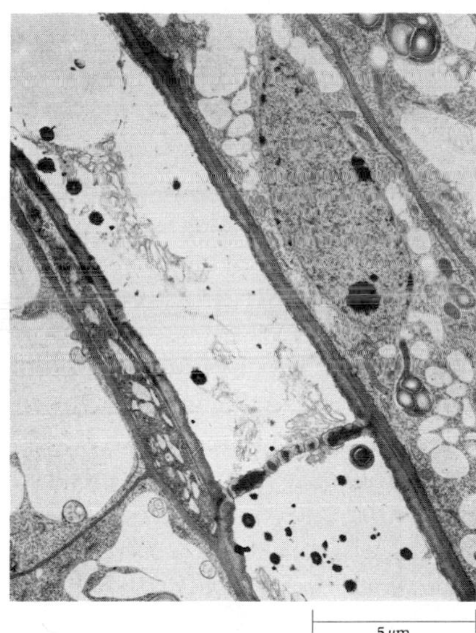

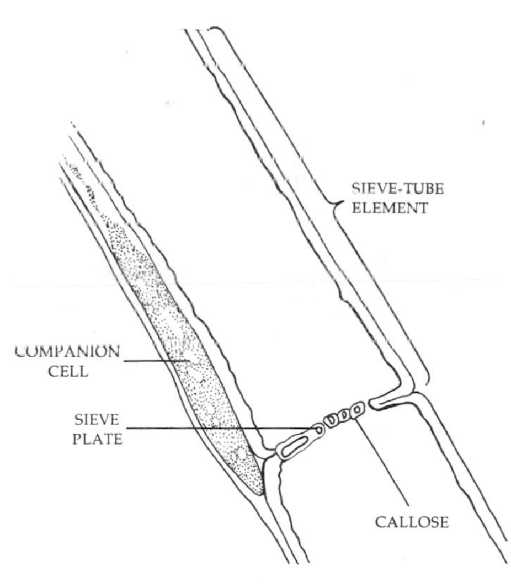

27–11
In angiosperms, the conducting elements of the phloem are sieve tubes, made up of individual cells, the sieve-tube elements. These cells, which lack nuclei at maturity, are usually found in close association with companion cells, which do have nuclei. Sieve-tube elements are joined by a sieve plate. The light-colored (electron-transparent) material near the pore is callose, a polysaccharide characteristically associated with sieve plates, especially in old or injured cells. Callose plugs the pores when a sieve-tube element is injured, thus preventing leakage. Immediately to the left of the upper sieve-tube element, identifiable by its dense cytoplasm, is a portion of a companion cell. A parenchyma cell is to the right.

27-12
Tracheids and vessel elements are the conducting cells of the xylem in angiosperms. (a) *Tracheids are a more primitive and less efficient type of conducting cell. Water passing from one tracheid to another passes through pits. Pits are not perforations but areas in which there is no secondary cell wall. Water moving from one tracheid to another passes through two primary cell walls and the middle lamella. Vessel elements differ from tracheids in that the primary walls of vessel elements are perforated at the end where they are joined with other vessel elements.* (b) *There may be numerous small perforations in adjoining walls of vessel elements, or* (c) *the adjoining walls may break down completely as the cells mature, forming a single opening. Vessel elements are also characteristically shorter and wider than tracheids and their adjoining walls are less oblique.*

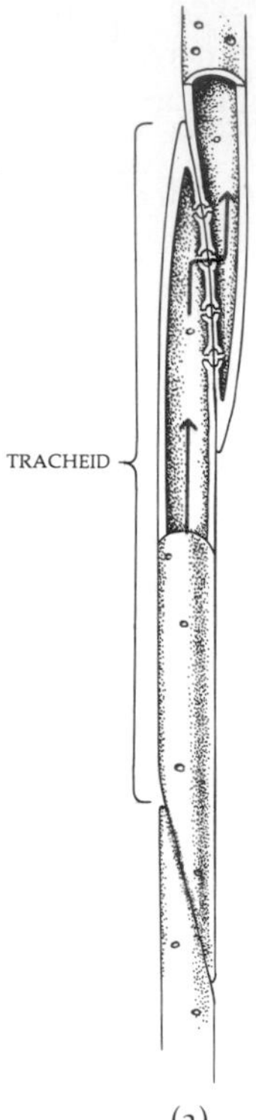

(a)

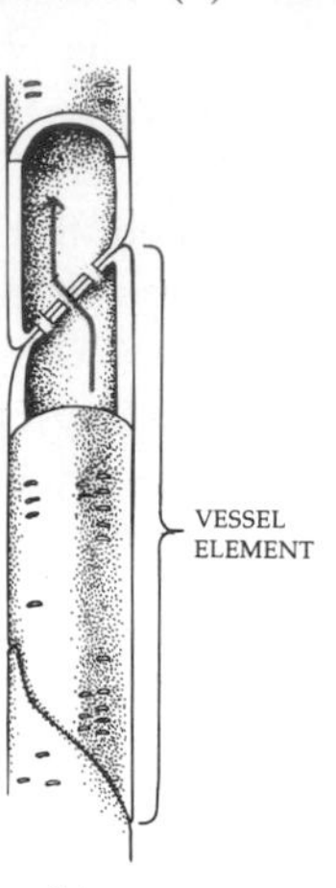

(b)

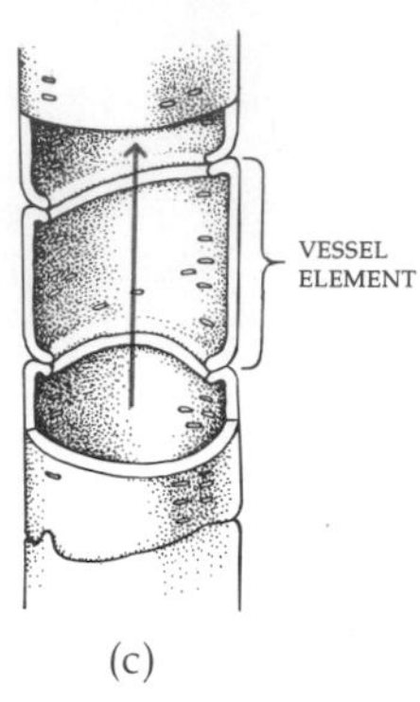

(c)

In angiosperms, the conducting cells of the xylem are *tracheids* and *vessel elements*. Both of these cell types have thick secondary walls impregnated with lignin, and both are dead at maturity. Tracheids are long, thin cells that overlap one another on their tapered ends (Figure 27-12a). These overlapping surfaces contain thin areas, pits, where no secondary wall has been deposited. Water passes from one tracheid to the next through the pits. Vessels, which are much larger, also differ from tracheids in that their end walls contain one or more perforations or are broken down entirely (Figure 27-12b and c). Thus, the vessel elements form a continuous *vessel*, which is a more effective conduit than a series of tracheids. Lower vascular plants and most gymnosperms have only tracheids. Most angiosperms have both tracheids and vessels.

Both phloem and xylem tissue contain parenchyma cells, which store food and water, and supporting fibers.

Stem Patterns

In green stems, the xylem and the phloem are usually arranged in longitudinal parallel strands, the vascular bundles, which are embedded in the ground tissue. In young dicot stems, the vascular bundles form a ring, the vascular cylinder, around a central area of ground tissue called the *pith*. The cylinder of ground tissue outside the vascular bundles is called the *cortex*. Within each bundle, the xylem is characteristically on the inside, adjacent to the pith, and the phloem is on the outside, adjacent to the cortex. In monocots the vascular bundles are scattered throughout the ground tissue (see Figure 27-14c).

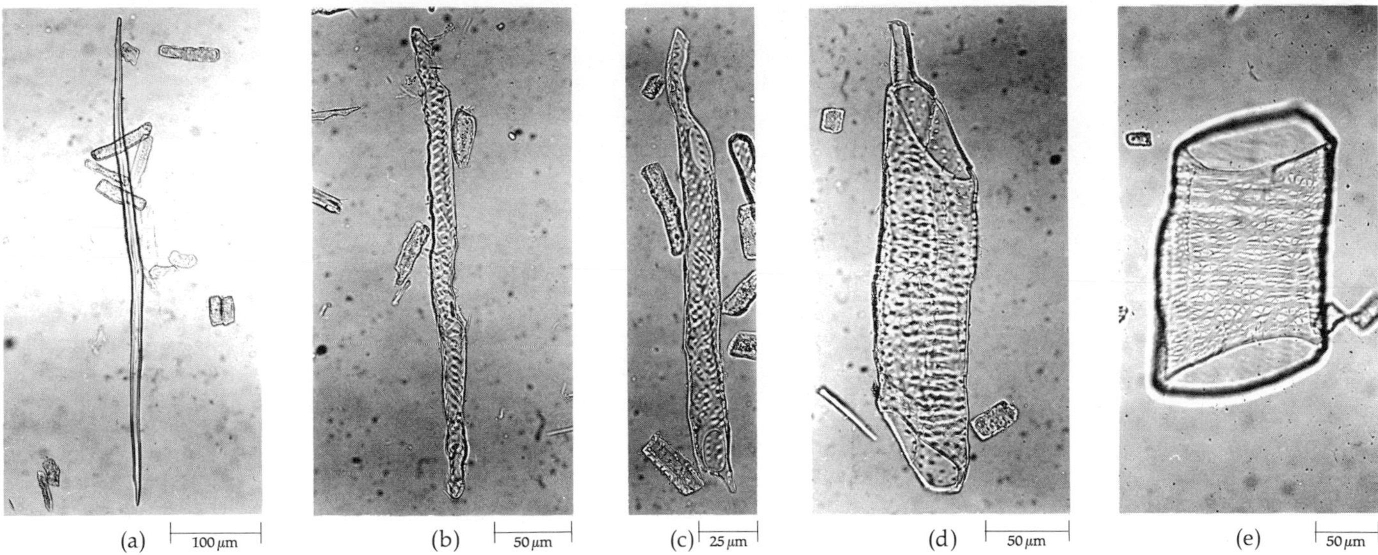

27–13
Cell types in the xylem of an oak: (a) *a fiber with some parenchyma cells;* (b) *a tracheid (the spots in the wall are pits);* (c) *a narrow vessel element;* (d, e) *wide vessel elements. Both fibers and vessel elements evolved from primitive tracheids.*

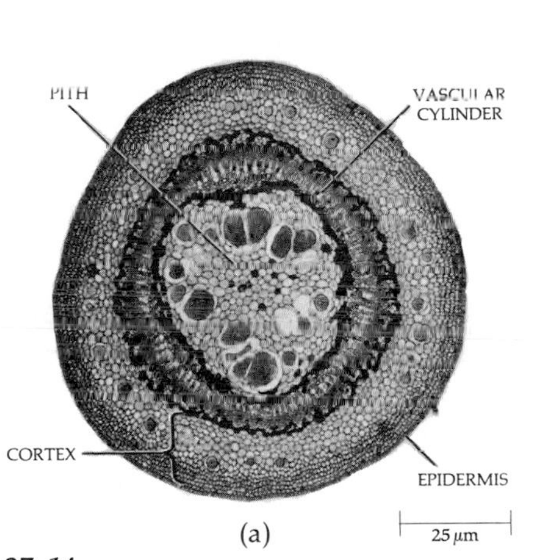

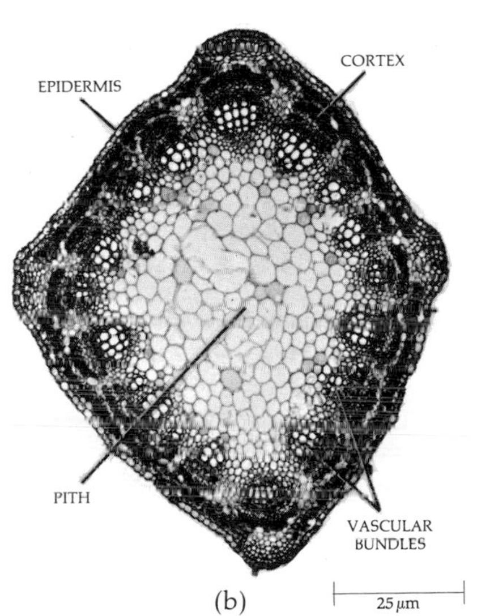

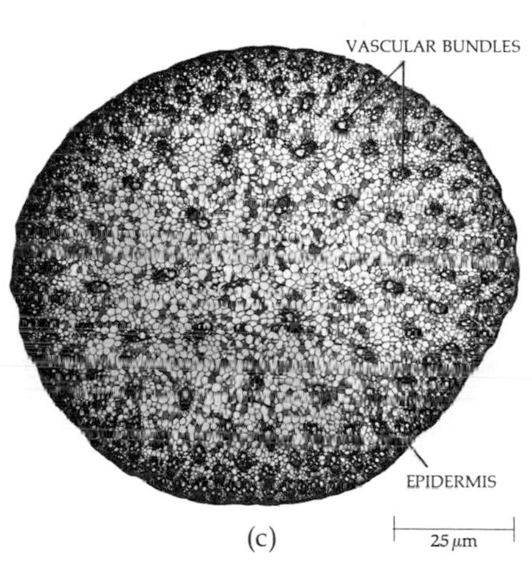

27–14
Cross sections of two dicot stems and one monocot stem. (a) *In the linden, a dicot, the vascular tissue forms a continuous cylinder.* (b) *In alfalfa, also a dicot, the cylinder is made up of separate vascular bundles.* (c) *In corn, a monocot, numerous vascular bundles are scattered throughout the ground tissue.*

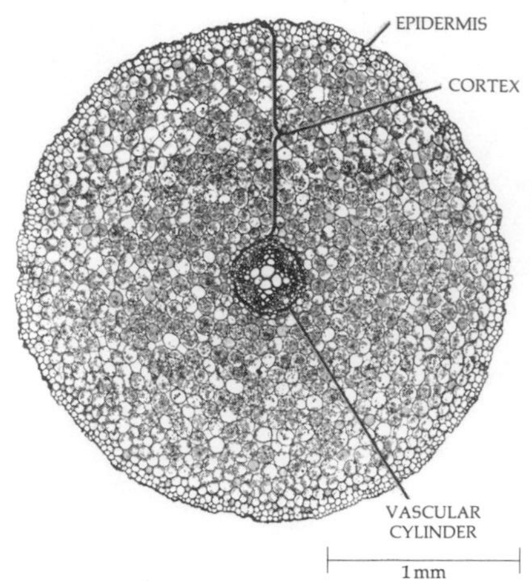

27–15
Root of a buttercup in cross section. The vascular cylinder is shown in more detail in Figure 27–19.

Special Adaptations of the Stem

The stems of some climbing plants coil themselves around the structures on which they are growing. Others produce modified branches in the form of tendrils. (As we saw previously, tendrils may also be modified leaves.) The tendrils of English ivy, grape, and Virginia creeper are all modified stems.

Runners, such as those found in most varieties of strawberry, are long, slender stems that grow along the surface of the soil. Rhizomes are underground stems that often, as in the grasses, form buds and from these buds produce upright stems bearing leaves and flowers.

Stems also may be adapted for food storage. White potatoes are enlarged rhizomes known as tubers. In some species of plants that grow in arid environments, stems are adapted for water storage. The water-storing tissues consist of large parenchyma cells that lack chloroplasts; the stem of a cactus may be 98 percent water by weight.

ROOTS

Roots anchor the plant and, as we noted at the beginning of the chapter, are specialized for taking up water and essential minerals. In an older plant, the root system may make up more than half of the plant body. The lateral spread of tree roots is usually greater than the spread of the crown of the tree. In a study made on a four-month-old rye plant, the total surface area of the root system, including root hairs, was calculated by extrapolation to be 639 square meters, 130 times the surface area of the leaves and stem. Root growth is affected by soil conditions and availability of water. The deepest known root was that of a pine growing on sandy, porous soil. It had penetrated to a depth of about 6.5 meters.

27–16
Root hairs of a radish seedling. Most of the uptake of water and minerals takes place through the root hairs, which begin to form just behind the growing tip of the root.

The Structure of the Root

The internal structure of the root is comparatively simple. There are three concentric layers: the *epidermis*, the *cortex*, and the *vascular cylinder* (Figure 27–15).

The Epidermis

The epidermal cells of the root, which enclose the entire surface of the root, absorb water and minerals from the soil. As you would expect, they lack the cuticle found on the surface of the epidermal cells of the leaf. They are also characterized by fine, threadlike outgrowths, known as root hairs (Figure 27–16). Root hairs are slender extensions of the epidermal cells themselves; the nucleus of the epidermal cell is often found within the root hair.

In the study of the rye plant previously mentioned, the roots were estimated to have some 14 billion root hairs; placed end to end, they would have extended more than 10,000 kilometers. Much of the water and minerals that enter the root are taken up through these delicate outgrowths of the epidermal cells. (However, as we shall see in Chapter 29, in many species mycorrhizal associations seem to substitute for root hairs.

The Cortex

The cortex occupies by far the greatest volume of the young root (Figure 27–15). The cells of the cortex are parenchyma cells, like those of the mesophyll of the leaf

27–17
Diagrammatic cross section of a root, showing the two pathways of uptake of water and minerals. Along pathway A, water moves by osmosis and solutes by active transport through the cell membranes and plasmodesmata of a series of living cells. Along pathway B, water flows through the cell walls and along their surfaces, and the solutes flow with the water or by diffusion. Notice the location of the Casparian strip and how it blocks off pathway B all around the vascular cylinder of the root. In order to pass the Casparian strip, the solutes must be transported through the cell membranes of the endodermal cells, as in pathway A.

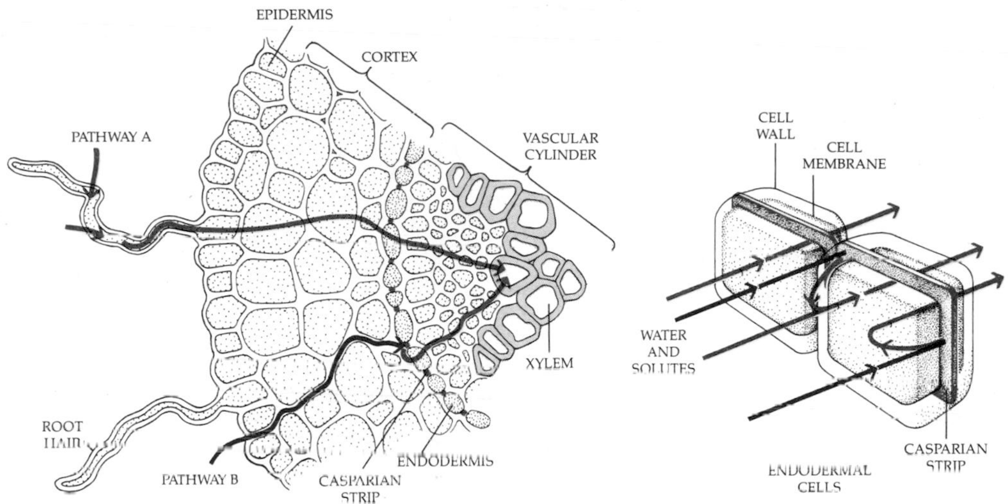

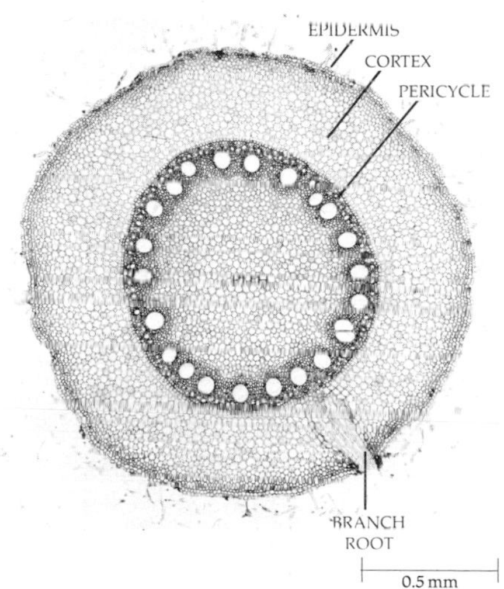

27–18
Cross section of the root of a corn plant, showing the vascular cylinder enclosing the pith. Part of a branch root can be seen at the lower right.

and the ground tissue of the stem; however, as you would expect, they lack chloroplasts. They store starch and other organic substances. The tissue of the cortex contains many air spaces. Oxygen-containing air enters these spaces through the epidermal cells and is used by the cortical cells in respiration.

Unlike the rest of the cortex, the cells of the innermost layer, the *endodermis*, are compact and have no spaces between them. Each endodermal cell is encircled by a *Casparian strip*, a fatty band within the cell wall (see Figure 27–17). The strip is continuous and is not permeable to water. Therefore, water and dissolved substances, which pass freely around the other cortical cells and through their cell walls, must pass through the cell walls and, more important, the cell membranes of endodermal cells. As you will recall (page 128), water, oxygen, and carbon dioxide pass freely through cell membranes, but many ions and other substances do not. Therefore, the membranes of the endodermal cells regulate the passage of such substances into the vascular tissues of the root and so determine what enters the plant body, as we shall see in Chapter 29.

The Vascular Cylinder

The vascular cylinder of the root consists of the vascular tissues (xylem and phloem) completely surrounded by one or more layers of cells, the *pericycle*. Branch roots arise from the pericycle. In most species, the vascular tissues are grouped in a solid cylinder, as shown in Figure 27–15. In some, however, they form a hollow cylinder around a pith, a central core of ground tissue (Figure 27–18). Figure 27–19 shows the details of the vascular cylinder of a buttercup.

27–19
Details of the vascular cylinder shown in Figure 27–15. The endodermis is considered part of the cortex. The outermost layer of the vascular cylinder is the pericycle, from which branch roots arise. The conducting tissues, the xylem and the phloem, are within the pericycle.

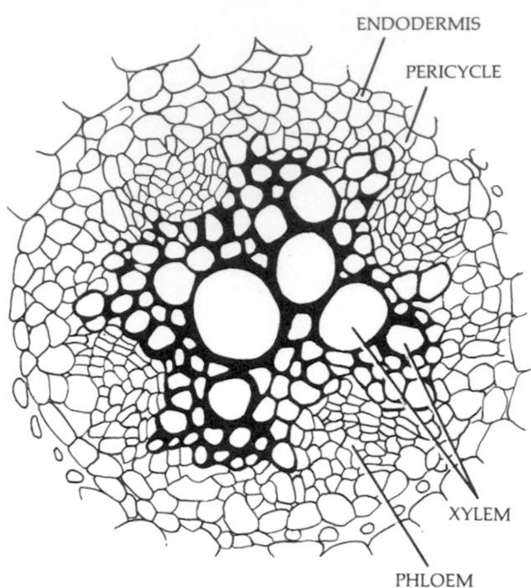

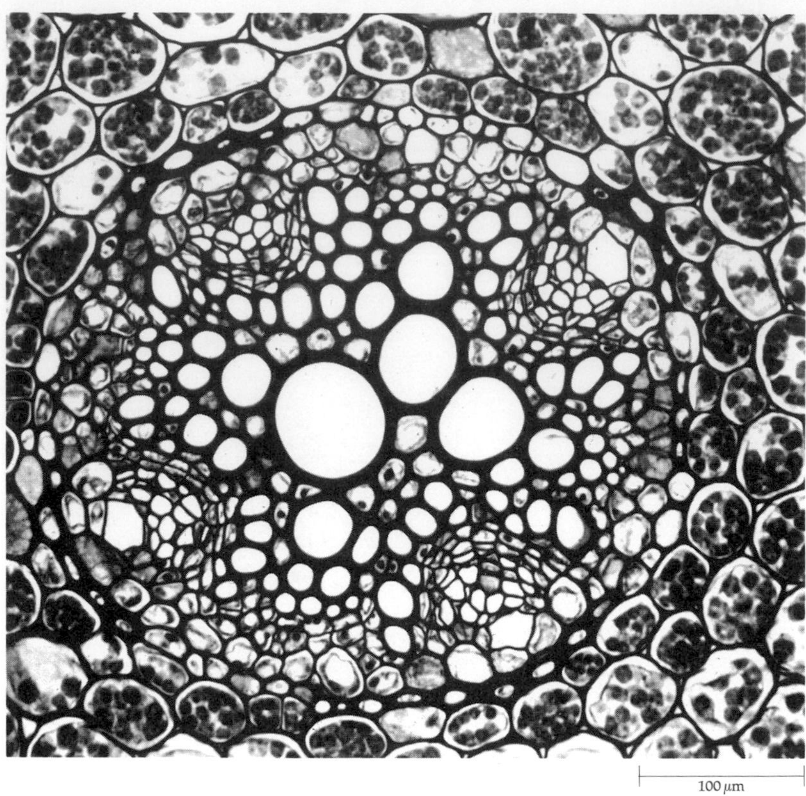

Patterns of Root Growth

The first root of a plant, which originates in the embryo, is called the radicle. In many dicots, this root develops into a *taproot* that, in turn, gives rise to lateral or branch roots (Figure 27–20a). In monocots, the primary root is usually short-lived and the final root system develops from the base of the stem; such roots are called adventitious roots. (Adventitious describes any structure growing from an unusual place.) The adventitious roots and their branches develop into a fibrous root system (Figure 27–20b).

Special Adaptations of Roots

Aerial roots are adventitious roots produced from aboveground structures. Some aerial roots, such as those of English ivy, cling to vertical surfaces and thus provide support for the climbing stem. In some plants, such as corn, aerial roots serve as prop roots (Figure 27–21a). Trees that grow in swamps, such as the red mangrove and the bald cypress (Figure 27–21b), often have prop roots. In swampy areas the soil is low in oxygen, and, in some species, the prop roots are believed not only to anchor the plant but also to supply the root cells with the oxygen needed for respiration. Black mangroves have roots whose tips grow upward out of the mud and serve this aerating function.

Many special root adaptations are found among epiphytes, such as the flowerpot plant (Figure 27–8). In some species of orchids, for instance, the root is the only photosynthetic organ.

27–20
Types of root systems. (a) *Taproot and branch roots of a carrot (a dicot).* (b) *Fibrous root system characteristic of grasses (monocots).*

Most roots are storage organs, and in some species the roots are highly specialized for this function, with an abundance of storage parenchyma. Beets, carrots, and sweet potatoes are examples of such roots.

SUMMARY

Plants are multicellular photosynthetic organisms adapted to life on land. The largest and most recently evolved class of plants is the angiosperms, the flowering plants. Angiosperms are divided into two subclasses: monocots and dicots. Together they number about 235,000 species.

The plant body has specialized photosynthetic areas (leaves), conducting and supporting structures (stems), and organs that anchor the plant in the soil and absorb water and minerals from it (roots).

The blade of the leaf is composed principally of photosynthetic parenchyma cells enclosed by epidermis. Parenchyma cells, which are thin-walled and many-sided, are the most common cells of the plant body. In the leaf, they include palisade cells and spongy parenchyma. Palisade cells, in which most of the photosynthesis takes place, are elongated cells with large central vacuoles. Spongy parenchyma cells, also photosynthetic, are surrounded by large air spaces. These spaces are normally filled with carbon dioxide, oxygen, water vapor, and other gases. The palisade and spongy parenchyma make up the mesophyll.

Veins, the vascular bundles of the leaf, conduct water and minerals to the leaf cells (through the xylem) and remove sugars from the leaf (through the phloem). The upper and lower surfaces of the leaf are covered with a layer of epidermal cells, which are transparent. The outer surfaces are covered with a fatty layer, the cuticle.

(a)

(b)

27–21
(a) *Prop roots of corn. These are adventitious roots, arising from the stem.* (b) *Aerial roots of bald cypress. Such roots serve to anchor the plant in a marshy habitat and also, perhaps, provide aeration.*

The epidermis contains specialized pores, the stomata, which open and close, regulating the exchange of gases and the release of water vapor. The leaf blade is attached to the stem by the petiole. The conducting tissues of the leaf are continuous with the vascular bundles of the stem.

Green stems, like leaves, have an outer layer of epidermal cells covered with a cuticle. Within the epidermis is the ground tissue of the stem, which in young dicot stems is divided into cortex and pith. The ground tissue is made up largely of turgid parenchyma cells. Also present may be collenchyma cells, which are living cells with irregularly thickened primary cell walls, and sclerenchyma cells, which have both primary and secondary cell walls, often containing lignin. Sclerenchyma cells may be dead at maturity. There are two kinds of sclerenchyma cells: fibers and sclereids. Fibers are elongated and somewhat elastic and are frequently found in various types of stems. Sclereids are irregularly shaped cells sometimes found scattered in stems and commonly found in dense layers in seed coats, nuts, and fruit stones.

The vascular tissues consist of phloem and xylem. In angiosperms, the conducting cells of the phloem are the sieve-tube elements, living cells with their end walls perforated (sievelike), which form continuous sieve tubes. Associated closely with each sieve-tube element are companion cells. The conducting tissues of the xylem are made up of a series of vessel elements and tracheids. Vessel elements and tracheids characteristically have thick secondary walls and are dead at maturity. Phloem and xylem also contain parenchyma cells and fibers.

Roots have an outer layer of epidermis but, unlike stems, have no cuticle. Extensions of the epidermal cells form root hairs, which greatly increase the absorptive surface of the root. Beneath the epidermis is a cortex composed mostly of parenchyma cells, often with large storage capacities. The innermost layer of the cortex is the endodermis, a single layer of specialized cells whose walls contain a waterproof zone, the Casparian strip. Just inside the endodermis is another layer of cells, the pericycle, from which branch roots arise. Within the pericycle are the xylem and phloem. The xylem and phloem of the root resemble those of the stem in composition but not in arrangement. In the root, xylem characteristically fills the central core and the phloem is arranged in strands outside the core of xylem.

There are characteristic differences between monocots and dicots. The arrangement of the veins of dicot leaves usually is netted, whereas the major veins of monocots are parallel. In monocot stems, strands of xylem and phloem, paired in vascular bundles, are scattered throughout the ground tissue. In dicot stems, the strands of xylem and phloem form a ring around a central zone of ground tissue, the pith. The ground tissue between the vascular bundles and the epidermis is the cortex. Other differences between dicots and monocots include: two seed leaves (cotyledons) in dicots, one in monocots; and commonly, a main taproot in dicots, numerous fibrous roots in monocots.

Leaves, stems, and roots are highly modified in some species. Special adaptations among leaves include wide variations in leaf shape, leaf abscission, and spines and tendrils, which are modified leaves. Leaves may also be specialized for storage of food, as in a bulb, or water, as in succulents. Stems may run along the surface of the ground (runners) or underground (rhizomes), and they may be specialized for water storage, as in cacti, or food storage, as in white potatoes. Roots may be modified as props, as aerators, and for clinging or climbing. Frequently they are specialized for food storage.

QUESTIONS

1. Sketch the interior of a leaf, labeling the principal cells or tissues. Describe the function of each of the labeled parts.

2. Sketch cross sections of (a) a dicot stem, (b) a monocot stem, and (c) a root. Label the principal cells or tissues in each sketch, and describe the function of each of the labeled parts.

3. The epidermis of stems and leaves is covered by a cuticle. This layer is absent in roots. What differences in function between stems and leaves on the one hand, and roots on the other, make this structural difference valuable to plants?

4. What two tissues are present in roots, but absent in stems? What functions do these tissues perform in the roots, and why are they necessary for the well-being of the plant?

5. We have seen that the tissue of the root cortex contains many air spaces. Also, we noted that some plants growing in swampy areas have aerial roots. You may have discovered the hard way that over-watering can kill house plants. What need of the plant root is indicated by these data? Why is this need not met by the leaves of the plant?

6. It has often been said that the principal differences among root, stem, and leaf are quantitative in nature rather than qualitative. Explain.

7. If an orchid root is green and photosynthetic, why is it called a root?

CHAPTER 28

Plant Reproduction, Development, and Growth

Unlike the higher animals, whose reproduction is almost exclusively sexual, many plants reproduce both sexually and asexually. (Asexual reproduction is often referred to as vegetative reproduction.) Organisms produced by asexual reproduction are genetically identical to their single parent, whereas organisms produced by sexual reproduction, which involves meiosis and fertilization, are different from both parents.

ASEXUAL REPRODUCTION

There are many forms of asexual reproduction among plants. One of the most familiar occurs by means of runners or rhizomes (Figure 28–1). Strawberries are a common example of plants that propagate by runners, as are spider plants, commonly grown as hanging plants. Plants that reproduce by rhizomes include potatoes, many flowering garden perennials, such as lilies-of-the-valley, irises, and dahlias, and the sod-forming grasses of lawns and pastures. Both runners and rhizomes develop adventitious roots.

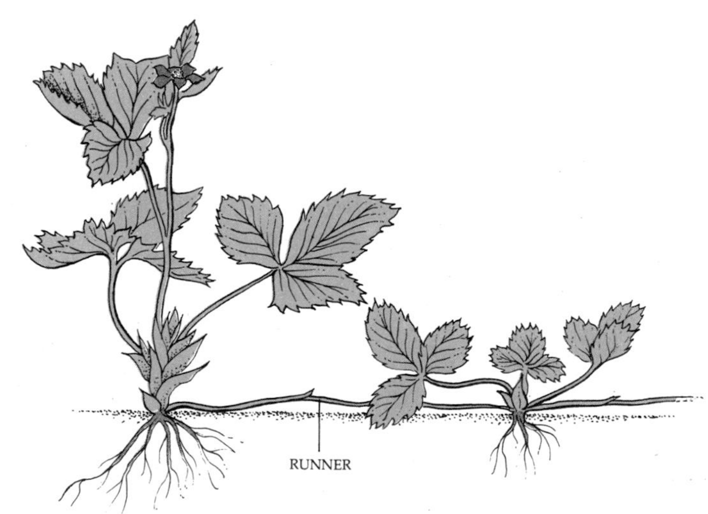

28–1
Wild strawberry plants reproduce asexually by means of aboveground stems, or runners. Roots and leaves develop at every second node along the modified stems. These plants also form flowers and reproduce sexually.

28–2
Tiny plantlets grow along the leaf margins of Kalanchoe, *the "maternity plant." When mature, they drop to the soil, take root, and form new plants.*

Many members of the lily family, which includes onions and tulips as well as lilies, reproduce asexually by bulbs. Some species of plants with arching stems, such as raspberries, develop new roots from stem tips that touch the soil. New plants may form if the stem is subsequently broken, separating it from the parent plant. The leaves of some plants, such as African violets, develop adventitious roots when they are detached from the parent plant and may give rise to new individuals in that way. Some species of *Kalanchoe* produce plantlets in the margins of leaves, which later drop to the ground and develop into separate plants. If the taproot of a dandelion is injured or broken near the soil line, a callus (a mass of undifferentiated cells) forms, plugging the wound. Eventually several new plants grow from this callus tissue. Thus both lawn mowers and grazing animals have highly beneficial effects on the dandelion population.

The capacity of many species of plants to reproduce asexually has been exploited in the development of cultivated varieties of plants for food or ornamental use. Such plants are, of course, genetically identical to the parent stock, and so vegetative reproduction is a way of preserving uniformity. Many plants are reproduced by stem cuttings, which simply involves sticking young stems in the ground and protecting them from drying out until adventitious roots appear. Rooting can often be facilitated by hormone treatment (see page 548). Another artificial form of plant propagation is grafting, in which a stem cutting is attached to the main stem of a rooted woody plant. Most fruit trees and roses are propagated in this way.

Many economically important plants are sterile and can only be propagated vegetatively; these include pineapples, bananas, seedless grapes, navel oranges, and numerous ornamental plants.

SEXUAL REPRODUCTION

28–3
Flower of a tulip. In the center of the corolla are the fused carpels surrounded by the stamens with their dark brown anthers, within which the pollen forms. This is a perfect flower.

The flower is the structure of sexual reproduction of the angiosperms. Unlike the gonads of animals, which are permanent organs that develop in the embryo, flowers are transitory, developing seasonally. A perfect flower, as it is called, consists of four sets of floral appendages, which grow in spirals or whorls; each floral part, evolutionarily speaking, is a modified leaf. The outermost parts of the flower are the *sepals*, which are commonly green and obviously leaflike in structure. The sepals, collectively known as the *calyx*, enclose and protect the flower bud. Next are the *petals*, collectively called the *corolla*; these are also usually leaf-shaped but are often brightly colored. They advertise the presence of the flower among the green leaves, attracting insects or other animals that visit flowers for their nectar or for other edible substances and so carry pollen from flower to flower. Calyx and corolla together are known as the *perianth*.

Within the corolla are the *stamens*. Each stamen consists of a single elongated stalk, the *filament*, and at the end of the filament, the *anther*. The pollen grains, when they are ripe, are released—often in large numbers—from the anther, usually through narrow slits or pores. The pollen grains are immature male gametophytes.*

* For a review of the concept of alternation of generations, see page 210.

The centermost appendages of the flower are the *carpels*, which contain the female gametophytes. Typically a single carpel consists of a *stigma*, which is a sticky surface specialized to receive the pollen; a slender stalk, the *style*, down which the pollen tube grows; and a base, the *ovary*. Within the ovary are the *ovules*, each of which encloses a female gametophyte with a single egg cell. When the egg cell is fertilized, the ovule develops into a seed. A single flower may have one carpel or several carpels. The carpels may be separate or fused together.

In some species, flowers are either male (staminate) or female (carpellate). Male and female flowers may be present on the same plant, as in corn, squash, oaks, and birches, or on different plants, such as the tree of heaven *(Ailanthus)*, the date palm, and the American mistletoe. Species in which both male and female flowers are borne on the same plant are known as monoecious ("in one house"); species in which the male and female flowers are on separate plants are known as dioecious ("in two houses").

POLLINATION OF THE FLOWER

For flowering plants, a new cycle of life begins when a grain of pollen—often brushed from the body of a foraging insect—comes into contact with the stigma of a flower of the same species. By the time this pollen grain is released from its parent flower, it usually consists of three haploid cells (two sperm cells and a "vegetative," or tube, cell), enclosed by the thickened outer wall of the pollen grain.

Pollen is commonly produced in great quantities; the probability of any particular pollen grain reaching the stigma of an appropriate flower is very small. The pollen grain contains its own nutrients and has so tough an outer coating that intact grains thousands of years old have been found in peat bogs.

In lower plants, you will recall, there is a distinct cycle of alternation of generations in which the sporophyte produces spores that produce gametophytes that produce gametes, with gametophyte and sporophyte having separate existences. In the course of plant evolution, the gametophyte stage has been steadily reduced and, in the angiosperms, all that remains of the male gametophyte is the tough, tiny pollen grain and the pollen tube. The sperm cells are the gametes.

28–4
Pollen grains. The walls of the pollen grain protect the male gametophyte on the journey between the anther and the stigma. These outer surfaces, which are remarkably tough and resistant, are often elaborately sculptured. As you can see, the pollen grains of different species are distinctly different: (a) *a chrysanthemum (spiny pollen grains such as these are common among composites),* (b) *a morning glory,* (c) *a lily.*

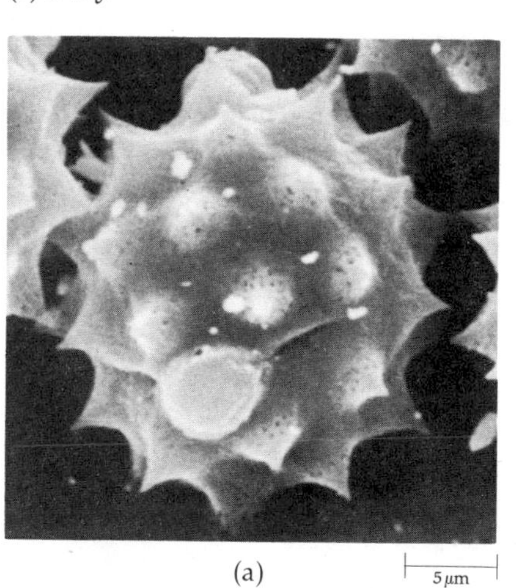

(a) 5 μm

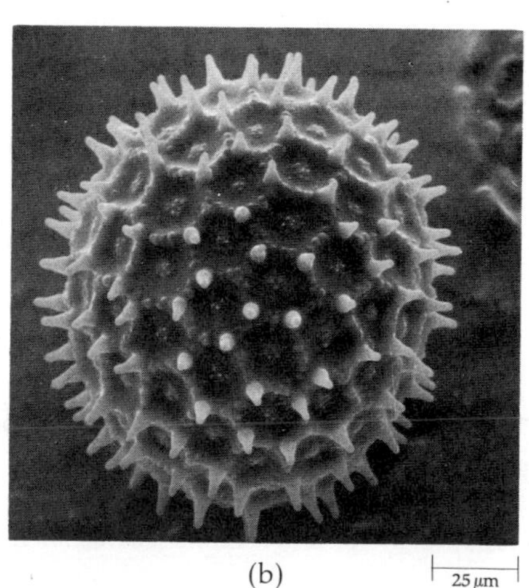

(b) 25 μm

(c) 50 μm

28–5

(a) Fertilization in angiosperms. The pollen tube of the male gametophyte, or pollen grain, grows down through the style and enters the ovule, in which the female gametophyte has developed to a seven-cell stage. One of the sperm nuclei unites with the egg nucleus, and the zygote is formed. The other sperm nucleus fuses with the two polar nuclei that are present in a single large cell (which in the drawing fills most of the ovule). From the resulting triploid (3n) cell, the endosperm will develop. The carpel shown here contains a single ovule. (b) Cross section of a pollen tube. Two lobes of the tube nucleus are visible. This nucleus directs the formation of the pollen tube and eventually disintegrates. At the right is a sperm cell. Numerous mitochondria are visible, as are several plastids. The sperm cell is actually a cell within a cell.

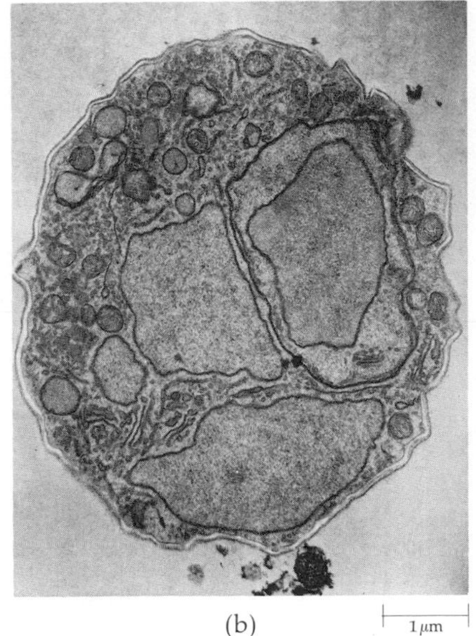

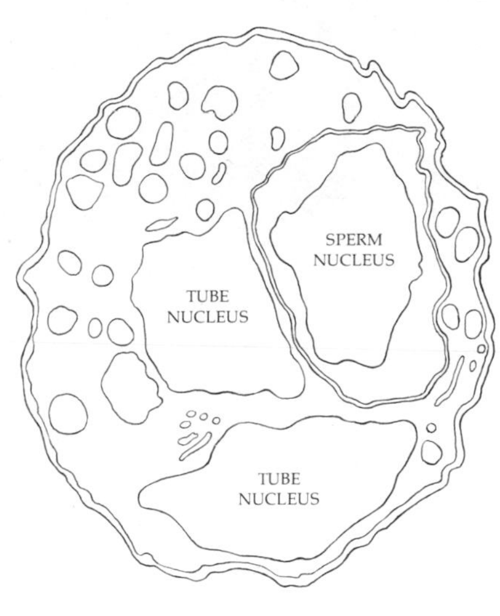

(b)

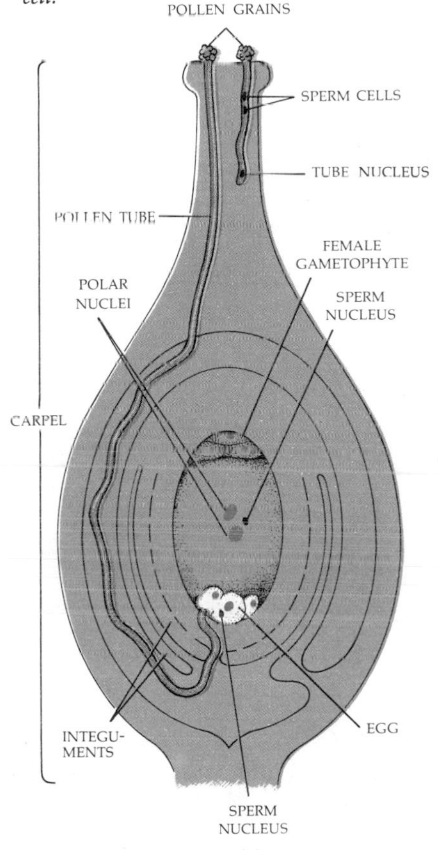

(a)

Once on the stigma, the pollen grain germinates, and, under the influence of the tube nucleus, a pollen tube grows down through the style into an ovule (Figure 28–5). The ovule contains the female gametophyte, which has also become reduced in size in the course of evolution. In many species it consists of seven cells, with a total of eight haploid nuclei. One of the seven cells is the egg, containing a single haploid nucleus. On either side of the egg are two small cells known as synergids. At the opposite end of the gametophyte are three small cells, the antipodal cells, whose function, if any, is unknown. The large central cell contains two haploid nuclei, called the polar nuclei because they move to the center from each end, or pole, of the gametophyte.

FERTILIZATION

One of the sperm cells carried by the pollen tube unites with the egg. This fertilized cell, the zygote, develops into the young sporophyte, or embryo. The nucleus of the second sperm cell unites with the two nuclei of the central cell. From the resultant $3n$ cell, a specialized tissue called the *endosperm* develops. It completely surrounds and nourishes the embryo. These extraordinary phenomena of fertilization and triple fusion—together called "double fertilization"—take place, in all the natural world, only among the flowering plants.

The seed consists of the embryo, which develops from the fertilized egg; the stored food, which consists of or derives from the endosperm; and the seed coat, which develops from the outermost layer or layers (integuments) of the ovule. The ovule or ovules are contained within the ovary, which is the enlarged base of either a single carpel or fused carpels.

(a)

(b)

28–6
Flowers—forms and variations. (a) *Stamens and stigma of a white crocus.* (b) *In* Hibiscus, *a column of stamens is fused around the style.* (c) *Corn* (Zea mays), *a monoecious flower. The tassels, at the top of the stem, are the male (pollen-producing) flowers. Each thread of "silk," seen emerging from the ear of corn, is the combined stigma and style of a female flower.* (d) Helianthus annuus, *"annual flower of the sun," is composed of numerous separate florets, each comprising a carpel and fused anthers enclosed in a small corolla of fused petals.* (e) *Shootingstars* (Dodecatheon pauciflorum). *The anthers encircle the style, at the tip of which is the stigma. At the top of the stem is a withered flower showing two anthers and a style.* (f) *In the southern magnolia, the carpels form a cone-shaped structure from which curved styles emerge. The yellow stamens, which surround the carpels, have dropped off.* (g) *Water hemlock, a member of the carrot family, an example of an inflorescence.*

(c)

(d)

(e)

(f)

(g)

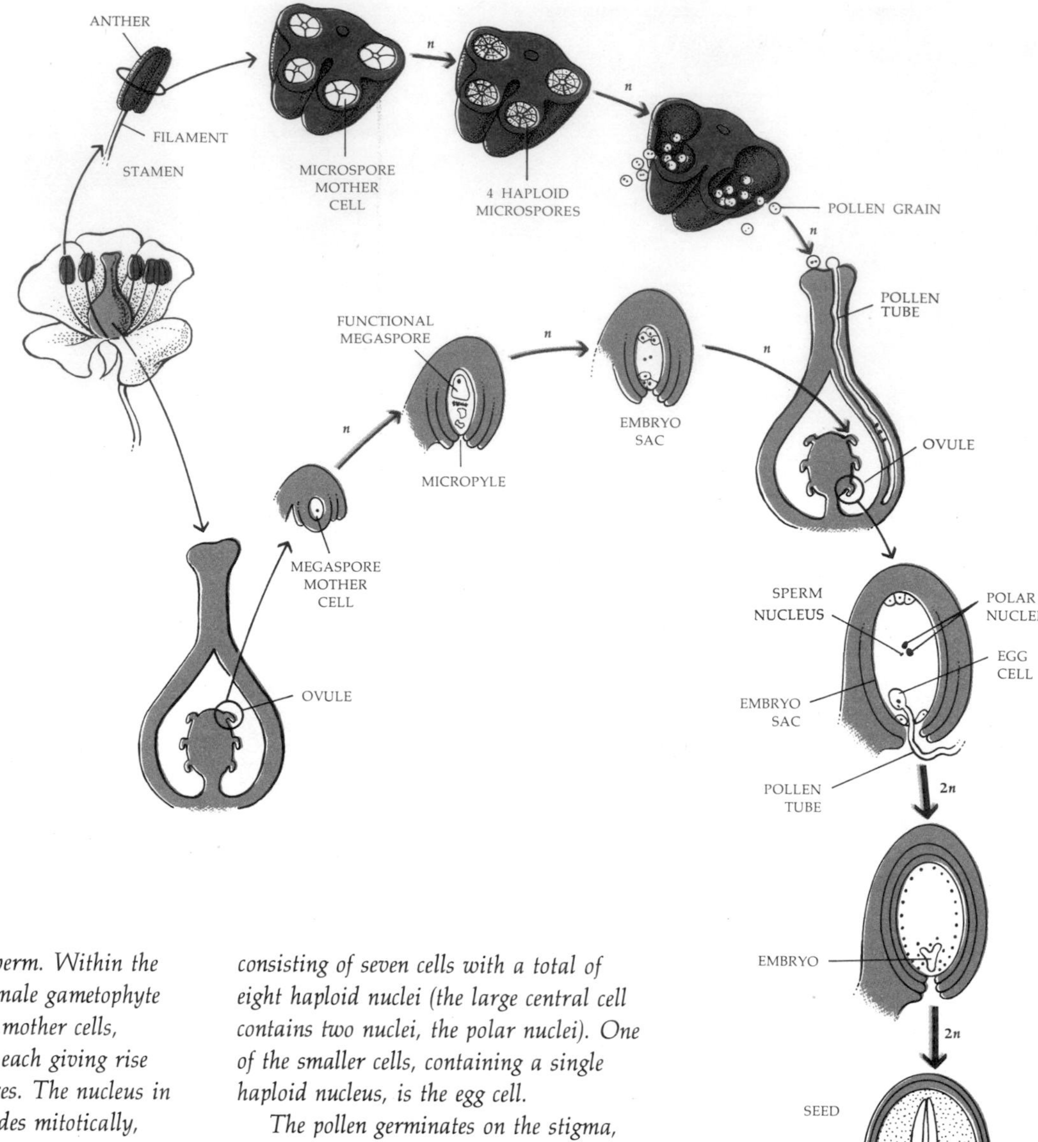

28–7

Life history of an angiosperm. Within the anther of the flower, the male gametophyte develops from microspore mother cells, which divide meiotically, each giving rise to four haploid microspores. The nucleus in each microspore then divides mitotically, and the microspore develops into a two-celled pollen grain. One of the cells subsequently divides again, usually upon germination, resulting in three cells per pollen grain: two sperm cells and the tube (pollen tube) cell.

Within the ovule, the female gametophyte develops from a megaspore mother cell, which divides meiotically to produce four haploid megaspores. Three of the megaspores disintegrate; the fourth divides mitotically, developing into an embryo sac consisting of seven cells with a total of eight haploid nuclei (the large central cell contains two nuclei, the polar nuclei). One of the smaller cells, containing a single haploid nucleus, is the egg cell.

The pollen germinates on the stigma, producing a pollen tube that grows down the style into the ovary. The two sperm cells enter the embryo sac through the tube; one sperm nucleus fertilizes the egg cell, the other merges with the polar nuclei, forming the triploid (3n) *endosperm. The embryo undergoes its first stages of development while still within the ovary of the flower, and the ovary itself matures to become a fruit. The seed, released from the mother sporophyte in a dormant form, eventually germinates, forming a seedling.*

28-8
A simple fruit, an aggregate fruit, and, on the right, a multiple fruit.

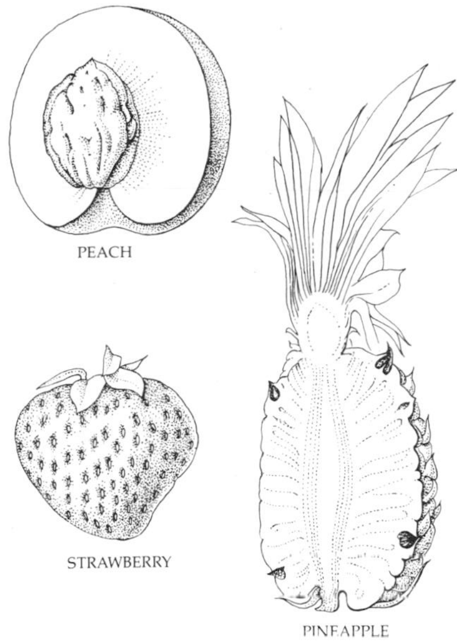

THE FRUIT

As the embryo develops, the ovary wall (pericarp) develops into the fruit. As you know from your own observations, fruits have many forms. They are generally classified as simple, aggregate, or multiple, depending on the arrangement of the carpels from which they develop. Simple fruits develop from one carpel or the united carpels of a single flower. Aggregate fruits, such as magnolia, raspberry, and strawberry, are formed from a number of separate carpels of one flower. Multiple fruits consist of the carpels of more than one flower. A pineapple, for example, is a multiple fruit formed from a flower cluster of many separate flowers. The ovaries of these flowers become fused as they mature.

Simple fruits are by far the most diverse of the three groups. When ripe, they may be soft and fleshy or dry. There are three main types of fleshy fruit—the berry, the drupe, and the pome. In the berry type, examples of which are tomatoes, dates, and grapes, there are one to several carpels, each of which may have one or many seeds. The inner layer of the pericarp is usually fleshy.

In the drupe, there are also one to several carpels, but each usually contains only a single seed. The inner wall of the fruit is stony and usually tightly adherent to the seed. Some familiar drupes are the peach, cherry, olive, and plum. The peach is a typical drupe; the skin, the succulent, edible portion of the fruit, and the stone are three layers of the mature ovary wall. The almond-shaped structure within the stone is the seed. The coconut is a drupe whose outer covering is fibrous rather than fleshy.

A highly specialized sort of fleshy fruit is the pome, which is characteristic of the subfamily of roses that produces rose hips. The pome is derived from an inferior ovary (page 415) in which the fleshy portion comes largely from the perianth. Apples and pears are pomes.

Dry fruits are classified into two groups, dehiscent and indehiscent. In dehiscent fruits, the tissues of the mature ovary wall break open, releasing the seeds. In indehiscent fruits, the seeds remain in the fruit after the fruit is shed from the parent plant.

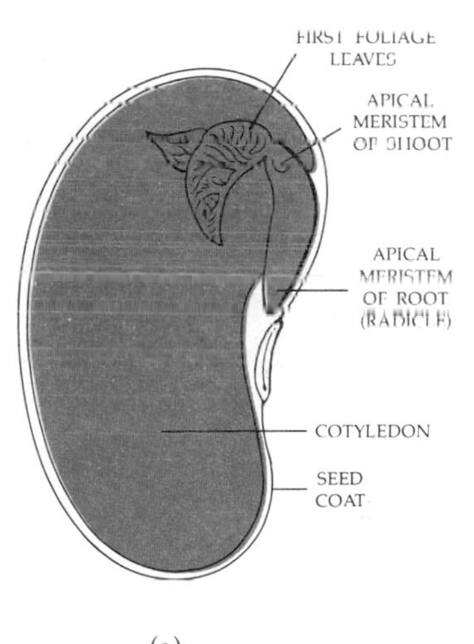

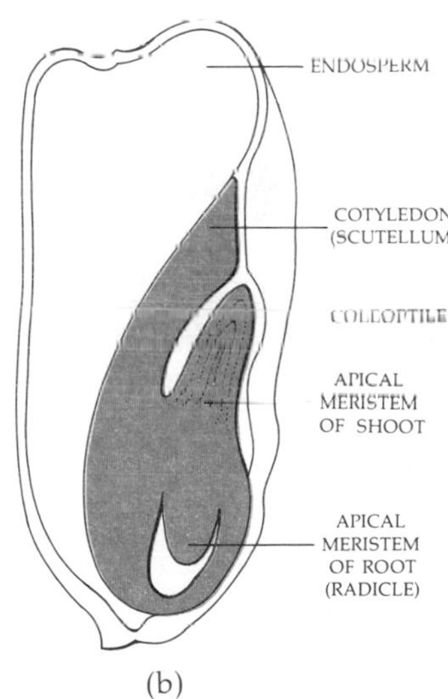

28-9
Seeds. (a) *In dicots such as the common bean, the endosperm is digested as the embryo (shown in color) grows and the food reserve is stored in the fleshy cotyledons.* (b) *In corn and other monocots, the single cotyledon, known as the scutellum in corn and other grains, absorbs food reserves from the endosperm*

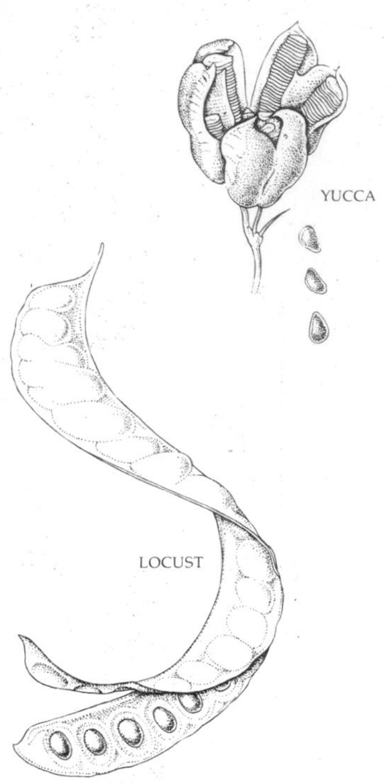

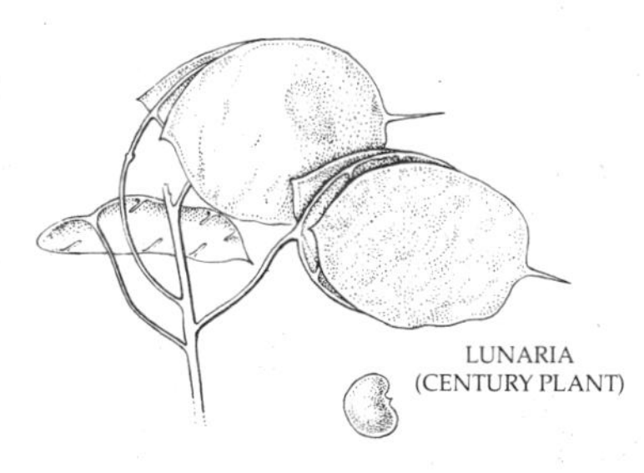

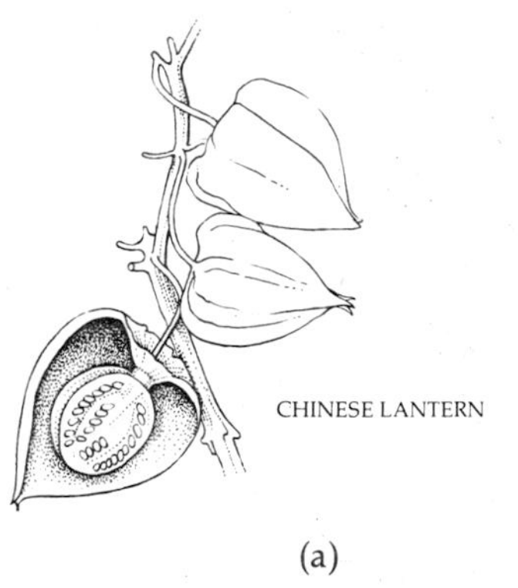

(a)

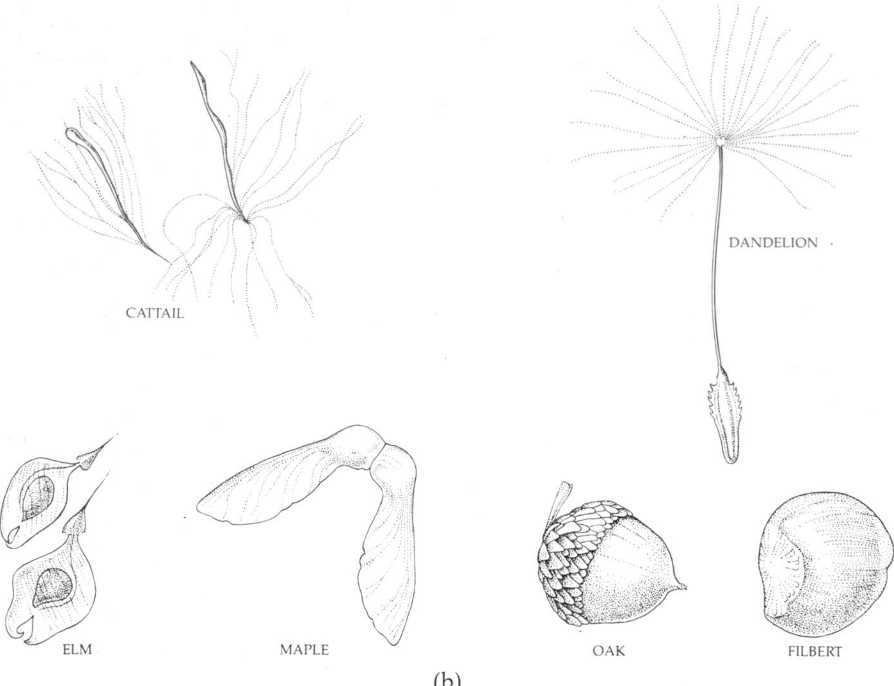

(b)

28–10
Examples of (a) *dehiscent and* (b) *indehiscent fruits.*

There are several sorts of dehiscent fruits. A fruit derived from a single carpel in which the ovary wall splits down one side is known as a follicle; the fruits of columbines and milkweeds are examples. In the pea family (legumes), the ovary splits down two sides; the pod is the mature pericarp and the peas are the seeds. Other examples of dehiscent fruit are shown in Figure 28–10a.

Indehiscent fruits (Figure 28–10b) are characteristic of a great variety of plant families. The most common is the achene, a small, single-seeded fruit. Winged achenes, such as those found in the elm and the ash, are called samaras. The most familiar kind of indehiscent fruit is the nut, which resembles the achene but has a stony coat and is derived from a compound ovary (an ovary formed from fused carpels). Examples of nuts are acorns and hazelnuts. Note that the word nut is used very indiscriminately in common speech: Peanuts are legumes; pine nuts are seeds; almonds and coconuts are drupes.

SEED DORMANCY

The seeds of most wild plants require a period of dormancy before they will germinate. This genetic requirement ensures that the seed will "wait" at least until the next favorable growth period. Seeds can remain dormant and yet viable—with the embryo in a state of suspended animation—for hundreds of years. The record for dormancy, as far as is known, has been set by some seeds of Arctic tundra lupine found in a lemming burrow in the Yukon. Deeply buried in the permanently frozen silt, they are estimated to be at least 10,000 years old. But when a sample was planted, the seeds germinated within 48 hours.

THE STAFF OF LIFE

Grains are the small, one-seeded fruits of grasses. Because they are dry, they can be stored as food. The collecting and storing of grains from wild grasses was believed to be an important impetus to the agricultural revolution of some 10,000 years ago (see Chapter 49). Today we are heavily dependent on cultivated wheat, rice, corn, rye, and other grains. In many countries, they constitute the chief food resource.

The fruit of wheat, sometimes known as the kernel, is made up of the embryo, the endosperm, and the surrounding seed coat. More than 80 percent of the bulk of the wheat kernel and 70 to 75 percent of its protein are in the endosperm. White flour is made from the endosperm. Wheat germ, the embryo, forms about 3 percent of the kernel. It is usually removed as wheat is processed because it contains oil, which makes the grain more likely to spoil. Bran is the seed coat plus the aleurone layer (outer part of endosperm); it constitutes about 14 percent of the kernel. The bran is also removed when wheat is milled to make white flour. Actually, the bran somewhat decreases the caloric value of the wheat kernel. We are unable to digest bran because it is mostly cellulose. Bran therefore tends to speed the passage of food through our intestinal tracts, resulting in decreased absorption. The wheat germ and the bran, which contain most of the vitamins, are sometimes used for human consumption but more often are fed to livestock.

Wheat is about 9 to 14 percent protein, most of which, as we noted, is contained in the endosperm. Its protein value is diminished, however, by its lack of certain essential amino acids, notably lysine.

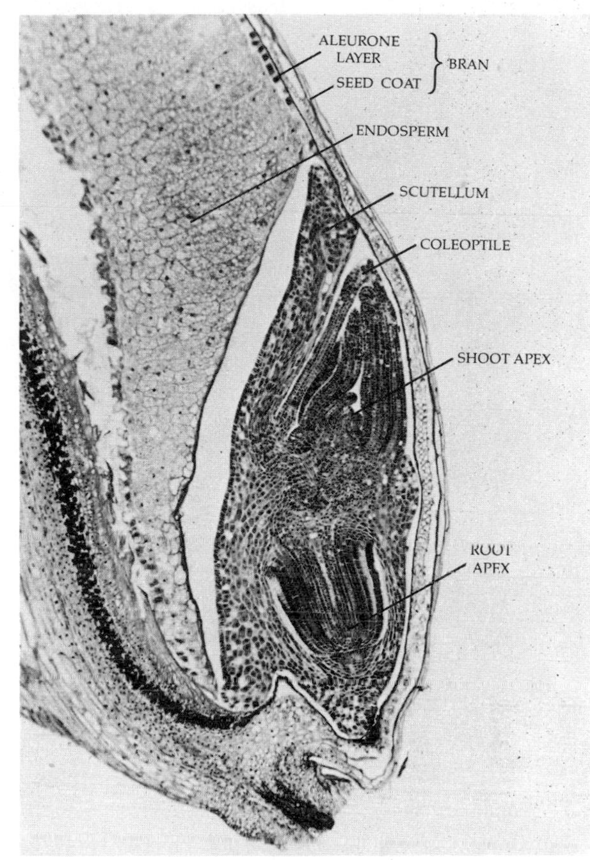

Longitudinal section of a wheat kernel.

The seed coat apparently plays a major role in maintaining dormancy. In some species, the seed coat seems to act primarily as a mechanical barrier, preventing the entry of water and gases, without which growth is not possible. In these cases, growth is initiated by the seed coat's being worn away in various ways—such as being washed by rainfall, abraded by sand or soil, burned away by a forest fire, decomposed by microbial action, or partially digested as it passes through the digestive tract of a bird or other animal. In other species, dormancy seems to be maintained chiefly by chemical inhibitors in the seed coat. These inhibitors undergo chemical changes in response to various environmental factors, such as light or prolonged cold or a sudden rise in temperature, which neutralize their effects, or they may be washed or eroded away. Eventually, the embryo resumes growth.

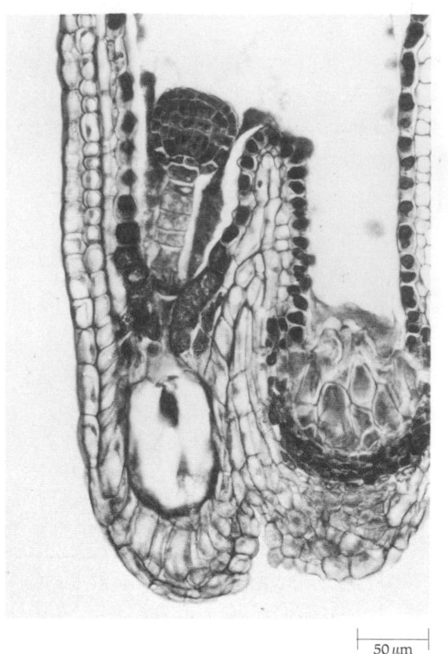

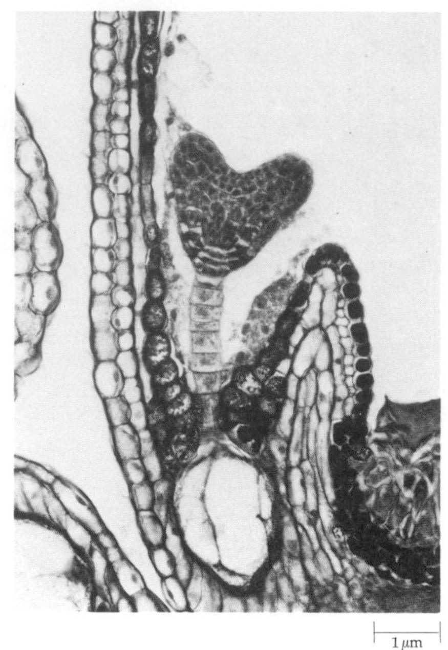

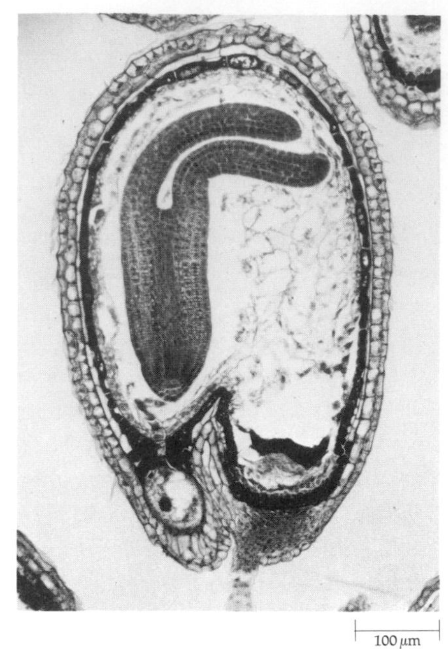

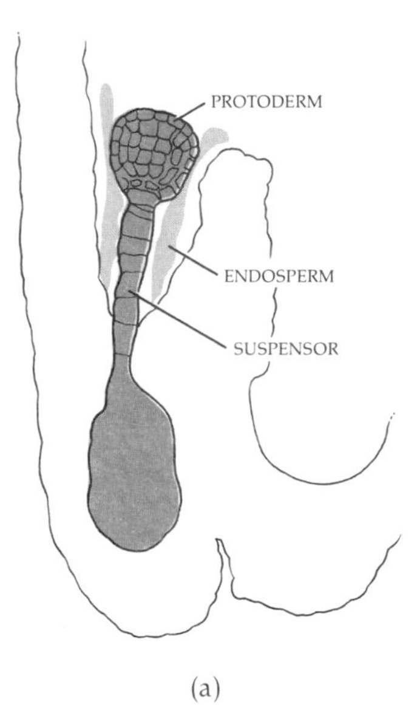

(a)

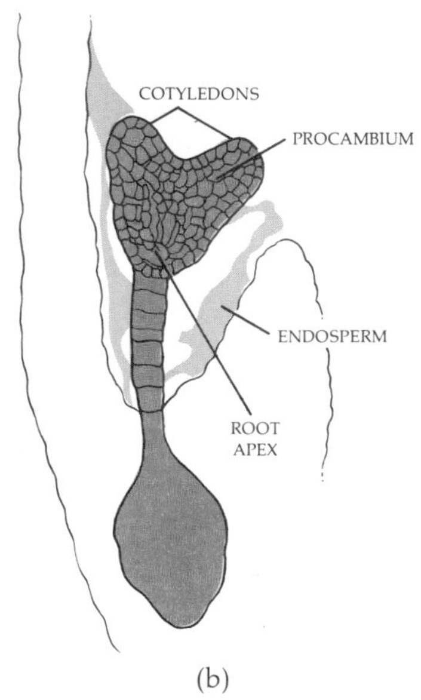

(b)

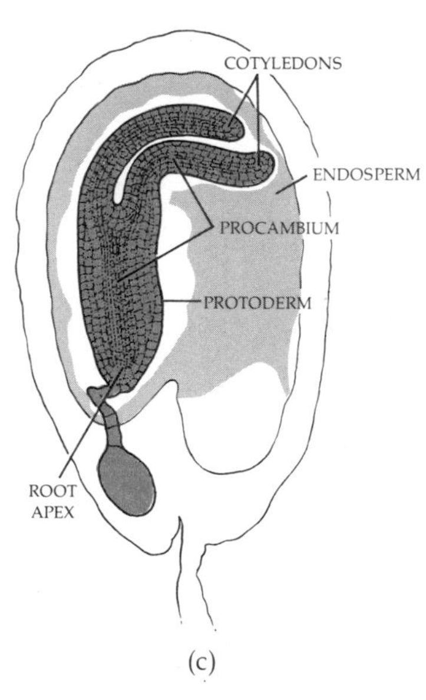

(c)

28–11
Some stages in the development of the embryo (shown in color) of shepherd's purse (Capsella bursa-pastoris), *a dicot.* (a) *The embryo proper is globular, and its protoderm is the future epidermis. The large cell at the bottom is the basal cell of the suspensor.* (b) *The cotyledons are beginning to emerge. The procambium will later give rise to the conducting tissues of the plant.* (c) *The cotyledons have developed further, and additional tissue differentiation has occurred.* (d) *The mature embryo. The apical meristems of the root and the shoot—the root apex and the shoot apex—are clearly differentiated.*

The dormancy requirement in seeds apparently evolved only recently, geologically speaking, among groups of plants subjected to the environmental stress of increasing winter cold characteristic of the most recent Ice Age. By this time—only $1\frac{1}{2}$ to 2 million years ago—the angiosperms were already a highly diversified group, and different populations responded to these pressures in different ways. This explains why even closely related plants have different mechanisms for maintaining and breaking dormancy.

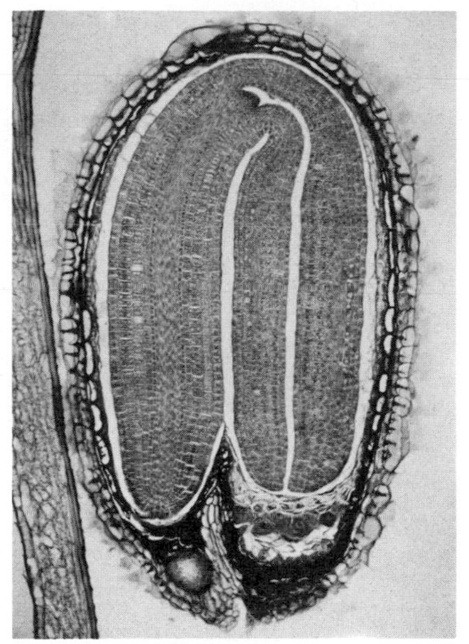

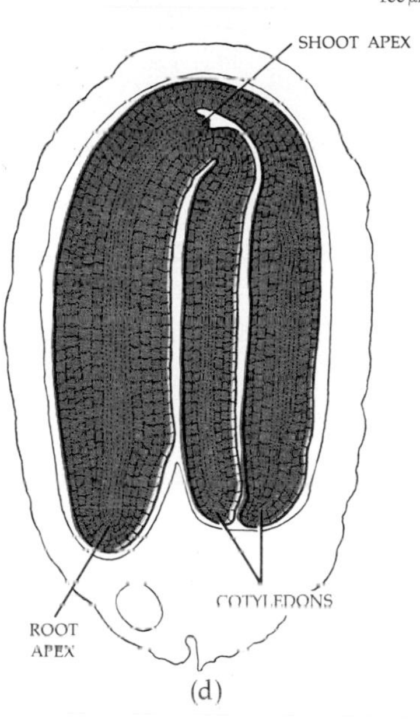

(d)

GROWTH OF THE PLANT

As we saw earlier, the "double fertilization" of angiosperms produces a $3n$ cell and the zygote (the fertilized egg cell). The $3n$ cell divides mitotically to produce endosperm, the tissue that nourishes the developing embryo and, in many cases, the young seedling. The fertilized egg cell also divides mitotically, and as the embryo grows, its cells begin to *differentiate*—become different from one another—and the embryo begins to take on a characteristic form, a process known as *morphogenesis.*

In its earliest stages, the embryo proper consists of a globular mass of cells. Suspensor cells, also formed from divisions of the fertilized egg cell, are believed to push the developing embryo into the endosperm and to be actively involved in the absorption of nutrients from the endosperm. As development of the embryo proceeds, changes in its internal structure result in the beginnings of the tissue systems of the plant. At the same time, or slightly later, emergence of the one cotyledon in monocots, or the two in dicots, occurs. The stages in the development of a dicot embryo are shown in Figure 28-11.

In the earliest stages of embryonic growth, cell division takes place throughout the body of the young plant. As the embryo grows older, however, the addition of new cells becomes gradually restricted to certain parts of the plant body: the *apical meristems* of the root and the shoot. During the rest of the life of the plant, all the primary growth—which chiefly involves the elongation of the plant body—originates in these meristems.

The existence of such meristematic areas, which add to the plant body throughout the life of the plant, is one of the principal differences between plants and animals. Higher animals stop growing when they reach maturity, although the cells of certain "turnover" tissues, such as nails, hair, skin, or the lining of the intestine, continue to divide.

Plants, however, continue to grow during their entire life span. Growth in plants is the counterpart, to some extent, of mobility in animals. Plants "move" by extending their roots and shoots, both of which involve changes in size and form. By growth, a plant modifies its relationship with the environment, turning toward the light and extending its roots. The sequence of growth stages in plants thus corresponds to a whole series of motor acts in animals, especially those concerned with the search for food and water. In fact, growth in plants serves many of the functions that we group under the term "behavior" in animals.

PRIMARY GROWTH

A seed characteristically contains very little moisture (only about 5 to 10 percent of its total weight). Germination begins with a massive entry of water into the seed.

The seed coat ruptures and the young sporophyte emerges.

Primary growth, which begins immediately, involves the elongation of stems and roots, the formation of branches, and the differentiation of the conducting tissues and other specialized tissues of the young shoot and root. All primary growth originates in the apical meristems of the shoot and root.

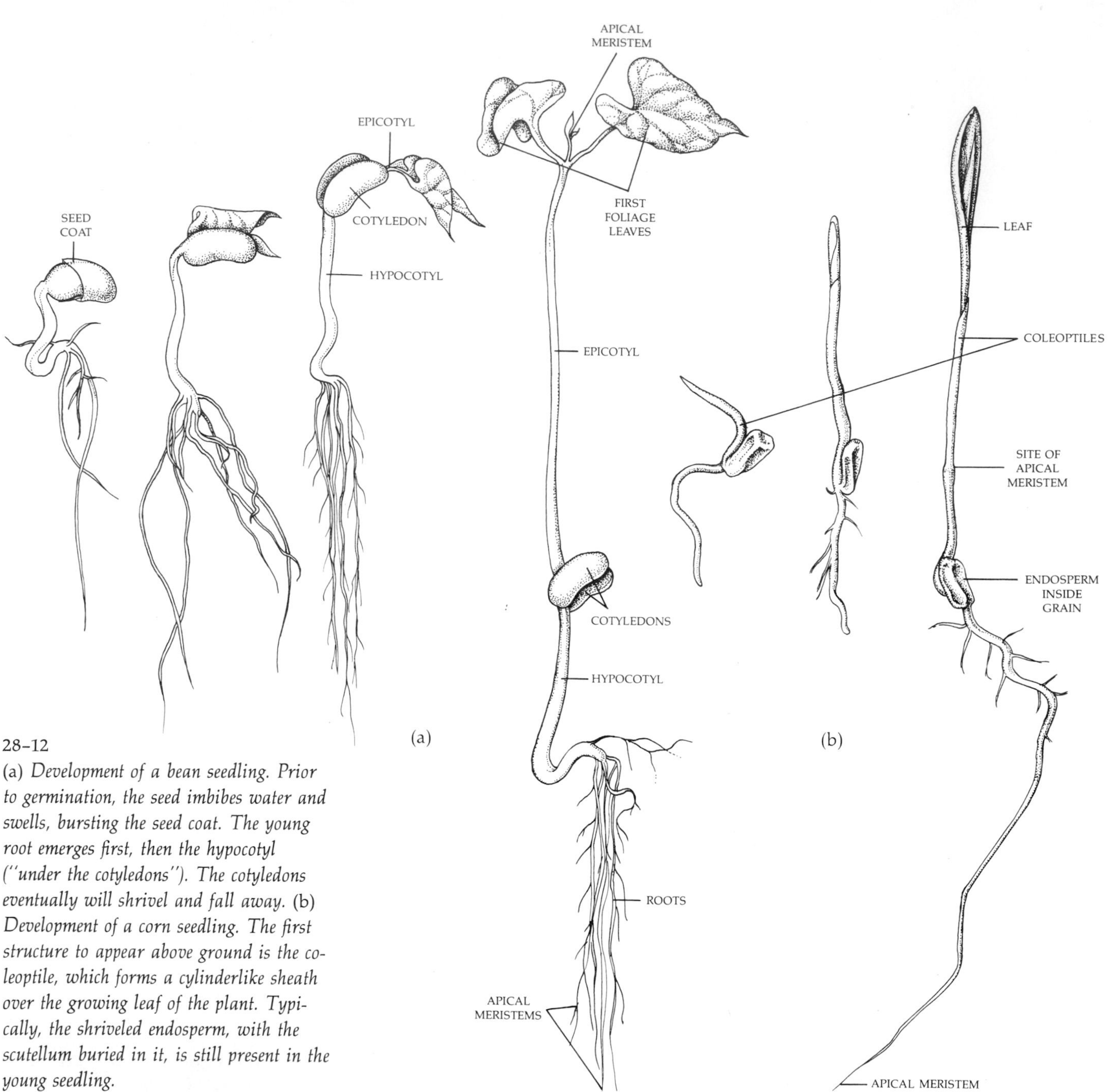

28–12
(a) *Development of a bean seedling. Prior to germination, the seed imbibes water and swells, bursting the seed coat. The young root emerges first, then the hypocotyl ("under the cotyledons"). The cotyledons eventually will shrivel and fall away.* (b) *Development of a corn seedling. The first structure to appear above ground is the coleoptile, which forms a cylinderlike sheath over the growing leaf of the plant. Typically, the shriveled endosperm, with the scutellum buried in it, is still present in the young seedling.*

28–13
The growth regions of a dicot root. New cells are produced by the division of cells within the apical meristem. The cells above the meristem undergo a characteristic series of changes as the distance increases between them and the root tip. First, there is a maximum rate of cell division, followed by cell elongation without further division. The latter accounts for most of the lengthening of the root. As the cells elongate, they differentiate into various specialized tissues of the root. The protoderm becomes the epidermis, the ground meristem becomes the cortex, and the procambium becomes the xylem and phloem. Some of the cells produced in the apical meristem form the protective root cap.

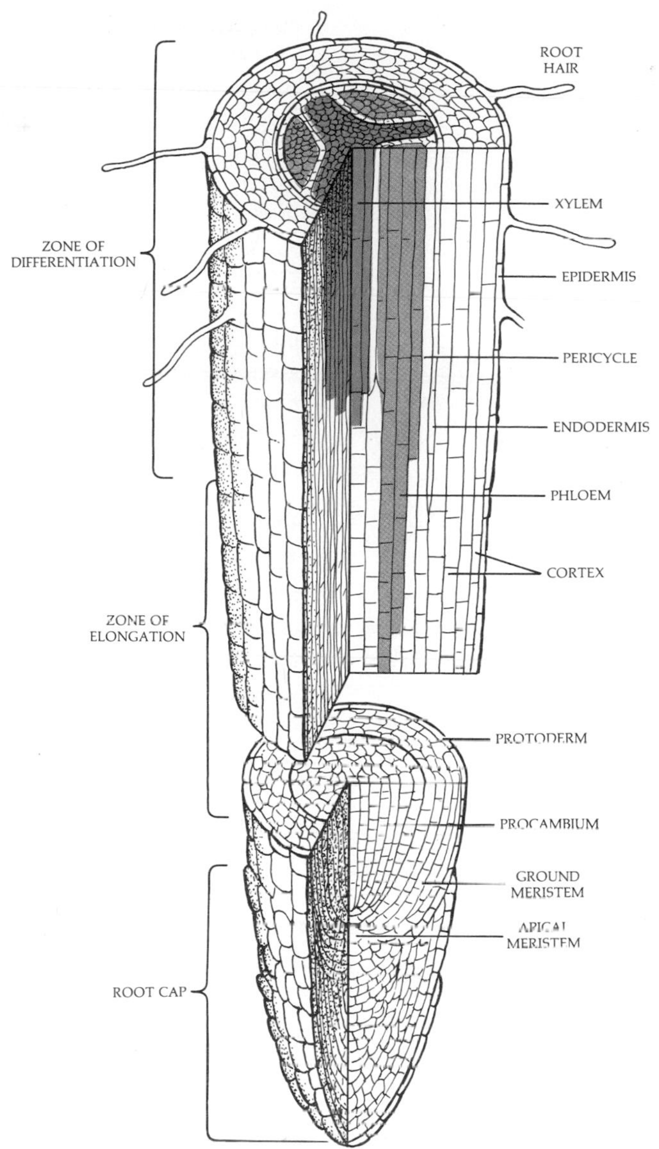

Primary Growth of the Root

The first part of the embryo to break through the seed coat, in nearly all seed plants, is the embryonic root, the radicle. Figure 28–13 diagrams the growing zone of the young root of a dicot. At the very tip is the root cap, which protects the apical meristem as the root is pushed through the soil. The cells of the root cap wear away and are constantly replaced by new cells from the meristem. The cells in the meristem designated as *apical initials* are those that produce new cells. All the other cells in the root are the progeny of these relatively few meristematic cells. The apical initials are always capable of dividing. Some of the daughter cells remain in the tip of the meristem to maintain the apical initial population. Others differentiate as they divide, some becoming cells of the root cap and others forming the

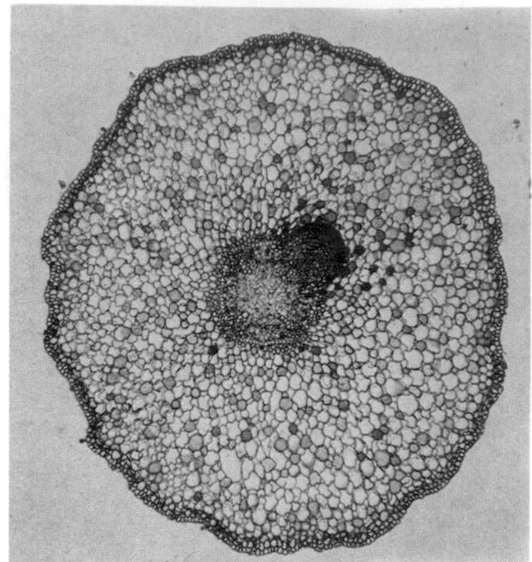

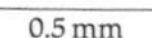

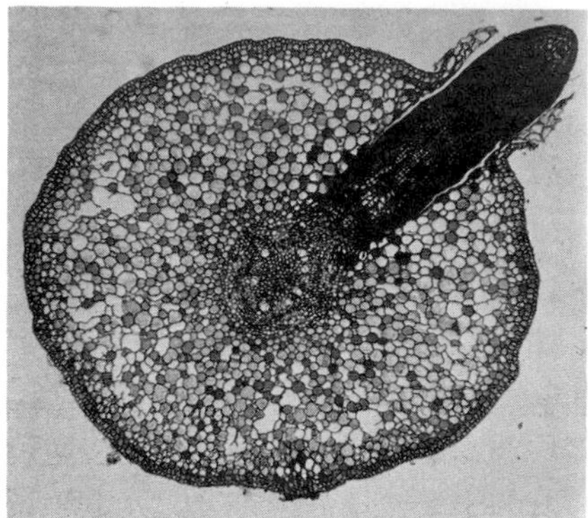

0.5 mm

28–14
Two stages in the development of a branch root in a willow. Such roots originate in the pericycle and grow out laterally through the endodermis, cortex, and epidermis. They destroy the tissues in their path, partly by crushing, partly by digestion by multiple enzymes.

complex tissues of the root. The maximum rate of cell division occurs at a point well above the tip of the meristem. Then, just above the point where cell division ceases, the cells gradually elongate, growing to 10 or more times their previous length, often within the span of a few hours. This process of elongation without cell division is the principal cause of root growth, although, of course, growth is ultimately dependent on the production of the new cells that become part of the zone of elongation.

As the cells stop dividing, some begin to differentiate, forming first the sieve-tube elements, the conducting cells of the phloem, and then vessel elements, the conducting cells of xylem. At the level of xylem differentiation, the endodermis takes shape, and to the inside of the endodermis, the pericycle forms. This tissue gives rise to branch roots. At about this same stage, the epidermal cells differentiate and begin to extend root hairs into the crevices between the soil grains.

This same basic pattern of growth is seen in the first root of a seedling and is repeated over and over again in the growing root tips of a tree 50 meters tall.

Primary Growth of the Shoot

The organization of the developing shoot tip is somewhat similar to that seen in the root: first, a zone in which most of the cell division takes place; next, a zone of cell elongation; and finally, a zone of differentiation. These zones are not as distinct in the shoot as they are in the root, however, because of the regular occurrence of nodes and their appendages. Also, no covering analogous to the root cap is produced over the shoot tip.

As in the root, the outermost layer of cells develops into the epidermis. In the shoot, these cells have an outer covering, the cuticle. Other cells differentiate to form ground tissue, which in dicots includes the cortex and the pith, and the primary vascular tissues—the primary xylem and the primary phloem. The pattern of development is more complicated, however, than in the root tip since the apical meristem of the shoot is the source of tissues that give rise to new leaves, branches, and flowers. (At the time of flowering the apical meristem forms the floral parts and ceases to exist.)

Figure 28–15 shows the shoot tip of a lilac. Here you can see the apical meristem, which is very small, and the beginnings—primordia—of leaves. As you can see, leaves are formed in an orderly sequence at the shoot tip. The leaf originates by the division of cells in a localized area along the side of the shoot apex. The vascular tissue differentiates upward, becoming part of the general vascular system that connects the plant from root to leaf tip. In some species, leaves arise simultaneously in pairs opposite one another, as in our figure. In other species, the leaves occur spirally or in circles (whorls) at the nodes. As the internodes elongate, the young leaves become separated so that the leaves clustered so tightly together around the apex in Figure 28–15 will eventually be spaced out along the stem of the plant.

Buds, Flowers, and Branches

As the growing tip of the shoot elongates, small masses of meristematic tissue are left just above the point at which the leaf joins the stem (the leaf axil). These new meristematic regions, the buds, remain dormant until after the growth of the adjacent leaf and internode is complete. In many species, the buds do not develop at all unless the apical meristem of the shoot is damaged or removed. (This phe-

28–15
Longitudinal section of the shoot tip of a lilac (Syringa vulgaris). *At the tip of the stem is the apical meristem, the central zone of cell division. The leaf primordia, from which new leaves will form, originate along the sides of the shoot apex. In this picture, two leaf primordia can be seen arising from the apical meristem. Successively older leaf primordia had formed on both sides. The developing stem below the apical meristem and the lateral buds on either side are also regions of active cell division.*

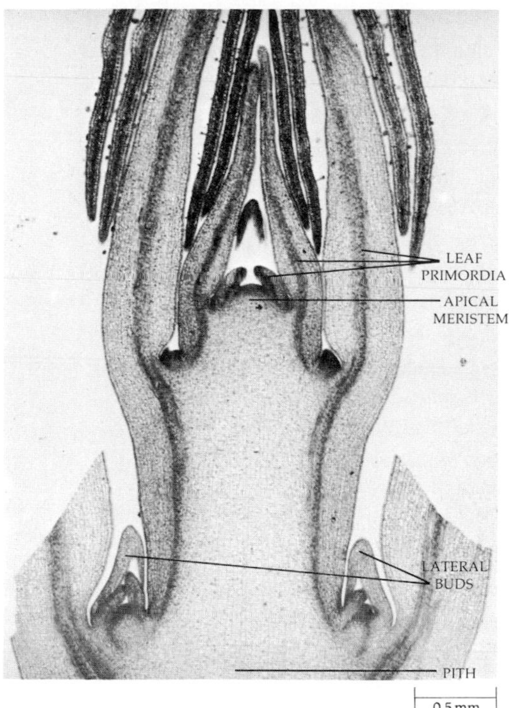

nomenon, known as apical dominance, will be discussed further in Chapter 30.) In some species, some buds are destined to become lateral branches, or a specialized shoot, such as a rhizome or a tuber or a flower. In other species, whether or not a particular bud will become a flower or a branch is determined by environmental conditions, particularly day length. (The hormonal regulation of flowering is discussed on page 552.)

SECONDARY GROWTH

Secondary growth is the process by which woody plants increase the thickness of trunks, stems, branches, and roots after primary growth has ceased. The so-called "secondary tissues" are not derived from the apical meristems; they are the result of the production of new cells by the vascular cambium and cork cambium. The cambiums are called *lateral meristems*.

The *vascular cambium* is a thin, cylindrical sheath of tissue in between the xylem and the phloem. In plants with secondary growth, the cambium cells divide continually during the growing season, adding new xylem cells—that is, secondary xylem—on the outside of the primary xylem, and secondary phloem on the inside of the primary phloem. Some daughter cells remain as a cylinder of undifferentiated cambium. As the tree grows older, the living parenchyma cells of the xylem in the center of the trunk die, and those vessels cease to function. This nonfunctional wood is called heartwood, as distinct from sapwood, which consists of living parenchyma cells and functional vessels.

As the girth of stems and roots increases by secondary growth, the epidermis becomes stretched and torn. In response to this tearing process, a new type of

28–16

(a) *Stem of a dicot before the onset of secondary growth.* (b) *Beginnings of secondary growth. Secondary xylem and secondary phloem are produced by the vascular cambium, a meristematic tissue formed late in primary growth. As the trunk increases in diameter, the epidermis is stretched and torn, which apparently triggers the formation of the cork cambium, from which cork is formed, replacing the epidermis.* (c) *Cross section of a three-year-old stem, showing annual growth rings. Rays are strands of living cells that transport nutrients and water laterally (across the trunk). On the perimeter of the outermost growth ring of xylem is the vascular cambium, encircled by a band of secondary phloem. The primary phloem and also the cortex will eventually disappear. In an older stem, the thin cylinder of active secondary phloem is immediately adjacent to the vascular cambium.*

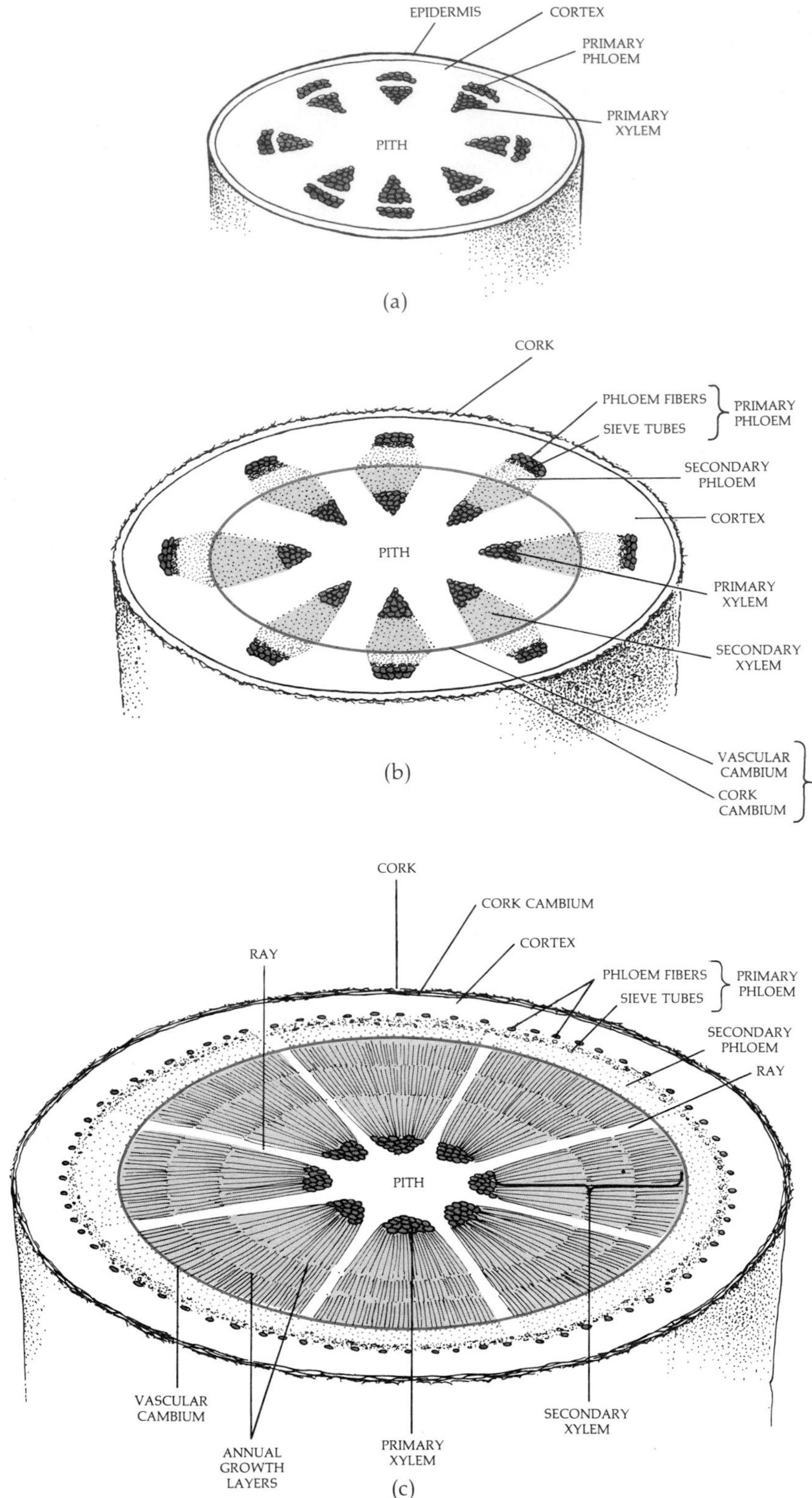

cambium, the *cork cambium,* forms from the cortex. Cork (phellem), which is a dead tissue at maturity, is produced from the cork cambium. The cork cambium forms anew each year, moving further inward until, finally, there is no cortex left. The tissues outside the vascular cambium, including the phloem, constitute the bark.

Figure 28–16 shows the cross section of the stem of a young dicot, in which some secondary growth has taken place. In the center of the stem is the pith, composed of loosely packed parenchyma cells. The first cylinders of tissues around the pith are layers of xylem, composed of vessels and other cell types. Around the outermost layer of xylem is a layer of meristematic tissue, the vascular cambium. Each year, during the growing season, the cambium undergoes a number of mitotic divisions, forming new xylem (secondary xylem) on its inner surface and new phloem (secondary phloem) on its outer surface. By this continuous formation of layers of xylem and (to a small extent) phloem, tree trunks increase their diameter as primary growth increases their height. Season after season the new xylem forms visible growth layers, or rings. Each growing season leaves its trace, so that the age of a tree can be estimated by counting the number of growth rings in a section near its base. Since the rate of growth of a tree depends on climatic conditions, it is possible to determine from the width of the annual growth layers of ancient trees fluctuations in temperature and rainfall that occurred hundreds of years ago.

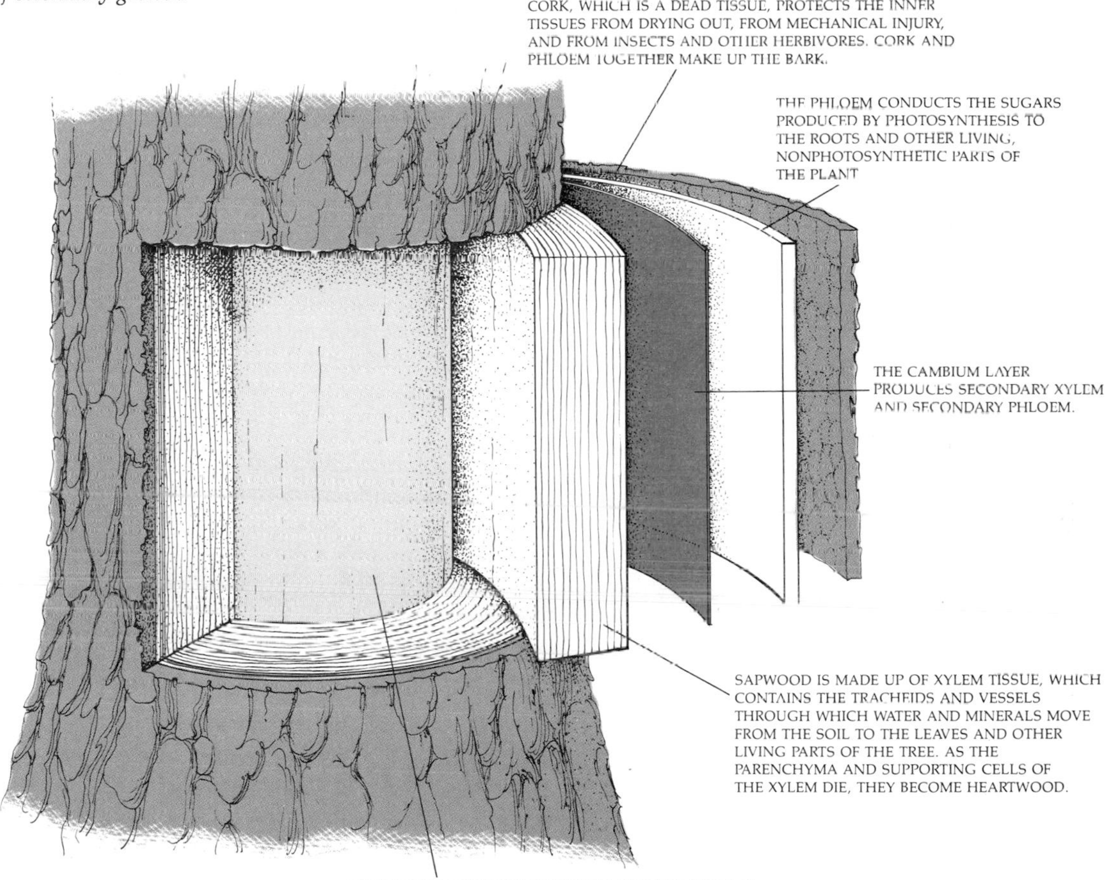

28–17
A tree trunk showing the relationships of the successive concentric layers. The heartwood is composed entirely of dead xylem cells. The cortex, which is outside the phloem in a green stem, is sloughed off in the process of secondary growth.

Table 28–1 *Summary of Main Cell Types in Angiosperms*

CELL TYPE	ORIGIN	LOCATION	CHARACTERISTICS	FUNCTION
Apical meristem	Embryonic cells	Apices of shoots and roots	Many-sided, small, thin-walled cells; vacuoles usually small	Origin of primary meristematic tissues
Epidermis	Protoderm	Surface of entire primary plant body	Flattened, variable in shape, overlaid by cuticle in shoot; specialized guard cells	Protective covering
Parenchyma	Protoderm, ground meristem, procambium, vascular cambium, cork cambium, wound tissues	Everywhere, usually dominant in pith, cortex, mesophyll, cork, phelloderm	Many-sided, usually thin-walled; abundant air spaces between cells	Photosynthesis, storage, wound healing, among others
Collenchyma	Ground meristem of leaf and stem	Peripheral in cortex of stem and in leaves	Elongate, with irregularly thickened primary walls	Support for young stems and leaves
Sclereid	Protoderm, ground meristem, procambium, vascular cambium, and cork cambium	Anywhere; in pith and cortex of stems; in leaves and flesh of fruits; seed coats	Irregular; massive secondary wall; alive or dead at maturity	Produces hard texture, mechanical support
Fiber	Procambium or vascular cambium; ground meristem	Primary and secondary xylem and phloem; cortex	Very long, narrow cell, with secondary cell walls; alive or dead at maturity	Support
Tracheid	Procambium or vascular cambium	Primary or secondary xylem	Elongate, with pits in walls; dead at maturity	Conduction of water and solutes
Vessel element	Procambium or vascular cambium	Interconnected series (= vessels) in primary or secondary xylem	Elongate, with pits in walls and end walls perforated; dead at maturity	Conduction of water and solutes
Sieve-tube element	Procambium or vascular cambium	Primary or secondary phloem, usually with companion cells; form interconnected series (= sieve tubes)	Elongate, with specialized sieve plates; nucleus lacking at maturity	Conduction of organic solutes
Vascular cambium	Procambium and parenchyma of pith rays in stems; procambium and pericycle in roots	Lateral, between secondary phloem and xylem tissues	Elongate fusiform (spindle-shaped) and short ray initials	Produces secondary xylem and phloem
Cork (phellem)	Cork cambium	Surface of stems and roots with secondary growth	Flattened cells, compactly arranged; dead at maturity, the cells often air-filled	Restricts gas exchange and water loss

ADAPTATIONS TO CLIMATE CHANGE

As we noted in Chapter 24, the angiosperms evolved during a relatively mild period in the earth's history. As the climate became colder, water became locked in snow and ice for part of the year. As a consequence, the angiosperms, which already possessed some adaptations to drought (perhaps because of highland origins), were placed under new environmental stress. Some did not survive, and some were pushed toward the Equator. Those that did survive in colder, drier areas did so because of selection for existing characteristics that offered advantages in these relatively unfavorable environments. Chief among such characteristics is the capacity to remain dormant during periods when water is in short supply and when climatic conditions are unfavorable for delicate growing buds, shoots, new leaves, and root tips. Depending on their characteristic patterns of active growth, dormancy, and death, modern plants are classified as annuals, biennials, and perennials.

Annuals, Biennials, and Perennials

28–18
Cross section of the trunk of a locust tree, showing annual growth rings. The rings are visible because of differences in the density of the wood produced early in the spring and that produced late in the summer. The early wood has large cells with thin walls, while the late wood has smaller cells with proportionately thicker walls. Within a given growth layer, the change from early wood to late wood is quite gradual, but a distinct change is visible where the small, thick-walled cells of the late wood of one growing season abut the larger, thin-walled cells of the next spring. In the locust, the heartwood is much darker than the sapwood.

Among annual plants, the entire cycle from seed to vegetative plant to flower to seed again takes place within a single growing season. Annuals include many of our weeds, wild flowers, garden flowers, vegetables, and grasses and most other monocots. Such plants usually show little or no secondary growth. All vegetative organs (roots, stems, and leaves) die, and only the dormant seed bridges the gap between one generation and the next. Plants with nonwoody stems, such as most annuals, are known as herbs.

In biennial plants, the period from seed germination to seed formation spans two growing seasons. The first season of growth often results in a rosette of leaves near the soil surface, a short stem, and a root, sometimes a storage root, such as a carrot. In the second growing season, extensive stem elongation, flowering, fruiting, and seed formation occur. This completes the life cycle and the vegetative organs die. Biennials may be herbaceous or woody.

Perennials are characteristically woody plants that live for many seasons and whose vegetative structures persist from year to year. In a woody plant, the stem increases in diameter and becomes firm due to the xylem inside. On the outside, a protective layer of dead tissue, the cork, accumulates.

The ancestral angiosperm is believed to have been a woody plant. Modern plants in favorable climates, such as a tropical rain forest, may live year after year with little change during the annual cycle, just as did the ancestral plants. Perennials that live in areas where part of the year is unfavorable to growth show a variety of adaptations. Some, such as cacti and most conifers, undergo little apparent change, although their rates of metabolism, and therefore of growth, change with the seasons.

Most dicots living in the temperate zones undergo both structural and functional changes that allow them to take maximum advantage of the growing season and still survive the rest of the year. In some, all of the aboveground structures die, and the plants overwinter as underground roots, stems (rhizomes or tubers), or buds (bulbs). Others, including many of the common vines, shrubs, and trees, are deciduous (page 495).

SUMMARY

The flower is the reproductive structure of the angiosperms. The anthers of the flower produce pollen grains, the male gametophytes; each anther is attached to a slender filament. The entire structure, anther plus filament, is known as the stamen. The carpel typically consists of the stigma, an area on which the pollen grains germinate, a slender style, and at its base, the ovary. The ovary contains one or more ovules. Within each ovule, a female gametophyte containing an egg cell develops. The ovule later becomes the seed, and the ovary becomes the fruit.

In flowering plants, a new life cycle begins when a pollen grain germinates upon the stigma of a flower of the same species, sending a pollen tube down the style and into an ovule. Within the ovule is the female gametophyte, which usually consists of seven cells, with a total of eight haploid nuclei, including the egg nucleus. One of the three haploid nuclei in the pollen grain (a sperm nucleus) unites with the egg cell nucleus, and the other sperm nucleus unites with the two polar nuclei of the female gametophyte. The plant embryo develops from the first union, and the nutritive endosperm, which is a triploid ($3n$) tissue, from the second. These phenomena of fertilization and triple fusion are found only among the flowering plants, and together are known as "double fertilization."

The fruit, which is formed from the ovary wall, is an adaptation for seed dispersal.

The angiosperm seed, or mature ovule, consists of the embryo, the seed coat, and stored food, which may be present in the form of endosperm. The petals, stamens, and other floral parts of the parent plant may fall away as the ovary ripens into fruit and the seed forms. The seed commonly enters a dormant period.

The embryo has two apical meristems—the apical meristem of the shoot and the apical meristem of the root or radicle—and either one or two cotyledons ("seed leaves"). From the endosperm, the cotyledons absorb nutrients that nourish the growing embryo. Early in embryonic life, cell division becomes confined to limited areas in the plant body: the apical meristems of the root and the shoot.

When the seed germinates, growth of the root and shoot proceeds from the apical meristems of the embryo. Certain cells within the meristems divide continuously. Some of these cells remain meristematic, continuing to divide, while others elongate and then differentiate, forming, according to their position, various specialized cells, including those of the xylem and the phloem. In the root, the apical meristem forms a root cap, which protects the root tip as it is pushed through the soil. Root hairs, which are the principal pathways of absorption by the roots, are outgrowths of the epidermal cells.

Leaf primordia arise from the apical meristem of the shoot. As the nodes are separated by elongation of the internodes, small apical meristems (buds) are formed in the axils of the leaves. These buds may remain dormant or may give rise to branches or specialized shoots, such as runners, rhizomes, tubers, or flowers.

Secondary growth is the process by which the woody perennials increase their girth. Such growth arises primarily from the vascular cambium, a sheath of meristematic tissue completely surrounding the xylem and completely surrounded by phloem. The cambium cells divide during the growing season, adding new xylem cells (secondary xylem) on their inner surfaces and new phloem cells (secondary phloem) on their outer surfaces. As the trunk increases in girth, the epidermis is eventually ruptured and destroyed and is replaced by cork.

Among the most useful adaptations of angiosperms, and probably one of the chief reasons for their evolutionary success, is the capacity, found in many species, to remain dormant during climatically unfavorable periods. Angiosperms are classified as annuals, biennials, or perennials, depending on whether the plant body dies at the end of one growing season (annual) or after two years (biennial), leaving only the seeds, or whether the plant body persists (perennial). In areas where there is marked seasonal variation in the availability of water, perennials are often deciduous, losing their leaves during periods when water is in short supply.

QUESTIONS

1. Distinguish among the following: runner/rhizome; ovary/ovule/egg cell/seed; pollination/fertilization; epidermis/endodermis/endosperm; apical initial/apical meristem/lateral meristem; primary growth/secondary growth.

2. Sketch a flower. Give the function of each floral part.

3. Identify the floral parts visible in the photograph at the left. Can you name the flower?

4. In many plants, pollen production in the anthers occurs prior to or after the full development of the carpel of the same flower. What are the consequences of this shift in time frames?

5. Which of the following is a berry: blackberry, strawberry, mulberry, grape?

6. The jack pine produces seeds with very tough seed coats. These seed coats rupture, and the seed germinates, only after exposure to intense heat, as in a forest fire. What might be the advantages of such seeds for the jack pine?

7. Sketch a dicot embryo at the time of seed release. Identify each part in terms of the future development of the plant body.

8. Suppose you carve your initials 1 meter above the ground on the trunk of a mature tree that is growing vertically at an average rate of 15 centimeters a year. How high will your initials be at the end of 2 years? At the end of 20 years?

9. As a result of secondary growth, most of our common trees increase in girth as they increase in height. What limitations would a lack of secondary growth impose on the form and mechanical support of a tree?

10. Sketch a tree trunk with secondary growth. Compare your sketch with Figure 28–16c.

11. Consider a mature tree. Which parts of the tree are composed of living cells? Why might it be advantageous for living cells to make up such a small fraction of the mass of the tree? Similarly, why might it be advantageous for a deciduous tree to invest as little material as possible in making leaves?

12. Describe the evolutionary advantages to a plant of the following: (a) the ripening of a fruit; (b) seed dormancy; (c) a biennial life cycle.

13. Many features deliberately bred into cultivated seeds and plants would be detrimental or fatal to wild varieties. Give four examples of such features, and explain why each would be disadvantageous to the plant in the wild.

CHAPTER 29 Transport Systems in Plants

29-1
Guttation droplets on the edge of a rose leaf. Guttation, the loss of liquid water, is a result of root pressure. The water escapes through specialized pores located near the ends of the principal veins of the leaf. Guttation, which is restricted to relatively small plants, usually occurs at night when the air is moist.

After William Harvey (1578–1657) discovered the circulation of the blood, botanists began an eager search for the analogous pathways and pumping mechanisms of plants. It was to no avail. Although plants do have a vascular system, as we noted in Chapter 27, the ways that fluids move in the plant body are quite different from the circulation of fluids in the bodies of animals. As long ago as 1727, Stephen Hales, in *Vegetable Staticks,* showed that fluids do not recirculate in plants, as blood does in animals, but that water moves predominantly toward the leaves. However, despite these remarkable early studies, it is only comparatively recently that plant physiologists have begun to understand the principles of the transport of water, sugar, and minerals in plants. Some aspects of the problem are, as you will see, still controversial.

One striking point of difference between plants and animals is that animals have a single major transport system, the blood vessels, and plants have two, the xylem and the phloem. In this chapter, we shall discuss first the movement of water in the xylem; second, the transport of sugars and other nutrients in the phloem; and, at the end of the chapter, a related subject, the uptake by the plant of nutrients from the soil.

THE MOVEMENT OF WATER: TRANSPIRATION

In order to carry out photosynthesis, plants must exchange gases with the surrounding air. In the course of this exchange, as we noted in Chapter 27, they lose water, which diffuses as vapor into the surrounding air. The quantity of water passing through a plant is enormous—far greater than that used by an animal of comparable weight. An animal requires less water because, as we noted, most of its water remains in its body and recirculates again. By contrast, in plants, more than 90 percent of the water that enters the roots is given off into the air as water vapor. A single corn plant needs 160 to 200 liters of water from seed to harvest, and a hectare of corn requires almost 5 million liters of water a season. The loss of water vapor from the plant body is known as *transpiration*.

Water enters the body of most plants almost entirely through the roots. During periods of rapid transpiration, water may be removed from around the roots so quickly that the soil in the vicinity of the roots becomes depleted; water will then move slowly by diffusion and capillary action through the soil toward the depleted region near the roots. (For a review of the principles of water movement, see pages 34 to 35, and 125 to 130. An understanding of these processes is necessary for understanding the rest of this chapter.)

Root cells, like other living parts of the plant, contain a higher concentration of solutes (both organic and inorganic) than does soil water, and so water from the soil enters the roots by osmosis. Osmosis is sufficient to move water a short distance up the stem, a phenomenon known as root pressure. Guttation (Figure 29–1) is a visible consequence of root pressure. But how can water reach 20 meters high to the top of an oak tree, travel three stories up the stem of a vine, or move 125 meters up in a tall redwood?

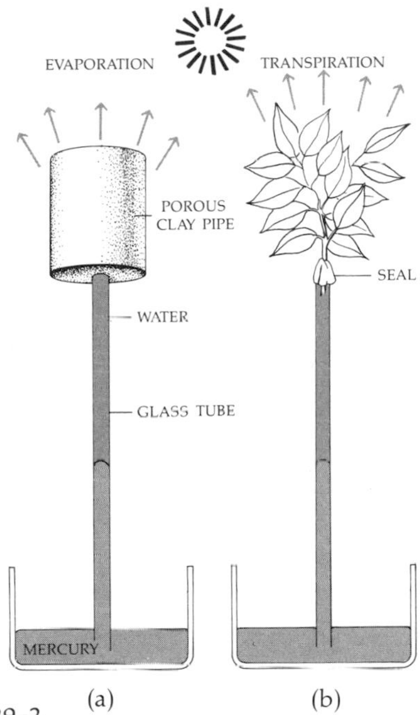

29–2
(a) *A simple model that illustrates the cohesion-tension theory. A piece of porous clay pipe, closed at both ends, is filled with water and attached to the end of a long, narrow glass tube also filled with water. The water-filled tube is placed with its lower end below the surface of a volume of mercury contained in a beaker. As water molecules evaporate from the pores in the pipe, they are replaced by water "pulled up" through the narrow glass tube in a continuous column. As the water evaporates, mercury rises in the tube to replace it.* (b) *Transpiration from plant leaves results in sufficient water loss to create a similar negative pressure.*

One important clue is the observation that during times when the most rapid transpiration is taking place—which is, of course, when the flow of water up the stem must be the greatest—xylem pressures are characteristically negative (less than atmospheric pressure). The existence of negative pressure can be demonstrated readily. If you peel a piece of bark from a transpiring tree and make a cut in the xylem, no sap runs out. In fact, if you place a drop of water on the cut, the drop will be drawn in.

What is the pulling force? It is not simple suction, as the negative pressure might indicate. Suction simply removes air from a system so that the water (or other liquid) is pushed up by atmospheric pressure. But atmospheric pressure is only enough to raise water (against no resistance) about 11 meters at sea level, and many trees are much taller than 11 meters.

According to the now generally accepted theory, the explanation is to be found not in the properties of the plant but in the remarkable properties of water, to which the plant has become exquisitely adapted. As we pointed out in Chapter 2, in every water molecule, the pair of hydrogen atoms is linked to a single oxygen atom. The hydrogen atoms are also held to the oxygen atoms of the nearest water molecules by hydrogen bonds. This secondary attraction can produce a tensile strength of as much as 140 kilograms per square centimeter (2,000 pounds per square inch) in a thin column of water. In the leaf, water evaporates, molecule by molecule, from the cells into the intercellular spaces within the leaves. The water potential of the leaf cell falls, and water from the vessels or tracheids moves, molecule by molecule, into the leaf cells. But each molecule in the xylem vessel is linked to other molecules in the vessel, and they, in turn, are linked to others, forming one long, narrow, continuous thread of water reaching right down to a root tip. As the molecule of water moves into the leaf, it tugs the next molecule along behind it.

Because the diameter of the vessels is very small and because the water molecules adhere to the walls, even as they are cohering to one another, gas bubbles, which could rupture the column, do not usually form. The pulling action, molecule by molecule, causes the negative pressure observed in the xylem. The technical term for a negative pressure is *tension*, and this theory of water movement is known as the *cohesion-tension theory*.

The principle of cohesion-tension is illustrated in Figure 29–2. As indicated in the diagram, the power for this process comes not from the plant, which plays only a passive role in transpiration, but directly from the energy of the sun.

29–3
Measurements in ash trees show that a rise in water uptake follows a rise in transpiration. These data suggest that the loss of water generates forces for its uptake.

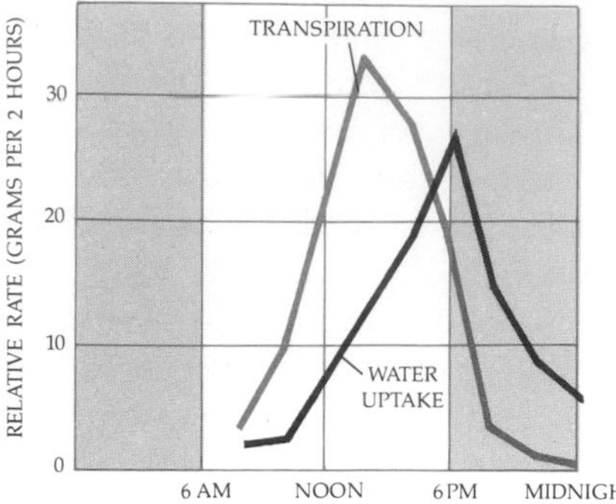

FACTORS INFLUENCING TRANSPIRATION

Transpiration, as we noted, is costly, especially when water supply is limited. Various factors affect the rate of water loss. One is temperature; the rate of evaporation doubles for every temperature rise of about 10°C. Humidity is also important. Water is lost much more slowly into air already laden with water vapor. Air currents also affect transpiration. A breeze cools your skin on a hot day because it blows away the water vapor that has accumulated near the skin surface and so facilitates evaporation. Similarly, wind blows away the water vapor from leaf surfaces, which makes the concentration gradient steeper and so hastens the diffusion of water molecules. Leaves of plants that grow in exposed, windy areas are often hairy; these hairs are believed to protect the leaf surface from the wind and so retard transpiration (Figure 29–4). By far the most important factor affecting transpiration, however, is the regulatory effect exercised by the opening and closing of the stomata.

Mechanism of Stomatal Movements

As shown in Figure 29–5, each stoma has two surrounding guard cells. Stomatal movements are caused by changes in the turgor of these cells. When the guard cells are turgid, they bow out, opening the stoma; when they lose water, they collapse, and the stoma closes.

Turgor, as we saw in Chapter 6 (page 129), is maintained or lost due to the osmotic movement of water into or out of cells along a gradient of solute concentration. Solute concentration in the guard cells can change as a result of the loss of water, the production of more solutes within the cells, or the transport of solutes into the guard cells.

Factors Influencing Stomatal Movements

A variety of environmental factors influence the opening and closing of the stomata. The principal one is the availability of water: When the plant body begins to lose water, the stomata close. This is not a direct effect; closing of the stomata occurs before the leaf begins to wilt. It is, however, an overriding one.

Stomatal changes also occur quite independently of the loss or gain of water.

29–4
Leaf hairs create a layer of still air over the epidermis, restricting gas exchange and water loss from the stomata (visible in the lower portion of the micrograph).

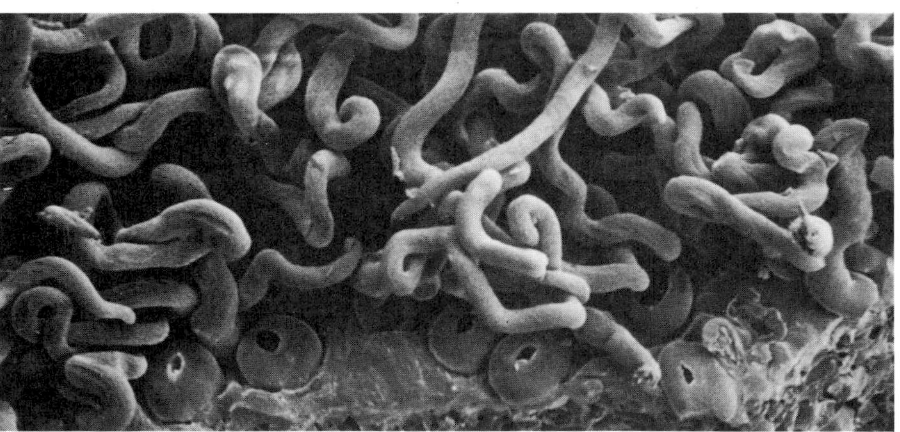

29–5
Mechanism of stomatal movement. A surface view is shown in (a) *and a cross section in* (b)*. Each stoma has two guard cells that open the stoma when they are turgid and collapse and close it when they lose turgor. In many species, the guard cells have thickened walls adjacent to the stomatal opening. As turgor pressure increases, the thinner parts of the cell wall are stretched more than the thicker parts, causing the cells to bow out and the stoma to open.*

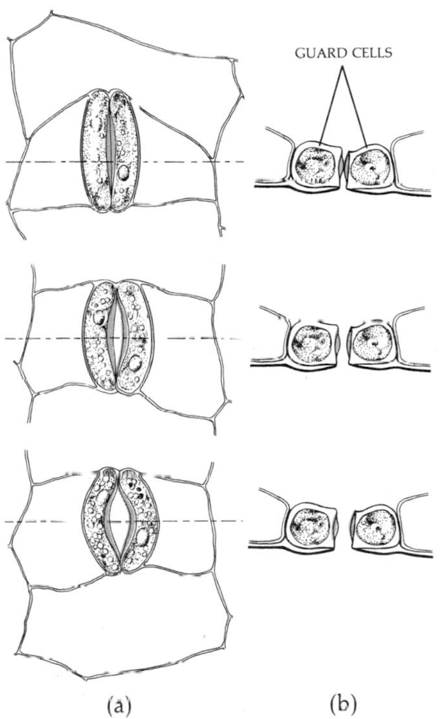

The simplest and most common example is found in the many species in which the stomata close regularly in the evening, when photosynthesis is no longer possible, and open in the morning. These stomatal movements in response to light occur even when there are no changes in the water available to the plant. An increase in temperature causes stomatal closing in many species. Carbon dioxide concentration also affects the stomata. An increase in carbon dioxide concentration in the intercellular spaces of the leaf causes the stomata to close. Conversely, in some species, exposure of the leaf to CO_2-free air causes the stomata to open. These changes occur even though the water supply and the light remain constant.

As long ago as the 1920s the so-called classic theory of stomatal function was formulated. According to this hypothesis, a decrease in carbon dioxide in the guard cells causes a slight increase in pH as a consequence of the decrease in the carbonic acid (H_2CO_3) concentration. In response to this increase in pH, the enzyme (amylase) that catalyzes the hydrolysis of starch to sugar becomes more active. As we noted in Chapter 6, the size of the molecules of a solute has little effect on osmosis; it is the number of molecules that counts. Therefore the breakdown of large starch molecules into many smaller molecules rapidly changes the osmotic equilibrium, causing water to move rapidly into the guard cells and the guard cells to open. Carbon dioxide decreases during photosynthesis and increases during respiration. Thus opening in the light and closing in the dark may be a result of the changes in carbon dioxide concentration. Similarly, heat increases the rate of respiration, so the heat effect may also be mediated by the CO_2-H_2CO_3 effect.

More recent data indicate, however, that this pleasing theory does not explain stomatal movements in all species. For example, some plants—onions, for instance—do not form starch, yet their stomata open and close. Also, the action spectrum for photosynthesis (Chapter 9, page 177) does not match that for stomatal opening; the greatest effect for the latter is in the blue region of the spectrum. As a consequence of such findings, a search has been underway for an alternative explanation. New techniques that permit the measurement of ion concentrations within single guard cells show that potassium (K^+) ions are accumulated in guard cells in the light and lost from these cells in the dark. Apparently, the K^+ ions are brought in by active transport, and the water follows.

The relative roles of these two mechanisms are still under debate. It is possible, of course, that different plants have evolved different mechanisms of stomatal regulation.

CAM Photosynthesis

As we mentioned, the stomata of most plants are open during the day, when photosynthesis takes place, and closed at night. This is not true of all plants, however. Many succulents—including cacti, pineapples, and members of the stonecrop family—close their stomata during the hot, dry days and open them at night. The carbon dioxide taken in during the night is converted to malic and isocitric acids. During the day, when the stomata are closed, the carbon dioxide is released from these organic acids and used in photosynthesis. This process is known as Crassulacean acid metabolism (from Crassulaceae, the plant family to which the stonecrops belong), and is generally referred to as CAM photosynthesis. It is analogous to the C_4 pathway in photosynthesis described in Chapter 9, although it apparently evolved independently. (As you can see, this is another instance of stomatal movement not explained by the CO_2 mechanism.)

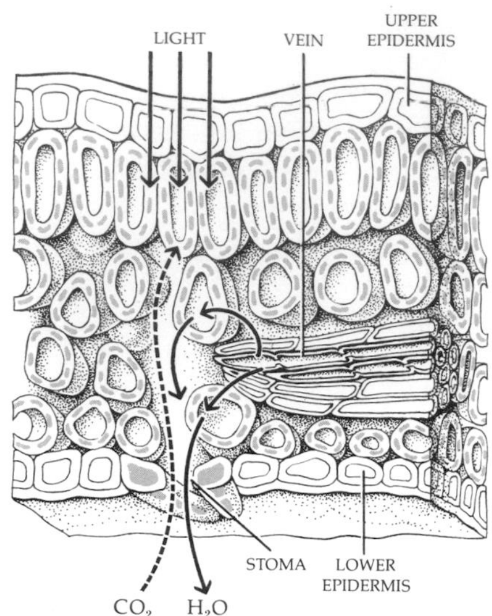

29–6
As carbon dioxide, essential for photosynthesis, enters the leaf through the stomata, water vapor is lost.

Photosynthesis and Water Loss

In short, photosynthesis requires carbon dioxide. Uptake of carbon dioxide involves loss of water. Evolution has not yet produced a way that a plant can maintain a free flow of carbon dioxide without water loss. It has, however, produced a number of ingenious ways of reducing water loss, through structural modifications and by regulation of stomatal movements. The way in which various external and internal stimuli act upon the membranes of guard cells to produce closing and opening of the stomata is not known.

THE MOVEMENT OF SUGARS: TRANSLOCATION

The fact that water is transported by vessels in the xylem was recognized by botanists 300 years ago. The role of phloem in the movement of sugar was not generally agreed upon, however, until well into the twentieth century. One reason for this long delay is that the sieve-tube elements are so delicate. When they are disturbed, callose and P-protein plug the pores in the sieve plates, impeding or preventing the movement of substances through them. Many botanists of the eighteenth and nineteenth centuries believed that sugars, like water, move through the xylem, and they developed elaborate theories involving the properties of water to support this contention. The direction of science has often been determined by brilliant, articulate, strong-willed men whose unproved hypotheses dominated their fields, even until after their deaths. We have seen examples of this both in genetics and in the development of evolutionary theory. It is likely that this phenomenon is still taking place in various fields of science.

THE EVIDENCE FOR THE PHLOEM

Early evidence supporting the role of the phloem in transport came from observations of trees that had had a complete ring of bark removed from them. As we saw in the last chapter, bark contains the phloem but not the xylem. When a photosynthesizing tree is "girdled" in this manner, the tissue above the ring becomes swollen, indicating the accumulation of fluid moving downward in the phloem from the photosynthesizing leaves.

Much more convincing evidence was obtained when radioactive tracers became available. If plants carry out photosynthesis in air containing carbon dioxide made with carbon 14 ($^{14}CO_2$), the sugars produced will include the radioactive ^{14}C label. If a tissue sample is taken from a plant containing ^{14}C-labeled sugars and placed on a photographic emulsion, emissions from the radioactive isotope will make tracks on the emulsion. These tracks reveal the concentration of radioactive material in the plant (see essay on facing page). Such studies have shown conclusively that sugars are transported in the sieve tubes of plants.

ASSISTANCE BY APHIDS

Cutting into a sieve tube with even the most delicate dissecting blade results in a surging of sap from the severed sieve tubes and the plugging up of the sieve-plate pores (page 499) with slime and callose. This technical problem has been cleverly circumvented with the help of aphids, which are very small sap-sucking insects.

RADIOACTIVE TRACERS IN PLANTS

Radioactive tracers can be used in a number of ways to study the synthesis, transport, and use of materials within plants. Initially, in any tracer study, the radioactive isotope must be incorporated into the plant. Radioactive carbon, for example, will be taken up by a plant if its leaves are exposed to carbon dioxide containing carbon 14. Or, radioactive phosphorus will be taken up if the roots are exposed to a solution containing phosphorus-32 ions.

The length of time the plant is exposed to the radioactive material is determined by the information the investigators hope to obtain. For example, in studies to determine the time required for carbon dioxide to be incorporated into the various products of photosynthesis, a sequence of exposure times would be used. In studies focusing on the location of a particular product formed from the radioactive substance, the length of exposure would depend on the time required for the chemical reactions being studied.

After exposure to the radioactive substance, the plant is quick-frozen and freeze-dried. In whole-plant autoradiography, the plant is flattened and then pressed against a sheet of x-ray film. Radiation given off by the isotope exposes the film adjacent to the portions of the plant in which the tracer is located. By comparing the flattened plant with the developed film, investigators can determine the location of the radioactive substance within the plant.

In tissue autoradiography (histo-autoradiography), the freeze-dried plant tissues are embedded in paraffin, resin, or a similar material. Next, they are sliced into very thin sections, which are mounted on microscope slides. The tissue sections are then placed in contact with a photographic emulsion or film. As in whole-plant autoradiography, the radiation from the isotope exposes the film in contact with the portions of the tissue section containing the radioactive material. After an appropriate interval of time, the film is developed. Comparison, under the microscope, of the developed film and the underlying tissue section reveals the exact location of the radioactive substance in the plant tissues.

In the study illustrated here, three leaves of a broad bean plant were enclosed in a flask and were exposed to $^{14}CO_2$ and light for 35 minutes (a). *During that time, the radioactive carbon dioxide was incorporated into sugars, which were then being transported to other parts of the plant. A cross section* (b) *and a longitudinal section* (c) *from the stem were placed in contact with autoradiographic film for 32 days. When the film was developed and compared with the underlying tissue sections, it was apparent that the radioactivity (visible as dark specks on the film) was confined almost entirely to the sieve tubes.*

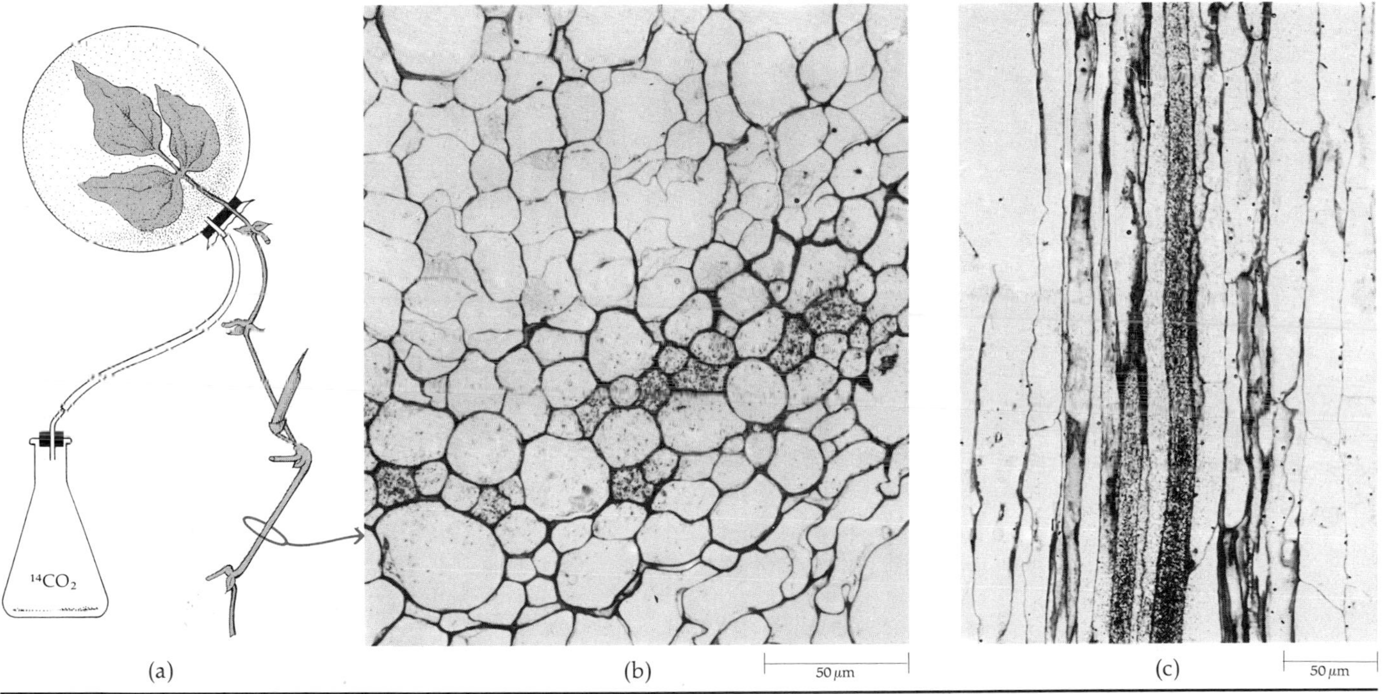

(a) (b) (c)

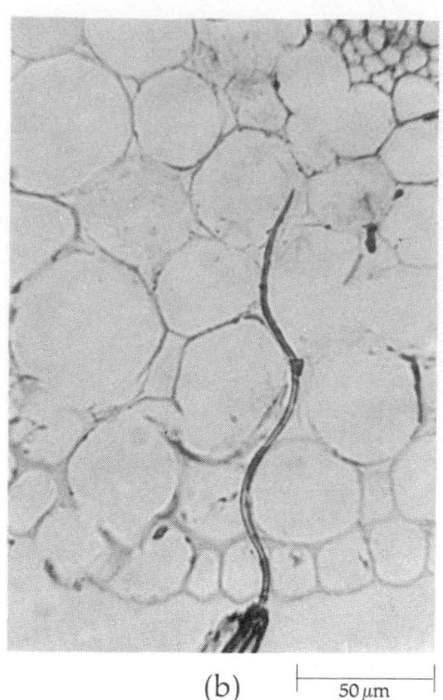

29–7
Assistance by aphids. (a) Aphids are very small insects that feed on plants, sucking out their juices; you probably have seen them on rose bushes. (b) The aphid, as the micrograph reveals, drives its sharp mouthparts, or stylets, like a hypodermic needle between the epidermal cells and then taps the contents of a single sieve-tube element. If the aphid is anesthetized, it is possible to sever the stylets and leave them undisturbed in the cell. The fluid will continue to exude through the stylets from the sieve tube for several days, and pure samples of the fluid flowing through the sieve tube can be collected for analysis without damaging the sieve tube or interfering with its function.

The aphid drives its sharp mouthparts, or stylets, through the epidermis of a plant, traversing cortical cells (Figure 29–7), and then taps the contents of a single sieve-tube element. If the aphid is anesthetized, it is possible to sever the stylets, leaving them in place in the cell. The solution will continue to exude through the stylets from the sieve tube for several days, and pure samples of phloem fluid can be collected for analysis. Many of the findings reported here were made possible by this simple and ingenious technique.

Data gathered by this technique and from radioactive tracers indicate that sieve-tube sap contains (by weight) 10 to 25 percent solutes, 90 percent or more of which are sugars, mostly sucrose. Low concentrations of amino acids and other nitrogen-containing substances are also present. The rate of movement of these substances along the sieve tube is remarkably fast: In one set of experiments, for example, it was estimated that the sap was moving at a rate of about 100 centimeters an hour, far faster than could be accounted for by diffusion.

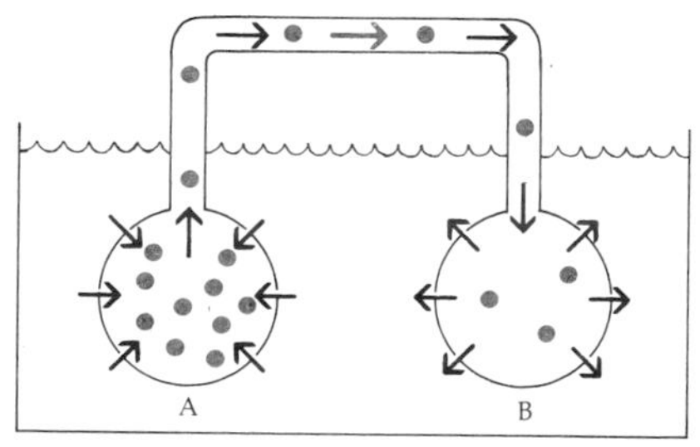

29–8
Model of pressure-flow hypothesis. Bulbs A and B, which are interconnected and are permeable to water, are placed in a bath of distilled water. Bulb A contains a higher concentration of sucrose than bulb B. Water enters bulb A from the medium, increasing the hydrostatic pressure and pushing the solution to bulb B. If B were connected to a third bulb, C, with a still lower concentration of sucrose, as sieve-tube elements are connected in a series, hydrostatic pressure building up in B would push the solution on to C, and so on.

PRESSURE-FLOW HYPOTHESIS

The most popular current explanation for translocation is the *pressure-flow hypothesis,* which proposes that solutes move in solutions that, in turn, move because of differences in water potential. The principle underlying this theory can be illustrated by a simple physical model consisting of bulbs permeable only to water and connected by glass tubes (Figure 29–8). The first bulb contains a solution in which there is a dissolved material, such as sucrose, and the second, to make the example as simple as possible, contains only water. When these interconnected bulbs are placed in distilled water, water will enter the first by osmosis. The entry of water will increase hydrostatic pressure (turgor) within this bulb and cause the water and the solutes in it to move along the tube to the second bulb, where the pressure again builds up. If the second bulb is connected with a third one containing water or a sucrose concentration lower than that now in the second one, the solution will flow from the second to the third by the same process, and so on indefinitely down

29-9
The pressure-flow mechanism as it is believed to occur in the plant body.

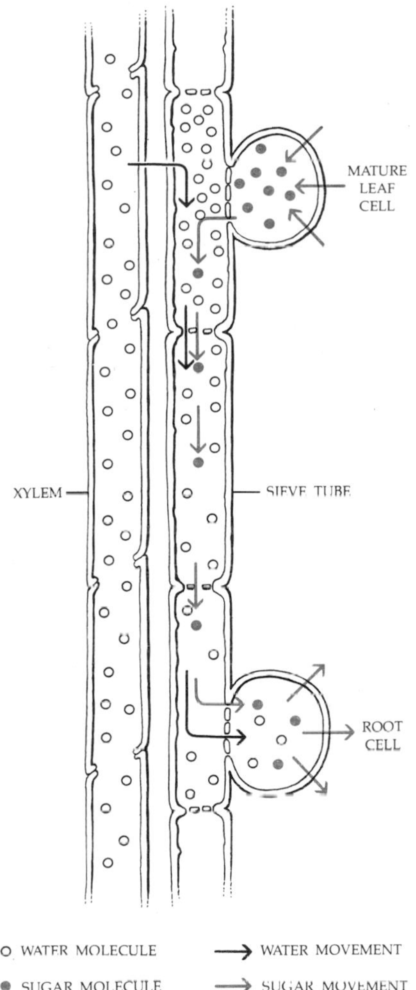

the sucrose gradient. This hypothesis is supported by the fact that distinct gradients in the concentrations of sucrose and other sugars have been demonstrated along the phloem tissues during the summer months. Moreover, it can account for the known rates of movement in the phloem.

According to the proponents of the pressure-flow mechanism, the sieve tubes play a passive role in the movement of the sugar solution through them. Active transport is also involved, however; the active transport is not directly involved with long-distance transport through the sieve tube, but with the movement of sugars and other substances into and out of the sieve tubes. In addition, the energy required for this active transport is expended not by the sieve tubes, but by other phloem cells (parenchyma cells and companion cells) bordering the sieve tubes. The energy requirements of this active transport would account for the comparatively high rate of respiration often found in phloem.

Some alternative hypotheses are under consideration. However, all of them apparently have serious deficiencies, leaving the pressure-flow hypothesis by far the most creditable explanation for the movement of substances through the phloem.

SOILS AND PLANT NUTRITION

MINERALS

The earth is composed of about 100 different elements; oxygen and silicon are by far the most abundant of these, followed by aluminum and iron. Elements are found in the earth in the form of *minerals*. A mineral is a naturally occurring inorganic substance, usually solid, with a definite chemical composition. Some minerals, such as diamond, sulfur, and copper, consist of single elements. Others, such as quartz (SiO_2) and calcite ($CaCO_3$), are compounds.

The mineral elements required by plants are usually classified either as macronutrients, which are generally expressed in percentage of dry weight of the plant and which range from about 0.5 to 3 or 4 percent, or as micronutrients, of which as little as a few parts per million may be present. Determination of many of the elements is generally made by burning the plant completely, which permits the carbon, hydrogen, oxygen, nitrogen, and sulfur to escape as gases; the ash is then analyzed. Proportions of each element vary in different species and in the same species grown under different conditions. Also, the ash will often contain elements, such as silicon, which are present in the soil and are taken up by plants but which are not generally thought to be required for plant growth.

Uptake of Minerals

Minerals are taken up in solution by the roots and travel through plants with the transpiration stream. As you saw in Figure 27-17 (page 503), the cells of the endodermis play a major role in determining which elements enter the xylem.

The mineral composition of plant cells is far different from the mineral composition of the medium in which the plant grows. For example, in one study, cells of pea roots were found to have a concentration of potassium ions 75 times greater than that of the nutrient solution. Similarly, in another study, the vacuoles of rutabaga cells were shown to contain 10,000 times more potassium than the external solution.

Thus it seems evident that minerals are brought into plant cells by active transport. Support for this hypothesis comes from observations indicating that the uptake of minerals is an energy-requiring process. For instance, if roots are deprived of oxygen or poisoned so that respiration is curtailed, mineral uptake is drastically decreased. Also, if a plant is deprived of light, it will cease to absorb minerals and eventually will release them back into the soil.

Ion Pumps

One way that roots move ions such as potassium against a concentration gradient is by the action of carrier proteins in the cell membrane (see page 131). Such a protein combines with the ion and then, utilizing the energy of ATP, carries it to the other side of the membrane and releases it. Because the membrane is partially impermeable to ions such as potassium, the continued action of the "pump proteins" results in differences in the concentration of the transported ions on one side of the membrane as compared to the other.

As a result of these differences in ion concentrations, there is often a difference of electric charge on either side of the membrane. In the nerve cells of animals, as we shall see, stimulation of the cell membrane results in an abrupt change in its permeability to ions. The rapid movement of ions across the membrane is the basis for transmission of nerve impulses. Similar events apparently involved in certain plant responses will be described in Chapter 30.

Functions of Minerals in Plants

29–10
Unlike animals, which require sodium, plants, in general, do not, and most cannot live in brackish waters, because of osmoregulatory problems. Exceptions are salt-marsh species, which not only can survive in salty water but some of which require salt as a micronutrient.

Plants require mineral elements for many different functions. For example, mineral ions affect osmosis (page 128) and so regulate water balance. Because in many plants several mineral ions can serve interchangeably in this role, this requirement is described as nonspecific. On the other hand, a mineral element may be a functional part of an essential biological molecule, and in this case the requirement is highly specific. An example is the magnesium atom in the chlorophyll molecule. Some minerals are required constituents of cell membranes, and some control membrane permeability. Others are indispensable components of a variety of enzyme systems that catalyze biological reactions in the cell. Still others provide a proper ionic environment in which biological reactions can occur.

Because mineral elements are involved in many fundamental processes, the effects of mineral deficiencies are typically very wide-ranging, affecting a number of structures and functions in the plant body.

Table 29–1 lists the mineral elements required by plants, the form in which they usually are absorbed, and some of the uses plants make of them. You might anticipate that organisms make use of what is most readily available; indeed, they seem to have done this to a considerable extent when life originated from elements in the gases of the primitive atmosphere. But the table reveals some findings that you might not expect. Sodium, for instance, which is one of the most abundant of the elements, is apparently not required at all by most plants. The fact that plants evolved with no functions requiring sodium is even more striking when you consider that sodium is vital to animals. (Sodium is the principal osmoregulator in animals and, as we shall see in the next section, is necessary for transmission of the nerve impulse.) In the seas, where both plant and animal life seem to have originated, sodium is the most abundant mineral element and is far more readily available than potassium, which it closely resembles in its essential properties. Potas-

(a)

(b)

29–11
(a) *Plants of the mustard family, such as the wintercress shown here, use sulfur in the synthesis of the mustard oils that give the plants both their name and their characteristic sharp taste.* (b) *Horsetails incorporate silicon into their cell walls, making them indigestible to most herbivores but useful, at least in colonial America, for scouring pots and pans.*

sium, however, is the principal osmoregulator in plants. Similarly, although silicon and aluminum are almost always present in large amounts in soils, few plants require silicon and apparently none requires aluminum. On the other hand, most plants need molybdenum, which is relatively rare.

Symbioses and Mineral Utilization

Mycorrhizae

Two types of symbiotic relationships play important roles in plant nutrition. One of these is *mycorrhizae* ("fungus-roots"), which are associations between plant roots and fungi. The importance of mycorrhizae was first recognized in connection with efforts to grow orchids in greenhouses. Orchids have microscopic seeds that germinate to form a tiny pad of tissue called a protocorm. Cultivators of orchids found that the plants seldom developed beyond the protocorm stage unless they were infected with a particular kind of fungus. Subsequently it was found that if seedlings of many forest trees are grown in nutrient solutions and then transplanted to prairie and other grassland soils, they fail to grow. Eventually they die from malnutrition despite the fact that soil analysis shows that there are abundant nutrients in the soil (Figure 29–12). If a small amount (0.1 percent by volume) of forest soil containing fungi is added to the soil around the roots of the seedlings, however, they will grow promptly and normally. Mycorrhizae are now thought to occur in more than 90 percent of all families of plants.

29–12
Effects of mycorrhizae on tree nutrition. Nine-month-old seedlings of white pine were grown for two months in a sterile nutrient solution and then transplanted to prairie soil. The seedlings on the left were transplanted directly. The seedlings on the right were placed in forest soil for two weeks before being transplanted to the prairie.

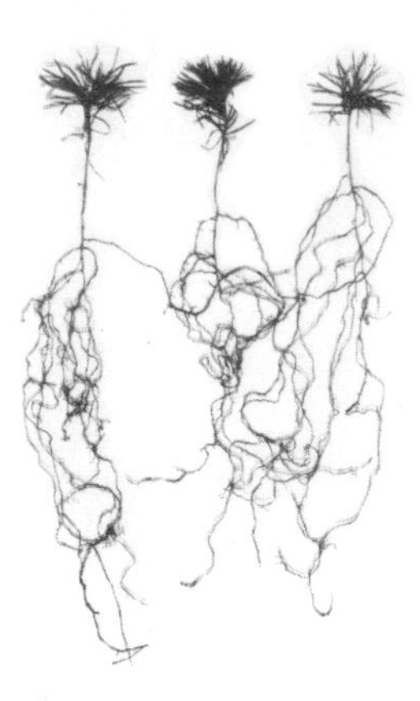

In mycorrhizal associations, the fungus may form a sheath of hyphae, or a fungal mantle, around the root. Only the cortex is actually invaded by the fungus. Roots with mycorrhizae usually lack root hairs, the role of water and mineral uptake from the soil evidently being at least partially assumed by the fungus.

The exact relationship between roots and fungi is not known. Apparently the roots secrete sugars, amino acids, and possibly some other organic substances that are utilized by the fungi. Although the evidence is just now accumulating, it seems that the chief contribution of the fungi is in water uptake and in converting minerals in the soil and in decaying material into an available form.

Rhizobia and Nitrogen Fixation

All but one of the elements listed in Table 29–1 are derived from the weathering of rocks. The exception is nitrogen, which enters the soil—and the plants that grow in the soil—by way of the atmosphere.

Molecular nitrogen (N_2), which constitutes about 78 percent of the air, is held together by an exceptionally strong bond. A large quantity of energy is required to break it. The process by which atmospheric nitrogen is incorporated into nitrogen compounds that can be utilized by plants is known as *nitrogen fixation*. It is carried out to a small extent by abiotic (nonliving) processes, such as the production of nitrogen oxides and ammonia by interactions between nitrogen and other atmospheric gases, with lightning as an energy source.

Nitrogen fixation is carried out commercially by combining 1 mole of nitrogen gas (N_2) and 3 moles of hydrogen gas (H_2) to produce 2 moles of ammonia (NH_3). The required energy is supplied by fossil fuels. As the costs of these fuels increase, so do the prices of the nitrogen-containing fertilizers produced. At present, the

Table 29-1 *A Summary of Mineral Elements Required by Plants*

ELEMENT	FORM IN WHICH ABSORBED	APPROXIMATE CONCENTRATION IN WHOLE PLANT (AS % OF DRY WEIGHT)	SOME FUNCTIONS
Macronutrients			
Nitrogen	NO_3^- (or NH_4^+)	1–3%	Component of amino acids, proteins, nucleotides, nucleic acids, chlorophyll, and coenzymes.
Potassium	K^+	0.3–6%	Osmoregulation. Activator of many enzymes. Involved in opening and closing of stomata.
Calcium	Ca^{2+}	0.1–3.5%	Component of cell walls. Enzyme cofactor. Involved in cell membrane permeability.
Phosphorus	$H_2PO_4^-$ or HPO_4^{2-}	0.05–1.0%	Formation of "high-energy" phosphate compounds (ATP and ADP). Component of nucleic acids and of several essential coenzymes.
Magnesium	Mg^{2+}	0.05–0.7%	Part of the chlorophyll molecule. Activator of many enzymes.
Sulfur	SO_4^{2-}	0.05–1.5%	Component of some amino acids, proteins, and coenzyme A.
Micronutrients			
Iron	Fe^{2+}, Fe^{3+}	10–1,500 parts per million (ppm)	Required for chloroplast development. Component of cytochromes.
Chlorine	Cl^-	100–10,000 ppm	Involved in osmosis and ionic balance; probably essential in photosynthesis in the reactions in which oxygen is produced.
Copper	Cu^{2+}	2–75 ppm	Activator of some enzymes.
Manganese	Mn^{2+}	5–1,500 ppm	Activator of some enzymes. Required for oxygen release in photosynthesis.
Zinc	Zn^{2+}	3–150 ppm	Activator of many enzymes.
Molybdenum	MoO_4^{2-}	0.1–5.0 ppm	Nitrate reduction and nitrogen fixation.
Boron	BO^{3-} or $B_4O_7^{2-}$ (borate or tetraborate)	2–75 ppm	Functions unknown. Possibly involved in carbohydrate transport.
Elements Essential to Some Plants or Organisms			
Cobalt	Co^{2+}	Trace	Required by nitrogen-fixing microorganisms.
Sodium	Na^+	Trace	Involved in osmotic and ionic balance, probably for many plants not essential. Required by some desert and salt-marsh species and may be required by all plants that utilize C_4 photosynthesis.

equivalent of 2 million barrels of oil a day is required for the production of nitrogen-containing fertilizers.

On a worldwide basis, most nitrogen fixation is carried out by a few types of free-living microorganisms, including blue-green algae, some free-living bacteria, and some species of bacteria that live in symbiosis with plants. Of these various classes of nitrogen-fixing organisms, the symbiotic bacteria are by far the most important in terms of total amounts of nitrogen fixed.

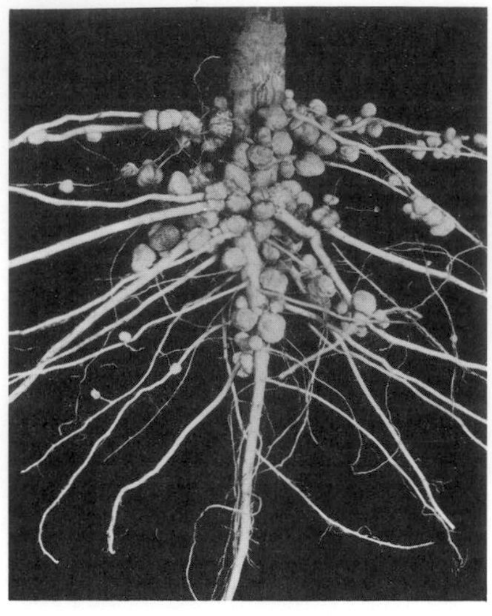

(a)

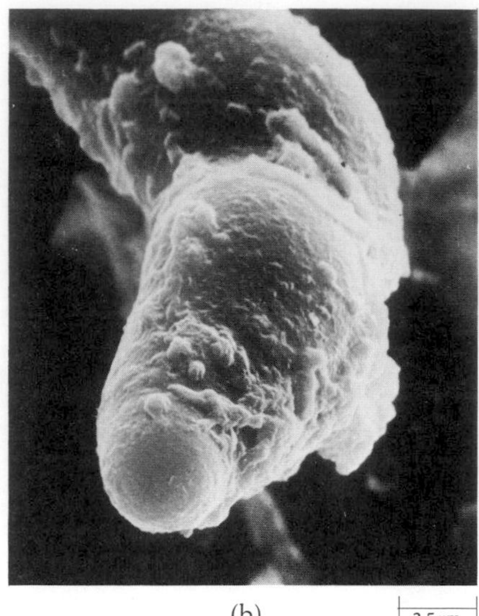

(b)

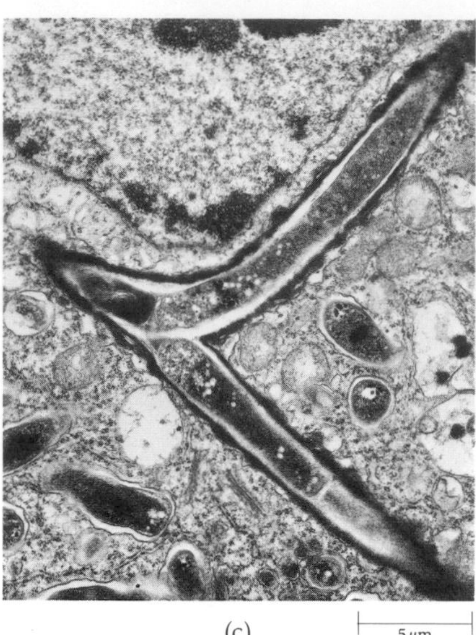

(c)

29–13
(a) *Nitrogen-fixing nodules on the roots of a soybean, a legume. These nodules are the result of a symbiotic relationship between a soil bacterium* (Rhizobium) *and root cells.* (b) *Tip of a root hair of a clover seedling, with several rhizobia and some soil particles.* (c) *Two branches of an infection thread (which, as you can see, is passing near the cell nucleus, at the top of the micrograph).* (d) *Cross section of an infected nodule, showing bacteria.*

The most common of the nitrogen-fixing symbiotic bacteria is *Rhizobium,* a type of bacterium that invades the roots of leguminous plants, such as clover, peas, beans, vetches, and alfalfa (Figure 29–13a). Each species of legume is associated with a distinct species of *Rhizobium.*

Rhizobia enter the roots of seedlings by way of the root-hair tips (Figure 29–13b). The cell membrane turns inward, forming an infection thread (Figure 29–13c). The bacteria at this stage are not actually in contact with the protoplasm, but are separated from it by the cell membrane. Once started, the infection thread branches, becoming populated with the multiplying bacteria. These threads typically pass close to the cell nuclei and seem to cause the nuclei to degenerate. The infection threads pass from one cell to another and finally invade the cortex of the root. Soon there is a proliferation of the surrounding cortical cells, presumably owing to the effects of hormones released during the growth of the rhizobia. Finally, a vesicle forms in the thread, near a cell nucleus. The vesicle breaks open, and the bacteria enter the cytoplasm of the host cell (Figure 29–13d).

Soon after their release, the bacteria begin to grow, increasing in size some tenfold. Ammonium (NH_4^+) produced by the bacteria is combined with carbon compounds synthesized by the photosynthetic cells of the plant to produce amino acids. A key molecule in the nitrogen-fixation process is an enzyme called nitrogenase, which is composed of two different polypeptides, one containing iron and the other molybdenum. The enzyme is inhibited by oxygen.

The beneficial effects to the soil of growing leguminous plants are so obvious that they have been recognized for hundreds of years. Where leguminous plants are grown, some of the "extra" nitrogen may be released into the soil; it then becomes available to other plants. In modern agriculture, it is common practice to rotate a nonleguminous crop, such as corn, with a leguminous one, such as alfalfa. The leguminous plants are then either harvested, leaving behind the nitrogen-rich roots, or, better still, plowed back into the field. A good crop of alfalfa that is plowed back into the soil may add as much as 450 kilograms of nitrogen to the soil per hectare, frequently enough to grow a crop of a nonleguminous plant.

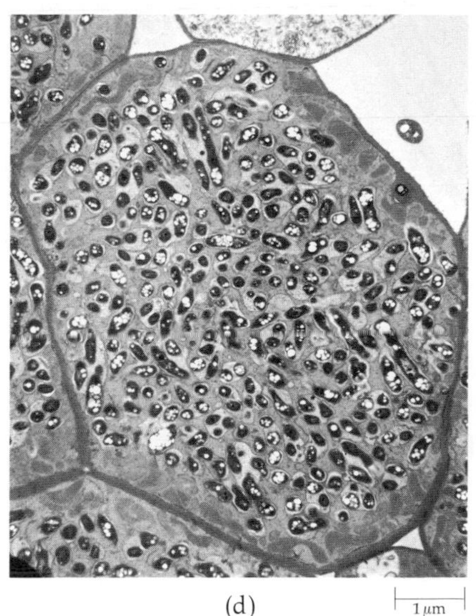

(d)

Genetic mapping (page 248) of a free-living nitrogen-fixing bacterium has shown that the genes involved in nitrogen fixation are clustered on one portion of the chromosome. In experiments at the University of Sussex, biologists have succeeded in transferring this gene cluster to a plasmid and then introducing the plasmid into *Escherichia coli* cells. The *E. coli* cells then were found to synthesize nitrogenase and to fix nitrogen. These experiments raise the hope, of course, that nitrogen-fixing genes can be transferred to the cells of plants such as corn, which could then, in effect, make their own nitrogen-containing compounds. However, it is generally agreed that the technical problems involved are so formidable that this prospect is remote.

SUMMARY

Transpiration is the loss of water vapor by plants. As a consequence of transpiration, plants require large amounts of water. Water enters the plant from the soil through the roots and travels through the plant body by means of the conducting cells of the xylem—vessel elements and tracheids. Water vapor is lost from the leaf largely through openings called stomata during the course of gas exchange.

The current and widely accepted theory of how water moves through the xylem is the cohesion-tension theory. According to this theory, water within the vessels is under negative pressure (tension) because the water molecules cling together in continuous columns as they are pulled by evaporation from above.

Diffusion of gases, including water vapor, into and out of the leaf is regulated by the stomata. The stomata are opened and closed by guard cells as a consequence of changes in turgor. Turgor is increased or decreased by the movement of potassium ions in and out of guard cells.

The movement of organic carbon compounds (mostly sucrose) from the photosynthetic parts of the plant is known as translocation. It takes place in the phloem. According to the pressure-flow hypothesis, sugars are carried along by the flow of water from an area of higher solute concentration in the sieve tubes of the leaves to areas of lower solute concentration in the sieve tubes of the roots and other plant parts. Active transport is involved in the movement of sugar into or out of the sieve tubes against the concentration gradient.

Minerals are naturally occurring inorganic substances. They are brought into the plant through the root by active transport across the endodermis and are carried through the plant body by the xylem. Active transport is also involved in the uptake of elements from the xylem by individual cells. Minerals fulfill a variety of functions. Some of the functions are nonspecific, such as the effects on osmosis. Others are specific, such as the presence of magnesium in the chlorophyll molecule. Many minerals are involved in enzyme reactions.

Two types of symbiotic relationships are important in mineral utilization by plants: those involving mycorrhizae and nitrogen-fixing bacteria. Mycorrhizae, "fungus-roots," are associations between plant roots and soil-dwelling fungi that facilitate the uptake of minerals by the roots. The association between nitrogen-fixing bacteria, such as rhizobia, and the roots of certain plants, particularly legumes, makes it possible for some plants to utilize atmospheric nitrogen directly.

QUESTIONS

1. Distinguish between transpiration/translocation; cohesion-tension/pressure-flow; phloem/xylem; rhizobia/mycorrhizae.

2. What properties of water discussed in Section 1 are important to the movement of water and solutes through plants?

3. Notice in the drawing of the guard cells that their inner walls are greatly thickened. What is the functional significance of this thickening?

4. Consider a tree transpiring most rapidly at midday and an investigator with a sensitive instrument for measuring changes in the diameter of the trunk. If water is pulled up from the top (cohesion-tension theory), what changes in diameter should be observed from night to day? (The change was, in fact, one of the early bits of evidence for the theory.)

5. In Figure 29–2, why doesn't air enter the tube through the top of the enclosed porous pipe or the leaf, even though water vapor can easily escape? How can the porous pipe (and, by analogy, the leaf) be permeable to air or water but effectively impermeable to an air-water interface?

6. Transpiration has often been described as a "necessary evil" to the plant. Why is it necessary? Why is it evil?

7. Gardeners advise removing many leaves of a plant after transplanting. Why does this help the plant survive?

8. Using the techniques described in this chapter for the analysis of phloem sap, how would you measure the rate of movement? (*Hint:* Aphids are plentiful.)

9. Rabbits often eat all of the bark off young trees, at the height they can reach. When they do this, the tree dies. Why?

10. (a) Experts in flower arrangements advise recutting the stems of flowers while holding them under water. Explain why. (b) Some florists advise adding ordinary table sugar to the water in which cut flowers are placed. When this is done, the flowers often remain fresh for several weeks, although any leaves remaining on the stems wilt and die. What is the explanation for these phenomena?

11. Most plants cannot live in areas in which there is a high salt concentration, such as salt marshes. Explain.

12. Identify the cells and tissues through which a molecule of water travels from the time it enters the root until it is used in photosynthesis.

CHAPTER 30

Plant Hormones and Plant Responses

30–1
An example of the effects of hormone treatment. In this case, gibberellin has caused bolting and flowering in cabbage plants. The plants on the left were not treated.

We saw in Chapter 28 that as a plant grows it does far more than increase its mass and volume. It differentiates, forming a variety of cells, tissues, and organs, and undergoes morphogenesis, taking on the shape characteristic of the adult sporophyte. Moreover, many of its activities are finely tuned to its environment and to the changing pattern of the seasons. Many of the details of how these processes are regulated are not known, but it is clear that plant development and growth depend on the interplay of a number of internal and external factors.

Chief among the internal factors are the plant hormones. Hormones, by definition, are substances that are produced in one tissue and transported to another, where they exert highly specific effects. Hormones help the plant integrate the growth, development, and metabolic activities of its various tissues. Typically they are active in very small quantities. In the shoot of a pineapple plant, for example, only 6 micrograms of auxin, a common growth hormone, are found per kilogram of plant material. One enterprising plant physiologist calculated that the weight of the hormone in relation to 1 kilogram of shoot is comparable to the weight of a needle in a 22-ton haystack. The term "hormone" comes from the Greek word meaning "to excite." It is now clear, however, that many hormones have inhibitory influences. So, rather than thinking of hormones as stimulators, it is perhaps more useful to consider them as chemical messengers. But this term also needs qualification. As we shall see, the response to the particular "message" depends not only on its content but also upon how it is "read" by its recipient. All known plant hormones have a multiplicity of effects.

HORMONES AND REGULATION OF PLANT GROWTH

THE AUXINS

The first plant hormones to be isolated were the auxins. The effects of these hormones were observed by Charles Darwin and his son Francis and first reported in *The Power of Movement in Plants,* published in 1881. The Darwins were studying the bending toward light, or *phototropism,* of grass seedlings. They noted that the bending takes place below the tip, in the lower part of the coleoptile. Then they showed that if they covered just the terminal portion of the coleoptile with a cylinder of metal foil or a hollow tube of glass blackened with India ink and

30–2
The Darwins' experiment. (a) *Light striking a growing coleoptile (such as the tip of this oat seedling) causes it to bend toward the light.* (b) *Placing an opaque cover over the tip of the seedling inhibits this bending response, but* (c) *an opaque collar placed below the tip does not. These experiments indicate that something produced in the tip of the seedling and transmitted down the stem causes the bending.*

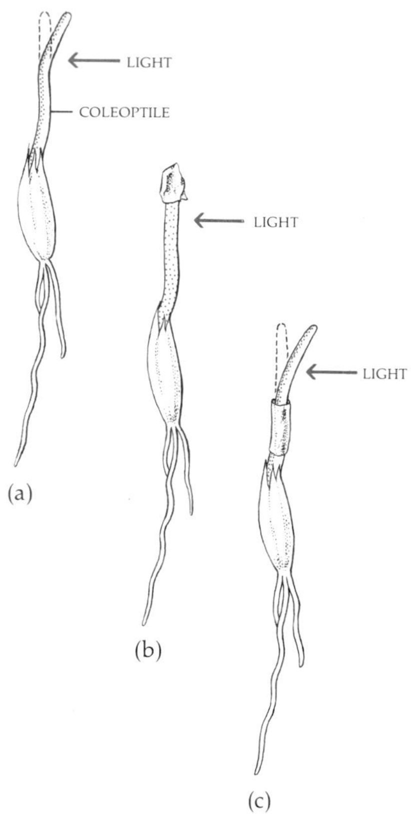

exposed the plant to a light coming from the side, the characteristic bending of the seedling did not occur. If, however, the tip was enclosed in a transparent glass tube, bending occurred normally. Bending also occurred normally when the lightproof cylinder was placed below the tip (Figure 30–2). "We must therefore conclude," they stated, "that when seedlings are freely exposed to a lateral light some influence is transmitted from the upper to the lower part, causing the material to bend."

In 1926, the Dutch plant physiologist Frits W. Went succeeded in separating this "influence" from the plants that produced it. Went cut off the coleoptile tips from a number of oat seedlings. He placed the tips on a slice of agar (a gelatinlike substance), with their cut surfaces in contact with the agar, and left them there for about an hour. He then cut the agar into small blocks and placed a block off-center on each stump of the decapitated plants, which were kept in the dark during the entire experiment. Within one hour, he observed a distinct bending *away* from the side on which the agar block was placed (Figure 30–3).

Agar blocks that had not been exposed to a coleoptile tip produced either no bending or only a slight bending toward the side on which the block had been placed. Agar blocks that had been exposed to a section of coleoptile lower on the shoot produced no physiological effect.

Went interpreted these experiments as showing that the coleoptile tip exerted its effects by means of a chemical stimulus (in short, a hormone) rather than a physical stimulus, such as an electric impulse. In response to this chemical stimulus, the side of the coleoptile adjacent to the agar block grew more rapidly, causing the coleoptile to bend away from that side. The hormone was named auxin, a term coined by Went from the Greek word *auxein*, "to increase."

Several different substances with auxin activity have now been isolated from plant tissues, and others have been synthesized in the laboratory; they are all known as auxins (Figure 30–4). The most common of the natural auxins is indoleacetic acid, abbreviated IAA. One of the synthetic auxins, known as 2,4-D, is

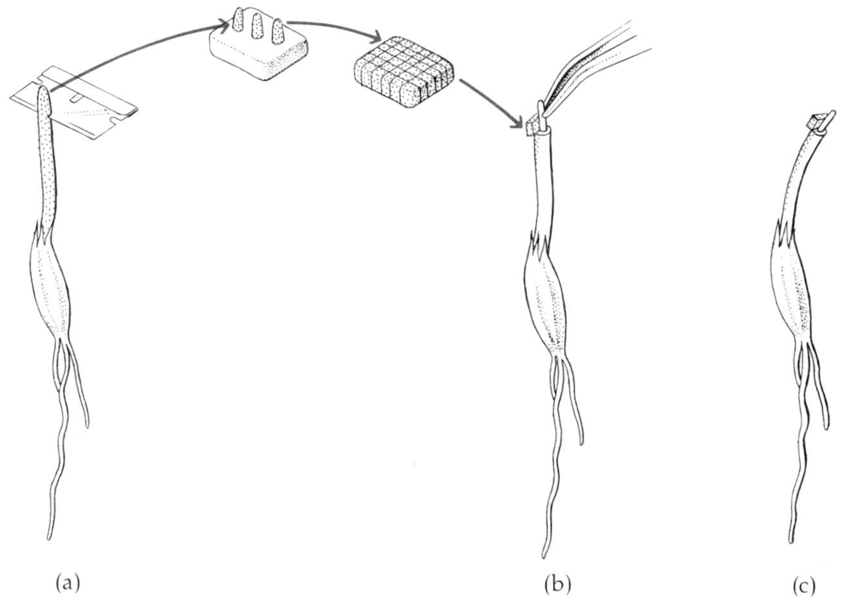

30–3
Went's experiment. (a) *He cut the coleoptile tips from oat seedlings and placed them on a slice of agar. After about an hour, he cut the agar in small blocks and* (b) *placed each block, off-center, on a decapitated seedling (the leaf is pulled up).* (c) *The seedlings bent away from the side on which the block was placed.*

30–4

Auxins. IAA (indoleacetic acid) is a natural auxin, isolated from plant tissues. 2,4-D is a synthetic auxin used as a weed killer (herbicide). Naphthalenacetic acid, also a synthetic auxin, is commonly used to induce the formation of adventitious roots in cuttings and to reduce fruit drop in orchard crops.

CH_2COOH N

IAA

$O—CH_2COOH$ Cl Cl

2,4-D

CH_2COOH

NAPHTHALENACETIC ACID

commonly used as an herbicide (like many other physiologically active compounds, auxins are toxic in high doses). For reasons not known, 2,4-D and related compounds are effective on broad-leaved plants at concentrations not harmful to grasses and so are commonly used on lawns to control broad-leaved weeds. Auxins, in massive doses, were among the herbicides used in Southeast Asia during United States' participation in the Vietnam War. The synthetic auxins, unlike IAA, are not readily broken down by natural plant enzymes or by bacteria. Their longer effective life makes them better suited for commercial purposes. Their destructive effects on vegetation are long-lasting.

The phototropism observed by the Darwins results from the fact that under the influence of light, auxins migrate from the light side to the dark side of the tip. The cells on the dark side, having more auxin, elongate more rapidly than those on the light side, causing the plant to bend toward the light. Auxin effects on stems and coleoptiles can be detected within 15 minutes of application of the hormone, suggesting that auxin exerts its effects by producing changes in the permeability of the cell membrane. The action spectrum for phototropism is not the same as that for photosynthesis; only light in the blue region of the spectrum (less than 500 nanometers in wavelength) produces the phototropic response.

Other Tissue Responses to Auxin

Various plant tissues show other responses to auxin. Shoots elongate in response to auxins produced in the meristem. Apparently, under the influence of auxin, the cell wall becomes more plastic and thus the cell takes up more water and elongates. Because of the construction of the cell walls (page 104), expansion of the cell is unidirectional.

When large concentrations of auxins are present, the growth of the main roots is inhibited, and this inhibition is presumed responsible for the capacity of seedlings to orient themselves in the ground (Figure 30–5). In low concentrations, auxins induce the formation of adventitious roots (Figure 30–6); rooting preparations used by gardeners contain auxin.

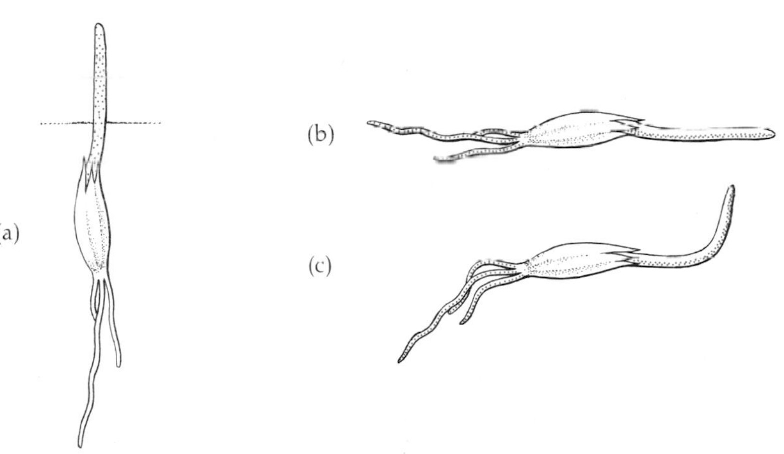

30–5

Why shoots grow upward and roots downward. (a) *When a seedling is perpendicular, the auxin produced in the tip is distributed evenly and the plant grows erect.* (b) *However, when a seedling is placed on its side, auxin is transported laterally and accumulates in cells on the lower side of the shoot;* (c) *these cells then grow more rapidly and so the shoot grows upward. It has been hypothesized that auxin similarly accumulates in the cells on the lower side of the root, causing the cells to grow more slowly, so that the root tip turns down. However, many investigators believe other inhibiting factors to be involved in the geotropism of roots.*

30–6
The upper row of holly cuttings was treated with a synthetic auxin 21 days before the picture was taken. The cuttings in the lower row were not. Note the growth of adventitious roots on the plants in the upper row.

In most dicot species, the growth of lateral buds is inhibited by auxins. If you pinch off the growing tip (meristem) of the stem of the houseplant *Coleus,* for example, the lateral buds begin to grow vigorously, producing a plant with a bushier, more compact body. If you treat the "eyes" (actually lateral buds) of a potato with auxin, they will be inhibited from sprouting and so the tubers can be stored longer. The formation of the abscission layer (page 496) has been correlated with diminished production of auxin in the leaf. Auxins are also involved in the growth of fruits (Figure 30–7).

Mechanism of Action

As we noted earlier, auxin increases the plasticity of the cell wall. This effect is brought about by a complex series of interactions that are not yet fully understood. During continued growth, auxin stimulates specific RNA and protein biosyntheses, and under the control of this newly forming protein, certain of the old bonds holding the cell wall together are broken as new carbohydrate material is incorporated into the structure. However, response to auxin by coleoptiles, as we noted, is very rapid, occurring in too short a time for new protein synthesis to be initiated. Current studies indicate that, in these reactions, auxin exerts its effects by binding to the cell membrane and changing its permeability. Auxin responses are associated with a rapid movement of hydrogen ions across the membrane.

THE GIBBERELLINS

The gibberellins were first discovered by a Japanese scientist who was studying a disease of rice plants called "foolish seedling disease." The diseased plants grew rapidly but were spindly and tended to fall over under the weight of the developing seeds. The cause of the symptoms, it was found, was a chemical produced by a fungus, *Gibberella fujikuroi,* which infected the seedlings. The substance, which was named gibberellin, was subsequently isolated not only from the fungus but from many species of plants (Figure 30–9). These hormones are produced in apical meristems, leaves, and plant embryos.

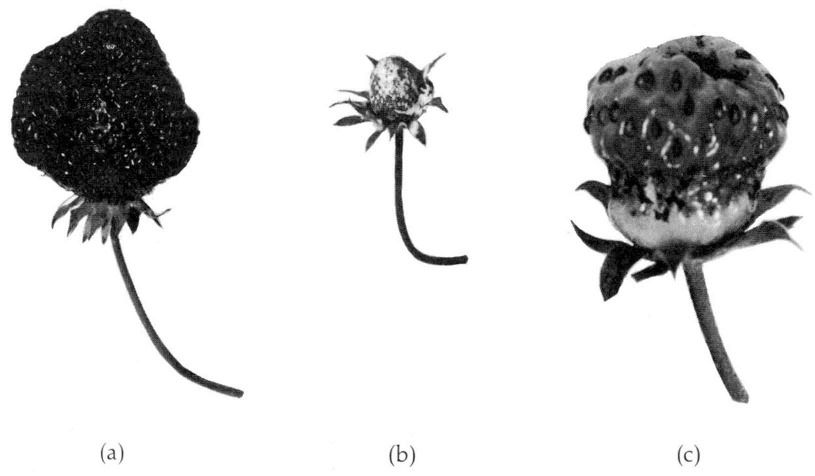

30–7
Auxin, apparently produced by developing seeds, promotes the growth of fruit. (a) *Normal strawberry,* (b) *strawberry from which all seeds have been removed, and* (c) *strawberry in which three horizontal rows of seeds were left. If a paste containing auxin is applied to* (b), *the strawberry grows normally.*

30-8

An experiment investigating the action of gibberellin in barley seeds. Forty-eight hours before the picture was taken, each of these seeds was cut in half and the embryo removed. The seed at the bottom was treated with plain water, the seed in the center was treated with a solution of 1 part per billion of gibberellin, and the seed at the top was treated with 100 parts per billion of gibberellin. As you can see, digestion of the starchy storage tissue has begun to take place in the seeds treated with gibberellin.

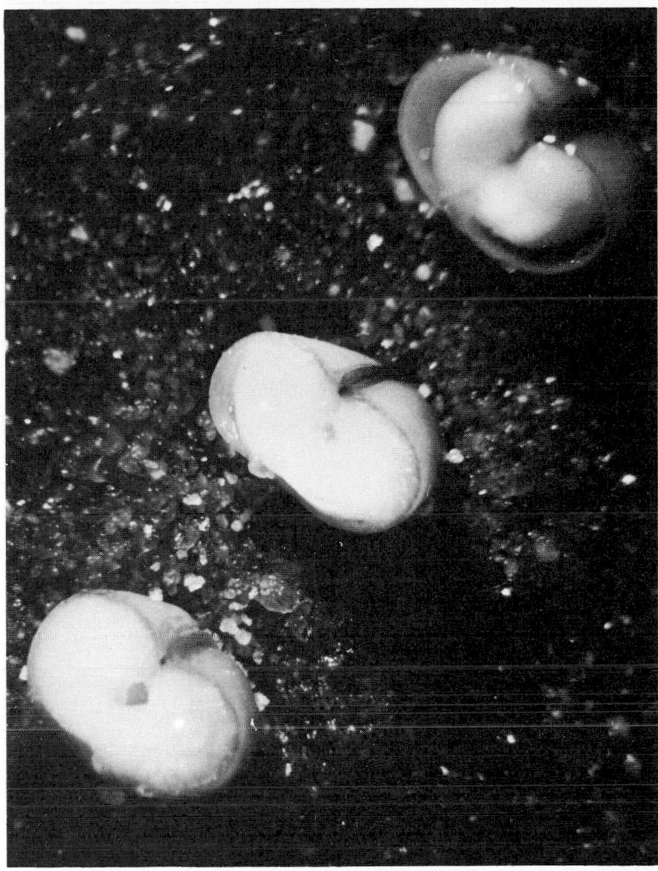

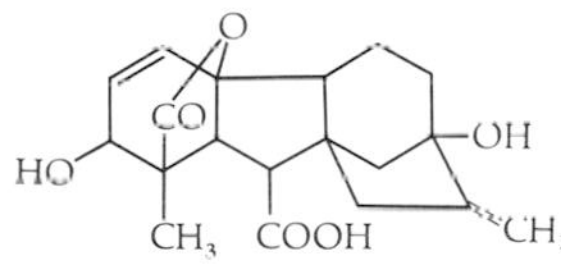

GIBBERELLIC ACID (GA_3)

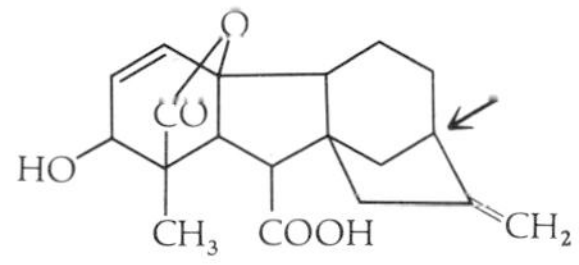

GA_7

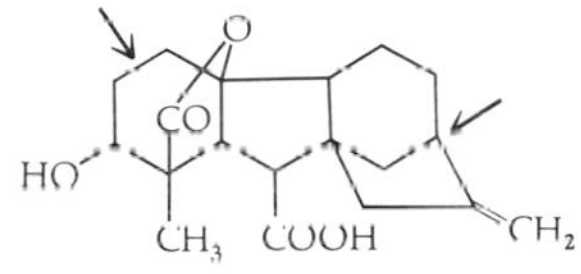

GA_4

30-9

Three of the more than 90 gibberellins that have been isolated from natural sources. Gibberellic acid (GA_3) is the most abundant in fungi and the most biologically active in many tests. The minor structural differences that distinguish the other two gibberellins are indicated by arrows.

The most remarkable results are seen when gibberellins are applied to plants that are genetic dwarfs. Under gibberellin treatment, these dwarfs become indistinguishable from normal tall plants.

Some plants—lettuce and cabbage are common examples—first grow as rosettes; the leaves develop but the internodes do not elongate until just before flowering. At that time the stem elongates rapidly, a phenomenon known as bolting. In the case of cabbage and other biennials, bolting and flowering do not normally occur until after a period of cold; however, bolting and flowering can be induced by the application of gibberellin (see Figure 30-1).

Gibberellins have also been shown to play a role in the growth of plant embryos and seedlings. In grass seeds, there is a specialized layer of cells, the aleurone layer, just inside the seed coat. These cells are rich in protein. During the early stages of germination, the embryo produces gibberellin, which diffuses to the aleurone layer. In response to the gibberellin, the aleurone cells produce enzymes that hydrolyze the starch, proteins, and other storage products to soluble sugars, amino acids, and other small molecules that the embryo and then the seedling uses for its growth as it pushes up through the soil.

Recent studies by Joseph Varner of Washington University show that, in the germinating barley grain, gibberellin acts directly at the nuclear level, activating formerly repressed genes and so resulting in new messenger RNA formation.

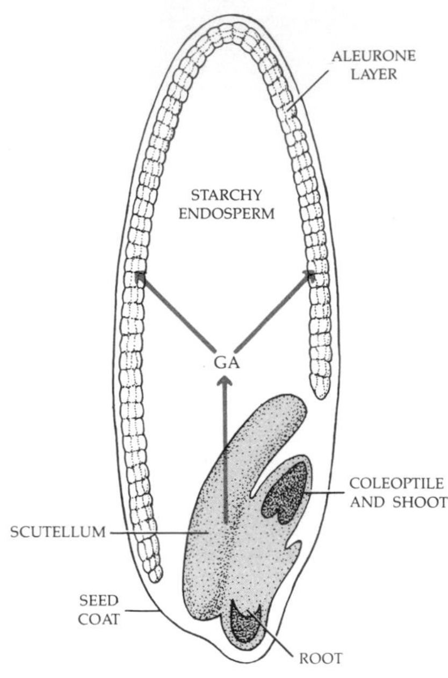

30–10
Mechanism of action of gibberellin in a barley seed. The embryo, shown in color, releases gibberellic acid (GA), which diffuses to the aleurone layer. The gibberellin induces the aleurone cells to synthesize enzymes that digest starch and proteins stored in the endosperm. The polymers are broken down into smaller molecules that the embryo and seedling use for growth.

THE CYTOKININS

The cytokinins are a group of hormones originally detected in coconut "milk," which is a liquid endosperm. They were found to promote the division of plant cells isolated in test tubes, and their name is derived from cytokinesis, meaning cell division. They have now been found in numerous plants, largely in actively dividing tissues, including germinating seeds, fruits, and roots (Figure 30–11).

Responses to Cytokinin and Auxin Combinations

Studies of responses to combinations of auxin and cytokinins are helping physiologists understand how plant hormones work to produce the total growth pattern of the plant. Apparently, the undifferentiated plant cell—such as the meristematic cell—has two courses open to it: Either it can enlarge, divide, enlarge, and divide again, or it can elongate without cell division. The cell that divides repeatedly remains essentially undifferentiated, or embryonic, whereas the elongating cell tends to differentiate and become specialized. In studies of tobacco stem callus, the addition of an auxin to the tissue culture produced rapid cell expansion, so that giant cells were formed. A cytokinin alone had little or no effect. Auxin plus cytokinin resulted in rapid cell division, so that large numbers of relatively small cells were formed.

By slight alterations in the relative concentrations of auxin and cytokinin, investigators have been able to affect the development of undifferentiated cells growing in tissue culture. When a high concentration of auxin is present, undifferentiated tissue gives rise to organized roots. With higher concentrations of cytokinins, buds appear (Figure 30–12).

KINETIN BAP

(a)

ZEATIN 2iP ADENINE

(b)

30–11
Cytokinins. (a) *Kinetin, the first of this group to be discovered, and 6-benzylamino purine (BAP) are commonly used synthetic cytokinins.* (b) *Zeatin and 2iP have been isolated from plant material. Note the resemblances between the purine adenine and a portion of the molecule of each of these cytokinins. The significance of the resemblance between adenine and the hormones of this group is not known. It may merely be another example of biological economy: the use of a single major biosynthetic pathway to produce a number of functionally different products.*

PLANTS IN TEST TUBES

As long ago as the 1930s, scientists developed techniques for the growing of plant cells in test tubes. Tiny fragments of meristem are implanted under sterile conditions in a medium containing minerals and various combinations of organic compounds. Under these conditions, the meristematic cells proliferate to form clumps of undifferentiated cells. Subsequently, it was discovered that, by adjusting the hormone balance in the medium, it was possible to make these cells differentiate and grow into mature plants. Using these techniques, hundreds or even thousands of subcultures of meristematic tissue can be produced in a relatively short time and in a small space. Then, by altering the hormone balance, each of these can be turned into a small but perfect plant. This technique has already been employed in the culture of exotic orchids and is being extended to other hard-to-grow varieties.

Within the last decade, it has become possible, using similar techniques, to grow isolated protoplasts—plant cells without cell walls—in the laboratory. These protoplasts can be grown as isolated cells, like cultures of bacteria. Alternatively, if they are grown in a suitable medium, they will regenerate cell walls, multiply, and differentiate into whole plants.

Cultured cells are being used in a variety of ways. For instance, it has been found that they can be used in rapid screening tests to determine resistance to infectious diseases or to detect nutritional requirements. In this way, a scientist not only can work much more rapidly, but also can do assays with millions of cells growing in a very small space, as compared to the far fewer number of plants that can be grown in a field or a greenhouse. Moreover, it has been found possible to fuse protoplasts from two different species of plant and create a hybrid. The only plant created so far by a fusion of protoplasts is a hybrid of tobacco that had already been produced by conventional crossbreeding techniques. (Producing a plant whose characteristics were known was a necessary preliminary test for the method.) However, this initial success has, of course, raised the hopes of producing entirely new superplants—organisms that might be capable, for instance, of simultaneously carrying out C_4 photosynthesis and fixing nitrogen. So far, fusion of protoplasts from cells as distantly related as corn and soybeans has been successful, but it has not been possible to induce hybrid cells to develop into plants.

30–12
Two buds forming on undifferentiated tissue from a geranium following treatment with both an auxin and a cytokinin. Callus from some types of plants will continue to grow as undifferentiated tissue, or roots, or buds, depending on the relative proportions of auxins and cytokinins.

However, lest you think this is simple, we shall describe another tissue culture study, in which tuber tissue of the Jerusalem artichoke was used. In this study, it was shown that a third substance, the calcium ion, can modify the action of the auxin-cytokinin combination. Auxin plus low concentrations of cytokinin was shown to favor cell enlargement, but as calcium ion was added to the culture, there was a steady shift in the growth pattern from cell enlargement to cell division. High concentrations of calcium ion apparently prevent the cell wall from expanding, and at such concentrations the cell switches course and divides. Thus, not only do hormones modify the effects of hormones, but these combined effects may be, in turn, modified by nonhormonal factors, such as calcium ions and, undoubtedly, many others.

ABSCISIC ACID

Soon after the discovery of the growth-promoting hormones, plant physiologists began to speculate that growth-inhibiting hormones would be found, since it is clearly advantageous to the plant not to grow at certain times and in certain seasons. Not long afterward, an inhibitory hormone was isolated from dormant buds. Subsequently, the same hormone was discovered in cotton bolls (the fruit of the cotton plant), where it was found to promote abscission. The hormone is called abscisic acid.

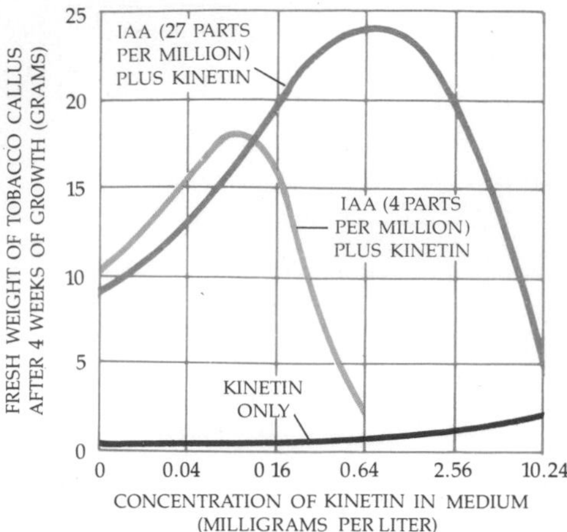

30–13

The response of tobacco cells in tissue culture to combinations of an auxin and a cytokinin. Kinetin alone has little effect on the growth of tobacco callus (a mass of undifferentiated cells). IAA alone (not shown), regardless of the concentration used, causes the culture to grow to a weight of about 10 grams. When both hormones are present, growth is greatly increased. Notice, however, that when optimum concentrations are exceeded, the growth rate declines.

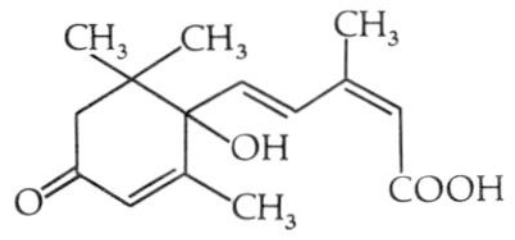

30–14

Abscisic acid (dormin), an inhibitor that blocks the action of the growth-promoting hormones.

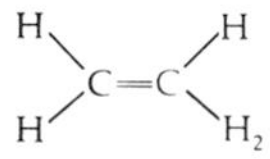

30–15

Ethylene, a simple hydrocarbon involved in the ripening of fruit. Unlike any other known hormone, it is a gas, raising the question of whether it should properly be considered a hormone at all.

Abscisic acid accelerates the dropping of leaves and of fruit; the presence of auxin inhibits abscission. Application of abscisic acid to vegetative buds changes them to winter buds by converting the outermost leaf primordia to bud scales; these inhibitory effects can be overcome by gibberellin. The appearance of hydrolyzing enzymes induced by gibberellin in barley seeds is inhibited by abscisic acid. Abscisic acid, though it has little effect on dwarf plants, reduces the growth of normal plants; this inhibition can be counteracted by gibberellin. Kinetin inhibits yellowing in excised leaves; abscisic acid causes green leaves to yellow. In short, abscisic acid, in many of its effects, serves as an inhibitor of the growth-promoting effects of other hormones, emphasizing the concept that growth is the result of a balance of different factors.

Finally, as if the subject of stomatal movements was not already sufficiently complex, abscisic acid has been shown to be a potent effector of stomatal closure. When leaves lose moisture, abscisic acid synthesis increases and the stomata close. A variety of tomato is known, called a "wilty mutant," that cannot make abscisic acid and cannot close its stomata.

ETHYLENE

Ethylene is an unusual growth regulator in that it is a gas, a simple hydrocarbon, $H_2C{=}CH_2$. Its effects have been known for a long time. In the early 1900s, many fruit growers made a practice of improving the color and flavor of citrus fruits by "curing" them in a room with a kerosene stove. (Long before this, the Chinese used to ripen fruits in rooms where incense was being burned.) It was long believed that it was the heat that ripened the fruits. Ambitious fruit growers, who went to the expense of installing more modern heating equipment, found to their sorrow that this was not the case. As experiments showed, it was actually the incomplete combustion products of kerosene that were responsible for ripening the fruits. The most active gas was identified as ethylene. As little as 1 part per million of ethylene in the air will speed the ripening process. Subsequently, it was found that ethylene is produced by plants, as well as by kerosene stoves, that it appears just before and also during fruit ripening, and that it is responsible for a number of changes in color, texture, and chemical composition that take place as fruits mature.

Auxin at certain concentrations causes a burst of ethylene production in some plants. It is now believed that some of the effects on fruits and flowers generally attributed to auxin are related to the release of ethylene. Similarly, it is now believed that abscisic acid is not the direct cause of abscission in most cases; rather, it is hypothesized, abscisic acid induces ethylene formation, which accelerates senescence, which stimulates abscission.

HORMONAL CONTROL OF FLOWERING

Flowering also appears to be under the control of a hormone or hormones, although to date no specific one has been isolated and identified. Some of the earliest experiments on this hypothetical flowering hormone were carried out by a Russian scientist, M. H. Chailakhyan, in the 1930s. Working with a species of chrysanthemum, Chailakhyan found that if the upper portion of the stem was stripped of its leaves and the leaves on the lower stem were exposed to an appropriate light cycle, the plant would flower, a phenomenon known as *photoinduction*. (We shall have

30–16
Experiments that indicate the existence of a flowering hormone. (a) *When certain plants, such as the cocklebur shown here, are exposed to an appropriate light cycle, those with leaves flower and those without leaves do not. When even one-eighth of a leaf remains on a plant, flowering occurs, and the illumination of a single leaf—not necessarily the whole plant—suffices. These experiments indicate that a chemical originating in the leaves causes the plant to flower.* (b) *This conclusion is supported by experiments on branched plants. Exposure of one branch to the light induces flowering on the other branch as well, even when only a portion of a leaf is present on the lighted branch.* (c) *When two plants are grafted together, exposure of one of the plants to the light cycle induces flowering in both the lighted plant and the grafted one.*

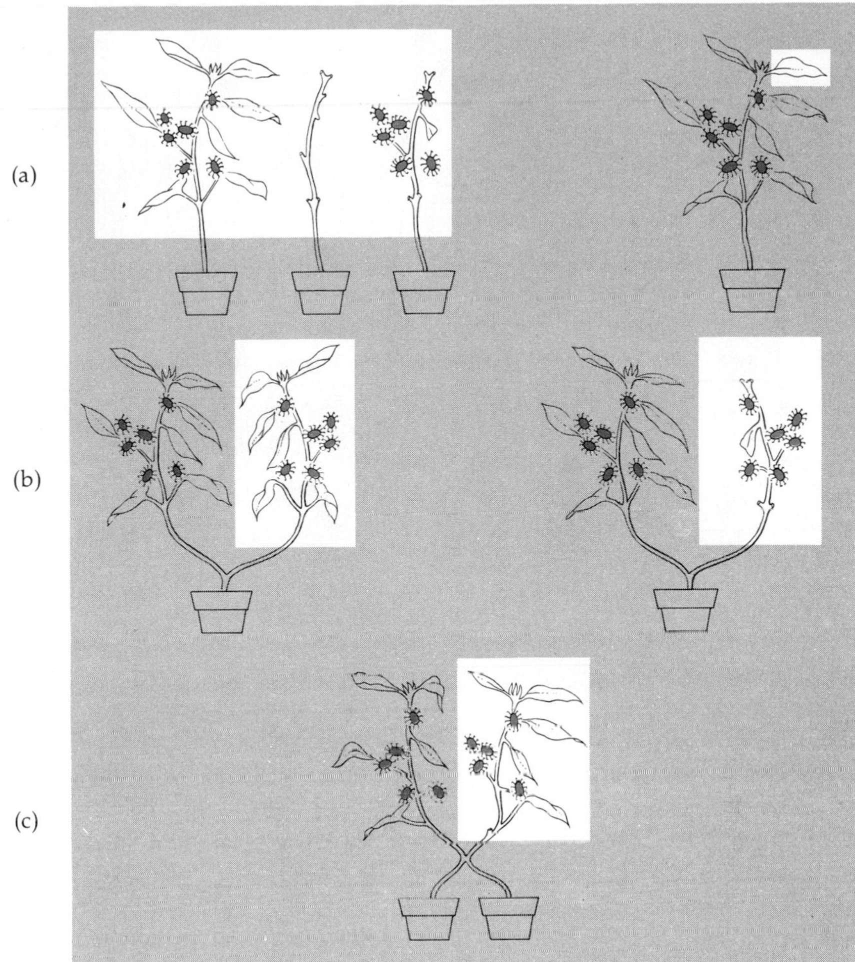

more to say about light cycles and flowering later in this chapter.) If, however, only the upper, leafless stem and its buds were exposed, no flowering occurred. He interpreted these results as indicating that the leaves form a hormone that moves to the apical meristem of the plant and initiates flowering. He named this hypothetical hormone florigen, the "flower maker."

Subsequent experiments showed that the flowering response does not take place if the leaf is removed immediately after photoinduction. But if the leaf is left on the plant for a few hours after the induction cycle is complete, it can then be removed without stopping flowering. The flowering hormone can pass through a graft from a photoinduced plant to a noninduced plant. If a branch is girdled, florigen movement ceases. This led to the conclusion that florigen moves by way of the phloem system, the means by which most organic substances are transported. However, despite this strong evidence for the existence of florigen, the hormone has never been isolated and, also, more recent evidence indicates that other, inhibitory factors are also involved in flowering (page 556).

30–17
Photoperiodism is the phenomenon that brings the plants of many species into flower simultaneously, year after year. This is a cherry tree, blooming in Brooklyn.

PLANTS AND PHOTOPERIODISM

In many regions of the biosphere, the most important environmental changes affecting plants (and indeed, land organisms, in general) are those that result from the changing seasons. Plants are able to accommodate themselves to these changes because of their capacity to sense and, more important, to anticipate the yearly calendar of events—the first frost, the spring rains, long dry periods, long growing spells, and even the time that nearby plants of the same species will be in flower. For many plants, all of these determinations are made in the same way: by measuring the relative periods of light and darkness. This phenomenon is known as *photoperiodism.*

PHOTOPERIODISM AND FLOWERING

The effects of photoperiodism on flowering are particularly striking (Figure 30–18). Plants are of three general types: day-neutral, short-day, and long-day. Day-neutral plants flower without regard to day length. Short-day plants flower in early spring or fall; they must have a light period shorter than a critical length—for instance, the cocklebur flowers when exposed to 15½ hours or *less* of light. Other short-day plants are poinsettias, strawberries, primroses, ragweed, and some chrysanthemums.

Long-day plants, which flower chiefly in the summer, will flower only if the light periods are *longer* than a critical length. Spinach, potatoes, clover, henbane, and lettuce are examples of long-day plants.

The discovery of photoperiodism explained some puzzling data about the distribution of common plants. Why, for example, is there no ragweed in northern Maine? The answer, investigators found, is that ragweed starts producing flowers when the day is less than 14½ hours long. The long summer days do not shorten to 14½ hours in northern Maine until August, and then there is not enough time for ragweed seed to mature before the frost. For similar reasons, spinach cannot produce seeds in the tropics. Spinach needs 14 hours of light a day for a period of at least two weeks in order to flower, and days are not this long in the tropics.

Note that the cocklebur and spinach will both bloom if exposed to 14 hours of daylight, yet one is designated as short-day and one as long-day. The important factor is not the absolute length of the photoperiod but rather whether it is longer or shorter than a particular critical interval for that variety. And in some varieties, 5 or 10 minutes' difference in exposure can determine whether or not a plant will flower.

These early field studies on photoperiodism in plants were carried out by W. W. Garner, H. A. Allard, and co-workers at the U.S. Department of Agriculture, known as the Beltsville group, for the small town in Maryland where they carried out their studies.

Photoperiodism has now been demonstrated in many species of insects, fish, birds, and mammals. It influences such diverse phenomena as the metamorphosis from caterpillar to butterfly, sexual behavior, migration, molting, and seasonal changes in coat or plumage.

30–18

(a) *The relative length of day and night determines when plants flower. The four curves depict the annual change in day length in four North American cities at four different latitudes. The black lines indicate the effective photoperiod of three different short-day plants. The cocklebur, for instance, requires* $15\frac{1}{2}$ *hours or less of light. In Miami, it can flower as soon as it matures, but in Winnipeg the buds do not appear until early in August, so late that frost usually kills the plants before the seed is mature.* (b) *Relationship between day length and the developmental cycle of plants in the temperate zone.*

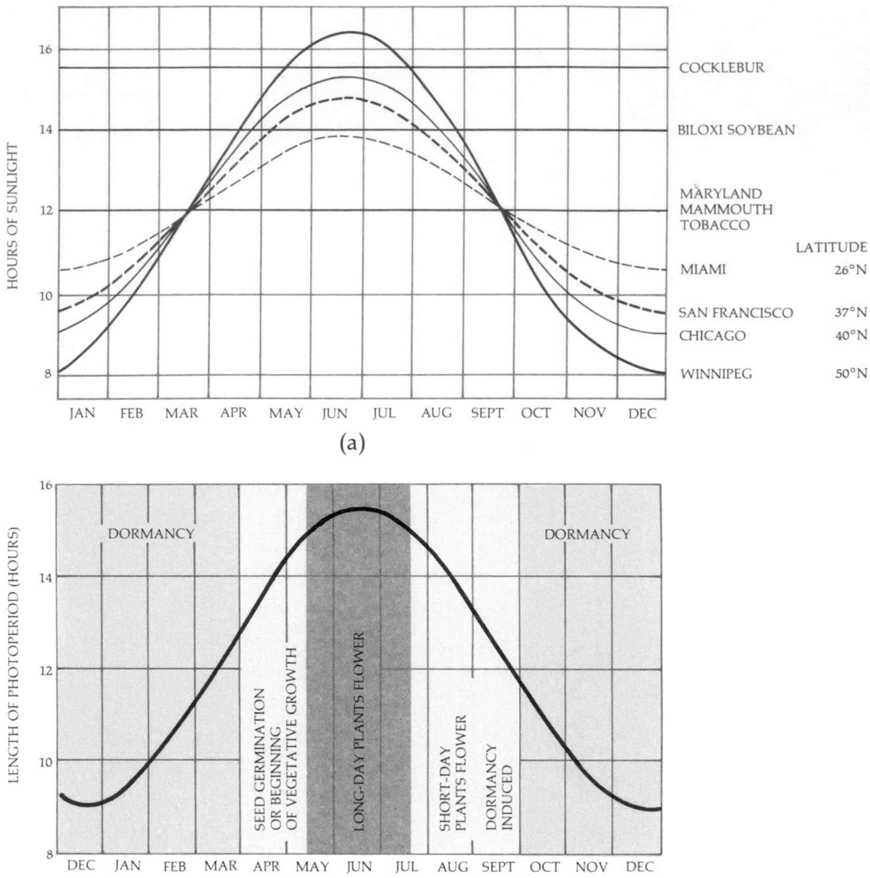

Measuring the Dark

Following the early studies by the Beltsville group, other investigators, Karl C. Hamner and James Bonner, began a laboratory study of photoperiodism. They also used the cocklebur as the experimental organism. As we mentioned previously, the cocklebur is a short-day plant, requiring $15\frac{1}{2}$ hours or less of light per 24-hour cycle to flower. It is particularly useful for experimental purposes because a single exposure under laboratory conditions to a short-day cycle will induce flowering two weeks later, even if the plant is immediately returned to long-day conditions. Also, the cocklebur can withstand a good deal of rough treatment. With the cocklebur it was possible to demonstrate that flowering will not occur if the leaves are removed but will occur if even a portion of one leaf remains on the plant. In other words, it is the leaf that "perceives" the light, and this function can, in some cases, be assumed by a fraction of the total leaf surface.

In the course of these studies, in which they tested a variety of experimental conditions, the investigators made a crucial and totally unexpected discovery. If the period of darkness is interrupted by as little as a 1-minute exposure to the light of a 25-watt bulb, flowering does not occur. Interruption of the light period by darkness has no effect whatsoever on flowering. Subsequent experiments with other short-

30–19
Experiments on photoperiodism showed that plants measure the darkness rather than the light. Short-day plants flower only when the darkness exceeds some critical value. Thus, the cocklebur, for instance, will flower on 8 hours of light and 16 hours of darkness. If the 16-hour period of darkness is interrupted even very briefly, as shown on the right, the plant will not flower. The long-day plant, on the other hand, which will not flower on 16 hours of darkness, will flower if the darkness is interrupted. Long-day plants flower only when the darkness is less than some critical value.

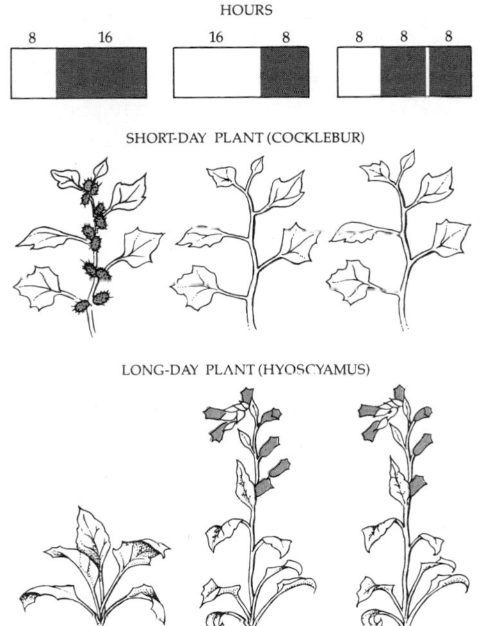

day plants showed that they, too, required periods not of uninterrupted light but of uninterrupted darkness (Figure 30–19).

What about long-day plants? They also measure darkness. A long-day plant that will flower if it is kept in a laboratory in which there is light for 16 hours and dark for 8 hours will also flower on 8 hours of light and 16 hours of dark if the dark is interrupted by even a brief exposure to light.

A. Lang of Michigan State University has carried out studies involving grafts of day-neutral, long-day, and short-day plants that strongly indicate the presence of a hormone or hormones other than the hypothetical florigen. For instance, he and his co-workers have shown that if the long-day plant *Hyoscyamus niger* is completely defoliated, flower buds will form irrespective of photoperiod; however, if even one leaf remains, it will flower only on long days. If a day-neutral plant is grafted to a long-day plant, flowering of the day-neutral plant is accelerated by exposure to long days. Exposure to short days, however, inhibits flowering of a day-neutral plant grafted to a long-day plant. In short, flower-inhibitors as well as flower-promoters are involved.

PHOTOPERIODISM AND PHYTOCHROME

Following up on the clues from the Hamner and Bonner experiments, the Beltsville group was able to detect and eventually to isolate the pigment involved in photoperiodism. This pigment, which they called *phytochrome*, exists in two different forms: One is P_{660} and one is P_{730}. The numbers refer to the wavelengths of light the two forms absorb; 660 nanometers is red light and 730 nanometers is far-red. P_{660} absorbs red light and is converted to P_{730}, which is the active form. This conversion takes place in daylight or in incandescent light; in both of these types of light, red wavelengths predominate over far-red. When P_{730} absorbs far-red light,

30–20
P_{660} changes to P_{730}, the active form, when exposed to red light, which is present in sunlight. P_{730} reverts to P_{660} when exposed to far-red light. In darkness, P_{730} slowly reverts to P_{660}.

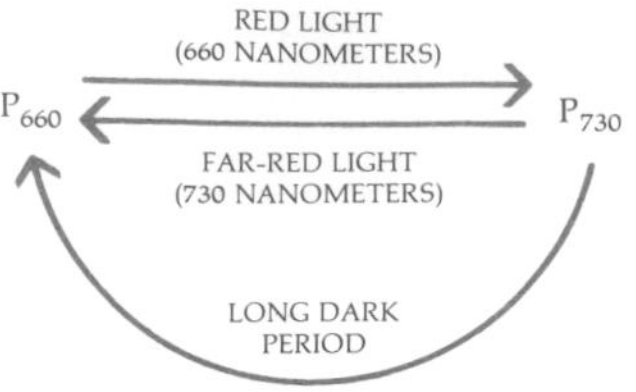

it is converted back to P_{660}. The P_{730} to P_{660} conversion can also take place in the dark, which is how it usually occurs in nature (Figure 30–20). P_{730}, the active form of the pigment, promotes flowering in long-day plants and inhibits flowering in short-day plants.

Other Phytochrome Responses

Many types of small seeds, such as lettuce, germinate only when they are in loose soil, near the surface. Otherwise the seedling would have little chance of reaching the light. Red light, a sign that sunlight is present, stimulates seed germination by converting phytochrome to the active form. Similarly, phytochrome is involved in the early development of seedlings. When a seedling develops in the dark, as it normally does underground, the stem elongates rapidly, pushing the shoot up through the soil layers. Any seedling grown in the dark will be elongated and spindly with small leaves (Figure 30–21). It will also be almost colorless, because the chloroplasts do not synthesize chlorophyll until they are exposed to light. Such a seedling is said to be etiolated. When the seedling tip reaches the light, normal growth begins. Phytochrome is involved in the switching of etiolated to normal growth. If a dark-grown bean seedling is exposed to only one minute of red (660 nanometers) light, it will respond with normal growth. If, however, the exposure to red light is followed by a one-minute exposure to far-red light, etiolated growth continues, thus negating the original exposure. The exposures can be alternated repeatedly and the plant responds only to the last one perceived.

The way in which phytochrome acts is not known. One recent suggestion is that it alters the permeability of the cell membrane, permitting particular substances to enter the cell, or, perhaps, inhibiting their entry, and that these substances, which probably include hormones, regulate the cell's activities.

30–21
Dark-grown seedlings, such as the ones on the right in each of these photographs, are thin and pale with longer internodes and smaller leaves than the normal seedlings on the left. This group of characteristics, known as etiolation, has survival value for the seedling because it increases its chances of reaching light before its stored energy supplies are used up. The plants in the photograph on the left are corn seedlings; on the right are bean seedlings.

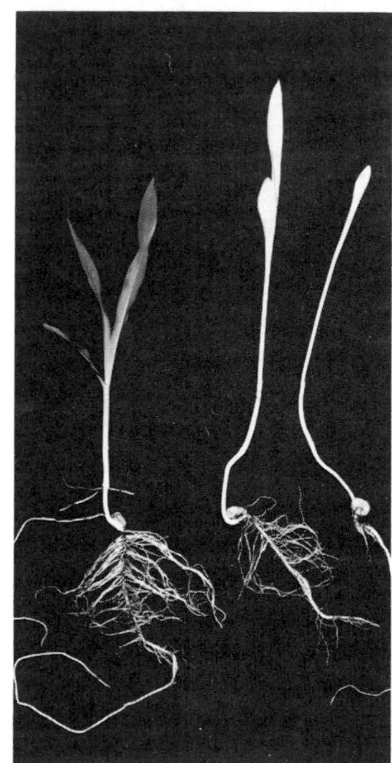

30–22
Leaves of the wood sorrel, day (a) *and night* (b). *One hypothesis concerning the function of such "sleep" movements is that they protect the leaves from absorbing moonlight on bright nights, thus protecting photoperiodic phenomena. Another theory, proposed by Darwin almost 100 years ago, is that the folding protects against heat loss from the leaves by night.*

(a)

Richard F. Trump, OMIKRON

(b)

CIRCADIAN RHYTHMS

Some species of plants have flowers that open in the morning and close at dusk or they spread their leaves in the sunlight and fold them toward the stem at night (Figure 30–22). As long ago as 1729, the French scientist Jean-Jacques de Mairan noticed that these diurnal (daily) movements continue even when the plants are kept in dim light. More recent studies have shown that less evident activities, such as photosynthesis, auxin production, and the rate of cell division, also have daily rhythms. The rhythms continue even when all environmental conditions are kept constant. These regular day-night cycles have come to be called *circadian rhythms*, from the Latin words *circa*, meaning "about," and *dies*, "day." Circadian rhythms now have been found throughout the plant and animal kingdoms.

BIOLOGICAL CLOCKS

Are these rhythms internal—that is, caused by factors within the plant or animal itself—or is the organism keeping itself in tune with some external factor? For a number of years, biologists debated whether it might not be some environmental force, such as cosmic rays, the magnetic field of the earth, or the earth's rotation, that was setting the rhythms. Attempts to settle this recurrent controversy have led to numerous experiments under an extraordinary variety of conditions. Organisms have been taken down into salt mines, shipped to the South Pole, flown halfway around the world in airplanes, and, most recently, orbited in satellites.

Although there is still a vocal minority that believes that circadian rhythms are under the influence of a subtle geophysical factor, most workers now agree that the rhythms are endogenous—that is, they originate within the organism. Strong evidence in support of this belief is that the rhythms are not exact. Different species and different individuals of the same species often have slightly different, but consistent, rhythms, often as much as an hour or two longer or shorter than 24 hours. Nothing, however, is known about the physical or chemical nature of this internal timing device, which is often referred to as a *biological clock*.

30-23
"Sleep movement" rhythms in the bean plant. Many legumes, such as the bean, orient their leaves perpendicular to the rays of the sun during the day and fold them up at night. These "sleep movements" can easily be transcribed on a rotating chart by a delicately balanced pen-and-lever system attached to the leaf by a fine thread (a). *The rhythm will persist for several days in continuous dim illumination. A representative recording is seen in* (b).

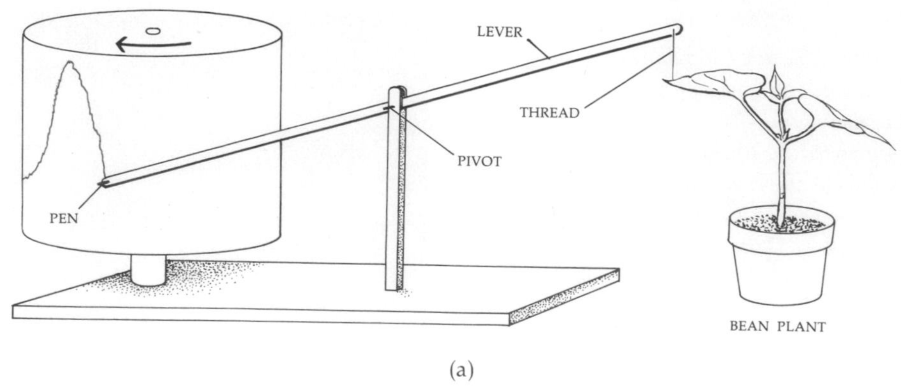

(a)

Another important feature of these rhythms is that they do not speed up as the temperature rises, as you might expect since enzymatic activities must be involved and enzyme reactions take place more rapidly as temperatures increase (Figure 4–10, page 84). Therefore, the clock must contain within its workings some type of compensatory mechanism—a feedback system that adjusts it to temperature changes. This idea is supported by the observation that in some organisms circadian rhythms seem to slow down as the temperature rises, rather than speed up, suggesting that overcompensation may be taking place.

CHANGING THE RHYTHMS

Although circadian rhythms probably originate within the organisms themselves, they can be modified by external conditions—a fact that is, of course, important to the survival of both individuals and species. For instance, a plant whose natural daily rhythm shows a peak every 26 hours when grown under continuous dim light can adjust its rhythm to 14 hours of light and 10 hours of darkness. It can also adjust to 11 hours of light and 11 of dark (or 22 hours). Such adjustment to an externally imposed rhythm is known as entrainment. If the new rhythm is too far removed from the original one, however, the organism will "escape" the entrained rhythm and revert to its natural one. A plant that has been kept on an artificial or forced rhythm, even for a long period of time, will revert to its normal internal period when returned to continuous dim light.

CLOCK FUNCTIONS

Biological clocks are believed to play an essential role in many aspects of plant and animal physiology. For instance, insects are more active in the early evening hours. Bats that feed on insects begin to fly each evening just when the insects are most available. Moreover, caged bats under controlled and constant laboratory conditions continue to show this sort of activity about every 24 hours, indicating that they are following an internal rhythm, not merely responding to environmental cues.

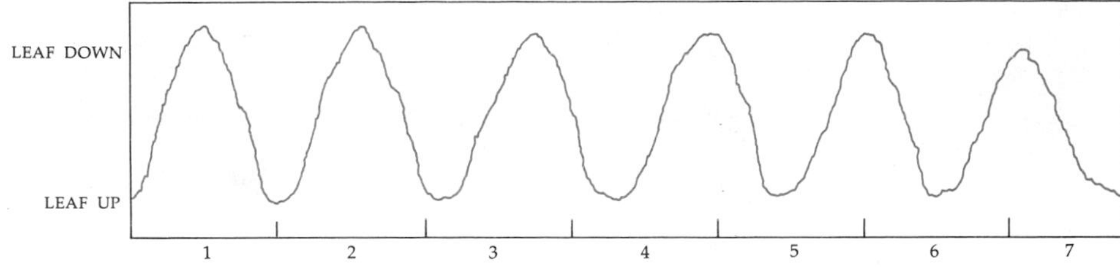

(b)

Some plants secrete nectar or perfume at certain specific times of the day. As a result, insects—which have their own biological clocks—become programmed to visit these flowers at these times, thereby ensuring maximum rewards for both the insects and the flowers.

The ability to tell time also appears to be involved in a number of complex and fascinating phenomena such as the extraordinary ability of migrating birds and turtles to navigate.

Biological clocks enable organisms to recognize the changing seasons of the year by "comparing" external rhythms of the environment, such as changes in day length, to their own relatively constant internal rhythms. This capacity is an important factor in regulating the growth cycle of many plants.

Finally, biological clocks are clearly involved in photoperiodism. In order for organisms to detect changes in day length—and remember that in some cases, they are accurate to within 15 minutes—they must have some constant, a clock, with which to compare them. The chemical nature of the biological clock—or indeed if there is just one kind of clock or many—is still not known.

30–24
Tendrils of Smilax *(greenbrier). Twisting is caused by varying growth rates on different sides of the tendril.*

TOUCH RESPONSES IN PLANTS

Many plants respond to touch. One of the most common examples is seen in tendrils or winding stems. They wrap around objects with which they come in contact (see Figure 30–24) and so enable the plant to cling and climb. The response can be rapid; a tendril may wrap around a support in less than a minute. Cells touching the support shrink slightly and those on the other side elongate. There is some evidence that auxin plays a role in this response.

A more spectacular response is seen in the sensitive plant, *Mimosa pudica*, in which the leaflets and sometimes entire leaves droop suddenly when touched. This response is a result of a sudden change in turgor pressure in cells at the base of the leaflets and leaves. The change in turgor pressure appears to involve the movement of potassium ions, as do the changes in stomata (page 533), but again, how the change is triggered is unknown. There is also some controversy about its survival

(a)

(b)

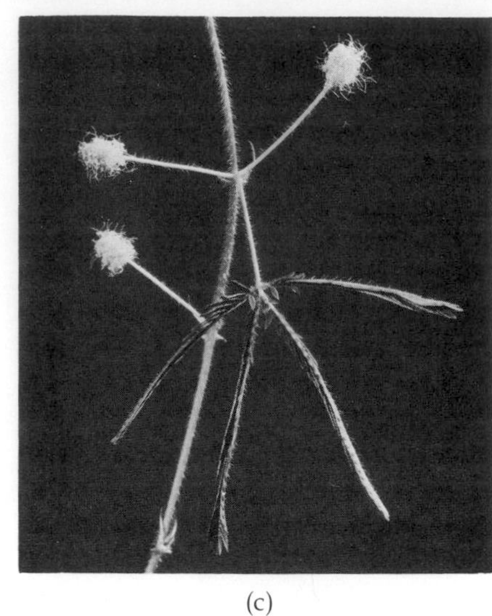

(c)

30–25
Sensitive plant (Mimosa pudica). (a) *Normal position of leaves and leaflets.* (b, c) *Successive responses to touch.*

value to the plant. *Mimosa pudica* often grows in dry, exposed areas where it may be subjected to drying winds; strong winds may shake the leaves enough to make them fold up, so conserving water. Another suggestion is that the wilting response makes the plant unattractive to grazing animals or startles chewing insects.

The triggering of turgor changes by touch is also involved in the capture of prey by the carnivorous Venus flytrap. The leaves of the Venus flytrap are hinged in the middle, and each leaf half is equipped with three sensitive hairs. When an insect walks on one of these leaves, attracted by the nectar on the leaf surface, it brushes against the hairs, triggering the traplike closing of the leaf. The toothed edges mesh, the leaf halves gradually squeeze closed, and the insect is pressed against digestive glands on the inner surface of the trap.

30–26
Touch response in the Venus flytrap.

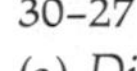

30–27
(a) *Diagram of a sundew tentacle, with electrodes in place.* (b) *A record of electric impulses produced by positioning a fruit fly so that its feet stroked the head of a tentacle. The recording ended when the contact was broken as the tentacle bent.*

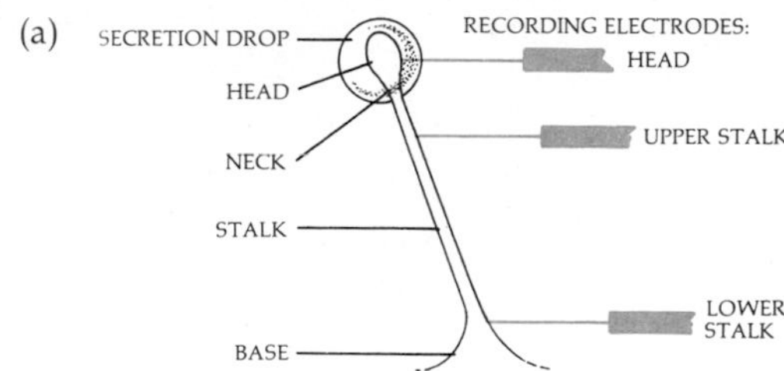

The trapping mechanism is so specialized that it can distinguish between living prey and inanimate objects, such as pebbles and small sticks, that fall on its leaves by chance: The leaf will not close unless two of its hairs are touched in succession or one hair is touched twice.

Studies have recently been made by investigators at Washington University in St. Louis on how the sundew traps insects. The club-shaped leaves of the plant are covered with tiny tentacles. A sticky droplet surrounding the tip of each tentacle attracts insects. When an insect is caught on the tip of a tentacle, the tentacle bends in rapidly, carrying the prey to the center of the leaf, where it is digested by enzymes. Microelectrodes placed in the tentacles (Figure 30–27) have revealed that the rapid response is accompanied by an electric impulse that moves down the tentacle. This impulse is similar, in principle, to the action potential that forms the basis of the nerve impulse in animals (see page 639).

The investigators predict that electric signals will be found to coordinate a variety of functions in plants.

SUMMARY

Plants respond to factors in both their internal and external environments. Such responses enable the plant to develop normally and to remain in touch with external conditions. Hormones are important factors in plant response. A hormone is a chemical produced in particular tissues of an organism and carried to other tissues of the organism, where it exerts a specific influence. Characteristically, it is active in extremely small amounts.

Auxins are hormones that are produced principally in physiologically active tissues, such as coleoptile tips and apical meristems. They cause lengthening of the shoot and the coleoptile, chiefly by promoting cell elongation. They often inhibit growth in lateral buds, thus restricting growth principally to the apex of the plant. The same concentration of auxin that promotes growth in the stem inhibits growth in the main root system. Auxins promote the initiation of branch roots and adventitious roots; however, they retard abscission in leaves and fruits. In fruits, auxin produced by seeds stimulates growth of the ovary wall. The capacity of auxins to produce such varied effects appears to result from different responses of the various target tissues.

The gibberellins were first isolated from a parasitic fungus that causes abnormal growth in rice seedlings. They were subsequently found to be natural growth hormones also present in plants. Gibberellins induce bolting and flowering in cabbage. In some dwarf plants, application of gibberellins restores normal growth. Gibberellins are also involved in embryo and seedling growth in grasses. They stimulate the production of hydrolyzing enzymes that act on the stored starch and proteins of the endosperm, converting them to sugar and amino acids, which nourish the seedling.

The cytokinins, a third class of hormone, promote cell division. It is possible by altering the concentrations of auxins and cytokinins to alter patterns of growth in undifferentiated plant tissue in tissue culture.

Abscisic acid, a growth-inhibiting hormone, induces dormancy in vegetative buds and accelerates abscission. Ethylene is a gas that is produced by fruits during the ripening process. Ethylene promotes the ripening of fruits and is now considered a natural growth regulator. There is evidence for the existence of a flowering hormone (or hormones), but it has not yet been isolated.

Plants respond to a number of environmental stimuli. Photoperiodism is the response of organisms to changing periods of light and darkness in the 24-hour day. Such a response controls the onset of flowering in many plants. Some plants will flower only when the periods of light exceed a critical length. Such plants are known as long-day plants. Other plants, short-day plants, flower only when the periods of light are less than some critical period. Day-neutral plants flower regardless of photoperiod. Interruption of the dark phase of the photoperiod, even by a brief exposure to light, can serve to reverse the photoperiod effects, indicating that the dark period rather than the light period is critical.

Phytochrome, a pigment commonly present in small amounts in the tissues of higher plants, is the receptor molecule for transitions between light and darkness. The pigment can exist in two forms, P_{660} and P_{730}. P_{660} absorbs red light of a wavelength of 660 nanometers (which is also ordinarily present in white, or visible, light). It is thereby converted to P_{730}, which absorbs far-red light of a 730-nanometer wavelength. With ordinary day-night sequences, P_{730} is converted back to P_{660} over a period of hours in the dark; it can also be converted to P_{660} by exposure to far-red light. P_{730} is the active form of the pigment; it promotes flowering in long-day plants and inhibits flowering in short-day plants. It promotes germination in lettuce seeds and normal growth in seedlings. Its mechanism of action is unknown.

Circadian rhythms are regular cycles of growth and activity occurring approximately on a 24-hour basis. Many of these rhythms have been shown to be independent of the organism's environment and to be controlled by some endogenous regulator—a biological clock. Its chemical nature is not known.

Some species of plants respond to touch. Examples include the winding of tendrils, the collapse of the leaves of the sensitive plant *(Mimosa)*, the triggering of the carnivorous Venus flytrap, and the bending of the tentacles of the sundew. The latter appears to involve an electric impulse.

QUESTIONS

1. Describe Went's experiments and the conclusions that can be drawn from them.

2. Describe the principal roles played by each of the following hormones: auxin, gibberellin, cytokinin, abscisic acid, and ethylene.

3. For many years it has been known that oranges hasten the ripening of bananas. Can you think of an explanation for this phenomenon?

4. Auxins have been "blamed" for a lot of responses in plants. Can these very different reactions be attributed to the sole action of indoleacetic acid? Explain.

5. Speculate on why a gas has been selected to mediate processes such as fruit ripening and senescence.

6. The distinctions we make among the plant hormones refer to general classes of their effects rather than to the precise chemical names of the substances actually produced by particular plants. Why might it be relatively easy to test chemicals for hormonal activity but be particularly difficult to determine just which chemical a plant actually produces and uses?

7. If you prune many shrubs, they grow bushier. Devise a simple experiment to show that the change in overall shape of the plant is not simply a result of reducing apical elongation by cutting. Suggest a hormonal mechanism for the phenomenon of bushiness induced by pruning.

8. Distinguish among the following: phototropism/photoperiodism/photoinduction; circadian rhythm/biological clock.

9. Plants must, in some manner, synchronize their activities with the seasons. Of the clues that they might use, day (or night) length seems to have been selected. What might be the advantage of using photoperiod rather than, say, temperature as a season detector?

10. Traveling from north to south, one can find varieties of the same species with different photoperiodic requirements. How would you expect them to differ?

11. Photoperiodic systems are not nearly as sensitive to low levels of illumination as are many visual systems. Why might extreme sensitivity be a positive disadvantage in a photoperiodic system? What havoc might be wrought by the wide spread use of bright lights for street illumination?

12. Suppose you were given a chrysanthemum plant, in bloom, one autumn, and you decided to keep it indoors as a houseplant. What precautions would you need to take the following autumn to ensure that it would bloom again?

13. Carefully examine the graph in Figure 30–23. What is happening to the rhythm of the plant's "sleep movements" toward the end of the week? Speculate as to why this might be happening. It has also been observed that as the plant is maintained over time under the experimental conditions, it becomes rather sickly. What factors do you suppose might be involved in this change?

SUGGESTIONS FOR FURTHER READING

Books

ESAU, KATHERINE: *Anatomy of Seed Plants,* 2d ed., John Wiley & Sons, Inc., New York, 1977.

An up-to-date and profusely illustrated text on plant anatomy.

GALSTON, A. W.: *The Green Plant,* Prentice-Hall, Inc., Englewood Cliffs, N.J., 1968.*

A convenient and concise summary of plant growth and development.

GREULACH, VICTOR: *Plant Structure and Function,* The Macmillan Company, New York, 1973.

A short, lucid introduction to botany, with emphasis on physiology.

LEDBETTER, M. C., and KEITH PORTER: *Introduction to the Fine Structure of Plant Cells,* Springer-Verlag, New York, 1970.

An excellent atlas of electron micrographs of plant cells, with a detailed explanation of each.

RAVEN, PETER, RAY EVERT, and HELENA CURTIS: *Biology of Plants,* 2d ed., Worth Publishers, Inc., New York, 1975.

An up-to-date and handsomely illustrated general botany text.

RAY, PETER M.: *The Living Plant,* 2d ed., Holt, Rinehart and Winston, Inc., New York, 1972.*

An outstanding, short text.

RICHARDSON, MICHAEL: *Translocation in Plants,* St. Martin's Press, Inc., New York, 1968.*

An extremely useful, brief review of experimental work on the movement of water in plants.

SALISBURY, FRANK B., and CLEON ROSS: *Plant Physiology,* Wadsworth Publishing Co., Inc., Belmont, Calif., 1969.

A good, modern plant physiology text for more advanced students.

TORREY, JOHN G.: *Development in Flowering Plants,* The Macmillan Company, New York, 1967.*

How flowering plants develop, with emphasis on the underlying physiological process.

Articles

BRILL, WINSTON T.: "Biological Nitrogen Fixation," *Scientific American,* March 1977, pages 68–81.

---

* Available in paperback.

SECTION 6 **BIOLOGY OF ANIMALS**

31–1
Genus, Homo; *species,* sapiens. *An immature form.*

CHAPTER 31 The Human Animal: An Introduction

Even those of us who have most marveled at the exquisite architecture of an orchid flower, or who are as content as Leeuwenhoek to watch the intricate and varied movements of a *Paramecium* and its neighbors, approach the subject of the biology of our own systems with a quickened interest. Indeed for some, student and scientist alike, the principal and perhaps the only reason for studying "lower forms" is the extent to which such studies bear directly on human welfare. But as *Escherichia coli*, T2 and T4 bacteriophages, and *Drosophila* remind us, there is no way to study only the human species, any more than it is possible to study only a fruit fly.

In this section of the book, which deals with vertebrate physiology, we shall examine *Homo sapiens* as an example of an animal organism. This one species, however, will not command our full attention. First, we shall rely heavily on your understanding of the principles governing such matters as surface and volume, diffusion and osmosis, and gas exchange and water balance, which were discussed and exemplified in earlier chapters. Second, in the chapters that follow, we shall introduce numerous examples from other organisms that illuminate our central theme—human physiology. Finally, we shall continue to emphasize broad principles rather than lose our way in a mass of details and intricate terminology. Because we are part of a continuum of nature, it is logical to study other animals in order to understand the human animal; it is equally logical to study the human animal (now one of the best understood) in order to gain a deeper understanding of animal life in general.

In 1966, George Gaylord Simpson, one of the leading modern students of evolution, wrote:

> The question "What is man?" is probably the most profound that can be asked by man. It has always been central to any system of philosophy or of theology. We know that it was being asked by the most learned humans 2,000 years ago, and it is just possible that it was being asked by the most brilliant australopithecines 2 million years ago. The point I want to make now is that all attempts to answer that question before 1859 are worthless and that we will be better off if we ignore them completely.*

* G. G. Simpson: "The Biological Nature of Man," *Science*, vol. 152, pages 472–478, 1966.

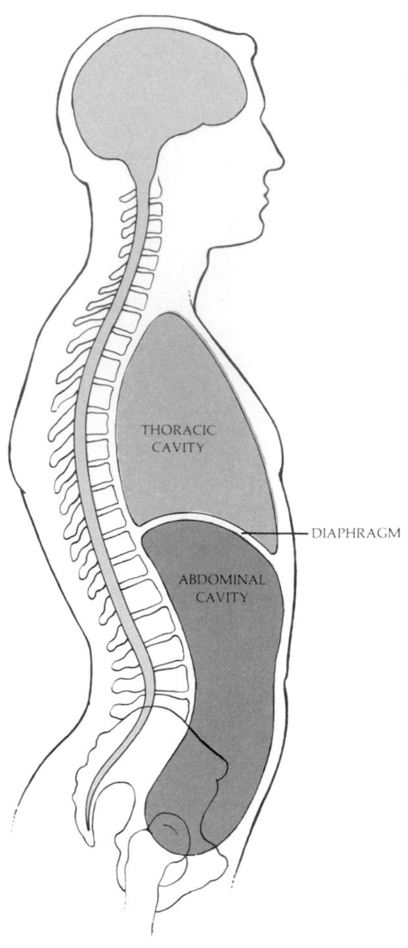

31–2
Humans, like other vertebrates, are characterized by a dorsal nervous system enclosed in vertebrae and the skull. As in other mammals, a muscular diaphragm divides the coelom into the thoracic cavity and the abdominal cavity.

Some of us may not accept Simpson's dismissal of pre-Darwinian philosophy, but there are few who would not agree about the relevance of his statement to human biology.

The human being is a vertebrate, and as such has a bony, articulated (jointed) endoskeleton that supports the body and grows as it grows. The spinal cord, which is dorsal, is surrounded by bony segments, the vertebrae, and the brain is enclosed in a protective casing, the skull. The skeleton is moved by means of muscles.

As in other vertebrates, and most invertebrates as well, the human body contains a cavity, or coelom. In humans, the coelom is divided into two parts, the thoracic cavity and the abdominal cavity. These are separated by a dome-shaped muscle, the diaphragm. The thoracic cavity contains the heart, the lungs, and the upper portion of the digestive tract. The abdominal cavity contains the stomach, intestines, liver, and other organs.

Human beings are also mammals. One of the most important characteristics of mammals is that they are warm-blooded. More precisely, they are homeotherms; that is, they maintain a high and relatively constant body temperature. As a consequence, mammals (and birds, which are also homeotherms) are able to achieve and sustain levels of physical activity and mental alertness generally far greater than those of animals whose temperatures rise and fall with those of their external environment. A concomitant of homeothermy is a high metabolic rate, which requires relatively large and constant supplies of food (fuel) molecules and oxygen.

Mammals have other important characteristics. They have hair or fur rather than scales or feathers. All mammals (except the monotremes) give birth to live young, as distinct from laying eggs, which all birds and most fish, amphibians, and reptiles do. They nurse their offspring, which involves a relatively long period of parental care. This degree of parental care is correlated with a relatively long learning period (as contrasted, for example, with insects, most species of fish, and amphibians and reptiles, most of which are independent from the moment they hatch from the egg). There is a tendency among the large mammals, in particular, toward fewer young per litter and an even longer period of parental care. Humans, for instance, rarely have more than two surviving young per birth, only two mammary glands with which to nurse them, and an extraordinarily long period of infancy and childhood, with dependency on parents often lasting well past physical maturity.

As a consequence, at least in part, of homeothermy and the long learning period, mammals are, in general, by far the most intelligent of all groups of organisms. They have the most highly developed systems for receiving, processing, and correlating information from the environment (although, as we noted in Chapter 26, in complexity and variety of sense organs, they are in some cases surpassed by some of the invertebrates).

The class Mammalia comprises a large number of different species, ranging from whales and dolphins, bats, moles and hedgehogs, hippopotamuses, zebras and rhinoceroses, to gorillas and marmosets. Among the members of this class, the human being is distinguished first by being versatile rather than specialized. A human cannot run as fast as a deer, swim as gracefully as a seal, or swing from branch to branch with the agility of a gibbon. But he or she can, if adequately motivated, run several kilometers, swim a river, and climb a tree—and few other mammals can do all three. Finally, for better or worse, *Homo sapiens* is by far the most intelligent of all mammalian species.

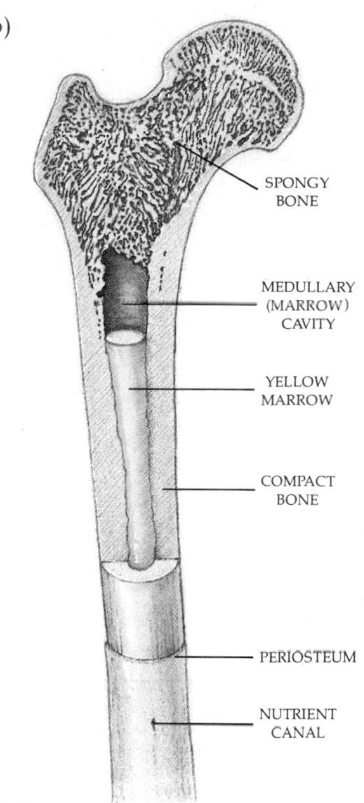

31–3
(a) The skeleton of a human adult contains 206 bones. Twenty-nine are in the skull, including 14 face bones and six small bones (ossicles) of the ears. There are 27 bones in each hand and 26 in each foot. Bones are living organs, made up not only of osseous tissue but many other tissue types as well, including nerve, muscle, vascular tissues, and cartilage. Calcium compounds, secreted by osteoblasts (immature bone cells), give bone its characteristic hardness. (b) The ends of long bones such as this femur consist of spongy bone, in which there are large spaces, surrounded by compact bone. The shaft, which is hollow, is composed of compact bone. A central cavity containing bone marrow extends through the center of the shaft. The marrow of long bones is yellow because of stored fat. The periosteum is a fibrous sheath, which contains blood vessels that nourish the bone tissues. Blood vessels emerge from bone through openings known as nutrient canals.

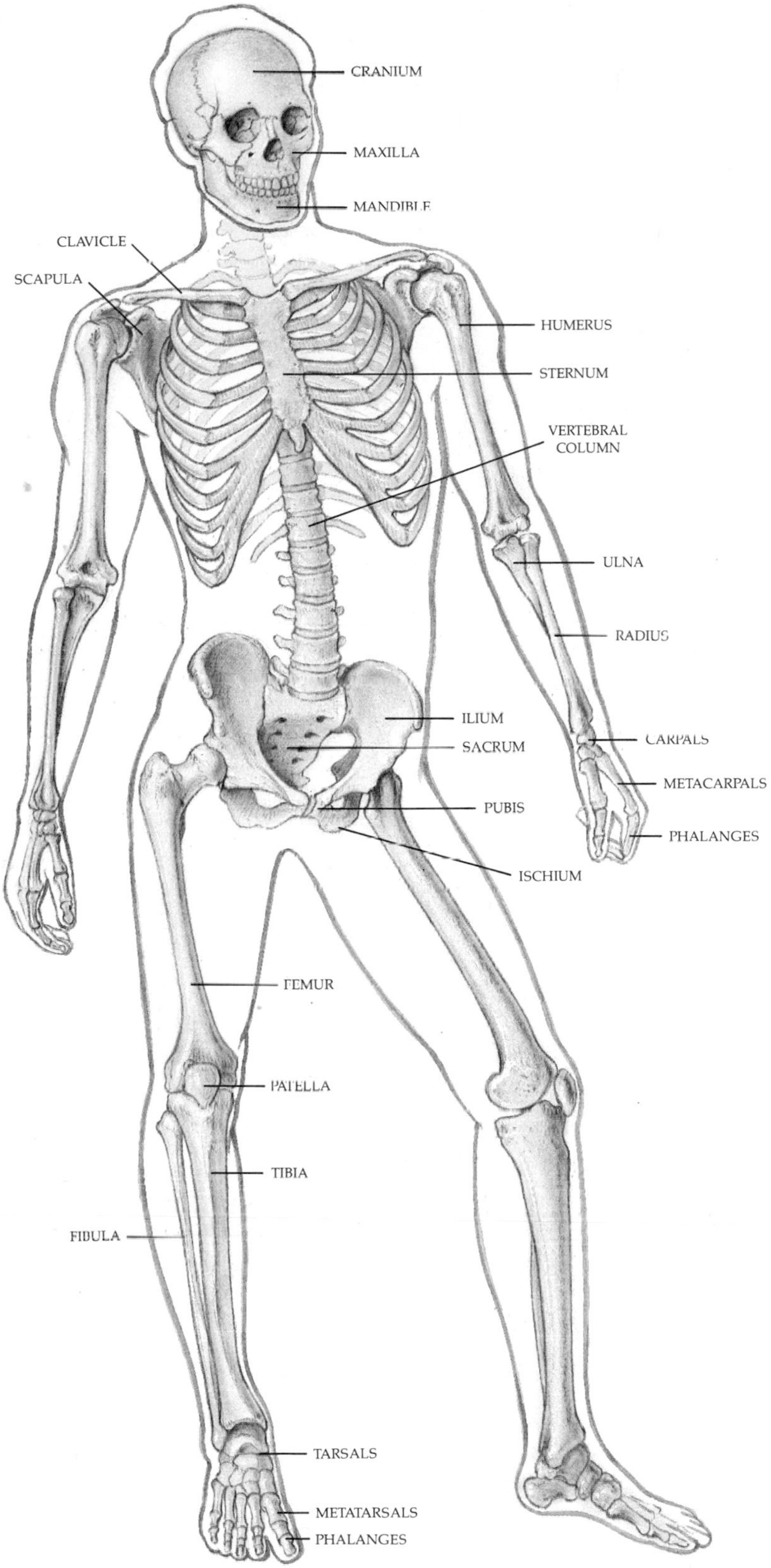

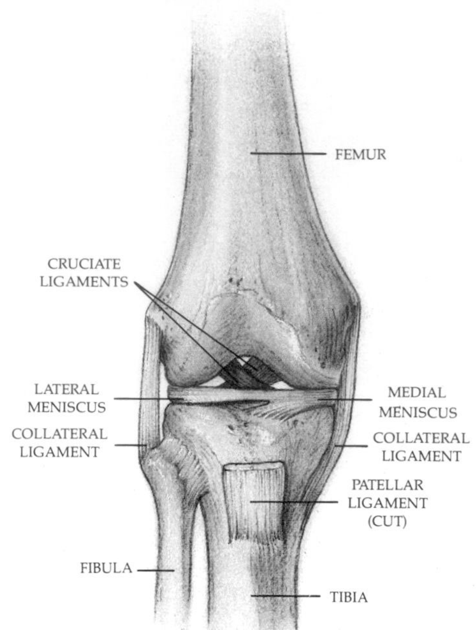

31–4
A knee is a hinge joint between the femur and the tibia, covering which is the "knee cap," or patella (removed in this drawing). As with all bones, bones of the knee joint are bound together by ligaments. The collateral ligaments bind the joint externally; they are slack when the knee is flexed and taut when the knee is extended. The cruciate ligaments cross within the joint, stabilizing it. The menisci are C-shaped wedges of cartilage, resting on the upper surface of the tibia and cushioning the joint.

Knees are particularly susceptible to injury, as Joe Namath, Bobby Orr, Wilt Chamberlain, and others will sadly attest. The reason for the knee's vulnerability becomes clear when you contemplate the analogous problem of fastening together two match sticks with several rubber bands, to produce a hinge that is both strong and flexible and can not only swing back and forth but also twist, bend, and rotate. Common knee injuries are torn ligaments, especially the collateral ligaments (as a result of the knee's being hit from the side), and crushed menisci. When the shock-absorbing wedges of menisci are lost, bone chips begin to accumulate in the joint.

CELLS AND TISSUES

The human body, like that of all other complex animals, is made up of a great variety of different cells. Each of these cells is, to some extent, an independent entity. Each needs oxygen, glucose (or some other source of carbon and energy), amino acids, and small amounts of other substances (such as iron). Each needs to dispose of carbon dioxide and other wastes. Human physiological processes, like those of any other organism, are concerned with providing second-by-second, 24-hour-a-day service to each of these billions of cells.

Although they greatly resemble one-celled organisms in their requirements, the cells of an animal body differ from one-celled organisms in that they develop and function as part of an organized whole. As we shall see, cells are organized into *tissues*, groups of cells similar in structure and function. Different kinds of tissues, united structurally and coordinated in their activities, form *organs*. The eye (Figure 31–5), with its many different types of fibers, is an example of an organ. Organs that function together in an integrated and organized way make up *organ systems*. The digestive system, for example, is composed of a number of different organs, each of which carries out a specific activity that contributes toward the overall process. Because the focus of this section is on the various functions of the body, we shall be examining it principally in terms of organ systems.

Although experts can distinguish more than 100 different types of cells in the human body, they customarily classify them in terms of only four tissue types: (1) epithelial, (2) connective, (3) muscle, and (4) nerve.

Epithelial Tissue

Epithelial tissue consists of a continuous sheet of cells that provides a protective covering of the whole body and contains various sensory nerve endings. It also forms the lining membranes of internal organs, cavities, and passageways. Hence, as a moment's reflection will reveal, everything that goes into and out of the body must pass through epithelial cells.

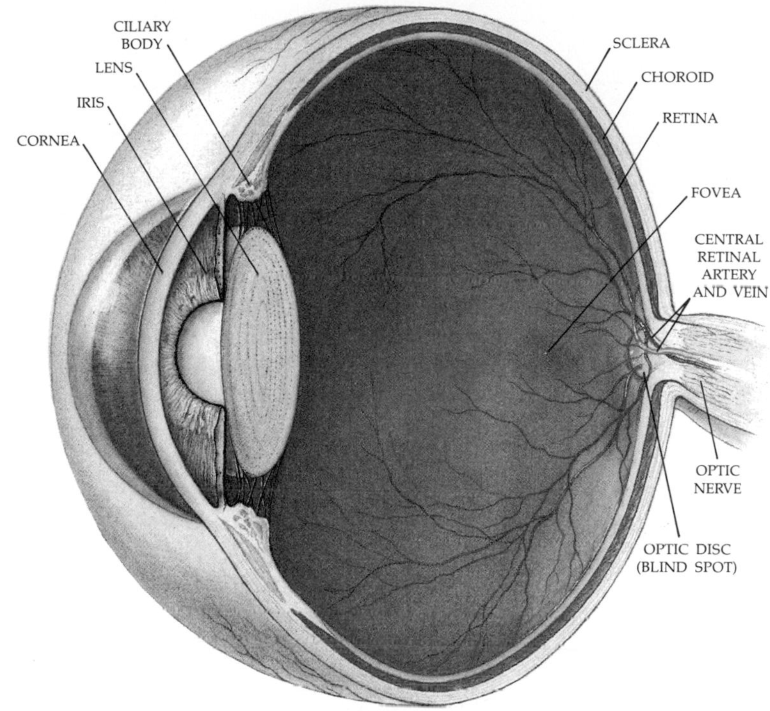

31–5
The eye is a complex organ composed of three layers of tissue, which form a fluid-filled sphere. The outer layer, the sclera, is white fibrous connective tissue that serves a protective function. The anterior portion of the sclera, the cornea, is transparent. The middle layer, the choroid, contains blood vessels. Its anterior portion is modified into the ciliary body, the suspensory ligament, and the iris. The ciliary body is a circle of smooth muscle from which extend the suspensory ligaments that hold the lens in place. The colored part of the eye, the iris, is a circular structure attached to the ciliary body. The pupil is a hole in the center of the iris. The innermost layer, the retina, contains the light-sensitive cells, the rods and cones. It has no anterior portion. Most of the eyeball is recessed in the orbit, protected by the bony socket of the skull.

One surface of the epithelial sheet is always attached to an underlying layer, called the basement membrane, composed of a fibrous polysaccharide material produced by the epithelial cells themselves. Epithelium is classified according to the shape of the individual cells as squamous, cuboidal, or columnar (Figure 31–6). It may consist of only a single layer of cells (simple epithelium), as found in the inner lining of the circulatory system, or several layers (stratified epithelium), as found in the epidermis of the skin.

The epithelium of the body cavities and passageways frequently contains modified epithelial cells that secrete mucus, which lubricates the surfaces. Glands are special types of epithelial tissue; gland cells produce specific substances, such as perspiration, saliva, hormones, or digestive enzymes. Glandular epithelium is composed of cuboidal or columnar epithelial cells.

31–6
The three types of epithelial cells that cover the inner and outer surfaces of the body. Squamous cells, which usually perform a protective function, make up the outer layers of the skin and the lining of the mouth and other mucous membranes. There are usually several layers of these flat cells piled on top of one another. Cuboidal and columnar cells, in addition to lining various passageways, perform much of the chemical work of the body.

SQUAMOUS

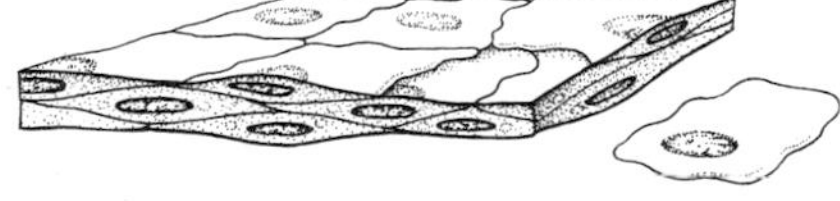

CUBOIDAL

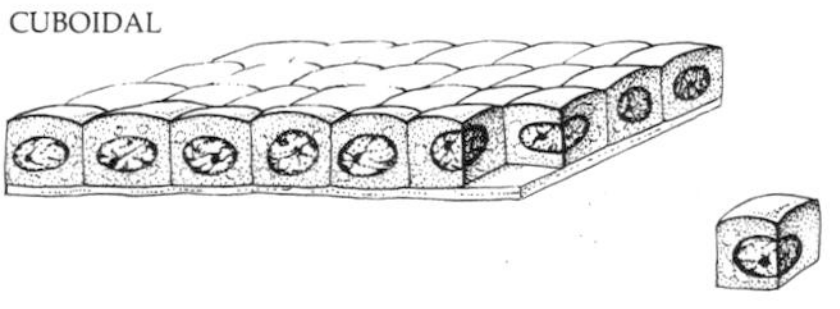

COLUMNAR

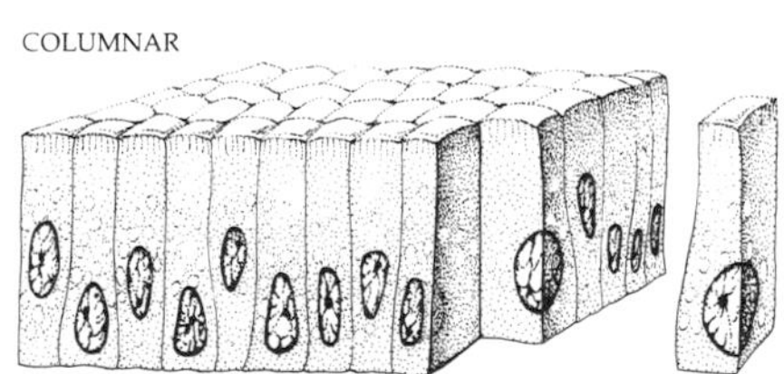

Connective Tissue

Connective tissue binds together and supports the other three kinds of tissue. Unlike epithelial tissue cells, the cells of connective tissue are widely separated from one another by large amounts of intercellular substances. The intercellular substances include: (1) connecting and supporting fibers, such as collagen, which is a major component of skin, tendons, ligaments, and bones; (2) elastic fibers, which are often found in the walls of hollow, distensible organs, such as the stomach and the uterus; and (3) reticular fibers, which form networks inside solid organs, such as the liver. All of these fibers are embedded in a matrix, or ground substance, that is more or less fluid and amorphous (without any shape or form).

Bone, like other connective tissues, consists of cells, fibers, and ground substance (Figure 31–7). It is unlike the others in that the collagen fibers are impregnated with hard crystals. Bone tissue is amazingly strong and light; our bones make up only about 18 percent of our weight.

Blood and lymph are connective tissues in which the matrix is plasma. The characteristic cells of blood and lymph are described in Chapter 35.

Muscle Tissue

About 40 percent of our body's weight is made up of muscle tissue. There are two general types: smooth muscle and striated muscle, so called because of its striped appearance (Figure 31–8). Striated muscle is the type of muscle that moves the

31–7
Electron micrograph and diagram of a bone cell (osteocyte). Young bone cells (osteoblasts) produce the intercellular bone matrix, an organic material consisting of collagen fibers and an unstructured ground substance. This matrix gradually calcifies and hardens.

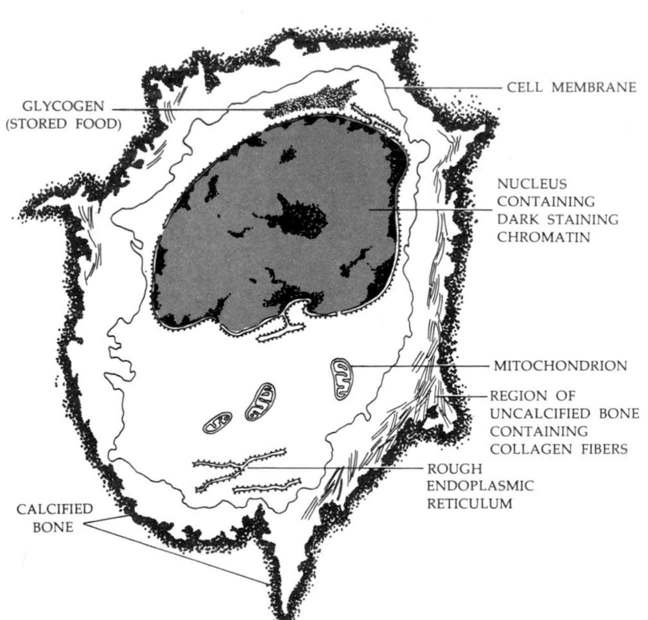

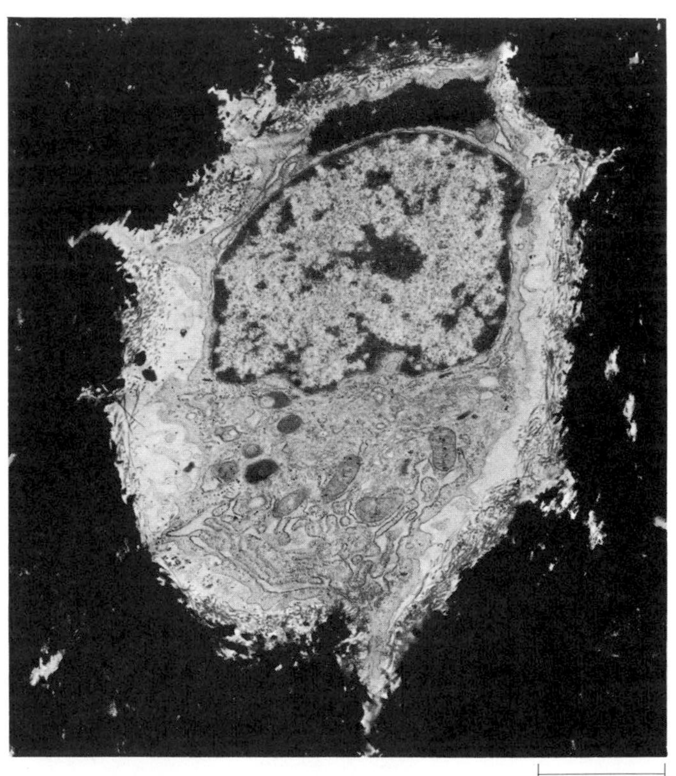

31–8
Photomicrograph of skeletal muscle fibers, showing striated pattern. The arrow indicates a small blood vessel, a capillary. Within it, you can see red blood cells, which carry oxygen to the muscle fibers, each of which is a single cell.

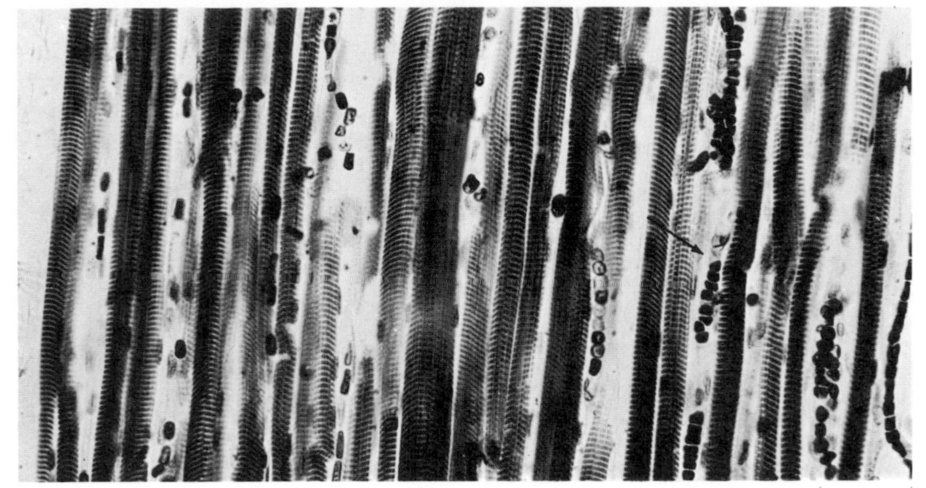

31–9
Muscles attached to bone move the vertebrate endoskeleton. They often work in antagonistic pairs, with one relaxing as the other contracts. Muscles cannot lengthen spontaneously; they lengthen only when pulled by antagonistic muscles. For example, when you move your hand toward your shoulder, as shown here, the biceps contracts and the triceps relaxes. When you move your hand down again, the triceps contracts, while the biceps relaxes. The muscles that move the skeleton, such as those diagrammed here, are known as skeletal muscles. They are striated, as shown in Figure 31–8.

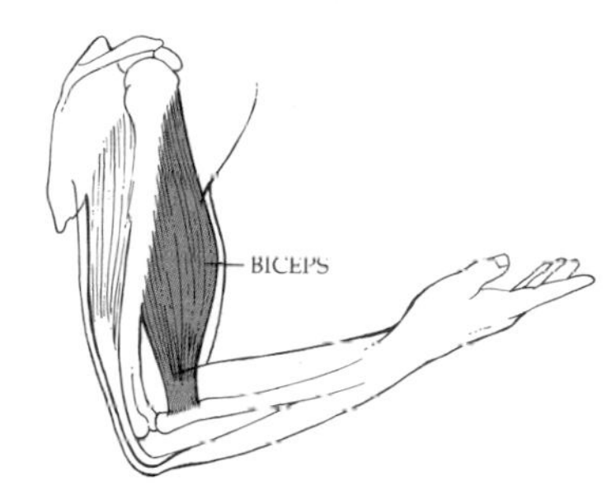

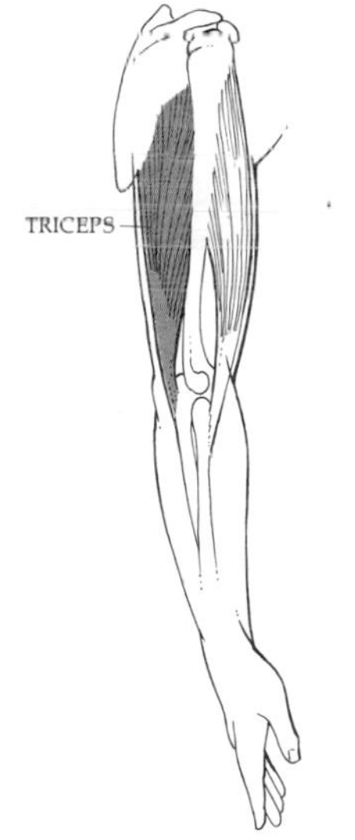

skeleton; skeletal striated muscle is sometimes called voluntary muscle, since we can move it at will. Cardiac (heart) muscle is also striated muscle. Smooth muscle surrounds the walls of the internal organs, such as the digestive organs, uterus, and bladder; it is sometimes called involuntary muscle.

In this discussion, we shall focus on skeletal muscle because it is the subject of much current research, and much of what we know about the other forms of muscle is inferred from what has been learned about skeletal muscle. Also, studies of the machinery of skeletal muscle, involving as they do both electron microscopy and biochemistry, have brought scientists tantalizingly close to visualizing the precise role of the individual molecules.

Muscle Action

Skeletal muscles, like all muscles, act by contracting. A skeletal muscle is typically attached to two or more bones, either directly or by means of the tough strands of connective tissue known as tendons. Some of these tendons, such as those that connect the finger bones and their muscles in the forearm, may be very long. When the muscle contracts, the bones move around a joint, which is held together by ligaments and contains a lubricating fluid. Most of the skeletal muscles of the body work in antagonistic pairs, one muscle flexing, or bending, the joint, and the other extending, or straightening, it (Figure 31–9).

A muscle, such as the biceps, consists of bundles of muscle fibers—often hundreds of thousands of fibers—held together by connective tissue. Each fiber is a single cell with many nuclei. These fibers are very large cells—50 to 100 micrometers in diameter and, often, several centimeters long. They are surrounded by an outer cell membrane that has been given the special name of sarcolemma.

Embedded in the cytoplasm of each muscle cell (fiber) are some 1,000 to 2,000 smaller structural units; they appear ribbonlike in electron micrographs but they are actually cylindrical strands. These strands, which are called *myofibrils* (from *myo*, the prefix for "muscle"), run parallel for the length of the cell. The nuclei are crowded to the periphery of the cytoplasm by the myofibrils and can typically be found at the cell surface, just beneath the sarcolemma.

Each myofibril is, in turn, composed of units called sarcomeres. The repetition of these units gives the muscle its characteristic striated pattern.

Figure 31–10 shows a sarcomere as seen in a longitudinal section of muscle. Each sarcomere is about 2 or 3 micrometers in length. The Z line is the dense black line seen in the electron micrograph; the I band is the relatively clear, broad stripe that the Z line bisects; and the A band is the large, dense stripe in the center of the sarcomere bisected by the central H zone. As the diagram shows, each sarcomere is composed of two types of filaments running parallel to one another. The thicker filaments in the central portion of the sarcomere are composed of a protein known as myosin; the thinner filaments are made of actin, also a protein. The Z line is where the actin filaments from adjacent sarcomeres interweave.

Muscles function by contracting. From the cell physiologist's point of view, the exciting feature of the sarcomere is that it is the contractile machine. When muscle is stimulated, the thin (actin) filaments of the sarcomere slide past the thick (myosin) filaments. Since the thin filaments are anchored into each Z line, this causes each sarcomere to shorten, and thus the myofibril as a whole contracts. According to the current hypothesis, cross bridges between the thick and thin filaments form, break, and re-form rapidly, as one filament "walks" along the other (Figure 31–11).

Actin and Myosin

The actin strands in muscle are composed, it has been found, of many smaller globular subunits assembled in a long chain. As shown in Figure 31–11e, each thin filament is composed primarily of two such actin chains wound around each other.

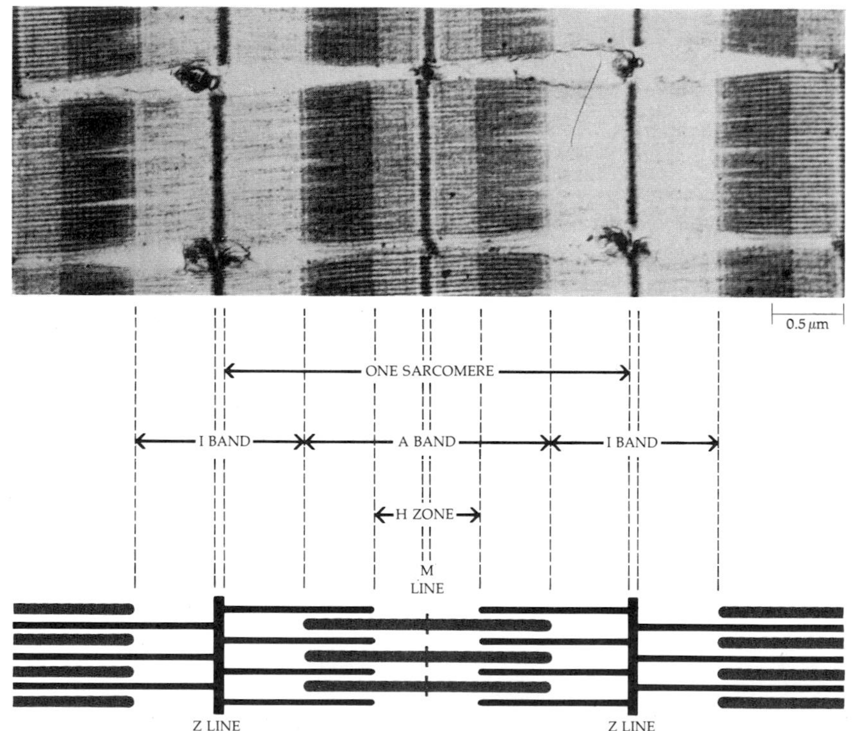

31–10
Electron micrograph and diagram of a sarcomere, the contractile unit of muscle. Each sarcomere is composed of an array of thick and thin protein filaments arranged longitudinally. The way the thin and thick filaments overlap makes the muscle fiber appear striated (banded). The Z line is where thin filaments from adjoining sarcomeres interweave. The I band is a region that contains only thin fibers. The A band marks the extent of the thick fibers. The part of the A band where there are no thin filaments is called the H zone. The thick filaments are interconnected and held in place at the M line. Muscle contraction involves the sliding of the thin filaments between the thick ones.

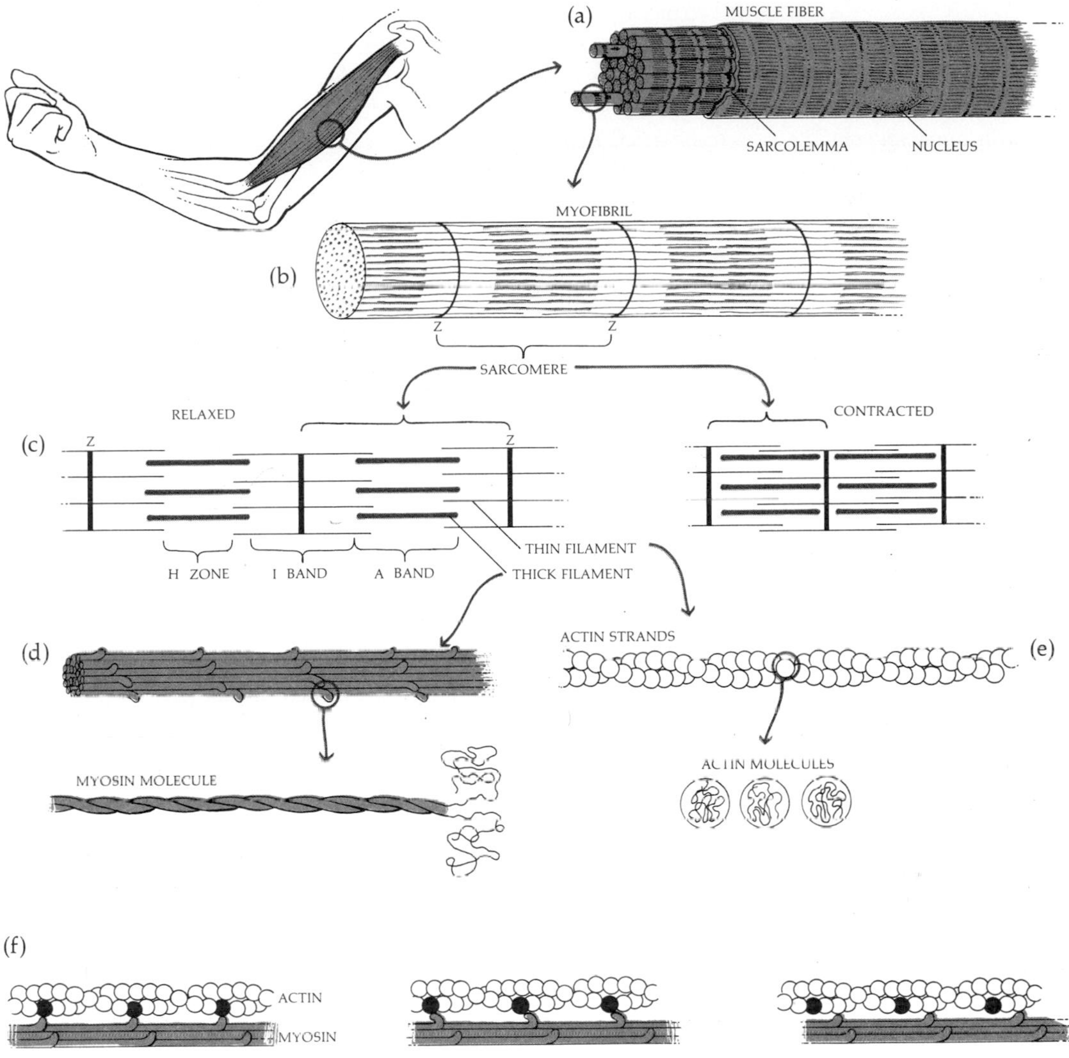

31–11
(a) *Skeletal muscle is composed of individual muscle cells, the muscle fibers. These are cylindrical cells, often many centimeters long, with numerous nuclei.* (b) *Each muscle fiber is made up of many cylindrical subunits, the myofibrils. These are rods of contractile proteins that run from one end of the cell to the other.* (c) *The myofibril is divided into segments, sarcomeres, by thin, dark partitions, the Z lines. The Z lines of adjacent myofibrils are in line with each other and appear to run from myofibril to myofibril across the fiber, giving the muscle cell its striated appearance.* (d) *Each sarcomere is made up of thick and thin filaments. Chemical analysis shows that the thick filaments consist of bundles of a protein called myosin. Each individual myosin molecule is composed of two protein chains wound in a helix; the end of each chain is folded into a globular structure.* (e) *Each thin filament consists of two actin strands coiled about one another in a helical chain. Each strand is composed of globular actin molecules.* (f) *According to current hypothesis, the globular myosin heads protruding from the thick filaments serve as hooks or levers, attaching to the actin molecules of the thin filament and pulling them toward the center of the H zone, shortening the sarcomere and contracting the myofibril. When the myofibrils contract, the fiber shortens, and when enough fibers shorten, the entire muscle shortens, producing skeletal movements.*

31-12
Electron micrograph of smooth muscle from the epididymis (part of the spermatic duct) of a mouse. Smooth muscle is made up of long, spindle-shaped cells containing contractile proteins. The proteins are not organized into sarcomeres, and the cells are therefore not striated (hence their name, smooth). Portions of a number of smooth muscle cells can be seen in the picture. The nucleus of one cell, distinguishable by its granular texture, can be seen in the middle of the micrograph. Very thin fibrils run lengthwise through the cells and make up the bulk of the cytoplasm. A number of mitochondria are visible.

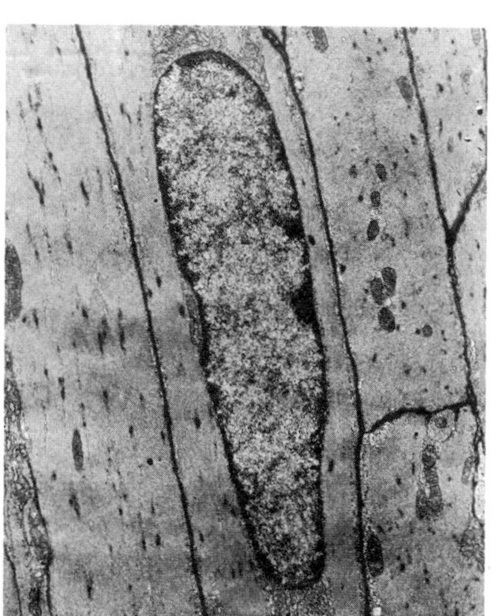

The myosin molecule has the longest protein chains known, each one consisting of some 1,800 amino acid units. Each long protein chain has a globular structure at one end. The myosin molecule consists of two of these chains wound around each other, with the globular "heads" free. The thick filaments, in turn, are composed of bundles of myosin molecules. These globular heads have two crucial functions: They are the binding sites that link the actin and myosin molecules, and they also act as enzymes to split ATP to ADP, thus providing the energy for muscle contraction.

To sum up, when a muscle fiber is stimulated, the thin (actin) filaments slide between the thick (myosin) filaments, and ATP is hydrolyzed to ADP in the process. The sarcomeres shorten and the fibers contract. If enough fibers contract, a whole muscle contracts, and the organism or part of the organism moves. Thus the intricate machinery of the sarcomere is, in essence, a mechanism for the conversion of chemical energy into kinetic energy, the energy of motion.

Smooth Muscle

Smooth muscle contracts much less rapidly than striated muscle, and its contractions are more prolonged. Smooth muscle is responsible for the movement of food along the intestinal tract, for example, and the emptying of the bladder and the constriction of small blood vessels to regulate blood flow.

Unlike striated muscle, smooth muscle consists of cells with single nuclei (Figure 31-12). These cells contain numerous fibrils that run lengthwise through the cell but are not arranged in any discernible pattern. Thus no striations are apparent. Like striated muscle, smooth muscle cells contain large amounts of actin and myosin, which are presumed, but not proved, to be in the fibrils and to play a role in the contraction of smooth muscle. Smooth muscle also requires ATP for contraction.

Nerve Tissue

The fourth major tissue type is nerve tissue. The basic units of nerve tissue are the neurons, or nerve cells, which make up about 10 percent of nerve tissue cells. Nerve cells receive and transmit signals from either the external or the internal environment. They also process this information and are responsible for the complex functions of consciousness, memory, and thought. The other 90 percent of nerve tissue cells are glial cells, which support the neurons physically and also nourish, protect, and insulate them. (In terms of the bulk of nerve tissue, each cell type contributes about half.)

A typical neuron consists of a cell body, which contains the nucleus and most of the metabolic machinery of the cell, and one or more long extensions, or processes (axons and dendrites) (Figure 31-13). Along these neuron processes, nerve impulses are transmitted from one part of the organism to another. These cells are the most morphologically spectacular of the body's cells. Neurons may reach astonishing lengths. For example, the axon of a single motor neuron—a nerve cell that activates muscle—may extend from the spinal cord down the whole length of the leg to the toe. Or a sensory neuron—one that transmits sensations to the brain—based near the spinal cord may send a dendrite down to the toe and an axon up the entire length of the spinal cord to terminate in the lower part of the brain. In an adult human, such a cell might be close to 2 meters long (5 meters in a giraffe).

31–13
Four examples of neurons. Dendrites, which are characteristically multiple, transmit signals toward the neuron cell body, whereas axons, of which there is only one per cell, transmit signals away from the cell body of the neuron.

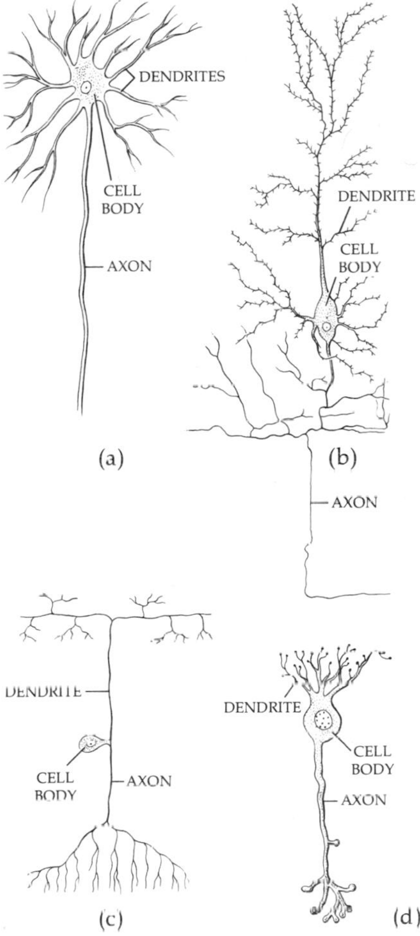

The cell bodies of neurons are often found in clusters. Clusters of neuron cell bodies within the brain and spinal cord are called *nuclei;* clusters of neuron cell bodies outside the brain and spinal cord are called *ganglia* (singular, ganglion).

Nerve bundles are composed of many nerve fibers from many neurons—usually hundreds and sometimes thousands. Each fiber is capable of transmitting separate messages, like the wires in a telephone cable.

ORGANS AND ORGAN SYSTEMS

Organs are composed of functionally related groups of tissue. The stomach, for example, is made up of layers of glandular epithelium (the stomach lining), connective tissue, nerves, blood vessels, and smooth muscle. Organ systems, in turn, generally comprise a number of organs that work together to accomplish a function necessary to maintain or continue the life of the organism. The stomach, for instance, is part of the digestive system. One of the major organ systems, the urinary system, is shown in Figure 31–14.

INTEGRATION AND CONTROL

The adult human body is composed of several hundred trillion cells of more than 100 kinds, grouped together in many different structural and functional units.

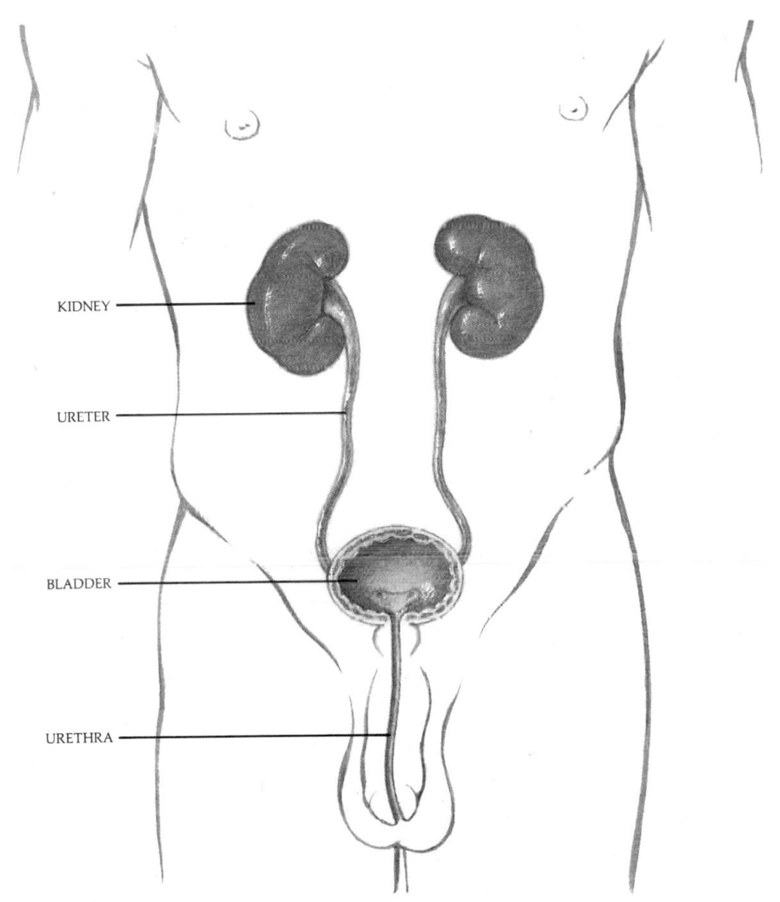

31–14
The urinary system is one of the major organ systems of the body. Fluids are processed in the kidneys, and waste products and water—the urine—pass along a pair of tubes, the ureters, to the bladder, where they are stored. Urine leaves the body by way of the urethra, which, in the male mammal, also serves as the passageway for semen.

Moreover, each cell is also working for itself, breaking down glucose, making ATP, building and maintaining membranes and organelles, producing enzymes and other protein molecules, dividing or not dividing. Yet in order for the organism as a whole to survive, all of these cells, tissues, and organs have to function as a whole.

For example, do you recall what it feels like to be in an absolute rage? The physical characteristics of a rage result from a simultaneous discharge of many nerve fibers. Certain of these cause the blood vessels in the skin and intestinal tract to contract; this contraction increases the return of the blood to the heart, raising the blood pressure and sending more blood to the muscles. The heart beats both faster and stronger, and the respiratory rate increases. The pupils dilate. The muscles underlying the hair follicles in the skin contract; this is probably a legacy from our furry forebears, which looked larger and more ferocious with their hair standing on end. The rhythmic movement of the intestine stops, and sphincters, the muscles at the end of the intestines and the opening of the bladder, relax; these reactions inhibit digestive operations, but the relaxing of the sphincters may also have the disconcerting consequence of causing involuntary defecation or urination. Hormones are produced that cause the release of large quantities of sugar from the liver into the bloodstream; this sugar provides an extra energy source for the muscles. The adrenal glands pour out epinephrine (adrenaline). As a consequence, the body as a whole is prepared for "fight or flight"—or, at least, action that would have been appropriate at some stage in our cultural evolution.

However, response to the external environment is only a small part of the organism's need for integration of its many individual functional units. Equally complex and vital, although less obvious, activities are required for the constant regulation of the internal environment. This maintenance of a relatively constant internal environment is known as *homeostasis*. As we noted in Chapter 1, homeostasis—"staying the same"—is one of the chief characteristics of living organisms. It is by regulating the internal environment that organisms have been able to free themselves to an increasing degree from the external environment, moving from salt water to fresh water, and from the water to the land and air.

Homeostasis is characteristic even of single cells and very simple organisms. In an organism as complex as *Homo sapiens,* it involves constant monitoring and regulation of a large number of different factors, such as oxygen and carbon dioxide, glucose and other nutrients, hormones, ions, pH, and temperature. Despite changes in the external environment and the organism's response to it, the concentrations of these substances in the body fluids remain relatively unchanged.

Virtually every cell, tissue, and organ in the body contributes in some way to this remarkable stability. The liver acts as a metabolic factory, removing organic molecules or adding them as they are needed. Oxygen for ATP production may be used rapidly, as during muscle exertion, or slowly, as during sleep, but the rates at which the lungs take in oxygen and the heart pumps blood are regulated so that the supply in the bloodstream available to the individual cells remains constant. The kidney processes and removes just the right amount of wastes, water, and salts—and so on, down the long list of organs, tissues, and cells. Therefore, although we shall examine the various organ systems of the body one by one—as it is conventional to do—it is important to remember that all are acting in concert and that the function of each is affected by all the others. In Chapter 34 we shall discuss the two major control agencies—the nervous system and the endocrine system—in more detail.

SUMMARY

We are vertebrates. As such, we have a bony, jointed supporting endoskeleton that includes a vertebral column enclosing the spinal cord and a skull. Our bodies contain a coelom that is divided by a muscular diaphragm into two compartments, the abdominal cavity and the thoracic cavity.

Our bodies are made up of cells of four principal tissue types: epithelial tissue, connective tissue, muscle, and nerve. Epithelium serves as a covering or lining for the body and its cavities. It also secretes mucus, perspiration, saliva, hormones, and digestive enzymes. Connective tissue strengthens and supports the other tissues of the body. It is characterized by intercellular substances. Muscle is composed of assemblies of contractile fibers. It converts chemical energy to kinetic energy. Nerve tissue includes neurons and glial cells. Neurons carry signals from one part of the body to another.

Tissues are groups of cells that are structurally and functionally similar. Various types of tissues are grouped in different ways to form organs, and organs are grouped to form organ systems.

The various cells, tissues, and organs of the body operate in an integrated manner so that the organism is capable of responding as a whole to its external environment and of regulating its internal environment (homeostasis).

QUESTIONS

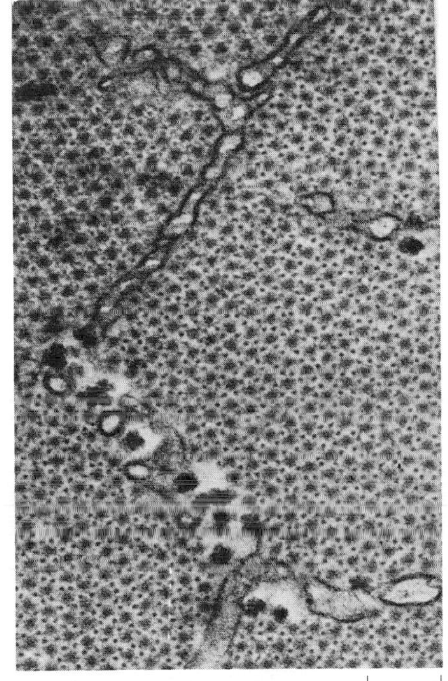

1. Distinguish among the following terms: animal/vertebrate/mammal/*Homo sapiens;* diaphragm/coelom; endoskeleton/exoskeleton; tendon/ligament; homeothermy/homeostasis; smooth muscle/striated muscle; muscle cell/myofibril; actin/myosin; sarcomere/sarcolemma; Z line/A band; nerve cell/neuron/glial cell.

2. What is the functional significance of each of the four tissue types? Give an example of each type.

3. What are the three types of epithelial tissue? What is the basis for classification of epithelial cells?

4. In addition to protection, what is the other major function of epithelial tissue?

5. What is the major structural difference between connective tissue and epithelial tissue? What functions of connective tissue account for this structural difference?

6. Discuss what is meant by antagonistic paired muscles.

7. Diagram a sarcomere, in a relaxed state, and label your drawing.

8. How does the structure you diagrammed in Question 7 enable muscle to perform its primary function?

9. Give some examples of homeostasis. Why is it important?

10. Can you interpret this electron micrograph? (You may be able to find the clue you need on page 576.)

CHAPTER 32

The Continuity of Life: Reproduction

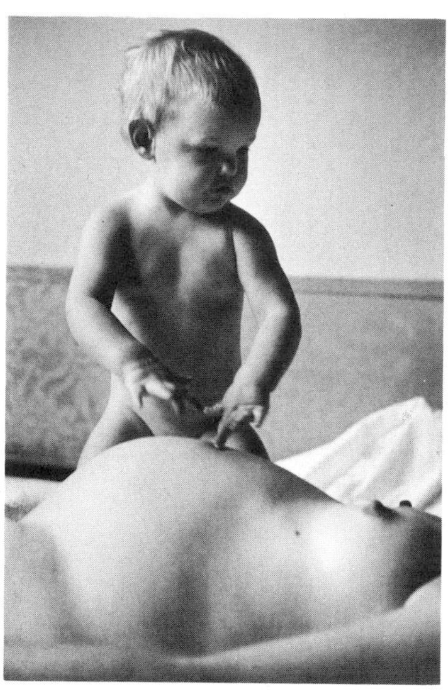

32–1
Mother and child.

When does a human life begin? When sperm encounters egg? When the fetus "quickens" in the womb? When the infant becomes viable as an independent entity? In the past, these matters were discussed by philosophers and theologians who were concerned with the question of when the soul enters the body. These issues have been revived in the ethical and legal controversies concerning abortion. In the evolutionary sense, however, none of these events marks the beginning of life. Life began more than 3 billion years ago and has been passed on since that time from organism to organism, generation after generation, to the present, and stretches on into the future, farther than the mind's eye can see. Each new organism is thus merely a temporary participant in the continuum of life. So is each sperm, each egg, indeed, in a sense, each living cell. From the viewpoint of the species, the continuum is shorter, of course—perhaps as brief as 10,000 years. And as for the individual, he or she is a unique blend of heredity and experience, never to be duplicated and therefore irreplaceable; but from the perspective of the biological continuum, an individual life lasts no longer than the blink of an eye.

In this and the following chapter, we are going to discuss the process by which parent organisms produce new individuals, maintaining the continuity of life and species, and how these new individuals develop to become reasonable facsimiles of their parent organisms. In the remaining chapters, we shall discuss how the individual organism maintains itself, solving, as an organism, the problems of survival by utilizing the solutions bequeathed it by its parents and other ancestors, even to its most distant forebears.

PATTERNS OF ANIMAL REPRODUCTION

Asexual reproduction, as we have seen, is the only form of reproduction among prokaryotes and some protists. Vertebrates, however, usually reproduce sexually. As you will recall,* sexual reproduction involves two events: meiosis and fertilization. In vertebrates, which are almost always diploid, meiosis produces gametes, the only haploid forms in the life cycle. The gametes are specialized for motility (sperm) or for storage of food (eggs) and are produced by separate individuals. In many lower organisms (insects, in particular), the generations are nonoverlapping,

* Or turn back to page 210 for a reminder.

32–2
Sexual reproduction always involves meiosis and fertilization. Fertilization may be external, as in frogs (a), *or internal, as in reptiles* (b), *birds* (c), *and mammals* (d).

as is the case in annual plants; but in vertebrates, parents not only survive after their young are produced but often are essential to the rearing of the young. This evolutionary trend toward increasing parental care becomes pronounced among birds and reaches its fullest expression among the mammals, as we noted in the previous chapter.

In most species of fish and in amphibians, as in many invertebrates, fertilization is external. Among organisms that produce amniote eggs (reptiles, birds, and monotreme mammals), fertilization is internal. The outer protective shell, produced as the egg moves down through the female reproductive tract, is laid down after the egg cell is fertilized, enclosing the embryo and its membranes. Fertilization is also internal among marsupial and placental mammals, in which the embryo develops within the mother and is nourished by her.

In the following pages, we shall describe sexual reproduction in humans. First, we shall trace the development of the male gametes, then of the female gametes, and we shall describe the special structures and activities that make fertilization and the subsequent implantation of the fertilized egg possible.

THE MALE REPRODUCTIVE SYSTEM

The primary sexual organs are the *gonads,* the organs in which the gametes are formed. The male gonads are the *testes* (singular, testis). The testes develop in the abdominal cavity of the male embryo and, in the human male, descend into an external sac, the *scrotum*. This descent usually occurs before birth. The function of the scrotum, apparently, is to keep the testes cool. A temperature 3°C lower than that of the body is necessary for the sperm to develop. Sperm are not produced in an undescended testis, and even temporary immersion of the testes in warm water—as in a hot bath—has been known to produce temporary sterility. (This effect is not reliable enough, however, to recommend its use as a birth control measure.) When the temperature outside the scrotal sac is warm, the sac is thin and hangs loose in multiple folds. When the outside temperature is cold, the muscles under the skin of the scrotum contract, drawing the testes close to the body. In this way, a fairly constant testicular temperature is maintained.

Spermatogenesis

Each testis is subdivided into about 250 compartments (lobes), and each of these is packed with tightly coiled seminiferous ("seed-bearing") tubules. These are the sperm-producing regions of the testes. Between the tubules are the interstitial cells that produce androgens, the male hormones, of which testosterone is the most important. Each seminiferous tubule is about 80 centimeters long, and the two testes contain a total of about 250 meters of tubules. The sperm are produced continuously within the tubules (Figure 32–4).

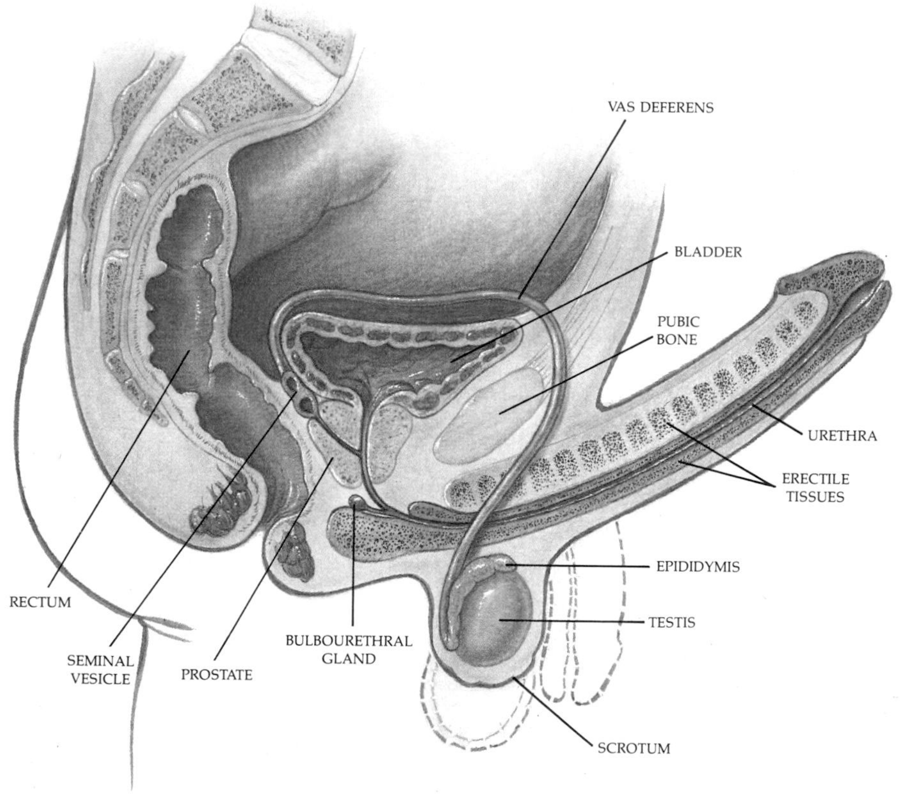

32–3
Diagram of the human male reproductive tract, showing the penis and scrotum before (dashed lines) and during erection. Sperm cells formed in the seminiferous tubules of the testis enter the vas deferens, which connects with a duct from the seminal vesicle and then within the prostate gland joins the urethra. The sperm cells are mixed with fluids, mostly from the seminal vesicles and prostate gland. The resulting mixture, the semen, is released through the urethra of the penis. The urethra is also the passageway for urine, which is stored in the bladder.

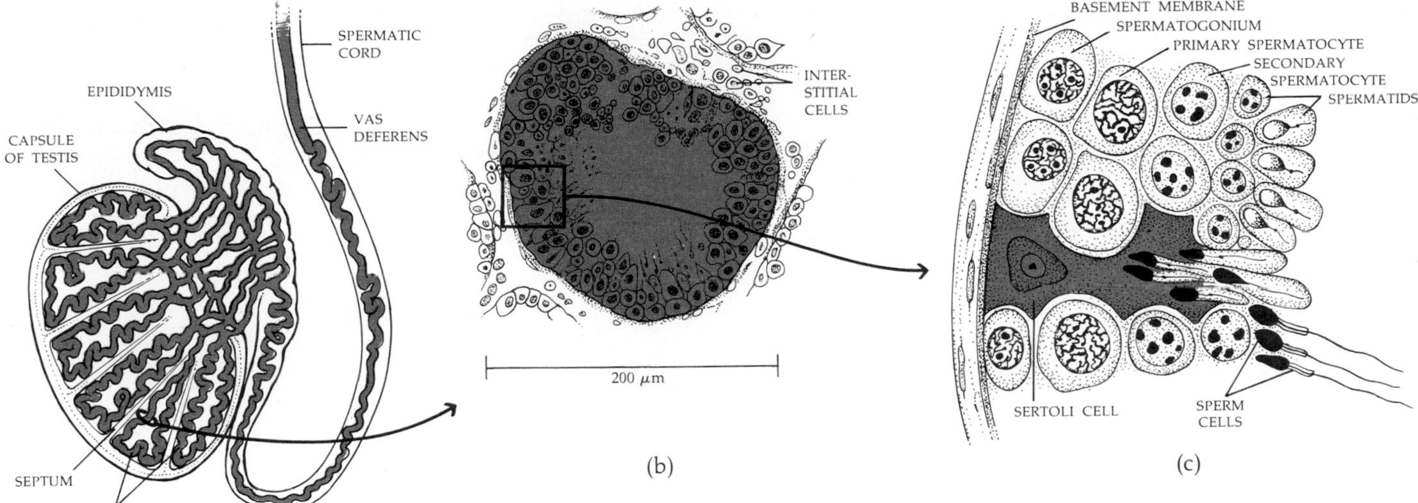

32–4
The testis (a) is made of tightly packed coils of seminiferous tubules (b) containing sperm cells in various stages of development. The entire developmental sequence takes eight to nine weeks. As shown in the idealized cross section (c), spermatogonia develop into cells known as primary spermatocytes. These divide (first meiotic division) into two equal-sized cells, the secondary spermatocytes. In the second meiotic division, four equal-sized spermatids are formed. These differentiate into functional sperm. (It is unusual to find all these stages in a single cross section.) The Sertoli cells support and nourish the developing sperm. The interstitial cells, which are found in connective tissues between the tubules, are the source of the male hormone testosterone. The sperm cells leave the testis through the epididymis and the vas deferens.

The tubules contain two types of cells: spermatogenic (sperm-producing) cells and Sertoli cells. The spermatogenic cells pass through several stages of differentiation. Once production of sperm begins at puberty in the human male, it goes on continuously and so, in a single seminiferous tubule, it is possible to find cells in all the different stages of spermatogenesis. It is unusual, however, to find all of the stages in a single cross section, because spermatogenesis characteristically occurs in waves that travel down the tubules.

The earliest (first-stage) cells are spermatogonia, which line the basement membrane of the tubules. Cells in the bottom layer divide continuously; some of the cells produced by these divisions remain undifferentiated in this first layer, while others, in the course of successive divisions, move away from the basement membrane and begin to differentiate, giving rise to primary spermatocytes. Spermatogonia are diploid and have, in the human, 44 autosomes and 2 sex chromosomes, an X and a Y; they divide by mitosis to produce primary spermatocytes. Primary spermatocytes undergo the first meiotic division* to produce two secondary spermatocytes, each of which contains 22 autosomes and either an X chromosome or a Y chromosome; each of the 23 chromosomes consists of two chromatids. The secondary spermatocytes undergo the second meiotic division to produce spermatids, each of which contains the haploid number of single chromosomes. It takes eight to nine weeks for a spermatogonium to differentiate into four sperm cells (spermatozoa). During this time the developing cells receive nutrient material from adjacent Sertoli cells.

Differentiation of Spermatids

A spermatid is a small spherical or polygonal ("many-sided") cell that develops, without further division, into a sperm cell. The sequence of changes by which the quite unremarkable spermatid becomes the highly specialized, very extraordinary sperm cell is an excellent example of cell differentiation. Differentiation is a process that, as we shall see, plays an important role in many stages of embryonic development.

* For a review of meiosis, see pages 213 to 215.

32–5
A mammalian spermatid in the process of differentiation. The nucleus almost fills the right half of the electron micrograph. The two centrioles have moved to a position just behind the nucleus, and the flagellum has begun to form from one of them. To the left of the nucleus are several mitochondria.

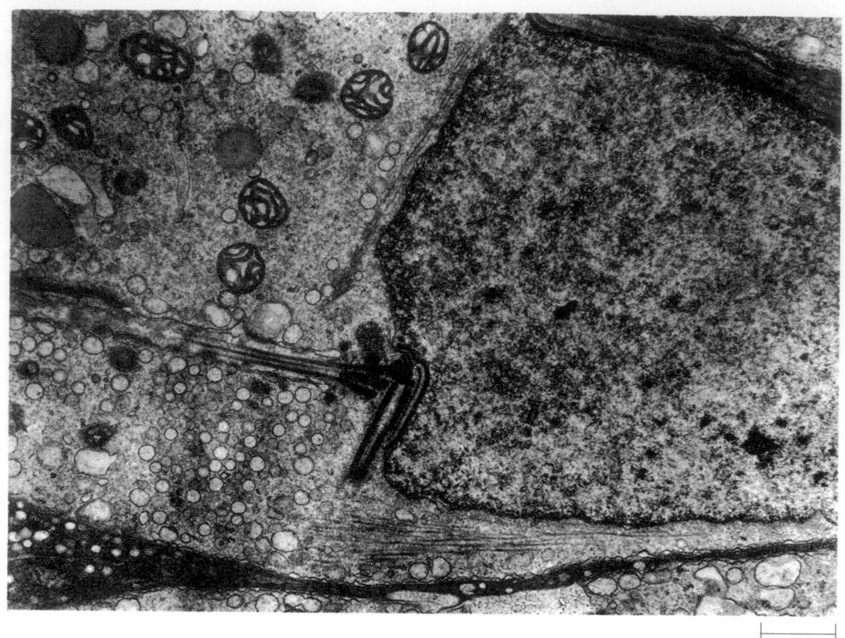

The first visible sign of differentiation of a spermatid is the appearance of vesicles containing small, dark granules within the Golgi body, which, as in most cells, lies close to the nuclear envelope. These vesicles enlarge and coalesce into a single vesicle, the acrosome. The acrosome contains enzymes that will help the sperm to penetrate the protective layer surrounding the egg. The position of the acrosomal vesicle determines the polarity of the sperm; that is, it establishes where the anterior end, or "head," is going to be.

During the early stages of acrosome formation, the cell's pair of centrioles moves to the outer cell membrane at the end of the cell opposite the acrosome. From one of these centrioles a fine thread grows out, the beginning of the sperm flagellum. The centrioles then move back to the nucleus, carrying the cell membrane inward with them. One centriole eventually lodges in a notch formed at the posterior end of the nucleus, while the other, the one associated with the flagellum, lies at right angles to the first, as shown in Figure 32–5.

As the tail grows, it becomes apparent that its axial filament has the characteristic 9 + 2 structure found in cilia and flagella generally (Figure 32–6). Mitochondria aggregate about its basal end, forming a continuous spiral, providing a ready energy source (ATP) for the flagellar movement. The rest of the axial filament, almost to its tip, is surrounded by nine additional protein fibers tightly coiled in a helix. These fibers, which are somewhat thicker than the microtubules within the flagellum, presumably play some role in sperm motility.

During this period, the nucleus condenses, apparently by eliminating water. Once the tail has formed, the cell rapidly elongates. Longitudinal bundles of microtubules can be seen in the cell at this time, and they may play a role in changing cell shape. As the cell lengthens, the bulk of the cytoplasm, together with the Golgi body, is sloughed away.

32–6

Diagram of a human sperm cell. The mature cell consists primarily of the nucleus, carrying the "payload" of tightly condensed DNA and associated protein, the very powerful tail (flagellum), and mitochondria, which provide the power for sperm movement. The acrosomal vesicle is a specialized lysosome (see page 111).

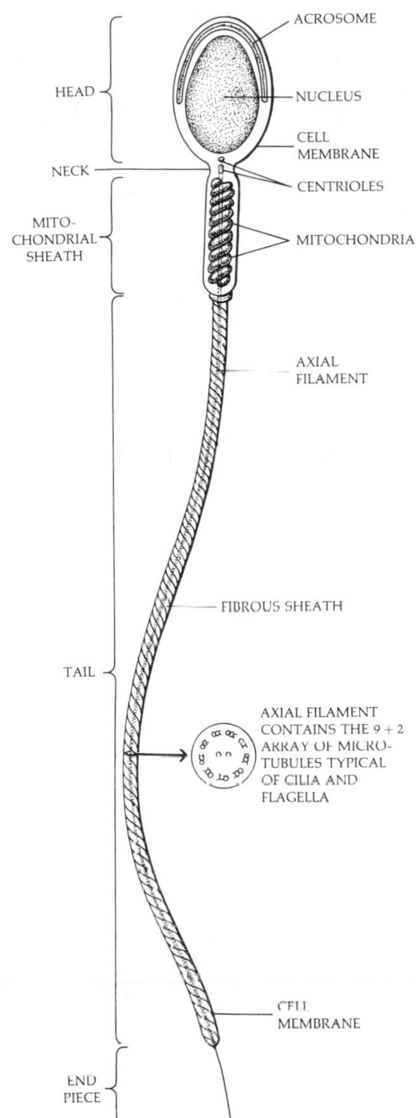

In its final form, the sperm cell consists of the acrosome, the tightly condensed nucleus, the mitochondria, a pair of centrioles, one of which is now serving as the basal body for the flagellum, and the long, powerful flagellum itself, all bounded by the cell membrane. In the fully differentiated sperm, all other functions have been subordinated to the task of providing motility for delivery of the "payload," the DNA and its associated protein, condensed and coiled in the sperm head. A young adult human male may produce several hundred million sperm per day; a ram may produce several billion.

Pathway of the Sperm

From the testis, the sperm are carried to the *epididymis,* which consists of a coiled tube 7 meters long, overlying the testis. It is surrounded by a thin, circular layer of smooth muscle fibers. (Testis and epididymis together constitute the testicle.) The sperm are immotile when they enter the epididymis and gain motility only after some 18 hours there. From the epididymis, the sperm pass to the *vas deferens,* where most of them are stored. The vas deferens (plural, vasa deferentia), an extension of the tightly coiled tubules of the epididymis, leads from the testicle into the abdominal cavity. Together with its accompanying nerves, arteries, and wrapping tissues, the vas deferens constitutes the spermatic cord. The spermatic cords, one from each testicle, are found along the same path the testes took during their descent in the embryo.

Within the posterior wall of the abdominal cavity, the vasa deferentia lead around the bladder where they merge with the ducts of the seminal vesicles and then enter the urethra. Each vas deferens is covered with a heavy, three-layered coat of smooth muscle whose contractions propel the sperm along it. Vasectomy (Figure 32–7) is gaining popular acceptance as a form of birth control. This is a relatively safe and almost painless procedure that involves cutting and tying off the vasa deferentia. The nerves and blood vessels of the sperm cord are left intact.

The urethra, which runs through the prostate gland and terminates at the tip of the penis, serves both for the excretion of urine and the ejaculation of sperm.

32–7

During a vasectomy, the vas deferens on each side is severed, and the cut ends are folded back and tied off, preventing the release of sperm from the testicle. The sperm cells are reabsorbed by the body, and the semen is normal except for the absence of sperm. The procedure seems to produce no effect at all on testosterone levels, sexual potency, or performance.

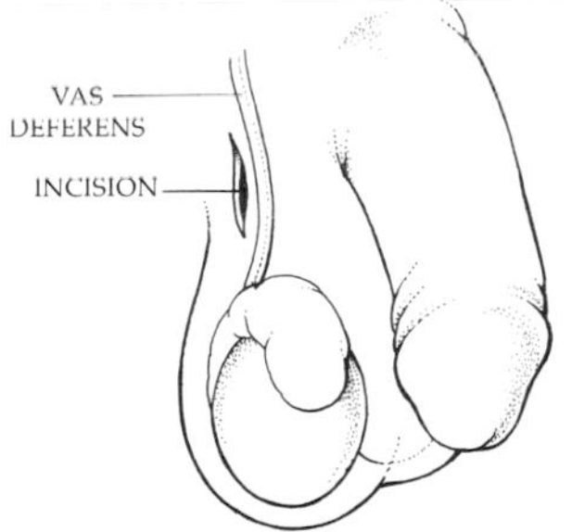

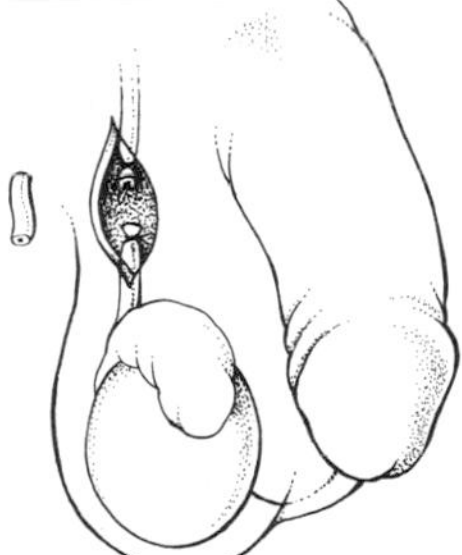

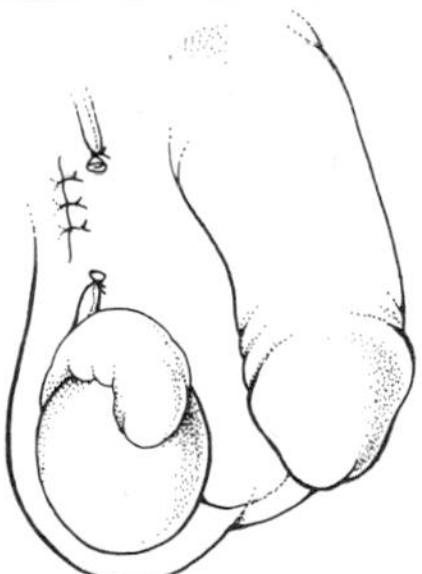

32–8

A cross section of a human penis. The penis is formed of three cylindrical masses of spongy erectile tissue that contain a large number of small spaces, each about the size of a pinhead. Erection of the penis is caused by dilation of the blood vessels carrying blood to the spongy tissues.

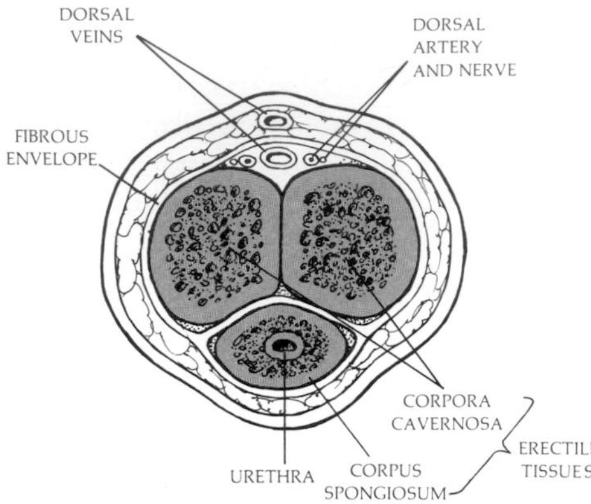

The Penis

The function of the penis is to deposit sperm cells within the reproductive tract of the female. The penis, in various forms, has evolved independently in a number of species of insects and in other invertebrates. It is found among some reptiles and birds; all flightless birds have penes and so do all ducks, flightless or not. In most reptiles and birds, however, one opening, the cloaca, serves as the passage for eggs or sperm and also for the elimination of wastes. These animals mate by juxtaposition of their cloacae. Only among mammals is the penis found in all species.

The human penis is formed of three cylindrical masses of spongy erectile tissue, each of which contains a large number of small spaces, each about the size of a pinhead. Two of these erectile masses form the upper two-thirds of the penis; the third lies beneath them, surrounding the urethra (Figure 32–8). This latter enlarges at the distal (far) end to form the glans penis, which forms a smooth protective cap over the spongy tissues. At the basal end, the mass enlarges to form the bulb of the penis, which is embedded below the pelvic cavity and is surrounded by muscles that participate in orgasm. The exterior portion of the penis is covered by a loose, thin layer of skin, which at the end forms an encircling fold over the glans. This fold, the foreskin, is sometimes surgically removed (circumcision). The urethra terminates in a slitlike opening in the glans.

Erection of the penis, which can be elicited by a variety of stimuli, is caused by dilation of the blood vessels carrying blood to the spongy tissues, resulting in the collection of blood within these spaces. As the tissues become distended, they compress the veins and so inhibit the flow of blood out of the tissues. With continued stimulation, the penis and the underlying bulb become hard and enlarged.

As the sperm move toward the urethra, the seminal vesicles secrete a fructose-rich fluid that nourishes the sperm cells and contains a high concentration of the fatty acid compounds prostaglandins. The prostaglandins cause contractions in the musculature of the uterus and oviducts and so may assist the sperm in reaching the egg. The prostate gland adds a thin, milky, alkaline fluid that helps neutralize the normally acid pH of the female reproductive tract. Sperm and the secretions of the testes make up only about 10 percent of the semen that is ejaculated.

Erection is accompanied by discharge from the bulbourethral glands (pea-shaped organs at the base of the penis) of a small amount of fluid. This serves as a lubricant to facilitate the movement of spermatozoa along the male urethra and to aid penetration of the penis into the female. Continued stimulation, such as may be produced by repeated thrusting of the penis in the vagina, characteristically leads to contraction of the muscles in the scrotum (raising the testes close to the body) and of the muscles encircling the epididymis and vas deferens. These contractions move the semen into the urethra. Finally, the muscles surrounding the bulb are stimulated. These muscle contractions propel the semen out through the urethra and produce some of the sensations associated with orgasm.

The sperm, along with the secretions from the seminal vesicles, the prostate gland, and the bulbourethral glands, constitute semen. The volume of semen measures from 3 to 6 milliliters per ejaculation.

Even though sperm constitute less than 10 percent of the semen, about 300 to 400 million of them are present in each ejaculate of a normal young adult male. Of those, only one can fertilize each egg cell. The rest die within three days, retaining

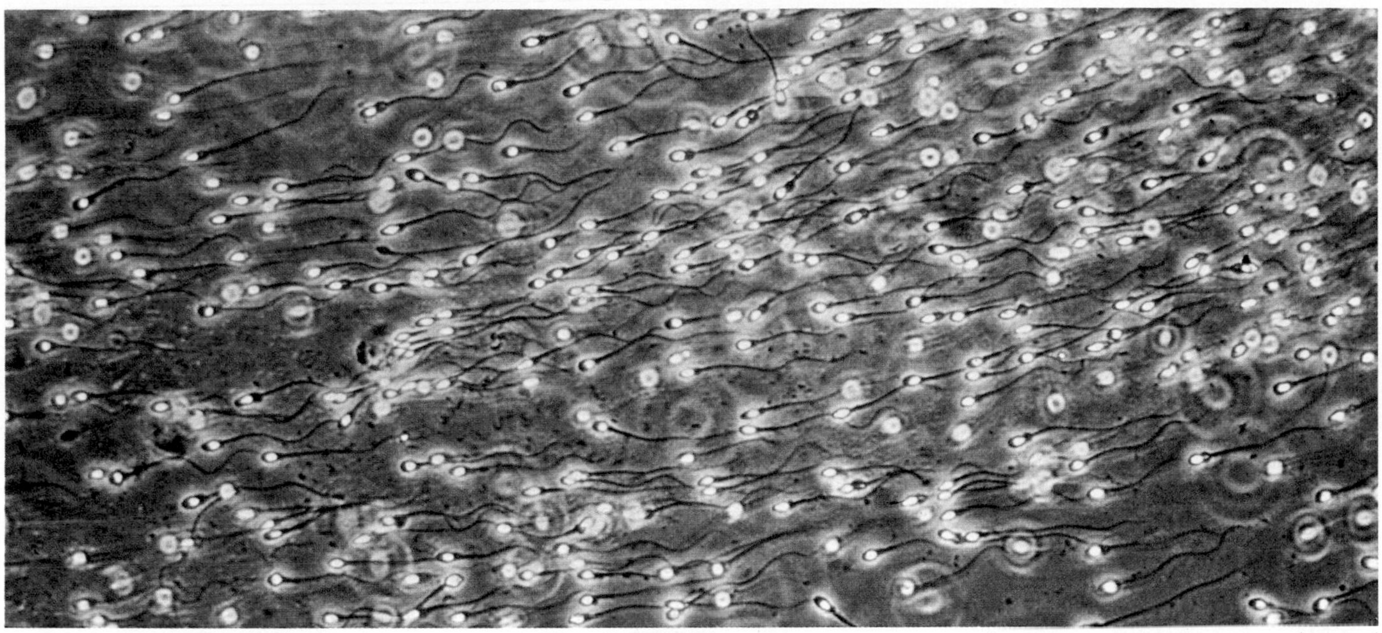

32-9
Human sperm. About 300 to 400 million sperm cells are present in the ejaculate of an average, healthy adult male.

their fertilizing ability for only about 48 hours. The other sperm cells apparently play an accessory role, however, perhaps by bringing about chemical changes necessary for fertilization. Males that produce fewer than 20 million sperm per milliliter of fluid are generally sterile.

The Role of Hormones

Hormones, as you will recall, are substances that are produced in one tissue and transported to another, where they exert highly specific effects. The principal male sex hormone is testosterone, produced primarily by the interstitial cells of the testes. Male hormones, including testosterone, are known collectively as *androgens.* Androgens produced in the male fetus are important in the development of the external genitalia. When the human male is about 10 years old, increased androgen production is associated with the enlargement of the penis and testes and also the prostate, the seminal vesicles, and other accessory organs. Androgens also affect other parts of the body not directly involved in the production and deposition of sperm. In the human male, these effects include growth of the larynx and an accompanying deepening of the voice, muscle development, skeletal size, and distribution of body hair. Androgens stimulate the biosynthesis of proteins and so of muscle tissue. They stimulate the apocrine sweat glands, whose secretion attracts bacteria and so produces body odors associated with sweat after puberty. And they may cause the sebaceous glands of the skin to become overactive, resulting in acne. Such characteristics, associated with sex hormones but not directly involved with reproduction, are known as secondary sex characteristics. Puberty in the male is generally considered as the time of onset of sperm production.

32–10
Secondary sex characteristics.

In other animals, the androgens are responsible for the lion's mane, the powerful musculature and fiery disposition of the stallion, the cock's comb and spurs, and the bright plumage of many adult male birds. They are also responsible for a variety of behavior patterns such as the scent marking of dogs, the courting behavior of sage grouse, and various forms of aggression toward other males found in a great many vertebrate species.

Since almost the beginnings of agriculture, domestic animals have been castrated in order to make them fatter, less tough, and more manageable. Eunuchs (human castrates) were traditionally, and for obvious reasons, used as harem guards, and as recently as the eighteenth century, selected boys were castrated before puberty in order to retain the purity of their soprano voices for church and opera choirs. As a consequence of these early practices, the effects of testosterone—or more precisely of its absence—were the first hormonal influences to be recognized and studied.

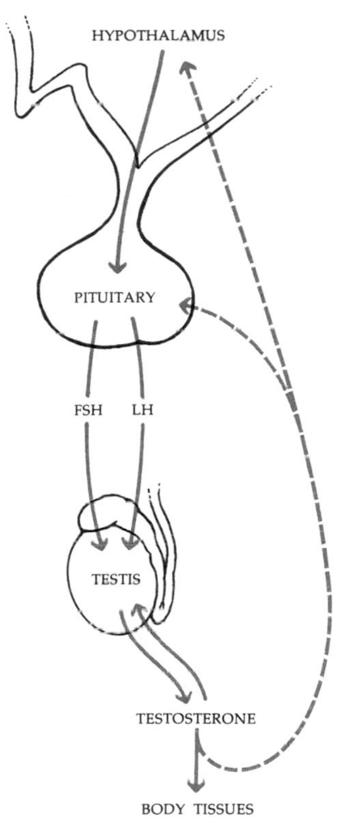

32–11
Production of the male sex hormone testosterone is controlled by a negative feedback system. The hypothalamus, a brain center, regulates the pituitary gland's production of the gonadotropin LH. LH stimulates production of testosterone by the interstitial cells of the testes. As testosterone increases to a certain level of concentration in the blood, it inhibits the production of LH. Decreased levels of LH result in decreased production of testosterone. FSH, another gonadotropin secreted by the pituitary under the influence of the hypothalamus, stimulates sperm production.

Regulation of Hormone Production

The production of testosterone (Table 32–1) is regulated by a gonadotropic (gonad-stimulating) hormone called luteinizing hormone (LH), produced by the pituitary gland. The pituitary, which is about the size and shape of a kidney bean, is located at the geometric center of the human skull. It is lodged just beneath an area of the brain known as the hypothalamus, and the pituitary's production of gonadotropic hormones is controlled by this brain center. LH is carried by the blood to the interstitial tissues of the testes, where it stimulates the output of testosterone. When the blood level of testosterone reaches a certain concentration, it, in turn, inhibits the release of LH. This type of regulatory system, in which the system is shut off when the products of its operation reach a certain level, is known as negative feedback. The most familiar example of a negative feedback system is a thermostat that turns off a furnace when the temperature rises. Many examples of negative feedback are known, both in biological systems and in those designed by engineers.

Another pituitary hormone, also under the control of the hypothalamus, is follicle-stimulating hormone (FSH). It acts on the seminiferous tubules, stimulating the development of sperm cells. Testosterone is also required for sperm development.

In man the rates of hormone release and, therefore, the rates of spermatogenesis are fairly constant. In many animals, however, hormone production is triggered by changes in temperature or daylight or other environmental cues that presumably

Table 32–1 *Major Mammalian Gonadotropic and Sex Hormones in Males*

HORMONE	PRINCIPAL SOURCE	PRINCIPAL EFFECTS	CONTROL
FSH	Pituitary	Stimulates sperm production	Hypothalamus
LH	Pituitary	Stimulates testosterone production	Hypothalamus
Testosterone	Testes	Produces and maintains male sex characteristics, stimulates sperm production	LH

32–12
Rutting is the annual period of intense sexual activity characteristic of some animals, such as the red deer stags shown here. It is triggered by male hormones. During the period from spring to midsummer, while new antlers are growing, testosterone is almost undetectable. As soon as testosterone levels in the blood begin to rise, the antlers die, losing their nerve and blood supply and their "velvet" cover. In the hard antler stages that follow, stags have pushing contests in which they lock antlers. Antler size is a determinant of hierarchy within a group. In early fall, testosterone levels begin to soar, the stag's voice deepens, his neck and shoulder muscles thicken, and he sets out to corral a harem, which he defends until rutting season is over. Then he returns to the "stag line," his antlers drop off, and the testosterone in his blood becomes once more undetectable.

act upon some receptive center in the nervous system. The nervous system in turn relays the data to the hypothalamus, which sets in motion the chain of hormonal events just described. Recent studies on bulls, for example, showed that after a bull sees a cow, his blood level of LH rises as much as seventeenfold; within about half an hour, the blood level of testosterone reaches its peak. Human testosterone production has also been shown to fluctuate measurably in response to environmental stimuli.

THE FEMALE REPRODUCTIVE SYSTEM

The gamete-producing organs in the female are the *ovaries*, each a solid mass of cells about 3 centimeters long. They are suspended in the abdominal cavity by ligaments (bands of connective tissue) and mesenteries. The oocytes, from which the eggs develop, are in the outer layer of the ovary.

Oogenesis

In human females, the primary oocytes begin to form about the third month of fetal development. By the time of birth, the two ovaries contain some 2 million primary oocytes, which have reached prophase of the first meiotic division. These primary oocytes remain in prophase until the female matures sexually. Then, under the influence of hormones, the first meiotic division resumes and is completed at about the time of the release of the oocyte from the ovary (ovulation). Of these 2 million primary oocytes, about 300 to 400 reach maturity, usually one at a time about every 28 days after puberty, to become secondary oocytes, which develop into mature egg cells (ova). Therefore, as many as 50 years may elapse

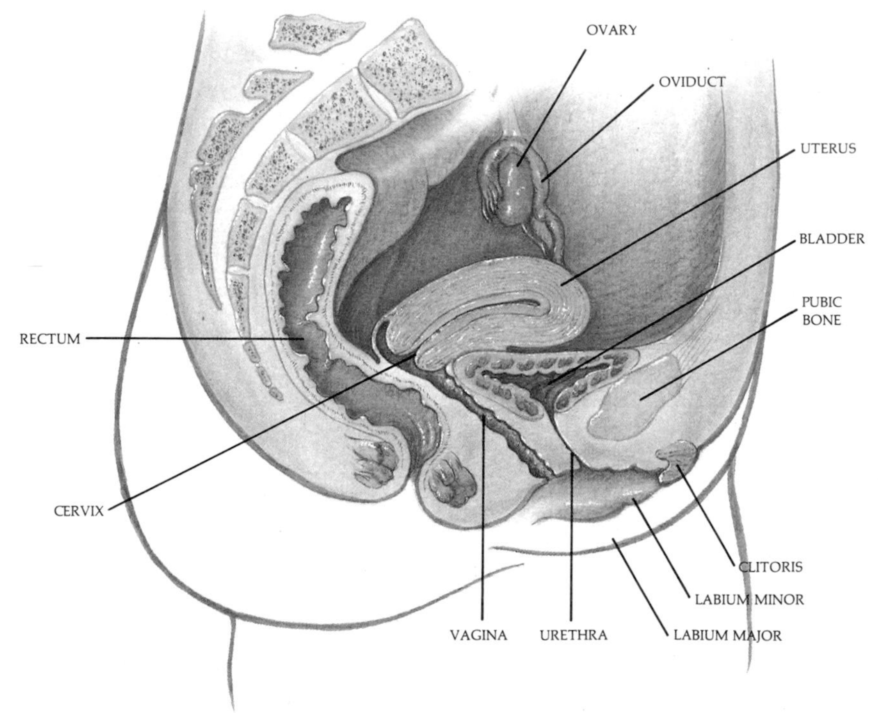

32–13
The female reproductive organs. Notice that the uterus lies at right angles to the vagina. This is one of the consequences of the bipedalism and upright posture of Homo sapiens *and one of the reasons that childbirth is more difficult for the human female than for other mammals.*

between the beginning and the end of the first meiotic division in a particular oocyte.

Maturation of the primary oocyte involves both meiosis and a great increase in size. This size increase reflects the accumulation of stored food reserves and metabolic machinery, such as messenger RNA and enzymes, required for the early stages of development. At meiosis an oocyte does not divide to form four ova. Instead, a single ovum and one to three polar bodies are formed (see Figure 11–12, page 217). When the oocyte is ready to complete meiosis, the nuclear envelope fragments, and the chromosomes move to the surface of the cell. As the nucleus divides, the cytoplasm of the oocyte bulges out. One set of chromosomes moves into the bulge, which then pinches off into a small cell, the first polar body. The rest of the cellular material forms the large secondary oocyte. The first meiotic division is completed a few hours before ovulation.

The second meiotic division does not take place until after fertilization. This division produces the ovum and another small polar body.

As a consequence of these unequal cell divisions, all the accumulated food reserves of the oocyte are passed on to a single ovum. The first polar body may also divide, although there is no functional reason for it to do so. All the polar bodies eventually die.

Oocytes develop near the surface of the ovary. An oocyte and the specialized cells surrounding it are known as an ovarian follicle. The cells of the follicle supply food to the growing oocyte and also secrete estrogens, hormones that initiate the buildup of the lining of the uterus (the endometrium). During the final stages of its growth, the follicle moves to the surface and produces a thin, blisterlike elevation that eventually bursts, releasing the oocyte (ovulation).

32–14
(a) *Oocytes develop near the surface of the ovary within follicles. After an oocyte is discharged from a follicle (ovulation), the remaining cells of the ruptured follicle give rise to the corpus luteum, which secretes estrogens and progesterone. If the ovum is not fertilized, the corpus luteum is reabsorbed in two to three weeks. If the ovum is fertilized, the corpus luteum persists, sustaining the production of estrogens and progesterone, which maintain the uterus during pregnancy.* (b) *Ovulation, as it occurs in the ovary of a rabbit.*

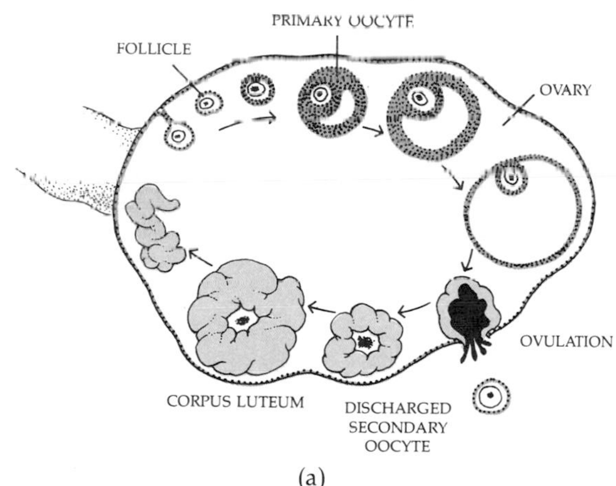

(a)

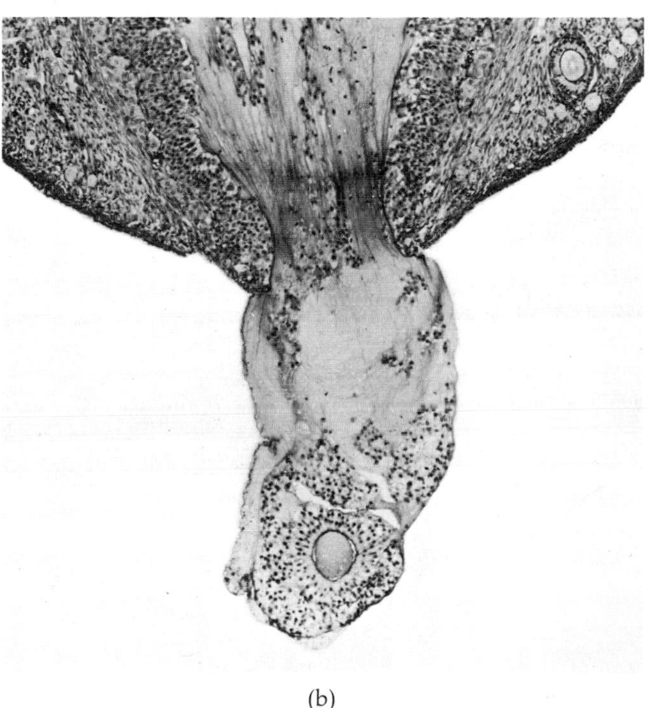

(b)

32–15
A human egg. The dark strands within the nucleus are chromosomes. A polar body is at the lower right. (Roberts Rugh and Landrum B. Shettles, M.D., From Conception to Birth: The Drama of Life's Beginnings, *Harper & Row, Publishers, Inc., New York, 1971.)*

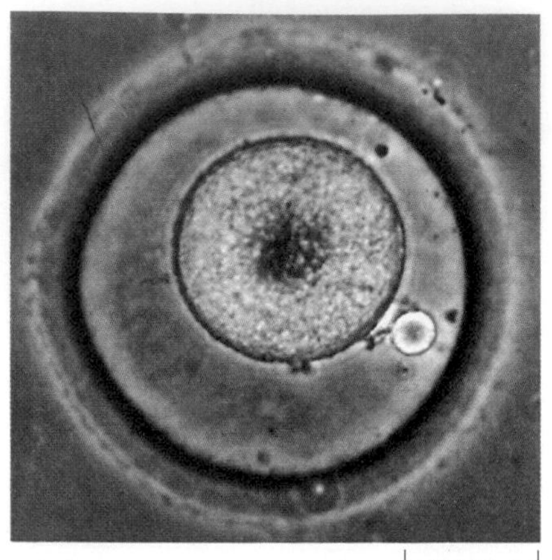

A human oocyte is about 100 micrometers in diameter, which is very large for a cell. It contains an unusually large supply of ribosomes, enzymes, amino acids, and all the other cellular machinery that will be used in the early, rapid stages of biosynthesis characteristic of embryonic cells.

Pathway of the Egg

When the oocyte is released from the follicle at ovulation, it is swept into the adjacent oviduct (sometimes called a Fallopian tube) by the movement of the funnel-shaped opening of the oviduct over the surface of the ovary and by the beating of cilia that line the fingerlike projections surrounding this opening.

The smooth muscles in the walls of the oviducts contract in continuous peristaltic waves that move the oocyte from the ovary to the uterus in about three days. An unfertilized oocyte, however, lives only about 24 hours after it is ejected from the follicle. So, fertilization, if it is to occur, must occur in an oviduct. If the egg cell is fertilized, it becomes implanted in the endometrium three to four days after the young embryo reaches the uterus, six or seven days after the egg was fertilized. If the egg cell is not fertilized, it dies, and the endometrial lining of the uterus is shed at menstruation. Fertilized eggs implanted in the endometrium are sometimes lost in abnormal menstrual flow, but it is difficult to estimate the number of such very short-lived pregnancies.

Sterilization in women is usually carried out by severing the oviducts, thus preventing sperm from meeting the egg and the passage of the egg to the uterus. Tubal ligation ("tying off the tubes"), as the procedure is called, usually does not produce hormone changes or prevent the menstrual cycle.

32–16
Fertilization of the egg by sperm. Once a month in the nonpregnant female of reproductive age, an oocyte is ejected from the ovary and is swept into one of the oviducts. Fertilization, when it occurs, normally takes place within an oviduct, after which the fertilized egg becomes implanted in the lining of the uterus. Muscular movements of the oviduct, plus the beating of the cilia that line it, propel the egg cell down the tube toward the uterus. If the egg cell is not fertilized, it dies, usually within 12 to 24 hours. A sperm cell has an average life of 48 hours within the female reproductive tract.

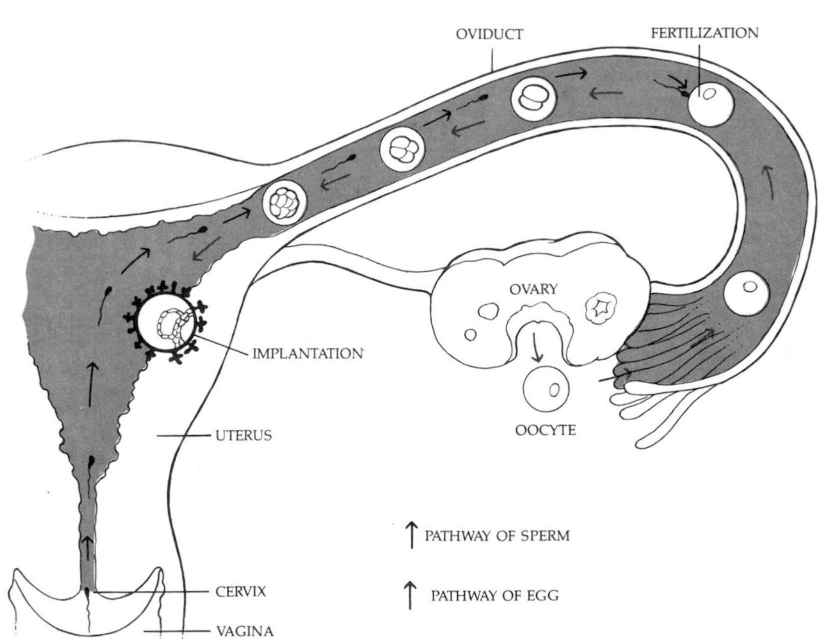

32–17

Implantation. The tiny embryo invades the lining of the uterus within a week after fertilization. Subsequently, the placenta begins to form; this organ is the source of hormones that help to maintain pregnancy. Implantation usually occurs three to four days after the young embryo reaches the uterus.

Hormonal Regulation in Females

Like spermatogenesis, oogenesis is under hormonal control. Unlike spermatogenesis in the human male, however, oogenesis in all vertebrate females is cyclic. It involves an interplay of hormones, including estrogens, progesterone, and the two gonadotropic (gonad-stimulating) hormones, follicle-stimulating hormone (FSH) and luteinizing hormone (LH). The timing and control of the cycle rest in the hypothalamus.

Estrogens are the female sex hormones. A variety of them, of which estradiol is probably the most important, are produced by the ovarian follicles under the stimulation of FSH. The production of estrogens inhibits the secretion of FSH in a feedback system similar to that which controls the production of sex hormones in males. Estrogens stimulate the development of the breasts and the external genitalia, and the distribution of body fat. Both estrogens and progesterone are required to prepare the endometrium for the implantation of the embryo; neither can do the job alone. This cooperation, or synergism, in which two or more agents act together to produce an effect that is greater in magnitude than the sum of the effects produced separately, is characteristic of many hormonal responses.

Synthetic estrogens have been prepared and are usually more potent than the natural product. For example, DES, diethylstilbestrol, which is used in the "morning-after pill," is a potent synthetic estrogen. It is believed to prevent implantation of the embryo in the endometrium. When administered late in pregnancy, however, it helps prevent spontaneous abortion. Recently this use has been discontinued because of the increased incidence of cervical cancer in women whose mothers were treated with DES late in pregnancy. For a time DES was used commercially to fatten cattle, but this use is now forbidden by government regulation since the hormone was found to cause cancer in experimental animals when administered in high doses.

The Menstrual Cycle

The beginning of the first menstrual cycle (menarche) marks the onset of puberty in human females. The average age of onset is $13\frac{1}{2}$, but the normal range is very wide. Puberty in the female begins on the average about a year and a half before puberty in the male. Puberty in the female is usually preceded by the appearance of the secondary sex characteristics, such as pubic and axillary hair and enlargement of the breasts.

Although the menstrual cycle does not require an environmental cue, as do reproductive cycles in many other vertebrates, it is clearly under the influence of external factors, to some extent. For example, some women find that emotional

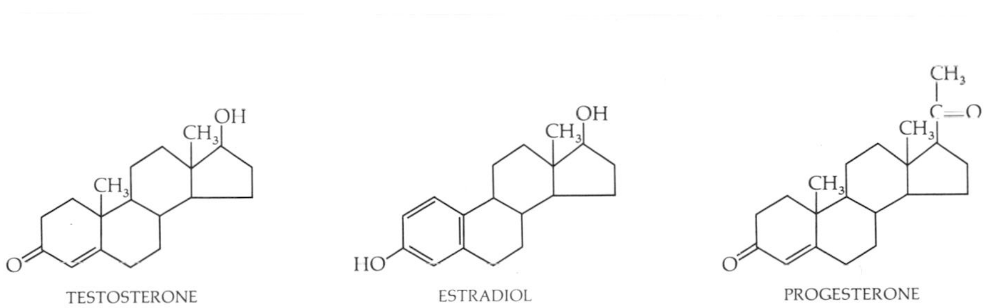

32–18

The chemical structures of testosterone, estradiol, and progesterone. Note that the hormones differ only very slightly chemically, in contrast to the great differences in their physiological effects—another example of the extreme specificity of biochemical actions. All of these belong to a group of chemicals known as steroids, characterized by the four-ring structure shown here.

TESTOSTERONE ESTRADIOL PROGESTERONE

upset delays a menstrual period or eliminates it completely. A similar mechanism may be responsible for the sterility that occurs among rodents, for example, when living in very crowded conditions or under other forms of stress. Human females living in groups—as in college dormitories—are familiar with the tendency of the menstrual cycles of the group to become synchronized. The mechanism for this is unknown, but it has been suggested that it may be a result of the exchange of a pheromone* among the individuals involved.

The menstrual cycle (Figure 32–19) begins with the casting off of the outer layer of endometrium (menstruation). After the menstrual flow ceases and under the influence of FSH and LH, another egg cell and its follicle begin to mature, and the follicle secretes increased amounts of estrogens. (Usually a number of follicles begin to enlarge simultaneously, but only one becomes mature enough to release its ovum, and the others regress.) The estrogens stimulate the regrowth of the endometrium. The rapid rise in estrogens near the midpoint of the cycle triggers a sharply increased production of LH by the pituitary gland (an example of positive feedback). Paradoxically, toward the end of the cycle, LH and FSH production decline as a result of the increased concentration of progesterone and estrogens (negative feedback).

* For another discussion of pheromones, see page 470.

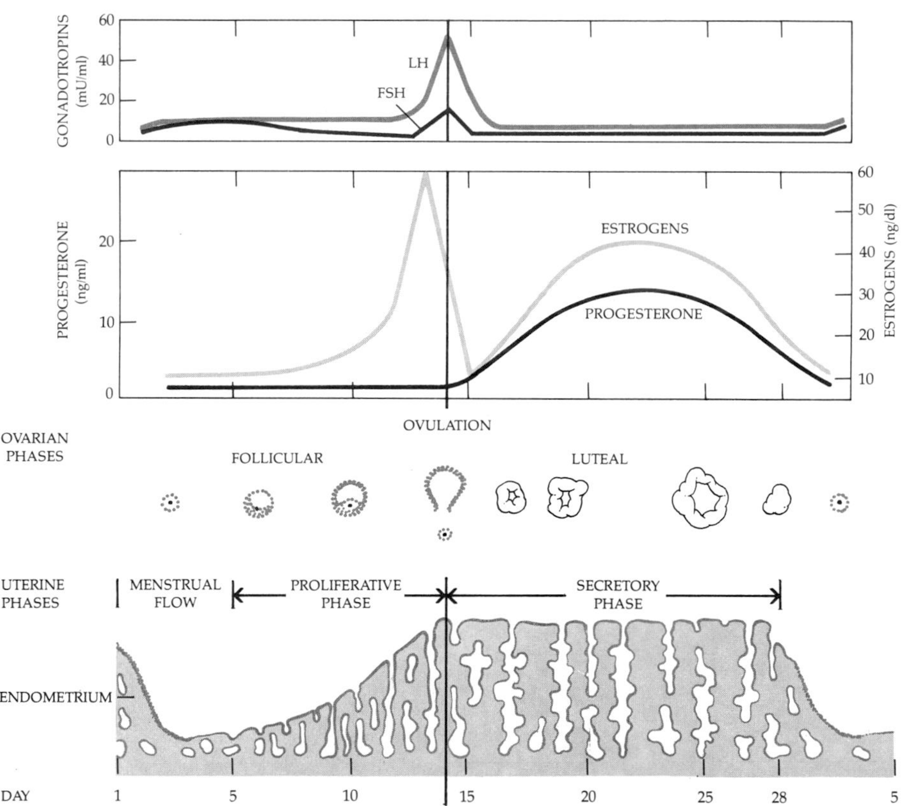

32–19
Diagram of events taking place during the menstrual cycle. The cycle begins with the first day of menstrual flow, which is caused by the shedding of the endometrium, the lining of the uterine wall. The increase of FSH and LH during the first week promotes the growth of the ovarian follicle and its secretion of estrogens. Under the influence of estrogens, the endometrium regrows. A sharp increase of LH from the pituitary about midcycle stimulates the release of the egg cell (ovulation). (It is not known what role, if any, is played by the simultaneous increase in FSH.) Following ovulation, LH and FSH levels drop. The follicle is converted to the corpus luteum, which secretes estrogens and also progesterone. Progesterone further stimulates the endometrium, preparing it for implantation. If pregnancy does not occur, the corpus luteum degenerates, the production of progesterone and estrogens falls, the endometrium begins to slough off, FSH and LH concentrations increase once more, and the cycle begins anew.

Table 32–2 *Major Mammalian Gonadotropic and Sex Hormones in Females*

HORMONE	PRINCIPAL SOURCE	PRINCIPAL EFFECTS	CONTROL
FSH	Pituitary	Stimulates growth of ovarian follicle, stimulates estrogen production	Hypothalamus
LH	Pituitary	Stimulates release of egg cell, stimulates progesterone production	Hypothalamus
Estrogens	Ovary, placenta	Produce and maintain female sex characteristics, thicken lining of uterus	FSH
Progesterone	Ovary, placenta	Further prepares uterine lining for pregnancy, inhibits uterine movements, promotes development of milk ducts	LH

All these events take place as a result of a shifting balance of hormones. The spurt of high LH stimulates the follicle to release the egg cell, which begins its passage to the uterus. Under the continued stimulus of LH, the cells of the emptied follicle grow larger and fill the cavity, producing the corpus luteum ("yellow body"). The cells of the corpus luteum, as they increase in size, begin to synthesize progesterone as well as estrogens. As the estrogen and progesterone levels increase, they inhibit the production of the gonadotropic hormones from the pituitary. Production of ovarian hormones then drops. The lining of the uterus can no longer sustain itself without hormonal support, and a portion of it is sloughed off in the menstrual fluid. Then, in response to the low level of ovarian hormones, the level of pituitary gonadotropic hormones begins to rise again, followed by development of a new follicle and a rise in estrogens as the next monthly cycle begins.

The cycle usually lasts about 28 days, but individual variation is common. Moreover, even in women with cycles of average length, ovulation does not always occur at the same time in the cycle.

Fertility pills for women contain either gonadotropins or a synthetic compound that decreases estrogen concentrations in the blood, stimulating the production of FSH by the pituitary. Multiple births may occur when several ova are released simultaneously as the result of such treatments.

Uterus, Vagina, and Vulva

The uterus is a hollow, muscular, pear-shaped organ about 7.5 centimeters long and 5 centimeters wide. It is lined by the endometrium, which has two principal layers, one of which is shed at menstruation and another from which the shed layer is regenerated. The smooth muscles in the walls of the uterus move in continuous waves that are more frequent at the broad fundus, or base, and less active at the narrower cervical end of the uterus. This motion possibly increases the motility of both the sperm on its journey to the oviduct and the oocyte as it passes from the oviduct to the uterus. The contractions increase when the endometrium is shed during a menstrual period and reach greatest strength when a woman is in labor.

The uterus lies almost horizontally in the abdominal cavity and is on top of the bladder (see Figure 32–13). The muscular sphincter guarding the opening of the uterus is the cervix. The sperm pass through this opening on their way toward the oocyte. The cervix dilates to allow the fetus to emerge at the time of birth.

The vagina is a muscular tube about 7.5 centimeters long that leads from the cervix of the uterus to the outside of the body. It is the receptive organ for the penis and also the birth canal. Its opening is between the urethra, the tube leading from the bladder, and the anus. The lining of the vagina is rich in glycogen, which bacteria normally present in the vagina convert to lactic acid. As a consequence the vaginal tract is mildly acidic, with a pH between 4 and 5.

The external genital organs of the female are collectively known as the vulva. The clitoris, which corresponds to the penis in the male,* is about 2 centimeters long and, like the penis, is composed chiefly of erectile tissue. The clitoris has two bulbs (analogous to the penile bulb of the male) that lie on either side of the lower third of the vagina. The labia (singular, labium) are folds of skin. The labia majora are fleshy and, in the adult, covered with pubic hair. They enclose and protect the underlying, more delicate structures. (Embryologically, they are homologous with the scrotum in the male.) The labia minora are thin and membranous.

* In the early embryo, the structures are identical.

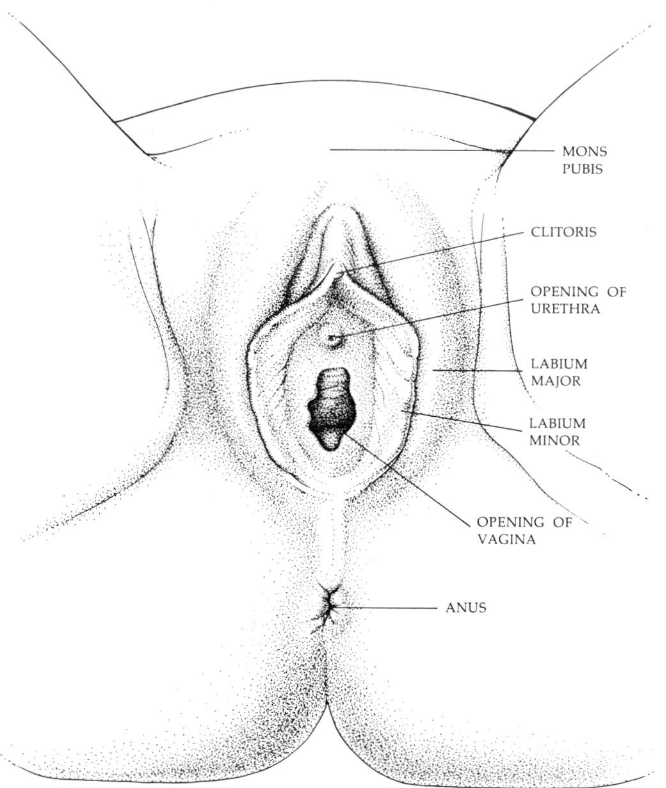

32–20
The external genitalia of the human female. The labia, clitoris, vaginal opening, and the mons pubis constitute the vulva. The mons pubis is a pad of fatty tissue overlying the pubic symphysis, where the two pubic bones join.

Orgasm in the Female

Under the influence of a variety of stimuli, the clitoris and its bulbs become engorged and distended with blood, as does the penis of the male. This process is somewhat slower in women than in men, largely because the valves trapping the blood in the sexual structures are not as efficient as in men. The distension of the tissues is accompanied by the secretion into the vagina of a fluid that both lubricates the walls of the vagina and neutralizes its highly acidic, and therefore spermicidal, secretion.

The thrusting of the penis into the vagina stimulates the lower third of the vagina and also causes the prepuce of the female (which covers the clitoris just as the male foreskin covers the glans of the uncircumcised penis) to move back and forth over the clitoris, stimulating it. Orgasm in the female, as in the male, is marked by rhythmic muscular contractions, followed by expulsion into the veins of the blood trapped in the engorged tissues. Homologous muscles produce orgasm in the two sexes, but in the female there is no ejaculation of fluid through the urethra.

At orgasm, the cervix drops down into the upper portion of the vagina, where the semen tends to form a pool. The female orgasm also may produce contractions in the oviducts that propel the sperm upward. It has been calculated that it would take a sperm cell at least two hours to make its way up the oviducts under its own power, and sperm have been found in the oviducts as soon as five minutes after intercourse. Orgasm in the female, however, is not necessary for conception.

CONTRACEPTIVE TECHNIQUES

A variety of contraceptive techniques is now available for couples who wish to prevent or defer pregnancy. In Table 32–3 they are rated in order of effectiveness, in terms of average number of pregnancies per year among the women of child-bearing age using the techniques. In most cases, two figures are given for effectiveness. The first, lower, figure is an "ideal" figure, obtainable when the method is used consistently and correctly. The second figure is an average figure, reflecting actual experience.

"The pill" consists of a combination of estrogen and progesterone. When taken daily, it keeps the level of these ovarian hormones in the blood high enough to shut off production of the pituitary hormones FSH and LH. Without FSH the ovarian follicles do not ripen, and in the absence of LH no ovulation occurs. Under the influence of the artificial hormones, the lining of the uterus thickens. The pills taken toward the end of the cycle no longer contain hormones. Without the synthetic estrogen and progesterone, the endometrium can no longer maintain itself and sloughs off. Although this produces a menstrual flow, ovulation has not occurred, so pregnancy is impossible.

Despite the availability of a number of contraceptive methods, there are many unwanted pregnancies each year, especially among teen-agers, as conservatively estimated by the numbers seeking abortions. In a survey of 4,611 unmarried pregnant girls 15 to 19 years old seeking abortion, it was found that more than half (53 percent) had failed to use any form of contraception. Of those who did use contraception, the majority used one of the less reliable methods. Those most fre-

Table 32-3 *Methods of Birth Control Currently Available*

METHOD	MODE OF ACTION	EFFECTIVENESS (PREGNANCIES PER 100 WOMEN PER YEAR)	ACTION NEEDED AT TIME OF INTERCOURSE	REQUIRES INSTRUCTION IN USE	POSSIBLE UNDESIRABLE EFFECTS
Vasectomy	Prevents release of sperm	0	None	No	Usually produces irreversible sterility
Tubal ligation	Prevents passage of egg cell to uterus	0	None	No	Usually produces irreversible sterility
"The pill" (estrogen and progesterone)	Prevents follicle maturation and ovulation	0–10	None	Yes, timing	Early—some water retention, breast tenderness, nausea; late—increased risk of cardiovascular disease
Intrauterine device (coil, loop, IUD)	Possibly prevents implantation	1–5	None	No	Menstrual discomfort, displacement or loss of device, uterine infection
"Minipill" (progesterone alone)	Probably prevents sperm from entering uterus	1–10	None	Yes, timing	?
"Morning-after pill" (50 × normal dose of estrogen)	Arrests pregnancy, probably by preventing implantation	?	None	Yes, timing	Breast swelling, nausea, water retention, cancer (?)
Condom (worn by male)	Prevents sperm from entering vagina	3–10	Yes, male must put on after erection	Not usually	Some loss of sensation in male
Diaphragm with spermicidal jelly	Prevents sperm from entering uterus, jelly kills sperm	3–17	Yes, insertion before intercourse	Yes, must be inserted correctly each time	None known
Vaginal foam, jelly alone	Spermicidal, mechanical barrier to sperm	3–22	Yes, requires application before intercourse	Yes, must use within 30 minutes of intercourse; leave in at least 6 hours after	None usually, may cause irritation
Withdrawal	Removes penis from vagina before ejaculation	9–25	Yes, withdrawal	No	Frustration in some
Rhythm	Abstinence during probable time of ovulation	13–21	None	Yes, must know when to abstain	Requires abstinence during part of cycle
Douche	Washes out sperm that are still in the vagina	?–40	Yes, immediately after	No	None

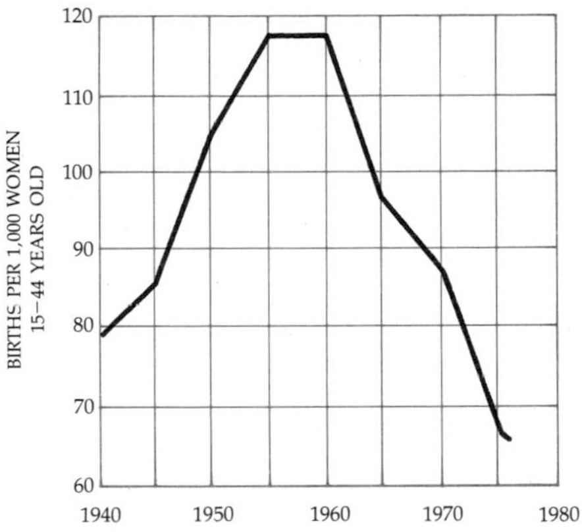

32–21
Birth rates among women in the United States, since 1940.

quently used were the condom (27 percent) and withdrawal (24 percent). Twenty-one percent used the pill. Among the reasons that teen-agers failed to use contraception were the beliefs that pregnancy could not occur because they were too young, or because they had sexual intercourse too infrequently, or because it was the "wrong time of the month." (Only about two-fifths of the teen-agers surveyed knew when in the menstrual cycle a woman is most likely to become pregnant.) Only 19 percent reported that they always used contraceptives; however, the percentage using medically prescribed contraceptive measures was considerably higher among those who had previously been pregnant.

Among the approximately 10 million girls in the United States aged 15 to 19, about 40 percent have had premarital sexual experience. Of these, 4 million, about 28 percent, have become pregnant one or more times.

HORMONES AND PREGNANCY

If the ovum is fertilized, it begins to develop in the oviduct, becomes implanted in the lining of the uterus, and the placenta begins to form. Almost immediately, the placenta begins to produce a gonadotropic hormone of its own in large quantities. This gonadotropic hormone causes the corpus luteum to continue its production of estrogens and progesterone even after the pituitary gonadotropic hormone production has ceased under the negative feedback system. Pregnancy tests often involve testing for the gonadotropic hormone from the placenta in either blood or urine. As the placenta matures, it also begins its own production of estrogens and progesterone. By the end of the third month of pregnancy, the placenta has taken over completely the production of estrogens and progesterone; placental gonadotropin is no longer produced and the corpus luteum regresses.

ESTRUS

Females of almost all species except *Homo sapiens* will mate only during their fertile period, which is known as *estrus,* or heat. Estrus may occur only once a year (as in wolves and deer), about once a month (as in cows and horses), or every few days (as in rats and mice).

The periods of estrus may last from only a few hours to three or four weeks. In animals, such as dogs, that produce eggs continuously during estrus, the eggs may be fertilized at different times and by different males, which explains in part why a mixed-breed litter can contain such an astonishing variety of siblings. Also, there may be a great size variation even among purebred pups, because many of them are actually of different ages at birth. In some mammals, such as cats, rabbits, and minks, although the egg is mature and the female receptive during estrus, ovulation occurs only under the stimulus of copulation, obviously a very efficient system, ensuring maximum economy in the utilization of gametes. There is suggestive evidence that in some women, also, ovulation may be triggered by sexual intercourse.

The human female appears to be one of the few female animals receptive to mating during unfertile periods. Anthropologists speculate that this receptivity co-evolved with the establishment of strong pair-bond relationships between human or prehuman males and females. A consequence of this pair-bond relationship is a society based on a family unit, in contrast to many other primate groups, in which the social and breeding unit is a troop or band. The establishment of such family units is seen, in turn, as the basis for the traditional division of labor among the sexes, with the female concentrating on childbearing and the home and the male on hunting, defending territory, and waging war against enemies. If the anthropologists are right, this behavioral and hormonal adaptation on the part of the human female has profoundly influenced the shape of human civilization.

SUMMARY

In vertebrates, reproduction is characteristically sexual and involves two parents, one of which produces sperm and the other eggs. Sperm and eggs are formed by meiosis in the gonads (the testes and the ovaries). The male gametes are produced by meiosis in the seminiferous tubules of the testes. The spermatogonia become primary spermatocytes; then, after the first meiotic division, secondary spermatocytes; and, following the second meiotic division, spermatids, which then differentiate into sperm cells. These sperm cells enter the epididymis, a tightly coiled tubule overlying the testis, where they are partially mobilized. (Full motility occurs only after they have entered the female reproductive tract.) The epididymis is continuous with the vas deferens, which carries sperm through the spermatic cord along the posterior wall of the abdominal cavity and around the bladder. Below the bladder in the region of the prostate gland, the two vasa deferentia merge with the urethra, which leads through the penis.

The penis is composed largely of spongy erectile tissue that can become engorged with blood, enlarging and hardening it. At the time of ejaculation, sperm are propelled along the vas deferens by contractions of a surrounding coat of smooth muscle. Secretions from the seminal vesicles, the prostate, and the bulbourethral glands are added to the sperm as they pass to the urethra. The resulting mixture, the semen, is expelled from the urethra by muscular contractions involving, among other structures, the base of the penis. These muscular contractions also contribute to the sensations of orgasm.

Production of sperm and the development of characteristics associated with masculinity are under the control of hormones, including testosterone (an androgen) and two gonadotropins, luteinizing hormone (LH) and follicle-stimulating hormone (FSH). LH acts on the interstitial cells (located between the seminiferous tubules) to stimulate the production of testosterone. FSH and testosterone stimulate the production of sperm. The gonadotropins are produced by the pituitary gland under the regulation of the hypothalamus, a brain center. Production of LH is controlled by the levels of sex hormones in the blood through a negative feedback system.

The female gamete-producing organs are the ovaries. The primary oocytes develop within nests of cells called follicles. The first meiotic division begins in the female fetus and is completed at ovulation. The second meiotic division is completed at fertilization. On the average, one potential ovum is produced every 28 days; it travels down the oviduct to the uterus. If it is fertilized (which usually takes place in an oviduct), it becomes implanted in the lining of the uterus (the endometrium). If it is not fertilized, it degenerates and the endometrial lining is shed at menstruation.

The production of ova, the menstrual cycle, and the development at puberty of the uterus, vulva, breasts, and the secondary sex characteristics of the female are all controlled by hormones. Before ovulation, FSH from the pituitary stimulates the ripening of the follicles and the secretion of estrogens. After ovulation, the corpus luteum, which forms from the emptied follicle, produces both estrogens and progesterone. Progesterone and estrogens both stimulate the growth of the endometrium.

In many vertebrates, gamete production and mating are cyclic. The part of the cycle during which mating occurs in the nonprimate female is known as estrus, or heat. In the human male, spermatogenesis is continuous. Although ovulation is cyclic in human females, they, unlike most other animals, are continuously receptive to sexual intercourse, a behavior pattern that may have evolved to strengthen pair bonding.

QUESTIONS

1. What are the advantages to the organism of asexual reproduction? Of sexual reproduction?

2. Consider the kinds of organisms that generally have external fertilization and those that generally have internal fertilization. What differences in their "life styles" require these differing mechanisms?

3. How is the shelled egg of a hen fertilized?

4. Would a vasectomy affect the structures associated with orgasm?

5. During which days in the menstrual cycle is a woman most likely to become pregnant? (Include data on longevity of eggs and sperm in making this calculation.) Why, in your opinion, is the use of the calendar method of birth control much less effective than other methods?

6. Describe the effects of luteinizing hormone (LH) and follicle-stimulating hormone (FSH) in the human male and female.

7. As you will recall from Chapter 18, the frequency of Down's syndrome increases with maternal age. On the basis of your knowledge of oogenesis, suggest a factor that may contribute to this increased frequency, and relate it to the chromosomal abnormalities of Down's syndrome.

8. The pregnancy rate among teen-aged girls increased 33 percent from 1971 to 1976. What do you think is the reason for this?

9. What would be an appropriate method of contraception for a couple who have infrequent intercourse (once a month)? For a couple who have frequent intercourse (three times a week), with plans to have children? For a couple who have frequent intercourse, but do not wish to have children? Under what circumstances, if any, would you elect to have a vasectomy or a tubal ligation?

10. Under what circumstances, if any, would you consider involuntary sterilization acceptable?

CHAPTER 33

The Continuity of Life: Development

Embryology first commanded the attention of biologists more than 100 years ago—long before genetics was an established scientific discipline. The course of development in many species has been minutely observed and meticulously documented. Yet the underlying processes of development are almost as mysterious as they were a century ago. As a biological challenge, development ranks second only to the functioning of the brain.

Most of the experimental work on development has been carried out on animals other than humans. (Even though abortion is now legal, experimentation on human fetuses is strictly forbidden.) However, the basic patterns of development are remarkably the same throughout the animal kingdom and so, as in genetics, for example, or cytology, we understand our own species by observations of others.

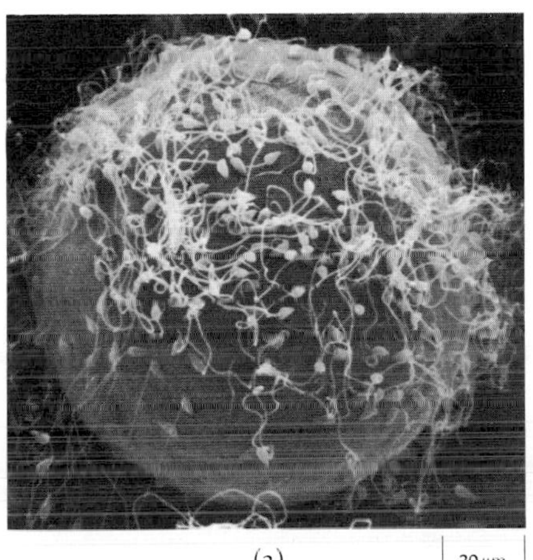

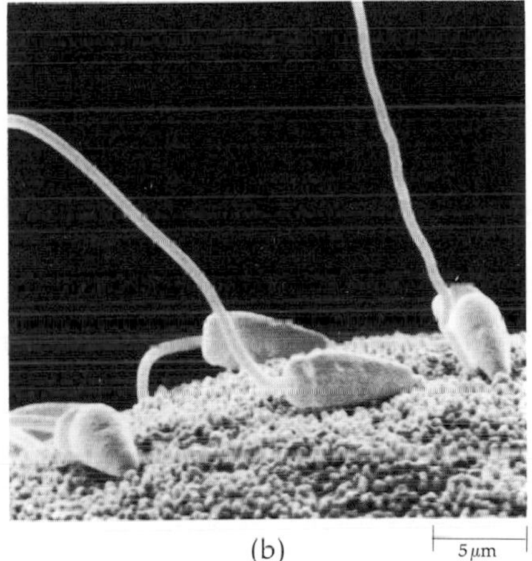

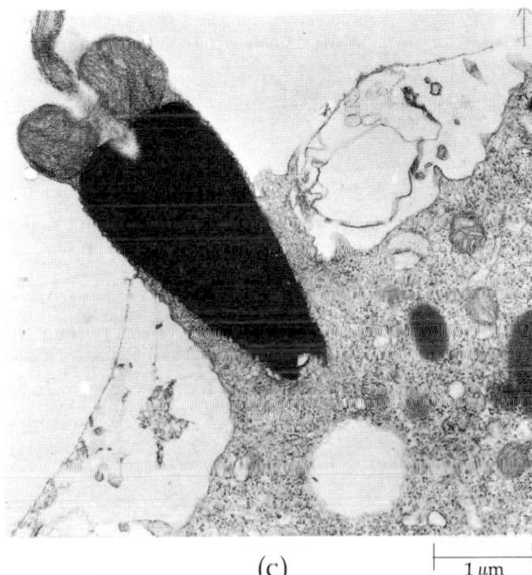

33–1
Despite the great differences in size of egg and sperm, both contribute equally to the hereditary characteristics of the individual. Since the nucleus is approximately the same size in both cells, the early microscopists postulated that this part of the cell must be the carrier of the hereditary determinants. The egg and sperm cells shown in (a) *and* (b) *are from the sea urchin, which has been used in many studies because sea urchins are relatively easy to obtain and fertilization, which is external, can be easily observed in the laboratory. The electron micrograph at the right* (c) *shows a sperm penetrating an egg cell. The torpedo-shaped sperm nucleus is at the left in the micrograph.*

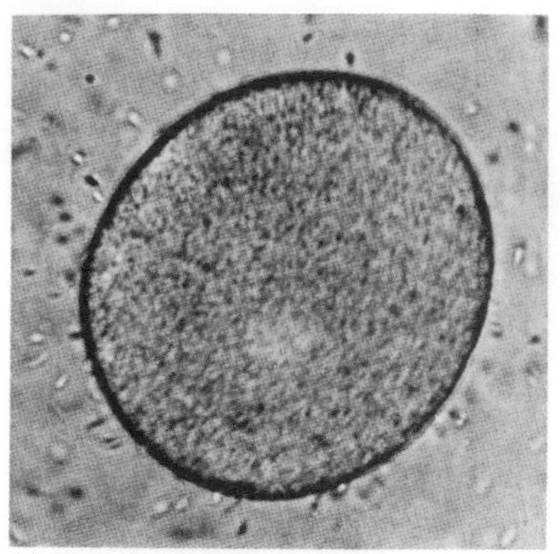

(a) *Numerous spermatozoa can be seen surrounding an unfertilized egg.*

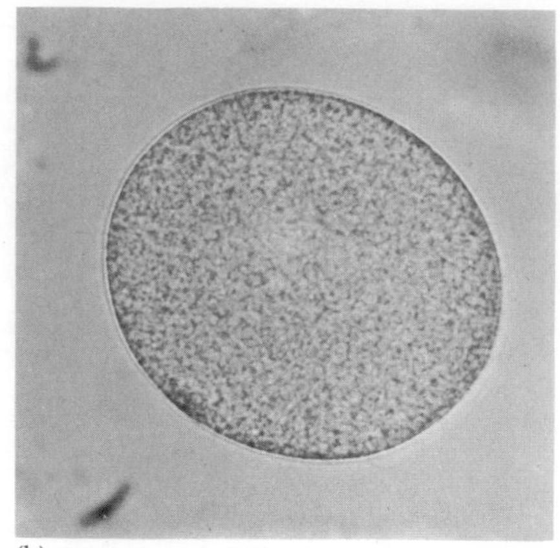

(b) *Fertilized egg; the fertilization membrane has just begun to form. The light area slightly to the right of center is the diploid nucleus.*

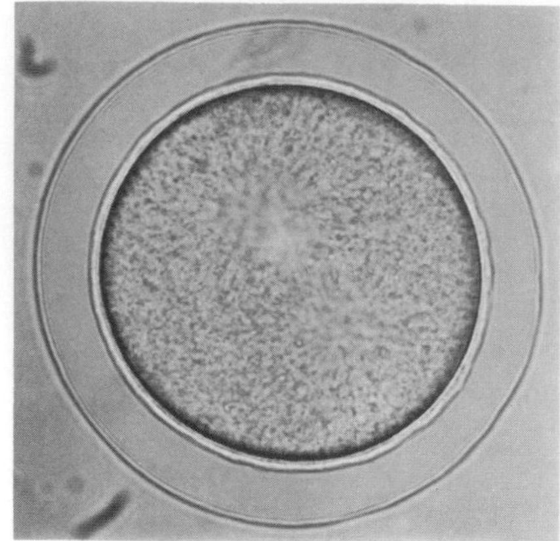

(c) *The fertilization membrane is fully formed. The egg has begun to divide; if you look closely, you can see that there are two nuclei.*

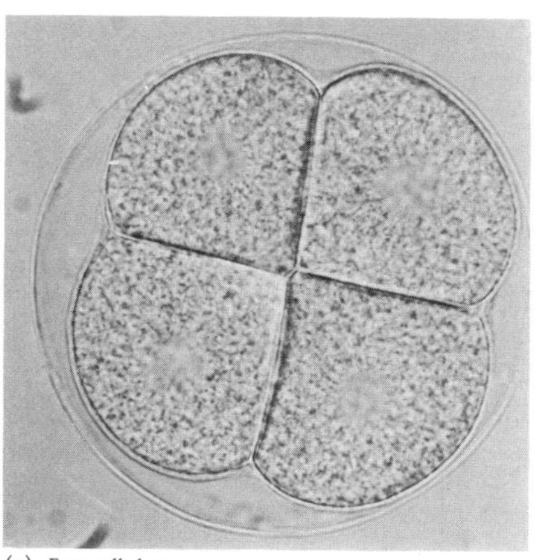

(e) *Four-celled stage.*

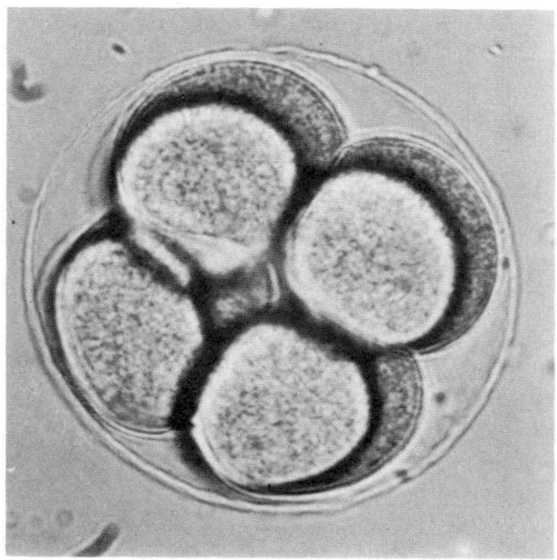

(f) *Eight-celled stage; the upper four blastomeres are smaller than the lower four.*

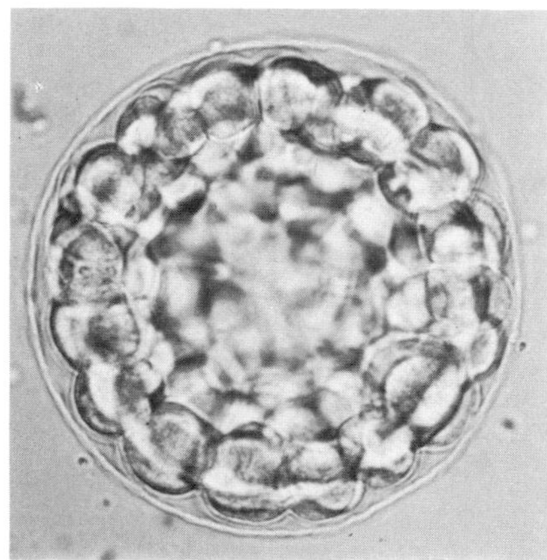

(g) *The blastocoel forms.*

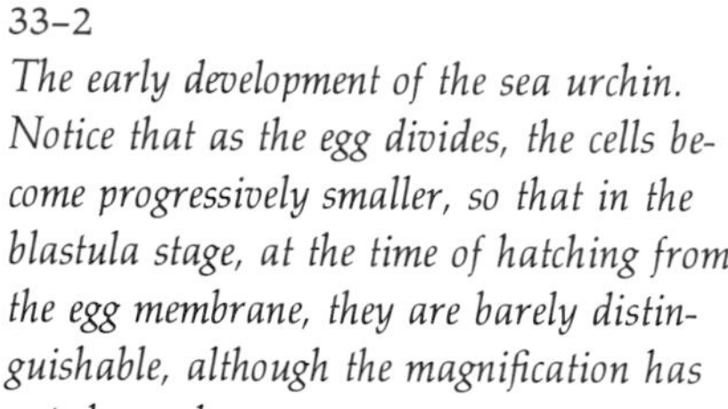

33–2
The early development of the sea urchin. Notice that as the egg divides, the cells become progressively smaller, so that in the blastula stage, at the time of hatching from the egg membrane, they are barely distinguishable, although the magnification has not changed.

DEVELOPMENT OF THE SEA URCHIN

Let us begin by following the course of development in a sea urchin.

The sea urchin has long been a favorite of embryologists because (1) the eggs, which are produced in large numbers, are fertilized and develop externally, making it possible to study their development under relatively simple laboratory conditions; (2) both egg and developing embryo are almost transparent, so that it is possible to observe many of the early events without disrupting them; and (3) the process is rapid. In about 48 hours, the zygote develops into a larval form, known as the pluteus. Moreover, sea urchins are abundant in pleasant places such as Woods Hole, Massachusetts, where biologists like to spend the summer.

Development begins with fertilization of the egg cell by the sperm (Figure 33–1).

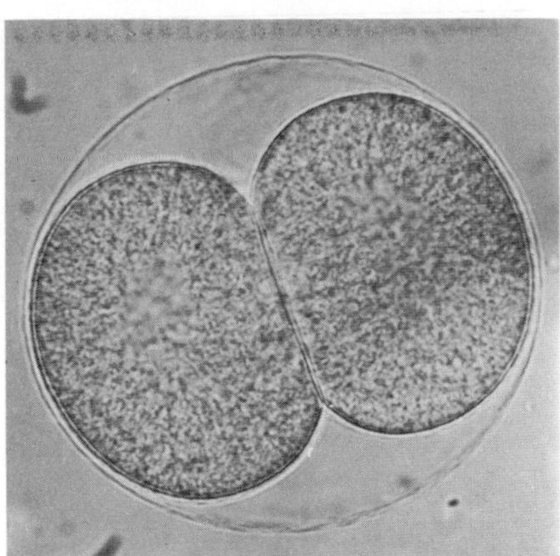

(d) *The first division.*

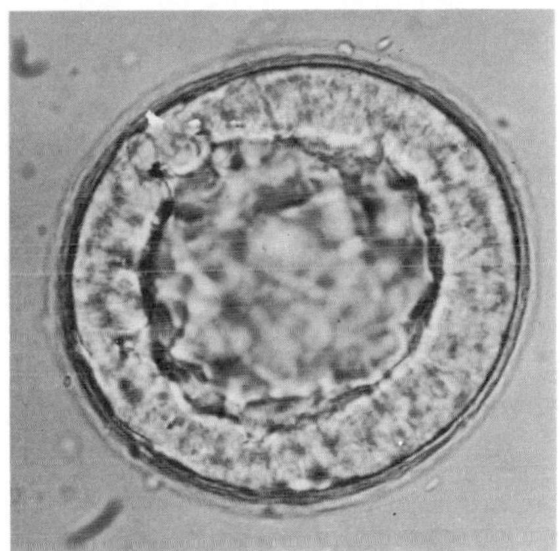

(h) *The mature blastula.* 0.1 mm

Like the mammalian sperm cell, the sperm cell of the sea urchin consists of a highly condensed, tightly packed nucleus, a small amount of cytoplasm, and a long flagellum, all surrounded by a cell membrane. The egg cell is much larger than the sperm cell (Figure 33–2a). Fertilization—the fusion of sperm and egg—has at least four consequences: (1) Changes take place in or on the outer membrane of the egg that prevent entry of other sperm (Figure 33–2b and c). (2) The genetic material of the male is introduced into the female gamete, and the two nuclei fuse, producing a diploid nucleus. The genotype of the new individual is thus established. (3) The egg is activated metabolically, as evidenced by a dramatic increase in protein synthesis. (4) The egg begins to divide by mitosis, and the developmental chain of events is set in motion.

In many species, although activation and mitosis follow fertilization, they can also proceed without it. In the sea urchin egg, for instance, even exposure to a hypertonic solution can start the developmental chain of events; frogs' eggs can be activated and will divide after being pricked with a glass needle or given a mild electric shock.

In fact, in some species it has been possible to demonstrate that activation does not require any nucleus at all, male or female. Sea urchin eggs can be divided in two, half with a nucleus, half without. Each half can then be activated artificially. Protein production in the half without a nucleus is as great initially, on a gram-for-gram basis, as in the half with a nucleus or in the whole egg (Figure 33–3), although, of course, the enucleated cell does not long survive. In other words, the egg in the course of its differentiation transcribes all the information needed for the early rounds of biosynthetic activity, and the transcribed mRNA is inactive in the egg until fertilization initiates translation.

The egg divides about once an hour for 10 hours. This process is known as cleavage, and the embryo at this stage is called a morula. As the cells divide, a fluid-filled cavity forms in the center. This cavity is the blastocoel, and the entire group of cells, the developing organism, is called the *blastula* (Figure 33–2h). The sea urchin blastula is about the same size as the egg cell from which it developed. Cleavage has not changed the total volume but has greatly altered the surface/volume ratio (page 99) and also the ratio of nucleus to cytoplasm of the individual cells.

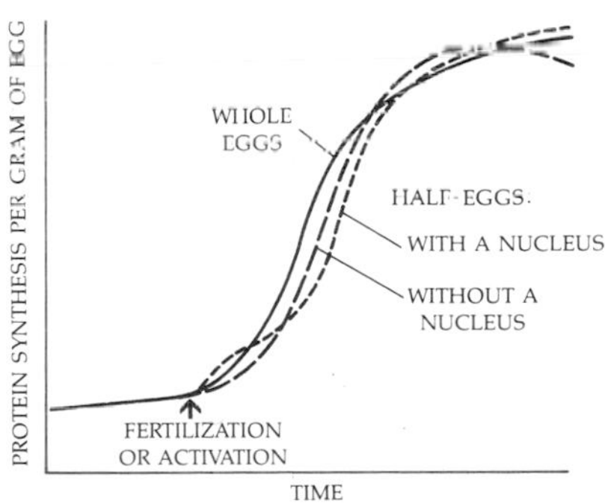

33–3
Rates of protein synthesis in activated whole eggs, half-eggs with a nucleus, and half-eggs without a nucleus. The initial pattern of protein synthesis is the same whether the activated egg contains a nucleus or not, indicating that the mRNA for early protein synthesis is present in the egg before fertilization.

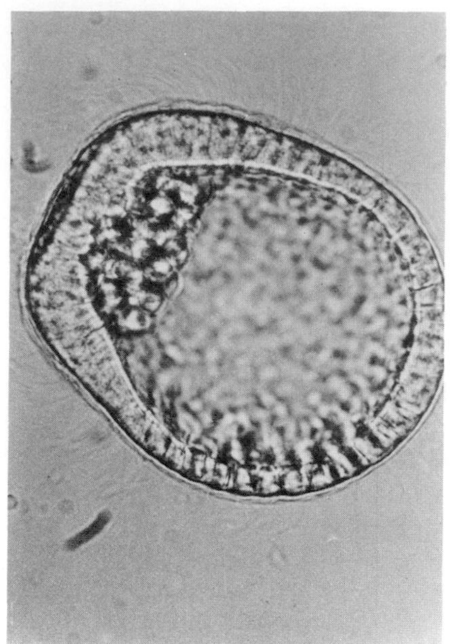

(a) *The beginning of gastrulation; the blastopore has begun to form at the upper left, and cells near the blastopore have begun to migrate across the blastocoel.*

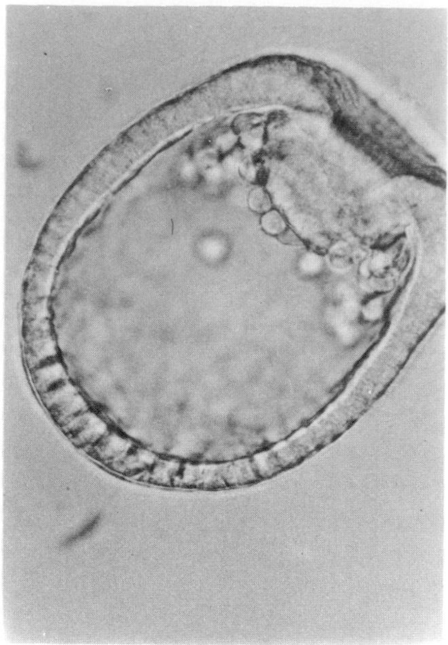

(b) *The outer cell layer begins to fold inward at the blastopore, forming the archenteron.*

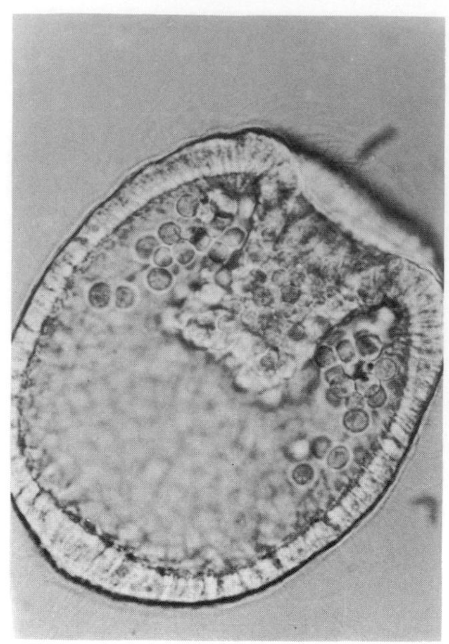

(c) *The outer layer of cells continues to move across the blastocoel.*

33–4
Gastrulation in the sea urchin.

The formation of the blastula is followed by a process known as *gastrulation* (Figure 33–4). In the sea urchin, gastrulation begins when cells on the inner surface of the blastula near the blastopore break loose and move toward the opposite pole. These cells are the primary mesenchyme. Next, the entire cell layer at one pole turns inward, moving through the blastocoel to the opposite pole, forming a new cavity, the gastrocoel, or archenteron, the primitive gut. The opening into the archenteron is called the blastopore. It will develop into the anus. (Sea urchins, you will recall—page 450—are deuterostomes, like other enchinoderms and like vertebrates.)

As a result of the movements that take place at gastrulation, three layers have been formed: an outside layer, the *ectoderm*; a middle layer, the *mesoderm*, which formed from the primary mesenchyme cells; and an inner layer, the *endoderm* (Figure 33–5). Also, the axis of the embryo has been established.

Once gastrulation is completed, evidence of cell differentiation can be observed. Cells in the mesoderm begin to secrete calcium-containing granules that develop into tiny three-armed spicules. These become the supporting skeleton for the pluteus. At the point at which the endoderm touches the opposite surface of the blastocoel, the ectodermal cells curve inward to form the mouth of the larva. The archenteron subdivides into mouth, stomach, intestine, and anus, and a ring of long cilia forms around the mouth region. The single fertilized egg cell has become a number of specialized, differentiated cells, performing specific functions, such as digestion; producing new organelles, such as cilia; and secreting new products, such as the calcium-containing skeleton.

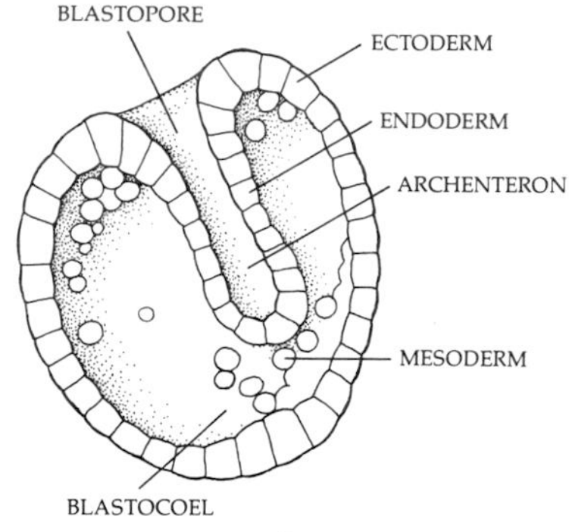

33–5
The sea urchin gastrula. Gastrulation produces a three-layered embryo. The archenteron becomes the gastrointestinal tract, and the blastopore becomes the anus. The blastocoel is almost entirely obliterated.

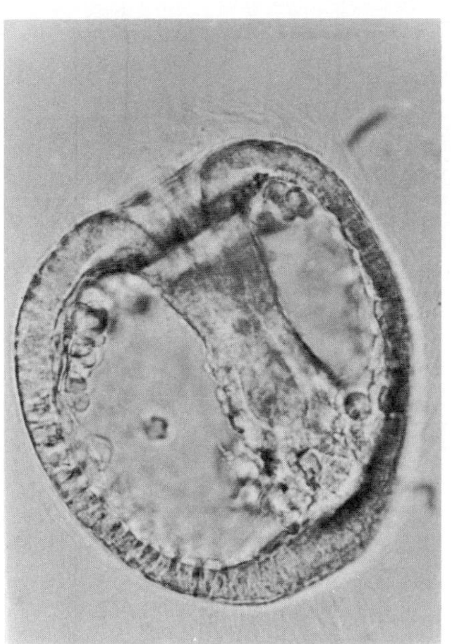

(d) *The mature gastrula.*

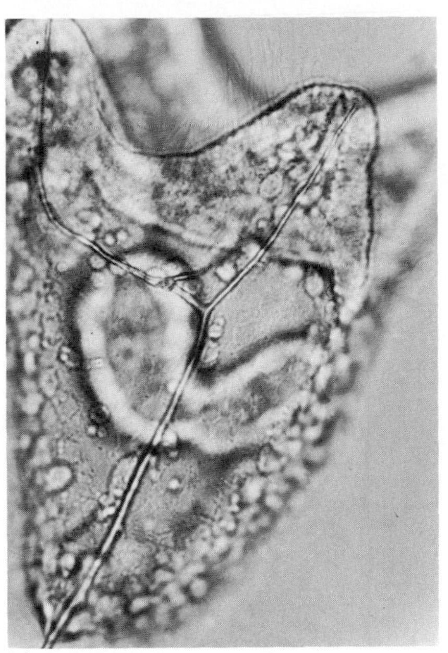

(e) *Gastrula cells differentiate and organize to form the pluteus larva.*

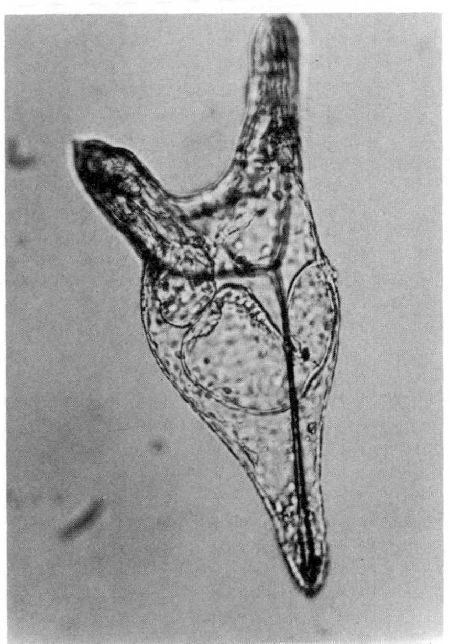

(f) *Within 48 hours after fertilization, the egg has developed into this multicellular organism, the pluteus.*

The Influence of the Cytoplasm

The sea urchin egg contains a relatively small amount of stored food (yolk), which, because of its weight, settles in the lower half of the egg. This half is called the vegetal half; the upper half is called the animal half. At cleavage, the first two cell divisions run from the pole of the animal half to the pole of the vegetal half, perpendicular to one another—the way you would quarter an apple. If these four cells are separated, each can develop into a normal pluteus. If, however, the embryo is divided in half along the third cleavage line, which splits the embryo across the equator, the two halves develop abnormally. The top half has large tufts of cilia but no gut. The bottom half is almost all gut, with no mouth, no arms, no cilia, and only a few, if any, spicules.

In other words, it is not just the nucleus that controls the differentiation of cells. The cytoplasm surrounding the nucleus also has a profound effect.

Let us describe another experiment. If you put the early blastula of a sea urchin in calcium-free water and agitate the water, the cells will separate. The cells still look the same, and as we shall see, neither their genetic material nor their cytoplasm has been altered in any crucial way. But they simply stop dividing and very shortly they deteriorate and die. Returning the cells to normal sea water does not restore their developmental potential. Suppose, however, the cells are returned to normal sea water and, by gently stirring, are brought into contact with one another. Under these conditions the cells reaggregate, arrange themselves in their former pattern, and once more begin to divide.

In short, development is determined by the nucleus, by the cytoplasm, and by interactions among cells.

33–6
The formation of the gray crescent in a frog's egg. (a) *Before fertilization, the upper two-thirds of the egg is covered by a heavily pigmented layer, and the greater part of the yolk is massed in the lower hemisphere. The nucleus of the egg is near the pole of the upper hemisphere.* (b) *The egg has been fertilized; the sperm has entered at the right and is moving toward the center of the egg. The trail it leaves behind it is caused by the disruption of pigment granules clustered near the egg surface. The whole pigment cap has rotated toward the point of sperm penetration. The cytoplasmic region on the other side of the egg, from which the pigment layer has moved away, becomes the gray crescent.*

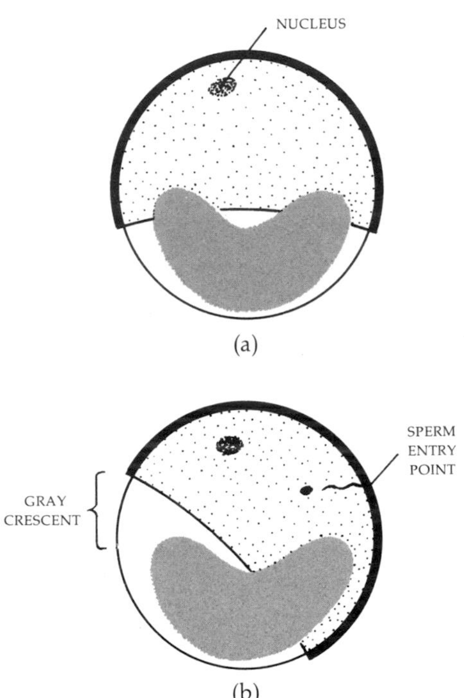

DEVELOPMENT OF AN AMPHIBIAN

The eggs of frogs and many other amphibians are laid in shallow water and, like sea urchin eggs, are fertilized externally so they are readily observed. The amphibian egg, however, contains a much larger amount of yolk. The animal half and the vegetal half of the egg differ markedly in appearance. For example, in frogs of the genus *Rana*, the yolk is massed in the lower hemisphere of the unfertilized egg, and the upper two-thirds of the egg is covered by a heavily pigmented layer (Figure 33–6a). Following fertilization, there is a massive reorganization of the cytoplasm. When the sperm penetrates the egg, the pigment cap rotates toward the point of sperm penetration and a gray crescent appears on the side of the egg, opposite the point of sperm entry (Figure 33–6b).

In a series of brilliant experiments by Hans Spemann in the early 1900s, it was shown that the cytoplasm associated with the crescent is of critical importance in the later development of the embryo. For instance, if an amphibian egg is divided at the two-cell stage, each blastomere may develop into a normal embryo. Whether or not both cells develop depends on whether or not each has received some of the cytoplasm containing the gray crescent (Figure 33–7).

These experiments add further support to the hypotheses that in the early stages of development, the nuclei are equivalent in their potential for directing

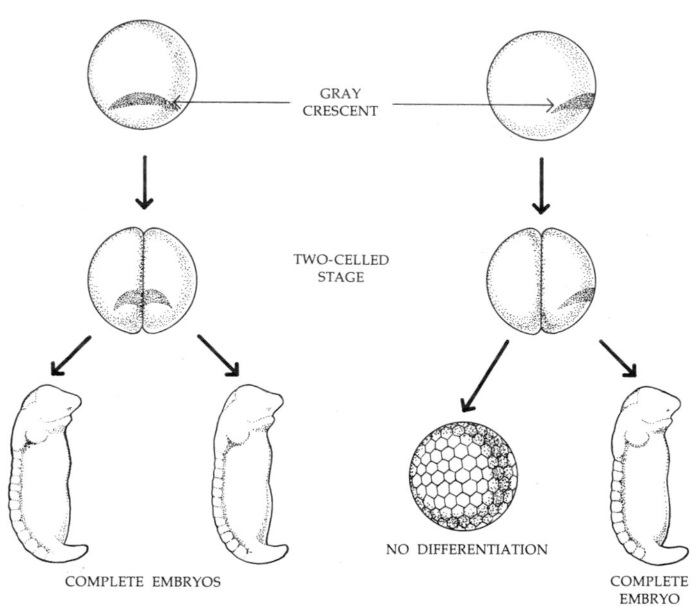

33–7
The importance of the gray crescent in development was demonstrated by separating the two cells formed by the first division of the egg. When the egg on the left divided, half of the gray crescent passed into each of the two new cells. When these cells were separated from each other, each formed a complete embryo. The first division of the egg on the right resulted in all the gray crescent going to one cell and none to the other. When these cells were separated, the one without the crescent did not develop.

33–8
Eggs that contain a large amount of yolk concentrated in one hemisphere, such as those of the frog, cleave unequally. The first two cleavages split the frog's egg through the poles to produce four cells shaped like the segments of an orange. The third cleavage separates the yolkier (vegetal) part from the less yolky (animal) part. As you can see, the four cells in the animal hemisphere are much smaller than the four in the yolky vegetal hemisphere. Subsequently, the yolkier cells cleave much more slowly than the less yolky ones.

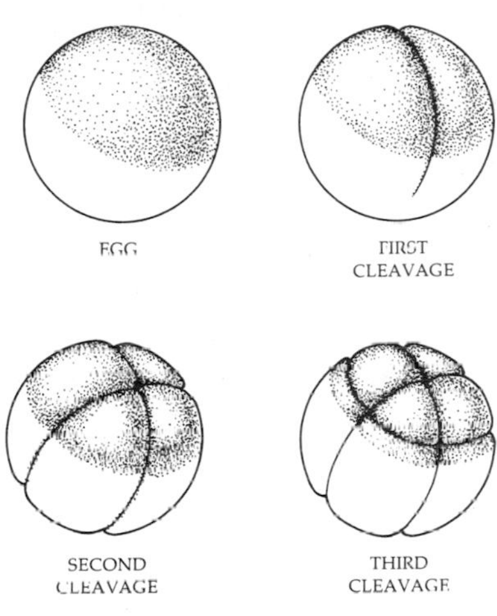

future development and that differences in the cytoplasm, rather than in the nuclei, determine the course of development.

The Amphibian Egg: Cleavage and Gastrulation

Cleavage in the amphibian egg differs from that in the sea urchin egg chiefly because of differences in the yolk content. When yolk is absent or present only in small amounts, as in the sea urchin, cleavage involves the whole egg. When larger amounts of yolk are present, as in frogs, the egg divides unevenly (Figure 33–8), with fewer cell divisions in the vegetal hemisphere.

As with the sea urchin egg, cleavage results in the formation of a blastula. The blastopore is a crescent-shaped slit; it always forms at the boundary between the gray crescent and the vegetal hemisphere (Figure 33–9a).

Gastrulation in the two types of egg differs in detail but not in principle. In the sea urchin, one wall of the gastrula forms a tubular extension toward the other wall. In the yolky frog's egg, the cells migrate inward, moving over the lip of the blastopore as if they were being hauled over a pulley (Figure 33–9b and c). The direction in which they move is the future main axis of the animal. These mesodermal cells along the axis are called the chordamesoderm.

At the end of gastrulation in the amphibian egg, as in the sea urchin egg, there are three tissue layers: endoderm, mesoderm, and ectoderm.

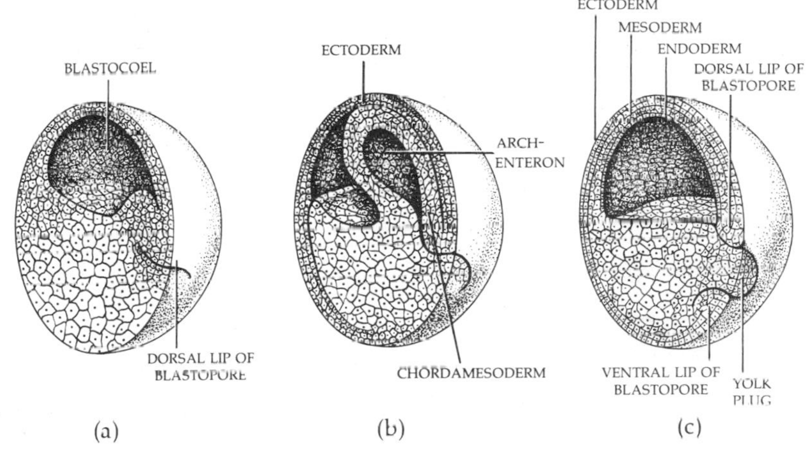

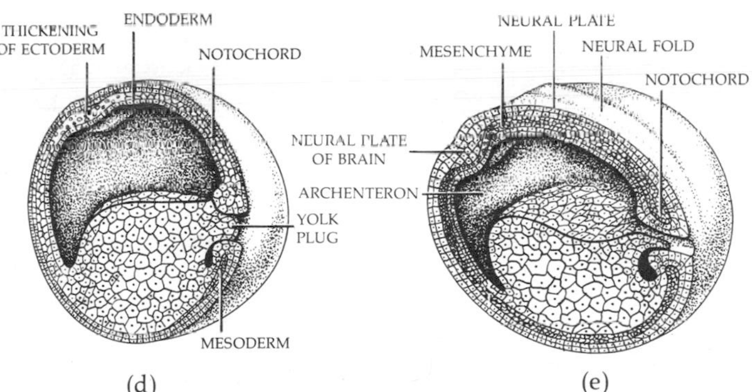

33–9
Development of the frog gastrula. (a) *The blastopore forms in the blastula.* (b) *Cells from the outer surface move into the blastopore, obliterating the blastocoel and creating the archenteron.* (c) *Three layers of tissues have been produced: ectoderm, endoderm, and mesoderm.* (d) *The ectoderm on the dorsal surface, overlying the notochord, thickens and flattens.* (e) *The neural plate and neural folds begin to form.*

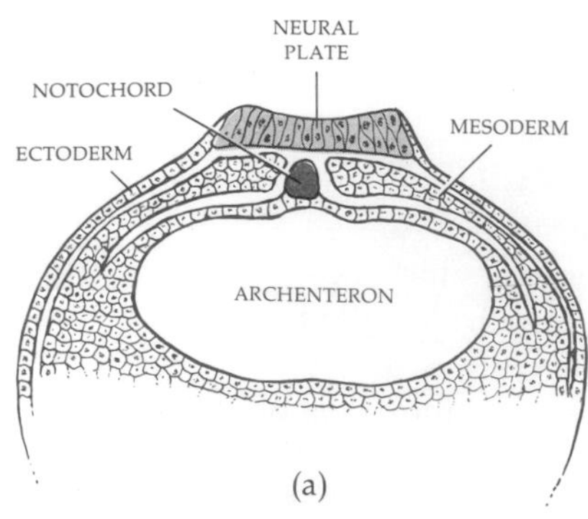

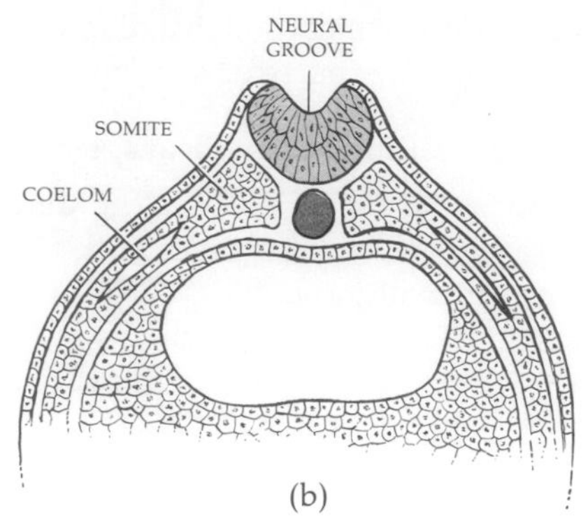

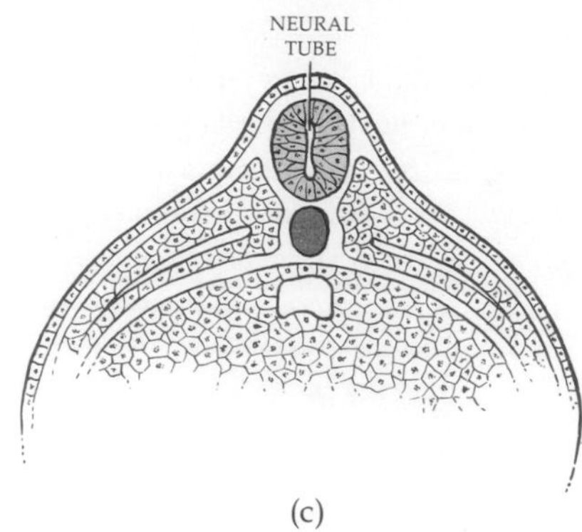

33–10
The formation of the neural tube from the neural plate in the frog. The thickened elevations of ectoderm on the right and left sides of the neural plate curve inward, forming the neural groove. The ridges bordering the neural groove then meet and fuse. Finally, the resulting neural tube pinches off from the rest of the ectodermal tissue.

The Amphibian: Neural Tube Formation

At the end of gastrulation, the first visible signs of differentiation appear. The chordamesoderm becomes the notochord. The ectoderm above the chordamesoderm begins to thicken, forming the neural plate (Figure 33–10a). The ridges of the neural plate meet and fuse to form the neural tube, which pinches off from the rest of the ectodermal tissue (Figure 33–10b and c). Thus the principal features of the vertebrate have been established.

The mesodermal tissue on either side of the notochord breaks up into two longitudinal masses that segment, forming blocks of tissue called somites. In the adjacent mesoderm the coelomic cavity forms. The endoderm, which contains the yolk, continues to take up much of the body mass until the stored nutrients are finally absorbed by the larva, the tadpole. The movements of the cells are so regular and predictable that it is possible to chart the fate of various cells of the late blastula (Figure 33–12).

The Organizer

The tissue that lies at the dorsal lip of the blastopore of the amphibian is one of the most thoroughly investigated in all of embryology. Spemann, in a continuation of the studies described earlier (Figure 33–7), discovered that he could divide the developing blastula in two and sometimes obtain two normal embryos. Once blastulation was completed and gastrulation had begun, however, he found that he could obtain only one normal embryo from the operation. The embryo that was normal always developed from the half that contained the dorsal lip of the blastopore. (This was actually an extension of his previous experiment of separating cells in the first cleavage stage. The dorsal lip forms in the blastula at the position once occupied in the egg by the gray crescent.)

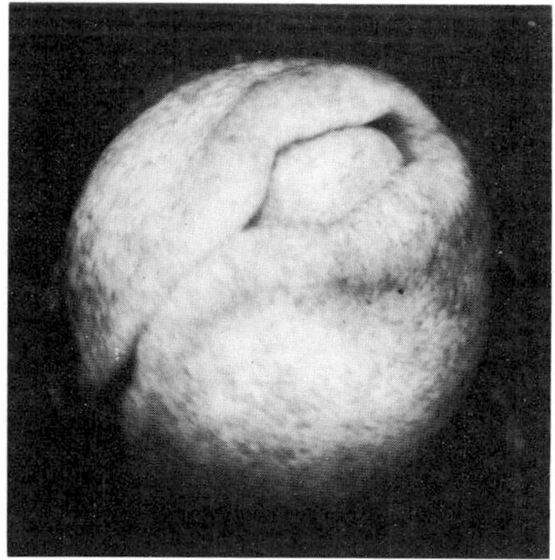

33–11
An amphibian embryo, showing the neural folds and neural groove.

In 1924, Hilde Mangold, a student of Spemann, excised the dorsal lip from one salamander embryo and grafted it into the belly region of another embryo. When the embryo that had received the graft developed, it was found that a second embryo had formed within its tissues, a sort of Siamese twin to the first. Because the two salamanders differed in color, it was possible to distinguish the tissues of the host embryo from those of the transplant. The transplanted dorsal lip had developed into notochord, as it would have had it not been removed in the first place. This in itself was noteworthy because previous experiments had shown that other portions of an embryo of similar age, if transplanted, would develop strictly according to the site to which they were transplanted. Even more remarkable, however, was the fact that the rest of the secondary embryo was made up of tissues of the host. As countless repetitions of this experiment have shown, the transplanted dorsal lip has the effect of organizing the tissues of the host into a second embryo (Figure 33–13). This is how the dorsal lip of the blastopore (the prospective chordamesoderm) came to be called the *organizer*.

We do not know the nature of the organizer or how it exerts its effect. However, it is known that in chordate embryos of every species, including *Homo sapiens*, the chordamesoderm serves this same role of primary organizer, inducing differentiation of the overlying ectoderm and setting in motion the other events of development.

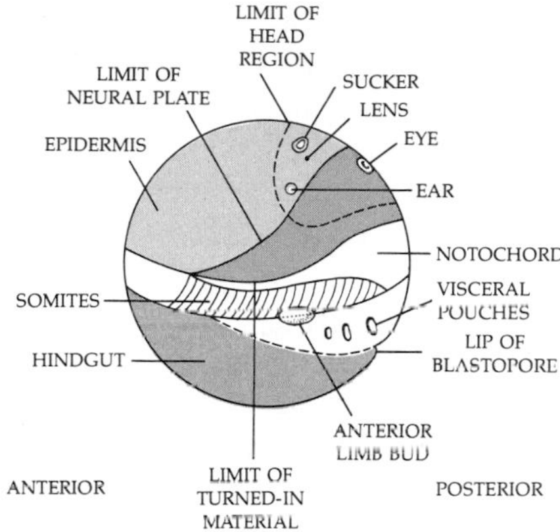

33–12
By the late blastula or early gastrula stage, the developmental fate of groups of cells in the embryo has already been determined. Experiments to learn what tissues and organs will be formed from which embryonic regions are carried out by the application of small amounts of harmless dye. The results of such experiments are used to construct a "fate map," such as the one shown here.

Induction and Inducers

It is now known that the action of Spemann's organizer is just one of many tissue interactions that take place in development. This general process is known as *embryonic induction*. Embryonic induction occurs when two different types of tissues come in contact with one another, as the result either of morphogenetic movement, as in the induction of the neural plate, or of growth.

Numerous experiments have shown that induction is mediated through physical contact and that some chemical material passes from the inducing tissue to the tissue that is induced. Induction takes place even when direct contact is prevented between the two tissues, as by placing a porous filter between them, but not if chemical exchanges are prevented, as by a piece of metal foil. The substance or substances must be of some very general nature. For example, the capacity of the organizer to induce neural plate formation is not species specific; thus one can provoke the formation of a secondary neural tube in a chick embryo by means of a

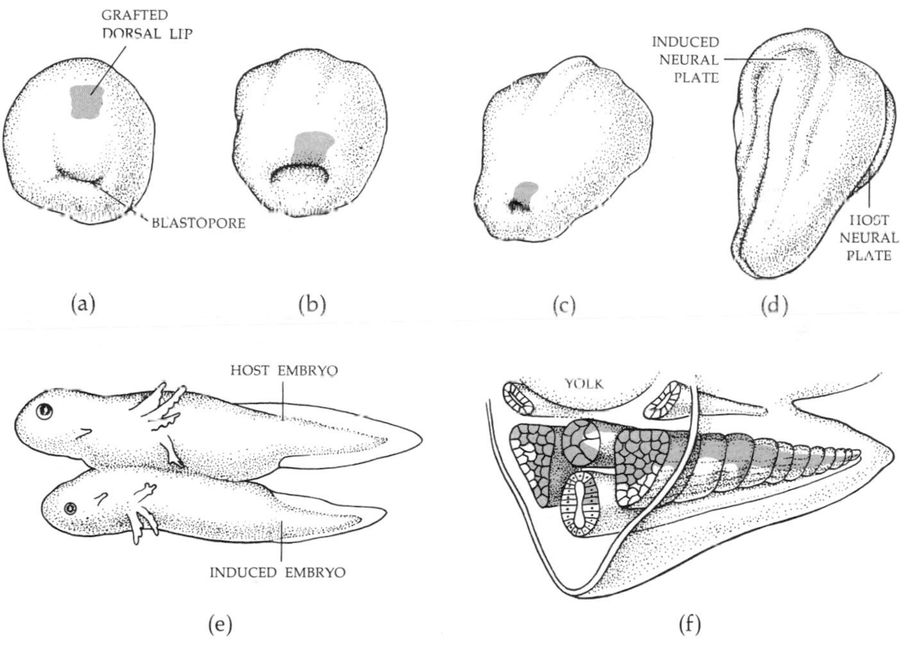

33–13
Experiment showing the activity of the organizer. (a) *The dorsal lip of the blastopore from one amphibian embryo is grafted onto another embryo. At gastrulation the grafted tissue moves into the interior of the embryo* (b, c) *through a second blastopore created in the host. In* (b) *and* (c) *the embryo is viewed from a different angle so the host blastopore is not visible. Two neural plates are formed* (d), *and a double embryo develops* (e). *In* (f) *you see the structure of the secondary embryo under the yolk of the host embryo. The colored cells are derived from the graft, and the light cells are host material that has been induced to undergo these differentiations.*

graft of tissue from a rabbit embryo. Moreover, experiments have shown that a number of chemicals, some of which never occur in embryos, will induce neural plate formation in embryonic ectoderm. It is not, then, that the inducing substance *endows* the ectoderm with the capacity to form nerve tissue; rather, it merely *evokes* a potentiality that is already present in the responding tissue. The organizing substances, in entering the responding cell, change the nature of its cytoplasm in such a way as to provoke a new pattern of gene activity.

DEVELOPMENT OF THE CHICK

The chick egg, familiar to us all, is different in many respects from the eggs previously examined. First, and most obvious, it is surrounded by a shell, which permits it to develop on the land. (Such eggs, as noted on page 476, are called amniote eggs.) Second, related to its terrestrial development, it is surrounded by a system of membranes, about which we shall have more to say later in this chapter. Third, it contains a large amount of yolk. In accord with this large amount of stored food, development in the chick can take much longer than in the sea urchin or the frog and does not stop with the emergence of a still immature, larval form, such as the pluteus or the tadpole.

The yolk of the chick egg, like that of other bird and also reptile eggs, is so large and dense that cleavage does not involve most of the egg mass. The only part of the egg that cleaves is a thin layer on top. These cleaving cells sit like a cap on the top of the yolk mass. A blastula like this is known as a *blastodisc*. Birds, reptiles, and mammals develop from a blastodisc.

If you break open a fertilized chick egg when it is first laid, you can see the blastodisc as a white mass about two millimeters in diameter on top of the yolk (Figure 33–14). Microscopic examination will show that this mass is made up of

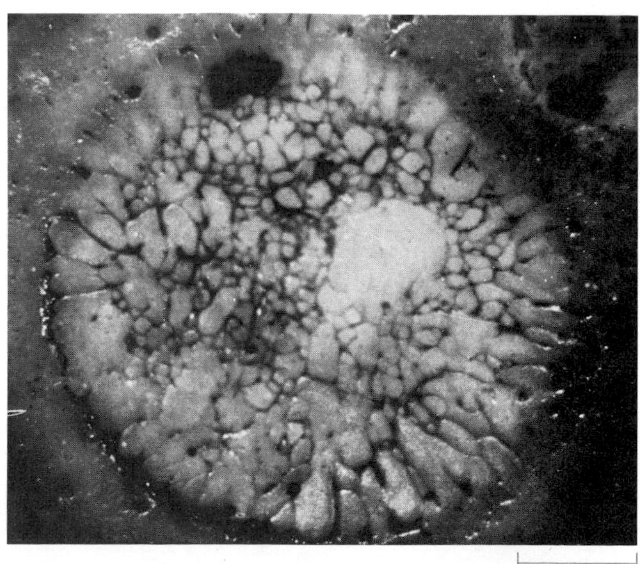

33–14
Blastodisc of early chick.

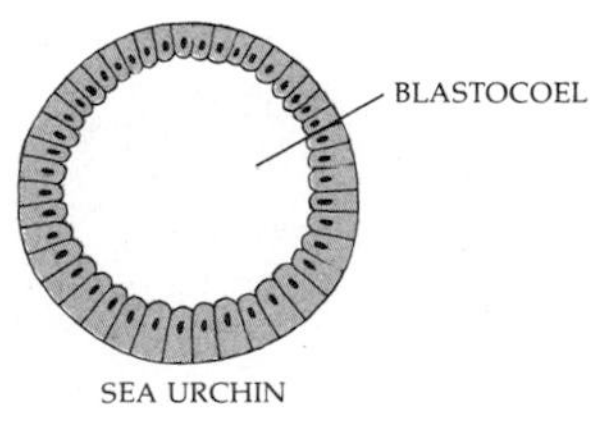

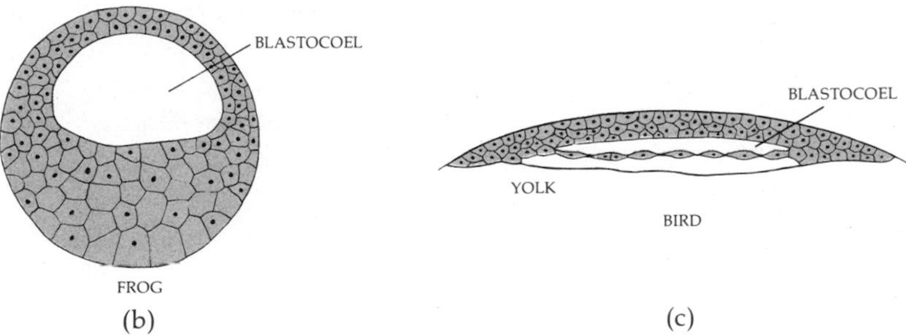

33–15
Blastulas of (a) *a sea urchin,* (b) *a frog, and* (c) *a bird. In the sea urchin and the frog, the blastula is a hollow sphere of cells. In the bird, it forms a blastodisc that sits on the surface of the yolk.*

many cells (close to 100,000) and that the cells are in two layers: The upper layer will give rise to the ectoderm and the lower layer the endoderm. The space between them is blastocoel (Figure 33–15). If the egg is left intact and development continues, within a short period of time, a visible line, known as the *primitive streak*, appears on the surface of the blastodisc; the primitive streak is, in effect, an elongated blastopore. Cells migrate through the primitive streak to form the mesoderm and some endoderm, thereby establishing the three characteristic tissue layers: ectoderm, mesoderm, and endoderm (Figure 33–16). On the completion of gastrulation, the neural plate forms and, subsequently the neural tube, as in the frog.

Extraembryonic Membranes of the Chick

Eggs develop in water. In the amniote egg, this water is contained within a set of membranes formed from the tissues of the embryo, the extraembryonic membranes. These membranes begin as extensions of the blastodisc. Each is formed from a combination of two of the basic tissue types.

As Figure 33–17 shows, the yolk gradually becomes surrounded by a membrane of mesoderm and endoderm, the *yolk sac membrane*. The function of this membrane is nutritive; the endodermal cells digest the yolk, and the blood vessels formed in the mesodermal component carry food molecules from the yolk into the embryo proper.

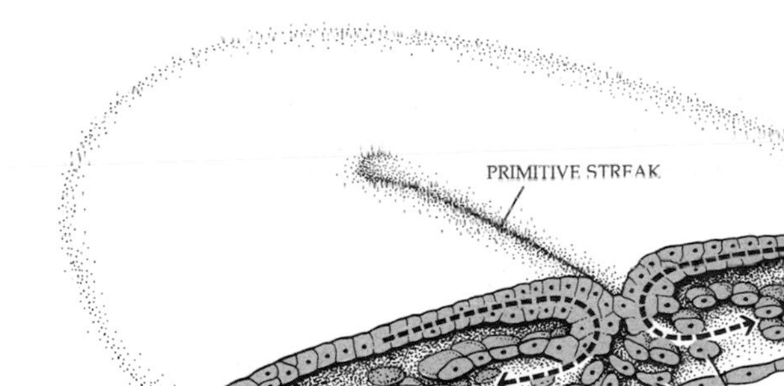

33–16
Gastrulation of the chick blastodisc. Cells migrate into and through the primitive streak to form the mesoderm and some endoderm. The result is a thin embryo with the three characteristic tissue layers.

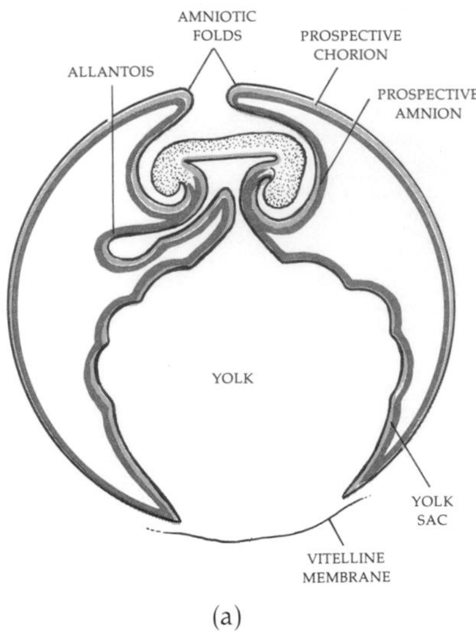

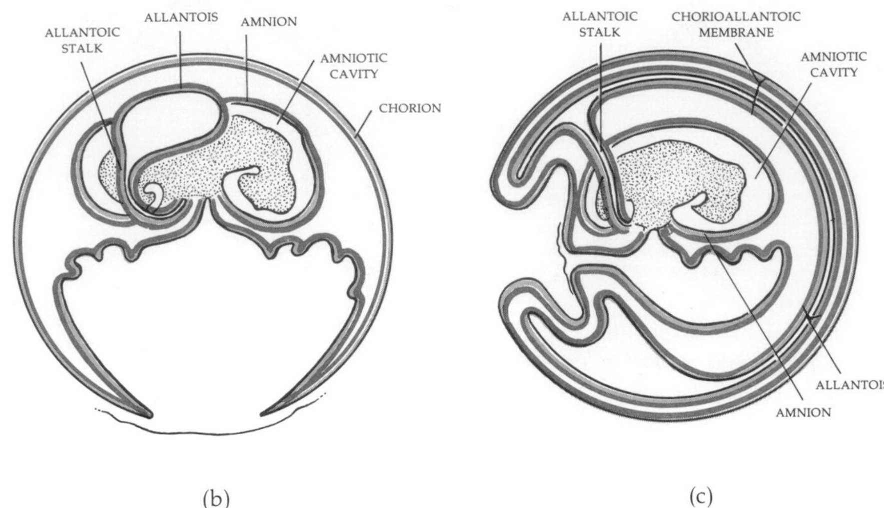

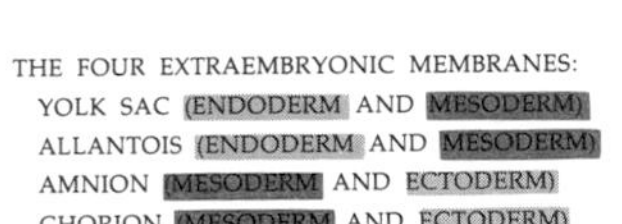

33–17
Development of the extraembryonic membranes of the chick. In early development, body folds of the embryo begin to separate from the underlying yolk. One membrane, the yolk sac, grows around and almost completely encloses the yolk. A second, the allantois, arises as an outgrowth of the rear of the gut. The third and fourth are elevated over the embryo by a folding process during which the membrane is doubled. When the folds fuse, two separate membranes are formed. The inner one is the amnion, and the outer one is the chorion. The chorion eventually fuses with the allantois to form the chorioallantoic membrane, which, in the later stages of development, encloses embryo, yolk, and all the other structures. Note that each membrane is composed of two types of primary tissue.

The extraembryonic ectoderm and mesoderm on the outside of the coelomic space fold upward around and over the embryonic region. These so-called amniotic folds eventually fuse with each other over the embryo, enclosing it inside a fluid-filled cavity, the *amniotic cavity*. The saline fluid in the cavity serves to buffer the embryo against mechanical and thermal shocks. As Figure 33–17a shows, the fusion of the amniotic folds creates two membranes, both composed of a layer of ectoderm and mesoderm, separated by coelomic space, which is continuous with the extraembryonic coelom. The inner membrane is the *amnion*, and the outer membrane is the *chorion*.

As development proceeds and the embryo becomes increasingly demarcated from the yolk mass (although always attached to it by the yolk sac stalk), a ventral and posterior pouch is formed from the primitive hindgut. This is the *allantois*, and, as you can see in Figure 33–17b, its wall is composed of endoderm and mesoderm. At first, the function of the allantois is excretion. Since bird and reptile eggs are closed off from their environment by the shell, there is no way to dispose of the products of nitrogen metabolism, most of which are toxic. Evolution found a dual answer to this problem. First, the nitrogenous wastes are converted to uric acid, which is insoluble and of low toxicity. The uric acid is then voided from the embryo into the allantois, which essentially serves as an embryonic garbage bag. (When a chick hatches, the accumulated uric acid can be found adhering to the abandoned shell.)

As development proceeds further, the allantois expands in volume and pushes out into the extraembryonic coelom, gradually enveloping the embryonic region as it does so. Eventually, as Figure 33–17c shows, the allantois effectively obliterates the extraembryonic coelom, and its walls fuse with the chorionic membrane lying under the eggshell. Since the allantois is plentifully supplied with blood vessels, the *chorioallantoic membrane*, formed by the fusion of allantoic wall with the chorion, acts as an efficient respiratory membrane for the embryo during its later development.

Organogenesis

While the membranes are forming, the development of the embryo itself continues. These later stages of development, following cleavage and gastrulation, are generally termed organogenesis. Organogenesis begins with the inductive interaction between ectoderm and underlying chordamesoderm. Each of the three primary tissues formed during gastrulation now proceeds to undergo growth, differentiation, and morphogenesis. The process is essentially the same in all vertebrates.

Ectoderm: Nervous System and Sensory Organs

The neural tube stretches out along the dorsal surface, growing and becoming longer and thinner as the embryo increases in size. The cells in the neural tube at first appear to be all alike, but some of them, the future motor neurons, begin to extend long processes that grow out beyond the neural tube and invade the peripheral organs and tissues.

When the neural tube is formed, some of the ectodermal cells at the crests of the neural folds are left behind in the surrounding tissue, as shown in Figure 33–18. Some of these cells from the neural crest now migrate into positions around the neural tube and form connections between the dorsal part of the tube and the surrounding tissues. Meanwhile, mesodermal cells have begun to migrate toward the neural tube and underlying notochord and to aggregate around them. Here they differentiate into cartilage and give rise to the hollow vertebral column. If a portion of neural tube is transplanted to another region of the mesoderm, mesodermal cells will aggregate similarly around it as a result of tissue interaction.

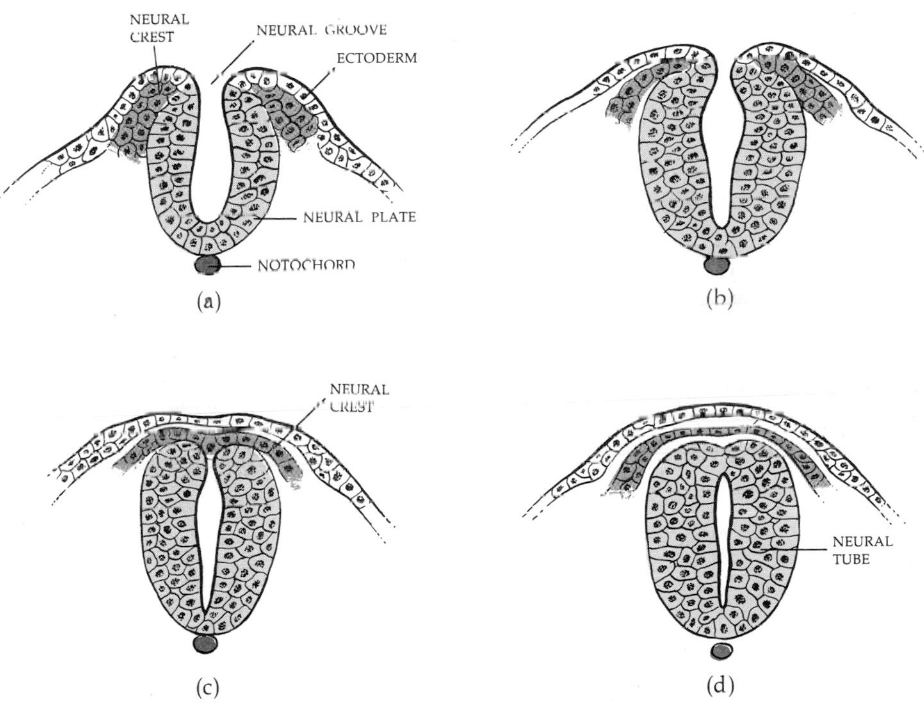

33–18
During the formation of the neural tube, some of the ectodermal cells at the crests of the neural folds are pinched out as the folds come together. Eventually, some of these neural crest cells will migrate down toward the notochord and aggregate into ganglia. These cells become sensory neurons that send extensions up to connect with the dorsal part of the spinal cord and axons into the surrounding tissues. Others become Schwann cells, which surround and insulate the nerve fibers, some become the pigment cells (melanocytes) found at the base of the epidermis, and some form the adrenal medulla.

33–19

The development of the brain and associated sense organs. At the front end of the neural tube, local swellings produce three distinct bulges—the forebrain, the midbrain, and the hindbrain. The forebrain then bulges laterally, and two optic vesicles appear and develop a cuplike shape. At the same time, the surface epidermis folds inward to meet the optic cups. The ears and nostrils also appear at first as epidermal infoldings.

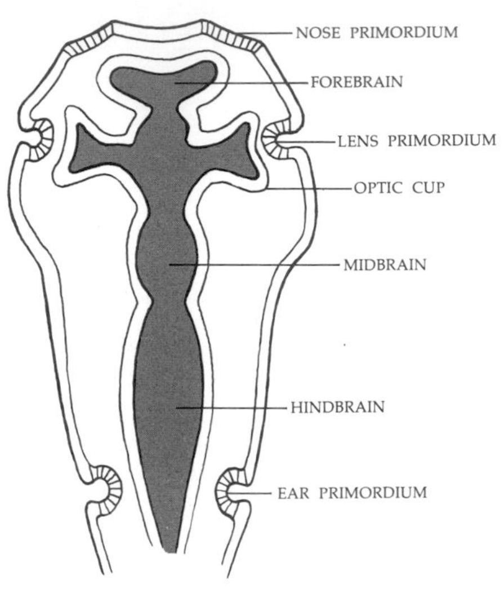

The brain begins as three bulges in the foremost part of the neural tube (Figure 33–19). Almost immediately, two saclike protrusions, the optic vesicles, appear on both sides of the forming brain. These spherical vesicles enlarge and come into contact with the epidermis, while still remaining connected to the neural tube by the narrowing optic stalk. Within this stalk, the optic nerve will develop.

When the optic vesicle comes into contact with the inner surface of the epidermis, the external surface of the vesicle flattens out and pushes inward. The vesicle thus becomes a double-walled cup. The rim of the cup becomes the edge of the pupil. The opening of the cup is large at first, but the rims bend inward and converge, so that the opening of the pupil becomes smaller. At the point where the optic vesicle touches the epidermis, the epidermis begins to differentiate into a lens, becoming transparent (Figure 33–20).

The differentiation of the eye lens is an example of a *secondary induction*. The inducing tissue (the optic vesicle) is itself the result of the primary induction of dorsal ectoderm by dorsal mesoderm. In fact, the formation of the complete eye involves at least six separate inductions, occurring in an orderly sequence, each one linked to the one before, with primary induction by the organizer as the starting point.

Other infoldings and differentiations of head ectoderm, associated with the presence of the embryonic brain, lead to the elaboration of the organs of smell and hearing.

Mesoderm

The series of somites that came to lie on each side of the notochord shortly after gastrulation now differentiate into three kinds of cells: (1) sclerotome cells, which later form skeletal elements; (2) dermatome cells, which become part of the developing skin; and (3) myotome cells, which form most of the musculature. In the lower vertebrates, the same pattern, with its simple double series of back muscles, is retained in adulthood; it is well suited to the side-to-side motion of the body in

33–20

The eye develops as a result of interactions between the epidermis and a lateral outgrowth of the forebrain, the optic vesicle. When the optic vesicle reaches the epidermis, the overlying epidermis begins to thicken and differentiate. The optic vesicle flattens out and invaginates, becoming the double-walled optic cup. The invaginated wall will become the retina of the eye; the outer, thinner wall develops into the pigment layer. The rim of the optic cup will later form the edge of the pupil. The thickened epidermal layer pinches off to become the transparent lens, and the overlying epidermis, which also becomes transparent, forms the cornea. The connection with the brain remains as the optic stalk, within which are the optic nerves. Note that the eye is thus a differentiated extension of the brain.

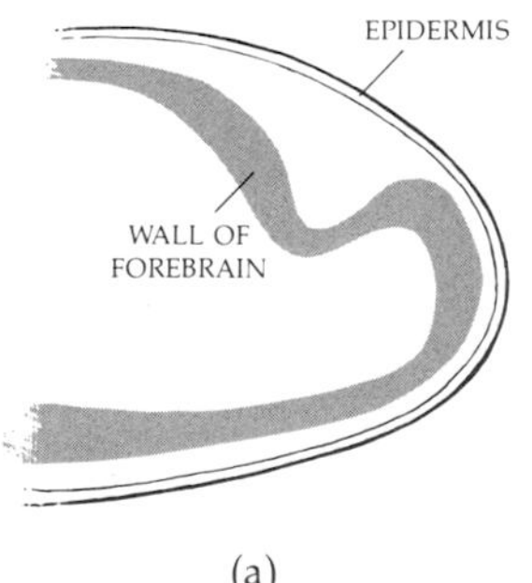

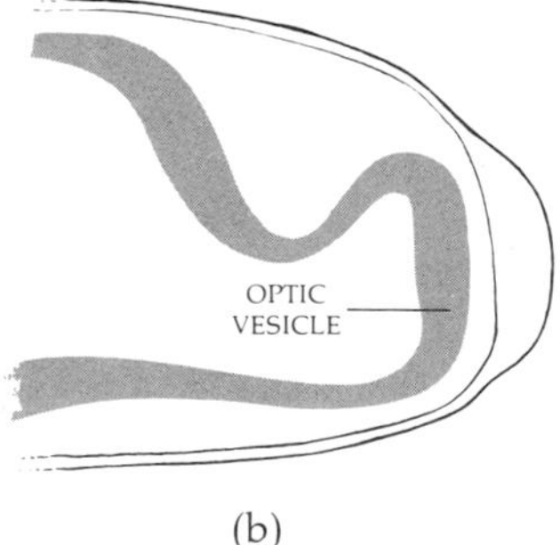

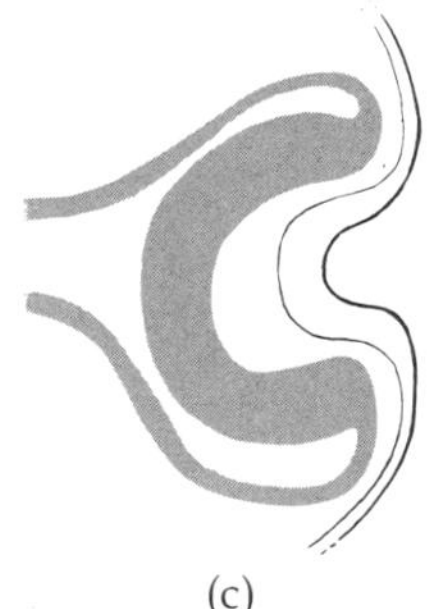

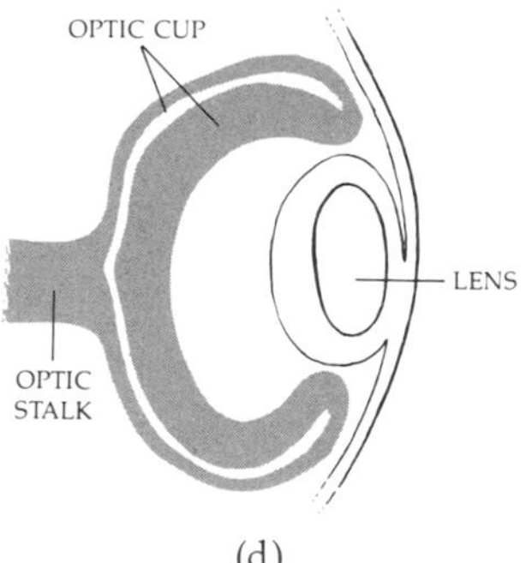

swimming and crawling. In the higher vertebrates, however, this initial segmented arrangement is almost obliterated by modifications and insertions of the muscles associated with terrestrial locomotion.

Mesoderm lateral to the somites forms the kidneys, the gonads, and the ducts of the excretory and reproductive systems. The ventral mesoderm splits into two layers. One sheet of mesoderm lines the thoracic and abdominal cavities, and another becomes the outer layer of the internal organs. The space between the two sheets is the coelom.

Endoderm

The endoderm differentiates into tissues of the respiratory and digestive tracts and a number of related organs. Early in development, endodermal pouches develop at the anterior end of the endodermal tube (the gut). They push laterally until they meet the ectoderm, which folds inward to meet the pouches, producing a series of grooves on the surface of the embryo. In lower vertebrates, endoderm and ectoderm fuse and a perforation forms, around which gill filaments develop. In terrestrial vertebrates, including humans, portions of these endodermal pouches develop into eustachian tubes (which connect the pharynx and the middle ear), tonsils, parathyroid glands, and the thymus gland. Posterior to the gill region, the lungs develop as similar outpocketings, branching into two sacs; more posterior still, outpushings from the primitive gut begin to differentiate into liver, gallbladder, and pancreas.

No organ system is derived from only one type of tissue. For example, the lining of the intestine is of endodermal origin; these lining cells secrete the digestive juices and absorb the digested materials. (Secretion and absorption are the principal functions of the intestine.) However, the functional structure of the intestine also includes muscles, connective tissue, blood vessels, nerves, and an outer wrapping, which are made up of tissues derived from mesoderm and ectoderm.

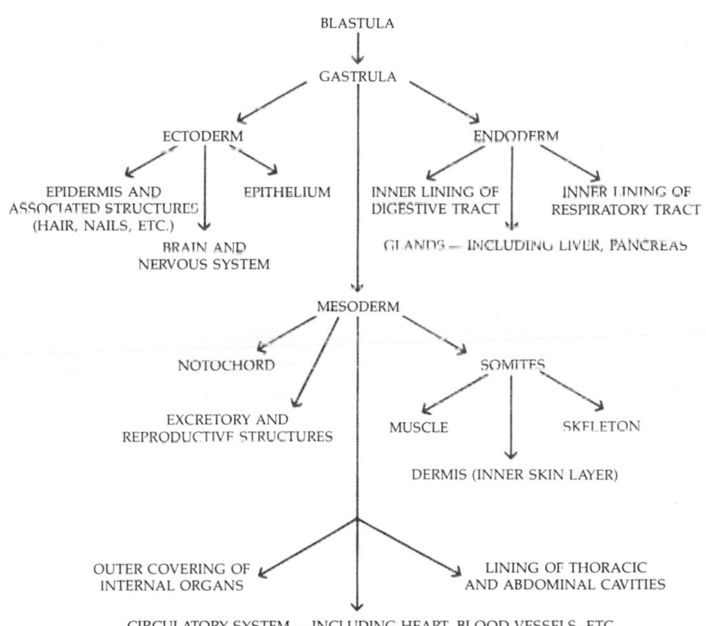

33–21
The principal channels of vertebrate development.

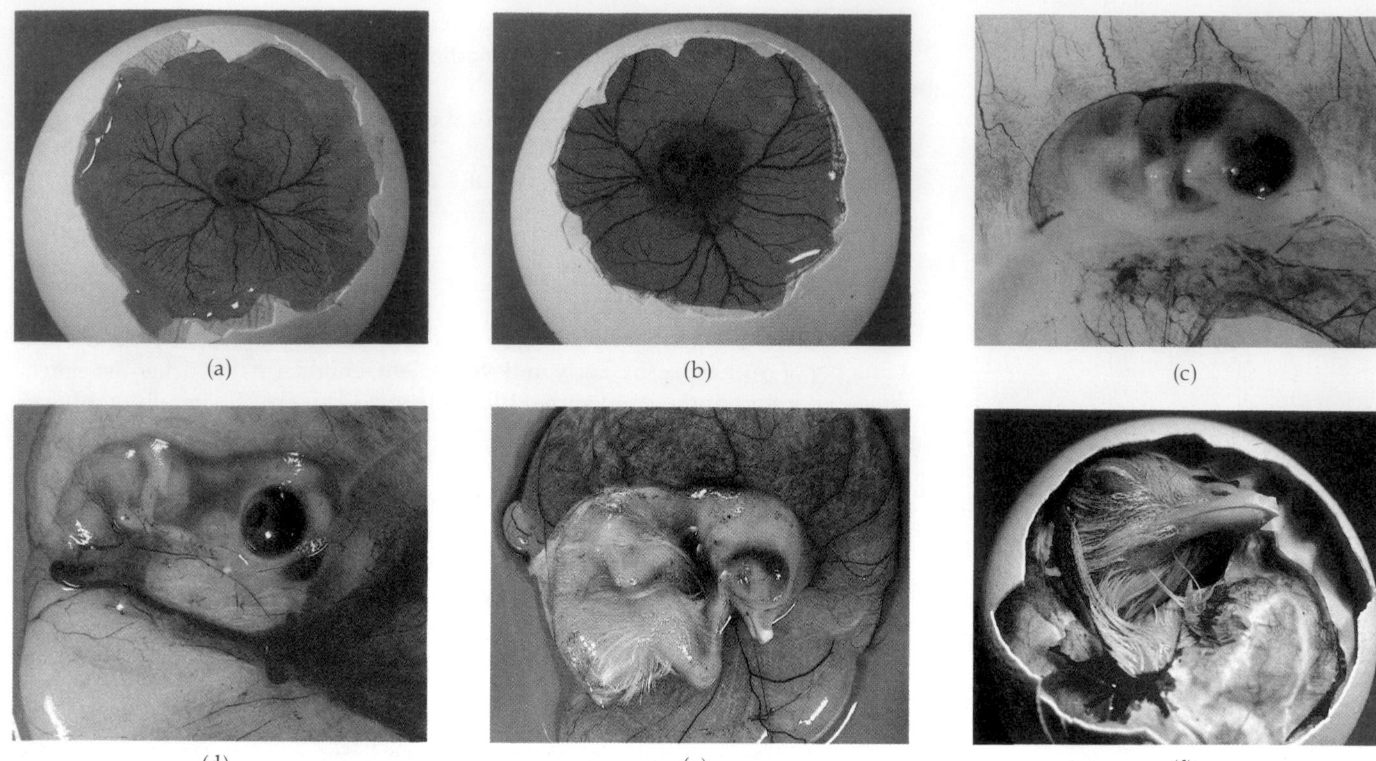

33–22
Chick embryo at (a) *4 days,* (b) *5 days,* (c) *8 days,* (d) *11 days,* (e) *14 days, and* (f) *21 days, just prior to hatching.*

EMBRYOLOGY AND EVOLUTION

Not long after the establishment of evolutionary theory, the German biologist E. H. Haeckel, on the basis of his studies of the embryos of a number of organisms, proposed that each organism as it grows from the one-celled egg to the multicelled individual passes through all the evolutionary stages that preceded it—that is, "ontogeny [development] recapitulates phylogeny," or, to put it more simply, "each animal climbs up its family tree." According to this theory, the human embryo might first resemble a one-celled amoeba, then a one-layered *Volvox,* then a two-layered coelenterate, and so on, right through fish and amphibian. This intriguing idea gained widespread popularity, since much evidence, such as the gill pouches of mammalian embryos, seemed to provide strong support. If the evidence hidden in the embryo could be deciphered, Haeckel's supporters believed, the complete evolutionary history of the species would be revealed. As a consequence, painstaking descriptions of embryonic development were the major biological works of the late nineteenth century—the type of accomplishment that earned a biologist a full professorship. Haeckel was wrong; embryos do not resemble any mature organism—they resemble, as we have seen so far, other embryos. However, these resemblances are indeed evidence of evolutionary relationships. Perhaps most important, structures such as gill pouches persist, not as signposts conveniently conserved for the biologist, but because they have selective value for the embryo, either in terms of its immediate survival or as a necessary pathway in development.

33–23
A human egg surrounded by sperm cells. At the top are two polar bodies. Surrounding the ovum is a layer of mucoprotein, the zona pellucida.

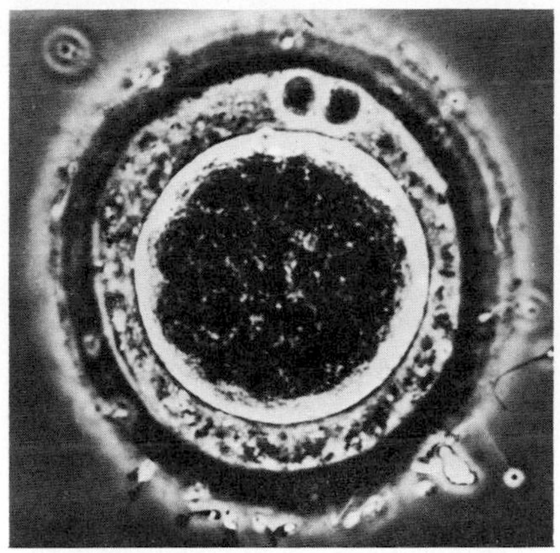

DEVELOPMENT OF THE HUMAN EMBRYO

Human egg cells, as we noted in the preceding chapter, mature in the ovary and are released at approximately 28-day intervals. Fertilization usually takes place in the oviduct. As in other species, fertilization results in introduction of the genetic material of the father, changes in the outer surface of the egg that prevent entry of other sperm cells, metabolic activation of the egg, and cleavage. After fertilization, the egg continues its passage down the oviduct, where the first cell divisions take place. At about 36 hours after fertilization, the fertilized egg divides to form two cells (Figure 33–24a); at 60 hours, the two cells divide to form four cells. At three days, the four cells divide to form eight. In this early stage, all of the cells are of equal size, as they are in the sea urchin.

As the blastula develops further, a large cavity forms; a cross section of the cell mass would look somewhat like a ring with the stone on the inside. The inner cell mass (the stone) is a ball of cells at one pole of the blastula and will develop into the embryo proper. The ring is called the trophoblast (from the Greek word *trophe*, "to nourish"). It is composed of a double layer of cells and completely encloses the developing embryo (Figure 33–25). The trophoblast is the precursor of the placenta and will give rise to the chorion. (In the chick, the surrounding membranes, you will recall, grew out from cells at the edge of the embryo early in development. It is as if the human embryo has skipped this step entirely, arriving, as it were, pre-packaged.)

At about the sixth day, the trophoblast makes contact with the tissues of the uterus and releases a hormone, chorionic gonadotropin. This hormone prevents menstruation by stimulating the corpus luteum to continue its production of estrogens and progesterone in large amounts. Pregnancy tests involve the detection of chorionic gonadotropin in blood or urine. The trophoblast cells, multiplying rapidly by now, chemically induce changes in the endometrium and invade it. As the embryo penetrates the endometrial tissues, it becomes surrounded by ruptured blood vessels and the nutrient-filled blood escaping from them. The trophoblast thickens and develops amoebalike projections that invade the uterine lining to develop into extraembryonic membranes.

33–24
(a) *A human embryo at the two-cell stage. The cells are still surrounded by an outer membrane. Sometimes they separate at this stage, resulting in identical twins.* (b) *The cells continue to divide, but, since the volume of the embryo does not increase, they are still easily contained within the same membrane.*

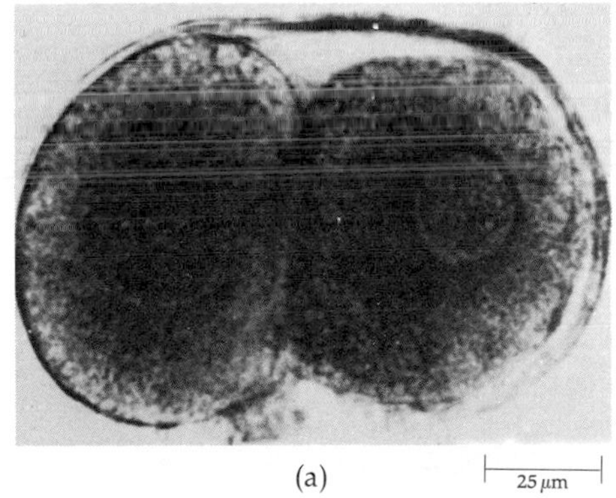

(a)

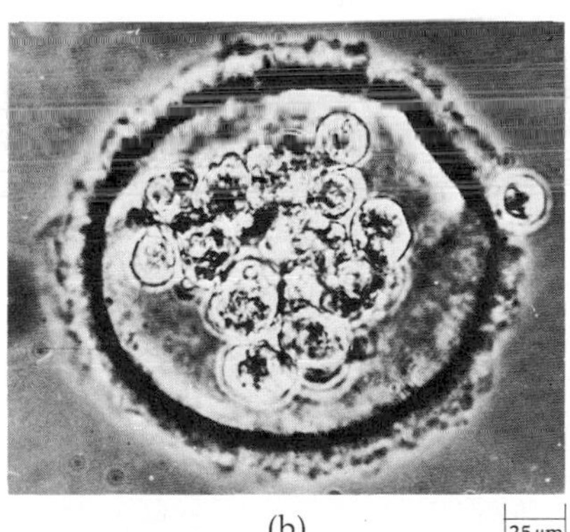

(b)

33–25
Human embryo at 13 days, showing the amniotic sac and the yolk sac.

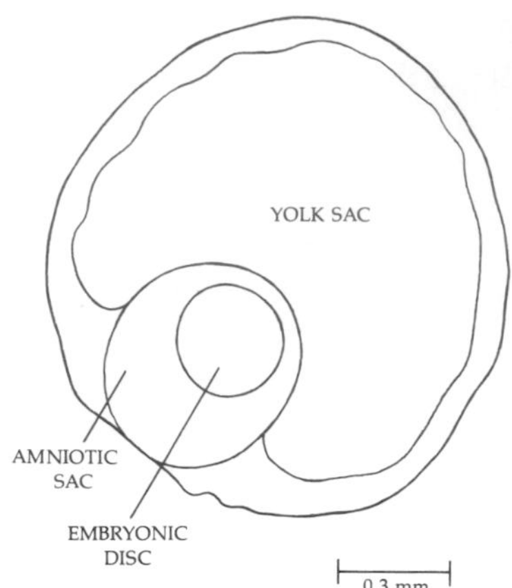

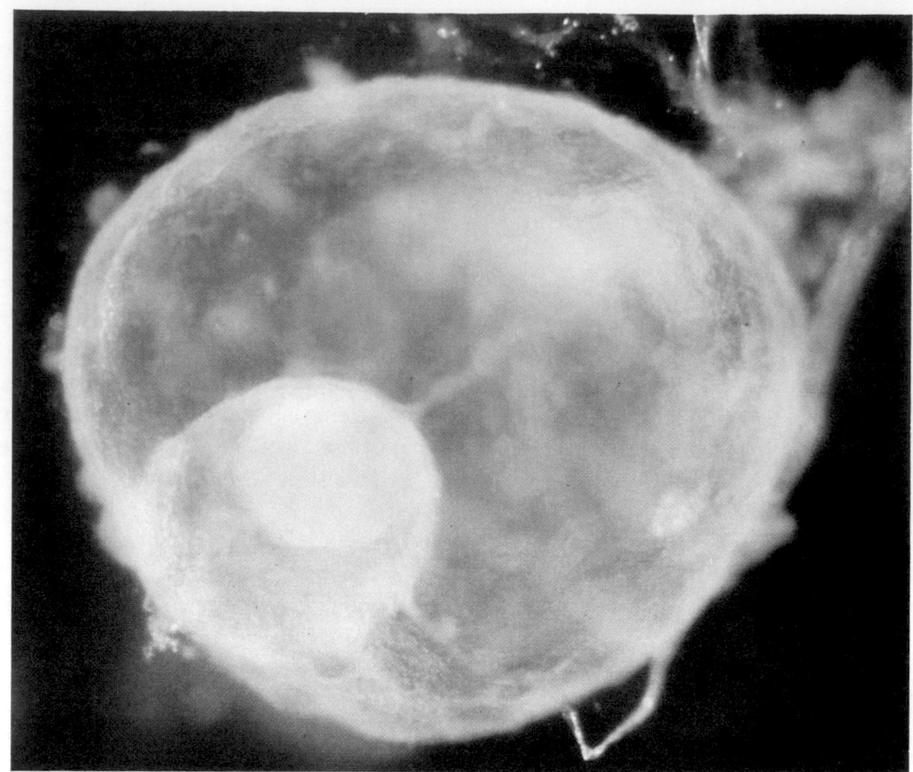

Human Embryonic Membranes and the Placenta

As the embryo becomes implanted, the embryonic membranes begin to develop. These have interesting similarities to and differences from the avian-reptilian membranes discussed previously. In the first place, the yolk sac, which develops first, has no yolk. This lack of yolk is a secondary evolutionary development. The monotremes (egg-laying mammals) produce eggs with yolks. The marsupials produce eggs with yolks, but the yolks are discarded at the first cleavage. The placentals produce eggs with no yolks but a prominent cavity where the yolk "used to be." Cleavage and the cellular migrations at gastrulation proceed as if yolk were still there.

The allantois connects with the hindgut and eventually becomes a primitive urinary bladder. In humans, wastes are not segregated within the fetus, as they are in the chicken, but are transported to the maternal bloodstream. In the course of development, the endodermal component of the allantois (the actual sac) becomes greatly reduced, but the allantoic mesoderm spreads out; it is the source of the major blood vessels on the embryonic side of the placenta. The allantoic stalk eventually becomes the umbilical cord, the embryo's principal connection with the uterine tissues. The blood vessels of the allantois transport oxygen and nutrients that have diffused in from the mother's circulatory system and carry off carbon dioxide and other wastes.

33–26
Levels of estrogen, progesterone, and chorionic gonadotropin excreted in the urine during pregnancy. Urinary excretion rates indicate the concentrations of these hormones in the blood.

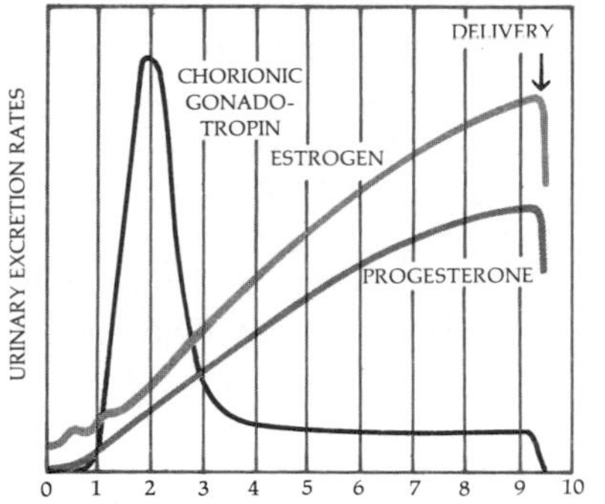

The third membrane is the amnion. The space between the amnion and the embryo is the amniotic cavity. This cavity fills with a saline fluid. For the genetic testing procedures described in Chapter 18, fluid samples are taken from the amniotic cavity. The samples obtained by amniocentesis contain cells that have been sloughed off by the embryo (see page 311).

The fourth membrane is the chorion, which is a combination of ectoderm from the trophoblast and mesoderm grown out from the embryo itself. By about the fourteenth day, chorionic villi begin to form; the formation of the villi represents the beginning of the mature placenta. The chorion (the outer membrane of the trophoblast), as we mentioned previously, is a source of gonadotropic hormone (Figure 33–26). By this time (about the time of the first missed menstrual period), a pregnancy test based on detection of this hormone will probably be positive.

By the end of the third week after conception, the placenta covers 20 percent of the uterus. It is a disc-shaped mass of spongy tissue through which oxygen, food molecules, and wastes are exchanged between mother and embryo. It is formed from a maternal tissue, the endometrium, as well as from the chorion of the fetus, and has a rich blood supply from both. Chorionic villi dip into pools of maternal blood; however, the fetal and maternal circulatory systems are not directly connected, so maternal and fetal blood cells do not mix. Molecules, including food and oxygen, diffuse from the maternal bloodstream through the placental tissue and into the blood vessels that carry them into the fetus. Similarly, carbon dioxide and other waste products from the fetus are picked up from the placenta by the maternal bloodstream and carried away for disposal through the mother's lungs and kidneys.

33–27
From the placenta, numerous fingerlike chorionic villi project into the maternal blood space in the wall of the uterus. This space is kept charged with blood from branches of the uterine artery. Across the thin barrier separating fetal from maternal blood, exchange of materials takes place: soluble food substances, oxygen, water, and salts pass into the umbilical vein from the mother's blood; carbon dioxide and nitrogenous waste, brought to the placenta in the umbilical artery, pass into the mother's blood. The placenta is thus the excretory organ of the fetus as well as its respiratory surface and its source of nourishment.

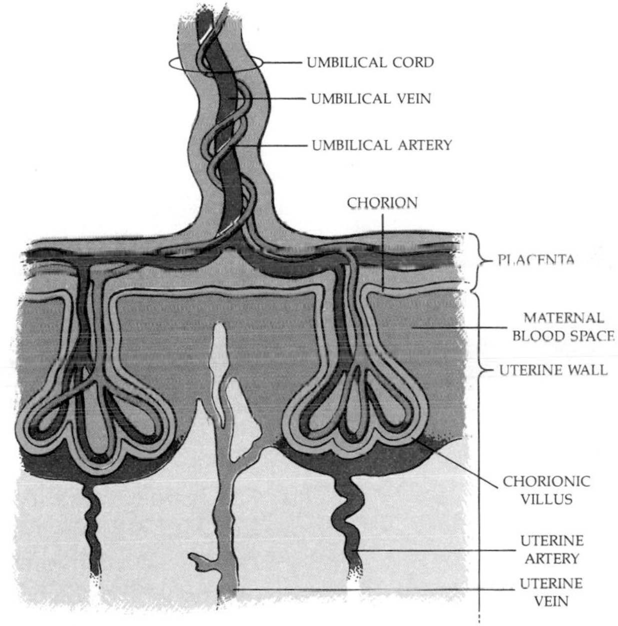

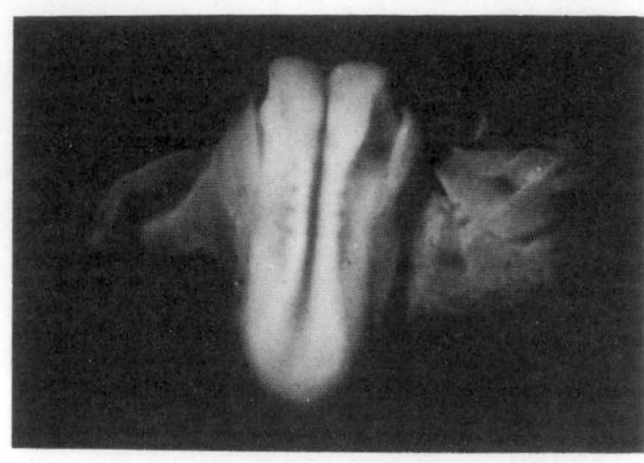

33–28
As the embryo develops, two ridges form and enlarge. The groove between them is called the neural groove. From the neural groove, the dorsal nerve cord—one of the "trademarks" of the chordates—develops. These two folds soon close. This embryo is 18 days old.

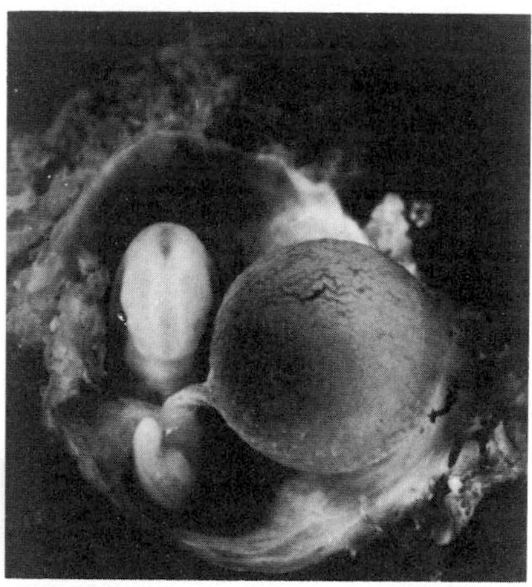

33–29
A human embryo at 28 days. The balloon-like structure is the yolk sac. The embryo is curved toward you. At the top you can see the bulge of the rudimentary brain. The "seam" where the neural ridges closed is still visible.

As the embryo grows larger, it remains attached to the placenta by the umbilical cord, which permits it to float freely in its sac of amniotic fluid.

It is the existence of the placenta that makes possible the long gestation period characteristic of the placental mammals.

The First Trimester

During the second week of its life, the embryo grows to 1.5 millimeters in length, and its major body axis begins to develop. (In this and subsequent measurements, the fetus is measured from crown to rump.) As it elongates, a primitive streak forms, very similar in appearance to the primitive streak of the chick. Cells migrating through the primitive streak form the mesoderm, establishing the three-layered embryo. The dorsal part of the body can be seen to be divided into somites.

During the third week, the embryo grows to 2.3 millimeters long, and most of its major organ systems begin to form: the neural groove (Figure 33–28), which is the beginning of the central nervous system (spinal cord and brain), the first organ system to develop; the heart and blood vessels; the primitive gut; and the muscle rudiments.

By 22 days, the very rudimentary heart, still only a tube, begins to flutter and then to pulsate. From this time on, the heart will not stop its 100,000 beats per day until the death of the individual. Soon after, the eyes begin to form. Also, by this time, about 100 cells have been set aside (in the yolk sac) as germ cells, from which the ova or sperm cells of the individual will eventually develop. These cells begin to crawl, amoebalike, toward the site at which the reproductive organs will develop.

By the end of the first month, the embryo is 5 millimeters in length and has increased its mass 7,000 times. The neural groove has closed, and the embryo is now C-shaped (Figure 33–29). At this stage, it can be clearly seen that the tissues lateral to the notochord are arranged in paired somites. Each embryo has 40 pairs of somites, from which muscles, bones, and connective tissues will develop. This segmentation of the muscles persists in the adult forms of lower vertebrates—fish, particularly—but not in the higher, terrestrial vertebrates. The heart, even as it beats, develops from a simple contracting set of paired tubes to a four-chambered vessel.

By 38 days, the germ cells reach their destination, the developing gonads. Although the primitive reproductive organs have begun to form by this time, male and female embryos are still morphologically identical. Development of male or female characteristics depends on whether or not the embryo is exposed to androgen. Without such an exposure, the embryo, regardless of its chromosomes, remains female. Gallbladder and pancreas are present and there is clear differentiation of the divisions of the intestinal tract. The liver now constitutes about 10 percent of the body of the fetus and is its main blood-forming organ. Arms, legs, elbows, knees, fingers, and toes are all forming during this time. As another reminder of our ancestry, there is a temporary tail.

By the end of the second month, the major steps in organ development are more or less complete, and the embryo, despite its very small size (30 millimeters), is almost human-looking, and from this time on it is generally referred to as a fetus. Its head is still relatively large, because of the early and rapid development of the brain, but the size of the head in proportion to the body will continue to be reduced throughout gestation (and throughout childhood as well).

33–30
A human embryo at 40 days, front and back views. Notice the spinal cord, brain, and paddlelike feet. The embryo is now about 16 millimeters long, little more than half an inch. (Roberts Rugh and Landrum B. Shettles, M.D., From Conception to Birth: The Drama of Life's Beginnings, *Harper & Row, Publishers, Inc., New York, 1971.)*

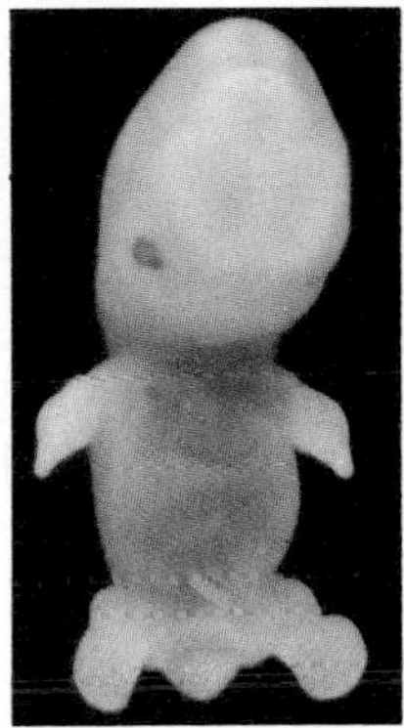

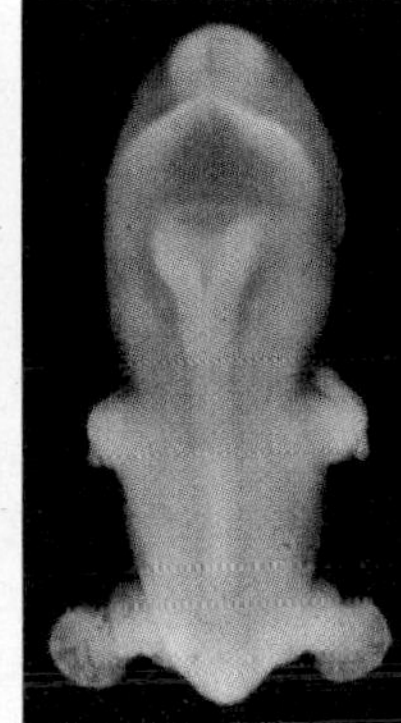

The first two months are the most sensitive period as far as the possible influence of external factors is concerned. For example, when the arms and legs are mere rudiments (fourth and fifth weeks), a number of substances can upset the normal course of events and result in limb abnormalities. The experience with the presumably safe tranquilizer thalidomide in the 1960s is a tragic and familiar example. Because of the widespread publicity about the "thalidomide babies," some of whom are now becoming thalidomide adults, there was an increase in research into teratogens (substances that cause fetal deformities). A large number of drugs have been found to cause birth defects, and pregnant women and their doctors have become far more cautious about the use of any drug during these critical first months. Heavy alcohol consumption also affects normal organogenesis; frequently, alcoholic mothers abort during the first trimester.

Infections may also affect the development of the embryo. Rubella (German measles) is a very mild disease in children and adults. Yet when contracted by the mother during the fourth through the twelfth weeks of pregnancy, it can have damaging effects on the formation of the heart, the lens of the eye, the inner ear, and the brain, depending on exactly when the infection occurs in relation to embryonic development. The correlation between congenital heart defects among newborn children and a previous German measles epidemic was reported by two Australian physicians in 1942. Some studies have since shown that more than 75 percent of the infants born to mothers having had German measles during the first few months of pregnancy develop heart defects. Similarly, exposure to x-rays at doses that would not affect an adult or even an older fetus may produce permanent abnormalities.

During the third month, the fetus begins to move its arms and kick its legs, and the mother may become aware of its movements. Reflexes, such as the startle reflex and (by the end of the third month) sucking, first appear at this time. Its face becomes expressive; the fetus can squint, frown, or look surprised. Its respiratory organs are fairly well formed by this time but, of course, are not yet functional. The external sexual organs begin to develop.

33–31
(a) *Embryo at 2 months. The yolk sac is now smaller in comparison to the embryo but still persists. The umbilical cord, connecting the embryo to the placenta, contains both veins and arteries. Portions of the skeleton are visible.* (b) *Human fetus at 2 months, 1 week, now almost 4 centimeters long. The eyes have lenses but are covered by lids that will fuse during the third month and remain closed for the next three months. (Rugh and Shettles, op. cit.)*

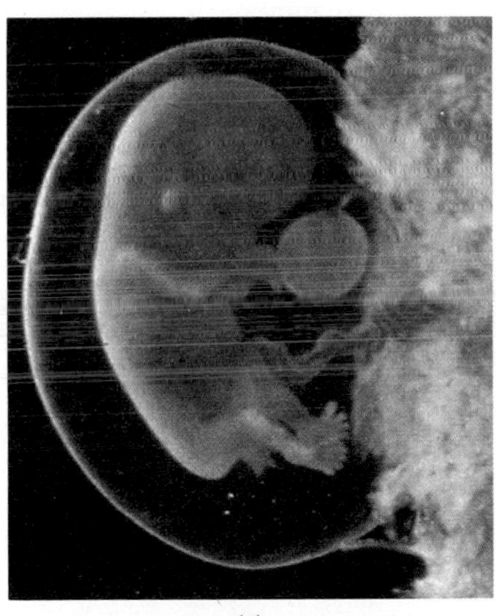

(a)

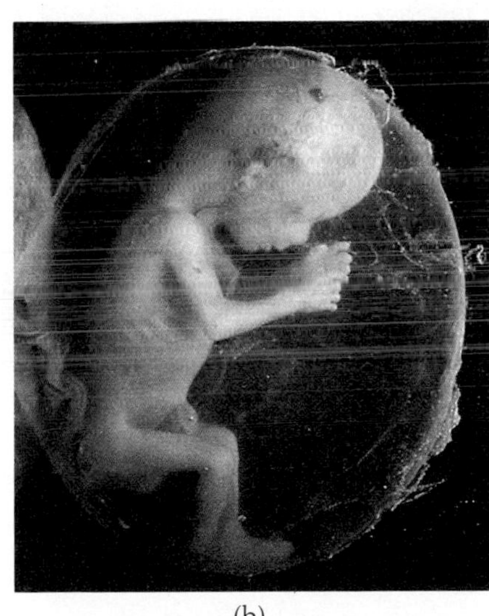

(b)

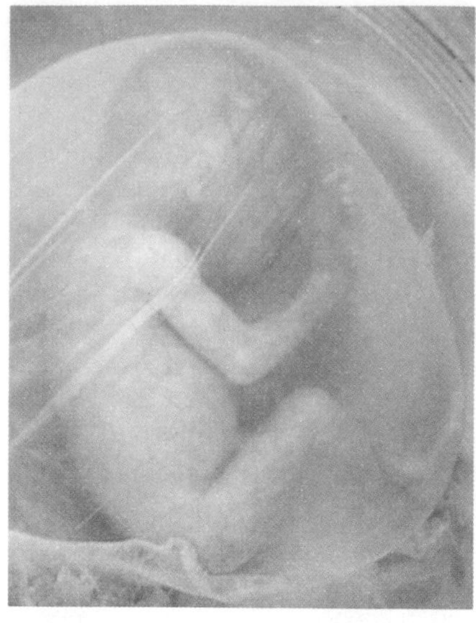

(a)

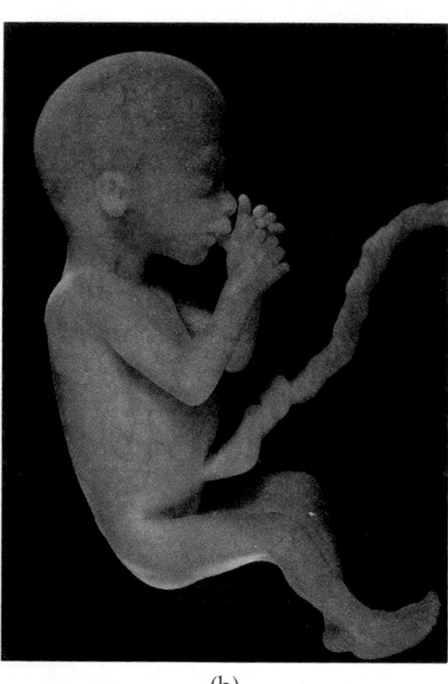

(b)

33–32
(a) *Fetus at 16 weeks. Blood vessels are visible through the translucent skin. The hands and feet are well formed; even the fingernails are clearly visible. The fetus is now about 13 centimeters long and fills the uterus, which is expanding as the fetus grows.* (b) *A 17-week-old fetus, sucking its thumb. (Rugh and Shettles, op. cit.)*

By the end of this month, the fetus is about 9 centimeters long from the top of its head to its buttocks and weighs about 15 grams ($\frac{1}{2}$ ounce). It can suck and swallow, and occasionally does swallow some of the fluid that surrounds it in the amniotic sac. The finger, palm, and toe prints are now so well developed that they can be clearly distinguished by ordinary fingerprinting methods. The kidneys and other structures of the urinary system develop rapidly during this period, although waste products are still disposed of through the placenta. By the end of this period—the first trimester of development—all the major organ systems have been laid down.

The Second Trimester

During the fourth month, movements of the fetus become obvious to the mother. Its bony skeleton is forming and can be seen with x-rays. The body is becoming covered with a protective cheesy coating. The four-month-old fetus is about 14 centimeters long and weighs about 115 grams (4 ounces).

By the end of the fifth month, the placenta covers about 50 percent of the uterus. The fetus has grown to almost 20 centimeters and now weighs 250 grams. It has acquired hair on its head, and its body is covered with a fuzzy, soft hair called the lanugo, from the Latin word for "down." Its heart, which beats between 120 and 160 times per minute, can be heard with a stethoscope. The five-month-old fetus is already discarding some of its cells and replacing them with new ones, a process that will continue throughout its lifetime. Nevertheless, a five-month-old fetus cannot yet survive outside of the uterus. The youngest fetus on record to survive was about 23 weeks old and required continuous assistance in breathing, taking food, and maintaining its body temperature.

During the sixth month, the fetus has a sitting height of 30 to 36 centimeters and weighs about 680 grams. By the end of the sixth month, it could survive outside the mother's body, although probably only with respiratory assistance in an incubator. Its skin is red and wrinkled, and although teeth are only rarely visible at birth, they are already forming dentine. The cheesy body covering, which helps protect the fetus against abrasions, is now abundant. Reflexes are more vigorous. In the intestines is a pasty green mass of dead cells and bile, known as meconium, which will remain there until birth.

The Final Trimester

During the final trimester, the fetus increases greatly in size and weight. In fact, the fetus normally doubles in size just during the last two months. During this period, many new nerve tracts are forming and new brain cells are being produced at a very rapid rate. By the seventh month, brain waves can be recorded, through the abdomen of the mother, from the cerebral cortex of the fetus. Some recent research indicates that the protein intake of the mother is important during this period if the child is to have full development of its brain. Infants weighing less than 2,000 grams (4 pounds, 6 ounces) at birth are at high risk of death or severe brain damage. Their risk of subtle intellectual deficits, often undetected until they enter school, may be even higher but has been studied much less thoroughly.

As the fetal period progresses, the physiology of the fetus becomes increasingly like the physiology of the adult, and so agents that affect the mother also threaten the late fetus. An increasingly familiar example is found in the infants being born with heroin or methadone addiction.

33–33

A human fetus, shortly before birth, showing the protective membranes surrounding it and the uterine tissues. The cervical plug is composed largely of mucus. It develops under the influence of progesterone and serves to exclude bacteria and other infectious agents from the uterus. In 95 percent of all births, the fetus is in this head-down position.

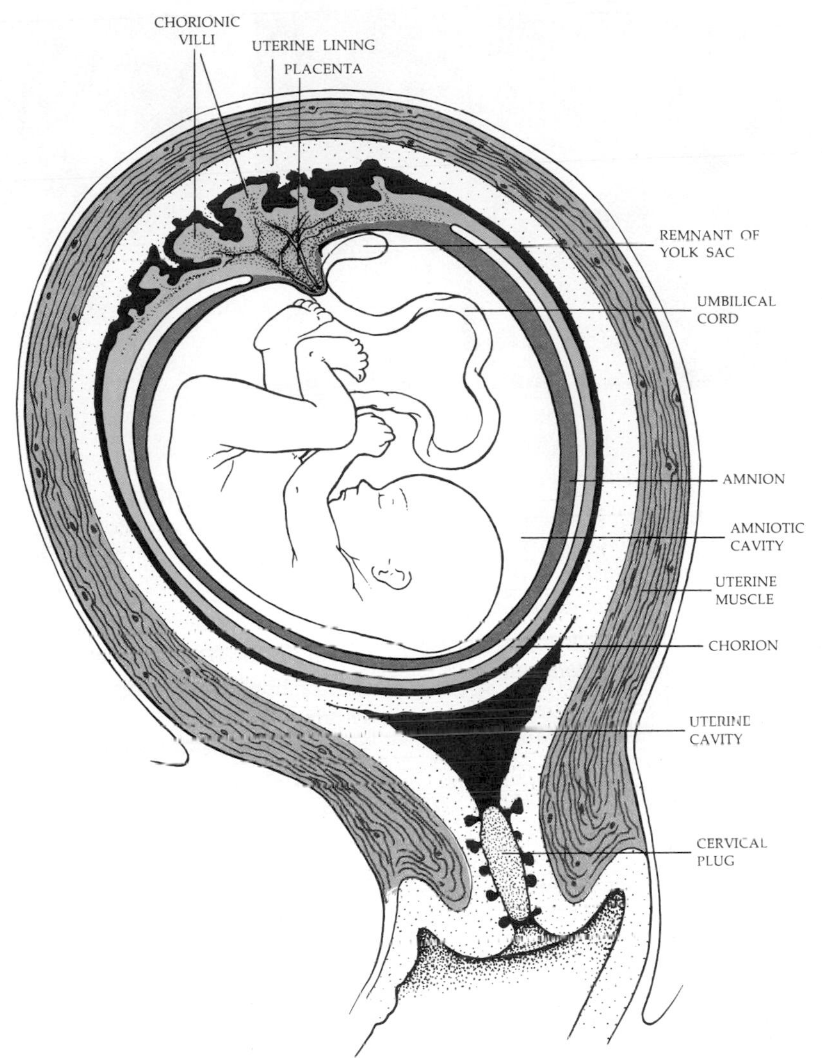

During the last month of pregnancy, the baby usually acquires antibodies from its mother. These are globular proteins—formed against bacteria, viruses, or other foreign invaders—that defend the body against attack by such microorganisms. The baby becomes immune to whatever the mother is immune. This immunity is only temporary. Within one to two months after birth, the maternal antibodies will be gradually replaced by antibodies manufactured by the baby's own immune system.

During this last month, the growth rate of the baby begins to slow down. (If it continued at the same rate, the child would weigh 90 kilograms—about 200 pounds—by its first birthday.) The placenta begins to regress and becomes tough and fibrous.

Birth

The date of birth is calculated as about 266 days after conception or 280 days after the beginning of the last regular menstrual period, but only some 75 percent of

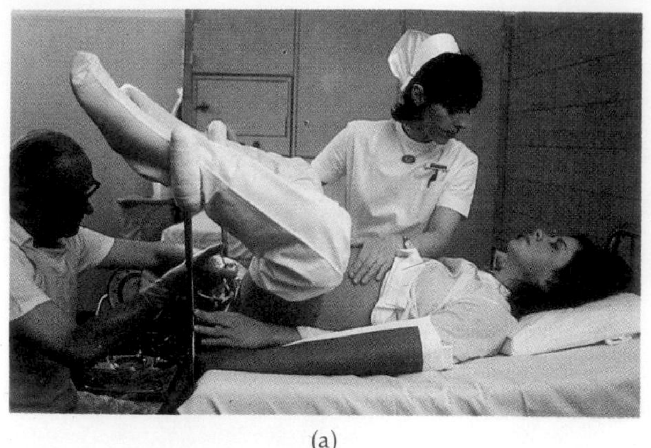

(a)

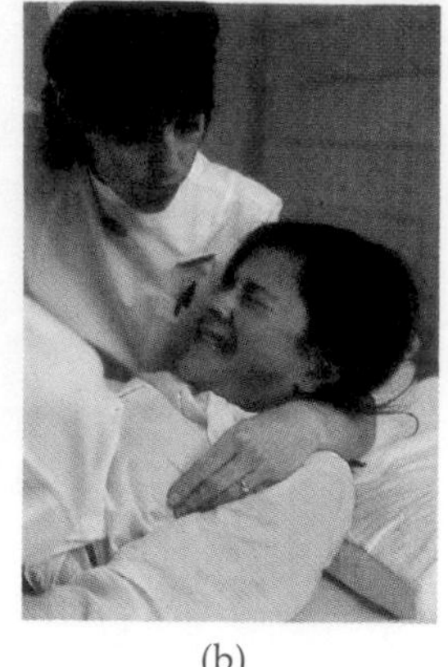

(b)

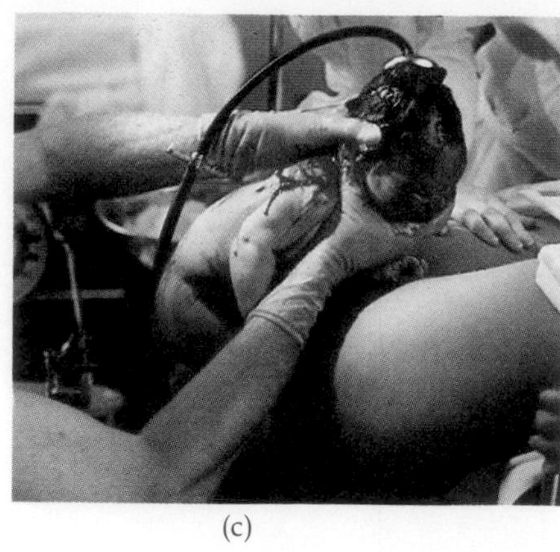

(c)

33–34
Birth of a baby. (a) *Mother in delivery room just before the beginning of the second stage of labor.* (b) *A strong contraction.* (c) *The baby appears. Visible on the top of the baby's head is a suction device (much safer than forceps) that has been placed there to aid in maneuvering the infant into position for delivery.* (d) *Doctor holding newborn infant.*

babies are born within two weeks of the scheduled time. Labor is divided into three stages: dilation, expulsion, and placental stages. Dilation, which lasts from 2 to 16 hours (it is longer with the first baby than with subsequent births), begins with the onset of contractions of the uterus and ends with the full dilation, or opening, of the cervix. At the beginning of the stage, uterine contractions occur at intervals of about 15 to 20 minutes and are relatively mild. By the end of the dilation stage, contractions are stronger and occur about every 1 to 2 minutes. At this point, the cervix is dilated to about 10 centimeters in diameter. Rupture of the amniotic sac, with the expulsion of fluids, usually occurs during this stage.

The second, or expulsion, stage lasts 2 to 60 minutes. It begins with the full dilation of the cervix and the appearance of the head in the cervix, called crowning. Contractions at this stage last from 50 to 90 seconds and are 1 or 2 minutes apart.

The third, or placental, stage begins immediately after the baby is born. It involves some contractions of the uterus and the expelling from the vagina of fluid, blood, and finally the placenta (also called the afterbirth), with the clamped umbilical cord attached. Minor uterine contractions continue; they help to stop the flow of blood and to return the uterus to its prepregnancy size and condition.

The baby emerges from the warm, protective enclosure where it has been nourished and permitted to grow for nine months. The umbilical cord—until that moment, its lifeline—is severed. The baby cries as it takes its first breath and then starts to breathe regularly. Its separate existence has begun.

Human beings are relatively helpless at birth. It has, in fact, been argued that they are born "too soon," and that the gestation period should be twice as long. Evidence for this point of view includes the fact that rapid growth continues in humans far beyond its cessation in other primates. At birth, a human fetus is made up of about ten trillion (10^{13}) cells consisting of more than 100 distinctly different types. By adulthood, the number of cells in the human body is several hundred trillion (3 to 5×10^{14}). (Such large numbers are hard to comprehend. It might help to realize that there are this many seconds in 100,000 years.) Most notable, our brains, at birth, are still growing at fetal rates. (Cranial capacity is only 23 percent that of the adult.) Unlike other placental mammals, the ends of our long bones are still entirely cartilaginous at birth, and we are born toothless.

The time at which we are born is, apparently, a result of at least two conflicting selective pressures. One is the steady increase in brain size that has marked homi-

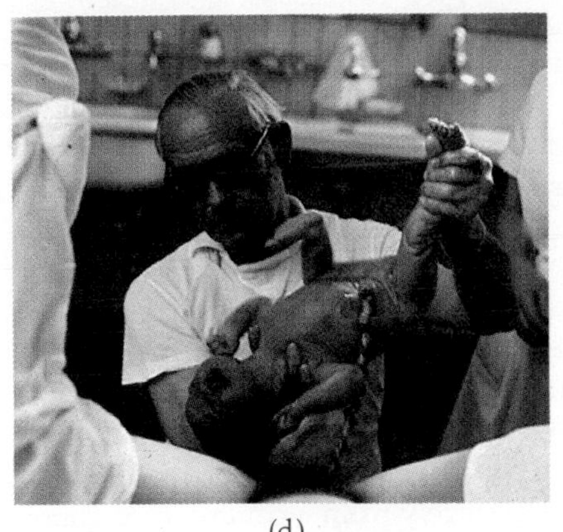

(d)

nid evolution and the other the tendency among the primates toward fewer young and longer periods of gestation. The human embryo remains in the uterus as long as is practicable, given its outsized brain. On the other hand, though we may be evicted prematurely, it is more than a decade before we could survive on our own, and often more than two decades before we have to. The process of physical growth and maturation requires almost one-third of the average life span.

CONTINUOUS DEVELOPMENT: THE SKIN

The processes of development—growth, differentiation, morphogenesis—do not cease with maturity. Cells and tissues constantly undergo replacement and repair. The extent varies with the type of cell. Nerve cells, for example, cannot be replaced, nor can they regenerate if the cell body is destroyed (though an axon can regrow if the cell body is intact). Red blood cells, epithelial cells, and sperm, however, are continuously discarded and replenished.

Consider, for example, the skin, the body's largest organ. As shown in Figure 33–35, the skin has three main layers, or zones. The innermost layer, lying below the skin, is made up largely of fatty tissue, which serves as a cushion and an insulator. Above the fatty layer is the lowest layer of the skin itself, the *dermis*. The dermis is a meshwork of fibrocytes surrounded by collagen, other proteins, and mucopolysaccharides. Smooth-muscle fibers lie in the dermis and are attached to the hair follicles; it is when these muscles contract that your hair stands on end. Small blood vessels twist through the dermis, bringing oxygen and chemical energy to the cells. The tiny nerve endings in the skin register sensations of touch, pain, heat, or cold—different receptors for each sensation.

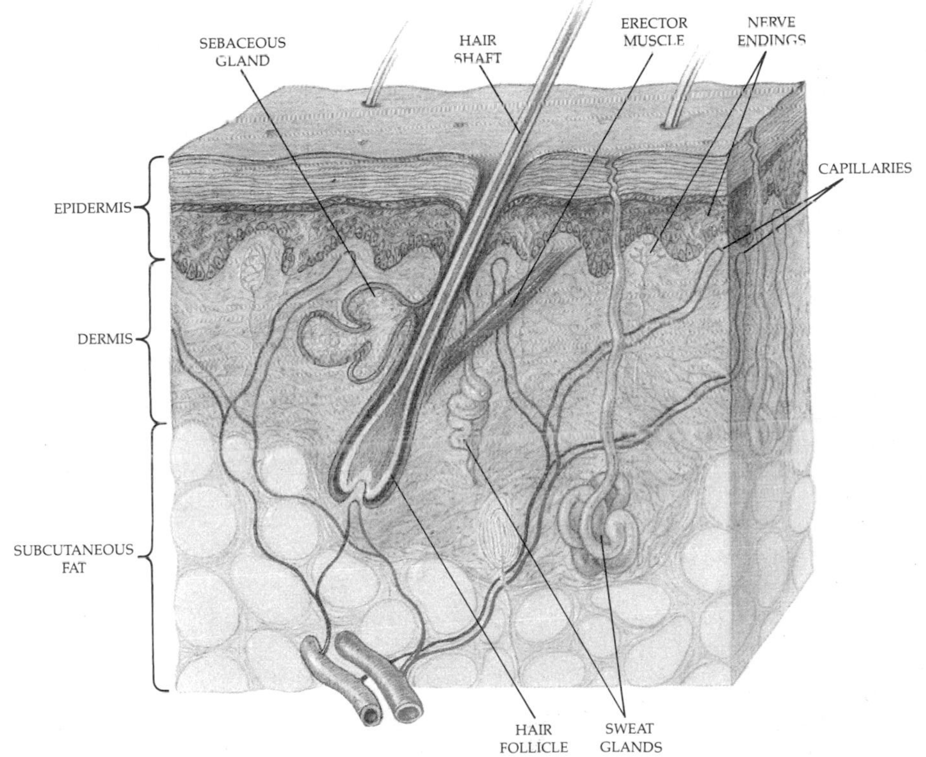

33–35
A section of human skin, showing the epidermis and the dermis, which together make up the skin, and the underlying subcutaneous fat. The dermis, consisting mostly of connective tissue, contains sensory receptors, hair follicles, erector muscles, which raise the hair when contracted, and sweat and sebaceous glands. The latter produce a fatty substance that lubricates the skin surface. Above the dermis are the two layers of epidermis: a lower layer of living cells (the basal layer) and an upper layer of dead cells (the stratum corneum) filled with keratin. The epidermis is a turnover system of epithelial cells: cells produced in the basal layer migrate toward the surface and die. At the base of the epidermis are melanocytes, the pigment organs that produce the granules responsible for skin colors. The epidermis and dermis are separated by the basement membrane.

The surface of the dermis forms hills and valleys, papillae, into which the overlying layer of skin, the *epidermis*, fits very exactly. The epidermis can be separated from the dermis by special treatments that disrupt or dissolve the thin layer of cementing substance holding them together. When the two tissues are separated, each retains the same papillary pattern, one an exact complement, or negative mold, of the other. It is the papillary pattern that determines the texture of the skin.

The epidermis, which has no blood supply, depends upon the underlying dermis for food and oxygen. Directly above the dermis is a layer of living epidermal cells, the basal layer, which divide constantly to replace cells lost from the skin surface. Among these cells are the melanocytes, the pigment-producing cells.

Once an epidermal cell is free of the basal layer, it begins to differentiate. As the cell moves toward the surface, it changes progressively from a columnar or cuboidal basal cell to a flattened, tile-shaped squamous cell. Also in the course of this migration it takes on new synthetic functions, producing keratin, which is a mixture of insoluble proteins. The production of keratin continues until the death of the cell.

The cells of the basal layer are only loosely attached to one another, but once they move out and begin their migrations, desmosomes (page 119) form between adjacent cells, linking them together. Thus the epidermal cells migrate to the surface bonded together into a continuous sheet.

The outer layer of epidermis is the stratum corneum, or "horny layer," of dead tissue that provides the waterproof outer wrapping on which our survival as terrestrial animals depends. The principal component of the stratum corneum is keratin, the substance from which nails, scales, feathers, claws, horns, hooves—all epidermal derivatives—are formed. The keratin deposited in dead cells makes up about half of the total substance of the dead epidermal layer.

The stratum corneum also forms a continuous sheet. In tissue sections prepared for the microscope, it usually looks like a loosely packed pile of dead cells, but it has been shown recently that this appearance is misleading. By appropriate treatments, the stratum corneum can be removed intact. It is a waterproof, pliable, semitransparent film, thin and tough, like plastic wrap. This horny, dead layer of cells is shed in fragments from the human skin (and is a major component of ordinary house dust), but in animals such as lizards, it peels off every few days almost intact.

Differentiation of the epidermis, like embryonic organogenesis, depends upon secondary induction. Epidermis will differentiate only under the influence of the underlying dermis. (This is a major factor in the healing of a burn or other wound. If the dermal layer of skin is destroyed, it cannot regenerate the epidermal layer.) However, as studies in tissue culture have shown (Figure 33–36), actual contact between the tissues is not essential.

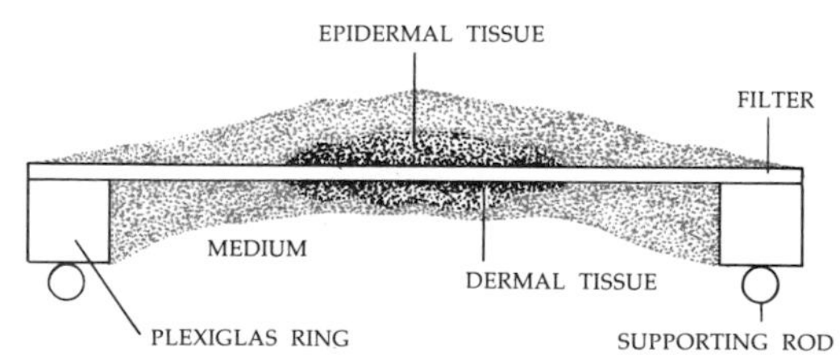

33–36
The differentiation and proliferation of epidermal cells depend on a continued interaction between epidermis and dermis. However, the two tissues do not have to be in direct contact for the necessary interactions to occur. This has been demonstrated by experiments using the filter apparatus shown here. Epidermis and dermis were grown in culture separated by a thin, porous membrane that permitted the passage of molecules but not cells. The epidermal cells directly opposite the dermis continued to divide and differentiate.

AGING

During the last 100 years, the average human life span has greatly increased. However, once an individual reaches the age of 70 or so, he or she will probably not live much longer than a 70-year-old of 100 or 200 years ago. In other words, death appears to be an internally programmed event. This program does not exist only for human beings; each species of mammal appears to have its characteristic life span, although few wild animals make it to the end of their allotted time.

The biological nature of this program has long intrigued both scientists and aging laymen. At one time it was believed that cells were immortal and that, therefore, aging was an "organismic" property. A leading theory of aging, in keeping with the cellular immortality concept, was that senescence was the result of the accumulation of toxins, particularly bacterial toxins in the gut. In this view aging was curable.

More recently the cellular immortality concept has been proved false. Studies of human cells in tissue culture by Leonard Hayflick of Children's Hospital Medical Center, Oakland, California, have shown that not only are they not immortal but that they have a very finite life span. (An interesting exception is found in cancer cells, which do keep dividing indefinitely in a nutrient medium.) Moreover, the life span of normal human cells is related to the age of the individual from whom the tissue sample was taken. Cells from a human embryo, for example, double about 50 times before they die, whereas cells from a middle-aged man or woman double only about 20 times. When nuclei are transplanted between "young" cells and "old" cells, Hayflick found the old cell lives out the life span of the young cell (from which it has received the nucleus) and vice versa. In short, these experiments suggest that cell nuclei have only a determinate life span and it is this that sets the limit to longevity. There is, however, some doubt whether or not the life and death of cells in tissue culture reflects exactly what takes place in the environment of the whole organism.

SUMMARY

Development is the process by which a fertilized egg becomes a complete organism, consisting of billions of cells and closely resembling its parent organism.

Development in most species of animals begins with fertilization—the fusion of egg and sperm. Fertilization results in: (1) changes in the egg cell membrane that prevent fertilization by another sperm cell; (2) introduction of the genetic material of the male parent into the egg; (3) metabolic activation of the egg cell; and (4) cell division.

Development takes place in three stages: cleavage, gastrulation, and organogenesis. During cleavage, the original egg cell divides, with very slight, if any, change in overall volume. In eggs containing little or no yolk, such as those of the sea urchin, cleavage results in cells of approximately equal size. In yolkier eggs, such as those of the frog, cleavage is unequal, with fewer and larger cells in the yolky (vegetal) hemisphere. In eggs with a great deal of yolk, such as the hen's egg, cleavage is limited to a small, nonyolky disc at the top of the egg. In all cases, when cleavage is complete, the embryo consists of a sphere of cells, the blastula, with a central cavity, the blastocoel.

Gastrulation involves the movement of cells into new relative positions and results in the establishment of three body layers: endoderm, mesoderm, and ectoderm. In the course of gastrulation, an opening forms, the blastopore. Cells from the outer surface of the embryo move through the blastopore into the blastocoel and form a new cavity, the archenteron, which will be the primitive gut. In the chick and the human, the homologue of the blastopore is the primitive streak. When the movement of cells is complete, the central part of the dorsal surface of the gastrula flattens, becoming the neural plate, which folds to form the neural tube. The cells underlying the neural plate become notochord (a mesodermal tissue), and the somites become embryonic muscle tissue. The endodermal cells, on the interior of the embryo on the ventral side, become gut.

The small area of tissue on the dorsal lip of the blastopore prior to gastrulation is known as the organizer because, as shown by experiments on amphibians, it induces the cells overlying it to form the neural tube. (It itself forms the notochord.) If it is implanted in another embryo, it will induce the formation of a second neural tube in that embryo.

Other organ systems also begin to develop by this same process of embryonic induction, in which one tissue induces changes in the growth or migration of an adjacent tissue with which it comes in contact. The nature of the inducing substances is not known.

Each of the three primary cell layers established during gastrulation gives rise to particular tissues and organs. The epidermis (outer skin) and the nervous system arise from ectoderm. The notochord, muscles, bones and cartilage, heart and blood vessels, excretory organs and gonads, the inner lining of the skin, and the outer linings of the digestive and respiratory tracts arise from mesoderm. The lungs, salivary glands, pancreas, liver, and inner lining of the digestive tract develop from endoderm. The body cavity, or coelom, is a space, or discontinuity, between two layers of mesoderm.

In amniote eggs, such as those of reptiles, birds, and mammals, the embryo, as it develops, forms four embryonic membranes: the yolk sac, amnion, chorion, and allantois.

The human embryo develops as a cell mass, a blastodisc, and an extraembryonic layer of cells, the trophoblast. During these early stages, there is a large increase in the number of cells but little or no increase in the total size of the embryo. When the embryo descends into the uterus, at about its sixth day of development, the trophoblast develops rapidly and invades the maternal tissues. The trophoblast becomes the chorion and eventually the fetal part of the placenta. Like the chick embryo, the human embryo is surrounded by an amnion filled with amniotic fluid. Also present are a yolk sac with no yolk and an allantoic membrane, which develops to become part of the umbilical cord, linking the embryo to the placenta.

When the embryo is about 2 weeks old, a primitive streak forms, followed by the development of a neural plate and neural ridges, which fold to form a neural tube. Although the embryo is still very small (about 2.5 millimeters long), most of the major organs have begun to form in these very early weeks, which is why damage caused to the embryo by viral infection, x-rays, or drugs during this period can be widespread. By the end of the second month, the embryo, now called a fetus, is almost human-looking, although it only weighs about 1 gram. By the end of the third month, all of the organ systems have been laid down. Birth occurs, on the average, 266 days after conception. Labor consists of dilation, expulsion, and placental stages.

Developmental processes continue after birth. The skin is an example of continuous development that depends on interaction between dermis and epidermis. Recent experiments indicate that aging occurs because the nuclei of cells have only a determinate life span.

QUESTIONS

1. Distinguish among the following: blastula/gastrula; blastocoel/archenteron; ectoderm/endoderm; primary induction/secondary induction; fetus/embryo.

2. Describe, in general terms, the end results of each of the following events: fertilization, cleavage, gastrulation, organogenesis.

3. Follow the course of a single cell from its place of origin in the fertilized egg of a frog to its position in the eye cup of an early frog embryo. List, for each stage, all of the influences on this cell that might affect its history.

4. Suppose you wanted to make an exact copy—in popular language, a clone—of yourself. How would you go about it? (*Hint:* You would need the cooperation of your mother.)

CHAPTER 34 Integration and Control

In the two previous chapters, we examined the processes by which organisms reproduce themselves. In the chapters that follow, we shall be concerned with how they maintain themselves. As you will see, the requirements of multicellular organisms are remarkably similar to those of a single-celled organism—a *Paramecium,* for instance, or an amoeba. Each heterotrophic cell requires energy-yielding molecules; an oxidizing agent, usually oxygen; and a means for disposal of waste products, such as carbon dioxide. One of the outstanding differences, however, between a *Paramecium* and a multicelled animal is the problem of coordinating the activities of all the various cells, tissues, and organs in order to promote the survival of the whole organism. This problem of the integration and control of the body's many activities is the subject of this chapter.

The principal regulatory systems of the body are the nervous system and the endocrine system, the hormone-secreting glands and their products. In Chapter 32, we saw some examples of the ways these two systems interact: for instance, in the regulation of estrogen concentration, and in the synchrony of copulation and ovulation in cats and rabbits. In this chapter, we are going to examine these vast communication networks in order to provide a background for understanding the control of such functions as circulation, digestion, excretion, and temperature regulation, which are the subjects of the chapters that follow. Discussion of the brain and of some of the current research on brain functions is reserved for Chapter 40.

THE NERVOUS SYSTEM

Functionally, the nervous system differs from the endocrine system chiefly in its capacity for rapid, transient response—a nerve impulse can travel through the body in a matter of milliseconds. Hormones, however, move at a somewhat slower rate (by way of the bloodstream) and, characteristically, elicit a slower and longer-lasting response. Plants, which are not noted for their liveliness, rely largely on an elaborate interplay of hormones to coordinate their activities. In general, animals, while also relying on hormones, are characterized by the presence of networks of specialized nerve cells. As we have seen, many have highly developed and finely tuned nervous systems that monitor internal and external changes and initiate responses to them.

34–1
Diagram of a motor neuron. The stimulus is received at any point on the naked nerve surface, usually by the dendrites, which transmit the signal to the cell body and to the axon. The signal travels along the axon, which is insulated by a myelin sheath composed of Schwann cells. The nodes of Ranvier are gaps in the axon sheath that occur at the junctions of adjacent Schwann cells. This motor neuron branches to form a neuromuscular junction with muscle fibers.

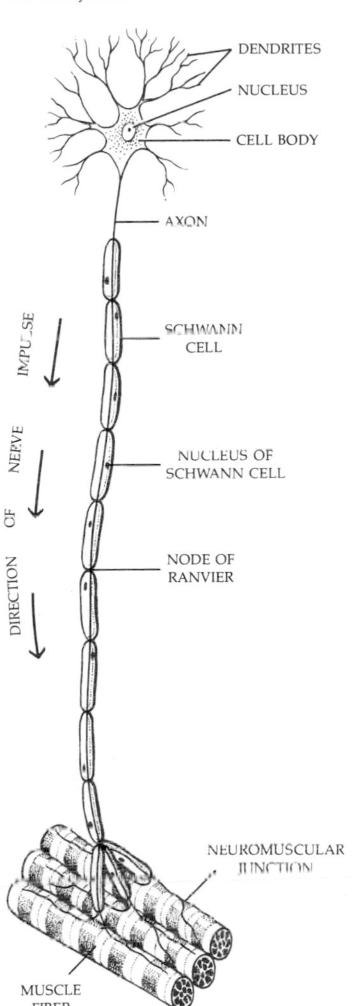

NEURONS

The functional unit of the nervous system, as we noted in Chapter 31, is the neuron. A neuron has three basic parts: the cell body itself, which contains the nucleus; the *dendrites,* which receive signals and transmit them to the cell body (the cell body itself may also receive signals); and the *axon,* which transmits impulses away from the dendrites and cell body (Figure 34–1). Axons are commonly referred to as nerve fibers. A single neuron usually has many dendrites but has only one axon, although the axon may be branched. Vertebrate nerve fibers are often enveloped in a myelin sheath formed by Schwann cells (Figure 34–2), which are a type of glial cell. As we shall see, myelinated fibers transmit messages more rapidly than fibers without a myelin sheath.

There are three types of neurons in the nervous system: (1) afferent, or sensory, neurons, which either act as receptors of sensory stimuli or connect to receptors; they carry impulses to the central nervous system; (2) interneurons, which connect two or more neurons within the central nervous system; and (3) efferent, or motor, neurons, which carry impulses away from the central nervous system to the effectors (glands or muscles).

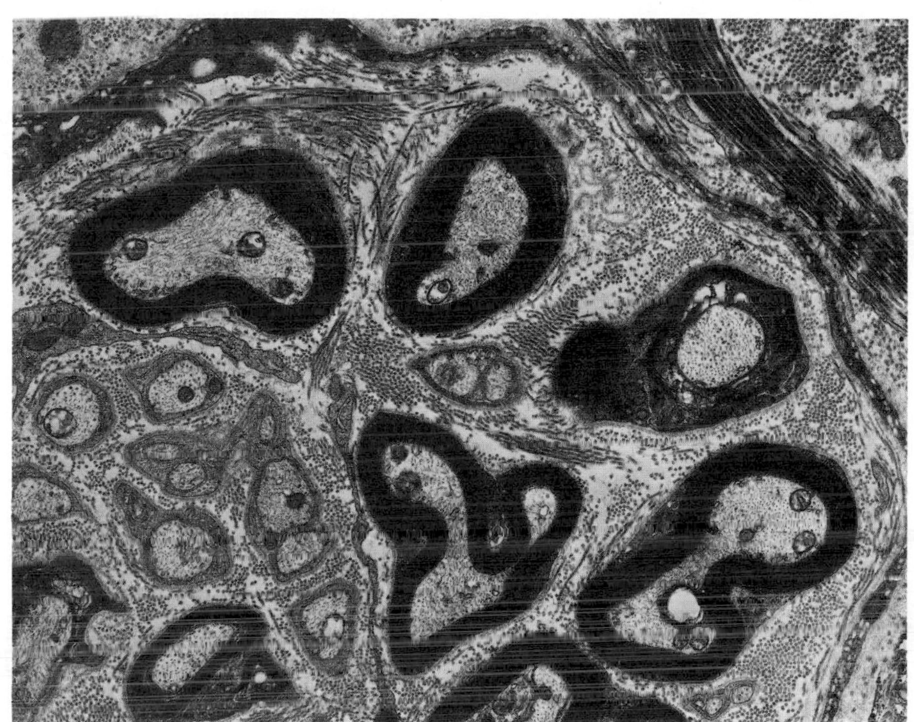

34–2
Cross section of a nerve showing myelinated and unmyelinated fibers. The myelinated fibers are those surrounded by the dark borders, myelin sheaths. These sheaths are formed from Schwann cells by the process illustrated in Figure 34–3a. The cytoplasm and nuclei of Schwann cells can be seen on the outer edges of some of the myelin sheaths.

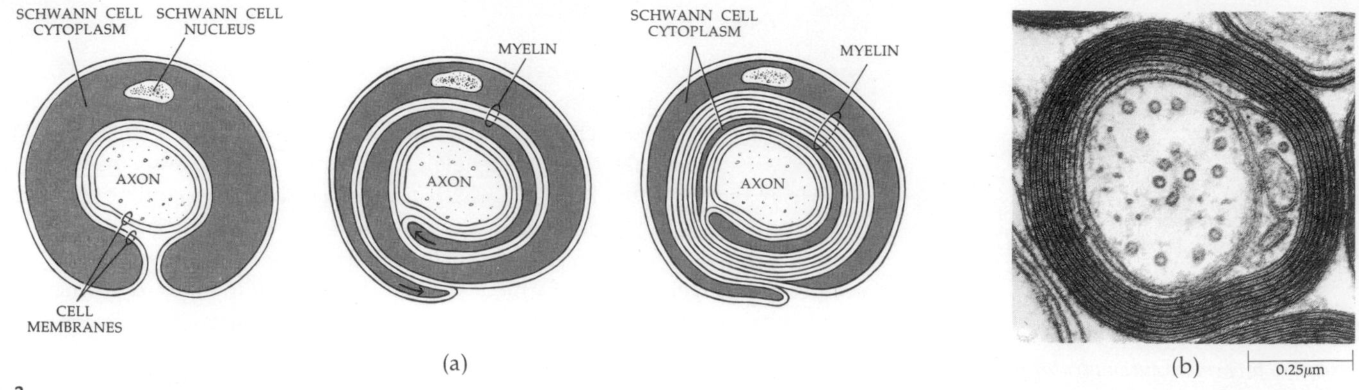

34-3
(a) *Formation of a myelin sheath by a Schwann cell. As the Schwann cell grows, it extends itself around and around the axon and gradually extrudes its cytoplasm from between the layers.* (b) *Electron micrograph of a cross section of a mature myelin sheath.*

Sensory neurons are activated by sensory stimuli. These stimuli may be received by naked nerve endings; some of the pain receptors in the human skin appear to be of this sort. Others, such as those in muscles, joints, and skin, have terminals enclosed in specialized capsules. More often sensory receptors are found within sensory organs, such as the eye or ear. The impulses generated by the receptors are carried along nerve fibers to the brain or spinal cord. (Note that in these sensory cells with very long processes, as shown in Figure 31-13c, the cell body is off to the side, and the formal definition of dendrite and axon does not apply.) The signals are then transmitted to *interneurons* in the central nervous system or, in the case of a simple reflex, to dendrites or cell bodies of *motor neurons,* from which they pass along the axons of the motor neurons to effectors, either muscle cells or gland cells. These events are summarized in Figure 34-4.

The Nerve Impulse

The major function of the nervous system is internal communication. For almost 200 years it has been known that nerve conduction is somehow associated with electrical phenomena. Thus, before we go any further, it may be worthwhile to review some simple principles of electricity. One basic concept is that there are two types of electric charge, positive and negative; like charges repel one another and unlike charges attract. Thus, negatively charged particles tend to move toward an area of positive charge, and vice versa. In a metal, such as copper, only negative charges move, but the principle is the same. Materials that permit the movement of charged particles, such as a solution containing ions or a copper wire, are known as conductors. Materials that do not permit the movement of charged particles, such as fat and rubber, are insulators.

The difference in the amount of electric charge between an area of positive charge and an area of negative charge is called the *electric potential.* An electric potential is a form of potential energy, like a boulder on the top of a hill or water behind a dam. This potential energy is converted to electrical energy (electricity) when charged particles are allowed to move through a solution or along the wire between two areas of different charge. The force with which the charged particles move, which is analogous to the force of water running downhill, is measured in volts or, if it is very small, in millivolts.

Major advances in understanding the nature of the nerve impulse occurred when it became possible to measure electrical impulses in an individual neuron.

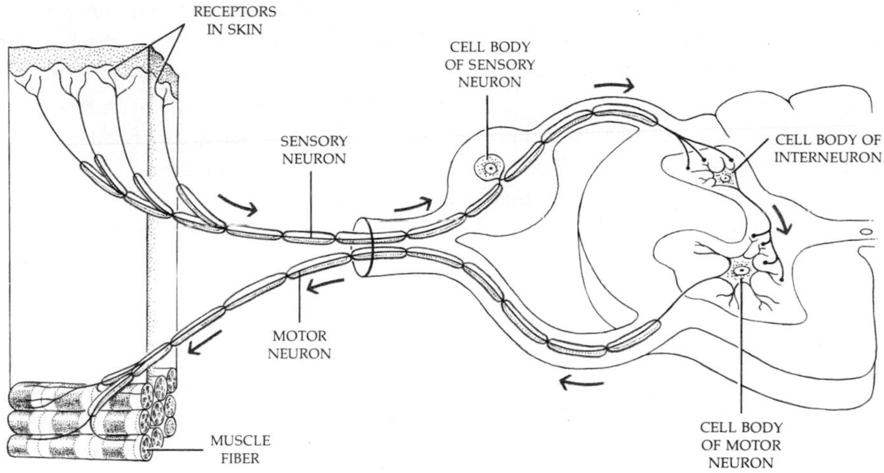

34–4
Sensory receptors in the skin, when appropriately stimulated, transmit signals to an interneuron in the spinal cord. As a result, a motor neuron is stimulated and muscle fibers contract.

Such measurements are made with microelectrodes, tiny enough to penetrate a living cell without injuring it. The microelectrodes are connected with a very sensitive voltmeter called an oscilloscope, which measures voltage (in millivolts) in relation to time (in milliseconds).

When both electrodes are outside the neuron, no voltage difference is recorded (Figure 34–5a). When one electrode penetrates the axon, a voltage difference of about 70 millivolts between the outside and the inside of the axon can be detected. The interior of the membrane is negative with respect to the exterior (Figure 34–5b). This is the resting potential of the membrane. When the axon is stimulated, the oscilloscope records a very brief reversal of polarity—that is, the interior becomes positive in relation to the exterior (Figure 34–5c). This reversal in polarity is called the *action potential*. The action potential is the nerve impulse.

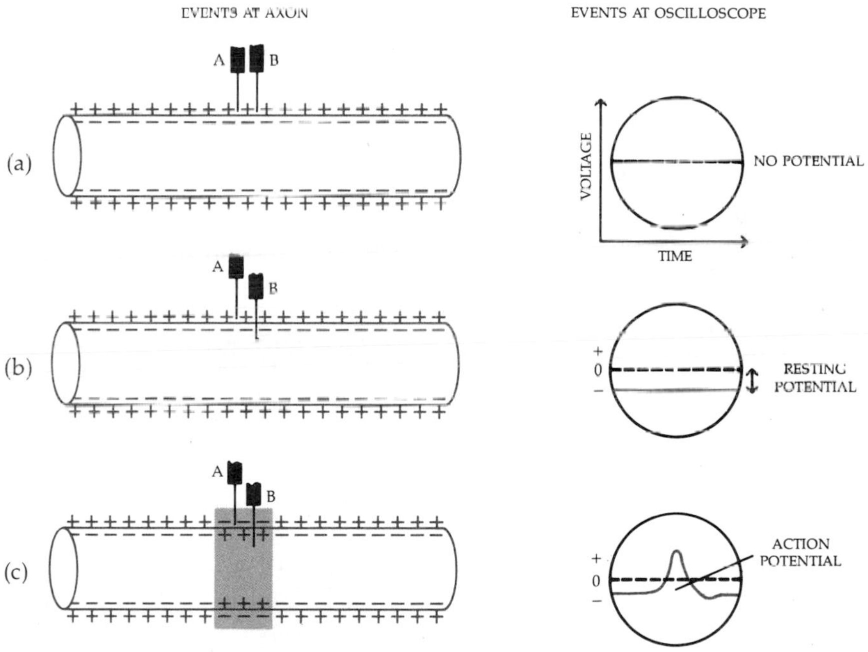

EVENTS AT AXON

EVENTS AT OSCILLOSCOPE

(a) NO POTENTIAL

(b) RESTING POTENTIAL

(c) ACTION POTENTIAL

34–5
The electric potential of the membrane of the axon is measured by microelectrodes connected to an oscilloscope. (a) *When both electrodes are outside the membrane, no potential is recorded.* (b) *When one electrode penetrates the membrane, the oscilloscope shows that the interior is negative with respect to the exterior and that the difference between the two is about 70 millivolts. This is the resting potential.* (c) *When the axon is stimulated and a nerve impulse passes along it, the oscilloscope shows a brief reversal of polarity—that is, the interior becomes positive in relation to the exterior. This reversal in polarity is the action potential.*

34–6
(a) *Nerve impulses can be monitored by electronic recording instruments. The impulses from any one neuron are all the same; that is, each impulse has the same duration and voltage change as any other.* (b) *Nerve impulses from a sensory neuron (touch receptor) in cat skin. The figures to the left indicate the skin indentation. The individual vertical lines represent action potentials on a compressed time scale.*

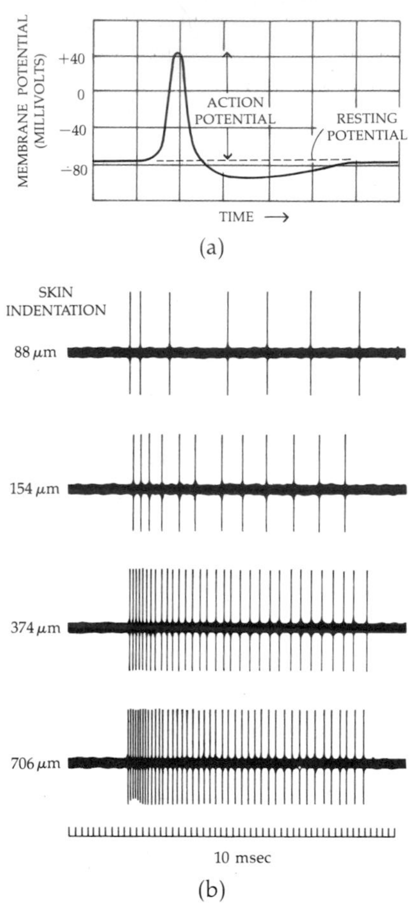

The nerve impulses (action potentials) from any one neuron are always the same. The only variation is the frequency at which impulses are produced. Figure 34–6b shows, for example, the action potentials produced by a single sensory neuron in the skin of a cat in response to pressure. All the impulses are the same size. This is known as an all-or-nothing response; either the neuron does not respond at all to the stimulus, or it responds maximally.

The Ionic Basis of the Action Potential

To understand the action potential, we must reexamine the chemistry of the cell and, in particular, the nature of the cell membrane. A difference in total electric charge between the inside and the outside of the membrane—the electric potential—is a characteristic common to all cells. It is produced by differences in the distribution of ions on either side of the membrane.

Let us look first at the axon of a motor neuron in its resting state. (Most of the studies on which these data are based were carried out on the giant axon of the squid, which has thus taken its place in biological history along with the fruit fly, the pea plant, and the red bread mold.) In this resting state, the concentration of potassium (K^+) ions on the inner surface of the membrane is 30 times as great as the concentration outside, and the concentration of sodium (Na^+) ions is about 10 times as great on the outside as on the inside. This concentration difference is maintained by the sodium-potassium pump (page 133), which pumps sodium out and potassium in.

The distribution of these ions is governed by three factors: (1) diffusion, which results in the movement of particles along a concentration gradient (page 127); (2) the attraction of particles with opposite charges and the repulsion of particles with like charges; and (3) the properties of the membrane itself. The membrane is not permeable to large anions (negatively charged ions) or to the sodium ion, Na^+. It is, however, somewhat permeable to the K^+ ions.

Now let us examine the forces at work on the K^+ ions, which, according to current hypotheses, are responsible for the electric potential of the membrane at rest, or the *resting potential*. Because of the concentration gradient, they tend to move out of the cell; in fact, if no other influence were at work, the K^+ ions would, of course, move down the concentration gradient until their distribution was equal on either side of the membrane. However, because of the impermeability of the membrane, anions cannot follow the K^+ ions out of the cell. This imbalance of positive and negative charges stops the movement of K^+ ions just as, for example, hydrostatic pressure can stop the osmotic movement of water molecules. As a result, an equilibrium is reached at which there is no net movement of K^+ ions across the membrane. At the point of equilibrium, because of the movement of K^+ ions down the concentration gradient, there is a slight excess of positive charge outside the membrane. This is the resting potential (Figure 34–7).

When the membrane is stimulated, it suddenly becomes highly permeable to Na^+ at the site of the stimulation. The Na^+ ions rush in, moving down their concentration gradient and attracted, at first, by the negative charge inside the cell. This influx of positively charged ions momentarily reverses the polarity of the membrane so that it becomes more positive on the inside than on the outside, producing the action potential (Figure 34–8). The change in Na^+ permeability lasts for only about half a millisecond; then the membrane regains its previous impermeability to Na^+. During this time, the permeability to K^+ increases, and there is

34–7
Resting axon, with membrane polarized. There is a slight excess of positive charge outside the membrane caused by the outward diffusion of K^+ ions.

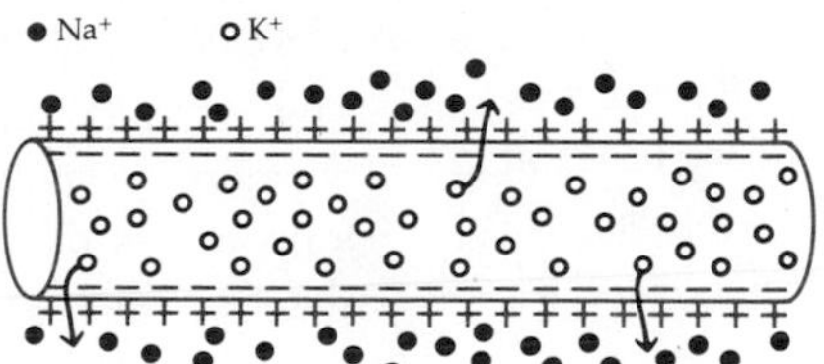

34–8
Action potential. A portion of the membrane becomes momentarily permeable to Na^+ ions. Na^+ ions rush in, and the polarity of the membrane is reversed.

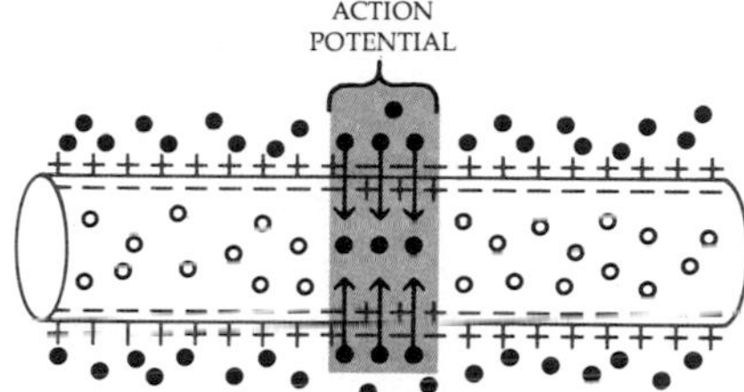

an outward flow of K^+ ions due to the concentration gradient and also the positive charge inside the cell at the peak of the action potential. This outward flow of positive K^+ ions counteracts the previous inward flow of positive Na^+ ions, and the resting potential is quickly restored. (The actual number of ions involved is very small. Only a very few Na^+ ions need enter to cause the depolarization of the membrane, and only the same small number of K^+ ions need move out of the cell to restore the resting potential.) Subsequently, the sodium-potassium pump restores the Na^+ and K^+ concentrations to their original levels.

Propagation of the Impulse

An important feature of the nerve impulse is that this transient depolarization continues to move along the axon, renewing itself indefinitely. The action potential is self-propagating because at the peak of the action potential, when the inside of the membrane at the active region is comparatively positive, positively charged ions move from this region to adjacent areas inside the axon, which are still comparatively negative. As a consequence, the adjacent area, in turn, becomes depolarized—that is, less negative (Figure 34–9). When it becomes slightly depolarized, its permeability to Na^+ increases. Na^+ ions rush in and create a new action potential, which, in turn, depolarizes the next adjacent area of the membrane. Thus the nerve impulse travels along the axon. (The segment of the axon behind the nerve impulse has a brief refractory period during which it cannot be reexcited and which keeps the action potential from going backwards.) Because of this renewal process, repeating itself along the length of the membrane, an axon, which would be a very poor conductor of an ordinary electric current, is capable of transmitting a nerve impulse over a considerable distance with absolutely undiminished strength.

34–9
Propagation of the action potential. In advance of the action impulse, a small segment of the membrane becomes slightly depolarized, owing to the movement of positively charged ions along the inside of the membrane. When the membrane becomes depolarized in this way, its permeability to Na^+ increases, and Na^+ ions rush in, creating a new action potential and depolarizing another segment of membrane.

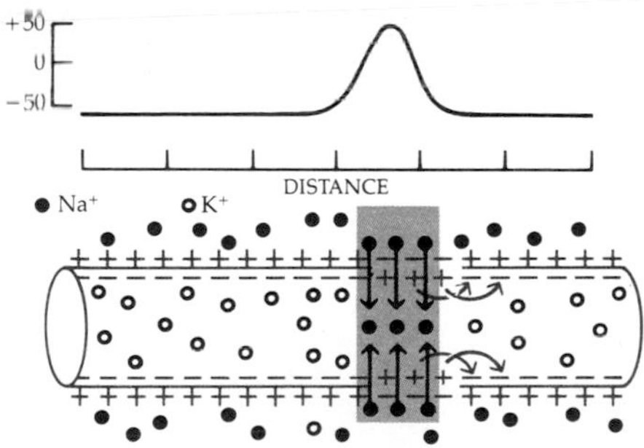

34–10
In an unmyelinated fiber (a), *the action potential travels continuously along the axon, whereas in a myelinated fiber* (b), *the impulse jumps from node to node, greatly accelerating the transmission.*

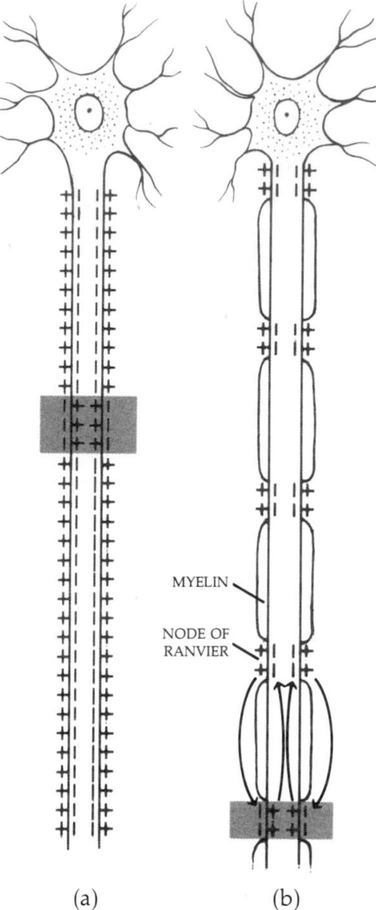

The Function of the Myelin Sheath

The propagation of the nerve impulse in this fashion, while fast compared to hormonal response, is relatively slow. Vertebrates solved this problem with what John Eccles describes as "a brilliant innovation." The innovation is the myelin sheath, which greatly speeds conduction of the nerve impulse. It is not a simple insulator, as one would logically expect. The important feature of the myelin sheath is that it is interrupted at regular intervals by openings, or nodes (see Figure 34–1). In myelinated fibers—which include all the large nerve fibers of vertebrates—the impulse, rather than traveling from point to point all along the membrane, jumps from node to node (Figure 34–10). This saltatory (leaping) transmission increases the velocity severalfold. Some large, myelinated nerve fibers, for instance, conduct impulses as rapidly as 200 meters per second, compared with velocities of only a few millimeters per second in small, unmyelinated fibers.

The Synapse

Signals travel from one nerve cell to another across a specialized junction known as a *synapse*. The neurons do not touch at the synapse; there is a space of about 20 nanometers between them known as the synaptic cleft. Transmission across most synapses in mammals is by chemical means. In synaptic knobs at the end of the axon are numerous small vesicles, visible in electron micrographs, which contain a chemical transmitter substance. Arrival of the action potential at the axon terminal causes these vesicles to empty their contents into the synaptic cleft. The transmitter substance crosses the cleft and combines with receptor molecules on the membrane of the postsynaptic cell, changing the permeability of the membrane. A number of these chemical transmitters have been tentatively identified. Two principal ones are acetylcholine and norepinephrine.

Unlike the nerve impulse along the fiber—which is an all-or-nothing proposition—signals transmitted by chemicals across a synaptic junction can modulate one another. That is, some may excite and some may inhibit. A single neuron may receive signals from hundreds, even thousands, of synapses, and, based on the summation of excitatory and inhibitory signals, it will or will not be excited and initiate an impulse. Synapses are therefore relay and control points that are extremely important in the functioning of the nervous system.

Each excitatory transmitter-receptor reaction can cause a very small decrease in membrane voltage. If many such reactions occur at about the same time, they can cause a depolarization of the axon and so initiate the nerve impulse. The chemicals released at inhibitory synapses, however, have a reverse effect. They stabilize the membrane near the resting potential or even increase the membrane voltage and thereby decrease the likelihood of an impulse's being generated.

The chemicals that transmit signals across nerve junctions are rapidly destroyed by specific enzymes after their release, diffuse away, or are taken up again by the synaptic knob. Such removal or destruction of the transmitter puts a halt to its effect. This in itself is an essential feature in the control of the activities of the nervous system. The activity of many drugs, such as LSD and psilocybin, is thought to depend on their interference with chemical transmission across synapses (see Table 40–1).

The Neuromuscular Junction

At the beginning of this chapter, we showed a diagram of a motor neuron, a nerve cell that stimulates striated skeletal ("voluntary") muscle. The cell bodies and dendrites of motor neurons are, as we shall see, in the central nervous system. In vertebrates, interneurons synapse with the dendrites and cell bodies of motor neurons and excite (or inhibit) them by means of transmitter chemicals. Nerve impulses, if they are initiated, travel the length of the motor neurons, leaping from node to node of the myelin sheath. The axon of each motor neuron branches as it reaches the muscle, and each branch terminates at a muscle fiber. As you will recall, each fiber of striated muscle is an elongated multinucleate cell.

A single motor neuron axon and all the muscle fibers it supplies are known as a motor unit. The number of muscle fibers in a motor unit varies according to the fineness of control that is required. In a muscle that moves the eyeball, for instance, a motor unit contains only about 10 muscle fibers, whereas in the biceps, each motor unit contains more than a thousand. Within a given muscle, fibers of different motor units are intermingled. A slight movement may involve the contraction of only a few motor units. The strength of contraction of a muscle as a whole depends on the number of motor units that have been activated and the frequency with which they are stimulated.

At the end of each branch, the nerve fiber emerges from the myelin sheath and becomes embedded in a groove on the surface of the muscle fiber, forming the neuromuscular junction. Transmission across this junction, as across the synapses between neurons, is by means of a chemical transmitter. The transmitter is acetylcholine. However, unlike synaptic transmission between neurons, transmission across the neuromuscular junction in vertebrate skeletal muscles is never inhibitory; either the muscle is signaled to contract or it is not signaled at all. (In arthropods, by contrast, integration of excitatory and inhibitory impulses takes place at individual muscle fibers rather than centrally in the nervous system.)

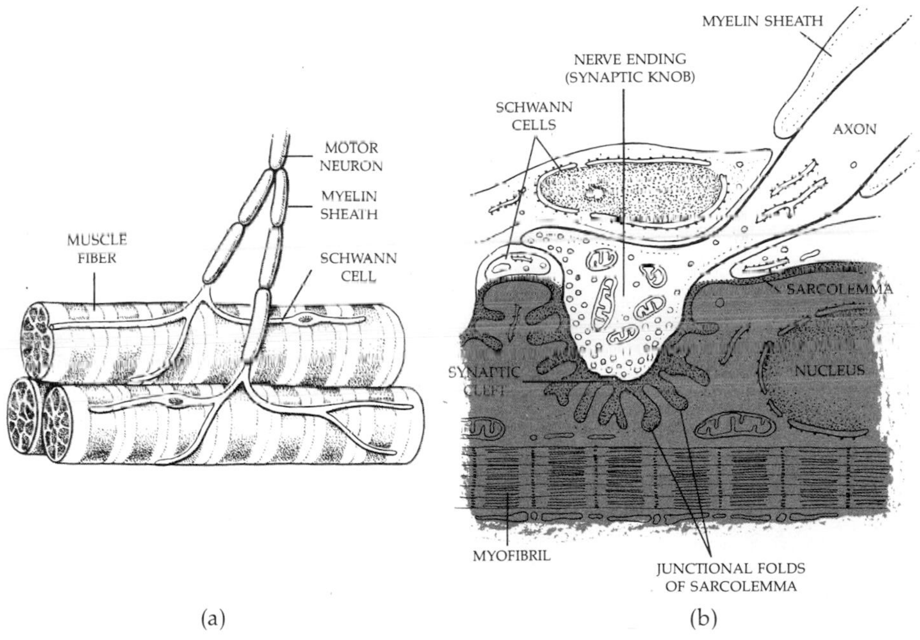

34-11
(a) *The axon of each motor neuron divides into branches, each forming a neuromuscular junction with a different muscle fiber (cell). The motor neuron and the numerous muscle fibers that it innervates are known as a motor unit. Stimulation of a motor axon stimulates all of the fibers in that motor unit. Within a given muscle, fibers of different motor units are intermingled.*
(b) *A neuromuscular junction. An action potential conducted along the axon of a motor neuron releases acetylcholine from synaptic vesicles into the synaptic cleft. This transmitter agent combines with reactive sites on the sarcolemma, altering the membrane permeability and initiating an action potential in the muscle fiber.*

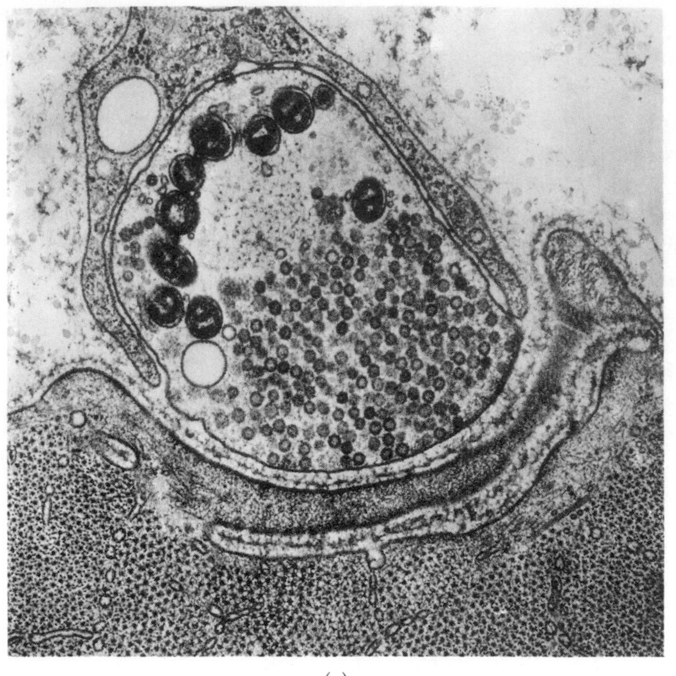

(a)

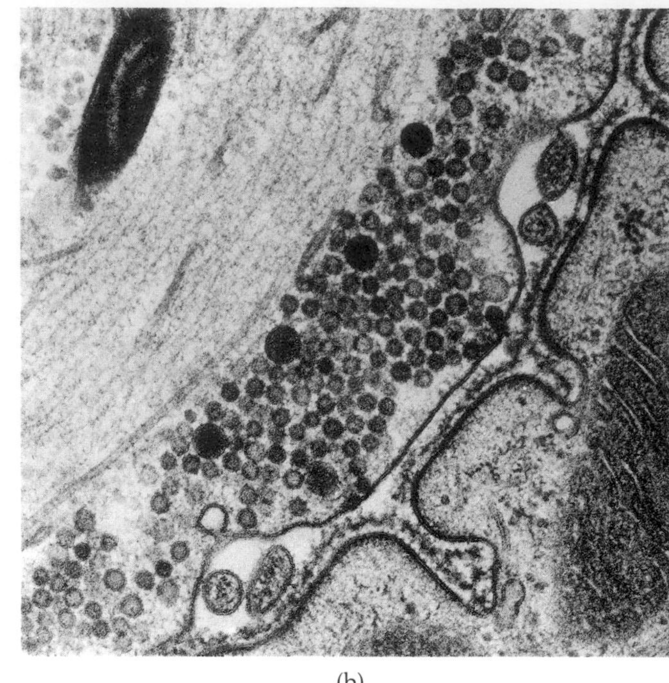

(b)

34–12
(a) *A motor neuron terminal. The larger, darker bodies are mitochondria. Below them are the spherical synaptic vesicles. At the bottom of the micrograph is a portion of a muscle cell. The light area immediately below the neuron terminal is the synaptic cleft, and below that is a junctional fold in the sarcolemma.* (b) *Synaptic vesicles discharging into the synaptic cleft.*

Many drugs act specifically on the neuromuscular junction. Curare, for instance, a plant extract that Indians of South America used to poison their arrow tips, inhibits transmission across the junction and produces paralysis. It is used medically as a muscle relaxant during surgery. The bacterial toxin of botulism, one of the most poisonous substances known, prevents nerve endings from liberating acetylcholine and so kills by paralysis of muscles controlling breathing.

THE CENTRAL NERVOUS SYSTEM

The brain and the spinal cord form the central nervous system. In vertebrates, they are protected by the bones of the skull and the vertebrae. In addition, they are wrapped in three layers of connective tissue, the meninges. The space between two of the layers is filled with cerebrospinal fluid, so that the entire central nervous system is cushioned within its bony enclosure. This degree of protection reflects not only the importance of central nervous system function but also the fact that neurons, once destroyed by aging, disease, injury, or poisons (such as ethyl alcohol), are never replaced. It is estimated that 10,000 neurons are lost naturally each day, a sobering thought.

Anatomically, the human brain consists of a number of different structures (Figure 34–13). The three principal divisions are the *brainstem*, the *cerebellum*, and the *cerebrum*. The brainstem is the old brain, evolutionarily speaking, and is surprisingly similar from fish to *Homo sapiens.* The cerebrum is divided into two cerebral hemispheres, each of which, in humans, is covered by the *cerebral cortex*. Both the brainstem and the cerebrum are made up of a number of anatomically distinct structures (see Table 34–1).

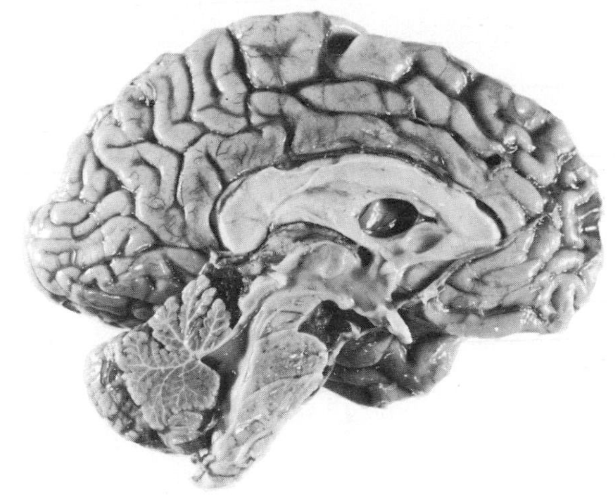

34–13
Longitudinal section of a human brain. The gray matter, characteristic of the cortex, consists of nerve cell bodies, glial cells, and unmyelinated fibers. The white matter is made up predominantly of myelinated fibers.

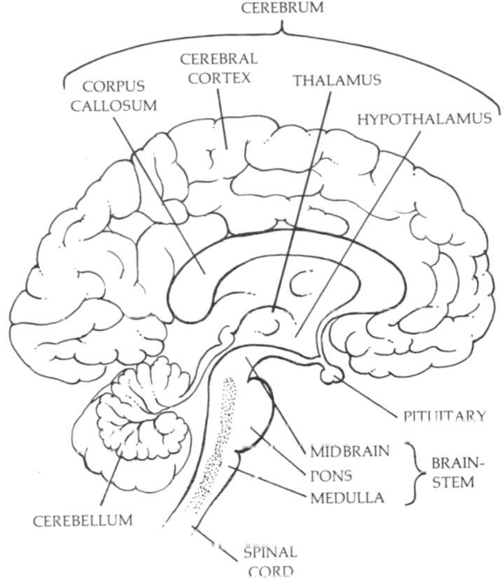

The brain and spinal cord are divided into gray matter and white matter. The white matter contains nerve fibers carrying information between various regions of the body and the brain and spinal cord. Its whiteness is due to the fatty layers of the myelin sheath. The gray matter contains interneurons, glial cells, and, in the brainstem and spinal cord, the cell bodies of motor neurons as well.

Figure 34–14 shows a segment of the human spinal cord, with pairs of spinal nerves entering and emerging from the cord through spaces between the vertebrae. Each of these pairs innervates the skeletal muscles and sensory receptors of a different and distinct area of the body. In mammals, there are 31 such pairs.

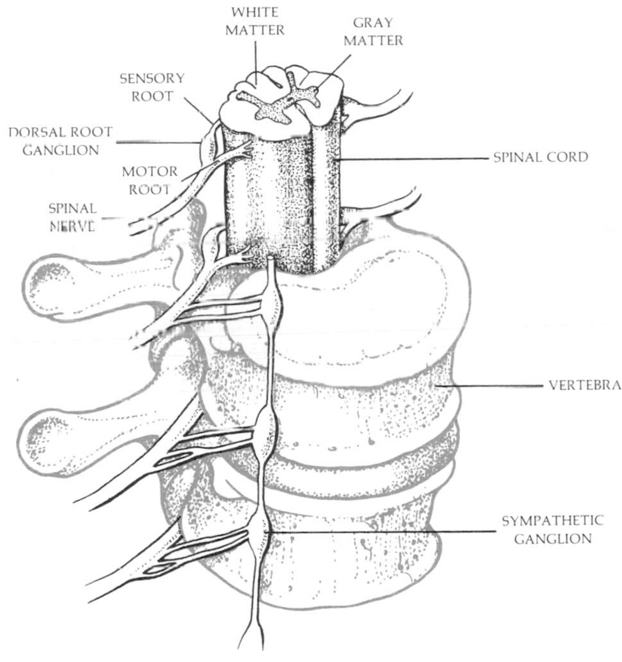

34–14
A segment of the human spinal cord. Each spinal nerve divides into two fiber bundles, the sensory root and the motor root, at the vertebral column. The sensory bundle connects with the cord dorsally; the cell bodies of the sensory neurons are in the dorsal root ganglia. The motor bundle connects ventrally with the spinal cord. The sympathetic ganglia are part of the autonomic nervous system (page 646). The butterfly-shaped gray matter within the spinal cord is composed of groups of neuron and glial cell bodies, and the surrounding white matter consists of ascending and descending tracts of fibers.

Table 34-1 *Structural Divisions of the Human Brain*

STRUCTURE	DESCRIPTION	FUNCTION
Brainstem	Stalk of brain; enlarged, knobby extension of spinal cord; consists of medulla, pons (bridge), and midbrain. Contains all afferent and efferent nerve fibers that pass between spinal cord and higher brain centers. Also contains nuclei involved with many autonomic reflexes.	Transmission of signals between spinal cord and higher brain centers; control of cardiovascular functions including heartbeat, respiration, other autonomic functions (see page 645)
Cerebellum	Bulbous, convoluted mass of neural tissue that, in humans, lies in back of and under the cerebrum	Unconscious coordination of muscular activities; reaches largest size, proportionately, in birds
Cerebrum	Largest, most prominent part of human brain, divided by groove into right and left hemispheres; each hemisphere consisting of four lobes: frontal, parietal, temporal, occipital	Receives and integrates incoming (afferent) information; makes associations between new data and stored information in cortex; coordinates responses
Cerebral cortex	Thin (2 millimeter) layer of gray matter, wrinkled and folded, overlying surface of cerebrum; reaches greatest development in higher primates, especially humans	Sensory, motor, and association functions (see page 757)
Inner cerebrum	Nerve tracts (white matter) connecting areas of cortex to each other and to rest of brain	Transmission of information to and from cerebral cortex
Corpus callosum	Nerve fibers between hemispheres	Exchange of information between hemispheres
Thalamus	Two egg-shaped masses of gray matter below cerebral hemispheres	Main relay center between brainstem and cerebrum; nuclei sort and process sensory information before transmission to cortex
Hypothalamus	Small mass (about 4 grams) below thalamus, above pituitary	Source of ADH and oxytocin (see page 650); controls pituitary; integration center for sex drive, anger, hunger, thirst, pleasure, temperature regulation
Hippocampus ("sea horse"), amygdala ("almond"), and fornix ("arch")	Nuclei and fiber bundles of "old brain" (rhinencephalon) (see page 754), located below cerebral hemispheres	Part of limbic system (see page 760)

THE PERIPHERAL NERVOUS SYSTEM

The neurons and fibers outside the central nervous system constitute the *peripheral nervous system*. (Note, however, that fibers of the peripheral nervous system extend into the central nervous system.) The neurons of the peripheral system include both motor (efferent) neurons and sensory (afferent) neurons. The fibers of motor and sensory neurons are bundled together into nerves, which are classified as *cranial nerves*, nerves connecting directly to the brain (such as the optic nerve), and *spinal nerves*, those that connect with the spinal cord.

As they reach the central nervous system, the motor and sensory components of the spinal nerves separate from each other. The sensory fibers feed into the dorsal (back) side of the spinal cord. They may synapse immediately with interneurons or motor neurons as in a simple reflex, they may turn and ascend toward the brain, or they may send processes in both directions. The cell bodies of these neurons are in the dorsal root ganglia outside the spinal cord. Motor fibers emerge from the spinal cord on the ventral (front) side. The cell bodies of these fibers are in the spinal cord, where they synapse with interneurons in the spinal cord.

The Reflex Arc

Figure 34-15 is a diagram of some of the neural connections in a reflex arc. In a simple reflex action—the withdrawal of your finger from a hot surface, for example—the stimulus is received by a sensory neuron (a pain-sensitive cell, in this case), and a signal is transmitted to the spinal cord. Here the sensory neuron synapses directly with a motor neuron or with an interneuron, which, in turn, relays the signal to a motor neuron. The motor neuron causes the appropriate muscle to contract, which moves your arm.

Thus you move your hand away "automatically" before your brain has had a chance to process the information. However, your brain has been informed and can call forth the next appropriate activity—such as muttering or turning off the stove.

34-15
Diagram of the reflex arc. Impulses from a receptor cell travel along the sensory fiber to the spinal cord. The cell body of the sensory neuron is located in a ganglion lying just outside the spinal cord. The sensory axon enters the cord and synapses with an interneuron in the gray matter of the cord. The interneuron relays the impulse to a motor neuron, which stimulates a muscle. The response is automatic and does not involve the brain, although the brain is informed about what is taking place.

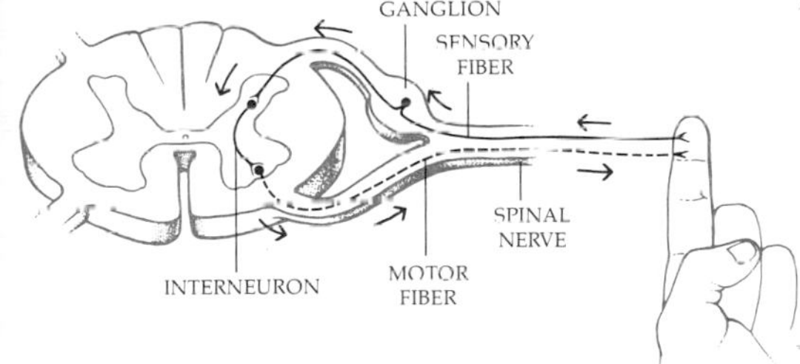

The Somatic and Autonomic Nervous Systems

There are two important subdivisions of the peripheral nervous system: the *somatic* and the *autonomic*. The autonomic nervous system consists of the nerves that control heart muscle, glands, and smooth muscle (the type of muscle found in the walls of blood vessels and in the digestive, respiratory, and reproductive tracts). The autonomic nervous system is generally categorized as an "involuntary" system, in contrast to the somatic system, which controls the muscles that we can move at will, that is, the skeletal muscles.

You will readily recognize that the distinction here between "voluntary" and "involuntary" is not clear-cut. Skeletal muscles often move involuntarily, as in a reflex action, and it is reported that some individuals, particularly practitioners of yoga, or those who have had biofeedback training, can control their rate of heartbeat and the contractions of some smooth muscle.

Anatomically, the motor neurons of the somatic system are distinct and separate from those of the autonomic nervous system, although fibers of both types may be carried in the same nerve. Also, the cell bodies of the motor neurons of the somatic system are located within the central nervous system, with long axons running without interruption all the way to the skeletal muscles. The efferent fibers of the autonomic nervous system also originate in cell bodies inside the central system; however, they do not usually travel all the way to their target organ, or effector, but instead form a synapse outside the brain or spinal cord with a second neuron, which then innervates the cell of the muscle or gland (Figure 34-16).

In other words, the autonomic nervous system consists of efferent neurons leaving the brain and spinal cord, and of peripheral motor neurons entirely outside the central nervous system.

The autonomic nervous system receives its sensory input from some of the same neurons as the somatic nervous system and also from sensory neurons monitoring changes in the interior of the body (visceral afferent nerves). An example is the group of neurons signaling changes in blood pressure. Such stimuli are involved in reflexes similar to the reflex arc just described, but we are not usually conscious of the sensations.

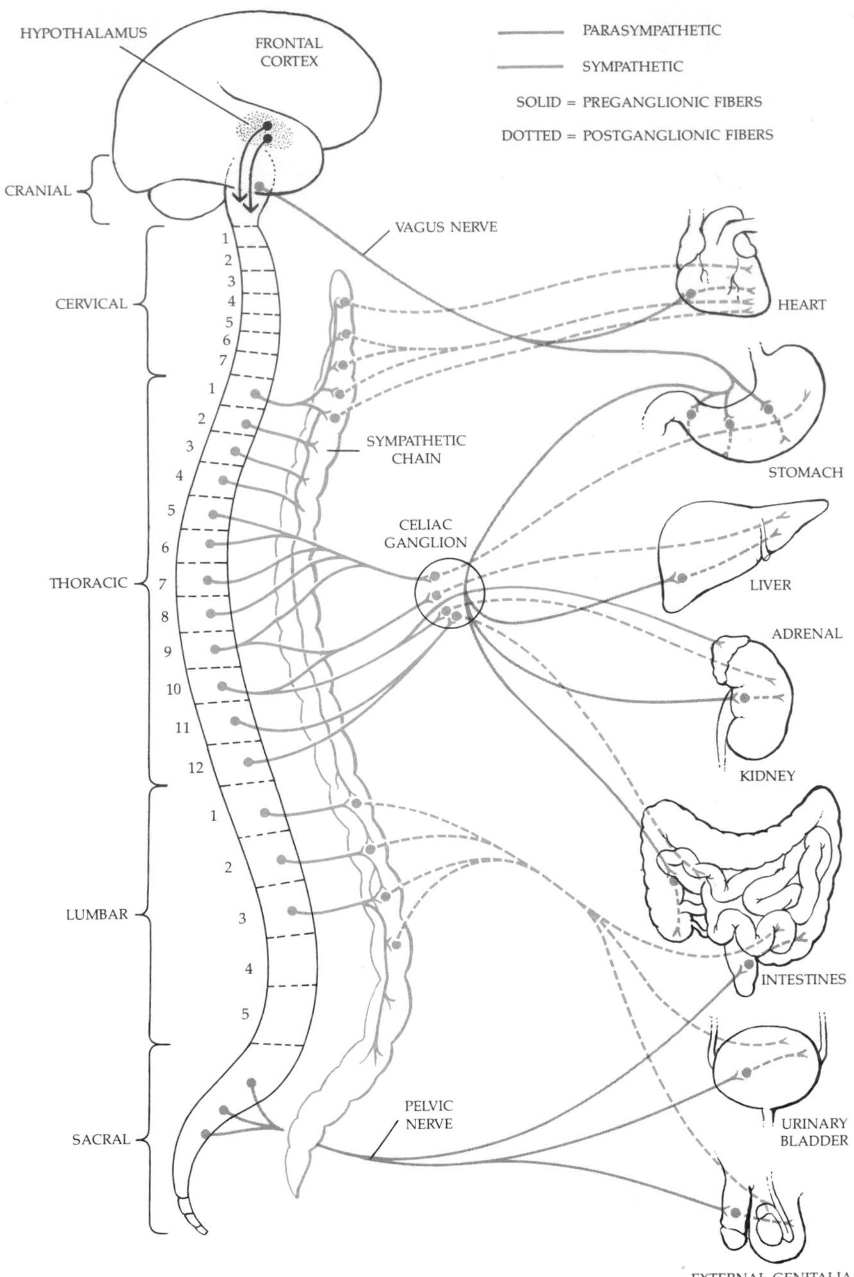

34–16
The autonomic nervous system. It differs from the somatic system anatomically in that its motor neurons are entirely outside the central nervous system and synapse with nerve fibers (axons) emerging from the brain and spinal cord. The fibers emerging from the central nervous system are known as preganglionic fibers, and those terminating in the effector organs are known as postganglionic fibers.

The autonomic nervous system consists of the sympathetic and the parasympathetic divisions. The preganglionic fibers of the parasympathetic division exit from the base of the brain and from the sacral region of the spinal cord and synapse with the postganglionic neurons at or near the target organs. The sympathetic division originates in the thoracic and lumbar regions; preganglionic fibers of the sympathetic division synapse with postganglionic neurons in the sympathetic chain or in other ganglia, such as the celiac ganglion, part of the solar plexus. The chemical transmitter of the sympathetic division is usually norepinephrine and of the parasympathetic, acetylcholine.

Most, but not all, internal organs are innervated by both divisions, which usually function in opposition to each other. In general, the sympathetic division produces the effect of exciting organs involved in "fight-or-flight" reactions, and the parasympathetic division stimulates more tranquil functions, such as digestion.

The two-neuron pathway with synapses outside the central nervous system constitutes a characteristic difference between the autonomic and somatic systems. Another major difference is that, in vertebrates, the somatic system can either stimulate or not stimulate its effector, whereas the autonomic nervous system, as we shall see, can stimulate or inhibit the activity of its target organ.

Divisions of the Autonomic Nervous System: Sympathetic and Parasympathetic

Neuroanatomists divide the autonomic nervous system into two divisions: the *sympathetic division* and the *parasympathetic division*. These two divisions are distinct anatomically, physiologically, and functionally. A particular effector is often under the influence of neurons from each of these divisions, and usually the two will have different and opposite effects.

There are three major physical differences between the sympathetic and parasympathetic divisions:

1. They differ in their sites of emergence from the central nervous system. The sympathetic division originates in the thoracic (chest) and lumbar (lower back) regions of the spinal cord. The parasympathetic division emerges through cranial (head) nerves, such as the vagus nerve, or from the sacral ("tail") region.
2. As previously stated, in the autonomic nervous system, there are always two neurons extending between the central nervous system and the effector organ. These neurons synapse at a ganglion. In the sympathetic division, this synapse is usually close to the central nervous system. Thus, characteristically, the preganglionic axon is short, and the postganglionic axon is long. In the parasympathetic division, the opposite is true: The synapse is close to or embedded in the target organ; the preganglionic axon is long, and the postganglionic axon is short.
3. Postganglionic sympathetic nerve endings release norepinephrine. All parasympathetic endings release acetylcholine. (Acetylcholine is also the transmitter substance at the first synapse outside the central nervous system for both divisions.)

34–17
Divisions of the nervous system.

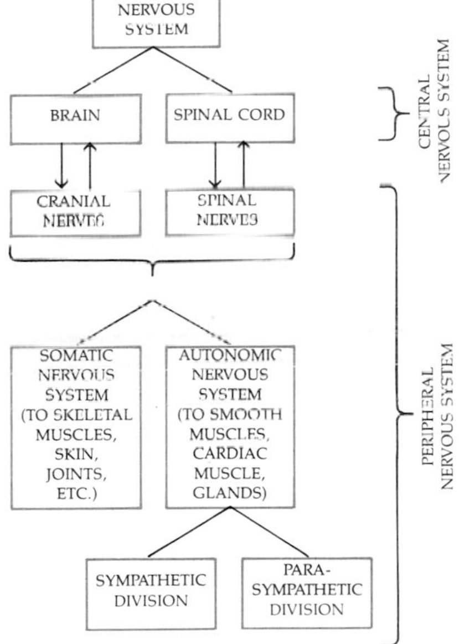

Generally speaking, the effects of the two divisions are antagonistic. The symptoms of rage, described on page 580, are the results of discharge by the neurons of the sympathetic division. The parasympathetic division, on the other hand, is more concerned with the restorative activities of the body, that is, with rest and rumination. It is not totally mobilized except, perhaps, after a heavy meal or following orgasm. Parasympathetic stimulation slows down the heartbeat, increases the movements of the wall of smooth muscles surrounding the intestine, and stimulates secretions of the salivary glands. As mentioned earlier, most large organs, such as the heart, are under the control of both sympathetic and parasympathetic nerves, and these work in close cooperation for the ultimate homeostatic regulation of the body's functions.

THE ENDOCRINE SYSTEM

Neurons establish one-to-one connections, like calling on the telephone. Hormones broadcast their message. Usually only certain groups of cells "listen," but some hormones, such as insulin, affect virtually every cell in the body. The effects mediated by the endocrine system are, generally speaking, slower, more diffuse, and more prolonged than those of the nervous system. They are also, of course, closely integrated with nervous system activities. We have seen some examples of ways in which the two systems interact, as, for instance, in the synchronization of

34–18
A cross section of pancreatic tissue. The pancreas is both an exocrine and an endocrine gland. The group of small cells in the center of the micrograph are islet cells, endocrine cells that secrete the hormones insulin and glucagon into the bloodstream. The surrounding exocrine cells produce digestive enzymes, which are carried through the pancreatic duct to the small intestine.

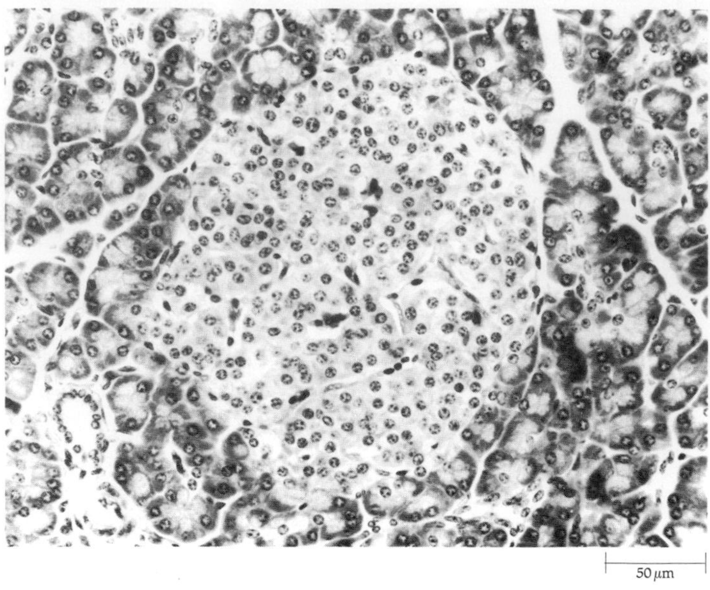

copulation and ovulation in the cat and rabbit or the release of epinephrine and norepinephrine from the adrenal medulla in "flight-or-fight" situations. We shall see much more evidence, both anatomical and functional, of the links between these two systems.

Hormones are secreted by glands—epithelial tissues specialized for secretion. Glands are classified as exocrine or endocrine. *Exocrine glands* secrete their product into ducts; examples are the sweat glands of the human skin and digestive glands. *Endocrine glands* are glands that secrete hormones. They secrete them into the bloodstream (or, more precisely, into the extracellular space from which the substances diffuse into the bloodstream). They are sometimes referred to as "ductless" glands. "Endocrine" is generally used as a synonym for "hormone-secreting," and endocrinology means the study of hormones.

THE DISCOVERY OF HORMONES

In 1849, A. A. Berthold, a physician and professor in Göttingen, carried out the first scientific experiment in endocrinology. Using six young cockerels, he set up three experimental groups: (1) Two birds were castrated; these became typical capons in which the combs, wattles, plumage, aggressiveness, crow, and sexual urge characteristic of the mature cock failed to develop. (2) Two additional birds were castrated, and the removed testes were reimplanted in the same animal but at a new site, distant from ducts, nerves, or any possible previous connections except with the circulatory system. These birds developed into normal cocks. (3) Two birds were subjected to "sham" surgery; that is, surgical incisions were made, but no tissues were removed or transplanted. They, too, developed normally. (These last animals served as controls; without them, Berthold could not be sure that the results obtained in the second group were related to removal and reimplantation of the testes and not to surgical stress alone.)

This experiment initiated a wholly new concept of controlling mechanisms:

Substances produced by particular tissues of the body could be carried by the bloodstream and could exert specific effects on distant tissues. Such substances came to be known as *hormones*, from the Greek word meaning "excite." It is now known, however, that many hormones act as inhibitors rather than exciters. (The juvenile hormone of insects, page 466, is an example.) The hormones are better described as chemical messengers, establishing communications between the various parts of the body. Some of the communications are concerned with homeostatic regulation, the perpetual adjustment of the internal environment. Others are involved with change—the changes that occur with sexual maturation or responses to emergencies.

Hormones exert their effects on particular cells. Because hormones travel through the bloodstream, most tissues in the body are equally exposed to them. Thus their specificity of action depends on the receptivity of the target tissue as well as on the chemical characteristics of the hormone. Moreover, target tissues may be receptive under some circumstances and not under others. For example, prolactin causes the production of milk—but only in a mammary gland also under the influence of estrogen, progesterone, thyroxine, adrenal steroids, and growth hormone.

In the following pages, we shall discuss some of the principal endocrine glands of mammals and the hormones they secrete. The hormones involved in digestion, which are not included here, will be discussed in Chapter 37.

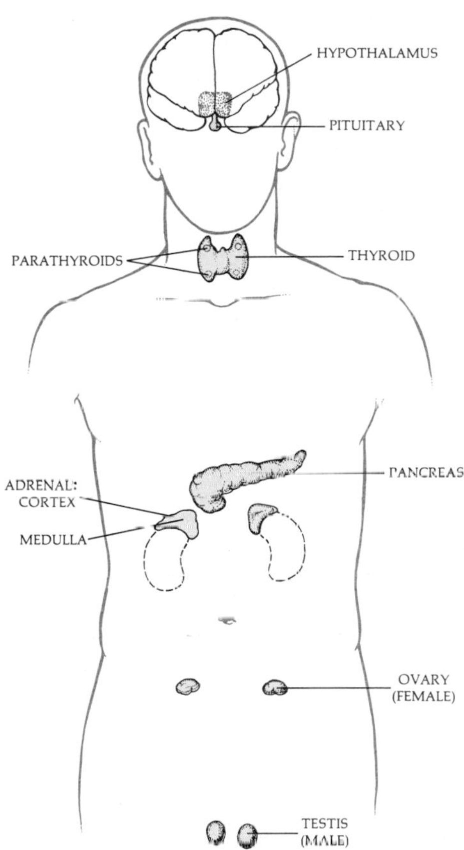

34–19
Some of the hormone-producing (endocrine) organs. The pituitary releases hormones that, in turn, regulate the hormone secretions of the sex glands, the thyroid, and the adrenal cortex (the outer layer of the adrenal gland). The pituitary is itself under the regulatory control of an area in the brain known as the hypothalamus, indicated in color. The hypothalamus thus is the major link between the two integrating and control centers.

THE PITUITARY GLAND

The pituitary gland is situated at the base of the brain in the geometric center of the skull. It is about the size of a kidney bean and has three lobes: the anterior, the intermediate, and the posterior.

The anterior lobe of the pituitary gland is the source of at least six different hormones. Four of these are tropic (from *trope*, "a turning") hormones—hormones that act upon other glands to regulate their secretions. Two of the tropic hormones, FSH and LH, which act upon the gonads, were discussed in Chapter 32. TSH, the thyroid-stimulating hormone, influences the thyroid. Like the gonadotropic hormones, TSH is regulated by negative feedback, responding to the concentration of thyroid hormone in the blood. (As we shall see, this negative feedback system involves the hypothalamus as well as the pituitary.) Adrenocorticotropic hormone (ACTH) has a similar regulatory relationship with cortisol, one of the hormones produced by the adrenal cortex.

Growth hormone, sometimes called somatotropin, is also produced by the anterior pituitary. It promotes the growth of bone and muscle. As is the case with most of the hormones, growth hormone is best known by the effects caused by too much or too little. If there is a deficit in somatotropin production in childhood, a midget results, the so-called "pituitary dwarf." An excess of somatotropin results in a giant; most circus giants are the result of an excessive growth hormone production during childhood. Excessive growth hormone in the adult does not lead to giantism, since growth of the long bones no longer occurs, but to acromegaly, an increase in the size of the jaw and the hands and feet, adult tissues that are still sensitive to the effects of growth hormone. Growth hormone also has effects on glucose metabolism, inhibiting the uptake and oxidation of glucose by many types of cells. It also stimulates the breakdown of fatty acids, thus conserving glucose.

34–20
"General" Tom Thumb with P. T. Barnum. Members of the sideshow, at one time a major circus attraction, were often persons with endocrine disorders. Giants and midgets are the result of too much or too little growth hormone.

Prolactin is also produced by the anterior pituitary. It stimulates secretion of milk in mammals. As long as the infant continues to nurse, the impulses produced by the suckling of the breast are transmitted by way of the central nervous system to the pituitary and cause it to produce prolactin; the prolactin, in turn, acts upon the breast to maintain the production of milk. Once suckling ceases, the synthesis and release of prolactin stops and so does milk production. Thus supply is regulated by demand. In some birds, prolactin stimulates the production of crop milk. (See the essay on page 652.)

The intermediate lobe of the pituitary gland is the source of melanocyte-stimulating hormone. In reptiles and amphibians, this hormone stimulates color changes associated with camouflage or with behavior patterns such as aggression or courtship. It has no known function in humans. The posterior lobe of the pituitary stores hormones produced by the hypothalamus.

HYPOTHALAMUS

The pituitary gland lies beneath an area of the brain known as the hypothalamus and is directly under its influence. The pituitary is also under the influence, by way of the hypothalamus, of other brain centers. Some nine hormones have now been isolated from the hypothalamus that act either to stimulate or inhibit the secretion of hormones by the anterior pituitary. These "brain hormones," as they are called, are small peptides, one only three amino acids in length. Released into the circulatory system, they travel through the pituitary stalk to their target cells in the anterior pituitary gland, thus technically meeting the criterion of hormones (Figure 34–22).

The first of these hormones to be discovered was TRH, thyrotropin-releasing hormone. As its name implies, it stimulates the release of thyrotropin from the pituitary gland. The second was LRH, which controls the release of luteinizing hormone. These discoveries, which were made by a group of scientists headed by Roger Guillemin at the Salk Institute, prompted the search for a releasing factor controlling somatotropin, the growth hormone. To their surprise, the investigators found not a releaser but an inhibitor, which has been given the name of somatostatin.

The hypothalamus is also the source of two hormones stored in and released from the posterior pituitary: oxytocin and antidiuretic hormone, or ADH. (ADH is sometimes called vasopressin because it increases blood pressure in many vertebrates; it has no such effect in humans, however, except in very high doses.)

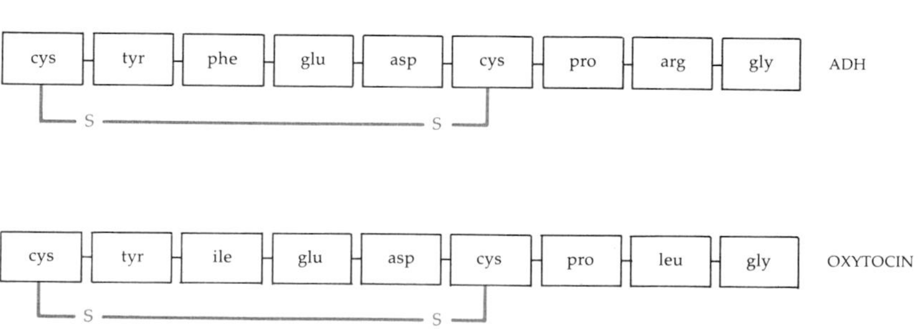

34–21
Primary structures of antidiuretic hormone (ADH) and oxytocin, hormones produced in the hypothalamus and released from the posterior pituitary. Note that each consists of only nine amino acids and that they differ by only two.

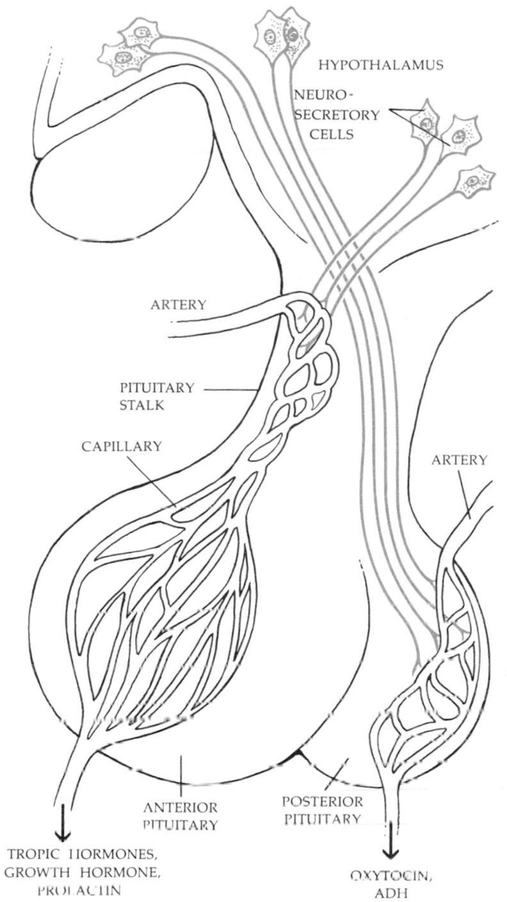

34–22
Relationships between the pituitary and the hypothalamus. Nerve fibers (in color) connecting the hypothalamus and posterior lobe of the pituitary transmit oxytocin and ADH, which are stored in and released from the posterior lobe. The anterior lobe is connected with the hypothalamus by a network of capillaries. Neurosecretory cells of the hypothalamus secrete hormones directly into the capillaries. They are carried by the blood to the anterior pituitary where they affect the production of other hormones.

Oxytocin is responsible for the "letting down" of milk that occurs when the infant begins to suckle. The hormone promotes contraction of the muscle fibrils around the milk-secreting cells of the mammary glands. It also accelerates childbirth by increasing uterine contractions. These contractions also cause the uterus to regain its normal shape after delivery. The release of oxytocin is under the control of the nervous system and may be triggered by increasing pressure within the uterine wall or by movements of the fetus. In experimental animals, labor can be induced by mechanical stimulation of the uterus or electrical stimulation of the hypothalamus. The hormone is also present in males, but its function in the male, if any, is unknown.

ADH decreases the excretion of water by the kidneys. It achieves this effect by increasing the permeability of the membranes of cells in the collecting ducts of the kidneys so that more water passes through them and back into the blood from the urine. (We shall examine this mechanism in more detail in Chapter 38.) Oxytocin has some ADH effect and ADH some oxytocin effect. This cross action is not surprising because each one of these hormones consists of only nine amino acids, and differences in the two hormones involve differences of only two amino acids among the nine (see Figure 34–21).

REPRODUCTIVE BEHAVIOR IN RINGDOVES

The ringdove is a small relative of the domestic pigeon. The late Daniel S. Lehrman at Rutgers University carried out a series of studies with these birds on the relationships between behavior and endocrine activity.

As is the case with many species of birds, the reproductive cycle in ringdoves is initiated by seasonal changes, such as temperature and day length. Although it is always springtime in the temperature- and light-controlled laboratory, the solitary male or female dove will show no signs of mating or nesting behavior. If a male and female are placed together, however, they will carry out repeated cycles of reproductive behavior, each lasting six to seven weeks and each accompanied by profound physiological changes.

The first day of the cycle is spent in courtship, with the male bowing, cooing, and strutting. A nest site is selected—an empty bowl provided by the experimenters—and during the following week both birds cooperate in gathering material for the nest. During this period, they mate. Between 7 and 11 days after the beginning of courtship, the female lays the eggs, which the male and female sit on in turn. In about 14 days, the eggs hatch, and the parents begin to feed their young crop milk, a liquid secreted by the lining of the adult's crop (a pouch in the bird's gullet) under the influence of prolactin from the pituitary. The young birds are fed for about two weeks, after which the parents lose interest in them. The young birds by then are able to peck grain on the floor of the cage. When the young are 15 to 25 days old, the adult male begins once again to bow and coo and the cycle starts over.

During the period from the beginning of courtship to the egg-laying, the oviducts of the female bird increase in size some 500 percent, from 800 to 4,000 milligrams. This increase in size can also be seen in a female bird that, although in the same cage as a male bird, is separated from the male by a glass partition. If the male bird is castrated, however, and does not behave like a male, the female's oviducts do not enlarge.

Physiological changes and behavior changes develop in parallel. For instance, if a nest containing eggs is put into the cage at the same time as the male and female are placed in it, the nest is absolutely ignored. When the time comes for the birds to build their nest, they usually build it right on top of the already present eggs. If, however, the birds are allowed to court for 7 days, they will sit on a newly introduced nest and eggs quite promptly. Presumably, the courtship prepares them for nesting by stimulating hormone production. To test this hypothesis, isolated birds were injected with the hormone progesterone for one week. They were then put into cages together and immediately given nests and eggs; the great majority sat promptly on the nest.

During the 14 days when the doves are sitting on their eggs, their crops increase enormously in weight, from 900 to as much as 3,000 milligrams. This increase in crop weight, which contributes to the parents' production of "milk," is stimulated by production of the hormone prolactin. If male birds are removed from the cage during the incubation period, their crops do not increase in weight. If they are merely separated from the female by a glass partition, however, and are permitted to watch her incubate the eggs, the visual stimulus alone is sufficient to induce the secretion of prolactin and the consequent enlargement of the crop.

Thus, as Lehrman indicated, there is a constant interplay between the higher brain centers and the endocrine system. The result of this interplay is a sequence of profound physiological changes accompanied by the appropriate behavioral changes necessary to complete the reproductive cycle.

Male ringdove bowing and cooing in courtship.

34-23
Thyroxine, the principal hormone produced by the thyroid gland. Note the four iodine atoms in its structure. Where iodine is present in the soil, it is available in minute quantities in drinking water and in plants. Sea salts are rich in iodine, and, in the United States, table salt is ordinarily iodized or must be specifically labeled as being uniodized. Triiodothyronine differs from thyroxine by having one less iodine atom on the OH-bearing ring.

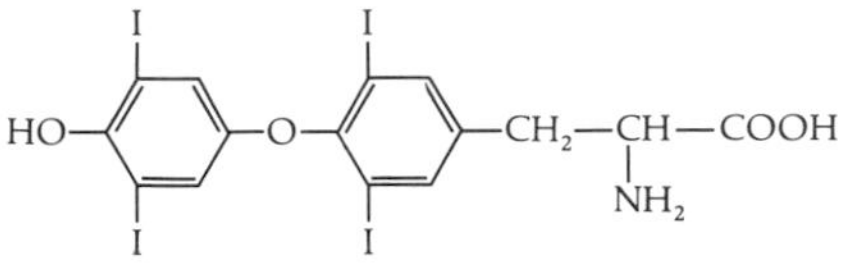

THYROXINE

THE THYROID GLAND

The thyroid, under the influence of the thyroid-stimulating hormone from the pituitary, produces thyroxine, which is an amino acid combined with four atoms of iodine (Figure 34-23). Thyroxine (or its metabolic product, triiodothyronine) accelerates the rate of cellular respiration. Hyperthyroidism, the overproduction of thyroxine, produces nervousness and excessive excitability, increased heart rate and blood pressure, and weight loss. Hypothyroidism in infancy affects development, particularly of the brain cells, and if not treated in time, can lead to permanent mental deficiency and dwarfism. In adults, hypothyroidism is associated with dry skin, intolerance to cold, and lack of energy.

Hypothyroidism may be caused by insufficient iodine, which is needed to make thyroxine, and, in these cases, it is often associated with goiter, an enlargement of the thyroid gland.

The thyroid gland is also the source of the hormone calcitonin, whose major action is to inhibit the release of calcium from bone. Its secretion is controlled directly by the calcium concentration of the fluid surrounding the thyroid cells.

ADRENAL CORTEX

The adrenal glands are on top of the kidneys. The outer layer of the adrenal gland, the adrenal cortex, is the source of a large number of steroids (see page 60) with hormonal activity. About 50 different steroids, all very similar in structure, have so far been isolated from the adrenal cortex of various mammals. Some of these undoubtedly represent steps in the synthesis of various hormones, though most of them have some hormonal activity. In humans, there are two major groups, the glucocorticoids and the mineralocorticoids.

34-24
The chemical structures of two steroid hormones secreted by the adrenal cortex. Cortisol is a glucocorticoid, and aldosterone is a mineralocorticoid. Note, however, the very minor structural differences between the two.

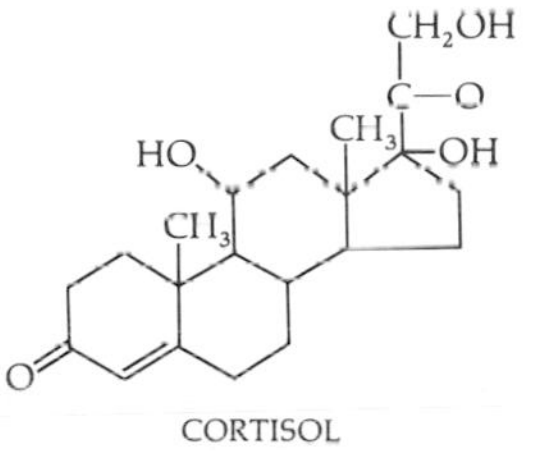

CORTISOL

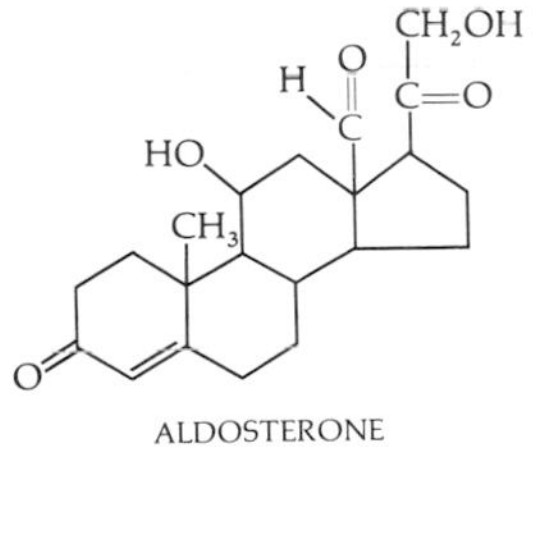

ALDOSTERONE

Cortisol is thought to be the most important glucocorticoid in humans. Cortisol and similar compounds are concerned primarily with the formation of glucose from protein and fat. Only the glucose present in the bloodstream immediately following a meal is derived directly from the diet. After this immediate supply is used up, glycogen stored in the liver is converted to glucose, and the glucose is released into the bloodstream. The liver's supply of glucose is sufficient for a few hours. Subsequently, protein and fat from the body tissues are converted into glucose. In this way, a fairly stable level of glucose is maintained in the bloodstream despite an irregular intake of food.

The glucocorticoids also act to suppress inflammation and are sometimes used medically as anti-inflammatory agents in the treatment of such diseases as arthritis. However, their severe side effects limit their usefulness. (Side effects of these drugs include reduced ability to combat infection, injury, or stress; redistribution of body fat, resulting in "buffalo hump" and "moon face"; and mental disturbance.) The most familiar of the glucocorticoids is cortisone, which is not itself secreted by the human adrenal gland. Instead, cortisone is produced in the liver by the breakdown of cortisol and is used to form other compounds. Little, if any, cortisone is actually released into the bloodstream.

Cortisol and the cortisol-like hormones are secreted in response to the tropic hormone ACTH, and they inhibit secretion of ACTH by negative feedback exerted principally on neurosecreting cells in the hypothalamus.

A second group, the mineralocorticoids, of which aldosterone is the primary

34-25
The production of many major hormones is regulated by a complex negative feedback system involving the pituitary and the hypothalamus. The hypothalamus stimulates the pituitary to secrete tropic hormones, and these, in turn, stimulate the secretion of hormones from the thyroid, adrenal cortex, and gonads (the testes or the ovaries). When the hormones produced by these target glands reach a certain concentration in the blood, the hypothalamus decreases its production of stimulating hormones, the pituitary stops producing hormones, and production of hormones by the target glands also stops. By way of the hypothalamus, which receives information from many other parts of the brain, hormone production is regulated also in response to changes in the external and internal environments.

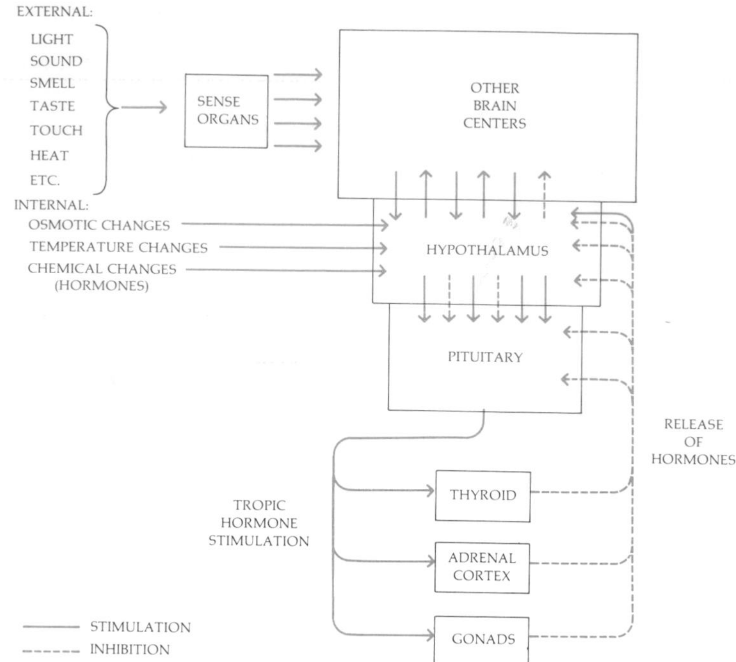

EPINEPHRINE

(a)

NOREPINEPHRINE

(b)

34-26
Epinephrine (or adrenalin) and norepinephrine are secreted by cells of the adrenal medulla. The ring structure to the left is called a catechol, and epinephrine and norepinephrine are characterized chemically as catecholamines, nitrogen-containing catechols.

example, is concerned with the regulation of ions, particularly sodium and potassium ions. The mineralocorticoids affect the transport of ions across the cell membranes of the kidneys and, as a consequence, have major effects both on ion concentrations in the blood and on water retention and loss. An increase in aldosterone secretion results in greater resorption of sodium and chloride ions and increases the excretion of potassium. A deficiency in mineralocorticoids precipitates a critical loss of sodium from the body in the urine and, with it, a loss of water by osmosis. This, in turn, leads to a reduction in blood pressure, which is dependent, in part, upon the water levels of the body.

The adrenal cortex is also a source of male sex hormones, which is why an adrenal tumor may result in increased production of these hormones and the production of facial hair and other masculine characteristics in a woman. Bearded ladies in the circus are often victims of such tumors.

ADRENAL MEDULLA

The adrenal medulla, the central portion of the adrenal gland, actually does not conform to the definition of a gland in that it is not made up of glandular epithelium. It is composed, instead, of nervous tissue; in fact, it is, in essence, a large ganglion whose nerve endings secrete epinephrine and norepinephrine. These hormones raise blood pressure, stimulate respiration, dilate the respiratory passages, and promote the activity of the enzyme that breaks down glycogen to glucose 6-phosphate, thus increasing concentrations of glucose in the bloodstream. They stimulate the general metabolic activity of cells. The adrenal medulla is stimulated by a nerve fiber of the sympathetic division and acts as an enforcer of sympathetic activity. Epinephrine and norepinephrine are destroyed by enzymes

34–27
Hormonal regulation of blood glucose. When blood sugar concentrations are low (hypoglycemia), the pancreas releases glucagon, which stimulates the breakdown of glycogen and the release of glucose from the liver and decreases the entry of glucose into cells. When blood sugar concentrations are high (hyperglycemia), the pancreas releases insulin, which removes glucose from the bloodstream by stimulating its passage into cells and promoting its conversion to glycogen, the storage form. ACTH, produced by the pituitary, stimulates the adrenal cortex to produce glucocorticoids, which cause the liver to convert stored proteins to glucose. Under conditions of stress, the adrenal medulla releases epinephrine and norepinephrine, which also raise blood sugar, and the production of ACTH is increased.

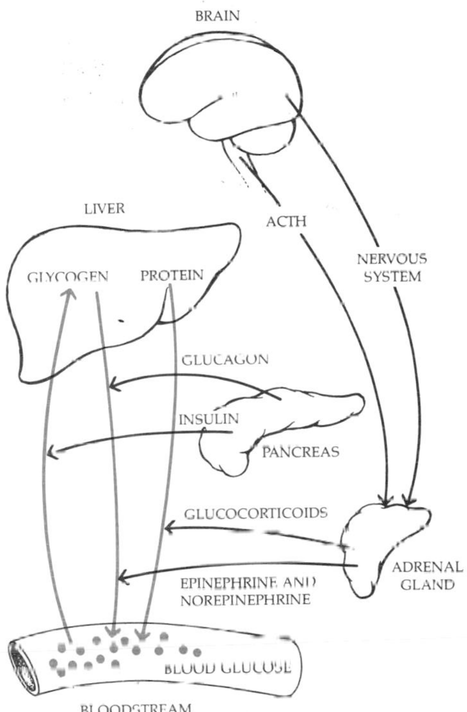

in the blood within minutes after their release, which is another mechanism for tight regulatory control. (The steroid and protein hormones are also broken down, although more slowly and principally by the liver.)

PANCREAS

The islet cells of the pancreas are the source of two hormones concerned with the metabolism of glucose: insulin and glucagon. Insulin is secreted in response to a rise in blood sugar or amino acid concentration (as after a meal). It lowers the blood sugar by stimulating cellular uptake of glucose and by stimulating the conversion of glucose to glycogen.

When there is an insulin deficiency, as in persons with diabetes mellitus, the concentration of blood sugar rises so high that glucose entering the kidney cannot be resorbed; the presence of glucose in the urine forms the basis of simple tests for diabetes. The loss of glucose is accompanied by loss of water by osmotic pressure. Dehydration, which can lead to collapse of circulation, is one of the causes of death in an untreated diabetic.

Mild diabetes can be controlled by diet. Severe diabetes is treated by insulin injections. (Because insulin is a protein, it cannot be given by mouth. It would simply be digested. Certain other oral drugs, which affect sugar metabolism in different ways, can be used in diabetes therapy.)

Glucagon, produced by different islet cells of the pancreas, increases blood sugar by stimulating the breakdown of glycogen to glucose in the liver and by stimulating the breakdown of fats and protein, which decreases glucose utilization.

Thus, as we have seen, at least five different hormones are involved in regulating blood sugar: growth hormone, cortisol, epinephrine, insulin, and glucagon. This tight control over blood glucose is particularly important for brain cells. Unlike other cells in the body, which can derive energy from the breakdown of amino acids and fats, brain cells can utilize only glucose and so are immediately affected by low blood sugar. Fainting from hunger is an indication of the sensitivity of the brain cells to glucose deprivation.

THE PARATHYROID GLANDS

The parathyroid glands were first discovered in 1850 by a British anatomist dissecting an Indian rhinoceros. They play an essential role in mineral metabolism, specifically in the regulation of calcium and phosphate, which exist in a reciprocal relationship in the blood. Calcium is normally present in mammalian blood in concentrations of about 6 milligrams per 100 milliliters of whole blood. A rise or fall of more than 2 or 3 milligrams per 100 milliliters can lead to such severe disturbances in blood coagulation, muscle contraction, and nerve function that death may follow within hours. Parathyroid hormone maintains blood calcium levels by increasing its absorption in the intestine and by reducing its excretion by the kidneys. It also stimulates the release into the bloodstream of calcium from bone, which contains 99 percent of the body's total calcium. This release occurs as the result of bone-destroying cells known as osteoclasts. Other cells are simultaneously engaged in producing new bone. By slightly shifting the balance between these two activities, the parathyroids exercise a very tight control over blood calcium levels (another example of homeostasis).

HUMAN CIRCADIAN RHYTHMS

Circadian rhythms serve to integrate bodily functions by bringing them into synchrony with one another and with the external environment. Human organisms, like plants and other animals, are governed by such rhythms. It has long been known that we are more likely to be born between 3 and 4 A.M. *and also to die in these same early morning hours. Body temperature fluctuates as much as 1°C (about 2°F) during the course of the day, usually reaching a high about 4* P.M. *and a low about 4* A.M. *Alcohol tolerance is greatest at 5* P.M. *Many people have an automatic internal alarm system that wakes them at the same hour every morning—whether they want to or not. Secretion of some hormones, heart rate, blood pressure, and urinary excretion of potassium, sodium, and calcium all vary according to a circadian rhythm. A recent study by the Federal Aviation Agency showed that pilots flying from one time zone to another—from New York to Europe, for instance—exhibit "jet lag," a general decrease in mental alertness and ability to concentrate and an increase in decision time and physiological reaction time. Measurements of circadian rhythms show that the body may be "out of sync" for as much as a week after such a flight. This brings into question present policies of diplomats speeding to foreign capitals at times of international crisis or of troops being air-transported into combat.*

Elaborate studies in which individuals have been kept isolated in constant conditions for long periods of time—in underground bunkers, for example—have indicated that circadian rhythms are under the control of a biological clock or clocks. Such internal clocks presumably are another mechanism for integrating and coordinating the body's multitude of physiological functions.

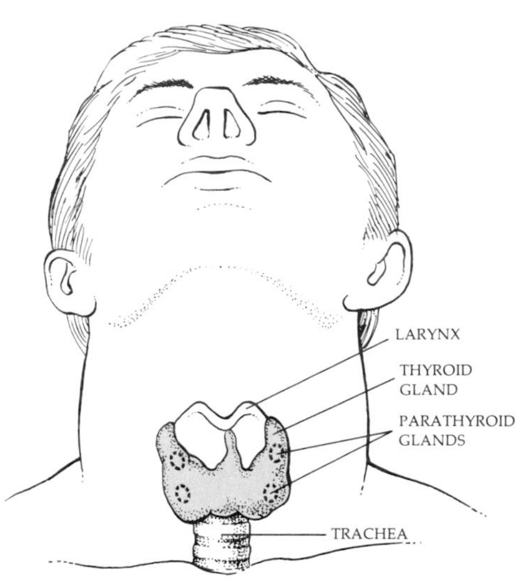

34–28
The pea-sized parathyroid glands, the smallest of the known endocrine glands, are located behind or within the thyroid gland. They produce parathyroid hormone (parathormone), which increases concentrations of blood calcium. Calcitonin, which decreases blood calcium, was also thought to be produced by the parathyroids. It is now known to be a thyroid hormone and, in order to stress its true status, is sometimes called thyrocalcitonin.

Hyperparathyroidism, caused by tumors of the parathyroids, occasionally occurs in humans. When there is too much parathyroid hormone, the bones lose too much calcium and become soft and fragile. Removal of the parathyroid glands without replacement therapy results in violent muscular contractions and spasms, leading to death.

PROSTAGLANDINS

The most recently discovered hormone-like chemicals are the prostaglandins. They were given this name because they were first detected in seminal fluid and were thought to be produced by the prostate gland. Actually, most of the prostaglandins in semen are synthesized in the seminal vesicles. These prostaglandins, which are found in the female reproductive tract after sexual intercourse, induce rhythmic contractions in the muscular wall of the uterus. The semen of some infertile males has been found to be poor in prostaglandins, and the uterus of infertile females is often unresponsive to these hormones. For these reasons, prostaglandins are believed to play a role in fertilization.

Since their initial discovery, 16 natural prostaglandins have been found. One major source is menstrual fluid. The prostaglandins in menstrual fluid also induce contractions in the uterus, which presumably aid in dispelling the endometrial lining. Because of their effects on the uterus, they are being studied for their possible use in birth control (by inducing menstruation in women in very early stages of pregnancy and so producing early abortions) and in facilitating labor. They may play a role in the effectiveness of IUDs.

More recently, prostaglandins have been implicated in inflammatory and immune reactions (see Chapter 35) and, by extension, in diseases such as rheumatoid arthritis and asthma. Aspirin inhibits prostaglandin activity, and this is thought to be the reason that it is so effective in reducing inflammation and fever. Some prostaglandins have antagonistic effects; for example, one causes smooth muscle to relax, and another causes the same muscle to contract.

Prostaglandins differ from hormones in a number of ways. (1) They are fatty acids. (2) Their target tissues are those of another individual (if their role in fertilization has been correctly interpreted). (3) They are among the most potent of all known biological materials, producing marked effects in very small doses. A concomitant of their extraordinary potency is the fact that they are produced in very small amounts and are rapidly broken down by enzyme systems in the body. If it were not for their presence in unusually large amounts in semen, they might never have been discovered. (4) They appear to be produced by cell membranes. (5) They often exert their effect in the tissue that produces them.

MECHANISM OF ACTION OF HORMONES

Virtually all cells of the body are equally exposed to the hormones released into the bloodstream, yet not all respond. Recent research has indicated the mechanisms underlying specificity of action of two major groups of hormones, the steroids and the protein hormones.

Steroid hormones are relatively small molecules. These molecules, it has been found, pass easily through cell membranes and so freely enter all the cells of the body. However, in their target cells, and only in their target cells, the hormones encounter a specific receptor molecule in the cytoplasm that combines with them. The receptor, a protein, and the steroid together act directly on the DNA of the cell to promote the synthesis of messenger RNA and so of specific enzymes and other proteins. (Note that this is similar to the operation of the operon, described on page 291. However, it differs in that hormone and receptor apparently serve to stimulate RNA synthesis in these cells, rather than to remove a repressor molecule, as occurs in the operon.) These findings explain how the very slight differences in configuration among steroid molecules can be correlated with such drastically different effects: The protein receptors, like enzymes, are highly specific in their combining properties.

By contrast, it is more likely that protein hormones, which are much larger molecules, do not enter cells but rather combine with receptor molecules on the membrane surface, setting in motion a "second messenger" that is responsible for the sequence of events inside the cell. These findings may have important medical implications. For example, it has long been known that juvenile diabetes—diabetes in young persons—is caused by a deficiency of the hormone insulin, which as we noted, promotes the uptake of glucose by cells. It was assumed by analogy that the diabetes found commonly in older persons and associated with obesity had the same cause. It has now been found, however, that adult diabetes results from a decrease in the number of insulin binding sites on receptor cells. Such patients are treated most effectively by diet.

Another group of studies, for which Earl W. Sutherland was awarded the Nobel Prize in 1971, has shown that a chemical known as cyclic AMP (Figure 34–29) is the second messenger in a number of target cells. Hormones that trigger the action of cyclic AMP include ACTH, TSH, LH, ADH, epinephrine, and glucagon. The differences in their effects are due to the presence of different enzyme systems within the cells that respond to cyclic AMP.

At almost the same time that cyclic AMP was identified by these research

CYCLIC AMP

(a)

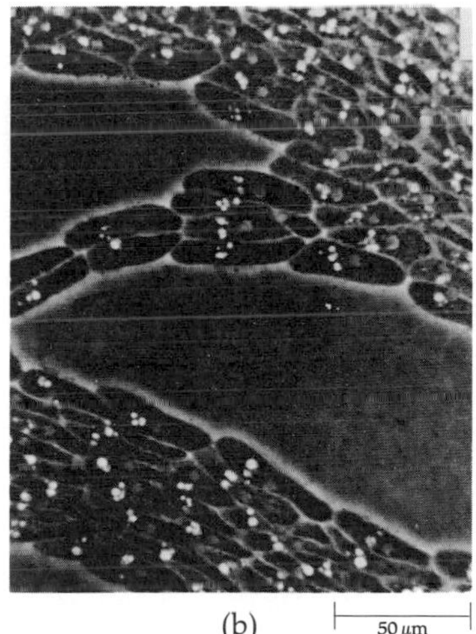

(b)

34–29

(a) Cyclic AMP (adenosine monophosphate) acts as a "second messenger" within the cells of vertebrates. Following stimulation by various hormones—the "first messengers"—cyclic AMP is formed from ATP. "Cyclic" refers to the fact that the atoms of the phosphate group form a ring. (b) Cyclic AMP is also the chemical that attracts the amoebas of the cellular slime molds, causing them to aggregate into a sluglike body, which then behaves like a multicellular organism. The arrow shows the direction in which the cells are moving.

workers in mammalian physiology, biologists studying a peculiar group of organisms known as the cellular slime molds isolated a chemical of great importance in this biological system. As you will recall, the cells of the cellular slime mold begin as individual amoebas and then come together to form a single organism (see pages 121 and 383). The chemical that calls them together was named acrasin, and has now been identified as cyclic AMP, another example of the long thread of evolutionary history linking all organisms.

SUMMARY

The nervous and endocrine systems provide the communication and integration necessary to coordinate the numerous internal functions that enable an animal to regulate its internal environment.

The unit of the nervous system is the neuron, or nerve cell, which consists of dendrites and a cell body, which receive impulses, and an axon, or nerve fiber, which relays the impulses to other cells. The nerve impulse travels along the fiber in the form of an action potential, which is a transient reversal in the polarity of the membrane of the fiber. Because the action potential is always the same, the message carried by a particular fiber can be varied only by a change in frequency and pattern. In myelinated cells, the nerve impulse leaps from node to node of the myelin sheath, thereby speeding conduction.

Nerve cells transmit signals to other neurons across a junction called a synapse. The signal crosses the synaptic cleft in the form of a chemical transmitter that stimulates or inhibits the production of an action potential in the target cell. Signals to muscle fibers are similarly transmitted across the neuromuscular junction.

The central nervous system consists of the brain and the spinal cord, which are encased, in vertebrates, in the skull and vertebral column. The nervous system outside the central nervous system constitutes the peripheral nervous system. Spinal nerves emerge from the vertebral column in pairs, each pair consisting of motor fibers and sensory fibers. The motor fibers of each pair innervate the muscles of a different area of the body, and the sensory fibers receive impulses from sensory receptors in the same area.

In vertebrates, the peripheral nervous system has two major divisions: (1) the somatic nervous system, and (2) the autonomic nervous system, which controls the muscles and glands involved in the digestive, circulatory, urinary, and reproductive functions. In the autonomic system, the efferent neurons arising from the spinal cord synapse outside the central nervous system with second neurons. These postganglionic neurons stimulate or inhibit the effectors. The autonomic system has two divisions: (1) the sympathetic division, which is largely responsible for reactions to the external environment; and (2) the parasympathetic division, which controls restorative activities, such as digestion and rest. The two divisions are anatomically distinct. The terminal chemical transmitter for the sympathetic division is norepinephrine; for the parasympathetic division, acetylcholine.

The endocrine system consists of glands that secrete chemicals (hormones) that are taken up into the bloodstream. These chemicals exert specific effects on certain organs and tissues. The effects of hormones on their target tissues depend on the presence in the target cells of specific receptors for that hormone. Table 34–2 lists the principal glands of vertebrates and their hormones. The nervous and endocrine systems influence one another's activities profoundly.

Table 34-2 *Some of the Principal Endocrine Glands of Vertebrates and the Hormones They Produce*

GLAND	HORMONE	PRINCIPAL ACTION	MECHANISM CONTROLLING SECRETION	CHEMICAL COMPOSITION
Pituitary, anterior lobe	Thyroid-stimulating hormone (TSH)	Stimulates thyroid	Thyroxine in blood; hypothalamic-releasing hormone	Glycoprotein
	Follicle-stimulating hormone (FSH)	Stimulates ovarian follicle, spermatogenesis	Estrogen in blood; hypothalamic-releasing hormone	Glycoprotein
	Luteinizing hormone (LH)	Stimulates interstitial cells in male, corpus luteum and ovulation in female	Testosterone or progesterone in blood; hypothalamic-releasing hormone	Glycoprotein
	Adrenocorticotropic hormone (ACTH)	Stimulates adrenal cortex	Adrenal cortical hormone in blood; hypothalamic-releasing hormone	Protein
	Growth hormone (somatotropin)	Stimulates bone and muscle growth, inhibits oxidation of glucose, promotes breakdown of fatty acids	Hypothalamic-inhibiting hormone (somatostatin)	Protein
	Prolactin	Stimulates milk production and secretion in "prepared" gland	Hypothalamic-inhibiting hormone	Protein
Thyroid	Thyroxine, other thyroxinelike hormones	Stimulate and maintain metabolic activities	TSH	Iodinated amino acids
	Calcitonin	Inhibits release of calcium from bone	Concentration of calcium in blood	Peptide (32 amino acids)
Parathyroid	Parathyroid hormone (parathormone)	Stimulates release of calcium from bone, promotes calcium uptake from gastrointestinal tract, inhibits calcium excretion	Concentration of calcium in blood	Protein
Ovary, follicle	Estrogens	Develop and maintain sex characteristics in females, initiate buildup of endometrium	FSH	Steroids
Ovary, corpus luteum	Progesterone and estrogens	Promote continued growth of endometrium	LH	Steroids
Testis	Testosterone	Supports spermatogenesis, develops and maintains sex characteristics of males	LH	Steroid
Hypothalamus (via posterior pituitary)	Oxytocin	Stimulates uterine contractions, milk ejection	Nervous system	Peptide (9 amino acids)
	Antidiuretic hormone (vasopressin)	Controls water excretion	Osmotic concentration of blood; nervous system	Peptide (9 amino acids)
Adrenal cortex	Cortisol, other cortisol-like hormones	Affect carbohydrate, protein, and lipid metabolism	ACTH	Steroids
	Aldosterone	Affects salt and water balance	Renin from kidney, K^+ ions in blood	Steroid
Adrenal medulla	Epinephrine and norepinephrine	Increase blood sugar, dilate some blood vessels, increase rate of heartbeat	Nervous system	Catecholamines
Pancreas	Insulin	Lowers blood sugar, increases storage of glycogen	Concentration of glucose in blood	Protein
	Glucagon	Stimulates breakdown of glycogen to glucose in the liver	Concentration of glucose and amino acids in blood	Protein

Hormone production is characteristically regulated by a negative feedback system. In the case of thyroxine and the steroid hormones, the concentration of a particular hormone in the blood affects the release from the pituitary of the tropic hormone that stimulates that particular gland. In the case of other hormones, such as parathyroid hormone, which increases calcium concentration in the blood, the increased calcium acts directly on the gland to inhibit secretion of the hormone. Such feedback systems are important mechanisms of homeostasis.

QUESTIONS

1. Distinguish between the following: neuron/nerve; afferent/efferent; somatic/autonomic; sympathetic/parasympathetic; axon/dendrite; resting potential/action potential; gray matter/white matter; pituitary/hypothalamus; somatostatin/somatotropin.

2. Describe the way in which the nerve impulse propagates itself. Draw a diagram of this process.

3. If a neuron is placed in a medium where ionic concentrations are the same as those of its own cytoplasm, what effect will this have on the resting potential?

4. What are four significant differences between the somatic and autonomic nervous systems?

5. What are three significant differences between the sympathetic and parasympathetic divisions?

6. Assume that three preganglionic neurons, A, B, and C, make adjacent synapses on the same postganglionic neuron, D. No postganglionic impulse is initiated as a result of single preganglionic impulses in A, B, or C, nor are they initiated if impulses arrive simultaneously in all three, in A and B, or in A and C. Only if impulses arrive together at the synapses of B and C will D fire an impulse. Explain these results in terms of excitatory and inhibitory transmitter substances and their effects on the postganglionic membrane voltage.

7. What are the classical experimental steps by which an unknown organ or tissue can be diagnosed as endocrine in function? How might this procedure prove deceptive?

8. Draw a diagram of the negative feedback system regulating the production of thyroxine.

9. What do you think a positive feedback system might be? Can you give a nonbiological example? Why is negative feedback more appropriate as a biological control system? When is positive feedback more effective?

10. Ideally, what types of functions should be controlled by hormones, as compared to nerves? Does the reality, as described in this chapter, fulfill your expectations?

CHAPTER 35

Circulation of the Blood

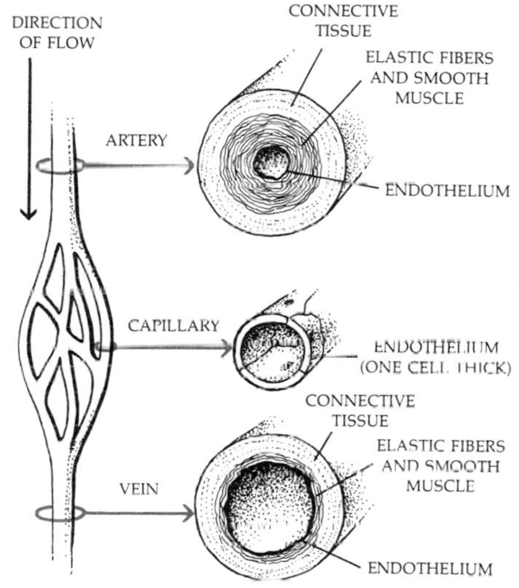

35–1
Structure of blood vessels. Arteries have thick, tough, elastic walls that can withstand the pressure of the blood as it leaves the heart. Capillaries have walls only one cell thick. Exchange of gases, nutrients, and wastes between the blood and the cells of the body takes place through these thin capillary walls. Veins have larger lumens (passageways) and thinner, more readily extensible walls that minimize resistance to the flow of blood on its return to the heart.

As we discussed earlier in this book, every cell in the body of a complex organism builds its own membranes and organelles, makes its own ATP, and assembles its own enzymes and other proteins. To engage in these activities, cells need oxygen and nutrients and must dispose of carbon dioxide and other wastes.

In single-celled organisms and very small multicellular ones, the needs of each cell are supplied directly by the medium in which the organism lives. In larger, more complex animals, these needs are supplied by a transport system. In humans and other vertebrates, gases (oxygen and carbon dioxide), nutrient molecules, hormones, and wastes are transported in the bloodstream. The blood circulates through a closed circuit of continuous vessels, propelled by the contractions of a specialized muscle, the heart. This system is known as the *cardiovascular system* (from *cardio,* meaning "heart," and *vascular,* meaning "vessel"). In this system, the heart pumps the blood into the large arteries, from which it travels to branching, smaller arteries (the *arterioles*) and then into very small vessels, the *capillaries.* Through the thin walls of the capillaries, nutrients, oxygen, carbon dioxide, and other molecules are exchanged between the blood and the fluids surrounding the body cells (the interstitial fluids). From the capillaries, the blood passes into small veins, the *venules,* then into larger veins, and through them, back to the heart. Thus the heart, arteries, and veins are, in essence, the means for getting the blood to and from the capillaries, where the actual function of the circulatory system is carried out.

THE BLOOD VESSELS

In humans, the diameter of the opening of the largest artery, the *aorta,* is 2.5 centimeters, that of the smallest capillary only 8 micrometers, and that of the largest vein, the *vena cava,* 3 centimeters. Arteries, veins, and capillaries differ not only in their diameters but also in the structure of their walls (Figure 35–1). Of the three types of vessels, the arteries have the thickest, strongest walls, made up of three layers. The inner layer, or endothelium (a type of epithelial tissue), forms the lining of the vessels; the middle layer contains smooth muscles and elastic tissues; the outer layer, also elastic, is made of collagen and other supporting tissues. Because of their elasticity, the arteries expand when the blood is pumped into them, and then relax slowly; as a consequence, by the time the blood leaves the

35–2
A large vein and a small artery. The artery can be identified by its thick, muscular wall.

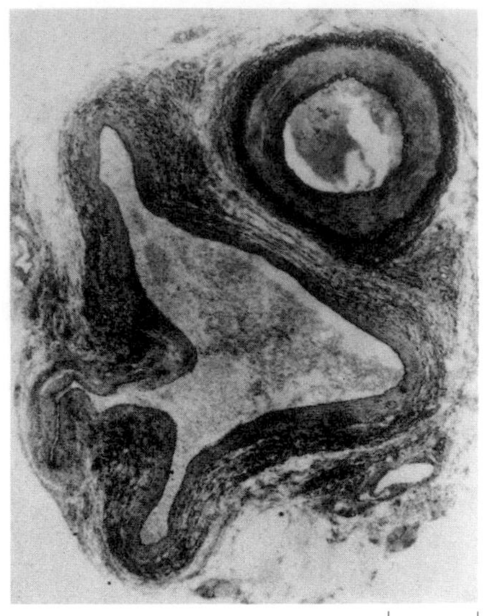

arteries it is flowing smoothly through the vessels, rather than in spurts, as it does when it leaves the heart. The pulsation felt when the fingertips are placed over an artery close to the body surface—as in the wrist—represents the alternating expansion and recoil of an elastic arterial wall.

Capillaries and Diffusion

The walls of the capillaries consist of only one layer of cells (Figure 35–3), the endothelium, and the passageway (lumen) of the smallest capillary is about 6 micrometers in diameter, just wide enough for red blood cells to move in single file. The total length of the capillaries in a human adult is more than 80,000 kilometers. Because of the total cross-sectional area, blood moves slowly through the capillary system. As it moves, gases (oxygen and carbon dioxide), hormones, and other materials are exchanged with the surrounding tissues by diffusion through junctions between the endothelial cells and through the cytoplasm of the endothelial cells. Organic molecules, such as glucose, are probably moved by transport systems of the cells. Pinocytotic vesicles have been observed in the capillary endothelial cells; they presumably serve to ferry dissolved materials in or out of the capillaries.

At the arteriolar end of the capillaries, the hydrostatic pressure of the blood is greater than its osmotic pressure, and the watery component of the blood leaves the capillaries along with nutrient molecules and oxygen. By the time the blood reaches the venous end of the capillaries, the hydrostatic pressure has dropped to a point where it now is less than the osmotic pressure of the blood, and fluids move back into the capillaries. As a consequence, there is little total fluid loss from the blood as it moves through the capillaries. (The relatively small amount that is not returned by osmosis is channeled back into the bloodstream by the lymphatic system, as we shall see later in this chapter.) When loss of water into tissues does

35–3
Electron micrograph of a capillary from cardiac muscle. Portions of two endothelial cells can be seen, fitting together to form the capillary lumen.

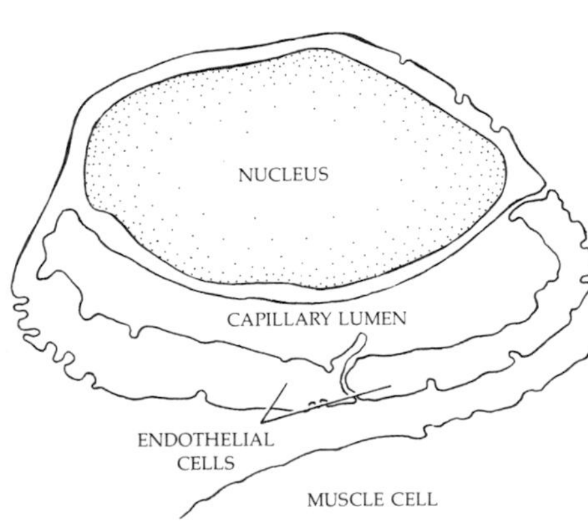

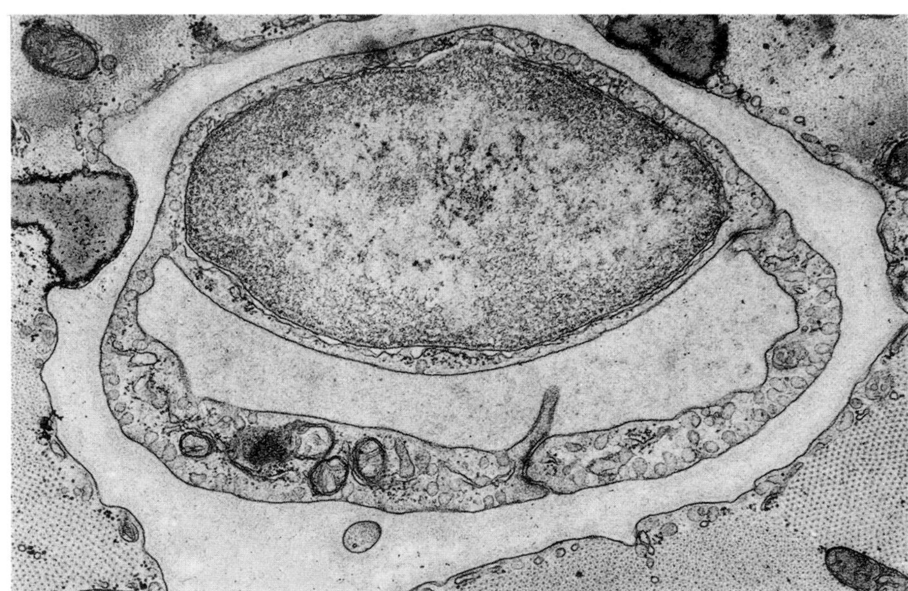

35-4
A child with kwashiorkor, a West African word that means "the sickness a child develops when another child is born." The swelling characteristic of this disease is caused by a deficit in blood proteins; because of the deficit, fluids are not drawn back in by osmotic forces to the extent that they are in well-nourished persons. Children with kwashiorkor are usually also weak and apathetic and have skin discoloration. The syndrome, which is the result of a deficiency in protein but not in total calories, usually develops after a child is weaned. Children with kwashiorkor can be found in the United States as well as in developing countries.

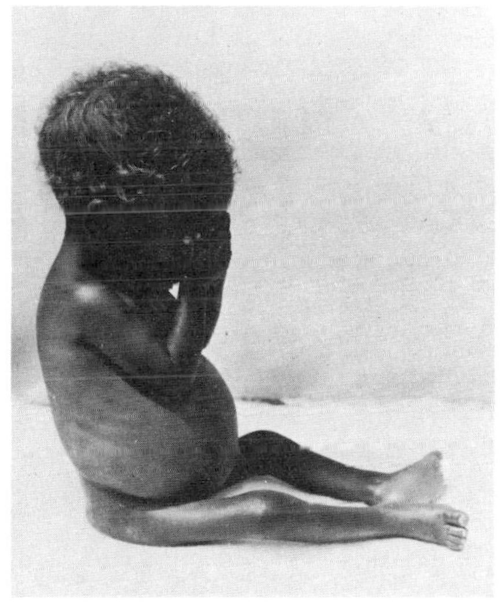

occur, as, for example, when the endothelium is damaged by a blow, swelling (edema) is produced. The edema associated with certain forms of malnutrition is also due to the movement of fluids out of the capillaries (Figure 35–4).

No cell in the human body is farther than 130 micrometers—a distance short enough for rapid diffusion—from a capillary. Even the cells in the walls of the large veins and arteries depend on this capillary system for their blood supply, as does the heart itself, like all the other organs of the body.

At the ends of the capillary beds, blood passes into the venules, smallest of the veins, then into larger veins, and finally back into the heart either through the superior (anterior) or inferior (posterior) vena cava. Like those of the arteries, the walls of the veins are three-layered, but they are less elastic and more pliable. An empty vein collapses, whereas an empty artery remains open. The veins, with their thin walls and relatively large diameters, offer little resistance to flow, facilitating the return of the blood to the heart.

Blood Pressure

Blood pressure is the force per unit area with which blood pushes against the walls of the blood vessels. It is conventionally described in terms of how high it can push a column of mercury. For medical purposes, it is usually measured at the artery of the left upper arm. Normal blood pressure in a young man is 120 millimeters of mercury (120 mm Hg) when the heart is contracting (the systolic blood pressure) and 80 mm Hg when the heart relaxes (diastolic pressure); this is stated as a blood pressure of 120/80. The blood pressure of a young woman is 4 to 8 mm Hg less.

The pressure is generated by the pumping action of the heart and changes with the rate at which it contracts. Blood flow is directly proportional to blood pressure; the greater the pressure, the greater the flow.

Constriction of the arteries by loss of elasticity or by the formation of fatty deposits within their walls increases the systolic blood pressure and also increases the work load on the heart.

Blood pressure is not the same in the various parts of the cardiovascular system. As the blood gets farther along the system of vessels, the pressure drops, becoming lower in the veins and lowest in the right atrium (upper chamber of the heart) (Figure 35–5). Thus the blood, moving toward an area of lower pressure, returns toward the heart.

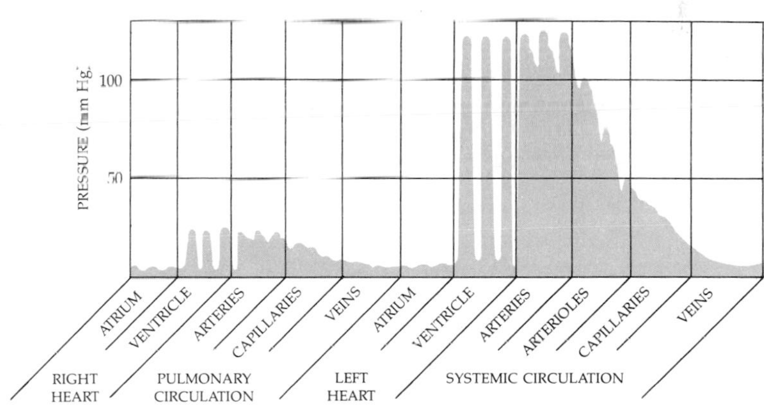

35–5
In mammals, blood goes from the right heart to the lungs, from the lungs to the left heart, and, from the left heart, it enters the systemic circulation, moving from arteries to arterioles to capillaries to veins. Blood pressures vary in the different areas of the cardiovascular system. The fluctuations in blood pressure produced by three heartbeats are shown in each section of the diagram. Note the fall in pressure as blood traverses the arterioles of the systemic circulation.

Gravity affects blood pressure, decreasing the flow of blood to the brain, as you know if you have experienced dizziness on standing suddenly. Fainting and falling down are protective responses that prevent serious damage to the brain cells as a result of inadequate blood supply. These responses are often thwarted by well-meaning bystanders anxious to get the affected individual "back on his feet." In fact, holding a fainting person upright can lead to severe shock and even death.

Regulating Blood Flow

Peripheral Resistance

The flow of any fluid is directly proportional to the difference in pressure between the two ends of the tube through which it flows and inversely proportional to the resistance. Resistance is essentially a measure of friction, the friction between the tube wall and the fluid flowing through it. We know from common experience that the narrower a tube, the more resistance it offers to the flow of a given volume of fluid. The exact relationship is given by a simple formula that states that resistance to flow of a given fluid is inversely proportional to the fourth power of the radius (r^4) of the vessel, or flow through the tube is proportional to the fourth power of the radius. Thus, if you double the radius of a tube, the flow through it increases sixteenfold. Conversely, sixteen times less fluid will flow through a tube of half the radius. As the blood reaches the branching arterioles, the radius of the vessels decreases, resistance increases greatly, and the rate of flow of blood decreases. Peripheral resistance is the term applied to the resistance that the vessels offer to the flow of blood from the arterial system to the heart.

35–6
Diagram of a capillary bed. The sphincter muscle at the end of the arteriole controls the blood flow through the capillaries. These muscles, which are innervated by the autonomic nervous system, make it possible to regulate the blood supply to different regions of the body depending on the physiological state of the organism.

Blood flow through capillary beds is regulated by contraction and relaxation of the arteriolar smooth muscles (Figure 35–6). These muscles are always in a state of partial contraction, producing what is called arterial tonus, or tone. These smooth muscles are controlled by autonomic nerves (both sympathetic and parasympathetic), the adrenal hormones epinephrine and norepinephrine, and chemicals produced locally in the tissues themselves. Sympathetic control tends to close down the blood supply to internal organs and open that to appendages and skeletal muscle, and parasympathetic has the opposite effect. Thus the oxygen supply can be distributed according to the requirements of different parts of the body at different times, with skeletal muscles getting more during exercise, for example, or the intestinal tract more during digestion. Psychological events also influence vasomotor activities. Familiar examples are blushing, turning pale with fear, angina pectoris (pain in the chest and left arm caused by a shortage of oxygen to the heart muscle) precipitated by emotion, and erection of the penis or clitoris following erotic stimulation.

THE HEART

Evolution of the Heart

In its simplest form—as in the earthworm—the heart is a muscular contractile part of a circulatory vessel. In the course of vertebrate evolution the heart has undergone some structural adaptations, as shown in Figure 35–7.

Fish have a single heart divided into the *atrium,* which is the receiving area for the blood, and the *ventricle,* which is the pumping area from which the blood is

35–7
Vertebrate circulatory systems. Oxygenated blood is indicated in color. (a) *In the fish, the heart has only one atrium (A) and one ventricle (V). Blood oxygenated in the gill capillaries goes straight to the systemic capillaries without first returning to the heart.* (b) *In amphibians, the single primitive atrium has been divided into two separate chambers. Oxygenated blood from the lungs enters one atrium, and oxygen-poor blood from the tissues enters the other. The blood is mixed somewhat in the single ventricle and then pumped again through the lungs and body tissues.* (c) *In birds and mammals, the ventricle is divided into two separate chambers, so that there are, in effect, two hearts—one for pumping oxygen-poor blood through the lungs and one for pumping oxygen-rich blood through the body tissues.*

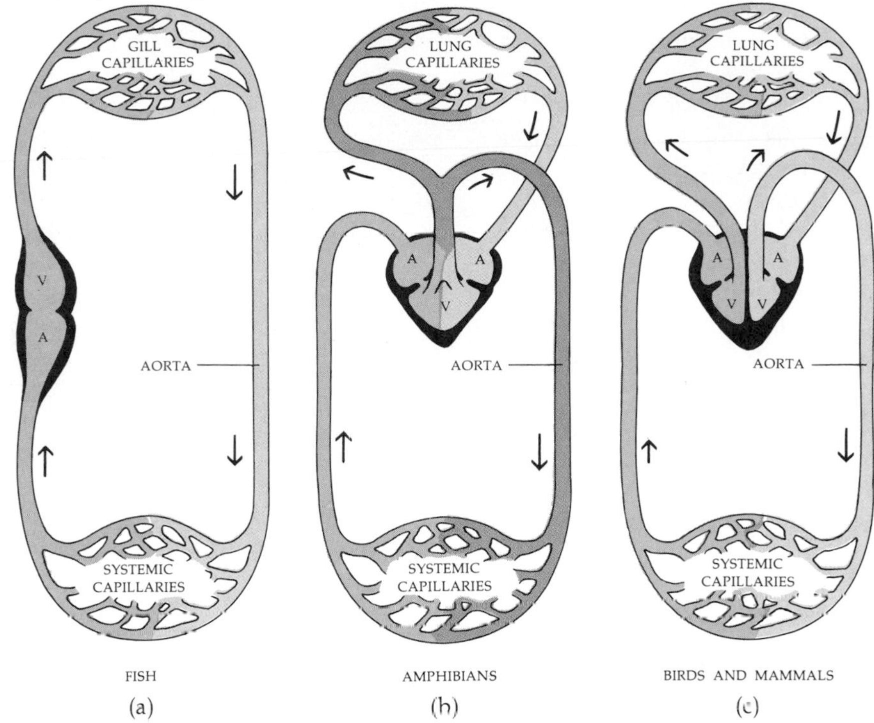

expelled into the vessels. The ventricle of the fish heart pumps blood directly into the capillaries of the gills, where it picks up oxygen and releases carbon dioxide. From the gills, oxygenated blood is carried to the tissues. By this time, however, most of the propulsive force of the heartbeat has been dissipated by the resistance of the capillaries in the gills, so that the blood flow through the rest of the tissues (the systemic circulation) is relatively sluggish.

In amphibians, there are two atria; one receives oxygenated blood from the lungs and the other receives deoxygenated blood from the systemic circulation. Both atria empty into the single ventricle, which pumps the mixed blood simultaneously through the lungs and the systemic circulation. By this arrangement, the blood enters the systemic circulation under high pressure, but not highly oxygenated.

In the birds and mammals, the heart is functionally separated longitudinally into two organs. The right heart receives blood from the tissues and pumps it into the lungs, where it becomes oxygenated. From the lungs it returns to the left heart, from which it is pumped into the body tissues. This efficient, high-pressure circulation system with its separation of oxygenated and deoxygenated blood is necessary to maintain the high metabolic rate of both birds and mammals, with their constant body temperature and their generally high level of physical and mental activity.

35–8
The human heart. Blood returning from the systemic circulation through the superior and inferior venae cavae enters the right atrium and is pumped to the right ventricle, which propels it through the pulmonary arteries to the lungs, where it is oxygenated. Blood from the lungs enters the left atrium through the pulmonary veins and is pumped to the left ventricle and then through the aorta to the body tissues.

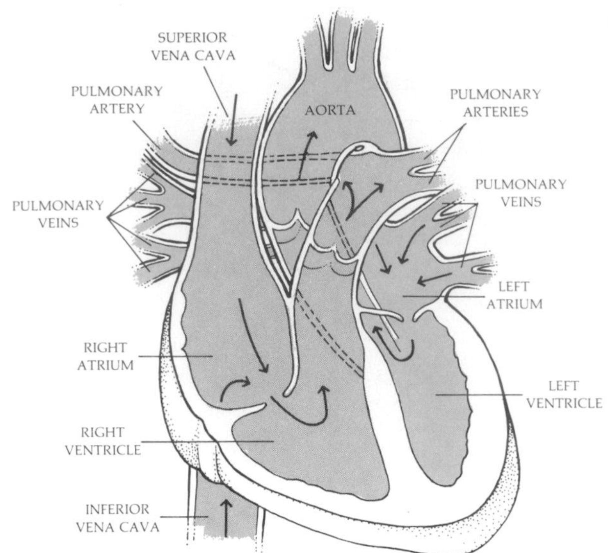

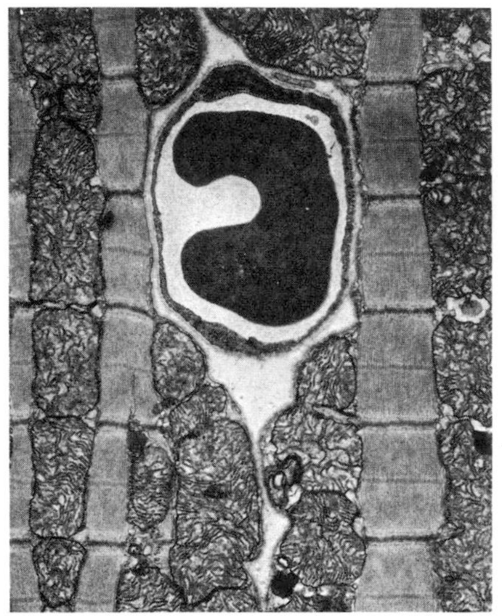

35–9
Electron micrograph of cardiac muscle, showing a red blood cell within a capillary in the muscle. The striated pattern of the cardiac muscle is clearly visible. Note also the large and numerous mitochondria and the extremely thin wall of the capillary. The heart, like all the other tissues of the body, is dependent on the fine capillary network for its supply of oxygen and other blood-borne materials.

The Heart and Regulation of Its Beat

Figure 35–8 shows a diagram of the human heart. Its walls are made up of connective tissue and a specialized muscle—the cardiac muscle. Blood returning from the body tissues enters the right atrium through two large veins, the superior and inferior venae cavae. Blood returning from the lungs enters the left atrium through the pulmonary veins. The atria, which are thin-walled compared with the ventricles, expand as they receive the blood. Both atria then contract simultaneously, assisting the flow of blood through the open valves into the ventricles. Then the ventricles contract simultaneously; the valves between the atria and ventricles are closed by the pressure of the blood in the ventricles. The right ventricle propels the blood into the lungs through the pulmonary arteries; the left ventricle propels it into the aorta, from which it travels to the other body tissues. Valves between the ventricles and the pulmonary artery and the aorta close after the ventricles contract, thus preventing backflow of blood.

Most muscle contracts only when stimulated by a motor nerve, but the contractions of cardiac muscle originate in the muscle itself. A vertebrate heart will continue to contract even after it is removed from the body if it is kept in a nutrient, oxygenated solution. In vertebrate embryos, as we saw, the heart begins to beat very early in development, before the appearance of any nerve supply. In fact, embryonic heart cells isolated in a test tube will beat.

The beat of the cardiac muscle is initiated by a special area of the heart, the *sinoatrial node,* which is located in the right atrium and functions as the pacemaker (Figure 35–10). It is composed of specialized cardiac muscle cells, which can depolarize spontaneously, initiating their own action potential and contraction. From the pacemaker the action potential spreads throughout the left and right atria. As it passes along the surface of the individual heart muscle cells, it activates their contractile machinery, and they contract. The action potential travels very quickly, so many cells are activated almost simultaneously.

35–10
The beat of the mammalian heart is controlled by a region of specialized muscle tissue in the right atrium, the sinoatrial node, which functions as the heart's pacemaker. Some of the nerves regulating the heart have their endings in this region. Excitation spreads from the pacemaker through the atrial muscle cells, causing both atria to contract almost simultaneously. When the wave of excitation reaches the atrioventricular node, its conducting fibers pass the stimulation to the bundle of His, which triggers simultaneous contraction of the ventricles. Because the fibers of the atrioventricular node conduct relatively slowly, the ventricles do not contract until the atrial beat has been completed.

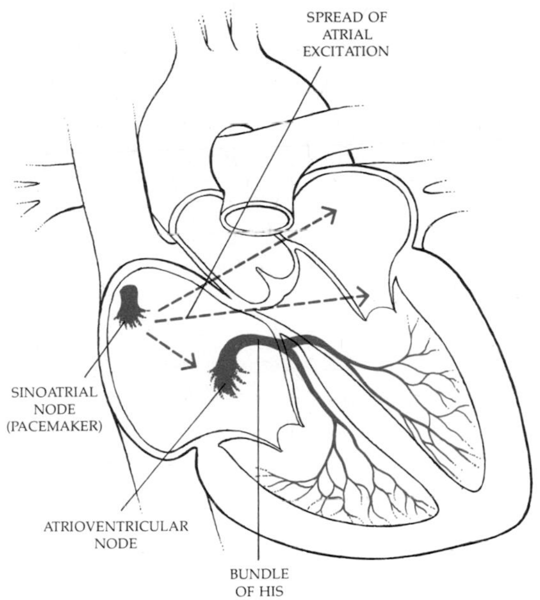

About 100 milliseconds after the pacemaker fires, impulses traveling through special conducting fibers stimulate a second area of specialized tissue, the *atrioventricular node*. The atrioventricular node is the only electrical bridge between the atria and the ventricles. It consists of slow-conducting cardiac muscle cells. Thus the atrioventricular node imposes a delay between the atrial and ventricular contractions, so that the atrial beat is completed before the beat of the ventricles begins. From the atrioventricular node, impulses are carried by special muscle fibers, the *bundle of His* (named after its discoverer), to the walls of the right and left ventricles, which then contract simultaneously.

If you listen to a heartbeat, you hear "lubb-dup, lubb-dup." The deeper, first sound ("lubb") is the closing of the valves between the atria and the ventricles; the second sound ("dup") is the closing of the valves leading from the ventricles to the arteries. If any one of the four valves is damaged, as from rheumatic fever, blood may leak back through the valve, producing the noise characterized as a "heart murmur" (a "ph-f-f-t" sound).

35–11
An electrocardiogram showing seven normal heartbeats. Each beat is denoted by a series of waves that record the electrical activity of the heart during contraction. Analysis of the rate and character of these waves can reveal abnormalities in heart function. Atrial relaxation occurs during ventricular excitation; therefore its record is obscured.

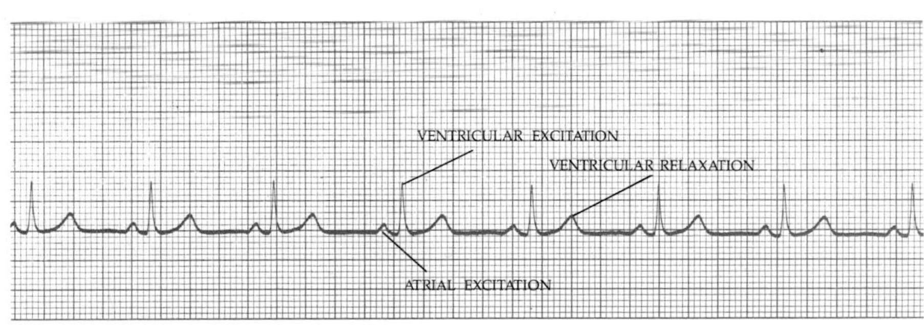

DISEASES OF THE HEART AND BLOOD VESSELS

Cardiovascular diseases cause more deaths in the United States than accidents and all other diseases combined. According to recent estimates, more than 28 million people in this country—13 percent of the total population—have some form of cardiovascular disease.

About 65 percent of the cardiovascular deaths are caused by heart attacks. A heart attack is the result of an insufficient supply of blood to an area of heart muscle; with their oxygen supply cut off, the muscle cells die. It can be caused by a blood clot—a thrombus—that forms in the blood vessels of the heart itself or by a clot that forms elsewhere in the body and travels to the heart and lodges in a vessel there. (A wandering clot such as this is known as an embolus.) A heart attack can also result from a blocking of a blood vessel due to atherosclerosis (see below). Recovery from a heart attack depends on how much of the heart tissue is damaged and whether or not other blood vessels in the heart can enlarge their capacity and supply these tissues, which then may recover to some extent. Angina pectoris is a related condition in which the heart muscle receives an insufficient blood supply—often a result of a narrowing of the vessels. The symptoms are a pain in the center of the chest and, often, in the left arm and shoulder.

A stroke is caused by interference with the blood supply to the brain. This may be the result of a thrombus (cerebral thrombosis), an embolus, or the bursting of a blood vessel in the brain. Its effects depend on the severity of the damage and where it occurs in the brain.

Atherosclerosis contributes to both heart attacks and strokes. In this disease, the linings of the arteries thicken and their surface becomes roughened by deposits of cholesterol, fibrin (clotting material), and cellular debris. The arteries, becoming inelastic, no longer expand and contract, and blood moves with increasing difficulty through the narrowed vessels. (Hardening of the arteries—arteriosclerosis—may subsequently occur if calcium becomes deposited in the artery walls). Thrombi and emboli thus form more easily and are more likely to block the vessel. The causes are not clear-cut. Arteriosclerosis is associated with high levels of cholesterol (page 60) in the blood, and also with cigarette smoking, stress, and lack of exercise. Female hormones seem to protect against arteriosclerosis; the condition is much less common in premenopausal women than it is among men of the same age. This difference in susceptibility to arteriosclerosis is a major reason that, in the United States, women now live almost 10 years longer than men.

High blood pressure—hypertension—also contributes significantly to cardiovascular disease. It places additional strain on the arterial walls and increases the chances of emboli. Despite lack of knowledge about the causes, hypertension can be treated and blood pressure reduced to safer levels. In the United States, high blood pressure is twice as common among black males as among white males and almost twice as common among black females as among white females. Related to this disturbing statistic is the fact that the incidence of deaths among black men and women from cardiovascular diseases is even higher than in the white population.

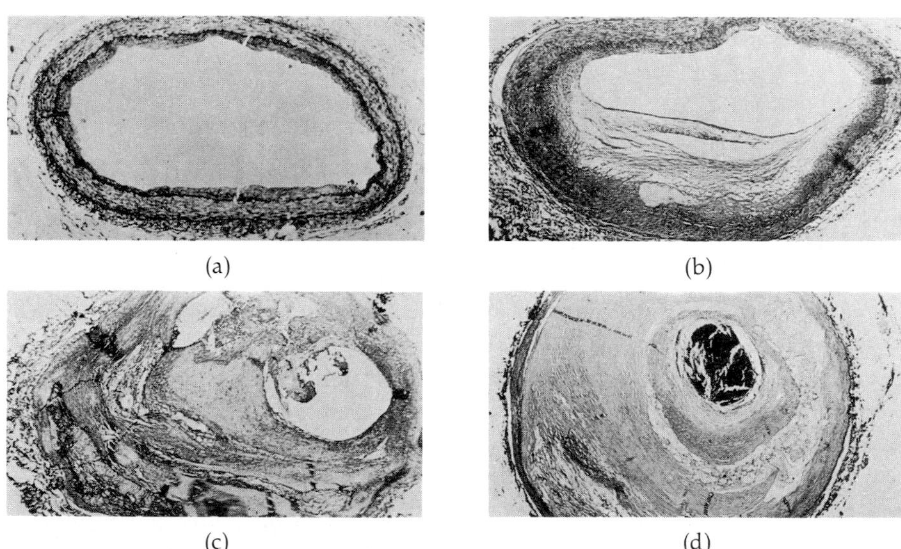

Atherosclerosis is the accumulation of fatty substances (mostly cholesterol) and the growth of fibrous coronary tissue in the wall of the artery. (a) *A normal artery in cross section.* (b) *Fatty deposits have begun to form, and the space left for blood flow has decreased.* (c) *The deposits have increased, and the diameter of the lumen is greatly reduced.* (d) *A clot has formed, blocking the vessel completely (coronary thrombosis, or heart attack).*

Cardiac Output

The total volume of blood pumped by the heart depends on the heart rate (beats per minute) and on the amount of blood ejected at each beat, the stroke volume. This total volume is called the cardiac output, which is defined as the amount (usually expressed in liters) of blood pumped per minute: cardiac output (liters per minute) = heart rate (beats per minute) × stroke volume (liters per beat). Thus if the heart beats 72 times per minute and ejects 70 milliliters into the aorta with each beat, the cardiac output is about 5 liters per minute (0.07 liter per beat times 72 beats per minute).

Control of Cardiac Output

Cardiac output is affected by both the endocrine and nervous systems. Epinephrine increases both the rate of heartbeat and the stroke volume of the heart, by increasing the force of contraction of the ventricles. Both sympathetic and parasympathetic nerves act on the sinoatrial (pacemaker) and atrioventricular nodes. Sympathetic stimulation increases the discharge rate of the pacemaker. Parasympathetic stimulation (by way of the vagus nerve) decreases the discharge rate of the pacemaker and increases the refractory period for conduction along the bundle of His (Figure 35-12).

35-12
Autonomic regulation of the rate of heartbeat. Sympathetic fibers stimulate the sinoatrial node, whereas parasympathetic fibers, which are contained in the vagus nerve, inhibit it.

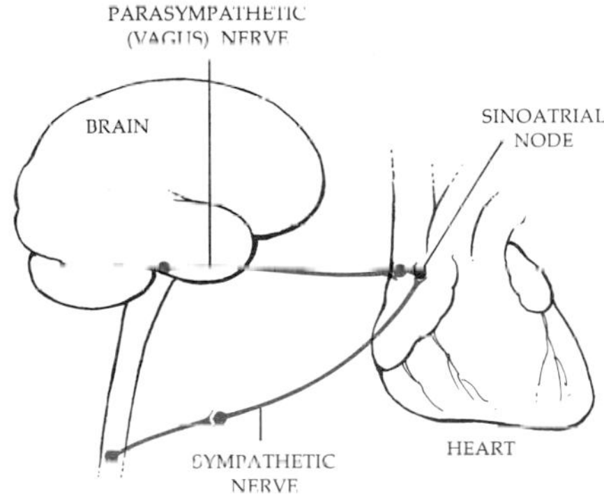

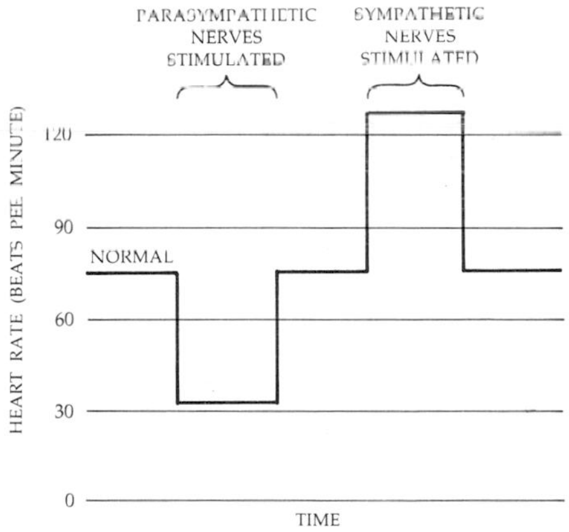

Cardiovascular Regulating Center

The sympathetic and parasympathetic nerves to the heart are controlled by the *cardiovascular regulating center,* which is located in the medulla. The cardiovascular regulating center is also the control center for the nerves controlling smooth muscle in the arterioles. Thus, if blood flow through a particular body area is increasing owing to dilation of blood vessels, the heart is simultaneously activated to develop greater pressure to support the greater flow.

The cardiovascular regulating center integrates the reflexes that control blood pressure. It receives information about existing blood pressure from specialized stretch receptors in large arteries in the neck (the carotid arteries), the venae cavae, the aorta, and the heart. The effector organs of the reflex are, as we have indicated, the heart and blood vessels.

The blood-pressure reflex is another example of negative-feedback control. When pressure falls, the activity of the heart is increased and the blood vessels are constricted, which raises the pressure again. Conversely, heart activity is decreased and the blood vessels dilated in response to high pressure.

VASCULAR CIRCUITRY

There are two principal circuits of the vascular system: the pulmonary circuit and the systemic circuit (Figure 35-13). In the pulmonary circuit, blood leaves the right ventricle of the heart through the pulmonary artery. This artery divides into right and left branches, which carry the blood to the right and left lungs, respectively. Within the lungs, the arteries divide into smaller and smaller vessels that finally become capillaries through which oxygen and carbon dioxide are exchanged. Blood flows from the capillaries into small venules and then into larger and larger veins, finally draining into the four pulmonary veins that carry the blood, now oxygenated, to the left atrium of the heart. The pulmonary arteries are the only

35–13
Diagram of the human circulatory system. Oxygenated blood is shown in color. Every tissue of the body contains capillaries that supply oxygen and nutrients to every cell of these tissues and carry off carbon dioxide and other wastes.

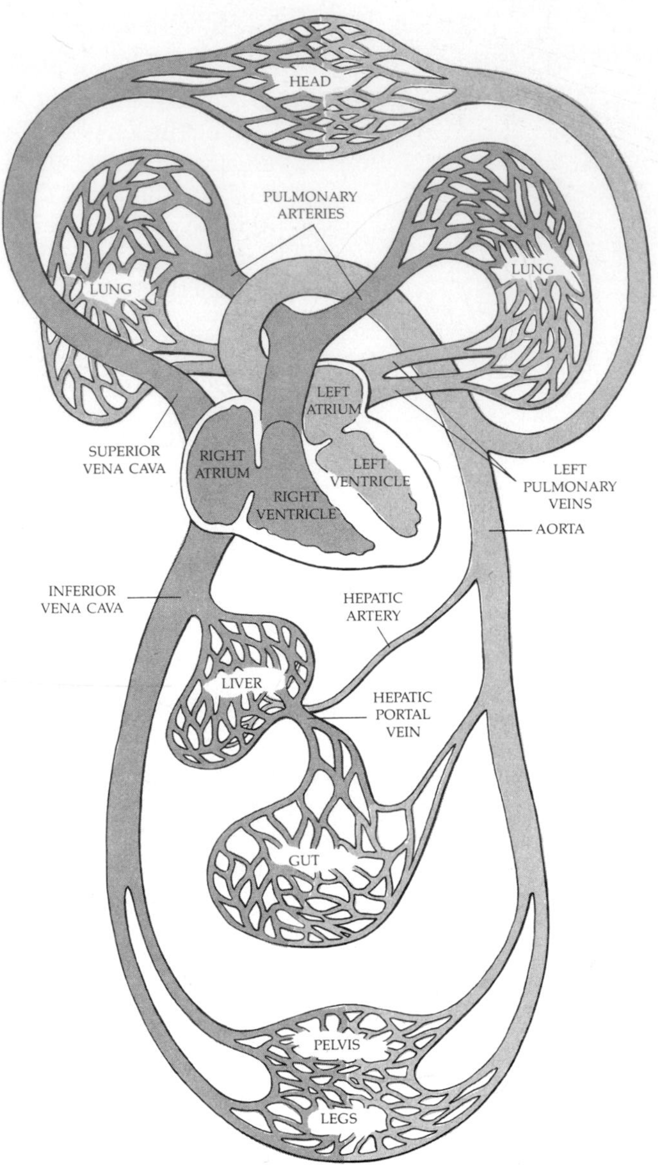

arteries that carry deoxygenated blood, and the pulmonary veins are the only veins that carry fully oxygenated blood.

The systemic circuit is much larger. Many arteries branch off the aorta after it leaves the left ventricle. The first two branches are the right and left coronary arteries, which bring oxygenated blood to the heart muscle itself. Another major subdivision of the systemic circulation supplies the brain. If the circulation of freshly oxygenated blood to the brain is cut off for even five seconds, unconsciousness results; after four to six minutes, brain cells are damaged irreversibly.

The hepatic portal system is a special subdivision of the systemic circulation.

HARVEY AND THE CIRCULATION OF THE BLOOD

William Harvey's discovery of the circulation of the blood is often cited as a model of biological research, not only because of its great importance to medicine but also as an example of how reason can push its way against the general current of scientific opinion. The stubbornness of Harvey's logic is even more impressive when one realizes that the microscopic observations that offered the final proof of Harvey's theory were not made until after his death.

Harvey lived from 1578 to 1657. By this time, the function of the heart as a pump had been recognized, but it was thought that the blood ebbed and flowed in the veins, like tides in the ocean, being gradually used up and replaced. The arteries were thought to carry only air, because the arteries of dissected cadavers were empty.

Harvey, working with animals, observed that a beating heart expelled the blood within it. By calculating the amount of blood expelled from the heart at each beat as compared with the total amount in the human body, he convinced himself that all of the blood in the body must return to the heart every few minutes. (His figures were not entirely accurate, but were accurate enough for the purpose. Actually, the amount of blood pumped per minute by the heart—5 liters—is equal to the entire amount of blood in the human body.)

One of the simple demonstrations Harvey used to support his theory is shown in the drawings below. If you tie a ligature around the upper arm, the valves in the veins show up as small swellings. If blood is pushed up toward the heart to a point above one of these swellings, it fails to flow back again, even if pushed. Concerning these observations, Harvey wrote "so provident a cause as nature had not so plac'd many valves without design."

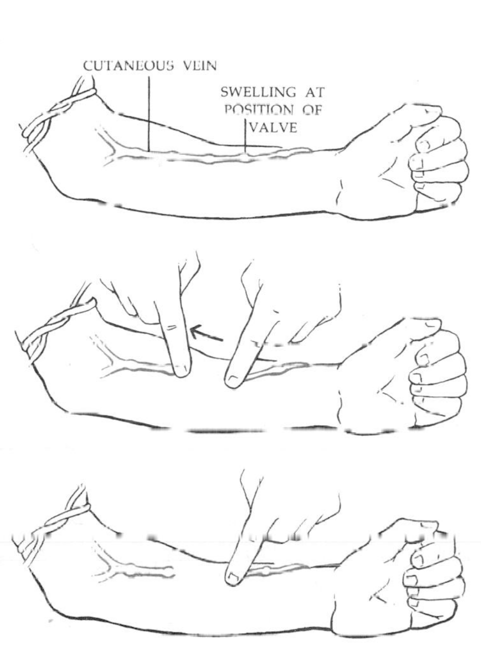

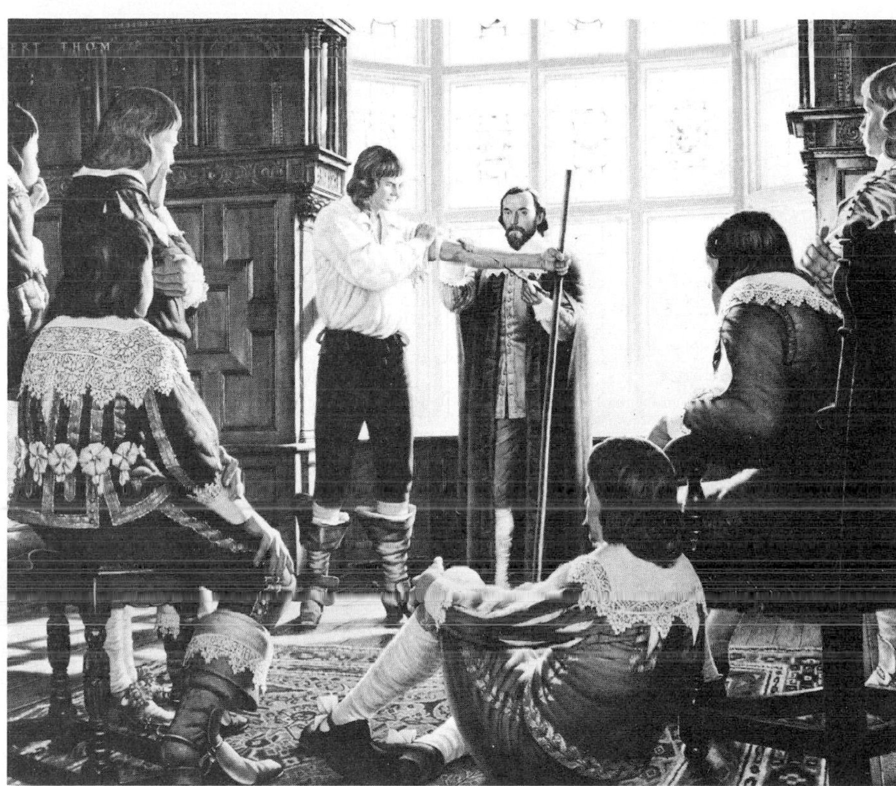

Venous blood collected from the digestive organs is shunted via the hepatic portal vein through the liver. There it goes through a second capillary system before it is emptied into the inferior vena cava. (This passage from capillaries to veins to capillaries is called a portal system.) In this way, the products of digestion can be directly processed by the liver. The liver also receives freshly oxygenated blood directly from a major artery, the hepatic artery.

There is a large drop in pressure as the blood goes through the capillaries. The return of blood to the heart through the veins is enhanced by body movements, which squeeze the veins between contracting muscles and force the blood upward. (If you have to stand still for long periods of time, try contracting your leg muscles periodically to move blood back toward the heart and prevent blood pooling.) Valves in the veins prevent backflow. Also, the pressure in the thoracic cavity is negative (less than atmospheric pressure), and as the thorax expands in respiration, the elastic walls of the veins in the thorax dilate, so that blood is both pulled (in the thorax) and pushed (in the veins) back into the heart.

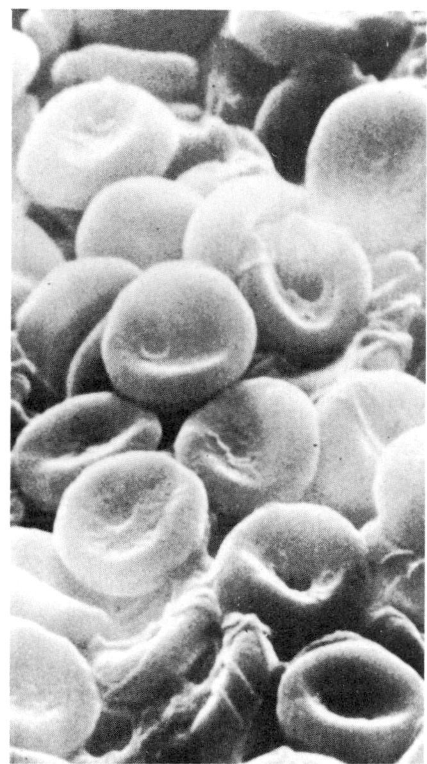

35–14

In vertebrates, oxygen is transported in red blood cells, shown here in a scanning electron micrograph. Red blood cells, which are only about 7 or 8 micrometers in diameter, are flexible and so are able to twist and turn in their passage through the capillaries (see Figure 35–9).

THE BLOOD

An individual weighing 75 kilograms (165 pounds) has about 5 liters of blood. About 60 percent of it is a straw-colored liquid called *plasma*. The plasma, which is more than 90 percent water, carries a large number of different kinds of ions and molecules. They include fibrinogen, the protein from which clots are formed; nutrients, such as glucose, fats, and amino acids; gases, including CO_2; various ions; antibodies, hormones, and enzymes; and waste materials, such as urea and uric acid.

The other 40 percent of the blood is made up of *erythrocytes,* or red blood cells; *leukocytes,* or white blood cells; and *platelets,* which are cytoplasmic fragments of cells. The relative amount of plasma varies among species, among individuals who live under different conditions, and between the sexes. For example, persons living at high altitudes (low oxygen pressure) have relatively less plasma and more red blood cells, and women have fewer red blood cells per milliliter than men.

Erythrocytes

Red blood cells transport oxygen. There are about 5 million of them per cubic millimeter of blood—some 25 trillion (25×10^{12}) in the whole body. In human beings, an individual red blood cell has a life span of about 120 to 130 days. New ones are produced in the bone marrow of adults at the rate of about 2 million per second.

Red blood cells are among the most highly specialized of all cells. As a mammalian red blood cell matures, it extrudes its nucleus and mitochondria, and its other cellular structures dissolve. Almost the entire volume of a mature red blood cell is filled with hemoglobin, about 300 million molecules per cell. Anemia is a deficiency either in the number of red blood cells or in total hemoglobin.

Leukocytes

There are about 6,000 to 9,000 white blood cells per cubic millimeter of blood—1 or 2 white blood cells for every 1,000 red blood cells. These cells are nearly colorless, are larger than red blood cells, contain no hemoglobin, and have a nucleus. Unlike red blood cells, leukocytes are not confined to the vascular system but can migrate

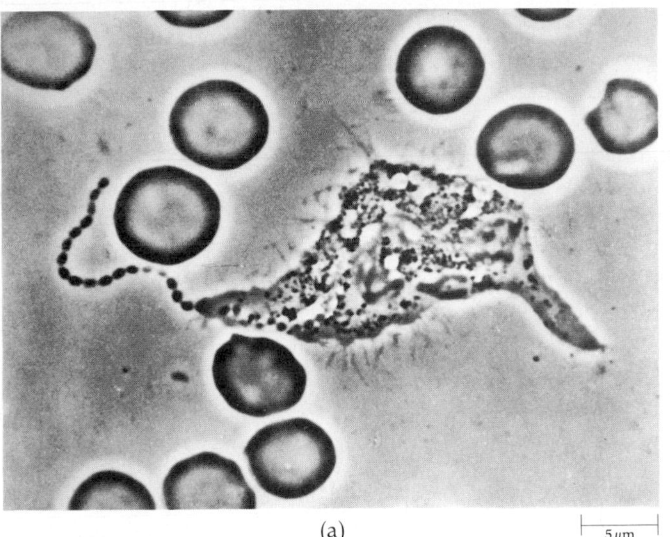

(a) 5 μm

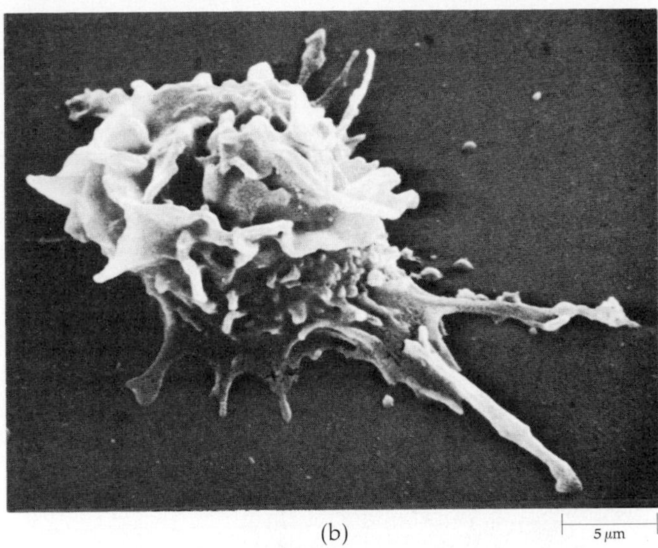

(b) 5 μm

35–15
(a) *A neutrophil (a white blood cell) phagocytizing a chain of streptococci. Numerous red blood cells are also visible.* (b) *A scanning electron micrograph of a neutrophil from a rabbit.*

between endothelial cells out into the tissues. They appear spherical in the bloodstream but in the tissues become flattened and amoebalike; like amoebas, they move by means of pseudopodia and many are phagocytic (Figure 35–15).

There are several different types of white blood cells, which are classified in a number of different ways according to their appearance under the microscope. Perhaps the simplest method of classification is the one that separates leukocytes into mononuclear and polymorphonuclear cells (Table 35–1). A mononuclear cell has a single rounded or kidney-shaped nucleus that fills most of the cell. A polymorphonuclear cell is so called because its nucleus assumes many *(poly)* shapes *(morphe)*. Polymorphonuclear cells are also called granulocytes because of granules present in the cytoplasm. The terms neutrophil, eosinophil, and basophil refer to different staining properties of these granules (Figure 35–15). Polymorphonuclear cells, like red cells, are produced in bone marrow.

Leukocytes are involved in defending the body against viruses, bacteria, and other foreign intruders; they do so in part by phagocytosis and also by the production of antibodies. Neutrophils are the chief phagocytic cells of the bloodstream. Monocytes, which are large cells (three times as large as an erythrocyte), migrate out through the capillary walls into the tissues. There they develop into macrophages, large phagocytic cells that colonize the tissues. Lymphocytes are of two main types: B-cells and T-cells. Both arise from cells that originate in the yolk sac and migrate into the fetus. B-cells, which develop in the fetal liver and spleen, produce antibodies that circulate in blood and lymph. T-cells, which mature in the thymus gland, are involved in tissue-transplant rejections and other immune reactions. In contrast to most other blood cells, lymphocytes have life spans of 10 years or more. Their longevity, as we shall see, is in keeping with their function.

Table 35–1 *Leukocytes*

CELL TYPE	% IN BLOOD
Polymorphonuclear cells (granulocytes)	
Neutrophils	55–65
Eosinophils	2–3
Basophils	1–2
Mononuclear cells	
Lymphocytes	20–30
Monocytes	4–7

Platelets

Platelets, so called because they look like little plates, are colorless, round or biconcave disks smaller than erythrocytes (about 3 micrometers in diameter). They are cytoplasmic fragments of unusually large cells, megakaryocytes, found in the bone marrow. Platelets play an important role in the formation of clots.

35–16
Beginning of the formation of a clot. Red blood cell enmeshed in fibrin.

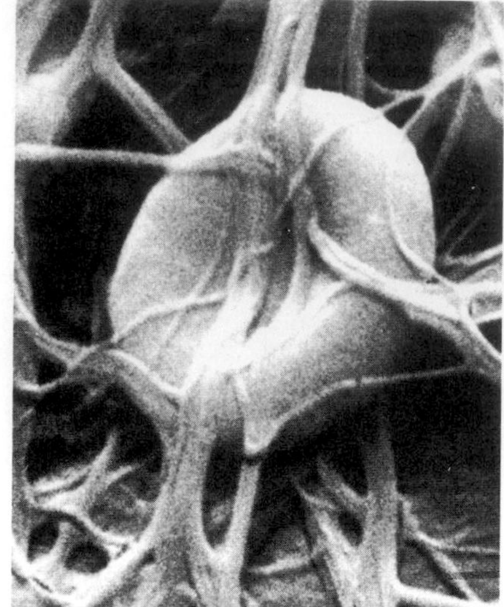

Blood Clotting

Blood clotting is a complex phenomenon; at least 15 factors involved in the process have been identified. It is not initiated when blood encounters air, as was reasonably assumed for a long period of time. The sequence of events begins when platelets encounter a rough surface, such as a torn tissue. This stimulates the platelets to release substances called thromboplastins. (The same effect occurs when platelets touch a glass surface, such as that of a test tube, which is one of the reasons the natural process was so difficult to analyze.) Thromboplastin acts to convert the enzyme prothrombin, a plasma protein produced in the liver, to its active form, thrombin:

$$\text{prothrombin} \xrightarrow{\text{thromboplastin}} \text{thrombin}$$

Several other factors normally present in the bloodstream are also required for this reaction, which involves several enzymatic steps. Thrombin converts fibrinogen, a soluble plasma protein, to fibrin:

$$\text{fibrinogen} \xrightarrow{\text{thrombin}} \text{fibrin}$$

The fibrin molecules clump together, forming an insoluble network that enmeshes red blood cells and platelets to form a clot (Figure 35–16). The clot contracts, pulling together the edges of the wound. When blood is removed from the body and placed in a test tube, it congeals to form a clot. The clot eventually contracts, leaving a clear fluid, the serum.

Clot formation, although essential to the survival of the organism, also poses a threat because clots can block the circulatory system. A heart attack, for example, can be caused by a clot in a blood vessel supplying heart tissue (Figure 35–17). The clot-forming process is very sensitive; just a few molecules of thromboplastin can set it off. On the other hand, its many steps provide numerous opportunities for interrupting the process. A number of natural inhibitors of clotting are known. For example, heparin, produced by cells of the liver, lungs, intestines, and other tissues, inhibits the conversion of prothrombin to thrombin. Heparin is used medi-

35–17
A heart attack. When a clot forms in a blood vessel, the cells in the area supplied by this vessel are deprived of oxygen and die. The severity of the heart attack depends, in part, on the extent of damage to the heart muscle.

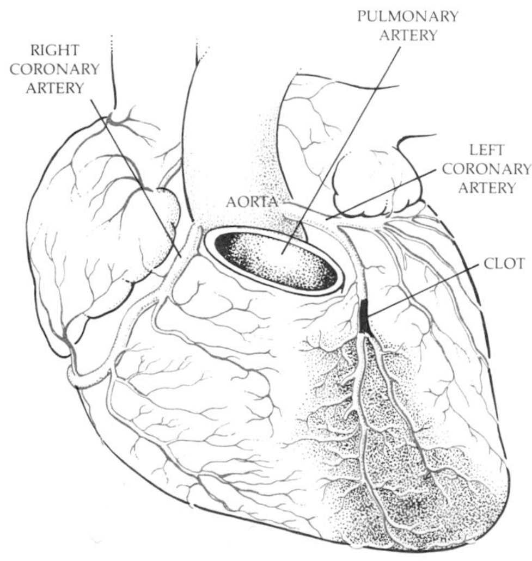

cally to limit clot formation after surgery, following a heart attack, or during coronary bypass operations.

Hemophilia is an inherited deficiency in blood clotting (see pages 306–307). The most common type of hemophilia involves a defective protein (known as Factor VIII) involved in the production of prothrombin. Hemophiliacs can be treated by administration of plasma concentrates of Factor VIII from normal blood.

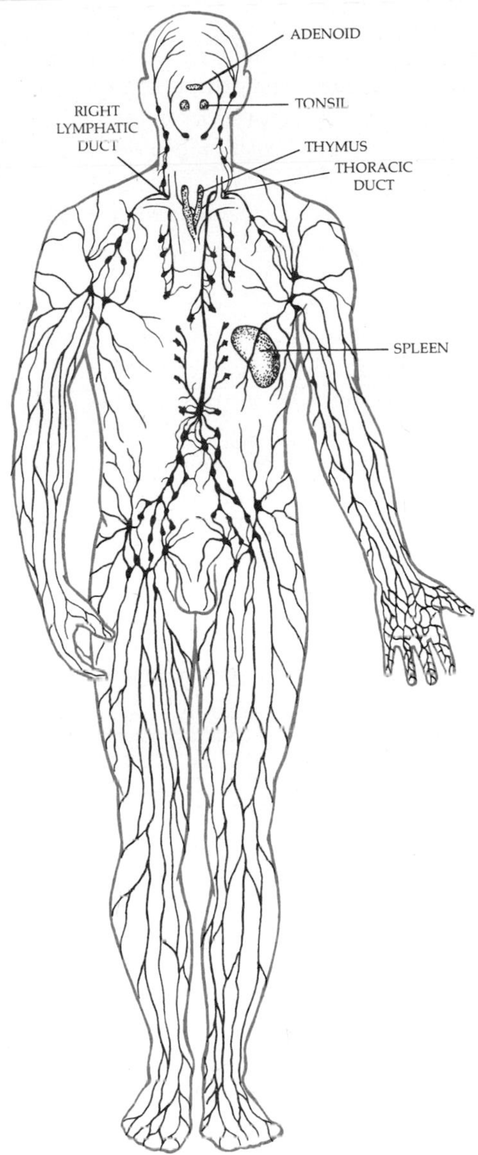

35-18
Human lymphatic system. Lymphatic fluid reenters the bloodstream through the thoracic duct and right lymphatic duct.

LYMPHATIC SYSTEM

As we noted previously, not quite all of the fluid forced out of the capillaries by the pressure of the circulating blood is returned to the capillaries by osmosis. In higher vertebrates, this fluid is collected by the lymphatic system, which routes it back to the bloodstream. The lymph also serves to transport some nutrients, particularly fats, absorbed from the digestive tract and picked up by the lymph capillaries. The lymphatic system is like the venous system in that it consists of an interconnecting network of progressively larger vessels. The larger vessels are, in fact, similar to veins in their structure. The small vessels are much like the blood capillaries; the most important difference is that, rather than forming part of a continuous circuit, the lymph capillaries end blindly in the tissues. Interstitial fluid seeps into the lymph capillaries, from which it travels in the form of lymph to large ducts from which it is emptied into the vena cava. Some nonmammalian vertebrates have lymph "hearts," which help to move the fluid. In mammals, lymph is moved by contractions of the body muscles with valves preventing backflow, as in the venous system of the blood. Also, recent studies have shown, lymph vessels contract rhythmically; these contractions may be the principal factor propelling the lymph.

Lymph Nodes

As the fluid travels through the lymphatic system, it passes through lymph nodes (sometimes incorrectly referred to as lymph glands). As you can see in Figure 35-18, single lymph nodes are distributed throughout the body, but most are found clustered in particular areas, such as the neck, armpits, and groin. The nodes range from the size of a pea to that of a lima bean. A lymph node is a mass of spongy tissue separated into compartments by connective tissue.

Lymph nodes have two functions: They remove foreign particles from the lymph before it enters the blood, and they are the sites of proliferation of lymphocytes, which are produced in lymph nodes at a rate of about 10 billion per day. The removal system is in part mechanical filtration; the nodes catch dead cells and particles that may have entered the body fluids. Lymph nodes near the respiratory system, for example, are often filled with carbon particles from soot or tobacco smoke. The lymph nodes also contain leukocytes that attack bacteria and other invaders. Cancer cells that have broken loose from the principal site of cancer growth are often caught in the lymph nodes, and hence in major cancer operations, surgeons often routinely remove lymph nodes adjacent to the cancer growth. If lymphatic vessels and nodes are damaged by disease or injury or are removed by surgery, fluids that would have been drained off may collect in the tissues, producing the condition known as lymphedema. Elephantiasis, which causes severe, often grotesque, swelling of the appendages, is a form of lymphedema caused by a parasitic worm that infects and blocks lymphatic vessels.

Other Lymphatic Organs

The spleen is largely lymphoid tissue. One of its principal functions is the culling of red blood cells from the circulation. As aged, damaged, and otherwise defective erythrocytes are broken down in the spleen, the iron is removed from the hemoglobin and recycled by the bone marrow for use in the production of new hemoglobin.

In many vertebrates, the spleen also serves as a reservoir of red blood cells that can be squeezed out into the circulation following hemorrhage. Whether the human spleen serves this same function is a matter of controversy.

The human thymus is a spongy, pinkish-gray, two-lobed organ that lies high in the chest, in front of the aorta and behind the breastbone. It is a relatively large organ in infancy, continues to grow until puberty, and then becomes smaller. The thymus is the source of hormones believed to be involved in the maturation of T-lymphocytes.

Tonsils and adenoids are also lymphoid tissues, which play a role in immunity to microorganisms entering the body through the nose or mouth.

DEFENSE MECHANISMS

The body's first line of defense is its outer wrapping of skin and mucous membranes. The skin with its tough layer of keratin is an effective barrier as long as it is intact. Mucous membranes are more fragile, but they are constantly flushed and cleansed with fluids, such as mucus and saliva, that contain antimicrobial substances, including lysozyme (page 80).

Once a microorganism penetrates this barrier, it encounters a second line of defense, consisting of a variety of agents carried by the circulating blood and lymph. Suppose, for example, you nick your skin. The injured cells immediately release histamine and other chemicals that lead to the distension of the nearby capillaries, thus increasing their permeability. Circulating white blood cells make their way through the distended capillary walls, crowding into the site of the injury. They appear to move by chemotaxis, the same sort of sensory mechanism that directs bacteria toward food sources (page 899). The neutrophils, which are the first to arrive, engulf any foreign invaders by phagocytosis. They literally eat themselves to death. Next on the scene are the lymphocytes. Blood clots begin to form, walling off the injured area. Pus may accumulate; it is made up chiefly of white blood cells, dead and alive, combined with tissue debris. The local temperature in the area often rises, creating an environment unfavorable to the multiplication of microorganisms while accelerating the motion of the white blood cells.

Development of Immunity

If this inflammatory response, as it is called, is insufficient, a third defense mechanism comes into play—the immune system. It differs from the other defenses of the body in that it is highly specific, involving recognition of a particular invader and the tailoring of an attack against it.

We noted in Chapter 21 (page 365) that an organism that has been infected with a particular bacterium or virus is often protected against reinfection by that pathogen. Common examples of diseases that confer immunity, as this phenomenon is called, are measles, mumps, and chicken pox. This protection against reinfection is caused by the formation of globular proteins known as *antibodies* (humoral immunity) and by sensitized lymphocytes (cellular immunity). This phenomenon is known as the *immune response,* and understanding the details of the process is of great biological interest and, of course, of great medical importance.

Like enzymes, antibodies are complex globular proteins. They are highly specific, and their specificity is based on their combining in a very precise way with another molecule. The molecule with which an antibody combines or that a lymphocyte recognizes is known as an *antigen.* Virtually all proteins and some polysaccharides and lipids can act as antigens. A single cell, such as a bacterial cell, may carry a number of antigens, each of which can elicit a specific antibody.

Antibodies act against invaders in one of three ways: (1) They may coat the foreign particle in such a way that it can be taken up by the phagocytic cells; (2) they may combine with it in such a way that they interfere with some vital activity—for example, covering the protein coat of a virus at the site where the virus attaches to the cell membrane; or (3) they may themselves, in combination with another blood component known as *complement,* actually lyse and destroy the foreign cell.

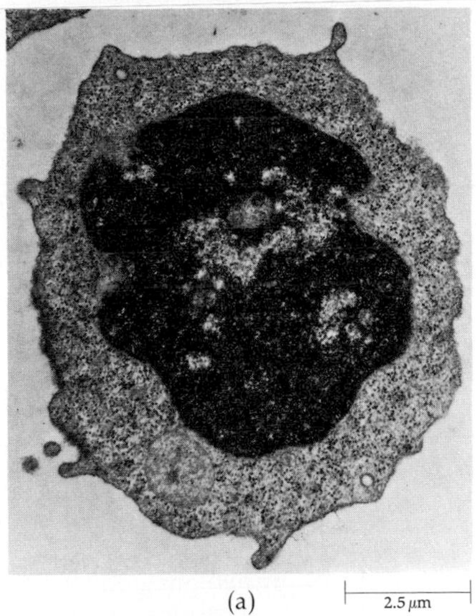

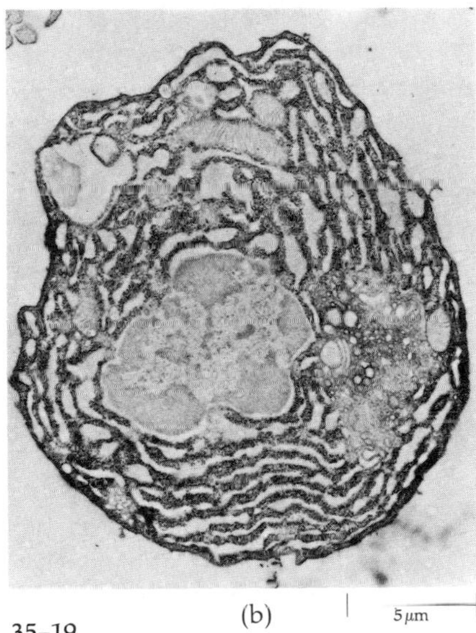

35–19
As a B-lymphocyte (a) *differentiates into a plasma cell* (b), *it grows larger, its nucleus becomes relatively smaller and less dense, and there is a large increase in endoplasmic reticulum and ribosomes. Note the mitochondrion above the nucleus and the large Golgi body to the right.*

Antibody synthesis begins with an encounter between an antigen and the antigen-specific antibody on the surface of a B-lymphocyte. Following this interaction, the B-cell differentiates into a plasma cell (Figure 35–19). For some antigens, interaction with B-cells is sufficient to cause the cells to differentiate into plasma cells. For most antigens, T-cells must also recognize the antigen and cooperate with the B-cells before the B-cells can differentiate. Plasma cells have the capacity to make large amounts of antibody against the particular antigen (and probably against no other). The antigen also stimulates the cell to divide, thereby greatly increasing antibody production. Division of a plasma cell takes about 10 hours, and the process is repeated nine times in four or five days, at the end of which time antibody production often catches up with multiplication of the infectious organism.

After the first bout of infection, the circulating antibodies disappear, but memory cells sensitized to the particular antigen persist indefinitely in the circulation. Memory cells are primed to begin immediate antibody production against that particular antigen and are responsible for long-term immunity.

The Structure of Antibodies

There are five major classes of antibodies (or immunoglobulins, as they are called by biochemists): IgG, IgM, IgA, IgE, and IgD. The latter two are present only in trace amounts. IgA is found principally in external secretions, such as saliva. IgM is formed early in the immune response and seems to be active in reactions involving complement. IgG is by far the most abundant in serum and is the best understood of the immunoglobulins.

In 1972, Gerald Edelman of Rockefeller University and R. R. Porter of Oxford were awarded the Nobel Prize for their work on antibody structures. These researchers elucidated the primary structure (the amino acid sequence) of an antibody for the first time. Each IgG antibody, as their work has shown, consists of four subunits, two "light" chains and two "heavy" chains. (The heavy chains have more amino acids.) Each of the chains has a region where the sequence of amino acids is apparently common to all IgG antibodies, the constant region, and another, variable region, where the sequence is specific for each particular antibody. The

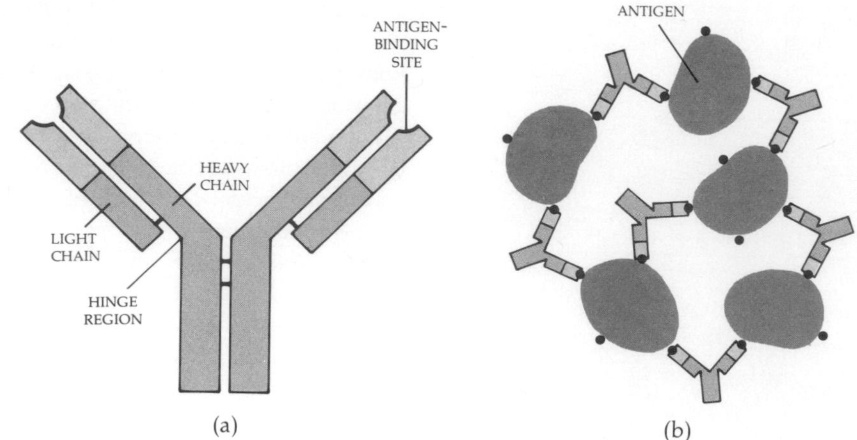

35–20
(a) *Diagram of antibody molecule. The two heavy chains are connected by two disulfide bridges, and each heavy chain has a flexible hinge region. The light and heavy chains both have constant (gray) and variable (colored) regions.* (b) *Because each antibody molecule has two antigen-binding sites, one antibody can bind to antigenic determinants on different cells or molecules, thus causing them to stick together (agglutination).*

combining sites, it is not surprising to find out, involve the variable regions. Each antibody has at least two such sites, so it can attach to two antigens at the same time. The antigens may be on the same microorganism, or on two different ones. Cross-linking (Figure 35–20) takes place when the two antigens are on two different microorganisms. Each heavy chain contains a flexible hinge region, so the two antigen-binding sites need not be a fixed distance apart.

How do the constant and variable portions of the polypeptide chains become welded into one molecule? Molecular genetic studies in mice using DNA-RNA recombination techniques have shown that in the embryo the C genes (which code for the constant chains) and the V genes (which code for the variable chains) are separate. However, in the course of development, the genes relocate so that they are adjacent and make one continuous strand of RNA. This is the first discovery that genes can change position on a chromosome and may mark a major advance in the understanding of other events during development and differentiation.

Theories of Antibody Formation

The most intriguing—and bewildering—fact about the immune response is the great number and variety of antigens for which a single individual can make antibodies. An early model of antibody formation proposed that each antigen served as a specific template or mold for the antibody and the imprint of this template persisted, "instructing" the cell in its antibody formation. This proposal, known as the instructive theory, has foundered as the result of the discoveries that (1) antibodies to different antigens have different amino acid sequences, and (2) the amino acid sequences of a protein and not any external template determine its three-dimensional shape.

The most widely accepted theory is that of clonal selection. According to this concept, large numbers of different cell lines (clones) of lymphocytes exist. Each makes only a single type of antibody specific for a certain antigen. In the absence of antigen, each clone exists only in small numbers and makes only small amounts of its particular antibody. When an antigen is introduced into the organism, it stimulates one specific clone of lymphocytes to multiply and differentiate (Figure 35–21). This model can be summarized as the one cell–one antibody theory. In terms of genetics, the model postulates that all the genes for making all the antibodies exist in each lymphocyte (indeed, if you follow the argument further, in every cell in the body) and that all but those coding for a single antibody are repressed in each clone of lymphocytes. (See page 291 for a further discussion of gene repression.)

This theory has been in some difficulty, as a matter of simple mathematics. A single individual is apparently capable of making antibodies for up to a million (10^6) different antigens. However, a human cell contains only about 10^4 genes in its entire genome, and many of these code for other proteins. Moreover, the antibody response occurs even when the antigen is a synthetic molecule never before encountered in nature. It is difficult to explain how or why evolution might have provided a specific antibody-producing lymphocyte for an antigen it could hardly have anticipated.

Genetic analyses of lymphocytes indicate that there are only a few genes coding for the constant regions and less than a hundred for the variable ones. Thus, although most investigators still favor the clonal selection hypothesis, pointing out the possibilities of gene combinations, the problem is far from solved.

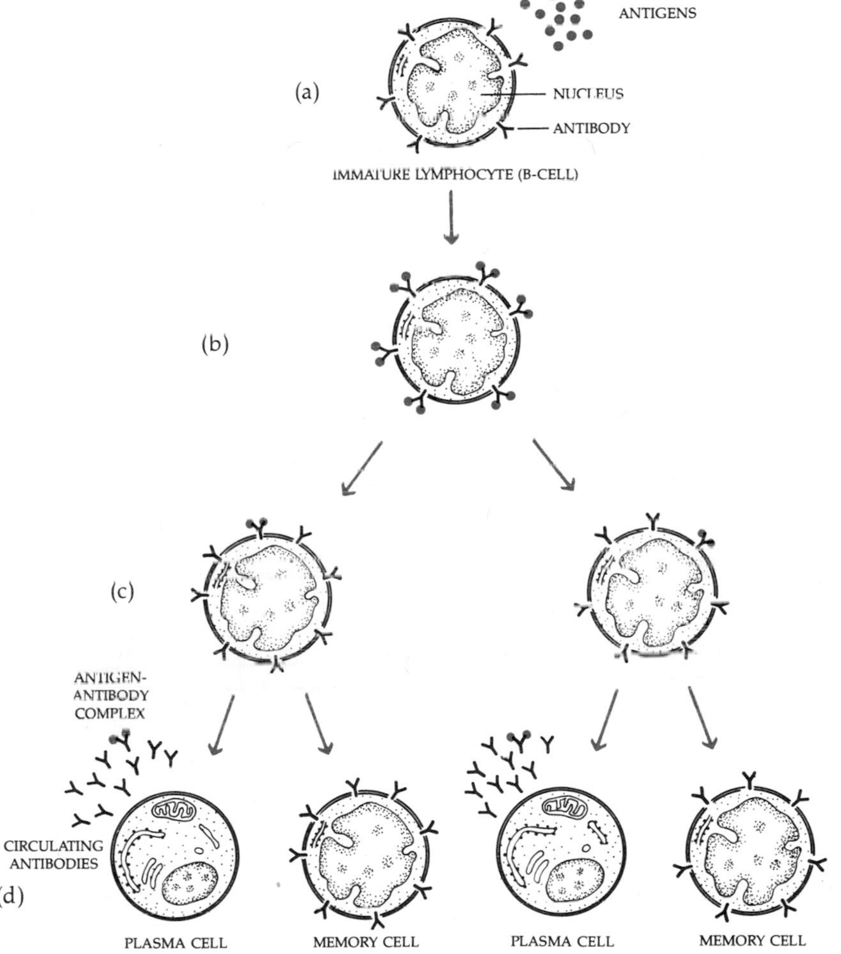

35–21
A hypothetical mechanism for the immune response. (a) *An immature lymphocyte (B-cell) with antibodies on its surface encounters the antigen molecules for which the antibodies are specific, and* (b) *antibodies and antigens bind together.* (c) *The B-cell then begins to divide and differentiate* (d), *forming plasma cells and memory cells. Plasma cells secrete circulating antibodies. Memory cells persist in the circulation, secreting antibodies only following an encounter with the specific antigen.*

B-Cells and T-Cells

As we mentioned previously, it is now known that there are two types of lymphocytes involved in immune responses, B-cells and T-cells.

The B-cells are the lymphocytes that mature into plasma cells and make circulating antibodies. The T-cell response is called cellular immunity, since it is characterized by the reaction of the T-lymphocyte directly with the foreign cell or antigen. Cellular immunity to an antigen may conveniently be recognized by carrying out a skin test with the antigen; the skin test used by physicians in detecting tuberculosis is an example. It has recently been realized that there are many types of T-lymphocytes having different functions. Some T-cells are involved in the rejection of tissue transplants. Other types of T-cells function in the regulation of the immune response. Another class of T-cells (called helper cells) apparently release a chemical that influences the B-cells in their antibody-producing activities. Macrophages (a class of monocytes) play an important role in immunity in addition to their phagocytic function mentioned earlier. They are important in initiating the immune response by presenting antigens to both B- and T-cells in a highly immunogenic form; this is called antigen processing.

Disorders Related to the Immune System

The immune response is a powerful bulwark against disease, but it sometimes goes awry. Hayfever and other allergies are the result of immune responses to pollen, dust, or other substances that are weak antigens to which most people do not react. These weak antigens interact with a special kind of antibody (lgE) that fixes onto mast cells, a type of white blood cell found in body fluids and on mucous membranes.

Another medical problem caused by the immune system is hemolytic anemia of the newborn. During the last month in the uterus, the human baby usually acquires antibodies from its mother. Most of these antibodies are beneficial. An important exception, however, is found in the antibodies formed against a blood

35–22
(a) *A T-lymphocyte (the smaller cell at the lower right) has attached itself to a tumor cell.* (b) *The cytolytic process has begun, as manifested by the many vesicles protruding from and being shed by the tumor cell surface. Note that the T-lymphocyte remains attached to the target cell.*

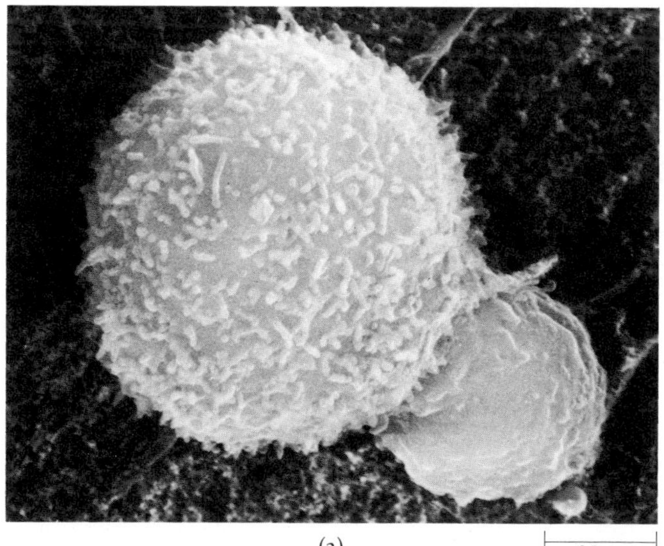

(a)

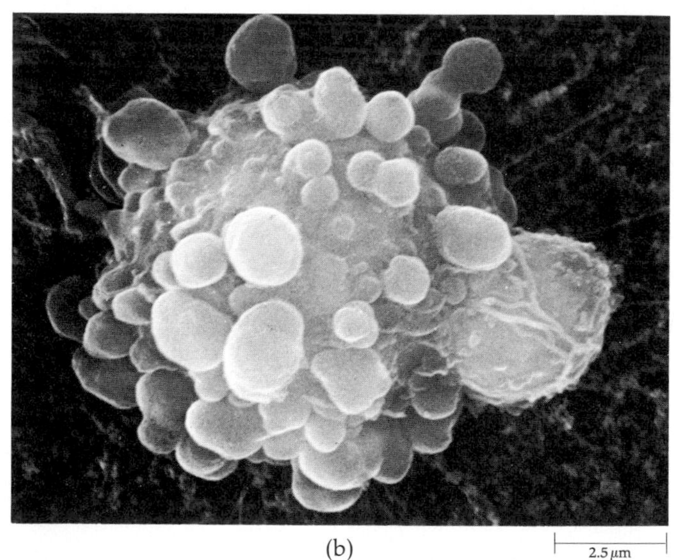

(b)

factor, the Rh factor (named after the rhesus monkeys in which the research leading to its discovery was carried out). The Rh factor is a genetically determined substance found on the surface of red blood cells. If a woman who lacks the Rh factor (that is, an Rh-negative woman) has children fathered by a man homozygous for the Rh factor, all the children will be Rh positive; if he is a heterozygote, about half of the children will be Rh positive. During her first pregnancy carrying an Rh-positive fetus, the mother is likely to produce antibodies against Rh antigens contained in fetal blood that enters her bloodstream. This generally occurs at birth, so the first Rh-positive child normally is not threatened. In subsequent pregnancies, these antibodies may be transferred to the fetus, causing erythroblastosis—a destruction of the red blood cells—which can be fatal, either before or just after birth. One method of treatment, which is quite an intricate procedure, is to transfuse the infant completely, replacing all its blood with Rh-negative blood so that the maternal antibodies do not cause clumping of the red blood cells. Such transfusions save about 35 percent of the babies who would have died of erythroblastosis. Recently, two medical scientists at Columbia University have developed a substance called RhoGAM, that contains antibodies against the Rh factor. RhoGAM, if injected into an Rh-negative woman at the birth of her first Rh-positive child, destroys the fetal Rh-positive cells that have entered her bloodstream, thus preventing the development of anti-Rh antibodies.

Self and Not Self

The immune system can ordinarily distinguish between "self" and "not self." Substances that are present during embryonic life, when the immune system is developing, will not be antigenic in later life. This recognition occasionally breaks down, however, and the immune system attacks the body. One type of anemia, myasthenia gravis, and certain other disorders have been identified as autoimmune diseases—that is, diseases in which an individual makes antibodies against his or her own cells. It is possible that other disorders, such as rheumatoid arthritis, the causes of which are not yet known, may prove to have the same basis.

The Immune Response and Tissue Transplant

Because it is programmed to act against foreign materials of all kinds, the immune system works vigorously against tissues—such as skin, kidney, or heart—transplanted from another individual (except an identical twin). The blood transfusion reactions described in Chapter 18 are examples of this same phenomenon. Surgeons and medical research workers concerned with extending the use of tissue transplants are seeking ways to suppress or paralyze the immune response in such a way that these foreign but potentially lifesaving tissues can survive. Rejection of transplanted tissue appears to involve T-cells. These white blood cells can be seen aggregating around a tissue transplant before and during its rejection. X-ray treatment and certain chemicals that inhibit white blood cell production suppress immune reactions. Such treatments, of course, also render the patient more susceptible to disease.

Immunity and Cancer

Cancer cells resemble the host's own cells in many ways. Yet, within the host, they act like foreign organisms, invading and "choking off" or competing with normal tissues. Moreover, they can be shown to have antigens on their cell surfaces that can be distinguished from the antigens of the normal host cells. Does this mean that people can mount an immune response against their own cancers? A growing number of cancer researchers believe that not only can cancer induce an immune response but also that it usually does so. In fact it usually does so successfully, overwhelming the cancer before it is ever detected by the patient or the physician. The cancers that are discovered represent occasional failures of the immune system. This conclusion suggests that bolstering the patient's immune defense may provide a means for cancer prevention or control.

SUMMARY

Transport of oxygen, nutrients, hormones, and wastes is accomplished in vertebrates by closed circulatory systems in which the blood is pumped by the muscular contractions of the heart into a vast circuit of arteries, arterioles, capillaries, venules, and veins. This network ultimately services every cell in the body. The essential function of the circulatory system is performed by the capillaries, through which substances are exchanged with the interstitial fluid surrounding the individual cells.

Heart structure in vertebrates varies with the demands of their differing metabolic rates. The fish has a two-chambered heart; the atrium receives deoxygenated blood and the ventricle expels blood through the oxygenating gill capillaries into the systemic circulation. In amphibians, both oxygenated and deoxygenated blood are received in a two-chambered atrium, are mingled somewhat in the ventricle, and are then pumped simultaneously through the lungs and through the systemic circulation. Birds and mammals have a double circulation system, made possible by a four-chambered heart that functions as two separate pumping organs. One side pumps deoxygenated blood to the lungs, and the other pumps oxygenated blood to the body tissues.

Synchronization of the heartbeat is controlled by the sinoatrial node (the pacemaker) located in the right atrium, and by the atrioventricular node, which delays the stimulation of ventricular contraction until the atrial contraction is completed. The rate of heartbeat is secondarily under neural and hormonal control. Acetylcholine inhibits the pacemaker; epinephrine and norepinephrine stimulate it.

The blood is composed of plasma, white blood cells (leukocytes), red blood cells (erythrocytes), and platelets. The fluid part of the blood is plasma, chiefly composed of water, in which are dissolved or suspended the nutrients, antibodies, enzymes, waste substances, and other specialized compounds necessary to the life functions. Leukocytes defend the body against invaders by the manufacture of antibodies and by phagocytosis. Erythrocytes are the vehicles of oxygen-bearing hemoglobin. Platelets serve in the formation of clots. Leukocytes, erythrocytes, and platelets are produced in bone marrow.

Fluid that seeps out of the capillaries as a consequence of hydrostatic pressure and does not return by osmosis is returned to the blood by the lymphatic system. The lymph also picks up bacteria, cellular debris, and other foreign particles. These are filtered out of the lymph by the lymph nodes. The lymphatic system is also the site of maturation of lymphocytes, the white blood cells involved in the immune reaction. The spleen, thymus, tonsils, and adenoids are also components of the lymphatic system.

The immune response is a defensive reaction that involves recognizing a substance entering the circulatory system as foreign and developing a specific defense against it. One important part of the system involves the production of antibodies by plasma cells, which are B-lymphocytes that have differentiated after contact with an antigen. Antibodies are complex proteins that combine with specific antigens and inactivate them. The immune system is also involved in reactions against blood transfusions and tissue transplants. Cellular immunity is mediated by cells derived from T-lymphocytes, which mature in the thymus gland. Disorders involving the immune system include allergies, Rh disease of infants, autoimmune diseases, and possibly cancer.

QUESTIONS

1. Distinguish between the following: blood/plasma; systolic/diastolic; aorta/vena cava; atrium/ventricle; right heart/left heart; sinoatrial node/cardiovascular regulating center; instruction theory/clonal selection theory.

2. How does the radius of a blood vessel affect the blood flow through it?

3. If a tube full of water is connected to one of your arteries, how high will the water go if the tube is vertical?

4. Explain the reasons for the changes in blood pressure shown in Figure 35–5.

5. When you are very frightened, you turn quite pale. What is occurring to cause this change? Why is such an adaptation useful?

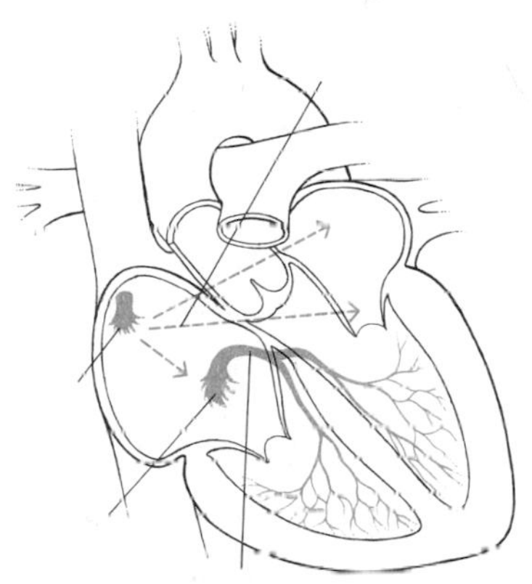

6. Label the diagram at the left.

7. The valves of the heart are not directly controlled by nerves. Yet, in most individuals, they open and shut at precisely the right points in the cardiac cycle for efficient heart operation. How is this precise timing possible? What does determine just when the valves will open and shut?

8. What is the advantage of the atrioventricular fibers being slow-conducting?

9. Trace the course of a single red blood cell from the right ventricle to the right atrium in a mammal.

10. Blood serum is the portion of the plasma remaining after a clot is formed. Name some of the components of the plasma that would not be present or would be present in a lesser amount in serum.

11. When an accident victim suffers blood loss, he or she is transfused with plasma rather than with whole blood. Why is plasma so effective in meeting the immediate threat to life?

12. Give two reasons why atherosclerotic arteries are much more susceptible to clot formation and blockage than are normal, healthy arteries.

13. What is probably the immediate cause of death by crucifixion? (If you need a clue, see page 664.)

14. If a live bacterium enters the body, what are the ways in which it can be destroyed?

CHAPTER 36

Respiration

Respiration has two meanings in biology. At the biochemical level, "respiration" refers to the oxygen-requiring chemical reactions that take place in the mitochondria and are the chief source of energy for the cell. At the level of a whole organism, "respiration" refers to the process of taking in oxygen from the environment and returning carbon dioxide to it. The latter process—which is, of course, essential for the former—is the subject of this chapter. In every organism from an amoeba to an elephant, exchange of gases takes place by diffusion. So we are going to begin this chapter with a discussion of the principles of diffusion as they refer to the movement of gases into and out of liquids.

AIR UNDER PRESSURE

Diffusion, you will recall (page 126), is the net movement of particles from a region of higher concentration to a region of lower concentration as a result of their random motion. In describing gases, scientists speak of the pressure of a gas rather than its concentration. At sea level, the air around us exerts a pressure on our skin of 1 atmosphere (about 15 pounds per square inch). This pressure is enough to raise a column of water about 10 meters high or a column of mercury 760 millimeters (29.91 inches) in the air (Figure 36–1). Atmospheric pressure is generally measured in terms of mercury simply because mercury is relatively heavy—so the column will not be inconveniently tall.

The total pressure of a mixture of gases, such as air, is the sum of the pressures of the separate gases in the mixture. The pressure of each gas is proportional to its volume. Oxygen, for instance, makes up about 21 percent by volume of dry air. (See Table 36–1.) Thus the pressure of the oxygen component of dry air is 21 percent of 760, or about 159 millimeters of mercury. This is known as the partial pressure of oxygen and is abbreviated Po_2. In air that contains water vapor (also a gas), the volume of O_2 is proportionately less and so the Po_2 is less.

If a liquid containing no gases is exposed to air at atmospheric pressure, each of the gases in the air diffuses into the liquid until the partial pressure of each gas in the liquid is equal to the partial pressure of the gas in the air. When one speaks of the Po_2 of the blood, one means the pressure of dry gas with which the dissolved O_2 is in equilibrium. The total amount of gas in the liquid may be very small. For instance, 100 ml of fully aerated sea water at 15°C contain only 0.5 ml of oxygen,

Table 36–1 *Composition of Dry Air*

GAS	% OF VOLUME
Oxygen	21
Nitrogen	77
Argon	1
Carbon dioxide	0.03
Other gases*	0.97

* Includes hydrogen, neon, krypton, helium, ozone, and xenon.

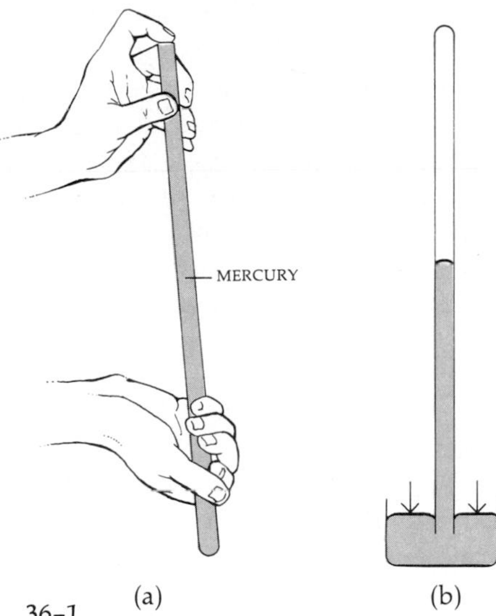

36–1
Atmospheric pressure is usually measured by means of a mercury barometer. (a) *To make a simple mercury barometer, fill a long glass tube, open at one end only, with mercury. Closing the tube with your finger,* (b) *invert it into a dish of mercury. Remove your finger. The mercury level will drop until the pressure of its weight inside the tube is equal to the atmospheric pressure outside. At sea level, the height of the column will be about 76 centimeters.*

or only 0.5 percent by volume. Thus when we speak of venous blood with a Po_2 of 40 mm Hg, we mean that it would be in equilibrium with air in which the partial pressure of oxygen were 40 mm Hg and that, if it were exposed to the usual mixture of air, oxygen would move from the air into the venous blood until the blood $Po_2 = 155$ mm Hg and equilibrium was reached. Gases move from a region of higher partial pressure to a region of lower partial pressure.

We are so accustomed to the pressure of the air around us that we are unaware of its presence or of its effects on us. However, if you visit a place—such as Mexico City—which is at a comparatively high altitude and therefore has a lower atmospheric pressure (and of course, a lower Po_2), you will feel lightheaded at first and will tire easily. Regular inhabitants of such places breathe more deeply, have enlarged hearts that circulate oxygen-carrying blood more rapidly, and have more red blood cells for oxygen transport.

The consequences of increased gas pressures are seen in deep-sea divers. Early in the history of deep-sea diving, it was found that when divers come up from the bottom too quickly, they get the "bends," which are always painful and sometimes fatal. The bends develop as a result of breathing compressed air. High pressures force more nitrogen from the air in the lungs into solution in the blood and tissues. If the body is rapidly decompressed, the gas expands and nitrogen bubbles out of the blood, like the carbon dioxide bubbles that appear in a bottle of soda when you first remove the top. The nitrogen bubbles lodge in the capillaries, stopping blood flow, or invade nerves or other tissues.

EVOLUTION OF RESPIRATORY SYSTEMS

The biochemical means by which organisms obtain energy in the absence of oxygen are not very efficient, as we saw in Chapter 8. Before oxygen came to be present in the atmosphere, the only forms of life were one-celled organisms; the only present-day forms that can carry on life processes without oxygen are a few types of bacteria and yeasts. The transition to an atmosphere containing free oxygen was responsible for one of the giant steps in evolution (page 338).

Oxygen enters and moves within cells by diffusion. Within the cell, it takes part in the oxidation of breakdown products of glucose and other carbon-containing compounds that serve as cellular energy sources. In this process, carbon dioxide is produced, which then diffuses out of the cell along the concentration (partial pressure) gradient. This is true of all cells, whether an amoeba, a *Paramecium,* a liver cell, or a brain cell. However, substances can move effectively by diffusion only for very short distances, less than 1 millimeter. These limits pose no problem for very small animals, in which each cell is quite close to the surface, or for animals in which much of the body mass is not metabolically active—like the "jelly" (mesoglea) of jellyfish. Many eggs and embryos also respire in this simple way, particularly in the early stages of development.

However, diffusion cannot possibly meet the needs of large organisms in which cells in the animal's interior may be many centimeters from the air or water serving as the oxygen source. As organisms increased in size in the course of evolution, there also evolved circulatory and respiratory systems that transport large numbers of gas molecules by bulk flow. (Remember that whereas diffusion is the result of the random movement of individual molecules, bulk flow is the overall movement of molecules in response to pressure or gravity.)

An early stage in the evolution of gas-transport systems is exemplified by the earthworm (page 440). As is the case with most other kinds of worms, earthworms have a network of capillaries just one cell layer beneath the surface of their bodies. Oxygen and carbon dioxide diffuse directly through the body surface into and out of the blood as it travels through these capillaries. The blood picks up oxygen by diffusion as it travels near the surface of the animal and releases oxygen by diffusion as it travels past the oxygen-poor cells in the interior of the earthworm's body. Conversely, blood picks up carbon dioxide from the cells and releases CO_2 by diffusion as it travels near the surface of the animal. Thus the gases move into and out of the earthworm by diffusion but are transported within the animal by bulk flow.

This system is particularly suitable for worms because their tube shape exposes a proportionately large surface area. Some worms can adjust their surface area in relation to oxygen supply. If you have an aquarium at home, you may be familiar with tubifex worms, which are frequently sold as fish food. When these worms are placed in water low in oxygen, such as a poorly aerated aquarium, they stretch out as much as 10 times their normal length and thus increase the surface area through which oxygen diffusion occurs.

Insects and some other arthropods, as we have seen (page 463), have evolved a different strategy. Air is piped directly into the tissues by a network of chitin-lined tubules (Figure 36–2). In large insects, in particular, diffusion is assisted by body movements, which propel the air into and out of the spiracles. This system is fine for small organisms but is the principal limitation on the size that can be obtained by insects and other tracheal-system breathers. The planet may eventually be taken over by cockroaches or ants, but it is safe to bet they will not be the giant forms of science fiction.

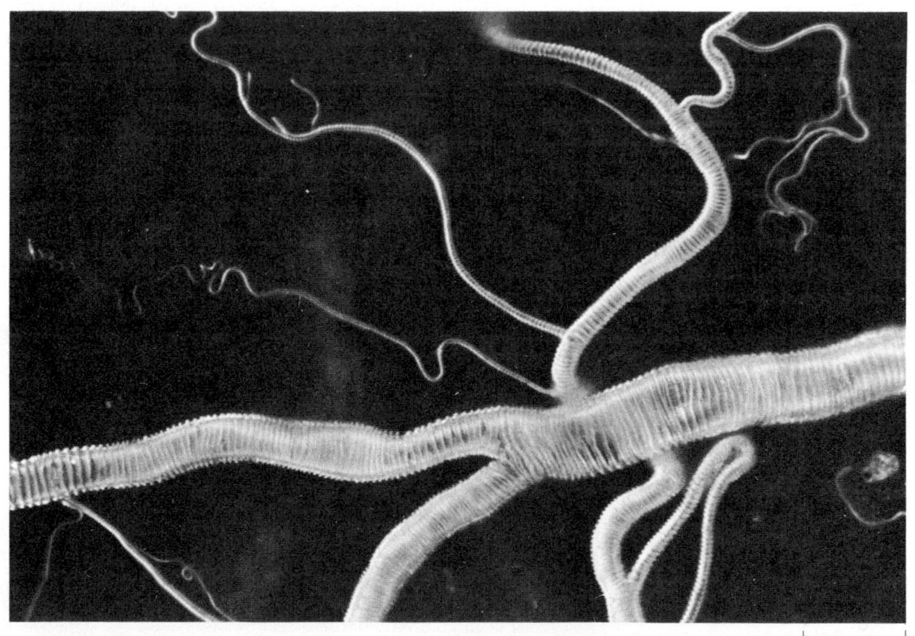

36–2
Insects and some other terrestrial arthropods breathe by means of tracheae. These are inward tubular extensions of the exoskeleton, which are thickened in places, forming spiral supports that hold the tubes open. The trachea is lined with chitin, which is molted when a new exoskeleton is formed. Gas exchange takes place at the thin, moist terminal ends of the tubules. This micrograph shows part of the tracheal system of a cockroach.

36-3
Gills are essentially outpocketings of the epithelium that increase the surface area exposed to water. Often the gills are covered by an exoskeleton, as in crustaceans, or a flaplike shield (the operculum), as in fish. In the axolotl, the amphibian shown here, the external nature of the gills is more clearly evident.

Evolution of the Gill

Gills and lungs are other ways of increasing the respiratory surface. Gills are usually outgrowths (Figure 36-3), whereas lungs are ingrowths, or cavities. The respiratory surface of the gill, like that of the earthworm, is a layer of cells, one cell thick, exposed to the environment on one side and to circulatory vessels on the other. The layers of gill tissue may be spread out flat, stacked, or convoluted in various ways. The gill of a clam, for instance, is shaped like a steam-heat radiator (which is also designed to provide a maximum surface-to-volume ratio).

The vertebrate gill is believed to have originated primarily as a feeding device. Primitive vertebrates respired mostly through their skin. They filtered water into their mouths and out of what we now call their gill slits, extracting bits of organic matter from the water as it went through. (*Branchiostoma*—page 472—which is believed to resemble closely the ancestral vertebrate, feeds in this way.)

In the course of time, numerous selection pressures, chiefly involved with predation, came into operation. As one consequence, there was a trend toward an increasingly thick skin, even one armored or covered with scales. Such a skin is not, of course, useful for respiratory purposes. At the same time, related forces were operating to produce animals that were larger and swifter and so more efficient at capturing prey and escaping predators. Such animals also had larger energy requirements—and consequently larger oxygen requirements. These problems were solved by the "capture" of the gill for a new purpose: respiratory exchange. The surface area and blood supply of the epithelium beneath the gill have slowly increased over the millennia. The modern gill is the result of this evolutionary process.

36–4

Rates of diffusion are proportional not only to the surface areas exposed but also to differences in concentration of the diffusing molecules. The greater the difference in concentration of a molecule, the more rapid its diffusion. In the gill of the fish, the circulatory vessels are arranged so that the blood is pumped through them in a direction opposite to that of the oxygen-bearing water. This arrangement results in a far more complete transfer of oxygen to the blood than if the blood flowed in the same direction as the water.

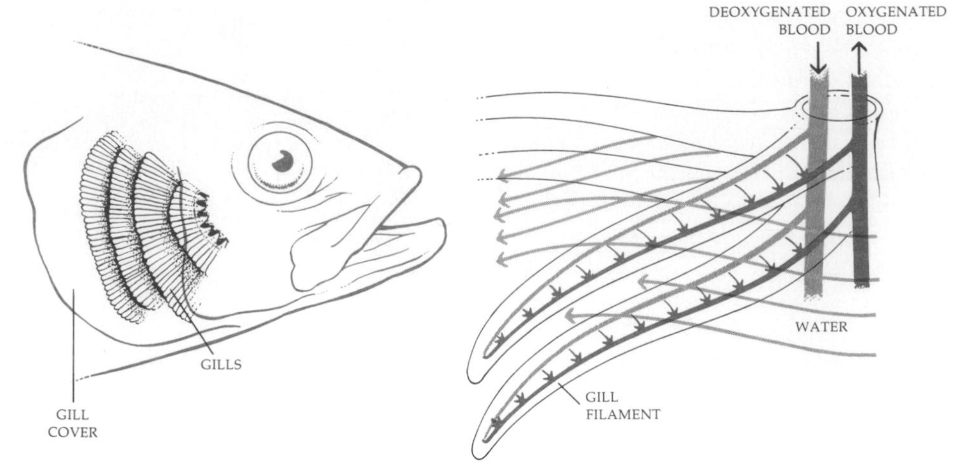

In most fish, the water (in which oxygen is dissolved) is pumped in at the mouth by oscillations of the bony gill cover and flows out across the gills (Figure 36–4). In the gill of the fish, the circulatory vessels are arranged so that the blood is pumped through them in a direction opposite to that of the oxygen-bearing water. This counter-current arrangement results in a far more efficient transfer of oxygen to the blood than if the blood flowed in the same direction as the water. Also the fish can regulate the rate of water flow, and sometimes assist it, by opening and closing its mouth. Fast swimmers, such as mackerel, can obtain enough oxygen only by keeping perpetually on the move. Such fish cannot be kept in an aquarium or any other space where their motion is limited because they will suffocate.

Evolution of the Lung

Lungs are internal cavities into which oxygen-containing air is taken. They have certain disadvantages as compared with gills; it is more efficient from the point of view of diffusion to have a continuous flow across the respiratory surface. Air, however, is a far better source of oxygen than is water; one-fifth of the air of the modern atmosphere is free oxygen. Not only must more water be processed to obtain a given amount of oxygen, but water also weighs a great deal more. A fish spends up to 20 percent of its energy in the muscular work associated with respiration, whereas an air breather expends only 1 or 2 percent of its energy in respiration. Also, oxygen diffuses about 300,000 times more rapidly through air than through water, and so can be replenished much more quickly as it is used up by respiring organisms. All the higher vertebrates—the birds and the mammals—are air breathers, even those that live in the water.

Lungs are not essential for air breathing. As we saw, earthworms are air breathers. The overwhelming advantage of gas exchange across an internal surface, however, is that the respiratory areas can be kept moist without a large loss of water by evaporation. Although lungs are largely a vertebrate "invention," they are found in some lower animals. Land-dwelling snails, for example, have independently evolved lungs that are remarkably similar to the lungs of some amphibians.

ENERGY AND OXYGEN

Oxygen consumption is directly related to energy expenditure, and, in fact, energy requirements are usually calculated by measuring oxygen intake or the release of carbon dioxide. The energy expenditure at rest is known as basal metabolism. The basal metabolism of a warm-blooded (endothermic) animal is much higher than that of an ectotherm, with 10 to 30 percent of the energy budget being expended on maintaining a constant temperature.

Metabolic rates increase sharply with exercise. A person exercising consumes 15 to 20 times the amount of oxygen that he or she consumes sitting still. Oxygen consumption of a running animal increases linearly with its speed. If you are an aquatic animal, such as a fish or a porpoise, swimming is far less expensive, in terms of energy requirements, than is running. However, a human, whose body is basically unfit for moving through the water, uses five times as much energy—as measured by oxygen consumption—in swimming as in running.

In terms of energy consumption, flying is less expensive than walking and running. Many birds can fly more than 1,000 kilometers nonstop, whereas an animal of similar weight could not travel on the ground for that distance without stopping for food. (Migratory birds may also hitchhike on atmospheric currents.) However, flying, for a bird, is more expensive, in energy-consumption terms, than swimming is for a fish.

Oxygen is required for the energy-producing reactions that take place in the mitochondria. The oxygen consumption of running animals—such as these gazelles—increases directly in proportion to their speed.

Some primitive fish had lungs as well as gills, although the lungs were not efficient enough to serve as more than accessory structures. These lungs were probably a special adaptation to life in fresh water, which, unlike ocean water, may stagnate (become depleted of oxygen) because of decay or algal bloom. A few species of lungfish still exist, which, by coming to the surface and gulping air into their lungs, can live in water that does not have sufficient oxygen to support other fish life.

36–5
Frogs have relatively small and simple lungs, and a major part of their respiration takes place through the skin. The chart shows the results of a study of a frog in which oxygen and carbon dioxide exchanges through the skin and lungs were measured simultaneously for one year. Can you explain why the respiratory activity peaks in April and May and is lowest in December and January?

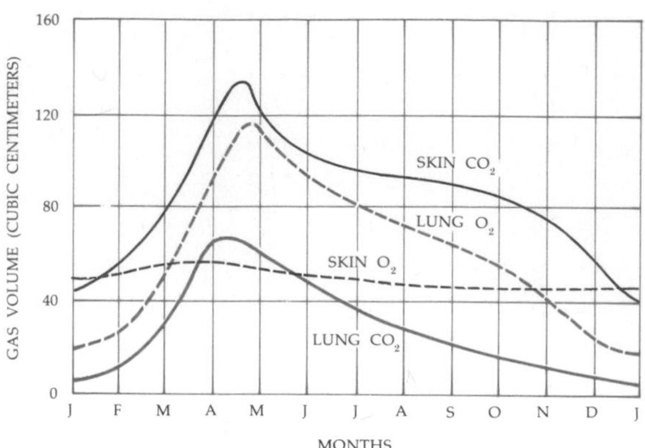

Amphibians and reptiles have relatively simple lungs, with small internal surface areas, although their lungs are far larger and more complex than those of the lungfish. The lungs of lungfish developed directly from the pharynx, the posterior portion of the mouth cavity, which leads to the digestive tract. In amphibians, reptiles, and other air-breathing vertebrates, we see the evolution of the windpipe, or *trachea*, guarded by a valve mechanism, the glottis, and nostrils, which make it possible for the animal to breathe with its mouth closed. Amphibians still rely largely on their skin for gas exchange (Figure 36–5), but reptiles breathe almost entirely through their lungs.

An important feature of all vertebrate lungs is that the exchange of air with the atmosphere takes place as a result of changes in lung volume. Such lungs are known as ventilation lungs or air sacs. Frogs gulp air and force it into their lungs in a swallowing motion, thus expanding their lung volume; then they open the glottis and let the air out again. In reptiles, birds, and mammals, air is sucked into the lungs as a consequence of changes in the size of the lung cavity, brought about by activity of the chest muscles. The various types of respiratory systems are summarized in Figure 36–6.

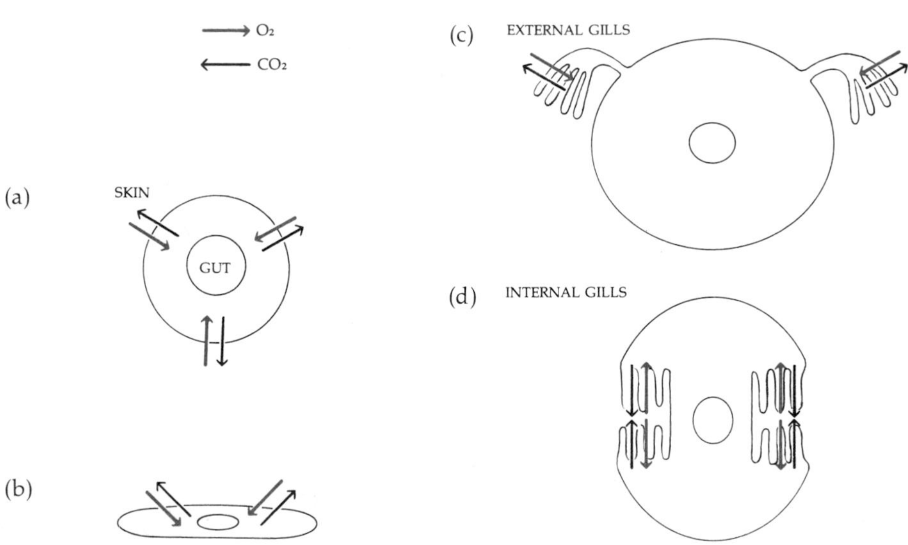

36–6
Respiratory systems. (a) *Gas exchange across the entire surface of the body is found in a wide range of small animals from protozoans to earthworms.* (b) *Gas exchange across the surface of a flattened body. Flattening increases the surface-to-volume ratio and also decreases the distance over which diffusion has to occur within the body. Flatworms are examples.* (c) *External gills increase the surface area but are unprotected and therefore easily damaged. External gills are found in polychaete worms and some amphibians. Gas exchange usually takes place across the rest of the body surface as well as the gills.* (d) *With internal gills, a ventilation mechanism draws water over the highly vascularized gill surfaces, as in fish.* (e) *Gas exchange at the terminal ends of fine tracheal tubes that branch through the body and penetrate all the tissues is characteristic of insects and some other arthropods.* (f) *Lungs are highly vascularized sacs into which air is drawn by a ventilation mechanism. Lungs are found in all air-breathing vertebrates.*

36-7

(a) *The lungs of birds are extraordinarily efficient. They are small and are expanded and compressed by movements of the body wall. Each lung has several air sacs attached to it, which empty and fill like balloons at each breath. No gas exchange takes place in the sacs. They appear rather to act as bellows. Thus, the lung is almost completely flushed with fresh air at every breath; there is little residual "dead" air left in the lungs, as there is in mammals.* (b) *Scanning electron micrograph of lung tissue from a domestic fowl. The tubes visible here are ventilated by air sacs, which serve to draw in fresh air. Gas exchange takes place in the broad meshwork of air capillaries and blood capillaries, which make up the spongelike respiratory tissue seen here surrounding the tubes.*

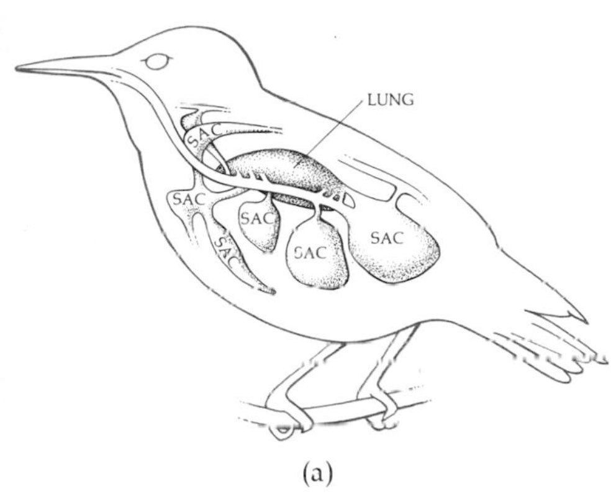

(a)

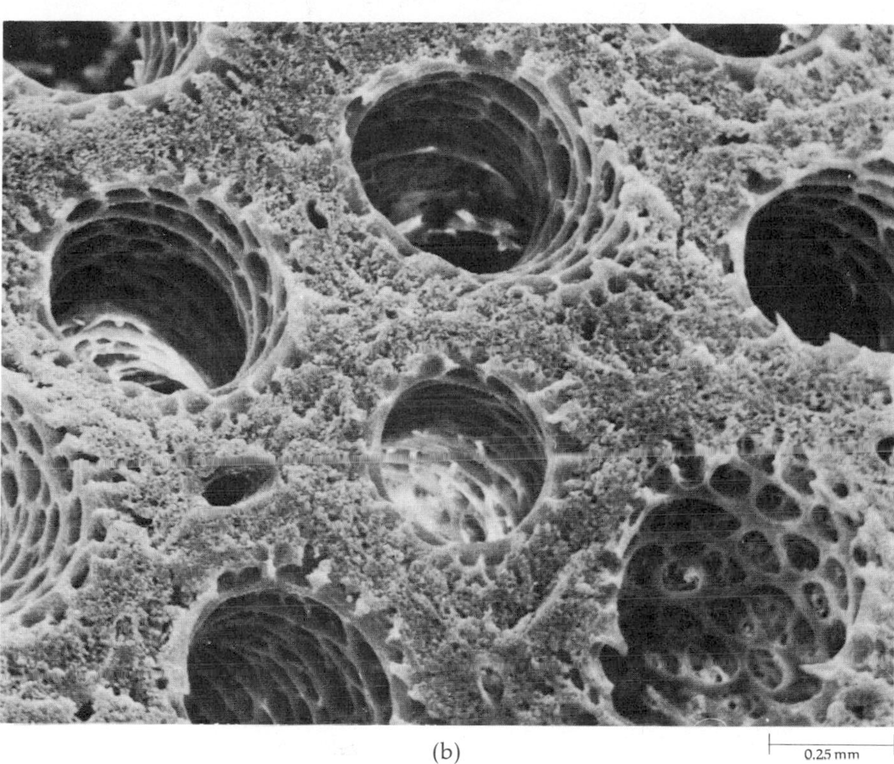

(b)

0.25 mm

(e) TRACHEAE

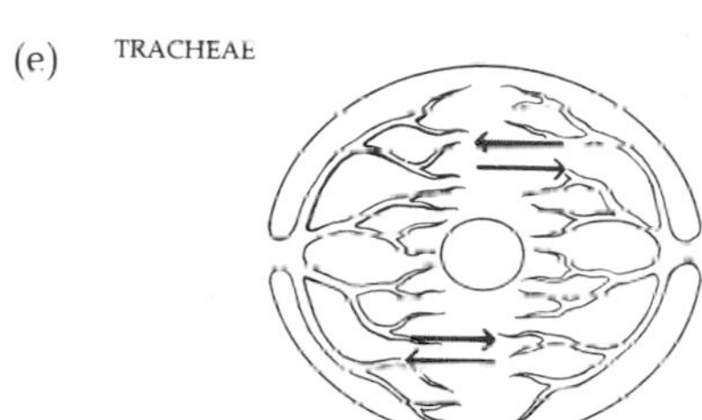

(f) LUNGS

RESPIRATORY PIGMENTS

Blood plasma is an even less efficient carrier of oxygen than sea water; at atmospheric pressure, only about 0.3 ml of oxygen will dissolve per 100 ml of plasma. All active animals—and this includes even the earthworm—have in their blood special oxygen-carrying protein molecules, known as respiratory pigments, that raise the oxygen transporting capacity of blood as much as seventyfold. (The only exception is some Antarctic fishes that have no respiratory pigments at all.) Hemoglobin is the carrier found in all vertebrates and in a wide variety of invertebrate species representing many different phyla. A form of hemoglobin (hemocyanin), which contains copper rather than iron, is the most common respiratory pigment of mollusks and arthropods. Unlike hemoglobin, which is red, hemocyanin is blue when combined with oxygen; other respiratory pigments are known, all of which are a combination of a metal and a large protein. In most invertebrates, respiratory pigments are simply dissolved in the blood plasma. In vertebrates and echinoderms the pigments are carried in red blood cells.

36-8
Oxygen-hemoglobin dissociation curve. This curve represents figures for normal adult human hemoglobin at 38° C and at a normal pH. As the partial pressure of oxygen drops, the oxygen and the hemoglobin dissociate.

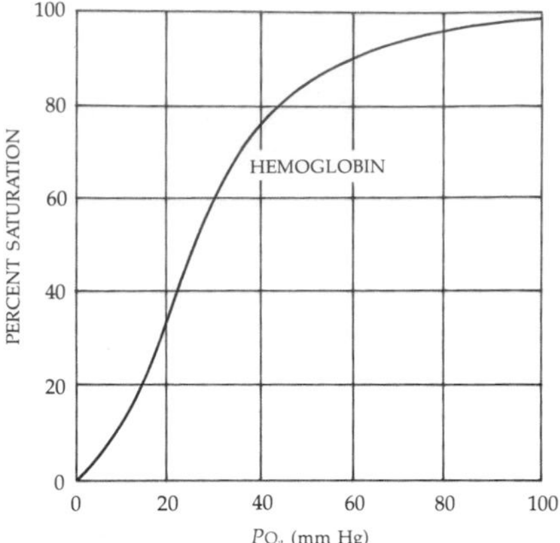

Hemoglobin, you will recall, is made up of four subunits, each of which comprises a heme unit and a protein chain (Figure 3-28, page 69). The heme unit consists of a porphyrin ring and one atom of iron. Each of the four atoms of iron in one hemoglobin molecule can bind reversibly with one molecule of oxygen. The oxygen molecules are added one at a time:

$$Hb_4 + O_2 \rightleftharpoons Hb_4O_2$$
$$Hb_4O_2 + O_2 \rightleftharpoons Hb_4O_4$$
$$Hb_4O_4 + O_2 \rightleftharpoons Hb_4O_6$$
$$Hb_4O_6 + O_2 \rightleftharpoons Hb_4O_8$$

Combination of the first Hb with O_2 increases the affinity of the second Hb for O_2, and oxygenation of the second increases the affinity of the third and so on. (As O_2 is taken up, the two beta chains move closer together, and this movement is apparently necessary for the shift in affinity to take place.) As a consequence of the change in the hemoglobin molecule, the curve relating the uptake of oxygen to Po_2 is not a straight line, but has a characteristic sigmoid shape (Figure 36-8). When hemoglobin is fully oxygenated, the oxygen content of human blood is as high as 20 percent.

HUMAN RESPIRATORY SYSTEM

In *Homo sapiens*, inspiration (breathing in) and expiration (breathing out) usually take place through the nose. The nasal cavities are lined with hairs and cilia, both of which trap dust and other foreign particles. The epithelial cells that line the cavities secrete mucus, which humidifies the air and collects debris that can be removed by swallowing, sneezing, or spitting. The cavities have a rich blood supply, which keeps their temperature high, warming the air before it reaches the lungs.

From the nasal passages, the air goes to the *pharynx* and from there to the *larynx*, located in the upper front part of the neck. An adult human larynx is shaped somewhat like a triangular box, with its point downward. Across it are stretched the vocal cords, which are two ligaments drawn taut across the lumen of the respiratory tract. Vibrations of these cords, by expired air, cause the sounds made in speech. The vocal cords are influenced by the male hormone. At puberty, the cords in males become longer and thicker, sometimes so rapidly that the adolescent male temporarily loses control over them, occasionally emitting embarrassing squeaks. Laryngitis, which is simply an inflammation of the vocal cords, interferes with their vibration, so you "lose your voice."

From the larynx, inspired air travels through the trachea, which is a long membranous tube, also lined with ciliated, mucus-producing epithelial cells (see Figure 5-2c, page 97). The walls of the trachea are strengthened by rings of cartilage that prevent it from collapsing during inspiration. The trachea leads into the *bronchi* (singular, *bronchus*), which subdivide into smaller and smaller passageways, the *bronchioles*.

The bronchi and bronchioles are surrounded by thin layers of smooth muscle. Epinephrine causes the bronchioles to dilate; histamine causes them to contract. Asthma is a spasm of these muscles resulting in labored breathing.

Cilia along the trachea, bronchi, and bronchioles beat continuously, pushing mucus and foreign particles embedded in mucus up toward the pharynx, from which it is generally swallowed. We are usually aware of this production of mucus only when it is increased above normal as a result of an irritation of the membranes.

The actual exchange of gases takes place in small air sacs, the *alveoli*, which are clustered in bunches like grapes around the ends of the smallest bronchioles. Each alveolus is about 1 or 2 millimeters in diameter, and each is surrounded by capillaries. The walls of the capillaries and of the alveoli each consist of only a single layer of flattened epithelial cells separated from one another only by a thin (basement) membrane; thus the barrier between an alveolus and its capillaries is only about 0.3 millimeter. Gases are exchanged between the air in the alveoli and the blood in the capillaries by diffusion.

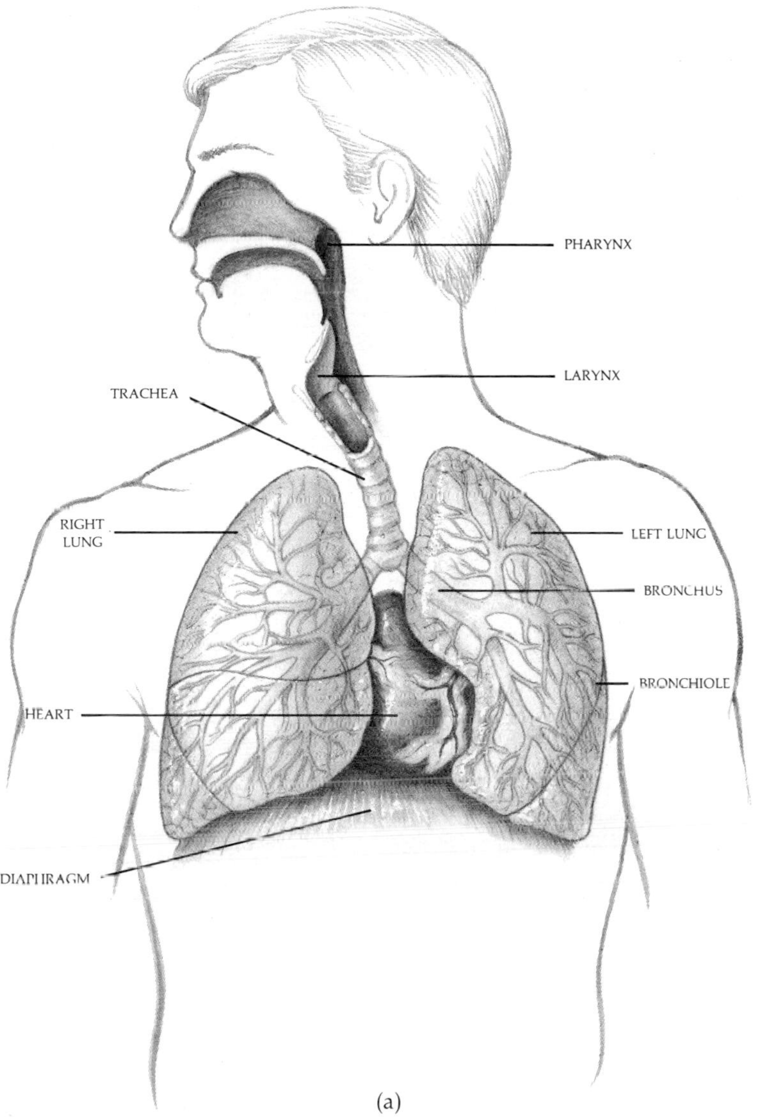

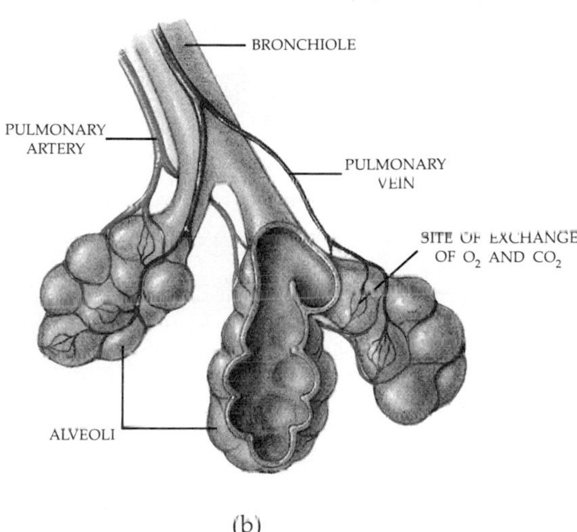

36–9
The human respiratory system. (a) *Air enters through the nose or mouth and passes into the pharynx and down the trachea, bronchi, and bronchioles to the alveoli* (b) *in the lungs. Within each alveolus, of which there are some 300 million in a pair of lungs, oxygen and carbon dioxide diffuse into and out of the bloodstream, through the capillary walls.*

CANCER OF THE LUNG

Lung cancer is the most rapidly increasing form of cancer in the United States and the most common cause of death from cancer in men. About 100,000 men and women will develop lung cancer in the United States this year, and more than 90 percent of them will die in less than three years. Most of these patients will be cigarette smokers.

The middle and lower lobes of a cancerous lung are shown in (a). *The cancer is the solid grayish-white mass. Its rapid growth has replaced most of the normal lung tissue in the middle lobe. It has also probably begun to spread to other parts of the body. Notice the bronchial branch that leads into the cancer and is destroyed by it. The lung tissue remaining around the cancer is compressed, airless, and dark red. Away from the cancer, in the lower lobe, the lung is filled with air and is light pink.*

The lung is normally protected by ciliated cells lining the tracheobronchial tree. The cilia catch and sweep out particulates in the respired air. Cancer cells arising in the bronchus generally do not have cilia. The scanning electron micrographs show (b) *the normal ciliated surface of the bronchus and* (c) *the surface of a bronchus with cancer.*

(a) (b) (c)

36-10
(a) *Scanning electron micrograph of lung tissue from a hamster, showing numerous alveoli.* (b) *Longitudinal section of an alveolar capillary from a dog. The large, dark, irregularly shaped structures are red blood cells. A portion of an alveolar cell is visible at the top.*

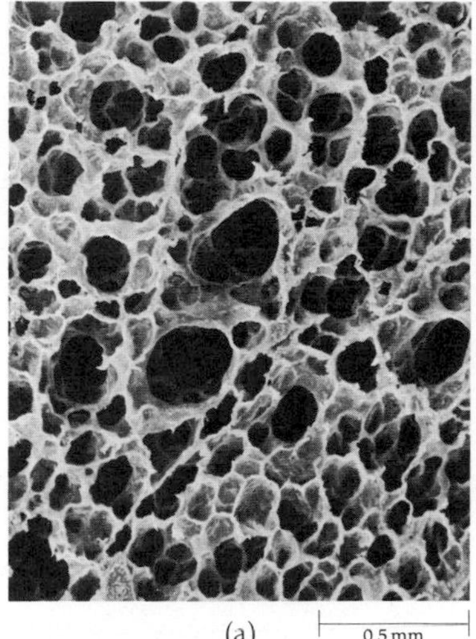

(a)

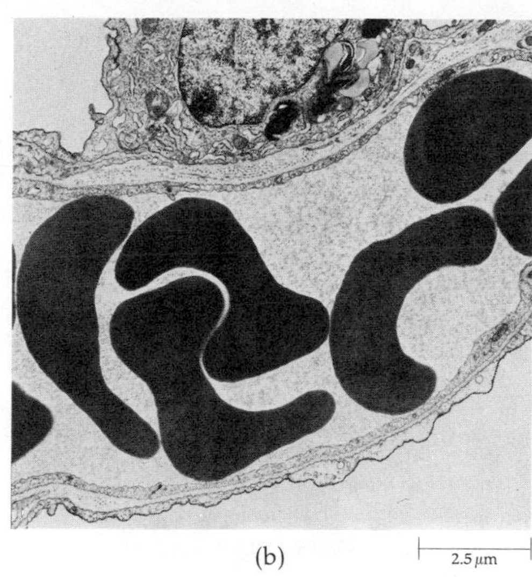

(b)

36-11
A model illustrating the way air is taken into and expelled from the lungs.

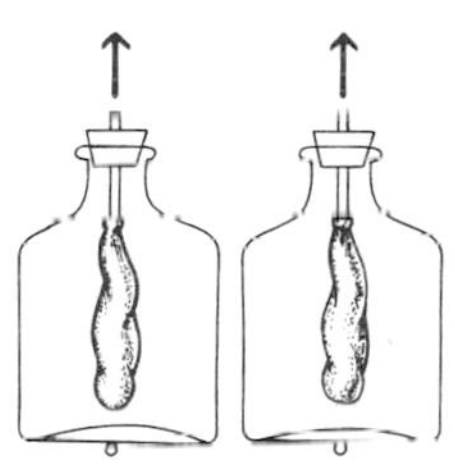

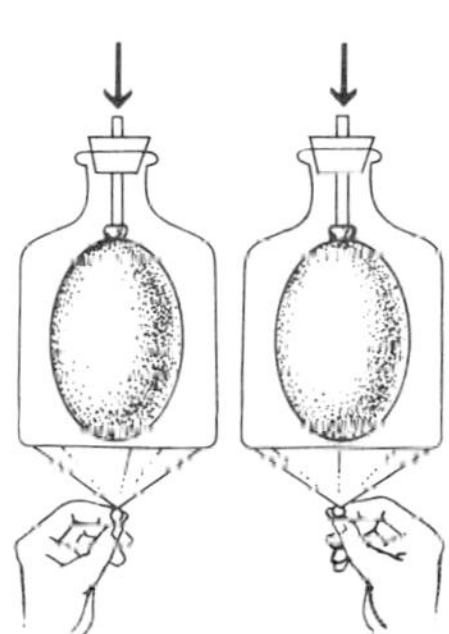

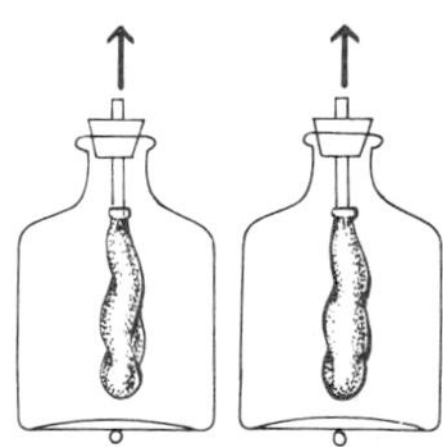

A pair of human lungs has about 300 million alveoli, providing a respiratory surface of some 70 square meters, or about 750 square feet—approximately 40 times the surface area of the entire human body.

The lungs are surrounded by a thin membrane and the thoracic cavity (part of the coelom) is lined by a similar membrane. These are known as the pleura. These membranes secrete a small amount of fluid that lubricates them so they slide past one another as the lungs expand and contract. Pleurisy is an inflammation of these membranes that causes them to secrete excess fluid that collects in the thoracic cavity.

Mechanics of Respiration

Inspiration and expiration are the results of changes in the volume of the thoracic cavity, brought about by the contraction and relaxation of the muscular diaphragm separating the thoracic and abdominal cavities and of the intercostal ("between-the-ribs") muscles. We inhale by contracting the dome-shaped diaphragm, which flattens it and increases its diameter, and by contracting the intercostal muscles, pulling the rib cage up and out. These movements enlarge the thoracic cavity, increasing its volume. This lowers the pressure in the cavity so that the pressure becomes less than atmospheric pressure. Air—which is, of course, at atmospheric pressure—enters the lungs. Air leaves the lungs as the muscles relax. Usually, only about 10 percent of the air in the lung cavity is exchanged at every breath, but as much as 80 percent can be exchanged by deliberate deep breathing. Whales and other large aquatic mammals suffocate on land because of the inability of their intercostal muscles to expand their massive chests when compressed under the weight of their bodies.

Gas Exchange

Gases are exchanged during the respiratory process (Table 36-2). Oxygen diffuses from the air in the alveoli into the capillaries. As it enters the bloodstream, most of it combines with hemoglobin. The amount of oxygen carried by the hemoglobin molecules is related to the partial pressure of oxygen (Po_2) in the blood. In adult

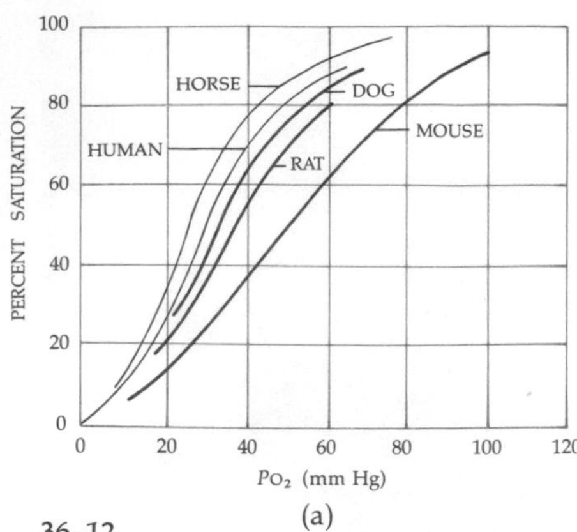

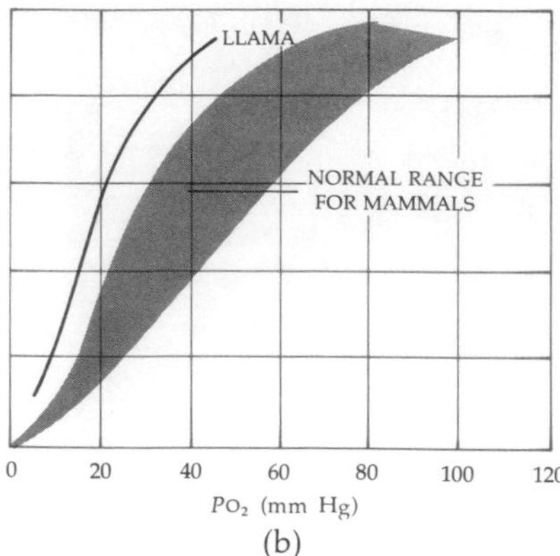

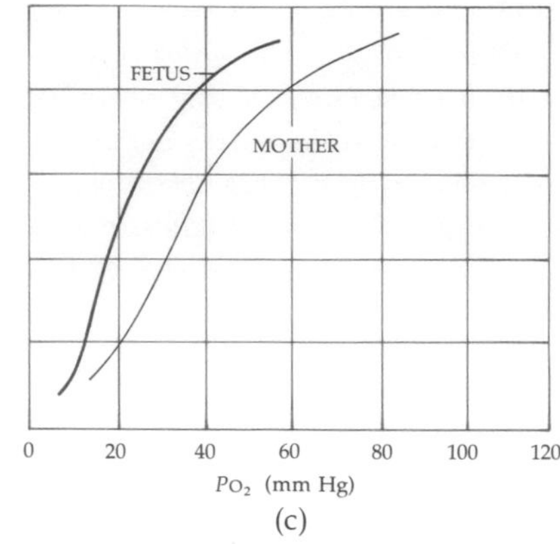

36–12
These curves show how the amount of oxygen carried by the hemoglobin is related to oxygen pressure. When oxygen pressure reaches 100 mm Hg—the pressure usually present in the human lung—the hemoglobin becomes totally saturated with oxygen. As the pressure drops, the oxygen bound to the hemoglobin molecule is given up. Therefore, when blood carrying oxygen reaches the capillaries, where pressure is only about 40 mm Hg or less, it gives up some of its oxygen to the tissues. A curve located to the right signifies that the oxygen is given up more readily at a given pressure. (a) *Small animals have higher metabolic rates and so need more oxygen per gram of tissue than larger animals. Therefore, they have blood that gives up oxygen more readily.* (b) *The llama, which lives in the high Andes of South America, has a hemoglobin that enables its blood to take up oxygen more readily at the low atmospheric pressures.* (c) *The fetus must take up all its oxygen from the maternal blood. The hemoglobin of mammalian fetuses has a greater affinity for oxygen than does the hemoglobin of adult mammals, and so the oxygen tends to leave the maternal blood and enter the fetal blood.*

humans, the partial pressure of oxygen in the blood as it leaves the lungs is about 100 millimeters of mercury (100 mm Hg); at this pressure, the hemoglobin is saturated with oxygen. As the hemoglobin molecules travel through the bloodstream, the Po_2 drops, and as it drops, the oxygen bound to the hemoglobin molecules is given up. Little oxygen is yielded as the Po_2 drops from 100 mm Hg to 60 mm Hg. This is a built-in safety factor that protects individuals at high altitudes or those who have heart or lung diseases that decrease blood Po_2. However, as the partial pressure drops below 60 mm Hg, oxygen is given up much more readily (Figure 36–12).

The Po_2 of the blood in the tissue capillaries is normally about 40 mm Hg. As a consequence, when the blood leaves the capillaries, its hemoglobin is still usually 70 percent saturated. This extra O_2 represents a reserve supply of oxygen should the demand increase—as a result, for example, of exercise.

Myoglobin and Its Function

Myoglobin is a protein molecule with an iron-containing (heme) group (Figure 36–13); in its structure, it resembles a single unit of the hemoglobin molecule (Figure 3–27, page 69). Myoglobin is found in skeletal muscle. It has a greater affinity for oxygen than hemoglobin does and begins to release significant amounts

Table 36–2 *Composition of Respiratory Gas at Standard Atmospheric Pressure*

GAS	INSPIRED AIR		EXPIRED AIR		ALVEOLAR AIR	
	% OF VOLUME	PARTIAL PRESSURE (mm OF Hg)	% OF VOLUME	PARTIAL PRESSURE (mm OF Hg)	% OF VOLUME	PARTIAL PRESSURE (mm OF Hg)
O_2	20.71	157	14.6	111	13.2	100
CO_2	0.04	0.3	4.0	30	5.3	40
H_2O	1.25	9.5	5.9	45	5.9	45
N_2	78.00	593	75.5	574	75.6	574

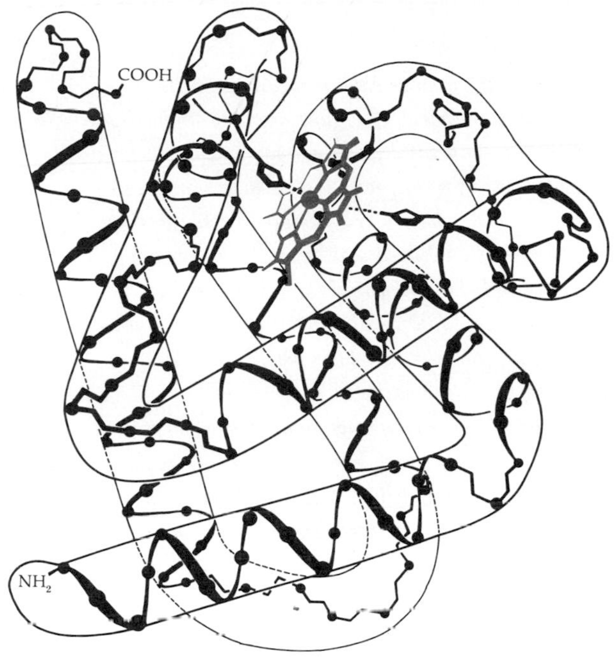

36–13
Tertiary structure of myoglobin, as deduced from x-ray diffraction analyses. Myoglobin closely resembles a single chain of the four-chain hemoglobin molecule. The heme portion is shown in color.

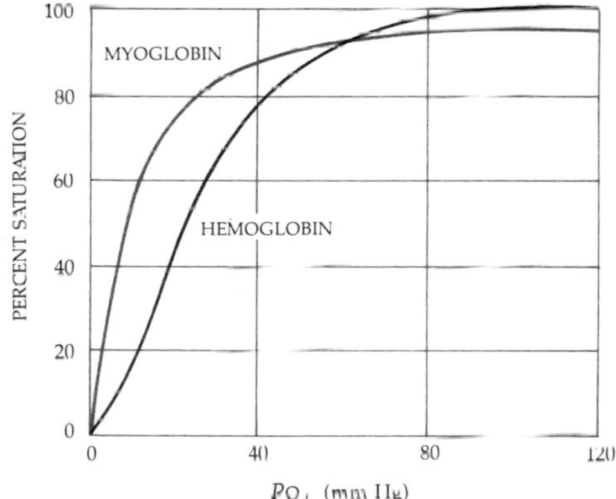

36–14
Comparison of the oxygen dissociation curves of myoglobin and hemoglobin. Note that myoglobin remains 80 percent saturated with oxygen until the partial pressure of oxygen falls below 20 mm Hg. Therefore, myoglobin retains its oxygen in the resting cell and relinquishes it only when strenuous muscle activity uses up the available oxygen provided by hemoglobin.

of oxygen only when the Po_2 falls below 20 mm Hg (Figure 36–14). Thus, when the muscle is at rest or engaged in only moderate activity, the myoglobin holds on to its oxygen. During strenuous exercise, however, when muscle cells are using oxygen rapidly and the partial pressure of oxygen in the muscle cells drops toward zero, myoglobin gives up its oxygen. Thus myoglobin provides an additional reserve of oxygen for active muscles.

Carbon Dioxide

A small amount of carbon dioxide is carried in the blood in the form of dissolved CO_2. Some (about 25 percent) is bound to hemoglobin molecules. Carbon dioxide does not combine with the heme units of the hemoglobin molecule, as oxygen does, but rather with the amino groups of the hemoglobin molecule. However, most of the carbon dioxide (about 65 percent) is carried in the blood as bicarbonate. Bicarbonate is produced in a two-stage reaction. First, carbon dioxide combines with water to form carbonic acid. This reaction is catalyzed by the enzyme carbonic anhydrase found in red blood cells. Carbonic acid, a weak acid, dissociates to yield bicarbonate and hydrogen ions:

$$CO_2 + H_2O \xrightleftharpoons{\text{carbonic anhydrase}} H_2CO_3 \rightleftharpoons HCO_3^- + H^+$$

As more carbon dioxide is taken up by the blood, the blood becomes increasingly acidic. As the acidity increases, hemoglobin gives up its oxygen more readily. Thus, as carbon dioxide enters the capillaries, the acidity of the blood increases and the yield of oxygen increases.

DIVING MAMMALS

Most mammals have about the same oxygen requirements as humans. Structurally, their respiratory organs are very similar; yet marine mammals can dive to great depths and stay submerged for long periods of time. A sperm whale, for example, has been found at a depth of more than 1,000 meters and has been known to stay submerged for 75 minutes. A Weddell seal, a much smaller animal, can dive to 600 meters and stay submerged for 70 minutes, or can swim submerged for between 2,000 and 4,400 meters—1 to 2 miles. Are their lungs different from ours? Or their blood? Do they have special oxygen reserves, like a submarine?

Studies in a variety of diving mammals have shown that none has lungs significantly larger, in proportion, than ours. In fact, seals (and perhaps other diving mammals also) exhale before diving or early in the dive. Blood volume is increased, however. In humans, blood is about 7 percent of the body weight, whereas in diving marine mammals, it is 10 to 15 percent. The blood vessels are proportionately enlarged, and they appear to serve as a reservoir of oxygenated blood. The proportion of red blood cells is higher and myoglobin is more concentrated, giving the muscles a very dark color. Still, physiologists calculate, these adaptations would not provide enough oxygen for long dives.

A major survival factor, studies in diving mammals have shown, is a group of automatic reactions known, collectively, as the diving reflex. During a dive, the heart rate slows and blood supply is reduced to tissues that are tolerant of oxygen deprivation, such as the digestive organs, skin, and muscles. Muscles obtain energy by anaerobic glycolysis, producing large amounts of lactic acid and building up an oxygen debt (see page 158). Most of the oxygen is shunted to the heart and brain, whose cells would begin to die after about four minutes without oxygen. In a pregnant female, the fetus similarly is given high priority for the available oxygen.

Recent studies by Martin Nemiroff of the University of Michigan Medical School indicate that the diving reflex is also present in human beings. In humans, its primary survival value is during birth, when the infant may be cut off from an oxygen supply during the later stages of labor. Operation of the reflex can be demonstrated in a very simple way: Immersing the face in cold water causes the heart rate to slow down.

These observations by Nemiroff came in the course of investigating about 60 near-drownings. It has been generally assumed that a person who is submerged for four minutes or more will die, or, at best, suffer irreversible brain damage. In the group of 60, 15 had been rescued after four minutes or more in the chilly waters of Michigan lakes. Of these 15, 11 survived without brain damage. One of these, a college student who crashed through the ice in an automobile accident, was under water for 38 minutes; he finished the college semester with a 3.2 average. Another survivor resumed his career as a practicing physician. All the survivors had been submerged in cold (below 21° C) water, which slowed the metabolic demands of their cells. Some required assisted breathing for as long as 13 hours. Moral: Even if a drowning victim is cold, blue, with no pulse, no heartbeat, and the eyes fixed in a glassy stare—as was the case with the student—if he or she has been in cold water, efforts at resuscitation should be started immediately and continued indefinitely. We are not Weddell seals, but owing to an ancient survival strategy, we have unexpected powers of underwater survival.

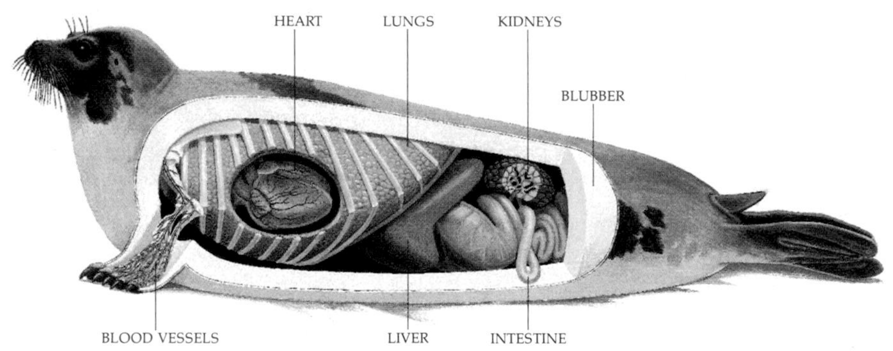

A seal exhales as it dives, reducing the likelihood of decompression sickness (the bends). It has $1\frac{1}{2}$ times as much blood as a land mammal of the same size and, when it dives, can reduce its heartbeat from 120 to 30 beats per minute. Other adaptations of the seal include an intestine more than twice as long as a human's, a kidney composed of lobules that can each perform all the functions of the whole kidney, special arrangements of blood vessels that allow regulation of body temperature, and a thick insulating layer of blubber.

36–15
The partial pressure of oxygen in an alveolus and the alveolar capillary. The figures represent millimeters of mercury.

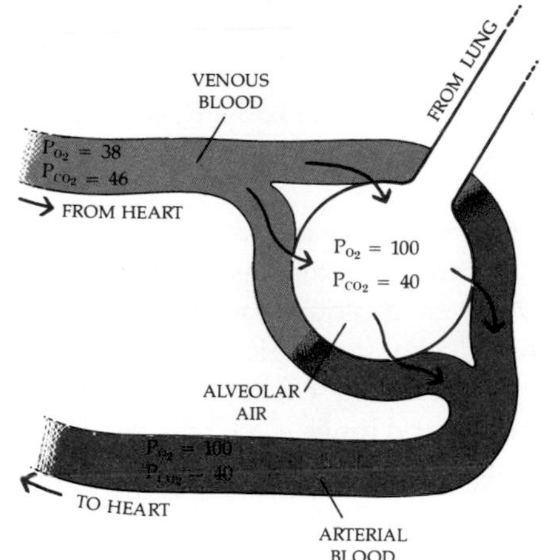

Control of Respiration

The rate and depth of respiration are controlled by respiratory neurons in the brainstem. These neurons, which are in the medulla (page 644), are responsible for normal breathing, which is rhythmic and involuntary, like the beating of the heart. Unlike the beating of the heart, however, which few of us can control voluntarily, breathing may be brought under voluntary control within certain limits.

The respiratory neurons in the brain become active spontaneously. They activate the motor neurons in the spinal cord that cause the diaphragm and intercostal muscles to contract. Periodically the respiratory neurons are inhibited to allow expiration. In addition to their own spontaneous activity, the respiratory neurons receive signals from receptors sensitive to carbon dioxide, oxygen, and hydrogen ions as well as to the degree of stretch of the lungs. Chemoreceptor cells located in the carotid arteries (which supply oxygen to the brain) signal the respiratory neurons when the concentration of oxygen in the blood decreases. These cells also monitor the concentration of dissolved carbon dioxide (carbonic acid), which is simultaneously monitored by centers in the brain. Thus, a number of regulatory mechanisms are at work.

Control of P_{CO_2} is of overriding importance. If the concentration of CO_2 increases only slightly, breathing immediately becomes deeper and faster, permitting more carbon dioxide to leave the blood until the carbon dioxide level has returned to normal. If you deliberately hyperventilate (breathe deeply and rapidly) for a few moments, you will feel faint and dizzy because of the blood's (and, therefore, the brain's) increased alkalinity. You can, as we noted, deliberately increase your breathing rate by contracting and relaxing your chest muscles, but breathing is normally under involuntary control. It is impossible to commit suicide by deliberately holding your breath; as soon as you lose consciousness, the involuntary controls take over once more.

The receptor cells sensitive to oxygen concentration provide a kind of back-up system for the carbon dioxide sensors. In cases of drug poisoning—for example, by morphine or barbiturates—the brainstem cells sensitive to carbon dioxide become depressed. This causes a decrease in the breathing rate, leading ultimately to a

36–16
Location of the carotid body, one of the receptors that monitors the concentration of dissolved oxygen (P_{O_2}) in the blood and also, to a lesser extent, monitors P_{CO_2}. Carbon dioxide concentrations are also measured directly by neurons in the brain.

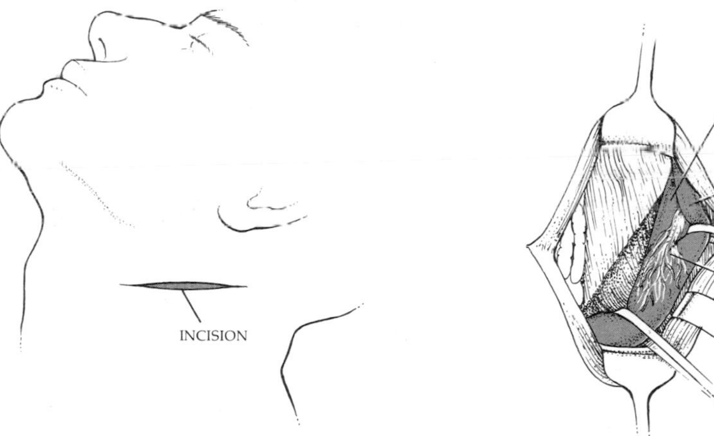

reduction in the oxygen concentration in the blood. The oxygen sensors are then stimulated, and they maintain breathing. Massive overdoses of these drugs, however, depress the activity of the oxygen sensors as well.

SUMMARY

Organisms obtain most of their energy from oxidation of carbon-containing compounds. This process requires oxygen and releases carbon dioxide. Respiration is the means by which an organism obtains the oxygen required by its cells and rids itself of carbon dioxide.

Oxygen is available in both water and air. It enters into cells and body tissues by diffusion, moving from regions of higher partial pressure to regions of lower partial pressure. However, movement of oxygen by diffusion requires a relatively large surface area exposed to the source of oxygen and a short distance over which the oxygen has to diffuse. Selection pressures for increasingly efficient means of gas exchange led to the evolution in vertebrates of gills and lungs. Both gills and lungs present enormously increased surface areas for the exchange of gases. They also have a rich blood supply for transporting these gases to and from other parts of the animal's body. Respiration in large animals involves both diffusion and bulk flow. Bulk flow brings air or water to the lungs or gills and circulates oxygen and carbon dioxide in the bloodstream. Gases are exchanged between the blood and the air in the lungs and between the blood and the tissues by diffusion.

Respiratory pigments increase the oxygen-carrying capacity of the blood. In vertebrates, the respiratory pigment is hemoglobin. Each hemoglobin molecule has four subunits, each of which can combine with one molecule of oxygen. The addition of each molecule of oxygen increases the affinity of the molecule for each subsequent molecule of oxygen. Conversely, the loss of each molecule of oxygen facilitates the loss of the subsequent one.

In humans, air enters the lungs through the trachea, or windpipe, and goes from there into a network of increasingly smaller tubules, the bronchi and bronchioles, which terminate in small air sacs, the alveoli. Gas exchange actually takes place across the alveolar walls. The lungs empty and fill as a result of changes in the pressure within the thoracic cavity, which, in turn, result from changes in the size of the thoracic cavity. The muscles that bring about respiration are under the control of nerves that originate in the spinal cord. These nerves are activated by respiratory neurons in the medulla of the brain that respond to signals caused by very slight changes in the carbon dioxide or hydrogen ion concentration of the blood. They are responsive to larger changes in oxygen concentration.

QUESTIONS

1. Distinguish among the following: gills/lungs; hemoglobin/myoglobin/hemocyanin; trachea/pharynx/larynx; bronchi/bronchioles/alveoli.

2. Discuss the importance of free oxygen in the earth's atmosphere.

3. Why are calculations of partial pressure used in determining the movement of gases between liquids and the atmosphere rather than concentrations? When does a gas move from an area of lower concentration to one of higher concentration?

4. Sketch the human respiratory system. When you have finished, compare your drawing to Figure 36–9.

5. Trace the path of an oxygen atom from the air into a body cell.

6. What are the advantages and disadvantages of obtaining oxygen from air rather than from water? You might be able to think of several of each besides those mentioned in the text.

7. Carbon monoxide (CO) attaches to hemoglobin; the resulting compound does not readily dissociate, can no longer combine with oxygen, and is a brighter red than normal hemoglobin. From these facts, suggest how you might recognize and give assistance to a victim of carbon monoxide inhalation.

8. One of the results of long-term smoking is the loss of bronchial cilia. What effects would you expect this to have on normal lung function?

CHAPTER 37

Digestion

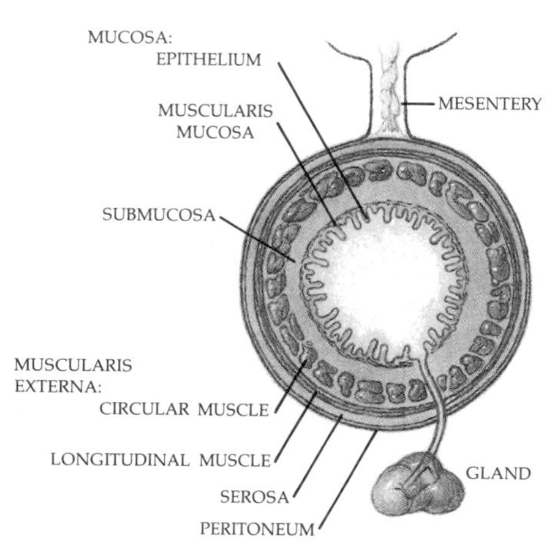

37–1
The layers of the digestive tract include (1) the mucosa, (2) the submucosa, which contains nerves and blood and lymph vessels, (3) the muscularis externa, and (4) the serosa, an outer coating. All of the outer surfaces are covered by peritoneum, a tissue of mesodermal origin. Peritoneum also makes up the mesenteries, which support the various organs, and it lines all of the abdominal cavity. Glands outside the digestive tract, principally the pancreas and liver, discharge digestive enzymes and bile into the tract through various ducts.

Digestion is the process by which heterotrophs break down the tissues of other organisms to molecules that can be used by their own cells. Such molecules include monosaccharides (such as glucose), amino acids, fatty acids, and iron and other inorganic substances. Once within the cells, these molecules can be processed for entry into the Krebs cycle, where they are oxidized and their energy is eventually packaged in the high-energy bonds of ATP. Or, after suitable alterations, they can be synthesized into proteins and other molecules required for cellular structures and activities. Digestion thus provides both the atoms and molecules needed by the individual cells and the cellular fuel for their various activities.

DIGESTIVE TRACT IN VERTEBRATES

In the course of digestion, food substances of various kinds are broken down into molecules that pass into the bloodstream, travel along it, and enter cells. This process takes place in a long, convoluted tube, extending from mouth to anus, known as the digestive tract or, rather less elegantly but just as accurately, as the gut. The surface of the gut is continuous with the surface of the body, and so, technically, the cavity of the gut is exterior to the body. The food molecules actually enter the body only when they pass through the walls of the digestive tract. Therefore, the process of digestion involves two components: breakdown of food molecules and their absorption into the body.

The digestive tube begins with the oral cavity and includes the mouth, pharynx, esophagus, stomach, small intestine, large intestine, and anus. Each of these areas of the gut is specialized for a particular phase in the overall process of digestion, but the fundamental structure of each is similar. The tube, from beginning to end, has four layers: (1) the *mucosa,* which is made up of glandular epithelial tissue, an underlying basement membrane, and connective tissue, with a thin outer coating of smooth muscle in some places; (2) the *submucosa,* which is made up of connective tissue and contains nerves and blood and lymph vessels; (3) the *muscularis externa,* muscle tissue; and (4) the *serosa,* an outer fibrous coating (Figure 37–1).

Along most of the digestive tract, the muscularis is made up of two layers of smooth muscle, the inner layer, in which the orientation is circular, and the outer layer, in which the cells are longitudinally arranged. Coordinated movements of these muscles produce ringlike constrictions that mix the food, as well as the wavelike motions, known as *peristalsis,* that move food along the digestive tract. At several points the circular layer of muscle thickens into heavy bands, called *sphincters.* These sphincters, by relaxing or contracting, control passage of food from one area of the digestive tract to another.

The Oral Cavity

The mechanical breakdown of food begins in the mouth. In humans, as in most other mammals, the first tearing and grinding of food is done by the teeth. (The gizzards of birds and earthworms serve the same function.) Children have 20 teeth, which are gradually lost and replaced by a second set of 32 teeth as the jaw grows larger. Of the 16 adult teeth in each jaw, four are incisors, flat chisel-like structures specialized for cutting; two are canines; four are premolars ("in front of the molars"), each of which has two cusps, or protuberances, and so are also called bicuspids; and six are molars, each of which has four or five cusps. The premolars and molars are used for grinding. Molars do not replace temporary teeth but are added as the jaw grows. The crown of the tooth—the visible part—is covered with enamel, the hardest substance in the body (mostly calcium phosphate); the root, within the gum, is covered with cement, a substance similar to bone. The bulk of the tooth is composed of dentine, another bonelike material, which forms slowly during the life of the tooth. The pulp cavity within the tooth contains the cells that produce dentine and also the nerves and blood vessels.

In mammals, the tongue serves largely to move and manipulate food. Some vertebrates, though, such as the hagfish and lampreys, have tongues equipped with horny "teeth." The tongues of frogs and toads flip out (they are attached at the front, not the back) to catch insects. Mammalian tongues carry the taste buds (Figure 37–3). In humans, the tongue has developed a secondary function of formulating sounds for communication.

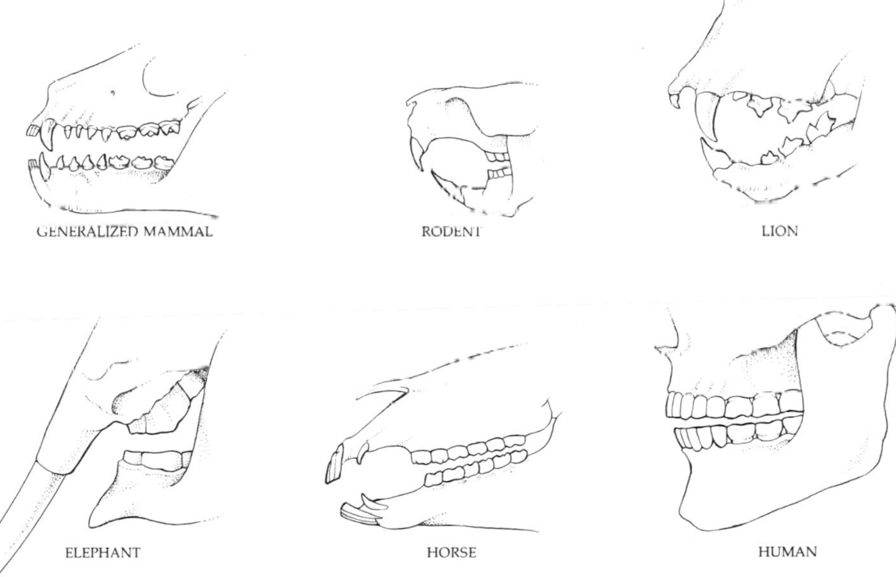

37–2
Teeth of various mammals. Rodents (whose name comes from the Latin rodere, *to gnaw) are characterized by their sharp, chisel-shaped incisors. Most rodents have no canines. Large predatory carnivores, such as the lion, have canines adapted for stabbing and slicing and large molars, which can easily crush the bones of large herbivores. In elephants, tusks are modified incisors; they are used for attack and defense and for rooting food from the ground or breaking branches. (The tusks of other mammals, such as walruses, are modified canines.) The modern horse is a grazing animal; its incisors clip off grasses, which are then ground by the large, flat molars. Humans, with a relatively unspecialized diet, have a correspondingly unspecialized dentition.*

37-3
(a) *The outer pore of a taste bud. The receptor cells of the taste bud are just visible within the pore. This scanning electron micrograph shows the surface of the human tongue. Other taste buds are found on the roof of the mouth, the pharynx, and the larynx.* (b) *Longitudinal section of a taste receptor.*

(a)

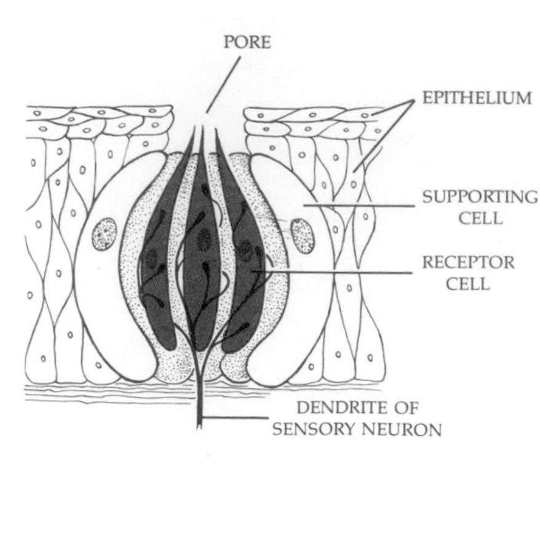

(b)

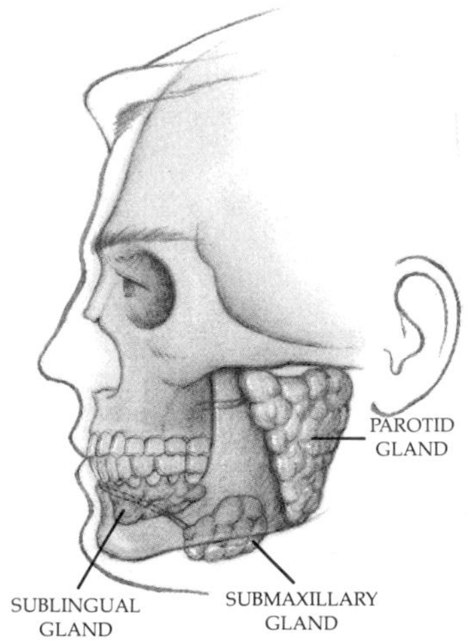

37-4
The bulk of the saliva is produced by three pairs of salivary glands. Additional amounts are supplied by minute glands in the mucous membrane lining the mouth. The parotid glands are the sites of infection of the mumps virus.

While the food is being chewed in the mouth, it is moistened by saliva, a watery secretion produced by three pairs of large salivary glands plus numerous minute glands, the buccal glands, which are located in the mucosa lining the mouth. On the average, we produce 1 to $1\frac{1}{2}$ liters of saliva every 24 hours. The saliva, which also contains mucus, lubricates the food so that it moves down easily. The saliva is slightly alkaline, owing to the presence of sodium bicarbonate. In humans, saliva also contains a digestive enzyme, amylase, which begins the breakdown of starches. (Carnivores, such as dogs, which characteristically tear and gulp their food, have no digestive enzymes in their saliva.) Like all digestive enzymes, amylase works by hydrolysis; that is, the breaking of each bond involves addition of a molecule of water.

The secretion of saliva is controlled by the autonomic nervous system. It can be initiated by reflexes originating in taste buds and in the walls of the mouth, and also by the mere smell or anticipation of food. (Think hard, for a moment, about eating a lemon.) Fear inhibits salivation; at times of great danger or stress, the mouth may become so dry that speech is difficult.

The Pharynx and Esophagus: Swallowing

From the mouth, food is propelled backward toward the pharynx, a passageway at the back of the mouth that connects with the larynx and the esophagus. The pharynx also serves as a resonating chamber in sound production. Swallowing is the passing of food to the esophagus, a muscular tube (about 25 centimeters long in adults), and through the esophagus to the stomach. Swallowing begins as a voluntary action and, once underway in humans, continues involuntarily. In humans, the upper part of the esophagus is striated muscle, but the lower part is smooth muscle. (Dogs, cats, and other animals that gulp their food have striated muscle along the whole length of the esophagus.) Both liquids and solids are propelled along the esophagus by peristalsis.

37–5

Separation of the digestive and respiratory systems in mammals makes it possible for them to breathe while eating. The pharynx, the common passageway of the two systems, is at the back of the mouth and connects with the larynx and the esophagus. (a) *Face and neck, showing parts of the respiratory and digestive systems.* (b) *Swallowing. As the food mass descends, the epiglottis tips downward, and the food mass passes into the esophagus.*

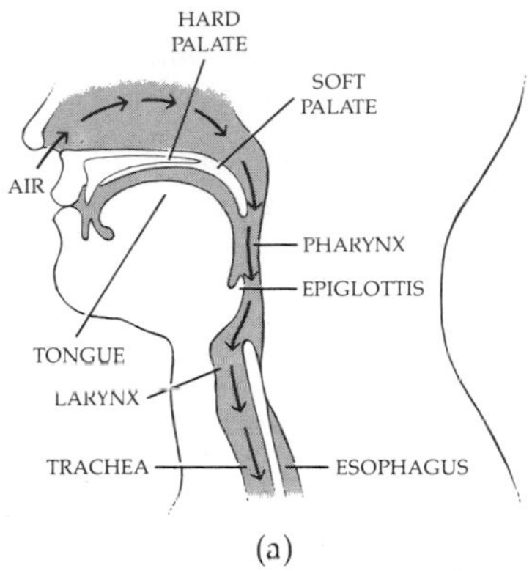

(a)

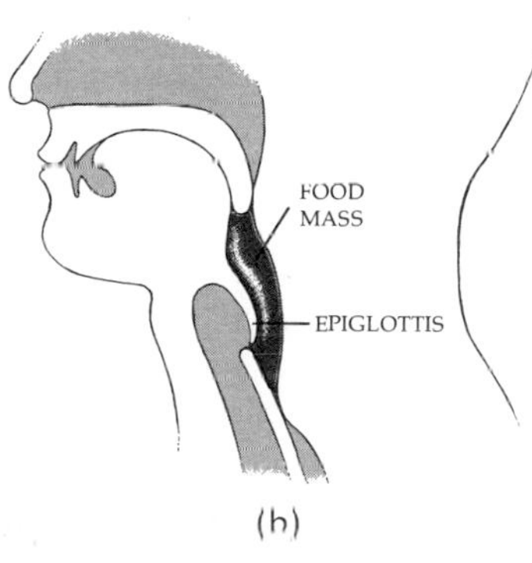

(b)

37–6

The human digestive tract.

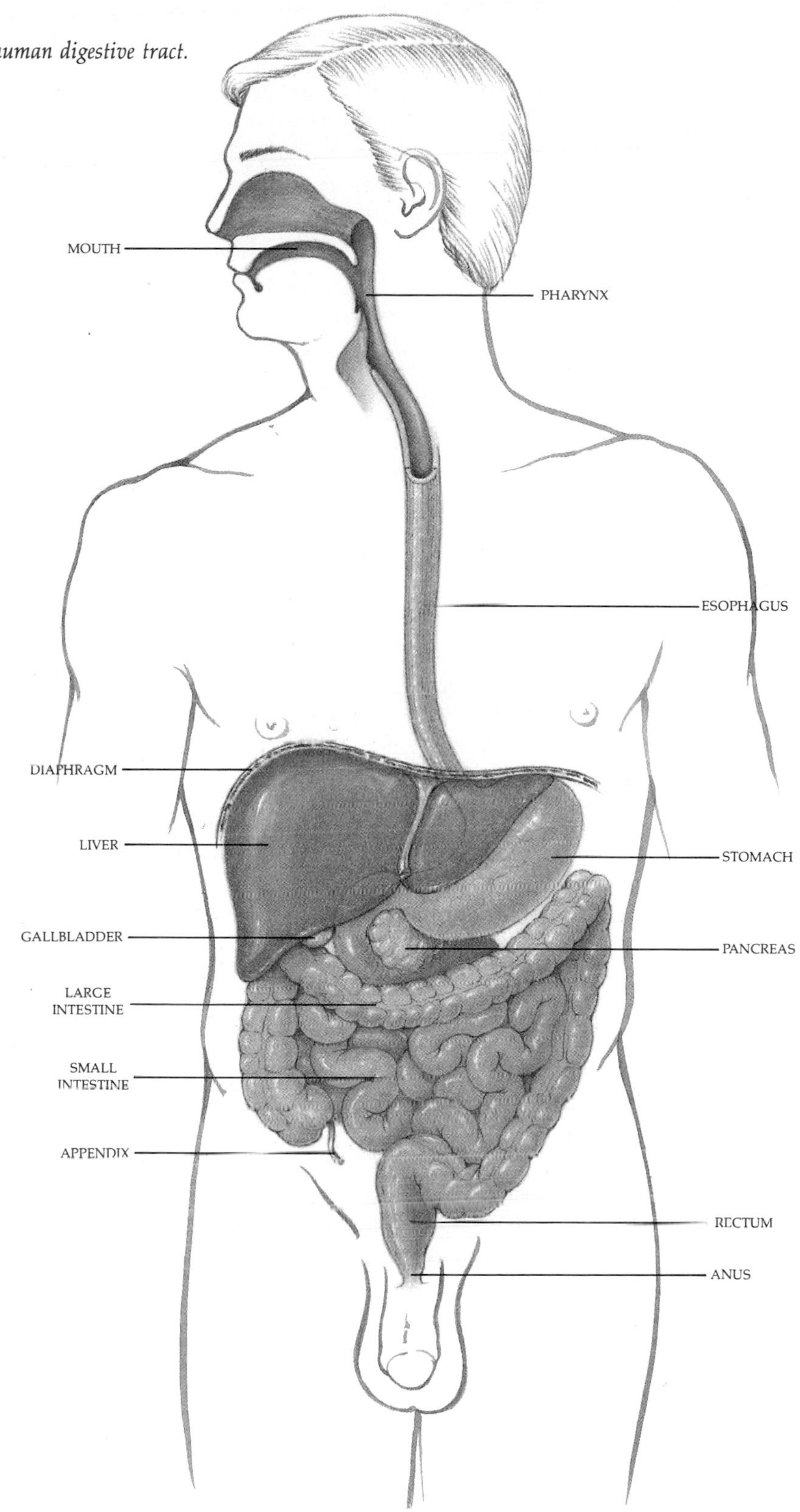

MOUTH
PHARYNX
ESOPHAGUS
DIAPHRAGM
LIVER
STOMACH
GALLBLADDER
PANCREAS
LARGE INTESTINE
SMALL INTESTINE
APPENDIX
RECTUM
ANUS

THE HEIMLICH MANEUVER

Food strangulation claims the lives of almost 4,000 persons a year in this country alone, more than accidents involving firearms or airplanes. It occurs when a food mass enters the trachea rather than the esophagus (Figure 37–5). If the food becomes lodged, the victim cannot speak or breathe and, if the airway is completely blocked, will die in four or five minutes. (The fact that the victim cannot speak helps onlookers to distinguish it from a heart attack; heart attack sufferers, although the symptoms are similar, can talk.)

The food can nearly always be dislodged by the Heimlich maneuver, a procedure so simple that it has been carried out successfully in at least two instances by eight-year-olds. There are three steps: (1) Stand behind the victim and wrap your arms around his or her waist. (2) Grasp your fist with your other hand, and place the fist against the victim's abdomen, slightly above the navel and below the rib cage. (3) Press your fist into the victim's abdomen with a quick upward thrust. The sudden elevation of the diaphragm compresses the lungs and forces air up the trachea, pushing the food out. Repeat several times, if necessary. If the victim is sitting, the procedure can be carried out in the same way. If the victim is lying on his back, face him, and kneel astride his hips. Put the heel of one hand on the abdomen above the navel and below the rib cage, put the other hand on top of the first hand, and press with a quick upward thrust. You can even do it on yourself: Place your hands at your waist, make a fist, and press quickly upward.

The victim should be seen by a physician immediately after the emergency treatment because it is possible to break a rib or cause other internal injuries, especially if the movements are performed incorrectly, but it is well worth the risk considering the alternative.

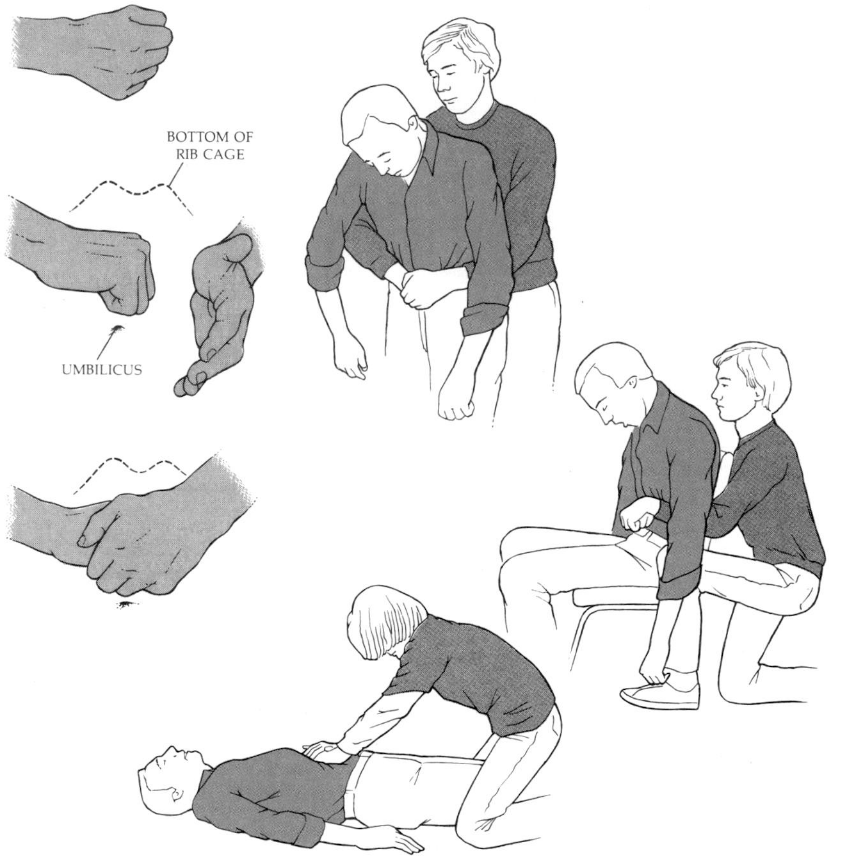

The maneuver can be carried out with the victim in a standing, sitting, or lying position. If the person is larger than you are, it is preferable to have him sitting or lying down. Note how the fist is made (you might want to practice this) and the position of the hands. These are both very important.

The Stomach: Secondary Processing

The esophagus passes through the diaphragm separating the thoracic and abdominal cavities and opens into the stomach, which, with the rest of the digestive organs, lies in the abdomen. The abdominal cavity is completely lined by the *peritoneum,* a thin layer of connective tissue covered by moist epithelium. The stomach, intestines, and other organs are suspended by folds of peritoneum known as *mesenteries* (Figure 37–1). Mesenteries consist of double layers of tissue, with blood vessels, lymph vessels, and nerves lying between the two layers.

The stomach is essentially a collapsible, elastic bag, which, unless it is fully distended, lies in folds. Distended, it holds 2 to 4 liters of food. The mucosal layer is very thick and contains numerous gastric pits (Figure 37–7). Mucus-secreting epithelial cells cover the surface of the stomach and line the gastric pits. Within the pits are the parietal cells, which produce hydrochloric acid, and the chief cells, which produce pepsinogen (see page 87).

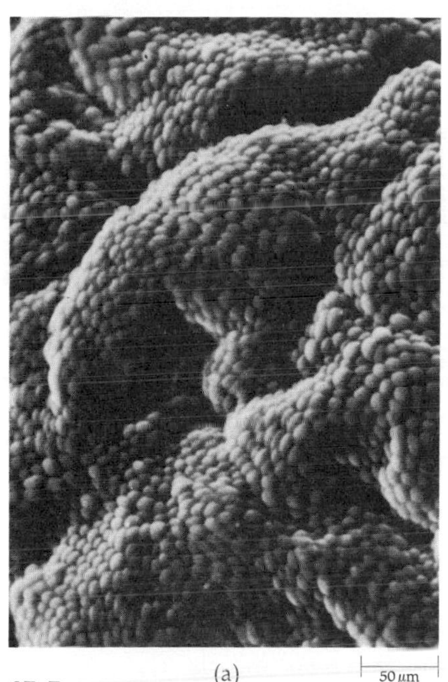

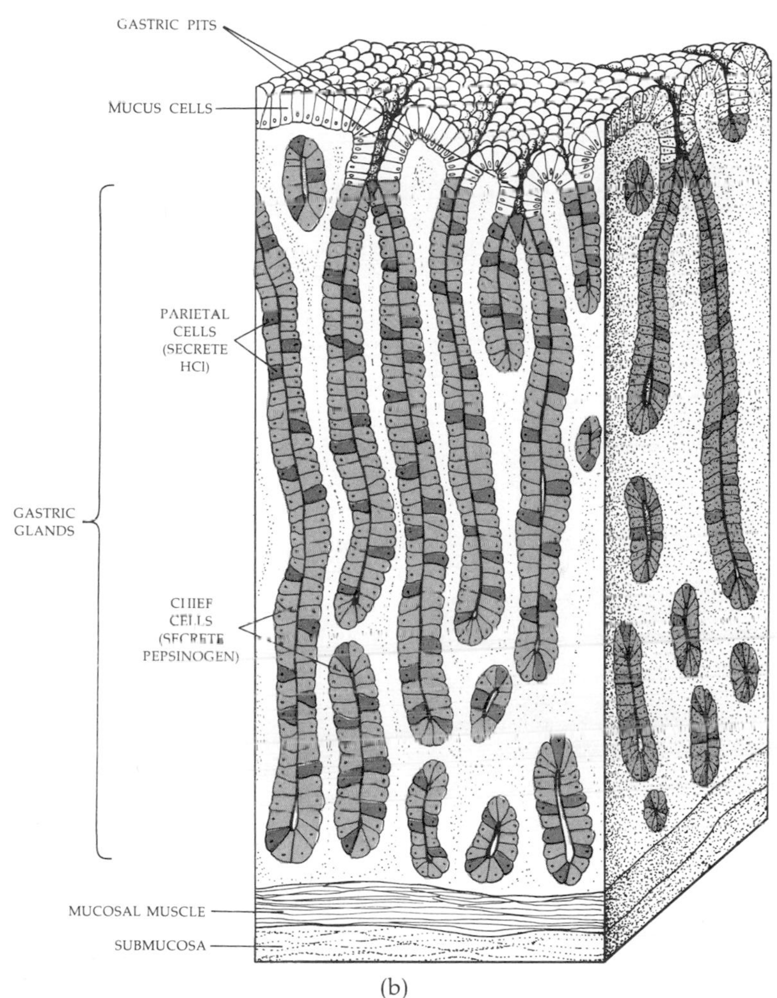

37–7
(a) *Surface of the stomach, as shown in a scanning electron micrograph. The numerous indentations are gastric pits.* (b) *A cross section of stomach mucosa. The parietal cells secrete hydrochloric acid and the chief cells produce pepsinogen. Mucus, secreted by other cells of the epithelium, coats the surface of the stomach and lines the gastric pits, protecting the stomach surface.*

As a consequence of the HCl secretion, the pH of gastric juice is normally between 1.5 and 2.5, far more acidic than any other body fluid. The burning sensation you feel if you vomit is caused by the acidity of gastric juice acting on unprotected membranes. Normally, the mucus in the stomach forms a barrier between the epithelium and the gastric juices and so prevents the stomach from digesting itself. The HCl kills most bacteria and other living cells in the ingested food. It also loosens the tough, fibrous components of tissues and erodes the cementing substances between cells. HCl initiates the conversion of pepsinogen to its active form, pepsin, by splitting off a small portion of the molecule. Once pepsin is formed, it acts on other molecules of pepsinogen to form more pepsin. Pepsin, which breaks proteins down into peptides, is active only at the low pH of the normal stomach.

The stomach is under both neural and hormonal control. Anticipation of food and the presence of food in the mouth stimulate churning movements of the stomach and the production of gastric juices. Fear and anger decrease the stomach's motility. When food reaches the stomach, its presence causes the release of a hormone, gastrin, from gastric cells into the bloodstream. This hormone acts on the cells of the stomach to increase their secretion of gastric juices.

In the stomach, as a result of the action of the mucus, the pepsin, the hydrochloric acid, and the churning motions, food is converted from a solid into a semiliquid mass. The stomach empties gradually through the pyloric sphincter, which separates the stomach and small intestine; the food mass is forced out, little by little, by peristalsis. The stomach is usually empty four hours after ingestion of a meal.

37-8
Microvilli on the surface of an intestinal epithelial cell. These cellular extensions greatly increase the absorptive surface of the intestine. Several mitochondria are visible at the bottom of this electron micrograph.

The Small Intestine: Digestion

In the small intestine, the breakdown of food molecules begun in the mouth and stomach is completed. The food molecules are then absorbed from the digestive tract into the circulatory system of the body, from which they are delivered to the individual cells.

Anatomically, the small intestine is characterized by circular folds in the submucosa; numerous fingerlike projections, *villi*, on the mucosa; and tiny cytoplasmic projections, *microvilli*, on the surface of the individual epithelial cells. All of these structural features increase the surface area of the small intestine. The small intestine is 300 to 400 centimeters (10 to 13 feet) long in the living adult; the total surface area of the human small intestine is about 180 square meters, the area of a singles tennis court. The upper 20 centimeters, known as the duodenum, is the most active in the digestive process; the rest is principally concerned with absorption of food nutrients.

The small intestine contains a variety of digestive enzymes, some produced by the intestinal cells and some by the pancreas (see Table 37-1). The epithelial cells of the mucosa also secrete mucus. The pancreatic enzymes enter the small intestine through the pancreatic duct, about 10 centimeters below the pyloric sphincter.

In the small intestine, amylases continue the breakdown of starch begun in the mouth. Lipases hydrolyze fats into glycerol and fatty acids. Three types of enzymes break down proteins. One group breaks apart the long protein chains. Each enzyme in this group acts only on the bonds linking particular amino acids, so that several enzymes are required to break a single large protein into shorter peptide

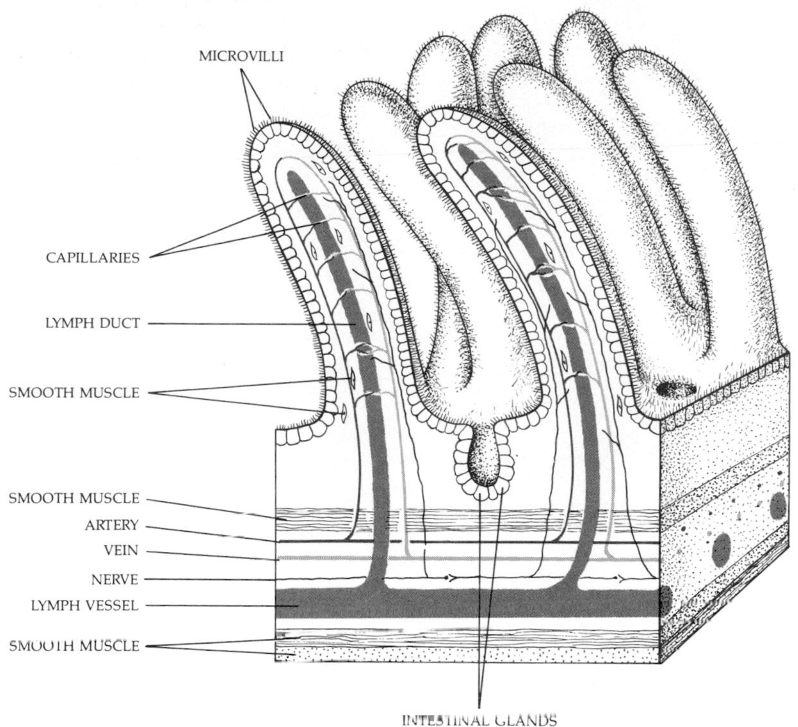

37–9
Longitudinal section of villi of small intestine. Food molecules are absorbed through the walls of the villi and, with the exception of fat molecules, enter the bloodstream by means of the capillaries. Fats, hydrolyzed to fatty acids and glycerol and packaged into chylomicrons, are taken up into the lymphatic system. The villi can move independently of one another; their motion increases after a meal.

Table 37–1 *Principal Digestive Enzymes*

ENZYME	SOURCE	SUBSTRATE	SITE OF ACTION
Salivary amylase	Salivary glands	Starches	Mouth
Pepsin	Stomach mucosa	Polypeptides, pepsinogen	Stomach
Pancreatic amylase	Pancreas	Starches	Small intestine
Lipase	Pancreas	Fats	Small intestine
Trypsin	Pancreas	Polypeptides, chymotrypsinogen	Small intestine
Chymotrypsin	Pancreas	Polypeptides	Small intestine
Carboxypeptidase	Pancreas	Polypeptides	Small intestine
Deoxyribonuclease	Pancreas	DNA	Small intestine
Aminopeptidase	Small intestine	Polypeptides	Small intestine
Dipeptidase	Small intestine	Dipeptides	Small intestine
Maltase	Small intestine	Maltose	Small intestine
Lactase*	Small intestine	Lactose	Small intestine
Sucrase	Small intestine	Sucrose	Small intestine
Enterokinase	Small intestine	Trypsinogen	Small intestine
Phosphatases	Small intestine	Nucleotides	Small intestine

* Often absent in adults, especially those of African origin.

chains. A second type of enzyme acts only on the end of a chain, splitting off dipeptides (pairs of amino acids), some on the amino end and some on the carboxyl end. A third group of enzymes comes into action breaking the remaining dipeptides into single amino acids. The amino acids and dipeptides are absorbed through the epithelial cells of the intestine and finally enter the bloodstream.

In addition to these many different enzymes, the small intestine receives an alkaline fluid from the pancreas, which neutralizes the stomach acid, and bile, which is produced in the liver and stored in the gallbladder. Bile contains a mixture of salts that, like laundry detergents, emulsify fats, breaking them apart into droplets. In this form they can be attacked by the lipases. Bile contains some sodium bicarbonate, which, with the pancreatic fluid, neutralizes the stomach acid. Neutralizing the acidity is essential because the intestinal enzymes are optimally active at a pH of 7 to 8 and would be denatured by the acid pH of the gastric juices that enter the intestine.

Bile also contains pigments that are derived from the breakdown of hemoglobin; the chief pigment is bilirubin. If the liver is not functioning properly, owing to damage, infection, or some other disease process, the bile pigments may not be removed from the blood. Abnormal amounts of bilirubin in blood cause the eyes and skin to become yellow, a condition known as jaundice.

The digestive activities of the small intestine are coordinated and regulated by hormones (Table 37–2). In the presence of food, the duodenum releases secretin, a hormone that stimulates the pancreas and biliary tract to secrete alkaline fluid, and cholecystokinin (also called pancreozymin), a hormone that acts on the pancreas to stimulate the secretion of enzymes and that triggers the emptying of the gallbladder. At least eight other substances are suspected of being gastrointestinal hormones. A complex interplay of stimuli and checks and balances serves to activate and inactivate digestive enzymes and to adjust the chemical environment. In addition to hormonal influences, the intestinal tract is also regulated by the autonomic nervous system. Stimulating the vagus nerve, which carries parasympathetic nerve fibers, causes the pancreas to secrete digestive enzymes.

Table 37–2 *Gastrointestinal Hormones*

HORMONE	SOURCE	STIMULUS FOR PRODUCTION	ACTION
Gastrin	Stomach	Protein-containing food in stomach, also parasympathetic nerves to stomach	Stimulates secretion of gastric juices
Secretin	Duodenum	HCl in duodenum	Stimulates secretion of alkaline bile and pancreatic fluids
Cholecystokinin	Duodenum	Fats and amino acids in duodenum	Stimulates release of pancreatic enzymes and stimulates release of bile from gallbladder

37–10

Portion of pancreatic cell involved in the production of digestive enzymes. Notice the extensive rough endoplasmic reticulum characteristic of cells that synthesize proteins for export. The cell nucleus can be seen on the right-hand side of the micrograph, and, just outside the nuclear envelope, are two mitochondria.

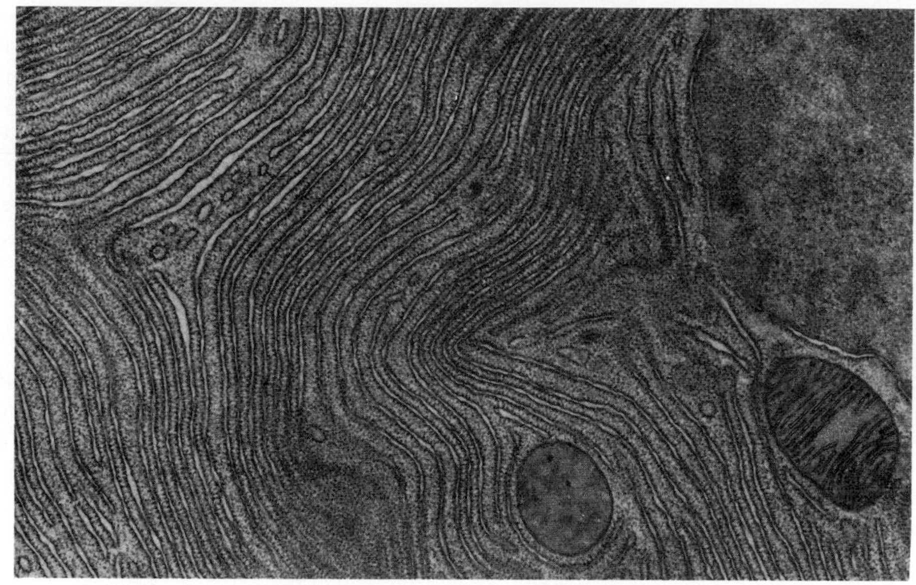

The Pancreas and Liver

The pancreas and the liver are specialized digestive organs, developing, both in the embryo and in the course of evolutionary history, from the digestive tract. The bulk of the tissues of the pancreas resemble salivary gland tissue. In addition, little clusters of cells in the pancreas, known as the islets of Langerhans (named after their discoverer, not some romantic Pacific atoll), secrete insulin and glucagon, which are released into the bloodstream rather than into the gut, as are the digestive enzymes.

The liver carries out an extraordinary variety of functions. It is the source of bile and also of prothrombin and other plasma proteins involved in blood clotting (page 674). The liver breaks down a variety of toxic substances, including alcohol and other drugs, and inactivates a number of hormones, thus playing an important role in hormone regulation. Liver cells remove the nitrogen group from amino acids, converting the nitrogenous waste to urea (page 719), which is excreted by the kidney. They are storage sites for fat-soluble vitamins, such as A, D, and E, and minerals. Moreover, as we shall see on the next page, the liver plays the central role in the regulation of blood glucose.

ABSORPTION

The Small Intestine: Absorption of Food Molecules

The food moves by peristalsis along the small intestine. Here most of the absorption of food molecules takes place. Monosaccharides are rapidly absorbed from the small intestine into the blood vessels of the villi, as are amino acids and dipeptides, the breakdown products of proteins. Fats are hydrolyzed to glycerol and fatty acids; these are taken up by the epithelial cells of the villi and resynthesized to fats.

Proteins present in the absorptive cells coat the fat molecules and so prevent their coalescing. These lipoprotein droplets (called chylomicrons) enter the lymphatic vessels and are transferred to the venous circulation via the thoracic duct. After you consume a meal high in fats, your blood plasma actually turns a milky color owing to the presence of these fat droplets. While they are in the bloodstream, the fat molecules are hydrolyzed once again, this time by a plasma enzyme. The fatty acids and glycerol then enter tissue cells where they are broken down by cellular respiration (see page 168), or they are stored as fat in subcutaneous tissue, mesenteries, muscle, or liver.

The Large Intestine: Absorption of Water and Elimination

The chief function of the large intestine is the resorption of water, sodium, and other minerals. Its epithelial cells secrete mucus, which lubricates the drying mass of the undigested food residue. In the course of digestion, large amounts of water—some 2 or 3 liters—enter the stomach and small intestines by osmosis from the body fluids or as secretions of the glands lining the digestive tract. When the resorption process is interfered with, as in diarrhea, severe dehydration can result. Infants dying from infectious diarrhea, still the chief cause of infant deaths in many countries, die principally as a consequence of water loss.

The large intestine also harbors a considerable population of symbiotic bacteria (including the familiar *Escherichia coli*), which break down additional food substances. Living on these food substances, largely materials we lack the enzymes to digest, they synthesize amino acids and vitamins, which are absorbed into the bloodstream. These bacteria are the chief source of vitamin K.

The appendix is a blind pouch off the large intestine, which may become irritated, inflamed, and then infected. If it ruptures, as a consequence of inflammation and swelling, it spills its bacterial contents into the abdominal cavity. The appendix has no known function in humans.

The pigments that result from hemoglobin breakdown in the liver pass into the intestine, giving the feces their characteristic color. However, the bulk of the fecal matter consists of bacteria (mostly dead cells) and cellulose fibers, along with other indigestible substances. It is stored briefly in the rectum and eliminated through the anus as feces.

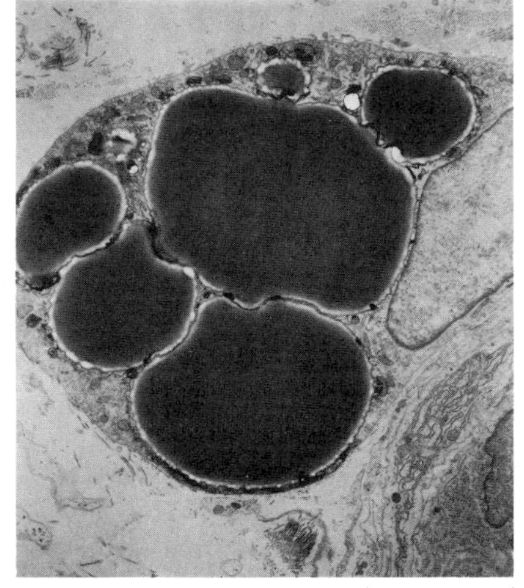

37-11
A fat cell from the skin of a newborn rat. The nucleus is to the right. The large stored droplets of fat that nearly fill the cell have been stained with osmium for electron microscopy. When excess calories are taken in in the diet, fat accumulates in these specialized cells, and when caloric intake is less than sufficient, fat is mobilized, broken down to glycerol and fatty acids, and released into the bloodstream.

REGULATION OF BLOOD GLUCOSE

As we noted at the beginning of this chapter, a major function of digestion is to provide an energy source for each of the individual cells of which the body is composed. Although vertebrates rarely eat 24 hours a day, their blood sugar—the cellular energy supply—remains extraordinarily constant. The liver plays the central role in this critical process. Glucose and other monosaccharides are absorbed into the blood from the intestinal tract and are passed directly to the liver by way of the portal vein. The liver converts most of these monosaccharides to glycogen (the liver stores enough glycogen to satisfy the body's needs for about four hours) and to fat. The fat is stored in the fat cells (fat cells also form fat from glucose). The liver similarly breaks down amino acids (excess amino acids are not stored) and converts them to glucose. The nitrogen from the amino acids is excreted in the form of urea, and the glucose is stored as glycogen.

AIDS TO DIGESTION

Compared to the microscopic photosynthetic cells of the open waters, plant life is large and tough. By logic, one would expect terrestrial herbivores to have sets of enzymes capable of breaking down cellulose and other structural polysaccharides. Not so. Terrestrial animals have evolved another solution: symbiotic relationships with bacteria and protozoa that perform these activities in their behalf, receiving, in return, a free food supply, protection (except from the host itself), and good housing. Digestion by symbionts is the major adaptation of plant-eaters to life on land.

In the warm, wet, airless interior of a mammalian digestive tract, symbionts break down plant products anaerobically. The glucose they use themselves; the host's share is mainly fatty acids, which are absorbed through the host intestine. In addition, the microorganisms synthesize many vitamins, especially those of the B group, which are utilized by the host.

The fermentation process takes place in different areas in the intestinal tract of different species. In the horse family, it occurs in the colon, which is enlarged, and in the cecum, a blind sac at the junction of the small and large intestines. (The cecum persists in humans in a vestigial form as the trouble-making appendix.) Rabbits and other lagomorphs have an enlarged cecum in which bacterial fermentation takes place and also a curious adaptation, peculiar to this group. At night, in their burrows, rabbits produce, from the cecum, a special type of feces, different from those excreted in the daytime, that consists almost entirely of bacteria. These feces they eat, thus recycling their beneficial symbionts.

The most elaborate specialization, anatomically speaking, is found among the ruminants. It is probably not a coincidence that the ruminants are among the most successful terrestrial herbivores, with their rise in evolutionary prominence taking place at the time the grasslands were undergoing rapid expansion. Ruminants hastily shred grass blades and other plant parts and store them, regurgitating them as cud and completing the chewing process at leisure. Rumination is an adaptation that reduced the time the wild herbivore needed to remain exposed on the open plains. The actual fermentation takes place in the rumen, a large esophageal pouch. The contents of the rumen are then passed to the stomach where absorption begins. The bacteria in the rumen not only break down the cellulose and other polysaccharides, releasing fatty acids, but also synthesize proteins, using ammonia and urea as nitrogen sources. Some of these bacteria are also digested by the host, thus providing it with protein.

Ruminant digestion is a major activity. In cattle, the rumen-stomach complex makes up about 15 percent of the weight of the animal. To keep the mixture moist, a cow secretes 60 liters of saliva a day. She also produces about 2 liters of gas a minute, most of which escapes in sweet, chlorophyll-scented belches.

Omnivorous humans have no such symbionts and so are totally unable to utilize cellulose, which is eliminated as roughage. However, like other mammals, we are dependent on the bacteria of our digestive tracts for the synthesis of vitamins, and, as in the rabbit, a major component of our feces is symbiotic bacteria.

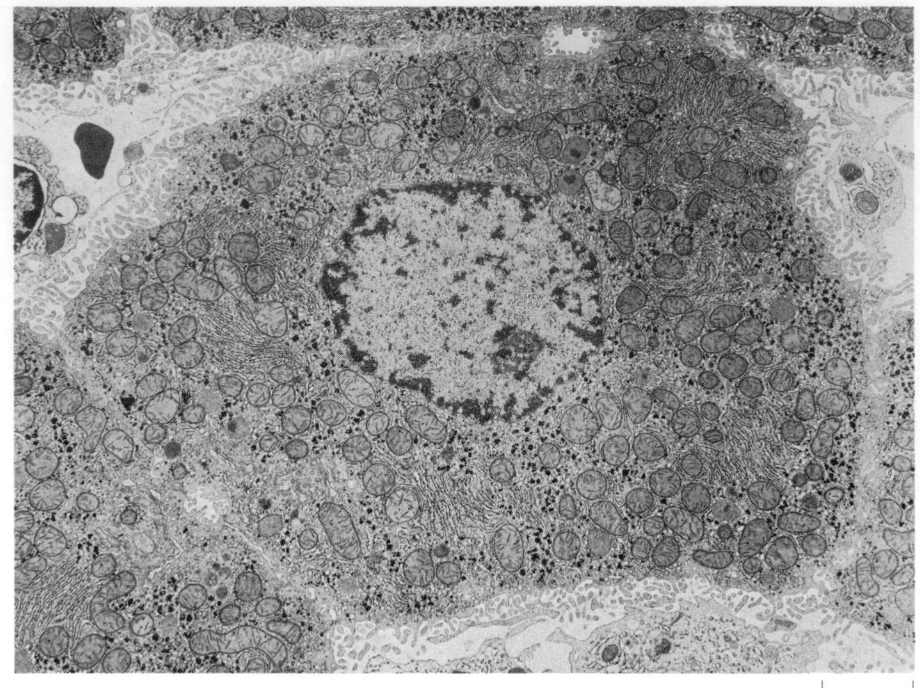

37–12
Electron micrograph of a liver cell. The dark granules are glycogen. Despite the variety of activities carried out by the liver, the cells of the liver all resemble one another in appearance and, apparently, in function. There seems to be no division of labor among cell groups, as there is in the pancreas, for example (page 648). Liver cells, as you would expect, are rich in mitochondria and in rough endoplasmic reticulum.

Whether the liver takes up or releases glucose and the amount it takes up or releases are determined primarily by the concentration of glucose in the blood. The concentration of glucose is, in turn, regulated by a number of hormones and is influenced by the autonomic nervous system (see Figure 34–27, page 655).

Insulin, as we have mentioned previously, stimulates the uptake of glucose by cells, thus decreasing blood glucose. Glucagon, also produced by the pancreas, promotes the breakdown of glycogen, thus increasing blood glucose. Recently, it has been found that the pancreas also synthesizes somatostatin, which lowers the production of both insulin and glucagon. In juvenile diabetics, who suffer from an insulin deficiency, administration of somatostatin, which would inhibit glucagon, may offer a means for controlling the disease. (Somatostatin is also produced by certain cells of the gastrointestinal tract, where it acts to inhibit production of gastrin, secretin, and other gastrointestinal hormones. In addition, it is produced by the hypothalamus, where it inhibits the pituitary release of growth hormone. A picture is now emerging of somatostatin as a hormone with localized, short-lived inhibitory activities, effective against a number of other hormones.)

SOME NUTRITIONAL REQUIREMENTS

Because of the liver's activities in converting various types of food molecules into glucose, the energy requirements of the body can be met by carbohydrates, proteins, or fats—the three principal types of food molecules. Energy requirements are ordinarily met by a combination of the three. Carbohydrates and proteins supply about the same number of calories per gram, and fats about twice as much as either of them.

37–13
(a) *Vitamin D is a steroid that is produced in the skin by the action of ultraviolet rays from the sun on a cholesterol-like compound (the absorbed energy opens the second ring of the molecule). According to one current hypothesis, the ancestors of modern* Homo sapiens *originated in the tropics and were all dark-skinned. In those populations that moved northward, however, the screening effects of the darker pigment, which inhibits the production of vitamin D, caused selection for lighter skin color, whereas no such selection pressures occurred in the sun-drenched areas nearer the equator. Nor did such selection occur among the Eskimos* (b), *who eat a diet rich in fish oils, a major source of vitamin D.*

VITAMIN D

(a)

(b)

In addition to the need for calories, the cells of the body need the 20 amino acids required for assembling proteins. When any one of the amino acids necessary for the synthesis of a particular protein is unavailable, the protein cannot be made and the other amino acids are broken down and excreted. Vertebrates are not able to synthesize all 20 amino acids. Humans (and albino rats) can synthesize 12, either from a simple carbon skeleton or from another amino acid. The other eight, which must be obtained in the diet, are known as essential amino acids.* Plants are the ultimate source of these essential amino acids, but it is difficult (although by no means impossible) to obtain sufficient quantities of them by eating a completely vegetable diet, largely because plant proteins are relatively deficient in lysine and tryptophan.

Mammals also require but cannot synthesize certain polyunsaturated fats that provide fatty acids needed for fat synthesis. These can be obtained by eating plants or insects (or eating other animals that have eaten plants or insects).

Vitamins are an additional group of molecules required by living cells that cannot be synthesized by animal cells. They are characteristically required only in small amounts. Many of them function as coenzymes. Table 37–3 indicates some of the principal vitamins required in the human diet. There is no clear evidence that the ingestion of amounts of any particular vitamin in excess of the amounts available in well-balanced diets has any beneficial effects on a normally healthy individual, and some, including vitamins A, D, and K, are toxic in large dosages.

SUMMARY

Digestion is the process by which food is broken down into molecules that can be taken up by the bloodstream and so distributed to the individual cells of the body. It occurs in successive stages, regulated by an interplay of hormones and nervous stimuli. In mammals food is processed initially in the mouth, where the breakdown of starch begins. It moves through the esophagus to the stomach, where gastric juices destroy bacteria and begin to break down proteins.

Most of the digestion occurs in the duodenum, the upper portion of the small intestine; here, digestive activity, which is performed by enzymes, is almost completely under hormonal regulation. The breakdown of starch by amylases continues, fats are hydrolyzed by lipases, and proteins are reduced to single amino acids. Monosaccharides, amino acids, and dipeptides are absorbed into the blood vessels of the villi; fats are absorbed into the lymphatic vessels. Hormones (secretin and cholecystokinin) are secreted by duodenal cells, and these, in turn, stimulate the functions of the pancreas and the liver. The pancreas releases an alkalizing fluid containing digestive enzymes; the liver produces bile, which is also alkaline.

Water is resorbed from the residue of the food mass as it passes through the large intestine. The large intestine contains symbiotic bacteria, which are the source of certain vitamins. Undigested remains are excreted from the large intestine.

The chief energy source for cells in the mammalian body is glucose circulating in the blood. The organ principally responsible for maintaining a steady supply of

* The essential amino acids are isoleucine, leucine, lysine, methionine, phenylalanine, threonine, tryptophan, and valine.

Table 37-3 *Vitamins*

DESIGNATION: LETTER AND NAME	MAJOR SOURCES	FUNCTION	DEFICIENCY SYMPTOMS
A, carotene	Egg yolk, green or yellow vegetables, fruits, liver, butter	Formation of visual pigments, maintenance of normal epithelial structure	Night blindness; dry, flaky skin
B-complex vitamins:			
B_1, thiamine	Brain, liver, kidney, heart, whole grains	Formation of cofactor involved in Krebs cycle	Beri-beri, neuritis, heart failure, mental disturbance
B_2, riboflavin	Milk, eggs, liver, whole grains	Part of electron carrier FAD	Photophobia, fissuring of skin
B_6, pyridoxine	Whole grains, liver, kidney, fish, yeast	Coenzyme for amino acid metabolism and fatty acid metabolism	Dermatitis, nervous disorder
B_{12}, cyanocobalamin	Liver, kidney, brain	Maturation of erythrocytes, coenzyme in amino acid metabolism	Anemia, malformed erythrocytes
Biotin	Egg yolk, synthesis by intestinal bacteria	Concerned with fatty acid synthesis, CO_2 fixation, and transamination	Scaly dermatitis, muscle pains, weakness
Folic acid	Liver, leafy vegetables	Nucleic acid synthesis, formation of erythrocytes	Failure of erythrocytes to mature, anemia
B_3, niacin (nicotinic acid)	Whole grains, liver and other meats, yeast	Part of electron carriers NAD, NADP, and of CoA	Pellagra, skin lesions, digestive disturbances
Pantothenic acid	Present in most foods	Forms part of CoA	Neuromotor disorders, cardiovascular disorders, GI distress
C, ascorbic acid	Citrus fruits, tomatoes, green leafy vegetables	Vital to collagen and ground (intercellular) substance	Scurvy, failure to form connective tissue fibers
D_3, calciferol	Fish oils, liver, milk and other dairy products, action of sunlight on lipids in the skin	Increases Ca^{2+} absorption from gut, important in bone and tooth formation	Rickets (defective bone formation)
E, tocopherol	Green leafy vegetables	Maintains resistance of red cells to hemolysis, cofactor in electron transport chain	Increased red blood cell fragility
K, naphthoquinone	Synthesis by intestinal bacteria, leafy vegetables	Enables synthesis of clotting factors by liver	Failure of blood coagulation

glucose is the liver, which stores glucose (in the form of glycogen) when glucose levels in the blood are high and breaks down glycogen, amino acids, or fats, releasing glucose, when the levels drop. These activities of the liver are regulated by at least five different hormones.

Requirements for good nutrition include calories (which can be obtained from carbohydrates, fats, or proteins), essential amino acids, certain essential fatty acids, vitamins, and minerals.

QUESTIONS

1. Distinguish, in terms of function, between: gastrointestinal hormones/gastrointestinal enzymes; vitamins/essential amino acids; digestion/absorption; rumen/cecum; villi/microvilli; cholecystokinin/chylomicron.

2. Diagram the human digestive tract, including all the major organs. Check your drawing against Figure 37–6.

3. Trace the chemical process of a hamburger on a bun through your digestive tract.

4. If you oxidize a pound of fat, whether butter or the fat in your own body cells, about 3,500 kilocalories are released. Suppose that you go on a weight-reducing diet, limiting yourself to 1,000 kilocalories a day. You spend most of your time sitting in a well-heated library (studying, of course) and so you expend only about 3,000 kilocalories a day. How many pounds will you lose each week?

5. What effects would you expect anger at the beginning of a meal—or during a meal—to have on your digestive processes?

6. Humans are omnivores rather than strict herbivores or carnivores or specialists on a single organism as a food source. Humans have an unusually large number of substances specifically required in their diets. How might these two observations be related? Can you make an educated guess as to which way cause-and-effect operates in this case?

CHAPTER 38

Some Homeostatic Mechanisms

One of the identifying properties of living systems is that they are homeostatic, meaning simply that they are able to maintain a relatively constant internal environment despite constant exchanges with and changes in the external environment. This capacity exists even in the simplest living organisms—a bacterial cell, for example, has a chemical composition quite distinct from the medium in which it lives.

In animals, many activities contribute toward homeostasis. We have seen numerous examples in the previous chapters—the homeostatic regulation of blood sugar, for example, or the fine-tuning of hormone levels by feedback systems. In this chapter, we are going to examine three particularly noteworthy homeostatic functions: regulating blood chemistry, maintaining water balance, and controlling temperature.

REGULATION OF BLOOD CHEMISTRY

Animals are about 60 percent water. About two-thirds of this water is within the cells, and one-third, the extracellular fluid, surrounds, bathes, and nourishes the cells (Figure 38–1). Thus this extracellular fluid serves the same purpose for the body cells that the pre-Cambrian seas served for the earliest cells that arose in them. As they became multicellular in the course of evolution, animals began to produce their own extracellular fluid, similar in composition to the salty fluid of the sea. As they did so, they also evolved mechanisms for regulating its composition.

In many invertebrates and all vertebrates, the internal chemical environment is to a large extent regulated by a special organ system. Among terrestrial vertebrates, the most important components of this system are the kidneys. It is possible to relate major advances in vertebrate evolution—such as the transition to land—to increasing efficiency of kidney function. In fact, one can trace vertebrate evolution in terms of the evolution of the kidney, as Homer Smith did in his classic book *From Fish to Philosopher.*

THE KIDNEY

The kidneys can be viewed as excretory organs. The two chief metabolic wastes that cells release into the bloodstream are carbon dioxide and the nitrogen com-

38-1
The fluid compartments of the human body, showing the main routes of exchange among them.

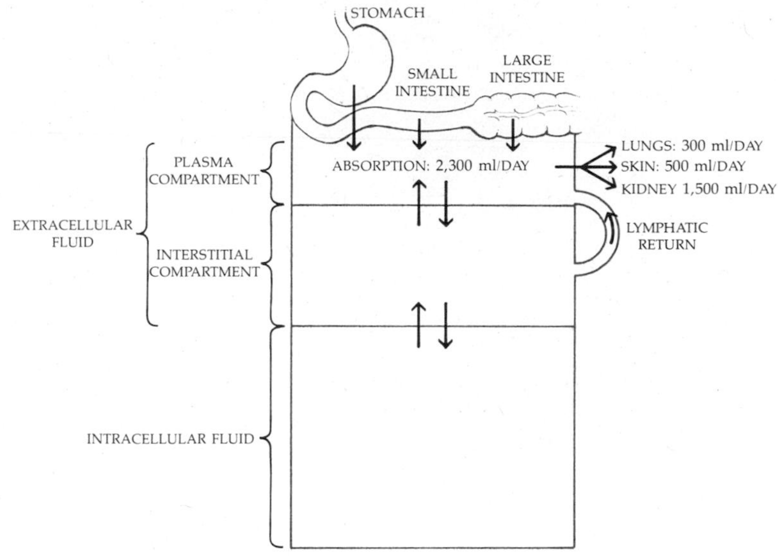

$$NH_2-\underset{COOH}{\overset{CH_3}{C}}-H + \tfrac{1}{2}O_2 \longrightarrow \underset{COOH}{\overset{CH_3}{C}}=O + NH_3$$

ALANINE — PYRUVIC ACID

(a)

$$NH_2-\underset{\|\atop O}{C}-NH_2$$

UREA

(b)

38-2
(a) Deamination—the removal of the amino group—is the first step in the breakdown of amino acids. The products of the reaction are ammonia and a carbon skeleton, which can be broken down for energy or converted to sugar or starch. (b) Urea, the principal form in which nitrogen is excreted in most mammals. It is formed in the liver by the combination of two molecules of ammonia with one of carbon dioxide through a complex series of energy-requiring reactions. What would be the other product of this reaction?

pounds, such as ammonia, produced by the breakdown of amino acids. Carbon dioxide is eliminated from the lungs, or it diffuses out into water through the skin or the gills. Animals that have unlimited access to water most often excrete nitrogenous wastes in the form of ammonia through the body surface, the gills, or the kidneys. In land-dwelling organisms, which do not have this unlimited water supply, nitrogenous wastes must be converted to some other form since ammonia is highly toxic, even in low concentrations.

Many birds, terrestrial reptiles, and insects eliminate waste nitrogen in the form of uric acid or uric acid salts (urates), which can be excreted as crystals. In birds, the uric acid and urates are mixed with the undigested wastes in the cloaca (the common exit chamber for the digestive and urinary tracts), and the combination is dropped as a semisolid paste, familiar to frequenters of public parks and admirers of outdoor statuary. This nitrogen-laden substance forms a rich natural fertilizer. Guano, the excreta of seabirds, accumulates in such quantities on the small islands where the birds gather in great numbers that at one time it was harvested commercially. Mammals excrete nitrogenous wastes mostly in the form of *urea,* which is produced by deamination of amino acids in the liver (Figure 38-2).

In a broader view, however, kidneys are also regulatory organs. Excretion is highly selective. For instance, although most of the urea in the blood that enters a normal mammalian kidney is usually excreted, almost all the amino acids are retained. Glucose is not excreted unless it is present in high amounts, as in diabetes mellitus. The presence of sugar in the urine is, in fact, a basis for the diagnosis of this form of diabetes and the evaluation of its treatment. Chemical regulation also involves maintaining closely controlled concentrations of ions such as Na^+, K^+, H^+, Mg^{2+}, Ca^{2+}, Cl^-, and HCO_3^-. These ions are important for their specific roles in various chemical processes, such as maintenance of protein structure, membrane permeability, action potentials, and muscle contraction. Some are important in the regulation of the pH of the blood, which is also monitored by the kidneys.

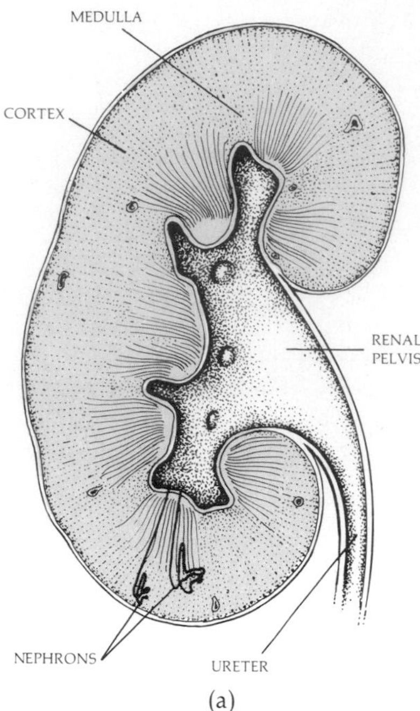

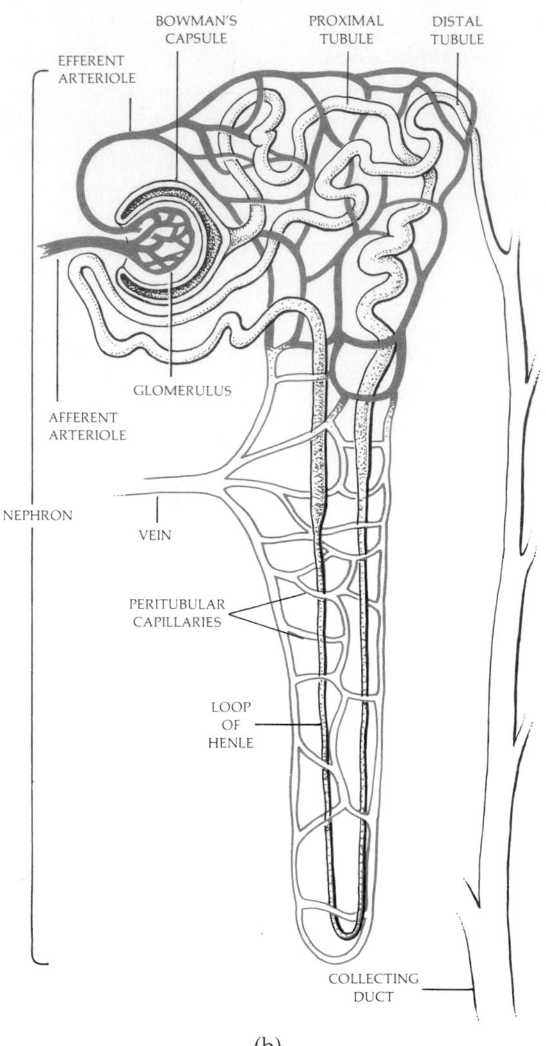

38–3

(a) *In longitudinal section, the human kidney is seen to be made up of an outer region, the cortex, which contains the fluid-filtering mechanisms, and an inner region, the medulla, through which collecting ducts carrying the urine merge and empty into the funnel-shaped renal pelvis, which enters into the ureter.* (b) *The nephron is the functional unit of the kidney. Blood enters the nephron through the afferent arteriole leading into the glomerulus. Fluid is forced out by the pressure of the blood through the thin capillary walls of the glomerulus into Bowman's capsule. The capsule connects with the long renal tubule, which loops down into the medulla and up again. As the fluid travels through the tubule, almost all the water, ions, and other useful substances are resorbed into the bloodstream through the capillaries surrounding the tubule. Other substances are secreted from the capillaries into the tubules. Waste materials and some water pass along the entire length of the tubules into the collecting duct and are excreted from the body.*

Structure of the Kidney

In humans and other mammals, the functional unit of the kidney is the *nephron.* It consists of a cluster of capillaries known as the *glomerulus,* a bulb called *Bowman's capsule,* and a long, narrow tube, the *renal tubule* (Figure 38–3). Each of the two human kidneys contains about a million nephrons with a total length of some 80 kilometers (50 miles) in an adult male.

Urine is formed in the nephrons and passed into the renal pelvis, which is, in essence, a funnel. From this funnel, the urine trickles continuously through the ureters to the bladder, which stores the urine until it is excreted through the urethra. (The urinary system is shown in Figure 31–14, on page 579.)

Function of the Kidney

Blood enters the kidney through the renal artery, which divides into progressively smaller branches, the arterioles, which each lead, in turn, to a glomerulus. Unlike other capillary beds, a glomerulus lies between two arterioles. The efferent arteriole—the one leading out—then divides again into capillaries, the peritubular ("around-the-tubes") capillaries, that then merge to form the renal vein.

The efferent vessel—the one leading out of the glomerulus—constricts, keeping the blood within the glomerulus at a pressure about twice that in other capillaries. As a consequence, fluid is forced out of the glomerular capillaries into Bowman's capsule. About one-fifth of the blood plasma that enters the kidney is thus forced into the nephron through the capillary walls of the glomerulus. This crucial first stage of kidney function is called *filtration*, and the fluid entering the capsule is referred to as the *filtrate* (Figure 38–4). Except for the absence of large molecules, the filtrate has the same chemical composition as the plasma.

The second stage in formation of the urine is secretion, in which molecules left in the plasma are selectively removed and actively secreted by the cells of the tubular walls into the filtrate. Penicillin, for example, is removed from the circulation in this way.

Another major step that occurs simultaneously with secretion is resorption. During the passage through the renal tubule, most of the water and solutes that entered the tubule in the first place are transported back into the bloodstream by the cells of the tubular walls through the capillary walls (Figure 38–5). Secretion and resorption of most solutes take place by active transport. As a consequence, the kidney has a high energy requirement, higher on a per-gram basis than even the heart.

Finally, the remaining fluid leaves the nephron; it is now the urine. You will note that excretion by the kidney is quite different from elimination by the intestinal tract. In the latter, the residue that is eliminated consists almost entirely of material that never entered the body.

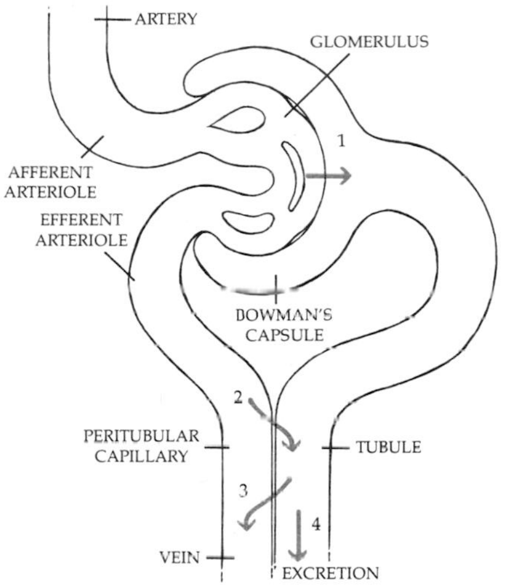

38–4
The four basic components of renal function: (1) filtration through the glomerular capillaries into Bowman's capsule; (2) secretion from the peritubular capillary into the tubule; (3) resorption from the tubule to the peritubular capillary; (4) excretion. The pressure in the glomerular capillaries is regulated by constriction of the efferent arteriole.

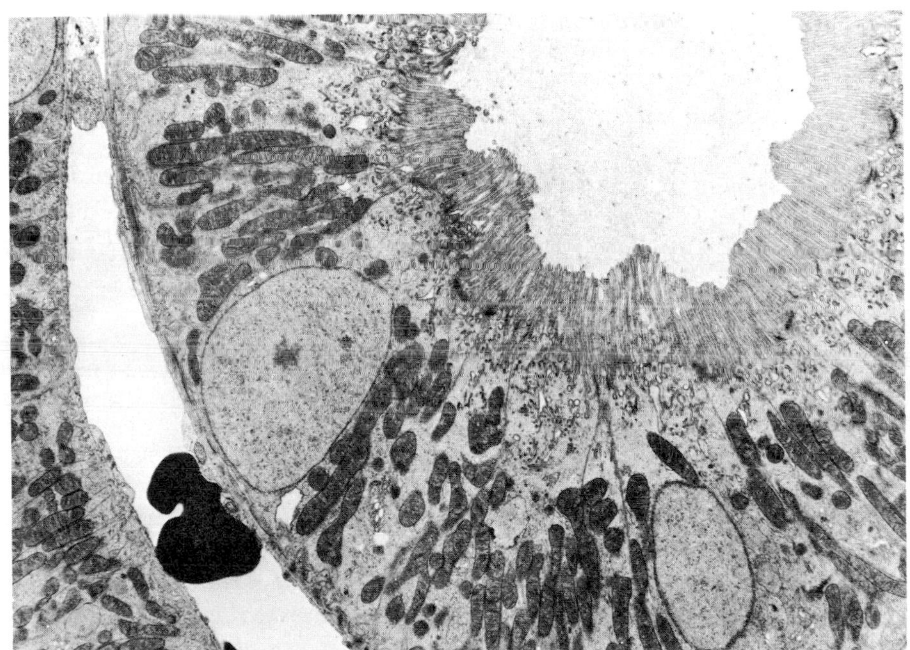

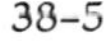
38–5
An electron micrograph of the cells of the proximal tubule of the kidney. The walls of renal tubules are made up of only a single layer of cells. Notice the many mitochondria and the brush border on the surface of the cells bordering the lumen of the tubule (the light area at the upper right of the micrograph). This border is made up of many very fine, hairlike extensions that, like microvilli, greatly increase the absorptive capacity of the cell. A dark, irregularly shaped red blood cell can be seen in the lumen of the peritubular capillary, on the left-hand side of the micrograph.

38–6
An American black bear ready for hibernation. During its three to five months of profound winter sleep, the black bear does not defecate, urinate, or require food or water. Investigators at the Mayo Clinic have shown that, contrary to their original expectations, protein synthesis increases in the dormant bears. The bears have some mechanism for recycling amino acids that conserves proteins, eliminates the production of urea, and probably contributes to body heat. Knowledge of such a mechanism might help in the management of patients with kidney failure.

Aldosterone and ADH

In mammals, two hormones act on the nephron to affect the composition of the urine. One of these is aldosterone, which is produced by the adrenal cortex. Aldosterone stimulates resorption of sodium from the distal tubule. When the adrenal glands are removed or when they function poorly (as in Addison's disease), sodium chloride and water are lost in the urine, and the tissues of the body become depleted of them. Generalized weakness results, and if a patient with Addison's disease is not given hormone replacement therapy, the fluid loss can eventually be fatal. Aldosterone production is controlled by a complex feedback circuit involving, as you would expect, sodium levels in the bloodstream.

The second hormone, ADH (antidiuretic hormone), is formed in the hypothalamus and secreted by the posterior pituitary. It affects the amount of water excreted in the urine. ADH acts on the membranes of the collecting ducts of the nephrons and increases their permeability to water, so more water moves, by diffusion, back into the blood from the nephron. The amount of ADH released depends on the osmotic concentration of the blood and also on the blood pressure. Osmotic receptors that monitor the solute content of the blood are located in the hypothalamus and thus influence ADH release directly. Pressure receptors that detect changes in blood volume are found in the walls of the heart, in the aorta, and in the carotid arteries. Stimuli received by these receptors are transmitted to the hypothalamus. Factors that increase the concentration of solutes in the blood or decrease blood pressure, or both, stimulate the production of ADH and conservation of water in the body. Such factors include dehydration and hemorrhage. Factors that decrease blood concentration, such as the ingestion of large amounts of water, or that increase blood pressure—epinephrine, for instance—signal the hypothalamus to stop producing ADH, and so more water is excreted. Cold stress inhibits ADH secretion and so increases urinary flow. Alcohol also suppresses ADH secretion and increases urinary flow, a phenomenon familiar to imbibers of beer and other alcoholic beverages. Pain and emotional stress trigger ADH secretion and thus decrease urinary flow.

The ways in which the cells of the walls of the kidney tubules act to select and reject particular molecules and the ways in which hormones alter this selection and rejection process are a part of the larger, and largely unsolved, problem of how substances are transported across cell membranes.

WATER BALANCE

AN EVOLUTIONARY PERSPECTIVE

The earliest organisms probably had a salt and mineral composition much like that of the environment in which they lived. The early organisms and their surroundings were probably also isotonic; that is, each had the same total effective concentration of dissolved substances, so that there was no net movement of water into or out of the cell. When organisms moved to fresh water (a hypotonic—less concentrated—environment), they had to develop systems for "bailing themselves out," since fresh water tended to move into their bodies by osmosis. The contractile vacuole of *Paramecium* is an example of such a bailing device (see page 129).

38–7
Some marine animals, such as the turtle, have special glands in their heads that can excrete sodium chloride at a concentration about twice that of sea water. Since ancient times, turtle watchers have reported that these great armored reptiles come ashore to lay their eggs with tears in their eyes, but it is only recently that biologists have learned that this is not caused by an excess of sentiment—as is the case with Lewis Carroll's mock turtle—but is, rather, a useful solution to the problem of excess salt. Marine birds have similar salt-secreting glands.

If the vertebrates evolved in fresh water, as is generally believed, the first function of the kidneys, phylogenetically speaking, was probably to pump water out and to keep salt and other desirable solutes, such as glucose, in. In freshwater fish, the kidney works primarily as a filter and resorber, and the urine is hypotonic—that is, it has a concentration of solutes lower than that of body fluids.

Saltwater fish have a different problem. Their body fluids are generally less concentrated than their environment, and so they tend to lose water by osmosis. Their need is to conserve water and thereby keep their body fluids from becoming too concentrated. This problem has been solved in different ways by different groups of fish. In hagfish, for example, body fluids are about as salty as the salt waters of the surrounding ocean and so are isotonic with them.

Cartilaginous fish such as the shark are also isotonic to sea water, but they achieve their isotonicity in a different way. In the course of evolution, they developed an unusual tolerance for urea—the breakdown product of nitrogen-containing molecules—so instead of constantly excreting it, as do most other fish, they retain a high concentration of it in the blood. This high concentration of urea makes their body fluids almost isotonic in relation to sea water; hence, they do not tend to lose water by osmosis.

The bony fish, the most abundant group of marine fish, have body fluids that are hypotonic in relation to the environment, having an osmotic pressure only about one-third that of sea water. Thus, like terrestrial animals, they have the potential problem of losing so much water to their environment that the solutes in the body fluids could become too concentrated and the cells could die. In bony marine fish, this problem has been solved by the evolution of special gland cells in the gills that excrete excess salt. Hence these fish can drink salt water and still remain hypotonic. (Freshwater fish, conversely, have salt-absorbing cells in their gills.)

Since terrestrial animals do not always have automatic access to either fresh or salt water, they must regulate water content in other ways, balancing off gains and expenditures.

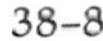

SOURCES OF WATER GAIN AND LOSS

Terrestrial animals gain water by drinking fluids, by eating water-containing foods, and as an end product of the oxidative processes that take place in the mitochondria, as we saw in Chapter 8. When 1 gram of glucose is oxidized, 0.6 gram of water is formed. When 1 gram of protein is oxidized, only about 0.3 gram of water is produced. Oxidation of 1 gram of fat, however, produces 1.1 grams of water

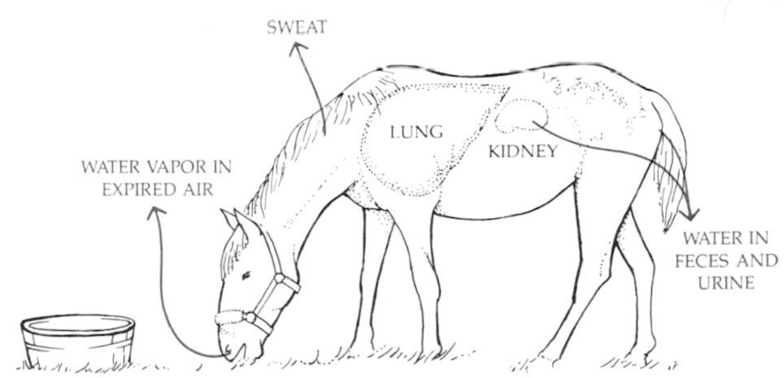

38–8
A mammal is in water balance when the total amount of water lost in expired air, evaporation from skin, and in urine and feces equals the total amount of water gained by the intake of food and fluids and by the oxidation of food molecules.

38–9
The kangaroo rat, a common inhabitant of the American desert, may spend its entire life without liquid water. It lives on seeds and other dry plant materials, and in the laboratory, it can be kept for months on a diet of only fatty seeds such as barley or rolled oats. Analysis shows the kangaroo rat is highly conservative in its water expenditures. It has no sweat glands, and being nocturnal, it searches for food only when the external temperature is relatively cool. Its feces have a very small water content, and its urine is highly concentrated. Its major water loss is through respiration, and even this loss is reduced by the animal's long nose in which some cooling of the expired air takes place, with condensation of water from it.

because of the high hydrogen concentration in fat (the extra oxygen comes from the air).

Some animals can derive all their water from food and do not require fluids. The kangaroo rat of the American desert, for example, can spend its entire life without drinking water if it eats the right type of food. It is not surprising that it selects a diet of fatty seeds. If it is fed high-protein seeds, such as soybeans—the oxidation of which produces a large amount of nitrogen waste and a relatively small amount of water—it will die of thirst unless some other source of water is available.

On the average, a human takes in about 2.3 liters of water a day in food and drink and gains an additional 200 milliliters a day by oxidation of food molecules. Water is removed from the blood and excreted as urine, is lost from the lungs in the form of the moisture in exhaled air, is eliminated in the feces, and is lost by evaporation from the skin.

Excretion of Water

In humans, urine is usually the major route of water loss. In a normal human adult, the rate of water excretion in urine averages 1.2 liters a day. Although the actual amount of urine produced may vary from a few hundred milliliters a day to several thousand, there will be a variation of less than 1 percent in the fluid content of the body. However, a minimum output of 500 milliliters of water is necessary for health since this much water is needed to remove potentially toxic waste products.

WATER CONSERVATION: THE LOOP OF HENLE

The control of urinary water loss is a major mechanism by which body water is regulated. Animals with free access to fresh water typically excrete the wastes from their bloodstream in a copious urine that is hypotonic in relation to their blood. The daily urinary output of a frog, for instance, totals 25 percent of its body weight. Most terrestrial animals cannot afford to be so profligate with water, however. In response to evolutionary pressures, birds and mammals have developed the ability to excrete hypertonic urines—that is, urines that are more concentrated than their body fluids. This ability is associated with a hairpin-shaped section of the nephron known as the *loop of Henle.* As is the case with the renal glomerulus, structure is the key to function. By sampling the fluids in and around the nephron and analyzing the ions present in each sample, physiologists have been able to determine how such a deceptively simple-looking structure makes this function possible.

As shown in the diagram (Figure 38–10), the filtrate first enters the proximal (near) convoluted tubule, descends the loop of Henle, ascends it, and then passes through the distal (far) convoluted tubule into the collecting duct.

The fluid entering the proximal tubule is isotonic with the blood plasma; that is, it has the same solute concentration as does the blood plasma. As the urine descends the loop of Henle, it becomes increasingly concentrated. In the ascending branch, it becomes less and less concentrated, and as it enters the distal tubule, it is hypotonic. By the end of the distal tubule, it is once more isotonic or still hypotonic in relation to the blood. In birds and mammals needing to conserve water, the fluid descending the collecting duct becomes increasingly concentrated, and when it is excreted, it is distinctly hypertonic in relation to the circulating blood. (Note that ADH alone, because of its mechanism of action, could not bring about the production of a hypertonic urine, only an isotonic one.)

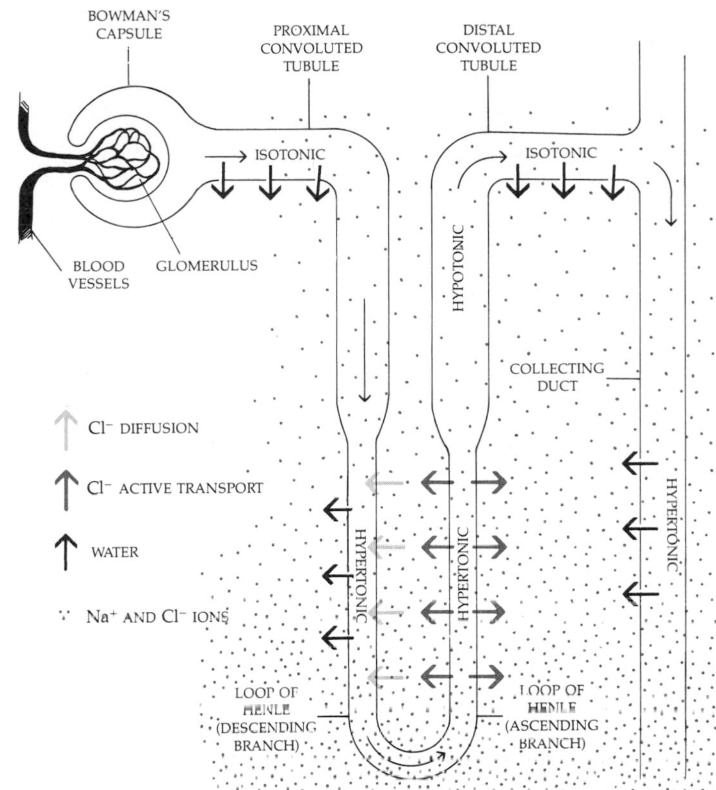

38–10
The formation of hypertonic urine in the human nephron. As the urine passes up the ascending branch of the loop of Henle, the cells lining the tubule pump out chloride (Cl^-) into the surrounding fluid. The chloride ions diffuse passively into the descending branch and are recirculated in the urine to the ascending branch, where they are pumped out again. Sodium ions (Na^+) follow passively, maintaining the balance of positive and negative charges. As a consequence of this recirculation of chloride, the loop of Henle and the lower section of the collecting duct are constantly bathed in a salty fluid. The cells of the ascending loop are impermeable to water, so it does not diffuse out when the chloride is pumped out. The membrane of the collecting duct, however, is permeable to water if ADH is present, so water diffuses out into the surrounding salty fluid, and a hypertonic urine is passed down the duct to the renal pelvis, the ureter, the bladder, and finally out the urethra.

This is how these findings are explained: Water diffuses freely through the cells lining the walls of the nephron, with the exception of the ascending branch of the loop of Henle, and, in the absence of ADH, the collecting duct. Chloride ions are actively pumped from the ascending loop of Henle back to the interstitial fluid with sodium ions following passively after, thus maintaining electrical balance. The ions pumped out of the ascending loop pass back into the descending loop by simple diffusion since the fluids surrounding the descending loop are more concentrated than those within it. Thus the salt recirculates to the ascending loop, where the chloride is pumped out again. This recirculation of NaCl has two consequences: (1) As the urine passes through the loop of Henle, much salt but little water is removed, and (2) the lower part of the loop of Henle and also the lower part of the collecting duct are bathed in a fluid containing many times the level of salt normally found in tissue or blood. The urine—which has become isotonic again by the time it leaves the distal tubule—then descends through the collecting duct, which traverses the zone of high salt concentration. If ADH is present, water is removed from the collecting duct by osmosis, leaving within the duct a urine that is isotonic with the surrounding briny fluid but hypertonic in relation to the body fluids as a whole. In this way, the body is able to excrete a fluid, the urine, far more concentrated than the plasma from which it is derived, and water is extracted from the urine and returned to the bloodstream.

SURVIVAL AT SEA

Seals and some whales get their water by eating fish, most of which have a high water content and an osmotic concentration only about one-third that of sea water. A person on a life raft cannot survive by eating fish, however, rumor to the contrary. Humans cannot form urine with a salt concentration higher than about 2.2 percent. Although a fish's fluids are less concentrated than this, a fish contains a large amount of protein; so our hypothetical shipwreck survivor is placed in the same position as that of the kangaroo rat eating soybeans. Drinking sea water would, of course, make the situation much worse, since the salt concentration of sea water is about 3.5 percent, and precious body water would be expended to dispose of the excess salt. (A kangaroo rat, however, which excretes a highly concentrated urine, can maintain water balance by drinking sea water.) Our would-be survivor could, however, squeeze tissue fluid from a fish. The best bet, however, would be a small solar still.

TEMPERATURE REGULATION

Some 200 years ago, Dr. Charles Blagden, then Secretary of the Royal Society of London, went into a room that had been heated to 126°C (260°F), taking with him a few friends, a small dog in a basket, and a raw steak. The entire group remained there for 45 minutes. Dr. Blagden and his friends emerged unaffected. So did the dog. (The basket had kept the dog's feet from being burned by the floor.) But the steak was cooked.

Temperature regulation is, of course, another example of homeostasis. The control of body heat is an important factor in regulation of the internal environment. (For example, the rate of enzyme activity doubles, approximately, with every 10°C rise in temperature.)

Temperature regulation, like water conservation, is a problem faced by animals as they become terrestrial. In general, large bodies of water, for the reasons discussed in Chapter 2, maintain a very stable temperature. The animals that live in them, with few exceptions, maintain a body temperature that is the same as the temperature of the water. Because the temperatures of large bodies of water are low (seldom exceeding about 5°C), the pace of life in the water, physically and intellectually, is generally a slow one. The few exceptions include the aquatic mammals and certain large fish, such as tuna, that retain inside their bodies heat generated by muscular activities.

Terrestrial reptiles—snakes, lizards, and tortoises—are able to maintain remarkably stable body temperatures during their active hours by absorbing solar radiation. By careful selection of suitable sites, such as the slope of a hill facing the sun, and by orienting their bodies with a maximum surface exposed to sunlight, they can heat themselves rapidly. Such heating can occur as quickly as 1°C per minute, even on mornings when the air temperature is close to 0°C. By frequently changing

38-11
In laboratory studies, the internal temperature of reptiles was shown to be almost the same as the temperature of the surrounding air. It was not until observations were made of these animals in their own environment that it was found that they have behavioral means for temperature regulation that give them a surprising degree of temperature independence. By absorbing solar heat, reptiles can raise their temperature well above that of the air around them. This Colorado collared lizard has become quite warm from the sun and has raised its body to allow cooling air currents to circulate across its belly.

position, these reptiles are able to keep their temperatures within quite a narrow range as long as the sun is shining. When it is not, they seek the safety of their shelters so that they are not immobilized in an exposed position and thus vulnerable to attack. Such animals, which obtain their heat from the outside, are known as *ectotherms.*

"WARM-BLOODED" ANIMALS

The so-called "warm-blooded" animals, primarily mammals and birds, are *endotherms.* With some exceptions, endotherms are also *homeotherms,* maintaining a quite constant internal body temperature. The primary source of heat in endotherms is the oxidation of glucose and other energy-containing molecules within the body cells. Endotherms have a much higher metabolic activity than ectotherms, with a large proportion of their energy budget allocated to heat production. Also, small endotherms have a proportionally larger heat budget than large ones, because of surface-volume ratios (page 98).

A warm-blooded mammal is warmer at the core than at the periphery. (Our temperature usually does not reach 37°C, or 98.6°F, until some distance below the skin surface). Heat is transported from the core to the periphery by the circulatory system. At the surface of the body, the heat is transferred to the air, as long as air temperature is less than body temperature. Temperature regulation involves increasing or decreasing heat production and increasing or decreasing heat loss at the body surface. Endotherms are characterized by layers of insulating material—fur, feathers, fat—and by mechanisms for disposing of excess heat, such as panting in dogs and sweating in humans.

The remarkable constancy of temperature characteristic of humans and many other mammals is maintained by an automatic system—a thermostat—in the hypothalamus (Figure 38-12), which precisely measures the body temperature and triggers the appropriate control mechanisms. Receptor cells in this center monitor the temperature of the blood flowing through the hypothalamus. Although the surface of the skin is covered with receptors for hot and for cold, these are not the

38–12
Body temperature in mammals, regulated by a complex network of activities involving both the nervous and endocrine systems, is controlled largely by the hypothalamus. (The production of TSH is not thought to be part of the response in humans.) Many behavioral responses are also involved.

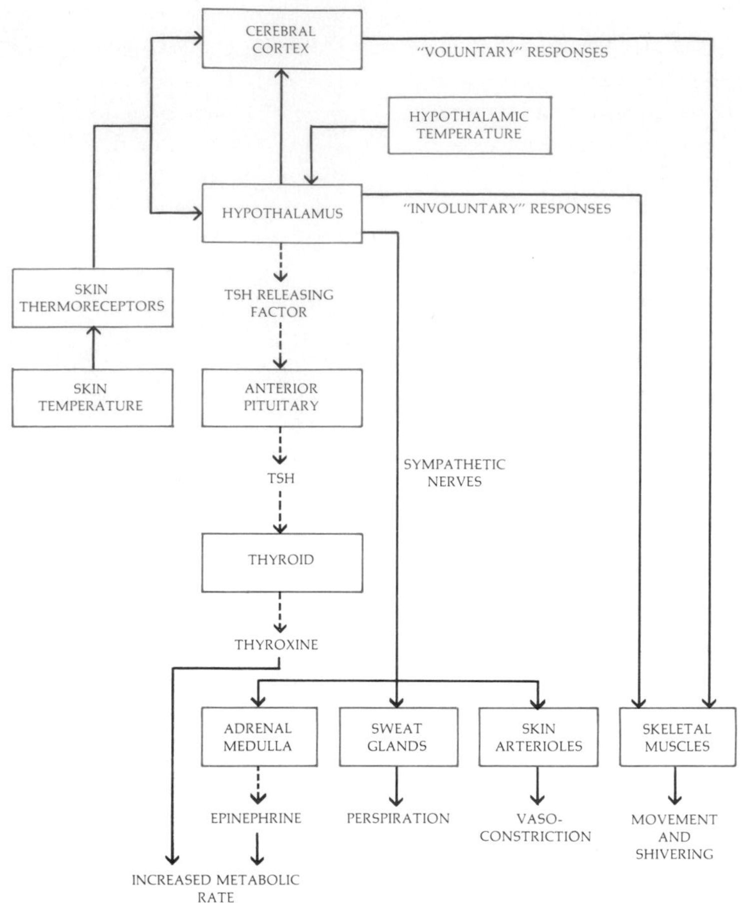

principal agents in the regulation of internal temperature, as has been demonstrated by some interesting experiments with human subjects. For example, in a room in which the air is warmer than body temperature, if the blood circulating through a person's hypothalamus is cooled, he or she will stop perspiring, even though the skin temperature continues to rise.

The elevation of temperature known as fever is due not to a malfunction of the thermostat but to a resetting. Thus, at the onset of fever, an individual typically feels cold and often has chills; although the body temperature is rising, it is still lower than the new thermostat setting. The substance primarily responsible for the resetting of the thermostat is a protein released by the white blood cells in response to a foreign organism. The adaptive value of fever, if any—and we can only presume that there is one—is a subject of current research.

Some endotherms, under certain difficult conditions, conserve energy by permitting their body temperatures to drop. Some hummingbirds and bats, for instance, allow their body temperature to drop when they are at rest. Some animals hibernate, turning their thermostats way down; others—bears, for instance—sleep through most of the winter, reducing their metabolic activities and living off stored fat.

38–13
Some adaptations to cold. (a) *Prior to hibernating, mammals, such as the dormouse shown here, store up food reserves in body fat.* (b) *Note the heat-conserving reduction in surface-to-volume ratio in this hibernating chipmunk.* (c) *Brown fat cells store fats whose energy is channeled directly into heat production. The mitochondria of these cells produce relatively little ATP. Brown fat is found in newborn human infants (but not in human adults) and also in hibernating animals. This electron micrograph shows the nucleus of a brown fat cell from a bat recently aroused from hibernation. Note the large, spherical mitochondria surrounding the nucleus. The fat of these cells has been metabolized in the course of hibernation, leaving only a compact mass of cells. Their color is due to the cytochromes of the many mitochondria.*

(a)

(b)

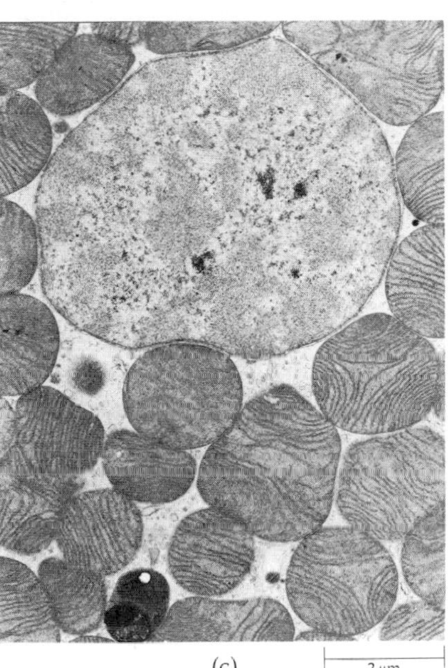

(c)

Regulation at Low Temperatures

As external temperatures go down, blood vessels near the skin surface constrict, limiting heat loss from the skin. Metabolic processes increase. Part of this increase is due to increased muscle activity (shivering, shifting from foot to foot), and part is due to hormones that stimulate metabolism. Muscular activity raises heat production by mobilizing glucose and ATP and by friction.

Prime mobilizers of chemical energy are the autonomic nerves to fat, which increase its metabolic breakdown, and the hormone epinephrine, which is released from the adrenal medulla. In some animals, the thyroid gland increases its release of thyroxine in response to cold. Thyroxine increases the metabolic rate; it apparently exerts its effects directly on the mitochondria. Cortisol production by the adrenal cortex may also increase under the stress of exposure to cold. Both of these effects, you will recall (page 651), are mediated by the hypothalamus's acting on the pituitary, which, via tropic hormones, stimulates the target gland. So again we have an example of integration and control in which the nervous and endocrine systems function as a single unit.

In general, the rate of heat loss depends both on the amount of surface area and on temperature differences between the body surface and the surroundings. Ma-

38-14
Aquatic mammals, such as this bull fur seal, have a heavy layer of fat beneath the skin that acts, in effect, like a diver's wet suit, insulating them against heat loss. On land, they have a problem unloading heat. One way they do it is by keeping their fur moist; another is by sleeping, which, in some species, reduces heat production by almost 25 percent.

rine animals that can survive in extremely cold water are able to tolerate a very great drop in their skin surface temperature; measurements of skin temperature have shown that it is only a degree or so above that of the surrounding water. By permitting the skin temperature to drop, these animals, which maintain an internal temperature as warm as ours, expend very little heat outside of the insulating fat layer and so keep warm—like a diver in a wet suit.

Animals adapt to extremely cold climates by increasing their insulation. Fur and feathers, both of which trap an insulating layer of air, are characteristic of Arctic land animals. Whales, walruses, seals, and penguins are insulated with fat (neither fur nor feathers serve as effective insulators when wet).

Regulation at High Temperatures

In endotherms, as the external temperature rises, the blood vessels near the skin surface dilate, and the supply of blood to the skin increases. If the air is cooler than the body surface, heat can be transferred directly to the air. Animals that live in hot climates characteristically have larger exposed surface areas than animals that live in the cold (Figure 38-15).

Heat can also be lost from the surface by the evaporation of perspiration or saliva. In humans and most other large animals, when the external temperature rises above body temperature, perspiration begins. Evaporation requires heat energy (see page 36), which comes from the body surface, and the blood just below the skin surface cools as this heat energy is used. Dogs and some animals pant, so that air passes rapidly over their large, moist tongues, evaporating saliva. Cats lick themselves all over as the temperature rises, and evaporation of this water cools their body surface. Horses and human beings sweat from all over their body surfaces. For animals that dissipate heat by water evaporation, temperature regulation at high temperatures necessarily involves water loss, which, in turn, stimulates thirst and water conservation by the kidneys. Dr. Blagden and his friends were probably very thirsty.

38-15
The size of the extremities in a particular type of animal can often be correlated with the climate in which it lives. (a) *The fennec fox of the North African desert has large ears that help it to dissipate body heat.* (b) *The red fox of the eastern United States has ears of intermediate size, and* (c) *the Arctic fox has relatively small ears. Similar correlations between size, weight, color, etc., and environment can also sometimes be made among animals of a single species living over an extended geographic range.*

(a)

(b)

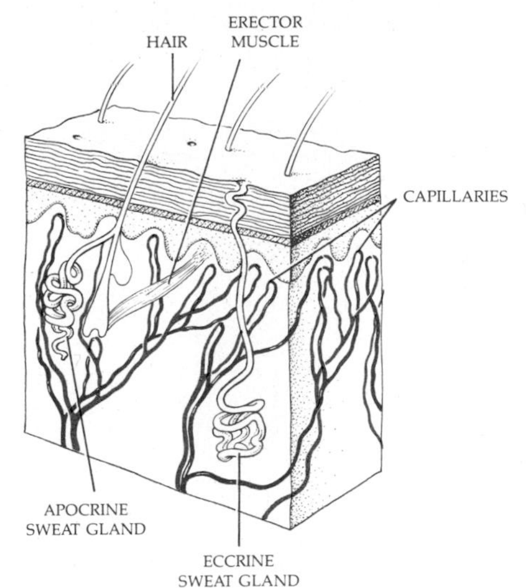

38–16

*Cross section of human skin showing structures involved in temperature regulation. In cold, the arterioles constrict, reducing the flow of blood through the capillaries, and the hairs, each of which has a small erector muscle under nervous control, stand upright. In animals with body hair (*Homo sapiens *is one of the few terrestrial mammals that is virtually hairless), air trapped between the hairs insulates the skin surface, conserving heat. In heat, the arterioles dilate and the eccrine glands secrete a salty liquid. Evaporation of this liquid cools the skin surface, dissipating heat (approximately 540 calories for every gram of* H_2O*). The apocrine glands are scent glands, producing a milky, odorous fluid. They are abundant in the armpit, navel, and anogenital areas. Beneath the skin is a layer of subcutaneous fat that serves as insulation, retaining the heat in the underlying body tissues.*

ADAPTATION OF DESERT ANIMALS

We have seen in the course of this chapter, first, that water balance is essential to the smooth functioning of the internal environment, and, second, that water evaporation is a major mechanism for dissipating heat. How do animals in the desert resolve these two conflicting demands? There are a variety of answers to this question.

First, desert animals tend to excrete an extremely concentrated urine. Some desert animals have loops of Henle that are so long they dip down into the ureter. As mentioned earlier, some animals, such as the kangaroo rat, can derive all their water from food even during pregnancy and lactation. Many desert animals regulate their temperature and thus their demands on their water supply by behavior, as do the reptiles mentioned earlier in this chapter. One of the most interesting examples is found in the architecture of the burrows built by prairie dogs, which follow some basic principles of aerodynamics to air-condition their tunnels.

As you may know, airplane wings are shaped so that air will move along the top surface faster than along the bottom surface. This creates lower pressure above the wings so that the plane is, in effect, sucked upward. A prairie dog constructs a burrow about 18 meters long, with an opening at either end. Then, following the principle illustrated by the airplane wing, it constructs a high mound around one end and a low mound around the other, often gathering dirt from some distance around and working and reworking the structure. Air moves faster over the higher mound, and the pressure is less than at the other, lower opening of the burrow. As a consequence, air is pulled through the burrow from the lower end. Duke University scientists, who have studied the prairie dog's engineering accomplishments, calculate that when air is moving over the mounded openings at the rate of 1.6 kilometers (1 mile) an hour, a prairie dog burrow gets a complete change of air every 10 minutes.

(c)

As a final example, let us consider the camel, for so long the companion of desert travelers. A camel has several advantages over a human in the desert. For one thing, the camel excretes a much more concentrated urine; in other words, it does not need to use so much water to dissolve its waste products. (In fact, we are very uneconomical with our water supply; even dogs and cats excrete a urine twice as concentrated as ours.)

COUNTERCURRENT EXCHANGE

In many Arctic animals, the veins and arteries leading to and from the extremities are juxtaposed in such a way that the chilled blood returning from the legs (or fins or tail) through the veins picks up heat from the blood entering the extremities through the arteries. The veins and arteries are closely apposed to give maximum surface for heat transfer. This arrangement, which serves to keep heat in the body and away from the extremities, is accomplished, largely, by countercurrent exchange.

The principle of countercurrent exchange is illustrated by a hot-water pipe and a cold-water pipe placed side by side to achieve maximum temperature transfer under ideal conditions. In (a), *the hot water and cold water flow in the same direction. Heat from the hot-water pipe warms the cold water until both temperatures equalize, at 5. Thereafter, no further exchange takes place and the water in both pipes remains lukewarm. In* (b), *the flow is in opposite directions, so that heat transfer continues for the length of the pipes. The result is that the hot water transfers most of its heat as it travels through the pipe, and the cold water is warmed to almost the initial temperature of the hot water. The principle of countercurrent exchange applies equally well to oxygen exchange (Figure 36–4) or to the exchange of solutes such as* Cl^-*, as into the loop of Henle.*

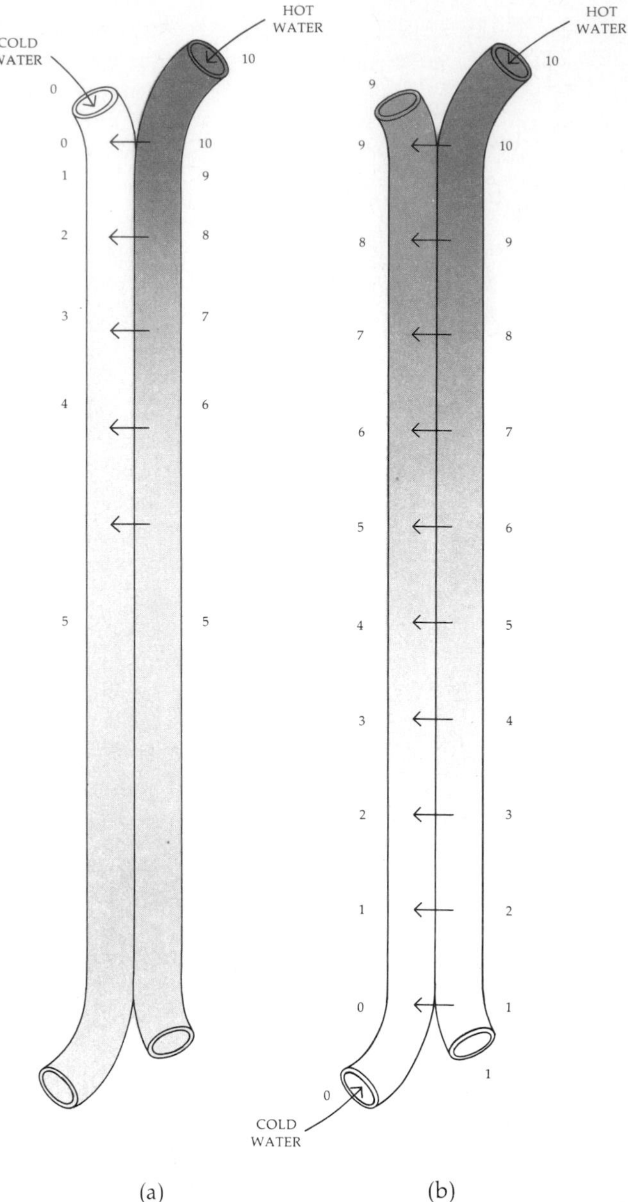

38-17

By facing the sun, a camel exposes as small an area of body surface as possible to the sun's radiation. Its body is insulated by fat on top, which minimizes heat gain by radiation. The underpart of its body, which has much less insulation, radiates heat out to the ground cooled by the animal's shadow. Other adaptations of the camel to desert life include long eyelashes, which protect its eyes from the stinging sand, and flattened nostrils, which retard water loss.

Also, a camel can lose more water proportionally than a human and still continue to function. If a person loses 10 percent of the body weight in water, he or she becomes delirious, deaf, and insensitive to pain. If as much as 12 percent is lost, the individual is unable to swallow and so cannot recover without assistance. Laboratory rats and many other common animals can tolerate dehydration of up to 12 to 14 percent of body weight. Camels can tolerate the loss of more than 25 percent of their body weight. They can go without drinking for one week in the summer months, three weeks in the winter.

Probably most important, the camel can tolerate a fluctuation in internal temperature of 5° to 6°C. This tolerance means that it can let its temperature rise during the daytime (which the human thermostat would never permit) and cool during the night. The camel begins the next day at below its normal temperature—storing up coolness, in effect. It is estimated that the camel saves as much as 5 liters of water a day as a result of being able to tolerate these internal temperature fluctuations.

Finally, a camel orients itself to the sun in such a way as to expose as small an area as possible to the sun's radiation. Its fatty hump on top insulates the animal against the sun's rays striking its back. The relatively uninsulated underpart of its body radiates heat out to the ground cooled by the animal's shadow. As you might guess, when it goes to sleep on cold desert nights, it curls up.

SUMMARY

Homeostasis—the capacity to maintain a constant internal environment—is a characteristic of living systems. Three major homeostatically controlled factors are the chemical composition of the blood, water balance, and body temperature.

The urinary system plays a major role in the capacity of an animal to regulate its internal chemical environment. Regulation is accomplished by the selective excretion of toxic waste products, especially the nitrogenous compounds produced in the breakdown of amino acids; the selective retention of nutrient molecules, such as glucose and amino acids, and of ions; and the control of the osmotic pressure of the blood by maintaining water balance.

The excretory unit of the kidneys is the nephron; each nephron consists of a long tubule attached to a closed bulb (Bowman's capsule), which encloses a twisted cluster of capillaries, the glomerulus. Fluids are forced from the glomerulus into the Bowman's capsule (filtration). By a combination of filtration, secretion, and resorption, the circulating body fluids are processed during their passage along the tubules and converted into urine.

Water and salt excretion are subject to hormonal regulation. In mammals, aldosterone, produced by the adrenal cortex, increases the resorption of sodium by the renal tubules. ADH, a hormone secreted from the pituitary gland, acts on the collecting ducts of the nephrons, increasing their permeability to water and thus decreasing water loss by returning urine water to the circulating blood.

The problems of water balance are different for animals living in salt water (hypertonic), fresh water, and terrestrial environments. In terrestrial mammals, water is gained from food, fluids, and cellular respiration. It is lost in feces, exhaled air, sweat, and, to the largest extent, the urine. An important means of water conservation is the capacity for excreting a urine that is hypertonic in relation to the blood. The loop of Henle is the portion of the mammalian nephron that makes possible the production of a hypertonic urine.

The control of body temperature is an essential feature of homeostasis. Ectotherms are animals that adjust their body temperatures principally by regulating the amount of heat taken in from the external environment. For endotherms, such as *Homo sapiens*, the source of heat is the oxidation of glucose and other fuel molecules by the body cells and friction from muscular activity.

Heat is carried by the bloodstream from the core of the body to the surface, where it is dissipated. In hot weather, the blood vessels near the surface of the skin dilate, increasing the flow of blood to the skin. As external temperatures rise above body temperatures, humans and most other large animals begin to sweat or pant. Evaporation of sweat or saliva from body surfaces requires heat and cools the surface. In the cold, blood vessels supplying the skin surface constrict. As temperatures drop, shivering begins; the increased cellular metabolism and muscular activity involved in shivering produce heat. Temperature is regulated by an automatic system, or thermostat, in the hypothalamus that measures the body temperature and sets in motion the response mechanisms.

Desert animals have a number of solutions to the dual problem of temperature regulation and water conservation. These include dietary habits, the capacity to excrete a very concentrated urine, a comparatively wide tolerance for fluctuations in total water and temperature, and a number of behavioral adaptations.

QUESTIONS

1. Distinguish among the following: efferent/afferent; Bowman's capsule/glomerulus; endotherm/ectotherm/homeotherm; extracellular fluid/intracellular fluid; aldosterone/ADH.

2. Explain the following terms in relation to kidney function: filtration, secretion, resorption, and excretion.

3. Sketch a nephron, indicating the pathways of the blood and the glomerular filtrate.

4. Diagram the paths of (a) glucose and (b) urea through the human kidney.

5. Sketch the urinary tract of a human female, labeling the various components.

6. Reexamine the micrograph in Figure 38–5. Describe the function of each structure visible in it in terms of the overall function of the nephron.

7. Why does a high-protein diet require an increased intake of water? (You should be able to think of two different reasons.) Why does a person lose some weight after shifting to a low-salt diet, even without reducing the caloric intake? Given the fact that amino acids in excess of the body's requirements are broken down by the liver, not stored, what is the advantage of a high-protein diet? What might be a disadvantage?

8. Viewed as a whole, the kidney selectively removes substances from the bloodstream. A closer look, however, reveals that it does so by filtering out all small molecules through the glomerulus and then resorbing those that are needed in the

tubules. (We'll neglect tubular secretion for the moment.) Thus, the system need not identify wastes as such; rather, it must identify useful substances. Why would such an arrangement have been beneficial to early mammals? Why is such an arrangement especially beneficial to modern mammals, including ourselves?

9. Describe what happens to the human organism as the external temperature rises. As it falls.

10. Compare the surface-to-volume ratio of an Eskimo igloo with that of a California ranch house. In what way is the igloo well suited to the environment in which it is found?

11. The primary temperature-measuring device in our bodies is located in the hypothalamus, at the base of the brain. Why is this a better location than some more peripheral point, such as the skin?

12. Unlike nocturnal, burrowing rodents, small desert birds cannot evade the effects of summer daytime heat and dryness. What predictions can you make about how they regulate body temperature and maintain water balance?

13. Compare and contrast the mechanisms used by marine mollusks, freshwater arthropods, and terrestrial amphibians to maintain water and ion balance.

14. Why is it an advantage for animals in cold climates to keep heat away from the extremities? (You may be able to think of two reasons.)

CHAPTER 39

The Senses

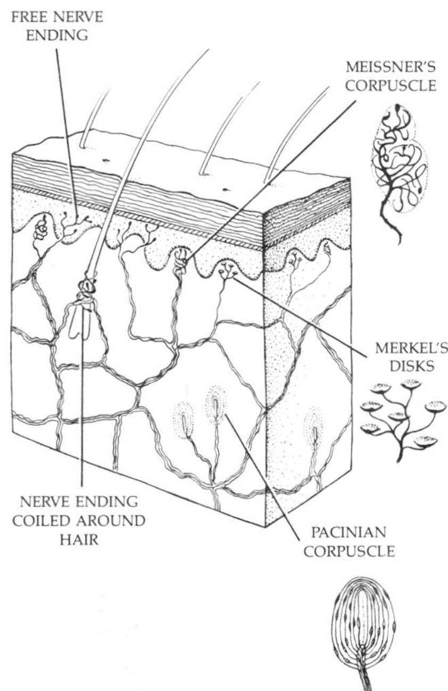

39–1

Some sensory receptors present in human skin. Merkel's disks and Meissner's corpuscles respond to touch, as do the nerve endings surrounding the hair follicle. (Actually, when many hairs are present, the other touch receptors are sparse or absent.) Pacinian corpuscles register pressure, and the free nerve endings are largely pain receptors.

Our sensory equipment is the means by which we stay aware of the world around us. It provides the mechanisms by which we and other animals are informed of predator and prey, friend and foe, whether something is good to eat or not, and changes in the weather or the seasons. It identifies infant to mother, mother to infant, and mate to mate. It is also our source of pleasure and of esthetic enjoyment and provides our tools for learning. Our sense organs, like the other organs of the body of any animal, are the products of adaptation and evolution, and by these long processes, they have been tailored to our specific requirements. For example, we think we see what is visible and hear what is audible, but visible and audible are actually indefinite and relative. What we see is very different from what an insect sees, and a bat or a fish, acoustically speaking, might as well be living on another planet. Moreover, it is likely that our insensitivity to a great number of the potential stimuli that surround us is also as useful as our sensitivity to others.

TYPES OF SENSORY RECEPTORS

Our sensory receptors are many and varied. Like most animals, we have mechanoreceptors (touch, hearing, position), chemoreceptors (taste and smell), photoreceptors (vision), temperature receptors, and pain receptors. (We do not have electroreceptors, but some fish do.)

Some sensory receptors are small and relatively simple in structure. Look again at the human skin (Figure 39–1) as an example. The simplest receptors are the free nerve endings, which are receptors for pain and perhaps for other sensations as well. Slightly more complex are the combinations of free nerve endings with a hair and its follicle. Each of these little organs is an exquisitely sensitive mechanoreceptor. When the hair is touched or bent, it sets off an action potential in the nerve endings of a sensory neuron that is carried directly to the central nervous system. Three other types of mechanoreceptors are also shown, each a combination of one or more free nerve endings and an outer layer or layers of connective tissue. Meissner's corpuscles and Merkel's disks are both involved with touch. They are found in particularly sensitive areas of the skin, such as the fingertips, palms, lips, and nipples, and are especially abundant where hairs are not present. They are responsible for the extraordinary cutaneous sensitivity of these parts of the body and are associated with the ability, for example, to read Braille, do certain magic

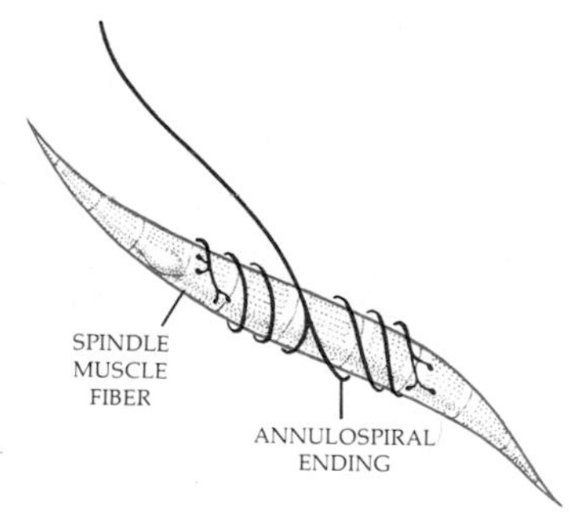

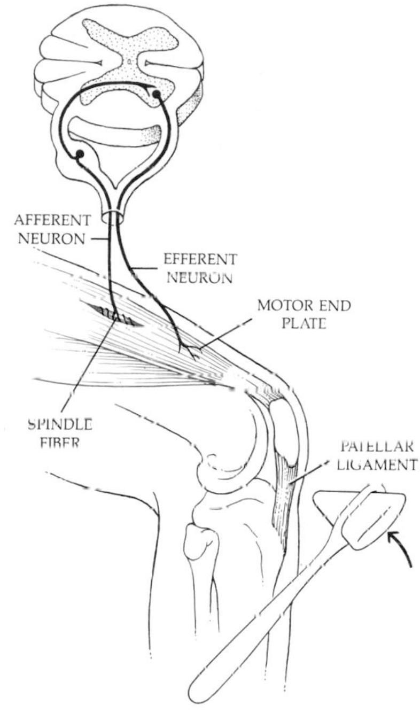

39–2
A light tap on the knee, as with a rubber-headed hammer, causes the lower leg to fly forward in the knee-jerk reflex. This reaction occurs because the tap pulls the patellar ligament, which, in turn, gives a sudden tug to a muscle of the upper leg. The stretching of this muscle is sensed by the nerve fibers of the muscle spindle, which send a message, encoded in action potentials, to the spinal cord. There the afferent (sensory) neuron synapses directly with a motor neuron that causes the muscle to contract. This reflex helps to maintain balance and posture.

tricks, crack a safe, or respond to a kiss. The Pacinian corpuscles, lying deeper within the tissues, respond to pressure. The free nerve ending of the corpuscle is surrounded by layers of connective tissue and fluid. This onion-shaped structure is easily deformed, so it responds to even very slight pressure changes, but it also, because of its structure, quickly adjusts so that the nerve ending stops firing when pressure is sustained. This is one reason that we can adapt rapidly to continuous pressure, such as that produced by sitting in a chair or leaning against a solid object. Of all the sensory receptors shown here, the simplest in structure—the free nerve endings associated with pain—are the least understood in function. It seems likely that some of them are not mechanoreceptors, as are the others, but rather chemoreceptors, responding to very small amounts of some chemical released by cells when they are injured.

Sensory receptors are categorized as interoceptors, exteroceptors, and proprioceptors. *Interoceptors* include the mechanoreceptors and chemoreceptors that measure blood pressure and CO_2 concentration in the carotid arteries. The temperature sensors of the hypothalamus are also interoceptors. We are usually not conscious of signals from these receptors, although sometimes the signals result in perceptions of, for example, pain, hunger, thirst, nausea, or the sensations, produced by stretch receptors, of having a full bladder or bowel.

Exteroceptors are the most familiar, including those in the ear, the eye, the olfactory epithelium, and the touch receptors and temperature receptors of the skin (see page 727).

Proprioceptors (from the Latin *proprius*, meaning "self") inform us about the orientation of our body in space and the position of our arms, legs, and other body parts. Because of proprioceptors you can, with your eyes shut, touch your nose with your fingers or tie your shoelace in the dark. The semicircular canals of the ear are major proprioceptive organs in many animals. Among the most common and the most active are muscle spindles, which act as gauges monitoring muscle length. The actual sensory receptors are nerve endings that are wrapped around specialized muscle fibers located deep within the muscle. They fire when the muscle is stretched. The muscle spindle is the bundle of specialized muscle fibers, and the free nerve endings are wrapped around them. It is involved in the well-known knee-jerk reflex (Figure 39–2).

All sense organs, as different as they may be in structure and function, are essentially similar. They are all transducers. That is, they respond to energy in one form and convert it to another form, like a telephone, which converts mechanical energy (sound waves) to electrical energy and then back again. Also, as in the telephone system, all of the information is carried in the same coded form, the all-or-nothing action potential. The differences lie not in the signals as they are transmitted, but in their reception and interpretation in the brain. We shall discuss this further in the following chapter.

LIGHT RECEPTION: VISION

Eyes have evolved independently several times in the course of evolutionary history. Among the most highly developed of modern light-receptor systems are the compound eye of the arthropods (see page 467), the eye of the octopus (see page 446), and the vertebrate eye, of which the human eye, shown in Figure 39–3, is an example. (See also Figure 31–5 on page 573.)

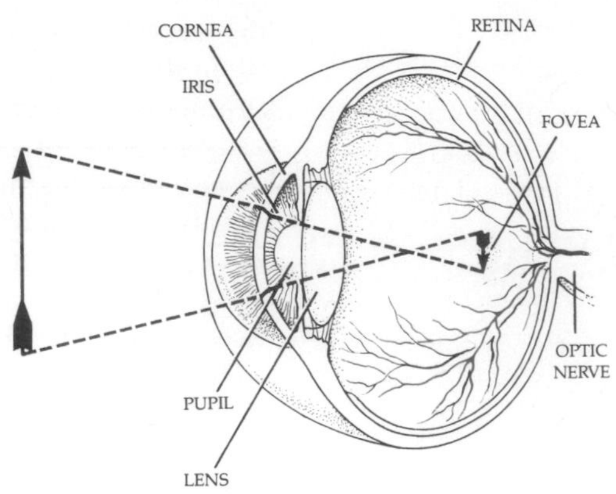

39–3
The eyeball is a fluid-filled sphere composed of three layers of tissue. The outer layer, the sclera, is white fibrous connective tissue that serves a protective function. The anterior portion of the sclera, the cornea, is transparent. The middle layer, the choroid, contains blood vessels. Its anterior portion is modified into the ciliary body, the suspensory ligament, and the iris. The ciliary body is a circle of smooth muscle from which the suspensory ligaments extend that hold the lens in place. The colored part of the eye, the iris, is a circular structure attached to the ciliary body. The pupil is a hole in the center of the iris. The innermost layer, the retina, coats the posterior hemisphere of the eyeball. It consists of the rods, the cones, and other nerve cells, and is adjacent to a non-neural pigmented layer of epithelium that adheres to the choroid layer. Only the front of the eye is exposed; the rest of the eyeball is recessed in the orbit, protected by the bony socket of the skull.

The vertebrate type of eye is often called a camera eye. In fact, it has a number of features in common with an ordinary camera equipped with several expensive accessories, such as a built-in cleaning and lubricating system, an exposure meter, and an automatic field finder. Light from the object being viewed passes through the transparent cornea and lens, which focus an inverted image of the object on the light-sensitive retina in the back of the eyeball. (Though the eye may be like a camera, the retina is not like a piece of film. As we shall see, much processing of information takes place in the retina itself.)

In mammals, fine focusing of the image on the retina is brought about by contracting or relaxing the ciliary muscles, thus changing the curvature, and hence the focal length, of the lens (Figure 39–4). (Fish and amphibians, which lack ciliary muscles, focus their eyes the same way one focuses a camera with a lens of fixed focal length. Muscles within the eye change the position of the lens, drawing it back toward the retina in order to focus more distant objects.) The iris controls the amount of light entering the eye by regulating the size of the pupil.

Stereoscopic (three-dimensional) vision depends on viewing the same visual field with both eyes simultaneously. When, under these conditions, both eyes are trained on a distant object, then both eyes see almost the same image. A nearby object, however, will present a slightly different image to each eye (parallax). On the basis of the disparity of the two images of the nearby object, we automatically compute the distance between us and the object. When the eyes are trained on a very distant object, however, the disparity is too small; hence, we are not able to make such judgments for faraway objects. Instead, we estimate their distance on the basis of known sizes of the objects or such visual clues as houses, people, or cars. A young child viewing a distant object—an airplane, for example—may see it as little rather than far away.

Tree-dwelling animals, such as the probable forerunners of *Homo sapiens*, usually have overlapping visual fields in the front that provide them with the stereoscopic vision essential for the distance judgments involved in moving from branch to branch. Predatory animals also tend to have stereoscopic vision, but animals that are more likely to be hunted than to hunt usually have an eye placed on each side of the head, giving a wide total visual field. Some birds with laterally placed eyes—the woodcock, the cuckoo, and certain species of crow—have binocular fields of vision both in front of them and behind them. Some of these and other specializations are shown in Figure 39–5.

39–4

Focusing the eye. Preliminary focusing is done by the cornea, and fine focusing by the lens, which is like a jelly-filled, transparent rubber balloon. The lens is held in place by suspensory ligaments attached to the ciliary body, which encircles the lens. The ciliary body contains the ciliary muscle, which is the muscle of accommodation. When it contracts, the lens is released, and, as it returns to its more nearly spherical shape, its curvature is increased. (a) *The normal eye views distant objects with the lens stretched and flattened;* (b) *for viewing near objects, the lens is relaxed and more convex, bending the light rays more sharply.* (c) *Nearsightedness occurs when the eyeball is too long for the lens to focus a distant image on the retina.* (d) *It is corrected by a concave lens that bends the light rays out enough so that the image is focused.* (e) *Farsightedness is the result of an eyeball too short for the lens to focus a near object.* (f) *It is corrected by a convex lens that bends the light rays in before they reach the lens of the eye.*

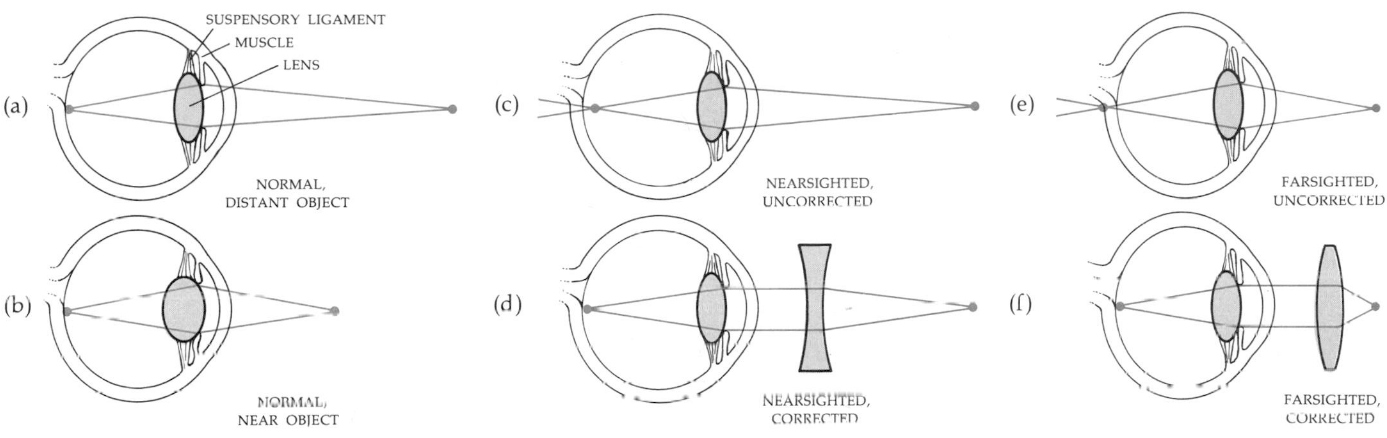

39–5

Some vertebrate adaptations. (a) *Mud skipper eyes are directed upward like periscopes.* (b) *Eagles and other predatory birds often have eyes as large as ours (although the skull and brain are much smaller), two foveas (see page 741) in each retina, and strong powers of near and far accommodation.* (c) *Rabbits have eyes set on each side of their skulls, an adaptation that allows them to watch on both sides for predators while they are feeding.* (d) *This ring-tailed cat is nocturnal. Like many other nocturnal animals, it has large eyes. Nocturnal animals characteristically have only rod cells, which are more light-sensitive than cone cells, and so they lack any color vision.*

(a)

(b)

(c)

(d)

The Retina

The retina of the vertebrate eye is anatomically inside out—that is, the photoreceptors of the eye are pointed toward the back of the eyeball—and light must reach the light-sensitive area by passing through layers of nerve cell bodies (see Figure 39–6). Light that is not captured by the photoreceptors is absorbed by pigmented epithelium that lines the back of the eyeball, just behind the photoreceptors. Some nocturnal vertebrates have a reflecting layer (tapetum) behind the photoreceptors that increases the likelihood of dim light stimulating the photoreceptors. This reflecting layer makes the animal's eyes seem to shine at night.

The inside-out arrangement of the retina appears to be an historical accident. One would expect, of course, that light would strike the receptor cells before traveling through all the other cell layers and blood vessels of the retina. Only about 10 percent of the light falling on the cornea reaches the retina. The great nineteenth-century physiologist Hermann von Helmholtz is reported to have said that if he had given one of his technicians the task of designing an image-forming sense organ and that functionary had brought him the human eye, the man would have been fired on the spot. However, as you will recall from Chapter 33, the retina is actually an outpocketing of the brain. The double folding to form the optic cup (page 618) is the anatomical reason that the various retinal cell layers are between the light source and the receptors.

The photoreceptor cells transmit their electrical signals to neurons known as bipolar cells, which pass them on to the ganglion cells, whose axons form the optic nerve. However, these neurons and others—horizontal and amacrine cells—have varied and elaborate interconnections made among them even before nerve im-

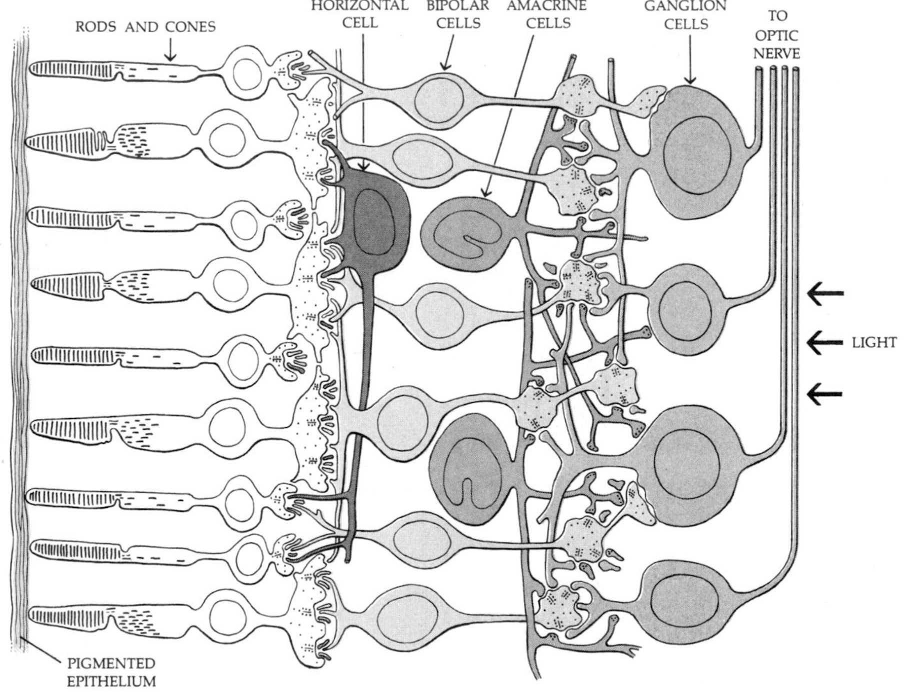

39–6
The retina of the vertebrate eye. Light (shown here as entering from the right) must pass through a layer of cells to reach the photoreceptors (the rods and cones) at the back of the eye. Signals from the photoreceptor cells are then transmitted through the neurons known as bipolar cells to the ganglion cells, whose axons converge to become the optic nerve. The transmission paths involve elaborate interconnections.

39–7
With this diagram, you can prove for yourself that you have a blind spot in each eye. Hold the book about 30 centimeters (12 inches) from your face, cover your left eye, and gaze steadily at the X while slowly moving the book toward your face. Note that at a certain distance, the image of the dot becomes invisible. Then cover your right eye and gaze steadily at the dot while you move the book toward your face. What happens to the X?

X ● / /

pulses leave the retina. This circuitry is involved in some preliminary analysis of visual information that takes place in the retina itself. The human retina contains about 125 million photoreceptors, and the optic nerve about 1 million axons.

Ganglion cell axons from all over the retina converge at the rear of the eyeball and bundle together like a cable to form the optic nerve. The point at which the axons pass out of the retina is a blind spot, since photoreceptor cells are absent there. We generally are not aware of the existence of the blind spot since we usually see the same object with both eyes and the "missing piece" is always supplied by the other eye. If you are not aware of your own blind spot, you can demonstrate its existence with the help of Figure 39–7.

As shown in Figure 39–3, there is an area in the human retina, just beside the blind spot, known as the *fovea*. This is the only area in the retina at which the imperfect lens of our eye forms a really sharp image. In this area, the photoreceptor cells are packed much more closely together, and many have a one-to-one connection with the bipolar cells. Birds, which rely on vision above all the other senses, may have two or three foveas. Most vertebrates, however, have no fovea, and their vision is comparable in acuity to what we see out of the corners of our eyes.

Rods and Cones

The 125 million photoreceptors of the human retina are of two types, named, because of their shapes, rods and cones (Figure 39–8). The cones, of which the human eye contains 5 to 10 million, provide greater resolution, giving a "crisper" picture. They are generally concentrated in the center of the retina; the fovea consists entirely of cones. We have about 160,000 cones per square millimeter of eye. A hawk, with an eye of about the same size, has some 1 million cones per square millimeter and therefore a visual acuity about eight times that of a human being.

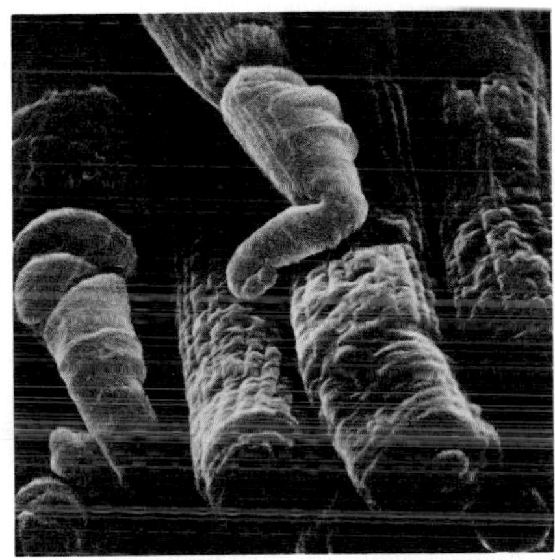

39–8
Rods and cones as shown by the scanning electron microscope. A single photon is sufficient to cause a response in a rod cell; as few as five to eight photons in the blue-green frequency range can result in conscious perception.

Rods do not provide as great a degree of resolution as cones do, but they are more light sensitive than cones. In dim light, our vision of objects depends entirely on rods. In humans, rods are concentrated outside of the fovea. Some nocturnal animals, such as toads, mice, rats, and bats, have retinas made up almost entirely of rods, and some diurnal animals, such as some reptiles and squirrels, have almost entirely cones.

Cones are responsible for color vision, which is why the world becomes colorless to us at night—the light is too dim to stimulate the cone system and we must depend on the rods.

Both rods and cones contain light-sensitive compounds consisting of a protein, called an opsin, and a carotenoid. As we noted previously, the eyes of arthropods, mollusks, and vertebrates all evolved separately, and yet each of the three types contains almost identical visual pigments. Moreover, since no animal can synthesize carotenoids, all vision depends on substances derived in the diet directly or indirectly from plants.

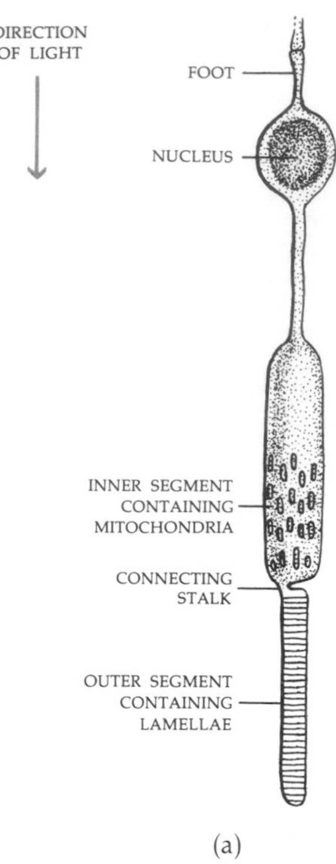

(a)

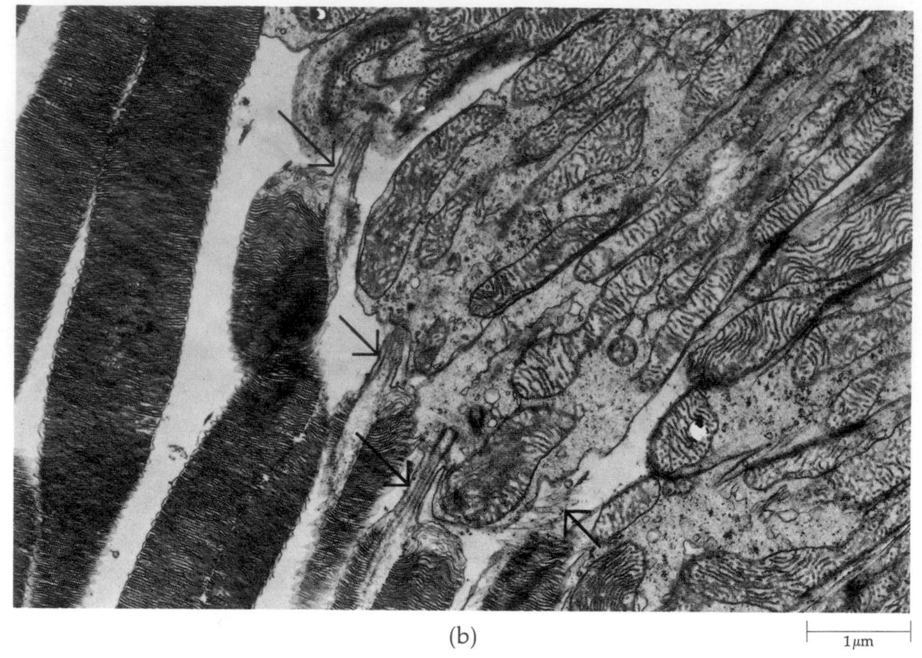

(b)

39–9
(a) *Diagram of a rod from the human retina. The molecules of light-sensitive pigment, rhodopsin, are located on the lamellae. The foot makes synaptic contact with bipolar and horizontal cells.* (b) *An electron micrograph of photoreceptor cells. The ends pointing toward the back of the eyeball consist of a stack of membranes, piled up one on top of the other like poker chips. The light-sensitive pigment is built into these membranes. In the adjoining part of the cell, new pigment molecules and other substances required by the light-receptor area are synthesized and transported through a narrow stalk (indicated by the arrows), separating the two portions of the cell, to the membrane-packed portion, where the actual work of this highly specialized cell is done. A cross section of the stalk reveals, surprisingly, that it has the internal structure of a cilium, lacking only the two central microtubules. The cells shown here are rods, but cones are built along the same general principles.*

The light-sensitive compound in the rods of the vertebrates is rhodopsin, sometimes called visual purple. The carotenoid portion of the rhodopsin molecule is called retinal. When rhodopsin absorbs light, retinal changes shape (Figure 39–10). This change in shape causes retinal to dissociate from its opsin. The opsin then also changes shape. In some way, as yet not fully understood, this breakdown of rhodopsin results in a change in the membrane of the rod cell that causes a signal to be transmitted to the bipolar cell and then along pathways to the brain. An enzymatic reaction changes retinal back to its original shape and then it recombines with opsin, restoring rhodopsin and so completing the cycle.

Eyes become adapted to the dark by a slow buildup of rhodopsin synthesis from vitamin A stored in the pigmented epithelium of the eye. Persons who have a vitamin A deficiency cannot build up sufficient rhodopsin and therefore suffer "night blindness."

In humans, there are three different types of cones, each containing one of three different visual pigments. Each of these pigments is made up of retinal and a slightly different opsin. As long ago as 1802, a three-receptor hypothesis for color vision was proposed; this hypothesis, supported by a number of psychological studies, has been greatly bolstered by the identification of the three separate cone types and their pigments. Each pigment is most sensitive to the wavelength of one of the three fundamental colors—blue, green, or red. Different shades of color stimulate different combinations of these cones, and the signals, after being processed in the retina, are further processed by the brain and so become what we perceive as color.

Visible light is only a small portion of the vast electromagnetic spectrum (see Figure 9–3, page 174). For the human eye, the visible spectrum ranges from violet

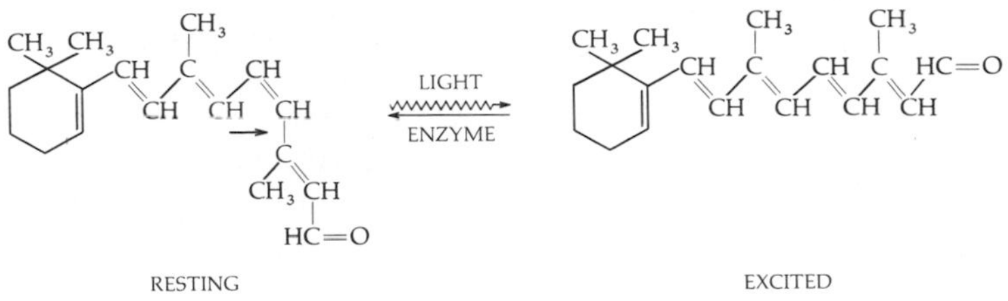

39-10
When the retinal portion of the rhodopsin molecule is excited by light, it changes shape, setting in motion a series of events that result in the sensation of light. Retinal is changed back to its original conformation by an enzyme, completing the visual cycle. The arrow indicates the bond around which the side chain of the "resting" retinal rotates to form the "excited" retinal.

light, which is made up of comparatively short rays, to red light, the longest rays visible to us. Fortunately, we cannot see in the infrared (heatwave) spectrum. Otherwise, we would see everything through an infrared glow emitted by our own bodies. Our photoreceptors are sensitive to ultraviolet. Ordinarily, these waves are filtered out by yellow pigment in the lens, but persons who have had cataracts removed can see well enough to read in ultraviolet light.

Vision and Behavior

Vision is a direct stimulus to action in many of the vertebrates. Predatory fish are vision-stimulated, and many such fish literally "cannot help" striking at a moving object of appropriate size. The frog can learn not to strike at a small moving object, but only with difficulty. If it comes to associate an unpleasant experience with a particular shape, it can restrain itself, although apparently with effort, from striking at that shape. On the other hand, it can never learn to take an object that is not moving. Even a frog that has eaten thousands of live flies will starve to death surrounded by motionless ones.

Investigators have found that certain ganglion cell axons leaving the retina of the frog produce impulses only when certain light receptors in the retina are stimulated by a small, rounded, moving object. In the laboratory, the visual stimulus for these ganglion cells can be simulated by a variety of objects—such as paper clips and pencil erasers—provided they are in motion. In a frog's natural environment, however, the objects likely to elicit this response most frequently are flying insects. A large, moving object or a small, still one provokes no response in these ganglion cells. These ganglion cell axons terminate in a special layer of cells in the frog's midbrain, and, although no further connections have been traced, it is tempting to speculate that signals are relayed from this part of the midbrain to areas containing motor nerves that control the muscles involved in striking.

Colors often play an important role in eliciting behavior in vertebrates. Studies of the three-spined stickleback, a small freshwater fish, have shown that fighting behavior among the males is elicited by the red color of their underbellies, which develops during the mating season. Niko Tinbergen, who kept a laboratory in England full of aquariums containing sticklebacks, found that the male sticklebacks would rush to the sides of their tanks and assume threatening postures every time a red mailtruck passed on the street outside. Similarly, it has been found that fighting behavior among English robins can be evoked by a small tuft of red feathers. Among birds, bright plumage, especially of the males, plays an important part in attracting the females and in courtship ceremonies.

In mammals, in general, vision is less important than it is in many of the other vertebrates. Many of the smaller mammals are nocturnal, and most of them live close to the ground in habitats in which their highly developed sense of smell provides the greatest amount of information about their environment. Presumably, it was the tree-dwelling habits of our immediate ancestors that made the sense of smell less useful for the primates. Most mammals do not have color vision, and, among the mammals, only primates have a central fovea for sharp vision. In *Homo sapiens,* highly developed vision is undoubtedly closely correlated with the development of the use of the hands for fine manipulative movements. This reasonable hypothesis is supported by the fact that the human eye is so constructed that the sharpest images—images within the fovea—can be made of objects that are within the reach of the hand.

CHEMORECEPTION: TASTE AND SMELL

If we were to lose our sense of taste or smell, our lives would lose many of their pleasures but we would not be greatly handicapped in our essential activities. Blindness or deafness is a far more serious threat. For many animals, however, the sense of smell is their window on the world, much as vision is ours. Watch a dog the next time it is let out of the house in the morning: It does not look around, human fashion; instead it sniffs the air and then explores the ground with its nose in order to bring itself up to date on the events of the previous night.

Although the organs of chemoreception seem much simpler than those of vision or hearing, we know less about how they work. We do not know exactly how a chemical can give rise to an electrical signal in a nerve cell or how the various different chemicals that we can distinguish by taste or smell manage to identify themselves to the brain.

Taste

Fish, particularly bottom feeders, have taste cells scattered over the surface of their bodies, and these cells play a major role in determining their behavior. In the carp, for example, the part of the brain that receives information from these taste cells is larger than all the other sensory centers combined. The catfish, also a bottom feeder, has taste cells on its body and, in addition, trails barbels, or "whiskers," along the bottom. These are richly supplied with gustatory nerve endings and when they are stimulated, the fish turns and snaps at the source of the stimulus.

39–11
Amphibians and reptiles have, in addition to nostrils, a special olfactory organ, the vomeronasal organ, in the roof of the mouth. Food substances are apparently tested in this organ. When a snake lashes its forked tongue in the air, like the Eastern hognose snake shown here, it is collecting samples for testing in its vomeronasal cavities.

In terrestrial animals, taste cells are located inside the mouth, where they act as sentinels, providing a final judgment on what is and what is not to be swallowed. The taste receptors and the supporting cells around them form the lemon-shaped clusters known as taste buds (see Figure 37–3, page 704). We are able to distinguish four primary tastes: sweet, sour, salty, and bitter. While each primary taste appears to stimulate a different type of taste bud, there are indications that a single taste bud may respond to more than one category of substance.

Most animals seem to have the same general range of taste discrimination, although there is some difference in what animals "like" and "don't like." Birds, for example, will readily eat seeds and insects with a bitter taste, whereas for most other vertebrates, bitterness serves as a warning signal. Cats are among the few animals that do not prefer substances with a sweet taste, and very few taste receptors that respond to sugar can be found in the cat's tongue. Most animals will work for a sweet reward even though they are not hungry, and the cat's indifference to sweetness is one of the reasons it is difficult to train—although its general social independence is probably a more significant one.

Smell

In fish, taste and smell are operationally very similar since both involve the detection of substances dissolved in the surrounding water. Taste receptors and smell receptors differ anatomically in fish as in higher vertebrates, however, and the centers of taste and of smell within the brain are entirely distinct.

In terrestrial animals, smell can be defined as the chemoreception of airborne substances. To be detected, however, these substances must first be dissolved in the watery layer of mucus overlying the olfactory epithelium. In humans, this epithelium, which is located high within the nasal passages, is comparatively small. The part within each nostril is only about as large as a postage stamp. Each of these areas contains some 600,000 receptor cells (Figure 39–12). Even with our relatively insensitive olfactory equipment, we are able to discriminate some 10,000 different odors. Much of what we call flavor in food is actually a result of volatile substances reaching the olfactory epithelium.

One of the current theories of odor discrimination was developed in the 1960s by John Amoore while he was still an undergraduate at Oxford University in England. According to the present version of Amoore's theory, all scents are made up of combinations of "primary" odors: The seven proposed by Amoore are camphoric, musky, floral, pepperminty, etherlike, pungent, and putrid. There are also, according to this hypothesis, different types of olfactory receptors corresponding to the primary odors. Molecules of a certain shape fit into correspondingly shaped receptors, just as a piece of a jigsaw puzzle fits into its own particular space. All substances that have a pungent odor, for example, have the same general shape and fit into the same type of receptor. Some molecules, depending on which way they are oriented, can fit into more than one receptor and so can evoke in the brain two different types of signals. From the signals received from the various different types of cells, the brain constructs a "picture" of an odor.

Smell plays a large role in the behavior of most mammals. The males of many species—including domestic dogs and cats—scent-mark their territories with urine as a warning signal to other males. Males are attracted to females by special odors associated with estrus. Such chemical messages among members of the same species are known as pheromones. (Insect pheromones are discussed on page 470.)

39–12

(a) *A patch of special tissue, the olfactory epithelium, arching over the roof of each nasal cavity, is responsible for our sense of smell.* (b) *The olfactory epithelium is composed of three types of cells: supporting cells, basal cells, and olfactory cells. Supporting cells are tall and columnar, wider near the surface than they are deep within the tissue, and their outer surfaces are covered with microvilli, similar to the microvilli found on the surface of intestinal cells. The basal cells, triangular in shape, are found along the deepest layer of the epithelium. Their function is unknown. They may give rise to new supporting cells when these are needed. The olfactory cells are the sensory receptors.* (c) *The olfactory cell is long and narrow. The cilia protruding from its upper surface are believed to be the odor receptors, although the way in which they function is not known. This outermost part of the cell is connected with the cell body by a long stalk containing microtubules arising from deep inside the cell. From the deepest portion of the cell body, a nerve fiber extends through the underlying tissue, transporting the sensory message to the brain.*

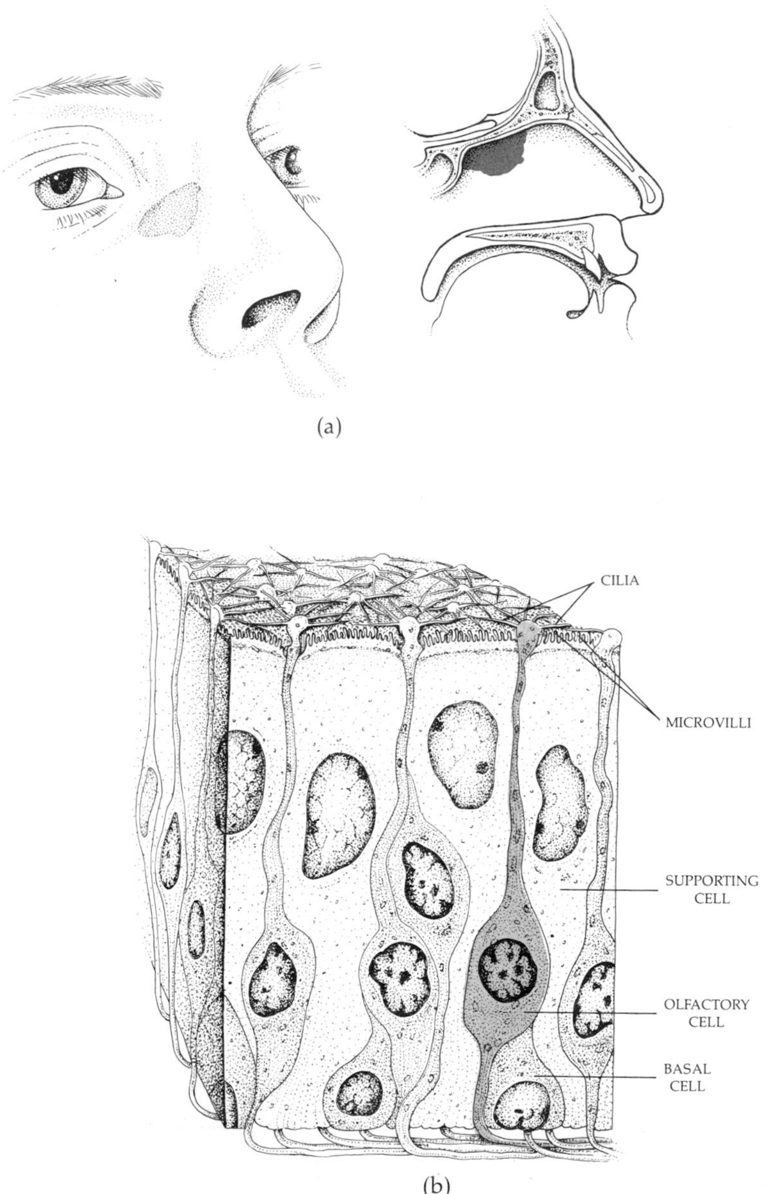

(a)

(b)

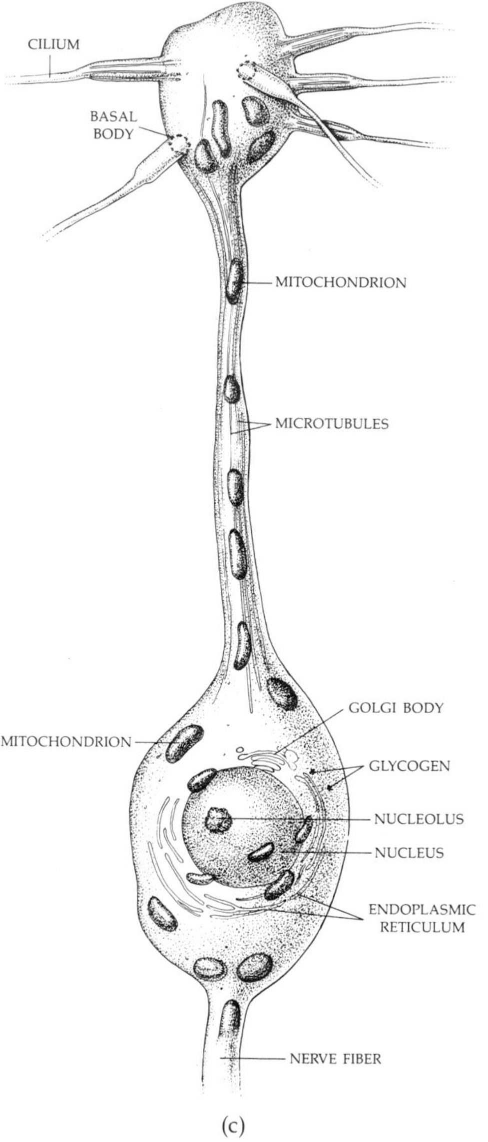

(c)

Among mice, juvenile females reach sexual maturity earlier when exposed to the odor of an adult male. Pregnant females may resorb their fetuses in the presence of the odor of a strange male. Both of these effects are mediated by pheromones in the urine of the male.

Recently a sex attractant has been isolated from female rhesus monkeys and identified as a fatty acid. Whether or not sex attractants and other chemical signals are exchanged among members of the human species is not known. However, the fact that the apocrine sweat glands, the sources of "body odor," begin to function at puberty strongly suggests that the chemicals they produce (also fatty acids) originally played such a role. The pubic and axillary hair that develops at about the same time would appear to have the function of retaining and amplifying these odors. It may be that the current obsession with soaps, deodorants, and feminine hygiene products represents another device by which society attempts to shape and curb natural human proclivities.

SOUND RECEPTION: HEARING

Figure 39–13 diagrams the structure of the human ear. Sound travels through the ear canal to the tympanic membrane (the eardrum), in which it sets up a vibration. This vibration is transferred to a series of three very small and delicate bones, which are called, because of their shapes, hammer (malleus), anvil (incus), and stirrup (stapes). Vibrations in the tympanic membrane cause the stirrup to push gently and rapidly against a membrane, the oval window, leading to the inner ear, and from the inner ear, the signal is transmitted by nerve fibers to the brain.

The middle ear, as the area between the eardrum and the inner ear is called, is connected with the upper pharynx by the eustachian tube. This connection makes it possible to equalize the air pressure in the middle ear with atmospheric pressure;

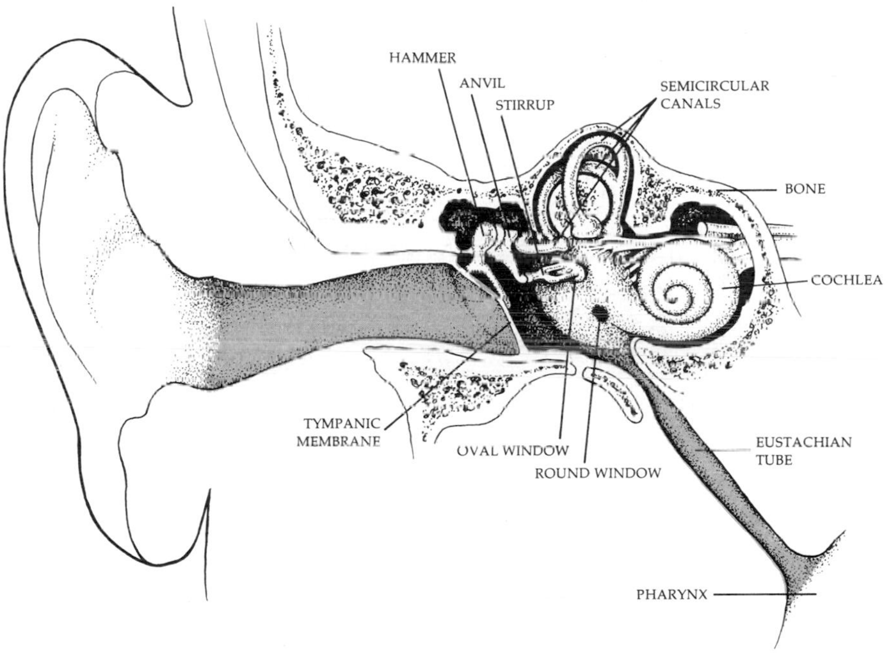

39–13
The structure of the human ear. Sound waves enter the outer ear and pass to the tympanic membrane, which they cause to vibrate. These vibrations are transmitted through the hammer and anvil to the stirrup. The space containing these three small bones (and which connects with the eustachian tube) is the middle ear. The stirrup is attached to another membrane (the oval window), and as the stirrup vibrates, it pushes against the oval window, leading to the cochlea, which is in the inner ear. Fluid vibrations in the inner ear activate hair cells, which send action potentials to the brain. The three semicircular canals are additional fluid-filled chambers within the bony labyrinth of the inner ear. With each one being in a plane perpendicular to the other two, rotational movements of the head set fluids in motion, activating their sensitive hair cells, which inform the brain.

it also unfortunately makes the middle ear a fertile breeding ground for infectious microorganisms that enter the body through the nose or throat.

The hair cells, which are the sensory receptors, are in the inner ear. They lie along a long membrane, the basilar membrane. In mammals, the membrane is coiled in a bony, spiral passageway, the *cochlea* (snail). Figure 39–14 shows a diagram of the cochlea partially uncoiled. It consists essentially of three fluid-filled canals separated by membranes. The upper and lower canals are connected with one another at the far end of the spiral. At the base of each of these two canals are the movable membranes known as the oval and round window membranes. The stirrup vibrates against the oval window membrane at the base of the upper canal, and the sound waves cause pressure waves, which travel the length of the cochlea along the basilar membrane, around the far end, and back again to the membrane at the base of the lower canal, the round window. As the oval window moves in, the round window moves out, keeping the pressure equalized. You will notice that, although hearing is generally considered to be the detection of airborne sounds, our hearing cells are ultimately stimulated by movements in fluid.

The central canal contains the organ of Corti, which rests on the basilar membrane of the canal. The movement of the fluid along the outer surface of the central canal causes vibrations in the basilar membrane, which, in turn, stimulate individual sensory cells, the hair cells, within the organ of Corti.

39–14

The part of the inner ear concerned with hearing is the cochlea, a coiled tube of $2\frac{1}{2}$ *turns—shown here* (a) *as if it were partially uncoiled. Vibrations transmitted from the tympanic membrane to the stirrup cause the stirrup to push against the oval window, resulting in pressure waves in the fluid that fills the cochlear canals. Short pressure waves in the fluid set up vibrations in the basilar membrane, stimulating the sensory cells in the organ of Corti, which rests in the basilar membrane. Sounds at different frequencies (or pitch) have their maximum effect on different areas of the membrane. The round window prevents the pressure from building up in the cochlea. A cross section of the cochlea is shown in* (b), *and a close-up of the organ of Corti in* (c).

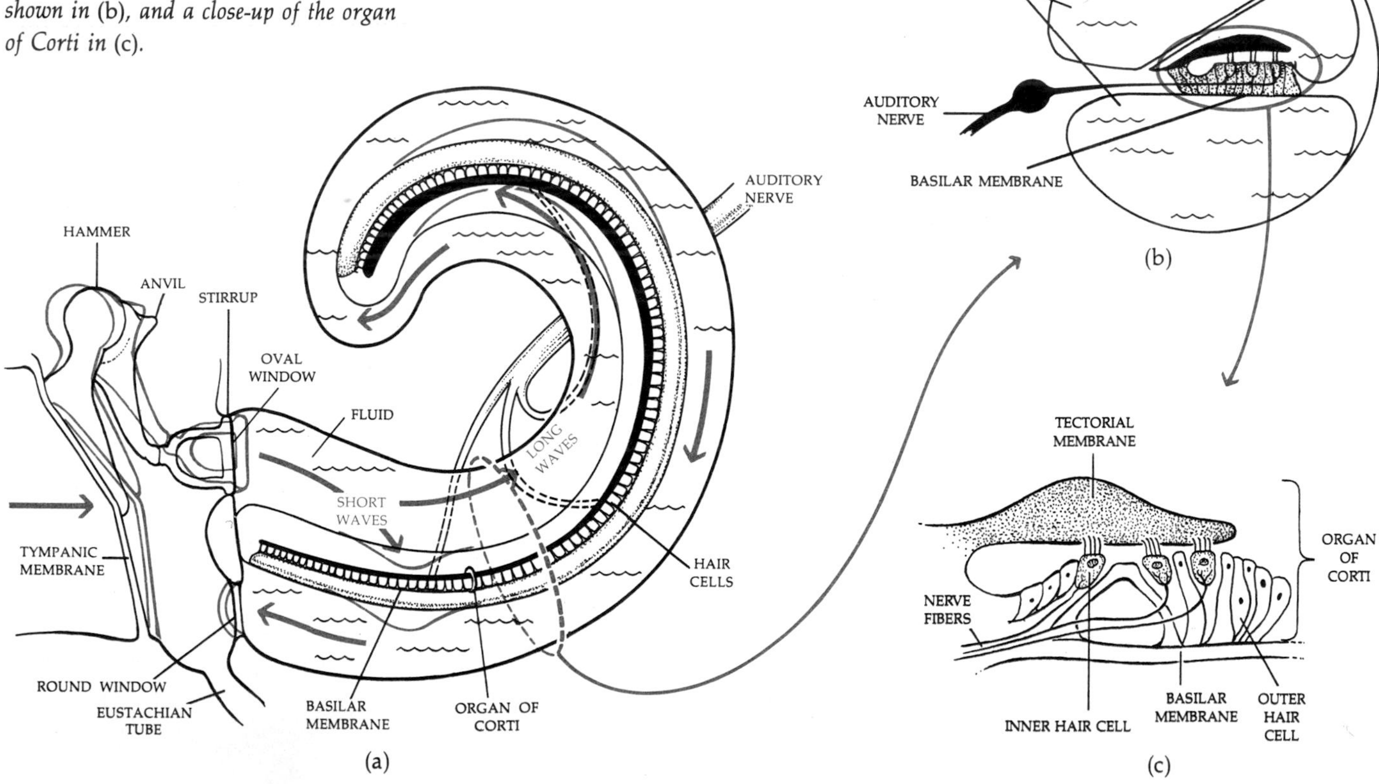

39-15
Scanning electron micrographs of hair cells from the organ of Corti of a guinea pig before (a) *and after* (b) *exposure to 24 hours of loud noise, comparable to that of a loud rock concert. On some of the hair cells, the orderly arrangement of cilia has been disrupted. Other hair cells have degenerated, losing their cilia.*

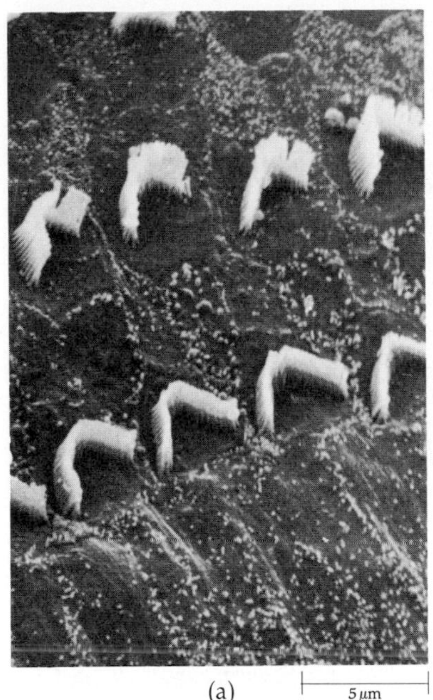

(a)

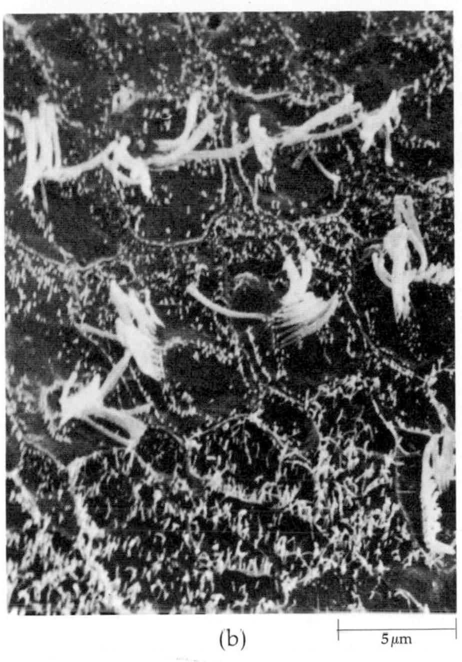

(b)

The basilar membrane is narrower and less plastic at the end nearer the middle ear. Hence it does not vibrate uniformly along its length; instead different areas of the membrane respond to different frequencies of sound. Thus, different hair cells are stimulated by different frequencies. Human beings are capable of very fine discriminations of sound; such discriminations are made by the brain on the basis of the signals received from the various hair cells.

In general, the human ear can detect sounds ranging from 16 to 20,000 cycles per second, although children can hear up to 25,000 cycles per second. (Middle C is 256 cycles per second.) From middle age onward, there is a progressive loss in ability to hear the higher frequencies.

Our inability to hear very low-pitched sounds (below 16 cycles per second) is a useful adaptation since if we could hear at low pitch, we would be barraged with sounds caused by the movements of our own bodies, our moving blood, and those conducted through our bones. We do hear some sounds conducted directly through the skull—for example, the noise we make when we chew celery. We also hear our own voices, mainly through the skull. This is why a recording of one's own voice sounds so startlingly unfamiliar.

Hearing and Behavior

Throughout the vertebrate world, hearing plays a major role in communication and the determination of behavior. Consider, for example, the mating calls of frogs, the songs of birds, the chirp of the baby chick, the challenging bellow of the male gorilla. Many of the auditory signals used by animals extend into ranges we cannot hear. It has been found, for instance, that infant rats and mice are constantly squeaking at frequencies so high that we cannot detect them without special recording devices. The world around us is doubtless filled with noises of which we are unaware.

(a)

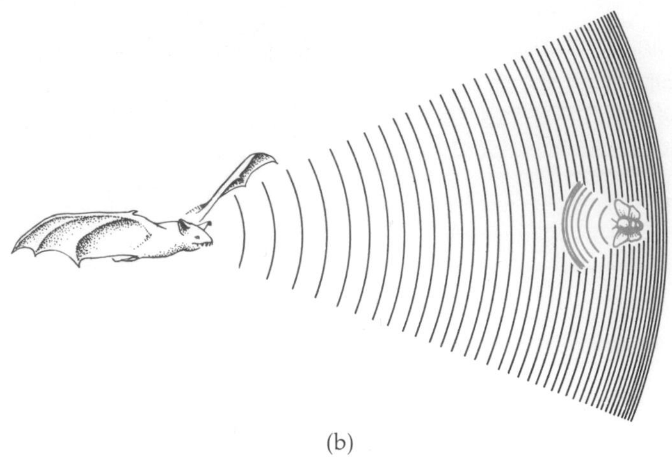

(b)

39–16
(a) *The long ears of this bat are adapted for picking up high-frequency echolocation signals. As the bat flies, it emits shrill squeaks, as many as 30 every second. When it comes closer to some object, it may increase this frequency to about 60 per second. These sounds do not travel far—about 6 meters is the maximum—and they bounce back only about 4½ meters. The bat is guided by these echoes in flying and in hunting prey.* (b) *Echolocation of a moth by a flying bat.*

One animal in which hearing plays a major and unusual role is the bat. In 1793, Lazzaro Spallanzani, a brilliant Italian scientist, became curious about how animals find their way around in darkness. Owls and other nocturnal creatures, he discovered, rely on their large eyes and become helpless in complete darkness. Bats, however, do not depend on vision at all. Spallanzani, in one of the first biological experiments, captured bats in the bell tower of the Cathedral of Pavia, blinded them, and turned them loose. Several days later, he captured the same bats again; not only had the blind animals found their way back to the bell tower but their stomachs were full of freshly caught moths and the other flying insects that were their normal diet. Guessing that the bats might "hear" their way through the darkness, Spallanzani plugged the ear canals of some captive bats and found that the deafened animals were wholly disoriented and bumped into obstacles at random.

Little more was known about the navigation of bats until electronic devices were developed that could record the presence of sounds outside the range of human hearing. Donald Griffin, while still an undergraduate student at Harvard, brought some bats to a laboratory that contained such an apparatus and found that the bats emitted shrill cries above the frequency range of human hearing. When these sound waves hit a solid object, they echo back to the ears of the bat. On the basis of these echoes, bats can navigate skillfully through a dark room strung with wires little thicker than a human hair or can catch an insect as small as a mosquito.

Few other animals can hear the shrill shrieks of bats. An exception is the noctuid moth. Just as the bat's cries have been honed by evolution to locate prey, such as noctuid moths, the tympanic organs of the moth are fine-tuned to detect the predatory cries of the bat.

SUMMARY

Sensory receptors are cells or groups of cells particularly adapted to detecting changes in the environment and so allowing the animal to initiate appropriate responses. Receptors can be classified as interoceptors that monitor internal conditions, proprioceptors that respond to joint position or stretch within the body and provide our positional sense, and exteroceptors that respond to conditions of

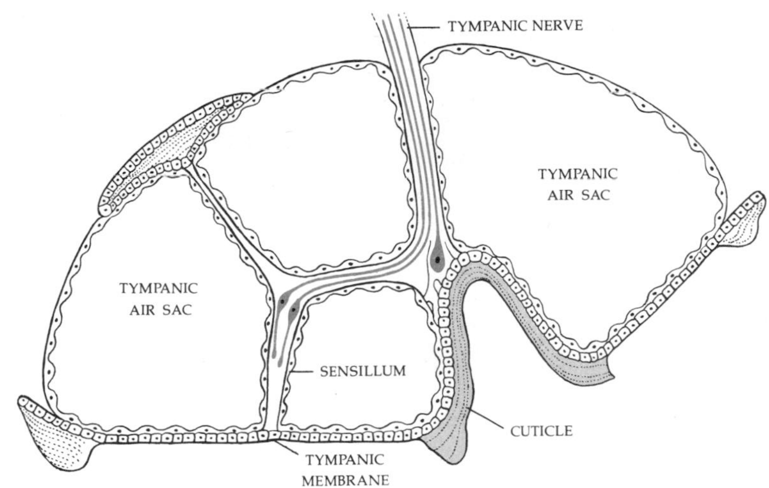

39–17
The most elaborate of the insect sound receptors is the tympanic organ. The tympanic air sacs are covered by a membranous drum, and the sensory cells are so arranged in the organ that they are stimulated by movements of the drum or the air-sac walls. Tympanic organs respond to vibrations in the air or in water.

the external environment, such as eyes, ears, and receptors for touch, pressure, heat, and cold in the skin.

The major human sense organ is the eye, which provides most of our environmental information. The outer, transparent layer of the eye is the cornea, behind which lies the lens. The image is focused on the light-sensitive retina by changes in the shape of the lens, produced by the ciliary muscles. Light passes through the eyeball to the retina where it is received by specialized cells, the rods and the cones. These point toward the back of the retina, toward a pigmented layer that lines the back of the eyeball and captures light not absorbed by the photoreceptor cells. Cones provide greater resolution and are responsible for color vision. Rods are involved in night vision. Some vertebrates, notably birds and humans, have areas specialized for acute vision; such an area is known as a fovea. The photoreceptor cells transmit signals to overlying bipolar cells from which they are relayed to a network of ganglion cells whose axons form the optic nerve. Where the optic nerve passes through the retina on its way to the brain, it creates a blind spot.

Chemoreception involves the detection of specific molecules. In humans, taste buds in the oral cavity are specialized for the discrimination of four primary tastes: sweet, sour, salty, and bitter. Detection of airborne molecules takes place in the olfactory epithelium. The mechanism of chemoreception is unknown. According to current hypotheses, chemoreceptors respond to the three-dimensional, hand-in-glove fit of particular molecules.

Hearing is the function of specialized hair cells in the inner ear. The hairlike processes are bent in response to sound waves, and the bending triggers nerve impulses that are transmitted to the brain. In humans, the outer ear functions as a funnel for capturing and routing sound waves to the tympanic membrane (eardrum). Vibrations in the tympanic membrane are transmitted via the bones of the middle ear (the hammer, anvil, and stirrup). The motion of the stirrup against the oval window, a membrane leading to the cochlea, causes vibrations in the fluid of the cochlea. The resulting movements of the basilar membrane stimulate the hair cells of the organ of Corti. The pattern of stimulation is then relayed to the brain.

QUESTIONS

1. Distinguish among the following terms: exteroceptor/interoceptor/proprioceptor; cochlea/organ of Corti; taste/smell; rod/cone; hormone/pheromone.

2. Label this drawing:

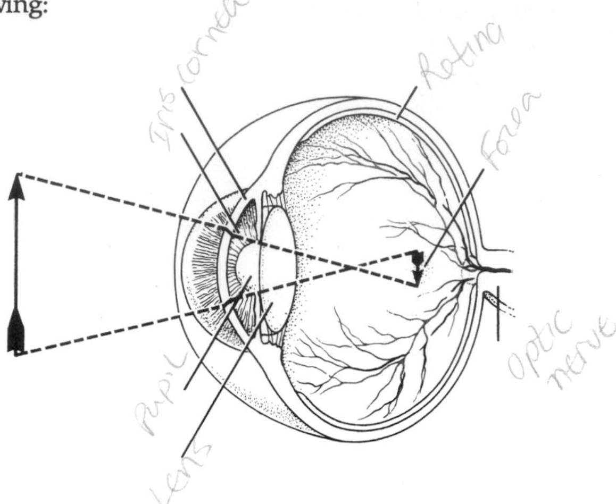

3. In most states, people who are blind in one eye are not issued driver's licenses. Why is this a wise restriction? Most people who are blind in one eye, however, function quite well in ordinary life. How do they manage to do so?

4. The visual acuity of most vertebrates is comparable to our peripheral vision. Think about what you can see out of the corners of your eyes. How can such limited vision be adequate for most vertebrates? Give at least two reasons.

5. The visual acuity of birds is appreciably sharper than ours. Why do birds need such enhanced vision?

6. Consider the process that occurs when another person speaks a word and you then hear the word. Describe the steps involved from the time the air leaves the lungs of the speaker until you are aware of the word spoken.

7. Why is hearing considered a form of mechanoreception?

8. What advantage do mammals gain by having such a long basilar membrane in the organ of Corti?

9. Why does the bat in Figure 39–16 roll up its ears when it sleeps?

CHAPTER 40 Brain and Behavior

The human brain, which weighs about 1,400 grams (3 pounds), has the consistency of semisoft cheese. Like the spinal cord, the brain is made up of both "white matter"—the fiber tracts—and gray matter. The gray matter consists of nerve cell bodies and glial ("glue") cells, which apparently support and nourish the neurons. The glial cells may also play a role in ionic balance and thus in electric potentials. In some areas of the brain, neurons and glial cells are so densely packed that a single cubic centimeter of gray matter contains some 6 million cell bodies, with each neuron connected to as many as 80,000 others.

For at least 2,000 years, people have wondered about the relationship between the brain—this mass of semisoft substance—and the mind, the center of consciousness, thought, and emotion. René Descartes—"I think; therefore, I am"—believed that the body is essentially a complex machine. He conceived of mind and body as two separate entities, whose meeting place was a small gland, the pineal, located

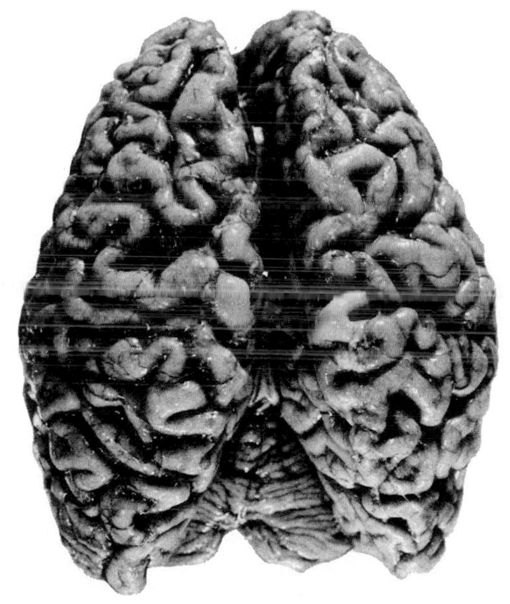

40–1
A brain, viewed from above. The corpus callosum is visible within the deep groove separating the two cerebral hemispheres, and the cerebellum can be seen below the hemispheres at the base of the brain (see also page 643). Note the many convolutions of the cerebral cortex. By these, you can immediately distinguish this brain as human.

within the skull. The persistence of this dualistic concept is reflected in such phrases, still in use today, as "mind over matter." Many years after the cell doctrine was accepted for all other parts of the body, the brain was considered to be an exception. One reason was that it was not possible to see neurons as readily as other types of cells because special staining techniques are required to reveal them. Another and undoubtedly more important reason is the feeling that mental processes are somehow special, different from other physiological functions.

Even today, however, many of us who accept without difficulty the fact that the tissues and cells of the digestive tract are responsible for digestion or that respiratory gases are exchanged across the surface of the lungs find it less easy to comprehend—to *really* believe—that the ideas, ideals, dreams, fantasies, thoughts, hypotheses, loves, hates, fears, and aspirations that make up the mind somehow can be explained by the interactions of the cells within our heads. Nor, indeed, can scientists yet fully explain "mind" in terms of brain, despite the many intriguing discoveries that have been made in this field. Thus, the brain is the last remaining stronghold of the vitalists.

It is also, by far, the most complex and highly organized structure on this planet.

EVOLUTION OF THE VERTEBRATE BRAIN

The human brain is, of course, the product of millions of years of evolution. As we noted in Chapter 33, the vertebrate brain begins embryologically as swellings at the anterior end of the neural tube.

Primitive vertebrate brains have the same linear arrangement (Figure 40–2). In the hindbrain, the most conspicuous changes take place in the cerebellum, which is an outgrowth of the medulla and which is involved with balance and the coordination of movement. The cerebellum is far larger and more convoluted in warm-blooded, fast-moving animals than in fish, amphibians, or reptiles. It reaches its greatest relative size in birds (Figure 40–3).

In the course of vertebrate evolution the midbrain has changed little in size, but it has changed in function. In lower vertebrates, it serves as the principal integra-

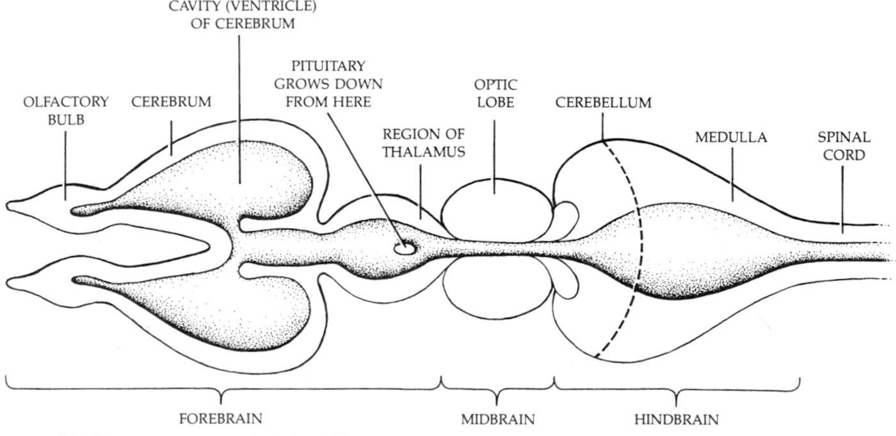

40–2
A dorsal view of the general pattern of the vertebrate brain. The brain has been cut horizontally to show the ventricles (cavities within the brain). The brain ventricles are continuous with the spinal canal.

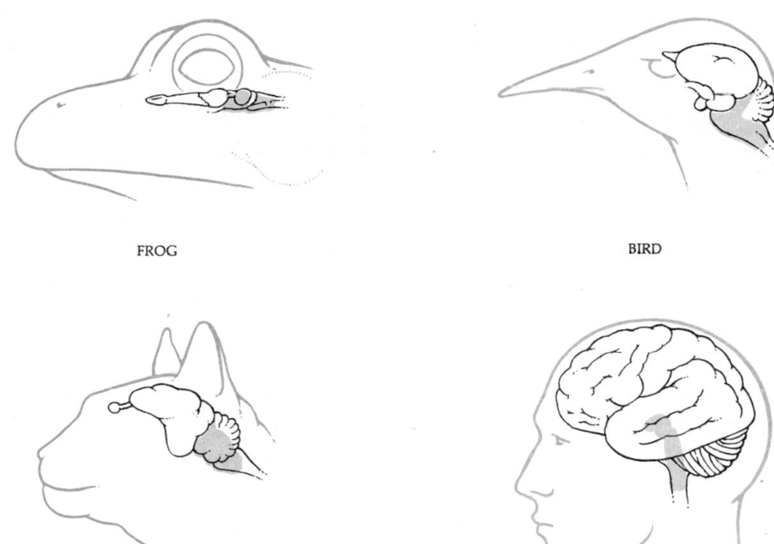

40–3
The brains of various vertebrates. The brainstems (indicated in color) include the pons, medulla, and midbrain. They are approximately the same in the various species. However, the cerebral hemispheres have become progressively larger in the course of evolution. The cerebral cortex, the upper surface of the cerebral hemispheres, reaches its greatest development in the primates, particularly Homo sapiens.

tion center. With the evolutionary development of the forebrain, however, the midbrain has become more of a relay center. One region of the midbrain, the optic lobes, is particularly well developed in nonmammalian vertebrates with good vision, such as birds and fish. In mammals, however, the visual centers are primarily in the forebrain, and the midbrain functions principally as a relay station and reflex center.

The forebrain shows the most dramatic changes. In fish, the olfactory bulb, which processes information from chemosensory receptors, is the dominant part of the cerebrum. This primitive forebrain is often designated as the rhinencephalon ("smell brain"). In the course of evolution, the olfactory bulbs remain about the same size relative to the size of the animal, but the rest of the cerebrum increases greatly and is the dominant structure of the mammalian brain. In primitive mammals, such as the opossum, the cerebral hemispheres are smooth, but in the course of mammalian evolution, the cortex, the outer layer of the cerebrum, has become thicker and increasingly convoluted. Among the primates, the olfactory bulbs are diminished in size (in keeping with the diminished importance of the sense of smell in this mammalian order), and the cerebral cortex has expanded to cover most of the rest of the brain.

Although it is not evident in Figure 40–3, which is not drawn to scale, not only have there been relative changes in various structures of the brain, but also the size of the entire brain has increased enormously in relation to the total body weight of the animal. For example, the brain of an adult gorilla, an animal close to *Homo sapiens* on the evolutionary ladder and larger in body size, weighs only about 400 grams, less than a third the weight of a human brain.

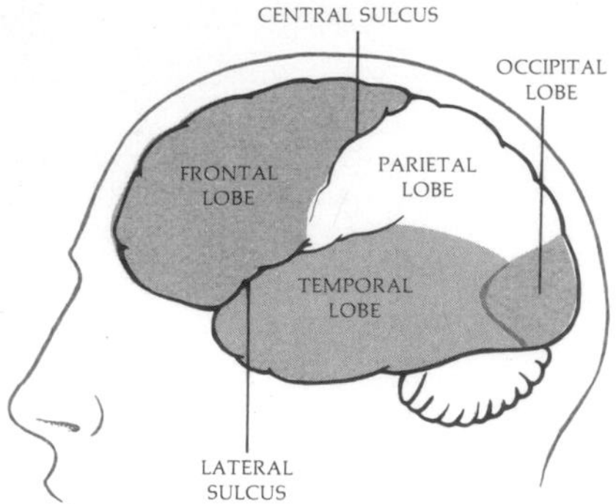

40–4
The principal sulci (fissures) and lobes of the human cerebral cortex.

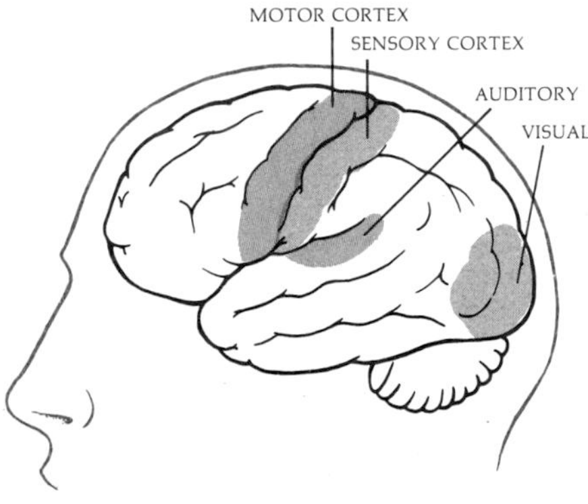

40–5
The human cerebral cortex, showing the location of the motor and sensory areas, on either side of the central sulcus, and the auditory and visual zones.

PROBING BRAIN FUNCTIONS

The Cerebral Cortex

Of the approximately 12 billion nerve cells in the human brain, about 9 billion are in the cerebral cortex. In *Homo sapiens*, as in other primates, each of the cerebral hemispheres is divided into lobes by two deep fissures, or grooves, in the surface. The principal fissures are the central sulcus, which runs down the side of each hemisphere, and the lateral sulcus (Figure 40–4). There are four lobes on each side: frontal, parietal, temporal, and occipital.

Besides these anatomical designations, certain areas of the cortex have been mapped functionally (Figure 40–5). Some of the information is based upon observation of human patients and experimental animals that have had particular areas of the cortex destroyed. Other studies involve stimulating particular areas of the cortex and observing what takes place in various parts of the body, or stimulating various sensory receptors and noting electrical discharges in parts of the cortex. Such investigations have been carried out in experimental animals as well as in humans undergoing brain surgery. (Touching the brain produces no sensation of pain because it has no sensory receptors.)

These studies have shown that the area just anterior to the central sulcus, in the frontal lobe, contains neurons concerned with integration of muscular activities. Here, apparently, is where the electrical signals producing muscular movements originate. Each point on the motor cortex, as it is called, controls the movement of a different part of the body. The relative amount of cortex allocated to a particular group of muscles varies from animal to animal. For instance, in humans, the area controlling the hand and fingers is very large (Figure 40–6), whereas the area controlling a cat's paw is somewhat smaller and that of a horse's hoof is undoubtedly very small.

Posterior to the central sulcus, in the parietal lobe, is the sensory cortex, which is involved with the reception of tactile stimuli and taste. It shows comparable emphasis in those parts of the body that are most richly endowed with sensory receptors: fingertips, tongue, lips, face, genitalia.

The auditory cortex is in the temporal lobe, partially buried within the lateral sulcus (Figure 40–5). Different regions of the auditory cortex respond to different frequencies of sound.

The visual cortex occupies the occipital lobe. By using a tiny point of light to stimulate very small regions of the retina, one after the other, investigators have been able to show that each region of the retina is represented by a corresponding but larger region on the visual cortex. This cortical region contains a variety of cells, different groups of which respond to different types of visual stimuli (see the essay on page 759).

In both experimental animals and humans, damage to specific areas of the parietal lobes seems to result in problems in discriminating specific stimuli. For example, if particular cortical areas of the parietal lobes are damaged, monkeys can no longer tell the difference between patterns, objects, or sounds that they were previously able to distinguish. In humans, damage to specific areas of the cortex on the left side of the brain (in most left-handed as well as all right-handed people) results in impairment of speech (Figure 40–7). The more anterior area (Broca's area) is adjacent to the region of the motor cortex that controls the movements of the

40–6
A combined cross section of one hemisphere of the human cerebrum, indicating the functional areas of the motor and sensory cortices. The motor cortex is indicated in black; stimulation of these areas causes responses in corresponding parts of the body. The sensory cortex is in color; stimulation of various parts of the body produces electrical activity in corresponding parts of this cortex. Notice the relatively huge motor and sensory areas associated with the hand and the mouth. This map is based largely on studies done by neurosurgeon Wilder Penfield on patients undergoing surgical treatment for epilepsy. The motor and sensory cortices are located on either side of the central sulcus. Because nerve fibers cross in the brainstem, the motor and sensory cortices of the right hemisphere are associated with the left side of the body, and vice versa.

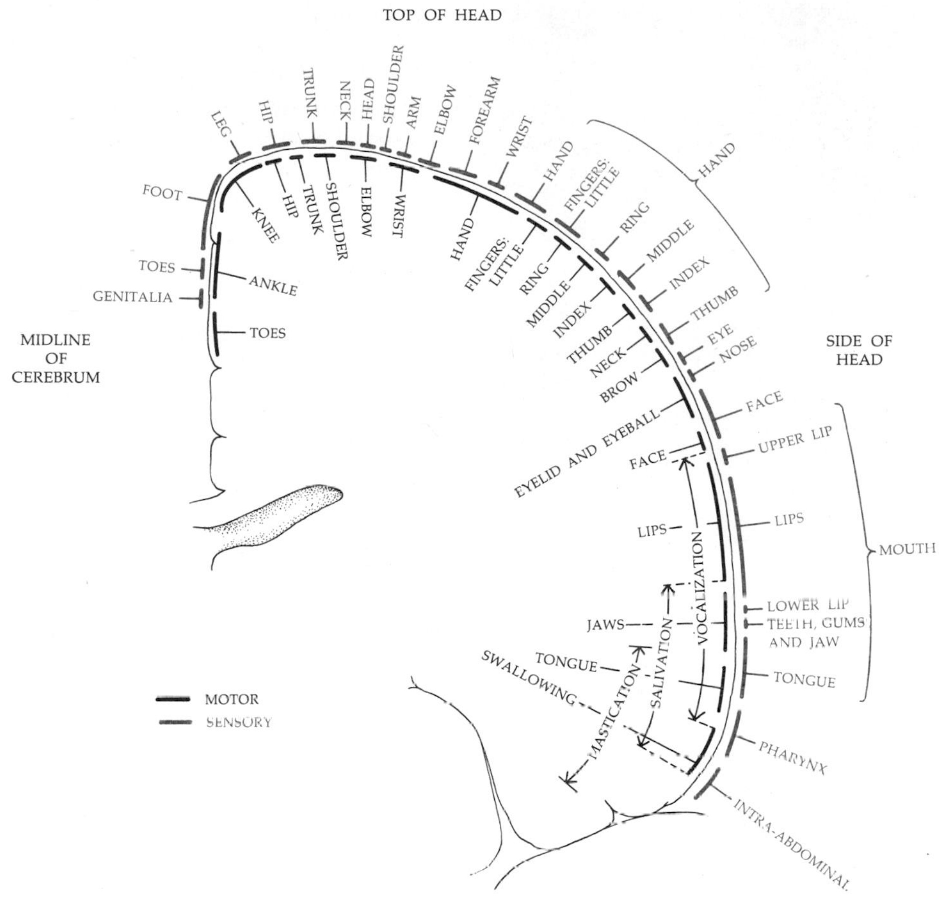

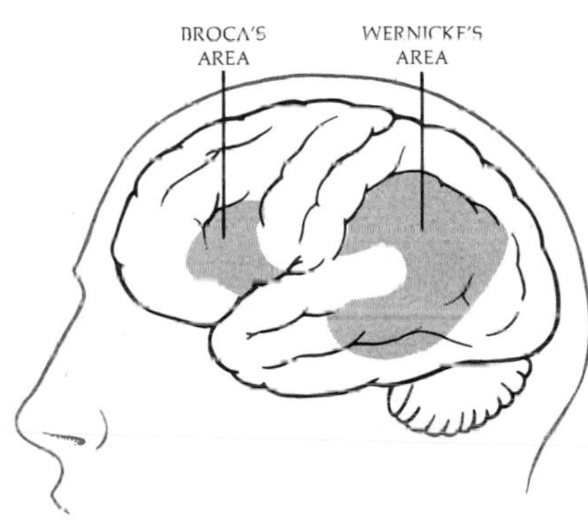

40–7
The human cerebral cortex, showing the areas associated with speech. Damage to Broca's area—the more anterior area—affects motor control of speech. Damage to Wernicke's area affects conceptual aspects of language.

muscles of the lips, tongue, jaw, and vocal cords. Damage to this area results in slow and labored speech but does not affect comprehension. When the more posterior area (Wernicke's area) is damaged, speech is fluent but often meaningless, and comprehension of both spoken and written words is impaired. The two speech areas are joined by a nerve bundle. The corresponding areas in the right brain are associated with music and spatial relations.

The Association Cortex

The unmapped areas of the cortex are sometimes known as the association, or "silent," cortex. The term "association" was introduced when this part of the cortex was thought to function as a sort of giant switchboard, interconnecting the motor and sensory zones. More recent studies suggest, however, that organization of the brain is more vertical than horizontal; for example, severing the motor cortex from the sensory areas by deep vertical cuts appears to have no effect at all on an animal's behavior. Anatomical studies support the interpretation that communication between the sensory and motor areas of the cortex takes place primarily via lower brain centers, particularly the thalamus.

(a)

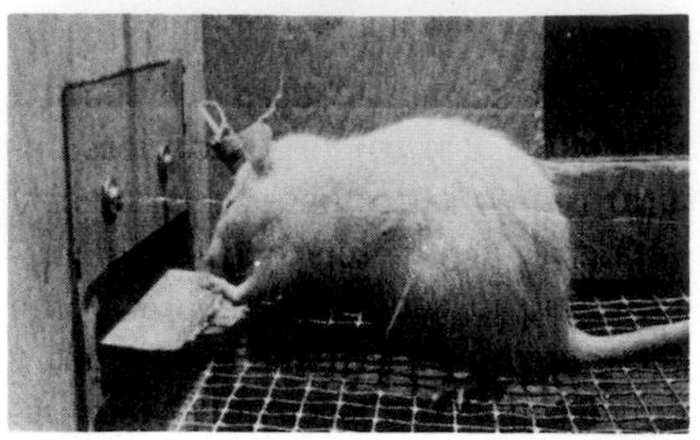

(b)

40–8
(a) *Rat with an electrode implanted in its hypothalamus.* (b) *By pressing a lever, the rat is able to stimulate particular cells in the hypothalamus with a weak electric current. Some animals press the lever as often as 8,000 times an hour, and even when hungry, will prefer the pleasure lever to one giving a food reward.*

Nor are the association areas actually "silent." Although sensory stimuli do not evoke local responses in these areas, and direct electrical stimulation of these areas does not elicit muscle movements, the association cortex shows steady background activity, apparent (but nonspecific) responses to stimuli, and increased activity following electrical stimulation of other areas of the brain. Clearly, something is taking place in these "silent" areas. The proportion of association area to motor and sensory areas is much higher in primates than in other mammals (even mammals such as cats), and is very large in humans. This suggests that these areas have something to do with what is special about the human mind.

About half of the association area of the cortex is in the frontal lobes, the part of the brain that has developed most rapidly during the recent evolution of *Homo sapiens.* The frontal lobes are responsible for our high forehead, as compared with the beetle brow of our most immediate ancestors. Public appraisal of their function is reflected in the terms "high brow" and "low brow."

The Hypothalamus

The hypothalamus is a small (about 4 grams or 1/10 ounce) but functionally mighty area of the brain. We noted in Chapter 34 that the hypothalamus produces oxytocin and ADH (antidiuretic hormone), which are then stored in and released from the posterior lobe of the pituitary. Moreover, cells of the hypothalamus secrete short polypeptide "releasing factors" into a specialized vascular network that carries these substances to the anterior pituitary where they control the secretion of all of the anterior pituitary hormones. In this way the hypothalamus modulates and integrates much of the endocrine function of the body. Factors affecting the internal environment, such as blood pressure and heart rate, are also modified by particular hypothalamic nuclei. The hypothalamus is also a control center for sexual drive, anger, hunger, thirst, and pleasure; in experimental animals, stimulation of particular areas in the hypothalamus can result in sexual arousal, anger, food-seeking behavior, overeating, overdrinking, and so on. Rats with electrodes implanted in the hypothalamus will spend much of their time pressing levers that stimulate these electrodes (Figure 40–8). This phenomenon can be localized in particular nerve cells that seem to form relay stations for several "pleasure pathways" through the brain.

Action Systems of the Brain

Behavior, especially complex behavior, involves circuits connecting many different areas of the brain. These circuits are often difficult to identify anatomically. Two in particular, the reticular activating system and the limbic system, have commanded the attention of investigators.

The Reticular Activating System

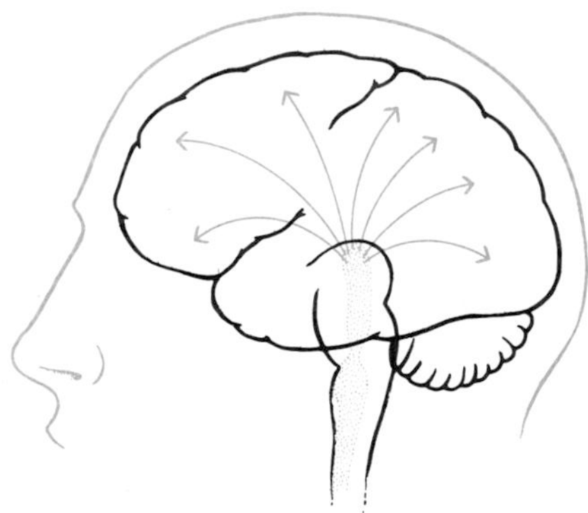

40–9
The reticular system is a diffuse network of neurons in the brainstem. Here, incoming stimuli are monitored, analyzed, and relayed to other areas of the brain.

The reticular system, a diffuse network of neurons in the brainstem, is of particular interest to physiological psychologists because it is involved with arousal of that hard-to-define state we know as consciousness. All the sensory systems have fibers that feed into the reticular system, which apparently filters incoming stimuli and "decides" whether or not they may be important. Electrical stimulation of this system, either artificially or by the incoming sensory impulses, results in increased electrical activity in other areas of the brain.

THE EYE AND THE BRAIN

When we look at an object, as we noted in Chapter 39, the lens of our eye, like the lens of a camera, projects the object onto the retina, stimulating the photoreceptor cells. Impulses are carried from the retina by the fibers of the optic nerve. These fibers lead to a visual relay center in the thalamus, where some of the fibers from the two eyes come together on each side of the brain and synapse with fibers leading to the visual cortex. Various experiments have shown that the spatial arrangement of neurons on the cortex corresponds topographically with the spatial arrangement of the image as it is received on the retina, except for "over-representation" of the fovea. (The fovea, which occupies only about 1 percent of the area of the retina, projects to nearly 50 percent of the visual cortex.)

How might it be that we perceive form? One answer is that such perception may have to do with the way the visual image is arranged on the visual cortex, allowing for the foveal distortion. However, studies conducted over the past 15 years on cats and monkeys by David H. Hubel and Torsten N. Wiesel of Harvard University reveal that quite a different mechanism may be involved. They used microelectrodes to record from single cells of the visual cortex, recorded the responses of individual cells to visual stimuli, and came up with some very surprising results. One class of cortical neurons, for example, was found to respond only to a horizontal bar; as the bar was tipped away from the horizontal, discharges from the cells slowed and ceased. Another class responded best to a vertical bar, ceasing to fire as the bar was oriented toward the horizontal. A third class fired only when the bar was moved from left to right; another only to a right-to-left movement. Another class of cells turned out to be a right-angle detector; such cells responded only to an angle moving across the visual field and were most excited when the angle was a right angle. Thus an important component of the visual information-processing system appears to be a set of very specifically tuned neurons, each responsive to one aspect of the shape of an object in the visual field. If this hypothesis is correct, then there must be some further information-processing system synthesizing the final perception by integrating the information received from its individual elements.

You will remember that neurons in the frog's retina carried out much this same kind of data processing. These very fundamental discoveries would seem to be leading to a new and far deeper level of understanding of the ancient dilemma about the relationship between the human mind and the outside world.

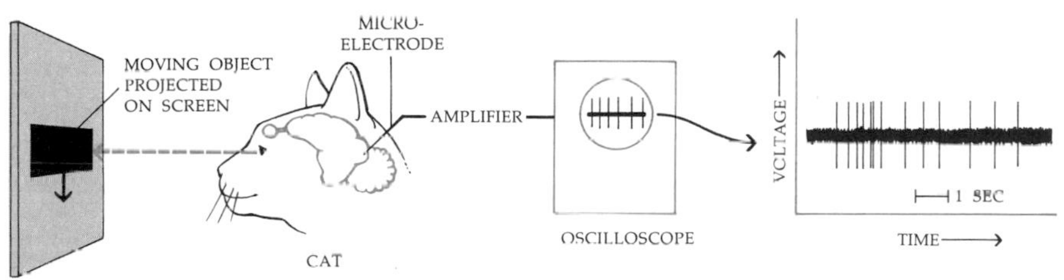

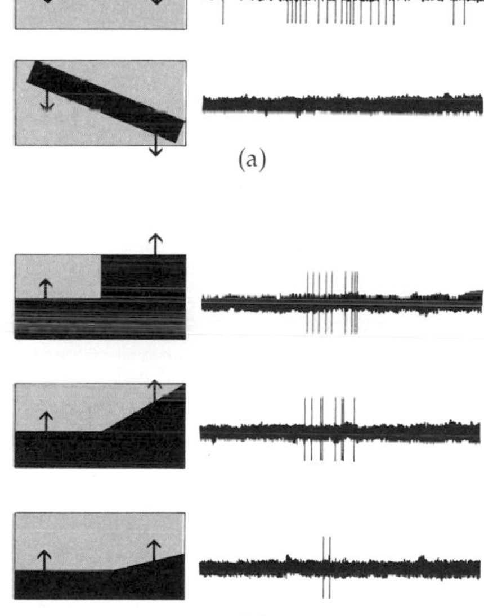

The experimental method of Hubel and Wiesel. A cat with a small electrode recording from a single cell in its visual cortex observes a screen on which a moving shape is projected. (a) *and* (b) *represent tracings from two different neurons. The first neuron* (a) *responds to a horizontal bar; as the bar is turned toward the vertical, the action potential ceases. The second neuron* (b) *responds to a right angle. Note the reduction in response as the angle increases. The arrows indicate the direction of movement of the shape on the screen.*

40-10

The limbic system, a group of interconnected structures within the cerebrum, is associated with emotional and motivational aspects of behavior. The structures include the hippocampus, amygdala, and fornix, and correspond to the forebrain of primitive vertebrates.

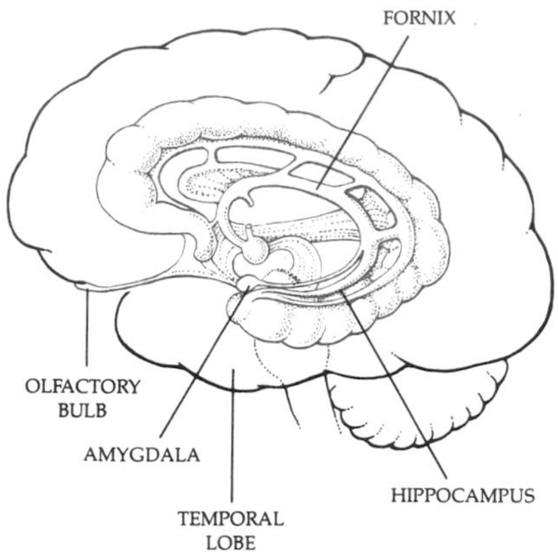

The existence of such a filtering system is well verified by ordinary experience. A person may sleep through the familiar blare of a subway train or a loud radio or TV program but wake instantly at the cry of the baby or the stealthy turn of a doorknob. Similarly, we may be unaware of the contents of a dimly overheard conversation until something important—our own name, for instance—is mentioned, and then our degree of attention increases.

The Limbic System

The limbic system is a neuron network that forms a loop around the inside of the brain, connecting the hypothalamus to the cerebral cortex. It is thought to be the circuit by which drives and emotions, such as hunger, thirst, and desire for pleasure, are translated into complex actions, such as seeking food, drinking water, or pressing a lever. The brain structure of the limbic system includes the hippocampus and other primitive regions of the brain, and corresponds, in effect, to the forebrain (rhinencephalon) of reptiles.

CHEMICAL ACTIVITY IN THE BRAIN

As we noted in Chapter 34, transmission of nerve impulses across synapses in the human nervous system involves the release of chemicals into the synaptic gap. In the peripheral nervous system, the chemical transmitters are acetylcholine and norepinephrine. Acetylcholine and norepinephrine are also transmitter substances in the brain. At least three additional substances have been found in the central nervous system: dopamine, serotonin (5-hydroxytryptamine), and gamma-aminobutyric acid. These are closely related chemically to norepinephrine, and together they constitute a group of chemicals known as the biogenic amines.

Norepinephrine is a transmitter substance of the reticular activating system and thus may play a role in arousal and attention. Severe depression, for example, may be related to an abnormally low level of norepinephrine in the synaptic gap. Two types of medications have been discovered that can increase the concentration of norepinephrine at the nerve ending. One works by inhibiting monoamine oxidase, the enzyme that normally destroys norepinephrine after its release. Such compounds are referred to as monoamine oxidase inhibitors. The other method of increasing levels of norepinephrine in the synaptic gap is by preventing its resorption by the synaptic knob. Antidepressants like amitriptilyne (Elavil) and imipramine (Tofranil) act in this way. Norepinephrine is found in particularly large concentrations in the hypothalamus and the limbic system.

Dopamine is the transmitter substance for a relatively small group of neurons involved with muscular activity. It has been discovered that Parkinson's disease, which is characterized by muscular tremor, is associated with a decrease in dopa-

40-11

The structures of some biogenic amines that function as neurotransmitter substances in the brain.

HO, HO—[ring]—$CHCH_2NH_2$ (OH) — NOREPINEPHRINE

HO, HO—[ring]—$CH_2CH_2NH_2$ — DOPAMINE

HO—[indole ring, N, H]—$CH_2CH_2NH_2$ — SEROTONIN (5-HYDROXYTRYPTAMINE)

O, HO—C—$CH_2CH_2CH_2NH_2$ — GAMMA-AMINOBUTYRIC ACID

Table 40–1 *Drugs That Affect Mood and Behavior*

DRUG	EFFECTS	MECHANISM OF ACTION
Caffeine (coffee, tea, cola drinks)	Stimulant; promotes alertness; increases motor activities; insomnia	Facilitates synaptic transmission
Nicotine (tobacco)	Stimulant	Mimics acetylcholine, facilitates synaptic transmission
Amphetamines (Benzedrine, Methedrine, Dexedrine)	Stimulants; lessen fatigue and depression; suppress appetite; cause malnutrition, exhaustion, impairment of judgment, psychoses; strongly addictive	Increase activity of biogenic amines in brain
Alcohol	Sedative (after preliminary excitatory effect); loss of motor coordination and alertness; euphoria; cirrhosis of liver	Depresses central nervous system (exact mechanism unknown)
Chlorpromazine (Thorazine, Compazine, Stelazine)	Tranquilizer; depresses reticular-system activity; suppresses hallucinations and delusions; antipsychotic	Blocks receptors of epinephrine and acetylcholine
Barbiturates (Seconal, Amytal, Nembutal, Tuinal)	Sedatives; induce sleep, euphoria, irritability, hallucinations; highly addictive	Interfere with synthesis and release of norepinephrine and serotonin
Opium, morphine, heroin, methadone, Demerol, codeine	Relieve pain; induce muscle relaxation, drowsiness, lethargy, euphoria, sleep; produce constipation, loss of appetite, kill by depressing respiratory center; highly addictive	Depress central nervous system
Cocaine	Local anesthesia; euphoria; alertness; addiction	Inhibits uptake of norepinephrine
MAO inhibitors	Antidepressants	Increases brain levels of norepinephrine
Psilocybin, mescaline	Increased sensory awareness; hallucinations; psychoses	Chemicals similar to biogenic amines; may enhance their effects
LSD	Rich visual imagery; sensory awareness; visual hallucinations	Inhibits brain serotonin, perhaps by decreasing activity of serotonin-producing cells in reticular system
Marijuana, hashish	Mild euphoria; enhanced sensory perception	Unknown

mine. L-Dopa, a closely related compound, has been used with great success in treating some persons with this fairly common condition. L-Dopa also appears to be of benefit in the treatment of some cases of depression.

Serotonin is found in many portions of the central nervous system, including, in particular, the hypothalamus and the limbic system. All of the cells producing serotonin, however, have their cell bodies in a single small area in the brainstem that is part of the reticular system. It thus appears that the serotonin synapses may characterize a very old portion of the brain dealing with basic functions such as sleep, consciousness, and emotions.

Table 40–1 lists drugs that act on the brain to alter mood and behavior. As you can see, many of these agents appear to exert their effects by enhancing or inhibiting the effects of chemical transmitters.

Internal Opiates

The word opium comes from the Greek *opion,* meaning "poppy juice." Since the time of the Greeks, poppy juice and its derivatives have been used for control of pain. They are the most potent pain-killers known, and their psychological effects are greatly enhanced by the fact that they produce euphoria. They also are addictive, and although the pharmaceutical industry, urged on by the possibility of great profits, has made repeated attempts to develop an opium derivative that is nonaddictive, their efforts have been uniformly unsuccessful. (Demerol was a recent failure.) Further, in the tests of all of these agents, it has not been possible to determine the mechanism by which the opiates decrease pain, produce euphoria, and result in addiction.

It has long been suspected that the opiates act upon the brain by binding to specific receptors. All substances with opiate-like action are related chemically and have similarities in their three-dimensional structures. Moreover, chemicals are known that block the effect of opiates and can, for instance, act as antidotes in cases of overdoses of opiates. These chemicals are apparently similar enough in structure to block the receptors but not similar enough to produce the opiate effects.

Using opium derivatives labeled with radioactive isotopes, investigators have been able to show that the brain does, indeed, have receptors for opiates. These receptors are located primarily in the limbic system and in the spinal cord; their effect is to inhibit the production of nerve impulses by neurons in this circuit. Such receptors have been found not just in humans, but in all other vertebrates tested.

Why would vertebrate brains have opiate receptors? Only one answer seemed logical: Because vertebrate brains themselves must produce opiates. This rather startling conclusion has now been found to be true. Not one but four naturally occurring substances have now been isolated that bind to the opiate receptors. All four have analgesic (pain-relieving) effects. The two found in brain tissue, known as enkephalins, are believed to function as neurotransmitters with inhibitory effects. It is hypothesized that, under normal conditions, our natural opiate receptors are exposed to a relatively low level of enkephalins. Under conditions of stress or pain, the release of enkephalins increases, producing the analgesic effect and, probably, the mood-elevating state. Administering morphine or other opiates enhances these effects. However, by a negative feedback mechanism, administered opiates reduce the normal production of enkephalins, resulting in dependency on the artificial source (addiction).

Substances isolated from the pituitary gland, which have been called endorphins, also have opiate activity. In fact, one has been found to be three times as powerful as morphine when injected intravenously, and 48 times as powerful when injected directly into the brain. Because endorphins are abundant in the pituitary

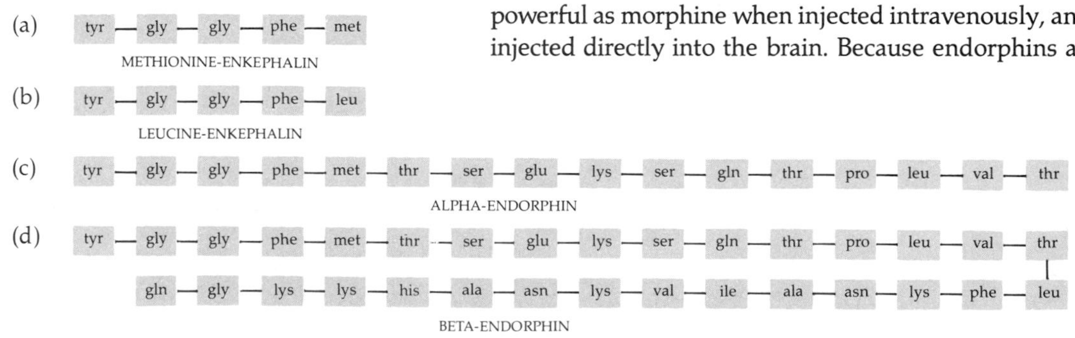

40–12

The amino acid sequences of four naturally occurring substances with opiate activity: (a) *methionine-enkephalin,* (b) *leucine-enkephalin,* (c) *alpha-endorphin, and* (d) *beta-endorphin. The enkephalins are found in the brain tissue of vertebrates, and the endorphins are in the pituitary gland. The first four amino acids of each sequence are identical.*

(a) tyr — gly — gly — phe — met

METHIONINE-ENKEPHALIN

(b) tyr — gly — gly — phe — leu

LEUCINE-ENKEPHALIN

(c) tyr — gly — gly — phe — met — thr — ser — glu — lys — ser — gln — thr — pro — leu — val — thr

ALPHA-ENDORPHIN

(d) tyr — gly — gly — phe — met — thr — ser — glu — lys — ser — gln — thr — pro — leu — val — thr — leu — phe — lys — asn — ala — ile — val — lys — asn — ala — his — lys — lys — gly — gln

BETA-ENDORPHIN

40–13
A continuous discharge of alpha waves seems to be associated with tranquility and well-being. By concentrating on a musical tone that sounds when alpha waves are produced, it is possible to learn to turn on those waves at will. This technique is known as biofeedback training, or BFT.

and because they are not found in the brain, where the effects of pain are clearly mediated, it has been suggested that the endorphins may be hormones whose primary effects and target tissues are as yet unknown.

ELECTRICAL ACTIVITY OF THE BRAIN

Another way in which the brain is studied is by electroencephalography. An electroencephalogram (EEG) is a record of continuous electrical activity of the brain as measured by the difference in electric potential between an electrode placed on a specific area of the scalp and a "neutral" electrode placed elsewhere on the body, or the difference between pairs of electrodes on the head. The voltages that arise are very weak—about 300 microvolts is the maximum in a normal adult—and extremely sensitive recording equipment is required. (A microvolt is a millionth of a volt, or a thousandth of a millivolt.)

EEG recordings obviously reflect only gross electrical activity. Donald Kennedy of Stanford University, one of the leading investigators in the field of nerve cell function, compares the electroencephalogram to measurements that an observer on Mars might make of the roars emanating from the Houston Astrodome during a baseball game. They would certainly indicate that something was going on in Houston, but it would be very difficult to infer the rules or determine the progress or the nature of the game from such recordings.

Nevertheless, there are characteristic EEG patterns that can be correlated with certain levels and types of brain activity.

Alpha and Beta Waves

One characteristic EEG pattern is the alpha wave, a slow, fairly irregular variation in electric potential, with a frequency of about 8 to 12 cycles per second, recorded at the back of the head. Alpha waves are usually produced during periods of relaxation; in most persons they are conspicuous only when the eyes are shut. About two-thirds of the general population have alpha waves that are disrupted by attention. Of the remaining third, about one-half have almost no alpha rhythm at all and about one-half have persistent alpha rhythms that are not easily disrupted by attention.

Another EEG pattern is the beta wave. Beta waves are of smaller amplitude (lower voltage) than alpha waves, but their frequency is greater, being about 18 to 32 cycles per second. Beta waves occur in bursts in the anterior part of the brain and are associated with mental activity and excitement.

Waves of lower frequencies (delta waves) are normally seen in infants or in adults during sleep. Delta waves in waking adults are signs of mental disturbance or brain injury.

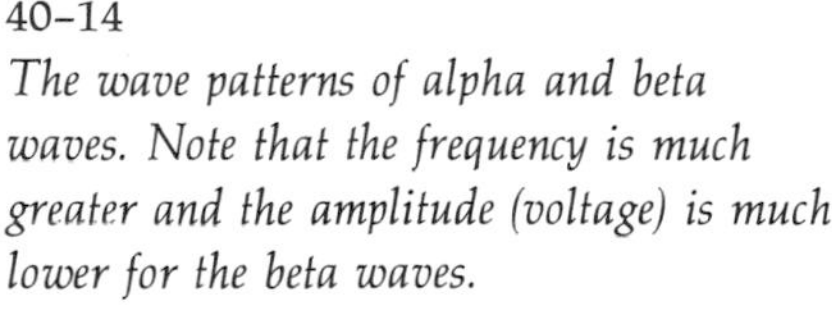

40–14
The wave patterns of alpha and beta waves. Note that the frequency is much greater and the amplitude (voltage) is much lower for the beta waves.

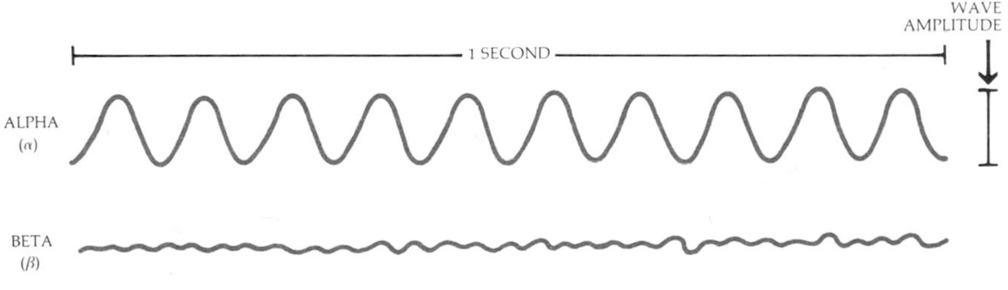

40-15
(a) *Brain wave recordings of a student with conspicuous alpha rhythm. Electrodes were placed at four positions on the head. When the student was asked to multiply 18 × 14, the alpha rhythm was suppressed and then resumed after the problem was solved (10 seconds of the recording are omitted in order to keep it on the page).*
(b) *Brain waves from a student with a complete absence of alpha rhythms. According to a recent study, people who have almost no alpha rhythms under any conditions think almost exclusively by visual imagery, whereas people with a persistent alpha rhythm tend to be abstract thinkers.*

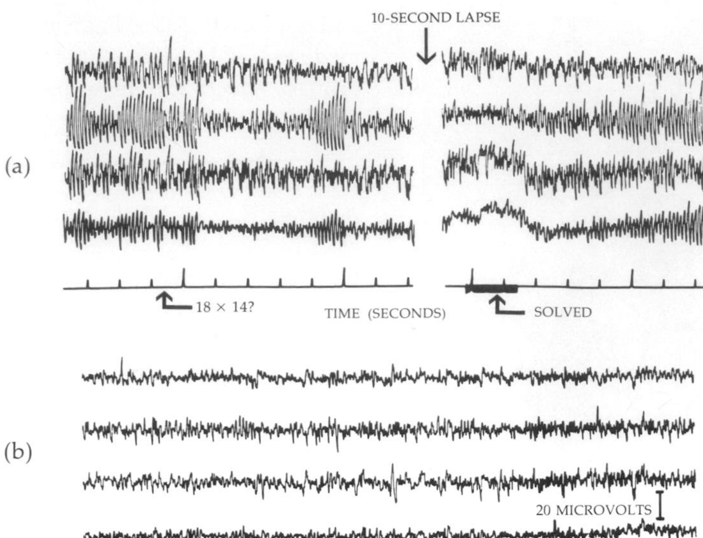

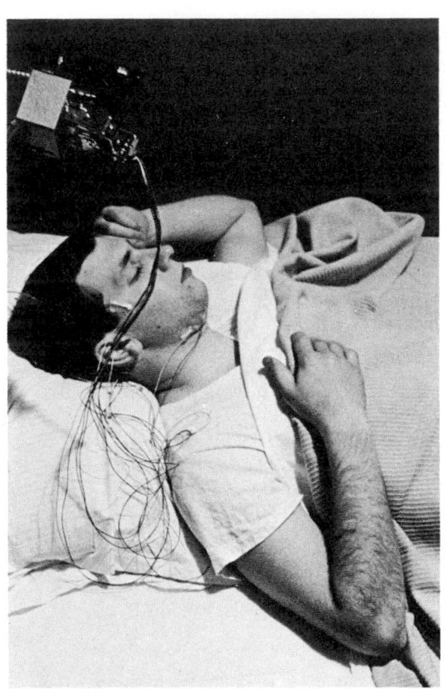
40-16
Volunteer wired to record brain waves, eye movements, breathing rate, and other physical functions.

SLEEP AND DREAMS

Considering the prominence of sleep in our daily existence, it is remarkable how little we know about it. Why does it occupy one-third of our lives? What happens during sleep to restore our capacity to function? Why does sleep, or the lack of it, have such profound effects on our alertness, judgment, good temper, coordination, and emotional stability? Guinea pigs have survived, never sleeping, after destruction of areas in the midbrain. There are also two reported cases (perhaps apocryphal) of persons, one in Italy and one in Australia, who never sleep, presumably because of some brain dysfunction, such as that inflicted on the guinea pigs. Thus, sleep is certainly not necessary for survival in the same way as oxygen, food, or water.

One of the oldest hypotheses concerning sleep is that it results from the buildup of some chemical in the brain that induces sleep and is then altered or destroyed during the sleep process. Certain chemicals, including serotonin, have been discovered to induce sleep, but whether any of these is the natural sleep promoter—or if, indeed, there is a sleep-promoting chemical—is not known. In fact, some studies seem to refute this attractive hypothesis; for example, Siamese twins sharing a common circulation may sleep at different times.

During sleep, heart rate and arterial pressure decrease, breathing becomes more shallow, the CO_2 concentration of the blood increases, and rectal temperature may drop more than 1°C. There are also obvious changes in the electroencephalogram, on the basis of which sleep can be divided into five stages.

Stage 1. Drowsiness, in which alpha waves are slowed slightly. This stage is not classified as true sleep but rather as a transition period between wakefulness and sleep. Sustained Stage 1 EEG patterns do not fulfill sleep requirements in experiments with human volunteers.

Stage 2. Light sleep, which is a mixture of slow (3 to 6 per second) low-amplitude waves interrupted by short bursts of faster (14 to 15 per second) waves known as sleep spindles.

Stage 3. Intermediate sleep with high-amplitude waves at about 2 per second (delta) and with occasional sleep spindles.

Stage 4. Deep sleep with very slow waves of comparatively high voltage and without spindles.

REM sleep. Every 80 to 120 minutes, there occurs a period of rapid, low-voltage activity similar to that seen in alert persons. This is also called paradoxical sleep and is associated with rapid eye movements (REMs), in which the eyeballs move jerkily as if the sleeper were watching some scene of intense activity.

As shown in Figure 40–17, sleepers pass through a regular sleep cycle, moving from one stage of sleep to another. The average young adult spends 50 percent of the sleeping time in Stage 2 and about 20 percent in REM sleep, divided into three to five episodes.

REM sleep differs from the other sleep stages. For example, although there is electroencephalographic evidence of alertness, the muscles are more relaxed than in ordinary Stage 4 sleep. Further, although the sleeper is harder to awaken from REM than from Stage 4 sleep, once awakened, he or she is more alert. Fluctuations in heart rate, blood pressure, and respiration are also seen during REM sleep. Many male subjects experience erections of the penis during REM sleep, and women have similar erections of the clitoris with secretions of vaginal fluid. A person awakened during a period of REM sleep nearly always reports that he or she has been dreaming. At one time, REM sleep was thought to be the period when all dreaming occurred.

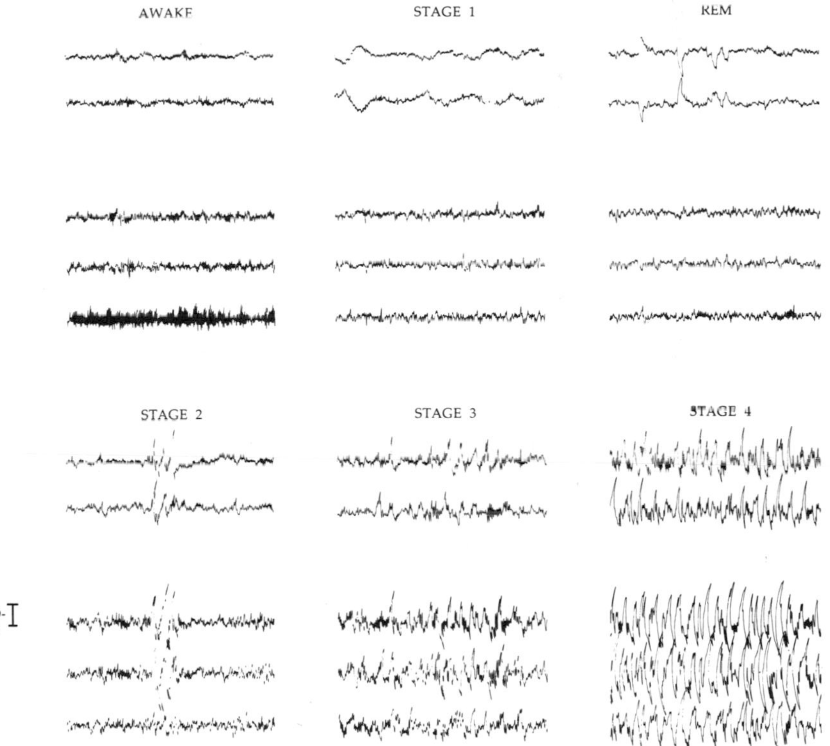

40–17
Stages of sleep as shown by an electroencephalogram. Five tracings are shown for each stage; the upper two are from electrodes attached beside the eye, and the lower three from electrodes attached on the scalp. These recordings represent 20 seconds of each sleep stage.

On the basis of more recent work, it has been suggested that dreaming can occur at any time during sleep but that conditions for recalling dreams are most favorable when subjects are awakened during REM sleep. Some alternative hypotheses have been suggested. One proposal is that REM sleep serves as an information-processing period during which data from the previous waking cycle are sorted, processed, and stored. So this stage of sleep is essential to the memory process. Laboratory rats forget tasks they have learned if they are deprived of REM sleep. Similarly, evidence indicates that the student who stays up all night cramming for an exam will not have as good a recollection of the material as the student who studies and then sleeps.

According to other hypotheses, only Stage 4 sleep of the total sleep cycle is necessary for the restorative functions of sleep. REM sleep is thought to be a partial arousal, serving the same sort of sentinel function as the occasional arousals seen in hibernating animals. These studies and the controversies they engender are continuing.

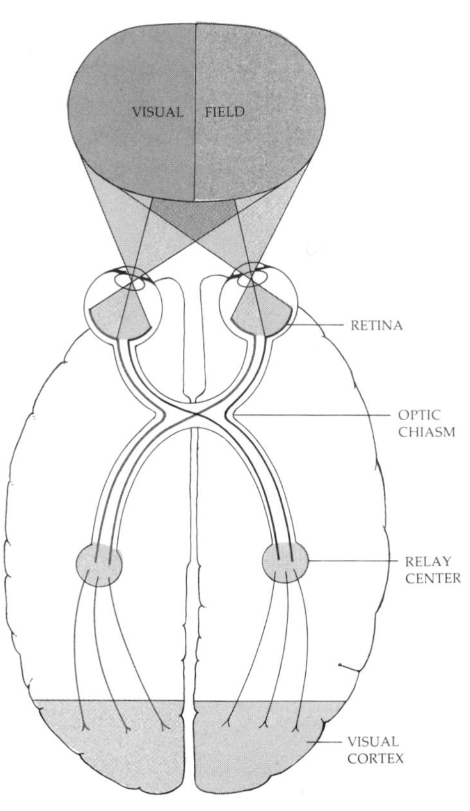

40–18
The optic chiasm viewed from below. The chiasm is the structure formed by the crossing-over of fibers traveling from the retina to the visual cortex. Because of this crossing-over, the right and left cortical areas each "see" only the opposite half of the visual field.

SPLIT-BRAIN STUDIES

Additional interesting implications concerning processing of information in the cortex have come from split-brain studies carried out by Roger Sperry and coworkers at the California Institute of Technology. Sperry severed the connections between the two cerebral hemispheres in cats and monkeys by cutting through the corpus callosum and the optic chiasm (see Figure 40–18). The animals' everyday behavior was normal, but by using specially contrived testing techniques, the investigators were able to show some remarkable and unexpected effects. The animals responded to test procedures as if they had two brains. When the right eye of an animal was covered, for instance, it could learn tasks and discriminations using only the left eye. But then if the left eye were covered and the same tests were presented to the right eye, the animal had to learn all over again. In fact, it was possible, without creating any confusion at all, to have both sides of the brain learn opposite solutions to the same problem.

Similar observations have been made in humans in whom the corpus callosum has been severed for the treatment of epilepsy. The patients are able to carry out their normal activities, but specially devised tests indicate the presence of what Sperry calls "two realms of consciousness." If such patients are asked to identify by touch objects that they cannot see, they can name those they can feel with the right hand but not those they can feel only with the left hand. This is apparently because information from the right hand goes to the left cerebral hemisphere and the speech centers of the brain are located in the left hemisphere; the right brain is mute. If, for instance, a patient's left hand is given a plastic object shaped like the number 2, which the patient cannot see, he or she is unable to identify the object verbally but can readily tell the experimenter what it is by extending two fingers.

In human patients with the optic chiasm intact, if a picture is briefly flashed before the eyes while the head is held in a fixed position, the right brain sees only the left side of the picture and the left brain sees only the right side. If the patients are then asked what they saw, they report only what they saw on the right side of the picture; that is, they report what the left brain saw. However, the left side of the picture can have an emotional effect on them. A patient may laugh, for instance, but cannot explain why.

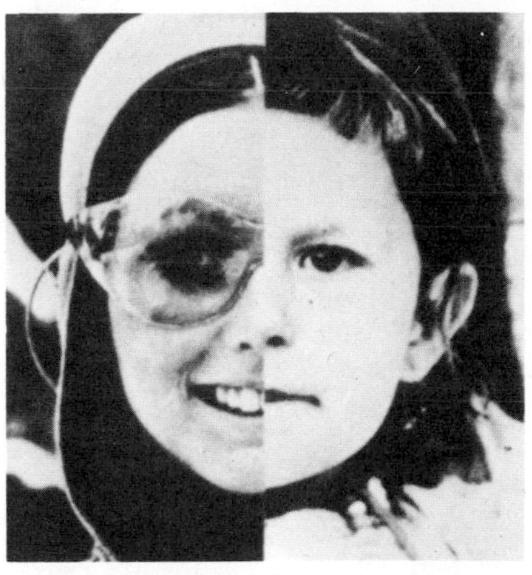

(a)

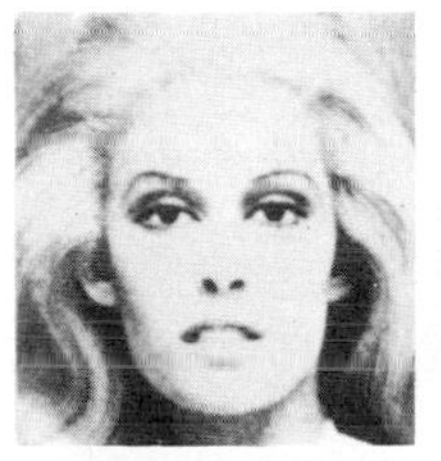
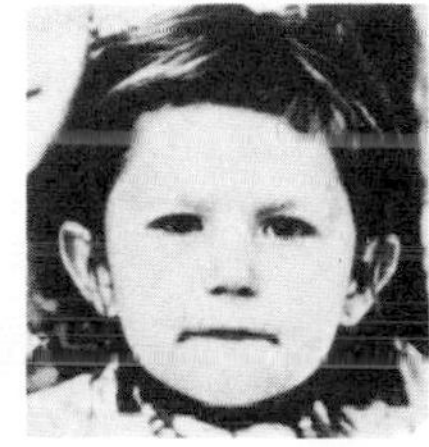

(b)

40–19
A split face (a) *used in tests of patients whose brains have been surgically divided is made up from two of the faces shown in the selection* (b). *The patient, wearing a headgear that restrains eye movements, sees the picture projected briefly on a screen. The left side of the brain recognizes the child; the right side sees the woman with glasses.*

Under normal circumstances, of course, the two cerebral hemispheres are integrated in their activities, and although one may be dominant in certain functions—as in language—the other retains at least the potential for this same function. The capacity of one hemisphere to compensate for the dysfunction of the other is well known to neurosurgeons who have observed patients functioning normally after removal of or damage to extensive areas of the right temporal lobes. Further, in a number of cases, children with left temporal lobe damage have, after an interval, learned to speak again, as a result of the right lobe's taking over the speech functions. The outlook for complete recovery varies with age. Apparently, lateralization of function *is* a developmental process.

More recent studies have revealed further differences in mental characteristics between the two hemispheres, with the left brain excelling in speech, logic, writing, and mathematics. The right brain excels in discriminations involving shape, form, tune, and texture, in nonverbal communication, and in general creativity. Some researchers have reported that some youngsters classified as retarded in the left-brain functions are normal or above average in right-brain functions. Similarly, according to one study, persons with left-brain damage frequently are more expressive nonverbally than normal individuals. These results suggest that the left brain may dominate the activity of the right brain.

The many convolutions of the cerebral cortex represent one way of increasing the working area of the brain. Specialization of function of the two cerebral hemispheres is another solution to the problem of increasing the capacity of the brain without enlarging its volume.

LEARNING AND MEMORY

For scientists interested in brain research, the biggest present challenge is to understand the mechanisms of memory and learning. If we define learning as a change in behavior based on experience, and if the functions of the brain—the mind—are to be explained in terms of the atoms and molecules and the structures composed of them, then learning must involve changes in these atoms or molecules or structures. But what is this change, and where and how does it take place?

Almost 50 years ago, the late Karl Lashley set out to locate this physical change, the trace of memory, which he called the engram. He taught rats and other animals to solve particular problems and then performed operations to see whether he could remove the portion of the cortex containing that particular engram. But he never found the engram. As long as he left enough brain tissue to enable the animal to respond to the test procedures at all, he left memory as well. And the amount of memory that remained was generally proportional to the amount of remaining brain tissue. Lashley concluded that memory is "nowhere and everywhere present."

Although much more sophisticated techniques are now available for brain research, modern neurobiologists have had little more success than Lashley in finding memory's hiding place. Moreover, the problem has become, if anything, more complex. It is now clear that there are two types of memory, short-term and long-term.

A simple example of short-term memory is looking up an unfamiliar number in a phone book; you usually remember it just long enough to dial it. Short-term memory can be obliterated experimentally by electric shock or other insults to the brain. Persons who have suffered a concussion may lose their memory of the events immediately preceding the blow, but there is rarely any loss of the memory of earlier experiences.

The transfer of information from short-term to long-term memory involves, according to the most widely held hypotheses, changes in synaptic pathways in the brain. However, there is as yet no good evidence as to the sorts of changes they might be. The process can be compared to the establishment of a footpath. The more frequently the path is traveled, the better established it becomes. This concept is supported by the familiar experience of consciously retrieving a name, for instance, by seeking out related information that puts one "on the right track."

The consolidation of memory—as this transfer process is called—requires protein synthesis. At least this is the case in rats and goldfish, in which most of these investigations have been carried out. The antibiotic puromycin stops protein synthesis by detaching the polypeptide chain from the ribosome. The administration of puromycin prevents memory consolidation without affecting already established long-term memory. Memory consolidation is similarly inhibited by drugs that block RNA synthesis, such as actinomycin D. (However, since the experimental animals are often sick as a side effect of these drugs, the results are difficult to interpret and the conclusions controversial.) Further, the brains of experimental animals that have been subjected to intensive learning experiences show increases in RNA and protein.

The hippocampus appears to play an important role in memory consolidation. Persons whose hippocampus has been severely damaged or surgically removed cannot consolidate memory. Their short-term memory is unimpaired, and their previous store of long-term memory is untouched, but the transfer of one to another can no longer take place. The hippocampus is also one of the parts of the brain in which increased RNA and protein are found after learning experiences, thus further implicating this part of the rhinencephalon in the transfer process.

One interesting hypothesis about memory is that it may involve changes in the glial cells. Remember, these make up about 90 percent of the cells of the nervous system. Unlike neurons, glial cells continue to grow and divide. In addition to their presumed function of nourishing the neurons, glial cells might affect the neurons' synthesis of neurotransmitter or the establishment of synapses. But this is only speculation.

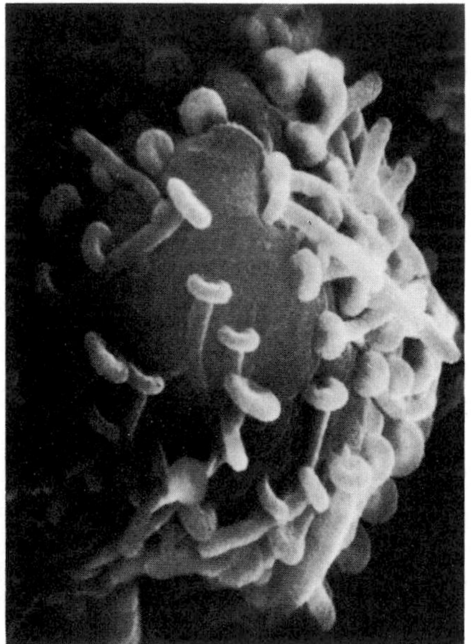

40–20
Synaptic knobs of the sea slug Aplysia *as shown by the scanning electron microscope. Much current research in neurobiology is being carried out in invertebrate systems.*

INVERTEBRATE MODELS

Many investigators studying the nervous system are using invertebrate models for a number of reasons. First, the nerve cells of invertebrates are larger and unmyelinated. This is because unmyelinated cells must be larger than myelinated ones if they are to conduct nerve impulses at comparable velocities. Second, invertebrate nervous systems have considerably fewer neurons, numbering in the thousands rather than in the billions. Each segmental ganglion in a leech, for example, contains only about 350 neurons. Single neurons can be seen under a microscope, teased apart with dissecting instruments, and probed with microelectrodes.

Third, many invertebrate behaviors are mediated by circuits that are relatively easy to trace. For example, if you stroke a leech gently, sensory cells respond, carry nerve impulses to the ganglion, where synapses are made with motor neurons, resulting in complex rhythmic movements. After you stroke the leech for a while, it ceases to respond; it has become habituated to the stimulus. This is a very simple form of learning. What kind of changes have taken place in the cells to produce this modification of behavior? By answering this sort of question, researchers are hoping to gain insight into the more complex problems of mind and matter in vertebrate nervous systems.

Earlier in this century of remarkable biological progress, the work of a number of brilliant scientists—including physicists, physicians, and biochemists, as well as biologists—focused with great success on the problem of the molecular basis of heredity. In the last several years, a number of outstanding research workers, including some prominent molecular geneticists, have abandoned their previous lines of research and turned their attention to the tantalizing subject of the cellular and molecular basis of learning and memory. It will be interesting to see if their ventures in neurobiology meet with the same spectacular success as the earlier efforts to discover the nature of the gene.

SUMMARY

The human brain is the most complex structure known. It has evolved from the primitive vertebrate brain. In the course of its evolution, the brainstem (medulla, pons, and midbrain) has remained much the same, but the cerebellum, which is involved with equilibrium and the coordination of motion, has increased greatly in size. The cerebral cortex, which is the outer layer of the cerebral hemispheres, has also greatly increased in size. The cerebral cortex is the area of the brain in which complex, conscious motor activities originate and to which sensory information is ultimately transmitted. It appears to be the site of memory and learning. In both experimental animals and humans, it has been possible to correlate particular areas of the brain with particular functions. These areas include the motor cortex, sensory cortex, and parts of the cortex concerned with vision, hearing, and speech. Most of the human cortex is "silent"; that is, it has no direct sensory or motor function. It is known as the association cortex, about half of which is in the frontal lobes, and is the part of the brain that has developed most rapidly in human evolution.

The hypothalamus is a small group of nuclei (nerve cell bodies) with a great variety of functions. It produces oxytocin and ADH and regulates the secretion of the various hormones from the anterior pituitary gland. It is also apparently involved in basic drives and emotions, such as hunger, pleasure, and sex. Two important systems of the brain are the reticular system, concerned with arousal and attention, and the limbic system, which connects the hypothalamus and other primitive brain structures with parts of the cerebral cortex.

Chemical transmitters in the brain include acetylcholine and norepinephrine, which are also found in the peripheral nervous system, and, in addition, serotonin, dopamine, and gamma-aminobutyric acid. A newly isolated group of chemical substances, the enkephalins and endorphins, resemble opiates in their structure and function and may also act as inhibitory transmitters. Many drugs appear to act on the brain by increasing or decreasing the concentration of particular neurotransmitters.

Wavelike variations in electric potential can be recorded from the brain surface (electroencephalography). Three principal types of brain waves can be recorded in normal subjects: beta waves, of relatively high frequency, associated with alertness and attention; alpha waves, of lower frequency, associated with relaxation; and delta waves, of still lower frequency, which normally appear only during sleep.

Sleep involves metabolic changes, such as decreases in heart rate, ventilation, and body temperature, and also changes in the electroencephalogram. The slow, rhythmic waves characteristic of deep sleep are interrupted by periods of rapid, high-frequency waves (REM, or paradoxical, sleep) typically accompanied by rapid eye movements. The function of sleep is not understood.

Studies on the eye and the brain reveal that visual form is coded by single neurons in the brain, each of which responds to one or two specific features of the visual image.

Split-brain studies have been carried out on experimental animals and human patients in which connections have been severed between the cerebral hemispheres. Results of such studies indicate that the two cerebral hemispheres of the same individual can function independently and that they differ somewhat in their capacities.

Three stages in the acquisition of memory have been identified: short-term memory, consolidation of memory, and long-term memory. Studies on the mechanisms of memory and learning involve the search for changes in the physical or chemical structure of brain cells or the connections among them.

QUESTIONS

1. Distinguish among the following: white matter/gray matter; forebrain/midbrain/hindbrain; reticular system/limbic system; olfactory lobe/optic lobe; cerebellum/cerebrum; Broca's area/Wernicke's area; alpha waves/beta waves/delta waves; left brain/right brain.

2. Sketch the human cerebral cortex and indicate on it the areas that have been "mapped."

3. On the basis of your own experience, define the differences between learning, memory, and habit.

4. What functions might be affected by damage (from stroke, accident, or disease) to the following structures: Cerebellum? Reticular system? Dorsal portion of the cerebral cortex anterior to the central sulcus?

5. Can you relate the relative sizes of various portions of the brains of the animals in Figure 40–3 with the habitats in which they live and the ways in which they live, obtain food, and escape predators?

SUGGESTIONS FOR FURTHER READING

Books

BECK, WILLIAM S.: *Human Design,* Harcourt Brace Jovanovich, Inc., New York, 1971.

A comprehensive, modern treatise, designed for the undergraduate student, covering "molecular, cellular, and systematic physiology" of humans.

BRECHER, EDWARD M., et al.: *Licit and Illicit Drugs: The Consumers Union Report on Narcotics, Stimulants, Depressants, Inhalants, Hallucinogens, and Marijuana—Including Caffeine, Nicotine, and Alcohol,* Little, Brown and Company, Boston, 1972.

A well-balanced, straightforward, and nonsensational review.

CALDER, NIGEL: *The Mind of Man,* 2d ed., Viking Press, Inc., New York, 1973.

A well-written, fast-moving, journalistic report on the "drama of brain research."

ECCLES, JOHN C.: *The Understanding of the Brain,* McGraw-Hill Book Company, New York, 1977.

A general survey, written for the layman, by one of the great leaders in the field of neurophysiology.

ECKERT, ROGER, and DAVID RANDALL: *Animal Physiology,* W. H. Freeman and Company, San Francisco, 1978.

The emphasis is on basic principles. Well-written and handsomely illustrated.

GANONG, W. T.: *Review of Medical Physiology,* 8th ed., Lange Medical Publications, Los Angeles, 1977.

A concise, accurate, up-to-date summary of human physiology. An excellent reference book.

GORDON, MALCOLM S., et al.: *Animal Physiology: Principles and Adaptations,* 3d ed., The Macmillan Company, New York, 1977.

A good animal physiology textbook, intended for an advanced course. The authors emphasize function as it relates to the survival of organisms in their natural environments.

GREGORY, R. L.: *Eye and Brain: The Psychology of Seeing,* 2d ed., McGraw-Hill Book Company, New York, 1973.*

A vivid introduction to the science of vision.

KUFFLER, STEPHEN W., and JOHN G. NICHOLLS: *From Neuron to Brain,* Sinauer Associates, Inc., Sunderland, Mass., 1976.

A lively account of the cellular basis of the workings of the nervous system, based largely on work with which these two distinguished neurobiologists have had first-hand experience.

NOSSAL, G. J. V.: *Antibodies and Immunity,* 2d ed., Basic Books, Inc., New York, 1978.

An account for the general public of research in this important field by a scientist who has himself made many significant contributions.

OATLEY, KEITH: *Brain Mechanisms and Mind,* E. P. Dutton & Co., Inc., New York, 1972.

A readable, up-to-date, well-illustrated survey written by an experimental psychologist for the general reader.

ROMER, ALFRED: *The Vertebrate Story,* 4th ed., The University of Chicago Press, Chicago, 1971.*

The history of vertebrate evolution, written by an expert but as readable as a novel.

RUGH, ROBERTS, and LANDRUM B. SHETTLES: *From Conception to Birth: The Drama of Life's Beginnings,* Harper & Row, Publishers, Inc., New York, 1971.

This is an account of the history of life before birth. The book describes in detail the development of the unborn child from the moment of fertilization and also the changes in the mother during pregnancy. It discusses such related topics as birth control, congenital malformation, labor, and delivery of the baby. There are a large number of illustrations, including a group of magnificent color photographs of the developing fetus.

* Available in paperback.

SCHMIDT-NIELSEN, KNUT: *Animal Physiology: Adaptations and Environment,* Cambridge University Press, New York, 1975.

Schmidt-Nielsen is concerned with underlying principles of animal physiology—the problems animals have to solve in order to survive. The emphasis is on comparative physiology, and the lucid exposition is illuminated by many interesting examples.

SCHMIDT-NIELSEN, KNUT: *Desert Animals,* Oxford University Press, New York, 1964.

Although considered the definitive work on the physiological problems relating to heat and water, this readable book also contains numerous anecdotes—such as that about Dr. Blagden—and many fascinating personal observations.

SMITH, HOMER W.: *From Fish to Philosopher,* Doubleday & Company, Inc., Garden City, N.Y., 1959.*

Smith was an eminent specialist in the physiology of the kidney. Writing for the general public, he explains the role of this remarkable organ in the story of how, in the course of evolution, organisms have increasingly freed themselves from their environments.

THOMPSON, R. F.: *Introduction to Physiological Psychology,* Harper & Row, Publishers, Inc., New York, 1975.

Intended for the undergraduate student, this text presents an up-to-date survey of the biological foundations of psychology.

VANDER, A. J., J. H. SHERMAN, and DOROTHY S. LUCIANO: *Human Physiology,* 2d ed., McGraw-Hill Book Company, New York, 1975.

Most highly recommended. The text is a model of clarity, and the diagrams, many of which we have borrowed, are splendid.

Articles

BARCHAS, JACK D., et al.: "Behavioral Neurochemistry: Neuroregulators and Behavioral States," *Science,* vol. 200, pages 964–973, 1978.

EPEL, DAVID: "The Program of Fertilization," *Scientific American,* November 1977, pages 128–138.

GUILLEMIN, ROGER: "Peptides in the Brain: The New Endocrinology of the Neuron," *Science,* vol. 202, pages 390–402, 1978.

LESTER, HENRY A.: "The Response to Acetylcholine," *Scientific American,* February 1977, pages 106–118.

LIEBER, CHARLES S.: "The Metabolism of Alcohol," *Scientific American,* March 1976, pages 25–33.

MCEWEN, BRUCE S.: "Interactions between Hormones and Nerve Tissue," *Scientific American,* July 1976, pages 48–58.

OLD, LLOYD J.: "Cancer Immunology," *Scientific American,* May 1977, pages 62–79.

PERUTZ, M.F.: "Hemoglobin Structure and Respiratory Transport," *Scientific American,* December 1978, pages 92–125.

ROUTTENBERG, AREH: "The Reward System of the Brain," *Scientific American,* November 1978, pages 154–164.

SNYDER, SOLOMON H.: "Opiate Receptors and Internal Opiates," *Scientific American,* March 1977, pages 44–56.

* Available in paperback.

PART III

Biology of Populations

SECTION 7

EVOLUTION

41–1
In any given population, there are variations among individual organisms.

CHAPTER 41

The Modern Theory of Evolution

In the Introduction, we traced the beginnings of Darwin's theory of evolution and noted the revolution that it brought about, not only in biology but also in almost every aspect of Western culture. Now, the theory of evolution is so much a part of modern biological thinking that today's discussions center around details of how the process takes place rather than the theory itself. As a result, one tends to lose the grand design. Before we begin our own detailed discussion, let us take another brief look at the theory, beginning with the body of evidence from which the theory emerged.

First, the earth has a long history. It has been in existence, by present calculations, for 4.6 billion years, and during this time it has experienced major geographic and climatic changes.

Second, in the course of the earth's history there has been a succession of living forms, as recorded in the fossil record, which now dates back more than 3 billion years. In this record, generally speaking, simpler forms precede more complex ones. In the case of many groups of organisms—vascular plants and vertebrates, for example—fossils can be found that exhibit a graded series of changes in morphological characteristics, linking older forms with the modern forms and revealing pathways diverging from common ancestors.

Third, living members of the various groups of organisms share the same basic plans of organization, although the actual structures may vary considerably, both in form and function. For example, Figure 41-2 shows the forelimbs of various vertebrates. The bones in each are the same, but they have been modified in each different type of organism. Such homologies are not surprising from our perspective, but consider what a powerful argument they were for Darwin against the concept he himself long held that each species was created separately and specifically.

Further evidence for evolution is seen in the great variety of existing organisms. Remember it was not until comparatively recently in the history of human civilization that naturalists began to explore other countries, much less other continents. Such explorers, Darwin and Wallace among them, began to see that species were not distinct, but that as one traveled—up the western coast of South America, for instance—one could observe slight differences in various characteristics of the plants and animals. These were interpreted as evidence against separate creation

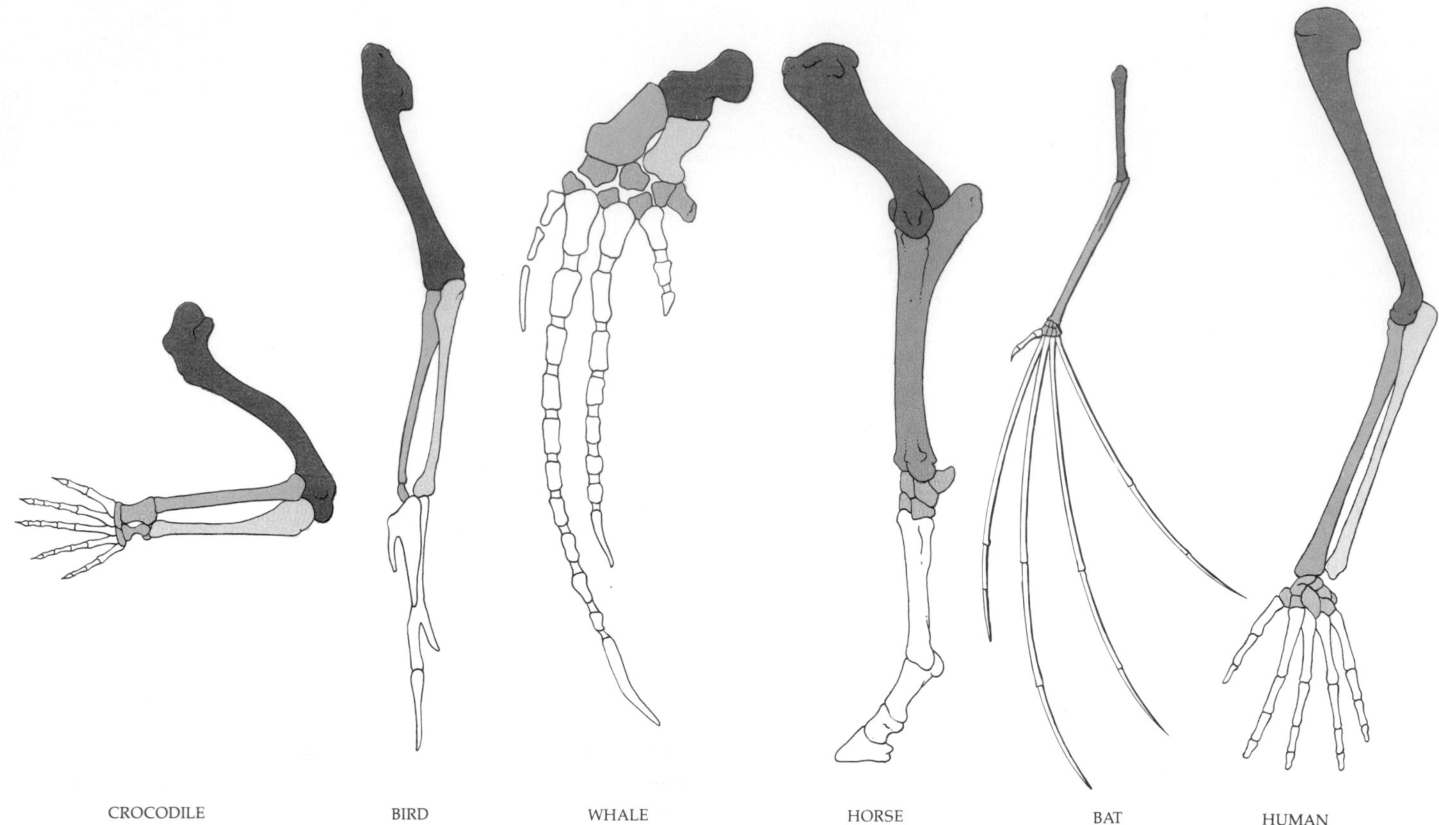

41–2
The bones in these forelimbs are color-coded to indicate fundamental similarities of structure and organization. The crocodile is placed first because it is the closest to the ancestral type—the form from which all the others arose.

Such structures, which have a common origin but not necessarily a common function, are known as homologues.

Analogous structures, by contrast, are superficially similar but have an entirely different evolutionary background—the wing of a bird and the wing of an insect, for example, or the spine of a cactus (a modified leaf) and the thorn of a rose (a modified branch).

and for the concept that organisms became modified with time, according to the different environments in which they lived.

The principle of evolution gains additional support, as we have noted on previous occasions, from the remarkable similarity of living things. The cellular basis of life was demonstrated within a decade of the publication of the first edition of *The Origin of Species.* Since that time, almost every major discovery in biochemistry—the glycolytic pathway, the electron transport chain, the nature of ATP—has emphasized the fundamental similarity of living cells. Additional evidence for the unity of life has been uncovered by advances, first in light microscopy and more recently in electron microscopy, that have revealed the two-ply structure of cell membranes, the existence of ribosomes, and the internal organization of eukaryotic cilia and flagella.

Finally, there is the exquisite adaptedness of living organisms to their environment. As we remarked in the Introduction, this seeming perfection of adaptation appeared to many as strong support for the doctrine of special creation. Now, however, it is possible to show that such adaptations have taken place over a long period of time as a consequence of very small changes and, moreover, are still taking place at this very moment. The co-evolution of insects and flowers, which we described previously, is one example, and there are many others, some of which will be cited in the chapters that follow.

THE THEORY TODAY

Although it is now more than a hundred years since the first publication of *The Origin of Species,* Darwin's original concept of how evolution comes about still provides the basic framework for our understanding of the process. His concept rested on four premises:

1. Like begets like—in other words, there is stability in the process of reproduction.
2. In any given population, there are variations among individual organisms, and some of the variations are inheritable.
3. In most species, the number of individuals that survive to reproductive age is small compared with the number produced.
4. Which individuals will survive and reproduce and which will not are determined to a significant degree by these variations. Some variations enable individuals to produce more offspring than other individuals. Darwin termed these "favorable" variations and argued that inherited favorable variations tended to become more and more common from one generation to the next. This is the process that Darwin called *natural selection.*

Modern evolutionary theory conceives of evolution as Darwin conceived of it, with the important addition of an understanding of the mechanisms of inheritance. Twentieth-century genetics answers two questions that Darwin was never able to resolve: (1) why genetic traits are not "blended out" but can disappear and reappear (like whiteness in pea flowers), and (2) how the variations appear on which natural selection acts. This combination of evolutionary theory and genetics is known as the synthetic theory of evolution. (Here "synthetic" does not mean artificial, which is the connotation it has for us in these days of synthetic fabrics and artificial colors and flavors, but has its original meaning of the putting together of two or more different elements.)

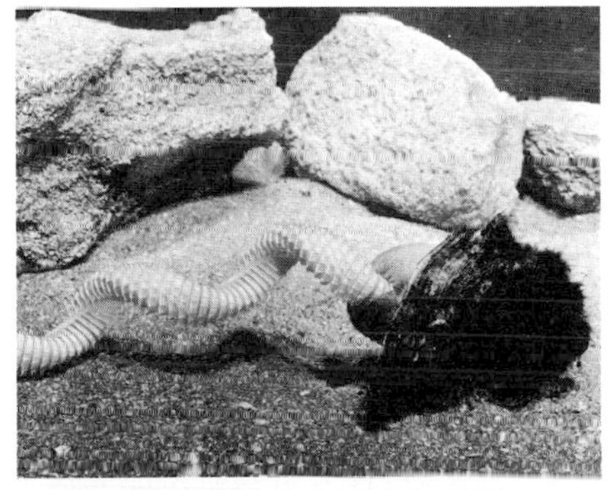

41–3
Within this long egg case, composed of a parchmentlike material, a whelk is depositing hundreds of eggs. When the baby mollusks hatch, they will break through thin areas in the individual capsules and emerge as miniature copies of their parents. Once the egg case is produced and the eggs laid, the mother whelk does not participate further in raising the young. Despite the large number of young produced, there is, in general, no increase in the number of whelks from generation to generation.

The branch of biology that emerged from this synthesis of Darwinian evolution and Mendelian principles is known as *population genetics.* A population, for the geneticist, is an interbreeding group of organisms. For instance, all the fish of one particular species in a pond are a population and so are all the fruit flies in one bottle. The population is defined and united by its *gene pool,* which is simply the total of all the alleles of all the genes of all the individuals in a population.

The relative proportion of alleles of certain genes in a gene pool may change from one generation to the next. In population genetics, evolution is defined as any change in the composition of the gene pool. In this view the individual is only a temporary vessel, holding a small portion of the gene pool for a short time, testing a particular combination. Fitness, in the Darwinian sense, is measured by only one criterion: the reproductive success of one genotype measured relative to the reproductive success of another.

Let us take an example. There are small insects known as fungus-eating gall midges that can reproduce both sexually and asexually. They are found on mushroom beds. When abundant food is available, a female can produce parthenogenetic eggs from which larvae develop within her body, devour her tissues, and emerge with eggs inside their own tissues ready to repeat the process while they are still larvae. The numbers of parthenogenetic young produced by these larval forms are not large compared with normal insect broods, but the generation time is so short that the rate of increase can be very rapid. For example, a female of one species can produce up to 47 larvae in 4 days. Thus, it would take only 20 days for well over 200 million offspring to be produced. The original mother would long since have disappeared, but she could have left 200 million offspring, and, as an additional bonus, each genotype would be an exact copy of her own. If she did this, by Darwinian standards she would be considered very fit indeed.

THE HARDY-WEINBERG PRINCIPLE

Mendel's great contribution was to demonstrate that traits—even when they seem to disappear—are actually preserved, "hidden" in the heterozygote. But, as put forth by Mendel and other classical geneticists, the laws of heredity had been demonstrated only for ideal conditions, in which simple, controlled matings were performed. For the purpose of evolutionary theory, it was necessary to establish how genes behave under the much more complex conditions that occur in nature. One needed to be able to detect and follow changes in the genetic makeup of a population—to analyze, for example, the fate of a new allele under natural breeding conditions. Population genetics is concerned with the sources and dynamics of variations in populations. Just as Mendel's laws form the cornerstone of classical genetics, the cornerstone of population genetics is a principle known as the Hardy-Weinberg law.*

G. H. Hardy was an English mathematician and G. Weinberg, a German physician. Working independently, they examined the behavior of alleles in an idealized population in which matings occur at random—and so are governed by the laws of probability—and in which individuals do not move in or out of the population (taking their genes with them) and in which every allele is equally viable. Let us suppose that it is possible to identify every allele of every gene of every organism in this imaginary population. Each of these alleles is a single, discrete unit, as Mendel demonstrated, as discrete as a colored bead, for instance. These beads are the gene pool of the imaginary population. If matings occur at random and all alleles are equally viable, these same alleles will simply be redistributed again and again in every generation. None will be lost and none will be gained from the gene pool, the composition of which will, under these imaginary circumstances, stay the same forever.

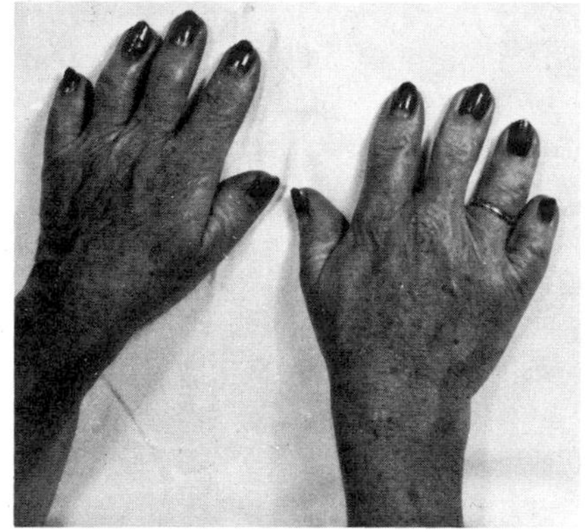

41-4
A dominant gene is responsible for the trait known as brachydactylism (short fingers). In the brachydactylous hands shown here, the first bones of the fingers are of normal length, but the second and third bones are abnormally short. If brachydactylism is caused by a dominant gene, why is it such a rare trait?

Now let us look at just one gene, selecting one that has only two alleles, which we will call *A* and *a*. We are interested in the relative proportions of *A* and *a*. These proportions could be expressed in terms of fractions, as Mendel did, but since the proportions of the two alleles will not be equal, as they were in Mendel's carefully controlled experiments, it is more convenient to express the numbers as decimals. For example, suppose 0.8 of the alleles of the gene under study are allele *A*. Given that there are only two alleles, we then know that 0.2 is allele *a*. These figures, 0.8

* The Hardy-Weinberg law was formulated in 1908 in response to a question raised by biologists: How could both dominant and recessive genes remain in populations? Why didn't dominants simply drive out recessives? If brachydactylism is a dominant gene, why didn't most or all of the population have short, fat fingers? If you asked this question yourself, here is the answer.

and 0.2, represent the frequency of the allele. By convention, the letter p is used to designate the frequency of one allele and q the frequency of the other. When there are only two alleles, p and q together must equal one: $p + q = 1$.

In diploid organisms, which is what Hardy and Weinberg (like Mendel before them) were considering, alleles come in pairs. What is the probability that any given individual in our imaginary population will have one *A* allele? The probability is 8 in 10, or 0.8. What is the probability that the second allele will be an *A*? That is also 0.8. What, then, is the probability for any one individual that both alleles will be *A*? To find this, as we noted on page 227, we simply multiply the two probabilities: $0.8 \times 0.8 = 0.64$, or $p^2 = 0.64$. Following the same line of reasoning, the probability that both alleles will be *a* is 0.2×0.2, or $q^2 = 0.04$. What is the probability that an individual will be *Aa*? It is $p \times q$, or 0.8×0.2, or 0.16; $pq = 0.16$. Similarly, the combination *aA* (which is, in effect, the same as *Aa*) has an equal chance of appearing. Now one has a population in which the homozygotes for *A* are 64 percent of the population (with a frequency of 0.64), the homozygotes for *a* are 4 percent (with a frequency of 0.04), and the heterozygotes, *Aa*, are 32 percent (with a frequency of 0.32). In other words, $p^2 + 2pq + q^2 = 1$. Remembering your algebra, you will recognize that this equation can also be expressed as $(p + q)^2 = 1$.

Let us imagine that another round of mating takes place. What is now the frequency of the alleles in the gene pool? We know from our calculations that the frequency of *AA*s is 0.64. In addition, half of the alleles in the heterozygotes ($2pq$) are *A*s, so the total of allele *A* is 0.64 plus $\frac{1}{2}$ of 0.32, or 0.8. Similarly, the total of allele *a* is $0.04 + 0.16$, or 0.2. As you can readily calculate, the probability that both alleles will be *AA* in the next generation is 0.8×0.8, and so on, and so on, generation after generation. The Hardy-Weinberg formula applies equally well when there are more than two alleles of the same gene, although the calculations then become more complex.

Now let us move a little closer to reality and suppose that the *a* allele in its homozygous state has a damaging effect. We can estimate the number of *aa*s in the population, perhaps by screening tests on newborn infants. Suppose, for instance, that the condition shows up in 1 in 10,000 infants. In other words, $q^2 = 1/10{,}000$ or 0.0001, and $q = \sqrt{0.0001}$ or 0.01. If $q = 0.01$, $p = 0.99$ and $2pq = 0.0198$, or almost 0.02. Thus about 2 percent of the population—one person in every 50—can be estimated to be a carrier of this allele.

Now suppose that we do the same screening tests five years later and find that q is not 0.01 but 0.009, and then we repeat it a few years later and find it has once again decreased very slightly, perhaps to 0.008. In other words, evolution has occurred. One allele is decreasing at the expense of another. Moreover, using Hardy-Weinberg as our yardstick, we know not only that the change has taken place but also that there must be some reason for it.

What the Hardy-Weinberg law says, in effect, is that the genetic recombination that occurs as a result of meiosis and fertilization does not by itself change the overall composition of the gene pool. A dominant allele, for instance, does not tend to increase in a population, and a recessive one does not tend to decrease. A rare variant will continue to persist in the population, despite the shuffling and reshuffling of genes that take place every generation. The interbreeding of organisms does not change the frequencies of alleles. Now we shall examine the various factors that do change the Hardy-Weinberg equilibrium—the agents of evolution.

THE AGENTS OF EVOLUTION

As we noted at the beginning of this discussion, the Hardy-Weinberg principle applies only to an ideal population. Moreover the population must be large enough for the laws of probability to operate. Also mating must be at random. Absolutely random mating seldom takes place, partly because of the fact that individuals of many species (trees and grasses with windborne pollen, for instance) are likely to breed with those geographically nearer to them, and also because of characteristics that make some individuals more successful in the competition for mates than others.

We shall now discuss four additional important factors that change the frequency of alleles in populations: mutations, gene flow, genetic drift, and, above all, natural selection.

Mutations

Mutations are defined as inheritable changes in the genotype (see Chapter 19). The rate of spontaneous mutation is low; for humans, the chance of a mutation occurring in any given gene, as we noted, has been estimated at less than 1 in 10,000. However, given the large number of genes in every gamete, there is an excellent possibility that any given gamete will carry a new mutation in at least one gene. Different genes have different rates of mutation. The reason for these differences is not known but probably has to do with the chemical nature of the gene itself or with its position in the chromosome.

At one time, many students of evolution viewed mutations as an important element in determining the *direction* of evolutionary change (Figure 41–5). Today, mutations are seen to be important only as the original source of the small cumulative variations upon which selection acts. Mutations alone do not effect major changes in the frequencies of alleles.

Gene Flow

Changes in the Hardy-Weinberg equilibrium of a population can, of course, be produced by the movement of new breeding individuals into the population or, as in the case of plants, the introduction of gametes (in the form of pollen) from other populations. This phenomenon is often referred to as *gene flow*. As we shall see, gene flow into a population often counteracts the effects of natural selection.

41–5
An example of the sometimes dramatic effects of mutation. The ewe in the middle is an Ancon, an unusually short-legged strain of sheep. The first Ancon on record was born in the late nineteenth century into the flock of a New England farmer. By inbreeding (the trait was transmitted as a recessive), it was possible to produce a strain of animals with legs too short to jump the low stone walls that traditionally enclosed New England sheep pastures. A similar strain was produced in northern Europe as a result of an independent mutation. At one time, it was thought that evolution took place in sudden, large jumps such as this—a concept sometimes referred to as the "hopeful monster" theory. One reason this concept was abandoned is that nearly all mutations producing dramatic changes in the phenotype are harmful, as this one would be in a wild population.

Genetic Drift

As we stated previously, the Hardy-Weinberg principle holds true only if the population is large. This qualification is necessary because the Hardy-Weinberg equilibrium depends on the laws of probability. These laws—the laws of chance—apply equally well to flipping coins, rolling dice, or betting at roulette. In flipping coins, it is possible for heads to show up five times in a row, but, on the average, heads will show up half the time and tails half the time. The more times the coin is flipped, the more closely the expected frequencies of half (0.50) and half (0.50) are approached.

Consider, for example, an allele, say *a*, that is present in 2 percent of the individuals. In a population of 1 million, 20,000 *a* alleles would be present in the gene pool. But in a population of 50, only one individual would carry this allele. If this individual failed to mate or were destroyed by chance before leaving offspring, allele *a* would be completely lost. Similarly, if 10 of the 49 without allele *a* were lost, the frequency of *a* would jump from 1 in 50 to 1 in 40. This phenomenon, a change in the gene pool that takes place as a result of chance, is *genetic drift*.

The Founder Principle

A small population that branches off from a larger one may or may not be genetically representative of the larger population from which it was derived. Some rare alleles may be overrepresented or may be lost completely. As a consequence, even when and if the small population increases in size, it will have a different genetic composition—a different gene pool—from that of the parent group. This phenomenon, a type of genetic drift, is known as the *founder principle*. In a small population, there is likely to be less variation in the population. Thus, even though mating may be random, the mates will be more closely related to each other—more similar genetically—with a consequent increase in homozygosity.

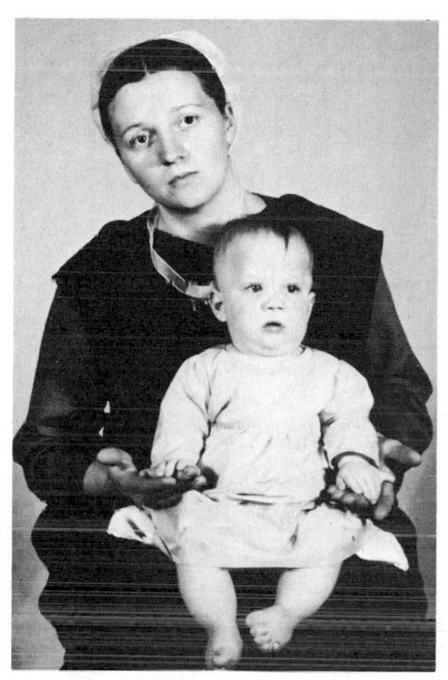

41–6
Among the Old Order Amish, a group founded by only a few couples some 200 years ago, there is an unusually high frequency of a rare allele. In its homozygous state, the allele results in extra fingers and dwarfism. This Amish child is a six-fingered dwarf.

An example of the founder principle is found in the Old Order Amish of Lancaster, Pennsylvania (Figure 41–6). Among these people, there is an unprecedented frequency of a recessive allele that, in the homozygous state, causes a combination of dwarfism and polydactylism (extra fingers). Since the group was founded in the early 1770s, some 61 cases of this rare congenital deformity have been reported, about as many as in all the rest of the world's population. Approximately 13 percent of the persons in the group, which numbers some 17,000, are estimated to carry this rare allele. The entire colony, which has kept itself virtually isolated from the rest of the world, is descended from only a few individuals. By chance, one of those must have been a carrier of the gene.

Genetic Drift and Non-Darwinian Evolution

Changes in the gene pool that originate as a result of random mutations and that increase or decrease as a result of genetic drift are sometimes referred to as non-Darwinian evolution. Now that scientists are beginning to analyze the genotype at the molecular level, more and more variations are being found that might be "neutral"—that is, that might have no effect on the fitness of the organism. Such variations will be described in more detail in the following chapter. Whether or not any variation can, in fact, be neutral and the extent to which changing frequencies of variation can be caused merely by random sampling (genetic drift) are matters currently under debate by population biologists.

Natural Selection

Natural selection is the fourth and by far the most important agent of evolution. Darwin coined the term "natural selection" as an analogy to what he called "artificial selection," the process by which breeders of domestic animals or crops select certain individuals and discard others as they seek to develop desirable characteristics in their stock. In Darwin's time, pigeon fanciers had produced a number of different and often very exotic breeds by means of artificial selection. In our own time, familiar examples are found among "purebred" dogs (Figure 41–7) and cultivated plants (Figure 41–8).

It is important to emphasize that selection acts upon the phenotype, the physical expression of the genotype. The phenotype, you will recall, is the result of the interaction of the genes of an individual with one another and with the external environment in the course of the individual's life (Figure 41–9). Natural selection is the process of interaction between an organism and its environment that results in the differential rate of reproduction of different phenotypes in a population. Thus differential reproduction may result in changes in relative frequencies of alleles in the gene pool—that is, in evolution.

41–7
All breeds of dogs have been produced by artificial selection, a demonstration of the tremendous potential for variability in a single species.

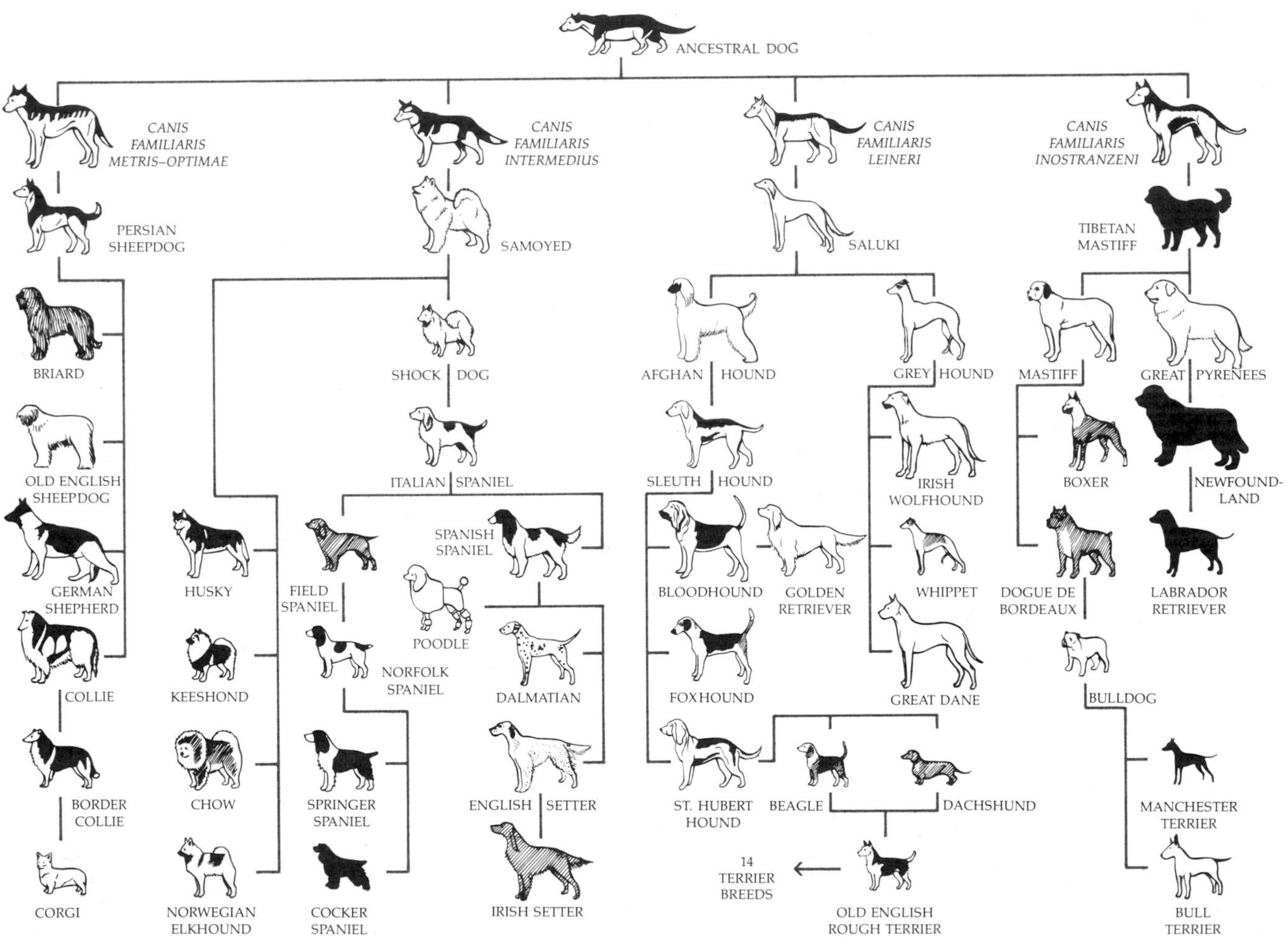

41–8
*Six vegetables produced from a single species of plant (*Brassica oleracea*, a member of the mustard family). They are the result of selection for leaves (kale), lateral buds (brussels sprouts), flowers and stem (broccoli), stem (kohlrabi), enlarged terminal buds (cabbage), and flower clusters (cauliflower). Kale most resembles the wild plant.*

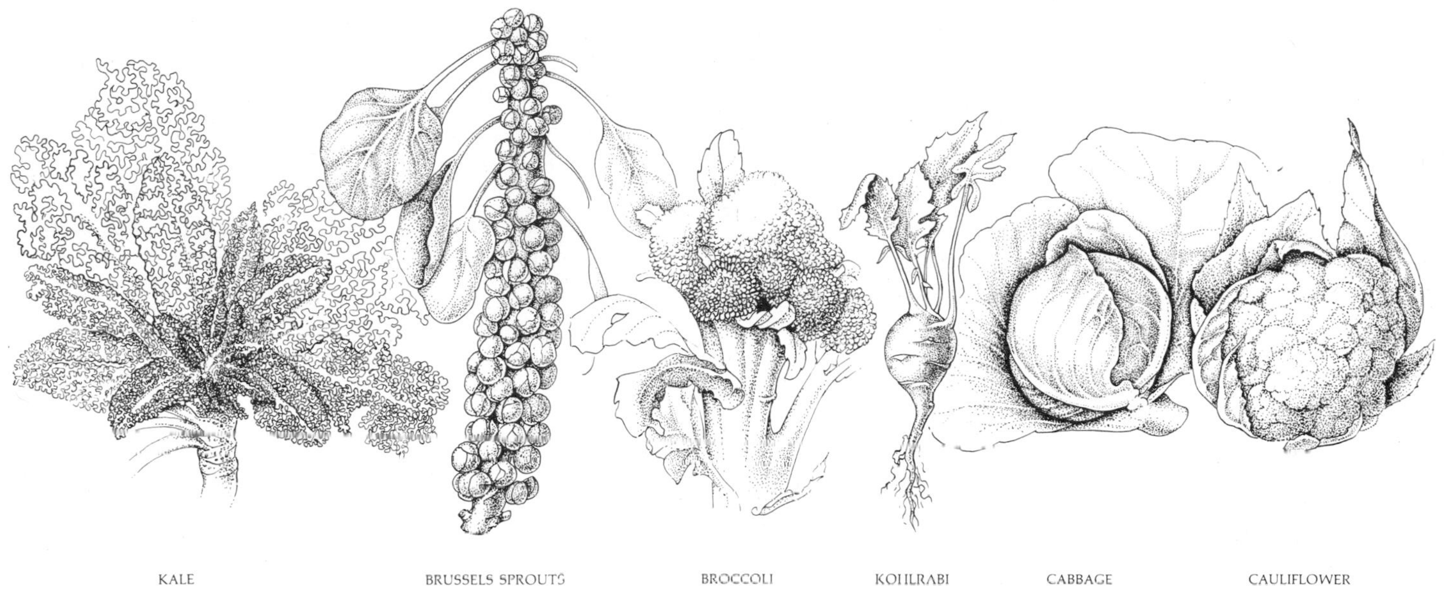

41–9
A pinyon pine growing out of a crack in a rock in California. Genotypically the tree is tall and straight; however, the constant, strong winds in which it has grown have produced this phenotype.

Table 41–1 *Survival in Relation to Number of Young in Swiss Starling*

Number of young in brood	1	2	3	4	5	6	7	8
Number of young marked	65	328	1,278	3,956	6,175	3,156	651	120
Percentage of marked birds recaptured after 3 months	–	1.8	2.0	2.1	2.1	1.7	1.5	0.8

Types of Selection

Three general types of selection operate within populations: stabilizing, disruptive, and directional (Figure 41–10).

Stabilizing selection, a process that goes on at all times in all populations, is the continual elimination of extreme individuals. Most mutant forms are probably immediately weeded out in this way, many of them in the zygote or the embryo. Clutch size in birds, for instance, is a result of stabilizing selection. Clutch size (the number of eggs a bird lays) is determined genetically. As you can see in Table 41–1, it is a disadvantage, in terms of surviving young, for the Swiss starling to have a clutch size of less than 4 or more than 5. Females whose genotypes dictate a clutch size of 4 or 5 will have more surviving young, on the average, than members of the same species that lay more or fewer eggs.

A somewhat similar example is seen in a study of the correlation between birth weight and survival for 13,730 babies born in London over a ten-year period

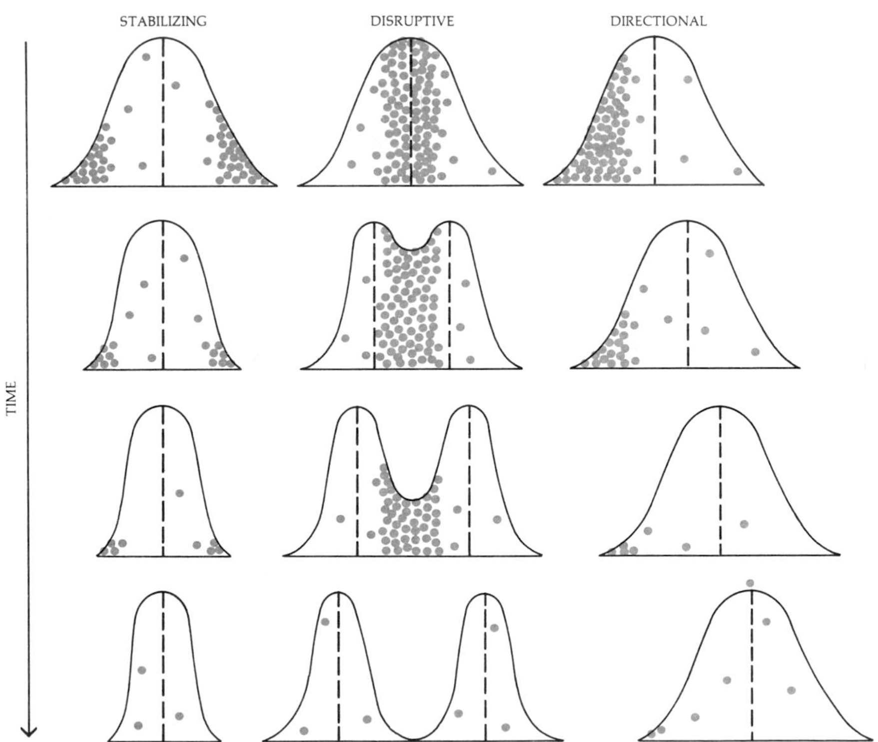

41–10
The three different types of natural selection. The dots represent individuals that failed to reproduce or that left less than the average number of offspring. Stabilizing selection involves the elimination of extremes. In disruptive selection, intermediate forms are eliminated, producing two divergent gene pools. Directional selection, which is the gradual elimination of one phenotype in favor of another, produces adaptive change. In these graphs, the vertical axes denote the proportion of individuals in a population with a particular characteristic, and the horizontal axes, the varying dimensions of whatever characteristic is being considered.

41–11
Relationship between weight at birth and survival in a particular group of human babies. The death rate is higher for babies weighing either more or less than about 3.8 kilograms.

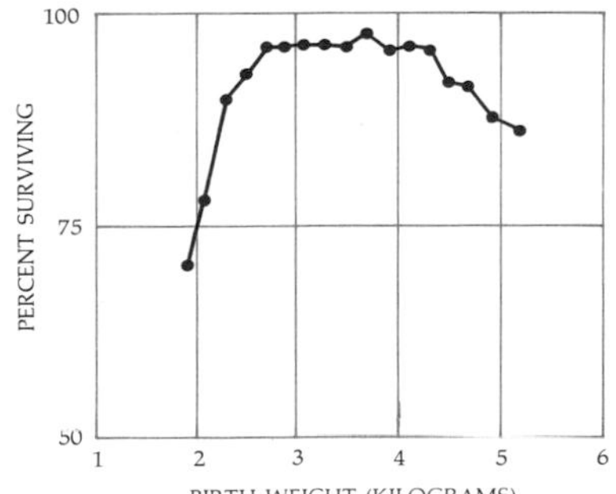

(1935–1945). As you can see in Figure 41–11, the optimum weight for this population of babies was about 3.8 kilograms. At this birth weight, 98.2 percent of the babies survived. The overall survival rate was 95.5 percent. To the extent that birth weight of offspring is determined by heredity, stabilizing selection was at work in this population to maintain this birth weight.

The second type of selection, *disruptive selection,* increases the two extreme types in a population at the expense of intermediate forms. Disruptive selection is probably a major factor in sexual dimorphism—the differences between males and females of the same species.

The third type, which is the one we shall be most concerned with, is *directional selection.* Directional selection acts either for or against an extreme phenotypic characteristic and so is likely to result in the gradual replacement of one allele or group of alleles by another in the gene pool.

The Peppered Moth

One of the best-studied examples of natural selection is that of *Biston betularia,* the peppered moth. These moths were well known to British naturalists of the nineteenth century, who remarked that they were usually found on lichen-covered trees and rocks. Against this background, the light coloring of these moths made them practically invisible, concealing them from predatory birds. Until 1845, all reported specimens of *Biston betularia* had been light-colored, but in that year one black moth of this species was captured at the growing industrial center of Manchester.

With the increasing industrialization of England, smoke particles began to pollute the foliage in the vicinity of industrial towns, killing the lichens and leaving the tree trunks bare. In heavily polluted districts, the trunks and even the rocks and ground became black. During this period, more and more black *Biston betularia* were found. This mutant black form spread through the population until black moths made up 99 percent of the Manchester population. Where did the black moths come from? Eventually, it was demonstrated that the black color was the result of a rare, recurring mutation.

Why were they increasing so rapidly? H. B. D. Kettlewell was among those who hypothesized that the color of the moths protected them from predators, notably birds. In the face of a number of strongly opposed entomologists—all of whom claimed they had never seen a bird eat a *Biston betularia* of any color—he set out to test his hypothesis. He marked a sample of moths of each color by carefully putting a spot on the underside of the wings, where it could not be seen by a predator. Then he released known numbers of marked individuals into a bird reserve near Birmingham, an industrial area where 90 percent of the local population consisted of black moths. Another sample was released into an unpolluted Dorset countryside, where no black moths ordinarily occurred. He returned at night with light traps to recapture his marked moths. From the area around Birmingham, he recovered 40 percent of the black moths, but only 19 percent of the light ones. From the area around Dorset, 6 percent of the black moths and 12.5 percent of the light moths were retaken. Clearly, if you are a moth, it is advantageous to be black near Birmingham but light around Dorset.

To clinch the argument, Kettlewell placed moths on tree trunks in both locations, focused hidden movie cameras on them, and was able to record birds actually selecting and eating the moths, which they do so rapidly that it is not surprising that it was not previously observed. Near Birmingham, when equal numbers of dark and light moths were available, the birds seized 43 light-colored moths and only 15 black ones; near Dorset, they took 164 black moths but only 26 light-colored forms.

A similar tendency for dark-colored forms to replace light-colored forms has been found among some 70 other moth species in England and some 100 species of moths in the Pittsburgh area of Pennsylvania.

41–12
Biston betularia, *the peppered moth, and its melanic form,* Biston betularia carbonaria, *at rest on a lichen-covered tree trunk in unpolluted countryside near Birmingham, England. A striking example of an evolutionary change resulting from a drastic environmental change, the black form of the moth began to appear in the latter part of the nineteenth century as the English countryside became increasingly polluted by industrial smoke.*

"Survival of the Fittest"

The phrase "survival of the fittest" is often used in describing the Darwinian theory. In the early twentieth century, the doctrine of survival of the fittest in natural populations was used by some individuals to defend gross social inequalities and ruthless competitive tactics in industry on the grounds that they were merely in accord with the "laws of nature." This philosophy was referred to by some as social Darwinism. However, in actual fact, very little in the process of change fits the concept of "nature red in tooth and claw." One fuchsia flowering a little brighter than its neighbors and better able to catch the attention of a passing hummingbird is a more pertinent model of the struggle for survival. There is only one criterion of fitness as it is measured by population geneticists: the relative number of descendants of an individual in a future population.

SUMMARY

Population genetics is a synthesis of the Darwinian theory of evolution with the Mendelian principles of genetics. A population, for the population geneticists, is an interbreeding group of organisms, defined and united by its gene pool (the sum of all the alleles of all the genes of all the individuals in the population). Evolution is, by definition, a change in the composition of the gene pool.

The Hardy-Weinberg law is the application of Mendelian genetics to population biology. It states that in a large population of individuals in which random mating occurs, genetic recombinations alone do not change the frequencies of alleles in the gene pool. For a gene with two alleles, the mathematical expression of the Hardy-Weinberg equilibrium is $(p + q)^2 = 1$, where p equals the frequency of one allele and q equals the frequency of the other allele. Thus, in any given population, the homozygotes for each allele are represented by p^2 and q^2, and the heterozygotes by $2pq$. Changes in the Hardy-Weinberg equilibrium are caused by mutations, gene flow, genetic drift, and, particularly, by natural selection.

The Hardy-Weinberg equilibrium applies only if the population is large enough. In a small population, chance may cause the frequency of certain alleles to increase or to decrease and perhaps even disappear; this phenomenon is known as genetic drift. A small population that has branched off from a larger population may not be a representative sample of the larger population; this phenomenon is known as the founder principle.

Natural selection is the cause of changes in the Hardy-Weinberg equilibrium resulting from differences in the number of offspring left by different organisms because of variations in their phenotypes. It may take the form of stabilizing selection (elimination of extreme individuals), disruptive selection (elimination of intermediate forms), or directional selection (replacement of one phenotype by another). Of these three, the first is the most frequent and the third is primarily responsible for the changes we usually associate with evolution.

QUESTIONS

1. Distinguish between the following terms: genotype/phenotype; natural selection/evolution; population/gene pool.

2. Why does not everyone in the human population have short fingers and a cleft chin?

3. Suppose, in a breeding experiment, you combined 7,000 *AA*s and 3,000 *aa*s. After the first generation, what would be the value of p^2? Of q^2? Of $2pq$? What would be the values after the second generation?

4. Among black Americans, the frequency of sickle cell anemia is about 0.0025. What is the frequency of heterozygotes? When one black American marries another, what is the probability that both will be heterozygotes? If both are heterozygotes, what is the probability that one of their children will have sickle cell anemia?

5. What is the difference between gene flow and genetic drift? How does each affect the gene pool of a population?

6. How does stabilizing selection affect the gene pool? Directional selection?

7. What characteristics in the human population are probably being maintained by stabilizing selection? Which might be subject to directional selection?

8. Following the example of the fungus-eating midge, why do not all mothers similarly sacrifice themselves for their offspring?

CHAPTER 42

Variability: Its Extent, Preservation, and Promotion

Like begets like, we know now, because of the remarkable precision with which the DNA is copied and transmitted from cell to cell, so that the DNA in the cells of any individual is, except for occasional mutations, a true replica of the DNA that individual received from its father and mother. This constancy is, of course, essential to the survival of the individual organisms of which the population is composed. However, if evolution is to occur, there must be variations among individuals. Such variations make it possible for species to change as conditions do; they provide the raw material on which natural selection acts.

As we mentioned in the Introduction, the notion of evolution—changes in organisms taking place through time—was part of the intellectual climate for some hundred years before Darwin published *The Origin of Species.* Darwin was able to propose his simple, coherent hypothesis because he recognized, where others did not, the importance of widespread, inheritable variations in the process of evolution. Much of the research in modern population genetics is concerned with the extent of such variability (far greater even than Darwin could have realized) and with the way variations are preserved and fostered in gene pools.

42–1
Even in populations in which the individuals appear almost identical, such as the members of this group of Adelie penguins, we know that variations exist because individuals have no difficulty in identifying their own mates or offspring. To these penguins, all people probably look alike.

THE EXTENT OF VARIABILITY

Darwin's awareness that variations exist among individuals in a population was based largely on his observations as a naturalist, both in England and during his voyage on the *Beagle.* He also recognized that the traits that emerged in the course of selective breeding of domestic animals and plants reflected variability somehow latent in the gene pool. This important hypothesis of Darwin has now been confirmed by a variety of laboratory investigations.

Bristle Number in Drosophila

In one group of studies, for example, the extent of latent variability in a natural population was demonstrated in the laboratory by experiments with the fruit fly *Drosophila melanogaster.* An easily observable hereditary trait, the number of bristles on the ventral surface of the fourth and fifth abdominal segments, was chosen for study. In the starting stock, the average number of bristles was 36. Two groups were interbred, one selected for increase of bristles and one for decrease. In every generation, individuals with the fewest bristles were selected and crossbred, and so were individuals with the highest number of bristles. Selection for low bristle number resulted in a drop after 30 generations from 36 to an average of about 30 bristles. In the high-bristle-number line, progress was at first rapid and steady. In 21 generations, bristle number rose steadily from 36 to an average of about 56 (Figure 42–2). No new genetic material had been introduced; within the single population the potential for a wide range of bristle numbers already existed. Subsequent experiments in *Drosophila* and other organisms have shown that the choice of bristle number was not merely a fortunate accident. No matter what single characteristic is selected for in breeding experiments, all reveal a comparable range of natural variability.

There is a second part to the bristle-number story. The low-bristle-number line soon died out, owing to sterility. Presumably, changes in factors affecting fertility had also taken place during selection. When sterility became severe in the high-bristle line, a mass culture was started; members of the high-bristle line were

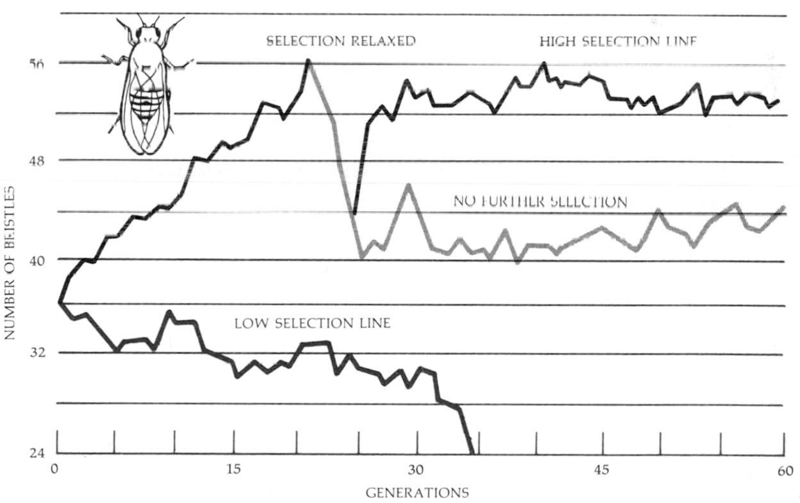

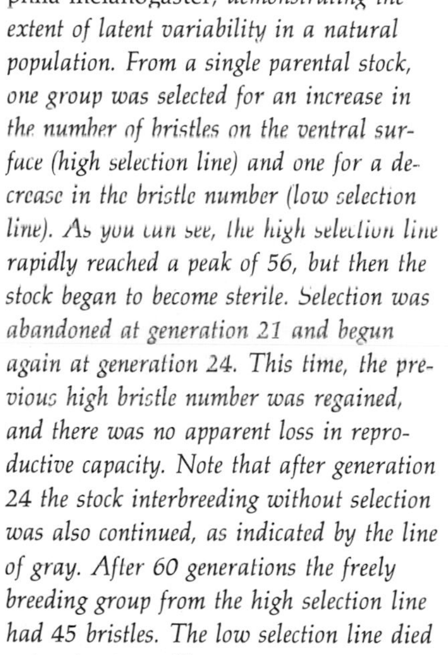

42–2
The results of an experiment with Drosophila melanogaster, *demonstrating the extent of latent variability in a natural population. From a single parental stock, one group was selected for an increase in the number of bristles on the ventral surface (high selection line) and one for a decrease in the bristle number (low selection line). As you can see, the high selection line rapidly reached a peak of 56, but then the stock began to become sterile. Selection was abandoned at generation 21 and begun again at generation 24. This time, the previous high bristle number was regained, and there was no apparent loss in reproductive capacity. Note that after generation 24 the stock interbreeding without selection was also continued, as indicated by the line of gray. After 60 generations the freely breeding group from the high selection line had 45 bristles. The low selection line died out owing to sterility.*

permitted to interbreed without selection. The average number of bristles fell sharply, and in five generations went from 56 to 40. Thereafter, as this line continued to breed without selection, the bristle number fluctuated up and down, usually between 40 and 45, which still was higher than the original 36. At generation 24, selection for high bristle number was begun again for a portion of this line. The previous high bristle number of 56 was regained, and this time there was no loss in reproductive capacity. Apparently, the genotype had rearranged and reintegrated itself so that genes controlling bristle number were present in more favorable combinations.

Mapping studies have shown that bristle number is controlled by a large number of genes, at least one on every chromosome and sometimes several at different sites on the same chromosome. Therefore, although we do not know how important bristle number is to the survival of the animal, selection for this trait in some way disrupted the entire genotype. Livestock breeders are well aware of this consequence of artificial selection. Loss of fertility is a major problem in virtually all circumstances in which animals have been purposely inbred for particular traits. This result emphasizes the fact that in natural selection it is the entire phenotype that is selected rather than certain isolated traits, as is often the case in artificial selection.

VARIABILITY IN GENE PRODUCTS

Molecular biology provides another method for studying latent variability. Genes, as we saw in Section 3, code for and regulate the production of proteins. Proteins therefore provide clues about the genes coding for them. J. L. Hubby and R. C. Lewontin ground up a population of fruit flies and extracted proteins from them. From these proteins they were able to isolate 18 different enzymes, which were detected on the basis of their activity. (The method used was electrophoresis, the same method used by Pauling to separate the variants of hemoglobin.) Then they analyzed each of the 18 enzymes separately to see if it was composed of a single protein or of proteins that were slightly different structurally.

Of the 18 enzymes studied in this way, nine were found to be composed of structurally indistinguishable proteins; in other words, the genes that had produced these proteins were the same throughout the entire population of fruit flies studied. However, nine of the groups contained two or more structurally different enzymes. Therefore, without any direct analysis of the genes themselves, the investigators were able to conclude that among the fruit flies studied, there were two or more alleles of the gene responsible for each of the nine enzyme groups. In one group of enzymes, there were as many as six slightly different structural forms; that is, six alleles for the gene coding for that enzyme group were shown to exist in the population as a whole.

Each fruit-fly population tested was heterozygous for almost a third of the genes so tested. Each individual, it was estimated, was probably heterozygous for about 12 percent of its genes. Similar studies on humans, using an accessible tissue such as blood or placenta, indicate that at least 25 percent of the genes in any given population are represented by two or more alleles, and individuals are heterozygous for at least 7 percent of their genes, on the average.

Explaining the Extent of Variation

The experiments of Hubby and Lewontin and those patterned after them disclosed a far greater degree of variability than had been previously imagined and, like most important scientific discoveries, raised major new questions. Most geneticists had previously thought that the individuals of a population should be close to genetic uniformity, as a result of a long history of selection for "optimal" genes. Yet, as these studies revealed, there is a great deal of genetic variation present in natural populations. One school of geneticists, the "selectionists," claim that even such small variations as those in enzyme structure are maintained by a balance of forces of natural selection that favor some genotypes at some times and in some areas and others at other times or in other localities. An opposing school, the "neutralists," claim that the observed variations in the protein molecules are so slight that they do not make any difference in the function of the organism and so are not affected by natural selection. This latter group believes that such variations result from the accumulation over a long period of time of random mutations, many of which are lost from the population by genetic drift, but some of which persist. (For further evidence in this dispute see pages 796 and 819.)

PRESERVATION OF VARIABILITY

Diploidy

The most important factor in the preservation of variability in eukaryotes is diploidy. In a haploid organism, any genetic variations are immediately exposed to the selection process, whereas in the diploid organism, such variations may be stored as recessives, as with the allele for white flowers in Mendel's pea plants. The extent to which a rare allele is protected is revealed by the following table:

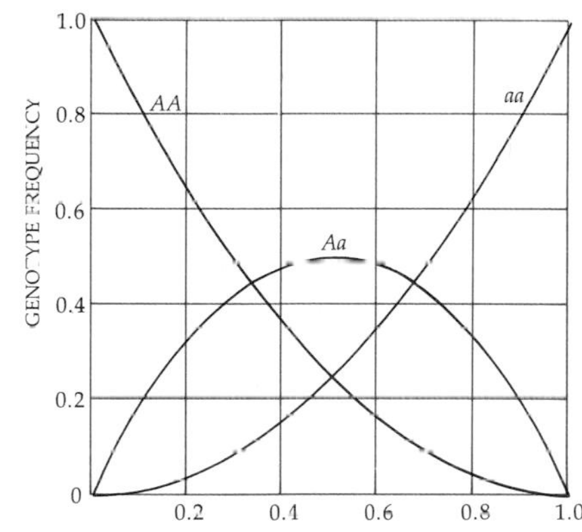

42–3
The relationship between the frequency of allele a *in the population and the frequencies of the genotypes* AA, Aa, *and* aa. *Naturally, the more* AA *genotypes there are, the lower the frequency of* a *alleles. Because of the interrelationships among* AA, Aa, *and* aa, *a change in the frequency of one allele, say* a, *results in a corresponding and symmetrical change in the frequencies of the other allele,* A, *and of the three genotypes.*

FREQUENCY OF ALLELE *a* IN GENE POOL	GENOTYPE FREQUENCIES *AA*	*Aa*	*aa*	PERCENTAGE OF *a* IN HETEROZYGOTES
0.9	0.01	0.18	0.81	10
0.1	0.81	0.18	0.01	90
0.01	0.9801	0.0198	0.0001	99

As you can see, the lower the frequency of allele *a*, the smaller the proportion of it exposed in the *aa* homozygote becomes, and the removal of the allele by natural selection slows down accordingly (see also Figure 42–3). This result is of special interest to students of eugenics, the study of improvements in the human gene pool through controlled breeding. For instance, consider a genetic disorder expressed only in the homozygous recessive with a frequency of about 0.01 in the human population. Individuals with the *aa* genotype make up 0.0001 of the population (1 child for every 10,000 born). It would take 100 generations, roughly 2,500 years, of a program of sterilization of defective homozygote individuals to halve the allele frequency (to 0.005) and reduce the number born with this genetic disorder to 1 in 40,000.

Suppose, however, that the recessive allele in the homozygous state conferred an advantage under certain unusual environmental circumstances or in conjunction with another particular set of genes. By appearing mostly in the heterozygous state, it would be protected against removal by selection. The allele would therefore persist in the gene pool and so by chance might be expressed under favorable selective circumstances.

The Evolution of Diploidy

Once sexual reproduction was established among the unicellular eukaryotes, the stage was set for the evolution of diploidy. By "accident"—an accident that apparently took place in a number of separate evolutionary lines—a zygote, originally the only diploid stage of the organism's life cycle, divided mitotically instead of meiotically, producing an organism with two complete sets of chromosomes. The fact that all the organisms we consider most highly evolved are diploid (or polyploid) is an indication of the important role of diploidy in the evolutionary process.

Heterozygote Superiority

Recessive alleles, even though they may be harmful in the homozygous state, may not only be sheltered in the heterozygous state, but may actually be selected for. This phenomenon, known as *heterozygote superiority,* is another way that variability is preserved. A dramatic example of heterozygote superiority is found in association with sickle cell anemia. Individuals homozygous for sickling almost never become parents; in Darwinian terms, their fitness is zero. Therefore, almost every time one sickling allele encounters another in a homozygous individual, two sickling alleles are removed from the population. At one time, it was thought that the sickling allele was maintained in the population by a steady influx of new mutations. Yet in some African tribes, as much as 45 percent of the population is heterozygous for sickling, and to replace the loss of sickle alleles by mutations alone would require a rate of about 1 mutation per 100 genes. This is about 5,000 times greater than any other known human mutation rate.

In the search for an alternative explanation, it was discovered that the sickling allele is maintained at high frequencies because the heterozygote has a selective advantage. In many regions of Africa, malaria is one of the leading causes of illness and death, especially among young children. Studies of the incidence of malaria among young children showed that susceptibility to malaria is significantly lower in individuals heterozygous for sickling. Moreover, for reasons that are not known, women who carry the sickling allele are more fertile.

Among blacks in the United States, only about 9 percent are heterozygous for the sickling allele. Since no more than half of this loss of the sickling allele can be explained by the black-white admixture in America, the conclusion is that once selection pressure in favor of the heterozygote is relaxed, the mutant will tend to be eliminated slowly from the population.

Other alleles that are harmful in the homozygous state appear to be maintained in the population by heterozygote superiority. For example, again for reasons that are not known, women who are carriers of hemophilia (heterozygotes) have a fertility rate about 20 percent greater than that of other women.

42–4
Most species of bamboo flower very seldom. Few wait less than 15 years between flowerings. In the meantime, like other grasses, it reproduces asexually by producing new shoots from underground rhizomes.

PROMOTING VARIATION

Sexual Reproduction

By far the most important method by which eukaryotic organisms promote variation in the phenotype of their offspring is by sexual reproduction. As we saw in Chapter 11, sexual reproduction combines parental genes in new ways so that wholly new genotypes appear in the next generation. Genes are recombined at meiosis, by crossing over and by the segregation of homologous chromosomes, and at fertilization, when new pairs of alleles are formed. By these processes, variations arising from mutations are reshuffled and so brought into a series of new combinations with other alleles, giving rise to new phenotypes upon which natural selection can operate. The effect of a particular allele varies, depending both on the other allele and on the rest of the genotype. A favorable allele is only favorable if it is in good company, and most traits that are selected for are undoubtedly the result of a particular winning combination of alleles.

Assuming the hypothesis is correct that efficient utilization of energy and other resources is under selective pressure, consider the expenditure of energy and resources involved in sex. In the words of Edward O. Wilson:

> Sexual reproduction is in every way a consuming biological activity. Reproductive organs tend to be elaborate in structure, courtship activities lengthy and energetically expensive, and genetic mechanisms of sex determination finely tuned and easily wrecked. But more important, a female that elects to reproduce by sex cuts her genetic contribution to each gamete by one-half, without, in the vast majority of species, receiving any material aid from the male. If an egg develops parthenogenetically without meiosis—the simplest way to proceed—all of the genes in the resulting offspring will be identical to those of the mother. In sexual reproduction, which entails reduction division during meiosis, only half are identical. The female, in other words, has thrown away half of her investment.*

Note that the only selective advantage conferred by sexual reproduction is the production of totally new phenotypes that can be tested in the next generation.

Some eukaryotic organisms reproduce only asexually. In some cases, it appears that the capacity for sexual reproduction was lost in the course of their evolutionary history. In effect, in balancing the costs, these organisms opted for the stability conferred by asexual reproduction. Many modern organisms, particularly members of the plant kingdom, hedge their bets, reproducing both sexually and asexually (Figure 42–4).

Mechanisms That Promote Outbreeding

Many ways have evolved by which new genetic combinations are promoted in sexually reproducing populations. Among animals, even in those invertebrates that are hermaphrodites (each organism produces both egg and sperm cells), such as earthworms, slugs, and snails (Figure 42–5), an individual seldom fertilizes its own eggs. Among plants, there is a variety of mechanisms that ensure that the sperm-

* E. O. Wilson, *Science*, vol. 188, page 1139, 1975.

(a)

(b)

42–5

(a) *Two Roman snails* (Helix pomatia) *mating. Like earthworms and slugs, snails are hermaphrodites. Hermaphroditism is advantageous for slow, solitary species because it doubles the chances of finding a mate, since every mature member of the same species a hermaphrodite meets will always be of the opposite sex—and of the same sex too.* (b) *Each individual can produce new offspring.*

42–6

The willowherb (Epilobium) *is an example of pollination "insurance." If the flower is not pollinated by insects the styles curl back to self-pollinate. No new alleles are introduced when the flower self-pollinates, but because of the reshuffling of chromosomes and genetic cross-overs at meiosis, the offspring may be both genotypically and phenotypically different from its single parent.*

bearing pollen is from a different individual than the stigma it lights upon. Many plants, such as the holly and the ginkgo, have male flowers on one tree and female on the other. In others, such as the avocado, the pollen of a particular plant matures at a time when its own stigma is not receptive. In some species, anatomical arrangements inhibit self-pollination (Figure 42–7).

Some plants have genes for self-sterility. Typically, such a gene has multiple alleles—s^1, s^2, s^3, and so on. A plant carrying the allele s^1 cannot pollinate a plant with an s^1 allele, one with an s^1/s^2 genotype cannot pollinate any plant with either of those alleles, and so forth. In one population of about 500 evening primrose plants, 37 different self-sterility alleles were found, and it has been estimated that there are more than 200 alleles for self-sterility in red clover. As a consequence, a plant with a rare self-sterility allele is more likely to be able to pollinate another plant than is a plant with a common self-sterility allele. Such a system strongly encourages variability in a population; selection for the rare allele makes it more common, whereas more common alleles become rarer.

The existence of such a wide variety of mechanisms, some of them very elaborate, reaffirms the selective value of genetic variability.

TYPES OF VARIABILITY

Variations within populations may be continuous or polymorphic. Continuous variation, as we noted on page 235, is characteristic of a trait, such as height, that is governed by several or more genes (polygenic inheritance). *Polymorphism* is the coexistence within a population of two or more phenotypically different forms with no intermediate forms connecting them.

42–7
Diagrams of a pin type and a thrum type of a primrose. Notice that the pollen-bearing anthers of the pin flower and the pollen-receiving stigma of the thrum flower are both situated about halfway up the length of the flower and that the pin stigma is level with the thrum anthers. An insect foraging for nectar in these plants would collect pollen in different areas of its body, so that thrum pollen would be deposited on pin stigmas, and vice versa.

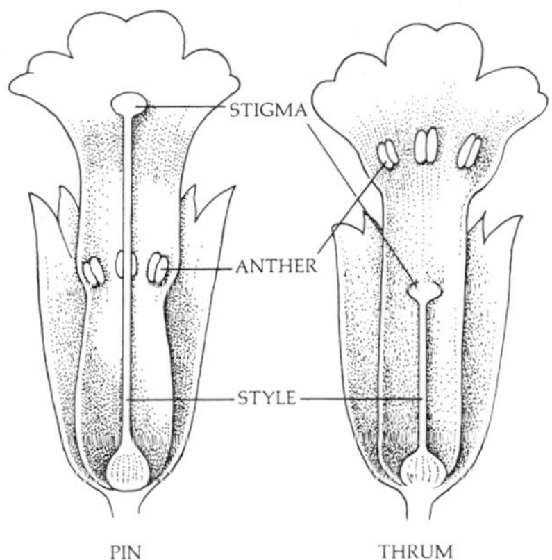

Polymorphism

Polymorphism may be transient or balanced. The coexistence of the light and melanic forms of *Biston betularia* offers an example of transient polymorphism, in which individuals with one trait are increasing as individuals with another are decreasing. Polymorphism may also be balanced, with neither form changing markedly in frequency. Cases of balanced polymorphism are of special interest because defenders of evolutionary theory are called upon in such instances to explain how natural selection permits the existence of two such dissimilar forms. There are many examples of balanced polymorphism in nature, of which we shall cite only three.

Color and Banding in Snails

One of the best-studied cases of balanced polymorphism is found among land snails of the genus *Cepaea.* In one species (*Cepaea nemoralis*), for instance, the shell of the snail may be yellow, brown, or any shade from pale fawn through pink and orange to red. The lip of the shell may be black or dark brown (normally) or pink or white (rarely), and up to five black or dark-brown longitudinal bands may decorate it (Figure 42–8). Fossil evidence shows that these different types of shells have coexisted for more than 12,000 years.

Studies among English colonies of *Cepaea nemoralis* have revealed selective forces at work in some of the colonies. An important enemy of the snail is the song thrush. Song thrushes select snails from the colonies and take them to nearby rocks, where they break them open, eating the soft parts and leaving the shells. By comparing the proportions of types of shells around the thrush "anvils" with the proportions in the nearby colony, the investigators have been able to show correlations between the shell patterns of snails seized by the thrushes and the habitats of the snails. For instance, of 560 individuals collected from a small bog near Oxford, 296 (52.8 percent) were unbanded; whereas of 863 broken shells collected from around the rocks, only 377 (43.7 percent) were unbanded.* In other words, in bogs, where the background is fairly uniform, unbanded snails are less likely to be preyed upon than banded ones.

* This difference (about 10 percent) may seem too small to be significant, but actually the number of shells counted was sufficiently large that there is only 1 chance in 1,000 that this difference could have occurred by chance.

(a)

(b)

42–8
(a) *Polymorphism in land snails: a banded and an unbanded snail.* (b) *An "anvil," where song thrushes break land snails open in order to obtain the soft, edible parts. From the evidence left by the empty shells, investigators have been able to show that in areas where the background is fairly uniform, unbanded snails have a survival advantage over the banded type. Conversely, in colonies of snails living on dark, mottled backgrounds (such as woodland floors), banded snails are preyed upon less frequently.*

Studies of many different colonies have confirmed these correlations. In uniform environments, a higher proportion of snails is unbanded, whereas in rough, tangled habitats, such as woodland floors, far more tend to be banded. Similarly, the greenest habitats have the highest proportion of yellow shells, but among snails living on dark backgrounds, the yellow shells are much more visible and are clearly disadvantageous, judging from the evidence conveniently assembled by thrushes.

Many of the snail colonies studied were at distances so great from one another that the possibility of immigration between populations could be ruled out. Why, then, are both shell types still present in these communities? One would expect populations living on uniform backgrounds to be composed almost entirely of unbanded snails, and colonies on dark, mottled backgrounds to lose most of their yellow-shelled individuals. The answer to this problem is not fully worked out, but it seems that there are physiological factors that are correlated with the particular shell patterns and that form a part of the same group of genes that control color and banding. Experiments have shown, for instance, that unbanded snails (especially yellow ones) are more heat-resistant and cold-resistant than banded snails. In other words, other strong selection pressures besides predation are at work, and they may maintain the genetic variations of color and banding.

42–9
Blue and white snow geese migrate from the Arctic, where they breed, across the Great Plains to Texas and Louisiana, where they winter along the coast. These geese were photographed en route at a National Wildlife Refuge in South Dakota.

Blue and White Snow Geese

At one time the lesser white snow goose (*Chen hyperborea,* "the goose from beyond the north wind") and the blue goose (*Chen caerulescens*) were believed to represent distinct species. More recently it has been found that they actually represent one species, polymorphic for color. The white is recessive; heterozygotes and homozygotes for blue are both the same dark blue color. However, birds mate preferentially with animals of their own color. As a result, there are more homozygotes than if mating were random.

Blue and white geese differ significantly in their physiological adaptations for breeding. One major difference is that the white geese, which breed earlier, are less visible on snow, whereas the later-breeding blue geese are better camouflaged on the muddy ground that follows. Apparently, the chances for the survival of young are approximately equal, so both the blue and white varieties are maintained.

However, as the snow geese make their long fall migration from the Arctic to the Gulf of Mexico, they are subject to human predation. The white-plumaged birds, which are more conspicuous and also more highly prized as trophies, are shot in much larger numbers than those with blue plumage, thus introducing a new selection pressure.

The Tippling Fruit Flies

Fruit flies, you may recall, live on a heady brew of fermenting fruit juices. In *Drosophila melanogaster,* there are two common forms of the enzyme alcohol dehydrogenase, which converts alcohols to sugars. These two forms are commonly known as fast (F) and slow (S) for their electrophoretic behavior. On a substrate of ethanol, F is almost twice as active as S. On the other hand, F is less stable than S at high temperature, and its activity declines more rapidly than that of S. On other alcohol substrates the activities of the two are very similar. Figure 42–10 shows the results of breeding experiments with the flies. Homozygotes and heterozygotes for

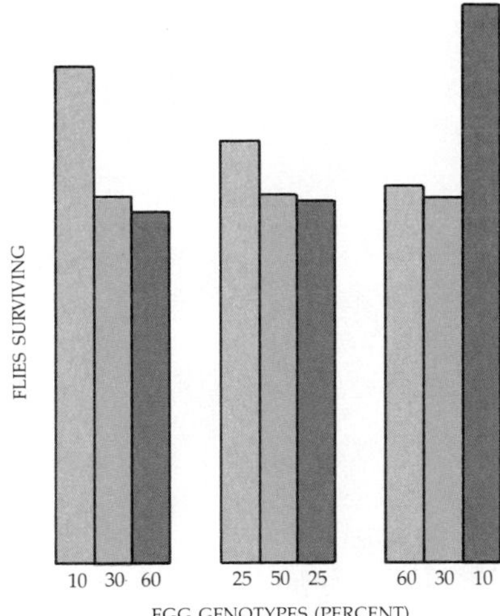

42–10
Polymorphism in fruit flies. There are two common forms of the enzyme alcohol dehydrogenase, which converts alcohol to sugar, fast (F) and slow (S). Eggs homozygous for the S gene (blue), homozygous for the F gene (yellow), and heterozygous (green) were placed in vials with a fixed supply of nutrients. The proportion of each genotype surviving was dependent upon its initial frequency.

the two alleles were placed in a glass jar with a limited supply of nutrients. When the homozygotes were equal in number, the FF flies showed a slightly higher survival rate than the SS flies, and when the FF homozygotes were rare, their survival rate was much higher. However, when the SS homozygotes were rare, they had a better survival rate than either the FF homozygotes or the heterozygotes (FS). These experiments are interpreted as showing (1) that the polymorphism is maintained by natural selection, and (2) that the selective force probably reflects the relative capacities of the different groups to use some form of alcohol under some particular environmental circumstance. When the fruit flies that can utilize this resource are rare, the comparative abundance of the resource is great and the group prospers. Thus, in this case at any rate, the enzyme variants are not neutral, but, as with the snail shells, one form is advantageous under some circumstances and the other better suited to other circumstances. Such findings lead biologists to question the activities of agricultural engineers who seek purebred varieties representing the "best" genotype. Frequency-dependent selection, in which the fitness of a phenotype depends upon its relative abundance in the local population, is a common mechanism by which polymorphism is maintained.

Human Blood Groups

The blood types A, B, AB, and O are a familiar reflection of polymorphism. Apparently, the three alleles associated with these blood types are a part of our ancestral legacy, since the same blood types are also found in other primates. Some population biologists regard the blood types as probably neutral in selective value. Others maintain that polymorphism in human blood groups is a result of selection. For example, among Caucasian males, the life expectancy is greatest for those with group O and least with group B; exactly the opposite is true for Caucasian females with these blood types. People with type A blood run a relatively higher risk of cancer of the stomach and of pernicious anemia. Men with type B blood are about 5 centimeters taller, on the average, than those with other blood types. People with type O blood have a higher risk of duodenal ulcers and are more likely to contract Asian flu. Thus selective pressures at work on each of the groups apparently serve to keep them in balance. The geographic distributions of the A, B, AB, and O groups are irregular. For instance, there is a large predominance of O in the western hemisphere and an increase in the B allele as one moves from Europe toward Central Asia. These differences may reflect some differences in the selective forces favoring particular blood groups under particular conditions.

Geographic Variation: Clines

Sometimes variations within the same species follow a geographic distribution and are correlated with temperature, humidity, or some other environmental condition. Such a graded variation in a trait or a complex of traits is known as a *cline*.

Many species exhibit north-south clines of various characteristics. House sparrows, for example, tend to have a smaller body size in the warmer parts of the range of the species and a larger body size in the cooler parts (see Figure 42–11). In cooler regions, a larger body size is advantageous for heat conservation. Conversely, tails, ears, bills, and other extremities of animals are relatively longer in the warmer areas of a species' range; such adaptations allow for heat radiation from the

42–11
The distribution by size of male house sparrows in North America. The higher numbers indicate larger sizes, based on a composite of sixteen different measurements of the birds' skeletons. This map was generated and drawn by a computer.

animal. Plants growing in the south often have slightly different requirements for flowering or for ending dormancy than the same plants growing in the north, although they all may belong to the same species. Most geographic variations, however, are too irregular to be classified into clines.

A species that occupies many different habitats may appear to be slightly different in each one. Each group of distinct phenotypes is known as an *ecotype*. Are the differences among ecotypes determined entirely by the environment or are there genetic differences too? Figure 42–12 illustrates a series of experiments carried out with the perennial plant *Potentilla glandulosa*, a relative of the strawberry. Experimental gardens were established at various altitudes, and wild plants collected from near each of the experimental sites were grown in all the gardens. (*Potentilla glandulosa*, like the strawberry, reproduces asexually by runners, making it possible to study genetically uniform replicates.) Under these conditions it was possible to demonstrate that many of the phenotypic differences among the ecotypes of *Potentilla glandulosa* were due to genetic variations. It is not surprising that in their very different environments, different traits were selected for and so, over time, genetic differences arose.

Variations and the Origin of Species

Darwin believed, as do modern students of evolution, that the differences between individuals within a population could, with time, be translated into the differences between species. We shall explore this concept further in the next chapter.

42–12

Ecotypes of Potentilla glandulosa, *a relative of the strawberry. Notice the correlation between the height of the plant and the altitude at which it grows; there are other phenotypic differences among the plants as well. When plants of the four geographic races are grown under identical conditions, many of these phenotypic differences persist and are passed on to the next generation, indicating that these plants are genotypically, as well as phenotypically, different.*

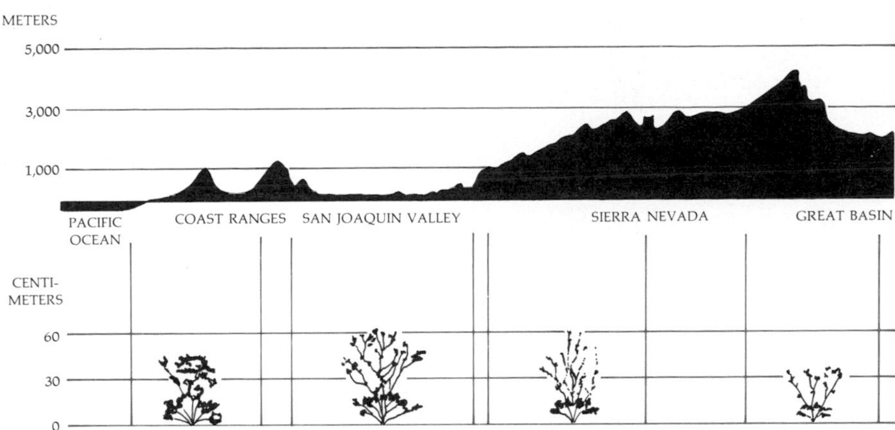

SUMMARY

Natural populations can be shown, by a variety of methods, to harbor a wide spectrum of latent variations. Variability is preserved in populations by diploidy (in which an organism has two sets of chromosomes) and by heterozygote superiority (in which a recessive allele, detrimental in its homozygous state, confers an advantage in its heterozygous state). Variability is promoted by sexual reproduction and by mechanisms, such as self-sterility alleles and anatomic adaptations in plants, which discourage self-fertilizing and promote interbreeding.

Variability may be continuous, as in traits that are governed by several or more genes (polygenic inheritance), or polymorphic. Balanced polymorphism is the coexistence within a population of two or more distinct phenotypes; it is the result of a balance of selection pressures. Variations that follow a geographic distribution are known as clines. Ecotypes are distinct groups of phenotypes of a species occupying different habitats.

QUESTIONS

1. Distinguish between the following terms: homozygote/heterozygote; clone/cline; polygenic inheritance/polymorphism.

2. In what sense does a female reproducing sexually throw away half of her investment? If you are not sure of the answer, review the diagram of meiosis, page 216. Does the male also throw away part of his "investment"?

3. In the experiment on bristle number in *Drosophila,* the average number of flies at the end of the experiment had 56 bristles. No fly at the beginning of the experiment had as many as 56, however. How do you explain this fact?

4. Explain how frequency-dependent selection maintains polymorphism.

5. Describe in your own words the steps in the Hubby-Lewontin experiment and the significance of the results.

CHAPTER 43

The Results of Natural Selection

Darwin believed evolution to be such a slow process that it could never be observed directly. However, the effects of human civilization have produced such extremely strong selection pressures on some organisms that it has been possible to observe not only the results but also the actual process of selection.

EVOLUTION OBSERVED

One example of evolution in action is that of the peppered moth (page 785). Replacement of light-colored moths by the melanic forms proceeded so briskly that eventually only a few of the light-colored population could be found, and these were far from industrial centers. Because of the prevailing westerly wind in England, the moths to the east of industrial towns tended to be of the black variety right up to the east coast of England, and the few light-colored populations were concentrated in the west, where lichens still grow.

Most recently, strong controls have been instituted in Great Britain on the particulate content of smoke, and the heavy soot pollution has begun to decrease. The light-colored moths are already increasing in proportion to the black forms, but it is not yet known whether a complete reversal either in pollution or in adaptive coloration will come about.

43–1
A scale insect on an orchid.

Insecticide Resistance

Another example of evolution in process is the development of insecticide resistance. Chemicals poisonous to insects, such as DDT, were originally hailed as major saviors of human health and property. They have since fallen into disfavor not only because of their tendency to accumulate in the environment, but also because of the extraordinary increase in resistant strains of insects. In fact, one insect is known to be able to remove a chlorine atom from a DDT molecule and use the remainder as food.

A particularly striking example of insecticide resistance has been found in the scale insects that attack citrus trees in California. In the early 1900s, a concentration of hydrocyanic gas sufficient to kill nearly 100 percent of the insects was applied to orange groves at regular intervals with great success. By 1914, orange growers near Corona, California, began to notice that the standard dose of the fumigant was no longer sufficient to destroy one type of scale insect, the red scale. A concentration of the gas that had left fewer than 1 in 100 survivors in the nonresistant strain left 22 survivors out of 100 in the resistant strain. By crossing resistant and nonresistant strains, it was possible to show that the two differed in a single gene. The mechanism for this resistance is not known, but one group of experiments has shown that the resistant individual can keep its spiracles closed for 30 minutes under unfavorable conditions whereas the nonresistant insect can do so for only 60 seconds.

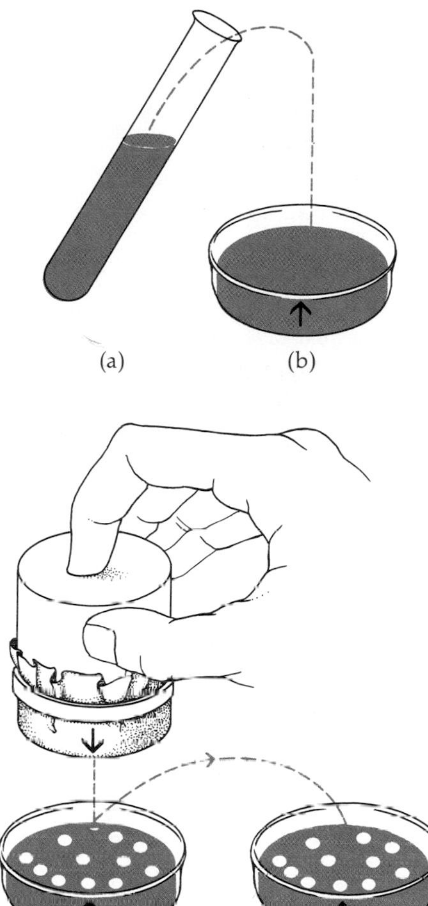

43-2
The Lederbergs' replica-plating method for detecting and isolating drug-resistant bacteria. (a) *The bacteria are cultured in a liquid medium containing a diverse collection of organic molecules referred to as a "broth."* (b) *A sample of the cell suspension is spread over the surface of a petri dish containing broth solidified with agar.* (c) *The plate is incubated until colonies appear. A piece of velveteen held by a ring that fits snugly around a cylindrical block is used to transfer a sample of each colony to a replicate plate, another petri dish containing a solidified medium* (d).

Bacterial Resistance to Antibiotics

The most familiar example of directional selection has been seen among bacteria following the rapid development of antibiotic drugs, which selectively destroy many types of bacteria. It was not long before bacteriologists and physicians began to notice that bacterial cells developed resistance to these drugs. Was this resistance a result of changes produced in the individual bacterium by the effects of the drugs—analogous to the changes produced in heroin addicts, which make them tolerant of doses that would be fatal to a nonaddict—or was bacterial resistance another example of adaptation by natural selection?

The question was important medically, and it was also a serious issue for molecular biologists. Work in molecular biology was just beginning at this time, and, as you will recall, much of it was being carried out with bacteria and other microorganisms. For the purposes of these studies, geneticists were assuming that bacteria were representative biological systems, a supposition that was already under attack since bacteria are not eukaryotes. If the resistance was the result of changes produced by the effects of the drugs in individual bacterial cells and if such changes were inheritable, this Lamarckian concept would violate the genetic principles governing higher organisms, giving the question broad significance.

The dilemma was resolved by a beautifully simple experiment performed by Joshua and Esther Lederberg. Bacterial cells grown in a broth were spread thinly on a dish containing agar, a type of jelly that provides a firm medium for bacterial growth. Once visible colonies began to appear—a matter of about 24 hours—the cells were transferred to different dishes containing different kinds of media, including, for example, doses of penicillin or streptomycin.

The method the Lederbergs developed for transferring the cells was ingenious. As you can see in Figure 43-2, a piece of velveteen was held on a cylindrical block in the same way that a piece of material is held on an embroidery hoop. The block was marked by an arrow, and so were all the culture dishes. The velveteen was pressed gently against the original culture dish and then onto the new medium, with the arrows in careful alignment each time. As a consequence, it was possible to relate each colony that grew on the new culture to a colony in the original culture.

After transfer of cells to a penicillin-containing agar, a small colony sometimes appeared. But how? Were the cells random mutants, or did some cells "adapt themselves" to the new environment? To answer this question, the Lederbergs took more cells from the spot on the original colony corresponding to the spot from which the penicillin-resistant cells arose and transplanted them to a new agar dish that contained no penicillin. Free from the competition of other bacterial cells, these cells grew and covered the entire dish. When members of this dish were tested, it was found that they were all penicillin-resistant—yet they had never been exposed to penicillin. The mutation had occurred independently of the exposure.

Penicillin-resistant cells had arisen spontaneously in the original bacterial colony, and when placed in an environment for which they were particularly well suited, they flourished and multiplied. They were, therefore, genetic mutants, selected by the environment but not produced by it.

This finding restored the status of bacteria as experimental models for medical genetics. It also led to the search for new drugs and drug combinations.

However, despite increased knowledge of the mechanisms of bacterial resistance and the ease with which it can be induced, antibiotics are still being employed indiscriminantly. In the winter of 1977–1978, United States' physicians began seeing increased numbers of cases of gonorrhea resistant to penicillin. Epidemiologists believe they have traced one of these resistant strains to the Philippines, where prostitutes, who are subject to governmental regulation, are routinely given low doses of penicillin. Nor should we dismiss this incident as simply an example of the less developed world's inability to cope with modern technology. Many physicians in the United States still routinely administer antibiotics for the treatment of colds, flu, and other virus-caused diseases against which they are completely ineffective. Antibiotics are mixed with commercial feeds for beef cattle and other livestock intended for human consumption. (This practice, which adds to the market weight of the livestock, has now been banned in Great Britain.)

THE FOSSIL RECORD OF EVOLUTION

Can relatively small changes, such as those observed in populations of peppered moths and scale insects, add up over the eons to the major changes associated with the emergence of entirely new forms? The leap is a large one, but the evidence that this is indeed the case is found in the fossil record.

Phyletic Evolution

Paleontologists—students of fossils—recognize three principal patterns of organic evolution: One of these is phyletic evolution. Phyletic evolution describes the changes taking place in a single line of organisms over a long period of time. A well-known example is the evolution of the horse (Figure 43–3), whose history is abundantly represented in the fossil record. At the beginning of the Tertiary period, some 65 million years ago, the horse lineage was represented by Eohippus ("dawn horse"). It was a small herbivore (25 to 50 centimeters high at the shoulders) with three toes on its hind feet, four toes on its front feet, and doglike footpads on which its weight was carried. Its eyes were halfway between the top of its head and the tip of its nose, and its teeth had small grinding surfaces and low crowns, which probably could not have stood much wear. Its teeth indicate that it did not eat grass; in fact, there was probably not much grass to eat. It probably lived on succulent leaves. Eohippus evolved by stages into a group of larger, three-toed herbivores. Some of these lines survived and some did not. The molar teeth of some of the successful lines had large crowns that continued to grow as they were worn away (as do those of modern horses).

During the Oligocene, horses, of which there were still several distinct species, became still larger. The middle or third digit of each foot expanded and became the weight-bearing part of the foot. Some of the species remained three-toed, while in others the non-weight-bearing digits became greatly reduced. The three-toed spe-

43–3
Phyletic evolution: the modern horse and some of its ancestors. Over the past 60 million years, small several-toed browsers, such as Eohippus, were replaced in gradual stages by members of the genus Equus, *characterized by, among other features, a larger size, broad molars adapted to grinding coarse grass blades, a single toe surrounded by a tough, protective keratin hoof, and a leg in which the bones of the lower leg had fused, with joints becoming more pulleylike and motion restricted to a single plane. Only a few of the evolutionary lines are shown here.*

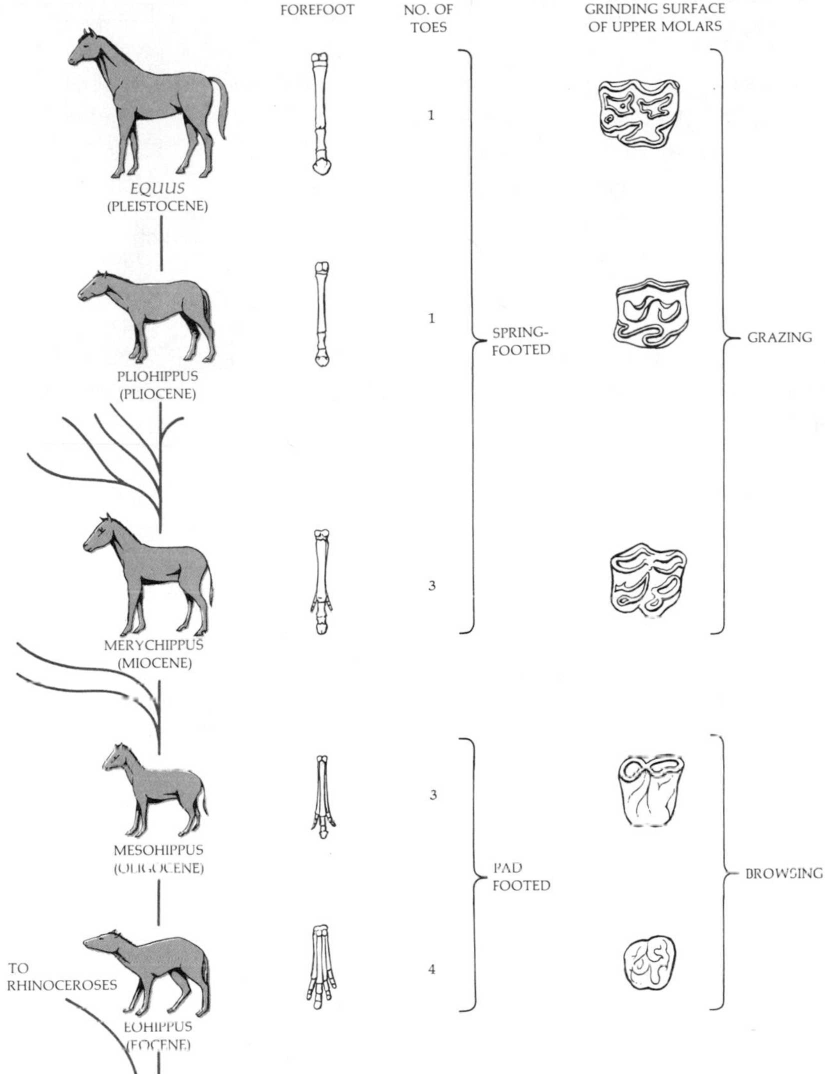

cies did not survive, but those in which the digits were greatly reduced—to one—did, and were the direct ancestors of the modern horse.

By hindsight, these changes can be correlated with changes in the environment and so interpreted as adaptive. In the time of little Eohippus, the land was marshy and the chief vegetation was leaves; the teeth of Eohippus were adapted for browsing. By the Miocene, the grasslands began to spread; groups of horses whose teeth became adapted to grinding grasses (which have much coarser blades than most dicots do) survived, whereas those who remained browsers did not. The placement of the eye higher in the head may have facilitated watching for predators while grazing. The climate became drier and the ground became harder; reduction of the number of toes, with the development of the spring-footed gait characteristic of the modern horse, was an adaptation to harder ground and to larger size. An

THE RECORD IN THE ROCKS

The earth's long history is recorded in the rocks that lie at or near its surface, layer piled upon layer, like the chapters in a book. These layers, or strata, are formed as rocks in upland areas are broken down to pebbles, sand, and clay and are carried to the lowlands and the seas. Once deposited, they slowly become compacted and cemented into a solid form as new material is deposited above them. As continents and ocean basins change shape, some strata sink below the surface of an ocean or a lake, others are forced upward into mountain ranges, and some are worn away, in turn, by water, wind, or ice or are deformed by heat or pressure.

Individual strata may be paper-thin or many meters thick. They can be distinguished from one another by the types of parent material from which they were laid down, the way the material was transported, and the environmental conditions under which the strata were formed, all of which leave their traces in the rock. They can be distinguished, moreover, by the types of fossils they contain. Small marine fossils, in particular, can be associated with specific periods in the earth's history. The fossil record is never complete in any one place, but because of the identifying characteristics of the strata, it is possible to piece together the evidence from many different sources. It is somewhat like having many copies of the same book, all with chapters missing—but different chapters, so it is possible to reconstruct the whole.

The geologic eras—Precambrian, Paleozoic, Mesozoic, and Cenozoic—which are the major subdivisions of the geologic "book," were established and named in the early nineteenth century. These eras were subdivided into periods, each named, quite simply, for the areas in which the particular strata were first studied or studied most completely: the Devonian for Devonshire in southern England, the Permian for the province of Perm in Russia, the Jurassic for the Jura Mountains between France and Switzerland, and so on.

Early attempts to date the various eras and periods were based simply on their relative ages compared to the age of the earth; obviously, a stratum occurring regularly above another was younger than the one below it. The first scientific estimate was made in the mid-1800s by the famous British physicist Lord Kelvin. On the basis of his calculations of the time necessary for the earth to have cooled from its original molten state, Kelvin maintained that the planet could not have been more than 25 million years old. This estimate greatly troubled Charles Darwin, who knew that it was too little time for evolution to have taken place. (Kelvin was not aware of the existence under the earth's surface of radioactive materials that heat the planet from within.) In the last 20 years, however, new methods for determining the ages of strata have been developed involving measurements of the decay of radioactive isotopes. As a result, the estimated age of the earth has increased, in little more than a century, from 25 million years to more than 4.5 billion.

Darwin's concept of the time necessary for major evolutionary changes to take place has been more than borne out by subsequent studies. For instance, in the evolution of the horse (page 803), a major index of change is relation of tooth width to tooth height as horses gradually changed from browsers (leaf eaters) to grazers (grass eaters). The largest actual change—and this is a fast evolutionary line—was a difference of 1.61 millimeters in the height of a particular tooth in 1 million years. In the entire sequence of dawn horse to Equus, *there were more than eight successive genera over 60 million years, or about 7.5 million years per genus.*

Geologic strata (and the fossils within them) are now dated whenever possible by analysis of the radioactive isotopes they contain. Many naturally occurring isotopes are radioactive. All the heavier elements—atoms that have 84 or more protons in their nuclei—are unstable and, therefore, radioactive. All radioactive elements emit radiation (as particles or rays) at a fixed rate; this process is known as radioactive "decay." The rate of decay is measured in terms of half-life: The half-life of a radioactive element is defined as the time in which half the atoms in a sample of the element lose their radioactivity and become stable. Since the half-life of an element is constant, it is possible to calculate the fraction of decay that will take place for a given isotope in a given period of time.

Half-lives vary widely, depending on the isotope. The radioactive nitrogen isotope ^{13}N has a half-life of 10 minutes, and tritium has a half-life of 12¼ years. The most common isotope of uranium (^{238}U) has a half-life of 4½ billion years. The uranium atom undergoes a series of decays, eventually being transformed to an isotope of lead (^{206}Pb).

The proportion of ^{238}U to ^{206}Pb in a given rock sample, for example, is a good indication of how long ago that rock was formed. However, because the daughter atoms (^{206}Pb, for instance) are fewer and so more likely than parent atoms (^{238}U, in this case) to go undetected, dates set by radioactive isotopes are more likely to be too recent. Hence, as methods become more precise and more samples are analyzed, geologic time stretches farther and farther backward.

animal twice as high as another tends to weigh about eight times as much. Little Eohippus was probably as fast as the modern horse, but a larger, heavier horse with the foot and leg structure of Eohippus would have been too slow to escape from predators. (During this same period predators were developing adaptations that rendered them better able to catch large herbivores, including horses.)

Thus the evolution of *Equus,* viewed from the long retrospective of the geologic record, represents a fairly straightforward accumulation of adaptive changes related to pressure for increased size and for grazing.

Splitting Evolution

A second mode of evolution involves the splitting of phylogenetic lineages. For example, *Ursus arctos*—the brown bear, a species that also includes the Kodiak and grizzly bears—is distributed throughout the northern hemisphere, ranging through the deciduous forests up through the taiga and into the tundra, as it was some 1½ million years ago. As is characteristic of such widespread species, there are many local ecotypes or races. During one of the massive glaciations of the Pleistocene, a population of *Ursus arctos* was split off from the main group, and, according to fossil evidence, this group, under extreme pressure from the harsh environment, evolved into the polar bear, *Ursus maritimus* (Figure 43–4). Brown bears, although they are members of the order of carnivores and closely related to dogs, are mostly vegetarians, supplementing their diet only occasionally with fish and game. The polar bear, however, is almost entirely carnivorous, with seals being its staple diet. It spends most of its time in the water and, unlike the brown bear, does not hibernate. The polar bear differs physically from the other bears in a number of ways, including, besides its white color, its carnivore-type teeth, its streamlined head and shoulders, and the stiff bristles that cover the soles of its feet, providing insulation and traction on the slippery ice.

43–4
The white coloring of the polar bear is probably related to its need for camouflage while hunting rather than a need to hide from predators since it is one of the world's largest carnivores. (Its greatest enemy of all time is modern man, who, prizing the skins for trophies, has taken to hunting the bears by helicopter.) Unlike their southern cousins, the brown bears, polar bears do not hibernate (except for females with young) but roam the ice and icy waters all winter long, in search of fish, seals, young walruses, and other game.

(a)

(b)

43–5
While the placental mammals were undergoing adaptive radiation on the other continents, the marsupials, geographically isolated in Australia, went through a parallel pattern of speciation. (a) *The ecological role of large herbivore, usually filled by hooved mammals, is occupied in Australia by kangaroos. There are some 45 herbivorous species of kangaroo; adults of the larger species weigh up to 90 kilograms and can jump 9 meters in a single leap.* (b) *Though the koala is called a bear, it has no true placental equivalent. It is arboreal and lives exclusively on the oily aromatic leaves of certain* Eucalyptus *trees. As for the* Eucalyptus, *more than 600 different species are known in Australia; none are native to other continents.* (c) *Wombats, of which there are four species, are the marsupial equivalent of large rodents. Nocturnal animals, they spend the day in tunnels and underground burrows dug with their shovel-like paws.*

The cave bear (*Ursus spelaeus*), which was twice as large as any existing bear, is pictured in ancient cave drawings but is now extinct. It also appears from fossil remains to have originated by splitting evolution from *Ursus arctos.*

Ernst Mayr of Harvard believes that the formation of new species by the splitting off of small populations from the parent stock (the founder principle) is responsible for almost all major evolutionary change. In such small populations, favorable genetic combinations, if present, can increase rapidly in number and frequency without being diluted out by gene flow. Hence, evolution probably does not occur steadily and gradually but in spurts, accounting for the sudden appearance of new species in the fossil record.

Adaptive Radiation

Adaptive radiation is believed by many experts to be the major pattern of evolution. It involves the sudden (in geologic time) diversification of a group of organisms that share a common ancestor, often itself newly evolved. It also involves the opening up of a new biological frontier that may be as vast as the land or the air, or, as in the case of the Galapagos finches (page 813), as small as an archipelago.

The fossil record contains many examples of adaptive radiation. For example, some 300 million years ago, the reptiles, liberated from an amphibian existence by the "invention" of the amniote egg (page 476), diversified rapidly into terrestrial environments. A similar, even more rapid burst of evolution later gave rise to the birds (see page 477). As the dinosaurs became extinct, the mammals similarly burst forth on the evolutionary scene, with many different kinds appearing simultaneously in the fossil record.

(c)

Common Features in Patterns of Evolution

Actually, as we noted earlier, these three major patterns of evolution—phyletic evolution, splitting evolution, and adaptive radiation—have a great deal in common. Although we can trace an overall pattern of evolution in *Equus* and its progenitors, the evolution of the horse did not proceed in a straight line, and it involved many branches. Conversely, although a branching off was the crucial event in the creation of *Ursus maritimus,* the modern polar bear is itself a product of phyletic evolution, as is every modern organism. Adaptive radiation is a combination of both phyletic and splitting evolution.

ORIGIN OF SPECIES

The crucial event in splitting evolution and in adaptive radiation is the branching of a single population into two or more populations that do not interbreed. (Some population biologists contend that such splitting also plays a major role in phyletic evolution, holding that major changes take place primarily in small, separate populations.) The polar bear could never have developed its distinctive characteristics had gene flow continued between it and the brown bear. Despite the title of Darwin's major work, he never dealt directly with the problem of speciation. It has been, however, a chief concern of twentieth-century evolutionists.

What is a species? In the words of Terrell H. Hamilton, "A species may be envisioned as an isolated pool of genes flowing through space and time, constantly adapting to changes in its environment as well as to the new environments encountered by its extension into other geographic regions."*

As a working definition, a species is a group of populations whose members can interbreed with one another but cannot (or at least usually do not) interbreed with members of other species. The key concept in both these definitions is that of genetic isolation. How does one pool of genes split off from another and begin its solitary journey? How do two species, often very similar to one another, inhabit the same place at the same time and yet maintain the genetic isolation essential to their identity?

Population biologists agree that new species can arise only when populations are genetically isolated from one another. If, under such circumstances of genetic isolation, the populations are subject to different selective pressures, they will begin to diverge genetically. As a consequence of such genetic divergence, the populations will eventually be unable to interbreed. At this point, speciation is said to have occurred.

Speciation with Geographic Isolation

Geographic Races

Every widespread species that has been carefully studied has been found to contain geographically representative populations that differ from each other to a greater or lesser extent, as with the ecotypes of *Potentilla glandulosa* (page 799). A species composed of geographic races of this sort is obviously particularly susceptible to speciation if geographic barriers arise, as with the polar bear.

* Terrell H. Hamilton, *Process and Pattern in Evolution,* The Macmillan Company, New York, 1967.

Because of the work of John Moore, the leopard frog (*Rana pipiens*) has become one of the best-known examples of geographic races and speciation. Leopard frogs are found in North America as far north as Quebec and as far south as Mexico. Moore has studied and compared a large number of characteristics in 29 separate populations of this species. He has found that there are marked variations, as might be expected, from population to population. Crossbreeding of individuals from different populations produces normal offspring when parents are drawn from populations that are geographically adjacent, such as central and southern Florida, or that lie at roughly the same latitude, such as Texas and central Florida. However, the greater the north-south gap separating the home populations of the parents, the greater also is the proportion of defective, often inviable, offspring. In Texas-Vermont hybrids, for example, mortality among the developing embryos may reach 100 percent. Thus what was clearly once a single species is becoming genetically fragmented.

Geographic barriers are of many different types. Islands are frequent sites for the development of unusual species. The breakup of Pangaea (see essay on facing page) profoundly altered the course of evolution, setting the island continent of Australia adrift, for example, as a veritable Noah's ark of marsupials. Populations of many organisms can become cut off from one another by barriers less obvious than oceans. For a plant, an island may be a mist-veiled mountaintop, and for a fish, a freshwater lake. A forest grove may be an island for a small mammal. A few meters of dry ground can isolate two populations of snails. Islands of this sort usually form by the creation of barriers between formerly contiguous geographic zones. The Isthmus of Panama, for instance, has repeatedly submerged and re-emerged in the course of geologic time. With each new emergence, the Atlantic and Pacific oceans became "islands," populations of marine organisms were isolated, and some new species formed. Then, when the oceans joined again (with the submergence of the Isthmus), the continents separated and became, in turn, the "islands."

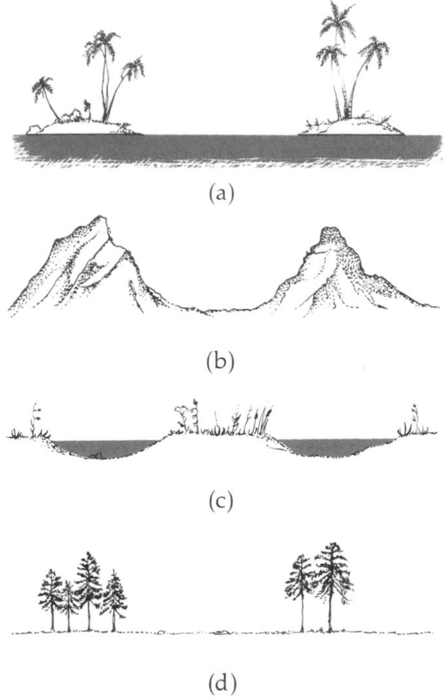

43–6
Depending on the species, islands, genetically speaking, may be (a) *true islands,* (b) *mountaintops,* (c) *ponds, lakes, or even oceans, or* (d) *isolated clumps of vegetation.*

(a) (b)

43–7
Two subspecies that have formed on either side of a natural geographic barrier, the Tana River in Kenya, (a) *the common giraffe,* Giraffa camelopardalis rothschildi, *and* (b) *the reticulated giraffe,* Giraffa camelopardalis reticulata. *The two are considered subspecies rather than species because they interbreed readily in zoos, producing fertile hybrids, and hybrids are also found in nature.*

THE BREAKUP OF PANGAEA

In the past 15 years, the theory of continental drift—viewed at one time with as much suspicion as reports of flying saucers and extrasensory perception—has become firmly established. According to this theory, the outermost layer of the earth is divided into a number of segments, or plates. These plates, on which the continents rest, slide across the surface of the earth, moving in relation to one another. The geologic expression of this relative motion occurs mainly at the boundaries between the adjacent plates. Where plates collide, volcanic islands such as the Aleutians may be formed or mountain belts such as the Andes or Himalayas may be uplifted. At the boundaries where plates are separating, volcanic material wells up to fill the void. It is here that ocean basins are created. Plates may also move parallel to the boundary that joins them but in opposite directions. Such is the case of the San Andreas fault.

About 200 million years ago, all the major continents were locked together in a supercontinent, Pangaea. Several reconstructions of Pangaea have been proposed, one of which is shown here. It is generally agreed that Pangaea began to break up about 190 million years ago, about the time the dinosaurs approached their zenith and the first mammals began to appear. First, the northern group of continents (Laurasia) split apart from the southern group (Gondwana). Subsequently, Gondwana broke into three parts: Africa–South America, Australia–Antarctica, and India. India drifted northward and collided with Asia about 50 million years ago. This collision initiated the uplift of the Himalayas, which continue to rise today as India still pushes northward into Asia.

By the end of the Cretaceous period—according to current reconstructions, about 65 million years ago—South America and Africa had separated sufficiently to have formed half the South Atlantic, and Europe, North America, and Greenland had begun to drift apart; however, final separation between Europe and North America–Greenland did not occur until the Eocene (43 million years ago). During the Cenozoic era, Australia finally split from Antarctica and moved northward to its present position, and the two Americas were joined by the Isthmus of Panama, which was created by volcanic action.

Geology now finds itself in somewhat the same position as astronomy did at the time of Galileo and Copernicus, and geology texts are being hastily rewritten in accordance with the new doctrines. From the outset, of course, the theory of continental drift has been closely interwoven with that of evolution. One of the earliest and most impressive pieces of evidence in favor of the new theory was the discovery of fossil remains of a small, snaggle-toothed reptile, Mesosaurus, *found in coastal regions of Brazil and South Africa but nowhere else. Students of evolution have not yet felt the full impact of the new geology; however, it is already beginning to call forth some important new interpretations of fossil history.*

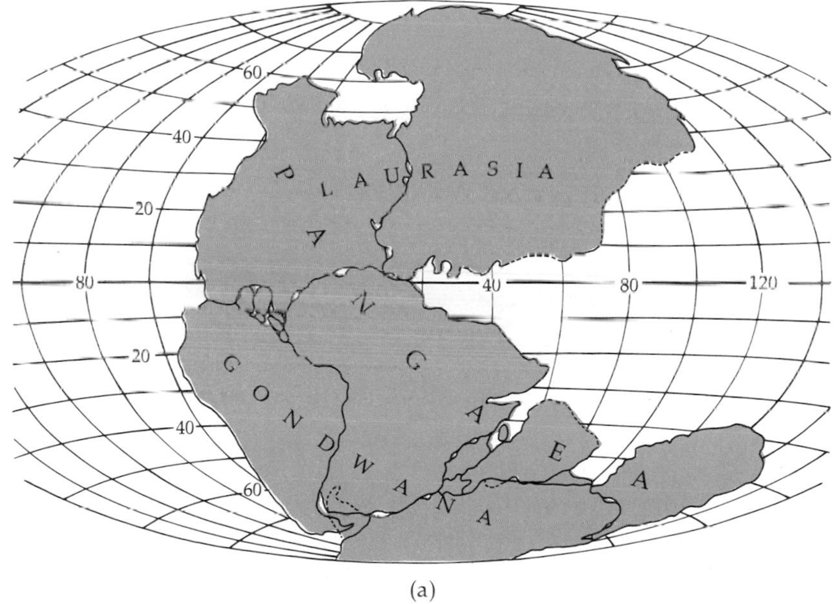

(a)

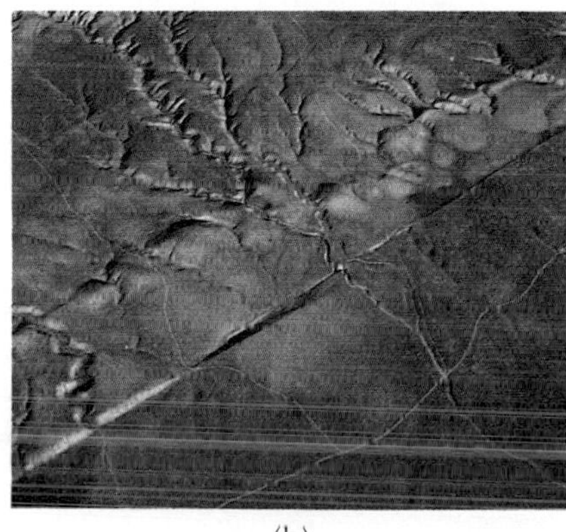

(b)

(a) Pangaea. (b) The San Andreas fault is a boundary between two giant plates that are moving past one another. The fault, running through San Francisco and continuing southeast of Los Angeles, is responsible for California's notorious earthquakes.

Speciation without Geographic Isolation

The only proven means of producing new species without geographic isolation occurs by hybridization or polyploidy. Hybrids, the offspring of parents of different species, occur in both animals and plants. In nature, they are far more frequent in plants. Some hybrid plants spread by runners and eventually cover large areas. Kentucky bluegrass, for instance, occurs widely throughout the United States, is a promiscuous hybridizer, and has formed crosses with many related species. Occasional offspring of these matings have been better adapted than either of the parents and have become successful in their own right.

Hybrids in both plants and animals are nearly always sterile because the chromosomes cannot pair at meiosis (having no homologues), a necessary step for producing viable gametes (Figure 43–8a). In plants, however, fertile forms may arise from these infertile hybrids by the process of polyploidy, in which the cells acquire more than two sets of chromosomes. Polyploid cells arise at a low frequency as the result of a "mistake" in mitosis, so that the chromosomes divide but the cell does not. If such cells divide by further mitosis so that they eventually produce a new individual, either sexually or asexually, that individual will have twice the number of chromosomes of its parent. Polyploid individuals can be produced deliberately in the laboratory by the use of the drug colchicine, which prevents separation of chromosomes during mitosis.

If polyploidy occurs in a sterile hybrid, each chromosome from both parents will be present in duplicate. It then can pair with its duplicate chromosome, meiosis will be normal, and fertility restored (Figure 43–8b). Approximately half of the 235,000 kinds of flowering plants have had a polyploid origin, and many important agricultural species, including wheat, are hybrid polyploids.

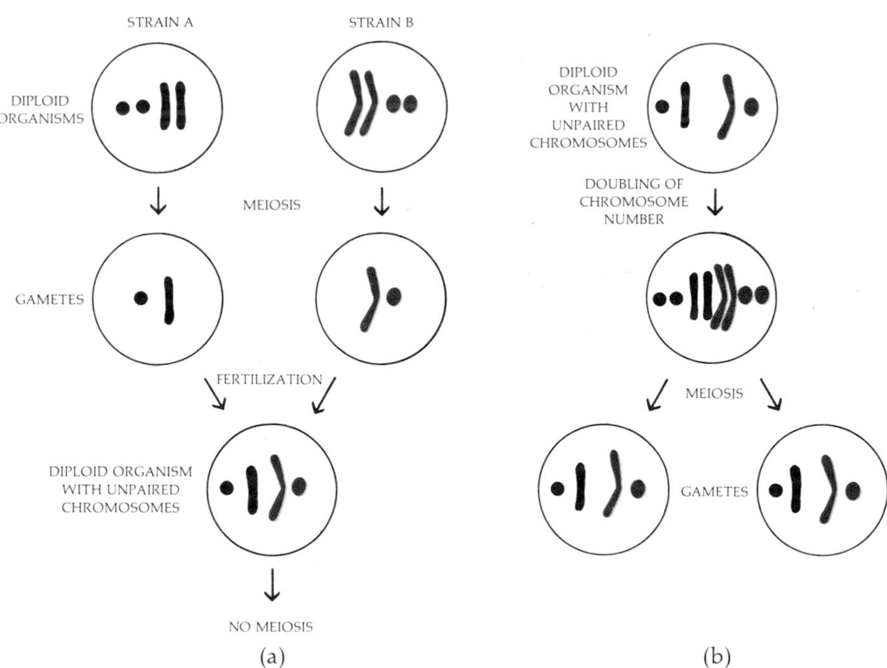

43–8
(a) *A hybrid organism (such as a mule) produced from two haploid* (n) *gametes can grow normally because mitosis is normal, but it cannot reproduce because the chromosomes cannot pair at meiosis.* (b) *However, a hybrid organism produced from two diploid* (2n) *gametes can both divide normally mitotically and undergo meiosis. Since each chromosome will have a partner, the chromosomes can pair at meiosis.*

Isolating Mechanisms

Some species of organisms are so similar phenotypically that only an expert with a microscope can tell them apart—*Drosophila* offers several examples. Yet such species may live together in the same area without interbreeding. What factors operate to maintain genetic isolation of closely related species? Isolating mechanisms may be conveniently divided into two categories: those distinguishable on an anatomical or physiological basis, and those that can be identified in terms of behavior.

Physiological Isolating Mechanisms

These barriers are due to physiological incompatibilities between members of the two species. There are a number of different types. For instance:

1. Differences in the shape of the genitalia may prevent insemination.
2. The sperm may not be able to survive in the reproductive tract of the female.
3. The pollen tube may not be able to grow on the stigma.
4. The sperm cell may not fuse with the ovum.
5. The ovum, once fertilized, may not be able to develop.
6. The young, or some of the young, may survive but may not become reproductively mature.
7. The offspring may be hardy but sterile—the mule, for instance. Sterility in such cases is apparently often caused by the fact that the dissimilar chromosomes cannot pair at meiosis.

Behavioral Isolating Mechanisms

One of the most significant things about the postmating physiological isolating mechanisms is that, in nature, they are rarely tested. In animals, physiologically incompatible genotypes are usually prevented from mating by behavioral isolating mechanisms. These premating behavioral mechanisms can take many forms.

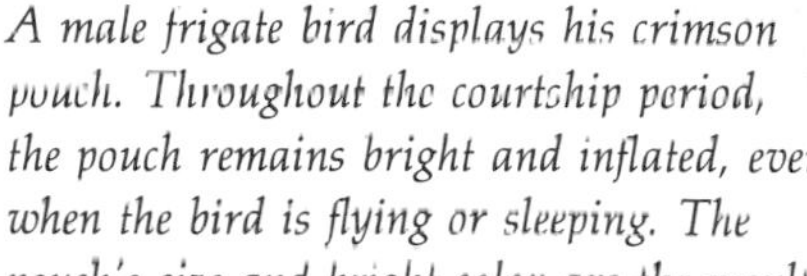

43–9
A male frigate bird displays his crimson pouch. Throughout the courtship period, the pouch remains bright and inflated, even when the bird is flying or sleeping. The pouch's size and bright color are the result of sexual selection.

43–10
Flashing patterns of various species of fireflies found in Delaware. Each division of the horizontal lines represents one second, and the height of each curve represents the brilliance of the flash. The black curves are male flashes, and the colored ones are the female responses. The male flashes first and the female answers, returning the species-specific signal. Firefly flashes differ from species to species, not only in duration, intensity, and timing, but also in color.

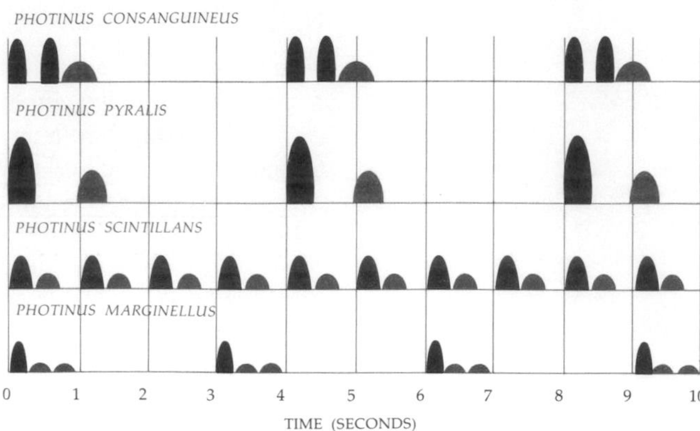

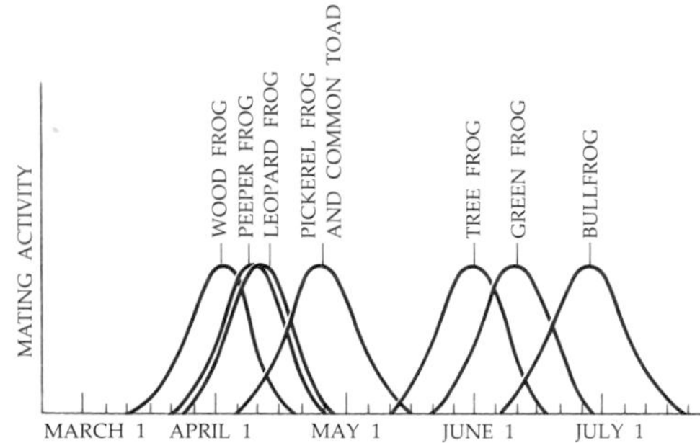

43–11
Mating timetable for various frogs and toads that live near Ithaca, New York. In the two cases where two different species have mating seasons that coincide, the breeding sites differ. Peepers prefer woodland ponds and shallow water; leopard frogs breed in swamps; pickerel frogs mate in upland streams and ponds; and common toads use any ditch or puddle.

Visual recognition is important, particularly among birds, which are visually oriented in all their behavior. According to observers, Galapagos finches recognize each other by their beaks. A male finch may mistakenly pursue a finch of another species—the finches often look very much alike from the rear—only to lose interest as soon as he sees the beak. He will not court a female of another species. The beak is a conspicuous feature in courtship, during which food is passed from the beak of the male to that of the female.

Among many invertebrates, pheromones (page 470) are the chief isolating mechanisms. They may serve as signals, as in the case of the Cecropia moth, to attract the male or, with oysters, for instance, to trigger the release of gametes by the female.

Bird songs, frog calls, and the strident love notes of cicadas and crickets all serve to identify members of a species to one another. For example, field studies of leopard frogs have recently revealed that some of the populations that can crossbreed under laboratory conditions do not. They are effectively isolated from one another by differences in their mating calls. Thus, new species have evolved.

The flashings of fireflies (which are actually beetles) are sexual signals. Each species of firefly has its own particular flashing pattern, which is different from that of other species both in duration of flashes and in intervals between them (Figure 43–10). For example, one common flash pattern consists of two short pulses of light separated by about 2 seconds, with the phrase repeated every 4 to 7 seconds. The lights also vary among the species in intensity and color. The male flashes first, and the female answers, returning the species-specific signal. By mimicking signals with a flashlight, it is possible to attract the males of a given species.

Temporal mechanisms also play an important role in sexual isolation. Species differences in flowering times are important isolating mechanisms in plants. Most mammals—we are a notable exception—have seasons for mating, often controlled by temperature or by day length. Figure 43–11 shows the mating calendar of species of frogs near Ithaca, New York. The spawning of freshwater fishes is often regulated by water temperatures. Members of the species *Drosophila pseudoobscura* and *Drosophila persimilis,* which often are found in the same areas, congregate and mate almost every day but differ in their mating hour.

43–12
The Galapagos Islands, some 950 kilometers west of the coast of Ecuador, have been called "a living laboratory of evolution." Species and subspecies of plants and animals that have been found nowhere else in the world inhabit these islands. "One is astonished," wrote Charles Darwin in 1837, "at the amount of creative force . . . displayed on these small, barren, and rocky islands. . . ."

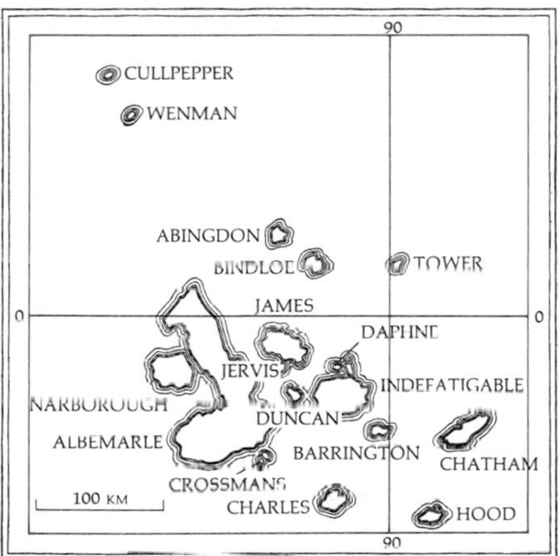

The fact that hybrid progeny are often less viable serves to reinforce behavioral isolating mechanisms in the species. A female cricket that answers to the wrong song or a frog whose individual calendar is not synchronized with that of the rest of the species will contribute less to the gene pool. As a consequence, there will be a steady selection for behavioral isolating mechanisms.

An Example: Darwin's Finches

Admirers of Charles Darwin find it particularly appropriate that one of the best examples of speciation is provided by the finches observed by Darwin on his voyage to the Galapagos Islands. All the Galapagos finches are believed to have arisen from one common ancestral group—perhaps either a single pair or even a single pregnant female—transported from the South American mainland, some 950 kilometers away. How she or they got there is, of course, not known, but it may have been the result of some particularly severe storm. (Every year, for instance, some American birds and insects appear on the coasts of Ireland and England after having been blown across the North Atlantic.) It is very likely that finches were the first land birds to colonize the islands.

We do have an idea of what greeted these unwilling adventurers. The Galapagos archipelago consists of 13 main volcanic islands (Figure 43–12) with many smaller islets and rocks. On some of the islands, craters rise to heights of more than a kilometer. The islands were pushed up from the sea more than a million years ago, and most of them are still covered with black basaltic lava.

The major vegetation is a dreary grayish-brown thornbush, making up vast areas of dense, leafless thicket, and a few tall tree cactuses. Inland and high up on the larger islands, the land is more humid, and there one can find rich, black soil and tall trees covered with ferns, orchids, lichens, and mosses, kept damp by a mist that forms around the volcanic peaks. During the rainy season, the area is dotted by sparkling shallow crater lakes. Thus the habitat available to the finches was a highly diversified one.

From the small ancestral group, 13 different species arose (plus one to the northeast on the Cocos Islands, 1,000 kilometers away). Apparently the various islands were near enough to one another so that, over the years, small founding groups could land. They were far enough apart, however, so that once a group was established, there would be little or no gene flow between it and the parent group for a period of time long enough for genetic barriers to develop. (Finches are not very good fliers; if they had been, this natural experiment in evolution would have been a failure.) Development of a new species requires, it is estimated, at least 10,000 years of genetic isolation, but since there are many islands, several species could have been evolving at the same time.

The ancestral type was a finch, a smallish bird with a short, stout, conical bill especially adapted for seed-crushing; canaries and sparrows are finches and so are cardinals. The ancestor is believed to have been a ground-feeding finch, and six of the Galapagos finches are ground finches. Four species of ground finches live together on most of the islands. Three of them eat seeds and differ from one another mainly in the size of their beaks, which in turn, of course, influences the size of seeds they eat. The fourth lives largely on the prickly pear and has a much longer and more pointed beak. The two other species of ground finch are usually found only on outlying islands, where some supplement their diet with cactus.

43–13
The woodpecker finch (Camarhynchus pallidus) *is a rare phenomenon in the bird world because it is a tool user. Like the true woodpecker, it feeds on grubs, which it digs out of trees using its beak for a chisel. Lacking the woodpecker's long, barbed tongue, it resorts to an artificial probe, a twig or cactus spine, to dislodge the grub. The woodpecker finch shown has selected a cactus spine, which it inserts into the grub hole. The bird has succeeded in prying out the grub, which it eats. If the pick selected turns out to be an efficient tool, the bird will carry it from tree to tree in its search for grubs.*

In addition to the ground finches, there are six species of tree finches, also differing from one another mainly in beak size and shape. One has a parrotlike beak, suited to its diet of buds and fruit. Four of these tree finches have insect-eating beaks, each adapted to a different size range of insects. The sixth, and most remarkable of the insect eaters, is the woodpecker finch, which has a beak like a woodpecker's that it uses as a chisel. Lacking the woodpecker's long, prying tongue, it carries about a twig or cactus spine to dislodge insects from crevices in the bark (Figure 43–13).

By all ordinary standards of external appearance and behavior, the thirteenth species of Galapagos finch would be classified as a warbler. Its beak is thin and pointed like a warbler's, it even has the warblerlike habit of flicking its wings partly open, and, warblerlike, it searches leaves, twigs, and ground vegetation for small insects. However, its internal anatomy and other characteristics clearly place it among the finches, and there is general agreement that it, too, is a descendant of the common ancestor or ancestors.

Notice also that the evolution of the Galapagos finches—the diversification of a single type to fill vacant ecological spaces—provides, as we noted previously, a good example of adaptive radiation.

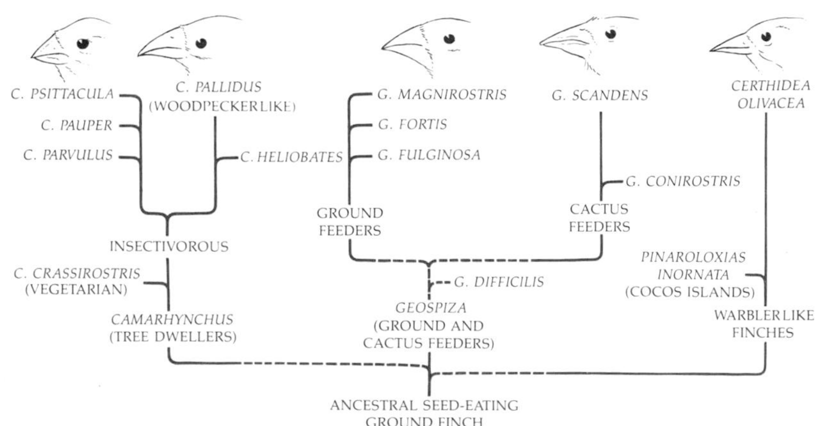

43–14
Evolutionary tree of the Galapagos finches. According to this hypothesis, all are derived from a single ancestral species. The birds are small and dusky-brown or blackish, with stubby tails. They differ anatomically, primarily in the shapes of their beaks. There are six ground finches, including the two cactus-feeding finches (Geospiza *species); six tree finches* (Camarhynchus *species), including the woodpecker finch; and one warbler finch* (Certhidea). *Another warblerlike finch is found 1,000 kilometers away on the Cocos Islands.*

ADAPTATION

Adaptation is the tangible result of evolution. As we noted, all organisms are adapted. For example, note how a squirrel's tail serves as counterbalance as the squirrel leaps and turns; in addition, the same marvelous structure serves as a parasol, a blanket, an aerial rudder, and, in case of mishap, a parachute. Pluck a burr from your clothing and consider the ingenuity with which it clings there. Consider the love and devotion characteristic of the domesticated dog. These are adaptations related to procurement of food and shelter as stringently selected for as the beak of a hawk. Thread a needle; your capacity to do so represents the cumulative effect of millions of years of selective pressures for digital dexterity and eye-hand coordination. (The needle itself made its appearance a mere 10,000 years ago.)

Convergent Evolution

Organisms that occupy similar environments often come to resemble one another even though they may be only very distantly related phylogenetically. When they are subjected to similar selection pressures, they show similar adaptations. The whales, a group that includes the dolphin and porpoises, are similar in many exterior features to sharks and other large fish, but the fins of whales conceal the remnants of a vertebrate hand. Whales are warm-blooded, like their land-dwelling ancestors, and they have lungs rather than gills. Similarly, two families of plants invaded the desert, giving rise to the cacti and the euphorbs. Both evolved large fleshy stems with water-storage tissues and protective spines, and they appear superficially similar. However, their quite different flowers reveal their widely separate evolutionary origins.

The biomes—the world's major groupings of organisms—contain some of the best examples of convergent evolution. We shall discuss these in Chapter 46.

43–15
An example of convergent evolution is provided by (a) *the bull fur seal and* (b) *the king penguin shown here. Although one is a mammal and the other a bird, both have streamlined, fishlike bodies and a layer of insulating fat below the skin. The penguin swims by means of its flipperlike wings, using its webbed feet as rudders. The fur seal also swims primarily with its webbed forelimbs, using them like oars.*

(a)

(b)

43–16
Some unusual adaptations. (a) *Among a number of different species of ants that live in arid habitats, certain individuals, known as repletes, are used as storage containers for honey.* (b) *A cave-dwelling crayfish, like other animals adapted for life in the lightless interiors of caves, is sightless and unpigmented. Its antennae are longer than those of an ordinary crayfish, and, moving slowly and silently, it is able to detect even slight disturbances in the water.* (c) *A red-eyed tree frog of Central America. These animals, like many others that dwell in the canopy of the tropical rain forest, never descend to the forest floor. The frogs' eggs are laid on leaves that overhang water. After they hatch, the tadpoles drop into the water where they complete their development. The bulges at the ends of the toes are suction pads.* (d) *Anteaters include marsupials and placentals, Old World and New World forms, closely related to one another only by certain special adaptations. They all have long noses, very long, wormlike tongues, and stout digging claws. This is the giant anteater of tropical America and, like most anteaters, it eats termites.*

Co-evolution

When two or more populations interact so closely that each is a strong selective force on the other, simultaneous adjustments occur that result in coadaptation, or, as it is more commonly called, *co-evolution*.

We have previously mentioned several examples of co-evolution. One of the most important, in terms of sheer numbers of species and individuals involved, was the co-evolution of flowers and their pollinators, described in Chapter 24. Others will be described in the next section. Here we shall give just one example.

Mimicry

Predation is a strong selective force, and many adaptations are related to escaping from or deterring predators. Some animals, insects in particular, have evolved natural defenses that inflict insult or injury on the predator, such as a sting, a revolting smell, or a bad-tasting or poisonous secretion. Suppose, however, that you were a delicate butterfly with a bad taste; you would sincerely hope that your would-be predator—a bird probably—*knew* you tasted bad without taking a bite. In such situations, it pays to advertise, and if you (reader now, not butterfly) think of animals with such defenses—a skunk, for example—you will note that they are generally conspicuously marked or colored. Obviously, the advertising campaign will pay off best for each individual if many individuals are involved, since the chances of any one having to make the supreme sacrifice are correspondingly reduced. Hence, members of different, quite unrelated species have come to resemble one another.

Müllerian mimics, named after F. Müller, who first described the phenomenon, are groups of organisms that, although not closely related phylogenetically, all have effective obnoxious defenses and all resemble one another. Bees, wasps, and hornets probably offer the most familiar example; even if we cannot tell which is which, we recognize them immediately as stinging insects and keep a respectful distance. Similarly, large numbers of bad-tasting butterflies are look alikes. Müllerian mimicry is adaptive for all the individuals involved because each prospers from a predator's experience with another.

43–17
Müllerian mimics: (a) *a yellowjacket,* (b) *a sand wasp, and* (c) *a masarid wasp. All share the same warning coloration, and all sting.*

(a)

(b)

(c)

43–18
A monarch butterfly (below) and one of its mimics, a viceroy butterfly. A Batesian mimic has been compared to an unscrupulous retailer who copies the advertisement of a successful firm. Müllerian mimics, by contrast, are reputable tradesmen who share a common advertisement and divide its costs. Viceroys are Batesian.

Batesian mimicry, first described by the British naturalist H. W. Bates in 1862, is deceptive mimicry. In Batesian mimicry, the innocuous mimic fools its predator by resembling a stinging or bad-tasting model that the predator has learned to avoid. Some species of stingless flies resemble bees or hornets, and many species of butterflies resemble monarchs or other unpalatable butterflies or moths.

Laboratory experiments have clearly demonstrated Batesian mimicry in operation. Jane Brower, working at Oxford, made artificial models by dipping mealworms in a solution of quinine, to give them a bitter taste, and then marking each one with a band of green cellulose paint. Other mealworms, which had first been dipped in distilled water, were painted green like the models, so as to produce mimics, and still others were painted orange to indicate another species. These colors were chosen deliberately: orange is a warning color, since it is clearly distinguishable, and green is usually found in species that are not repellent and for whom, therefore, advertising would be unwise.

The painted mealworms were fed to caged starlings, which ordinarily eat mealworms voraciously. Each of the nine birds tested received models and mimics in varying proportions. After initial tasting and violent rejection, the models were generally recognized by their appearance and avoided. In consequence, their mimics were protected also. Even when mimics made up as much as 60 percent of the green-banded worms, 80 percent of the mimics escaped the predators.

Batesian mimicry obviously works to the advantage only of the mimic. The model, on the other hand, suffers from attacks not only by inexperienced predators but by predators who have had their first experience with mimic rather than with model. The mimetic pattern will be at its greatest advantage if the mimic is rare—that is, less likely to be encountered than the model—and also if the mimic makes its seasonal appearance after the model, thus reducing its chances of being encountered first. However, if the model is sufficiently distasteful—as in the case of the quinine-soaked mealworms—it may protect mimics even if the latter are very common.

EXTINCTION

Only a small fraction of all the species that have ever lived are presently in existence—certainly less than 1/10 of 1 percent, perhaps less than 1/1,000 of 1 percent. Extinction is very much a part of the evolutionary process, with the elimination of species making room for new ones. Certain times in the earth's history have seen the elimination of major life forms: the trilobites that dominated the early Paleozoic seas; the dinosaurs that disappeared at the end of the Mesozoic era; and the great land mammals that became extinct during the Pleistocene, victims of its harsh climatic changes or, perhaps, of the predations of a new life form, *Homo sapiens.*

At present, it has been estimated, extinction is proceeding at a far more rapid rate than ever before in evolutionary history, owing partly to human predation but mostly to the wholesale destruction of natural habitats. Just as those of us who want to protect a child from the symptoms of the PKU syndrome try to place a monetary value on this enterprise (page 312), conservationists try to indicate the value of these endangered resources, pointing out that seemingly worthless plants have been proven to contain chemicals with medicinal uses or that large mammals have "recreational" value and are the basis of revenue to national parks. The arguments are easier when the animal is photogenic, like the magnificent Bengal

MOLECULAR CLOCKS

Evolutionary relationships may be established by comparisons of living species and by examination of fossil forms. Most recently a new method has become available: the study of macromolecules. In different species some of the amino acids that make up similar proteins may be different, and the degree of difference between the proteins appears to correspond well with the degree of evolutionary divergence between the species. For example, the protein chain of cytochrome c *is exactly the same in human beings and chimpanzees, but it differs from the cytochrome* c *of* Neurospora *by 44 out of 100 amino acids. The composition of cytochrome* c *from more than 60 species has now been determined, and it appears that about 20 million years is required to produce a change in 1 percent of the amino acids in this particular protein.*

The altered protein is able to function because the changes involve only certain amino acids; 35 of the amino acids are the same for all species. These are presumably the amino acids of the active site. Moreover, when a substitution is made, it almost always involves an amino acid of similar chemical properties, so that the conformation of the chain is altered little if at all. Presumably mutations occur that involve other amino acids, but they are weeded out as disadvantageous.

Other proteins change at different rates. Hemoglobin, for example, appears to change at a rate of 1 percent in 6 million years, presumably because it is more tolerant of change than cytochrome c.

Although macromolecules may evolve at different rates, they are consistent in providing the same answer concerning the relative divergence of species. Moreover, these answers are generally, but not always, in accord with accepted interpretations of the fossil evidence. Perhaps most important, these ancient proteins may hold the key to unsolved evolutionary problems, such as the relationships of the invertebrate phyla to one another and to vertebrate origins.

43–19
The last known surviving member of the subspecies of tortoise found on the Galapagos island of Culpepper. With his death, this group will be extinct.

tiger, or pathetically appealing, like the dodo. We suggest, however, that as in the case of the endangered child, the conservationists compile these figures not to persuade themselves but because they feel that these terms are the only ones "others" will understand and accept. Most endangered species, however, have no monetary value. Nor, as we shall see in the next section, has it been proven that the "web of life" will be threatened by their disappearance. We know only that a species takes a very long time to evolve, and once it is gone it is gone forever. We shall end our discussion of evolution on a religious note, based on reverence. As David W. Ehrenfeld of Rutgers University points out, there has been, in all of Western culture, only one similar conservation effort, and not a single species was excluded as unworthy. "Of clean beasts, and of beasts that are not clean, and of fowls, and of everything that creepeth on the earth, there went in two and two . . ." Not a bad precedent.

SUMMARY

Examples of the effects of natural selection on contemporary populations can be seen in the peppered moth (*Biston betularia*) and in the phenomenon of insecticide and antibiotic resistance.

The effects of natural selection on past populations can be seen in the fossil record. Three principal patterns of evolution are observed: phyletic, or "straight line," evolution; splitting, or "branching," evolution; and adaptive radiation. These are not distinct processes; phyletic evolution always involves branching, splitting evolution always involves phyletic evolution, and adaptive radiation involves a combination of both forms.

A species is defined in population biology as an isolated gene pool moving through time and space. Genetic isolation is required for a new species to originate. Among animals, genetic isolation appears usually to require geographic isolation. Among plants, it may also be achieved by hybridization or polyploidy. Polyploidy is often a mechanism whereby hybrids become fertile. Genetic barriers among species are maintained by a variety of physiological and behavioral isolating mechanisms.

Adaptation is the result of interactions between the organism and its environment, including other organisms.

QUESTIONS

1. Distinguish between the following: phyletic evolution/splitting evolution; physiological isolation/behavioral isolation; hybridization/polyploidy; adaptation/evolution; Batesian mimicry/Müllerian mimicry.

2. In your own words, define "species."

3. Describe the separate steps involved in the formation of the distinct species of Galapagos finches.

4. What physical features of the Galapagos made possible the evolution of Darwin's finches? Name them.

5. In your opinion, why is there at present only one species in the genus *Homo*? Is a second *Homo* species likely to arise?

SUGGESTIONS FOR FURTHER READING

Books

CALDER, NIGEL: *The Restless Earth*, Viking Press, Inc., New York, 1972.

A handsomely illustrated report on the new geology, written for the general reader.

CARLQUIST, SHERWIN: *Island Life: A Natural History of the Islands of the World*, Natural History Press, Garden City, N.Y., 1965.

An exploration of the nature of island life and of the intricate and unexpected evolutionary patterns found in island plants and animals.

DAWKINS, RICHARD: *The Selfish Gene*, Oxford University Press, New York, 1976.*

Dawkins demonstrates that we are "survival machines" programmed to preserve the "selfish molecules" known as genes. A wit-sharpening account and also an entertaining one, although not to be taken too seriously.

EHRLICH, PAUL R., RICHARD W. HOLM, and DENNIS R. PARNELL: *The Process of Evolution*, McGraw-Hill Book Company, New York, 1974.

A good, short text for an undergraduate course in evolution theory, appropriate for students who have completed a year of general biology.

EICHER, DON L.: *Geologic Time*, Prentice-Hall, Inc., Englewood Cliffs, N.J., 1968.*

A short, supplementary text that describes the fossil strata and their significance.

* Available in paperback.

GOULD, STEPHEN JAY: *Ever Since Darwin,* W. W. Norton & Company, New York, 1977.*

A collection of thoughtful, witty, and well-written essays from Natural History.

HAMILTON, TERRELL H.: *Process and Pattern in Evolution,* The Macmillan Company, New York, 1967.*

This short book, designed as a supplementary text, is outstanding for its clarity of definition and presentation of evolutionary concepts.

LACK, DAVID: *Darwin's Finches,* Harper & Row, Publishers, Inc., New York, 1961.*

This short, readable book, first published in 1947, gives a marvelous account of the Galapagos, its finches and other inhabitants, and of the general process of evolution.

LEWONTIN, R. C.: *The Genetic Basis of Evolutionary Change,* Columbia University Press, New York, 1974.

This book is intended for the advanced student, but it is also highly recommended for any reader interested in the problems and perspectives of population genetics.

MAYR, E.: *Animal Species and Evolution,* Harvard University Press, Cambridge, Mass., 1963.

A masterly, authoritative, and illuminating statement of contemporary thinking about species—how they arise and their role as units of evolution.

SIMPSON, GEORGE GAYLORD: *Horses,* Oxford University Press, New York, 1951.

The story of the horse family in the modern world and through 60 million years of history, as recorded by a leading paleontologist.

SMITH, JOHN MAYNARD: *The Theory of Evolution,* 3d ed., Penguin Books, Inc., Baltimore, Md., 1975.*

Written for the general public, this book is notable for its many concrete examples of evolution, past and present.

WILSON, E. O., and WILLIAM H. BOSSERT: *A Primer of Population Biology,* Sinauer Associates, Inc., Sunderland, Mass., 1971.*

An introduction to the mathematics of population biology.

Articles

AYALA, FRANCISCO J.: "The Mechanisms of Evolution," *Scientific American,* September 1978, pages 56–69.

CALDER, WILLIAM A., III: "The Kiwi," *Scientific American,* July 1978, pages 132–142.

CLARKE, BRYAN: "The Causes of Biological Diversity," *Scientific American,* August 1975, pages 50–60.

DICKERSON, RICHARD E.: "The Structure and History of an Ancient Protein," *Scientific American,* April 1972, pages 58–72.

EHRENFELD, DAVID W.: "The Conservation of Non-Resources," *American Scientist,* vol. 64, pages 648–656, 1976.

LEWONTIN, RICHARD C.: "Adaptation," *Scientific American,* September 1978, pages 212–230.

MAYR, ERNST: "Evolution," *Scientific American,* September 1978, pages 46–55.

* Available in paperback.

SECTION 8

ORGANISMS IN NATURE

CHAPTER 44

Ecology: The Biosphere and the Ecosystem

44–1
The energy on which life depends enters the biosphere in the form of light. It is converted to chemical energy by photosynthetic organisms, such as these sunlit corn plants. Such photosynthetic organisms are, in turn, the energy source for all heterotrophs, including ourselves.

Ecology is the study of the interactions of organisms with their physical environment and with each other. As a science, it seeks to discover how an organism affects, and is affected by, its living and nonliving environment and to define how these interactions determine the kinds and numbers of organisms found in a particular place at a particular time.

Although few will quarrel with this conventional definition, closer analysis would show that, in fact, ecology means different things to different people. At one time it was synonymous with natural history and with the observation—and also, the early ecologists made clear, the enjoyment—of organisms in their natural environment. Darwin was an ecologist in this sense, as were many of the biologists of his age, although they would have been more comfortable with the term naturalist. (The term ecology was coined by the German biologist Ernst Haeckel, from the Greek word *oikos,* meaning "house" or "home," also the stem of the word "economy." Haeckel, who was also responsible for ontogeny, phylogeny, and many other words equally resonant but no longer apposite, was a younger contemporary of Darwin.)

In 1962, Rachel Carson published her book *The Silent Spring* and heralded a new era in which the word ecology came to be associated with the preservation of wildlife and the conservation of natural resources. The movement became both popular and productive, and, as a consequence, Lake Erie was brought back to life, the Hudson River is once more producing edible fish, strip mining has been put under federal regulations, the smog has been lifted from many of our major cities, DDT has been banned, and our springs are no longer silent. All the problems have not, of course, been solved, but these and other noteworthy achievements indicate that similar problems are soluble.

Now, however, the word ecology is taking on a less favorable and new shade of meaning. Young people, who were in the forefront of the earlier ecology movement, are perhaps more concerned, in these harder times, about their own immediate futures and less about the planet's more distant one. In many rural communities, local residents seeking "progress"—by which they usually mean, quite simply, jobs—find themselves in conflict with conservationists, whom they sometimes regard as wealthy elitists, issuing edicts from an Ivory Tower and little in touch with the real needs of the people. The questions are not easy ones to resolve.

44-2
A plastic cylinder is used to enclose a portion of a Puerto Rican rain forest. Carbon dioxide in the air, which flows from top to bottom in the "sleeve," is measured. Gross primary production of this portion of the rain forest is calculated by adding the amount of carbon dioxide gained at night to that lost during the day.

During this period of ecological activism, another ecological revolution was in the making. The objective of these revolutionists was to place ecology on a quantitative rather than a descriptive basis. Pioneers in this undertaking were the Odums—Eugene, Howard, and William—who devised and utilized methods for measuring productivity and energy flow in ecological systems. Another group of studies was initiated by Robert MacArthur and E. O. Wilson. Their book *The Theory of Island Biogeography* makes a direct assault on what many consider the most interesting question of population biology: Why are there so many species? The study is noteworthy both for its brilliance of conception and for the fact that it is written almost entirely in the language of mathematics.

These new ecologists are an audacious lot. They seek not only to quantify the many variables that may affect an organism in nature but even to manipulate these variables—that is, to perform experiments and to make predictions and construct testable hypotheses. Their tool is the computer.

THE BIOSPHERE

Within our own lifetimes, an unprecedented event has taken place. For the first time, inhabitants of this planet have been able to look back on earth from space. From this vantage point, earth, once known to us as Terra Firma, appears as a frail, small wanderer in the vastness of the universe.

The biosphere, that part of the earth on which life exists, is only a thin film on the surface of this little planet. It extends about 8 or 10 kilometers above sea level and a few meters down into the soil, as far as roots penetrate and microorganisms are found. It includes all of the surface waters and some of the ocean depths. It is patchy, differing in both depth and density.

The Sun

The sun powers the biosphere. It is responsible for the wind and the weather as well as for the energy flow that characterizes life.

44-3
The earth's atmosphere is seen as a thin, bright shell in this photograph taken from an orbiting spacecraft.

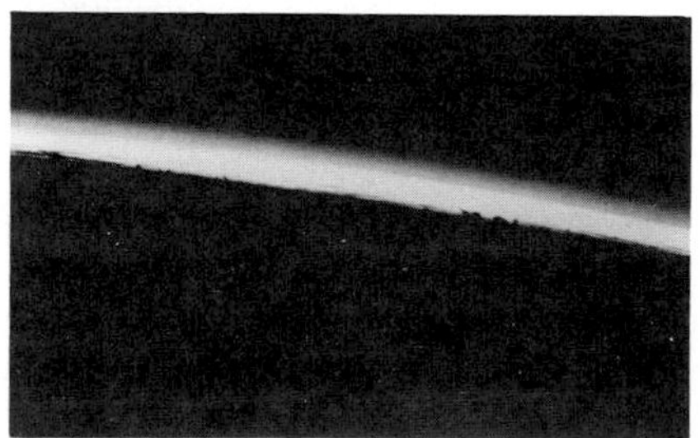

Every day, year in and year out, energy from the sun arrives at the upper surface of the earth's atmosphere at an average rate of 1.94 calories per square centimeter per minute, a total of about 10^{24} calories per year. This is known as the solar constant, and although it is only a tiny fraction of the total energy radiated by the sun, it is a tremendous quantity of energy. Of this total amount of energy, about 30 percent is reflected back into space from clouds and dust; because of these reflections, earth, as seen from outer space, is a shining planet, as bright as Venus. About 20 percent is absorbed in the various layers of the earth's atmosphere. Of this 20 percent, about 1 to 3 percent is absorbed by oxygen and ozone (O_3) in the outermost layers of the atmosphere. This percentage, though small, is very important because it represents most of the radiation in the ultraviolet range. These high-energy radiations damage organic molecules; until photosynthesis produced this protective oxygen-ozone layer, living systems could probably have survived only below the surface of the waters. Of the 20 percent of incoming radiation absorbed in the atmosphere, 17 percent is taken up in the lower layers of the atmosphere, mostly by water vapor, dust, and water droplets. This absorption of radiation warms the atmosphere slightly, though much of the energy is stored as latent heat in the water vapor.

44–4

The greenhouse effect. (a) *Light rays penetrate the glass of the greenhouse, are absorbed by the plants and soil, and are then reradiated as longer-waved heat rays. The glass does not permit these rays to escape, and so the heat remains within the greenhouse.* (b) *Carbon dioxide, like glass, is transparent to light but absorbs the infrared (heat) rays, preventing their escape. An increase in atmospheric carbon dioxide, such as may be caused by the burning of fossil fuels, could lead to an increase in the temperature of the biosphere as a whole.*

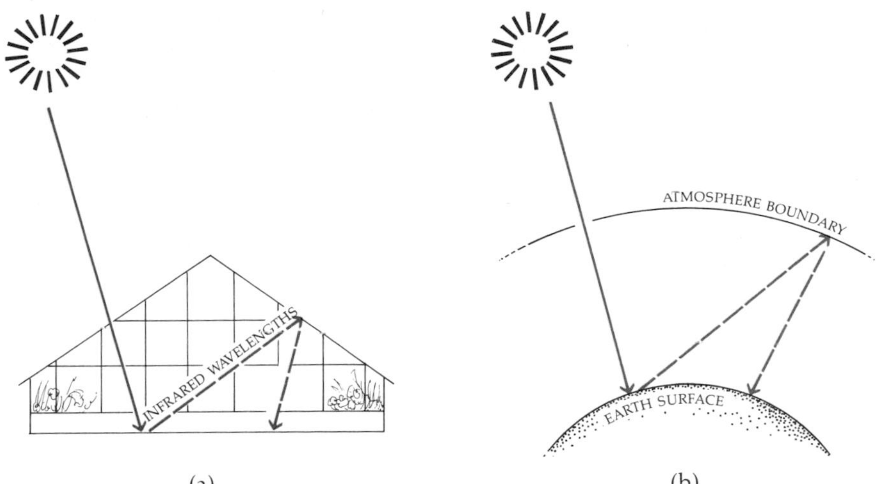

The remaining 50 percent of the incoming radiation reaches the surface. A small amount of this is reflected from bright areas of the earth's surface, but most is absorbed. Energy absorbed by the oceans warms the surface of the water, evaporating water molecules, forming clouds, and providing the rainfall that is a source of water for the dry land areas, powering the water cycle (page 38). The unequal heating of the waters of the ocean combined with the rotation of the earth cause the movements of the ocean currents. Energy absorbed into the earth as light is reradiated from the earth's surface in longer (infrared) wavelengths, that is, as heat. The components of the atmosphere that are transparent to light are not transparent to infrared rays, and so the heat is held in the atmosphere, warming the surface of the earth. This is known as the greenhouse effect; the floor of the greenhouse absorbs incoming light and reradiates it as heat, and the glass of the greenhouse prevents the heat from escaping, so that the air inside a greenhouse on a sunny day is warmer than the air outside (Figure 44–4).

All of these factors balance out; heat loss and heat gain are in equilibrium. If they were not, life would be in trouble. Yet the balance is a delicate one. Increase the earth's reflectivity, thicken its cloud cover, change the gas content of its atmosphere, or decrease its ozone layer, and the entire system is in hazard.

Climate, Wind, and Weather

The amount of energy received by various parts of the earth's surface is not uniform. At the equator, the sun's rays are almost perpendicular to the earth's surface, and this sector receives more energy from the sun than the areas to the north and south, with the polar regions receiving the least (Figure 44–5a). Moreover, because the earth, which is tilted on its axis, rotates once every 24 hours and completes an orbit around the sun about once every 365 days, the amount of energy reaching different parts of the surface changes hour by hour and season by season (Figure 44–5b).

Temperature variations over the surface of the earth and the earth's rotation establish major patterns of air circulation and rainfall. These patterns depend, to a large extent, on the fact that cold air is denser than warm air. As a consequence, hot air rises and cold air falls. As air rises, it encounters lower pressure and conse-

44-5
(a) *A beam of solar energy striking the earth near one of the poles is spread over a wider area of the earth's surface than is a similar beam striking the earth near the equator.* (b) *In the northern and southern hemispheres, temperatures change in an annual cycle because the earth is slightly tilted on its axis in relation to its pathway around the sun. In winter, the northern hemisphere tilts away from the sun, which decreases the angle at which the sun's rays strike the surface and also decreases the duration of daylight, both of which result in lower temperatures. In the summer, the northern hemisphere tilts toward the sun. Note that the polar region of the northern hemisphere is continuously dark during the winter and continuously light during the summer.*

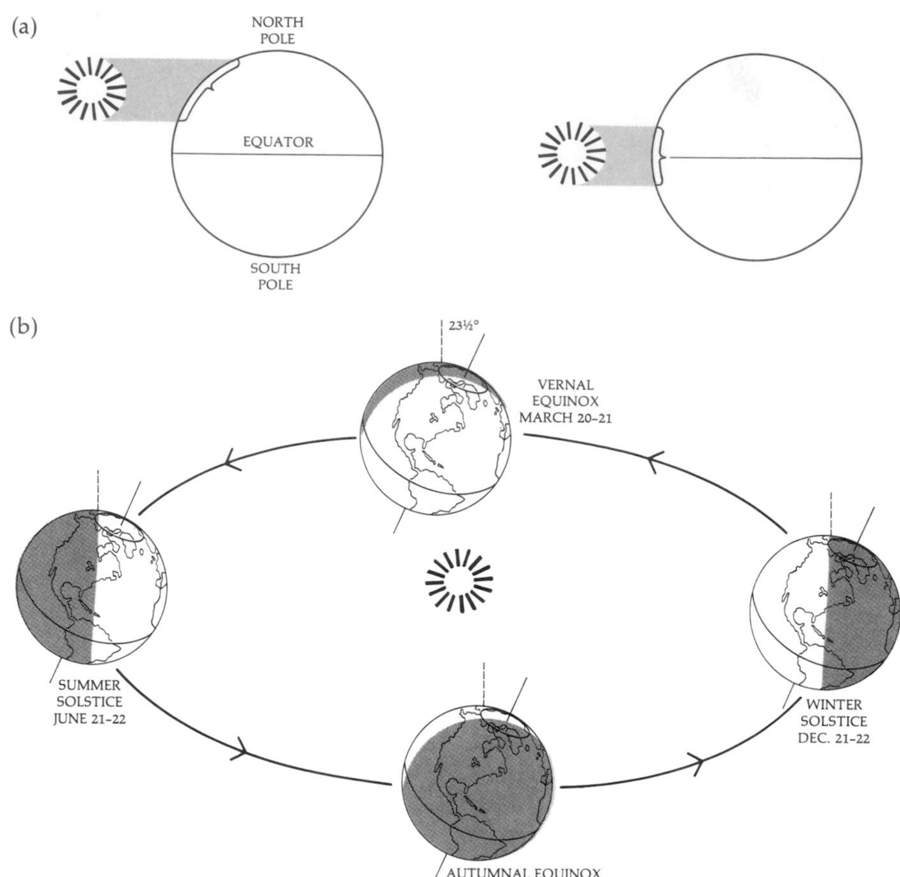

44-6
The earth's surface is covered by many belts of air currents, which determine the major patterns of rainfall and wind. Air rising at the equator loses moisture in the form of rain, and falling air at latitudes of 30° north and south is responsible for the great deserts found at these latitudes.

quently expands, and as a gas expands, it cools. Cooler air holds less moisture, so rising, cooling air tends to lose its moisture in the form of rain or snow.

The air is hottest along the equator, the region heated most by the sun. This air rises, creating a low-pressure area (the doldrums), moves away from the equator, cools, loses some of its moisture, and falls again at latitudes of about 30° north and south, the regions where most of the great deserts of the world are found. This air warms, picks up moisture, and rises again at about 60° latitude (north and south); this is the polar front, another low-pressure area. A third, weaker belt rising at the polar front descends again at the poles, producing a region in which again, as in other areas of descending air, there is virtually no rainfall. The earth's spinning movement twists the winds caused by the heat transfer from equator to poles, creating the major wind patterns (Figure 44-6).

The worldwide patterns are modified locally by a variety of factors. For example, along our own West Coast, where the winds are prevailing westerlies, the western slopes of the Sierra Nevadas have abundant rainfall, while the eastern slopes are dry and desertlike (Figure 44-7). As the air from the ocean hits the western slope, it rises, is cooled, and releases the water. Then, after passing the crest of the mountain range, the air descends again, becomes warmer, and its water-holding capacity increases, resulting in a so-called "rain shadow" on the eastern slope.

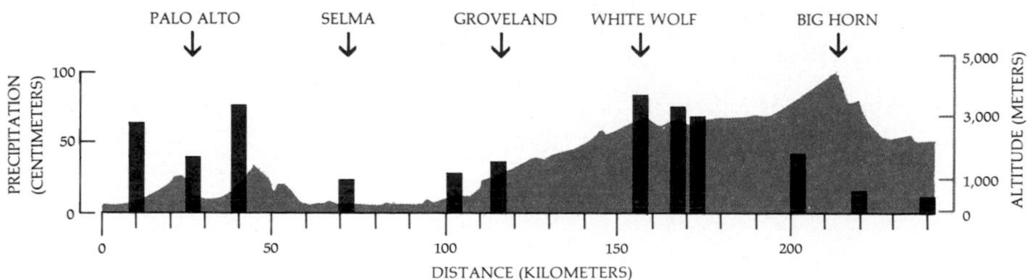

44-7
The mean annual rainfall (vertical columns) in relation to altitude at a series of stations from Palo Alto on the Pacific Coast across the Coastal Range and the Sierra Nevada mountains. The prevailing winds are from the west, and there are rain shadows on the eastern slopes of the two mountain ranges.

The Earth's Surface

Much of the iron and other heavier materials of which the earth is composed is collected in a dense core in the center. Surrounding the core is a lighter layer of solid and molten rock. The outermost layer is made up of overlapping, mobile plates on which the continents rest.

The continents themselves are composed of granite, which is an igneous rock (from the Latin word *ignis*, for "fire"); igneous rock is formed directly from molten material. The surfaces of the continents change constantly. They are crumpled by contractions and collisions as the continents rise, sink, and collide because of the motion of the plates on which they are carried (page 809). As a consequence, the earth's surface is not at all uniform but varies widely from place to place in its composition and in its height above sea level. Both the mineral content of the earth's surface and the altitude affect the growth of plants and other living organisms, and as we saw in the example of the Sierra Nevadas, the mountain ranges of the continents do much to determine the patterns of rainfall.

These varying patterns of temperature, seasonal variation, and precipitation are responsible for the local variations in the patterns of life on the earth's surface. We shall return to this subject at the end of Chapter 46. First, however, let us establish some general principles about the relationships of living things to the atmosphere and to the environment.

ECOSYSTEMS

In all cases, the interactions of the various organisms in a natural setting have two consequences: (1) a flow of energy through photosynthetic autotrophs (green plants or algae) to heterotrophs, which eat either autotrophs or other heterotrophs; and (2) a cycling of materials, which move from the abiotic (nonliving) environment through the bodies of living organisms and back to the abiotic environment.

Such a combination of living and nonliving—biotic and abiotic—components through which energy flows and minerals recycle is known as an *ecosystem*. Taking a large, astronautical view, the entire biosphere can be seen as a single ecosystem. This view is useful when studying materials that are circulated on a worldwide basis, such as carbon dioxide, oxygen, and water. A suitably stocked aquarium or terrarium is also an ecosystem, and such models may be useful in studying certain ecological problems, such as the details of transfer of a particular mineral element. Most studies of ecosystems have been made, however, on more or less self-contained natural units—on a pond, for example, or a swamp, or a meadow.

THE ATMOSPHERE

The atmosphere that lies between the earth and outer space forms four concentric shells, or layers, distinguished by temperature differences. The lowest layer, closest to the earth, is called the troposphere. It extends out about 10 kilometers (6 miles) and so covers all points on the earth's surface, although the tallest mountain peaks extend almost to its limits. About 75 percent of all the molecules in the atmosphere are contained in this layer. At its outer boundary, the temperature is below −50° C. It derives its name from tropos, *"to turn"; it is the layer that turns over (see Figure 44–6). Almost all of the phenomena included under the broad general heading of weather take place within the troposphere.*

The layer above the troposphere is the stratosphere, which extends to an altitude of about 50 kilometers. In the stratosphere, the temperature increases with altitude; the temperature at the outer boundary of the stratosphere is only slightly cooler than the temperature at the earth's surface. The major reason for the warming of the stratosphere is its layer of ozone, which is proportionately densest near the outer boundary. Ozone (O_3) is formed when molecules of oxygen gas (O_2) are broken apart by radiant energy and recombine. Ozone molecules absorb most of the ultraviolet rays of the sun, rays that would be lethal to most forms of life should they reach the earth's surface. The solar energy absorbed by the ozone is changed to heat and so the temperature rises. About 99 percent of the atmosphere lies below the outer boundary of the stratosphere.

In the mesosphere, the third layer, there is a gradual decrease in temperature as the concentration of ozone decreases in the steadily thinning atmosphere. Then, above the mesosphere, we encounter another of those surprising temperature reversals: The temperature increases after we enter the thermosphere. This increase is, in fact, easy to understand if you recall that temperature and heat are not the same; temperature is the average kinetic energy of molecules, whereas heat also involves volume (page 35). The individual molecules of the mesosphere, unshielded from the sun's rays, move at great speed, but there are few molecules, so that there is little heat. At the outer boundary of the thermosphere, however, the thin atmosphere blends with the hydrogen and helium atoms in the space between the stars.

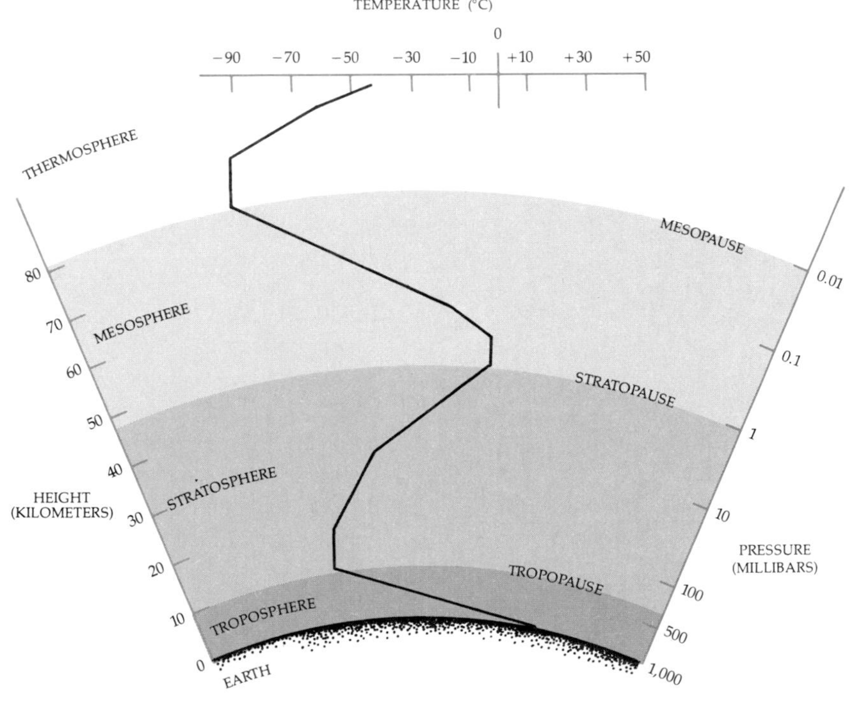

There are four main divisions of the atmosphere: troposphere, stratosphere, mesosphere, and thermosphere. The boundaries between them are determined by abrupt changes in average temperature.

The portion of an ecosystem in which a particular organism lives is known as the organism's *habitat*. Thus, within a pond ecosystem, an amoeba's habitat would be a few square millimeters of bottom ooze, the surface of the pond would be the habitat of a water strider, and the open water the habitat of a largemouth bass.

The position occupied by each species of organism within an ecosystem is known as its *ecological niche*. The ecological niche is multidimensional and includes not only an organism's position in the food web but also all its other relationships with its environment, including other living organisms in the niche. We shall have more to say about the concept of the niche in Chapter 45.

Energy for the Ecosystem

Of the energy that reaches the earth's surface, only a very small fraction is diverted into living systems—an estimated 1/10 of 1 percent on a worldwide basis.

Even when the light falls where vegetation is abundant, as in a forest, a cornfield, or a marsh, only 1 or 2 percent of that light (calculated on an annual basis) is used in photosynthesis. Yet this fraction, as small as it is, may result in the production—from carbon, oxygen, water, and a few minerals—of several thousand grams (dry weight) of organic matter per year in a single square meter of field or forest, a total of about 90 billion metric tons of such organic matter per year on a worldwide basis.

TROPHIC LEVELS

The passage of energy from one organism to another takes place along a particular food chain. A *food chain* is a sequence of organisms related to one another as prey and predator; the first is eaten by the second, the second by the third, and so on, in a series of *trophic levels*, or feeding levels. In most ecosystems, food chains are linked together in complex *food webs*, with many branches and lateral connections and involving many different species (Figure 44–8).

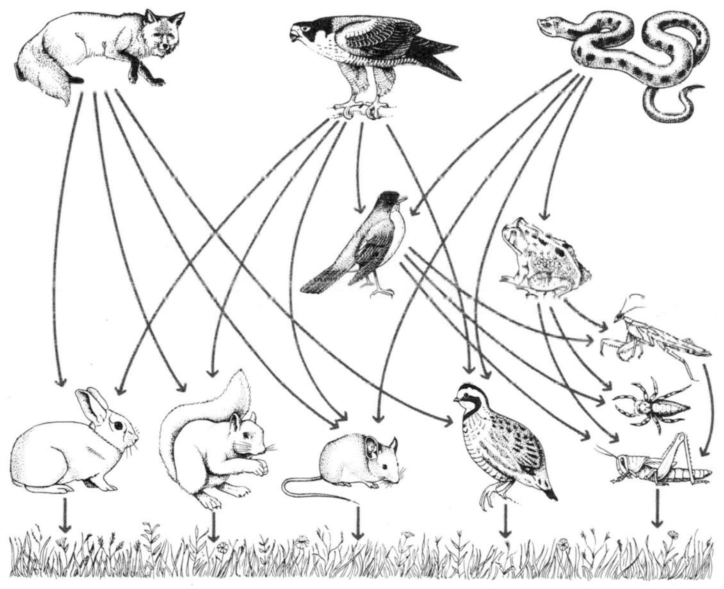

44–8
Diagram of a food web. The arrows point from consumers to their energy source. This food web is much simplified; in reality, many more animals and plants would be involved.

44–9
Calculation of the productivity of a field in Michigan in which the vegetation was mostly perennial grasses and herbs. Measurements are in terms of calories per square meter per year. In this field, the net production—the amount of chemical energy stored in plant material—was 4,950,000 calories per square meter per year. Thus, slightly more than 1 percent of the 471,000,000 calories per square meter per year of sunlight reaching the field was converted to chemical energy and stored in the plant bodies.

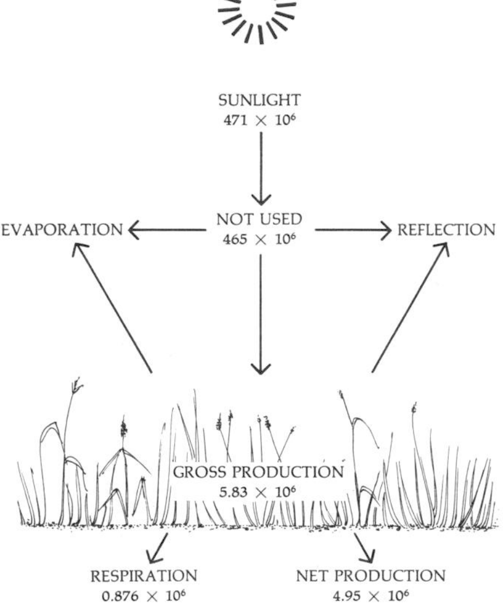

The Producers

A primary producer is always at the first trophic level of a food web. On land, the primary producer is usually a plant; in aquatic ecosystems, it is usually a photosynthetic alga. These photosynthetic organisms use light energy to make carbohydrates and other compounds, which then become sources of chemical energy. Producers far outweigh consumers; 99 percent of all the organic matter in the biosphere is made up of plants and algae. All heterotrophs combined account for only 1 percent.

Ecologists speak of the *productivity* of ecosystems; productivity is measured by the amount of energy (measured in calories) stored in chemical compounds or by the increase in biomass (measured in grams or metric tons) in a particular length of time. (*Biomass* is a convenient shorthand term meaning the weight of all the living organisms in a given area.) *Net productivity* represents the amount of energy stored in chemical compounds less the amount of glucose and other compounds used in respiration; in agricultural communities, the standing crop at the end of the season represents the net primary production.

The Consumers

Energy enters the animal world through the activities of the herbivores, animals that eat plants and algae. An herbivore may be a caterpillar, an elephant, a sea urchin, a snail, or a field mouse; each type of ecosystem has its characteristic complement of herbivores. Of the organic material consumed by herbivores, much is eliminated undigested. Some of the chemical energy from digested food is transformed to other types of energy—heat or motion—or used in the digestive process itself. A fraction of the material is converted to new animal biomass. Increase in animal biomass is the sum of the increase in weight of individual animals plus the weight of new offspring.

The next level in the food web, the secondary consumer level, involves a carnivore, a meat-eating animal that devours herbivores. The carnivore may be a lion, a minnow, a starfish, a robin, or a spider, but in each of these cases, only a small

44–10
Two consumers. (a) *A pika gathering plant material, which it will stack and dry in the sun before storing the "hay" among the rocks of its mountain habitat.* (b) *Leopard with dead prey in tree.*

(a)

(b)

44–11
The fate of food energy ingested by an organism. Only about 10 percent is converted to animal biomass, available for consumption by an organism at the next trophic level.

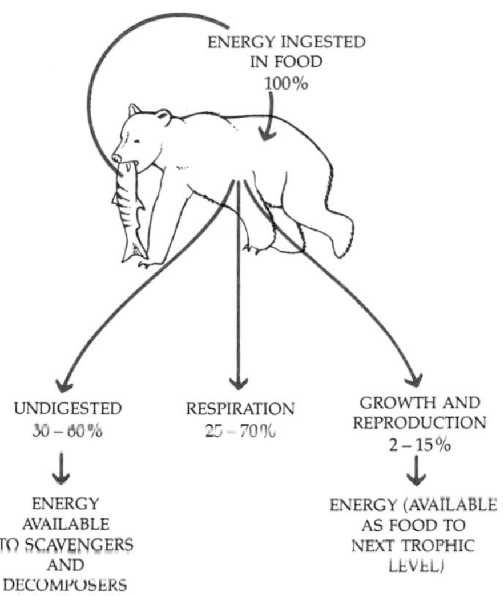

part of the organic substance present in the body of the herbivore becomes incorporated into the body of the consumer. Some webs have third and fourth consumer levels, but five links are usually the limit, largely because of the waste involved in the transfer of energy from one trophic level to another. Other important consumers include the decomposers, whose role will be described in more detail later in this chapter.

Ecological Pyramids

The flow of energy through a food web is often represented by a graph of quantitative relationships among the various trophic levels. Because large amounts of energy and biomass are dissipated at every trophic level, so that each level retains a much smaller amount than the preceding level, these diagrams nearly always take the form of pyramids. An *ecological pyramid,* as such a diagram is called, may be (1) a pyramid of numbers showing the numbers of individual organisms at each level; (2) a pyramid of biomass, based either on the total weight of the organisms at each level or on the number of calories at each level; or (3) a pyramid of energy flow showing the productivity of the different trophic levels.

The shape of any particular pyramid tells a great deal about the ecosystem it represents. Figure 44–12a, for example, shows a pyramid of numbers for a grassland ecosystem. In this type of ecosystem, the primary producers (the individual grass plants) are small, and so a large quantity of them is required to support the primary consumers (the herbivores). In an ecosystem in which the primary producers are large (for instance, trees), one primary producer may support many herbivores, as indicated in Figure 44–12b.

A pyramid of biomass for a grassland ecosystem, like the pyramid of numbers for that system, takes the form of an upright pyramid, as shown in Figure 44–13a. Pyramids of biomass are inverted only when the producers have very high reproduction rates. For example, in the ocean, the standing crop of phytoplankton (plant plankton) may be smaller than the biomass of the zooplankton (animal plankton) that feeds upon it (Figure 44–13b). Because the growth rate of the phytoplankton is much more rapid than that of the zooplankton, a small biomass of phytoplankton can supply food for a larger biomass of zooplankton. Like pyramids of numbers, pyramids of biomass indicate only the quantity of organic material present at one time; they do not give the total amount of material produced or, as do pyramids of energy, the rate at which it is produced.

44–12
Pyramids of numbers for (a) *a grassland ecosystem, in which the number of primary producers (grass plants) is large, and* (b) *a temperate forest, in which a single primary producer, a tree, can support a large number of herbivores.*

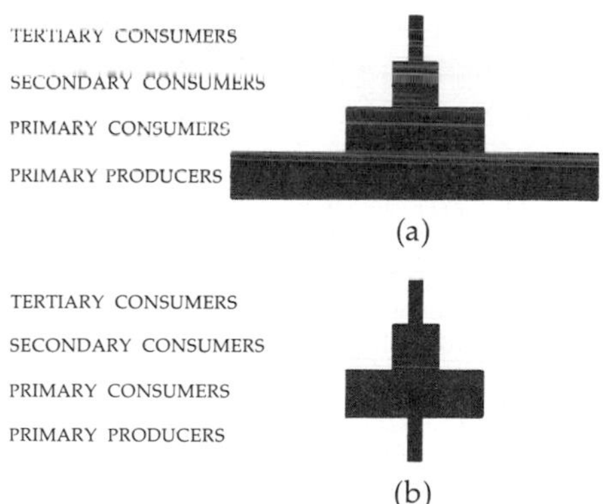

44–13
Pyramids of biomass for (a) *a field in Georgia and* (b) *the English Channel. Such pyramids reflect the mass present at any one time, hence the seemingly paradoxical relationship between phytoplankton and zooplankton.*

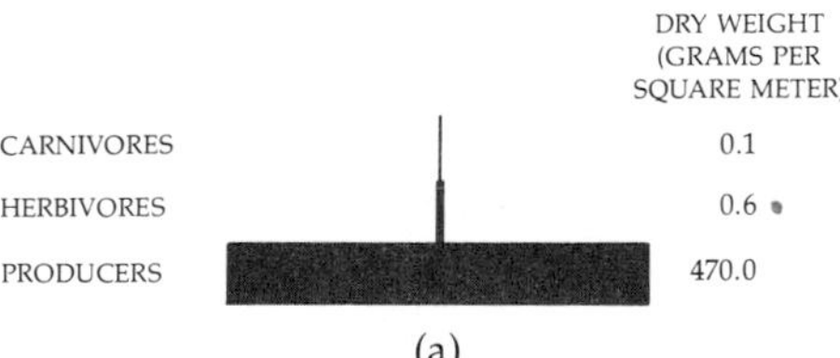

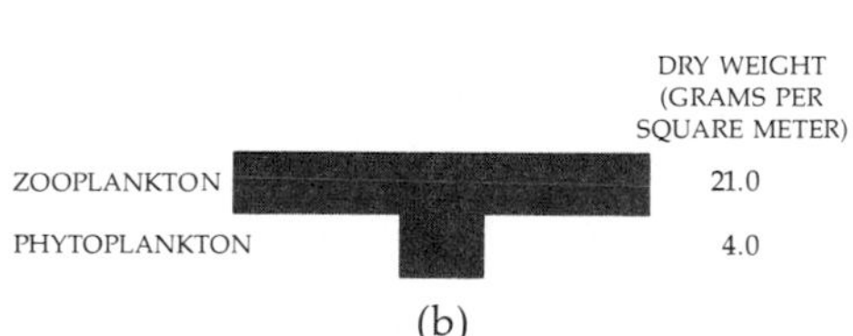

A CHEMOSYNTHETIC ECOSYSTEM

Throughout this text, we have been informing you that life on earth depends on radiant energy from the sun, which is converted by plants or algae to chemical energy by the process of photosynthesis. Not so! In 1977, a new type of ecosystem was discovered based on chemosynthesis.

Its discoverer was Alvin, *the little research submarine from Woods Hole Oceanographic Institution, and its crew of scientific investigators, led by John B. Corliss of Oregon State University. The place was the Galapagos Rift, a boundary between tectonic plates. Sea water seeps into the fissures in the volcanic rocks erupting along this boundary, becomes heated, and rises again. As a consequence, oases of warmth are created in the near-freezing waters 2.5 kilometers below the surface. More important, the water reacts with the rocks deep within the crust. Here, under extreme heat and pressure (300° C, 280 kilograms per square centimeter), chemical reactions take place. The crucial one, according to early evidence, is the conversion of sulfate in the sea water to hydrogen sulfide. Chemosynthetic bacteria use the hydrogen sulfide as an energy source, becoming the primary producers in a complex food web.*

Five such oases have now been located, each with a somewhat different animal community. (According to the investigators, the types of animals that are present are probably the result of whatever floating larvae colonized the area in its early stages of development.) Among the most spectacular inhabitants are clams, measuring some 20 centimeters in diameter, which cover the lava floor. Other permanent inhabitants include crabs, mussels, and octopods. One oasis, known as the Garden of Eden, is dominated by huge tube worms. The unusual size of the organisms found at these warm springs of the Rift is attributed to the high productivity of the chemosynthetic bacteria. Thus, although most life on earth is, indeed, based on radiant energy from the sun, there are clearly highly viable alternatives not only on this planet, but perhaps on others as well.

Deep in the Pacific Ocean, at the Garden of Eden oasis in the Galapagos Rift, giant tube worms, never before seen by man, flourish in the warmth of a sea-floor spring. Cancroid crabs hide in the crevices of the limpet-encrusted lava, while a pink brotulid fish swims by. Compared to the near-freezing temperatures normally found at these depths, it is 17° C (63° F).

44–14
Pyramid of energy flow for a river system in Florida. This type of pyramid, regardless of the ecosystem under analysis, is never inverted.

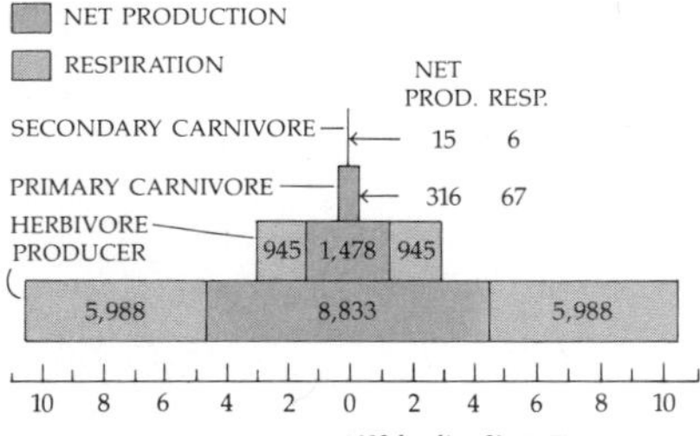

Pyramids of energy, in keeping with the second law of thermodynamics, have the same basic shape for every ecosystem (Figure 44–14). In general, of the total amount of energy in the biomass that passes from one trophic level to another in a food chain, only about 10 percent is stored in body tissue. Approximately 90 percent of the total calories is either unassimilated or "burned" in respiration by the animals at that level. Hence, the tertiary consumers, the carnivores that eat the carnivores that eat the herbivores, are reduced to approximately $1/10 \times 1/10 \times 1/10 = 1/1{,}000$ the energy stored in the plants that are eaten. Supercarnivores, which eat these tertiary carnivores, are reduced to one-tenth of this, or 1/10,000 the energy in the plant material. Individuals at the top of the ecological pyramid, such as humans, may have to be quite large, as individuals, to capture and eat other animals, but even if they are larger as individuals, they are usually smaller in total number and total biomass. And they are always lower in total captured energy than the organisms they eat.

This is the basis of the argument for living lower on the food chain. Beef, for instance, is energetically expensive. In the case of steer, only 10 percent of the energy present in the grasses or grain that fed the steer is converted to animal biomass, and even less to edible meat. Consequently, in terms of energy, those who eat meat are consuming 10 times as much of the world's energy resources as those who eat grains and other plant products.

BIOGEOCHEMICAL CYCLES

Energy takes a one-way course through an ecosystem, but many substances cycle through the system. Such substances include water, nitrogen, carbon, phosphorus, potassium, sulfur, magnesium, calcium, sodium, chlorine, and also a number of minerals, such as iron and cobalt, that are required by living systems in only very small amounts. The water cycle is shown on page 38, and the carbon cycle is on page 145. You may want to look at them again at this time.

Movements of inorganic substances are referred to as *biogeochemical cycles* because they involve geological as well as biological components of the ecosystem. The geological components are the atmosphere, which is made up largely of gases, including water vapor; the lithosphere, the solid crust of the earth; and the hydrosphere, comprising the oceans, lakes, and rivers, which cover three-fourths of the earth's surface.

The biological components of biogeochemical cycles include the producers, consumers, and decomposers. The decomposers, which are primarily bacteria and fungi, break down dead and discarded organic matter, completing the oxidation of the energy-rich compounds formed by photosynthesis. As a result of the metabolic work of the decomposers, waste products—dead leaves and branches, the roots of annual plants, feces, carcasses, even the discarded exoskeletons of insects—are broken down to inorganic substances that are returned to the soil or water. As we shall see in the case of the nitrogen cycle, the role of each decomposer may be very specialized—the performance of a single energy-yielding metabolic step. It has been said that for almost every reaction in the biosphere in which the conversion of one compound to another yields at least 15 kilocalories per mole, some organism has evolved that can exploit this energy to survive. From the soil or water, inorganic substances are again taken into the bodies of plants and enter their cycles again.

44–15
The phosphorus cycle. Phosphorus is essential to all living systems as a component of the nucleotides of DNA and RNA and also of the energy-carrier molecules ATP and ADP. Like other minerals, it is released from dead tissues by the activities of decomposers, taken up from soil and water by plants and algae, and recycled through the ecosystem.

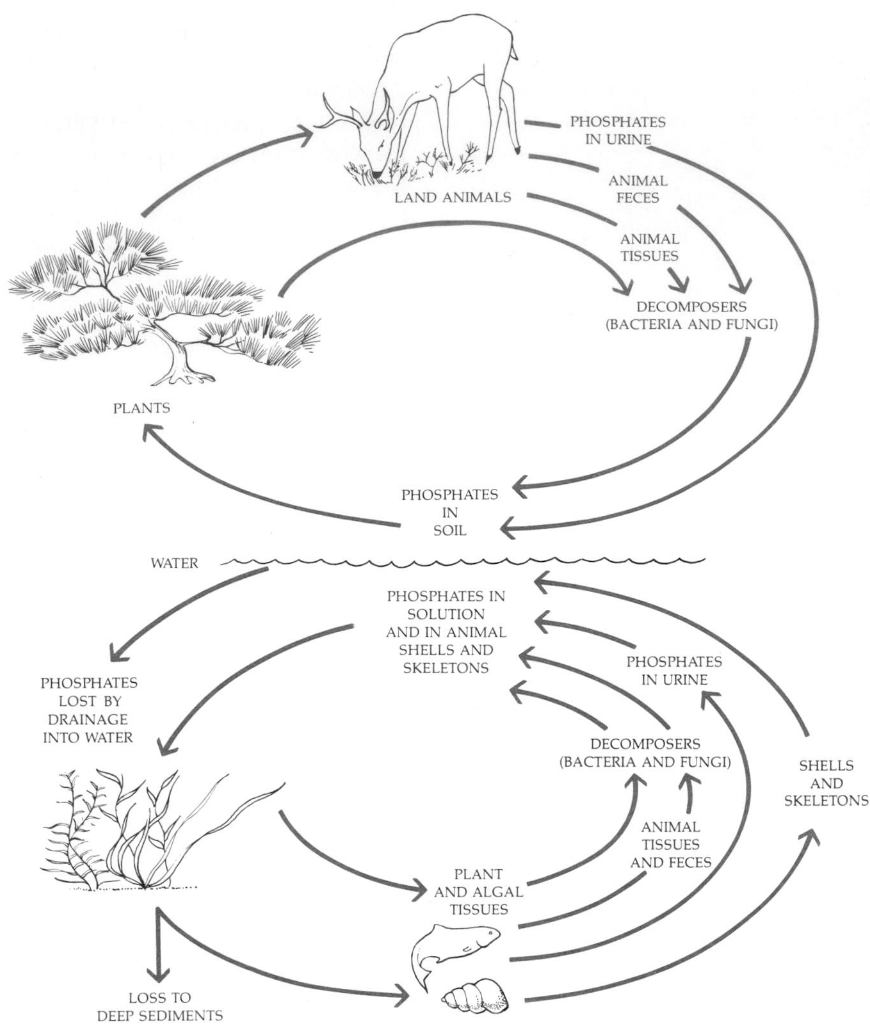

The cycling of one important mineral, phosphorus, through the ecosystem is shown in Figure 44–15. Other minerals in the ecosystem undergo similar cycles, differing only in details.

The Nitrogen Cycle

The chief reservoir of nitrogen is the atmosphere; in fact, nitrogen makes up 78 percent of the gases in the atmosphere. Since most living things, however, cannot use elemental atmospheric nitrogen to make amino acids and other nitrogen-containing compounds, they must depend on nitrogen present in soil minerals. So, despite the abundance of nitrogen in the biosphere, a shortage of nitrogen in the soil is often the major limiting factor in plant growth. The process by which this limited amount of nitrogen is circulated and recirculated throughout the world of living organisms is known as the *nitrogen cycle*. The three principal stages of this cycle are (1) ammonification, (2) nitrification, and (3) assimilation.

Much of the nitrogen found in the soil, a result of the decomposition of organic materials, is in the form of complex organic compounds, such as proteins, amino acids, nucleic acids, and nucleotides. However, these nitrogenous compounds are usually rapidly decomposed into simple compounds by soil-dwelling organisms. Certain soil bacteria and fungi are mainly responsible for the decomposition of dead organic materials. These microorganisms use the proteins and amino acids as a source of their own needed proteins and release the excess nitrogen in the form of ammonia (NH_3) or ammonium (NH_4^+). This process is known as *ammonification.*

Several species of bacteria common in soils are able to oxidize ammonia (NH_3) or ammonium (NH_4^+). This oxidation of ammonia or ammonium, known as *nitrification,* is an energy-yielding process, and the energy released in the process is used by these bacteria as their primary energy source. One group of bacteria oxidizes ammonia (or ammonium) to nitrite (NO_2^-):

$$2NH_3 + 3O_2 \longrightarrow 2NO_2^- + 2H^+ + 2H_2O$$

Nitrite is toxic to higher plants, but it rarely accumulates. Members of another genus of bacteria oxidize the nitrite to nitrate, again with a release of energy:

$$2NO_2^- + O_2 \longrightarrow 2NO_3^-$$

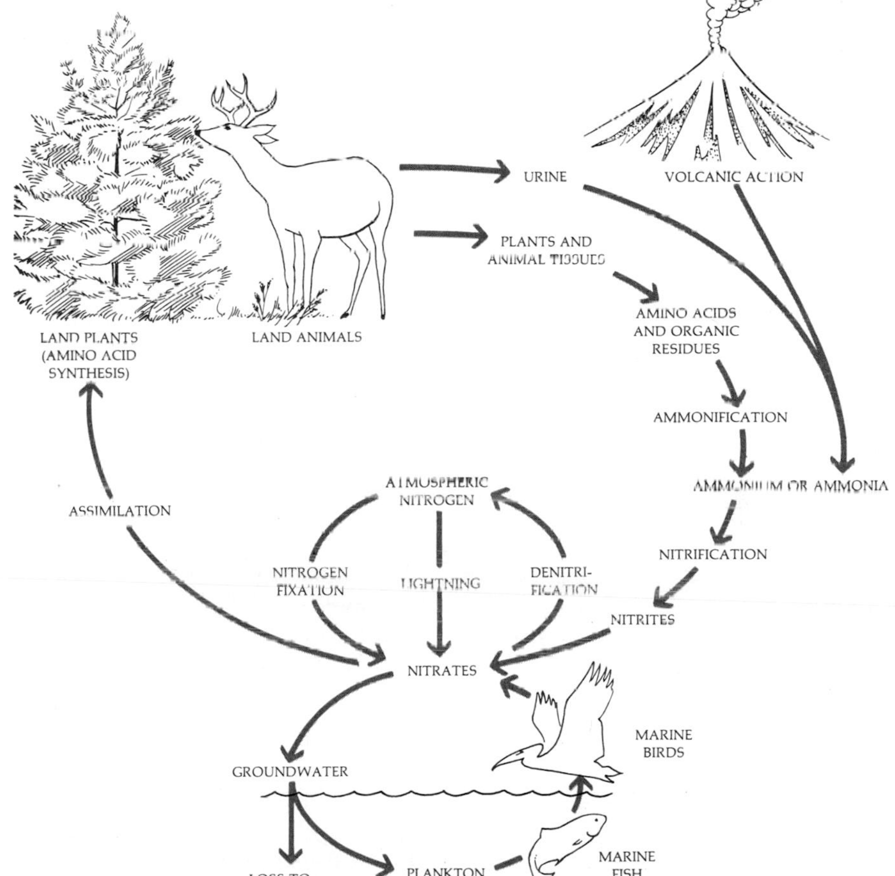

44-16
The nitrogen cycle. Although the reservoir of nitrogen is in the atmosphere, where it makes up 78 percent of dry air, the movement of nitrogen in the ecosystem is more like that of a mineral than of a gas. Only a few microorganisms are capable of nitrogen fixation.

Although plants can utilize ammonium directly, nitrate is the form in which most nitrogen moves from the soil into the roots.

Once the nitrate is within the plant cell, it is reduced back to ammonium. In contrast to nitrification, this assimilation process requires energy. The ammonium ions thus formed are transferred to carbon-containing compounds to produce amino acids and other nitrogenous organic compounds needed by the plant.

The nitrogen-containing compounds of green plants are returned to the soil with the death of the plants (or of the animals that have eaten the plants) and are reprocessed by soil organisms and microorganisms, taken up by the plant roots in the form of nitrate dissolved in the soil water, and reconverted to organic compounds. In the course of this cycle, a certain amount of nitrogen is always "lost," in the sense that it becomes unavailable to plants.

The main source of nitrogen loss is the removal of plants from the soil. Soils under cultivation often show a steady decline of nitrogen content. Nitrogen may also be lost when topsoil is carried off by soil erosion or when ground cover is destroyed by fire. Nitrogen is also leached away by water percolating down through the soil to the groundwater. In addition, numerous types of bacteria are present in the soil that, when oxygen is not present, can break down nitrates, releasing nitrogen into the air and using the oxygen for the oxidation of carbon compounds (respiration). This process, known as *denitrification,* takes place in poorly drained (hence, poorly aerated) soils.

As you can see, if the nitrogen lost from the soil were not steadily replaced, virtually all life on this planet would finally flicker out. The "lost" nitrogen is returned to the soil by nitrogen fixation, the process by which gaseous nitrogen from the air is incorporated into organic nitrogen-containing compounds and thereby brought into the nitrogen cycle. It is carried out to a small extent by abiotic processes; most nitrogen fixation, however, is carried out by a few types of microorganisms, the most important of which are species of bacteria that live in symbiosis with plants (see pages 540–542). Fifty million metric tons of nitrogen are added to the soil each year, of which 45 million tons are biological in origin. (The other 10 percent is largely in the form of chemical fertilizers.) Just as all organisms are ultimately dependent on photosynthesis for energy, they all depend on nitrogen fixation for their nitrogen.

Concentration of Elements

The elements needed by living systems often are present in the tissues of these systems in higher concentrations than in the surrounding air, soil, or water. This concentration of elements comes about as a result of selective uptake of such compounds by living cells, amplified by the channeling effects of the ecological pyramids of biomass and energy. Under natural circumstances, this concentration effect is usually valuable; animals generally have a greater requirement for minerals than do plants because so much of the biomass of plants is cellulose. In some special cases, the effect of the accumulation may be quite dramatic. As we noted previously, foraminiferans built the white cliffs of Dover, and the thousands of kilometers of coral reefs in the world's warmer waters are composed of calcium extracted from sea water over the millennia by one tiny polyp after another.

Foreign substances can get caught up in these biogeochemical cycles, and, as they are passed from one living organism to another, reach high concentrations as they approach the top of the food chain.

ENERGY COSTS OF FOOD GATHERING

How much does a calorie cost, in terms of calories? For the expenditure of 1 calorie, an organism in most natural populations obtains from 2 to 20 calories in food energy. This is true for organisms whose expenditures are very high—such as the hummingbird, which spends up to 330 calories a minute—as well as for organisms whose expenditures are very low—such as the damselfly, which uses less than a calorie a day.

In simple human societies, in which individuals obtain their food without fossil-fuel energy subsidies, the ratio of food calories gained to calories invested is similar to that which prevails for the rest of the animal kingdom. Hunter-gatherers average 5 to 10 calories for each calorie spent; shifting agriculture (which requires no fertilizer) yields about 20 calories per calorie spent.

As is true in most societies (the social insects and civilized man are among the exceptions), almost the entire adult population has a share in the business of getting food. In the United States, about 20 percent of the population is involved in the food supply system. (Only about 2 percent are actually farmers; the rest are involved in food processing, transportation, and marketing.) Thus 80 percent are, for better or worse, free for other pursuits.

On the surface, it would seem that we expend less energy than most animals on the mundane work of food supply. Not so. At the turn of the century, for each calorie expended, including human labor, fuel for farm machinery and food transport, and the energy cost of fertilizer, we received about a calorie in return. Today, in the United States, as in other "advanced" technological societies, for every calorie invested we get a return of 0.1 calorie. This cost figure does not include the energy used for heating or lighting or running private automobiles (even those that bring the food home from the market) or electric can openers.

Obviously, other animals cannot live so profligately; their energy income must exceed their expenditures. Like these other organisms, we also are dependent almost exclusively on solar energy. There is an important difference, however; because of our technology, we have been able to draw upon energy stored millions of years ago. It is only in the last decade that we have come to realize that not only are these resources finite but also they may soon be expended.

From Robert M. May, "Energy Costs of Food Gathering," Nature, vol. 255, page 669, 1975.

A Bushman, a hunter-gatherer, returns triumphantly with a porcupine he has killed.

44–17

(a) *Concentration of DDT residues being passed along a simple food chain. As organic matter is transferred from one link to the next in the chain, about nine-tenths of it is usually respired or excreted; only the remaining 10 percent forms new biomass. The losses of DDT residue, however, are small in proportion to the respiration and excretion losses. Consequently, the concentration of DDT increases as the material passes along the chain, and high concentrations occur in the carnivores.* (b) *An eaglet and an egg that will never hatch, photographed in a nest near the Muskegon River in Michigan. DDT causes a bird's liver to break down the hormones that mobilize calcium at the time of egg production, resulting in thin-shelled, fragile eggs, like the one shown here. Birds at the top of food chains, such as the osprey, peregrine falcon, and bald eagle, were principal victims.*

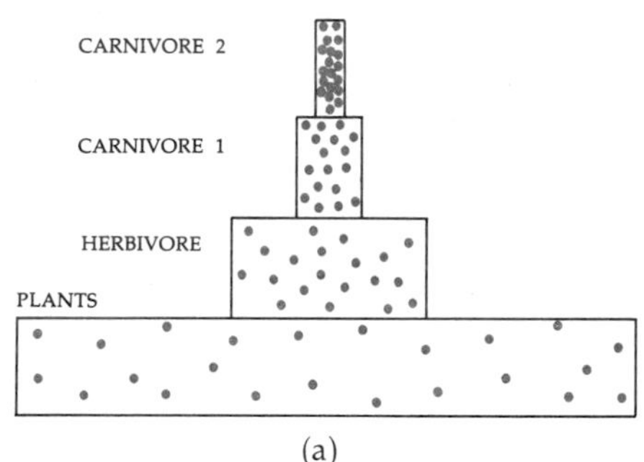

(a)

(b)

DDT is probably the best known of the toxic substances whose effects were amplified in this way (Figure 44–17). Another was strontium-90, a radioactive element produced by nuclear testing in the 1950s. It is closely related to calcium and can take its place in many biochemical reactions. In dairy lands, strontium-90 made its way through grasses into dairy cows and into milk; from there it became concentrated in the bones and teeth of children.

By 1959 the bones of children in North America and Europe averaged an estimated 2.6 picocuries of strontium-90 per gram of bone calcium, compared with 0.4 picocurie per gram of calcium in the bones of adults. This amount of radioactivity has not been proved dangerous, but, as we mentioned in Chapter 19, exposure to radioactive elements is known to cause leukemia, bone cancers, and genetic abnormalities and generally to shorten the life span; and the minimum exposure that can produce such effects is not established. The half-life of strontium-90 is 28 years; in other words, it takes 28 years for half of the element to lose its radioactivity, so this exposure still continues.

The channeling of strontium-90 along the calcium food chain might have been anticipated. Other results were less predictable. For example, in the Arctic, only slight radiation exposure from the atomic tests was expected because the amount of fallout that reaches the ground at the poles is much less than it is in the temperate zones. However, Eskimos in the Arctic were discovered to have concentrations of radioactivity in their bodies that were much higher than those found in the inhabitants of the temperate regions. The link in the chain was the lichens. Lichens, which absorb their minerals largely from the rain, had absorbed a large amount of fallout material, little of which had time to decay and none of which was dissipated by absorption into the soil. In the winter, caribou live almost exclusively on lichens, and at the top of the food chain, Eskimos live largely on caribou.

Table 44-1 *Soil Classification*

	DIAMETER OF FRAGMENTS (MICROMETERS)
Coarse sand	200–2,000 (0.2–2 millimeters)
Sand	20–200
Silt	2–20
Clay	Less than 2

Soils and Mineral Cycles

The composition of the soil influences the availability of minerals in an ecosystem. Soil, the uppermost layer of the earth's crust, is composed of weathered rock associated with organic material, both living and in various stages of decomposition. It typically has three layers: the A horizon, the B horizon, and the C horizon. The A horizon, or topsoil, is the zone of maximum organic accumulation (humus). The B horizon, or subsoil, consists of inorganic particles in combination with mineral nutrients that have leached down from the A horizon. The C horizon is made up of loose rock that extends down to the bedrock beneath it. Figure 44–18 shows profiles of three common soil types.

The mineral content of the soil depends in part on the parent rock from which the soil is formed. These differences in mineral content can be extremely localized, with sharp lines of demarcation. Geologists sometimes use types of vegetation or changes in color or growth patterns of plants as indicators of mineral deposits.

In most soils, however, the mineral content is more dependent on the biotic component of the ecosystem. In an undisturbed environment, most of the mineral nutrients stay within the ecosystem. If, however, the vegetation is repeatedly removed, as when crops are harvested or grasslands are overgrazed, or the top, humus-rich layer of the A horizon is eroded, the soil rapidly becomes depleted and can be used for agricultural purposes only if it is heavily fertilized.

Another factor influencing the mineral content of soils is the soil composition. The smaller fragments of rock are classified as sand, silt, or clay, according to size (see Table 44–1). Water and minerals drain rapidly through soil composed of large particles (sandy soil). Soil composed of small particles (clay) holds the water against gravity. Moreover, the small clay particles are negatively charged and so hold positively charged ions, such as calcium (Ca^{2+}), potassium (K^+), and magnesium (Mg^{2+}). However, a pure clay soil is not suitable for plant growth because it is usually too tightly packed to let in enough oxygen for the respiration of plant roots, soil animals, and most soil microorganisms. Clay soils that contain enough large particles to keep the soil from packing are known as loams, and these are generally the best soils for plant growth.

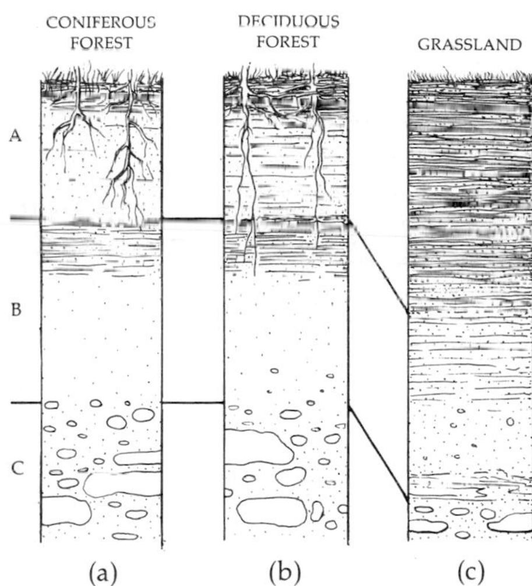

44–18
Diagrams of soil layers of three major soil types. (a) *The litter of the northern coniferous forest is acidic and slow to decay, and the soil has little accumulation of humus, is very acidic, and is leached of minerals.* (b) *In the cool, temperate deciduous forest, decay is somewhat more rapid, leaching less extensive, and the soil more fertile. Such soils have been used extensively for agriculture, but they need to be prepared by adding lime (to reduce acidity) and fertilizer.* (c) *In the grasslands, almost all of the plant material above the ground dies each year as do many of the roots, and so much organic matter is constantly returned to the soil. In addition, the finely divided roots penetrate the soil extensively. The result is highly fertile soil, often black in color, with a topsoil sometimes more than a meter in depth.*

44–19
Prairie soil. Note how the roots of the grasses bind the topsoil.

The pH of the soil also affects its capacity to retain minerals. In an acidic soil, hydrogen ions replace other positively charged ions clinging to clay particles, and these nutrient ions easily leach out of the soil. Soil pH also affects the solubility of certain nutrient elements. Calcium, for example, increases in solubility (and therefore in its availability to plant roots) as pH increases. Crops such as alfalfa, sweet clover, and other legumes have high calcium requirements and so grow well only in alkaline soil. Rhododendrons and azaleas, on the other hand, have high requirements for iron, which is abundant only when the soil is acidic.

Soils and plant life interact. Plants constantly add to the humus, thereby changing not only the content of the soil but also its texture and its capacity to hold minerals and water. In turn, the plants are dependent upon the mineral content of the soil and its holding capacity. As these improve, plants increase in biomass and also often change in kind, thereby producing further changes in the soil. Thus, under natural conditions, the soil is constantly changing its composition.

Recycling in a Forest Ecosystem

Recent studies of an ecosystem in a deciduous forest have shown that the plant life of the community plays a major role in retaining the nutrient elements. The studies were made in the Hubbard Brook Experimental Forest in the White Mountain National Forest in New Hampshire. The investigators first established a procedure for determining the mineral budget—input and output, "profit" and "loss"—of areas in the forest. By analyzing the content of rain and snow, they were able to estimate input, and by constructing concrete weirs that channeled the water flowing out of selected areas, they were able to calculate output. (A particular advantage of the site is that bedrock is present just below the soil surface, so that little material leaches downward.) They discovered, first, that the natural forest was extremely efficient in conserving its mineral elements. For example, the annual net loss of calcium from the ecosystem was 9.2 kilograms per hectare (a hectare is about 2½ acres). This represents only about 0.3 percent of the calcium in the system. In the case of nitrogen, the ecosystem was actually accumulating this element at a rate of about 2 kilograms per hectare per year. There was a similar, though somewhat smaller, net gain of potassium.

44–20
Weir at Hubbard Brook Experimental Forest in New Hampshire. Water from each of six experimental ecosystems was channeled through a weir, such as this one built where the water leaves the watershed, and was analyzed for chemical elements. The watershed behind the weir in this photograph has been stripped of vegetation. The experiments showed that such deforestation greatly increased the loss of elements from the system.

In the winter of 1965–1966, all of the trees, saplings, and shrubs in one 15.6-hectare area of the forest were cut down completely. No organic materials were removed, however, and the soil was undisturbed. During the following summer, the area was sprayed with a herbicide to inhibit regrowth. During the four months from June through September, 1966, the runoff of water from the area was four times higher than in previous years. Net losses of calcium were 10 times higher than in the undisturbed forest, for example, and of potassium, 21 times higher. The most severe disturbance was seen in the nitrogen cycle. Plant and animal tissues continued to be decomposed to ammonia or ammonium, which then were acted upon by nitrifying bacteria to produce nitrates, the form in which nitrogen is usually assimilated by plants. However, no plants were present, and the nitrate, a negatively charged ion, was not held in the soil. Net losses of nitrate nitrogen averaged 120 kilograms per hectare per year from 1966 to 1968. As a side effect, the stream draining the area has become polluted with algal blooms, and its nitrate concentration now exceeds the levels established by the U.S. Public Health Service for drinking water.

ECOLOGICAL SUCCESSION

If land is laid bare as a result of a landslide, erosion, excavation, the eruption of a volcano, the retreat of a glacier, the rising of a new island from the sea, or some other such phenomenon, it will, if the environment is not too harsh, slowly become covered with vegetation and its accompanying animal life. The vegetation that initially colonizes the bare land is usually replaced in the course of time by a second type, which gradually crowds out the first and which itself may eventually be replaced. In areas where these stages of replacement have been studied over a number of years, it has been found that communities of living organisms replace one another in a predictable and orderly sequence. This process of community change is known as *ecological succession*.

The process of ecological succession is carried out by the living organisms themselves. According to the conventional model, each temporary community changes the local conditions of temperature, light, humidity, soil content, and other abiotic factors, and so sets up favorable conditions for the next temporary community. When the site has been modified as much as possible (here the limits are set by the environment), succession ceases, or at least slows down considerably. The final community is known as the *climax community*. In a sense, the climax community, too, is in a state of constant change; individual organisms die, and their places are taken by new individuals. However, these individuals will probably be of the same sort and of the same species as the ones being replaced.

The physical characteristics of the environment determine the nature of the mature community. In regions where conditions are particularly unfavorable, such as the tundra, the process of succession involves relatively few stages and the climax community is correspondingly simple. Where physical conditions are less limiting, the mature community is rich and diverse.

Primary Succession

The colonization by plants of an area not previously covered by vegetation is known as *primary succession*. Rocks and cliffs are common sites of primary succession. The first stage in such areas is the formation of soil. The solid rock is broken

(a)

(b)

44–21
(a) *Yapoah Crater, a volcanic cinder cone east of the Cascade Mountains in central Oregon. Succession leading to the establishment of climax forest on such a cone may take centuries and may often be interrupted by further volcanic activity long before it is complete.* (b) *An early stage of succession on a rocky slope. Lichens have begun to accumulate soil, and a bladder fern has sprung up in a small crevice.*

down by weathering processes, such as freezing and thawing or heating and cooling, which cause the rocks to expand and contract, thus splitting them. Water and wind exert a scouring action that breaks the fragmented rock into smaller particles, often carrying the fragments great distances. Water enters between the particles, and soluble materials such as rock salt dissolve in the water. Water in combination with carbon dioxide from the air forms a mild acid that dissolves substances that will not dissolve in water alone. Chemical reactions that contribute to the disintegration of the rock begin to take place. Soon, if other conditions such as light and temperature permit, bacteria, fungi, lichens, and then small plants begin to gain a foothold. Growing roots further split the rocks, and the disintegrating bodies of the plants and of the animals associated with them add to the accumulating material. Finally, larger plants move in, anchoring the soil in place with their root systems, and a new community has begun. The patterns are similar worldwide, although the species differ from place to place.

Lake Michigan Sand Dune Succession

Primary succession has been particularly well studied on the sand dunes of Lake Michigan. The dunes studied are part of beach systems running parallel to the shoreline above the level of the lake, which has been falling with the retreat of the glaciers. Because the soil, the climate, and the plant and animal life of the surrounding area are basically the same, these dunes have offered an unusual opportunity to study changes associated with time alone.

Bare sand surface is colonized at first by grasses, such as Marram grass, which propagates almost exclusively by rhizomes. It can stabilize a bare area in as little as six years. Before a sand dune is so stabilized, it shifts continually and no other plant life can gain a foothold. Once the sand dune is stabilized, the Marram grass dies out for some unknown reason.

44-22
Sand dune near Lake Michigan with beach grass in the foreground. Later successional stages can be seen on the more remote dunes.

The first tree to appear is usually the cottonwood, after which the dune is likely to be invaded quickly—in about 50 to 100 years—by pines. Black oak replaces the pines, appearing in about 100 to 150 years. As this occurs, the understory changes, with shade-tolerant shrubs replacing the light-requiring shrubs associated with the open pine forests. The oldest dunes (about 12,000 years old) are still populated with black oaks and associated vegetation, and many ecologists doubt that they will ever reach the beech-maple stage, which represents the climax community for that area.

During the early stages of succession, the originally very alkaline pH drops rapidly as carbonates are leached from the soil very quickly. Nitrogen and organic carbon increase markedly in these early stages. Most of the soil changes in the original barren dune sand occur within about 1,000 years.

Secondary Succession

When succession occurs on a site previously occupied by vegetation, it is known as *secondary succession.* Secondary succession commonly occurs on areas laid bare by human intervention, such as abandoned farmland and strip mines, roadsides, and landfills. (As you can see, the distinctions between primary and secondary succession are not biologically clear-cut, and some ecologists recommend that the terms be dropped.) Such an open area is bombarded by the seeds of numerous plants and is captured by those that can germinate most quickly. In an open field, these are the plants that can survive the sunlight and drying winds—weeds and grasses and such trees, for example, as cedars, white pines, poplars, and birch. For a while, these plants are dominant, but eventually they eliminate themselves because their

44-23
Secondary succession. A young stand of loblolly pine is taking over an abandoned field in Wake County, North Carolina. The plow furrows are still clearly visible. The field was abandoned perhaps 20 years ago.

(a)

(b)

44–24
(a) *The seedling of a red maple rising above needles of white pine. Mature white pines filter the light so that their own seedlings cannot survive and only those tolerant of shade, such as maple and oak seedlings, can gain a foothold.* (b) *Seedling trees of balsam fir growing up under and replacing quaking aspen in northern Minnesota—a stage in forest succession leading to a climax community of white spruce and balsam fir.*

seedlings cannot compete in the shade cast by the parent trees. Some seedlings, however—for instance, oaks, red maples, white ash, and tulip trees—can thrive in partial shade, and these species become established next. As the forest matures, these trees grow tall and finally they shut out so much light that even *their* seedlings cannot grow. Eventually, the only young trees that can grow in the forest are those that can survive the dimmest light, such as the hemlock, beech, and sugar maple, and these ultimately take over the forest. Nothing else can compete with them in the conditions that have been established. This, in temperate regions, is the climax forest.

Fire

In some areas, fire plays an important part in determining the final stage in forest succession. Young seedlings of deciduous trees are very susceptible to fire, while pines are resistant to it. Jack pines, in fact, open their cones and release their seeds only after they have been heated, so they tend to spread most readily after a fire. In the southern pine forests and the northern lake area of the United States, recurrent ground fires maintain the pinewoods, keeping the forest perpetually young. Similarly, the forests of sugar pines and giant sequoias on the western coast of the United States are maintained by fires. The fires destroy competitive trees that are faster growing and less fire-resistant than the giant sequoias, and also lay bare the ground so the small sequoia seeds can germinate. These ground fires, characteristic of the fire-type ecosystem, are distinctly different from the uncontrolled crown fires of northern forests, which spread through the treetops, destroying entire communities of plants and animals. Also, there is now good evidence that, in prehistory, grass fires played a significant role in maintaining grassland ecosystems by eliminating the early stages of forest succession.

44–25

When fire sweeps through a forest, secondary succession—with regeneration from nearby unburned stands of vegetation—is initiated. Some plants produce sprouts from the stumps; others seed abundantly on the burn. In one group of pines, the closed-cone pines, the cones do not open to release their seeds until they have been heated by fire.

Immature and Mature Ecosystems

As ecosystems pass through the various stages of succession, one type of community is replaced by another, distinctly different type. Although the changes that take place as the ecosystem matures differ in detail as to the exact species involved and the rates of change, many ecologists believe that they all have certain results in common.

First, there is an increase in total biomass. Compare, for example, a recently abandoned field, which is an immature ecosystem, with a deciduous forest, which is a more mature one.

Second, there is a decrease in net productivity in relation to biomass (standing crop). In other words, the biomass of a mature system does not tend to increase with time, as does that of a system in earlier stages of succession, since it is dominated by slower-growing species.

Third, as we would expect from the Hubbard Brook experiments, mature systems have a greater capacity to entrap and hold nutrients within the system.

Fourth, during the early stages of succession, the diversity of species increases. There is also a general increase in the size of organisms, the length of their lives, and the complexity of their life cycles.

Fifth, it is hypothesized that the ecosystem becomes more stable as it matures. Mature ecosystems, according to this hypothesis, show a greater capacity for homeostasis and are less likely to be drastically affected by biological changes—such as population explosions.

Agriculture, stressing net productivity, exploits the features of the immature ecosystem.

As is the case with many "classic" models in ecology, the concept of each series of plants making way for the next is now under question. Other factors are believed to play major roles in ecological succession. These factors include predation (grazing) and competition between opportunistic species and equilibrium species, a topic that will be considered in the next chapter.

SUMMARY

The biosphere is the part of the earth that contains living organisms. It is a thin film on the surface of the planet, irregular in its thickness and its density. Its source of energy is the sun; at the upper surface of the earth's atmosphere, energy from the sun arrives at a rate of 1.94 calories per square centimeter per minute (the solar constant). The atmosphere is a blanket of gas and dust molecules held close to the surface by gravity. The atmosphere filters out harmful ultraviolet rays from the sun and impedes the escape of heat from the earth's surface. The biosphere is affected by the position and movements of the earth in relation to the sun and by the movements of air and water. These conditions cause wide differences in temperature and rainfall from place to place and season to season on the earth's surface.

An ecosystem, or ecological system, is a unit of biological organization made up of all the organisms in a given area and the environment in which they live. It is characterized by interactions between the living (biotic) and nonliving (abiotic) components that result in (1) a flow of energy from the sun through autotrophs to heterotrophs, and (2) a cycling of minerals and other inorganic materials.

Within an ecosystem, there are trophic (feeding) levels. All ecosystems have at least two such levels: autotrophs, which are plants or photosynthetic algae, and herbivores, which are usually animals. The autotrophs, the primary producers in the ecosystem, convert a small proportion (above 1 percent) of the sun's energy into chemical energy. The herbivores, which eat the autotrophs, are the primary consumers. A carnivore that eats the herbivore is a secondary consumer, and so on. About 10 percent of the energy transferred at each trophic level is stored in body tissue. There are seldom more than five links in a food chain.

The movements of water, carbon, nitrogen, and minerals through ecosystems are known as biogeochemical cycles. In such cycles, inorganic materials from the air, water, or soil are taken up by primary producers, passed on to consumers, and eventually transferred to decomposers, chiefly bacteria and fungi. The decomposers break down dead and discarded organic material and return it to the soil or water in a form in which it can be used again by the primary producers. Characteristics of soil affect the presence, retention, and recycling of minerals. These characteristics include the rock from which the soil was formed, the presence of humus on the soil surface, and soil composition and pH. Soils and plant life interact. Plants and decomposers increase the availability of minerals in the soil and affect soil texture; these changes, in turn, improve the soil's capacity to maintain plant life, leading to a further increase in the humus content.

Ecological succession is an orderly sequence of changes in the type of vegetation and other organisms on a particular site. Primary succession occurs in areas previously unoccupied by living organisms, whereas secondary succession occurs in disturbed areas where vegetation has been removed. Both types of succession result from modifications produced by the living organisms themselves and culminate in the establishment of a climax community, which is the characteristic community for that area.

Maturation in ecosystems is accompanied by increases in biomass and in number of species and decrease in net productivity.

QUESTIONS

1. Distinguish among the following: equinox/solstice; biomass/net productivity; pyramid of numbers/pyramid of energy; A horizon/B horizon/C horizon.

2. Describe what happens to the energy in light striking a temperate forest ecosystem. What happens when it strikes a cornfield? A pond? A field on which cattle are grazing?

3. Describe what happens to a nutrient mineral in each of those areas.

4. What are the characteristics of a mature ecosystem? Using these characteristics as criteria, how would you classify land under agriculture in terms of ecosystem maturity? On the basis of this answer, what would you expect to be some of the consequences of intensive agriculture?

CHAPTER 45

Population Dynamics: The Numbers of Organisms

In the previous chapter, we defined ecology as the scientific study of the interactions that determine the distribution and abundance of organisms. We saw how the flow of energy and recycling of minerals that define an ecosystem call for the presence of organisms of various categories and in certain numbers. In other words, an ecosystem will probably contain herbivores and first-level carnivores, and the biomass of the former will usually be at least 10 times that of the latter.

In this chapter, we shall look more closely at the question of numbers and investigate the factors that affect the sizes of the various populations within an ecosystem.

THE RATE OF INCREASE

One factor affecting numbers of individuals in a population is its rate of increase, or reproductive capacity. As Darwin noted more than 100 years ago, the reproductive capacities of almost all species are very high.

It can be calculated, for instance, that a single female housefly producing an average of 120 eggs per laying, with half developing into females and with seven generations per year per female, could produce 6,182,442,727,320 houseflies in the space of one year. At the other end of the reproductive scale is a pair of elephants, which have a gestation period of 20 to 22 months and usually give birth to only one offspring at a time. If all offspring and their offspring, in turn, survived and reproduced, this one pair of elephants would have 19 million descendants at the end of 750 years, as noted by Darwin.

Reduced to its simplest elements, the rate of increase of a population is the difference in any given time interval between the rate of addition of individuals due to birth and immigration and the rate of subtraction due to death and emigration. If we assume immigration and emigration balance out, the rate of increase (r) becomes simply the births minus the deaths, the reproduction rate less the death rate. This rate—births minus deaths—is the *intrinsic rate of increase* of a population and is symbolized as r.

The algebraic formula is a simple one:

$$\frac{\triangle N}{\triangle t} = rN$$

It states that the change in number ($\triangle N$) over a given period of time ($\triangle t$) is equal to the rate of increase *times the number already present.* This type of growth, known as exponential growth, is the same principle as that of compound interest; the more you have, the more you get. A graph of the equation is shown in Figure 45–1a. As you can see, it starts off slowly, and then it shoots up very rapidly. A curve such as this is characteristic, for example, of bacteria or protists cultivated in the laboratory or of seasonal blooms of algae. Such growth curves are also typical of insect invasions, such as the gypsy moth infestations that recurrently plague the northeastern United States. Often, in nature, the populations grow either until they are overcome by an epidemic of disease organisms or until they reduce their food supply so that they are not able to complete their metamorphoses and reach the reproductive stages. (Spraying of the caterpillars with insecticides often reduces the size of the population just sufficiently to delay the population crash.)

Figure 45–1b is a model of the simplest sort of growth pattern seen for most populations in nature. It is a graph of the equation:

$$\frac{\triangle N}{\triangle t} = rN\left(\frac{K-N}{K}\right)$$

In this equation, r is the intrinsic rate of increase, as it is in the previous exponential equation, and r is again multiplied by the number present at any one time; but this figure is modified by the introduction of a new factor, K, which is the *carrying capacity* of the environment. The carrying capacity is defined simply as the number of individuals in a particular population that the environment can support over an indefinite period of time. It often is determined by food resources, but some other factors, such as light, water, or nesting space, may be the limiting ones. Note that when N is very small, $(K - N)/K$ is close to 1, and the curve approximates the exponential growth curve of Figure 45–1a. As N increases, $(K - N)$ decreases, and growth slows down. Finally, if N exceeds K, the growth rate becomes negative, and the population declines. This model holds true for microorganisms in a laboratory, for instance, or for organisms that have found their way to a new and favorable environment not being exploited by any other organism. This type of growth is called logistic. Note that this model has possible practical utility. If, for instance, one wants to control a population of rats, killing half of them may merely reduce the population to the point at which it increases most rapidly. A more effective approach would be to reduce the carrying capacity, which, in the case of rats,

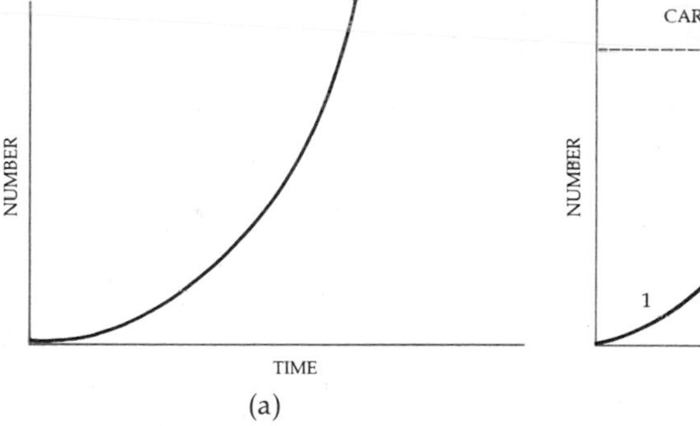

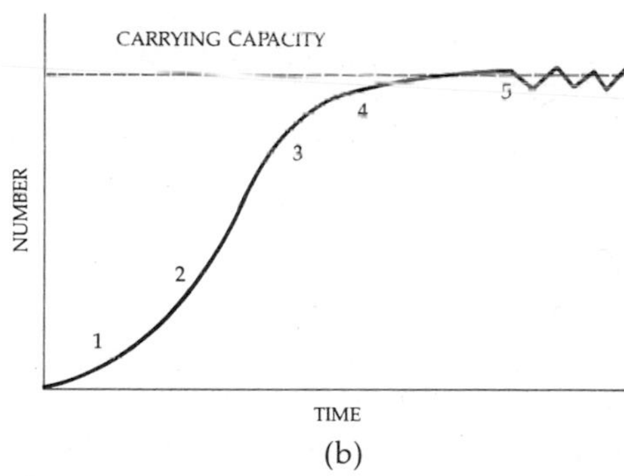

45–1
(a) *Exponential growth curve. After an initial establishment phase, the population increases compound-interest fashion.* (b) *The more usual logistic growth curve is sigmoid, or S-shaped. As with exponential growth, there is an establishment phase (1), and a phase of rapid acceleration (2). Then, as the population approaches environmental limits, the growth rate slows down (3 and 4), and finally stabilizes (5).*

45–2
Charting populations by age and sex permits predictions about future growth rates. In India, for example, nearly 45 percent of the population is under 15 years of age. Even if these young men and women limit their family size enough to produce only enough children to replace themselves (which means cutting the current birth rate in half), population growth will not level off until about the year 2040—and at a level of well over a billion. The population of Sweden, by contrast, will remain the same unless the birth rate is dramatically increased.

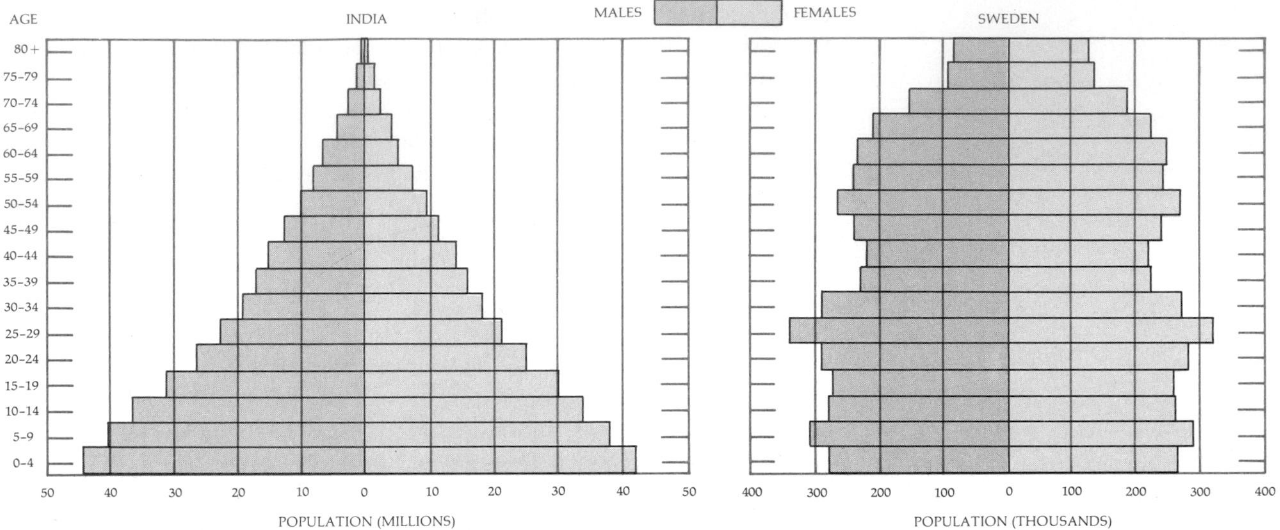

45–3
In many tropical countries, the death rate has fallen rapidly since 1940, resulting in a rapid growth of the population. The drop in death rate is the result of increased medical services, control of malaria by DDT, and the availability of new antibacterial drugs, especially the antibiotics. Note that the birth rate has also begun to fall. The area on the graph marked in color indicates population growth. The data shown here are from Sri Lanka (formerly Ceylon).

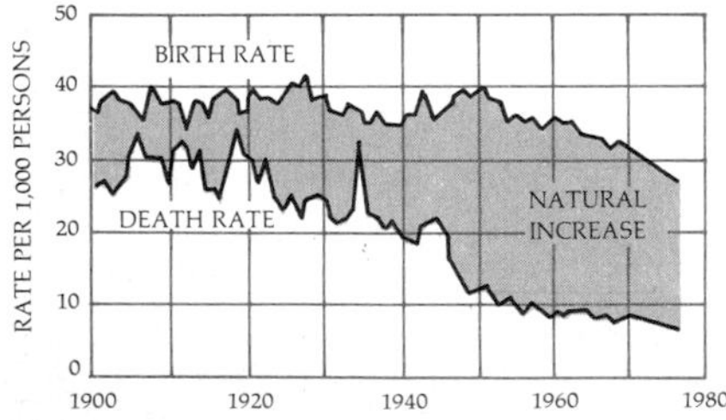

usually means tighter control of garbage disposal. Similarly, if one wants to control the cropping of a particular type of economically valuable fish, the fish should not be cropped below the level of rapid growth unless one is willing to wait a long time for the population to recover.

Age and Birth Rate

In the simple models presented in Figure 45–1, no allowances were made for changes in the ages of individuals in the population. In organisms in which the life span exceeds the reproductive span, the average age of the population at any given time may profoundly affect the rate of increase (Figure 45–2). Considerations such as this help to explain why the population of the United States has continued to increase despite the fact that young couples are having slightly less, on the average, than two children per couple. (Another reason is continued immigration.)

Mortality

Another factor in population growth is, of course, mortality. In the so-called developing nations—the nations showing the most rapid population increases—gains in population are the result more of the decrease in mortality, especially among infants and young people, than of any increase in birth rates (Figure 45–3).

The rate of increase or decrease of a population is the difference between the birth rate and the death rate. Death rates are high in natural populations. For example, in a study made of the saltmarsh song sparrows of San Francisco Bay, it

45–4
Representative survivorship curves, plotted from life-table data on the basis of survivors per 1,000 (log scale, vertical coordinate) and age in relative units of mean life span (horizontal coordinate). As you can see, the oyster's chances of survival greatly improve with age, while mortality among Hydra *remains constant throughout the life span. The curve at the top of the graph is for a hypothetical population in which all individuals live out the average life span of the species—a population, in other words, in which all individuals die at about the same age. The fact that the curve for humans approaches this hypothetical curve indicates that the human population as a whole is reaching a uniform age of mortality.*

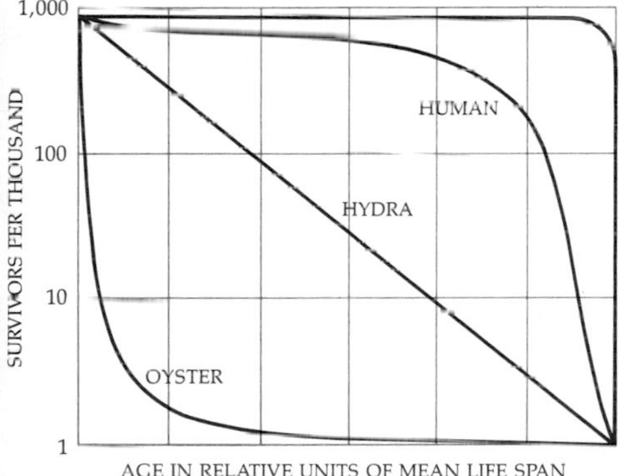

was estimated that of every 100 eggs laid, 26 are lost before hatching. Of the 74 live nestlings, only 52 leave the nest, and of these 52, 80 percent die the first year. The remaining 10 breed the following season, but during the next year, 43 percent of these die, leaving only 6 out of the original 100. Each subsequent year, mortality among the survivors amounts to 43 percent.

Once a bird survives its first, risk-laden year, the mortality rate for the years before old age remains more or less constant. In many species, the probability of survival for one more year (or one more period of time) seems to be highest during the middle of life and lowest for both very young and very old individuals. Patterns of mortality differ from species to species, however. Figure 45–4 shows the basic types of survivorship curves. In the oyster, for example, mortality is extremely high during the free-swimming larval stage, but once the individual attaches itself to a favorable substrate, life expectancy levels off. Among *Hydra,* the mortality rate is the same at all ages.

If all individuals in a population lived out the average life span of their species, the survivorship curve would resemble the hypothetical curve shown at the top of the graph in Figure 45–4. The mortality curve for humans, as you can see, approaches this curve, indicating that the human population as a whole is reaching a uniform age of mortality.

Patterns of mortality affect the structure of populations and may be important also in population management.

Emigration

Emigration is the departure of individuals from a population. Arctic lemmings, among the most famous emigrants, increase in number and then periodically emigrate in large numbers, apparently in response to crowding. In northern Europe, they have been known to march by the millions across the countryside and, on reaching the coasts, reportedly drown in the sea. Similar spectacular emigrations occur among locusts, resulting in the plagues of locusts recorded from Biblical times. When a locust population increases in number, changes in hormone production begin to take place among members of the new generations. These changes result in longer wings, more slender bodies, darker colors, and more active behavior. The locust swarms are made up of these individuals. Emigration may also occur continuously but less spectacularly, involving only small numbers of individuals at a time.

45–5
Greenland collared lemmings. Lemmings are inhabitants of the North American and Eurasian tundra. Lemming populations, like those of many other small rodents, undergo drastic cyclic fluctuations in numbers, resulting in mass emigrations.

The stimuli that cause mass emigration undoubtedly differ from species to species, but the result is always the same: a reduction in the size of the local population. (Note that emigration and migration are two quite different phenomena. Emigration is a one-way trip, whereas migration refers to periodic, seasonal movements.)

THE LIMITS TO GROWTH

The size of a population depends on a number of factors that differ for different populations. If any of these essential requirements is in short supply, even though all the other necessities are met, growth of the population is not possible. Whatever it is that determines whether or not a population can grow in a given environment is known as a *limiting factor.*

The way limiting factors affect growth of organisms is illustrated by a common pollution problem. When phosphorus is the limiting factor in the growth of freshwater algae in a lake or slow-moving stream, phosphate-containing detergents added to the water from sewage systems will produce a spectacular bloom of algae. This period of rapid growth continues until the available supply of another essential element—perhaps calcium—is consumed. Then the algae begin to die, the decomposing bacteria take over, the respiration of the bacteria begins to use up the oxygen in the water, and oxygen becomes the limiting factor. Eventually the concentration of oxygen drops below the tolerance level of fish and other organisms, and these, too, begin to die. If the process is not interrupted, the lake becomes completely stagnant, and only bacteria and other microorganisms can survive in it.

Ecologists often divide the factors that affect the growth of a population into those that are density-dependent and those that are not. Density-independent factors include extremes of temperature, elements present in or absent from soils, day length, rainfall, and so forth. Density-dependent factors exert effects that vary in proportion to the size of the population. These are the factors that are defined as K in our previous equation and that are responsible for the S-shaped curve in Figure 45-1b. In our previous example of water pollution, the limits on growth imposed by shortages of phosphorus, calcium, and then oxygen are all density-

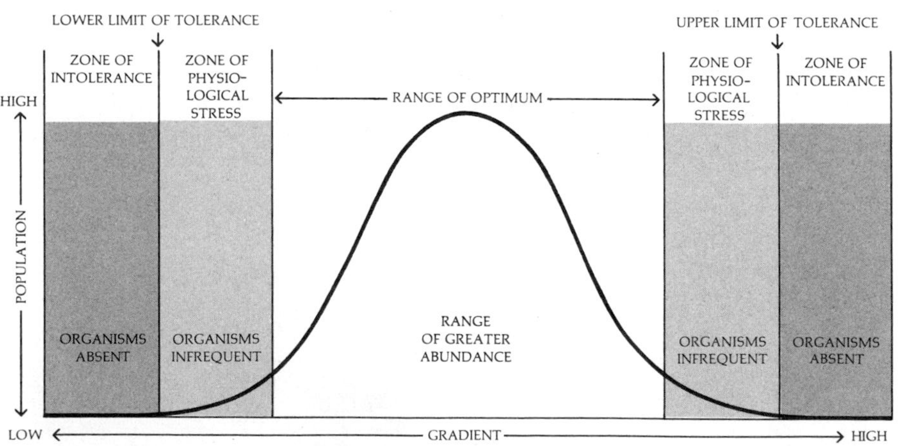

45-6
The principle of limiting factors. Every species has a characteristic limiting-factor curve for each factor in its environment. The three critical points on the curve are the lower limit of tolerance, the point of greatest abundance, and the upper limit of tolerance.

dependent. Competition among members of the same population for a requirement in short supply is one of the major forces of natural selection.

Interspecific Interactions

Interactions between organisms of different species also affect the size of local populations. Two types of interaction are of paramount importance: competition and predation. Competition generally involves organisms on the same trophic level, whereas predation occurs between organisms on different trophic levels.

Interspecific Competition: The Ecological Niche

Competition affects both the kinds and numbers of organisms found in a community. Closely linked to the phenomenon of competition is the concept of the ecological niche, which, as we noted earlier, is defined as an organism's position in the structure of the ecosystem. According to the principle of *competitive exclusion,* formulated by the Russian biologist G. F. Gause, in any given community, only one species can occupy any given ecological niche for an extended period of time.

Gause bolstered his hypothesis by a number of laboratory experiments. His simplest, now classic, experiment involved laboratory cultures of two species of *Paramecium, Paramecium aurelia* and *Paramecium caudatum.* When the two species were grown under identical conditions in separate containers, *P. aurelia* grew much more rapidly than *P. caudatum,* indicating that the former used the available food supply more efficiently than the latter. When the two were grown together, the former rapidly outmultiplied the latter, which soon died out (Figure 45–7).

A similar situation occurs with two species of duckweed, *Lemna gibba* and *Lemna polyrrhiza. Lemna gibba* grows more slowly in pure culture than *L. polyrrhiza; L. gibba,* however, always replaces *L. polyrrhiza* when they are grown together. Again, evolution has provided one with an advantage. The plant bodies of *L. gibba* have air-filled sacs that serve as little pontoons, so that these plants form a mass over the other species, cutting off the light. As a consequence, the shaded *L. polyrrhiza* dies out (Figure 45–8).

It is possible to devise different culture conditions under which the outcomes of both the *Paramecium* and *Lemna* experiments can be reversed. However, regardless of conditions, one species always wins.

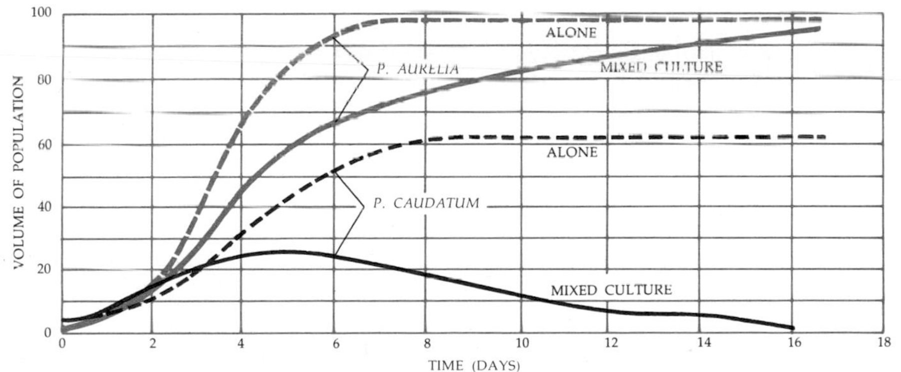

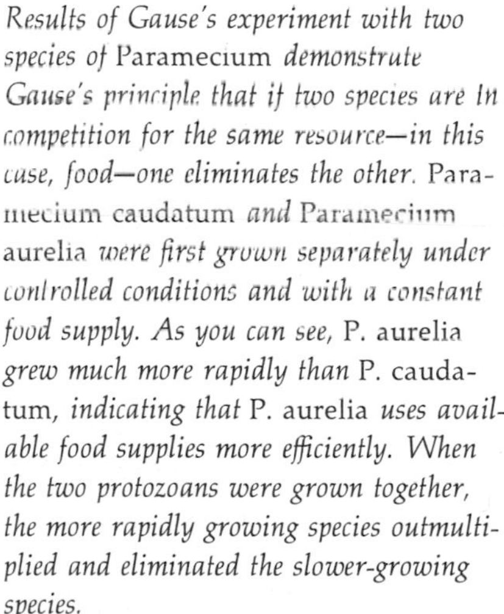

45–7
Results of Gause's experiment with two species of Paramecium *demonstrate Gause's principle that if two species are in competition for the same resource—in this case, food—one eliminates the other.* Paramecium caudatum *and* Paramecium aurelia *were first grown separately under controlled conditions and with a constant food supply. As you can see,* P. aurelia *grew much more rapidly than* P. caudatum, *indicating that* P. aurelia *uses available food supplies more efficiently. When the two protozoans were grown together, the more rapidly growing species outmultiplied and eliminated the slower-growing species.*

45–8
An experiment with two species of floating duckweed, tiny angiosperms found in ponds and lakes. One species, Lemna polyrrhiza, *grows more rapidly in pure culture than the other species,* Lemna gibba. *But* L. gibba *has tiny air-filled sacs, which, like pontoons, float it on the surface, and so it shades the other species, making it the victor in the competition for light.*

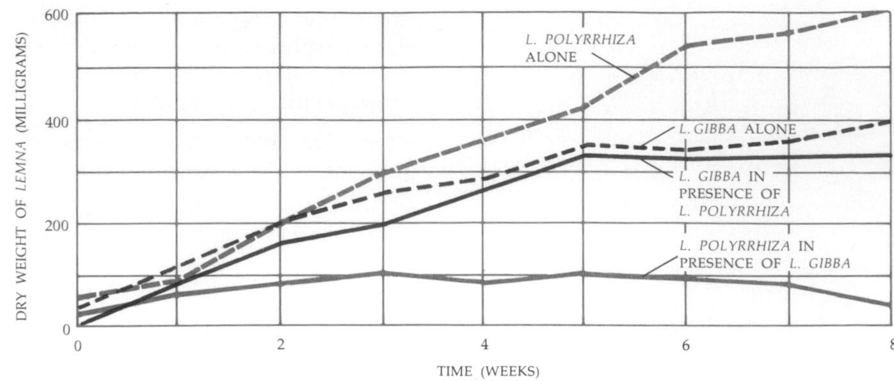

The Niche in Nature

Gause's principle leads one to predict that in a situation in which two very similar species coexist in nature, it will be possible to demonstrate differences in the niches they inhabit. The prediction has been fulfilled by a number of observations. For example, some New England forests are inhabited by five closely related species of warbler, all about the same size and all insect eaters. Why do these birds not eliminate one another by competitive exclusions? An analysis by Robert MacArthur showed that these warblers have different feeding zones in the canopy (Figure 45–9), and, because they exploit slightly different resources, they are able to coexist without direct competition.

J. H. Connell studied competition between two genera of barnacles in Scotland. Barnacles are crustaceans. Before they change from their immature, larval forms, in which they are free-swimming, into their adult, feeding forms, they cement themselves to rocks and secrete shells. One species of barnacle, *Chthamalus stellatus*, occurs in the high part of the intertidal seashore, and another, *Balanus balanoides*, occurs lower down. Although *Chthamalus* larvae, after their period of drifting in the plankton, often attach to rocks in the lower, *Balanus*-occupied zone, no adults are ever found there.

The history of a population of barnacles can be recorded very accurately by holding a pane of glass over a patch of barnacles and noting with glass-marking ink each spot where a barnacle is. Once attached, barnacles remain fixed, so that by returning later, one can check exactly which barnacles have died and which new ones have arrived.

45–9
A demonstration of Gause's principle: the feeding zones in a spruce tree of five species of North American warblers. The colored areas in the tree indicate where each species spends at least half its feeding time. In this way, all five species feed in the same trees with diminished competition.

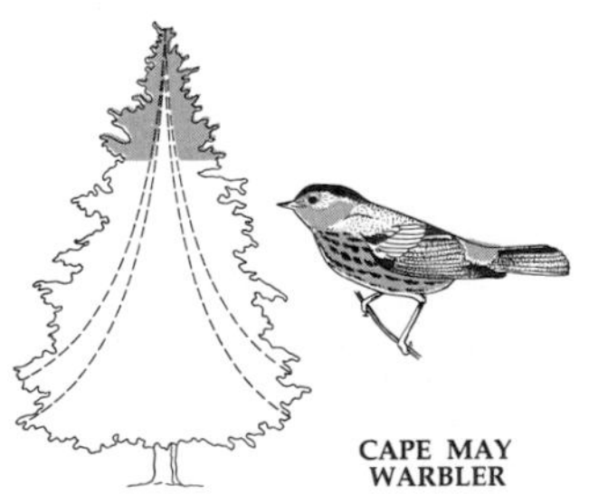

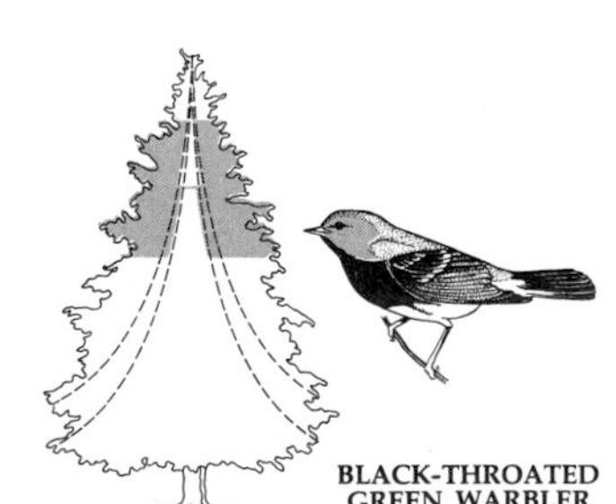

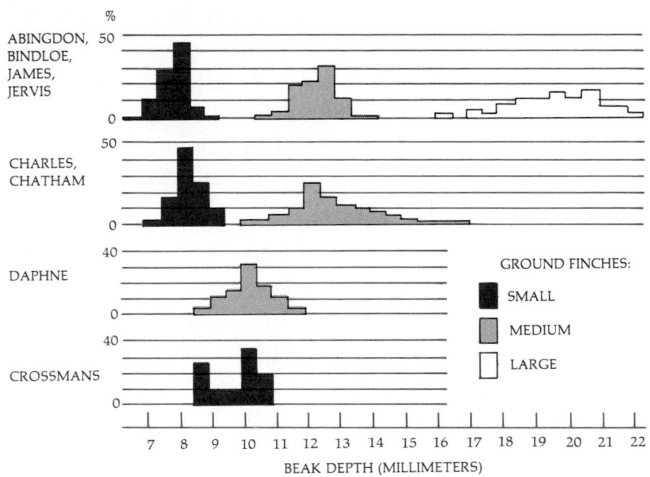

45–10
Beak sizes in three species of ground finch found on the Galapagos Islands. Beak measurements are plotted horizontally, and the percentage of specimens of each species is shown vertically. Daphne and Crossmans, which are very small islands, each have only one species of ground finch. These species have beak sizes halfway between those of the medium-sized and small finches on the larger islands.

By doing this, Connell was able to show that in the lower zone, *Balanus,* which grows faster, ousts *Chthamalus* by crowding it off the rocks or growing over it. When *Chthamalus* was isolated from contact with *Balanus,* it lived with no difficulty in the lower zone, showing that competition with *Balanus* restricted *Chthamalus* to the higher, less favorable zone. However, when *Chthamalus* was removed, *Balanus* was unable to live in its range. *Chthamalus* survives in the intertidal community by special physiological adaptations that enable it to inhabit a marginal area.

A final example is afforded by Darwin's finches. As we noted in Chapter 43, the large, medium, and small ground finches are very similar except for differences in overall body size and in the sizes of their beaks. These differences in beak size are correlated with the fact that they eat seeds of different sizes. Because of these different eating habits, competition for food is reduced. On islands such as Abingdon and Bindloe, where all three species of ground finch exist together, there are clear-cut differences in beak size. On Charles and Chatham Islands (see page 813), the large species is not found, and the beak sizes of the medium ground finches found on these islands overlap the beak sizes of the large finches found on Abingdon and Bindloe. Daphne and Crossmans, which are very small islands, each have only one species; Daphne has the medium-sized finch and Crossmans the small finch. These two populations have similar beak sizes, which are intermediate between those of the medium-sized and small finches on the larger islands (Figure 45–10).

Thus it would appear that although competing species may eliminate one another in the laboratory, in natural environments, which are far more complex, the pressure of competition may result in simply subdividing the niche; that is, the two species evolve in such a way as to minimize competition between them.

A habitat has been defined as an organism's address, a niche as its profession, and these definitions offer an easy way to remember the difference between the two. However, the modern concept of the ecological niche includes much more than what the organism does for a living. In the words of Richard Lewontin, who grapples with this problem of defining the niche:

The ecological niche is a multidimensional description of the total environment and way of life of an organism. Its description includes physical factors, such as temperature and moisture; biological factors, such as the nature and quantity of food sources and of predators, and factors of the behavior of the organism itself, such as its social organization, its pattern of movement, and its daily and seasonal activity cycles.*

Predation

Predation is the eating of live organisms, including plants by animals, animals by animals, and even, as we have seen, animals by plants. As we saw in the previous chapter, predation is the principal route of energy flow through an ecosystem. Here we are going to examine the effects of predation on population size. The classic example is that of the lynx and the snowshoe hare; the data (Figure 45–11) are based on pelts received yearly by the Hudson's Bay Company over a period of almost 100 years. As you can see, there are oscillations in population density that occur about every 10 years. Generally speaking, a rise in the hare population is followed by a rise in the lynx population; the hare population then plummets, and the lynx population follows. Note that other factors also may be involved that were not recorded by the Hudson's Bay Company. For example, the fluctuation in the population numbers of hares may have resulted from overcropping of the vegetation. Nevertheless, even those ecologists most skeptical of the "classic" models agree that for many populations, predation is by far the most important factor in the regulation of population size.

Most predators have more than one prey species, as indicated by the food web shown on page 831. Characteristically, when one prey species becomes less abundant, predation on other, more abundant species increases so the proportions fluctuate. Thus, although oscillations occur in the population size of a prey species, they are usually not as striking as those seen with the lynx and the hare.

* Richard Lewontin, "Adaptation," *Scientific American,* September 1978, page 215.

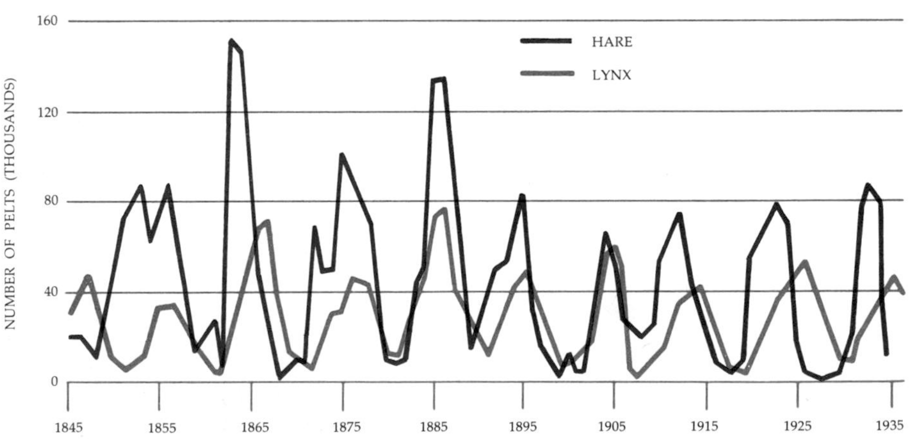

45–11
The number of lynx and snowshoe hare pelts received yearly by the Hudson's Bay Company over a period of almost 100 years, indicating a pattern of 10-year oscillations in population density. The lynx reaches a population peak every 9 or 10 years, and these peaks are followed in each case by several years of sharp decline. The snowshoe hare follows the same cycle, with a peak abundance generally preceding that of the lynx by a year or more. Presumably a surge in the population of snowshoe hares, the prey, results in a corresponding increase in the predator population, with a resultant decrease in prey followed by a crash in the predator population. This type of pattern is characteristic of very simple or immature ecosystems.

45-12
Prickly-pear cactus on a homestead in Australia. Such rapid and environmentally destructive spread is often seen among alien organisms introduced into a region where they have no natural enemies.

Predation on one species may increase the numbers of another. For instance, in England on chalk soils, grasses grow more vigorously than the annual flowering plants, but the annuals were able to maintain their populations in the area because rabbits cropped the grasslands. However, a virus epidemic severely reduced the rabbit population, the grasses took over, and the wildflowers disappeared.

Similarly, Jane Lubchenko of Harvard University was able to show in a series of experiments that the herbivorous marine snail *Littorina littorea* controls the abundance and type of algae in higher intertidal pools on the New England coast. In such pools, the snails' preferred food (the green alga *Enteromorpha*) is competitively superior, and its removal by the snail permits the growth of other algal species. (However, in exposed areas, where *Enteromorpha* is competitively inferior, its removal by the snail facilitates the growth of the dominant species and decreases the number of species in the area.)

The introduction of a new species in a community where it has no natural predators may sometimes have disastrous results. When prickly-pear cactus, for instance, was brought to Australia from South America, it escaped from the garden of the gentleman who imported it and spread into fields and pastureland until more than 12 million hectares were so densely covered with prickly pears that they could support almost no other vegetation (Figure 45–12). The cactus then began to take over the rest of Australia at the rate of about 400,000 hectares a year. It was not brought under control until a natural predator was imported—a South American moth, whose caterpillars live only on the cactus. Now only an occasional cactus and a few moths can be found. (Note, however, that introduction of the moth was a risky experiment.)

STRATEGIES FOR GROWTH

Given the goal of maximizing one's contribution to the gene pool, is it better to produce large numbers of offspring or to invest the same amount of energy in caring for a relative few? The answer seems to be that it depends. If one is a pioneer in a new environment, a rapid reproduction rate would appear to be the best strategy, capturing the available space for one's own genes. Examples would include a new mushroom for a gall midge (see page 777) or an early stage of ecological succession in the case of grass seeds. Species that adopt such a life-history tactic are called fugitive or opportunistic species, and their tactic is sometimes called *r*-selection. Opportunistic species are characterized, among other things, by high intrinsic rates of increase, rapid development, early reproduction, small body size, and uncertain adult survival. Under more constant conditions, when the population is at or near its carrying capacity, the advantage appears to go to species with lower reproduction rates, longer development times, larger body size, and greater capacities to compete. Such species are known as equilibrium species, and their tactic is called *K*-selection.

According to this model, ecological succession involves a shift in the balance between the fugitive or opportunistic species that originally invade a plowed field, for instance, and the equilibrium species that eventually become established in the climax community. Once a mature community is established, the opportunistic species are at a disadvantage, but enough survive to exploit their *r* tactics in disturbed areas, as in the growth of wild flowers along the side of a road or the spread of weeds in a garden plot.

(a) (b) (c) (d)

45–13

Reproductive strategies. (a) *A leatherback turtle returning to the sea after laying her eggs in a hole dug in the sand. The eggs, which she has covered over with sand, will be incubated by the warmth of the sun. Once the eggs are laid, the mother turtle does not participate further in raising the young.* (b) *A female water bug lays her eggs on the back of the male. The male then carries the eggs, aerating them, until they hatch. In the photograph, one egg can be seen hatching at the left.* (c) *Baby Canada geese remain with their parents through the summer after they are hatched and migrate with them to the winter feeding grounds. In proportion to life span, the time allotted to parental care by birds is longer than for any other animal group.* (d) *Elephant calves are suckled by their mothers for at least two years, and the young are zealously guarded by the mother and the sisters and the aunts.*

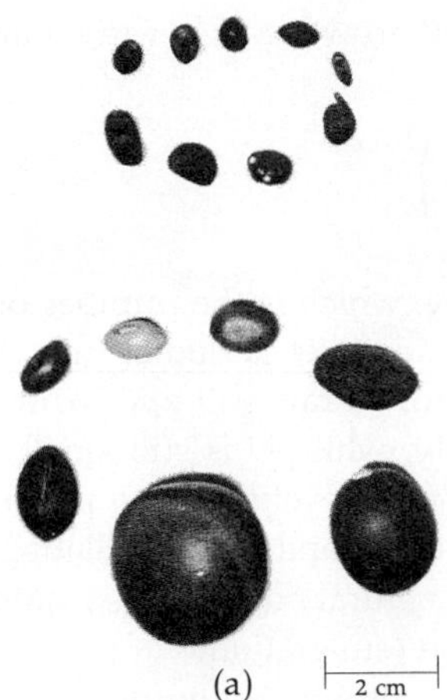

(a)

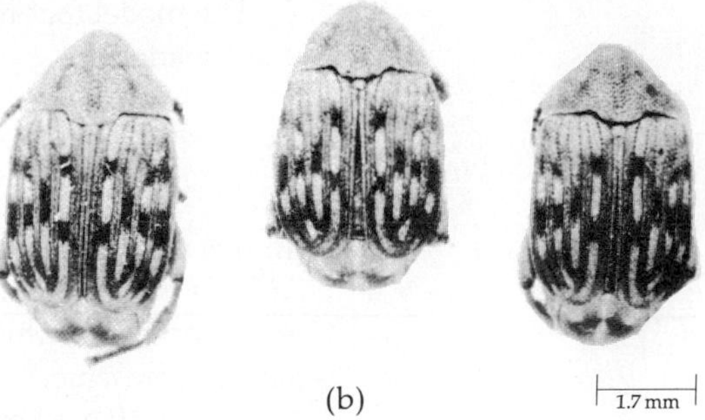

(b)

45–14
(a) *Seeds of Central American legumes. Each of the seeds is derived from a different species of tree or shrub belonging to this plant family. All of the seeds in the lower group are presumably poisonous, and none is attacked by seed-eating insects. Those in the upper group are edible and heavily attacked by seed-eating beetles of the family Bruchidae* (b), *which complete their larval development within the seeds. Exit holes may be seen in several of the seeds of the upper group.*

Whether a species is opportunistic or equilibrium is a result, of course, of natural selection. Daniel Janzen, now at the University of Pennsylvania, observed interactions between some tropical woody members of the pea family growing in Central America, and a family of beetles (bruchids) that inhabited the same area and lived on the seeds. The seeds' only chance for germination was being whisked away by some dispersal agent—such as a bird—before the beetles could devour them. Some species produced numerous, very small seeds, with a reduction in food reserves per seed, on which the beetles preyed heavily; their strategy was an increase in numbers. The other group produced fewer and larger seeds, which contained not only additional food but also a chemical substance that deterred predation by the beetles and so protected both seed and food supply from the predators. The small-seeded species were following an r-tactic course, and the large-seeded species were following a K-tactic course.

K tactics, as we noted, are good under some circumstances and r tactics under others. Judith Meyers and Charles Krebs studied emigration in a population of meadow mice. As with many other groups of rodents, an increase in population size culminates in the movement of large numbers of individuals out of the population. What triggers the dispersal? What decides which individuals leave and which remain behind? The investigators found that the population was polymorphic for r and K strategies. One group of mice had a higher birth rate, began to reproduce at an earlier age, and tended to disperse when population density increased. Thus, these individuals were the ones that migrated, leaving the equilibrium species behind.

SUMMARY

Population size is determined by a number of factors. One of these is the intrinsic rate of increase, symbolized as r; r is equal to the difference in any given time span between the number being born plus the number of immigrants and the number dying plus the number of emigrants. This type of growth, known as exponential growth, is represented by the equation:

$$\frac{\triangle N}{\triangle t} = rN$$

In this equation, N stands for the number of individuals in the population at any given time (t). Exponential growth is approached very seldom in natural populations.

The model that more nearly represents a natural growth rate is represented by the equation:

$$\frac{\Delta N}{\Delta t} = rN\left(\frac{K - N}{K}\right)$$

In this equation, K represents the carrying capacity, which is the number of individuals that the environment can support over an indefinite period of time. In any environment in which resources are limited, the growth rate will approximate the curve represented by this equation: starting slowly, while N is still small; rising rapidly during a period of almost exponential growth; leveling off to produce an S-shaped curve as the carrying capacity is approached; and then oscillating at its upper level. Factors that affect the rate of growth in natural populations include age of population, patterns of mortality, and dispersion (emigration).

A complex of biotic and abiotic factors limit the growth of a population. Density-independent factors include extremes of temperature, moisture, day length, and local soil conditions. Density-dependent factors are often carbon and energy sources, minerals, and other nutrients.

Population size is affected by interactions with organisms of other species. Two principal types of interspecific interactions are competition and predation.

The ecological niche is a multidimensional description of an organism's way of life. It is determined by abiotic factors and also biotic ones, including competition from other organisms. Where niches overlap, one population may be eliminated, or one or both populations may evolve in such a way as to reduce or eliminate the competition.

Predation is the most powerful influence on size of many types of populations, often keeping the population well below the carrying capacity in terms of other resources. In complex food webs, predation is usually density-dependent; when a population decreases in size, predation pressure relaxes and thus the population usually has a chance to recover.

Species may be categorized as opportunistic or equilibrium, depending on whether species members (a) produce large numbers of young with small food supplies or little or no parental care and a high dispersal rate, or (b) produce few young and invest their energies in improved care of existing young, better competition, and avoidance of predators.

QUESTIONS

1. Distinguish between the following: intrinsic rate of increase/growth rate; exponential growth/logistic growth; predation/competition.

2. An old French riddle: "The pond lilies in a certain pond grow at a rate such that each day they cover twice as much of the pond as they did the day before. The pond is of a size that it will be completely covered at the end of 30 days. On what day is the pond half covered? One-tenth covered? One-hundredth covered?" What is the relevance of this riddle to human ecology?

3. Compare the results of MacArthur's experiment with the warblers with Connell's experiment with barnacles. What step did Connell perform that MacArthur did not? Why is the step important?

4. How could the following affect the growth rate of a population (explain your answers): age at first reproduction; generation time; pre-reproductive mortality; post-reproductive mortality; length of period of parental care?

5. Name 10 characteristics of opportunistic species as compared to equilibrium species.

6. Remembering the logistic growth equation, if a population increased by 10 individuals over two days and the initial population size was 100, what is r? Calculate r if the same increase occurred over 20 days and the initial population size was 10.

CHAPTER 46 Interactions in Communities

46–1
The redwood forest along the coast of northern California and southern Oregon is, like every biological community, the result of the interaction of numerous factors, including temperature, altitude, rainfall, seasonal changes, and a particular evolutionary history.

A community is a group of organisms inhabiting a common environment and interacting with one another. It is made up of populations, and it is the biotic component of ecosystems. In the previous chapter, we examined how these interactions affected the numbers of organisms in the community. In this chapter, we shall concentrate more on the qualitative aspects of these relationships and examine how these interactions have shaped the community and guided the evolution of its members. As we shall see, ecology and evolution are intertwined, in what G. E. Hutchinson of Yale University aptly called "the ecological theater and the evolutionary play."

Figure 46–1 shows a view from within a redwood forest and also demonstrates one of the principal ecological factors determining the existence of this particular community—fog. Because of the fog, there is ample moisture along the coast of northern California and southern Oregon during the otherwise dry summer. As in all terrestrial communities, the basic structure is determined by the plant life; here the redwoods dominate the plant community. Not only are they responsible for the gradients of shade and moisture, but also their rotting leaves and branches condition the soil and change its characteristics so that only certain other plants can live in it. At the foot of the redwoods characteristically grow swordfern and redwood sorrel (Figure 46–2).

46–2
At the foot of the redwood trees characteristically grow (a) *swordfern* (Polystichum munitum) *and* (b) *redwood sorrel* (Oxalis oregana). *We call this particular association of plants the redwood community, rather than the swordfern community, because the redwoods so thoroughly dominate it. In fact, it can be defined only in terms of the redwoods themselves and does not have any other sort of unity.*

(a)

(b)

46–3
A birch forest, showing layers of plant growth. The shrub layer is made up of thimbleberry and mountain maple, a favorite browse of deer.

Other forests of the temperate zones resemble the redwood forest in having one or two dominant species of trees—trees that are the most conspicuous features of the landscape, that overshadow the other vegetation, and that alter the habitat. In deciduous woodlands there are up to four layers of plant growth.

1. The tree layer, in which the crowns form a continuous canopy. The canopy is usually between 10 and 35 meters high.
2. The shrub layer, which grows to a height of about 5 meters. Shrubs and bushes resemble trees in that they are woody and deciduous, but they branch at or close to the ground.
3. The field layer, made up of grasses and other herbaceous (nonwoody) plants, including the wild flowers, which typically bloom in the spring before the trees regain their leaves. Bracken and other ferns, whose large leaf areas make them efficient interceptors of light, are also often conspicuous members of the field layer.
4. The ground layer, which consists of mosses and liverworts. The ground is also often covered with leaf litter.

COMPETITION AMONG PRODUCERS

A forest looks like a stable community, and in many ways it is; the redwoods, for example, have held domain on the foggy northwest coast of America since the early Tertiary period, more than 100 million years ago. But silent competition still continues among the plant life. The intertwined roots compete for water and minerals. When a tree falls, fugitive, opportunistic species try to claim the land. And wherever the environment changes, the competitive equilibrium shifts.

On the African savanna, the dominant species are grasses. These compete with

46–4
Grazing by herbivores has encouraged the growth of mesquite, at the expense of grasses, in many areas of the southwestern United States.

acacia trees, and factors such as grazing, habitat destruction, and annual fluctuations in rainfall shift the balance. Similarly in the American southwest, grasses and mesquite bushes vie with one another for dominance. The deciding factor here is grazing, mostly by cattle. Mesquite were rare before cattle were introduced into the western United States.

In a bog, mosses of the genus *Sphagnum* may form a continuous cover, with several species growing together. Does this violate Gause's principle? Closer examination reveals that each species is actually confined to its own characteristic microenvironment, one growing in wet hollows, others in drier places on the sides of hummocks, and others only on the tops of hummocks where they are in uneasy balance with species of flowering plants. As these microenvironments change, one species replaces another.

Sometimes the competition is more aggressive, even on the first trophic level. The leaves of many trees and shrubs contain chemical compounds that are leached out by rain after the leaves fall to the ground. The compounds inhibit the germination and establishment of seedlings that might compete with the plant. This phenomenon, known as allelopathy, has been observed for a number of species, among them, eucalyptus and pine trees.

COMPETITION AMONG CONSUMERS

We have already mentioned several examples of the competition among animals. We have also seen that the effects, as we can observe them after the fact, do not seem to be the competitive exclusion of one population (as they are in the laboratory) but rather the restriction of the population to a narrower ecological niche. We shall present one more example.

When you first visit the grasslands of East Africa, no matter how much you have thought or read about it or how many pictures you have seen, you are bound to be overwhelmed by the sheer numbers of herbivores. The plains surround you, stretching from horizon to horizon, and there are animals as far as you can see. Moreover, these herds are composed of many different kinds of animals, and they are all dependent upon the same vegetation, which, to the inexperienced eye, looks far more homogeneous than the animals.

The sharing of resources by the browsers is the more obvious process. Giraffes reach the taller branches, 3 or 4 meters off the ground. Gerenuks, long-legged, long-necked antelopes, stand on their hind legs to nibble delicately in the middle branches. The lower branches, a meter or so from ground level, are browsed upon by the rhinos, and the smaller antelopes, such as the dik-dik, hide in the thickets and browse on the lowest branches.

The grazers apportion the resources just as equably, although their adaptations are less visible. Zebras, which, unlike the other grazers, have front teeth in both their upper and lower jaws, eat the longer, tougher grass stems. The wildebeest graze younger stems and blades nearer the ground level, and the gazelles, with their tiny muzzles, nip off the smallest green shoots of grass and also crop the little dicots that grow among the grasses and are almost entirely ignored by the larger herbivores. Thus, not only do the zebras leave enough for the wildebeests and the wildebeests for the gazelles, but the tramping and cropping of the larger animals actually encourage the growth of the younger vegetation on which the smaller grazers depend.

46–5
A carnivorous predator, a barn owl, and its prey, a field vole.

PREDATION

As we noted earlier, predation is the channel by which energy flows through the ecosystem. It is often the most important factor in determining the size of populations and so may be properly regarded, in such populations, as the most powerful agent of natural selection. Many predators feed upon the most common prey species available, switching to another diet only when the plant or animal becomes rare. Thus, in some cases, very similar species can coexist in the same ecosystem because of predation.

Looking at a natural population of herbivores, whether aphids or antelopes, the observer cannot help but note the uniformly good condition of the animals. It is very rare to observe a sick, starving, or injured animal. Under normal circumstances, predators tend to keep prey well within the carrying capacity of their environment. Moreover, predation, especially of large herbivores, tends to cull animals in poor physical condition. Wolves, for instance, have great difficulty overtaking healthy adult caribou or even healthy calves. A study of Isle Royale, an island in Lake Superior, showed that in some seasons more than 50 percent of the animals the wolves killed had lung disease, although the incidence of such individuals in the population was less than 2 percent. (Human hunters, however, with their superior weapons and their desire for a "prize" specimen, are more likely to injure or destroy strong, healthy animals.)

Predation also may profoundly affect the characteristics of both predator and prey. Natural selection strongly favors the most efficient predator. Thus, over the course of the generations, the big cats become swifter and more cunning, the necks of the giraffes become progressively longer, the grosbeaks' bills grow thicker and stronger, and eagles become more eagle-eyed. Some examples follow.

Bats and Moths: A Paradigm

In addition to strength, swiftness, cunning, and acuity, evolution has produced some unusual predatory stratagems. Bats, for example, hunt by sonar, emitting shrill cries well above the range of human hearing. When these sound waves hit a solid object, they echo back to the ears of the bat. On the basis of these echoes, bats can navigate skillfully through a dark room strung with wires little thicker than a human hair or can catch an insect as small as a fly.

Most animal species that have hearing organs have organs tuned to receive sounds from members of their own species. A notable exception is the noctuid moth. These moths are among the principal prey of bats, and their tympanic organs have been precisely tuned in the course of evolution to receive, you will not be surprised to hear, the shrill cries of bats, and so, with increasing skill, to avoid their predators.

Parasitism

Parasitism is a special form of predation in which the predator is considerably smaller than the host. The plants and animals in a natural community support hundreds of parasites of many species—in fact, perhaps millions, if one were to count viruses.

As with other forms of predation, infectious diseases are most likely to wipe out the very young, the very old, and the disabled—either directly or, often, indirectly, by making them more susceptible to other predators or to the effects of climate or food shortages. It is logical that a parasite-caused disease should not be too virulent or too efficient. If a parasite were to kill all the hosts for which it is adapted, it too would perish. This principle is particularly well illustrated by a series of misadventures on the continent of Australia.

46–6
Rabbits crowd a water hole in Australia. Imported from Europe as potentially valuable herbivores, they soon overran the countryside. The myxoma virus was introduced to control them, but now host and parasite are coexisting.

There were no rabbits in Australia until 1859, when an English gentleman imported a dozen from Europe to grace his estate. Six years later he had killed a total of 20,000 on his own property and estimated he had 10,000 remaining. In 1887, in New South Wales alone, Australians killed 20 million rabbits. By 1950, Australia was being stripped of its vegetation by giant rabbit hordes. In that year, rabbits infected with myxoma virus were released on the continent. (Myxoma virus causes only a mild disease in the South American rabbits, its normal hosts, but is usually fatal to the European rabbit.) At first, the effects were spectacular, and the rabbit population steadily declined, yielding a share of pastureland once more to the sheep herds, on which much of the economy of the country depends. But then occasional rabbits began to survive, and their litters also showed resistance to the myxoma virus.

A double process of selection had taken place. The original virus was so rapidly fatal that often a rabbit died before it could be bitten by a mosquito and thereby infect another rabbit; the virus strain then died with the rabbit. Strains less drastic in their effects, on the other hand, had a better chance of survival since they had a greater opportunity to spread to a new host. (After an initial infection, a rabbit is immune to the virus, just as human beings usually become immune to mumps or measles after one infection.) So, first, selection began to work in favor of a less virulent strain of myxoma virus. Almost simultaneously, rabbits that were resistant to the original virus began to appear. Now, as a result of co-evolution, the two are reaching a commensal equilibrium, like most hosts and parasites.

46–7
Like many other plants of arid regions, this species of acacia has thorns that protect it against herbivores. The graceful gerenuk, however, which has co-evolved in the savanna, is, in turn, specially adapted to circumvent these defenses. Acacias and other trees and shrubs provide gerenuks with water as well as food; these giraffe-necked gazelles never need to drink.

Natural Defenses in Plants

Plants, the producers of terrestrial ecosystems, are under heavy selection pressure from predators and have many adaptations—natural defenses—to deter herbivores. Some natural defenses in plants are structural, such as the sharp-toothed edges of the holly leaf, the thorns of the rosebush, the spines of the cactus, and the stings of the nettle. Plants also produce a number of chemical substances for which the plant itself has no apparent physiological use and which seem to function as defenses against leaf-eating insects and other predators. Eating foxgloves *(Digitalis purpurea)* can cause convulsive heart attacks in vertebrates. Other plants containing digitalis-like toxins include oleander *(Nerium oleander),* of which a single leaf may be fatal to humans, and the members of a large family of plants known as the milkweeds.

Some plants contain chemicals with insect-hormone activities. A number of gymnosperms have been found to contain chemicals with molting-hormone (ecdysone) activity that fatally accelerate insect metamorphosis. The balsam fir has been found to produce a chemical that resembles juvenile hormone (page 466) and arrests the development of certain insect species. R. H. Whittaker of Cornell describes these chemicals as "a fiendishly subtle mechanism of defense."

Plant defense chemicals include many substances presently useful to us, among them digitalis, quinine, castor oil, and peppercorns and other spices; some substances of more questionable value, such as nicotine, caffeine, and morphine; and others, such as the active principles in marijuana, mescaline, and peyote, whose desirability for the animal world is a matter of opinion. Plants appear to have been waging such chemical warfare long before the coming of *Homo sapiens,* and it is possible that some small leafhopper was the first of all animals to have its mind expanded in a psychedelic experience.

In cultivating plants for agricultural purposes, we have carefully bred out their toxic and bitter-tasting chemicals and so, having destroyed their natural defenses, must protect them with insecticides. (Another reason for the vulnerability of cultivated plants as compared to wild ones is that the former are often grown in large pure stands that attract and support large populations of species-specific insects and other parasites and predators.)

Natural Defenses in Animals

Natural defenses in animals include structural adaptations and the capacity to produce noxious chemicals, both of which are seen in plants, and behavioral strategies as well. Some animals are formidably armored—for example, the armadillo, the porcupine, and the sea urchin. Some have ingenious behavioral devices. The armadillo and the pillbug roll themselves up in tight armor-plated balls. Some species of cephalopods vanish, jet-propelled, leaving behind only an ink cloud. Many lizards have brightly colored tails that break off when they are attacked and, conspicuous and wriggling, divert the predator from the prey—which is escaping, tailless but with all its vital organs intact. Hermit crabs often carry sea anemones as passengers on their shells. The anemone protects the crab from predators while enjoying the benefits of mobility (and perhaps some stray morsels of food). When the crab changes shells, it often coaxes its sea anemone from its old home to the new one.

(a)

(b)

(c)

46–8
(a) When attacked, a blue-tailed skink leaves its tail to distract predators while it escapes. (b) When threatened, porcupines release barbed quills. (c) An arrow-poison frog from South America. Skin secretions from these frogs are used by local Indians to tip their arrows with a poison that produces a curare-like paralysis and death.

Concealment and Camouflage

Hiding is one of the chief means of escape from predators. The young of many birds and of some other vertebrates respond to the warning cries of their parents by "freezing" or by running to cover. Small mammals often have nests or burrows or makeshift residences in hollow logs or beneath tree roots in which they conceal themselves from their enemies.

Protective coloration is common. Mice, lizards, and arthropods that live on the sand are often light-colored, and such light-colored species, if removed from their home territories, will immediately return to them. Snails that live on mottled backgrounds are often banded. Grass snakes are grass colored, as are many of the insects that live among the grasses.

Some animals are countershaded for camouflage. The next time you pass a fish market, look at the specimens laid out on view. Fish are nearly always darker on the top than on the bottom. Countershading reduces the contrast between the shaded and unshaded areas of the body when the sun is shining on the organism from overhead. A fish was once found—the Nile catfish—that was reverse-countershaded; that is, its dorsal surface was light and its ventral surface dark. The selective theory of camouflage was momentarily threatened, but scientific order was restored when it was discovered that the Nile catfish characteristically swims upside down.

Other organisms hide by looking like something else. One type of insect, the treehopper, looks like a thorn; another, the walkingstick, like a twig. Young larvae of some swallowtail butterflies look like bird droppings. In order for such disguises to work, animals must behave appropriately. *Biston betularia,* the peppered moth, lies very flat and motionless on its tree trunk, so that its colors blend with those of the tree. These and other moths that sit exposed on the bark of trees even habitually orient themselves so that the dark markings on their wings lie parallel to the dark cracks in the bark. Some desert succulent plants look like smooth stones, revealing their vegetable nature only once a year when they flower.

(a)
(b)
(c)
(d)
(e)
(f)

46–9
Concealment and camouflage. (a) *Bark katydid;* (b) *stone grasshopper in the desert;* (c) *horned lizard;* (d) *a francolin, an East African bird;* (e) *viper buried in sand;* (f) *a leaflike insect* (Anaea). *The physical adaptation of these animals is dependent, in large part, on a crucial behavioral adaptation. In times of danger, all of these animals remain absolutely still.*

46–10
Deception and disguise. (a) *Frog fish.* (b) *Owl butterfly. With wings folded, the butterfly is camouflaged. The startling eyespots are a second line of defense.*

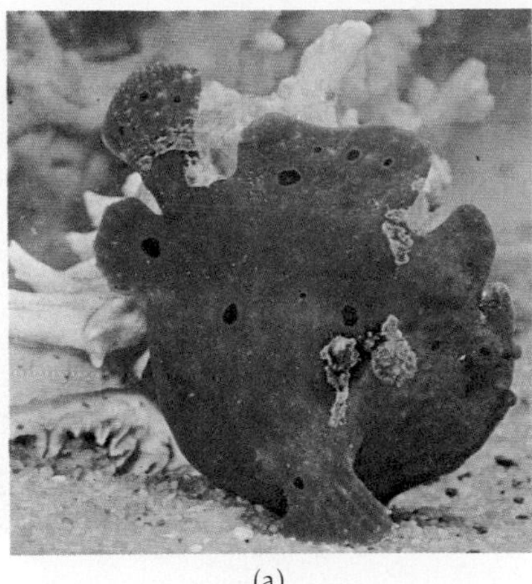
(a)

(b)

Some insects manage through warning coloration to frighten off their would-be predators. Large spots that look like eyes are commonly found on the backs of butterflies or the bodies of caterpillars, where they will suddenly appear when the insect spreads its wings or arches its body. Small birds reported to flee at the sight of these insects are probably themselves the prey of larger, large-eyed birds, such as owls or hawks. Smaller eyespots, while probably not frightening, seem to have the effect of deflecting the point of attack away from the head. Examination of wounded butterflies has shown that if a part of the wings bears beak marks or is missing, it is most often the part that contains the eyespots. One investigator has tested and confirmed this conclusion by painting eyespots on the wings of living insects, releasing them, and later recapturing them for examination.

On Being Obnoxious

Some animals are protected from predators by a disagreeable taste, odor, or spray, often derived from distasteful chemicals in the plants they eat. Such animals are, generally, carefully avoided. Monarch and other related butterflies feed in their larval stages on milkweeds, and the digitalis-like compounds of the plants are concentrated in their tissues. The caterpillars are immune to these poisons, but the birds that eat the caterpillars become violently ill and learn to avoid these butterflies.

Obviously, tasting bad, while useful, is not an ideal defense from the point of view of the individual, since making this fact known may demand a certain amount of personal sacrifice. (Actually, birds drop the monarch butterfly after the first bite, but they often inflict fatal injury in the process.) Obnoxious sprays and odors have the advantage of warding off predators before they harm the prey. To us, the most familiar of such animals is the skunk, which advertises its malodorous threat by its distinctive coloration. Many insects and other arthropods have developed defense secretions. In some millipedes, for example, the secretion oozes out of a gland onto the surface of the animal's body. Other arthropods are able to spray their secretion over a distance, in some instances even aiming it precisely at the attacker. The caterpillar *Schizura concinna,* whose single spray gland opens ventrally just behind its head, directs the spray simply by aiming its front end. In the beetle *Eleodes longicollis,* the spray glands are in the rear, and the beetle, when disturbed, does a

46–11
Newly hatched stinkbugs grouped next to their empty egg shells. This clumping behavior and their bright coloration, combined with a repellent smell, warn predators of the stinkbug's unpleasant taste. They harbor symbiotic bacteria in their intestines that aid digestion of plant juices. These they obtain shortly after hatching by drinking moisture adhering to the egg shells contaminated by maternal feces.

quick headstand and sprays a secretion from its abdominal tip. The soldiers of certain termite species possess a pointed cephalic nozzle from which their defensive spray is ejected. This spray not only can incapacitate a small predator but also acts as an attractant to summon more troops. In the whip scorpion, two glands open at the tip of a short knob that moves like a gun turret. Many arthropods possess a number of glands but discharge only from those closest to the point of attack, thus saving ammunition and gaining efficiency. The secretions usually act as topical irritants, especially to the mouth, nose, and eyes of the predator. Because of their relatively permeable skin, frogs and toads, common predators of arthropods, are sensitive to these irritants over their entire bodies.

Vertebrates learn quickly to recognize an obnoxious prey. Blue jays, for instance, having once been sprayed by a walkingstick, will remain aloof from it even when the insect is presented two or three weeks later. Predators also may learn to counter these defenses. Grasshopper mice feed on beetles like *Eleodes longicollis* by jamming them butt-end into the earth, so that their chemical arsenal is harmlessly expended in the soil, and then eating them head first, another example of co-evolution.

SYMBIOSIS: MUTUALISM

Symbiosis ("living together") is a close and permanent association between organisms of different species. There are several types of symbiotic relationships. If the relationship is beneficial to both, it is called mutualism. If one species benefits from the association while one is neither harmed nor benefited, it is called commensalism. If one species benefits and the other is harmed, the relationship is known as parasitism. However, since all the details of symbiotic relationships between species are not fully understood, these distinctions are not always useful. The examples of mutualism in nature are many and varied (see Figure 46–12). One, which we have mentioned previously, is the association between a ruminant animal, such as a cow, and the bacteria and protozoa that inhabit its stomach and break down cellulose. *Trichonympha,* a protist, which is the source of the spectacular cross section of flagella shown on page 116, has a similar association with wood termites.

46–12

Symbioses. (a) *Sea anemones on the back of a snail shell occupied by a hermit crab. The anemone protects and camouflages the crab and, in turn, gains mobility—and so a wider feeding range—from its association with the crab. Hermit crabs, which periodically move into new, larger shells, will coax their anemones to move with them.* (b) *Cleaner fish are permitted to approach larger fish with impunity because they feed off the algae, fungi, and other microorganisms on the fish's body. The fish recognize the cleaners by their distinctive markings. Other species of fish, by closely resembling the cleaners, are able to get close enough to the large fish to remove large bites of flesh. What would probably happen if cleaner mimics began to outnumber cleaners?* (c) *Aphids suck phloem, removing certain amino acids, sugars, and other nutrients from it and excreting most of it as "honey dew," or "sugar-lerp," as it is called in Australia where it is harvested as food by the aborigines. Some species of aphids have been domesticated by some species of ants. These aphids do not excrete their honeydew at random, but only in response to caressing movements of the ant's antennae and forelimbs. The aphids involved in this symbiotic association have lost all their own natural defenses, including even their hard outer skeletons, relying upon their hosts for protection.* (d) *Oxpeckers live on the ticks they remove from their hosts. An oxpecker forms an association with one particular animal, such as the wart hog shown here, conducting most of its activities, including courtship and mating, on the back of its host.*

Similarly, symbiotic algae, such as those found in the coral reef, provide their hosts with the products of photosynthesis and receive protection in return.

A classic example of symbiosis is provided by the lichens, which are part alga, part fungus. The body of the lichen is composed largely of fungal mycelium, and, held within this mycelium, are numerous photosynthetic algal cells. The two organisms together form a closely integrated unit that can grow under conditions where neither the fungus nor the alga alone could survive. Lichens occur from arid desert regions to the Arctic; they grow on bare soil, tree trunks, sunbaked rocks, and windswept alpine peaks all over the world. They are often the first colonists of bare rocky areas.

Ants and Acacias

Trees and shrubs of the genus *Acacia* grow throughout the tropical and subtropical regions of the world. In Africa and tropical America, acacia species are protected by thorns. (The acacias of Australia, which has few large grazing animals except recently introduced species, have no thorns.)

On one of the African species of *Acacia,* ants of the genus *Crematogaster* gnaw entrance holes in the walls of the thorns and live permanently in them. Each colony of ants inhabits the thorns on one or more trees. The ants obtain food from nectar-secreting glands on the leaves of the acacias and eat caterpillars and other herbivores that they find on the trees. Both the ants and the acacias appear to benefit from this symbiotic association.

In the lowlands of Mexico and Central America, the ant-acacia relationship has been extended to even greater lengths. The bull's-horn acacia, a common plant of that area, is found particularly frequently in cutover or disturbed areas, where the competition for light is intense. It grows extremely rapidly. This species of acacia has a pair of greatly swollen thorns several centimeters in length at the base of most leaves. The petioles bear nectaries, and at the very tip of each leaflet is a small structure rich in oils and proteins known as a Beltian body. Thomas Belt, the

46–13
Ants and acacias. (a) *The beginning. A queen ant cuts an entrance into a thorn on a seedling bull's-horn acacia. She will hollow out the thorn and raise her first brood inside it.* (b) *The tip of an acacia leaf. The orange structures at the tips of the leaves are Beltian bodies. They are a source of food for the ants.* (c) *A worker ant drinking from the nectary.* (d) *Warriors in a battle for possession of an acacia. Such battles occur when the branches of acacias grow to touch each other. The largest colony usually wins.*

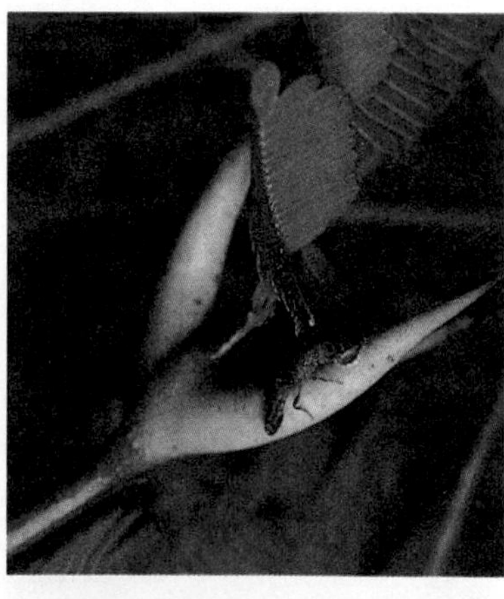
(a)

(b)

(c)

(d)

naturalist who first described these bodies, noted that their only apparent function was to nourish the ants. Ants live in the thorns, obtain sugars from the nectaries, eat the Beltian bodies, and feed them to their larvae.

Worker ants, which swarm over the surface of the plant, are very aggressive toward other insects, and indeed toward animals of all sizes. They become alert at the mere smell of a person or a cow, and when their tree is brushed by an animal, they swarm out and attack at once, inflicting painful burning stings. The effect has been described as similar to walking into a large nettle. Moreover, and even more surprisingly, alien plants sprouting within as much as a meter of occupied acacias are chewed and mauled, and twigs and branches of other trees that touched an occupied acacia are similarly destroyed. Not so surprisingly, acacias inhabited by these ants grow very rapidly, soon overtopping other vegetation.

Janzen, who first analyzed the ant-acacia relationship in detail, removed ants from acacias artificially, by insecticides or by removing thorns or entire occupied branches. Acacias without their ants grew slowly and usually suffered severe damage from insect herbivores. Their stunted bodies were soon overshadowed by competing species of plants and vines. As for the ants, according to Janzen, these particular species can live only on acacias.

(a)

(b)

46-14
Sand dunes of (a) *Padre Island, Texas, and* (b) *the Sahara Desert in El Golea, Algeria.*

BIOMES

In Chapter 44, we observed that the biosphere is not uniform in depth and density, but patchy. *Biomes,* the big patches into which the biosphere is divided, are large collections of communities. The patches are discontinuous, but a patch may closely resemble another patch on the other side of the planet (Figure 46-14). For instance, the deserts of the world look remarkably the same. However, when one looks more closely, one will see that although the physical features of the environment—temperature and rainfall—are the same, the organisms are not the same. But they look and act alike. The same is true of the chaparral of California and the Mediterranean, the grasslands of North and South America, and so on.

A biome is thus an abstraction. When we speak of the tropical forest biome, we are not speaking of a particular geographical region, but rather of all the tropical forests on the planet. As with most abstractions, important details are omitted. For example, the boundaries are not so sharp as shown on maps nor are all areas of the world easy to categorize. However, the biome concept emphasizes one important truth: Where the climate is the same, the organisms are also very similar, even though these organisms are not related genetically and are far apart in their evolutionary history. This phenomenon, as we have noted previously, is called convergent evolution.

TERRESTRIAL BIOMES

Tropical Forests

In the equatorial zone, where most of the tropical forest formations are found, the mean daily temperature is the same throughout the year, and the length of day varies by less than one hour. Rainfall is seasonal, however, with maxima at the time of the equinoxes (see page 828). Variations in total rainfall from one area to another within these zones are caused principally by mountains and their rain shadows.

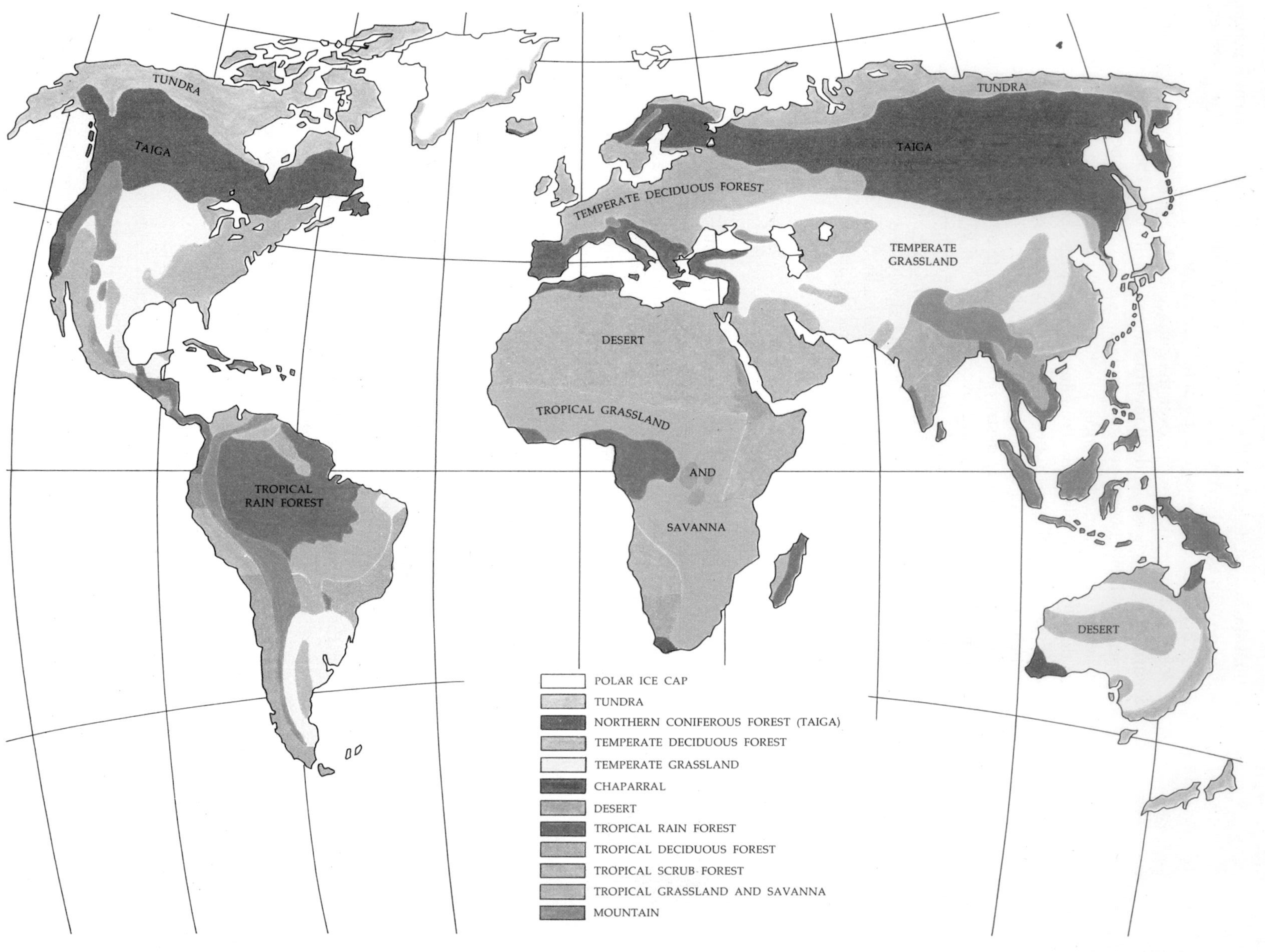

46–15
Biomes of the world.

46–16
A tropical rain forest, Rancho Grande National Park, in Venezuela. Notice the height of the trees and the huge, buttressed trunk of the tree in the center. The vine growing up it is a philodendron.

Tropical Rain Forest

In the tropical rain forest, rainfall is abundant all year around; total rainfall is between 200 and 400 centimeters per year, and a month with less than 10 centimeters of rain is considered relatively dry. More species of plants and animals live in the tropical rain forest than in all the rest of the biomes of the world combined. As many as 100 species of trees can be counted on 1 hectare (2.47 acres). Conversely, in contrast to the deciduous forests of the northeastern United States, where, typically, only a few species are represented in any particular area, in the tropical rain forest a species may be represented only once per hectare.

The competition among plants of the tropical forest is for light. About 70 percent of all species of plants are trees. The upper tree story consists of solitary giants 50 to 60 meters tall. A lower story of trees characteristically forms a solid canopy. They are remarkably similar in appearance. Their trunks are usually slender and branch only near the crown. The crowns are high up and relatively small as a result of crowding. Because the soil is perpetually wet, their roots do not reach deep into it, and trunks often end in thick buttresses that provide firm, broad anchorage. Their leaves are large, leathery, and dark green; their bark is thin and smooth; and their flowers are generally inconspicuous and greenish or whitish in color.

Woody vines, or lianas, are abundant, especially where an opening has appeared in the forest, as a result, for example, of a tree's falling; vines as long as 240 meters have been measured. There are also many *epiphytes*, which are plants that grow on other plants, often high above the forest floor (Figure 46–17). The epiphytes of the tropical rain forest germinate in the branches of trees and obtain water and minerals from the humid air of the canopy. Unlike the plants that have contact with the moist floor, epiphytes need to conserve water between rainfalls. Some epiphytes resemble desert succulents, having fleshy water-storing leaves and stems. Others have spongy roots or cup-shaped leaves that capture moisture and organic debris; many of these epiphytes can take up nutrients from decaying organisms in these storage tanks. A variety of plants, including ferns, orchids, mosses, and bromeliads, have exploited this life style.

(a)

(b)

46–17
Inhabitants of the tropical rain forest. (a) *Epiphytes, such as this bromeliad, grow in the canopy of the tropical rain forest, obtaining water and minerals from the moist air. Bromeliads are members of the pineapple family that are especially adapted to the tropical forest biome.* (b) *A splendid parakeet* (Neophema splendida).

An extraordinary variety of insects, birds, and other animals, including mammals, have moved into the treetops along with the vines and epiphytes to make it the most abundantly and diversely populated area of the tropical rain forest.

Little light reaches the forest floor (from 0.1 to 1 percent of the total), and the few plants that are found there are adapted to growing at low light intensities. Many of these, such as the African violet, are familiar to us as house plants.

There is almost no accumulation of leaf litter on the forest floor such as we find in our northern forests; decomposition is too rapid. Everything that touches the ground disappears almost immediately—carried off, consumed, or rapidly decomposed. In many places the ground is bare.

The soils of tropical rain forests are relatively infertile. Many are chiefly composed of a red clay; these red soils are known as laterites, from the Latin word *later,* or "brick." When laterite soils are cleared, in many cases they either erode rapidly or form thick, impenetrable crusts that cannot be cultivated after a season or two. Tropical soils are generally deficient in minerals; most of the nitrogen, phosphorus, calcium, and other nutrients are found in the plants rather than the soil. Also, those minerals that are present in the soil are leached out by the heavy rainfall. As a consequence, soils where tropical forests have stood are very poor agriculturally and will support crops for only a few years. The rapidly expanding human population in the tropics has made the traditional tropical agricultural practices of clearing and short-term cultivation immensely destructive because they are now carried out on such a wide scale. Although the tropical rain forest now forms about half of the forested area of the earth, some ecologists predict that, at the present rate of destruction, it may almost all have disappeared by the year 2000, and with it thousands of species of plants and animals found nowhere else in the world.

One of the questions ecologists would like to answer, before the tropical forests disappear, concerns their enormous diversity. One hypothesis holds that it is the result of past and present competition, which has produced many small ecological niches, each held by the best competitor. Another, more recent concept, is that the diversity is the result of nonequilibrium, with disruptions and changes in habitat preventing the competition that will eliminate species and decrease diversity.

Other Tropical Forests

Where the rainfall is more seasonal, tropical rain forests give way to tropical deciduous forests, which are dominated by trees that lose their leaves during the dry seasons. Many of the species flower before they put out new leaves (transpiration, or water loss, is extremely low in leafless, flowering trees because petals have few stomata). Tropical scrub forests, which have trees with small, water-conserving leaves, are found where rainfall is limited all year.

46–18
A pack of hyenas eating a zebra they have killed.

Savannas

Savannas are tropical grasslands with scattered clumps of trees (Figure 46–19). The transition from open forest with grassy undergrowth to savanna is gradual and is determined by the duration and severity of the dry season and, often, by fire and grazing and browsing animals.

In the savanna, the competition is for water, and grasses are the victors. Grasses are well suited to a fine, sandy soil with seasonal rain because their roots form a dense root system capable of extracting the maximum amount of water during the rainy period. During dry seasons, the aboveground portions of the plants die, but

46–19
A savanna in Kenya, with a giraffe, zebras, and an impala. The trees in the background are acacias.

the roots are able to survive even many months of desiccation. The balance between woody plants and grasses is a delicate one. If rainfall decreases, the trees die entirely. If rainfall increases, the trees increase in number until they shade the grasses, which, in turn, die. If the grasses are overgrazed (which often happens when people begin to use the savanna for agricultural purposes and introduce livestock), enough water is left in the soil so that the woody plants can increase in number, and the grassland is eventually destroyed.

The best-known savannas are those of Africa, which are inhabited by the most abundant and diverse group of large herbivores in the world, including the gazelle, the impala, the eland, the buffalo, the giraffe, the zebra, and the wildebeest.

The Desert

The great deserts of the world are located at latitudes of about 30°, both north and south, and extend poleward in the interiors of the continents. These are areas of falling, warming air and, consequently, little rainfall. Only about 5 percent of North America is desert. The Sahara Desert, which stretches all the way from the Atlantic coast of Africa to Arabia, is the largest in the world, almost equal to the size of the United States, and is increasing in size, spreading along its southern boundaries. This spread is due in large part to the growth of the human population, resulting, in turn, in intensified grazing by domestic animals along the margins of the desert.

Desert regions are characterized by less than 25 centimeters of rain a year. Because there is little water vapor in the air to moderate the temperature, the nights are often extremely cold. The temperature may drop as much as 30°C at night, in comparison with the humid tropics, where day and night temperatures vary by only a few degrees.

(a)

(b)

(c)

(d)

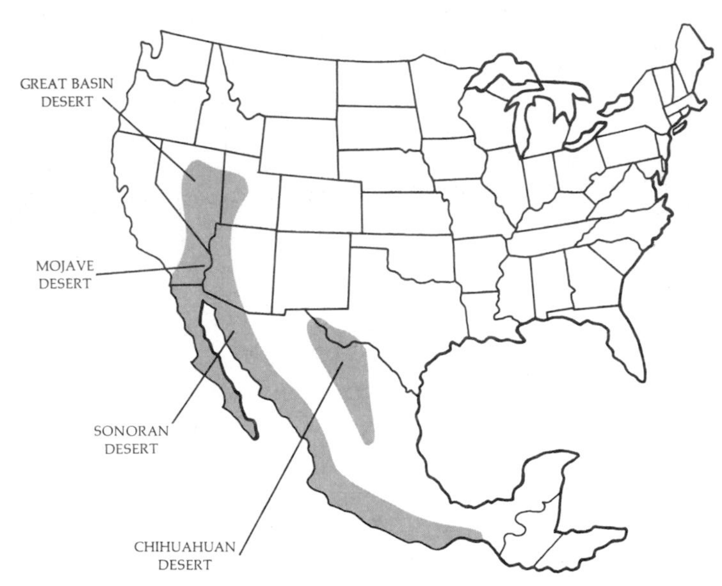

46–20

The North American deserts. The Sonoran stretches from southern California to western Arizona and down into Mexico. A dominant plant, the giant saguaro cactus (a) *is often as much as 15 meters high, with a widespreading network of shallow roots. Water is stored in a thickened stem, which expands, accordionlike, after a rainfall.*

To the southeast is the Chihuahuan desert, one of whose principal plants is the agave (b)*, or century plant, a monocot.*

North of the Sonoran is the Mojave, whose characteristic plant is the grotesque Joshua tree (c)*. This plant was named by early Mormon colonists who thought that its strange, awkward form resembled a bearded patriarch gesticulating in prayer. The Mojave contains Death Valley, the lowest point on the continent (90 meters below sea level), only 130 kilometers from Mt. Whitney, whose elevation is more than 4,000 meters.*

The Mojave blends into the Great Basin, a cold desert bounded by the Sierra Nevada to the west and the Rockies to the east. It is the largest and bleakest of the American deserts. The dominant plant form is sagebrush (d)*, shown here in the background. The large green plant in the foreground is shadscale, and the yellow-flowered plant to the left is rabbitbrush.*

(a)
(b)
(c)
(d)
(e)
(f)
(g)

46–21
Inhabitants of the North American deserts: (a) Golden carpet, found only (and rarely) in Death Valley; (b) Gamble quail; (c) giant hairy scorpion; (d) roadrunner with a whiptail lizard; (e) horned lizard with a grasshopper; (f) bighorn sheep of the southwestern desert mountains; (g) gila monster.

Many of the plants in the desert are annuals that race from seed to flower to seed during periods when water is available, and during the brief growing seasons, the desert may be carpeted with flowers. Many of the perennials are succulents (adapted for water storage). Some, like cacti, have no leaves, some are drought-deciduous (dropping their leaves in dry seasons), and some have small, leathery water-conserving leaves. C_4 and CAM photosynthesis (pages 187 and 533), both of which conserve water, are common among desert plants, as are extensive root systems that are able to trap maximum amounts of water during the brief periods it is available.

The animals that live in the desert are also specially adapted to this extreme climate. Reptiles and insects, for example, have waterproof outer coverings and dry, and therefore water-conserving, excretions. The few mammals of the desert are small and nocturnal and get what little water they require from the plants they eat.

Chaparral

Regions with mild, rainy winters and long, hot, dry summers, such as the southern coast of California, are dominated by small trees or often by spiny shrubs with broad, thick evergreen leaves. In the United States, such areas are known as *chaparral*. Similar communities are found in areas of the Mediterranean (where they are called the *maquis*, which became the name of the French underground in World War II), in Chile (where they are the *matorral*), in southern Africa, and along portions of the coast of Australia. Although the plants of these various areas are phylogenetically unrelated, they closely resemble one another in their growth patterns and characteristic appearance.

46–22
The North American chaparral. (a) *A cacomistle, or ring-tailed cat, a common inhabitant of the chaparral.* (b) *The bushy vegetation that characterizes the chaparral grows as dense as a mat on the foothills of southern California. It is the result of long, dry summers, during which much of the plant life is semidormant, followed by a brief, cool rainy season. The name comes from* chaparro, *the Indian word for the scrub oak that is one of the prominent components of the chaparral. Chaps, the leather leggings worn by the cowboys making their way through this dense, dry growth, has the same derivation.*

(a)

(b)

Mule deer live in the North American chaparral during the spring growing season, moving out to cooler regions during the summer. The resident vertebrates—brush rabbits, lizards, wren-tits, and brown towhees—are generally small and dull-colored, matching the dull-colored vegetation.

Grasslands

Grasslands, which are transitional areas between the deserts and the temperate forests, are found in the interior areas of continents. They are characterized also by periodic droughts, rolling to flat terrain, and hot-cold seasons (rather than the wet-dry seasons of the savanna). The great grasslands of the world include the plains and prairies of North America, the steppes of Russia, the veld of South Africa, and the pampas of Argentina.

The vegetation is largely bunch or sod-forming grasses, often mixed with legumes (clover and alfalfa) and various annuals. In North America, there is a transition from the more desertlike, western short-grass prairie (the Great Plains), through the moister, richer tall-grass prairie (the Corn Belt), to the eastern temperate deciduous woodland. Grasslands become drier and drier at increasing distances from the Atlantic Ocean and the Gulf of Mexico, which are the major sources of moisture-bearing winds in the eastern half of the continent.

The grasslands of the world support small, seed-eating rodents and also large herbivores, such as the bison of early America, the gazelles and zebras of the African veld, the wild horses, wild sheep, and ibex of the Asiatic steppes, and now the domestic herbivores. These large, grass-eating mammals, in turn, support the carnivores, such as lions, tigers, and wolves, as well as omnivorous humans. The herds of grazing animals serve, as do ground fires, to maintain the nature of the grassland landscape, destroying tree seedlings and preventing their encroachment.

46–23
The grasslands of North America. (a) *Short-grass prairie with two familiar inhabitants, a black-tailed prairie dog peering out of its burrow and a bull snake.* (b) *A June day on a tall-grass prairie in North Dakota. The cottonwood grove by the prairie creek is characteristic of this biome. A thunderstorm is gathering on the horizon.*

(a)

(b)

(a)
(b)
(c)
(d)

46–24
The deciduous forest of North America. (a) *Aspens growing on the Wasatch Mountains in Utah.* (b) *A raccoon fishing and* (c) *a red fox killing a cottontail rabbit.* (d) *Large-flowered trillium* (Trillium grandiflorum). *All are common inhabitants of the deciduous forests.*

Temperate Deciduous Forest

Deciduous forests occupy areas where there is a warm, mild growing season with moderate precipitation, followed by a colder period less suited to plant growth. Leaf-shedding in the deciduous forest, as in the tropical forest, evolved as a protection against water loss.

The dominant trees of the deciduous forests vary from region to region, depending largely on the local rainfall. In the northern and upland regions, oak, birch, beech, and maple are the most prominent trees. Before the chestnut blight struck North America, an oak-chestnut forest extended from Cape Ann, Massachusetts, and the Mohawk River valley of New York to the southern end of the Appalachian highland. Maple and basswood predominate in Wisconsin and Minnesota, and maple and beech in southern Michigan, becoming mixed with hemlock and white pine as the forest moves northward. The southern and lowland regions have forests of oak and hickory. Along the southeastern coast of the United States, the wet, warm climate supports an evergreen forest of live oak and magnolia.

The deciduous forest supports an abundant animal life. Smaller mammals, such as chipmunks, voles, squirrels, raccoons, opossums, and white-footed mice, live mainly on nuts and other fruits, mushrooms, and insects. The wolves, bobcats, gray foxes, and mountain lions, in the areas where they have not been driven out by the encroachments of civilization, feed on these smaller mammals. Deer live mainly on the forest borders, where they browse on shrubs and seedlings.

Beneath the ground layer is the soil of the forest, often a rich, gray-brown topsoil. Such soil is composed largely of organic material—decomposing leaves and other plant parts and decaying insects and other animals—and the bacteria, protozoans, fungi, worms, and arthropods that live on this organic matter. The roots of plants penetrate the soil to depths measurable in meters and add organic matter to the soil when they die. Carnivorous arthropods carry fragments of their prey to considerable depths in the soil. The myriad passageways left by dead roots and fungi and by the earthworms and other small animals that inhabit the forest make the soil into a sponge that holds the water and minerals. The land where deciduous forests have stood makes good farmland.

Temperate deciduous forests once covered most of eastern North America, as well as most of Europe, part of Japan and Australia, and the tip of South America. In the United States, only scattered patches of the original forest still remain.

Coniferous Forests: The Taiga

Most conifers are evergreens, with small, compact leaves protected by a thick cuticle that guards against water loss. The dropping of leaves is a more efficient adaptation to a hot-cold season, and conifers generally cannot compete with deciduous trees in temperate zones with adequate summer rainfall. However, deciduous trees require a warm summer period of at least four months to permit their leaves to regrow. The northern boundary between the deciduous forest and the coniferous forest occurs where the summers become too short and the winters too long for deciduous trees to grow well.

The northern coniferous forest, called the taiga, is characterized by long, severe winters and a constant cover of winter snow. It is composed chiefly of evergreen needle-leaved trees such as pine, fir, spruce, and hemlock. A thick layer of needles

46–25

Taiga of North America. (a) *A bull moose, with strands of velvet hanging from his newly polished antlers, in Mount McKinley National Park, Alaska. The moose is browsing on willow, its staple food in this region.* (b) *A coniferous forest of balsam spruce and white pine, photographed near the Canadian border.* (c) *Floor of a virgin northern coniferous forest, carpeted with red pine needles. Decay is slower than on the warmer, wetter floors of the deciduous forest, and there is no undergrowth due to the shade cast by the mature trees.*

(a)

(b)

(c)

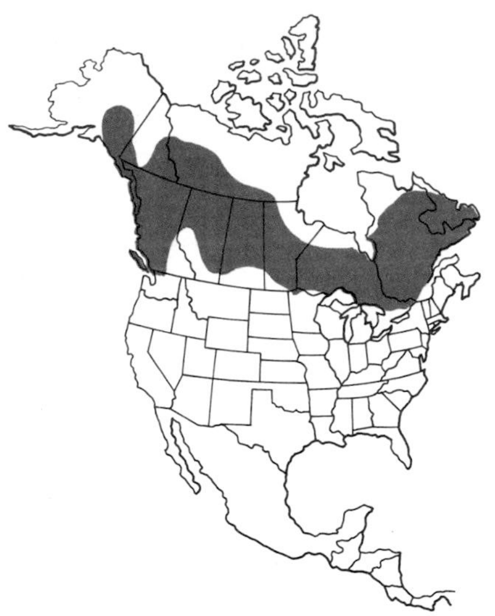

and dead twigs, matted together by fungus mycelium, covers the ground. Along the stream banks grow deciduous trees, such as tamarack, willow, birch, alder, and poplar. There are virtually no annual plants.

The principal large animals of the North American coniferous forest are elk, moose, mule deer, black bears, and grizzlies. Among the smaller animals are porcupines, red-backed mice, snowshoe hares, shrews, wolverines, lynxes, warblers, and grouse. The small animals use the dense growths of the evergreens for breeding and for shelter. Wolves feed upon these mammals, particularly the larger ones. The black bear and grizzly bear eat everything—leaves, buds, fruits, nuts, berries, fish, the supplies of campers, and occasionally the flesh of other mammals. Porcupines are bark eaters, and many seriously damage trees by girdling them. Moose and mule deer are largely browsers. The ground layer of the coniferous forest is less richly populated by invertebrates than that of the deciduous forest because the accumulated litter is slower to decompose.

The Tundra

Where the climate is too cold and the winters too long even for the conifers, the taiga grades into the tundra. The tundra is a form of grassland that occupies one-tenth of the earth's land surface, forming a continuous belt across northern North America, Europe, and Asia. Its most characteristic feature is permafrost, a layer of permanently frozen subsoil. During the summer the ground thaws to a depth of a few centimeters and becomes wet and soggy; in winter it freezes again.

46–26
(a) Tundra of North America on a long Arctic day. Cotton grass surrounds a kettle pond formed from a chunk of glacial ice. (b) The Arctic tern represents one of a number of bird species that breed in the tundra, taking advantage of the long summer days to gather food for their nestlings. The terns winter in the Antarctic, following migration routes of 13,000 to 18,000 kilometers. Within three months after hatching, the young must be ready to migrate. (c) A snowshoe, or varying, hare in its winter coat. Extra hair on its feet enables it to travel easily over the snow.

(a)

(b)

(c)

This freeze-thaw process, which tears and crushes the roots, keeps the plants small and stunted. Drying winter winds and abrasive driven snow further reduce the growth of the Arctic tundra.

The virtually treeless vegetation of the tundra is dominated by herbaceous plants, such as grasses, sedges, rushes, and heather, beneath which is a well-developed ground layer of mosses and lichens, particularly the lichen known as reindeer moss. All the flowering plants are perennials.

The largest animals of the Arctic tundra are the musk oxen and caribou of North America, and the reindeer of the Old World. Lemmings (small, furry rodents with short tails) and ptarmigans (pigeon-sized grouse) feed off the tundra. The white fox and the snowy owl of the Arctic live largely on the lemmings.

During the brief Arctic summer, insects emerge in great numbers, and migratory birds visit, taking advantage of the insect hordes and the long periods of daylight. The growing season in many areas of the tundra is less than two months.

Altitude and Terrestrial Biomes

The mean temperature of the atmosphere decreases about 0.5°C for each degree of latitude. Increases in elevation produce a similar effect, and, in general, a change in mean atmospheric temperature corresponding to 1° in latitude occurs with a rise of about 100 meters (about 100 yards) in elevation. Climbing a mountain in the tropics is analogous to a trek toward the poles. Thus, plants and animals of almost Arctic characteristics are sometimes found near the equator.

FRESHWATER BIOMES

In general, limnologists—scientists who study natural fresh waters in all their aspects—classify freshwater habitats into two groups: standing water (lakes and ponds) and running water (rivers and streams).

Lakes and Ponds

Lakes vary in size from large seas, covering thousands of square kilometers, to small ponds. Lakes contain three distinct zones: littoral, limnetic, and profundal.

The *littoral zone,* at the edge of the lake, is the most richly inhabited. Here the most conspicuous plants are angiosperms rooted to the bottom, such as cattails and rushes. Water lilies grow farther out from shore. There is often a green blanket of duckweed, a small, free-floating angiosperm. Other pond weeds grow entirely beneath the water. These plants lack a waxy outer cuticle and so can absorb minerals through their epidermis as well as through their roots. Their submerged plant surfaces harbor large numbers of small organisms. Snails, small arthropods, and mosquito larvae feed upon the plants. Other insects that live among the submerged plants, such as the larvae of the dragonfly and damselfly, and the water scorpion, are carnivorous. Clams, worms, snails, and still other insect larvae burrow in the mud. Frogs, salamanders, water turtles, and water snakes are found almost exclusively in these littoral zones, where they feed primarily upon the insects. Fish, too, are found in greater numbers along the lake margins. Ducks, geese, and herons feed on the plants, insects, mollusks, fish, and amphibians abundant in this zone. The shallow margins of some lakes and ponds are marshy. Among the inhabitants of the marshes are such invertebrates and vertebrates as snails, frogs, ducks, herons, bitterns, muskrats, otters, and beavers.

In the *limnetic zone,* the zone of open water, small, floating algae—phytoplankton—are usually the only photosynthetic organisms found. This zone, which extends down to the limits of light penetration, is the habitat, for example, of smallmouth bass, bluegills, and, in colder waters, of trout.

The deepwater *profundal zone,* extending down from the limnetic zone, has no plant life. Its principal occupants are bacteria and fungi that decompose the organic debris filtering down from the overlying water.

As we noted in Chapter 2, water increases in density as its temperature drops until it reaches 4°C. As a consequence, a lake during the summer will consist of an

46–27
On the left, a great blue heron at the edge of a lake. The water is covered with duckweed. On the right, a pond with water lilies.

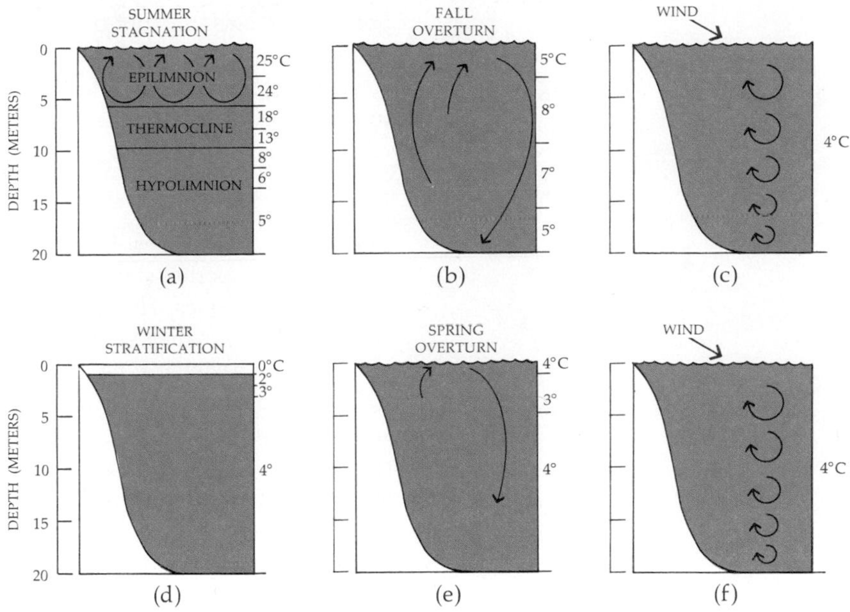

46–28
Seasonal cycle of temperature changes in a temperate-zone lake. Water, like air, increases in density as it cools, reaching a maximum density at 4°C. In the summer, the top layer of water, called the epilimnion, becomes warmer than the lower layers and therefore remains on the surface. Only the water in this warm, oxygen-rich layer circulates. In the middle layer, which is called the thermocline, there is an abrupt drop in temperature. Since the thermocline water does not mix, it cuts off oxygen from the third layer, the hypolimnion, producing summer stagnation (a). *In the fall, the temperature of the epilimnion drops until it is the same as that of the hypolimnion. The warmer water of the thermocline then rises to the surface, producing the fall overturn* (b). *Aided by the fall winds, all the water of the lake begins to circulate* (c), *and oxygen is returned to the depths. As the surface water cools below 4°C, it expands, becoming lighter, and remains on the surface; in many areas, it freezes. The result is winter stratification* (d). *In spring, as ice melts and the water on the surface warms to 4°C, it sinks to the bottom, producing the spring overturn* (e), *after which the water again circulates freely* (f).

upper layer of water that has been heated by the sun and a lower layer of denser, cold water. In winter, the coldest, least dense water (below 4°C) will remain on top, and the slightly warmer water will sink to the bottom. This stratification effectively seals off the depths from any contact with the air. Also, nutrients derived from the organic debris that collects on the lake bottom cannot rise to the surface. Consequently, during the summer and winter, the supply of oxygen in the profundal zone and of nutrients in the limnetic zone may run short, imposing special stresses on the inhabitants of these zones. In the spring and fall, when the entire body of water approaches the same temperature, mixing occurs, redistributing oxygen and nutrients and often causing "blooms" of phytoplankton to appear.

Rivers and Streams

Rivers and streams are characterized by continuously moving water. They may begin as outlets of ponds or lakes, as runoffs from melting ice or snow, or they may arise from springs (flows of groundwater emerging from bedrock).

The character of life in a stream is determined to a large extent by the swiftness of the current, which characteristically changes as a stream moves downward from its source and, fed by tributaries, increases in volume and decreases in speed. In swift streams, most organisms live in the riffles, or shallows, where small photosynthetic algae and mosses cling to rock surfaces. Many insects, both adult and immature forms, live on the underside of rocks and gravel in the riffles. For those small organisms that can survive the swiftly moving current, there is an abundance of oxygen and of nutrients, such as leaves, needles, and twigs, which fall from the river banks and are swept along by the flowing waters.

As the stream travels along its course, the riffles are often interrupted by quieter pools, where organic materials may collect and be decomposed. Few plants can gain footholds on the shifting bottoms of stream pools, but some invertebrates, such as dragonflies and water striders, are typically found in or about the pools.

Some organisms, notably trout, move back and forth between the riffles and the pools.

As the streams broaden and become slower, they begin to take on the characteristics of lakes and ponds.

THE OCEANS

The oceans cover almost three-fourths of the surface of the earth. Life extends to their deepest portions, but photosynthesizing organisms are restricted to the upper, lighted zones. The sea has an average depth of more than 3 kilometers and, except for a very small surface fraction, is dark and cold. Most of it, therefore, is inhabited by bacteria, fungi, and animals, rather than by plants.

The sea absorbs light readily. Even in clear water, less than 40 percent of the sunlight reaches a depth of 1 meter, and less than 1 percent of the sunlight that reaches the surface penetrates below 50 meters. Red, orange, and yellow wavelengths of the light are absorbed first, so that only the shorter wavelengths, specifically blue and green, penetrate deeply. Thus, below depths of a few meters, only those photosynthesizing organisms capable of utilizing the short wavelengths of light can grow.

There are two main divisions of life in the open ocean: *pelagic* (free-floating) and *benthic* (bottom-dwelling). A major component of the pelagic division is plankton. It is composed of photosynthetic algae (phytoplankton), intermingled with small shrimp and other crustaceans and the eggs and larval forms of many fish and invertebrates (zooplankton). These planktonic forms provide food for the fish and other animals that are also part of the pelagic division. The benthic division contains the bottom dwellers, the sessile animals, such as sponges, sea anemones, and clams, and many mobile animals, such as worms, starfish, snails, crustaceans, and deepwater fish. A variety of fungi and bacteria also inhabit the benthic zone, subsisting on the sparse but steady accumulation of debris drifting from the more populated levels of the ocean.

46–29
Waves breaking at Cape Kiwanda, on the coast of Oregon.

The major ocean currents, which are produced by a combination of winds and the earth's rotation, profoundly affect life in the oceans and alter the climate along the ocean coasts. These patterns of water circulation—clockwise in the northern hemisphere and counterclockwise in the southern hemisphere—move currents of warm water north and south from the equator. One such current, the Gulf Stream, warms a portion of the eastern coast of North America and the western shores of Europe, and another warms the eastern coast of South America. The same patterns of circulation bring cold waters to the western coasts of North and South America. Where the winds move the water continuously away from the shores, as off the coasts of Portugal and Peru, cold water rich in nutrients is brought to the surface in upwellings. Such areas traditionally have highly profitable fishing industries.

Despite the fact that oceans cover three times more surface area of the planet than does the land, the total productivity of the open ocean—as measured by the amount of carbon converted to organic compounds by photosynthesis—is only about one-third as great. In fact, the open ocean is only slightly more productive per square meter than the desert, presumably because of the low concentration of minerals in the areas of ocean where light penetrates and photosynthetic organisms can survive.

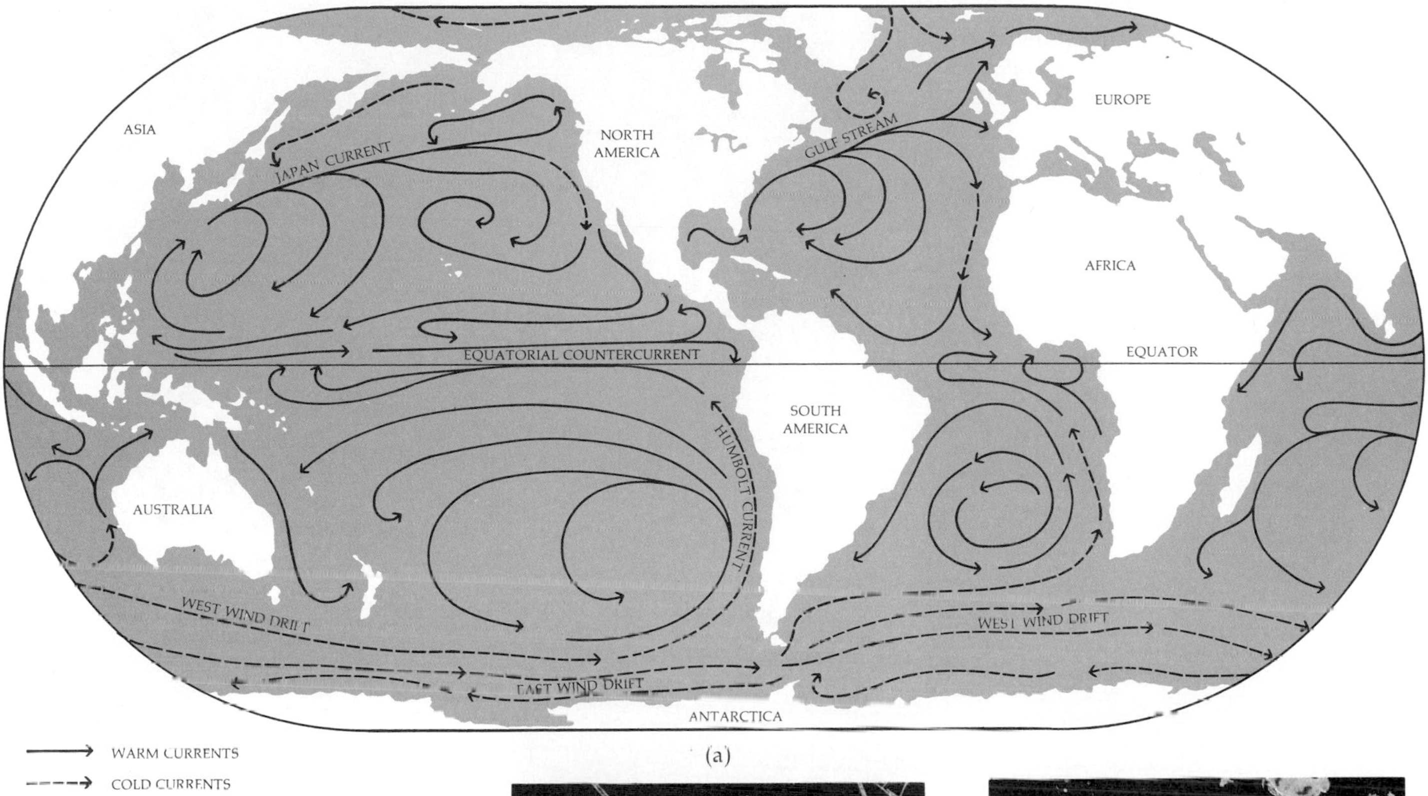

ASIA
NORTH AMERICA
EUROPE
JAPAN CURRENT
GULF STREAM
AFRICA
EQUATORIAL COUNTERCURRENT
EQUATOR
SOUTH AMERICA
HUMBOLDT CURRENT
AUSTRALIA
WEST WIND DRIFT
WEST WIND DRIFT
EAST WIND DRIFT
ANTARCTICA

(a)

→ WARM CURRENTS
- - -→ COLD CURRENTS

(b) (c)

(d) (e)

46–30
(a) *The major currents of the ocean have profound effects on climate. Because of the warming effects of the Gulf Stream, Europe is milder in temperature than is North America at similar latitudes. The eastern coast of South America is warmed by water from the equator, and the Humboldt Current brings cooler weather to the western coast of South America.* (b) *A sample of living phytoplankton showing several species of diatoms and dinoflagellates.* (c) *Zooplankton, the primary consumers in the sea, eat phytoplankton. Plankton are the main component of the pelagic division of the sea.* (d) *Starfishes eating soft coral and* (e) *an amphipod, both photographed at McMurdo Sound, Antarctica.*

46-31
A reef in the New Hebrides, islands in the Pacific, east of Australia. Staghorn coral is in the foreground.

The Coral Reef

The coral reef is the most diverse of all marine communities. The reef structure itself is formed by colonial coelenterates. Each polyp in the colony secretes its own calcium-containing skeleton, which then becomes part of the reef. The photosynthetic activity of the reef is carried out almost entirely by symbiotic algae living within the corals. Carbon, oxygen, and dissolved minerals flow over the reef as a result of the movement of waves and the ocean currents. The reef furnishes both food and shelter for other sea animals, including numerous species of reef fishes and a tremendous variety of invertebrates, such as sponges, sea urchins, polychaetes, and crustaceans. The coelenterates and algae that form the reef can grow only in warm, well-lighted surface water, where the temperature seldom falls below 21°C.

The longest reef in the world is the Great Barrier Reef of Australia, which extends some 2,000 kilometers. Other reefs are found throughout tropical waters and as far north as Bermuda, which is warmed by the Gulf Stream.

The Seashore

The edges of the continents extend 10 to 20 kilometers out into the sea. Along these edges, known as the continental shelves, nutrients are washed out from the land, and life is much denser than in the open seas. In temperate latitudes at the edge of the sea, where the primary producers are rockweeds (such as *Ficus*) and kelps (*Laminaria*), net primary productivity is as high as anywhere else in the biosphere. Sessile animals, such as sponges and corals, are found all over the ocean bottom, but they are abundant only in areas that are close to the shores. They are especially numerous in warmer waters. Predators, such as mollusks, echinoderms, crustaceans, and many kinds of fish, roam over the bottoms of the continental shelves. Eel grasses, turtle grasses, and seaweeds provide shelter for many animals and increase the supply of oxygen. Snails, slugs, and worms crawl over the surfaces of the plants and algae, eating off the encrusted growth.

Seashores—where the sea and the land join—are of three general types along most of the shores of the temperate zones: rocky, sandy, and muddy. The organisms that live on rocky coasts, like those that live in the riffles of fast-moving streams, have special adaptations for clinging to rocks. The algae have strong holdfasts. The starfish of the rocky coasts lies spread-eagled on the rocks, clinging with its suction cups. The abalone holds tight with its well-developed muscular foot. Mussels secrete coarse, ropelike strands that anchor them to rocky surfaces.

The organisms of the rocky coast face the additional problem of the rising and falling tides that threaten them alternately with too much and too little water. Life along the rocky coasts is highly stratified. The supratidal zone, which is flooded only occasionally by high tides, is a zone of dark algal and lichen growth. The intertidal zone, submerged and exposed daily, is characterized by rockweed (a brown alga), often intermixed with Irish moss (a red alga) and other seaweeds. Animal life includes barnacles, oysters, blue mussels, limpets, and periwinkles. The sublittoral zone, exposed only occasionally, contains forests of the large brown alga *Laminaria*, sea squirts, starfish, and various other invertebrates. Kelp beds in shallow subtidal zones have been shown to be some of the most productive areas of the biosphere. The characteristic stratification is due, in part, to gradients of light, temperature, and wave actions, partly to competitive interactions, and also to predation by herbivores, such as sea urchins.

Sandy beaches have fewer bottom dwellers because of the constantly shifting sands. Clams, numerous ghost crabs, sand fleas, lugworms, and other small invertebrates live below the surface of the sand, sometimes emerging at low tide to feed on the debris washed in and out by the tide. Along the sandy beaches, beach grasses, which spread by means of underground stems, are important for stabilizing the shifting dunes.

46–32
(a) *Sea otter in a kelp bed off the coast of California.* (b) *Tidal pools at low tide on Alaska's rocky southeast coast.* (c) *Common inhabitants of sandy tropical beaches are ghost crabs. These crabs, also known as racing crabs, run sideways at high speeds across the sand.*

(a)

(b)

(c)

(a)

(b)

46–33
(a) *A salt marsh. The grasses are* Spartina. (b) *Shore birds on a mud flat, San Francisco Bay.*

The mud flat, while not so rich or diverse in species as the rocky coast, supports a large number of organisms with many animals living not only on but also beneath its surface. A mud flat can support tens of thousands of individuals per cubic meter.

Mud flats, salt marshes, and estuaries (areas where the fresh water of streams and rivers drains into the sea) are the receiving grounds for a constant flow of nutrients drained off from the land and so are extremely rich in animal life. They serve an important role as the spawning places and nurseries for many forms of marine life.

In the tropics and subtropics (including parts of Florida, Puerto Rico, and Hawaii), mangrove forests are important tideland communities, spawning and nourishing marine organisms and exporting minerals and nutrients.

Because they are often located in prime recreational and commercial areas and because they cannot be directly exploited for agriculture or lumbering, thousands of kilometers of mud flats, salt marshes, and mangrove forests are destroyed each year as these wetlands are filled and paved and rendered sterile for human occupation. Their protection is of special importance because of their role in nurturing life in the oceans.

SUMMARY

A community is a group of organisms inhabiting a common environment and interacting with one another. These interactions involve competition, predation, and symbioses, all of which shape and affect the characteristics of the organisms that make up the community and of the community as a whole. The principal observable effect of competition is an allotment of the physical resources of the community among the populations that inhabit it. Predation and symbioses result in continuous directional selection of both members of the association.

As a result of variations in temperature and precipitation over the surface of the planet, the biosphere is divided into different types of communities, called biomes. The major terrestrial biomes include the tropical rain forests, the savannas, the deserts, the chaparrals, the grasslands, the temperate deciduous forests, the taiga, and the tundra. Life formations of fresh water include the standing waters (lakes and ponds) and running waters (rivers and streams). Marine environments comprise the oceans, the coral reefs, and the shores.

QUESTIONS

1. Distinguish among the following: ecosystem/community/biome; tundra/taiga; mutualism/commensalism; the ecological theater/the evolutionary play.

2. The zebras, in their grazing habits, would appear to be especially helpful toward other herbivores. How do you explain their seeming generosity?

3. What are the eight principal biomes? Name their principal abiotic features.

4. Name a plant and an animal associated with each biome, and describe the special adaptations of each.

5. Describe the conditions resulting in spring and fall overturn in a lake.

6. Introducing a new species into a community can have a number of possible effects. Name some of these possible consequences both to the community and to the species. What types of studies should be made before the importing of an "alien" organism? Some states and many countries have laws restricting such importations. Has your own state adopted any such laws? Are they, in your opinion, ecologically sound?

CHAPTER 47

Introduction to Ethology

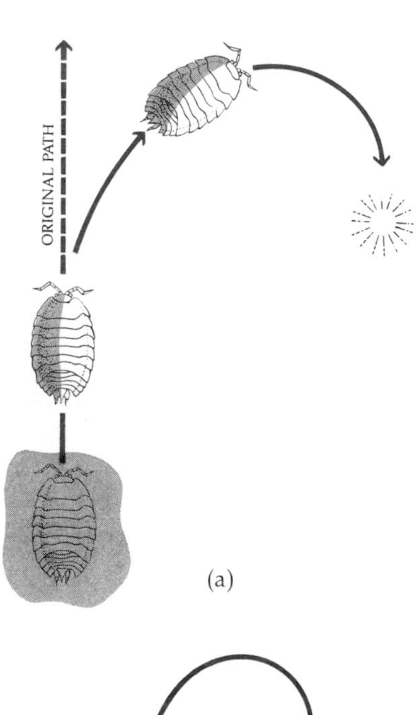

47–1
(a) *The phototactic wood louse is deflected by light from the straight path in which it normally moves. This is an example of what Jacques Loeb referred to as forced movement.* (b) *With its right eye covered, the wood louse moves constantly toward the left.*

The study of animal behavior has its historical origins in several different sources. Among the first to attempt a scientific study of behavior was physiologist Jacques Loeb. Loeb, who worked in the United States in the late nineteenth century, was a mechanist, in the tradition of Descartes (page 28). He worked on the assumption that all motions of an animal were, in his words, "determined by internal or external forces," and attempted to define these forces. Figure 47–1 shows the movement of a wood louse, which is positively phototactic (meaning that it turns toward the light). Loeb explained such movements in the following way:

> Normally the processes inducing locomotion are equal in both halves of the central nervous system, and the tension of the symmetrical muscles being equal, the animal moves as in a straight line. If, however, the velocity of chemical reactions in one side of the body, e.g., in one eye of the insect, is increased, the physiological symmetry of both sides of the brain and, as a consequence, the equality of tension of the symmetrical muscles no longer exists. The animal will thus be compelled to change the direction of its motion. . . . The conduct of animals consists of forced movements.

The intellectual descendants of Loeb are the neurophysiologists who are attempting, for example, to relate changes in behavior to changes in the nervous systems of invertebrates such as the leech or the sea slug (page 768). Although the oversimplified concepts of "forced motions" have long since been abandoned, the neurophysiologists seek, like Loeb, to find the physical and chemical foundations of behavior.

A second group of students of animal behavior comprises the so-called behaviorists, of whom B. F. Skinner, an experimental psychologist at Harvard University, is the best known. Skinner has trained hundreds of individual animals, largely rats and pigeons, to carry out a variety of unfamiliar tasks. Skinner has been interested in studying general principles of behavior, especially as they apply to learning. His work may be considered analogous to that of Mendel, who used the garden pea, or of Morgan, who worked exclusively with *Drosophila,* in order to develop general principles of genetics. On the basis of his studies, Skinner has worked out theories of learning that are now widely applied in teaching, particularly in programmed instruction; the chief principle involved is that of immediate reward, or positive reinforcement.

A third school of behavioral studies had its origins in the work of the Austrian zoologist Konrad Lorenz and his former student, Niko Tinbergen. Lorenz, Tinbergen, and their followers call themselves *ethologists*—from the Greek word *ethos,* meaning "character" or "custom"—and ethology has come to mean the study of behavior characteristic of species, with emphasis on the evolution and adaptive value of such behavior.

In this chapter and the next, we shall be following largely in the path of the ethologists, examining behavior as it takes place in the ecological theater.

BEHAVIOR IN PROKARYOTES

Even organisms as seemingly simple as bacteria respond to stimuli. Some photosynthetic bacteria, for example, respond to light, moving toward it. One such species is *Halobacterium halobium* (see page 186), the salt-loving bacterium with rhodopsin in its cell membrane. It is tempting to speculate that rhodopsin is the light-detector for *H. halobium,* just as it is for the light receptors of the vertebrate eye.

Escherichia coli and some other bacteria show chemotactic responses, responding either positively or negatively to the presence of chemicals in their environment by changing the direction of flagellar rotation. When presented simultaneously with an attractant and a repellent, they respond to whichever one is present in the more effective concentration. Such a response indicates that these bacteria have a processing mechanism that compares opposing signals from different chemoreceptors, sums these signals up, and then communicates the sum to the flagella. The processing mechanism has been shown to involve proteins in the cell membranes.

Recently, scientists at the University of California, working with the bacterium *Salmonella,* showed that the cells could move up a very slight concentration gradient of a nutrient amino acid (serine)—an obviously adaptive response. How are the bacteria able to distinguish an area of lower concentration from an area of only very slightly higher concentration? There seem to be only two possible answers to this question. One is that they sense the gradient instantaneously by comparing the concentration of serine at their "head" with that at their "tail." This possibility has been rejected because the bacteria are so small that the differences in concentration from head to tail would be too minute and too variable to be detected. The second hypothesis, supported by preliminary data, is that they are able to compare concentrations at points many cell lengths apart. In other words, according to this hypothesis, information about the concentration at one time and place is stored by the bacteria and later compared with information received at another time and place—a primitive form of memory.

PROTIST BEHAVIOR PATTERNS

Slightly more complex patterns of behavior are seen among the protists, at the eukaryotic level of organization. Amoebas, for example, are also chemotactic. When something edible, an algal cell or a fellow protozoan, is in the vicinity of an amoeba, the amoeba can sense it at some distance, at least the length of its own body away. It then sends out a pseudopodium that is shaped quite specifically for its intended victim. A fine pincerlike projection will be formed for a small, quiet morsel; a much stronger and more massive pseudopodium will reach for a large

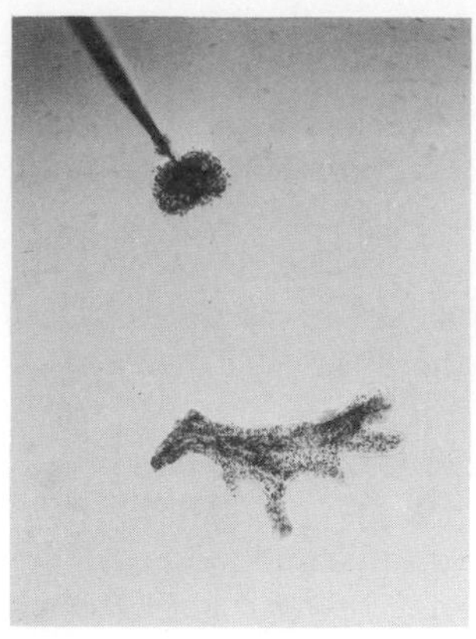
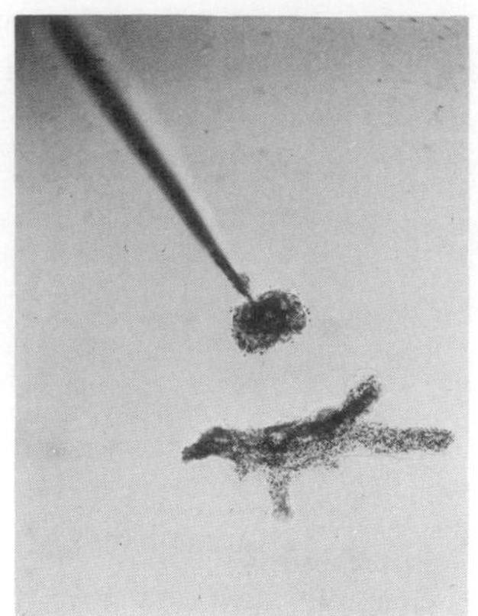
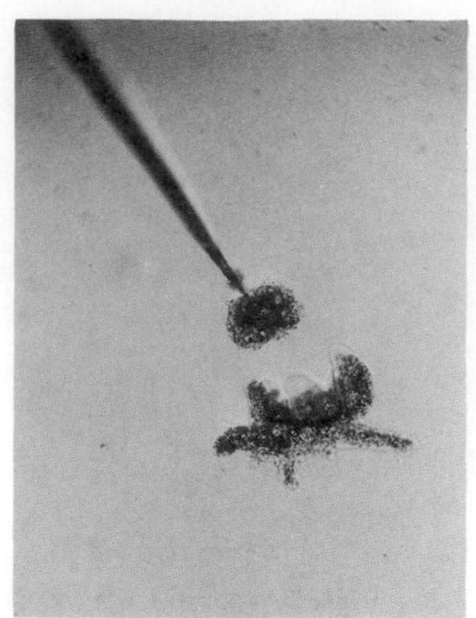
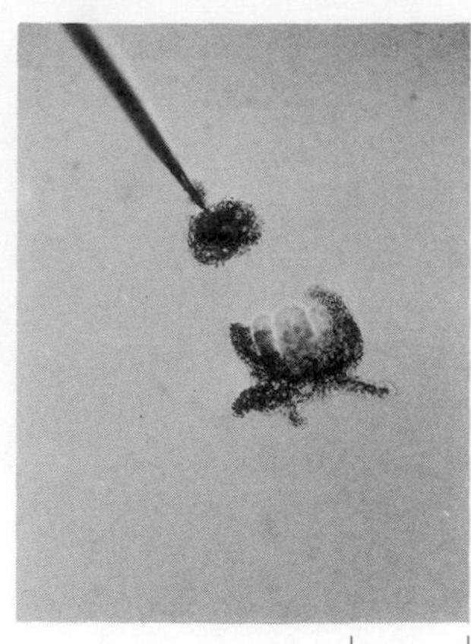

47–2
An amoeba pursues its prey. The initial stimulus produced by the "prey"—a fragment of Hydra *tentacle—induces pseudopodia from the amoeba and causes the amoeba to move toward the* Hydra *fragment. As the piece of tentacle is moved away, the amoeba moves after it and will remain in pursuit so long as the prey is close enough to continue to produce the stimulus.*

ciliate or vigorously moving object. If the intended victim moves away, the amoeba will remain in pursuit so long as it is close enough to the prey to receive stimuli from it (Figure 47–2).

Amoebas also react to light. If a bright pinpoint of light is focused on the advancing pseudopodium of an amoeba, the pseudopodium withdraws. If the entire body of the amoeba is exposed to bright light, the cell contracts suddenly and will even extrude any half-eaten food. However, if it cannot escape from the light, the amoeba, after a few moments' hesitation, will resume its normal activities. This latter response, by which the stimulus comes to be ignored and a previous behavior pattern restored, is known as *habituation*. It is a very important component of animal behavior. If our own cells were not capable of habituation, we would be continuously responding to and distracted by, for example, the touch of our clothing or background noises.

Avoidance in Paramecium

Paramecium shows behavior that appears more complex than that of amoebas but is just as simple in principle. These protists are responsive to a variety of stimuli, including very subtle changes in temperature and chemistry that we can detect only by finely calibrated instruments. But though the detection methods are highly sensitive, the responses are fixed (stereotyped).

As a *Paramecium* swims, the strongly beating cilia around the cytostome create currents that constantly bring a sample of the water to the oral zone, which seems to be the testing area. In this way, the *Paramecium* continuously explores the environment that lies ahead, shifting and sampling (Figure 47–3).

In general, *Paramecium* responds to changes in its environment by avoidance reactions. If it receives a negative stimulus, it turns away. If it turns, it always turns in the same direction, to its left, which is the side away from the cytostome; this occurs because of the relaxation of the beat of the body cilia. It does not matter which side the stimulus comes from. If the microscopist takes a blunt needle and jabs a *Paramecium* on the aboral side, it still turns toward that side; and once it has turned, it will continue in this new direction indefinitely.

As shown in Figure 47–4, if the negative stimulus is powerful, such as a poisonous chemical, the *Paramecium* will stop short, reverse its ciliary beat and back up, turn toward the aboral side about 30°, and then start forward again, testing. If

47-3
A Paramecium *samples a drop of India ink. Currents created by the strongly beating cilia draw particles of the ink toward the animal's cytostome.*

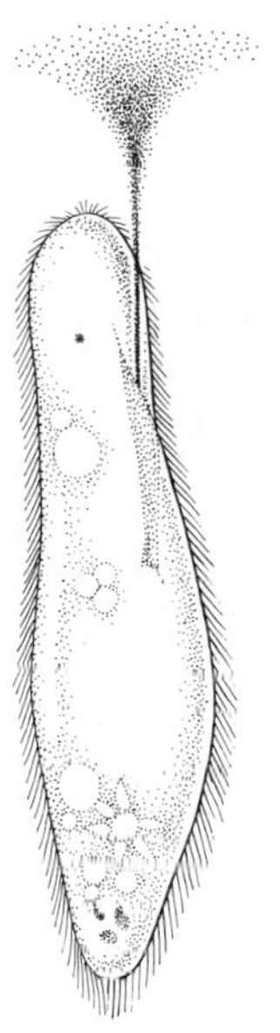

necessary, it will repeat this performance. Under a strong negative stimulus, it will turn a full 360° and will continue to turn until an avenue of escape is found. It can find its way around solid objects in the same way.

The end result of this, of course, is about the same as if the *Paramecium* were "attracted" to a favorable situation. For example, *Paramecium* is extremely temperature-sensitive. If you place the cells in a culture dish that can be made warmer at one end than another, you will find, by trying different combinations of temperatures, that they will tend to congregate in water that is about 27°C (80°F). If you place them on a slide with water that is below 27°C and then warm one little portion of it gently (you can do this by placing a hot needle on the cover slip), the cells will all gather in this warm spot. They reach it by chance as they move about through the water. But once they get into it, they literally cannot get out again, since as soon as they reach the edge of it, they receive a sample of cooler water, which will turn them right back again. Similarly, on a slide that contains water that is too hot, they will trap themselves in an area that has been cooled slightly, even by only a few degrees.

They have also been shown to congregate in the same way in a spot that is slightly acidic, by avoiding the neutral or alkaline areas. Bacteria, which are a primary food of *Paramecium*, create a slightly acidic environment by their metabolism, and indeed, the *Paramecium* itself creates a slightly acidic environment by giving off carbon dioxide; so their avoidance of neutral or alkaline areas enhances the tendency of these protists to group together and also to gather where there is food. This tendency is strengthened further by the fact that they are more likely to cling to some other object—such as the leaf or stalk of a plant, a pile of debris, or even a shred of filter paper—if they are in a slightly acidic medium.

As described here, the behavior of a *Paramecium*, helpless to escape from a drop of warm water or forced to cling by the presence of carbon dioxide, seems very unlifelike. Yet, as one watches the cell negotiating the changes in its environment, its actions seem purposeful. Both of these impressions are true. The individual *Paramecium* has no choice as to its behavior, but the behavior of these protists, in general, is a result of millions of years of choice, the choice by natural selection.

47-4
Avoidance behavior in Paramecium. *The colored area at the top of the figure represents a drop of a toxic substance, and the arrows indicate the direction of movement. The* Paramecium *tests the substance, backs up, turns 30°, and starts forward again in a new direction. Many protists are capable of this sort of simple behavior.*

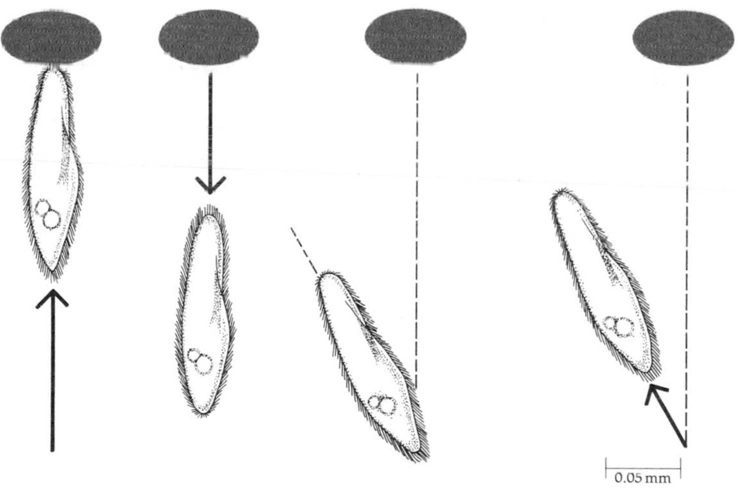

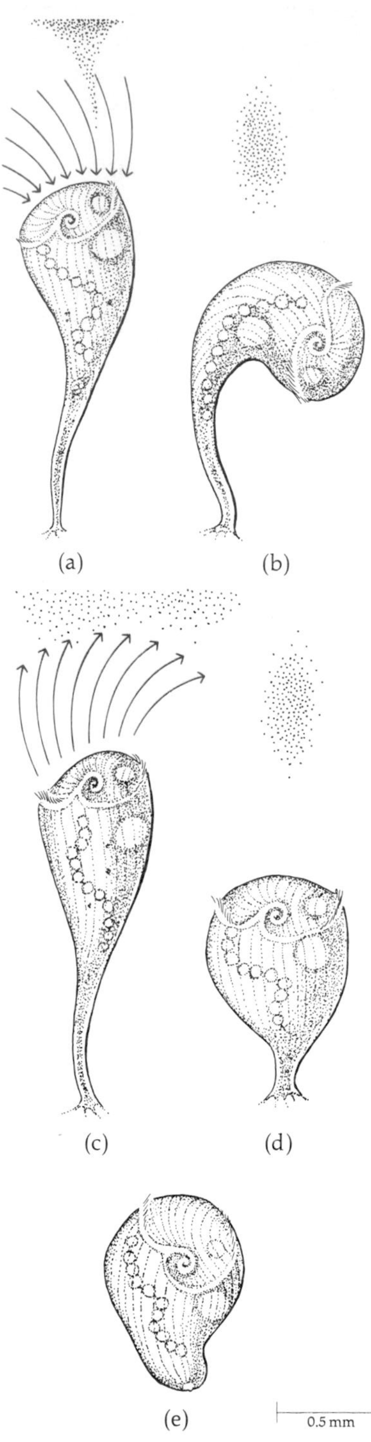

47–5
Response of Stentor *to a cloud of ink particles.*

Learning in Stentor

If a *Stentor* is poked with a needle, it gives a standard avoidance response of the sort described for *Paramecium*—bending, always toward the aboral side, regardless of where the poke comes from. If the *Stentor* is poked repeatedly, it will contract its body and then extend it again. If the poking continues, the *Stentor* will probably contract again, perhaps several times, but then either it will ignore the jabbings and continue feeding, which it cannot do in the contracted position, or it will detach itself and swim away. Whether or not it leaves appears to depend at least to some extent on the richness of the food supply.

Similarly, if a *Stentor* is annoyed—by a cloud of ink particles, for instance—it will at first bend, perhaps repeatedly, to the aboral side (Figure 47–5). If the offensive stimulus continues, the *Stentor* will reverse its cilia and start blowing the particles away. If bending and blowing are not successful, it contracts and waits a minute before it stretches forth again. Once it has contracted, it does not bend or blow again, but it may draw back several times and reach out again to sample the water before it withdraws. Again, the length of time that it tolerates the noxious stimulus depends on whether or not its site of attachment had previously proved to be a good feeding area. In short, these reactions of *Stentor* are based, at least to some degree, on previous experience and so represent a simple form of learning, which is defined, most simply, as the process by which experience produces adaptive change in the behavior of the individual.

Thus, even at the one-celled level it is possible to discern many of the basic patterns of behavior seen throughout the animal kingdom and how they serve the organism in its relationship to its environment.

INSTINCT AND ITS MODIFICATION

Instinct

Some of the liveliest research and controversy in the field of behavior has centered around the word "instinct." If you think of some of the common ways in which this word is used, you will begin to glimpse the problems associated with it. What would you give as examples of instinct? Mother love? Dog-and-cat fights? Fear of snakes? What do you mean when you say "I took an instinctive dislike to that person"?

As used by biologists in the nineteenth century, the word "instinct" meant an inherited pattern of behavior. This is still a useful working definition. On the other hand, we know now that all that can be inherited is that which is contained in the nucleic acid of the egg and the sperm. On the basis of this knowledge, it is possible to describe quite precisely how certain traits, such as sickle cell anemia, are inherited. The problem of inheritance of other, multifactored characteristics, such as the shape of a jawbone, is far more complicated. And the unknown links between, for example, the three-day ritual of a bower bird adorning his nuptial palace and a particular sequence of purines and pyrimidines are many indeed. Modern ethologists tend to refer to *innate behavior patterns* rather than to use the older, more ambiguous term of instinct. As a useful working definition, innate behavior patterns can be defined as patterns of behavior that emerge in a roughly complete form the first time an animal responds to certain biologically relevant stimuli.

GENETIC STUDIES OF BEHAVIOR

One approach to the study of genetically programmed behavior is the crossbreeding of species with different behavior patterns. Lovebirds of one species, for instance, cut long strips of paper, bark, or leaves with which to build their nests. The bird then grasps one end of the strip in its bill, tucks the strip into the feathers of the lower back, and so carries it off to its nest. Another species cuts materials in the same way but carries the strips back to the nest in their bills. Hybrids between the two species always try to tuck the nesting material into their feathers but invariably fail to do so. Sometimes the tucking movements are incomplete, or sometimes the bird will hold the strip by the middle or the end, will fail to let go at the right moment, or will tuck it into some inappropriate part of the body. The F_1 *hybrids, which were studied over a three-year period, never entirely gave up trying the tucking movements, but by the end of the period, the futile activities related to this method of carrying had decreased considerably.*

The chief obstacle to the wider application of studies of this type is that, since behavioral differences between species that are closely enough related to interbreed are usually not so clear-cut as in this example, interpretation of the results is difficult.

Other genetic experiments involve selective breeding for particular traits, as in the experiments on bristle number in Drosophila *(page 789). In laboratory experiments, rats were divided into three groups: "bright," "intermediate," and "dull," on the basis of their capacity to run a standard laboratory maze made up of blind ends and detours. The brightest of the bright were bred together, and the dullest of the dull. After seven generations, two groups were produced that were far brighter or duller on an average than the original bright or dull group, and between which there was almost no overlap. However, when the rats were placed in a maze based on visual clues, the "bright" rats performed no better than the "dull" ones. The experimenter had been selecting for the ability to run a certain kind of maze, not for intelligence, or even for general maze-running ability.*

(a)

(b)

Instinctive behavior in lovebirds. (a) *Bird with a strip of paper it has just cut for nesting material.* (b) *Tucking a strip of paper into its flank feathers, the bird is getting ready to fly back to its partially completed nest. Lovebirds are small parrots.*

Various investigators, including Seymour Benzer and his associates at the California Institute of Technology, are studying behavioral mutants in Drosophila. *A number of single-gene mutations have been found that affect a fruit fly's response to light, or rate of wing beat, or capacity to respond to particular stimuli. A single mutation such as this may have multiple physical effects: For instance, the mutation that slows the wing beat also results in yellow color in the exoskeleton. It may also have far-reaching behavioral consequences; the slowing of the wing beat so alters the ritual courtship dance that females refuse to mate with the mutant male. The goal of these studies—as yet unrealized—is to trace the physical links between a particular genetic defect and an observed aberration in behavior.*

47-6

A male bowerbird in his bower. The male builds the bower, which is the focus of his territory, and mating almost always takes place in it. The bower may be pulled down and rebuilt several times a year.

In studying such patterns, ethologists seek to determine the stimuli, pinpoint the responses, and identify the adaptive value of the behavioral pattern. For instance, if you saw a female tick drop from a bush onto a warm stone and ram her proboscis into it, thereby killing herself, it would not seem she had exhibited adaptive behavior. Perhaps, however, if you found out that the stone had recently been vacated by a hot, perspiring human being, you might piece together the reasons for the tick's "suicide." The female tick, like the mosquito and certain other insects, requires a blood meal before laying her eggs. After mating, she climbs to a branch where she remains for months or even years, until a warm-blooded animal passes by. She senses its presence by one stimulus alone: the smell of butyric acid, a characteristic animal odor. Her biting is determined by a second stimulus: the temperature of a mammalian body surface. Thus two simple responses to two simple stimuli—odor and temperature—trigger an adaptive behavior pattern.

Dissection of Instinct

As another example, why does the mother hen, long a symbol of maternal concern, behave the way she does, sheltering her brood and literally risking her life to defend it against aggressors? Experiments with the turkey hen have shown that she responds innately to one stimulus only: the peeping of the chick. It is no biological accident that a chick peeps from the moment it is hatched "as if its life depended on it." If a baby chick is placed under a glass jar where the mother can see it struggling to escape, she will pay no attention to it. On the other hand, if she can hear it peep, she will charge to its rescue, seeking it frantically. Any small object that peeps will be accepted by a hen that is a new mother—for instance, a small stuffed polecat mounted on wheels and equipped with a loudspeaker. Any small object that does not peep will be destroyed if it comes near the nest.

Behavior of the Stickleback

One of the favorite subjects of study by a number of ethologists (Niko Tinbergen, in particular) is the three-spined stickleback, a small, brightly colored freshwater fish that thrives in aquariums and will readily carry out its normal cycles of behavior in captivity. In the springtime, in their natural habitats, male sticklebacks isolate themselves from the school in which they ordinarily live, move to shallow water, and select a territory in which to breed. This behavior, it has been found, is a response to hormonal changes brought on by the increasingly long days (or short nights) of springtime. Arrival at the nesting site triggers another sequence of activ-

47-7

Colors elicit behavioral responses in many animal species. In experiments by Niko Tinbergen with sticklebacks, the crude models, colored red, caused much more positive reactions in both male and female sticklebacks (aggressiveness in the males and attraction in the females) than did the exact replica of a male, which was not colored. Tinbergen shared a Nobel Prize in 1973 for his work in animal behavior.

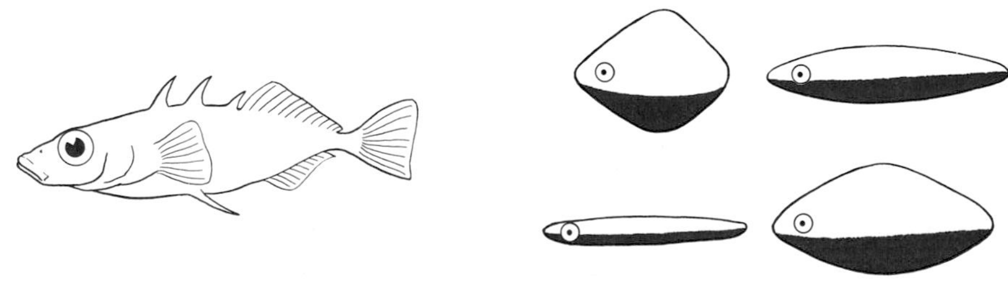

47-8
Sticklebacks, small freshwater fish, have elaborate mating behavior. The male at breeding time, in response to increasing periods of sunlight, changes from dull brown to the radiant colors shown here. He builds a nest and begins to court females, zigging toward them and zagging away from them. A female ready to lay eggs responds by displaying her swollen belly. The male leads her down to the tunnel-like nest. He prods her tail and, in response, she lays her eggs and swims off. He follows her through, fertilizes the eggs, and stays to tend the brood. If either partner fails in any step of this quite elaborate ritual, no young are produced.

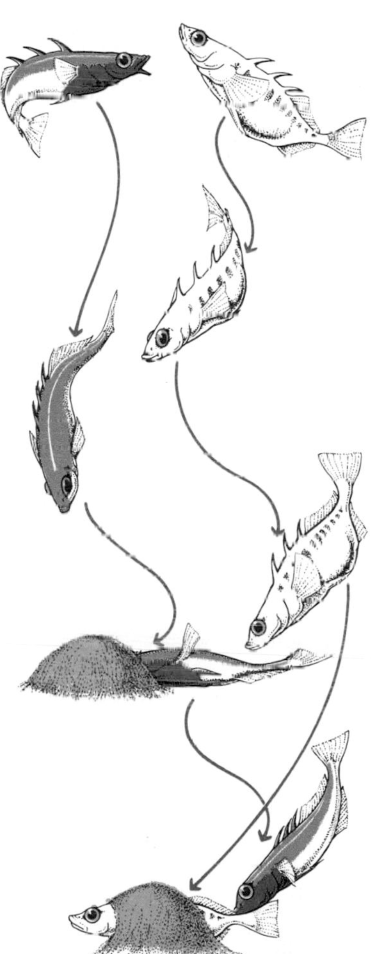

ities. How does the stickleback know he has arrived? Apparently the environmental clue in this case is the presence of certain types of vegetation.

At the time the male begins seeking his breeding site, he changes color—his eyes become a shining blue; his back, instead of dull brownish, becomes greenish; and his underbelly becomes red. Once the site is chosen, the now radiant male begins to build the nest, picking up mouthfuls of sand from the bottom, which he carries and drops some 10 to 15 centimeters away. When he has made a shallow pit, he gathers nest material, usually threads of algae, and presses it down into the pit. Occasionally, he creeps over the material with slow quivering movements, secreting a sticky substance from the kidney that glues the threads together. In this way, in a matter of a few days, a green mound is completed. The male then bores a tunnel in it by wriggling through the mound.

As nesting activities proceed, the colors of the male become even more intense. The underside becomes a brighter red, and the black color cells on the back contract to minute dots, exposing a deeper layer of shiny whitish-blue. In this brilliant nuptial costume, the male parades up and down near his nest. During this period, he is extremely aggressive toward other male sticklebacks, driving them away from the nesting site.

The female sticklebacks, meanwhile, have become swollen with unfertilized eggs. They cruise about in schools, passing through the territories of the males. Each male, if ready to receive a female, reacts to the group by performing a zigzag dance consisting of a series of leaps that take him alternately toward the females and away from them. Most of the females retreat, but one that is ready to lay her eggs turns toward the male, her swollen belly in full view, and begins to follow him as he moves away. The male then heads straight for the nest and thrusts his head inside. He lies flat on his side with his colorful flank on display. The female wriggles past him, enters the nest, and remains in it, her head protruding from one end and her tail from the other. The male now begins to prod the base of her tail with his snout. Quickly, the female spawns. Once her eggs are laid, she slips through the nest and swims away. The male then enters the nest and fertilizes the eggs. He also repairs any damage that has been done to the nest.

The male may lead as many as three or four females to his nest to deposit eggs, which he fertilizes. The presence of the eggs sets in motion a new pattern of behavior. The male ceases to court passing females and devotes himself to parenthood, ventilating the eggs by fanning them and protecting the young once they have hatched. This long and complex sequence of activities can be performed to absolute perfection by a male stickleback raised in complete isolation—that is, it is an innate behavior pattern.

Innate Releasing Mechanism

Studies of stereotyped behavior patterns such as these have led to the concept of the *innate releasing mechanism* (IRM). This concept explains instinctive behavior by assuming that within each animal there are a number of inborn movement forms, which ethologists call *fixed action patterns.* The fixed action patterns of a given species are as specific and constant as its anatomical characteristics. Every specific action pattern remains blocked until the releasing stimulus activates the IRM, which causes the specific act to be performed. A given releaser is related to a given IRM, according to this theory, as a key is to a lock. Such a releaser may be a visual clue, a

47–9
The male stickleback is attracted to a female model because of its swollen abdomen.

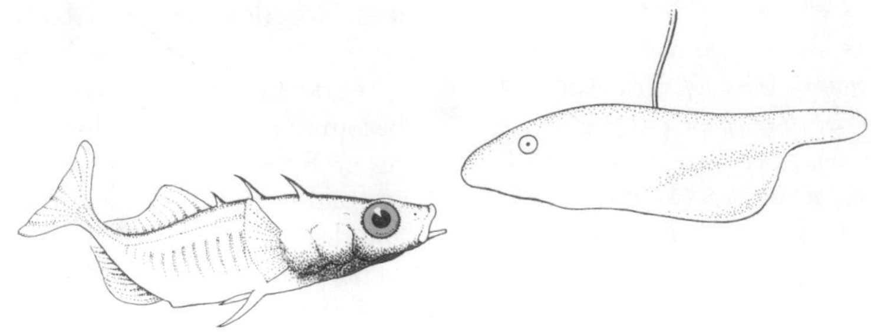

scent, or an activity performed by another animal. For example, to test the IRM that elicits aggressiveness between male fish, Tinbergen constructed a series of models (Figure 47–7), by which he was able to show that a male stickleback reacts much more aggressively to a very crude model of a male with a red underbelly than to an exact replica without the red coloration.

Females similarly are attracted to males because of their red underbellies, and males are attracted to females on the basis of their swollen abdomens (Figure 47–9). A male will court a female stickleback with a swollen abdomen, a crude model of such a female, and also similarly gravid females of other small species; but he will only lead a female to the nest if she responds appropriately to his dance. Her response is the innate releasing mechanism for the second phase of his courtship, leading her to the nest. It is his prodding that causes her to release the eggs; without such prodding she will not spawn, while gentle prodding with a glass rod by the experimenter can induce egg laying. Finally, if the eggs are removed, the male will continue courting, but if they are present, their presence releases parental behavior.

In a stream system on the Olympic Peninsula, courting sticklebacks are jet black. This population of jet black fish evolved some 8,000 years ago, it is estimated, in a glacial lake under the pressures of predation by black fish of another species. Since that time, presumably, the females have been courted by black males. However, one of those black female sticklebacks, when given a choice between a black and a red-bellied male, chooses the red-bellied male five out of six times. Even after 8,000 years, the genetic program persists.

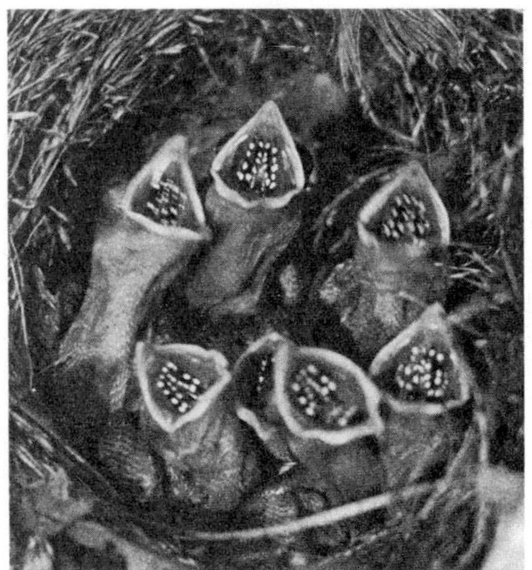

47–10
Bearded-tit nestlings have very conspicuous mouth markings. The innate releasing mechanism of the parent is adapted to these species-specific markings, and only those young are fed that have the right markings.

Releasers in Bird Behavior

Thrushes

Tinbergen has made similar studies of behavior in birds. Young thrushes, like the young of many other birds, thrust their heads upwards toward their parents to beg for food, revealing the orange-colored lining of their mouths. The baby that gapes the most vigorously is fed first; this ensures that each healthy baby will be fed in turn, since the hungrier the baby is, the harder it gapes. A weak fledgling, however, will sometimes not be able to gape vigorously enough to elicit the feeding response in the parents and so will die. This, too, though cruel by anthropomorphic standards, is adaptive.

When the nestlings first hatch, they are blind, and their gaping response is released by a light touch on the nest or a puff of air. A heavy touch makes the baby

47-11
Response of baby thrushes to Tinbergen's models of parents. Fledglings gape toward a model if it is within certain size limits, is above them, and shows movement. The key stimulus is the size of the head in proportion to the body. To be recognized as a "head," the protrusion must be about one-third the size of the "body." On the left, the young birds are gaping at the smaller of the two protrusions; on the right, they have selected the larger protrusion.

birds crouch. When they are about one week old, the gaping response is released by a visual stimulus, but the birds still gape straight up. At the age of about 10 days, the gaping is directed toward the head of the parent. Tinbergen tested various models of parents to see what the fledglings would gape toward. He found that for a "parent" to release the response, it must be within certain limits of size, be above the nestling, and show movement. The response is directed toward the "head," a protrusion on the object's outline; the protrusion is recognized as a "head" if it is about one-third the size of the "body" (Figure 47-11).

Herring Gulls

The herring gull does not gape to be fed but pecks at the tip of its parent's bill. The parent then regurgitates the food onto the ground, picks some of it up in its bill tip, and holds it near the young, which pecks the food from its parent's bill. The herring gull has a yellow bill with a red patch near the end of the lower mandible. The importance of the various characteristics of the adult's bill as stimuli eliciting pecking has also been studied by using a series of models (Figure 47-12). Neither the color of the bill nor the color of the head was found to influence the response. The model obtaining the highest number of responses was long and thin, pointing downward, with a red patch contrasting with the bill near the tip. In fact, the greatest response of all was obtained to a red knitting needle with three white bands near the tip.

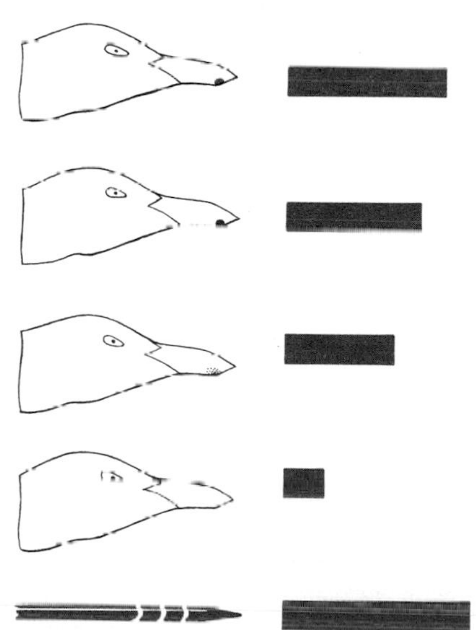

47-12
Herring-gull models and a knitting needle are used to determine the character of the stimulus that elicits pecking responses from the young gulls. The bar graph on the right indicates the relative number of responses obtained to each model. Note that the knitting needle elicited the highest number of responses.

The knitting needle is an example of what is known as a *supernormal stimulus,* of which a number have been found. Tinbergen, using dummy eggs made of wood, showed that the female herring gull, when faced with a choice between an egg the size of her usual eggs and one much larger, will select the larger—even one so large that she cannot sit on it without falling off.

Herring gulls, like some other species of birds, will retrieve eggs that have rolled out of the nest, rolling them back in again with their bills. G. P. Baerends of the University of Groningen in the Netherlands, experimenting with dummy eggs, found that the "egg" that stimulated this egg-retrieving response most effectively was pale green, $1\frac{1}{2}$ times the normal diameter, and covered with many small, dark speckles. (The normal egg is olive brown with both small and large speckles.)

47-13
A bedraggled hedge sparrow feeding a young cuckoo. Only the English species of cuckoo lays its eggs in the nests of other birds. In North America similar behavior is seen in the cowbird. The hedge sparrow's own egg, visible at the bottom left, was ejected from the nest by the cuckoo when it hatched.

Cuckoos

Responses to supernormal stimuli are not usually harmful under natural conditions; the likelihood of a herring-gull chick encountering a knitting needle is very slight. In some cases, however, such responses have been exploited. The English cuckoo, for instance, lays its eggs in the nests of other birds. The baby cuckoo has a larger head than the young of the songbirds whose nests it usually parasitizes, and since it has a wider gape, it is fed in preference to other fledglings.

The behavior of the cuckoo and its foster parent also emphasizes both the extreme adaptiveness of genetically determined behavior and its limitations. The cuckoo almost immediately upon hatching makes movements that result in the ejection of any other occupants, whether unhatched eggs or fledglings, from the nest. If a baby bird is pushed out of the nest and left clinging to the rim, the parent, which under other circumstances will die to preserve her young, will not push the baby bird back into the nest, nor will it feed it so long as it is outside. But the parent will continue to feed the murderous young cuckoo, now gaping alone.

MODIFICATIONS OF BEHAVIOR

All patterns of behavior, even those that seem relatively complete on their first appearance, depend not only on environmental cues but also on the normal physiological development of the animal. Also, many innate patterns can be altered to some extent by experience, and many types of behavior, even some that we think of as "instinctive"—such as mating behavior or mother love—may develop only if the animal has had particular experiences. Thus, although such experiences are conventionally categorized as forms of learning, "instinct" and "learning" are no longer considered as alternatives but as part of a continuum.

One of the simplest forms of behavior modification is habituation, which we discussed in our description of the behavior of protists at the beginning of the chapter.

Another simple and familiar type of learning is known as *associative learning,* in which one object comes to be linked, through experience, with another one. If you keep pets in your home, you will be able to cite many examples of associative learning, such as goldfish coming to the corner of the aquarium to be fed as you walk toward the tank or your dog becoming excited at the sight of a leash. The famous conditioned reflex of Pavlov is a form of associative learning.

Another way in which behavior is modified by experience is categorized as *trial-and-error learning.* A newly hatched chick will peck at any spots that contrast with the background; as it grows older, it will learn which of these spots represent edible objects. Also, its pecking accuracy will improve. When young chicks first peck at grain, only about 15 percent of the pecks lead to obtaining a morsel worth swallowing. Accuracy improves with practice and also with maturity. Chicks that have been kept in the dark and fed by hand for a number of days will peck more accurately than younger chicks, but they will not peck as accurately as chicks their own age that have had pecking experience.

Closely related to these types of learning is the development of *discrimination.* For example, young blackbirds gape in response to auditory stimulation. Initially, a very wide range of noises will elicit gaping, but this range becomes narrower and narrower until eventually, under natural circumstances, the young bird will gape only when it hears the call notes of its own mother. Similarly, in the human baby that has begun to smile, there is a period during which two dots on a white oval background evoke the smile. Later, other features, such as eyebrows and mouth, become necessary. Later still, masks become ineffective and only a human face will elicit smiling. By the middle of the first year, the baby will smile only when it sees the face of its mother and father and perhaps one or two other individuals whom it knows.

Imitation is often a component of learning. We think of a cat killing a mouse or a cheetah hunting a gazelle as "instinctive behavior," but the appropriate behavior patterns do not develop in kittens or cubs reared in isolation. Imitation is perhaps most obvious in the higher primates ("monkey see, monkey do"), but there are also many examples in birds.

During the 1950s, for example, householders in England became surprised and sometimes angry witnesses of the development by imitation of an entirely new type of behavior in birds. Titmice began opening milk bottles left on doorsteps and drinking off the cream. The tearing of paper and paperlike bark is a common natural trait of this species, so one can guess that the first opened bottle may have been a happy accident. At least 11 species of birds took up the practice. The milk bottles were usually attacked within a few minutes after they were left at the door, and it was reported that troops of birds were following milk carts down the street.

In some cases, learning by imitation appears to be part of a bird's essential education, as it is in the higher mammals. Many types of birds engage in a practice known as *mobbing,* in which all the members of the flock descend upon a fox or owl or some other natural predator. This often drives the enemy away or, at the very least, reveals its presence and so interferes with its hunting. In many species, young birds do not recognize their enemies instinctively and only learn to identify them as a result of the mobbing activity of other birds. They themselves then join in the mobbing and become part of the educational system.

In the case of mobbing, there is no intention on the part of the older birds to teach the younger ones. Mobbing takes place even if young birds are not present, just as warning cries are given by birds even if no companions are present for them to warn.

Another category of behavior modification is *insight learning*—in other words, "seeing through" a problem. If a dog sees a dish of food through a screened hole in a short piece of fence, it may scratch helplessly at the screen or sit down and cry. Eventually, by running back and forth, it may find its way around the fence. Some puppies will immediately run around the fence the next time they are confronted with the same problem. If the fence is then made longer, however, some of these puppies will run only as far as the length of the original fence, while others will follow the fence to its end. The latter group has demonstrated insight learning. They understand not just the specific problem but the nature of the problem.

Similarly, if a monkey is taught that it can find a food reward under the lid marked *X* when the other two lids are marked *O*, or under a yellow lid when the other two lids are red, it may then look first under a round lid when the other two are square. This is also an example of insight learning; members of only a very few species are able to solve the oddity problem, as this is called.

Imprinting

Birds, such as swans, chickens, and turkeys, that are physiologically mature enough to leave the nest soon after they are hatched follow the first moving object they see. For ducklings, the effective objects can range from a matchbox on a string to a walking man. Once one of these newly hatched birds has become attached to a particular object, it will follow only that object. Under natural circumstances, of course, this object will be a parent, and it is this following response that keeps the young birds close behind and well within the protective range of the parent until the end of their juvenile period, when the response is lost.

The learning pattern involved in this act of recognition is called *imprinting*, and it differs from most other types of learning in that the period in which it can occur is very limited. Studies of the newly hatched mallard duckling, using a mechanically operated decoy, showed that imprinting was most effective between 12 and 16 hours after hatching. After 24 hours, 80 percent of the ducklings failed to follow moving objects; and after 30 hours, all the ducklings avoided them. Similarly, chicks will not follow a moving object when they are only a few hours old nor when they are several days old but only during the intervening period. It is believed that the end of the imprinting period is marked by the development of fear responses, which seem to arise in normally reared birds at about this time.

The more difficult it is for the young bird to follow the moving object, the more firmly it becomes imprinted, according to experiments made with ducklings and a moving decoy. The ducklings that had to waddle rapidly, climb over obstacles, and even endure mild punishment by electric shock to follow the decoy were more persistent on subsequent following trials than those that were able to keep up with the decoy more easily.

The gosling is especially susceptible to imprinting. The only way to make an incubator-hatched gosling follow other geese is to bundle it up in a bag and rush it to the barnyard as it is hatching, thus ensuring that the first object it sees is a goose.

Imprinting also seems to influence mate selection in the adult birds; Lorenz reports many examples of ducks and geese becoming sexually fixated on objects or on members of other species, including Lorenz himself. Controlled experiments with ducklings support these observations. Ducklings raised with females of another species courted females of that species when they reached maturity, while ducklings raised with females of their own species did not.

The term "imprinting" is usually applied only to phenomena associated with the following response in birds, but similar effects are seen in other animals. For instance, if a lamb is taken from its mother right after birth and bottle-raised, it will not follow other sheep when it is returned to pasture with the rest of the flock.

First impressions seem to determine the choice of a mate even in birds that do not show the following response. In order to crossbreed two strains of pigeons of different colors, breeders find that the future fathers of the crossbreeds must be raised by foster parents of the opposite species; the male pigeon will not pay court to a female that does not resemble "Mom."

Similar effects occur in parents. The parent jewel fish, a species of cichlid, tend their large brood very solicitously despite the fact that their babies are the same size as the small animals that make up their regular diet. If the first batch of eggs is taken from them and replaced by the eggs of another fish, they will tend these babies equally well, but when they have a second brood of their own, they will eat them. The parents apparently recognize their young by imprinting, in the way that

Nina Leen, Time-Life Picture Agency

47–14

Many species of precocial birds (birds that are able to walk as soon as they are born) will follow the first moving object they see after hatching and will continue to show this following response as they mature. The phenomenon is known as imprinting. The birds shown here are goslings, and the object of their affection is animal behaviorist Konrad Lorenz.

a young bird will become imprinted to the first object or animal presented to it. At any time in the reproductive cycle, substitutions of eggs or young can be made if they are of the same species as the first brood. Analogously, if a female lovebird is given foster fledglings of a different color than her own to raise from eggs, she will subsequently raise only nestlings of the species of her foster young.

Song Learning in Birds

About 150 days after hatching, a juvenile male white-crowned sparrow begins to utter a tentative, twittering call that has only a vague resemblance to the full song of the mature male. During the next 50 days or so, the juvenile's song gradually becomes more complex and sophisticated until by the time the bird is 200 days old, he is singing a full song. His song not only identifies him as a white-crowned sparrow but also often provides information as to the specific locality from which he comes; in other words, the song patterns differ slightly from place to place, forming characteristic dialects. These dialects can be detected by the human observer and reproduced by good whistlers such as ethologist Peter Marler of Rockefeller University.

Laboratory experiments have been undertaken by Marler and others to dissect the sequence of events in song learning in this species. Figure 47–15 shows sound spectograms of (a) a wild bird, and (b) an adult, hand-reared bird that had never heard a song of its own species. Obviously some form of learning is required for production of full song. Exposing birds to tape-recorded songs at various stages of their development produced some surprising results. If the bird is exposed to the full song of a white crowned male in the period of 10 to 50 days after hatching, he will eventually develop the same song (and even the same dialect) five to six months later. This is true even if the bird is kept in complete sound isolation after two months of age. If, however, he does not hear the song until about the time he begins to sing himself, he can never produce a full song. Exposure to the song of other birds has no effect.

If the bird is deafened after he has heard songs of his own species during the

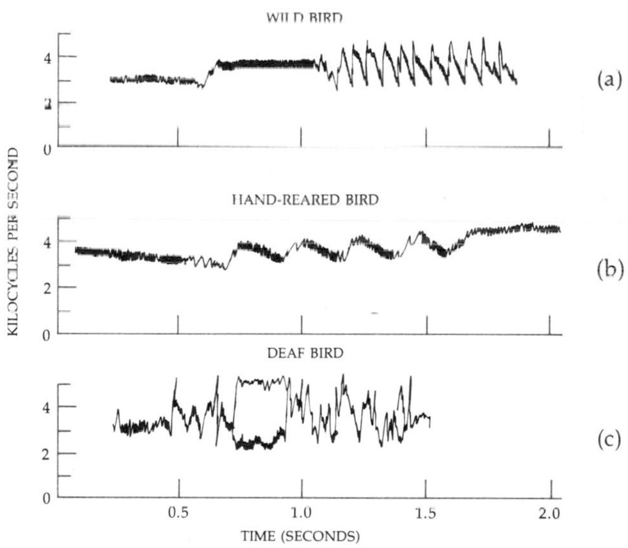

47–15
Sound spectograms for white-crowned sparrows: (a) wild; (b) reared by hand but isolated from songs of other white-crowned sparrows; (c) deafened prior to five months of age, after having heard normal white-crowned sparrow songs.

47-16
These well-trained rats have learned to work together to obtain a food reward. Each rat was placed in a different position—on the platform at the left, on the chamber floor, and in a box suspended beyond the chamber. The first rat has descended from the platform using a ladder held by the second rat. (a) *By pulling chains, the two rats draw the box closer so the third rat can join them.* (b) *The rats then step on three buttons sequentially,* (c) *opening the door to their food reward.*

(a)

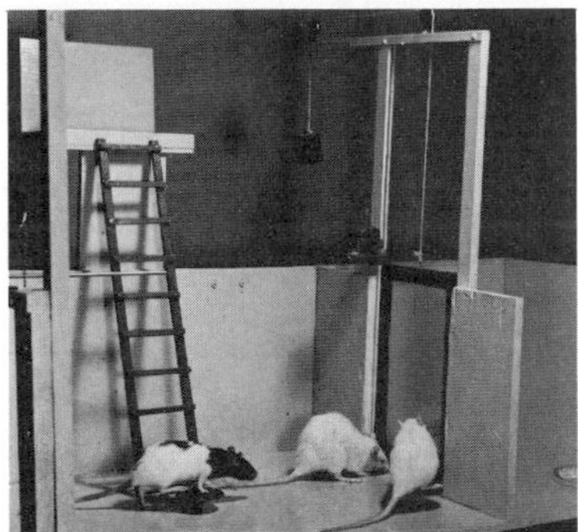

(b)

(c)

critical learning period (Figure 47-15c), the full song is not normal, though it is more complex than that of the sound-isolated bird. Apparently, then, for normal song, three requirements must be met: First, the bird must have the genetic capacity to recognize and reproduce the song; second, he must hear the song during a critical imprinting period; and third, he must be able to hear himself sing the song. Apparently as he sings, he compares his own song to the song stored in his memory during the critical period.

This study is of interest to students of behavior for two reasons. The first is that it demonstrates the complex interplay between the innate and the learned, bringing us a little closer to an understanding of these two ambiguous terms. Second, song learning in the white-crowned sparrow may provide a model for the human learning of language. Clearly there are some analogies. We have the capacity to learn to understand and to speak a complex language; this capacity is species-specific and clearly a part of our genetic endowment. Also, obviously, we learn the language we are exposed to. The question of a critical period is one which has interested psychologists for many years and is part of the reason for the excitement engendered by the discovery of a "wild child," who presumably had not been exposed to language during the normal learning period. So far, however, this question has not been resolved.

OPEN AND CLOSED PROGRAMS

As we have seen from these examples, instinctive is not the opposite of learned; what can be learned is a part of an organism's genotype just as much as an "instinct" is. Ernst Mayr suggests that we replace these overworked terms with the concepts of open and closed genetic programs. An open program is one that is subject to modification by experience during the life of the individual. Little or nothing can be inserted into a closed program. Like *K* and *r* strategies, each offers advantages under particular circumstances and as part of a number of other components of the life cycle. For example, in many types of organisms behavioral patterns associated with courtship response tend to be the result of closed programs; there is high selective value in being able to execute those responses flawlessly. An organism may have both closed and open programs. A bird, for instance, may perform the courtship ritual in a highly stereotyped way but still learn to take tops off of milk bottles.

Closed behavior patterns are more advantageous for organisms that have a short life span and therefore little opportunity for input. Closed behavior is probably also more advantageous for very small animals; there is evidence, as we mentioned in Chapter 40, that the capacity to learn with experience—that is, intelligence—is generally correlated with brain size. Finally, animals that depend largely on an open program not only require a long time in which to acquire experience but must also be protected by experienced animals during the period in which learning takes place.

SUMMARY

Ethology is the study of the behavior of animals in their natural environment, with emphasis on evolutionary factors and the proximate causes, or releasers, of the behavior.

47-17
This rhesus monkey is being raised in isolation. When she is an adult, she will probably refuse to mate. If she does become pregnant, she will not exhibit normal maternal behavior and will probably neglect and may possibly abuse her infant. In short, these "instinctive activities" are, in fact, part of an open program of behavior that depends not only on genetic programming but also on individual experience and development.

Bacteria and protists exhibit a variety of simple behavioral responses including phototaxis, chemotaxis, avoidance, and habituation. They also show a limited capacity to modify behavior on the basis of stored information.

Instinct is defined as genetically programmed behavior. Some of the programs are quite simple (the feeding strategy of the tick). Some are complex (the courtship of the stickleback). Analysis of the behavioral patterns shows that they can be broken down into a series of reactions to different stimuli, which may involve environmental clues (such as day length), hormonal changes, and recognition of specific releasers, which may take the form of particular shapes, colors, or movements. According to the ethologists, organisms possess an innate releasing mechanism (IRM), which, in response to the appropriate releaser, results in the animal's exhibiting a particular behavior. The IRM is in the genotype; the behavior is its phenotypic expression.

Learning is defined as modification of behavior as a result of experience. Various categories of learning are recognized, including associative learning, in which one stimulus comes to be associated with another; trial-and-error learning; discrimination; and imitation. Imprinting is a special form of learning that is very limited both in what can be learned and in the time interval in which the learning can take place.

A closed program of behavior is one in which responses are fixed—stereotyped—and in which little or no modification can take place. An open program is one that is variable, although it is limited, since it is dependent on other factors—such as perceptual development, hormones, neural circuits, and so forth. Closed programs are associated with organisms that have a short life span and so little opportunity for learning. Open programs are advantageous for organisms with a long life span and a long period of parental care.

QUESTIONS

1. Distinguish between the following: behaviorist/ethologist; releasing stimulus/IRM; associative learning/insight learning; imitation/imprinting.

2. Domestic dogs often turn in a small circle several times before they lie down. Can you present any evidence that this behavior is the result of a closed genetic program?

3. What are the advantages to an organism of a closed genetic program? Of an open one?

4. If a bee larva dies, some strains of honey bee, known as hygienic, uncap the cell and remove the corpse. This behavior is genetically determined; hygienic queens produce only hygienic daughters, and nonhygienic queens do not. A breeding experiment was performed in which unhygienic bees were crossed with hygienic ones. All of the F_1 generation was unhygienic. A backcross was then made between the F_1 unhygienic hybrid and the original hygienic strain. Twenty-nine colonies of bees resulted. In eight of the colonies, the bees left cells capped and did not remove dead larvae. In six colonies, workers uncapped cells and removed dead larvae. In nine colonies, workers uncapped cells but left larvae untouched. In six colonies, workers did not uncap cells but would remove larvae in cells that were uncapped by experimenters. Interpret these results in Mendelian terms.

CHAPTER 48

Social Behavior

A *society* is a group of individuals belonging to the same species and organized in a cooperative manner. It is thus something more than an aggregation of individuals. Often, similar organisms—bacteria, for instance, or *Paramecium,* or mealworms—are found gathered in the same place because of environmental conditions, such as humidity, shelter, temperature, food supply, and so on. These aggregations do not represent societies in the sense that the word is used by students of animal behavior. In a society, stimuli exchanged among members of the group hold the group together. These exchanges of stimuli, which are essentially forms of communication, result in what is defined as social behavior.

INSECT SOCIETIES

We shall begin with a description of insect societies, since they are by far the most ancient of all societies, and also, with the single exception of modern human societies, by far the most complex. *Eusocial,* or "truly social," insects include wasps, bees, and ants (all of which belong to the order Hymenoptera) and termites.

48–1
Insect societies are held together by a complex network of chemical, tactile, and visual communications. Here honey-bee workers exchange nectar and perhaps pheromones with one another.

Of all animal organizations, the insect societies are probably the best understood, in terms both of their evolution and of the interplay of forces that keep them together. As with other animals, the social insects evolved from forms that were originally solitary. In fact, eusociality has evolved on at least eight separate occasions in bees and four times among wasps. The type and degree of care provided to the eggs and larvae by the wasp or bee is coming to be used as a taxonomic characteristic for determining relationships among the many species, just as slight differences in the shape of the body or the color of the wing might be used.

STAGES OF SOCIALIZATION

Most species of bees and wasps are solitary. Among the solitary species, the female builds a small nest, lays her eggs in it, stocks it with a food supply, seals it off, and leaves it forever. She usually dies before the larvae mature.

Among subsocial or presocial species, the mother returns to feed the larvae for some period of time, and the emerging young may subsequently lay their eggs in the same nest or comb. However, the colony is not permanent (usually being destroyed over the winter), there is no division of labor, and all females are fertile.

Eusocial insects are characterized by cooperation in caring for the young and a division of labor, with sterile individuals working on behalf of reproductive ones. The honey bees are the most familiar example.

HONEY BEES

A honey-bee society usually has a population of 30,000 to 40,000 workers and one adult queen. The life span of a worker is approximately six weeks. Each worker, always a female, begins life as a fertilized egg deposited by the queen in a separate wax cell. (Drones, or male bees, develop from unfertilized eggs.) The fertilized egg hatches to produce a white, grublike larva that is fed almost continuously by the nurse workers; each larval bee eats about 1,300 meals a day. After the larva has grown until it fills the cell, a matter of about six days, the nurses cover the cell with a wax lid, sealing it in. It pupates for about 12 days, after which an adult emerges.

(a)

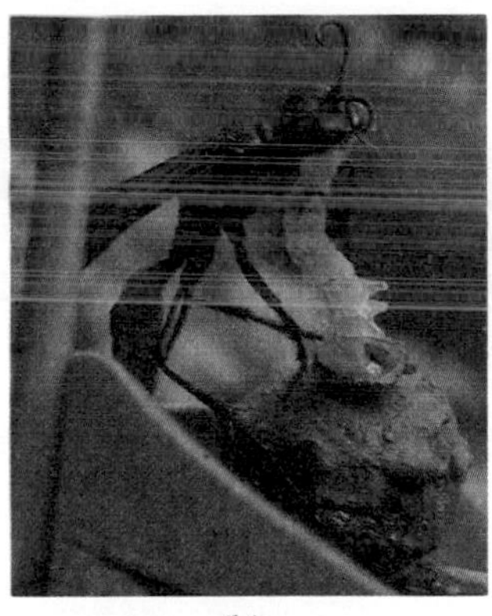

(b)

48–2
Wasps, most of which are not truly social insects, do not tend their young, but often provide for them. (a) *Wasps of the* Apanteles, *a solitary and parasitic genus, inject their eggs under the skins of caterpillars. The resultant larvae eat the internal tissues of the caterpillar, chew their way to the surface and spin the cocoons shown here. Adults emerge from these cocoons.* (b) *The potter wasp builds a graceful clay vase, lays the eggs inside, and stocks it with caterpillars for the larvae. A caterpillar, paralyzed but not dead, provides up to 40 days' food.*

48–3
Bumblebees are large, hairy bees, primarily adapted to colder climates. They are social bees, but their societies are smaller and simpler than those of the highest social bees, the honey bees. The life cycle is annual; only the fertilized queen survives the winter.

The newly emerged adult worker rests for a day or two and then begins successive phases of employment. She is first a nurse, bringing honey and pollen from storage cells to the queen, drones, and larvae. This occupation usually lasts about a week, but it may be extended or shortened, depending on the conditions of the community. Then she begins to produce wax, which is exuded from the abdomen, passed forward by the hind legs to the front legs, chewed thoroughly, and then used to enlarge the comb. During this stage of employment as a houseworking bee, she may also remove sick or dead comrades from the hive, clean emptied cells for reuse, or serve as a guard at the hive entrance. During this period, she begins to make brief trips outside, seemingly to become familiar with the immediate neighborhood of the hive. It is only in the third and final phase of her existence that the worker honey bee forages for nectar and pollen.

The Queen

The queen begins life as an egg genetically identical to that of the worker. The differences between the two depend on the substance fed the queen-to-be in the larval stage and on the pheromonal influences she, in turn, exerts upon her subjects.

Queens are raised in special cells larger than the ordinary cells and shaped somewhat like a peanut shell. Larvae in queen cells are fed special glandular secretions, known as royal jelly. Attempts have been made to identify the substance in royal jelly that confers queenhood, but so far they have not been successful.

If a hive loses its queen, workers will notice her absence very quickly and will become quite agitated. Very shortly, they begin enlarging worker cells to form emergency queen cells, and the larvae in the enlarged cells are then fed royal jelly. Any diploid larva so treated will become a queen.

The queen exerts influences on her subjects by means of pheromones, of which there appear to be several. As shown by the British entomologist C. G. Butler and his co-workers, the influence of one of the pheromones, known as queen substance, inhibits ovarian development in the worker bees and prevents them from becoming queens or producing rival queens. This pheromone passes among the workers of the hive orally. As the workers meet, they often exchange the contents of their stomachs. Studies in which queen substance has been tagged with a

(a)

(b)

48–4
Honey-bee workers. (a) *The first segment of each of the three pairs of legs (tarsi) has a patch of bristles on its inner surface. Those of the first and second pairs are pollen brushes, which gather the pollen that sticks to the bee's hairy body. On the third pair of legs, the bristles form a pollen comb that collects pollen from the brushes and the abdomen. From the comb, the pollen is forced up into the pollen basket, a concave surface fringed with hairs on the upper segment of the third pair of legs. Transfer of pollen to the pollen basket occurs in midflight. The sting is at the tip of the abdomen.* (b) *The mouthparts are fused into a sucking tube containing a tongue with which the bee obtains nectar. The antennae, attached to the head by a ball-and-socket joint, contain both touch receptors and olfactory organs. The large compound eyes cannot see red (which is black, or colorless, to them) but can see ultraviolet, which is invisible to human eyes.*

radioactive label have shown that, as a result of this activity, the pheromone travels through the hive with remarkable speed. Within only half an hour after removal of the queen, the shortage of the queen substance is already noticed, and the hive begins to grow restless. It is difficult to understand how a single queen can produce enough pheromone to influence the entire hive of as many as 30,000 to 40,000 workers, as well as tend to her stupendous egg-laying chore (commonly more than 1,000 per day). It has been suggested that, after the pheromone is passed among the workers, it is fed back to the queen in a reduced form and she need simply oxidize it to reactivate it.

Note that the words "queen" and "royal" imply, by analogy, that the queen bee's life is more desirable than that of her sisters. However, she can also be viewed, from a slightly different perspective, as an egg-laying machine held captive and operated by the workers.

Winter Organization

The honey-bee colony differs from that of subsocial bees in that it survives the winter. This means that the bees must stay warm despite the cold. Honey bees cannot fly if the temperature falls below 10°C and cannot walk if the temperature is below 7°C. Within the wintering hive, bees maintain their temperature by clustering together in a dense ball; the lower the temperature, the denser the cluster. The clustered bees produce heat by constant muscular movements of their wings, legs, and abdomens. In very cold weather, the bees on the outside of the cluster keep moving toward the center, while those in the center move to the colder outside periphery. The entire cluster moves slowly about on the combs, eating the stored honey from the combs as it moves.

New Colonies

In the spring, when the nectar supplies are at their peak, so many young are raised that the group separates into colonies. The first new colony is always founded by the old queen, who leaves the hive, taking about half of the workers with her. This helps to ensure survival of the new colony since this queen is of proven fertility. The group stays together in a swarm for a few days, gathered around the queen, after which the swarm will settle in some suitable hollow tree or other shelter found by its scouts.

As the old queen is preparing to leave the hive, the new queens are getting ready to emerge. These two events are synchronized by sound signals transmitted through the comb. As these signals are exchanged, the workers remain motionless. During this period, ovarian development begins in some of the workers, a few of which lay eggs. The unfertilized eggs develop into drones. After the old queen leaves the hive, a new young queen emerges, and any other developing queens are destroyed. The young queen then goes on her nuptial flight, exuding a pheromone (apparently also the queen substance) that entices the drones of neighboring colonies. She mates only on this one occasion (although she may mate more than once) and then returns to the hive to settle down to a life devoted to egg production.

During her nuptial flight, the queen receives enough sperm to last her entire life, which may be some five to seven years. These are stored in her spermatheca and are released, one at a time, to fertilize each egg as it is being laid. The queen usually lays unfertilized eggs only in the spring, at the time males are required to inseminate the new queens.

THE DANCING BEES

Probably the most remarkable example of communication among insects is the "language of the bees" discovered by Karl von Frisch. The honey bee returning to the hive performs a dance upon the comb, which signals quite precisely to her hivemates the location of the food she has found.

When the nectar source is near to home, the bee does a round dance, which arouses other workers to seek the food source. From the scent of the bee's body, the workers are informed at the same time about the type of flower she has found. When the food source is some distance from the hive (more than 85 meters, in the case of an Austrian strain of honey bee, Apis mellifera carnica, *with which von Frisch did his original studies), the bee does the waggle dance, in which she runs a short distance in a straight line, waggling her abdomen from side to side, and returns in a semicircle to the starting point. Then she repeats the straight run and comes back in a semicircle on the opposite side. This cycle is repeated many times. The time taken by the straight run indicates the distance of the source from the hive. If the source is 350 meters away from the hive, the worker bee performs a straight run lasting ½ second; if the source is twice as far away, the straight run takes about 1 second, or twice as long.*

More remarkable, the direction of the straight run of the waggle dance indicates the direction of the food source. The bee locates the food source for herself by reference to the position of the sun, and this is how she finds the food again. Sometimes when she returns, she dances first on a horizontal platform outside the hive. In this case, the straight run of the waggle dance will point directly toward the food source. When she reenters the hive, however, to share and store the nectar, her only dancing spot is on the vertical surface of the comb in the dark interior of the hive. Here she orients herself, not by the sun, but by gravity, and the point directly overhead takes the place of the sun in the bee's map, just as our maps are generally oriented so that North Pole is at the top. If the food source lies at an angle of 40° to the left of the sun, for example, the waggling run points 40° to the left of the vertical. The bees that follow the dancer translate the information back into a direction for orientation to the sun and start off toward the food source. A young bee that has been raised outside the hive away from other bees is able to respond to the dance the first time she is exposed to it in the hive; no learning is required.

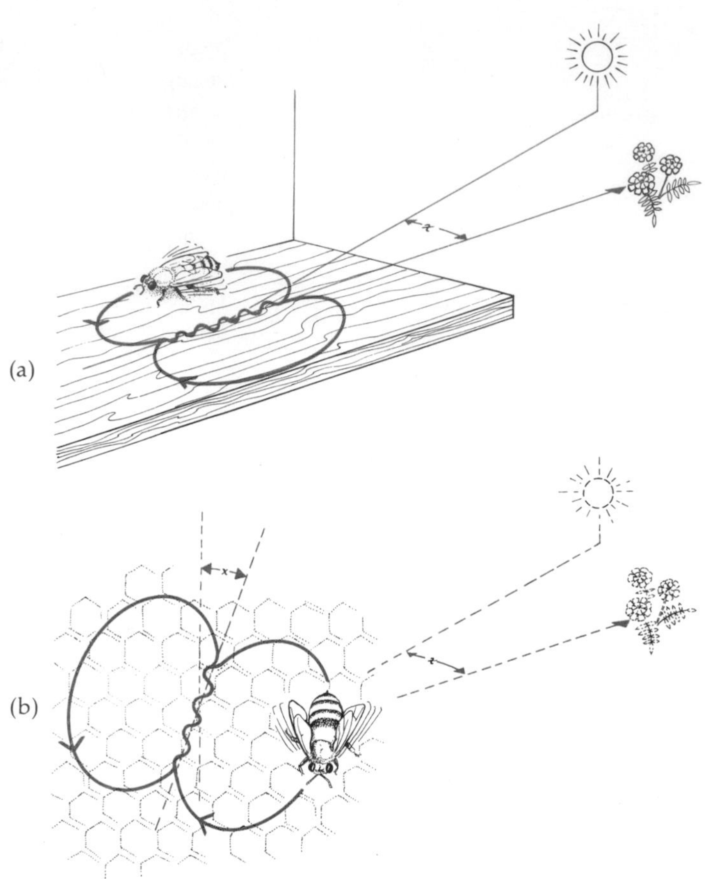

The waggle dance of the honey bee communicates to her fellow workers the location of a food source she has found. The bee locates the source with reference to the position of the sun. (a) *If she performs her dance outside the hive on a horizontal platform, the straight run of the dance will point directly toward the food source.* (b) *If she performs her dance inside the hive on the vertical wall of the comb, she orients herself by gravity and the point directly overhead takes the place of the sun. The angle* X *is the same for both dances. The distance of the food source from the hive is indicated by how fast she performs the straight-run portion of the dance pattern.*

(a)

(b)

(c)

48–5
Honey bees. (a) *Workers tending honey and pollen storage cells. The honey is made from nectar processed by special enzymes in the workers' bodies.* (b) *Workers feed the queen (top center) and lick queen substance, from her body. Queen substance, a pheromone, prevents sexual maturation in the workers.* (c) *Four pupae in different stages of development. A fully developed worker at the far left has shed her pupal skin and is ready to emerge, which she will do by gnawing through the wax cap.*

The drones' only service to the hive is their participation in the nuptial flight. Since they are unable to feed themselves, they become an increasing liability to the hive. As nectar supplies decrease in the fall, they are stung to death by their sisters or are driven out to starve.

THE EVOLUTION OF INSECT SOCIETIES

Eusociality, as we noted previously, has arisen many times among the Hymenoptera. A clue as to why this might occur was offered on page 242 in our discussion of sex determination. Among bees and other hymenopterans, a fertilized (and therefore diploid) egg develops into a female, and an unfertilized, haploid egg develops into a male. Haplo-diploidy has some unusual ramifications in terms of Darwinian fitness. Let us look first at the more familiar situation in which both sexes are diploid, as in our own species. First calculate your degree of relatedness to your mother and then to your siblings. Half your alleles came from your mother and half from your father, making your chance of having any particular gene from either one ½. You and your sibling (ignoring the sex chromosome), each received half of both of your alleles from your mother, and you each have half a chance of getting a particular allele, so your total chance is ½ times ½, or ¼. Your genetic relationship with your father is the same, so the total relatedness between two human siblings is ¼ plus ¼, or ½. Note that human offspring have the same degree of relatedness to each other as to their parents.

Now consider the hymenopterans. The total relationship is the same on the maternal side, but on the paternal side, since the father is haploid, each sister will get the full share of the father's genes. As a consequence, the relationship between sisters is not ½ but ¾ (¼ + ½). Thus, sister hymenopterans are more closely related to one another than they are to their own daughters or than they are to their mother. (The queen, however, may mate with one of several males, which presents some difficulty for the theory.)

Hence, from this perspective, the hive of workers is not laboring for the queen; they are working for one another, and the queen thus becomes merely the machine by which they produce more of their own genotype (¾ at a time).

48-6
Haplo-diploidy in honey bees (for simplicity, only one pair of chromosomes is shown). Females, including the workers, are diploid; males, the drones, are haploid. Every worker therefore receives all the chromosomes of her father and half the chromosomes of her mother. Thus, workers with the same father share, on the average, three-fourths of their genes with their sisters.

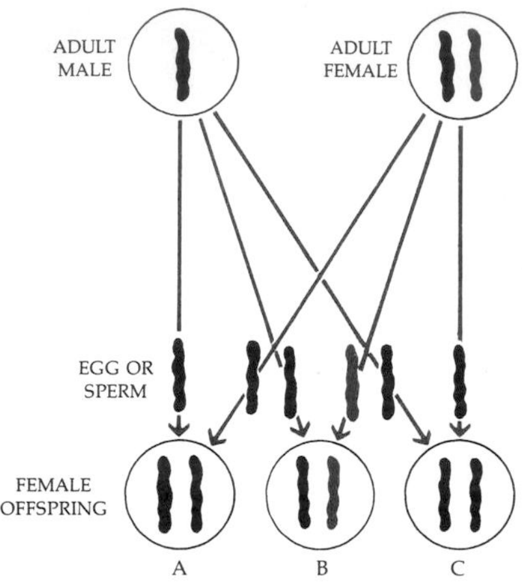

Their relationship with their brothers is quite different, however. The brother gets no genes from his father; he gets half his mother's chromosomes, and that is all. His chances of getting the same half as any given sister are ½ × ½, or 1 in 4. On this basis one would predict that the male bee would be "selfish," performing little in the way of service to the hive, and that the females would invest little energy in caring for their brothers, both of which predictions turn out to be true.

However, haplo-diploidy is not an essential component of eusociality in insects. Termites have societies as complex as those of the honey bees, and all individuals are diploid. Among termites, the sterile workers include males as well as females. Moreover, a queen bee, as we have noted, may mate more than once, thus reducing the relatedness. However, the question of the correlation between relatedness (kin selection) and altruism (self-sacrifice) is still very much alive.

VERTEBRATE SOCIETIES

Vertebrate societies range from small, often transient groups of closely related individuals, in which the social focus is on the raising of the young, to larger, permanent, quite stable groups, such as those found among some of the primates. Some of the societies—such as flocks of birds, schools of fish, and large migratory herds of herbivores—may number in the thousands, but they are not usually as large as the societies of most eusocial insects.

Homo sapiens is among the most social of the social animals, and so, from the point of view of the human observer, social living would appear to be the norm. However, in fact, living in a society has decided disadvantages for the individual organism. The most important of these is the sharing of food resources. Another is the much greater vulnerability to disease. Also a large group may attract a large predator, as when sharks attack a school of fish. Finally, an individual in a society may have to compete for a mate, thereby diminishing his or her chances of leaving viable offspring. The advantage of social living seems to be related almost entirely to avoidance of predation.

48-7
Note that zebras standing together like this can not only scratch each other's back and whisk flies from each other but can also command a 360° view of the savanna.

Grazing animals feed in herds, and birds, when threatened, tend to tighten their flock. It is almost impossible for a hawk to catch a bird within the flock, since the hawk will be subjected to a constant bombardment of other bird bodies moving at high speeds. Computer simulation models indicate (1) the likelihood of predation for any one individual depends on its distance from other individuals, and (2) as each individual decreases its distance from other individuals, herding or schooling behavior automatically results.

COMMUNICATION AMONG VERTEBRATES

As we noted at the beginning of this chapter, societies are held together by reciprocal stimuli—communications. In the social insects, these stimuli are principally (though not exclusively) in the form of pheromones, an extremely efficient system for societies where members number in the thousands, whose life span is short, and whose lives are spent mostly in darkness.

Among vertebrates, communication is often visual, involving, as we saw in the previous chapter, signals of color and shape. Such signals may take the form of behavior patterns, often quite complex, which ethologists call rituals or ceremonies.

For instance, most vertebrates—fish and birds in particular—have a distinct preference for maintaining discreet distances between themselves and other animals, whether of their own or another species. Birds lined up on a telephone wire, for instance, will be found almost equidistant from each other, as though the spaces were measured off by an invisible meterstick. Even those animals that fly in flocks or swim in schools tend not to touch one another.

One function of the mating ceremonies of these animals is to allay those highly adaptive feelings of hostility so that mating can take place. As a consequence, many of these ceremonies incorporate both aggressive and appeasing behavior. In the mating behavior of the stickleback (page 905), for example, the male's "zig" toward the female is identical to a motion of attack, and when the male "zags," the motion entices her toward the nest. Most females flee the attack motion, but the

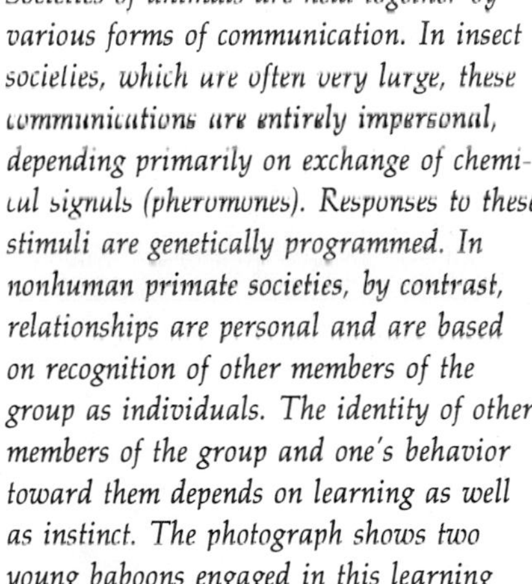

48–8
Societies of animals are held together by various forms of communication. In insect societies, which are often very large, these communications are entirely impersonal, depending primarily on exchange of chemical signals (pheromones). Responses to these stimuli are genetically programmed. In nonhuman primate societies, by contrast, relationships are personal and are based on recognition of other members of the group as individuals. The identity of other members of the group and one's behavior toward them depends on learning as well as instinct. The photograph shows two young baboons engaged in this learning process.

one whose need to lay her eggs is sufficiently great stands still, turns sideways (an appeasing gesture on her part since an antagonist would zig back), and then follows the male's zag.

Mating ceremonies also serve to maintain genetic isolation between species (page 811) and to synchronize reproductive readiness (page 652).

The Ceremony of the Graylag Goose

In mating ceremonies between birds, feelings of fear and aggression brought into play by the closeness of another individual may be handled by being redirected at either a real or an imaginary antagonist. Konrad Lorenz describes the so-called "triumph ceremony" in the graylag goose, one of his favorite research subjects and companions. As a form of greeting to his partner, the gander proceeds to attack an "enemy." This attack is performed, as is a real attack, with the head and neck pointing obliquely forward and upward and is accompanied by a raucous trumpeting. After the "enemy" is routed or defeated, the gander returns to his partner. On his return, the gander holds his head lowered and pointed forward, but instead of pointing directly at the goose, as he would at an enemy, he points obliquely past her. She comes forward to meet him, her head inverted submissively, and he cackles to her triumphantly.

This ceremony, which probably had a purely sexual origin, now serves to hold entire flocks together, and even small goslings participate in elements of the triumph ceremony. When a young male performs the ceremony with a strange young female, it usually marks the beginning of a mating bond that may last for the entire lifetime of the individuals. The ceremony is typically performed between young geese the year before mating and breeding begin, and it will continue to be performed by the partners throughout their entire lives whenever they encounter one another even after a short separation. By the intensity with which the ceremony is performed, an experienced observer can judge the length and strength of the bond between the partners.

Although few animals—except perhaps humans of ambassadorial rank—have such elaborate social rituals, higher animals of many species exhibit obligate social formalities upon meeting. Think, for example, of the greeting ceremonies between domestic dogs, and how they vary with sex, rank, and familiarity. Rituals serve the function both of allaying the anxieties of the individuals involved and of identifying them to one another as members of an "in" group.

48–9
Rituals and ceremonies are common among birds. (a) *Greeting ceremonies between storks at the nest. Pair bonding between storks characteristically endures for a lifetime, based, in part, on the repeated performance of this ritual.* (b) *Courtship display between two albatrosses of the Galapagos.* (c) *Female tern begging for a fish from a male tern. If he is ready to mate, he will give her the fish (although she may have to ask several times). If she is ready, she will eat it; otherwise she returns it. Such mating ceremonies serve to promote genetic isolation.*

(a)

(b)

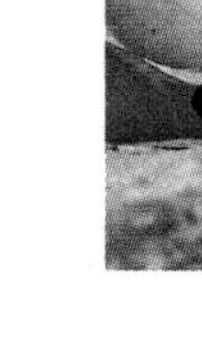

(c)

48-10
Two Thomson's gazelles, sparring strictly according to the rules.

Rituals and Aggression

Rituals also serve an important function in settling disputes. They are a way of deciding who has won without holding the actual battle. For instance, male iguanas of the Galapagos Islands fight by pushing their heads against one another; the one that drops to its belly in submission is no longer attacked. Male cichlid fish of one species first display, presenting themselves head on and then side on, with their dorsal fins erected, and then beat water at each other with their tails. If this does not bring about a decision, each grasps the other by its thick, strong lips. They then pull and push with great force until one lets go and, unharmed but defeated, swims away. Rattlesnakes, which could kill each other with a single bite, never bite when they fight but instead glide along side by side, each pushing its head against the head of the other, trying to push it to the ground in a form of Indian wrestling. Many antlered animals, such as stags of the fallow deer, which have long and vicious horns, follow an equally careful ceremony and attack only when they are facing each other, so that their antlers are used only for dueling and not for goring. These prohibitions against killing and maiming make sense biologically; the winner is not injured, and the loser lives to fight another day.

SOCIAL STRUCTURE IN VERTEBRATE SOCIETIES

Social Dominance

Vertebrate societies are often arranged in hierarchies. One type of social dominance that has been studied in some detail is the *pecking order* in chickens. A pecking order is established whenever a flock of hens is kept together over any period of time. In any one flock, one hen usually dominates all the others; she can peck any other hen without being pecked in return. A second hen can peck all hens but the first one; a third, all hens but the first two; and so on through the flock, down to the unfortunate pullet that is pecked by all and can peck none in return.

48-11
A subordinate baboon turns his buttocks toward a superior. This gesture, known as presenting and used by females to indicate their readiness to mate, is also used between males and between females to signify submission or conciliation or to beg for special favors. The superior is reassuring the subordinate with a pat on the back.

Hens that rank high in pecking order have privileges such as first chance at the food trough, the roost, and the nest boxes. As a consequence, they can usually be recognized at sight by their sleek appearance and confident demeanor. Low-ranking hens tend to look dowdy and unpreened and to hover timidly on the fringes of the group.

During the period when a pecking order is being established, frequent and sometimes bloody battles may ensue, but once rank is fixed in the group, a mere raising or lowering of the head is sufficient to acknowledge the dominance or submission of one hen in relation to another. Life then proceeds in harmony. If a number of new members are added to a flock, the entire pecking order must be reestablished, and the subsequent disorganization results in more fighting, less eating, and less tending to the essential business, from the poultry dealer's point of view, of growth and egg laying.

Pecking orders reduce the breeding population. Cocks and hens low in the pecking order copulate much less frequently than socially superior chickens. Thus the final outcome is the same as if the social structure did not exist: The stronger and otherwise superior animals eat better, sleep better, and leave the most offspring. However, because of the social hierarchy, this comes about with a minimum expenditure of lives and energy.

Territories and Territoriality

Most vertebrates stay close to their birthplaces, occupying a home range that is likely to be the same home range occupied by their parents. Even migratory birds that travel great distances are likely to return year after year to the same areas. Often these home ranges are defended, either by individuals (or more likely, mating pairs) or by groups against other individuals or groups of the same species. Areas so defended are known as *territories,* and the behavior of defending an area against intruders of the same species is known as *territoriality.*

Territoriality in some form has now been recognized in animals as widely diverse as crickets, howler monkeys, fur seals, dragonflies, red deer, beaver, prairie dogs, many types of lizards, and a large number of species of birds and fish.

A territory may be an area of a plain, a corner of a small wood, or a few meters at the bottom of a pond. Sometimes the territory consists of little more than the nest itself and the immediate area around it. For the male bitterling, a small fish, the territory is an area immediately surrounding a freshwater mussel. The bitterling admits only egg-laden females into his territory, then leads them to the mussel, where they lay their eggs within its gills. The male then injects his sperm while swimming over the siphon of the mussel.

Territoriality in Birds

Territoriality was first recognized by an English amateur naturalist and bird watcher, Eliot Howard, who observed that the spring songs of male birds served not only to court the females but also to warn other males of the same species away from the terrain that the prospective father had selected for his own. In general, a territory is established by a male. Courtship of the female, nest building, raising of the young, and often feeding are carried out within this territory. Frequently the female also participates in territory defense.

48–12
Territories come in many shapes and sizes. (a) The male Uganda kob displays on his stamping ground, which is about 15 meters in diameter and is surrounded by similar stamping grounds on which other males display. A female signifies her choice by entering one of the stamping grounds and grazing there. Only a small proportion of males possess stamping grounds, and those that do are the only ones that breed. For a fiddler crab (b), it is a burrow, from which he signals with his large claw, beckoning females and warning off other males. (c) Howler monkeys shift their territories as they move through the jungle canopy, but maintain spacing between groups by chorusing. (d) Territoriality is common among reef fish. For many species, a territory is a crevice in the coral, but for others, such as the skunk clown fish shown here, the territory is a sea anemone. The fish is covered by thick slime that partially protects it from the poison of the tentacles, but its acceptance by the anemone is chiefly a consequence of behavioral adaptation of the fish, which even mates and raises its brood among the tentacles.

(a) (b) (c) (d)

SOCIAL DOMINANCE AND SEX REVERSAL

An unusual form of social dominance is seen among one species of wrasse, Labroides dimidiatus, *a small, brilliant blue fish that inhabits the waters around Australia's Great Barrier Reef. Each group of wrasses consists of a male and a number of females; the group is territorial, with territoriality exhibited largely among males. The single male and his females are organized in a hierarchy; the largest, oldest individual is the male, who dominates all the females of the group. Next in rank is the largest, oldest female, who occupies the center of the territory with the male and is dominant over the other smaller and younger females. When the male dies, either neighboring males invade the territory, or, as occurs in the majority of cases, the ranking female becomes dominant. Within 1½ to 2 hours after the male's death, she begins to show male aggressive behavior, and within a few hours she is exerting dominance over the other females and patrolling the borders of the territory. Within a few days, the new dominant has begun male courtship and spawning behavior, and small regions of dormant testicular tissue within her ovaries have begun to develop. By the end of about two weeks, the ovaries have become testes and "she" is fully male, able to produce sperm and fertilize the eggs of the subordinate, female family members.*

Labroides dimidiatus, *a cleaner wrasse, at the Great Barrier Reef.*

In some species of birds, the usual sex roles are reversed. The female red-necked phalarope establishes and defends the territory, while the drab male defends the nest, incubates the eggs, and takes care of the young. Some birds, such as the American robin and the European swallow, tend to return to the same territory year after year, often with the same mate. Storks and night herons, however, are attached to a nesting place rather than to a mate, and if the mate of the previous year does not return the following year, he or she is readily exchanged for a new incumbent.

By virtue of territoriality, a mating pair is assured of a monopoly of food and nesting materials in the area and of a safe place to carry on all the activities associated with reproduction and care of the young. Some pairs carry out all their domestic activities within the territory. Others perform the mating and nesting activities in the territories, which are defended vigorously by the males, but do their food gathering on a nearby communal feeding ground, where the birds congregate amicably together. A third type of territory functions only for courtship and mating, as in the bower of the bowerbird or the arena of the prairie chicken. In these territories, the males prance, strut, and posture—but very rarely fight—while the females look on and eventually indicate their choice of a mate by entering his territory. Males that have not been able to secure a territory for themselves are not able to reproduce; in fact, there is evidence from studies of some territorial species, such as the Australian magpie, that adults that do not secure territories do not mature sexually.

Bird territories vary greatly in size. The golden eagle, for instance, defends a territory of about 90 square kilometers; the European robin, a territory of about 6 square kilometers; and the king penguin, a territory of only ½ square meter. Individuals of the same species often have territories of somewhat different sizes, depending partly upon the density of food and shelter in a given area, partly upon population pressure, and partly upon the aggressiveness of the individual.

Territorial Defense

Even though territorial boundaries may be invisible, they are clearly defined and recognized by the territory owner. With birds, for example, it is not the mere proximity of another bird of the same species that elicits aggression, but his presence within a part of a particular area. The territory owner patrols his territory by flying from tree to tree. He will ignore a nearby rival outside his territory, but he will fly off to attack a more distant one that has crossed the border. Animals of other species are generally ignored unless they are prey or predators.

Once an animal has taken possession of a territory, he is virtually undefeatable on it. Among territory owners, prancing, posturing, scent marking, and singing and other types of calls usually suffice to dispel intruders, which are at a great psychological disadvantage. If a male stickleback, for example, is placed in a test tube and moved into the territory of a rival male, he visibly wilts, his posture becoming less and less aggressive the farther within the territory he is transported. Similarly, a male cichlid will dart toward a rival male within his territory but as he chases the rival back into its own territory, he begins to swim more slowly, the caudal fin seemingly working harder and harder, just as if he were making his way against a current that increases in strength the farther he pushes into the other male's home ground. The fish know just where the boundaries are and, after chasing each other back and forth across them, will usually end up with each one trembling and victorious on his own side of the truce line.

48–13
An Alaskan fur seal harem; the bull is in the upper right-hand corner of the picture. The only breeding males are those adult bulls large and strong enough to secure a beach territory and round up a harem for the summer. Females arrive at the breeding grounds in June, about a month after the males have established their territories. There is a period of almost constant conflict while each bull herds as many females as he can into his own territory. In a very short time, the females give birth to pups conceived during the preceding summer. There is then a period of mating, and for the rest of the summer the bull will continue to guard the harem, neither eating nor leaving his territory for several months. In the fall, the males and females separate, the females and young going south and the bulls moving to the Gulf of Alaska. In the spring, the bull will gather another family together again, but his harem may consist of an entirely new group of females.

(a)

(b)

48–14

An encounter between two Grant's gazelles. (a) *With head upraised and muscles tensed, two bucks circle one another, displaying their horns.* (b) *One gazelle lowers his head in submission, and the contest is over.*

Similarly, the expulsion from communal territories is typically accomplished by ritual rather than by force. For example, among the red grouse of Scotland, the males crow and threaten only very early in the morning, and then only when the weather is good. This ceremony may become so threatening that weaker members of the group leave the moor. Those that leave often starve or are killed by predators. Once the early-morning contest is over, the remaining birds flock together and feed side by side for the rest of the day.

Territories and Population Regulation

As we mentioned previously, animals that do not gain possession of territories or that are excluded from the "home range" or breeding area do not produce young. As older animals die, younger ones will contend for their places, keeping the breeding population stabilized. If conditions are particularly favorable, or if the range is enlarged, more animals can gain access to the breeding community. However, many never gain territories.

The submissive acceptance of an inferior social position would seem to offer little reward in terms of "fitness" to those deprived of a territory and therefore of an opportunity to reproduce. Would not it be advisable for such individuals to fight with their last calorie of strength rather than to give up the struggle? What, after all, have they got to lose? This question is such a seductive one that it led to the formulation of the hypothesis that populations regulate their own size and that individuals that do not reproduce do so for the good of the group. We shall discuss this largely discarded hypothesis at the end of the chapter, but for the moment, note only that submission here serves the same function as it does in ritual combat—the loser survives and so has a second chance. For territorial animals, waiting may be worthwhile.

Some years ago, for example, a group of scientists undertook a study in Maine to determine whether or not the warblers nesting in the spruce trees exerted a controlling effect on the caterpillars of the spruce bud worm. They mapped the position of the singing males within an area of 16 hectares (40 acres), found 148 pairs, and began to shoot them all. After 3 weeks, they had shot 302 male warblers (and a lesser number of females), and there were still male warblers singing throughout the 16 hectares. The moral of this story, among others, is that animals that do not have territories are always ready to claim them and sometimes succeed in doing so. Thus, in the long run, waiting may be a better evolutionary bet than engaging in a probably unsuccessful combat. Territoriality is a successful strategy, in short, because it ritualizes competition and secures for the winner adequate resources for the breeding of young. The limitation of the size of the population is a side effect of these activities.

SOCIAL BEHAVIOR AMONG WOLVES

Wolves are large members of the genus *Canis*, closely related to domestic dogs. Adult males average about 43 to 45 kilograms (95 to 100 pounds, about the size of a good-sized German Shepherd). A few individuals may weigh more; an 80-kilogram male was reported from Alaska. Wolves are carnivorous and predatory; their usual prey consists of large animals, such as moose, caribou, and deer, although they sometimes eat smaller herbivores, such as rabbits and mice. Their only predators are human, but they are host to many parasites, such as ticks and tapeworms, and diseases, including rabies. The habitat of the wolf once included all of the biomes of the Northern Hemisphere, excluding only tropical rain forests and deserts. At present, the only substantial numbers of wolves in North America are found in Canada (the estimated number there is 17,000 to 28,000) and Alaska (5,000 to 10,000). The only population left in the rest of the United States, in northern Minnesota, numbers about 500 to 1,000 individuals. The state of Wisconsin placed the wolf under legal protection in 1957, but that was too late; there are no more wolves in Wisconsin. An additional 20 to 30 wolves live on Isle Royale in Lake Superior. There are also a few wolves left in upper Michigan, and there may be a small population in the Rocky Mountains. Now the Endangered Species Act of 1973 provides legal protection for all the wolves in the 48 contiguous states.

The Wolf Pack

Carnivorous animals that prey on large mammals either weigh as much as their prey or hunt in packs; wolves belong to the latter category. Wolf packs are usually made up of two to eight members, although much larger groups have been reported occasionally, as have solitary wolves. A pack in this size range seems to represent a maximally efficient hunting unit; L. David Mech, author of what is considered the definitive study of wolves, reports that only about six wolves of a large pack of 15 actually hunt at any one time. Natural selection would also work against tendencies to form very large packs since a single kill would not feed all members of such a pack. However, since packs vary greatly in size—as compared to clutch size in birds, for instance—selection pressures must also be either varied or not very strong.

Although a number of field studies of wolves have been done, no one has been able to observe a single wild wolf pack over a long enough period of time to determine how a pack originates. Most packs are made up of a breeding pair, pups, and extra adults old enough to breed (but which usually do not). The social bonds among members of a wolf pack are very strong. In both strength and expression, they closely resemble the bonds between a dog and its master, and, indeed, it seems certain that the existence of a strong bonding system among the evolutionary ancestors of the domestic dog is the reason why the dog is so uniquely domesticable.

Wolf packs travel, sleep, and hunt together. They share their kills, either by regurgitating food, by carrying home part of their prey, or by leading other members of the pack to a fresh kill. This sharing of food among adult members of a group is rare among wild animals; for example, among the primates, organized sharing of food occurs only in our own species.

48–15
Wolf courtship. It is the female that is making the advances. A breeding pair, such as this one, usually consists of the dominant male and the dominant female.

Studies made of animals in captivity indicate a strong inhibition among wolves against taking food away from an animal already in possession of it, even by wolves that are both superior in strength and hungry.

One of the strong bonds that holds the wolf pack together is the sexual bond between the breeding pair. Wolves may begin choosing mates when they are about one year old, although a wolf is not sexually mature until the age of 22 months, and there is some evidence that these pair bonds, once formed, are maintained until one member of the pair dies.

48–16
Evolution of the wolf. Other members of the genus Canis *(dogs) include the domestic dog, the coyote, the jackal, and the dingo. Creodonts ("flesh-teeth") were early carnivorous mammals, ancestors of the modern carnivores and descendants of insectivores.*

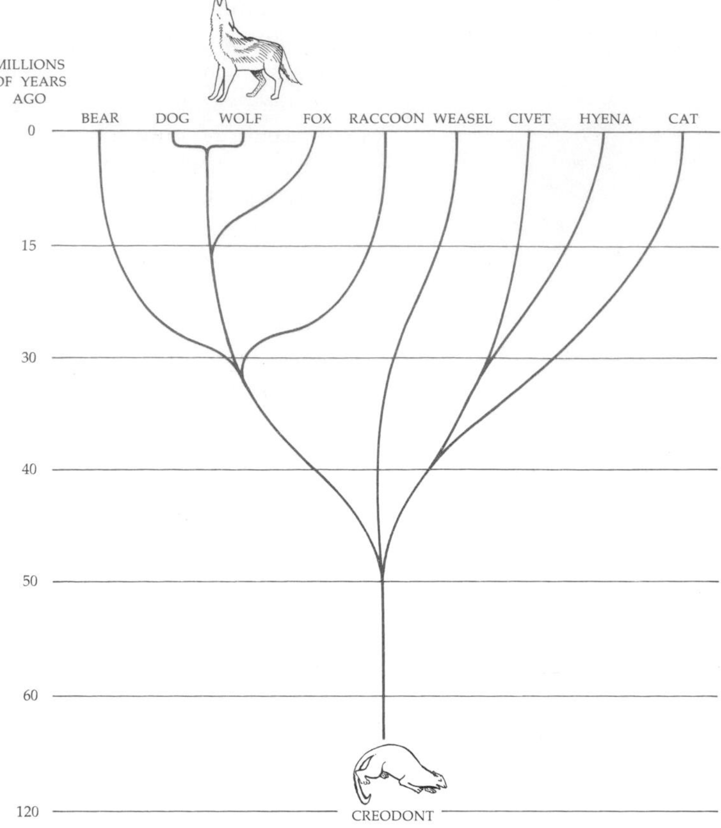

48–17
A mother wolf nursing her pups outside the den. Dens are dug a few weeks before the birth of the young, and the young remain in the dens until they are about eight weeks old. Care of the pups is shared among members of the pack. An average litter contains five or six young.

The bond between a pair of wolves is not related to the sexual availability of the female. Female wolves come into heat only once a year, and although courtship (which closely resembles the play behavior of younger animals) is carried out over a period of several weeks, copulation occurs only during the one week of estrus, the time when ovulation takes place.

It has been suggested that this bonding is reinforced by the copulatory tie, a phenomenon unique to members of the dog family. In the copulatory tie, the female's vaginal sphincter muscles constrict about the swollen penis of the male after intromission, and, as a consequence, the two animals remain attached by their genitals for 15 minutes or more following ejaculation, during which time they stand or lie rump to rump. The copulatory tie immobilizes both animals in an extremely awkward position, highly vulnerable to harassment and attack. To the casual observer it would not seem likely to increase the affection of the individuals involved toward one another. Recent studies, however, indicate that it may increase fertility.

Strong bonds are formed between mothers and pups and between the pups and other members of the pack. During the pups' first three weeks of life, they remain in the den, in close contact with one another, and are nursed by their mother. After this time, they begin to eat predigested food that they obtain by touching their mouths to the mouths of adult wolves, which stimulates the adults to regurgitate meat for the pups. Wolf pups are fed in this manner not only by their mother but by any adult member of the pack. As the pups grow older, they are played with by all members of the pack, and numerous instances have been reported of wolf mothers going off to hunt with other members of the pack while leaving another adult behind as a baby sitter. The adult that remains behind is probably fed regurgitated food, as are the nursing mothers.

Persons who have raised wolves in captivity report that it is easy to establish friendly relationships with wolf pups less than 20 weeks old, but that after this time it becomes increasingly difficult. Wild wolf pups would not normally encounter any other animals except members of their own pack during this early socialization period. Hence this behavioral characteristic keeps members of a wolf pack closely united with one another and unfriendly or hostile toward outsiders.

Social Dominance in Wolves

There are separate dominance ("pecking") orders within each pack, one for males and one for females. The dominant male usually leads the pack when it travels, has first choice of food, takes the initiative in hunting, and is most aggressive in defending the pups.

In this case, as with other social dominance orders, there is little fighting within established wolf packs. Among dominant males, the most frequent threat is a fixed stare; in more extreme situations, dominants may bare their teeth with the corners of their mouths pulled forward, raise their ears, wrinkle their brows, and emit rumbling growls.

Subordinates show submission toward dominants in much the same way as domestic dogs do, by lowering their heads and tails, flattening their ears, and, under extreme threat, rolling over on their backs. Subordinate members also often express affection for the dominant wolf, nuzzling the mouth of the pack leader, using the food-begging gesture of the pups. Sometimes all the members of a pack will surround the leader and lick his face and poke his mouth with their muzzles. This ceremony often occurs when wolves first wake up, when the group has been separated from one another for some time, and when the pack has scented prey and is ready to start on a hunt.

48–18
(a) *Dominant male (center) surrounded by pack in greeting ceremony.* (b) *A subordinant female licking the muzzle of the dominant female. This same muzzle-nuzzle gesture is used by pups begging for food.*

(a)

(b)

Usually the dominant male and the dominant female are the only mating pair. Since all the male members of the pack prefer the dominant female, and all the female members prefer the dominant male, courtship usually does not occur between other pack members. If it does, it is disrupted by the dominant animals. Hence social dominance has the effect of limiting the size of the group. According to some studies, as many as 40 percent of all adult female wolves do not breed in any given year, and fewer than half—in some studies as few as 6 percent—of the pups survive for one year. When packs are heavily hunted, the ratio of pups to adults increases, indicating that more pups are produced or survive, or both, under these conditions. In short, wolf packs tend to remain stable in size. This concept is supported, for example, by actual population counts from Isle Royale.

When the dominant male or female dies, subordinate wolves will then mate and bear pups. It has been suggested that new packs are formed at times of social unrest after a leader has died, leaving two adult males of almost equal rank.

Territoriality

Wolf packs are territorial, each pack occupying a home range of 130 square kilometers or more. A wolf or wolf pack intruding on another's home range is attacked aggressively. Two adjacent ranges may overlap, but since the areas of overlap are small and the territories very large, the chance of two packs encountering one another is not great. The boundaries of territories are scent marked with urine; scent marking probably conveys information about both the extent of the territory and the amount of time that has passed since the pack passed by, thus making it easy for packs to avoid one another.

Howling may also be related to territoriality, analogous to singing in male birds. When a pack howls, one wolf begins and the others join in, finally working up to a full chorus. A session lasts about a minute and a half and is obviously pleasurable for the wolves, as are howling sessions for domestic dogs—somewhat like a community sing. Howling can be detected by the human ear up to 6 kilometers away, and probably for a greater distance by other wolves. The howl of each wolf in the pack is different. One observer reported, in fact, that if he howled on the same note as his pet wolf, she would shift her howl by a few notes. Thus a howl could provide information to a neighboring pack not only of the presence of another wolf pack but also of the number of members in it. Howling also seems to serve as a way to regroup pack members that may have been separated during a hunt.

48–19
Howling in wolves, which communicates information about both location and individual identity, is probably related to territoriality.

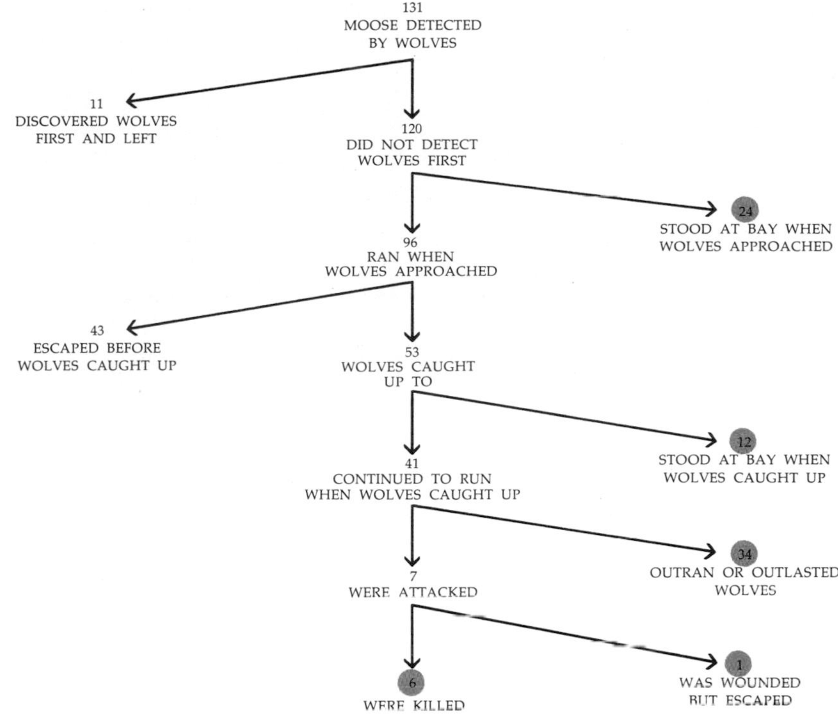

48–20
Results of interactions between all or part of a large wolf pack (15 to 16 wolves) and 131 moose on Isle Royale. The figures indicated by color are the numbers of animals actually "tested" by the wolves.

The Wolf as Predator

Wolves prey mostly on large animals, including deer, moose, caribou, elk, and bighorn sheep. Characteristically, they consume almost all of their prey, including hair, hide, and all but the largest bones. If the entire carcass cannot be eaten on the spot, the remains are either carried away or revisited.

Wolves often locate prey by scent, sometimes from a distance of more than 1½ kilometers. Staying downwind, they first stalk the animal slowly. They begin to run only when the quarry runs; apparently the movement of the prey animal serves as a stimulus. The beginning of the chase is the crucial part of the hunt; this is actually a testing period. Generally, the only prey animals that are chased for more than a few minutes are those that falter because of immaturity, old age, or illness, or calves that have been deserted by their mothers. Healthy adult animals of all prey species can easily outrun wolves. Larger animals often simply stand their ground; a full-grown moose can weigh over 450 kilograms and is well able to kill a wolf. Domestic animals such as sheep and cattle have no such advantages, however. Having evolved under human protection, they are typically unable to defend themselves either by fight or flight and so are an easy target for a wolf pack or even for a lone animal.

There have been occasional reports of wolves using cunning and complex strategies in their hunts—such as driving the prey to a predetermined ambush—but if such tactics occur at all, they would appear to be rare.

48–21
Moose under attack by a pack of wolves. The moose stood its ground for 5 minutes, after which the wolves gave up and left.

48–22
Wolves feeding on an elk carcass. The two wolves at the rear on the left are subordinate to the male on the right.

It is difficult to state with certainty whether predation by wolves serves as a form of population control for prey species. The principal limitation on the population size of large herbivores appears to be the food supply, yet, as we noted previously, only rarely does a population under natural conditions expand to the point where large numbers of individuals die from starvation. Therefore it can be argued that the prey population remains at about the size that resources can support, whether or not wolves are present. (In this regard, it is interesting that before wolves arrived on Isle Royale, almost no twin calves were ever seen among the moose population; by 1963, the twinning rate had reached 38 percent.) Others contend that the "culling" of the population by wolves benefits the prey, not only by removing ill and unfit individuals but also by holding down the population. In Yellowstone National Park, where wolves have been exterminated, great herds of elk have ravaged the vegetation and starve periodically. Clearly wolves and their prey can coexist for long periods of time, and regardless of the presence or absence of wolves, populations of large herbivores cannot expand indefinitely. Moreover, the very young, the old, the ill, and the inept will be the least likely to survive against almost any selective force.

The wolf is in the unfortunate position of being in direct competition with humans for the position of top-level carnivore. Every place the competition has been intense, as throughout the contiguous United States, the wolf has been exterminated, and, with the new hunting methods available, the problem of wiping out the remaining relatively small wolf populations in Alaska and Canada is not a difficult one should humans, pursuing Gause's principle, choose to do so.

SOCIAL ORGANIZATION AMONG BABOONS

Baboons are large, quadrupedal African monkeys with protruding, doglike muzzles. One genus (*Papio*) and four species are commonly recognized: a savanna species (*Papio anubis*), of which there are at least two, and probably more, distinct races; two forest species (the drill and the mandrill), both short-tailed baboons living in West Africa; and a desert species, found in North Africa and Arabia, *Papio hamadryas*, the hamadryas baboon.

The baboons are of particular ecological interest for two reasons. First, unlike most monkeys, they live on the ground and have done so for several million years. For this reason, differences between baboons and other monkeys can be related to some degree to their environments. Second, there is a marked difference in social organization between the savanna and forest species, on the one hand, and the desert species, on the other. Since the groups are closely related genetically, obviously sharing a common ancestor in the not-too-distant past, it is very likely that the differences can be explained ecologically rather than genetically. Both of these statements must be tentative, however; field studies of baboons have been made only since the late 1950s, and most other monkey groups have been studied inadequately or not at all.

The Genus Papio

Baboons are large, with adult males weighing about 55 kilograms. Their size is clearly related to a terrestrial existence; among the primates, only large animals, such as baboons, some of the great apes, and *Homo sapiens*, are terrestrial.

Among all baboons, adult males are very different in appearance from adult females (Figure 48–23). This phenomenon, known as sexual dimorphism, is much less pronounced among other primates and particularly other monkeys. The males have mantles and prominent canines and, most obviously, they are about twice the size of the females. Sexual dimorphism, which is much less pronounced among tree-dwelling primates, appears to be related to the increased hazards of life on the ground and to the male role of defender.

48–23
A male baboon and a female with an infant. Males are much larger than the females. Other conspicuous differences are their bright hindquarters (pale pink in the nonestrous female) and, in this species, the long mane, or mantle. The young are born black and their coats become lighter as they mature.

48–24
Baboons sleep in groups, either in tall trees, as shown here, or on ledges or steep cliffs. The size of the group and the choice of a sleeping site appear to depend more on the terrain than on the species.

Like most primates, baboons are almost exclusively vegetarian, living on leaves, fruits, flowers, and young stems, all of which they pick and consume on the spot. They also dig for roots, bulbs, and tubers. This vegetable diet is supplemented by insects and occasionally by meat, such as snakes, lizards, fledgling birds, or young gazelles, when they are encountered accidentally. There is no organized hunting for prey animals and no sharing of prey. Baboons sleep in tall trees or on cliffs, returning to a nesting site every night. Hence they must often travel to find food, sometimes for great distances across open country. Baboons have a harder time finding food than do tree-dwelling primates, which are more likely to live in the lush tropical rain forests and which sleep where they eat. The difference in size between the two sexes of baboons is probably related to the difficulties of the animals in finding food. The male is large enough to act as protector, whereas the female's smaller size reduces the food requirements of the family unit and so promotes survival of the young.

The Structure of the Band

Baboons are organized into multimale bands; those observed have ranged in size from eight to more than 185 individuals. Adult females outnumber adult males by about 2 to 1. Much of this difference can be accounted for by the much slower maturation rate of the males. Although an adult male can probably produce viable sperm by the time he is about four years old, he does not reach full size nor are his canines fully erupted until he is eight or more, and therefore is not counted as an adult male either in the social organization of the band or by the field observer peering through binoculars. Females, on the other hand, are adults by the age of two or three.

The bands are territorial, with fairly distinct home ranges, and bands rarely meet. When they do, the one farthest from the center of its customary range generally moves away, sometimes after an exchange of vocalizations, canine displays, and gestures, but more often with no visible reaction at all, except perhaps a slight display of nervousness. Bands that know each other may share large sleeping groves. Overt fighting has been reported once; it occurred between two groups trying to sleep in the same clump of trees.

When the bands move in open country, females and infants characteristically are in the middle of the group, close to adult males. Mass counterattacks by adult males, who placed themselves between the group and the attackers, have been observed to rout a leopard and disperse a large dog pack, killing or wounding several of its members.

48–25
A dominant male, threatening. The huge canines are found only in the males, indicating that they are related less to diet (male and female eating habits are the same) than to dominance and aggression. However, altercations among baboons seldom result in serious injury.

Another function of the band is the pooling of information. Hans Kummer, one of the most experienced and imaginative observers of baboons, describes a band at the beginning of a day's march as resembling a giant amoeba, stretching out and then withdrawing one pseudopod after another, as small segments begin to move off in one direction, only to rejoin the band if the others do not follow. Finally, a movement is initiated that meets with general approval, and the day's march begins.

The size of the group seems to be somewhat related to its habitat. Very small groups are correlated with small areas of vegetation, quite widely separated from one another, whereas larger groups may be found where a typical feeding site might be a clump of trees. Logically, troop size must be a compromise between the availability of food resources and the need of the troop to defend itself.

(a)

(b)

48-26
(a) *Mother and nursing infant.* (b) *A female baboon presenting, in social deference, to a mother and newborn infant because she wants to approach and touch the infant.*

Social Dominance

In all baboon species except the hamadryas, the band is organized around a dominance hierarchy of adult males. As in other dominance hierarchies, the band is maintained mostly by behavioral conventions, with only occasional fights, which rarely result in injury. Dominant males threaten subordinates by staring, yawning (with the ears laid back), raising the eyebrows, and, at close range, by grinding their teeth. Subordinate males demonstrate submissiveness by looking away, grinning, turning their backs, and presenting their hindquarters. The dominant male may mount the presenting male briefly. The gesture of presentation, also used by females in estrus as an invitation to copulate, is the most reliable index of hierarchical position. The top-ranking male presents to no other individual, the second-ranking male only to the top one, and so on down the social ladder. (Presentation as a gesture of submission or appeasement is seen among many primate species.) Adult females are subordinate to males but also have a separate dominance hierarchy, although not as clearly structured as that of the males. Adult females present to superior females.

Socially superior males mate more often than inferior males. In some groups, it was observed that a number of males copulated with females when they were not in full estrus, when the likelihood of fertilization is considerably less. But when a superior female is in full estrus, as revealed by maximum swelling and coloring of the genital area, only the highest-ranking male mates with her. Thus the highest-ranking male is the one most likely to father offspring. Dominant males also have first choice of nesting sites. However, because there is no shared food supply, dominants are not allocated a larger share of food than subordinates, as they are in some other pecking orders.

Dominant males play a leading role in territorial disputes, thus maintaining the spacing among bands. They are also the most aggressive in defending the band against intraspecific aggression. If another member of the group barks a warning or screams in fear, the dominant male steps forth and inspects the danger.

Social Bonds

As among almost all higher primates, grooming is a prominent form of social behavior among baboons. It involves a brisk parting of the fur of another individual with the fingers of both hands and the picking off of small particles, including insects; these items are often swallowed. Dominant males, females in estrus, and females with infants are groomed most frequently, but all members of the group receive some grooming attention. The function of grooming is clearly not only hygienic but also a continual reinforcement of social bonds.

Infants are another clear social bond. The infant, which is all black and therefore very conspicuous for its first few weeks of life (perhaps so it can be guarded more zealously), usually clings to its mother, who may support it with her arm. With the mother's permission, it may be touched or even held briefly by other members of the troop. The infant or infant-mother pair are clearly attractive to other members of the group, who tend to cluster around, especially when the infant is newborn. Mothers with infants are groomed frequently, particularly by other females, and dominant males attend them closely as the troop moves. An infant that loses its mother will be adopted, sometimes by a childless female, but often by a young male.

48–27
As among all social primates (except Homo sapiens), *grooming is one of the primary bonds among group members. The amount of grooming an individual receives is closely related to his or her position in the social hierarchy. Here an adult female grooms a dominant male. Being groomed is not only (obviously) pleasurable but also removes insects and other parasites from the skin.*

The Hamadryas Baboon

The hamadryas baboon is unlike the other members of the genus *Papio* in that there are three distinct levels of social organization. The principal unit is a male with one or more females and their young, totaling up to seven or eight members. These one-male units group together into bands, similar in size and organization to the bands of the other baboon species. Finally, the bands come together in troops, some of which have been counted as containing 750 members.

Students of primate behavior hypothesize a direct relationship between these social organizations and the environmental resources of the hamadryas. The habitat, which is on the edges of deserts, is typically arid grassland, interspersed with thorny acacias and other small trees and bushes. There are no tall trees, and the baboons of this area sleep on ledges on the vertical slopes of steep cliffs. (This is learned behavior. The hamadryas will sleep in tall trees if they are available, and other species will sleep on cliffs if there are no tall trees.) The baboons move across open country, often for long distances, between their sleeping cliffs and their feeding sites. They travel in bands, break up into one-male units for feeding—typically one unit to a tree—regroup into bands, return to the cliffs, where they often sleep in troops, and then recongregate each morning in the band. Thus each social unit serves a clear and important function. The one-male unit is an optimal foraging unit, provided with a protector; the band provides for mutual defense when traveling; and the troop makes possible maximum utilization of safe sleeping sites within the habitat.

The strongest bonds are those that hold the one-male unit together. The females mate exclusively with the unit leader, and almost all social interactions, such as grooming, are carried out within the unit. Juveniles sometimes leave to play with those of other units within the band.

48–28
A hamadryas band. Note the grouping by one-male units.

The structure of the hamadryas band is generally similar to that of other baboon species. There is a male dominance hierarchy, and males act as defenders. About 20 percent of the adult males are not part of any family unit, but, as with other baboon species, these males are part of the band. The band moves as a whole, apparently acting on information from knowledgeable members of the group.

The ties that hold the band together are less strong than those that hold the one-male unit of the hamadryas together. If a member, because of age, illness, or injury, drops out of a nonhamadryas band, he or she is left behind. If a disabled female drops out of a hamadryas band, her male leaves the band to stay with her. The band, however, like the one-male unit, is a stable social group, always composed of the same members. The troop, on the other hand, is made up of varying groups of bands from the same territory that apparently recognize one another and are mutually tolerant.

Because the male-dominated band is the primary social structure of the other *Papio* species (and also of the macaques, arboreal monkeys who are their closest relatives), the hamadryas band appears to be a legacy from their common ancestors. Similarly, the troop seems easy to understand as a direct outgrowth of environmental pressures; baboons of other species share water holes and large sleeping groves when they have to.

What, however, are the bonds that hold the one-male unit together? The cohesion of the unit turns out to depend primarily on two behavioral patterns, both those of the male. The first is a herding instinct toward the females. The females are taught to follow the male. He constantly watches them over his shoulder and, if one drops back or attempts to slip away, he stares, threatens, and sometimes nips her on the back of the neck, after which she immediately follows. If a male hamadryas is presented with a female olive baboon (*Papio anubis*), he accepts her quite readily and immediately trains her to follow him.

48–29
A young male hamadryas with a juvenile female that he has stolen from another one-male unit. New family groups originate in this way.

The second behavioral pattern is a strong inhibition among baboon males against taking another male's females. This, too, was tested experimentally. Two adult males, who knew each other, and a strange adult female were trapped. One male was enclosed with the female while the other was permitted to watch from a cage 10 meters away. The male caged with the female immediately made grooming, mounting, and herding advances toward the female, as he would toward a member of his harem. Fifteen minutes later, the second male was put in the same enclosure with the pair. He not only refrained from fighting over the female, but he avoided even looking at them. The animals were separated, and two days later the experiment was repeated again, this time switching the males. Once again, the male that had watched the activities between the male-female pair was strongly inhibited in his behavior.

Another adaptive feature of this inhibition is that it applies only to adult males and females. Young males are attracted to juvenile females, which they "kidnap" from their family groups before they are sexually mature and train to follow them. Thus the younger male has a detour around the inhibition barrier, and new family units can come into existence.

Lest you be tempted to extrapolate to human behavior from these accounts of baboon behavior, it is of some interest that, among the baboonlike geladas, groups are formed and held together by activities of both sexes. Dominant males pair with dominant females, wife number one is dominant over wife number two, and so on. The geladas are large monkeys that live in the mountainous grasslands of Ethiopia. Although they are referred to as baboons in common usage, they are different enough to be recognized as a separate genus (*Theropithecus gelada*).

The baboons remind us that social behavior in animals is ecologically adaptive—that is, that it promotes the survival of members of the social organization (and particularly of their young) within a given habitat or range of environmental conditions.

SOCIOBIOLOGY

Sociobiology means, quite simply, the biology of social behavior and, as such, is a normal outgrowth of ethology. The evolution of social behavior began to command serious attention in 1962 when Scottish biologist V. C. Wynne-Edwards set off an academic furor by proposing the previously mentioned hypothesis that natural selection might act on groups as well as individuals. This concept offered satisfying explanations for many hitherto puzzling phenomena—such as altruism in social insects and territorial exclusion—which seemed to serve the group but whose advantage to the individual was not immediately clear. In 1964, W. D. Hamilton offered a cogently argued response to Wynne-Edwards, based mainly on the genetic relationship of the social insects, in which he presented the ideas of kin selection we discussed on page 920. Hamilton demonstrated, in these cases at any rate, that the concept of group selection was neither necessary nor useful—since no mechanism has been suggested as to how it might operate. Since that time numerous publications have appeared demonstrating the benefit in evolutionary terms to the individual of various types of group behavior, even, paradoxically, of those forms of behavior that lead to the failure to reproduce by the individual exhibiting the behavior. Wynne-Edwards has modified his original position, and most biologists are in agreement that group selection, if it exists at all, is not a major adaptive force in social behavior.

The demonstrations by Hamilton and others of the extent to which evolutionary theory could explain social behavior, such as territoriality and altruistic acts, gave impetus to another group of studies and controversies that are still continuing. These center around attempts to explain human behavior on the same evolutionary bases used so successfully to explain the behavior of other social organisms. The most important serious publication in this area is the brilliant and beautiful *Sociobiology* by E. O. Wilson, a Harvard scientist whose work on insect societies, in particular, has won high esteem. The last chapter of *Sociobiology* is given over to speculations on the genetic basis of altruism, ethical values, "ease of doctrination," and other characteristics attributed to humankind.

A number of prominent biologists have launched attacks on the application of the theories of sociobiology to the human species. These actions may seem contrary to the spirit of biology in general; over and over again, for instance, we have had occasion to point out that *Homo sapiens* is a member of the biological continuum, even sharing genes and enzymes with our fellow prokaryotes. There are two reasons for the objections.

First, with the exception of some very simple reflexes—such as the suckling response in newborn infants—there is no evidence for a genetic basis for any human behavior. In fact, as we have noted, little is known about the genetic basis of any complex behavior. There is, on the other hand, much evidence that the behavior of members of the human species is determined by open programs (page 912). This concept is supported by numerous observations that social behavior varies widely with culture and, within any culture, individual behavior varies widely with personal experience. Whatever may be programmed into our genes—and it is likely we shall never know what this is—would appear to be overridden by our cultural evolution and personal history.

Second—and this is the reason for all the excitement—biological determinism is a politically dangerous philosophy. In the past, it has offered a justification for slavery, for the "final solution" of the Jewish "problem," and for all the ugly remnants of racism that persist today. The assignment of particular behavioral characteristics to particular races has proved economically very convenient, as convenient as social Darwinism (page 786). Note particularly that such analyses always work out to the benefit of whoever is doing the analysis. As Kate Millet says, in *Sexual Politics:* "Patriarchy has a tenacious or powerful hold through its successful habit of passing itself off as nature." Similarly, how simple it is, again using biology as our ally, for all of us to forgive ourselves for and even to justify violence, aggressiveness, docility, and greed.

SUMMARY

A society is a group of individuals of the same species organized in a cooperative manner and communicating with one another in some way.

Among the most complex societies are those of insects. These societies are matriarchies, centering around the care of the queen and the raising of the brood. The behavior of members of the society is determined by chemical substances—pheromones—exchanged among members of the society. In eusocial insects, only one female is reproductive; the other females are workers, caring for the reproducing individuals and the young. Among hymenopterans males are haploid and females diploid. Hence the female workers are more closely related to one another than they are to their mother or would be to their own daughters. Thus haplodiploidy is believed to play an important role in eusociality in hymenopterans.

Among vertebrates, the principal advantage of social living appears to involve either avoidance of predators or the communal hunting of prey.

Some animal societies are organized in terms of social hierarchies. Socially inferior animals—those low in the pecking order—reproduce less frequently than their social superiors. They are also the first to starve if food is limited or to be driven out if shelter is limited.

Territories are areas defended by an individual or a society against others of the same species. Territories may be "real"—that is, they may be actual areas of land containing food and nesting material to support a mating pair and young—or they may be symbolic, such as an arena. In either case, the only animals that breed are those with territories. Animals without territories provide replacements for territory owners and a reserve for population expansion if the range or food supply of the population is increased.

Acts of aggression are restricted in animal societies by social dominance, territoriality, and other forms of social behavior.

Wolf societies (packs) are organized in a dominance hierarchy. Usually a pack contains only one breeding pair, the socially dominant male and female. Because all females prefer the dominant male and all males prefer the dominant female, mating rarely occurs among other members of the pack, and so the size of the pack is limited. All members of the pack share in caring for the pups, and strong social bonds develop between pups and other members of the band, which also appear to contribute to the coherence of the pack. Wolves are territorial, using scent marking, and perhaps howling, to mark their territories.

Baboons are large, quadrupedal terrestrial monkeys, with pronounced sexual dimorphism. They are organized into multimale bands. The males have a pronounced social hierarchy; superior males mate more often, have first choice of nesting sites, and play a more important role in defending the band than their social inferiors. The hamadryas baboon is unusual in that the band is subdivided into one-male family units. These units group into bands, and the bands come together in large troops, which utilize communal sleeping cliffs. Thus, social organization is hypothesized to be related to the habitat. The one-male family is an optimal foraging unit, the band is the traveling unit, and the troop provides for the most efficient use of the sleeping resources.

Sociobiology is the biology of social behavior.

QUESTIONS

1. Distinguish between the following: community/society; eusocial/subsocial; worker/drone; altruism/kin selection.

2. Territoriality and social dominance achieve the same results. What are they? How do the results differ in a population in which these forms of behavior are not present?

3. In what ways are human societies different from insect societies? How are they similar?

4. What behavioral conventions limit intraspecific aggression in *Homo sapiens*? Do any foster it?

5. J. B. S. Haldane, arguing once in a pub about the nature of altruism, did some quick calculations on the back of an envelope and announced: "I will lay down my life for two brothers or eight cousins." What did he mean?

(a)

(b)

(c)

(d)

49–1

Until about 10,000 years ago, Homo sapiens *were all hunter-gatherers; now only a few remain, among them the Bushmen of the Kalahari Desert of South Africa.* (a) *Women and children beside a water hole. The groups, which average 15 to 25 members, move five or six times a year, setting up a base camp near a water hole. The basic social unit is the family, and marriages are very stable. The Bushmen have no easily detectable protein or vitamin deficiencies, few chronic diseases, low mortality, and a life span comparable to that of urban man.* (b) *Man drinking from the shell of an ostrich egg. Bushmen make small holes in emptied egg shells, fill them with water, and bury them at strategic spots along their customary routes.* (c) *Men setting out from camp to hunt, using the same weapons and hunting the same game as they have from prehistoric times.* (d) *Woman surrounded by mongongo nuts. High in calories and protein and drought-resistant because of their hard shells, they are the staple diet of the Bushmen. Women, who are the chief gatherers of the plant products, supply more than 50 percent of the food, and their social status in the group is high.*

CHAPTER 49

Human Evolution and Ecology

Where does the story of human evolution begin? We might start with a chance combination of chemicals in some warm Precambrian sea. Or perhaps even with the formation of a small planet 150 million kilometers from a star. Or it might begin more than 4½ billion years later, when some little tribe of man-apes found they could sharpen a digging stick or hone the flat edge of a stone. In any case, it is a very long story, measured in human terms, and many of its details are lost to us, probably forever.

For present purposes, let us start the story about 200 million years ago, in the early Mesozoic era, at about the time of the first dinosaurs. In this same period of time—give or take a few million years—the first mammals appeared, arising from a primitive reptilian stock. Our information about these mammals is very slight. The entire length of the Jurassic and Cretaceous periods has left us with only a few fragments of skulls and some occasional teeth and jaws. From these scraps of evidence, we know that the first mammals were about the size of a house cat. They had sharp teeth, indicating that they were basically carnivorous, but since they were too small to attack most other vertebrates, they are assumed to have lived on insects and worms, supplementing their diet with tender buds, fruits, and perhaps eggs. These first mammals were probably nocturnal, judging by the large size of their eye sockets, and they were almost certainly warm-blooded. If such an animal were alive today, it would be classified as an insectivore, something like a ground shrew.

For about 130 million years, these small mammals led furtive existences in a land dominated by reptiles. Then suddenly, as geologic time is measured, the giant reptiles, the dinosaurs, disappeared. The cause of the dinosaur extinction is one of the great biological mysteries. No species of land animals weighing more than 9 kilograms survived. The disappearance of the dinosaurs occurred at a time when, geologists believe, there was a drop in the average temperature and, perhaps more important, a marked increase in seasonal temperature fluctuations. In any case, by the end of the Cretaceous period all of the dinosaurs had disappeared forever, and about 65 million years ago an explosive radiation of the mammals began.

The early mammals immediately diverged into the two dozen or so different lines that included (1) the monotremes, or egg-laying mammals, of which the duckbilled platypus is one of the few remaining examples; (2) the marsupials, such

THE ICE AGES

During most of our planet's history, its climate appears to have been warmer than it is at the present time. However, these long periods of milder temperatures have been interrupted periodically by Ice Ages, so called because they are characterized by glaciations, or persistent accumulations of ice and snow. Such glaciations occur whenever the summers are not hot enough and long enough to melt ice that has accumulated during the winter. In many parts of the world, an alteration of only a few degrees in temperature is enough to begin or end a glaciation.

An early Ice Age appears to have occurred at the beginning of the Paleozoic era, some 600 million years ago. Another, marked by extensive glaciations in the Southern Hemisphere, closed the Paleozoic, some 250 million years ago. The conifers evolved during this period and possibly the angiosperms, as older forest types disappeared. A more recent, less severe cold, dry period occurred at the end of the Mesozoic, about 65 million years ago, and was perhaps a principal cause of the extinction of the dinosaurs.

The most recent Ice Age began during the Pleistocene epoch, about $1\frac{1}{2}$ million years ago. The Pleistocene has been marked by four extensive glaciations that have covered large areas of North America, England, and northern Europe. Between the glaciations there have been intervals, called interglacials, during which the climate has become warmer. In each of these four Pleistocene glaciations, sheets of ice, thicker than 3 kilometers in some regions, spread out locally from the poles, scraped their way over much of the continents—reaching as far south as southern Illinois in North America and covering Scandinavia, most of Great Britain, northern Germany, and northern Russia—and then receded again. We are living at the end of the fourth glaciation, which began its retreat only some 10,000 years ago.

Glaciers are accumulations of snow and ice that flow across a land surface as a result of their own weight. This glacier is flowing into the sea. As the glacier moves forward, its edges melt; the rocks carried within it have become concentrated in the dark band at the right. This is part of the Greenland ice sheet, which has an area of about 1,726,000 square kilometers and at some places is more than 3 kilometers thick. Glaciers covered much of North America and Europe during the period of the evolution of the genus Homo.

During these periods of violent climatic changes, the fossil record shows that the populations of these regions were under extraordinary evolutionary pressures. Plant and animal populations moved, changed, or became extinct. In the interglacial periods, during which the average temperatures were at times warmer than those of today, the tropical forests and their inhabitants spread up through today's temperate zones. During the periods of glaciations, only animals of the northern tundra could survive in these same locations. Rhinoceroses, great herds of horses, large bears, and lions roamed Europe in the interglacial periods, and in North America, as the fossil record shows, there were camels and horses, saber-toothed cats, and great ground sloths, one species as large as an elephant. In the colder periods, reindeer ranged as far south as southern France, while during the warmer periods, the hippopotamus reached England. It was during this most recent period of glaciations and interglacials that hominids of the genus Homo *evolved.*

The reason for these large changes in temperature is one of the most controversial issues in modern science. They have been variously ascribed to changes in the earth's orbit, variations in the earth's angle of inclination toward the sun, migration of the magnetic poles, fluctuations in the solar energy, migration of the continents, higher elevations of the continental masses, continental drift, and combinations of these and other causes. Also under debate is the question of whether this period of glaciations is over—perhaps for another 200 million years—or whether we are merely enjoying a brief interglacial before the ice sheets begin to creep toward the equator once more.

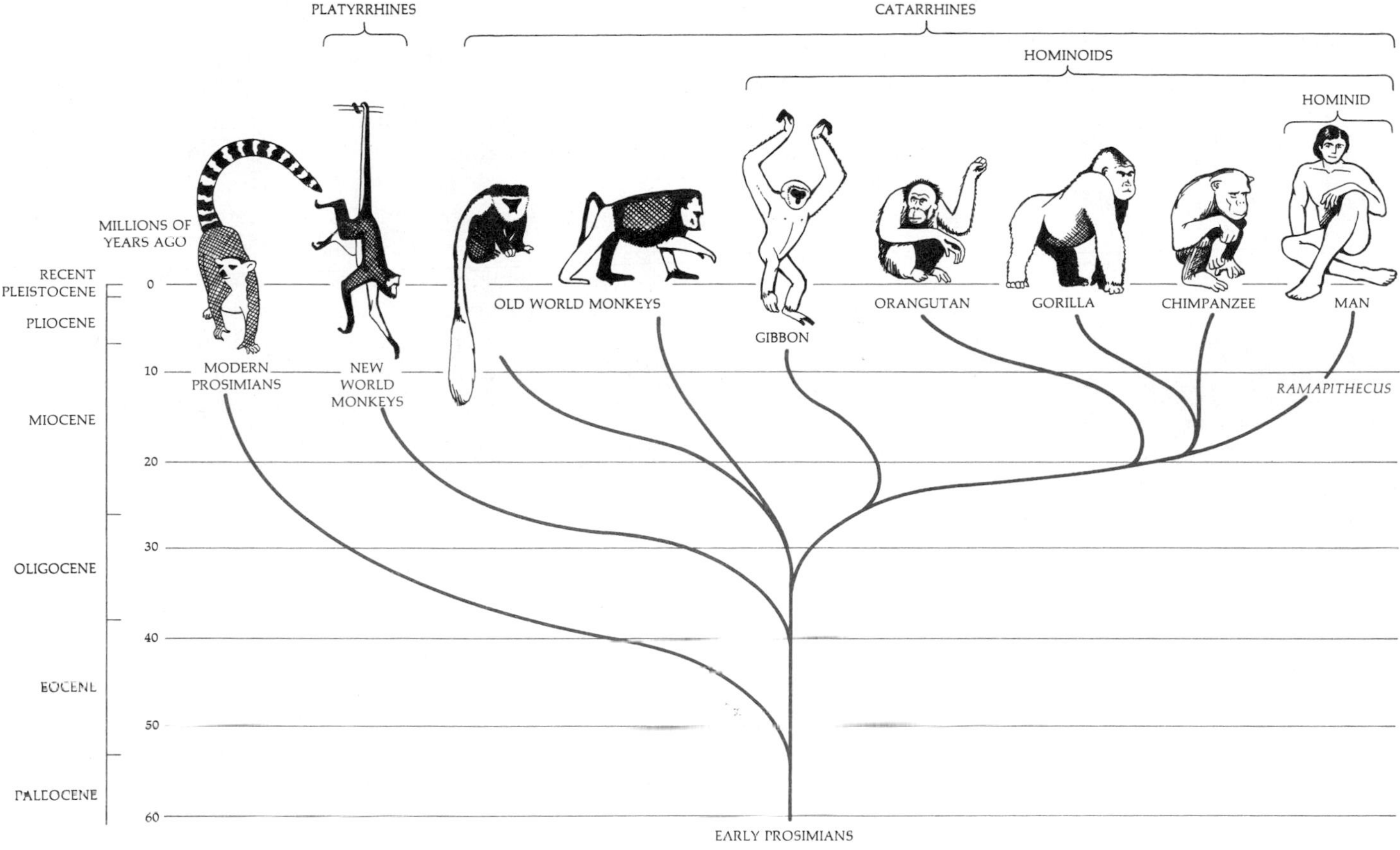

49–2
A tentative phylogenetic tree of the primates, based on fossil evidence. The monkeys, apes, and man are anthropoids.

as the kangaroos, opossums, koala bears, and others whose young are born in embryonic form and continue their development in pouches; and (3) the placentals, by far the largest group. Among the placentals are carnivores, ranging in size from the saber-toothed tiger down to small, weasel-like creatures; herbivores, which include not only the many wild grazing animals but also most of our domesticated farm animals; the omnipresent rodents; and such odd groups as the whales and dolphins, the bats, the modern insectivores, and the primates. Man is a placental mammal and a member of the primate order, as are tarsiers, lemurs, monkeys, and apes, among others.

TRENDS IN PRIMATE EVOLUTION

Primate evolution began when a group of the small, shrewlike mammals took to the trees; fossils of early primates indicate that they closely resembled modern-day tree shrews (Figure 49–3). Most trends in primate evolution seem to be related to various adaptations to arboreal life.

49–3
A modern tree shrew, which the earliest primates probably resembled. If you look closely you will see five-digited paws. Although clawed, they can be spread out and used for grasping. In some classification systems, tree shrews are grouped with the primates, and in others, with the insectivores, which indicates the closeness of the two evolutionary lines.

The Primate Hand and Arm

With a few exceptions, primates have five digits and a divergent thumb. The divergent thumb, which can be brought into opposition to the forefinger, greatly increases gripping powers and dexterity. There is an evolutionary trend among the primates toward finer manipulative ability that reaches its culmination in man (Figure 49–4).

Compared to other mammals, however, primates are relatively unspecialized. Their extremities resemble those of the primitive mammals—indeed, of the reptiles—more closely than do the extremities of mammals of most of the other major orders (see page 776). The first four-legged mammals all had five separate digits on each hand and foot, and each digit except the thumb and the first toe had three separate segments that made it flexible and capable of independent movement. In the course of evolution, most mammals developed hooves and paws more suited for running, seizing prey, and digging; other mammals developed flippers for swimming. The primates retained and elaborated on the primitive five-digited pattern.

In the basic quadrupedal structure of the early mammals and reptiles, the forelimb is supported by two bones (the radius and the ulna), a pattern that provides for flexibility. Among mammals, it is the primates, in particular, that have retained the ability to twist the radius, the bone on the thumb side, over the ulna so that the hand can be rotated through a full semicircle without moving the elbow or the upper arm.

Similarly, only a few mammals have retained the ability to move the upper arm freely in the shoulder socket. A dog or horse, for instance, usually moves its legs in only one plane, forward and backward; some South American monkeys, apes, and man are among the few higher mammals that can rotate the arm widely in the socket.

Primates also have nails rather than claws. Nails leave the tactile surface of the digit free and so greatly increase the sensitivity of the digits for exploration and manipulation.

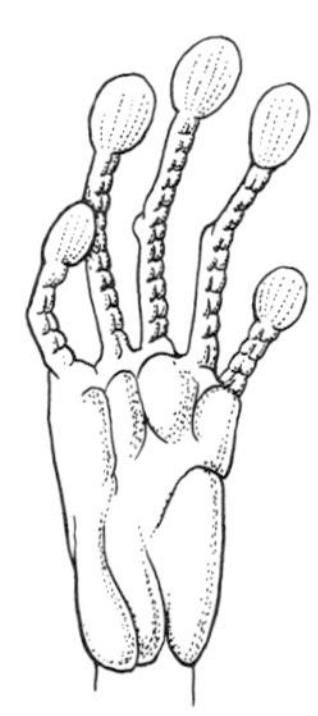

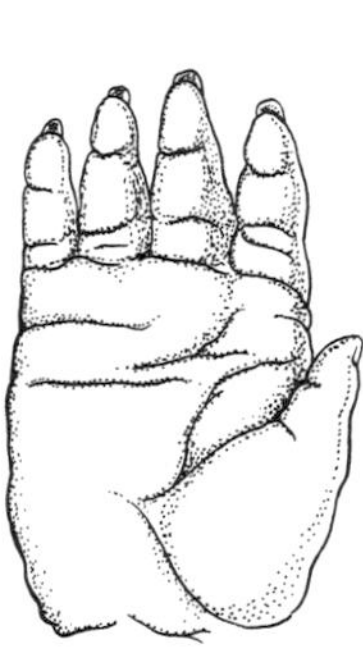

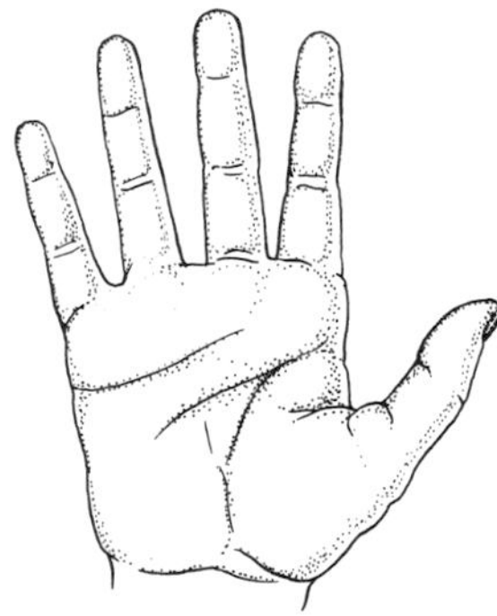

49–4
Some primate hands. The hand of the tarsier has enlarged skin pads for grasping branches. In the orangutan, the fingers are lengthened and the thumb reduced, which provide for efficient brachiating. The gorilla's hand, which is used in walking as well as handling, has shortened fingers. Man's thumb is larger proportionately than that of any other primate, and opposition of thumb and fingers, on which the handling ability depends, is greatest in man.

49–5
Life in the treetops made maternal care a major factor in infant survival. Also the necessity for carrying the young for long periods resulted in strong selection pressures for reduced numbers of offspring. (a) *Anthropoids, such as this vervet monkey, usually have single births.* (b) *A mother chimpanzee with her infant. Field studies suggest that bonds between mother and offspring and perhaps also among siblings last well into adulthood, perhaps for a lifetime.*

(a)

(b)

Visual Acuity

Another result of the move to the trees is the high premium placed on visual acuity, with a decreasing emphasis on the role of olfaction, the most important of the senses among many of the other mammalian orders. (Flying produced similar evolutionary pressures among the birds, which were also evolving rapidly during this same period.) This shift from dependence on smell to dependence on sight has anatomical consequences. Tree shrews, like many other animals, have eyes that are directed laterally, but among the other primates there can be traced a steady evolutionary trend toward frontally directed eyes and stereoscopic vision.

Almost all primate retinas have cones as well as rods; cones, as we discussed on page 741, are concerned with color vision and with fine visual discrimination. All primate retinas have foveas, areas of closely packed cones that produce sharp visual images.

Care of the Young

Another principal trend in primate evolution is toward increased care of the young. Because mammals, by definition, nurse their young, they tend to have longer, stronger mother-child relationships than other vertebrates (with the exception, in some cases, of birds). In the larger primates, the young mature slowly and have long periods of dependency and learning (Figure 49–5).

Uprightness

Another adaptation to arboreal life is an upright posture. Even quadrupedal primates, such as monkeys, sit upright. One consequence of this posture is a change in the orientation of the head, allowing the animal to look straight ahead while in a vertical position; it is this characteristic, above all others, that makes primates look so "human" to us. Vertical posture was an important preadaptation for the upright stance of modern man.

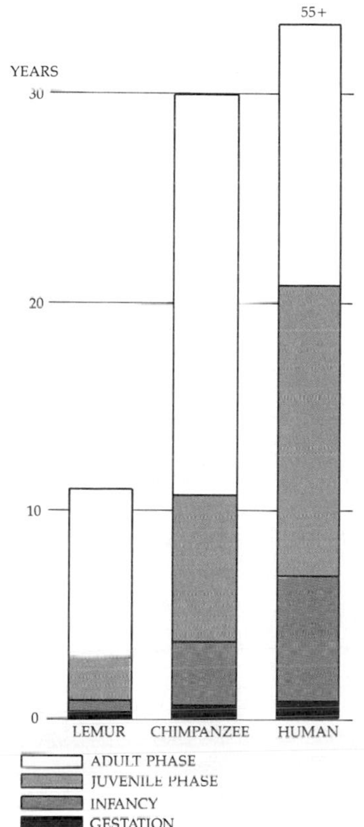

49–6
In the course of primate evolution there has been a trend toward longer periods of juvenile dependency and thus longer periods of learning.

MAJOR LINES OF PRIMATE EVOLUTION

Prosimians

Primates are divided into two major groups: the *prosimians* (lorises, bush babies, tarsiers, and lemurs) and the higher primates, the *anthropoids* (monkeys, apes, and man). During the Paleocene and the Eocene (about 65 to 38 million years ago), a great abundance and variety of prosimians inhabited the tropical and subtropical forests that spread much farther north and south of the equator than they do today. Modern prosimians (Figure 49–7) are small, arboreal, furry, eat fruit and insects, and are almost all nocturnal. The prosimians are generally considered an unnatural group (as are the protists, for example) in the sense that they are not closely related.

Monkeys

Monkeys, along with the apes and man, make up the higher primates, the anthropoids. The monkeys are generally larger than prosimians, their skulls are more rounded, and they are more intelligent. Like tarsiers, but unlike other prosimians, they have full stereoscopic vision, and also color vision. They are all diurnal.

Monkeys generally move in bands composed of adult males and females, infants, and juveniles. The females protect and otherwise care for the young, and the males often serve protective functions for the group, ranging from observing and warning to attacking predators outright. In some species, males join together to mob a predator, harassing it by hooting and calling, jumping on branches until they fall on enemy heads, sometimes throwing sticks and branches, and, in some species, by group defecation and urination.

49–7
Some prosimians. (a) *A lesser bush baby. Bush babies move over level ground by hopping like kangaroos, and sometimes leap 3 or 4 meters in the air from one tree to another. The fingers work together, not independently like human fingers.* (b) *A ring-tailed lemur, combing his tail. The second digit of each foot is a special grooming claw.* (c) *A native of Indonesia, the little tarsier (about the size of a kitten) has existed relatively unchanged for some 50 million years. Its upper lip, like ours, is free from the gum below it, giving it the ability to make faces. It has stereoscopic vision and a larger brain than the lemur's. Living entirely in trees, it has hands and feet with enlarged skin pads for grasping branches. As you may have guessed from the owl-like eyes, tarsiers are nocturnal.*

(a)

(b)

(c)

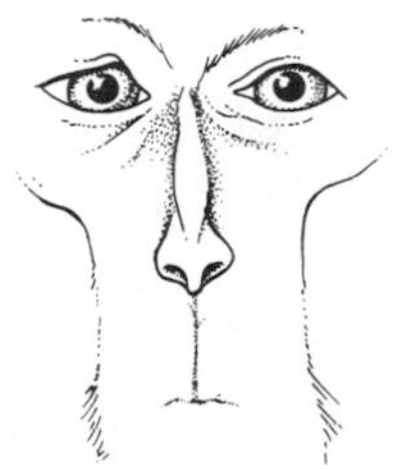

49-8
Early in their evolution, the anthropoids split into two main lines, the platyrrhine, or flat-nosed (left), and the catarrhine, or downward-nosed (right). New World monkeys are platyrrhines; Old World monkeys and the hominoids are catarrhines. There are other characteristic anatomical differences between the two groups.

The monkeys arose from prosimian stock apparently during the Eocene period; related fossil forms are found in abundance in the Oligocene. There are two principal groups, the New World monkeys, also known as platyrrhines (meaning flat-nosed), and the Old World monkeys, the catarrhines (downward-nosed). The separation of these groups took place very early in their history, with the platyrrhines evolving in South America and the catarrhines in Africa, both during the Oligocene, some 26 to 38 million years ago.

Despite their long chronological separation, members of the two groups closely resemble each other, an example of parallel evolution.

The New World monkeys, from South and Central America, are all strictly arboreal in their habits, and many of them use their tails as a fifth prehensile limb, which none of the Old World monkeys can do. Any monkey you see hanging by its tail is definitely a New World monkey. The New World monkeys include marmosets, howler monkeys, the spider monkey, and the capuchin—the familiar monkey of the organ-grinder.

The Old World monkeys are both arboreal and terrestrial. The tree dwellers include the colobus monkeys, langurs, the mangabeys, and the guenons. Their tails are used for balance rather than as prehensile organs. The ground dwellers, which came from tree-dwelling ancestors, include the baboons and macaques, or rhesus monkeys, all of which walk on all fours. (The social behavior of baboons is described in Chapter 48.)

49-9
Only New World monkeys, such as the spider monkey, can hang by their tails. The platyrrhine monkeys are the only primates native to the Americas.

Apes

Hominoids—now represented by the apes and us—first appeared in the fossil record some 35 million years ago. The modern apes comprise four principal genera: *Hylobates* (gibbons), *Pongo* (orangutans), *Pan* (chimpanzees), and *Gorilla* (gorillas). Apes, with the exception of the gibbons, are larger than monkeys. Their brain case is larger in proportion to their size. Like the Old World monkeys, they are catarrhines. They are all capable of brachiation—that is, swinging from one arm and then the other with their bodies upright. Although, among modern genera, only the gibbons move primarily in this way, brachiation may have played a role in the transition from the body structures associated with the crouching position characteristic of the lower primates to the body structure that makes possible our erect posture. Apes have relatively long arms and short legs. As a result, even when they are on all fours, their bodies are partially erect.

The social structures of the genera vary. Gibbons form permanent pairs, each couple living by itself with its offspring. Male orangutans are solitary; adult females are usually accompanied by one or two youngsters. Gorillas and chimpanzees are highly social.

Gorillas, the largest of living primates, are among the shyest and the gentlest, according to recent field studies. Gorillas have almost completely abandoned trees,

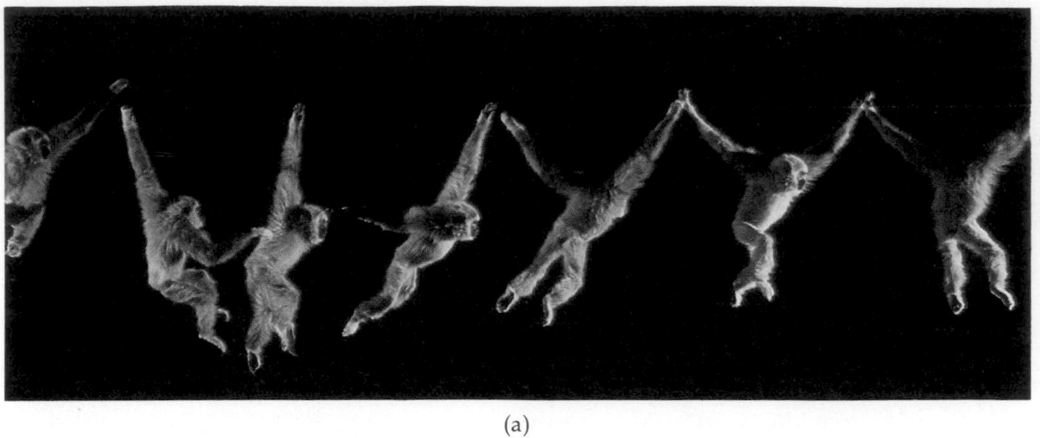

(a)

(b)

ANTHROPOID APES

The anthropoid apes include gibbons, orangutans, gorillas, and man's closest living relatives, chimpanzees. Gibbons are the smallest of the apes. They live in pairs or in small family groups, and, along with some human beings, they are the only monogamous anthropoids. Gibbons can stand and walk upright, but usually they move through the trees, swinging arm over arm (brachiating) as shown above (a). *The rarest and least studied of the large primates, the modern orangutan* (b), *is found only in the tropical rain forests of Borneo and Sumatra. Males, which can weigh as much as 100 kilograms (220 pounds), are usually found alone.*

A male gorilla is about as tall as a tall man but weighs more than three times as much (some 180 to 275 kilograms). (c) *Gorillas have a beetling brow, a strong, heavy jaw, and, on top of the skulls of adult males, a bony crest to which the strong jaw muscles are attached. Its legs are shorter in proportion to its body than are man's, but its arms are much longer—with a span of about 3 meters—and extremely powerful.*

Gorillas usually walk on all fours but they can stand erect, as they do when challenging an enemy, and they also walk erect for short distances. They live principally on the ground and feed on ground plants. In (d), *a mature male and family are feeding on a banana plant.*

Each gorilla troop, consisting of from 8 to 24 individuals, is led by a mature male. Although reportedly placid by nature, when threatened, males will give fearsome exhibitions of power, thrashing with broken branches, standing erect, beating their chests, barking and roaring, even hitting themselves under the chin to make their teeth rattle. Given the massive size of an adult male, such as the silver-backed male lunging forward to charge (e), *the display is impressive.*

(c)

(d)

(e)

(f)

(g)

(h)

(i)

(j)

Chimpanzees are somewhat smaller than man; an adult male weighs about 45 kilograms. They feed and sleep in the trees, but spend a large part of their waking hours on the ground. They walk on all fours, on the flats of their feet and the knuckles of their hands. Although their fingers are shorter than man's, they have a fine precision grip. They use simple tools, such as a stick to pry out termites, a leaf as a blotter, or a stick as a weapon. Whereas the gorilla is shy by nature, chimps are gregarious, curious, boisterous, and extroverted. (f) *A juvenile and infant wrestle in play;* (g) *a male pant-hoots;* (h) *a mother shares food with her young;* (i) *a young male relaxes in a tree; and* (j) *a chimpanzee group sits and grooms.*

CULTURAL EVOLUTION AMONG MACAQUES

The little Japanese island of Koshima, a high, wooded mountain surrounded by a beach, is inhabited by macaques. The monkeys lived and fed in the forest until a decade or so ago, when a group of Japanese researchers began throwing sweet potatoes on the beach for them. The group quickly got used to venturing onto the beach, brushing sand off the potatoes, and eating them. One year after the feeding started, a two-year-old female the scientists had named Imo was observed carrying a sweet potato to the water, dipping it in with one hand, and brushing off the sand with the other. Soon, other macaques began to wash their potatoes, too. Only macaques that had close associations with a potato washer took up the practice themselves. Thus it spread among close companions, siblings, and their mothers, but adult males, which were rarely part of these intimate groups, did not acquire the habit. However, when the young females that learned potato washing matured and had offspring of their own, all of them learned potato washing from their mothers. Today all of the macaques of Koshima dip their potatoes in the salt water to rinse them off, and many of them, having acquired a taste for salt, dip them between bites.

This was only the beginning. Later, the scientists began scattering wheat kernels on the beach. Like the others, Imo, now four years old, had been picking the grains one by one out of the sand. One day she began carrying handfuls of sand and wheat to the shore and throwing them into the water. The sand sank, the wheat kernels floated to the top, and Imo collected the wheat and ate it. The researchers were particularly intrigued by this new behavior since it involved throwing away food once collected, much less a part of the macaques' normal behavioral repertory than holding on to food and cleaning it off. Washing wheat spread through the group in much the same way that washing potatoes had. Now the macaques, which had never even been seen on the beaches before the feeding program began, have taken up swimming. The youngsters splash in the water on hot days. Some of them dive and bring up seaweed, and at least one has left Koshima and swum to a neighboring island, perhaps as a cultural missionary.

with the exception of some of the smaller animals that sleep in nests in the lower branches. The larger gorillas sleep on the ground, and all feed on ground plants. They live in groups ranging from eight to 24 individuals, with about twice as many females as males, and a number of juveniles and infants. Each gorilla troop has a large, mature (silver-backed) male as a leader.

Chimpanzees, the primates generally considered to be the closest to humans, usually move in groups, feeding on wild fruits, seeds, and pods of a large variety, and occasionally killing smaller animals for meat. Within a group there is a male dominance hierarchy; females are usually subordinate to adult males, but there is also a female dominance hierarchy. Temporary (two- or three-day) bonds may form between a male and a female in estrus, but otherwise there are no pair bonds, and receptive females typically mate with a series of males. However, there are often friendships among members of the group, and friends may spend as much as two hours a day in mutual grooming. Males cooperate in capturing prey, which consists most often of other primates, such as infant baboons and colobus monkeys. Prey may be shared among the group. Offspring maintain strong bonds with their siblings and their mothers, apparently throughout their lifetimes. Chimps are gregarious, curious, boisterous, and extroverted. Although groups move through home ranges, there is not a strong sense of territoriality, and the bands are not well knit and exclusive as they are in some other primate groups. The composition of the group changes as newcomers are taken in and other members wander off.

Thus, as you can see, it is difficult to make correlations between human behavior and the behavior of other higher primates.

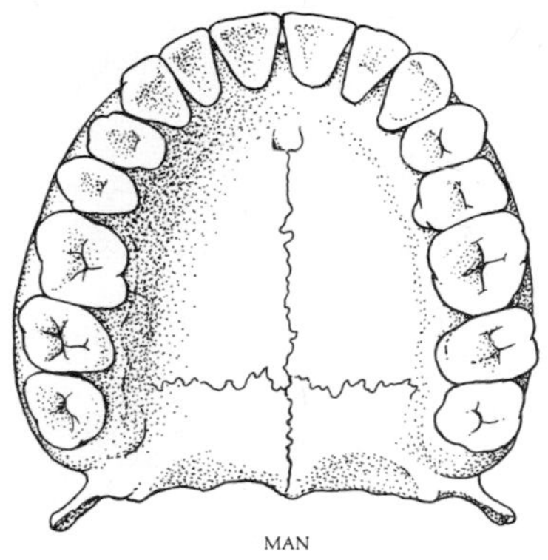

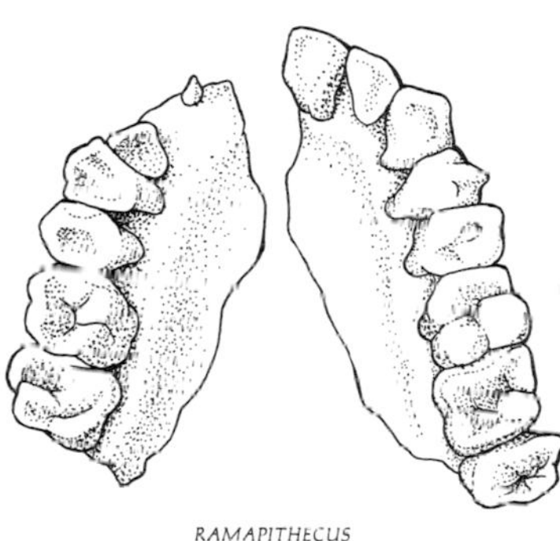

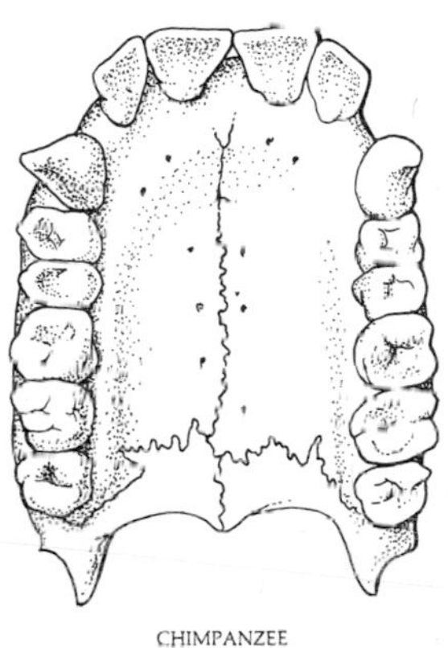

49–10
A comparison of the upper jaws of man, Ramapithecus, *and a chimpanzee. The jaw of* Ramapithecus *is more rounded and thus more manlike than that of the ape.*

Ramapithecus

When the great apes were just beginning their separate lines of evolution, there lived a primate that has been called *Ramapithecus. Ramapithecus* is known only from fossil fragments of upper and lower jaws from India and Africa (Figure 49–10). The fossils are dated about 12 to 14 million years ago. What can these few fossils tell us? First, when the bits of jawbone are pieced together, it is clear that *Ramapithecus* had a smaller and broader dental arch, as compared to other large contemporary primates or, indeed, to modern apes. The jaw fragments also indicate that *Ramapithecus* was comparatively small, about the size of a chimpanzee. Second, adjacent molars on the jaw fragments of *Ramapithecus* show marked differences in wear in comparison to contemporary primitive primates, in which no sharp variations in wear are evident. This suggests that the molar eruption sequence of *Ramapithecus* was much slower; as a consequence, there is reason to believe that *Ramapithecus* matured more slowly than its hominoid contemporaries—as does modern man.

Finally, the teeth and their condition indicate that tasks such as biting off and tearing up vegetation, for which apes use their front teeth, were not carried out to the same extent by the teeth of *Ramapithecus.* Physical anthropologists hypothesize that he (or it, as you prefer) used his forelimbs for these purposes, and, on the basis of this hypothesis, they further hypothesize a trend toward bipedalism. Additional support for bipedalism is found in the fact that the back teeth are specialized for more powerful chewing, which may indicate that *Ramapithecus* had abandoned the fruits and other moist vegetation of the forest for the harder, drier food (such as tough rhizomes and dried seeds) of the bush and savanna.

And then the curtain falls for another 10 to 11 million years, leaving many questions unanswered.

49–11
The skull of a child, found in a limestone quarry in Taung, South Africa, in 1924, was the first specimen discovered of the earliest known hominid, an australopithecine. The oldest australopithecine fossils, according to radioisotope dating methods, are specimens found near Lake Omo in eastern Africa, representing remains of creatures that lived about 3½ million years ago. Shown here are (a) *the skull and* (b) *a reconstruction drawing of the head of the child, which was about five or six years old when it died. The australopithecines were erect-walking creatures with brains somewhat larger proportionately than the gorilla's and with teeth that are more similar to ours than to the ape's.* Australopithecus *lived on the ground, and some of them probably ate meat and used tools of their own making.*

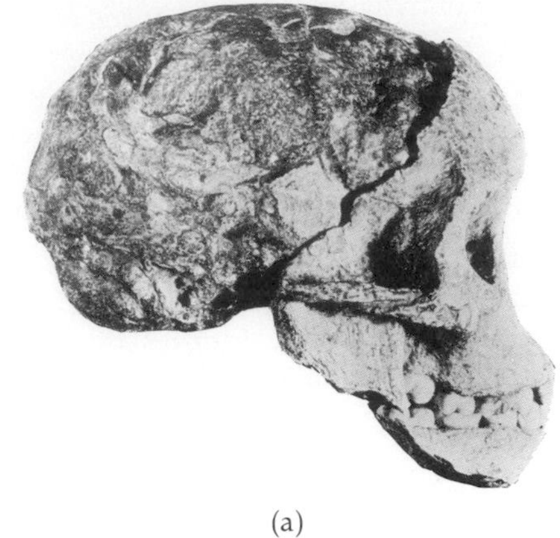

(a)

(b)

The First Hominids

In 1924 an explosion in a quarry that was being mined in South Africa loosened a piece of rock containing the skull of a child. This specimen, along with many others, was crated up and sent to anatomist Raymond Dart in Johannesburg. Dart was quick to see that the little creature had many humanlike features that distinguished it from both modern apes and their ancestors. These included the rounded appearance of the skull, the size of the fossilized brain within it, and the roundness of the jaw. Also, the point of attachment of the vertebral column to the skull indicated that the young animal was a biped.

Dart reported his find in the British journal *Nature,* naming the new species *Australopithecus* ("southern ape"). His announcement was largely ignored. The scientific community had not yet recovered from the effects of the Piltdown forgery.* Besides, it was generally agreed at that time that Europe, with all its cultural advantages, was the cradle of mankind, not darkest Africa.

Subsequent fossil finds, however, confirmed Dart's findings and their interpretation. Gradually a picture began to emerge of a group of small, lightly built hominids, not more than 1½ meters (5 feet) tall, with cranial capacities ranging from 450 to 700 cubic centimeters, as compared to 340 to 750 for male gorillas (a much larger animal) and 1,000 to 2,000 for modern man. Their hands had a broad, flat thumb and the beginnings of a human-style grip. The teeth were very much like our own, although the molars were larger. Australopithecines walked upright, probably covering the ground with quick, short steps in a sort of jog trot, rather than by our far more efficient heel-and-toe striding movement.

They used simple tools (Figure 49–12) and ate meat; collections of bones of small animals—frogs, lizards, rats, and mice—and of the young of larger animals, such as antelopes, have been found at various australopithecine sites. The evidence indicates that they lived in small groups, like most anthropoids. In the past decade, owing largely to the efforts and extraordinary talents of the Leakey family, discov-

49–12
Pebble tools such as these have been found at Olduvai and other sites in eastern Africa in fossil strata of 2 million or more years ago. Australopithecines (presumably) took pebbles of lava and quartz and either struck off flakes in two directions at one end, making a somewhat pointed implement, or in a row on one side, making a chopper. They measure up to 10 centimeters in length and were probably used to prepare plant food and to butcher game. The flakes were also used apparently as scrapers.

* In the Piltdown forgery, the skull of a man and the jaw of an ape, carefully doctored to fit prevailing ideas concerning early hominids, were planted as fossils and, embarrassingly enough, widely accepted by anthropologists.

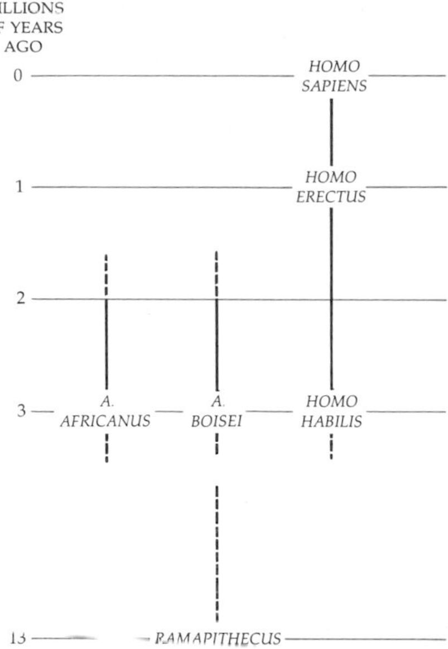

49–13
The evolution of the genus Homo, *according to some hypotheses.*

eries of australopithecine fossils have increased in number and in antiquity. According to radioisotope dating, the oldest remains were deposited about $3\frac{1}{2}$ million years ago. The picture has also become much more complicated.

It is now beginning to appear that there was not one but several lines of hominid evolution (which, of course, is in accord with evolutionary patterns of other organisms). The Leakeys now recognize at least two species of australopithecines, *Australopithecus africanus* and *Australopithecus boisei*, and a third species, which they call *Homo habilis. Homo habilis*, according to this interpretation, was the species that used tools, had the larger brain, and is our ancestor.

How Did It Happen?

What prompted one group of primitive apes to make the first fateful step to bipedalism? We know from pollen specimens that grasslands were spreading and forests contracting during the Miocene, the period during which *Ramapithecus* made its first appearance. The grasslands may have represented a new adaptive zone. The horse and other grazing animals evolved rapidly during this time, filling the new niches available to herbivores. Baboons also seem to have become terrestrial in this period. We do not know why one group of apes and not another moved into the grasslands. One possible explanation is that this group was driven out of the woodland by competition with other primates. The fact that *Ramapithecus* was smaller than other contemporary apes is in keeping with this speculation. Perhaps they possessed some important preadaptation that we do not know about.

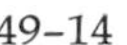
49–14
Louis and Mary Leakey, the British anthropologists, with one of their sons at Olduvai Gorge in the Serengeti plain of Tanzania. Olduvai, from which many important australopithecine fossils have been recovered, represents a geologic sequence that is almost continuous from the present to nearly 2 million years ago. Like our Grand Canyon, it was cut by river action. Because it was by the side of a lake, it was visited by many animals and was apparently a popular hominid camping ground. Louis Leakey died in 1975, but Mary Leakey and their son Richard are still leading figures in the search for early man.

49-15
The fossil jaw of a child, discovered by Mary Leakey in 1975 and dated at about $3\frac{1}{2}$ million years ago. A canine tooth is pushing up through the milk teeth, indicating that the young hominid had a long period of maturation (and hence of learning). The small size of the teeth is interpreted to mean use of tools and perhaps weapons and a diet of meat.

Standing upright on the ground would have greatly increased the range of vision of a ground-dwelling species with highly developed eyesight; other mammals, including bears, squirrels, and prairie dogs, often raise themselves on their two hind legs to look around. Standing upright also, of course, frees the hands, which then can be employed for picking protein-rich grains from wheat and other grasses, for tool making and using, for carrying food, and for brandishing weapons. Use of weapons for defense correlates well with the reduction in the size of the canines among manlike fossil primates, as compared with baboons, for example, or gorillas, among which the males have large canines that are used in aggressive display. The advantages of bipedalism must have been very great for natural selection to have operated in its favor, because the first bipedal primates must have moved slowly and inefficiently compared with the swift movement of other large animals, either on the ground or in the trees. Among the consequences of bipedalism is the enormous human brain, the most characteristic feature of the genus *Homo.*

THE EMERGENCE OF MAN

By about 1 million years ago, a group of hominids had evolved that are placed, without dispute, into the genus *Homo.* Within this genus are two species, the now-extinct *Homo erectus* and our own species, *Homo sapiens.*

Homo erectus

Homo erectus was distinctly different from even the most advanced australopithecine. Members of this species had body skeletons much like our own and were about the same size as modern man. The bones of their legs indicate that they had

(a)

(b)

49-16
(a) *Eye-hand coordination is another characteristic of higher primates, such as the female olive baboon shown here with her young.* (b) *Bipedalism frees the upper extremities for a variety of activities. An adult female chimpanzee "fishes" for termites with a twig.*

49–17
A rock painting from Spain. Such paintings date from postglacial times, about 8000 to 3000 B.C. In contrast to the earlier cave paintings, paintings such as this were executed on sheltered rock surfaces, frequently show human figures, and are small in scale. These paintings typically show forest animals rather than the great plains animals that stalk the caves of Lascaux, France, and Altamira, Spain. The deer are apparently being driven toward the archers by beaters. The hunting of small, nonmigratory animals such as these is believed to mark the transition between the nomadic, hunting-gathering existence of early man and the agricultural revolution.

a stride similar to our own. The chief differences between *H. erectus* and modern man are in the skull. Skull specimens of *H. erectus* are thick and massive, with low foreheads. Their jaws are large and chinless, with large teeth. Brain capacity ranges from 700 to 1,100 cubic centimeters, overlapping that of modern man.

Also, unlike the australopithecines, which have only been found in eastern and southern Africa, members of *H. erectus* seem to have occupied a much larger range. Fossil specimens have been found in Java (Java man), in the Choukoutien caves in China (Peking man), and in various sites in Africa, including Olduvai. There are also a jaw from Heidelberg and some skull fragments from Vérteszöllös, Hungary, that are believed to represent this species. The fossils cover a period of about 700,000 years, from about 1 million to 300,000 years ago. Along with the skeletal fragments have been uncovered a number of clues about the social life of *H. erectus.*

Hunting

Probably the most significant factor in the evolution of *H. erectus* was that they were hunters. As we mentioned earlier, recent field studies have shown that baboons

49–18
A reconstruction of the skull of H. erectus *from the Solo River, Java.* Homo erectus *was much larger than earlier hominids and had a much larger brain. The skull walls are thick and heavy, and the brow ridges are prominent. The protruding occipital crest (cheekbone) was the point of attachment of strong, heavy neck muscles. The jaw is prognathous (protruding) and chinless.*

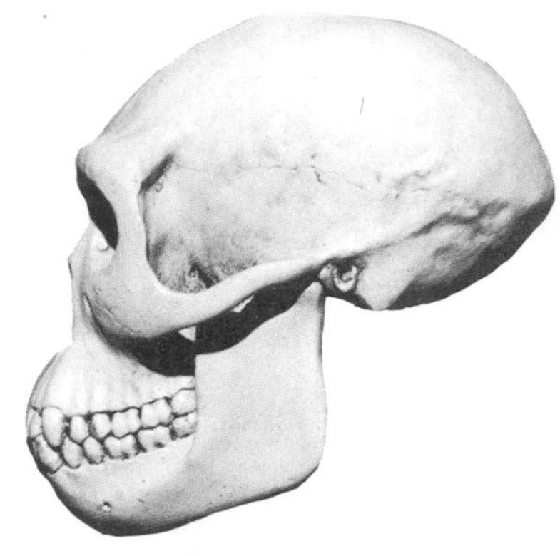

and chimpanzees eat meat occasionally and obviously enjoy it (Figure 49–19), and the australopithecines were also meat eaters. Members of the species we call *H. erectus* became hunters on a different scale, however. During various stages of the Pleistocene era, there seem to have been far more grasslands than now exist. Some areas that are now deserts were savannas, and to the north, much land that is now forest was prairie or steppe, which was abundantly able to support large numbers of grazing and browsing animals. Man was a rare species then, greatly outnumbered by these large herbivores, and so would have had little difficulty finding game. And indeed, from sites as far apart as eastern Africa and China and dating back from 1 million to 350,000 years ago, the bones of very large animals—elephants, rhinoceroses, antelopes, bears, hippopotamuses, and giant baboons—are found in the debris of human campsites.

We do not know how these early hunters killed animals so much larger and more ferocious than themselves. Stone implements found at the sites were apparently made to be held in the hand and were used for chopping, cutting, or pounding. These tools therefore appear to have been used for butchering or preparing food rather than for killing game. There is some evidence that early man killed animals by stampeding them into marshes or over cliffs, perhaps with the help of fire to frighten them. American Indians slaughtered game in this way until very recent times.

Certainly hundreds of thousands of years passed before men developed weapons as sophisticated as spears with stone or metal tips or bows and arrows. Anthropologists who have studied primitive tribes have been impressed by the fact that, in general, their weapons are not efficient and that hunters take little interest in perfecting them. These modern hunters emphasize patience, endurance, skill in stalking, and highly specialized knowledge of animals and terrain, and perhaps prehistoric man did also. Clearly the capacity to cooperate with other members of the hunting party would have been of extreme importance. It is not known whether or not *H. erectus* was capable of speech, but it would have been of great advantage to him ("Head off that hippopotamus!").

49–19
The chimpanzee shows some behavioral traits also attributed to early man. (a) *Chimpanzee preparing to eat the forequarters of a young baboon.* (b) *Male chimpanzee throwing rocks and shouting at adult baboons.*

(a)

(b)

49–20
The hand ax is a stone that has been worked on all its surfaces to provide what appears to be a gripping surface and various combinations of cutting edges, sometimes with a more or less sharp point. Making such an implement requires both skill and time, as reported by anthropologists who have tried. Hand axes came into use about 1 million years ago and are associated with H. erectus.

Tools

In the hands of *H. erectus,* the pebble tool (see Figure 49–12) became a new and highly distinctive implement, the hand ax (Figure 49–20). Hand axes were used extensively throughout Africa, India, and much of the Near East and Europe; tens of thousands have been found in the valleys of the Somme and Thames alone. All these hand axes closely resemble one another. Unlike the pebble tools, they were clearly fashioned according to a formal pattern. Thus we see the emergence of a clear cultural tradition, with skills and learning passed from one generation to another. Also, the wide distribution of the hand ax indicates communication and exchange among the groups of men roaming over these vast territories at that time.

Home and Hearth

All higher primates except modern man are continuously on the move, and though they range back and forth over the same territory, they sleep in a different place every night. Any member of the group that cannot keep up is left behind. Early men were also nomads, as are most modern hunter-gatherers, but there is evidence that they maintained base camps. Some of these camps appear to have been fairly permanent, in the sense that they were returned to year after year.

The australopithecines and the earliest men probably lived mostly in the open, although one australopithecine site is in the natural shelter of overhanging rocks. The campsites found in southeastern Africa are near streams or lakes, which would not only have provided water for the campers but would also have attracted game.

The Choukoutien caves, near Peking, China, inhabited 350,000 years ago, are among the earliest caves known to have been used for human habitation. It is probably no coincidence that within these same caves are also the first traces of the human use of fire. Before men had fires, caves were inhabited by bears and other large carnivores and were too dangerous for humans to occupy. Fire, however, made them available for human habitation, and they have been used by man up to modern times. The caves near Shanidar (Figure 49–21) in the remote mountains of Iraq have been occupied continuously—to the present day—for some 100,000 years.

49–21
This large cave near the little village of Shanidar in northern Iraq has been continuously inhabited for more than 100,000 years. Nine Neanderthal skeletons have been found in the cave, including one who was, according to analysis of fossil pollen, buried on a bed of woody branches and June flowers gathered from the hillside.

49–22
Increase in brain volume in the course of hominid evolution. Some of the increase can be correlated with the increase in body size that was also taking place during this period, but most of it is believed to represent the results of strong selection pressures for intelligence.

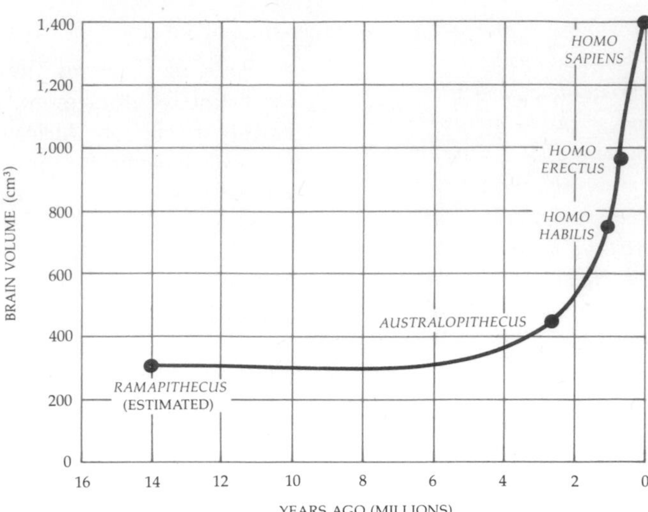

FLAKE STRUCK FROM CORE

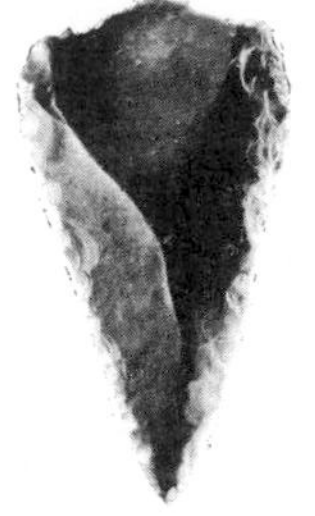
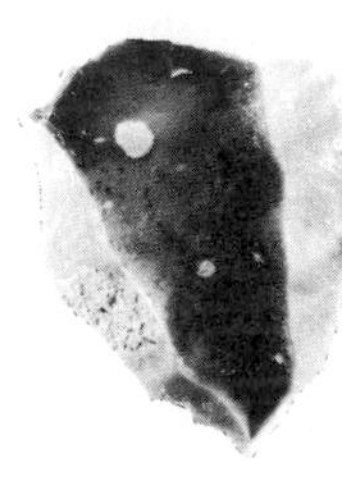

49–23
Fossils of Neanderthal man are associated with a decrease in the use of core tools, such as the hand ax, and a predominance of flake tools. Multiple flakes were struck off from a central core and then retouched. They are known as points and scrapers, which also presumably describes their functions. They are usually 5 to 8 centimeters long. Some may have been used as points for spears or javelins, but, in general, it appears that they were hand-held. Spears, which were in use at that time, were of fire-hardened wood.

Increase in Intelligence

In mammals, the proportion of brain size to body size can be correlated with the intelligence of the species under consideration. It is therefore reasonable to assume that the enormous increase in cranial capacity that took place in the course of human evolution signifies strong selection pressures for intelligence. In fact, insofar as one can compare the evolution of the human brain with the evolution, for instance, of the foreleg of the horse (which is considered very rapid), the human brain evolved 100 times more quickly. The increase in cranial capacity (Figure 49–22) took place well after bipedalism and the early use of tools and well before the development of sophisticated tools or weapons, which did not take place until a mere 30,000 to 40,000 years ago, as we shall see.

Homo sapiens

The earliest fossils that are classified as *Homo sapiens* come from Swanscombe in England and Steinheim in Germany. They consist only of some skull fragments, but these clearly indicate that the brains of these men were larger than those of *H. erectus* and that their skulls were less massive. These skull fragments are dated at about 250,000 to 150,000 years ago, during an interglacial period when Europe was warm, probably warmer than it is today.

Neanderthal Man

Then we have another of those large and frustrating gaps in the fossil record. The next specimens that are known to us date from around 80,000 to about 40,000 years ago, the period of the last glaciation. This period abounds with specimens of what we have come to call Neanderthal man. They have been found largely in Europe but also in the Near East and Central Asia. Neanderthal man stood erect, had a brain capacity somewhat larger than modern man, a thick skull, a prognathous (protruding) muzzle, a low forehead, and heavy brow ridges. He is now classified as a variety of *H. sapiens.*

Neanderthals also used simple, hand-held stone tools (Figure 49–23). Some of the stone tools appear to have been used for scraping hides, suggesting that Neanderthals wore clothing made of animal skins, which would certainly have been in keeping with the climate in which they lived.

Neanderthals buried their dead, often with food and weapons and, in at least one instance, with spring flowers. Formal burials such as these suggest a belief in life after death. Whether or not Neanderthals were able to speak is a matter of current debate, but it is difficult to understand how a society could hold such abstract concepts without some means of exchanging ideas among its members.

Cro-Magnon Man

About 30,000 to 40,000 years ago, specimens of Neanderthal man disappear abruptly from the fossil record and are replaced by what is known as Cro-Magnon man, or sometimes as Upper Pleistocene man, who is physically indistinguishable from modern man. We do not know what became of the Neanderthals. Perhaps they were exterminated in warfare, although there is no evidence of this. Perhaps they were simply unable to compete for food and living space with the better-equipped Cro-Magnon type. Perhaps they interbred, although there are no clear traces of intermediate forms. Some have suggested that modern men brought with them some disease to which they themselves were resistant and the Neanderthals were not. In any case, soon after the appearance of Cro-Magnon man, there were no other hominids in Europe, and within the course of 10,000 to 20,000 years, this new variety of primate had spread over the face of the planet.

Cro-Magnon man, when he first appeared in Europe, came bearing a new, quite different, and far better tool kit (Figure 49–24). His stone tools were essentially flakes—which, of course, had been in use for more than 2½ million years—but they were struck from a carefully prepared core with the aid of a punch (a tool made to make another tool). These flakes, usually referred to as blades, were smaller, flatter, and narrower, and, most important, they could be and were shaped in a large variety of ways. They included, from the beginning, various scraping and piercing tools, flat backed knives, awls, chisels, and a number of different engraving tools. Using these tools to work other materials, especially bone and ivory, Cro-Magnon man made a variety of projectile points, barbed points for spears and harpoons, fishing hooks, and needles. Thus, although Cro-Magnon man lived much the same sort of existence as that of his forebears, he apparently lived it with more possessions, more comfort, and more style.

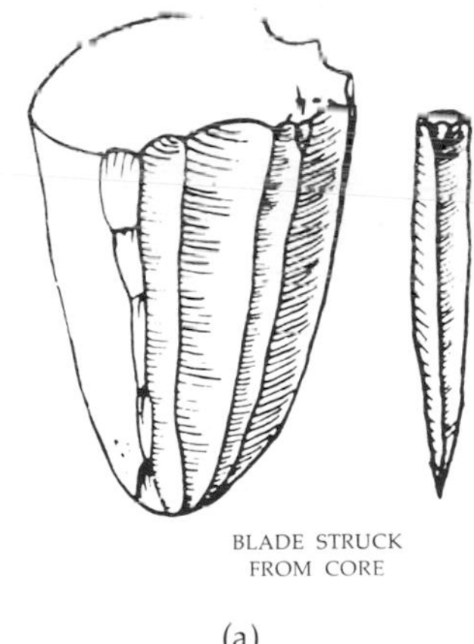

(a)

(b)

49–24
(a) *Cro-Magnon culture was characterized by a great increase in the types of specialized tools made from long and relatively thin flakes with parallel sides, called blades.* (b) *These beautifully worked "laurel leaf" blades are often more than 30 centimeters long and only ½ centimeter thick.*

49–25
In almost every cave, a small number of the animals show wounds, as seen in this steppe wisent from the Cave of Niaux in France. Although the animals are shown in realistic detail, the weapons and the hunters, if shown at all, are abstractions, perhaps to keep their identity a secret. The illusion of movement is greatly enhanced by the patterns of light and shadow in the dark, narrow recesses of the cave.

Perhaps our closest emotional links to this most immediate ancestor are the cave paintings of western Spain and southern France (Figure 49–25). The examples that remain to us, many surprisingly untouched by time, clearly form a part of a rich artistic tradition that has endured for at least 10,000 years. The cave drawings are almost entirely of animals, nearly all game animals, and they are deep within the caves, so they must have been viewed (as they must have been painted) by the light of crude lamps or torches.

The meaning of these drawings and paintings has long been a matter of debate. Some of the animals are marked with darts or wounds (although very few appear to be seriously injured or dying). Such markings have led to the suggestion that the figures are examples of sympathetic magic, in which there is the notion that one can do harm to one's enemy by sticking needles in his image. The fact that many of the animals appear to be pregnant suggests that they may symbolize fertility. Many appear also to be in motion. Perhaps these animals, so vital to the hunters' welfare, were migratory in these areas, and they may have seemed to vanish at certain times in the year, mysteriously returning, heavy with young, in the springtime. This return of the animals might have been an event to be solicited or celebrated in much the same spirit as the rites of spring of more recent peoples or Easter are celebrated.

Cave art came to an end perhaps 8,000 or 10,000 years ago. Not only were the tools and pigments laid aside, but the sacred places—for such they seem to have been—were no longer visited. New forces were at work to mold the course of human existence, and the end of this first great era in human art is our clearest line of demarcation between the old life of man the hunter and the new life that was to come.

The Agricultural Revolution

The most important event in the cultural evolution of man was the change to an agricultural way of life. The reasons for such a transition are not clear, but one factor seems to have been a change in the climate. The most recent of the glaciations began to retreat about 18,000 years ago, withdrawing slowly for about 10,000 years. As the glaciers retreated, the plains of northern Europe and of North America, once cold grasslands, or steppes, gave way to forest. Many of the great herbivores that roamed these steppes retreated northward and eventually vanished; the woolly mammoth was last seen in Siberia more than 10,000 years ago. While some animals became extinct, humans adapted, as they had during the entire period of violent climatic changes that have marked their evolutionary history. With the migratory animals gone, man shifted his attention to smaller game. The hunting and gathering of small animals—instead of large migratory herbivores—undoubtedly resulted in a less nomadic existence for the hunters and formed a prelude to the agricultural revolution. (However, it should be noted that similar fluctuations had occurred previously in the history of man without leading to a revolution in the way of life.)

The earliest traces of agriculture, dating back about 11,000 years, are found in areas of the Near East, which are now parts of Iran, Iraq, and Turkey. Here were the raw materials required by an agricultural economy: cereals, which are grasses with seeds capable of being stored for long periods without serious deterioration, and herbivorous herd animals, which can be readily domesticated. The grasses in these areas were wild wheats and barley, which still grow wild in the foothills. The animals were wild sheep and goats.

By 8,100 years ago, agricultural communities were established in eastern Europe. By 7,000 years ago (about 5000 B.C.), agriculture had spread to the western Mediterranean and up the Danube River into central Europe and, by 4000 B.C., to Britain. During this same period, agriculture originated separately in Central and South America, and perhaps slightly later in the Far East.

THE POPULATION EXPLOSION

One immediate and direct consequence of the agricultural revolution was an increase in population. It is estimated that about 25,000 years ago there may have been as many as 3 million people. By the close of the Pleistocene epoch, some 10,000 years ago, the human population probably numbered a little more than 5 million, spread over the entire world. By 4000 B.C., about 6,000 years ago, the population had increased enormously, to about 90 million, and by the time of Christ, it is estimated, there were 130 million people. In other words, the population increased more than 25 times between 10,000 and 2,000 years ago.

By 1650, the world population had reached 500 million, many people lived in urban centers, and the development of science, technology, and industrialization had begun, bringing about further profound changes in human life and its relationship to nature.

In 1980, it is estimated that there will be about 4½ billion people on our planet (Figure 49–26), and the number is increasing at a rapid rate. Although the birth rate in the United States plunged dramatically from 18.4 babies born per 1,000 people in the population in 1970 to 15 in 1973, and has remained there ever since, the world's population is growing at 2 percent per year. This means that about 171 people are added to the world population every minute, about 246,500 each day, and 90 million every year. If this rate of increase is sustained, there will be 6½ billion people on earth by the year 2000.

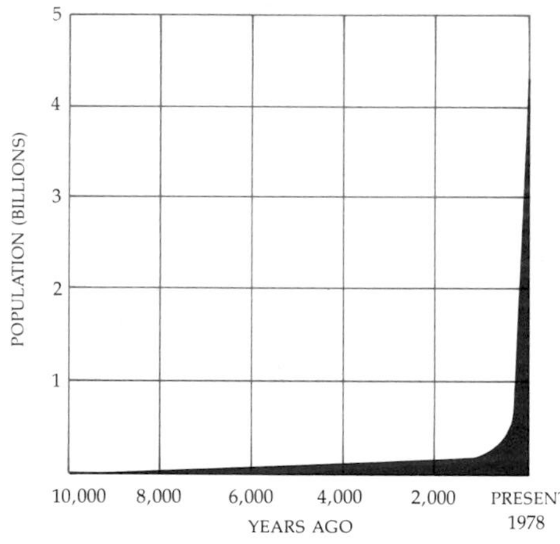

49–26
Growth of the human population. The total biomass of human beings is now estimated at 135 million metric tons, far greater than that of any other animal, with 1 million metric tons added every year.

Life versus Death

The size of a population is, of course, the result both of its birth rate and its death rate (see page 850), and the enormous recent increase in the world's population is actually more a product of the decline in the death rates, especially among the young, than an increase in births. For example, in Pakistan, the population was 20 million in 1911, 35 million in 1950, and 78 million in 1978. During this period, birth rates seem to have remained the same—about 45 per 1,000—but the death rate is one-third of what it was only 20 years ago.

Population growth also depends on the age of the population. In Pakistan, again, almost half of the population is under 15, so if the ideal of a two-child family were to be obtained in the near future, Pakistan's population could still double in the next half century. (The effects of population structure on population growth are diagrammed on page 852.)

The Rich and the Poor

The unequal distribution of growth makes the problem more difficult. First, births are occurring at the greatest rates in precisely those areas where the new arrivals have the least chance of an adequate diet, good housing, schools, medical care, or future occupations. Moreover, because the affluent citizens of developed countries are not constantly reminded of the soaring population rate by the problems of hunger and crowding, so evident to the poor of other countries, they may feel that it is not their problem. Yet a child born to a middle-class American will consume, in his or her lifetime, a far greater amount of the limited resources of the world—more than twice the amount of food, for instance—than a child born in a less developed country. In the words of Jean Mayer, chairman of the National Council on Hunger and Malnutrition in the United States:

> Rich people occupy much more space, consume more of each natural resource, disturb the ecology more, and create more land, air, water, chemical, thermal and radioactive pollution than poor people. So it can be argued that from many viewpoints it is even more urgent to control the number of the rich than it is to control the numbers of the poor.

Birth Rates, Death Rates, and Social Security

Despite the fact that the most recent phase of the population increase is related to a reduction in death rates, particularly from infectious diseases, some experts believe that further reductions in death rates and a general increase in the standard of living will reduce the rate of population growth. There seems to be a correlation between economic deprivation and high birth rates. It is in the developing countries that the greatest rates of population growth are found. In India, for example, where large numbers of people are now starving, the annual rate of population increase is 2.1 percent. Yet, most Indian women do not seek help in birth control until they have three or four children. The desire for large families is deeply rooted in the Indian culture. Moreover, in Indian tradition, children provide security for their parents in old age. Two recent studies of life expectancy in India show that, with the high death rate among children, it is necessary for a mother to bear five children if the couple is to be 95 percent certain that one son will survive the father's sixty-fifth birthday.

Barry Commoner of Washington University in St. Louis, who has long been in the front ranks of our environmental protectors, is of the opinion that population pressures can be significantly reduced by reducing death rates, particularly among infants. Wherever income rises and death rates drop below a critical point, birth rates begin to fall too, he points out. This was true historically in industrialized nations, and it is becoming true in some countries where living standards are just beginning to rise. However, for many countries, a significant rise in the standard of living seems an almost unattainable goal and, in some cases, a rapidly receding one.

It is clear that a reduction in birth rate is not directly correlated with the availability of birth-control measures. United States' birth rates are very low now, 15 per 1,000, but they were equally low in the 1930s during the Depression. The Caribbean island of Jamaica has a Family Planning Center in almost every small town, dispensing pills and contraceptive devices at very low cost, yet the current birth rate is 29 per 1,000 population.

OUR HUNGRY PLANET

The consequences of the increase in the human population include the problems of pollution, depletion of fossil fuels, destruction of natural resources, and extinction of other species. By far the most difficult problem and also the most urgent is

49–27
Field workers weeding a rice plot at the International Rice Research Institute in the Philippines. The new strains of rice are dwarf varieties with short, stiff stalks that respond well to heavy fertilization. Older, nondwarf strains become too tall when heavily fertilized and tend to fall over, spilling the grain.

hunger and starvation. Of the world's 4½ billion people, at least 1 billion are inadequately nourished. It is estimated that about one-third of the deaths that now occur worldwide are due directly or indirectly to malnutrition. Perhaps even more important, in terms of the world's future, are the effects, both physical and psychological, of the prolonged chronic hunger of so large a proportion of our population.

The effort to increase agriculture by the development of new crop plants—especially grains—is called the Green Revolution. Enormous progress is being made. From 1950 to 1970, the production of wheat in Mexico increased from 270,000 metric tons per year to 2.35 million; the corn harvest increased a more modest 250 percent, with yields per hectare almost doubling. Between 1950 and the present, India has increased its production of food grains about 2.8 percent a year. (Its population during this period has increased about 2.1 percent a year, which is significantly less.) Since 1971, China, the most populous nation in the world, has become agriculturally self-sustaining. Most of this success has come about as a result of improved varieties of crop plants, combined with better techniques of irrigation and fertilization. Moreover, and perhaps most important, the full potential of the Green Revolution has not yet been realized.

Despite its acknowledged success, this massive effort has come under criticism in recent years. One reason is the increasing cost of fertilizer; these new grains require intensive cultivation. Because the large landowners are able to afford the investment in fertilizer and farming equipment that the small-scale farmers cannot, these new agricultural developments are seen as accelerating the consolidation of farm lands into a few large holdings by the very wealthy. Most serious of all, although food production is still outstripping population growth, the Green Revolution, at its most productive, will not be able to keep pace indefinitely with the rapid growth of the world's population. (If the population had remained at its 1950 levels, there would be enough food to feed everyone on the planet adequately today.)

Finally, there is a more fundamental though more elusive reason for the dissatisfaction with the Green Revolution. When it was first introduced, it appeared to many to be an almost magical solution to problems so enormous and distressing that they had seemed insoluble. It is now clear, however, that poverty and famine and the unrest and violence they may bring will not be solved by a "technological fix." The Green Revolution must of course go forward. At the same time, we must recognize that the broader solutions are social, political, and ethical, involving not only the growth of crops but their distribution, not only the limiting of populations but the raising of living standards of these populations to tolerable levels.

SCIENCE AND HUMAN VALUES

In the Introduction, we drew an analogy between the goals of science and those of art, religion, or philosophy. There is an important difference, however. The raw materials of science are our observations of the phenomena of the natural universe. Science is limited to what is observable and measurable and, in this sense, is rightly categorized as materialistic. Hunches are abandoned, hypotheses superseded, theories shattered, but the observations endure, and, moreover, they are used over and over again, sometimes in wholly new ways. It is for this reason that scientists stress and seek objectivity. (In the arts, by contrast, the emphasis is on subjectivity—experience as filtered through the individual consciousness.)

Because of this emphasis on objectivity, value judgments cannot be made in science in the way that such judgments are made in philosophy, religion, and the arts, and indeed in our daily lives. Whether or not something is good or beautiful or right in a moral sense, for example, cannot be determined by the scientific method. Such judgments, even though they may be supported by a broad consensus, are not subject to proof by the scientific method.

At one time, sciences, like the arts, were pursued for their own sake, for pleasure and excitement and satisfaction of the insatiable curiosity with which we are both cursed and blessed. In the twentieth century, however, the sciences have spawned a host of giant technological achievements—the H-bomb, the Salk vaccine, DDT, indestructible plastics, nuclear energy plants, perhaps even ways to manipulate our genetic heritage—but have not given us any clues about how to use them wisely. Moreover, science, as a result of these very achievements, appears enormously powerful. It is thus little wonder that among members of the present generation, as perhaps not in any previous one, there are many individuals who are angry at science, as one would be angry at an omnipotent authority who apparently has the power to grant one's wishes but who refuses to do so.

The reason that science cannot and does not solve the problems we want it to is inherent in its nature. Most of the problems we now confront can be solved only by value judgments. For example, science gave us nuclear weapons and also some predictions as to the extent of the biological damage that their use might cause. Yet it could not help us, as citizens, in weighing the risk of damage from fallout against the desire for a strong national defense. In fact, in the 1950s, the most outspoken advocate of atomic testing was one of the country's most prominent physicists, as was the most ardent opponent.

Science has produced the information for the construction of the artificial kidney and can explain how it works. Yet the application of scientific methods cannot help us decide who, given limited availability of such complex and expensive machines, should be treated by them and who should be abandoned. Similarly, science can predict the possible extent of damage to a particular area from the use of pesticides, but it cannot weigh this damage against the reduction of food crops or the increase in malaria that would occur were they prohibited.

At one time, it was traditional for scientists to abstain from such political and ethical controversies, maintaining quite rightly that scientists have no special qualifications in such matters. "Living in an ivory tower" became a well-worn phrase. Now there is an increasing tendency for scientists as citizens to take public stands, particularly on matters involving the applications of science. For example, a large number of distinguished scientists, including several Nobel laureates, recently issued a public statement questioning the safety of nuclear reactors. Also, in recent years, a small group of molecular geneticists took the unprecedented move of calling for a general moratorium on the research in which they themselves were involved.

It is one of the ironies of this so-called "age of science and materialism" that probably never before have ordinary individual men and women, including scientists, been confronted with so many moral and ethical problems.

In the age of the dinosaurs, the earliest primates survived, it would appear, largely by their wits; now, if we are to survive the monsters of our own creating, we will have to do it, again, by the contents of our own skulls. For within the human mind—that complex collection of neurons and synapses—resides the uniquely human capacity to accumulate knowledge, to plan with foresight, and so to act with enlightened self-interest and even, on occasion, with compassion.

SUMMARY

The first mammals arose from primitive reptilian stock about 200 million years ago and coexisted with the dinosaurs for 130 million years. The extinction of the dinosaurs was followed by a rapid adaptive radiation of the mammals. The primates are an order of mammals that became adapted to arboreal life. Primates are characterized by five-digited extremities adapted for grasping with nails rather than claws and by freely movable limbs. They are dependent more upon vision than upon smell, and the higher primates all have stereoscopic vision with foveas for fine focus and cones for color vision.

The two principal groups of living primates are the prosimians and the anthropoids. Fossil prosimians were widespread and abundant during the Paleocene and Eocene epochs, some 65 to 38 million years ago. Modern prosimians include lorises, bush babies, lemurs, and tarsiers.

The anthropoids include the New World monkeys, the Old World monkeys, and the hominoids (apes and man). Early in hominoid evolution, in the Miocene epoch, the hominoids diverged. *Ramapithecus,* known only from some jaw fragments dated 12 to 14 million years ago, is believed to be ancestral to man. *Ramapithecus* was smaller than its contemporaries, with a broader dental arch and teeth adapted for powerful chewing. Evidence suggests that *Ramapithecus* was bipedal.

From $3\frac{1}{2}$ million to 1 million years ago, groups of hominids lived that were bipedal, had larger brains than apes, and, some of whom at least, used simple tools and ate meat. Evidence now indicates the coexistence of at least three species, which have been called *Australopithecus africanus, Australopithecus boisei,* and *Homo habilis.*

Homo erectus lived about 1 million to 300,000 years ago. Individuals of the species were successful hunters of very large animals, indicating that they lived in fairly large groups, shared food, and worked together cooperatively. The hand ax is associated with *H. erectus,* and its widespread use indicates cultural exchanges between groups. Some groups lived in caves and had fire, two developments that are probably related.

Neanderthal man is considered a variety of *Homo sapiens.* Neanderthal fossils date from about 80,000 to 30,000 years ago, the period of the last glacial advance. The majority of specimens have been found in Europe. Neanderthals had fire, inhabited caves, hunted large animals, and probably wore clothing. They used stone tools of a characteristic type. They buried their dead, sometimes with food and weapons. Neanderthals disappeared 30,000 to 40,000 years ago.

Cro-Magnon men, or Upper Pleistocene men, who replaced the Neanderthal, were nomads and hunters, like their predecessors. They had tools (blade tools) distinctly different, smaller, and far more varied than had been seen previously. Their art, which covers a period of some 10,000 years, is evidence of a culture rich not only in beauty but also in mystery, magic, and tradition.

The most important event in the cultural evolution of man was the advent of agriculture, about 10,000 years ago.

About 8000 B.C., when agriculture had its beginning, there were probably about 5 million people in the world. By 4000 B.C., the population had increased to about 90 million; by the time of Christ, there were probably around 130 million people in the world; and by 1650, there were about 500 million. In 1980, there will be 4½ billion, and the population continues to increase, especially in the developing countries. The most serious consequence of this rapid increase in world population is widespread malnutrition and starvation.

QUESTIONS

1. Distinguish among the following: primate/prosimian; monkey/ape; hominid/hominoid/anthropoid; *Homo habilis/Australopithecus.*

2. If you were to meet one of the following, how would you distinguish him (or her): *Australopithecus? Homo habilis? Homo erectus?* Cro-Magnon man?

3. Name five evolutionary trends among primates, and discuss the probable selective value for each.

4. Can you offer a possible relation between these two facts: (1) the only modern lemurs are a widely diversified group found on the island of Madagascar; (2) the only other primate on Madagascar is man, and he is a late arrival.

5. Buckminster Fuller, the architect, has said, "Pollution is resources we are not harvesting." Give some examples to support this statement. Do you agree with his definition?

6. In an editorial entitled "Food, Overpopulation, and Irresponsibility" that appeared recently in a scientific journal, the authors concluded: "Because it creates a vicious cycle that compounds human suffering at a high rate, the provision of food to the malnourished nations of the world that cannot, or will not, take very substantial measures to control their own reproductive rates is inhuman, immoral and irresponsible." What is your opinion?

SUGGESTIONS FOR FURTHER READING

Books

ALCOCK, JOHN: *Animal Behavior: An Evolutionary Approach,* Sinauer Associates, Inc., Sunderland, Mass., 1975.

A general text on animal behavior.

"The Biosphere," *Scientific American,* September 1970.*

A reprint of the September 1970 issue of Scientific American. *Its 11 chapters by different authors are devoted entirely to energy flow and biogeochemical cycles and their relationship to current human problems of food consumption and pollution.*

BROWN, JERRAM: *The Evolution of Behavior,* W. W. Norton & Company, New York, 1975.

A text for advanced undergraduates and graduates, with emphasis on ethology and evolution.

* Available in paperback.

CAMPBELL, BERNARD: *Human Evolution*, Heinemann Educational Books, Ltd., London, 1967.

Campbell emphasizes the way that the special adaptations characterizing the hominids can be related to their environment, ecology, and culture.

CARSON, RACHEL: *Silent Spring*, Houghton Mifflin Company, Boston, 1962.*

This is the book that awakened the nation to the dangers of pesticides. Once seen as highly controversial, it is now regarded as a classic of modern ecology.

COLINVAUX, PAUL: *Introduction to Ecology*, John Wiley & Sons, Inc., New York, 1973.

A good, general text by a teacher who clearly enjoys his subject, designed primarily for second-year students and science majors. It is notable for its stress on and criticism of the concepts of ecology.

COLINVAUX, PAUL: *Why Big Fierce Animals Are Rare*, Princeton University Press, Princeton, N.J., 1978.

A series of essays that form an excellent introduction to modern ecology.

COMMONER, BARRY: *The Closing Circle*, Alfred A. Knopf, Inc., New York, 1971.*

The author traces the relationship between the present ecological crisis and technological and social factors in our society. His title, The Closing Circle, *refers to his conviction that if we are to survive, "we must learn how to restore to nature the wealth we borrow from it." Commoner's book is criticized by Ehrlich and some other ecologists and demographers as presenting an oversimplified and too optimistic solution to environmental problems.*

EHRLICH, PAUL, and ANNE EHRLICH: *Population/Resources/Environment*, 2d ed., W. H. Freeman and Company, San Francisco, 1972.

Required, although not cheerful, reading for all concerned with the ecological crisis. The Ehrlichs present a wealth of useful data, even though you may not agree with all their conclusions.

EMMEL, THOMAS C.: *An Introduction to Ecology and Populations*, W. W. Norton & Company, New York, 1973.*

A short, sound text for the beginner, logically presented and sensibly organized.

HUTCHINSON, G. EVELYN: *The Ecological Theater and the Evolutionary Play*, Yale University Press, New Haven, Conn., 1965.

By one of the great modern experts on freshwater ecology, this is a charming and sophisticated collection of essays on the influence of environment in evolution—and also on an astonishing variety of other subjects.

KREBS, CHARLES J.: *Ecology: The Experimental Analysis of Distribution and Abundance*, Harper & Row, Publishers, Inc., New York, 1972.

This excellent, modern text focuses on populations and on interactions among them. Requires some knowledge both of biology and of mathematics.

KRUTCH, J. W.: *The Desert Year*, Viking Press, Inc., New York, 1960.*

A description by one of the best contemporary American nature writers of the animal and plant life of the American desert.

KUMMER, HANS: *Primate Societies: Group Techniques of Ecological Adaptation*, Aldine, Atherton, Inc., Chicago, 1971.

Most of this book is concerned with behavior in baboons. This book is greatly enriched both by the author's personal experience in observing baboons and other primates and by his interpretation of

* Available in paperback.

primate behavior patterns in terms of evolutionary and ecological principles. Some wonderful and unusual photographs, also by the author.

LAPPÉ, FRANCES MOORE: *Diet for a Small Planet,* Ballantine Books, Inc., New York, 1971.*

Lappé clearly establishes the feasibility of eating "low on the food chain," thereby greatly increasing the availability of proteins to other human populations.

LEAKEY, RICHARD, and ROGER LEWIN: *Origins,* E. P. Dutton & Co., Inc., New York, 1977.

A beautifully illustrated, thoughtful, and imaginative book about 3 million years of human history, and how we know what we do about it, with some speculations on human nature.

LE GROS CLARK, W. E.: *Antecedents of Man: An Introduction to the Evolution of the Primates,* Harper & Row, Publishers, Inc., New York, 1963.*

Le Gros Clark's emphasis is on the anatomy and physiology of the entire primate order.

MECH, L. DAVID: *The Wolf: The Ecology and Behavior of an Endangered Species,* Natural History Press, Garden City, N.Y., 1970.

The definitive book on the social history of wolves.

MORAN, JOSEPH M., MICHAEL D. MORGAN, and JAMES H. WIERSMA: *An Introduction to Environmental Sciences,* Little, Brown and Company, Boston, 1973.

A clear, matter-of-fact presentation of the basic features of our planet and, in particular, of the biosphere. It provides a good background for the study of ecology in general and problems of environmental pollution and technological change in particular.

PILBEAM, DAVID: *The Ascent of Man: An Introduction to Human Evolution,* The Macmillan Company, New York, 1972.

A thorough survey of primate evolution, including evolution of the hominids.

SANDERS, N. K.: *Prehistoric Art in Europe,* Penguin Books, Ltd., Harmondsworth, England, 1968.

An illustrated history of art's first 30,000 years.

SMITH, ROBERT L.: *Ecology and Field Biology,* 2d ed., Harper & Row, Publishers, Inc., New York, 1974.

The outstanding sections of this text are those that deal with descriptions of biomes and communities and the plants and animals within them. Although meant to accompany a course in field biology, this book would be an asset and a pleasure to the amateur naturalist.

STORER, JOHN H.: *The Web of Life,* New American Library, Inc., New York, 1966.*

One of the first books ever written on ecology for the layman. In its simple presentation of the interdependence of living things, it remains a classic.

TINBERGEN, NIKO: *Curious Naturalists,* Natural History Library, Doubleday & Company, Inc., Garden City, N.Y., 1968.*

Some charming descriptions of the activities and discoveries of scientists studying the behavior of animals in their natural environment.

UCKO, PETER J., and ANDRÉE ROSENFELD: *Paleolithic Cave Art,* McGraw-Hill Book Company, New York, 1967.*

This book not only presents the cave art, in photos and drawings, but also examines its contents and context and discusses the various interpretations of it.

* Available in paperback.

VAN LAWICK-GOODALL, JANE: *In the Shadow of Man,* Houghton Mifflin Company, Boston, 1971.

An absorbing personal account of 11 years spent observing the complex social organization of a single chimpanzee community in Tanzania.

WICKLER, WOLFGANG: *Mimicry in Plants and Animals,* World University Library, London, 1968.*

Many examples and illustrations of a delightful subject.

WILSON, E. O.: *The Insect Societies,* Harvard University Press, Cambridge, Mass., 1971.

A comprehensive and fascinating account of the social insects. An unusual nominee for the National Book Award.

WILSON, E. O.: *Sociobiology: The New Synthesis,* Harvard University Press, Cambridge, Mass., 1975.

In this extremely interesting, beautifully written, and beautifully illustrated book, Wilson undertakes to set forth the biological principles that govern social behavior in all kinds of animals. The last chapter, which concerns the sociobiology of man, has become a focus of the current controversy about the inheritance of behavioral traits in Homo sapiens.

Articles

ISAAC, GLYNN: "The Food-Sharing Behavior of Protohuman Hominids," *Scientific American,* April 1978, pages 90–119.

MAY, ROBERT M.: "The Evolution of Ecological Systems," *Scientific American,* September 1978, pages 160–175.

MAYR, ERNST: "Behavior Programs and Evolutionary Strategies," *American Scientist,* vol. 62, pages 650–659, 1974.

SHABECOFF, PHILIP: "New Battles over Endangered Species," *New York Times Magazine,* June 4, 1978.

SIMONS, ELWYN L.: *"Ramapithecus," Scientific American,* May 1977, pages 28–35.

SMITH, JOHN MAYNARD: "The Evolution of Behavior," *Scientific American,* September 1978, pages 176–192.

WASHBURN, SHERWOOD L.: "The Evolution of Man," *Scientific American,* September 1978, pages 194–208.

WOODWELL, GEORGE M.: "The Carbon Dioxide Question," *Scientific American,* January 1978, pages 34–43.

---

* Available in paperback.

Appendixes

APPENDIX A METRIC TABLE

	QUANTITY	NUMERICAL VALUE	ENGLISH EQUIVALENT	CONVERTING ENGLISH TO METRIC
Length	kilometer (km)	1,000 (10^3) meters	1 km = 0.62 mile	1 mile = 1.609 km
	meter (m)	100 centimeters	1 m = 3.28 feet	1 yard = 0.914 m
			= 1.09 yards	1 foot = 0.305 m
	centimeter (cm)	0.01 (10^{-2}) meter	1 cm = 0.394 inch	= 30.5 cm
	millimeter (mm)	0.001 (10^{-3}) meter	1 mm = 0.039 inch	1 inch = 2.54 cm
	micrometer (μm)	0.000001 (10^{-6}) meter		
	nanometer (nm)	0.000000001 (10^{-9}) meter		
	angstrom (Å)	0.0000000001 (10^{-10}) meter		
Area	square kilometer (km^2)	100 hectares	1 km^2 = 0.3861 square mile	1 square inch = 6.4516 cm^2
	hectare (ha)	10,000 square meters	1 ha = 2.471 acres	1 square foot = 0.0929 m^2
	square meter (m^2)	10,000 square centimeters	1 m^2 = 1.1960 square yards	1 square yard = 0.8361 m^2
			= 10.764 square feet	1 acre = 0.4047 ha
	square centimeter (cm^2)	100 square millimeters	1 cm^2 = 0.155 square inch	1 square mile = 2.590 km^2
Mass	metric ton (t)	1,000 kilograms =	1 t = 0.9842 ton	1 ton = 1.0160 t
		1,000,000 grams		1 pound = 0.4536 kg
	kilogram (kg)	1,000 grams	1 kg = 2.205 pounds	1 ounce = 28.35 g
	gram (g)	1,000 milligrams	1 g = 0.0353 ounce	
	milligram (mg)	0.001 gram		
	microgram (μg)	0.000001 gram		
Time	second (sec)	1,000 milliseconds		
	millisecond	0.001 second		
	microsecond	0.000001 second		
Volume (solids)	1 cubic meter (m^3)	1,000,000 cubic centimeters	1 m^3 = 35.315 cubic feet	1 cubic yard = 0.7646 m^3
			= 1.3080 cubic yards	1 cubic foot = 0.0283 m^3
	1 cubic centimeter (cm^3)	1,000 cubic millimeters	1 cm^3 = 0.0610 cubic inch	1 cubic inch = 16.387 cm^3
Volume (liquids)	kiloliter (kl)	1,000 liters	1 kl = 264.17 gallons	1 gal = 3.785 l
	liter (l)	1,000 milliliters	1 l = 1.06 quarts	1 qt = 0.94 l
	milliliter (ml)	0.001 liter	1 ml = 0.034 fluid ounce	1 pt = 0.47 l
	microliter (μl)	0.000001 liter		1 fluid ounce = 29.57 ml

APPENDIX B

TEMPERATURE CONVERSION SCALE

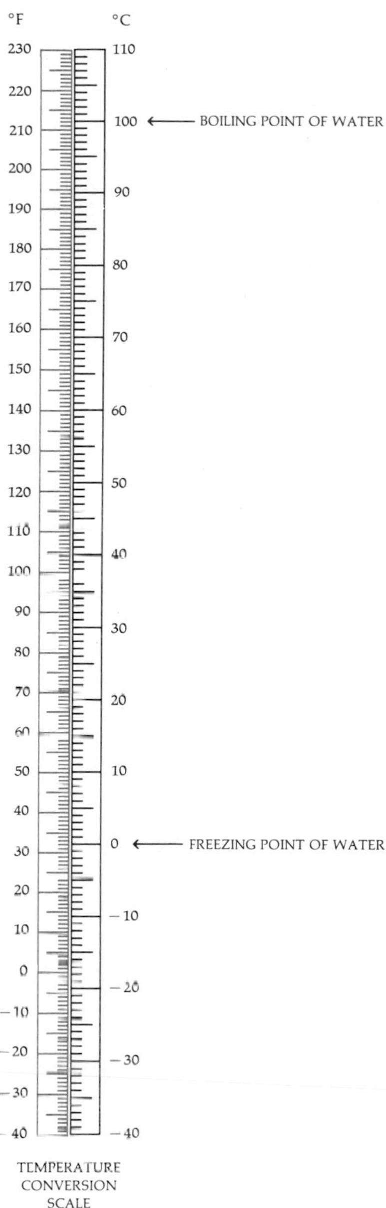

TEMPERATURE CONVERSION SCALE

FOR CONVERSION OF FAHRENHEIT TO CELSIUS, THE FOLLOWING FORMULA CAN BE USED:

$$°C = \frac{5}{9}(°F - 32)$$

FOR CONVERSION OF CELSIUS TO FAHRENHEIT, THE FOLLOWING FORMULA CAN BE USED:

$$°F = \frac{9}{5}°C + 32$$

APPENDIX C

CLASSIFICATION OF ORGANISMS

There are several ways to classify organisms. The one presented here is based closely upon that of Whittaker (in *Science*, vol. 163, pages 150–160, 1969). Organisms are divided into five major groups, or kingdoms: Prokaryotae, Protista, Fungi, Plantae, and Animalia.

The chief taxonomic divisions are kingdom, phylum, class, order, family, genus, species. The following classification includes all of the major phyla. Certain classes and orders, particularly those mentioned in this book, are also included, but the listing is far from complete. The number of species given for each group is the estimated number of living species described and named.

KINGDOM PROKARYOTAE

Prokaryotes are cells that lack a nuclear envelope, chloroplasts and other plastids, mitochondria, and 9 + 2 flagella. Prokaryotes are unicellular but sometimes aggregate into filaments or other superficially multicellular bodies. Their predominant mode of nutrition is absorption, but some groups are photosynthetic or chemosynthetic. Reproduction is primarily asexual, by fission or budding, but conjugation occurs in some species.

PHYLUM

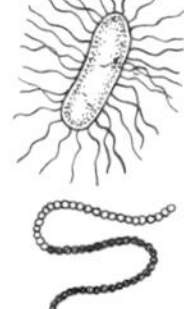

PHYLUM SCHIZOPHYTA: the bacteria. Unicellular prokaryotes; reproduction usually asexual by cell division; nutrition usually heterotrophic. About 1,600 species.

PHYLUM CYANOPHYTA: blue-green algae. Unicellular or colonial; prokaryotic; chlorophyll, but no chloroplasts; nutrition usually autotrophic; reproduction by fission. Common on damp soil and rocks and in fresh and salt water. About 200 distinct, nonsymbiotic species.

KINGDOM PROTISTA

Eukaryotic organisms, including unicellular heterotrophs (protozoans) and unicellular or multicellular algae. Their modes of nutrition include ingestion, photosynthesis, and sometimes absorption. Reproduction both sexual (in some forms) and asexual. They move by 9 + 2 flagella or pseudopods or are nonmotile.

PHYLUM

PHYLUM PROTOZOA: microscopic, unicellular or simple colonial heterotrophic organisms; reproduction usually asexual by mitotic division; classified by type of locomotion. About 30,000 species.

CLASS

Class Mastigophora: protozoans with flagella, including a number of symbiotic forms such as *Trichonympha* and *Trypanosoma,* the cause of sleeping sickness.

Class Sarcodina: protozoans with pseudopods, such as amoebas. No stiffening pellicle; some secrete shells.

Class Ciliophora: protozoans with cilia, including *Paramecium* and *Stentor.*

Class Sporozoa: parasitic protozoans; usually without locomotive organs during a major part of their life cycle. Includes *Plasmodium,* several species of which cause malaria.

PHYLUM

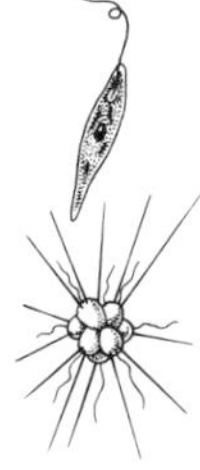

PHYLUM EUGLENOPHYTA: euglenoids. Unicellular photosynthetic (or sometimes secondarily heterotrophic) organisms with chlorophylls *a* and *b*. They store food as paramylon, an unusual carbohydrate. Euglenoids usually have a single apical flagellum and a contractile vacuole. Sexual reproduction is unknown. Euglenoids occur mostly in fresh water. There are some 800 species.

PHYLUM CHRYSOPHYTA: golden algae and diatoms. Unicellular photosynthetic organisms with chlorophylls *a* and *c* and the accessory pigment fucoxanthin. Food stored as the carbohydrate leucosin or as large oil droplets. Cell walls consisting mainly of pectic compounds, sometimes heavily impregnated with siliceous materials. Some 6,000 to 10,000 living species.

CLASS

Class Bacillariophyceae: diatoms. Chrysophyta with double siliceous shells, the two halves of which fit together like a pillbox. They are sometimes motile by the secretion of mucilage fibrils along a specialized groove, the raphe. There are many extinct and 5,000 to 9,000 living species.

Class Chrysophyceae: golden algae. A diverse group of organisms, including flagellated, amoeboid, and nonmotile forms, some naked and others with a cell wall that may be ornamented with siliceous scales. At least 1,000 species.

PHYLUM

PHYLUM PYRROPHYTA: "fire" algae, sometimes called golden-brown algae. Unicellular photosynthetic organisms with chlorophylls *a* and *c*. Food is stored as starch. Cell walls contain cellulose. The phylum contains some 1,100 species, mostly biflagellated organisms, of which the great majority belong to the following class:

CLASS

Class Dinophyceae: dinoflagellates. Pyrrophyta with lateral flagella, one of which beats in a groove that encircles the organism. They probably have no form of sexual reproduction, and their mitosis is unlike that in any other organism. More than 1,000 species.

PHYLUM

PHYLUM XANTHOPHYTA: Autotrophic unicellular, amoeboid, and colonial organisms with chlorophyll *a* and carotenoids. Food stored as leucosin. Cell walls consisting primarily of cellulose and pectin. About 450 species.

PHYLUM CHLOROPHYTA: green algae. Unicellular or multicellular, characterized by chlorophylls *a* and *b* and various carotenoids. The carbohydrate food reserve is starch. Motile cells have two whiplash flagella at the apical end. True multicellular genera do not exhibit complex patterns of differentiation. Multicellularity has arisen at least three times, and quite possibly more often. There are about 7,000 known species and possibly many more.

PHYLUM PHAEOPHYTA: brown algae. Multicellular marine organisms characterized by the presence of chlorophylls *a* and *c* and the pigment fucoxanthin. Their food reserve is a carbohydrate called laminarin. Motile cells are biflagellate, with one forward flagellum of the tinsel type and one trailing one of the whiplash type. A considerable amount of differentiation is found in some of the kelps, with specialized conducting cells for transporting photosynthate to dimly lighted regions of the alga present in some genera. There is, however, no differentiation into leaves, roots, and stem, as in the plants. About 1,100 species.

PHYLUM RHODOPHYTA: red algae. Primarily marine organisms characterized by the presence of chlorophyll *a* and pigments known as phycobilins. Their carbohydrate reserve is a special type of starch (floridian). No motile cells are present at any stage in the complex life cycle. The algal body is built up of closely packed filaments in a gelatinous matrix and is not differentiated into leaves, roots, and stem. It lacks specialized conducting cells. There are some 4,000 species.

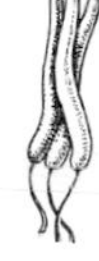

PHYLUM GYMNOMYCOTA: the slime molds. Heterotrophic amoeboid organisms that mostly lack a cell wall but form sporangia at some stage in their life cycle. Predominant mode of nutrition is by ingestion. There are three classes:

CLASS

Class Myxomycetes: plasmodial slime molds. Slime molds with multinucleate plasmodium that creeps along as a mass and eventually differentiates into sporangia, each of which is multinucleate and eventually gives rise to many spores. About 450 species.

Class Acrasiomycetes: cellular slime molds. Slime molds in which there are separate amoebas that eventually swarm together to form a mass but retain their identity within this mass, which eventually differentiates into a compound sporangium. Seven genera and about 26 species.

Class Protostelidomycetes: In this recently discovered group, the amoebas may remain separate or mass together, but each one eventually differentiates into a simple stalked sporangium with one or two spores at its apex. Five genera and more than a dozen species.

KINGDOM FUNGI

Eukaryotic unicellular or multinucleate organisms in which the nuclei occur in a basically continuous mycelium; this mycelium becomes septate (partitioned off) in certain groups and at certain stages of the life cycle. They are heterotrophic, with nutrition by absorption. Reproductive cycles often include both sexual and asexual phases. Some 100,000 species of fungi have been named.

CLASS

Class Oomycetes: Mostly aquatic fungi with motile cells characteristic of certain stages of the life cycle; their cell walls are composed of glucose polymers, including cellulose. There are several hundred species.

Class Zygomycetes: terrestrial fungi, such as black bread mold, with the hyphae septate only during the formation of reproductive bodies; chitin predominant in the cell walls. The class includes several hundred species.

Class Ascomycetes: terrestrial and aquatic fungi, including *Neurospora,* powdery mildews, morels, and truffles. The hyphae are septate but the septa perforated; complete septa cut off the reproductive bodies, such as spores or gametangia. Chitin is predominant in the cell walls. Sexual reproduction involves the formation of a characteristic cell, the ascus, in which meiosis takes place and within which spores are formed. The hyphae in many Ascomycetes are packed together into complex "fruiting bodies." Yeasts are unicellular Ascomycetes that reproduce asexually by budding. About 30,000 species.

Class Basidiomycetes: terrestrial fungi, including the mushrooms and toadstools, with the hyphae septate but the septa perforated; complete septa cut off reproductive bodies, such as spores or gametangia. Chitin is predominant in the cell walls. Sexual reproduction involves formation of basidia, in which meiosis takes place and on which the spores are borne. There are some 25,000 species.

Fungi Imperfecti: Mainly fungi with the characteristics of Ascomycetes but in which the sexual cycle has not been observed; a few probably belong to other classes. The Fungi Imperfecti are classified by their asexual spore-bearing organs. There are some 25,000 species, including a *Penicillium,* the original source of penicillin, fungi which cause athlete's foot and other skin diseases, and many of the molds which give cheese, such as Roquefort and Camembert, their special flavor.

KINGDOM PLANTAE

Multicellular photosynthetic terrestrial eukaryotes and closely related forms. The photosynthetic pigment is chlorophyll *a*, with chlorophyll *b* and a number of carotenoids serving as accessory pigments. The cell walls contain cellulose. There is considerable differentiation of organs and tissues. Their reproduction is primarily sexual with alternating gametophytic and sporophytic phases; the gametophytic phase has been progressively reduced in the course of evolution.

PHYLUM

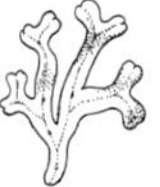

PHYLUM BRYOPHYTA: mosses, hornworts, and liverworts. Multicellular plants with photosynthetic pigments and food reserves similar to those of the green algae. They have gametangia with a multicellular sterile jacket one cell layer thick. The sperm are biflagellate and motile. Gametophytes and sporophytes both exhibit complex multicellular patterns of development, but the conducting tissues are usually completely absent and not well differentiated when present. Most of the photosynthesis in these primarily terrestrial plants is carried out by the gametophyte, upon which the sporophyte is initially dependent. More than 23,500 species.

CLASS

Class Hepaticae: liverworts. The gametophytes are either thallose (not differentiated into roots, leaves, and stem) or leafy, and the sporophytes relatively simple in construction. About 9,000 species.

Class Antherocerotae: hornworts. The gametophytes are thallose. The sporophyte grows from a basal meristem for as long as conditions are favorable. Stomata are present on the sporophyte. About 100 species.

Class Musci: mosses. The gametophytes are leafy. The sporophytes have complex patterns of spore discharge. Stomata are present on the sporophyte. About 14,500 species.

PHYLUM

PHYLUM TRACHEOPHYTA: vascular plants. Plants with complex differentiation of organs into leaves, roots, and stems. The only motile cells are the male gametes of some species, which are propelled by many cilia. The vascular plants have well-developed strands of conducting tissue for the transport of water and organic materials. The main trends of evolution in the vascular plants involve a progressive reduction in the gametophyte, which is green and free-living in ferns but heterotrophic and more or less enclosed by sporophytic tissue in the others; the loss of multicellular gametangia and motile sperm; and the evolution of the seed. The phylum includes the following subphyla with living representatives:

SUBPHYLUM

SUBPHYLUM LYCOPHYTINA: lycophytes. Homosporous and heterosporous vascular plants with microphylls; extremely diverse in appearance. All lycophytes have motile sperm. There are five genera and about 1,000 species.

SUBPHYLUM SPHENOPHYTINA: horsetails. Homosporous vascular plants with jointed stems marked by conspicuous nodes and elevated siliceous ribs and sporangia borne in a strobilus at the apex of the stem. Leaves are scalelike. Sperm are motile. Although now thought to have evolved from a megaphyll, the leaves of the horsetails are structurally indistinguishable from microphylls. There is one genus, *Equisetum,* with about two dozen living species.

SUBPHYLUM PTEROPHYTINA: ferns, gymnosperms, and flowering plants. Although diverse, these groups possess in common the megaphyll, which in certain genera has become much reduced. About 260,000 species.

CLASS

Class Filicineae: the ferns. They are mostly homosporous, although some are heterosporous. The gametophyte is more or less free-living and usually photosynthetic. Multicellular gametangia and free-swimming sperm are present. About 11,000 species.

Class Coniferinae: the conifers. Seed plants with active cambial growth and simple leaves, in which the ovules are not enclosed and the sperm are not flagellated. There are some 50 genera and about 550 species, the most familiar group of gymnosperms.

Class Cycadinae: cycads. Seed plants with sluggish cambial growth and pinnately compound, palmlike or fernlike leaves. The ovules are not enclosed. The sperm are flagellated and motile, but are carried to the vicinity of the ovule in a pollen tube. Cycads are gymnosperms. There are nine genera and about 100 species.

Class Ginkgoinae: ginkgo. Seed plants with active cambial growth and fan-shaped leaves with open dichotomous venation. The ovules are not enclosed and are fleshy at maturity. Sperm are carried to the vicinity of the ovule in a pollen tube, but are flagellated and motile. They are gymnosperms. There is one species only.

Class Angiospermae: flowering plants. Seed plants in which the ovules are enclosed in a carpel (in all but a very few genera), and the seeds at maturity are borne within fruits. They are extremely diverse vegetatively but characterized by the flower, which is basically insect-pollinated. Other modes of pollination, such as wind pollination, have been derived in a number of different lines. The gametophytes are much reduced, with the female gametophyte often consisting of only eight cells or nuclei at maturity. Double fertilization involving two of the three nuclei from the mature microgametophyte gives rise to the zygote and to the primary endosperm nucleus; the former becomes the embryo and the latter a special nutritive tissue, the endosperm. About 250,000 species.

SUBCLASS

Subclass Dicotyledonae: dicots. Flower parts are usually in fours or fives; leaf venation is usually netlike, pinnate, or palmate; there is true secondary growth with vascular cambium commonly present; there are two cotyledons; and the vascular bundles in the stem are in a ring. About 190,000 species.

Subclass Monocotyledonae: monocots. Flower parts are usually in threes, leaf venation is usually parallel, true secondary growth is not present, there is one cotyledon, and vascular bundles in the stem are scattered. About 60,000 species.

KINGDOM ANIMALIA

Eukaryotic multicellular organisms. Their principal mode of nutrition is by ingestion. Many animals are motile, and they generally lack the rigid cell walls characteristic of plants. Considerable cellular migration and reorganization of tissues often occur during the course of embryonic development. Their reproduction is primarily sexual, with male and female diploid organisms producing haploid gametes which fuse to form the zygote. More than a million species have been described and the actual number may be close to 10 million.

PHYLUM

PHYLUM PORIFERA: sponges. Simple multicellular animals, largely marine, with stiff skeletons, and bodies perforated by many pores that admit water containing food particles. All have choanocytes, "collar cells." About 4,200 species.

PHYLUM COELENTERATA: coelenterates. Animals with radially symmetrical, "two-layered" bodies of a jellylike consistency. Reproduction is asexual or sexual. They are the only organisms with cnidoblasts, special stinging cells. All are aquatic and most are marine. About 11,000 species.

CLASS

Class Hydrozoa: Hydra, Obelia, and other *Hydra*-like animals. They are often colonial and frequently have a regular alternation of asexual and sexual generations. The polyp form is dominant.

Class Scyphozoa: marine jellyfishes or "cup animals," including *Aurelia.* The medusa form is dominant. They have true muscle cells.

Class Anthozoa: sea anemones ("flower animals") and colonial corals. They have no medusa stage.

PHYLUM

PHYLUM CTENOPHORA: comb jellies and sea walnuts. They are free-swimming, often almost spherical animals. They are translucent, gelatinous, delicately colored, and often bioluminescent. They possess eight bands of cilia, for locomotion. About 80 species.

PHYLUM PLATYHELMINTHES: flatworms. Bilaterally symmetrical with three tissue layers. The gut has only one opening. They have no coelom or circulatory system. They have complex hermaphroditic reproductive systems and excrete by means of special (flame) cells. About 15,000 species.

CLASS

Class Turbellaria: planaria and other nonparasitic flatworms. They are ciliated, carnivorous, and have ocelli ("eyespots").

Class Trematoda: flukes. They are parasitic flatworms with digestive tracts.

Class Cestoidea: tapeworms. They are parasitic flatworms with no digestive tracts; they absorb nourishment through body surfaces.

PHYLUM

PHYLUM RHYNCHOCOELA: proboscis, nemertine, or ribbon worms. They are nonparasitic, usually marine, and have a tubelike gut with mouth and anus, a protrusible proboscis armed with a hook for capturing prey, and simple circulatory and reproductive systems. About 600 species.

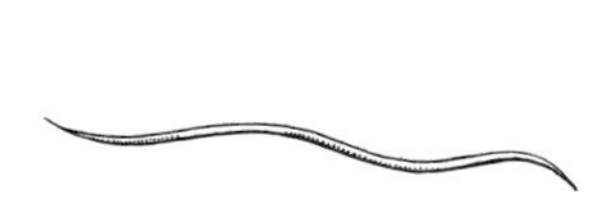

PHYLUM NEMATODA: roundworms. The phylum includes minute free-living forms, such as vinegar eels, and plant and animal parasites, such as hookworms. They are characterized by elongated, cylindrical, bilaterally symmetrical bodies. About 80,000 species.

PHYLUM ACANTHOCEPHALA: spiny-headed worms. They are parasitic worms with no digestive tract and a head armed with many recurved spines. About 300 species.

PHYLUM CHAETOGNATHA: arrow worms. Free-swimming, planktonic marine worms, they have a coelom, a complete digestive tract, and a mouth with strong sickle-shaped hooks on each side. About 50 species.

PHYLUM NEMATOMORPHA: horsehair worms. They are extremely slender, brown or black worms up to 1 meter long. Adults are free-living, but the larvae are parasitic in insects. About 250 species.

PHYLUM ROTIFERA: microscopic, wormlike or spherical animals, "wheel animalcules." They have a complete digestive tract, flame cells, and a circle of cilia on the head, the beating of which suggests a wheel; males are minute and either degenerate or unknown in many species. About 1,500 species.

PHYLUM GASTROTRICHA: These are microscopic, wormlike animals that move by longitudinal bands of cilia. About 140 species.

PHYLUM BRYOZOA: "moss" animals. These microscopic aquatic organisms are characterized by a U-shaped row of ciliated tentacles, with which they feed. They usually form fixed and branching colonies. Bryozoans superficially resemble hydroid coelenterates but are much more complex; having anus and coelom, they retain larvae in special brood pouch. About 4,000 species.

PHYLUM BRACHIOPODA: lamp shells. Marine animals with two hard shells (one dorsal and one ventral), they superficially resemble clams. Fixed by a stalk or one shell in adult life, they obtain food by means of ciliated tentacles. About 260 living species; 3,000 extinct.

PHYLUM PHORONIDEA: sedentary, elongated, wormlike animals that secrete and live in a leathery tube. They have a U-shaped digestive tract and a ring of ciliated tentacles with which they feed. Marine. About 15 species.

PHYLUM ANNELIDA: ringed or segmented worms. They usually have a well-developed coelom, a one-way digestive tract, head, and circulatory system, nephridia, and well-defined nervous system. About 8,800 species.

CLASS

Class Archiannelida: small, simple, probably primitive, marine worms. About 35 species.

Class Polychaeta: mainly marine worms, such as *Nereis*. They have a distinct head with tentacles, antennae, and specialized mouthparts. Parapodia are often brightly colored. About 4,000 species.

Class Oligochaeta: soil, freshwater, and marine annelids, including the earthworm (*Lumbricus*). They have scanty bristles and usually a poorly differentiated head. About 2,500 species.

Class Hirudinea: leeches. They have a posterior sucker and usually an anterior sucker surrounding the mouth. They are freshwater, marine, and terrestrial; either free-living or parasitic. About 300 species.

PHYLUM

PHYLUM MOLLUSCA: unsegmented animals, with a head, a mantle, and a muscular foot, variously modified. They are mostly aquatic; soft-bodied, often with one or more hard shells, and a three-chambered heart. All mollusks, except bivalves, have a radula (rasplike organ used for scraping or marine drilling). About 110,000 species.

CLASS

Class Amphineura: chitons. The simplest type of mollusks, they have an elongated body covered with a mantle in which are embedded eight dorsal shell plates. About 700 species.

Class Pelecypoda: two-shelled mollusks, including clams, oysters, mussels, scallops. They usually have a hatchet-shaped foot and no distinct head. Generally sessile. About 15,000 species.

Class Scaphopoda: tooth or tusk shells. They are marine mollusks with a conical tubular shell. About 350 species.

Class Gastropoda: asymmetrical mollusks, including snails, whelks, slugs. They usually have a spiral shell and a head with one or two pairs of tentacles. About 80,000 species.

Class Cephalopoda: octopus, squid, *Nautilus*. They are characterized by a "head-foot" with eight or ten arms or many tentacles, mouth with two horny jaws, and well-developed eyes and nervous system. The shell is external (*Nautilus*), internal (squid), or absent (octopus). All except *Nautilus* have ink glands. About 400 species.

PHYLUM

PHYLUM ARTHROPODA: The largest phylum in the animal kingdom, arthropods are segmented animals with paired jointed appendages, a hard jointed exoskeleton, a complete digestive tract, reduced coelom, no nephridia, a dorsal brain, and a ventral nerve cord with paired ganglia in each segment. About 765,257 species.

CLASS

Class Merostomata: horseshoe crabs. They are aquatic, with book gills. About 5 species.

Class Crustacea: lobsters, crabs, crayfish, shrimps. Crustaceans are mostly aquatic, with two pairs of antennae, one pair of mandibles, and typically two pairs of maxillae. The thoracic segments have appendages, and the abdominal segments are with or without appendages. About 30,000 species.

Class Arachnida: spiders, mites, scorpions. Most members are terrestrial, air-breathing; usually have four or five pairs of legs; first pair of appendages used for grasping; have chelicerae (pincers or fangs) in place of jaws or antennae. About 35,000 species.

Class Onychophora: simple, terrestrial arthropods. All belong to one genus, *Peripatus*. They have many short, unjointed pairs of legs. About 73 species.

Class Insecta: insects, including bees, ants, beetles, butterflies, fleas, lice, flies, etc. Most insects are terrestrial, and most breathe by means of tracheae. They have one pair of antennae, three pairs of legs, and three distinct parts of the body (head, thorax, and abdomen). Most have two pairs of wings. About 700,000 species.

Class Chilopoda: centipedes. They have 15 to 173 trunk segments, each with one pair of jointed appendages. About 2,000 species.

Class Diplopoda: millipedes. They have an abdomen with 20 to 100 segments, each with two pairs of appendages. About 7,000 species.

PHYLUM

PHYLUM ECHINODERMATA: starfish and sea urchins. Echinoderms are radially symmetrical in adult stage, with a well-developed coelom, an endoskeleton of calcareous ossicles and spines, and a unique water vascular system. They have tube feet and are marine. About 6,000 species.

CLASS

Class Crinoidea: sea lilies and feather stars. Sessile animals, they often have a jointed stalk for attachment, and they have ten arms bearing many slender lateral branches. Most species are fossils.

Class Asteroidae: starfish. They have five to fifty arms, an oral surface directed downward, and two to four tube feet.

Class Ophiuroidea: brittle stars, serpent stars. They are greatly elongated, with highly flexible slender arms and rapid horizontal locomotion.

Class Echinoidea: sea urchins and sand dollars. Skeletal plates form rigid external covering that bears many movable spines.

Class Holothuroidea: sea cucumbers. They have a sausage-shaped or wormlike elongated body.

PHYLUM

PHYLUM HEMICHORDATA: small group of wormlike marine animals, including the acorn worms. They have a notochordlike structure in the head end, gill slits, and a solid nerve cord. About 91 species.

PHYLUM CHORDATA: animals having at some stage a notochord, pharyngeal gill slits, and a hollow nerve cord on the dorsal side. About 44,794 species.

SUBPHYLUM

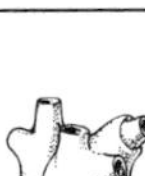

SUBPHYLUM TUNICATA: tunicates or ascidians. Adults are saclike, usually sessile, often forming branching colonies. They feed by ciliary currents, have gill slits, a reduced nervous system, and no notochord. Larvae are active, with well-developed nervous system and notochord. They are marine. About 1,600 species.

SUBPHYLUM CEPHALOCHORDATA: lancelets. This small subphylum contains only *Branchiostoma* and related forms. They are somewhat fishlike marine animals with a permanent notochord the whole length of the body, a nerve cord, a pharynx with gill slits, and no cartilage or bone. About 13 species.

SUBPHYLUM VERTEBRATA: the vertebrates. The most important subphylum of Chordata. In the vertebrates the notochord is replaced by cartilage or bone, forming the segmented vertebral column or backbone. A skull surrounds a well-developed brain. They usually have a tail. About 43,090 species.

CLASS

Class Agnatha: lampreys and hagfish. These are eel-like aquatic vertebrates without limbs, with a jawless sucking mouth, and no bones, scales, or fins.*

Class Chondrichthyes: sharks, rays, skates, and other cartilaginous fish. They have complicated copulatory organs, scales, and no air bladders. They are almost exclusively marine.*

Class Osteichthyes: the bony fish, including nearly all modern freshwater fish, such as sturgeon, trout, perch, anglerfish, lungfish, and some almost extinct groups. They usually have an air bladder or (rarely) a lung.*

Class Amphibia: salamanders, frogs, and toads. They usually breathe by gills in the larval stage and by lungs in the adult stage. They have incomplete double circulation and a usually naked skin. The limbs are legs. They were the first vertebrates to inhabit the land and are ancestors of the reptiles. Their eggs are unprotected by a shell and embryonic membranes. About 2,000 species.

Class Reptilia: turtles, lizards, snakes, crocodiles; includes extinct species such as the dinosaurs. Reptiles breathe by lungs and have incomplete double circulation. Their skin is usually covered with scales. The four limbs are legs (absent in snakes). They are ectotherms. Most live and reproduce on land, although some are aquatic. Their embryo is enclosed in an egg shell and has protective membranes. About 5,000 species.

Class Aves: birds. Birds are homeothermic animals with complete double circulation and a skin covered with feathers. The forelimbs are wings. Their embryo is enclosed in an egg shell with protective membranes. Includes the extinct *Archaeopteryx.* About 8,590 species.

Class Mammalia: mammals. Mammals are homeothermic animals with complete double circulation. Their skin is usually covered with hair. The young are nourished with milk secreted by the mother. They have four limbs, usually legs (forelimbs sometimes arms, wings, or fins), a diaphragm used in respiration, a lower jaw made up of a single pair of bones, three bones in each middle ear connecting eardrum and inner ear, and seven vertebrae in the neck. About 4,500 species.

* The total number of species of fish is estimated to be about 23,000.

SUBCLASS

Subclass Prototheria: monotremes. These are the oviparous (egg-laying) mammals with imperfect temperature regulation. There are only two living species: the duckbill platypus and spiny anteater of Australia and New Guinea.

Subclass Metatheria: marsupials, including kangaroos, opossums, and others. Marsupials are viviparous mammals without a placenta (or with a poorly developed one); the young are born in an undeveloped state and are carried in an external pouch of the mother for some time after birth. They are found chiefly in Australia.

Subclass Eutheria: mammals with a well-developed placenta. This subclass comprises the great majority of living mammals. There are 12 principal orders of Eutheria:

Order

Insectivora: shrews, moles, hedgehogs, etc.
Edentata: toothless mammals—anteaters, sloths, armadillos, etc.
Rodentia: the rodents—rats, mice, squirrels, etc.
Artiodactyla: even-toed ungulates (hoofed mammals)—cattle, deer, camels, hippopotamuses, etc.
Perissodactyla: odd-toed ungulates—horses, zebras, rhinoceroses, etc.
Proboscidea: elephants
Lagomorpha: rabbits and hares
Sirenia: the manatee, dugong, and sea cows. Large aquatic mammals with the forelimbs finlike, the hind limbs absent.
Carnivora: carnivorous animals—cats, dogs, bears, weasels, seals, etc.
Cetacea: the whales, dolphins, and porpoises. Aquatic mammals with the forelimbs fins, the hind limbs absent.
Chiroptera: the bats. Aerial mammals with the forelimbs wings.
Primates: the lemurs, monkeys, apes, and humans.

GLOSSARY

This list does not include units of measure or names of taxonomic groups, which can be found in Appendixes A and C, or terms that are used only once in the text and defined there.

abdomen: In vertebrates, the portion of the trunk containing visceral organs other than heart and lungs; in arthropods, the posterior portion of the body, made up of similar segments and containing the reproductive organs and part of the digestive tract.

abscission [L. *ab,* away, off + *scissio,* dividing]: In plants, the dropping of leaves, flowers, fruits, or stems at the end of a growing season, as the result of formation of an abscission layer, a layer of specialized cells, and the action of a hormone (abscisic acid).

absorption [L. *absorbere,* to swallow down]: The movement of water and dissolved substances into a cell, tissue, or organism.

absorption spectrum: A graph or other display of the wavelengths (colors) of light absorbed by a particular pigment.

acetylcholine (a-**sea**-tell-**co**-leen): One of the chemicals (neurotransmitters) responsible for the passing of nerve impulses across synaptic junctions.

acid [L. *acidus,* sour]: A substance that, on dissociation, releases hydrogen ions (H^+) but not hydroxide ions (OH^-); having a pH of less than 7; the opposite of a base.

actin [Gk. *aktis,* a ray]: One of the two major proteins of muscle (the other is myosin); makes up the Z-band of a sarcomere.

action potential: A transient change in electric potential across a membrane; in nerve cells, results in transmission of a nerve impulse; in muscle cells, results in contraction.

action spectrum: A graph or other display of the wavelengths (colors) of light that elicit a particular reaction or response.

activation energy: The minimum amount of energy that must be supplied from an outside source before a chemical reaction will proceed.

active site: That part of an enzyme molecule into which the substrate fits during the reaction catalyzed by the enzyme.

active transport: The pumping of a substance across a cell membrane from a point of lower concentration to a point of higher concentration (against a concentration gradient); an energy-requiring process.

adaptation [L. *adaptare,* to fit]: (1) The evolution of features that make a group of organisms better suited to live and reproduce in their environment. (2) A peculiarity of structure, physiology, or behavior that aids the organism in its environment.

adaptive radiation: The evolution from a primitive and unspecialized ancestor to several divergent forms, each specialized to fit a distinct niche.

adenosine diphosphate (ADP): The compound formed by hydrolysis of an ATP molecule.

adenosine triphosphate (ATP): The nucleotide that provides the energy currency for cell metabolism; composed of adenine, ribose, and three phosphate groups. On hydrolysis, ATP loses one phosphate and one hydrogen to become adenosine diphosphate (ADP), releasing energy in the process. ATP is formed from ADP + P_i in an enzymatic reaction that traps energy released by catabolism or energy captured in photosynthesis.

ADH: Abbreviation of antidiuretic hormone.

adhesion [L. *adhaerrere,* to stick to]: A sticking together of unlike substances.

ADP: Abbreviation of adenosine diphosphate.

adrenal gland [L. *ad,* near + *renes,* kidney]: A vertebrate endocrine gland. The cortex (outer surface) is the source of cortisol, aldosterone, and other steroid hormones; the medulla (inner core) secretes epinephrine.

adrenaline: *See* Epinephrine.

aerobic [Gk. *aēr,* air + *bios,* life]: Any biological process that can occur in the presence of molecular oxygen (O_2).

afferent [L. *ad,* near + *ferre,* to carry]: Bringing inward to a central part, applied to nerves and blood vessels.

agar: A gelatinous material prepared from certain red algae that is used to solidify nutrient media for growing microorganisms.

aldosterone [Gk. *aldainō,* to nourish + *stereō,* solid]: A hormone produced by the adrenal cortex that stimulates the resorption of sodium from the kidney.

alga, *pl.* **algae** (al-gah, al-jee): A photosynthetic organism lacking multicellular sex organs.

alkaline: Pertaining to substances that release hydroxide (OH^-) ions in water; having a pH greater than 7; basic; opposite of acidic.

alleles (al-**eels**) [Gk. *allelon,* of one another]: Two or more genes that occupy the same position (locus) on homologous chromosomes (and so are separated from each other at meiosis); alleles can mutate one to the other.

alternation of generations: A sexual life cycle in which a haploid (n) phase alternates with a diploid ($2n$) phase. The gametophyte (n) produces gametes (n) by mitosis. Fertilization of gametes yields zygotes ($2n$). Each zygote develops into a sporophyte ($2n$) that forms haploid spores (n) by meiosis. Each haploid spore forms a new gametophyte, completing the cycle.

altruism: Self-sacrifice for the benefit of others.

amino acids (am-**ee**-no) [Gk. *Ammon,* referring to the Egyptian sun god, near whose temple ammonium salts were first prepared from camel dung]: Organic molecules containing nitrogen in the form of NH_2 and an acidic carbon atom, COOH, bonded to the same carbon atom; the "building blocks" of protein molecules.

amnion (**am**-neon) [Gk. dim. of *amnos,* lamb]: Fluid-filled sac that surrounds the embryo in reptiles, birds, and mammals.

amniote egg: An egg that is isolated and protected from the environment by a more or less impervious shell during the period of its development and that is completely self-sufficient, requiring only oxygen from the outside.

amoeboid [Gk. *amoibē,* change]: Moving or eating by means of pseudopodia (temporary cytoplasmic protrusions from the cell body).

AMP: Abbreviation of adenosine monophosphate.

amphibian [Gk. *amphibios,* living a double life]: A class of vertebrates intermediate in many characteristics between fish and reptiles, which live part or most of the time on land but must return to water to reproduce because fertilization is external.

anaerobe [Gk. *an,* without + *aēr,* air + *bios,* life]: Cell that can live without free oxygen; obligate anaerobes cannot live in the presence of oxygen; facultative anaerobes can live with or without oxygen.

anaerobic [Gk. *an,* without + *aēr,* air + *bios,* life]: Applied to a process that can occur without oxygen, as anaerobic fermentation.

analogous [Gk. *analogos,* proportionate]: Applied to structures similar in function but different in evolutionary origin, such as the wing of a bird and the wing of an insect.

anaphase (**anna**-phase) [Gk. *ana,* up + *phasis,* form]: A stage in mitosis or meiosis in which the chromatids of each chromosome separate and move to opposite poles.

androgen [Gk. *andros,* man + *genos,* origin, descent]: Male sex hormone.

angiosperms (**an**-jee-o-sperms) [Gk. *angeion,* vessel + *sperma,* seed]: The flowering plants. Literally, a seed borne in a vessel; thus, any plant whose seeds are borne within a matured ovary (fruit).

annual plant [L. *annus,* year]: A flowering plant that completes its life cycle (from seed germination to seed production) and dies within a single growing season.

antennae: Long, paired sensory appendages on the head of many arthropods.

anterior [L. *ante,* before, toward, in front of]: The front end of an organism; in human anatomy, the ventral surface.

anther [Gk. *anthos,* flower]: In flowering plants, the pollen-bearing portion of a stamen.

anthropoid [Gk. *anthropos,* man, human]: A higher primate; includes monkeys, apes, and humans.

anthropomorphism [Gk. *anthropos,* human + *morphe,* form]: Assignment of human characteristics, abilities, or feelings to nonhuman organisms.

antibiotic [Gk. *anti,* against + *bios,* life]: An organic compound formed and secreted by an organism that is inhibitory or toxic to other species.

antibody [Gk. *anti,* against]: A globular protein that is produced in response to a foreign substance (antigen), with which it reacts specifically.

antidiuretic hormone (ADH) [Gk. *anti,* against + *diurgos,* thoroughly wet; Gk. *hormon,* excite, stimulate]: A hormone secreted by the hypothalamus that inhibits urine excretion by inducing the resorption of water from the nephrons of the kidneys; also called vasopressin.

antigen [Gk. *anti,* against + *genos,* origin, descent]: A foreign substance, usually a protein or polysaccharide, that stimulates the formation of specific antibodies.

aorta (a-**ore**-ta) [Gk. *aeirein,* to lift, heave]: The major artery in blood-circulating systems; the aorta sends blood to the other body tissues.

apical dominance [L. *apex,* top]: In plants, the influence of a terminal bud in suppressing the growth of lateral buds, sometimes the result of the release of growth-regulating hormones.

apical meristem [L. *apex,* top + Gk. *meristos,* divided]: In vascular plants, the growing point at the tip of the root or stem.

arboreal [L. *arbor,* tree]: Tree-dwelling.

archegonium, *pl.* **archegonia** [Gk. *archegeonos,* first of a race]: In embryo-forming plants, the multicellular egg-producing organ.

archenteron [Gk. *arch,* first, or main + *enteron,* gut]: The main cavity within the early embryo (gastrula) of many animals; lined with endoderm, it opens to the outside by means of the blastopore and ultimately becomes the digestive tract.

artery: A vessel carrying blood from the heart to the tissues; arteries are usually thick-walled, elastic, and muscular.

arthropod [Gk. *arthron,* joint + *pous, podos,* foot]: Any invertebrate animal with jointed appendages; a member of the phylum Arthropoda.

artificial selection: The breeding of special strains of organisms for the purpose of enhancing desired traits.

ascus, *pl.* **asci** (**as**-kus, **as**-i) [Gk. *askos,* wineskin, bladder]: In the Ascomycetes, a specialized cell within which two haploid nuclei fuse to produce a zygote that immediately divides by meiosis; at maturity, an ascus contains ascospores.

asexual reproduction: Any reproductive process, such as fission or budding, that does not involve the union of gametes.

atmospheric pressure [Gk. *atmos,* vapor + *sphaira,* globe]: The weight of the earth's atmosphere over a unit area of the earth's surface.

atom [Gk. *atomos,* indivisible]: The smallest particle into which a chemical element can be divided and still retain the properties characteristic of the element.

atomic nucleus [Gk. *atomos,* indivisible; L. *nucleus,* a little nut]: The central core of an atom, containing protons and neutrons, around which electrons move.

atomic number: The number of protons in the nucleus of an atom; equal to the number of electrons in the neutral atom.

atomic weight: The average weight of all the isotopes of an element relative to the weight of an atom of carbon (^{12}C), which is by convention assigned the integral value of 12.

ATP: Abbreviation of adenosine triphosphate, the principal energy-storage compound of the cell.

atrium, *pl.* **atria** (a-tree-um) [L., yard, court, hall]: A thin-walled chamber of the heart that receives blood and passes it on to a thick, muscular ventricle.

autonomic [Gk. *autos,* self + *nomos,* usage, law]: Self-controlling, independent of outside influences.

autonomic nervous system [Gk. *autos,* self + *nomos,* usage, law]: A system of motor nerves and ganglia in vertebrates that is not under voluntary control; innervates the heart, glands, visceral organs, and smooth muscle; subdivided into the sympathetic and the parasympathetic divisions.

autopolyploid [Gk. *autos,* self + *polus,* many + *ploion,* a vessel]: A polyploid in which the chromosomes all come from the same species.

autosome [Gk. *autos,* self + *soma,* body]: Any chromosome other than the sex chromosomes. Humans have 22 pairs of autosomes and 1 pair of sex chromosomes.

autotroph [Gk. *autos,* self + *trophos,* feeder]: An organism that is able to synthesize all needed organic molecules from simple inorganic substances (e.g., H_2O, CO_2, NH_3, and some energy source, e.g., sunlight) in contrast to heterotroph. Plants, algae, and some bacteria are autotrophs.

auxin [Gk. *auxein,* to increase + *in,* of, or belonging to]: One of a group of plant hormones with a variety of growth-regulating effects, including promotion of cell elongation.

axis: An imaginary line passing through a body or organ around which parts are symmetrically aligned.

axon [Gk. *axon,* axle]: The part of a neuron that carries impulses away from the cell body.

backcross: A cross between a hybrid and one of its parents. More generally, a cross between an F_1 heterozygote and an individual with a parental genotype.

bacteriophage [L. *bacterium* + Gk. *phagein,* to eat]: A virus that parasitizes a bacterial cell.

bacterium [Gk. dim of *baktron,* staff]: A unicellular prokaryote.

bark: In higher plants, all tissues outside the cambium in a woody stem.

basal body [Gk. *basis,* foundation]: A cytoplasmic organelle of protozoa and animals that organizes cilia or flagella; identical in structure to the centriole, which is involved in mitosis and meiosis in most protozoa and animals and some plants.

basal metabolism [Gk. *basis,* foundation; *metabolē,* change, changing]: The metabolism of an organism when it is using just enough energy to maintain vital processes; measured as the amount of heat (in terms of quantity of oxygen consumed or carbon dioxide given off) produced by an animal at rest (but not asleep), determined at least 14 hours after eating and expressed as kilocalories per square meter of body surface (or per unit weight) per hour.

base: A substance that, on dissociation, releases hydroxide (OH^-) ions but not hydrogen (H^+) ions; having a pH of more than 7; the opposite of an acid. *See* Alkaline.

base-pairing rule: In the formation of nucleic acids, the requirement that adenine must always pair with thymine (or uracil) and guanine with cytosine.

basidium, *pl.* **basidia** (ba-sid-i-um): A specialized reproductive cell of the Basidiomycetes, often club-shaped, in which nuclear fusion and meiosis occur; homologous with the ascus.

behavioral isolating mechanisms: Modes of behavior, such as display rituals of courtship patterns, that serve to prevent mating between species.

biennial [L. *biennium,* a space of two years; *bi,* twice + *annus,* year]: Occurring once in two years; a plant that requires two years to complete its reproductive cycle: vegetable growth occurs in the first year, sexual reproduction and death in the second.

bilateral symmetry [L. *bi,* twice, two + *lateris,* side; Gk. *summetria,* symmetry]: A body form in which the right and left halves of an organism are approximate mirror images of each other.

bile: A yellow secretion of the vertebrate liver, temporarily stored in the gall bladder and composed of organic salts that emulsify fats in the small intestine.

binomial system [L. *bi,* twice, two + Gk. *nomos,* usage, law]: A system of naming organisms in which the "scientific" name consists of two parts (the first designates genus and the second, species); originated by Linnaeus and sometimes called the Linnaean system.

biological clock [Gk. *bios,* life + *logos,* discourse]: Proposed internal factor(s) in plants and animals that governs what seem to be the innate biological rhythms (growth and activity patterns) of some organisms.

biomass [Gk. *bios,* life]: Total weight of all organisms (or some group of organisms) living in a particular habitat or place.

biomes: Communities of organisms recognized by common patterns of distinctive vegetation and climate; for example, the grassland biome, the tropical rain forest biome, etc.

biosphere [Gk. *bios,* life + *sphaira,* globe]: The zones of air, land, and water at the surface of the earth occupied by living things.

biosynthesis [Gk. *bios,* life + *sunthesis,* a putting together]: Formation by living organisms of organic compounds from elements or simple compounds.

blade: The broad, expanded part of a leaf or leaflike organ.

blastodisc [Gk. *blastos,* sprout + *discos,* a round plate]: Disklike area on the surface of a large, yolky egg that undergoes cleavage and gives rise to the embryo.

blastomere: One of many cells produced by division during the cleavage stage of development of the egg.

blastopore [Gk. *blastos,* sprout + *poros,* a way, means, path]: In the gastrula stage of an embryo, the opening that connects the archenteron with the outside; represents the future mouth in some animals (protostomes), the future anus in others (deuterostomes).

blastula [Gk. *blastos,* sprout]: An animal embryo after cleavage and before gastrulation; usually consists of a hollow sphere, the walls of which are composed of a single layer of cells.

bond energy: The energy required to break a particular chemical bond.

Bowman's capsule: In the vertebrate kidney, the bulbous unit of the nephron, including the glomerulus. It is the site of filtration of the renal fluid from the blood, the initial process in urine formation.

brainstem: The most posterior portion of the vertebrate brain; includes medulla, pons, and midbrain.

bronchus, *pl.* **bronchi** (**bronk**-us, **bronk**-eye) [Gk. *bronchos,* windpipe]: One of a pair of respiratory tubes branching into either lung at the lower end of the trachea; it subdivides into progressively finer passageways, the bronchioles, culminating in the alveoli.

bud: (1) In plants, an embryonic shoot, including rudimentary leaves, often protected by specialized bud scales. (2) In animals, an asexually produced outgrowth that develops into a new individual.

buffer: A substance that prevents appreciable changes of pH in solutions to which small amounts of acids or bases are added.

bulb: A modified bud with thickened leaves adapted for underground food storage.

bulk flow: The overall movement of a liquid induced by gravity, pressure, or an interplay of both.

calorie [L. *calor,* heat]: The amount of energy in the form of heat required to raise the temperature of 1 gram of water 1°C; in making metabolic measurements the kilocalorie (Calorie) is generally used. A Calorie is the amount of heat required to raise the temperature of 1 kilogram of water 1°C.

Calvin cycle: The set of reactions through which carbon dioxide is reduced to carbohydrates during the dark phase of photosynthesis.

capillaries [L. *capillaris,* relating to hair]: Smallest thin-walled blood vessels through which exchanges between blood and the tissues occur; connect arteries with veins.

capillary action: The movement of water or any liquid along a surface; results from the combined effect of cohesion and adhesion.

capsule (**kap**-sul) [L. *capsula,* a little chest]: (1) A slimy layer around the cells of certain bacteria. (2) The sporangium of Bryophyta. (3) A dehiscent, dry fruit that develops from two or more carpels.

carbohydrate [L. *carbo,* charcoal + *hydōr,* water]: An organic compound consisting of a chain of carbon atoms to which hydrogen and oxygen are attached in a 2:1 ratio; includes sugars, starch, glycogen, cellulose, etc.

carnivore [L. *caro, carnis,* flesh + *voro,* to devour]: Predator that obtains its nutrients and energy by eating animals.

carotene [L. *carota,* carrot]: A yellow, orange, or red pigment found in plants and some algae; converted into vitamin A in the vertebrate liver.

carotenoids [L. *carota,* carrot]: A class of pigments that includes the carotenes (yellows, oranges, and reds) and the xanthophylls (yellow); accessory pigments in photosynthesis.

carpel: A leaflike floral structure enclosing the ovule or ovules of angiosperms, typically divided into ovary, style, and stigma; a flower may have one or more carpels, either single or fused.

carrying capacity: In ecology, the largest number of organisms of a particular species that can be maintained indefinitely in a given part of the environment.

cartilage [L. *cartilago,* gristle]: The connective tissue in skeletons of higher vertebrates; forms much of the skeleton of adult lower vertebrates and immature higher vertebrates.

Casparian strip (after Robert Caspary, German botanist): A thickened, waxy strip that extends around and seals the walls of endodermal root cells in plants, thus restricting the diffusion of solutes across the endodermis into the vascular tissues of the root.

catalyst [Gk. *katalysis,* dissolution]: A substance that accelerates the rate of a chemical reaction but is not used up in the reaction; enzymes are catalysts.

cell [L. *cella,* a chamber]: The structural unit of organisms, composed of cytoplasm and one or more nuclei and surrounded by a membrane. In most plants, fungi, and bacteria, there is a cell wall outside the membrane.

cell cycle: The timed sequence of events occurring in a cell in the period between mitotic divisions.

cell membrane: The outermost membrane of the cell; also called the plasma membrane.

cell plate: In the dividing cells of most plants (and in some fungi and algae), a flattened structure that forms at the equator of the mitotic spindle during early telophase; gives rise to the middle lamella.

cell theory: All living things are composed of cells; cells arise only from other cells. No exception has been found to these two rules since they were first proposed well over a century ago.

cellulose [L. *cellula,* a little cell]: The chief constituent of the cell wall in all green plants; an insoluble complex carbohydrate formed of microfibrils of glucose molecules.

cell wall: A plastic or rigid structure, produced by the cell and located outside the cell membrane in most plants, algae, fungi, and bacteria; in plant cells, it consists mostly of cellulose.

central nervous system: In vertebrates, the brain and spinal cord; in invertebrates it usually consists of one or more cords of nervous tissue plus their associated ganglia.

centriole (**sen**-tree-ole) [Gk. *kentron,* center]: A cytoplasmic organelle generally found near the base of each flagellum or cilium and identical in structure to a basal body; nonflagellated animal cells have centrioles at the spindle poles during division.

centromere (**sen**-tro-mere) [Gk. *kentron,* center + *meros,* a part]: Region of constriction of chromosome that holds sister chromatids together.

cerebellum [L. dim. of *cerebrum,* brain]: An enlarged part of the dorsal side of the vertebrate brain; functions in coordinating muscular activities and maintaining equilibrium.

cerebral cortex [L. *cerebrum,* brain]: A layer of neurons (gray matter) forming the upper surface of the cerebrum, well developed only in mammals; the seat of conscious sensations and voluntary muscular activity.

cerebrum [L., brain]: The portion of the vertebrate brain occupying the upper part of the skull, consisting of two cerebral hemispheres united by the corpus callosum; coordinates most activities.

chelicera [Gk. *cheilos,* the edge, lips + *cheir,* arm]: First pair of appendages in arachnids; used for seizing and crushing prey.

chemical reaction: An interaction among atoms, ions, or molecules that results in the formation of new combinations of atoms, ions, or molecules; the making or breaking of chemical bonds.

chemiosmotic theory: Model proposing that ATP formation in mitochondria and chloroplasts is the result of a proton gradient produced by electron transport.

chemoreceptor: A sensory cell or organ that responds to the presence of a specific chemical stimulus; includes smell and taste receptors.

chemotropism [Gk. *cheimōn,* storm + *trope,* a turning]: A behavior in which an organism responds to a chemical stimulus by turning toward or away from it.

chiasma, *pl.* **chiasmata** (**kye**-az-ma) [Gk., a cross]: The X-shaped figure formed by the junction of nonsister chromatids during meiosis; thought to be the site of crossing over.

chitin (**kye**-tin) [Gk. *chitōn,* a tunic, undergarment]: A tough, resistant, nitrogen-containing polysaccharide present in the exoskeleton of arthropods, the epidermal cuticle or other surface structures of many other invertebrates, and in the cell walls of certain fungi.

chlorophyll [Gk. *chloros,* green + *phyllon,* leaf]: A class of green pigments found in chloroplasts of plant cells; necessary for photosynthesis.

chloroplast [Gk. *chloros,* green + *plastos,* formed]: A membrane-bound, chlorophyll-containing organelle in eukaryotes (algae and green plants) that is the site of photosynthesis.

chordate [L. *chorda,* cord, string]: Member of the animal phylum Chordata of which all members possess a notochord, dorsal nerve cord, pharyngeal gill slits, and tail, at least at some stage of the life cycle.

chorion (**core**-ee-on) [Gk. *chorion,* skin, leather]: The outermost membrane of the embryos of reptiles, birds, and mammals; in placental mammals it contributes to the structure of the placenta.

chromatid (**crow**-ma-tid) [Gk. *chrōma,* color]: Either of the two strands of a duplicated chromosome, which are joined at the centromere.

chromatin [Gk. *chrōma,* color]: The deeply staining nucleoprotein complex of the chromosomes.

chromosome [Gk. *chrōma,* color + *soma,* body]: One of the bodies in the cell nucleus along which the genes are located; visualized as threads or rods of chromatin, which appear in a contracted form during mitosis and meiosis.

chromosome map: A diagram of the linear order of the genes on a chromosome; determined from the frequency of crossing over between pairs of genes.

cilium, *pl.* **cilia** (**silly**-um) [L., eyelash]: A short hairlike structure embedded in the surface of some eukaryotic cells, usually in large numbers and arranged in rows; has a highly characteristic internal structure of two inner microtubules surrounded by nine pairs of outer microtubules; involved in locomotion.

circadian rhythms [L. *circa,* about + *dies,* day]: Regular rhythms of growth or other activity that occur on an approximately 24-hour cycle.

class: A taxonomic grouping of related, similar orders; category above order and below phylum.

cleavage: The successive cell divisions of a fertilized egg to form a multicellular blastula.

climax community: Final, relatively stable community in a successional series.

cline [Gk. *klinein,* to lean]: A graded series of changes in some characteristic within a species, correlated with some gradual change in climate or another geographical factor.

clitoris (**klit**-o-ris) [Gk. *kleitoris,* a small hill]: A small erectile body at the anterior part of the vulva; homologous to the penis.

cloaca [L., sewer]: The exit chamber from the digestive system in lower vertebrates; also may serve as the exit for the reproductive and urinary systems.

clone [Gk. *klon,* twig]: A line of cells, all of which have arisen from the same single cell by mitotic division; a population of individuals descended by asexual reproduction from a single ancestor.

cnidoblast (**ni**-do-blast) [Gk. *knide,* nettle + *blastos,* sprout]: A stinging cell containing a nematocyst; characteristic of coelenterates.

cochlea [Gk. *kochlias,* snail]: Part of the inner ear of mammals; concerned with hearing.

codon (**code**-on): Basic unit ("letter") of the genetic code; three adjacent nucleotides in a molecule of mRNA that form the code for a specific amino acid.

coelenteron (see-**len**-t-ron) [Gk. *koilos,* hollow + *enteron,* gut]: A digestive cavity with only one opening, characteristic of the phylum Coelenterata (jellyfish, hydra, corals, etc.).

coelom (**seal**-um) [Gk. *koilos,* a hollow]: A body cavity within which the viscera of higher animals are suspended; formed between layers of mesoderm.

coenzyme [L. *co,* together + Gk. *en,* in + *zyme,* leaven]: An organic molecule that plays an accessory role in enzyme-catalyzed processes, often by acting as a donor or acceptor of a substance involved in the reaction: NAD^+, $NADP^+$, and FAD are common coenzymes.

co-evolution [L. *co,* together + *e-,* out + *volvere,* to roll]: The simultaneous development of adjustments in two or more populations that interact so closely that each is a strong selective force on the other.

cohesion [L. *cohaerere,* to stick together]: The attraction or holding together of like molecules or like substances.

cohesion-tension theory: A theory accounting for the upward movement of water in plants. According to this theory, transpiration of a water molecule results in a negative (below 1 atmosphere) pressure in the leaf cells, inducing the entrance from the vascular tissue of another water molecule, which, because of the cohesive property of water, pulls with it a chain of water molecules extending up from the cells of the root tip.

coleoptile (coal-ee-**op**-tile) [Gk. *koleon,* sheath + *ptilon,* feather]: The sheath enclosing the apical meristem and leaf primordia of the germinating monocot. Often interpreted as the first leaf.

collagen [Gk. *kolla,* glue]: A fibrous protein in bones, tendons, and connective tissues.

colon: The large intestine of vertebrates leading to the rectum or cloaca.

colony: A group of unicellular or multicellular organisms living together in close association.

community: The organisms inhabiting a common environment and interacting with one another.

competition: Interaction between members of the same population or of two or more populations in order to obtain a mutually required resource.

compound [L. *componere,* to put together]: A molecule composed of two or more kinds of atoms in definite ratios, held together by chemical bonds.

compound eye: In arthropods, a complex eye composed of many separate elements (ommatidia), each with light-sensitive cells and a lens that can form an image.

cone: (1) In plants, the reproductive structure of a conifer. (2) In vertebrates, a type of light-sensitive neuron in the retina, concerned with the perception of color and with the most acute discrimination of detail.

conifer [Gk. *konos,* cone + *phero,* carry]: One of the cone-bearing plants; includes mostly trees, such as pines and firs.

conjugation [L. *conjugatio,* a joining, connection]: The sexual process in some unicellular organisms by which genetic material is passed from one cell to another across a cytoplasmic bridge.

connective tissues: Supporting or packing tissues that lie between groups of nerves, glands, and muscle cells, and beneath epithelial cells, in which the cells are irregularly distributed through a relatively large amount of intercellular material; include bone, cartilage, blood, and lymph.

continental drift: The gradual drifting of the world's continents that has occurred over hundreds of millions of years.

convergent evolution [L. *convergere,* to turn together; *evolutio,* to unfold]: The independent development of similarities between unrelated groups, such as porpoises and sharks, resulting from adaptation to similar environments.

corolla (ko-**role**-a) [L. dim. of *corona,* wreath, crown]: Petals, collectively; usually the conspicuously colored flower parts.

cortex [L., bark]: (1) The outer layer. (2) In a stem or root, the primary tissue bounded externally by the epidermis and internally by the central cylinder of vascular tissue.

cotyledon (cottle-**ee**-don) [Gk. *kotyledon,* a cup-shaped hollow]: A leaflike structure of the embryo of a seed plant; contains stored food used during germination.

covalent bond [L. *con,* together + *valere,* to be strong]: A chemical bond formed between atoms as a result of the sharing of one or more pairs of electrons.

cross-fertilization: Fusion of gametes formed by different individuals; as opposed to self-fertilization.

crossing over: During meiosis, the exchange of genetic material between paired chromatids of homologous chromosomes.

cuticle (**ku**-tik-l) [L. *cuticula,* dim. of *cutis,* the skin]: Layer of waxy substance (cutin) on outer surface of plant cell walls; noncellular, outermost layer of many invertebrates.

cyclosis (si-**klo**-sis) [Gk. *kyklosis,* circulation]: The circulation of cytoplasm within a cell.

cytochromes [Gk. *kytos,* vessel + *chroma,* color]: Heme-containing proteins that participate in electron transport chains; involved in cellular respiration and photosynthesis.

cytokinesis [Gk. *kytos,* vessel + *kinesis,* motion]: Division of the cytoplasm of a cell following nuclear division.

cytokinin [Gk. *kytos,* vessel + *kinesis,* motion]: One of a group of chemically related plant hormones that promotes cell division, among other effects.

cytoplasm (**sight**-o-plazm) [Gk. *kytos,* vessel + *plasma,* anything molded]: The living matter within a cell, excluding the nucleus.

cytostome [Gk. *kytos,* vessel + *stoma,* mouth]: In protozoans, the "mouth" opening.

deciduous [L. *decidere,* to fall off]: Refers to plants that shed their leaves at a certain season.

decomposers: Organisms (bacteria, fungi) in an ecosystem or community that convert dead organic material into plant nutrients.

denaturation: The loss of the native configuration of a macromolecule resulting, for instance, from heat treatment, extreme pH changes, chemical treatment, or other denaturing agents. It is usually accompanied by loss of biological activity.

dendrite [Gk. *dendron,* tree]: Nerve fiber, typically branched, that conducts impulses toward a nerve body.

deoxyribonucleic acid (DNA) (dee-ox-y-rye-bo-new-**klee**-ick): The carrier of genetic information in cells, composed of two complementary chains of nucleotides wound in a double helix; capable of self-replication as well as coding for RNA synthesis.

dermis [Gk. *derma,* skin]: The outer layer of the skin beneath the epidermis.

development: The progressive production of the phenotypic characteristics of a multicellular organism, beginning with the fertilization of an egg.

diaphragm [Gk. *diaphrassein,* to barricade]: In mammals, a sheetlike tissue (tendon and muscle) forming the partition between the abdominal and thoracic cavities; functions in breathing.

dicotyledon (dye-cottle-**ee**-don) [Gk. *di,* double, two + *kotyledon,* a cup-shaped hollow]: A subclass of angiosperms having two seed leaves or cotyledons, among other distinguishing features; often abbreviated as dicot.

differentiation: The developmental process by which a relatively unspecialized cell or tissue undergoes a progressive (usually irreversible) change to a more specialized cell or tissue.

diffusion [L. *diffundere,* to pour out]: The net movement of suspended or dissolved particles down a concentration gradient as a result of the random spontaneous movements of individual particles; the process tends to distribute the particles uniformly throughout a medium.

digestion [L. *digestio,* separating out, dividing]: The breakdown of complex, usually insoluble foods into simple, usually soluble forms by means of enzymes.

dikaryon (dye-**care**-ee-on) [Gk. *di,* two + *karyon,* kernel]: A cell or organism with paired but not fused nuclei derived from different parents. Found among fungi (Ascomycetes and Basidiomycetes).

dioecious (dye-**ee**-shus) [Gk. *di,* two + *oikos,* house]: In plants, having the male and female (or staminate and ovulate) flowers on different individuals of the same species.

diploid [Gk. *di,* double, two + *ploion,* vessel]: The condition in which each autosome is represented twice ($2n$); in contrast to haploid (n).

DNA: Abbreviation of deoxyribonucleic acid.

dominance: In genetic terminology, the ability of one allelic form of a gene to determine the phenotype of a heterozygous individual. The homologous chromosome carries a different allele, which is said to be recessive.

dominant allele: An allele whose phenotypic effect is the same in both the heterozygous and homozygous conditions.

dormancy [L. *dormire,* to sleep]: A period during which growth ceases and is resumed only if certain requirements, as of temperature or day length, have been fulfilled.

dorsal [L. *dorsum,* the back]: Pertaining to or situated near the back; opposite of ventral.

double fertilization: The fusion of the egg and the sperm nucleus (resulting in a $2n$ fertilized egg, the zygote) and the simultaneous fusion of a second male gamete nucleus with the polar nuclei (resulting in a $3n$ endosperm nucleus); a unique characteristic of angiosperms.

duodenum (duo-**dee**-num) [L. *duodeni,* twelve each—from its length, about 12 fingers' breadth]: The upper portion of the small intestine in vertebrates, where food is digested into molecules that can be absorbed by intestinal cells.

ecological dominant: A species that, by virtue of size, number, or behavior, exerts a controlling influence on its environment and, as a result, determines what other kinds of organisms exist in that ecosystem.

ecological niche: A description of the roles and associations of a particular species in the community of which it is a part; the way in which an organism interacts with the biotic and abiotic parts of its environment.

ecological pyramid: A graphic representation of the quantitative relationships between the trophic levels of a food chain. Because large amounts of energy and biomass are dissipated at every trophic level, these diagrams nearly always take the form of pyramids.

ecological succession: The gradual process by which the species composition of a community changes, culminating in a stable composition, known as a climax community.

ecology [Gk. *oikos,* home + *logos,* a discourse]: The study of the interactions of organisms with their physical environment and with each other and of the results of such interactions.

ecosystem [Gk. *oikos,* home + *systema,* that which is put together]: All organisms in a community plus the associated abiotic environmental factors with which they interact.

ectoderm [Gk. *ecto,* outside + *derma,* skin]: (1) The outermost layer of body tissue. (2) One of the three embryonic germ layers, it gives rise to the outer epithelium of the body (skin, hair, nails), the sense organs, the brain and spinal cord, etc.

effector [L. *ex,* out of + *facere,* to make]: Cell, tissue, or organ (such as muscle or gland) capable of producing a response to a stimulus.

efferent [L. *ex,* out of + *ferre,* to bear]: Carrying away from a center, applied to nerves and blood vessels.

egg: A female gamete, or germ cell, which usually contains abundant cytoplasm and yolk; usually immotile, often larger than a male gamete.

electron: A subatomic particle with a negative electric charge equal in magnitude to the positive charge of the proton but with a mass 1/1,837 that of the proton; normally found within orbitals surrounding the atom's positively charged nucleus.

electron acceptor: Substance that accepts or receives electrons in an oxidation-reduction reaction, becoming reduced in the process.

electron carrier: A specialized molecule, such as a cytochrome, that can lose and gain electrons reversibly, alternately becoming oxidized and reduced.

electron donor: Substance that donates or gives up electrons in an oxidation-reduction reaction, becoming oxidized in the process.

element: A substance composed only of atoms of the same atomic number and which cannot be decomposed by ordinary chemical means; one of about 105 distinct natural or synthetic types of matter that, singly or in combination, compose all materials of the universe.

embryo [Gk. *en,* in + *bryein,* to swell]: The early developmental stage of an organism produced from a fertilized egg; a young organism before it emerges from the seed, egg, or the body of its mother. In humans, the embryo is called a fetus following the third month of pregnancy.

endergonic [Gk. *endon,* within + *ergon,* work]: Energy-requiring, as in a chemical reaction; applied to an "uphill" process.

endocrine gland [Gk. *endon,* within + *krinein,* to separate]: Ductless gland whose secretions (hormones) are released into the circulatory system; in vertebrates, includes pituitary, sex glands, adrenal, thyroid, and others.

endoderm [Gk. *endon,* within + *derma,* skin]: (1) The innermost layer of body tissue. (2) One of the three embryonic germ layers, it gives rise to the epithelium that lines certain internal structures, such as most of the digestive tract and its outgrowths, most of the respiratory tract, and the urinary bladder, liver, pancreas, and some endocrine glands.

endodermis [Gk. *endon,* within + *derma,* skin]: In plants, a one-celled layer of specialized cells that lies between the cortex and the vascular tissues in young roots. The Casparian strip of the endodermis prevents diffusion of solutes across the root.

endometrium [Gk. *endon,* within + *metrios,* of the womb]: The glandular lining of the uterus in mammals; thickens in response to progesterone secretion during ovulation and is sloughed off in menstruation.

endoplasmic reticulum [Gk. *endon,* within + *plasma,* from cytoplasm; L. *reticulum,* network]: An extensive system of membranes present in most eukaryotic cells, dividing the cytoplasm into compartments and channels, often coated with ribosomes.

endosperm [Gk. *sperma,* seed]: In plants, a $3n$ tissue containing stored food that develops from the union of a male nucleus and the polar nuclei of the egg; found only in angiosperms.

entropy [Gk. *en,* in + *trope,* turning]: A measure of the randomness or disorder of a system.

enzyme [Gk. *en,* in + *zyme,* leaven]: A protein molecule that catalyzes a chemical reaction.

epidermis [Gk. *epi,* on or over + *derma,* skin]: In plants and animals, the outermost layers of cells.

epinephrine: A hormone produced by the medulla of the adrenal gland, which increases the concentration of sugar in the blood, raises blood pressure and heartbeat rate, and increases muscular power and resistance to fatigue; also a chemical transmitter across synaptic junctions. Also called adrenaline.

episome: A genetic element found in bacteria that can exist either free (as a plasmid) or as part of the chromosome.

epithelial tissue [Gk. *epi,* on or over + *thele,* nipple]: In animals, a type of tissue that covers a body or structure or lines a cavity; epithelial cells form one or more regular layers with little intercellular material.

equilibrium [L. *aequus,* equal + *libra,* balance]: The state of a system in which no further net change is occurring; result of counterbalancing forward and backward processes.

erythrocyte (eh-**rith**-ro-site) [Gk. *erythros,* red + *kytos,* vessel]: Red blood cell, the carrier of hemoglobin.

estrogens: Female sex hormones, which are the predominant secretions of the ovarian follicle during the preovulatory phase of the menstrual cycle.

estrus [Gk. *oistros,* frenzy]: The mating period in female mammals, characterized by ovulation and intensified sexual activity.

ethology [Gk. *ethos,* habit, custom + *logos,* discourse]: The study of whole patterns of animal behavior in natural environments, stressing adaptation and evolution.

eukaryote (you-**car**-ry-oat) [Gk. *eu,* good + *karyon,* nut, kernel]: A cell having a membrane-bound nucleus, membrane-bound organelles, and chromosomes in which DNA is combined with special proteins; an organism composed of such cells.

eusocial [Gk. *eu,* good + L. *socius,* companion]: Applied to insect societies in which sterile individuals work on behalf of individuals involved in reproduction.

evolution [L. *e-*, out + *volvere,* to roll]: Any change in the gene pool from one generation to the next; Darwinian evolution is the result of natural selection operating on random genetic variations.

exergonic [Gk. *ex,* out of + *ergon,* work]: Energy-producing, as in a chemical reaction; applied to a "downhill" process.

exoskeleton: The outer supporting covering of the body; common in arthropods.

exponential growth: In populations, the increasingly accelerated rate of growth due to the increasing number of individuals being added to the reproductive base. In the absence of control factors, all populations experience exponential growth.

extraembryonic membranes: Membranes formed of embryonic tissues that lie outside the embryo and are concerned with its protection and metabolism; include amnion, chorion, allantois, and yolk sac.

F_1 (first filial generation): The offspring resulting from the crossing of plants or animals of a parental generation.

F_2 (second filial generation): The offspring resulting from crossing members of the F_1 generation among themselves.

Fallopian tubes: *See* Oviduct.

family: A taxonomic grouping of related, similar genera, the category below order and above genus.

feedback inhibition: A control mechanism whereby an increase in some substance inhibits the process leading to the increase.

fertilization: The fusion of two gamete nuclei to form a diploid (or polyploid) zygote nucleus.

fetus [L., pregnant]: An unborn or unhatched vertebrate that has passed through the earliest developmental stages; a developing human from about the third month after conception until birth.

fibril [L. *fibra,* fiber]: Any minute, threadlike organelle within a cell.

fibrous protein: Insoluble structural protein in which the polypeptide chain is coiled along one dimension; constitutes the main structural elements of many animal tissues.

filament [L. *filare,* to spin]: (1) A chain of cells. (2) In plants, the stalk of a stamen.

fission [L. *fissus,* split]: Asexual reproduction by division of the cell or body into two or more equal, or nearly equal, parts.

fitness: The relative ability to leave offspring.

flagellum, *pl.* **flagella** (fla-**jell**-um) [L. *flagellum,* whip]: A long, threadlike organelle found in eukaryotes and used in locomotion and feeding; has an internal structure of nine pairs of microtubules encircling two central microtubules.

flower: The reproductive structure of angiosperms; a complete (perfect) flower includes sepals, petals, stamens (male sex organs), and carpels (female sex organs).

food chain, food web: A set of interactions among organisms, including producers, herbivores, and carnivores, through which energy and materials move within a community or ecosystem.

fossil [L. *fossilis,* dug up]: The remains of an organism, or direct evidence of its presence (such as tracks). May be an unaltered hard part (tooth or bone), a mold in a rock, petrification (wood or bone), unaltered or partially altered soft parts (a frozen mammoth).

fossil fuels: The remains of once-living organisms that are burned to release energy. Examples are coal, oil, and natural gas.

founder principle: A variety of genetic drift in which an altered gene pool, of lower heterozygosity, results in a small population that has branched off from a larger one.

free energy: Energy that has the ability to do work.

fruit [L. *fructus,* fruit]: In angiosperms, a matured, ripened ovary or group of ovaries and associated structures; contains the seeds.

function [L. *fungor,* to busy oneself]: Characteristic role or action of a structure or process in the normal metabolism or behavior of an organism.

gamete (**gam**-meet) [Gk., wife]: The haploid reproductive cell whose nucleus fuses with that of another gamete of an opposite sex (fertilization); the resulting cell (zygote) may develop into a new diploid individual, or, in some species, may undergo meiosis to form haploid somatic cells.

gametophyte: In plants having alternation of haploid and diploid generations, the haploid (n) gamete-producing generation.

ganglion, *pl.* **ganglia** (**gang**-lee-on) [Gk. *ganglion,* a swelling]: Aggregate of nerve cell bodies.

gastrula [Gk. *gaster,* stomach]: An embryo in the process of gastrulation, the stage of development during which the blastula with its single layer of cells turns into a three-layered embryo, made up of ectoderm, mesoderm, and endoderm, often enclosing an archenteron.

Gause's principle (after G. F. Gause, German geneticist): The hypothesis that two species with identical ecological requirements cannot coexist stably in the same locality and the species that is more efficient in utilizing the available resources will exclude the other.

gene [Gk. *genos,* birth, race; L. *genus,* birth, race, origin]: A unit of heredity in the chromosome; a sequence of nucleotides in a DNA molecule that codes for a polypeptide.

gene flow: The exchange of genes between different populations.

gene frequency: In a population, the proportion of a particular allele at a given genetic locus (at which two or more allelic forms may occur).

gene pool: All the alleles of all the genes in a population.

genetic code: The system of nucleotide triplets in DNA and RNA that carries genetic information; referred to as a code because it determines the amino acid sequence in the enzymes and other protein components synthesized by the organism.

genetic drift: Random fluctuations in the relative allele frequencies of a small breeding population.

genetic isolation: The absence of genetic exchange between populations or species as a result of morphological, behavioral, or physiological mechanisms that prevent reproduction, or as a result of geographic separation.

genome: The haploid set of chromosomes, with their associated genes.

genotype (**jean**-o-type): The genetic constitution, latent or expressed, of a cell or an organism, as contrasted with the phenotype; the sum total of all the genes present in an individual.

genus, *pl.* **genera** (**jean**-us) [L. *genus,* race, origin]: A taxonomic grouping of closely related species.

geologic eras: See Table 26-2, page 478.

germ cells: Gametes or the cells that give rise directly to gametes.

germination [L. *germinare,* to bud]: In plants, the resumption of growth or the development from seed or spore.

germ layer: A layer of distinctive cells in an embryo; each germ layer always gives rise to certain tissues or structures as the organism develops. The majority of multicellular animals have three embryonic layers: ectoderm, mesoderm, and endoderm.

gibberellins (jibb-e-**rell**-ins) [Fr. *gibberella,* genus of fungi]: A group of chemically related plant growth hormones, whose most characteristic effect is stem elongation in dwarf plants and bolting.

gill: The respiratory organ of aquatic animals, usually a thin-walled projection from some part of the external body surface or, in vertebrates, from some part of the digestive tract.

gland [L. *glans, glandis,* acorn]: An organ specialized to produce one or more secretions that are discharged to the outside of the gland.

globular protein [L. dim. of *globus,* a ball]: A polypeptide chain folded into a roughly spherical shape.

glomerulus (glom-**mare**-u-lus) [L. *glomus,* ball]: In the vertebrate kidney, a cluster of capillaries enclosed by the Bowman's capsule; blood plasma minus large molecules filters through the walls of the glomerular capillaries into the renal tubules.

glucagon [Gk. *glukus,* sweet + *agō,* to lead toward]: Hormone produced in the pancreas that acts to raise the concentration of blood sugar.

glucose [Gk. *glukus,* sweet]: A six-carbon sugar ($C_6H_{12}O_6$); the most common monosaccharide in animals.

glycogen [Gk. *glukus,* sweet + *genos,* race or descent]: A complex carbohydrate (polysaccharide); one of the main stored food substances of most animals and fungi; it is converted into glucose by hydrolysis.

glycolysis (gly-**coll**-y-sis) [Gk. *glukus,* sweet + *lysis,* loosening]: The process by which a glucose molecule is changed anaerobically to two molecules of pyruvic acid with the liberation of a small amount of useful energy; catalyzed by soluble cytoplasmic enzymes.

Golgi body (**goal**-jee): An organelle present in many eukaryotic cells consisting of flat, disk-shaped sacs, tubules, and vesicles. It functions as a collecting and packaging center for substances that the cell manufactures for export.

gonad [Gk. *gone,* seed]: Gamete-producing organ of multicellular animals; ovary or testis.

granum, *pl.* **grana** [L., grain or seed]: In chloroplasts, stacked membrane-bound disks (thylakoids) that contain chlorophylls and carotenoids and are the sites of the light reactions of photosynthesis.

groundwater: Water in the zone of saturation where all openings in rocks and soil are filled, the upper surface of which forms the water table.

growth, exponential: *See* Exponential growth.

guard cells: Specialized epidermal cells surrounding a pore, or stoma; changes in turgor of a pair of guard cells causes opening and closing of the pore.

habitat [L. *habitare,* to live in]: The surroundings in which individuals of a particular species can usually be found.

half-life: The average time required for one-half of any amount of a radioactive element to undergo radioactive disappearance or decay.

haploid [Gk. *haploos,* single + *ploion,* vessel]: Having only one set of chromosomes (n), in contrast to diploid ($2n$); characteristic of all eukaryotic gametes, gametophytes in plants, and of some microorganisms.

Hardy-Weinberg law: The mathematical expression of the relationship between relative frequencies of two or more alleles in an idealized population; it states that both the allele frequencies and the genotype frequencies will remain constant in the absence of evolutionary forces.

heme: The iron-porphyrin group of heme proteins.

hemoglobin [Gk. *haima,* blood + L. *globus,* a ball]: The iron-containing protein in vertebrate blood that carries oxygen.

hemophilia [Gk. *haima,* blood + *philios,* friendly]: A hereditary disease characterized by failure of the blood to clot and consequent excessive bleeding from even minor wounds.

herbaceous (her-**bay**-shus) [L. *herba,* grass]: In plants, nonwoody.

herbivore [L. *herba,* grass + *vorare,* to devour]: A consumer that eats plants or other photosynthetic organisms to obtain its food and energy.

heredity [L. *herres, heredis,* heir]: The transmission of characteristics from parent to offspring through the germ cells.

hermaphrodite [Gk. *Hermes* and *Aphrodite*]: An organism possessing both male and female reproductive organs; hermaphrodites may or may not be self-fertilizing.

heterogamy [Gk. *heteros,* different + *gamos,* union or reproduction]: Reproduction involving two types of gametes.

heterosis [Gk. *heteros,* other, different]: Hybrid vigor, the overall superiority of the hybrid over either parent.

heterotroph [Gk. *heteros,* other, different + *trophos,* feeder]: An organism that cannot manufacture all its needed organic compounds and so must feed on organic materials formed by other organisms; in contrast to autotroph.

heterozygote [Gk. *heteros,* other + *zugōtos,* a pair]: A diploid organism that carries two different alleles at one or more genetic loci.

heterozygote superiority: The greater fitness of a heterozygote as compared with the two homozygotes at a given genetic locus; also called hybrid vigor.

hibernation [L. *hiberna,* winter]: A period of dormancy and inactivity, varying in length depending on the species and occurring in dry or cold seasons. During hibernation, metabolic processes are greatly slowed and, even in mammals, body temperature may drop to just above freezing.

homeostasis (home-e-o-**stay**-sis) [Gk. *homos,* same or similar + *stasis,* standing]: Maintenance of a relatively stable internal physiological environment or internal equilibrium in an organism, population, or ecosystem.

homeotherm [Gk. *homos,* same or similar + *therme,* heat]: Organism capable of maintaining a stable body temperature independent of the environment; warm-blooded. Birds and mammals are homeotherms.

hominid [L. *homo,* man]: Humans and closely related primates; includes modern and fossil hominoids but not the apes.

hominoid [L. *homo,* man]: Hominids and the great apes.

homologues [Gk. *homologia,* agreement]: Chromosomes that carry corresponding genes and associate in pairs in the first stage of meiosis; each member of the pair is derived from a different parent.

homology [Gk. *homologia,* agreement]: Similarity in structure and/or position, assumed to result from a common ancestry, regardless of function, such as the wing of a bird and the foreleg of a mammal.

homozygote [Gk. *homos,* same or similar + *zugōtos,* a pair]: A diploid organism that carries identical alleles at one or more genetic loci.

hormone [Gk. *hormaein,* to excite]: An organic molecule secreted, usually in minute amounts, in one part of an organism and transported to another part of that organism where it has a specific effect on some target organ or tissue.

host: An organism on or in which a parasite lives; recipient of grafted tissue.

hybrid [L. *hybrida,* the offspring of a tame sow and a wild boar]: Offspring of two parents that differ in one or more heritable characteristics; offspring of two different varieties or of two different species.

hydrogen bond: A weak molecular bond linking a hydrogen atom that is covalently bonded to another atom (usually oxygen, nitrogen, or fluorine) to another oxygen, nitrogen, or fluorine atom of the same or another molecule.

hydrolysis [Gk. *hydro,* water + *lysis,* loosening]: Splitting of one molecule into two by addition of H^+ and OH^- ions of water.

hypertonic [Gk. *hyper,* above + *tonos,* tension]: Having a concentration of solutes high enough to gain water across a semipermeable membrane from another solution.

hypha [Gk. *hyphe,* web]: A single tubular filament of a fungus; the hyphae together make up the mycelium, the matlike "body" of a fungus.

hypothalamus [Gk. *hypo,* under + *thalamos,* inner room]: The floor and sides of the vertebrate brain just below the cerebral hemispheres; controls the autonomic nervous system and the pituitary gland and contains centers that regulate body temperature and appetite.

hypothesis [Gk. *hypo,* under + *tithenai,* to put]: A temporary working explanation or supposition based on accumulated facts and suggesting some general principle or relation of cause and effect; a postulated solution to a scientific problem that must be tested by experimentation and, if not validated, discarded.

hypotonic [Gk. *hypo,* under + *tonos,* tension]: Having a concentration of solutes low enough to lose water across a semipermeable membrane to another solution.

imprinting: A rapid and extremely narrow form of learning to respond to a specific stimulus, which occurs during a very short, critical period in the early life of an organism, such as the following response in certain birds.

inbreeding: The mating of individuals closely related genetically.

independent assortment: *See* Mendel's second law.

induction [L. *inducere,* to induce]: The process in an embryo in which one tissue or body part causes the differentiation of another tissue or body part.

instinct: An inflexible, unlearned behavior or response triggered by specific stimuli; thought to be genetically determined.

insulin: A peptide hormone produced by the vertebrate pancreas, the action of which results in the lowering of the concentration of sugar in the blood.

intelligence: The capacity to profit by experience, through reasoning and the analysis and association of ideas.

interphase: The stage between two mitotic or meiotic cycles.

invagination [L. *in,* in + *vagina,* sheath]: The local infolding of a layer of tissue, especially in animal embryos, so as to form a depression or pocket opening to the outside.

ion (eye-on): An atom or molecule that has lost or gained one or more electrons. By the process known as ionization, the atom or molecule becomes electrically charged.

irritability: The ability to respond to stimuli or changes in environment; a general property of all living organisms.

isogamy [Gk. *isos,* equal + *gamos,* wedding]: A type of sexual reproduction in algae and fungi in which the gametes are alike in size.

isolating mechanisms: Mechanisms that prevent gene exchange between different populations or species; may be morphological, behavioral, or physiological.

isotonic [Gk. *isos,* equal + *tonos,* tension]: Having the same concentration of solutes as that of another solution. If two isotonic solutions are separated by a semipermeable membrane, there will be no net flow of water across the membrane.

isotope [Gk. *isos,* equal + *topos,* place]: Atom of an element that differs from other atoms of the same element in the number of neutrons in the atomic nucleus; isotopes thus differ in atomic weight. Some isotopes are unstable and emit radioactivity.

karyotype [Gk. *kara,* the head + *typos,* stamp or print]: The general appearance of the chromosomes in a genome with regard to number, size, and shape.

keratin [Gk. *karas,* horn]: One of a group of tough, fibrous proteins formed by certain epidermal tissues and especially abundant in skin, claws, hair, feathers, and hooves.

kidney: In vertebrates, the organ that regulates the balance of water and solutes in the blood and the excretion of nitrogen wastes in the form of urine.

kinetic energy: Energy of motion.

kinetochore [Gk. *kinetikos,* putting in motion + *choros,* chorus]: Region within the centromere to which the spindle fibers attach.

Krebs cycle: Stage of cellular respiration in which pyruvate fragments are completely broken down into carbon dioxide; molecules reduced in the process can be used in ATP formation.

lamella (lah-**mell**-ah) [L. dim. of *lamina,* plate or leaf]: Layer, thin sheet.

larva [L., ghost]: An immature animal that is morphologically very different from the adult; examples are caterpillars and tadpoles.

laterite [L. *latus, lateris,* side]: A leached tropical soil with a high content of iron and aluminum.

leaching: The dissolving of minerals and other elements in soil or rocks by the downward movement of water.

learning: The process that leads to adaptive change in individual behavior as the result of experience.

leucoplast [Gk. *leukos,* white + *plastes,* molder]: In plants, a colorless cell organelle that serves as a starch repository; usually found in cells not exposed to light, such as roots and internal stem tissue.

leukocyte [Gk. *leukos,* white + *kytos,* vessel]: White blood cell. Two of the principal types are neutrophils, active in phagocytosis, and lymphocytes, involved in immune reactions.

lichen: Organism composed of an alga and a fungus that are symbiotically associated.

life cycle: The entire span of existence of any organism from time of zygote formation (or asexual reproduction) until it itself reproduces.

limiting factor: Any environmental factor the level or concentration of which limits the growth or some other activity of an organism or population.

linkage: The tendency for certain genes to be inherited together because they are located on the same chromosome.

lipid [Gk. *lipos,* fat]: One of a large variety of organic fats or fatlike compounds; includes waxes, steroids, phospholipids, and carotenes.

locus, *pl.* **loci** [L., place]: In genetics, the position of a gene in a chromosome. There may be one or several alleles of the gene at any locus.

loop of Henle (after F. G. J. Henle, German pathologist): A hairpin-shaped portion of the kidney tubule system found in birds and mammals in which a hypertonic urine is formed by processes of osmosis and active transport.

lymph: Colorless fluid occurring in special lymph ducts, derived from blood by filtration through capillary walls.

lysis [Gk. *lysis,* a loosening]: Disintegration of a cell by breaking its cell membrane.

lysogenic bacteria (lye-so-jenn-ick) [Gk. *lysis,* a loosening + *genos,* race or descent]: Bacteria carrying a bacteriophage genome integrated into the bacterial chromosome. The viral genome may subsequently begin to replicate independently and set up an active cycle of infection, causing lysis of the bacterial cells.

lysosome [Gk. *lysis,* loosening + *soma,* body]: A membrane-bound organelle in which hydrolytic enzymes are segregated.

macromolecule [Gk. *makros,* large + L. dim. of *moles,* mass]: A molecule of very high molecular weight; refers specifically to proteins, nucleic acids, polysaccharides, and complexes of these.

mantle: (1) In mollusks, the outermost layer of the body wall or a soft extension of it; usually secretes a shell. (2) In geology, the region of earth below the crust extending down to the molten outer core.

marine [L. *marini(us),* from *mare,* the sea]: Living in salt water.

marsupial [Gk. *marsypos,* pouch, little bag]: A nonplacental mammal in which the female has a ventral pouch or folds surrounding the nipples; the premature young leave the uterus and crawl into the pouch, where each one attaches itself by the mouth to a nipple until development is completed.

mechanoreceptor: A sensory cell or organ that receives mechanical stimuli such as those involved in touch, pressure, hearing, and balance.

medulla (med-**dull**-a) [L., the innermost part]: (1) The inner as opposed to the outer part of an organ, as in the adrenal gland. (2) The most posterior region of the vertebrate brain; connects with the spinal cord.

medusa: The free-swimming, bell- or umbrella-shaped stage in the life cycle of many coelenterates; a jellyfish.

megaspore [Gk. *megas,* great, large + *spora,* a sowing]: In plants, a haploid (n) spore that develops into a female gametophyte.

meiosis (my-**o**-sis) [Gk. *meioun,* to make smaller]: The two successive nuclear divisions in which a single diploid ($2n$) cell forms four haploid (n) cells, and segregation, crossing over, and reassortment of the genes occur; gametes or spores may be produced as a result of meiosis.

Mendel's first law: The factors for a pair of alternative characters are separate and only one may be carried in a particular gamete (genetic segregation). In modern form: Alleles segregate in meiosis.

Mendel's second law: The inheritance of a pair of factors for one trait is independent of the simultaneous inheritance of factors for other traits, such factors "assorting independently" as though there were no other factors present (later modified by the discovery of linkage). Modern form: Unlinked genes assort independently.

menstrual cycle [L. *mensis,* month]: In certain primates, the cyclic, hormone regulated changes in the condition of the uterine lining; marked by the periodic discharge of blood and disintegrated uterine lining through the vagina. Mammals with a menstrual cycle lack a well-defined period of estrus.

meristem [Gk. *meris,* part of, portion + *stamon,* the warp of a loom]: The undifferentiated plant tissue, including a mass of rapidly dividing cells, from which new tissues will arise.

mesoderm [Gk. *mesos,* middle + *derma,* skin]: One of the three embryonic germ layers, it gives rise to muscle, connective tissue, the circulatory system, and most of the excretory and reproductive systems.

messenger RNA (mRNA): The single-stranded RNA that carries genetic information from the gene to the ribosome, where it determines the order of the amino acids in the formation of a polypeptide; mRNA is formed by transcription and functions in translation.

metabolism [Gk. *metabole,* change]: The sum of all chemical reactions occurring within a cell or organism.

metamorphosis [Gk. *metamorphoun,* to transform]: Abrupt transition from larval to adult form, such as the transition from tadpole to adult frog.

metaphase [Gk. *meta,* middle + *phasis,* form]: The stage of mitosis or meiosis during which the chromosomes lie in the equatorial plane of the spindle.

microbe [Gk. *mikros,* small + *bios,* life]: A microscopic organism.

microfilament [Gk. *mikros,* small + L. *filare,* to spin]: A small, fibrous structure found in the cytoplasm of eukaryotic cells; it appears to play a role in cellular and cytoplasmic motion.

micronutrient [Gk. *mikros,* small + L. *nutrire,* to nourish]: A mineral required in only minute amounts for plant growth, such as iron, chlorine, copper, manganese, zinc, molybdenum, and boron.

microspore [Gk. *mikros,* small + *spora,* a sowing of seeds]: In plants, a spore that develops into a male gametophyte; in seed plants, it becomes a pollen grain.

mimicry [Gk. *mimos,* mime]: The superficial resemblance in form, color, or behavior of certain organisms (mimics) to other more powerful or more protected ones (models), resulting in protection, concealment, or some other advantage for the mimic.

mineral: A naturally occurring element or inorganic compound.

mitochondrion, *pl.* **mitochondria** [Gk. *mitos,* thread + *chondros,* cartilage or grain]: An organelle bound by a double membrane in which the reactions of the Krebs cycle and electron transport chain take place, resulting in the formation of ATP, CO_2, and H_2O from acetyl CoA and ADP. Mitochondria are the organelles in which most of the ATP of the eukaryotic cell is produced.

mitosis [Gk. *mitos,* thread]: Nuclear division characterized by chromosome duplication and the formation of two identical daughter nuclei.

molecular weight: The sum of the atomic weights of the constituent atoms in a molecule.

molecule [L. dim. of *moles,* mass]: Smallest unit of a compound (consisting of two or more atoms) that displays the properties of the compound.

molting: Shedding of all or part of an organism's outer covering; in arthropods, periodic shedding of the exoskeleton to permit an increase in size.

monocotyledon [Gk. *monos,* single + *kotyledon,* a cup-shaped hollow]: A subclass of angiosperms, characterized by a variety of features, among which is the presence of a single seed leaf (cotyledon); abbreviated as monocot.

monoecious (mo-**nee**-shuss) [Gk. *monos,* single + *oikos,* house]: In higher plants, having the male and female organs (the anthers and the carpels, respectively) on the same individual but on different flowers.

monomer [Gk. *monos,* single + *meros,* part]: A simple molecule of relatively low molecular weight that can be linked to others to form a polymer.

monosaccharide [Gk. *monos,* single + *sakcharon,* sugar]: A simple sugar, such as glucose, fructose, ribose.

morphogenesis [Gk. *morphe,* form + *genesis,* origin]: The development of size, form, and other structural features of organisms.

morphology [Gk. *morphe,* form + *logos,* discourse]: The study of form and structure, at any level of organization.

motor neuron: Neuron that transmits nerve impulses from the central nervous system to skeletal muscle; efferent neuron.

muscle fiber: Muscle cell; a long, cylindrical, multinucleated cell containing numerous myofibrils, which is capable of contraction when stimulated.

mutagen [L. *mutare,* to change + *genus,* source or origin]: A chemical or physical agent that increases the mutation rate.

mutant [L. *mutare,* to change]: A mutated gene or an organism carrying a gene that has undergone a mutation.

mutation [L. *mutare,* change]: The change of a gene from one allelic form to another; an inheritable change in the DNA sequence of a chromosome not resulting from recombination.

mycelium [Gk. *mykes,* fungus]: The mass of hyphae forming the body of a fungus.

myelin sheath [Gk. *myelinos,* full of marrow]: A fatty layer surrounding the axons of some nerve cells in vertebrates; made up of the membranes of Schwann cells.

myofibril [Gk. *mys,* muscle + L. *fibra,* fiber]: Contractile element of a muscle fiber, made up of thick and thin filaments arranged in sarcomeres.

myoglobin [Gk. *mys,* muscle + L. *globus,* a ball]: An oxygen-binding, heme-containing globular protein found in muscles.

myosin [Gk. *mys,* muscle]: One of the principal proteins in muscle; makes up the thick filaments.

NAD: Abbreviation of nicotinamide adenine dinucleotide, a coenzyme that functions as an electron acceptor.

natural selection: The differential reproduction of genotypes, resulting from interactions among a variety of phenotypes and the environment.

nectar [Gk. *nektar,* the drink of the gods]: A sugary fluid that attracts insects to plants.

nematocyst [Gk. *nema, nematos,* thread + *kyst,* bladder]: A threadlike stinger, containing a poisonous or paralyzing substance, found in the cnidoblast of coelenterates.

nephridium [Gk. *nephros,* the kidneys]: A type of excretory organ found in many invertebrates.

nephron [Gk. *nephros,* kidney]: The functional unit of the kidney structure in reptiles, birds, and mammals; a human kidney contains about 1 million nephrons.

nerve: A group or bundle of nerve fibers with accompanying connective tissue.

nerve fiber: A filamentous process of a neuron; either dendrite or axon.

nerve impulse: A rapid, transient change in electric potential across the membrane of a nerve cell; it is propagated along a nerve fiber from one part of an animal to another.

nervous system: All the nerve cells of an animal; the receptor-conductor-effector system; in humans, the nervous system consists of brain, spinal cord, and all nerves.

neural groove: Dorsal, longitudinal groove that forms in a vertebrate embryo; bordered by two neural folds; preceded by the neural-plate stage and followed by the neural-tube stage.

neural plate: Thickened strip of ectoderm in early vertebrate embryos that forms along the dorsal side of the body and gives rise to the central nervous system.

neural tube: Primitive, hollow, dorsal nervous system of the early vertebrate embryo formed by fusion of neural folds around the neural groove.

neurohormone [Gk. *neuron,* nerve]: A hormone secreted by a preganglionic nerve ending; includes acetylcholine, epinephrine, and norepinephrine.

neuron [Gk. *neuron,* nerve]: Nerve cell, including cell body, dendrites, and axon.

neurosecretory cell [Gk. *neuron,* nerve]: A nervelike cell that releases one or more neurohormones into the circulatory system.

neutron (**new**-tron): An uncharged particle with a mass slightly greater than that of a proton. Found in the atomic nucleus of all elements except hydrogen, in which the nucleus consists of a single proton.

niche: *See* Ecological niche.

nitrify: To convert organic nitrogen compounds into nitrites and nitrates, as by nitrifying bacteria and fungi.

nitrogenous base: A nitrogen-containing molecule having basic properties (tendency to acquire an H^+ ion); a purine or pyrimidine.

nitrogen cycle: Worldwide circulation and reutilization of nitrogen atoms, chiefly due to metabolic processes of living organisms; plants take up inorganic nitrogen and convert it into organic compounds (chiefly proteins), which are assimilated into the bodies of one or more animals; excretion and bacterial and fungal action on dead organisms return nitrogen atoms to the inorganic state.

nitrogen fixation: Incorporation of atmospheric nitrogen into inorganic nitrogen compounds available to plants, a process that can be carried out only by certain soil bacteria and many blue-green algae, or by certain symbiotic microorganisms in association with legumes.

node [L. *nodus,* knot]: In plants, a joint of a stem; the place where branches and leaves are joined to the stem.

nondisjunction [L. *non,* not + *disjungere,* to separate]: The failure of homologous chromosomes to separate during meiosis, resulting in one or more extra chromosomes in the cells of some offspring and correspondingly fewer in the cells of others.

nuclear envelope [L. *nucleus,* a little nut]: The double membrane surrounding the nucleus within a eukaryotic cell.

nucleic acid: A macromolecule consisting of nucleotides; the principal types are deoxyribonucleic acid (DNA) and ribonucleic acid (RNA).

nucleolus (new-**klee**-o-lus) [L. *nucleolus,* a small kernel]: A small, dense body, containing DNA, RNA, and protein, present in the nucleus of eukaryotic cells; site of production of ribosomal RNA.

nucleotide [L. *nucleus,* a little nut]: A building block of nucleic acid composed of phosphate, a five-carbon sugar (either ribose or deoxyribose), and a purine or a pyrimidine base.

nucleus [L. *nucleus,* a kernel]: (1) The cellular organelle that contains the genetic information in the form of DNA. In the eukaryotic cell, a specialized body bound by a double membrane, containing the chromosomes. (2) The central part of an atom. (3) A group of nerve cell bodies in the central nervous system.

ocellus, *pl.* **ocelli** [L. dim. of *oculus,* eye]: A simple light receptor common among invertebrates.

olfactory [L. *olfacere,* to smell]: Pertaining to smell.

ommatidium, *pl.* **ommatidia** [Gk. *ommos,* eye]: The single visual unit in the compound eye of arthropods; contains light-sensitive cells and a lens able to form an image.

omnivorous [L. *omnis,* all + *vorare,* devour]: Eating "everything," for example, using both plants and animals as food.

ontogeny [Gk. *on,* being + *genesis,* origin]: The developmental history of an individual organism from zygote to maturity.

oocyte (o-uh-sight) [Gk. *oion,* egg + *kytos,* vessel]: A cell that gives rise by meiosis to an ovum.

oogamy (oh-**og**-amy) [Gk. *oion,* egg + *gamos,* marriage]: Kind of sexual reproduction in which one of the gametes (the egg) is large and nonmotile, while the other gamete (the sperm) is smaller and motile.

operator gene: A segment of DNA to which a repressor protein may bind, thus regulating the transcription of an operon.

operon [L. *opus, operis,* work]: A unit of genetic transcription; a group of adjacent structural genes whose functions are related to a particular biochemical pathway and which are regulated by a single repressor protein.

opposable thumb: Thumb that rotates at the joint so that the tip of the thumb can be placed opposite the tip of any one of the four fingers.

order: A taxonomic grouping of related, similar families; the category below class and above family.

organ [Gk. *organon,* tool]: A body part composed of several tissues grouped together in a structural and functional unit.

organelle [Gk. *organon,* instrument, tool]: A formed body in the cytoplasm of a cell.

organic [Gk. *organon,* instrument, tool]: Pertaining to (1) organisms or living things generally, or (2) compounds formed by living organisms, or (3) the chemistry of compounds containing carbon.

organism [Gk. *organon,* instrument, tool]: Any living creature, either unicellular or multicellular.

organizer [Gk. *organon,* instrument, tool]: The part of an embryo capable of inducing undifferentiated cells to follow a specific course of development; in particular, the dorsal lip of the blastopore in the amphibian.

osmosis [Gk. *osmos,* impulse, thrust]: The net movement of water across a selectively permeable membrane (a membrane that permits the free passage of water but prevents or retards the passage of a solute). The net movement of water is from the side containing a lesser concentration of solute to the side containing a greater concentration.

osmotic pressure [Gk. *osmos,* impulse, thrust]: A measure of the difference in solute concentrations on either side of a semipermeable membrane; pressure generated by the osmotic flow of water.

ovary [L. *ovum,* egg]: (1) In animals, the egg-producing organ. (2) In flowering plants, the enlarged basal portion of a carpel or a fused carpel, containing the ovule or ovules; the ovary matures to become the fruit.

oviduct [L. *ovum,* egg + *ductus,* duct]: The tube serving to transport the eggs to the uterus or to the outside; Fallopian tubes (in humans).

ovulation: In animals, release of an egg or eggs from the ovary.

ovule [L. dim. of *ovum,* egg]: In seed plants, the megasporangium, a structure composed of a protective outer coat, a tissue specialized for food storage, and a female gametophyte with egg cell; becomes a seed after fertilization.

ovum, *pl.* **ova** [L., egg]: The egg cell; female gamete.

oxidation: Gain of oxygen, loss of hydrogen, or loss of an electron by an atom, ion, or molecule. Oxidation and reduction take place simultaneously, with the electron lost by one reactant being transferred to another reactant. Oxidation-reduction reactions are an important means of energy transfer within living systems.

oxidative phosphorylation: The use of the electron energy released during cellular respiration to phosphorylate (add a phosphate group onto) ADP, thereby yielding energy-rich ATP molecules.

pacemaker: Area of the vertebrate heart that initiates the heartbeat; located where the superior vena cava enters the right atrium; sinoatrial node.

paleontology [Gk. *palaios,* old + *onta,* things which exist + *logos,* discourse]: The study of the life of past geologic times, principally by means of fossils.

palisade cells: In plant leaves, the columnar, chloroplast-containing cells of the mesophyll.

pancreas (pang-kree-us) [Gk. *pan,* all + *kreas,* meat, flesh]: In vertebrates, a small complex gland located between the stomach and the duodenum, which produces digestive enzymes and the hormones insulin and glucagon; sweetbread.

parasite [Gk. *para,* beside, akin to + *sitos,* food]: An organism that lives on or in an organism of a different species and derives nutrients from it. *See* Symbiosis.

parasympathetic division [Gk. *para,* beside, akin to]: A subdivision of the autonomic nervous system of vertebrates with centers located in the brain and in the most anterior part and in the most posterior parts of the spinal cord; stimulates digestion; generally inhibits other functions and restores the body to normal following emergencies. Chemical transmitter: acetylcholine.

parenchyma (pah-**renk**-ee-ma) [Gk. *para,* beside, akin to + *en,* in + *chein,* to pour]: A plant tissue composed of living, thin-walled, randomly arranged cells with large vacuoles; usually photosynthetic or storage tissue.

parthenogenesis [Gk. *parthenon,* virgin + *genesis,* birth]: The development of an egg without fertilization.

pecking order: A social ranking in poultry flocks in which each bird dominates (can peck) others lower in rank, but is dominated by birds higher in rank.

pellicle [L. dim. of *pellis,* skin]: A thin, translucent envelope located outside the cell membrane in many protozoans.

peptide [Gk. *pepto,* to soften, digest]: Two or more amino acids linked together. Molecules made up of a relatively small number of amino acids (2 to about 100) are called peptides, while those formed of a larger number of amino acids are called polypeptides or proteins.

peptide bond [Gk. *pepto,* to soften, digest]: The type of bond formed when two amino acids are joined end to end; the acidic group ($-COOH$) of one amino acid is linked chemically to the basic group ($-NH_2$) of the next, and a molecule of water (H_2O) is removed.

perennial [L. *per,* through + *annus,* year]: A plant that persists in whole or in part from year to year and usually produces reproductive structures in more than one year.

peritoneum [Gk. *peritonos,* stretched over]: A membrane that lines the body cavity and forms the external covering of the visceral organs.

permeable [L. *permeare,* to pass through]: Penetrable by molecules, ions, or atoms; usually applied to membranes that let given solutes pass through.

petiole (pet-ee-ole) [Fr., from L. *petiolus,* dim. of *pes, pedis,* a foot]: The stalk of a leaf, connecting the blade of the leaf with the branch or stem.

pH: A symbol denoting the relative concentration of hydrogen ions in a solution; pH values range from 0 to 14; the lower the value, the more acidic a solution, that is, the more hydrogen ions it contains; pH 7 is neutral, less than 7 is acidic, more than 7 is alkaline.

phagocyte (fag-o-sight) [Gk. *phagein,* to eat + *kytos,* cell]: Any cell that engulfs foreign particles.

phagocytosis [Gk. *phagein,* to eat + *kytos,* cell]: Cell "eating"; the intake of solid particles by a cell, by flowing over and engulfing them; characteristic of amoebas, the digestive cells of some invertebrates, and vertebrate white blood cells.

phenotype [Gk. *phainein,* to show + *typos,* stamp, print]: The observable properties of an organism (as distinguished from its genotype).

pheromone (fair-o-moan) [Gk. *phero,* to bear, carry]: Substance secreted by an animal that influences the behavior or morphological development of other animals of the same species, such as the sex attractants of moths, the odor trail of ants.

phloem (flow-em) [Gk. *phloos,* bark]: Vascular tissue of higher plants; conducts sugars and other organic molecules from the leaves to other parts of the plant; composed of sieve cells (in gymnosperms) or sieve tubes and companion cells (in angiosperms), parenchyma, and fibers.

phospholipids: Cellular materials that contain phosphorus and are soluble in organic solvents; important components of cellular membranes.

phosphorylation: Addition of a phosphate group or groups to a molecule.

photon [Gk. *photos,* light]: The elementary particle of light and other electromagnetic radiations.

photoperiodism [Gk. *photos,* light]: The response to relative day and night length, a mechanism by which organisms measure seasonal change.

photophosphorylation [Gk. *photos,* light + *phosphoros,* bringing light]: Formation of ATP in the chloroplast during photosynthesis.

photoreceptor [Gk. *photos,* light]: A cell or organ capable of detecting light.

photosynthesis [Gk. *photos,* light + *syn,* together + *tithenai,* to place]: The conversion of light energy to chemical energy; the synthesis of organic compounds from carbon dioxide and water in the presence of chlorophyll, using light energy.

phototropism [Gk. *photos,* light + *trope,* turning]: Movement in which the direction of the light is the determining factor, such as the growth of a plant toward a light source; turning or bending response to light.

phylogeny [Gk. *phylon,* race, tribe]: Evolutionary history of a taxonomic group. Phylogenies are often depicted as "evolutionary trees."

phylum [Gk. *phylon,* tribe, stock]: A taxonomic grouping of related, similar classes; a high-level category beneath kingdom and above class.

physiology [Gk. *physis,* nature + *logos,* a discourse]: The study of function in cells, organs, or entire organisms; the processes of life.

phytochrome [Gk. *phyton,* plant + *chroma,* color]: A plant pigment that is a photoreceptor for red or far-red light and is involved with a number of developmental processes, such as flowering, dormancy, leaf formation, and seed germination.

phytoplankton [Gk. *phyton,* plant + *planktos,* wandering]: Aquatic, free-floating, microscopic, photosynthetic organisms.

pigment [L. *pigmentum,* paint]: A colored substance that absorbs light over a narrow band of wavelengths.

pinocytosis [Gk. *pinein,* to drink + *kytos,* a hollow vessel]: Cell "drinking"; the intake of fluid droplets by a cell, probably by a mechanism similar to that of phagocytosis but triggered by different stimuli.

pituitary [L. *pituita,* phlegm]: Endocrine gland in vertebrates; the anterior lobe is the source of tropic hormones, growth hormone, and prolactin and is stimulated by neurosecretory cells in the hypothalamus; the posterior lobe stores and releases oxytocin and ADH produced by the hypothalamus.

placenta [Gk. *plax,* a flat object]: A tissue formed in part from the inner lining of the mammalian uterus and in part from the extraembryonic membranes; serves as the connection through which exchanges of nutrients and wastes occur between the blood of the mother and that of the embryo.

plankton [Gk. *planktos,* wandering]: Free-floating, mostly microscopic, aquatic or marine organisms; includes both photosynthetic (phytoplankton) and heterotrophic (zooplankton) forms.

planula [L. dim. of *planus,* a wanderer]: The ciliated, free-swimming type of larva formed by many coelenterates.

plasma [Gk. *plasma,* form or mold]: The clear, colorless fluid component of vertebrate blood, containing dissolved salts and proteins; blood minus the blood cells.

plasmid *See* Episome.

plasmodesmata [Gk. *plassein,* to mold + *desmos,* band, bond]: In plants, minute, cytoplasmic threads that extend through pores in cell walls and connect the protoplasts of adjacent cells.

plastid [Gk. *plastos,* formed or molded]: A cytoplasmic, often pigmented, organelle in plant cells; includes leucoplasts, chromoplasts, and chloroplasts.

platelet (**plate**-let) [Gk. *platus,* flat]: In mammals, a minute, granular body suspended in the blood and involved in the formation of blood clots.

pleiotropy (**plee**-o-trope-ee) [Gk. *pleios,* more + *trope,* a turning]: The capacity of a gene to affect a number of different phenotypic characteristics.

polar [L. *polus,* end of axis]: Having parts or areas with opposed or contrasting properties, such as positive and negative charges, head and tail.

polar body: Minute, nonfunctioning cell produced during those meiotic divisions that lead to egg cells; contains a nucleus but very little cytoplasm.

pollen [L., fine dust]: Spores that form the male gametophytes of seed plants.

pollination [L. *pollen,* fine dust]: The transfer of pollen from the anther to a receptive surface of a flower.

polygenic inheritance [Gk. *polus,* many + *genos,* race, descent]: The determination of a given characteristic, such as weight or height, by the interaction of many genes.

polymer [Gk. *polus,* many + *meris,* part or portion]: A large molecule composed of many molecular subunits.

polymorphism [Gk. *polus,* many + *morphe,* form]: Occurrence together in a single population of two or more morphologically distinct forms.

polyp [Gk. *polus,* many + *pous,* foot]: The sessile stage in the life cycle of coelenterates.

polypeptide [Gk. *polus,* many + *pepto,* to soften, digest]: A molecule consisting of many (about 100 or more) amino acids linked together by peptide bonds.

polyploid [Gk. *polus,* many + *ploion,* vessel]: Cell with more than two complete sets of chromosomes per nucleus.

polyribosome: A threadlike molecule of messenger RNA together with a number of ribosomes. The ribosomes move along the mRNA, directing the synthesis of polypeptide chains.

polysaccharide [Gk. *polus,* many + *sakcharon,* sugar]: A carbohydrate polymer composed of monosaccharide monomers in long chains; includes starch, cellulose.

population: Any group of individuals of one species that occupy a given area at the same time; in genetic terms, an interbreeding group of organisms.

posterior: Of or pertaining to the rear end. In humans, the back of the body is said to be posterior.

potential energy: Energy in a potentially usable form that is not, for the moment, being used.

preadapted [L. *pre,* before + *adaptare,* to fit]: A feature of an organism that suits the organism for survival in new circumstances.

pressure-flow hypothesis: A hypothesis accounting for sap flow through the phloem system. According to this hypothesis, the solution containing nutrient sugars is moved by osmotic pressure from the sieve tubes to the cells of the rest of the plant.

primary growth: In plants, growth originating in the apical meristem of the shoots and roots, as contrasted with secondary growth; results in an increase in length.

primary structure of proteins: The amino acid sequence of a protein.

primate: A member of the order of mammals that includes anthropoids and prosimians.

primitive [L. *primus,* first]: Not specialized; at an early stage of evolution or development.

primitive streak [L. *primus,* first]: The thickened, dorsal, longitudinal strip of ectoderm and mesoderm in early avian, reptilian, and mammalian embryos; equivalent to blastopore in other forms.

procambium [L. *pro,* before + *cambium,* exchange]: In plants, a primary meristematic tissue; gives rise to vascular tissues of the primary plant body and to the vascular cambium.

producer, in ecological pyramids: An autotrophic organism, usually a photosynthesizer, which contributes to the net primary productivity of a community.

progesterone [L. *progerere,* to carry forth or out + *steiras,* barren]: In mammals, a steroid hormone produced by the corpus luteum that prepares the uterus for implantation of the ovum.

prokaryote [L. *pro,* before + Gk. *karyon,* nut, kernel]: A cell lacking a membrane-bound nucleus or membrane-bound organelles; a bacterium or a blue-green alga.

promoter: Specific segment of DNA to which RNA polymerase attaches to initiate transcription of mRNA from an operon.

prophase [Gk. *pro,* before + *phasis,* form]: An early stage in nuclear division, characterized by the condensing of the chromosomes and their movement toward the equator of the spindle. Homologous chromosomes pair up (synapse) during meiotic prophase.

proprioceptor [L. *proprius,* one's own]: Receptor that senses movements, position of the body, or muscle strength.

prosimian [L. *pro,* before + *simia,* ape]: A lower primate; includes lemurs, lorises, tarsiers, and tree shrews, as well as many fossil forms.

prostaglandins [Gk. *prostas,* a porch or vestibule + L. *glaus,* acorn]: A group of fatty acid hormones discovered in semen, now found to be present in many other tissues; believed to play a role in fertilization.

prostate gland [Gk. *prostas,* a porch or vestibule + L. *glaus,* acorn]: A mass of muscle and glandular tissue surrounding the base of the urethra in male mammals, through which the seminal vesicle passes and which secretes an alkaline fluid that has a stimulating effect on the sperm as they are released.

protein [Gk. *proteios,* primary]: A complex organic compound composed of one or more polypeptide chains, each made up of many (about 100 or more) amino acids linked together by peptide bonds.

proton: A subatomic particle with a single positive charge equal in magnitude to the charge of an electron and with a mass of 1 atomic mass unit; a component of every atomic nucleus.

protoplasm: *See* Cytoplasm.

protoplast [Gk. *protos,* first + *plastos,* molded]: The cell membrane and its contents, not including the cell wall.

pseudopod [Gk. *pseudes,* false + *pous, pod-,* foot]: A temporary cytoplasmic protrusion from an amoeboid cell, which functions in locomotion or in feeding by phagocytosis.

pulmonary artery [L. *pulmonis,* lung]: In vertebrates, the artery carrying blood from the heart to the lungs, where it is oxygenated.

pulmonary vein: In vertebrates, a vein carrying oxygenated blood from the lungs to the left atrium of the heart, from which blood is pumped into the left ventricle and from there to the body tissues.

Punnett square: The checkerboard diagram used for analysis of gene segregation.

pupa [L., girl, doll]: A developmental stage of some insects, in which the organism is nonfeeding, immotile, and sometimes encapsulated or in a cocoon; the pupal stage occurs between the larval and adult phases.

purine [Gk. *purinos,* fiery, sparkling]: A nitrogenous base such as adenine or guanine; one of the components of nucleic acids.

pyramid, ecological: *See* Ecological pyramid.

pyramid of energy: A diagram of the energy relationships among various levels in a food chain; plants (at the base of the pyramid) represent the largest accumulation of energy, herbivores next, then primary carnivores, secondary carnivores, etc.

pyrimidine: A nitrogenous base such as cytosine, thymine, or uracil; one of the components of nucleic acids.

queen: The fertile, or fully developed, female of social bees, ants, and termites whose function is to lay eggs.

radial symmetry [L. *radius,* a staff, beam, or ray + Gk. *summetros,* symmetry]: The regular arrangement of parts around two or more planes that would divide an organism into halves that are approximately mirror images; seen in coelenterates and adult echinoderms.

radiation [L. *radius,* a spoke of a wheel, hence, a ray]: The movement of energy from one place to another.

radioactive isotope: An isotope with an unstable nucleus that stabilizes itself by emitting radiation.

recessive allele [L. *recedere,* to recede]: An allele that is not expressed phenotypically in the heterozygous condition.

recombination: In genetics, the formation of gene combinations that differ from the combinations present in the parents; may be accomplished through assortment of unlinked genes during sexual reproduction, crossing over between linked genes, or both.

reduction [L. *reducere,* to lead back]: Loss of oxygen, gain of hydrogen, or gain of an electron by an atom, ion, or molecule; oxidation and reduction take place simultaneously with the electron lost by one reactant being transferred to another. Oxidation-reduction reactions are an important means of energy transfer within living systems.

reflex [L. *reflectere,* to bend back]: Unit of action of the nervous system involving a sensory neuron, often an interneuron or -neurons, and one or more motor neurons.

regulator gene: A gene that functions to control the rate of transcription of other genes.

renal [L. *renes,* kidneys]: Pertaining to the kidney.

repressor [L. *reprimere,* to press back, keep back]: The substance produced by a regulator gene that inhibits messenger-RNA formation.

resolving power [L. *resolvere,* to loosen, unbind]: The ability of a lens to distinguish two lines as separate.

respiration [L. *respirare,* to breathe]: (1) In aerobic organisms, the intake of oxygen and the liberation of carbon dioxide. (2) In cells, the oxygen-requiring stage in the breakdown and release of energy from fuel molecules.

reticular activating system [L. *reticulum,* a network]: A core of tissue that runs centrally through the entire brainstem; a weblike network of fibers and neurons; involved with consciousness.

reticulum [L., network]: Any fine network, as in endoplasmic reticulum.

retina [L. dim. of *rete,* net]: The photosensitive layer of the vertebrate eye; contains several layers of neurons and light-receptors (rods and cones); receives the image formed by the lens and transmits it to the brain via the optic nerve.

rhizoid [Gk. *rhiza,* root]: Rootlike anchoring tissue in nonvascular plants.

rhizome [Gk. *rhizoma,* mass of roots]: In vascular plants, an underground stem; may be enlarged for storage, or may function in vegetative propagation.

ribonucleic acid (RNA) (rye-bo-new-**klee**-ick): A class of nucleic acids characterized by the presence of the sugar ribose and the pyrimidine uracil; includes mRNA, tRNA, and rRNA. RNA is the genetic material of many viruses.

ribosome: A small organelle composed of protein and ribonucleic acid; the site of translation in protein synthesis; in eukaryotic cells, often bound to the endoplasmic reticulum. Many ribosomes attached to a single strand of mRNA comprise a polyribosome.

RNA: Abbreviation of ribonucleic acid.

rod: Light-sensitive nerve cell found in the vertebrate retina; sensitive to very dim light, responsible for "night vision."

root: The descending axis of a plant, normally below ground and serving both to anchor the plant and to take up and conduct water and minerals.

saprophyte [Gk. *sapros,* rotten]: An organism that secures its food directly from dead and decaying tissues of other organisms; includes many fungi and most nonparasitic bacteria.

sarcomere [Gk. *sarx,* the flesh + *meris,* part of, portion]: Functional and structural unit of contraction in striated muscle.

secondary sex characteristics: Characteristics of animals that distinguish between the two sexes but that do not produce or convey gametes; includes facial hair of the human male and body-fat distribution of the human female.

secretion [L. *secermere,* to sever, separate]: Product of any cell, gland, or tissue that is released through the cell membrane and that performs its function outside the cell that produced it.

seed: A complex organ formed by the maturation of the ovule of seed plants following fertilization; upon germination, a seed develops into a new sporophyte; generally consists of seed coat, embryo, and a food reserve.

segregation: The separation of alleles into different gametes during meiosis. *See* Mendel's first law.

self-fertilization: The union of egg and sperm produced by a single hermaphroditic organism.

self-pollination: The transfer of pollen from anther to stigma in the same flower or to another flower of the same plant, leading to self-fertilization.

semen [L., seed]: Product of the male reproductive system; includes sperm and the sperm-carrying fluid.

sensillum, *pl.* **sensilla:** Sensory-receptor unit on the body surface of arthropods; responsive to touch, smell, taste, and vibration (sound).

sensory neuron: A neuron that carries impulses from a receptor to the central nervous system or central ganglion.

sensory receptor: A cell, tissue, or organ that detects internal or external stimuli.

sessile [L. *sedere,* to sit]: Attached; not free to move about.

sex chromosomes: Pairs of chromosomes that are similar in one sex but not in the other—for example, human females are *XX,* males are *XY;* chromosomes that determine sexual characteristics.

sex-linked characteristic: A genetic characteristic, such as color blindness, determined by a gene located either on the *X* or *Y* chromosome.

sexual reproduction: Reproduction involving meiosis and fertilization.

shoot: The aboveground portions, such as the stem and leaves, of a vascular plant.

sieve cell: A long, slender cell of the phloem of gymnosperms; involved in transport of materials synthesized in the leaves to other parts of the plant.

sieve tube: A series of sugar-conducting cells (sieve-tube elements) found in the phloem of angiosperms.

smooth muscle: Nonstriated muscle; lines the walls of internal organs and arteries and is under involuntary control.

social dominance: A hierarchical pattern of social organization involving domination of some members of a group by other members in a relatively orderly and long-lasting pattern.

sociobiology: The study of the biological basis of social behavior.

solution: A homogeneous mixture (usually liquid) in which one or more substances (the solutes) are uniformly dispersed throughout a solvent.

somatic cells [Gk. *soma,* body]: The differentiated cells composing body tissues of multicellular plants and animals; all body cells except the germ cells.

somatic nervous system [Gk. *soma,* body]: In vertebrates, the motor, sensory, and interneurons; the "voluntary" system, as contrasted with the "involuntary," or autonomic, nervous system.

somite: One of the segments into which the body of many animals is divided, especially arthropods, annelids, and vertebrate embryos.

specialized: (1) Of organisms, having special adaptations to a particular habitat or mode of life. (2) Of cells, having particular functions in a multicellular organism.

species, *pl.* **species** [L., kind, sort]: A group of organisms that actually (or potentially) interbreed in nature and are reproductively isolated from all other such groups; a taxonomic grouping of morphologically similar individuals (the category beneath genus).

species-specific: Characteristic of (and limited to) a particular species.

specific heat: The amount of heat (in calories) required to raise the temperature of 1 gram of a substance 1° C. The specific heat of water is 1 calorie per gram.

specificity: Uniqueness, as in proteins in a given organism and of enzymes in given reactions.

sperm [Gk. *sperma,* seed]: A mature male sex cell or gamete, usually motile and smaller than the female gamete.

spermatheca [Gk. *sperma,* seed + *theke,* box, coffin, vault]: Receptacle for sperm storage; found in many species of female invertebrates.

spermatid [Gk. *sperma,* seed]: Each of four haploid (n) cells resulting from the meiotic divisions of a spermatocyte; each spermatid becomes differentiated into a sperm cell.

spermatocytes [Gk. *sperma,* seed + *kytos,* vessel]: The diploid ($2n$) cells formed by the enlargement of the spermatogonia; they give rise by meiotic division to the spermatids.

spermatogenesis [Gk. *sperma,* seed + *genesis,* origin]: The process by which spermatogonia develop into sperm.

spermatogonia [Gk. *sperma,* seed + *gonos,* a child, the young]: The unspecialized diploid ($2n$) germ cells on the walls of the testes, which, by meiotic division, become spermatocytes, then spermatids, then spermatozoa, or sperm cells.

spermatozoon, *pl.* **spermatozoa** [Gk. *sperma,* seed + *zoos,* alive, living]: A sperm cell.

spinal cord: Part of the vertebrate central nervous system; consists of a thick, dorsal, longitudinal bundle of nerve fibers extending posteriorly from the brain.

spiracle [L. *spirare,* to breathe]: One of the external openings of the respiratory system in terrestrial arthropods.

sporangiophore (spo-**ran**-ji-o-for) [Gk. *spora,* seed + *phore,* from *phorein,* go bear]: A branch bearing one or more sporangia.

sporangium, *pl.* **sporangia** [Gk. *spora,* seed]: A unicellular or multicellular structure in which spores are produced.

spore [Gk. *spora,* seed]: An asexual reproductive cell capable of developing into an adult without fusion with another cell; in contrast to a gamete.

sporophyll [Gk. *spora,* seed + *phyllon,* the leaves]: Spore-bearing leaf. The carpels and stamens of flowers are modified sporophylls.

sporophyte [Gk. *spora,* seed + *phytos,* growing]: The spore-producing diploid ($2n$) phase in the life cycle of a plant having alternation of generations.

stamen [L., a thread]: The male organ of a flower, which produces microspores or pollen; usually consists of a stalk, the filament, bearing a pollen-producing anther at its tip.

starch [M.E. *sterchen,* to stiffen]: A class of complex, insoluble carbohydrates, the chief food-storage substances of plants; composed of up to several hundred sugar units and readily broken down enzymatically into these units.

statocyst [Gk. *statos,* standing + *kystis,* sac]: An organ of balance, consisting of a vesicle containing granules of sand (statoliths) or some other material that stimulates sensory cells when the animal moves.

stem: The aboveground part of the axis of vascular plants, as well as anatomically similar portions below ground (such as rhizomes).

stereoscopic vision [Gk. *stereos,* solid + *optikos,* pertaining to the eye]: Ability to perceive a single, three-dimensional image from the simultaneous but separate images delivered to the brain by each eye.

stigma [Gk. *stigme,* a prick mark, puncture]: In plants, the region of a carpel serving as a receptive surface for pollen grains, which germinate on it.

stimulus [L., goad, incentive]: Any internal or external change or signal that influences the activity of an organism or of part of an organism.

stoma, *pl.* **stomata** (**sto**-ma) [Gk. *stoma,* mouth]: A minute opening bordered by guard cells in the epidermis of leaves and stems through which gases pass. Also used to refer to the entire stomatal apparatus, the guard cells plus their included pore.

striated muscle [L., from *striare,* to groove]: Skeletal voluntary muscle and cardiac muscle. The name derives from the striped appearance, which reflects the arrangement of contractile elements.

stroma [Gk. *stroma,* a bed, from *stronnymi,* to spread out]: The ground substance that makes up the interior of the chloroplast in which the grana are dispersed.

structural gene: Any gene that produces a protein in distinction to other genes of an operon (operator and regulator genes).

substrate [L. *substratus,* strewn under]: (1) The foundation to which an organism is attached. (2) A substance acted on by an enzyme.

succession [L. *succedo,* go up to]: In ecology, the slow, orderly progression of biotic changes in the composition of a community.

sucrose: Cane sugar; a common disaccharide found in many plants; a molecule of glucose linked to a molecule of fructose.

sugar: Any monosaccharide or disaccharide.

symbiosis [Gk. *syn,* together with + *bioonai,* to live]: An intimate and protracted association between two or more organisms of different species. Includes mutualism, in which the association is beneficial to both; commensalism, in which one benefits and the other is neither harmed nor benefitted; and parasitism, in which one benefits and the other is harmed.

sympathetic division: A subdivision of the autonomic nervous system, with centers in the midportion of the spinal cord; slows digestion; generally excites other functions. Chemical transmitters: epinephrine or norepinephrine.

synapse [Gk. *synapsis,* a union]: The region of nerve-impulse transfer between two neurons.

synapsis [Gk., a union]: The pairing of homologous chromosomes that occurs prior to the first meiotic division.

syngamy (**sin**-gamy) [Gk. *syn,* with + *gamos,* a marriage]: The union of gametes in sexual reproduction; fertilization.

synthesis [Gk. *syntheke,* a putting together]: The formation of a more complex substance from simpler ones.

systematics [Gk. *systema,* that which is put together]: Scientific study of the kinds and diversity of organisms and of the relationships among them.

taxonomy [Gk. *taxis,* arrange, put in order + *nomos,* law]: The study of the classification of organisms, the ordering of organisms into a hierarchy that reflects their essential similarities and differences.

telophase [Gk. *telos,* end + *phasis,* form]: The last stage in mitosis and meiosis, during which the chromosomes become reorganized into two new nuclei.

template: A pattern or mold guiding the formation of a negative or complement.

tentacles [L. *tentare,* to touch]: Long, flexible protrusions located about the mouth in many invertebrates; usually prehensile or tactile.

territory: An area or space occupied and defended by an individual or a group; trespassers are attacked (and usually defeated); may be the site of breeding, nesting, food gathering, or any combination thereof.

testcross: A mating between a phenotypically dominant individual and a homozygous recessive "tester" to determine the genetic constitution of the dominant phenotype, that is, whether it is homozygous or heterozygous for the relevant gene.

testis, *pl.* **testes** [L., witness]: The sperm-producing organ; also the source of male sex hormone.

testosterone [Gk. *testis,* testicle + *steiras,* barren]: A hormone secreted by the testes in higher vertebrates and stimulating the development and maintenance of male sex characteristics and the production of sperm; an androgen.

tetrad [Gk. *tetras,* four]: In plants, a group of four haploid spores formed by meiosis within a mother cell.

thalamus [Gk. *thalamos,* chamber]: A part of the vertebrate forebrain just posterior to the cerebrum; an important intermediary between all other parts of the nervous system and the cerebrum.

thallus [Gk. *thallos,* a young twig]: A simple plant body without true roots, leaves, or stems.

theory [Gk. *theorein,* to look at]: A generalization based on many observations and experiments.

thermodynamics [Gk. *therme,* heat + *dynamis,* power]: The study of transformations of energy, especially heat. The first law of thermodynamics states that in all processes, the total energy of an isolated system remains constant. The second law states that in all processes the entropy, or degree of randomness, tends to increase.

thorax [Gk. *thorax,* breastplate]: (1) In vertebrates, that portion of the trunk containing the heart and lungs. (2) In insects, the three leg-bearing segments between head and abdomen.

thylakoid [Gk. *thylakos,* a small bag]: A saclike membrane-limited organelle within the chloroplast; stacks of thylakoids collectively form the grana.

thyroid [Gk. *thyra,* a door]: An endocrine gland of vertebrates, located in the neck; source of an iodine-containing hormone (thyroxine) that increases the metabolic rate and affects growth.

tissue [L. *texere,* to weave]: A group of similar cells organized into a structural and functional unit.

trachea, *pl.* **tracheae** (**trake**-ee-a) [Gk. *tracheia,* rough]: An air-conducting tube, such as the windpipe of mammals and the breathing systems of insects.

tracheid (**tray**-key-idd) [Gk. *tracheia,* rough]: In vascular plants, an elongated, thick-walled conducting and supporting cell of xylem, characterized by tapering ends and pitted walls without true perforations.

transcription [L. *trans,* across + *scribere,* to write]: The process by which the genetic information in DNA specifies a complementary RNA molecule.

transduction [L. *trans,* across + *ducere,* to lead]: The transfer of genetic material (DNA) from one bacterium to another by a lysogenic bacteriophage.

transfer RNA (tRNA) [L. *trans,* across + *ferre,* to bear or carry]: A class of small RNAs (about 80 nucleotides) with two functional sites; one recognizes a specific activated amino acid; the other carries the nucleotide triplet (anticodon) for that amino acid. Each type of tRNA accepts a specific activated amino acid and transfers it to a growing polypeptide chain as specified by the nucleotide sequence of the messenger RNA being transcribed.

transformation [L. *trans,* across + *formare,* to shape]: A genetic change produced by the incorporation into a cell of DNA from another cell.

translation [L. *trans,* across + *latus,* that which is carried]: The synthesis of a polypeptide from amino acids carried by tRNAs to the ribosome-bound mRNA specifying the polypeptide.

translocation [L. *trans,* across + *locare,* to put or place]: (1) In plants, the transport of the products of photosynthesis from a leaf to another part of the plant. (2) In genetics, the breaking off of a piece of a chromosome and its attachment to a nonhomologous chromosome.

transpiration [L. *trans,* across + *spirare,* to breathe]: In plants, the loss of water vapor from the stomata.

tritium: A radioactive isotope (^{3}H) of hydrogen with a half-life of 12.5 years.

trophic level [Gk. *trophos,* feeder]: The position of a species in the food web or chain; a step in the movement of biomass or energy through a system.

tropic [Gk. *trope,* a turning]: Pertaining to behavior or action brought about by specific stimuli, for example, phototropic ("light-oriented") motion, gonadotropic ("stimulating the gonads") hormone.

tuber [L. *tuber,* bump, swelling]: A much-enlarged, short, fleshy underground stem, such as that of the potato.

turgor [L. *turgere,* to swell]: The distension of a plant cell wall by the fluid contents of the cell.

ultracentrifuge: A high-speed centrifuge that can attain speeds up to 60,000 rpm and centrifugal fields up to 500,000 times gravity and thus is capable of rapidly sedimenting macromolecules.

urea [Gk. *ouron,* urine]: An organic compound formed in the vertebrate liver; the principal form of disposal of nitrogenous wastes by mammals.

ureter [Gk. from *ourein,* to urinate]: The tube carrying urine from the kidney to the cloaca (in reptiles and birds) or to the bladder (in amphibians and mammals).

urethra [Gk. from *ourein,* to urinate]: The tube carrying urine from the bladder to the exterior of mammals.

uric acid [Gk. *ouron,* urine]: An insoluble nitrogenous waste product that is the principal excretory product in birds, reptiles, and insects.

urine [Gk. *ouron,* urine]: The liquid waste filtered from the blood by the kidney and stored in the bladder pending elimination through the urethra.

uterus [L., womb]: The muscular, expanded portion of the female reproductive tract modified for the storage of eggs or for housing and nourishing the developing embryo.

vacuole [L. *vacuus,* empty]: A fluid-filled cellular organelle, bound by a membrane and containing a solution isotonic with the cytoplasm.

vagina [L., a sheath]: The part of the female reproductive duct in mammals that receives the male penis during copulation; leads to the uterus.

vagus nerve [L. *vagus,* wandering]: A nerve arising from the medulla of the vertebrate brain that innervates the heart and visceral organs; carries parasympathetic fibers.

valence [L. *valere,* to have power]: A measure of the bonding capacity of an atom, based on the number of electrons in its outer energy level.

vascular [L. *vasculum,* a small vessel]: Containing or concerning vessels that conduct fluid.

vascular bundle: In plants, a group of longitudinal supporting and conducting tissues (xylem and phloem).

vascular cambium [L. *vasculum,* a small vessel + *cambium,* exchange]: A cylindrical sheath of meristematic cells that divide mitotically, producing secondary phloem to one side and secondary xylem to the other, but always with a cambial cell remaining.

vas deferens (vass **deff**-er-ens) [L., *vas,* a vessel + *deferre,* to carry down]: In mammals, the tube carrying sperm from the testes to the urethra.

vein [L. *vena,* a blood vessel]: (1) In plants, a vascular bundle forming a part of the framework of the conducting and supporting tissue of a leaf or other expanded organ. (2) In animals, a blood vessel carrying blood from the tissues to the heart.

vena cava (**vee**-na **cah**-va) [L., blood vessel + hollow]: A large vein that brings blood from the tissues to the right atrium of the four-chambered vertebrate heart. The superior vena cava collects blood from the forelimbs, head, and anterior or upper trunk; the inferior vena cava collects blood from the posterior body region.

ventral [L. *venter,* belly]: Pertaining to the undersurface of an animal that moves on all fours; to the front surface of an animal that holds its body erect.

ventricle [L. *ventriculus,* the stomach]: A muscular chamber of the heart that receives blood from an atrium and pumps blood out of the heart, either to the lungs or to the body tissues.

vertebral column [L. *vertebra,* joint]: The backbone; in nearly all vertebrates, it forms the supporting axis of the body and it protects the spinal cord.

vesicle [L. *vesicula,* a little bladder]: A small, intracellular membrane-bound sac.

vessel [L. *vas,* a vessel]: A tubelike element of the xylem of angiosperms; composed of dead cells (vessel elements) arranged end to end. Its function is to conduct water and minerals from the soil.

viability [L. *vita,* life]: Ability to live.

villus, *pl.* **villi** [L., a tuft of hair]: In vertebrates, one of the minute, fingerlike projections lining the small intestine that serve to increase the absorptive surface area of the intestine.

virus [L., slimy, liquid, poison]: A submicroscopic, noncellular particle, composed of a nucleic acid core and a protein coat; parasitic; reproduces only within a host cell.

viscera [L., internal organs]: The collective term for the internal organs of an animal.

vitamin [L. *vita,* life]: Any of a number of unrelated organic substances that cannot be synthesized by a particular organism and are essential in minute quantities for normal growth and function.

vulva [L., the womb]: External genitalia of the mammalian female; in humans, includes the clitoris and the labia.

water table: The upper limit of permanently water-saturated soil; the top layer of the groundwater.

wild type: In genetics, the phenotype that is characteristic of the vast majority of individuals of a species in a natural environment.

worker: A member of the nonreproductive laboring caste of social insects.

xanthophyll [Gk. *xanthos,* yellow + *phyllon,* leaf]: In algae and green plants, one of a group of yellow pigments; a member of the carotenoid group.

xylem [Gk. *xylon,* wood]: A complex vascular tissue through which most of the water and minerals are conducted from the roots to other parts of the plant; consists of tracheids or vessel elements, parenchyma cells, and fibers; constitutes the wood of trees and shrubs.

yolk: The stored food material in egg cells that nourishes the embryo.

zoology [Gk. *zoe,* life + *logos,* a discourse]: The study of animals.

zooplankton [Gk. *zoe,* life + *plankton,* wanderer]: A collective term for the nonphotosynthetic organisms present in plankton.

zygote (**zi**-got) [Gk. *zygon,* yolk, pair]: The diploid ($2n$) cell resulting from the fusion of male and female gametes (fertilization); a zygote may either develop into a diploid individual by mitotic divisions, or may undergo meiosis to form haploid (n) individuals that divide mitotically to form a population of cells.

ILLUSTRATION ACKNOWLEDGMENTS

I–1 Kenneth W. Fink, Photo Researchers

I–2 (left) F. Erize, Bruce Coleman; (right) Hans Reinhard, Bruce Coleman; (bottom) Robert L. Dunne, Bruce Coleman

I–3 (a) Grant Heilman; (b) Frank M. Carpenter; (c) R. Levi-Setti

I–4 The American Museum of Natural History

I–5 M. P. L. Fogden, Bruce Coleman

I–6 The Royal College of Surgeons of England

I–9 (a) Oronce Finé, Théorique de la huitième sphère et sept planètes, 1528; (b) Ann Ronan Picture Library

Page 10 The Royal College of Surgeons of England

I–10 Treat Davidson, National Audubon Society Collection/PR

1–1 © California Institute of Technology and the Carnegie Institution of Washington

Page 20 (top left) Runk/Schoenberger, Grant Heilman Photography; (top right) Grant Heilman; (bottom left) G. Ronald Austing, Bruce Coleman; (bottom right) Douglas P. Wilson

Page 21 (top left) Roberts Rugh and Landrum B. Shettles, M. D., *From Conception to Birth: The Drama of Life's Beginnings*, Harper & Row, Publishers, Inc., New York, 1971; (top right) Stephen Dalton, Photo Researchers; (bottom) Grant Heilman

1–6 Grant Heilman

1–7 (b) Grant Heilman

Page 28 The Bettmann Archive

2–1 E. R. Degginger, Bruce Coleman

2–2 NASA

2–4 Stephen Dalton

2–5 Bill Ratcliffe

2–6 Heather Angel

2–8 (b) Bill Ratcliffe

2–11 Jeanne M. Riddle

Page 44 Grant Heilman

Page 48 John M. Sieburth

3–4 After E. J. DuPraw, *Cell and Molecular Biology*, Academic Press, Inc., New York, 1968

Page 51 Photograph from J. D. Watson, *The Double Helix*, Atheneum, New York, 1968

3–8 (c) After Albert L. Lehninger, *Biochemistry*, Worth Publishers, Inc., New York, 1975; (d) L. M. Beidler; (e) Don Fawcett

3–9 (c) R. D. Preston

3–10 (b) Larry West

3–12 Russ Kinne, Photo Researchers

3–15 B. E. Juniper

3–20 Linus Pauling

3–21 (b) After E. O. Wilson *et al.*, *Life: Cells, Organisms, Populations*, Sinauer Associates, Inc., Sunderland, Mass., 1977

3–22 (b) Richard Dickerson

3–24 Dan Friend

3–25 (a) Emil Bernstein and Eila Kairinen, Gillette Company Research Institute, Rockville, Md., *Science*, **173**: 1971. © 1971 by The American Association for the Advancement of Science. (b) Jack Dermid

3–26 (b) Eugene B. Small and Douglas S. Marszalek, *Science*, **163**: 1064–1065, 1969. © 1969 by the American Association for the Advancement of Science

3–30 Patricia Farnsworth

4–2 After Lehninger, *op. cit.*

4–4 J. T. Pickett-Heaps

4–7 After Lehninger, *op. cit.*

4–11 Danish Information Office

4–12 Don Fawcett, *An Atlas of Fine Structure. The Cell: Its Organelles and Inclusions*, W. B. Saunders Company, Philadelphia, 1966

4–13 Ylla, Rapho/PR

4–14 After L. Stryer, *Biochemistry*, W. H. Freeman and Company, San Francisco, 1975

4–15 After J. D. Watson, *Molecular Biology of the Gene*, 2d ed., The Benjamin/Cummings Publishing Company, Menlo Park, Calif., 1970

4–17 After Lehninger, *op. cit.*

4–18 After Peter H. Raven, Ray F. Evert, and Helena Curtis, *Biology of Plants*, 2d ed., Worth Publishers, Inc., New York, 1976

5–1 Michael A. Walsh

5–2 Micrographs by (a) A. Ryter; (b) Ursula Goodenough; (c) Keith R. Porter

5–4 (b) L. E. Roth, D. J. Pihlaja, and Y. Shigenaka, Journal of Ultrastructure Research, **30**:7, 1970

Pages 100 and 101 David M. Phillips

5–5 (a) J. D. Robertson

Page 103 Rare Book Division, The New York Public Library

5–6 (a) Micrograph by Myron C. Ledbetter; (b) after Peter Albersham, *Scientific American*, April, 1975

5–8 Keith R. Porter

5–9 Ursula Goodenough

5–10 (a) Daniel Branton; (b) Don Fawcett, *op. cit.*

5–11 Don Fawcett, from William Bloom and D. W. Fawcett, *A Textbook of Histology*, 10th ed., W. B. Saunders Company, Philadelphia, 1975

5–12 Don Fawcett

5–13 (left) Don Fawcett, from Bloom and Fawcett, *op. cit.*; (right) C. J. Flickinger, *Journal of Cell Biology*, **49**:221, 1975

5–14 After Stephen Wolfe, *Biology of the Cell*, Wadsworth Publishing Co., Inc., Belmont, Calif., 1972. By permission of the publisher

5–15 Don Fawcett

5–16 Myron C. Ledbetter

5–17 David Stetler

5–18 Don Fawcett

5–19 (a) E. Lazarides, *Journal of Cell Biology*, **65**:549, 1975; (b) K. Weber and U. Groeschel-Stewart, *Proceedings of National Academy of Sciences U.S.*, **171**:4561, 1974

5–20 Gregory Antipa

5–21 (a) After DuPraw, *op. cit.*; (b) A. V. Grimstone

5–22 Étienne de Harven, *The Nucleus*, Academic Press, Inc., New York, 1968

5–23 J. F. M. Hoeniger, *Journal of General Microbiology,* **40**:29, 1965

5–24 James W. Perry

5–25 (a) Keith R. Porter; (b) L. A. Staehelin

5–26 David W. Buck

5–27 J. T. Bonner

6–1 W. G. Whaley, The University of Texas Electron Microscope Laboratory

6–3 P. Berger, National Audubon Society Collection/PR

6–5 After Lehninger, *op. cit.*

6–6 (a), (b) Thomas Eisner

6–8 After S. J. Singer and A. L. Nicholson, *Science,* **175**:720, 1972

6–9 (a) P. Pinto da Silva and D. Branton, *Journal of Cell Biology,* **45**:598, 1970

6–10 Don Fawcett

6–13 Gregory Antipa

6–14 Gregory Antipa

6–15 Birgit H. Satir

7–1 R. D. Estes

7–2 The Bettmann Archive

7–3 The Bettmann Archive

7–4 The Bettmann Archive

7–6 Gene Ahrens, Bruce Coleman

Page 150 Lotte Jacobi

7–8 (a) Larry West; (b) Jane Burton, Bruce Coleman; (c) Tom Brakefield, Bruce Coleman; (d) John Dominis, LIFE © 1971

8–1 (a) Bob Evans, Peter Arnold; (b) Gary Robinson; (c) Richard Bray, *Science,* **200**:333–334, 1978. © by the American Association for the Advancement of Science

8–4 (b) Grant Heilman

8–6 Keith R. Porter

Page 161 R. H. Kirschner

8–7 (b) Lester J. Reed, from P. D. Boyer, ed., *The Enzymes,* vol. 1, Academic Press, Inc., New York, 1970

8–8 After A. J. Vander, J. H. Sherman, and Dorothy S. Luciano, *Human Physiology,* McGraw-Hill Book Company, New York, 1969

8–9 After Lehninger, *op. cit.*

8–12 After T. Takano, O. B. Kallai, R. Swanson, and R. E. Dickerson, *Journal of Biological Chemistry,* **248**:5244, 1973

8–13 After P. C. Hinkle and R. E. McCarty, *Scientific American,* March, 1978

8–17 After Lehninger, *op. cit.*

9–1 Stephen B. Gough and William J. Woelkering

9–8 Photograph by Oxford Scientific Films, Bruce Coleman

9–9 E. Gantt and S. F. Conti, *Journal of Bacteriology,* **97**:1486, 1969

9–10 (a) Daniel Branton; (b) J. Greenwood; (c) L. K. Shumway; (d) Myron C. Ledbetter

9–14 After Lehninger, *op. cit.*

Page 185 N. Pfeiffer

9–16 After Hinkle and McCarty, *op. cit.*

9–17 J. Heslop-Harrison

Page 186 Walther Stoeckenius, *Scientific American,* June, 1976

9–18 After Stryer, *op. cit.*

9–22 W. M. Laetsch, *Scientific Progress,* **57**:323, 1969

10–1 Kurt Hirschhorn, M. D., from *Birth Defects: Original Article Series,* vol. **4**, no. 4, *Guide to Human Chromosome Defects* by Audrey Redding and Kurt Hirschhorn, M. D., The National Foundation

10–2 Carnegie Institution of Washington

10–4 (a) A. Ryter

10–6 William Tai

10–7 William Tai

10–8 Andrew Bajer

10–9 William Tai

10–10 General Biological Supply Company

10–11 (a) S. Inoué, *Chromosome Bd.,* **5**:487–500, Springer, 1953; (b) J. Pickett-Heaps

10–12 (a) Étienne de Harven, *op. cit.*; (b) Michael Friedlander

10–14 H. W. Beams and R. G. Kessel, *American Scientist,* **64**:279, 1976

10–15 James Cronshaw

10–16 Micrographs by William Tai

11–1 Arnold Sparrow

11–7 B. John

11–8 William Tai

11–9 Mary E. Clutter

11–10 After DuPraw, *op. cit.*

11–13 Hugh Spencer, National Audubon Society Collection/PR

11–14 After John L. Moore, *Heredity and Development,* Oxford University Press, New York, 1963

Page 220 William Tai

12–1 The Bettmann Archive

12–2 The Bettmann Archive

Page 224 Pasteur Institute and The Rockefeller University Press

12–5 After Karl von Frisch, *Biology,* translated by Jane Oppenheimer, Harper & Row, Publishers, Inc., New York, 1964

Page 232 C. D. K. Cook, *A Monographic Study of Ranunculus Subgenus Batrachium (DC) A. Gray,* **6**:47–237, October, 1966

12–11 Dr. V. Orel, The Moravian Museum, Brno

12–12 C. G. G. J. van Steenis

12–14 Photograph by F. B. Hutt, *Journal of Genetics,* **22**:126, 1930

12–15 (a) After E. D. Merrell, 1964, (b) *Journal of Heredity,* vol. 5, 1914

13–2 Mary E. Clutter

Page 243 S. Baty

13–7 Micrograph by B. John

13–11 B. P. Kaufmann

14–1 Drawing by B. O. Dodge, from E. L. Tatum

14–4 After Linus Pauling *et al., Science,* **110**:543–548, 1949

14–5 Robert Austrian, *Journal of Experimental Medicine,* **92**:21, 1953

14–6 After Derry D. Koob and William E. Boggs, *The Nature of Life,* Addison-Wesley Publishing Company, Inc., Reading, Mass., 1972

15–1 Lee D. Simon

15–3 From J. Cairns, G. S. Stent, and J. D. Watson, eds., *Phage and the Origins of*

Molecular Biology, Cold Spring Harbor Laboratory of Quantitative Biology, Cold Spring Harbor, N.Y., 1966

15-4 Lee D. Simon

15-5 A. K. Kleinschmidt, D. Land, D. Jacherts, and R. K. Zahn, *Biochemica Biophysica Acta*, **61**:857-864, 1962

15-6 From Watson, *op. cit.*, 1968

15-8 After Lehninger, *op. cit.*

15-10 From Watson, *op. cit.*, 1968

15-12 After Lehninger, *op. cit.*

15-13 After Lehninger, *op. cit.*

16-1 After Wendell Stanley and Evan G. Valens, *Viruses and the Nature of Life*, E. P. Dutton & Co., Inc., New York, 1961

16-4 O. L. Miller, Jr., and Barbara R. Beatty, Biology Division, Oak Ridge National Laboratory

16-5 (b) Dr. Sung-Hou Kim

16-6 Ursula Goodenough

16-8 After Keith Roberts

16-10 Hans Ris

16-11 O. L. Miller, Jr., Barbara A. Hamkalo, and C. A. Thomas, Jr., *Science*, **169**:392-395, 1970. © 1970 by the American Association for the Advancement of Science

17-1 John Cairns

17-2 Sunil Palchaudhuri, E. Bell, and M. R. J. Salton, *Infection* and *Immunity*, **11**:1141, 1975

17-3 Judith Carnahan and Charles Brinton

Page 290 Stanley N. Cohen

17-9 Jack Griffith

17-10 George T. Rudkin

17-11 J. Gall

17-13 Donald E. Olins and Ada L. Olins, University of Tennessee, Oak Ridge Graduate School of Biomedical Sciences and Oak Ridge National Laboratory

18-2 (a) Herbert A. Fischler, Isaac Albert Research Institute of the Kingsbrook Jewish Medical Center; (b) John S. O'Brien

18-3 Patricia Farnsworth

18-4 UPI

18-7 After I. Michael Lerner, *Heredity, Evolution, and Society*, W. H. Freeman and Company, San Francisco, 1968

18-8 Gernsheim Collection, Humanities Research Center, The University of Texas

18-9 The Bettmann Archive

18-10 Paediatric Research Unit, Guy's Hospital Medical School

Page 312 The National Foundation

18-14 George Ancona

19-2 After W. F. Bodmer and L. L. Cavalli-Sforza, *Genetics, Evolution, and Man*, W. H. Freeman and Company, San Francisco, 1976

19-4 After E. Noritski, *Human Genetics*, The Macmillan Company, New York, 1977

19-5 After Bodmer, *op. cit.*

19-9 Joseph L. Melnick, Baylor College of Medicine, Houston

19-11 Tsuyoshi Kakefuda

19-12 After *City of Hope Quarterly*, **7**:2, Winter, 1978

19-13 *Ibid.*

19-14 David A. Jackson

20-1 Edward S. Ross

Page 334 NASA

20-2 S. Moorbath

20-3 S. Jonasson, from R. Anderson *et al.*, *Science*, **148**:1179-1190, 1965. © 1965 by the American Association for the Advancement of Science

20-4 Photograph by C. Ponnamperuma

20-5 (a) J. W. Schopf and E. S. Barghoorn, *Science*, **156**:508-512, 1967. © 1967 by the American Association for the Advancement of Science. (b) E. S. Barghoorn and J. W. Schopf, *Science*, **152**:758-763, 1966. © 1966 by the American Association for the Advancement of Science

20-6 (a) William J. Larsen, *Journal of Cell Biology*, **47**:373, 1970; (b) D. A. Stetler and W. M. Laetsch, *American Journal of Botany*, **56**:260, 1969

20-7 (a) Michael D. Lagios, Children's Hospital of San Francisco; (b) Leonard Muscatine

20-10 (a), (b) Eric V. Gravé; (c) Michael P. Godomski, National Audubon Society Collection/PR; (d) Les Blacklock, (e) Alvin E. Staffan

20-11 Burndy Library

Page 344 (left) John H. Gerard, DPI; (right) H. Reinhard, Bruce Coleman

20-12 (a) John S. Flannery, Bruce Coleman; (b) Ken Brate, Photo Researchers; (c) H. Reinhard, Bruce Coleman

21-1 M. E. Edwards

21-2 Lee D. Simon

21-3 Leon J. Le Beau, American Society for Microbiology

21-4 (a) J. Adler; (b) after M. L. DePamphilis and J. Adler, *Journal of Bacteriology*, **105**:395, 1971

21-5 J. F. M. Hoeniger

21-6 David Greenwood

21-7 J. F. M. Hoeniger

21-8 (a) Center for Disease Control; (b), (c) M. A. Listgarten and S. S. Socransky, "Electron Microscopy of Axial Fibrils, Outer Envelope and Cell Division of Oral Spirochetes," *Journal of Bacteriology*, **88**:108, 1964

21-10 R. S. Wolfe

21-11 Micrograph by J. F. M. Hoeniger and C. L. Headley, *Journal of Bacteriology*, **96**:1835-1847, 1968

21-12 P. L. Grilione and J. Pangborn, *Journal of Bacteriology*, **124**:1558, 1975

21-13 E. Boatman

21-14 Micrograph by M. Jost

21-15 (a) H. Forest; (b) Eric V. Gravé

21-16 (a), (b) J. D. Almeida and A. F. Howatson, *Journal of Cell Biology*, **16**:616, 1963; (c) Frederick A. Murphy; (e) Lee D. Simon

21-17 Ny Carlsberg Glyptotek

22-1 Eric V. Gravé

22-2 Eric V. Gravé

22-3 D. S. Marszalek and E. B. Small

22-4 Eric V. Gravé

22-5 Eric V. Gravé, Photo Researchers

22-6 (a) G. A. Horridge and S. L. Tamm,

Science, **163**:817–818, 1969. © 1969 by the American Association for the Advancement of Science. (b) Eric V. Gravé

22–8 Jerry Lesser, Bruce Coleman

22–12 H. N. Guttmann

22–13 (a) Eric V. Gravé, Photo Researchers; (b) Douglas P. Wilson

22–14 Kent Cambridge Scientific Company

22–15 Douglas P. Wilson

Page 378 Micrographs by D. Kubai and H. Ris

22–16 Edward S. Ross

22–18 (a) Valerie Taylor, Ron Taylor Film Productions; (b) Oxford Scientific Films, Bruce Coleman; (c) Douglas P. Wilson

22–19 (a) M. I. Walker, Photo Researchers; (b), (d) M. I. Walker; (c) George I. Schwartz

22–20 Douglas P. Wilson

22–21 Larry West

23–1 Les Blacklock

23–3 H. C. Hoch and D. P. Maxwell

23–4 Jack Dermid

23–5 John H. Troughton

23–6 After John S. Niederhauser and William Cobb, *Scientific American,* May, 1959

23–7 Alma W. Barkdale, *Mycologia,* **55**:493–501, 1973

23–9 (a) Alvin E. Staffan; (b) John Shaw

23–10 C. Bracker

23–12 (a) Agenzia Fotografica, Luisa Ricciarini, Milan; (b) Craig Blacklock; (c) Jane Burton, Bruce Coleman

Page 394 David Pramer

23–14 (a) Jack Dermid; (b), (c) Grant Heilman

24–1 R. Carr, Bruce Coleman

24–2 Edward S. Ross

24–3 Ray F. Evert

24–4 Micrograph by Trond Braten

24–7 Ray F. Evert

24–9 Turtox

24–11 Alvin E. Staffan

24–12 (a) Edward S. Ross; (b), (c) Grant Heilman

24–13 After Raven *et al., op. cit.*

24–14 After Raven *et al., op. cit.*

24–15 (a) Bruce Coleman; (b) John Shaw; (c) Robert Carr

24–16 Grant Heilman

24–17 (a) Richard B. Peacock, Photo Researchers; (b) Gene Ahrens, Bruce Coleman

24–18 © Field Museum of Natural History, painting by Charles R. Knight

24–19 Grant Heilman

24–20 (a) William Harlow; (b) Larry West

24–23 Grant Heilman

24–26 (b) Larry West

24–27 (b) Grant Heilman

24–28 (b) Edward S. Ross

24–30 After Grover C. Stephens and Barbara Best North, *Biology,* John Wiley & Sons, Inc., New York, 1974

24–31 (a), (e) Edward S. Ross; (b) W. G. Calder; (c) David Overcash, Bruce Coleman; (d) D. J. Howell; (f) Jack Dermid

24–33 (a) H. N. Darrow, Bruce Coleman; (b) Mike Godfrey; (c), (d) Heather Angel

Page 421 R. Borneman, Photo Researchers

25–1 Douglas P. Wilson

25–2 Runk/Schoenberger, Grant Heilman Photography

25–3 Nat Fain

25–4 Maria Wimmer

25–5 Douglas P. Wilson

25–7 The American Museum of Natural History

25–8 Nat Fain

25–12 After Ralph Buchsbaum and Lorus J. Milne, *The Lower Animals: Living Invertebrates of the World,* Doubleday & Company, Inc., Garden City, N.Y., 1962

25–14 Nat Fain

25–15 (a) Douglas P. Wilson; (b) Runk/Schoenberger, Grant Heilman Photography

25–16 Douglas Faulkner

25–18 Tom McHugh, Photo Researchers

25–20 Walter E. Harvey, National Audubon Society Collection/PR

25–23 After M. B. V. Roberts, *Biology: A Functional Approach,* The Ronald Press Company, New York, 1971

25–25 (a) British Museum (Natural History); (b) Carolina Biological Supply Company

25–26 Maria Wimmer

25–27 J. E. Sulston

25–32 Runk/Schoenberger, Grant Heilman Photography

25–33 Maria Wimmer

25–34 Russ Kinne, Photo Researchers

25–36 After Roberts, *op. cit.*

25–37 (a), (c) Runk/Schoenberger, Grant Heilman Photography; (b) Peter Ward, Bruce Coleman; (d) Jane Burton, Bruce Coleman

25–38 (b) M. J. Wells and Heinemann Educational Books, Ltd.

25–39 Douglas P. Wilson

25–40 (a) Jane Burton, Bruce Coleman; (b) Russ Kinne, Photo Researchers; (c) Runk/Schoenberger, Grant Heilman Photography; (d) Paul E. Taylor, National Audubon Society Collection/PR

26–1 Grant Heilman

26–2 (a), (c) Edward S. Ross; (b) Jack Dermid

26–3 After Wilson *et al., op. cit.* 1977

26–4 Runk/Schoenberger, Grant Heilman Photography

26–6 Peter Ward, Bruce Coleman

26–9 (a), (b) Edward S. Ross; (c) S. C. Bisserôt, Bruce Coleman

26–11 (a)–(c) Edward S. Ross; (d) W. Clifford Healey, National Audubon Society Collection/PR

26–12 (a) Edward S. Ross; (b) Larry West

26–15 (a), (c), (d) Edward S. Ross; (b) Larry West

26–16 (a) Oxford Scientific Films, Bruce Coleman; (b)–(d) David Overcash, Bruce Coleman; (e)–(i) E. R. Degginger, Bruce Coleman

26–17 After Malcolm S. Gordon, *Animal Physiology: Principles and Adaptations,* 2d ed., The Macmillan Company, New York, 1972

26–18 (a) L. M. Beidler; (b) after Roberts, *op. cit.*

26–19 Larry West

26–21 Edward S. Ross

26–22 Ross Hutchins

26–23 Edward S. Ross

26–27 Donald H. Fritts

26–28 (a) The American Museum of Natural History; (b) John S. Flannery, Bruce Coleman

26–29 H. Reinhard, Bruce Coleman

26–30 Runk/Schoenberger, Grant Heilman Photography

26–31 Lynwood M. Chace, National Audubon Society Collection/PR

26–32 After Cleveland P. Hickman and Cleveland P. Hickman, Jr., *Biology of Animals,* C. V. Mosby Company, St. Louis, Mo., 1972

26–33 (a) Jack Dermid; (b) R. E. Pelham, Bruce Coleman

26–34 The American Museum of Natural History

26–35 Anthony Mercieca, Photo Researchers

26–36 (a) The New York Zoological Society; (b) Jack Dermid

26–38 (a) Tom McHugh, Photo Researchers; (b) Karl Maslowski, Photo Researchers; (c) Jeff Foote, Bruce Coleman; (d) Mark N. Boulton, National Audubon Society Collection/PR; (e) Catherine Bray, National Audubon Society Collection/PR; (f) Mary M. Thatcher, Photo Researchers

27–1 Jack Dermid

27–5 (a) L. M. Beidler; (b) John H. Troughton

27–6 Ray F. Evert

27–7 Ray F. Evert

27–8 Ray F. Evert

27–9 (a) Dan Clark, Grant Heilman Photography; (b) John Colwell, Grant Heilman Photography; (c), (d) Grant Heilman; (e) Larry West

27–10 Myron C. Ledbetter and Keith R. Porter, *Introduction to the Fine Structure of Plant Cells,* Springer-Verlag, New York, 1970

27–11 Ledbetter and Porter, *op. cit.*

27–13 Ray F. Evert

27–14 (a), (b) Ray F. Evert; (c) Carolina Biological Supply Company

27–15 Ray F. Evert

27–16 Jack Dermid

27–17 After Peter Ray, *The Living Plant,* 2d ed., Holt, Rinehart and Winston, Inc., New York, 1971

27–18 Ray F. Evert

27–19 Micrograph by Ray F. Evert

27–21 (a) John Colwell, Grant Heilman Photography; (b) Grant Heilman

28–2 Heather Angel

28–3 Jane Burton, Bruce Coleman

28–4 (a), (b) Patrick Echlin; (c) J. Heslop-Harrison

28–5 (b) Micrograph by Myron Ledbetter

28–6 (a) Jane Burton, Bruce Coleman; (b), (d) Anita Sabarese; (c) Edward Ross; (e) Larry West; (f) Jack Dermid; (g) John H. Gerard

Page 517 Ray F. Evert

28–11 Micrographs by Ray F. Evert

28–13 After Ray, *op. cit.*

28–14 Howard Towner

28–15 Peter Ray

28–17 From Richard Ketchum, *The Secret Life of the Forest,* American Heritage Press, New York, 1970

28–18 Grant Heilman

Page 529 Ron Lurbow, National Audubon Society Collection/PR

29–1 George Whiteley, Photo Researchers

29–2 After M. Richardson, *Translocation in Plants,* St. Martin's Press, Inc., New York, 1968 (after Stout and Hoagland, 1939)

29–3 After A. C. Leopold, *Plant Growth and Development,* McGraw-Hill Book Company, New York, 1964

29–4 John H. Troughton

Page 535 (b), (c) Walter Eschrich and Eberhard Fritz

29–7 (a) Martin H. Zimmermann; (b) George A. Schaefers

29–9 After W. A. Jensen and F. B. Salisbury, *Botany: An Ecological Approach,* Wadsworth Publishing Co., Inc., Belmont, Calif., 1972

29–10 Jack Dermid

29–11 Grant Heilman

29–12 S. A. Wilde

29–13 (a) Raymond C. Valentine; (b) P. J. Dart; (c), (d) R. R. Hebert, J. G. Criswell, and R. W. F. Hardy, Central Research and Development Department, E. I. du Pont de Nemours & Company, Wilmington, Del.

30–1 S. H. Wittwer and The Michigan Agricultural Station

30–6 U.S. Department of Agriculture

30–7 J. P. Nitsch, *American Journal of Botany,* **37**:3, 1950

30–8 J. E. Varner

30–10 After Wilson *et al., op. cit.,* 1977

30–12 H. R. Chen

30–17 Marvin B. Winter

30–18 After Aubrey W. Naylor, *Scientific American,* May, 1952

30–21 Frank Salisbury

30–24 Jack Dermid

30–25 Jack Dermid

30–26 Grant Heilman

31–1 Ken Heyman

31–7 Micrograph by Keith R. Porter

31–8 M. H. Ross and Edward J. Reith

31–10 Micrograph by Hugh E. Huxley

31–12 Don Fawcett

Page 581 Micrograph by Hugh E. Huxley

32–1 Ken Heyman

32–2 (a) George Porter, Photo Researchers; (b) Tom McHugh, Photo Researchers; (c) Robert W. Hernandez, Photo Researchers; (d) George Holton, Photo Researchers

32–5 David M. Phillips

32–6 After Roberts, *op. cit.*

32–10 (top left) Tom McHugh, Photo Researchers; (top right) N. DeVore, III, Bruce Coleman; (middle left) Charles G.

Summers, Jr., Bruce Coleman; (middle) Jane Burton, Bruce Coleman; (middle right) Russ Kinne, Photo Researchers; (bottom) L. L. Rue, III, Bruce Coleman

32-12 L. L. Rue, III, National Audubon Society Collection/PR

32-14 (b) R. J. Blandau

32-17 Carnegie Institution of Washington

32-21 National Center for Health Statistics

33-1 (a) David Epel and Mia Tegner; (b) William Byrd; (c) Everett Anderson

33-2 Tryggve Gustafson

33-3 After J. J. W. Baker and G. E. Allen, *The Study of Biology*, 3d. ed., Addison-Wesley Publishing Company, Inc., Reading, Mass., 1977

33-4 Tryggve Gustafson

33-5 After E. O. Wilson *et al.*, *Life on Earth*, 2d ed., Sinauer Associates, Inc., Sunderland, Mass., 1978

33-6 After Waddington, 1966

33-8 After Huettner, 1949

33-9 After Huettner, 1949

33-10 After Huettner, 1949

33-11 John A. Moore

33-12 After Baker and Allen, *op. cit.*

33-13 After Waddington, 1966

33-14 M. P. Olson

33-15 After Wilson *et al.*, *op. cit.*, 1978

33-16 *Ibid.*

33-17 After Torrey, 1962

33-19 After Maurice Sussman, *Animal Growth and Development*, 2d ed., Prentice-Hall, Inc., Englewood Cliffs, N.J., 1964

33-20 After Charles W. Bodemer, *Modern Embryology*, Holt, Rinehart and Winston, Inc., New York, 1968

33-21 After Sussman, *op. cit.*

33-22 Runk/Schoenberger, Grant Heilman Photography

33-23 Rugh and Shettles, *op. cit.*

33-24 Rugh and Shettles, *op. cit.*

33-27 After Roberts, *op. cit.*

33-28 Carnegie Institution of Washington

33-29 Rugh and Shettles, *op. cit.*

33-34 Paoli Koch, Photo Researchers

33-36 Rugh and Shettles, *op. cit.*

34-2 Keith R. Porter

34-3 (b) Cedric S. Raine, from P. Morell, ed., *Myelin*, Plenum Publishing Corp., New York, 1977

34-5 After Wilson *et al.*, *op. cit.*, 1978

34-6 (b) David Hubel

34-8 After S. W. Kuffler and J. G. Nicholls, *From Neuron to Brain*, Sinauer Associates, Inc., Sunderland, Mass., 1976

34-9 *Ibid.*

34-10 After C. R. Noback, *The Human Nervous System: Basic Principles of Neurobiology*, McGraw-Hill Book Company, New York, 1975

34-11 After Keith R. Porter and Mary Bonneville, *Fine Structure of Cells and Tissues*, Lee and Febiger, Philadelphia, 1968

34-12 John Heuser

34-13 Photograph by Arthur Jacques

34-17 After S. S. Mader, *Inquiry into Life*, William C. Brown Company, Publishers, Dubuque, Iowa, 1976

34-18 Don Fawcett

34-20 The Bettmann Archive

Page 652 Rae Silver

34-27 After Wilson *et al.*, *op. cit.*, 1978

34-29 (b) J. T. Bonner

35-1 After Roberts, *op. cit.*

35-2 Thomas Eisner

35-3 Micrograph by Don Fawcett

35-4 U.N. Food and Agriculture Organization

35-6 After Roberts, *op. cit.*

35-9 Keith R. Porter

35-11 Robert Rosenblum

Page 668 American Heart Association

35-12 After Vander, Sherman, and Luciano, *op. cit.*

Page 671 Photograph by Parke Davis Company

35-14 Thomas L. Hayes and L. McDonald, *Experimental and Molecular Pathology*, vol. 10, Academic Press, Inc., New York, 1969

35-15 (a) Pfizer, Inc.; (b) Peter B. Armstrong

35-16 Emil Bernstein, The Gillette Company Research Institute

35-19 Joseph Feldman

35-20 (a) After Lehninger, *op. cit.*

35-21 After Richard A. Lerner and Frank J. Dixon, *Scientific American*, June, 1973

35-22 A. Liepins, Sloan-Kettering Institute for Cancer Research, New York

36-2 Thomas Eisner

36-3 Russ Kinne, Photo Researchers

Page 689 Alan Root

36-7 (b) Preparation and micrograph by H. R. Duncker, Department of Anatomy, Justus-Liebig-University, Giessen, Germany

Page 694 Myron Melamed

36-10 (a) Keith R. Porter; (b) E. R. Weibel

36-12 After Knut Schmidt-Nielsen, *How Animals Work*, Cambridge University Press, New York, 1972

Page 698 Anker Odum, from *Scenic Wonders of Canada*. © 1976 by Reader's Digest Association (Canada) Ltd.

36-15 After W. S. Beck, *Human Design*, Harcourt Brace Jovanovich, Inc., New York, 1971

37-2 After W. W. Howells, *Mankind in the Making*, rev. ed., Doubleday & Company, Inc., Garden City, N.Y., 1967

37-3 (a) I. Kaufman Arenberg

37-7 (a) Jeanne M. Riddle

37-8 Don Fawcett

37-10 From Bloom and Fawcett, *op. cit.*

37-11 Keith R. Porter

Page 713 U.S. Department of Agriculture

37-12 G. Decker

37-13 (b) L. L. Rue, III, Photo Researchers

38-1 After J. E. Crouch and R. McClintic, *Human Anatomy and Physiology*, John Wiley & Sons, Inc., New York, 1971

38-5 From William Bloom and D. F. Fawcett, *Histology*, 9th ed., W. B. Saunders Company, Philadelphia, 1968

38-6 Allan D. Cruickshank, Photo Researchers

38-7 From Lewis Carroll, *Alice's Adventures in Wonderland,* illustrated by John Tenniel

38-9 Bob Leatherman, National Audubon Society Collection/PR

38-11 Jack Dermid, Photo Researchers

38-12 After Vander *et al., op. cit.*

38-13 (a) Nicholas Mrosovsky; (b) L. L. Rue, III, National Audubon Society Collection/PR; (c) Bloom and Fawcett, *op. cit.,* 1975

38-14 Phyllis McCutcheon, Photo Researchers

38-15 (a) Ron Garrison, San Diego Zoo; (b) L. L. Rue, III, Bruce Coleman; (d) L. L. Rue, III, National Audubon Society Collection/PR

38-17 Pat Kirkpatrick, Photo Researchers

39-1 After Beck, *op. cit.*

39-5 (a) Russ Kinne, Photo Researchers; (b) Gordon S. Smith, Photo Researchers; (c) Jerry Focht, Photo Researchers; (d) A. W. Ambler, National Audubon Society Collection/PR

39-8 E. R. Lewis

39-9 (b) Keith R. Porter

39-11 Jack Dermid

39-15 Robert E. Preston, courtesy of Professor J. E. Hawkins, Kresge Hearing Research Institute, University of Michigan Medical School

39-16 (a) S. C. Bisserôt, Bruce Coleman

40-1 Arthur Jacques

40-2 After P. W. Davis and E. P. Solomon, *The World of Biology,* McGraw-Hill Book Company, New York, 1974

40-8 James Olds

40-11 After Lehninger, *op. cit.*

40-13 Aquarius Electronics, Mendocino, Calif., © 1971

40-15 W. R. Klemm, *Science, The Brain and Our Future,* Pegasus, New York, 1972

40-16 Jerry Hecht, National Institute of Mental Health

40-17 Laverne C. Johnson, *American Scientist,* **61**:3, 1973

40-19 Jerre Levy, Colwyn Trevarthen, and Roger W. Sperry, *Brain: The Journal of Neurology,* 1972

40-20 E. R. Lewis

41-1 Baron Wolman

41-3 Marineland of Florida

41-4 Victor McKusick

41-5 Edmund B. Gerrard

41-6 Victor McKusick

41-8 After Jack R. Harlow, *Scientific American,* September, 1976; vegetables courtesy of the Green Thumb, Water Mill, New York

41-9 Tony Gauba, National Audubon Society Collection/PR

41-10 After Charles W. Brown, Santa Rosa Junior College

41-12 H. B. D. Kettlewell

42-1 F. Erize, Bruce Coleman

42-2 After Mather and Harrison, *Heredity,* vol. 3, 1949

42-4 P. Ward, Bruce Coleman

42-5 (a) Hans Reinhard, Bruce Coleman; (b) Jane Burton, Bruce Coleman

42-6 Heather Angel

42-8 P. M. Sheppard

42-9 Jen and Des Bartlett, Bruce Coleman

42-10 After Bryan Clarke, *Scientific American,* August, 1975

42-11 From S. J. Gould, *Ever Since Darwin,* W. W. Norton & Company, New York, 1977

42-12 After J. Clauser and W. M. Hiesey, Carnegie Institution of Washington, publication 615, 1958

43-1 Jerome Wexler, National Audubon Society Collection/PR

43-3 After Stephens and North, *op. cit.*

43-4 Department of Indian Affairs and Northern Development

43-5 (a) J. R. Brownlie, Bruce Coleman; (b) Fritz Prenzel, Bruce Coleman; (c) J. Wallis, Bruce Coleman

43-6 After E. O. Wilson and William H. Bossert, *A Primer of Population Biology,* Sinauer Associates, Sunderland, Mass., 1971

43-7 (a) M. Philip Kahl, Bruce Coleman; (b) N. Myers, Bruce Coleman

Page 809 (b) John S. Shelton

43-9 Jen and Des Bartlett, Bruce Coleman

43-10 After Bruce Wallace and Adrian Srb, *Adaptation,* Prentice-Hall, Inc., Englewood Cliffs, N.J., 1964

43-11 *Ibid.*

43-13 I. Eibl-Eibesfeldt

43-14 After David Lack, *Darwin's Finches,* Harper & Row, Publishers, Inc., New York, 1961

43-15 (a) C. Haagner, Bruce Coleman; (b) Robert Clark, Photo Researchers

43-16 (a) M. W. Larson, Bruce Coleman; (b) Robert Mitchell; (c) Alan Blank, Bruce Coleman; (d) Jeanne White, National Audubon Society Collection/PR

43-17 Edward S. Ross

43-18 Edward S. Ross

43-19 George Holton, Photo Researchers

44-1 Elihu Blotnick, BBM Associates

44-2 Howard T. Odum

44-3 NASA

44-10 (a) Craig Blacklock; (b) N. Myers, Bruce Coleman

44-11 From Maarten J. Chrispeels and David Sadava, *Plants, Food and People,* W. H. Freeman and Company, San Francisco, 1977

Page 834 Photograph by John M. Edmond, from John B. Corliss and Robert D. Ballard, National Geographic, vol. 152, no. 4, October, 1977

Page 839 Irven deVore, Anthro/Photos

44-17 (b) Michigan Department of Natural Resources

44-19 R. H. Wright, National Audubon Society Collection/PR

44-20 G. E. Likens

44-21 (a) U.S. Forest Service; (b) Larry West

44-22 Larry West

44-23 Jack Dermid

44-24 (a) Jack Dermid; (b) U.S. Forest Service

44–25 Jack Dermid

45–5 Tom McHugh, Photo Researchers

45–6 After S. Charles Kendeigh, *Animal Ecology*, Prentice-Hall, Inc., Englewood Cliffs, N.J., 1961 (after Shelford, 1911)

45–8 After John L. Harper, *Symposia Society for Experimental Biology*, **15**:25, 1961

45–12 Biological Section Laboratory, Australian Department of Lands

45–13 (a) Jane Burton, Bruce Coleman; (b) Cosmos Blank, Photo Researchers; (c) Eric Hosking, Bruce Coleman; (d) H. Reinhard, Bruce Coleman

45–14 Photographs by Howard Towner; seeds and beetles supplied by Daniel Janzen

46–1 U.S. Forest Service

46–2 (a) Craig Blacklock; (b) J. Spurr, Bruce Coleman

46–3 Les Blacklock

46–4 Grant Heilman

46–5 Graham F. Date

46–6 Australian News and Information Service

46–7 N. Myers, Bruce Coleman

46–8 (a) Edward S. Ross; (b) Doug Fulton, Photo Researchers; (c) S. C. Bisserôt, Bruce Coleman

46–9 Edward S. Ross

46–10 (a) S. C. Bisserôt, Bruce Coleman; (b) Alan Blank, Bruce Coleman

46–11 Edward S. Ross

46–12 (a) R. Marischal, Bruce Coleman; (b) Douglas Faulkner; (c) Edward S. Ross; (d) L. L. Rue, III, Bruce Coleman

46–13 Daniel Janzen

46–14 (a) S. J. Krasemann, National Audubon Society Collection/PR; (b) Tomas D. W. Friedman, Photo Researchers

46–16 Karl Weidmann, National Audubon Society Collection/PR

46–17 (a) Edward S. Ross; (b) John Markam, Bruce Coleman

46–18 Sylvio Fresco, Bruce Coleman

46–19 Joe Van Wormer, Bruce Coleman

46–20 (a) L. L. Rue, III, Bruce Coleman; (b) Max Thompson, National Audubon Society Collection/PR; (c) N. Myers, Bruce Coleman; (d) Verna R. Johnston, Photo Researchers

46–21 (a) Bill Ratcliffe; (b) Jen and Des Bartlett, Bruce Coleman; (c) Robert H. Wright, National Audubon Society Collection/PR; (d) Grace Thompson, Photo Researchers; (e) Anthony Mercieca, National Audubon Society Collection/PR; (f) Willis Peterson; (g) Alan Blank, Bruce Coleman

46–22 (a) Jen and Des Bartlett, Bruce Coleman; (b) Dennis Brokaw

46–23 (a) Tom McHugh, Photo Researchers; (b) Patricia Caulfield

46–24 (a) Ed Cooper; (b) Les Blacklock; (c) L. L. Rue, III, Photo Researchers; (d) Craig Blacklock

46–25 (a) Craig Blacklock; (b), (c) Les Blacklock

46–26 (a) Les Blacklock; (b) Alvin E. Staffan; (c) Bill Ruth, Bruce Coleman

46–27 Les Blacklock

46–29 Ed Cooper

46–30 (b), (c) Douglas P. Wilson; (d) William R. Curtsinger, Rapho/PR; (e) Kjell B. Sandved, Photo Researchers

46–31 Allan Power, Bruce Coleman

46–32 (a) Jeff Foote, Bruce Coleman; (b) Charlie Ott, National Audubon Society Collection/PR; (c) Jack Dermid

46–33 (a) Jack Dermid; (b) Al Lowry, Photo Researchers

47–2 K. W. Jeon

Page 903 William C. Dilger

47–6 A. W. Ambler, Photo Researchers

47–8 After Tinbergen, 1951

47–10 Eric Hosking

47–13 John Markam, Bruce Coleman

47–15 After Mazudazo Konish

47–16 © Frank Lotz Miller, Black Star

47–17 Wisconsin Regional Primate Research Center

48–1 Stephen Dalton, Photo Researchers

48–2 (a) Edward S. Ross; (b) Alan Blank, Bruce Coleman

48–3 Edward S. Ross

48–4 (a) Oxford Scientific Films, Bruce Coleman; (b) Treat Davidson, National Audubon Society Collection/PR

48–5 (a) Edward S. Ross; (b) Colin G. Butler, Bruce Coleman; (c) Stephen Dalton, Photo Researchers

48–6 From David P. Barash, *Sociobiology and Behavior*, Elsevier North-Holland, Inc., New York, 1977

48–7 S. Baty

48–8 Masud Quraishy, Bruce Coleman

48–9 (a) M. P. Kahl, Photo Researchers; (b) George Holton, Photo Researchers; (d) Jeff Foote, Bruce Coleman

48–10 Tanaka Kojo, Animals Animals Enterprises

48–11 T. W. Ransom

48–12 (a) L. L. Rue, III, Bruce Coleman; (b) R. R. Pawlaski, Bruce Coleman; (c) S. Gaulin, Anthro/Photos; (d) Douglas Faulkner

Page 926 Jane Burton, Bruce Coleman

48–13 V. B. Scheffer, U.S. Fish and Wildlife Service

48–14 Richard D. Estes

48–15 Rod Allin, Bruce Coleman

48–17 Russ Kinne, Photo Researchers

48–18 (a) Patricia Caulfield; (b) Robert L. Dunne, Bruce Coleman

48–19 Kenneth Fink, Bruce Coleman

48–20 After L. David Mech, *The Wolf: The Ecology and Behavior of an Endangered Species*, Natural History Press, Garden City, N. Y., 1970

48–21 L. David Mech

48–22 Tom McHugh, Photo Researchers

48–23 Ron Garrison, San Diego Zoo

48–24 Irven deVore, Anthro/Photos

48–24 Irven deVore, Anthro/Photos

48–26 (a) Tom McHugh, Photo Researchers; (b) Irven deVore, Anthro/Photos

48–27 Irven deVore, Anthro/Photos

48–28 J. Popp, Anthro/Photos

48–29 J. Popp, Anthro/Photos

49–1 (a), (c) S. Trevor, Bruce Coleman; (b) Norman Myers, Bruce Coleman; (d) Irven deVore, Anthro/Photos

Page 946 George Herben, Photo Researchers

49-3 Ron Garrison, San Diego Zoo

49-5 (a) Edward S. Ross; (b) H. Albrecht, Bruce Coleman

49-6 After R. E. Leakey and R. E. Lewin, *Origins*, E. P. Dutton & Co., Inc., New York, 1977

49-7 (a) George Holton, Photo Researchers; (b) Tom McHugh, Photo Researchers; (c) M. P. L. Fogden, Bruce Coleman

49-8 After W. W. Howells, *op. cit.*

49-9 L. L. Rue, III, Photo Researchers

Page 952 (a) Ralph Morse, Time-Life; (b) Irven deVore, Anthro/Photos; (c) George Holton, Photo Researchers; (d), (e) Lee Lyon, Bruce Coleman

Page 953 (f), (h) Teleki/Baldwin; (g), (i), (j) S. Wrangham, Anthro/Photos

Page 954 C. R. Carpenter

49-11 The American Museum of Natural History

49-12 The American Museum of Natural History

49-13 After Leakey and Lewin, *op. cit.*

49-14 Robert F. Sisson, © National Geographic Society

49-15 © National Geographic Society

49-16 (a) L. L. Rue, III, National Audubon Society Collection/PR; (b) Geza Teleki

49-17 The American Museum of Natural History

49-18 Tom McHugh, Photo Researchers

49-19 Geza Teleki

49-20 Lee Boltin

49-21 Ralph S. Solecki

49-22 From David Pilbeam, *The Ascent of Man: An Introduction to Human Evolution*, The Macmillan Company, New York, 1972

49-23 Drawings from Jacques Bordaz, *Tools of the Old and New Stone Age*, The Natural History Press, Garden City, N.Y., 1970; photographs by Lee Boltin

49-24 Drawings from Bordaz, *op. cit.*; photographs from Peabody Museum

49-25 Tom McHugh, Photo Researchers

49-27 International Rice Research Institute

Front endpaper Barren ground caribou migrating; Charlie Ott, National Audubon Society Collection/PR

Back endpaper Peltate scales from Russian olive fruit; John N. A. Lott

Page vi Cave painting of wild ox from Lascaux, France; Archives Photographique, Ministère des Affaires Culturelles Caisse Nationale des Monuments Historiques et des Sites

Page xii Nuclei of grasshopper spermatids; David M. Phillips

Contents: page xiv E. R. Degginger, Bruce Coleman; page xv Jane Burton, Bruce Coleman; page xvi Arnold Sparrow; page xvii A. K. Kleinschmidt, D. Land, D. Jacherts, and R. K. Zahn, *Biochemica Biophysica Acta*, **61**:857-864, 1962; page xviii John S. Flannery, Bruce Coleman; page xix Eric V. Gravé, Photo Researchers; page xx Catherine Bray, National Audubon Society Collection/PR; page xxi John H. Gerard; page xxii Tom McHugh, Photo Researchers; page xxiii H. R. Duncker, Department of Anatomy, Justus-Leibig-University, Giessen, Germany; page xxiv F. Erize, Bruce Coleman; page xxv Edward S. Ross; page xxvi Robert L. Dunne, Bruce Coleman; page xxvii S. Wrangham, Anthro/Photos

INDEX

Page numbers in italic type indicate references to illustrations and tables; boldface type is used to distinguish major text references from minor text references.

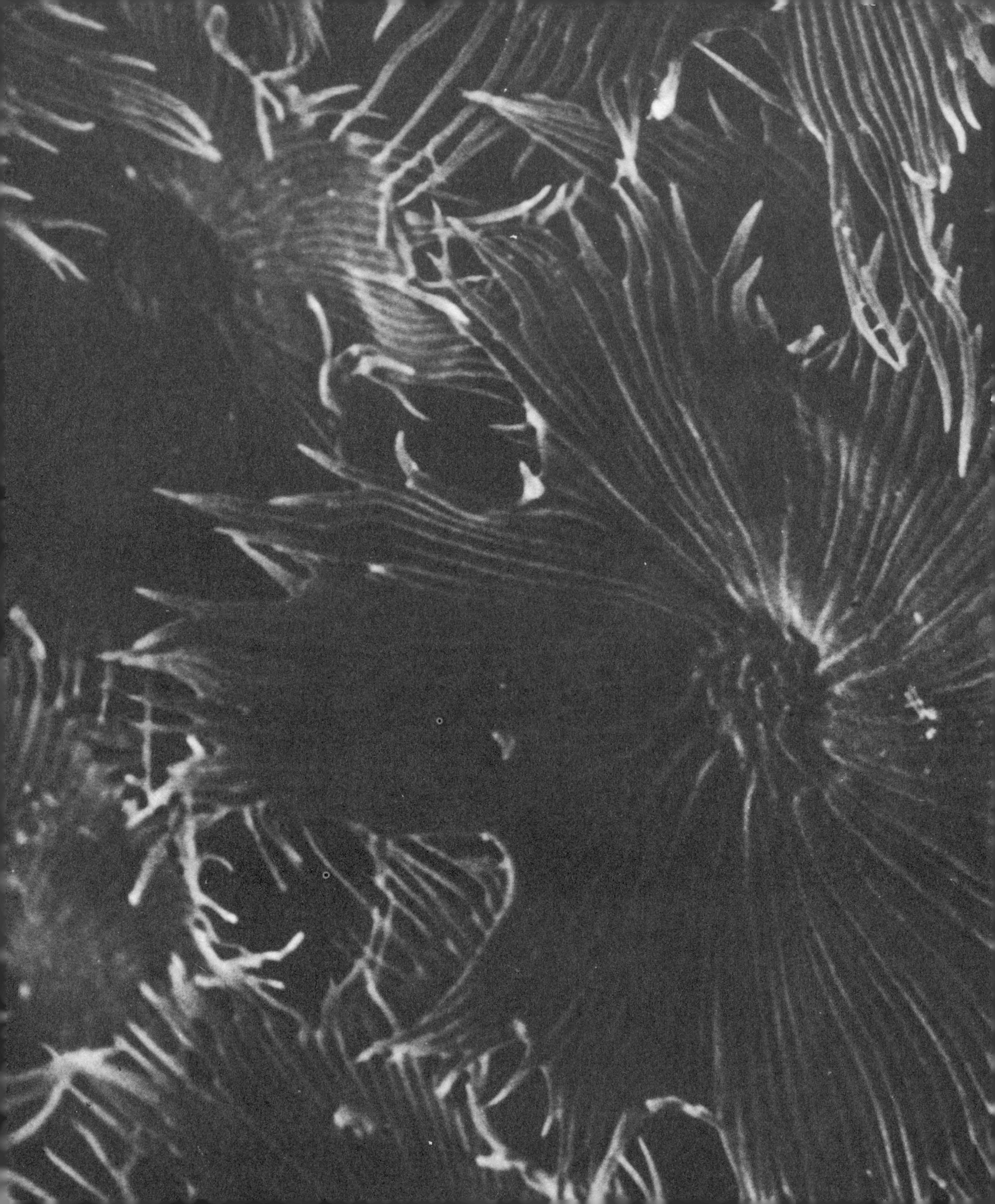